FLORA OF CYPRUS

VOLUME TWO

BOSEA CYPRIA

FLORA OF CYPRUS

VOLUME TWO

R. D. MEIKLE
The Herbarium, Royal Botanic Gardens, Kew

1985
Published by The Bentham-Moxon Trust
Royal Botanic Gardens, Kew

Vol. 1. ISBN 0 9504876 3 5

Vol. 2. ISBN 0 9504876 4 3

PREFACE

VOLUME 2 completes this survey of the flora of Cyprus. In the main it follows closely the format of Volume 1, but one minor change from previous usage has been made in deference to the views expressed by Prof. W. T. Stearn in Taxon, 16: 168–178 (1967): names published in *Florae Graecae Prodromus* and *Flora Graeca*, and previously attributed to "Sibthorp et Smith" are now attributed to Smith alone, or rarely to John Lindley, his successor as editor of the *Flora Graeca*.

A key for the identification of plant families represented in the Flora will be found on pp. 1899–1907, but the full list of cited literature promised in Vol. 1 has had to be omitted, since its inclusion would have further distended a volume already rather obese. Moreover the provision of such information has been rendered to a great extent supererogatory by the publication of: *World List of Scientific Periodicals* (Butterworths, London etc., 1963–1980, continued as *British Union — Catalogue of Periodicals* 1980–); *Flora Europaea* (5 vols., Cambridge University Press, 1964–1980); *B-P-H* (*Botanico-Periodicum-Huntianum*; Hunt Botanical Library, Pittsburgh, 1968); *The Kew Record of Taxonomic Literature* (H.M. Stationery Office, London, 1971–); *Serial Publications in the British Museum* (*Natural History*) *Library* (3rd Edition, British Museum (Nat. Hist.) Publication No. 778, 1980), and F. A. Stafleu and R. S. Cowan's comprehensive and scholarly *Taxonomic Literature* (4 vols. to date, Bohn, Scheltema and Holkema, Utrecht, 1976–1983).

R. D. M.

ACKNOWLEDGEMENTS

MAY I again take the opportunity of expressing my warmest thanks to the Government of Cyprus for consistent, patient and generous support of a project which has consumed many years and which, at times, seemed as if it might never reach completion. To the Forestry Department of the Ministry of Agriculture and Natural Resources, and to the Director of Forests, Mr. L. I. Leontiades, I am particularly indebted for information, and for specimens of plants otherwise known to me only by report, also for the fresh material of *Bosea cypria* used in the preparation of the frontispiece for Volume 2.

Apart from those whose contributions to the Flora are acknowledged in the body of the text, many of my former colleagues on the staff of the Royal Botanic Gardens, Kew, have assisted me with advice and information; it is not possible to thank them individually here, but I am sure they will not object to the individual mention of Mr. J. L. Gilbert, Mrs. J. Kennedy O'Byrne (Miss P. Z. Scammell), Mrs. P. Middleton (Miss P. Carman) and Mrs. L. Booth (Miss L. Lehmanova), who, from 1955 onward, were engaged as official assistants for the Flora, and whose cheerful help (and criticism) considerably eased the task of compilation. Here I must also acknowledge the very practical, though unofficial, assistance given to me by Mr. Edwards Casey, whose glossary of Cyprus place-names, received at the outset of the project, has proved an invaluable stand-by, saving many hours of tedious search. Nor should I fail to acknowledge the important contribution of the Kew typing staff in making sense of some very testing manuscripts.

Finally, a particular debt of gratitude is owed to Miss P. Halliday, not only for her contributions to the Flora, but for her untiring assistance as joint proof-reader for Volume 2, after my retirement from the Royal Botanic Gardens. No doubt some errors will be found to have escaped our detection, but, without her additional scrutiny, I am certain the blemishes would have been many times more numerous and more serious.

R. D. M.

CONTENTS

PREFACE *page* v

ACKNOWLEDGEMENTS vii

LIST OF ILLUSTRATIONS xi

FAMILIES IN VOL. 2 833

FLORA 835

APPENDIX IV   Revised List of Collectors 1893

APPENDIX V   List of new names, new combinations, etc. 1897

APPENDIX VI   Key to Families 1899

INDEX TO VOLS. 1 and 2 1909

> Long worke it were,
> Here to account the endlesse progenie
> Of all the weedes, that bud and blossom there;
> But so much as doth need, must needs be counted here.

"The Garden of Adonis"
SPENSER, *The Faerie Queene*, Book III, Canto VI, Stanza 30.

ILLUSTRATIONS

Bosea cypria
| | Boiss. ex Schinz et Autran | *frontispiece* |

Plate 53	*Centranthus calcitrapa* (L.) Dufr. ssp.	
	orbiculatus (Sm.) Meikle	*page* 839
Plate 54	*Scabiosa cyprica* Post	855
Plate 55	*S. brachiata* Sm.	857
Plate 56, figs. 1–2	*Bellis annua* L. ssp. *minuta* (DC.) Meikle	870
Plate 56, figs. 3–4	*B. sylvestris* Cyr.	870
Plate 57, figs. 1–2	*Phagnalon rupestre* (L.) DC. ssp.	
	rupestre	884
Plate 57, figs. 3–4	*P. rupestre* (L.) DC. ssp. *graecum* (Boiss.	
	et Heldr.) Hayek	884
Plate 58, figs. 1–4	*Achillea santolina* L.	905
Plate 58, figs. 5–8	*A. cretica* L.	905
Plate 59, figs. 1–5	*Anthemis plutonia* Meikle	911
Plate 59, figs. 6–11	*A. palaestina* (Reuter ex Kotschy)	
	Boiss.	911
Plate 60, figs. 1–3	*Senecio glaucus* L. ssp. *cyprius* Meikle	933
Plate 60, figs. 4–9	*S. leucanthemifolius* Poir. var. *vernalis*	
	(Waldst. et Kit.) J. Alexander	933
Plate 61	*Echinops spinosissimus* Turra	941
Plate 62	*Catananche lutea* L.	987
Plate 63	*Taraxacum aphrogenes* Meikle	1022
Plate 64, figs. 1–3	*Campanula delicatula* Boiss.	1051
Plate 64, figs. 4–6	*C. podocarpa* Boiss.	1051
Plate 65	*Solenopsis minuta* (L.) C. Presl ssp.	
	nobilis (E. Wimmer) Meikle	1058
Plate 66	*Cyclamen cyprium* Kotschy	1081
Plate 67	*Centaurium erythraea* Rafn ssp.	
	rhodense (Boiss. et Reuter) Melderis	1113
Plate 68, figs. 1–3	*Cynoglossum troodi* Lindberg f.	1127
Plate 68, figs. 4–6	*C. montanum* L. ssp. *extraeuropaeum*	
	Brand	1127
Plate 69	*Convolvulus oleifolius* Desr.: figs. 1–3	
	var. *oleifolius*; fig. 4 var. *deserti*	
	Pamp.; fig. 5 var. *pumilus* Pamp.	1169
Plate 70	*Odontites cypria* Boiss.	1231
Plate 71	*Orobanche cypria* Reuter	1239
Plate 72, figs. 1–5	*Pinguicula crystallina* Sm.	1245
Plate 72, figs. 6–11	*Acinos troodi* (Post) Leblebici	1245
Plate 73	*Origanum cordifolium* (Aucher-Eloy et	
	Montbret ex Benth.) Vogel	1263
Plate 74	*Nepeta troodi* Holmboe	1302
Plate 75	*Sideritis cypria* Post	1305
Plate 76	*Ballota integrifolia* Benth.	1318

xi

ILLUSTRATIONS

Plate 77, figs. 1–2a   *Phlomis cypria* Post var. *cypria*   1322

Plate 77, figs. 3–5a   *P. cypria* Post var. *occidentalis* Meikle   1322

Plate 77, figs. 6–7a   *P. brevibracteata* Turrill   1322

Plate 78, figs. 1–3   *Teucrium cyprium* Boiss. ssp. *kyreniae* P. H. Davis   1340

Plate 78, figs. 4–7   *T. cyprium* Boiss. ssp. *cyprium*   1340

Plate 79, figs. 1–6   *Plantago cretica* L.   1352

Plate 79, figs. 7–11   *P. amplexicaulis* Cav.   1352

Plate 80   *Phytolacca pruinosa* Fenzl   1397

Plate 81   *Rumex cyprius* Murb.   1407

Plate 82   *Aristolochia sempervirens* L.   1418

Plate 83   *Euphorbia dimorphocaulon* P. H. Davis   1438

Plate 84   *Alnus orientalis* Decne.   1475

Plate 85   *Quercus alnifolia* Poech   1483

Plate 86   *Epipactis troodi* Lindberg f.   1500

Plate 87   *Platanthera chlorantha* (Custer) Reichb. ssp. *holmboei* (Lindberg f.) J. J. Wood   1512

Plate 88   Flowers of 1. *Ophrys sphegodes* Mill. ssp. *mammosa* (Desf.) Soó; 2. *O. bornmuelleri* M. Schulze ex Bornm.; 3. *O. umbilicata* Desf. ssp. *attica* (Boiss. et Orph.) Soó; 4. *O. sphegodes* Mill. ssp. *transhyrcana* (Czernjak.) Soó; 5. *O. apifera* Huds.; 6. *O. argolica* H. Fleischm. ssp. *elegans* (Renz) E. Nelson   1521

Plate 89   *Ophrys sphegodes* Mill. ssp. *mammosa* (Desf.) Soó: Variation in speculum markings   1522

Plate 90   *Orchis anatolica* Boiss.   1544

Plate 91, figs. 1–5   *Crocus cyprius* Kotschy   1565

Plate 91, fig. 6   *C. hartmannianus* Holmboe   1565

Plate 91, fig. 7   *C. veneris* Tappeiner   1565

Plate 92   *Gladiolus triphyllus* Sm.   1572

Plate 93, figs. 1–3a   *Narcissus tazetta* L.   1575

Plate 93, figs. 4–5   *N. serotinus* L.   1575

Plate 94   *Tulipa cypria* Stapf   1598

Plate 95, figs. 1–2   *Gagea juliae* Pascher   1604

Plate 95, figs. 3–4   *G. peduncularis* (J. et C. Presl) Pascher   1604

Plate 95, figs. 5–6   *G. fibrosa* (Desf.) J. A. et J. H. Schultes   1604

Plate 95, figs. 7–8   *G. chlorantha* (M. Bieb.) J. A. et J. H. Schultes   1604

Plate 96   *Allium rubrovittatum* Boiss. et Heldr.   1622

Plate 97, figs. 1–2   *Scilla morrisii* Meikle   1638

Plate 97, figs. 3–6   *S. cilicica* Siehe   1638

Plate 98   *Arum hygrophilum* Boiss.   1668

Plate 99   *Triglochin bulbosa* L.   1675

Plate 100   *Carex illegitima* Cesati   1710

Plate 101   *Briza humilis* M. Bieb.   1718

ILLUSTRATIONS

Plate 102, figs. 1–8	*Brachypodium firmifolium* Lindberg f.	1748
Plate 102, figs. 9–13	*Lindbergella sintenisii* (Lindberg f.) Bor	1748
Plate 103	*Bromus diandrus* Roth	1801
Plate 104	*Taeniatherum crinitum* (Schreb.) Nevski	1829
Plate 105	*Cheilanthes marantae* (L.) Domin	1881

FAMILIES IN VOL. 2

51.	Valerianaceae	85.	Ulmaceae
52.	Dipsacaceae	85a.	Cannabaceae
53.	Compositae	86.	Moraceae
54.	Campanulaceae	87.	Platanaceae
55.	Ericaceae	88.	Juglandaceae
56.	Monotropaceae	88a.	Casuarinaceae
57.	Plumbaginaceae	89.	Betulaceae
58.	Primulaceae	90.	Corylaceae
58a.	Ebenaceae	91.	Fagaceae
59.	Styracaceae	92.	Salicaceae
60.	Oleaceae	93.	Orchidaceae
61.	Apocynaceae	93a.	Musaceae
62.	Asclepiadaceae	94.	Iridaceae
62a.	Loganiaceae	95.	Amaryllidaceae
63.	Gentianaceae	95a.	Agavaceae
63.	Hydrophyllaceae	96.	Dioscoreaceae
64.	Boraginaceae	97.	Liliaceae
65.	Convolvulaceae	98.	Juncaceae
66.	Solanaceae	98a.	Palmae
67.	Scrophulariaceae	99.	Typhaceae
68.	Orobanchaceae	100.	Sparganiaceae
69.	Lentibulariaceae	101.	Araceae
69a.	Bignoniaceae	102.	Alismataceae
69b.	Pedaliaceae	103.	Juncaginaceae
69c.	Acanthaceae	104.	Potamogetonaceae
70.	Verbenaceae	105.	Ruppiaceae
71.	Labiatae	105a.	Najadaceae
72.	Plantaginaceae	106.	Posidoniaceae
72a.	Nyctaginaceae	107.	Zannichelliaceae
73.	Amaranthaceae	107a.	Zosteraceae
74.	Chenopodiaceae	108.	Cymodoceaceae
74a.	Basellaceae	109.	Cyperaceae
75.	Phytolaccaceae	110.	Gramineae
76.	Polygonaceae	111.	Pteridophyta
77.	Rafflesiaceae	111a.	Selaginellaceae
78.	Aristolochiaceae	111b.	Equisetaceae
79.	Lauraceae	111c.	Ophioglossaceae
79a.	Proteaceae	111d.	Pteridaceae
80.	Thymelaeaceae	111e.	Adiantaceae
81.	Elaeagnaceae	111f.	Dennstaedtiaceae
82.	Santalaceae	111g.	Aspidiaceae
83.	Euphorbiaceae	111h.	Aspleniaceae
84.	Urticaceae	111i.	Polypodiaceae

51. VALERIANACEAE

Annual or perennial herbs, or rarely shrubs, frequently malodorous; leaves opposite or whorled, occasionally all basal, exstipulate, entire or variously toothed or dissected; flowers in bracteate cymes, hermaphrodite or unisexual, generally zygomorphic; calyx-tube adnate to ovary, calyx-limb at first inconspicuous, frequently accrescent after anthesis and crowning the fruit; corolla infundibuliform, sometimes spurred or saccate at base, lobes 3–5, unequal, imbricate; stamens 1–4, inserted in the upper or lower part of the corolla-tube, alternate with the corolla-lobes, often exserted; filaments free; anthers versatile or basifixed, introrse, the 2 thecae dehiscing longitudinally; ovary with 1 fertile loculus and 2 (often very reduced) sterile, empty ones; ovule solitary, pendulous, anatropous; style simple, slender, stigma truncate or shortly lobed. Fruit dry, indehiscent; seed with or without endosperm; testa membranous; embryo straight.

Fifteen genera widely distributed in Europe, Asia, Africa and America.

Stamens 3; corolla-tube slightly gibbous near the base:
  Rhizomatous or tuberous perennials; calyx-teeth accrescent and plumose in fruit, forming a
    pappus - - - - - - - - - - - - - **1. Valeriana**
  Annuals; calyx-teeth not plumose, nor forming a pappus in the fruit - **3. Valerianella**
Stamen 1; corolla-tube distinctly gibbous or sometimes distinctly spurred near the base; calyx
  forming a plumose pappus in the fruit - - - - - - - **2. Centranthus**

1. VALERIANA *L.*

Sp. Plant., ed. 1, 31 (1753).
Gen. Plant., ed. 5, 19 (1754).

Rhizomatous or tuberous, commonly malodorous, perennial herbs; stems generally fistulose and unbranched; leaves simple or pinnate; inflorescence consisting of one or more, dense or lax, many-flowered cymes; calyx-teeth 5–15, involute at anthesis, spreading, accrescent and plumose in fruit; corolla narrowly infundibuliform, slightly gibbous at base, lobes (3–)5, unequal, spreading; stamens 3; stigma shortly 3-lobed. Fruit compressed, apparently 1-locular, the sterile loculi barely distinguishable, apex crowned with a conspicuous, plumose pappus.

About 200 species in Europe, Asia, Africa and America.

1. **V. italica** *Lam.*, Tabl. Encycl., 1: 92 (1791); Post, Fl. Pal., ed. 2, 1: 604 (1932); Osorio-Tafall et Seraphim, List Vasc. Plants Cyprus, 99 (1973).
   V. dioscoridis Sm. in Sibth. et Sm., Fl. Graec. Prodr., 1: 21 (1806), Fl. Graec., 1: 24, t. 33 (1806); Unger et Kotschy, Die Insel Cypern, 232 (1865); Boiss., Fl. Orient., 3: 90 (1875); Holmboe, Veg. Cypr., 174 (1914); Davis, Fl. Turkey, 4: 557 (1972).
   [*V. tuberosa* (non L.) Sm. in Sibth. et Sm., Fl. Graec. Prodr., 1: 21 (1806); Poech, Enum. Plant. Ins. Cypr., 15 (1842); Boiss., Fl. Orient., 3: 91 (1875) pro parte quoad plant. cypr.; Holmboe, Veg. Cypr., 174 (1914); Osorio-Tafall et Seraphim, List Vasc. Plants Cyprus, 99 (1973).]
   [*V. sisymbriifolia* (non Vahl) H. Stuart Thompson in Journ. Bot., 44: 309 (1906); Osorio-Tafall et Seraphim, List Vasc. Plants Cyprus, 99 (1973).]
  TYPE: "On le cultive depuis longtemps au jardin du Roi" (P, in herb. Jussieu et herb. Lam.). Erroneously thought to have come from Liguria. Perhaps one of the plants originally collected in Crete by Tournefort.

Erect perennial 30–60 (–100) cm. high; roots tuberous, fusiform, clustered, crowned with a fibrous collar of decayed petiole-bases; stems unbranched, glabrous, fistulose, obscurely striate, sometimes purplish and pruinose; leaves glabrous, bright or yellowish green, those of non-flowering, current year's growths simple, broadly ovate, 3–6 cm. long, 2·5–4 cm. wide, entire or obscurely undulate, apex obtuse or rounded, base abruptly cuneate; petiole slender, 6–10 (or more) cm. long; leaves of the flowering stems normally imparipinnate, occasionally simple or almost so, the lowermost 7–15 (–23) cm. long, 1–7 cm. wide, with (0–) 1–11 irregularly opposite or alternate, widely spaced leaflets; terminal leaflet broadly ovate, obovate or suborbicular, 1–5 cm. long, 0·8–4.5 cm. wide, entire or variously toothed or undulate, lateral leaflets usually much smaller than the terminal, asymmetrically ovate or oblong, 0·5–3·5 cm. wide; obtuse or acute, entire or variously toothed; petioles up to 15 cm. long; cauline leaves usually in 2–3 remote pairs, much smaller than the basal and more shortly petiolate, with narrow, often lanceolate or linear-lanceolate, entire or toothed leaflets, the uppermost very much reduced, the leaflets sometimes almost filiform; inflorescence a crowded terminal cyme 1·5–3·5 cm. wide, sometimes accompanied by 2 (rarely 4) remote, long-stalked, lateral cymes; bracts filiform, 6–8 mm. long, 1–2·5 mm. wide at base, margins sparsely glandular-ciliate; bracteoles similar to the bracts, but smaller and often proportionately wider; flowers sessile or subsessile, fragrant; calyx tubular, about 2·5 mm. long, 0·8 mm. wide, calyx-teeth strongly incurved and barely visible; corolla whitish, pale pink or purplish, about 5 mm. long, lobes 5, subequal, broadly and bluntly oblong, about 1·5 mm. long, 1·3 mm. wide; filaments inserted about half-way down the corolla-tube, subulate, terete, about 3·5–4 mm. long; anthers yellow, exserted, oblong, 1·2 mm. long, 0·8 mm. wide; style about 5 mm. long, stigma obscurely 3-lobed. Infructescence a rather lax, spreading, compound cyme; fruit narrowly and bluntly deltoid, about 5 mm. long, 2 mm. wide at base, pale brown, glabrous adaxially with a prominent median nerve, longitudinally 5-nerved abaxially and shortly pilose between the nerves; pappus-rays 12, about 4 mm. long, white-plumose; testa pale ochraceous, minutely and regularly papillose.

HAB.: In clefts of rock and in garigue on rocky, calcareous and non-calcareous slopes; sea-level to over 5,000 ft. alt.; fl. Febr.–May.

DISTR.: Divisions 1–3, 7, 8, locally common. Eastern Mediterranean region from S. Yugoslavia to Palestine.

1. Toxeftera near Ayios Yeoryios (Akamas), 1962, *Meikle* 2095!
2. Near Prodhromos, 1862, *Kotschy* 712! Philani, 1862, *Kotschy*; Khionistra S.W. side, 1937, *Kennedy* 907! Selladhi tou Mavrou Dasous above Spilia, 1962, *Meikle* 2816!
3. Stavrovouni, 1787, *Sibthorp*; Kato Lefkara, 1937, *Syngrassides* 1524! also, 1941, *Davis* 2729!, and, 1967, *Merton* in ARI 649!
7. Kantara, 1880, *Sintenis & Rigo* 92! Pentadaktylos, 1901, *A. G. & M. E. Lascelles* s.n.! Kyrenia, 1936, *Kennedy* 424! St. Hilarion, 1937, *Syngrassides* 1536! Kornos, 1941, *Davis* 2961! Larnaka tis Lapithou, 1941, *Davis* 2975! Vasilia, 1941, *Davis* 3027! Lefkoniko Pass, 1949, *Casey* 614!
8. Ayios Theodhoros, 1905, *Holmboe* 495; Koma tou Yialou, 1935, *Syngrassides* 703! and, 1950, *Chapman* 254! Apostolos Andreas, 1938, *Syngrassides* 1780! Ronnas Bay, 1941, *Davis* 2381! Platanisso, 1950, *Chapman* 204!

NOTES: The Cyprus material departs from the type, and approaches *V. sisymbriifolia* Vahl in its lower stature, compact, usually unbranched, inflorescences and relatively broad, entire or subentire leaflets. It agrees, however, with much of the material of the species from Turkey and the Aegean area, and cannot be regarded as distinct from *V. italica* at any rank.

2. CENTRANTHUS *Necker ex Lam. et DC.*

Fl. Franç., 4: 238 (1805).
Kentranthus Necker, Elem. Bot., 1: 122 (1790) nom. invalid.
I. B. K. Richardson in Bot. Journ. Linn. Soc., 71: 211–234 (1975).

Glabrous annual or perennial herbs; flowering stems usually erect and undivided; leaves simple or compound, sessile or petiolate; inflorescence generally a branched, elongate compound cyme; flowers hermaphrodite or unisexual; calyx adnate to ovary, teeth 5–25, involute at anthesis, spreading, accrescent and plumose in fruit; corolla ± zygmorphic, usually with 5 somewhat unequal lobes, tube gibbous, or spurred at the base with an internal longitudinal membrane; stamen 1; stigma entire, subclavate or shortly lobed. Fruit compressed, apparently 1-locular, the sterile loculi much reduced, apex crowned with a conspicuous, plumose pappus.

Nine species widespread in S. Europe and the Mediterranean region.

Perennial 30–80 (–100) cm. high; leaves ovate or lanceolate, entire or subentire, glaucous
1. **C. ruber**
Annual 6–30 (–50) cm. high; leaves usually toothed or lobed, not glaucous - 2. **C. calcitrapa**

SECT. 1. **Centranthus** Perennial herbs; leaves simple; corolla-tube slender, cylindrical, spurred at the base, with an internal membrane.

1. **C. ruber** (*L.*) *DC.* in Lam. et DC., Fl. Franç., 4: 239 (1805); Boiss., Fl. Orient., 3: 91 (1875); Post, Fl. Pal., ed. 2, 1: 604 (1932); Davis, Fl. Turkey, 4: 558 (1972); I. B. K. Richardson in Bot. Journ. Linn. Soc., 71: 220 (1975).
Valeriana rubra L., Sp. Plant., ed. 1, 31 (1753).

Erect rhizomatous perennial 30–80 (–100) cm. high with a branched rootstock and fleshy roots; stems with poorly developed branches, fistulose, glaucous, finely striatulate and usually somewhat thickened at the nodes; leaves ovate or broadly lanceolate 3–8 (–12) cm. long, 1–6 cm. wide, glaucous, rather fleshy, apex acute or obtuse, margins entire or dentate, base tapering to a short flattened petiole in the lower half of the stem, often rounded or broadly cuneate above and sessile or subsessile, the uppermost leaves sometimes with a wide, toothed amplexicaul base, and a long, acuminate apex; inflorescence elongate, consisting of 2 or more, approximate or distant pairs of crowded dichasia subtended by narrowly oblong or subulate, glabrous or sparsely pubescent bracts 2–10 mm. long, 1–3 mm. wide at base; flowers sessile or subsessile; calyx tubular, about 1·5 mm. long, 0·5 mm. wide, calyx-teeth involute and barely visible; corolla bright reddish pink or occasionally white, tube very slender, about 8–10 mm. long, produced basally into a slender pointed spur commonly 4–5 mm. long, corolla-lobes 5, spreading, subequal, bluntly oblong, about 2 mm. long, 1·3 mm. wide; style slender, about 12 mm. long; stigma subclavate, very obscurely lobed; filaments about 3 mm. long, inserted at the mouth of the corolla-tube; anthers oblong, yellow, about 2 mm. long, 1·3 mm. wide. Infructescence rather lax and spreading; fruits narrowly ovate, 3·5 mm. long, 1·5 mm. wide, dull brown, minutely pustulate, adaxial surface with an obscure median nerve, abaxial surface prominently 3-nerved, margins of fruit blunt, somewhat recurved; pappus-rays 15–18, about 4·5 mm. long, white-plumose; testa pale brown, minutely and regularly papillose.

var. **sibthorpii** (*Heldr. et Sart. ex Boiss.*) *Baldacci* in Nuov. Giorn. Bot. Ital., n.s., 5: 13 (1898).
Valeriana angustifolia Sm. in Sibth. et Sm., Fl. Graec. Prodr., 1: 20 (1806), Fl. Graec., 1: 22, t. 29 (1806) non Miller (1768) nom. illeg.
Centranthus sibthorpii Heldr. et Sart. ex Boiss., Diagn., 2, 2: 119 (1856).
C. ruber (L.) DC. ssp. *sibthorpii* (Heldr. et Sart. ex Boiss.) Hal., Consp. Fl. Graec., 1: 747 (1901); Hayek, Prodr. Fl. Pen. Balc., 2: 491 (1930); I. B. K. Richardson in Bot. Journ. Linn. Soc., 71: 221 (1975).
TYPE: Greece; "In montibus elatioribus circa Athenas". *Sibthorp* (OXF).

Leaves narrowly ovate or lanceolate, all entire, the uppermost less than 1 cm. wide, tapering to apex and base and not at all amplexicaul.

HAB.: Gardens and old walls; sea-level to 4,000 ft. alt.; fl. June–July.

DISTR.: Divisions 2, 7, almost certainly an escape from cultivation. S. and W. parts of Balkan Peninsula, possibly also Aegean area.

2. Trikoukkia Church, 1970, *A. Genneou* in ARI 1557! Platres, 1981, *Meikle*!
7. Kyrenia, 1955, *G. E. Atherton* 105!

NOTES: The distinctions between *C. ruber* var. *ruber* and *C. ruber* var. *sibthorpii*, though fairly obvious, are sometimes misleading, since lateral or secondary growths of the typical species can be very similar to normal growths of var. *sibthorpii*. It is possible that the Cyprus plant is simply such a depauperate state of *C. ruber* var. *ruber*, and the identification should be checked against material of normal, robust, primary stems.

SECT. 2. **Calcitrapa** *Lange* Annual herbs; leaves generally toothed or pinnatisect; corolla gibbous or shortly spurred.

2. **C. calcitrapa** (*L.*) *Dufr.*, Hist. Nat. Méd. Fam. Valér., 39 (1811); Poech, Enum. Plant. Ins. Cypr., 15 (1842); Unger et Kotschy, Die Insel Cypern, 231 (1865); Boiss., Fl. Orient., 3: 93 (1875); Holmboe, Veg. Cypr., 174 (1914); Post, Fl. Pal., ed. 2, 1: 605 (1932); Rechinger f. in Arkiv för Bot., ser. 2, 1: 431 (1950); Davis, Fl. Turkey, 4: 559 (1972); Osorio-Tafall et Seraphim, List Vasc. Plants Cyprus, 99 (1973); I. B. K. Richardson in Bot. Journ. Linn. Soc., 71: 231 (1975) — as "*C. calcitrapae*".
    Valeriana calcitrapa L., Sp. Plant., ed. 1, 31 (1753) — as *V.* "*calcitrapæ*", Fl. Monsp., 8 (1756) — as *V.* "calcitrapa", Sp. Plant., ed. 2, 1: 44 (1762) — as *V.* "*calcitrapa*"; Sibth. et Sm., Fl. Graec. Prodr., 1: 20 (1806), Fl. Graec., 1: 22, t. 30 (1806).

Erect, commonly unbranched, glabrous annual 6–30 (–50) cm. high; stems terete, fistulose, striatulate, frequently tinged purple; basal leaves obovate, spathulate or suborbicular, 0·5–3·5 cm. long, 0·5–3 cm. wide, pale green or tinged purple, obscurely nerved, apex rounded or obtuse, margins subentire, coarsely and bluntly serrate or pinnatisect, base rounded or broadly cuneate; petiole flattened, up to 3 cm. long; cauline leaves in 2–6 remote pairs, the lowermost resembling the basal, the upper sessile or subsessile, often narrow, oblong, acute or subacute, coarsely toothed or deeply pinnatisect; inflorescence a compact or lax, branched or un-branched, corymbose cyme; bracts persistent, narrowly oblong or linear, 2–3 mm. long, 0·8–1 mm. wide, keeled, sparsely glandular, apex acute or acuminate; flowers sessile or subsessile, calyx about 2 mm. long, 1 mm. wide, strongly compressed with the involute calyx-teeth barely visible; corolla pink, narrowly infundibuliform, about 2 mm. long, 1·5 mm. wide, tube gibbous or shortly spurred near the base, lobes 5, rather unequal, spreading, oblong, about 0·7 mm. long, 0·5 mm. wide, apex rounded or shallowly emarginate; filaments about 0·3 mm. long, inserted at mouth of corolla-tube; anthers broadly oblong, about 0·4 mm. long, 0·3 mm. wide; style about 2 mm. long; stigma distinctly 3-lobed. Infructescence a lax, spreading, or rather crowded and much-branched dichasium; fruits caducous, narrowly ovate, strongly dorsiventrally compressed, about 2·5 mm. long, 1·5 mm. wide, dull brown, glabrous (or rarely pilose) often thinly pustulate; adaxial surface with an obscure median nerve, abaxial surface prominently 3-nerved, with an incurved, minutely papillose rim; pappus-rays about 14, white-plumose, 2·5–3 mm. long, united into a membranous basal cup about 0·8 mm. long; testa pallid, minutely and regularly papillose; embryo becoming very oily when scraped or crushed.

ssp. **orbiculatus** (*Sm.*) *Meikle* **stat. nov.**
    Valeriana orbiculata Sm. in Sibth. et Sm., Fl. Graec. Prodr., 1: 21 (1806), Fl. Graec., 1: 23 (as *V. orbiculata*), t. 31 (as *V. rotundifolia*) (1806).
    Centranthus orbiculatus (Sm.) Dufr., Hist. Nat. Méd. Fam. Valér., 39 (1811).
    C. calcitrapa (L.) Dufr. var. *orbiculatus* (Sm.) DC., Prodr., 4 (2): 632 (1830); Unger et Kotschy, Die Insel Cypern, 231 (1865); Boiss., Fl. Orient., 3: 93 (1875); Holmboe, Veg. Cypr., 174 (1914).
    TYPE: Cyprus; "In monte Crucis [Stavrovouni] insulae Cypri, *D. F. Bauer.*"

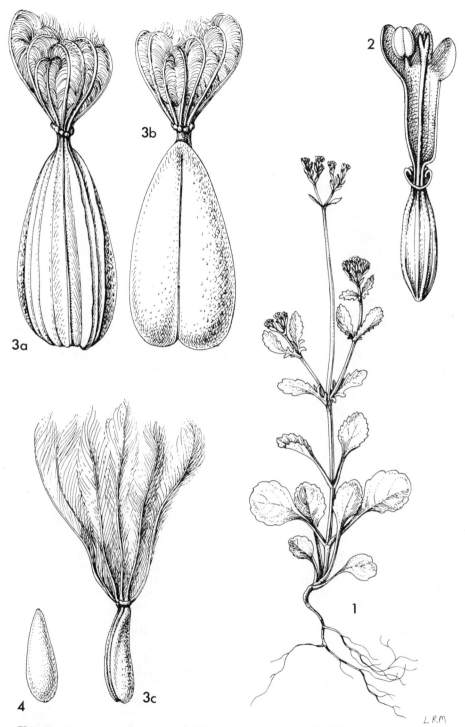

Plate 53. CENTRANTHUS CALCITRAPA (L.) Dufr. ssp. ORBICULATUS (Sm.) Meikle **1,** habit, × 1; **2,** section of flower, × 30; **3a, b, c,** fruit, × 30; **4,** seed, × 12. (**1, 3, 4** from *Syngrassides* 1572, **2** from *Davis* 3109.)

Plant slender, usually less than 30 cm. high; basal leaves spathulate or suborbicular, subentire or bluntly toothed; cauline leaves usually in 2–3 (–4) remote pairs, subentire, toothed or lobed, but not pinnatisèct; corolla-tube gibbous near the base but not spurred. Infructescence a lax, sparingly branched dichasium. *Plate 53*.

HAB.: Rocky slopes and moist screes, sometimes under Pines on calcareous or non-calcareous formations; 300–5,500 ft. alt.; fl. March–June.

DISTR.: Divisions 2, 3, 7, locally abundant. Type species widespread in the Mediterranean region and Atlantic Islands, ssp. *orbiculatus* perhaps endemic in Cyprus.

2. Near Phini, 1862, *Kotschy* 735! 3 m. above Lefka, 1927, *Rev. A. Huddle* 65! 66! Kykko Monastery, 1937, *Syngrassides* 1572! Platres, 1938, *Kennedy* 908! Papoutsa, 1941, *Davis* 3109! Above Palekhori, 1957, *Merton* 2997! Ayia Irini near Platres, 1960, *N. Macdonald* 141! Prodhromos, 1961, *D. P. Young* 7277! 7370! Ayia valley, Paphos Forest, 1962, *Meikle* 2658!

3. Stavrovouni, 1787, *F. L. Bauer sec. Sibthorp et Kotschy*, also, 1862, *Kotschy* 200! 5 m. W. of Malounda, 1963, *J. B. Suart* 57!

7. Antiphonitis Monastery, 1787, *Sibthorp*; Bellapais, 1880, *Sintenis & Rigo* 517! Armenian Monastery, 1912, *Haradjian* 393; Yaïla, 1941, *Davis* 2845! ¾ mile E. of St. Hilarion, 1949, *Casey* 373! Trypa Vouno, 1956, *G. E. Atherton* 1386!

NOTES: Frequently dismissed as a minor variant of the polymorphic *C. calcitrapa*, but significantly uniform in Cyprus, where the typical species has yet to be collected.

[FEDIA CORNUCOPIAE (*L.*) *Gaertn.*, Fruct. Sem. Plant., 2: 37 (1790); *Valeriana cornucopiae* L., Sp. Plant., ed. 1, 31 (1753), an erect, usually branched annual, not unlike *Centranthus calcitrapa* (L.) Dufr., but with the inflorescence branches conspicuously thickened and inflated in fruit, the flowers with a 2–4-dentate calyx, a strongly zygomorphic, 2-lipped corolla, and 2 stamens. Listed by Osorio-Tafall & Seraphim (List Vasc. Plants Cyprus, 99; 1973). The source of this record has not been traced, but as *C. calcitrapa*, and some *Valerianella* species with thickened infructescence branches, have been misidentified as *Fedia*, there is reason to suspect an error. *Fedia cornucopiae* is locally frequent in the western Mediterranean region eastward to Greece and Crete; it is apparently absent from Turkey, Syria and Palestine, and is an unlikely occurrence in Cyprus, even if the record cannot be positively rejected.]

3. VALERIANELLA *Moench* *
Meth., 493 (1794).

Erect annual herbs, with or without opposite branches, dichotomously forked above, often with some retrorse indumentum below; basal leaves forming a rosette, cauline leaves opposite, diminishing upwards; bracts often leaf-like; inflorescence terminal, pedunculate, always condensed, capitate or with some ± scorpioidal cymose branching just detectable, bracteolate; flowers sessile; calyx various, usually developing in fruit; corolla blue to pink to whitish; stamens 3; ovary with 2 clearly developed sterile loculi, visible in fruit, as well as the seed-bearing loculus. Fruit hairy or glabrous, of various forms with calyx 3–12-coronate or represented by hooks or horns or a "fan" of greenish reticulate-veined tissue atop the fertile loculus, or calyx much reduced or even absent.

A genus of about 50 species, one group centred in the Mediterranean and Middle East, extending to N.W. Europe, Central Asia, Pakistan and Kenya; the other group North American. The Old World species all seem to produce

* by M. J. E. Coode

a strong and characteristic stale odour (attracting cats) when dried; the New World species seem to be odourless.

The 10 species found in Cyprus are all very similar vegetatively and belong to 4 sections, as currently accepted: spp. 1–4 belong to sect. *Coronatae* Boiss. (calyx inflated or coronate, no blocks of spongy tissue present); 5–6 to sect. *Platycoele* DC. (calyx absent or reduced to a tooth or rim; no blocks of spongy tissue present); 7–9 to sect. *Cornigerae* Soy.-Willem. emend. Krok (calyx represented by 3 horns; blocks or ridges of spongy tissue present) and 10 to sect. *Siphonocoele* Soy.-Willem. (calyx a low rim or oblique tube or fan; blocks of spongy tissue absent). For a fuller understanding of the fruit structure see Coode in Notes Roy. Bot. Gard. Edinb., 27:219–256 (1967), where details are discussed which seem inappropriate here. All the Cyprus species are included in this 1967 paper.

Fruits all of one kind, surmounted by a ± symmetrical coronate or inflated persistent calyx, or calyx represented by an oblique tube or fan or simple, flattened, greenish tooth or rim; or calyx ± absent; spongy tissue absent; inflorescence branches not inflated in fruit (except occasionally in *V. muricata*); bracteoles various:

Fruiting calyx an inflated bladder-like vesicle with about 6 inflexed teeth around the ostiole, developing very early and detectable in flower; bracteoles very broad and very membranous and white-ciliate - - - - - - - - - **4. V. vesicaria**

Fruiting calyx otherwise; bracteoles narrower, ciliate or glabrous:

Fruiting calyx ± coronate; ovary in fruit densely hairy, squarish-wedge-shaped, tapering to base in face view (i.e. with fertile loculus at the back); bracteoles obviously ciliate and with wide membranous margins:

Fruit not more than 2 mm. long overall, lobes of fruiting calyx weakly developed, not hooked, sometimes only ± 3 - - - - - - - **3. V. lasiocarpa**

Fruit larger, lobes of fruiting calyx strongly developed, usually terminated by a smooth indurate hooked point:

Interior of calyx-cup glabrous, lobes 6; these distinctions generally visible in flower
2. V. coronata

Interior of calyx-cup pubescent, lobes up to 12, of different widths, sometimes only 6 but then with subsidiary hooks - - - - - - **1. V. discoidea**

Fruiting calyx ± absent or reduced to a rim or fan, best developed on the fertile loculus, rarely an oblique tube; ovary in fruit glabrous or weakly hairy, rarely obviously hairy, and, if so, then ovary ovoid; bracteoles mostly green and ± glabrous, except for *V. pumila* with ciliate bracteoles:

Fruits longer than broad, incurved, with a distinct channel between the sterile loculi; fruiting calyx absent, fertile loculus projecting slightly beyond the sterile
5. V. carinata

Fruits about as long as broad or only slightly longer; fruiting calyx present as a distinct tooth or fan or rarely an oblique tube:

Fruits almost globose, glabrous, with two large sterile loculi and calyx reduced to a tooth on the fertile loculus, sometimes spreading forwards to the sterile loculi as an irregular rim; bracteoles ciliate - - - - - - - **6. V. pumila**

Fruits ovoid, usually hairy, the two sterile loculi reduced to curving (often pale) ridges on the anterior surface, surmounted by a fan or oblique tube with venation clearly visible; bracteoles glabrous - - - - - - - **10. V. muricata**

Fruits usually of two kinds, those of the lowermost forking of the inflorescence much narrower (probably seedless) than the rest, surmounted by 3 short or long horns or projections, and with conspicuous blocks or ridges of spongy tissue; branches of inflorescence in fruit usually inflating; bracteoles narrow and mostly green:

Fruits with one or more long horns, visible early in flower:

Fruits asymmetric, with one long and two short horns; bracteole margins ± entire
7. V. echinata

Fruits ± symmetrical with 3 long horns; bracteole margins clearly repand-dentate with a few rather glandular teeth - - - - - - - - **8. V. triceras**

Fruits with 3 short horns - - - - - - - - - **9. V. orientalis**

1. **V. discoidea** (*L.*) *Lois.*, Not. Plantes Fl. France, 148 (1810); Boiss., Fl. Orient., 3: 111 (1875); Holmboe, Veg. Cypr., 173–4 (1914); Davis, Fl. Turkey, 4: 577 (1972); Osorio-Tafall et Seraphim, List Vasc. Plants Cyprus, 99 (1973).

    Valeriana locusta L. var. *discoidea* L., Syst. Nat., ed. 10, 2: 860 (1759).

    [*Valerianella chlorostephana* (non Coss. et Dur.) H. Stuart Thompson in Journ. Bot., 44: 309 (1906).]

    TYPE: None designated. There is a specimen in LINN 48.17 (*Valeriana locusta* 14δ)!

Annual herb, erect, 3–27 cm. tall, single-stemmed to richly branched from the base, generally with short, simple, fairly stout and often reflexed hairs below, glabrous above; basal leaves very narrowly oblanceolate, (2–) 4–7 cm. long, 5–10 mm. wide, gradually attenuate at base, entire to sinuous to very distantly ± repand-dentate or -serrate; lower stem leaves similar but smaller and more oblong than oblanceolate, with an abruptly narrowed base; upper stem leaves smaller and narrower, usually with 1–3 pairs of linear laciniae or teeth near the base; inflorescence a subglobose head, often 2–3 aggregated together, internodes in the head ± nil; bracts like the uppermost leaves; uppermost node hairy (like the base of the plant); outermost bracteoles green, oblong, 3–5 mm. long, c. 1 mm. wide, inner bracteoles much more membranous, .ovate, c. 2 mm. long, 1 mm. wide, transparent with a midrib and a thicker acuminate green tip and relatively long soft white hairs particularly on the margins; innermost bracteoles even smaller and more delicate; corolla pale mauve to bluish or white. Fruit: ovary densely white-hairy, squarish in section and side-view, wedge-shaped in face view, 1·5–2 mm. long; calyx coronate, c. 5 mm. diam., lobes ± 6 with subsidiary or additional lobes, ovate, acuminate and strongly hooked at apex, subsidiary hooks on lobe margins sometimes present, ± glabrous outside and sometimes on part of the inside, pubescent on margins and ± densely so in the calyx cup, with some strong reticulating venation.

HAB.: Fields, pastures, banks, open stony places, olive groves, coastal maquis, eroded slopes; usually at low altitudes from sea-level to 600 (–1,800) ft. alt.; fl. Febr.–March, rarely April.

DISTR.: Divisions 1, 3, 4, 6–8, apparently commoner in the north of the island; N. Africa, S. Europe to Turkey, W. Syria and N. Iran.

1. Avgas R. near Ayios Yeoryios (Akamas), 1962, *Meikle* 2011! Lachi to Neokhorio, west of Polis, 1978, *Chesterman* 48!
3. 11 miles S.W. of Larnaca between Mazotos and Alaminos, 1978, *Chesterman* 43 pro parte!
4. Ayia Napa, 1905, *Holmboe* 50; Famagusta, 1957, *Merton* 2891! Sotira, Paralimni and Ayia Napa, 1979, *Chesterman* 132!
6. Ayia Irini (Morphou), 1941, *Davis* 2557! W. of Orga, 1956, *Poore* in *Merton* 2608! 48½ miles between Xeros and Polis, *Harris* 3104 pro parte! W. of Panagra Gorge, 3 m. W. of Orga, 1967, *Merton* in ARI 106!
7. Kythrea, *Sintenis & Rigo* 96 pro parte! also, 1950, *Chapman* 86! and, 1967, *Merton* in ARI 332! St. Hilarion, 1902, *A. G. & M. E. Lascelles* s.n.! Myrtou, 1936, *Syngrassides* 914! Tjiklos Cliff, Kyrenia, 1949, *Casey* 523! Above Kyrenia Pass, 1953, *Casey* 1301! Kyrenia, 1956, *G. E. Atherton* 1105!
8. Apostolos Andreas, 1941, *Davis* 2318! Kantara, 1941, *Davis* 2462!

NOTES: It is curious that on the adjacent mainland in Turkey, *V. coronata* is the commonest *Valerianella*, but on Cyprus it seems that *V. discoidea* and *V. vesicaria* are far more frequent than *V. coronata*.

2. **V. coronata** (*L.*) *DC.* in Lam. et DC., Fl. Franç., 4: 421 (1805); Boiss., Fl. Orient., 3: 110 (1875); Sintenis in Oesterr. Bot. Zeitschr., 32: 51, 193 (1882); Post, Fl. Pal., ed. 2, 1: 609 (1932); Davis, Fl. Turkey, 4: 576 (1972); Feinbrun, Fl. Palaest., 3: 261, t. 447 (1978).
　　Valeriana locusta L. var. *coronata* L., Sp. Plant., ed. 1, 34 (1753).
TYPE: Portugal: "*in arvis* Lusitaniae" Hort. Cliff., 16.

Very similar to *V. discoidea* in habit, inflorescence and bracts but inflorescence in well-grown specimens often completely globose, and the branching laxer, with short internodes between the flowers (and abortive flower-buds at the tips of these very short branches, with very small bracteoles); bracteoles webbed down the branchlets at base; outermost bracteoles similar; inner bracteoles much wider, mostly hyaline but with a central band rather thicker and greener, sometimes with a distinct midrib, margins white-ciliate and "midribs" ± sparsely white-hairy outside; innermost bracteoles narrower than inner, otherwise similar; corolla rosy-lilac. Fruit detaching separately; ovary c. 2 mm. long, densely pilose, squarish and ribbed in side-view, tapering in front view; calyx coronate, 5–6 mm. diam., the whole glabrous or sparsely hairy on margins and outside only, with pronounced reticulate venation, usually rather regularly 6-lobed, each lobe triangular-ovate, acuminate, terminated by an indurate hook.

HAB.: Stony slopes, bare soil, [or as a weed in adjacent cultivated ground]; from low levels to 4,000 ft. alt.; fl. March–May.

DISTR.: Divisions 2, 6 and the borders of 5 and 7; widespread throughout S. Europe, N. Africa, Turkey to Iraq, Iran and Central Asia to Tien Shan, Caucasus, Crimea.

2. Phini, 1880, *Sintenis & Rigo* 1009 pro parte! Papoutsa, 1962, *Meikle* 2935!
6. Nicosia, 1973, *P. Laukkonen* 168!
7. Above Kythrea, 1880, *Sintenis & Rigo* 96 pro parte!

3. V. lasiocarpa (*Stev.*) *Betcke*, Animadv. Bot. Valer., 26 (1826); Boiss., Fl. Orient., 3: 108 (1875); Holmboe, Veg. Cypr., 173 (1914); Davis, Fl. Turkey, 4: 580 (1972); Osorio-Tafall et Seraphim, List Vasc. Plants Cyprus, 99 (1973).
Fedia lasiocarpa Stev. in Mém. Soc. Nat. Mosc., 5: 350 (1817).
TYPE: U.S.S.R.; "In Tauria et Caucaso orientali rarior", *Steven* (LE).

Like *V. discoidea* but often smaller; inflorescences slightly laxer, so that branching pattern can be observed; bracts 2–2·5 mm. long, 0·5 mm. wide, green with a narrow hyaline ± ciliate margin and expanded, membranous base; outermost bracteoles ovate, c. 2·5 mm. long, 1 mm. wide, membranous with a green tip and greenish central band and midrib, margins ciliate; inner bracteoles similar but slightly smaller and sometimes obovate, webbed at base down the inflorescence branches; colour of corolla not indicated, but probably pink rather than blue. Fruit: ovary not more than 1·5 mm. long, 1 mm. wide, ± oblong or sometimes slightly ovate, densely pilose, the hairs white, ± spreading and often slightly incurved; calyx remaining green, weakly coronate with scarcely developed lobes, sometimes reduced almost to a mere sinuous rim, not hooked in Cyprus (see Notes Roy. Bot. Gard. Edinb. 27: 244; 1967), the whole fruit not more than 2 mm. long.

HAB.: Weed in old fallow; 500–1,000 ft. alt.; fl. c. March.

DISTR.: Divisions 5, 6. South and central Russia, Crimea, Turkey, Iraq, Iran, Rumania, Bulgaria — perhaps introduced in Cyprus?

5. Anayia, 1950, *Chapman* 321!
6. Nicosia, 1905, *Holmboe* 308.

NOTE: Description based partly on *Sintenis* 350 and *Davis* D41656 from Turkey.

4. V. vesicaria (*L.*) *Moench*, Meth., 493 (1794); Unger et Kotschy, Die Insel Cypern, 231 (1865); Boiss., Fl. Orient., 3: 112 (1875); Sintenis in Oesterr. Bot. Zeitschr., 31: 288–9, 393 (1881); 32: 51, 259 (1882); Holmboe, Veg. Cypr., 174 (1914); Post, Fl. Pal., ed. 2, 1: 610 (1932); Davis, Fl. Turkey, 4: 578 (1972); Osorio-Tafall et Seraphim, List Vasc. Plants Cyprus, 99 (1973); Feinbrun, Fl. Palaest., 3: 262, t. 449 (1978).
Valeriana locusta L. var. *vesicaria* L., Sp. Plant., ed. 1, 33 (1753).
TYPE: "*in* Creta & Halepo" (Hort. Cliff., BM?).

Very similar to *V. discoidea* in habit, inflorescence and bracts, except that the leaves are often slightly more toothed; outermost bracteoles green, oblong, 2–3 mm. long, 0·5 mm. wide, whitish-ciliate, and with a few hairs on the midrib; inner bracteoles ± ovate, c. 3 mm. long, 1–2 mm. wide, green at tip and often down the middle band, the rest hyaline-ciliate; innermost bracteoles broad-ovate to subcircular, c. 4 mm. long and almost as wide, entirely hyaline and very delicate except for a greenish patch at the tip, white-ciliate; corolla purple, pale purple or lavender. Fruits often detaching in pairs with bracteoles attached, relatively firmly joined down the fertile carpels by the interlocked, dense, white, lanate indumentum; ovary c. 2 mm. long, wedge-shaped or obconical; calyx ± coronate in flower with 6 triangular lobes each ending in an indurate tip, quickly inflating by intercalary growth in the fused bases of the lobes to become a vesicle-like, hollow sphere 4–5 mm. diam., with the lobe-tips inflexed around an ostiole, outer surface pubescent and reticulate-veined.

HAB.: Pathsides, edges of cultivation, fields, rocky waste places, sandy soil under Olives, grassy slopes; low altitudes, from sea-level to 1,000 (–3,800) ft. alt.; fl. February–April.

DISTR.: Divisions 1–8. South Europe from France eastwards to Turkey; Syria, Palestine, Iraq, Iran; Central Asia; Algeria.

1. Limni Mine, 1957, *Merton* 3022!
2. Platres, 1938, *Kennedy* 1431!
3. Yermasoyia River, 1948, *Mavromoustakis* s.n.! Near Souni, 1960, *N. Macdonald* 152! 11 miles S.W. of Larnaca between Mazotos and Alaminos, 1978, *Chesterman* 43 pro parte!
4. Phaneromene near Larnaca, 1905, *Holmboe* 152; Dhekelia, 1981, *Hewer* 4716!
5. Athalassa, 1933, *Syngrassides* 9! 5 miles from Nicosia towards Famagusta, 1950, *Chapman* 47! Above Dheftera, Klirou, 1952, *F. M. Probyn* 66!
6. Nicosia, 1905, *Holmboe* 307; 48½ miles between Xeros and Polis, *Harris* 3104 pro parte! Morphou to Myrtou, (undated), *A. Genneou* in ARI 1405! Mesaoria, near Kykko Metokhi, 1952, *F. M. Probyn* 19!
7. Ayios Khrysostomos Monastery, 1862, *Kotschy* 435; Ardhana, 1880, *Sintenis & Rigo* 94! Kyrenia, *Syngrassides* 275! Masari, 1968, *Economides* in ARI 1076 pro parte! St. Hilarion, 1940, *Mavromoustakis* 139! also, 1949, *Casey* 326! Ayios Epiktitos, *N. Macdonald* 68!
8. Ayios Philon near Rizokarpaso, 1941, *Davis* 2255! Ayios Dhimitrios, 1956, *G. E. Atherton* 1179!

NOTES: With *V. discoidea*, evidently the commonest *Valerianella* in Cyprus. Identifiable even in flower by the very broad, very delicate ciliate bracteoles and calyx showing signs of inflation very early.

[V. LOCUSTA (*L.*) *Laterrade*, Fl. Bordelaise, ed. 2, 93 (1821); Osorio-Tafall et Seraphim, List. Vasc. Plants Cyprus, 99 (1973).
    Valeriana locusta L., Sp. Plant., ed. 1, 33 (1753) with var. *olitoria* L.; Sibth. et Sm., Fl. Graec. Prodr., 1: 22 (1806).
    Valerianella olitoria (L.) Poll., Hist. Fl. Palat., 1: 30 (1776); Holmboe, Veg. Cypr., 173 (1914).

This species is not represented among any Cyprus collections so far seen, although it would not be surprising if it were found, since it occurs on the adjacent mainland in Turkey and is also the species most commonly used as a salad vegetable. It differs from *V. carinata* in having a straight slightly shorter fruit, laterally flattened, with a conspicuous block of spongy tissue on the fertile loculus. All literature references to this species appear to be based on the original Sibthorp and Smith report.]

5. **V. carinata** *Lois.*, Not. Plantes Fl. France, 149 (1810); Boiss., Fl. Orient., 3: 106 (1875); Post, Fl. Pal., ed. 2, 1: 608 (1932); Davis, Fl. Turkey, 4: 571 (1972); Feinbrun, Fl. Palaest., 3: 260, t. 444 (1978).
    TYPE: France; "dans les champs aux environs de Paris", *Loiseleur* (P).

Differs from *V. discoidea* in having less divided leaves, a laxer inflorescence in which the branching is detectable, and particularly in the much more foliaceous bracts and bracteoles, all narrowly ± oblong-obovate and obtuse, non-membranous and green (except for the bases of innermost bracts) and glabrous; corolla pale blue. Fruit: ovary oblong, incurved, the fertile loculus slightly wider and distinctly longer than the sterile loculi, a deep groove present between the sterile loculi, all 3 loculi approximately similar in cross-section, glabrous or very sparsely short-hairy at most; calyx absent.

HAB.: Roadsides, stony mountainsides; 4,000 ft. alt.; fl. May.

DISTR.: Division 2; Europe and North Africa to Transcaucasia, Iran and Iraq.

2. Selladhi tou Mavrou Dasous, above Spilia, 1962, *Meikle* 2823d! Below Madhari (Yironas River), 1962, *Meikle* 2827!

6. **V. pumila** (*Willd.*) *DC.* in Lam. et DC., Fl. Franç., 4: 242 (1805); Davis, Fl. Turkey, 4: 572 (1972); Feinbrun, Fl. Palaest., 3: 260, t. 445 (1978).
    Valeriana pumila Willd., Sp. Plant., 1: 184 (1797).
    Valeriana locusta L. η *pumila* L., Syst. Nat., ed. 12, 2: 73 (1767), nom. illegit., superfl.
    (≡ *Valeriana locusta* L. δ *multifida* Gouan, Hort. Reg. Monsp., 23; 1762).
    Fedia tridentata Steven in Mém. Soc. Imp. Nat. Moscou, 2: 178 (1809), 5: 346 (1817).

Valerianella tridentata (Stev.) Reichb., Fl. Germ. Excurs., 198 (1831); Krok in K. Svensk. Vet.-Akad. Handb., n.s., 5: 73 (1864); Boiss., Fl. Orient., 3: 109 (1875); Post, Fl. Pal., ed. 2, 1: 608 (1932).
TYPE: France; Montpellier.

Similar to *V. discoidea* but differing in the slightly laxer inflorescence, showing the branching; bracteoles narrower throughout, only the inner bracteoles ovate, webbed at base; corolla white, pinkish or lilac. Fruit c. 1·5 mm. long, 1·5 mm. wide, glabrous, ovary ± hemispherical or flattened-globose, the sterile loculi obvious and as large as or larger than the fertile, contiguous at base and apex but with a depression or deeper gap between them in the middle, slightly shorter than the fertile; calyx reduced (in Cyprus) to a greenish tooth on the tip of the fertile loculus with an obscure rim forwards on to the top of the sterile loculi.

HAB.: Fields; 500–c. 3,000 ft. alt.; fl. March–May.

DISTR.: Divisions 2 and 6; S. Europe, E. Mediterranean, Middle East.

2. Phini, Troödos, 1880, *Sintenis & Rigo* 1009 pro parte!
6. Masari, 1968, *Economides* in ARI 1076 pro parte!

NOTES: Only known from 2 specimens; a common species in Turkey and elsewhere, showing great variation in size of fruit and in size and division of calyx.

7. V. echinata (*L.*) *DC.* in Lam. et DC., Fl. Franç., 4: 242 (1805); Unger et Kotschy, Die Insel Cypern, 231 (1865); Boiss., Fl. Orient., 3: 102 (1875); Sintenis in Oesterr. Bot. Zeitschr., 31: 393 (1881); 32: 51, 193 (1882); Holmboe, Veg. Cypr., 173 (1914); Post, Fl. Pal., ed. 2, 1: 607 (1932); Davis, Fl. Turkey, 4: 567 (1972); Osorio-Tafall et Seraphim, List Vasc. Plants Cyprus, 99 (1973); Feinbrun, Fl. Palaest., 3: 259, t. 441 (1978).
Valeriana echinata L., Syst. Nat., ed. 10, 2: 861 (1759).
TYPE: France; Montpellier, "*a D. Sauvages*". A specimen in Herb. LINN — 48.19!

Differs vegetatively from *V. discoidea* in having the hairs on the lower parts reduced to scarcely more than papillae in a few lines on the stem and on leaf margins near the base, and in having the uppermost leaves divided at base into a few triangular, though sometimes linear, teeth; lowermost branch-forks of the inflorescence usually held rather close together, and with a flower between them; in fruit these branches and often the upper part of the stem inflate perceptibly; above the lowest inflorescence-fork the inflorescences rather few-flowered but branching very condensed, all flowers ± erect (so head not globular); bracts oblong, green, c. 5 mm. long, 1 mm. wide, margins ± entire and not membranous; bracteoles similar but getting smaller towards the innermost; corolla lavender or bluish-mauve, calyx perceptibly 3-horned (one horn longer) even in bud. Fruits of two kinds; those in lowermost fork and often those in central fork of the heads thinner and apparently longer, probably abortive, the long horn not outcurving and the total length of ovary (including horn) c. 6 mm.; normal fruits falcate, asymmetric, glabrous; ovary up to 4 mm. long and long horn 2–3 mm. long, usually with blocks of spongy tissue on the outside, sometimes apparently ± smooth especially when not fully ripe; calyx represented by 3 horns, the longest much thicker and bigger than the others and terminating one of the sterile carpels. Finally the inflorescence branches disarticulate with the fruits tending to be persistent in the head.

HAB.: Rocky places, shady places under rocks; gregarious; 1,100–2,700 ft. alt.; fl. March–April.

DISTR.: Seen from Division 7 only, but Kotschy records the species from 2 and 3; throughout the Mediterranean region and eastwards to Iraq, N. Iran, Crimea.

2. Lemithou near Prodhromos, 1862, *Kotschy* 873!
3. Stavrovouni [Santa Croce], *Kotschy.*
7. Ayios Khrysostomos Monastery, 1862, *Kotschy* 437; Buffavento Castle, 1880, *Sintenis & Rigo* 93! Yaïla, 1941, *Davis* 2841! Kyrenia, 1949, *Casey* 641!

NOTES: There is insufficient Cyprus material at Kew to know whether the full range of variation of spongy tissue-blocks and inflation of the upper branches found in Turkey also occurs in Cyprus. With both spongy tissue and hooks it seems that the plant is able to utilise both water and animals for dispersal.

8. V. triceras *Bornm.* in Mitt. Thür. Bot. Verein., 34: 66 (1908); Meikle in Hooker's Icon. Plant., 7: t. 3654 (1969); Davis, Fl. Turkey, 4: 568 (1972); Osorio-Tafall et Seraphim, List Vasc. Plants Cyprus, 99 (1973).

TYPE: Turkey: "in rupestribus cacuminis montis Dyo Adelphia" (Iki Kardaş), 8–900 m. 15 May 1906, *Bornmueller* 9623 (holotype B, isotype K! Herb. Hub.-Mor.).

Like *V. discoidea* vegetatively; the inflorescence, however, much fewer-flowered (7–10 flowers) and usually with a flower in the lowermost fork; bracts narrow-oblong, delicate but probably green, c. 3 mm. long, 0·5 mm. wide, ± entire; outer bracteoles similar but with distinct, repand-dentate, membranous margins, each of the teeth ending in a gland-like papilla, inner and innermost bracteoles similar but smaller; corolla lilac-blue. Fruits of two kinds: those of the lower fork thinner than the rest, probably infertile, ovary 4 mm. long, horns 1·5 mm. long; normal fruits: ovary 2·5–3 mm. long, ornamented with longitudinal papillose blocks or ridges of spongy tissue, calyx represented by 3 equal spreading uncinate horns 2–2·5 mm. long. Fruits packed densely (not easy to see inflorescence branching), perhaps persistent and like *V. echinata* in dispersal.

HAB.: Among *Quercus alnifolia* on igneous, stony, north-facing slopes; 4,000–4,200 ft. alt.; fl. April–May.

DISTR.: Division 2; recently this species has been found in S. and W. Turkey and on Samos, always at 2,700 ft. or above.

2. Papoutsa, 1941, *Davis* 3110! and, 1962, *Meikle* 2932!

9. V. orientalis *(Schlechtd.) Boiss. et Bal.* in Boiss., Diagn., 2, 2: 120 (1856); Boiss., Fl. Orient., 3: 103 (1875); Post, Fl. Pal., ed. 2, 1: 607 (1932); Davis, Fl. Turkey, 4: 568 (1972); Feinbrun, Fl. Palaest., 3: 259, t. 442 (1978).

Fedia orientalis Schlechtd. in Linnaea, 17: 126 (1843).

TYPE: Syria: "in lapidosis collinum pr. Aleppo", 19 April, *Kotschy* 116 (isotypes E! BM!).

Very similar to *V. echinata* in all respects, except in the laxer inflorescence, and the form of the fruit, and the ease with which the ripe fruits detach, leaving the inflated inflorescence branches and bracteoles visible. Fruits: thinner, sterile fruits of lower forks with ovary 3 mm. long plus horns less than 1 mm., sometimes the whole fruit reduced even further; normal fruits symmetrical, ovoid-globose to semiglobose, c. 2·5 mm. long, 2·5 mm. wide, with big thick blocks or ridges of spongy tissue arranged longitudinally; calyx represented by 3 equal, short, outcurved horns, c. 1 mm. long.

HAB.: Rocky igneous (or rarely calcareous) slopes, streamsides in grassy turf, rocky roadside banks, among summit rocks; 1,200–4,650 ft. alt.; fl. April–May.

DISTR.: Divisions 2, 3, 7. Probably not as weedy as most other species of *Valerianella*; S. & W. Turkey, Syria, Palestine.

2. Polystipos 1957, *Merton* 3116! Stavros tis Psokas, 1962, *Meikle* 2749! Kionia, summit rocks, 1967, *Merton* in ARI 515! Palekhori, Agros, *Economides* in ARI 1338!
3. Stavrovouni, near Monastery, 1974, *Meikle* 4062!
7. Between Klepini and Pentadaktylos, 1954, *Casey* 1331!

10. V. muricata *(Stev.) Baxt.* in Loudon, Hort. Brit. Suppl., 3: 654 (1839); Davis, Fl. Turkey, 4: 581 (1972); Feinbrun, Fl. Palaest., 3: 260, t. 443 (1978).

Fedia muricata Stev. in Roem. et Schult., Syst. Veg., 1: 366 (1817).

F. truncata Reichb., Plant. Crit., 2: 7, no. 225, t. 115 (1824).

Valerianella truncata (Reichb.) Betcke, Animadv. Bot. Valer., 22 (1826); Boiss., Fl. Orient., 3: 105 (1875); Sintenis in Oesterr. Bot. Zeitschr., 32, 193 (1882); Holmboe, Veg. Cypr., 173 (1914); Post, Fl. Pal., ed. 2. 1: 608 (1932); Osorio-Tafall et Seraphim, List Vasc. Plants Cyprus, 99 (1973).

V. truncata (Reichb.) Betcke var. *muricata* (Stev.) Boiss., Fl. Orient., 3: 106 (1875).
[*V. eriocarpa* (non Desv.) Unger et Kotschy, Die Insel Cypern, 231 (1865); H. Stuart Thompson in Journ. Bot., 44: 309 (1906).]
TYPE: Not indicated; probably from Crimea or Caucasus.

Like *V. discoidea* vegetatively but leaves and bracts lacking basal teeth (though uppermost leaves and bracts have membranous auricles at base); inflorescence becoming much laxer with rather flattened branching clearly visible at least in fruit, lower branches and upper stem sometimes inflating slightly in fruit; bracts oblong, acute, 2–3 mm. long, 0·5 mm. wide, green, glabrous or with occasional asperities on the margin, basal auricles of each opposite pair apparently fused into a cup, midrib and leaf margins decurrent as narrow wings down the main forkings of the inflorescence; outer bracteoles similar but broader and more membranous, margins strongly webbed at base, inner and innermost bracteoles similar but smaller, all glabrous throughout; corolla mauve-lilac to pale mauve. Fruits overall 1·5– 2 mm. long, of which the ovary is slightly more than half; ovary ovoid, in Cyprus with obvious short ascending simple hairs particularly on the fertile carpel, usually red-brown with the 2 sterile loculi visible as paler curving ridges on the inner face meeting above and below but separated in the middle by an elliptic area; calyx a reticulate-veined fan, longest and acute above the fertile loculus, curved forward and meeting above the sterile loculi as a low ridge, fan margin entire or rather irregular, rarely the anterior ridge either developed so that the calyx is a complete oblique tube or so reduced as to leave a gap in the middle; fruits falling readily leaving (in well-grown specimens) the rather characteristic inflorescence branches straight, divaricate and clothed with persistent, imbricate, narrow bracteoles.

HAB.: Wheatfields, rocky slopes, on steep hillside in shade of *Pinus*, coastal maquis, roadside banks; sea-level to 2,800 ft. alt.; fl. March–April.

DISTR.: Divisions 1, 3, 5–8; locally abundant. Widespread in Europe, North Africa, Middle East to Central Asia and Pakistan.

1. Wheatfields between Lachi and Neokhorio, 1978, *Chesterman* 47!
3. Limassol, 1951, *Mavromoustakis* 38! Amathus, 1976, *Holub* s.n.!
5. Athalassa, 1936, *Syngrassides* 1171!
6. Orga, 1956, *Poore* 1! Dhiorios, 1962, *Meikle* 2325!
7. S. of Kourtella Peak above Karmi, 1953, *Casey* 1314! Kyrenia, 1956, *G. E. Atherton* 1307! Ayios Khrysostomos Monastery, 1862, *Kotschy* 436.
8. Valia near Ayios Theodoros, 1905, *Holmboe* 500; Koma tou Yialou, Karpas, 1950, *Chapman* 221! Kantara, 1880, *Sintenis & Rigo* s.n. (given as 95 in Holmboe)!

NOTES: In Cyprus this is apparently the sole representative of the variable group including *V. eriocarpa* and *V. dentata*, found on the mainland in Turkey, but the Cyprus plants do some-times show the characters of these species — e.g. there are fruits with tubular calyces on *Mavromoustakis* 38 and *Syngrassides* 1171 — the only character used in the Flora of Turkey account to separate *V. eriocarpa*, and fruits with very small calyces occur on many of the collections which would key out in the Flora of Turkey to *V. dentata*. It is suspicious that all three names are used over a wide geographical range; the group as a whole is found virtually throughout the full range of the genus in the Old World — from the area bounded by North Africa to Kenya to Pakistan to Central Asia, Scandinavia, Britain and Spain.

52. DIPSACACEAE

Annual or perennial herbs, or rarely shrubs or subshrubs; leaves opposite (rarely verticillate), exstipulate, simple, entire or variously toothed or lobed; inflorescence cymose, commonly an involucrate capitulum or spike; receptacle often paleaceous; flowers 4- or 5-merous, generally hermaphrodite; ovary usually enveloped in a persistent, cup-shaped involucel, often expanded apically into a membranous, scarious or bristly corona; calyx cupuliform or almost reduced to teeth or bristles; corolla sympetalous, 4–5-lobed, sometimes zygomorphic and 2-lipped; corolla-lobes imbricate; stamens 4 (or 2) inserted at the mouth of the corolla-tube and alternating with the lobes; filaments usually free, exserted; anthers versatile, 2-thecous, introrse, dehiscing longitudinally; ovary inferior, 1-locular; ovule solitary, pendulous, anatropous; style slender; stigma simple or occasionally 2-lobed. Fruit an achene, generally enclosed in the persistent, indurated involucel, sometimes crowned with the persistent calyx; seed with a membranous testa; endosperm present; embryo straight.

About 10 genera and more than 200 species, chiefly in Europe, Asia and the Mediterranean region, with outliers in tropical Africa and South Africa.

Corolla 4-lobed; receptacular scales large, scarious, with a conspicuous, purple-tinged arista
1. Cephalaria
Corolla 5-lobed; receptacular scales small or wanting:
Involucel with a well-developed corona; calyx-bristles 5 or 10, sometimes scabrid or
pectinate-ciliate, but not plumose -   -   -   -   -   -   **2. Scabiosa**
Involucel without, or with a very inconspicuous, corona; calyx-bristles 10–30 or more,
plumose with greyish or whitish hairs-   -   -   -   -   -   **3. Pterocephalus**

1. CEPHALARIA *Schrader ex Roemer et Schultes*

Syst. Veg., 3: 1, 43 (1818) nom. cons.

Z. Szabó, Mon. Gen. Cephalaria in Mat. Term. Közlem., 38: 1–352 (1940).

Annual, biennial or perennial herbs, or rarely suffruticose; leaves simple, entire, toothed or pinnatipartite; inflorescence an ovoid or subglobose capitulum; involucral bracts in several rows; receptacle paleaceous, the scales often conspicuous; calyx cupuliform, fimbriate; corolla white, yellow, mauve or blue, 4-lobed, the outermost florets sometimes radiant. Involucel (in fruiting specimens) 4- or 8-angled, with 4 or 8 apical bristles.

About 65 species, chiefly in S. Europe, the Mediterranean region and western Asia, also in southern Africa.

1. **C. syriaca** (*L.*) *Schrader*, Ind. Sem. Hort. Gött. (1814); Poech, Enum. Plant. Ins. Cypr., 15 (1842); Boiss., Fl. Orient., 3: 120 (1875); E. G. Bobrov in Bull. Appl. Bot. Leningrad, 21: 311 (1929); Post, Fl. Pal., ed. 2, 1: 612 (1932); Szabó in Mat. Term. Közlem., 38: 176 (1940); Lindberg f., Iter Cypr., 33 (1946); Davis, Fl. Turkey, 4: 590 (1972); Osorio-Tafall et Seraphim, List Vasc. Plants Cyprus, 99 (1973).

Scabiosa syriaca L., Sp. Plant., ed. 1, 98 (1753); Sibth. et Sm., Fl. Graec. Prodr., 1: 81 (1806), Fl. Graec., 2: 6, t. 105 (1813).

Cephalaria syriaca (L.) Schrader var. pedunculata DC., Prodr., 4: 648 (1830); Unger et Kotschy, Die Insel Cypern, 232 (1865).

C. syriaca (L.) Schrader f. pedunculata (DC.) Boiss., Fl. Orient., 3: 120 (1875); Holmboe, Veg. Cypr., 174 (1914).

Erect annual up to 1 m. high; stems usually unbranched in the lower part, dichotomously branched above, rather pallid, thinly or densely setose, terete near base, ribbed above; leaves subsessile, lanceolate or narrowly oblong, occasionally pinnatipartite, 3–15 cm. long, 0·5–3 cm. wide, dark green, glabrous or thinly setose, apex obtuse or acute, margins entire, serrate or crenate, bases shortly connate to form a tube around the stem; inflorescence a dense, ovoid capitulum 2·5 cm. long and 0·8–2 cm. wide; peduncle of terminal capitulum at each dichotomy sometimes very short (var. *sessilis* DC.), or up to 15 cm. long, scabrid-setose and conspicuously ribbed, erect, or the lateral peduncles and branches strongly divaricate (var. *sessilis* DC.); involucral bracts scarious, broadly ovate, 5–8 mm. long, 4–5 mm. wide, acuminate, glabrous or pubescent externally; receptacular scales similar, but often rather larger, obovate, concave, ciliolate, the apex abruptly narrowed into a conspicuous, purplish arista, 3–7 mm. long; outer flowers not radiant; calyx cupuliform, about 1·5 mm. long, 2 mm. wide, pubescent externally and internally, apex fimbriate with pubescent setae 1–1·5 mm. long; corolla mauve or blue, narrowly infundibuliform, 6–8 mm. long, 2·5–4 mm. wide at mouth, pubescent externally; lobes 4, subequal, oblong, obtuse or subacute, about 2·5 mm. long, 1·8–2 mm. wide; stamens 4, exserted; filaments 4–5 mm. long, slender, anthers yellow, narrowly oblong, about 1·5 mm. long, 0·5 mm. wide; style filiform, about 7 mm. long; stigma lateral, about 1 mm. long. Fruiting involucel narrowly oblong-obovoid, obscurely 4-angled, about 5 mm. long, 2·5 mm. wide, shortly pilose, 8-ribbed, the apex crowned with 8 unequal, pubescent setae, those at the angles about 4–5 mm. long, the intermediate about 2·5 mm. long; achene fusiform, pale brown, closely invested by the persistent involucel.

ssp. **syriaca**
TYPE: "*in* Syria."

Fruit-involucel with the intermediate setae 4–5-times shorter than those at the angles; calyx shortly fimbriate.

HAB.: Cultivated fields and field-borders; sea-level to c. 5,000 ft. alt.; fl. April–May.

DISTR.: Divisions 2, 3, 6. Turkey, Syria, Lebanon, Palestine, ? N. Africa, east to Iraq and Iran, a casual or naturalized elsewhere.

2. Phiti, 1905, *Holmboe* 774; Prodhromos, 1939, *Lindberg f.* s.n.
3. Amathus, 1787, *Sibthorp*; Episkopi, 1862, *Kotschy*; Kolossi, 1862, *Kotschy*.
6. Nicosia, 1880, *Sintenis & Rigo* sec. Szabó.

NOTES: All these records may be more correctly referable to *C. syriaca* ssp. *phoeniciaca* Bobr. *infra*.

ssp. **phoeniciaca** *Bobr.* in Bull. Appl. Bot. Leningrad, 21: 314 (1929); Rechinger f. in Arkiv för Bot., ser. 2, 1: 432 (1950); Osorio-Tafall et Seraphim, List Vasc. Plants Cyprus, 99 (1973).
TYPE: Not indicated; to be selected from amongst the syntypes cited by Bobrov.

Fruit-involucel with the intermediate setae about half as long as those at the angles; calyx fimbriate with rather long setae.

HAB.: Cultivated fields and field-borders; sea-level to c. 2,000 ft. alt.; fl. April–May.

DISTR.: Divisions 1, 2, 5, 7. Also Turkey, Syria, Lebanon, Palestine, Egypt, N. Africa.

1. Stroumbi, 1913, *Haradjian* 727 !
2. Between Karavostasi and Lefka, 1937, *Syngrassides* 1548 !
5. Prastio, 1936, *Syngrassides* 998 !
7. St. Hilarion, 1934, *Syngrassides* 680 ! Kyrenia, 1937, *Casey* s.n. !

NOTES: All the Cyprus material examined is referable to *C. syriaca* ssp. *phoeniciaca*, and it is not unlikely that it is the only subspecies represented on the island. *C. syriaca* is evidently rather rare in Cyprus, which is probably fortunate, for a note on one of the specimens in herb. Kew. (Fl. Palaestina Exsicc. 187) reads: "A very troublesome weed in crop-fields of Mediterr. Palestine. It has poisonous properties, and a small quantity of it is sufficient to spoil the taste of the flour." *C. syriaca* occurs as an adventive in many parts of Europe.

2. SCABIOSA *L.*

Sp. Plant., ed. 1, 98 (1753).
Gen. Plant., ed. 5, 43 (1754).
Tremastelma Raf., Fl. Telluriana, 4: 96 (1838).
Callistemma (Mertens et Koch) Boiss., Fl. Orient., 3: 146 (1875).

Annual, biennial or perennial herbs, or occasionally shrubs or subshrubs; leaves simple or pinnatipartite; inflorescence an ovoid or hemispherical capitulum; involucral bracts herbaceous, in 1–3 rows; receptacular scales linear-lanceolate; calyx cupuliform at base, produced into 5 or 10, often conspicuous, bristles; corolla white, yellowish, pink, blue or purple, 5-lobed, the outermost florets sometimes radiant. Involucel (in fruiting specimens) with an 8-sulcate tube, or sometimes with 8 pits at the apex of the tube, expanded above into a membranous or scarious, veined corona.

About 100 species in temperate Europe and Asia, the Mediterranean region, the mountains of East Africa, and South Africa.

Calyx bristles 10, pectinate-ciliate, dorsiventrally compressed   - - - **5. S. brachiata**
Calyx bristles 5, scabridulous or hispidulous, but not pectinate-ciliate, nor compressed:
  Capitula subsessile or very shortly pedunculate, much overtopped by the subtending leaves
                                                         3. S. prolifera
  Capitula long-pedunculate, much exceeding the subtending leaves:
    Perennials:
      Leaves usually entire, ovate, obovate or spathulate, rarely dentate or with a few
      irregular lobes; a shrub or subshrub with erect stems; inflorescence unbranched or
      sparingly branched; flowers mauve or purplish - - - - **4. S. cyprica**
      Leaves pinnatisect; herbaceous perennials, woody only at the base:
        Plants caespitose; leaves crowded towards base of stems; capitula 2–4 cm. diam.;
        flowers pinkish-lilac - - - - - - S. CRENATA (p. 850)
        Plants lax, not caespitose:
          Peduncles slender, divaricate; involucel-tube deeply pitted at apex; rootstock
          woody, branched; flowers milky-white - - - - **1. S. argentea**
          Peduncles not noticeably slender or divaricate; involucel-tube not pitted at apex;
          flowers purple, mauve or pink , rarely white -   - S. ATROPURPUREA (p. 850)
    Annuals or biennials:
      Capitula 1–1·5 cm. diam., much exceeded by the involucral bracts; leaves entire or
      remotely lobed; involucel-tube pitted at apex - - - - **2. S. sicula**
      Capitula 1·5–3 cm. diam., not exceeded by involucral bracts; upper leaves narrowly
      pinnatipartite; involucel-tube not pitted at apex -   - S. ATROPURPUREA (p. 850)

S. ATROPURPUREA *L.*, Sp. Plant., ed. 1, 100 (1753). Erect, glabrous or pubescent annual, biennial or sometimes perennating, 20–60 cm. high; upper cauline leaves pinnatipartite with narrowly oblong or linear lobes; capitula depressed-hemispherical, 1·5–3 cm. diam., long-pedunculate; florets dark purple (var. *atropurpurea*), lavender, pink or rarely white (var. *setifera* (Lam.) DC., *S. maritima* L., *S. atropurpurea* L. ssp. *maritima* (L.) Arcang.), sometimes with the outermost markedly radiant. Fruiting involucel with a deeply sulcate-plicate tube and a short, crispate, inrolled corona.

S. atropurpurea L. var. *setifera* (Lam.) DC. has been collected in an old garden just outside the walls of Kyrenia Castle (1955, *G. E. Atherton* 202 !), but is neither native nor naturalized in Cyprus. It has a wide distribution in the Mediterranean region, whereas typical *S. atropurpurea* is scattered (Portugal, Spain, Italy, Sardinia, Sicily, Malta) and may be a garden escape in some of its stations; it is normally a more robust plant than var. *setifera*, with larger capitula of dark purple, radiant florets, but type and variety are connected by intermediates. Both are popular garden plants.

S. CRENATA *Cyr.*, Plant. Rar. Neap., 1: 11 (1788); Unger et Kotschy, Die Insel Cypern, 233 (1865); H. Stuart Thompson in Journ. Bot., 44: 309 (1906); Holmboe, Veg. Cypr., 174 (1914).

A caespitose perennial, woody at the base; stems 3–25 (–80) cm. long; leaves often crowded towards the base of the stem, oblong, deeply pinnatisect or bipinnatisect, with numerous, bluntly oblong, hairy or subglabrous lobes; capitula depressed-hemispherical, 2–4 cm. diam., generally long-pedunculate; involucral bracts inconspicuous; florets pinkish-lilac, adpressed-hairy externally, the outermost with a long tube. Fruiting involucel with a conspicuous, membranous, many (26–29)-veined, infundibuliform corona, and a shorter tube which is densely setose towards the base and deeply 8-pitted above; calyx-bristles 2–3-times as long as corona of involucel.

Recorded by Kotschy "auf sonnigen Abhängen gegen Sta. Croce" [Stavrovouni, Div. 3], but without collector's number. It has not since been seen in this frequently visited part of the island, and the record must be regarded as dubious. Sibthorp's *Scabiosa coronopifolia* (Fl. Graec. Prodr., 1: 85; 1806, Fl. Graec., 2: 13, t. 114; 1813), now regarded as synonymous with *S. crenata*, is unlocalized, but probably came from Greece. H. Stuart Thompson (*loc. cit.*) attributes the Cyprus record to Sibthorp, but I can find no evidence for this. *S. crenata* is found in Italy, Sicily, Yugoslavia, Albania and Greece, but does not occur in Turkey, Syria or Palestine, so that it is not likely to be found in Cyprus.

1. **S. argentea** L., Sp. Plant., ed. 1, 100 (1753); Sibth. et Sm., Fl. Graec. Prodr., 1: 83 (1806), Fl. Graec., 2: 9, t. 108 (1813); Holmboe, Veg. Cypr., 174 (1914); Post, Fl. Pal., ed. 2, 1: 616 (1932); Davis, Fl. Turkey, 4: 613 (1972); Osorio-Tafall et Seraphim, List Vasc. Plants Cyprus, 99 (1973).
   S. ucranica L., Sp. Plant., ed. 2, 144 (1762); Boiss., Fl. Orient., 3: 139 (1875) pro parte excl. var. *β eburnea*.
   TYPE: "*in* Oriente".

Erect or spreading perennial, 30–80 (–100) cm. high, rootstock woody, usually branched and scaly; stems slender, subterete, densely white-hirsute at the base, becoming glabrous upwards, sparingly branched except in the region of the inflorescences; leaves oblong, 2–4 cm. long, 1–2 cm. wide, deeply pinnatisect into 9 or more lanceolate, linear or narrowly oblong, acute lobes, the lower leaves densely white-hirsute, the upper more sparsely so; petioles up to 1·5 cm. long, flattened and dilated to form a sheathing base; inflorescence a lax, divaricate, repeatedly dichotomous cyme, with linear, entire or sparsely pinnatisect, glabrescent bracts (or modified leaves) 1–2·5 cm. long, 0·2 cm. wide, at each dichotomy; peduncles very slender, glossy, 4–15 cm. long; capitula flattened-hemispherical 1·5–3 cm. diam.; involucral bracts in 2 series, the outermost linear, 5–10 mm. long, 1·5–3 mm. wide, thinly hairy, the inner often a little shorter; receptacular scales narrowly ovate-acuminate, 3–4 mm. long, 1·5–2 mm. wide, ciliate; florets milky-white, the outermost distinctly radiant, about 1 cm. long, with an infundibuliform, pubescent tube and 5 very unequal lobes, the 3 lower oblong, truncate, about 3 mm. long, 2·5 mm. wide, the 2 upper ovate, acute or subacute, 1–1·5 mm. long and almost as wide at the base; inner florets more equally lobed, with short, ovate lobes; filaments glabrous, slightly flattened, about 5 mm. long; anthers narrowly oblong, about 2 mm. long, 0·5 mm. wide; style relatively thick, about 6 mm. long; stigma obscurely 2-lobed. Fruiting involucel with a conspicuous, membranous, flat or shallow-concave, 20–26-veined corona, about 5 mm. diam., the veins produced beyond the margin as small apiculi; involucel-tube about 2 mm. long and almost as wide, densely setose in the lower half, deeply 8-pitted above; calyx-bristles 6–7 mm. long, much exceeding corona, thickened and shortly connate at base, with a central tuft of hairs.

HAB.: Dry hillsides and vineyards on chalk and limestone, 2,000–3,600 ft. alt.; fl. June–August.

DISTR.: Division 2, rare, S. Europe eastwards from Italy, and western Asia from Turkey to Iran.

2. Omodhos, 1880, *Sintenis & Rigo* 692! Aphamis, 1937, *Kennedy* 913! Perapedhi, 1940, *Davis* 1854!

2. S. sicula L., Mantissa Altera, 196 (1771); Sibth. et Sm., Fl. Graec. Prodr., 1: 82 (1806); Boiss., Fl. Orient., 3: 142 (1875); Sintenis in Oesterr. Bot. Zeitschr., 32: 124, 364 (1882); Holmboe, Veg. Cypr., 174 (1914); Davis, Fl. Turkey, 4: 617 (1972); Osorio-Tafall et Seraphim, List Vasc. Plants Cyprus, 99 (1973).

S. divaricata Jacq., Hort. Vindob., 1: 5, t. 15 (1772); Post, Fl. Pal., ed. 2, 1: 616 (1932).

S. ucranica L. var. γ *Sicula* (L.) Coult., Mém. Dipsac., 35 (1823); Unger et Kotschy, Die Insel Cypern, 233 (1865) [as *S. "Ukrainica"* var. *sicula*].

[*S. ucranica* (non L.) Poech, Enum. Plant. Ins. Cypr., 17 (1842); H. Stuart Thompson in Journ. Bot., 44: 309 (1906) — as *S. "ukranica"*.]

TYPE: *"in* Sicilia*"*.

Erect or sprawling annual 10–40 cm. high with a slender taproot; stems usually much branched, subterete or obscurely angular, thinly crisped-pilose or subglabrous; basal leaves oblanceolate-spathulate, 2–6 cm. long, 0·5–2 cm. wide, thinly pilose, obtuse, entire or remotely lobulate, tapering to a short, flattened, sheathing petiole; upper leaves 2·5–7 cm. long, 0·2–3 cm. wide, narrowly oblanceolate and entire, or commonly pinnatisect into 3–7 (or more), irregular, oblanceolate, subacute lobes; inflorescence a lax, divaricate cyme; peduncles slender, wiry, subglabrous or thinly hairy, often purplish, 5–15 (–20) cm. long; capitula hemispherical, 1–1·5 cm. diam.; involucral bracts in 2 series, oblong-lanceolate, subacute or obtuse, thinly strigillose, the outer up to 40 mm. long, 7 mm. wide, much exceeding the florets, the inner usually shorter; receptacular scales filiform-subulate, 4–5 mm. long, 0·5 mm. wide, hispidulous; florets mauve or pinkish-purple, not markedly radiant, about 4–5 mm. long, with a narrowly infundibuliform, hairy tube and 5 unequal lobes, the 3 lower oblong, truncate 1–1·5 mm. long, 1 mm. wide, the 2 upper less than 1 mm. long and subacute; inner florets more equally lobed, with short, ovate lobes; filaments glabrous, about 2 mm. long, shortly exserted; anthers yellow, oblong, about 7 mm. long, 5 mm. wide; style relatively thick, about 4·5 mm. long; stigma slightly clavate, obscurely 2-lobed. Fruiting involucel with a conspicuous, membranous, flattish or shallow-concave 25-veined corona, 5-9 mm. diam., the veins produced beyond the margin as distinct and conspicuous spinules; involucel-tube about 2·5 mm. long and almost as wide, shortly adpressed-hairy in the lower half, deeply 8-pitted above; calyx-bristles 6–7 mm. long, much exceeding the corona, reddish-purple, scabridulous-hispid, shortly connate at base, but without a central tuft of hairs.

HAB.: Dry stony slopes; roadsides; margins of cultivated fields; near sea-level to 3,800 ft. alt.; fl. March–May.

DISTR.: Divisions 1–5, 7, 8. Spain, Sicily and the eastern Mediterranean region from Yugoslavia to Syria, eastwards to Iraq and N. Iran.

1. Between Smyies and Neokhorio (Akamas), 1941, *Davis* 3304!
2. Mandria, 1938, *Kennedy* 1435! Platres, 1938, *Kennedy* 1436!
3. Kophinou, 1905, *Holmboe* 583; near Michael Arkhangelos Monastery, Stavrovouni, 1957, *Merton* 3162! Near Souni, 1960, *N. Macdonald* 189! Episkopi, 1967, *Merton* in ARI 631!
4. Athna Forest, 1952, *Merton* 1444!
5. Between Lefkoniko and Kythrea, 1880, *Sintenis & Rigo* 43! Near Apostolos Varnavas Monastery, 1948, *Mavromoustakis* s.n.! and, 1962, *Meikle* 2636!
7. Ayios Khrysostomos Monastery, 1859, *Kotschy*; Larnaka tis Lapithou, 1936, *Syngrassides* 917! Yaïla, 1941, *Davis* 2831! 2831b! Lakovounara Forest, 1950, *Chapman* 356! and, 1974,

Osorio-Tafall & Meikle in *Meikle* 4077! Above Kyrenia Pass, 1953, *Casey* 1300! Kalyvakia, 1957, *Merton* 3095! Ayios Epiktitos, 1960, *N. Macdonald* 69!
8. Near Eleousa, 1880, *Sintenis & Rigo*; Valia, 1905, *Holmboe* 481.

3. S. prolifera *L.*, Syst. Nat., ed. 10, 889 (1759), Sp. Plant., ed. 2, 144 (1762); Sibth. et Sm., Fl. Graec. Prodr., 1: 82 (1806), Fl. Graec., 2: 8, t. 107 (1813); Poech, Enum. Plant. Ins. Cypr., 17 (1842); Unger et Kotschy, Die Insel Cypern, 233 (1865); Boiss., Fl. Orient., 3: 144 (1875); Holmboe, Veg. Cypr., 175 (1914); Post, Fl. Pal., ed. 2, 1: 617 (1932); Lindberg f., Iter Cypr., 17 (1946); Davis, Fl. Turkey, 4: 618 (1972); Osorio-Tafall et Seraphim, List Vasc. Plants Cyprus, 99 (1973); Electra Megaw, Wild Flowers of Cyprus, 9, t. 10 (1973).
TYPE: *"in* India" [sic].

Robust, erect annual, (10–) 15–30 (–50) cm. high with a slender taproot; stems usually unbranched except in the region of the inflorescence, subterete or obscurely ribbed, thinly crispate-pilose, often stained purple; basal leaves usually withered at anthesis, obovate-spathulate, 3–10 cm. long, 1·5–4 cm. wide, pale green, thinly pilose, obtuse, entire or remotely and irregularly dentate, the base tapering gradually to a short, flattened petiole; upper leaves similar, but usually rather smaller and narrower; inflorescence a rigid, divaricate or spreading cyme, with 2–4 (rarely more) primary branches arising from just below a central capitulum, these branches often again divided, usually dichotomously, at their apex; bracts foliaceous; peduncles short, usually less than 1 cm. long (rarely up to 2·5 cm.), sometimes almost wanting, densely clothed with long, spreading, white hairs; capitula flattened-hemispherical, 2–4 cm. diam.; involucral bracts in 2–3 series, lanceolate or narrowly oblong, 0·5–1·5 cm. long, 0·1–0·5 cm. wide, subacute or obtuse, densely pilose, usually rather shorter than, or equalling the outermost florets; receptacular scales filiform, 8–9 mm. long, glabrous below, pilose near apex; florets creamy-white or yellowish, conspicuously radiant, the corolla-tube rather broadly infundibuliform, 2–3 mm. long, 1·5–2 mm. wide at apex, densely sericeous-hairy, 5-lobed, the lower 3 lobes of the outermost florets oblong-truncate, the median about 10 mm. long, 6–7 mm. wide, the 2 lateral about 6–7 mm. long, 2–2·5 mm. wide, all shortly crenate at the apex, upper 2 lobes ovate, about 1·5–2 mm. long and almost as wide, obtuse or subacute, entire; inner florets with subequal, ovate lobes, 1–2 mm. long and almost as wide at the base; filaments glabrous, 3–4 mm. long, distinctly exserted; anthers yellow, narrowly oblong, about 2·5 mm. long, 0·5 mm. wide; style relatively thick, about 5 mm. long; stigma slightly clavate, obscurely 2-lobed or sulcate. Fruiting involucel with a conspicuous, membranous, cup-shaped, 30–35-veined corona, 10–12 mm. diam., the veins produced beyond the margin of the corona as short cusps or spinules, involucel-tube about 7 mm. long, 3·5 mm. wide, densely pilose in the lower half, with 8 deep, elongate pits above; calyx-bristles about 5 mm. long, not much exceeding the corona, pallid, scabridulous, shortly connate and purplish at the base, and ciliate with long white hairs.

HAB.: Roadsides, waste ground, fallow and cultivated fields; sea-level to 1,800 (to 5,000) ft. alt.; fl. Febr.–May.

DISTR.: Divisions 1–8, locally very common. Also S.E. Turkey (Amanus), Syria, Palestine.

1. Polis area, 1962, *Meikle*!
2. Prodhromos, 1929, *C. B. Ussher* 10!
3. Frequent about Limassol, *Kennedy* 1559! *Chapman* 562! *Mavromoustakis* 1/57! *N. Macdonald*, 71! Pissouri, 1959, *C. E. H. Sparrow* 12! Episkopi, 1959, *C. E. H. Sparrow* 13!
4. Ormidhia, 17 April 1787, *Sibthorp*; and 1862, *Kotschy* 176! Near Larnaca, *Deschamps* 224 b; also, 1955, *Merton* 2010! Sotira, 1905, *Holmboe* 410.
5. Near Kythrea, 1880, *Sintenis & Rigo* 41! Salamis, 1962, *Meikle*!
6. About Nicosia, 1913, *Miss Godman* 13! 1927, *Rev. A. Huddle* 67! 1930, *F. A. Rogers*! Dhiorios, 1962, *Meikle*!
7. Common about Kyrenia, Ayios Epiktitos, Panagra, Larnaka tis Lapithou, Dhikomo, etc. *A. G. & M. E. Lascelles*! *J. A. Tracey*! *Syngrassides* 269! 423! *Davis* 3018! *Casey* 572!

Miss Mapple 43! *G. E. Atherton* 1166! 1395! *Merton* in ARI 116! *I. M. Hecker* 31! *P. Laukkonen* 279! etc.

8. Near Leonarisso, 1880, *Sintenis & Rigo*; Rizokarpaso, 1912, *Haradjian* 218! Platanisso, 1950, *Chapman* 241! Between Boghazi and Gastria, 1941, *Davis* 2431! Akradhes, 1962, *Meikle*!

NOTES: Although now one of the commonest and most conspicuous plants of lowland Cyprus, there is some reason for supposing that *S. prolifera* has greatly increased in the past 100 years. Sibthorp and Kotschy note it only from Ormidhia, and though, by 1880, Sintenis and Rigo could report it from several additional localities, it would appear that it was much less familiar to them than it would be if they were botanizing in Cyprus at the present time.

4. S. cyprica *Post* in Bull. Herb. Boiss., 5: 757 (1897); H. Stuart Thompson in Journ. Bot., 44: 309 (1906); Holmboe, Veg. Cypr., 174, fig. 57 (1914); Lindberg f., Iter Cypr., 33 (1946); Chapman, Cyprus Trees and Shrubs, 78 (1949); Osorio-Tafall et Seraphim, List Vasc. Plants Cyprus, 99 (1973).

TYPE: Cyprus; "prope Perapedia (Cypri); floret aestate. [*Post*] No. 14" (BEI).

Shrub 30–100 cm. high; stems erect, branched towards base, covered with rough, fuscous, pubescent bark in the lower part, terete or subterete above, often purplish, glabrescent, or clothed with an indumentum of crispate pubescence intermixed with short, scattered, white bristles; leaves ovate, obovate or spathulate, often rather crowded towards the base of the stems; lamina 0·8–3 cm. long, 0·3–1·5 cm. wide, densely adpressed silvery-silky when young, greyish-green at maturity, apex obtuse, acute or mucronate, margins commonly entire, sometimes remotely serrate or even with 1–2 deeply cut lobes, base tapering to a very short petiole which forms a shortly connate, pubescent sheath around the stem; inflorescence erect, very sparingly branched; peduncle often very long (sometimes exceeding 25 cm.), sulcate, thinly adpressed-pilose; capitulum flattened-hemispherical, 1·5–3·5 cm. diam.; involucral bracts in 2–3 series, narrowly lanceolate, 5–8 mm. long, 1–2 mm. wide, much shorter than florets, adpressed grey-pubescent; receptacular scales linear-subulate, about 8 mm. long, 1 mm. wide, membranous towards base, with an herbaceous, adpressed-pubescent apex; florets conspicuously radiant, mauve or purple, pubescent externally; corolla-tube infundibuliform, 3–4 mm. long; lobes of outermost florets very unequal, the 3 lower oblong, truncate, the median 5–6 mm. long, 3–4 mm. wide, dentate-laciniate at apex, the laterals similar but distinctly shorter and narrower, 2 upper lobes shortly oblong or suborbicular, blunt, entire, about 1·5 mm. long and almost as wide; inner florets with subequal, oblong or suborbicular lobes, 1–1·5 mm. long; filaments slender, glabrous, about 7 mm. long; anthers (? yellow), narrowly oblong, about 2·5 mm. long, 0·5 mm. wide, long-exserted; style 8–9 mm. long; stigma distinctly thickened, obscurely 2-lobed or sulcate. Fruiting involucel with a conspicuous, often purple, membranous, cup-shaped, 28–33-veined corona, 5–7 mm. diam., the veins scarcely produced beyond the plicate margin; involucel-tube about 4 mm. long, 3 mm. wide, densely white hirsute all over, so that the 8 apical pits are scarcely visible; calyx bristles short, about 2 mm. long, shorter than corona, scabridulous, connate at the base and forming a small, puberulous, greenish disc. *Plate 54*.

HAB.: Dry hillsides amongst garigue, on chalk; 1,000–3,800 ft. alt.; fl. June–July.

DISTR.: Divisions 2, 3. Endemic.

2. Perapedhi, 1894, *Post* 14; between Platres and Phini, 1898, *herb. Post* s.n.!, also, 1937, *Kennedy* 910! 911! Perapedhi, 1905, *Holmboe* 1091; Mandria-Omodhos, 1937, *Kennedy* 912! Kato Platres, 1939, *Lindberg* f. s.n.!

3. Oritaes Forest, 1937, *Chapman* 302!, and, 1956, *Merton* 3531! Below Trimiklini, and from there to Perapedhi, 1940, *Davis* 1834! Between Kividhes and Ezimi, 1941, *Davis* 3531!

NOTES: Restricted to one area in S.W. Cyprus. Post suggests a relationship with *S. kurdica* Post from the Amanus area (S.E. Turkey–Syria), and the allied *S. paucidentata* Hub.-Mor. is

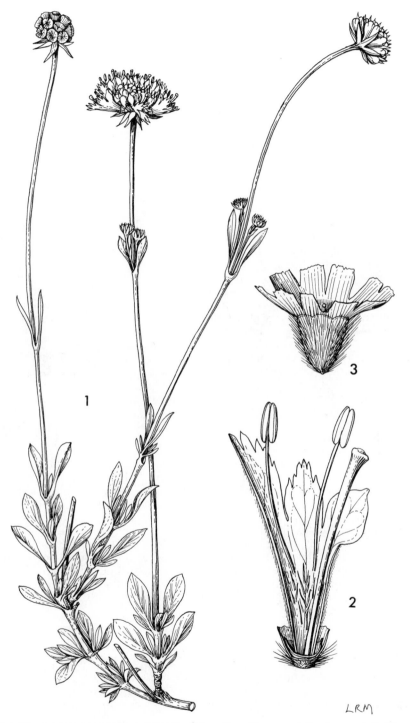

LRM

Plate 54. SCABIOSA CYPRICA Post   **1**, habit ×⅔; **2**, section of flower, ×4; **3**, achene, ×4. (**1**, **2** from *Kennedy* 911; **3** from *Kennedy* 912.)

another affinity suggested in Fl. Turkey (4: 612). It is also evident that *S. cyprica* is allied to *S. cretica* L., *S. minoana* Greuter and *S. albocincta* Greuter, the first from the Balearics, S. Italy and Sicily, the other two endemic in Crete; these last agree with *S. cyprica* in having the calyx bristles distinctly shorter than the corona of the involucel.

5. **S. brachiata** *Sm.* in Sibth. et Sm., Fl. Graec. Prodr., 1: 83 (1806), Fl. Graec., 2: 9, t. 109 (1813).
   Knautia palaestina L., Mantissa Altera, 197 (1771) non *Scabiosa palaestina* L., Mant. I, 37 (1767).
   Pterocephalus palaestinus (L.) Coult., Mém. Dipsac., 81, t. 1, fig. 14 (1823); Unger et Kotschy, Die Insel Cypern, 233 (1865).
   Callistemma brachiatum (Sm.) Boiss., Fl. Orient., 3: 146 (1875).
   Callistemma palaestinum (L.) Heldr., Herb. Graec. Norm., no. 1148 (1891); Holmboe, Veg. Cypr., 175 (1914).
   Tremastelma palaestinum (L.) Janchen in Oesterr. Bot. Zeitschr., 66: 395 (1916); Davis, Fl. Turkey, 4: 621 (1972); Osorio-Tafall et Seraphim, List Vasc. Plants Cyprus, 100 (1973).
   TYPE: "In arvis insulae Cypri", 1787, *Sibthorp* (OXF).

Erect annual 5–20 (–30) cm. high with a slender taproot; stems usually unbranched below, divaricately branched in the region of the inflorescence, subterete or obscurely striate, hirsute with a mixed indumentum of long and short, spreading, white hairs; basal leaves obovate or broadly oblanceolate, 3–7 cm. long, 1–3 cm. wide, adpressed-sericeous, obtuse or subacute, entire, or crenate-serrate towards apex, or sometimes irregularly lobed, base tapering to a flattened petiole up to 3 cm. long; upper leaves similar, but usually narrower and more acute, subsessile or shortly petiolate, sometimes pinnatisect with a few, blunt lobes; inflorescence unbranched or a lax, divaricate cyme; peduncles 3–12 (–20) cm. long, pilose with erect, subadpressed hairs; capitula depressed-hemispherical, 1·5–2·5 cm. diam.; involucral bracts in 2 series, lanceolate-acuminate, 1–1·5 cm. long, 0·3–0·5 cm. wide, pale at the base, darker green above, densely strigose-hirsute, equalling the florets or a little shorter, sharply reflexed in fruit; receptacular scales linear, 7–10 mm. long, 0·5 mm. wide, strigose-hirsute; florets pale mauve, the outer shortly but distinctly radiant, tube narrowly infundibuliform, about 4 mm. long, densely hirsute externally, lobes of outermost florets very unequal, the 3 lower oblong, truncate, 3–4 mm. long, 1·5–2·5 mm. wide, the 2 upper broadly oblong or suborbicular, 1·5–2 mm. long and about as wide; inner florets with suborbicular or bluntly oblong, subequal lobes, 1·5–2 mm. long; filaments about 2 mm. long, whitish, glabrous; anthers yellow, narrowly oblong, about 1·5 mm. long, 0·5 mm. wide; style 5–6 mm. long; stigma clavate, obscurely sulcate. Fruiting involucel with a conspicuous, flattish or shallow-concave, c. 30-veined corona 7–10 mm. diam., the veins produced beyond the margin as short, scabridulous cusps; involucel-tube about 4–5 mm. long and almost as wide, clothed below with blunt papillose hairs, deeply and conspicuously 8-pitted above, calyx-bristles 10, dorsiventrally compressed, about 7 mm. long, much exceeding the corona, reddish, pectinate-ciliate, shortly connate, greenish and glandular at base. *Plate 55.*

HAB.: On rocky calcareous hillsides in garigue; 650–3,000 ft. alt.; fl. April–June.

DISTR.: Divisions 2, 7. Eastern Mediterranean region from Greece to Syria.

2. Between Mandria and Perapedhi, 1965, *Meikle* 2865!
7. [Near Kyrenia, 1787, *Sibthorp*]; Pentadaktylos, 1880, *Sintenis & Rigo* 580! Lakovounara Forest, 1955, *Merton* 2210! Kalyvakia, 1957, *Merton* 3094!

NOTES: Generally referred to an independent genus, *Tremastelma* Raf., but the distinction appears to rest solely on the 10, not 5, calyx-bristles, and seems to be a sectional (*Scabiosa* L. sect. *Callistemma* Mertens et Koch in Röhlings Deutschl. Fl., 1: 758; 1823) rather than a generic one. In all other respects *S. brachiata* conforms with the genus *Scabiosa*.

I do not know by what authority Kotschy refers Sibthorp's *Scabiosa cerignensis* to *Pterocephalus palaestinus* (L.) Coult. (= *S. brachiata* Sm.). Sibthorp (Walpole, Travels, 18;

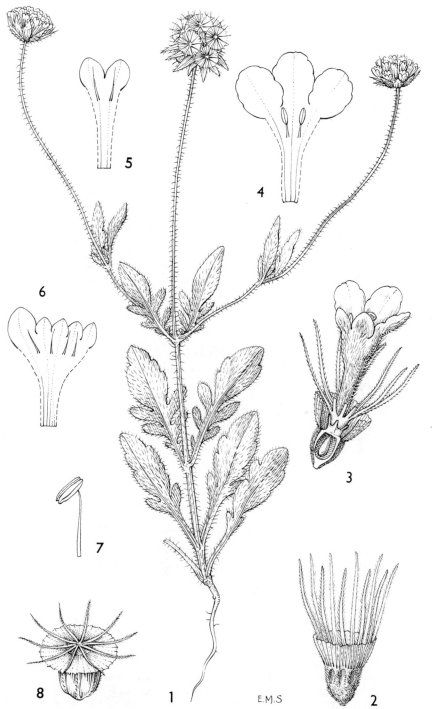

Plate 55. SCABIOSA BRACHIATA Sm. **1**, habit, $\times \frac{2}{3}$; **2**, ovary and involucre, $\times 4$; **3**, ovary and involucre in section, $\times 4$; **4**, corolla, abaxial section, $\times 4$; **5**, corolla, adaxial section, $\times 4$; **6**, inner corolla expanded, $\times 5$; **7**, stamen, $\times 8$; **8**, fruit, $\times 2$. (**1–8** from *Meikle* 2865.)

1820) collected *S. cerignensis* by the shore near Kyrenia on 20 April, 1787. There is no subsequent record for *S. brachiata* from this locality.

[S. SIBTHORPIANA *Sm.* in Sibth. et Sm., Fl. Graec. Prodr., 1: 84 (1806), Fl. Graec., 2: 10, t. 110 (1813).
     S.multiseta Vis., Stirp. Dalmat. Spec., 1, t. 1 (1826).
     Pterocephalus palaestinus (L.) Coult. var. *dalmaticus* DC., Prodr., 4: 653 (1830).
     Tremastelma palaestinum (L.) Janch. var. *lyratum* (Vis.) Hayek, Prodr. Fl. Pen. Balc., 2: 509 (1930).

The type of *S. sibthorpiana* Sm. is cited: "In insulâ Cypro", but this is almost certainly an error, despite the fact that Smith (in Rees, Cyclopedia, 31; 1819) says it was "Gathered by Dr. Sibthorp, in Cyprus, with the preceding [*S. brachiata* Sm.]". *S. sibthorpiana* has not been found on the island by any other collector, and almost certainly came from Greece. It differs from *S. brachiata* (*Tremastelma palaestinum* (L.) Janchen) in its more robust habit, pinnatisect cauline leaves, cut into narrow segments, sparse, not sericeous, indumentum, and capitula with the outermost florets conspicuously enlarged and radiant. Although often regarded as conspecific with *S. brachiata*, it seems to be permanently distinct.

     S. sibthorpiana is widely distributed in the Balkan peninsula, from Yugoslavia and Albania to S. Greece. I have not seen any specimens from the Aegean Islands or western Asia.]

3. PTEROCEPHALUS *Adanson*
Fam. Plantes, 2: 152 (1763).

Annual or perennial herbs or subshrubs; leaves simple or variously lobed, usually hairy; inflorescence a hemispherical capitulum; involucral bracts herbaceous, in several series; receptacular scales hairy, sometimes wanting; calyx stipitate, sometimes cup-shaped, produced into 5–24 plumose bristles; corolla white, pink, or mauve, rarely yellow, 5-lobed, the outermost florets sometimes radiant. Involucel (in fruiting specimens) with an 8-sulcate tube, apex sometimes aristate, toothed or with a short corona.
     About 25 species in Europe, Asia and Africa.

Annual herbs; leaves commonly lobed or pinnatisect:
     Involucel without a corona or internal collar; ribs of involucel-tube united at apex
                                                P. PLUMOSUS (p. 858)
     Involucel with a narrow corona and internal collar; ribs of involucel-tube not united at apex
                                                  1. P. brevis
Perennial shrubs with woody branches; leaves shortly toothed, not lobed   **2. P. multiflorus**

[P. PLUMOSUS (*L.*) *Coult.*, Mém. Dipsac., 31 (1823); Unger et Kotschy, Die Insel Cypern, 232 (1865); Boiss., Fl. Orient., 3: 147 (1875); Holmboe, Veg. Cypr., 175 (1914); Post, Fl. Pal., ed. 2, 1: 618 (1932); Davis, Fl. Turkey, 4: 622 (1972); Osorio-Tafall et Seraphim, List Vasc. Plants Cyprus, 100 (1973).
     Knautia plumosa L., Mantissa Altera, 197 (1771).
     Scabiosa papposa L., Sp. Plant., ed. 1, 101 (1753) nec *S. papposa* L., Syst. Nat., ed. 12, 112 (1767) nec *Pterocephalus papposus* Coult., Mém. Dipsac., 32 (1823) nom. rejic.
     Scabiosa plumosa (L.) Sm. in Sibth. et Sm., Fl. Graec. Prodr., 1: 84 (1806), Fl. Graec., 2: 11, t. 111 (1813); Clarke, Travels, 2 (3): 723 (1816).
TYPE: Not localized.

An erect annual to about 50 cm. high; stem and leaves clothed with a mixed indumentum of short, glandular and long, eglandular hairs; lower leaves oblong or obovate, crenate-dentate or pinnatisect with a large terminal lobe, upper leaves usually pinnatisect; involucral bracts lanceolate, entire, acute, about as long as the outermost florets; corolla mauve or pinkish, the lower lobes of the outermost florets somewhat enlarged and radiant. Fruiting involucel minutely dentate at apex, but without a corona or internal collar; involucel-tube ribbed, the ribs joined above; calyx-bristles 10–15, about 9–11 mm. long.

The occurrence of *P. plumosus* on Cyprus is very questionable. It was first noted by Clarke in 1801 (Travels 2 (3): 723; 1816) and subsequently by Kotschy (Die Insel Cypern, 232; 1865) from (7) Ayios Khrysostomos (1859, *Kotschy* Suppl. 409) and (1) Paphos (1862, *Kotschy* 660). Sintenis (Oesterr. Bot. Zeitschr., 32: 124, 364; 1882) records it from near Lefkoniko and Cape Andreas, but in both instances the plant collected was *P. brevis* Coult. H. Stuart Thompson (Journ. Bot., 44: 309; 1906) and Holmboe (Veg. Cypr., 175; 1914) repeat the Kotschy records. Setting aside Clarke's unlocalized reference, evidence for the occurrence of *P. plumosus* on Cyprus is based solely on Kotschy's citations, and, in view of the fact that no one else has found the plant, it may be wondered if these too are not misidentifications of *P. brevis*. *P. plumosus* is found in the Balkan peninsula, Aegean Islands, Turkey, W. Syria and eastwards to N. Iran.]

1. **P. brevis** *Coult.*, Mém. Dipsac., 32, t. 1, fig. 16 (1823); B. L. Burtt in Notes Roy. Bot. Gard. Edinb., 22: 280 (1957); Davis, Fl. Turkey, 4: 623 (1972); Osorio-Tafall et Seraphim, List Vasc. Plants Cyprus, 100 (1973).

Scabiosa involucrata Sm. in Sibth. et Sm., Fl. Graec. Prodr., 1: 84 (1806), Fl. Graec., 2: 11, t. 112 (1813) nom. illeg.

Pterocephalus involucratus Spreng., Syst. Veg., 1: 384 (1824); Boiss., Fl. Orient., 3: 148 (1875); Post, Fl. Pal., ed. 2, 1: 619 (1932).

P. papposus (L.) Coult. var. *luteiflorus* Lindberg f., Iter Cypr., 33 (1946).

[*P. papposus* (nec *Scabiosa papposa* L.) Poech, Enum. Plant. Ins. Cypr., 16 (1842); Unger et Kotschy, Die Insel Cypern, 232 (1865); Holmboe, Veg. Cypr., 175 (1914).]

[*P. plumosus* (non (L.) Coult.) Sintenis in Oesterr. Bot. Zeitschr., 32: 124, 364 (1882).]

TYPE: "Patria ignota" (G).

Erect or spreading annual, 5–30 (–40) cm. high; stems usually much branched, subterete, clothed with a mixed indumentum of long, eglandular, and short, glandular hairs; leaves oblong in outline, 2–7 cm. long, 0·5–2 cm. wide, greyish, pubescent, the lowermost irregularly serrate-laciniate, the upper pinnatisect, with numerous oblong, acute or obtuse, serrate or subentire lobes, base tapering to a short, flattened petiole; inflorescence a spreading cyme, or sometimes reduced to a solitary capitulum in starved specimens; peduncles up to 25 cm. long, rather slender, thinly hirsute; capitula depressed-hemispherical, 1·5–2·5 cm. diam.; involucral bracts oblong-lanceolate, 1–2·5 cm. long, 0·3–0·6 cm. wide, herbaceous, pubescent, pinnatifid with narrow lobes, or often entire, usually equalling or exceeding the florets; receptacle strongly convex, thinly hairy, florets attached to peg-like protuberances, scales absent; corolla mauve or dirty white, pilose externally, glandular internally, tube 4–5 mm. long, 1 mm. wide, lobes unequal, ovate, acute, the lowermost (often stained purple) 1–2 mm. long, 0·7 mm. wide, the other 4 about 0·8–1 mm. long, 0·5 mm. wide; filaments slender, glabrous, attached near base of tube, about 3–4 mm. long; anthers yellow, oblong, about 0·7 mm. long, 0·4 mm. wide; style 5–6 mm. long, thickening towards apex; stigma narrowly clavate, obscurely 2-lobed. Fruiting involucel about 5 mm. long, 2 mm. wide, conspicuously 8-ribbed, pallid or tinged purplish, with strigose ribs, apex with a very narrow, barely visible corona; interior glabrous, with a narrow collar surrounding the apex of ovary; fruit fusiform, about 2 mm. long, 0·7 mm. wide, pubescent; calyx shortly stipitate, calyx-bristles 10–12, white-plumose, about 6–7 mm. long.

HAB.: On dry calcareous slopes and rocky pastures, in garigue; sea-level to 2,600 ft. alt.; fl. March–May.

DISTR.: Divisions 1, 3–8. Also Karpathos, S.E. Turkey, Syria, Palestine, Iraq, Iran.

1. Between Neokhorio and Smyies, 1941, *Davis* 3304A! Akamas, 1948, *Kennedy* in *Casey* 173!
3. Kophinou, 1905, *Holmboe* 584; Parekklisha, 1948, *Kennedy* 1616! Mile 22, Nicosia–Limassol road, 1950, *Chapman* 551! Romios Rocks, 1950, *F. J. F. Barrington* 18!
4. Akhyritou, 1962, *Meikle* 2631!
5. Between Lefkoniko and Kythrea, 1880, *Sintenis & Rigo* 42! Pera, 1908, *Clement Reid* s.n.! Athalassa, 1939, *Lindberg f.* s.n.; Ayia Katerina, Salamis, 1948, *Mavromoustakis* s.n.! Mia Milea, 1950, *Chapman* 670!
6. Vizakia–Koutraphas, 1936, *Syngrassides* 1072! Mile 3, Nicosia–Myrtou road, 1950, *Chapman* 709!
7. Buffavento, 1859, *Kotschy* 409; Lapithos Pass, 1902, *A. G. & M. E. Lascelles* s.n.! Near Kyrenia, 1936, *M. E. Dray* 6! St. Hilarion, 1939, *Lindberg f.* s.n.! Buffavento, 1941, *Davis* 2815! Yaïla, 1941, *Davis* 2831A! St. Hilarion, 1949, *Casey* 694! Melounda, 1963, *J. B. Suart* 93!
8. Cape Andreas, 1880, *Sintenis & Rigo*; Ephtakomi, 1905, *Holmboe* 530.

NOTES: Lindberg f. (Iter Cypr., 33; 1946) distinguishes his var. *luteiflorus* from the type by its entire involucral bracts and yellow florets. Entire or pinnatifid bracts are equally frequent in Cyprus specimens, and one collector (*M. E. Dray* 6) has noted "Flower whitish-mauve. It has turned yellowish in drying". In the circumstances, it seems best to reduce var. *luteiflorus* to synonymy under *P. brevis*.

2. P. multiflorus Poech, Enum. Plant. Ins. Cypr., 16 (1842); Unger et Kotschy, Die Insel Cypern, 233 (1865); Boiss., Fl. Orient., 3: 150 (1875); Holmboe, Veg. Cypr., 176 (1914); Lindberg f., Iter Cypr., 33 (1946); Chapman, Cyprus Trees and Shrubs, 78 (1949).
P. cyprius Boiss., Diagn., 1, 2: 110 (1843).

Much-branched shrub 30–100 cm. high; branches erect or spreading, the old ones covered with rough fuscous bark, the young shoots 4-angled, clothed with long, eglandular and short, glandular hairs; leaves elliptic, obovate or subspathulate, 1–4 cm. long, 0·5–1·5 cm. wide, thinly or rather densely canescent with a mixed indumentum of glandular and eglandular hairs, apex obtuse or acute, margins serrate or crenate, base tapering to or abruptly narrowed to an indistinct, flattened petiole 0·5–1·5 cm. long; inflorescence a lax, spreading cyme, or capitula sometimes solitary in depauperate specimens; peduncles to 17 cm. long; capitula depressed-hemispherical, 1·2–3 cm. diam.; involucral bracts lanceolate, in several series, 7–15 mm. long, 2–5 mm. wide, shorter than or exceeding the florets, herbaceous, ± densely pilose with eglandular and glandular hairs, apex subacute to acuminate, margins entire; receptacular scales wanting, florets attached to short, peg-like, shortly pilose protuberances; florets pinkish-mauve, hairy externally, tube cylindrical 6–8 mm. long, 1–2 mm. wide, lobes unequal, the 3 lower about 2–3 mm. long, 1–1·5 mm. wide, oblong, acute, the 2 upper about 1·5 mm. long, 0·8 mm. wide; filaments slender, glabrous, about 4 mm. long, attached just below the middle of the corolla-tube; anthers yellow, oblong, about 0·6 mm. long, 0·4 mm. wide; style 8–10 mm. long; stigma clavate or almost spathulate, oblique, obscurely 2-lobed. Fruiting involucel urceolate, closely enveloping fruit, about 4 mm. long, 1·8 mm. wide, subterete or obscurely 8-costate, hairy; corona short, undulate at apex, not clearly differentiated from the rest of involucel; calyx shortly stipitate, calyx-bristles about 20-28, white plumose, 6–7 mm. long, free almost to base; fruit fusiform, about 3 mm. long, 1·4 mm. wide, very thinly adpressed-hairy.

ssp. **multiflorus**
TYPE: Cyprus; "in monte 'Olympo' insulae Cypri (Kotschy *pl. Cypr*!)", 1840 (W).

Leaves elliptic, tapering at apex and base, often densely pilose; inflorescences usually much branched, bearing 5–9 (or more) capitula; individual capitula rather small, 1–2 cm. diam.

HAB.: In garigue on rocky igneous or calcareous slopes; in open Pine forest; in vineyards and by roadsides; 1,500–5,800 ft. alt.; fl. July–Nov.

DISTR.: Divisions 2, 3. Endemic.

2. Locally abundant; Prodhromos, Troödos, Platania, Platres, Selladhi tou Petrou, etc. *Aucher-Éloy* 756; *Kotschy* 747! *Post* s.n.! *Haradjian* 453! 526! 950! *C. B. Ussher* 74! *Chapman* 97! 298! *Kennedy* 909! 914! 915! 1433! *Wyatt* 10! *Lindberg f.* s.n.! *Davis* 1828! *Mavromoustakis* 46! *Casey* 899! *F. M. Probyn* s.n.! *Mrs. H. R. P. Dickson* 09! *G. E. Atherton* 526! *N. Macdonald* 201! *D. P. Young* 7416! *Economides* in ARI 1218! *Barclay* 1106! *A. Genneou* in ARI 1560! *P. Laukkonen* 921!

3. Trimiklini, 1935, *Syngrassides* 736! Lefkara, 1941, *Davis* 2735! S. of Kalokhorio, 1959, *P. H. Oswald* 144!

NOTES: The type of *P. multiflorus* is a Kotschy specimen collected in Cyprus in the autumn of 1840; it would appear, however, that Aucher-Éloy was the first to find the plant, probably high on the Troödos Range, in August 1831.

ssp. **obtusifolius** *Holmboe*, Veg. Cypr., 176 (1914); Chapman, Cyprus Trees and Shrubs, 78 (1949).

　　P. obtusifolius (Holmboe) C. E. Gresham in New Flora & Silva, 5: 271 (1933).

TYPE: Cyprus; "Kythrea and Lapithos", *Sintenis & Rigo* 579! *Holmboe* 837 (K! O).

Leaves obovate or subspathulate, often rather thinly hairy, with a blunt or subacute apex and a suddenly narrowed base; inflorescences often (but not always) rather sparingly branched, bearing (1–) 3–5 capitula; individual capitula rather large, frequently 2–3 cm. diam.

HAB.: Dry calcareous slopes and rocky ground; sea-level to 2,500 ft. alt.; fl. April–Nov.

DISTR.: Divisions 3, 6, 7. Endemic.

3. Akrotiri Forest, 1939, *Mavromoustakis* 99! Cape Gata, 1941, *Davis* 3572!

6. Ayia Irini, 1941, *Davis* 2564!

7. Common on the Northern Range; *Sintenis & Rigo* 579! *A. G. & M. E. Lascelles* s.n.! *Holmboe* 837; *Kennedy* 916! 1432! *Lindberg f.* s.n.! *Davis* 1735! 2196! 2940! 3639! *Casey* 214! *Miss Mapple* 45! *G. E. Atherton* 46! 146! 747! *I. M. Hecker* 8!

NOTES: Distinguishable from ssp. *multiflorus* only by its proportionately broader, blunter leaves. The other characteristics mentioned by Holmboe (i.e. the toothing of the leaves, number and size of capitula) are subject to frequent exception when a large range of specimens is available. However, such slight differences as do exist deserve recognition, if only because *P. multiflorus* ssp. *obtusifolius* is evidently in the first stages of speciation, a process completed, or nearly completed, in many vicariads with a Troödos/Northern Range distribution.

Kew material of *Sintenis & Rigo* 579 (a syntype of *P. multiflorus* Poech ssp. *obtusifolius* Holmboe) is variously labelled "In mont. supra Lapithos. 31/5 [1880]" or "In vineis pr. Galata. 24/6 [1880]", though it is clear that the specimens themselves all belong to the same subspecies, that is, to ssp. *obtusifolius*. One must assume that all the Kew specimens came from Lapithos; it is possible, however, that typical *P. multiflorus* from Galata (2) may have been distributed elsewhere under the same number.

53. COMPOSITAE

Annual or perennial herbs or subshrubs, rarely trees or climbers; leaves alternate, opposite or basal, generally exstipulate, simple or compound, frequently lobed or dissected; inflorescences branched or unbranched, sometimes scapose, terminal, axillary or leaf-opposed, consisting of small flowers (*florets*) arranged in heads (*capitula*) surrounded by few or many, overlapping bracts (*phyllaries*) upon a foveolate, glabrous, hairy or scaly receptacle; florets unisexual, hermaphrodite or neuter, all alike (*homogamous*) or sexually diverse (*heterogamous*), actinomorphic and tubular or campanulate, or variously zygomorphic, the marginal in each capitulum frequently ligulate (*ray-florets*) the central tubular (*disk-florets*); calyx wholly absent, or reduced to hairs, bristles or scales, commonly accrescent after fertilization and crowning the fruit with a *pappus*; corolla generally 5- or 3-toothed or lobed at the apex; stamens 5, inserted on the corolla, and alternating with the corolla-lobes; filaments short; anthers introrse, usually cohering marginally into a tube around the style, commonly with apical and basal appendages; ovary inferior, 1-locular; ovule solitary, erect, basal, anatropous; style solitary, divided into 2 stigmatiferous branches. Fruit an achene (or rarely fleshy); seed without endosperm; embryo straight.

One of the largest of flowering plant families, reputed to include more than 900 genera and 13,000 species, distributed throughout temperate and tropical regions, and particularly well represented in the Mediterranean area and western Asia.

The family includes many ornamental plants, but in proportion to its size, relatively few culinary or medicinal herbs or vegetables, though the Lettuce (*Lactuca sativa* L.), Sunflower (*Helianthus annuus* L.), Globe Artichoke (*Cynara scolymus* L.) and Chicory (*Cichorium intybus* L.) are of considerable economic importance.

In the following account, the term *inflorescence* refers to the whole flowering region of the plant, including all the true inflorescences, or capitula.

Male and female florets in separate unisexual capitula, the male (staminate) capitula terminal; the female (pistillate) lower down, in the axils of the upper leaves:
    Fruiting involucre subglobose or broadly ovoid, about 5 mm. long, tuberculate but not spinose; male involucre with phyllaries united into a cup - - - **14. Ambrosia**
    Fruiting involucre oblong-ellipsoid or ovoid, about 13 mm. long, rigidly spinose; male involucre with free phyllaries - - - - - - - **15. Xanthium**
Florets all hermaphrodite, or hermaphrodite, female or neuter together in the same capitulum:
    Florets all ligulate, ligules generally with 5 small apical teeth; plants generally bearing latex (*Lactucoideae*):
        Florets blue, purple, pink or white:
            Achenes obovoid, not tapering apically or beaked; pappus of short scales; florets generally sky-blue (rarely white); phyllaries generally in 2 series - **55. Cichorium**
            Achenes cylindric or fusiform, tapering apically or beaked; pappus of hairs or bristles:
                Pappus of scabridulous or barbellate hairs; phyllaries in several series
                        73. Prenanthes
                Pappus at least in part of plumose hairs; phyllaries apparently in 1 series:
                    Achenes uniform; all the pappus of plumose hairs; receptacle naked
                        77. Tragopogon
                    Achenes dimorphic, the outer with a pappus of scabridulous bristles, the inner with plumose hairs; receptacle with a few, filiform scales - **78. Geropogon**
        Florets yellow or yellowish:
            Leaves and involucres spinose; stems spinose-winged; plants Thistle-like **53. Scolymus**
            Leaves and involucre not spinose; stems not spinose-winged; plants not Thistle-like:
                Leaves linear, entire, grass-like:
                    Inflorescence branched; capitula small; achenes linear, strongly curved, spinulose dorsally; pappus much reduced and inconspicuous - - **58. Koelpinia**
                    Inflorescence not branched; capitula large, solitary, terminal; achenes neither strongly curved nor spinulose dorsally; pappus of conspicuous, plumose hairs
                        77. Tragopogon
                Leaves not linear or grass-like:
                  Phyllaries wholly scarious or membranous - - - - **54. Catananche**
                  Phyllaries wholly herbaceous, or, at most, with scarious margins:
                    Pappus minute or wanting; marginal achenes subulate or narrowly fusiform, partly enveloped by the persistent, accrescent phyllaries, usually stellately patent - - - - - - - **59. Rhagadiolus**
                  Pappus present; achenes not as above:
                    Leaves all basal, forming a rosette:
                      Inflorescence branched; receptacle scaly - - **65. Hypochaeris**
                      Inflorescence unbranched, capitula solitary, terminal on a leafless scape; receptacle naked:
                      Scape distinctly swollen towards apex; pappus wholly of small scales
                        56. Hyoseris
                      Scape not swollen towards apex; pappus at least in part of hairs:
                      Leaves hispidulous with compound, 2–3-branched bristles; pappus of outer achenes scaly; pappus of inner achenes of barbellate or plumose hairs - - - - - - - **66. Leontodon**

Leaves glabrous or almost glabrous; pappus uniform, of scabridulous
hairs   -   -   -   -   -   -   -   - **67. Taraxacum**
Leaves extending up the stems, not all basal:
Indumentum of glochidiate, apically branched hairs or bristles:
Outer phyllaries foliaceous, broadly ovate or cordate, wholly or partly
concealing the inner phyllaries   -   -   - **62. Helminthotheca**
Outer phyllaries small, short, not concealing the inner phyllaries
61. Picris
Indumentum of simple, not glochidiate, hairs or plants glabrous:
Receptacle scaly, with deciduous, membranous scales **65. Hypochaeris**
Receptacle naked:
Pappus of short scales; outer achenes strongly incurved, partly
concealed by the persistent, erect phyllaries   - **60. Hedypnois**
Pappus of hairs or bristles:
Achenes strongly compressed or flattened:
Cauline leaves much reduced, the bases strongly decurrent,
forming interrupted green wings down the stem
72. Scariola
Cauline leaves sometimes auriculate, but not forming wings
down the stem:
Involucre 1·5–2·5 cm. long at anthesis (up to 3 cm. long in
fruit); achenes almost flattened with a narrow marginal
wing; root tuberous   -   -   - **70. Streptorhamphus**
Involucre less than 1·5 cm. long at anthesis; root not tuberous:
Involucre conical after anthesis; achenes without a beak
75. Sonchus
Involucre cylindrical or campanulate after anthesis;
achenes with a slender, abruptly differentiated beak
71. Lactuca
Achenes not strongly compressed or flattened:
Plants stoloniferous, stems creeping and rooting at the nodes;
roots bearing small, globose tubers   - **64. Aetheorhiza**
Plants not stoloniferous, stems not creeping:
Achenes with a distinct (sometimes deciduous) apical beak:
Base of stem setose with fulvous bristles; cauline leaves
sparse, often much reduced   -   - **68. Chondrilla**
Base of stem not setose; cauline leaves usually well
developed:
Beak of achene oblique, hollow at base; phyllaries rather
fleshy, apparently in one series; plants not tuberous
79. Urospermum
Beak of achene not oblique or hollowed at base; phyllaries
in several series, not fleshy; tuberous perennials
69. Cephalorrhynchus
Achenes without an apical beak:
Pappus-hairs wholly or partly plumose   - **80. Scorzonera**
Pappus-hairs smooth, scabridulous or barbellate, but not
plumose:
Peduncles bracteate, the bracts sometimes merging with
the outer phyllaries:
Phyllaries   broad,   with   conspicuous,   scarious-
membranous margins:
Pappus deciduous; pappus-hairs uniform, connate at
base into a ring   -   -   - **74. Reichardia**
Pappus persistent; pappus-hairs dimorphic, not
connate at base into a ring   - **76. Launaea**
Phyllaries   narrow,   without   distinct   scarious-
membranous margins -   -   -   - **57. Tolpis**
Peduncles not bracteate; outer phyllaries commonly
distinct from inner, forming an ill-defined calyculus
63. Crepis
Florets not all ligulate, the inner (disk-florets) tubular, the outer (ray-florets) sometimes
ligulate, or sometimes all the florets tubular; plants seldom bearing latex (*Asteroideae*):
Capitula aggregated into a globose or ovoid, spinose head:
Phyllaries without a metallic sheen, fused into a spine-tipped, turbinate cupule; plant
lactiferous -   -   -   -   -   -   -   -   -   -   -   -   - **28. Gundelia**
Phyllaries with a metallic sheen; not fused into a cupule; plant not lactiferous
29. Echinops

Capitula not aggregated into a head:
Leaves broadly ovate or suborbicular, cordate at base, almost as broad as long:
Phyllaries terminating in a slender, hooked spine; florets purple or purplish

    34. Arctium

Phyllaries not spine-tipped; florets yellow - - - - - **25. Tussilago**

Leaves not broadly ovate, suborbicular or cordate at base, generally much longer than broad:
Involucre of numerous, imbricate, spine-tipped phyllaries ("Thistles"):
Stems with spinose wings:
Pappus-hairs plumose:
Receptacle without scales or bristles - - - - **42. Onopordum**
Receptacle scaly or bristly:
Median phyllaries with pectinate-spinose appendages; leaves and stem-wings armed with golden spines; capitula surrounded and overtopped by uppermost leaves - - - - - - - **38. Picnomon**
Median phyllaries simply spinose; leaves and stems without golden spines; capitula not overtopped by uppermost leaves - - **39. Cirsium**
Pappus-hairs smooth, scabridulous or barbellate, not plumose:
Filaments free; anthers caudate-sagittate at base; achenes somewhat compressed, without mucilage-channels in the furrows - **35. Carduus**
Filaments united into a ring around the style; anthers not cordate-sagittate at base; achenes 4-angled, with mucilage-channels in the furrows

    36. Tyrimnus

Stems without spinose wings:
Leaves linear-subulate (resembling Pine-needles); shrubs or subshrubs

    40. Ptilostemon

Leaves not linear-subulate; plants herbaceous:
Leaves spiny:
Pappus scaly:
Achenes sericeous-hairy:
Innermost phyllaries conspicuously enlarged, scarious, coloured, radiating like ray-florets - - - - - **32. Carlina**
Innermost phyllaries not conspicuously enlarged, not coloured or radiating; florets bright blue - - - **30. Cardopatium**
Achenes glabrous - - - - - - - - **51. Carthamus**
Pappus hairy or bristly:
Achenes sericeous-hairy; outermost phyllaries forming a pectinate-spinulose "cage" around the inner phyllaries - - **33. Atractylis**
Achenes glabrous:
Pappus-hairs (or some of them) plumose:
Uppermost leaves forming a coloured involucre around the capitula

    37. Notobasis

Uppermost leaves not forming a coloured involucre   **43. Cynara**
Pappus-hairs smooth, scabridulous or barbellate, not plumose:
Leaves white-veined; phyllaries terminating in a rigid, recurved spine

    44. Silybum

Leaves not white-veined; phyllaries erect, the innermost with a pectinate-fimbriate appendage   - - **52. Carduncellus**
Leaves not spiny:
Capitula with ligulate ray-florets - - - - **13. Pallenis**
Capitula without ligulate ray-florets - - - - **48. Centaurea**
Involucre smooth or almost smooth, not spine-tipped or prickly to the touch:
Capitula with ligulate ray-florets:
Ligules blue or purplish - - - - - - - - ASTER (p. 871)
Ligules white or yellow:
Leaves deeply dissected into narrow segments:
Leaves opposite; phyllaries united into a cylindrical involucre

    16. Tagetes

Leaves alternate, phyllaries not united:
Capitula numerous, small in flattish, dense, corymbose heads

    17. Achillea

Capitula not in dense corymbose heads:
Receptacle scaly - - - - - - **19. Anthemis**
Receptacle naked:
Achenes crowned with a hairy pappus - - - **26. Senecio**
Achenes without a hairy pappus:
Ultimate leaf-divisions very narrowly linear, almost filiform

    22. Matricaria

Ultimate leaf-divisions broader, flattened, often dentate, not filiform:
    Leaves strongly aromatic; plant perennial   - **21. Tanacetum**
    Leaves not strongly aromatic; plant annual
        20. Chrysanthemum
Leaves entire or shortly dentate, not dissected or lobed:
  Achenes without a pappus:
    Ray-florets white or pinkish; achenes uniform:
      Capitula solitary; leaves not aromatic -   -   - **2. Bellis**
      Capitula numerous forming a branched, corymbose inflorescence; leaves aromatic - - - - - - **21. Tanacetum**
    Ray-florets yellow; achenes polymorphic, elongate, often incurved and dorsally tuberculate or echinate -   -   - **27. Calendula**
  Achenes with a hairy or scaly pappus; ray-florets yellow:
    Receptacle scaly or bristly:
      Outer phyllaries blunt, not spine-tipped -   -   **12. Asteriscus**
      Outer phyllaries spine-tipped -   -   -   - **13. Pallenis**
    Receptacle naked:
      Pappus consisting wholly of hairs:
        Phyllaries in several, imbricate series, free -   -   **10. Inula**
        Phyllaries in 1 series, coherent laterally to form a tube or cup
          26. Senecio
      Pappus with an outer row of short, partly connate scales **11. Pulicaria**
Capitula without ligulate ray-florets:
  Subshrubs with stems woody towards base:
    Phyllaries scarious, glossy:
      Capitula in corymbose heads; phyllaries yellow - - **9. Helichrysum**
      Capitula solitary; phyllaries brown -   -   -   - **7. Phagnalon**
    Phyllaries not scarious or glossy:
      Leaves and stems densely tomentose; capitula subglobose - **18. Otanthus**
      Leaves and stems not tomentose; capitula cylindrical or narrowly ovoid
        45. Staehelina
  Herbaceous annuals, biennials or perennials:
    Florets blue, purple, mauve or pink:
      Inflorescence crowded, corymbose, consisting of numerous small capitula
        1. Eupatorium
      Inflorescence not crowded or corymbose; capitula usually rather large:
        Phyllaries with a pectinate apical appendage - -   - **48. Centaurea**
        Phyllaries without a pectinate apical appendage:
          Pappus-hairs plumose -   -   -   -   - **43. Cynara**
          Pappus-hairs smooth, scabridulous or barbellate, not plumose, or pappus scaly:
            Leaves entire or almost entire:
              Innermost phyllaries conspicuously enlarged, scarious, pinkish or purplish, resembling ligulate ray-florets **31. Xeranthemum**
              Innermost phyllaries not conspicuously enlarged or coloured
                41. Jurinea
            Leaves deeply lobed or pinnatisect:
              Achenes sericeous; inner phyllaries scarious, lanceolate-acuminate; slender annual -   -   -   -   - **47. Crupina**
              Achenes glabrous, inner phyllaries indurated, black-tipped, obtuse or shortly acute; perennial -   -   - **49. Mantisalca**
    Florets yellow, orange or brownish:
      Receptacle naked:
        Leaves (or at least the basal ones) lobed, deeply toothed or dissected:
          Achenes with a hairy pappus, capitula small, numerous in panicles
            3. Conyza
          Achenes without a hairy pappus:
            Capitula solitary, terminal:
              Leaves rather fleshy, sparingly lobed - - **23. Chlamydophora**
              Leaves not fleshy, dissected into numerous, slender, filiform segments -   -   -   -   - **22. Matricaria**
            Capitula in racemes or panicles, individually very small
              24. Artemisia
        Leaves entire or shortly dentate:
          Phyllaries scarious or membranous, yellowish or brownish; leaves entire, lanuginose or tomentose - - **8. Pseudognaphalium**
          Phyllaries not scarious or membranous; leaves shortly dentate, gland-dotted, aromatic, sparsely pubescent -   -   **21. Tanacetum**

Receptacle scaly or bristly:
Pappus absent:
Upper cauline leaves forming a rosette around the subglobose clusters
of sessile capitula -   -   -   -   -   -   -   **4. Evax**
Upper cauline leaves not forming a rosette around a sessile cluster of
capitula:
Annuals:
Capitula large, solitary; phyllaries subglabrous or thinly arach-
noid; receptacle bristly   -   -   -   - **51. Carthamus**
Capitula small, clustered, woolly all over; receptacle scaly, the
scales saccate, deciduous with achene -   **5. Bombycilaena**
Perennial; leaves and stems densely white-tomentose; capitula in
small terminal corymbs -   -   -   -   - **18. Otanthus**
Pappus present:
Capitula large, generally solitary; plants glabrous or thinly villous:
Stems 20–50 cm. long; phyllaries scarious, yellowish, the inner
acuminate, without an apical appendage; perennial
46. Serratula
Stems generally less than 20 cm. long, often very short; phyllaries not
scarious or yellowish, the inner with a pectinate-spinulose apical
appendage; annual -   -   -   -   -   - **50. Cnicus**
Capitula very small, clustered; tomentose annuals   -   - **6. Filago**

SUBFAMILY 1. **Asteroideae** Plants generally without latex; corolla of disk-
florets not ligulate; pollen grains usually with uniformly distributed spines.

TRIBE 1. **Eupatorieae** *Cass.* Leaves usually opposite; capitula homogamous, discoid, usually in corymbs; florets all tubular; receptacle naked;
corolla purple, pink or white, never yellow; anther-base obtuse; style-
branches elongate, obtuse or clavate. Achenes without a beak; pappus of
scabridulous hairs.

1. EUPATORIUM *L.*

Sp. Plant., ed. 1, 836 (1753).
Gen. Plant., ed. 5, 363 (1754).

Perennial herbs, shrubs and subshrubs; leaves opposite or whorled, simple
or ternatisect; capitula homogamous, discoid, in terminal corymbs;
involucre cylindrical or campanulate, with few or many imbricate phyllaries
in 2 or more series; corolla whitish, pink or purplish, tubular, shortly 5-
lobed; anthers with obtuse bases. Achenes cylindrical or fusiform, truncate,
5-angled, sparingly glandular; pappus-hairs as long as corollas, minutely
scabridulous, often tinged brown or purple.

About 1,200 species, chiefly in America, with relatively few representatives in Europe, Africa and Asia.

1. **E. cannabinum** *L.*, Sp. Plant., ed. 1, 838 (1753); Boiss., Fl. Orient., 3: 154 (1875); Post, Fl.
Pal., ed. 2, 2: 16 (1933); Lindberg f., Iter Cypr., 35 (1946); Osorio-Tafall et Seraphim, List
Vasc. Plants Cyprus, 105 (1973); Davis, Fl. Turkey, 5: 173 (1975); Feinbrun, Fl. Palaest., 3:
295, t. 490 (1978).

Erect perennial herb up to 1·5 m. high; stems longitudinally sulcate,
sparsely or densely crispate-pubescent, often purplish; leaves ternatisect or
the upper ones simple, divisions ovate-elliptic or lanceolate, 2–15 cm. long,
1–4 cm. wide, dull green on both sides, sparsely or rather densely crisped-
pubescent, the undersurface thinly dotted over with shining, sessile glands,
apex acute or acuminate, margins remotely serrate; petiole short, usually
less than 2 cm. long; inflorescence usually much branched, the partial
inflorescences densely corymbose; capitula small, 5–7-flowered; involucre
cylindrical-campanulate, about 4–6 mm. long, 2·5 mm. wide, phyllaries few
(about 10), rather loose, in 3 series, oblong, 1–6 mm. long, about 1 mm. wide,

pubescent externally, the inner scarious, often purple-tinged; corolla purplish, about 6 mm. long, 1·5 mm. wide at apex, lobes deltoid, erect, about 0·8 mm. long; anthers about 1·5 mm. long, 0·5 mm. wide, apex subacutely appendiculate, base obtuse; style-base about 3 mm. long, arms linear, obtuse, about 4·5 mm. long, minutely papillose in the upper half. Achene narrowly fusiform, angled, fuscous, about 3 mm. long, 0·6 mm. wide, sparsely glandular; pappus-hairs whitish, about 4·5 mm. long, minutely scabridulous.

var. syriacum (*Jacq.*) *Boiss.*, Fl. Orient., 3: 154 (1875); Post, Fl. Pal., ed. 2, 2: 16 (1933).
    E. syriacum Jacq., Misc. Austr., 2: 349 (1781–82), Icones Plant. Rar., 1: 17, t. 170 (1782).
    E. cannabinum L. ssp. *syriacum* (Jacq.) Lindberg f., Iter Cypr., 35 (1946); Osorio-Tafall et Seraphim, List Vasc. Plants Cyprus, 105 (1973).
    TYPE: Cultivated in Vienna from material "ex horto Argentoratensi [Strasbourg]", probably sent by J. R. Spielmann, who sent Jacquin *Cistus* (*Helianthemum*) *syriacus* and other Levantine plants.

Plant generally less robust than typical *E. cannabinum*; leaf-divisions shorter and broader, often ovate-elliptic, acute, usually less than 7 cm. long, often 3 cm. wide, together with the young stems often rather densely white-pubescent; corymbs rather small, with relatively few capitula.

HAB.: Moist ground by springs; c. 500 ft. alt.; fl. Aug.–Oct.

DISTR.: Division 7, rare. Syria, Lebanon, Palestine.

7. Near the spring at Lapithos, 1939, *Lindberg f.* s.n.!

NOTES: Only one immature, sterile specimen seen, and should be re-collected in better condition. Perhaps overlooked by other collectors because of its late flowering, but evidently rare. The variety is dismissed by most authors as scarcely deserving of recognition.

TRIBE 2. **Astereae** Leaves alternate or basal, usually simple; capitula heterogamous and radiate, or homogamous and discoid; involucre campanulate or shallow-concave; receptacle naked; corolla white, yellow, pink, purple or blue; anther-bases acute or obtuse; style-branches compressed or flattened with a sterile apical appendage. Achenes compressed, not beaked; pappus of scabridulous hairs, or wanting.

2. BELLIS *L.*

Sp. Plant., ed. 1, 886 (1753).
Gen. Plant., ed. 5, 378 (1754).

Annual or perennial herbs; stems sometimes very short or almost wanting; leaves alternate or sub-basal, simple, oblanceolate or spathulate, entire or serrate; capitula solitary, pedunculate, often scapose, heterogamous, radiate; involucre campanulate, phyllaries herbaceous, in 1–2 series, often blackish-green; ray-florets female, ligulate, in 1 series, reddish, pink, white or bluish; disk-florets cylindrical, 5-lobed, hermaphrodite, yellow. Achenes compressed, with a thickened margin; pappus absent or of short, inconspicuous bristles.

About 15 species chiefly in Europe and the Mediterranean region.

Plants perennial, acaulescent or with very short stems:
    Phyllaries ovate-oblong, 3–7 mm. long, with convex sides, subglabrous in the upper half; leaves usually spathulate, narrowing rather abruptly to a distinct petiole; ray-florets 5–8 mm. long - - - - - - - - - - - **1. B. perennis**
    Phyllaries narrowly oblong, (3–) 5–12 mm. long, with sub-parallel sides, thinly pilose all over or ciliate all round; leaves tapering gradually to base, sessile or subsessile; ray-florets 10–17 mm. long- - - - - - - - - - - **2. B. sylvestris**
Plant annual, stems well developed, slender - - - - - **3. B. annua** ssp. **minuta**

1. B. perennis *L.*, Sp. Plant., ed. 1, 886 (1753); Unger et Kotschy, Die Insel Cypern, 234 (1865); Boiss., Fl. Orient., 3: 173 (1875); Holmboe, Veg. Cypr., 177 (1914); Post, Fl. Pal., ed. 2, 2: 19 (1933); Osorio-Tafall et Seraphim, List Vasc. Plants Cyprus, 101 (1973); Davis, Fl. Turkey, 5: 135 (1975); Feinbrun, Fl. Palaest., 3: 296, t. 491 (1978).

TYPE: "*in Europae apricis pascuis*".

Perennial, sub-rosulate or caespitose herb; rootstock short, premorse, roots numerous, thick, fibrous; stems very short or almost wanting; leaves generally spathulate, crowded, 1–9 cm. long, 0·6–2 cm. wide, thinly hairy or subglabrous on both surfaces, indistinctly 3-nerved, apex rounded or obtuse, margins subentire or remotely and unequally serrate or serrulate, base tapering rather abruptly to a ± well-marked, flattened petiole 1–5 cm. long; inflorescences scapose; peduncles up to 25 cm. long, terete, thinly pilose, obscurely sulcate, often purplish, commonly rather densely adpressed-pubescent at apex; capitula solitary; involucre widely campanulate or saucer-shaped; phyllaries ovate-oblong, 3–7 mm. long, 1–2 mm. wide, dark blackish-green, pilose near base, glabrous or subglabrous above, and usually with an apical tuft of hairs, apex obtuse or rounded; ray-florets ligulate, white or with the underside reddish or pinkish, 5–8 mm. long, 1·5–2 mm. wide towards apex, 3-nerved, apex rounded or sometimes emarginate; disk-florets narrowly tubular-campanulate, yellow, about 2·5 mm. long, 1 mm. wide, with 5 erect, ovate-deltoid, papillose lobes about 0·5 mm. long; anthers about 1 mm. long, 0·3 mm. wide, base obtuse, apex shortly cuspidate; style about 1·5 mm. long, divided above into 2 erect branches about 0·5 mm. long, forming a clavate stigma; receptacle prominently conical, naked, up to 1 cm. long. Achenes strongly compressed, obovate, pale brown, thinly pubescent, with a conspicuous thickened margin, about 2 mm. long, 1 mm. wide.

HAB.: In moist turf, usually by springs or streams; 600–4,500 ft. alt.; fl. Febr.–June.

DISTR.: Divisions 2, 6, 7, not common. Widespread in Europe and east to the borders of Iran; an introduced weed in many temperate areas of the world.

2. Near Prodhromos, 1862, *Kotschy* 706; Pedhoulas, 1932, *Syngrassides* 41 ! Trikoukkia, 1937, *Kennedy* 933 ! Platres, 1937, *Kennedy* 934 ! and, 1941, *Davis* 3179 ! Moniatis, 1937, *Kennedy* 935 ! Platania, 1947, *Mavromoustakis* 30 ! also, 1949, *Casey* 840 ! Stavros tis Psokas, 1962, *Meikle* in K. 5441 ! Above Askas, 1979, *Edmondson & McClintock* E 2900 !
6. Before Myrtou on the Nicosia-Myrtou road, 1957, *Merton* 2849 !
7. Dhikomo, 1936, *Syngrassides* 1187 ! Kambyli, 1954, *Merton* 1857 ! 2521 !

2. B. sylvestris *Cyr.*, Plant. Rar. Neap., fasc. 2: 12, t. 4 (1792); Kotschy in Oesterr. Bot. Zeitschr., 12: 279 (1862) et in Unger et Kotschy, Die Insel Cypern, 234 (1865); Boiss., Fl. Orient., 3: 174 (1875); Holmboe, Veg. Cypr., 177 (1914); Post, Fl. Pal., ed. 2, 2: 20 (1933); Osorio-Tafall et Seraphim, List Vasc. Plants Cyprus, 101 (1973); Davis, Fl. Turkey, 5: 136 (1975); Feinbrun, Fl. Palaest., 3: 296, t. 492 (1978).

TYPE: Italy; Naples, "*nascitur in pratis, & secus margines agrorum, locis non apricis*" (? NAP).

Superficially very similar to *B. perennis*, but generally more robust with larger leaves and capitula; plants rosulate, generally solitary, not branched or caespitose as in *B. perennis*, and probably not persistently leafy as in the latter species; leaves oblanceolate, 3·5–11 cm. long, 0·5–2 cm. wide, dull and often rather densely crispate-pubescent on both surfaces, apex commonly acute or mucronate, but sometimes rounded, margins frequently serrate or serrulate but sometimes subentire, base tapering very gradually to an indistinct petiole, or leaves apparently sessile; scapes often few, frequently 17–25 (–30) cm. long, normally densely adpressed-pubescent and conspicuously white barbate-villose towards apex; capitula conspicuously large; phyllaries narrowly oblong with sub-parallel margins (not ovate-oblong with convex margins as in *B. perennis*), (3–) 5–12 mm. long, 2–3 mm. wide, blackish-green, thinly pilose all over, or at least with the margins distinctly ciliate throughout; ray-florets white or with a red or pink undersurface, often 10–17 mm. long, 2–2·5 mm. wide, irregularly 2–3-toothed at apex, conspicuously 3–4-nerved; disk-florets yellow, about 3 mm. long, 1 mm. wide, with 5, ovate-deltoid, erect apical lobes 0·5 mm. long; stamens and styles as in *B. perennis*; receptacle prominently conical, subglabrous or densely white pilose. Achenes strongly compressed, obovate, pale brown, thinly pubescent, with conspicuously thickened margins, apex somewhat depressed or even indented, about 2 mm. long, 1–1·5 mm. wide. *Plate 56, figs. 3, 5.*

HAB.: On hillsides in garigue; roadsides; grassy banks; streamsides; edges of fields; often in shade; sea-level to 3,000 ft. alt., usually lowland; fl. Oct–March.

DISTR.: Divisions 1–3, 7, 8. Mediterranean region.

1. Avgas near Ayios Yeoryios (Akamas), 1962, *Meikle* 2023!
2. Prodhromos, 1862, *Kotschy* 706a; between Pyrgos and Polis, 1935, *Syngrassides* 864! Omodhos, 1937, *Kennedy* 932!
3. Mazotos, 1905, *Holmboe* 197; Episkopi, 1963, *J. B. Suart* 134!
7. Ayios Epiktitos, 1880, *Sintenis & Rigo* 323! Kyrenia Pass, 1901, *A. G. & M. E. Lascelles* s.n.! Aghirda, 1901, *A. G. & M. E. Lascelles* s.n.! Kyrenia, 1936, *Kennedy* 936! also, 1955, *G. E. Atherton* 754! 768! Vasilia, 1940, *Davis* 2012! Bellapais, 1949, *Casey* 201! and, 1970, *I. M. Hecker* 5! Ayios Panteleimon near Myrtou, 1956, *Merton* 2463!
8. Kantara, 1880, *Sintenis & Rigo* 323a! also, 1905, *Holmboe* 547; Koma tou Yialou, 1935, *Syngrassides* 661!

NOTES: Many authors have commented on the difficulties of distinguishing *Bellis perennis* and *B. sylvestris*. In their typical forms both are easily recognised and separated, but whatever criteria one applies, one is inevitably left with a residue of connecting forms which might be referred to either species with equal justification. Leaf-shape, phyllary-shape and flower size seem to furnish the most useful distinctions, but further observation in the field, and experimental work, is called for as a basis for sound taxonomic assessment; meantime I have followed tradition in allowing both to rank as species.

3. B. annua L., Sp. Plant., ed. 1, 887 (1753); Sibth. et Sm., Fl. Graec. Prodr., 2: 184 (1813), Fl. Graec., 9: 59, t. 876 (1837); Boiss., Fl. Orient., 3: 175 (1875) pro parte; Post, Fl. Pal., ed. 2, 2: 20 (1963) pro parte.

Erect or ascending annual 3–14 cm. high; roots very slender, fibrous; stems usually well developed, slender, subglabrous or thinly pilose, obscurely sulcate; leaves alternate, sometimes crowded, bright green, obovate-spathulate, subglabrous or thinly pilose, 1–3 cm. long, 0·4–1 cm. wide, apex subacute, obtuse or rounded, margins irregularly serrate or shortly lobed or entire, base tapering gradually to an indistinct petiole; inflorescences sub-scapose, leaf-opposed; capitula solitary; peduncles very slender, 2–6 cm. long, thinly adpressed-pubescent; involucre widely campanulate or saucer-shaped; phyllaries ovate or lanceolate, acute or subacute, sometimes acuminate, subglabrous or thinly pilose, often edged blackish-green, about 2–3 mm. long, 1·5 mm. wide; ray-florets white, or tinged reddish-pink below, 3–8 mm. long, 0·8–1·5 mm. wide, obscurely nerved, apex shortly emarginate; disk-florets yellow, about 1–1·5 mm. long, 0·8 mm. wide, deeply divided into 4–5, erect, concave lobes; anthers 0·3–0·5 mm. long, 0·1–0·2 mm. wide, apex shortly cuspidate; style about 0·5–0·6 mm. long, branches erect, forming a small broadly clavate stigma; receptacle prominently conical. Achenes strongly compressed, obovate, about 1·5 mm. long, 1 mm. wide, pale brown, thinly pilose, apex very slightly emarginate, margins distinctly thickened.

ssp. **minuta** (*DC.*) *Meikle* **stat. nov.**
B. annua L. var. *minuta* DC., Prodr., 5: 304 (1836); Halácsy, Consp. Fl. Graec., 2: 13 (1902); Hayek, Prodr. Fl. Pen. Balc., 2: 581 (1931); Rechinger f., Fl. Aegaea, 608 (1943).
[*B. annua* (vix L.) Poech, Enum. Plant. Ins. Cypr., 17 (1842); Unger et Kotschy, Die Insel Cypern, 234 (1865); H. Stuart Thompson in Journ. Bot., 44: 332 (1906); Holmboe, Veg. Cypr., 177 (1914); Osorio-Tafall et Seraphim, List Vasc. Plants Cyprus, 101 (1973); Davis, Fl. Turkey, 5: 135 (1975) pro parte; Feinbrun, Fl. Palaest., 3: 297, t. 493 (1978).]

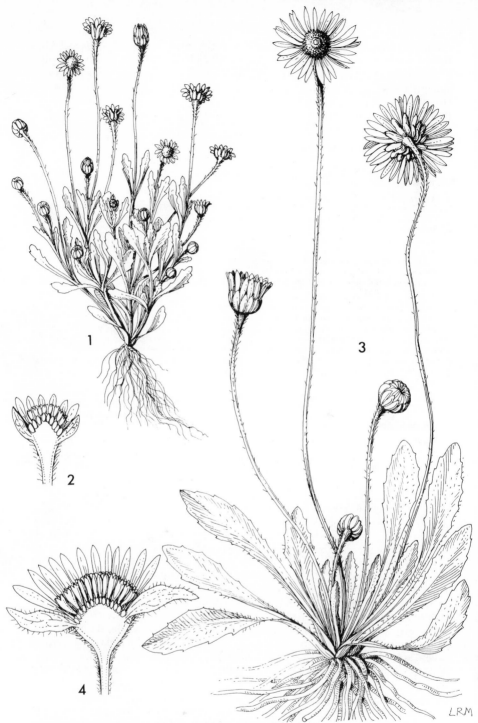

Plate 56. Figs. **1, 2.** BELLIS ANNUA L. ssp. MINUTA (DC.) Meikle **1,** habit, × 1; **2,** capitulum, longitudinal section, × 3; figs. **3, 4.** BELLIS SYLVESTRIS Cyr. **3,** habit, × 1; **4,** capitulum, longitudinal section, × 3. (**1,** from *Davis* 2378; **2** from *Hewer* 4724; **3, 4** from *A. G. & M. E. Lascelles* s.n.)

TYPE: "circa Byzantium locis humidis legit cl. L. Castagne et in Graeciâ (h. acad. Monac.) (G, K!).

Leaves in basal tufts or scattered along slender, branched stems, obovate-spathulate, irregularly serrate; involucre dark blackish-green, usually less than 7 mm. diam., phyllaries acute, 2–2·5 mm. long, 0·8–1 mm. wide; ray-florets about 3 mm. long, white, the ligule equalling or shorter than the involucre. *Plate 56, figs. 1, 2.*

HAB.: Moist depressions where water has stood during the winter; sea-level to 1,000 ft. alt.; fl. Febr.–March.

DISTR.: Divisions 1, 3, 4, 6–8. Eastern Mediterranean region from Greece and Crete to Palestine, and further east to Iran.

1. Near Paphos, 1840, *Kotschy* 63; Erimidhes near Ayios Yeoryios (Akamas), 1962, *Meikle* 2140!
3. Zakaki marshes, 1939, *Mavromoustakis* 9! 15!
4. Ayia Napa, 1905, *Holmboe* 37a; also, 1940, *Davis* 2051!
6. Marsh near Dhikomo, 1952, *Casey* 1220!
7. Near Ayios Khrysostomos Monastery, 1880, *Sintenis & Rigo* 322! Above Ayios Amvrosios, 1941, *Davis* 2165! Kambyli, 1956, *Merton* 2518!
8. Between Gastria and Áyios Theodhoros, 1936, *Syngrassides* 930! Ronnas Bay near Rizokarpaso, 1941, *Davis* 2378!

NOTES: All the Cyprus specimens are uniformly small-flowered, with the ray-florets shortly exceeding the involucre, whereas in typical *B. annua* the capitula are twice as large, often rivalling in size those of *B. perennis*. The distinction between the typical plant and ssp. *minuta* is well demonstrated if one compares the illustration in *Flora Graeca* 9: t. 876 with that in Feinbrun, *Fl. Palaest.*, 3: t. 493.

Bellis microcephala Lange (*B. annua* L. ssp. *microcephala* (Lange) Nyman) is similarly small-flowered, but is at once distinguished by its narrow, pallid, acuminate phyllaries, which are often distinctly hispidulous-pilose. *B. microcephala* is found in S. Spain and N. Africa, but is absent from our area.

ASTER SQUAMATUS (*Spreng.*) *Hieron.* in Engl. Bot. Jahrb., 29: 19 (1900) (*Conyza squamata* Spreng., Syst. Veg., 3: 515; 1826). An erect, glabrous perennial, with narrowly lanceolate or oblanceolate leaves 3–10 (–15) cm. long; inflorescence lax, branched, paniculate; involucre narrowly campanulate, about 8 mm. long, 6 mm. wide; phyllaries lanceolate- or oblanceolate-acuminate, in several series, often purplish towards apex; florets inconspicuous, whitish, pink or bluish, the rays largely concealed by the copious, dirty-white, scabridulous pappus. Achenes narrowly oblong, somewhat compressed, light brown, thinly pilose, about 2 mm. long, 0·4 mm. wide.

A native of South America, with a scattered distribution as an alien in the Mediterranean region. It has been once collected in Cyprus, in the garden at (3) Kolossi Castle, 1981, *Hewer* 4751! but, as yet, can scarcely be considered an established element in the flora. [Now, 1984, well established in (4.) Larnaca area, *A. Hansen*!]

3. CONYZA *Less.*

Syn. Gen. Comp., 203 (1832) nom. cons.

Annual or perennial herbs or subshrubs; leaves alternate, simple, entire or variously toothed or lobed, usually narrow; inflorescence many-branched, paniculate; capitula heterogamous, small, numerous; involucre campanulate; phyllaries 2–3-seriate, imbricate; receptacle flat, without scales; central, hermaphrodite florets few; marginal, pistillate (or female) florets numerous, in several rows, with filiform corollas, ligules inconspicuous or wanting, shorter than the corolla-tubes and shorter than, or at most slightly exceeding, the pappus. Achenes obovate-oblanceolate, compressed; pappus of scabridulous (often brown- or purple-tinged) hairs.

About 60 species widely distributed in temperate and tropical regions, often as introduced weeds.

1. **C. bonariensis** (*L.*) *Cronq.* in Bull. Torrey Bot. Club, 70: 632 (1943); Osorio-Tafall et Seraphim, List Vasc. Plants Cyprus, 101 (1973); Davis, Fl. Turkey, 5: 133 (1975); Feinbrun, Fl. Palaest., 3: 299, t. 497 (1978).
Erigeron bonariense L., Sp. Plant., ed. 1, 863 (1753).
E. crispus Pourr. in Hist. Mém. Acad. Sci. Toulouse, 3: 318 (1788); Post, Fl. Pal., ed. 2, 2: 18 (1933); Lindberg. f., Iter Cypr., 35 (1946); Osorio-Tafall et Seraphim, List Vasc. Plants Cypr., 101 (1973).
E. linifolius Willd., Sp. Plant., 3: 1955 (1803); Boiss., Fl. Orient., 3: 169 (1875); Holmboe, Veg. Cypr., 177 (1914).
[*E. canadensis* (non. L.) H. Stuart Thompson in Journ. Bot., 44: 332 (1906); Holmboe, Veg. Cypr., 177 (1914).]
TYPE: "*in* America *australi*".

Erect annual 10–60 cm. high; stems angled, usually not much branched below the inflorescence, thinly or densely pilose with a mixture of long, spreading, and short, subadpressed hairs; leaves variable, usually linear, lanceolate or oblanceolate, 1–12 cm. long, 0·2–2 cm. wide, scabridulous, adpressed-pubescent, the lower often irregularly serrate or lobed and acute or obtuse, the upper usually entire, acute; petiole obscure or wanting; inflorescence a many-branched panicle, the lateral branches often ultimately exceeding the central one; capitula very numerous, small; peduncles slender, pubescent, 5–20 mm. long; bracts subulate or filiform, foliaceous, 8–12 mm. long, 1–1·5 mm. wide, pubescent; phyllaries in 2 series, linear, acuminate, 2–5 mm. long, 0·3–0·5 mm. wide, herbaceous, hispidulous, distinctly costate; florets purplish, almost concealed by the pappus; corolla of hermaphrodite, central florets about 5 mm. long, 0·4 mm. wide at apex, narrowly tubular, apical lobes 5, erect, ovate-deltoid, about 0·3 mm. long, 0·2 mm. wide, minutely papillose-pilose; stamens exserted; anthers narrowly oblong, about 0·8 mm. long, 0·2 mm. wide, with a blunt base, and acute apical appendage; style concealed by anthers, branches erect, forming a slightly clavate stigma about 0·2 mm. long; female (marginal) florets filiform, about 4–5 mm. long, 0·1–0·2 mm. wide, glabrous, minutely and sharply lobed or ligulate at apex; style as long as corolla-tube; stigma narrowly clavate; receptacle flattish, naked, the phyllaries strongly reflexed with age. Achenes narrowly oblong, pale brown, sparsely hairy, compressed, about 1·5 mm. long, 0·4 mm. wide; pappus sessile, brownish, of 15–20 slender, scabridulous hairs about 3·5 mm. long.

HAB.: Gardens, roadsides, waste ground; sea-level to 3,500 ft. alt.; fl. Jan.–Dec.

DISTR.: Divisions 1–4, 6–8, probably common everywhere in the lowlands; a native of S. America now widespread in S. Europe and the Mediterranean region.

1. Ktima, 1913, *Haradjian* 695! Abundant about Paphos, 1981, *Meikle*!
2. Chakistra, 1905, *Holmboe* 1034; Platres, 1938, *Kennedy* 1437! also, 1948, *S. G. Cowper* s.n.!
3. Limassol, ? 1905, *Michaelides* teste *Holmboe*.
4. Larnaca, 1880, *Sintenis & Rigo* 898!
6. Nicosia, 1900, *A. G. & M. E. Lascelles* s.n.! also, 1905–1959, *Holmboe* 291; *Syngrassides* 797! 1281! *Lindberg f.* s.n.! *P. H. Oswald* 29!
7. Armenian Monastery, 1912, *Haradjian* 372! Kephalovryso, Kythrea, 1941, *Davis* 3635! Kyrenia, 1948, *Casey 22! and, 1955–1956, G. E. Atherton* 6! 393! 826! Myrtou, 1955, *G. E. Atherton* 410!
8. Ayios Andronikos, 1934, *Syngrassides* 538!

NOTES: Not mentioned by Kotschy (1865), and apparently introduced into Cyprus sometime between that date and 1880, when it was first recorded from Larnaca. By the early part of the succeeding century it had spread to Limassol, Nicosia and Paphos, and thence to most parts of the lowlands.

C. CANADENSIS (*L.*) *Cronq.* in Bull. Torrey Bot. Club, 70: 632 (1943); *Erigeron canadense* L., Sp. Plant., ed. 1, 863 (1753) superficially similar to *C. bonariensis* but with a sparse stem-indumentum solely of spreading hairs, smaller capitula with about 25 female flowers, and a glabrous or subglabrous involucre, is represented by a single specimen (*F. A. Rogers* 0646!) said to have been collected near Nicosia, 28 March, 1930. In view of the unreliability of this collector's localizations, further evidence is required before *C. canadensis* can be admitted to the flora. Earlier records (H. Stuart Thompson, Holmboe) are based on misidentifications of *C. bonariensis*.

TRIBE 3. **Inuleae** *Cass.* Leaves generally alternate, simple, entire; inflorescence usually branched, terminal or axillary, or capitula rarely solitary; capitula heterogamous and radiate, or homogamous, the outer florets usually female, the inner hermaphrodite or functionally male; corolla generally yellow, rarely blue or purplish; receptacle with or without scales; anthers sagittate-caudate at base; style-branches flattened, rounded, acute or truncate at apex, but without appendages. Achenes terete or subterete, occasionally rostrate; pappus of hairs or scales, sometimes absent.

PLUCHEA DIOSCORIDIS *(L.) DC.*, Prodr., 5: 450 (1836) (*Baccharis dioscoridis* L., Cent. Plant., 1, 27; 1755; Hume in Walpole, Mem. Europ. Asiatic Turkey, 254; 1817), a branched, glandular-puberulous shrub 1–3 m. high, with acute, ovate-oblong, serrated leaves, and crowded terminal corymbs of small capitula with pinkish or purplish florets, has been recorded by Hume from "Larnica and Limosol", June–July 1801. If not a misidentification, then the record was almost certainly based on cultivated plants, since the species, a native of Egypt, Arabia and Palestine, is otherwise unknown from Cyprus.

4. **EVAX** *Gaertn.*

Fruct. Sem. Plant. 2: 393 (1791).

J. Chrtek et J. Holub in Preslia, 35: 1–17 (1963).

G. Wagenitz in Willdenowia, 5: 395–444 (1969).

R. B. Fernandes et I. Nogueira in Bol. Soc. Brot., ser. 2, 45: 317–347 (1971).

Dwarf, solitary or caespitose, woolly or tomentose annuals; stems simple or branched; leaves alternate, entire, sessile, forming more or less distinct rosettes; inflorescence a terminal, subglobose cluster of sessile capitula, immersed in a nest of leaves; phyllaries few, small, scarious; receptacle conical, scaly, the scales conspicuous, scarious, yellow or brownish, sometimes lanate dorsally, concave or keeled, subtending the female florets; hermaphrodite or functionally male, central florets few, usually without subtending scales; corollas of female florets filiform, with 2–4 apical teeth; corollas of hermaphrodite (or functionally male) florets tubular, with 4–5 apical teeth; anthers caudate-sagittate at base. Achenes free, dorsally compressed, usually hairy or papillose; pappus absent.

About 20 species in Europe, the Mediterranean region east to Central Asia, and in N. America.

Rosette-leaves distinctly longer than inflorescence; plants unbranched or sparingly branched:
   Rosette-leaves 7–10 mm. long, 2·5–4 mm. wide, not much longer than inflorescence, apex blunt or subacute, not mucronate, base scarcely tapering - - - **1. E. pygmaea**
   Rosette-leaves 10–40 mm. long, 3–7 mm. wide, much exceeding inflorescence:
      Apex of rosette-leaves obtuse or subacute, generally mucronate, base tapering, quasi-petiolate; stem-leaves few, inconspicuous - - - - - - **2. E. contracta**
      Apex of rosette-leaves acute, scarcely mucronate, base not tapering; stem-leaves well developed, erect, imbricate - - - - - E. ASTERISCIFLORA (p. 875)
Rosette-leaves shorter than inflorescence; plants with very numerous spreading branches, forming mats of contiguous rosettes - - - - - **3. E. eriosphaera**

1. **E. pygmaea** *(L.) Brot.*, Fl. Lusit., 1: 363 (1804); Boiss., Fl. Orient., 3: 242 (1875); Post, Fl. Pal., ed. 2, 2: 38 (1933); Davis, Fl. Turkey, 5: 111 (1975).
   Filago pygmaea L., Sp. Plant., ed. 1, 927 (1753).
   TYPE: "*in* Europa *australi*".

Dwarf, acaulescent annual, seldom more than 4 cm. high; stems usually unbranched; leaves 20 or more, the lower 5–6 mm. long, 1·5–3 mm. wide, the upper oblong or narrowly spathulate, 7–10 mm. long, 2·5–4 mm. wide, whitish-tomentose, obtuse or subacute, crowded, sessile, forming a rosette around the aggregated capitula; inflorescence a dense, dome-shaped cluster, 5–12 mm. diam.; receptacular scales (*paleae*) thinly scarious, brownish, obovate, cuspidate-acuminate, thinly tomentose dorsally, about 5 mm. long, 2·5 mm. wide; female flowers filiform, about 2 mm. long, apex minutely but sharply 4-toothed; style as long as corolla-tube; hermaphrodite (or functionally male) flowers crowded at the centre of the capitulum; corolla narrowly tubular, about 2·5 mm. long, 0·6 mm. wide at apex, lobes 4, ovate-deltoid, about 0·5 mm. long, 0·2 mm. wide; anthers narrowly linear, about 1 mm. long, less than 0·1 mm. wide, apex acute, base caudate-sagittate; style slightly exceeding corolla, branches linear-acute, about 0·2 mm. long. Achenes narrowly oblong-fusiform, about 1 mm. long, 0·3 mm. wide, dark brown, minutely papillose.

HAB.: Dry, open, stony hillsides; pathsides; fallows; c. 1,000 ft. alt.; fl. April–May.

DISTR.: Division 7, evidently rare. S. Europe, Aegean Islands, Turkey, Lebanon.

7. "In montibus circa Kythraea", 24 May, 1880, *Sintenis & Rigo* 319, partly!

NOTES: One plant only, removed from a sheet of *Evax contracta* Boiss., and determined by J. Holub. In view of the uncertainty that attends Sintenis & Rigo localizations, some doubt must attach to the provenance of the specimen, and one hopes the plant will be re-found. Since these dowdy little Composites are often overlooked, *Evax pygmaea* may not be so rare in Cyprus as would appear.

2. E. contracta *Boiss.*, Diagn., 1, 11: 3 (1849); Fl. Orient., 3: 243 (1875); Unger et Kotschy, Die Insel Cypern, 235 (1865); Holmboe, Veg. Cypr., 177 (1914); Post, Fl. Pal., ed. 2, 2: 38 (1933); Osorio-Tafall et Seraphim, List Vasc. Plants Cyprus, 103 (1973); Davis, Fl. Turkey, 5: 111 (1975).

    Filago contracta (Boiss.) Chrtek et Holub in Preslia, 35: 3 (1963); Feinbrun, Fl. Palaest., 3: 308, t. 511 (1978).

    [*Evax asterisciflora* (non (Lam.) Pers.) Sintenis in Oesterr. Bot. Zeitschr., 31: 392 (1881); 32: 122, 259 (1882).]

TYPE: "in *Arabiâ petreâ* ad fines Palaestinae (Boiss.) ad fontes petroli propè *Dalechi* Persiae australis Kotschy No. 968".

Dwarf, acaulescent annual, seldom more than 5 cm. high; stems sometimes branched at the base, often unbranched; leaves crowded into a rosette, thinly greyish-tomentose, oblanceolate-subspathulate, 10–40 mm. long, 3–7 mm. wide, thinly tomentose on both surfaces, apex obtuse or subacute, generally mucronate, base tapering to an apparent petiole, shortly sheathing; inflorescence a dense, dome-shaped cluster, 6–20 mm. diam.; receptacular scales (paleae) thinly scarious, pale brown, sparsely lanuginose externally, obovate-cuspidate, about 4 mm. long, 2 mm. wide; female flowers filiform, about 2 mm. long, apex very minutely but sharply 4-toothed; style-branches shortly exserted; hermaphrodite flowers very few at centre of capitulum, narrowly tubular, about 2 mm. long, 0·4 mm. wide at apex, lobes 4, ovate-deltoid, tinged orange-brown, about 0·3 mm. long, 0·2 mm. wide; anthers narrowly linear, about 0·8 mm. long, less than 0·1 mm. wide, apex acute, base caudate-sagittate; style slightly exceeding corolla, branches erect, linear-acute, about 0·2 mm. long. Achenes narrowly obovate, strongly compressed, pale brown, about 1·2 mm. long, 0·8 mm. wide, conspicuously white-papillose.

HAB.: Dry, trampled ground, old fallow-land; dry hillsides; sea-level to 1,000 ft. alt.; fl. March–May.

DISTR.: Divisions 4, 5, 7, locally common. Eastern Mediterranean region and eastwards to Iran.

4. Larnaca, 1859, *Kotschy* 476.
5. Mile 3, Nicosia-Famagusta road, 1950, *Chapman* 400! Mia Milea, 1950, *Chapman* 491!
   Kythrea, 1950, *Chapman* 107!
7. Mountains above Kythrea, 1880, *Sintenis & Rigo* 319! Near Ardhana, 1880, *Sintenis & Rigo*
   319a! Below Tjiklos cliff, Kyrenia, 1949, *Casey* 522!

E. ASTERISCIFLORA (*Lam.*) *Pers.*, Syn. Plant., 2: 422 (1807); Boiss., Fl. Orient., 3: 243 (1875); Holmboe, Veg. Cypr., 178 (1914); Osorio-Tafall et Seraphim, List Vasc. Plants Cyprus, 103 (1973) (*Gnaphalium asterisciflorum* Lam., Encycl. Méth., 2: 760; 1788), from the western Mediterranean region, is recorded from Cyprus by Boissier on the strength of an unlocalized specimen said to have been collected there by Labillardière. The specimen is in herb. Boissier (G) and seems to have been correctly identified, but it may be questioned if it came from Cyprus. *Evax asterisciflora* extends from Spain and western North Africa as far east as Sicily, but is absent from the eastern Mediterranean area. It resembles *E. contracta* Boiss., but is usually more robust, with a well-developed leafy stem, and larger, acute or acuminate rosette-leaves about 4 times as long as the inflorescence-cluster.

Records for *Evax asterisciflora* published by Sintenis in Oesterr. Bot. Zeitschr. (1881, 1882) are erroneous, and based upon misidentifications of *E. contracta* Boiss.

3. E. eriosphaera *Boiss. et Heldr.* in Boiss., Diagn., 1, 11: 3 (1849), Fl. Orient., 3: 244 (1875); Unger et Kotschy, Die Insel Cypern, 235 (1865); Holmboe, Veg. Cypr., 178 (1914); Post, Fl. Pal., ed. 2, 2: 39 (1933); Osorio-Tafall et Seraphim, List Vasc. Plants Cyprus, 103 (1973); Davis, Fl. Turkey, 5: 113 (1975).
TYPE: Turkey; "in aridis planitiei *Adalia* in Pamphyliâ (Heldr.)" (G, K!).

Dwarf annual, usually less than 4 cm. high, commonly with very numerous spreading branches, each ending in a leaf-rosette, and forming a dense, caespitose mat often 12 cm. diam.; stem leaves sparse, narrowly spathulate, usually withered by anthesis; rosette-leaves narrowly spathulate, usually less than 7 mm. long, 3 mm. wide, and shorter than the inflorescence, thinly or densely lanuginose, apex subacute or obtuse, base tapering to an apparent petiole; inflorescences dense, white-woolly, subglobose, 5–15 mm. diam.; receptacular scales (paleae) oblong-acuminate, somewhat keeled, densely lanuginose dorsally, about 3 mm. long, 0·8 mm. wide; female flowers filiform, apex minutely 2–4-toothed; style-branches not exserted; hermaphrodite flowers narrowly tubular, about 2 mm. long, 0·4 mm. wide at apex, with 4 ovate-deltoid, slightly recurved lobes about 0·2 mm. long; anthers narrowly oblong, about 0·8 mm. long, 0·1 mm. wide, base caudate-sagittate, apex subacute; styles included, style-branches about 0·2 mm. long, subacute. Achenes narrowly obovoid, compressed, dark olive-brown, minutely and thinly papillose, about 0·8 mm. long, 0·4 mm. wide.

HAB.: Bare, hard ground and caked mud by paths and roadsides; sea-level to 1,000 ft. alt.; fl. March–May.

DISTR.: Divisions 1, 3–8. Aegean Islands, Turkey, Lebanon.

1. Between Neokhorio and Smyies, 1948, *Mercy Casey* in *Casey* 3!
3. Kophinou, 1905, *Holmboe* 590!
4. Larnaca, 1862, *Kotschy* 309!
5. Near Lefkoniko, 1880, *Sintenis & Rigo* 320!
6. Nicosia, 1950, *Chapman* 535! Between Skylloura and Pyrga, 1956, *Merton* 2762!
7. Mountains near Kythrea, 1880, *Sintenis & Rigo* 320a!
8. Akradhes, 1962, *Meikle* 2516!

5. BOMBYCILAENA (*DC.*) *Smolj.*

Notulae Syst. Herb. Inst. Bot. Acad. Sci. URSS, 17: 448 (1955).
Micropus L. sect. *Bombycilaena* DC., Prodr., 5: 460 (1836).

Decumbent or erect annuals; stems usually well developed; leaves alternate, entire, sessile, tomentose or lanuginose; inflorescences consisting of 2–5 capitula in terminal or axillary, conspicuously lanuginose clusters; phyllaries few, small, scarious; receptacle cylindrical; receptacular scales compressed laterally, saccate, enveloping the female florets and falling off with the fruit; female florets filiform, attached laterally to ovary, apex of corolla-tube 2-dentate; style-branches linear; hermaphrodite (or functionally male) florets few at the centre of each capitulum, corolla tubular, 5-dentate at apex. Achenes obovoid, compressed; pappus absent.

Two species in S. Europe, the Mediterranean region and western Asia, east to the Caucasus.

1. **B. discolor** (*Pers.*) *Lainz* in Bol. Inst. Estud. Astur. (Supl. Ci.), 16: 194 (1973); Davis, Fl. Turkey, 5: 114 (1975).
   Micropus discolor Pers., Syn. Plant., 2: 423 (1807).
   M. bombicinus Lag., Gen. Sp. Nov., 32 (1816); Unger et Kotschy, Die Insel Cypern, 235 (1865) as *M. bombycinus*; Boiss., Fl. Orient., 3: 241 (1875); Holmboe, Veg. Cypr., 177 (1914); Post, Fl. Pal., ed. 2, 2: 37 (1933); Osorio-Tafall et Seraphim, List Vasc. Plants Cyprus, 103 (1973), all as *M. bombycinus*.
   Bombycilaena bombycina (Lag.) Smolj. in Notulae Syst. Herb. Inst. Bot. Acad. Sci. URSS, 17: 450 (1955) in adnot.
   [*Micropus erectus* (non L.) Sm. in Sibth. et Sm., Fl. Graec. Prodr., 2: 208 (1813); Poech, Enum. Plant. Ins. Cypr., 17 (1842); Unger et Kotschy, Die Insel Cypern, 235 (1865); H. Stuart Thompson in Journ. Bot., 44: 332 (1906); Holmboe, Veg. Cypr., 177 (1914); Osorio-Tafall et Seraphim, List Vasc. Plants Cyprus, 103 (1973).]
   TYPE: Spain; "prope Matritum. (Lagasca)" (L).

Erect or decumbent, branched or unbranched annual (1–) 3–14 cm. high; stems tomentose-lanuginose; leaves oblong, subacute, 5–20 mm. long, 3–7 mm. wide, densely white-tomentose; inflorescences terminal and axillary, sessile, subglobose, 8–13 mm. diam., conspicuously white-lanuginose, like tufts of cotton-wool; receptacular scales strongly compressed laterally, cucullate-saccate, densely lanuginose externally, about 2·5 mm. long, 2 mm. wide; female florets filiform, minute; corolla-tube about 1 mm. long, less than 0·1 mm. wide, concealed within the dorsal keel of the receptacular scale and attached obliquely to the ovary; style-branches filiform, about 0·5 mm. long, exserted from receptacular scale; hermaphrodite (or functionally male) florets 2–4 at centre of capitulum, not subtended by scales, corolla-tube narrow, about 1·5 mm. long, 0·4 mm. wide at apex, lobes ovate, subacute, erect, about 0·2 mm. long and almost as wide at base, fringed with conspicuous stalked glands; anthers linear, about 0·4 mm. long, 0·1 mm. wide, apex subacute, base caudate-sagittate; style shortly exserted, style-branches connivent. Achenes closely and permanently enveloped in the indurated, lanuginose receptacular scale, lunate-obovate, about 2·5 mm. long, 2 mm. wide.

HAB.: Dry, stony hillsides; fallows and cultivated fields; tracks and roadsides; 300–3,000 ft. alt.; fl. March–May.

DISTR.: Divisions 2–5, 7, 8. S. Europe and Mediterranean region east to Iraq.

2. Philani, 1908, *Clement Reid* s.n.! Mandria, 1939, *Kennedy* 1440!
3. Kophinou, 1905, *Holmboe* 589; Apsiou, 1962, *Meikle* 2887!
4. Larnaca, 1862, *Kotschy* 266a.
5. Lefkoniko, 1880, *Sintenis & Rigo*; Nisou, 1933, *Syngrassides* 31! and, 1936, *Syngrassides* 1134! Kythrea, 1950, *Chapman* 108!
7. Ayios Khrysostomos Monastery, 1862, *Kotschy* 439; mountains above Kythrea, 1880, *Sintenis & Rigo* 318! Near Kantara, 1880, *Sintenis & Rigo* 318a! St. Hilarion, 1949, *Casey*

687! and, 1950, *Casey* 1027! Lakovounara, 1950, *Chapman* 359! 508! Kyrenia Pass, 1956, *G. E. Atherton* 1376!

8. Cape Andreas, 1880, *Sintenis & Rigo.*

NOTES: Commonly confused with *B. erecta* (L.) Smolj., which is superficially very similar, but with a shorter indumentum than in *B. discolor*, so that the capitula appear to be smaller, and often overtopped by the subtending foliage.

6. FILAGO *L.*

Sp. Plant., ed. 1, 927, 1199 (1753).
Gen. Plant., ed. 5, 397 (1754) nom. cons.
J. Chrtek et J. Holub in Preslia, 35: 1–17 (1963).
G. Wagenitz in Willdenowia 5 (3): 395–444 (1969).

Tomentose annuals; leaves alternate, entire; capitula small, conical, ovoid or ellipsoid, generally in axillary or terminal, subglobose clusters often surrounded by conspicuous floral leaves; phyllaries few, small, scarious; receptacle filiform, cylindrical or obconical; receptacular scales in several series, imbricate, acute or obtuse, scarious or partly herbaceous, concave or keeled and subtending or enveloping the outer, female florets, erect or spreading in fruit; florets yellowish, the outer female, epappose, with filiform, minutely 2–4-toothed corollas; inner partly female and partly hermaphrodite; or wholly hermaphrodite, pappose, with a tubular, 4–5-toothed corolla. Achenes cylindrical, oblong or obovate, often somewhat compressed, smooth or papillose, sometimes dimorphic.

About 30 species in Europe, Mediterranean region, western Asia and America.

Leaves oblong or lanceolate-spathulate, apex obtuse, rounded, subacute or abruptly acute:
  Capitula crowded into subglobose heads:
    Heads terminal, solitary or terminating dichotomous lateral branches; receptacular scales not reflexing after anthesis:
      Receptacular scales superposed in fives in a regular peripheral series of five, forming a sub-pentagonal capitulum; tips of scales conspicuous, shining yellow
        2. F. pyramidata
      Receptacular scales not in a regular radial or peripheral series; tips of scales not conspicuously shining:
        Leaves narrowly oblong, acute   -   -   -   -   -   **1. F. eriocephala**
        Leaves lanceolate-spathulate, obtuse or rounded at apex -   **3. F. aegaea** ssp. **aristata**
    Heads terminal and lateral, sessile or subsessile, forming narrow spiciform inflorescences or partial inflorescences; receptacular scales strongly reflexed after anthesis
        5. F. arvensis
  Capitula solitary or few together in open cymes, not in subglobose heads   - **4. F. mareotica**
  Leaves linear-subulate, tapering to an acute apex; capitula forming small, irregular clusters, usually overtopped by the floral leaves -   -   -   -   -   -   **6. F. gallica**

SECT. 1. **Filago** Capitula crowded into subglobose heads; receptacular scales concave or keeled, acute or acuminate, remaining erect or suberect in fruit; inner florets female and hermaphrodite, or occasionally all hermaphrodite, pappose.

1. **F. eriocephala** *Guss.*, Plant. Rar., 344, t. 59 (1826); Unger et Kotschy, Die Insel Cypern, 241 (1865); Davis, Fl. Turkey, 5: 103 (1975); Feinbrun, Fl. Palaest., 3: 306, t. 506 (1978).
  F. germanica L. var. *eriocephala* (Guss.) Parl., Plant. Nov., 9 (1842); Boiss., Fl. Orient., 3: 245 (1875); Post, Fl. Pal., ed. 2, 2: 39 (1933); Lindberg f., Iter Cypr., 35 (1946).
  F. germanica L. ssp. *eriocephala* (Guss.) Holmboe, Veg. Cypr., 178 (1914).
  Gifola eriocephala (Guss.) Chrtek et Holub in Preslia, 35: 6 (1963).
  Filago vulgaris Lam. ssp. *eriocephala* (Guss.) Osorio-Tafall et Seraphim, List Vasc. Plants Cyprus, 103 (1973) nom. invalid.
  TYPE: Italy; "In collibus aridis maritimis Lucaniae; *Policoro* una cum *F. germanica*, ac *F. pyramidata*" *Gussone* (? NAP).

Erect or decumbent annuals, 4–25 cm. high, sometimes unbranched, or sometimes with numerous basal branches; stems terete, lanuginose-tomentose; leaves erect, sessile, narrowly oblong, acute, 5–20 mm. long, 2–5 mm. wide, thinly lanuginose-tomentose on both sides, or subglabrous above; inflorescences terminal on dichotomous branches, subglobose, 6–12 mm. diam.; capitula numerous, congested, exceeding the floral leaves; receptacular scales erect, ovate-acuminate, concave-carinate, 3–3·5 mm. long, 1·5 mm. wide, herbaceous and lanuginose dorsally, with broad yellowish-brown, scarious margins; female florets with a filiform corolla-tube, 2·5–3 mm. long, apex sparsely glandular, minutely toothed; style-branches filiform, about 0·5 mm. long, distinctly exserted from corolla-tube; central florets mixed female and hermaphrodite, pappose; hermaphrodite (or functionally male) florets narrowly tubular, about 1·5–2 mm. long, 0·3 mm. wide at the minutely 4-lobed, glandular apex; anthers linear, subacute, about 0·8 mm. long, 0·2 mm. wide; style-branches included in corolla-tube, connivent; pappus of numerous, very sparsely scabridulous bristles about 1·5 mm. long. Achenes fusiform, plano-convex, with acute margins, mid-brown, minutely and sparsely papillose, about 0·9 mm. long, 0·3 mm. wide; pappus-hairs about 3 mm. long, recurved, minutely scabridulous.

HAB.: Amongst garigue on dry, open, calcareous hillsides, in thin Cypress forest; vineyards; or in rock-crevices by the sea; sea-level to 2,000 ft. alt.; fl. April–June (to Sept.).

DISTR.: Divisions 1–4, 7, 8. Widespread in the Mediterranean region.

1. Ayios Neophytos Monastery, 1939, *Lindberg f.* s.n.; Dhrousha, 1941, *Davis* 3230!
2. Near Prodhromos, 1862, *Kotschy* 821 sec. *Wagenitz*; Galata, 1880, *Sintenis & Rigo* 303 sec. *Wagenitz*; Kannaviou, 1939, *Lindberg f.* s.n.; between Ambelikou and Kambos, 1939, *Lindberg f.* s.n.; Kakopetria, 1939, *Lindberg f.* s.n.
3. Kophinou, 1905, *Holmboe* 594.
4. Akhna Forest, 1952, *Merton* 737!
7. On mountains near Kantara, 1880, *Sintenis & Rigo* 303a (see Notes); frequent about Kyrenia, 1948–1973, *Casey* 6! *G. E. Atherton* 173! 224! 1351! 1368! *Chapman* 615! Glykyotissa (Snake) Island, 1955, *G. E. Atherton* 338! Melounda, 1963, *J. B. Suart* 89! Tjiklos, 1973, *P. Laukkonen* 343! St. Hilarion, 1939, *Lindberg f.* s.n.
8. Cape Andreas, 1880, *Sintenis & Rigo* sec. *Wagenitz*.

NOTES: *Sintenis* 303, said by Wagenitz (in Willdenowia, 5: 402; 1969) to have been collected on "Troodos, Weinberge bei Galata" (as recorded on the label of the specimen at Lund) is represented at Kew by a sheet bearing the same number, but labelled "In montibus prope Kantara", which is in quite a different area of the island. Similar confusions are common with these collectors, and detract from the value of their otherwise estimable collections.

2. F. pyramidata *L.*, Sp. Plant., ed. 1, 1199 (1753); Wagenitz in Willdenowia, 5: 403 (1969); Osorio-Tafall et Seraphim, List Vasc. Plants Cyprus, 104 (1973); Davis, Fl. Turkey, 5: 104 (1975); Feinbrun, Fl. Palaest., 3: 306, t. 507 (1978).

    F. spathulata C. Presl, Delic. Prag., 1: 99 (1822); Boiss., Fl. Orient., 3: 246 (1875); Post, Fl. Pal., ed. 2, 2: 39 (1933); Rechinger f. in Arkiv för Bot., ser. 2, 1: 432 (1950).

    F. prostrata Parl. in Ann. Sci. Nat., ser. 2, 15: 302 (1841) non DC. (1837) nom. illeg.

    Gifola spathulata (C. Presl) Reichb., Icones Fl. Germ., 16: 26 (1853).

    Filago spathulata C. Presl var. *prostrata* Heldr., Herb. Graec. Norm. no. 531 (1856).

    F. germanica L. var. *prostrata* (Heldr.) Fiori in Fiori et Paoletti, Fl. Anal. Ital., 3: 274 (1904).

    F. germanica L. ssp. *spathulata* (C. Presl) Holmboe, Veg. Cypr., 178 (1914).

    F. germanica L. ssp. *decumbens* Holmboe, Veg. Cypr., 178 (1914).

    F. vulgaris Lam. ssp. *prostrata* (Heldr.) Osorio-Tafall et Seraphim, List Vasc. Plants Cyprus, 103 (1973) nom. invalid.

    [*F. germanica* (non. L.) Kotschy in Unger et Kotschy, Die Insel Cypern, 241 (1865).]

TYPE: "*in* Hispania. *Loefling.*"

Erect or decumbent annual 4–18 (–30) cm. high; stems unbranched, or, more frequently much branched from the base, thinly or densely tomentose; leaves erect, sessile, oblong, 8–25 mm. long, 2–7 mm. wide, obtuse or subacute, thinly or rather densely grey-tomentose; inflorescences terminal on dichotomous branches, subglobose, about 1 cm. diam.; capitula numerous, congested, somewhat echinate in appearance, often equalled by or sometimes exceeded by the floral leaves; receptacular scales symmetrically arranged in a series of 5, both along the radius of the capitulum and around its periphery, narrowly ovate, caudate-acuminate, about 5 mm. long, 1·5 mm. wide, deeply concave-carinate, lanuginose dorsally with broad, shining, yellowish, scarious margins, the long, slender acumen distinctly recurved; female florets with a filiform corolla-tube about 3 mm. long, apex sparsely glandular, minutely toothed; style-branches filiform, about 0·4 mm. long, distinctly exserted from the corolla-tube; central florets pappose, consisting of a few female mixed with hermaphrodite (or functionally male) florets, the latter narrowly tubular, about 2·5 mm. long, 0·2 mm. wide, the corolla-tube shortly 4-lobed at apex; anthers narrowly linear, about 0·8 mm. long, 0·1 mm. wide, apex subacute; style-branches connivent, subtruncate, included in corolla-tube, pappus-hairs about 2·5 mm. long, sparsely scabridulous. Achenes obovoid, about 0·8 mm. long, 0·4 mm. wide, plano-convex, olive-brown, smooth or thinly papillose.

HAB.: Dry rocky slopes; waste ground; sandy fields; vineyards; thin Cypress forest; sea-level to 5,400 ft. alt.; fl. March–June.

DISTR.: Divisions 1–8. Widespread in the Mediterranean region and eastwards to Central Asia.

1. Stroumbi, 1913, *Haradjian* 743.
2. Livadhi tou Pasha, 1937, *Kennedy* 956! Platres, 1938, *Kennedy* 1439! Perapedhi, 1962, *Meikle* 2801!
3. Kophinou, 1905, *Holmboe* 595; Amathus, 1978, *Holub* s.n.!
4. Larnaca, 1862, *Kotschy* 266; and, 1905, *Holmboe* 254; Akhna, 1952, *Merton* 738!
5. Kythrea, 1950, *Chapman* 173! Mia Milea, 1952, *Merton* 775! Athalassa, 1967, *Merton* in ARI 151! Mile 10, Nicosia-Famagusta road, 1967, *Merton* in ARI 475!
6. Ayia Irini, 1941, *Davis* 2532! Nicosia, 1972, *W. R. Price* 1021!
7. Kyrenia, 1933, *Syngrassides* 188! also, 1956, *G. E. Atherton* 1350! Lakovounara, 1950, *Chapman* 363! Kyrenia Pass, 1956, *G. E. Atherton* 1375!
8. Apostolos Andreas, 1941, *Davis* 2332!

NOTES: *Kotschy* 238a from (2) Makheras Monastery, listed as *F. germanica* L. in *Die Insel Cypern*, 241, very probably belongs here. *F. vulgaris* Lam. (*F. germanica* L. pro parte) does not occur in Cyprus.

3. F. aegaea *Wagenitz* in Willdenowia, 6 (1): 126 (1970).

Prostrate or erect, grey-tomentose annual; primary stem almost wanting (ssp. *aegaea*) or up to 6 cm. long, lateral branches generally numerous, sometimes with secondary branchlets, occasionally rooting; leaves greyish-tomentose (especially on the leaf undersurface), spathulate, lanceolate-spathulate, or almost orbicular, with a rounded apex, crowded under the inflorescences and equalling or a little shorter than them; capitula 5–15 in each inflorescence-cluster; receptacular scales superposed in 4–5 series, but not in 5 distinct radial series, rather rigid, acute or acuminate from a broadly ovate base, glabrous and with a broad hyaline margin towards the base, adpressed-tomentose in the upper part almost to the apex, convex, the outer somewhat keeled towards the base, with a slightly recurved apex at anthesis, all more or less inflexed in fruit; outer, female florets with a filiform corolla-tube; inner florets 4–6, hermaphrodite. Achenes oblong, 0·7–0·8 mm. long, brown, papillose; pappus (of central florets) white, deciduous, consisting of one series of scabridulous bristles about 1·5 mm. long.

ssp. **aristata** *Wagenitz* in Willdenowia, 6 (1): 129 (1970).

TYPE: Cyclades; Naxos, Koronos, 900–960 m. alt., 1958, *Runemark et Snogerup* 9739 (LD).

Stem usually developed, up to 6 cm. long; leaves lanceolate-spathulate; receptacular scales distinctly acuminate, with an acumen 0·5–1 mm. long.

HAB.: Dry calcareous hillsides; c. 1,000 ft. alt.; fl. March–June.

DISTR.: Division 7, apparently rare. Also Crete and Aegean Islands; ssp. *aegaea* in Crete, Karpathos and Cyclades.

7. Mountains above Phlamoudhi, 1880, *Sintenis & Rigo* 305!

NOTES: I have not seen any type material of *F. aegaea* or its subspecies *aristata*, and the above description is taken solely from that given by Wagenitz: the Cyprus specimen of *F. aegaea* ssp. *aristata* (kindly loaned by the Botanical Museum, University, Lund) bears a very close resemblance to some forms of the common *F. pyramidata* L., and indicates the need for further investigation.

SECT. 2. **Gifolaria** *Coss. et Kralik*   Capitula solitary; receptacular scales in a series of 5, divergent in fruit; innermost florets hermaphrodite (or functionally male); pappus present or absent.

4. **F. mareotica** *Del.*, Descr. Égypte Hist. Nat., 2: 274, t. 47, fig. 2 (1813); Boiss., Fl. Orient., 3: 246 (1875).
TYPE: Egypt; "Cette plant croît auprès des anciennes carrières d'Alexandrie et du lac Mareotis" (MPU).

Erect annual 1–5 cm. high; stems adpressed-canescent, sometimes sparingly branched, but generally with numerous, crowded, erecto-patent, dichotomous branches; leaves oblong-oblanceolate, small, inconspicuous, densely adpressed-canescent, obtuse or subacute, sessile, generally less than 5 mm. long, 2·5 mm. wide; inflorescence a dichotomous cyme; capitula solitary, sessile; receptacular scales superposed in a series of 3–5, ovate, acute or acuminate, distinctly keeled, about 2·5 mm. long, 1·5 mm. wide, lanuginose towards the base dorsally, with wide, brownish, scarious margins; female florets filiform, about 1·3 mm. long; style-branches exserted for about 0·3 mm.; central hermaphrodite (or functionally male) florets narrowly tubular, about 1 mm. long, 0·4 mm. wide, apex with 4 short, erect, rather blunt, ovate-deltoid, brownish lobes less than 0·2 mm. long and wide; anthers linear, about 0·7 mm. long, 0·1 mm. wide, apex acute; style included; style-branches erect, connivent, subtruncate. Achenes oblong, a little compressed, about 0·8 mm. long, 0·4 mm. wide, brownish, sparsely papillose; pappus absent.

HAB.: Sandy margin of salt-lake; near sea-level; fl. March–May.

DISTR.: Division 4, very rare. Also Libya and Egypt.

4. Side of Larnaca Salt Lake near the Hala Sultan Tekké, 1978, *J. Holub* s.n.!

NOTES: The Cyprus material (admittedly in early fruiting condition) is evidently without pappus. This, according to Wagenitz, is characteristic of typical *F. mareotica*, while pappus hairs are present in *F. mareotica* var. *floribunda* (Pomel) Maire, which has a wider distribution than the type, extending westwards through N. Africa to S. Spain.

SECT. 3. **Oglifa** *(Cass.)* DC.   Capitula in small, subglobose clusters or sometimes solitary; receptacular scales 15–20, obtuse or acute, but not acuminate, strongly and stellately divergent in fruit; innermost florets mixed female and hermaphrodite, all with pappus.

5. **F. arvensis** *L.*, Sp. Plant., ed. 1, addend. post indic. (1753); Boiss., Fl. Orient., 3: 247 (1875); Post, Fl. Pal., ed. 2, 2: 40 (1933); Wagenitz in Willdenowia, 5 (3): 426 (1969); Feinbrun, Fl. Palaest., 3: 308, t. 513 (1978).
Gnaphalium arvense L., Sp. Plant., ed. 1, 856 (1753).
G. lagopus Stephan ex Willd., Sp. Plant., 3: 1897 (1803).
Filago arvensis L. var. *lagopus* (Stephan ex Willd.) DC., Prodr., 6: 249 (1838); Unger et Kotschy, Die Insel Cypern, 241 (1865); Boiss., Fl. Orient., 3: 247 (1875); H. Stuart Thompson in Journ. Bot., 44: 332 (1906); Post, Fl. Pal., ed. 2, 2: 40 (1933); Lindberg f., Iter Cypr., 35 (1946).
F. lagopus (Stephan ex Willd.) Parl. in Giorn. Tosc. Sci. Med. Fis. Nat., 1 (2): 182 (1841).
F. arvensis L. ssp. *lagopus* (Stephan ex Willd.) Nyman, Consp. Fl. Europ., 385 (1879);

Holmboe, Veg. Cypr., 178 (1914); Osorio-Tafall et Seraphim, List Vasc. Plants Cyprus, 104 (1973).

Logfia arvensis (L.) Holub in Notes Roy. Bot. Gard. Edinb., 33: 432 (1975); Davis, Fl. Turkey, 5: 108 (1975).

TYPE: "*in* Europae *campis aridis, arvisque sabulosis*".

Erect or decumbent, white-lanuginose annual 2–17 cm. high; stems simple or branched at the base; leaves numerous, erect, sessile, generally overlapping, oblong, 6–20 mm. long, 2–5 mm. wide, apex obtuse or shortly acute; capitula crowded into a spiciform inflorescence towards the apex of the stem, or inflorescences sometimes with numerous short, erecto-patent, spiciform branches; capitula densely lanuginose, in subglobose clusters of 3–10 or more, generally exceeding the floral leaves; receptacular scales ovate-acute, deeply carinate, about 4–5 mm. long, 2–2·5 mm. wide, densely lanuginose almost to apex, spreading stellately in fruit and exposing a flattened, pitted receptacle apex more than 1 mm. diam.; female florets filiform, 3–3·5 mm. long, enfolded within receptacle-scale, apex minutely toothed, glandular; style-branches exserted, linear, recurved, about 0·4 mm. long; central florets mixed female and hermaphrodite, the latter with a narrow corolla-tube, about 3 mm. long, 0·3 mm. wide at apex, lobes 4, ovate-deltoid, spreading, orange-brown, glandular, about 0·2 mm. long, 0·2 mm. wide at base; anthers narrowly linear, about 0·8 mm. long, 0·1 mm. wide, apex subacute; style included, branches linear-subtruncate, erect, about 0·3 mm. long. Achenes narrowly ovoid-ellipsoid, compressed, olive-brown, papillose, about 0·8 mm. long, 0·4 mm. wide; pappus-hairs silky, remotely scabridulous, white, about 3 mm. long.

HAB.: Stony, igneous mountainsides; 2,500–5,000 ft. alt.; fl. May–June.

DISTR.: Division 2, rather rare. Europe and Mediterranean region east to the Himalayas.

2. Prodhromos, 1862, *Kotschy* 845; Philani, 1908, *Clement Reid* s.n.! Platres, 1938, *Kennedy* 1438! and Kato Platres, 1948, *Mavromoustakis* s.n.! Kakopetria, 1939, *Lindberg f.* s.n.; Platania, 1939, *Lindberg f.* s.n.

NOTES: In common with much of the material from the eastern Mediterranean area, Cyprus specimens of *F. arvensis* are densely lanuginose, with rather larger heads of capitula. They were formerly referred to *F. arvensis* L. var. *lagopus* (Stephan ex Willd.) DC. or *F. arvensis* L. ssp. *lagopus* (Stephan ex Willd.) Nyman, but recent authors (Wagenitz in Willdenowia, 5 (3): 427; 1969, J. Holub in Davis, Fl. Turkey, 5: 108; 1978) question the application of this name, nor is it easy to draw any clear distinction between such plants and typical *F. arvensis* from Europe. Their resemblance to *Logfia davisii* Holub ex Grierson (from E. Anatolia, Syria, Lebanon and Palestine) has also been noted, but they differ from the latter in their more robust overall habit, and in the structure and composition of their capitula.

The nomenclature of *Filago pyramidata* L., *F. arvensis* L. and *F. gallica* L. furnishes problems of a different order (see Greuter et Rechinger f. in Boissiera, 13: 136–138; 1967); it is possible to argue that none of these combinations is validly published in Sp. Plant., ed. 1 (1753), and that their validation must be sought in Fl. Anglica, 23 (1754) or in Fl. Suecica, ed. 2, 302 (1755). Linnaeus has undoubtedly contributed to the muddle by misrepresenting his own intentions, but the situation is one where finesse is out of place, and I am accepting all the *Filago* names published in Sp. Plant., ed. 1 as valid.

6. F. gallica *L.*, Sp. Plant., ed. 1, addend. post indic. (1753); Unger et Kotschy, Die Insel Cypern, 241 (1865); Boiss., Fl. Orient., 3: 248 (1875); H. Stuart Thompson in Journ. Bot., 44: 332 (1906); Holmboe, Veg. Cypr., 178 (1914); Post, Fl. Pal., ed. 2, 2: 40 (1933); Wagenitz in Willdenowia, 5 (3): 428 (1969); Osorio-Tafall et Seraphim, List Vasc. Plants Cyprus, 104 (1973); Feinbrun, Fl. Palaest., 3: 309, t. 514 (1978).

Gnaphalium gallicum L., Sp. Plant., ed. 1, 857 (1753).

Logfia gallica (L.) Coss. et Germ. in Ann. Sci. Nat., ser. 2, 20: 291 (1843); Davis, Fl. Turkey, 5: 109 (1975).

[*F. gallica* L. ssp. *tenuifolia* (non. *F. tenuifolia* C. Presl) Holmboe, Veg. Cypr., 178 (1914); Osorio-Tafall et Seraphim, List Vasc. Plants Cyprus, 104 (1973).]

TYPE: "*in* Anglia, Gallia".

Erect or decumbent, silvery-tomentose annual 3–18 (–25) cm. high; stems simple, unbranched, or much branched from the base; leaves numerous, erect, sessile, overlapping, linear-oblong or subulate, 5–25 mm. long, 1–2 mm. wide at base, tapering to an acute apex; inflorescences cymose, repeatedly dichotomous; capitula 2–14, forming small, irregular clusters generally exceeded by the floral leaves; receptacular scales erect, pentagonally arranged, superposed in 2–3 series, narrowly ovate, 4–4·5 mm. long, 1·5 mm. wide, deeply saccate-carinate, lanuginose towards the base dorsally, tapering gradually to a scarious, brownish, subacute apex; female florets filiform, very slender, about 2 mm. long, closely enveloped by the receptacular scales; style-branches exserted from corolla-tube, linear, brownish, about 0·4 mm. long; central florets mixed female and hermaphrodite, the latter with narrow corolla-tubes about 2·5 mm. long, 0·3 mm. wide at apex, lobes 4, ovate-deltoid, slightly recurved, yellowish-brown, about 0·2 mm. long and almost as wide at base; anthers narrowly linear, about 0·8 mm. long, less than 0·1 mm. wide, apex subacute; style included; stigmas linear, subtruncate, erect, brownish, about 0·2 mm. long; receptacular scales stellately spreading in fruit to expose a very small, pitted receptacle, less than 0·5 mm. diam. Achenes narrowly oblong, about 0·8 mm. long, 0·3 mm. wide, somewhat compressed, light brown, papillose, pappus rather silky, white, hairs about 2·5 mm. long, sparingly scabridulous.

HAB.: In low, open grass steppe; in garigue on dry mountainsides; in vineyards; on banks by roadsides or in light, open Cypress forest; sea-level to 3,900 ft. alt.; fl. April–May.

DISTR.: 1–4, 7, 8. S. Europe, Mediterranean region, Atlantic Islands.

1. Dhrousha, 1941, *Davis* 3257!
2. Platres, 1938, *Kennedy* 1441! Galata, 1941, *Davis* 3203! Ayios Nikolaos near Kakopetria, 1962, *Meikle* 2851!
3. Kalavasos, 1905, *Holmboe* 614a; 614b; Cape Gata, 1905, *Holmboe* 675.
4. Larnaca aerodrome, 1950, *Chapman* 696! Akhna Forest, 1952, *Merton* 736!
7. Near Ayios Khrysostomos Monastery, 1862, *Kotschy* 439a.
8. Mountains near Eleousa, 1880, *Sintenis & Rigo* 307!

NOTES: *Filago tenuifolia* C. Presl (Delic. Prag., 101 (1822); *F. gallica* L. ssp. *tenuifolia* (C. Presl) Holmboe quoad nomen) is disregarded by most modern authors as unsatisfactorily distinguished from typical *F. gallica*. Plants from Cyprus agree well with western European material of the latter.

Similar nomenclature problems as mentioned under *F. arvensis* are attached to the name *F. gallica*; and the notes under the former species are equally applicable here.

7. PHAGNALON *Cass.*
Bull. Sci. Soc. Philom. Paris, 1819: 174 (1819).

Small shrubs and subshrubs generally with white-tomentose stems; leaves alternate, entire or sinuate-dentate; capitula generally solitary, terminating long, slender peduncles, medium-sized, heterogamous, disciform (all the florets tubular); involucre rather broadly campanulate; phyllaries in several series, scarious, at least at the margins, often shining and brownish; receptacle flattish, naked; outer florets female, filiform, inner narrowly tubular, 5-lobed at apex; corolla-tube yellow; anthers minutely caudate-sagittate at base; style-branches slender, obtuse or subtruncate at apex. Achenes oblong or nearly cylindrical, somewhat compressed, often pubescent; pappus a single series of relatively few, deciduous, scabridulous hairs.

About 30 species in the Mediterranean region, North Africa and western Asia.

All the phyllaries, except the uppermost row, rounded, obtuse or subacute
　　　　　　　　　　　　　　　　　　　　　　1. P. rupestre ssp. **rupestre**
All the phyllaries acute or subacute, the lowermost narrowly deltoid, the uppermost linear-subulate　-　-　-　-　-　-　-　-　-　**1. P. rupestre** ssp. **graecum**

1. **P. rupestre** (*L.*) *DC.*, Prodr., 5: 396 (1836); Unger et Kotschy, Die Insel Cypern, 234 (1865); Boiss., Fl. Orient., 3: 221 (1875); Holmboe, Veg. Cypr., 178 (1914); Post, Fl. Pal., ed. 2, 2: 31 (1933); Lindberg f., Iter Cypr., 35 (1946); Chapman, Cyprus Trees and Shrubs, 79 (1949); Rechinger f. in Arkiv för Bot., ser. 2, 1: 433 (1950); Osorio-Tafall et Seraphim, List Vasc. Plants Cyprus, 104 (1973); Davis, Fl. Turkey, 5: 78 (1975); Feinbrun, Fl. Palaest., 3: 311, t. 517 (1978).

 Conyza rupestris L., Mantissa 1, 113 (1767).

Erect or sprawling, much branched subshrub 6–40 (–50) cm. high; stems terete, at first conspicuously white-tomentose, becoming glabrous and dark brown on old growths; leaves numerous, sessile, semi-amplexicaul, linear, linear-oblong or linear-oblanceolate, 10–30 mm. long, 2–5 mm. wide, bright or dark, rather glossy green above, densely white-tomentose below, apex acute, margins undulate-sinuate, often distinctly revolute; capitula solitary, terminal, peduncles about 6 cm. long, arachnoid-tomentose, merging imperceptibly into the stem-apices and generally clothed with several small, modified leaves; involucre campanulate, 8–15 mm. diam.; phyllaries numerous, in several series, light brown or mid-brown, generally with a fuscous-brown margin or centre, erect, scarious, narrowly oblong, with rounded, obtuse, acute or acuminate-subulate apices, the lowermost series 1–2 mm. long, 1–1·5 mm. wide, the uppermost often 10 mm. long, but rarely more than 1·5 mm. wide, all remaining erect and subadpressed in fruit; female florets filiform, corolla-tube about 6 mm. long, divided at apex into 4–5 minute, narrow lobes barely 0·2 mm. long; style-branches exserted, about 0·4 mm. long, linear, blunt; hermaphrodite flowers narrowly tubular, corolla-tube 5–6 mm. long, 0·4 mm. wide at apex, lobes narrowly ovate or lanceolate, blunt, about 0·5 mm. long, 0·2 mm. wide at base, often fuscous-tipped, and minutely papillose dorsally; anthers narrowly linear, about 2 mm. long, 0·1 mm. wide, apical appendage blunt or subacute; style included, branches about 0·8 mm. long, connivent, slightly swollen at apex; receptacle up to 7 mm. diam., slightly convex. Achenes narrowly oblong or subcylindrical, about 1 mm. long. 0·3 mm. wide, dark brown, subadpressed-pubescent; pappus whitish or pale brown, usually composed of 5–7 minutely scabridulous hairs 6–7 mm. long, in a single series.

ssp. **rupestre**
 TYPE: "*in* Arabia. *Forskåhl* H [ortus] U [psaliensis]" (LINN).

Phyllaries, excepting the uppermost row, rounded, blunt or at most subacute, generally fuscous-brown dorsally with paler, stramineous margins. *Plate 57, figs. 1, 2.*

 HAB.: In garigue on dry, rocky ground; sea-level to 2,700 ft. alt.; fl. March–May.

 DISTR.: Divisions 4–7, locally common. Eastern Aegean Islands, Turkey, Syria, Lebanon, Palestine, Egypt, Libya, and eastwards to Iran.

4. Larnaca, 1912, *Haradjian 66*! Famagusta, 1912, *Haradjian 118*! Cape Greco, 1958, *N. Macdonald 23*! Coast E. of Dhekelia, 1981, *Hewer 4713*!
5. Kythrea, 1927, *Rev. A. Huddle 68*! also, 1941, *Davis 2950*! and, 1950, *Chapman 179*! Between Nisou and Stavrovouni, 1933, *Syngrassides 21*! Salamis, 1934, *Chapman 137*! Athalassa, 1939, *Lindberg f. s.n.*!
6. Nicosia, 1950, *Chapman 605*! also, 1973, *P. Laukkonen 117*!
7. Buffavento, 1862, *Kotschy 386*! Near Kantara, 1880, *Sintenis & Rigo 321*! St. Hilarion, 1898, *Post s.n.*! also, 1901, *A. G. & M. E. Lascelles s.n.*! Kyrenia, 1908–1972, *Clement Reid s.n.*! *Syngrassides 271*! *Casey 778*! Miss Mapple 107! *G. E. Atherton 1304*! *W. R. Price 966*! Koronia, *Chapman 249*! Yaïla, 1941, *Davis 2902*!

ssp. **graecum** (*Boiss. et Heldr.*) *Hayek*, Prodr. Fl. Pen. Balc., 2: 600 (1931); Chapman, Cyprus Trees and Shrubs, 79 (1949).
 Phagnalon graecum Boiss. et Heldr. in Boiss., Diagn., 1, 11: 6 (1849); Boiss., Fl. Orient., 3: 221 (1875); Post in Mém. Herb. Boiss., 18: 94 (1900); H. Stuart Thompson in Journ. Bot., 44: 332 (1906); Holmboe, Veg. Cypr., 178 (1914); Lindberg f., Iter Cypr., 35 (1946); Osorio-Tafall et Seraphim, List Vasc. Plants Cyprus, 104 (1973); Davis, Fl. Turkey, 5: 79 (1975).

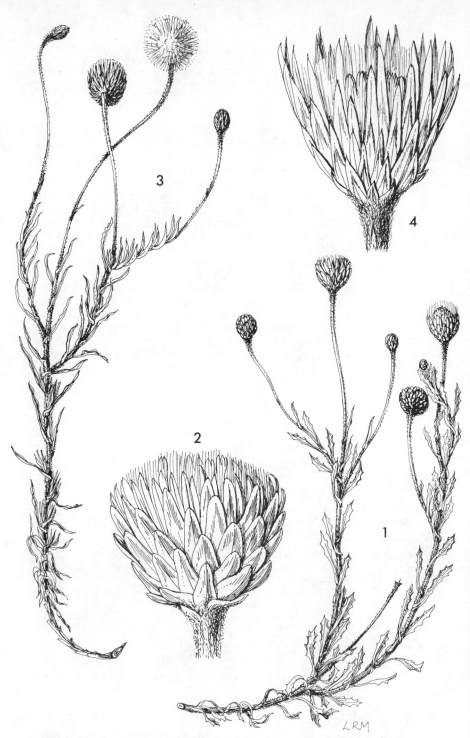

Plate 57. Figs. **1, 2.** PHAGNALON RUPESTRE (L.) DC. ssp. RUPESTRE **1,** habit, × 1; **2,** capitulum, × 6; figs. **3, 4.** PHAGNALON RUPESTRE (L.) DC. ssp. GRAECUM (Boiss. et Heldr.) Hayek **3,** habit, × 1; **4,** capitulum, × 6. (**1, 2** from *G. E. Atherton* 1304; **3, 4** from *Edmondson & McClintock* E 2699.)

TYPE: "in rupestribus regionis calidae totius Graeciae ad *Patras, Nauplia* (Boiss.), Atticâ ad *Acropolim* (Sprun.), Messeniâ ad *Pylos* (Heldr.), *Cretâ* (Sieb. sub *Con. saxatili*), *Zacyntho* (Marg.)" (G, K !).

Phyllaries all acute or subacute, the lowermost narrowly ovate-deltoid, the uppermost linear-subulate, all often stramineous, or sometimes with darker fuscous-brown tips and margins. *Plate 57, figs. 3, 4.*

HAB.: In garigue on dry, rocky ground; on walls; sea-level to 2,900 ft. alt.; fl. March–May.

DISTR.: Divisions 1–3, [7]. Greece, Crete, Aegean Islands, Malta, W. & S. Turkey.

1. Ayios Neophytos Monastery, 1939, *Lindberg f.* s.n.!
2. Amiandos R., 2,600 ft. alt., 1938, *Kennedy* 1442! Kryos Potamos, 2,900 ft. alt., 1938, *Kennedy* 1443!
3. Mazotos, 1905, *Holmboe* 176; Kophinou, 1905, *Holmboe* 569; Limassol, 1913, *Haradjian* 560! Kolossi, 1939, *Lindberg f.* s.n.! Amathus, 1964, *J. B. Suart* 153! Lefkara, 1967, *Merton* in ARI 644! Near Apsiou, 1979, *Edmondson & McClintock* E 2699!
[7. St. Hilarion, 1898, *Post* s.n.; Kyrenia mountains, roadside, 2,000 ft. alt., 1972, *W. R. Price* 978! See Notes.]

NOTES: Evidently rarer than *P. rupestre* ssp. *rupestre*, and perhaps restricted to the southern and western part of the island; both records from the Northern Range may be expunged — the Post specimen from St. Hilarion is correctly *P. rupestre* ssp. *rupestre*, while *W. R. Price* 978 is almost certainly mislabelled, since it bears the inapplicable data "9 ins. Fl. pink".
Phagnalon rupestre (L.) DC. (*Conyza rupestris* L.) has been consistently misunderstood. The type (Linn 993/8) was grown at Uppsala from seeds sent to Linnaeus by Forsskaal, and originally collected in Egypt (Alexandria); there are four specimens of the same plant in herb. Forsskaal, which were named *Conyza tomentosa* Forssk. (Fl. Aegypt.–Arab., 148; 1775). True *P. rupestre* differs from the western Mediterranean plant so named in having the inflorescences terminating a well-developed, elongate branch or lateral branchlet, the peduncle merging imperceptibly with the branch and bearing a number of small leaves at its base, the latter sometimes extending (like bracts) to the middle or upper part of the apparent peduncle. Exactly the same structure is to be seen in ssp. *graecum*. In western Mediterranean specimens (from Portugal, Spain, Balearics, Corsica, Sardinia, Italy, Sicily, Yugoslavia, Malta, western N. Africa and Atlantic Islands) the flower-bearing shoots are so condensed that most inflorescences appear to be axillary, or to arise abruptly from a terminal tuft of normal-sized leaves; the peduncles are either naked or with a minute, barely visible, bract. The western species (or subspecies) is I think correctly *Phagnalon tenorii* (Spreng.) C. Presl, Fl. Sic., XXIX (1826) (*Conyza tenorii* Spreng., Plant. Minus Cognit. Pugillus, 1: 55 (1813); *Conyza geminiflora* Tenore; *Phagnalon rupestre* (L.) DC. var. *illyricum* Lindberg f. or *P. rupestre* (L.) DC. ssp. *illyricum* (Lindberg f.) Ginzberger).
While the relationship between *P. rupestre* and *P. tenorii* is obvious, I do not think it is as close as that between *P. rupestre* and *P. graecum*, thus, while I am prepared to accept *P. rupestre* and *P. tenorii* as distinct species, I feel certain that subspecific rank (under *P. rupestre*) is correct for *P. graecum*. So far as I can judge, the only area where *P. tenorii* and *P. rupestre* ssp. *graecum* meet is in Malta, though I suspect the two must also come together in at least some parts of S. Yugoslavia, Albania or N. Greece. I doubt if *P. tenorii* and *P. rupestre* ssp. *rupestre* ever meet in nature.

8. PSEUDOGNAPHALIUM *Kirpiczn.**

in Act. Inst. Bot. Acad. Sci. SSSR, ser. 1, 9: 33 (1950).

O. M. Hilliard et B. L. Burtt in Journ. Linn. Soc. Bot., 82: 202 (1981).

Annual, biennial or perennial herbs; leaves lanceolate, linear-lanceolate, oblanceolate or spathulate, lanuginose or more or less glandular, often decurrent, margins entire, flat or undulate; capitula heterogamous, in small corymbs; involucre scarious, whitish or yellowish; phyllaries 3–4-seriate, equalling or slightly exceeding the florets; receptacle naked, smooth or faveolate; female florets filiform or narrowly tubular, peripheral in 1 or several series, exceeding in number the hermaphrodite; central florets hermaphrodite, tubular, scarcely expanded at apex; corolla-lobes glandular abaxially. Achenes with scattered, 3-celled, gelatinous idioblasts; pappus-bristles scabridulous.
About 40 species in the New World, Africa and Eurasia.

* by E. Georgiadou

1. P. luteo-album (*L.*) *Hilliard et B. L. Burtt* in Journ. Linn. Soc. Bot., 82: 206 (1981).

    Gnaphalium luteo-album L., Sp. Plant., ed. 1, 851 (1753); Boiss., Fl. Orient., 3: 224 (1875); Holmboe, Veg. Cypr., 178 (1914); Post, Fl. Pal., ed. 2, 2: 33 (1933); Lindberg f., Iter Cypr., 35 (1946); Rechinger f. in Arkiv för Bot., ser. 2, 1: 433 (1950); Osorio-Tafall et Seraphim, List Vasc. Plants Cyprus, 104 (1973); Davis, Fl. Turkey, 5: 98 (1975); Feinbrun, Fl. Palaest., 3: 311, t. 519 (1978).

    TYPE: "*in* Helvetia, G. Narbonensi, Hispania, Lusitania" (lectotype in herb. van Royen sheet 900, 286–294 L).

    Erect or spreading annual; stems up to 50 cm. long, simple or branched, arachnoid or whitish-lanate, sometimes woody towards base; basal leaves oblanceolate to spathulate, occasionally rather crowded or sub-rosulate, 0·8–8 cm. long, 0·3–1·5 cm. wide, arachnoid-floccose or whitish-lanate, apex obtuse or rounded, margins flat or somewhat undulate, base tapering to an ill-defined petiole; cauline leaves usually narrowly oblong or almost linear, acute or obtuse, sessile, sometimes amplexicaul or shortly decurrent, becoming progressively smaller upwards, the uppermost sometimes terminating in a short, scarious acumen; inflorescence a simple or branched corymb; peduncles very short or wanting; capitula (2–) 5–40 or more, campanulate; phyllaries in about 4 series, membranous-scarious, stramineous, the outer ovate, acute, about 3·5 mm. long, 2 mm. wide, the median about 4 mm. long, 1·3 mm. wide, the innermost about 4 mm. long, 1 mm. wide; female florets greatly outnumbering the hermaphrodite, filiform, about 2·2 mm. long, 0·1 mm. wide, corolla-lobes 2–3, very short, glandular, tinged purplish; hermaphrodite florets tubular, about 2·8 mm. long, 0·4 mm. wide, corolla-lobes usually 5, glandular, tinged purple. Achene shortly cylindrical, 0·5 mm. long, 0·2 mm. wide, brownish, with scattered, 3-celled, gelatinous idioblasts; pappus whitish, hairs 2·3–2·8 mm. long, partly coherent at base, caducous, scabridulous.

    HAB.: Rocky or sandy ground, often near water; sea-level to 5,500 ft. alt.; fl. April–July.

    DISTR.: Divisions 2–4, uncommon. Almost throughout temperate and tropical regions of the world.

    2. Troödos, 1913, *Haradjian* 993! Platres, 1930, *E. Wall* s.n., also, 1937, *Kennedy* 957! and, 1971, *G. Joscht* 6439! Mesapotamos Monastery, 1939, *Lindberg f.* s.n.; Kambos, 1939, *Lindberg f.* s.n.; Agros, 1941, *Davis* 3059! Ayios Merkourios, 1962, *Meikle* 2280!

    3. Kolossi, 1939, *Lindberg f.* s.n.

    4. Larnaca, 1905, *Holmboe* 252; also, 1938, *Syngrassides* 1838!

9. HELICHRYSUM *Mill.**

Gard. Dict. Abridg., ed. 4 (1754) as "*Elichrysum*", corr. Pers., Syn. Plant., 2: 414 (1807) nom. et orth. cons.

    Suffruticose ± rhizomatous chamaephytes; stems erect, ascending or decumbent, usually unbranched, leafy for their whole length; leaves simple, alternate or rarely opposite, linear, oblong, oblanceolate or spathulate, sessile or petiolate, subamplexicaul, subglabrous to lanate, often glandular and aromatic, margins entire, inrolled, revolute or rarely flat, apex obtuse or acute; inflorescence corymbose, of several or many capitula; capitula at first cylindrical, ovoid or subglobose, becoming turbinate or campanulate at anthesis, homogamous or heterogamous, the female florets fewer than the hermaphrodite; involucre scarious, bright yellow or straw-coloured (in our area), the phyllaries closely or loosely imbricate, sometimes reflexed after anthesis, stereome arachnoid or arachnoid-lanate, more or less glandular, divided; receptacle concave, naked or irregularly dentate; florets yellow, the females narrowly tubular, 3–5-dentate, the hermaphrodite tubular or widening towards apex 5(4)-dentate; anthers with lanceolate apical

                  * by E. Georgiadou

appendages, basal appendages caudate, sometimes ciliate; style-arms truncate, penicillate. Achenes cylindrical, more or less terete, glabrous, pilose or glandular-papillose, not beaked; pappus-hairs generally in 1 series, whitish or yellowish, scabrid or barbellate.

About 500 species in Europe, Africa, S.W. Asia and Australia. The persistent, often brightly coloured, scarious phyllaries are decorative, and some of the species are cultivated as "everlastings".

Phyllaries glossy, bright yellow; capitula campanulate to broadly campanulate at anthesis, 4–9·5 (–11·3) mm. wide; plant not, or scarcely, aromatic -   -   - **1. H. conglobatum**
Phyllaries rather dull, straw-coloured; capitula cylindric-campanulate or obpyramidal at anthesis, 3–4 (–5) mm. wide; plant strongly aromatic   -   -   - **2. H. italicum**

1. **H. conglobatum** (*Viv.*) *Steudel*, Nomencl. Bot., ed. 2, 738 (1840); Unger et Kotschy, Die Insel Cypern, 240 (1865); Sintenis in Oesterr. Bot. Zeitschr., 31: 326 (1881).
    Gnaphalium conglobatum Viv., Fl. Libycae Spec., 54, t. 3, fig. 5 (1824).
    G. siculum Sprengel, Syst. Veg., 3: 476 (1826) nom. illeg.
    Helichrysum rupicola Pomel, Nouv. Mat. Fl. Atlant., 1: 47 (1874) as *H. "rupicolum"*; Holmboe, Veg. Cypr., 178 (1914); Chapman, Cyprus Trees and Shrubs, 79 (1949); Rechinger f. in Arkiv. för Bot., ser. 2, 1: 433 (1950).
    H. siculum Boiss., Fl. Orient., 3: 229 (1875); Post, Fl. Pal., ed. 2, 2: 33 (1933); Osorio-Tafall et Seraphim, List Vasc. Plants Cyprus, 104 (1973).
    H. siculum Boiss. var. *brachyphyllum* Boiss., Fl. Orient., 3: 230 (1875); Post, Fl. Pal., ed. 2, 2: 34 (1933).
    H. stoechas (L.) Moench ssp. *barrelieri* (Ten.) Nyman, Consp. Fl. Europ., 381 (1879); Davis, Fl. Turkey, 5: 83 (1975).
    H. rupicola Pomel ssp. *brachyphyllum* (Boiss.) Holmboe, Veg. Cypr., 179 (1914); Chapman, Cyprus Trees and Shrubs, 79 (1949).
    H. siculum Boiss. ssp. *brachyphyllum* (Boiss.) Lindberg f., Iter Cypr., 35 (1946); Osorio-Tafall et Seraphim, List Vasc. Plants Cyprus, 104 (1973).
    TYPE: Libya; "in Magna Syrteos aggeribus arenosis", figured in *Fl. Libycae Spec.*, t. 3, fig. 5!

Caespitose, suffruticose perennials (5–) 10–50 (–80) cm. high, not aromatic or weakly so, freely branched with grey-green vegetative shoots; flowering stems (6–) 8–42 (–46) cm. long, ascending, flexuous or erect, rarely decumbent, arachnoid-lanate or sometimes lanate; leaves grey-green or green, puberulent to arachnoid above, arachnoid-lanate below, or frequently lanate on both sides, eglandular or sparsely glandular (densely so in ssp. *decipiens* Georgiadou), lower cauline leaves (5–) 10–53 (–77) mm. long, 0·8–6 (–7·3) mm. wide (narrower in ssp. *decipiens* Georgiadou), usually oblanceolate, sometimes broadly ovate, obovate or suborbicular, rarely linear, those on the lowest third of the stem usually the largest, upper leaves reduced, linear, (1·5–) 5–16 mm. long; inflorescence simple and compact- or branched-corymbose, (6–) 25–64 (–86) mm. wide, comprising (2–) 3–47 (–60) capitula; peduncles 1·5–6 (–10) mm. long, arachnoid-lanate; capitula ovoid in bud, campanulate to broadly campanulate at anthesis, 4·2–7·8 (–8·2) mm. long, 4–9·5 (–11·3) mm. wide; involucre bright, glossy yellow, phyllaries 20–38 (–41), laxly imbricate in 5–7 series, the outermost elliptical, obtuse to subacute, arachnoid at base; median phyllaries 1½–2 times as long as outermost, obovate to oblanceolate, obtuse or subobtuse, the stereome extending up to 1/3 of total length, arachnoid and glandular; innermost phyllaries twice as long as outermost, glandular and sparingly arachnoid, narrowly oblong-spathulate to linear-oblong, subacute to acute, stereome extending to 2/3 of total length; florets 22–52 (–55), the female (6–) 8–19 (–21), hermaphrodite (13–) 14–38 (–41), tube (3–) 3·5–4 (–5) mm. long. Achenes cylindrical, 0·5–0·8 (–1) mm. long, with scattered, small, white, gelatinous, 3-celled idioblasts; pappus white, bristles 11–21 per floret, scabrid.

    HAB.: On rocky or sandy shores or sand-dunes; edges of Pine forest, or in garigue or maquis on stony hillsides; sea-level to 2,000 ft. alt.; fl. March–May.

DISTR.: Divisions 1–8, locally common. Widespread in the Mediterranean region. *H. conglobatum* ssp. *conglobatum* is distributed from S. Italy and Sicily to the E. Mediterranean region and in Africa from Egypt to Algeria; ssp. *decipiens* Georgiadou occurs in the W. Mediterranean region.

1. Common; Yiolou, Dhrousha, Cape Arnauti, Lara, Coral Bay, Tsadha, Stroumbi, Baths of Aphrodite, Ayios Neophytos Monastery, Peyia, etc. *Syngrassides* 1849! *Davis* 3266! *Meikle* 2300! *Georgiadou* 552! 585! 1380! 1386! 1412! 1428! 1453! etc.
2. Near Evrykhou, 1859, *Kotschy* 479; Kambia, 1908, *Clement Reid* s.n.! Saïttas, 1932, *Syngrassides* 351! Mallia, 1950, *F. J. F. Barrington* s.n.! 5 km. N.E. of Pano Panayia, 1973, *Bauer & Spitzenberger* s.n.! Near Moniatis, 1976, *Georgiadou* 465!
3. Common; Limassol, Akrotiri Forest, Stavrovouni, Zagala, Amathus, Apsiou, Mari, Curium, Episkopi, Pissouri, etc. *Haradjian* 558! *Mavromoustakis* 110! *Davis* 3138! *Georgiadou* 410! 411! 514! 586! 592! 1353! 1355! 1368! etc.
4. Ayia Napa, 1905, *Holmboe* 36; between Larnaca and Nicosia, 1976, *Georgiadou.*
5. Kythrea, 1880, *Sintenis*; also, 1941, *Davis* 2951! Athalassa, 1939, *Lindberg f.* s.n.! Mile 5, Nicosia — Limassol road, 1976, *Georgiadou.*
6. Ayia Irini, 1936, *Syngrassides* 1251! also, 1953, *Kennedy* 1793! and, 1968, *Barclay* 1072! Cape Kormakiti, 1956, *Merton* 2594!
7. Common; Buffavento, Kantara, Akanthou, Koronia, Kyrenia, Lapithos, Lakovounara, Tjiklos, Dhavlos, etc. *Sintenis & Rigo* 308! *Holmboe* 560; *Syngrassides* 26! *Chapman* 255! *Casey* 582! *Miss Mapple* 1! *Merton* 2238! *G. E. Atherton* 255! *P. Laukkonen* 236! *H. Painter* 28! 29! etc.
8. Valia, 1905, *Holmboe* 492; Rizokarpaso, 1912, *Haradjian* 158! Eleousa, 1937, *Syngrassides* 1641! Apostolos Andreas, 1941, *Davis* 2314! Yialousa, 1970, *A. Genneou* in ARI 1455! 1460!

NOTES: Plants growing in sand or rock-fissures in exposed coastal areas are often very dwarf and condensed, with numerous, woody branches and a close dome of densely lanate, small, broadly ovate, obovate or suborbicular leaves. These extreme forms are, however, connected to the typical plant by a series of intermediates, not infrequently to be found nearby, but in more sheltered situations.

2. H. italicum (*Roth*) *Don* in Loudon, Hort. Brit., 342 (1830); Boiss., Fl. Orient., 3: 234 (1875); Chapman, Cyprus Trees and Shrubs, 80 (1949); Osorio-Tafall et Seraphim, List Vasc. Plants Cyprus, 104 (1973); Davis, Fl. Turkey, 5: 83 (1975).
   Gnaphalium italicum Roth in Roemer et Usteri, Mag. für Bot., 4 (10): 19 (1790), Catalecta Bot., 1: 115 (1797).
   Helichrysum italicum (Roth) Don var. *canum* Boiss. ex Holmboe, Veg. Cypr., 179 (1914).
   [*H. microphyllum* (non Willd.) Camb.) Kotschy in Unger et Kotschy, Die Insel Cypern, 241 (1865); Lindberg f., Iter Cypr., 35 (1946); Rechinger f. in Oesterr. Bot. Zeitschr., 94: 186 (1948); Osorio-Tafall et Seraphim, List Vasc. Plants Cyprus, 104 (1973).]
   [*H. italicum* (Roth) Don var. *microphyllum* (non *H. microphyllum* (Willd.) Camb.) Boiss., Fl. Orient., 3: 234 (1875) pro parte quoad plant cypr.]
   TYPE: *Gnaphalium italicum* B–W no. 15445, neotype. Photo!

Subshrubby or shrubby perennial 10–60 (–70) cm. high, grey-green to yellow-green, arachnoid-lanate or sericeous-lanate, glandular and aromatic (smelling of curry); vegetative shoots numerous, usually slender, often bearing axillary fascicles of leaves, the fascicles 2–10 mm. long; flowering stems (1–) 3–40 (–52) cm. long, erect, ascending or flexuose, occasionally branched; leaves linear, rarely oblanceolate, inrolled, the margins straight, lower to mid-cauline leaves (2–) 4–43 mm. long, 0·5–2·5 mm. wide, often closely leafy, the lower deflexed to patent, the upper reflexed, gradually becoming more distantly arranged, the uppermost reduced to 1–10 mm. just below the inflorescence; inflorescences (5–) 9–60 (–90) mm. wide, simple to compound-corymbose, lax or compact with 4–100 capitula; peduncles 1–5 (–10) mm. long, arachnoid and glandular; capitula (3·5–) 4–6·5 (–7) mm. long, (2·5–) 3–6·8 (–7·5) mm. wide, narrowly ovoid to cylindrical in bud, cylindrical-campanulate to obpyramidal at anthesis, sometimes constricted just above the middle, heterogamous; involucre straw-coloured; involucral bracts (19–) 21–38 (–41), very closely imbricate in 5–6 series, acute, flat, outermost ovate to narrowly ovate, often coriaceous, arachnoid to tomentose and glandular, middle twice as long as the outermost, narrowly obovate to narrowly elliptical, the stereome extending up to the middle of the bract, arachnoid and glandular, innermost three times as long as the outermost, narrowly lanceolate to narrowly oblong, acute, glandular and sparsely arachnoid; florets (12–) 14–31 (–39), female 3–12, hermaphrodite 9–27, 2·5–3·5 mm. long. Achenes 0·5–0·8 mm. long, cylindrical, with scattered short, 3-celled idioblasts; pappus white, scabrid to subscabrid; bristles 16–24 per floret with acute apices.

HAB.: Rocky, igneous mountainsides [200–] 1,000–5,600 ft. alt.; fl. June–Sept.

DISTR.: Divisions 2, 3. Widely distributed in the Mediterranean region east to Cyprus.

2. Locally abundant, 1831–1980; Prodhromos, Kykko, Platres, Troödos, Galata, Makheras, Kambos, etc. *Aucher-Eloy* 3545! *Kotschy* 839; *Sintenis & Rigo* 797! *Haradjian* 527! *C. B. Ussher* 15! 39! 73! 76! *Kennedy* 959! 961! *Lindberg f.* s.n.! *Davis* 1838! *S. G. Cowper* 2! *Casey* 918! *G. E. Atherton* 540! *Georgiadou* 401! 3003! *Georgiadou & Vassiliou* 448! 456! 460! etc.
3. Mazotos and Moni, 1862, *Kotschy* 577! Stavrovouni area, 1932–1977, *Syngrassides* 395! *Lindberg f.* s.n.; *Georgiadou* 406! 1352! 1338!

NOTES: In Cyprus normally confined to igneous mountainsides at altitudes above 1,000 ft. Kotschy's specimens from Moni and Mazotos were found in river-beds, and may have been carried down from higher elevations.

10. INULA L.

Sp. Plant., ed. 1, 881 (1753).
Gen. Plant., ed. 5, 375 (1754).

Perennial, or sometimes annual, herbs and subshrubs; stems usually erect, or plants acaulescent; leaves entire or toothed, petiolate or sessile, frequently amplexicaul; inflorescences corymbose or paniculate, or capitula solitary; capitula heterogamous, radiate or disciform; involucre generally hemispherical or broadly campanulate, phyllaries in several series, imbricate, the outer often herbaceous, or even foliaceous, the inner frequently scarious; receptacle naked, flat or convex; female florets tubular or ligulate, few or numerous, or rarely absent, yellow or orange; hermaphrodite disk-florets usually numerous, tubular, 5-lobed at apex, yellow or orange; anthers with sagittate, often fimbriate basal appendages; style-branches obtuse, somewhat compressed, papillose dorsally towards apex. Achenes oblong or fusiform, smooth or longitudinally ribbed, often pilose with a truncate or abruptly constricted apex; pappus hairs in 1 or more series, free or connate at base, smooth, scabridulous or barbellate.

The genus comprises 200 or more species widely distributed in Europe, Asia and Africa; because of its heterogeneous character, authors have divided it into smaller groupings, here, pending a satisfactory revision of the genus *sensu lato*, regarded as sections.

Capitula disciform, without ligulate ray-florets; phyllaries generally with squarrose, herbaceous apices - - - - - - - - - - - **1. I. conyzae**
Capitula radiate, with ligulate ray-florets; apices of phyllaries not squarrose:
  Leaves succulent, linear or narrowly oblanceolate; capitula solitary or few together
      2. I. crithmoides
  Leaves not succulent; flowers in branched panicles:
    Leaves and stems viscid-glandular, odorous; achenes constricted at apex; pappus-hairs connate at base:
      Capitula 1–2 cm. diam., ligules 6–7 mm. long; perennials with a woody base
          3. I. viscosa
      Capitula less than 1 cm. diam.; ligules 3–3·5 mm. long; slender annuals **4. I. graveolens**
    Leaves and stems not viscid-glandular or odorous; achenes truncate at apex; pappus-hairs free or almost free - - - - - - - - - [I. BRITANNICA p. 890]

SECT. 1. **Bubonium** *DC.* Perennial, erect herbs; inflorescences corymbose or paniculate, or capitula sometimes solitary; involucres broadly campanulate, phyllaries commonly herbaceous and squarrose at apex.

Achenes ± pilose, truncate at apex, angled or longitudinally ribbed; pappus hairs free to base.

1. **I. conyzae**(*Griesselich*) *Meikle* **comb. nov.**
  Conyza squarrosa L., Sp. Plant., ed. 1, 861 (1753), non *Inula squarrosa* L.
  C. vulgaris Lam., Fl. Franç., 2: 73 (1778) nom. illeg.
  Aster conyzae Griesselich, Kleine Bot. Schrift., 122 (before July 1836).
  Inula conyza DC., Prodr., 5: 464 (Oct. 1836); Boiss., Fl. Orient., 3: 190 (1875); Post in Mém. Herb. Boiss., 18: 95 (1900); H. Stuart Thompson in Journ. Bot., 44: 332 (1906); Osorio-Tafall et Seraphim, List Vasc. Plants Cyprus, 104 (1973).
  I. vulgaris Trevisan, Fl. Euganea, 29 (1842); Holmboe, Veg. Cypr., 179 (1914); Post, Fl. Pal., ed. 2, 2: 24 (1933); Davis, Fl. Turkey, 5: 67 (1975).
  TYPE: "*in* Germaniae, Belgii, Angliae, Galliae *siccis*".

Erect biennial or short-lived perennial with a thick, oblique rootstock; stems 20–130 cm. high, pubescent or thinly tomentose, often branched above, basal leaves ovate-oblong, 8–12 cm. long, 5–7 cm. wide, tapering basally to a flattened petiole 5–7 cm. long, apex acute or obtuse, margins irregularly dentate, upper leaves sessile, narrowly ovate-oblong or ovate-lanceolate, 3–7 cm. long, 1–2·5 cm. wide, denticulate or entire, pubescent on both surfaces; inflorescence a dense, corymbose panicle; involucre widely campanulate, about 1 cm. diam.; phyllaries imbricate in 4–6 series, the lower (and outer) shortly oblong, 2–5 mm. long, 1–1·5 mm. wide, shortly adpressed-pilose dorsally, with an obtuse, squarrose, herbaceous apex, upper (and inner) phyllaries narrowly linear-oblong, up to 12 mm. long, 1 mm. wide, scarious, ciliate, sometimes purplish, with an acute apex; female florets in 1 series, narrowly tubular, corolla 6–7 mm. long, 0·2 mm. wide, apex very shortly and obliquely ligulate, 3-lobed, lobes about 0·3 mm. long, style-branches shortly exserted, linear, compressed, obtuse, about 1 mm. long; hermaphrodite florets about 6 mm. long, 0·8 mm. wide at apex, lobes erect, subacute, concave, about 0·4 mm. long; anthers linear, about 3 mm. long, 0·1 mm. wide, apex subacute, base caudate-sagittate; style-branches connivent, about 1 mm. long, compressed, obtuse. Achenes narrowly oblong-cylindrical, about 2 mm. long, 0·4 mm. wide, closely ribbed longitudinally, brownish, glabrous towards base, shortly pilose towards the truncate apex; pappus hairs about 20 in a single series, free to base, whitish, shortly barbellate, subequal, 5–6 mm. long.

HAB.: Dry, igneous mountainsides; 2,900–5,000 ft. alt.; fl. Aug.–Sept.

DISTR.: Division 2, very rare. Western, central and S. Europe eastwards to N. Iran; Algeria.

2.   "Au pied du mont Machaira" [Makheras = Kionia], Aug. 1898, *Post* s.n.; road from Troödos to Prodhromos, 1900, *A. G. & M. E. Lascelles* s.n. !

NOTES: Evidently very rare, and not seen by any recent collectors. Its nearest occurrence outside Cyprus is in the Amanus mountains of S. Turkey, but it is absent from Palestine.

I. BRITANNICA *L.*, Sp. Plant., ed. 1, 882 (1753) (non "*britanica*" — *pace* DC. in Lam. et DC., Fl. Franç., 4: 149 (1805) adnot.; vide Briquet in Burnat, Fl. Alpes-Marit., 6: 240; 1917) has been recorded "Bei Prodromo [Prodhromos] in Schluchten im Schatten", 1840, by Kotschy (in Unger et Kotschy, Die Insel Cypern, 235; 1865), but has not since been seen in Cyprus. Kotschy does not cite a specimen, and the plant seen was most probably *Pulicaria dysenterica* (L.) Bernh., which occurs in the area, and is not infrequently misidentified as *Inula britannica*. Later Cyprus records for the latter (H. Stuart Thompson in Journ. Bot., 44: 332; 1906; Holmboe, Veg. Cypr., 179; 1914) all refer back to Kotschy's statement.

Inula britannica has larger flowers than those of *Pulicaria dysenterica*, with broader phyllaries, the inner fringed with brown, glandular hairs. It is at once distinguished by its pappus, consisting of free hairs in a single series.

In *Pulicaria* the pappus consists of an outer cup of scales surrounding an inner ring of hairs.

SECT. 2. **Limbarda** (*Adans.*) *DC.* Herbs and subshrubs with linear leaves; capitula solitary or in panicles; involucres broadly campanulate; phyllaries narrow, adpressed, not squarrose at apex. Achenes pilose, truncate, angled; pappus hairs free to base.

2. I. crithmoides *L.*, Sp. Plant., ed. 1, 883 (1753); Unger et Kotschy, Die Insel Cypern, 235 (1865); Boiss., Fl. Orient., 3: 195 (1875); Holmboe, Veg. Cypr., 179 (1914); Post, Fl. Pal., ed. 2, 2: 24 (1933); Lindberg f., Iter Cypr., 35 (1946); Chapman, Cyprus Trees and Shrubs, 80 (1949); Osorio-Tafall et Seraphim, List Vasc. Plants Cyprus, 104 (1973); Davis, Fl. Turkey, 5: 72 (1975); Feinbrun, Fl. Palaest., 3: 314, t. 522 (1978).

TYPE: "*in* Angliae, Galliae, Lusitaniae, Hispaniae *maritimis*".

Sprawling woody-based herb or subshrub; stems 30–50 cm. long, glabrous, sulcate, usually branched; leaves numerous, linear or narrowly oblanceolate, 10–40 mm. long, 1–5 mm. wide, glabrous, succulent, widest near the entire or shortly lobed, acute or subacute apex, tapering to a slender base; capitula solitary, or sometimes 4–5 on short axillary branchlets forming a loose terminal corymb; involucre very broadly campanulate or hemispherical, 1–1·5 cm. diam.; phyllaries in 3–4 series, oblong or linear, the lower 1–2 mm. long, 1–1·5 mm. wide, the upper linear, acute, about 6 mm. long, 1 mm. wide, all glabrous with scarious margins; ray-florets yellow, ligulate, in a single series, about 10–12 mm. long, ligule about 1·5 mm. wide, shortly 3-dentate at apex; style about 6 mm. long, branches exserted, linear, obtuse, recurved, about 0·8 mm. long; disk-florets narrowly tubular, yellow, about 6 mm. long, 0·8 mm. wide at apex, lobes ovate-deltoid, concave, erect, subacute, about 0·6 mm. long; anthers linear, about 2·5 mm. long, 0·1 mm. wide, apex acute or subacute, base caudate-sagittate; style exserted, about 6·5 mm. long, branches linear, obtuse, slightly recurved, about 0·8 mm. long. Achene oblong, truncate, about 2·5 mm. long, 0·8 mm. wide, angled, densely adpressed-pilose; pappus-hairs numerous, whitish or brownish, free to base in a single series, barbellate, unequal, 2·5–6·5 mm. long.

HAB.: By salt-lakes, salt-marshes and seashores; near sea-level; fl. June–Aug.

DISTR., Divisions 3, 4, 6. W. European and Mediterranean coasts; Atlantic Islands.

3. Near Akrotiri, 1862, *Kotschy* 601; near Limassol, 1939, *Lindberg f.* s.n.; Limassol Salt Lake, 1941, *Davis* 3547!

4. Near Livadhia, 1936, *Syngrassides* 1011! Famagusta, 1938, *Chapman* 332! and, 1939, *Lindberg f.* s.n.! Larnaca aerodrome, 1950, *Chapman* 412!

6. Syrianokhori swamp, 1953, *Casey* 1322!

SECT. 3. **Cupularia** *Willk.* Annual and perennial herbs and subshrubs; leaves and stems viscid-glandular, malodorous; capitula numerous in panicles; phyllaries in 3–5 series, not squarrose at apex. Achenes not angled or ribbed, sparsely pilose, constricted at apex; pappus hairs connate at base, forming a shallow cup.

3. I. viscosa (*L.*) *Ait.*, Hort. Kew., ed. 1, 3: 223 (1789); Unger et Kotschy, Die Insel Cypern, 235 (1865); Boiss., Fl. Orient., 3: 198 (1875); Holmboe, Veg. Cypr., 179 (1914); Post, Fl. Pal., ed. 2, 2: 25 (1933); Lindberg f., Iter Cypr., 35 (1946); Chapman, Cyprus Trees and Shrubs, 80 (1949); Osorio-Tafall et Seraphim, List Vasc. Plants Cyprus, 104 (1973); Davis, Fl. Turkey, 5: 73 (1975); Feinbrun, Fl. Palaest., 3: 314, t. 523 (1978).

Erigeron viscosum L., Sp. Plant., ed. 1, 863 (1753); Hume in Walpole, Mem. Europ. Asiatic Turkey, 254 (1817).

Dittrichia viscosa (L.) Greuter in Exsicc. Genav., fasc. 4: 71 (1973).

TYPE: "*in* Narbonensi, Hispania, Italia".

Bushy perennial with a woody rootstock; branches numerous, erect, 50–150 cm. high; stems terete, glandular, glutinous, longitudinally sulcate;

leaves numerous, lanceolate, sessile or somewhat amplexicaul at base, 2·5–9 cm. long, 0·3–2 cm. wide, viscid-glandular on both surfaces, rank-smelling, apex acute or acuminate, margins irregularly serrate or serrulate; inflorescence consisting of numerous capitula forming a loose, broadly or narrowly pyramidal panicle; capitula 1–2 cm. diam., on slender, glandular peduncles 1–2 cm. long; involucre broadly campanulate, phyllaries oblong-linear, the outer about 2 mm. long, 1 mm. wide, the inner up to 9 mm. long, 1 mm. wide, all subadpressed, obtuse or shortly acute, glandular, the inner with wide, shortly ciliate, scarious margins; florets bright yellow; ray-florets about 10 in one series, basal tube about 3 mm. long, ligule 6–7 mm. long, 1·5 mm. wide, shortly 3-dentate at apex; style distinctly exserted, branches linear, obtuse, recurved, about 1 mm. long; disk-florets narrowly tubular, 6–7 mm. long, 1·5 mm. wide at apex, lobes 5, ovate-deltoid, about 0·8 mm. long, 0·5 mm. wide at base, subacute, slightly recurved; anthers linear, about 3 mm. long, 0·2 mm. wide, apex acute, base caudate-sagittate with filiform appendages up to 2 mm. long; style 7 mm. long, branches exserted, linear, obtuse, about 0·8 mm. long, slightly recurved. Achene fusiform, about 2 mm. long, 0·8 mm. wide, pale brown, thinly pilose, not angled or ribbed, abruptly constricted at apex; pappus-hairs about 20, subequal, whitish, barbellate, brittle, about 4·5 mm. long, united basally to form a distinct, shallow cup about 0·7 mm. diam.

HAB.: Roadsides, gardens, waste ground, stony hillsides, often in moist situations by rivulets or springs; sea-level to 5,000 ft. alt.; fl. Aug.–Nov.

DISTR.: Divisions 1–5, 7, probably common everywhere. Mediterranean region; Atlantic Islands.

1. Common about Kato Paphos, 1981, *Meikle*!
2. Common; Prodhromos, Galata, Makheras Monastery, Pyrgos, Platres, Milikouri, Kambi, Yialia, Platania, etc., 1862–1981, *Kotschy*; *Holmboe* 802! 1137; *Syngrassides* 862! *Kennedy* 937! *Lindberg f.* s.n.! *S. G. Cowper* s.n.! *Chapman* 110; *G. E. Atherton* 506! etc.
3. Limassol, 1801, *Hume*; Episkopi, 1963, *J. B. Suart* 132!
4. Larnaca, 1801, *Hume*; Kouklia, 1905, *Holmboe* 401.
5. Kakoradjia, 1934, *Syngrassides* 530! Dheftera, 1937, *Nattrass* 868.
7. Ayios Khrysostomos Monastery, 1862, *Kotschy*; near Myrtou, 1862, *Kotschy*; Kantara, 1880, *Sintenis*; Kyrenia, 1936–1970, *Kennedy* 938! *C. H. Wyatt* 61! *Casey* 40! *G. E. Atherton* 328! 329! 391A! 492! *I. M. Hecker* 12! Vasilia, 1968, *Barclay* 1051!

NOTES: So common, and so late in flowering, that it is seldom collected; it often forms a continuous yellow band along roadsides in the autumn.

4. I. graveolens (*L.*) *Desf.*, Fl. Atlant., 2: 275 (1799); Boiss., Fl. Orient., 3: 199 (1875) pro parte; H. Stuart Thompson in Journ. Bot., 44: 332 (1906); Holmboe, Veg. Cypr., 179 (1914); Post, Fl. Pal., ed. 2, 2: 25 (1933); Lindberg f., Iter Cypr., 35 (1946); Osorio-Tafall et Seraphim, List Vasc. Plants Cyprus, 104 (1973); Davis, Fl. Turkey, 5: 72 (1975); Feinbrun, Fl. Palaest., 3: 314, t. 524 (1978).

Erigeron graveolens L., Cent. Plant. 1, 28 (1755).

Dittrichia graveolens (L.) Greuter, Exsicc. Genav., fasc. 4: 71 (1973).

TYPE: France; "Monspelii".

Erect annual 25–100 cm. high; stems branched or unbranched, sulcate, viscid-glandular, sometimes with long, spreading, eglandular hairs intermixed; leaves linear-oblanceolate or linear, 2–6 cm. long, 0·2–0·8 cm. wide, sessile, viscid-glandular, smelling of camphor, the lowermost obtuse, minutely denticulate, the upper acute, entire or subentire, capitula very numerous, in lax, much-branched pyramidal panicles; peduncles 3–15 mm. long; capitula small, usually less than 10 mm. diam.; involucre widely campanulate; phyllaries oblong or linear-oblong, in 3-4 series, the outermost about 2 mm. long, 1 mm. wide, herbaceous, glandular, the inner 6 mm. long, 1 mm. wide, with an herbaceous midrib and broad, scarious margins, all subadpressed at anthesis, becoming reflexed after the fruit is shed; florets pale yellow (often reddish on drying); ray-florets about 8 in one series, basal tube about 2·5 mm. long, ligule about 3–3·5 mm. long, 0·8 mm. wide, generally revolute and shortly 3-dentate at apex; style distinctly exserted, branches linear, obtuse, about 0·6 mm. long; disk-florets narrowly tubular, 3·5–4 mm. long, 0·6 mm. wide at apex, lobes 4, erect, ovate-deltoid, about 0·4 mm. long, 0·3 mm. wide at base; anthers linear, about 1 mm. long, 0·1 mm. wide, apex acute, base caudate-sagittate, with filiform appendages about 0·8 mm. long; style about 3 mm. long, branches linear, obtuse, about 0·6 mm. long, erect or somewhat recurved. Achene fusiform, about 2 mm. long, 0·8 mm. wide, pale brown, thinly pilose, not angled or ribbed, abruptly constricted at apex; pappus-hairs 20–25, subequal, whitish, barbellate, brittle, 3–4 mm. long, very shortly united basally to form an indistinct, shallow cup about 0·5 mm. diam.

HAB.: Roadsides, gardens, waste ground, stony hillsides, often with *I. viscosa* (L.) Ait.; sea-level to 3,800 ft. alt.; fl. Oct.–Nov.

DISTR.: Divisions 1, 2, 5–7. Mediterranean region and eastwards to Afghanistan and N.W. India.

1. Paphos, 1939, *Lindberg f.* s.n.; Skoulli, 1939, *Lindberg f.* s.n.!
2. Platres, 1937, *Kennedy* 939! and, 1955, *Kennedy* 1851! Pano Pyrgos, 1937, *Syngrassides* 1720!
5. Dheftera, 1931, *Nattrass* 156; Athalassa, 1966, *Merton* in ARI 25!
6. Nicosia, 1935, *Nattrass* 693; Makhedonitissa Monastery, 1935, *Syngrassides* 872!
7. Kyrenia, 1948, *Casey* 87! also, 1955, *G. E. Atherton* 275! Kazaphani, 1955, *G. E. Atherton* 783!

NOTES: Distinctly less common than *I. viscosa* (L.) Ait., and readily distinguished in the field by its slender habit, narrow leaves and small capitula.

11. PULICARIA *Gaertn.*
Fruct. Sem. Plant., 2: 461 (1791).

Annual and perennial herbs resembling *Inula*; leaves alternate, sessile, somewhat amplexicaul or auriculate at base; inflorescence generally branched, paniculate or corymbose; capitula heterogamous, radiate or disciform; involucre hemispherical; phyllaries narrow, imbricate in several series; receptacle naked; ray-florets female, often ligulate, yellow, in 1-2 series; disk-florets hermaphrodite, tubular, 5-lobed at apex; style-branches linear, obtuse; anthers linear, caudate-sagittate at base. Achenes terete or somewhat compressed, smooth or ribbed; pappus in 2 rows, the outer consisting of short, broad, partly connate, persistent scales, the inner of 7–25, or more, brittle, barbellate hairs.

About 40 species, chiefly in Europe and the Mediterranean region, also in Africa.

Rhizomatous perennials, 50–130 cm. high - - - - **1. P. dysenterica** ssp. **uliginosa**
Annuals, 5–50 cm. high:
  Capitula terminal and lateral; peduncles generally elongate, slender; leaves flat or flattish
2. P. arabica
  Capitula terminal on leafy branches and branchlets; peduncles short, not very distinct; leaf-margins strongly recurved or revolute - - - - - - **3. P. sicula**

1. **P. dysenterica** (*L.*) *Bernh.*, Syst. Verz. Pfl., 153 (1800); Boiss., Fl. Orient., 3: 201 (1875); Post, Fl. Pal., ed. 2, 2: 25 (1933); Davis, Fl. Turkey, 5: 75 (1975); Feinbrun, Fl. Palaest., 3: 317, t. 530 (1978).
  Inula dysenterica L., Sp. Plant., ed. 1, 882 (1753).

Rhizomatous perennial 50–130 cm. high; stem erect, terete, longitudinally sulcate, pubescent or arachnoid-tomentose, usually unbranched in the lower half, sparingly or repeatedly branched above; leaves thin, oblong or oblong-lanceolate, 1–8 cm. long, 0·3–2 cm. wide, pubescent, tomentose or subglabrous, apex generally acute, base semi-amplexicaul with short or conspicuous, obtuse or acute auricles, margins entire or varyingly undulate-serrate; inflorescence sparsely branched and subcorymbose, or much branched, with numerous, lax, divaricated branches, forming a diffuse panicle; peduncles stout or slender, pubescent or tomentose, up to 6 cm. long; involucre hemispherical, 0·8–2 cm. diam.; phyllaries in 4–5 (or more) series, linear-subulate, barely 1 mm. wide, the lowermost about 2mm. long, the uppermost to 10 mm. long, pubescent or arachnoid-tomentose, often dark-tipped; florets clear yellow; ray-florets numerous, ligules narrow, 3–10 mm. long, barely 1 mm. wide, apex sharply 3-dentate; style exserted, 2–3 mm. long, branches recurved, linear, obtuse, 0·4–0·5 mm. long; disk-florets numerous forming a low dome, narrowly tubular, 2·5–4·5 mm. long, 0·6–0·8 mm. wide at apex, lobes erect, deltoid, acute, densely papillose dorsally, about 0·4–0·7 mm. long, 0·4–0·6 mm. wide at base; anthers linear, acute, 1·5–2 mm. long, 0·1–0·2 mm. wide, base caudate-sagittate, the appendages sometimes 1·5 mm. long; style about as long as corolla-tube, branches recurved, linear, obtuse, 0·4–0·6 mm. long. Achene oblong-cylindrical, subterete, longitudinally ribbed, mid-brown, thinly pilose, apex truncate, about 1 mm. long, 0·4 mm. wide; pappus in 2 rows, the outer consisting of connate, fimbriate-erose, whitish scales, about 0·1–0·2 mm. long, the inner of about 20 barbellate, whitish hairs 2–3 mm. long.

ssp. **uliginosa** *Nyman*, Consp. Fl. Europ., 394 (1879).

    Inula dentata Sm. in Sibth. et Sm., Fl. Graec. Prodr., 2: 181 (1813), Fl. Graec. 9: 57, t. 874 (1837).

    Pulicaria uliginosa Stev. ex DC., Prodr., 5: 478 (1836) non S. F. Gray (1821) nom. illeg.

    P. dentata (Sm.) DC., Prodr., 5: 480 (1836).

    P. dysenterica (L.) Bernh. var. *microcephala* Boiss., Fl. Orient., 3: 202 (1875).

    P. dysenterica (L.) Bernh. ssp. *dentata* (Sm.) Holmboe, Veg. Cypr., 180 (1914); Lindberg f., Iter. Cypr., 36 (1946); Osorio-Tafall et Seraphim, List Vasc. Plants Cyprus, 104 (1973).

    [*P. dysenterica* (non (L.) Bernh.) Poech, Enum. Plant. Ins. Cypr., 17 (1842); Unger et Kotschy, Die Insel Cypern, 236 (1865); H. Stuart Thompson in Journ. Bot., 44: 332 (1906).]

    TYPE: U.S.S.R. (Crimea); "In Tauriâ meridionali [Nikita] ad fossas aquosas legit cl. Steven" (G, K!).

Plants as tall as, or taller than *P. dysenterica* (L.) Bernh. ssp. *dysenterica*, but generally much more slender; stems and leaves shortly pubescent, viscid, rarely arachnoid-tomentose; capitula numerous, rather small, generally less than 1 cm. diam., in lax, spreading panicles; peduncles up to 6 cm. long, slender, pubescent (not tomentose); involucre pubescent; ray-florets with ligules generally less than 5 mm. long; corolla-tube of disk-florets about 3 mm. long. Achenes and pappus as in ssp. *dysenterica*.

HAB.: Swamps, marshes, sides of streams and irrigation channels, roadside ditches, moist shaded hollows on mountains; sea-level to 4,500 ft. alt.; fl. Aug.–Nov.

DISTR.: Divisions 1–3, 5, 7. Widespread in the Mediterranean region and eastwards to Afghanistan.

1. Ayios Neophytos Monastery, 1939, *Lindberg f.* s.n.; Polis, 1969, *A. Genneou* in ARI 1359!
2. Troödhitissa Monastery, 1840, *Kotschy* 10; Platres [?1905] *Michaelides* sec. *Holmboe*; also, 1937, *Kennedy* 940! Pharmakas, 1905, *Holmboe* 1129; Kryos Potamos, 3,900 ft. alt., 1937, *Kennedy* 941! and Kephalovryson, Platres, 1937, *Kennedy* 942! Kambos, 1939, *Lindberg f.* s.n.! Galata, 1939, *Lindberg f.* s.n.; Prodhromos, 1955, *G. E. Atherton* 568!
3. Asomatos Marshes, 1939, *Mavromoustakis* 61! Yermasoyia River, 1940, and, 1947, *Mavromoustakis* 39!
5. Kakoradjia, 1936, *Syngrassides* 531!
7. Kyrenia, 1932-1955, *Syngrassides* 630! *Kennedy* 943! 944! *Casey* 51! *G. E. Atherton* 331! 348! 377! Below Bellapais, 1948, *Casey* 117! Upper Thermia, 1955, *G. E. Atherton* 493! Lapithos, 1955, *G. E. Atherton* 692! Karavas, 1966, *Merton* in ARI 37!

NOTES: Although most authors note that this intergrades with typical *P. dysenterica*, it is nonetheless very distinct over most of its range, and uniformly so in Cyprus, where it looks very different from *P. dysenterica* of western Europe.

2. P. arabica (*L.*) *Cass.* in Dict. Sci. Nat., 44: 94 (1826); Unger et Kotschy, Die Insel Cypern, 236 (1865); Boiss., Fl. Orient., 3: 205 (1875); Holmboe, Veg. Cypr., 180 (1914); Post, Fl. Pal., ed. 2, 2: 27 (1933); Lindberg f., Iter Cypr., 36 (1946); Rechinger f. in Arkiv för Bot., ser. 2, 1: 433 (1950); Osorio-Tafall et Seraphim, List Vasc. Plants Cyprus, 104 (1973); Davis, Fl. Turkey, 5: 76 (1975); Feinbrun, Fl. Palaest., 3: 318, t. 532 (1978).

Inula arabica L., Mantissa 1, 114 (1767).

[*?I. pulicaria* (non L.) Hume in Walpole, Mem. Europ. Asiatic Turkey, 254 (1817).]

TYPE: Egypt; "*in* Aegypto"; cultivated at Uppsala.

Erect or diffuse, usually much-branched annual, 5–30 (–50) cm. high; stems terete, shallowly sulcate, often purplish, thinly viscid-glandular with scattered long hairs, or occasionally thinly arachnoid-lanuginose; leaves narrowly oblong or strap-shaped, often recurved, 1–6 cm. long, 0·2–1 cm. wide, thinly glandular-pubescent or pilose, apex obtuse, mucronate or shortly acute, base semi-amplexicaul, sometimes shortly auriculate, margins entire or almost so, occasionally a little undulate; inflorescence lax, spreading, paniculate; peduncles slender, divaricate, glandular-pubescent, up to 5 cm. long; capitula small, usually less than 1 cm. diam.; involucre hemispherical; phyllaries subadpressed, in 3–4 series, linear, the lower 1–3 mm. long, less than 1 mm. wide, acute, herbaceous, glandular, the upper 7–8 mm. long, 1 mm. wide, acuminate and often with a purple acumen, and with broad scarious margins; florets yellow, the marginal (12–20) radiate, ligulate, ligules about 3 mm. long, 1 mm. wide, recurved, apex shortly 3-toothed; style exserted, branches recurved, linear, obtuse, about 0·5 mm. long; disk-florets narrowly tubular, 4–5 mm. long, 0·6 mm. wide at apex, lobes deltoid-acuminate, about 0·6 mm. long, 0·4 mm. wide at base, distinctly papillose dorsally; anthers linear, acute, about 2 mm. long, 0·2 mm. wide, with caudate-sagittate basal appendages; style about as long as corolla-tube, branches connivent, linear, obtuse, about 0·8 mm. long. Achene fusiform, about 1 mm. long, 0·4 mm. wide, mid-brown, subterete, not angled or sulcate, thinly subadpressed-pilose; outer pappus an erose cup, about 0·1 mm. long, of connate, whitish scales, inner pappus a single series of 10–12 brittle, faintly scabridulous hairs, 3–4 mm. long.

HAB.: Pastures and fallows, generally in hollows where water has lain during the winter; sea-level to 2,000 ft. alt.; fl. May–July.

DISTR.: Divisions 2–5, 8, rather rare. Turkey, Lebanon, Palestine, Egypt and east to Afghanistan.

2. Near Galata, 1913, *Haradjian* 1001 !
3. Akrotiri Forest, 1939, *Mavromoustakis* 54 !
4. Phaneromene near Larnaca, 1862, *Kotschy* 979 ! Famagusta, 1955, *Lindberg f.* s.n. ! Sotira, Ayios Antonios, 1955, *Merton* 2264 !
5. Athalassa, 1939, *Lindberg f.* s.n. !
8. Boghaz near Trikomo, 1937, *Syngrassides* 1650 !

NOTES: In the absence of specimens the identity of *Inula pulicaria*, recorded by Hume from Larnaca and Limassol, must remain uncertain, though it was most probably *Pulicaria arabica*, which occurs in both areas. True *Inula pulicaria* L. (now, correctly, *Pulicaria vulgaris* Gaertn.) has not been collected in Cyprus.

3. P. sicula (*L.*) *Moris*, Fl. Sard., 2: 363 (1843); Boiss., Fl. Orient., 3: 205 (1875); H. Stuart Thompson in Journ. Bot., 44: 332 (1906); Holmboe, Veg. Cypr., 179 (1914); Post, Fl. Pal., ed. 2, 2: 27 (1933); Osorio-Tafall et Seraphim, List Vasc. Plants Cyprus, 104 (1973); Davis, Fl. Turkey, 5: 76 (1975); Feinbrun, Fl. Palaest., 3: 319, t. 533 (1978).

Erigeron siculum L., Sp. Plant., ed. 1, 864 (1753).

Jasonia sicula (L.) DC. ex Decaisne in Ann. Sci. Nat., sér. 2, 2: 261 (1834); Unger et Kotschy, Die Insel Cypern, 236 (1865).

TYPE: Sicily; "*in* Siciliae *paludosis*".

Erect annual 20–50 cm. high; stems terete, faintly sulcate longitudinally, often purplish, usually branched towards apex, unbranched below, but sometimes branched from base; leaves linear, entire, with recurved or revolute margins, sparingly scabridulous-pubescent, often small, 8–30 (–50)

mm. long, 1–5 mm. wide, apex obtuse or subacute, base very shortly and
obscurely amplexicaul-auriculate; inflorescence a lax panicle, comprising
many, slender, erecto-patent branches; capitula terminal on the branches
and branchlets, the latter minutely leafy, with bract-like leaves almost to
apex; peduncles very short, slightly incrassate, scaberulous-pubescent;
involucre hemispherical, 8–12 mm. diam.; phyllaries linear-lanceolate,
acuminate, subadpressed in 2–3 series, pubescent, often purplish, the lower
about 2 mm. long, 1 mm. wide, the upper (inner) to 6 mm. long, 1 mm. wide;
florets yellow, not much exceeding the involucre; ray-florets few (about 12),
ligule small, barely 4 mm. long, 0·8 mm. wide, shortly 3-toothed at apex;
style shortly exserted from tube, branches spreading, linear, obtuse, about
0·5 mm. long; disk-florets narrowly tubular, about 4 mm. long, 0·8 mm. wide
at apex, lobes 5, narrowly deltoid, about 0·4 mm. long, 0·3 mm. wide at base,
erect or slightly recurved, a little papillose dorsally near apex; anthers
narrowly linear, about 1·5 mm. long, 0·1 mm. wide, apex acute, base
caudate-sagittate; style included or almost included, branches suberect,
linear, obtuse, about 0·4 mm. long. Achene fusiform, about 1·5 mm. long, 0·4
mm. wide, slightly compressed, without longitudinal ribs, mid-brown,
rather densely subadpressed-pilose; outer pappus of numerous, shortly
connate, laciniate scales about 0·2 mm. long, inner pappus of 18–25 smooth
or slightly scabridulous hairs 3–3·5 mm. long.

HAB.: In brackish marshes; near sea-level; fl. July–Oct.

DISTR.: Division 4, very rare. Mediterranean region from Spain to Palestine.

4. "Auf der Ebene bei Larnaca unweit Phaneromene, 28 Mai noch nicht in Blüthe", 1862,
   Kotschy 978.

NOTES: Not seen since, and perhaps an error, though Kotschy must have known the plant,
and may have identified it correctly, even from non-flowering material. It should be looked for
in the numerous moist, brackish places and seasonally marshy hollows to be found in the
neighbourhood of Larnaca.

12. ASTERISCUS *Mill.**

Gard. Dict. Abridg., ed. 4 (1754).
Odontospermum Necker ex Sch. Bip. in Webb et Berth., Hist. Nat. Iles
Canar., 3 (2), 2: 233 (1844).

Annual or perennial herbs; leaves simple, alternate; capitula many-
flowered, heterogamous, ligulate, subsessile, axillary and terminal;
involucre 2–3-seriate, bracts (phyllaries) imbricate, outermost spreading,
foliaceous, appendiciform, with indurated bases and leafy apices; receptacular scales concave or shallowly boat-shaped; flowers yellow; ligulate
florets uniseriate, marginal, 3-toothed; disk-florets regular, tubular, 5-
toothed. Achenes turbinate, ribbed; marginal 3-angled, inner 4-angled or ±
terete.

A genus comprising about 10 species, mainly eastern Mediterranean in
distribution, but also extending to Europe, Canary Is., Cape Verde Is. and
N. Africa S. to Central Sahara.

1. **A. aquaticus** (*L.*) *Less.*, Syn. Comp., 210 (1832); Schkuhr, Handbk., t. 257 (1791–1808);
   Lam., Tableau Encycl. et Méth. (Bot.), 4, 1: t. 682 (1796); DC., Prodr., 5: 486 (1836); Unger
   et Kotschy, Die Insel Cypern, 236 (1865); Boiss., Fl. Orient., 3: 179 (1875); Post, Fl. Pal.,
   ed. 2, 2: 20 (1933); Osorio-Tafall et Seraphim, List Vasc. Plants Cyprus, 105 (1973); Davis,
   Fl. Turkey, 5: 50 (1975); Feinbrun, Fl. Palaest., 3: 321, t. 537 (1978).
       Buphthalmum aquaticum L., Sp. Plant., ed. 1, 903 (1753); Willd., Sp. Plant., 3: 2232
   (1804); Sibth. et Sm., Fl. Graec. Prodr., 2: 196 (1813), Fl. Graec., 9: 76, t. 899 (1837); Clarke,
   Travels, 8: 441 (1818).

* by P. Halliday

Odontospermum aquaticum Sch. Bip. in Webb et Berth., Hist. Nat. Iles Canar., 3 (2), 2: 233 (1844); Holmboe, Veg. Cypr., 180 (1914); Lindberg f., Iter Cypr., 35 (1946).
TYPE: "*in* Creta, Lusitania, Massiliae".

Erect, aromatic annual 4·5–28 (–35) cm. high, usually branched either from base or in the upper half, leafy, glandular and pilose throughout, said to be slightly viscid; leaves bright green, 3–6·5 (–8·5) cm. long, 0·9–1·8 (–2·2) cm. wide, decreasing in size upward, basal sometimes withered at anthesis, lower cauline obovate, attenuate, sub-amplexicaul, obtuse, upper oblong to linear-lanceolate, sub-amplexicaul, obtuse to acute; capitula 1·3–1·4 cm. diam., rarely up to 1·8 cm., (excluding outermost bracts), radiate, solitary, shortly pedunculate, in angles of branches or terminating branches; involucre ± hemispherical, outermost phyllaries exceeding capitulum, oblong, occasionally linear-lanceolate, 1·8–2·2 (–3) cm. long, with a somewhat indurated base; inner phyllaries erect, ± oblong, up to 20 mm. long, ± 5 mm. wide, acute; receptacular scales ± 3 mm. long, obovate, shallowly boat-shaped, sub-membranous, acute, hairy and glandular on outer surface, ciliate in upper half, each enclosing a floret; ligules rather pale golden-yellow, female, spreading, oblong, ± 6 mm. long, 2 mm. wide, with 3 apical teeth and 6 veins, outer surface of floret glandular and long-hairy; disk-florets hermaphrodite, regular, tubular, 3 mm. long, 1 mm. wide, 5-dentate, glandular-hairy on upper outer surface; anthers 1–1·5 mm. long, with a well-defined lanceolate appendage and shortly sagittate, subcaudate base; filaments very short; style bifid, branches compressed, spathulate or acute. Achenes turbinate, 2 mm. long, 1 mm. wide, sericeous, marginals 3-angled, inner 4 (–5)-angled; pappus-scales lanceolate, 1–1·5 mm. long, free, scarious, lacerate, obtuse to rounded, subulate.

HAB.: Cultivated and waste land; in garigue on dry hillsides; by roadsides; on sandy seashores; sea-level to 650 ft. alt.; fl. April–June.

DISTR.: Divisions 1, 3–8. Widespread in the Mediterranean region and east to Azerbaijan; Atlantic Islands.

1. Between Neokhorio and Polis, 1941, *Davis* 3341! Coral Bay, 1978, *J. Holub* s.n.! Kato Paphos, 1978, *J. Holub* s.n.!
3. Kannaviou, 1939, *Lindberg f.* s.n.; ½ mile S. of Pyrgos, 1963, *J. B. Suart* 36! Amathus, 1978, *J. Holub* s.n.!
4. Larnaca, 1862, *Kotschy* 327; also, 1939, *Lindberg f.* s.n.; near Paralimni, 1958, *N. Macdonald* 49! Kouklia reservoir, 1905, *Holmboe* 379! and, 1967, *Merton* in ARI 449!
5. Between Kythrea and Nicosia, 1880, *Sintenis & Rigo* 554! North of Pyroï, 1934, *Syngrassides* 468! Athalassa, 1939, *Lindberg f.* s.n.
6. Nicosia, 1901, *A. G. & M. E. Lascelles* s.n.! Mile 3, Nicosia-Myrtou road, 1950, *Chapman* 573!
7. Kyrenia, 1949, *Casey* 781! also, 1956, *G. E. Atherton* 1353! Lakovounara Forest, 1950, *Chapman* 471!
8. Shore S. of Galinoporni, 1962, *Meikle* 2503!

13. PALLENIS *Cass.* *

Dict. Sci. Nat., 37: 275 (1825) nom. cons.
Benth. et Hook. f., Gen. Plant., 2: 340 (1873).

Erect ± hairy annual herbs with simple, alternate leaves; capitula heterogamous, ligulate, solitary in angles of branches or terminal; involucre ± hemispherical, 2–3-seriate, outermost phyllaries foliaceous, much-exceeding capitulum, spreading, spine-tipped; inner ± erect; receptacle convex with shallowly boat-shaped, keeled scales; florets yellow, all fertile; ray-florets female, 1–3-seriate, apical teeth 3; disk-florets hermaphrodite,

* by P. Halliday

regular, tubular with single wing along inner face, and with 5 deeply cut, ±
erect apical teeth. Achenes of ligulate florets dorsiventrally compressed,
with 2 lateral wings; those of disk-florets laterally compressed, with a single
narrow wing along inner face; pappus-scales minute, free, scarious, lacerate.
A monotypic genus from eastern Mediterranean countries.

P. spinosa (*L.*) *Cass.* in Dict. Sci. Nat., 37: 276 (1825); DC., Prodr., 5: 487 (1836); Sibth. et Sm.,
Fl. Graec., 9: 75, t. 898 (1837); Unger et Kotschy, Die Insel Cypern, 236 (1865); Boiss., Fl.
Orient., 3: 180 (1875); H. Stuart Thompson in Journ. Bot., 44: 332 (1906); Holmboe, Veg.
Cypr., 194 (1914); Mrs Frank Tracey in Journ. R. H. S., 58: 305 (1933); Lindberg f., Iter
Cypr., 35 (1946); Osorio-Tafall et Seraphim, List Vasc. Plants Cyprus, 105 (1973); Davis,
Fl. Turkey, 5: 51 (1975); Feinbrun, Fl. Palaest., 3: 320, t. 536 (1978).
    Buphthalmum spinosum L., Sp. Plant., ed. 1, 903 (1753); Reichb., Icones Fl. Germ., 16,
tt. 938, 939 (1853).
    B. asteroideum Viv., Fl. Lyb. Spec., 57: t. 25, f. 2 (1824).
    Asteriscus spinosus Sch. Bip. in Webb et Berth., Hist. Nat. Iles Canar., 2, 2: 230
(1842–60).
    TYPE: "G. Narbonensi, Hispania, Italia, *ad margines agrorum*".

Stiffly erect annual herb to 48 cm. high (in Cyprus), usually less, variably
long-hairy; stems leafy throughout, usually branched, often only in upper
half; basal and lower cauline leaves oblanceolate, 2·5–11·8 cm. long, 5–20
mm. wide, petiolate, decurrent, with a ± clasping base, spine-tipped,
shallowly serrate; upper cauline leaves usually oblong-lanceolate, occasionally as basal, 2–7 cm. long, 6–20 mm. wide, ± amplexicaul, auricled, usually
sessile, occasionally sub-petiolate, obtuse to acute, spine-tipped, entire to
very shallowly serrate; capitula ± 2·8 cm. diam.; receptacular scales oblong-
oblanceolate, keeled, 6 mm. long, 1 mm. wide, acute to obtuse, cuspidate;
outermost phyllaries up to 4·5 cm. long, with conspicuous nerves, abruptly
contracted into a stiff stalk, spine-tipped and spreading in a star around
capitulum, inner phyllaries stiff, oblong, 6–8 mm. long, 2–2·5 mm. wide,
cuspidate, 1- or 3-veined, ciliate, the upper part of outer surface hairy;
florets yellow, ligules of ray-florets oblong, 8 mm. long, ± 1 mm. wide, with 3
apical teeth and 3 veins, tube expanded, 5 mm. wide, compressed, 2-winged,
± ⅓ of floret-length, ciliate, whole floret pubescent on outer surface; disk-
florets ± terete and thickened at base, 4–5 mm. long, ± 1 mm. wide,
glabrous; anthers ± 2 mm. long, cylindrical, base sagittate-caudate,
appendage lanceolate; style bifid, branches compressed, ± rounded at apex.
Achenes of ray-florets oblong-obovate in outline, 3 mm. long, 2·5 mm. wide,
ciliate, those of disk-florets turbinate, 1·5 mm. long, ± 1 mm. wide; pappus-
scales less than 1 mm. long.

    HAB.: Cultivated and waste land; roadsides; in garigue or open Pine forest on dry hillsides;
edges of salt-marsh; sea-level to 2,000 ft. alt.; fl. March–July.

    DISTR.: Divisions 1, 3–8. Widespread in the Mediterranean region east to Iran; Atlantic
Islands.

1. Paphos, 1862, *Kotschy* 661; Dhrousha, 1941, *Davis* 3248!
3. Near Mazotos, 1862, *Kotschy* 554! Kophinou, 1905, *Holmboe* 586! Yerasa, 1941, *Davis* 3073!
Akrotiri Forest, 1947, *Mavromoustakis* s.n.! Cherkez, 1954, *Mavromoustakis* 22! 45!
4. Larnaca aerodrome, 1950, *Chapman* 416! Cape Greco, 1958, *N. Macdonald* 32! Dhekelia,
1973, *P. Laukkonen* 173!
5. Athalassa, 1939, *Lindberg f.* s.n.
6. Nicosia, 1929, *C. B. Ussher* 6!
7. Buffavento, 1937, *Miss Godman* 15! Lakovounara Forest, 1950, *Chapman* 160! Kyrenia,
1955, *Miss Mapple* 10! and, 1955, *G. E. Atherton* 197! Karakoumi, 1956, *G. E. Atherton*
1253! Kazaphani, 1955, *G. E. Atherton* 112!
8. Ayios Andronikos, 1948, *Mavromoustakis* s.n.! Yialousa, 1970, *A. Genneou* in ARI 1450!

TRIBE 4. **Heliantheae** *Cass.* Herbs or shrubs; all the leaves, or at least the
lower ones, usually opposite, entire, toothed or lobed; phyllaries imbricate,
in several series; receptacle usually scaly; capitula usually heterogamous,

generally radiate; florets mostly yellow; ray-florets female or neuter; disk-florets tubular, hermaphrodite or staminate; anthers with an acute or obtuse base, not caudate-sagittate; style-branches with marginal stigmatic surfaces. Achenes compressed or angular; pappus generally present, of bristles or scales.

14. AMBROSIA L.*

Sp. Plant., ed. 1, 987 (1753).
Gen. Plant., ed. 5, 425 (1754).

Annual or perennial herbs; leaves opposite and alternate or all alternate, lobed or pinnatisect; capitula homogamous, monoecious, the males in leafless terminal racemes, the female in the axils of the uppermost leaves; receptacle scaly; involucre of male flowers gamophyllous, forming a dentate cup holding numerous florets; corolla regular, tubular, 5-toothed at apex; filaments free; anthers coherent; gynoecium rudimentary; involucre of female flowers gamophyllous, subglobose, tuberculate, with an apical beak partly enveloping the style of the single floret; corolla wanting; style-branches deeply 2-cleft, exserted. Achene ovoid or sub-globose, naked, enveloped in the persistent involucre.

About 30 species with a cosmopolitan distribution, but chiefly North American.

1. **A. maritima** *L.*, Sp. Plant., ed. 1, 988 (1753); Boiss., Fl. Orient., 3: 252 (1875); Post, Fl. Pal., ed. 2, 2: 43 (1933); Lindberg f., Iter Cypr., 33 (1946); Osorio-Tafall et Seraphim, List Vasc. Plants Cyprus, 101 (1973); Davis, Fl. Turkey, 5: 47 (1975); Feinbrun, Fl. Palaest., 3: 322, t. 540 (1978).
    TYPE: *"in* Hetruriae, Cappadociae *maritimis arenosis"*.

Aromatic, adpressed-canescent perennial; stems woody at the base, erect, 15–35 (–100) cm. high; leaves ovate in outline, 1·5–6 cm. long, 3–4·5 (–5) cm. wide, lobed or pinnatisect, the pinnae shallowly or rather deeply, obtusely dentate; uppermost cauline leaves simple or lobed; petioles up to 15 mm. long; receptacular scales filiform, 1–1·3 mm. long; male capitula nodding, 3–4 mm. diam., containing 7–20 florets; corolla tubular, about 3 mm. long, 1–1·5 mm. wide at apex, yellow, pubescent externally; anthers oblong, 1–1·5 mm. long, 0·4–0·5 mm. wide, each with a conspicuous ovate-acuminate appendage; style exserted, unbranched, apex rounded, papillose; female florets erect, greenish, involucre about 5 mm. long, 3–4 mm. diam., conspicuously tuberculate. Achene subglobose or broadly ovoid, 4–5 mm. long, 3–4 mm. diam., dark brown, glabrous, with a shortly acute apex.

HAB.: Sandy seashores; at or near sea-level; fl. July–Dec.

DISTR.: Divisions 4, 8, rare. Widespread in the Mediterranean region.

4. Famagusta, 1939, *Lindberg f.* s.n.!
8. Boghaz, 1956, *Poore* 39!

15. XANTHIUM L.*

Sp. Plant., ed. 1, 987 (1753).
Gen. Plant., ed. 5, 424 (1754).

Annual herbs; leaves alternate, entire or lobed; inflorescences terminal and axillary; capitula unisexual, monoecious, solitary or in clusters; male capitula terminal, many-flowered; phyllaries free, uniseriate; receptacle

* by P. Halliday

scaly; corolla greenish, tubular, regular, 5-dentate; filaments mona-delphous; anthers free, exserted, sagittate at base, minutely appendiculate at apex; gynoecium rudimentary; style undivided; female capitula at base of inflorescence, ovoid or ellipsoid, 2-locular, with 1 floret in each loculus; involucre clothed with straight or hooked spines, persistent and indurated in fruit, and crowned with 2 short or long apical beaks; corolla and stamens absent; style persistent, bifid, exserted from a hole on the inner face of the apical beak. Achene obovate or elliptic, compressed, epappose, concealed within the persistent involucre.

Stems armed with branched, yellow spines; leaves lanceolate in outline, about 3 times as long as broad, grey-white canescent beneath, green above; capitula generally solitary
1. **X. spinosum**
Stems unarmed; leaves ovate in outline, about 1½ times as long as broad, green on both surfaces; capitula generally in clusters-   -   -   -   -   -   -   -   2. **X. strumarium**

1. **X. spinosum** *L.*, Sp. Plant., ed. 1, 987 (1753); Boiss., Fl. Orient., 3: 252 (1875); Holmboe, Veg. Cypr., 180 (1914); Post, Fl. Pal., ed. 2, 2: 43 (1933); Lindberg f., Iter Cypr., 36 (1946); Osorio-Tafall et Seraphim, List Vasc. Plants Cyprus, 101 (1973); Davis, Fl. Turkey, 5: 48 (1975); Feinbrun, Fl. Palaest., 3: 324, t. 543 (1978).
   TYPE: "*in* Lusitania".

Much-branched, spiny annual with slightly ribbed stems to at least 42 cm. high in our area (up to 100 cm. elsewhere); spines conspicuous, yellow, solitary or in pairs at the base of the petioles, usually 3-branched, the branches about 15-25 mm. long; leaves numerous, or sometimes small and inconspicuous, lanceolate in outline, 1·8–7 cm. long, 1–2·5 cm. wide, acute, with a tapering base, the upper often entire, the lower generally with 3–5 entire lobes, the mid-lobe usually much longer than the laterals, green above, densely grey-white canescent beneath and on veins above; petiole up to 2 cm. long, canescent; capitula shortly pedunculate, the males about 5 mm. diam.; phyllaries oblanceolate, about 2 mm. long, 1 mm. wide, pilose; receptacular scales oblanceolate, pilose, ciliate, about 2·5 mm. long; florets infundibuliform, externally glandular in the upper half; anthers about 1 mm. long; female capitula often solitary, at first subglobose, about 2 mm. diam., without any basal phyllaries; apical beaks slender, straight, one usually more developed than the other. Fruiting involucre oblong-ellipsoid, 12–13 mm. long, about 4 mm. wide; spines hooked, 4–6 mm. long; achene narrowly oblong-cylindrical, slightly compressed, about 6–7 mm. long, 1 mm. wide, fuscous.

HAB.: Roadsides; cultivated and waste ground; sea-level to 50 ft. alt.; fl. June–Sept.

DISTR.: Divisions 3, 4, 6. Distributed as a weed almost throughout the warmer regions of the world.

3. Between Kividhes and Limassol, 1934, *Syngrassides* 514!
4. Larnaca, 1905, *Holmboe* 1094; also, 1913, *Haradjian* 1015! and 1939, *Lindberg f.* s.n.!
6. Syrianokhori, 1936, *Syngrassides* 1246! Morphou, 1949, *Casey* 938!

2. **X. strumarium** *L.*, Sp. Plant., ed. 1, 987, (1753); Boiss., Fl. Orient., 3: 251 (1875); Druce in Rep. B.E.C., 9: 470 (1931); Post, Fl. Pal., ed. 2, 2: 43 (1933); Osorio-Tafall et Seraphim, List Vasc. Plants Cyprus, 101 (1973); Davis, Fl. Turkey, 5: 48 (1975); Feinbrun, Fl. Palaest., 3: 323, t. 541 (1978).
   X. brasilicum Vell., Fl. Flum. Icones, 10, t. 23 (1825); Lindberg f., Iter Cypr., 36 (1946); Osorio-Tafall et Seraphim, List Vasc. Plants Cyprus, 101 (1973).
   TYPE: "*in* Europa, Canada, Virginia, Jamaica, Zeylona, Japonia".

Erect, scabridulous, glandular annual to 1 m. high; stems and petioles often (? always) purple-mottled; branches few; leaves ovate in outline, 4–13 cm. long, 5–14 cm. wide, obtuse, (3–) 5–7-lobed with rather wide basal sinuses, clothed with short white bristles and sessile, yellow glands, lobes shallowly toothed; petiole 1–10 cm. long; capitula generally clustered, sometimes solitary, shortly pedunculate; male capitula subglobose, 5–7 mm. diam., phyllaries linear-lanceolate, pilose, glandular, 2–3 mm. long, 0·5–1 mm. wide; receptacular scales obovate, about 3 mm. long; florets infundibuliform, 3 mm. long, glandular in the upper half; anthers 1·5 mm. long; female capitulum with a basal involucel of numerous uniseriate, free, lanceolate, pilose-glandular phyllaries about 3 mm. long, 1 mm. wide; apical beaks erect, incurved or divergent; style filiform, deeply bifid, papillose. Fruiting involucre oblong-ellipsoid or ovoid, 13 mm. long, 5–6 mm. wide (excluding spines), glandular-pubescent; spines hooked, 4–6 mm. long, glandular-pubescent at base; achene narrowly elliptic, 9–15 mm. long, 4–5 mm. wide, strongly compressed dorsiventrally and flattened-hemispherical in transverse section, fuscous, longitudinally ribbed, and sometimes with 2 narrow wings; style persisting as a rostrate appendage about 2–5 mm. long; seed with a pale brown, minutely shagreened testa.

HAB.: Waste and cultivated ground; irrigation ditches; seashores; sea-level to 5,000 ft. alt.; fl. July–Oct.

DISTR.: Divisions 2, 3, 6, 7. Distributed as a weed almost throughout the warmer regions of the world.

2. Prodhromos, 1928 or 1930, *Druce*; also, 1949, *Casey* 932! 954! and, 1955, *G. E. Atherton* 607! Platres, 1937, *Kennedy* 598! Kryos Potamos, 1937, *Kennedy* 599! Kakopetria, 1955, *Chiotellis* 435!
3. Akhelia, 1939, *Lindberg f.* s.n.!
6. Makhedonitissa Monastery, 1935, *Syngrassides* 871! Kokkini Trimithia, 1966, *Merton* in ARI 3!
7. Lapithos, 1955, *G. E. Atherton* 683!

Apart from these native or naturalized species, *Compositae* tribe *Heliantheae* is represented in Cyprus by a number of ornamental or useful plants belonging to several, distinct genera:

Zinnia elegans Jacq. (Collect. Suppl., 152; 1796), the most commonly cultivated of the Central American Zinnias, is a robust, erect annual, about 50–100 cm. high, with opposite, sessile, entire leaves and solitary, terminal capitula; the ray-florets are fertile, conspicuous and persistent, in shades of red, pink, yellow or white.

Helianthus tuberosus L. (Sp. Plant., ed. 1, 905; 1753), the Jerusalem Artichoke and *H. annuus* L. (Sp. Plant., ed. 1, 904; 1753), the Sunflower, are primarily of economic importance, though the latter is frequently grown as an ornamental; in both, the upper cauline leaves are alternate, scabrid, simple or sometimes lobed, and the golden-yellow ray-florets are sterile; *H. tuberosus* is a perennial, grown for its edible, knobby-fusiform tubers; it has been collected "Auf den Aeckern bei Prodromo [Prodhromos] fast wild, 1840" (*Kotschy* 32 in Unger et Kotschy, Die Insel Cypern, 237; 1865), but not since recorded by any other collector. *H. tuberosus* is a native of North America, the "Jerusalem" in its popular name being a corruption of the French *Girasol*. *H. annuus*, a tall annual (often 2–3 m. high), with very large capitula, is valued for its oil-bearing achenes. It too is a native of North America.

Dahlia pinnata Cav. (Icones, 1: 57, t. 80; 1791) and *D. coccinea* Cav. (Icones, 3: 33, t. 266; 1796) are the ancestors or parents of the garden Dahlias, often regarded as of hybrid origin (*D.* × *cultorum* Thorsr. et Reis.); they are natives of Central America. The genus (closely allied to *Cosmos*, *Coreopsis* and *Bidens*) has thick fistulose stems and compound fleshy leaves; the inner phyllaries of the involucre are large and membranous, and the richly coloured ray-florets are generally sterile.

Cosmos bipinnatus Cav. (Icones, 1: 10, t. 14; 1791) a tall annual, sometimes 2–3 m. high, has the leaves dissected into narrow, linear segments, and long-pedunculate capitula with broad, white, pink or purplish, sterile ray-florets. It is a native of Mexico, now cultivated (and locally naturalized) in most temperate and tropical regions of the world.

TRIBE 5. **Helenieae** *Benth*. Herbs or shrubs; leaves opposite or alternate, entire, toothed or lobed; phyllaries usually in 1–3 series; receptacle naked; capitula heterogamous, generally radiate; ray-florets female or neuter, generally ligulate, entire or 3-dentate at apex; disk-florets hermaphrodite, or sometimes sterile, tubular, 4–5-lobed at apex; anther with an obtuse or acute base, rarely sagittate; style-branches truncate at apex or appendiculate. Achenes often elongate or turbinate, glabrous or hairy, pappus generally present, of scales or rarely bristles.

16. TAGETES *L.*

Sp. Plant., ed. 1, 887 (1753).
Gen. Plant., ed. 5, 378 (1754).

Erect or diffuse, pungent-aromatic herbs; leaves opposite, generally pinnatisect; inflorescence corymbose or capitula solitary, pedunculate; involucre cylindrical or infundibuliform, consisting of almost wholly connate phyllaries in one series, sometimes with the addition of 1 or more small basal bracts; receptacle usually small; ray-florets yellow or orange, female, ligulate, usually spreading, sometimes very short or wanting; disk-florets tubular, 5-lobed at apex, hermaphrodite; anthers obtuse at base; style-branches slender, sometimes truncate-penicillate, or with an apical appendage. Achenes narrow, linear, compressed or angular, tapering to the base; pappus scaly, heteromorphous, the scales free or connate, acute or truncate, ciliate or awned.

About 60 species, mostly from Central and South America; the so-called African Marigold, *Tagetes erecta* L. (Sp. Plant., ed. 1, 887; 1753) and the smaller French Marigold, *T. patula* L. (Sp. Plant., ed. 1, 887; 1753), are popular annuals for summer flower-beds.

1. **T. minuta** *L.*, Sp. Plant., ed. 1, 887 (1753); A. Hansen in Tutin et al., Fl. Europ., 4: 144 (1976). TYPE: "*in* Chili".

Erect annual up to 1 m. high; stem glabrous, strongly sulcate, usually unbranched below, much branched, with erect, almost fastigiate, branches in the neighbourhood of the inflorescence; leaves opposite or the upper cauline alternate, imparipinnatisect, 2–8 cm. long, 1·5–5 cm. wide, with 5–11 (or more), rather distant, linear-lanceolate, sharply serrulate, acuminate segments, about 5–40 mm. long, 1–4 mm. wide, lamina dark green, dotted over with large, irregular, pellucid oil-glands; petiole usually indistinct, dilated at base; inflorescence densely fastigiate-corymbose; capitula very numerous, shortly pedunculate, crowded; involucre cylindrical, 10 mm. long, 2–3 mm. diam., shining yellow-green, with small glanduligerous streaks, apical lobes semicircular, about 0·5 mm. long, 0·8 mm. wide; ray-florets 3–5, pale yellow, about 4·5 mm. long, with a recurved, flabelliform, bluntly 3-toothed ligule, about 1·5 mm. long, 2 mm. wide, and a slender, pilose basal tube about 3 mm. long; style-branches linear, recurved, blunt, about 0·8 mm. long; disk-florets 3–5, narrowly tubular-infundibuliform, about 4 mm. long, 1 mm. diam. at apex, with a pilose tube and 5, erect, narrowly ovate-acute apical lobes about 0·8 mm. long, 0·3 mm. wide at base; anthers loosely coherent, linear, about 1·2 mm. long, 0·3 mm.

wide, base blunt, apex with a conspicuous, fuscous, narrow-lanceolate, subacute appendage about 0·2 mm. long; style-branches slightly exserted, recurved, about 0·8 mm. long, linear, truncate, penicillate. Achene linear-fusiform, angled, fuscous, about 5–7 mm. long, 0·8 mm. wide at apex, tapering to a narrow base, with a conspicuous whitish basal callus; pappus-scales about 5, subulate, whitish, scabridulous, shortly connate at base, unequal, 1–3 mm. long, 0·2 mm. wide near base.

HAB.: Waste and cultivated ground, roadsides; sea-level to 4,000 ft. alt.; fl. Aug.–Nov.

DISTR.: Divisions 1, 2. A native of S. America, now found as a weed in most of the warmer parts of the world.

1. Not infrequent about Kato Paphos, 1981, *Meikle*! also, 1982, *G. Godfrey* s.n.!
2. Between Platania and Kakopetria, 1970, *A. Genneou* in ARI 1611! Platres, 1981, *Meikle*!

NOTES: Beginning to spread on the island, and will probably be common before long, as in many other countries. The reasons for its introduction are obscure; the strong-smelling oil is said to serve as an insecticide and larvicide, and the plant, dug in as green manure, is reputed to eliminate nematodes, but it is unlikely to have been imported for these reasons.

TRIBE 6. **Anthemideae** *Cass.* Herbs and small shrubs, often aromatic; leaves alternate, commonly dissected; capitula heterogamous, radiate or sometimes homogamous and discoid; involucre campanulate or saucer-shaped; phyllaries in several series, imbricate, herbaceous or scarious; receptacle scaly or naked; ray-florets yellow or white (rarely purplish) female or neuter; disk-florets yellow or purple, tubular, hermaphrodite or occasionally male; anther-bases obtuse; style-branches blunt or truncate. Achenes terete, angular or compressed, not beaked; pappus absent or reduced to a crown of scales, not setose or pilose.

17. ACHILLEA *L.*

Sp. Plant., ed. 1, 896 (1753).
Gen. Plant., ed. 5, 382 (1754).

Perennial herbs and subshrubs; leaves generally pinnatisect, often finely so, rarely entire or subentire; inflorescence generally a terminal, many-flowered corymb; involucre oblong-campanulate, ovoid or hemispherical; phyllaries in several series, imbricate, usually with scarious margins; receptacle convex or flattish, scaly, with persistent, scarious scales; florets white or yellow (rarely pink or purplish); ray-florets female, in 1 series, ligules usually very short, 3-toothed; disk-florets hermaphrodite, 5-dentate at apex, tube more or less compressed, dilated at base and covering the apex of the achene; style-branches truncate, penicillate at apex. Achenes oblong or obovate, dorsiventrally compressed; pappus absent.

About 200 species, chiefly in temperate regions of the Old World.

Stem thinly pilose; leaves 10–15 mm. wide, with spreading linear-subulate segments; ray-florets yellow; perennial herb - - - - - - - - - - **1. A. biebersteinii**
Stems white-tomentose; leaves 1–3 mm. wide, with minute, overlapping, transverse segments; small shrubs or subshrubs with a woody base:
   Ligules of ray-florets yellow, about 1 mm. diam., scarcely exceeding involucre
                           2. A. santolina
   Ligules of ray-florets white, about 4–5 mm. diam., much exceeding the involucre and sharply reflexed - - - - - - - - - - - **3. A. cretica**

1. **A. biebersteinii** *Afan.* in Notul. Syst., 19: 361 (1959); Davis, Fl. Turkey, 5: 250 (1975); Feinbrun, Fl. Palaest., 3: 341, t. 574 (1978).
   A. micrantha Willd., Sp. Plant., 3 (3): 2209 (1803) non Willd. (1789) nom. illeg.
   [? *A. aegyptiaca* (non L.) Sm. in Sibth. et Sm., Fl. Graec. Prodr., 2: 193 (1813), Fl. Graec., 9: 71 (1842) pro parte quoad plant. cypr.; Holmboe, Veg. Cypr., 182 (1914); Osorio-Tafall et Seraphim, List Vasc. Plants Cyprus, 101 (1973).]
   [? *A. tournefortii* DC., Prodr., 6: 28 (1837) pro parte quoad plant. cypr.; Poech, Enum.

Plant. Ins. Cypr., 18 (1842); Unger et Kotschy, Die Insel Cypern, 238 (1865); Boiss., Fl. Orient., 3: 260 (1875) pro parte quoad plant. cypr.]
TYPE: Turkey, "*in* Cappadocia" (B-W, lectotype).

Robust perennial 25–50 (–75) cm. high; stems unbranched or branched at base, subterete or slightly angled, longitudinally sulcate, softly pilose with spreading hairs, and thinly glandular with shining yellow glands; leaves oblong-linear in outline, the lower 6–10 cm. long, 1–1·5 cm. wide, 2–3-pinnatisect, the ultimate divisions very numerous, linear-subulate, softly glandular-hirsute, base distinctly dilated and sheathing; upper cauline leaves becoming progressively shorter, 2–4 cm. long, 0·5–1 cm. wide, sessile and distinctly amplexicaul at base; inflorescence densely corymbose, with 2–5 main branches divided above into numerous, short, pilose branchlets; capitula very numerous; peduncles usually less than 2 mm. long; involucre ovoid-campanulate, about 4 mm. long, 3·5 mm. wide; phyllaries narrowly oblong, 1–4 mm. long, 0·5–1·5 mm. wide, obtuse, glandular-pubescent dorsally with wide scarious margins; ray-florets 5–6, tube about 2 mm. long, 0·4 mm. wide, ligule orange-yellow, patent or reflexed, sub-reniform, about 2 mm. long, 2–2·5 mm. wide, apex bluntly 3-toothed; style-branches shortly exserted, truncate, about 0·2 mm. long; disk-florets about 4 mm. long, 0·5 mm. wide at apex, apical lobes ovate-deltoid, erect, about 0·5 mm. long, 0·3 mm. wide at base; anthers linear, subacute, about 1 mm. long, 0·2 mm. wide, generally exceeding the style; style-branches erect, about 0·5 mm. long, penicillate; receptacle-scales oblong, obtuse, a little concave, about 4 mm. long, 1·2 mm. wide, pilose dorsally. Achenes narrowly oblong-obovate, about 1 mm. long, 0·5 mm. wide, strongly compressed, brown, glabrous, apex rounded, margins narrowly cartilaginous.

HAB.: Roadsides, waste ground; sea-level to 500 ft. alt.; fl. April–June.

DISTR.: Divisions 4, 6, sometimes as a garden escape; perhaps not strictly indigenous anywhere. Widely distributed from S.E. Europe to Central Asia.

4. Famagusta, a garden escape, 1970, *A. Hansen* 336!
6. Morphou, 1972, *W. R. Price* 1059! Nicosia, 1973, *P. Laukkonen* 248!

NOTES: The identity of the plant recorded by Sibthorp as *Achillea aegyptiaca* is doubtful, though — if the Cyprus record is not merely an erroneous localization — it might possibly have been *A. biebersteinii* Afan.

2. A. santolina *L.*, Sp. Plant., ed. 1, 896 (1753); Unger et Kotschy, Die Insel Cypern, 239 (1865); Boiss., Fl. Orient., 3: 266 (1875); Holmboe, Veg. Cypr., 182 (1914); Post, Fl. Pal., ed. 2, 2: 46 (1933); Osorio-Tafall et Seraphim, List Vasc. Plants Cyprus, 102 (1973); Feinbrun, Fl. Palaest., 3: 341, t. 575 (1978).
A. wilhelmsii C. Koch in Linnaea, 24: 328 (1851); Davis, Fl. Turkey, 5: 232 (1975).
[*A. cretica* (non L.) H. Stuart Thompson in Journ. Bot., 44: 332 (1906).]
TYPE: "*in* Oriente".

Erect subshrub 10–30 (–35) cm. high; stems much branched, densely white-tomentose or sometimes glabrescent; leaves numerous, rather crowded, linear-vermicular, sessile, 10–25 mm. long, 1–2 mm. wide, thinly tomentellous or glabrescent, pinnatisect, with numerous, very small, overlapping, blunt, transverse segments; inflorescence corymbose, terminal, rather dense, 2–5 cm. diam., capitula numerous, very shortly pedunculate, with white-tomentose peduncles; involucre campanulate-hemispherical, 4–5 mm. diam.; phyllaries oblong, obtuse or subacute, 1·5–2·5 mm. long, 0·8–1 mm. wide, tomentose dorsally with narrow scarious margins; ray-florets yellow, very small, tube about 1·5 mm. long, ligule suborbicular, about 1 mm. diam., with 2-3 rounded lobes; style exserted; style-branches oblong, truncate, about 0·4 mm. long, divergent; disk-florets about 2 mm. long, 1 mm. wide at apex, with 5 ovate-deltoid, reflexed lobes about 0·4 mm.

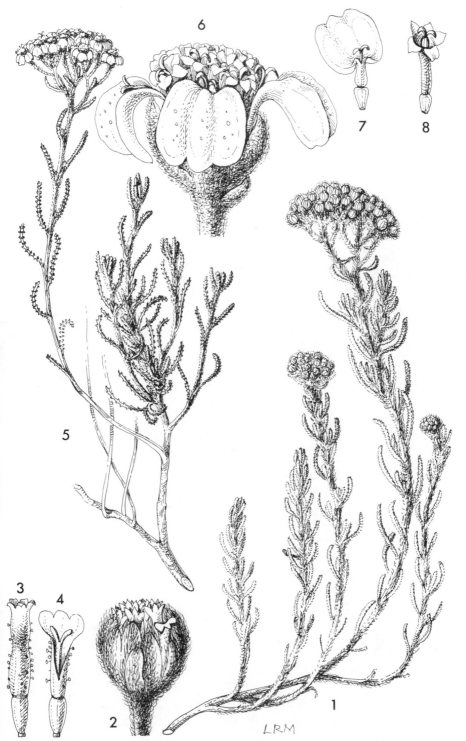

Plate 58. Figs. **1–4.** ACHILLEA SANTOLINA L. **1,** habit, × 1; **2,** capitulum, × 9; **3,** disk-floret, × 15; **4,** ray-floret, × 15; figs. **5–8.** ACHILLEA CRETICA L. **5,** habit, × 1; **6,** capitulum, × 9; **7,** disk-floret, × 6; **8,** ray-floret, × 6. (**1–4** from *A. G. & M. E. Lascelles* s.n.; **5–8** from *Casey 1570*.)

long, 0·3 mm. wide at base; anthers linear, acute, about 1·3 mm. long, 0·2 mm. wide; style shorter than anthers, branches erect, oblong, truncate, penicillate, about 0·5 mm. long; receptacle-scales narrowly ovate, about 2 mm. long, 1 mm. wide, concave, tomentellous dorsally. Achenes oblong-ovate, about 1·5 mm. long, 0·5 mm. wide, brown, glabrous, strongly compressed. *Plate 58, figs. 1–4.*

HAB.: Dry fields; roadsides; sea-level to 150 ft. alt.; fl. March–May.

DISTR.: Divisions 4, 5, rare. Also Turkey, Syria, Palestine and eastwards to Transcaucasia, Afghanistan and Pakistan.

4. Famagusta, 1901, *A. G. & M. E. Lascelles* s.n. !
5. Syngrasis, 1862, *Kotschy* 543 ! Gypsos, 1905, *Holmboe* 563; between Milea and Piyi, 1936, *Syngrassides* 939 !

NOTES: Apparently restricted to the eastern Mesaoria, whereas *A. cretica* L., with which it has been confused, is confined to the west of Cyprus.

3. A. cretica *L.*, Sp. Plant., ed. 1, 899 (1753); Boiss., Fl. Orient., 3: 269 (1875); Holmboe, Veg. Cypr., 182 (1914); Osorio-Tafall et Seraphim, List Vasc. Plants Cyprus, 102 (1973); Davis, Fl. Turkey, 5: 241 (1975).
  [*A. santolina* (non L.) Sm. in Sibth. et Sm., Fl. Graec., 9: t. 891 (1837).]
  TYPE: "*in* Creta".

Much-branched shrub 7–30 cm. high, forming neat rounded bushes, old wood greyish, coarsely fissured; branches longitudinally sulcate, white-tomentose; leaves numerous, rather crowded, sessile, tomentellous, linear-vermicular, 10–25 mm. long, 2–3 mm. wide, pinnatisect, with numerous, usually close and overlapping, very small, blunt, transverse segments; inflorescence a compact or rather open corymb; capitula 5–50, with tomentellous peduncles up to 10 mm. long; involucre very broadly ovoid-campanulate or subglobose, 4–5 mm. diam.; phyllaries broadly ovate or suborbicular, shallowly concave, 2·5–3·5 mm. long, 2·5–3 mm. wide, tomentellous dorsally, with very wide scarious margins; ray-florets white, tube about 1·5 mm. long, ligule sharply reflexed, suborbicular, 4–5 mm. diam., with 3 rounded lobes; style distinctly exserted, branches linear, truncate, recurved, about 0·5 mm. long; disk-florets 3 mm. long, tube about 1 mm. wide at apex, with 5, recurved, ovate-deltoid lobes about 0·5 mm. long, 0·4 mm. wide at base; anthers linear, about 1·2 mm. long, 0·1 mm. wide, subacute; style shortly exserted, branches recurved, linear, truncate, penicillate, about 0·5 mm. long; receptacle-scales oblong, subacute, concave, about 3 mm. long, 1·2 mm. wide, scarious, the narrow midrib tomentellous dorsally. Achene oblong-obovate, about 1·8 mm. long, 1 mm. wide, strongly compressed, with a narrow, pallid wing. *Plate 58, figs. 5–8.*

HAB.: Rocky slopes near the sea, often in tufa near springs; 50–500 ft. alt.; fl. April–July (–Oct.).

DISTR.: Divisions 2, 6. Crete, Aegean Islands, S.W. Turkey.

2. Pyrgos, 1905, *Holmboe* 809; also, 1935, *Syngrassides* 861 ! and, 1953, *Kennedy* 1790 ! 1791 ! also, 1954, *Casey* 1570 !
6. Orga, 1953, *Kennedy* 1792 ! also, 1956, *Merton* 2602 ! 2653 ! and, 1956, *Poore* 32 ! Ayios Yeoryios near Liveras, 1962, *Meikle* 2409 !

NOTES: The Cyprus plant agrees with most of the material from Rhodes and the eastern Aegean area in having relatively few, rather large capitula. It is the *Achillea santolina* of Flora Graeca, t. 891 and *Achillea erioclada* DC., Prodr., 6: 296 (1838). Typical *A. cretica*, from Crete, has very numerous, crowded, small capitula, but there are transitions, and present evidence does not provide a satisfactory basis for segregation, even at infraspecific rank.

[A. ARMENORUM *Boiss. et Hausskn.* in Boiss., Fl. Orient., 3: 269 (1875), listed by Osorio-Tafall et Seraphim, List Vasc. Plants Cyprus, 102 (1973), is most probably a misidentification of *A. cretica* L. True *A. armenorum* is an extremely rare alpine, apparently confined to Beryt Dağ in S. Turkey. It is not likely to be found in Cyprus.]

18. OTANTHUS *Hoffsgg. et Link*

Fl. Portug., 2: 364 (1809).

Diotis Desf., Fl. Atlant., 2: 260 (1798) non Schreber (1791).

Tomentose perennials; stems much branched; leaves sessile, alternate, entire or subentire; inflorescence terminal, corymbose; capitula homogamous, discoid, subglobose; involucre hemispherical; phyllaries 1–2-seriate; receptacle convex, scaly; florets yellow, tubular, strongly compressed dorsiventrally, apex shortly 5-lobed; anthers obtuse at base; style-branches truncate, penicillate. Achenes 3–4-ribbed, partly enveloped by the persistent, accrescent corolla-base; pappus absent.

One species in south-west Europe and the Mediterranean region.

1. **O. maritimus** (*L.*) *Hoffsgg. et Link*, Fl. Portug., 2: 364 (1809); Osorio-Tafall et Seraphim, List Vasc. Plants Cyprus, 102 (1973); Davis, Fl. Turkey, 5: 253 (1975); Feinbrun, Fl. Palaest., 3: 343, t. 580 (1978).
   Filago maritima L., Sp. Plant., ed. 1, 927 (1753).
   Athanasia maritima (L.) L., Sp. Plant., ed. 2, 2: 1182 (1763).
   Diotis maritima (L.) Desf., Tabl. École Bot., 99 (1804); Boiss., Fl. Orient., 3: 253 (1875); Holmboe, Veg. Cypr., 182 (1914); Druce in Rep. B.E.C., 9: 470 (1931); Post, Fl. Pal., ed. 2, 2: 44 (1933).
   TYPE: "*in* Europae *australis, marisque mediterranei littoribus*".

Erect or sprawling, subshrubby perennial 15–30 cm. high; stems usually much branched, densely white-tomentose; leaves oblong, subacute, sessile, tomentose, 5–20 mm. long, 4–5 mm. wide; capitula subglobose, about 6–8 mm. diam., forming small (3–10-flowered) terminal corymbs; peduncles tomentose, up to 15 mm. long; phyllaries ovate, blunt, about 4 mm. long, 2 mm. wide, tomentose dorsally; corolla-tube about 3 mm. long, 0·8 mm. wide at apex, glabrous, sparsely glandular, strongly compressed and narrowly winged towards base, lobes ovate-deltoid, about 0·5 mm. long, 0·4 mm. wide at base; anthers shortly exserted, linear, about 1·5 mm. long, 0·2 mm. wide, with a conspicuous apical appendage; style slightly shorter than anthers, branches linear, truncate, penicillate, recurved, about 0·5 mm. long; receptacle-scales carinate, oblong, subacute, about 4 mm. long, 2 mm. wide, tomentose dorsally towards apex. Achenes obovate, strongly compressed, yellow-glandular, pale brown with a narrow cartilaginous margin, about 1·8 mm. long, 1 mm. wide, apex partly buried in the accrescent, compressed-alate, spongy, brownish base of the corolla.

HAB.: Sand dunes and sandy seashores; near sea-level; fl. June–Aug.

DISTR.: Divisions 3, 4, 6. Distribution that of the genus.

3. Limassol, ? 1905, *Michaelides* sec. *Holmboe*; Cape Gata, 1941, *Davis* 3570! S. of Mari, 1956, *Poore* 26!
4. Between Famagusta and Salamis, 1928 or 1930, *Druce*; mouth of Yialias River, 1968, *Economides* in ARI 1192!
6. Ayia Irini, 1951, *Merton* 877!

19. ANTHEMIS *L.*

Sp. Plant., ed. 1, 893 (1753).
Gen. Plant., ed. 5, 381 (1754).

Annual or perennial herbs or subshrubs; leaves alternate, often deeply 1–3-pinnatisect, rarely subentire or dentate; stems generally branched; capitula terminal, usually pedunculate, solitary, mostly heterogamous and radiate, occasionally homogamous, discoid; involucre hemispherical, or flattish, phyllaries closely imbricate in several series, commonly with pallid or dark, scarious or membranous margins; receptacle conical or hemispherical, wholly or partly clothed with scales; ray-florets in 1 row, fertile or neuter, white, yellow or occasionally pinkish or purplish; disk-florets hermaphrodite, tubular, 5-lobed, yellow or rarely purplish, tube often compressed, sometimes dilated towards base; anthers obtuse at base; style-branches usually truncate, penicillate. Achenes falling readily at maturity, or remaining attached to the receptacle and falling with the capitulum, obpyriform, subcylindric or angled, sometimes dorsiventrally compressed, smooth, costate or tuberculate, sometimes with a short apical crown or a unilateral tooth or auricle, but without a distinct pappus.

About 150 species in Europe, the Mediterranean region and western Asia.

Plants perennial with a tough, woody rootstock; disk purplish or pinkish; rays (when present) with a purplish base:
  Cauline leaves simply pinnatisect, with entire or occasionally dentate, oblong lobes; achenes with a very short, inconspicuous corona - - - - - - **3. A. tricolor**
  Cauline leaves bipinnatisect; achene with a conspicuous, well-developed corona
                                                    4. A. plutonia
Plants annual; disk yellow; rays (when present) white:
  Receptacular scales linear-filiform or linear-subulate, 0·2–0·3 mm. wide:
    Achenes conspicuously tuberculate:
      Achenes caducous at maturity, soon falling from the receptacle - **7. A. cotula**
      Achenes persistent at maturity, forming a dense subglobose head - **8. A. parvifolia**
    Achenes longitudinally ribbed, not or scarcely tuberculate:
      Achenes caducous at maturity; peduncles not noticeably thickened after anthesis; plant erect or ascending - - - - - - A. BORNMUELLERI (p. 916)
      Achenes persistent at maturity, forming a dense head; peduncle distinctly thickened after anthesis; plant often prostrate or decumbent with radiating branches
                                                9. A. pseudocotula
  Receptacular scales oblong-obovate or obcuneate, 1–4 mm. wide:
    Phyllaries with a conspicuous, fuscous margin; tube of disk-florets goblet-shaped, abruptly dilated above the middle - - - - - - - - - **6. A. chia**
    Phyllaries with a pale, hyaline or stramineous margin; tube of disk-florets not abruptly dilated above the middle:
      Capitula radiate; receptacular scales cartilaginous, obcuneate; achenes strongly compressed dorsiventrally:
        Receptacular scales truncate or obscurely muticous at apex, usually pallid in fruit
                                           1. A. amblyolepis
        Receptacular scales conspicuously cuspidate or mucronate, often dark bronze-purple in fruit - - - - - - - - - - **2. A. palaestina**
      Capitula discoid; receptacular scales membranous, oblong-obovate; achenes clavate-obconical not compressed dorsiventrally - - - - **5. A. rigida**

SECT. 1. **Cota** *(Gay) Rupr.* Annual or perennial herbs; receptacle usually convex, not conical, scales oblong, obovate or obcuneate, sometimes aristate or cuspidate, persistent; ligules white, yellow or purplish. Achenes strongly compressed dorsiventrally, rhomboid in cross-section, epappose and without an apical auricle.

1. **A. amblyolepis** *Eig* in Pal. Journ. Bot., J. series, 1: 208 (1938); Osorio-Tafall et Seraphim, List Vasc. Plants Cyprus, 102 (1973).
    A. palaestina (Reuter ex Kotschy) Boiss. ssp. *amblyolepis* (Eig) Feinbrun in Israel Journ. Bot., 25: 81 (1976), Fl. Palaest., 3: 336, t. 567 (1978).
    [? *Cota altissima* (non (L.) Gay) Kotschy in Unger et Kotschy, Die Insel Cypern, 240 (1865).]
    [*A. palaestina* (non (Reuter ex Kotschy) Boiss.) Boiss., Fl. Orient., 3: 283 (1875) pro parte; Sintenis in Oesterr. Bot. Zeitschr., 31: 193 (1881) pro parte; Post, Fl. Pal., ed. 2, 2: 50 (1933); Lindberg f., Iter Cypr., 34 (1946); Davis, Fl. Turkey, 5: 219 (1975).]
    [*A. altissima* (non L.) C. et W. Barbey, Herborisations au Levant, 99 (1882) pro parte; Sintenis in Oesterr. Bot. Zeitschr., 32: 121 (1882) pro parte.]
    [*A. cota* L. ssp. *palaestina* (non *A. palaestina* (Reuter ex Kotschy) Boiss.) Holmboe, Veg. Cypr., 181 (1914) pro parte quoad *Sintenis et Rigo* 314.]

TYPE: Not designated; probably best lectotypified by *Gaillardot* 2416 from Lebanon, which has already been selected by Grierson (Notes Roy. Bot. Gard. Edinb., 33: 216; 1974) as the lectotype of his *A. palaestina*.

Erect or ascending annual (2–) 10–35 (–50) cm. high; stems commonly branched from base, usually not much branched above, or occasionally unbranched, glabrous or thinly hispidulous, obscurely angled or sulcate; leaves numerous, oblong in outline, 15–30 mm. long, 5–20 mm. wide, bipinnatisect, the ultimate pinnules small, oblong, 1–2 mm. long, 0·5–1·5 mm. wide, flattish, subglabrous, pungent-acuminate, sometimes with 1–2 pungent-acuminate lateral lobes; petiole flattened, often indistinct or wanting; peduncles usually less than 6 cm. long, distinctly ribbed, stout and often conspicuously thickened after anthesis; capitulum rather large, commonly 2·5–4 cm. in overall diameter; involucre depressed-hemispherical; phyllaries in 3 series, the outermost ovate-acuminate, about 4 mm. long, 1·5 mm. wide, thinly arachnoid-lanuginose, with an obscure, greenish or purplish midrib and membranous margin, the innermost oblong, blunt, about 7 mm. long, 2·5 mm. wide, with a brownish or hyaline, erose, membranous margin; ray-florets white, ligules spreading, oblong, about 10–13 mm. long, 3–4 mm. wide, apex shortly and bluntly 3-dentate; style about 3 mm. long, scarcely exserted from corolla-tube, branches about 0·3 mm. long, blunt; disk convex, yellow; disk-florets narrowly tubular, compressed, about 4 mm. long, 1·2 mm. wide at apex, sparingly glandular externally, apical lobes 5, erect or recurved, about 0·8 mm. long, 0·6 mm. wide; anthers about 2 mm. long, 0·2 mm. wide, apex acute; style about 2·5 mm. long, included, branches recurved, about 0·6 mm. long, truncate-penicillate; receptacular scales obcuneate, the outer 2·5–4 mm. long, the inner much smaller, all glabrous, subcartilaginous and brownish, with an undulate, truncate or obscurely muticous apex. Achenes strongly compressed, narrowly obcuneate, about 3 mm. long, 1·5 mm. wide at apex, brownish, obscurely nerved, minutely glandular all over, with a narrow apical rim and sub-alate margins.

HAB.: Cultivated and fallow fields; occasionally in coastal garigue; sea-level to 3,750 ft. alt.; fl. Febr.–May.

DISTR.: Divisions 2, 4, 5, 7, 8. Also W. and S. Turkey, Syria, Lebanon, Palestine.

2. Platres, 1938, *Kennedy* 1444 ! and, 1941, *Davis* 3198 !
4. Larnaca, 1880, *Sintenis & Rigo* 804 partly ! Between Ormidhia and Dhekelia, 1936, *Syngrassides* 1078 ! also, 1981, *Hewer* 4704 !
5. Kythrea, 1939, *Lindberg f.* s.n.
7. Dhavlos, 1880, *Sintenis & Rigo* 314 ! Kyrenia, 1949, *Casey* 368 ! and, 1956, *G. E. Atherton* 940 ! 970 ! Ayios Epiktitos, 1959, *P. H. Oswald* 72 ! and, 1967, *Merton* in ARI 583 !
8. Rizokarpaso, 1941, *Davis* 2360 !

NOTES: *Anthemis amblyolepis* Eig is evidently much less common in Cyprus than its close ally *A. palaestina* (Reuter ex Kotschy) Boiss. (*A. melanolepis* Boiss.), and seems to occur chiefly in the northern part of the island, about Kyrenia, though it may have been overlooked elsewhere.

Holmboe regarded it as a subspecies of *A. altissima* L. (*A. cota* L. pro parte), and Feinbrun as a subspecies of *A. palaestina*; both views are very reasonable, but, pending investigation of the *A. altissima* complex in its widest sense, I prefer to treat it as a species.

2. A. palaestina (*Reuter ex Kotschy*) *Boiss.*, Fl. Orient., 3: 283 (1875) pro parte quoad nomen; Sintenis in Oesterr. Bot. Zeitschr., 31: 193 (1881) pro parte; Post, Fl. Pal., ed. 2, 2: 50 (1933) pro parte.
Cota palaestina Reuter ex Kotschy in Unger et Kotschy, Die Insel Cypern, 240 (1865).
Anthemis melanolepis Boiss., Fl. Orient., Suppl., 297 (1888); Post, Fl. Pal., ed. 2, 2: addenda p. 819 (1933); Eig in Pal. Journ. Bot., J. series, 1: 206 (1938); Lindberg f., Iter Cypr., 34 (1946); Osorio-Tafall et Seraphim, List Vasc. Plants Cyprus, 102 (1973); Davis, Fl. Turkey, 5: 219 (1975).
A. cota L. ssp. *melanolepis* (Boiss.) Holmboe, Veg. Cypr., 181 (1914).
A. cota L. ssp. *melanolepis* (Boiss.) Holmboe f. *apiculata* Holmboe, Veg. Cypr., 181 (1914).

A. melanolepis Boiss. var. *genuina* Eig in Pal. Journ. Bot., J. series, 1: 206 (1938).
A. melanolepis Boiss. var. *cypria* Eig in Pal. Journ. Bot., J. series, 1: 206 (1938).
A. palaestina (Reuter ex Kotschy) Boiss. ssp. *palaestina* [*"palaestina"*] Feinbrun, Fl. Palaest., 3: 336, t. 556 (1978).
   [*A. altissima* (non L.) C. et W. Barbey, Herborisations au Levant, 99 (1880) pro parte; Sintenis in Oesterr. Bot. Zeitschr., 32: 121 (1882) pro parte.]
   [*A. cota* (non L.) Holmboe, Veg. Cypr., 180 (1914).]
   TYPE: Cyprus; "Bei Larnaca östlich von der Marine auf Meeresgerölle [*Kotschy*] n. 304. Um Prodromo [*Kotschy*] n. 870" (? W; see Notes).

Very similar superficially to *A. amblyolepis* Eig; stems commonly more branched, both at the base and above, often decumbent and tinged purplish; leaves bipinnatisect, the rhachis and ultimate divisions narrower and more conspicuously pungent-serrate or -lobed than in *A. amblyolepis*; peduncles less noticeably thickened after anthesis; receptacular scales with a distinct, sharp median cusp or mucro, the outer scales often conspicuously aristate-cuspidate, and exceeding the florets at or soon after anthesis, all the scales commonly becoming dark bronze-purple in fruit. Achenes rather narrow, about 3 mm. long, 1·2 mm. wide, sometimes with a distinct apical rim or corona, often (but not always) distinctly 8–10-ribbed on either face, with sub-alate margins. *Plate 59, figs. 6–11.*

HAB.: Cultivated and fallow fields; roadsides; coastal or riverine shingle; occasionally on stony hillsides; sea-level to 5,000 ft. alt.; fl. March–June.

DISTR.: Divisions 1–6. Also Rhodes, Syria, Lebanon, Palestine.

1. Dhrousha, 1941, *Davis* 3259! Polis, 1962, *Meikle* 2308!
2. Prodhromos, 1862, *Kotschy* 305 partly! also, 1939, *Lindberg f.* s.n.! Kryos Potamos, 3,500 ft. alt., 1937, *Kennedy* 946! Platres, 1955, *Kennedy* 1850! Near Palekhori, 1957, *Merton* 2998! and, 1979, *Edmondson & McClintock* E 2899! Kapedhes, 1967, *Merton* in ARI 297!
3. Pyrgos, 1939, *Mavromoustakis* 14! Lefkara, 1967, *Merton* in ARI 636!
4. Larnaca, 1880, *Sintenis & Rigo* 804 partly! Ormidhia, 1954, *Merton* 1871! Athna, *Merton* 2672!
5. Kythrea, 1880, *Sintenis & Rigo* 313! Nisou, 1934, *Syngrassides* 1400! Lakatamia, 1939, *Syngrassides* 369! Mora, 1956, *Merton* 2536! Kokkini Trimithia, 1957, *Merton* 2959! Near Pyroï, 1974, *Meikle* 4090!
6. Nicosia, 1902, *A. G. & M. E. Lascelles* s.n.! also, 1905, *Holmboe* 268; Dhali, 1937, *Syngrassides* 1516!

NOTES: Grierson (in Notes Roy. Bot. Gard. Edinb., 33: 216; 1974) has selected *Gaillardot* 2416 from Beirut (here suggested as an appropriate lectotype of *A. amblyolepis* Eig) as the lectotype of *A. palaestina*, apparently overlooking the prior claims of *Kotschy* 304 and 870, the only specimens referred by this author to his *Cota palaestina*, which must be regarded as the basionym of *Anthemis palaestina*, for which Boissier, albeit unwittingly, provided the new combination. *Kotschy* 305, cited by Boisser, consists of duplicates evidently distributed from an amalgamation of *Kotschy* 304 and 870, and expressly labelled as coming from two localities ("In maritimis et montanis insulae prope Larnaca et ad Prodromo divulgata") though under a single number. The Kew sheet consists of plants with aristate-cuspidate receptacular scales, answering exactly to Kotschy's description of *Cota palaestina*, and Eig reports that material from Berlin and Vienna is (or was) referable to the same plant, whereas the sheet of *Kotschy* 305 in herb. Boissier (G) is said by Eig to be *A. amblyolepis*.
   Why Kotschy should add the comment "Cypern eigen" to the description of *Cota palaestina* is inexplicable, particularly as he acknowledges that the name was taken from a Reuter manuscript [attached to a Gaillardot, Lebanese specimen] in the Boissier herbarium.

SECT. 2. **Anthemis** Annual or perennial herbs or subshrubs; receptacle convex or shortly conical, scales oblanceolate or oblong-obovate, acute or shortly acuminate, often persistent; ligules white, yellow or pink, sometimes persistent, disk-florets yellow or purplish. Achenes obconical or turbinate, sometimes angled, not compressed, epappose with an apical corona sometimes elevated on one side or produced into an auricle.

3. A. tricolor *Boiss.*, Fl. Orient., 3: 288 (1875); Holmboe, Veg. Cypr., 181 (1914); Osorio-Tafall et Seraphim, List Vasc. Plants Cyprus, 102 (1973).

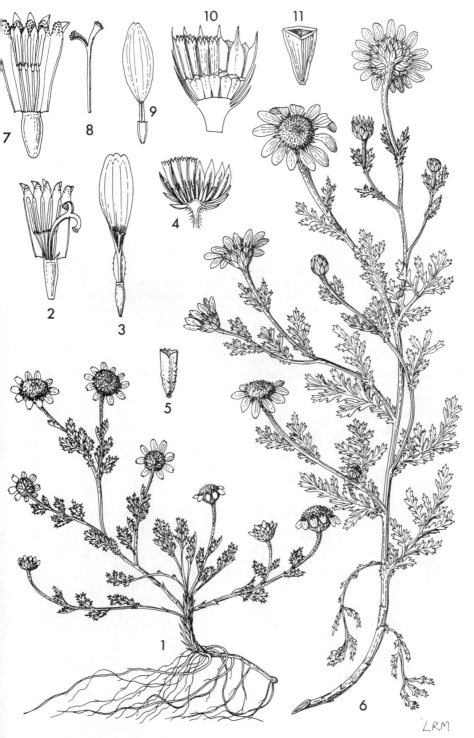

Plate 59. Figs. **1–5.** ANTHEMIS PLUTONIA Meikle **1,** habit, ×⅔; **2,** disk-floret, ×6; **3,** ray-floret, ×6; **4,** capitulum, longitudinal section, ×2; **5,** achene, ×6; figs. **6–11.** ANTHEMIS PALAESTINA (Reuter ex Kotschy) Boiss. **6,** habit, ×⅔; **7,** disk-floret, ×6; **8,** style, ×6; **9,** ray-floret, ×2; **10,** capitulum, longitudinal section, ×2; **11,** achene, ×6. (**1–4** from *Kennedy* 951; **5** from *Kennedy* 953; **6–11** from *A. G. & M. E. Lascelles* s.n.)

A. cypricola Lindberg f., Iter Cypr., 34 (1946); Osorio-Tafall et Seraphim, List Vasc. Plants Cyprus, 102 (1973).

[*A. pontica* (non Willd.) Sm. in Sibth. et Sm., Fl. Graec. Prodr., 2: 190 (1813), Fl. Graec., 9: 66, t. 885 (1839); Poech, Enum. Plant. Ins. Cypr., 18 (1842); Unger et Kotschy, Die Insel Cypern, 237 (1865) pro parte.]

[*A. rosea* (non Sm.) C. et W. Barbey, Herborisations au Levant, 99 (1882).]

TYPE: The illustration in Sibthorp & Smith, *Flora Graeca*, 9: 66, t. 885!

Prostrate or decumbent, canescent perennial; stems 6–45 cm. long, somewhat angular, purplish, thinly adpressed-pilose or glabrescent, radiating from a woody rootstock, unbranched or sparingly branched; basal leaves forming a loose tuft, oblong-ovate in outline, 1–4 cm. long, 0·8–1·5 cm. wide, bipinnatisect, greyish, adpressed-lanuginose, ultimate divisions acute or obtuse, oblong-lanceolate; petiole slender, up to 3·5 cm. long; cauline leaves subsessile or shortly petiolate, rather remote, seldom much more than 1 cm. long, simply pinnatisect, or with a few pinnae bipinnatisect, the ultimate segments rather broad, obtuse or shortly acute; capitula solitary, terminal; peduncles slender, 2–8 cm. long; involucre hemispherical 6–13 mm. diam.; phyllaries closely imbricate in 3 series, oblong, the lowermost about 1·5 mm. long, 1 mm. wide, subdeltoid, subacute, the upper to 4 mm. long, 1·5 mm. wide, oblong, obtuse or rounded at apex, all thinly lanuginose with a raised greenish midrib and wide scarious margins; receptacle conical, scales obovate, concave, about 3 mm. long, 1·5 mm. wide, hyaline-membranous, with a faint midrib and shortly cuspidate, often reddish, apex, margins erose; florets persistent; ligules of ray-florets white or pinkish with a darker purplish base, broadly oblong, about 5–6 mm. long, 3·5–4·5 mm. wide, often recurved, apex with 3 rounded lobes, tube 1–2 mm. long; style about as long as tube, reddish-purple, branches recurved, linear, truncate, about 0·5 mm. long; disk 8–13 mm. diam., strongly convex, florets pink or purple, tubular, compressed or even sub-alate, about 3 mm. long, 1 mm. wide at apex, lobes 5, ovate, recurved, acute, about 0·5 mm. long, 0·3 mm. wide; anthers linear, subacute, about 1·5 mm. long, 0·2 mm. wide; style about 2 mm. long, branches about 0·3 mm. long, truncate, penicillate. Achenes narrowly oblong-obconical, about 1–1·8 mm. long, 0·8 mm. wide at apex, pale brown, obscurely 4-angled, glandular-verruculose, with a narrow, lop-sided, erose, apical corona scarcely 0·3 mm. long.

HAB.: Dry, rocky or stony ground, usually (but not always) on chalk or limestone; sea-level to 2,700 ft. alt.; fl. Febr.–May.

DISTR.: Divisions 1–8. Endemic.

1. Dhrousha, 1941, *Davis* 3238! 3252! Coral Bay, 1978, *J. Holub* s.n.! S. of Paphos, 1980, *J. E. De Langhe* 111/80!
2. Lavramis near Yialia, 1905, *Holmboe* 808; Perapedhi, 1941, *Davis* 3085! and, 1962, *Meikle* 2800! near Palekhori, 1979, *Edmondson & McClintock* E 2907! Pano Panayia, 1979, *Hewer* 4593!
3. Limassol, 1913, *Haradjian* 559! Between Nisou and Stavrovouni, 1933, *Syngrassides* 8! Stavrovouni, 1937, *Syngrassides* 1474! Akrotiri, 1939, *Mavromoustakis* 106! Skarinou, 1941, *Davis* 2693! Yerasa, 1941, *Davis* 3074! Cape Gata, 1954, *Scott-Moncrieff* in *Casey* 1651! Near Limassol, 1957, *Mavromoustakis* 5!
4. Larnaca, 1880, *C. & W. Barbey* s.n.; also, 1912, *Haradjian* 36! Ayia Napa, 1905, *Holmboe* 378; Dhekelia, 1936, *Syngrassides* 1058! and, 1973, *P. Laukkonen* 217! Famagusta, 1936, *Syngrassides* 1074! Cape Greco, 1958, *N. Macdonald* 3! also, 1960, *N. Macdonald* 109! and, 1962, *Meikle* 2598! Dhekelia, 1981, *Hewer* 4705!
5. Kythrea, 1927, *Rev. A. Huddle* 70! Athalassa, 1932, *Syngrassides* 222! also, 1939, *Lindberg f.*, s.n.! and, 1949, *Casey* 552! Mia Milea, 1933, *Syngrassides* 306! Ayios Varnavas Monastery, 1948, *Mavromoustakis* s.n.! S of Koutsovendis, 1959, *C. E. H. Sparrow* 7!
6. Nicosia, 1930, *F. A. Rogers* 0658! 0732! Makhedonitissa Monastery, 1936, *Syngrassides* 1112! Near Nicosia, 1941, *Davis* 2145! Kykko metokhi, 1952, *F. M. Probyn* 37! also, 1973, *P. Laukkonen* 252!
7. Mountains above Kythrea, 1880, *Sintenis & Rigo* 316! Ayios Khrysostomos Monastery, 1889, *Pichler* s.n.! Kyrenia, 1931–1967, *J. A. Tracey* 63! *C. H. Wyatt* 4! 16! *G. E. Atherton* 1306! 1336! *Polunin* 8556! Panagra, 1936, *Mrs. Dray* 4! Between Kythrea and Yaïla, 1940,

Davis 2850! Larnaka tis Lapithou, 1941, *Davis* 3009! Lakovounara Forest, 1950, *Chapman* 50!
8. Valia, 1905, *Holmboe* 498; Karpas peninsula, 1930, *C. B. Ussher* 104!

4. A. plutonia *Meikle* in Ann. Musei Goulandris, 6: 88 (1983). See App. V, p. 1898.
    A. tricolor Boiss. var. *artemisioides* Holmboe, Veg. Cypr., 181 (1914).
    A. tricolor Boiss. f. *artemisioides* (Holmboe) Lindberg f., Iter Cypr., 34 (1946); Rechinger f. in Arkiv för Bot., ser. 2, 1: 433 (1950).
    ? *A. topaliana* Beauverd in Bull. Soc. Bot. Genève, ser. 2, 26: 156 (1936) nom. invalid. sine descr. lat.
    [*A. pontica* (non Willd.) Kotschy, Die Insel Cypern, 237 (1865) pro parte.]
    TYPE: Cyprus; Platres, 3,400 ft. alt., 7 April, 1937, *Kennedy* 950 (K!).

Prostrate or decumbent, sericeous-pilose perennial; stems 6–21 cm. long, obscurely angular, often purplish, generally densely adpressed-pilose or lanuginose, radiating from a woody rootstock, and often forming intricate mats, unbranched or sparingly branched; basal leaves forming a loose tuft, oblong in outline, 0·8–2·5 cm. long, 0·5–1·8 cm. wide, bipinnatisect, sericeous-pilose, ultimate divisions generally narrow, acute or acuminate; petiole slender, 1–2·5 cm. long; cauline leaves subsessile or shortly petiolate, bipinnatisect, closely resembling the basal, but usually smaller, the ultimate segments acute or acuminate; capitula solitary, terminal; peduncles slender, 1–10 cm. long, purplish, adpressed-pilose; involucre hemispherical, 6–13 mm. diam.; phyllaries closely imbricate in 3 series, narrowly oblong, the lowermost about 1·2 mm. long, narrowing upwards to an acute apex from a base about 0·8 mm. wide, the upper about 3·5 mm. long, 1–1·2 mm. wide, acute or subacute, all thinly lanuginose or subglabrous externally, commonly reddish or purplish, with a narrow scarious, erose-ciliate margin; receptacle strongly convex or conical, scales narrowly obovate, concave-carinate, about 3–3·5 mm. long, 1–1·2 mm. wide, hyaline-membranous, with a faint midrib, acuminate or rarely shortly cuspidate apex, and erose margins; florets persistent; ligules of ray-florets white or pinkish, generally less than 5 mm. long and 2·5 mm. wide, often almost obsolete or absent, apex with 3 obscure, rounded lobes; tube 1–2 mm. long; style as long as tube, reddish-purple, branches recurved, linear, truncate, about 0·3–0·5 mm. long; disk 6–12 mm. diam., strongly convex, pink or dark purplish-red, sometimes with a yellow centre or margin; florets tubular, compressed or sub-alate, about 2 mm. long, 0·8 mm. wide at apex, lobes 5, recurved, ovate, acute, about 0·5 mm. long, 0·3 mm. wide at base, papillose and commonly dark-tipped; anthers about 1·2 mm. long, 0·2 mm. wide, subacute; style about 1·8 mm. long; stigmas recurved, truncate-penicillate, about 0·3 mm. long. Achenes oblong-obconical, about 2 mm. long, 0·8 mm. wide at apex, pale brown or stained purple, obscurely 4-angled, glandular-verruculose, with a conspicuous, somewhat lop-sided, erose-undulate apical corona about 0·8 mm. long. *Plate 59, figs. 1–5.*

HAB.: Rocky igneous mountainsides; 800–6,400 ft. alt.; fl. March–July.

DISTR.: Divisions 2, 3, locally abundant. Endemic.

2. Abundant on the Troödos Range, 2,500–6,400 ft. alt.: Prodhromos, Khionistra, Kryos Potamos, Xerokolymbos, Platres, Platania, Pedhoulas, Papoutsa, Palekhori, Makheras, Stavros tis Psokas, etc. *Kotschy* 339! *Sintenis & Rigo* 316b! *Holmboe* 984; *Haradjian* 398! *Feilden* s.n.! *C. B. Ussher* 65! *Syngrassides* 720! *Wyatt* 4! *Kennedy* 949–953! 1447–1449! *Lindberg f.*, s.n.! *Davis* 1799! 1872! 3103! 3511A! *Casey* 740! 823! *Mavromoustakis* 29! *F. M. Probyn* 16! *Meikle* 2763! *Economides* in ARI 955! 1214! *Edmondson & McClintock* E 2907a! etc.
3. Ayia Varvara (Stavrovouni), 1937, *Syngrassides* 1497! 1498!

NOTES: Closely allied to *A. tricolor* Boiss., but consistently distinguished by its hairiness, bipinnatisect cauline leaves and relatively large, conspicuously coronate achenes. Both species have been found together, near Palekhori, above Askas (*Edmondson & McClintock* 2907! 2907a!) and seem as distinct here as elsewhere. The presence or absence of ligulate florets, commented upon by Holmboe and others in distinguishing *A. tricolor* var. *tricolor* from *A. tricolor* var. *artemisioides*, is of minor taxonomic significance. Within *A. plutonia* (which includes *A. tricolor* var. *artemisioides*) some forms, especially from very high altitudes, are destitute of ligules; most plants have very small ligules, and a few, from the lowest limits of distribution, have ligules almost, but not quite, as large as in *A. tricolor*.

Anthemis plutonia appears to be confined to the Plutonic rocks of the Troödos massif and its outliers; a specimen collected in April 1902 by A. G. & M. E. Lascelles, and purporting to come from St. Hilarion (7), is almost certainly mislabelled.

5. **A. rigida** *Boiss. ex Heldr.*, Herb. Graec. Norm. 1856, no. 503 (? 1857); Greuter in Candollea, 23: 263 (1968); Davis, Fl. Turkey, 5: 204 (1975).

  Anacyclus creticus L., Sp. Plant., ed. 1, 892 (1753); Sibth. et Sm., Fl. Graec. Prodr., 2: 188 (1813).

  Santolina rigida Sm. in Sibth. et Sm., Fl. Graec. Prodr., 2: 166 (1813), Fl. Graec., 9: 40, t. 853 (1839) nom. superfl. illeg.

  Lyonnetia pusilla Cass. in Dict. Sci. Nat., 34: 106 (1825); Poech, Enum. Plant. Ins. Cypr., 18 (1842); Unger et Kotschy, Die Insel Cypern, 238 (1865).

  Lyonnetia rigida DC., Prodr., 6: 14 (1838).

  Anthemis cretica (L.) Nyman, Sylloge Fl. Europ., 7 (1855); Boiss., Fl. Orient., 3: 299 (1875); Holmboe, Veg. Cypr., 181 (1914) non L. (1753) nom. illeg.

  A. pusilla Greuter in Boissiera, 13: 142 (1967); Osorio-Tafall et Seraphim, List Vasc. Plants Cyprus, 102 (1973).

  [*Lyonnetia abrotanifolia* (non (Willd.) Less.) Kotschy in Unger et Kotschy, Die Insel Cypern, 238 (1865).]

  TYPE: Cyprus; the illustration in Sibthorp & Smith, Fl. Graec., 9: 40, t. 853!

Prostrate annual; stems usually numerous, radiating from a central rootstock, occasionally simple in starved specimens, 2–13 cm. long, sulcate, thinly adpressed-pilose or glabrescent, often purplish, generally shortly branched towards apex, unbranched below; leaves bright or glaucous-green, oblong in outline, 0·5–3 cm. long, 0·3–2 cm. wide, bipinnatisect, subglabrous or very thinly lanuginose, ultimate segments lanceolate or narrowly oblong, acute or obtuse; petiole slender, flattened, 0·5–2·5 (–4) cm. long; capitula terminal on the branches, solitary, generally discoid, subglobose, 7–10 (–13) mm. diam.; peduncle usually short, 1–3 (–4) cm. long, stout and thickening conspicuously after anthesis, rather densely subadpressed-hirsute, often arcuate; involucre broadly campanulate, 3·5–10 mm. diam.; phyllaries adpressed-imbricate, usually in 2 series, oblong, the outer 2–2·5 mm. long, 1·2 mm. wide, the inner to 3·5 mm. long, all membranous or submembranous, with a conspicuous or ill-defined herbaceous midrib, thinly hirsute dorsally, with an obtuse or rounded, somewhat erose apex; receptacle much thickened, strongly convex, scales oblong-obovate, persistent, hyaline-membranous, concave, acute or subacute, about 3–3·5 mm. long, 1·2–1·5 mm. wide; disk-florets golden-yellow, tubular, about 3 mm. long, 1–1·5 mm. wide at apex, thinly glandular externally, lobes ovate, acute, somewhat recurved, about 0·6 mm. long, 0·4 mm. wide at base; anthers linear, subacute, about 1·2 mm. long, 0·2 mm. wide; style 2 mm. long, branches recurved, truncate-penicillate, about 0·8 mm. long. Achenes clavate-obconical, about 1·4 mm. long, 0·5 mm. wide at apex, pallid, faintly ribbed longitudinally, minutely glandular, with a very narrow, lop-sided, erose, apical corona.

HAB.: Sandy or stony ground near the sea; sea-level to 500 ft. alt.; fl. Febr.–May.

DISTR.: Divisions 1, 3, 4, 6–8. Also Greece, Crete, Aegean Islands, W. & S. Turkey.

1. Peyia Forest, Kalifes, 1962, *Meikle* 2044! Neokhorio, 1962, *Meikle* 2209! Paphos, 1978, *Holub* s.n.! and, 1980, *J. E. De Langhe* 76/80!
3. Petra tou Romiou, 1950, *F. J. F. Barrington* 63!
4. Larnaca, 1862, *Kotschy* 250! also, 1889, *Pichler* s.n.! 1905, *Holmboe* 101! 1912, *Haradjian* 33! 1950, *Chapman* 24! 456! Famagusta, 1862, *Kotschy*; Cape Pyla, 1880, *Sintenis & Rigo*; Ayia Napa, 1905, *Holmboe* 27; Cape Kiti, 1955, *Merton* 2018! Cape Greco, 1958, *N. Macdonald* 7! Dhekelia, 1981, *Hewer* 4701!

6. Between Pendayia and Xeros, 1934, *Syngrassides* 489! Xeros, 1970, *A. Genneou* in ARI 1396!
7. Dhavlos, 1880, *Sintenis & Rigo* 311! also, 1957, *H. Painter* 30! Kyrenia, 1932–1973, *Syngrassides* 9! *C. H. Wyatt* 50! *Casey* 471! *G. E. Atherton* 1302! *P. H. Oswald* 156! *P. Laukkonen* 272! Ayios Epiktitos, 1951, *Casey* 1119! Karakoumi, 1956, *G. E. Atherton* 1225!
8. Cape Andreas, 1880, *Sintenis & Rigo*; Apostolos Andreas Monastery, 1941, *Davis* 2283!

NOTES: The Cyprus populations of *A. rigida* are very uniform, consisting of typical discoid plants, and varying only to a limited degree in the width and sharpness of the ultimate divisions of the leaves. The species occurs in extraordinary abundance in many coastal areas.

SECT. 3. **Chia** *Yavin* Annual herbs; receptacle conical, scales oblong, deciduous; ligules white; disk yellow. Achenes cylindrical-obconical, the outer auriculate, the inner with a short rim or corona.

6. A. chia *L.*, Sp. Plant., ed. 1, 894 (1753); Boiss., Fl. Orient., 3: 311 (1875); Post, Fl. Pal., ed. 2, 2: 34 (1933); Davis, Fl. Turkey, 5: 207 (1975); Feinbrun, Fl. Palaest., 3: 335, t. 564 (1978). TYPE: Khios; "*in* Chio".

Erect annual 2–24 (–30) cm. high; stems thinly adpressed-pilose, sulcate, often purplish, branched or unbranched; leaves bluntly oblong in outline, 0·5–3 cm. long, 0·5–2·5 cm. wide, deeply bipinnatisect, glabrous or very thinly pubescent, the ultimate divisions flattish, with narrowly revolute margins, obtuse or acute; petiole slender, up to 4 cm. long on basal leaves, the cauline usually subsessile or shortly petiolate; capitula solitary, terminal, often rather large and showy; peduncle 2–10 (–15) cm. long, sulcate, adpressed-pubescent, often rather densely so towards apex; involucre depressed-hemispherical or flattish, 10–15 mm. diam.; phyllaries narrowly oblong in 3 series, closely imbricate, 2–5 mm. long, 1·2–1·8 mm. wide, greenish, glabrous or subglabrous, with a conspicuous, fuscous, membranous margin, apex obtuse, erose; ligules of ray-florets white, 8–15 mm. long, 3–5 mm. wide, apex bluntly 2–3-lobed; disk convex, yellow, florets goblet-shaped, about 2 mm. long, abruptly dilated about the middle, 1·5 mm. diam. at apex, lobes 5, ovate, spreading or recurved, about 0·5 mm. long, 0·3 mm. wide at base; anthers linear, subacute, about 1 mm. long, 0·2 mm. wide, apical appendage conspicuous; style about 1·8 mm. long, branches suberect, about 0·4 mm. long, truncate-penicillate; receptacle conical, scales oblong, 2–2·5 mm. long, 1–1·5 mm. wide, concave-carinate, membranous, fuscous towards apex, cuspidate-erose, deciduous. Achenes subcylindric, about 2–2·5 mm. long, 1–1·2 mm. wide at apex, often curved, pallid, strongly and bluntly 10-ribbed, with a short apical corona or auricle.

HAB.: Bare rocky soil over igneous rocks; 2,000 ft. alt.; fl. March–April.

DISTR.: Division 2, very rare. S. Italy eastwards to Palestine.

2. Galata, 1953, *Merton* 1662!

NOTES: The only record for this species, which is usually easily recognizable by its conspicuous dark-margined phyllaries and ornamental *Leucanthemum*-like capitula.

SECT. 4. **Maruta** *(Cass.) Reichb.* Annual herbs; receptacle convex or conical, scales subulate or setiform, commonly deciduous; ligules usually white, disk yellow. Achenes obconical, clavate or turbinate, round in cross-section.

7. A. cotula *L.*, Sp. Plant., ed. 1, 894 (1753); Boiss., Fl. Orient., 3: 316 (1875); Holmboe, Veg. Cypr., 182 (1914); Post, Fl. Pal., ed. 2, 2: 36 (1933); Lindberg f., Iter Cypr., 33 (1946); Osorio-Tafall et Seraphim, List Vasc. Plants Cyprus, 102 (1973); Davis, Fl. Turkey, 5: 208 (1975); Feinbrun, Fl. Palaest., 3: 337, t. 569 (1978).
Maruta cotula (L.) DC., Prodr., 6: 13 (1838); Unger et Kotschy, Die Insel Cypern, 238 (1865).
TYPE: "*in* Europae *ruderatis, praecipue* in Ucrania".

Erect annual 15–50 (–80) cm. high; stems shallowly sulcate, glabrous or subglabrous towards base, thinly adpressed-pubescent above, often purplish, commonly branched only in the upper half, or sometimes from the base; leaves oblong, 2–7 cm. long, 1–3 cm. wide, bipinnatisect, plumulose, the ultimate divisions narrowly linear or filiform, thinly setulose-pubescent, with revolute margins and a sharply apiculate tip; petioles of basal leaves up to 2 cm. long, flattened, sheathing at base, upper leaves sessile or subsessile; inflorescence loose, corymbiform; capitula terminal on the branches, rather small; peduncles slender, 2–7 cm. long, thinly pilose towards apex; involucre depressed-hemispherical or flattened, 6–8 mm. diam.; phyllaries closely imbricate in 3 series, narrowly oblong, 2–4 mm. long, about 1 mm. wide, thinly lanuginose dorsally, with an ill-defined midrib, hyaline-membranous margins, and an obtuse or rounded, erose apex; receptacle elongate-conical, to about 4 mm. long, naked near the base, scaly in the upper part, scales linear-filiform, slightly carinate, pallid, 2–2·5 mm. long, about 0·3 mm. wide; ray-florets sterile, ligules white, 4–8 mm. long, 3–5 mm. wide, entire or shallowly 3-lobed at apex; disk very prominently convex or subglobose, yellow, florets set somewhat obliquely on top of immature achene, tube 1·8 mm. long, 1 mm. wide at apex, somewhat dilated just above middle, 5-lobed, lobes ovate, acute, about 0·4 mm. long, 0·2 mm. wide at base, recurved; anthers linear, subacute, about 0·8 mm. long, 0·2 mm. wide; style 1·8 mm. long, branches recurved, about 0·5 mm. long, truncate-penicillate. Achenes caducous, subcylindric-obconical, about 1 mm. long, 0·5 mm. wide at apex, brown, conspicuously tuberculate from a papillose surface, with scattered, shining, sessile glands, apex truncate or obscurely auriculate.

HAB.: Cultivated and fallow fields, roadsides, occasionally in garigue, or on river gravels, or on the drier edges of salt-marshes; sea-level to 5,000 ft. alt.; fl. April–Sept.

DISTR.: Divisions 2–8. Widespread in Europe and the Mediterranean region; Atlantic Islands, and as a weed in many temperate parts of the world.

2. Between Galata and Spilia, 1880, *Sintenis & Rigo* 795! Near Prodhromos, 1880, *Sintenis & Rigo* 796! also, 1955, *G. E. Atherton* 556! 630! and 1961, *D. P. Young* 7316! Platres, 1937, *Kennedy* 947! 948! also, 1938, *Kennedy* 1445! and, 1941, *Davis* 3186! Kambos, 1939, *Lindberg f.* s.n.! Kykko Monastery, 1939, *Lindberg f.* s.n.; Troödos area, 1950, *F. M. Probyn* s.n.! Yialia, 1955, *Merton* 2103!
3. Stavrovouni, 1939, *Lindberg f.* s.n.; Limassol, 1947, *Mavromoustakis* s.n.!
4. Chali near Famagusta, 1962, *Meikle* 2622!
5. Kythrea, 1880, *Sintenis & Rigo* 315! 549b! also, 1927, *Rev. A. Huddle* 69! Vatili, 1905, *Holmboe* 336! Dhikomo, 1936, *Syngrassides* 991! Salamis, 1962, *Meikle* 2589!
6. Peristerona, 1936, *Syngrassides* 1215!
7. Kyrenia, 1955, *Miss Mapple* 31! also, 1955, *G. E. Atherton* 67!
8. Between Ephtakomi and Trikomo, 1865, *Kotschy* 533! Ayios Theodhoros, 1970, *A. Genneou* in ARI 1469!

A. BORNMUELLERI *Stoy. et Acht.* in Notizbl. Bot. Gart. Berlin, 13: 522 (1937); Feinbrun, Fl. Palaest., 3: 336, t. 568 (1978).
    A. galilaea Eig in Pal. Journ. Bot., J. series, 1: 201 (1938).
    A. cotula L. ssp. *paleacea* Eig var. *hierosolymitana* Eig in Pal. Journ. Bot., J. series, 1: 195 (1938).
    A. galilaea Eig var. *hierosolymitana* (Eig) Yavin in Israel Journ. Bot., 19: 142 (1970).
TYPE: Israel; "in arenosis siccis ad Jaffam", 1897, *Bornmüller* 873 (PR, K!).

Indistinguishable in habit and overall appearance from *A. cotula* L., but receptacle dome-shaped or broadly conical, covered nearly all over with linear-subulate scales. Achenes caducous, subcylindric-obconical, about 1–1·5 mm. long, 0·5–0·8 mm. wide, pale brown, smooth or obscurely tuberculate, with conspicuous, blunt, longitudinal ribs; apex rounded, sometimes auriculate.

HAB.: Cultivated and fallow-fields; 2,000 ft. alt.; fl. April–May.

DISTR.: Division 1, rare. Also Samos, Syria, Lebanon, Palestine.

1. Dhrousha, 1941, *Davis* 3274!

NOTES: The specimen is cited by Z. Yavin in Israel Journ. Bot., 19: 142 (1970) under *A. galilaea* Eig var. *hierosolymitana* (Eig.) Yavin, the same author noting that this differs from *A. cotula* not only in the smooth (or almost smooth), ribbed achenes, but also in the shape and clothing of the receptacle and the fertile ray-florets. I am not, however, convinced that the Cyprus specimen, which is unfortunately without mature, fruiting capitula, is satisfactorily distinguished from *A. cotula*. It is included here, partly to draw attention to the record, and partly in the hope that future investigation may provide a clearer analysis of *A. cotula* populations on the island.

8. A. parvifolia *Eig* in Pal. Journ. Bot., J. series, 1: 198 (1938); Z. Yavin in Israel Journ. Bot., 19: 147 (1970); Feinbrun, Fl. Palaest., 3: 339, t. 571 (1978).
  [*A. pseudocotula* (non Boiss.) Lindberg f., Iter Cypr., 34 (1946) pro parte.]
  TYPE: S. Turkey; "Env. of Jemele, between Mersina and Fundukpinar, fields in a Pinetum", 1931, *Eig & Zohary* s.n. (HUJ).

Erect annual 20–35 cm. high; stems sulcate, subglabrous or thinly adpressed-pilose, usually much branched; cauline leaves (basal leaves not seen) sessile, oblong in outline, deeply bipinnatisect, 1–2 cm. long, 0·3–1·3 cm. wide, thinly pilose, ultimate segments lanceolate, 0·5–1·5 mm. long, sharply mucronulate; inflorescence lax, subcorymbiform; capitula rather small, terminal on the branches; peduncles rigid, sulcate, to 5 cm. long, not noticeably thickened in fruit, thinly pubescent; involucre depressed-hemispherical, 6–8 mm. diam.; phyllaries closely imbricate in 3 series, the outermost subdeltoid, about 2 mm. long, 1 mm. wide at base, the innermost oblong, obtuse, 3–3·5 mm. long, all greenish or brownish and thinly pilose externally, with conspicuous erose-membranous margins; receptacle conical, about 4 mm. long, clothed nearly all over with linear-subulate, subcarinate scales about 2–2·5 mm. long, 0·2 mm. wide; ray-florets apparently sterile on specimens examined ("sterile" Eig, "fertile" Feinbrun), ligule white, oblong, 4–4·5 mm. long, 3–3·5 mm. wide, apex obscurely 3-lobed, tube about 1 mm. long, distinctly compressed and narrowly winged; style equalling tube, branches linear, truncate, recurved, about 0·3 mm. long; disk subglobose; florets tubular, about 1·5 mm. long, 1 mm. wide at apex, somewhat dilated above middle, glandular externally, lobes ovate, acute, recurved, about 0·4 mm. long, 0·3 mm. wide at base; anthers linear, subacute, about 0·8 mm. long, 0·2 mm. wide; style 1·5 mm. long, branches strongly recurved, about 0·5 mm. long, truncate-penicillate. Achenes persistent, cylindrical-obconical, about 1·2 mm. long, 1 mm. diam. at the truncate apex, pale brown, with conspicuous longitudinal rows of tubercles arising from the glandular-papillose surface, corona reduced to a minute unilateral auricle.

HAB.: Cultivated and fallow fields; sea-level to 300 ft. alt.; fl. April–June.

DISTR.: Divisions 4, 5, rare. Also in S. Turkey, Rhodes, Israel [Tripoli; S. France, ? introduced.]

4. Famagusta, 1906, *B.V.D. Post* 837.
5. Trikomo, 1939, *Lindberg f.* s.n.!

NOTES: Similar to *A. cotula* L., but distinguished by its persistent achenes, attached to the receptacle even after maturity, as in *A. pseudocotula* Boiss., from which it differs in its slender, not incrassate, peduncles and small foliage with neatly arranged, short pinnules.

9. A. pseudocotula *Boiss.*, Diagn., 1, 6: 86 (1846), Fl. Orient., 3: 317 (1875); C. et W. Barbey, Herborisations au Levant, 139 (1880); Holmboe, Veg. Cypr., 182 (1914); Post, Fl. Pal., ed. 2, 2: 56 (1933); Lindberg f., Iter Cypr., 34 (1946); Rechinger f. in Arkiv för Bot., ser. 2, 1: 433 (1950); Osorio-Tafall et Seraphim, List Vasc. Plants Cyprus, 102 (1973); Davis, Fl. Turkey, 5: 209 (1975); Feinbrun, Fl. Palaest., 3: 338, t. 570 (1978).

Erect or decumbent annual; stems 8–30 cm. long, subglabrous or distinctly adpressed-pubescent, simple or branched above, or with numerous branches radiating from the rootstock; leaves bipinnatisect, thinly or rather densely subadpressed-pilose, 1–4 cm. long, 0·4–1·5 cm. wide, the ultimate segments filiform, linear or narrow-lanceolate, mucronulate; petioles of basal leaves slender, 1–2 cm. long, cauline leaves sessile or subsessile; capitula terminal on long branches, usually rather larger than those of *A. cotula* L.; peduncles 1–8 cm. long, thickening conspicuously after anthesis; involucre depressed-hemispherical or flattish, 8–12 mm. diam.; phyllaries imbricate in 3 series, oblong, obtuse, 2–4 mm. long, about 1·5 mm. wide, thinly or rather densely pilose dorsally, with a greenish midrib and broad hyaline-membranous margins; receptacle elongate-conical clothed in upper part with linear-subulate, sub-carinate, sharply acuminate scales; ray-florets with a white, broadly oblong, obscurely 3-lobed ligule 0·5-1 cm. long, 0·4 cm. wide, tube about 2 mm. long, strongly compressed, narrowly winged; style about as long as tube, branches linear, truncate, about 0·3 mm. long; disk strongly convex or subglobose, florets tubular, about 3 mm. long, 1 mm. wide, dilated from about middle to apex and base, thinly glandular externally, lobes 5, ovate, acute, erect or recurved, about 0·4 mm. long, 0·3 mm. wide at base; anthers linear, subacute, about 1 mm. long, 0·2 mm. wide; style about 2·5 mm. long, branches recurved, about 0·5 mm. long, truncate-penicillate. Achenes persistent, forming a dense head (like a minute pineapple), subcylindric-obconical or indistinctly tetragonal, 1–1·5 mm. long, 0·8–1 mm. wide, brownish, strongly ribbed longitudinally and minutely glandular-papillose, but not or scarcely tuberculate, apex truncate, oblique, with a very narrow, spreading corona, sometimes extended to form a short, unilateral auricle.

ssp. **rotata** (*Boiss.*) *Eig* in Pal. Journ. Bot., J. series, 1: 197 (1938); Z. Yavin in Israel Journ. Bot., 19: 147 (1970); Osorio-Tafall et Seraphim, List Vasc. Plants Cypr., 102 (1973); Feinbrun, Fl. Palaest., 3: 338 (1978).
    A. rotata Boiss., Fl. Orient., 3: 318 (1875); Holmboe, Veg. Cypr., 182 (1914); Post, Fl. Pal., ed. 2, 2: 56 (1933).
    [*A. arvensis* (non L.) Sm. in Sibth. et Sm., Fl. Graec. Prodr., 2: 189 (1813) pro parte quoad plant. cypr.; Poech, Enum. Plant. Ins. Cypr., 17 (1842); Sintenis in Oesterr. Bot. Zeitschr., 32: 123 (1882); H. Stuart Thompson in Journ. Bot., 44: 332 (1906); Holmboe, Veg. Cypr., 181 (1914).]
    [*A. pamphylica* (non Boiss. et Heldr.) Holmboe, Veg. Cypr. 182 (1914).]
    TYPE: "in Arabiâ petreâ (Boiss!) in Aegypto prope Alexandriam (Samar.!), Cypro (Kotschy!)" (G).

Plants prostrate or decumbent, with many stems radiating from the rootstock; leaves usually small, with lanceolate-apiculate divisions, commonly greyish-pubescent; peduncles often densely pilose towards apex; phyllaries pilose dorsally. Achenes obliquely truncate at apex, generally without an auricle.

HAB.: Dry grassy ground; dune-slacks; roadsides; cultivated and fallow fields; coastal garigue; vineyards; forest clearings; sea-level to 4,000 ft. alt.; fl. March–July.

DISTR.: Divisions 2–8. E. Aegean Islands, Turkey, Syria, Lebanon, Palestine, Egypt and eastwards to Iraq and Iran.

2. Kyperounda, 1940, *Davis* 1880! Kato Platres, 1958, *N. Macdonald 64*! Below Alithinou near Polystipos, 1974, *Meikle* 4054!
3. Kolossi, 1862, *Kotschy*; Limassol, 1930, *E. Wall* s.n.! Ayia Varvara (Stavrovouni) 1937, *Syngrassides* 1486!
4. Larnaca, 1862, *Kotschy*; Cape Kiti, 1955, *Merton* 2018!
5. Near Kythrea, 1880, *Sintenis & Rigo* 312! Mile 7, Nicosia-Famagusta road, 1950, *Chapman* 373! Mia Milea, 1967, *Merton* in ARI 198!
6. Ayia Irini, 1941, *Davis* 2541! Orga, 1956, *Poore* 2655! Cape Kormakiti, 1967, *Merton* in ARI 131!

7. Lakovounara Forest, 1950, *Chapman* 161! Mile 6, Kyrenia-Akanthou road, 1950, *Chapman* 613! Ayios Epiktitos, 1951, *Casey* 1118! 7½ miles N.É. of Halevga, 1959, *P. H. Oswald* 113! 8. Valia, 1905, *Holmboe* 489; Apostolos Andreas Forest, 1962, *Meikle* 2481!

NOTES: Very uniform in Cyprus; normally with very numerous (often purplish) radiating, decumbent branches, and small, canescent leaves.

In addition to the above, the following *Anthemis* species have been recorded from Cyprus, but the records remain unsubstantiated, and must be discarded pending confirmation.

A. TINCTORIA *L.*, Sp. Plant., ed. 1, 896 (1753) (sect. *Cota* (Gay) Rupr.), an erect perennial 40–70 cm. high with greyish, bipinnatisect leaves, long peduncles and relatively large capitula with (or rarely without) yellow rays. Recorded by Hume (in Walpole, Mem. Europ. Asiatic Turkey, 254; 1817) from Larnaca and Limassol, but probably a misidentification of *Chrysanthemum coronarium* L. A widespread species in southern Europe and the Mediterranean area, often cultivated as an ornamental elsewhere.

A. ROSEA *Sm.* in Sibth. et Sm., Fl. Graec. Prodr., 2: 191 (1813), Fl. Graec., 9: 67, t. 887 (1839) (sect. *Anthemis*), a slender, erect or decumbent annual 3–20 cm. high, with finely dissected, bi-or tripinnatisect leaves, and small capitula with pink rays. Recorded by Sibthorp (or Smith) "In insulae Cypri collibus siccis, vulgaris", but (as Holmboe points out) the plant illustrated under this name in *Flora Graeca* most probably came from the island of Samos. It has not since been found in Cyprus. Later records for *A. rosea* in Poech, Unger & Kotschy and Boissier are based on the statements in *Fl. Graec. Prodr.* and *Flora Graeca*; references to the species by Kotschy (in Oesterr. Bot. Zeitschr., 12: 278; 1862) and C. & W. Barbey (Herborisations au Levant, 99; 1882) are respectively misidentifications of *A. plutonia* Meikle and *A. tricolor* Boiss.

A. PEREGRINA *L.*, Syst. Nat., ed. 10, 2: 122 (1759); Sibth. et Sm., Fl. Graec., 9: 64, t. 883 (1839) (sect. *Anthemis*), now often regarded as synonymous with *A. tomentosa* L., Sp. Plant., ed. 1, 893 (1753), a grey-lanuginose annual, 2–25 cm. high, with pinnatisect or bipinnatisect leaves, short, white rays and a prominent convex or conical, yellow disk, is recorded by Kotschy (in Unger et Kotschy, Die Insel Cypern, 237; 1865) as "In Cypern nicht selten", but evidently in error, as the plant has not been seen there by anyone else. Later records by H. Stuart Thompson and Holmboe are based solely on Kotschy's statement. *A. tomentosa* has a wide distribution in the eastern Mediterranean, from Italy and Sicily eastwards, and might occur in Cyprus, but, in the absence of a precise localization, it cannot be included in the flora.

A. AUSTRALIS *Sm.* in Sibth. et Sm., Fl. Graec. Prodr., 2: 190 (1813), Fl. Graec., 9: 67, t. 886 (1839) non Willd. (1804); *A. cypria* Boiss., Fl. Orient., 3: 300 (1875); Holmboe, Veg. Cypr., 181 (1914) (sect. *Anthemis*), recorded "In insulae Cypri maritimis", is considered by Lindley to be synonymous with *A. secundiramea* Biv. It is a prostrate or decumbent annual, less than 10 cm. high, with numerous radiating branches, finely bipinnatisect leaves, arcuate peduncles thickening conspicuously after anthesis, and small capitula, with a campanulate involucre, short, white rays and a yellow disk. There are no specimens in the Sibthorp herbarium (OXF) and, apart from the possibility of it being a radiate form of *A. rigida* Boiss. ex Heldr., one may assume an erroneous localization. *A. secundiramea* Biv. is found in the central Mediterranean region, from S. France through Italy to Sicily, Malta and Pantellaria. A specimen in herb. Kew (herb. Gay) labelled "In insulâ Cypro (Labill!) – Webb dedit Januar. 1841 (anonymos in herb. Labill.)" has been confirmed as *A. secundiramea* by Stoyanoff and Achtaroff, and may be part of Sibthorp's collecting. Its existence does not, however, add much weight to the evidence for the occurrence of this species on the island.

20. CHRYSANTHEMUM L.
Sp. Plant., ed. 1, 887 (1753).
Gen. Plant., ed. 5, 379 (1754).

Annual herbs; leaves toothed or pinnatisect; capitula heterogamous, radiate, solitary, terminal; involucre depressed-hemispherical; phyllaries imbricate in 3–4 series, obtuse, with broad membranous or scarious margins; receptacle convex, naked; ray-florets female or neuter, with yellow, white or particoloured ligules; disk-florets hermaphrodite, yellow, tubular, 5-lobed at apex; anthers obtuse at base; style-branches truncate-penicillate. Achenes dimorphic, the marginal triquetrous, sometimes winged, the inner rounded or compressed, sometimes angled, 10-ribbed; pappus absent.

In its modern, restricted sense comprises 3 species distributed in Europe and the Mediterranean region.

Leaves simply pinnatisect, lobed, or sometimes subentire; achenes terete or angular, but not winged - - - - - - - - - - - - - - - - **1. C. segetum**
Leaves bipinnatisect; achenes sharply angular or compressed, conspicuously winged
2. C. coronarium

1. C. segetum L., Sp. Plant., ed. 1, 889 (1753); Unger et Kotschy, Die Insel Cypern, 239 (1865); Boiss., Fl. Orient., 3: 336 (1875); Holmboe, Veg. Cypr., 182 (1914); Post, Fl. Pal., ed. 2, 2: 61 (1933); Osorio-Tafall et Seraphim, List Vasc. Plants Cyprus, 102 (1973); Davis, Fl. Turkey, 5: 253 (1975); Feinbrun, Fl. Palaest., 3: 349, t. 588 (1978).
TYPE: "*in* Scaniae, Germaniae, Belgii, Angliae, Galliae *agris*".

Erect, glabrous, glaucous annual (10–) 15–50 (–70) cm. high; stems usually branched (except in starved specimens), obscurely sulcate; leaves oblong or obcuneate in outline, 1–10 cm. long, 0·5–5 cm. wide, the basal deeply lobed or coarsely pinnatisect, with oblong, irregularly serrate lobes, apex acute or obtuse, base tapering to an indistinct, flattened petiole; cauline leaves sessile or subsessile, becoming progressively smaller upwards, the uppermost often oblong, subentire and amplexicaul, peduncles 2–10 cm. long, distinctly swollen below the capitulum, becoming thickened in fruit, involucre 1–2·5 cm. diam.; phyllaries loosely imbricate in 3 series, oblong, 3–8 mm. long, 2–4 mm. wide, apex rounded, with a wide, brownish, membranous margin; receptacle shallowly convex, accrescent after anthesis; ray-florets female, ligule oblong, 10–15 mm. long, 6–7 mm. wide, yellow, apex rounded or obscurely 3-lobed, tube about 2·5 mm. long; style about 2 mm. long, branches oblong, obtuse or truncate, about 0·4 mm. long; disk-florets yellow, 3–4 mm. long, tubular, dilated above the middle and 1·5 mm. wide at apex, base expanded and partly covering apex of immature achene, lobes 5, recurved, ovate, acute, papillose, about 0·5 mm. long, 0·4 mm. wide at base, with a brownish-yellow median resin-canal continued down the inside of the tube for 0·8–1 mm. as a raised line; anthers linear, about 1·5 mm. long, 0·3 mm. wide, apex subacute; style 2 mm. long, branches truncate-penicillate, 0·6 mm. long. Achenes of ray-florets bluntly triquetrous, oblong, pale brown, prominently ribbed, about 2 mm. long, 1·5 mm. wide, truncate at apex and base; achenes of disk-florets shortly subcylindrical, about 2·4 mm. long, 1·2 mm. wide, strongly 10-ribbed.

HAB.: Cultivated fields; occasionally on open ground in garigue; sea-level to 1,000 ft. alt.; fl. Febr.–May.

DISTR.: Divisions 1–8. Widely distributed as a weed in Europe and the Mediterranean region.

1. Near Khrysokhou, 1862, *Kotschy*; near Ayios Neophytos Monastery, 1981, *Hewer* 4757!
2. Near Anadhiou ["Slewra"], 1862, *Kotschy*.
3. Lefkara, 1941, *Davis* 2723! Mile 22, Nicosia–Limassol road, 1950, *Chapman* 548!
4. Larnaca, 1880, *Sintenis & Rigo* 806! Phaneromene near Larnaca, 1905, *Holmboe* 145.
5. Kythrea, 1932, Syngrassides 260! Mia Milea, 1957, *Merton* 2967!
6. Near Kykko metokhi, Nicosia, 1952, *F. M. Probyn* 13! Near Klirou, 1967, *Merton* in ARI 278!
7. Kyrenia, 1952, *Casey* 1240! also 1956, *G. E. Atherton* 971! Kazaphani, 1956, *G. E. Atherton* 1152! Dhavlos, 1957, *H. Painter* 3!
8. Near Komi Kebir, 1912, *Haradjian* 282! 283! Rizokarpaso, 1941, *Davis* 2362! Apostolos Andreas Monastery, 1948, *Mavromoustakis* s.n.!

2. C. coronarium *L.*, Sp. Plant., ed. 1, 890 (1753); Hume in Walpole, Mem. Europ. Asiatic Turkey, 254 (1817); Unger et Kotschy, Die Insel Cypern, 239 (1865); Boiss., Fl. Orient., 3: 336 (1875); Holmboe, Veg. Cypr., 182 (1914); Post, Fl. Pal., ed. 2, 2: 61 (1933); Lindberg f., Iter Cypr., 34 (1946); Osorio-Tafall et Seraphim, List Vasc. Plants Cyprus, 103 (1973); Davis, Fl. Turkey, 5: 254 (1975); Feinbrun, Fl. Palaest., 3: 349, t. 589 (1978).

Erect, glabrous, glaucescent annual up to 1 m. high; stems obscurely sulcate, branched or unbranched; leaves subsessile or shortly petiolate, oblong in outline, 1·5–9 cm. long, 0·6–5 cm. wide, deeply bipinnatisect, the ultimate divisions narrowly oblong, irregularly serrate-lobed, the apices sharply apiculate; capitula terminal on peduncles 2–10 cm. long; involucre 1·5–2·5 cm. diam.; phyllaries loosely imbricate in 4 series, broadly ovate or oblong, 2–10 mm. long, 2–5 mm. wide, apex obtuse or rounded, margins widely membranous, brownish, usually fissured; receptacle strongly convex, not much accrescent after anthesis; ray-florets female, ligules broadly oblong, 10–15 mm. long, 6–7 mm. wide, yellow or sometimes creamy-white with a yellow base, apex rounded, emarginate or obscurely 3-lobed, tube 3–4 mm. long; style about 3 mm. long, branches oblong, truncate, about 0·6 mm. long; disk-florets yellow, tubular, 3–4 mm. long, gradually dilated above the middle and about 1·5 mm. wide at apex, base slightly expanded but not covering top of immature achene, lobes 5, recurved, ovate, acute, papillose, about 0·5 mm. long, 0·4 mm. wide at base, without an internal resin-canal; anthers linear, subacute, about 1·8 mm. long, 0·3 mm. wide; style 2·8 mm. long, branches linear, truncate-penicillate, about 0·8 mm. long. Achenes of ray-florets sharply triquetrous, with excavate sides and narrowly winged angles, brown, about 3 mm. long, 2·5 mm. wide; achenes of disk-florets laterally compressed, about 3 mm. long, 2·5 mm. wide, abaxial face with a wide, curved wing, lateral faces bluntly 3-ribbed, adaxial face with a curved wing terminating apically in a sharp angle or beak.

var. **coronarium**
TYPE: "*in* Creta, Sicilia".

Ray- and disk-florets uniformly bright yellow.

HAB.: Cultivated fields; waste ground; roadsides; sea-level to 2,000 ft. alt.; fl. Jan.–Dec.

DISTR.: Divisions 1, 3–8; abundant throughout the lowlands; widespread in the Mediterranean region and eastward to Iran.

1. Yeroskipos, 1939, *Lindberg f.* s.n.; Ayios Yeoryios (Akamas), 1962, *Meikle*! Polis, 1962, *Meikle*! Kissonerga, 1967, *Economides* in ARI 994!
3. Limassol, 1801, *Hume*; also 1862–1913, *Kotschy* s.n.; *Michaelides*; *Haradjian* 606! Kouklia, 1939, *Lindberg f.* s.n.; Yerasa, 1947, *Mavromoustakis* s.n.! Pyrgos, 1963, *J. B. Suart* 44! Amathus, 1978, *Holubová* s.n.!
4. Larnaca, 1801, *Hume*; also, 1862–1905, *Kotschy* s.n.; *Sintenis & Rigo* s.n.; *Holmboe* 144; Paralimni, 1940, *Davis* 2067! Cape Greco, 1958, *N. Macdonald* 33! Near Dhekelia, 1981, *Hewer* 4712!
5. Kythrea, 1880 *Sintenis & Rigo* s.n.; also, 1950, *Chapman* 104! Athalassa, 1933, *Syngrassides* 1375! Salamis, 1962, *Meikle*!
6. Nicosia, 1935, *Syngrassides* 805! also, 1939, *Lindberg f.* s.n.; 1950, *Chapman* 524! 1973, *P. Laukkonen* 101! Neapolis, 1973, *P. Laukkonen* 202! Dhiorios, 1962, *Meikle*!

7. Dhavlos, 1880, *Sintenis & Rigo* s.n.; also 1957, *H. Painter* 1! Kyrenia, 1932–1955, *Syngrassides* 301! *Casey* 468! *Miss Mapple* 30! *G. E. Atherton* 64! 201! 317! Ayios Epiktitos, 1968, *Economides* 1053!
8. Rizokarpaso, 1880, *Sintenis & Rigo* s.n.; also, 1912, *Haradjian* 261! Akradhes, 1962, *Meikle*!

var. **discolor** *Urv.* in Mém. Soc. Linn. Paris, 1: 368 (1822); Feinbrun, Fl. Palaest., 3: 349 (1978).
  C. coronarium L. ssp. *discolor* (Urv.) Rechinger f. in Beih. Bot. Centralbl., 54B: 634 (1936); Osorio-Tafall et Seraphim, List Vasc. Plants Cyprus, 103 (1973).
  TYPE: Aegean; "in insulis Archipelagi", *Dumont d'Urville* (P, K !).

Ray-florets creamy-white with a yellow base.

HAB.: Cultivated and waste ground; roadsides; fl. Jan.–Dec.

DISTR.: Division 6. Sporadic in populations of *C. coronarium* var. *coronarium*.

6. Nicosia, 1974, *Meikle* 4074!

NOTES: Recorded from the Aegean area, Turkey and Israel, and not infrequently cultivated elsewhere as an ornamental.

[CHRYSANTHEMUM MYCONIS *L.*, Sp. Plant., ed. 2, 1254 (1763) (*Leucanthemum myconis* (L.) Giraud in Ann. Univ. Grenoble, sect. Sci.-Med., 11: 195; 1935) has been recorded by Kotschy (in Oesterr. Bot. Zeitschr., 12: 278; 1862) as common in the valleys of the Troödos range, but clearly in error, and the plant seen was almost certainly *C. segetum* L., which Kotschy (Die Insel Cypern, 239; 1865) notes from this area. *C. myconis* has oblong or obovate, sessile, serrate leaves, yellow florets and conspicuously coronate achenes. It occurs (rarely) in S. Turkey and more abundantly in Palestine, and might yet be found in Cyprus.]

21. TANACETUM *L.*

Sp. Plant., ed. 1, 843 (1753).
Gen. Plant., ed. 5, 366 (1754).
V. H. Heywood in Anal. Inst. Bot. Cavanilles, 12: 313–325 (1954).

Perennial herbs, or rarely suffruticose, generally rhizomatous; leaves various, entire, toothed, lobed or pinnatisect; capitula solitary, or in lax or dense corymbs, heterogamous or homogamous; involucre hemispherical or campanulate; phyllaries imbricate in 3–4 series, herbaceous or scarious, usually with membranous margins; receptacle flattish, naked; ray-florets female, with short, white, yellow (or pinkish) ligules; disk-florets yellow, tubular, 5-lobed at apex, hermaphrodite; anthers blunt at base; style-branches truncate. Achenes cylindrical, prismatic-cylindrical or subclavate, 5–10-ribbed, glabrous, sometimes glandular; corona short, often irregularly dentate or lobed.

More than 50 species in temperate regions of Europe, Asia and North America.

Leaves oblong, margins closely and bluntly serrate, inflorescence rather densely corymbose
                                                                                    1. T. balsamita
Leaves deeply bipinnatisect; inflorescence lax   -   -   -   -   -   **2. T. parthenium**

1. **T. balsamita** *L.*, Sp. Plant., ed. 1, 845 (1753); Davis, Fl. Turkey, 5: 264 (1975).
  Pyrethrum balsamita (L.) Willd. var. *tanacetoides* Boiss., Fl. Orient., 3: 346 (1875); A. K. Jackson et Turrill in Kew Bull., 1938: 466 (1938).
  [*Pyrethrum balsamita* (non (L.) Willd.) Kotschy in Unger et Kotschy, Die Insel Cypern, 239 (1865).]
  [*Chrysanthemum balsamita* (non L.) Osorio-Tafall et Seraphim, List Vasc. Plants Cyprus, 103 (1973).]
  TYPE: "*in* Hetruria, Narbona".

Erect, rhizomatous perennial up to 130 cm. high; stems stout, subglabrous or thinly adpressed-pubescent, obscurely ridged, usually unbranched except in the region of the inflorescence; leaves simple, oblong, pleasantly aromatic, closely gland-dotted, adpressed-pubescent on both surfaces, apex rounded or obtuse, margins closely and bluntly serrate, the basal up to 12 cm. long, 5 cm. wide, with slender, flattened, or canaliculate petioles up to 12 cm. long, the cauline becoming progressively smaller and more shortly petiolate upwards, the uppermost subsessile, 3–4 cm. long, 1–1·5 cm. wide; inflorescence rather densely corymbose with numerous capitula, few- or many-branched with erect branches; involucre widely campanulate, 5–7 mm. diam.; phyllaries closely imbricate in 4 series, 1·5–3·5 mm. long, 0·5–1 mm. wide, oblong, glandular and thinly pubescent externally, the outermost wholly herbaceous, obtuse or subacute, the innermost with conspicuous, rounded, lacerate-membranous apices; florets all tubular (ssp. *balsamita*; *T. balsamita* L. var. *tanacetoides* Boiss.) or the outermost white, ligulate, radiate (ssp. *balsamitoides* (Sch. Bip.) Grierson), the ray-florets oblong, 3·5–7 mm. long, 1·7–3·5 mm. wide; tubular disk-florets about 2·5 mm. long, 0·8 mm. wide, thinly glandular externally, lobes 5, small, bluntly ovate-deltoid, erect, about 0·3 mm. long, and almost as wide at base; anthers subacute, about 1 mm. long, 0·2 mm. wide; style about 2 mm. long, with a globose basal swelling, branches recurved, truncate-penicillate, about 0·5 mm. long; receptacle flat, about 2·5 mm. diam. Achenes apparently not maturing in our area, oblong-cylindrical, about 1·5 mm. long, 0·5 mm. wide, pale brown, glandular externally, 7-ribbed, with a short, crenulate corona.

HAB.: Roadsides or near streams in Pine forest; 2,200–5,500 ft. alt.; fl. Sept.–Oct.

DISTR.: Division 2, probably an escape from cultivation. Throughout Europe and eastwards to Central Asia, but often cultivated or an escape from cultivation.

2. Kryos Potamos, 5,500 ft. alt., 1937, *Kennedy* 945 ! Troödos, 1940, *Davis* 1945 ! also, 1950, *F. M. Probyn* s.n. ! and, 1970, *A. Genneou* in ARI 1627 ! Kakopetria, 1955, *Chiotellis* 460 ! 46th milestone above Pano Amiandos, 1959, *P. H. Oswald* 146 !

NOTES: An old garden plant, long grown for its mint-scented foliage, and almost certainly not indigenous in Cyprus, where Kotschy (Die Insel Cypern, 239; 1865) says it is (or was) widely grown. Only the discoid, typical ssp. *balsamita* has been collected.

2. T. parthenium (*L.*) *Sch. Bip.*, Ueber die Tanaceteen, 55 (1844); Davis, Fl. Turkey, 5: 268 (1975).
  Matricaria parthenium L., Sp. Plant., ed. 1, 890 (1753).
  Chrysanthemum parthenium (L.) Bernh., Syst. Verz., 145 (1800); A. K. Jackson in Kew Bull., 1937: 343 (1937); Osorio-Tafall et Seraphim, List Vasc. Plants Cyprus, 103 (1973).
  Pyrethrum parthenium (L.) Sm., Fl. Brit., 2: 900 (1800); Boiss., Fl. Orient., 3: 344 (1875); Post, Fl. Pal., ed. 2, 2: 62 (1933).
  TYPE: "*in* Europae *cultis, ruderatis*".

Erect, pungent-aromatic perennial 20–60 cm. high, sometimes with a persistent woody base; stems puberulous, angled, usually much branched and bushy; leaves ovate-oblong in outline, 2–7 cm. long, 1·5–5 cm. wide, glandular, puberulous, bipinnatisect, the ultimate divisions 2–6 mm. wide, oblong, obtuse or shortly acute, petioles slender, canaliculate, those of the lower leaves up to 8 cm. long, those of the upper progressively shorter; inflorescence a lax terminal corymb; involucre hemispherical, 5–6 mm. diam.; phyllaries imbricate in 3 series, oblong, acute, 1·5–3 mm. long, 0·8–1 mm. wide, the outer herbaceous, pubescent, the innermost with a short lacerate-membranous apex; capitula heterogamous, radiate, ray-florets with white, oblong, bluntly 3-lobed ligules about 5–6 mm. long, 2–3 mm. wide, tube about 1 mm. long; style 1·5 mm. long, branches linear, truncate, 0·4 mm. long; disk-florets yellow, tubular-infundibuliform, about 3 mm. long, 1·5 mm. wide at apex, lobes 5, recurved, deltoid-acute, about 0·6 mm. long, 0·5 mm. wide at base; androecium and gynoecium small, (?)

imperfectly developed in specimens examined; anthers linear, subacute, about 0·8 mm. long, 0·1 mm. wide; style 0·8 mm. long, branches recurved, truncate-penicillate, about 0·4 mm. long; receptacle convex, pitted, 3–4 mm. diam. Achenes fusiform, somewhat curved, pale grey-brown, about 1·5 mm. long, 0·5 mm. wide at apex, glabrous, glandular, prominently 6-ribbed, corona minute, crenate or lobed.

HAB.: Gardens, waste ground, roadsides; 3,700–5,200 ft. alt.; fl. June–Sept.

DISTR.: Division 2, a garden escape. Widespread in most temperate regions of the world, and formerly valued as a febrifuge.

2. Platres, 1934, *Syngrassides* 521 ! Kryos Potamos, 5,200 ft. alt., 1938, *Kennedy* 1446 !

22. MATRICARIA *L.*

Sp. Plant., ed. 1, 890 (1753).
Gen. Plant., ed. 5, 380 (1754).

Annual herbs; leaves 2–3-pinnatisect, the ultimate divisions narrowly linear or filiform; capitula generally solitary, terminal, heterogamous and radiate, or homogamous and discoid; phyllaries imbricate in 2–3 series, with membranous margins; receptacle naked, conical; ray-florets (when present) with white ligules, female; disk-florets yellow, tubular, 4–5-lobed, hermaphrodite; style-branches truncate. Achenes oblong-cylindrical, often curved, glabrous, longitudinally ribbed; corona present or absent, or reduced to an auricle.

About 40 species widely distributed in the northern hemisphere; also in South Africa.

Capitula radiate, with white ray-florets; phyllaries greenish or with pale brown, membranous
    margins -   -   -   -   -   -   -   -   -   **1. M. recutita** var. **coronata**
Capitula discoid, yellow; phyllaries with a conspicuous fuscous margin -   -   **2. M. aurea**

1. **M. recutita** *L.*, Sp. Plant., ed. 1, 891 (1753); Dandy, List British Vasc. Plants, 115 (1958); Feinbrun, Fl. Palaest., 3: 344, t. 581 (1978).
    M. chamomilla L., Sp. Plant., ed. 2, 1256 (1763); Unger et Kotschy, Die Insel Cypern, 239 (1865); Boiss., Fl. Orient., 3: 323 (1875); Post, Fl. Pal., ed. 2, 2: 58 (1933); Davis, Fl. Turkey, 5: 293 (1975) non L. (1753) nom. illeg.

Erect or spreading annual, 6–35 (–50) cm. high; stems glabrous, sulcate or bluntly angled, usually much branched, slender; leaves finely bipinnatisect, oblong in outline, 1–7 cm. long, 0·6–2 cm. wide, ultimate divisions narrowly linear or almost filiform, 1–1·5 mm. wide, glabrous, subacute or shortly mucronate; petioles short or wanting, dilated at base and sheathing the stem; capitula solitary, terminal on the branches, sweet-smelling; involucre depressed-hemispherical, 3–10 mm. diam.; phyllaries imbricate in 2 (–3) series, oblong, about 1·5–3 mm. long, 1–1·5 mm. wide, greenish with a brownish, membranous margin; ray-florets with a white, oblong, shortly 3-lobed ligule about 6–8 mm. long, 3–3·5 mm. wide, tube about 1 mm. long; style about 1·5 mm. long, branches linear, truncate, a little recurved, about 0·4 mm. long; disk very strongly convex or subglobose, 4–7 mm. diam., florets yellow, tube about 1·5 mm. long, expanded rather suddenly in the upper half, and about 0·7 mm. wide at apex, thinly glandular externally, lobes 5, deltoid-acute, about 0·5 mm. long, 0·4 mm. wide at base; anthers linear, about 0·8 mm. long, 0·1 mm. wide, subacute, style about 1 mm. long, branches truncate-penicillate, about 0·5 mm. long; receptacle sharply conical, up to 5 mm. long, regularly foveolate. Achenes oblong-cylindrical, about 1 mm. long. 0·4 mm. wide, slightly curved, brownish, thinly glandular, with 5 pallid longitudinal ribs on the adaxial face, apex truncate, or with an irregularly lacerate-lobed, membranous, whitish corona about 1 mm. long.

var. **coronata** (*Gay ex Boiss.*) *Gruenberg-Fertig* in Feinbrun, Fl. Palaest., 3: 344 (1978).
 M. pusilla Willd., Enum. Plant. Hort. Reg. Bot. Berol., 2: 907 (1809).
 M. chamomilla L. var. *coronata* Gay ex Boiss., Voy. Bot. Espagne, 2: 316 (1840), Fl. Orient., 3: 324 (1875).
 M. chamomilla L. var. *pappulosa* Margot et Reuter, Fl. Zante, 96 (1841); Davis, Fl. Turkey, 5: 294 (1975).
 M. coronata (Gay ex Boiss.) Gay ex Koch, Syn. Fl. Germ. Helv., ed. 2, 416 (1843).
 M. chamomilla L. ssp. *pusilla* (Willd.) Holmboe, Veg. Cypr., 182 (1914); Osorio-Tafall et Seraphim, List Vasc. Plants Cyprus, 103 (1973).
 M. chamomilla L. var. *pusilla* (Willd.) Fiori et Paoletti, Nuov. Fl. Anal. Ital., 2: 620 (1927).
 TYPE: "in Zacyntho (Marg.) Hispaniâ australi, Canariis (Courant), Louisianâ (Teinturier), Mexico (Andrieux)". (G).

Achenes of ray-florets with a conspicuous, irregularly lacerate-lobed, membranous corona 1–1·5 mm. long, frequently longer than the achene itself; achenes of disk-florets either ecoronate, or with a corona similar to that of the ray-achenes, but usually shorter.

HAB.: Cultivated and fallow fields; waste ground, roadsides; sea-level to 500 ft. alt.; fl. March–May.

DISTR.: Divisions 3–7, locally common. Widespread in the Mediterranean region and east to N.W. India, Atlantic Islands.

3. Between Limassol and Kolossi, 1859, *Kotschy* 433; Phasouri, 1966, *A. Matthews* 6!
4. Larnaca, 1862, *Kotschy* 44; also, 1877, *herb. Post* 612! and, 1880, *C. & W. Barbey* s.n.
5. Kythrea, 1932, *Syngrassides* 209! Near Nisou, 1934, *Syngrassides* 1385! Mile 3, Nicosia-Famagusta road, 1950, *Chapman* 384! Mora, 1956, *Merton* 2537! Athalassa, 1957, *Merton* 2898! Mia Milea, 1967, *Merton* in ARI 179! Dheftera, 1967, *Merton* in ARI 302!
6. Nicosia, 1930–1973, *F. A. Rogers* 0720! *Chapman* 533! *Merton* 1436! 3080! *P. Laukkonen* 60!
7. Kyrenia, 1945, *Casey* 594! 600! also, 1956, *G. E. Atherton* 1094!

NOTES: All the specimens examined have the achenes of the ray-florets coronate; in most the disk-florets have ecoronate achenes, but in two sheets, (*Chapman* 533! *Merton* in ARI 179!) one of two specimens mounted has all the achenes coronate, the other only those of the ray-florets. I cannot believe that the distinction between *M. chamomilla* (= *M. recutita*) var. *chamomilla* (var. *kochiana* (Sch. Bip.) Fiori et Paoletti — with achenes of ray-florets alone coronate — and var. *coronata* Gay ex Boiss. — with all the achenes coronate — is worth maintaining. *M. recutita* var. *recutita*, with all the achenes ecoronate, has not yet been satisfactorily recorded from Cyprus; it has a more northerly distribution, and is the prevailing variety over much of Europe.

2. M. aurea (*Loefl.*) *Sch. Bip.* in Bonplandia, 8: 369 (1860); Boiss., Fl. Orient., 3: 324 (1875); Post, Fl. Pal., 2: 58 (1933); Davis, Fl. Turkey, 5: 295 (1975); Feinbrun, Fl. Palaest., 3: 345, t. 582 (1978).
 Cotula aurea Loefl., Iter Hisp., 163 (1758); L., Sp. Plant., ed. 2, 2: 1257 (1763); Sibth. et Sm., Fl. Graec. Prodr., 2: 187 (1813), Fl. Graec., 9: 61, t. 878 (1837).
 C. complanata Sm. in Sibth. et Sm., Fl. Graec. Prodr., 2: 187 (1813), Fl. Graec., 9: 62, t. 879 (1837) quoad plant. cypr.; Poech, Enum. Plant. Ins. Cypr., 18 (1842).
 Anthemis complanata (Sm.) Halácsy, Consp. Fl. Graec., 2: 58 (1902); Holmboe, Veg. Cypr., 181 (1914).
 [*Anacyclus orientalis* (non L.) DC., Prodr., 6: 17 (1837) pro parte quoad plant. cypr.; Unger et Kotschy, Die Insel Cypern, 238 (1865).]
 [*Anthemis montana* L. var. *tenuiloba* (non (DC.) Boiss.) H. Stuart Thompson in Journ. Bot., 44: 332 (1906).]
 TYPE: Spain, without precise localization.

Erect or spreading, glabrous annual, 5–8 (–25) cm. high; stems slender, sulcate, usually much branched at base; leaves finely bipinnatisect, oblong in outline, 1–2·5 cm. long, 0·2–0·8 cm. wide, ultimate divisions linear-subulate, 0·5–3 mm. long; petioles short or wanting; capitula solitary, terminal, discoid, on slender peduncles up to 2 cm. long; involucre depressed-hemispherical, 4–8 mm. diam.; phyllaries imbricate in 2 series, oblong, about 2–3 mm. long, 1–1·5 mm. wide, apex rounded, margins fuscous-membranous, erose; disk-florets tubular, bright yellow, about 1 mm. long, 0·3 mm. wide, lobes 4, broadly deltoid, about 0·2 mm. long; anthers linear, subacute, about 0·4 mm. long, 0·1 mm. wide; style 0·6 mm. long, branches erect, truncate-penicillate, about 0·2 mm. long; receptacle conical, about 3 mm. long. Achenes minute, oblong-cylindrical, about 1 mm. long, 0·4 mm. wide, brown, longitudinally striate, apex with a whitish, membranous, erose or irregularly lobed corona about 0·4 mm. long.

HAB.: Roadsides, walls, waste ground; sea-level to 500 ft. alt.; fl. Febr.–April.

DISTR.: Divisions 4, 6, 7. Mediterranean region and eastwards to the Himalayas.

4. Waste ground N. of old city, Famagusta, 1959, *P. H. Oswald* 45!
6. Walls of Nicosia, 1880, *Sintenis & Rigo* 310!
7. Roadside, Kyrenia, 1956, *G. E. Atherton* 1107!

23. **CHLAMYDOPHORA** *Ehrenb. ex Less.*

Syn. Gen. Compos., 265 (1832).

Annual herbs; stems rather fleshy; lower leaves opposite, upper alternate, entire or lobed; capitula solitary, terminal, homogamous, discoid; involucre depressed-hemispherical, phyllaries imbricate, blunt, in 2–3 series; receptacle naked, convex; florets yellow, hermaphrodite; corolla tubular, apex 4–5-lobed; anthers obtuse at base; style-branches truncate or obtuse. Achenes fusiform, conspicuously 10-ribbed, apex crowned with a well-developed, membranous, spathaceous corona often as long as, or longer than the achene.

One species in the eastern Mediterranean region.

1. **C. tridentata** (*Del.*) *Ehrenb. ex Less.*, Syn. Gen. Compos., 266 (1832); Boiss., Fl. Orient., 3: 359 (1875); Holmboe, Veg. Cypr., 183 (1914); Davis, Fl. Turkey, 5: 293 (1975); Feinbrun, Fl. Palaest., 3: 348, t. 587 (1978).
    Balsamita tridentata Del., Fl. d'Égypte, 129, t. 47, fig. 1 (1813).
    Tanacetum uliginosum Sm. in Sibth. et Sm., Fl. Graec. Prodr., 2: 167 (1813), Fl. Graec., 9: 42, t. 855 (1837).
    Cotula tridentata (Del.) Dinsmore in Post, Fl. Pal., ed. 2, 2: 65 (1933); Osorio-Tafall et Seraphim, List Vasc. Plants Cyprus, 103 (1973); Meikle, Fl. Cypr., 1: 2 (1977).
    [*Cotula coronopifolia* (non L.) Kotschy in Unger et Kotschy, Die Insel Cypern, 240 (1865).]
    TYPE: Egypt; "près d'Alexandrie, aux environs de la colonne de Pompée et du lac *Mareotis*" (MPU).

Erect or spreading, glabrous annual 2–30 cm. high; stems sulcate, usually branched; leaves linear-oblong, 1–5 cm. long, 0·2–0·8 cm. wide, bright green, rather fleshy, entire or commonly with 3 subacute or obtuse apical lobes, sometimes irregularly pinnatilobed; petiole absent, base of lamina amplexicaul, the bases of the lower, opposite leaves connate to form a cup around the stem; peduncles slender 1·5–10 cm. long; involucre 7–10 mm. diam.; phyllaries loosely imbricate, oblong, 2–4 mm. long, 1·5–2 mm. wide, glabrous, with a conspicuous, brown, erose-membranous margin, receptacle glabrous, convex, irregularly dotted over with small, blunt prominences; florets yellow, tubular, about 1·8 mm. long, 0·4 mm. wide, apex 5-lobed, the lobes deltoid, about 0·3 mm. long and almost as wide at base; anthers minute, oblong, apiculate, about 0·4 mm. long, 0·1 mm. wide; style about 1 mm. long, branches erect, connivent, truncate-penicillate, about 0·3 mm. long. Achenes fusiform, slightly curved, about 1·5 mm. long, 0·5 mm. wide, dark brown with 10 conspicuous whitish longitudinal ribs and a short basal stipe; corona white, membranous, spathaceous, about 2·3 mm. long, 1·3 mm. wide.

HAB.: Salt-marshes, and by brackish springs and runnels inland, often on Kythrean marls and sandstones; sea-level to 800 ft. alt.; fl. Febr.–May.

DISTR.: Divisions 3–6, 8. Eastern Mediterranean region, from Crete and the Aegean Islands to Egypt and Tunisia.

3. Mazotos, 1862, *Kotschy*.
4. Larnaca, 1859, *Kotschy* 254! also, 1862, *Kotschy* 208 (? 308)! and, 1967, *Merton* in ARI 389! Ayia Napa, 1905, *Holmboe* 33; Larnaca Salt Lake, 1905, *Holmboe* 107.
5. Kythrea, 1880, *Sintenis & Rigo* 553! also, 1941, *Davis* 2948! Near Pyroï bridge, 1936, *Syngrassides* 1046! Lakovounara Forest, 1950, *Chapman* 74! Near Kato Dhikomo, 1962, *Meikle* 2968!
6. Near Syrianokhori, 1941, *Davis* 2606!
8. Boghaz, 1941, *Davis* 2410!

24. ARTEMISIA L.

Sp. Plant., ed. 1, 845 (1753).
Gen. Plant., ed. 5, 367 (1754).

Annual or perennial herbs, shrubs or subshrubs; leaves usually alternate and pinnatisect, often aromatic; inflorescence generally paniculate or racemose; capitula often very numerous, small, homogamous and discoid, or heterogamous and disciform; involucre generally ovoid, oblong or campanulate; phyllaries in 2–4 series, imbricate, with membranous or scarious margins; receptacle naked, convex, glabrous or hairy; florets narrowly tubular, all hermaphrodite with a shortly 5-lobed corolla in homogamous capitula, or the outer female, with a 2–3-lobed corolla, and the inner functionally male with a 5-lobed corolla, in heterogamous capitula; anthers blunt at base; style-branches truncate, recurved, often penicillate in female or hermaphrodite florets, connivent in functionally male florets. Achenes obovoid or fusiform, smooth, epappose, or with a minute, ill-defined corona. Wind-pollinated.

About 400 species widely distributed in the northern hemisphere; Tarragon (*A. dracunculus* L.) and Absinth (*A. absinthium* L.) are used for flavouring food or drink; Southernwood (*A. abrotanum* L.) is valued for its fragrant leaves.

Leaves and stems silvery-tomentellous; plant perennial; capitula 5–7 mm. diam.

1. A. arborescens

Leaves and stems glabrous or thinly pubescent; plant annual; capitula about 2 mm. diam.

2. A. annua

1. **A. arborescens** *L.*, Sp. Plant., ed. 2, 1188 (1763); Boiss., Fl. Orient., 3: 372 (1875), Suppl., 301 (1888); Holmboe, Veg. Cypr., 183 (1914); Post, Fl. Pal., ed. 2, 2: 66 (1933); Rechinger f. in Arkiv för Bot., ser. 2, 1: 433 (1950); Osorio-Tafall et Seraphim, List Vasc. Plants Cyprus, 103 (1973); Davis, Fl. Turkey, 5: 318 (1975); Feinbrun, Fl. Palaest., 3: 353, t. 596 (1978). TYPE: "*in* Italia, Oriente".

Erect shrub, to 1 m. high, with pale brown, thinly tomentellous, shallowly sulcate branches; leaves oblong-deltoid in outline, 3–10 cm. long, 1·5–8 cm. wide, bipinnatisect, silvery-canescent, the ultimate divisions linear, subacute, 5–20 mm. long, 1–1·5 mm. wide; petiole subterete, to 3 cm. long; inflorescence terminal, narrowly paniculate, to 30 cm. long, 15 cm. wide, consisting of numerous erect, leafy branches bearing racemes of subglobose capitula; bracts herbaceous, similar to the leaves but often simply pinnatisect or undivided and linear; peduncles slender, tomentellous, up to 8 mm. long; involucre hemispherical, 5–7 mm. diam.; phyllaries imbricate in 2–3 series, oblong, rounded at apex, 2–2·5 mm. long, 1–1·5 mm. wide, tomentellous dorsally with a narrow membranous margin; capitula homogamous; florets brownish-yellow; corolla tubular, papillose externally, 2–2·5 mm. long, 1–1·5 mm. wide, apex with 5 short, acute, deltoid lobes; anthers oblong, shortly acute, about 0·8 mm. long, 0·3 mm. wide; style 1·5 mm. long, branches spreading, truncate-penicillate, about 0·4 mm. long;

receptacle convex, hairy, about 1·5 mm. diam. Achenes fusiform, about 1·3 mm. long, 0·6 mm. wide at apex, obscurely angled, dark brown, rather densely golden-glandular, generally crowned with the persistent, spongy, corolla-tube.

HAB.: Hedges and dry-stone walls near dwellings; 200–3,000 ft. alt.; fl. May–June.

DISTR.: Divisions 1–3, 5, 7, 8, probably always a relic of cultivation. Widespread in the Mediterranean region, but often as an introduction.

1. Kritou Terra, 1905, *Holmboe* 765.
2. Makheras Monastery, 1937, *Syngrassides* 1625!
3. Salamiou, 1941, *Davis* 3425!
5. Marathovouno, 1936, *Syngrassides* 1258!
7. Between Pano Dhikomo and Ayios Khrysostomos Monastery, 1880, *Sintenis & Rigo* 668!
8. Near Komi Kebir, 1912, *Haradjian* 312!

2. A. annua *L.*, Sp. Plant., ed. 1, 847 (1753); Boiss., Fl. Orient., 3: 371 (1875); Post, Fl. Pal., ed. 2, 2: 66 (1933); Davis, Fl. Turkey, 5: 317 (1975).
[*A. campestris* (non L.) H. Stuart Thompson in Journ. Bot., 44: 332 (1906); Holmboe, Veg. Cypr., 183 (1914).]
TYPE: "*in* Sibiriae *montosis*".

Erect annual to 1 m. high; stems usually branched, glabrous or thinly pubescent, sulcate, often purplish; basal leaves usually withered by anthesis, broadly oblong-deltoid, 2–5 cm. long. 1·5–4 cm. wide, bipinnatisect, the ultimate divisions linear, 2–5 mm. long, 0·5–1·5 mm. wide, glabrous or thinly puberulous, often serrulate; petioles flattened, to 3 cm. long; cauline leaves usually sessile or subsessile, often deltoid in outline, elegantly bipinnatisect, becoming progressively smaller upwards; inflorescence a lax, pyramidal, nebulous panicle, often 30 cm. long and 15 cm. wide at base, consisting of innumerable, very small, subglobose capitula, rather densely aggregated in secondary, lateral panicles; peduncles very short, filiform, usually less than 2 mm. long; involucre hemispherical, about 2 mm. diam.; phyllaries rather loosely imbricate in 2 series, oblong, blunt, the outer about 0·8 mm. long, 0·4 mm. wide, the inner about 1–1·2 mm. long, 0·8 mm. wide, shining green, with a narrow membranous margin; florets 10–12, homogamous; corolla tubular, about 1 mm. long, 0·5 mm. wide, glandular externally, apex with 5, minute, deltoid-acute, recurved lobes; stamens oblong, about 0·4 mm. long, 0·2 mm. wide, acute at apex; style about 0·3 mm. long, branches spreading, truncate-penicillate, about 0·3 mm. long; receptacle minute, convex, about 0·5 mm. diam. Achenes obovoid, somewhat compressed, about 0·8 mm. long, 0·4 mm. wide, mid-brown, minutely glandular-granulose.

HAB.: Cultivated ground; about 500 ft. alt.; fl. Sept.–Nov.

DISTR.: Division 6. Widespread in central and southern Europe, and eastwards to Central Asia.

6. Nicosia, "garden, Sept. 1900", *A. G. & M. E. Lascelles* s.n.!

NOTES: The only record, and perhaps a fleeting occurrence, though the fact that *A. annua* flowers late in the year may have caused it to be overlooked by other collectors.

TRIBE 7. **Senecionideae** *Cass.* Herbs, shrubs or rarely trees; leaves generally alternate, simple or pinnatisect; capitula mostly heterogamous and radiate, sometimes homogamous and discoid; involucre herbaceous; phyllaries commonly in 1 series with a basal calyculus of small bracts; florets generally yellow, sometimes white or purplish; ray-florets (when present) female, in one series; disk-florets hermaphrodite or sometimes functionally male; anthers rounded or sagittate at base, but not caudate; style-branches truncate or with a sterile apiculus. Achenes various; pappus of hairs or bristles, rarely absent.

25. TUSSILAGO *L.* *

Sp. Plant., ed. 1, 865 (1753).
Gen. Plant., ed. 5, 372 (1754).

Perennial, rhizomatous herb; leaves mostly basal, appearing after the flowers; scapes several, each bearing numerous scale-leaves and a solitary, terminal, heterogamous, radiate capitulum; involucre 2-seriate, campanulate, phyllaries of each series equal; receptacle flat, naked; florets yellow, rays numerous, female, ligulate; disk-florets few, functionally male, tubular, with 5 apical lobes. Achenes narrowly cylindrical, ribbed, glabrous; pappus of scabrid hairs.

A monotypic genus, widely distributed in Europe, the Mediterranean area and western Asia, and introduced elsewhere.

1. **T. farfara** *L.*, Sp. Plant., ed. 1, 2: 865 (1753); Boiss., Fl. Orient., 3: 377 (1875); Holmboe, Veg. Cypr., 183 (1914); Post, Fl. Pal., ed. 2, 2: 67 (1933); Osorio-Tafall et Seraphim, List Vasc. Plants Cyprus, 105 (1973); Davis, Fl. Turkey, 5: 168 (1975).
   TYPE: "*in* Europae *argillosis subtus humidis*".

Perennial herb to 12 cm. high in Cyprus, to 20–30 cm. elsewhere; leaves often in pseudo-rosettes; broadly ovate to suborbicular in outline, cordate, 3·3–10 cm. long, 3·2–17 cm. wide, angled or shallowly lobed with irregularly toothed margins, white-felted at first, upper surface becoming glabrous; petiole 1·5–12·5 cm. long, deeply grooved above; inflorescence unbranched, scapes 8–11 (–15) cm. long, sometimes lengthening to 30 cm. or more in fruit, more or less white-felted; scale-leaves reddish, lanceolate, often with ciliate margins, 11–14 mm. long, to 6 mm. wide; capitula 1·5–2·5 cm. diam., nodding after anthesis; involucre reflexing in fruit; outer phyllaries linear, acute, ciliate, 3–4 (–7) mm. long, 1–2 mm. wide, inner phyllaries 7–8 (–14) mm. long, 1–2 mm. wide, lanceolate-oblong, with scattered, black, stipitate, glandular hairs externally; florets golden-yellow; ray-florets in several series, very narrowly oblanceolate, almost filiform, 8–9 (–10) mm. long, 0·5–0·75 mm. wide, with 1–3 veins; style slender, very shortly bifid at apex; disk-florets 8 mm. long, tube slender below, expanded to 1·5 mm. wide at apex, lobes linear, acute, 2 mm. long, 0·4–0·5 mm. wide; anthers linear, 2–2·5 mm. long, 0·2 mm. wide, with a lanceolate appendage, and a bluntly sagittate base; style imperfect, sub-clavate, very slightly divided, minutely papillose and slightly swollen in the upper part. Achenes linear, glabrous, brown, with distinct longitudinal ribs, 3–4 mm. long, 0·5 mm. wide; pappus-hairs white, scabridulous, 10–15 mm. long.

HAB.: Moist, open ground on banks, and in damp fields, usually on clay, 2,000–5,500 ft. alt.; fl. March–April.

DISTR.: Division 2, rare. General distribution that of the genus.

2. Ayios Ioannis, 1905, *Holmboe* 1146; above Pedhoulas, 1952, *Casey* 1280! Side of Prodhromos Dam, 1974, *Osorio-Tafall & Meikle* in *Meikle* 4043!

NOTES: Apparently indigenous in Cyprus, but perhaps originally introduced as a medicinal plant; Coltsfoot has long had a reputation for the treatment of coughs and ailments of the chest.

26. SENECIO *L.*

Sp. Plant., ed. 1, 886 (1753).
Gen. Plant., ed. 5, 373 (1754).

Annual, biennial or perennial herbs, shrubs or climbers; leaves alternate, entire, toothed, lobed or pinnatisect; inflorescence paniculate or corymbose,

* by P. Halliday

or flowers solitary; capitula heterogamous and radiate, or homogamous and discoid; ray-florets female; disk-florets hermaphrodite; receptacle flat or convex, naked; involucre campanulate or cylindrical; phyllaries in 1 series, generally with a conspicuous or inconspicuous basal calyculus of supplementary bracts; ray-florets mostly yellow, ligulate in 1 series; disk-florets tubular, apex 4–5-lobed; anthers usually rounded at base, or very shortly sagittate; style-branches generally truncate-penicillate, sometimes shortly appendiculate. Achenes mostly terete, cylindrical, glabrous or pilose, 5–10-ribbed, truncate; pappus of silky, scabridulous hairs; involucre generally reflexing after achenes are shed.

A huge genus, with an estimated, 1,500–2,000 species, distributed throughout temperate and tropical regions of the world. Many of the species are poisonous to stock; a few are valued as ornamentals.

Leaves white-tomentose; shrubby perennial - - - - - S. CINERARIA (p. 934)
Leaves not tomentose; plant annual:
   Capitula without rays, or ray-florets with very minute ligules:
     Pappus-hairs twice as long as achene; capitula in small terminal cymes or corymbs, or sometimes solitary- - - - - - - - - - **1. S. vulgaris**
     Pappus-hairs equalling or shorter than achene; capitula numerous in rather large terminal cymes or corymbs - - - - - - - - - **2. S. aegyptius**
   Capitula conspicuously radiate:
     Leaves and stems arachnoid; involucre with a calyculus of blackish (sphacelate)-tipped bracts - - - - - - - - **3. S. leucanthemifolius** var. **vernalis**
     Leaves and stems glabrous or glabrescent; involucre without a calyculus
                                 4. S. glaucus ssp. **cyprius**

1. S. vulgaris *L.*, Sp. Plant., ed. 1, 867 (1753); Unger et Kotschy, Die Insel Cypern, 242 (1865); Boiss., Fl. Orient., 3: 386 (1875); Holmboe, Veg. Cypr., 183 (1914); Post, Fl. Pal., ed. 2, 2: 68 (1933); Lindberg f., Iter Cypr., 36 (1946); Osorio-Tafall et Seraphim, List Vasc. Plants Cyprus, 105 (1973); Davis, Fl. Turkey, 5: 165 (1975); Feinbrun, Fl. Palaest., 3: 355, t. 598 (1978).

TYPE: "*in* Europae *cultis, ruderatis, succulentis*".

Erect, branched or unbranched, glabrous or sparingly arachnoid annual 3–30 (–50) cm. high; stems sulcate, rather fleshy; leaves oblong or obovate in outline, 1–5 cm. long, 0·5–2·5 cm. wide, margins dentate, sinuate, lobed or irregularly lacerate-pinnatisect, the basal often shortly petiolate, the upper sessile, commonly amplexicaul and auriculate at base; capitula in small terminal corymbs or sometimes solitary, subsessile or shortly pedunculate, discoid; involucre cylindrical-campanulate, 4–8 mm. long, 3–6 mm. wide; phyllaries linear, acuminate, about 0·8 mm. wide, closely connivent in a single series, 3-nerved, apex usually fuscous, margins narrowly scarious; calyculus of 6 or more small, subulate, fuscous-tipped bracts; florets yellow, narrowly tubular, about 5 mm. long, about 0·6 mm. wide in the upper part, lobes 5, deltoid, recurved, about 0·3 mm. long, 0·2 mm. wide at base; anthers linear, about 0·8 mm. long, 0·1 mm. wide, with a lanceolate appendage, and rounded base; style slender, 3–4 mm. long, branches narrowly oblong, truncate, about 0·4 mm. long. Achenes cylindrical-fusiform, about 2·3 mm. long, 0·8 mm. wide, closely adpressed-pubescent, brown, 8–10-ribbed, apex expanding into a shallow cup; pappus-hairs white, silky, scabridulous, 5–6 mm. long, deciduous.

HAB.: Cultivated and waste ground; roadsides; bare, rocky ground; sand-dunes and seashores; sea-level to 3,500 ft. alt.; fl. Jan.–Dec.

DISTR.: Divisions 2, 4–8. Europe, N. Africa and Asia; found as a weed throughout temperate regions of the world.

2. Milikouri, 1939, *Lindberg f.* s.n.; Kato Platres, 1958, *N. Macdonald* 63! Above Statos, 1981, *Hewer* 4772!

4. Larnaca, 1862, *Kotschy* 250a; Famagusta, 1969, *A. Genneou* in ARI 1374! Cape Greco, 1958, *N. Macdonald* 43!

5. Near Nisou, 1934, *Syngrassides* 1399!
6. Nicosia, 1957, *Merton* 3077! Xeros, 1970, *A. Genneou* in ARI 1401!
7. Yaïla, 1941, *Davis* 2838! Kyrenia, 1949, Casey 242! also, 1952, *Casey* 1216! and, 1956, *G. E. Atherton* 810! Bellapais, 1955, *G. E. Atherton* 394!
8. Ayios Philon near Rizokarpaso, 1941, *Davis* 2219!

2. S. aegyptius *L.*, Sp. Plant., ed. 1, 867 (1753); Boiss., Fl. Orient., 3; 387 (1875); Post, Fl. Pal., ed. 2, 2: 68 (1933); Osorio-Tafall et Seraphim, List Vasc. Plants Cyprus, 105 (1973).

Erect, branched or unbranched, glabrous or very sparsely arachnoid annual 30–60 cm. high; stems sulcate; leaves very variable, obovate, oblong or oblanceolate in outline, 2–10 cm. long, 0·8–5 cm. wide, the basal usually petiolate, the upper sessile or subsessile, coarsely and irregularly dentate-laciniate or pinnatisect, with blunt or acute divisions; capitula in lax, terminal cymes or corymbs; peduncles slender, 2–20 mm. long; involucre shortly cylindrical, 4–6 mm. long, 3–4 mm. wide, the base indurated in fruit; phyllaries linear, acuminate, 4–5 mm. long, about 0·5 mm. wide, closely connivent in a single series, midrib prominent, margins narrowly scarious, apex scarcely fuscous (or sphacelate); calyculus of a few, scattered, subulate, greenish or brownish-tipped bracts; florets yellow, all tubular or the marginal shortly revolute-ligulate; ligulate florets female, tube slender, 2–3 mm. long, ligule 1·5–2 mm. long, shortly 3-lobed at apex; style 2·5 mm. long, branches linear, truncate, about 0·5 mm. long, recurved; disk-florets narrowly tubular, about 3·5 mm. long, 0·5 mm. wide at the expanded apex, lobes 5, very short, deltoid; anthers linear, about 0·8–1 mm. long, 0·1 mm. wide, with a lanceolate appendage and blunt base; style about 2·5 mm. long, branches connivent, linear, truncate, about 0·4 mm. long. Achenes cylindrical, about 2 mm. long, 0·5 mm. wide, obscurely ribbed, dark brown, closely adpressed-pubescent, apex truncate; pappus-hairs less than 2 mm. long, white, silky, distinctly scabridulous, deciduous.

var. **discoideus** *Boiss.*, Fl. Orient., 3: 388 (1875), Suppl., 302 (1888).
    S. arabicus L., Mantissa 1, 114 (1767).
    S. aegyptius L. var. *arabicus* (L.) Holmboe, Veg. Cypr., 183 (1914).
    TYPE: "*in* Aegypto", cult. Uppsala.

Capitula discoid, without marginal ligulate florets.

HAB.: Roadsides; 400–500 ft. alt.; fl. Jan.–Dec.

DISTR.: Division 5, very rare. Also Egypt.

5. By roadsides, Kythrea, 5 June, 1880, *Sintenis & Rigo* 550!

NOTES: Not collected since, though probably still extant. Resembles *Senecio vulgaris*, but much larger in all its vegetative parts, less succulent, and with proportionately small capitula. The pappus-hairs are slightly shorter than the achene.

3. S. leucanthemifolius *Poir.*, Voy. en Barbarie, 2: 238 (1789); Boiss., Fl. Orient., 3: 388 (1875); Post, Fl. Pal., ed. 2, 2: 69 (1933); J. C. M. Alexander in Notes Roy. Bot. Gard. Edinb., 37: 399 (1979).

Erect or decumbent, glabrous or arachnoid, branched or unbranched annual 5–30 (–60) cm. high; stems sulcate, often purplish; leaves very variable, obovate, oblong, spathulate or oblanceolate in outline, dentate, lobed or irregularly laciniate-pinnatifid or pinnatisect, 1–5 (–8) cm. long, 0·5–2 cm. wide, the basal shortly petiolate, the cauline sessile, often with projecting basal auricles; capitula few or numerous in lax terminal cymes or corymbs, or often solitary at the ends of inflorescence-branches; peduncles up to 7 cm. long, sparingly bracteate; involucre campanulate or shortly cylindrical, 4–8 mm. long and almost as wide; phyllaries linear, acuminate, about 0·8 mm. wide, closely connivent in a single series, reflexed after fruits are shed, commonly with a fuscescent apex, midrib prominent, margins scarious; calyculus of 6 or more small, subulate, generally fuscous-tipped (or sphacelate) bracts; florets golden-yellow, the marginal conspicuous ligulate and female (rarely wanting), the inner tubular and hermaphrodite; ligule of ray-florets 4–10 mm. long, 2–3·5 mm. wide, apex rounded or bluntly 3-lobed, tube slender, thinly puberulous externally, 3–5 mm. long; style about as long as tube, branches oblong, truncate, about 0·4 mm. long; disk-florets narrowly tubular-infundibuliform, about 4·5 mm. long, 1 mm. wide at apex, lobes 5, ovate-deltoid, papillose, about 0·3 mm. long, 0·3 mm. wide at base; anthers linear, about 1·8 mm. long, 0·2 mm. wide, with a blunt appendage and a slightly, bluntly, sagittate base; style about 3 mm. long, branches divergent, oblong, truncate-penicillate, about 0·5 mm. long. Achenes cylindrical-fusiform, 2–3 mm. long, 0·5 mm. wide, ribbed, brown, densely or sparsely adpressed-strigulose; pappus-hairs white, scabridulous, deciduous, 4–5 (–7) mm. long.

var. **vernalis** (*Waldst. et Kit.*) *J. Alexander* in Notes Roy. Bot. Gard. Edinb., 37: 403 (1979).

   S. vernalis Waldst. et Kit., Plant. Rar. Hung., 1: 23, t. 24 (1802); Unger et Kotschy, Die Insel Cypern, 242 (1865); Boiss., Fl. Orient., 3: 389 (1875); Holmboe, Veg. Cypr., 183 (1914); Post, Fl. Pal., ed. 2, 2: 69 (1933); Rechinger f. in Arkiv för Bot., ser. 2, 1: 433 (1950); Osorio-Tafall et Seraphim, List Vasc. Plants Cyprus, 105 (1973); Davis, Fl. Turkey, 5: 166 (1975); Feinbrun, Fl. Palaest., 3: 355, t. 599 (1978).

   [*S. crassifolius* (vix Willd.) Sintenis in Oesterr. Bot. Zeitschr., 32: 398 (1882).]

   TYPE: Yugoslavia; Srem, "Crescit ad sepes vinearum & in aggeribus in Comitatu Syrmiensi" (PR).

Plant thinly arachnoid-pubescent with tangled, white, septate hairs; stems unbranched or branched from the base, usually not much branched above, except in the region of the inflorescence, generally less than 25 cm. high; basal leaves subspathulate, crenate-dentate, often purple below; stem leaves sessile, amplexicaul-auriculate, narrowly oblong or oblanceolate, margins coarsely and irregularly dentate, or lacerate-pinnatifid with short, toothed lobes; inflorescence few(2–6)-flowered, lax, spreading, with long, slender peduncles; capitula (including rays) 1–1·5 cm. diam., the ligules usually well developed and conspicuous. *Plate 60, figs. 4–9.*

   HAB.: Bare igneous slopes and screes; roadsides; sand-dunes; sea-level to 4,000 ft. alt.; fl. Febr.–May.

   DISTR.: Divisions 2, [6], 8. Europe and Mediterranean region eastwards to Afghanistan; said to have been introduced into western and central Europe in the 19th century, and to be spreading there.

2. Prodhromos, 1859 and 1862, *Kotschy* 465; 876; Kykko Monastery, 1913, *Haradjian* 928; also, 1941, *Davis* 3473! and, 1970, *A. Hansen* 783; Stavros valley, 1933, *Foggie* 93; Kryos Potamos, 2,800 ft. alt., 1937, *Kennedy* 954! Above Vavatsinia, 1941, *Davis* 2702! Papoutsa, 1941, *Davis* 3112! Lower Marathasa valley, 1953, *Casey* 1285! Above Palekhori, 1957, *Merton* 3004! Near Makheras Monastery, 1967, *Merton* in ARI 520! Vroisha, 1959 or 1960, *P. H. Oswald* 95! Palekhori, 1969, *Economides* in ARI 1343! also, 1979, *Edmondson & McClintock* E 2889!

[6. Near Nicosia; 1930, *F. A. Rogers* 0723; See Notes.]

8. Cape Andreas, 1880, *Sintenis & Rigo* 301! also, 1957, *Merton* 2916!

   NOTES: Largely confined to the Troödos massif, with an unexpected outlier in the extreme N.E. of the island at Cape Andreas, a distribution echoing that of *Hedera*. The Nicosia record is unlikely, and almost certainly an error; few of Rogers' specimens are reliably localized.

   Cyprus specimens agree closely with the illustration of *S. vernalis* in *Plant. Rar. Hung.*, t. 24; they are often small, but this may indicate only that they are normally found on poor, stony soils.

   The identity of *Senecio crassifolius* (? Willd.) Sm. in Sibth. et Sm., Fl. Graec. Prodr., 2: 177 (1813), Fl. Graec., 9: 53, t. 868 (1837); Poech, Enum. Plant. Ins. Cypr., 18 (1842) is doubtful; no such plant grows "in insulae Cypri maritimis", but it resembles some variants of *S. leucanthemifolius* from other parts of the eastern Mediterranean.

   The plant distributed as *S. crassifolius* by Sintenis & Rigo is not the same as that figured in *Fl. Graeca* t. 868, and is here identified as *S. leucanthemifolius* var. *vernalis*.

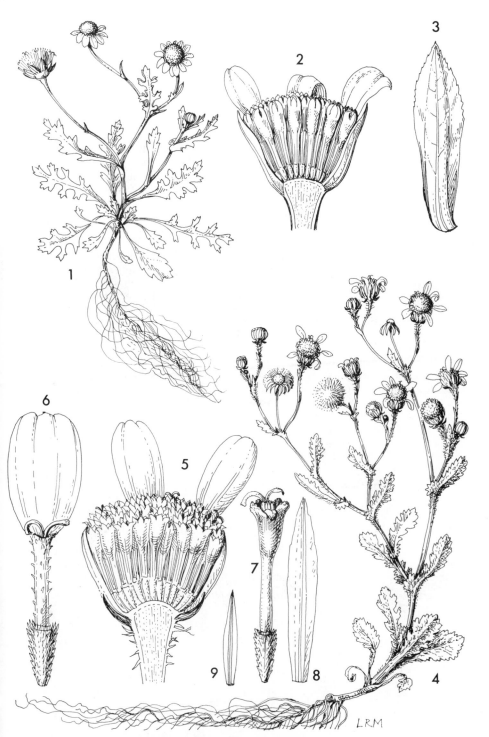

Plate 60. Figs. **1–3.** SENECIO GLAUCUS L. ssp. CYPRIUS Meikle **1,** habit, × 1; **2,** capitulum, longitudinal section, × 6; **3,** phyllary, × 12; figs. **4–9.** SENECIO LEUCANTHEMIFOLIUS Poir. var. VERNALIS (Waldst. et Kit.) J. Alexander **4,** habit, × 1; **5,** capitulum, longitudinal section, × 6; **6,** ray-floret, × 12; **7,** disk-floret, × 12; **8,** phyllary, × 12; **9,** calyculus-bract, × 12. (**1** from *Casey* 1282 & *Meikle* 2064; **2, 3** from *Meikle* 2064; **4** from *Kennedy* 955 & *Edmondson & McClintock* E 2889; **5–9** from *Kennedy* 955.)

4. S. glaucus *L.*, Sp. Plant., ed. 1, 868 (1753); Feinbrun, Fl. Palaest., 3: 356, t. 601 (1978); J. C. M. Alexander in Notes Roy. Bot. Gard. Edinb., 37: 411 (1979).

　　S. joppensis Dinsmore in Post, Fl. Pal., ed. 2, 2: 69 (1933); Feinbrun, Fl. Palaest., 3: 356, t. 600 (1978).

　　S. glaucus L. ssp. *joppensis* (Dinsmore) Feinbrun, Fl. Palaest., 3: 356 (1978).

　　Erect or decumbent, glabrous or sparsely arachnoid, glaucous, sometimes fleshy, branched or unbranched annual, (2–) 8–40 (–60) cm. high; stems sulcate; leaves oblong or lanceolate in outline, 1–5 (–13) cm. long, 0·5–3 (–8) cm. wide, the basal obscurely or distinctly petiolate, the upper usually sessile, amplexicaul or auriculate, all deeply lobed, or more commonly pinnatisect or pinnatipartite into a varying number of entire or pinnatilobed or dentate, linear divisions; capitula in lax terminal cymes or corymbs; peduncles up to 10 cm. long, sparsely bracteate; involucre campanulate, 4·5–8 mm. long, and as wide or a little wider, often with a distinct, orbicular, indurated base in fruiting specimens; phyllaries linear, acuminate, 0·8–1 mm. wide, connivent in a single series, reflexed after the fruits are shed, midrib prominent, margins narrowly scarious, apex sometimes sphacelate, but not always so; calyculus absent or of 1–12 small, subulate, herbaceous, or sometimes sphacelate bracts; ray-florets normally ligulate, female; ligules bright yellow, 6–12 mm. long, 2–3·5 mm. wide, apex rounded or bluntly 3-lobed, tube slender, glabrous or thinly puberulous externally, about 3 mm. long; style about as long as tube, branches narrowly oblong, truncate, divergent, about 0·5 mm. long; disk-florets narrowly tubular-infundibuliform, about 4 mm. long, 0·8–1 mm. wide at apex, lobes 5, ovate-deltoid, papillose, about 0·4 mm. long, 0·3 mm. wide at base; anthers linear, about 1·4 mm. long, 0·2 mm. wide, with a subacute appendage, and an obscurely and bluntly sagittate base; style 3·5 mm. long, branches recurved, narrowly oblong, truncate-penicillate, about 0·8 mm. long. Achenes cylindrical-fusiform, 1·8–2·2 mm. long, 0·5 mm. wide, ribbed, fuscous, densely or sparsely strigillose, apex truncate; pappus-hairs white, scabridulous, deciduous, (4–) 5–6 (–7) mm. long.

ssp. **cyprius** *Meikle* in Ann. Musei Goulandris, 6: 89 (1983). See App. V, p. 1898.

　　TYPE: Cyprus; Xeros, on kafkalla by the sea, 1970, *A. Genneou* in ARI 1399 (K!).

　　Plants usually dwarf, tufted, less than 15 cm. high, but occasionally up to 45 cm.; leaves narrowly oblong, 1–5 cm. long, 0·4–2 cm. wide, bright or dull green, glabrous or at first thinly arachnoid, sometimes rather fleshy, irregularly lacerate-pinnatifid, with a broad rhachis and sharply dentate, deltoid or shortly oblong, often trifurcate, lateral lobes; involucre without a calyculus, basal part clearly indurated in fruit. *Plate 60, figs. 1–3.*

　　HAB.: Sandy or stony ground by the sea; near sea-level; fl. Jan.–March.

　　DISTR.: Divisions 1, 6. Endemic.

　1. Karavopetres near Ayios Yeoryios (Akamas), 1962, *Meikle* 2064! Ayios Yeoryios Island (Akamas), 1962, *Meikle* 2160! Paphos, 1981, *Hewer* 4780!
　6. Ayia Irini, 1941, *Davis* 2121! 2562! also, 1952, *Merton* 1429! and, 1953, *Casey* 1282! Xeros, 1970, *A. Genneou* in ARI 1399!

　　NOTES: Clearly allied to *S. joppensis* Dinsmore, which J. C. M. Alexander (Notes Roy. Bot. Gard. Edinb., 37: 412; 1979) regards as synonymous with *S. glaucus* L. ssp. *glaucus*. It differs from this, however, in the uniform absence of a calyculus, and from *S. glaucus* L. ssp. *coronopifolius* (Maire) J. Alexander additionally in its coarsely dentate-lobed leaves and indurated involucre-base. I have not seen any matching material from outside Cyprus.

S. CINERARIA *DC.*, Prodr., 6: 355 (1838) (*S. bicolor* (Willd.) Todaro ssp. *cineraria* (DC.) Chater in Journ. Linn. Soc. Bot., 68: 273; 1974), a shrubby perennial to 60 cm. high; leaves densely white-tomentose, ovate-oblong in outline, 4–8 cm. long, 2·5–5·5 cm. wide, pinnatisect or bipinnatisect, with rather broad, blunt divisions; capitula in dense terminal corymbs, shortly radiate or occasionally discoid; involucre campanulate, white-tomentose; florets yellow.

A native of the western and central Mediterranean area, popular as an ornamental foliage plant, and frequently cultivated in Cyprus. It has been recorded from the foreshore near Kyrenia (Glykyotissa), 1955, *N. Chiotellis* in *G. E. Atherton* 213!, where it is almost certainly a garden escape.

For a note on the nomenclature of the species, see D. J. Mabberley in Watsonia, 14 (3): 279 (1982).

[S. SQUALIDUS *L.*, Sp. Plant., ed. 1, 869 (1753), an erect or diffuse, glabrous or subglabrous, short-lived perennial, up to 60 cm. high, with variously toothed, lobed or dissected leaves, and lax corymbs of rather large (2–3 cm. diam.) capitula with bright yellow ray- and disk-florets, is represented in the Kew herbarium (K) by two specimens made from plants cultivated (1938) in the herbarium ground, and purporting to have been raised from seeds collected by Mrs. Kennedy in Cyprus, by the Kryos Potamos, 5,850 ft. alt. There are no specimens from which the seed could have been taken, and since *S. squalidus* was a common weed in the herbarium ground, one must suspect that it was accidentally substituted for a Cyprus plant, which may not have germinated.

S. squalidus is widely distributed in central and southern Europe, and occurs as a weed or aggressive colonist elsewhere. It may yet be found in Cyprus.]

TRIBE 8. **Calenduleae** *Cass.* Herbs or shrubs; leaves generally alternate, glandular; capitula heterogamous, radiate; ray-florets female, fertile; disk-florets tubular, functionally male or sterile; receptacle naked; phyllaries in 1–2 series, herbaceous; anthers sagittate-caudate at base; style bifid in female florets, clavate in functionally male or sterile florets. Achenes epappose, heteromorphic.

27. CALENDULA *L.**

Sp. Plant., ed. 1, 921 (1753).
Gen. Plant., ed. 5, 393 (1754).
D. Lanza, Mon. Gen. Calendula L., Palermo (1919) et in Atti Reale Accad. Palermo, ser. 3, 12: 1–166 (1923).
C. C. Heyn, O. Dagan & B. Nachman in Israel Journ. Bot., 23: 163–201 (1974).

Annual or perennial, aromatic, often straggling herbs, sometimes woody at the base, usually clothed with glandular, multicellular, sometimes arachnoid hairs; leaves entire or shallowly toothed, sessile, often more or less amplexicaul; capitula many-flowered, heterogamous, radiate, solitary, terminal and axillary; involucre 2-seriate; phyllaries imbricate, subequal, linear-lanceolate, acuminate, with narrow scarious margins; receptacle flat, naked; ray-florets ligulate, female, fertile, yellow or orange; disk-florets actinomorphic, tubular, 5-dentate, functionally male, yellow, orange, brown or purple. Achenes epappose, 2–3-seriate, heteromorphic, usually incurved, dorsally rugose, tuberculate or echinate; marginal achenes beaked, cymbiform or winged, inner achenes curved or annulate.

About 9 species with a wide distribution in the Mediterranean region and

* by P. Halliday

western Asia. *Calendula officinalis* L., the Pot Marigold, formerly grown as a culinary or medicinal herb, is popular as an ornamental, and has been found as an escape from cultivation at (3) Curium (1982, *A. Fell* s.n. !)

1. **C. arvensis** *L.*, Sp. Plant., ed. 2, 2: 1303 (1763); Sibth. et Sm., Fl. Graec. Prodr., 2: 207 (1813), Fl. Graec., 10: 14, t. 920 (1840); Hume in Walpole, Mem. Europ. Asiatic Turkey, 254 (1817); Poech, Enum. Plant. Ins. Cypr., 18 (1842); Unger et Kotschy, Die Insel Cypern, 242 (1865); Boiss., Fl. Orient., 3: 418 (1875); H. Stuart Thompson in Journ. Bot., 44: 332 (1906); Post, Fl. Pal., ed. 2, 2: 72 (1933); Davis, Fl. Turkey, 5: 171 (1975); Feinbrun, Fl. Palaest., 3: 359, t. 604 (1978).

　　C. aegyptiaca Pers., Syn. Plant., 2: 492 (1807); Boiss., Fl. Orient., 3: 419 (1875); Post, Fl. Pal., ed. 2, 2: 72 (1933).

　　C. persica C. A. Mey., Verz. Pflanz. Cauc., 72 (1831); Boiss., Fl. Orient., 3: 418 (1875); Nattrass et Pappaiannou in Cypr. Agric. Journ., 34 (1): 26 (1939); Lindberg f., Iter Cypr., 34 (1946); Rechinger f. in Arkiv för Bot., ser. 2, 1: 433 (1950).

　　C. gracilis DC., Prodr., 6: 453 (1837).

　　C. micrantha Tineo et Guss. in Guss., Fl. Sic. Syn., 2: 874 (1845); Freyn in Bull. Herb. Boiss., 5: 782 (1897).

　　C. persica C. A. Mey. var. *gracilis* (DC.) Boiss., Fl. Orient., 3: 419 (1875); C. et W. Barbey, Herborisations au Levant, 99 (1882); H. Stuart Thompson in Journ. Bot., 44: 332 (1906).

　　C. persica C. A. Mey. ssp. *gracilis* (DC.) Holmboe, Veg. Cypr., 183 (1914); Osorio-Tafall et Seraphim, List Vasc. Plants Cyprus, 105 (1973).

　　C. arvensis L. ssp. *micrantha* (Tineo et Guss.) Holmboe, Veg. Cypr., 183 (1914); Osorio-Tafall et Seraphim, List Vasc. Plants Cyprus, 105 (1973).

TYPE: "*in* Europae *arvis*".

Erect, spreading or decumbent, much-branched annual (3–) 20–30 (–60) cm. high, pubescent with glandular, multicellular, sometimes arachnoid hairs; leaves bright or greyish green, (2·2–) 3·5–15 (–22) cm. long, (0·3–) 0·6–2·7 (–3·7) cm. wide, obtuse to acute, apiculate, subentire or sinuate-dentate to denticulate, lower leaves attenuate to base, decurrent, upper leaves amplexicaul; capitula terminal, 1·1–2·7 cm. diam. (including ligules); phyllaries 5–9 mm. long, about 1–2 mm. wide; ray-florets yellow to deep orange, ligulate or oblanceolate, 4–11 mm. long, 1·5–3 mm. wide, usually 4-veined, with 2-3 apical lobes, base clothed with long, multicellular hairs; disk-florets yellow, orange, brown or purple, 3–4 mm. long, about 1·5 mm. wide at apex; anthers 1·5–2 mm. long, base caudate-sagittate, appendage ovate; style of ray-florets bifid, branches compressed, acute; style of disk-florets imperfect, clavate, papillose. Fruiting capitula 1–1·4 (–2·2) cm. diam.; achenes polymorphic, all annulate, or the inner annulate, the marginal beaked and cymbiform, or all the marginal cymbiform; annulate achenes 2–4 mm. diam., variably rugose dorsally; beaked achenes 8–18 mm. long, variably incurved, sometimes roughly L-shaped, rugose, or often spinose, or sometimes laciniate-winged dorsally; cymbiform achenes 6–9 mm. long, inflated, transversely rugose dorsally.

HAB.: Cultivated and waste ground, roadsides; sometimes in open garigue or batha; sea-level to 500 (–3,900) ft. alt.; fl. March–May (and sporadically throughout the year).

DISTR.: Divisions 1–8, common. S.Europe and Mediterranean region eastwards to Afghanistan and India; an introduced weed in Australia, South Africa, North and South America.

1. Yeroskipos, 1939, *Lindberg f.* s.n. ! Near Lyso, 1979, *Edmondson & McClintock* E 2825 !
2. Philani, 1908, *Clement Reid* s.n. ! Pedhoulas, 1932, *Syngrassides* 40 ! Platres, 1939, *Kennedy* 1450 ! Kapedhes, 1967, *Merton* in ARI 296 ! Near Statos, 1981, *Hewer* 4770 !
3. Limassol, 1913, *Haradjian* 607 ! also, 1947, *Mavromoustakis* 20 ! Near Zakaki, 1947, *Mavromoustakis* s.n. ! Curium, 1961, *Polunin* 6688 ! Episkopi, 1964, *J. B. Suart* 142 ! Akhelia, 1967, *Economides* in ARI 1010 !
4. Larnaca, 1862, *Kotschy* 251 ! also, 1950, *Chapman* 8 ! Paralimni, 1940, *Davis* 2053 ! 2056 ! Cape Greco, 1958, *N. Macdonald* 1 ! Near Famagusta, 1969, *A. Genneou* in ARI 1373 ! Dhekelia, 1981, *Hewer* 4711 !
5. Athalassa, 1933, *Syngrassides* 1374 ! and, 1936, *Syngrassides* 867 ! Kythrea, 1950, *Chapman* 341 ! also, 1966, *O. Huovila* 46 ! and, 1967, *Merton* in ARI 338 ! Mile 3, Nicosia-Famagusta road, 1950, *Chapman* 380 ! 601 ! Lakovounara Forest, 1950, *Chapman* 49 ! Salamis, 1962, *Meikle* 2575 ! 2578 ! Mia Milea, 1967, *Merton* in ARI 189 !

6. Nicosia, 1930–1973, *F. A. Rogers* 0667! *Merton* 9! 40! 3083! *P. Laukkonen* 49! 208! Orta Keuy, 1936, *Syngrassides* 1166! Xeros, 1970, *A. Genneou* in ARI 1403!
7. Pentadaktylos, 1880, *Sintenis & Rigo* 309! Kyrenia, 1949–1972, *Casey* 406! 506! 520! *G. E. Atherton* 378! 390! 794! 811! 871! 1291! *W. R. Price* 982! Kharcha, 1950, *Chapman* 271! St. Hilarion, 1952, *F. M. Probyn* 47! Karavas, 1970, *A. Genneou* in ARI 1385!
8. Ayios Andronikos, 1880, *Sintenis & Rigo* 309a! Komi Kebir, 1912, *Haradjian* 284! 323! Ronnas Bay, 1941, *Davis* 2355!

NOTES: The Cyprus population of this polymorphic species includes the three main races, or variants, namely: (1) plants with a thinly arachnoid indumentum, uniformly yellow or orange florets, and trimorphic (rostrate, alate or cymbiform, and inner annulate) achenes — *C. arvensis* L. sens. strict.; (2) plants with a short, glandular, non-arachnoid indumentum, orange ray-florets and orange, brown or purple disk-florets; achenes usually dimorphic or trimorphic — *C. aegyptiaca* Pers. (3) plants with a thinly arachnoid indumentum, uniformly yellow or orange florets; achenes all annular and rugose dorsally — *C. persica* C. A. Mey. The third variant appears to be rather more frequent than the other two, but all are evidently well represented on the island, and joined by intermediates into a complex which defies further analysis.

TRIBE 9. **Arctoteae** *Cass.* Herbs or shrubs, sometimes spinescent and lactiferous; leaves alternate or basal; capitula heterogamous or homogamous, radiate or discoid, solitary or aggregated into compound heads; receptacle scaly; phyllaries free or connate at base, sometimes becoming indurated; anthers blunt or shortly sagittate at base; style-branches short, style thickened and papillose below bifurcation. Achenes with a scaly pappus or epappose, angular, without a beak.

28. GUNDELIA *L.**

Sp. Plant., ed. 1, 814 (1753).
Gen. Plant., ed. 5, 356 (1754).

Erect perennial herb with a stout rootstock and milky latex; leaves alternate, pinnatifid, spinose-dentate, decreasing in size upwards, the uppermost involucrate; inflorescence sub-corymbose; capitula individually few-flowered, aggregated into *Eryngium*-like compound heads, each subtended by a spiny, sub-cymbiform bract; phyllaries and receptacular scales fused into an indurated, turbinate cupule, comprising several 1-flowered compartments, the spines largely confined to the apex, but sometimes reduced to scale-like processes on the sides of the cupule as well; florets 5-7, white, yellow, pink, mauve, purple or red, sometimes maroon outside, yellow within, the central alone hermaphrodite and fertile. Achene turbinate, or obovoid, large, smooth, glabrous; pappus a small crown of persistent, uniseriate, denticulate, slightly indurated scales.

A monotypic genus, distributed from Transcaucasia through Turkey to Cyprus, Syria, Palestine and Israel, and eastwards to Iran and Transcaspia. A tumbleweed, each cupule finally dispersing as a 1-seeded unit.

1. **G. tournefortii** *L.*, Sp. Plant., ed. 1, 814 (1753); Kotschy in Oesterr. Bot. Zeitschr., 12: 277 (1862); Unger et Kotschy, Die Insel Cypern, 234 (1865); Boiss., Fl. Orient., 3: 421 (1875); H. Stuart Thompson in Journ. Bot., 44: 332 (1906); Holmboe, Veg. Cypr., 183 (1914); Post, Fl. Pal., ed. 2, 2: 74 (1933); Rechinger f. in Arkiv för Bot., ser. 2, 1: 433 (1950); Osorio-Tafall et Seraphim, List Vasc. Plants Cyprus, 103 (1973); Davis, Fl. Turkey, 5: 325 (1975); Feinbrun, Fl. Palaest., 3: 360, t. 607 (1978).
TYPE: "*in* Armenia, Syria".

Leafy, Thistle-like herb to 45-50 cm. or more in height; rootstock clothed with remains of old leaves and membranous scales; stems erect, simple or branched, stout, smooth, glabrous or thinly clothed with multicellular,

*by P. Halliday

arachnoid hairs; leaves lanceolate-elliptic in outline, pinnatifid to pinnatisect, the basal petiolate, up to 42 cm. long, 21 cm. wide, the cauline sessile,
often somewhat keeled, up to 36 cm. long, 19 cm. wide, acute, base auricled
or sometimes decurrent for a short distance as a narrow wing, venation
conspicuous, raised on both surfaces, sometimes white, or sometimes with a
purple midrib, margins toothed, yellow-spinose, thinly or densely
arachnoid-pilose in the sinuses between the spines, surface of lamina
glabrous or thinly arachnoid-pilose; aggregated heads of capitula terminal,
subglobose or ovoid, 5-7 cm. long, to 4·5 cm. wide; individual capitula to
about 2 cm. wide, subtending bracts equalling or exceeding involucre, with
an apical spine up to 5 cm. long; phyllaries fused except for the spinescent
tips; florets 7–10 (–13) mm. long, apex of tube minutely pubescent
externally, lobes linear, acute, 3–5 mm. long, about 1 mm. wide; anthers
linear, yellow or brown, 4–6 mm. long, appendage ovate-lanceolate, 0·7–1
mm. long, base shortly sagittate, tails adpressed to filament, obtuse; style
bifid, thickened and papillose towards apex, branches recurved, somewhat
compressed, slender, linear or somewhat dilated towards the acute apex.
Fruiting cupule turbinate and fluted, or obovoid and smooth, 1·2–1·3 cm.
long, about 7 mm. wide, becoming woody as achene develops, compartments 5–7 with horizontal or oblique apical aperture, fringed with spines or
spinules of varying length, the central (fertile) compartment almost
unarmed or surmounted by a tuft of fused spinules, 0·5–7 mm. long. Achene
somewhat compressed, sub-tetragonal, 6 mm. long, 3 mm. wide; pappus-
scales 1·5–2 mm. long, 1 mm. wide at base, forming a corona about 3–3·5
mm. diam.

HAB.: Seashores; roadsides; non-calcareous batha amongst *Sarcopoterium*; sea-level to 1,500
ft. alt.; fl. May–June.

DISTR.: Divisions 1, 3, 8, rare. General distribution that of genus.

1. Paphos, 1901, *A. G. & M. E. Lascelles* s.n.!
3. Between Limassol and Kolossi, 1862, *Kotschy*; Kholetria, Kryanera, 1941, *Davis* 3456!
8. Komi Kebir, 1912, *Haradjian* 280; between Leonarisso and Tavros, 1948, *Mavromoustakis*
    s.n.!

NOTES: Davis notes that the plant is popularly known as *Silifa*, and that its heads are eaten
like Globe Artichokes (*Cynara scolymus* L.); this suggests that *Gundelia tournefortii* is
commoner in Cyprus than the very few records would indicate.
Several varieties have been recognized, but, in Cyprus at least, it is impossible, on present
evidence, to judge the taxonomic significance of this variation.

TRIBE 10. **Cardueae** *Cass.* Herbs, rarely shrubs; leaves alternate,
commonly lobed and spinose; capitula mostly homogamous and discoid,
sometimes heterogamous with the neuter or female florets often enlarged,
zygomorphic and radiant, but rarely ligulate; receptacle generally scaly,
setose or pilose; phyllaries in several series, imbricate, herbaceous or
scarious, commonly spinose; florets mostly yellow or purple; anthers
caudate-sagittate at base; style frequently swollen and shortly pilose below
point of bifurcation of the short, obtuse or subacute, papillose style-
branches. Achenes compressed or terete, sometimes heteromorphic; pappus
of bristles and/or scales, sometimes of hairs, occasionally obsolete.

29. ECHINOPS *L.*

Sp. Plant., ed. 1, 814 (1753).
Gen. Plant., ed. 5, 356 (1754).

Mostly robust, Thistle-like perennials; leaves undivided or 1–3-pinnati-
sect, generally spinose; inflorescences branched or unbranched; capitula 1-
flowered, crowded into a globose head on a swollen common receptacle, each

head subtended by small, often concealed, bracts and falling apart in fruit; individual capitula with an involucre consisting of an outer (or basal) series of white bristles or pallid, narrow scales (the *"pencil"*) and an inner (or upper) series of rigid, imbricate, free or partly fused, often metallic-blue phyllaries; florets tubular, hermaphrodite, blue, cream or white (rarely crimson or green), apex of corolla divided into 5 linear lobes. Achenes enveloped by the persistent involucre, elongate-oblong, often densely adpressed-pilose; pappus coroniform of short, free or connate bristles or scales.

About 100 species widely distributed in Europe, Asia and Africa; a few are cultivated as ornamentals.

1. **E. spinosissimus** *Turra*, *Farsetia* Nov. Plant. Gen., 13 (1765); Osorio-Tafall et Seraphim, List Vasc. Plants Cyprus, 105 (1973) excl. syn.; Kožuharov in Tutin et al., Fl. Europaea, 4: 213 (1976).

   E. viscosus DC., Prodr., 6: 525 (1838); Boiss., Fl. Orient., 3: 429 (1875); Holmboe, Veg. Cypr., 183 (1914); Post, Fl. Pal., ed. 2, 2: 75 (1933); Rechinger f. in Arkiv för Bot., ser. 2, 1: 433 (1950); Davis in Notes Roy. Bot. Gard. Edinb., 21: 131 (1953); Fl. Turkey, 5: 618 (1975) non Schrader ex Reichb. (1832) nom. illeg.

   E. creticus Boiss. et Heldr. in Boiss., Diagn., 1, 10: 87 (1849).

   E. glandulosus Weiss in Verh. Zool.-Bot. Gesellsch. Wien, 18: 433 (1868).

   E. viscosus DC. ssp. *creticus* (Boiss. et Heldr.) Rechinger f., Fl. Aegaea, 641 (1943); Lindberg f., Iter Cypr., 35 (1946); Osorio-Tafall et Seraphim, List Vasc. Plants Cyprus, 105 (1973).

   E. viscosus DC. ssp. *glandulosus* (Weiss) Rechinger f., Fl. Aegaea, 641 (1943); Lindberg f., Iter Cypr., 35 (1946); Osorio-Tafall et Seraphim, List Vasc. Plants Cyprus, 105 (1973).

   [*E. spinosus* (non L.) Sm. in Sibth. et Sm., Fl. Graec. Prodr., 2: 209 (1813) pro parte, Fl. Graec. 10: 18, t. 924 (1840) pro parte; Poech, Enum. Plant. Ins. Cypr., 18 (1842); Gaudry, Recherches Sci. en Orient, 188 (1855); Unger et Kotschy, Die Insel Cypern, 242 (1865); Boiss., Fl. Orient., 3: 429 (1875), Suppl., 304 (1888); Holmboe, Veg. Cyr., 183 (1914).]
   TYPE: *"in Creta"*.

Robust erect perennial to 2 m. high; stems usually branched, strongly sulcate, tomentose, arachnoid-floccose or glabrescent, thinly or densely clothed with glandular or eglandular, multicellular, long or short, often purplish hairs, or occasionally very shortly glandular or subglabrous; leaves oblong in outline, 5–30 cm. long, 3–12 cm. wide, sessile, green and glandular-pubescent above, white-tomentose below, 2–3-pinnatisect, the ultimate divisions elongate-subulate, spinose, margins strongly recurved, nervation prominent below, glandular; leaf-bases dilated, amplexicaul and sub-auriculate, the auricles strongly spinose; inflorescence lax, branched; heads of capitula usually solitary, terminal, 2·5–5·5 cm. diam., subtending bracts small, concealed by head; basal pencil of bristles about as long as the lowermost phyllaries, whitish, the innermost bristles flattened, with a small, spathulate, ciliolate apex; phyllaries generally pale or bright metallic-blue, scarious, the outermost concave, narrowly rhomboid-subspathulate, about 7–9 mm. long, 2–3 mm. wide, glabrous, with a brown dorsal stripe and a sharply acuminate, brownish apex, the middle linear, bluntly carinate, about 2 cm. long, 0·2–0·3 cm. wide, spinose, some, especially of capitula near top of head, produced into a conspicuous elongate spine, frequently 3–4 cm. long, innermost phyllaries translucent, linear-lanceolate, acuminate, about 12–15 mm. long, 1–2 mm. wide, connate basally into a tube 4–5 mm. long; corolla blue (or occasionally white), tube slender, 6–9 mm. long, 0·8 mm. wide, shortly glandular externally, apex abruptly dilated, to 2 mm. wide, lobes 5, linear, 6–9 mm. long, 0·8 mm. wide, apex subacute, recurved, base with a shallow dorsal concavity, generally with a small apical hair-tuft; anthers linear, about 7 mm. long, 0·6 mm. wide, base sharply sagittate-caudate, apex acute; style 15–16 mm. long, 0·4 mm. wide, apex with a swollen, pilose annulus, branches erect, connivent, dorsally compressed, obtuse, about 3 mm. long. Achenes oblong-fusiform, about 4 mm. long, 1·5 mm. wide, brown, densely pilose externally with subadpressed, fulvous, shining bristles; pappus a small corona, about 1 mm. long, of fused, whitish or brown, strigillose bristles. *Plate 61.*

HAB.: Roadsides, waste ground, or in garigue on rocky hillsides; sea-level to 3,800 ft. alt.; fl. July–Oct.

DISTR.: Divisions 1, 2, 5–7, locally abundant. Eastern Mediterranean region from Sicily to Palestine, and eastwards to Iran.

1. Between Kouklia and Yeroskipos, 1862, *Kotschy*; above Khrysokhou, 1862, *Kotschy*; Ayios Neophytos Monastery, 1939, *Lindberg f.* s.n. ! common about Paphos, 1981, *Meikle* !
2. Phiti, 1905, *Holmboe* 1048; near Platres, 1930, *E. Wall* s.n., also, 1937, *Kennedy* 962 ! Between Ambelikou and Kambos, 1939, *Lindberg f.* s.n.; Makheras Monastery 1940, *Davis* 1894 !
5. Pedieos R. basin, 1935, *Syngrassides* 664 !
6. Kykko metokhi, Nicosia, 1966, *Merton* in ARI 27 !
7. Pentadaktylos, 1880, *Sintenis & Rigo* 790 ! Kyrenia, 1955, *G. E. Atherton* 90 !

NOTES: Probably much more common than the records indicate, but prickly and late-flowering, so that collectors tend to ignore it.

The Cyprus material is not at all uniform, and while much of it probably falls within the range of variation of *E. spinosissimus* Turra ssp. *bithynicus* (Boiss.) Kožuharov, the residue falls somewhere between this and typical *spinosissimus*. Davis (Notes Roy. Bot. Gard. Edinb., 21: 131; 1953) comments, "Although it [the Cyprus plant] approaches that Egyptian species [*E. spinosus* L.] in leaf-shape, the habit . . . . . is essentially that of *E. viscosus* [DC. non Schrader ex Reichb.]. It is either a new subspecies of the latter or a closely related new species". In view of the taxonomic uncertainties posed by the *E. spinosissimus* complex, and the paucity of herbarium material, I think it best to avoid further additions to, or subdivision of, the aggregate species.

A pathological condition, whereby the inflorescence-heads become very numerous and crowded, is not infrequent in Cyprus populations of the species.

30. CARDOPATIUM *A. L. Juss.*

Ann. Mus. Hist. Nat. Paris, 6: 324 (1805).

Spinose perennials; root thick, fleshy; stems erect, short, much divided near apex, forming a bushy, corymbose inflorescence; leaves pinnatisect with spinose lobes; capitula homogamous, discoid, few-flowered, sessile, crowded into spinose terminal glomerules; involucre ovoid-oblong, phyllaries imbricate in several series, the outer foliaceous, pectinate-spinose; florets tubular, apex deeply 5-lobed; anthers caudate-sagittate at base, with fimbriate tails; style-branches erect, blunt, with a pilose annulus; receptacle small, with fimbriate scales. Achenes ovoid or fusiform, sericeous-pilose; pappus persistent, of 1–2 rows of fimbriate, acuminate scales.

Four species in the Mediterranean region and eastwards to Central Asia.

1. **C. corymbosum** (*L.*) *Pers.*, Syn. Plant., 2: 500 (1807) ["*Cardopatum*"]; Boiss., Fl. Orient., 3: 442 (1875); Post, Fl. Pal., ed. 2, 2: 77 (1933); Lindberg f., Iter Cypr., 33 (1946); Osorio-Tafall et Seraphim, List Vasc. Plants Cyprus, 105 (1973); Davis, Fl. Turkey, 5: 597 (1975); Feinbrun, Fl. Palaest., 3: 366, t. 614 (1978).
   Echinops corymbosus L., Sp. Plant., ed. 1, 815 (1753).
   Carthamus corymbosus (L.) L., Sp. Plant., ed. 2, 2: 164 (1763).
   Brotera corymbosa Willd., Sp. Plant., 3: 2399 (1803); Holmboe, Veg. Cypr., 183 (1914) — as "*Broteroa*".
   Cardopatium orientale Spach in Ann. Sci. Nat., ser. 3, 5: 237 (1846); Gaudry, Recherches Sci. en Orient, 187 (1855); Unger et Kotschy, Die Insel Cypern, 243 (1865).
   TYPE: "*in* Apulia, Hellesponto, Lemno, Tracia, *in campis*".

Low-growing perennial, usually less than 35 cm. high, with a thick tap root; rootstock clothed with the fibrous remains of decayed petioles; basal leaves rosulate, oblong-obovate or oblanceolate in outline, 7–20 cm. long, 3·5–8 cm. wide, pinnatisect, the divisions flat, green, subglabrous, coarsely and irregularly spinose-lobed; petiole short, dilated, often purplish; stems

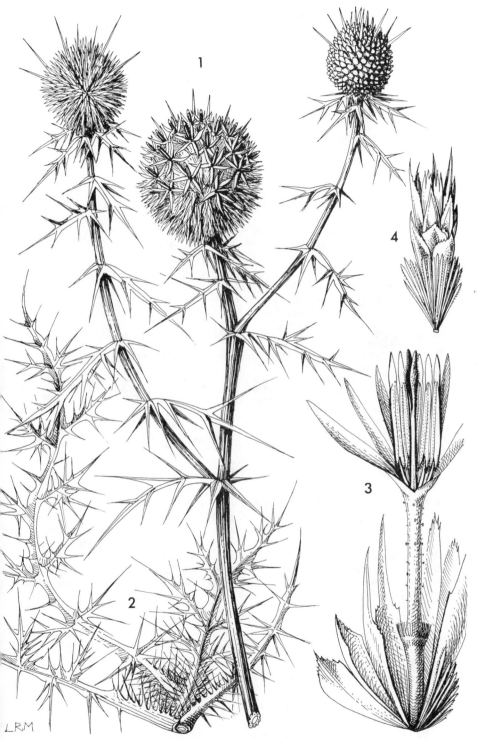

Plate 61. ECHINOPS SPINOSISSIMUS Turra **1,** flowering stem, × 1; **2,** leaves, × 1; **3,** capitulum, × 6; **4,** capitulum, showing "pencil" scales, × 3. (**1, 2** from *Sintenis & Rigo* 790 & *Kennedy* 962; **3, 4** from *Kennedy* 962.)

LRM

short, ridged, spinose, glabrous or subglabrous; cauline leaves few, similar to basal, but usually smaller and subsessile; inflorescence much branched, forming a low dome often more than 30 cm. diam.; capitula very numerous, crowded, subtended by spinose-pinnatisect, foliaceous bracts, 1·5–3 cm. long, 0·5–1 cm. wide; involucre narrowly ovoid, about 1 cm. long, 0·5 cm. wide; phyllaries erect, loosely imbricate, 6–12 mm. long, 2–3·5 mm. wide, the outer tough, herbaceous, pectinate-spinose, the inner entire or sub-entire, scarious with an acuminate-spinose apex, all subglabrous or thinly arachnoid-pilose; florets bright cobalt-blue, usually about 6 in each capitulum; corolla-tube 4–5 mm. long, 0·8–1 mm. wide, glabrous, sulcate, abruptly dilated to 1·5 mm. wide at apex, lobes 5, linear, obtuse, about 5–6 mm. long, 1·5 mm. wide at base; anthers about 6 mm. long, 0·5 mm. wide, apex acute, base sagittate-caudate, the tails fimbriate; style stout, about 7 mm. long, branches blunt, erect-connivent, about 0·7 mm. long with a basal, pilose annulus. Achenes fusiform, 4–5 mm. long, 1·5–2 mm. wide, densely fulvous-sericeous; pappus of 10 or more, erect, subulate, lacerate, membranous scales about 3·5 mm. wide at base.

HAB.: Pastures, fallow fields, waste ground; in garigue on rocky slopes; rocky or sandy seashores; sea-level to 1,000 ft. alt.; fl. May–August.

DISTR.: Divisions 1, 3–8. S. Italy and eastwards to Palestine.

1. Near Yeroskipos, 1862, *Kotschy*; Paphos, 1905, *Holmboe* 726; abundant about Kato Paphos, 1981, *Meikle*!
3. Kolossi, 1862, *Kotschy*; Limassol, 1939, *Lindberg f.* s.n.; also, 1941, *Davis* 3543! and, 1963, *J. B. Suart* 118!
4. Larnaca, 1939, *Lindberg f.* s.n.; Kouklia reservoir, 1968, *Economides* in ARI 1255!
5. Athalassa, 1939, *Lindberg f.* s.n.
6. Nicosia, 1929, *C. B. Ussher* 4! Between Yerolakkos and Myrtou, 1932, *Syngrassides* 194!
7. Mountains above Kythrea, 1880, *Sintenis & Rigo* 546! Kyrenia, 1938, *C. H. Wyatt* 37! also, 1939, *Lindberg f.* s.n., and, 1949, *Casey* 807! 1955, *G. E. Atherton* 198!
8. Near Leonarisso, 1880, *Sintenis & Rigo*.

31. XERANTHEMUM *L.*

Sp. Plant., ed. 1, 857 (1753).
Gen. Plant., ed. 5, 369 (1754).

Erect, annual, tomentose herbs; leaves small, narrow, entire, sessile; capitula solitary, terminal, pedunculate, heterogamous, disciform; involucre ovoid or fusiform; phyllaries in several series, scarious, imbricate, the innermost often much enlarged and brightly coloured; receptacle flat, scaly; marginal florets sterile, corolla unequally 2-lipped; disk-florets actinomorphic, hermaphrodite, tubular, apex shortly 5-lobed; anthers sagittate at base; style with 2 very short branches. Achenes fusiform, adpressed-pilose, hilum basal; pappus of 5 or 10–15, aristate-acuminate scales, in 1 series.

1. **X. inapertum** (*L.*) *Mill.*, Gard. Dict., ed. 8, no. 2 (1768); Unger et Kotschy, Die Insel Cypern, 243 (1865); Boiss., Fl. Orient., 3: 445 (1875), Suppl., 304 (1888); H. Stuart Thompson in Journ. Bot, 44: 333 (1906); Holmboe, Veg. Cypr., 184 (1914); Post, Fl. Pal., ed. 2, 2: 78 (1933); Osorio-Tafall et Seraphim, List Vasc. Plants Cyprus, 106, (1973); Davis, Fl. Turkey, 5: 605 (1975).
    X. annuum L. var. *inapertum* L., Sp. Plant., ed. 1, 858 (1753).
    [*X. squarrosum* (non Boiss.) H. Stuart Thompson in Journ. Bot., 44: 333 (1906).]
    [*X. annuum* L. ssp. *annettae* (non (Kalen.) Holmboe) Holmboe, Veg. Cypr., 184 (1914) quoad spec. cypr.; Osorio-Tafall et Seraphim, List Vasc. Plants Cyprus, 106 (1973).]
    TYPE: "*in* Italia".

Erect grey-tomentose annual 5–30 cm. high; stems branched or unbranched, distinctly angled; basal leaves commonly forming a rosette or tuft, oblanceolate, 1·5–7 cm. long, 0·5–1·3 cm. wide, entire, apex acute or apiculate, base gradually tapering; cauline leaves lanceolate, oblanceolate, oblong or narrowly obovate, decreasing in size upwards, 0·5–4 cm. long, 0·2–1 cm. wide, often distinctly apiculate; capitula terminal on leafless stems or peduncles up to 12 cm. long; involucre ovoid or subglobose, 1–1·5 cm. long, 0·6–1·4 cm. wide; phyllaries in 5 series, loosely imbricate, scarious, the outermost ovate-apiculate, 3–4 mm. long, 2–3 mm. wide, glabrous, stramineous, inner progressively longer, narrower and more acuminate, the innermost narrowly oblong or oblong-obovate, 12–15 mm. long, 3–4 mm. wide, erect, stained pink, lilac or purple dorsally; marginal florets purplish, narrowly tubular, tube about 3 mm. long, less than 0·5 mm. wide, apical lobes erect, very unequal, 0·5–1·5 mm. long; style exserted, about 9–10 mm. long, apex clavate, base imbedded in a yellow, fleshy (nectariferous ?) collar, branches short, erect, connivent; disk-florets tubular, 3–4 mm. long, 0·5 mm. wide, apical lobes 5, narrowly deltoid, 0·5–0·6 mm. long, 0·2 mm. wide at base, erect; anthers linear, 2 mm. long, 0·2 mm. wide, apex acute, base sharply sagittate; style 3·5 mm. long, branches divergent, narrowly oblong, truncate, about 0·4 mm. long; receptacular scales linear-subulate, about 6–9 mm. long, 1–2 mm. wide at base. Achenes fusiform, 3·5–4 mm. long, 1–1·3 mm. wide at apex, dark brown, rather densely sericeous; pappus of 5 stramineous, erect, subulate-filiform, strigillose scales, 4–4·5 mm. long, 0·5 mm. wide at the dilated base.

HAB.: Rocky fields and hillsides on calcareous or igneous formations; 600–5,000 ft. alt.; fl. Febr.–May.

DISTR.: Divisions 2, 3, 5–7. Southern Europe and Mediterranean region east to the Caucasus.

2. Prodhromos, 1862, *Kotschy* 810a; Makheras, 1952, *F. M. Probyn* 103! Perapedhi, 1960, *N. Macdonald* 147! Between Mandria and Perapedhi, 1962, *Meikle* 2866!
3. Four miles N. of Kandou, 1963, *J. B. Suart* 64!
5. Athalassa, 1936, *Syngrassides* 1329!
6. Mile 30, Nicosia-Troödos road, 1955, *Merton* 1989!
7. Pentadaktylos 1880, *Sintenis & Rigo* 547! Lapithos Pass, 1902, *A. G. & M. E. Lascelles* s.n.! Buffavento, 1941, *Davis* 2182! 2802! Yaïla below Halevga, 1941, *Davis* 2832! Kyrenia Range, 2,000 ft. alt., 1948, *Kennedy* 1660!

32. CARLINA *L*.
Sp. Plant., ed. 1, 828 (1753).
Gen. Plant., ed. 5, 360 (1754).
H. Meusel et A. Kästner in Feddes Repert., 83: 213–232 (1972).

Annual, biennial or perennial, spinose herbs or subshrubs; stems sometimes much reduced or apparently wanting; leaves toothed, lobed or pinnatisect; capitula homogamous, discoid; phyllaries polymorphic, the outermost foliaceous, pinnatifid or pinnatisect, spinose, the median erect, herbaceous, linear or narrowly oblong, the innermost much enlarged and exceeding the florets, scarious, variously coloured, radiating (and simulating ligulate ray-florets) when dry; receptacle flat, faveolate, each floret surrounded by partly fused, laciniate, scarious scales; florets tubular, yellow, apex 5-lobed; anthers caudate-sagittate at base; style-branches short. Achenes terete or subterete, truncate, sericeous; hilum basal; pappus of 1–2 rows of plumose-laciniate scales.

About 30 species in Europe, the Mediterranean region and temperate Asia.

Innermost series of phyllaries (resembling ligulate ray-florets) golden-yellow; plant perennial, robust, very spiny, usually more than 15 cm. high - - **1. C. involucrata** ssp. **cyprica**
Innermost series of phyllaries (resembling ligulate ray-florets) pink or purple; plants generally less than 15 cm. high, not very spiny:

Plants perennial with a thick, fusiform taproot; florets and receptacular scales generally
purple towards apex -   -   -   -   -   -   -   -   -   -   **2. C. pygmaea**
Plants annual with a slender taproot; florets and receptacular scales yellowish or pallid
3. C. lanata

1. **C. involucrata** *Poir.*, Voy. en Barbarie, 2: 234 (1789); Post, Fl. Pal., ed. 2, 2: 80 (1933); Meusel
et Kästner in Feddes Repert., 83: 229 (1972).
    C. corymbosa L. var. *involucrata* (Poir.) Boiss., Fl. Orient., 3: 449 (1875).
    C. corymbosa L. ssp. *involucrata* (Poir.) Holmboe, Veg. Cypr., 185 (1914); Osorio-Tafall et
Seraphim, List Vasc. Plants Cyprus, 106 (1973) quoad nomen.

Erect perennial (7–) 15–50 (–80) cm. high; rootstock woody, often
branched; stem stout, glabrous, smooth, simple or sparingly branched; basal
and cauline leaves mostly shed by, or soon after, anthesis, oblong-lanceolate
in outline, 5–20 cm. long, 2–7 cm. wide, deeply pinnatisect with narrow,
spinose divisions, glabrous or thinly arachnoid-pilose, venation often
distinct, closely reticulate; capitula solitary or few, forming an open,
branched inflorescence; uppermost cauline leaves forming a conspicuous,
spiny ruff (4–) 10–15 cm. diam. below each capitulum; outermost phyllaries
foliaceous, spinose, similar to the uppermost leaves, but smaller and
narrower, commonly arachnoid-pilose externally, with an indurated,
gibbous base, inner phyllaries about 2 cm. long, with an expanded,
pectinate-spinose apex about 0·6 cm. wide, the innermost linear-oblong,
acuminate, stramineous, golden-yellow, about 1·5–2 cm. long, 0·2–0·4 cm.
wide, forming what appears to be a row of ligules around a central, compact,
somewhat convex, yellow disk, 2–4 cm. diam.; corolla narrowly tubular,
about 8 mm. long, 1·2 mm. wide, lobes 5, erect, narrowly lanceolate-
acuminate, about 3 mm. long, 0·8 mm. wide; anthers linear, about 7·5 mm.
long, 0·5 mm. wide, tapering to apex, base caudate-sagittate, the tails about
2 mm. long; style stout, exserted, about 11–12 mm. long, branches erect,
connivent, about 1 mm. long. Achenes narrowly fusiform, about 3 mm. long,
1 mm. wide at apex, densely adpressed-sericeous; pappus of about 10
principal rays, pinnately subdivided into numerous, filiform, sericeous-
plumose branches.

ssp. **cyprica** *Meusel et Kästner* in Feddes Repert., 88: 402 (1977).
    [*C. vulgaris*? (non L.) Kotschy in Unger et Kotschy, Die Insel Cypern, 243 (1865).]
    [*C. curetum* (non Hal.) H. Stuart Thompson in Journ. Bot., 44: 333 (1906).]
    [*C. graeca* (non Heldr. et Sart.) Lindberg f., Iter Cypr. 34 (1946).]
    [*C. corymbosa* L. ssp. *graeca* (non (Heldr. et Sart.) Rechinger f.) Osorio-Tafall et
Seraphim, List Vasc. Plants Cyprus, 106 (1973).]
    [*C. corymbosa* L. ssp. *graeca* (Heldr. et Sart.) Rechinger f. × *C. pygmaea* (Post) Holmboe
teste Burtt et Davis in Kew Bull., 1949: 104 (1949).]
    TYPE: Cyprus; Mia Milea, 500 ft. alt.; 1 July, 1954, *Merton* 1913 (K !).

Middle and upper leaves broader than in type; cauline leaves and outer
phyllaries closely and acutely denticulate right up to the base of the spines;
capitula usually few, 1–3, large, (2–4 cm. diam.)

HAB.: Roadsides, waste ground, rough pastures; sometimes in garigue on dry mountainsides;
sea-level to 5,000 ft. alt.; fl. July–Oct.

DISTR.: Divisions 1–7. Endemic.

1. Between Kouklia and Yeroskipos, 1862, *Kotschy*. Abundant about Paphos, 1981, *Meikle* !
2. Between Pano Platres and Perapedhi, 1900, *A. G. & M. E. Lascelles* s.n. ! Vasa near
Omodhos, 1905, *Holmboe* 1156; Platres, 1937, *Kennedy* 963 ! Palekhori, 1940, *Davis* 1988 !
Perapedhi, 1970, *A. Genneou* in ARI 1583 ! Prodhromos, 1955, *G. E. Atherton* 537 !
3. Ephtagonia, 1956, *Merton* 2812a partly !
4. Ayia Napa, 1939, *Lindberg f.* s.n.; Famagusta, 1939, *Lindberg f.* s.n.
5. Kythrea, 1940, *Davis* 1931 ! 1953 ! 1954 ! 1961 ! Mia Milea, 1954, *Merton* 1913 ! Limnia, 1956,
Poore 11 !
6. Nicosia, 1939, *C. B. Ussher* 3 !
7. Kyrenia, 1955, *G. E. Atherton* 129 ! 372 !

NOTES: Much commoner than records indicate, but flowering late, and too prickly for most
collectors.

2. C. pygmaea (*Post*) Holmboe, Veg. Cypr., 184 (1914); Osorio-Tafall et Seraphim, List Vasc. Plants Cyprus, 106 (1973).

C. lanata L. var. *pygmaea* Post in Mém. Herb. Boiss., 18: 95 (1900); H. Stuart Thompson in Journ. Bot., 44: 333 (1906).

TYPE: "in montosis Cypri; floret augusto et septembro. [Post] No. 909" (BEI).

Erect, dwarf perennial 2–10 cm. high (or, under exceptional circumstances, up to 30 cm. high); taproot thick, fleshy, fusiform; rootstock sometimes branched; stems arachnoid-tomentose; leaves crowded, persistent, the lower oblanceolate in outline, 3–8 cm. long, 0·8–2·5 cm. wide, thinly arachnoid-tomentellous, apex blunt or shortly acute, margins pinnatifid, with short, blunt or acute, spinulose-denticulate lobes, upper leaves similar, but becoming progressively smaller; capitula solitary or 2–3 together in a crowded cluster; involucre 1·3–2 (–3) cm. wide; outer phyllaries foliaceous, oblanceolate, pinnatisect, with spinose-acuminate lobes, spinulose-denticulate margins and a shortly stipitate base, inner phyllaries smaller, densely arachnoid-tomentose, spinose-acuminate, the innermost series linear, acute, 15–25 mm. long, 1·5–2 mm. wide, scarious, shining pink or vinous-purple (rarely white), spreading when dry and resembling ligulate ray-florets; receptacular bristles linear-acuminate, 15–20 mm. long, 1–1·5 mm. wide, purple at apex, exceeding the pappus-hairs and simulating florets; florets narrowly tubular, purple, about 10 mm. long, 1 mm. wide at apex, longitudinally 5-costate, lobes erect, narrowly deltoid, about 1·5 mm. long, 0·5 mm. wide at base; anthers linear, acuminate, 6–6·5 mm. long, 0·4 mm. wide, base caudate-sagittate, with fimbriate tails about 0·8 mm. long; style purple, about 10 mm. long, distinctly thickened in the upper half, branches erect, connivent, about 0·8 mm. long. Achenes narrowly fusiform, 3·5–4 mm. long, less than 1 mm. wide at apex, densely brownish adpressed-sericeous; pappus-hairs of 10 principal rays, pinnately subdivided into numerous, filiform, sericeous-plumose, pallid branches.

HAB.: Dry, bare pastures and hillsides on sandstone or igneous rocks; sea-level to 3,500 ft. alt.; fl. June–Oct.

DISTR.: Divisions 1–6. Endemic.

1. Akamas, on serpentine [probably near Smyies], 1941, *Davis* 3313! Smyies, 1948, *Kennedy* 1626! 1627!
2. Above Elias Bridge, 1901, *A. G. & M. E. Lascelles* s.n.! Between Mesapotamos and Kato Amiandos, 1940, *Batten-Poole* in herb. *Davis* s.n.! Roudhkias valley between Pano Panayia and Kykko Monastery, 1941, *Davis* 3389! Troödos area, 1950, *F. M. Probyn* s.n.!
3. Akrotiri Bay, 1939, *Mavromoustakis* s.n.! Near Ephtagonia, 1956, *Merton* 2812! 2812a! 2813! (see Notes).
4. Dhekelia, Gibraltar Line, 1970, *A. Genneou* in ARI 1585!
5. Koutsovendis, 1936, *Lady Loch* 3! Between Vouno and Mia Milea 1937, *Syngrassides* 1733! Kythrea, 1940, *Davis* 1952! 1960! also, 1941, *Davis* 2159! 2797! Vouno, 1941, *Davis* 2804!
6. Between Nicosia and Aghirda, 1905, *Holmboe* 1165!

NOTES: Specimens (*Merton* 2812, 2812a) collected near Ephtagonia, and thought by the collector to be hybrids between *Carlina involucrata* Poir. ssp. *cyprica* Meusel et Kästner ("*C. corymbosa*" auct.) and *C. pygmaea* (Post) Holmboe, appear to be a mixture of these two species, but not hybrids; *Merton* 2812 has been re-determined as *C. pygmaea* by Meusel & Kästner. Specimens (from Kythrea) cited as the same hybrid by Burtt & Davis (Kew Bull., 1949: 104), have been referred to *C. involucrata* ssp. *cyprica* by Meusel & Kästner.

3. C. lanata L., Sp. Plant., ed. 1, 828 (1753); Boiss., Fl. Orient., 3: 451 (1875); Holmboe, Veg. Cypr., 184 (1914); Post, Fl. Pal., ed. 2, 2: 81 (1933); Lindberg f., Iter Cypr., 34 (1946); Osorio-Tafall et Seraphim, List Vasc. Plants Cyprus, 106 (1973); Davis, Fl. Turkey, 5: 600 (1975); Feinbrun, Fl. Palaest., 3: 369, t. 620 (1978).

TYPE: "*in* Italia, G. Narbonensi".

Erect annual 5–15 cm. high, with a slender taproot; leaves oblong-lanceolate, 1·5–8 cm. long, 0·5–1·5 cm. wide, thinly arachnoid-tomentellous, apex acute-spinose, margins obscurely spinose-pinnatifid, with scattered, slender spinules; venation often conspicuously raised and reticulate;

capitula generally solitary, 2–3·5 cm. wide (including scarious phyllaries); outer phyllaries foliaceous, arachnoid-lanuginose, lanceolate, spinose, 1–4 cm. long, 0·4–1 cm. wide, inner phyllaries lanceolate, entire, spinose-acuminate, 0·8–2 cm. long, 0·2–0·8 cm. wide, innermost series linear, acute, scarious, shining pink or purplish, 1·3–2 cm. long, 0·1–0·3 cm. wide; receptacular bristles filiform, minutely papillose, about 8–10 mm. long, considerably exceeding pappus-hairs; florets narrow-tubular, yellowish, 6–8 mm. long, 0·8 mm. wide at apex, lobes narrowly deltoid, about 1·5 mm. long, 0·4 mm. wide at base, erect, connivent; anthers linear, acuminate, 3·5 mm. long, 0·3 mm. wide, base caudate-sagittate, with fimbriate tails 0·8–1 mm. long; style 8–9 mm. long, branches blunt, erect, connivent. Achenes fusiform-cylindrical, 2·5–3 mm. long, 0·8 mm. wide at apex, densely adpressed-sericeous, dull brown; pappus of 10 principal rays, sub-palmately divided into 2–3, pallid, filiform, sericeous-plumose branches.

HAB.: On dry stony ground amongst *Cistus* and *Sarcopoterium*; sea-level to 1,000 ft. alt.; fl. June–July.

DISTR.: Divisions 1, 3, 4, 7, 8. S. Europe and Mediterranean region east to Iraq and Iran.

1. Smyies, 1948, *Kennedy* 1625!
3. Ayios Theodhoros (Limassol), 1905, *Holmboe* 1069; near Limassol, 1939, *Lindberg f.* s.n.
4. Near Larnaca, 1939, *Lindberg f.* s.n.!
7. Kyrenia, 1948, *Casey* 138! and, 1948, *Kennedy* 1589!
8. Boghaz, 1937, *Syngrassides* 1646! Ayios Theodhoros, 1939, *Lindberg f.* s.n.

33. ATRACTYLIS *L.*

Sp. Plant., ed. 1, 830 (1753).
Gen. Plant., ed. 5, 360 (1754).

Annual or perennial herbs; leaves dentate or pinnatisect, generally spinose; capitula mostly homogamous and discoid, rarely heterogamous and radiate; involucre ovoid or campanulate; phyllaries in several series, dimorphic, the outer herbaceous, often pectinate-spinose, loosely suberect, the inner obovate or linear-oblanceolate, acuminate, erect, imbricate, often scarious; receptacle flattish, with rigid receptacular scales split into long bristles; florets tubular, hermaphrodite, 5-lobed at apex, or the marginal female and ligulate; anthers caudate-sagittate at base, with fimbriate tails; style with very short, erect branches. Achenes terete or subterete, sericeous; hilum basal; pappus in 1–3 series, divided above into slender, plumose bristles.

About 20 species in Europe, the Mediterranean region and Asia.

1. **A. cancellata** *L.*, Sp. Plant., ed. 1, 830 (1753); Unger et Kotschy, Die Insel Cypern, 243 (1865); Boiss., Fl. Orient., 3: 453 (1875); Holmboe, Veg. Cypr., 185 (1914); Post, Fl. Pal., ed. 2, 2: 82 (1933); Lindberg f., Iter Cypr., 34 (1946); Osorio-Tafall et Seraphim, List Vasc. Plants Cyprus, 106 (1973); Davis, Fl. Turkey, 5: 604 (1975); Feinbrun, Fl. Palaest., 3: 372, t. 623 (1978).
    Acarna cancellata (L.) Willd., Sp. Plant., 3: 1701 (1803); Sibth. et Sm., Fl. Graec. Prodr., 2: 159 (1813); Fl. Graec., 9: 29, t. 839 (1837); Poech, Enum. Plant. Ins. Cypr., 18 (1842).
    TYPE: "*in* Hispaniae, Siciliae, Cretae *agris*."

Erect or spreading annual 2–25 cm. high; stems slender, branched or unbranched, glabrous or thinly arachnoid-villose, subterete or obscurely angled, often purple; basal leaves in a rosette, oblong-oblanceolate, 10–50 mm. long, 3–15 mm. wide, thinly arachnoid-pubescent, apex acute or obtuse, margins spinulose, subentire or slightly lobed; cauline leaves similar, but usually narrower, remote, tapering to a semi-amplexicaul base; inflorescence generally branched except in starved specimens, capitula solitary, terminal; involucre ovoid-campanulate, 1–1·5 cm. long and almost as wide; outer phyllaries loosely erect, linear, 10–15 mm. long, less than 1 mm. wide, conspicuously pectinate-spinulose, forming as it were a cage around the capitulum, inner phyllaries closely adpressed-imbricate, the outermost ovate, greenish, about 3 mm. long, 2–3 mm. wide, acute or apiculate, the inner progressively longer and narrower, the innermost about 10–15 mm. long, 2–3 mm. wide, linear-oblong, with an acuminate-cuspidate, ciliate, scarious, often purplish apex; florets all narrowly tubular, hermaphrodite, about 7 mm. long, 0·5 mm. wide at apex, purple, lobes lanceolate, acuminate, about 1·3 mm. long, 0·3 mm. wide at base; anthers linear, about 3·5 mm. long, 0·2 mm. wide, apex acuminate, base caudate-sagittate, with fimbriate tails about 1·5 mm. long; style about 6·5 mm. long, branches erect, connivent, about 0·3 mm. long. Achenes oblong-fusiform, about 3 mm. long, 1·3 mm. wide at apex, dark brown, densely covered with long, silky, transversely undulate, whitish or pale brown hairs; pappus in one series of 20 plumose, pallid bristles 8–9 mm. long, shortly connate at base.

HAB.: Dry igneous or calcareous slopes; roadsides; field margins; sea-level to 2,500 ft. alt.; fl. March–June.

DISTR.: Divisions 1–3, 5–8. Widespread in the Mediterranean region and east to Iran.

1. Near Paphos, 1840, *Kotschy* 62; Lyso, 1913, *Haradjian* 827!
2. Perapedhi, 1938, *Kennedy* 1451! Soli, 1963, *Townsend* 63/67!
3. Amathus, 1862, *Kotschy* 590; and 1978, *Holub* s.n·.! Kophinou, 1905, *Holmboe* 585; Lefkara, 1940, *Davis* 1914! Between Perapedhi and Trimiklini, 1941 *Davis* 3447! Polemidhia hills, 1947, *Mavromoustakis* s.n.! Evdhimou, 1950, *F. J. F. Barrington* 54! Near Mamonia, 1960, *N. Macdonald* 156!
5. Between Nisou and Stavrovouni, 1933, *Syngrassides* 20! Near Pyroï, 1934, *Syngrassides* 464! Kythrea, 1941, *Davis* 2952! Lakovounara Forest, 1950, *Chapman* 81! 472! 655! Vizakia, 1936, *Syngrassides* 1017!
6. Nicosia, 1972, *W. R. Price* 1041! also, 1973, *P. Laukkonen* 254!
7. Near Melandrina, 1862, *Kotschy* 517; near Panteleimon Monastery, 1862, *Kotschy* 939! Mountains above Kythrea, 1880, *Sintenis & Rigo* 299! Kyrenia Pass, 1939, *Lindberg f.* s.n.! Kyrenia, 1948, *Casey* 779! and, 1958, *P. H. Oswald* 24!
8. Near Apostolos Andreas Monastery, 1880, *Sintenis & Rigo*; near Rizokarpaso, 1912, *Haradjian* 138!

34. ARCTIUM *L.*

Sp. Plant., ed. 1, 816 (1753).
Gen. Plant., ed. 5, 357 (1754).
J. Arènes in Bull. Jard. Bot. Brux., 20: 67–156 (1950).

Erect, robust biennials; stems sulcate, often purplish, subglabrous or arachnoid-pilose; leaves large, papery, cordate, entire or denticulate; petioles usually well developed, sometimes fistulose, inflorescences branched, racemose or corymbose; capitula sessile or pedunculate, subglobose, homogamous, discoid; phyllaries in several series, imbricate, produced into a slender, subulate, generally uncinate appendage; receptacle flat, densely setose; florets purple (or rarely white), tubular, 5-lobed at apex; anthers sagittate at base, the tails ciliate or fimbriate; style usually exserted, branches linear, often shortly pilose at base. Achenes oblong, truncate, somewhat compressed; pappus-hairs short, scabridulous, deciduous.

About 4–5 species widely distributed in Europe, the Mediterranean region and temperate Asia, or as introduced weeds elsewhere.

The whole capitulum, at maturity, becomes attached by the hooked phyllaries to the fur or fleece of animals, the achenes being shed subsequently as the capitulum disintegrates.

1. **A. lappa** *L.*, Sp. Plant., ed. 1, 816 (1753) emend. Mill., Gard. Dict., ed. 8, no. 1 (1768); Arènes in Bull. Jard. Bot. Brux., 20: 74 (1950); Tutin et al., Fl. Europaea, 4: 215 (1976).
Lappa vulgaris Hill, Veg. Syst., 4: 28, t. 25, fig. 1 (1762).

L. major Gaertn., De Fruct., 2: 379, t. 162, fig. 3 (1791); Boiss., Fl. Orient., 3: 457 (1875); H. Stuart Thompson in Journ. Bot., 44: 333 (1906); Holmboe, Veg. Cypr., 185 (1914).
Arctium majus (Gaertn.) Bernh., Syst. Verz. Pflanz. Erfurt, 154 (1800).
A. vulgare (Hill) Druce in Ann. Scott. Nat. Hist., no. 60: 222 (1906); Osorio-Tafall et Seraphim, List Vasc. Plants Cyprus, 106 (1973).
TYPE: "*in Europae cultis ruderatis*".

Erect, branching biennial 60–150 cm. high; stems subglabrous or thinly arachnoid, sulcate, often purplish; leaves very large, bluntly cordate, 16–40 cm. or more long, 13–30 cm. or more wide, green and sparsely pubescent above, grey-tomentellous below, apex usually rounded, margins undulate-sinuate, the obscure lobes sometimes apiculate, basal sinus usually rather wide; nervation conspicuous; petiole thick, canaliculate above, arachnoid, solid, 15–30 cm. or more long; inflorescence a lax or dense corymb, the capitula terminal on peduncles 4–10 cm. long; involucre subglobose, 2·5–3 cm. diam.; phyllaries in many series, closely imbricate at the base, with a spreading, slender, subulate appendage, usually more than 1 cm. long and always hooked at the apex, in our area glabrous or thinly ciliolate, sometimes arachnoid elsewhere; florets purple, only slightly exceeding involucre; corolla narrowly tubular-infundibuliform, 14–15 mm. long, about 1·3 mm. wide at apex, glabrous, lobes linear-deltoid, about 2·5 mm. long, 0·5 mm. wide at base, erect, slightly concave; anthers linear, subacute, about 5 mm. long, 0·5 mm. wide, base caudate-sagittate, the tails more than 1 mm. long; style about as long as corolla-tube, branches linear, truncate, dorsiventrally compressed, about 1 mm. long. Achene compressed-fusiform, truncate, about 6 mm. long, 2·5 mm. wide, dull brown, obscurely rugulose; pappus soon deciduous, consisting of unequal, pallid, scabridulous bristles 0·5–2 mm. long.

HAB.: Rocky streamsides at high altitudes; 4,000–4,500 ft. alt.; fl. July–Oct.

DISTR.: Division 2, rare. Widespread in Europe and temperate Asia, but becoming rare in the eastern Mediterranean, though perhaps more widespread than records indicate.

2. Troödhitissa, 1901, *A. G. & M. E. Lascelles* s.n. ! Platres, rather common by Kryos Potamos, above village, 1957, *Merton* 3202 !

NOTES: F. K. Kupicha (in Davis, Fl. Turkey, 5: 356; 1975) discards the name *Arctium lappa* L. as a *nomen confusum*, since the apparent type material is referable to the plant subsequently named *A. tomentosum* by Miller, and not to the species described above. Such a break with the traditional identification of *A. lappa* L. would almost certainly necessitate rejection of the name under Art. 69 of the Code. But it seems much more likely that the material in herb. Linn. is referable to *Arctium lappa* L. β of Sp. Plant., ed. 1, 816 (1753) and not to the plant which Linnaeus regarded as typical *A. lappa*. This is also Miller's interpretation of the situation, and one which obviates a number of name changes. I do not agree with Dr. Kupicha that *A. lappa* L. (*A. majus* (Gaertn.) Bernh; *A. vulgare* (Hill) Druce) and *A. minus* (Hill) Bernh. are distinguishable only as subspecies, and again view Miller's analysis of the situation as the more satisfactory one.

35. CARDUUS *L.*

Sp. Plant., ed. 1, 820 (1753).
Gen. Plant., ed. 5, 358 (1754).
S. M. A. Kazmi in Mitt. Bot. Staatsamml. München, 5: 279–550 (1964).

Erect, annual, biennial or perennial herbs; stems commonly winged and spiny, leaves lobed or pinnatisect, the margins and apices of lobes spinose or spinulose; capitula homogamous, discoid, solitary and terminal or sometimes clustered; involucre ovoid or subglobose; phyllaries in several series, imbricate with a spine-tipped, erect, patent or reflexed, entire, apical appendage; receptacle flat or convex, densely setose; florets purple, pink or white; corolla tubular, equally or unequally lobed at apex; filaments free, pilose; anthers caudate-sagittate at base; style-branches short or rather long, obtuse. Achenes oblong-obovoid, compressed, glabrous, truncate or umbonate, hilum basal or sub-basal; pappus in several series, bristles simple, scabridulous or barbellate, connate basally.

About 120 species widely distributed in Europe, the Mediterranean region and Asia.

Capitula solitary on well-developed, often elongate peduncles; outer and median phyllaries narrowed rather abruptly above and terminating in a conspicuous, linear-subulate appendage - - - - - - - - - **1. C. argentatus** ssp. **acicularis**
Capitula in sessile or very shortly pedunculate clusters; phyllaries lanceolate-subulate, tapering from near base to a spine-tipped apex - - - - **2. C. pycnocephalus** ssp. **albidus**

[C. ACANTHOIDES *L.*, Sp. Plant., ed. 1, 821 (1753), an erect biennial 30–100 cm. high, stems prickly-winged and leafy almost to apex; peduncles short; capitula solitary or clustered, subglobose, and corollas purple. Recorded by Sibthorp "In Peloponneso; nec non in Cypro et Cretâ insulis" (Fl. Graec. Prodr., 2: 149; 1813), but has not been found since, and the record is almost certainly an error. *C. acanthoides* has a wide distribution in Europe, but is absent from areas adjacent to Cyprus; all subsequent records for its occurrence on the island (H. Stuart Thompson, Holmboe, etc.) are based on the statement in Fl. Graec. Prodr.]

1. C. argentatus *L.*, Mantissa 1, 280 (1771); Boiss., Fl. Orient., 3: 522 (1875) pro parte; Post, Fl. Pal., ed. 2, 2: 88 (1933); Davis, Fl. Turkey, 5: 432 (1975); Feinbrun, Fl. Palaest., 3: 376, t. 633, 634 (1978).

Erect annual, 20–60 cm. high; stems branched or unbranched, winged for the greater part of their length, with prickly, coarsely toothed or lobed wings, thinly arachnoid-tomentellous; leaves oblong in outline, 4–22 cm. long, 1–12 cm. wide, greenish and thinly arachnoid above, whitish-tomentose below, margins coarsely and irregularly lacerate-pinnatifid, with broad, spinose-denticulate or spinose-lobulate lobes; capitula terminal on slender, floccose-tomentose peduncles 4–30 cm. long, ovoid-oblong, 1–1·8 cm. long, 0·6–1·3 cm. wide; phyllaries closely imbricate in 5–6 series, indurated-scarious, narrowly lanceolate or lanceolate-subulate, 2–16 mm. long, 1·5–3 mm. wide at base, sparingly arachnoid-floccose externally, apex of outer and median phyllaries sometimes produced into an elongate, subulate, herbaceous, 3-nerved, strigillose, erect or somewhat spreading appendage, apex of innermost phyllaries acuminate with narrow, membranous margins; corolla pink-purple, tube slender, 8–9 mm. long, distinctly swollen, and about 1·2–1·5 mm. wide, just below apex, lobes linear, acute or subacute, erect or spreading, 4–5 mm. long, 0·5 mm. wide; filaments 1–1·5 mm. long, lanuginose; anthers linear, acute, about 6 mm. long, 0·4 mm. wide; style 13–15 mm. long, branches erect, connivent, obtuse, with a basal ring of hairs. Achene compressed, oblong, truncate, about 4–5 mm. long, 2 mm. wide, the apex produced centrally into a short, cylindrical stump-like anthophore; ventral and dorsal faces glabrous, shallowly 5–8-sulcate, margins shortly scabridulous; pappus tinged brownish, deciduous, bristles numerous, about 12 mm. long, scabridulous.

ssp. **acicularis** (*Bert.*) Meikle **stat. nov.**
    C. acicularis Bert. in Ann. Stor. Nat. Bologna, 1: 274 (1829); Kazmi in Mitt. Bot. Staatsamml. München, 5: 454 (1964); Osorio-Tafall et Seraphim, List. Vasc. Plants Cyprus, 106 (1973); Davis, Fl. Turkey 5: 434 (1975).
    [*C. argentatus* (non L.) Sm. in Sibth. et Sm., Fl. Graec. Prodr., 2: 149 (1813); Poech, Enum. Plant. Ins. Cypr., 19 (1842); Boiss., Fl. Orient., 3: 522 (1875) pro parte quoad plant. cypr.; Holmboe, Veg. Cypr., 185 (1914); Rechinger f. in Arkiv för Bot., ser. 2, 1: 433 (1950); Osorio-Tafall et Seraphim, List Vasc. Plants Cyprus, 106 (1973).]
    [*C. argenteus* Sintenis in Oesterr. Bot. Zeitschr., 31: 193, 326 (1881); 32: 262 (1882) nomen.]
    TYPE: Italy; "Nascitur in agro Ravennati", *Bertoloni* (BOLO, K !).

Outer and median phyllaries produced into a long, subulate, spinescent, erect or spreading, 3-nerved, subglabrous or thinly arachnoid-pubescent appendage, often equalling or exceeding the florets; peduncles commonly shorter than in *C. argentatus* L. ssp. *argentatus*.

HAB.: Cultivated and fallow fields; waste ground, roadsides; in garigue on stony hillsides; sea-level to 2,000 ft. alt.; fl. March–May.

DISTR.: Divisions 1–8. N. Mediterranean region, from S. France through Italy and Greece to Palestine.

1. Stroumbi near Paphos, 1913, *Haradjian* 748!
2. Paliambela, 1962, *Meikle* 2251! Near Omodhos, 1979, *Hewer* 4656!
3. Cherkez, 1950, *Mavromoustakis* 21! Near Limassol, 1950–51, *Mavromoustakis* 40! 53! N.E. of Apsiou, 1979, *Edmondson & McClintock* E 2702! Pissouri, 1981, *Hewer* 4760! 4762!
4. Larnaca, 1880, *Sintenis*.
5. Kythrea, 1880, *Sintenis & Rigo*.
6. Myrtou, 1932, *Syngrassides* 202! Near Dhiorios, 1962, *Meikle* 2354!
7. Pentadaktylos, 1880, *Sintenis & Rigo* 294 partly! Armenian Monastery, 1912, *Haradjian* 394! St. Hilarion, 1937, *Miss Godman* 17! Kyrenia Range, 1938, *C. H. Wyatt* 9! Bellapais, 1951, *Casey* 1154! Templos, 1956, *G. E. Atherton* 1116! Kazaphani, 1956, *G. E. Atherton* 1154! Kyrenia, 1956, *G. E. Atherton* 1283! Between Myrtou and Panagra, 1967, *Merton* in ARI 115!
8. Cape Andreas, 1938, *Syngrassides* 1781! also, 1948, *Mavromoustakis* s.n.!

NOTES: The distinctions between *Carduus argentatus* L. and *C. acicularis* Bert. break down in Cyprus, where most specimens are intermediate in character between the two, with a definite leaning towards *C. acicularis*, though occasionally (cf. *Edmondson & McClintock* E 2702!) nearer to *C. argentatus*. Since the extremes of the aggregate are deserving of recognition, I have compromised by treating them as subspecies, and have, after some hesitation, referred all the Cyprus material to *C. argentatus* ssp. *acicularis*.

Three specimens: *Chapman* 525! *P. H. Oswald* 34!, both from (6) Nicosia, and *Sintenis & Rigo* 294a (part of *S. & R.* 294)! from (7) Pentadaktylos, are intermediate between *C. argentatus* L. ssp. *acicularis* (Bert.) Meikle and *C. pycnocephalus* L., and are probably hybrids. *Chapman* 525 has in fact been so determined by S. M. A. Kazmi.

2. C. pycnocephalus *L.*, Sp. Plant., ed. 2, 2: 1151 (1763); Boiss., Fl. Orient., 3: 520 (1875); H. Stuart Thompson in Journ. Bot., 44: 333 (1906); Holmboe, Veg. Cypr., 185 (1914); Druce in Proc. Linn. Soc., 141st session: 51 (1930); Post, Fl. Pal., ed. 2, 2: 87 (1933); Osorio-Tafall et Seraphim, List Vasc. Plants Cyprus, 106 (1973); Davis, Fl. Turkey, 5: 435 (1975).

Erect branched or unbranched perennial 15–75 (–100) cm. high; stems broadly winged to apex, with irregular-deltoid, spinose and spinulose lobes, thinly arachnoid-tomentellous; leaves oblong in outline, 7–20 cm. long, 1·5–8 cm. wide, greenish, minutely pustulate and very sparsely strigillose above, greyish-tomentose below, deeply and irregularly lacerate-pinnatilobed, with acute, spinose or spinulose lobes and lobules, occasionally undulate-lobed or shallowly pinnatifid; capitula generally in terminal or lateral clusters at the ends of leafy, spinose branches, ovoid-oblong, 1–1·5 (–2) cm. long, 0·8–1·5 cm. wide, glabrescent or arachnoid-tomentose; phyllaries in 4–6 series, closely imbricate, erect or somewhat recurved, greenish or purplish, lanceolate-subulate, spine-tipped, 3–20 mm. long, 1–3 mm. wide, the outer 3-nerved, with thickened margins, the inner 1-nerved, with scarious, ciliate margins; florets rosy-purple; corolla narrowly tubular, 13–16 mm. long, about 1–1·5 mm. wide at the slightly expanded apex, lobes erect or slightly spreading, linear, acuminate, 4–4·5 mm. long, about 0·4 mm. wide; filaments about 1·5 mm. long, thinly pilose; anthers linear, acuminate, about 4 mm. long, 0·3 mm. wide, base caudate-sagittate, tails about 0·8 mm. long; style 13–14 mm. long, branches erect, connivent, blunt, about 1·5 mm. long, with a pilose basal ring. Achene oblong, about 5 mm. long, 2 mm. wide, strongly compressed, truncate, with an anthophore-stump, glabrous, scabridulous, with 16 shallow longitudinal furrows; pappus tinged brownish, deciduous, bristles numerous, in several series, shortly barbellate.

ssp. **albidus** (*M. Bieb.*) *Kazmi* in Mitt. Bot. Staatsamml. München, 5: 446 (1964); Davis, Fl. Turkey, 5: 436 (1975).

C. albidus M. Bieb., Fl. Taur.-Cauc., 2: 269 (1808).
C. pycnocephalus L. var. *albidus* (M. Bieb.) Boiss., Fl. Orient., 3: 521 (1875).
TYPE: "Frequens in incultis Tauriae et ad Caucasum, etiam in Iberiâ", *Bieberstein* (LE).

Capitula smaller than in ssp. *pycnocephalus*, seldom much more than 15 mm. long, often rather densely arachnoid-tomentellous, with narrower, erect or slightly spreading phyllaries, the terminal appendage commonly short and inconspicuous.

HAB.: Cultivated and fallow fields; waste ground, roadsides; in garigue on stony hillsides; sea-level to 2,000 ft. alt.; fl. March–May.

DISTR.: Divisions 1, 4, 5–8. Eastern Mediterranean region, and eastwards to Afghanistan and Pakistan.

1. Near Dhrousha, 1941, *Davis* 3284!
4. Coast W. of Ayia Napa, 1981, *Hewer* 4710! 4723!
5. Kythrea, 1938, *Syngrassides* 1797! Lakovounara Forest, 1950, *Chapman* 141! Vatili, 1967, *Merton* in ARI 250!
6. Syrianokhori, 1935, *Syngrassides* 628! Nicosia, 1973, *P. Laukkonen* 211!
7. Pentadaktylos, 1880, *Sintenis & Rigo* 294 partly! also, 1974, *Meikle* 4040!
8. Near Yialousa, 1880, *Sintenis & Rigo* 502! Ronnas Bay near Rizocarpaso, 1941, *Davis* 2353!

NOTES: Probably all records of *C. pycnocephalus* L. from Cyprus are referable to ssp. *albidus* (M. Bieb.) Kazmi, but I have cited only those specimens which I have been able to examine. The differences between the type subspecies and ssp. *albidus* are slight, and of less account because of the notorious variability of the species. *Meikle* 4040 from Pentadaktylos is unusual in having marbled leaves with pallid venation; in this respect it resembles *C. marmoratus* Boiss. et Heldr. (*C. arabicus* Jacq. ssp. *marmoratus* (Boiss. et Heldr.) Kazmi); in every other respect, however, it falls within *C. pycnocephalus* ssp. *albidus*.

36. TYRIMNUS (*Cass.*) *Cass.*
Dict. Sci. Nat., 41: 314, 335 (1826).
Carduus L. subgenus *Tyrimnus* Cass. in Bull. Sci. Soc. Philom. Paris, 1818: 168 (1818).

Erect annual or biennial; stems winged, prickly; leaves pinnatifid, with spinose lobes; inflorescence generally branched; capitula terminal, solitary, long-pedunculate, heterogamous, discoid; involucre broadly ovoid-campanulate or subglobose; phyllaries in many series, closely imbricate, tapering to an elongate, weakly spinose acumen; receptacle bristly; florets mauve-pink (sometimes white), the marginal sterile; corolla narrowly tubular, 5-lobed; filaments united, subglabrous; anthers not caudate at base. Achenes compressed-obovoid, obtusely 4-angled, 4-sulcate with mucilage-bearing furrows; pappus deciduous, hairs in several series, the outer long, scabridulous, the inner short and slender.

One species widely distributed in the Mediterranean region.

1. **T. leucographus** (*L.*) *Cass.* in Dict. Sci. Nat., 41: 335 (1826); Boiss., Fl. Orient., 3: 555 (1875); Post, Fl. Pal., ed. 2, 2: 93 (1933); A. K. Jackson et Turrill in Kew Bull., 1939: 478 (1939); Osorio-Tafall et Seraphim, List Vasc. Plants Cyprus, 106 (1973); Davis, Fl. Turkey, 5: 439 (1975).
    Carduus leucographus L., Sp. Plant., ed. 1, 820 (1753).
    TYPE: Italy; "*in* Campania".

Erect annual or biennial up to 2 m. high; stem glabrous or subglabrous, sulcate, narrowly winged with spinose wings; leaves oblong, or the basal obovate, subglabrous above, thinly arachnoid-tomentellous below, pinnatifid with spine-tipped lobes and spinulose or denticulate margins; often marbled with pallid venation; inflorescence loosely branched; peduncles 15–56 cm. long, sulcate, thinly arachnoid-floccose; capitula solitary, 1·3–2 cm. long and almost as wide; phyllaries closely imbricate, subglabrous or thinly arachnoid, lanceolate-subulate, about 3–18 mm. long, 1–3 mm. wide at base, sometimes tinged purple, apex weakly spinose; corolla pinkish-mauve or white, considerably exceeding involucre, tube 14–15 mm. long, about 1·5 mm. wide at apex, lobes 5, linear, subacute, about 3 mm. long, 0·4 mm. wide at base; filaments united in a tube around the style, glabrous or subglabrous; anthers linear, about 5 mm. long, 0·3 mm. wide, apex aristate-acuminate, base obtuse; style 15–16 mm. long, branches 0·8 mm. long, erect, connate except at apex. Achenes narrowly oblong, 5–6 mm. long, 2 mm. wide, compressed, umbonate, 4-angled with strong, pallid ribs at the angles, the intervening furrows filled with brown mucilage; pappus deciduous, hairs white, silky, minutely scabridulous, up to 17 mm. long, the innermost series about 0·8 mm. long, very slender.

HAB.: Margins of dried-up streams and rivers; vineyards; all records from non-calcareous rocks; 100–2,950 ft. alt.; fl. May–June.

DISTR.: Divisions 1, 2, rare. General distribution that of the genus.

1. Dhrousha, 1941, *Davis* 3267!
2. Pomos, 1938, *Syngrassides* 1843! Mandria, 1938, *Kennedy* 1456!

37. NOTOBASIS (*Cass.*) *Cass.*
Dict. Sci. Nat., 35: 170 (1825).
Lamyra (Cass.) Cass. subgenus *Notobasis* Cass. in Dict. Sci. Nat., 25: 225 (1822).

Erect, spiny annual; stems not winged, branched or unbranched; leaves pinnatifid or pinnatilobed, the lobes spine-tipped; inflorescence a racemose cluster; capitula homogamous, or the marginal florets neuter, discoid, sessile or shortly pedunculate, subtended by deeply pinnatisect, rigidly spinose, often purplish, bract-like floral leaves; involucre broadly ovoid; phyllaries in several series, imbricate, lanceolate, tapering to a spinose apex; receptacle bristly; corolla purple, tubular, obliquely 5-lobed; filaments free, pilose; anthers not caudate-sagittate at base; style-branches erect. Achenes compressed-obovoid, smooth, hilum sub-lateral, very narrow and slit-like; apical collar and stump-like anthophore wanting; pappus deciduous, hairs in several series, scabridulous in the marginal achenes, plumose in the inner.

One species widely distributed in the Mediterranean region and eastwards to Iran.

1. **N. syriaca** (*L.*) *Cass.* in Dict. Sci. Nat., 35: 171 (1825); Poech, Enum. Plant. Ins. Cypr.; 20 (1842); Unger et Kotschy, Die Insel Cypern, 248 (1865); Boiss., Fl. Orient., 3: 553 (1875); Post, Fl. Pal., ed. 2, 2: 92 (1933); Davis, Fl. Turkey, 5: 419 (1975); Feinbrun, Fl. Palaest., 3: 377, t. 635 (1978).
    Carduus syriacus L., Sp. Plant., ed. 1, 823 (1753).
    Cirsium syriacum (L.) Gaertn., De Fruct., 2: 283, t. 163, fig. 2 (1791); Holmboe, Veg. Cypr., 185 (1914); Osorio-Tafall et Seraphim, List Vasc. Plants Cyprus, 106 (1973).
    Cnicus syriacus Willd., Sp. Plant., 3: 1683 (1803); Sibth. et Sm., Fl. Graec. Prodr., 2: 155 (1813), Fl. Graec., 9: 22, t. 831 (1837).
    TYPE: "*in* Syria, Creta, Hispania".

Robust annual (8–) 50–100 (–150) cm. high; stems often simple, sub-glabrous or thinly arachnoid, deeply sulcate; basal-leaves oblong-obovate, 20–30 cm. long, 4–9 cm. wide, shortly pinnatifid, the lobes and lobules ending in rigid spines, lamina glabrous, subglabrous or thinly arachnoid; cauline leaves progressively smaller and more deeply spinose-pinnatisect upwards, those in the region of the inflorescence pinnatisect almost to midrib, often purplish, with narrow, subulate, rigidly spinose divisions; capitula terminal and lateral, sessile or shortly pedunculate, frequently clustered, forming an irregular raceme in well-developed specimens; involucre broadly ovoid, 1–1·5 cm. long and about as wide; phyllaries closely imbricate, lanceolate, scarious, scabridulous-glandular and sometimes arachnoid externally, 5–15 mm. long, 2–3 mm. wide at base, tapering to a spinose acumen; corolla narrowly tubular, about 12 mm. long, 1 mm. wide at apex, lobes erect, linear, the adaxial about 5–6 mm. long, 0·5 mm. wide at base, the abaxial about 3 mm. long, 0·4 mm. wide at base; filaments about 2·5 mm. long, shortly pilose; anthers exserted, linear, acuminate, about 6 mm. long, 0·4 mm. wide, very shortly sagittate at base; style about 16–18 mm. long, branches erect and connate almost to apex. Achene strongly compressed, obovoid (shaped like a sheep-tick), 5–6 mm. long, 4 mm. wide, smooth, brown with darker speckling; hilum slit-like, oblique, apex with a conspicuous, circular, oblique scar at point of attachment of the corolla; pappus-hairs white, silky, those of the outermost florets scabridulous, those of the inner long-plumose.

HAB.: Cultivated and fallow fields; waste ground, roadsides; sometimes on dry, stony hillsides; sea-level to 1,000 ft. alt.; fl. Febr.–May.

DISTR.: Divisions 1, 3–8; abundant in the lowlands. General distribution that of genus.

1. Dhrousha, 1941, *Davis* 3285!
3. Kophinou, 1905, *Holmboe* 602; Kolossi, 1961, *Polunin* 6670! Amathus, 1978, *Holub* s.n.!
4. Larnaca, 1862, *Kotschy* 259! Coast E. of Dhekelia, 1981, *Hewer* 4706!
5. Kythrea, 1880–1941, *Sintenis & Rigo* 500! *Syngrassides* 1798! *Davis* 2910! Mia Milea, 1950, *Chapman* 517! Vatili, 1967, *Merton* in ARI 249!
6. Myrtou, 1932, *Syngrassides* 458!
7. Dhavlos, 1880, *Sintenis & Rigo*; Kyrenia, 1949–1956, *Casey* 651! *G. E. Atherton* 717! 1164! 1257!
8. Near Apostolos Andreas Monastery, 1948, *Mavromoustakis* s.n.!

38. PICNOMON *Adans.*
Fam. Plantes, 2: 116, 590 (1763) ("Piknomon")

Erect, spiny annual; stems conspicuously winged, commonly unbranched except in the region of the inflorescence; leaves linear-oblong, sessile, obscurely pinnatifid, the lobes tipped with golden spines; inflorescence a dense terminal cluster, subtended by conspicuously spinose, bract-like leaves; involucre narrowly ovoid or subcylindrical; phyllaries imbricate in several series, arachnoid-tomentellous, apex shortly spinose or with a pectinately compound, spinose appendage; receptacle bristly; florets homogamous, discoid, purplish, unequally 5-lobed; filaments free, glabrous or glandular, sagittate at base; style-branches erect, connivent, blunt. Achenes compressed-obovoid, smooth, with a slit-like hilum, apical collar and anthophore-stump; pappus deciduous, connate at base, hairs long, plumose.

One species widely distributed in the Mediterranean and eastward to Iran and Afghanistan.

1. **P. acarna** (*L.*) *Cass.* in Dict. Sci. Nat., 40: 188 (1826); Gaudry, Recherches Sci. en Orient, 188 (1855); Lindberg f., Iter. Cypr., 35 (1946) ["*Picnemon*"]; Davis, Fl. Turkey, 5: 412 (1975); Feinbrun, Fl. Palaest., 3: 380, t. 639 (1978).
   Carduus acarna L., Sp. Plant., ed. 1, 820 (1753).
   Cirsium acarna (L.) Moench, Meth. Suppl., 226 (1802); Boiss., Fl. Orient., 3: 549 (1875); H. Stuart Thompson in Journ. Bot., 44: 333 (1906); Holmboe, Veg. Cypr., 186 (1914); Post, Fl. Pal., ed. 2, 2: 91 (1933); Osorio-Tafall et Seraphim, List Vasc. Plants Cyprus, 106 (1973).
   TYPE: "*in* Hispania".

Erect annual (12–) 20–60 (–150) cm. high; basal parts of stem and basal leaves thinly arachnoid or subglabrous, upper parts of stem and upper leaves densely arachnoid-tomentellous; basal leaves narrowly oblong-obovate, rather coarsely pinnatifid with spinose lobes; stem-leaves linear-oblong, 4–20 cm. long, 0·5–1·5 cm. wide, obscurely lobed, the lobes marked by clusters of rigid, golden spines, often 1–2 cm. long, sinuses between lobes shortly spinulose-ciliate, base of leaf continued down stem as a very conspicuous, spinose wing; uppermost leaves surrounding the clusters of capitula, narrowly linear, recurved-arcuate, the apex and lobes rigidly golden-spinose; capitula 2–8 in terminal clusters at the apex of divaricate, winged, inflorescence-branches, the whole forming a broadly domed corymb; involucre 13–25 mm. long, 6–10 mm. wide; phyllaries narrowly oblong, the outer 9–13 mm. long, about 2 mm. wide, densely lanuginose dorsally, with a pectinate-spinose apical appendage 3–7 mm. long, innermost up to 20 mm. or more long, apex shortly spinulose; florets pink-purple; corolla-tube about 10 mm. long, very slender towards base, narrowly infundibuliform, and about 1 mm. wide towards apex, lobes erect, linear, acute, sub-cucullate, about 3–5 mm. long, 0·3 mm. wide; filaments about 2 mm. long, thinly lanuginose; anthers linear, about 6 mm. long, 0·3 mm. wide, with a slender acumen, and shortly sagittate base; style 15–16 mm. long, branches erect, closely connivent or connate, blunt. Achenes compressed-obovoid or almost oblong, about 6 mm. long, 2·5 mm. wide, shining pale brown, obscurely striatulate longitudinally, with a distinct apical collar; anthophore-stump less than 0·3 mm. long; pappus-hairs white, plumose, about 2 cm. long.

HAB.: Roadsides, waste ground; dry rocky hill-sides; 500–5,000 ft. alt.; fl. July–Sept.

DISTR.: Divisions 2, 5–7. General distribution that of genus.

2. Near Kykko Monastery, 1905, *Holmboe* 1031; also, 1939, *Lindberg f.* s.n.! Prodhromos, 1937, *Syngrassides* 1744! and, 1955, *G. E. Atherton* 538! Agros, 1937, *Syngrassides*; Platres, 1937, *Kennedy* 964! Amiandos, 1970, *A. Genneou* in ARI 1563!
5. Alambra, 1932, *Syngrassides* 382!
6. Between Panteleimon Monastery and Nicosia, 1853–54, *Gaudry* sec. H. Stuart Thompson; Peristerona, 1936, *Syngrassides* 1250!
7. Kornos, 1934, *Syngrassides* 433! Kyrenia, 1948, *Casey* 69! also, 1955, *G. E. Atherton* 128! Lapithos, 1966, *Merton* in ARI 43!

39. CIRSIUM Mill.
Gard. Dict. Abridg., ed. 4 (1754).

Annual, biennial or perennial herbs, commonly with spinose-winged stems and spinose leaves; stems branched or unbranched, or plants acaulescent; leaves subentire or pinnatilobed, often decurrent; capitula homogamous, discoid, sometimes unisexual (dioecious), solitary or clustered; involucre ovoid or subglobose; phyllaries numerous, imbricate in several series, spine-tipped; receptacle bristly; florets purple, white or yellowish, tubular, unequally 5-lobed; filaments free, often pilose; anthers sagittate at base; style-branches erect, connivent or connate. Achenes oblong, compressed, smooth, generally with an elevated apical rim (or collar) and an anthophore-stump (or umbo), hilum ovate, oblique; pappus deciduous, or rarely persistent, hairs in several series, plumose (rarely scabrid), connate at base, the inner sometimes thickened at apex.

About 150 species widely distributed in temperate regions of the northern hemisphere.

1. **C. vulgare** (*Savi*) *Ten.*, Fl. Nap., 5: 209 (1835–36); Davis, Fl. Turkey, 5: 397 (1975).
   Carduus lanceolatus L., Sp. Plant., ed. 1, 821 (1753).
   Cirsium lanceolatum (L.) Scop., Fl. Carn., ed. 2, 2: 130 (1772); Boiss., Fl. Orient., 3: 538 (1875); Post in Mém. Herb. Boiss., 18: 95 (1900); H. Stuart Thompson in Journ. Bot., 44: 333 (1906); Holmboe, Veg. Cypr., 185 (1914); Post, Fl. Pal., ed. 2, 2: 90 (1933); Osorio-Tafall et Seraphim, List Vasc. Plants Cyprus, 106 (1973) non Hill (1769) nom. illeg.
   Carduus vulgaris Savi, Fl. Pisana, 2: 241 (1798).
   TYPE: Not indicated, presumably from the neighbourhood of Pisa (?PI).

Erect, robust biennial 50–150 cm. high; stems branched or unbranched, angled, sulcate, thinly pilose with mixed arachnoid and multicellular hairs, generally spinose-winged, at least in the upper part; basal leaves forming a loose rosette, often 30 cm. or more long, 8–12 cm. wide, oblong, pinnatilobed, with acute, spine-tipped lobes and lobules, upper surface green, coarsely strigose, lower surface densely greyish-tomentellous, petioles often well developed, 7–10 cm. long, spinulose, arachnoid-tomentose; cauline leaves becoming progressively smaller and more deeply and acutely lobed, sessile, the bases decurrent as wings along part of the stem; inflorescence loosely branched; capitula solitary, or sometimes in small clusters, terminal, without an outer involucre of bract-like leaves; involucre subglobose, 2·5–5 cm. diam.; phyllaries numerous, in many series, suberect, subulate, 5–40 mm. long, 1–2·5 mm. wide at base, arachnoid dorsally, with entire margins and a rigidly spinose apex; corolla purple, tube very slender, 20–25 mm. long, less than 0·5 mm. wide, except at the expanded apex and base, lobes linear, erect, subacute, 6–9 mm. long, 0·3 mm. wide; filaments about 5 mm. long, thinly pilose; anthers linear, 6–7 mm. long, 0·3 mm. wide; style 20–25 mm. long, branches erect, connate except at the very apex, blunt. Achenes oblong, 4–5 mm. long, 1·8–2 mm. wide, compressed, obscurely angled, pale brown with darker, blackish marbling; pappus-hairs silky, long-plumose, whitish, about 2 cm. long.

HAB.: Gardens, orchards, waste ground, roadsides; [sea-level] to 5,000 ft. alt.; fl. July–Oct.

DISTR.: Divisions 2, [3]. Throughout Europe, Mediterranean region and eastwards to Central Asia; an introduced weed in many temperate regions.

2. Kyperounda ["Kippalunga"]. ? 1898, *Post* s.n.; Platania, 1936, *Syngrassides* 1094! also 1956, *Merton* 2807! Trikoukkia, 1937, *Syngrassides* 1691! Prodhromos, 1949, *Casey* 947! also, 1955, *G. E. Atherton* 563! Near Spilia, 1966, *Merton* in ARI 77!

[3. Limassol, ? 1905, *Michaelides* teste *Holmboe*. See Notes.]

NOTES: In view of the fact that all other records for *Cirsium vulgare* are from high altitudes on the Troödos Range, one must question the accuracy of the Limassol localization cited by Holmboe, also his comment that the Thistle is "not rare" in Cyprus. It is, in fact, distinctly uncommon, and most probably a recent introduction.

40. PTILOSTEMON *Cass.*

Bull. Sci. Soc. Philom. Paris, 1816: 200 (1816).
W. Greuter in Boissiera, 22: 1–215 (1973).

Biennial or perennial herbs or subshrubs; leaves lanceolate or linear, entire, sinuate or pinnatilobed, glabrescent or thinly arachnoid-tomentellous above, densely white-tomentose below, the cauline alternate, sessile, not decurrent, margins sometimes with spinose lobes; inflorescence generally branched, corymbose or paniculate; capitula discoid, the outermost series of florets functionally male, the inner hermaphrodite, or occasionally all hermaphrodite and fertile; involucre ovoid or subglobose; phyllaries in several series, imbricate, coriaceous, acute or spine-tipped, generally entire; receptacle bristly; corolla purple, pink or white, narrowly tubular, lobes 5, usually unequal, erect or rarely patent; filaments free, generally pilose; anthers with a sword-shaped apical appendage, caudate-sagittate base, and a translucent median vitta; style-branches short, erect or divergent, usually blunt. Achenes obliquely obovoid, scarcely compressed, smooth, apex with an obsolete collar, and without a median anthophore-stump; pappus deciduous, hairs in several series, plumose (or the outermost scabridulous), connate at base.

Fourteen species widely distributed in the Mediterranean region and eastward to Caucasus and Iraq.

1. P. chamaepeuce (*L.*) *Less.*, Gen. Cynaroceph. Spec. Arctot., 5 (1832); Greuter in Boissiera, 22: 105 (1973); Davis, Fl. Turkey, 5: 417 (1975); Feinbrun, Fl. Palaest., 3: 380, t. 640 (1978).

　　Serratula chamaepeuce L., Sp. Plant., ed. 1, 819 (1753).

　　Staehelina chamaepeuce (L.) L., Syst. Nat., ed. 12, 2: 538 (1767); Sibth. et Sm., Fl. Graec. Prodr., 2: 163 (1813), Fl. Graec., 9: 34, t. 847 (1837).

　　Ptilostemon muticus Cass. in Dict. Sci. Nat., 44: 59 (1826) nom. illeg.

　　Cirsium chamaepeuce (L.) Ten., Fl. Nap., 5: 211 (1835–36) quoad nomen; Holmboe, Veg. Cypr., 186 (1914); Chapman, Cyprus Trees and Shrubs, 80 (1949).

　　Chamaepeuce mutica DC., Prodr., 6: 657 (1838); Post, Fl. Pal., ed. 2, 2: 93 (1933).

　　Chamaepeuce alpini Jaub. et Spach, Ill. Plant. Orient., 5: 26, t. 425 (1854); Boiss., Fl. Orient., 3: 554 (1875) nom. illeg.

Lax, evergreen shrub 30–80 cm. high; old branches greyish-floccose, prominently scarred, young branches densely white-tomentose; leaves crowded towards base of stems, more remote towards apices, linear, 3–10 cm. long, 0·1–0·3 cm. wide, dark green above, densely white-tomentose below, apex acute or shortly spinulose, margins strongly revolute, base tapering, sometimes with small stipuloid lobes; inflorescence loosely and irregularly corymbose, or capitula sometimes solitary and terminal on long, sparsely leafy flowering shoots; involucre ovoid, subglobose or campanulate, 14–22 mm. long, 12–20 mm. wide, glabrescent or tomentose; phyllaries narrowly ovate or lanceolate, closely imbricate in several series, the outermost 4–5 mm. long, 2–3 mm. wide, sometimes with an herbaceous, linear-subulate appendage more than 5 mm. long, the inner often tinged purple, with a short, erect or reflexed, herbaceous acumen, the innermost elongate-lanceolate, with a long, tapering acumen; florets purple; corolla-tube slender, (6–) 7–12 (–14) mm. long, about 0·5 mm. wide, limb (8–) 9–13 mm. long, distinctly dilated towards base and about 1 mm. wide, lobes erect, linear, acuminate, 3–5 mm. long, about 0·5–0·7 mm. wide at base; filaments 4–6 mm. long, conspicuously pilose about the middle; anthers shortly exserted, 7–12 mm. long, 0·7 mm. wide, apical appendage 2–3 mm. long, basal tails 1·5–3 mm. long, laciniate-fimbriate; apex of style a little thickened, about 2–3 mm. long, with a pilose basal ring, branches very short, erect, blunt. Achenes oblique-obovoid, 4–5 mm. long, 2·5–3·5 mm. wide, smooth, brown with darker marbling, apical collar very short and inconspicuous; pappus-hairs whitish, 12–18 mm. long, plumose with barbellate apices.

var. **cyprius** *Greuter* in Boissiera, 22: 113 (1973).

　　[*Chamaepeuce mutica* (non DC.) Poech, Enum. Plant. Ins. Cypr., 19 (1842); Kotschy in Unger et Kotschy, Die Insel Cypern, 248 (1865); Sintenis in Oesterr. Bot. Zeitschr., 31: 395 (1881).]

　　[*Chamaepeuce alpini* Jaub. et Spach var. *camptolepis* (non Boiss.) Boiss., Fl. Orient., 3: 554 (1875) quoad plant. cypr.]

　　[*Chamaepeuce fruticosa* (non *Cnicus fruticosus* Desf.) Boiss., Fl. Orient., Suppl., 309 (1888).]

　　[*Cirsium chamaepeuce* (L.) Ten. ssp. *camptolepis* (non *Chamaepeuce alpini* Jaub. et Spach var. *camptolepis* Boiss.) Holmboe, Veg. Cypr., 186 (1914); Lindberg f., Iter Cypr., 34 (1946); Rechinger f. in Arkiv för Bot., ser. 2, 1: 433 (1950).]

　　[*Cirsium chamaepeuce* (L.) Ten. ssp. *fruticosum* (non *Cnicus fruticosus* Desf.) Holmboe, Veg. Cypr., 186 (1914).]

　　[*Ptilostemon chamaepeuce* (L.) Less. ssp. *polycephalus* (non *Chamaepeuce polycephala* DC.) Osorio-Tafall et Seraphim, List Vasc. Plants Cyprus, 107 (1973) nom. invalid.]

　　[*Ptilostemon fruticosus* (non *Cnicus fruticosus* Desf.) Osorio-Tafall et Seraphim, List Vasc. Plants Cyprus, 107 (1973) nom. invalid.]

TYPE: Cyprus; "Troodos: Felsen am Flusse bei Galata, d. 25./6., *Sintenis*, Reise auf Cypern 1880, n. 548 "(LD, K! etc.).

Outermost phyllaries with a long, foliaceous, deflexed or spreading, linear-subulate appendage; median phyllaries with a shorter, strongly recurved, spinulose-acuminate appendage; involucre commonly floccose-tomentose externally, but sometimes glabrescent.

HAB.: Rocky slopes and rock-fissures on igneous or calcareous formations; 500–5,500 ft. alt.; fl. May–July.

DISTR.: Divisions 1–3, 7, 8. Endemic.

1. Ayios Neophytos, 1939, *Lindberg f.* s.n.! Yioulou near Baths of Aphrodite, 1962, *Meikle* 2230!
2. Near Galata, 1880, *Sintenis & Rigo* 548! Prodhromos, 1898, *Post* s.n.; Agros, 1905, *Holmboe* 1143; Kakopetria, 1930, *C. B. Ussher* 126! also, 1935, *Syngrassides* 735! and, 1939, *Lindberg f.* s.n.; Limnitis, 1938, *Chapman* 355! Vouni near Makheras Monastery, 1940, *Davis* 1895! Troödos, 1948, *Mavromoustakis* s.n.! also, 1950, *F. M. Probyn* s.n.! Stavros tis Psokas, 1979, *Hewer* 4647!
3. Between Limassol and Silikou, 1901, *A. G. & M. E. Lascelles* s.n.! Kalavasos, 1905, *Holmboe* 610! Lefkara, 1940, *Davis* 1900! Between Moni and Kellaki, 1956, *Merton* 2553! Khalassa, 1970, *A. Genneou* in ARI 1510!
7. Buffavento, 1859, *Kotschy* 475, and 1889, *Pichler* s.n.; Pentadaktylos, 1880, *Sintenis & Rigo* 548a; Kyrenia, 1936, *Kennedy* 958! also, 1955, *Miss Mapple* 53! and, 1955, *G. E. Atherton* 142! 254! St. Hilarion, 1939–1952, *Lindberg f.* s.n.; *Davis* 3627! *F. M. Probyn* 141! also, 1955, *G. E. Atherton* 746! Kyrenia Pass, 1939, *Lindberg f.* s.n.; Koronia, 1937, *Chapman* 239! Halevga, 1940, *Davis* 1752! Karmi, 1948, *Kennedy* in *Casey* 177! Sina Oros, 1970, *A. Hansen* 401!
8. Near Komi Kebir, 1912, *Haradjian* 300!

41. JURINEA *Cass.*

Bull. Sci. Soc. Philom. Paris, 1821: 140 (1821).
Dict. Sci. Nat., 24: 287 (1822).

Perennial herbs and subshrubs; stems branched or unbranched, or plants occasionally acaulescent; leaves unarmed, entire, sinuate, pinnatilobed or pinnatisect, the uppermost sometimes forming an outer involucre around the capitula; inflorescence branched, corymbose, or capitula solitary, terminal; involucre cylindrical, campanulate or subglobose, phyllaries in many (4–8) series, loosely or closely imbricate, narrow, apex unarmed or spinulose, erect or recurved; receptacle bristly; florets purplish, homogamous, discoid; corolla with a long, slender tube, apex deeply 5-lobed, lobes unequal, linear; filaments glabrous, free; anthers caudate-sagittate at base; style-branches usually very short, erect or divergent. Achenes obpyramidal, subtetragonal or rarely compressed, with a small membranous apical corona, hilum basal; pappus persistent or ultimately deciduous together with the anthophore-stump, hairs in several series, scabrid, barbellate or plumose, the innermost sometimes conspicuously longer and broader than the others.

About 100 species, chiefly in western and central Asia.

1. **J. cypria** *Boiss.*, Fl. Orient., Suppl., 311 (1888); Post in Mém. Herb. Boiss., 18: 95 (1900); Holmboe, Veg. Cypr., 185 (1914); Lindberg f., Iter Cypr., 35 (1946); Osorio-Tafall et Seraphim, List Vasc. Plants Cyprus, 106 (1973).
    TYPE: Cyprus; "in vineis ad Spielia [Spilia] montis Adelphe Cypri", 26 June, 1880, *Sintenis & Rigo* 788 (G, K!).

Erect perennial 20–70 cm. high, usually with several stems arising from a thick woody rootstock; stems generally unbranched except in region of inflorescence, angular, sulcate, often densely white-tomentose towards base, floccose or arachnoid-tomentellous above; basal leaves mostly withered by anthesis, oblanceolate, 15–25 cm. long, 1·5–4 cm. wide, green and glabrescent or thinly floccose above, usually thinly canescent-tomentellous below, apex obtuse, sometimes apiculate, margins entire or sinuate-lobed, narrowly recurved, base tapering to inconspicuous, semi-amplexicaul, basal auricles; cauline leaves progressively smaller, narrower, more acuminate, and often more glabrous upwards, the uppermost sometimes forming a loose involucre around the capitula; inflorescence usually much branched, corymbose; capitula cylindrical-campanulate 2–4 cm. long, 1·5–3 cm. wide; phyllaries erect, lanceolate-acuminate, the outermost 7–10 mm. long, 2–3 mm. wide at base, the innermost up to 30 mm. long, all thinly tomentellous dorsally; florets dull purplish-pink; corolla-tube 16–20 mm. long, about 0·5–0·8 mm. wide, glabrous; limb about 15 mm. long, 1·5–2 mm. wide, glandular-pubescent at area of junction with tube, lobes linear, 6–8 mm. long, about 1 mm. wide at base, apex subacute, shortly cucullate; filaments 5–6 mm. long; anthers exserted, linear, about 16 mm. long, 0·5 mm. wide, apical appendage 5–6 mm. long, sharply acuminate, basal tails entire, about 3 mm. long; style shortly exceeding stamens, branches compressed dorsiventrally, divergent, blunt, about 3 mm. long, with a glandular-pilose ring at the base. Achenes fusiform-truncate, pale brown, about 5–6 mm. long, 2·5–3 mm. wide, compressed, obscurely angled, longitudinally sulcate, crowned with a neatly denticulate corona; pappus persistent, ultimately breaking off with the hemispherical anthophore-stump; pappus-hairs whitish, unequal, 2–20 mm. long, connate at base, plumose.

HAB.: In vineyards and stony hillsides on calcareous or igneous rocks; 3,000–5,700 ft. alt.; fl. June–Sept.

DISTR.: Division 2 only. Endemic.

2. Fairly common; Spilia, Khrysorroyiatissa, Khandria, Agros, Troödos, Platres, Platania, Aphamis, Perapedhi, etc. *Sintenis & Rigo* 788! *Post* s.n.; *A. G. & M. E. Lascelles* s.n.! *Holmboe* 1144! *Haradjian* 500! *C. B. Ussher* 13! *Kennedy* 968! 969! 1852! 1853! *Lindberg* f. s.n.! *Davis* 1830! *Mavromoustakis* 34! *F. M. Probyn* 102! *P. Laukkonen* 430! etc.

NOTES: Allied to *J. aucheriana* DC. and *J. brevicaulis* Boiss. from eastern Anatolia.

42. ONOPORDUM *L.*

Sp. Plant., ed. 1, 827 (1753).
Gen. Plant., ed. 5, 359 (1754).
Eig in Pal. Journ. Bot., J. series, 2: 185–199 (1942).

Spinose, commonly arachnoid-tomentose biennials and perennials; stems generally with spinose wings, often robust; leaves spinose, subentire, pinnatifid or pinnatisect, decurrent, the basal often forming a rosette; inflorescence branched or capitula solitary, usually large; involucre subglobose; phyllaries entire, erect or recurved, acuminate, or with an abruptly narrowed, subulate, spine-tipped acumen; florets discoid, homogamous, hermaphrodite, generally purple; corolla-tube slender, limb oblique, deeply 5-lobed, with subequal, narrow, erect lobes; filaments pilose; anthers sagittate at base, with entire or fimbriate tails; style-branches short, obtuse; receptacle fleshy, flat, foveolate or alveolate, not bristly. Achenes obovoid or oblong, smooth or rugulose, compressed or 4-angled, apex truncate; hilum basal; pappus deciduous, hairs in several series, scabridulous, barbellate or plumose, connate at base.

About 50 species in Europe, the Mediterranean region, and western Asia.

Stems thinly arachnoid or subglabrous; leaves green above, thinly arachnoid-tomentellous below; plants generally less than 1·5 m. high - - - - - - **1. O. cyprium**
Stems densely white-tomentose; leaves white-tomentose on both surfaces; plants robust, often 2 (–2·5) m. high - - - - - - - - - - **2. O. bracteatum**

1. **O. cyprium** *Eig* in Pal. Journ. Bot., J. series, 2: 190 (1942); Rechinger f. in Arkiv för Bot., 2: 190 (1942); Osorio-Tafall et Seraphim, List Vasc. Plants Cyprus, 107 (1973).
   O. insigne Holmboe f. *pallida* Lindberg f., Iter Cypr., 35 (1946).
   [*O. graecum* (non Gouan) Sm. in Sibth. et Sm., Fl. Graec. Prodr., 156 (1813); Poech, Enum. Plant. Ins. Cypr., 19 (1842); Unger et Kotschy, Die Insel Cypern, 246 (1865).]
   [*O. virens* (non DC.) Kotschy in Unger et Kotschy, Die Insel Cypern, 246 (1865); H. Stuart Thompson in Journ. Bot., 44: 333 (1906).]

[*O. sibthorpianum* (non Boiss. et Heldr.) Sintenis in Oesterr. Bot. Zeitschr., 32: 397 (1881); Post in Mém. Herb. Boiss., 18: 95 (1900); H. Stuart Thompson in Journ. Bot., 44: 333 (1906); Holmboe, Veg. Cypr., 186 (1914).]

    [*O. sibthorpianum* Boiss. et Heldr. ssp. *anatolicum* (non *O. sibthorpianum* Boiss. et Heldr. var. *anatolicum* Boiss.) Holmboe, Veg. Cypr., 186 (1914).]

    [*O. illyricum* (non L.) Holmboe, Veg. Cypr., 186 (1914); Osorio-Tafall et Seraphim, List Vasc. Plants Cyprus, 107 (1973).]

    [*O. tauricum* Willd. ssp. *elatum* (non *O. elatum* Sm.) Holmboe, Veg. Cypr., 186 (1914); Osorio-Tafall et Seraphim, List Vasc. Plants Cyprus, 107 (1973).]

    TYPE: Cyprus; "Inter Bellepais et Cerignia [Kyrenia]. 28 Majo 1880", *Sintenis & Rigo* 545 (G, K!).

Erect biennial or perennial to 1·2 m. high; stems unbranched or sparingly branched, thinly arachnoid or glabrescent, narrowly winged with spinose-lobed wings; leaves deeply pinnatisect, the basal lanceolate, 20–40 cm. long, 6–12 cm. wide, green and subglabrous above, glandular and arachnoid-tomentellous below, the divisions oblong, acute, with a spinose apex and spinose-deltoid marginal lobes, cauline leaves becoming progressively smaller upward, the uppermost commonly narrow-oblong with spinose-deltoid lobes; inflorescence lax, sparingly branched (or sometimes unbranched); capitula terminal on closely spinose, winged branches; involucre subglobose, 2·5–3·5 cm. diam. (excluding phyllary-appendages), subglabrous or thinly arachnoid; phyllaries closely imbricate, 2–3·5 cm. long, 0·3–0·5 mm. wide at the dilated base, rather abruptly narrowed above into a rigidly spinose, subulate, often purplish appendage, the appendages of the outer phyllaries often sharply recurved, the inner spreading or erect; florets generally exceeding the involucre, purple or sometimes whitish; corolla-tube 13–15 mm. long, limb narrowly tubular, about as long as tube, 2·5 mm. wide, lobes linear, erect, about 7–9 mm. long, 0·5 mm. wide; filaments shortly papillose-pubescent; anthers linear, about 10 mm. long, 0·5 mm. wide, with a slender, rostrate apical appendage and a shortly sagittate base; style slender, about 25 mm. long, branches about 5 mm. long, united almost to apex, with a shortly pilose basal annulus, and blunt obscurely bifid tips; receptacle deeply foveolate, the pits with conspicuous, erose-fimbriate walls. Achenes oblong, 4-angled, somewhat compressed, glabrous, fuscous-brown, transversely rugulose, about 4–5 mm. long, 2 mm. wide; pappus-hairs brownish, shortly barbellate, to about 11 mm. long, distinctly connate at base.

HAB.: Roadsides, field borders, fallows and waste land; sea-level to 4,000 ft. alt.; fl. April–July.

DISTR.: Divisions 1, 2, 4–7. Endemic.

1. Cape Arnauti, 1913, *Haradjian* 797! Khlorakas, 1967, *Economides* in ARI 963!
2. Base of Troödos, ? 1898, *Post* s.n.; Kryos Potamos, 2,900 ft. alt., 1937, *Kennedy* 967! Near Pedhoulas, 1940, *Davis* 1789! Evrykhou, 1941, *Davis* 3199!
4. Larnaca, 1877, [? *J. Ball* in] *Herb. Post* 631!
5. Athalassa, 1939, *Lindberg f.* s.n.! Road from Nicosia to Famagusta, 1972, *W. R. Price* 996!
6. Kykko fields near Nicosia, 1936, *Syngrassides* 1121! Nicosia, Dept. of Agriculture, 1937, *Syngrassides* 1620! Nicosia, 1939, *Lindberg f.* s.n.; also, 1950, *Chapman* 624! Above Peristerona, 1957, *Merton* 2994!
7. Near Kephalovryso, Kythrea, 1862, *Kotschy* 346; between Bellapais and Kyrenia, 1880, *Sintenis & Rigo* 545! Kyrenia, 1949, *Casey* 783! also, 1955, *Miss Mapple* 33!

NOTES: Probably common in all parts of lowland Cyprus. The white-flowered form, recorded by Lindberg f. from (5) Athalassa as *O. insigne* Holmboe f. *pallida* [sic] is scarcely worth taxonomic recognition, since similar albino forms are to be seen in many species of Thistles, often in company with normal, purple-flowered plants.

O. cyprium belongs to the taxonomically complex aggregate that includes *O. boissieri* Willk. (non Freyn et Sint.) and *O. carduiforme* Boiss. It comes nearest to the latter, differing chiefly in its very deeply pinnatisect leaves (the divisions reaching almost to the midrib) and in its very long, abruptly subulate, phyllary-appendages. The dense glandulosity on the leaf-undersurface is visible only where the arachnoid tomentum is thin, or where it has been removed.

2. O. bracteatum Boiss. et Heldr. in Boiss., Diagn., 1, 10: 91 (1849); Fl. Orient., 3: 561 (1875); Rechinger f. in Arkiv för Bot., ser. 2, 1: 433 (1950); Osorio-Tafall et Seraphim, List Vasc. Plants Cyprus, 107 (1973); Davis, Fl. Turkey, 5: 363 (1975).
    O. boissieri Freyn et Sint. in Bull. Herb. Boiss., 3: 470 (1895); Holmboe, Veg. Cypr., 186 (1914) non Willk. (1870) nom. illeg.
    O. insigne Holmboe, Veg. Cypr., 187 (1914); Rechinger f. in Arkiv för Bot., ser. 2, 1: 433 (1950).
    TYPE: Turkey; "in saxosis prope Aglansoun ad radices montis Boudroun Pisidiae. Fl. Aug. (Heldreich)" (G).

Erect biennial or monocarpic perennial 50–200 (–250) cm. high; stems robust, conspicuously winged with white-tomentose, lobulate wings densely armed with numerous rigid, yellow-brown spines; basal leaves lying flat on the ground and forming a conspicuous rosette in winter and spring, lanceolate, densely white-tomentose, 12–60 (–70) cm. long, 4–20 cm. wide, margins irregularly and not very deeply lobed, armed with yellowish spines often 2 cm. or more in length; nervation obscure or distinct and reticulate; petiole 3–10 cm. long; cauline leaves few, sessile, much reduced, linear-lanceolate or narrowly deltoid, the upper generally less than 10 cm. long, 1·5 cm. wide, plicate-lobed, with viciously spinose lobes; inflorescence terminal and subterminal, shortly branched with rigid, divaricate, spinose-winged, white-tomentose branches; capitula terminal on the branches, depressed-globose; involucre 6–8 cm. diam.; phyllaries lanceolate, 2–6 cm. long, 0·5–0·8 mm. wide near base, tapering to an acuminate or subulate, rigidly spinose acumen, the outer generally recurved or patent, the inner erect or slightly recurved, apex often purplish, margins thinly arachnoid, often conspicuously ciliate with short, fragile, deciduous hairs; florets purple, slightly exceeding involucre; corolla-tube slender, about 30 mm. long, limb about 15 mm. long, 2–2·5 mm. wide, lobes erect, linear, subacute, about 9 mm. long, 0·8 mm. wide; filaments about 2 mm. long, pubescent-papillose; anthers linear, about 12 mm. long, 0·7 mm. wide, apex rostrate-acuminate, base shortly sagittate with fimbriate tails; style about 40 mm. long, branches erect, about 5 mm. long, united almost to apex, blunt, with a distinct, shortly pilose, basal annulus; receptacle deeply foveolate, with membranous, erose walls. Achene oblong, 5–6 mm. long, 2·5–3 mm. wide, 4-angled, somewhat compressed, glabrous, fuscous-brown, longitudinally nerved and transversely rugulose; pappus deciduous, hairs brownish, barbellate, unequal, up to 20 mm. long, distinctly connate at base.

HAB.: Dry, stony hillsides on chalk or igneous rocks; 2,500–4,000 ft. alt.; fl. June–Aug.

DISTR.: Divisions 2, 3. Also S. Balkan peninsula, Crete, Karpathos, Turkey, Khios, Rhodes.

2. Between Omodhos and Prodhromos, 1880, Sintenis & Rigo 794; Ayios Theodhoros near Agros, 1905, Holmboe 1068; Aphamis, 1937, Kennedy 965! 966! Perapedhi, 1940, Davis 1853! Agros, 1940, Davis 1891!
3. Valley S. of Yerasa, on limestone, 1981, Hewer 4767!

NOTES: Particularly abundant on the chalk foothills S. of the main Troödos range.

43. CYNARA L.

Sp. Plant., ed. 1, 827 (1753).
Gen. Plant., ed. 5, 359 (1754).

Perennial herbs; stems erect, sulcate or winged, sometimes wanting; leaves often large, 1–3-pinnatisect, sometimes with spinose divisions; capitula large, broadly ovoid or subglobose, homogamous, discoid, solitary or in lax terminal corymbs; phyllaries in several series, imbricate, coriaceous, sometimes terminating in a spine; florets purple or whitish, narrowly tubular, limb unequally 5-lobed; filaments free, glandular or pubescent; anthers with a blunt apical appendage and sagittate base; style exserted, branches erect, shortly bifid; receptacle flattish, fleshy, foveolate, the pits with setose walls. Achenes oblong-obovoid, compressed, 4-angled, smooth, truncate; hilum basal; pappus deciduous, hairs in several series, plumose, connate at base.

About 10 species in the Mediterranean region and western Asia.

Florets dirty white; plant less than 30 cm. high; leaves coriaceous, rather regularly spinose-
pinnatisect - - - - - - - - - - - - - - **2. C. cornigera**
Florets violet-purple; plant usually more than 30 cm. high; leaves irregularly lobate-
pinnatisect, spinose or unarmed:
Leaves spinose; outer phyllaries with a long terminal spine - - **1. C. cardunculus**
Leaves unarmed; outer phyllaries without a terminal spine, or with a very short spine
C. SCOLYMUS (p. 962)

1. C. cardunculus *L.*, Sp. Plant., ed. 1, 827 (1753); Unger et Kotschy, Die Insel Cypern, 246 (1865); Boiss., Fl. Orient., 3: 557 (1875); Post in Mém. Herb. Boiss., 18: 95 (1900); H. Stuart Thompson in Journ. Bot., 44: 333 (1906); Holmboe, Veg. Cypr., 186 (1914); Post, Fl. Pal., ed. 2, 2: 94 (1933); Osorio-Tafall et Seraphim, List Vasc. Plants Cyprus, 107 (1973); Davis, Fl. Turkey, 5: 328 (1975).
    C. horrida Aiton, Hort. Kew., ed. 1, 3: 148 (1789); Gaudry, Recherches Sci. en Orient, 187 (1855); H. Stuart Thompson in Journ. Bot., 44: 333 (1906).
    TYPE: "*in* Creta".

Perennial to 1 m. high; stems robust, conspicuously sulcate, thinly arachnoid-tomentose; leaves numerous, extending up the stems almost to the capitulum, oblong-lanceolate in outline, 30–40 cm. or more long, 10–15 cm. wide, green and glabrescent above, thinly tomentose below, deeply pinnatisect, the divisions narrowly oblong, up to 7 cm. long, 1·5 cm. wide, apex caudate-acuminate, margins with deltoid, spine-tipped lobes, rhachis and leaf-base with digitate spine-clusters; capitula solitary, terminal, or in sparsely branched corymbs; involucre broadly ovoid-subconical, 3–4·5 cm. long, 3–4 cm. wide at base; phyllaries closely imbricate, the outer 3–4 series with a broad flattish base, about 1 cm. long and almost as wide, abruptly constricted above into a rigid, suberect terminal spine 2–3 cm. long, the inner 3 series with a broad, ovate, acute, mucronate or very shortly spinose, purplish, reflexed apical appendage, 10–13 mm. long, 5–7 mm. wide; florets purple-violet, slightly exceeding the involucre; corolla-tube slender, glabrous, 30–40 mm. long, limb narrow, about 14 mm. long, 2 mm. wide, lobes linear, erect, about 12 mm. long, 0·6 mm. wide; filaments about 2 mm. long, free, glandular; anthers linear, about 11 mm. long, 0·6 mm. wide, exserted, apical appendage blunt, base shortly sagittate; style to 50 mm. long, branches erect, about 15 mm. long, united almost to the blunt, bifid apex, base slightly swollen and spongy, shortly papillose with short longitudinal furrows; receptacle foveolate, the walls densely setose with long, white bristles. Immature achene oblong, 4-angled, truncate, about 4 mm. long, 2 mm. wide; pappus copious, tinged brown towards base, hairs free or almost free, rigid, elegantly plumose.

HAB.: Roadsides, field borders; dry open hillsides; near sea-level to 1,000 ft. alt.; fl. May–July.

DISTR.: Divisions 4–7. Widespread in the Mediterranean region, but often as an introduction or escape from cultivation.

4. Neighbourhood of Larnaca, 1853–54, *Gaudry.*
5. Athalassa, 1933, *Mrs Tracey*; Syngrasis, 1936, *Syngrassides* 1241 !
6. Neighbourhood of Nicosia, 1853–54, *Gaudry*; and, 1898, *Post* s.n.; Aghirda, 1905, *Holmboe* 829; near Skylloura, 1953, *Merton* s.n. !
7. Between Panteleimon Monastery and Paleomylos, 1862, *Kotschy* 942.

NOTES: Gaudry, Holmboe and Merton all remark on the abundance of *C. cardunculus* in the region of the Mesaoria, where it would appear to be, if not wholly indigenous, at least completely naturalized. Unfortunately comments on its abundance are not borne out by a proportionate number of collectings, perhaps because Cynaras are tedious plants to press. *C. cardunculus* is said to provide the Cardoon of gardens, grown, like Celery, for its blanched petiole, but less popular in cultivation than *C. scolymus* L., the Globe Artichoke. Gaudry says of the wild *C. cardunculus* of Cyprus: "La partie comestible de son calice est plus délicate et plus tendre que dans les artichauts cultivés".

C. SCOLYMUS *L.*, Sp. Plant., ed. 1, 827 (1753); Unger et Kotschy, Die Insel Cypern, 246 (1865); Post, Fl. Pal., ed. 2, 2: 94 (1933); Davis, Fl. Turkey, 5: 328 (1975), the Globe Artichoke, generally supposed to be a cultivated derivative of *C. cardunculus* L., with unarmed, deeply 2–3-lobed leaves and large capitula with blunt or shortly acute, or at most very briefly spinose phyllaries, is widely grown in Cyprus for the succulent receptacles and phyllary-bases. Plants grown from seed are said to revert to the spiny *C. cardunculus*.

2. C. cornigera *Lindley* in Sibth. et Sm., Fl. Graec., 9: 25 (1837), 10: 74 (1840); Davis, Fl. Turkey, 5: 329 (1975).

C. sibthorpiana Boiss. et Heldr. in Boiss., Diagn., 1, 10: 94 (1849); Unger et Kotschy, Die Insel Cypern, 247 (1865); Boiss., Fl. Orient., 3: 557 (1875), Suppl., 309 (1888); Holmboe, Veg. Cypr., 186 (1914); Post, Fl. Pal., ed. 2, 2: 94 (1933); Osorio-Tafall et Seraphim, List Vasc. Plants Cyprus, 107 (1973).

[*C. humilis* (non L.) Sm. in Sibth. et Sm., Fl. Graec. Prodr., 2: 157 (1813), Fl. Graec., 9: 25, t. 835 (1837); Poech, Enum. Plant. Ins. Cypr., 19 (1842).]

TYPE: "In insulâ Cypro, et in Peloponneso". (OXF).

Erect perennial 15–30 cm. high; stems sulcate, arachnoid-floccose, unbranched or sparingly branched, sometimes much reduced; leaves mostly basal or sub-basal, spreading, lanceolate in outline, 15–50 cm. long, 5–20 cm. wide, coriaceous, glabrescent and bright green above with paler nervation, closely greyish-tomentose below, pinnatisect almost to midrib into narrowly oblong, acute, coarsely and rigidly spinose-dentate divisions 3–10 cm. long, 1–3·5 cm. wide; petiole obscure, spinose; capitula solitary or 2–3 in lax corymbs; involucre depressed-globose or hemispherical, 3–5 cm. diam.; phyllaries closely imbricate in many series, glabrous or subglabrous, the outermost small, broadly oblong, 5–10 mm. long, and about as wide, with an abruptly acute or shortly spinose, erect or reflexed apex, inner progressively larger, with a broad, oblong, verrucose or smooth base 1–2 cm. long, 0·8–1·5 cm. wide, abruptly constricted into a stout, erect or spreading terminal spine 1·5–4·5 cm. long; florets dirty white, slightly exceeding, or overtopped by phyllary-spines; corolla-tube slender, about 3 cm. long, limb oblique, narrowly tubular, 12–13 mm. long, 2 mm. wide, lobes linear, erect, subacute, about 0·8 mm. wide; filaments free, glandular, about 2·5 mm. long; anthers not exserted, linear, about 8–9 mm. long, 0·5 mm. wide, apical appendage blunt, base shortly sagittate; style exserted, about 5 cm. long, branches erect, about 10 mm. long, united almost to shortly bifid apex, base indistinctly dilated, minutely papillose-pilose. Achenes oblong, slightly compressed, smooth, light brown, about 4 mm. long, 2·5 mm. wide; pappus copious, hairs free to, or almost to base, silky, rather rigid, white, about 1·5 cm. long, elegantly plumose.

HAB.: Rocky pastures or rock-crevices, usually on chalk or limestone; sea-level to 1,000 ft. alt.; fl. April–May.

DISTR.: Divisions 1, 3–6. Also S. Greece, Aegean Islands, Egypt, Libya.

1. Side of Sotira, near Fontana Amorosa, 1787, *Sibthorp*.
3. Between Mazotos and Moni, 1862, *Kotschy* 571! Near Mazotos, 1981, *Hewer* 4732A!
4. Pergamos, 1955, *Merton* 2214! Cape Greco, 1962, *Meikle* 2597! Cape Kiti, 1979, *Edmondson & McClintock* E 2966!
5. Kythrea, 1880, *Sintenis & Rigo* 544!
6. Mile 3, Nicosia–Myrtou road, 1950, *Chapman* 582!

44. SILYBUM *Adans.*

Fam. Plantes, 2: 116, 605 (1763) nom. cons.

Robust spinose biennials; stems erect, simple or branched; leaves alternate, sinuate-lobed or pinnatisect with spinose margins, commonly blotched with white, the basal forming a rosette, the cauline decurrent;

capitula large, solitary, homogamous, discoid; involucre subglobose; phyllaries in several series, imbricate, coriaceous, apex dilated, spinulose-pectinate, generally produced into a long, rigid, recurved spine; florets pink or purple; corolla-tube slender, limb deeply cleft into 5 narrow lobes; filaments glandular-papillose, partly united; anthers with an acute apical appendage, base shortly sagittate; style exserted, branches united almost to apex; receptacle flat, fleshy, densely setose. Achenes narrowly oblong-obovate, somewhat compressed, glabrous, with a raised apical rim and anthophore-stump; hilum narrow, sub-basal; pappus deciduous, hairs in several series, scabridulous, connate at base.

Two species in the Mediterranean region and eastwards to Afghanistan; *S. marianum* (L.) Gaertner cultivated and naturalized in western and central Europe.

1. **S. marianum** (*L.*) *Gaertner*, De Fruct. et Sem., 2: 378 (1791); Poech, Enum. Plant. Ins. Cypr., 19 (1842) as "*Sylibum*"; Unger et Kotschy, Die Insel Cypern, 246 (1865) as "*Sylibum*"; Boiss., Fl. Orient., 3: 556 (1875); H. Stuart Thompson in Journ. Bot., 44: 333 (1906); Holmboe, Veg. Cypr., 186 (1914); Post, Fl. Pal., ed. 2, 2: 93 (1933); Osorio-Tafall et Seraphim, List Vasc. Plants Cyprus, 107 (1973); Davis, Fl. Turkey, 5: 369 (1975); Feinbrun, Fl. Palaest., 3: 382, t. 642 (1978).
   Carduus marianus L., Sp. Plant., ed. 1, 823 (1753); Sibth. et Sm., Fl. Graec. Prodr., 2: 150 (1813).
   TYPE: "*in* Angliae, Galliae, Italiae *aggeribus ruderatis*".

Erect, glabrous biennial, 25–200 cm. high; stems commonly unbranched except in region of inflorescence, conspicuously sulcate; leaves oblanceolate, coarsely sinuate-lobed with spinose margins, generally white-blotched, the basal 9–50 cm. long, 5–25 cm. wide, the cauline diminishing in size upwards, sessile, shortly decurrent, with distinct, spinose basal auricles; capitula solitary at the ends of the inflorescence-branches; involucre subglobose, about 4–5 cm. diam.; phyllaries imbricate, glabrous, with a broadly oblong base about 1–1·5 cm. long, 0·8–1 cm. wide, apex rigidly pectinate-spinose, produced into a stout spine 2·5–7 cm. long; florets purplish; corolla-tube slender, about 20 mm. long, limb narrowly infundibuliform, about 8 mm. long, with erecto-patent, linear, subacute lobes, about 6–7 mm. long, 0·8 mm. wide; filaments glandular-papillose, connate into a tube about 1·5 mm. long; anthers linear, about 6 mm. long, 0·7 mm. wide, with a sharply acute apical appendage; style about 30 mm. long, branches united almost to the subacute apex, without a pilose basal annulus. Achene 6–7 mm. long, about 3 mm. wide, oblong, distinctly compressed, streaked fuscous-brown, apical rim pallid, very narrow, anthophore-stump about 0·8 mm. long; pappus-hairs rather rigid, white, up to 15 mm. long, scabridulous or rather remotely barbellate.

HAB.: Field-margins, roadsides, waste land; sea-level to 3,000 ft. alt.; fl. March–May.

DISTR.: Divisions 2, 4, 5, 6, 8. Widespread in the Mediterranean region eastwards to Afghanistan; naturalized in western Europe and elsewhere.

2. Khrysorroyiatissa Monastery, 1862, *Kotschy* 697.
4. Cape Kiti, 1981, *Hewer* 4729!
5. Kythrea, 1862, *Kotschy* 346; also, 1941, *Davis* 2909! Mile 3, Nicosia–Famagusta road, 1950, *Chapman* 602!
6. Syrianokhori, 1935, *Syngrassides* 634! Kykko metokhi, Nicosia, 1936, *Syngrassides* 1122!
8. Near Cape Andreas, 1880, *Sintenis*.

NOTES: Holmboe (Veg. Cypr., 186) says, "common in lower regions", but, like other prickly plants, seldom collected.

45. STAEHELINA *L.*

Sp. Plant., ed. 1, 840 (1753).
Gen. Plant., ed. 5, 364 (1754).

Glandular, unarmed, shrubs and subshrubs; leaves alternate, simple, entire or pinnatifid; capitula homogamous, discoid, solitary or in clusters; involucre oblong or cylindrical; phyllaries in several series, closely imbricate, entire, obtuse or acute; florets whitish; corolla-tube slender, limb slightly oblique, narrow, deeply 5-lobed, with narrow, erect lobes; filaments free, glabrous; anthers linear with an acute apical appendage, base caudate-sagittate; style shortly exserted, branches erect, or somewhat recurved, with a pilose basal annulus; receptacle flat, sparingly scaly or naked. Achenes narrowly oblong-cylindrical, compressed or subterete, generally sericeous with a narrow apical rim; hilum basal; pappus in one series, scaly, the scales divided into slender, scabridulous or barbellate bristles.

About 6 species widely distributed in the Mediterranean region.

1. **S. lobelii** *DC.* in Ann. Mus. Hist. Nat. Paris, 16: 194, t. 5, fig. 30 (1810); Post, Fl. Pal., ed. 2, 2: 83 (1933); B. L. Burtt et Davis in Kew Bull., 1949: 107 (1949); Chapman, Cyprus Trees and Shrubs, 81 (1949); Osorio-Tafall et Seraphim, List Vasc. Plants Cyprus, 107 (1973); Davis, Fl. Turkey, 5: 595 (1975).

　　S. apiculata Labill., Icones Plant. Syr., 4: 3, t. 1 (1812); Boiss., Fl. Orient., 3: 455 (1875).

TYPE: "in Syria *Labillardière*" (G, FI).

Caespitose subshrub 13–30 (–60) cm. high, with numerous stems arising from a stout woody rootstock; stems slender, erect, glandular-hispidulous, generally unbranched except in region of inflorescence; leaves numerous, alternate, linear-lanceolate, sessile, acute, bright green, glandular and viscid when young, 10–20 mm. long, 1·5–3 mm. wide, margins entire; inflorescence loosely branched; capitula solitary, terminal on the branches; involucre cylindrical, about 10 mm. long, 4 mm. wide, phyllaries closely imbricate, thinly floccose dorsally, glandular, the outermost ovate, subacute, about 2·5 mm. long, 2 mm. wide, the inner progressively longer and more acute, the innermost about 10 mm. long, 2 mm. wide; florets whitish, exserted; corolla-tube slender, about 5 mm. long, limb oblique, narrowly infundibuliform, deeply divided into 5 erecto-patent, linear, obtuse lobes about 4–5 mm. long, 0·5 mm. wide; filaments about 1·5 mm. long; anthers linear, about 6 mm. long, 0·5 mm. wide, apical appendage acute, base caudate-sagittate, tails minutely fimbriate; style 13–14 mm. long, branches about 1 mm. long, slightly recurved, blunt, with a distinct, pilose basal annulus. Achenes narrowly oblong-fusiform, about 5 mm. long, 1 mm. wide, dark grey-brown, densely whitish-sericeous, apex truncate with a narrow rim; pappus persistent, pale whitish-brown, about 9 mm. long, the individual scales divided pinnately almost to base into slender, barbellate bristles.

HAB.: Vertical fissures of limestone rocks; 1,600–2,500 ft. alt.; fl. July–Sept.

DISTR.: Division 7, rare. Also S. Turkey and Lebanon.

7. Between Halevga and summit of Yaïla, 1941, *Davis* 2834! ½ mile W. of St. Hilarion Castle, in a gorge on Karmi clay, 1949, *Casey* 217! Above Boghazi Farm, Bellapais, 1957, *Poore* 48!

46. SERRATULA *L.*

Sp. Plant., ed. 1, 816 (1753).
Gen. Plant., ed. 5, 357 (1754).

Unarmed perennial herbs; stems simple or branched; leaves alternate, simple or lyrate-pinnatisect; inflorescence corymbose or capitula solitary, terminal on inflorescence branches; capitula homogamous, discoid (or plants rarely dioecious); involucre cylindrical to subglobose; phyllaries in several series, closely imbricate, scarious, yellowish or purplish; florets purple, pink, white or yellow; corolla-tube slender, limb often slightly oblique, deeply 5-lobed, with narrow lobes; filaments free, glabrous; anthers with an acute apical appendage and a sagittate or caudate-sagittate base; style shortly branched, the branches often with a pilose basal annulus; receptacle scaly or setaceous. Achenes subterete or compressed, glabrous, with a narrow apical rim; hilum oblique; pappus in several series, persistent or deciduous, bristles unequal, barbellate or shortly plumose.

About 70 species widely distributed in the northern hemisphere.

1. **S. cerinthifolia** (*Sm.*) *Boiss.*, Fl. Orient, 3: 585 (1875); Holmboe, Veg. Cypr., 187 (1914); Post, Fl. Pal., ed. 2, 2: 98 (1933); Lindberg f., Iter Cypr., 36 (1946); Osorio-Tafall et Seraphim, List Vasc. Plants Cyprus, 107 (1973); Davis, Fl. Turkey, 5: 453 (1975); Feinbrun, Fl. Palaest., 3: 388, t. 654 (1978).
Centaurea cerinthifolia Sm. in Sibth. et Sm., Fl. Graec. Prodr., 2: 197 (1813).
Serratula cordata Cass. in Dict. Sci. Nat., 50: 468 (1827); Unger et Kotschy, Die Insel Cypern, 248 (1865).
[*Centaurea behen* (non L.) Sm. in Sibth. et Sm., Fl. Graec. Prodr., 2: 200 (1813); Sibth. in Walpole, Travels, 26 (1820); Poech, Enum. Plant. Ins. Cypr., 19 (1942); Unger et Kotschy, Die Insel Cypern, 244 (1865); H. Stuart Thompson in Journ. Bot., 44: 333 (1906).]
TYPE: "In herbario Sibthorpiano, ut et Linnaeano, sine nomine aut loco natali invenitur". (? OXF).

Erect perennial 20–50 cm. high, with a thick woody rootstock; stems glabrous, rather slender, sulcate, generally branched a little way above base; leaves glabrous, glaucescent, the basal simple, elliptic, entire, 5–12 cm. long, 3–4 cm. wide, or lyrate-pinnatisect with 5–8 oblong lateral divisions and a much larger, obovate or elliptic, blunt or acute, terminal lobe; petioles slender, glabrous, 8–12 cm. long; cauline leaves sessile, amplexicaul-auriculate, oblong, 2–12 cm. long, 1–2 cm. wide, apex rounded or shortly acute, basal auricles rounded; capitula solitary at the ends of the leafy branches; involucre oblong-cylindric, about 2·5 cm. long, 1·2–1·5 cm. wide; phyllaries scarious, yellowish, glabrous, scabridulous dorsally, acute, erect, the outermost ovate, about 2·5 mm. long, 2 mm. wide, the inner linear, up to 25 mm. long, 2 mm. wide; florets yellow, distinctly exserted from involucre; corolla-tube slender, 15–16 mm. long, limb about 12 mm. long, 2 mm. wide, lobes linear, erect, about 7–8 mm. long, 0·8 mm. wide, subacute; filaments about 2 mm. long; anthers linear, about 9 mm. long, 0·8 mm. wide, apical appendage acute, base shortly sagittate; style about 25 mm. long, branches about 6·5 mm. long, acute, united almost to apex, basal pilose annulus indistinct. Achenes oblong-fusiform, slightly compressed, glabrous, longitudinally ribbed, about 6 mm. long, 2 mm. wide, apical rim obscure; pappus persistent, copious, bristles up to 12 mm. long, whitish, barbellate or shortly plumose, free to the base.

HAB.: Dry stony mountainsides on limestone or igneous rocks; field-margins, roadsides; sea-level to 4,000 ft. alt.; fl. June–Aug.

DISTR.: Divisions 1–3, 7. Also Turkey and Palestine east to Iran.

1. Slopes of Sotira, 1787, *Sibthorp*; Akamas, 1905, *Holmboe* 780; Dhrousha, 1941, *Davis* 3227!
2. Pano Panayia, 1905, *Holmboe* 1057; Ayios Nikolaos, 1914, *Feilden* s.n.! Platres, 1929, *C. B. Ussher* 20! also, 1937, *Kennedy* 970! and, 1955, *Kennedy* 1854! Perapedhi, 1940, *Davis* 1833! and, 1970, *A. Genneou* in ARI 1519!
3. Lefkara, 1967, *Merton* in ARI 645!
7. Near Ayios Khrysostomos Monastery, 1862, *Kotschy* 392; near Paleomylos, 1862, *Kotschy* 938; St. Hilarion, 1880–1939, *Sintenis & Rigo* 541! *Syngrassides* 631! *Lindberg f.* s.n.! Above Kythrea, 1903, *A. G. & M. E. Lascelles* s.n.! Armenian Monastery, 1912, *Haradjian* 385! Lapithos, 1940, *Davis* 1737! Kyrenia, 1931–1973, *J. A. Tracey* 67! *Casey* 54! 775! *G. E. Atherton* 141! 159! *P. Laukkonen* 345! Koronia, 137, *Chapman* 296! Ayios Yeoryios (Kyrenia), 1970, *A. Genneou* in ARI 1693!

47. CRUPINA (Pers.) DC.

Ann. Mus. Hist. Nat. Paris, 16: 157 (1810).
Centaurea L. sect. *Crupina* Pers., Syn. Plant., 2: 488 (1807); M. Le Vaillant in
Rev. Gén. Bot., 77: 111–124 (1970).

Erect annuals; leaves alternate, generally pinnatifid or pinnatisect, the margins ciliate with multicellular, asperate hairs; inflorescences lax, branched; capitula heterogamous, disciform, the marginal florets sterile; involucre narrowly ovoid or fusiform; phyllaries in several series, loosely imbricate, scarious, unarmed; florets pink or purple; corolla-tube hairy, limb regularly and deeply 5-lobed; filaments free, papillose; anthers with an acute apical appendage and very shortly sagittate base; style-branches united almost to apex; receptacle flat, bristly. Achenes cylindrical, terete or somewhat compressed, sericeous, apex truncate; pappus persistent, setaceous, generally fuscous or brownish, consisting of several outer rows of barbellate bristles, and an inner row of shorter scales.

Four species in S. Europe, the Mediterranean region, and western Asia.

1. **C. crupinastrum** (*Moris*) *Vis.*, Fl. Dalm., 2: 42, t. 51, fig. 3 (1847); Boiss., Fl. Orient., 3: 699 (1875); Holmboe, Veg. Cypr., 187 (1914); Post, Fl. Pal., ed. 2, 2: 118 (1933); Lindberg f., Iter Cypr., 35 (1946); Osorio-Tafall et Seraphim, List Vasc. Plants Cyprus, 107 (1973); Davis, Fl. Turkey, 5: 587 (1975); Feinbrun, Fl. Palaest., 3: 388, t. 653 (1978).
   Centaurea crupinastrum Moris, Fl. Sard., 2: 443 (1840–43); Sintenis in Oesterr. Bot. Zeitschr., 31: 392 (1881), 32: 192, 398 (1882).
   [*Crupina vulgaris* (non Cass.) Kotschy in Unger et Kotschy, Die Insel Cypern, 244 (1865); H. Stuart Thompson in Journ. Bot., 44: 333 (1906).]
   TYPE: Sardinia; "In apricis maritimis montanisque: frequens". (TO).

Slender, erect annual, (8–) 15–70 cm. high; stems angled or deeply sulcate, glabrescent or thinly pilose, unbranched or much branched with erecto-patent branches; cotyledons broadly and bluntly obovate, often persisting until anthesis; basal leaves coarsely and bluntly pinnatisect, with an enlarged terminal lobe, crenate-serrate, glabrous or thinly pilose; lower cauline leaves oblong in outline, 3–10 cm. long, 0·8–3 cm. wide, sessile, deeply pinnatisect, divisions narrowly oblong or linear, obtuse or acute, irregularly and bluntly serrulate, fringed with rigid, glochidiate bristles; upper cauline leaves smaller, with narrower, linear divisions; rhachis thinly subadpressed-villose; inflorescence lax, with slender, leafless branches up to 20 cm. long; capitula solitary, terminal; involucre narrowly ovoid, 1·2–2 cm. long, 0·5–1·4 cm. wide; phyllaries loosely imbricate, minutely glandular-scabridulous dorsally, scarious, often purplish, with narrow, hyaline margins, the outer ovate, acute, 2–4 mm. long, 1·5–3 mm. wide, the inner lanceolate-acuminate, up to 20 mm. long, 3 mm. wide; florets bright purple, often distinctly exceeding involucre, the outer sterile florets somewhat enlarged and radiant, with spreading lobes; corolla-tube slender, 5–7 mm. long, thinly clothed with spreading, smooth-walled, hyaline hairs, limb narrowly infundibuliform, 6–9 mm. long, divided into linear, subacute lobes, 5–7 mm. long, 0·8–1 mm. wide; filaments about 1·8 mm. long, with a brownish (? glandular) apex; anthers linear-acuminate, about 6 mm. long, 0·5 mm. wide; style about 15 mm. long, branches erect, about 2 mm. long, connate almost to the subacute apex, with a pilose basal annulus. Achenes cylindrical-turbinate, not much compressed, about 3–3·5 mm. long, 2·5 mm. wide, closely velutinous-sericeous, apex truncate, base glabrous, shining dark brown, sharply keeled, with a narrow lateral hilum; pappus generally fuscous-brown, the outer bristles barbellate or scabridulous, about 6–7 mm. long, the outermost rows reduced to short scales, the innermost row conspicuously distinct, consisting of 5 acute, erect, scales about 1 mm. long, 0·4 mm. wide surrounding the flat top of the achene.

HAB.: On rocky coastal pastures or on dry hillsides in garigue; sometimes in open Pine or Cypress forest; sea-level to 4,000 ft. alt.; fl. April–May.

DISTR.: Divisions 1–8, common. Mediterranean region eastwards to Iran.
1. Stroumbi, 1913, *Haradjian* 785! Dhrousha, 1941, *Davis* 3289! Near Lyso, 1979, *Edmondson & McClintock* E 2827!
2. Platres, 1938, *Kennedy* 1452! 1453! Kryos Potamos, 3,200 ft. alt., 1938, *Kennedy* 1454! Mandria, 1938, *Kennedy* 1455! Mesapotamos Monastery, 1939, *Lindberg f.* s.n.
3. Polemidhia hills, 1948, *Mavromoustakis* s.n.! Episkopi, 1960, *N. Macdonald* 82!
4. Kalopsidha, 1905, *Holmboe* 402; Famagusta, 1912, *Haradjian* 120! Cape Greco, 1958, *N. Macdonald* 22!
5. Lakovounara forest, 1950, *Chapman* 79! Athalassa, 1967, *Merton* in ARI 154!
6. Nicosia, 1905, *Holmboe* 286; also, 1927, *Rev. A. Huddle* 71! Dhiorios, 1952, *Probyn* 85!
7. Common; Kyrenia, Pentadaktylos, Kantara, Lapithos, St. Hilarion, Halevga, Vasilia, Melounda, etc. *Sintenis & Rigo* 298! *A. G. & M. E. Lascelles* s.n.! *Clement Reid* s.n.! *Syngrassides* 8! *C. H. Wyatt* 30! *Lindberg f.* s.n.! *Mavromoustakis* 28! *Davis* 2961A! *Kennedy* 1661! *Mavromoustakis* s.n.! *Merton* 2231! *Miss Mapple* 62! *G. E. Atherton* 1300! *J. B. Suart* 76! *P. Laukkonen* 295! etc.
8. Cape Andreas, 1880, *Sintenis & Rigo*; Yioti, 1962, *Meikle* 2522!

NOTES: Some specimens with small capitula (e.g. *Kennedy* 1452, 1453; *Chapman* 79; *Merton* 2231) have been determined as *Crupina vulgaris* Cass., but I think in error, since all have the achenes of *C. crupinastrum*, with a sharply keeled base, and narrow lateral hilum. In *C. vulgaris*, which otherwise much resembles *C. crupinastrum*, the achene is bullet-shaped, with a blunt, rounded base and a basal, orbicular hilum. Its distribution is generally more western and northern than that of *C. crupinastrum*, but it might occur in Cyprus.

One sheet of *Kennedy* 1453 has achenes with the fuscous pappus usual in *C. crupinastrum*; in a second sheet of the same number the pappus is almost white.

48. CENTAUREA L.

Sp. Plant., ed. 1, 909 (1753).
Gen. Plant., ed. 5, 389 (1754).
G. Wagenitz in Flora, 142: 213–279 (1955).

Annual, biennial or perennial herbs or subshrubs; sometimes acaulescent; leaves alternate, or occasionally all radical, commonly pinnatifid or pinnatisect, not spiny; inflorescences branched or capitula solitary, terminal; capitula heterogamous, disciform, or the marginal florets enlarged and radiant; involucre fusiform, ovoid or subglobose; phyllaries imbricate in several series, commonly scarious or papyraceous, sometimes indurated, with a terminal appendage varying greatly in shape and texture, entire or fimbriate, blunt or sometimes produced into a rigid spine; florets purple, pink, blue, yellow or white, the marginal often enlarged and sterile, the central hermaphrodite; corolla tubular or infundibuliform, 4–8-lobed, the lobes generally narrow; filaments free; anthers with a sagittate or caudate-sagittate base; style-branches erect or spreading, acute or obtuse; receptacle bristly. Achenes oblong, usually glabrous and somewhat compressed, with a truncate apex and lateral hilum; pappus present or absent, persistent or caducous, consisting of several series of unequal scabridulous, barbellate or plumose bristles, the innermost row often reduced and scale-like.

More than 600 species widely distributed in the northern hemisphere; a few are cultivated as ornamentals.

Involucre conspicuously spinose:
  Florets yellow:
    Stems winged with decurrent leaf-bases; involucral spines slender -   - **2. C. solstitialis**
    Stems not winged; involucral spines stout, rigid   -   -   -   -   - **5. C. hyalolepis**
  Florets purple, pink or white:
    Pappus present; involucre broadly ovoid or subglobose, 10–15 mm. diam.   **3. C. iberica**
    Pappus absent; involucre narrowly ovoid or fusiform, 4–6 mm. diam.
                     4. C. calcitrapa ssp. **angusticeps**

Involucre unarmed or phyllaries with a very short apical spinule:
 Annuals with conspicuously radiant, bright blue marginal florets; pappus absent
                                                                    7. C. cyanoides
 Perennials with thick taproots; marginal florets pink or mauve; pappus present:
  Stems elongate, prostrate; capitula numerous, small, forming sub-racemose inflorescences;
   phyllaries without a conspicuous hyaline margin   -   -   -   -   **1. C. veneris**
  Stems short, erect, or plants commonly acaulescent; capitula few, rather large, solitary or
   clustered; phyllaries with a conspicuous hyaline margin   -   -   **6. C. aegialophila**

SECT. 1. **Acrolophus** (*Cass.*) *DC.* Biennial or perennial herbs or
subshrubs; lower leaves lyrate, pinnatisect or bipinnatisect, upper often
simple; inflorescence usually branched, bearing numerous small capitula;
phyllaries deltoid or subdeltoid, fimbriate or denticulate, often with a
terminal mucro or short spine; florets pink, purple or occasionally yellow,
the marginal somewhat radiant, without staminodes. Achenes with a short
scabridulous pappus, the innermost row of bristles generally short, or
pappus sometimes absent.

1. C. veneris *B. L. Burtt et Davis* in Kew Bull., 1949: 105 (1949); Osorio-Tafall et Seraphim, List
  Vasc. Plants Cyprus, 108 (1973).
  TYPE: Cyprus; Akamas, in Pine forest on serpentine, 1941, *Davis* 3315 (K !).

Prostrate, greyish-tomentose perennial with a tough woody rootstock;
stems numerous, 8–25 cm. long, radiating to form a loose mat; leaves very
variable, the basal broadly spathulate and entire, or oblong and pinnatisect,
2–5 cm. long, 1–2 cm. wide, the divisions broadly obovate, entire or
denticulate, rounded or subacute, greenish and glandular-punctate above,
grey-tomentose below; petiole up to 2·5 cm. long, sheathing at the base;
cauline leaves usually simple, spathulate, entire, denticulate or sometimes
lobed, 0·5–1·5 cm. long, 0·4–1 cm. wide, apex obtuse or rounded, base
tapering to an obscure petiole; inflorescence sub-racemose, the capitula
terminal on short shoots towards the tips of the branches; involucre ovoid,
8–10 mm. long, 5–8 mm. wide; phyllaries rather loosely imbricate,
canescent, the outer ovate, about 4–6 mm. long, 2–3 mm. wide, brownish at
apex, with a blunt, decurrent, fimbriate-hyaline appendage, the innermost
oblong, about 7–9 mm. long, 2–2·5 mm. wide, with an entire or shortly
denticulate appendage; florets mauve-pink, the marginal radiant with a
slender tube 9–10 mm. long and an infundibuliform limb, divided almost to
base into 5 linear-lanceolate, acuminate, unequal lobes 3–5 mm. long, 0·5–1
mm. wide; inner florets narrowly tubular, with a tube about 6 mm. long, and
a glandular limb about 5 mm. long, 2 mm. wide, divided into 5, subequal,
erect, linear, acute lobes; filaments about 2 mm. long, long-papillose;
anthers linear, about 5 mm. long, 0·3 mm. wide, with a very long, acute
apical appendage; style about 12–13 mm. long; branches about 1·5 mm.
long, erect and connate almost to apex, with a conspicuous, pilose basal
annulus. Achenes oblong, compressed, about 3–3·5 mm. long, 1·8–2 mm.
wide, shining fuscous-brown, glabrous or very thinly pilose, longitudinally
ribbed; pappus whitish, the bristles unequal, barbellate, up to 2 mm. long,
the innermost series short.

HAB.: In open Pine forest and gravelly clearings on serpentine; 1,000–2,500 ft. alt.; fl.
April–May.

DISTR.: Divisions 1, 3, local and confined to serpentine. Endemic.
1. Akamas, central ridge above Smyies, 1937–1956, *Chapman* 283 ! *Davis* 3315 ! *Kennedy* 1597 !
  1630 ! *Merton* 3014 !
3. Limassol Forest, 1,500 ft. alt., 1956, *Poore* 29 ! Kakomallis above Louvaras, 1962, *Meikle*
  2884 !
NOTES: Probably allied to *C. cuneifolia* Sm., but only very remotely so, and with no more
obvious relationships. Its distribution recalls that of *Alyssum akamasicum* and *A.*

chondrogynum, but no clear distinction can be made between the populations found on the two widely separated areas of serpentine.

C. BEHEN *L.*, Sp. Plant., ed. 1, 914 (1753) (sect. *Microlophus* (Cass.) DC.). A single, very large, oblong, basal leaf, with conspicuous reticulate venation, collected 19 April, 1962, close to Engomi road junction near (5) Salamis (*Meikle* 2633 !) is probably referable to this species, otherwise unrecorded from Cyprus. *C. behen* is a robust perennial, often more than 1 m. high with branched inflorescences of yellow-flowered, ovoid-conical capitula. It is widely distributed in the eastern Mediterranean region, and eastwards to Iran.

SECT. 2. **Mesocentron** (*Cass.*) *DC.* Erect, branched annuals or biennials; median and upper cauline leaves decurrent; capitula ovoid or subglobose; phyllaries with spinose (and basally spinulose-ciliate) appendages; florets generally yellow. Achenes with a scabridulous pappus, the innermost row of bristles shorter than the outer.

2. **C. solstitialis** *L.*, Sp. Plant., ed. 1, 917 (1753); Unger et Kotschy, Die Insel Cypern, 244 (1865) – as *C. "solsticialis"*; Boiss., Fl. Orient., 3: 685 (1875); H. Stuart Thompson in Journ. Bot., 44: 333 (1906); Holmboe, Veg. Cypr., 188 (1914); Post, Fl. Pal., ed. 2, 2: 114 (1933); Osorio-Tafall et Seraphim, List Vasc. Plants Cyprus, 108 (1973); Davis, Fl. Turkey, 5: 542 (1975). TYPE: "*in* Gallia, Anglia, Italia".

Erect annual or biennial 15–100 cm. high; stems usually branched, thinly arachnoid-floccose, sulcate, conspicuously winged with long-decurrent leaf-bases; leaves grey-green, thinly floccose-tomentose, the basal generally lyrate-pinnatifid, to 15 cm. long, 6 cm. wide, the upper linear, entire, 1–6 cm. long, 0·2–0·8 cm. wide, apex cuspidate or apiculate, base strongly decurrent; inflorescence loosely branched; capitula terminal on long erecto-patent branches; involucre broadly ovoid or subglobose, 1–1·5 cm. long and almost as wide; phyllaries closely imbricate, thinly floccose, ovate, the outer tipped with a slender yellow (or reddish) spine, 0·5–4 cm. long, bordered basally on either side with 1–2 much shorter lateral spinules, the inner elongate, with a rounded or emarginate, hyaline-membranous apical appendage; florets usually yellow (rarely pink), the marginal not radiant; corolla-tube glabrous, 12–13 mm. long, limb tubular-infundibuliform, about 7 mm. long, 1·8 mm. wide, lobes erect, linear, subequal, subacute, about 4 mm. long, 0·4 mm. wide; filaments about 1·8 mm. long, glandular-papillose; anthers linear, 6–7 mm. long, 0·3 mm. wide, with a very long, acute apical appendage; style included, about 19-20 mm. long, branches erect, blunt, about 2·5 mm. long, connate to apex. Achenes dimorphic, oblong-fusiform, 2–2·5 mm. long, 1–1·3 mm. wide, the marginal blackish, without pappus, the inner brown with a copious pappus of white, scabridulous bristles up to 4 mm. long.

HAB.: Cultivated fields and waste ground; ? sea-level to 1,000 ft. alt.; fl. June–Aug.

DISTR.: Divisions 3, 5, evidently rare. Widespread as a weed in Europe and the Mediterranean region, western Asia and North America.

3. Limassol, ? 1905, *Michaelides* teste *Holmboe*.
5. Syngrasis, 1862, *Kotschy* 541a.

NOTES: I admit this widespread weed to the flora of Cyprus solely on the authority of Kotschy and Holmboe; the latter's statement that it is "common in lower regions" is clearly untrue, but it is difficult to think what Holmboe had in mind. Most likely *C. solstitialis* is no more than a casual occurrence on the island. It is also missing from most of Palestine.

SECT. 3. **Calcitrapa** *DC.* Annuals or biennials (rarely perennials); stems usually much branched; leaves lobed or pinnatisect, not decurrent; involucre ovoid or subglobose; phyllaries often with a hyaline margin and a spinose appendage; florets pink, purple or yellow, the marginal not radiant. Achenes sometimes epappose, pappus, when present, of scabridulous bristles, the innermost row shorter than the outer.

3. C. iberica *Trev. ex Spreng.*, Syst. Veg., 3: 406 (1826); Poech, Enum. Plant. Ins. Cypr., 19 (1842); Boiss., Fl. Orient., 3: 690 (1875); Holmboe, Veg. Cypr., 190 (1914); Post, Fl. Pal., ed. 2, 2: 115 (1933); Osorio-Tafall et Seraphim, List Vasc. Plants Cyprus, 108 (1973); Plitmann in Israel Journ. Bot., 22: 48 (1973), 24: 19 (1975); Davis, Fl. Turkey, 5: 543 (1975); Feinbrun, Fl. Palaest., 3: 399, t. 673 (1978).
TYPE: "Armenia, Caucas.".

Erect or spreading annual or biennial 20–60 (–100) cm. high; stems glabrous or subglabrous, longitudinally sulcate, repeatedly branched with divaricate branches; leaves sparsely crispate-pilose or glabrescent, the lower petiolate, oblong in outline, 10–15 (or more) cm. long, 4–6 (or more) cm. wide, deeply pinnatisect into narrowly oblong, lacerate-dentate or irregularly lobulate divisions, the upper smaller and sessile, the uppermost frequently simple, entire, bluntly oblanceolate, or irregularly lobed or pinnatisect; inflorescence lax, spreading; capitula terminal on short floral branchlets commonly leafy to apex; involucre coriaceous, glossy yellowish, glabrous, broadly ovoid or subglobose, 1–2 cm. long, 1–1·5 cm. wide; phyllaries closely imbricate, the outer broadly ovate or suborbicular, 5–8 mm. long, 4–6 mm. wide, with a narrow hyaline margin and a stout, yellowish spinose appendage 1·5–3 cm. long, flanked on either side at the base by 2–3 slender spinules 2–4 mm. long; inner phyllaries oblong, up to 20 mm. long, 6 mm. wide, with a broad, rounded or emarginate, erose, hyaline-edged appendage; florets purple, mauve, pink or almost white, conspicuously exserted from involucre; corolla-tube about 9 mm. long, limb subcylindrical, thinly glandular externally, about 9 mm. long, divided almost half-way into 5 linear, acute, erect lobes about 0·5 mm. wide; filaments shortly pilose-papillose, about 4 mm. long; anthers linear, about 6·5 mm. long, 0·5 mm. wide, with a very long, acute apical appendage; style included, 16–17 mm. long, branches erect, about 1·5 mm. long, connate almost to the blunt apex, pilose basal annulus conspicuous. Achenes oblong, 3–3·5 mm. long, 1·5 mm. wide, brown, glabrous; pappus-bristles whitish, scabridulous or minutely barbellate, up to 4 mm. long, the innermost row much shorter than the outer.

HAB.: Cultivated and waste ground; roadsides; 500–3,500 ft. alt.; fl. May–July.

DISTR.: Division 2, apparently very local.

2. Galata, 1880, *Sintenis & Rigo* 791 ! Evrykhou, 1840, *Kotschy* 2, also, 1880, *Sintenis & Rigo* 792 ! also, 1955, *Merton* 2361 ! Lefka, 1905, *Holmboe* 811; between Galata and Kakopetria, 1936, *Syngrassides* 1269 ! Kakopetria, 1949, *Casey* 857 ! Platres, 1938, *Kennedy* 1457 !

NOTES: Holmboe's comment that *C. iberica* is common in the western part of the Mesaoria is possibly true, but is not reflected by collectings, which are few in number, and almost wholly from the Troödos massif.

4. C. calcitrapa *L.*, Sp. Plant., ed. 1, 917 (1753); Clarke, Travels, 2 (3): 718 (1816); Unger et Kotschy, Die Insel Cypern, 244 (1865); Boiss., Fl. Orient., 3: 689 (1875); Holmboe, Veg. Cypr., 188 (1914); Post, Fl. Pal., ed. 2, 2: 115 (1933); Osorio-Tafall et Seraphim, List Vasc. Plants Cyprus, 108 (1973); Plitmann in Israel Journ. Bot., 24: 16 (1975); Davis, Fl. Turkey, 5: 544 (1975).

Spreading annual or biennial 15–60 cm. high; stems sulcate, thinly pilose with multicellular hairs, glandular in the upper part, branches numerous, divaricate; basal leaves oblong in outline, glandular-pubescent, 12–20 cm. long, 4–7 cm. wide, deeply pinnatisect into narrowly oblong or oblanceolate, blunt or acute, entire or irregularly serrulate divisions; petiole short, often obscure; upper leaves rather sparse, much smaller than the basal, with narrower, almost linear divisions; petiole generally wanting; inflorescence a repeatedly branched, open dichasium; capitula usually solitary, sometimes paired, subsessile or very shortly pedunculate; involucre ovoid or almost fusiform, 8–20 (–25) mm. long, 5–8 (–10) mm. wide; phyllaries coriaceous, yellowish, glabrous, closely imbricate, the outer ovate or broadly oblong, 3–8 mm. long, 2·5–6 mm. wide, with an apical, yellowish spine 0·6–2·5 cm. long, flanked basally on either side by 2–3 short (1–4 mm. long) spinules, margins of phyllaries perceptibly, but very narrowly, hyaline; innermost phyllaries narrowly oblong, appendage broad, rounded or emarginate, erose-denticulate, hyaline or brownish; florets well exserted beyond involucre, mauve or purple-pink; corolla-tube about 10 mm. long, limb cylindrical, about 10 mm. long, 1·5 mm. wide, glandular externally, lobes erect, acute, about 3 mm. long, 0·5 mm. wide; filaments about 2 mm. long, glandular-papillose; anthers linear, acute, 6–8 mm. long, 0·4 mm. wide; style included, about 20 mm. long, branches about 1·5–2 mm. long, erect and united to the blunt apex, with a distinct, pilose basal annulus. Achenes oblong, about 2·5 mm. long, 1·3 mm. wide, brown, glabrous, apex truncate, puberulous but without pappus.

ssp. **angusticeps** (*Lindberg f.*) *Meikle* **stat. nov.**
  C. angusticeps Lindberg f. in Act. Soc. Sci. Fenn. [Iter Cypr.], n.s. B., 2: 34 (1946); Osorio-Tafall et Seraphim, List Vasc. Plants Cyprus, 108 (1973).
  C. angusticeps Lindberg f. forma *albiflora* Lindberg f. in Act. Soc. Sci. Fenn. [Iter Cypr.], n.s. B., 2: 34 (1946).
  TYPE: Cyprus; "In campo sterili juxta opp. Nicosia", 8 June 1939, *Lindberg f.* s.n. (H, K!).

Capitula numerous; involucre small, narrowly ovoid or fusiform, 8–14 mm. long, 4–6 mm. wide; spines rather slender, generally less than 20 mm. long and 2·5 mm. wide at base; upper cauline leaves generally divided into narrow, linear, entire or subentire lobes.

HAB.: Cultivated and fallow fields, waste ground, roadsides; sea-level to 500 ft. alt.; fl. June–July.

DISTR.: Divisions 4–7, locally abundant. ? Endemic.

4. Between Prastio and Kouklia, 1939, *Lindberg f.* (form with white florets).
5. Kythrea, 1936, *Syngrassides* 1284! 1286!
6. Near Peristerona, 1880, *Sintenis & Rigo* 793! also, 1955, *Merton* 2447! Yerolakkos, 1932, *Syngrassides* 197! Nicosia, 1939, *Lindberg f.* s.n.!
7. Abundant around Kyrenia, 1880–1955, *Sintenis & Rigo* 543 partly! *M. E. Dray* 4! *Casey* 797! *G. E. Atherton* 63! 162! 184!

NOTES: This is almost certainly the plant cited by Plitmann (Israel Journ. Bot., 24: 16; 1975) as "a) Heads many, dense, small (narrow); spines 10–16 per head, 9–18 mm. long. Cyprus: Larnaka, 26. viii. 1931, *Eig* s.n. (HUJ)". G. Wagenitz (in Davis, Fl. Turkey, 5: 545; 1975) correctly remarks that it is very near to *C. calcitrapa* L. ssp. *cilicica* (Boiss. et Bal.) Wagenitz, but, to judge from the very few specimens seen, this has even smaller capitula, with shorter, more slender involucral spines. The Cyprus specimens are, with a few exceptions, very uniform, and certainly distinct from European material of *C. calcitrapa*. The exceptions are *Davis* 3542 from (3) Ayios Amvrosios, near Pakhna, 21 May 1941, with very stout spines and large capitula, superficially resembling *C. iberica* Trev. ex Spreng., but with the achenes apparently epappose. It is unfortunately immature, and unsatisfactory for critical examination. *S. Economides* in ARI 1015 from (?3) "Ashielia [? Akhelia] to the sea", 25 July 1967, comes closer to typical *C. calcitrapa*, and might pass for such, were it not for the fact that *some* of the achenes have a well-developed pappus. It probably belongs to a group of puzzling specimens mentioned in the notes under the following species, *C. hyalolepis* Boiss., and may be of hybrid origin.

5. C. hyalolepis *Boiss.*, Diagn., 1, 6: 133 (1845); Unger et Kotschy, Die Insel Cypern, 244 (1865); Post, Fl. Pal., ed. 2, 2: 116 (1933); Plitmann in Israel Journ. Bot., 22: 49 (1973), 24: 18 (1975); Davis, Fl. Turkey, 5: 545 (1975); Feinbrun, Fl. Palaest., 3: 399, t. 674, 675 (1978).
  C. pallescens Del. var. *hyalolepis* (Boiss.) Boiss., Fl. Orient., 3: 691 (1875); C. et W. Barbey, Herborisations au Levant, 99 (1880).
  C. pallescens Del. ssp. *hyalolepis* (Boiss.) Holmboe, Veg. Cypr., 190 (1914); Lindberg f., Iter Cypr., 34 (1946); Osorio-Tafall et Seraphim, List Vasc. Plants Cyprus, 108 (1973).

? *C. monacantha* Clarke, Travels, 2 (1): 354 (1812); Poech, Enum. Plant. Ins. Cypr., 19 (1842).
[*C. pallescens* (non Del.) Sintenis in Oesterr. Bot. Zeitschr., 32: 292 (1882); R. M. Nattrass, First List Cyprus Fungi, 19 (1937).]
TYPE: "in *Syriâ* Aucher No. 3136, inter *Bagdad* et *Alep* Olivier, inter *Alep* et *Mossoul* Kotschy herb. Mus. Vindob." (G).

Erect annual or biennial 20–60 cm. high; stems sulcate or angled, thinly pilose with white multicellular hairs, glabrescent and glandular above, usually much branched, with spreading branches; basal leaves oblong in outline, 4–15 cm. long, 2·5–7 cm. wide, thinly scabrid-hispidulous above and below, coarsely lacerate-pinnatisect or bipinnatisect into a varying number of narrow or rather broad, irregularly serrate divisions, the serrations commonly terminating in a minute hard spine or apiculus; petiole flat, winged, up to 7 cm. long; cauline leaves sessile, coarsely pinnatisect, or sometimes simple, oblanceolate, with spinulose-serrulate margins, diminishing in size upwards, the topmost almost subtending the capitula; inflorescence spreading, much branched with rather slender angled or subterete branches; capitula solitary, subsessile; involucre broadly ovoid or subglobose, yellowish-green, glabrous, 6–14 mm. long and often almost as wide; phyllaries closely imbricate, broadly ovate or suborbicular, 3–10 mm. long, with a conspicuous hyaline-membranous border, often 2–3 mm. wide; outer phyllaries terminating in a spinose appendage, the spine usually stout and rigid, 1–3 cm. long, yellowish, generally (but not always) flanked with 2–3 short spinules on either side at the base; innermost phyllaries with a broad, rounded or emarginate, denticulate, decurrent, membranous, hyaline or brownish appendage; florets yellow, exserted beyond apex of involucre; corolla-tube very slender, 7–8 mm. long, limb narrowly cylindrical, 7–8 mm. long, thinly glandular externally, lobes linear, acute, erect, about 4–5 mm. long, 0·5 mm. wide; filaments about 2 mm. long, thinly glandular-papillose; anthers linear, about 7 mm. long, 0·4 mm. wide, apex acute; style included, about 15 mm. long, branches erect, united to blunt apex, with a distinct pilose basal annulus. Achenes oblong, brown, glabrous, about 2·5 mm. long, 1·2 mm. wide; pappus whitish, of unequal scabridulous bristles up to 3 mm. long, the innermost row much shorter than the others.

HAB.: Cultivated and fallow fields, waste ground, roadsides, sometimes on bare, rocky ground; sea-level to 2,200 ft. alt.; fl. April–July.

DISTR.: Divisions 1–8, common. S. Greece and Crete eastwards to Iran.

1. Paphos, 1913, *Haradjian* 632! Polis, 1955, *Merton* 2167! Khlorakas, 1967, *Economides* in ARI 965!
2. Kykko, 1905, *Holmboe* 1030; Kakopetria, 1955, *Merton* 2319!
3. Between Kolossi and Limassol, 1862, *Kotschy* 548; Kophinou, 1905, *Holmboe* 572; Alaminos, 1937, *Syngrassides* 1503! Kolossi, 1939, *Lindberg f.* s.n.! Kannaviou, 1939, *Lindberg f.* s.n.; near Limassol, 1957, *Mavromoustakis* 6/57!
4. Larnaca, 1877, ? *J. Ball in herb. Post* 611! also *J. Ball* s.n.! Between Dherinia and Paralimni, 1936, *Syngrassides* 882! Between Prastio and Kouklia, 1939, *Lindberg f.* s.n.
5. Kythrea, 1880, *Sintenis & Rigo* 297! Near Lefkoniko, 1880, *Sintenis & Rigo* 297a! Near Athalassa, 1933, *Syngrassides* 1369! also, 1967, *Merton* in ARI 814! Road to Famagusta, 1972, *W. R. Price* 997!
6. Nicosia, 1929, *C. B. Ussher* 5! also, 1930, *F. A. Rogers* 0652! 1939, *Lindberg f.* s.n.! and, 1973, *P. Laukkonen* 353! Mile 3, Nicosia–Myrtou road, 1950, *Chapman* 581! Morphou, 1972, *W. R. Price* 1044!
7. Kyrenia, 1880, *Sintenis & Rigo* 543 partly! Between Vasilia and Larnaka tis Lapithou, 1937, *Syngrassides* 1541! Kyrenia area, 1955, *G. E. Atherton* 91! 102!
8. Between Rizokarpaso and Apostolos Andreas Forest, 1957, *Merton* 3074! Koma tou Yialou, 1970, *A. Genneou* in ARI 1464!

NOTES: *Kotschy* 2, collected in 1840 near Evrykhou, and originally identified by Poech as *C. iberica* Trev. ex Spreng., but later (Die Insel Cypern, 245) transferred to *C. hyalolepis*, has been returned to *C. iberica*, since it comes from a locality where this species is locally plentiful. Kotschy may, however, have been correct in his re-identification.

Centaurea hyalolepis, one of the more distinct members of the *calcitrapa* group, is usually easy to identify. Two collectings are, however, puzzling: *Lindberg f.* s.n. collected at (3) Kolossi on 3 July 1939, and named "*Centaurea iberica* Trev. ex Spreng. var. *hermonis* Boiss. f. *alba*", closely resembles *C. calcitrapa* L. ssp. *angusticeps* (Lindberg f.) Meikle in habit and shape of capitulum, but the phyllaries have a relatively wide scarious margin, and the achenes have a pappus. *Merton* 1948, collected, 1954, at (2) Ayios Nikolaos, west of Platres, is similar, but with more globose involucres; two of the specimens have pappose achenes, two are epappose. The identity of these puzzling plants is uncertain, but it would not be unreasonable to suggest that they are hybrids between *C. calcitrapa* and *C. hyalolepis*, this would at least explain their "intermediate" characteristics. It may be of interest to note that the Lindberg f. plants had white florets, as did also some of the Merton plants, though others are noted as having a purple corolla.

[C. ACICULARIS *Sm.* in Sibth. et Sm., Fl. Graec. Prodr., 2: 203 (1813), Fl. Graec. 10: 8, t. 911 (1840) (sect. *Acrocentron* (Cass.) DC.), a rosulate, acaulescent perennial with a thick taproot, arachnoid-tomentose, lacerate-pinnatisect leaves, yellow florets and an oblong-ovoid involucre armed with slender, yellow spines and spinules, is recorded by Smith "In Liro [Leros] et Cypro insulis", but has not since been collected in Cyprus, and probably never occurred there. All subsequent records (Unger et Kotschy, Die Insel Cypern, 244 (1865); Boiss., Fl. Orient., 3: 678 (1875); Holmboe, Veg. Cypr., 188 (1914) and Osorio-Tafall et Seraphim, List Vasc. Plants Cyprus, 108; 1973) are based on the statement in Fl. Graec. Prodr. and Fl. Graec. *C. acicularis* is a native of Greece, the Aegean Islands and western Turkey.]

SECT. 4. **Aegialophila** (*Boiss. et Heldr.*) *O. Hoffm.* Acaulescent perennials, with a thick, often tuberous, taproot; leaves simple or lyrate-pinnatisect; florets pink, the marginal sometimes radiant; involucre ovoid with a decurrent, entire, scarious appendage usually terminating in a short spinule. Achenes densely pilose; pappus deciduous, sericeous, hairs plumose or barbellate, the innermost row shorter than the others.

6. C. aegialophila *Wagenitz* in Notes Roy. Bot. Gard. Edinb., 33: 230 (1974); Davis, Fl. Turkey, 5: 558 (1975).

Aegialophila cretica Boiss. et Heldr. in Boiss., Diagn., 1, 10: 106 (1849); Fl. Orient., 3: 704 (1875); Unger et Kotschy, Die Insel Cypern, 245 (1865); Rechinger f. in Oesterr. Bot. Zeitschr., 94: 191 (1948).

Centaurea cretica (Boiss. et Heldr.) Nyman, Sylloge, 34 (1854); Holmboe, Veg. Cypr., 187 (1914); Lindberg f., Iter Cypr., 34 (1946); Osorio-Tafall et Seraphim, List Vasc. Plants Cyprus, 107 (1973) non *C. cretica* (L.) Spreng. (1826) nom. illeg.

Aegialophila cretica Boiss. et Heldr. var. *alpina* Post in Mém. Herb. Boiss., 18: 96 (1900).

[*Centaurea pumila* (non *C. pumilio* L.) Clarke, Travels 2 (3): 718 (1816); Poech, Enum. Plant. Ins. Cypr., 19 (1842); Gaudry, Recherches Sci. en Orient, 188 (1855).]

[*Aegialophila pumila* (non *Ae. pumilio* ["*pumila*"] (L.) Boiss.) Kotschy in Unger et Kotschy, Die Insel Cypern, 245 (1865); Sintenis in Oesterr. Bot. Zeitschr., 32: 364 (1882); Boiss., Fl. Orient., Suppl., 317 (1888).]

TYPE: Crete; "in arenosis maritimis Cretae ad partum *Sitia* infrà *Piskokephali* Eparchiae *Sitia*", *Heldreich* (G).

Perennial with a thick, tuberous taproot; stems usually very short or wanting, occasionally up to 20 cm. long, sulcate, arachnoid-floccose; leaves alternate, usually spreading and rosulate, simple and ovate, or oblong in outline, lyrate with a few, much reduced lateral lobes, or deeply and irregularly pinnatisect (or bipinnatisect) with blunt lobes and lobules, 1–10 (–15) cm. long, 0·8–6 cm. wide, subglabrous or shortly puberulous above, sometimes thinly arachnoid-tomentose below; petiole often well developed, up to 10 cm. long; capitula solitary or in dense, subsessile clusters of 4–6 or more; involucre broadly ovoid or oblong-ovoid, 13–30 mm. long, 9–28 mm. wide; phyllaries rather loosely imbricate in several series, chartaceous, greenish or purplish, subglabrous, the outer ovate, 3–20 mm. long, 3–6 mm.

wide, with a distinct, but narrow, hyaline-membranous, entire, or erose
margin, and a rigid apical spinule seldom more than 2 mm. long, innermost
phyllaries narrowly oblong, up to 30 mm. long, 5 mm. wide, with a brownish,
scarious, rounded, pectinate-denticulate apical appendage; florets pink or
pale mauve-purple, the marginal neuter, but not or scarcely radiant; corolla-
tube slender, about 12–15 mm. long, limb of neuter marginal florets
infundibuliform, about 12 mm. long, divided into 5 lanceolate, acuminate,
erect lobes about 5–7 mm. long, 1–2 mm. wide at base, limb of fertile florets
narrowly cylindrical, lobes erect, about 6 mm. long, less than 1 mm. wide at
base; filaments pilose, 3–4·5 mm. long; anthers linear, about 8 mm. long, 0·8
mm. wide, apical appendage short, subacute, base very shortly glandular-
sagittate; style to 30 mm. long, shortly exserted, branches divergent,
subacute, about 1·5 mm. long, basal annulus obsolete. Achenes fusiform,
about 6 mm. long, 2 mm. wide at truncate apex, densely subadpressed
sericeous-pilose; pappus up to 12 mm. long, hairs barbellate or shortly
plumose, whitish with a purplish base, innermost row rigid, scaly, oblong,
about 1–1·5 mm. long.

HAB.: Sandy seashores, sand-dunes; dry stony hillsides; open Pine forest; sea-level to 6,000 ft.
alt.; fl. March–Aug.

DISTR.: Divisions 1–8, common. Also Crete and S. Turkey.

1. Ktima, 1862, *Kotschy* 673; Dhrousha, 1941, *Davis* s.n.! Akamas, 1941, *Davis* 3317! Sotira,
   Kako Skala, near Baths of Aphrodite, 1949, *Casey* 763! Paphos, Tombs of the Kings, 1978,
   J. Holub s.n.! Near Polis, 1979, *Edmondson & McClintock* E 2778!
2. Troödos, ? 1898, *Post* s.n.; above Prodhromos, 1905, *Holmboe* 958; also, 1940, *Davis* 1810!
   Khionistra, 5,000–6,000 ft. alt., *C. B. Ussher* 72! *Kennedy* 972! *Wyatt* 14! *Lindberg f.* s.n.;
   Davis 1871! *H. R. P. Dickson* 02! Platres, 1937, *Kennedy* 971! and, 1938, *Kennedy* 1458!
   Kryos Potamos, 5,700 ft. alt., 1938, *Kennedy* 1459! Stavros tis Psokas, 1940, *Davis* 1758!
   Yialia, 1955, *Merton* 2114! Prodhromos, 1955, *G. E. Atherton* 577! Perapedhi, 1970, *A.
   Genneou* in ARI 1505!
3. Limassol, 1859, *Kotschy* 490; also, 1977, *J. J. Wood* 13! Between Nisou and Stavrovouni,
   1933, *Syngrassides* 3! Zakaki marshes, 1939, *Mavromoustakis* 103! Paramali, 1959, *N.
   Macdonald* 41! Amathus, 1963, *J. B. Suart* 75! also, 1978, *J. Holub* s.n.!
4. Near Livadhia, 1840, *Kotschy* s.n.; near Larnaca, 1877, *J. Ball* s.n.! Famagusta, 1903, *A. G.
   & M. E. Lascelles* s.n.! Ayia Napa, 1905, *Holmboe* 6; near Ayios Memnon, 1948,
   Mavromoustakis s.n.! Cape Greco, 1958, *N. Macdonald* 39! Cape Kiti, 1979, *Edmondson &
   McClintock* E 2965! Dhekelia, 1981, *Hewer* 4709!
5. Near Kythrea, 1880, *Sintenis & Rigo* 296! Salamis, 1905, *Holmboe* 456; Athalassa, 1936,
   Syngrassides 1234! also, 1939, *Lindberg f.* s.n.! Trikomo, 1939, *Lindberg f.* s.n.
6. Near Morphou, 1937, *Miss Godman* 16! Nicosia, 1939, *Lindberg f.* s.n.; Ayia Irini, 1967,
   Economides in ARI 935!
7. Kyrenia, 1931, *J. A. Tracey* 65! also, 1936, *Kennedy* 973! 974! 1939, *Lindberg f* s.n.! 1955,
   Miss Mapple 99! and, 1955, *G. E. Atherton* 102! Ayios Amvrosios, 1932, *Syngrassides* 449!
   Halevga, 1940, *Davis* 1741! Between Orga and Vasilia, 1941, *Davis* 2107! Mile 6,
   Kyrenia–Akanthou road, 1950, *Chapman* 614!
8. Yialousa, 1880, *Sintenis & Rigo* 296! Apostolos Andreas, 1940, *Davis* 2308! also, 1957,
   Merton 3088! and, 1973, *P. Laukkonen* 286! Koma tou Yialou, 1970, *A. Genneou* in ARI
   1477!

NOTES: Plants from high altitudes on the Troödos range are uniform in having simple ovate,
or very sparingly lobed lyrate leaves, but similar plants are to be found elsewhere, especially on
the Northern Range and in the Karpas peninsula, so that the taxonomic value of *Aegialophila
cretica* Boiss. et Heldr. var. *alpina* Post cannot be more than trifling. The Cyprus populations of
C. aegialophila are readily distinguished from Egyptian, Syrian and Palestine *C. pumilio* L. by
their glabrescent leaves, very shortly spinulose phyllaries, and inconspicuously radiant
capitula.

SECT. 5. **Cyanus** (*Mill.*) *DC.* Annuals or perennials; leaves tomentose or
glabrescent, simple, pinnatifid or pinnatisect, the upper often decurrent;
involucre ovoid or subglobose; phyllaries usually not very hard or
coriaceous, appendage more or less deltoid, scarious, hyaline or brownish,
decurrent, generally pectinate or ciliate, without a terminal spine or spinule;
florets blue, violet, purple or yellow, the marginal conspicuously radiant,

more than 5-lobed. Achenes barbate around the hilum, usually pappose, the bristles scabridulous, frequently shorter than the achene.

7. **C. cyanoides** *Berggren et Wahlenb.* in Isis, 21: 971 (1828); Boiss., Fl. Orient., 3: 635 (1875), Suppl. 313 (1888); Holmboe, Veg. Cypr., 188 (1914); Post, Fl. Pal., ed. 2, 2: 105 (1933); Osorio-Tafall et Seraphim, List Vasc. Plants Cyprus, 108 (1973); Feinbrun, Fl. Palaest., 3: 393, t. 659 (1978).

TYPE: Lebanon; on sandy hills by the roadside near Bsharri, 17 June 1821, *Berggren* (? UPS).

Erect annual 15–40 cm. high; stems usually branched, thinly arachnoid-floccose, angled or sulcate; basal leaves simple or lyrate, 2–5 cm. long, 0·8–2 cm. wide, arachnoid-canescent, margins entire or denticulate, petiole short but usually distinct; cauline leaves sessile, oblong or narrowly obovate, 1·5–6 cm. long, 0·3–2·5 cm. wide, entire or serrulate, thinly floccose, apex rounded or acute-apiculate; capitula solitary, terminal on naked peduncles 5–10 cm. long and distinctly incrassate towards apex; involucre ovoid, 1–1·5 cm. long, 0·8–1 cm. wide; phyllaries ovate, 3–10 mm. long, 2·5–3 mm. wide, glabrous, apical appendage 1·5–2·5 mm. long, dark brown at base, pectinate with narrow, hyaline teeth; marginal florets conspicuously radiant, bright blue, corolla-tube slender, 7–8 mm. long, limb narrowly infundibuliform, divided almost to base into 3 main divisions 7–8 mm. long, the 2 lateral divisions lanceolate, often lacerate, about 1·2 mm. wide, the median division 5–6 mm. wide, with 5 lanceolate, acute, apical lobes about 3–4 mm. long, 1 mm. wide at base; central florets purplish, tube about 7 mm. long, limb cylindrical, 6–7 mm. long, commonly fissured to base down one side, with 5, linear, acute, apical lobes about 3 mm. long, 0·5 mm. wide at base; filaments free, about 2 mm. long, glabrous with a conspicuous fimbriate zone about the middle; anthers about 6 mm. long, appendage acute, about 2 mm. long, base indistinctly sagittate; style exserted, about 16 mm. long, branches divaricate, subacute, about 0·5 mm. long, with a distinct basal annulus. Achenes oblong, about 5 mm. long, 2·2 mm. wide, thinly puberulous, basal part livid or fuscous, striatulate, apex yellowish with a marginal excavation surrounding a narrow crest (or ? elaiosome), base dome-shaped; hilum orbicular; pappus absent.

HAB.: Cultivated fields; ? c. 1,000 ft. alt.; fl. April–June.

DISTR.: Division 7, very rare. Also Syria, Lebanon, Palestine.

7. Between Kyrenia and Pano Dhikomo, in fields, 1 June 1880, *Sintenis & Rigo* 614!

NOTES: Only once collected, and perhaps, with improvements in agricultural practice, now extinct.

[C. REUTERIANA *Boiss.*, Diagn., 1, 4: 18 (1844), a Turkish endemic, is erroneously recorded from Cyprus by Druce (Rep. B.E.C., 9: 470; 1931). As noted by B. L. Burtt (in Kew Bull., 1954: 70; 1954), the record is most probably a slip of the pen for *Crepis reuteriana* Boiss., q.v.]

49. MANTISALCA *Cass.*

Bull. Sci. Soc. Philom. Paris, 1818: 142 (1818).

Erect, unarmed biennials and perennials, much resembling *Centaurea*; stems usually branched, basal leaves often crowded, sinuate or pinnatifid; cauline leaves sparse, remote, much reduced, the upper linear or narrowly oblong, serrulate or lobulate, sessile; capitula terminal, usually solitary, heterogamous, the marginal florets somewhat radiant; involucre ovoid; phyllaries yellowish, scarious or coriaceous, imbricate in several series, often blackish at the apex, with a very short spinulose appendage; florets pink or mauve; corolla with a slender tube and 5–8-lobed limb; filaments free; anthers caudate-sagittate at base; receptacle setose. Achenes sometimes dimorphic, slightly compressed, longitudinally costate, transversely rugulose; hilum lateral; pappus in several series, the outer bristly, scabridulous, the innermost row consisting of one or more, broader, acuminate scales.

Five species in the Mediterranean region.

1. M. salmantica (*L.*) *Briq. et Cavill.* in Arch. Sci. Phys. Nat. Genève, ser. 5, 12: 111 (1930); Osorio-Tafall et Seraphim, List Vasc. Plants Cyprus, 108 (1973); Davis, Fl. Turkey, 5: 464 (1975).

    Centaurea salmantica L., Sp. Plant., ed. 1, 918 (1753); Lindberg f., Iter Cypr., 34 (1946).

    Microlonchus salmanticus (L.) DC., Prodr., 6: 563 (1838); Boiss., Fl. Orient., 3: 700 (1875).

TYPE: "*in* Europa *australi*".

Erect perennial 15–100 cm. high; stems slender, wiry, longitudinally ribbed, glabrous or subglabrous, usually branched; basal leaves often crowded into a tuft, obovate in outline, 5–12 cm. long, 1·5–3 cm. wide, thinly pilose, generally lyrate-pinnatifid with a broad, subacute terminal lobe, base tapering to a more or less distinct petiole; cauline leaves sparse, remote, linear or oblong, 1–3 cm. long, 0·2–1 cm. wide, sessile, subglabrous, generally serrulate or denticulate with spinule-tipped teeth, sometimes pinnatifid; capitula terminal on slender peduncles; involucre ovoid, 1–2 cm. long, 0·8–1·3 cm. wide; phyllaries scarious-indurate, ovate or oblong, yellowish with minutely arachnoid-ciliate margins, the outermost shortly ovate, about 2–3 mm. long and almost as wide, the inner oblong, up to 18 mm. long, 4 mm. wide, all blackish (or sphacelate) at apex with a very short, erect or deflexed, deciduous, spinule-appendage; florets pinkish or mauve, the marginal distinctly radiant; corolla-tube slender, glabrous, about 7-8 mm. long, limb of marginal florets infundibuliform, divided almost to base into 8 lobes, 5 of them linear-acuminate, about 8 mm. long, 0·5 mm. wide, the remaining 3 filiform, about 6 mm. long; inner (fertile) florets cylindrical, limb about 12 mm. long, 1–1·5 mm. wide, lobes subequal, erect, acuminate, about 5 mm. long, 0·5 mm. wide at base; filaments about 4 mm. long, almost smooth; anthers linear, 6-7 mm. long, 0·4 mm. wide, apical appendage about 2·5 mm. long, acuminate; style exserted, about 20 mm. long, branches about 2·5 mm. long, united below, free and divaricate towards the obtuse apex, basal annulus distinct. Achenes oblong-fusiform, about 4–5 mm. long, 2 mm. wide, dark brown, glabrous, distinctly ribbed longitudinally and rugulose transversely, the lateral hilum with a much thickened, pallid waxy border, evidently with an adnate elaiosome; pappus-bristles whitish, scabridulous, about 3 mm. long, the innermost (solitary) bristle conspicuously wider than the others (about 0·3 mm. wide at base) rigid, and about 3·5 mm. long.

HAB.: Maritime sands; near sea-level; fl. May–August.

DISTR.: Division 5, very rare. Widely distributed in S. Europe and the Mediterranean region, but becoming rare in the eastern Mediterranean.

5. Shore near Trikomo, 1939, *Lindberg f.* s.n.

NOTES: The specimen has not been examined, but the species was known to Harald Lindberg, having been collected by him in Morocco in 1926, so there is no reason to suspect a misidentification.

50. CNICUS *L.*

Sp. Plant., ed. 1, 826 (1753) nom. cons.

Gen. Plant., ed. 5, 358 (1754).

Arachnoid-villose annuals; stems spreading, often very short; leaves alternate, often forming basal rosettes, sinuate-pinnatifid, with minutely spinulose marginal teeth, capitula large, terminal, heterogamous, disciform; involucre broadly ovoid or subglobose; phyllaries in several series, imbricate, papery, foliaceous, the inner with a pectinate-spinulose apical appendage; florets yellowish, the marginal sterile; corolla-limb 5-lobed; filaments free, papillose; anthers shortly sagittate-caudate at base; style-branches short, with a pilose basal annulus; receptacle flat, densely setose. Achenes cylindrical, glabrous, costate, with a denticulate apical rim; hilum lateral; pappus persistent in 2 rows each of 10 bristles, the outer long, aristate, the inner short, fimbriate.

One species widely distributed in S. Europe and the Mediterranean region eastwards to Iran and Afghanistan.

1. **C. benedictus** *L.*, Sp. Plant., ed. 1, 826 (1753); Poech, Enum. Plant. Ins. Cypr., 19 (1842); Unger et Kotschy, Die Insel Cypern, 245 (1865); Boiss., Fl. Orient., 3: 705 (1875); H. Stuart Thompson in Journ. Bot., 44: 333 (1906); Holmboe, Veg. Cypr., 190 (1914); Post, Fl. Pal., ed. 2, 2: 119 (1933); Osorio-Tafall et Seraphim, List Vasc. Plants Cyprus, 108 (1973); Davis, Fl. Turkey, 5: 589 (1975); Feinbrun, Fl. Palaest., 3: 405, t. 686 (1978).

   Centaurea benedicta (*L.*) L., Sp. Plant., ed. 2, 2: 1296 (1763); Sibth. et Sm., Fl. Graec. Prodr., 2: 201 (1813), Fl. Graec., 10: 4, t. 906 (1840).

   TYPE: "*in* Chio, Lemno, Hispania *ad versuras agrorum*".

Sprawling or tufted annual, seldom more than 20 cm. high, often dwarfed and acaulescent with rosulate leaves; stems reddish-purple, rather conspicuously villose or thinly lanuginose, generally branched only at base; leaves narrowly oblong in outline, 3–15 cm. long, 0·8–4 cm. wide, bright green and thinly villose with multicellular hairs, coarsely sinuate-pinnatifid, with acute, erose-denticulate lobes, the marginal teeth normally terminating in a short spinule, basal leaves with purplish petioles up to 8 cm. long, cauline normally sessile; inflorescence branched, or capitula commonly solitary, terminal, subsessile or very shortly pedunculate; involucre 1·5–2·5 cm. long, 1·5–2 cm. wide; phyllaries rather loosely imbricate, oblong, 6–10 mm. wide, arachnoid-villose, greenish, the outermost about 8–10 mm. long, generally with a simple spinulose apex, the spinule purple, about 4–5 mm. long, the innermost much longer, with a conspicuous, purple, spreading, pectinately spinulose appendage, sometimes 10 mm. long; florets pale straw-yellow; marginal, sterile florets with a slender corolla-tube about 15–16 mm. long, limb infundibuliform, divided almost to base into 3 (–5) linear, acute lobes about 4 mm. long, 0·4 mm. wide; fertile florets with a slender corolla-tube and cylindrical, slightly oblique limb, about 7 mm. long, divided into 5 linear, subacute, erect lobes about 3·5–4 mm. long, 0·5 mm. wide; anthers usually purplish, linear, 3·5–4 mm. long, 0·4 mm. wide, apical appendage subacute, about 1·5 mm. long, base very shortly sagittate; filaments about 1·5–2 mm. long, thinly papillose; style included, 18–20 mm. long, branches about 0·8 mm. long, erect or slightly divergent, blunt, with a distinct basal annulus. Achenes cylindrical-fusiform, 7–8 mm. long, 2–2·5 mm. wide, prominently ribbed, greyish, shining, with a pale, toothed apical corona and a very conspicuous lateral hilum; outer pappus-bristles spreading, about 7-8 mm. long, sparingly scabridulous, pallid, inner bristles erect, barely 2·5 mm. long, conspicuously fimbriate.

HAB.: Field margins, vineyards, or in garigue on rocky hillsides, on limestone or igneous rocks; sea-level to 4,000 ft. alt.; fl. Febr.–May.

DISTR.: Divisions 2–8. General distribution that of the genus.

2. Galata, 1880, *Sintenis & Rigo* 295! Platres, 1937, *Kennedy* 976! and, 1954, *Merton* 1946! Agros, 1940, *Davis* 1892! 3053! Papoutsa, 1979, *Edmondson & McClintock* E 2869!
3. Alethriko, 1905, *Holmboe* 235; near Mosphiloti, 1967, *Merton* in ARI 612!
4. Dhekelia, 1936, *Syngrassides* 1055!
5. Athalassa, 1966, *Y. Ioannou* s.n.!
6. Nicosia, Government House gardens, 1935, *Syngrassides* 635! Dhiorios, 1962, *Meikle* 2318!
7. Mountains above Kythrea, 1880, *Sintenis & Rigo* 295a! Between Akanthou and Kalogrea, 1880, *Sintenis & Rigo* 295b! 6 miles E. of Kyrenia, 1949, *Casey* 494!
8. Near Ayios Symeon, 1880, *Sintenis.*

51. CARTHAMUS *L.*

Sp. Plant., ed. 1, 830 (1753).
Gen. Plant., ed. 5, 361 (1754).
A. Ashri et P. F. Knowles in Agron. Journ. 52: 11–17 (1960); P. Hanelt in Fedde Repert., 67: 41–180 (1963).

Mostly spinose annuals, occasionally unarmed or perennial; leaves alternate, entire, pinnatifid or pinnatisect, often rigidly coriaceous, frequently glandular; venation generally prominent, reticulate; cauline leaves sessile, frequently amplexicaul, but not decurrent; inflorescence corymbose or capitula solitary, terminal; capitula homogamous, discoid; phyllaries in several series, the outer usually much enlarged, foliaceous, often spinose, the inner scarious, acute, spine-tipped or sometimes with a dilated, pectinate-laciniate apical appendage; florets yellow, orange, white, purple or mauve; corolla-limb cylindrical, deeply 5-lobed; filaments free, glabrous or pilose; anthers sagittate at base, apex with a blunt appendage; style-branches long or short, usually with an obtuse apex; receptacle densely setose. Achenes glabrous, usually angled and obpyramidal with a rounded or truncate apex, frequently with a denticulate or pitted rim, the marginal often transversely rugose and epappose, the inner pappose; hilum lateral; pappus in several series, scaly or of scabridulous or plumose bristles.

About 15 species chiefly in the Mediterranean region, a few extending eastwards to Afghanistan and India. The genus is taxonomically difficult, and insufficiently collected; whenever possible flower-colour should be noted, and mature achenes included (preferably separately) with flowering specimens.

Leaves entire or minutely spinulose-serrulate; involucre unarmed; florets orange or reddish
C. TINCTORIUS (p. 978)
Leaves pinnatifid or pinnatilobed with spinose lobes; involucre spinose:
Florets yellow or yellowish   -   -   -   -   -   -   4. C. lanatus ssp. baeticus
Florets purple or mauve:
Median phyllaries with an expanded, ovate, pectinate-fimbriate terminal appendage
1. C. dentatus ssp. ruber
Median phyllaries without an expanded terminal appendage:
Median phyllaries terminating in an abruptly acute, spinose-cuspidate apex
2. C. boissieri
Median phyllaries tapering gradually to a slender, attenuate apex
3. C. tenuis ssp. foliosus

SECT. 1. **Carthamus.** Middle phyllaries entire; florets yellow, orange-red or white; corolla-lobes pilose at apex; pappus, when present, of long, plumose bristles.

C. TINCTORIUS *L.*, Sp. Plant., ed. 1, 830 (1753); Unger et Kotschy, Die Insel Cypern, 245 (1865); Boiss., Fl. Orient., 3: 709 (1875); Post in Mém. Herb. Boiss., 18: 96 (1900); Hanelt in

Fedde Repert., 67: 90 (1963); Osorio-Tafall et Seraphim, List Vasc. Plants Cyprus, 108 (1973); Davis, Fl. Turkey, 5: 591 (1975).

An erect annual 40–150 cm. high; stems whitish, glabrous, angled or sulcate; leaves ovate, glabrous, green, 3–9 cm. long, 1–2 cm. wide, sessile, margins entire or finely spinulose-serrate; inflorescence a lax corymb; capitula subglobose, about 2 cm. long and almost as wide; phyllaries subglabrous or thinly arachnoid, the outer with ovate or elliptic, entire, foliaceous appendages, the inner shortly acuminate, closely-ribbed, scarious; florets exserted, orange-yellow or reddish. Achenes pallid, 4-angled, epappose, about 8 mm. long, 4–5 mm. wide at apex.

The Safflower, of uncertain origin, has been cultivated in the Mediterranean region, Africa and Asia since ancient times, the florets providing a yellow or reddish dye, often used as a cheap substitute for the true Saffron (made from the stigmas of *Crocus sativus* L.). The large achenes are edible, and are used as poultry or cattle-food, or as the source of a useful oil.

Carthamus tinctorius is recorded from Cyprus by several authors; Kotschy says it is grown in the Mesaoria, and it has been collected (*Syngrassides* 715) from a vegetable garden between Limassol and Yermasoyia (3), but it is not native or naturalized on the island.

SECT. 2. **Odontagnathius** (*DC.*) *Hanelt* Middle phyllaries constricted above, with a dilated, pectinate-laciniate terminal appendage; florets purple; tips of corolla-lobe hairy; pappus scaly.

1. **C. dentatus** *Vahl*, Symb. Bot., 1: 69, t. 17 (1790); Boiss., Fl. Orient., 3: 709 (1875); Post, Fl. Pal., ed. 2, 2: 121 (1933); Hanelt in Fedde Repert., 67: 92 (1963); Davis, Fl. Turkey, 5: 592 (1975).

Erect annual up to 100 cm. high; stems subterete, whitish or pale brown, glandular-pilose or glabrescent, commonly unbranched in the lower half; basal leaves withered at anthesis, usually narrowly lyrate-pinnatifid; cauline leaves sessile, amplexicaul, ovate or lanceolate, acuminate, 1–6 cm. long, 0·5–1·5 cm. wide, greyish-green, shortly glandular-pubescent, apex tapering to a rigid spine, base rounded, margins with 3–7 pairs of spinules, or sometimes spinulose-pinnatifid; nervation prominent, reticulate; inflorescence a lax, or occasionally dense, corymb; capitula ovoid, narrowing noticeably towards apex, 2–3 cm. long, 1–2 cm. wide; outer phyllaries foliaceous, erect, spreading or recurved, lanceolate or almost linear, 2–5·5 cm. long, 0·4–0·8 cm. wide, carinate, margins with 3–6 pairs of spines; middle phyllaries oblong-lanceolate, to 3 cm. long, 0·7 cm. wide, erect, glandular-pubescent externally, constricted in the upper part and terminating in an ovate, pectinate-fimbriate or laciniate, often purple-tinged, spine-tipped appendage up to 1 cm. long and 0·4–0·5 cm. wide; innermost phyllaries about as long as the middle, gradually acuminate to a spinose apex; florets purple; corolla-tube very slender, 17–18 mm. long, limb cylindrical 8–10 mm. long, 2 mm. wide, lobes linear, subacute, erect, 5–6 mm. long, 0·4 mm. wide; filaments about 1·5 mm. long, with a median pilose zone; anthers about 7 mm. long, 0·4 mm. wide; style 25 mm. long, branches erect, connivent, bluntish, about 7 mm. long, without a distinct basal annulus. Central achenes obpyramidal, bluntly angled, pallid, about 4 mm. long and almost as wide at apex; pappus copious, scales narrowly lanceolate-acuminate, subfuscous or greyish, margins shortly spinulose-serrate, 10–16 (–18) mm. long, about 1–1·3 mm. wide, the innermost row as long as, or shorter than the outer, sometimes erose-truncate at apex.

ssp. **ruber** (*Link*) *Hanelt* in Fedde Repert., 67: 98 (1963).
 C. ruber Link in Linnaea, 9: 580 (1835); Čelakovsky in Sitz. Ber. Böhm. Gesellsch. Wissensch., 1885: 79, 81 (1886); Holmboe, Veg. Cypr., 190 (1914); Osorio-Tafall et Seraphim, List Vasc. Plants Cyprus, 108 (1973).
 [*C. glaucus* (non M. Bieb.) Boiss., Fl. Orient., Suppl., 317 (1888) pro parte quoad plant. cypr.]
 TYPE: Greece, Peloponnese, without exact locality. (B, ? destroyed).

Leaves narrowly lanceolate or almost linear; outer phyllaries generally much longer than the inner, linear-lanceolate, tapering to a spinose apex, spreading or recurved.

 HAB.: Roadsides; vineyards; dry, stony hillsides; 900–4,500 ft. alt.; fl. July–Sept.

 DISTR.: Divisions 2, 6. Also Greece, Crete, Aegean Islands, Turkey.

 2. Near Evrykhou, 1880, *Sintenis & Rigo* 789 ! Platres, 1937, *Kennedy* 977 ! Agros, 1940, *Davis* 1893 ! Prodhromos, 1952, *Casey* 1279 ! and, 1955, *G. E. Atherton* 590 ! Kyperounda, 1954, *Merton* 1937 !
 6. Myrtou, 1970, *A. Genneou* in ARI 1581 !

 NOTES: Davis (in Notes Roy. Bot. Gard. Edinb., 21: 128, 1953) rejects the distinction between *C. dentatus* Vahl and *C. ruber* Link and argues that the latter should be regarded as a synonym of the former. While agreeing that the supposed differences in pappus-structure do not bear critical examination, it seems to me that the narrow-leaved ssp. *ruber*, which is tolerably uniform in Cyprus, calls for recognition, at least at infraspecific level. It is certainly very distinct from the lectotype of *C. dentatus* as portrayed in Fedde Repert., 63: t. VIII, fig. 13.

 SECT. 3. **Lepidopappus** *Hanelt* Middle phyllaries entire or occasionally serrulate-laciniate towards apex; florets purple, mauve or whitish; tips of corolla-lobes glabrous; pappus scaly.

 2. **C. boissieri** *Hal.* in Verh. Zool.-bot. Gesellsch. Wien, 49: 186 (1889); Linderg f., Iter Cypr., 34 (1946); Hanelt in Fedde Repert., 67: 116 (1963); Davis, Fl. Turkey, 5: 594 (1975).
 C. glaucus M. Bieb. ssp. *boissieri* (Hal.) Holmboe, Veg. Cypr., 34 (1946); Osorio-Tafall et Seraphim, List Vasc. Plants Cyprus, 108 (1973).
 TYPE: Crete; without precise locality, 1847, *Raulin* (G).

Erect annual 20–50 cm. high; stems slender, whitish or pale brown, glandular-pubescent, sometimes thinly arachnoid, sparingly branched usually in the upper part only; leaves somewhat glaucous, glandular-pubescent or thinly arachnoid, the basal oblanceolate or narrowly obovate, 8–15 cm. long, 1·5–3·5 cm. wide, regularly pinnatifid, with short, acute, spinulose-dentate lobes, base tapering to an indistinct petiole; cauline leaves sessile, amplexicaul, patent or somewhat recurved, pinnatifid, 2–5 cm. long, 0·5–2 cm. wide, with 5–6 pairs of narrow, rigid, spine-tipped lobes; nervation prominent; inflorescence loosely or rather densely corymbose; capitula ovoid, narrowing towards apex, 15–25 mm. long, 8–12 mm. wide; outer phyllaries foliaceous, spreading or recurved, much resembling the upper cauline leaves, and generally exceeding the florets, inner phyllaries scarious, glandular-pubescent, ovate-acuminate, 1–2 cm. long, 0·7–1 cm. wide, acumen rather abruptly narrowed at apex, with a short, exserted spine; florets purple or mauve; corolla-tube very slender, about 17 mm. long, limb cylindrical, about 10 mm. long, 1·8 mm. wide, lobes erect, linear, about 6 mm. long, 0·8 mm. wide, subacute and slightly thickened at the glabrous apex; filaments about 2·5 mm. long, with a very conspicuous, pilose, median zone; anthers about 7 mm. long, 0·4 mm. wide, apex subacute; style about 27 mm. long, exserted, branches 6–8 mm. long, erect, connate, dark purplish. Central achenes bluntly angled, pale brown, about 5 mm. long, 4 mm. wide at the minutely denticulate, truncate apex; outer pappus-scales linear-lanceolate, up to 8 mm. long, 0·8 mm. wide, brownish, ciliate, innermost row pallid, truncate, 1·5–2 mm. long.

 HAB.: Cultivated and waste ground, sometimes on bare hillsides; sea-level to 500 ft. alt.; fl. June–Aug.

DISTR.: Divisions 4, 5, 7, 8. Also Crete and Aegean Islands.

4. Near Larnaca, 1939, *Lindberg f.* s.n.; near Famagusta, 1939, *Lindberg f.* s.n.
5. Athalassa, 1939, *Lindberg f.* s.n.; Trikomo, 1939, *Lindberg f.* s.n.!
7. Kyrenia, 1949, *Casey* 798! also, 1955, *G. E. Atherton* 371!
8. Between Rizokarpaso and Apostolos Andreas Monastery, 1932, *Syngrassides* 303!

3. C. tenuis (*Boiss. et Blanche*) *Bornm.* in Verh. Zool.-bot. Gesellsch. Wien, 48: 605 (1898); Post, Fl. Pal., ed. 2, 2: 121 (1933); Hanelt in Fedde Repert., 67: 117 (1963); Davis, Fl. Turkey, 5: 594 (1975).

Kentrophyllum tenue Boiss. et Blanche in Boiss., Diagn., 2, 3: 51 (1856).

Carthamus glaucus M. Bieb. var. *tenuis* (Boiss. et Blanche) Boiss., Fl. Orient., 3: 707 (1875).

Erect annual 20–80 cm. high; stems whitish or pale brown, glandular, frequently arachnoid, sometimes glabrescent, commonly branched from near base; basal leaves oblanceolate, pinnatisect or pinnatifid; cauline leaves lanceolate, 2–5 (–10) cm. long, 0·4–0·8 (–1·5) cm. wide, greyish-green, densely glandular and often thinly arachnoid; margins dentate-lobate, or sometimes pinnatifid with 3–7 pairs of narrow, rigid, spine-tipped lobes; inflorescence lax, open or rather densely corymbose; capitula ovoid, often narrowly so, and tapering to apex, 2–3 cm. long, 0·8–1·5 cm. wide; outer phyllaries foliaceous, linear-lanceolate, 2–3·5 cm. long, 0·3–0·5 cm. wide, suberect or patent, sometimes recurved, equalling or exceeding the florets, glandular, occasionally arachnoid, tapering above to a long, slender, spine-tipped acumen, with 3–5 pairs of short lateral spinules; median and inner phyllaries scarious, lanceolate, glandular-pubescent externally, 1·8–2·5 cm. long, 0·4–0·6 cm. wide at base, gradually tapering above to a slender, spine-tipped apex; florets purple or mauve; corolla-tube slender, 17–20 mm. long, limb cylindrical, about 9 mm. long, 1·8–2 mm. wide, lobes linear, erect, about 6 mm. long, 0·5 mm. wide, tapering to a thickened, subacute apex; filaments about 1·5 mm. long, with a conspicuously pilose median zone; anthers linear, about 6 mm. long, 0·4 mm. wide, with a short, bluntish apical appendage; style exserted, about 27 mm. long, branches erect, united, about 6–7 mm. long, dark purple, with a blunt apex. Central achenes obpyramidal, bluntly angled, pale brownish, about 4 mm. long, 3·5 mm. wide across the truncate, entire apex; pappus pale brownish, the outer scales lanceolate, ciliate, 6–7 mm. long, 1 mm. wide, narrowed abruptly to a distinct apical cusp; inner scales pallid, ovate-elliptic, about 1·5 mm. long, 1 mm. wide.

ssp. **foliosus** *Hanelt* in Fedde Repert., 67: 122 (1963); Feinbrun, Fl. Palaest., 3: 104, t. 683 (1978).

[*C. ruber* (non Link) Lindberg f., Iter Cypr., 34 (1946).]

TYPE: Palestine; "ad radices Carmeli", *Gaillardot* 1987 (G, JE, W, LE).

Stems pallid, often arachnoid, leaves narrowly lanceolate, tapering to apical spine; capitula small, usually about 2–2·5 cm. long, 0·8–1 cm. wide, often rather crowded; outer, foliaceous phyllaries usually exceeding the florets and frequently twice as long as the median phyllaries, narrowly lanceolate, with shortly spinulose margins and a long, slender, spinose acumen; median pappus-scales often partly truncate, about twice as long as achene.

HAB.: Dry cultivated fields and waste ground; near sea-level; fl. July–Sept.

DISTR.: Divisions 7, 8. Also Palestine and Egypt.

7. Kyrenia, 1948, *Casey* 42! and, 1956, *Casey* 1731!
8. Ayios Theodhoros, near the sea, 1939, *Lindberg f.* s.n.!

SECT. 4. **Atractylis** *Reichb.* Middle phyllaries with an entire or irregularly fissured apex; florets yellow or whitish; tips of corolla-lobes sometimes with swollen, verruculose-papillose cells but without hairs; pappus scaly.

4. C. lanatus *L.*, Sp. Plant., ed. 1, 830 (1753); Sibth. et Sm., Fl. Graec. Prodr., 2: 160 (1813), Fl. Graec., 9: 31, t. 841 (1837); Boiss., Fl. Orient., 3: 706 (1875); Post, Fl. Pal., ed. 2, 2: 120 (1933); Hanelt in Fedde Repert., 67: 135 (1963); Davis, Fl. Turkey, 5: 592 (1975).
  Kentrophyllum lanatum (L.) DC. in DC. et Duby, Bot. Gall., ed. 2, 1: 293 (1828).

Erect annual 16–100 cm. high; stems whitish or pale brown, glandular-pilose and thinly arachnoid or sometimes almost glabrous, sulcate, unbranched except in the region of the inflorescence, or sparingly branched; basal leaves oblong in outline, 5–15 cm. long, 1·5–4 cm. wide, glandular-pilose or glabrescent, deeply pinnatisect into 5–8 or more pairs of narrow, spine-tipped, serrate and lobulate divisions; petiole short, ill-defined; cauline leaves sessile, amplexicaul, resembling the basal but usually smaller, or often ovate-lanceolate, 3–4 cm. long, 1·5–2 cm. wide, tapering from a broad base to an acuminate, spinose apex, the margins with 3–4 pairs of rigid, patent, or forward-pointing, spine-tipped lobes; inflorescence a lax or dense corymb; capitula ovoid, 2–3 cm. long, 0·8–2 cm. wide, sometimes up to 3 cm. wide in fruit; outer phyllaries foliaceous, erect, spreading or recurved, glandular, subglabrous or arachnoid, lanceolate or linear-lanceolate, 2·5–5·5 cm. long, 0·3–1·5 cm. wide, tapering from base to spine-tipped apex, generally with 2–4 rigid, patent, spinose lateral lobes; median phyllaries ovate-oblong, 1·5–2·5 cm. long, 0·3–0·8 cm. wide, scarious, pallid or purplish, puberulous, narrowing rather abruptly to an entire or irregularly fissured (and sometimes slightly dilated), spine-tipped apex; florets generally yellow, sometimes whitish, exserted; corolla-tube slender, 17–20 mm. long, limb cylindrical, about 9 mm. long, 2 mm. wide, lobes linear, erect, about 6 mm. long, 0·5 mm. wide at base, tapering to an acute, slightly cucullate, minutely papillose apex; filaments about 2 mm. long, with a conspicuous pilose zone about two-thirds way to apex; anthers linear, about 6 mm. long, 0·4 mm. wide, apical appendage short, rather blunt; style about 30 mm. long, exserted, branches about 6 mm. long, united almost to apex. Achenes turbinate, obscurely angled, pale brown, 4·5–6 mm. long, 3·5–5 mm. wide across the minutely crenulate apex; pappus whitish, scales linear, 6–9 (–12) mm. long, apex acuminate, rounded or truncate, margins ciliate, innermost row short, oblong, about 1·5 mm. long, 0·4 mm. wide.

ssp. **baeticus** (*Boiss. et Reut.*) *Nyman*, Consp. Fl. Europ., 419 (1879).
  Carthamus creticus L., Sp. Plant., ed. 2, 2: 1163 (1763); Hume in Walpole, Mem. Europ. Asiatic Turkey, 254 (1817).
  Kentrophyllum baeticum Boiss. et Reut., Pugill. Plant. Afr. Hisp., 65 (1852).
  Carthamus lanatus L. ssp. *creticus* (L.) Holmboe, Veg. Cypr., 190 (1914); Lindberg f., Iter Cypr., 34 (1946); Hanelt in Fedde Repert., 67: 143 (1963); Osorio-Tafall et Seraphim, List Vasc. Plants Cyprus, 108 (1973).
  [*Kentrophyllum lanatum* (non (L.) DC.) Poech, Enum. Plant. Ins. Cypr., 19 (1842); Unger et Kotschy, Die Insel Cypern, 245 (1865).]
  [*Carthamus lanatus* (non L.) Post in Mém. Herb. Boiss., 18: 96 (1900); H. Stuart Thompson in Journ. Bot., 44: 333 (1906).]
  TYPE: Spain; "in Hispaniâ prope *Malaga*, etc. (Boiss., Reut., Bourgeau!)" (G).

Stems usually pallid, often glabrescent; leaves dark green, subglabrous with spreading lateral spines; capitulum often rather small, generally less than 2 cm. wide; outer phyllaries narrow, attenuate, patent or recurved, generally exceeding florets, at least twice as long as the median phyllaries; corolla commonly pale yellow, sometimes whitish.

HAB.: Cultivated and fallow fields, roadsides, vineyards; sea-level to 4,000 ft. alt.; fl. April-July.

DISTR.: Divisions 1–5, 7. Widely distributed in the Mediterranean region, and predominant in the eastern half.

1. Paphos, 1905, *Holmboe* 725; also, 1913, *Haradjian* 640!
2. Pano Panayia, 1905, *Holmboe* 1047; Kakopetria, 1939, *Lindberg f*. s.n.; Kyperounda, 1954, *Merton* 1936!
3. Limassol, 1801, *Hume* s.n.
4. Larnaca, 1801, *Hume* s.n.; Pergamos, 1955, *Merton* 2217!
5. Syngrasis, 1936, *Syngrassides* 976! Athalassa, 1936, *Syngrassides* 1231! also, 1939, *Lindberg f*. s.n.
7. Between Ayios Khrysostomos Monastery and Kyrenia, 1862, *Kotschy*; near the Armenian Monastery, 1912, *Haradjian* 344! Near Kyrenia, 1939, *Lindberg f*. s.n.! Near Bellapais, 1939, *Lindberg f*. s.n.!

NOTES: Typical *C. lanatus* L. seems to be absent from Cyprus, where all the specimens seen have the small capitula and long, attenuate, spreading or recurved outer phyllaries characteristic of ssp. *baeticus*. *Merton* 1936 from Kyperounda is exceptional in having whitish florets, and could conceivably be a hybrid, possibly *C. dentatus* Vahl ssp. *ruber* (Link) Hanelt x *C. lanatus* L. ssp. *baeticus* (Boiss. et Reut.) Nyman.

52. CARDUNCELLUS *Adanson*
Fam. Plant., 2: 116, 532 (1763).

Perennial spinose or spinulose herbs; stems erect, generally unbranched or sparingly branched, glabrous or pilose, or plants rarely acaulescent; leaves alternate, spinose-dentate, lyrate, pinnatifid or pinnatisect, the basal usually petiolate, the cauline sessile, often amplexicaul; capitula solitary, terminal, homogamous, discoid; phyllaries in several series, imbricate, the outer foliaceous, with spinulose-serrate margins, the inner with a short fimbriate, or rarely spine-tipped appendage; florets blue or purple, all hermaphrodite; corolla-tube slender, limb deeply 5-lobed; filaments pilose in the middle part; anthers shortly caudate-sagittate; style long-exserted, shortly bifid at apex; receptacle densely bristly. Achenes glabrous, obpyramidal or cylindrical, often 4-angled; hilum lateral; pappus generally copious, bristles scabridulous, barbellate or shortly plumose.

About 14 species in southern Europe and the Mediterranean region.

1. **C. caeruleus** (*L.*) *C. Presl*, Fl. Sic., xxx (1826).
   Carthamus caeruleus L., Sp. Plant., ed. 1, 830 (1753); Sibth. et Sm., Fl. Graec. Prodr., 2: 161 (1813), Fl. Graec., 9: 32, t. 843 (1837); Boiss., Fl. Orient., 3: 710 (1875); Post, Fl. Pal., ed. 2, 2: 122 (1933).
   TYPE: "*in agro* Tingitano & Hispalensi *inter segetes*".

Erect perennial to about 1 m. high; rootstock thick, woody, oblique; stems sulcate, subglabrous or pilose with whitish, arachnoid hairs; basal leaves oblanceolate or narrowly obovate in outline, 10–20 cm. long, 2–3·5 cm. wide, scabridulous above and below and sometimes thinly pilose, apex usually acute, base tapering to an ill-defined petiole, margins coarsely spinulose-serrate, sometimes pinnatisect, rarely subentire; cauline leaves sessile, amplexicaul, the lower narrowly oblong, up to 20 cm. long, 3·5 cm. wide, the upper progressively smaller, the uppermost ovate-deltoid, 3–4 cm. long, 1·5–2·5 cm. wide, all rather rigid and coriaceous, prominently veined, with boldly spinose-serrate or spinose-lobulate margins; capitula solitary, terminal, shortly pedunculate, broadly oblong or subglobose, 2–4 cm. long, 2–3 cm. wide; outer phyllaries erect or suberect, ovate-lanceolate, 2–3·5 cm. long, 0·7–1·3 cm. wide at base, distinctly veined, subglabrous or thinly arachnoid, apex spinulose, acute or acuminate, margins closely spinulose-serrate; inner phyllaries scarious, oblong, 1·5–2·5 cm. long, 0·6–0·8 cm. wide, with a blunt or rounded, papery, brown, fimbriate terminal appendage; florets blue; corolla-tube to 3 cm. long, slender, limb cylindrical, lobes linear, 6–7 mm. long, 0·8 mm. wide, apex obtuse, slightly cucullate; filaments about

2·5 mm. long, with a conspicuously pilose median zone; anthers linear, blunt, 6–7 mm. long, 0·5 mm. wide; style 3–4·5 cm. long, exserted, branches united almost to apex. Achenes shortly cylindrical or obscurely 4-angled, about 5–6 mm. long, 4 mm. wide, pale brown, smooth below, distinctly rugose towards apex; pappus copious, bristles rigid, pale brownish, up to 10 mm. long, barbellate, the innermost row short.

HAB.: Cultivated fields; near sea-level; fl. May–June.

DISTR.: Division 7, very rare. Widespread in the Mediterranean region, but becoming rare in the eastern half.

7. Near Kyrenia, 1 June 1880, *Sintenis & Rigo* 613!

NOTES: The only record, not previously noted for the flora of Cyprus. The specimen has spinose-serrate (not lobulate) leaves, and agrees best with *C. caeruleus* (L.) C. Presl var. *dentatus* DC. (Prodr., 6: 615; 1837), a minor variant.

SUBFAMILY 2. **Lactucoideae.** Plants commonly lactiferous; corolla of all the florets ligulate; pollen grains with spines arranged in rows.

TRIBE 11. **Lactuceae** *Cass.* (*Cichorieae* Spreng.) Mostly unarmed herbs, rarely shrubs or trees; leaves alternate or basal; capitula homogamous, florets all ligulate and hermaphrodite, ligules commonly 5-toothed at apex; phyllaries in 1 or several series, herbaceous or membranous-scarious; anthers generally sagittate at base; style-branches usually slender, divergent; receptacle naked or scaly. Achenes often beaked; pappus present or absent, rarely scaly.

53. SCOLYMUS *L.*

Sp. Plant., ed. 1, 813 (1753).
Gen. Plant., ed. 5, 355 (1754).

Spinose annual, biennial or perennial herbs; stems branched or unbranched; leaves alternate, rigid, spinose-lobed or spinose-pinnatifid, the bases decurrent and forming wings along the stem; capitula sessile, axillary and terminal; outermost phyllaries patent or suberect, spinulose-pectinate, rigid, foliaceous, inner phyllaries imbricate, herbaceous with membranous margins and a mucronate or spinulose apex; florets all ligulate, yellow, apex 5-dentate; anthers sagittate at base, with acuminate-mucronate auricles; style-branches slender; receptacle conical or elongate, clothed with scales. Achenes dorsally compressed, partly enveloped by and adnate to the deciduous receptacular scales; pappus absent or reduced to a few, caducous bristles.

Three species widely distributed in the Mediterranean region.

Leaves with conspicuously thickened, white margins and nerves; corolla clothed with dark
　　brown hairs in the lower half; stems subglabrous; pappus absent　-　　- **1. S. maculatus**
Leaves without thickened, white margins; corolla whitish-pubescent in the lower half; stems
　　arachnoid-pubescent; pappus present　-　　-　　-　　-　　-　　- **2. S. hispanicus**

1. **S. maculatus** *L.*, Sp. Plant., ed. 1, 813 (1753); Boiss., Fl. Orient., 3: 713 (1875); Sintenis in Oesterr. Bot. Zeitschr., 32: 399 (1882); Holmboe, Veg. Cypr., 190 (1914); Post, Fl. Pal., ed. 2, 2: 123 (1933); Lindberg f., Iter Cypr., 36 (1946); Rechinger f. in Arkiv för Bot., ser. 2, 1: 434 (1950); Osorio-Tafall et Seraphim, List Vasc. Plants Cyprus, 103 (1973); Davis, Fl. Turkey, 5: 624 (1975); Feinbrun, Fl. Palaest., 3: 406, t. 687 (1978).
TYPE: "*in* G. Narbonensi, Italia".

Erect, glabrous or subglabrous annual 40–160 cm. high; stem stout, whitish, conspicuously winged with deltoid, denticulate, spine-tipped lobes, sparingly branched except in region of inflorescence; basal leaves pinnatifid, oblanceolate in outline, 10–20 cm. long, 2–5 cm. wide, with spinose-dentate margins, base tapering to an indistinct petiole; upper leaves sessile, very thick and rigid, oblong-deltoid, 2–7 cm. long, 2–5 cm. wide, coarsely spinose, with much thickened, white margins and nerves; inflorescence few-flowered, corymbose; capitula sessile; involucre ovoid, about 15 mm. long, 7 mm. wide, glabrous, outermost phyllaries spreading or loosely erect, up to 30 mm. long, coarsely and rigidly spinose; inner phyllaries 4–15 mm. long, 3–4 mm. wide, ovate or oblong with a spine-tipped or mucronate apex and narrow membranous margins; florets yellow, thinly clothed in the lower half with long, dark brown, crispate hairs; corolla-tube 5–6 mm. long, ligule 9–10 mm. long, about 1·8 mm. wide; filaments glabrous, about 0·8 mm. long; anthers linear, blunt, about 6 mm. long, 0·3 mm. wide; style 9–10 mm. long, shortly pilose, branches short, blunt, less than 0·4 mm. long. Achenes almost flattened, oblong-obovate, 4–5 mm. long, 3 mm. wide, almost totally enveloped by the pale, folded, adnate receptacular scales, and apparently winged all round; pappus absent.

HAB.: Cultivated and fallow fields, waste ground, roadsides; sea-level to 2,000 ft. alt.; fl. April–July.

DISTR.: Divisions 1–4, 6–8. Widespread in S. Europe and Mediterranean region.
1. Stroumbi, 1913, *Haradjian* 742; Ktima, 1913, *Haradjian* 700.
2. Phiti, 1905, *Holmboe* 775.
3. Between Limassol and Skarinou, 1941, *Davis* 3555!
4. Near Larnaca, 1939, *Lindberg f.* s.n.!
6. Myrtou, 1932, *Syngrassides* 215! Peristerona, 1955, *Merton* 2454!
7. Kyrenia, 1949, *Casey* 766! and, 1955, *G. E. Atherton* 100!
8. Near Rizokarpaso, 1880, *Sintenis & Rigo* 499! Near Komi Kebir, 1912, *Haradjian* 309!

NOTES: A second sheet of *Sintenis & Rigo* 499 (K!) is labelled "In montibus circa Kythrea". It is clearly from the same collecting as *Sintenis & Rigo* 499, "In campis pr. Rizo Carpasso", cited above, and, in fact, was probably collected between Ayios Symeon and Leonarisso (8) on 28 April, 1880.

2. S. hispanicus *L.*, Sp. Plant., ed. 1, 813 (1753); Gaudry, Recherches Sci. en Orient, 188 (1855); Kotschy in Oesterr. Bot. Zeitschr., 12: 277 (1862); Unger et Kotschy, Die Insel Cypern, 249 (1865); Boiss., Fl. Orient., 3: 713 (1875); H. Stuart Thompson in Journ. Bot., 44: 333 (1906); Holmboe, Veg. Cypr., 190 (1914); Post, Fl. Pal., ed. 2, 2: 123 (1933); Lindberg f., Iter Cypr., 36 (1946); Osorio-Tafall et Seraphim, List Vasc. Plants Cyprus, 103 (1973); Davis, Fl. Turkey, 5: 624 (1975); Feinbrun, Fl. Palaest., 3: 407, t. 688 (1978).
TYPE: "*in* Italia, Sicilia, G. Narbonensi".

Erect biennial or perennial (10–) 30–80 (–100) cm. high; stems pallid, arachnoid-pilose, narrowly winged with very unequal, sharply spinose lobes, generally branched; basal leaves narrowly oblong, 12–20 cm. long, 2·5–4 cm. wide, pinnatifid with broadly deltoid, spine-tipped and spinose-denticulate lobes, scabridulous-pilose or glabrescent on both surfaces, margins and nerves not conspicuously thickened or whitish; cauline leaves sessile, amplexicaul, diminishing in size upwards, rigidly spinose; inflorescence subspicate, the capitula usually solitary and sessile in the axils of the uppermost leaves; involucre oblong, 1–2 cm. long, about 1 cm. wide; outer phyllaries foliaceous, rigid, spinose, suberect, 1·5–4 cm. long, 0·5–1 cm. wide, inner phyllaries lanceolate-acuminate, spine-tipped, with a very narrow, shortly ciliolate, membranous margin, 8–20 mm. long, 3–7 mm. wide; florets golden-yellow, softly white-pubescent in the lower half; corolla-tube 8–10 mm. long, ligule 14–15 mm. long, 2·5 mm. wide, apex minutely 5-dentate; filaments glabrous, flattened, about 1 mm. long; anthers linear, about 6 mm. long, 0·4 mm. wide, apical appendage rounded; style 18–20 mm. long, shortly pilose in the upper half, branches filiform, blunt, about 1·5 mm. long. Achenes almost flattened, broadly ovate, 8–9 mm. long, 7–8 mm. wide, almost totally enveloped by the pale, folded, adnate receptacular scales, forming membranous-margined, laciniate, lateral wings; pappus of 3, pale, barbellate bristles up to 6 mm. long.

HAB.: Cultivated and fallow fields, waste ground, roadsides; sea-level to 3,000 ft. alt.; fl. May–Aug.

DISTR.: Divisions 1–7, common. Widespread in S. Europe and the Mediterranean region.

1. Paphos, 1905, *Holmboe* 724; also, 1913, *Haradjian* 634! Khlorakas, 1967, *Economides* in ARI 964!
2. Lemithou, 1939, *Kennedy* 1460!
3. Between Limassol and Kolossi, 1862, *Kotschy*; Yermasoyia, ? 1905, *Michaelides* sec. *Holmboe*; between Limassol and Skarinou, 1941, *Davis* 3554! Episkopi, 1950, *F. J. F. Barrington* 60!
4. Near Larnaca and Kiti, 1862, *Kotschy*; Larnaca–Famagusta road, 1950, *Chapman* 700!
5. Near Sykhari, 1901, *A. G. & M. E. Lascelles* s.n.! Athalassa, 1939, *Lindberg f.* s.n.! Nicosia–Famagusta road, 1972, *W. R. Price* 997A!
6. Between Yerolakkos and Myrtou, 1932, *Syngrassides* 174! Peristerona, 1955, *Merton* 2453!
7. Kyrenia, 1949, *Casey* 765! also, 1955, *Miss Mapple* 12! *G. E. Atherton* 62! 358!

54. CATANANCHE L.*

Sp. Plant., ed. 1, 812 (1753).
Gen. Plant., ed. 5, 354 (1754).

Erect annual or perennial herbs; leaves mostly basal, entire or sparingly dentate; capitula usually solitary, many-flowered; florets blue or yellow; involucre cylindrical-campanulate; phyllaries in several series, imbricate, scarious or membranous; receptacle flat, pitted, with long, filiform scales; anthers sagittate at base; style-branches very short, subacute or blunt. Achenes turbinate, 5-angled; pappus persistent, of 5–7 ovate-lanceolate, often aristate, scales.

Five species, mainly Mediterranean in distribution. *C. caerulea* L., from S.W. Europe and the western Mediterranean region, is sometimes cultivated as an ornamental.

1. **C. lutea** L., Sp. Plant., ed. 1, 812 (1753); Sibth. et Sm., Fl. Graec. Prodr., 2: 145 (1813), Fl. Graec., 9: 15, t. 821 (1837); Poech, Enum. Plant. Ins. Cypr., 20 (1842); Unger et Kotschy, Die Insel Cypern, 250 (1865); Boiss., Fl. Orient., 3: 714 (1875); Holmboe, Veg. Cypr., 190 (1914); Post, Fl. Pal., ed. 2, 2: 123 (1933); Osorio-Tafall et Seraphim, List Vasc. Plants Cyprus, 108 (1973); Davis, Fl. Turkey, 5: 626 (1975); Feinbrun, Fl. Palaest., 3: 407, t. 689 (1978).
TYPE: "*in* Creta".

Erect, sparingly branched, adpressed-villous annual, 9–27 (–58) cm. high; leaves mostly basal, pale green, lanceolate, 7–27 cm. long, 0·7–2 cm. wide, 3-nerved, base sometimes attenuate to a long, ill-defined petiole, apex acute or obtuse, margins entire or with a few, remote teeth; cauline leaves few, sessile, decreasing in size upwards; inflorescence usually branched, capitula 1–2 cm. diam., usually solitary and terminal on long naked or almost naked branches, several sessile, few-flowered capitula generally produced at ground-level in the axils of the basal leaves; receptacular bristles up to 5 mm. long; involucre cylindrical-campanulate, 1·7–2·2 cm. long, 0·8–1·5 cm. wide; phyllaries silvery, shining, scarious, often with a dark midrib and thickened base, the outer ovate, 4–5 mm. long, 2·5 mm. wide, the inner lanceolate, 0·9–2 cm. long, 0·3–0·6 cm. wide, acute, apex minutely serrulate; florets all hermaphrodite, dark lemon-yellow, 1·3–1·5 cm. long, 0·2–0·3 cm. wide, ligule oblong, truncate, 4–5-dentate at apex, 6-veined, corolla-tube hairy externally; anthers linear, 2–3 mm. long, 0·3 mm. wide, base caudate-sagittate, apical appendage oblong, blunt; style slender, pubescent above, shortly bifid at apex, branches linear or subspathulate, blunt or rounded at

* by P. Halliday

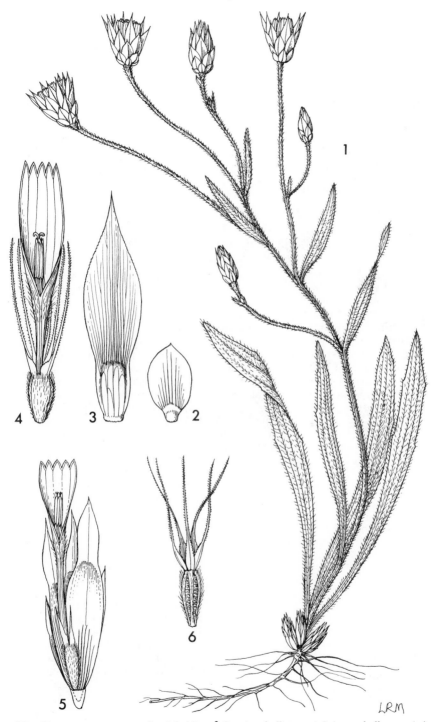

Plate 62. CATANANCHE LUTEA L. **1,** habit, × ⅔; **2,** outer phyllary, × 4; **3,** inner phyllary, × 4; **4,** floret, × 4; **5,** part of capitulum, longitudinal section, × 4; **6,** achene, × 4. (**1–5** from *Merton* 2015 & *Syngrassides* 916; **6** from *Syngrassides* 916.)

apex. Achenes turbinate, about 3 mm. long, 1–1·5 mm. wide, 5-angled, angles stiffly pilose; pappus of 5 (–7) aristate scales up to 8 mm. long, about 1 mm. wide at base. *Plate 62.*

HAB.: Pastures; hillsides; stabilized sand-dunes; sea-level to 2,000 ft. alt.; fl. March–May.

DISTR.: Divisions 1, 3, 4, 7, 8. Eastern Mediterranean region from Sardinia, Italy and Sicily eastwards to Turkey, Syria and Palestine.

1. Stroumbi, 1913, *Haradjian* 723! Dhrousha, 1941, *Davis* 3250!
3. Lemona, 1905, *Holmboe* 751.
4. Pergamos, 1955, *Merton* 1971! Cape Kiti, 1955, *Merton* 2015!
7. Ayios Khrysostomos Monastery, 1862, *Kotschy* 393; Dhavlos, 1880, *Sintenis & Rigo* 494! Ayios Amvrosio, 1932, *Syngrassides* 425! Larnaka tis Lapithou, 1936, *Syngrassides* 916! also, 1941, *Davis* 2974! Lakovounara, 1950, *Chapman* 518! also, 1957, *Merton* 2966! Near Krini, 1962, *Casey* 1740! 1741! Near Panagra, 1962, *Meikle* 2397!
8. Valia, 1905, *Holmboe* 491; near Rizokarpaso, 1912, *Haradjian* 189!

55. CICHORIUM *L.*

Sp. Plant., ed. 1, 813 (1753).
Gen. Plant., ed. 5, 354 (1754).

Annual, biennial or perennial herbs or subshrubs; branches generally rigid, divaricate, sometimes spinescent; leaves mostly basal or sub-basal, oblanceolate, entire, toothed, lyrate-runcinate or pinnatifid, glabrous or pilose; capitula sometimes terminal on thickened, pedunculoid branches, sometimes sessile, solitary or in clusters, in the axils of the branches; phyllaries usually in 2 series, the outer short, forming a calyculus, the inner long, narrow, erect, herbaceous; florets blue (rarely pink or white); ligules truncate and 5-toothed at apex; anthers sagittate at base; style-branches slender, obtuse; receptacle flattish, somewhat pitted, naked or almost so. Achenes obovoid, truncate, glabrous, obscurely 5-angled, or longitudinally ribbed; pappus of 1–3 rows of short, blunt or shortly aristate scales.

About 9 species in Europe, the Mediterranean region, and south to Ethiopia. *Cichorium intybus* L. (Chicory) and *C. endivia* L. (Endive) are cultivated as salad plants, or their roots used to flavour coffee.

Spinescent subshrubs, forming low, domed, intricately branched bushes; capitula 5–6-flowered
  3. C. spinosum
Unarmed, erect or spreading herbs, not intricately branched; capitula more than 6-flowered:
  Outer phyllaries erect, scarcely half as long as linear inner phyllaries, the latter 8–9 mm. long, 1·5–2 mm. wide at base; plant perennial with a thick, woody rootstock   **1. C. intybus**
  Outer phyllaries loosely erect or with somewhat recurved apices, almost as long as the inner phyllaries, the latter oblong, 10–12·5 mm. long, 2–3 mm. wide; plant annual with a slender taproot -   -   -   -   -   -   -   -   -   -   -   **2. C. endivia**

1. **C. intybus** *L.*, Sp. Plant., ed. 1, 813 (1753); Boiss., Fl. Orient., 3: 715 (1875); Post, Fl. Pal., ed. 2, 2: 124 (1933) pro parte; Davis, Fl. Turkey, 5: 627 (1975) pro parte.
TYPE: "*in* Europa *ad margines agrorum viarumque*".

Erect perennial to 100 cm. high with a thick, woody taproot; stems sulcate, glabrous or thinly hispidulous, repeatedly branched with erecto-patent branches; leaves mostly basal, lanceolate or oblanceolate in outline, 7–15 cm. long, 1–5 cm. wide, subglabrous above, thinly pilose below, entire, pinnatifid or lyrate-runcinate, base tapering to a short, ill-defined petiole; inflorescence lax, with numerous rigid, erecto-patent, almost leafless branches; capitula lateral and terminal on very slightly swollen branches, sessile, 3–4 cm. diam.; involucre subcylindric; outer phyllaries erect, ovate, about 4–5 mm. long, 2 mm. wide at base, with a gibbous, indurated pallid median zone, surrounded by a narrow, herbaceous, hispidulous-ciliate margin, apex acute, often shortly spinose; inner phyllaries linear, 8–9 mm. long, erect from a slightly dilated base 1·5–2 mm. wide, glabrous or thinly glandular-pilose, apex subacute, somewhat cucullate, midrib strigillose dorsally; florets bright, soft blue, ligules spreading, 15–19 mm. long, 4–5 mm. wide, apex sharply 5-dentate; filaments glabrous, about 1 mm. long; anthers linear, 5–6 mm. long, 0·3 mm. wide, apical appendage blunt or rounded; style 10–12 mm. long, pilose in the upper part, branches filiform, spreading, 2·5–3 mm. long. Achenes obovoid, about 2·5 mm. long, 1·5 mm. across the truncate apex, obscurely 5-angled, brownish with darker marbling, densely and minutely verrucose-papillose; pappus whitish, a single row of subacute, ovate, erose scales about 0·2 mm. long, 0·1 mm. wide.

HAB.: Sides of cultivated land, roadsides, waste ground; 900–3,500 ft. alt.; fl. May–Aug.

DISTR.: Divisions 2, 6, evidently uncommon. Widespread in Europe and western Asia, but becoming rare in many areas of the eastern Mediterranean.

2. Platres, 1929, *C. B. Ussher* 21! also, 1937, *Kennedy* 980! 981!
6. Myrtou, 1970, *A. Genneou* in ARI 1580!

NOTES: The only genuine records, all others being referable to the following, *C. endivia* L. f. *divaricatum* (Schousboe) Webb. The differences between *C. intybus* and *C. endivia* have been so poorly indicated that the two, distinct, species have been repeatedly and hopelessly confused. No record for either species, from the eastern Mediterranean region, should be accepted without examination of supporting specimens.

2. C. endivia *L.*, Sp. Plant., ed. 1, 813 (1753); Sibth. et Sm., Fl. Graec. Prodr., 2: 146 (1813); Poech, Enum. Plant. Ins. Cypr., 20 (1842); Unger et Kotschy, Die Insel Cypern, 250 (1865); Post, Fl. Pal., ed. 2, 2: 124 (1933); Davis, Fl. Turkey, 5: 628 (1975).

Erect or spreading annual, (5–) 10–60 (–100) cm. high, with a slender taproot; stems sulcate, glabrous, pubescent or sometimes thinly lanuginose, usually much branched, with divaricate branches; basal leaves very variable, generally oblong or oblanceolate in outline, 15–30 cm. long, 2–6 (or more) cm. wide, subentire, pinnatifid, lyrate-runcinate, or variously crisped and curled in cultivars, generally glabrescent or subglabrous, but sometimes persistently crispate-pilose; stem leaves generally more numerous and better developed than in *C. intybus*, sessile, and becoming progressively smaller upwards, entire, toothed or pinnatifid, with a rounded or acute apex, and an amplexicaul, often prominently auriculate base, the auricles commonly exceeding the stem-diameter; inflorescence as in *C. intybus*, but the lateral capitula generally in dense clusters, and the flowering branchlets often conspicuously swollen; capitula generally less than 2 cm. diam. (occasionally to 3 cm. in cultivated specimens); involucre campanulate, 8–10 mm. long, 6–8 mm. wide; outer phyllaries rather loosely erect, ovate-elliptic, 8–12 mm. long, 4–5 mm. wide, with a large, pallid, indurated, median zone, surrounded by a broad, herbaceous, ciliate margin, the cilia often remote, and 1–1·5 mm. long, apex acute or shortly spinulose; inner phyllaries oblong, 10–12·5 mm. long, 2–3 mm. wide, not much longer than the outer, loosely erect, apex obtuse or subacute, margins and midrib glabrous or long-ciliate; florets blue, ligules spreading, 9–14 mm. long, 3–4 mm. wide, puberulous dorsally, apex sharply 5-dentate; filaments glabrous, about 0·8 mm. long; anthers linear, about 3·5 mm. long, 0·3 mm. wide, apical appendage blunt; style 9–10 mm. long, pilose in the upper half, branches about 2·5 mm. long, filiform, divergent. Achenes obovoid or turbinate, about 3 mm. long, 1·3 mm. across the truncate apex, distinctly 5-angled and prominently ribbed, brownish with darker marbling, densely and minutely verrucose-papillose; pappus whitish, a single row of oblong or ovate, erose-laciniate scales, about 0·8–1 mm. long, 0·3–0·4 mm. wide.

f. **divaricatum** (*Schousboe*) *Webb* in Webb et Berth., Phytogr. Canar., 2: 391 (1847–48).
C. pumilum Jacq., Obs. Bot. 4: 3, t. 80 (1771); Sibth. et Sm., Fl. Graec. Prodr., 2: 146 (1813), Fl. Graec., 9: 16, t. 822 (1837); Holmboe, Veg. Cypr., 191 (1914); Post, Fl. Pal., ed.

2, 2: 124 (1933); Lindberg f., Iter Cypr., 34 (1946); Osorio-Tafall et Seraphim, List Vasc. Plants Cyprus, 108 (1973); Davis, Fl. Turkey, 5: 628 (1975); Feinbrun, Fl. Palaest., 3: 408, t. 690 (1978).
    C. divaricatum Schousboe, Vextriget i Marokko, 199–200 (1800); Boiss., Fl. Orient., 3: 716 (1875), Suppl., 318 (1888).
    C. intybus L. var. *divaricatum* (Schousboe) DC., Prodr., 7: 84 (1838); Unger et Kotschy, Die Insel Cypern, 250 (1865).
    C. endivia L. var. *pumilum* (Jacq.) Vis., Fl. Dalmat., 2: 97 (1847).
    C. intybus L. ssp. *divaricatum* (Schousboe) Bonnier et Layens, Tabl. Synopt. Fl. France, 183 (1894).
    C. endivia L. ssp. *pumilum* (Jacq.) Hegi, Illustr. Fl. Mitteleurop., 6 (2); 998 (1928).
    C. endivia L. ssp. *divaricatum* (Schousboe) P. D. Sell in Bot. Journ. Linn. Soc., 71: 240 (1976).
    TYPE: "In campis regionis tingitanae" [Tangier], cultivated Copenhagen (? C).

Basal leaves narrow, oblanceolate, generally less than 4 cm. wide, subentire, dentate, dentate-pinnatifid or lyrate-runcinate, often pubescent or shortly crispate-pilose.

HAB.: Cultivated and fallow fields, roadsides; occasionally in garigue on hillsides; sea-level to 4,000 ft. alt.; fl. March–June.

DISTR.: Divisions 2–8. Widespread in the Mediterranean region, becoming rare in the northern part of the area, and in southern Europe.

2. Kryos Potamos, 4,000 ft. alt., 1937, *Kennedy* 978! Mandria, 1937, *Kennedy* 979!
3. Kannaviou, 1939, *Lindberg f.*, s.n.; Akhelia, 1939, *Lindberg f.* s.n.; Limassol, 1941, *Davis* 3581!
4. Livadhia, 1905, *Holmboe* 515; Kondea, 1935, *Syngrassides* 627! Sotira, Ayios Antonios, 1950, *Chapman* 584! Cape Greco, 1958, *N. Macdonald* 53! 54! Kouklia reservoir, 1967, *Merton* in ARI 490!
5. Dhikomo, 1936, *Syngrassides* 994! Kythrea, 1936, *Syngrassides* 1259!
6. Nicosia, 1939, *Lindberg f.* s.n.
7. Between Ayios Khrysostomos Monastery and Kyrenia, 1862, *Kotschy*; Lakovounara Forest, 1950, *Chapman* 658! Kyrenia, 1949, *Casey* 794! Templos, 1956, *G. E. Atherton* 1397!
8. Near Galinoporni, 1880, *Sintenis & Rigo* 493! Ayios Theodhoros, 1970, *A. Genneou* in ARI 1468!

NOTES: Boissier (Fl. Orient., 3: 716) correctly notes that this is the wild form of the cultivated Endive, and indeed so little removed from it that cultivated plants, allowed to seed, will rapidly revert to their natural state. The situation is analogous to that of *Eruca sativa* L. var. *sativa* and var. *eriocarpa* (Boiss.) Post (see Fl. Cyprus, 1: 104), but the differences between the variants are even less substantial. Several authors refer to the glandulosity of the phyllaries in f. *endivia*, but the distinction, if genuine, is not very apparent.

3. **C. spinosum** *L.*, Sp. Plant., ed. 1, 813 (1753); Sibth. et Sm., Fl. Graec. Prodr., 2: 147 (1813), Fl. Graec., 9: 16, t. 823 (1837); Poech, Enum. Plant. Ins. Cypr., 21 (1842); Unger et Kotschy, Die Insel Cypern, 251 (1865); Boiss., Fl. Orient., 3: 71 (1875); Holmboe, Veg. Cypr., 191 (1914); Post, Fl. Pal., ed. 2, 2: 821 (1933); Lindberg f., Iter Cypr., 34 (1946); Chapman, Cyprus Trees and Shrubs, 81 (1949); Osorio-Tafall et Seraphim, List Vasc. Plants Cyprus, 108 (1973); Davis, Fl. Turkey, 5: 627 (1975).
    TYPE: "*in* Cretae, Siciliae *collibus arenosis maritimis*".

Intricately branched subshrub (3–) 5–15 (–30) cm. high; stems glabrous, sulcate, repeatedly branched with divaricate branches, the ultimate divisions rigid, leafless, bluntly spinose or spinescent; leaves generally (but not always) restricted to the bases of the stems, oblanceolate in outline., 3–10 cm. long, 0·3–1 cm. wide, usually lyrate-pinnatifid, with a blunt, oblong terminal lobe and small, deltoid, or bluntly oblong, lateral lobes, sometimes dentate or entire, glabrous, base tapering to a short petiole which is dilated and sheathing at the base; capitula generally solitary, lateral or in the axils of the branches, rarely terminal; involucre subcylindric; outer phyllaries ovate, erect, 2·5–4 mm. long, about 2 mm. wide, glabrous, with a large, pallid, indurated median zone, and a narrow acute or subacute margin; inner phyllaries oblong, 8–9 mm. long, 2–2·5 mm. wide, spreading or recurved at anthesis, glabrous, with a narrow hyaline margin and a subacute, slightly cucullate apex; florets few (usually 5–6), spreading, blue; ligules narrowly oblong, 9–12 mm. long, 3 mm. wide, glabrous or subglabrous, apex 5-dentate; filaments about 0·5 mm. long, glabrous; anthers about 4 mm. long, 0·3 mm. wide, apical appendage rounded; style about 8·5 mm. long, pilose in the upper half, branches filiform, divergent, about 2 mm. long. Achenes oblong-turbinate, 2–2·5 mm. long, about 1·3 mm. wide across truncate apex, glabrous, pale brown, bluntly angled or ribbed, very minutely verruculose-papillose; pappus pallid, scales minute, in 1 row, oblong, less than 0·3 mm. long, 0·2 mm. wide, erose-laciniate.

HAB.: Rock-crevices by the sea, occasionally in sandy fields; at or near sea-level; fl. June–Oct.

DISTR.: Divisions 1, 2, 6–8, locally common. Widely distributed, particularly on Mediterranean islands, from the Balearics east to Cyprus.

1. Paphos, Tombs of the Kings, 1787, *Sibthorp*; Khrysokhou, 1905, *Holmboe* 791; Paphos, 1939, *Lindberg* f. s.n.; Ayios Yeoryios (Akamas), 1962, *Meikle*!
2. Pomos Point, 1938, *Chapman* 374!
6. Ayia Irini, 1962, *Meikle*!
7. Between Kyrenia and Melandrina, 1862, *Kotschy* 504; Kyrenia, 1936–1973, *Kennedy* 982! *Lindberg* f. s.n.! *Casey* 867! *G. E. Atherton* 108! *A. Genneou* in ARI 1553! *P. Laukkonen* 442!
8. Near Apostolos Andreas Monastery, 1937, *Syngrassides* 1642! Cape Andreas, 1962, *Meikle*!

NOTES: *G. E. Atherton* 579 (K!) labelled "Prodhromos, 4,500 ft." is surely an error in labelling. There are no other records for *Cichorium spinosum* from the area, or from such an altitude.

56. HYOSERIS L.*

Sp. Plant., ed. 1, 809 (1753).
Gen. Plant., ed. 5, 351 (1754).

Annual or perennial acaulescent herbs; scapes erect or decumbent with solitary, homogamous, ligulate capitula; involucre cylindric-campanulate; phyllaries in 2 series, the outer small, the inner much longer, somewhat accrescent in fruit, and becoming hardened and thickened; receptacle flat, naked; florets yellow, often with a broad, purple, longitudinal stripe on the outer surface; anthers sagittate at base; style-branches slender, rather blunt. Achenes of 3 kinds: marginal somewhat compressed; median flat and winged; inner cylindrical; pappus of rigid hairs, or of rigid hairs and scales.

About 4 species in Europe and the Mediterranean region.

1. **H. scabra** L., Sp. Plant., ed. 1, 809 (1753); Sibth. et Sm., Fl. Graec. Prodr., 2: 141 (1813); Boiss., Fl. Orient., 3: 718 (1875); Sintenis in Oesterr. Bot. Zeitschr., 31: 194, 326 (1881); Holmboe, Veg. Cypr., 191 (1914); Post, Fl. Pal., ed. 2, 2: 125 (1933); Osorio-Tafall et Seraphim, List Vasc. Plants Cyprus, 108 (1973); Davis, Fl. Turkey, 5: 631 (1975); Feinbrun, Fl. Palaest., 3: 411, t. 695 (1978).
   H. microcephala Cass. in Dict. Sci. Nat., 22: 338 (1821); Poech, Enum. Plant. Ins. Cypr., 20 (1842); Unger et Kotschy, Die Insel Cypern, 250 (1865) [as *H.* "*microcepala*"].
   TYPE: "*in* Sicilia".

Compact, low-growing, glabrescent, rosulate annual, 6–10 (–12) cm. high, the whole plant thinly or densely scurfy with whitish, vesiculose papillae; leaves dull green, oblanceolate in outline, 4–10·5 cm. long, 1·5–2·4 cm. wide, pinnatilobed, terminal lobe 3-dentate at apex, lateral lobes broadly dentate, diminishing in size downwards; petiole short, winged, or sometimes almost wanting; scapes not branched, erect or prostrate, usually about as long as the leaves or a little longer, glabrescent, reddish or purplish, terete, and conspicuously swollen (and hollow) towards apex; capitulum cylindric-campanulate, 1–1·3 cm. long, 0·5 cm. diam. at base of involucre; outer phyllaries about 5, erect, subulate to ovate, 2·5–3 mm. long, 1–1·5 mm. wide; inner phyllaries about 7, erect, oblong-lanceolate, 8–15 mm. long, 2–3 mm.

* by P. Halliday

wide, with wide, scarious margins; florets about 12 in each capitulum; ligules yellow (occasionally whitish), oblong, slightly narrowed to base, 9–9·5 mm. long, 2 mm. wide, truncate with (3–) 5 apical teeth and 6 veins; corolla-tube about 3 mm. long, about ⅓ of the total length of floret, mouth fringed with multicellular hairs; anthers linear, about 2 mm. long, 0·2 mm. wide, apical appendage rounded, base sagittate with short, obtuse, adpressed tails; style about 7 mm. long, minutely pilose in the upper part, branches obtuse, about 1·5 mm. long. Marginal achenes enveloped by hardened, thickened phyllaries, 8–9 mm. long, 2 mm. wide, somewhat compressed and incurved, tapering upwards, with 2 longitudinal grooves on the concave face, and 2 corresponding ribs on the convex face, usually minutely hispid in the upper part, with a pappus of scabridulous scales, 0·5–1 mm. long; median achenes oblong in outline, 7–10 mm. long, 2·5–3 mm. wide, flat, winged, with a prominent midrib on the inner surface, and a corresponding groove flanked by 2 ribs on the outer surface, the margins of the wings hispid, at least towards apex; pappus 5–6 mm. long, of 5 narrowly lanceolate, attenuate scales, and several rigid, scabrid bristles; inner achenes 8–9 mm. long, about 1 mm. wide, cylindrical or bluntly angled, smooth, with 5 longitudinal grooves, glabrous or minutely hispid; pappus like that of median achenes.

HAB.: Fallow fields near the sea; in coastal (rarely inland) garigue, or in pockets of soil on maritime rocks; sea-level to 600 ft. alt.; fl. Febr.–April.

DISTR.: Divisions 1, 4, 5, 7, 8. Widely distributed in the Mediterranean region.

1. Erimidhes near Ayios Yeoryios (Akamas), 1962, *Meikle* 2142!
4. Larnaca, 1880, *Sintenis*.
5. Kythrea, 1880, *Sintenis*; Mile 3, Nicosia–Famagusta road, 1950, *Chapman* 387!
7. Kyrenia, 1952, *Casey* 1223! and, 1956, *G. E. Atherton* 947!
8. Near Apostolos Andreas Monastery, 1938, *Syngrassides* 1785! Near Rizokarpasao, 1941, *Davis* 2272!

57. **TOLPIS** *Adans.*
Fam. Plant., 2: 122, 612 (1763).

Annual or perennial herbs; stems usually branched, leafy in the lower part; leaves alternate, entire, toothed or lobed; inflorescence generally branched; capitula homogamous, ligulate; phyllaries in 2–3 series, sometimes overtopped by filiform bracts arising from near the apex of the peduncle; florets yellow or the central ones purplish, ligules 5-dentate at apex; anthers sagittate at base; style-branches slender, blunt; receptacle flat, naked, pitted. Achenes uniform or dimorphic, longitudinally ribbed, apex truncate; pappus of short or long bristles, or a mixture of the two, the marginal achenes sometimes with a few short bristles or epappose.

About 20 species in the Mediterranean region and Atlantic islands extending south to tropical Africa.

Bracts near apex of peduncle filiform, equalling or exceeding phyllaries and forming what appears to be an outer involucre; plant annual with a slender taproot   -   **1. T. barbata**
Bracts near apex of peduncle small, much shorter than the phyllaries and not forming an outer involucre; plant biennial or perennial with a thick, woody root   -   -   **2. T. virgata**

1. **T. barbata** (*L.*) *Gaertner*, De Fruct. et Sem., 2: 372, t. 160, fig. 1 (1791); Sibth. et Sm., Fl. Graec. Prodr., 2: 140 (1813); Poech, Enum. Plant. Ins. Cypr., 21 (1842); Unger et Kotschy, Die Insel Cypern, 251 (1865); Davis, Fl. Turkey, 5: 629 (1975); Feinbrun, Fl. Palaest., 3: 509, t. 692 (1978).
   Crepis barbata L., Sp. Plant., ed. 1, 805 (1753).
   Tolpis umbellata Bertol., Rar. Lig. Plant., 1: 13 (1803); Holmboe, Veg. Cypr., 191 (1914).
   TYPE: "*in* Monspelii, Vesuvii, Siciliae, Messanae *arenosis maritimis*".

Slender, erect, thinly arachnoid-tomentellous or glabrescent annual 14–60 cm. high; stems sulcate, usually branched; basal leaves lanceolate, or oblong-obovate, 7–12 cm. long, 1–4 cm. wide, toothed or shortly pinnatifid, base tapering to an ill-defined petiole; cauline leaves linear or narrowly lanceolate, sessile, entire or denticulate, 2–4 cm. long, 0·2–0·6 cm. wide; inflorescence branched, forming a lax, dichasial cyme; peduncles elongate, slender, slightly swollen near apex and bearing numerous, filiform, sub-apical bracts, 5–8 mm. long, equalling or exceeding the phyllaries; involucre campanulate, 5–6 mm. long, 4–5 mm. wide; phyllaries erect, linear, about 1–1·5 mm. wide at base, dark green, thinly arachnoid; florets lemon-yellow, the central ones often purplish; corolla-tube slender, pilose, about 4 mm. long, often purplish; ligule 4–5 mm. long, 1·5 mm. wide, pilose dorsally; filaments glabrous, about 0·3 mm. long; anthers linear, 1·5–2 mm. long, 0·2 mm. wide, apical appendage blunt; style about 8–9 mm. long, glabrous below, thinly pilose above, branches obtuse, about 0·2 mm. long, divergent. Achenes dimorphic, the outermost oblong, truncate, dark brown, sometimes shortly pilose and slightly curved, about 1·6 mm. long, 0·6 mm. wide, pappus of numerous short, whitish bristles about 0·3 mm. long, occasionally with an additional long bristle; inner achenes narrowly oblong, about 1·5 mm. long, 0·4 mm. wide, dark brown, longitudinally ribbed, minutely papillose-scabridulous, pappus-bristles 3–4, scabridulous, about 4 mm. long, much longer than the achene, with very short bristles in a row at their base.

HAB.: With *Cistus monspeliensis* in non-calcareous garigue; sea-level to 2,000 ft. alt.; fl. April–May.

DISTR.: Division 1. S. Europe, Mediterranean region.

1. Dhrousha, 1941, *Davis* 3245! 3280!

NOTES: Apart from Sibthorp's unlocalized (and rather doubtful) record, the only genuine evidence of the occurrence of *T. barbata* on the island. All other references seem to apply to the following, *T. virgata* (Desf.) Bertol., which is evidently much more common. Sintenis's record from the coast near (8) Cape Andreas (Oesterr. Bot. Zeitschr., 32: 398; 1882) may be correct, but I have seen no supporting specimens.

2. T. virgata (*Desf.*) *Bertol.*, Rar. Lig. Plant., 1: 15 (1803); Holmboe, Veg. Cypr., 191 (1914); Post, Fl. Pal., ed. 2, 2: 128 (1933); Lindberg f., Iter Cypr., 36 (1946); Osorio-Tafall et Seraphim, List Vasc. Plants Cyprus, 109 (1973); Davis, Fl. Turkey, 5: 630 (1975); Feinbrun, Fl. Palaest., 3: 410, t. 693 (1978).

    Crepis virgata Desf. in Act. Soc. Hist. Nat. Paris, 1: 37, t. 8 (1792); Fl. Atlant., 2: 230 (1799).

    Crepis altissima Balb., Cat. Plant. Hort. Taur., 15 (1804).

    Tolpis altissima (Balb.) Pers., Syn. Plant., 2: 377 (1807); Boiss., Fl. Orient., 3: 725 (1875); Post in Mém. Herb. Boiss., 18: 96 (1900); H. Stuart Thompson in Journ. Bot., 44: 333 (1906).

    Tolpis quadriaristata Biv., Mon. Tolpis, 9 (1809); Sibth. et Sm., Fl. Graec. Prodr., 2: 140 (1813), Fl. Graec., 9: 7, t. 810 (1837).

    [*T. umbellata* (non Bertol.) Poech, Enum. Plant. Ins. Cypr., 21 (1842); Kotschy in Unger et Kotschy, Die Insel Cypern, 251 (1865); Sintenis in Oesterr. Bot. Zeitschr., 32: 364 (1882).]

TYPE: "Cette plante croît dans les terreins sabloneux, et au bord des champs cultivés aux environs de Tunis et d'Alger" (P).

Erect biennial or perennial 50–100 cm. high, with a thick woody rootstock; stems sulcate, glabrescent or thinly pilose, usually branched in the upper half; basal leaves oblong-oblanceolate, 8–18 cm. long, 2–4 cm. wide, papery, thinly pilose, apex acute, margins subentire or remotely denticulate, base tapering to a flattened petiole 1·5–4 cm. long; stem leaves few, linear or elliptic-oblanceolate, 4–15 cm. long, 0·6–3 cm. wide, entire, denticulate or very shortly lobed, sessile or shortly petiolate; inflorescence lax, with long, erect or spreading, leafless, angular-sulcate branches; involucre campanulate; phyllaries linear, acuminate, thinly tomentose dorsally, the outer 3–4 mm. long, 1 mm. wide at base, longer than subapical peduncle-bracts, inner phyllaries to 10 mm. long, 1–1·5 mm. wide at base; florets yellow (turning greenish on drying); corolla-tube of outer florets 3–4 mm. long, shortly pilose, ligule 6–7 mm. long, 1·5–2 mm. wide, 5-toothed at apex; inner florets shorter, the ligule 4–5 mm. long; filaments very short, glabrous; anthers linear, 3–4 mm. long, 0·3 mm. wide, apical appendage blunt; styles about 9 mm. long, those of the central florets long-exserted, shortly pilose in the upper half, branches short, blunt, divergent, about 0·3 mm. long. Achenes oblong, about 1·8 mm. long, 0·4–0·5 mm. wide, dark brown, angular and longitudinally ribbed, minutely scabridulous-papillose; pappus of short, whitish scales, about 0·1 mm. long, interspersed with 4–8 (–14) scabridulous bristles up to 5 mm. long.

HAB.: Roadsides, field-borders, vineyards, waste ground, or open, stony hillsides; sea-level to 4,500 ft. alt.; fl. May–July.

DISTR.: Divisions 1–4, 7. S. Europe and Mediterranean region.

1. Ktima [? 1840, *Kotschy* 61], 1913, *Haradjian* 678!
2. Prodhromos, 1862, *Kotschy* 810; Chakistra, 1905, *Holmboe* 1033; Kykko Monastery, 1913, *Haradjian* 964! Kakopetria, 1936, *Syngrassides* 1261! Kryos Potamos, 3,000 ft. alt., 1937, *Kennedy* 983!
3. Near Khirokitia, 1901, *A. G. & M. E. Lascelles* s.n.! Ayia Nikola, near Arminou, 1941, *Davis* 3403!
4. Mile 14, Nicosia–Larnaca road, 1952, *Merton* 847!
7. Near Lapithos, 1880, *Sintenis & Rigo* 551! also, 1939, *Lindberg f.* s.n.; St Hilarion, 1939, *Lindberg f.* s.n.! Kyrenia Pass, 1939, *Lindberg f.* s.n.; Kyrenia, 1949, *Casey* 817! also, 1955, *G. E. Atherton* 136! 181! 324! *Miss Mapple* 69! Tjiklos, 1973, *P. Laukkonen* 348!

58. KOELPINIA *Pallas* *
Reise, 3, 2: 755, t. L1, fig. 2 (1776).

Erect or sprawling, branched annuals with linear, entire, grass-like leaves; flowering stems bearing several shortly pedunculate capitula; phyllaries in 2 series, the outer short, the inner much longer, equalling the florets, slightly accrescent in fruit, but not enveloping the achenes; florets inconspicuous, yellow; anthers sagittate at base; style-branches slender, rather obtuse; receptacle flat, naked. Achenes much exceeding the involucre, spreading, slightly to strongly incurved or arcuate, linear, terete, attenuate, ribbed, clothed apically and dorsally with glochidiate spinules or stout bristles; pappus of several minute, blunt bristles, evident on immature achenes and developing into a crown of radiating hooked spinules as the fruit matures.

About 3 or 4 species in S. Europe, Mediterranean region, western and central Asia, Baluchistan, India.

1. **K. linearis** *Pallas*, Reise, 3, 2: 755, t. L1, fig. 2 (1776); Poech, Enum. Plant. Ins. Cypr., 20 (1842); Unger et Kotschy, Die Insel Cypern, 249 (1865); Boiss., Fl. Orient., 3: 722 (1875); Holmboe, Veg. Cypr., 19 (1914); Post, Fl. Pal., ed. 2, 2: 127 (1933); Osorio-Tafall et Seraphim, List Vasc. Plants Cyprus, 109 (1973); Davis, Fl. Turkey, 5: 631 (1975); Feinbrun, Fl. Palaest., 3: 410, t. 694 (1978).
    Lapsana koelpinia L.f., Suppl. Plant., 348 (1781); Sibth. et Sm., Fl. Graec. Prodr., 2: 145 (1813), Fl. Graec., 9: 13, t. 819 (1837).
    Rhagadiolus koelpinia (L.f.) Willd., Sp. Plant., 3: 1626 (1804).
    TYPE: U.S.S.R.: "ad montem Bogdensem deserti Astrachanensis observata", 1773, *Pallas*.

Erect or sprawling, slender, sparsely pubescent, branched annual, with ribbed stems 3·5–22 cm. high; leaves grass-like, linear, entire, 3-nerved, glaucous-green, (1·5–) 4–11·5 cm. long, 0·5–3·5 mm. wide at base, apex acute, base sheathing; capitula homogamous, ligulate, 6–8 mm. diam.; involucre cylindrical; phyllaries in 2 series, the outer lanceolate, linear-lanceolate or subulate, 2–3 mm. long, 0·5 mm. wide, inner lanceolate or

* by P. Halliday

narrow-lanceolate, equalling florets, about 6 mm. long at anthesis, to 9 mm. long in fruit, about 1 mm. wide at base; florets yellow (often drying purplish), glabrous; ligules oblong, tapering slightly to base, 6 mm. long, 1·5–2 mm. wide, with 4–6 veins, apex truncate with 3–5 teeth; anthers linear, 1·5 mm. long, base sagittate, with long, slender, adpressed tails. Achenes like a bird's claws, up to 15 in a cluster, at first erect, straight and reddish, becoming stellate-patent, arcuate or incurved, and blackish at maturity, usually minutely pubescent, 10–14 mm. long, 1–1·5 mm. wide at base, strongly ribbed longitudinally, with few to many hooked spinules (up to 1 mm. long) mostly on the dorsal surface of the achene, except at the base, arising from the ribs, apex with a radiating crown of similar, unequal spinules; pappus (of immature achenes) of several minute, stout, blunt bristles.

HAB.: Roadsides, field margins; streamsides and stony hillsides; chiefly on Kythrean marls; 300–600 ft. alt.; fl. March–May.

DISTR.: Divisions 5–7. General distribution that of the genus.

5. Near Kythrea, 1880, *Sintenis & Rigo*; between Nicosia and Nisou, 1936, *Syngrassides* 1115!
Mia Milea–Koutsovendis road, 1974, *Osorio–Tafall & Meikle* in *Meikle* 4070!
6. Near Nicosia, 1880, *Sintenis & Rigo* 498! also, 1962, *Meikle* 2549!
7. Lakovounara Forest, 1950, *Chapman* 78!

59. RHAGADIOLUS *Juss.**

Gen. Plant., 168 (1789) nom. cons.
R. D. Meikle in Taxon, 28: 133–141 (1979).

Erect annuals; stems usually branched; leaves mostly basal, forming a rosette, variously toothed, lobed or pinnatisect, occasionally subentire; inflorescence loosely branched; capitula small, sessile or shortly pedunculate, homogamous, ligulate; involucre cylindrical; phyllaries in 2 series, the outer (c. 5) minute, the inner much longer, boat-shaped, accrescent, variably thickened in fruit, enclosing or enveloping the marginal achenes; florets yellow; anthers sagittate at base; style-branches slender, rather obtuse; receptacle flat, naked. Achenes of 2 kinds: the marginal subulate or narrowly fusiform, straight or curved, more or less persistent, erect or stellately patent, the inner strongly arcuate, often hispidulous, caducous, pappus minute, persistent, ciliolate, or of scabridulous hairs, or absent.

About 6 species (including those formerly assigned to *Garhadiolus* Jaub. et Spach) widely distributed in the Mediterranean region, Atlantic Islands, and eastwards to Transcaspia, Iran, Baluchistan and India.

Marginal achenes 5–6, short (1–1·3 cm. long), straight, fusiform, wholly enveloped by glabrous phyllary; leaf with large, suborbicular terminal lobe and distinct petiole -   1. R. edulis
Marginal achenes (7–) 8, about 1·6 cm. long, subulate, attenuate, incurved, only partially enveloped by dorsally hispid phyllary; leaf repand-dentate, terminal lobe not large or suborbicular; petiole indistinct   -   -   -   -   -   -   -   2 R. stellatus

1. **R. edulis** *Gaertner*, De Fruct. et Sem., 2: 354 (1791); Rechinger f. in Arkiv för Bot., ser. 2, 1: 434 (1950); Feinbrun, Fl. Palaest., 3: 412, t. 697 (1978).
Lapsana rhagadiolus L., Sp. Plant., ed. 1, 812 (1753); Sibth. et Sm., Fl. Graec. Prodr., 2: 144 (1813), Fl. Graec., 9: 12, t. 818 (1837).
Rhagadiolus stellatus (L.) Gaertner var. *edulis* (Gaertner) DC., Prodr., 7 (1): 77 (1838); Poech, Enum. Plant. Ins. Cypr., 20 (1842); Post, Fl. Pal., ed. 2, 2: 127 (1933); Lindberg f., Iter Cypr., 36 (1946); Davis, Fl. Turkey, 5: 688 (1975).
R. stellatus (L.) Gaertner ssp. *edulis* (Gaertner) Holmboe, Veg. Cypr., 191 (1914) quoad nomen; Osorio-Tafall et Seraphim, List Vasc. Plants Cyprus, 109 (1973).
TYPE: Herb. Stockholm; IDC microfiche 330/3, 330/4 "Herb. Alstroemerii" (lectotype, S!)

* by P. Halliday

Erect or spreading annual (10–) 20–50 (–70) cm. high; stems branched or unbranched at the base, thinly or densely hispidulous in the lower part, glabrous above; basal leaves bright green, usually forming a rosette, oblanceolate or obovate in outline, 2·9–25 cm. long, lyrate-pinnatisect (rarely without lateral lobes), the terminal lobe much enlarged, 1·1–4·5 cm. wide, suborbicular, cordate, or occasionally reniform, sinuate or bluntly angled, each angle shortly mucronate, lateral lobes much smaller, usually in 2 pairs, remote, upper pair oblong and rounded, lower pair deltoid and mucronate; petiole distinct, not winged, often equalling or exceeding the lamina; cauline leaves few, lanceolate or linear-lanceolate, entire or shallowly toothed, 2·3–10 cm. long, 0·5–3·6 cm. wide, sparsely pilose above, more densely so beneath, with longer hairs along the nerves; inflorescence paniculate, spreading, much branched; involucre subcylindrical, 6–8 (–11) mm. long, 2·5–3·5 mm. wide; outer phyllaries 1 mm. long, 0·6–0·8 mm. wide, ovate, with scarious, erose margins; inner phyllaries usually 5–6, occasionally up to 8, narrowly oblong, about 8 mm. long, less than 1 mm. wide, at first erect, later accrescent, spreading, somewhat keeled and thickened, and enclosing the achene, glabrous, with narrow scarious margins and an incurved, shortly beaked (or spurred) apex; ligules yellow, oblong, 9 mm. long, 2 mm. wide, 6-veined, basal tube about 1 mm. long, apex truncate with 4–5 teeth; anthers linear, 2 mm. long, 0·2 mm. wide, apical appendage ovate-oblong; style-branches linear. Marginal achenes 5–6 (–7), straight, fusiform, not exceeding the enveloping phyllary, 9–11 (–12) mm. long, 1–1·5 mm. wide, smooth, glabrous; inner achenes 1 or few, sometimes absent, 5–8 mm. long, 0·5–1 mm. wide, strongly arcuate, hispidulous, without enveloping phyllaries.

HAB.: Cultivated and waste ground, roadsides; in open garigue or on rocky ground below cliffs; sea-level to 4,800 ft. alt.; fl. March–June.

DISTR.: Divisions 1, 2, 6–8. Mediterranean region eastwards to Iraq and Iran.

1. Lyso, 1913, *Haradjian* 872.
2. Near Kykko Monastery, 1937, *Syngrassides* 1547! Prodhromos, 1939, *Lindberg f.* s.n.! also, 1951, *Casey* 1197!
6. Near Liveras, 1962, *Meikle* 2413!
7. Kyrenia, 1949, *Casey* 610! Templos, 1956, *G. E. Atherton* 1176! Kyrenia Pass, 1956, *G. E. Atherton* 1383!
8. Yioti, 1962, *Meikle* 2525!

NOTES: Apparently less common than *R. stellatus* (L.) Gaertner, but rarely distinguished from this by collectors, and perhaps overlooked.

2. R. stellatus (*L.*) *Gaertner*, De Fruct. et Sem., 2: 354, t. 157, fig. 2 (1791); Unger et Kotschy, Die Insel Cypern, 249 (1865) pro parte; Boiss., Fl. Orient., 3: 722 (1875) pro parte; Holmboe, Veg. Cypr., 191 (1914) pro parte; Post, Fl. Pal., ed. 2, 2: 127 (1933); Rechinger f. in Arkiv för Bot., ser. 2, 1: 434 (1950); Feinbrun, Fl. Palaest., 3: 412, t. 696 (1978).

Lapsana stellata L., Sp. Plant., ed. 1, 811 (1753); Sibth. et Sm., Fl. Graec. Prodr., 2: 144 (1813), Fl. Graec., 9: 12, t. 817 (1837).

Rhagadiolus stellatus (L.) Gaertner var. *leiocarpus* DC., Prodr., 7: 77 (1838); Poech, Enum. Plant. Ins. Cypr., 20 (1842).

R. stellatus (L.) Gaertner ssp. *stellatus*; Osorio-Tafall et Seraphim, List. Vasc. Plants Cyprus, 109 (1973).

R. stellatus (L.) Gaertner var. *stellatus*; Davis, Fl. Turkey, 5: 687 (1975).

TYPE: "Monspelii, Bononiae" (LINN!).

Erect or spreading annual (7·5–) 12–43 cm. high; stems usually branched, slightly ribbed, sparsely strigose, or usually densely so towards base, rarely the whole plant scabridulous; basal leaves usually forming a rosette, dull, pale green, oblanceolate or narrowly obovate in outline, (1–) 2·5–29·5 cm. long, (0·4–) 0·8–6·2 cm. wide, subentire or irregularly toothed or pinnatisect, the lobes minutely denticulate and mucronate, terminal lobe not conspicuously enlarged or suborbicular; petiole short, rather indistinct; lamina scabridulous or sparsely strigose above, sometimes glabrous below except on the veins; cauline leaves linear-lanceolate, lanceolate, oblanceolate or elliptic, usually sessile, entire, subentire or dentate, 1·6–17·4 (–23·4) cm. long, 0·5–5·4 (–7·6) cm. wide, diminishing in size upwards and finally becoming bractiform, usually more densely strigose than the basal leaves; inflorescence lax, paniculate; capitula 8–9 (–12) mm. long, base of involucre often purplish; outer phyllaries 1–2 mm. long, about 1 mm. wide, ovate to cordate, acute, with scarious, more or less erose margins; inner phyllaries 5–8 mm. long, much thickened in fruit, keeled, with a crest of (usually pale) bristles, and sometimes with narrow, scarious margins; ligules yellow, oblong, 6–8·5 mm. long, 1·5–2·5 mm. wide, base tubular for 1 mm., apex truncate, with (3–) 5 irregular teeth, veins 5–6, glabrous; anthers linear, 2–3 mm. long, 0·2 mm. wide, base sagittate with long slender tails, apical appendages oblong, rounded; style dark-coloured (in dried material), papillose for the top 1 mm. or for up to half its length, branches linear, rounded. Marginal achenes 7–8, slender, subulate, attenuate with a long narrow neck, (13–) 14–16 (–18) mm. long, 1–2·5 mm. wide, about 2–4 mm. longer than the indurated, enveloping phyllary; inner achenes 1–3 (–4), occasionally none, arcuate, dorsally hispidulous, 6–12 mm. long, 0·5–1 mm. wide; pappus absent.

HAB.: Cultivated and waste ground, roadsides; in open garigue or Pine forest; sea-level to 4,050 ft. alt.; fl. March–June.

DISTR.: Divisions 1–8. S. Europe, Mediterranean region and eastwards to Iran.

1. Ayios Yeoryios (Akamas), 1960, *N. Macdonald* 52! Polis, 1962, *Meikle* 2268!
2. Pano Panayia, 1913, *Haradjian* 909! Platres, 1938, *Kennedy* 1462! Between Platres and Phini, 1938, *Kennedy* 1463! Statos, 1981, *Hewer* 4771!
3. Episkopi, 1862, *Kotschy* 614! Malounda, 1950, *Chapman* 307! Amathus, 1978, *J. Holub* s.n.!
4. Larnaca, 1862, *Kotschy* 84; also, 1880, *Sintenis* 552; and, 1905, *Holmboe* 157b; between Dherinia and Paralimni, 1936, *Syngrassides* 888! E. of Dhekelia, 1981, *Hewer* 4720! Near Famagusta, 1912, *Haradjian* 104; Kouklia, 1905, *Holmboe* 372.
5. Near Kythrea, 1880, *Sintenis & Rigo* 552a! Mile 5, Nicosia–Famagusta road, 1950, *Chapman* 46! Salamis, 1962, *Meikle* 2576!
6. E. of Karpasha, 1959, *P. H. Oswald* 104!
7. Near Phlamoudhi, 1880, *Sintenis & Rigo* 552b! Kyrenia, 1949, *Casey* 371! 529! Vasilia, 1956, *G. E. Atherton* 1098! Above Kythrea, 1967, *Merton* in ARI 208!
8. Near Rizokarpaso, 1912, *Haradjian* 221; Ayios Philon near Rizokarpaso, 1941, *Davis* 2273!

60. HEDYPNOIS *Mill.* *
Gard. Dict. Abridg., ed. 4, 2 (1754).

Erect or straggling, variably hispid annuals; leaves alternate or in a basal rosette; capitula solitary, homogamous, many-flowered, terminal and axillary, pedunculate, the peduncles commonly somewhat swollen beneath the capitulum; involucre subcylindrical or campanulate; phyllaries in 2 series, the outer small, the inner much larger, boat-shaped, accrescent and becoming thickened in fruit, enveloping the marginal achenes; ligules yellow; anthers sagittate at base, with a blunt apical appendage; style-branches blunt; receptacle flat, naked. Achenes cylindrical, ribbed, strigose along ribs, incurved; pappus of marginal achenes coroniform, or, of inner achenes, of 5 free scales.

One variable species (sometimes split into 2–3 segregate species) widely distributed in the Mediterranean region and western Asia.

1. **H. rhagadioloides** (*L.*) *F. W. Schmidt* in Samm. Phys.-oek. Aufs., 1: 279 (1795); Spreng., Syst. Veg., 3: 670 (1826); Holmboe, Veg. Cypr., 191 (1914); Rechinger f. in Arkiv för Bot., ser. 2,

* by P. Halliday

1: 434 (1950); Osorio-Tafall et Seraphim, List Vasc. Plants Cypr., 109 (1973); Greuter in Candollea, 31: 230 (1976); Feinbrun, Fl. Palaest., 3: 413, t. 698 (1978).

Hyoseris rhagadioloides L., Sp. Plant., ed. 1, 809 (1753).

H. hedypnois L., Sp. Plant., ed. 1, 809 (1753).

H. cretica L., Sp. Plant., ed. 1, 810 (1753).

Hedypnois cretica (L.) Dum.-Cours., Le Botaniste Cultivateur, 2: 339 (1802); Sibth. et Sm., Fl. Graec. Prodr., 2: 142 (1813), Fl. Graec., 9: 9, t. 813 (1835); Unger et Kotschy, Die Insel Cypern, 249 (1865); Boiss., Fl. Orient., 3: 719 (1875); C. et W. Barbey, Herborisations au Levant, 99 (1880); H. Stuart Thomson in Journ. Bot., 44: 333 (1906); Post, Fl. Pal., ed. 2, 2: 126 (1933); Lindberg f., Iter Cypr., 35 (1946); Davis, Fl. Turkey, 5: 686 (1975).

H. monspeliensis Willd., Sp. Plant., 3: 1616 (1804); Sibth. et Sm., Fl. Graec. Prodr., 2: 142 (1813).

H. polymorpha DC., Prodr., 7: 81 (1838); Poech, Enum. Plant. Ins. Cypr., 20 (1842); Unger et Kotschy, Die Insel Cypern, 249 (1865); Sintenis in Oesterr. Bot. Zeitschr., 31: 194, 326 (1881), 32: 123 (1882).

H. rhagadioloides (L.) F. W. Schmidt ssp. *cretica* (L.) Holmboe, Veg. Cypr., 191 (1914); Osorio-Tafall et Seraphim, List Vasc. Plants Cyprus, 109 (1973).

TYPE: "*in* Europa *australi*".

Erect or decumbent herb (2–) 4–14 (–41) cm. high, with smooth or sparsely strigose, branching stems; leaves greyish-green or tinged purple, the basal obovate or narrowly obovate (2·5–) 5–11 (–23·5) cm. long, (0·6–) 1·2–1·8 (–3·5) cm. wide, sessile or shortly petiolate with a winged petiole, lamina hispid with 2–3-furcate bristles on both surfaces, sometimes densely so on the upper surface, or sometimes only along veins and margins of lower surface, apex obtuse, acute or 3-dentate, margins subentire or dentate, the teeth diminishing in size downwards, base attenuate, decurrent; cauline leaves similar to basal, or linear-lanceolate, lanceolate or oblong, (1·5–) 2·5–12·5 (–19·5) cm. long, (0·2–) 0·6–2·3 (–3·6) cm. wide, decreasing in size upwards, usually sessile, but the lowermost occasionally petiolate; inflorescence branched or unbranched; peduncles usually (but not always) swollen and hollow, to 7 mm. diam. just below capitulum, striate and often densely, minutely hispid, with forked bristles much shorter than those on the leaves; capitulum subcylindric at anthesis, 8–12 mm. long, (6–) 8–10 (–13) mm. diam., suborbicular or hemispherical in fruit, 1–1·5 (–1·7) cm. diam.; outer phyllaries few, linear-lanceolate or subulate, 2–5 mm. long, with an apical tuft of minute hairs, and with or without longer white bristles on surface and margins; inner phyllaries narrowly lanceolate or linear-lanceolate, 6–8 mm. long, 1–1·5 mm. wide, glabrous or variably hispid on surface and margins, with bristles of varying length, or sometimes hispid only on the upper, dorsal surface; florets dull golden-yellow, the outer surface of ligules purple in dried specimens; ligules elliptic-oblong, 5–7 mm. long, 1·5–2·5 mm. wide, truncate, 5–6-veined, apex with 3–5 teeth, tube about ⅓ the length of floret, glabrous except for some hairs on the outer surface of the mouth; anthers linear, 2–2·5 mm. long, 0·2 mm. wide, with an ovate-oblong, rounded apical appendage and sagittate base; style papillose towards apex, branches about 1 mm. long, rounded. Fruiting capitulum resembling a bunch ("hand") of bananas; achenes erect and incurved, not spreading, the marginal 5–7 mm. long, up to 1 mm. wide at base, each enveloped by an accrescent, indurated, boat-shaped phyllary 10 mm. long, 1·5 mm. wide; pappus coroniform, with erose scales about 1 mm. long, the inner naked, less curved, 5–7 mm. long, with 5 subulate pappus-scales produced into distinct, scabrid awns 4–6 mm. long, each scale usually with many short, hair-like teeth near the base.

HAB.: Cultivated and fallow fields, roadsides; sand-dunes; grass-steppe or in open garigue on hillsides; sea-level to 3,900 ft. alt.; fl. Febr.–June.

DISTR.: Divisions 1–8. S. Europe and Mediterranean region east to Iran; Atlantic Islands; an introduced weed in S. Africa, Australia, N. & S. America.

1. Stroumbi, 1913, *Haradjian* 725! Peyia Forest, 1962, *Meikle* 2048!

2. Near Philani, 1908, *Clement Reid* s.n.! Platres, 1938, *Kennedy* 1461! and, 1955, *P. Paolides* s.n.! Mesapotamos Monastery, 1937, *Lindberg f.* s.n.!
3. Near Stavrovouni, 1937, *Syngrassides* 1485! Cherkez, 1947, *Mavromoustakis* s.n.! Near Amathus, 1978, *J. Holub* s.n.!
4. Near Larnaca 1859 and 1862, *Kotschy* 275; 460; and, 1880, *Sintenis* 803! and, 1912, *Haradjian* 6; Ayia Napa, 1905, *Holmboe* 3; 13; Perivolia, 1938, *Syngrassides* 1751!
5. Near Kythrea, 1880, *Sintenis & Rigo* 496! Salamis, 1905, *Holmboe* 454; between Nicosia and Nisou, 1936, *Syngrassides* 1120! Mile 3, Nicosia–Famagusta road, 1950, *Chapman* 389! Mia Milea, 1967, *Merton* in ARI 346! Road to Klirou, 1967, *Merton* in ARI 497!
6. Makhedonitissa Monastery, 1936, *Syngrassides* 893! Nicosia, 1936, *Syngrassides* 943! Vizakia, 1936, *Syngrassides* 1238! Mile 3, Nicosia–Myrtou road, 1950, *Chapman* 572!
7. Mountains above Kythrea, 1880, *Sintenis & Rigo* 497! Kyrenia, 1949, *Casey* 591! 770! Lakovounara Forest, 1950, *Chapman* 140! 6 miles S.E. of Kyrenia, 1951, *Casey* 1117!
8. Near Rizokarpaso, 1912, *Haradjian* 221! Komi Kebir, 1912, *Haradjian* 304!

NOTES: Some authors subdivide the species into segregate species, subspecies or varieties. In Cyprus the degree of variability is such that, for the purposes of the Flora, I prefer to regard *H. rhagadioloides* as a single unit.

61. PICRIS *L.*

Sp. Plant., ed. 1, 792 (1753).

Gen. Plant., ed. 5, 347 (1754).

Erect or decumbent, annual, biennial or perennial herbs, generally hispid with glochidiate bristles; stems usually branched; basal leaves often rosulate, attenuate to base, apex acute or blunt, margins subentire, dentate or pinnatifid; cauline leaves sessile, often entire or subentire; inflorescence commonly lax, corymbose or paniculate; capitula homogamous, ligulate, solitary at the apex of a distinct, often swollen peduncle; involucre generally campanulate at anthesis, often becoming urceolate in fruit; phyllaries in several series, the outer usually short, spreading or reflexed, forming a calyculus, the inner longer, linear-lanceolate, erect, concave-carinate, often partly enveloping the outer achenes; florets yellow or purplish externally, generally much exceeding involucre, 5-dentate at apex; anthers sagittate at base; style-branches slender; receptacle flat, naked. Achenes uniform or dimorphic, fusiform, truncate or attenuate to apex, or sometimes beaked, generally conspicuously rugose transversely, straight or curved; pappus generally of copious, silky, plumose hairs, connate into a basal annulus and deciduous, pappus of marginal achenes sometimes short, scabridulous or of a few bristles, deciduous or persistent.

More than 40 species widely distributed in temperate regions of the Old World.

Glochidiate bristles on peduncle and involucre uniformly 2-barbed; pappus-hairs uniform, plumose; leaves and stems strigose or very shortly hispidulous, cauline leaves not or scarcely auriculate at base:
Achenes 4–6 mm. long, apex attenuate, substrostrate; fruiting peduncles conspicuously swollen, abruptly constricted at apex, just below involucre; basal leaves usually narrow, oblanceolate - - - - - - - - - - - **1. P. pauciflora**
Achenes 2–3 mm. long, apex shortly attenuate, not at all rostrate; fruiting peduncles not much swollen, not constricted at apex; basal leaves oblanceolate or narrowly obovate
2. P. cyprica
Glochidiate bristles on peduncle and involucre 2–4 (–5)-barbed; pappus of marginal achenes short, scabridulous or barbellate, pappus of inner achenes long, plumose; leaves and stems conspicuously hispidulous; cauline leaves distinctly auriculate at base - **3. P. altissima**

1. **P. pauciflora** Willd., Sp. Plant., 3: 1557 (1803); Boiss., Fl. Orient., 3: 737 (1875); Rechinger f. in Arkiv för Bot., ser. 2, 1: 434 (1950); Osorio-Tafall et Seraphim, List Vasc. Plants Cyprus, 109 (1973); Davis, Fl. Turkey, 5: 681 (1975).
   [*P. sprengeriana* ["*sprengeliana*"] (non L.) Kotschy in Unger et Kotschy, Die Insel Cypern, 254 (1865).]
   TYPE: "*in* Gallia *australi*", *Desfontaines* (B).

Erect annual (10–) 20–50 cm. high; stems usually branched, obscurely ribbed, thinly strigillose with spreading 2-barbed bristles and minute arachnoid hairs; basal leaves loosely rosulate, oblanceolate, attenuate to an ill-defined petiole, 2–10 cm. long, 0·5–1·3 cm. wide, thinly glochidiate-strigillose on both surfaces, apex blunt or acute, margins remotely denticulate or repand-pinnatifid; cauline leaves similar, but usually smaller and narrower, the uppermost often entire or subentire; inflorescence lax, with divaricate branches; peduncles to 15 cm. (or more in fruit), glochidiate-strigillose, becoming distinctly swollen above, but abruptly constricted at apex in fruit; involucre 0·8–1·2 cm. long, 0·8–1 cm. wide, becoming subglobose-urceolate in fruit; outer phyllaries loosely spreading or reflexed, linear-subulate, 3–6 mm. long, 0·8–1 mm. wide at base, glochidiate-strigillose dorsally, with blackish tips; inner phyllaries erect, linear, twice as long as outer, with a distinct, pallid, hispid-lanuginose keel and blackish-green margins; corolla-tube about 4 mm. long, glabrous below, pilose at apex; ligules yellow, purplish externally, about 8 mm. long, 1·5 mm. wide, apex 5-dentate; filaments glabrous, about 0·5 mm. long; anthers about 3 mm. long, 0·2 mm. wide, apical appendage blunt; style about 6 mm. long, minutely pilose in the upper half, branches filiform, divergent, blunt, about 1 mm. long. Achenes fusiform, curved, about 4–6 mm. long, 0·8 mm. wide, brown or fuscous, glabrous, conspicuously ribbed transversely, apex attenuate or subrostrate; pappus-hairs usually 17 or more, white, silky, plumose, deciduous, 6–8 mm. long.

HAB.: Roadsides; stony, igneous mountainsides and screes at high altitudes; 2,800–4,500 ft. alt.; fl. May–July.

DISTR.: Division 2. Also France, Italy, Balkan Peninsula, Turkey, Crimea, Iraq, Iran.

2. Prodhromos, 1862, *Kotschy* 728 ! also, 1951, *Casey* 1198 ! 1204 ! Kryos Potamos, 3,200 ft. alt., 1938, *Kennedy* 1464 partly ! Pano Platres, 1930, *E. Wall* s.n.; also, 1938, *Kennedy* 1465 partly ! 1468 partly ! Between Ayios Theodhoros and Zoopiyi, 1941, *Davis* 3078 ! Kykko Monastery, 1941, *Davis* 3474 ! Ayios Nikolaos near Kakopetria, 1962, *Meikle* 2849 ! Between Vavatsinia and Ayii Vavatsinias, 1967, *Merton* in ARI 717 !

2. P. cyprica *Lack* in Notes Roy. Bot. Gard. Edinb., 33: 426 (1975); Davis, Fl. Turkey, 5: 681 (1875).

　　[*P. pauciflora* (non Willd.) Sintenis in Oesterr. Bot. Zeitschr., 32: 364, 398 (1882); Holmboe, Veg. Cypr., 192 (1914).]

TYPE: Turkey; "montagne près Anamour", May–June 1872, *A. Péronin* 48 (PRC, K ! etc.).

Robust or slender, sparsely glochidiate-strigillose annual, usually less than 30 cm. high in our area, but up to 70 cm. in Turkey; glochidia 2-barbed; stems usually with divaricate branches, obscurely ribbed; basal leaves loosely rosulate, oblanceolate or narrowly obovate, 2–8 cm. long, 1–2 cm. wide, apex rounded or very shortly acute, margins entire or remotely sinuate-denticulate, base tapering to an ill-defined petiole; cauline leaves few, smaller and narrower than the basal, often entire; inflorescence lax, spreading; peduncles to 10 cm. long at anthesis, lengthening to 15 cm., and becoming indistinctly swollen (in Cyprus specimens), but not constricted at apex in fruit; involucre broadly campanulate at anthesis, usually less than 1 cm. long, and often as wide or even a little wider, becoming somewhat urceolate in fruit; outer phyllaries loosely spreading or reflexed, linear-subulate, 3–4 mm. long, 0·8–1 mm. wide at base, rather densely glochidiate-strigillose dorsally, with fulvous bristles; inner phyllaries rather loosely erect, linear, twice as long as the outer or more, apparently greenish, and rather densely fulvous-strigillose along the dorsal keel; florets relatively large and conspicuous; corolla-tube about 6 mm. long, glabrous below, densely pilose towards apex; ligules yellow or stained purple externally, 8–9 mm. long, 1·5–2 mm. wide, apex 5-dentate; filaments glabrous, about 0·4 mm. long; anthers 2·5–3 mm. long, 0·2 mm. wide, apical appendage blunt; style 7–9 mm. long, minutely pilose in the upper half, branches filiform, divergent, blunt, about 0·8 mm. long. Achenes fusiform, strongly curved, 2–3 mm. long, 0·8 mm. wide, brown, glabrous, conspicuously ribbed transversely; apex shortly attenuate; pappus-hairs 15 or more, white, silky, plumose, deciduous, 5–6 mm. long.

HAB.: Rocky, limestone and igneous slopes; sea-level to 2,500 ft. alt.; fl. April–July.

DISTR.: Divisions 1–3, 7, 8. Also Turkey.

1. Lyso, 1913, *Haradjian* 895!
2. Yialia, 1955, *Merton* 2113 partly!
3. Near Stavrovouni, 1937, *Syngrassides* 1473!
7. Near Dhavlos, 1880, *Sintenis & Rigo* 292! Near Pentadaktylos, 1880, *Sintenis & Rigo* 292a! Armenian Monastery, 1912, *Haradjian* 361! 363! Hillside above Kyrenia Pass, 1953, *Casey* 1312! W. end forest road above Kyrenia, 1955, *G. E. Atherton* 140! Antiphonitis Monastery, 1955, *Merton* 2274 partly!
8. Cape Andreas, 1880, *Sintenis & Rigo* 292b!

NOTES: Most of the specimens cited above have been examined and determined by Dr. H. W. Lack, the acknowledged authority on this difficult genus. Cyprus specimens of *P. cyprica* are uniformly slender, and much less robust than the type, with less noticeably incrassate peduncles; they resemble *P. pauciflora* Willd., but differ markedly in the larger "flowers" and much smaller, boomerang-shaped achenes.

3. P. altissima *Del.*, Fl. d'Egypte, 260, t. 41, fig. 2 (1813–14); Davis, Fl. Turkey, 5: 682 (1975).
P. sprengeriana Chaix in Vill., Hist. Dauph., 1: 369 (1786) quoad plant. vix *Hieracium sprengerianum* L., Sp. Plant., ed. 1, 804 (1753); Boiss., Fl. Orient., 3: 738 (1875); Sintenis in Oesterr. Bot. Zeitschr., 32: 291, 398 (1882); Holmboe, Veg. Cypr., 192 (1914); Post, Fl. Pal., ed. 2, 2: 131 (1933) — as *P.* "*sprengerina*"; Lindberg. f., Iter Cypr., 36 (1946); Rechinger f. in Arkiv. för Bot., ser. 2, 1: 434 (1950) pro parte; Osorio-Tafall et Seraphim, List Vasc. Plants Cyprus, 109 (1973); Feinbrun, Fl. Palaest., 3: 418, t. 705 (1978).
TYPE: Egypt; "Iles sèches et sablonneuses du Nil", *Delile* (MPU).

Erect annual 20–80 cm. high; stems simple or branched, inconspicuously ribbed, hispid with spreading 2–4 (–5)-barbed bristles intermixed with a much shorter, scrofulose, very sparse indumentum; basal leaves forming a rosette, narrowly obovate, 5–15 cm. long, 1·3–3 cm. wide, thinly but conspicuously hispidulous on both surfaces, apex rounded or obtuse, margins sinuate-dentate, base tapering to an indistinct flattened petiole; cauline leaves few, narrowly oblong or obovate, 1·5–8 cm. long, 0·3–2 cm. wide, sessile, hispidulous, apex acute or subacute, margins entire or obscurely dentate, base generally semi-amplexicaul, auriculate, with short, rounded, projecting auricles; inflorescence a lax, spreading, much-branched panicle; involucre broadly campanulate, about 8 mm. long, 6–7 mm. wide, becoming subglobose-urceolate in fruit; outer phyllaries narrowly oblong, about 2–4 mm. long, 1 mm. wide, spreading, often tinged purplish, hispid; inner phyllaries erect, oblong-acuminate, about 1–1·5 mm. wide at base, commonly blackish or purplish and hispid with 2–5-barbed bristles; florets yellow or often purplish externally; corolla-tube slender, about 4 mm. long, glabrous below, pilose near apex; ligule about 9 mm. long, 2 mm. wide, apex 5-dentate; filaments very short, glabrous; anthers linear, 3·5–4 mm. long, about 0·4 mm. wide, apical appendage short, blunt; style 7–8 mm. long, minutely pilose in the upper half, branches filiform, blunt, about 1·2 mm. long. Fruiting peduncles 3–15 cm. long, not much swollen; achenes fusiform, curved, strongly ribbed transversely, dark brown, about 3·5 mm. long, 0·8 mm. wide, apex attenuate but scarcely rostrate; pappus of marginal achenes short, about 2–2·5 mm. long, scabridulous or barbellate; pappus of inner achenes white, silky, plumose, 5–6 mm. long.

HAB.: Cultivated and fallow fields, roadsides; stony hillsides; near sea-level to 5,000 ft. alt.; fl. May–July.

DISTR.: Divisions 1–3, 7, 8. Widespread in the Mediterranean region.
1. Paphos, 1913, *Haradjian* 635; Ktima, 1913, *Haradjian* 690; 707! Ayios Neophytos Monastery, 1939, *Lindberg f.* s.n.
2. Panayia, 1913, *Haradjian* 915! Platres, 1938, *Kennedy* 1465 partly! 1466! Kannaviou, 1939, *Lindberg f.* s.n.; Troödos, 1939, *Lindberg f.* s.n.; Lefkara, 1940, *Davis* 1905!
3. Limassol, 1913, *Haradjian* 590; Kolossi, 1939, *Lindberg f.* s.n.
7. Between Ayios Amvrosios and Ayios Epiktitos, 1932, *Syngrassides* 191! Dhikomo, 1936, *Syngrassides* 995! St. Hilarion, 1939, *Lindberg f.* s.n.; Kyrenia Pass [Boghazi], 1939, *Lindberg f.* s.n.! also, 1955, *Merton* 2275! Kyrenia, 1949, *Casey* 776!
8. Ayios Andronikos, 1880, *Sintenis & Rigo* 293! Near Rizokarpaso, 1912, *Haradjian* 157! 242! Komi Kebir, 1912, *Haradjian* 275!

[P. KOTSCHYI *Boiss.*, Fl. Orient., 3: 738 (1875), recorded without precise locality by H. Stuart Thompson in Journ. Bot., 44: 333 (1906) as *P. longirostris* Sch. Bip. var. *kotschyi* Sch. Bip., is an error. The Lascelles' specimen upon which the record was based has since been identified as *Urospermum picroides* (L.) F. W. Schmidt. Subsequent records by Holmboe, Osorio-Tafall et Seraphim, etc. are based on the same Lascelles' specimen.]

62. HELMINTHOTHECA *Zinn*
Cat. Plant. Goetting. 430 (1757).
Helminthia A. L. Juss., Gen. Plant., 468, 170 (1789).

Annual or perennial herbs; stems and leaves hispid with glochidiate, 2–6-barbed bristles; basal leaves attenuate to a shortly petiolate base; cauline leaves sessile, sometimes amplexicaul and auriculate; capitula homogamous, ligulate, pedunculate; involucre ovoid or urceolate in fruit; phyllaries in 2 series, the outer foliaceous, ovate or cordate, wholly or partly concealing the linear or oblong inner phyllaries; florets yellow; stamens and styles as in *Picris* L.; receptacle naked. Achenes uniform or dimorphic, obovoid, transversely ribbed; pappus-hairs plumose, those of marginal achenes sometimes much shorter than those of the inner.

Six species in Europe and the Mediterranean region east to Iran; some of the species are introduced weeds elsewhere.

1. **H. echioides** (L.) Holub in Folia Geobot. Phytotax., 8: 176 (1973); Davis, Fl. Turkey, 5: 684 (1975); Feinbrun, Fl. Palaest., 3: 421, t. 711 (1978).
   Picris echioides L., Sp. Plant., ed. 1, 792 (1753); Holmboe, Veg. Cypr., 192 (1914); Lindberg f., Iter Cypr., 35 (1946); Osorio-Tafall et Seraphim., List Vasc. Plants Cyprus, 109 (1973).
   Helminthia echioides (L.) Gaertner, De Fruct. et Sem., 2: 368, t. 159, fig. 2 (1791); Boiss., Fl. Orient., 3: 742 (1875); Sintenis in Oesterr. Bot. Zeitschr., 32: 291 (1882); Post, Fl. Pal., ed. 2, 2: 133 (1933); Rechinger f. in Arkiv för Bot., ser. 2, 1: 434 (1950).
   TYPE: "*in* Germaniae, Angliae, Belgii, Galliae *versuris agrorum*".

Erect, hispid annual (5–) 10–18 cm. high; stems branched or unbranched, obscurely ribbed, generally armed with numerous rigid, 4-barbed, pallid bristles; basal leaves obovate or oblanceolate, 4–12 cm. long, 1–2 (or more) cm. wide, pustulate-hispid, apex acute or obtuse, base tapering to an ill-defined petiole, margins entire or dentate; cauline leaves usually smaller and narrower, sessile, semi-amplexicaul, with shortly projecting auricles; inflorescence usually branched, with rather rigid, spreading or erecto-patent branches; peduncles densely hispid, 1–3 cm. long at anthesis, sometimes lengthening to 5 cm. or more, but not conspicuously incrassate, in fruit; involucre broadly campanulate, 1–1·8 cm. long, 0·8–1·5 cm. wide; outer phyllaries erect, foliaceous, hispid, ovate-acuminate or subcordate, 1–1·8 cm. long, 0·6–1 cm. wide; inner phyllaries linear, 1–1·8 cm. long, 0·2 cm. wide, scarious-membranous, with a hispid midrib produced into a long, hispid apical, caudate appendage; florets inconspicuous, often purplish externally; corolla-tube very slender, about 6 mm. long, glabrous below, pilose at apex; ligule about 7 mm. long, 0·8 mm. wide, apex shortly 5-toothed; filaments glabrous, about 0·2 mm. long; anthers linear, about 2 mm. long, 0·2 mm. wide, with a blunt apical appendage; style about 10 mm. long, minutely pilose in the upper half, branches filiform, blunt, about 1 mm. long. Achenes oblong or narrowly obovoid, compressed, 3–3·5 mm. long, 1–1·5 mm. wide, light brown, glabrous, transversely rugose, apex narrowed abruptly to a slender stipe 5–6 mm. long, bearing a pappus consisting of many, white, plumose hairs 6–7 mm. long; marginal achenes a little longer than the inner ones, with a shorter pappus.

HAB.: Cultivated fields; brackish marshes near the sea; sea-level to 2,000 ft. alt.; fl. April–July.

DISTR.: Divisions 1, 2, 4, 7, uncommon. Widespread in Europe, the Mediterranean region and eastwards to Iran. An introduced weed elsewhere.

1. Stroumbi, 1913, *Haradjian* 722.
2. Kannaviou, 1939, *Lindberg f.* s.n.!
4. Mouth of Pedieos R. near Famagusta, 1955, *Merton* 2265!
7. Near Dhavlos, 1880, *Sintenis & Rigo* 501!

63. CREPIS *L.*

Sp. Plant., ed. 1, 805 (1753).
Gen. Plant. ed. 5, 350 (1754).
E. B. Babcock in Univ. Calif. Publ. 21: pp. XII + 1–198 (1947); 22: X + 199–1030 (1947).

Annual, biennial or perennial herbs; stems leafy or scapiform, branched or unbranched, glabrous, pubescent, glandular or eglandular, sometimes setose; basal leaves forming a loose rosette, variously lobed or pinnatisect, sometimes almost entire; cauline leaves usually much reduced; inflorescence generally branched and corymbose, or capitula sometimes solitary; involucre cylindrical or campanulate, sometimes becoming urceolate after anthesis; phyllaries in 2 or several series, glabrous, pilose, glandular or setose, the outer generally shorter than the inner and often forming a calyculus; florets generally yellow, rarely pink or white, ligule with 5 (rarely 4) apical teeth; anthers yellow or green (rarely pinkish or whitish), sagittate at base, usually with a blunt apical appendage; style-branches filiform, divergent, usually blunt, yellow or green. Achenes uniform or heteromorphic, usually cylindrical or fusiform, truncate or attenuate into a slender beak, longitudinally ribbed; pappus white, yellowish or brownish, caducous or persistent, hairs free or united at base, fine or coarse, scabridulous or barbellate.

About 200 species widely distributed in the northern hemisphere, and in Africa.

Plants perennial with a thickened, premorse rootstock and tough, fibrous roots:
  Pappus-hairs not tousled, rather rigid; apex of peduncle and base of involucre thinly tomentose, often glandular; basal leaves lyrate-pinnatisect with a broad, blunt or acute terminal lobe - - - - - - - - - - - **1. C. fraasii**
  Pappus-hairs tousled, softly lanuginose; apex of peduncle and base of involucre glabrous or very sparsely puberulous; basal leaves runcinate-pinnatisect with a small, deltoid-acute, terminal lobe - - - - - - - - - - - **2. C. reuteriana**
Plants annual with a slender taproot:
  Capitula sessile along the inflorescence-branches; inner phyllaries persistent, becoming swollen, indurated and incurved after anthesis, and covering the achenes
7. C. zacintha
  Capitula pedunculate:
    Stems scapose; leaves (or apparent leaves) all basal, forming a rosette; involucre usually clothed with blackish hairs; capitula small - - - - - **6. C. sancta**

Stems not scapose; leaves not all basal:
  Involucre glabrous or subglabrous:
    Leaves obovate or spathulate, shortly toothed or lobed, or subentire, shortly and
      densely glandular-pilose; achenes tapering to base -   -   -   **3. C. pulchra**
    Leaves lyrate-pinnatisect, subglabrous or thinly hispidulous-pilose; achenes
      conspicuously dilated and excavate at base -   -   -   - **4. C. palaestina**
  Involucre pilose, hispidulous or setose:
    Peduncles very short, 5–10 (–20) mm. long, becoming thick and rigid in fruit; stems
      usually setose with scattered, rigid, spreading bristles -   -   **9. C. aspera**
    Peduncles slender, or if not slender, up to 20 cm. long; stems not setose with spreading,
      rigid bristles:
      Capitula rather large and conspicuous; involucre 7–12 mm. long:
        Receptacle ciliate, but without scales; peduncles and upper parts of stems usually
          hispidulous -   -   -   -   -   -   -   **5. C. foetida** ssp. **foetida**
        Receptacle scaly, with long, attenuate-caudate membranous scales; peduncles
          and upper parts of stems usually glabrous or subglabrous
                                                        5. C. foetida ssp. **commutata**
      Capitula small, numerous; involucre 4–5 mm. long -   -   - **8. C. micrantha**

1. **C. fraasii** *Sch. Bip.* in Flora, 25: 173 (1842); Unger et Kotschy, Die Insel Cypern, 257 (1865);
   Holmboe, Veg. Cypr., 195 (1914); Lindberg f., Iter Cypr., 35 (1946); Davis, Fl. Turkey, 5:
   820 (1975).
      C. montana Urv., Enum. Plant. Ins. Pont. Eux., 101 (1822); Post, Fl. Pal., ed. 2, 2: 153
   (1933) quoad nomen; Babcock in Univ. Calif. Publ., 22: 288 (1947); Osorio-Tafall et
   Seraphim, List Vasc. Plants Cyprus, 109 (1973) non Bernh. (1800) nom. illeg.
      C. sieberi Boiss., Diagn. 1, 11: 53 (1849), Fl. Orient., 3: 845 (1875); Sintenis in Oesterr.
   Bot. Zeitschr., 32: 191, 193, 291, 398 (1882).
      [*C. raulinii* (non Boiss.) Kotschy in Unger et Kotschy, Die Insel Cypern, 257 (1865).]
      [*C. dioscoridis* (non L.) H. Stuart Thompson in Journ. Bot., 44: 334 (1906); Holmboe,
   Veg. Cypr., 195 (1914).]
      [*C. hierosolymitana* (non Boiss.) Babcock in Univ. Calif. Publ., 22: 286 (1947); Rechinger
   f. in Arkiv för Bot., ser. 2, 1: 436 (1950).]
   TYPE: "Hr. Fraas fand diese Pflanze in Griechenland bei Patadjik (nun Hypati)".

Erect perennial (8–) 20–50 (–70) cm. high, with a brown-lanuginose
(sometimes glabrescent), premorse rootstock and tough, fibrous roots; stems
slender, angled, often reddish-purple, usually thinly pilose towards base,
subglabrous above, with a few, erecto-patent branches in the upper part;
basal leaves lyrate-pinnatisect, oblong-oblanceolate in outline, 8–22 cm.
long, 1·5–5 cm. wide, subglabrous or hispidulous-pilose, especially along
midrib, terminal lobe usually large, broadly ovate, blunt or acute, with a few
shallow, broad-based teeth, lateral lobes numerous, close, oblong, acute or
obtuse, broadly dentate; petiole usually short, pilose, up to 5 cm. long;
cauline leaves few or wanting, sessile, narrowly oblong or lanceolate,
coarsely laciniate-dentate or pinnatilobed, rarely subentire, with an
acuminate apex and bluntly auriculate base; inflorescence lax, much-
branched, subcorymbose; peduncles slender, up to 5 cm. long; involucre
campanulate, 7–10 mm. long, 6–8 mm. wide, thinly tomentose towards base
(and apex of peduncle) frequently with a few, spreading, glandular hairs;
outer phyllaries ovate-acute, about 3 mm. long, 1·5 mm. wide towards base,
often purplish, with a narrow membranous margin; inner phyllaries erect,
linear-acuminate, often purplish, about 7–10 mm. long, 1·5 mm. wide,
spreading or reflexed in fruit; florets golden-yellow; corolla-tube about 4
mm. long, glabrous below, thinly pilose at apex; ligule about 9–10 mm. long,
2–3 mm. wide, apex with 5 small, papillose-incrassate teeth; filaments
glabrous, 0·8–1 mm. long; anthers yellow, about 4 mm. long, 0·3 mm. wide;
style about 8 mm. long, thinly setulose in the upper part, branches linear-
filiform, blunt, about 2 mm. long, setulose dorsally; receptacle glabrous.
Achenes fusiform, about 3·5–4 mm. long, 0·6 mm. wide, brown, slightly
curved, obscurely ribbed, attenuate above to a truncate apex, but not at all
beaked; pappus white, hairs slender (but not tousled) 3·5–4·5 mm. long,
scabridulous.

HAB.: In garigue on calcareous and igneous hillsides; 150–1,500 ft. alt.; fl. April–June.

DISTR.: Divisions 1–4, 7, 8, locally very common. Greece, Crete, Aegean Islands, Turkey.

1. Ridge between Inia and Smyies N.W. of Vlambouros (Akamas), 1979, *Edmondson & McClintock* E 2762!
2. Common; Khionistra, Galata, Kryos Potamos, Platres, Prodhromos, Tripylos, Troödos, Stavros tis Psokas, Klirou near Palekhori, Vavatsinia, etc. *Sintenis & Rigo* 909! 931! *Topali* s.n.; *Holmboe* 959; *Haradjian* 439; 464; *Kennedy* 985–987! 1472! 1474! *Lindberg f.* s.n.! *Davis* 3479! *Pavlides* s.n.! *Merton* 2370! 2371! *Meikle* 2760! *D. P. Young* 7794! *Merton* in ARI 293! *Edmondson & McClintock* E 2911! *Hewer* 4742!
3. Near Stavrovouni, 1937, *Syngrassides* 1476! Lefkara, 1941, *Davis* 2762! Near Yerasa, 1981, *Hewer* 4764! 4765!
4. On conglomerate near Larnaca, 1862, *Kotschy* 245. See notes.
7. Pentadaktylos, 1862, *Kotschy* 384! also, 1880, *Sintenis & Rigo* 282! St. Hilarion, 1901, *A. G. & M. E. Lascelles* s.n.! Panagra, 1936, *Syngrassides* 1028! Yaïla, 1941, *Davis* 2905! ¾ mile E. of St. Hilarion, 1949, *Casey* 457!
8. Rizokarpaso, 1912, *Haradjian* 146! also, 1938, *Druce* s.n.! Koma tou Yialou, 1950, *Chapman* 253!

NOTES: The Kew specimen of *Kotschy* 245, from near Larnaca, is evidently mislabelled; it comprises two plants of *Crepis kotschyana* Boiss. (*C. bureniana* Boiss.) a species distributed from Syria to Central Asia and N.W. India (and collected by Kotschy in Iran), but otherwise unknown from Cyprus.

Records from Cyprus for the closely related *C. hierosolymitana* Boiss. are, I am sure, referable here. *C. fraasii* has also been confused with the very similar *C. reuteriana* Boiss. (infra), which is, evidently, much rarer on the island.

2. C. reuteriana *Boiss.*, Diagn. 1, 11: 55 (1849), Fl. Orient., 3: 846 (1875); Post, Fl. Pal., ed. 2, 2: 154 (1933); Lindberg f., Iter Cypr., 35 (1946); Babcock in Univ. Calif. Publ., 22: 651 (1947); Rechinger f. in Arkiv för Bot., ser. 2, 1: 436 (1950); Osorio-Tafall et Seraphim, List Vasc. Plants Cyprus, 110 (1973); Davis, Fl. Turkey, 5: 826 (1975); Feinbrun, Fl. Palaest., 3: 442, t. 748 (1978).

[? *Centaurea reuteriana* Druce in Rep. B.E.C., 9: 470 (1931); B. L. Burtt in Kew Bull., 1954: 70 (1954).]

TYPE: "in montanis dumosis Anatoliae, *Smyrnae* in montibus suprà *Siclar* (Boiss.) in fauce *Tsimboukkan* Pamphyliae (Heldr.), in Antilibano *Rascheya* (Boiss.)". (G).

Erect perennial, superficially very similar to *C. fraasii* Sch. Bip. (supra), with a premorse rootstock and tough fibrous roots, but without the brown lanuginose indumentum characteristic of *C. fraasii*; stems slender, up to 70 cm. high, angular, often purplish, generally glabrous above, sometimes hispidulous-pilose towards base, branches few, erecto-patent, confined to the upper part of the stem; basal leaves rosulate, narrowly oblong or oblanceolate in outline, 5–20 cm. long, 1–4 cm. wide, runcinate-pinnatisect (very like those of a *Taraxacum*) with a deltoid-acute, sparingly dentate or entire terminal lobe, and numerous, often retrorse, acute, entire, dentate, lateral lobes decreasing in size downwards, lamina subglabrous or hispidulous, especially along midrib; petiole indistinct or up to 5 cm. long, glabrous or hispidulous, often purple; cauline leaves wanting or much reduced, linear or subulate, usually entire or with a few scattered teeth; inflorescence lax, subcorymbose, with numerous branches; peduncles slender, up to 10 cm. long, glabrous or sometimes very sparingly pubescent at the very apex; involucre cylindrical or narrowly campanulate, 10–13 mm. long, 6–8 mm. wide, spreading or reflexed in fruit; outer phyllaries narrowly ovate-acuminate, 3–4 mm. long, 1·5 mm. wide, subglabrous or thinly puberulent; inner phyllaries linear, subacute, often purplish, about 10–13 mm. long, 1–1·5 mm. wide, erect at anthesis; florets golden-yellow; corolla-tube about 4 mm. long, glabrous below, pilose towards apex, ligule about 8 mm. long, 1·5 mm. wide, apex with 5 papillose, incrassate teeth; filaments glabrous, about 0·5 mm. long; anthers linear, about 4 mm. long, 0·3 mm. wide, yellowish; style 10–11 mm. long, thinly setulose in the upper half, branches linear-filiform, greenish in dried specimens, about 2 mm. long. Achenes fusiform, almost straight, 4–5 mm. long, 0·8 mm. wide, light brown, indistinctly ribbed, somewhat attenuate above to a truncate apex; pappus white, tousled-lanuginose, hairs very slender, 5–6 mm. long.

HAB.: In garigue on igneous or calcareous hillsides; 500–5,000 ft. alt.; fl. April–June.

DISTR.: Divisions 1–3, 7, 8. Also Turkey, Syria and Palestine.

1. Stroumbi, 1913, *Haradjian* 758!
2. Troödos, 1912, *Haradjian* 437! 457; Xerokolymbos, 4,400 ft. alt., 1937, *Kennedy* 984! Platres, 1938, *Kennedy* 1476! Prodhromos, 1939, *Lindberg f.* s.n.! also, 1951, *Casey* 1205! Platania, 1939, *Lindberg f.* s.n.; S. flank Mt. Troödos, 3,700 ft. alt., 1955, *P. Pavlides* s.n.!
3. Stavrovouni, 1862, *Kotschy* 210! Lefkara, 1967, *Merton* in ARI 652!
7. Near Kantara, 1880, *Sintenis & Rigo* 282!
8. Komi Kebir, 1912, *Haradjian* 302!

NOTES: The soft woolly, tousled appearance (like cotton-wool) of the pappus hairs is unmistakable, and will alone distinguish *C. reuteriana* from its near ally *C. fraasii* Sch. Bip., though leaves and glabrescence of peduncles and involucres provide additional differences. The two species seem to have similar distributions and ecological requirements.

3. C. pulchra *L.*, Sp. Plant., ed. 1, 806 (1753); Sibth. et Sm., Fl. Graec. Prodr., 2: 139 (1813); Poech, Enum. Plant. Ins. Cypr., 22 (1842); Unger et Kotschy, Die Insel Cypern, 257 (1865); Boiss., Fl. Orient., 3: 846 (1875); Holmboe, Veg. Cypr., 195 (1914); Osorio-Tafall et Seraphim, List Vasc. Plants Cyprus, 109 (1973); Davis, Fl. Turkey, 5: 827 (1975).
TYPE: "*in* Gallia".

Erect annual (20–) 30–70 cm. high; stems and leaves glandular-pilose, viscid; stems angular, sulcate, generally branched in well-grown specimens; basal leaves obovate-spathulate, (2–) 7–16 cm. long, 1–5 cm. wide, shallowly and remotely dentate or lobulate, or subentire, apex rounded or shortly acute, base tapering to an indistinct, flattened petiole; cauline leaves usually well developed, sessile, amplexicaul-auriculate, oblong or obovate, 2–14 cm. long, 0·3–5 cm. wide, entire or denticulate; inflorescence lax, much-branched, paniculate, branches glabrous or glandular; peduncles 2–8 cm. long, erecto-patent; involucre cylindric-campanulate, 10–12 mm. long, 4–6 mm. wide, ultimately spreading or reflexed; outer phyllaries very small, ovate-deltoid, about 1·5–2 mm. long, 0·8 mm. wide, glabrous or sub-glabrous, inner phyllaries rigidly erect at anthesis, linear, subacute, about 10–12 mm. long, 0·8 mm. wide, glabrous, with a very strong, prominent midrib; florets yellow; corolla-tube slender, about 4 mm. long, somewhat lanuginose at apex, ligule about 7–8 mm. long, 0·8–1 mm. wide, apex with 5, rather irregular, papillose, incrassate teeth; filaments glabrous, about 0·5 mm. long; anthers yellowish, linear, about 3 mm. long, 0·2 mm. wide; style about 8 mm. long, thinly setulose in the upper part, branches linear-filiform, greenish-livid in dried specimens, about 1 mm. long. Achenes dimorphic, the marginal cylindrical-fusiform, curved or almost straight, sometimes epappose, about 7 mm. long, 0·8 mm. wide, truncate, scarcely attenuate, glabrous, obscurely ribbed; the inner a little longer, almost straight, scabridulous; pappus white, tousled-lanuginose, hairs 6–7 mm. long, very slender, minutely scabridulous.

HAB.: Screes and stony hillsides, usually on igneous rocks; 1,000–6,400 ft. alt.; fl. May–June.

DISTR.: Divisions 2, [4], 7, not common. Widespread in southern Europe, N. Africa and eastwards to Iraq and Iran.

2. Phini, 1862, *Kotschy* s.n.; Kato Platres, 1938, *Kennedy* 1475! Kryos Potamos, 4,200 ft. alt., 1939, *Kennedy* 1473! Summit of Khionistra, 1939, *Lindberg f.* s.n.! Kykko, 1941, *Davis* 3475! About 2 miles from Spilia on the Lagoudhera road, 1955, *Merton* 2416! Psilondendron, 1962, *Meikle* 2786!
[4. Near Larnaca, 1862, *Kotschy* 294. See Notes.]
7. Kalogrea, 1880, *Sintenis & Rigo* 555!

NOTES: The Kotschy record from Larnaca must be questioned, since no other collector has found *Crepis pulchra* in this area; the plant collected by Kotschy may have been the allied *C. palaestina* (Boiss.) Bornm., cited by Boissier as coming from Larnaca, but not listed in *Die Insel*

Cypern. Holmboe, however, states that this latter specimen is "Kotschy Suppl. 458! in Herb. Boiss." and not *Kotschy* 294.

All the Cyprus material examined has smooth and scabridulous achenes in the same capitulum, but both are uniformly pappose.

4. C. palaestina (*Boiss.*) *Bornm.* in Beih. Bot. Centralbl., 31 (2): 236 (1914); Holmboe, Veg. Cypr., 195 (post–Oct. 1914); Babcock in Univ. Calif. Publ., 22: 656 (1947); Osorio-Tafall et Seraphim, List Vasc. Plants Cyprus, 110 (1973); Davis, Fl. Turkey, 5: 829 (1975); Feinbrun, Fl. Palaest., 3: 442, t. 749 (1978).

Cymboseris palaestina Boiss., Diagn., 1, 11: 51 (1849), Fl. Orient., 3: 830 (1875); Post, Fl. Pal., ed. 2, 2: 152 (1933).

TYPE: Palestine; "in umbrosis Palaestinae, in *Carmelo*, monte *Ithabure*, prov. *Samariâ*. Legi Apr. et Maio" (G).

Erect annual 30–80 cm. high; stems usually rather slender, often purplish, sulcate, glabrous or thinly hispidulous-pilose especially towards base, branched or unbranched, generally rather leafy; basal leaves forming a loose rosette, oblong-obovate in outline, 6–15 cm. long, 2–6 cm. wide, subglabrous or thinly hispidulous-pilose, lyrate-pinnatisect, with a large, blunt or rounded terminal lobe, and short, oblong, blunt or subacute, entire or angled-denticulate lateral lobes, base tapering to a short, flattened, glabrous or pilose, often purplish, petiole; cauline leaves sessile, thinly papery, the lower similar to the basal leaves, the upper often oblong, amplexicaul-auriculate, up to 10 cm. long, 6 cm. wide, with a denticulate margin and broad, blunt auricles; inflorescence lax, branched, paniculate or corymbose; peduncles slender, up to 12 cm. long, subglabrous or thinly pilose, distinctly swollen at apex just below capitulum; involucre narrowly campanulate or subcylindric, 10–16 mm. long, 6–9 mm. wide, erect, spreading or reflexed in fruit; outer phyllaries ovate, acute, 2–4 mm. long, 1·5–3 mm. wide, glabrous or subglabrous with a pallid, membranous margin; inner phyllaries linear, subacute, to 16 mm. long, 1·5–2 mm. wide, subglabrous, the midrib becoming thick, prominent and pallid in fruit, the phyllaries enveloping the marginal achenes; florets yellow, conspicuous; corolla-tube slender, about 5 mm. long, thinly lanuginose towards apex, ligule 10–12 mm. long, about 1 mm. wide, apex with 5 short, papillose-incrassate teeth; filaments glabrous, about 1 mm. long; anthers yellowish, linear, about 4·5 mm. long, 0·3 mm. wide; style about 12 mm. long, thinly setulose towards apex, branches dark livid-green in dried specimens, linear-filiform, about 1·5 mm. long. Achenes dimorphic, the marginal (enveloped by innermost phyllaries) epappose, pale brown, fusiform, strongly dorsiventrally compressed, about 10 mm. long, 1·5 mm. wide at base, minutely scabridulous with narrow marginal wings, ventral surface finely ribbed, dorsal surface almost smooth; inner achenes pale brown, fusiform, almost straight, about 9–10 mm. long, 1·5 mm. wide at the conspicuously dilated, excavate base, terete or subterete, longitudinally ribbed, smooth or distinctly scabridulous, gradually tapering above to a slender, truncate apex, but not beaked; pappus white, tousled, lanuginose, hairs very slender, scabridulous, to about 6 mm. long.

HAB.: Amongst *Juncus* sp. (probably *J. rigidus* Desf.) by margin of salt-marsh; near sea-level; fl. April–May.

DISTR.: Division 4, very rare. Also Turkey, Syria and Palestine.

4. Larnaca (probably by Salt Lake), 1862, *Kotschy* Suppl. 458 teste Boissier and Holmboe.

NOTES: The only record, not cited in *Die Insel Cypern*, and included here solely on the authority of Boissier and Holmboe.

5. C. foetida *L.*, Sp. Plant., ed. 1, 807 (1753); Boiss., Fl. Orient., 3: 851 (1875); Holmboe, Veg. Cypr., 195 (1914); Post, Fl. Pal., ed. 2, 2: 155 (1933); Babcock in Univ. Calif. Publ., 22: 687 (1947); Rechinger f. in Arkiv för Bot., ser. 2, 1: 436 (1950); Osorio-Tafall et Seraphim, List

Vasc. Plants Cyprus, 110 (1973); Davis, Fl. Turkey, 5: 831 (1975); Feinbrun, Fl. Palaest., 3: 443, t. 751 (1978).

Barkhausia foetida (L.) F. W. Schmidt, Samml. Phys. Aufs., 1: 283 (1795); Sintenis in Oesterr. Bot. Zeitschr., 31: 392 (1881).

Crepis fallax Boiss., Fl. Orient., 3: 850 (1875); Druce in Rep. B.E.C., 9: 470 (1931).

Erect or spreading annual (10–) 15–70 cm. high; stems generally branched, obscurely or distinctly sulcate, hispidulous or subglabrous, especially in the upper part; basal leaves forming a loose rosette; oblanceolate or obovate in outline, 5–20 cm. long, 1–5 cm. wide, irregularly lyrate- or runcinate-pinnatisect, lacerate, pinnatifid or occasionally subentire with a few remote teeth, densely or sparsely hispidulous or subglabrous, apex acute or obtuse, base tapering to a distinct or ill-defined petiole; cauline leaves sessile, often shortly auriculate, irregularly lacerate or pinnatisect, or occasionally subentire, the lower oblong, the upper almost linear; inflorescence very lax; peduncles erect or erecto-patent, up to 20 cm. long, glabrous or hispidulous, not much swollen at apex; involucre at first broadly campanulate, 7–12 mm. long and almost as wide, becoming cylindrical or subglobose-urceolate in fruit; phyllaries pale green or purplish, shortly tomentellous with few or numerous, pale or dark, glandular or eglandular hairs and bristles, the outer ovate or linear-lanceolate, usually less than half as long as inner, but sometimes two-thirds as long, spreading or loosely erect and subadpressed in fruit, inner phyllaries linear, subacute, 1–2 mm. wide, remaining erect, and enveloping the marginal achenes in fruit, usually with a conspicuously thickened midrib; florets golden-yellow or reddish externally, sometimes large and conspicuous; corolla-tube slender, 3–4 mm. long, glabrous or puberulous; ligule 6–8 mm. long, 1·5–2 mm. wide, with 5 short, papillose-incrassate apical teeth; filaments glabrous, less than 0·5 mm. long; anthers yellowish, linear, 2·5–3 mm. long, 0·2 mm. wide; style 7–9 mm. long, thinly setulose towards apex, branches linear-filiform, about 1·5 mm. long, yellowish; receptacle pitted, the pits with ciliate walls, or sometimes furnished with linear-subulate, caudate-attenuate, simple or bifid, membranous scales, 5–7 mm. long, 0·3–0·5 mm. wide at base. Marginal achenes about 6 mm. long, 0·6 mm. wide, fusiform, somewhat curved, brown, longitudinally ribbed, distinctly scabridulous, attenuate to a shortly beaked apex, pappose or epappose; inner achenes 12–13 mm. long, 0·6 mm. wide, usually almost straight, brown, longitudinally ribbed, minutely scabridulous, attenuate to a long (6–8 mm.) filiform beak, always pappose; pappus white, silky, almost wholly exserted from fruiting involucre, hairs slender, scabridulous, about 5 mm. long.

ssp. **foetida**

C. rhoeadifolia M. Bieb., Fl. Taur.-Cauc., 3: 538 (1819).

C. glandulosa Guss., Plant. Rar., 329, t. 56 (1826) non Bast. (1814) nom. illeg.

C. zacynthia Marg. et Reut. ex DC., Prodr., 7: 158 (1838).

C. foetida L. ssp. *rhoeadifolia* (M. Bieb.) Čelak., Prodr. Fl. Böhm., 190, 785 (1871); Babcock in Univ. Calif. Publ., 22: 695 (1947).

C. foetida L. ssp. *vulgaris* Babcock in Journ. Bot., 76: 205 (1938); Univ. Calif. Publ., 22: 688 (1947).

C. foetida L. ssp. *vulgaris* Babcock f. *fallax* (Boiss.) Babcock in Journ. Bot., 76: 206 (1938); Univ. Calif. Publ., 22: 693 (1947).

C. foetida L. var. *glandulosa* Lindberg f., Iter Cypr., 34 (1946).

C. foetida L. var. *zacynthia* (Marg. et Reut.) Lindberg f., Iter Cypr., 35 (1946).

TYPE: "*in* Gallia".

Receptacle with ciliate pits, but not scaly; stems, leaves and peduncles generally hispidulous; involucre usually green, shortly glandular-pilose, or hispidulous with long, usually pale, eglandular bristles; capitula often relatively large and conspicuous.

HAB.: Roadsides, field-margins; sand-dunes; grass-steppe; in garigue on rocky hillsides, or in open Pine forest; sea-level to 5,000 ft. alt.; fl. March–June.

DISTR.: Divisions 1, 2, 4–6. West, central and southern Europe east to Iran. Introduced elsewhere.

1. Stroumbi, 1913, *Haradjian* 764; Lyso, 1913, *Haradjian* 882! 889.
2. Galata, 1880, *Sintenis & Rigo* 293b! Troödos, 1912, *Haradjian* 419; 502; 506; Platania, 1936, *Syngrassides* 1273! Makheras Monastery, 1937, *Syngrassides* 1631! Platres, 1937–1955, *Kennedy* 988! 1464 partly! 1468 partly! 1469! 1470! 1855! 1856! also *Pavlides* s.n.! Between Ambelikou and Kambos, 1939, *Lindberg f.* s.n.; Troödos, 1939, *Lindberg f.* s.n.; Kato Platres, 1939, *Lindberg f.* s.n.; Prodhromos, 1939, *Lindberg f.* s.n.; and, 1955, *G. E. Atherton* 566! Kakopetria, 1939, *Lindberg f.* s.n.; and 1962, *Meikle* 2850!
4. Larnaca, 1939, *Lindberg f.* s.n.! Cape Greco, 1958, *N. Macdonald* 81!
5. Athalassa, 1950, *Chapman* 641!
6. Ayia Irini, 1936, *Syngrassides* 1207! also, 1941, *Davis* 2550! 2582! and, 1954, *Casey* 1332! Syrianokhori, 1952, *Merton* 1758!

NOTES: Babcock records both *C. foetida* ssp. *foetida* and *C. foetida* ssp. *rhoeadifolia* from Cyprus, but once the distinct *C. foetida* ssp. *commutata* is removed from the aggregate Cyprus material, I do not think the residue can be further subdivided. Most of the specimens examined have involucres rather densely clothed with long, pale, eglandular bristles, and do not agree very exactly with either ssp. *foetida* or ssp. *rhoeadifolia*, though with some leaning towards the latter (see Babcock loc. cit., 692, notes on variant 18 — from Galata *not* Galatia).

Sand-dune plants from Ayia Irini and Syrianokhori are very dwarf with shortly pinnatifid leaves, and call for further study.

ssp. **commutata** (*Spreng.*) *Babcock* in Journ. Bot., 76: 207 (1938); Univ. Calif. Publ., 22: 697 (1947); Davis, Fl. Turkey, 5: 833 (1975).
  Rodigia commutata Spreng., Neue Entd., 1: 275 (1820); Unger et Kotschy, Die Insel Cypern, 256 (1865); Boiss., Fl. Orient., 3: 880 (1875); Sintenis in Oesterr. Bot. Zeitschr., 31: 392 (1881); 32: 124, 191, 193, 398 (1882); Holmboe, Veg. Cypr., 193 (1914); Post, Fl. Pal., ed. 2, 2: 160 (1933); Osorio-Tafall et Seraphim, List Vasc. Plants Cyprus, 109 (1973).
  Crepis commutata (Spreng.) Greuter in Candollea, 31: 232 (1976) nom. invalid. sine relat. loc.
TYPE: Greece, "in insulis maris Ionii unde uvae passae veniunt e quibus Rodigius Stolpensis largitus est".

Receptacle paleaceous, with linear-subulate, caudate-attenuate, simple or bifid, pale, membranous scales longer than the fertile part of the achene, and almost as long as the involucre; upper parts of stems and peduncles often glabrous or subglabrous; involucre commonly stained purplish-livid, tomentellous, with a sparse indumentum of long glandular or eglandular hairs; capitula often smaller than in ssp. *foetida*, the outer phyllaries forming a patent or slightly reflexed calyculus in fruit. Plant often very small and slender, never as robust as well-grown specimens of *C. foetida* ssp. *foetida*; peduncles often proportionately very long.

HAB.: Old pastures and grass-steppe; roadsides and grassy banks; sea-level to 1,500 ft. alt.; fl. March–May.

DISTR.: Divisions 5–8. Also Bulgaria, Greece, Crete, Turkey, Syria and east to Iran.

5. Between Mia Milea and Mandres, 1933, *Syngrassides* 324! Mile 7, Nicosia–Famagusta road, 1950, *Chapman* 376! 391! Mia Milea, 1950, *Chapman* 482! also 1951 and 1955, *Merton* 287! 2033! Athalassa, 1967, *Merton* in ARI 161! Near Klirou, 1967, *Merton* in ARI 274!
6. Dhiorios, 1962, *Meikle* 2348!
7. Melandrina, 1862, *Kotschy* 526; Pentadaktylos, 1880, *Sintenis & Rigo* 291! Kantara, 1880, *Sintenis*; Kyrenia, 1948–1952, *Casey* 496! 527! 1164! 1224! Kharcha, 1950, *Chapman* 279! Lakovounara, 1950, *Chapman* 357! Above Antiphonitis Monastery, 1955, *Merton* 2274 partly!
8. Cape Andreas, 1880, *Sintenis & Rigo* 291a!

NOTES: Very distinct and uniform in Cyprus, less robust than *C. foetida* ssp. *foetida*, with long glabrous or subglabrous peduncles, and rather small, thinly pilose involucres.

6. C. sancta (*L.*) *Bornm.* in Mitt. Thür. Bot. Ver., n.s., 30: 79 (1913); Babcock in Univ. Calif. Publ., 19: 403 (1941), ibid., 22: 730 (1947); Davis, Fl. Turkey, 5: 834 (1975); Feinbrun, Fl. Palaest., 3: 444, t. 753 (1978).

Hieracium sanctum L., Cent. Plant. 2: 30 (1756).

C. nemausensis Gouan, Illustr. Observ. Bot., 60 (1773).

Trichocrepis bifida Vis., Stirp. Dalmat., 19, t. 7 (1826).

Lagoseris bifida (Vis.) Koch, Syn. Fl. Germ., 435 (1837); Boiss., Fl. Orient., 3: 881 (1875), Suppl., 329 (1888); Sintenis in Oesterr. Bot. Zeitschr., 32: 191, 193 (1882); Holmboe, Veg. Cypr., 193 (1914).

Pterotheca bifida (Vis.) Fisch. et Meyer, Ind. Sem. Hort. Petrop., 1837: 43 (1837); Unger et Kotschy, Die Insel Cypern, 257 (1865); H. Stuart Thompson in Journ. Bot., 44: 334 (1906).

Lagoseris sancta (L.) K. Malý in Glasn. Muz. Bosni Herceg., 20: 556, 562 (1908); Post, Fl. Pal., ed. 2, 2: 160 (1933); Rechinger f. in Arkiv för Bot., ser. 2, 1: 434 (1950); Osorio-Tafall et Seraphim, List Vasc. Plants Cyprus, 110 (1973).

Crepis sancta (L.) Bornm. ssp. *nemausensis* (Gouan) Thell. in Mém. Soc. Nat. Sci. Nat. et Math. Cherbourg, ser. 4, 38: 577 (1912); Babcock in Univ. Calif. Publ., 22: 731 (1947).

C. sancta (L.) Bornm. ssp. *bifida* (Vis.) Thell. in Mém. Soc. Nat. Sci. Nat. et Math. Cherbourg, ser. 4, 38: 577 (1912); Babcock in Univ. Calif. Publ., 22: 736 (1947).

TYPE: "*in* Palaestina. *Hasselquist*".

Slender erect or spreading annual 4–25 cm. high; leaves all basal, forming a distinct rosette, polymorphic, usually obovate in outline, 3–18 cm. long, 0·5–4 cm. wide, glabrous or thinly hispidulous, especially along midrib and nerves, lyrate- or runcinate-pinnatisect, pinnatifid, dentate or subentire, apex rounded, blunt or shortly acute, base tapering to an indistinct petiole; scapes usually arcuate-ascending, sulcate, subglabrous, or distinctly (often blackish) glandular-pilose, unbranched or branched near apex; inflorescence lax, few (2–5)-flowered in our area, or capitula often solitary; peduncles slender, erect, 2–12 cm. long, glabrous or glandular-pilose; involucre campanulate or subcylindric, 4–10 mm. long, 2·5–7 mm. wide, usually subfuscous, strongly reflexed after the fruits have been shed; outer phyllaries ovate, 2–3 mm. long, 1·5 mm. wide at base, subglabrous or with a few blackish glandular hairs dorsally, margin membranous-hyaline; inner phyllaries linear, subacute, usually glandular-pilose dorsally, sometimes glabrous, concave in fruit and enfolding the outer achenes, but not in-crassate or with a conspicuously thickened midrib, margins membranous, hyaline; florets rather small, golden-yellow; corolla-tube very slender, about 2 mm. long, lanuginose at apex, ligule about 6 mm. long, 1–1·2 mm. wide, apex with 5 minute, papillose-incrassate teeth; filaments glabrous, about 1 mm. long, rather conspicuously exserted from corolla-tube; anthers linear, yellowish, about 3 mm. long, 0·2 mm. wide; style about 8 mm. long, thinly setulose near apex, branches linear-filiform, dark green when dried, about 1–1·5 mm. long. Achenes dimorphic, the marginal pallid, oblong, truncate, epappose or sparsely pappose, 3–5 mm. long, 0·6–1·5 mm. wide, with thick or thin, spongy lateral wings, sulcate or smooth dorsally; inner achenes narrowly fusiform, 3–4 mm. long, about 0·4 mm. wide, longitudinally ribbed, minutely asperulous or sometimes distinctly scabridulous, brown, tapering slightly above to a truncate apex, but not beaked; pappus white, silky, hairs scabridulous, to about 4 mm. long.

HAB.: Cultivated and fallow fields; coastal garigue; grassy banks and roadsides; sea-level to 4,500 ft. alt.; fl. Dec.–April.

DISTR.: Divisions 2–8. Widely distributed in S. Europe and the eastern Mediterranean region eastwards to N. India and Central Asia.

2. Between Prodhromos and Tris Elies, 1862, *Kotschy* 856.

3. Zakaki marshes, 1939, *Mavromoustakis* 22! Akrotiri, 1939, *Mavromoustakis* 10! 11!

4. Larnaca, 1862, *Kotschy* 85; Famagusta, 1912, *Haradjian* 80.

5. Vatili, 1905, *Holmboe* 331.

6. Court Garden (Nicosia), 1901, *A. G. & M. E. Lascelles* s.n.!

7. Kantara, 1880, *Sintenis & Rigo* 281! Kyrenia, 1949–1956, *Casey* 145! 239! 1062! *G. E. Atherton* 785! 938! 945! 1021! Kambyli, 1956, *Merton* 2519! Ayios Epiktitos, 1967, *Merton* in ARI 318!

8. Ayios Philon, 1941, *Davis* 2256! Cape Andreas, 1962, *Meikle* 2474!

NOTES: Variable in Cyprus, as elsewhere, but not readily divisible into recognizable infraspecific units. A few depauperate specimens from Athalassa (*Merton* in ARI 167) and Lakovounara (*Chapman* 197, 513) may belong here, but have cauline leaves, not normally to be found in this species; they are too abnormal and immature for precise identification, and may be forms of *Crepis foetida* L. ssp. *commutata* (Spreng.) Babcock.

7. C. zacintha (*L.*) *Babcock* in Univ. Calif. Publ., 19: 404 (1941), ibid., 22: 760 (1947); Davis, Fl. Turkey, 5: 836 (1975); Feinbrun, Fl. Palaest., 3: 445 (1978).

   Lapsana zacintha L., Sp. Plant., ed. 1, 811 (1753).

   Zacintha verrucosa Gaertn., De Fruct. et Sem., 2: 358, t. 157, fig. 7 (1791); Boiss., Fl. Orient., 3: 830 (1875); Sintenis in Oesterr. Bot. Zeitschr., 32: 399 (1882); Post, Fl. Pal., ed. 2, 2: 152 (1933); Rechinger f. in Arkiv för Bot., ser. 2, 1: 434 (1950); Osorio-Tafall et Seraphim, List Vasc. Plants Cyprus, 110 (1973).

   TYPE: "*in* Italia".

Erect or spreading, subglabrous or thinly hispidulous annual 6–30 cm. high; basal leaves forming an irregular rosette, oblong-oblanceolate in outline, 4–16 cm. long, 0·8–4 cm. wide, irregularly lacerate-pinnatisect, with an enlarged dentate terminal lobe and small, obtuse or acute lateral lobes, base tapering to an indistinct petiole; cauline leaves few, sessile or subsessile, often auriculate with acute auricles, the uppermost bract-like; stems obscurely sulcate, branched above to form a spreading, open, rigid, repeatedly bifurcate inflorescence; capitula small (less than 5 mm. diam. at anthesis), sessile in the bifurcations, and lateral or terminal on the ultimate branches, generally remote and becoming more so in fruit; involucre campanulate, becoming indurated and incurved in fruit, enveloping the marginal achenes; outer phyllaries ovate-oblong, subacute, about 3 mm. long, 1 mm. wide, glabrous or thinly hispidulous with a narrow scarious margin; inner phyllaries linear-oblong at anthesis, 5–6 mm. long, 1 mm. wide, blunt, cucullate and shortly papillose at apex; florets small, yellow; corolla-tube slender, 1–2 mm. long, thinly pilose, ligule 8–9 mm. long, 1–1·2 mm. wide, apex with 5, small, incrassate-papillose teeth; filaments glabrous, about 0·3 mm. long; anthers yellow, 2–2·5 mm. long, 0·2 mm. wide; style 4–5 mm. long, branches linear-filiform, green when dried, about 0·8 mm. long. Achenes enclosed in the persistent, indurated, 8-lobed involucre, dimorphic, the marginal pubescent, laterally compressed, rounded dorsally, with 2 narrow, longitudinal ventral wings, about 2–2·5 mm. long, 1–1·5 mm. wide, inner achenes glabrous, minutely papillose-scabridulous, longitudinally ribbed, brown, about 2–2·5 mm. long, 0·8–1 mm. wide, somewhat curved and laterally compressed, the apex often (but not always) conspicuously oblique; pappus very caducous, perhaps borne only by the central achenes, hairs 1–1·5 mm. long, white, coarsely scabridulous or barbellate.

HAB.: Cultivated and fallow fields, waste ground, roadsides; about 500 ft. alt.; fl. March–May.

DISTR.: Divisions 5, 8, very rare. Widely distributed in the Mediterranean region.

5. Kythrea, 1880, *Sintenis & Rigo* 492!
8. Near Rizokarpaso, 1912, *Haradjian* 256! Also noted by Sintenis from the fields in the neighbourhood of Galinoporni and Ayios Symeon.

8. C. micrantha *Czerep.* in Fl. U.R.S.S., 29: 684 (1964); Davis, Fl. Turkey, 5: 837 (1975); Feinbrun, Fl. Palaest., 3: 445, t. 754 (1978).

   C. parviflora Desf. ex Pers., Syn. Plant., 2: 376 (1807); Sintenis in Oesterr. Bot. Zeitschr., 32: 291 (1882); Boiss., Fl. Orient., Suppl., 325 (1888); Holmboe, Veg. Cypr., 195 (1914); Post, Fl. Pal., ed. 2, 2: 154 (1933); Lindberg f., Iter Cypr., 35 (1946); Rechinger f. in Arkiv för Bot., ser. 2, 1: 436 (1950); Osorio-Tafall et Seraphim, List Vasc. Plants Cyprus, 109 (1973) non Moench (1794) nom. illeg.

   C. muricata Sm. in Sibth. et Sm., Fl. Graec. Prodr., 2: 138 (1813), Fl. Graec., 9: 4, t. 807 (1837).

   TYPE: "in Oriente" (? LE).

Erect glabrous or hispidulous-pilose annual (14–) 30–80 (–100) cm. high; stems usually slender, often deeply sulcate, pinkish or purplish towards base, frequently with copious, spreading or erecto-patent branches; basal leaves forming a rosette, oblanceolate or obovate in outline, (3–) 5–20 cm. long, (0·5–) 1·5–8 cm. wide, papery, subglabrous or thinly hispidulous-pubescent, subentire, remotely dentate, irregularly lacerate or deeply pinnatisect, base tapering to an ill-defined petiole; cauline leaves sessile, conspicuously amplexicaul-auriculate, 1·5–14 cm. long, 0·3–4 cm. wide, tapering from a sagittate base to an acuminate apex, entire or remotely dentate; inflorescence lax, repeatedly branched, corymbose; peduncles slender, usually less than 3 cm. long; capitula numerous, small, 5–7 mm. diam.; involucre campanulate, about 4 mm. long and about as wide, erect or spreading in fruit; outer phyllaries erect, ovate, acute, about 2 mm. long, less than 1 mm. wide at base, thinly tomentellous with a dark acumen; inner phyllaries oblong, subacute, about 4–5 mm. long, 1·5 mm. wide at base, thinly tomentellous or sparsely hispidulous, with a narrow scarious margin and dark acumen; florets golden-yellow, the ligules often reddish externally; corolla-tube slender, about 2 mm. long, pilose at apex, ligule 4–4·5 mm. long, 1·5 mm. wide, shortly and bluntly 5-toothed at apex; filaments glabrous, about 0·2 mm. long; anthers yellowish, about 3 mm. long; style about 5 mm. long, thinly setulose in the upper part, branches linear-filiform, greenish, about 0·6 mm. long. Achenes straight or slightly curved, fusiform or subcylindric, about 1·8 mm. long, 0·4 mm. wide, tawny-brown, minutely scabridulous, with prominent longitudinal ribs and an abruptly constricted, truncate apex; pappus white, caducous, hairs silky, 3–3·5 mm. long, distinctly scabridulous or shortly barbellate.

HAB.: Cultivated and waste land; grassy banks, and grassy verges at edge of forest; sea-level to 5,000 ft. alt.; fl. May–Aug.

DISTR.: Divisions 1–3, 7, 8. Eastern Mediterranean region and eastwards to Iran.

1. Lyso, 1913, *Haradjian* 862!
2. Prodhromos, 1880, *Sintenis & Rigo* 808! also, 1952, *Casey* 1276! Pano Panayia, 1913, *Haradjian* 908! Near Galata, 1913, *Haradjian* 1010! Galata, 1936, *Syngrassides* 1271! Mesapotamos Monastery, 1939, *Lindberg f.* s.n.! Kakopetria, 1955, *Chiotellis* 433!
3. Between Yerasa and Kalokhorio, 1962, *Meikle* 2892!
7. St. Hilarion, 1880, *Sintenis & Rigo* 283! Above Antiphonitis Monastery, 1955, *Merton* 2273!
8. Cape Andreas, 1880, *Sintenis & Rigo* 283! Rizokarpaso, 1912, *Haradjian* 234! Komi Kebir, 1912, *Haradjian* 265!

9. C. aspera L., Sp. Plant., ed. 2, 1132 (1763); Sibth. et Sm., Fl. Graec. Prodr., 2: 137 (1813), Fl. Graec., 9: 2, t. 804 (1837); Boiss., Fl. Orient., 3: 857 (1875); Holmboe, Veg. Cypr., 195 (1914); Post, Fl. Pal., ed. 2, 2: 157 (1933); Babcock in Univ. Calif. Publ., 22: 878 (1947); Rechinger f. in Arkiv för Bot., ser. 2, 1: 436 (1950); Osorio-Tafall et Seraphim, List Vasc. Plants Cypr., 110 (1973); Davis, Fl. Turkey, 5: 840 (1975); Feinbrun, Fl. Palaest., 3: 446, t. 756 (1978).

Nemauchenes inermis Cass. in Dict. Sci. Nat., 34: 363 (1825).

N. aspera (L.) Steudel, Nomencl. Bot., ed. 2, 2: 189 (1841); Poech, Enum. Plant. Ins. Cypr., 22 (1842); Unger et Kotschy, Die Insel Cypern, 256 (1865).

Crepis aspera L. var. *inermis* (Cass.) Boiss., Fl. Orient., 3: 857 (1875); Rechinger f. in Arkiv för Bot., ser. 2, 1: 436 (1950).

[*C. aculeata* (non (DC.) Boiss.) Sintenis in Oesterr. Bot. Zeitschr., 32: 364 (1882); Boiss., Fl. Orient., Suppl., 326 (1888); Holmboe, Veg. Cypr., 195 (1914); Osorio-Tafall et Seraphim, List Vasc. Plants Cyprus, 110 (1973).]

TYPE: "*in* Oriente, Sicilia, Palaestina".

Erect, glaucous annual 15–50 (–70) cm. high; stems sulcate, often tinged purple towards base, sparsely or densely setose with yellowish or black-based bristles, rarely unarmed (var. *inermis* (Cass.) Boiss.), otherwise glabrous or clothed with a thin, sparse, whitish tomentum, much branched in luxuriant specimens, or simple in starved ones; basal leaves oblong-obovate, 4–14 cm. long, 1–4 cm. wide, glabrous with setose midrib and nerves, apex blunt or acute, margins coarsely dentate or lacerate, often setulose; cauline leaves sessile, amplexicaul-auriculate, with projecting, acute (rarely blunt) auricles, the upper very small and bract-like, margins often coarsely laciniate-pinnatisect or deeply and sharply dentate; inflorescence spreading, repeatedly dichotomous with rather rigid branches; capitula rather small, subsessile or very shortly pedunculate, the peduncles 5–10 (–20) mm. long, at first slender, but becoming thick and rigid in fruit; involucre campanulate at anthesis, 6–7 mm. long, and almost as wide or wider, becoming indurated and subglobose-urceolate in fruit; outer phyllaries loosely spreading, broadly ovate-acute, about 3 mm. long, 2 mm. wide, with a hispid midrib and wide membranous margins; inner phyllaries erect, becoming incrassate and indurated in fruit and enveloping the marginal achenes, carinate, about 6 mm. long, 2–2·5 mm. wide, with a blackish-setose keel and broad scarious margins; florets yellow; corolla-tube slender, about 3 mm. long, sparsely and shortly pilose, ligule about 7 mm. long, 2 mm. wide, apex rather deeply and irregularly 5-toothed; filaments glabrous, about 0·3 mm. long; anthers yellowish, about 3 mm. long, 0·2 mm. wide; style about 7–8 mm. long, sparsely setulose in the upper part, branches greenish, linear-filiform, about 1–1·5 mm. long. Achenes generally dimorphic, the marginal (enveloped by the inner phyllaries) pallid, strongly compressed laterally, about 3 mm. long, 1–1·2 mm. wide, with 2 narrow, longitudinal, dorsal wings, and a broader ventral wing, often notched or minutely toothed near its apex, surface distinctly, but minutely scabridulous, apex narrowed, but not at all beaked; inner achenes fusiform, straight, or slightly curved, brown, about 2–2·5 mm. long, 0·7 mm. wide, prominently ribbed, and distinctly scabridulous, apex attenuate to a slender beak about 2·5 mm. long; pappus white, silky, caducous, hairs slender, about 4–6 mm. long, minutely scabridulous.

HAB.: Cultivated and fallow fields, roadsides; on rocky ground by the coast or inland; sea-level to 3,800 ft. alt.; fl. March–June.

DISTR.: Divisions 1–4, 8. Also S. Turkey, Syria, Palestine, Egypt.

1. Paphos, 1913, *Haradjian* 641! Stroumbi, 1913, *Haradjian* 769! Cape Arnauti, 1913, *Haradjian* 804.
2. Pano Panayia, 1913, *Haradjian* 920; Platres, 1938, *Kennedy* 1471! Yialia, 1955, *Merton* 2107!
3. Limassol, 1913, *Haradjian* 609! also, 1930, *E. Wall* s.n.; and, 1957, *Mavromoustakis* 2/57! Cherkez, 1947, *Mavromoustakis* s.n.! Pissouri, 1967, *Merton* in ARI 677! Amathus, 1978, *Holub* s.n.!
4. Famagusta, 1912, *Haradjian* 115! Sotira, Ayios Antonios, 1950, *Chapman* 590! and, 1952, *Merton* 725! E. of Paralimni, 1958, *N. Macdonald* 48!
8. Yialousa, 1880, *Sintenis & Rigo* 284! Rizokarpaso, 1912, *Haradjian* 130; 163; 164; near Apostolos Andreas Monastery, 1962, *Meikle* 2469! also, 1973, *P. Laukkonen* 290!

NOTES: Babcock (Univ. Calif. Publ., 22: 881 (1947) accepts *Sintenis & Rigo* 284 as *C. aspera* L., and not *C. aculeata* (DC.) Boiss., but (p. 873) cites the Kew material of the same number under the latter name. This is an error; all the Kew material of *Sintenis & Rigo* 284 is *C. aspera*, and *C. aculeata* can be excluded from the flora.

64. AETHIORHIZA *Cass.*

Dict. Sci. Nat., 48: 425 (1827).
E. B. Babcock et G. L. Stebbins in Univ. Calif. Publ., 18: 235–240 (1943).
Rechinger f. in Phyton, 16: 211–220 (1974).

Stoloniferous perennial herbs with globose root-tubers; leaves entire, dentate or pinnatifid; inflorescence scapose, the scapes glabrous or glandular-pilose; capitula usually solitary, sometimes 2–4; involucre narrowly campanulate; phyllaries imbricate in several series, fuscescent, often glandular; florets yellow, or the ligules stained reddish externally;

anthers sagittate at base, with a blunt or subacute apical appendage; style-branches filiform, blunt, usually greenish in dried specimens; receptacle flat, naked. Achenes fusiform, 4-ribbed, base excavate, apex not beaked; pappus copious, white, silky, persistent, hairs almost smooth or scabridulous.

One species (comprising 3 subspecies) widely distributed in the Mediterranean region and in coastal areas of S.W. Europe.

1. A. bulbosa (*L.*) *Cass.* in Dict. Sci. Nat., 48: 425 (1827); Unger et Kotschy, Die Insel Cypern, 257 (1865); Sintenis in Oesterr. Bot. Zeitschr., 31: 194 (1881), 32: 291 (1882); Rechinger f. in Phyton, 16: 215, 216, 218, (1974); Davis, Fl. Turkey, 5: 696 (1975).

     Leontodon bulbosus L., Sp. Plant., ed. 1, 798 (1753).

     Crepis bulbosa (L.) Tausch in Flora, 11, Erg.-Bl., 1: 78 (1828); Boiss., Fl. Orient., 3: 832 (1875); Holmboe, Veg. Cypr., 195 (1914); Post, Fl. Pal., ed. 2, 2: 153 (1933); Rechinger f. in Arkiv för Bot., ser. 2, 1: 436 (1950); Osorio-Tafall et Seraphim, List Vasc. Plants Cyprus, 109 (1933).

     C. bulbosa (L.) Tausch var. *polycephala* Boiss., Fl. Orient., 3: 833 (1875).

     Aethiorhiza bulbosa (L.) Cass. ssp. *microcephala* Rechinger f. in Phyton, 16: 217 (1974) pro parte quoad plant. cypr.

     TYPE: "Monspelii, *inque* Italia".

Creeping perennial with long slender stolons rooting at the nodes and forming a low, leafy mat; root-tubers subglobose, 5–15 mm. diam.; leaves forming loose rosettes, oblanceolate in outline, 3–30 cm. long, 0·5–5 cm. wide, entire, toothed or runcinate-pinnatifid, thinly papery, glabrous, sometimes glaucous, apex blunt or acute, base tapering to a slender, ill-defined petiole; scape 13–50 (–60) cm. long, usually unbranched and terminating in a solitary capitulum, sometimes (*Crepis bulbosa* (L.) Tausch var. *polycephala* Boiss.) with 2–4, shortly pedunculate, terminal and lateral capitula; scapes often reddish, glabrous below, rather densely black-glandular in the upper half; involucre 9–15 mm. long, 6–12 mm. wide; phyllaries linear-oblong, glabrous or black-glandular, especially towards base, commonly fuscescent, linear-oblong, the outer about 5 mm. long, 1–1·5 mm. wide at base, the inner a little wider, up to 15 mm. long, all acute, blunt or rounded, and shortly ciliolate at apex; corolla-tube slender, about 6 mm. long, thinly lanuginose, ligule 10–12 mm. long, 1·5 mm. wide, deeply 5-lobulate at apex, with narrow, incrassate-tipped lobules; filaments about 0·8 mm. long, glabrous; anthers yellowish, about 3·5 mm. long, 0·3 mm. wide; style 9–10 mm. long, thinly setulose in the upper part, branches filiform, 2–2·5 mm. long, slender, greenish, ultimately reflexed. Achenes cylindric-fusiform, bluntly 4-ribbed, brown, minutely scabridulous, about 3–4 mm. long, 0·4 mm. wide, apex truncate, scarcely attenuate; pappus-hairs straight, shining white, to about 8 mm. long.

HAB.: Moist rocky ground by the sea; edges of salt-marshes and freshwater swamps; sometimes a weed in gardens and cultivated fields; sea-level to 3,000 ft. alt.; fl. March–May.

DISTR.: Divisions 1–4, 6–8. General distribution that of genus.

1. Ktima, 1913, *Haradjian* 657; Stroumbi, 1913, *Haradjian* 763.
2. Pano Panayia, 1913, *Haradjian* 903.
3. Akrotiri, 1905, *Holmboe* 687; also, 1967, *Merton* in ARI 703! Limassol, 1913, *Haradjian* 588!
4. Larnaca, 1859, *Kotschy* 425; also, 1862, *Kotschy* 253! and, 1905, *Holmboe* 96.
6. Makhedonitissa Monastery, 1936, *Syngrassides* 895!
7. Dhavlos, 1880, *Sintenis & Rigo* 280! Kyrenia, 1949, *Casey* 485! 490!
8. Near Yialousa, 1880, *Sintenis*; near Rizokarpaso, 1912, *Haradjian* 211, 241; near Leon-arisso, 1957, *Merton* 3071! Cape Andreas, 1962, *Meikle* 2473!

NOTES: Rechinger f. (in Phyton 16: 215–218; 1974) cites *Sintenis & Rigo* 280 and *Casey* 485 [488] as *Aethiorhiza bulbosa* ssp. *bulbosa*, referring *Kotschy* 253 pro parte and *Haradjian* 241, 588, 657, 903 to *Aethiorhiza bulbosa* ssp. *microcephala* Rechinger f. While the differences between the subspecies (based chiefly on size of involucre and achenes) are evident outside Cyprus, I do not find that any satisfactory distinctions can be drawn between the individuals which form the polymorphic population of the species on the island. In the circumstances, I have not attempted an infraspecific analysis. The Cyprus plants differ uniformly from material collected elsewhere in having very glandular scapes, the glands often extending far down the scape and not concentrated near the apex; they may represent yet another subspecies; polycephalous individuals appear to be fairly frequent.

65. HYPOCHAERIS L.

Sp. Plant., ed. 1, 810 (1753).
Gen. Plant., ed. 5, 352 (1754).

Annual or perennial herbs; leaves mostly basal, forming a rosette, entire or variously toothed or lobed, often hispidulous; inflorescence usually branched, sometimes simple, with all or some of the cauline leaves reduced to small bracts; capitula homogamous, ligulate, solitary at the ends of the branches; involucre narrowly campanulate or subcylindric; phyllaries usually imbricate in several series; florets yellow, 5-toothed at apex; anthers sagittate at base; style-branches slender, blunt; receptacle flat, paleaceous with narrow hyaline scales subtending the florets. Achenes uniform or dimorphic, the inner beaked, the marginal sometimes truncate, all narrowly oblong or subcylindric, angled or subterete, rugulose or muricate; pappus of plumose hairs or the outer row sometimes shorter and not plumose, marginal achenes sometimes epappose.

About 100 species, chiefly in South America; about 10 species in Europe and the Mediterranean region, but sparsely represented in our area.

Involucres and stems glabrous; phyllaries distinctly imbricate in several series   **1. H. glabra**
Involucres setose, stems hispid at least towards apex; phyllaries apparently in 1 series
  2. H. achyrophorus

1. **H. glabra** L., Sp. Plant., ed. 1, 811 (1753); Boiss., Fl. Orient., 3: 783 (1875); Post, Fl. Pal., ed. 2, 2: 140 (1933); Rechinger f. in Arkiv för Bot., ser. 2, 1: 434 (1950); Davis, Fl. Turkey, 5: 670 (1975); Feinbrun, Fl. Palaest., 3: 415, t. 700 (1978).

  Hypochaeris minima Cyrill., Plant. Rar. Neap., 1: XXIX, t. 10 (1788); Sibth. et Sm., Fl. Graec. Prodr., 2: 143 (1813), Fl. Graec., 9: 11, t. 810 (1837).

  H. glabra L. var. *erostris* Boiss., Fl. Orient., 3: 783 (1875).

  H. glabra L., ssp. *minima* (Cyrill.) Holmboe, Veg. Cypr., 192 (1914); Osorio-Tafall et Seraphim, List Vasc. Plants Cyprus, 110 (1973).

  [*Hyoseris minima* (non L.) Kotschy in Unger et Kotschy, Die Insel Cypern, 250 (1865).]

  [*Arnoseris pusilla* (non Gaertner) H. Stuart Thompson in Journ. Bot., 44: 333 (1906).]

  TYPE: "*in* Dania, Germania, Belgio".

Erect annual 6–25 cm. high; stems leafless, or with the leaves reduced to small, subulate bracts, glabrous, unbranched or sparingly branched in the upper part; leaves usually forming a neat basal rosette, obovate or oblanceolate, 1·5–5 cm. long, 0·4–1·5 cm. wide, subglabrous or rather densely hispidulous, apex rounded or subacute, margins subentire, dentate or sinuate-lobed, base tapering to a very short, indistinct petiole; capitula terminal on the stems or branches; involucre cylindric-campanulate, glabrous, about 13–15 mm. long, 6–8 mm. wide; outer phyllaries erect, unequal, 3–6 mm. long, about 1·5 mm. wide, apex subacute; inner phyllaries up to 15 mm. long, about 1·5 mm. wide, acute or subacute with hyaline-membranous margins; florets deep yellow; receptacular scales subulate-acuminate, 5–6 mm. long, 0·8 mm. wide at base; corolla-tube slender, glabrous except at fimbriate apex, 5–6 mm. long, ligule 4–5 mm. long, about 1·2 mm. wide, apex shortly and bluntly 5-toothed; filaments very short, glabrous; anthers yellowish, about 1·8 mm. long, 0·3 mm. wide; style 6–8 mm. long, minutely setulose in the upper part, branches filiform, blunt, about 0·4 mm. long. Marginal achenes (var. *glabra*) subcylindric, about 5 mm. long, 0·6 mm. wide, dark brown, somewhat pruinose, muricate, with numerous longitudinal ribs, apex truncate, base tapering to a slender point; inner achenes (var. *glabra*) or all the achenes (var. *loiseleuriana* Godr.)

similar but with a slender beak 4–6 mm. long, or occasionally (var. *erostris* Boiss.) all the achenes truncate; pappus tinged brown, the outer hairs 3–4 mm. long, scabridulous or shortly barbellate, the inner up to 10 mm. long, thinly plumose.

HAB.: Sandy shores and roadsides, or in garigue on stony hillsides; sea-level to 5,000 ft. alt.; fl. March–May.

DISTR.: Divisions 2, 3, 5. Widely distributed in Europe and the Mediterranean region.

2. Prodhromos, 1862, *Kotschy* 814; Platres, 1930, *E. Wall* s.n.; also, 1938, *Kennedy* 1477! Above Palekhori, 1957, *Merton* 3149! Selladhi tou Mavrou Dasous above Spilia, 1962, *Meikle* 2810!

3. Between Ayia Varvara and Stavrovouni, 1937, *Syngrassides* 1472!

5. Maritime sands E. of Trikomo, 1970, *A. Hansen* 646!

NOTES: All the specimens examined are referable to *H. glabra* L. ssp. *glabra*, with truncate marginal, and rostrate inner achenes. The achene-beak does not develop until after anthesis, and *H. glabra* L. var. *erostris* Boiss. is based, at least in part, on juvenile material.

2. H. achyrophorus *L.*, Sp. Plant., ed. 1, 810 (1753); Davis, Fl. Turkey, 5: 671 (1975); Feinbrun, Fl. Palaest., 3: 415, t. 701 (1978).
      Seriola aethnensis L., Sp. Plant., ed. 2, 2: 1139 (1763); Unger et Kotschy, Die Insel Cypern, 252 (1865) – as *S. "Aetnensis"*; Boiss., Fl. Orient., 3: 785 (1875); Sintenis in Oesterr. Bot. Zeitschr., 31: 392 (1881); 32: 123, 193, 398 (1882); H. Stuart Thompson in Journ. Bot., 44: 333 (1906); Post, Fl. Pal., ed. 2, 2: 822 (1933).
      Hypochaeris aethnensis (L.) Ball in Journ. Linn. Soc., 16: 542 (1878); Holmboe, Veg. Cypr., 191 (1914) – as *H. "aetnensis"*; Osorio-Tafall et Seraphim, List Vasc. Plants Cyprus, 110 (1973).
   TYPE: "*in Creta*".

Erect annual 7–40 cm. high; stems commonly with 1 well-developed leaf, the others reduced to small subulate bracts, subglabrous or thinly hispidulous, generally branched; leaves forming a loose basal rosette, obovate or spathulate, 1·5–10 cm. long, 0·5–3 cm. wide, thinly hispidulous, apex rounded or acute, base tapering to a very short, indistinct petiole, margins entire, subentire or dentate; capitula terminal on the stems or branches; involucre campanulate, 6–13 mm. long and almost as wide; phyllaries linear, fuscescent, densely setose dorsally, in 2–3 series, but the outer very inconspicuous, linear, less than 5 mm. long and 1 mm. wide, inner up to 13 mm. long, 2 mm. wide, acute or subacute, erect until after the fruits are shed, with broad, hyaline, membranous margins; florets deep yellow; receptacular scales linear-subulate, 8–10 mm. long, 0·8 mm. wide, membranous, keeled, with a distinct midrib produced as a slender terminal awn; corolla-tube 4–5 mm. long, glabrous with long, irregular fimbriae at its apex, ligule 4–5 mm. long, 1–1·3 mm. wide, irregularly incrassate-dentate at apex; filaments very short, glabrous; anthers yellowish, about 2 mm. long, 0·2 mm. wide; style about 7 mm. long, minutely setulose towards apex, branches about 0·5 mm. long, linear, blunt, yellowish. Achenes dimorphic, narrowly cylindrical-fusiform, bluntly 4-ribbed, dark brown, transversely rugose, the marginal remaining enveloped in the persistent inner phyllaries, usually epappose, somewhat curved, about 6 mm. long, 0·5 mm. wide, apex attenuate but not beaked, the inner pappose, almost straight, about 9 mm. long, 0·5 mm. wide, the apex produced into a long, slender beak; pappus tinged brownish, the outer hairs very short, slender, scabridulous, less than 2 mm. long, the inner 5–6 mm. long, plumose with fragile pinnules distinctly flattened and shortly connate at base.

HAB.: In garigue on dry rocky slopes; in sandy cultivated fields, or in dry grass-steppe; sea-level to 3,100 ft. alt.; fl. March–May.

DISTR.: Divisions 1–8. Mediterranean region, becoming rarer in the east.

1. Mirillis River near Ayios Yeoryios (Akamas), 1962, *Meikle* 2067! Near Polis, 1962, *Meikle* 2312!

2. Mandria, 1938, *Kennedy* 1467!
3. Limassol, 1859, *Kotschy* 978!
4. Cape Greco, 1862, *Kotschy* 156; Cape Kiti, 1981, *Hewer* 4730!
5. Near Athienou, 1880, *Sintenis*; Mia Milea, 1950, *Chapman* 485!
6. Kormakiti, 1936, *Syngrassides* 1223!
7. Pentadaktylos, 1880, *Sintenis*; Kantara, 1880, *Sintenis & Rigo* 489! Melounda, 1963, *J. B. Swart* 92!
8. Near Cape Andreas, 1880, *Sintenis*; 3½ miles E. of Kantara, 1952, *Merton* 1484! W. of Leonarisso, 1957, *Merton* 3073!

66. LEONTODON *L.*

Sp. Plant., ed. 1, 798 (1753).
Gen. Plant., ed. 5, 349 (1754) nom. cons.

Perennial (or rarely annual) rosulate herbs clothed with simple, branched or compound hairs or bristles; leaves dentate, sinuate, pinnatifid or subentire; inflorescence scapose, simple or sparingly branched, the cauline leaves reduced to small scale-like bracts or absent; capitulum terminal, homogamous, ligulate; involucre campanulate or subcylindric; phyllaries in 2 or more series, imbricate; florets yellow (rarely pink or purple); anthers sagittate at base, apical appendage blunt or subacute; style-branches slender, blunt; receptacle naked. Achenes uniform or dimorphic, subcylindric or fusiform, longitudinally ribbed and often transversely rugulose, apex truncate or beaked; pappus of plumose hairs or bristles, or the marginal achenes sometimes with a scaly corona.

About 50 species in Europe, the Mediterranean region and temperate Asia.

1. **L. tuberosus** *L.*, Sp. Plant., ed. 1, 799 (1753); Davis, Fl. Turkey, 5: 673 (1975).
   Apargia tuberosa (L.) Willd., Sp. Plant., 3: 1549 (1803); Sibth. et Sm., Fl. Graec. Prodr., 2: 130 (1813), Fl. Graec., 8: 72, t. 797 (1833).
   Thrincia tuberosa (L.) DC. in Lam. et DC., Fl. Franç., 4: 52 (1805); Poech, Enum. Plant. Ins. Cypr., 21 (1842); Boiss., Fl. Orient., 3: 726 (1875); Sintenis in Oesterr. Bot. Zeitschr., 31: 392 (1881); Post, Fl. Pal., ed. 2, 2: 129 (1933); Feinbrun, Fl. Palaest., 3: 417 (1978).
   T. tuberosa (L.) DC. var. *olivieri* DC., Prodr., 7: 100 (1849); Unger et Kotschy, Die Insel Cypern, 251 (1865).
   Leontodon tuberosus L. ssp. *olivieri* (DC.) Holmboe, Veg. Cypr., 192 (1914); Osorio-Tafall et Seraphim, List Vasc. Plants Cyprus, 110 (1973).
   TYPE: "*in* Hetruriae, Galloprovinciae, Narbonae *pratis*".

Erect perennial 15–30 (–50) cm. high; rootstock premorse; roots clustered, fusiform, tuberous; leaves all basal, rosulate, oblanceolate in outline, 6–30 cm. long, 1–4 cm. wide, runcinate-pinnatifid or dentate, thinly or rather densely hispidulous with glochidiately 2–3-branched bristles, apex acute or obtuse, base tapering to an ill-defined petiole; inflorescence scapose, unbranched; capitula solitary, nodding before anthesis; involucre campanulate, fuscescent, subglabrous, or rather densely hispidulous, 10–18 mm. long, 8–10 mm. wide; phyllaries erect, imbricate, unequal, linear-oblong, subacute or obtuse, the outer 3–4 mm. long, 1·3 mm. wide, the inner up to 18 mm. long, and 1–1·5 mm. wide; florets golden-yellow, the ligules often olive-green externally; corolla-tube slender, 6–7 mm. long, long-pilose at apex, ligule 10–12 mm. long, 2–3 mm. wide, apex sharply and rather irregularly 5-toothed; filaments glabrous, about 0·5 mm. long; anthers yellowish, about 4 mm. long, 0·3 mm. wide; style about 10 mm. long, branches yellowish, linear-filiform, blunt, 1–1·5 mm. long. Achenes dimorphic, the marginal partly enveloped by the persistent inner phyllaries, rather pallid, curved, cylindric-fusiform, 6–7 mm. long, 1 mm. wide, longitudinally ribbed and transversely rugulose, tapering slightly to a truncate apex, and crowned with a pappus of lacerate, yellowish scales about 0·8 mm. long; the inner achenes narrowly fusiform, slightly curved, brown, about 8 mm. long, 0·8 mm. wide, longitudinally ribbed and transversely muricate, tapering gradually to a slender beak, outer hairs of pappus about 3 mm. long, barbellate, inner 6–7 mm. long, yellowish-white, plumose, somewhat dilated into a scaly, stramineous base.

HAB.: Pastures and fallow fields; roadsides; sand-dunes, or in garigue on moist hillsides; sea-level to 4,000 ft. alt.; fl. Febr.–May.

DISTR.: Divisions 1–8. Widespread in the Mediterranean region.

1. Paphos, 1840, *Kotschy* 66; Stroumbi, 1913, *Haradjian* 756!
2. Platres, 1938, *Kennedy* 989A! Khrysorroyiatissa Monastery, 1978, *J. Holub* s.n.!
3. Amathus, 1947, *Mavromoustakis* s.n.!
4. Larnaca, 1862, *Kotschy* 59; Cape Greco, 1862, *Kotschy* 158; Ayia Napa, 1905, *Holmboe* 30; between Dhekelia and Ormidhia, 1936, *Syngrassides* 1059! Larnaca aerodrome, 1950, *Chapman* 27!
5. Near Kythrea, 1880, *Sintenis & Rigo* 490! Vatili, 1905, *Holmboe* 332! Dhikomo, 1936, *Syngrassides* 1184! 1188! Mia Milea, 1950, *Chapman* 484!
6. Kormakiti, 1937, *Nattrass* 838.
7. Kyrenia, 1935–1956, *Syngrassides* 626! *Kennedy* 989! *Casey* 218! *G. E. Atherton* 757! 767! 939! Lakovounara Forest, 1950, *Chapman* 200! 454! Near Panagra, 1962, *Meikle* 2356! Above Kythrea, 1967, *Merton* in ARI 218! Ayios Epiktitos, 1968, *Économides* in ARI 1182!
8. Rizokarpaso, 1912, *Haradjian* 244! Cape Andreas, 1962, *Meikle* 2475!

67. TARAXACUM *Weber*

in Wiggers, Prim. Fl. Holsat., 56 (1780) nom. cons.
H. Handel-Mazzetti, Mon. Gatt. Taraxacum, X + 175 pp. (1907).
H. Dahlstedt in Acta Hort. Berg., 9: 1–36 (1926).
R. Doll in Feddes Repert., 93 (7–8): 481–624 (1982).

Perennial, lactiferous, scapose herbs; leaves forming a basal rosette, synanthous or hysteranthous, polymorphic, but commonly runcinate-pinnatisect; scapes 1 or more per plant, simple, fistulose; capitula solitary, terminal, rarely 2 or more, terminal and lateral, homogamous, ligulate; involucre cylindric-campanulate; phyllaries in several, imbricate series, the outermost short, sometimes spreading or reflexed and forming a calyculus, apex often with a distinct gibbous or horned appendage; florets usually yellow, sometimes white or pink; anthers shortly sagittate at base; style-branches slender, blunt; receptacle naked. Achenes oblong-fusiform, 10-ribbed, muricate or tuberculate towards apex, usually with a slender beak arising from a narrow, conical base; pappus white or tinged brown or purple, hairs in several series, slender, scabridulous, persistent.

A problematic genus, estimated to include about 2,000 species, many of them apomicts. Little is known about the biology of the species occurring in Cyprus, but it is likely that at least some of them are normal, sexual amphimicts; most of the Cyprus species are hysteranthous, the flowers appearing in Oct.–Nov. before the leaves, with anthesis sometimes continuing into the succeeding spring. All have very thick roots, often containing copious latex and sometimes with a densely fulvous-lanuginose apical indumentum.

The names used by the older authors often embrace several distinct taxa, and cannot now be applied with certainty or cited precisely as synonyms, though an attempt has been made here to allocate the various names that have been used in past accounts of Cyprus plants.

Outer phyllaries patent or reflexed, with narrow, indistinct, membranous margins; petioles purplish; achenes rusty-red; pappus white or almost white; plants probably always synanthous, commencing to flower in spring - - - - - - **4. T. holmboei**
Outer phyllaries erect, subadpressed, generally with a distinct, membranous margin; achenes pale or dark brown; plants generally hysteranthous, commencing to flower in autumn in advance of the leaves:

Leaves runcinate-pinnatifid with broad, deltoid, entire or toothed lateral lobes, occasionally
   subentire or broadly and irregularly lobed; pappus tinged brown -   -   **1. T. cyprium**
Leaves pinnatisect almost to midrib, with numerous, small, oblong-acuminate or
   suborbicular lateral lobes, separated by smaller lobules:
   Lateral lobes acuminate, oblong; apex or scape pilose or lanuginose; pappus tinged brown
                                                                     2. T. hellenicum
   Lateral lobes blunt or suborbicular; apex of scape glabrous; pappus white
                                                                     3. T. aphrogenes

1. **T. cyprium** *Lindberg f.*, Iter Cypr., 7: 36 (1946); N. Haran in Pal. Journ. Bot., J. series, 5: 237
   (1952); Osorio-Tafall et Seraphim, List Vasc. Plants Cypr., 110 (1973); Feinbrun, Fl.
   Palaest., 3: 429, t. 725 (1978).
      [*T. gymnanthum* (non (Link) DC.) Poech, Enum. Plant. Ins. Cypr., 22 (1842); Unger et
   Kotschy, Die Insel Cypern, 256 (1865); H. Stuart Thompson in Journ. Bot., 44: 333
   (1906).]
      [*T. officinale* (non Weber) Kotschy in Unger et Kotschy, Die Insel Cypern, 256 (1865) pro
   parte.]
      [*T. officinale* Weber var. *genuinum* (non Koch) Boiss., Fl. Orient., 3: 787 (1875) pro parte
   quoad plant. cypr.]
      [*T. calocephalum* (non Hand.-Mazz.) Hand.-Mazz., Mon. Taraz., 107 (1907) pro parte
   quoad plant. cypr.; Holmboe, Veg. Cypr., 193 (1914); Osorio-Tafall et Seraphim, List Vasc.
   Plants Cyprus, 110 (1973).]
      [*T. megalorrhizon* (non (Forssk.) Hand.-Mazz.) Hand.-Mazz., Mon. Tarax., 38 (1907) pro
   parte quoad plant. cypr.; Holmboe, Veg. Cypr., 193 (1914) pro parte; Osorio-Tafall et
   Seraphim, List Vasc. Plants Cyprus, 110 (1973).]
      [*T. minimum* (non (Briganti ex Guss.) Terrac.) Dahlst. in Act. Hort. Berg., 9: 18 (1929)
   pro parte quoad plant. cypr.; Davis, Fl. Turkey, 5: 801 (1975) pro parte quoad plant. cypr.]
      [*T. aleppicum* (non Dahlst.) van Soest in Davis, Fl. Turkey, 5: 799 (1975) pro parte
   quoad plant. cypr.]
   TYPE: Cyprus; "Paradisi juxta monasterium Kykko, in muro in horto", 1939, *Lindberg f.* s.n.
(H, K!)

Rosulate perennial with a very thick, rugose, blackish taproot; leaves
variable, narrowly oblong-oblanceolate in outline, 3–8 cm. long, 0·8–2·5 cm.
wide, glabrous, runcinate-pinnatifid, or irregularly lobed, or occasionally
subentire, when pinnatifid, terminal lobe usually deltoid, subaequilateral,
entire or toothed, lateral lobes often broadly deltoid, entire or toothed;
petiole flattened, sometimes indistinct, up to 5 cm. long; capitula generally
appearing in autumn (Sept.–Nov.) in advance of the leaves; scapes up to 30
cm. long at anthesis, elongating in fruit, often densely lanuginose at apex;
involucre campanulate, 10–18 mm. long, 6–12 mm. wide, outer phyllaries
loosely erect, oblong-ovate, 2–4 mm. long, 1·5–2·5 mm. wide, greenish or
purplish, with a short lingulate apex and rather conspicuous, membranous
margins; inner phyllaries linear, about 1·5 mm. wide, generally with a dark
green or purplish apical appendage and distinct, dark midrib; ligules golden-
yellow above, often olive-green below, linear, 10–20 mm. long, 1·5–2·5 mm.
wide, apex with 5 small, irregular, dark teeth; anthers linear, subacute,
about 5 mm. long, 0·3 mm. wide; style about 11 mm. long, shortly pilose in
the upper part, branches filiform, slightly greenish, about 3 mm. long.
Achene (including apical cone) about 4·5 mm. long, 1·3 mm. wide, light
brown, conspicuously verruculose towards base, muricate or echinulate
towards apex, apical cone narrow, papillose, beak slender, about 8 mm. long,
smooth; pappus brownish- or yellowish-white, hairs fragile, 4–5 mm. long,
thinly scabridulous.

HAB.: Roadsides, waste ground; in garigue on rocky hillsides; sea-level to 4,500 ft. alt.; fl.
Sept.–Nov. (sometimes abnormally at other seasons).

DISTR.: Divisions 1, 2, 7. Also Palestine.

1. Ktima, 1840, *Kotschy* 57.
2. Near Prodhromos, 1862, *Kotschy* 903; also, 1955, *G. E. Atherton* 524! Platanistasa, 1905,
   Holmboe 1162; Platres, 1928, *Druce* s.n.; also, 1936–37, *Kennedy* 991! 993! Kryos Potamos,
   3,900 ft. alt., 1937, *Kennedy* 990! Xerokolymbos, 4,300 ft. alt., 1937, *Kennedy* 992!

Evrykhou, 1937, *Syngrassides* 1678! Kykko Monastery, 1939, *Lindberg f.* s.n.! Lefka, 1940, *Davis* 1978! Kakopetria, 1966, *Merton* in ARI 75! Saïttas, 1970, *A. Genneou* in ARI 1612!
7. Kyrenia, 1936, *Kennedy* 994! also, 1948, *Casey* 88! and, 1955, *G. E. Atherton* 724! Vasilia, 1940, *Davis* 2011!

NOTES: This, and the following, are jointly referred to *T. megalorrhizon* (Forssk.) Hand.-Mazz. by earlier authors, but the identity of the latter (said to have been collected in the area of the Dardanelles) is very doubtful.

Haran (Pal. Journ. Bot., J. series, 5: 237–247; 1952) has examined specimens of *T. cyprium* from Palestine, and concludes that it is a typical tetraploid apomict with 2n = 32.

2. T. hellenicum *Dahlst.* in Act. Hort. Berg., 9: 11, t. 1, figs. 16–19 (1929); Druce in Rep. B.E.C., 9: 470 (1931) pro parte; Davis, Fl. Turkey, 5: 800 (1975).

[*Leontodon laevigatus* (non Willd.) Sm. in Sibth. et Sm., Fl. Graec. Prodr., 2: 129 (1813) pro parte quoad plant. cypr.]

[*Taraxacum laevigatum* (non (Willd.) DC.) Poech, Enum. Plant. Ins. Cypr., 22 (1842); Unger et Kotschy, Die Insel Cypern, 256 (1865); Holmboe, Veg. Cypr., 193 (1914); Osorio-Tafall et Seraphim, List Vasc. Plants Cypr., 110 (1973).]

[*T. officinale* (non Weber) Sintenis in Oesterr. Bot. Zeitschr., 31: 288 (1881).]

[*T. megalorrhizon* (non (Forssk.) Hand.-Mazz.) Hand.-Mazz., Mon. Tarax., 38 (1907) pro parte quoad plant. cypr.; Holmboe, Veg. Cypr., 193 (1914) pro parte.]

[*T. pseudonigricans* (non Hand.-Mazz.) Hand.-Mazz., Mon. Tarax., 51–52 (1907) pro parte quoad plant. cypr.; Holmboe, Veg. Cypr., 193 (1914); Osorio-Tafall et Seraphim, List Vasc. Plants Cyprus, 110 (1973).]

TYPE: Not indicated.

Rosulate perennial with thick, rugose, blackish taproot, leaves narrowly oblong-lanceolate in outline, 3–10 cm. long, 1–3 cm. wide, glabrous, divided almost or quite to midrib into numerous narrowly lanceolate or oblong, acute, entire or denticulate, patent or retrorse lateral lobes, commonly alternating with smaller, narrower lobules; petiole indistinct, purplish, to about 2 cm. long; capitula generally appearing with the leaves in late autumn and continuing to develop until the succeeding April; scapes rather slender, up to 10 cm. long at anthesis, elongating and becoming horizontal in fruit, often densely lanuginose towards apex; involucre campanulate, 8–10 mm. long and almost as wide; outer phyllaries loosely erect, ovate, 3–5 mm. long, 2·5–3 mm. wide, dark greenish-violet with a broad, conspicuous, membranous margin; inner phyllaries linear, to about 10 mm. long, 2·5 mm. wide, violet or greenish with a narrow, membranous margin, and a slightly dilated, somewhat erose apex; ligules yellow with a broad, longitudinal greenish stripe on the outer (lower) surface, tube about 2·5 mm. long, 0·5 mm. wide, ligule about 11 mm. long, 2·5 mm. wide, apex with 5, small, irregular, dark teeth; anthers about 4 mm. long, 0·3 mm. wide, with a small, bluntish apical appendage [and a small amount of irregular pollen]; style about 9 mm. long, branches filiform, blunt, 2·5–3 mm. long, greenish. Achenes dorsiventrally compressed, a little curved, pale brown, about 5 mm. long (excluding beak), 1·5 mm. wide, base verruculose, apex distinctly muricate, beak slender, about 5 mm. long; pappus tinged brownish, hairs fragile, 5–6 mm. long, thinly scabridulous.

HAB.: Pastures and rocky slopes; pathsides; sea-level to 2,000 [–4,500] ft. alt.; fl. Oct.–April.

DISTR.: Divisions [2] 3, 5, 6, 8. Widely distributed in the Mediterranean region.

2. Prodhromos, 1859 and 1862, *Kotschy* Suppl. no. 1025; 903 partly; summit of Olympus, 1862, *Kotschy*. See Notes.
3. Stavrovouni, 1934, *Syngrassides* 1391!
5. Near Athienou, 1880, *Sintenis & Rigo* 285!
6. Nicosia, 1928, *Druce* s.n.! English School, Nicosia, 1958, *P. H. Oswald* 21!
8. Cape Andreas, 1962, *Meikle* 2472!

NOTES: To judge from the microfiche of the sheet in herb. Boiss., *Kotschy* 903, from Prodhromos, consists of a mixture, one part being *T. cyprium* Lindberg f., which certainly occurs there, the other *T. hellenicum* Dahlst., which no-one else has found at such an altitude. The record may, however, be correct, also that from the summit of Khionistra, where *T.*

holmboei Lindberg f. is the more likely occurrence. *T. hellenicum* is apparently lowland in distribution, and one suspects some error with these records from high altitudes on Troödos.

3. T. aphrogenes *Meikle* in Ann. Musei Goulandris, 6: 89 (1983). See App. V, p. 1898.
   TYPE: Cyprus; in rock crevices just above the sea S.E. of Paphos; 30 Oct. 1981, *Meikle 5044* (K!).

Rosulate perennial, with a thick, blackish taproot, generally densely fulvous-lanuginose at the apex, around the leaf-bases; leaves very small, glabrous, narrowly oblanceolate in outline, 3–5 cm. long, 0·3–1 cm. wide, dissected almost to midrib into numerous minute, bluntly oblong or suborbicular lobes and alternating lobules, terminal lobe suborbicular or bluntly deltoid, sometimes 3-fid; petiole short, indistinct, dilated and sheathing at base; capitula appearing with the leaves in autumn; scapes very slender, to about 11 cm. long at anthesis, not elongating much in fruit, tinged bronze-purple, glabrous; involucre campanulate, 8–10 mm. long, 8–9 mm. wide, glabrous, outer phyllaries erect, subadpressed, 3–5 mm. long, 1·5 mm. wide, ovate, with an abruptly narrowed, blunt acumen, greenish or bronze-tinged with a narrow, but distinct, membranous margin; inner phyllaries linear-oblong, blunt, about 10 mm. long, 1·5 mm. wide, purplish-bronze with a narrow membranous margin; florets golden-yellow, the marginal ligules tinged bronze on the outer (lower) surface; corolla-tube about 2 mm. long; ligules about 7 mm. long, 2 mm. wide, minutely and bluntly 5-toothed at apex; filaments glabrous, about 0·5 mm. long; anthers linear, yellowish, about 3·5 mm. long, 0·3 mm. wide with a small, subacute apical appendage; pollen evidently plentiful, with well-formed grains; style about 9 mm. long, sparingly setulose in the upper half, branches filiform, blunt, yellowish, about 3 mm. long. Achenes dorsiventrally compressed, oblong, dark fuscous-brown, about 4 mm. long, 1 mm. wide, basal part almost smooth, apical part conspicuously muricate, apical cone short, truncate, beak slender, about 4 mm. long, hairs white, about 3·5 mm. long, remotely scabridulous. *Plate 63.*

HAB.: Fissures of maritime rocks; just above sea-level; fl. Oct.–? Dec.

DISTR.: Division 1. Endemic.

1. Ayios Nikolaos (Akamas), 1957, *Merton 3049*! Abundant on rocks just S.E. of Paphos, 1981, *Meikle 5044* (holotype)!

NOTES: *Merton 3049*, collected on 6th April 1957, is without flowers, and suggests that *T. aphrogenes* finishes flowering before the spring. Merton notes on the label: "This appears to be the same species as that found on Petra tou Romiou". I have not been able to trace any specimen from the latter locality.
   The anthers appear to contain plenty of good pollen, and *T. aphrogenes* is evidently a normal, sexual, amphimict. Dr. J. Richards informs me (1982) that its nearest ally is *T. bithynicum* DC. (Prodr. 7: 149; 1838), a member of the section *Scariosae* Hand.-Mazz. found at high altitudes on the mountains of Crete, Turkey and Lebanon.

4. T. holmboei *Lindberg f.*, Iter Cypr., 36 (1946); R. Doll in Feddes Repert,, 83: 698 (1973), 84: 134 (1974); Osorio-Tafall et Seraphim, List Vasc. Plants Cyprus, 110 (1973).
   TYPE: Cyprus; "M. Troodos, Platania, in margine viae" 23 June, 1939, *Lindberg f.* s.n. (H, K!).

Rosulate perennial with a long, but not very thick taproot; leaves variable, oblong-oblanceolate in outline, 3·5–7 cm. long, 0·8–2 cm. wide, glabrous, coarsely runcinate-pinnatifid, with broad, deltoid-acute, entire or denticulate lateral lobes, and a rather small, acute-deltoid terminal lobe, sometimes obtuse and shallowly lobed, coarsely dentate or subentire; petiole short, slender, usually less than 2 cm. long, purplish, sometimes ill-defined; capitula hysteranthous, developing after the leaves have expanded in May and June; scapes short, generally less than 8 cm. long, bronze-tinged,

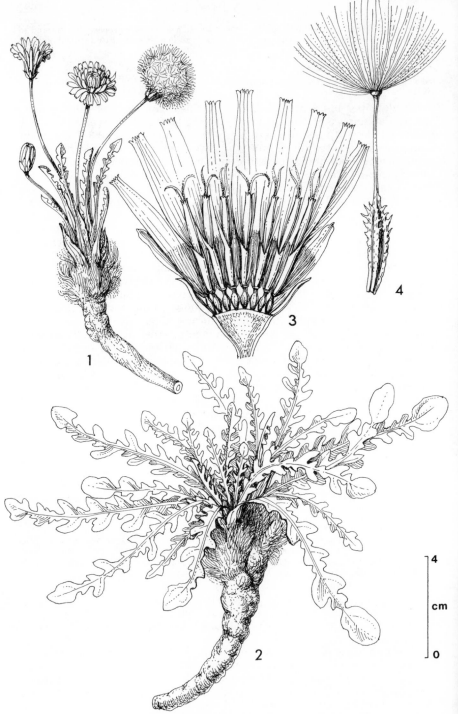

Plate 63. TARAXACUM APHROGENES Meikle **1,** habit, autumnal flowering stage, × 1; **2,** habit, spring post-flowering stage, × 1; **3,** capitulum, longitudinal section, × 6; **4,** achene, × 9. (**1, 3, 4** from *Meikle* 5044; **2,** from *Merton* 3049.)

glabrous or thinly pilose; involucre campanulate, 8–10 mm. long and almost as wide; outer phyllaries patent or reflexed, forming a basal calyculus, greenish, ovate-oblong, obtuse or subacute, 3–4 mm. long, about 2 mm. wide, with a very narrow, indistinct membranous margin; inner phyllaries greenish, linear-oblong, obtuse, about 10 mm. long, 2–3 mm. wide; ligules golden-yellow above, tinged inky-blue or purplish below; corolla-tube slender, about 4 mm. long, glabrous or very thinly pubescent, ligule linear-oblong, 9–10 mm. long, 1–1·2 mm. wide, apex minutely 5-toothed; filaments glabrous, 0·5 mm. long; anthers linear, yellowish, abut 3·5 mm. long, 0·3 mm. wide, apical appendage small, ovate, blunt; pollen sparse, the grains irregularly formed (teste R. Doll); styles 10 mm. long, branches filiform, blunt, greenish, about 2 mm. long. Achenes narrowly oblong, strongly compressed dorsiventrally, bright rusty-red, about 4 mm. long, 1 mm. wide, basal part strongly ribbed and angled, apical part conspicuously and sharply muricate, almost echinate, apical cone slender, about 0·7 mm. long, beak slender 6–7 mm. long; pappus white, hairs 5–6 mm. long, remotely scabridulous.

HAB.: Roadsides; Pine forests, or gravelly screes at high altitudes; 4,000–6,400 ft. alt.; fl. May–June.

DISTR.: Division 2. Endemic.

2. Platania, 1939, Lindberg f. s.n.! Prodhromos, 1957, Merton 3195! Khionistra, summit, 1962, Meikle 2779!

NOTES: Here also probably belongs the specimen recorded as Taraxacum officinale by Kotschy "Um die Spitze des Troodos", but in the absence of a specimen it is difficult to be certain.
R. Doll (Feddes Repert., 84: 134; 1973) refers T. holmboei to Taraxacum Weber sect. Erythrosperma Dahlst. It is certainly very distinct from other species found in Cyprus, with its reflexed outer phyllaries and rusty-red achenes, and more resembles the Dandelions of western Europe. Dr. J. Richards (1982) tells me that it is most probably indistinguishable from T. gracilens Dahlst. from Greece.

68. CHONDRILLA L.

Sp. Plant., ed. 1, 796 (1753).
Gen. Plant., ed. 5, 348 (1754).

Biennial or perennial herbs, often lactiferous, sometimes suffruticose; stems usually branched, ribbed or angled, often sparsely leafy; leaves alternate or mostly radical, irregularly lobed or pinnatisect, or entire; inflorescence often paniculate, capitula pedunculate or sessile, homogamous, ligulate; involucre cylindrical, phyllaries in 2–3 series, the outer very much shorter than the inner; florets yellow, the ligules shortly 5-dentate at apex; anthers sagittate at base; style-branches filiform; receptacle naked. Achenes oblong or fusiform, glabrous, subterete or angular, 5-ribbed, generally muricate towards apex, beak deciduous; pappus white, hairs slender, silky, scabridulous.

About 25 species in Europe, the Mediterranean region and temperate Asia.

1. C. juncea L., Sp. Plant., ed. 1, 796 (1753); Poech, Enum. Plant. Ins. Cypr., 22 (1842); Unger et Kotschy, Die Insel Cypern, 256 (1865); Boiss., Fl. Orient., 3: 792 (1875); Post in Mém. Herb. Boiss., 18: 96 (1900); H. Stuart Thompson in Journ. Bot., 44: 333 (1906); Holmboe, Veg. Cypr., 193 (1914); Post, Fl. Pal., ed. 2, 2: 143 (1933); Lindberg f., Iter Cypr., 34 (1946); Rechinger f. in Arkiv för Bot., ser. 2, 1: 434 (1950); Osorio-Tafall et Seraphim, List Vasc. Plants Cyprus, 111 (1973); Davis, Fl. Turkey, 5: 812 (1975); Feinbrun, Fl. Palaest., 3: 428, t. 724 (1978).
TYPE: "in Germania, Helvetia, Gallia ad agrorum margines".

Erect, glaucous biennial or perennial up to 1 m. high; lower part of main stem often densely fulvous-setose or scabridulous, upper part usually glabrous, angled, ribbed or subterete, branches generally numerous, ascending; basal leaves rosulate, narrowly oblong or obovate in outline, 4–10 cm. long, 1–3·5 cm. wide, lyrate-pinnatifid, glaucous, with denticulate or subentire lobes, apex usually acute, base tapering to an indistinct petiole; cauline leaves few, remote, sessile, linear, 2–8 cm. long, 0·2–0·5 cm. wide, entire or the lower sometimes closely denticulate; inflorescence very lax, the capitula sessile or subsessile, solitary or in small, remote clusters along the branches; involucre cylindric, about 8 mm. long, 2–3 mm. wide, thinly tomentellous; outer phyllaries very short and inconspicuous, ovate, 1·5–2 mm. long, about 1 mm. wide; inner phyllaries linear, about 8 mm. long, 1 mm. wide, with a conspicuous, pale midrib and narrow, membranous margins; florets few, yellow; corolla-tube slender, about 6 mm. long, minutely brownish-papillose; ligule about 8 mm. long, 1·5 mm. wide, apex shortly 5-dentate; anthers yellow, linear, about 5 mm. long, 0·3 mm. wide, apical appendage subacute; style about 12–14 mm. long, setulose towards apex, branches filiform, yellowish, about 2 mm. long. Achenes narrowly oblong, slightly curved, pale brown, about 3·5 mm. long, 1 mm. wide, longitudinally ribbed, lower two-thirds almost smooth, upper third conspicuously muricate, with an apical corona of small scales, beak slender, about 5 mm. long; pappus white, copious, hairs in a single series, about 6 mm. long, remotely scabridulous.

HAB.: In garigue on rocky hillsides; roadsides, edges of cultivated ground; sea-level to 6,300 ft. alt.; fl. June–Sept.

DISTR.: Divisions 1, 2, 6. Also South and Central Europe, the Mediterranean region and eastwards to Afghanistan and Central Asia.

1. Ktima, 1840, *Kotschy* 64.
2. Kambos, ? 1898, *Post* s.n.; Troödos, 1901, *A. G. & M. E. Lascelles* s.n.! Khionistra, 4,500–6,300 ft. alt., 1912–1950, *Haradjian* 513; *Kennedy* 996! *Lindberg f.* s.n.! *Davis* 1949! *F. M. Probyn* s.n.! Kykko Monastery, 1913, *Haradjian* 972! Platres, 1937, *Kennedy* 995! *N. Macdonald* 202! Mesapotamos Monastery, 1941, *Davis* 3453! Moniatis, 1976, *A. Della* in ARI 1692! Prodhromos, 1949, *Casey* 878! also, 1955, *G. E. Atherton* 522!
6. Government House Grounds, Nicosia, 1935, *Syngrassides* 798! Kykko metokhi, 1966, *Merton* in ARI 21!

69. CEPHALORRHYNCHUS *Boiss.*
Diagn., 1, 4: 28 (1844).

Biennial or perennial herbs with tuberous rhizomes; stems erect, glandular, sparingly branched; leaves alternate, lyrate-pinnatisect, the basal (arising from the tuber), long-petiolate, the cauline sessile or shortly petiolate, generally amplexicaul-auriculate; inflorescence lax, corymbose or paniculate; capitula small, pedunculate, homogamous, ligulate; involucre cylindric; outer phyllaries erect, subadpressed, very much shorter than the inner, glabrous or thinly glandular; florets yellowish (sometimes drying bluish), rather few; anthers sagittate at base; style-branches filiform, blunt; receptacle naked. Achenes fusiform, scarcely compressed, beak slender, with a conspicuously dilated apical disk; pappus white, deciduous, hairs in 2 series, the outer very short (forming a neat ruff), the inner long, slender, fragile, remotely scabridulous.

About 10 species in S.E. Europe and western Asia.

1. **C. cypricus** (*Beauverd*) *Rechinger f.* in Arkiv för Bot., ser. 2, 1: 435 (1950); Osorio-Tafall et Seraphim, List Vasc. Plants Cyprus, 111 (1973).
Cicerbita cyprica Beauverd in Bull. Soc. Bot. Genève, ser. 2, 26: 156 (1934).
[*Lactuca hispida* (non DC.) Kotschy in Unger et Kotschy, Die Insel Cypern, 255 (1865); Lindberg f. Iter Cypr., 35 (1946).]

[*Cephalorrhynchus candolleanus* (non Boiss.) Boiss., Fl. Orient., 3: 820 (1875) pro parte quoad Kotschy 784; nom. illeg.]

[*Lactuca hispida* DC. ssp. *candolleana* (non *Cephalorrhynchus candolleanus* Boiss.) Holmboe, Veg. Cypr., 194 (1914) pro parte quoad plant. cypr.]

[*Cephalorrhynchus tuberosus* (non (Stev.) Schchian) Davis, Fl. Turkey, 5: 775 (1975) quoad plant. cypr.]

TYPE: Cyprus; Troödos, 1931, *S. Topali* (G).

Erect biennial with a globose rhizome 2–4 cm. diam.; stems 10–40 (–50) cm. high, striate, glabrous in the lower part, glandular with long-stalked glands in the upper part, generally unbranched except in the region of the inflorescence; basal leaves with slender petioles up to 15 cm. long; lamina very variable, sometimes broadly cordate, but more often lyrate-pinnatisect, with a large, suborbicular, shallowly dentate terminal lobe and 2–10 narrowly oblong, subentire or dentate lateral lobes, 2–12 cm. long, 1·3–8 cm. wide, glabrous, pale green and white-spotted above, purple below; cauline leaves similarly variable, smaller than the basal and diminishing upwards, petiolate, with petioles up to 10 cm. long, or subsessile, the base of the lamina elongate and pseudo-petiolate, usually dilated towards the point of junction with the stem, and commonly amplexicaul-auriculate, with conspicuous, acute or rounded, entire or dentate auricles; inflorescence loosely corymbose or paniculate, sometimes with relatively few (6–10) capitula, but sometimes much branched, with erecto-patent branches and numerous capitula; peduncles slender, up to 3 cm. long, usually glandular; involucre cylindric, 8–12 mm. long, 2·5–3·5 mm. wide; outer phyllaries erect, lanceolate-acuminate, 3–5 mm. long, about 1 mm. wide, glabrous or thinly glandular, greenish or purplish with a narrow, membranous margin; inner phyllaries linear, subacute, glabrous or sparsely glandular, 8–12 mm. long, 1·5–2 mm. wide; florets pale yellow (often turning blue when dried); corolla-tube slender, 3–6 mm. long, minutely papillose and thinly pilose at apex, ligule narrowly oblong, 7–10 mm. long, 2·5–3·5 mm. wide, apex shortly 5-dentate; filaments very short; anthers yellowish, linear, about 3–4 mm. long, 0·3 mm. wide, apical appendage often erose-apiculate; style 8–10 mm. long, setulose in the upper part, branches filiform, about 1·5–2 mm. long. Achenes narrowly fusiform, 5-angled, not compressed, about 4 mm. long, 0·8–1 mm. wide, light brown with dark brown longitudinal stripes, minutely scabridulous with prominent longitudinal ribs, apex acute or acuminate, beak slender, about 3 mm. long, with a conspicuous, shortly ciliate, terminal disk; pappus white, the inner hairs very slender, fragile and caducous, about 5 mm. long, remotely scabridulous.

HAB.: Shady mountainsides or streamsides, under Pines, Cedars or Junipers, usually at high altitudes; 2,500–6,400 ft. alt.; fl. April–July.

DISTR.: Division 2. Endemic.

2. Khionistra (Troödos, Kryos Potamos, Xerokolymbos, Kharchis, etc.), 3,900–6,400 ft. alt., 1862–1955, *Kotschy 784*! *Sintenis & Rigo 798*! *Kennedy 1009–1015*! *Lindberg f.* s.n.! *Davis 3503*! *3158*! *Mavromoustakis* s.n.! *Casey 676*! *Merton 2372*! Tripylos, 1941, *Davis 3483*! Ayios Theodhoros River, near Stavros tis Psokas, 1962, *Meikle 2698*!

70. STEPTORHAMPHUS *Bunge*

Beitr. Kenntn. Fl. Russlands, 205 (1852).
Mem. Acad. Imp. Sci. St. Pétersb., 7: 381 (1854).

Erect perennial herb with a thick, fleshy rootstock; leaves alternate, lyrate-pinnatisect, glabrous or subglabrous, glaucous, the basal petiolate, the cauline sessile, amplexicaul-auriculate; inflorescence subcylindric or thyrsoid; capitula pedunculate, homogamous, ligulate; involucre cylindric; phyllaries erect, imbricate in several series, glabrous; florets pale yellow or tinged purplish; anthers sagittate at base; style-branches filiform; receptacle naked. Achenes strongly compressed, with flattened margins, beak slender, smooth; pappus deciduous, in 2 series, the outer of short bristles arising from a disk, the inner long, silky, white, sparingly scabridulous.

About 7 species in the eastern Mediterranean region eastwards to Central Asia.

1. S. tuberosus (*Jacq.*) *Grossh.*, Fl. Kavk., 4: 258 (1935); Davis, Fl. Turkey, 5: 775 (1975).
Lactuca tuberosa Jacq., Hort. Vindob., 1: 18, t. 47 (1772); Boiss., Fl. Orient., 3: 806 (1875); Post, Fl. Pal., ed. 2, 2: 145 (1933); Feinbrun, Fl. Palaest., 3: 436, t. 738 (1978).
L. cretica Desf. in Ann. Mus. Hist. Nat. Paris, 11: 160, t. 19 (1808); Boiss., Fl. Orient., 3: 805 (1875), Suppl., 322 (1888); Sintenis in Oesterr. Bot. Zeitschr., 32: 398 (1882); H. Stuart Thompson in Journ. Bot., 44: 333 (1906); Holmboe, Veg. Cypr., 194 (1914); Post, Fl. Pal., ed. 2, 2: 145 (1933); Lindberg f., Iter Cypr., 35 (1946); Rechinger f. in Arkiv för Bot., ser. 2, 1: 435 (1950); Osorio-Tafall et Seraphim, List Vasc. Plants Cyprus, 111 (1973).
L. leucophaea Sm. in Sibth. et Sm., Fl. Graec. Prodr., 2: 127 (1813), Fl. Graec., 8: 69, t. 794 (1833); Poech, Enum. Plant. Ins. Cypr., 21 (1842); Unger et Kotschy, Die Insel Cypern, 255 (1865).
TYPE: *Hortus Vindobonensis*, 1: t. 47! ("Sub titulo Lactucae creticae tuberosae accepi hujus plantae semina", *Jacquin*).

Erect glabrous perennial, 10–50 cm. high; rootstock a thick, fusiform tuber; stem usually unbranched, longitudinally sulcate; basal leaves oblanceolate, 10–17 cm. long, 1·5–3 cm. wide, glabrous, glaucous, lyrate-pinnatisect or often rather shallowly pinnatifid, apex acute or obtuse, base tapering to an ill-defined petiole; cauline leaves similar, but smaller, sessile, amplexicaul, generally with projecting blunt or acute auricles; inflorescence lax, usually consisting of 15 or fewer capitula in a cylindric or thyrsoid raceme or panicle; peduncles 2–4 cm. long, patent or sharply upcurved at apex; bracts usually numerous, squamiform; capitula erect or suberect, noticeably accrescent in fruit; involucre cylindric, 1·5–2 cm. long at anthesis (up to 3 cm. long in fruit), 0·5–0·7 cm. wide; phyllaries in several overlapping series, glabrous, glaucous, often purplish, the outermost ovate, acute, about 7 mm. long, 5 mm. wide, the inner becoming progressively longer and narrower, the innermost narrowly oblong, up to 20 mm. long, 3·5 mm. wide; florets pale creamy yellow, stained purplish or coppery-red externally; corolla-tube very slender, 12–13 mm. long, thinly pubescent at apex, ligule 14–15 mm. long, about 3 mm. wide, shortly 5-dentate at apex; anthers yellow or fuscous, about 3 mm. long, 0·3 mm. wide, apical appendage blunt; style 16–17 mm. long, branches slender, filiform, about 2 mm. long. Achenes almost flattened, elliptic, 3·5–4 mm. long, 2 mm. wide, dark reddish-brown with a narrowly winged margin, minutely rugulose, with a varying number of small, projecting papillae, beak very slender, thread-like, pale, smooth, 10–11 mm. long, terminating in a distinct disk or corona 0·4 mm. diam.; hairs white, silky, remotely scabridulous, about 12–13 mm. long.

HAB.: Bare, stony hillsides, sometimes in garigue or under bushes; in open Pine forest, or on muddy salt-flats; sea-level to 5,600 ft. alt.; fl. April–June.

DISTR.: Divisions 1–4, 7, 8. E. Mediterranean region and eastwards to Iraq and Iran.

1. Smyies, 1941, *Davis* 3307!
2. Lavramis, near Yialia, 1905, *Holmboe* 806; Troödos, 1912, *Haradjian* 404; Kryos Potamos, 5,100–5,600 ft. alt., 1937, *Kennedy* 1032! 1033!
3. Between Yerasa and Kalokhorio, 1962, *Meikle* 2896!
4. Larnaca Salt Lake, near Hala Sultan Tekke, 1979, *Edmondson & McClintock* E 2960!
7. Pentadaktylos, 1880, *Sintenis & Rigo* 290! St. Hilarion, 1933, *Syngrassides* 632! also, 1949, *Casey* 730! Near Kantara, 1950, *Casey* 1048! Kambyli, 1957, *Merton* 2972! N. side Sina Oros, 1967, *Merton* in ARI 809!
8. Near Rizokarpaso, 1912, *Haradjian* 250!

71. LACTUCA *L.*
Sp. Plant., ed. 1, 795 (1753).
Gen. Plant., ed. 5, 348 (1754).

Annual, biennial or perennial herbs, sometimes suffruticose; stems erect, lactiferous, often setulose especially towards the base; leaves alternate, subentire, dentate or deeply lyrate-pinnatisect, glabrous or setulose, especially on midrib and nerves below, often glaucous, the basal tapering to an ill-defined petiole, the cauline sessile, commonly auriculate or decurrent; inflorescence often much branched; capitula homogamous, ligulate, often rather small; involucre cylindrical; phyllaries erect, imbricate in 3–4 series, sometimes lengthening in fruit; florets yellow or bluish, rarely white; anthers sagittate at base; style-branches filiform; receptacle naked. Achenes dorsally compressed, beak long, slender, smooth; pappus of numerous slender, white or yellowish, scabridulous or almost smooth hairs, sometimes with an outer series of very short hairs arising from a small disk at the top of the achene-beak.

About 100 species widely distributed in the Old World or as introductions elsewhere.

Base of stem, and undersurface of leaf-midrib prickly; leaf-blade held in a vertical plane
  (edgewise) to stem - - - - - - - - - - **2. L. serriola**
Base of stem, and undersurface of leaf-midrib smooth; leaf-blade held horizontally:
  Stems slender, wiry, pallid; inflorescence elongate, spiciform - - - **1. L. saligna**
  Stems stout, succulent; inflorescence paniculate - - - - L. SATIVA (p. 1028)

1. L. saligna *L.*, Sp. Plant., ed. 1, 796 (1753); Boiss., Fl. Orient., 3: 810 (1875); Post in Mém. Herb. Boiss., 18: 96 (1900); H. Stuart Thompson in Journ. Bot., 44: 333 (1906); Holmboe, Veg. Cypr., 194 (1914); Post, Fl. Pal., ed. 2, 2: 146 (1933); Osorio-Tafall et Seraphim, List Vasc. Plants Cyprus, 111 (1973); Davis, Fl. Turkey, 5: 777 (1975); Feinbrun, Fl. Palaest., 3: 437, t. 740 (1978).
TYPE: "*in* Gallia, Lipsiae".

Slender erect or spreading perennial (or ? biennial) with a tough, woody rootstock, (7–) 10–65 cm. high; stems simple or branched from near the base, glabrous, pallid, longitudinally sulcate; leaves glabrous, glaucous, oblong or oblanceolate in outline 1–10 (–16) cm. long, 0·5–4·5 cm. wide, usually pinnatifid with a rhomboid-deltoid, acute apex and 2–6, broad, remote, deltoid lateral lobes, sometimes entire and oblanceolate, the base tapering to an ill-defined petiole, the upper sessile, generally with acute, projecting auricles; inflorescence narrow, spiciform, unbranched or with spreading, slender branches; capitula solitary or in remote clusters, with slender, glabrous, bracteate peduncles generally less than 13 mm. long; involucre narrowly cylindrical, 5–12 mm. long at anthesis, accrescent to 20 mm. in fruit, glabrous, glaucous, outermost phyllaries oblong, 3–5 mm. long, 1–2 mm. wide, erect; inner phyllaries linear-oblong, blunt or subacute, to 12 mm. long at anthesis, 1·5–3 mm. wide; florets 4–5, pale yellow (often drying blue); corolla-tube very slender, 4–5 mm. long, villose at apex, ligule 6–7 mm. long, 1–1·5 mm. wide, shortly toothed at apex, involute after anthesis; anthers about 3 mm. long, 0·2 mm. wide, apical appendage small, blunt; style 8–9 mm. long, shortly pilose in the upper half, branches linear, about 0·5 mm. long. Achene narrowly oblong-elliptic, strongly compressed, fuscous-brown, scabridulous, with 7–8 prominent longitudinal nerves on each face, apex acute, beak slender, smooth, pallid, about 6 mm. long, ending in a conspicuous disk 0·3 mm. diam.; pappus white, silky, hairs about 5 mm. long, fragile, remotely scabridulous.

HAB.: Roadsides, gardens, edges of fields and irrigation channels; 1,000–5,600 ft. alt.; fl. July–Oct.

DISTR.: Divisions 2, 5, 7. Widespread in Europe and the Mediterranean region.

2. Between Kato Mylos and Agros, 1937, *Syngrassides* 170! Troödos, 1940, *Davis* 1943!
   Prodhromos, 1949, *Casey* 877! 948! 953! also, 1955, *G. E. Atherton* 549! Nikos, 1956, *Merton*
   2828! Kakopetria, 1955, *N. Chiotellis* 454! 458! also, 1966, *Merton* in ARI 76!
5. Kythrea, 1940, *Davis* 1964!
7. Mylos Forest Station near Kyrenia, 1938, *Syngrassides* 1293!

NOTES: Post remarks that it is common everywhere, but fails to cite a single precise locality. It
is probable that *Lactuca saligna* is more frequent than the above records indicate; it is an
unprepossessing plant, and flowers late in the collecting season.

2. L. serriola *L.*, Cent. Plant. 2, 29 (1756); Osorio-Tafall et Seraphim, List Vasc. Plants Cyprus,
   111 (1973); Davis, Fl. Turkey, 5: 778 (1975); Feinbrun, Fl. Palaest., 3: 436, t. 739 (1978).
      L. scariola L., Sp. Piant., ed. 2, 2: 1119 (1763); Boiss., Fl. Orient., 3: 809 (1875); Holmboe,
   Veg. Cypr., 194 (1914); Post., Fl. Pal., ed. 2, 2: 145 (1933); Lindberg f., Iter Cypr., 35 (1946)
   nom. illeg.
      [*L. aculeata* (non Boiss. et Kotschy) Freyn in Bull. Herb. Boiss., 5: 785 (1897).]
   TYPE: "*in* Europa *australi*".

Erect biennial (25–) 50–100 (–150) cm. high; stem generally unbranched,
pallid, sulcate, setose in the lower part, glabrous above; leaves held in a
vertical plane (edgewise to the stem), at least in sunlight, oblong in outline,
4–10 cm. long, 2–6 cm. wide, pinnatifid or pinnatisect with 4–6, oblong,
subtruncate lateral lobes, and a deltoid or oblong, bluntish terminal lobe,
glaucous, glabrous except for the spinulose margins and undersurface of
midrib; basal leaves tapering to base, but scarcely petiolate, cauline leaves
sessile, auriculate, with projecting blunt or acute auricles; inflorescence
paniculate, much branched in luxuriant specimens, almost racemose in
depauperate specimens, lateral branches often equalling or exceeding the
main axis; capitula small, often rather remote; peduncles slender, bracteate,
up to 2·5 cm. long, but often very short; involucre narrowly ovoid or
cylindric, 8–10 mm. long, 3–4 mm. wide at anthesis, distinctly accrescent in
fruit; phyllaries densely papillose-asperulous, the outermost 2–3 mm. long,
1–1·5 mm. wide, ovate, subacute, the inner 8–10 mm. long, about 2 mm.
wide; florets 8–10 or more, yellow (drying blue); corolla-tube slender, about
4 mm. long, villose at apex, ligule about 5–7 mm. long, 1·5 mm. wide,
minutely dentate at apex; anthers about 2·5 mm. long, 0·3 mm. wide, apical
appendage blunt or subacute; style 6–7 mm. long, branches linear, recurved,
about 0·5 mm. long. Achenes narrowly oblong-obovate, 3·5 mm. long,
1·2–1·5 mm. wide, strongly compressed, light brown, scabridulous or shortly
pilose, with 7–8 prominent nerves on each face, apex acute, beak slender,
smooth, pallid, 3–3·5 mm. long, ending in a conspicuous disk 0·3 mm. diam.;
pappus white, silky, hairs about 5 mm. long, fragile, remotely scabridulous.

HAB.: Cultivated and waste ground, or on stony mountainsides and summits; sea-level to
6,400 ft. alt.; fl. July–Oct.

DISTR.: Divisions 2–7. Widespread in temperate Europe, Asia and the Mediterranean region;
introduced in many areas.

2. Pano Panayia, 1905, *Holmboe* 1064; between Platres and Perapedhi, 1935, *Syngrassides* 757!
   Platres, 1937, *Kennedy* 999! 1000! 1001! 1002! Agros, 1940, *Davis* 1885! Summit of
   Khionistra, 1940, *Davis* 1942! Prodhromos, 1949, *Casey* 937! and, 1955, *G. E. Atherton* 628!
3. Stavrovouni, 1939, *Lindberg f.* s.n.
4. Larnaca, 1893, *Deschamps* 315; Ayia Napa, 1939, *Lindberg f.* s.n.
5. Kythrea, 1939, *Lindberg f.* s.n.!
6. Morphou, 1955, *N. Chiotellis* 476!
7. Kyrenia, Tchingen, 1955, *G. E. Atherton* 299! Kyrenia, 3-mile Beach, 1955, *G. E. Atherton*
   351!

L. SATIVA *L.*, Sp. Plant., ed. 1, 795 (1753); Unger et Kotschy, Die Insel
Cypern, 255 (1865), the Lettuce, thought to be an ancient derivative of *L.
serriola* L., but differing in its smooth stems and leaves, and in the leaf-blade

being held in the normal, horizontal position, has long been cultivated as a salad plant, and is widely grown, and eaten, in Cyprus. The three principal cultivated variants are: var. *capitata* L., the Cabbage Lettuce, grown for its compact, globular "hearts"; var. *longifolia* Lam., the Cos Lettuce, with erect, narrow leaves forming a subcylindric "heart"; and var. *crispa* L., the Curled Lettuce, with crisped, toothed leaves, forming, at most, a loose, open "heart". Modern horticulture has, however, brought into existence many intermediate, connecting variants between these three sorts.

72. SCARIOLA *F. W. Schmidt*
Samml. Phys.-oekon. Aufsätze, 1: 270 (1795).

Biennial and perennial herbs similar to *Lactuca* L.; leaves runcinate-pinnatifid or -pinnatisect, the basal with an ill-defined petiole, the cauline sessile with conspicuously decurrent bases; inflorescence paniculate; capitula homogamous, ligulate; florets few, yellow or yellowish; receptacle naked. Achenes compressed, longitudinally ribbed, scabridulous-hispid, apex attenuate into a short, ill-defined, hispidulous beak; pappus white, silky, hairs scabridulous.

About 5 species in Europe, the Mediterranean region, and eastwards to Central Asia.

Capitula containing 5 (or more) florets; ligules 2·5–3 mm. wide; plant biennial, 35–150 cm. high; branches usually virgate; partial inflorescences spiciform -   -   - **1. S. viminea**
Capitula containing 4 florets; ligules 4 mm. wide; plant perennial, 7–30 cm. high; branches not virgate; inflorescence corymbose -   -   -   -   -   -   - **2. S. tetrantha**

1. **S. viminea** (*L.*) *F. W. Schmidt*, Samml. Phys.-oekon. Aufsätze, 1: 270 (1795); Davis, Fl. Turkey, 5: 783 (1975).
   Prenanthes viminea L., Sp. Plant., ed. 1, 797 (1753).
   Lactuca viminea (L.) J. et C. Presl, Fl. Čechica, 160 (1819); Boiss., Fl. Orient., 3: 818 (1875); Holmboe, Veg. Cypr., 194 (1914); Post, Fl. Pal., ed. 2, 2: 147 (1933); Lindberg f., Iter Cypr., 35 (1946); Osorio-Tafall et Seraphim, List Vasc. Plants Cyprus, 111 (1973); Feinbrun, Fl. Palaest., 3: 438, t. 742 (1978).
   TYPE: "*in* Gallia, Lusitania".

Biennial 35–150 cm. high, with a long, tough taproot; stems erect, slender, stramineous, glabrous, generally with numerous, simple, erecto-patent branches; leaves variable, subglabrous, or thinly or densely hispidulous-pilose, the basal oblong in outline, 6–15 cm. long, 1–6 cm. wide, runcinate-pinnatisect, with an oblong or deltoid, acute, terminal lobe and 8 or more, linear or oblong, dentate lateral lobes; cauline leaves diminishing rapidly upwards, soon reduced to linear or lingulate, bracteoid scales with long-decurrent bases forming interrupted green wings along the stem; overall inflorescence paniculate, partial inflorescences spiciform, the capitula solitary or in small sessile or subsessile groups along the flowering branches; involucre narrowly cylindrical, 10–12 mm. long, 2–2·5 mm. wide, distinctly accrescent in fruit; outer phyllaries ovate, 3–4 mm. long, 1–1·5 mm. wide, minutely ciliolate, apex often tinged purple; inner phyllaries linear, up to 12 mm. long, 1·5 mm. wide; florets yellow (or tinged red externally), opening at midday, rather few (about 5 in each capitulum); corolla-tube very slender, about 7 mm. long, glabrous, ligule 9–10 mm. long, 2·5–3 mm. wide, shortly 5-toothed at apex; anthers yellow, linear, 4–4·5 mm. long, 0·3 mm. wide, apical appendage small, blunt; style about 15 mm. long, setulose in the upper part; branches linear, blunt, about 1·5 mm. long. Achenes strongly compressed, lanceolate, about 12 mm. long, 1·3 mm. wide, blackish, scabridulous, prominently 7-ribbed on each face, apex gradually attenuate into a blackish, ill-defined beak, about as long as the fertile part of the achene, with a conspicuous apical disk; pappus-hairs white, fragile, about 7 mm. long, remotely scabridulous.

HAB.: Roadsides, earthy banks and in garigue on dry hillsides; 1,000–6,250 ft. alt.; fl. July–Oct.

DISTR.: Divisions 2, 7. Widely distributed in the Mediterranean region and east to N. Iraq.

2. Kakopetria, 1937, *Syngrassides* 1674! Platres, 1937, *Kennedy* 1003! 1004! 1005! 1478! 1528! Kryos Potamos, 4,300–5,600 ft. alt., 1937, *Kennedy* 1006! 1526! 1530! Khionistra, 6,250 ft. alt., 1938, *Kennedy* 1519! Damaskinari, Prodhromos, 1949, *Casey* 943! Troödos, 1952, *F. M. Probyn* 136! Platania, 1970, *A. Genneou* in ARI 1604!

7. Above Kythrea, 1,000 ft. alt., 1940, *Davis* 1965! 9 miles from W. end Forest road, Kyrenia, 1955, *G. E. Atherton* 143!

NOTES: All the above refer to typical *S. viminea*, with long, virgate branches and long-decurrent cauline leaves. The following are dwarfer and divaricately branched (agreeing with specimens from outside Cyprus identified as *Lactuca ramosissima* (All.) Gren. et Godr. or *L. viminea* (L.) J. et C. Presl var. *decumbens* Hal.) and sometimes approaching the following, *Scariola tetrantha* (B. L. Burtt et P. Davis) Soják.

2. Xerokolymbos, 5,400 ft. alt., 1937, *Kennedy* 1007! Kryos Potamos, 5,100–5,800 ft. alt., 1937–38, *Kennedy* 998 partly! 1479! 1485! 1523! 1524! Khionistra, 6,200–6,400 ft. alt., 1938–70, *Kennedy* 1484! 1486! 1522! *E. Chicken* 5! Troödos, 5,500–6,400 ft. alt., 1940–63, *Davis* 1940! *Casey* 893! *D. P. Young* 7826!

The status of these is doubtful, nor is there any certainty that they represent a single, homogeneous population. Further investigation of the *Scariola viminea* complex throughout its distribution is required before the problems relating to the Cyprus plants can be solved.

2. S. tetrantha (*B. L. Burtt et P. Davis*) *Soják* in Novit. Bot. et Del. Sem. Hort. Bot. Univ. Carol. Prag., 1962: 46 (1962).
  Lactuca tetrantha B. L. Burtt et P. Davis in Kew Bull., 1949: 106 (1949); Osorio-Tafall et Seraphim, List Vasc. Plants Cyprus, 111 (1973).
  TYPE: Cyprus; Khionistra, 5,000–6,200 ft. alt., on serpentine screes, 1940, *Davis* 1941 (K!).

Erect perennial 7–30 cm. high; taproot thick, tough, commonly branched, especially at apex; basal leaves forming a persistent rosette, commonly stained purple, especially on the undersurface, narrowly obovate in outline, 3–12 cm. long, 0·8–2·5 cm. wide, lyrate-pinnatisect, with a broad, blunt, subacute or rounded terminal lobe, and much smaller, broadly deltoid or oblong, patent or retrorse lateral lobes, glabrous, base tapering to an ill-defined flattened petiole; cauline leaves reduced to small lanceolate-oblong or lingulate, bracteoid scales, bases decurrent, but much less so than in *S. viminea*; inflorescence condensed, often subcorymbose, with numerous, crowded capitula, sometimes lax, with erecto-patent, dichotomous branches; involucre cylindric, often purplish, 14–15 mm. long, 4 mm. wide at anthesis; outermost phyllaries ovate, concave, acute, about 4 mm. long, 3·5 mm. wide, margins glabrous or minutely ciliolate, inner phyllaries narrowly oblong, blunt or subacute, glabrous, about 14–15 mm. long, 2–2·5 mm. wide; florets 4, yellow, the ligules stained coppery-red on the undersurface; corolla-tube glabrous, about 7 mm. long, ligule about 8 mm. long, 4 mm. wide, apex shortly dentate; anthers yellowish, about 5 mm. long, 0·3 mm. wide, apical appendage 0·5 mm. long, blunt or subacute; style about 13 mm. long, setulose in the upper part, branches linear, about 1·5 mm. long. Achenes strongly compressed, lanceolate, about 12 mm. long, 1–1·2 mm. wide, blackish, scabridulous, prominently 7-ribbed on each face, apex gradually attenuate into an ill-defined beak about as long as the fertile part of the achene, with a conspicuous apical disk; pappus-hairs white, fragile, about 7 mm. long, remotely scabridulous.

HAB.: Rocky mountainsides and rock-crevices on serpentine, often near streams; 5,000–6,300 ft. alt.; fl. July–Oct.

DISTR.: Division 2. Endemic.

2. Khionistra, 5,000–6,300 ft. alt., 1937–1957, *Kennedy* 997! 1008! 1483! 1527! *Syngrassides* 1688! *Davis* 1941! *Casey* 936! *F. M. Probyn* 135! *H. R. P. Dickson* 10! Kryos Potamos, 5,100–5,900 ft. alt., 1937–38, *Kennedy* 998! 1480! 1482! 1520! 1521! 1525! 1529! 1531! *Barclay* 1105!

73. PRENANTHES *L.*

Sp. Plant., ed. 1, 797 (1753).
Gen. Plant., ed. 5, 349 (1754) (*"Prenantes"*).

Perennial herbs, sometimes ligneous at base; stems erect, glabrous, or glandular-setose; leaves oblong or elliptic in outline, entire, cordate-sagittate or lyrate-pinnatisect, the basal usually with flattened petioles, the cauline sessile or subsessile, sometimes amplexicaul; inflorescence corymbose, paniculate or occasionally spiciform; capitula homogamous, ligulate; involucre usually narrow, cylindric; phyllaries unequal, in several series, not much accrescent in fruit; florets generally blue, mauve or white, seldom yellow; anthers sagittate at base; style-branches slender; receptacle naked. Achenes fusiform or subcylindric, sometimes a little compressed, ribbed, smooth or scabridulous, beak wanting or shorter than fertile part of the achene; pappus white or yellowish, the hairs fragile, scabridulous or barbellate.

About 40 species widely distributed in temperate and tropical regions of the Old World, also in temperate North America.

1. **P. triquetra** *Labill.*, Icon. Plant. Syr. Rar., 3: 4, t. 2 (1809).
   Lactuca triquetra (Labill.) Boiss., Fl. Orient., 3: 819 (1875); Post, Fl. Pal., ed. 2, 2: 147 (1933); A. K. Jackson et Turrill in Kew Bull., 1938: 465 (1938); Osorio-Tafall et Seraphim, List Vasc. Plants Cyprus, 111 (1973).
   TYPE: "in Libano ad montium radices", *Labillardière* (FI, K!).

Erect, scoparioid perennial 40–80 cm. high; rootstock tough, woody, stems sharply triquetrous, pith-filled, glaucous-green, dry and hard to the touch, soon becoming leafless; branches numerous, erect, slender; leaves (when present) sparse, glabrous, the basal narrowly oblong-obovate, 2–4 cm. long, 0·5–1 cm. wide, apex acute or rounded, margins entire or remotely dentate, base tapering to an ill-defined petiole; cauline leaves sessile, linear-oblanceolate, 1·5–7 cm. long, 0·3–0·8 cm. wide; inflorescence spiciform; capitula subsessile, usually solitary, remote; peduncles very short, covered with imbricate bracts; involucre cylindric, 12–14 mm. long, 2·5–3 mm. wide; outermost phyllaries ovate, subacute, about 3 mm. long, 2 mm. wide, glabrous with purplish, membranous margins; innermost phyllaries narrowly oblong, to 14 mm. long, 2·5–3 mm. wide, subacute, glabrous with purplish membranous margins; florets usually 5, lilac-blue; corolla-tube glabrous, slender, about 7 mm. long; ligule 10–11 mm. long, 3·5 mm. wide, deeply 5-dentate at apex; anthers yellow, 6–7 mm. long, 0·3 mm. wide, apical appendage short, blunt; style 17–18 mm. long, setulose in the upper part, branches filiform, 2·5 mm. long. Achene narrowly fusiform, about 5 mm. long, 1 mm. wide, almost smooth or minutely scabridulous, with numerous, prominent longitudinal ribs; apex tapering, but not beaked, ending in a distinct disk; pappus rather persistent, whitish or yellowish, hairs fragile, 6–7 mm. long, scabridulous or shortly barbellate.

HAB.: On chalky cliffs, or in flushes on serpentine; 1,500 ft. alt.; fl. Sept.–Oct.
DISTR.: Division 3, very rare. Also Lebanon.

3. Above Yerasa, 1937, *Syngrassides* 1698! Limassol Forest, 1,500 ft. alt., 1956, *Poore* 24!

NOTES: Apparently very local, but flowering late in the season, and perhaps less uncommon than these two records would suggest. The generic status of *P. triquetra* is obscure: it is certainly not a *Lactuca*, except in the widest sense, and fits uncomfortably into *Prenanthes*; further work on the tribe *Lactuceae* may find a better home for it.

74. REICHARDIA Roth*
Bot. Abhandl., 35 (1787).

Erect annual or perennial herbs; stems 1 to many; leaves alternate, entire to deeply pinnatisect, the lobes ciliate-denticulate, often with crisped-undulate margins; cauline leaves usually amplexicaul; capitula few to many, homogamous, ligulate, long-pedunculate, peduncle often swollen just beneath capitulum; phyllaries multiseriate, imbricate, the inner lanceolate, subequal, unaltered after anthesis, the outer shorter and broader with scarious margins, often grading into the bracts of the peduncle; receptacle flat, without scales; ligules yellow, often with purplish stripe on outer face, sometimes purplish at base, apex truncate, usually 5-toothed; anthers with sagittate bases; style-arms slender. Achenes of 2 types: the outer 4–5 angled and transversely rugose-verrucose, glabrous with a slightly contracted base, the apex slightly constricted beneath the projecting rim; the inner paler, usually smooth; pappus hairs white, copious, multiseriate, soft, simple, united at base into a ring, caducous.

About 6 species, mostly found in Europe, the Orient and N. Africa, but also extending to the Canary Is., India, Australia and Polynesia, Socotra and N.E. Tropical Africa.

Plant perennial; phyllaries with narrow margins (up to 0·5 mm. wide, usually less)
                    1. R. picroides

Plant annual; phyllaries with broad margins (1–1·5 mm.) wide:
Scape bracteate; plant up to 43 cm. tall - - - - - - **2. R. intermedia**
Scape ebracteate; plant small, usually less than 10 cm. tall- - - - **3. R. tingitana**

1. **R. picroides** (*L.*) *Roth*, Bot. Abhandl., 35 (1787); Post, Fl. Pal., ed. 2, 2: 151 (1933); Osorio-Tafall et Seraphim, List Vasc. Plants Cyprus, 111 (1973); Davis, Fl. Turkey, 5: 694 (1975).
    Scorzonera picroides L., Sp. Plant., ed. 1, 792 (1753).
    Sonchus picroides (L.) Lam., Encycl. Méth., 3: 398 (1791); Sibth. et Sm., Fl. Graec. Prodr., 2: 126 (1813), Fl. Graec., 8: 68, t. 793 (1833).
    Picridium vulgare Desf., Fl. Atlant., 2: 221 (1799); Poech, Enum. Plant. Ins. Cypr., 21 (1842); Unger et Kotschy, Die Insel Cypern, 255 (1865); Boiss., Fl. Orient., 3: 828 (1875); Rechinger f. in Arkiv för Bot., ser. 2, 1: 436 (1950).
    [*Sonchus chondrilloides* (non Desf.) Sm. in Sibth. et Sm., Fl. Graec. Prodr., 2: 125 (1813), Fl. Graec., 8: 67, t. 791 (1833).]
    TYPE: Not stated.

Erect perennial herb 6·5–31 cm. high, with a branched woody rootstock, growing in clumps, leafy for half the length of the stem; stems long, ribbed, slender, branched, sparsely papillose near nodes, with several bracts scattered along peduncle from topmost node to base of involucre, where they are crowded and grade into involucral bracts; basal leaves obovate in outline, 2·5–11·5 cm. long, 0·3–3·2 cm. wide; petiole broadly winged, causing leaf to appear sessile, or leaf tapering at base into petiole, the latter often equalling the length of the leaf-blade; apex obtuse to rounded, spinulose; margins inconspicuous, cartilaginous, often sparsely spinulose, regularly or irregularly dentate or pinnatifid; cauline leaves amplexicaul, ovate-lanceolate, 6·8–15·2 cm. long, 2·4–4·4 cm. wide, decreasing in size upwards, obtuse to acuminate, margins entire, spinulose, sometimes pinnatifid, especially towards base of stem; peduncles usually more than 5 cm. long, often as much as 15 cm., bracts ovate-cordate, acute to acuminate, 2–4 mm. long, with narrow scarious margins; capitula solitary, terminal, never subsessile nor clustered at centre of plant, the receptacle swollen in the fruiting state, the base spongy in appearance, 1·2–1·3 cm. diam.; phyllaries 3-seriate, overlapping in flowering state, but becoming spaced or barely

* by P. Halliday

overlapping as achenes develop, usually pouched at the tip; with a cushion of minute hairs inside the pouch, otherwise glabrous; outer and middle phyllaries ovate-lanceolate, 4–6 mm. long, 2–3 mm. wide with scarious margins usually less than 0·5 mm. wide, inner phyllaries lanceolate with narrower scarious margins, (8–) 12–13 mm. long, 2·5–3 mm. wide; florets yellow, exceeding involucre in length; tube slender, ⅓ to ½ the length of the floret, densely hairy near the mouth externally; ligule strap-shaped, often slightly widened at base, about 10 mm. long, 1·5–2 mm. wide, with 4–5 veins; anthers linear or oblong, 3·5–4 mm. long, with a well-developed collar at base, tails very short, adpressed, appendages oblong with rounded apex and no constriction at junction with thecae; style forked for at least 1·5 mm., with the upper 4–5 mm. minutely pubescent. Outer achenes subcylindrical or oblong, somewhat curved longitudinally, ± 3 mm. long, 1 mm. wide, transversely rugose-verrucose, truncate with distinct apical rim and scar, inner achenes subcylindrical, but broader at the base, 4·5 mm. long, 1 mm. wide; pappus 6·5–8 mm. long.

HAB.: Sea-cliffs and rocky coastal pastures; fallow fields near the sea, with *Sarcopoterium*; waste ground; sea-level to 150 ft. alt.; fl. Febr.–Nov.

DISTR.: Divisions 7, 8, rare. S. Europe and Mediterranean region east to Syria, Lebanon and Egypt.

7. Kharcha, 1950, *Chapman* 269! Kyrenia Castle and sea-cliffs 3 miles E. of Kyrenia, 1951–1958, *Casey* 1066! *G. E. Atherton* 203! *P. H. Oswald* 23! ¼ mile W. of Kyrenia, 1955, *G. E. Atherton* 749!

8. Rizokarpaso, 1912, *Haradjian* 248!

2. R. intermedia (*Sch. Bip.*) Coutinho, Fl. Port., 676 (1913); Post, Fl. Pal., ed. 2, 2: 151 (1933); Osorio-Tafall et Seraphim, List Vasc. Plants Cyprus, 111 (1973); Davis, Fl. Turkey, 5: 697 (1975); Feinbrun, Fl. Palaest., 3: 439, t. 744 (1978).
    Picridium intermedium Sch. Bip. in Webb et Berth., Phyt. Canar., 2, 2: 451 (1849–50); Boiss., Fl. Orient., 3: 828 (1875) pro parte; Holmboe, Veg. Cypr., 194 (1914) pro parte quoad nomen.
    [*P. vulgare* (non Desf.) Sintenis in Oesterr. Bot. Zeitschr., 31: 393 (1881).]
    TYPE: "vulgaris est in insulis Canariensibus: Webb! *Infierno Adexe, Fuenta del Orotava*. Teneriffae. Specimina magna et pulchra in *Barranco de las Augustias*. Palmae sec, Buch, l.c., p. 175. *Fuertaventura*: Bourgeau! n. 9"

Erect, somewhat glaucous, glabrous or sparsely papillose annual up to 43 cm. high, leafy near base only; stems long, ribbed, slender, branched, the ultimate divisions sparsely bracteate with top of scape usually naked; basal leaves as in *R. picroides* but more regularly toothed and less deeply divided, 3·2–9·1 cm. long, 1·4–3·2 cm. wide, tapering abruptly to a winged petiole; apex acute to obtuse, spinulose, margins spinulose, more noticeably so than in *R. picroides*; cauline leaves 4·5–7·2 cm. long, 0·6–1·4 cm. wide, decreasing in size upwards, lanceolate, acuminate, spinulose, sometimes minutely so, or the lower cauline leaves sometimes shallowly toothed with an obtuse apex, resembling basal leaves; peduncles long, swollen beneath capitulum, often up to 21 cm. long, with bracts 4–8 mm. long, fewer than in *R. picroides*, well spaced, never crowded beneath involucre; capitula as in *R. picroides*, but receptacle not particularly swollen in fruiting state; phyllaries 3-seriate, overlapping, entire, glabrous, with conspicuous (in dried material) scarious margins c. 1 mm. wide, sometimes more, the dark central zone forming a linear central stripe, pouched at the tip, with or without apical cushions of minute hairs; outer phyllaries ovate, cordate, 3–4 mm. long, 2 mm. wide; middle phyllaries oblong or obovate, 5–7 mm. long, 2 mm. wide; inner phyllaries linear-lanceolate, 12–13 mm. long, 2–3 mm. wide; florets yellow or purplish on outer surface or at apex; tube about ½ the length of the floret, long-hairy externally; ligule strap-shaped or elliptic, gradually tapering to base, 8–9 mm. long, 1–2 mm. wide, with (2–) 5 teeth at apex, and 4 (–5) veins;

anthers linear, oblong, 1·5–2·5 mm. long, with inconspicuous collars at the
sagittate base, the tails short, obtuse, appendages oblong-ovate without
constriction at base; style forked for up to 2 mm., the top 4 mm. pubescent.
Outer achenes cylindrical, transversely rugose-verrucose, 3–4 mm. long,
1–1·5 mm. wide; inner achenes occasionally with a few slight transverse
rugosities, curved and cylindrical, but with a slight beak and abruptly
widened at base, the extremity inrolled to a central cavity, 5–6 mm. long,
0·75–1·5 mm. wide (at base); pappus 9–10 mm. long.

HAB.: Rocky ground on calcareous hillsides; 500–2,000 ft. alt.; fl. March–May.

DISTR.: Divisions 5, 7, rare. Mediterranean region; Atlantic Islands.

5. Between Mia Milea and Mandres, 1933, *Syngrassides* 325!
7. Pentadaktylos, 1880, *Sintenis & Rigo* 287! N.W. of Bellapais, 1952, *Casey* 1273!

3. R. tingitana (*L.*) *Roth*, Bot. Abhandl., 35 (1787); Post, Fl. Pal., ed. 2, 2: 151 (1933); Feinbrun,
  Fl. Palaest., 3: 439, t. 794 (1978).
    Scorzonera tingitana L., Sp. Plant., ed. 1, 791 (1753).
    Scorzonera orientalis L., Sp. Plant., ed. 2, 1113 (1763).
    Sonchus tingitanus (L.) Lam., Encycl. Méth., 3: 397 (1791); Sibth. et Sm., Fl. Graec.
  Prodr., 2: 126 (1813), Fl. Graec., 8: 67, t. 792 (1833).
    Picridium tingitanum (L.) Desf., Fl. Atlant., 2: 220 (1799); Poech, Enum. Plant. Ins.
  Cypr., 21 (1842); Sch. Bip. in Webb et Berth., Phyt. Canar., 3: 2, 2: 451 (1849–50); Unger et
  Kotschy, Die Insel Cypern, 254 (1865); Boiss., Fl. Orient., 3: 827 (1875).
    P. orientale (L.) DC., in Lam. et DC., Fl. Franç., 4: 16 (1805); Holmboe, Veg. Cypr., 194
  (1914).
    P. hispanicum Poir. in Lam., Encycl. Méth., Suppl., 4: 410 (1816).
    P. tingitanum (L.) Desf. var. *minus* Boiss., Fl. Orient., 3: 828 (1875).
    Reichardia tingitana (L.) Roth var. *orientalis* (L.) Aschers. et Schweinf., Ill. Fl. Egypte,
  100 (1887).
    R. orientalis (L.) Hochr., in Ann. Cons. Jard. Bot. Genève, 7–8: 238 (1904); Osorio-Tafall
  et Seraphim, List Vasc. Plants Cypr., 111 (1973).
  TYPE: Tangier; "*in* Tingide".

Erect annual, 2·5–10 (–19·3) cm. high; stems 1 or more, ribbed, stout,
branched or unbranched, variably white-papillose or glabrous, or plant
almost stemless; leaves pale grey-green, usually forming a pseudo-rosette,
the basal most commonly obovate in outline, 15–80 mm. long, 4–16 mm.
wide; leaf-blade tapering to winged petiole or subpetiolate with a widened
clasping base (sometimes the petiole is so broadly winged that the leaf
appears to be sessile); apex obtuse to acute, spinulose, margins subentire or
more usually regularly or irregularly, shallowly or coarsely dentate to
pinnatifid, lobes toothed and spinulose, surrounded by a narrow, usually
white, cartilaginous border from which arise white spinules of varying
length; cauline leaves sessile, ovate, lanceolate or oblong, 5–14·2 cm. long,
2·2–6·5 cm. wide, the uppermost much smaller, with broad, clasping, often
auricled, bases; apex obtuse to acuminate, margins shallowly or more deeply
toothed or lobed, furnished with spinules like basal leaves; midrib often
densely white-spinulose, sometimes with scattered multicellular hairs
towards the base of the upper surface; peduncles 0·9–5·6 cm. long to first
node, 4–5 mm. diam. at widest point, ebracteate except for 1 or 2 bracts at
extreme tip; bracts with scarious margins, glabrous; capitula usually
solitary, on an unbranched peduncle, or at ends of branches; involucre
cylindrical, about 1 cm. long, phyllaries erect, commonly 3–4-seriate,
glabrous, or rarely the outer and middle series pubescent in the upper half,
all with conspicuous scarious margins to 1·5 mm. wide, and with a cushion of
minute hairs inside at base of mucro; outer phyllaries ovate-cordate, 5–6
mm. long, 4–6 mm. wide, acute; middle phyllaries oblong, emarginate,
mucronate or cuspidate, 8–10 mm. long, 4–6 mm. wide; inner phyllaries
lanceolate, 10–14 mm. long, 2–4 mm. wide; florets golden-yellow with purple
bases, tube of floret very slender, $\frac{1}{3}$ to $\frac{1}{2}$ the floret-length, apex of tube densely

hairy on outer surface; ligule elliptic, with 3–5 apical teeth and 5 veins, about 15 mm. long, 1·5–2 mm. wide; stamens inserted just below mouth of tube; anthers linear to narrowly elliptic, 2·5–3 mm. long, with well-developed anther-collars at base, tails closely adpressed, abruptly tapering, acute; appendages oblong, obtuse or rounded, not constricted at junction with thecae; style forked, the branches usually spreading, the apical 2·5–4 mm. slightly swollen and minutely pubescent. Outer achenes subcylindrical or oblong, somewhat curved lengthwise, 2·5–3 mm. long, 1 mm. wide, transversely rugose-verrucose, truncate, with slight but distinct collar or rim and scar; inner obconical, smooth to slightly wrinkled, 3·5–4 mm. long, 1–1·5 mm. wide; pappus 7·5–10 mm. long.

HAB.: Dry, rocky pastures; pathsides and streamsides; gravelly seashores; sea-level to 2,000 ft. alt.; fl. March–May.

DISTR.: Divisions, 1, 2, 4, 5, 7, 8. S. Europe and Mediterranean region east to India; also Socotra, Australia, Polynesia and E. tropical Africa S. to Tanzania.

1. Karavopetres near Ayios Yeoryios, 1962, *Meikle* 2066!
2. Karavostasi to Lefka, 1935, *Syngrassides* 812!
4. Shore at Ormidhia, 1787, *Sibthorp*; Cape Pyla, 1880, *Sintenis*.
5. Between Mia Milea and Mandres, 1933, *Syngrassides* 309! Nicosia–Kyrenia road, Mile 7, 1954, *Casey* 1328!
7. Above Kythrea, 1880, *Sintenis*; Pentadaktylos, 1880, *Sintenis & Rigo* 286! Between Photta and Lapithos, 1902, *A. G. & M. E. Lascelles* s.n.!
8. Cape Andreas, 1880, *Sintenis*.

NOTES: Examination of a wide range of specimens shows that the distinctions between the condensed, small-flowered *R. orientalis* (L.) Hochr. and the taller, large-flowered *R. tingitana* cannot be maintained.

75. SONCHUS *L.*
Sp. Plant., ed. 1, 793 (1753).
Gen. Plant., ed. 5, 347 (1754).
L. Boulos in Bot. Notiser, 125: 287–319 (1972); 126: 155–196 (1973); 127: 402–451 (1974).

Annual, biennial or perennial, lactiferous herbs and subshrubs; stems terete or somewhat angled, glabrous, tomentellous or glandular; leaves entire, pinnatifid or pinnatisect, generally auriculate at base, margins often denticulate or spinulose; inflorescence usually branched; capitula homogamous, ligulate; involucre campanulate or cylindric, often conical after anthesis; phyllaries in several series, the innermost often submembranous; florets yellow, occasionally stained blue externally; anthers deep yellow or brownish, sagittate at the base; style-branches slender; receptacle naked, alveolate. Achenes erostrate, oblanceolate or subelliptic, compressed, 1–5-ribbed; pappus dimorphic, some of the hairs caducous, relatively thick, scabridulous, the rest persistent, cottony, white.

About 54 species, in temperate Europe and Asia, also temperate and tropical Africa.

Capitula inconspicuous, usually less than 1·5 cm. diam.; plant with a vertical taproot:
  Achenes compressed but without a winged margin:
    Lateral lobes of pinnatisect leaf oblong or oblong-deltoid, or sometimes almost suppressed, commonly spinulose-denticulate - - - - - - - **1. S. oleraceus**
    Lateral lobes of pinnatisect leaf narrowly oblong, lorate or almost filiform, generally entire
      3. S. tenerrimus
  Achenes almost flattened, with a distinct marginal wing - - - - **2. S. asper**
Capitula showy, 4–5 cm. diam.; plant with a horizontal, creeping, brittle rhizome
S. ARVENSIS (p. 1039)

1. **S. oleraceus** *L.*, Sp. Plant., ed. 1, 794 (1753); Gaudry, Recherches Sci. en Orient., 198 (1855); Boiss., Fl. Orient., 3: 795 (1875); Sintenis in Oesterr. Bot. Zeitschr., 31: 193 (1881); H.

Stuart Thompson in Journ. Bot., 44: 333 (1906); Holmboe, Veg. Cypr., 194 (1914); Post, Fl. Pal., ed. 2, 2: 144 (1933); Lindberg. f., Iter Cypr., 36 (1946); Boulos in Bot. Notiser, 126: 155 (1973); Osorio-Tafall et Seraphim, List Vasc. Plants Cypr., 111 (1973); Davis, Fl. Turkey, 5: 691 (1975); Feinbrun, Fl. Palaest., 3: 434, t. 734 (1978).

S. ciliatus Lam., Fl. Franç., ed. 1, 2: 87 (1778); Unger et Kotschy, Die Insel Cypern, 254 (1865).

S. oleraceus L. var. *triangularis* Wallr., Sched. Crit. Plant. Fl. Hal., 1: 432 (1822); Druce in Rep. B.E.C., 9: 470 (1931).

S. oleraceus L. f. *albescens* Neuman, Sverig. Fl. 57 (1901).

S. oleraceus L. var. *ciliatus* (Lam.) Druce, British Plant List, ed. 2, 74 (1928), Rep. B.E.C., 9: 470 (1931).

S. oleraceus L. var. *albescens* (Neuman) Druce, British Plant List, ed. 2, 74 (1928), Rep. B.E.C., 9: 470 (1931).

TYPE: "*in Europae cultis*".

Erect annual or biennial 10–150 cm. high; stems fistulose, glabrous, obscurely sulcate, often glaucescent, the uppermost nodes sometimes glandular, branches generally few; basal leaves narrowly obovate or oblong in outline, 6–30 cm. long, 2–12 cm. wide, usually pinnatisect, with a broad deltoid or suborbicular, entire or spinulose-dentate terminal lobe and 4–6 (or more) oblong, acute or obtuse, entire or spinulose-dentate lateral lobes, or occasionally with the lateral lobes almost suppressed; petiole ill-defined, winged, to about 10 cm. long; cauline leaves similar, glabrous, glaucescent, sessile, the base generally dilated into conspicuous, amplexicaul, acute or obtuse, entire or denticulate, projecting auricles; inflorescence sparingly and irregularly branched, often subcorymbose; peduncles 0·5–2·5 cm. long at anthesis, lengthening to 6 cm. or more in fruit, glabrous or glandular, commonly floccose-tomentellous, especially near apex; involucre campanulate (or conical immediately after anthesis) 8–12 mm. long, 6–10 mm. wide; phyllaries at first erect, imbricate, becoming reflexed after the fruits are shed, glabrous or glandular, often somewhat livid, the outer narrowly deltoid, 3–5 mm. long, 1·5–2 mm. wide, the inner linear-lanceolate, up to 12 mm. long, 2 mm. wide; florets yellow (sometimes pale creamy-yellow); corolla-tube slender, about 7 mm. long, thinly lanuginose towards apex; ligule about 6 mm. long, 1 mm. wide, obscurely 3–5-dentate at apex; anthers about 2·5 mm. long, 0·3 mm. wide, apical appendage bluntly ovate, livid; style about 12 mm. long, minutely setulose towards apex, branches linear-filiform, livid, about 1·2 mm. long. Achene pale brown, narrowly obovate or oblanceolate, about 3 mm. long, 1 mm. wide, strongly compressed, scabridulous, 3-nerved on each face, not margined; pappus white, cottony, rather tousled, hairs up to 7 mm. long, minutely and remotely scabridulous.

HAB.: Gardens, cultivated fields, roadsides, often in moist situations; sea-level to 5,000 ft. alt.; fl. Febr.–Oct.

DISTR.: Divisions 2–8. A cosmopolitan weed.

2. Troödos, 1912, *Haradjian* 492! Sitjes, 1939, *Lindberg f.* s.n.; Kato Platres, 1939, *Lindberg f.* s.n.; Kakopetria, 1955, *N. Chiotellis* 440! Prodhromos, 1955, *G. E. Atherton* 606!
3. Limassol, 1859, *Kotschy* 459; Zakaki, 1955, *Merton* 2075!
4. Larnaca, 1859, *Kotschy* 459a; also, 1928 or 1930, *Druce* s.n.; Famagusta, 1928 or 1930, *Druce* s.n.; Ayia Napa, 1939, *Lindberg f.* s.n.
5. Kythrea, 1880, *Sintenis & Rigo* 288! also, 1932, *Nattrass* 226; and, 1936, *Syngrassides* 933! Between Dhikomo and Sykhari, 1936, *Syngrassides* 1189!
6. Nicosia, 1928 or 1930 *Druce* s.n.; also, 1950, *Chapman* 460! Myrtou, 1928 or 1930, *Druce* s.n.; and, 1955, *G. E. Atherton* 416!
7. Kyrenia, 1932–1956, *Nattrass* 191; *Casey* 124! *Miss Mapple* 40! *G. E. Atherton* 104! 321! 1001!
8. Rizokarpaso, 1941, *Davis* 2364!

NOTES: Boulos (Bot. Notiser, 126: 158: 1973) discusses the polymorphic nature of *S. oleraceus*, and concludes that the infraspecific taxa, recognized by Wallroth, Druce and others, are not worth maintaining; the variability supports the hypothesis that *S. oleraceus* originated as a hybrid between *S. asper* (L.) Hill and *S. tenerrimus* L.

2. S. asper (*L.*) *Hill*, Herb. Brit., 1: 47 (1769); Boiss., Fl. Orient., 3: 796 (1875); Druce in Rep. B.E.C., 9: 470 (1931); Post, Fl. Pal., ed. 2, 2: 144 (1933); Lindberg f., Iter Cypr., 36 (1946); Rechinger f. in Arkiv för Bot., ser. 2, 1: 434 (1950); Boulos in Bot. Notiser, 126: 164 (1973); Osorio-Tafall et Seraphim, List Vasc. Plants Cyprus, 111 (1973).

    S. oleraceus L. var. γ *asper* L., Sp. Plant., ed. 1, 794 (1753).

Erect annual or biennial 10–120 cm. high; stems sometimes angled, obscurely sulcate, glaucescent, glabrous or glandular in the upper part; basal leaves narrowly oblong-obovate in outline, 6–21 cm. long, 1·5–8 cm. wide, glabrous, glaucous or green, irregularly pinnatisect, pinnatifid or sinuate-dentate, apex blunt or acute, margins rigidly spinulose or sometimes merely hispidulous; petiole winged, ill-defined; cauline leaves similar, but often more rigid, with deeply lobed spinulose margins and an amplexicaul-auriculate base; inflorescence irregularly corymbose or subumbellate; capitula 6 or more, often rather crowded; peduncles 1–2 cm. long at anthesis, glabrous or glandular, lengthening to 4 cm. or more in fruit; involucre campanulate, about 10 mm. long, 8–10 mm. wide, becoming incrassate and spongy in fruit; outer phyllaries lanceolate, 3–4 mm. long, 0·8–1·5 mm. wide, glabrous or glandular; inner phyllaries linear-acuminate, to 10 mm. long, about 1·5 mm. wide; florets yellow, the ligules often stained purple externally; corolla-tube slender, about 6 mm. long, thinly lanuginose towards apex; ligule about 6 mm. long, 0·8 mm. wide, minutely 3–5-dentate at apex; anthers about 1·5 mm. long, 0·2 mm. wide, apical appendage acute, blackish; style about 8 mm. long, minutely setulose towards apex, branches linear-filiform, livid, about 0·8 mm. long. Achenes subelliptic, flattened, brown, about 3 mm. long, 1·8 mm. wide, almost smooth, with a conspicuously winged margin and 3 prominent longitudinal ribs on each face, the margins and ribs smooth or minutely, retrorsely ciliolate; pappus white, cottony, rather tousled, hairs 8–10 mm. long, sparsely and remotely scabridulous.

ssp. **asper**
    TYPE: "*in* Europae *cultis*".

Leaves often shining dark green, rarely glaucescent; margins and ribs of achene not retrorsely ciliolate. Annual (or ? biennial).

    HAB.: Cultivated and waste ground; 500–3,000 ft. alt.; fl. Febr.–Oct.

    DISTR.: Divisions 1, 2, 6. A cosmopolitan weed.

1. Stroumbi, 1913, *Haradjian* 731; Lyso, 1913, *Haradjian* 870; 891.
2. Platania, 1939, *Lindberg f.* s.n.
6. Nicosia, 1928 or 1930, *Druce* s.n.

    NOTES: No specimens of *S. asper* ssp. *asper* have been seen, and it is possible that some, or all, of the above records should be referred to *S. asper* ssp. *glaucescens* (Jordan) Ball (infra).

ssp. **glaucescens** (*Jordan*) *Ball* in Journ. Linn. Soc. Bot., 16: 548 (1876); Boulos in Bot. Notiser, 126: 165 (1973); Davis, Fl. Turkey, 5: 691 (1975).
    S. nymanii Tineo et Guss. in Guss., Fl. Sic. Syn., 2: 860 (1844); Post, Fl. Pal., ed. 2, 2: 144 (1933); Feinbrun, Fl. Palaest., 3: 435, t. 736 (1978).
    S. glaucescens Jordan, Obs. Plant. Crit., 5: 75, t. 5 (1847); Boiss., Fl. Orient., 3: 796 (1875).
    S. sp. (forsan *S. Nymani* Tin.) Rechinger f. in Arkiv för Bot., ser. 2, 1: 434 (1950).
    [*S. asper* (non (L.) Hill) A. K. Jackson in Kew Bull., 1937; 344 (1937).]
    TYPE: France; "il croît sur les rochers maritimes, aux îles d'Hyères, à Portquerolle, et à Ste.-Marguerite près Toulon (LY)".

Leaves generally glaucescent, often rigidly spinulose along the margins; margins and ribs of achenes ciliolate with minute retrorse bristles. Biennial.

    HAB.: Cultivated ground; water meadows; ditches, often in marshy places; sea-level to 2,000 ft. alt.; fl. Febr.–Oct.

DISTR.: Divisions 1, 2, 5, 7. S. Europe, Mediterranean region, Atlantic Islands; introduced elsewhere.

1. Prodhromi near Polis, 1937, *Syngrassides* 1712!
2. Near Galata, 1913, *Haradjian* 1000!
5. Dhikomo marshes, 1936, *Syngrassides* 1032! Kythrea, 1936, *Syngrassides* 1287!
7. 1 mile E. of Kyrenia, 1956, *Merton* 2769! Kambyli, 1957, *Merton* 2978!

S. asper (*L.*) *Hill* ssp. **glaucescens** (*Jordan*) *Ball* × **S. oleraceus** *L.*

Boulos has determined *Casey* 1206, collected, 1951, at Prodhromos (2) as possibly this hybrid. It has the spinulose-margined leaves of *S. asper* ssp. *glaucescens*, but the imperfectly formed (and probably sterile) achenes are narrow, and without the marginal wing of *S. asper*, much more resembling those of *S. oleraceus*.

3. S. tenerrimus *L.*, Sp. Plant., ed. 1, 794 (1753); Sibth. et Sm., Fl. Graec. Prodr., 2: 125 (1813), Fl. Graec., 8: 66, t. 790 (1833); Poech, Enum. Plant. Ins. Cypr., 21 (1842); Unger et Kotschy, Die Insel Cypern, 254 (1865); Boiss., Fl. Orient., 3: 797 (1875); Sintenis in Oesterr. Bot. Zeitschr., 31: 193 (1881); Holmboe, Veg. Cypr., 194 (1914); Post, Fl. Pal., ed. 2, 2: 144 (1933); Rechinger f. in Arkiv för Bot., ser. 2, 1: 434 (1950); Boulos in Bot. Notiser, 126: 158 (1973); Osorio-Tafall et Seraphim, List Vasc. Plants Cyprus, 111 (1973); Davis, Fl. Turkey, 5: 692 (1975); Feinbrun, Fl. Palaest., 3: 434, t. 735 (1978).
TYPE: "Monspelii, Florentiae".

Annual, biennial or perennial; stems 20–100 cm. high, terete or angled near the base, shallowly sulcate, glabrous or sparsely glandular towards apex, often purplish towards base, unbranched or branched from near the base; basal leaves often withered at anthesis, oblong in outline, 5–25 cm. long, 1–10 cm. wide, glabrous, glaucous, deeply and rather irregularly pinnatisect into numerous, oblong or lorate, entire or denticulate lobes; petiole ill-defined, flattened, sheathing; cauline leaves similar, but sometimes divided into very narrow almost filiform lobes, base amplexicaul-auriculate, with acute, projecting auricles; inflorescence rather irregularly branched, subcorymbose; capitula numerous, often rather crowded; peduncles 1–2 cm. long at anthesis, often glandular, elongating a little in fruit; involucre campanulate, 10–12 mm. long, 6–9 mm. wide, often floccose-tomentose at base; phyllaries reflexing after the fruits are shed, glabrous or thinly glandular, the outer about 4 mm. long, 1·5 mm. wide, the inner up to 12 mm. long, 1·5 mm. wide, rather blunt at apex; florets yellow; corolla-tube slender, very thinly lanuginose towards apex, about 7 mm. long, ligule about 4 mm. long, 0·8 mm. wide, minutely 5-dentate at apex, anthers 1·5 mm. long, 0·2 mm. wide, apical appendage subacute, pallid; style 9–10 mm. long, minutely setulose towards apex, branches livid, filiform, about 0·8 mm. long. Achenes oblanceolate, about 3·5 mm. long, 1 mm. wide, strongly compressed but without marginal wings, bright brown, conspicuously scabridulous-rugulose, tapering to a very narrow base; pappus white, cottony, rather tousled, hairs up to 8 mm. long, minutely and remotely scabridulous.

HAB.: In grassland; field borders; slacks in sand-dunes and in open, sandy Pine forest; shingly stream-beds; sea-level to 2,000 ft. alt.; fl. Febr.–May.

DISTR.: Divisions 1, 4, 5, 7, 8. Mediterranean region and eastwards to Iran; Atlantic Islands; Sudan; Ethiopia; probably an introduced weed elsewhere.

1. Between Androlikou and Prodhromi, 1957, *Merton* 3033! Ayios Yeoryios Island, 1962, *Meikle* 2153!
4. Larnaca, 1862, *Kotschy* 255, also, 1880, *Sintenis*; Famagusta, 1912, *Haradjian* 77.
5. Salamis, 1957, *Merton* 3207!
7. Armenian Monastery, 1912, *Haradjian* 345.
8. Cape Andreas, 1880, *Sintenis & Rigo* 289! also, 1938, *Syngrassides* 1782!

S. ARVENSIS *L.*, Sp. Plant., ed. 1, 793 (1753) a perennial with slender, brittle, creeping rhizomes, and large, showy capitula often 4–5 cm. diam., of bright golden-yellow florets, is recorded by Kotschy (in Unger et Kotschy, Die Insel Cypern, 254; 1865) "Am Capo Greco [1862, *Kotschy*] n. 159". No other collector has found *Sonchus arvensis* — primarily a plant of north-west Europe — in Cyprus, and an error is likely, though it is difficult to imagine with what other plant *S. arvensis* could be confused.

76. LAUNAEA *Cass.*

Dict. Sci. Nat., 25: 61, 321 (1822).

Annual or perennial herbs, or occasionally spiny subshrubs; stems usually branched; leaves polymorphic, frequently rosulate, toothed, lobed or pinnatisect; inflorescence usually lax, paniculate; capitula homogamous, ligulate; involucre campanulate or subcylindric; phyllaries in several series, erect, imbricate, often scarious-margined, glabrous; florets yellow or stained purplish externally; anthers sagittate at base; receptacle naked. Achenes cylindrical or 4-angled, sometimes compressed, without a beak, the outer pilose or tuberculate, the inner glabrous and smooth; pappus copious, white, caducous or persistent, sometimes dimorphic with long, filiform bristles interspersed amongst shorter, more slender hairs.

About 40 species widely distributed in the drier regions of the Mediterranean, western Asia, and Africa.

1. L. resedifolia (*L.*) *O. Kuntze*, Revis. Gen., 1: 351 (1891) — as "*Launaya*"; Osorio-Tafall et Seraphim, List Vasc. Plants Cyprus, 11 (1973); Feinbrun, Fl. Palaest., 3: 430, t. 726 (1978).
Scorzonera resedifolia L., Sp. Plant., ed. 1, 1198 (1753).
[*L. mucronata* (non (Forssk.) Muschler) Holmboe, Veg. Cypr., 194 (1914) — as "*Launaya*"; Lindberg f., Iter Cypr., 35 (1946); Osorio-Tafall et Seraphim, List Vasc. Plants Cyprus, 111 (1973).]
TYPE: "*in* Hispania. *Loefling*".

Erect, glabrous, perennial 5–30 cm. high; taproot tough, woody, slender-cylindrical, descending deeply; stems sulcate, branched or unbranched; leaves glaucous, mostly rosulate, variable, linear, lanceolate or narrowly oblong in outline, 3–10 cm. long, 0·5–3 cm. wide, deeply and irregularly pinnatisect into narrow oblong lobes or sometimes subentire, or irregularly dentate, the teeth usually terminating in a callused, whitish apiculus; petioles ill-defined, flattened, sheathing; cauline leaves few, remote, linear-lanceolate or oblong, subentire, dentate or pinnatisect, grading upwards into bracts; inflorescence loosely branched, with erecto-patent, bracteate branches; capitula usually few, terminating the branches, to about 3 cm. diam.; peduncles indistinct from branches; involucre campanulate, glabrous, about 10 mm. long, 8–10 mm. wide; phyllaries in 3–4, closely imbricate series, purplish or glaucous, usually with distinct scarious margins, the outer broadly ovate, about 3 mm. long and almost as wide, with a small, blunt, ciliate apical appendage, inner oblong, to 10 mm. long, 3 mm. wide, with a similar apical appendage; florets rich yellow, stained purplish or livid externally; corolla-tube about 4 mm. long, narrowly infundibuliform, rather densely pilose in the upper part, ligule 9–10 mm. long, about 2–2·5 mm. wide, deeply and narrowly 5-lobed at apex; anthers yellowish, about 3·5 mm. long, 0·3 mm. wide, apical appendage ovate, subacute; style about 9 mm. long, branches yellow, filiform, about 2 mm. long. Achenes slender, 4-angled, base shortly 4-lobed, apex truncate, the outer achenes about 5 mm. long, 0·5 mm. wide, fuscous, densely whitish papillose-puberulous, inner 6–7 mm. long, 0·5 mm. wide, brown, smooth, glabrous; pappus white, silky, persistent, hairs almost smooth, mixed with slightly longer, rigid, very slender, scabridulous or shortly barbellate bristles.

HAB.: Sand-dunes; near sea-level; fl. March–May.

DISTR.: Divisions 4–6. Widespread in the Mediterranean region.

4. Near Famagusta, 1939, *Lindberg f.* s.n. ! Ayios Memnon, 1948, *Mavromoustakis* s.n. !
5. Salamis, 1905, *Holmboe* 455; also, 1949, *Mercy Casey* in *Casey* 574 !
6. Ayia Irini, 1936, *Syngrassides* 1038 ! also, 1941, *Davis* 2548 ! 1954, *Casey* 1333 ! and, 1962, *Meikle* 2379 !

NOTES: Several of the Cyprus sheets have been determined as *Launaea resedifolia* (L.) O. Kuntze var. *aegyptiaca* Amin; I cannot, however, find any published description of this variety.

77. TRAGOPOGON *L.*

Sp. Plant., ed. 1, 789 (1753).
Gen. Plant., ed. 5, 346 (1754).

Annual, biennial or perennial herbs; stems often robust, glabrous or sometimes sparsely floccose-tomentose, branched or unbranched; leaves simple, entire, linear, lanceolate or narrowly ovate, with parallel nerves (like a monocotyledon), sessile, often amplexicaul or semi-amplexicaul; capitula solitary, terminal, homogamous, ligulate; peduncle often swollen at apex; phyllaries herbaceous, subequal, accrescent, apparently in 1 series, free to base or almost to base, often exceeding the florets with a long, attenuate acumen; florets yellow, lilac, pink or purple; anthers sagittate at base; style-branches slender; receptacle naked. Achenes cylindric or fusiform, usually 5- or 10-ribbed, the ribs scabrid-tuberculate, beak usually long, with an apical annulus below the pappus; pappus uniform, persistent, with 2 rows of long, rather rigid plumose hairs connate at the base.

About 50 species in Europe, the Mediterranean region and western Asia, found as introductions elsewhere. Salsify (*T. porrifolius* L.) is occasionally cultivated for its succulent roots; it is akin to *T. sinuatus* Avé-Lall. (infra), with purple florets, but has broader leaves, and more shortly beaked fruits, the beak slightly expanded, but not clavate, at its apex.

·1. T. sinuatus *Avé-Lall.*, Plant. Ital. Bor. Germ. Austr. Rar., 17 (1829).

　　T. australis Jord., Cat. Jard. Dijon, 1848: 32 (1848); Unger et Kotschy, Die Insel Cypern, 253 (1865); H. Stuart Thompson in Journ. Bot., 44: 333 (1906).
　　T. coelesyriacus Boiss., Diagn., 1, 11: 47 (1849); Feinbrun, Fl. Palaest., 3: 422, t. 712 (1978).
　　T. longirostris Bisch. ex Sch. Bip. in Webb et Berth., Phytogr. Canar., 2: 469 (1850); Boiss., Fl. Orient., 3: 745 (1875); Sintenis in Oesterr. Bot. Zeitschr., 32: 291, 364 (1882); Post, Fl. Pal., ed. 2, 2: 134 (1933); Lindberg f., Iter Cypr., 36 (1946); Osorio-Tafall et Seraphim, List Vasc. Plants Cyprus, 112 (1973); Davis, Fl. Turkey, 5: 659 (1975).
　　T. porrifolius L. ssp. *longirostris* (Bisch. ex Sch. Bip.) Holmboe, Veg. Cypr., 192 (1914).
　　[*T. crocifolius* (non L.) Sm. in Sibth. et Sm., Fl. Graec. Prodr., 2: 120 (1813), Fl. Graec., 8: 58, t. 779 (1833).]
　　[*T. porrifolius* (non L.) Poech, Enum. Plant. Ins. Cypr., 21 (1842); Osorio-Tafall et Seraphim, List Vasc. Plants Cyprus, 112 (1973).]
　　[*T. eriospermus* (non Ten.) Unger et Kotschy, Die Insel Cypern, 253 (1865).]

TYPE: France; "in vineis agri Nicaeensis, mense Majo florens" (? LE).

Erect, branched or unbranched biennial or perennial, 20–70 cm. high, with a thick, fleshy taproot; stems straight, terete, sulcate, distinctly swollen just below capitulum; basal leaves crowded, grass-like, 20–30 cm. long, 3–4 mm. wide, tapering to apex, base sheathing, usually subtending a mass of soft, brownish wool, otherwise glabrous; cauline leaves few, remote, 10–20 cm. long, base conspicuously dilated and amplexicaul, apex slender, tapering; capitula solitary, terminal; involucre widely campanulate, glabrous; phyllaries linear-lanceolate, glabrous, spreading and 3–4 cm. long, 3–4 mm. wide at anthesis, erect and accrescent to 6 cm. or more in fruit;

florets purple or mauve; corolla-tube slender, pilose at apex, 9–10 mm. long; ligule 14–15 mm. long, 3–4 mm. wide, minutely 5-dentate at apex; filaments glabrous, about 1 mm. long; anthers linear, purplish, about 5 mm. long, 0·3 mm. wide, apical appendage short, subacute; style 14–15 mm. long, glabrous, branches filiform, blunt, about 3 mm. long. Fertile part of achene narrowly fusiform, slightly curved, about 7 mm. long, 1·5 mm. wide, pale brown, 10-ribbed with muricate-tuberculate ribs, apex tapering into an angular beak about 17 mm. long, apex of beak abruptly swollen, subclavate, terminating in a lanuginose annulus; pappus-hairs stout, rigid, persistent, finely plumose, the outer series thicker than the inner, but otherwise similar, whitish or pale brown.

HAB.: Cultivated and waste ground, roadsides; in garigue on stony hillsides; sea-level to 6,400 ft. alt.; fl. March–June.

DISTR.: Divisions 1, 2, 4–8. Widespread in the Mediterranean region; Atlantic Islands.

1. Aspro Potamos near Ayios Yeoryios (Akamas), 1962, *Meikle* 2089!
2. Khionistra summit, 1862, *Kotschy* 776; also, 1880, *Sintenis & Rigo* 800! and, 1937, *Kennedy* 1021! Prodhromos, 1862, *Kotschy;* Panayia, 1913, *Haradjian* 913! Platres, 1937, *Kennedy* 1016! 1018! Kryos Potamos, 3,000 ft. alt. and 5,400 ft. alt., 1937, *Kennedy* 1017! 1020! Xerokolymbos, 4,300 ft. alt., 1937, *Kennedy* 1019! Platania, 1939, *Lindberg f.* s.n.; about Khionistra, 1939, *Lindberg f.* s.n.; Kalokhorio, Makheras, 1950, *Chapman* 308! Government Cottage, Troödos, 1952, *F. M. Probyn* 113! Pomos Point, 1962, *Meikle* 2254!
4. Kiti, 1862, *Kotschy* 506 sec. *Holmboe;* Dhekelia, 1973, *P. Laukkonen* 169.
5. Kythrea, 1880, *Sintenis & Rigo* 485! Xeri, 1932, *Syngrassides* 70! Mile 10, Nicosia–Famagusta old road, 1967, *Merton* in ARI 487!
6. Nicosia, 1901, *A. G. & M. E. Lascelles* s.n.! also, 1930, *F. A. Rogers* 0665! and, 1935, *Syngrassides* 633! 1966, *O. Huovila* 63! Ayia Irini, 1941, *Davis* 2580! Kykko metokhi, 1952, *F. M. Probyn* 40!
7. Kyrenia, 1949, *Casey* 588! and, 1955, *Miss Mapple* 74! Kharcha, 1950, *Chapman* 268! Lakovounara Forest, 1950, *Chapman* 359!
8. Near Eleousa, 1880, *Sintenis.*

[T. BUPHTHALMOIDES (*DC.*) *Boiss.*, Fl. Orient., 3: 750 (1875) (*Scorzonera buphthalmoides* DC., Prodr., 7: 121; 1838), with broader, linear-lanceolate leaves, yellow ligules exceeding the phyllaries, and shortly beaked achenes (the beak about as long as the fertile part of the achene) is recorded by Post (in Mém. Herb. Boiss., 18: 96; 1900), "Champs de Chypre, printemps", but without any precise locality. Apart from a reference to this record by Holmboe and Stuart Thompson, there is no other evidence of the occurrence of this species (which is Irano-Turanian in general distribution) in Cyprus. The fact that Post fails to mention the relatively common *T. sinuatus* suggests an error in identification.]

78. GEROPOGON *L.*

Sp. Plant., ed. 2, 2: 1109 (1763).

Annual herbs closely akin to *Tragopogon* L., but with a sparsely paleaceous, not naked receptacle, and dimorphic fruits, the outer achenes with a pappus consisting of (3–) 5 (–7) unequal, scabridulous awns, the inner achenes with plumose pappus-hairs as in *Tragopogon.*

One species widely distributed in Mediterranean region and western Asia.

1. **G. hybridus** (*L.*) *Sch. Bip.* in Webb et Berth., Phytogr. Canar., 2: 472 (1850); Davis, Fl. Turkey, 5: 668 (1975); Feinbrun, Fl. Palaest., 3: 423, t. 715 (1978).
   Tragopogon hybridum L., Sp. Plant., ed. 1, 789 (1753).
   Geropogon glabrum L., Sp. Plant., ed. 2, 2: 1109 (1763); Sibth. et Sm., Fl. Graec. Prodr., 2: 119 (1813) — as *G. glaber;* Poech, Enum. Plant. Ins. Cypr., 21 (1842) — as *G. glabrum;* Unger et Kotschy, Die Insel Cypern, 252 (1865); Boiss., Fl. Orient., 3: 744 (1875) — as *G. glabrum;* Sintenis in Oesterr. Bot. Zeitschr., 31: 393 (1881), 32: 398 (1882) — as *G. glaber;* Post, Fl. Pal., ed. 2, 2: 134 (1933) — as *G. glaber.*

G. hirsutum L., Sp. Plant., ed. 2, 1109 (1763); Sibth. et Sm., Fl. Graec. Prodr., 2: 119 (1813), Fl. Graec., 8: 57, t. 778 (1833) — as G. hirsutus.
Tragopogon hirsutum (L.) Kotschy in Unger et Kotschy, Die Insel Cypern, 253 (1865) non T. hirsutus Gouan (1765) nom. illeg.
T. glaber (L.) Hoffm. in Engl. et Prantl, Pflanzenfam., IV. 5: 365 (1893); Holmboe, Veg. Cypr., 192 (1914); Lindberg f., Iter Cypr., 36 (1946); Osorio-Tafall et Seraphim, List Vasc. Plants Cyprus, 112 (1973).
TYPE: "in Italia".

Erect annual 10–40 cm. high; stems usually glabrous, branched or unbranched; a little swollen under the capitulum, sulcate; leaves grass-like, sometimes (but not always) crowded towards the base, up to 18 cm. long, 0·8 cm. wide, sometimes very narrowly linear, generally glabrous, apex acuminate, base of cauline leaves amplexicaul-vaginate, but not conspicuously dilated; capitulum terminal; involucre campanulate, erect or spreading at anthesis, usually erect later; phyllaries usually 7–8, linear-subulate, glabrous, about 30 mm. long, 2·5–3 mm. wide at base, tapering to a fine acumen; florets pale mauve or pink; corolla-tube glabrous, about 3·5 mm. long, ligule 12–14 mm. long, about 3 mm. wide, apex shortly 5-dentate; filaments glabrous, about 0·8 mm. long; anthers dark purplish, about 2·5 mm. long, 0·2 mm. wide, apical appendage small, subacute; style about 7 mm. long, glabrous, branches filiform, about 2·5 mm. long; receptacular scales filiform, up to 2 cm. long. Achenes very narrowly fusiform or subcylindric, 2–2·5 cm. long, 1·5 mm. wide, pale brown, the outer almost smooth, the inner obscurely ribbed, with 10 minutely scabridulous ribs, apex of outer achenes tapering gradually to a scabridulous beak about 2·5 cm. long, beak of inner achenes a little shorter; pappus of outer achenes consisting of 5 unequal, scabridulous, erecto-patent awns, the shortest 2–3·5 mm. long, the longer up to 9 mm. long, minute, atrophied awns sometimes present between the developed ones; pappus of inner achenes consisting of about 20, thinly plumose, unequal bristles, the stouter about 15–18 mm. long, the more slender about 12 mm. long.

HAB.: Cultivated and fallow fields, waste ground, roadsides; occasionally with Tragopogon sinuatus on open, stony hillsides; sea-level to 1,000 ft. alt., usually lowland; fl. Febr.–April.

DISTR.: Divisions 1, 3–8. General distribution that of genus.

1. Ktima, 1913, Haradjian 659! Polis, 1962, Meikle 2307!
3. Mazotos, 1905, Holmboe 170; Pano Polemidhia, 1948, Mavromoustakis s.n.! Zyyi, 1956, Merton 2547! Pissouri, 1981, Hewer 4778!
4. Larnaca, 1787, Sibthorp; Dhekelia, 1977, A. Della in ARI 1682!
5. Between Famagusta and Salamis, 1934, Syngrassides 485!
6. Nicosia–Morphou road, Mile 15, 1973, P. Laukkonen 5!
7. Pentadaktylos, 1880, Sintenis & Rigo 484! Larnaka tis Lapithou, 1941, Davis 3006! Kyrenia, 1949, Casey 534! also, 1956, G. E. Atherton 1258! Lakovounara Forest, 1950, Chapman 360! Vasilia, 1956, G. E. Atherton 1103! Ayios Epiktitos, 1967, Merton in ARI 572!
8. Gastria, 1936, Syngrassides 928! Between Boghaz and Gastria, 1941, Davis 2414! Ovgoros, 1941, Davis 2474! Also noted by Sintenis, 1880, from near Cape Andreas.

79. UROSPERMUM Scop.

Introd. Hist. Nat., 122 (1777) — as Vrospermvm.

Annual, biennial or perennial herbs; leaves narrowly oblong or obovate, lyrate-pinnatisect or irregularly lacerate, the basal indistinctly petiolate, the cauline amplexicaul-auriculate; capitula solitary, terminal on long, stout peduncles, often rather large, homogamous, ligulate; involucre campanulate; phyllaries in 2 ill-defined series, connate below, herbaceous, often rather fleshy; florets yellow; anthers sagittate at base; style-branches slender; receptacle convex, foveolate, pubescent, but without scales. Achenes laterally compressed or subterete, cylindric-fusiform, apical beak long, oblique, hollow at the base, articulated at point of junction with achene, and with a distinct apical disk; pappus copious, deciduous, silky, white, plumose.

Two species in the Mediterranean region and western Asia.

1. **U. picroides** (*L.*) *F. W. Schmidt*, Samml. Phys.-oekon. Aufsätze, 1: 275 (1795); Unger et Kotschy, Die Insel Cypern, 252 (1865); Boiss., Fl. Orient., 3: 743 (1875); Sintenis in Oesterr. Bot. Zeitschr., 31: 259, 392, 393 (1881); Holmboe, Veg. Cypr., 192 (1914); Post, Fl. Pal., ed. 2, 2: 133 (1933); Lindberg f., Iter Cypr., 36 (1946); Osorio-Tafall et Seraphim, List Vasc. Plants Cyprus, 112 (1973); Davis, Fl. Turkey, 5: 685 (1975); Feinbrun, Fl. Palaest., 3: 416, t. 702 (1978).

Tragopogon picroides L., Sp. Plant., ed. 1, 790 (1753).

[*Picris longirostris* Sch. Bip. var. *kotschyi* (non (Boiss.) Sch. Bip.) H. Stuart Thompson in Journ. Bot., 44: 333 (1906).]

[*P. kotschyi* (non Boiss.) Holmboe, Veg. Cypr., 192 (1914); Osorio-Tafall et Seraphim, List Vasc. Plants Cyprus, 109 (1973).]

TYPE: "*in* Creta, Monspelii".

Erect, thinly hispid or hispidulous annual, 20–50 cm. high; stems sulcate, usually branched; leaves thin, papery, hispidulous below along the nerves, the basal oblong-obovate in outline, 4–30 cm. long, 1·5–8 cm. wide, lyrate-pinnatisect, irregularly dentate-lacerate or subentire, apex obtuse or acute, base tapering to an ill-defined, flattened petiole, the cauline sessile, amplexicaul-auriculate, with acute or rounded, projecting auricles; peduncles to 15 cm. long, not much thickened at apex; involucre conspicuously, but thinly, hispid, fleshy, 1·5–1·8 cm. long, 1–1·8 cm. wide; phyllaries usually 7–8, erect at anthesis, reflexed after fruits are shed, narrowly deltoid, 3–4 mm. wide near base, often with a dark violet margin; florets yellow, corolla-tube slender, about 7 mm. long, glabrous below, long-pilose towards apex, ligule 10–11 mm. long, 1·5 mm. wide, minutely 5-dentate (and often violet) at apex; filaments glabrous, about 0·5 mm. long; anthers about 2·5 mm. long, 0·2 mm. wide, apical appendage blunt; style 10–11 mm. long, minutely setulose above, branches linear, blunt, about 1 mm. long. Achenes subcylindric, slightly compressed, mid-brown, trans-versely rugose-lamellate, beak about 8 mm. long, smooth or scabridulous, conspicuously gibbous at base, apical disk about 0·3 mm. diam.; pappus-hairs in a single series, white, plumose, about 10 mm. long.

HAB.: Cultivated and waste ground, roadsides; sometimes on stony hillsides or in grassland; sea-level to 4,000 ft. alt., usually lowland; fl. March–June.

DISTR.: Divisions 1–8. Widespread in the eastern Mediterranean region and western Asia.

1. Paphos, 1913, *Haradjian* 637! Near Ayios Yeoryios (Akamas), 1962, *Meikle* 2118! Ayios Yeoryios Island, 1962, *Meikle* 2115! N.E. of Polis, 1979, *Edmondson & McClintock* E 2826!
2. Prodhromos, 1862, *Kotschy* 825; Lefka, 1908, *Clement Reid* s.n.!
3. Limassol, 1913, *Haradjian* 587! also, 1950, *Barrington* 9! and, 1954, *Mavromoustakis* 8! Kithasi, 1981, *Hewer* 4784!
4. Cape Pyla, 1880, *Sintenis & Rigo*; Stavrovouni, 1939, *Lindberg f.* s.n.! Cape Greco, 1958, *N. Macdonald* 79! Troulli, 1981, *Hewer* 4737!
5. Kythrea, 1880, *Sintenis & Rigo* 486! Athalassa, 1950, *Chapman* 641A!
6. Makhedonitissa Monastery, 1936, *Syngrassides* 1106! Ayia Irini, 1941, *Davis* 2573! 2596!
7. Kyrenia, 1932, *Syngrassides* 357! also, 1949, *Casey* 476! Near Thermia, 1955, *G. E. Atherton* 379! 388!
8. Komi Kebir, 1912, *Haradjian* 263! 269! 277! 335! Near Leonarisso, 1957, *Merton* 3072!

80. SCORZONERA *L.*

Sp. Plant., ed. 1, 790 (1753).
Gen. Plant., ed. 5, 346 (1754).

Mostly perennial herbs, rarely annuals or subshrubs; rootstock often thick, fleshy or tuberous, lactiferous; stem present or plants acaulescent; leaves various, alternate, simple and entire or pinnately lobed and dissected, petiolate or sessile; capitula homogamous, ligulate, often solitary and terminal, occasionally several in racemose or corymbose inflorescences, generally pedunculate; involucre narrowly campanulate or subcylindric; phyllaries herbaceous, in 2 series, somewhat accrescent; florets yellow, pink, purple or white; anthers sagittate at base; style-branches slender; receptacle naked. Achenes oblong-cylindric or prismatic, glabrous or pilose, erostrate, sometimes dilated and excavate at base; pappus-hairs in several, (usually 3) series, plumose or partly plumose, rarely wholly scabridulous.

About 150 species in temperate Europe and Asia.

Plants annual or biennial, 20–50 (or more) cm. high; stems glabrous; leaves often glabrous, deeply pinnatisect, with long, narrow, acuminate lobes; achenes glabrous   **1. S. laciniata**

Plants perennial with a thick taproot; generally less than 20 cm. high; stems and leaves canescent or pubescent, rarely glabrous; leaves very variable, entire, undulate or irregularly lobed or pinnatisect:

Involucre about 1 cm. long; achenes smooth, ribbed, thinly lanuginose towards apex, with a hollow, sterile base - - - - - -   **2. S. jacquiniana var. subintegra**

Involucre about 1·5–2 cm. long; achenes glabrous, some of them conspicuously, transversely rugose-verrucose, not hollow at base - - - - -   **3. S. troodea**

---

1. S. laciniata L., Sp. Plant., ed. 1, 791 (1753); Boiss., Fl. Orient., 3: 757 (1875); Post, Fl. Pal., ed. 2, 2: 821 (1933); A. K. Jackson in Kew Bull., 1937: 344 (1937); Osorio-Tafall et Seraphim, List Vasc. Plants Cyprus, 112 (1973); Davis, Fl. Turkey, 5: 634 (1975).

TYPE: "*in* Germania, Gallia".

Erect biennial with a slender taproot, often divided at its apex; stems 20–50 (or more) cm. high, glabrous, sulcate, generally with several erecto-patent branches; leaves lanceolate or oblong in outline, 8–20 cm. long, 0·6–5 cm. wide, glabrous, irregularly pinnatisect into narrow oblong or linear, entire, acuminate lobes; petiole ill-defined, flattened, to 7 cm. long; capitula terminal, solitary on stout, sulcate peduncles 6–15 cm. long; involucre narrowly campanulate, 12–15 mm. long, 8–10 mm. wide at anthesis, accrescent to 30 mm. long in fruit; phyllaries lanceolate or narrowly ovate, the outer about 4–6 mm. long, 3–4 mm. wide, the inner up to 15 mm. long, all herbaceous, glabrous or canescent, with a very narrow, scarious margin; florets yellow; corolla-tube slender, about 5–7 mm. long, pilose towards apex, ligule about 8 mm. long, 1·8 mm. wide, apex 5-dentate; filaments glabrous, about 0·5 mm. long; anthers 1·8–2 mm. long, 0·2 mm. wide, apical appendage blunt; style about 10 mm. long, papillose in the upper part, branches filiform, obtuse or acute, about 1 mm. long. Achenes cylindric, 12–13 mm. long, 1–1·3 mm. wide, glabrous, conspicuously 10-ribbed, the basal part slightly swollen and hollow for about 5 mm. the apex truncate; pappus copious, slightly brownish, hairs 18–20 mm. long, plumose almost to apex.

HAB.: Cultivated fields, gardens, roadsides; about 500 ft. alt.; fl. March–May.

DISTR.: Division 6, probably introduced. Widely distributed in Europe, the Mediterranean region and western Asia.

6. Makhedonitissa Monastery, 1936, *Syngrassides* 1108! Nicosia, 1950, *Chapman* 463! 628! also, 1957, *Merton* 3152!

NOTES: The very limited distribution of *S. laciniata* in Cyprus suggests that it is a recently introduced weed.

---

2. S. jacquiniana (*Koch*) Čelak., Prodr. Fl. Böhm., 1: 218 (1867); Boiss., Fl. Orient., 3: 757 (1875); Post in Mém. Herb. Boiss., 18: 96 (1900); H. Stuart Thompson in Journ. Bot., 44: 333 (1906); Post, Fl. Pal., ed. 2, 2: 136 (1933) — as "*S. Jacquiana*"; Osorio-Tafall et Seraphim, List Vasc. Plants Cyprus, 112 (1973).

Podospermum canum C. A. Meyer, Verz. Pflanz. Cauc., 62 (1831); Unger et Kotschy, Die Insel Cypern, 252 (1865) nom. superfl. illeg.

P. jacquinianum Koch, Syn. Fl. Germ. Helv., 425 (1837); Sintenis in Oesterr. Bot. Zeitschr., 31: 259, 326 (1881), 32: 123, 398 (1882).

Scorzonera cana Hoffm. in Engl. et Prantl, Pflanzenfam., IV. 5: 365 (1897); Holmboe, Veg. Cypr., 192 (1914); Davis, Fl. Turkey, 5: 635 (1975).

Perennial herb with a thick, fleshy, fuscous-brown rootstock, often divided into several branches at the apex; stems short, usually less than 15 cm. long, but (var. *jacquiniana*) sometimes attaining 30–50 cm., shortly lanuginose-canescent or subglabrous, branched or unbranched; leaves polymorphic, sometimes simple, entire or subentire, subulate, linear or lanceolate, or variously lobed and pinnatisect, with linear-acuminate, or oblong-obtuse lobes, floccose or subglabrous, up to 20 cm. long, 7 cm. wide, but usually very much smaller in our area; capitula solitary, terminal on stout or slender, shortly lanuginose or subglabrous peduncles 1–15 (or more) cm. long; involucre campanulate, 0·8–1·5 cm. long, 0·4–1·2 cm. wide; phyllaries rather loosely imbricate, often purplish, floccose or subglabrous, the outer ovate or ovate-lanceolate, 3–6 mm. long, 2–3 mm. wide, the inner lanceolate to 15 mm. long, about 3 mm. wide, all distinctly accrescent in fruit; florets yellow, much longer than the involucre at anthesis; corolla-tube slender, 4–7 mm. long, pilose at apex, ligule 6–10 mm. long, 2·5–3·5 mm. wide, conspicuously and somewhat irregularly 5-dentate at apex; filaments glabrous, about 0·5 mm. long; anthers about 4 mm. long, 0·3 mm. wide, with an obtuse apical appendage; style glabrous, about 8–9 mm. long, branches filiform, blunt, to 3 mm. long. Achenes oblong or subcylindric, about 6–8 mm. long, 1·5–2 mm. wide, with 5 prominent, whitish, and 5 obscure ribs, basal half hollow, pallid, slightly swollen, upper half sometimes thinly lanuginose; pappus copious, slightly brownish, hairs 9–10 mm. long, plumose almost to apex.

var. **subintegra** *Boiss.*, Fl. Orient., 3: 758 (1875); Post, Fl. Pal., ed. 2, 2: 136 (1933).
   Scorzonera subintegra (Boiss.) Thiébaut in Bull. Soc. Bot. Fr., 95: 18 (1948); Feinbrun, Fl. Palaest., 3: 424, t. 716 (1978).
   TYPE: Palestine; "in excelsioribus Libani et montis Hermon (Boiss! Bl! Gaill! Ky 180!)", etc. (G).

Plant dwarf, condensed, acaulescent, usually less than 15 cm. high; leaves lanceolate in outline, generally pinnatisect, but with relatively few, small, often elliptic-oblong, obtuse or acute lobes; capitula mostly solitary, usually rather small, on erect or arcuate, canescent, often purplish, peduncles; involucre campanulate, about 1 cm. long, 0·8 cm. wide; phyllaries loosely imbricate, somewhat canescent, often purplish, with narrow scarious margins and a small, rounded, lacerate, scarious, apical appendage, the outer phyllaries narrowly ovate, about 3 mm. long, 1·5 mm. wide, the inner oblong-lanceolate, about 10 mm. long, 2 mm. wide; florets yellow, tinged green or brownish externally, corolla-tube narrowly infundibuliform, about 5 mm. long, pilose at apex, ligule 8–9 mm. long, about 3 mm. wide, 6-nerved, apex rather bluntly and irregularly dentate; anthers yellow, about 4 mm. long, 0·3 mm. wide, apical appendage small, subacute; style 10–11 mm. long, minutely setulose towards apex, branches filiform, blunt, about 3 mm. long. Achenes narrowly oblong, compressed, angular, unilaterally 5-ribbed, 6–7 mm. long, 2 mm. wide, basal part sterile, hollow for about 3 mm., glabrous, upper, fertile part thinly lanuginose; pappus copious, whitish tinged brown, hairs about 9 mm. long, lanuginose-plumose.

HAB.: Stony (often eroded) hillsides; dry grass steppe; roadsides; sea-level to 1,500 ft. alt.; fl. March–May.

DISTR.: [2], 4–8. Also Lebanon and Palestine.
[2. Troödos, ? 1898, *Post* s.n. See Notes].
4. Larnaca, 1859, *Kotschy* 389; and, 1862, *Kotschy* 326; Cape Pyla, 1880, *Sintenis*; near Troulli Mine, 1981, *Hewer* 4735! 4736! 4738!
5. Kythrea, 1880, *Sintenis & Rigo* 487! also, 1941, *Davis* 2946! Lefkonico, 1880, *Sintenis*;

between Mia Milea and Mandres, 1933, *Syngrassides* 316! also, 1950, *Chapman* 489! Ayios Dhemetrianos near Kythrea, 1950, *Chapman* 106! Nisou, 1936, *Syngrassides* 1255! Road to Klirou, 1967, *Merton* in ARI 498!
6. Ayios Yeoryios near Liveras, 1962, *Meikle* 2406!
7. Between Orta Keuy and Kyrenia, 1936, *Syngrassides* 1163! Lakovounara, 1950, *Chapman* 75!
8. Cape Andreas, 1880, *Sintenis*.

NOTES: Post's record from Troödos is very doubtful; no other collector has found *S. jacquiniana* there, and *S. troodea* Boiss. was probably the plant seen.

Scorzonera jacquiniana is so polymorphic that one suspects it may be an aggregate of ill-defined species rather than a single taxon. The Cyprus plants are uniformly distinct in the much reduced, sparse lobing of the leaves.

3. S. troodea *Boiss.*, Fl. Orient., Suppl., 320 (1888); Holmboe, Veg. Cypr., 193 (1914); Lindberg f., Iter Cypr., 36 (1946); Osorio-Tafall et Seraphim, List Vasc. Plants Cyprus, 112 (1973).

[*Podospermum villosum* (non Stev.) Kotschy in Unger et Kotschy, Die Insel Cypern, 252 (1865).]

[*Scorzonera papposa* (non DC.) Post in Mém. Herb. Boiss., 18: 96 (1900); H. Stuart Thompson in Journ. Bot., 44: 333 (1906); Holmboe, Veg. Cypr., 193 (1914); Osorio-Tafall et Seraphim, List Vasc. Plants Cyprus, 112 (1973).]

[*S. mollis* (non M. Bieb.) H. Stuart Thompson in Journ. Bot., 44: 333 (1906); Holmboe, Veg. Cypr., 193 (1914); Osorio-Tafall et Seraphim, List Vasc. Plants Cyprus, 112 (1973).]

TYPE: Cyprus; "in monte Troodos Cypri (Sint. et Rigo 799!)" (G, K!).

Perennial herb with a long, tough, fleshy taproot, generally divided into several branches at the apex; stems usually short or plants subacaulescent, but sometimes up to 20 cm. long where plants are growing in shade, somewhat angled, thinly canescent; leaves polymorphic, obovate, oblong, elliptic, lanceolate or almost linear in outline, entire, undulate, lacerate, lobed or sometimes pinnatisect almost to the midrib into linear or subulate divisions, equally variable in overall size, from 3–20 cm. long, 0·3–3 cm. wide, the cauline much reduced and amplexicaul, the basal numerous and tapering to an indistinct petiole, lamina subglabrous, canescent or sometimes shortly grey-tomentose, especially on the undersurface; inflorescence subscapose; capitula generally solitary, peduncle up to 12 cm. long; involucre narrowly campanulate, about 1·5–2 cm. long, 1–1·5 cm. wide; phyllaries subglabrous or canescent, with a narrow scarious margin and a subacute, ciliolate apex, the outer ovate, 3–5 mm. long, 2–3 mm. wide, the inner up to 20 mm. long, 7 mm. wide; florets yellow, stained purplish externally; corolla-tube slender, 8–10 mm. long, pilose at apex, ligule to 15 mm. long, 4 mm. wide, 6-nerved, shortly 5-dentate at apex; filaments glabrous, about 0·5 mm. long; anthers about 6 mm. long, 0·3 mm. wide, apical appendage short, obtuse; style about 20 mm. long, minutely setulose towards apex, branches filiform, obtuse, 7–8 mm. long. Achenes dimorphic, pale brown, glabrous, the outer narrowly oblong, bluntly 5-angled, truncate, coarsely and transversely rugose-verrucose, up to 18 mm. long, 2·5–3 mm. wide, without a hollow, empty base, the inner subterete or obscurely rugose-verruculose, about 12 mm. long, 1·5 mm. wide; pappus whitish, brownish or greyish, hairs up to 12 mm. long, lanuginose-plumose below, barbellate towards apex.

HAB.: In Pine woods, and on screes and rocky slopes on serpentine or rarely on chalk; 2,500–6,400 ft. alt.; fl. May–July.

DISTR.: Divisions 2, 3, [4]. Endemic.

2. Common about Khionistra, Platres, Prodhromos, Omodhos, Ayios Nikolaos, etc. *Sintenis & Rigo* 799! *Post* s.n.; *A. G. & M. E. Lascelles* s.n.! *Haradjian* 407! 518! *Syngrassides* 723! *Kennedy* 1022–1031! 1704–1706! 1708–1720! 1722–1725! *C. Wyatt* 16! *Lindberg f.* s.n.! *Davis* 1804! 1827! *Casey* 832! *F. M. Probyn* 114! *H. R. P. Dickson* 07! *D. P. Young* s.n.! *S. Economides* in ARI 1227! *N. Macdonald* 163! *A. Genneou* in ARI 1511! *H. B. Larsen* s.n.!
3. Kakomallis above Louvaras, 1962, *Meikle* 2883!
[4. Near the Salt Lake, Larnaca, 1859, *Kotschy* Suppl. 831. See Notes.]

NOTES: Largely restricted to the serpentines of the Troödos massif, and formerly confused with *S. mollis* M. Bieb. from the eastern Mediterranean region and western Asia. The latter has not been found in Cyprus, though it is just possible that Kotschy's record (under *Podospermum villosum* Stev.) from Larnaca may be correct. In the absence of additional evidence, it is, however, more likely that a misidentification or mislabelling is responsible for it. *S. mollis* has numerous narrow-lanceolate or linear, grass-like leaves, and lamellate-scabridulous achenes.

[S. ARANEOSA *Sm.* in Sibth. et Sm., Fl. Graec. Prodr., 2: 123 (1813), Fl. Graec., 8: 63, t. 785 (1833); Unger et Kotschy, Die Insel Cypern, 253 (1865); Boiss., Fl. Orient., 3: 776 (1875); Holmboe, Veg. Cypr., 193 (1914); Rechinger f. in Oesterr. Bot. Zeitschr., 94: 194 (1948); Osorio-Tafall et Seraphim, List Vasc. Plants Cyprus, 112 (1973) is unknown in Cyprus, except as an erroneous localization in Fl. Graec. Prodr. and Fl. Graec. It was most probably collected in the Kikladhes, where it is now known to occur.]

S. PARVIFLORA *Jacq.*, Fl. Austr., 4: 3, t. 305 (1776); Boiss., Fl. Orient., 3: 771 (1875); Holmboe, Veg. Cypr., 193 (1914); Osorio-Tafall et Seraphim, List Vasc. Plants Cyprus, 112 (1973); Davis, Fl. Turkey, 5: 646 (1975). (*S. cypria* Kotschy in Unger et Kotschy, Die Insel Cypern, 25: 3; 1865), with glabrous, narrow-lanceolate leaves, subscapose inflorescences, solitary, terminal capitula, and bluntly ribbed, glabrous achenes, was reported (as *S. cypria*) from Cyprus, "Auf Anhöhen um Larnaca selten, April 1859. n.459"; all subsequent records are based upon this report, nor has any other collector found the plant at Larnaca or elsewhere on the island. The specimen numbered *Kotschy* Suppl. 459 in Herb. Boiss. (microfiche) seems to be *S. parviflora*; it is, however, labelled "in maritimis ad Larnaca", agreeing with Holmboe's "Near the sea-shore at Larnaca". One suspects mislabelling, with the possibility that the plant came from S. Turkey.

54. CAMPANULACEAE

Annual or perennial herbs, rarely woody, often with latex; leaves generally alternate and simple, exstipulate. Inflorescence racemose or cymose; flowers usually hermaphrodite, actinomorphic or zygomorphic, mostly 5-merous, sometimes resupinate; calyx with 3–5, open lobes, sometimes with appendages between the lobes; corolla tubular or campanulate with valvate lobes. Stamens 5, inserted near the base of the corolla; filaments free, often expanded at the base and almost covering a nectariferous disk at the base of the style; anthers free or coherent into a tube, introrse, dehiscing longitudinally. Ovary inferior or semi-inferior, 2–5 (–10)-locular, with axile placentation; ovules numerous; style terminal, simple; stigmas as many as carpels. Fruit a capsule, dehiscing in various ways, or a berry, often crowned with the persistent calyx-lobes; seeds numerous, with fleshy endosperm.

About 80 genera and 2,000 species widely distributed in temperate and tropical regions.

Flowers actinomorphic; stamens free; capsule dehiscing laterally by small valves or pores:
  Hypanthium and capsule hemispherical, turbinate or obovoid; corolla campanulate or
    tubular - - - - - - - - - - - - - - **1. Campanula**
  Hypanthium and capsule elongate; corolla rotate or subrotate, 5-lobed - **2. Legousia**
Flowers zygomorphic; stamens connate around style; capsule dehiscing apically **3. Solenopsis**

SUBFAMILY 1. **Campanuloideae**   Flowers actinomorphic; stamens free or rarely with coherent anthers.

1. CAMPANULA *L.*

Sp. Plant., ed. 1, 163 (1753).
Gen. Plant., ed. 5, 77 (1754).

Annual or perennial herbs; flowers solitary or in racemes, panicles or cymes; hypanthium hemispherical, turbinate or obovoid; calyx deeply 5-partite, sometimes with reflexed appendages between the teeth; corolla generally blue or purple, campanulate or tubular, rarely subrotate, the 5 lobes usually not extending below the middle of the tube; stamens free, the filaments normally dilated at the base and covering the nectariferous disk; anthers free, dehiscing in bud; ovary inferior or semi-inferior, 3- or 5-locular; ovules numerous; style rather thick, hairy or papillose; stigmas 3 or 5, recurving with age. Capsule crowned with the persistent calyx-teeth; dehiscing laterally by small valves or pores; seeds small, numerous.

About 300 species, chiefly in North Temperate regions and the Mediterranean.

Inflorescence narrow, spiciform; a robust perennial (or biennial) 15–80 cm. high
                                                           5. C. peregrina
Inflorescence cymose or corymbose, not narrow or spiciform; slender annuals:
  Calyx-lobes deltoid, patent in fruit and forming a 5-pointed star:
    Corolla minute, about 3·5 mm. long, scarcely exceeding the calyx-lobes -   **1. C. erinus**
    Corolla 7–9 mm. long, much exceeding the calyx-lobes -   -   -   **2. C. drabifolia**
  Calyx-lobes linear or narrowly deltoid, erect or suberect in fruit:
    Plants erect; flowers crowded in corymbose cymes -   -   -   - **3. C. podocarpa**
    Plants weak, diffuse; flowers in lax dichotomous cymes -   -   - **4. C. delicatula**

1. **C. erinus** *L.*, Sp. Plant., èd. 1, 169 (1753); Sibth. et Sm., Fl. Graec. Prodr., 1: 142 (1806), Fl. Graec., 3: 11, t. 214 (1819); Unger et Kotschy, Die Insel Cypern, 258 (1865); Boiss., Fl. Orient., 3: 932 (1875); H. Stuart Thompson in Journ. Bot., 44: 334 (1906); Holmboe, Veg. Cypr., 176 (1914); Post, Fl. Pal., ed. 2, 2: 167 (1933); Lindberg f., Iter Cypr., 33 (1946); Osorio-Tafall et Seraphim, List Vasc. Plants Cyprus, 100 (1973); Davis, Fl. Turkey, 6: 52 (1978).
TYPE: "*in* Italia, Galloprovincia, Monspelii".

A slender, erect or spreading annual 5–30 cm. high; stems usually much branched (except in starved specimens), longitudinally ridged, thinly strigose-hispidulous; leaves bright green, obovate, the basal 1–3 cm. long, 0·5–1 cm. wide, thinly strigose, the apex subacute or rounded, the margins somewhat irregularly toothed, lobed or crenate, the base tapering to an indistinct petiole; upper leaves progressively smaller, often obcuneate, with a few, rather coarse lobes; flowers in lax cymes, sometimes arising singly at dichotomies between pairs of leaf-like bracts, or opposite a single bract on an inflorescence-branch; pedicels very short, rarely exceeding 5 mm. (but sometimes lengthening to 10 mm. in fruit), thinly hispidulous, at first erect, later recurved; hypanthium broadly turbinate, about 2 mm. long, 3 mm. wide, strigose-hispidulous; calyx-lobes at first erect, narrowly deltoid, about 3 mm. long, 1 mm. wide at base, hispidulous especially along the margin, becoming patent in fruit and accrescent to 5 mm. long and 2·5 mm. wide at base; corolla inconspicuous, pale blue with a white tube, about 3·5 mm. long, 2·5 mm. wide, glabrous externally, lobes ovate, 1–2 mm. long, thinly strigillose externally; filaments dilated at base, about 1 mm. long, 0·3–0·5 mm. wide; style about 3·5 mm. long, pilose in the upper half; stigmas 3, about 0·5–1 mm. long. Fruit depressed-turbinate, about 3 mm. long, 6 mm. wide at the flattish apex, the spreading, persistent calyx-lobes forming a 5-pointed star, dehiscing by basal valves; seeds narrowly oblong-ellipsoid, about 0·6 mm. long, 0·3 mm. wide; testa pale brown, shining, stained darker at either extremity.

HAB.: On banks, cliffs and rocky hillsides, sometimes on old walls, usually on limestone, but also on igneous rocks.

DISTR.: Divisions 2–4, 6–8. Widespread in the Mediterranean region and eastwards to Arabia and Iran.

2. Near Kalopanayiotis, 1937, *Syngrassides* 1568! Khrysorroyiatissa Monastery, 1941, *Davis* 3441!
3. Cape Gata, 1862, *Kotschy* 604.
4. Near Larnaca, 1862, *Kotschy* 103.
6. Nicosia Walls, 1905, *Holmboe* 282; Famagusta Gate, Nicosia, 1956, *Merton* 718!
7. Common on the Northern Range; Lapithos, Pentadaktylos, St. Hilarion, Agirdha, Larnaka tis Lapithou, Kyrenia, Panagra, Phlamoudhi, etc. *Sintenis & Rigo* 27! *Lindberg f.*! *Davis* 2512! 2991! 3048! *Casey* 439! *Merton* 2224! *G. E. Atherton* 862! 1318! *P. H. Oswald* 123! *Meikle* 2369A! *Merton* in ARI 745! etc.
8. Cape Andreas, 1880, *Sintenis & Rigo*; 1938, *Syngrassides* 1760! Near Galinoporni, 1962, *Meikle* 2445!

2. C. drabifolia *Sm.* in Sibth. et Sm., Fl. Graec. Prodr., 1: 142 (1806), Fl. Graec., 3: 11, t. 215 (1819); Unger et Kotschy, Die Insel Cypern, 258 (1865) — as *C. drabaefolia*; Boiss., Fl. Orient., 3: 933 (1875); H. Stuart Thompson in Journ. Bot., 44: 333 (1906) — as *C. drabaefolia*; Osorio-Tafall et Seraphim, List Vasc. Plants Cyprus, 100 (1973); Davis, Fl. Turkey, 6: 53 (1978).
TYPE: "In vineis, et inter Gossypia, insulae Sami, et prope Athenas" *Sibthorp* (OXF).

Very similar in general appearance to *C. erinus* L., a strigose-hispidulous annual, often much branched with divaricately spreading branches; leaves as in *C. erinus*, sometimes subentire, or sometimes sharply and deeply toothed; flowers in lax, repeatedly branched cymes, shortly pedicellate, recurved in fruit; hypanthium as in *C. erinus*, but often rather densely strigose-hispidulous; calyx-lobes narrowly deltoid, hispidulous, especially along the margins, becoming patent in fruit; corolla violet-blue with a white tube, 7–9 mm. long, 5 mm. wide, often somewhat turbinate, much exceeding the calyx-lobes; corolla-lobes ovate, acute, 2–3 mm. long, spreading. Fruit and seeds as in *C. erinus*.

HAB.: Banks and slopes on igneous formations, 1,500–2,000 ft. alt.; fl. April–May.

DISTR.: Division 2, very rare. Greece, Crete, Aegean Islands, S.W. Turkey.

2. Between Prodhromos and Tris Elies, 1862, *Kotschy* s.n. Roudhkias valley between Pano Panayia and Kykko, 1941, *Davis* 3384! Mavres Sykies, 1962, *Meikle* 2734! Below Phterykoudhi, 1979, *Edmondson & McClintock* E 2926!

NOTES: The Kotschy record has not been verified, and may properly be referable to the following, *C. podocarpa* Boiss. The plant from Mavres Sykies has the large corollas of *C. drabifolia*, but is in other respects indistinguishable from *C. erinus*. For a further comment on the problems posed by this difficult group, see the notes under 3. *C. podocarpa* Boiss.

3. C. podocarpa *Boiss.*, Diagn., 1, 11: 68 (1849), Fl. Orient., 3: 934 (1875); Davis, Fl. Turkey, 6: 54 (1978).
C. cypria Rechinger f. in Arkiv för Bot., ser. 2, 1: 432 (1950); Osorio-Tafall et Seraphim, List Vasc. Plants Cyprus, 100 (1973).
TYPE: Turkey; "in Pamphyliâ ubi legit et mihi comm. amic. Pestalozza." (G!).

An erect, branched or unbranched annual 6–15 (–23) cm. high; stems longitudinally ridged, as in *C. erinus* L., thinly strigose-hispidulous; leaves oblong-obovate, 10–20 mm. long, 3–8 mm. wide, the basal distinctly petiolate, the upper sessile or subsessile, lamina rather densely strigose, apex obtuse or acute, margins subentire, crenate, or irregularly and bluntly toothed; inflorescence a rather congested corymbose cyme; flowers numerous, very shortly pedicellate or subsessile, clustered at the tips of the branches; hypanthium sparsely hispid, about 2 mm. long, 3 mm. wide; calyx-lobes narrowly deltoid, obtuse, about 3 mm. long, 1 mm. wide at base, permanently erect and adpressed to corolla-tube, hispid, the hairs often blackish at the base; corolla narrowly tubular, 5–7 mm. long, 2–2·5 mm. wide, tube glabrous, white, lobes violet-blue, narrowly ovate, 1·5–3 mm. long, 0·8–1·5 mm. wide at base, very sparsely hispidulous; filaments about 1 mm. long, dilated at the base; anthers linear, 3–4 mm. long, 0·4 mm. wide, very shortly auriculate at the base; style 6–6·5 mm. long, shortly protruding from mouth of corolla-tube; stigmas 3, bluntly linear-oblong, about 0·8 mm. long. Fruit depressed-turbinate, about 2 mm. long, 3–4 mm. wide, crowned with the erect calyx-lobes and persistent corolla; seeds oblong-ellipsoid, about 0·6 mm. long, 0·3 mm. wide; testa pale brown, shining, stained darker at either extremity. *Plate 64, figs. 4–6.*

HAB.: Rocky mountainsides, perhaps always on serpentine; 5,000–6,000 ft. alt.; fl. May–June.

DISTR.: Division 2, very rare. Also S.W. Turkey (Muğla Province).

2. Troödos, 5,000–6,000 ft. alt., 1912, *Haradjian* 432!

NOTES: The aggregate comprising *C. erinus* L. (1753), *C. drabifolia* Sm. (1806), *C. rhodensis* A. DC. (1830), *C. raveyi* Boiss. (1844), *C. attica* Boiss. et Heldr. (1849), *C. delicatula* Boiss. (1849), *C. podocarpa* Boiss. (1849), *C. cypria* Rechinger f. (1950), *C. creutzburgii* Greuter (1967) and *C. pinatzii* Greuter (1967) has been discussed by Greuter (Boissiera, 13: 132–135; 1967), and attention has been drawn to the very considerable problems posed by the taxonomy of the eastern Mediterranean segregates. So far as Cyprus is concerned, *C. erinus* is uniformly distinct, but is linked to *C. drabifolia* Sm. by the large-flowered plant (here identified as *C. drabifolia* Sm.) from Mavres Sykies (*Meikle* 2734) which resembles *C. erinus* in all respects save in size of corolla, and which does not exactly agree with the *C. drabifolia* of Fl. Graeca, t. 215. Likewise, *C. podocarpa* (incl. *C. cypria*) is linked to *C. drabifolia* on one side, and to *C. raveyi* Boiss. on the other, while *C. delicatula* Boiss. is linked to *C. drabifolia* through *C. pinatzii* Greuter. The problem, for the present, cannot be satisfactorily solved. To consider all the above microspecies as falling within the range of *C. erinus* would be an easy, but misleading, simplification of a complex situation; but to regard them as fully distinct taxa conceals the general tendency for each one to vary in the direction of another. I have chosen to err with the splitters, and to attach a significance, which may be illusory, to such characters as habit, indumentum, shape of inflorescence, shape and position of calyx-lobes, size and shape of corolla-tube and corolla-lobes. A better solution must await experimental work on the group.

4. **C. delicatula** *Boiss.*, Diagn., 1, 11: 67 (1849), Fl. Orient., 3: 933 (1875), Fl. Orient., Suppl., 333 (1888); Holmboe, Veg. Cypr., 176 (1914); Osorio-Tafall et Seraphim, List Vasc. Plants Cyprus, 100 (1973); Davis, Fl. Turkey, 6: 54 (1978).

TYPE: Turkey; "in *Cariâ* ad Moglah [Muğla] (Aucher No. 1838) in Lyciâ (Pestalozza)" (G, K!).

A weak, sprawling or decumbent annual, rarely more than 10 cm. high; stems much branched, obscurely angled or striate, thinly or rather densely patent-pubescent; leaves broadly ovate or obovate, 0·7–2 cm. long, 0·5–1 cm. wide, adpressed-pilose, apex blunt or rounded, base tapering to a short petiole, margins subentire or shortly toothed; inflorescence lax, repeatedly dichotomous, the flowers in the dichotomies, or terminal; pedicels rarely exceeding 2 mm., often very short or almost wanting; hypanthium semiglobose, about 2 mm. long and wide, densely white, hispidulous-pilose; calyx-lobes linear-acute, about 5 mm. long, less than 1 mm. wide at base, erect in flower and fruit, rather densely pilose; corolla pale lavender-blue, very small, 5–6 mm. long, 1·5 mm. wide, thinly hispidulous-pilose externally, lobes ovate, about 1 mm. long; filaments about 0·5 mm. long, not much dilated; anthers linear, about 1·4–1·5 mm. long, 0·3 mm. wide, style about 2 mm. long, thickened and papillose above; stigmas 3, about 0·5 mm. long. Fruit erect or slightly nodding, hemispherical, about 3 mm. long, 4 mm. wide, crowned with the persistent calyx; seeds very small, oblong-ellipsoid, about 0·4 mm. long, 0·2 mm. wide; testa pale brown, shining, darker at either extremity. *Plate 64, figs. 1–3.*

HAB.: On limestone rocks and cliffs; 1,000–2,000 ft. alt.; fl. April–June.

DISTR.: Division 7, rare. Also Aegean Islands and W. Turkey.

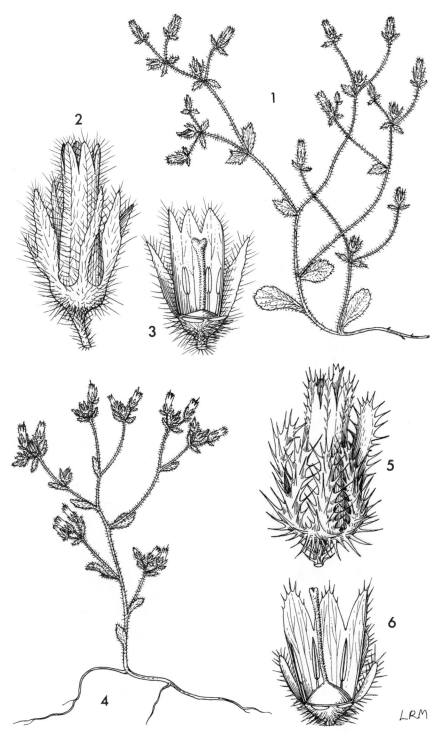

LRM

Plate 64. Figs. **1–3.** CAMPANULA DELICATULA Boiss. **1,** habit, $\times \frac{2}{3}$; **2,** flower, $\times 6$; **3,** flower, longitudinal section, $\times 4$; figs. **4–6.** CAMPANULA PODOCARPA Boiss. **4,** habit, $\times 2$; **5,** flower, $\times 6$; **6,** flower, longitudinal section, $\times 4$. (**1** from *Sintenis & Rigo* 516; **2, 3** from *Davis* 3022; **4–6** from *Haradjian* 432.)

7. Pentadaktylos, 1880, *Sintenis & Rigo* 516! Larnaka tis Lapithou, 1941, *Davis* 3022!

NOTES: One of the most readily recognizable of the *C. erinus* group, at once distinguished by its weak, sprawling stems, hairy calyces and large, rather flabby calyx-lobes.

5. C. peregrina *L.*, Mantissa Altera, 204 (1771); Poech, Enum. Plant. Ins. Cypr., 22 (1842); Kotschy in Oesterr. Bot. Zeitschr., 12: 299 (1862); Unger et Kotschy, Die Insel Cypern, 258 (1865); Boiss., Fl. Orient., 3: 939 (1875); Holmboe, Veg. Cypr., 176 (1914); Post, Fl. Pal., ed. 2, 2: 167 (1933); Lindberg f., Iter Cypr., 33 (1946); Osorio-Tafall et Seraphim, List Vasc. Plants Cyprus, 100 (1973); Davis, Fl. Turkey, 6: 62 (1978).

TYPE: "ad C.B.S. [Caput Bonae Spei — Cape of Good Hope]? inter semina capensia excrevit" (LINN!).

Erect, densely setose-hispidulous perennial (or biennial) 15–80 cm. high; stems usually simple or not much branched; basal leaves spathulate, up to 10 cm. long, 6 cm. wide, narrowing abruptly to a flattened or narrowly winged petiole sometimes 12–15 mm. long, apex rounded, margins crenate or bluntly dentate; cauline leaves obovate, obtuse or subacute, 8–15 cm. long, 4–6 cm. wide, crenate or shortly dentate, sessile or subsessile, or the base tapering to a short, winged petiole; inflorescence spiciform, 6–20 (–30) cm. long, 2–4 cm. wide, generally unbranched, the flowers solitary (or sometimes paired), sessile or very shortly pedicellate, in the axils of lanceolate, foliaceous bracts, 0·8–3 cm. long, 0·2–1 cm. wide; hypanthium obconical, 7–8 mm. long, 5–6 mm. wide, distinctly costate, conspicuously setose; corolla pale blue, widely infundibuliform, about 2 cm. long, 3–4 cm. wide, glabrous, divided almost half way into 5 ovate, acute lobes; filaments about 4·5 mm. long, strongly dilated and about 2·5 mm. wide at the base, margins densely ciliate; anthers linear, about 4 mm. long, 0·6 mm. wide; style 8–10 mm. long, slender, glabrous; stigmas 3, filiform, about 7 mm. long. Fruit turbinate, obconical, about 12 mm. long, 10 mm. wide at apex, pale brown, rather papery, distinctly costate, rupturing erratically when fully ripe; seeds ovoid-ellipsoid, distinctly compressed, about 0·5 mm. long, 0·4 mm. wide; testa bright shining brown, darker at the extremities.

HAB.: By springs and sides of streams, or in damp Pine forest, usually at high altitudes, but sometimes on wet ground at lower elevations; near sea-level to 4,500 ft. alt.; fl. July–Oct.

DISTR.: Divisions 2, 3. Also Rhodes, Turkey, Syria, Palestine.

2. Locally frequent: Galata, Troödhitissa Monastery, Stavros tis Psokas, Kakopetria, Platres, Kato Mylos, Perapedhi, Prodhromos, Platania, Kalopanayiotis, Madhari, etc. *Kotschy* (1840) 9! *Sintenis & Rigo* 739! *Syngrassides* 865! *Kennedy* 917–919! *Davis* 1844! *Casey* 922! *G. E. Atherton* 594! *P. H. Oswald* 150! *A. Genneou* in ARI 1285! 1564! etc.

3. Kolossi, 1939, *Lindberg f.* s.n.! Asomatos marshes, 1940, *Mavromoustakis* 79!

2. LEGOUSIA *J. F. Durande*

Fl. Bourgogne, 1: 37 (1782).

Annual herbs, inflorescences corymbose-cymose or occasionally spiciform; flowers subsessile or shortly pedicellate; hypanthium elongate; corolla rotate or subrotate, 5-lobed; stamens 5, free; filaments dilated towards base; style simple; stigmas 3, filiform, ultimately revolute. Capsule crowned with the persistent calyx, linear or narrowly oblong-ellipsoid, prismatic, 3-locular, dehiscing by sub-apical valves; seeds numerous, small.

About five species in Europe, western Asia, North Africa and the Atlantic Islands.

Inflorescence corymbose-cymose, flowers all chasmogamous:
Corolla 15–23 mm. diam., generally exceeding the calyx   -   -   **1. L. speculum-veneris**
Corolla generally less than 3 mm. diam., much shorter than the calyx -   -   **2. L. hybrida**
Inflorescence spiciform, some of the flowers frequently cleistogamous; corolla usually less than
    8 mm. diam., shorter than the calyx   -   -   -   -   -   -   **3. L. falcata**

1. L. speculum-veneris (*L.*) *Chaix* in Vill., Hist. Plant. Dauph., 1: 338 (1786) — as *Legouzia speculum veneris*; Osorio-Tafall et Seraphim, List Vasc. Plants Cyprus, 100 (1973); Davis, Fl. Turkey, 6: 85 (1978).

Campanula speculum-veneris L., Sp. Plant., ed. 1, 168 (1753).

Specularia speculum (L.) A.DC., Mon. Campan., 346 (1830); Boiss., Fl. Orient., 3: 958 (1875); H. Stuart Thompson in Journ. Bot., 44: 334 (1906); Holmboe, Veg. Cypr., 177 (1914) — all as *S. speculum*; Post, Fl. Pal., ed. 2, 2: 170 (1933) — as *S. speculum-veneris*; Rechinger f. in Arkiv för Bot., ser. 2, 1: 432 (1950) — as *S. speculum veneris*.

TYPE: "*inter segetes* Europae *australis*".

Erect or sprawling, usually much-branched annual 4–30 cm. high; stems angled, glabrous, smooth or thinly scabridulous, often tinged purplish; basal leaves oblong-spathulate, 2 cm. long, 1 cm. wide, often petiolate, with petioles up to 1·5 cm. long; cauline leaves sessile, oblong or obovate, 0·5–2·5 cm. long, 0·3–1 cm. wide, glabrous, pale green, apex obtuse, base abruptly cuneate or subamplexicaul, margins generally undulate, narrowly revolute, minutely scabridulous especially towards apex; flowers usually rather crowded in branched, terminal and subterminal, corymbose cymes, sessile; hypanthium narrowly prismatic, 10–15 mm. long, 1–1·5 mm. wide at anthesis, shortly and rather densely papillose-scabridulous; calyx-lobes linear, erect or spreading, 6–8 mm. long, 0·5–1 mm. wide, apex acute, margins recurved, minutely papillose-scabridulous especially towards apex; corolla rotate or very shallowly infundibuliform, 5-lobed, rich violet-purple or rarely pale blue or white, 1·5–2·3 cm. diam., generally exceeding the calyx, lobes obtuse or acute, usually with a small, papillose mucro; filaments glabrous, about 0·8–1 mm. long, somewhat dilated towards base; anthers linear, pallid, about 3·5 mm. long, 0·4 mm. wide; style 4–7 mm. long; stigmas linear, about 1·5 mm. long, at first erect and connivent, later strongly revolute. Capsule narrowly oblong, constricted at the apex, about 15 mm. long, 3·5 mm. wide, angled, minutely papillose-scabridulous; apical valve ovate-elliptic, about 2 mm. long, 1·3 mm. wide, deciduous; seeds broadly ovoid, about 1 mm. long, 0·8 mm. wide, strongly compressed; testa shining brown, darker around the rim.

HAB.: Cultivated and fallow fields, roadsides, sometimes amongst garigue on open stony hillsides; sea-level to 3,700 ft. alt.; fl. March–May.

DISTR.: Divisions 1–3, 5–8. Widespread in southern Europe and the Mediterranean region eastwards to Jordan.

1. Dhrousha, 1941, *Davis* 3275 !
2. Troödos, 1912, *Haradjian* 532; Karavostasi, 1932, *Syngrassides* 251 ! Platres, 1937, *Kennedy* 922 ! Perapedhi, 1937, *Kennedy* 923 ! Kakopetria, 1952, *F. M. Probyn* 94 ! Beyond Vouni towards Yialia, 1957, *Merton* 2950 ! Yialia, 1959, *C. E. H. Sparrow* 16 ! N. of Kykko Monastery, 1959, *P. H. Oswald* 111 ! Pomos Point, 1962, *Meikle* 2243 !
3. Limassol, 1913, *Haradjian* 562, 570, 571; between Yerasa and Limassol, 1941, *Davis* 3086 ! Limassol aerodrome, 1950, *Chapman* 561 ! Zakaki, 1955, *Merton* 2068 ! Near Souni, 1960, *N. Macdonald* 129 ! Moni, 1964, *J. B. Suart* 212 !
5. Near Alambra, 1967, *Merton* in ARI 614 !
6. Nicosia, Government House, 1927, *Rev. A. Huddle* 74 !
7. Near Kephalovryso, 1880, *Sintenis & Rigo* 604 ! St. Hilarion, 1901, *A. G. & M. E. Lascelles* s.n. ! Vasilia, 1936, *Syngrassides* 1021 !, also, same locality, 1941, *Davis* 2960 ! 1956, *G. E. Atherton* 1100 ! 1967, *Merton* in ARI 533 ! Kyrenia, 1949, *Casey* 567 !
8. Apostolos Andreas, 1957, *Merton* 3086 !

NOTES: Not collected by Sibthorp, Kotschy or Holmboe, and evidently on the increase, since it is a conspicuous plant, not likely to have been overlooked by these collectors.

[L. PENTAGONIA (*L.*) *Thell.* in Vierteljahr. Naturf. Ges. Zürich, 52: 465 (1907) — *Campanula pentagonia* L., Sp. Plant., ed. 1, 169 (1753) — is very similar to *L. speculum-veneris* (L.) Chaix, but with larger (up to 3 cm. diam.) flowers in less crowded, racemose or paniculate inflorescences, and with the filaments strongly dilated and hairy at the base. It occurs in all the countries adjacent to Cyprus, but has not yet been collected on the island. The filaments provide the most satisfactory distinction between the two species; it is also true that the capsules of *L. pentagonia* are less constricted at the apex than those of *L. speculum-veneris*.]

2. L. hybrida (*L.*) *Delarbre*, Fl. Auvergne, ed. 2, 47 (1800); Osorio-Tafall et Seraphim, List Vasc. Plants Cyprus, 101 (1973); Davis, Fl. Turkey, 6: 84 (1978).
   Campanula hybrida L., Sp. Plant., ed. 1, 168 (1753).
   Specularia hybrida (L.) A.DC., Mon. Campan., 348 (1830); Unger et Kotschy, Die Insel Cypern, 259 (1865); Boiss., Fl. Orient., 3: 960 (1875); H. Stuart Thompson in Journ. Bot., 44: 334 (1906); Holmboe, Veg. Cypr., 177 (1914).
   TYPE: "*in* Anglia, Gallia, *inter segetes*".

Erect or spreading, branched or unbranched annual, 4–15 cm. high; stems angled, glabrous, smooth or thinly scabridulous, sometimes tinged purple or bronze; basal leaves oblong-spathulate, about 10 mm. long, 0·8 mm. wide, often petiolate, with petioles up to 10 mm. long; cauline leaves sessile, oblong or obovate, 0·5–1·5 cm. long, 0·3–0·8 cm. wide, glabrous, pale green, apex obtuse, base abruptly cuneate or sub-amplexicaul, margins strongly undulate, narrowly revolute, minutely scabridulous; flowers usually rather crowded in branched, terminal and subterminal, corymbose cymes, sessile; hypanthium 10–15 mm. long, 2–3 mm. wide at anthesis, generally glabrous and smooth, or occasionally minutely and sparsely scabridulous; calyx-lobes lanceolate, acuminate, 5–8 mm. long, 1–1·5 mm. wide, slightly accrescent in fruit, glabrous, margins recurved, minutely papillose-scabridulous; corolla infundibuliform-rotate, very small, 5-lobed, whitish externally, violet-purple within, generally less than 3 mm. diam., much shorter than calyx, lobes obtuse, or acute, usually with a small, minutely papillose mucro; filaments glabrous, about 0·5–0·8 mm. long, somewhat dilated towards base; anthers linear, pallid, about 0·8–1 mm. long, 0·3–0·4 mm. wide, shortly acute or obtuse; style about 1·5 mm. long; stigmas linear, about 0·5 mm. long, at first erect and connivent, later revolute, usually concealed by the withered corolla. Capsule narrowly oblong, strongly constricted at apex, 15–30 mm. long, 3–4 mm. wide, angled, smooth or minutely and very sparsely scabridulous; apical valve ovate-elliptic, about 1·5 mm. long, 1 mm. wide, deciduous; seeds elliptic-ovate, about 1·3 mm. long, 0·7 mm. wide, strongly compressed; testa shining brown, not much darker around the rim.

HAB.: Cultivated and fallow fields; sea-level to 3,800 ft. alt.; fl. March–April.

DISTR.: Divisions 1–3, 6. W. and S. Europe, Mediterranean region, Atlantic Islands.

1. Yiolou, 1962, *Meikle* 2231!
2. Phini, 1862, *Kotschy* 800; Platres, 1938, *Kennedy* 921!
3. Two miles W. of Limassol, 1964, *J. B. Suart* 180!
6. Vizakia, 1936, *Syngrassides* 1013! Peristerona, 1936, *Syngrassides* 1214!

3. L. falcata (*Ten.*) *Fritsch ex Janchen* in Mitt. Naturw. Ver. Univ. Wien, 5: 100 (1907); Osorio-Tafall et Seraphim, List Vasc. Plants Cyprus, 101 (1973); Davis, Fl. Turkey, 6: 84 (1978).
   Prismatocarpus ["*Prysmatocarpus*"] *falcatus* Ten., Prodr. Fl. Nap., XVI (1811), Atlas, 1: t. XX (1811–1838).
   Specularia falcata (Ten.) A.DC., Mon. Campan., 345 (1830); Unger et Kotschy, Die Insel Cypern, 259 (1865); Boiss., Fl. Orient., 3: 960 (1875); Holmboe, Veg. Cypr., 177 (1914); Post, Fl. Pal., ed. 2, 2: 170 (1933).
   Triodanis falcata (Ten.) McVaugh in Wrightia, 1: 28 (1945).

Erect or spreading, branched or unbranched annual, (2–) 6–30 (–35) cm. high; stems angled, glabrous or hispid-scabridulous, commonly tinged purple; basal leaves oblong-spathulate, 10–20 mm. long, 8–15 mm. wide, usually petiolate, with petioles up to 20 mm. long; cauline leaves sessile, ovate or obovate, 1–3·5 cm. long, 0·8–1·5 (–2) cm. wide, glabrous or hispidulous, pale green, apex obtuse or subacute, base cuneate, margins flat or undulate, entire or crenate, narrowly revolute; inflorescence spiciform, the flowers apparently solitary and remote (the lateral branches suppressed) in the axils of the upper cauline leaves, sessile or apparently pedicellate, the atrophied lateral branches forming what appear to be bracteoles; hypanthium 8–15 mm. long, 2–2·5 mm. wide at anthesis, almost parallel-sided, or sometimes becoming fusiform in fruit, glabrous, sparsely papillose-scabridulous, or sometimes rather densely hispid-scabridulous; calyx-lobes linear-lanceolate or linear-subulate, 3–15 (–20) mm. long, 0·5–2 mm. wide, suberect or spreading, often falcately curved to one side, glabrous or hispidulous, distinctly accrescent in fruit, margins strongly recurved; corolla present or absent (in cleistogamous flowers), infundibuliform-rotate, 5-lobed, violet-purple, mauve or lavender-blue, generally less than 8 mm. diam. and distinctly shorter than the calyx, lobes acute or subacute, about 4–5 mm. long, not or indistinctly mucronate; filaments glabrous, about 0·8–1 mm. long, gradually dilated towards base; anthers linear, subacute, pallid, about 2 mm. long, 0·4 mm. wide; style 4–5 mm. long, thinly and shortly pilose above; stigmas linear, about 0·8 mm. long, at first erect and connivent, later revolute. Capsule narrowly oblong, or sometimes swollen and fusiform, not abruptly constricted at apex in fruit, 8–20 mm. long, 2·5–3 mm. wide, angled, smooth and glabrous, or papillose, or sometimes rather densely hispid-scabridulous, apical valve ovate-elliptic, about 1·5 mm. long, 1 mm. wide, deciduous; seeds suborbicular, lentil-shaped, about 0·8–1 mm. diam., with a well-marked, thickened rim; testa shining brown, rather paler around the rim.

var. falcata

TYPE: Italy; "presso Napoli può raccogliersi nelle vicinanze del lago di Agnano" (NAP? — figured in Fl. Nap. Atlas, 1: t. XX !).

Capsule, and plant as a whole, glabrous, subglabrous or thinly papillose-scabridulous.

HAB.: Stony pastures and amongst garigue on hillsides; sea-level to 500 ft. alt.; fl. April–May.

DISTR.: Division 1. Widespread in the Mediterranean region, Atlantic Islands, eastwards to Iran.

1. Akamas, 1950, *Casey* 1020!

NOTES: Apparently much less common in Cyprus than *L. falcata* var. *scabra* (*infra*).

var. scabra (*Lowe*) Meikle comb. nov.

Prismatocarpus scaber Lowe in Trans. Cambridge Phil. Soc., 6 (3): 16 (1838).
Specularia falcata (Ten.) A.DC. var. *scabra* (Lowe) A.DC. in DC., Prodr., 7 (2): 490 (1939).
TYPE: Madeira, June 1837, *C. Lemann* 1838 (K !).

Capsule, and plant as a whole, hispid-scabridulous.

HAB.: Stony pastures and amongst garigue on hillsides; by roadsides and streamsides in Pine forest; 800–5,400 ft. alt.; fl. April–May.

DISTR.: Divisions 2, 7. Madeira, Sardinia, Thasos.

2. Prodhromos, 1862, *Kotschy* 734! Kalopanayiotis, 1937, *Syngrassides* 1569! Platres, 1937, *Kennedy* 930! Kryos Potamos, 3,200 ft., 1937, *Kennedy* 924! Xerokolymbos, 1937, *Kennedy* 925! Livadhi tou Pasha, 1937, *Kennedy* 926! Tripylos, 1941, *Davis* 3489! Near Platres, 1960, *N. Macdonald* 158! Palaeokhori, Tripylos, 1962, *Meikle* 2669!
7. Mountains above Phlamoudhi, 1880, *Sintenis & Rigo* 23! Kantara, 1955, *J. Hughes* s.n.! Above Antiphonitis monastery, 1955, *Merton* 2272! Kyrenia Boghaz, 1956, *G. E. Atherton* 1382!

NOTES: The prevailing variant in Cyprus, almost confined to high altitudes on the Northern and Troödos Ranges, where it sometimes occurs in vast quantity on dry mountainsides, forming carpets consisting of innumerable, small, depauperate, largely cleistogamous individuals similar to *Specularia falcata* (Ten.) A.DC. var. *pusilla* Boiss., (Fl. Orient., 3: 960; 1875) but with the characteristic hispid-scabridulous hypanthia and capsules of var. *scabra*.

Such small plants appear to be no more than environmental states of typical var. *scabra*, which is generally to be found nearby.

The distribution of *L. falcata* var. *scabra* is very unusual: the Cyprus plants agree closely with type material from Madeira, otherwise plants with hispidulous capsules have been noted only from Sardinia (*Reverchon* 304!) and Thasos (*H. G. Tedd* 987!), though I suspect var. *scabra* has a wider distribution in the Mediterranean region. Contrary to some statements, var. *scabra* is distinct from *Legousia castellana* (Lange) Samp. This latter has a glabrous, or very sparsely scabridulous hypanthium and is questionably distinct, even as a variety, from typical *L. falcata*.

Legousia falcata, in the aggregate sense, has also been recorded from (2) Galata, *Kotschy*, (3) near Pissouri, *Kotschy*, (8) near Komi Kebir and Ephtakomi, *Sintenis, Holmboe* 517.

[L. PERFOLIATA (*L.*) *Britton* in Mem. Torrey Bot. Club, 5: 309 (1894) — *Triodanis perfoliata* (L.) Nieuwl. in Amer. Midland Nat., 3: 192 (1914) — a widespread North American species, is represented by a small fragment of an inflorescence collected "Nicosia, 600 ft. Apr. 1927, *Rev. Alfred Huddle* 73" (K!). The plant is otherwise unknown, either in Cyprus, or, it would appear, in the Mediterranean area, and one must assume that it was found as a casual on cultivated ground, possibly in the vicinity of Government House, since *Rev. A. Huddle* 74 (*Legousia speculum-veneris*), also collected in April 1927, is so localized.

Legousia perfoliata somewhat resembles *L. falcata* but has long, unbranched, spicate inflorescences, with the relatively short capsules concealed by the conspicuous cordate-amplexicaul bracts. The upper flowers in each inflorescence are chasmogamous, with blue, rotate-infundibuliform corollas; the lower flowers are generally cleistogamous.]

SUBFAMILY 2. **Lobelioideae** Flowers zygomorphic; anthers coherent, filaments connate almost to the base.

3. SOLENOPSIS *C. Presl*

Prodr. Mon. Lobel., 32 (1836).

Laurentia Adans., Fam. Plant., 2: 134, 568 (1763); F. E. Wimmer in Engl., Pflanzenr., 107 (IV. 276b): 386 (1953) nom. illeg.

Annual or perennial herbs, with erect or procumbent stems, or acaulescent; leaves generally alternate, simple; flowers usually axillary, solitary, pedunculate; hypanthium obovoid or turbinate; calyx-lobes 5, linear or subulate, persistent in fruit; corolla zygomorphic, persistent, bilabiate; tube entire (not split dorsally as in *Lobelia*) cylindrical or narrowly infundibuliform; corolla-lobes subequal or unequal, 2 upper erect, 3 lower pointing downward; stamens included or very shortly exserted; 2 lower anthers with pilose tufts at apex. Fruit a 2-locular capsule, dehiscence apical, loculicidal; seeds small, shining, numerous.

About 25 species, chiefly in S. Africa and Australia, also Mediterranean region, N. & S. America. Questionably distinct (as a genus) from *Lobelia*, into which genus Linnaeus placed the Mediterranean species.

1. S. **minuta** (*L.*) *C. Presl*, Prodr. Mon. Lobel., 32 (1836); Meikle in Kew Bull., 34: 373–375 (1979).

    Lobelia minuta L., Mantissa Altera, 292 (1771).

    Lobelia setacea Sm. in Sibth. et Sm., Fl. Graec. Prodr., 1: 145 (1806), Fl. Graec., 3: 16, t. 221 (1819) non Thunb.

    Laurentia minuta (L.) A.DC. in DC. Prodr., 7: 410 (1839); Lindberg f., Iter Cypr., 33 (1946); Wimmer in Engl., Pflanzenr., 107 (IV, 276b): 388 (1953); Osorio-Tafall et Seraphim, List Vasc. Plants Cyprus, 101 (1973).

    Laurentia tenella A.DC. in DC. Prodr., 7: 410 (1839); Poech, Enum. Plant. Ins. Cypr., 22 (1842); Kotschy in Oesterr. Bot. Zeitschr., 12: 279 (1862); Unger et Kotschy, Die Insel

Cypern, 258 (1865); Boiss., Fl. Orient., 3: 884 (1875); Holmboe, Veg. Cypr., 177 (1914); Post, Fl. Pal., ed. 2, 2: 161 (1933).

Acaulescent, rhizomatous perennial, forming dense leafy mats, the stems so condensed as to be barely visible below the foliage; leaves narrowly obovate-spathulate, lamina 0·8–3 cm. long, 0·5–1·5 cm. wide, bright green, glabrous or thinly clothed with short, thick, hyaline papillose hairs; apex rounded or subacute, margins entire, subentire or rarely distinctly lobed, nervation obscure, base tapering to a flattened petiole up to 5 cm. long; flowers solitary (or very rarely paired); peduncles 2–12 cm. long, generally much exceeding the leaves or bearing 1–3, alternate or opposite, linear-subulate bracts 2–8 mm. long; 0·3–0·6 mm. wide; hypanthium broadly turbinate or subglobose, 1·5–2 mm. long, about 1·5 mm. wide, glabrous, obscurely ridged; calyx-lobes linear or subulate, acute, erect, 1·5–3 (–4) mm. long, 0·3–1 mm. wide; corolla distinctly bilabiate, with a pallid throat and pale or bright blue lobes; corolla-tube very narrowly infundibuliform, 3–5 mm. long, 2 upper lobes lanceolate, acute, 2–4·5 mm. long, 0·8–1·5 mm. wide, 3 lower lobes oblong, obtuse or rounded at the apex, 1·5–4 mm. long, 1–3 mm. wide, the central lobe distinctly broader than the laterals, throat generally clothed with a varying number of short, white papillose hairs; stamens very shortly exserted from corolla-tube; filaments 2·5–3 mm. long, connate above, free or partly free towards base; anthers united in a ring around the stigma, dark fuscous-blue, each about 1–1·5 mm. long, 0·5 mm. wide, the 2 lower with tufts of white papillose hairs at the apex; style about 3 mm. long; stigma with 2 flat, subacute, lingulate lobes surrounded at the base by a ring of stigmatic hairs. Capsule subglobose, 2·5–4 mm. diam.; seeds ovoid-oblong, about 0·4–0·6 mm. long, 0·3–0·4 mm. wide; testa bright shining brown, minutely and closely, longitudinally striatulate.

ssp. **nobilis** (*E. Wimmer*) *Meikle* in Kew Bull., 34: 374 (1979).

    Laurentia minuta (L.) A.DC. f. *nobilis* E. Wimmer in Ann. Naturh. Mus. Wien, 56: 333 (1948), Engl. Pflanzenr., 107 (IV, 276b): 390 (1953) in adnot.

    Lobelia tenella Biv., Sic. Plant. Cent. 1, 53. t. 2 (1806) excl. syn. non L.

    Lobelia bivonae Tineo, Cat. Plant. Hort. Reg. Panor., 279 (1827).

    Solenopsis bivonaeana C. Presl, Prodr. Mon. Lobel., 32 (1836).

    TYPE: Not indicated by Wimmer. *Kotschy* 576 from Cyprus "Frequens ad fontes in pago Moni inter Larnaca et Limassol die 28 Aprilis [1862]" (W, K!), here chosen as lectotype.

Plants robust; leaves (including petiole) commonly 4–8 cm. long, 1–1·5 cm. wide, quite glabrous or rarely with a very few papillose hairs; peduncles often 5–12 cm. long; corolla large, commonly 10–12 mm. long, 8–10 mm. wide. Seeds 0·6 mm. long, 0·4 mm. wide. *Plate 65.*

HAB.: Moist, dripping banks by streams and springs; generally at 3,000–5,500 ft. alt., but sometimes descending to the plains; fl. March–Oct.

DISTR.: Divisions 2, 3, 6, locally abundant in Division 2. Also Sicily.

2. Locally abundant: Evrykhou, Troödhitissa, Makheras, Galata, Kykko, Platres, Amiandos, Moniatis, Khionistra, Platania, Prodhromos, Kakopetria, Phini, Tripylos, etc. *Sibthorp*; *Aucher-Eloy* 3854! *Sintenis & Rigo* 742! *Post*! *Holmboe* 1075; *Haradjian* 416! 944! *C. B. Ussher* 35! *Syngrassides* 520! 1257! *Kennedy* 927! 928! 929! 930! *Lindberg f.*, s.n.! *Davis* 1787! *S. G. Cowper* s.n.! *Casey* 828! *Merton* 2321! *G. E. Atherton* 557! *P. H. Oswald* 149! *D. P. Young* 7379! *Meikle* 2874! *Economides* in ARI 1229! *A. Genneou* in ARI 1571! *P. Laukkonen* 391! etc.

3. Moni, 1862, *Kotschy* 576!

6. Strovolos aqueduct, 1901, *A. G. & M. E. Lascelles* s.n.! Nicosia, 1905, *Holmboe* 292!

NOTES: Aggregate *Solenopsis minuta* is fairly readily divisible into a number of geographical races, which, though morphologically interconnected, warrant recognition at some rank. Wimmer preferred to regard them as forms, but since that rank is generally given to taxa of little or no phytogeographic significance, they have here been elevated to subspecies. Since Linnaeus supposed that the type of his *Lobelia minuta* came from the Cape of Good Hope ("*in* Cap. b. spei *subaquosis*"), and since no Linnaean material appears to survive, it is difficult to

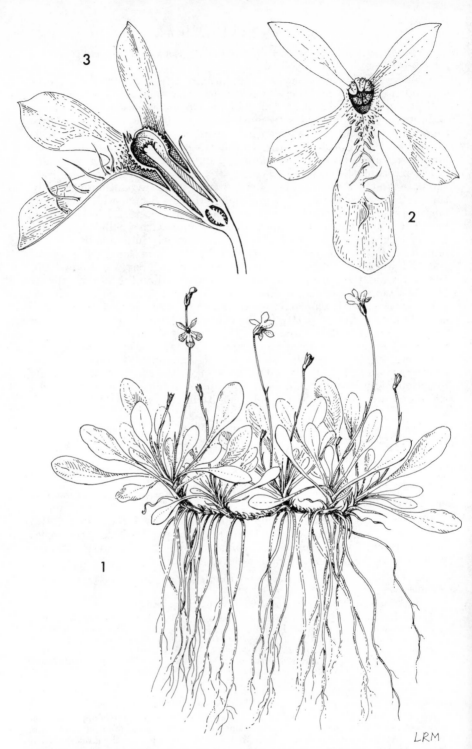

Plate 65. SOLENOPSIS MINUTA (L.) C. Presl ssp. NOBILIS (E. Wimmer) Meikle **1**, habit, ×1; **2**, flower, front view, ×9; **3**, flower, longitudinal section, ×9. (**1** from *Kennedy* 927, 928; **2, 3** from *Kennedy* 928.)

identify the species with exactitude. My own feeling is that the locality cited by Linnaeus is erroneous and that he was dealing with cultivated (perhaps mislabelled) plants, descended from the material collected by Tournefort in Crete. Accordingly I suggest that specimen no. 861 of *Rapuntium creticum, minimum, Bellidis folio, flore maculato* in the Tournefort Herbarium (P) is an appropriate neotype.

55. ERICACEAE
L. Watson in Journ. Linn. Soc. Bot., 59: 111–125 (1965).

Mostly shrubs or subshrubs, rarely trees; leaves alternate, rarely opposite or verticillate, simple, often evergreen, exstipulate, entire or toothed, frequently small, narrow or subulate, and thick with strongly recurved or revolute margins (*ericoid*); buds naked or perulate. Inflorescences various, terminal or lateral, generally with bracts or bracteoles; flowers hermaphrodite, actinomorphic or somewhat zygomorphic; calyx often persistent, 4–5 (–7)-lobed; corolla usually gamopetalous, often inserted below a nectariferous disk, cylindrical, urceolate, campanulate or hypocrateriform, with 3–7 imbricate, or occasionally valvate, lobes. Stamens usually twice as many as corolla–lobes, obdiplostemonous, free, inserted on the disk or receptacle, or sometimes with the filaments adnate to the base of the corolla-tube; anthers 2-thecous, frequently with awn-like or spur-like appendages, usually dehiscing by apical pores, or sometimes by longitudinal slits; pollen in tetrads, occasionally mixed with slender, viscid hairs. Ovary superior or inferior, (2–) 4–5 (–7)-locular, placentation axile, ovules generally numerous, anatropous; style simple, stigma usually capitate. Fruit commonly a loculicidal or septicidal capsule, sometimes a nut, berry or drupe; seeds with a straight, cylindrical embryo and copious endosperm.

About 125 genera and more than 1,300 species, chiefly in temperate regions or on mountains in the tropics, with concentrations of *Erica* species in S. Africa and *Rhododendron* species in W. China and the Himalayas.

Leaves whorled, small (less than 12 mm. long), linear; corolla persistent; fruit a capsule; small shrubs and subshrubs - - - - - - - - - - - - - - **1. Erica**
Leaves alternate, large (more than 3 cm. long), elliptic, oblong or oblong-lanceolate; corolla deciduous; fruit a fleshy berry; large shrubs or small trees - - - **2. Arbutus**

SUBFAMILY 1. **Ericoideae** Mostly shrubs without perulate buds, and with numerous small, coriaceous (*ericoid*) leaves; corolla often persistent; anthers usually appendaged; ovary superior. Fruit generally a loculicidal capsule; seeds not winged.

1. ERICA *L.*
Sp. Plant., ed. 1, 352 (1753).
Gen. Plant., ed. 5, 167 (1754).

Evergreen shrubs and subshrubs; leaves whorled, small, narrow, coriaceous with strongly revolute margins (*ericoid*); petioles short, erect; inflorescences various, lateral or terminal, umbellate, racemose, paniculate or clustered; flowers 4 (–5)-merous; calyx deeply lobed, shorter than corolla, not petaloid; corolla cylindrical, campanulate or urceolate, persistent, lobes generally shorter than tube; stamens 8 (–10), anthers with or without appendages, opening by pores. Fruit a loculicidal, many-seeded capsule.

About 500 species, chiefly S. African, with a much smaller secondary centre of distribution in S.W. Europe.

Flowers in terminal, umbellate clusters; corolla usually 5-lobed; anthers included in the cylindrical or urceolate corolla-tube - - - - - - - **1. E. sicula**
Flowers axillary, forming subterminal spiciform inflorescences; corolla 4-lobed; anthers exserted from campanulate corolla-tube - - - - - **2. E. manipuliflora**

1. **E. sicula** *Guss.*, Cat. Plant. Hort. Reg. Boccad., 74 (1821); Kotschy in Oesterr. Bot. Zeitschr., 12: 276 (1862); Chapman, Cyprus Trees and Shrubs, 60 (1949).
    Pentapera sicula (Guss.) Klotzsch in Linnaea, 12: 498 (1838); Unger et Kotschy, Die Insel Cypern, 297 (1865); Boiss., Fl. Orient., 3: 969 (1875); Post, Fl. Pal., ed. 2, 2: 173 (1933); Osorio-Tafall et Seraphim, List Vasc. Plants Cyprus, 81 (1973).
    Pentapera sicula (Guss.) Klotzsch var. *libanotica* C. et W. Barbey, Herb. au Levant, 144 (1882).
    Erica sicula Guss. var. *libanotica* (C. et W. Barbey) Holmboe, Veg. Cypr., 142 (1914).
    Pentapera sicula (Guss.) Klotzsch ssp. *libanotica* (C. et W. Barbey) Yaltirik in Notes Roy. Bot. Gard. Edinb., 28: 13 (1967).
    Erica sicula Guss. ssp. *libanotica* (C. et W. Barbey) P. F. Stevens in Davis, Fl. Turkey, 6: 97 (1978).
    TYPE: Sicily; "Monte Cofani presso Trapani". *Gussone* (NAP).

Erect or spreading shrub up to 65 cm. high, sometimes forming colonies up to 2 m. diam.; stems much branched, at first sulcate, densely pubescent or shortly tomentellous, later terete, brownish, with fuscous or greyish, flaking bark; leaves in whorls of 4, spreading or erect, usually dense and overlapping, linear, 8–12 mm. long, about 1 mm. wide, thick, glabrous or ± densely glandular-puberulous, apex obtuse or subacute, base rather abruptly narrowed, margins strongly revolute and meeting to form a well-marked median furrow running the length of the lower surface of the lamina; petiole erect, pallid, about 2 mm. long, usually pubescent; flowers 3–8 (or more) in rather dense terminal umbels; bracts pale pink, lanceolate, acuminate, about 3·5 mm. long, 1·5 mm. wide, puberulous, with glandular margins; pedicels angular, 5–8 mm. long, glandular-pubescent; bracteoles similar to bracts, but narrower, often almost linear, and more strongly concave; calyx usually deeply 5-lobed, pinkish, lobes narrowly deltoid-ovate, about 6 mm. long, 1–1·5 mm. wide at base, erect, glandular-puberulous; corolla pale or bright pink, cylindrical or narrowly urceolate, 7–8 mm. long, about 5 mm. wide, shortly pubescent, lobes 5, erect or recurved, suborbicular, about 1 mm. long and as wide or somewhat wider; filaments erect, flattened, linear, glabrous, about 3·5 mm. long; anthers included, narrowly ovate, connivent, dark purple, shortly puberulous, about 2 mm. long, 1 mm. wide; ovary globose, about 2 mm. diam., densely tomentose-villose; style stout, 4·5–5 mm. long, equalling or shortly exceeding corolla-tube, sparsely pilose; stigma capitate, about 0·7 mm. diam. Fruit depressed-globose, 4–5 mm. diam., shortly tomentose-villose, obscurely 10-sulcate; seeds oblong, about 0·8 mm. long, 0·4 mm. wide; testa dark brown, shining, minutely and sparsely, longitudinally sulcate.

HAB.: Fissures of limestone rocks; occasionally in garigue on limestone mountainsides; 900–3,200 ft. alt.; fl. March–August.

DISTR.: Divisions 2, 7, very local. Also Sicily, S. Turkey, Lebanon, Libya (Cyrenaica).

2. Aphamis, 3,200 ft. alt., in garigue between vineyards, 1937, *Kennedy* 665!
7. Pentadaktylos, 1859, *Kotschy* 253; Ayios Khrysostomos, 1862, *Kotschy* 429; near Akanthou, 1880, *Sintenis & Rigo* 222! St. Hilarion, 1904–1905, *Hartmann*; Buffavento, 1937, *Syngrassides* 1738! Yaïla, 1938–1941, *Kennedy* 1348! *Davis* 2796! 2903! Halevga, 1940–1949, *Davis* 1919! *Chapman* 126! second gorge W. of St. Hilarion, 1951, *Casey* 1202! Akanthou, 1967, *Merton* in ARI 793! Above Phlamoudhi, 1967, *Merton* in ARI 747!

NOTES: Holmboe (Veg. Cypr., 144) refers the Cyprus populations of this species to *Pentapera sicula* var. *libanotica* C. et W. Barbey, remarking: "Our plant agrees with the Syrian one in having more glabrescent leaves and thinner branches than the Sicilian type". The distinctions are, however, unsatisfactory, since Cyprus specimens vary considerably in both respects, depending, I suspect, on whether they come from open or shaded habitats. More recently the variety has been given subspecific rank, and P. F. Stevens (in Davis, Fl. Turkey, 6: 97) comments that it also differs in having "longer, narrower leaves and brighter pink flowers". The additional leaf character is again unsatisfactory as regards Cyprus specimens, but it would appear true that *E. sicula* from Sicily and Cyrenaica has very pallid flowers, or that the corollas may even be pure white (*teste* J. D. Hooker in Bot. Mag. t. 7030), while some of the Cyprus specimens are said to have bright pink flowers, though, for the most part, the flower-colour is not indicated, and cannot be judged from dried specimens. Until *E. sicula* has been more carefully investigated, throughout its distribution, I prefer not to subdivide the species, though it may be that future research will confirm the validity of such subdivision.

2. **E. manipuliflora** *Salisb.* in Trans. Linn. Soc., 6: 344 (1802); Osorio-Tafall et Seraphim, List Vasc. Plants Cyprus, 81 (1973); Davis, Fl. Turkey, 6: 96 (1978).

   E. verticillata Forssk., Fl. Aegypt.–Arab., 210 (1775); Boiss., Fl. Orient., 3: 970 (1875); H. Stuart Thompson in Journ. Bot., 44: 334 (1906); Holmboe, Veg. Cypr., 142 (1914); Post, Fl. Pal., ed. 2, 2: 173 (1933); Chapman, Cyprus Trees and Shrubs, 60 (1949) non Bergius (1767) nom. illeg.

   TYPE: Turkey; "Circa *Constantinop.*" *Forsskål* (C).

Erect shrub up to 1 m. high; stems much branched, often rather fastigiate, at first strongly sulcate or angled, shortly puberulous or subglabrous, later terete, greyish-brown or fuscous, with finely fissured bark; leaves in whorls of 4, erect or less often spreading, usually dense and overlapping, linear, 4–7 mm. long, about 0·7 mm. wide, thick, subglabrous or ± densely papillose-puberulous, apex obtuse, base rather abruptly narrowed, margins strongly revolute and meeting to form a well-marked median furrow running the length of the lower surface of the lamina; petiole erect, pallid, about 0·8 mm. long, usually puberulous; flowers fragrant, generally in pairs, sometimes solitary or in threes in the leaf-axils of the sub-apical leaves, forming elongate, cylindrical, spiciform or racemose inflorescences; bracts brownish, deltoid, about 0·5 mm. long, 0·5 mm. wide at base, subglabrous with ciliolate margins; pedicels filiform, purplish, 4–12 mm. long, angled, glabrous or subglabrous; bracteoles subulate-deltoid, concave, ciliolate, about 0·5 mm. long, 0·3 mm. wide at base; calyx 4-lobed, pallid, lobes ovate, about 1 mm. long, 0·8 mm. wide, erect, subglabrous, ciliolate; corolla pale pink, campanulate, about 2·5 mm. long, 3 mm. wide, glabrous, lobes 4, erect, broadly and bluntly ovoid-deltoid, about 1 mm. long, 2 mm. wide at base; filaments erect, flattened, linear, glabrous, about 2·5 mm. long, 0·2 mm. wide at base; anthers exserted, 0·8 mm. long, 0·6 mm. wide, dark purple, the 2 glabrous thecae free to the base; ovary tetragonous, glabrous, crimson, about 0·7 mm. long, 0·6 mm. wide, seated on a conspicuous, 8-lobed purple disk; style about 4 mm. long, glabrous, much exceeding corolla-tube; stigma truncate. Fruit tetragonous, about 2 mm. long, 1·6 mm. wide, dividing into 4 spreading valves; seeds broadly ovoid, about 0·5 mm. long, 0·4 mm. wide; testa pale brown, coarsely reticulate.

HAB.: Steep rocky slopes and hillsides, often with *Rhamnus alaternus*, usually on chalk or limestone; 300–600 ft. alt.; fl. (July–) Aug.–Dec.

DISTR.: Divisions 6, 7, rare. Eastern Mediterranean region from Italy to Lebanon.

6. Kormakiti, Dec. 1901, *P. Gennadius* in *A. G. & M. E. Lascelles* s.n.!
7. N.E. of Akanthou on steep north-facing slopes, 1956, *Merton* 2834! Between Akanthou and Phlamoudhi, on hills locally named "Makriarashi" or "Ktnassa" and "Stallos" *Waterer* ms.

SUBFAMILY 2. **Vaccinioideae** Shrubs or small trees with perulate buds; leaves broad, flat; inflorescence usually a leafless raceme or panicle; corolla deciduous, urceolate; anthers usually appendaged; ovary superior or inferior. Fruit a berry, drupe or loculicidal capsule; seeds not winged.

2. ARBUTUS L.

Sp. Plant., ed. 1, 395 (1753).
Gen. Plant., ed. 5, 187 (1754).

Evergreen shrubs or small trees; twigs terete; leaves alternate, petiolate; flowers 5-merous, in terminal panicles; calyx deeply lobed; corolla urceolate, waxy, with 5 short, recurved lobes; stamens 10, included; filaments pilose; anthers versatile, with 2 apical appendages, dehiscing by apical pores; ovary superior, sessile on a swollen 10-lobed disk; style simple; stigma capitate or obscurely lobed; ovules numerous. Fruit a globose, 5-locular, many-seeded berry; seeds small, compressed.

Eleven species in western Europe, Mediterranean region, Atlantic Islands, western N. America and Mexico.

Bark rough, fibrous; inflorescences subglabrous or pubescent; fruits 1·5–2 cm. diam., rough with projecting papillae; leaves usually serrate, not noticeably discolorous   **1. A. unedo**
Bark smooth, flaking; inflorescence glandular-pilose; fruits less than 1·3 cm. diam., verruculose with flattened papillae; leaves usually entire, markedly pallid below   **2. A. andrachne**

1. **A. unedo** L., Sp. Plant., ed. 1, 395 (1753); Sibth. et Sm., Fl. Graec. Prodr., 1: 274 (1809), Fl. Graec., 4: 66, t. 373 (1823); Boiss., Fl. Orient., 3: 966 (1875); Post, Fl. Pal., ed. 2, 2: 172 (1933); Chapman, Cyprus Trees and Shrubs, 59 (1949); J. R. Sealy et D. A. Webb in Journ. Ecol., 38: 223–236 (1950); Osorio-Tafall et Seraphim, List Vasc. Plants Cyprus, 81 (1973); Davis, Fl. Turkey, 6: 99 (1978).
TYPE: "*in* Europa *australi*".

Shrub or small bushy tree, 3–5 (–12) m. high; trunk short, with numerous divergent branches, forming a rounded crown; bark brown, tinged reddish, rough, sometimes coming away in irregular shreds; young shoots often reddish, very sparsely glandular-hispidulous or glabrous; leaves coriaceous, elliptic or oblong-lanceolate, 3–8 cm. long, 1–3 cm. wide, dark shining green above, rather paler and duller below, indistinctly reticulate-veined, glabrous, apex acute, margins usually sharply serrate, base tapering to a short channelled petiole generally less than 1 cm. long; inflorescence a rather dense, drooping, terminal panicle, 4–5 cm. long, 3–4 cm. wide, bracts ovate-deltoid, acuminate, concave, glabrous, 2–4 mm. long, 1–2 mm. wide, the lowermost often foliaceous and merging with the leaves; bracteoles wanting; pedicels short and relatively stout, 2–4 mm. long; calyx-lobes imbricate, broadly and bluntly ovate-deltoid, about 1 mm. long and a little wider at base, pallid and ciliate at the margin; corolla urceolate, waxy, white, tinged green or pink, glabrous externally, thinly white-pilose within, 7–10 mm. long, 4–7 mm. wide, lobes very short, rounded, imbricate, ciliolate; filaments flattened, lanceolate, villose, about 2 mm. long, 1–1·3 mm. wide; anthers oblong, about 1·8–2 mm. long, 1 mm. wide, with conspicuous apical caudate appendages; ovary subglobose, about 2 mm. diam., minutely verruculose; style about 3·5 mm. long; stigma capitate or obscurely lobed. Fruit a globose red or orange, yellow-fleshed berry 1·5–2 cm. diam.; exocarp rough with crowded, low pyramidal papillae; seeds irregularly ovoid, bluntly angled, about 3 mm. long, 1·6 mm. wide; testa light brown, closely reticulate with thick-walled, narrowly rectangular areoles.

HAB.: In low *Pinus brutia–Juniperus phoenicea* woodland, apparently on limestone; 100–300 ft. alt.; fl. Oct.–March.

DISTR.: Division 1, very rare. S.W. Europe and Mediterranean region, northwards to western Ireland.

1. Akamas, Psindron near Ayios Konon, 1949–1962, *Chapman* 291 ! *Poore* 58 ! *Meikle* 2120 ! In three spots in this area, namely, "Samakoudhi" near Ayios Konon, "Argaki tou Mouzouri" nearby, and in the "Argaki tou Exosirondes" valley, S. of Fontana Amorosa Bay, according to R. R. Waterer ms.

NOTES: Apparently indigenous here, but the very small population could possibly have arisen from introduced plants; R. R. Waterer (ms.) suggests it may have been introduced as a fruit tree; all the known localities are on the sites of Roman settlements.

2. **A. andrachne** *L.*, Sp. Plant., ed. 2, 566 (1762); Sibth. et Sm., Fl. Graec. Prodr., 1: 274 (1809), Fl. Graec., 4: 67, t. 374 (1823); Poech, Enum. Plant. Ins. Cypr., 27 (1842); Unger et Kotschy, Die Insel Cypern, 297 (1865); Boiss., Fl. Orient., 3: 966 (1875); Holmboe, Veg. Cypr., 142 (1914); Post, Fl. Pal., ed. 2, 2: 172 (1933); Lindberg f., Iter Cypr., 25 (1946); Chapman, Cyprus Trees and Shrubs, 59 (1949); Osorio-Tafall et Seraphim, List Vasc. Plants Cyprus, 81 (1973); Elektra Megaw, Wild Flowers of Cyprus, 9, t. 11 (1973); Davis, Fl. Turkey, 6: (1978).
   A. andrachne L. f. *arguteserrata* Lindberg f., Iter Cypr., 25 (1946).
   TYPE: "*in* Oriente".

Large bush or small tree 3–5 (–12) m. high; trunk short with numerous, divergent, contorted branches forming a broad, irregular crown; bark smooth, reddish, peeling off annually in early summer to expose a pale, greyish, or greenish underlayer; young shoots and leaves reddish-bronze, generally quite glabrous; leaves thickly coriaceous, broadly elliptic, 4–10 cm. long, 2–6 cm. wide, dark shining green above, distinctly glaucous and very finely reticulate-veined below, apex acute or subacute, margins usually entire, but sometimes serrate in juvenile or exceptionally robust specimens (f. *arguteserrata* Lindberg f.), base rounded or very broadly cuneate; petiole commonly 2–4 cm. long, shallowly channelled above; inflorescence an erect, much-branched panicle (3–) 6–9 (–12) cm. long, (3–) 5–8 (–10) cm. wide; bracts ovate-deltoid, acuminate, concave, glandular-ciliate, crimson, 2–4 mm. long, 2–3 mm. wide, the lowermost often foliaceous and merging with the leaves; bracteoles wanting, pedicels and rhachis densely glandular-pilose; calyx-lobes imbricate, broadly suborbicular, about 2 mm. wide, with a broad membranous, glandular-ciliate margin; corolla urceolate, waxy, white or pale green, glabrous externally, thinly white-pilose within, about 7 mm. long, 6 mm. wide, lobes very short, rounded, imbricate, ciliolate, strongly recurved; filaments flattened, with a broad ovate base narrowing abruptly to a short filiform apex, thinly pilose, about 2 mm. long, 1·5 mm. wide near the base; anthers shortly oblong, about 1·5 mm. long, 1·2 mm. wide, with conspicuous apical caudate appendages; ovary subglobose-conical, about 2 mm. diam., minutely verruculose and ± pilose with long hairs; style about 3 mm. long; stigma capitate. Fruit a globose, red, yellow-fleshed berry, exocarp verruculose with crowded, low, flattened papillae; seeds irregularly ovoid, bluntly angled, about 3 mm. long, 1·5 mm. wide; testa light brown, closely reticulate with thick-walled, narrowly rectangular areoles.

HAB.: In *Pinus brutia*, *Cupressus sempervirens* or *Juniperus phoenicea* woodland and maquis on rocky calcareous or igneous slopes; 300–5,000 ft. alt.; fl. Febr.–May.

DISTR.: Divisions 1–3, 7, 8, locally common. Eastern Mediterranean region from Albania to Palestine; Crimea; Caucasus.

1. Akamas, 1941, *Davis* 3334! Ayios Konon, 1962, *Meikle* 2119!
2. Common; Troödhitissa Monastery, Prodhromos, Kykko, Platres, Platania, Ora, Ayios Ioannis, etc. *Sibthorp*; *Sintenis & Rigo* 4a! *Haradjian* 945! *Kennedy* 663! *Lindberg f.* s.n.! *G. E. Atherton* 643! *Merton* in ARI 721! etc.
3. Common about Kakomallis near Louvaras, 1962, *Meikle*!
7. Common: Antiphonitis Monastery, Kantara, Pentadaktylos, Kyrenia, St. Hilarion, Halevga, etc. *Sibthorp*; *Sintenis & Rigo* 4! *A. G. & M. E. Lascelles* s.n.! *J. A. Tracey* 59! *Syngrassides* 1379! 1458! *Lindberg f.* s.n.! *Casey* 147! *Chapman* 468! *G. E. Atherton* 1078! *Barclay* 1080! *P. Laukkonen* 113!
8. Valia, 1905, *Holmboe* 475.

2a. **A. andrachne** *L.* × **A. unedo** *L.*
   A. × *andrachnoides* Link, Enum. Plant. Hort. Reg. Bot. Berol. Altera, 1: 395 (1821).

One Cyprus specimen (Akamas Forest, 28 March, 1938, *Kridhiotis* in *E. Chapman* 353) has been tentatively referred to this hybrid by J. R. Sealy. The specimen has leaves exactly intermediate between those of the putative parents, with subentire or obscurely serrate margins and short, but well-developed petioles. The inflorescence is immature, but lacks the glandulosity of *A. andrachne*, though the calyx-lobes closely resemble those of this species. On the whole it would seem that the cautious determination should be accepted as correct. At the time (20 April, 1939) Sealy was not aware that *A. unedo* occurred in Cyprus, nor that both *A. andrachne* and *A. unedo* grew in close proximity in the Akamas, where the specimen was collected, otherwise he might have been more confident.

A. × andrachnoides is said to be fertile, and to form introgressive populations. It is difficult to define, sometimes resembling one parent, sometimes the other; for the most part it has the flaking bark of *A. andrachne* and some of the glandular bristles on young shoots and twigs that characterize *A. unedo*. It should be looked for in the neighbourhood of Ayios Konon, since existing material is inadequate for certain proof of its occurrence in Cyprus.

56. MONOTROPACEAE

Ericaceae A. L. Jussieu subfam. *Monotropoideae* (Eichler) M. W. Henderson in Contr. Bot. Lab. Univ. Pennsylv., 5: 106 (1919); G. D. Wallace in Wasmann Journ. Biol., 33: 1–88 (1975).

Mycotrophic perennials without chlorophyll; roots brittle, succulent, much branched, creeping, or clustered; floral axes annual, arising from adventitious buds on the roots, without true leaves, but varyingly clothed with scales and sterile bracts; inflorescences racemose, bracteate, or sometimes reduced to a single flower; flowers hermaphrodite, actinomorphic, pedicellate; sepals 0–4 (–5), free, ± distinct from corolla; corolla sympetalous or polypetalous wth 3–5 (–6), commonly saccate lobes or divisions; stamens usually twice as many as corolla-lobes or divisions; filaments terete or flattened, glabrous or pubescent; anthers usually small, dehiscing by apical slits; pollen grains shed singly; nectaries usually present between the bases of the filaments; ovary superior, 1–5 (–6)-locular, placentation axile or intruded parietal; style simple, straight, sometimes articulated at the base; stigma capitate or infundibuliform, occasionally subtended by a ring of hairs; ovules anatropous. Fruit a berry or loculicidal (rarely indehiscent) capsule; seeds numerous, small, occasionally winged.

Ten genera chiefly in temperate N. America and Asia.

1. MONOTROPA L.
Sp. Plant., ed. 1, 387 (1753).
Gen. Plant., ed. 5, 183 (1754).

Inflorescences scaly, racemose or reduced to 1 flower, at first nodding, usually becoming erect in fruit; flowers pedicellate, bracteate; calyx ± differentiated, sepals (when present) 2–5, imbricate; corolla polypetalous; petals 3–6, imbricate, usually pubescent internally with ciliate or erose margins; stamens 6–10 (–12) in one series or 2 unequal series; nectaries of

8–10, paired lobes projecting downwards from between the bases of the stamens; ovary globose or elongate-ovoid, longitudinally sulcate, 5 (4 or 6)-locular, usually pubescent; placentation axile; style distinctly articulated at base; stigma capitate or infundibuliform, 4–6 lobed. Fruit a loculicidal capsule, dehiscing by longitudinal slits; seeds minute, fusiform; endosperm present.

Two species widely distributed in temperate regions of the northern hemisphere.

1. **M. hypopithys** *L.*, Sp. Plant., ed. 1, 387 (1753); Boiss., Fl. Orient., 3: 975 (1875); Post, Fl. Pal., ed. 2, 2: 175 (1933); G. D. Wallace in Wasmann Journ. Biol., 33: 31 (1975); Davis, Fl. Turkey, 6: 107 (1978).

TYPE: "*in* Sueciae, Germaniae, Angliae, Canadae *sylvis*".

Erect, unbranched perennial 7–30 cm. high; stems, scales and perianth-segments pale ochre-yellow, sometimes tinged red or pink; roots slender, forming a compact mass; stems at first rather succulent, becoming firm and dry with age, distinctly sulcate; cauline scales lingulate, acute or subacute, 1–2 cm. long, 0·5–1 cm. wide, usually small and rather dense near the base of the stem, larger and more remote above; inflorescence at first nodding, later erect, racemose, (1–) 3–12 (or more)-flowered, rather dense; bracts similar to cauline scales, but usually larger and blunter; pedicels 3–10 (–15) mm. long, glabrous or ± floccose-pubescent with scattered glands; sepals 1–2 (or more), not well differentiated from petals, elliptic or lanceolate-elliptic, 8–10 mm. long, 2–5 mm. wide, acute, tapering to a narrow base, usually glabrous; petals oblong-obovate, about 15 mm. long, 5–6 mm. wide, glabrous (var. *glabra* Roth) or shortly pilose within (var. *hypopithys*), apex rounded, somewhat plicate, often slightly recurved, base distinctly saccate; stamens 6–7 in specimens examined, apparently variable in number; filaments 8–10 mm. long, flattened, glabrous (var. *glabra* Roth) or pilose (var. *hypopithys*); anthers oblate-reniform, about 1 mm. long, 1·3 mm. wide, dehiscing along the upper margin; ovary broadly ovoid, about 7 mm. long, 6 mm. wide, glabrous or pubescent; style stout, 3–5 mm. long; stigma bright yellow, shallowly infundibuliform, about 2·5 mm. diam., obscurely lobed. Capsule about 8 mm. long, 6–7 mm. wide, deeply sulcate; seeds narrowly fusiform, about 0·8 mm. long, 0·2 mm. wide; testa pale brown, reticulate, transparent, extending from either end of the seed as narrow wings.

HAB.: In deep litter under *Pinus nigra* ssp. *pallasiana*; 5,500–6,000 ft. alt.; fl. July–August.

DISTR.: Division 2, very rare. Widespread in temperate Europe, Asia and N. America.

2. Troödos, 1950, *F. M. Probyn* s.n.! and, 1963, *D. P. Young* 7811!

NOTES: The two specimens seen have glabrous petals, filaments and ovaries, and may be referred to *M. hypopithys* L. var. *glabra* Roth, Tent. Fl. Germ., 2: 462 (1789), though in general robustness they more resemble var. *hypopithys*. The distinctions between the two varieties are not satisfactory, nor is there any obvious justification for treating var. *glabra* as a species (*M. hypophegea* Wallr., Sched. Crit., 1: 191; 1822). Further exploration may show that glabrous and pilose variants grow together in Cyprus, as elsewhere.

57. PLUMBAGINACEAE
F. Pax in Engl. et Prantl, Pflanzenfam., ed. 1, IV. 1: 116–125 (1890).

Perennial, or rarely annual, herbs and shrubs; leaves alternate or in basal rosettes, exstipulate, simple, entire or sometimes lobed; epidermis frequently with glands exuding water or calcium salts; inflorescences with bracts and bracteoles, usually cymose, sometimes capitate or spicate; flowers actinomorphic, 5-merous, often clustered in spikelets; calyx gamosepalous, tubular or infundibuliform, persistent, plicate, dentate or lobed, often showy, coloured or wholly or partly scarious; corolla lobed almost to base or with a distinct tube; stamens opposite the corolla-lobes; pollen sometimes dimorphic; ovary superior, 5-carpellate, 1-locular, 1-ovulate; styles 5, free or partly united, stigmas sometimes dimorphic. Fruit generally enveloped by persistent calyx, dry, indehiscent, circumscissile, or irregularly dehiscent; endosperm mealy; embryo straight.

About 10 genera with a cosmopolitan distribution; well represented in the Mediterranean region and frequently maritime or indicative of saline conditions.

Leaves all basal, often forming rosettes; corolla split to near the base, or with a short tube not
  exceeding the calyx - - - - - - - - - - - - **1. Limonium**
Leaves extending along the stems and branches; corolla-tube long, slender, much exceeding
  calyx - - - - - - - - - - - - - - - **2. Plumbago**

1. LIMONIUM *Mill.*
Gard. Dict. Abridg., ed. 4 (1754) nom. cons.
Statice L., Sp. Plant., ed. 1, 274 (1753) nom. rej.

Perennial, or rarely annual, herbs and subshrubs; leaves usually forming basal rosettes, entire or sometimes pinnately lobed or sinuate; inflorescence leafless, branched or unbranched, sterile flowerless branches sometimes present; bracts and bracteoles scale-like; flowers frequently forming dense or lax, secund spikes along the ultimate inflorescence-divisions, commonly in bracteolate, 1–5-flowered clusters; calyx persistent, infundibuliform, with a conspicuous, scarious or coloured, 5- (or 10-) costate limb; corolla often inconspicuous, divided almost to the base or with a basal tube; stamens usually inserted at base of corolla-tube; pollen and stigmas often dimorphic; styles 5, free or connate at the base; stigmas filiform. Fruit circumscissile or dehiscing irregularly.

About 300 species with a cosmopolitan distribution. In Europe and the Mediterranean region the occurrence of local, apomictic, morphologically distinct populations gives rise to taxonomic problems.

The corolla is generally said to be divided quite or almost to the base; in dried specimens it more often appears to consist of a basal tube and short, spreading lobes, but such is its consistency that it is difficult to say if the basal tube is composed of connate parts, or if the free (or almost free) lobes cohere in drying.

Leaves sinuately pinnatilobed; stem winged; calyx-limb conspicuous, brightly coloured
  1. L. sinuatum
Leaves simple, entire, stem not winged; calyx-limb membranous, pallid:
  All the inflorescence-branches bearing flowers:
    Plants annual, with a slender taproot; inner bracteoles tuberculate, involute, closely
      enveloping the flower - - - - - - - - - **7. L. echioides**

Plants perennial, with a tough scaly rootstock; inner bracteoles not conspicuously
    tuberculate or involute:
    Inflorescence repeatedly branched, large, diffuse, many-flowered:
        Leaves distinctly nerved, 8–30 cm. long, apex rounded, obtuse or emarginate
                                                    2. L. meyeri
        Leaves nerveless or indistinctly nerved, 2–10 cm. long, apex acute, terminating in a
           short, brownish awn - - - - - - - **3. L. narbonense**
    Inflorescence sparingly branched, not large or diffuse:
        Leaves subsessile, forming tight, crowded rosettes; inflorescence very slender, (2–)
           5–10 cm. high - - - - - - - - **4. L. albidum** ssp. **cyprium**
        Leaves tapering gradually to a petiole, forming loose, irregular rosettes; inflorescence
           18–35 cm. high - - - - - **5. L. ocymifolium** ssp. **bellidifolium**
Some of the inflorescence-branches sterile or without flowers; leaves narrow-spathulate
                                               6. L. virgatum

1. **L. sinuatum** (*L.*) *Mill.*, Gard. Dict., ed. 8, no. 6 (1768); Post, Fl. Pal., ed. 2, 2: 412 (1933);
    Osorio-Tafall et Seraphim, List Vasc. Plants Cyprus, 82 (1973); Feinbrun, Fl. Palaest., 3:
    9, t. 11 (1978); Davis, Fl. Turkey, 7: 467 (1982).
    Statice sinuata L., Sp. Plant., ed. 1, 276 (1753); Poech, Enum. Plant. Ins. Cypr., 15
    (1842); Unger et Kotschy, Die Insel Cypern, 230 (1865); Boiss., Fl. Orient., 4: 858 (1879);
    Holmboe, Veg. Cypr., 143 (1914); Lindberg f., Iter Cypr., 26 (1946).
    TYPE: "*in* Sicilia, Palaestina, Africa".

Erect or spreading perennial 15–60 cm. high with a tough, woody
rootstock; stems simple or branching at the base, subglabrous or setulose,
conspicuously 3–5-winged the wings terminating at the nodes as linear or
subulate, leaf-like appendages 3–7 mm. long, 1–5 mm. wide; leaves all basal,
narrowly oblong or obovate in outline, 1·5–17 cm. long, 0·5–4 cm. wide,
sinuately pinnatilobed with 9–18 or more blunt or rounded lobes, the
terminal usually reduced and aristate; petioles flattened, 0·5–4 cm. long;
inflorescence repeatedly and irregularly branched, the ultimate divisions
0·5–3 cm. long, strongly 3-winged, the wings broadening upward, one of
them produced into a relatively large, acute, leaf-like appendage, the others
with smaller appendages; bracts scarious, stramineous, long-ciliate,
bracteoles acuminate, the outer scarious, the innermost 2-keeled, rigid, with
2–3 short, spinescent, ciliate awns; flowers sessile in pairs, forming dense
corymbose clusters at the tips of the ultimate divisions of the branches;
calyx infundibuliform, about 16 mm. long, the lower half slender,
cylindrical, puberulous, about 2 mm. wide, the upper half lavender-blue,
pink or white, plicate, flaring to 6 mm. diam. at the truncate apex; corolla
hypocrateriform, pallid, crumpling after anthesis, tube slender, about as
long as lower half of calyx, limb about 4 mm. diam., divided into 5 blunt or
emarginate, spreading lobes; stamens included, filaments glabrous, inserted
about half way down corolla-tube; anthers yellow, oblong, about 1·3 mm.
long, 0·5 mm. wide; ovary glabrous, oblong, about 1·5 mm. long, 0·5 mm.
wide; styles about 2 mm. long, stigmas filiform, about 1 mm. long. Fruit
concealed by persistent calyx and indurated bracteoles, membranous,
narrowly oblong, about 5 mm. long, 1·3 mm. wide, circumscissile about
1·5–2 mm. from the 5-ribbed, bluntly conical apex; seed about 5 mm. long,
1·2 mm. wide, narrowly oblong-elliptic, compressed, longitudinally 5-
costate, testa dark brown, smooth.

HAB.: Sandy ground by the sea; near sea-level; fl. March–July.

DISTR.: Divisions 1, 3–8, locally abundant. Widespread in the Mediterranean region.

1. Ktima, 1840, *Kotschy* 47; Paphos, 1913, *Haradjian* 627! also, 1941 *Davis* 3345! and, 1955, *P.
    Pavlides* in *Kennedy* s.n.! Yeroskipos, 1939, *Lindberg f.*, s.n.; Ayios Yeoryios (Akamas),
    1962, *Meikle* 2134! Between Lachi and Baths of Aphrodite, 1979, *Edmondson & McClintock*
    E 2749!
3. Curium, 1961, *Polunin* 6696! Amathus, 1978, *J. Holub* s.n.!
4. Between Larnaca and Livadhia, 1862, *Kotschy* 173! Famagusta, 1900, *A. G. & M. E.
    Lascelles* s.n.! Larnaca, 1905, *Holmboe* 564! Famagusta 1912, *Haradjian* 95! also, 1931, *J. A.
    Tracey* 68! 1934, *Syngrassides* 474! 1939, *Lindberg f.*, s.n.! 1972, *W. R. Price* 1016!

5. Salamis, 1905, *Holmboe* 437; also, 1948, *Mavromoustakis* s.n.! and, 1958, *Merton* 3214!
6. Xeros, 1937, *Syngrassides* 1571! Near Morphou, 1908, *Clement Reid* s.n.!
7. Akanthou, 1880, *Sintenis & Rigo* 22! Kyrenia, 1929, *C. B. Ussher* 7! also, 1938, *C. H. Wyatt* 53! 1939, *Syngrassides* 6! 1949, *Casey* 818! 1955, *Miss Mapple* 32! 57! *G. E. Atherton* 223! Akanthou, 1967, *Merton* in AR1 792! Lambousa Beach, 1973, *P. Laukkonen* 267!
8. Cape Andreas, 1880 *Sintenis & Rigo* 22a! Koma tou Yialou, 1970, *A. Genneou* in AR1 1479!

2. L. meyeri (*Boiss.*) *O. Kuntze*, Rev. Gen. Plant., 2: 395 (1891); Feinbrun, Fl. Palaest., 3: 10, t. 13 (1978); Davis, Fl. Turkey, 7: 469 (1982).
    Statice meyeri Boiss. in DC., Prodr., 12: 645 (1848).
    S. gmelini Willd. var. *laxiflora* Boiss. in DC., Prodr., 646 (1848), Fl. Orient., 4: 859 (1879).
    [*S. limonium* L. var. *macroclada* (non Boiss.) Kotschy in Unger et Kotschy, Die Insel Cypern, 230 (1865); H. Stuart Thompson in Journ. Bot., 44: 334 (1906).]
    [*S. limonium* (non L.) Holmboe, Veg. Cypr., 143 (1914); R. M. Nattrass, First List Cypr. Fungi, 23 (1937).]
    [*Limonium vulgare* (non Mill.) Post, Fl. Pal., ed. 2, 2: 413 (1933) pro parte.]
    [*Statice gmelini* (non Willd.) Lindberg f., Iter Cypr., 26 (1946).]
    [*Limonium gmelinii* (non (Willd.) O. Kuntze) Osorio-Tafall et Seraphim, List Vasc. Plants Cyprus, 82 (1973).]
    TYPE: "In Tauriâ (herb. Fauché) provinciis cis et transcaucasicis (Fisch.! C. A. Mey.! Hohen.!) et in Atticae littoribus si specim. ab amiciss. Spruner missa huc recte spectant."

Erect perennial with a thick scaly rootstock; leaves all basal, forming a loose rosette, green or glaucous, narrowly obovate or subelliptic, 8–30 cm. long, 2–7 cm. wide, coriaceous, distinctly pinnate-veined, apex blunt or rounded, rarely emarginate, margins entire, narrowly cartilaginous, base subsessile or tapering gradually to an indistinct, flattened petiole 2–3 cm. long; inflorescence 30–60 (–100) cm. high, repeatedly, diffusely or erectly branched, forming a lax, open panicle, main stem and branches angled or sulcate, often glaucous and tuberculate; bracts scarious, brown, deltoid, acute; flowers usually paired in lax or crowded, secund, terminal spikelets 1–5 cm. long; bracteoles 8, broadly and bluntly ovate, 1–3 mm. long and about as wide, with a thick herbaceous keel and wide, brownish, papery margins; calyx infundibuliform, about 2·5 mm. long, 1·8 mm. wide at apex, plicate, thinly hairy or glabrous externally, with 5 thick nerves stopping short of the acute, erose, hyaline-membranous apical teeth, often with shorter, blunter teeth between the nerves, corolla lavender-blue, tube about 3·5 mm. long, lobes acute, concave, ovate, about 1·5 mm. long, 1 mm. wide; stamens inserted about 1 mm. from base of tube; filaments glabrous, about 2 mm. long; anthers oblong, about 0·8 mm. long, 0·5 mm. wide, yellow, ovary longitudinally ribbed, about 1 mm. long, 0·3 mm. wide, glabrous; styles 5, glabrous, about 2 mm. long; stigmas filiform, about 0·3 mm. long. Ripe fruit not seen.

HAB.: Seashores and salt-marshes, but not necessarily in moist situations; near sea-level; fl. July–Oct.

DISTR.: Divisions 3–5, 7. Eastern Mediterranean region from (Greece?) Turkey, Syria, Palestine and Egypt east to the Crimea, Caucasus and N. Iran.

3. Limassol Salt Lake near Akrotiri, 1941, *Davis* 3586!
4. Larnaca, 1860, *J. D. Hooker & D. Hanbury* s.n.! also, 1862, *Kotschy* s.n.; 1930, *C. B. Ussher* 115! Famagusta, 1939, *Lindberg f.* s.n.! Between Prastio and Kouklia, 1939, *Lindberg f.* s.n.!
5. Near Salamis, 1968, *A. Genneou* in AR1 1267!
7. Kyrenia, 1955, *G. E. Atherton* 334! 728!

3. L. narbonense *Mill.*, Gard. Dict., ed. 8, no. 2 (1768).
    Statice globulariifolia Desf. var. *glauca* Boiss., Voy. Bot. Espagne, 531 (1841), t. 155, figs. a, 1a–5a (1840).
    S. delicatula Girard in Ann. Sci. Nat., ser. 3, 2: 327 (1844).
    S. raddiana Boiss. in DC., Prodr., 12: 653 (1848).
    Limonium delicatulum (Girard) O. Kuntze, Rev. Gen. Plant., 2: 395 (1891); Täckholm, Students' Fl. Egypt, ed. 2, 403 (1974).

Statice mucronulata Lindberg f., Iter Cypr., 26 (1946).

Limonium mucronulatum (Lindberg f.) Osorio-Tafall et Seraphim, List Vasc. Plants Cyprus, 82 (1973) nom. invalid.

TYPE: Cultivated in Chelsea Physic Garden; originally from Narbonne (BM !)

Erect perennial from a tough rootstock; superficially very similar to *L. meyeri* (Boiss.) O. Kuntze; leaves all basal, forming a rather dense rosette, glaucous, coriaceous, obovate-acute or elliptical, 2–10 cm. long, 1–3 cm. wide, nervation very obscure or virtually invisible, apex acute, terminating in a short, brownish awn about 1–2 mm. long, margins entire, narrowly cartilaginous, base tapering to a short, flattened, indistinct petiole; inflorescence (14–) 30–70 cm. high, repeatedly and diffusely branched, the ultimate branches very slender and often somewhat pendulous, indistinctly angled and usually smooth; bracts scarious, brown with a pale marginal zone, acuminate, 3–5 mm. long, 2·5–3 mm. wide; flowers sometimes solitary (in Cyprus specimens), usually paired or in threes; bracteoles 3, the median ovate, acute, carinate, about 3 mm. long, 2 mm. wide, glabrous, brown with wide hyaline margins, lateral broadly ovate, concave, about 2 mm. long, 1·5 mm. wide; calyx narrowly infundibuliform, about 3 mm. long, 2 mm. wide at apex, plicate, thinly hairy towards the base externally, with 5 conspicuous nerves stopping short of the ovate, subacute, membranous, hyaline teeth; corolla lavender-blue, tube infundibuliform, about 3 mm. long, lobes oblong-emarginate, about 1·2 mm. long and almost as wide, apical sinus broad, conspicuous; stamens inserted about 1 mm. below apex of corolla-tube; filaments glabrous, about 1 mm. long, anthers oblong, about 0·7 mm. long, 0·4 mm. wide; ovary longitudinally ribbed, about 1 mm. long, 0·3 mm. wide; styles 5, glabrous, about 2 mm. long; stigmas filiform, about 0·3 mm. long. Ripe fruit not seen.

HAB.: Seashores; near sea-level; fl. June (?–Oct.).

DISTR.: Division 4. Also S. France, Spain, N. Africa, Egypt, Palestine.

4. Seashore near Larnaca, 1939, *Lindberg f.* s.n. !

4. L. albidum *(Guss.) Pignatti* in Bot. Journ. Linn. Soc., 64: 365 (1971).

Statice albida Guss., Suppl. Fl. Sic. Prodr., 88 (1832).

S. psiloclada Boiss. var. *albida* (Guss.) Boiss. in DC., Prodr., 12: 651 (1848) nom. illeg.

Tufted perennial with a much-branched, woody rootstock bearing numerous, crowded rosettes of grey-green, coriaceous, subsessile leaves; lamina obovate-spathulate, 0·5–3 cm. long, 0·4–1 cm. wide, minutely rugulose-papillose, apex shortly acute, obtuse or rounded, often somewhat recurved, margins narrowly cartilaginous, often undulate, base tapering to, at most, a very short and obscure petiole; nervation obscure or invisible; inflorescence slender, lax, paniculate, rather irregularly branched, (2–) 5–10 cm. high, branches glabrous, terete or obscurely angled; flowers 2–4 together, forming lax, often flexuous, terminal spikes; bracts small, deltoid, 1·5–2 mm. long and almost as wide, acute, brown with a wide hyaline-membranous margin; inner bracteole 4–7 mm. long, 3–4 mm. wide, concave-carinate, subacute, thick, herbaceous with a broad hyaline-membranous margin, outer bracteoles much smaller, 1·5–2·5 mm. long, almost as wide, obtuse, with wide membranous margins; calyx narrowly infundibuliform, 4–6 mm. long, 1·5–3 mm. wide at apex, tube glabrous or thinly pilose towards base externally, sometimes a little curved, nerves 5, prominent, falling a little short of the subacute, membranous teeth, the latter 0·8–1 mm. long; corolla lavender-blue, lobes slightly exceeding calyx, oblong, emarginate, about 1 mm. long and almost as wide; stamens inserted about 1·5 mm. down corolla (filaments continuing to base of corolla); filaments glabrous, the free part about 1 mm. long; anthers yellow, oblong, about 0·6 mm. long, 0·4 mm. wide; ovary glabrous, cylindrical, about 2 mm. long, 0·8 mm. wide; styles glabrous, about 2·5 mm. long; stigmas filiform, about 0·5 mm. long. Fruit narrowly ellipsoid, about 1·8 mm. long, 0·4 mm. wide; seed with a rich brown, glossy testa, distinctly furrowed longitudinally down one side.

ssp. **cyprium** *Meikle* in Ann. Musei Goulandris, 6: 88 (1983). See App. V, p. 1898.

TYPE: Cyprus; Cape Andreas, on rocks by the sea; 20 ft. alt.; 11 Nov. 1951, *Merton* 606 (K !)

Inner bracteoles less than 5 mm. long, straight; flowers small, scarcely exceeding bracteoles, with a straight calyx-tube; otherwise as in typical *L. albidum*. Leaves small, crowded.

HAB.: Calcareous maritime rocks; near sea-level, fl. June–Oct.

DISTR.: Divisions 7, 8, rare. Endemic.

7. Kyrenia, 1938, *C. H. Wyatt* 51 ! 15 miles E. of Kyrenia, 1956, *Poore* 15 ! near Panayia Glykyotissa Church, W. of Kyrenia, 1959, *P. H. Oswald* 153 !
8. Cape Andreas, in clefts of rocks at the tip of the Cape, 1937–1962, *Syngrassides* 1663 ! *Merton* 606 ! *Meikle* 2464 !

NOTES: Differing from typical *L. albidum* (Guss.) Pignatti from Lampione and Lampedusa only in its smaller flowers with straighter bracteoles and calyx-tubes. The specimens from the neighbourhood of Kyrenia appear to be less robust than those from Cape Andreas, but this may be fortuitous, since a very robust specimen on the type sheet of ssp. *cyprium* (*Merton* 606) is accompanied by one just as small and slender as any from Kyrenia.

5. L. ocymifolium (*Poir.*) *O. Kuntze*, Rev. Gen. Plant., 2: 396 (1891); Rechinger f., Fl. Aegaea, 427 (1943); Tutin et al., Fl. Europ., 3: 47 (1972); Davis, Fl. Turkey, 7: 470 (1982).
    Statice ocymifolia Poir. in Lam., Encycl. Suppl., 5: 238 (1817).

Erect, tufted perennial 18–35 cm. high with a tough, brown, woody rootstock; leaves basal or sub-basal, forming a loose, irregular rosette, apparently not very persistent when withered, lamina spathulate, 1·5–3·5 cm. long, 0·5–1·2 cm. wide, glaucescent, rather thick (becoming rugulose when dried), apex rounded or distinctly emarginate, base tapering gradually to an indistinct sheathing petiole; nervation very indistinct, consisting of an ill-defined midrib; inflorescence lax, with a naked, elongate main axis, becoming branched and flexuous above, with 5–8 or more, alternate, erecto-patent, stout, well-spaced, floriferous branches 3–8 cm. long, sometimes divided into short, lateral branchlets; bracts deltoid, acute, about 2 mm. long and almost as wide at the base, brown with a wide hyaline membranous margin; flowers in lax or rather dense, secund spikes, either solitary or in groups of 2–3; inner bracteole oblong or boat-shaped, acute, straight, about 5–6 mm. long, 3 mm. wide, thick and herbaceous dorsally, with a wide membranous margin; outer bracteoles much smaller, deltoid, acute, about 2–2·5 mm. long; calyx-tube straight, about 7 mm. long, 1·5–2 mm. wide at apex, subcylindric or narrowly infundibuliform, glabrous above, distinctly pilose towards base externally, 5-nerved, the nerves falling short of the oblong, shortly emarginate, hyaline-membranous teeth, the latter about 1 mm. long, 0·8 mm. wide; corolla lilac or pinkish, tube glabrous, about 5–6 mm. long, lobes suborbicular, shallowly emarginate, about 1·5 mm. diam.; filaments glabrous, about 1 mm. long; anthers yellow, oblong, about 1 mm. long, 0·5 mm. wide; styles 5, glabrous, filiform, about 3 mm. long; stigmas filiform, about 1 mm. long. Fruit and seeds not seen.

ssp. **bellidifolium** (*Sm.*) *Meikle* **comb. nov.**
    Statice bellidifolia Sm. in Sibth. et Sm., Fl. Graec. Prodr., 1: 211 (1806), Fl. Graec., 3: 90, t. 295 (1821) non *Limonium bellidifolium* (Gouan) Dumort. (1827).
    S. aucheri Girard in Ann. Sci. Nat., ser. 3, 2: 328 (1844).
    S. ocymifolia Poir. var. *bellidifolia* (Sm.) Boiss., Fl. Orient., 4: 861 (1879) nom. illeg.
    S. ocymifolia Poir. ssp. *bellidifolia* (Sm.) Nyman, Consp. Fl. Europ., 610 (1881) nom. illeg.

Limonium ocymifolium (Poir.) O. Kuntze var. *bellidifolium* Rechinger f., Fl. Aegaea, 427 (1943).

TYPE: "Ad littora maris in Archipelagi insulis. In Rhodo" (OXF), figured in *Flora Graeca*, 3, t. 295!

Flowers usually solitary, rather distant, forming lax, elongate, secund spikes.

HAB.: Rock crevices by the sea; near sea-level; fl. April–May.

DISTR.: Division 4. Also Aegean Islands.

4. Cape Greco, 1958, *N. Macdonald* 56! Same locality, 1973, *E. Chicken* 480!

NOTES: Evidently rare, but perhaps overlooked elsewhere; it appears to begin flowering earlier than most *Limonium* species.

6. L. virgatum (*Willd.*) *Fourr.* in Ann. Soc. Linn. Lyon, n.s., 17: 141 (1869); Post, Fl. Pal., ed. 2, 2: 413 (1933); Davis, Fl. Turkey, 7: 470 (1982).

Statice virgata Willd., Enum. Plant. Hort. Reg. Bot. Berol., 336 (1809); Boiss., Fl. Orient., 4: 863 (1879); Post in Mém. Herb. Boiss., 18: 99 (1900); H. Stuart Thompson in Journ. Bot., 44: 334 (1906); Holmboe, Veg. Cypr., 143 (1914); Lindberg f., Iter Cypr., 26 (1946); Rechinger f. in Årkiv för Bot., ser. 2, 1: 428 (1950).

[*S. oleifolia* (non Scop.) DC. in Lam. et DC., Fl. Franç., 3: 422 (1805); Sibth. et Sm., Fl. Graec. Prodr., 1: 212 (1806) — as *S. oleaefolia*.]

S. echioides (non L.) Sm. in Sibth. et Sm., Fl. Graec. Prodr., 1: 213 (1806) pro parte quoad plant. cypr.]

[*S. rorida* Sm. in Sibth. et Sm., Fl. Graec., 3: 91 (1819); Boiss., Fl. Orient., 4: 862 (1879) pro parte quoad plant. cypr.]

[*S. graeca* Boiss. var. *microphylla* (non Boiss.) Kotschy in Unger et Kotschy, Die Insel Cypern, 230 (1865).]

[*S. graeca* (non Poir.) Holmboe, Veg. Cypr., 144 (1914).]

[*Limonium oleifolium* (non Mill.) Osorio-Tafall, List Vasc. Plants Cypr., 82 (1973); Feinbrun, Fl. Palaest., 3: 11, t. 15 (1978).]

TYPE: "*in* Hispania" (B).

Erect, tufted perennial 10–35 cm. high, with a tough, branched woody rootstock; leaves all basal, forming irregular rosettes, persistent when withered, and clothing the stems, lamina narrowly spathulate, dark green, minutely pustulose, 1–4 (–9) cm. long, 0·2–0·6 (–2) cm. wide, apex rounded, margins entire, narrowly cartilaginous, base tapering very gradually to an indistinct, sheathing petiole; nervation very indistinct or invisible; inflorescence repeatedly branched almost to base, the branches terete, erecto-patent, the lower ones sterile, the uppermost floriferous; bracts deltoid, acute, about 2 mm. long and almost as wide at the base, brown with a membranous margin; flowers (1–) 2–3 (–5) together in dense or lax, secund, terminal spikes; inner bracteole oblong-lanceolate, acute, usually curved at anthesis, about 5–6 mm. long, 3 mm. wide, thick and herbaceous dorsally, with a wide membranous margin, outer bracteoles much smaller, deltoid-acute, usually less than 2·5 mm. long; calyx-tube curved at anthesis, 5–6 mm. long, 1·5–2 mm. wide at the mouth, narrowly infundibuliform, subglabrous or thinly hairy (especially towards the base) externally, distinctly 5-nerved, the nerves falling short of the ovate, subacute, membranous teeth, the latter 0·8–1 mm. long, 0·6–0·8 mm. wide; corolla lavender-blue, pink or purple, tube glabrous, 5–6 mm. long, lobes 1–1·5 mm. long, 1 mm. wide, shallowly emarginate; stamens inserted about 2 mm. down corolla-tube; filaments glabrous, about 1–2 mm. long; anthers yellow, oblong, about 1 mm. long, 0·5 mm. wide; ovary glabrous, about 1·5 mm. long, 0·3 mm. wide; styles 5, very slender, filiform, about 3 mm. long; stigmas filiform, about 0·3 mm. long. Seed narrowly ellipsoid, about 2·5 mm. long, 0·4 mm. wide, bluntly angled; testa rich brown, smooth.

HAB.: Maritime rocks, sandy shores, and margins of salt-lakes; occasionally inland on saline ground; sea-level to 500 ft. alt.; fl. May–August.

DISTR.: Divisions 1, 3, 4, 6–8. Widespread in the Mediterranean region.

1. Paphos, 1981, *Meikle* 5039!
3. Limassol, 1859, *Kotschy* 989; also, 1947, *Mavromoustakis* s.n.! Limassol Salt Lake, 1941, *Davis* 3577! Akrotiri, 1905, *Holmboe* 689.
4. Larnaca Salt Lake, 1880, *Sintenis & Rigo* 510! also, 1898, *Post* s.n.! 1905, *Holmboe* 100; near *Larnaca*, 1912, *Haradjian* 75! also, 1913, *Haradjian* 1016! and, 1939, *Lindberg f.* s.n.! Ayia Napa, 1939, *Lindberg f.* s.n.; between Larnaca and Famagusta 1939, *Lindberg f.* s.n.; Cape Greco, 1958, *N. Macdonald* 56a! Mouth of Yialia River, 1968, *Economides* in ARI 1191!
6. Between Skylloura and Yerolakkos, marshy ground by roadside, 1934, *Syngrassides* 532! Skylloura, 1952, *Merton* 860!
7. Kyrenia, 1929, *C. B. Ussher* 54! also, 1938–1968, *C. H. Wyatt* 54! *Casey* 790! *Merton* 1325! *G. E. Atherton* 12! *Poore* 16! *Barclay* 1053! Akhiropiitos, 1936, *Syngrassides* 1295! Ammos, 1939, *Lindberg f.* s.n.
8. Near Ayios Theodhoros, 1939, *Lindberg f.* s.n.; Cape Andreas, 1962, *Meikle* 2465!

7. L. echioides (*L.*) *Mill.*, Gard. Dict., ed. 8, no. 11 (1768) (as *L. "echoideum"*); Osorio-Tafall et Seraphim, List Vasc. Plants Cyprus, 82 (1973); Davis, Fl. Turkey, 7: 476 (1982).
     Statice echioides L., Sp. Plant., ed. 1, 275 (1753); Sibth. et Sm., Fl. Graec., 3: 92, t. 299 (1819); Poech, Enum. Plant. Ins. Cypr., 15 (1842); Unger et Kotschy, Die Insel Cypern, 230 (1865); Boiss., Fl. Orient., 4: 870 (1879); Holmboe, Veg. Cypr., 144 (1914); Lindberg f., Iter Cypr., 26 (1946).
     S. aristata Sm. in Sibth. et Sm., Fl. Graec. Prodr., 1: 213 (1806); Clarke, Travels, 2 (3): 723 (1816); Poech, Enum. Plant. Ins. Cypr., 15 (1842).

Erect annual 5–40 cm. high with a slender taproot; leaves all basal, sessile or subsessile forming a neat rosette, lamina obovate-spathulate, 0·4–6 cm. long, 0·3–2 cm. wide, glabrous, minutely pustulose, often reddish, apex rounded, emarginate or shortly apiculate, margins entire, very narrowly cartilaginous, tapering to base, nervation obscure or invisible; inflorescences solitary or several from each rosette, repeatedly dichotomously branched, with divaricate, glabrous, minutely pustulose branches; bracts deltoid, acute or acuminate, 3–4 mm. long, 3–4 mm. wide, reddish with a wide membranous margin; bracteoles very unequal, tuberculate externally, the inner 5–6 mm. long, 3 mm. wide, subacute, herbaceous dorsally with a membranous margin, strongly involute and closely enveloping the flowers, outer bracteoles ovate-deltoid, generally less than 3 mm. long and 2 mm. wide; flowers often solitary within the bracteoles, sometimes 2 or more, forming lax, secund spikes, or occasionally alternating on a flexuous rhachis; calyx-tube slightly curved, cylindrical, 3·5–4 mm. long, about 1 mm. wide at apex, glabrous or shortly pubescent, limb membranous, truncate, shallowly or deeply divided into broadly or narrowly deltoid lobes, with the 5 prominent nerves terminating at the apex of the limb, shortly excurrent, or sometimes largely free and produced into an uncinate-circinnate awn 2–2·5 mm. long; corolla lavender-blue or pinkish, tube slender, 5–6 mm. long, lobes spreading, oblong-emarginate, about 1 mm. long and almost as wide; free part of filaments about 1 mm. long, glabrous; anthers yellow, oblong, about 0·4 mm. long, 0·3 mm. wide; ovary narrowly ellipsoid, glabrous, about 0·8 mm. long, 0·3 mm. wide; styles slender, about 4 mm. long; stigmas filiform, about 0·4 mm. long. Seed narrowly ellipsoid or cigar-shaped, about 2·5 mm. long, 0·4 mm. wide; testa shining, rich brown, minutely granulose.

ssp. **echioides**
TYPE: France; "Monspelii".

Calyx-nerves produced far beyond the limb into spreading, uncinate-circinnate awns, calyx-limb divided almost to base into narrow lobes; calyx-tube generally glabrous; inflorescences often irregularly alternate-flowered, or secund, the flowers sometimes in clusters; leaves generally with rounded or marginate apices; plants commonly small and slender.

HAB.: Sandy ground inland or by the sea; damp saline ground often associated with Kythrean marls; sea-level to 700 ft. alt.; fl. April–June.

DISTR.: Divisions 3–5, 7, rather uncommon. Widespread in the Mediterranean region, but commoner in the west.

3. Akrotiri, 1905, *Holmboe* 672.
4. Paralimni, 1905, *Holmboe* 434; Perivolia, near Cape Kiti, 1936, *Syngrassides* 980! Larnaca Salt Lake, 1979, *J. Holub* s.n.!
5. Salamis, 1905, *Holmboe* 438; Athalassa, 1956, *Poore* 8! Mia Milea, 1967, *Economides* in ARI 933!
7. Kyrenia, 1939, *Lindberg f.* s.n.!

ssp. **exaristatum** (*Murb.*) *Maire* in Jahandiez et Maire, Cat. Plant. Maroc, 3: 571 (1934).
   Statice echioides L. ssp. *exaristata* Murb., Contrib. Fl. Nord-Ouest Afr., ser. 3: 1 (1899); Holmboe, Veg. Cypr., 144 (1914).
   Limonium exaristatum (Murb.) P. Fournier, Quatre Fl. France, 720 (1937); Osorio-Tafall et Seraphim, List Vasc. Plants Cyprus, 82 (1973).
   Statice echioides L. var. *exaristata* (Murb.) Lindberg f., Iter Cypr., 26 (1946).
   TYPE: Illustrated by Murbeck (*loc. cit.*, t.x, figs. 3, 4); type to be selected from the numerous specimens cited in this publication.

Calyx-nerves stopping at apex of limb, or shortly excurrent; calyx-limb subtruncate or divided into broadly ovate lobes; calyx-tube pubescent at least towards base; inflorescences regularly secund, the flowers generally solitary or in pairs; leaves often subacute or with a short apiculus; plants generally more robust than in ssp. *echioides*.

HAB.: Sandy ground inland or by the sea; salt flats and saline ground associated with Kythrean marls; sea-level to 600 ft. alt.; fl. April–June.

DISTR.: Divisions 4–8, locally common. Mediterranean region, chiefly from Sardinia, Sicily, Lampedusa and Tunis eastwards to Turkey and Cyprus.

4. Pyla, *Syngrassides* 429! Larnaca, 1880, *Sintenis & Rigo* 509 teste *Murbeck*, also 1939, *Lindberg f.* s.n.; 1950, *Chapman* 689! 1952, *Merton* 826! 1979, *Edmondson & McClintock* E 2950! Famagusta, 1939, *Lindberg f.* s.n.! Paralimni, 1948, *Mavromoustakis* s.n.! Cape Greco, 1962, *Meikle* 2602! Kouklia reservoir, 1968, *Economides* in ARI 1187!
5. Near Kythrea, 1880, *Sintenis & Rigo* 509! Apostolos Varnavas Monastery, 1948, *Mavromoustakis* s.n.! Lakovounara, 1950, *Chapman* 657! Near Gypsos, 1955, *Merton* 2433!
6. Nicosia, 1908, *Clement Reid* s.n.! Ayia Irini, 1962, *Meikle* 2388!
7. Akanthou, 1934, *Syngrassides* 1342!
8. Near Boghazi, 1862, *Kotschy* 536! Near Rizokarpaso, 1912, *Haradjian* 246!

NOTES: The differences between the two subspecies are, on the whole, well marked and consistent; the resemblances are, however, so close that *subspecies*, rather than *species*, seems the appropriate rank. From evidence available, there would not appear to be any obvious difference in ecological requirements between the two, but there is room for field investigation here. The species seems to reach its easternmost limit in Cyprus, where it is often very abundant. Material from (Div. 6) Skylloura ["Siluri"] cited by Kotschy (Die Insel Cypern, 230) cannot be referred to a subspecies. Murbeck lists *Sintenis & Rigo* 509 as coming from Larnaca Salt Lake. All the material so numbered at Kew is from the neighbourhood of Kythrea, but it is quite likely that these collectors distributed specimens from both localities under the same number.

[STATICE TATARICA *L.* (≡ *Goniolimon tataricum* (L.) Boiss.), a native of S.E. Europe and western Asia, is erroneously recorded from "Larnica and Limosol" by Hume (in Walpole, Mem. Europ. Asiatic Turkey, 254; 1817). The identity of the plant so named is uncertain, but it may have been *Limonium sinuatum* (L.) Mill.]

[STATICE SPICATA *Willd.* (≡ *Psylliostachys spicata* (Willd.) Nevski) is noted as the host of *Uromyces limonii* (DC.) Lév. in Nattrass, First List Cypr. Fungi, 23 (1937), but almost certainly in error. *Psylliostachys spicata*, an annual with sinuate leaves and small white flowers in densely crowded, cylindrical, terminal and lateral spikes, is a native of Central Asia westwards to Palestine, usually growing in saline deserts. It has not otherwise been recorded from Cyprus, nor is it very likely to occur there, though it is difficult to imagine what plant was intended under the name.]

2. PLUMBAGO L.

Sp. Plant., ed. 1, 151 (1753).
Gen. Plant., ed. 5, 75 (1754).

Perennial herbs and shrubs; leaves alternate, simple; flowers in terminal and lateral spikes; bracts and bracteoles foliaceous; calyx tubular, 5-nerved, 5-dentate, covered with conspicuous, stipitate glands; corolla hypocrateriform, with a long tube and 5 spreading lobes; stamens free, hypogynous; style divided into 5 filiform stigmatic lobes. Fruit oblong-ovoid, pericarp membranous, circumscissile near base and splitting upwards into 5 valves.

About 15 species widely distributed in the warmer regions of the world.

1. **P. europaea** *L.*, Sp. Plant., ed. 1, 151 (1753); Hume in Walpole, Mem. Europ. Asiatic Turkey, 254 (1817); Boiss., Fl. Orient., 4: 875 (1879); H. Stuart Thompson in Journ. Bot., 44: 334 (1906); Holmboe, Veg. Cypr., 144 (1914); Post, Fl. Pal., ed. 2, 2: 415 (1933); Lindberg f., Iter Cypr., 26 (1946); Osorio-Tafall et Seraphim, List Vasc. Plants Cyprus, 82 (1973); Feinbrun, Fl. Palaest., 3: 8, t. 10 (1978); Davis, Fl. Turkey, 7: 464 (1982).
TYPE: "*in* Europa *australi*".

A suffruticose, spreading perennial about 1 m. high; stems glabrous, conspicuously ridged or sulcate, much branched, with long internodes; leaves variable, the lower bluntly obovate-elliptic, to 8 cm. long, 3 cm. wide, glabrous, indistinctly nerved, base attenuate, amplexicaul-auriculate, margins entire or obscurely denticulate-asperulous; upper leaves progressively narrower and more acute, the uppermost lanceolate, 10–30 mm. long, 2–5 mm. wide, often distinctly tuberculate-asperulous; flowers in rather dense terminal and lateral spikes 1–3 cm. long; bracts foliaceous, oblong, obtuse about 3 mm. long, 2 mm. wide; bracteoles similar but smaller, sometimes suborbicular; calyx persistent, about 6 mm. long, 2 mm. diam., densely stipitate-glandular, with 5 broad herbaceous nerves joined by scarious interstices, teeth deltoid, acute, about 0·8 mm. long and about as wide at the base; corolla mauve, purplish or with the limb blue above, and magenta below and on the tube, the latter about 12 mm. long and 2·5 mm. wide at apex, limb spreading, divided into 5, oblong, obtuse, costate lobes, each about 4–5 mm. long, 2 mm. wide; filaments about 11 mm. long; anthers oblong, yellow, slightly exserted from corolla-tube, about 1·5 mm. long, 0·8 mm. wide; ovary ellipsoid, glabrous, about 1 mm. long, 0·4 mm. wide; style 7 mm. long, sparsely pilose; stigmatic lobes about 1·5 mm. long, papillose on the adaxial surface. Fruit ovoid-acuminate, about 5 mm. long, 2–2·5 mm. wide, the membranous pericarp splitting upwards from base at maturity; seed pyriform; testa rich brown, minutely rugulose.

HAB.: Roadsides, hedgerows, margins of vineyards, dry stone walls and ruins, sometimes on rocky hillsides or sand-dunes; sea-level to 3,500 ft. alt.; fl. July–Nov.

DISTR.: Divisions 1–5. Widespread in S. Europe and the Mediterranean region eastwards to Transcaucasia, Iran and Afghanistan.

1. Yeroskipos, 1939, *Lindberg f.* s.n.; Ayios Neophytos Monastery, 1939, *Lindberg f.* s.n.
2. Phini, 1937, *Kennedy* 686! Kilani, 1937, *Syngrassides* 1697! Milikouri, 1939, *Lindberg f.* s.n.; Moudhoulas near Pedhoulas, 1940, *Davis* 1975! Platres, 1952–1955, *F. M. Probyn* 140! *Kennedy* 1866!
3. Limassol, 1801, *Hume*; Kouklia, 1901, *A. G. & M. E. Lascelles* s.n.! also, 1936, *Lady Loch* 10! Between Limassol and Kividhes, 1935, *Syngrassides* 515! Kolossi, 1939, *Lindberg f.* s.n. and *Mavromoustakis* 66! Palea Paphos, 1952, *Merton* 1637!
4. Larnaca, 1801, *Hume*.
5. Salamis, 1939, *Lindberg f.* s.n.! and, 1968, *A. Genneou* in ARI 1212!

P. AURICULATA *Lam.*, Encycl., Méth., 2: 270 (1786) (*P. capensis* Thunb.) from South Africa, is widely cultivated in Cyprus gardens. It is a scandent shrub, similar in many respects to *P. europaea*, but with much larger, pale-or sky-blue corollas, the limb often more than 2 cm. diam., with broad, contiguous lobes. Although commonly grown, there are no records for it as a naturalized plant on the island.

58. PRIMULACEAE

F. Pax et R. Knuth in Engl., Pflanzenr., 22 (IV. 237): 1–386 (1905).

Annual or perennial herbs, or rarely subshrubs; leaves usually opposite, verticillate or basal, rarely alternate, simple or very occasionally dissected; inflorescences racemose, paniculate, umbellate, often scapose, or flowers solitary, axillary or apparently terminal; flowers hermaphrodite, usually 5-merous, actinomorphic or rarely zygomorphic; calyx free or adnate to ovary, commonly persistent; corolla generally present, hypogynous or sometimes perigynous, gamopetalous, tube short or long, lobes imbricate or contorted, sometimes reflexed; stamens as many as corolla-lobes and opposite them, sometimes with alternating staminodes; filaments free or connate at the base; anthers 2-thecous, introrse, dehiscing longitudinally; ovary superior or semi-inferior, ovoid or globose, 1-locular (apparently 5-carpellate), placentation free-central, placentas sessile or stipitate, often spongy; ovules few or many, semi-anatropous; style simple (heterostyly not uncommon); stigma usually capitate or truncate. Fruit a circumscissile or 5 (or 10)-toothed or 5-valved capsule, sometimes dehiscing irregularly; seeds usually compressed or angular, with copious fleshy or horny endosperm; embryo small, usually straight.

About 20 genera and 1,000 species, widely distributed, but most abundantly represented in temperate regions of the northern hemisphere.

Root a subglobose tuber; corolla-lobes sharply reflexed - - - - - - **3. Cyclamen**
Root not a subglobose tuber; corolla-lobes not reflexed:
  Leaves all basal:
    Rhizomatous perennial; leaves large, rugose-bullate; flowers apparently solitary; corolla
      conspicuous - - - - - - - - - - - - **1. Primula**
    Annuals; leaves small, smooth; flowers in scapose umbels; corolla inconspicuous
                                    2. Androsace
  Leaves not all basal:
    Leaves opposite; flowers solitary, axillary; ovary superior:
      Capsule dehiscing by 5 valves, corolla minute, barely visible; leaves narrowly linear-
      lanceolate - - - - - - - - - - - **4. Asterolinon**
      Capsule circumscissile; corolla conspicuous, usually red or blue; leaves ovate or ovate-
      lanceolate - - - - - - - - - - - - **5. Anagallis**
    Leaves alternate; flowers in terminal racemes or panicles; ovary semi-inferior **6. Samolus**

1. PRIMULA *L.*

Sp. Plant., ed. 1, 142 (1753).
Gen. Plant., ed. 5, 70 (1754).

Rhizomatous perennial (or rarely monocarpic) herbs; leaves basal, entire or sometimes lobed, sessile or petiolate; flowers usually rather large, dimorphic-heterostylous, in scapose umbels, heads, verticels or spikes, rarely solitary or apparently solitary through the suppression of the scape; involucral bracts foliaceous, often produced or saccate towards the base; calyx tubular, campanulate or infundibuliform, lobes imbricate; corolla hypognous, campanulate, infundibuliform or hypocrateriform with a long tube, lobes imbricate, entire or emarginate; stamens inserted in, or at the mouth of the corolla-tube; filaments short; anthers included; ovary globose or ovoid; style simple; stigma capitate; ovules numerous. Capsule dehiscing by 5–10 valves; seeds ± peltately attached.

About 500 species mostly in temperate regions of the northern hemisphere.

1. **P. vulgaris** *Huds.*, Fl. Angl., ed. 1, 70 (1762); W. Wright Smith et H. R. Fletcher in Trans. et Proc. Bot. Soc. Edinb., 34: 452 (1948); D. H. Valentine et A. Kress in Tutin et al., Fl. Europaea, 3: 16 (1972); Davis, Fl. Turkey, 6: 113 (1978).

    P. veris L. var. *acaulis* L., Sp. Plant., ed. 1, 143 (1753).

    P. acaulis (L.) Hill, Veg. Syst., 8: 25 (1765); Boiss., Fl. Orient., 4: 24 (1875); Pax in Engl., Pflanzenr., 22 (IV. 237): 54 (1905); Post, Fl. Pal., ed. 2, 2: 180 (1933).

TYPE: Not indicated.

Acaulescent herb with a thick oblique rhizome and fleshy roots; leaves rugose-bullate, obovate, 13–20 cm. long, 3·5–6 cm. wide, thinly crispate-pubescent, especially below, apex rounded, margins irregularly denticulate, base tapering to a flattened, winged petiole, about 3–5 cm. long; flowers apparently solitary (scape generally suppressed), arising in a loose cluster from the centre of the leaf-rosette, fragrant; pedicels rather fleshy, pubescent, pinkish, to 12 cm. long (or sometimes longer); calyx tubular, strongly 5-ribbed, softly crispate-hairy with multicellular hairs, about 15–20 mm. long, 4–5 mm. wide, lobes subulate-deltoid, about 8 mm. long, 2 mm. wide at base; corolla pale greenish-yellow, tube about 20 mm. long, equalling or slightly exceeding the calyx, lobes 5, spreading, broadly obovate-emarginate, about 12–15 mm. long, 10–12 mm. wide, with a darker basal zone; stamens inserted near apex of tube (thrum-eyed) in some plants, or half-way down tube in other (pin-eyed) plants; anthers subsessile, narrowly ovate, about 2·5 mm. long, 1·2 mm. wide; ovary subglobose, about 2·5 mm. diam., glabrous; style about 14 mm. long in pin-eyed flowers, about 8 mm. long in thrum-eyed flowers; stigma large, capitate. Capsule broadly ovoid, about 10 mm. long, 8·5 mm. wide, dehiscing by 8–10 short, sharply deltoid, strongly recurved teeth; seeds about 2 mm. long, 1·3 mm. wide, irregularly ovoid, angled; testa dark brown, rough, densely foveolate.

HAB.: Damp shady mountainside near stream, under Pines and Alders and associated with *Rosa* spp. and *Viola* spp.; 5,300 ft. alt.; fl. April.

DISTR.: Division 2 only, rare, perhaps introduced. Western and Central Europe and Mediterranean region eastwards to Turkey, Syria, Caucasus and Iran.

2. Kryos Potamos, 5,300 ft. alt.; 1952, *Kennedy* 1766 !

NOTES: Although there is no evidence of deliberate introduction, a measure of suspicion must attach to the above record. The plant agrees exactly with western European material of *P. vulgaris*, and differs in some respects from the usual eastern Mediterranean races, though, admittedly, these are by no means homogeneous. All the specimens seen have "pin-eyed" flowers, but appear to be fertile. In view of the extensive distribution of *P. vulgaris*, further investigation may confirm its claim to be considered a Cyprus plant.

2. ANDROSACE *L.*

Sp. Plant., ed. 1, 141 (1753).
Gen. Plant., ed. 5, 69 (1754).

Annual or, more commonly, perennial herbs; leaves generally all basal, sessile or petiolate, rosulate; flowers mostly rather small, not dimorphic-heterostylous, generally in scapose umbels; involucral bracts subulate or rarely foliaceous; calyx-tube campanulate or globose, lobes 5, deltoid or lanceolate; corolla hypogynous, infundibuliform or hypocrateriform, usually with a short tube, throat constricted by pleats, scales or a thickened rim, lobes 5, imbricate, generally entire; stamens included, inserted in corolla-tube; ovary globose or subglobose; style short; stigma capitate. Capsule globose, dehiscing by 5 teeth or frequently almost to the base into 5 valves; seeds few to many, peltately attached.

About 100 species mostly in mountainous regions of the northern hemisphere.

1. **A. maxima** *L.*, Sp. Plant., ed. 1, 141 (1753); Unger et Kotschy, Die Insel Cypern, 295 (1865); Boiss., Fl. Orient., 4: 18 (1875); R. Knuth in Engl., Pflanzenr., 22 (IV. 237): 212 (1905); Holmboe, Veg. Cypr., 142 (1914); Post, Fl. Pal., ed. 2, 2: 180 (1933); Osorio-Tafall et Seraphim, List Vasc. Plants Cyprus, 81 (1973); Davis, Fl. Turkey, 6: 123 (1978).
TYPE: "*inter* Austriae *segetes*".

Erect annual 3–12 cm. high; leaves all basal, forming a neat rosette, sessile or subsessile, elliptic or obovate-elliptic, subglabrous or thinly pilose with white multicellular hairs, 0·8–4 cm. long, 0·4–2 cm. wide, apex acute, margins glandular-serrulate or denticulate, base tapering to an indistinct flattened petiole; scapes solitary or numerous with a sparse indumentum of scattered white hairs and shorter glands; involucral bracts obovate-oblanceolate, foliaceous, 4–14 mm. long, 3–8 mm. wide; pedicels erect or spreading, usually about as long as the bracts; calyx-tube globose, about 3 mm. diam., lobes large, broadly lanceolate, acuminate, about 5–10 mm. long, 2–5 mm. wide; corolla white with a yellow throat, tube cylindrical, about 4 mm. long, 1 mm. wide, glabrous or very sparsely glandular, lobes oblong, about 1·5 mm. long, 0·8 mm. wide, apex rounded; stamens inserted near base of tube, filaments very short, about 0·3 mm. long; anthers ovate, yellow, about 0·6 mm. long, 0·4 mm. wide; ovary globose; style very short, about 0·3 mm. long; stigma capitate. Capsule globose, about 6 mm. diam., dehiscing by 5 bluntly deltoid, slightly recurved teeth about 2 mm. long, 2·5 mm. wide at base; seeds about 15, ellipsoid, sharply 3-angled, sunk in a spongy placenta, about 2·4 mm. long, 1·3 mm. wide; testa dark brown, rough, densely foveolate.

HAB.: Dry cornfields and fallows; 300–3,000 ft. alt.; fl. Febr.–April.

DISTR.: Divisions 2, 3, 5, 6, locally common. Central and southern Europe and Mediterranean region eastwards to Central Asia.

2. Phini, 1862, *Kotschy* 881; Lefkara, 1941, *Davis* 2746!
3. Trimiklini, 1941, *Davis* 3124!
5. Perakhorio near Nisou, 1936, *Syngrassides* 1129!
6. Nicosia, 1880, *Sintenis & Rigo*; *Holmboe* 263; *Casey* 1291! Skylloura, 1970, *A. Hansen* 525! Locally common in this division.

3. CYCLAMEN *L.*

Sp. Plant., ed. 1, 145 (1753).
Gen. Plant., ed. 5, 71 (1754).
O. Schwarz et L. Lepper in Feddes Repert., 69: 73–103 (1964).
D. E. Saunders, Cyclamen, An Alpine Garden Society Guide, 49 pp. revised ed. (1973).

Acaulescent perennials with globose or subglobose tubers; leaves rather fleshy, cordate or reniform, entire, toothed or lobed, commonly marbled above and reddish below, synanthous or hysteranthous; petioles long, tapering to the base; flowers solitary, axillary, nodding; pedicel long, without bracteoles, generally (except in *C. persicum* Mill.) twisting spirally in fruit; calyx deeply 5-lobed, the lobes narrowly subulate-deltoid or broadly deltoid; aestivation contorted; corolla hypogynous, deeply 5-lobed, tube short, hemispherical, lobes sharply reflexed or rarely spreading, sometimes auriculate at the base; stamens 5, inserted at the base of the corolla-tube; filaments very short, dilated; anthers large, subsagittate, acuminate, generally included in corolla-tube; ovary globose or conical, 1-locular; style straight, slender; stigma small, truncate, sometimes exserted; ovules numerous, semi-anatropous. Capsule globose or ovoid, dehiscing irregularly by 5 teeth or valves; seeds irregularly subglobose or angular; testa spongy, coated with a sweet secretion attractive to ants; embryo transverse; seedling with 1 developed cotyledon, the other suppressed.

About 17 species, chiefly in the Mediterranean region eastwards to Iran and the Caucasus.

Flowers appearing in spring, after the leaves; base of corolla-lobes with a darker zone, but not
  blotched; auricles absent; pedicels not coiling after anthesis   -   -   - **1. C. persicum**
Flowers appearing in autumn, usually in advance of the leaves; base of corolla-lobes blotched,
  with conspicuous auricles; pedicels coiling after anthesis:
    Calyx-lobes narrowly subulate-acuminate; flowers strongly sweet-scented; pedicels coiling
      from apex downward in fruit   -   -   -   -   -   -   -   - **2. C. cyprium**
    Calyx-lobes broadly ovate-deltoid; flowers not markedly fragrant; pedicels coiling
      irregularly in two directions from about the middle   -   -   -   - **3. C. graecum**

1. **C. persicum** *Mill.*, Gard. Dict., ed. 8, no. 3 (1768); R. Knuth in Engl., Pflanzenr., 22 (IV. 227):
    248 (1905); Holmboe, Veg. Cypr., 143 (1914); Post, Fl. Pal., ed. 2, 2: 179 (1933); Osorio-
    Tafall et Seraphim, List Vasc. Plants Cyprus, 81 (1973); Elektra Megaw, Wild Flowers of
    Cyprus, 10, t. 12 (1973); Davis, Fl. Turkey, 6: 131 (1978).
      C. latifolium Sm. in Sibth. et Sm., Fl. Graec., 2: 71, t. 185 (1813); Boiss., Fl. Orient., 4: 12
    (1875).
      C. cyprium Sibth. in Walpole, Travels, 25 (1820) nomen.
      [*C. hederifolium* (non Ait.) Kotschy in Unger et Kotschy, Die Insel Cypern, 295 (1865) —
    as *C. hederaefolium*.]
    TYPE: Cultivated in Chelsea Physic Garden (BM!).

Tuber large, depressed-globose, up to 20 cm. diam., covered with rough, fissured, corky bark; roots rather fleshy emerging from the base and lower sides of the tuber; leaves usually cordate, sometimes a little angled, long-petiolate, glabrous, rather fleshy, 4–12 cm. long, 3–10 cm. wide, apex acute or subacute, margins closely denticulate with thickened cartilaginous teeth, base with a deep, open or closed sinus; lamina generally marbled above, green or red below; flowers appearing in spring after the leaves, often sweetly fragrant; pedicels arising from leaf-axils on a condensed floral stem or "trunk" which is commonly rough with the bases of decayed petioles, up to 30 cm. long, at first erect or ascending, becoming downward-arcuate as the fruit matures but not coiling, commonly purplish-bronze and thinly glandular; calyx-lobes narrowly ovate, acuminate, 4–5 mm. long, 2–3 mm. wide, obscurely penninerved, sparsely glandular, with a pale membranous margin; corolla pink or white with a dark crimson basal zone, lobes oblong, acute or obtuse, about 3 cm. long, 1–1·5 cm. wide, usually entire and with an exauriculate base; filaments less than 0·5 mm. long, 1–1·3 mm. wide; anthers about 4 mm. long, 2·3 mm. wide, densely verruculose dorsally with small violet-purple warts; ovary subglobose-conical, about 3·5 mm. diam., thinly covered with brown, glandular hairs; style about 4·5 mm. long, glabrous, straight, tapering from base to apex; stigma truncate, shortly exserted. Capsule globose, 1·3–1·7 cm. diam., dehiscing to about half-way by 4–6 unequal, recurved teeth; seeds irregularly angular, 2–2·5 mm. long and about as wide; testa dark fuscous-brown, rough, densely foveolate.

HAB.: In maquis or garigue, or occasionally in the open, on rocky limestone or sandstone hillsides; sometimes in crevices of old walls; near sea-level to 3,000 ft. alt.; fl. end of Dec.–April.

DISTR.: Divisions 1, 3, 7, 8, locally abundant. Also Crete, Karpathos, E. Aegean Islands, Rhodes, S. Turkey, Syria, Palestine, Tunisia.

1. Near Paphos, 1787, *Sibthorp*, and, 1974, *Meikle*! Ayios Yeoryios Island, 1962, *Meikle* 2156! Near Baths of Aphrodite, 1962, *Meikle*!

3. Limassol, 1905, *Michaelides*.
7. Common and locally abundant along the Northern Range from Kornos to Kantara. *Kotschy* 514; *Sintenis & Rigo* 25 ! *Davis* 2031 ! *Casey* 250 ! *G. E. Atherton* 791 ! 802 ! *Economides* in ARI 1059 ! *I. M. Hecker* 44 ! *P. Laukkonen* 85 ! etc.
8. Valia, 1905, *Holmboe* 494 ! Koma tou Yialou, 1935, *Syngrassides* 702 ! Rizokarpaso, 1941, *Davis* 2385 ! Eleousa Forest, 1957, *Merton* 2875 ! Locally abundant in the Karpas.

[C. REPANDUM Sm. in Sibth. et Sm., Fl. Graec. Prodr., 1: 128 (1806), Fl. Graec., 2: 72, t. 186 (1813); Post in Mém. Herb. Boiss., 18: 96 (1900); Holmboe, Veg. Cypr., 143 (1905) is recorded from Evrykhou ["Ericon"] (Div. 2) by Post, but probably in error for *C. cyprium* Kotschy. *C. repandum* flowers with the leaves in spring, like *C. persicum*, but differs in its much smaller flowers, with a narrower corolla-tube, its pedicels coiling in fruit, and its leaves coarsely toothed and angular-lobed. The tuber is generally small, usually less than 5 cm. diam., depressed-globose, with a brown-pubescent skin. The corolla is typically bright magenta-pink with a darker basal zone, but white and pale pink races are recorded.]

[C. CRETICUM *Hildebr.* in Beih. Bot. Centralbl., 19: 367 (1906), from Crete, is closely allied to *C. repandum* Sm., but has a smaller, pure white or faintly pink-tinged corolla, and a stigma scarcely exserted from the corolla-tube. It occurs at Tchingen, near Kyrenia (Div. 7), but only as a relic of former cultivation.]

C. COUM *Mill.*, Gard. Dict., ed. 8, no. 6 (1768) has been reported on several occasions, but such reports have yet to be substantiated by specimens. It is probably on the basis of such unconfirmed records that the species is tentatively included in Osorio-Tafall et Seraphim, List Vasc. Plants Cyprus, 81 (1973). *C. coum* Mill. has a small, pubescent tuber, suborbicular or reniform, entire leaves, and small, purplish-pink or white flowers with short, blunt corolla-lobes, each with a clearly defined, dark purple basal blotch surrounding a small, white, central "eye". It flowers with the leaves in winter and early spring, and has spirally coiled fruiting pedicels. *C. coum* is widely distributed from Bulgaria eastwards to the Caucasus and Iran, and south to Lebanon. It may well occur in Cyprus, especially in moist, shaded situations in forests high on the Troödos Range. The most exact and promising records received to date are from the late Mr. Eliot Hodgkin, who, in a letter (dated 16th April, 1971), notes: "*Cyclamen* sp. ? *C. coum* 1. Kykko Monastery. In dense shade in punky, deep leafmould under bushes of *Quercus alnifolia* and other shrubs. Growing with *C. cyprium*. 2. Below Nikos. Near the road. Corm small and smooth, and roundish leaves unspotted. Roots springing from the base only. Growing with *C. cyprium*". Mr. Hodgkin collected some tubers and brought them into cultivation, but, with his death in March 1973, nothing more has been heard of this unidentified *Cyclamen*.

2. C. cyprium *Kotschy* in Unger et Kotschy, Die Insel Cypern, 295 (1865); R. Knuth in Engl., Pflanzenr., 22 (IV. 237); 254 (1905); Holmboe, Veg. Cypr., 143 (1914); Osorio-Tafall et Seraphim, List Vasc. Plants Cyprus, 81 (1973); Elektra Megaw, Wild Flowers of Cyprus, 10, t. 13 (1973).
[*C. neapolitanum* (non Ten.) Boiss., Fl.Orient., 4: 13 (1875) pro parte.]
TYPE: Cyprus; cultivated in Vienna from tubers collected near Galata, 1862, *Kotschy* s.n. (W).

Tuber subglobose, or more usually depressed-globose, normally less than 7 cm. diam., with a rough, greyish bark; roots emerging from one area, usually to the side of the lower surface of the tuber; leaves broadly cordate, 7–14 cm. long, 6–11 cm. wide, long-petiolate, glabrous and rather fleshy, apex acute, margins coarsely dentate or shallowly lobed, base with a deep, often narrow, sinus; lamina conspicuously marbled above, almost always rich purple or crimson-purple below; flowers appearing in autumn, generally a little in advance of the leaves, strongly and sweetly fragrant; pedicels arising from a slender floral "trunk", 7–15 cm. long, at first erect, but coiling tightly from the apex downwards soon after anthesis, usually bronze or purplish and rather densely brown-glandular; calyx-lobes narrowly subulate-acuminate, 5–7 mm. long, 1·5–2 mm. wide at base, with an indistinct midrib and an indumentum of short, brownish, glandular hairs; corolla white or very pale pink, lobes oblong, obtuse or subacute, 1·5–2 cm. long, 0·6–0·8 cm. wide, usually entire and distinctly auriculate at the base, with a conspicuous, M-shaped magenta blotch; filaments rather slender, to about 1 mm. long, less than 0·4 mm. wide; anthers about 4 mm. long, 1·4 mm. wide, closely and irregularly verruculose dorsally with small violet-purple warts; ovary subglobose-conical, about 2·5 mm. diam., densely brown-glandular; style slender, glabrous, about 8 mm. long, not tapering much from base to apex; stigma truncate, indistinctly exserted. Capsule globose, about 1–1·2 cm. diam., dehiscing to less than half-way by 5 deltoid, spreading or somewhat recurved teeth; seeds irregularly angular, 2–2·5 mm. long and about as wide; testa dark brown, rough, densely foveolate, very viscid when freshly shed. *Plate 66.*

HAB.: Shaded calcareous or igneous rocks, steep hillsides or streambanks, generally under shrubs or trees; 100–3,500 ft. alt.; fl. Sept.–Jan. (exceptionally to March).

DISTR.: Divisions 1, 2, 7, locally abundant. Endemic.

1. Toxeftera near Ayios Yeoryios (Akamas), 1962, *Meikle* 2099! Smyies, 1962, *Meikle* 2185!
2. Common; Galata, Evrykhou, Perapedhi, Platres, Vroisha Lagoudhera, Pyrgos Valley, Kalopanayiotis, Stavros tis Psokas, Kykko, Nikos, etc. *Kotschy; Hartmann; A. G. & M. E. Lascelles! Kennedy* 682! 683! 684! *Lady Loch* 4! 10! *Syngrassides* 1723! *Davis* 1982! 2016! *P. H. Oswald* 177! *A. Genneou* in ARI 1291! *I. M. Hecker* 27!, etc.
7. Common; St. Hilarion, Kyrenia, Ayios Amvrosios, Karmi, Vasilia, Bellapais; Kazan; Sisklipos, Pentadaktylos, etc. *C. B. Ussher* 66! *Kennedy* 685! *Lady Loch* 1! 19! *Syngrassides* 1469! *Davis* 2013! *Casey* 196! 198! *G. E. Atherton* 730! *P. H. Oswald* 158! 159!, etc.

NOTES: A very distinct taxon, with no obvious relatives, save possibly *C. cilicium* Boiss. et Heldr. from central and southern Turkey, which is a smaller plant, with narrow, exauriculate corollas.

3. C. graecum *Link* in Linnaea, 9: 573 (1834); Boiss., Fl. Orient.; 4: 13 (1875); R. Knuth in Engl., Pflanzenr., 22 (IV. 237): 254 (1905); Osorio-Tafall et Seraphim, List Vasc. Plants Cyprus, 81 (1973); Davis, Fl. Turkey, 6: 129 (1978).
　　C. cypro-graecum E. et N. Mutch in Quart. Bull. Alpine Gard. Soc., 23: 164 (1955) nomen.
TYPE: Greece; Nauplion, "Ad rupes circa Naupliam frequens florebat Septembri" *Berger* (M, ?B).

Tuber globose or subglobose, 4–10 cm. diam. with a rough, fissured, corky bark; roots fleshy, retractile, springing from the centre of the lower surface of the tuber; leaves broadly cordate, 6–14 cm. long, 3–14 cm. wide, long-petiolate, rather fleshy, apex acute or obtuse, margins undulate or obscurely angled, closely denticulate with small, blunt cartilaginous teeth; lamina distinctly marbled above, paler or purplish below; flowers appearing in autumn a little in advance of the leaves, not usually fragrant; pedicels arising from an elongate floral "trunk", 7–14 cm. long, at first erect, but coiling irregularly from the middle or base after anthesis, bronze or purplish, rather densely brownish-glandular; calyx-lobes ovate-deltoid, acute or subacute, about 4·5 mm. long, 2·5 mm. wide, obscurely penninerved, with pallid, membranous, commonly somewhat undulate margins, and a thin, brown, glandular indumentum; corolla white or less often pale pink, lobes oblong, 1·5–2 cm. long, 0·6–0·8 cm. wide, acute or obtuse, generally entire and distinctly auriculate at the base with a conspicuous, crescent-shaped magenta blotch; filaments short, about 0·5 mm. long and almost as wide;

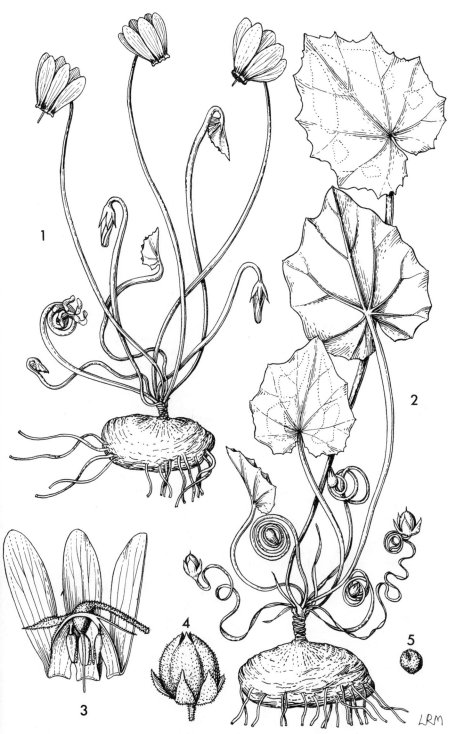

Plate 66. CYCLAMEN CYPRIUM Kotschy   **1,** autumnal, flowering plant, habit, ×⅔; **2,** spring, fruiting plant, habit, ×⅔; **3,** flower, longitudinal section, ×2; **4,** fruit, ×2; **5,** seed, ×2. (**1, 3** from *Kennedy* 683; **2** from *Kennedy* 682; **4, 5** from *Syngrassides* 1469.)

anthers about 3 mm. long, 1·5 mm. wide, conspicuously verruculose dorsally with numerous small, violet-purple warts; ovary subglobose-conical, about 2·5 mm. diam.; style rather stout, about 3 mm. long, tapering markedly from base to apex; stigma truncate, included or about equalling the corolla-tube. Capsule very large, sometimes slightly ovoid, about 1·5 cm. diam., dehiscing rather irregularly or by 5 deeply cut teeth; seeds irregularly angular, about 2·5 mm. long and wide; testa dark brown, rough, densely foveolate.

HAB.: In red soil on dry, rocky limestone slopes, frequently in the shade of dwarfed *Pistacia lentiscus* bushes; about 80–200 ft. above sea-level; fl. Sept.–Oct.

DISTR.: Division 6, rare. Greece, Crete, Aegean area, S.W. Turkey.

6. Near Liveras, 1953, *Kennedy* 1788! Also, same locality, 1954–1968, *Lady Loch* 2! *Meikle* 2422! *I. M. Hecker* 25!

NOTES: Locally plentiful over a very small area, and to date found nowhere else on the island, though one suspects it may have a more extensive distribution.

The Cyprus plant was at first thought distinct, but on further examination, agrees in all important respects with typical *C. graecum*. The leaves of the Cyprus plant tend to be rather duller and less attractively marbled than in specimens from Greece.

4. ASTEROLINON *Hoffmsgg. et Link*

Fl. Portug., 1: 332 (1820).

Small annual herbs; leaves opposite, sessile, entire, glabrous; flowers solitary, axillary, pedicellate; calyx persistent, spreading in fruit, divided almost to the base into 5 acuminate lobes; corolla minute, much shorter than calyx, campanulate-rotate, lobes entire or erose-denticulate; stamens 5, free, inserted near base of corolla; anthers obtuse, dorsifixed; ovary globose; style filiform; stigma minute, capitate. Capsule globose, 5-valved; seeds few, relatively large, corrugate.

Two species, one in the Mediterranean region and western Asia, the other in tropical East Africa.

1. **A. linum-stellatum** (*L.*) *Duby* in DC., Prodr., 8: 68 (1844); Unger et Kotschy, Die Insel Cypern, 286 (1865); Boiss., Fl. Orient., 4: 10 (1875); R. Knuth in Engl., Pflanzenr., 22 (IV. 237): 316 (1905); Holmboe, Veg. Cypr., 143 (1914); Post, Fl. Pal., ed. 2, 2: 178 (1933); Osorio-Tafall et Seraphim, List Vasc. Plants Cyprus, 81 (1973).
   Lysimachia linum-stellatum L., Sp. Plant., ed. 1, 148 (1753); Sibth. et Sm., Fl. Graec. Prodr., 1: 130 (1806), Fl. Graec., 2: 74, t. 189 (1813); Poech, Enum. Plant. Ins. Cypr., 27 (1842); Davis, Fl. Turkey, 6: 138 (1978).
   TYPE: "*in* Gallia".

Slender, erect or spreading, glabrous annual 2–8 cm. high; stems sharply 4-angled, usually much branched; leaves narrowly linear-lanceolate, 5–10 mm. long, 1–2·5 mm. wide, apex acuminate, margins minutely erose, base semi-amplexicaul; pedicels 2–3 mm. long, filiform, curved downwards; calyx-lobes lanceolate-subulate, 2–3 mm. long, 0·5–0·7 mm. wide, acuminate or shortly aristate; corolla-lobes whitish or pinkish, suborbicular, often shortly mucronate, about 0·5 mm. diam.; filaments narrow, about 0·4 mm. long; anthers suborbicular, minute, about 1 mm. diam.; ovary about 0·4 mm. diam., glabrous; style about 0·3 mm. long; stigma minute, capitate. Capsule 2–2·5 mm. diam., dividing almost to base into 5 lanceolate, boat-shaped valves; seeds broadly and bluntly ellipsoid, about 1 mm. long, 0·8 mm. wide, deeply excavate about hilum; testa dull, dark fuscous-brown, minutely foveolate, with a prominent dorsal ridge, on either side of which are 8–10 well-marked transverse corrugations.

HAB.: Well-drained gravelly or rocky ground, cliff-tops, dried-up river beds, sometimes in Pine forest or as a weed in cultivated or fallow fields; near sea-level to 5,000 ft. alt.; fl. March–April.

DISTR.: Divisions 1–4, 6–8. S.W. Europe and Mediterranean region east to Iran, Atlantic Islands.

1. Ayios Yeoryios (Akamas), 1962, *Meikle*! Polis, 1962, *Meikle*!
2. Prodhromos, 1862, *Kotschy*; valley above Pyrgos, 1957, *Merton* 2861!
3. Episkopi, 1960, *N. Macdonald* 37!
4. Near Cape Pyla, 1880, *Sintenis & Rigo*.
6. Dhiorios, 1962, *Meikle*!
7. Pentadaktylos, 1880, *Sintenis & Rigo* 29! and, 1974, *Meikle* 4026! St. Hilarion, 1941, *Davis* 2513! Yaïla, 1941, *Davis* 2891! Kyrenia, 1949, *Casey* 290!
8. Valia, 1905, *Holmboe* 479; Koma tou Yialou, 1950, *Chapman* 215! Akradhes, 1962, *Meikle*!

NOTES: Kotschy's record from Prodhromos is a little dubious in view of the fact that no other collector has recorded *Asterolinon* from high altitudes on the Troödos Range; normally the plant prefers lower ground, below 3,000 ft. alt. It is probably fairly general on the island, but, being insignificant, is not often collected.

5. ANAGALLIS L.

Sp. Plant., ed. 1, 148 (1753).
Gen. Plant., ed. 5, 73 (1754).

Annual or perennial herbs, sometimes tending to be suffruticose; leaves opposite or alternate, occasionally in threes, sessile or petiolate, entire; flowers axillary, pedicellate, the pedicels often longer than the leaves; calyx persistent, divided almost to the base into 5 patent, subulate or lanceolate lobes; corolla hypogynous, rotate or campanulate, deeply divided into 5 contorted, entire or erose-denticulate lobes; stamens 5, hypogynous, inserted near base of corolla; filaments ± connate at base, frequently pilose; ovary globose; ovules many, semi-anatropous; style slender; stigma truncate or capitate. Capsule globose, circumscissile; seeds angled; embryo transverse.

About 30 species distributed throughout the world, but chiefly in Europe, the Mediterranean region and Africa.

Corolla red or reddish - - - - - - - - **1. A. arvensis** var. **arvensis**
Corolla not red or reddish:
  Corolla white or whitish - - - - - - - **1. A. arvensis** var. **pallida**
  Corolla blue:
    Corolla-lobes overlapping, fringed with numerous short glands; pedicels at anthesis usually much exceeding the broad subtending leaves   **1. A. arvensis** var. **caerulea**
    Corolla lobes not overlapping, denticulate at apex, but eglandular or sparsely glandular; pedicels not, or only slightly exceeding the narrow subtending leaves
      1. A. arvensis ssp. **foemina**

1. A. arvensis *L.*, Sp. Plant., ed. 1, 148 (1753); Unger et Kotschy, Die Insel Cypern, 296 (1865); Boiss., Fl. Orient., 4: 6 (1875); Holmboe, Veg. Cypr., 143 (1914); Post, Fl. Pal., ed. 2, 2: 177 (1933); Osorio-Tafall et Seraphim, List Vasc. Plants Cyprus, 82 (1973); Davis, Fl. Turkey, 6: 139 (1978).

Sprawling or decumbent, glabrous annual; stems usually much branched, sharply 4-angled, up to 30 cm. long; leaves opposite, sessile, broadly ovate or ovate-lanceolate, 8–25 mm. long, 5–15 mm. wide, apex acute, acuminate or occasionally obtuse, margins bluntly papillose; flowers solitary; pedicels slender, 0·8–3·5 cm. long at anthesis, at first erect, lengthening and curving downwards in fruit; calyx-lobes lanceolate-acuminate, 5–7 mm. long, 1–1·5 mm. wide, with a keeled midrib and membranous, erose margins; corolla-lobes suborbicular or cuneate-obovate, 4–8 mm. long, 2–4 mm. wide, blue, red, pink or whitish, usually with a darker, purplish basal zone, apex rounded or subtruncate, margins subentire, erose or irregularly denticulate; filaments shortly connate at base, about 1·5–2·5 mm. long, densely pilose with purplish, shaggy, multicellular hairs; anthers pale yellow, narrowly ovate, about 0·8–1 mm. long, 0·5–0·8 mm. wide; ovary globose, glabrous, about 1·5 mm. diam.; style 2·5–3 mm. long; stigma capitate. Capsule pale brown, globose, about 5 mm. diam., usually crowned with the persistent style; seeds sharply trigonous-ellipsoid, about 1·5 mm. long, 1 mm. wide; testa dark fuscous-brown, dull, rough with paler, obtuse or subacute papillae.

ssp. **arvensis**

Corolla-lobes broadly obovate or suborbicular, overlapping, apical margin fringed with numerous, 3-celled stalked glands; hairs of the filaments 5–8-celled; pedicels at anthesis generally much exceeding the subtending leaf.

var. **arvensis**
> *A. arvensis* L. var. *phaenicea* Gouan, Fl. Monsp., 29 (1765); Boiss., Fl. Orient., 4: 6 (1875).
> *A. phoenicea* Scop., Fl. Carniol., ed. 2, 1: 139 (1772).
> *A. arvensis* L. var. *phoenicea* (Scop.) Gren. et Godr., Fl. France, 2: 467 (1852); R. Knuth in Engl., Pflanzenr., 22 (IV. 237): 323 (1905).
> *A. arvensis* L. ssp. *phoenicea* (Scop.) Vollmann in Ber. Bayer. Bot. Ges., 9: 44 (1904); Holmboe, Veg. Cypr., 143 (1914); Lindberg f., Iter Cypr., 26 (1946); Osorio-Tafall et Seraphim, List Vasc. Plants Cyprus, 82 (1973).
> TYPE: "*in* Europae *arvis*".

Corolla bright salmon-red with a darker basal zone.

HAB.: Cultivated and fallow fields, gardens, roadsides, waste ground; sea-level to 5,000 ft. alt.; fl. Feb.–Oct.

DISTR.: Divisions 2, 3, 5–8. A widespread weed in temperate regions.

2. Platres, 1905, *Michaelides*; Kykko, 1913, *Haradjian* 966! Mavrovouni, 1932, *Syngrassides* 91! Lefka, 1932, *Syngrassides* 245! Prodhromos, 1949, *Casey* 882! and, 1955, *G. E. Atherton* 529! Kakopetria, 1955, *Merton* 2313! and, 1955, *N. Chiotellis* 455! Pomos Point, 1962, *Meikle* 2241!
3. Akrotiri, 1905, *Holmboe* 681; Limassol, 1949, *Mavromoustakis* 59!
5. Vatili, 1905; *Holmboe* 349! Kythrea, 1880, *Sintenis & Rigo* 28!
6. Nicosia, by Pedios R., 1939, *Lindberg f.* s.n.! Dheftera, 1967, *Merton* in ARI 500!
7. Vasilia, 1941, *Davis* 3043! Kazaphani, 1948, *Casey* 115! Lapithos, 1955, *Miss Mapple* 87! Kambyli, 1956, *Merton* 2724!
8. Near Rizokarpaso, 1912, *Haradjian* 257! Ayios Theodoros, 1970, *A. Genneou* in ARI 1474!

NOTES: Less common in Cyprus than the following (var. *caerulea* Gouan), but, on this account, more often collected on the island, and disproportionately represented in collections.

var. **caerulea** *Gouan*, Fl. Monsp., 30 (1765); Davis, Fl. Turkey, 6: 140 (1978).
> *A. caerulea* L., Fl. Monsp., 11 (1756) nomen vix rite publ.
> *A. arvensis* L. f. *caerulea* Lüdi in Hegi, Illustr. Fl. Mittel-Europa, 5 (3): 1870 (1927).
> *A. arvensis* L. f. *azurea* Hyl. in Uppsala Univ. Arsskr., 1945 (7): 256 (1945).
> TYPE: France; unlocalized, in herb. Gouan (K!).

Corolla bright blue with a darker basal zone.

HAB.: Cultivated and fallow fields, gardens, roadsides, waste ground; sea-level to ? 600 ft. alt.; Feb.–May [? Oct.]

DISTR.: Divisions 2, 4–8, often abundant. Widely distributed as a weed, like var. *arvensis*, but tending to replace the latter in warm-temperate regions.

2. Mavrovouni, 1932, *Syngrassides* 88!
4. Near Larnaca, 1973, *P. Laukkonen* 32! Cape Greco, 1958, *N. Macdonald* 2!
5. Near Kythrea, 1880, *Sintenis & Rigo* 27!
6. Kykko Farm near Nicosia, 1927, Rev. *A. Huddle* 75!
7. Kyrenia, 1949, *Casey* 488! and, same locality, *G. E. Atherton* 919! 946!
8. Ayios Philon near Rizokarpaso, 1941, *Davis* 2261!

NOTES: Probably common throughout the lowlands, but so often confused with *A. arvensis* L. ssp. *foemina* (Mill.) Schinz et Thell. that records unsupported by specimens have had to be omitted.

var. **pallida** *Hook.f.*, Student's Fl. Brit. Isles, 303 (1870). *A. arvensis* var. *albiflora* Druce in Rep.
B.E.C., 9: 470 (1931) nomen.
TYPE: Not indicated.

Corolla white or whitish, with a darker basal zone.

HAB.: Unknown; probably on waste ground.

DISTR.: ? Division. Found here and there throughout the range of *A. arvensis* L.

? DIV. "Dconos Forest", 1928 or 1930, *Druce* s.n.

NOTES: The locality cannot be traced, and may be a misspelling, as with several other place names in Druce's list.

ssp. **foemina** (*Mill.*) *Schinz et Thell.* in Bull. Herb. Boiss., ser. 2, 7: 497 (1907); Marsden-Jones et
Weiss in Proc. Linn. Soc., 150: 146–155 (1938).
    A. foemina Mill., Gard. Dict., ed. 8, no. 2 (1768); Osorio-Tafall et Seraphim, List Vasc.
Plants Cyprus, 82 (1973); Davis, Fl. Turkey, 6: 141 (1978).
    A. caerulea Schreb., Spic. Fl. Lips., 5 (1771); Post, Fl. Pal., ed. 2, 2: 177 (1933).
    A. arvensis L. ssp. *caerulea* (Schreb.) Hartm., Svensk och Norsk Exc.-Fl., 32 (1846);
Holmboe, Veg. Cypr., 143 (1914); nom. illeg.
    A. arvensis L. var. *caerulea* (Schreb.) Gren. et Godr., Fl. France, 2: 467 (1852); Boiss.,
Fl. Orient., 4: 6 (1875) non *A. arvensis* L. var. *caerulea* Gouan (1765) nom. illeg.
    A. arvensis L. var. *gentianea* Beck in Ann. K. K. Naturhist. Hofmus. Wien, 13: 3 (1898).
TYPE: Cultivated in Chelsea Physic Garden (BM).

Corolla-lobes blue, rather narrowly obovate-cuneate, not overlapping, apical margin irregularly denticulate, with a few, scattered, 4-celled stalked glands, each with a rather elongate terminal cell; pedicels at anthesis shorter than, or only slightly exceeding the subtending, relatively narrow, ovate leaves. Plant said to be more erect than *A. arvensis* ssp. *arvensis*, and leaves ± glaucescent.

HAB.: Cultivated and fallow fields, gardens, roadsides, waste ground, margins of irrigation channels; sea-level to 4,500 ft. alt.; fl. (? Feb.–) May–Oct.

DISTR.: Divisions 2, 6, 7. A widespread weed, especially in warm-temperate regions.

2. Prodhromos, 1955, *G. E. Atherton* 530!
6. Nicosia airport, 1951, *Merton* 1211!
7. Kyrenia, 1955, *Miss Mapple* 9! also, same locality, 1955, *G. E. Atherton* 326! 370!

NOTES: Also collected, but without locality, by *Miss Samson* (1904) and *C. B. Ussher* 82 (1929). It is certainly less common than *A. arvensis* L. var. *caerulea* Gouan, but may have been passed over as a form of this latter. The narrow leaves, relatively short pedicels and very sparsely glandular corolla-lobe margins will distinguish ssp. *foemina*; Marsden-Jones and Weiss (*loc. cit.*) say that the flower colour of ssp. *foemina* is distinct from var. *caerulea*, being a greyish-blue rather than a clear gentian-blue, but this can be judged only from fresh, living specimens. It is noteworthy that all the Cyprus material of *A. arvensis* ssp. *foemina* appears to have been collected in summer or autumn; most of the material of *A. arvensis* var. *caerulea* has been collected between February and April.

6. SAMOLUS *L.*

Sp. Plant., ed. 1, 171 (1753).
Gen. Plant., ed. 5, 78 (1754).

Perennial herbs or rarely subshrubs; leaves alternate, sometimes forming a loose basal rosette, simple, entire; inflorescence terminal, racemose or corymbose, bracteate, the bracts adnate to the pedicels and simulating bracteoles; flowers 5-merous; calyx-tube adnate to lower half of ovary, lobes persistent; corolla perigynous, campanulate or rotate, lobes imbricate; stamens inserted near base of corolla-tube opposite the lobes and alternating with solitary or clustered, subulate or ligulate staminodes; ovary globose; style simple; stigma truncate or capitate; ovules numerous, semi-anatropous. Fruit an ovoid or globose capsule opening by 5 teeth or valves.

About 10 species with a cosmopolitan distribution.

1. S. valerandi *L.*, Sp. Plant., ed. 1, 171 (1753); Unger et Kotschy, Die Insel Cypern, 296 (1865); Boiss., Fl. Orient., 4: 5 (1875); R. Knuth in Engl., Pflanzenr., 22 (IV. 237): 337 (1905); Holmboe, Veg. Cypr., 142 (1914); Post, Fl. Pal., ed. 2, 2: 177 (1933); Lindberg f., Iter Cypr., 26 (1946); Osorio-Tafall et Seraphim, List Vasc. Plants Cyprus, 82 (1973); Davis, Fl. Turkey, 6: 142 (1978).

TYPE: *"in maritimis* Europae, Asiae & Americae *borealis"*.

Glabrous perennial 11–30 (–45) cm. high; stems longitudinally sulcate, generally erect and branched at the base or unbranched, occasionally decumbent with erect, lateral, flowering branches; leaves mostly basal, forming a loose rosette, obovate or spathulate, 3–11 cm. long, 0·8–3 cm. wide, pale green, obscurely nerved, apex rounded or subacute, base tapering gradually or abruptly to a flattened petiole up to 4 cm. long; inflorescence many-flowered, racemose, 6–12 cm. long, unbranched or with a few short lateral branches; bracts lanceolate, about 1·5 mm. long, 0·5 mm. wide, adnate to the pedicel for more than half its length; pedicels erect, 5–12 mm. long, geniculate at the bract, the upper part bent upwards at a sharper angle than the lower; calyx cup-shaped, about 2 mm. long, 2·5 mm. wide, divided amost half way into 5 bluntly deltoid lobes; corolla white, tube about 1 mm. long, lobes lingulate, spreading, about 1·5 mm. long, 1 mm. wide, apex rounded, truncate or slightly emarginate; filaments about 0·7 mm. long; ovary globose, about 1 mm. diam.; style about 0·5 mm. long; stigma capitate. Capsule 1·7 mm. diam., opening by 5 recurved teeth or valves; seeds polyhedral, about 0·6 mm. wide; testa dark fuscous-brown, dull, minutely rugulose-foveolate.

HAB.: Wet ground by streams and springs, sometimes in brackish marshes; sea-level to 4,500 ft. alt.; fl. April–August.

DISTR.: Divisions 1–3, 6–8. Cosmopolitan.

1. Yeroskipos, 1862, *Kotschy*; Tsadha, 1905, *Holmboe* 733; Ktima, 1913, *Haradjian* 680 ! Baths of Aphrodite, 1959, *C. E. H. Sparrow* 17 !
2. Prodhromos, 1862, *Kotschy*, and, 1955, *G. E. Atherton* 597 ! Troödos area, 1894, *Post*, also, 1900, *A. G. & M. E. Lascelles* s.n.! and, 1950, *F. M. Probyn* 118 ! Kakopetria, 1936, *Syngrassides* 1270 ! Xerokolymbos, 4,300 ft. alt., 1937, *Kennedy* 1352 ! Mesopotamos, 1939, *Lindberg f.* s.n.; Platania, 1947, *Mavromoustakis* s.n. ! Between Troödos and Pano Platres, 1973, *P. Laukkonen* 392 !
3. Frequent about Zakaki, Cherkez and Asomatos, 1939, *Mavromoustakis* 34 ! 40 ! 47 ! 54 ! 89 ! 114 ! One mile N. of Akrotiri, 1941, *Davis* 3576 ! Pano Livadhia, 1962, *Meikle* 2921 ! Near Trimiklini, 1963, *J. B. Suart* 113 !
6. Near Nicosia, by Pedieos R., 1939, *Lindberg f.* s.n. !
7. Kephalovryso, near Kythrea, 1862, *Kotschy* 322; also, 1880, *Sintenis & Rigo*; Kyrenia, 1948, *Casey* 49 ! also, 1955, *G. E. Atherton* 31 ! 310 ! 330 ! Lapithos, 1955, *Merton* 2249 ! Ayios Epiktitos, 1955, *G. E. Atherton* 237 ! Lambousa, 1955, *G. E. Atherton* 342 ! Bellapais, 1956, *G. E. Atherton* 1320 ! Vasilia, 1970, *A. Genneou* in ARI 1575 !
8. Ayios Andronikos, 1934, *Syngrassides* 528 !

58a. EBENACEAE

Trees and shrubs with hard, often blackish, heartwood; leaves alternate or rarely opposite, simple, generally entire, exstipulate. Flowers axillary, solitary or in cymes, actinomorphic, dioecious or occasionally hermaphrodite; calyx 3–6-lobed, persistent and often accrescent in fruit; corolla sympetalous, often urceolate with 3–7, spreading, contorted or sometimes imbricate lobes; stamens 2–7-times the number of the corolla-lobes, or rarely equal in number and alternate to them, usually attached at or near the base of the corolla-tube, reduced to staminodes in the female flower; filaments free or variously connate, often very short; anthers commonly long, narrow, basifixed, introrse, 2-thecous, dehiscing longitudinally. Ovary normally superior, (2–) 4–12 (–16)-locular; ovules pendulous, 1–2 in each loculus; placentation axile; styles free or ± connate; stigmas small, terminal. Fruit a fleshy or leathery berry; seeds with a thin testa and copious, cartilaginous endosperm; embryo straight or curved, cotyledons foliaceous.

Three genera with a wide distribution, chiefly in the tropics, some species being valued for their heavy, dark timber, notably Ebony (*Diospyros ebenum* J. G. Koenig), while others bear edible fruits, particularly the Kaki or Chinese Persimmon (*Diospyros kaki* Thunb.) and the Date Plum (*D. lotus* L.). The last-named, a round-headed tree with dark, lustrous green, oblong-elliptic, entire leaves, small axillary 4 (–5)-merous flowers, and brownish-yellow, globose fruits about 2 cm. diam., is reported from Salamis Plantation, and from gardens at Larnaca, Pedhoulas, Phini, Lapithos (Bovill, Rep. Plant. Work Cyprus, 14: 1915; Frangos in Cypr. Agric. Journ., 18: 86; 1923). It is sometimes grown as a stock for *D. kaki*, which, especially in selected specimens, has very large, attractive fruits, with a sweet (but insipid) flavour when fully ripe. *Diospyros lotus* is a native of temperate Asia, from Turkey to China. It cannot be considered native, or even naturalized, in Cyprus.

59. STYRACACEAE
Janet R. Perkins in Engl., Pflanzenr., 30 (IV. 241): 1–111 (1907).

Trees and shrubs with stellate or lepidote indumentum; leaves simple, alternate, exstipulate; flowers generally hermaphrodite; calyx cupuliform or tubular, persistent, free or adnate to ovary, lobes 4–5, valvate or open, often small or obsolete; corolla sympetalous, 4–5 (–7)-lobed, often with deeply cleft lobes or almost polypetalous, valvate or imbricate; stamens as many as or twice as many as corolla-lobes, in one series, inserted at or towards base of corolla; filaments united into a tube below or occasionally free; anthers 2-thecous, introrse, dehiscing longitudinally, often narrow; ovary superior to inferior, 3–5-locular below, sometimes 1-locular above; ovules 1 to several in each loculus, axile, erect or pendulous, anatropous; style simple, usually slender; stigma capitate or 3–5 lobed. Fruit a drupe or capsule; seeds with copious endosperm and a straight or slightly curved embryo; cotyledons oblong or broad and flattish.

Eleven genera, chiefly in E. Asia and America; only one genus, represented by a single species, in the Mediterranean region.

1. STYRAX L.
Sp. Plant., ed. 1, 444 (1753).
Gen. Plant., ed. 5, 203 (1754).

Inflorescence a lax, terminal or axillary, simple or branched raceme; calyx cupuliform, free from the superior ovary except at the base, lobes small or obsolete; corolla 5- (or rarely 7- or more) lobed, tube short; stamens 10 (rarely more or fewer), filaments attached near base of corolla, shortly connate; anthers linear; ovary conical, subglobose or depressed-globose, often hairy, at first 3-locular, later becoming 1-locular through the retraction of the septa; stigma small, capitate or obscurely lobed. Fruit globose or ovoid; pericarp indehiscent and fleshy, or dry and dehiscing regularly or irregularly from the apex into 3 valves; seeds usually 1 through abortion, large, subglobose or ellipsoid; embryo straight; cotyledons usually broad.

About 100 species, chiefly in tropical Asia and America.

1. **S. officinalis** *L.*, Sp. Plant., ed. 1, 444 (1753) as *S. officinale*; Sibth. et Sm., Fl. Graec. Prodr., 1: 275 (1809), Fl. Graec., 4: 68, t. 375 (1823); Poech, Enum. Plant. Ins. Cypr., 27 (1842); Unger et Kotschy, Die Insel Cypern, 297, 410–419 (1865); Boiss., Fl. Orient., 4: 35 (1875); J. R. Perkins in Engl., Pflanzenr., 30 (IV. 241): 79 (1907); Holmboe, Veg. Cypr., 144 (1914); Post, Fl. Pal., ed. 2, 2: 181 (1933); Chapman, Cyprus Trees and Shrubs, 60 (1949); Osorio-Tafall et Seraphim, List Vasc. Plants Cyprus, 82 (1973); Elektra Megaw, Wild Flowers of Cyprus, 10, t. 14 (1973); Davis, Fl. Turkey, 6: 144 (1978).

 S. officinarum Sintenis in Oesterr. Bot. Zeitschr., 32: 291 (1882) nomen.
 TYPE: "*in* Syria, Judaea, Italia".

Spreading deciduous shrub or small tree, 2–7 m. high; twigs at first scurfy with stellate hairs, soon glabrous, smooth, with dark, fuscous-brown bark; leaves broadly ovate, oblong or suborbicular, 3–5 cm. long, 2–5 cm. wide, thinly stellate-puberulous above, cinereous below with a mixed indumentum of very short and longer stellate hairs, nervation obscurely and openly reticulate, apex acute or obtuse, margins entire, base rounded or broadly cuneate; petiole short, usually 3–7 mm. long, channelled above, scurfy with stellate hairs; inflorescence a lax 2–7-flowered raceme, terminal on short lateral shoots; flowers fragrant; bracts 2–3 mm. long, 1 mm. wide, scurfy, caducous; pedicels 5–15 mm. long, densely scurfy; calyx cup-shaped, 5–7 mm. long, 6–8 mm. diam., apex truncate, or with 5–6 very short, apical teeth, densely scurfy; petals milky-white, united into a short glabrous tube, about 5 mm. long, at the base, divided above into 5–7 lanceolate, spreading lobes, about 15–17 mm. long, 5–7 mm. wide, stellate-pubescent externally, subglabrous within; stamens 10–14, inserted at apex of corolla-tube; filaments erect, flattened, about 7 mm. long, 0·5 mm. wide, thinly stellate-pubescent near base, running the length of the anther and forming a slightly dilated connective; anthers linear, slightly recurved, 4–5 mm. long, 1 mm. wide; ovary subglobose, about 2 mm. diam., stellate-pubescent; style 15–20 mm. long, stellate-pubescent and rather thickened towards base, glabrous above and tapering to a truncate apex. Fruit drupaceous, subglobose, about 1·5 cm. diam. with a greenish, scurfy, rather fleshy pericarp, the persistent calyx remaining attached at the base, dehiscing irregularly or into 3 valves when fully ripe; seed subglobose, about 1 cm. diam., with 2 conspicuous wrinkles near the hilum; testa bright brown, smooth.

HAB.: In maquis or open Pine forest, often in moist situations by streams; near sea-level to 5,500 ft. alt.; fl. March–June.

DISTR.: Divisions 1–3, 7, 8, locally common. Widespread in the eastern Mediterranean region, from Italy to Palestine, with a close ally (*S. officinalis* L. var. *californicus* (Torr.) Rehd.) in California.

1. Stroumbi, 1934, *Syngrassides* 844! Between Ayios Neophytos and Stephani, 1939, *Lindberg f.* s.n.! By Avgas River near Ayios Yeoryios, 1962, *Meikle* 2022!
2. Common; Platres, Amiandos, Evrykhou, Khrysorroyiatissa, Prodhromos, Omodhos, Kakopetria, etc. *Sibthorp*; *C. B. Ussher* 18! *Kennedy* 687! *Davis* 3439! *G. E. Atherton* 641! *Merton* 3169! *D. P. Young* 7364! *A. Genneou* in ARI 1273!
3. Near Lefkara, 1862, *Kotschy*, and, 1967, *Merton* in ARI 588! Near Limnatis, 1941, *Davis* 3069!
7. Common; Lapithos, Pentadaktylos, St. Hilarion, Kazaphani, Kyrenia Pass, etc. *Kotschy*; *Sintenis & Rigo* 1A! *A. G. & M. E. Lascelles*! *Syngrassides* 1457! *Lindberg f.*; *G. E. Atherton* 716! 1377! etc.
8. Between Dhavlos and Komi Kebir, 1880, *Sintenis & Rigo* 1! Near Cape Andreas, 1880, *Sintenis & Rigo*; Koronia, *Chapman* 50! Yialousa, 1958, *F. J. F. Barrington* s.n.!

NOTES: The mystery concerning the aromatic resin, known to pharmacists as Storax or Storax calamita, and supposed to have been obtained from *Styrax officinalis*, is not yet satisfactorily solved. The resin was known to Dioscorides and Pliny the Elder, and from what they say, it is evident that the source was *Styrax officinalis*. Furthermore Garidel (Hist. Plant., 450, t. 95; 1715) identifies the source beyond any doubt, remarking that *S. officinalis* grows in great quantity in the Forest of the Chartreuse de Montrieux (France; Dept. Var), where the resin is extracted by making incisions in the bark. It is at first liquid, and strongly aromatic, and is put into little phials by the monks, for distribution to their friends. Duhamel (Traité des Arbres, 2: 289; 1755) confirms this account, and says he has seen the balsam exuding from the trees at Chartreuse de Montrieux. The Abbé Mazéas (Journ. des Sçavans, 1769; 105) independently testifies to the existence of resin-producing *Styrax officinalis* near Rome. But I can find no evidence that anyone has seen or collected resin from *Styrax officinalis* since that date. Hanbury (in Pharm. Journ., 16: 417–423; 461–465; 1857) concludes: 1. That the original and classical Storax was produced by *Styrax officinalis* L. 2. That it was always scarce and valuable, and has since disappeared from commerce. 3. That liquid *Storax* is the produce of *Liquidambar orientalis* Mill. Professor Stamatios D. Krinos of Athens (quoted by Hanbury) was, however, of the opinion that all forms of Storax came from *Liquidambar*.

Is there any evidence that *Styrax officinalis* ever produced an aromatic resin in Cyprus? Mrs. Chapman (Cyprus Trees and Shrubs, 60) mentions only that the fruits are ground to a powder and used to stupefy eels. It is not without significance, however, that some other *Styrax* species, notably *S. benzoin* Dryand., are well known to contain aromatic resins.

60. OLEACEAE

A. Lingelsheim in Engl., Pflanzenr., 72 (IV. 243. 1–2) (1920)
Oleoideae-Fraxineae et *Syringeae*.

Trees, shrubs or woody climbers, often with supra-axillary accessory buds; leaves opposite (rarely alternate), exstipulate, simple or pinnately compound, entire or toothed; flowers in bracteolate cymes or panicles, or apparently solitary, generally hermaphrodite, actinomorphic; calyx often small, usually 4-lobed or 4-toothed, occasionally absent; petals usually 4 (sometimes more), free or united, sometimes absent, valvate or imbricate, rarely contorted; stamens 2 (rarely 4), epipetalous or hypogynous in apetalous genera, alternating with the carpels; filaments commonly short; anthers large, often with the connective produced into an apiculus, 2-thecous, the thecae placed back to back, dehiscing longitudinally; disk absent; ovary superior, 2-locular, ovules usually 2 (sometimes more) in each loculus; style simple; stigma generally 2-lobed. Fruit a capsule, berry, drupe or samara, generally 1–4 seeded; seeds with or without endosperm, sometimes polyembryonic; embryo straight.

About 29 genera widely distributed in tropical and temperate regions. The family includes many ornamental plants (*Syringa, Ligustrum, Osmanthus, Forsythia*) and some economically valuable ones (*Olea, Fraxinus*).

Corolla with a long, slender tube; fruit a berry - - - - - JASMINUM (p. 1089)
Corolla without a tube, or with a short tube, or corolla absent; fruit a drupe or a samara:
  Leaves compound, imparipinnate; fruit a samara - - - - FRAXINUS (p. 1091)
  Leaves simple; fruit a drupe:
    Leaves densely greyish-lepidote below; drupe large, with an oily pericarp - **2. Olea**
    Leaves not lepidote below; drupe small, pericarp not oily - - - **1. Phillyrea**

JASMINUM *L.*

Sp. Plant., ed. 1, 7 (1753).
Gen. Plant., ed. 5, 7 (1754).

Deciduous or evergreen, erect or climbing shrubs; branches usually angled; leaves opposite or sometimes alternate, generally imparipinnate with entire leaflets; flowers mostly in terminal cymes, occasionally solitary, actinomorphic, often fragrant; corolla hypocrateriform with a long, slender tube, and 4–9 convolute lobes; stamens 2, included; ovary 2-locular with 1–4 ovules in each loculus. Fruit a 2-lobed (or 1-lobed by abortion) berry, usually black and shining. Seed with scanty endosperm.

About 200 species chiefly in the tropics and subtropics of the Old World. Several Jasmines are popular as garden plants, and some of the fragrant species are highly valued in perfumery. Only three species are on record as cultivated in Cyprus, though almost certainly others are to be found in gardens there, and the following key may be of assistance in identifying them:

Flowers white or pink, fragrant:
  Leaves imparipinnately 5–9-foliolate:
    Calyx-lobes subulate-filiform, 4–6 mm. long:
      Corolla-lobes 15–20 mm. long; inflorescence lax, cymose, the pedicels of the outer flowers
        exceeding those of the inner -   -   -   -   J. GRANDIFLORUM (p. 1090)
      Corolla-lobes 9–12 mm. long; inflorescence rather condensed and subumbellate, the
        pedicels of the outer flowers not much exceeding those of the inner
                          J. OFFICINALE (p. 1090)
    Calyx-lobes minute, less than 2·5 mm. long -   -   -   - J. POLYANTHUM (p. 1091)
  Leaves 3-foliolate; calyx-lobes very short:
    Corolla-tube more than 2 cm. long -   -   -   -   -   - J. ANGULARE (p. 1091)
    Corolla-tube less than 2 cm. long -   -   -   -   - J. AZORICUM (p. 1091)
Flowers yellow, scentless; leaves 3-foliolate:
  Leaves persistent; flowers shortly stalked, corolla-lobes 1·5–2 cm. long   J. MESNYI (p. 1091)
  Leaves deciduous; flowers sessile or subsessile, corolla-lobes less than 1·5 cm. long
                          J. NUDIFLORUM (p. 1091)

J. GRANDIFLORUM *L.*, Sp. Plant., ed. 2, 9 (1762); Hume in Walpole, Mem. Europ. Asiatic Turkey, ed. 1, 253 (1817); P. S. Green in Baileya, 13: 146 (1965).

A straggling bush or climber with angled stems; leaves opposite, imparipinnate, 7–9-foliolate, subglabrous or thinly and shortly scabridulous; lateral leaflets obtuse or rounded, 7–18 mm. long, 6–15 mm. wide, terminal leaflet larger, elliptic, acute or acuminate, 1·5–3 cm. long, 0·7–1·5 cm. wide; inflorescence a lax, spreading cyme; pedicels of the outer flowers often 2 cm. long, those of the central flowers generally shorter, often less than 1 cm. long; calyx with 5, slender, subulate-filiform lobes, commonly 5–6 mm. long; corolla white or pinkish, very fragrant, tube about 1·5 cm. long, lobes 5, spreading, bluntly oblong-obovate, 15–20 mm. long, 8–14 mm. wide. Fruit a black, often 2-lobed, berry.

A plant of uncertain origin, now widely cultivated in the Mediterranean region, and in other warm-temperate and subtropical countries, and much valued in perfumery.

Hume reports it as in cultivation at Limassol in July 1801, and it has since been collected at Neapolis, Nicosia (1973, *P. Laukkonen* 432 !).

J. OFFICINALE *L.*, Sp. Plant., ed. 1, 7 (1753); Unger et Kotschy, Die Insel Cypern, 263 (1865); Boiss., Fl. Orient., 4: 43 (1879); P. S. Green in Baileya, 13: 154 (1965).

Similar to *J. grandiflorum* in general appearance but the leaves generally (5–) 7-foliolate, the terminal leaflet usually very much larger than the laterals, often 3–5 cm. long, with a slender, attenuate acumen; inflorescence rather condensed and sub-umbellate, the pedicels of outer and inner flowers more or less equal; corolla white or pinkish, fragrant, tube about 1·5 cm. long, lobes ovate, cuspidate-acuminate, 9–12 mm. long, 5–8 mm. wide. Fruit a black berry.

A native of China and the Himalayan region, now cultivated in temperate countries almost throughout the world, and distinctly hardier than *J.*

grandiflorum. Kotschy says it is widespread in Cyprus gardens, but cites no localities; he adds that the stems "are carefully cultivated to make pipe stems".

J. polyanthum Franch. (Rev. Hort., 1891: 270; 1891) is similar, but more slender, with more numerous flowers in lax cymes. It is at once distinguished by the very short, almost obsolete, calyx-lobes. *J. polyanthum* is a native of China, now popular in cultivation.

Two other white- or pinkish-flowered, fragrant species, also commonly cultivated, are distinguished from the above by their regularly 3-foliolate leaves. *J. angulare* Vahl (Symb. Bot., 3: 1; 1794), from S. Africa, has rather compact inflorescences of exceptionally long-tubed fowers; *J. azoricum* L. (Sp. Plant., ed. 1, 7; 1753), from Madeira, is very similar, but with laxer inflorescences of more shortly tubed flowers. Both species have the very short calyx-lobes of *J. polyanthum*.

J. MESNYI *Hance* in Journ. Bot., 20: 37 (1882); P. S. Green in Baileya, 13: 150 (1965).
    J. primulinum Hemsl. in Kew Bull., 1895: 109 (1895).

A scrambling evergreen shrub with strongly angled, green twigs; leaves opposite, 3-foliolate; the leaflets glabrous and dark green above, elliptic, acute, 1–4·5 cm. long, 0·6–2 cm. wide; flowers solitary on short, bracteate, axillary stalks, scentless; calyx with 5–6, spreading, narrowly lanceolate lobes 5–8 mm. long, 1·5–2·5 mm. wide; corolla bright yellow, tube generally less than 2 cm. long, lobes 6–10 or more, giving the flowers a "double" or "semi-double" appearance, broadly and bluntly ovate-oblong, 1·5–2 cm. long, about 1 cm. wide.

A native of western China, now frequently grown in the Mediterranean region, or as a greenhouse plant further north. It is to be seen in many gardens about Nicosia, and no doubt elsewhere, flowering profusely in early spring.

J. nudiflorum Lindl. (in Journ. Hort. Soc., 1: 153; 1846), the Winter Jasmine, is very similar to *J. mesnyi*, with opposite, 3-foliolate leaves, and solitary, yellow, scentless flowers. It is, however, deciduous, and the flowers are sessile or very shortly stalked, with smaller, "single", 6-lobed corollas. It is a native of China, very popular as a winter-flowering shrub in many cool-temperate countries. There are no records from Cyprus, but *J. nudiflorum* may well be grown there.

One other yellow-flowered Jasmine, *J. fruticans* L. (Sp. Plant., ed. 1, 7; 1753) is indigenous in many countries bordering the Mediterranean, from Spain to Palestine, and eastwards to Iran. It is a low, much-branched, evergreen shrub, with rather small, alternate, 3-foliolate leaves; the leaflets subglabrous, bluntly oblanceolate; the scentless or very weakly scented flowers are borne in small terminal cymes, and have 5-lobed corollas, the lobes bluntly oblong-ovate and usually less than 7 mm. long. *J. fruticans* (unlike *J. mesnyi* and *J. nudiflorum*) is copiously fertile, generally bearing an abundant crop of shining, black berries. It is not recorded for Cyprus, but may yet be found, either as a native, or as an escape from cultivation.

FRAXINUS *L.*

Sp. Plant., ed. 1, 1057 (1753).
Gen. Plant., ed. 5, 1026 (1754).
Lingelsheim in Engl., Pflanzenr., 72 (IV. 243): 1–65 (1920).

Deciduous trees; leaves opposite, generally imparipinnate with toothed leaflets; flowers hermaphrodite, unisexual or polygamous, often incon-spicuous, in dense terminal and lateral racemes or panicles; calyx campanulate and 4-lobed, or absent; petals (2–) 4 (–6), free or somewhat connate, often absent; stamens usually 2; ovary 2-locular; stigmas 2. Fruit a 1 (rarely 2)-seeded samara with an elongate, apical wing.

About 70 species widely distributed in the northern hemisphere, but not represented by any indigenous species in Cyprus.

Corolla present; flowers conspicuous in dense terminal panicles   -   - F. ORNUS (p. 1092)
Corolla absent; flowers inconspicuous in lateral clusters:
  Young twigs pubescent; leaflets generally 3–5 -   -   -   - F. VELUTINA (p. 1093)
  Young twigs glabrous; leaflets generally 5–11 (–15):
    Inflorescence a raceme; leaves remotely toothed -   - - F. ANGUSTIFOLIA (p. 1092)
    Inflorescence a panicle; leaves usually rather closely toothed - F. EXCELSIOR (p. 1092)

F. ORNUS *L.*, Sp. Plant., ed. 1, 1057 (1753), a dense, round-headed tree to 20 m. high, with glabrous twigs and greyish or brownish winter buds; leaves imparipinnate, oblong, 20–30 cm. long, 10–15 cm. wide, petiole and rhachis glabrous; leaflets 7–9, ovate-oblong, acute or caudate-cuspidate, glabrous except on the underside towards the base of the midrib, 5–10 cm. long, 2–5 cm. wide, margins usually rather obscurely serrate; flowers very numerous in showy terminal panicles; calyx with 4 deeply cut, ovate lobes about 1 mm. long; corolla of 4, free, ligulate, white petals, 8–9 mm. long, 1 mm. wide; ovary and style about 2 mm. long; stigma shortly oblong. Fruit a small, narrowly oblanceolate samara, about 2 cm. long, 0·4 cm. wide, apex rounded or subacute.

A native of S. Europe, the Mediterranean region and western Asia, cultivated for its ornamental inflorescences. Kotschy's record from the Northern Range about Pentadaktylos (Oesterr. Bot. Zeitschr., 12: 276; 1862) is undoubtedly an error, but *Fraxinus ornus*, the Manna Ash, has been noted in cultivation at (2) Pasha Livadhi (Bovill, Rep. Plant. Work Cyprus, 30; 1915) and at Nicosia (Frangos in Cypr. Agric. Journ., 18: 88; 1923). Manna, a sweet, nutritious and mildly laxative substance, is exuded from incisions in the bark.

F. EXCELSIOR *L.*, Sp. Plant., ed. 1, 1057 (1753), the Common Ash, a common species in many parts of Europe, is recorded by Frangos from a nursery at Lapithos (Cypr. Agric. Journ., 18: 100; 1923). It is a robust, open-crowned tree, up to 40 m. high, with glabrous, ashy-grey twigs, and black, velutinous winter buds; the leaves are imparipinnate, oblong, 20–30 cm. long, 14–16 cm. wide, with 7–13 (–15) narrowly ovate-oblong, acuminate-caudate leaflets, usually 8–10 cm. long and 1·5–3 cm. wide; the leaflets are normally glabrous except on the underside towards the base of the midrib, and have numerous short (sometimes almost obsolete) marginal teeth; the flowers, borne in lateral panicles, may be unisexual or hermaphrodite, and consist simply of 2 stamens and/or a pistil similar to that of *F. ornus*, but with longer, narrower stigmas. The fruit is a pale brown, narrowly oblong, slightly twisted samara, usually 3–4 cm. long and 0·5–0·7 cm. wide, with an acute, obtuse or emarginate apex.

F. ANGUSTIFOLIA *Vahl* (Enum. Plant., 1: 52; 1804) is similar to *F. excelsior*, but usually less robust, seldom exceeding 25 m. in height, with glabrous twigs and brownish or blackish winter buds; the leaves are normally 5–11-foliolate, with narrow leaflets 6–10 cm. long and 1–2 cm. wide, the sharply serrate leaflets are either completely glabrous on both sides (ssp. *angustifolia*), or pilose below towards the base of the midrib (ssp. *oxycarpa* (Willd.) Franco et Rocha Afonso); the flowers, like those of *F. excelsior*, are without calyx or petals, and consist solely of stamens and pistils; they are borne on slender, unbranched pedicels along an elongate rhachis, forming a simple raceme (visible in infructescences), whereas the inflorescences of *F. excelsior* are repeatedly branched, forming a compound panicle, with lateral clusters of flowers and fruits. The samara of *F. angustifolia* is very similar to that of *F. excelsior*.

F. angustifolia has a wide distribution in S. Europe, the Mediterranean region and western Asia. Its presence in Cyprus as a native would not be surprising, but trees seen (2) along the banks of the Ayios Theodhoros R. near Stavros tis Psokas (April 26, 1962, *Meikle* 2690!) were most probably planted by the Department of Forests, together with some other exotics noticed in the same area. All the trees had the glabrous leaflets of ssp. *angustifolia*, but the distinctions between this and ssp. *oxycarpa* are trifling.

F. VELUTINA *Torrey* (in Emory, Not. Reconn. Leavenworth to San Diego, 149; 1848), from south western U.S.A. and northern Mexico, a small tree usually less than 15 m. high, is reported as planted at Salamis (Bovill and Frangos) and Koronia (Bovill), but may not have survived. It differs from all the preceding in having more or less densely pubescent twigs and leaves, the latter generally with few (3–5 or less often 7) obscurely toothed or subentire leaflets; the flowers are in branched panicles, like those of *F. excelsior*, and are without calyx or petals. The fruit is a small samara, seldom more than 2 cm. long and 0·5 cm. wide, the seed-bearing portion of the samara is not dorsiventrally flattened or compressed as in the three other European and Asiatic species.

1. PHILLYREA *L.*
Sp. Plant., ed. 1, 7 (1753).
Gen. Plant., ed. 5, 8 (1754).

Evergreen shrubs and small trees; leaves opposite, simple, entire or variously toothed; flowers small, whitish, hermaphrodite, in short axillary racemes; calyx shortly 4-lobed; corolla 4-lobed, the lobes imbricate in bud and longer than the tube; stamens 2, epipetalous, exserted, with short filaments; ovary 2-locular, with 2 pendulous ovules in each loculus; style short; stigma 2-lobed. Fruit a black drupe.

Two closely allied species in the Mediterranean region, Atlantic Islands and western Asia.

Leaves ovate, usually distinctly nerved and often serrate or crenate - - **1. P. latifolia**
Leaves lanceolate or narrowly elliptic with very obscure nervation, normally entire
P. ANGUSTIFOLIA (p. 1094)

1. P. latifolia *L.*, Sp. Plant., ed. 1, 7 (1753); Unger et Kotschy, Die Insel Cypern, 264 (1865); Osorio-Tafall et Seraphim, List Vasc. Plants Cyprus, 83 (1973); Davis, Fl. Turkey, 6: 157 (1978).
P. media L., Sp. Plant., ed. 2, 1: 10 (1762); Boiss., Fl. Orient., 4: 36 (1875); Holmboe, Veg. Cypr., 144 (1914); Post, Fl. Pal., ed. 2, 2: 183 (1933); Chapman, Cyprus Trees and Shrubs, 61 (1949).
TYPE: "*in* Europa *australi*".

A rigid, much-branched shrub to 3 m. high, with pale grey, finely fissured bark and terete, shortly puberulous twigs; leaves ovate, 1·5–3 cm. long, 0·6–1·5 cm. wide, coriaceous, dark green above, slightly paler below, nervation usually distinct, apex acute or subacute, base rounded, margins narrowly cartilaginous, generally shortly and rather regularly serrate; petioles mostly less than 4 mm. long, rather stout, puberulous; flowers in short, rather congested racemes usually less than 2 cm. long; rhachis angled, puberulous; pedicels about 1·5 mm. long; bracts caducous, chartaceous, narrowly obovate, about 1·5 mm. long, 1 mm. wide, concave, puberulous;

calyx with 4 rounded, membranous lobes about 0·5 mm. long, 0·8 mm. wide;
corolla whitish or tinged purple externally, tube widely infundibuliform,
about 0·5 mm. long, lobes bluntly ovate, about 1·5 mm. long, 1 mm. wide,
strongly recurved at anthesis; filaments about 1 mm. long; anthers
relatively large, oblong, yellow, about 2 mm. long, 1·5 mm. wide; ovary
flask-shaped, about 1 mm. long, 0·8 mm. wide, glabrous, tapering to an
indistinct style; stigmas shortly oblong, about 0·3 mm. long, obscurely 2-
lobed. Fruit a blue-black, somewhat pruinose drupe 5–6 mm. diam., with a
short apiculus formed by the persistent style and stigmas; endocarp pale
brown, obscurely veiny, rather thin, separating readily from the exocarp;
seed globose, about 3·5 mm. diam.; testa pale brown, minutely shagreened;
endosperm oily.

HAB.: In *Arbutus-Pistacia* maquis on calcareous mountainsides; 1,900–2,000 ft. alt.; fl.
March–April.

DISTR.: Divisions [2], 7, rare. Widespread in the Mediterranean region.

[2]. "In den Vorbergen zwischen Lefkera [Lefkara] und Maschera [Makheras Mon.]". See Notes.

7. North side of Kyrenia Range E. of Lefkoniko Pass, 1952, *Merton* 1631! West of Kantara,
1952, *Merton* 1647! Olymbos Peak, 1957, *Merton* 2893! Near Kantara, 1971, *Guichard* CYP
13!

NOTES: The localization of the Kotschy record from Division 2 must be questioned. *Phillyrea
latifolia* has not since been found in this area, and Kotschy is known to have collected on the
Northern Range near the Lefkoniko Pass. There are, however, suitable habitats for the species
about Lefkara, and it may yet be refound.

P. ANGUSTIFOLIA *L.*, Sp. Plant., ed. 1, 7 (1753), with a wide distribution in
the central and western Mediterranean region, is grown in the nursery of the
Municipal Gardens, Nicosia (1974, *Meikle* 4099!) and may be found as a
cultivated plant elsewhere in Cyprus. It forms a rigid shrub up to 4–5 m.
high, with flowers and inflorescences virtually indistinguishable from those
of *P. latifolia*, from which, however, it differs in its lanceolate or narrowly
elliptic, entire, thickly coriaceous and obscurely nerved leaves. It may be
doubted if it is, at most, more than a subspecies of the protean *P. latifolia*.

2. OLEA *L.*

Sp. Plant., ed. 1, 8 (1753).
Gen. Plant., ed. 5, 8 (1754).

Evergreen trees and shrubs, with terete or angled twigs; leaves opposite,
simple, generally entire and coriaceous; flowers hermaphrodite, polygamous
or unisexual in axillary or terminal racemes or panicles; calyx cupuliform
with a truncate or shallowly 4-lobed apex; corolla 4-lobed, the lobes
induplicate-valvate, valvate or somewhat imbricate, longer than the
corolla-tube; stamens 2, epipetalous, with short filaments; ovary 2-locular,
generally with 2 pendulous ovules in each loculus; style short; stigma
obscurely 2-lobed. Fruit a drupe; seeds normally solitary; endosperm
present.

About 20 species with a wide distribution in temperate and tropical
regions of the Old World.

1. **O. europaea** *L.*, Sp. Plant., ed. 1, 8 (1753); Unger et Kotschy, Die Insel Cypern, 263 (1865);
Boiss., Fl. Orient., 4: 36 (1879); Holmboe, Veg. Cypr., 144 (1914); Post, Fl. Pal.,ed. 2, 2: 182
(1933); Lindberg f., Iter Cypr., 26 (1946); Chapman, Cyprus Trees and Shrubs, 61 (1949);
Osorio-Tafall et Seraphim, List Vasc. Plants Cyprus, 82 (1973); Davis, Fl. Turkey, 6: 155
(1978).

O. sylvestris Mill., Gard. Dict., ed. 8, no. 3 (1768).

O. europaea L. var. *sylvestris* (Mill.) Lehr, De Olea Europaea, 20 (1779).

O. oleaster Hoffmsgg. et Link, Fl. Port., 1: 387 (1813–20).

O. sativa Hoffmsgg. et Link, Fl. Port., 1: 388 (1813–20).

O. europaea L. var. *oleaster* (Hoffmsgg. et Link) DC., Prodr., 8: 284 (1844); Post, Fl. Pal., ed. 2, 2: 182 (1933).

O. europaea L. var. *sativa* (Hoffmsgg. et Link) DC., Prodr., 8: 284 (1844); Post, Fl. Pal., ed. 2, 2: 182 (1933).

O. europaea L. ssp. *sylvestris* [*silvestris*] (Mill.) Hegi, Ill. Fl. Mittel-Europa, 5 (3): 1937 (1927).

O. europaea L. ssp. *oleaster* (Hoffmsgg. et Link) Negodi in Béguinot, Arch. Bot., 3: 79 (March 1927); Osorio-Tafall et Seraphim, List Vasc. Plants Cyprus, 82 (1973).

TYPE: "*in* Europa *australi*".

A bush or tree up to 15 m. high with a gnarled trunk and spreading branches, young twigs ashy-grey, lepidote, bluntly angled; leaves coriaceous, oblanceolate, narrowly obovate or oblong, 1–6 cm. long, 0·4–1·5 cm. wide, dull green and sparingly lepidote above, ashy-grey and densely lepidote below, apex acute, obtuse or rounded, mucronate, base tapering or rounded, margins entire, very narrowly recurved; petiole short, usually less than 5 mm. long, densely lepidote; inflorescence a crowded, axillary raceme or panicle, generally less than 2 cm. long; peduncle short, angular, lepidote; pedicels very short or flowers subsessile; bracts caducous, narrowly ovate, concave, about 2–2·5 mm. long, 1 mm. wide, lepidote externally; calyx cupuliform, scarious, subglabrous, about 0·5 mm. long, 2 mm. wide, subtruncate, or with 4 very short, blunt lobes; corolla white, tube very short, not exceeding calyx, lobes 4, spreading, oblong-ovate, obtuse or shortly acute, 2–3 mm. long, 1·5 mm. wide; stamens 2, filaments less than 1 mm. long; anthers relatively large, oblong, yellow, about 2 mm. long, 1 mm. wide; ovary glabrous, flask-shaped, about 1·5 mm. long, 1 mm. wide; style about 0·5 mm. long; stigmas oblong, about 0·5 mm. long, 0·3 mm. wide, erect. Fruit an oily, ovoid drupe, 1–2·5 (–3·5) cm. long, 0·8–2 (–3) cm. wide, becoming black when mature, endocarp hard and bony, pale brown, narrowly ellipsoid, longitudinally rugulose, about 0·8–1·5 cm. long, 0·5–0·8 cm. wide; testa pale brown, longitudinally sulcate, minutely reticulate; endosperm copious, embryo large, wholly filling the centre of the seed.

HAB.: On hillsides in garigue or maquis, and extensively cultivated in the lowlands; sea-level to c. 3,000 ft. alt.; fl. April–June.

DISTR.: All Divisions as a cultivated tree; also recorded as spontaneous from Divisions 1–3, 7, 8. Widespread as a wild or cultivated plant throughout the Mediterranean region and Atlantic Islands, also, as a cultivated plant, throughout the world, wherever the climate is suitable.

1. Akamas, 1941, *Davis* 3297!
2. Adelphi, 1949, *Chapman*; near Vavatsinia, 1967, *Merton* in ARI 725! Near Kalopanayiotis, 1968, *A. Genneou* in ARI 1280! Between Asinou and Ayios Theodhoros, 1947, *Meikle*!
3. Near Prastio, 1956, *Merton* 2540!
7. St. Hilarion, 1939, *Lindberg f.* s.n.; coastal maquis west of Kyrenia, 1955, *G. E. Atherton* 697!
8. Cape Andreas, 1937, *Syngrassides* 1655! Matakos near Galinoporni, 1962, *Meikle*!

NOTES: The problems relating to the origin and status of the Olive are no less imponderable than those relating to the Carob (vol. 1: 589) or the Grape (vol. 1: 360). While there is no disputing that populations of "wild" Olives are to be found, often far from cultivation, in many parts of Cyprus, it is now impossible to ascertain whether these are, at least in part, remnants of an ancient indigenous stock, or whether, as some would aver, they are all descended from cultivated trees.

The "wild" Olive is generally identifiable by its shrubby, spinose habit, small leaves, and small, stony fruits. It is valued only as a stock upon which to graft cultivated varieties. But, as W. B. Turrill has noted (Kew Bull., 6: 441; 1951), "The 'wild olives' classified as *Olea europaea* var. *oleaster* vary greatly and in many characters parallel to variations of cultivated olives". Such "wild" Olives are, moreover, to be found in areas generally agreed to fall outside the likely geographical range of the Olive as an indigene. The names *O. europaea* var. *sylvestris* and *O. europaea* var. *oleaster* are therefore little more than convenient labels for distinguishing uncultivated from cultivated Olives. Their taxonomic significance is questionable, and they have not been given recognition here, however useful they may be in general discussion.

As to the origin of the cultivated Olive, it would seem that the researches of many authors (Newberry, Ciferri, Mazzolani, Chevalier, Turrill, etc.) have, as yet, failed to reach final conclusions. Despite undoubted affinities with the Asiatic and African *O. ferruginea* Royle and *O. indica* Burm. f. (*O. africana* Mill.; *O. chrysophylla* Lam.), I am not wholly convinced that the more obvious derivation, that is, from the "wild" *O. europaea* populations of the eastern Mediterranean, may not be the correct one, nor would I be prepared to assert, contrary to the evidence of my eyes, that the Olive is not an ancient and indigenous component of maquis in this region.

REFERENCES:

Newberry, P. E. On some African species of the genus Olea and the original home of the cultivated olive-tree, in Proc. Linn. Soc., Session 150: 3–16 (1937).

Mazzolani, G. La Patria dell'Olivo, in L'Olivicoltore, 18 (5): 141–149 (1941).

Chevalier, A. L'Origine de l'olivier cultivé et ses variations, in Rev. Internat. Bot. Appl. et Agric. Trop., 28: 1–25 (1948).

Ciferri, R. Dati et Ipotesi sull'Origine e l'Evoluzione dell'Olivo, in Olearia, 1950: 115–122 (1950).

Turrill, W. B. Wild and Cultivated Olives, in Kew Bull., 6: 437–442 (1951).

Wickens, G. E. Speculations on long distance dispersal and the flora of Jebel Marra, Sudan Republic, in Kew Bull., 31: 131–132 (1976).

61. APOCYNACEAE

K. Schum. in Engl. et Prantl, Pflanzenfam., ed. 1, IV. 2: 109–189 (1895).

Mostly lactiferous trees, shrubs and woody climbers, occasionally herbaceous; leaves simple, opposite or whorled, entire, generally exstipulate or with much modified stipules; inflorescence usually cymose or flowers sometimes solitary and axillary; bracts and bracteoles commonly present; flowers actinomorphic, (4–) 5-merous, hermaphrodite; calyx generally lobed almost to the base, the lobes often glandular towards the base internally; corolla hypocrateriform, infundibuliform, campanulate, urceolate or rotate, sympetalous, often hairy within; corolla-lobes contorted in bud or rarely valvate, frequently asymmetrical; stamens usually 5, inserted on the corolla, free or connivent around the stylar head (*clavuncle*); filaments very short or almost wanting; anthers introrse, dehiscing longitudinally, connective sometimes produced into an apical awn or flap, thecae sometimes forming caudate basal appendages; disk and nectaries often present; ovary superior, or occasionally partly sunk in the disk, generally consisting of 2 free carpels united above by the style; ovules usually pendulous, 1–many in each carpel; style simple, often with an elaborate, swollen, hairy-zoned apex. Fruit commonly of 2 divaricate follicles, sometimes of 2 indehiscent mericarps, or baccate; seeds generally compressed or flattened, often with an apical coma; endosperm present or absent; embryo straight.

About 200 genera chiefly in tropical regions. Some species are important drug plants; many are poisonous; a few are valued as ornamentals or as a source of rubber.

Flowers solitary, axillary; trailing subshrubs or perennials -   -   -   - VINCA (p. 1097)
Flowers several or numerous in corymbose or paniculate cymes:
  Robust shrub; flowers showy, 4–5 cm. diam.; follicles united until fully mature - **1. Nerium**
  Rhizomatous perennial; flowers small, 3–6 mm. diam.; follicles free and divaricate long
    before maturity -   -   -   -   -   -   -   -   - **2. Trachomitum**

VINCA L.

Sp. Plant., ed. 1, 209 (1753).
Gen. Plant., ed. 5, 98 (1754).
M. Pichon in Bull. Mus. Hist. Nat. Paris, ser. 2, 23; 439–444 (1951).
W. T. Stearn in W. I. Taylor et N. Farnsworth, The Vinca Alkaloids, 19–94 (1973).

Creeping subshrubs or herbaceous perennials; leaves usually evergreen, opposite; flowers solitary, axillary, pedicellate, generally borne on short, ascending shoots; calyx with 5, slender, deeply cut lobes; corolla hypocrateriform, usually blue or violet, tube dilated towards apex, with an internal zone of hairs above the insertion of the stamens, but without apical appendages, lobes 5, truncate or acute, strongly asymmetric, overlapping to the left in bud; stamens 5, inserted about the middle of the tube; filaments very short, sharply geniculate at the base; anthers with the connective produced into flap-like appendages which meet over the clavuncle; ovary of 2 free carpels joined only by the simple style; clavuncle (stylar head) pilose at the apex, stigmas forming a zone around the broadest part of the clavuncle. Fruit of 2 horn-like follicles, each containing 4–8, narrow, glabrous seeds.

About 6 species widely distributed in western, central and southern Europe and western Asia.

V. MAJOR L., Sp. Plant., ed. 1, 209 (1753); Boiss., Fl. Orient., 4: 45 (1875); Pichon in Bull. Mus. Hist. Nat. Paris, ser. 2, 23: 442 (1951) (excl. var. *difformis* (Pourr.) Pichon); Osorio-Tafall et Seraphim, List Vasc. Plants Cyprus, 83 (1973); W. T. Stearn in W. I. Taylor et Farnsworth, The Vinca Alkaloids, 79 (1973); Davis, Fl. Turkey, 6: 163 (1978).

Creeping perennial, with sterile shoots up to 1 m. long, rooting at the tips, and short, ascending flowering stems; leaves broadly ovate, 3·5–6 cm. long, 2–4·5 cm. wide, evergreen, subcoriaceous, glabrous with ciliate margins, apex subacute, base truncate or often shallowly cordate; petiole flattened, 1–2 cm. long, thinly pilose with 2 distinct apical glands; pedicels glabrous, 3–4 cm. long; calyx infundibuliform, tube 3–4 mm. long, lobes linear-subulate, ciliate, about 10 mm. long, 1–1·5 mm. wide, glandular laterally near the base; corolla mauve-blue (or sometimes white), tube 1·8–2 cm. long, cylindrical in the lower half, dilated and infundibuliform in the upper half, glabrous externally, densely barbate internally at and above the insertion of the stamens; corolla-lobes patent, about 2 cm. long and almost as wide, obliquely truncate, glabrous; filaments about 2·5 mm. long, geniculate at the base, dilated to more than 1 mm. wide near apex; anthers yellow, oblong, about 1·5 mm. long, connective-appendage tongue-shaped about 1·5 mm. long and almost as wide, pilose externally; ovaries glabrous, about 1·5 mm. long, 1 mm. wide, flanked by 2 very large nectary-glands; style about 1 cm. long, gradually swelling upwards to the much enlarged clavuncle, the latter about 2·5–3 mm. long and wide, with a glabrous, frilled base and a sulcate, densely hirsute apex. Fruit (seldom produced) consisting of 2, spreading, arcuate-acuminate, pale brown, longitudinally rugulose, glabrous follicles 3–6 cm. long, 0·5–0·8 cm. wide at the widest part; seeds narrowly oblong, about 8 mm. long, 2·5 mm. wide, rounded dorsally with a deep ventral furrow; testa dark brown, closely and regularly ornamented with short, longitudinal, rugulose ridges.

A popular garden plant, not native in Cyprus, but recorded as an escape from cultivation at (6) Neapolis, Nicosia (1973, *P. Laukkonen* 201!) and at (7) Kyrenia (1956, *G. E. Atherton* 902!). Dr. A. Hansen notes that it has run wild at (2) Pingos Hotel, Troödos (1970, *A. Hansen* 768) and in the western outskirts (6) of Nicosia (1970, *A. Hansen* 876). *Vinca major* is widely distributed in western and central Europe and the Mediterranean region as far east as the Caucasus, but, like other *Vinca* species, it is so often a naturalized introduction that its natural distribution is no longer easy to assess.

VINCA HERBACEA *Waldst. et Kit.*, Descr. Icon. Plant. Rar. Hung., 1: 8, t. 9 (1799) is also noted from Cyprus (Osorio-Tafall et Seraphim, List Vasc. Plants Cyprus, 83; 1973), but I have seen no specimens. It is a much smaller plant than *V. major*, the stems rarely exceeding 50 cm., and dying back to the base in winter; the leaves are ovate-elliptic or elliptic, sometimes narrowly so, about 2–3 cm. long and 0·6–1·5 cm. wide, with smooth or scabridulous margins and a very short petiole; the violet-blue flowers are about half as large as those of *V. major*, with ciliate calyx-lobes scarcely exceeding 6 mm., and narrow, frequently acuminate, corolla-lobes. It appears to be much less sterile than is usual in *V. major*.

Vinca herbacea is widely distributed in south-central and eastern Europe, and in the eastern Mediterranean region eastwards to Iran and Central Asia. It is known from all the countries adjacent to Cyprus, and might be expected to occur in our area, but precise localization is necessary before it can be fully admitted to the flora.

VINCA MINOR *L.*, Sp. Plant., ed. 1, 209 (1753) much resembles *V. herbacea*, but has persistent, trailing, rooting stems, evergreen subcoriaceous leaves and blue (rarely purple) flowers with very short, narrowly deltoid (not linear-subulate), glabrous calyx-lobes, rarely more than 4 mm. long. Like *V. major*, it is generally sterile. *V. minor* has an extensive distribution in western, central and eastern Europe as far east as the Crimea, but it is not really a Mediterranean plant, and a Cyprus record (not backed by specimen or locality) is suspect. It may have been based on a cultivated specimen, or possibly on a misidentification of *V. herbacea*.

1. NERIUM *L.*

Sp. Plant., ed. 1, 209 (1753).
Gen. Plant., ed. 5, 99 (1754).
R. E. Woodson in Ann. Missouri Bot. Gard., 17: 15–24 (1930).

Evergreen shrubs; leaves narrow, opposite or whorled; inflorescence a terminal, corymbose cyme; calyx deeply 5-lobed, fringed with gland-like *squamellae* at the base internally; corolla large, showy, infundibuliform with toothed appendages at the throat and 5, wide lobes overlapping to the right in bud; stamens 5; anthers sagittate, connivent around the clavuncle, with subulate basal appendages and a filiform apical awn; disk absent; carpels at first united, becoming free when mature. Fruit a follicle containing numerous hairy seeds, each with an apical coma.

Three closely related species (or perhaps subspecies) in the Mediterranean region eastwards to India. Widely cultivated in warm-temperate and tropical regions for its ornamental flowers. All parts of *Nerium oleander* are highly poisonous.

1. **N. oleander** *L.*, Sp. Plant., ed. 1, 209 (1753); Hume in Walpole, Mem. Europ. Asiatic Turkey, ed. 1, 254 (1817); Sibthorp in Walpole, Travels, 14 (1820); Unger et Kotschy, Die Insel Cypern, 264 (1865); Boiss., Fl. Orient., 4: 47 (1875); Holmboe, Veg. Cypr., 145 (1914); Post, Fl. Pal., ed. 2, 2: 188 (1933); Lindberg f., Iter Cypr., 27 (1946); Chapman, Cyprus Trees and Shrubs, 62 (1949); D. Zafrir, The Nerium oleander in Israel, 48 pp., col. pl. (1962); Osorio-Tafall et Seraphim, List Vasc. Plants Cyprus, 84 (1973); Davis, Fl. Turkey, 6: 159 (1978). TYPE: "*in* Creta, Palaestina, Syria, India".

Robust, leggy, erect shrub up to 4 m. high; branches subterete or obscurely angled, at first puberulous, soon glabrous with finely rugulose, pale grey-brown bark; leaves usually ternate, sometimes opposite or subopposite, linear-lanceolate, 7–14 cm. long, 1·3–3 cm. wide, coriaceous, glabrous above, rather paler below and dotted over with minute hair-capped stomatal pits; midrib very prominent below, lateral nervation close, parallel; apex sharply acute, margins narrowly revolute, base tapering to a short, stout petiole; inflorescence much branched, peduncles and pedicels closely pubescent or puberulous, often tinged bronze or purple; bracts broadly subulate, somewhat keeled, puberulous, caducous, often tinged purple, to about 7 mm. long and 2·5 mm. wide at the base; pedicels short, seldom exceeding 6 mm., stout, angled, puberulous; calyx with 5 erect or somewhat recurved, subulate, pubescent, purplish, concave lobes about 7 mm. long, 2 mm. wide, fringed internally at the base with a number of narrow, pallid, gland-like scales or *squamellae* which may, on occasion, be transformed into fully or partly developed supernumerary petals; corolla showy, rose-pink, crimson or white, 4–5 cm. diam., tube cylindrical for about 6 mm. from the base, widely infundibuliform above, glabrous externally, thinly hairy internally especially around the base of the free part of the filaments, throat furnished with a fringe of irregularly 3-lobed, laciniate appendages; corolla-lobes patent, broadly obovate or suborbicular, overlapping, 2–2·5 cm. wide; stamens inserted at base of corolla-tube; filaments thinly hairy, adherent to corolla-tube for about 7 mm., free for about 3–4 mm., geniculate towards apex; anthers sagittate, yellow, about 4 mm. long, 1·5 mm. wide, with basal barbs about 1 mm. long, and twisted, filiform, purple-hairy apical awns about 7 mm. long; style 8–9 mm. long, gradually swelling upwards to a rather small clavuncle; ovary oblong, about 1·5 mm. long and almost as wide, densely white-hairy. Fruit narrowly fusiform, subterete, 10–15 cm. long, 0·8–1·2 cm. wide, purplish, longitudinally striatulate, dividing into two follicles when fully ripe; seeds narrowly oblong, about 7 mm. long, 1·5 mm. wide, densely hairy, with an apical coma of brownish hairs about 1 cm. long.

HAB.: Wet, rocky gullies, sides of streams and rivers; sea-level to 3,000 ft. alt.; fl. May–July.

DISTR.: Divisions 1–8, locally common. Widespread in the Mediterranean region.

1. Evretou, 1937, *Syngrassides* 1715! Ayios Yeoryios (Akamas), 1962, *Meikle*! Polis, 1962, *Meikle*!
2. Makheras Monastery, 1862, *Kotschy*, and 1905, *Holmboe* 1123; Kryos Potamos, 3,000 ft. alt., 1938, *Kennedy* 1353! Ayia, *Chapman* 168; Saïttas, 1931, Mrs. *Tracey*; near Kakopetria, 1962, *Meikle*!
3. Stavrovouni, 1787, *Sibthorp*; Kouklia, 1862, *Kotschy*; Kalavasos, 1905, *Holmboe* 621.
4. Larnaca, 1801, *Hume*; Ayia Napa, 1853, *Gaudry*, also 1862, *Kotschy*; 1905, *Holmboe* 43, and 1958, *N. Macdonald* 84!
5. Athalassa, 1932, *Syngrassides* 414! and, 1939, *Lindberg f.* s.n.!
6. Near Cape Kormakiti, 1862, *Kotschy*; Dhiorios, 1962, *Meikle*!
7. Kyrenia, 1931, *Nattrass* 90! also, 1955, *Miss Mapple* 86! *G. E. Atherton* 189! 222! Near Ayios Amvrosios, 1955, *Merton* 2278!
8. North side of Karpas peninsula, 1853, *Gaudry*; Akradhes, 1962, *Meikle*!

N. INDICUM *Mill.*, Gard. Dict., ed. 8, no. 2 (1768) (*N. odorum* Aiton, Hort. Kew., ed. 1, 1: 297; 1789) is smaller in all its parts, with small, strongly fragrant flowers with deeply fimbriate, filiform corolla appendages. It is a native of Iran, Arabia and Pakistan, but is widely cultivated (more frequently than *N. oleander*) in tropical countries. *N. indicum* may be grown in Cyprus gardens, but there are no records. The distinctions between it and *N. oleander* (which may, *teste G. E. Atherton* 222! also be perfumed) are not very impressive, and one feels that subspecific rank would better suit it, and its very close allies *N. kotschyi* Boiss. and *N. mascatense* A.DC.

2. TRACHOMITUM *Woodson*
in Ann. Missouri Bot. Gard., 17: 157–164 (1930).

Herbaceous perennials arising from creeping, fibrous rhizomes; stems erect or ascending, terete, generally glabrous; leaves opposite or sub-opposite, petiolate or subsessile, entire, mucronate, usually glabrous or with the margins and underside of the midrib scabridulous; inflorescence a terminal, branched, monochasial, paniculate or corymbose, usually many-flowered, bracteate cyme; calyx deeply 5-lobed, eglandular within; corolla cylindrical, campanulate or infundibuliform, with 5 erect or spreading lobes, furfuraceous externally, lobes contorted to the right in bud, corolla-appendages adnate to the base of the corolla-tube, united into a ring, the 5 short lobes opposite the corolla-lobes; stamens 5; filaments short; anthers wth convergent basal appendages and a small apical appendage; receptacle raised above mouth of calyx-tube, subentire or somewhat lobed; nectaries 5, distinct, ovoid-cylindrical; carpels 2, united by the style and clavuncle; stigmas obscurely 2-lobed. Fruit comprising 2, free, divaricate, slender follicles; seeds with an apical coma, terete, without endosperm.

One polymorphic species, subdivided by some authors into numerous species, subspecies or varieties, and distributed from Italy and the Balkan peninsula eastwards across Asia to China. The genus was long included in *Apocynum* L. (type: *A. androsaemifolium* L.) but differs in having inflorescences with single-stalked (monochasial) bases, and united corolla-scales, as against a trichasial (3-stalked) base and free corolla-scales in *Apocynum*.

1. **T. venetum** (*L.*) *Woodson* in Ann. Missouri Bot. Gard., 17: 158 (1930); Osorio-Tafall et Seraphim, List Vasc. Plants Cyprus, 83 (1973).

    Apocynum venetum L., Sp. Plant., ed. 1, 213 (1753); Boiss., Fl. Orient., 4: 48 (1875); Post, Fl. Pal., ed. 2, 2: 188 (1933); Lindberg f., Iter Cypr., 27 (1946).

    Trachomitum sarmatiense Woodson in Ann. Missouri Bot. Gard., 17: 162 (1930).

    T. venetum (L.) Woodson ssp. *sarmatiense* (Woodson) Avetisian in Biol. Zhurn. Armen., ⸱20 (2): 104 (1967); Davis, Fl. Turkey, 6: 160 (1978).

    TYPE: "*in* Adriatici *maris insulis*". See Notes.

Erect, rhizomatous perennial up to 2·5 m. high; stems terete, glabrous, purplish, often somewhat pruinose, obscurely sulcate longitudinally, branches long, spreading, opposite or sometimes alternate; leaves opposite, oblong, elliptic or lanceolate, 1–8 cm. long, 0·3–2·5 cm. wide, obscurely penninerved, glabrous or scabridulous along margins, and on lower surface along midrib, apex acute, obtuse or rounded, generally apiculate, margins entire, narrowly recurved, base rounded or cuneate; petiole very short and indistinct, or up to 8 mm. long, sometimes flanked at the base by several small, finger-like glands; inflorescence a lax, spreading or compact, much-branched, paniculate or corymbose cyme; inflorescence branches glabrous or pubescent; bracts subulate, 2–3 mm. long, 1 mm. wide at base, puberulous with membranous margins; pedicels slender, puberulous, rarely more than 5 mm. long; calyx campanulate, deeply lobed, with 5 erect, oblong lobes 1–2 mm. long, 0·7–1 mm. wide, pubescent externally with membranous margins and obtuse or subacute apices; corolla subcylindrical or campanulate-infundibuliform, 4–7 mm. long, 3–6 mm. wide, bright or pale pink, furfuraceous externally and internally; tube 2·5–3·5 mm. long, lobes erect or spreading, oblong-lingulate or suborbicular, 1·5–3·5 mm. long, 1–3 mm. wide; filaments stout, about 1 mm. long, geniculate; anthers narrowly ovate, 1·5 mm. long, 0·7 mm. wide at base, apical appendage acuminate; carpels glabrous, 0·8 mm. long, 0·7 mm. wide; style broadly clavate, about 0·7 mm.

long; stigma bluntly conical, about 0·5 mm. long. Fruit consisting of 2 pendulous or spreading, straight or slightly falcate, subulate, glabrous follicles, 8–24 cm. long, 0·4–0·5 cm. wide; seeds narrowly oblong or fusiform, 2·5–3 mm. long, 0·5 mm. wide; testa dark brown, irregularly reticulate with narrow, longitudinally elongate sculpturing; comal hairs white, about 2·5 cm. long.

HAB.: Sandy seashores; near sea-level; fl. June–August.

DISTR.: Division 4 only. General distribution that of the genus.

4. Near Famagusta ("In litore arenoso inter opp. Famagusta et Salamis"), 1939, *Lindberg f.* s.n.!

NOTES: The taxonomy of the *Trachomitum venetum* complex is so involved, and so unsatisfactory, that the aggregate species has been left intact. It is true that typical *T. venetum* from the Adriatic represents a local, extreme variant, with subsessile leaves and small inflorescences, but when this is subtracted from the aggregate, one is left with a bewildering residue, clearly consisting of many local variants, but no less clearly including many intermediate, linking specimens. The Cyprus plant closely resembles specimens from Anatolia (Kizilirmak R. near Kalečik, *Bornmüller* 14373! *Guichard* T/126/60! *Albury, Cheese & Watson* 1761! also from the mouth of the Meander near Izmir, *Balansa* 246!) and from Afghanistan (Bahrak, *Grey-Wilson & Hewer* 1651! Banks of Hari Rud near Minaret of Djam, *Grey-Wilson & Hewer* 1222! etc.). It differs from *T. venetum* and *T. sarmatiense* in having large inflorescences of relatively large flowers with infundibuliform corollas. It falls within the orbit of *T. venetum* (L.) Woodson ssp. *scabrum* (Russan.) Rechinger f. (Fl. Iranica, no. 103: 5; 1974), but this appears to include more than one variant, and is typically so closely allied to *T. sarmatiense* (or *T. venetum* ssp. *sarmatiense*) that it is barely distinguishable save by the greater scabridulousness of the leaves.

Osorio-Tafall & Seraphim (List Vasc. Plants Cyprus, 83) asterisk *T. venetum*, to indicate that they regard it as "a taxon that is known or believed to have been introduced in Cyprus by the agency of man". This may well be so, though I have, to date, no positive evidence of introduction.

Two other exotic *Apocynaceae* are reported from Cyprus.

OCHROSIA ELLIPTICA *Labill.*, Sert. Austro.-Caled., 25, t. 30 (1824), an evergreen, lactiferous tree with thick, blunt, obovate leaves in whorls of 3 or 4 (or rarely opposite) and dense terminal corymbose cymes of small whitish flowers, followed by solitary or paired, hard, compressed-ellipsoid drupes, which turn bright red when ripe, has been collected in a nursery at Ktima (1935, *Syngrassides* 863!) and may still survive in cultivation on the island. *O. elliptica* is a native of Australia, New Caledonia and the New Hebrides.

HOLARRHENA ANTIDYSENTERICA (*Roth*) *A. DC.* in DC., Prodr., 8: 413 (1824) (*Echites antidysenterica* Roth) a native of India and Malaya, is recorded from Nicosia by Bovill (Rep. Plant. Work Cyprus, 8; 1915). It is a small, graceful tree, with opposite, relatively large, elliptic or ovate, papery, deciduous leaves, and attractive, white, scentless, Jasmine-like flowers in rather dense terminal cymes. The fruit consists of two slender follicles, often 30 cm. long, filled with silky-comose seeds. There are no recent records from Cyprus.

62. ASCLEPIADACEAE

K. Schum. in Engl. et Prantl, Pflanzenfam., ed. 1, IV. 2: 189–306 (1895).

Lactiferous perennial herbs, subshrubs and climbers, seldom shrubs or trees; leaves simple, opposite, rarely whorled or alternate, exstipulate, generally entire; flowers usually in cymes or umbels; calyx deeply 5-lobed; corolla gamopetalous, frequently rotate, less often campanulate or tubular, 5-lobed, the lobes contorted or rarely valvate in bud; appendages arising from backs of stamens or base of corolla and forming a conspicuous single or double, nectariferous corona; stamens 5; filaments usually very short or wanting; anthers introrse, with or without horny wings, usually adherent to the clavuncle (or stylar head) and forming a *gymnostegium*; pollen either granular in tetrads (*Periplocoideae*), discharging into a spathulate *pollen-carrier* with a glutinous base, or united into waxy *pollinia* (*Asclepiadoideae*), with 1, or occasionally 2, pollinia in the adjacent halves (or adjacent thecae) of 2 contiguous anthers, the pollinia united in pairs (or fours) by *arms*, *translators* or *caudicles* to a median, glutinous, sutured, pollen-carrier or *corpusculum*, attached to the clavuncle between adjacent anthers; stigma generally on the edge of, or at the base of the clavuncle; disk absent; ovary of 2 free carpels united by the style and clavuncle; ovules numerous, pendulous on ventral placentas. Fruit generally of 1 or 2 divaricate, horn-like follicles; seeds mostly with an apical tuft (coma) of silky hairs; endosperm sparse, cartilaginous; embryo straight.

About 280 genera and 1,800 species, widely distributed, but chiefly in warm-temperate and tropical regions.

Some authors divide the family into *Periplocaceae* (here subfamily *Periplocoideae*) and *Asclepiadacae* proper (subfamily *Asclepiadoideae*), but the connections with Apocynaceae are such that there would seem to be equal justification for widening the scope of this latter to include both *Asclepiadaceae* and *Periplocaceae* as subfamilies.

Flowers in umbels; erect herbs with linear or lanceolate leaves　　-　　-　　ASCLEPIAS (p. 1104)
Flowers in cymes or glomerules, not in umbels; sprawling or scandent herbs or shrubs:
　　Leaves and stems tomentellous; corolla-lobes broadly ovate-acuminate, flowers in dense
　　　　short-stalked clusters or glomerules　-　　-　　-　　-　　-　　-　　**2. Vincetoxicum**
　　Leaves and stems glabrous, or at most very sparsely pubescent; corolla-lobes elongate, strap-
　　　　shaped, lanceolate or narrowly oblong:
　　　　Leaves small, ovate-elliptic; flowers few in small, simple cymes　-　　-　　**1. Cyprinia**
　　　　Leaves large, cordate; flowers numerous in much-branched cymes　-　　-　　**3. Cionura**

SUBFAMILY 1. **Periplocoideae.** Pollen granular, in tetrads, discharged from anthers on to spathulate or infundibuliform pollen-carriers attached basally to the clavuncle by a glandular corpusculum; anthers free from each other, without horny wings.

1. CYPRINIA *Browicz*
in Feddes Repert., 72: 127 (1966).

Glabrous, climbing or scrambling shrubs; leaves entire, coriaceous, shortly petiolate; flowers in simple, few-flowered axillary cymes; calyx-lobes relatively broad, blunt; corolla rotate, deeply lobed, with 5, spreading, strap-shaped lobes; corona segments ribbon-like, deeply cleft into 2 slender, filiform lobes; stamens free with very short filaments; anthers awned; pollen-carriers spathulate; clavuncle discoid, crenate. Fruit consisting of 2, slender, divaricate follicles; seeds with an apical coma.

One species in Cyprus and southern Turkey.

1. **C. gracilis** (*Boiss.*) *Browicz* in Feddes Repert., 72: 128, fig. 1 (1966); Davis, Fl. Turkey, 6: 165 (1978).

    Periploca gracilis Boiss., Fl. Orient., 4: 50 (1875), Suppl., 344 (1888); Holmboe, Veg. Cypr., 145 (1914); Chapman, Cyprus Trees and Shrubs, 62 (1949); Osorio-Tafall et Seraphim, List Vasc. Plants Cyprus, 84 (1973).

    TYPE: Turkey: "Ravins encaissés situés près du village de Tchoupourlu, à 2 lieues au NO. de Mersina (Cilicie)", *Balansa* (Plantes d'Orient, 1855) 719 (G, K !).

Slender woody scrambler or climber with milky latex; stems much branched, terete, glabrous with conspicuous nodes, green in the first year, dark brown in older growths; leaves ovate-elliptic, elliptic or lanceolate, coriaceous, 1–5 cm. long, 0·3–2 cm. wide, dark glossy green above, rather paler below, turning purplish in autumn; nervation obscure; apex apiculate, acute or acuminate, base rounded or cuneate, margins entire, very narrowly recurved; petiole up to 3 mm. long, deeply canaliculate above with a swollen base which is barbellate adaxially; inflorescence a (1–) 3 (–5)-flowered cyme, usually as long as, or a little longer than, the subtending leaves; peduncle slender, 1–1·5 cm. long, tipped with 1–2 narrowly lanceolate, scarious, brownish, caducous bracts 1–1·3 mm. long, 0·5 mm. wide; pedicels to about 1 cm. long; calyx divided almost to the base, lobes broadly and bluntly obovate or suborbicular, about 1 mm. diam., with membranous margins; corolla white ("white as snow", *Kennedy*), greenish or greenish-yellow (*Chapman, Merton*), deeply divided into obtuse or subacute, strap-shaped or narrowly lanceolate lobes 6–7 mm. long, 1·5–2 mm. wide; corona-segments linear or ribbon-like, about 4 mm. long, 0·5 mm. wide, cleft almost to the middle into 2 filiform lobes which are continued downwards as distinct veins; stamens with a very short thick filament; anthers narrowly ovate, about 1 mm. long, 0·4 mm. wide, hispidulous externally, apex produced into a long, thread-like awn; pollen-carriers bluntly spade-shaped or spathulate, about 0·6 mm. long, brownish, membranous, the short basal stipe attached to a small, thickened corpusculum situated close to the clavuncle at the base of 2 longitudinal corolla-flanges; carpels about 0·5 mm. long; styles barely developed; clavuncle orbicular, about 0·5 mm. diam., with crenulate margins and a prominent central boss. Fruit consisting of 2, slender, divaricate, pale brown follicles, 5–7 cm. long, 0·3–0·5 cm. wide; seeds narrowly oblong, flattened, 5–7 mm. long, 1·5–2 mm. wide, longitudinally costate; testa dark brown, minutely verruculose; coma sessile, silky, about 2 cm. long.

    HAB.: Growing through *Rhus, Rubus, Pistacia, Rosa, Quercus alnifolia* and *Pinus nigra*, on igneous or (rarely) calcareous mountainsides; 2,000–4,000 ft. alt.; fl. May–July.

    DISTR.: Divisions 1, 2, 7. Also S. Turkey (near Mersin).

1. Near Lyso, 1905, *Holmboe* 778.
2. Between Omodhos and Paleomylos, 1880, *Sintenis & Rigo* 700 ! Prodhromos, 1905, *Holmboe* 954; Platania, without date, *Chapman* 75; above Kakopetria, 1936, *Syngrassides* 1262 ! Platres, 1930, *E. Wall* s.n.; and, 1937, *Kennedy* 690 ! Between Kakopetria and Platania, 1955, *Merton* 2307 ! Southern flank of Troödos, 3,800 ft. alt., 1955, *Kennedy* 1858 ! 1859 ! 1860 ! Kionia, above Makheras monastery, 1956, *Merton* 2837 ! Nicosia -Troödos road, at junction with Kyperounda road, 1973, *P. Laukkonen* 723 !
7. Rocky place S.W. of Kyrenia, 2,000 ft. alt., 1948, *Mercy Casey* in *Casey* 183 !

    NOTES: Mostly on igneous rocks, though the Kyrenia record (*Casey* 183) would appear to be on limestone.

SUBFAMILY 2. **Asclepiadoideae** Pollen united into pollinia attached by arms (translators, caudicles) to a median, sutured corpusculum; anthers united, with horny wings, their filaments coherent into a tube or *column*.

ASCLEPIAS *L.*

Sp. Plant., ed. 1, 214 (1753).

Gen. Plant., ed. 5, 102 (1754).

Gomphocarpus R. Br. in Mem. Wernerian Soc. Edinb., 1: 37 (1811).

Perennial or rarely annual herbs or subshrubs; rootstock often tuberous; leaves opposite or sometimes whorled, often with revolute margins; inflorescences extra-axillary, often arising between the petiole-bases, apparently umbellate, but strictly condensed cymes; corolla-lobes divided almost to base, seemingly valvate, but actually contorted and overlapping a little to the right in bud, patent or reflexed at anthesis, corona single, or double with an internal process or "horn", segments boat-shaped, often compressed laterally, with a thickened dorsal keel and apical appendages; pollinia pendulous; clavuncle depressed, pentagonal. Fruit an inflated follicle, or less often a pair of follicles; seeds numerous, flattened, with a silky coma.

About 150 species, chiefly in N. and Central America, South and tropical Africa. The distinction between *Asclepias* L. (with a single corona) and *Gomphocarpus* R. Br. (with a double corona, the outer segment enclosing an inner process or "horn"), is wholly artificial, and has been rejected by N. E. Brown (Fl. Trop. Afr., 4 (1): 314; 1902; Fl. Capensis, 4 (1): 664; 1907), R. E. Woodson (in Ann. Missouri Bot. Gard., 41: 1–211; 1954) and R. A. Dyer (Gen. S. Afr. Plants, 1: 483; 1975).

A. FRUTICOSA *L.*, Sp. Plant., ed. 1, 216 (1753); R. E. Woodson in Ann. Missouri Bot. Gard., 41: 151 (1954).

   Gomphocarpus fruticosus (L.) [R. Br. in] Ait. f., Hort. Kew., ed. 2, 2: 80 (1811); Boiss., Fl. Orient., 4: 61 (1875); Post, Fl. Pal., ed. 2, 2: 194 (1933); Osorio-Tafall et Seraphim, List Vasc. Plants Cyprus, 84 (1973); Davis, Fl. Turkey, 6: 166 (1978).

Erect, suffruticose perennial to about 1 m. high; stems unbranched or sparingly branched, terete, obscurely striatulate, shortly crispate-pubescent; leaves subsessile or shortly petiolate, linear-lanceolate, 3–10 cm. long, 0·4–1 cm. wide, shortly pubescent especially along midrib and margins, base narrowly cuneate, apex tapering to a fine pointed acumen, margins narrowly revolute, flowers usually 6–8 in a rather lax, nodding umbel; peduncle slender 2–2·5 cm. long; bracts filiform-subulate, about 10 mm. long, 1 mm. wide, pubescent; pedicels 1–2 cm. long; calyx divided almost to base, lobes subulate, pubescent, about 3 mm. long, 1 mm. wide at base; corolla-lobes spreading or reflexed, ovate, 5–6 mm. long, 4–5 mm. wide, white, shortly lanuginose along the margins; corona greenish-yellow or tinged purple, segments laterally compressed, about 3 mm. long, 1·5 mm. wide, dorsal keel rounded, thickened, apex truncate, with 2 short, falcate, incurved appendages, arising close to the free ventral margins; anthers about 3 mm. long, 1·4 mm. wide, margins horny, turned sharply outwards, apex produced into a short, blunt or emarginate, membranous wing; pollinia narrowly pyriform, about 1 mm. long, 0·4 mm. wide, tapering above to a short, slender caudicle attached to a small, oblong, brown corpusculum; clavuncle flattish or pulviniform, bluntly 5-lobed. Fruit consisting of 1 ovoid-acuminate, inflated follicle 5–6 cm. long, 2·5–3 cm. wide, rather densely echinate with soft, greenish spines, apex tapering to a short, pointed beak; seeds oblong, strongly compressed, about 4·5 mm. long, 2·5 mm. wide, margins strongly incurved; testa fuscous-brown, closely verruculose and rather coarsely and irregularly reticulate-ridged; coma about 2·5 cm. long, softly silky.

A native of South Africa, introduced into cultivation in 1714 by the Duchess of Beaufort, and now established, as an escape from cultivation, in the Mediterranean region, the West Indies, S. America and elsewhere. In Cyprus it has been recorded from nursery gardens at (6) Morphou, 1911 (Misc. Reports, Kew 1879–1920, letter no. 88) and collected at (1) Ktima, 1930, *C. B. Ussher* 93! (6) Astromeritis, 1936, *Syngrassides* 1082! and (7) Kyrenia, 1956, *Merton* 284! where the collector notes: "formerly cultivated; now persistent locally".

A. CURASSAVICA *L.*, Sp. Plant., ed. 1, 215 (1753); R. E. Woodson in Ann. Missouri Bot. Gard., 41: 59 (1954); Osorio-Tafall et Seraphim, List Vasc. Plants Cyprus, 84 (1973), an herbaceous annual (sec. Woodson) with a woody base; stems erect, 50–100 cm. high, thinly pubescent; leaves opposite or in whorls of 3, lanceolate, acuminate, 6–10 cm. long, 0·8–3 cm. wide, bright green, thinly pubescent; inflorescence an erect, 6–20-flowered umbel; peduncle 2–8 cm. long; bracts subulate-filiform; pedicels about 1·5–2 cm. long, calyx-lobes linear or narrowly oblong-ovate, 2–3 mm. long, blunt or subacute; corolla-lobes crimson or scarlet, narrowly ovate, about 7 mm. long, 4 mm. wide, strongly reflexed at anthesis; column conspicuous, slender, about 2 mm. long; corona golden-yellow, segments boat-shaped, about 4 mm. long, with an internal appendage or "horn" incurved over the flattened, pentagonal clavuncle; pollinia pyriform, about 1 mm. long, 0·4 mm. wide, tapering to flattened, brownish, geniculate caudicles about 0·5 mm. long; corpusculum large, fuscous, diamond-shaped. Fruit a smooth, narrow, ovoid-acuminate follicle 7–8 cm. long, 1·5 cm. wide; seeds flattened, ovate, 5–6 mm. long, 3·5–4 mm. wide, with a broad marginal wing; testa dark brown, minutely asperulous.

Thought to be a native of South America (the specific epithet refers to Curaçao), but now cultivated and naturalized almost throughout the tropics. It is recorded from (5) Kythrea, 1936, *Syngrassides* 1283!, but is probably cultivated elsewhere in Cyprus, where it may survive as a garden escape.

2. VINCETOXICUM *Wolf*

Gen. Plant. Vocab. Char. Def., 130 (1776).
R. Ross in Act. Bot. Neerl., 15: 161 (1966).

Rhizomatous perennials with erect or partly twining stems; leaves opposite, shortly petiolate; flowers in axillary or extra-axillary, pedunculate or subsessile cymes or umbels; calyx and corolla deeply lobed; corolla-lobes erect or spreading, contorted to the right in bud; corona single, fleshy, with 5 more or less connate, rounded or deltoid segments; pollinia pendulous; clavuncle pentagonal. Fruit a single fusiform, smooth follicle or rarely a pair of follicles; seeds flattened, with a silky coma.

About 20 species widely distributed in Europe and Asia.

1. **V. canescens** (*Willd.*) *Decne.* in DC., Prodr., 8: 523 (1844); Boiss., Fl. Orient., 4: 52 (1875); Post, Fl. Pal., ed. 2, 2: 191 (1933); Davis, Fl. Turkey, 6: 168 (1978).
    Asclepias canescens Willd. in Ges. Nat. Freunde Berlin Neue Schr., 3: 418 (1801).
    TYPE: Turkey; "in Cappadocien und Galatien" *Sestini* (B).

Sprawling or subscandent perennial up to 70 cm. high, rhizome short, woody; stems and leaves rather densely grey-tomentellous; lower leaves broadly and bluntly ovate, 3·5–6 cm. long, 3–5 cm. wide, becoming progressively narrower and more acuminate upwards, the uppermost lanceolate 2–4·5 cm. long, 0·5–2 cm. wide, nervation ascending, obscure above, rather prominent below; petiole very short and thick, seldom more than 4 mm. long; flowers in dense cymose, subsessile or shortly pedunculate glomerules, arising from the middle and upper nodes; pedicels short, thick, crispate-pilose, 2–4 mm. long; bracts narrowly deltoid, about 2 mm. long, 1 mm. wide at base, glabrous with long-ciliate margins; calyx-lobes deltoid, pilose, 1·5–2 mm. long, 1–1·5 mm. wide, shortly connate at the base, with solitary or paired, oblong glands on the internal surface of the sinuses; corolla-lobes yellowish, erect or spreading, glabrous externally, pilose or glabrous internally, broadly ovate-acuminate, about 4 mm. long, 3 mm. wide; corona segments bluntly deltoid, about 1 mm. long and wide, attached to a short, fleshy column; anthers about 1·3 mm. long, 1 mm. wide, with an ovate, membranous apical appendage inflexed over the clavuncle; pollinia shortly oblong-ovoid, not much compressed, about 0·4 mm. long, 0·3 mm. wide, attached by very short, gelatinous caudicles to a glossy, dark brown, narrowly oblong corpusculum about 0·3 mm. long, 0·1 mm. wide; clavuncle about 1·5 mm. diam., pentagonal, with a conspicuous, sulcate, central boss; ovaries squat, broadly and bluntly ovoid. Fruit usually a solitary, narrowly ovoid, rostrate-acuminate follicle (rarely a pair of follicles) 6–7 cm. long, 1·5–2 cm. wide towards the rather inflated base, pericarp greyish, shortly pubescent; seeds ovate, flattened with inflexed margins, 8–9 mm. long, 5–6 mm. wide; testa dark brown, minutely asperulous, but without ridges or reticulations; coma soft, silky.

HAB.: Igneous screes and stony slopes; 5,000–5,600 ft. alt.; fl. May–July.

DISTR.: Division 2, very rare. Also E. Aegean Islands, Turkey, N. Syria.

2. N.W. face of Moni tou Pellou (S.W. spur of Khionistra), 5,000 ft. alt., 1951, *Casey* 1211! Troödos, 5,600 ft. alt., 1952, *F. M. Probyn* 132!

NOTES: The Cyprus plant is typical *V. canescens*, with tomentellous stems and leaves, and short peduncles (see K. Browicz in Fragm. Flor. et Geobot., 21 (3): 261–271; 1975). E. C. Casey notes: "This plant has previously been collected by Mrs. Kennedy and Mrs. Chapman", but there are no specimens from either of these collectors at Kew. No flowering material has been seen, and the description of the species is partly drawn up from Turkish specimens.

3. CIONURA *Griseb.*

Spic. Fl. Rum. et Bith., 2: 69 (1844).

Shrub or subshrub; stems erect or scandent, glabrous or thinly pubescent; leaves opposite, broadly cordate, long-petiolate; flowers numerous, in terminal and lateral, extra-axillary, corymbose cymes; calyx deeply lobed; corolla white, lobes erect, narrowly oblong or strap-shaped, blunt; corona single, segments flat, tongue-shaped, blunt; column short; anthers with a fleshy connective produced apically into an elongate membranous appendage inflexed over the clavuncle; pollinia erect; clavuncle pentagonal at the base, produced above into an elongate, pillar-like appendage. Fruit a solitary, smooth, fusiform follicle; seeds flat, comose.

One species in the eastern Mediterranean region from Yugoslavia to Palestine and Sinai, and eastwards to Iran and Afghanistan.

1. **C. erecta** (*L.*) *Griseb.*, Spic. Fl. Rum. et Bith., 2: 69 (1844); Boiss., Fl. Orient., 4: 62 (1875); Davis, Fl. Turkey, 6: 174 (1978).

   Cynanchum erectum L., Sp. Plant., ed. 1, 213 (1753).
   Marsdenia erecta (L.) R. Br. in Mem. Werner. Soc. Edinb., 1: 28 (1809); Holmboe, Veg. Cypr., 145 (1914); Post, Fl. Pal., ed. 2, 2: 195 (1933); Osorio-Tafall et Seraphim, List Vasc. Plants Cyprus, 84 (1973).
   [*Vincetoxicum officinale* (non Moench) H. Stuart Thompson in Journ. Bot., 44: 334 (1906).]
   TYPE: "*in* Syria".

Low, branched shrub or subshrub, 0·5–1 m. high, or sometimes climbing up trees and shrubs to 5 m. or more; stems terete, glabrous or very sparsely pubescent; leaves broadly cordate, 3–8 cm. long, 2–7·5 cm. wide, subglabrous or shortly puberulous especially on the nerves, apex acute, often shortly cuspidate; petiole 1·5–5 cm. long, canaliculate above; flowers very numerous in lax, corymbose cymes, unpleasantly scented; calyx-lobes narrowly deltoid, blunt or subacute, about 1·8 mm. long, 1 mm. wide at the base, ciliolate, with 1 or 2 small, narrowly oblong glands internally at the sinuses; corolla-lobes milky white, narrowly oblong or strap-shaped, 7–8 mm. long, 2–2·5 mm. wide, with an obtuse, rounded or shortly emarginate, slightly expanded apex, convolute to the right in bud, erect or suberect at anthesis; corona segments tongue-shaped, glabrous, about 1 mm. long, 0·4 mm. wide; anthers about 0·5 mm. long, 0·4 mm. wide, pollen-cavities very small, connective produced to form a narrowly tongue-shaped, membranous appendage 0·5 mm. long, 0·3 mm. wide; pollinia clavate, about 0·3 mm. long, 0·1 mm. wide, not much compressed, apex tapering to a very short caudicle; corpusculum minute, linear, dark brown, about 0·1 mm. long; clavuncle barely 0·4 mm. diam., pentagonal, produced into a central column 1·5 mm. long and shortly bifid at the apex; ovaries squat, broadly and bluntly ovoid, glabrous. Fruit a smooth, fusiform follicle 8–9 cm. long, 1·5–2 cm. wide with a tapering apex; pericarp greenish or pale brown; seeds ovate, flattened, about 8 mm. long, 3·5 mm. wide, margins pallid, inflexed; testa dark brown, marginal part minutely verruculose, central part minutely rugulose, but not ridged or reticulate; coma white, silky, about 2 cm. long.

HAB.: Sand-dunes and sandy banks near the sea; about sea-level; fl. May–June.

DISTR.: Division 3, rare. Distribution that of the genus.

3. Sand banks near Limassol, 1901, *A. G. & M. E. Lascelles* s.n. ! Common on sand-dunes along the shore between Limassol and Amathus, 1905, *Holmboe* 643.

NOTES: The Kew sheet of the Lascelles specimen has been misidentified as *Vincetoxicum officinale* Moench (= *V. hirundinaria* Medik.). This is the basis for Stuart Thompson's record for the latter, and no doubt, for the inclusion of *V. hirundinaria* in Osorio-Tafall et Seraphim, List Vasc. Plants Cyprus, 84 (1973). So far as I am aware, *V. hirundinaria* does not grow in Cyprus, even as an introduction (cf. Holmboe, Veg. Cypr., 145, note).

62a. LOGANIACEAE

Trees and shrubs, or rarely herbs or climbers; leaves opposite (rarely alternate), simple, entire or toothed; stipules usually reduced to a line joining the bases of the petioles; flowers in cymes, panicles or racemes, occasionally solitary, hermaphrodite; calyx 4–5-lobed; corolla sympetalous, often tubular, lobes 4–5, imbricate, rarely contorted or valvate; stamens 4–5, inserted on the corolla-tube; filaments free, often short; anthers 2-thecous, opening longitudinally; ovary generally superior, 2 (–4)-locular; ovules usually numerous; placentation axile; style single with a capitate or 2-lobed stigma. Fruit generally a septicidal, 2-valved capsule, sometimes a drupe or a berry; seeds often winged; endosperm copious or scanty; embryo usually straight.

An ill-defined family, nowadays often subdivided into segregates, viz. *Buddlejaceae, Antoniaceae, Potaliaceae*, etc. It is not represented in Cyprus by any indigenous species, but *Buddleja madagascariensis* Lam., a native of Madagascar, has been collected in the Municipal Gardens, Nicosia (*Meikle* 4082 !) and may be grown elsewhere on the island. It is a spreading shrub, about 2–3 m. high, with the twigs and leaf-undersurface densely tomentose with pallid, ashy hairs, and with terminal panicles of tubular, orange, 4-merous flowers, which have a heavy, rather foetid smell. Two other species, *B. davidii* Franch., from China, with panicles of honey-scented, mauve or violet flowers, and *B. globosa* Hope, from Chile and Peru, with globose heads of orange flowers, are both commonly cultivated (and *B. davidii* often naturalized) in many parts of Europe, and may occur in Cyprus.

63. GENTIANACEAE

Gilg in Engl. et Prantl, Pflanzenfam., ed. 1, IV. 2: 50–108 (1895).

Annual and perennial herbs; leaves simple, exstipulate, usually opposite and entire, decussate, often sessile and generally glabrous; flowers actinomorphic, 4–5 (–12)-merous, usually hermaphrodite and in dichasial cymes; bracts and bracteoles present or absent; calyx gamosepalous usually with imbricate lobes; corolla gamopetalous, infundibuliform, campanulate, rotate, or sometimes hypocrateriform, lobes mostly convolute; stamens inserted on the corolla, as many as the corolla-lobes and alternate with them, sometimes declinate; anthers generally versatile, 2-thecous, introrse; disk often present; ovary superior, usually 1-locular, with 2, commonly extruded, parietal placentas or placentation rarely axile; ovules numerous, anatropous; style simple; stigma simple or 2-lobed. Fruit usually a septicidal capsule, rarely a berry; seeds numerous, small; endosperm fleshy; embryo very small.

About 10 genera and more than 1,000 species, found in a wide range of habitats almost throughout the world.

Corolla rotate, the tube much shorter than the 6–8 (–12) lobes - - - 1. **Blackstonia**
Corolla hypocrateriform, the tube equalling or longer than the (4–) 5 lobes - 2. **Centaurium**

1. BLACKSTONIA *Hudson*

Fl. Anglica, ed. 1, 146 (1762).
L. Zeltner in Bull. Soc. Neuchât. Sci. Nat., 93: 1–56 (1970).

Glabrous annuals; stems usually erect, often unbranched; leaves opposite, amplexicaul and more or less connate along the bases (perfoliate); flowers forming dichasia, 6–12-merous; calyx divided nearly to base into narrow lobes; corolla rotate, tube short, lobes narrowly elliptic, acute, yellow; epipetalous nectaries absent; stamens inserted at throat of corolla; anthers exserted, basifixed; style filiform, twice bifid towards apex; stigmas 4. Capsule 1-locular, 2-valved.

Four species in western and central Europe, the Mediterranean region, Crimea and eastwards to Iraq.

Calyx divided almost to base into linear-filiform lobes; pedicels slender, generally less than 2 cm. long; styles united for about 1·5 mm.; cauline leaves conspicuously perfoliate
1. B. perfoliata
Calyx with a distinct cupular base, lobes lanceolate-subulate; pedicels stout, commonly 2–3 (–8) cm. long; styles united for 2·5–3·5 mm.; cauline leaves shortly connate, not conspicuously perfoliate - - - - - - - - - - - **2. B. acuminata**

1. **B. perfoliata** (*L.*) *Hudson*, Fl. Anglica, ed. 1, 146 (1762); Zeltner in Bull. Soc. Neuchât. Sci. Nat., 93: 36 (1970); Osorio-Tafall et Seraphim, List Vasc. Plants Cyprus, 83 (1973); Davis, Fl. Turkey, 6: 177 (1978).
Gentiana perfoliata L., Sp. Plant., ed. 1, 232 (1753).

Chlora perfoliata (L.) L., Syst. Nat., ed. 12, 2: 267 (1767); Boiss., Fl. Orient., 4: 66 (1875); Holmboe, Veg. Cypr., 145 (1914); Post, Fl. Pal., ed. 2, 2: 197 (1933); Lindberg f., Iter Cypr., 26 (1946).

Erect annual 6–30 (–45) cm. high; stems terete, glabrous, usually unbranched; leaves glabrous, often glaucous, the basal crowded into an irregular, rather persistent rosette, sessile, ovate, obovate or oblong, acute or obtuse, entire, 1–5 cm. long, 0·5–2·5 cm. wide, the cauline perfoliate, remote, sharply acute or shortly cuspidate, 0·5–6 cm. long, 0·5–5 cm. wide, nervation obscure; inflorescence a lax or crowded, terminal dichasium; bracts foliaceous, sessile or with the bases of each pair somewhat connate, 2–10 mm. long, 2–8 mm. wide; pedicels slender, usually less than 2 cm. long at anthesis, elongating in fruit, but never conspicuously long; calyx generally divided almost to the base into 8 linear-filiform lobes 7–15 mm. long, 0·5 mm. wide at base, lobes glabrous, convex or obscurely keeled dorsally; corolla bright yellow or golden-yellow, tube 2·5–8 mm. long, lobes usually 8, narrowly obovate-elliptic, 4–14 mm. long, 2·5–6 mm. wide, acute, spreading at anthesis, afterwards erect and persisting until the fruits are ripe; filaments attached just below the sinuses, linear-filiform, 1·5–3 mm. long; anthers yellow, linear, 2–3·5 mm. long, 0·6 mm. wide; ovary ellipsoid, compressed, 4–7·5 mm. long, 2·5–3·5 mm. wide; styles united for 1·5 mm., then twice bifid into 4 arms 1–1·5 mm. long; stigmas decurrent along arms. Fruit compressed-ellipsoid, lustrous brown, 5–9 mm. long, 3·5–5 mm. wide; seeds oblong-ovoid or subglobose, 0·3–0·5 mm. long and almost as wide; testa dark brown, conspicuously reticulate-foveolate, the surfaces within the pits minutely and regularly verruculose.

ssp. **intermedia** *(Ten.) Zeltner* in Bull. Soc. Neuchât. Sci. Nat., 93: 45 (1970).
    Chlora perfoliata (L.) var. *minor* Ten., Fl. Nap., 1: 198 (1811–1815).
    C. intermedia Ten., Syll. Fl. Neap., 565 (1831).
    C. perfoliata (L.) L. ssp. *intermedia* (Ten.) Nyman, Consp. Fl. Europ., 501 (1881) quoad nomen.
    TYPE: Italy; "in collibus apricis et siccis circa Neapolim: *Valle delle Fontanelle*; *Valle di S. Rocco, Strada della storta dopo Poggio reale* et alibi", *Tenore* (NAP).

Plants very slender, generally less than 30 cm. high; inflorescences rather lax and often few-flowered; calyx-lobes usually less than 10 mm. long, 2·5–3 mm. wide; seeds about 0·3–0·4 mm. long.

HAB.: On hillsides amongst garigue, sometimes on moist ground by streams; by roadsides, or on sand-dunes; sea-level to 1,900 ft. alt.; fl. April–July.

DISTR.: Divisions 1, 3, 7, 8. Widespread in the Mediterranean region.

1. Tsadha, 1905, *Holmboe* 734; Yiolou, 1938, *Syngrassides* 1874! Ayios Neophytos monastery, 1939, *Lindberg f.* s.n.! Ayios Minas near Smyies, 1941, *Davis* 3302!
3. Kolossi, 1939, *Lindberg f.* s.n.; Episkopi, 1950, *F. J. F. Barrington* 48! Khalassa, 1970, *A. Genneou* in ARI 1514!
7. Above Lapithos, 1955, *Merton* 2241! Tchiklos, 1956, *G. E. Atherton* 1361! Dhavlos, 1957, *H. Painter* 26! Akanthou, 1967, *Merton* in ARI 810!
8. Valia, 1905, *Holmboe* 501; Cape Andreas, 1962, *Meikle* 2478!

NOTES: All the Cyprus material examined consists of slender small-flowered plants, agreeing with the taxon named *B. perfoliata* ssp. *intermedia* by Zeltner. Although it is difficult to find characters other than relative ones to distinguish the subspecies from type *perfoliata*, yet the differences between the Cyprus plant, and *Blackstonia perfoliata* of central and western Europe are such that some measure of recognition is demanded. I am not sure that the anther and filament characters cited by Zeltner invariably hold good, and there would appear to be little difference in the corolla-colour of ssp. *intermedia* and ssp. *perfoliata*.
When the Cyprus populations of *B. perfoliata* have been examined cytologically, I suspect it will be found that they are diploid (n = 10).

2. **B. acuminata** *(Koch et Ziz) Domin* in Bull. Int. Acad. Tchèque Sci. Cl. Nat. et Med., 34: 25 (1933); Zeltner in Bull. Soc. Neuchât. Sci. Nat., 93: 36 (1970); G. Zijlstra in Act. Bot. Neerl., 21: 587–597 (1972).

Chlora acuminata Koch et Ziz, Cat. Plant. Palat., 20 (1814); Reichb., Plant. Crit., 3: 6, t. CCVII (1825).

    C. serotina Koch ex Reichb., Plant. Crit., 3: 6, t. CCVIII (1825); Boiss., Fl. Orient., 4: 66 (1875); Post in Mém. Herb. Boiss., 18: 96 (1900); H. Stuart Thompson in Journ. Bot., 44: 334 (1906); Holmboe, Veg. Cypr., 145 (1914); Post, Fl. Pal., ed. 2, 2: 197 (1933); Lindberg f., Iter Cypr., 26 (1946); Osorio-Tafall et Seraphim, List Vasc. Plants Cyprus, 83 (1973).

    C. perfoliata (L.) L. var. *sessilifolia* Griseb., Gen. et Sp. Gent., 117 (1839); Unger et Kotschy, Die Insel Cypern, 265 (1865).

    C. perfoliata (L.) L. ssp. *serotina* (Koch ex Reichb.) Arcang., Fl. Ital., ed. 1, 475 (1882).

    Blackstonia perfoliata (L.) Huds. ssp. *serotina* (Koch ex Reichb.) Volmann, Fl. Bayern, 594 (1914); Davis, Fl. Turkey, 6: 177 (1978).

    TYPE: France; Montpellier, "circa Monspelium legit Salzmann" (ER?).

Erect annual 9–50 cm. high; stems terete, glabrous, often glaucous, usually unbranched; leaves glabrous, usually glaucous, the basal sometimes forming a loose rosette which is frequently withered before anthesis, oblong, obtuse, entire, 1–4 cm. long, 0·5–2·5 cm. wide, sessile, with a cuneate or tapering base, cauline leaves remote, amplexicaul or shortly connate (and perfoliate), narrowly or broadly ovate or oblong-ovate, acute or shortly cuspidate, 1–5 cm. long, 0·5–2 cm. wide, obscurely 3-nerved; inflorescence a lax or rather congested terminal, generally few (3–15)-flowered dichasium, or flowers often solitary in depauperate specimens; bracts as in *B. perfoliata*, but with the bases often very shortly connate; pedicels stout, often strictly erect and elongate, commonly 2–3 cm. long at anthesis, up to 8 cm. long in fruit; calyx generally divided from about 1·5–2 mm. from base into 8, lanceolate-subulate, acuminate lobes 6–11 mm. long, 1–1·5 mm. wide, and rather distinctly 3-nerved just above the connate, cup-shaped base, lobes glabrous, flattish or somewhat convex dorsally; corolla bright or golden yellow, tube 4–5 mm. long, lobes usually 8, ovate, often apiculate, 4–7 mm. long, 2–3·5 mm. wide, spreading at anthesis, erect and persistent in fruit; filaments attached just below the sinuses, linear-filiform, 2–2·5 mm. long; anthers yellow, oblong or linear, 0·8–2 mm. long, 0·4–0·6 mm. wide; ovary ellipsoid, compressed, 3·5–4 mm. long, 2·5–3·5 mm. wide; styles united for 2·5–3·5 mm., then twice bifid into 4 short arms, less than 1 mm. long; stigmas decurrent along arms. Fruit compressed-ellipsoid, dark brown, 6–9 mm. long, 3–5 mm. wide; seeds irregularly oblong or ovoid, about 0·4–0·5 mm. long, 0·3 mm. wide; testa dark brown, coarsely and irregularly reticulate-foveolate, with deep-set, membranous-walled pits, minutely and regularly verruculose at the base.

    HAB.: Damp ground by streams and springs; sea-level to 4,500 ft. alt.; fl. May–July.

    DISTR.: Divisions 2–8. Europe and Mediterranean region, from the Netherlands through south and central Europe to Turkey. See Notes.

2. Prodhromos, 1862, *Kotschy*; Panayia, 1862, *Kotschy* 615a; Galata, 1880, *Sintenis & Rigo* 605a! Phini, 1962, *Meikle* 2875!
3. Alekhtora [?"Alektriona"], 1894, *Post*; Limassol, 1905, *Holmboe* 665, and 1939, *Lindberg f.* s.n.; Kakoradjia, 1932, *Syngrassides* 394! Kolossi, 1939, *Lindberg f.* s.n.; Stavrovouni, 1939, *Lindberg f.* s.n.; Limassol Salt Lake, near Akrotiri, 1941, *Davis* 3559! Between Yerasa and Kalokhorio, 1962, *Meikle* 2891!
4. Larnaca Salt Lake, 1979, *Edmondson & McClintock* E 2956!
5. Above Dheftera*, 1956, *Merton* 2764!
6. Kykko (metokhi) Mill*, 1901, *A. G. & M. E. Lascelles* s.n.!
7. Kyrenia*, 1931, *J. A. Tracey* 43! also, 1939, *Lindberg* f. s.n.! and, 1949, *Casey* 811*! Lapithos, 1955, *Merton* 2252!
8. Cape Andreas, 1962, *Meikle* 2479!

    NOTES: Like *B. perfoliata* (L.) Huds. (with which it is frequently confused), *B. acuminata* comprises two chromosome races, ranked as subspecies by Zeltner (op. cit., 25–36), one, ssp. *acuminata*, relatively robust, with the anthers almost equalling the filaments, the other (ssp. *aestiva* (K. Malý) Zeltner), slender, with short anthers often less than half as long as filaments. Unlike *B. perfoliata*, however, the two subspecies of *B. acuminata* are very difficult to determine with certainty, in the absence of cytological data. I strongly suspect that the diploid ssp. *aestiva* (n = 10) is commoner in Cyprus than the tetraploid (n = 20) ssp. *acuminata*, but my suspicions await confirmation by cytological research into local populations. In the meantime, I have marked with an asterisk those collections which I have reason to believe are ssp. *acuminata*. The differences between *B. acuminata* and *B. perfoliata* are, in my opinion, quite ample to justify specific status for the two taxa. In addition to morphological characters, *B. acuminata* flowers later than *B. perfoliata*; plants of the latter (*Meikle* 2478) were in flower at Cape Andreas on 5th April, 1962, while plants of *B. acuminata* growing with them (*Meikle* 2479) were just beginning to develop flower buds. The two species often grow together, though it would appear that *B. acuminata* is more definitely restricted to moist localities than *B. perfoliata*.

2. CENTAURIUM *Hill*

British Herbal, 62 (1756).
L. Zeltner in Bull. Soc. Neuchât. Sci. Nat., 93: 57–164 (1970).
A. Melderis in Bot. Journ. Linn. Soc., 65: 224–250 (1972).

Annual, biennial or rarely perennial, glabrous herbs; stems usually erect, angled; leaves sessile, opposite, the basal often forming loose rosettes; inflorescences generally dichasial cymes, sometimes subcapitate or spiciform, or rarely flowers solitary; calyx deeply divided into (4–) 5 narrow lobes; corolla hypocrateriform, usually pink or purplish, sometimes white or yellow, (4–) 5 lobed; epipetalous nectaries absent; stamens inserted at apex of corolla-tube; anthers narrowly oblong or linear, twisting spirally on dehiscence; style slender, persistent, 2-fid at apex; stigmas capitate, oblong or spathulate. Capsule oblong or fusiform, 1-locular, 2-valved.

About 50 species with a cosmopolitan distribution, but chiefly in temperate regions of the world.

Corolla yellow, tube 15–20 mm. long - - - - - - - **5. C. maritimum**
Corolla pink or white, tube 7–12 mm. long:
    Inflorescence spiciform, the flowers spaced along one side of an elongate axis
        4. C. spicatum
    Inflorescence a dichasial, often corymbose cyme:
        Corolla-lobes 8–12 mm. long; upper parts of stems, bracts and calyces more or less
        scabridulous; leaves commonly forming a persistent basal rosette
            1. C. erythraea ssp. **rhodense**
        Corolla-lobes 3–5 mm. long; upper parts of stems, bracts and calyces not or scarcely
        scabridulous; leaves seldom forming a persistent basal rosette:
            Inflorescence an open, spreading dichasium; plants often (but not always) branched
            from near the base; pedicels usually well developed - - **2. C. pulchellum**
            Inflorescence an erect, crowded, fastigiate, many-flowered cyme; plants often (but not
            always) branched from near the top of a long, erect stem; pedicels generally wanting
            or very short - - - - - - - - - **3. C. tenuiflorum**

1. **C. erythraea** Rafn, Danm. Holst. Fl., 2: 75 (1800); Melderis in Bot. Journ. Linn. Soc., 65: 227
(1972); Osorio-Tafall et Seraphim, List Vasc. Plants Cyprus, 83 (1973); Davis, Fl. Turkey,
6: 178 (1978).
    Erythraea centaurium Borkh. in Roemer, Arch. Bot., 1 (1): 30 (1796); Unger et Kotschy,
Die Insel Cypern, 265 (1865); Boiss., Fl. Orient., 4: 68 (1875); Post in Mém. Herb. Boiss., 18:
96 (1900); H. Stuart Thompson in Journ. Bot., 44: 334 (1906); Holmboe, Veg. Cypr., 144
(1914); Post, Fl. Pal. ed. 2, 2: 198 (1933).

Erect biennial (or annual) 5–50 cm. high; stems sharply angled, glabrous or minutely scabridulous, unbranched or with a varying number of erect or ascending branches; basal leaves commonly forming a persistent rosette, obovate, oblanceolate or spathulate, 1–6 cm. long, 0·6–2 cm. wide, glabrous, sessile, 3–7-nerved, apex acute, obtuse or rounded, base gradually or abruptly cuneate; cauline leaves in remote pairs, broadly or narrowly oblong, ovate or obovate, 0·5–5 cm. long, 0·2–2 cm. wide, apex acute or obtuse, rarely acuminate, inflorescence a lax or congested, usually many-flowered, terminal, subcorymbose dichasium; bracts foliaceous, sometimes scabridulous, resembling the uppermost cauline leaves but generally shorter and narrower; flowers with short, stout, glabrous or scabridulous pedicels, generally less than 3 mm. long; calyx divided almost to base into 5, narrowly linear-subulate, acute or subacute, keeled, glabrous or scabridulous lobes 5–8 mm. long, less than 0·8 mm. wide at base; corolla-tube 7–12 mm. long, lobes usually 5, ovate or narrowly oblong-ovate, 4–12 mm. long, 2–3 mm. wide at base, generally bright pink, sometimes purplish or white, apex acute or subacute, sometimes appearing to be acuminate through the inrolling of the margins after anthesis; stamens inserted at apex of corolla-tube, filaments filiform, glabrous, 1·2–7 mm. long; anthers yellow, narrowly oblong, 0·8–5 mm. long, 0·4–0·5 mm. wide, twisting spirally after dehiscence; ovary fusiform 6–10 mm. long, 1–1·8 mm. wide, glabrous; style slender, 1·5–5 mm. long, arms 1–1·5 mm. long; stigmas subspathulate. Fruit a fusiform capsule, about 10 mm. long, 2 mm. wide; seeds irregularly oblong-subglobose, 0·3–0·5 mm. diam.; testa dark brown, coarsely and irregularly reticulate.

ssp. **rhodense** (*Boiss. et Reuter*) *Melderis* in Bot. Journ. Linn. Soc., 65: 234 (1972); Davis, Fl. Turkey, 6: 179 (1978).

    Erythraea rhodensis Boiss. et Reuter in Boiss., Diagn., 2, 6: 121 (1859).

    E. centaurium Pers. var. *laxa* Boiss., Fl. Orient., 4: 68 (1875); H. Stuart Thompson in Journ. Bot., 44: 334 (1906); Post, Fl. Pal., ed. 2, 2: 198 (1933) quoad nomen; Rechinger f. in Arkiv för Bot., ser. 2, 1: 428 (1950).

    Centaurium majus (Hoffmsgg. et Link) Ronniger ssp. *rhodense* (Boiss. et Reuter) Zeltner in Bull. Soc. Neuchât. Sci. Nat., 93: 109 (1970).

    C. erythraea Rafn ssp. *grandiflorum* (? non (Pers.) Melderis) Osorio-Tafall et Seraphim, List Vasc. Plants Cyprus, 83 (1973).

TYPE: Rhodes; "in insulâ *Rhodo* ubi legit cl. Cadet de Fontenay" (G).

Cauline leaves narrowly oblong, conspicuously smaller than the basal; inflorescence lax, often with long, spreading, lateral branches; upper parts of stems, bracts, pedicels and calyces scabridulous, sometimes densely so; flowers large, with narrow, acute corolla-lobes 8–12 mm. long; calyx-lobes 5–8 mm. long, usually just a little more than half as long as the corolla-tube; filaments 5–7 mm. long; anthers 4–5 mm. long. Seeds relatively large, almost 0·5 mm. diam., very coarsely reticulate. *Plate 67*.

HAB.: On dry open hillsides in garigue, usually on chalk or limestone; sea-level to 3,500 ft. alt.; fl. May–July.

DISTR.: Divisions 1–3, 7, 8. Eastern Mediterranean region, from S. Italy and Sicily eastwards to Palestine.

1. Tsadha, 1901, *A. G. & M. E. Lascelles* s.n.! Akamas, 1949, *Casey* 751!
2. Troödos, 1894, *Post*; Aphamis, 1937, *Kennedy* 688! Perapedhi, 1940, *Davis* 1840!
3. Limassol-Paphos road, 1903, *A. G. & M. E. Lascelles* s.n.! Akrotiri, 1905, *Holmboe* 673; Oritaes Forest, 1937, *Chapman* 301! Cape Gata, 1941, *Davis*, 3562! Polemidhia, 1947, *Mavromoustakis* s.n.! Cape Zevgari, 1950, *F. J. F. Barrington* 46! Near Trimiklini, 1954, *Mavromoustakis* 18! Lania, 1970, *A. Genneou* in ARI 1535!
7. Armenian Monastery, 1912, *Haradjian* 355! Between Halevga and the Armenian Monastery, 1940, *Davis* 1745! Lapithos, 1955, *Merton* 2225! Near Kyrenia, 1955, *Miss Mapple* 52! Akanthou, 1967, *Merton* in ARI 812!
8. Near Komi Kebir, 1912, *Haradjian* 272.

NOTES: The Cyprus populations of this plant are remarkably uniform and distinct, consisting of robust biennials with lax, spreading, scabridulous inflorescences of very large, narrow-lobed flowers. Occasionally, in exposed situations, the plants are small and condensed (e.g. *Davis* 3562! *F. J. F. Barrington* 46!), but these are fairly obvious habitat-states. Nowhere on the island, it would appear, does one find typical *C. erythraea*, with crowded inflorescences of small, usually bluntish-lobed flowers. Indeed, so marked are the differences, that one is tempted to follow Zeltner (*op. cit.*, 109) and to attach ssp. *rhodense* to *Centaurium majus* (Hoffmsgg. et Link) Ronniger (or to *C. grandiflorum* (Pers.) Ronniger). Apart from nomenclatural problems involving the identity and origin of *C. grandiflorum*, the taxonomy of the *C. erythraea* complex is so daunting, that Melderis's opinion "that none of these appears to be sufficiently distinct or constant to be recognized as a distinct species", is probably safest. While a few specimens from the western Mediterranean [especially from S. Spain] are (chromosome number apart?) indistinguishable from Cyprus material of *C. erythraea* ssp. *rhodense*, I find it difficult to accept Zeltner's and Melderis's conclusion that *Erythraea sanguinea* Mabille (from Corsica) is

2

1

4

3

LRM

Plate 67. CENTAURIUM ERYTHRAEA Rafn ssp. RHODENSE (Boiss. et Reuter) Melderis **1,** habit, ×2; **2,** flower, longitudinal section, ×3; **3,** dehisced capsule, ×4; **4,** seed, ×16. (**1, 2** from *Kennedy* 688; **3, 4** from *Davis* 1840.)

synonymous with *Erythraea rhodensis* Boiss. et Reuter. Isotype material of *E. sanguinea* (K!) looks very different from the Cyprus plant.

2. C. pulchellum *(Swartz) Druce*, Fl. Berks., 342 (1898); Zeltner in Bull. Soc. Neuchât. Sci. Nat., 93: 96–103 (1970); Melderis in Bot. Journ. Linn. Soc., 65: 245 (1972); Osorio-Tafall et Seraphim, List Vasc. Plants Cyprus, 83 (1973); Davis, Fl. Turkey, 6: 180 (1978).

  Gentiana pulchella Swartz in Kongl Svenska Vet. Akad. Nya Handl., 4: 85, t. III, figs. 8, 9 (1783).

  G. ramosissima Vill., Fl. Delph., 23 (1786).

  Erythraea ramosissima (Vill.) Pers., Syn. Plant., 1: 283 (1805); Boiss., Fl. Orient., 4: 67 (1875); H. Stuart Thompson in Journ. Bot., 44: 334 (1906).

  E. pulchella (Swartz) Fries, Novit. Fl. Suec., part 2: 30 (1814); Sintenis in Oesterr. Bot. Zeitschr., 32: 398 (1882); Holmboe, Veg. Cypr., 144 (1914); Post, Fl. Pal., ed. 2, 2: 198 (1933).

  E. ramosissima (Vill.) Pers. var. *pulchella* (Swartz) Griseb. in DC., Prodr., 9: 57 (1845); Unger et Kotschy, Die Insel Cypern, 264 (1865).

  Centaurium ramosissimum (Vill.) Druce in Rep. B.E.C., 4: 274 (1915).

  C. pulchellum (Swartz) Druce ssp. *ramosissimum* (Vill.) P. Fourn., Quatre Fl. France, 855 (1938).

  Erythraea pulchella (Swartz) Fries ssp. *ramosissimum* (Pers.) Lindberg f., Iter Cypr., 27 (1946).

  TYPE: Finland; Åland Islands, Finnstrom parish, *Swartz* (S, LINN).

Erect annual (3–) 5–20 (–42) cm. high; basal leaves small, oblong-spathulate, obtuse or rounded, tapering to the base, usually less than 1 cm. long and 0·8 cm. wide, sometimes forming loose rosettes, generally withering before anthesis; cauline leaves broadly or rather narrowly oblong, 0·5–3 cm. long, 0·3–1·5 cm. wide, obtuse or shortly acute, narrowing rather abruptly at the base, rather remote and much shorter than the internodes in well-developed specimens, crowded and overlapping in starved specimens; inflorescence normally a spreading, dichasial (1–) 3–70-flowered cyme, with erecto-patent branches often arising from near the base of the plant, but sometimes branching from near the apex of the main axis; bracts lanceolate or narrowly elliptic, smooth, 2·5–15 mm. long, 0·8–5 mm. wide; pedicels up to 5 mm. long at anthesis, occasionally elongating to 10 mm. in fruit, or flowers sometimes subsessile; calyx-lobes linear-subulate, 8–9 mm. long, 0·5–0·6 mm. wide at base, a little shorter than the corolla-tube, and more or less adherent to it, tapering to a fine acumen and sharply keeled dorsally; corolla bright or pale pink, or occasionally white (*C. meyeri* (Bunge) Druce; *Erythraea ramosissima* (Vill.) Pers. var. *albiflora* Boiss.); tube 8–10 mm. long, slightly constricted below the limb; lobes oblong-ovate, 3·5–5 mm. long, 2–2·5 mm. wide, concave, apex obtuse or subacute, generally minutely erose, sometimes irregularly emarginate; stamens inserted at apex of corolla-tube; filaments filiform, 3–4 mm. long; anthers yellow, oblong, 1–3 mm. long, 0·5 mm. wide; ovary narrowly fusiform, 5–6 mm. long, 0·8–1 mm. wide; style about 2·5 mm. long; stigmas oblong, blunt, about 0·5 mm. long. Fruit pale brown, 10–12 mm. long, about 2–2·5 mm. wide; seeds irregularly oblong or subglobose, about 0·3 mm. long, 0·2 mm. wide; testa brown, coarsely and conspicuously fuscous-reticulate.

  HAB.: Stony hillsides on limestone or igneous formations, open rocky pastures; roadsides; sand-dunes; sometimes in marshy situations by streams, irrigation channels or salt-lakes; sea-level to 3,800 (–5,000) ft. alt., but generally lowland; fl. March–May.

  DISTR.: Divisions 1–8. Widespread in Europe and Mediterranean region eastwards to China.

1. Between Evretou and Sarama, 1934, *Syngrassides* 968! Near Neokhorio, 1941, *Davis* 3338! Smyies, 1948, *Kennedy* 1629!
2. Prodhromos, 1862, *Kotschy* 615; Kambia, 1908, *Clement Reid* s.n.! Platres, 1929, *C. B. Ussher* 77! Mandria, 1937, *Kennedy* 689! North of Kykko Monastery, 1959, *P. H. Oswald* 109!
3. Limassol Salt Lake, 1948, *Kennedy* 1594! Parekklisha, 1948, *Kennedy* 617! Phasouri, Asomatos, 1962, *Meikle* 2928! Malounda, 1963, *J. B. Suart* 92! Episkopi, 1967, *Merton* in ARI 624!

4. Larnaca, 1939, *Lindberg f.* s.n.!, and, 1948, *Kennedy* 1619! Larnaca airport, 1952, *Merton* 827! Ayios Memnon, 1948, *Mavromoustakis* s.n.! Sotira, Ayios Antonios, 1951, *Merton* 732! Mile 14, Nicosia-Larnaca road, 1952, *Merton* 843! Kouklia, 1952, *Merton* 852! Pergamos, 1955, *Merton* 1969! 2215! Mouth of Pedieos River, 1955, *Merton* 2260! East of Paralimni, ? 1958, *N. Macdonald* 47!

5. Vatili, 1905, *Holmboe* 317; Kythrea, 1927, *Rev. A. Huddle* 78! Near Apostolos Varnavas Monastery, 1948, *Mavromoustakis* s.n.! Near Gypsos, 1955, *Merton* 2432! 4 miles E. of Nicosia, ? 1967, *Merton* in ARI 480!

6. Nicosia, 1927, *Rev. A. Huddle* 77! Syrianokhori, 1935, *Syngrassides* 695! Ayia Irini, 1941, *Davis* 2529! Peristerona, 1955, *Merton* 2410! 2421! Orga, 1956, *Poore* in *Merton* 2657! Near Liveras, 1962, *Meikle* 2414!

7. St. Hilarion, 1880, *Sintenis & Rigo* 606! Kyrenia, 1931, *J. A. Tracey* 44!, also, 1948, *Casey* 178! 187!, 1955, *Miss Mapple* 60! 83! and, 1955, *G. E. Atherton* 200! Larnaca tis Lapithou, 1941, *Davis* 3007! Halevga, 1948, *Mercy Casey* 133! Lakovounara, 1950, *Chapman* 514! 663! Kyrenia Pass, 1956, *G. E. Atherton* 1381 A! Glykyotissa Island, 1957, *P. H. Oswald* 140!

8. Cape Andreas, 1880, *Sintenis & Rigo*; Valia, 1905, *Holmboe* 488; Mutsefla, Khalosta Forest, 1962, *Meikle* 2457!

NOTES: Very polymorphic, and, in Cyprus, to judge from differences in size of corolla and anthers, very probably including more than one cytological infraspecies. For notes on the problem of distinguishing *C. pulchellum* from *C. tenuiflorum*, see under the latter species.

3. C. tenuiflorum (*Hoffmsgg. et Link*) *Fritsch* in Mitt. Naturw. Ver. Wien, 5: 97 (1907); Zeltner in Bull. Soc. Neuchât. Sci. Nat., 93: 83–96 (1970); Melderis in Bot. Journ. Linn. Soc., 65: 246 (1972); Osorio-Tafall et Seraphim, List Vasc. Plants Cyprus, 83 (1973); Davis, Fl. Turkey, 6: 181 (1978).

Erythraea tenuiflora Hoffmsgg. et Link, Fl. Port., 1: 354, t. 67 (1813–20); Holmboe, Veg. Cypr., 144 (1914); Post, Fl. Pal., ed. 2, 2: 198 (1933); Lindberg f., Iter Cypr., 27 (1946); Rechinger f. in Arkiv för Bot., ser. 2, 1: 428 (1950).

[*E. latifolia* (non Sm.) Boiss., Fl. Orient., 4: 67 (1875).]

TYPE: Portugal; "sur les bords de la mer, près d'A-Costa", *Hoffmannsegg & Link* (B).

Erect annual (15–) 20–40 cm. high; basal leaves as in *C. pulchellum* (Swartz) Druce, generally withering before anthesis; cauline leaves subequal, oblong, acute or obtuse, 0·5–2 cm. long, 0·3–0·8 cm. wide, often in numerous (5–15) pairs along the straight, unbranched lower part of the stem, commonly remote, and shorter than the internodes, but not always so; inflorescence a dense, many (60–180 or more)-flowered, repeatedly branched, fastigiate, corymbose cyme, usually (but not always) branched from near the apex of the stem; bracts lanceolate or almost linear, smooth or minutely papillose-scabridulous, 2–15 mm. long, 0·5–3 mm. wide; pedicels generally very short or wanting at anthesis and in fruit; calyx-lobes linear-subulate, 5–6 mm. long, 0·4–0·5 mm. wide at base, usually much shorter than corolla-tube, and more or less adherent to it, tapering to a fine acumen and sharply keeled dorsally; corolla bright or pale pink or occasionally white, tube 7–9 mm. long, distinctly constricted below the limb; lobes oblong, 3–4·5 mm. long, 1–1·5 mm. wide, concave (margins often inrolled after anthesis), apex obtuse, commonly a little erose or minutely emarginate; stamens inserted at apex of corolla-tube; filaments filiform, about 2–3 mm. long; anthers yellow, oblong, 1–3 mm. long, 0·4–0·5 mm. wide; ovary narrowly fusiform, 5–8 mm. long, 0·8–1 mm. wide; style 1·5–2·5 mm. long; stigmas oblong-spathulate, blunt, about 0·5 mm. long. Fruit pale or dark brown, 7–10 mm. long, 1·5–2 mm. wide; seeds irregularly oblong or subglobose, about 0·3 mm. long, 0·2–0·3 mm. wide; testa brown, coarsely and conspicuously fuscous-reticulate.

HAB.: Marshy ground by springs or salt-lakes; sea-level to 700 ft. alt.; fl. June–Aug.

DISTR.: Divisions 1, 3, 4, 6. S. Europe and Mediterranean region eastwards to Turkey, Syria and Palestine.

1. Ayios Neophytos Monastery, 1939, *Lindberg f.* s.n.; Smyies, 1948, *Kennedy* 1628!

3. Near Limassol, 1905, *Holmboe* 664; Zakaki marshes, 1930–1943, *Mavromoustakis* 38! 53! 73! 85–88! 91! Same area, 1939, *Lindberg f.* s.n.; near Akrotiri, 1939, *Mavromoustakis* 124! and, 1941, *Davis* 3563! Akhelia, 1939, *Lindberg f.* s.n.; Kolossi, 1939, *Lindberg f.* s.n.

4. Pyla marshes, 1935, *Syngrassides* 835! Near Larnaca Salt Lake, 1948, *Kennedy* 1618! 1620!
    Mile 14, Nicosia-Larnaca road, 1952, *Merton* 842! Akhyritou, 1948, *Mavromoustakis* s.n.!
    Kouklia, 1952, *Merton* 851!
6. By Pedieos R. in Nicosia, 1939, *Lindberg f.* s.n.

NOTES: Many authors (including Boissier, Litardière and Zeltner) have remarked upon the
difficulty of distinguishing *C. pulchellum* from *C. tenuiflorum*. In this respect the Cyprus
populations of the two species are no less perplexing than those from other parts of the
Mediterranean region. I can find no reliable differences in pedicel-length, relative lengths of
calyx and corolla, nor shape of corolla-lobes, and the traditional distinction, namely, that *C.
pulchellum* is branched from the lower part of the main axis, and *C. tenuiflorum* from the upper,
while generally true, is subject to such overlapping as to be useful only if taken into
consideration with other diagnostic features. Of these, I find the dense, erect-fastigiate
inflorescence of *C. tenuiflorum* quite the most serviceable, and have perhaps erred in rejecting
from the ambit of *C. tenuiflorum* all specimens which have lax, open or divaricate cymes. It
means that some specimens which would be considered *C. tenuiflorum* by many taxonomists,
are here to be found under *C. pulchellum*. It also means that, in Cyprus, *C. tenuiflorum* is very
largely restricted to brackish marshes about Limassol, Larnaca and the eastern Mesaoria. In
some of these areas it evidently grows with *C. pulchellum*, and yet remains distinct (viz. *Merton*
842 from Mile 14, Nicosia-Larnaca road (*C. tenuiflorum*) and *Merton* 843 (*C. pulchellum*) from
the same locality; also *Merton* 851 from Kouklia (*C. tenuiflorum*) and *Merton* 852 (*C. pulchellum*)
from the same locality). Were it not for the seeming validity of such distinctions, one would be
tempted to dismiss *C. tenuiflorum* as no more than an ecological race of *C. pulchellum*. At the
same time, it must be admitted that whatever morphological criteria are employed, one is
invariably left with a residue of specimens which might be referred, with equal justification, to
either species. Some of these puzzling plants may be hybrids, though Zeltner (*op. cit.*) does not
proffer any evidence of hybridization within the group. That there are different cytological
races of *C. pulchellum* and *C. tenuiflorum* in Cyprus is suggested by the occurrence, within both
species, of variants with large or small corollas and anthers. Such variants do not seem to differ
in other respects. Most of the Cyprus material of *C. tenuiflorum* resembles the diploid (n = 10)
ssp. *acutiflorum* (Schott) Zeltner, but I am not aware of any chromosome count based on Cyprus
specimens.

   Davis 3336, from (1) Ayios Yeoryios above Neokhorio (Akamas), found with *C. pulchellum*
(*Davis* 3338!) answers to the description of *C. tenuiflorum* var. *hermannii* (Sennen) Zeltner in
having exceptionally large flowers with corolla-lobes 5–7 mm. long; it is, however, said to have
"intense reddish-pink" flowers, whereas Zeltner says the flowers of var. *hermannii* are pale pink
or lilac-pink.The specimens are depauperate, and cannot be identified, even to species, with
complete certainty.

4. **C. spicatum** (*L.*) *Fritsch* in Mitt. Naturw. Ver. Wien, 5: 97 (1907); A. K. Jackson in Kew Bull.,
    1934: 272 (1934); Melderis in Bot. Journ. Linn. Soc., 65: 247 (1972); Osorio-Tafall et
    Seraphim, List Vasc. Plants Cyprus, 83 (1973); Davis, Fl. Turkey, 6: 182 (1978).
    Gentiana spicata L., Sp. Plant., ed. 1, 230 (1753).
    Erythraea spicata (L.) Pers., Syn. Plant., 1: 283 (1805); Boiss., Fl. Orient., 4: 69 (1875);
    Post, Fl. Pal., ed. 2, 2: 199 (1933); Lindberg f., Iter Cypr., 27 (1946).
    E. spicata (L.) Pers. f. *albiflora* Lindberg f., Iter Cypr., 27 (1946).
    TYPE: "*in montibus* Euganeis [Colli Euganei, N. Italy] & Monspelii."

Erect annual or biennial 18–40 cm. high; stems angled, usually much
branched, with erecto-patent branches arising from near the base, or
sometimes from near the apex of the central axis; basal leaves usually
withered before anthesis, oblong, obovate or suborbicular, 4–15 mm. long,
4–10 mm. wide, scarcely forming a rosette; cauline leaves rather thick and
fleshy, subequal, oblong, obtuse or subacute, 1–4 cm. long, 0·4–1·2 cm. wide,
usually almost equalling, or sometimes exceeding, the internodes; inflorescence monochasial, spiciform, the subsessile flowers spaced along one
side of a slightly arcuate axis 10–13 cm. or more long; bracts narrowly
lanceolate or subulate, 3–13 mm. long, 1–4 mm. wide, smooth, somewhat
keeled; calyx-lobes usually 5, unequal, linear-subulate, 5–8 mm. long,
0·4–0·6 mm. wide at base, apex acuminate, dorsal surface sharply keeled;
corolla pink or white (*Erythraea spicata* (L.) Pers. f. *albiflora* Lindberg f.);
tube 7–8 mm. long, equalling, and adherent to, the calyx-lobes, a little
constricted below the limb; corolla-lobes 4–5, oblong, about 4 mm. long, 1·2
mm. wide, concave, apex obtuse or subacute, commonly a little erose;

stamens inserted about half way down corolla-tube (and continuing, adnate to tube almost to its base); filaments filiform, the free part about 5 mm. long; anthers narrowly oblong, yellow, about 2·5 mm. long, 0·5 mm. wide; ovary fusiform, 4·5–5 mm. long, 1·2 mm. wide; style slender, 4–5 mm. long; stigmas oblong-spathulate, rounded, about 0·5 mm. long. Fruit dark brown, about 9 mm. long, 2·3 mm. wide; seeds irregularly subglobose, about 0·4 mm. diam.; testa brown, coarsely and conspicuously fuscous-reticulate.

HAB.: Marshy ground by springs or salt-lakes, often with *C. tenuiflorum* (Hoffmsgg. et Link) Fritsch; occasionally on damp walls; sea-level to 600 ft. alt.; fl. May–Aug.

DISTR.: Divisions 1, 3–8. S. Europe and Mediterranean region eastwards to Central Asia.

1. Skoulli near Polis, 1939, *Lindberg f.* s.n.
3. Zakaki marshes, 1930, *Mavromoustakis* 74! Akhelia, 1939, *Lindberg f.* s.n.; near Limassol Salt Lake, 1939, *Lindberg f.* s.n., and, same locality, 1941, *Davis* 3558!
4. Near Famagusta, 1934, *Syngrassides* 963! and, 1939, *Lindberg f.* s.n.; near Larnaca Salt Lake, 1939, *Lindberg f.* s.n.!, and, 1948, *Kennedy* 1621! 1622! Larnaca airport, 1952, *Merton* 844!
5. Between Prastio and Kouklia, 1939, *Lindberg f.* s.n.; below Aghirda, 1955, *Merton* 2380! Near Gypsos, 1955, *Merton* 2431!
6. Kykko metokhi, 1932, *Syngrassides* 398! Panagra-Orga road, 1967, *Economides* in ARI 1024!
7. Kyrenia, 1955, *G. E. Atherton* 32!
8. Boghaz, 1937, *Syngrassides* 1481!

5. C. maritimum (*L.*) *Fritsch* in Mitt. Naturw. Ver. Wien, 5: 97 (1907); Melderis in Bot. Journ. Linn. Soc., 65: 248 (1972); Osorio-Tafall et Seraphim, List Vasc. Plants Cyprus, 83 (1973); Davis, Fl. Turkey, 6: 182 (1978).
    Gentiana maritima L., Mantissa 1, 55 (1767).
    Chironia maritima (L.) Willd., Sp. Plant., 1 (2): 1069 (1798); Clarke, Travels Europ. Asiatic Turkey, 2 (3): 718 (1816).
    Erythraea maritima (L.) Pers., Syn. Plant., 1: 283 (1805); Boiss., Fl. Orient., 4: 68 (1875); Holmboe, Veg. Cypr., 144 (1914); Post, Fl. Pal., ed. 2, 2: 198 (1933).
    TYPE: "*in* Italiae, Galliae *australis maritimis*".

Erect annual (4–) 6–20 (–35) cm. high; stems angled, usually unbranched or sparingly branched; basal leaves very small, broadly ovate or sub-orbicular, often less than 5 mm. diam., not forming a rosette; cauline leaves ovate or oblong-ovate, 5–20 mm. long, 3–10 mm. wide, subequal, acute or obtuse, rather fleshy and glaucescent, generally very much shorter than the internodes; flowers solitary, terminal, or up to 20 in a lax, erecto-patent dichasium; bracts foliaceous, lanceolate or linear-lanceolate, 5–20 mm. long, 0·5–5 mm. wide at base, smooth, sometimes a little keeled; pedicels stout, 3–10 (–20) mm. long; calyx-lobes usually 5, linear-subulate, 8–15 mm. long 0·5–1 mm. wide at base, usually connate for about 2–3 mm. from base, free and gradually acuminate above, not strongly adherent to corolla-tube, dorsal surface sharply keeled; corolla pale yellow (or rarely tinged reddish), tube 15–20 mm. long, usually much exceeding the calyx-lobes, or rarely equalling them, slightly constricted below the limb; corolla-lobes normally 5, ovate, acute, 5–9 mm. long, 2–4 mm. wide, concave, apex acute, not or scarcely erose; stamens inserted at apex of corolla-tube; filaments filiform, 1–2 mm. long; anthers oblong, yellow, about 0·8–1 mm. long, 0·5 mm. wide; ovary narrowly fusiform, 10–15 mm. long, 1·5–2 mm. wide; style short, 1·5–2 mm. long; stigmas narrowly oblong, truncate, 1·5–2 mm. long, 0·5 mm. wide. Fruit pale brown, about 12–15 mm. long, 2 mm. wide; seeds irregularly subglobose, often angled, about 0·3 mm. diam.; testa brown, coarsely and conspicuously fuscous-reticulate.

HAB.: Rocky slopes and banks, in dry or moist situations; sea-level to 1,200 ft. alt.; fl. April–May.

DISTR.: Divisions 1, 2, 4, rare. W. and S. Europe and Mediterranean region to Turkey and Lebanon.

1. Ayios Nikolaos chiftlik (Akamas), 1950, *Casey* 1016! Between Inia and Smyies, 1979, *Edmondson & McClintock* E2757!
2. Jemali Bridge, 1962, *Meikle* 2767!
4. "Found by our companion, *Dr. John Hume*, at a ruined aqueduct near to *Larneca*. We never saw it in any other part of the island", 1801, *Clarke* (Travels, *loc. cit.*).

63a. HYDROPHYLLACEAE

A. Peter in Engl. et Prantl, Pflanzenfam., ed. 1, IV (3a): 54–71 (1895).
A. Brand in Engl., Pflanzenr., 59 (IV. 251): 1–210 (1913).

Annual or perennial herbs (or rarely subshrubs), commonly glandular or hispid; leaves opposite, alternate or basal, entire or pinnately (rarely palmately) divided, exstipulate; inflorescences commonly in scorpioid cymes without bracteoles, or flowers rarely solitary and axillary; flowers hermaphrodite, actinomorphic, usually 5-merous; calyx deeply lobed, the lobes usually imbricate, often with appendages in the sinuses; corolla rotate, campanulate or infundibuliform, mostly with 5 imbricate lobes; stamens 5, epipetalous, generally inserted at base of corolla and alternating with the petals; anthers 2-thecous, versatile, introrse, dehiscing longitudinally; scale-like appendages often present at the base of, and alternating with, the stamens; disk and hypogynous staminodes sometimes present; carpels 2; ovary superior (rarely half-inferior), 1-locular (or apparently 2-locular through the extrusion of the parietal placentas); ovules generally numerous, sessile or pendulous, anatropous or amphitropous; style 1, usually divided above; stigmas terminal, often capitate. Fruit a loculicidal 2 (or rarely 4)-valved capsule, rarely indehiscent; seeds sometimes carunculate; endosperm copious or thin; embryo small.

About 20 genera and 250 species widely distributed, but particularly abundant in North America. There are no indigenous representatives in Cyprus, but at least one species of *Phacelia* Juss., *P. tanacetifolia* Benth. (in Trans. Linn. Soc., 17: 280; 1835), from California, has been collected by *T. F. Hewer* (4743!) at (3) Layia (1981), and by *Merton* s.n. (unlocalized, but probably from gardens or as an escape from gardens, in the vicinity of Nicosia). *P. tanacetifolia* is a robust, hispidulous-glandular annual up to 80 cm. high, with oblong, bipinnate leaves, cut into numerous, jagged segments, and many-flowered terminal cymes of tubular-infundibuliform, lavender-blue flowers. Apart from being ornamental and easily grown, *P. tanacetifolia* is valued as a bee-plant, and, according to F. N. Howes (Plants and Beekeeping, 176–177; 1945) "Without doubt it deserves the reputation it enjoys, for it undoubtedly secretes nectar very freely and its bluish-pink flowers are often to be seen covered with bees, the flowers being visited at all hours of the day". The same author comments that the honey made from *Phacelia tanacetifolia* nectar is "light green and of fine flavour".

64. BORAGINACEAE

M. Gürke in Engl. et Prantl, Pflanzenfam., ed. 1, IV (3a): 71–131 (1897).

Herbs and subshrubs, or sometimes trees or woody climbers, often clothed with bristly hairs; leaves mostly alternate, simple, exstipulate; flowers generally in scorpioid cymes, rarely solitary, bracteolate or ebracteolate, usually hermaphrodite, actinomorphic or occasionally zygomorphic, mostly 5-merous; calyx with imbricate, or rarely valvate or open, teeth or lobes, occasionally irregular; corolla cylindrical, hypocrateriform, rotate or infundibuliform, lobes imbricate or contorted in bud, often with folds, swellings or scale-like appendages (faucal scales) at, and sometimes partly closing, the throat; stamens epipetalous, as many as the corolla-lobes and alternate with them, equal or sometimes unequal; filaments occasionally appendiculate; anthers 2-thecous, introrse, basifixed or dorsifixed near the base, dehiscing longitudinally; carpels 2; ovary generally seated on a nectariferous disk, superior, at first 2-locular, becoming 4-locular at maturity with 1 anatropous ovule in each loculus; placentation axile, sometimes apparently basal; style simple, gynobasic or occasionally terminal; stigma simple, capitate or conical, sometimes 2 (4)-lobed. Fruit commonly of 4 (less often) 2 nutlets, less often a 1–4-seeded nut or drupe; seed erect, oblique or horizontal; testa membranous; endosperm absent or scanty; embryo straight or curved.

About 100 genera and more than 2,000 species widely distributed in temperate and tropical regions, and a conspicuous element in the Mediterranean flora.

A natural group, nowadays often made still more homogeneous by the exclusion of the predominantly woody, drupaceous subfamilies *Cordioideae* and *Ehretioideae* as a distinct family (or families). *Wellstedia* Balf. f. from Somalia and Socotra, with 4-merous flowers and capsular fruits has also been excluded from *Boraginaceae*, and is referred to its own family, *Wellstediaceae*.

Tree with subglobose, drupaceous fruits - - - - - - CORDIA (p. 1120)
Herbs or subshrubs:
  Corolla cylindrical or fusiform, with a long tube and short, inconspicuous lobes; flowers
    yellow - - - - - - - - - - - **13. Onosma**
  Corolla rotate, hypocrateriform or infundibuliform, with well-developed lobes:
    Subshrubs with woody branches - - - - - - - **9. Lithodora**
    Herbs:
      Stamens exserted from corolla-tube:
        Corolla zygomorphic, infundibuliform - - - - - **12. Echium**
        Corolla actinomorphic, stellate-rotate - - - - - **4. Borago**
      Stamens included in corolla-tube:
        Nutlets concealed within a flattened cockscomb-like accrescent calyx **3. Asperugo**
        Nutlets not concealed within a flattened cockscomb-like accrescent calyx:
          Nutlets covered with barbed (glochidiate) spines - - - **2. Cynoglossum**
          Nutlets smooth or tuberculate, not spinose:
            Inflorescence ebracteate:
              Stigma large, conical; style very short, terminal; nutlets dull, rugulose-
                tuberculate or papillose - - - - - **1. Heliotropium**
              Stigma small, capitate or obscurely lobed; style well-developed, gynobasic;
                nutlets smooth, shining- - - - - - **8. Myosotis**
            Inflorescence bracteate:
               Calyx becoming inflated and ventricose in fruit - - - - **6. Nonea**
               Calyx not inflated or ventricose in fruit:
                  Corolla yellow or yellowish:
                    Leaves obovate or oblanceolate, erose-dentate, 0·8–4 cm. wide; faucal
                      scales well developed - - - - - - **5. Anchusa**

Leaves linear or narrowly oblanceolate, entire, 0·2–0·3 cm. wide; faucal
    scales wanting - - - - - - **11. Neatostema**
Corolla blue, purple, pink or whitish, not yellow or yellowish:
    Nutlets 1·5–2 mm. long, smooth and shining - - - **8. Myosotis**
    Nutlets 2–3 mm. or more long, rugulose-tuberculate or papillose:
      Corolla distinctly zygomorphic - - - - **12. Echium**
      Corolla actinomorphic:
        Faucal scales conspicuous - - - - **5. Anchusa**
        Faucal scales obscure or wanting:
          Perennials; apical half of nutlet bent horizontally or deflexed
                  7. Alkanna
          Annuals; apical half of nutlet erect or suberect **10. Buglossoides**

SUBFAMILY 1. **Cordioideae** Mostly trees and shrubs; style terminal; fruit a drupe.

CORDIA MYXA *L.*, Sp. Plant., ed. 1, 190 (1753); Unger et Kotschy, Die Insel Cypern, 278 (1865); Boiss., Fl. Orient., 4: 124 (1875); Holmboe, Veg. Cypr., 146 (1914); Post, Fl. Pal., ed. 2, 2: 219 (1933); Chapman, Cyprus Trees and Shrubs, 63 (1949); Osorio-Tafall et Seraphim, List Vasc. Plants Cyprus, 85 (1973) — as "*C. mixa*"; Davis, Fl. Turkey, 6: 246 (1978).

Tree to 8 m. high; leaves broadly ovate, obovate or suborbicular, 6–12 cm. diam., glabrous and rather glossy above, thinly pubescent below with prominent nerves and reticulate venation, apex rounded or obtuse, margins entire or obscurely lobulate, often distinctly recurved; petioles 1–3 cm. long; inflorescence a terminal, corymbose or thyrsoid, much-branched cyme; calyx campanulate, about 5 mm. long, 4 mm. wide, shortly lobed and pilose at apex, persistent, accrescent and becoming discoid in fruit; corolla white, shortly tubular, with 5 recurved, oblong-lanceolate lobes 3–5 mm. long, 1·5–2 mm. wide; stamens exserted; filaments thinly pilose; anthers oblong, yellow; ovary narrowly ovoid-acuminate; style about 1·5–2 mm. long, divided above into 4 linear or narrowly flabelliform stigmas. Fruit subglobose, apiculate, about 1·5 cm. diam., ochraceous or pinkish when ripe, persisting and turning dark brown during the winter; mesocarp fleshy, viscid; endocarp deeply rugose, broadly elliptic or suborbicular, like a small plum stone, about 1·2 cm. long, 1 cm. wide, initially 4-locular, but only one seed developing.

A native of tropical Asia, widely (and anciently) cultivated and sometimes naturalized in North Africa and western Asia, and recorded from or about gardens at Ayia Napa (*Kotschy; Syngrassides*), Larnaca (*Chapman*), Syngrasis (*Syngrassides* 1301 !), Kyrenia (*Mrs. Houston ! C. B. Ussher* 80 !), Lapithos (*Syngrassides*) and Komi Kebir (*Syngrassides, Chapman*).

Many authors and collectors note that the sticky mesocarp is used (sometimes mixed with honey) to make birdlime.

SUBFAMILY 2. **Heliotropoideae** Herbs or shrubs; style terminal; fruit a drupe, or comprising 4 (or fewer by abortion) nutlets.

1. HELIOTROPIUM *L.*

Sp. Plant., ed. 1, 130 (1753).
Gen. Plant., ed. 5, 63 (1754).
H. Riedl in Ann. Naturhist. Mus. Wien, 69: 81–93 (1966).
R. K. Brummitt in Bot. Journ. Linn. Soc., 64: 60–67 (1971).

Annual or perennial herbs, or sometimes subshrubs; leaves alternate, usually entire; flowers generally in branched, ebracteate, scorpioid, terminal cymes; calyx mostly divided almost to base; corolla hypocrateriform or infundibuliform, generally without faucal appendages, often with small teeth alternating with the lobes; stamens included, filaments very short; style very short; stigma (or stigmatic cone) depressed or elongate, entire or 2 (or 4)-lobed. Fruit of 4, free or connate nutlets, occasionally 1 nutlet through abortion.

About 200 species with a cosmopolitan distribution.

Calyx-lobes united almost to apex; nutlet 1, enclosed in the calyx and falling with it
                                                             4. H. supinum
Calyx-lobes free almost to base; nutlets 4, falling out of the persistent calyx when ripe:
   Corolla with 5 linear or filiform internal scales, limb 4–6 mm. diam.; stem clothed with short, subadpressed and long, spreading hairs -   -   -   -   **3. H. hirsutissimum**
   Corolla without internal scales, limb 2–3·5 mm. diam.; stem clothed with short, subadpressed hairs:
      Calyx clothed with long, spreading, shaggy hairs, lobes becoming patent soon after anthesis; anthers broadly lanceolate, about 1 mm. long; stigmatic cone produced into a long, subulate-filiform acumen - - - - - - **1. H. europaeum**
      Calyx clothed with subadpressed, upward-pointing hairs, lobes remaining erect and embracing the nutlets, not becoming patent until the nutlets have been shed; anthers narrowly lanceolate, about 1·5 mm. long; stigmatic cone sometimes narrowly conical, but not produced into a subulate-filiform acumen -   -   - **2. H. dolosum**

1. **H. europaeum** *L.*, Sp. Plant., ed. 1, 130 (1753); Hume in Walpole, Mem. Europ. Asiatic Turkey, ed. 1, 254 (1817) quoad nomen; Gaudry, Recherches Sci. en Orient, 187 (1855) quoad nomen; Unger et Kotschy, Die Insel Cypern, 278 (1865) quoad nomen; Boiss., Fl. Orient., 4: 130 (1875) quoad nomen; H. Stuart Thompson in Journ. Bot., 44: 335 (1906) quoad nomen; Post, Fl. Pal., ed. 2, 2: 220 (1933) quoad nomen; Lindberg f., Iter Cypr., 28 (1946) quoad nomen; Brummitt in Bot. Journ. Linn. Soc., 64: 60–67 (1971); Davis, Fl. Turkey, 6: 252 (1978).
   H. europaeum L. var. *tenuiflorum* Guss., Fl. Sic. Prodr., 1: 205 (1827).
   TYPE: "*in* Europa *australi*".

Erect or spreading annual, 15–40 cm. high; stems divaricately branched, terete, rather densely clothed with subadpressed hairs about 0·5 mm. long; leaves ovate, oblong or obovate, 2–8 cm. long, 0·8–3 cm. wide, canescent with tuberculate-based hairs above and, more densely so, below; apex obtuse or rounded, margins entire, base narrowing abruptly to a petiole 0·5–3 cm. long; nervation obscure above, rather prominent and densely pilose below; inflorescences terminal and lateral, at first dense, elongating greatly (to 10 cm.) after anthesis and the lowermost flowers becoming rather remote; flowers sessile, in 2 ranks, said to be scentless; calyx persistent, divided almost to base into 5 narrow-lanceolate lobes about 2·5 mm. long, 0·6 mm. wide near base, lobes densely and softly lanuginose with long, spreading, shaggy hairs, at first erect, but soon spreading and becoming stellate-patent after the nutlets are shed; corolla creamy-white with a greenish tube 2·5–3 mm. long, slightly exceeding the calyx, pilose externally with spreading hairs; lobes spreading, broadly ovate, about 1 mm. long, 0·8 mm. wide, translucent and distinctly veined, the sinuses plicate and often minutely toothed; stamens inserted about 0·5 mm. above base of corolla-tube; anthers subsessile, broadly lanceolate, about 1 mm. long, 0·5 mm. wide at base; ovary subglobose, about 0·5 mm. diam., glabrous or very sparsely glandular; style very short; stigmatic cone about 1·5 mm. long, broadly conical below with a basal glandular-papillose zone, produced above into a long, subulate-filiform, shortly bifid, glabrous or sparsely pilose acumen about 1 mm. long. Nutlets 4, free, ovoid, about 1·5 mm. long, 1 mm. wide, pale or dark brown, glabrous or thinly pilose, coarsely rugose-tuberculate, obscurely papillose.

HAB.: Cultivated and waste ground; sea-level to 500 ft. alt.; fl. April–November.

DISTR.: Divisions 6, 7. Widespread in South, West and Central Europe, Mediterranean region and Atlantic Islands.

6. Morphou, 1955, *N. Chiotellis* 474!
7. Near Kyrenia, 1880, *Sintenis & Rigo* 534 b (K!); Akanthou, 1940, *Davis* 2025!

NOTES: Apparently uncommon, but so often confused with the following (*H. dolosum* De Not.), that it may have been overlooked. Because of this confusion, checked specimens are alone cited. *H. europaeum* (perhaps misidentified *H. dolosum*) is also recorded from: 2. Above Prodhromos, 1862, *Kotschy* 835; Kakopetria, 1939, *Lindberg f.* s.n.; 3 (or 4). Larnaca and Limassol, 1801, *Hume* s.n.; 4. Larnaca, 1933, *Nattrass* 363; Perivolia, 1939, *Lindberg f.* s.n.; 6. Nicosia, 1936, *Nattrass* 834, and, 1939, *Lindberg f.* s.n.

2. H. dolosum *De Not.*, Repert. Fl. Ligust., 318 (1844); Brummitt in Bot. Journ. Linn. Soc., 64: 60–67 (1971); Davis, Fl. Turkey, 6: 253 (1978).

[*H. eichwaldii* (non Steud.) Boiss., Fl. Orient., 4: 131 (1875) pro parte.]

[*H. undulatum* (non Vahl) H. Stuart Thompson in Journ. Bot., 44: 335 (1906); Osorio-Tafall et Seraphim, List Vasc. Plants Cyprus, 85 (1973).]

[*H. europaeum* L. ssp. *tenuiflorum* (non (Guss.) Nyman) Holmboe, Veg. Cypr., 147 (1914); Osorio-Tafall et Seraphim, List Vasc. Plants Cyprus, 85 (1973).]

[*H. europaeum* L. var. *tenuiflorum* (non Guss.) Post, Fl. Pal., ed. 2, 2: 220 (1933) pro parte.]

TYPE: Italy, "In litore Liguriae occiduae prope *il Ceriale* legit hortulanus H. R. bot. genuensis" (GE).

Very similar to *H. europaeum* L. in stature, habit and inflorescence, but flowers said to be fragrant; calyx-lobes a little broader, about 0·8 mm. wide near base, subadpressed-pilose with upward-pointing hairs, remaining erect for some time after anthesis, and closely enveloping the base of the corolla until the nutlets are shed; corolla-tube less conspicuously pilose externally, with subadpressed hairs; corolla-lobes spreading, suborbicular, about 1·5 mm. long, rather opaque and obscurely veined, the sinuses plicate and often minutely toothed; anthers narrowly lanceolate, slightly hooked at apex, about 1·5 mm. long, 0·5 mm. wide at base; style very short; stigmatic cone 1–1·5 mm. long, conical, sometimes narrowly so, but without an attenuate, subulate-filiform acumen, shortly 2–4-fid, thinly pilose. Nutlets 4, free, ovoid, about 1·5 mm. long, 1 mm. wide, dark brown, glabrous, densely papillose, obscurely rugose-tuberculate, sometimes shortly pilose towards apex.

HAB.: Cultivated and waste ground; sea-level to 3,900 ft. alt.; fl. April–Nov.

DISTR.: Divisions 1, 2, 4–6, 8. Eastern Mediterranean region from Italy to Palestine and eastwards to Iran.

1. Lyso, 1913, *Haradjian* 828!
2. Platres, 1914, *G. St. C. Feilden* s.n.! Kyperounda, 1954, *Casey* 1350! Between Kalokhorio and Athrakos, 1970, *Edmondson & McClintock* E 2796!
4. Near Famagusta, 1947, *Mavromoustakis* 13! Dhekelia, 1982, *A. Fell* s.n.!
5. Between Prastio and Kouklia, 1939, *Lindberg f.* s.n.! Salamis, 1962, *Meikle* 2608!
6. Between Koutraphas and Astromeritis, 1955, *Merton* 2404! English School, Nicosia, 1958, *P. H. Oswald* 4!
8. Ayios Theodhoros, 1934, *Syngrassides* 509!

NOTES: As with *H. europaeum* L., because of confusion, only checked specimens are cited above, though Holmboe's record of *H. europaeum* L. ssp. *tenuiflorum* (Guss.) Nyman from 3. Khoulou (1905, *Holmboe* 750a) most probably belongs here, together with *P. Laukkonen* 441 from (? 6) "Kykko Camp", 1973, named as *H. dolosum* at Kew, but not available when this account was written. *H. dolosum* is also represented by an unlocalized sheet (*A. G. & M. E. Lascelles* s.n.) at Kew; the collectors remark "common everywhere".

H. europaeum L. var. *tenuiflorum* Guss. (Fl. Sic. Prodr., 1: 205; 1827) is, to judge from a Gussone specimen collected in Ischia (K!) scarcely more than a minor variant of *H. europaeum*.

3. H. hirsutissimum *Grauer*, Plant. Minus Cognit. Decuria, 1 (1784); Brummitt in Bot. Journ. Linn. Soc., 64: 60–67 (1971); Osorio-Tafall et Seraphim, List Vasc. Plants Cypr., 85 (1973); Davis, Fl. Turkey, 6: 254 (1978).

H. villosum Willd., Sp. Plant., 1: 741 (1797); Desf. in Ann. Mus. Hist. Nat. Paris, 10: 427, t. 10 (1807); Boiss., Fl. Orient., 4: 133 (1875); Holmboe, Veg. Cypr., 147 (1914); Post, Fl. Pal., ed. 2, 2: 221 (1933); Lindberg f., Iter Cypr., 28 (1946).

TYPE: Aegean Islands; "in Insula Melo" (possibly based on *Heliotropium majus, villosum, flore magno, inodoro* of Tournefort, Cor. Inst. Rei. Herb., 7; 1700 — Tournefort herb. no. 642 (P)).

Erect or sprawling annual, 10–50 cm. high; stems much branched, terete, rather densely clothed with short subadpressed (or crispate) hairs interspersed with conspicuous patent hairs up to 3 mm. long; leaves ovate or elliptic, 1–8 cm. long, 0·8–3·5 cm. wide, canescent with tuberculate-based hairs above and below, apex obtuse or rounded, margins entire, base narrowing to a petiole 0·5–3 cm. long; nervation obscure above, rather prominent and densely pilose below; inflorescences terminal and lateral, commonly simple, at first dense, elongating to 15 cm. or more after anthesis and the lowermost flowers often becoming remote; rhachis densely lanuginose-pilose; flowers sessile, alternate, commonly 1-ranked, said to be scented; calyx persistent, divided almost to base into 5 bluntly lanceolate lobes about 3·5 mm. long, 0·8 mm. wide near base, densely and softly pilose with long, straight, spreading hairs, remaining erect for some time after anthesis, but ultimately stellate-patent; corolla white with a yellow centre, or occasionally yellow all over, tube greenish, about 4 mm. long, thinly hairy externally, with 5 filiform or narrowly oblong, pilose scales, about 1 mm. long, attached internally about half-way down the tube below the sinuses; lobes spreading, suborbicular or broadly oblong, rounded, about 2 mm. long and almost as wide, translucent and distinctly veined, margins undulate-crispate, sinuses sometimes denticulate, but scarcely plicate; stamens inserted about 1·5 mm. above base of corolla-tube; anthers subsessile, lanceolate, about 1 mm. long, 0·5 mm. wide at base; ovary subglobose, about 0·8 mm. diam., glabrous, often distinctly 4-lobed; style very short; stigmatic cone depressed-conical or hemispherical, about 1 mm. diam., conspicuously pilose with long hairs. Nutlets 4, free, ovoid, about 1·5 mm. long, 1 mm. wide, fuscous-brown, glabrous, obscurely rugose-tuberculate, minutely papillose.

HAB.: Cultivated and waste ground; sea-level to 4,400 ft. alt.; fl. May–Nov.

DISTR.: Divisions 1–8, locally common. Eastern Mediterranean region from Greece to Egypt and Cyrenaica.

1. Polis, 1936, *Lady Loch* 4! Ayios Neophytos Monastery, 1939, *Lindberg f.* s.n.
2. Prodhromos, 1955, *G. E. Atherton* 637! Galata, 1970, *A. Genneou* in ARI 1606!
3. Kolossi, 1905, *Michaelides* teste *Holmboe*; Khoulou, 1905, *Holmboe* 750b; Akhelia, 1939, *Lindberg f.* s.n.; Skarinou, 1940, *Davis* 1985! Cherkez, 1946, *Mavromoustakis* 37! Paramali, 1959, *C. E. H. Sparrow* 44!
4. Dhekelia, 1982, *A. Fell* s.n.!
5. Athalassa, 1932, *Syngrassides* 211!
6. Nicosia, 1900, *A. G. & M. E. Lascelles* s.n.! English School, Nicosia, 1958, *P. H. Oswald* 1!
7. Common; Kyrenia, Bellapais, Halevga, Vasilia, Myrtou, etc. *Kennedy* 691! 716! *C. H. Wyatt* 62! *Lindberg f.* s.n.! *Davis* 1920! 2003! *Casey* 47! 48! *Miss Mapple* 78! *G. E. Atherton* 68! 166! 323! 407! *Barclay* 1088! etc.
8. Ayios Andronikos, 1934, *Syngrassides* 537!

4. H. supinum *L.*, Sp. Plant., ed. 1, 130 (1753); Boiss., Fl. Orient., 4: 127 (1875); Post, Fl. Pal., ed. 2, 2: 220 (1933); A. K. Jackson in Kew Bull., 1937: 344 (1937); Brummitt in Bot. Journ. Linn. Soc., 64: 60 (1971); Osorio-Tafall et Seraphim, List Vasc. Plants Cyprus, 85 (1973); Davis, Fl. Turkey, 6: 255 (1978).

TYPE: Spain; France; "Salmanticae *juxta agros*, Monspelii *in littore*".

Sprawling annual; stems much branched, terete, 7–40 cm. long, rather densely clothed with subadpressed hairs interspersed with a few, longer, spreading ones; leaves ovate, oblong or obovate, usually rather small, 0·7–4 cm. long, 0·5–2 cm. wide, densely canescent with adpressed and longer spreading hairs above and below, apex obtuse or rounded, margins entire, base tapering to a petiole 0·3–1·5 cm. long; nervation often impressed above,

prominent below; inflorescences terminal and lateral, generally simple, at
first dense, 1·5–2 cm. long, elongating to 8 cm. or more after anthesis; rhachis
densely pilose with a mixed indumentum of long and short hairs; flowers
sessile, commonly 2-ranked at first, becoming 1-ranked with the elongation
of the rhachis; calyx deciduous with the fruit, lobes 2·5–3 mm. long at
anthesis, united almost to apex, accrescent in fruit, densely clothed with
long spreading hairs, with bluntish apices; corolla white, tube about 2 mm.
long, rather shorter than the calyx, adpressed-hairy externally and
internally, lobes suberect, narrowly oblong, about 0·5 mm. long, 0·2 mm.
wide; stamens inserted near base of corolla-tube; filaments free from tube for
about 0·4 mm.; anthers narrowly lanceolate, about 1 mm. long, 0·4 mm.
wide at base; ovary ovoid, glabrous, about 1 mm. long, 0·5 mm. wide; style
attached laterally, about 0·4 mm. long, retrorse-hairy; stigmatic cone
narrowly conical, about 0·5 mm. long, 0·2 mm. diam. at base, shortly hairy.
Nutlet 1, enclosed within, and falling with, the accrescent calyx, broadly
ovoid, compressed, about 3 mm. long, 2·5 mm. wide with a narrow marginal
wing, glabrous, dark brown, obscurely rugulose-tuberculate, not papillose.

HAB.: Cultivated ground; near sea-level; fl. ? May–Nov.

DISTR.: Divisions 2, 8, rare. S. Europe and Mediterranean region eastwards to India.

2. Pomos, 1934, *Syngrassides* 972!
8. Ayios Andronikos, 1934, *Syngrassides* 539!

SUBFAMILY 3. **Boraginoideae** Mostly herbs; style gynobasic; fruit of 4
(rarely fewer) free nutlets.

2. CYNOGLOSSUM L.

Sp. Plant., ed. 1, 134 (1753).
Gen. Plant., ed. 5, 65 (1754).
A. Brand in Engl., Pflanzenr., 78 (IV. 252): 114–153 (1921).

Biennial, perennial or rarely annual herbs; cymes terminal and
subterminal, generally ebracteate, often lengthening conspicuously in fruit;
calyx 5-partite almost to the base, the lobes markedly accrescent in fruit;
corolla cylindrical-rotate or infundibuliform, 5-lobed, the lobes usually
broad and blunt, throat normally closed by quadrangular or crescent-
shaped appendages; stamens included; filaments short, inserted at or above
middle of corolla-tube; ovary 4-lobed; style frequently persistent; stigma
small, truncate or capitate. Nutlets 4, depressed-ovoid, often densely
echinulate with glochidiate spines, and with an apical rostrum adnate to the
style.

About 50 species found almost throughout temperate and tropical
regions.

Corolla blue with conspicuous, darker, reticulate venation; stems adpressed-pubescent
                                                                    3. C. creticum
Corolla uniformly violet, purple, pink or red, not conspicuously reticulate-veined; stems loosely
    clothed with spreading or subadpressed hairs:
    Corolla 6–8 mm. diam., violet or purple; basal leaves marcescent, 5–15 cm. long, 1–4 cm.
        wide; inflorescence 2–6-branched   -   -   - **1. C. montanum** ssp. **extraeuropaeum**
    Corolla about 4 mm. diam., pink or brick-red; basal leaves persistent, 2–5 cm. long, 0·5–1 cm.
        wide; inflorescence usually bifurcate -   -   -   -   -   -   **2. C. troodi**

1. **C. montanum** *L.*, Demonstr. Plant., 5 (1753), Amoen. Acad., 3: 402 (1756); Brand in Engl.,
    Pflanzenr., 78 (IV. 252): 126 (1921); Davis, Fl. Turkey, 6: 309 (1978).

Erect biennial 15–75 cm. high; stem generally unbranched or sometimes
branched from near the base, terete, striatulate, more or less densely clothed
with long spreading or subadpressed whitish hairs, or rarely subglabrous;

basal leaves crowded into a loose tuft, generally withering before anthesis, oblong or elliptic-obovate, 5–15 cm. long, 1–4 cm. wide, thinly or rather densely white-pilose above and below, rarely pustulose, apex obtuse or subacute, base tapering to a canaliculate petiole 1–10 cm. long; cauline leaves narrowly oblong, 2·5–15 cm. long, 0·5–2 cm. wide, sessile, amplexicaul or frequently dilated and subauriculate at the base, apex acute or obtuse, uppermost cauline leaves commonly very small, bract-like, ovate-oblong; inflorescences generally with 2–6, lax, erecto-patent, somewhat arcuate branches conspicuously elongating after anthesis so that the fruits are evenly spaced along the pilose rhachis; bracts absent; pedicels at first 2–3 mm. long, erecto-patent, densely adpressed-hairy, lengthening to 10 mm. (or sometimes more) and becoming recurved in fruit; calyx-lobes oblong, obtuse, 2–4 mm. long, 0·5–2 mm. wide at anthesis, densely hairy externally, subglabrous internally, accrescent in fruit and often reflexed; corolla uniformly violet, purple, maroon, crimson or dark red, glabrous, 6–8 mm. diam., tube campanulate, 3–4 mm. long, lobes spreading, hemispherical, 1·5–2·5 mm. long, 2–4 mm. wide at base, faucal scales oblong, emarginate, densely papillose or shortly pilose, about 1·2–1·5 mm. long, 0·7 mm. wide; stamens attached about 0·5 mm. above base of corolla-tube; filaments about 0·4 mm. long; anthers oblong, about 0·7–1 mm. long, 0·5–0·6 mm. wide; ovary glabrous; style glabrous, about 1 mm. long at anthesis, lengthening to 4 mm. in fruit; stigma truncate or domed. Nutlets depressed-ovoid, 4–6 mm. long, 3–5 mm. wide, pale brown when ripe, more or less uniformly covered with thick, rigid, glochidiate spinules, the intervening spaces smooth or densely verruculose-papillose.

ssp. **extraeuropaeum** Brand in Engl., Pflanzenr., 78 (IV. 252): 126 (1921).

    C. montanum L. ssp. *extraeuropaeum* Brand var. *asiaticum* Brand in Engl., Pflanzenr., 78 (IV. 252): 127 (1921).

    ? *C. teheranicum* H. Riedl in Oesterr. Bot. Zeitschr., 110: 512, t. 1 (1963) et Rechinger f., Fl. Iran., no. 48/15: 144 (1967).

    TYPE: Turkey; "Elmalu, in collibus", 11 May, 1860, *E. Bourgeau* s.n. (G, isolectotype K!).

Leaves and stems rather densely white-hirsute or thinly lanuginose; flowers shortly pedicellate. Nutlets rather small, often less than 5 mm. long, convex, immarginate (not noticeably flattened or depressed dorsally), uniformly covered with rigid glochidiate spinules, the spaces between the bases of the spinules closely and densely verruculose-papillose. Otherwise as in *C. montanum* L. ssp. *montanum*. Plate 68 figs. 4–6.

HAB.: Stony mountainsides on serpentine, under *Pinus nigra* ssp. *pallasiana*; 5,500–5,700 ft. alt.; fl. May–July.

DISTR.: Division 2. Also Turkey and ? Iran.

2. Troödos, 5,500–5,700 ft. alt., chiefly on roughly terraced ground between the Nicosia road and the site of the former Olympus Camp Hotel, 1939–1973; *Lindberg f.* s.n., *Mavromoustakis* s.n.! *Kennedy* 1768! 1809! 1812! 1814! 1816! 1880! 1881! *Merton* 2337! 2369! 2439! *P. Laukkonen* 356! *Ann Matthews* 2!

NOTES: The researches of C. C. Lacaita (in Bull. Ort. Bot. Nap., 3: 290; 1913) have established beyond reasonable doubt the identity of Columna's (or Colonna's) "*Cynoglossa media altera virente folio, rubro flore, montana frigidarum regionum*" (Ecphrasis, 176, t. 175; 1616) from central Italy, which is the basis of *Cynoglossum montanum* L. Unfortunately Lacaita does not unravel the difficult taxonomic tangle of *C. montanum* L., *C. nebrodense* Guss., *C. pustulatum* Boiss. and *C. hungaricum* Simonkai. Nor is Brand's treatment of the group (in Engl., Pflanzenr., *loc. cit.*) wholly satisfactory, since it involves the transfer of *C. hungaricum*, as a variety, to *C. germanicum* Jacq., and recognition of *C. nebrodense* as specifically distinct from *C. montanum*. Furthermore *C. montanum* is divided into two subspecies, separated only on size of corolla, a questionable distinction, since corolla-size can vary considerably on the same inflorescence, depending on whether the flowers are at the beginning, or at the close, of anthesis.

    From my own observations, *C. montanum*, *C. nebrodense* and *C. hungaricum* are conspecific, and I doubt if they can be satisfactorily distinguished even at varietal level. Spanish material labelled *C. pustulatum* Boiss., with glabrous or subglabrous stems and pustulose leaves, looks distinct, at least as a subspecies of *C. montanum*, as does Cyprus material of *C. montanum*, particularly in its small, densely verruculose-papillose nutlets, which closely resemble those of *C. creticum* Mill. The Cyprus plant matches well with *Bourgeau* (1860) specimens from Elmalu, Lycia, cited by Brand under his *C. montanum* L. ssp. *extraeuropaeum* var. *asiaticum*, and here chosen as the lectotype of the subspecies.

Most of the Cyprus specimens are said to have violet or purple flowers, while Turkish specimens are noted as having crimson, maroon or dark red flowers, otherwise I can find no difference between Cyprian and Turkish sheets.

I suspect that *C. teheranicum* H. Riedl is no more than a local variant of *C. montanum* ssp. *extraeuropaeum*, but have seen no authentic material of this species.

2. C. troodi *Lindberg f.* in Act. Soc. Sci. Fenn., n.s.B, 2: 27 (1946); Rechinger f. in Svensk Bot. Tidskr., 43: 38 (1949); Osorio-Tafall et Seraphim, List Vasc. Plants Cyprus, 86 (1973).

[*Paracaryum myosotoides* (non (Labill.) Boiss.) Kotschy in Oesterr. Bot. Zeitschr., 12: 279 (1862); Unger et Kotschy, Die Insel Cypern, 284 (1865); Boiss., Fl. Orient., 4: 257 (1875) quoad plant. cypr.; Holmboe, Veg. Cypr., 147 (1914).]

[*Mattiastrum lithospermifolium* (non (Lam.) Brand) Brand in Fedde Repert., 14: 155 (1915) et in Engl., Pflanzenr., 78 (IV. 252): 63 (1921) quoad plant. cypr.]

[*Cynoglossum montanum* L. ssp. *extraeuropaeum* Brand var. *asiaticum* (non Brand) Turrill in Kew Bull., 1929: 232 (1929).]

TYPE: Cyprus; "In monte Troodos. 18 Junio 1880, *Sintenis & Rigo* 828 (LD, lectotype; K, isolectotype!).

Erect perennial (or ? sometimes biennial) 15–25 cm. high; stem usually branched at base and covered with fuscous, withered leaf-bases, subterete, striatulate or obscurely angled above and thinly or rather densely clothed with long, spreading, white hairs; basal leaves crowded into a rather dense tuft, persisting until after anthesis, narrowly oblong, lanceolate, or oblong-spathulate, 2–5 cm. long, 0·5–1 cm. wide; cauline leaves rather sparse and usually stopping well below the inflorescences, narrowly oblong, 1·5–5 cm. long, 0·3–0·7 cm. wide, sessile, amplexicaul or frequently dilated and subauriculate at the base, apex acute or obtuse; inflorescences mostly bifurcate, at first dense and compact, lengthening to 15 cm. or more in fruit so that the fruits are widely spaced, erect or erecto-patent; pedicels usually less than 2 mm. long at anthesis, sometimes up to 10 mm. long and distinctly recurved in fruit, together with the rhachis densely adpressed-pilose; calyx-lobes oblong, obtuse, 2–2·5 mm. long, 0·5–1 mm. wide at anthesis, rather densely adpressed white-hairy externally, subglabrous internally, accrescent and lightly reflexed in fruit; corolla uniformly pinkish-red or brick-red, glabrous, about 4 mm. diam., tube cylindrical, about 3 mm. long, lobes suberect, oblong, rounded at apex, about 1·2 mm. long, 1 mm. wide at base, faucal scales narrowly oblong, about 1·5 mm. long, 0·5 mm. wide, minutely papillose, dilated at base, strongly recurved at the shortly emarginate apex; stamens attached about 1 mm. above base of corolla-tube; filaments about 0·3 mm. long; anthers oblong, about 0·8 mm. long, 0·6 mm. wide; ovary glabrous; style glabrous, about 1 mm. long at anthesis, lengthening to 3–4 mm. in fruit; stigma truncate or domed. Nutlets depressed-ovoid, convex dorsally, immarginate, 4–6 mm. long, 3–5 mm. wide, pale brown when ripe, more or less uniformly covered with thick, rigid, glochidiate spinules, the spaces between the bases of the spinules sparsely and obscurely verruculose-papillose. *Plate 68, figs. 1–3.*

HAB.: Stony mountainsides on serpentine, under *Pinus nigra* ssp. *pallasiana*; 5,000–6,400 ft. alt.; fl. May–Aug.

DISTR.: Division 2. Endemic.

2. Locally abundant Troödos–Prodhromos–Kryos Potamos–Xerokolymbos to summit of Khionistra, 1862–1955; *Kotschy* 741! *Sintenis & Rigo* 828! *Haradjian* 408! 987! *Kennedy* 693! 694! 695! 696! 697! 1769! 1813! 1815! *Lindberg f.* s.n.! *Davis* 1798! 3517! *Mavromoustakis* s.n.! *F. M. Probyn* 125! *Casey* 1348! *Merton* 2328! 2796! *N. Macdonald* 187! *D. P. Young* 7330! *Ann Matthews* 1!

Plate 68. Figs. **1–3.** CYNOGLOSSUM TROODI Lindberg f. **1,** habit, × ⅔; **2,** flower, longitudinal section, × 6; **3,** infructescence, × ⅔; figs. **4–6.** CYNOGLOSSUM MONTANUM L. ssp. EXTRA-EUROPAEUM Brand **4,** habit, × ⅔; **5,** infructescence, × ⅔; **6,** flower, longitudinal section, × 6. (**1, 2** from *Kennedy* 694; **3** from *Merton* 2796; **4, 6** from *Kennedy* 1768; **5,** from *Kennedy* 1814.)

NOTES: Closely allied to *C. montanum* L., but consistently distinguished by its perennial habit, small narrow leaves and bifurcate inflorescences of small reddish or pinkish flowers. Unfortunately both species are included in Harald Lindberg's original description, the specimen labelled "in pineto (*P. pallasiana*) juxta "Olympus Camp Hotel. 15·6 [1939]" being *C. troodi* in the sense of the above description, etc., that labelled "in ruderatis juxta Olympus Camp Hotel". 22·6 [1939] being *C. montanum* L. ssp. *extraeuropaeum* Brand. It is because of this confusion that I have chosen *Sintenis & Rigo* 828 as lectotype of the name *C. troodi* in preference to any of specimens collected by Lindberg f. and cited in his "Iter Cyprium".

3. C. creticum *Mill.*, Gard. Dict., ed. 8, no. 3 (1768); Brand in Engl., Pflanzenr., 78 (IV. 252): 129 (1921); Post, Fl. Pal., ed. 2, 2: 251 (1933); Lindberg f., Iter Cypr., 27 (1946); Osorio-Tafall et Seraphim, List Vasc. Plants Cyprus, 86 (1973); Davis, Fl. Turkey, 6: 308 (1978).
  C. pictum Ait., Hort. Kew., ed. 1, 1: 179 (1789); Unger et Kotschy, Die Insel Cypern, 284 (1865); Boiss., Fl. Orient., 4: 265 (1875); H. Stuart Thompson in Journ. Bot., 44: 335 (1906); Holmboe, Veg. Cypr., 147 (1914).
  TYPE: Cultivated in Chelsea Physic Garden; "in Andalusia, I received the seed of this from Gibraltar", *Miller* (? BM).

Robust, erect biennial up to 80 cm. high; stem usually unbranched, striate or angular, softly adpressed-pubescent; basal leaves forming loose rosettes, usually withering before anthesis, oblanceolate, 7–15 cm. long, 1·5–4 cm. wide, softly grey pubescent above and below, nervation obscure, apex obtuse or subacute, base gradually tapering to a petiole 5–15 cm. long; cauline leaves numerous along the whole length of the stem, narrowly oblong-lingulate, 2·5–15 (–28) cm. long, 0·6–3 (–5) cm. wide, sessile, amplexicaul or frequently dilated and auriculate at the base, apex acute or obtuse; inflorescence much branched, at first rather dense, soon becoming lax and lengthening to 20 cm. or more in fruit, so that the fruits are widely and evenly spaced, erect or erecto-patent; pedicels 2–7 mm. long at anthesis, lengthening to 15 mm. (or more) in fruit and becoming distinctly recurved, together with the rhachis densely and softly, subadpressed grey-pilose; calyx-lobes oblong, obtuse, 3–5 mm. long, 1–2 mm. wide at anthesis, rather densely, subadpressed grey-hairy on both surfaces, accrescent to 10 mm. (or more) and reflexed in fruit, and usually exceeding the nutlets; corolla purplish in bud, soon changing to blue with a conspicuous network of dark inky-blue or purple veins, glabrous, 10–15 mm. diam., tube pallid, shortly cylindrical, about 2 mm. long, lobes spreading, suborbicular, 4–5 mm. diam., faucal scales oblong, about 2–2·5 mm. long, 1·3 mm. wide, dark blue, densely papillose-pilose, apex subtruncate or shortly emarginate; stamens attached about 1 mm. above base of corolla-tube; filaments about 0·5 mm. long; anthers shortly oblong, about 1 mm. long, 0·7 mm. wide; ovary glabrous, seated on a conspicuous, pallid disk; style glabrous, about 1·5 mm. long at anthesis, lengthening to 4 mm. in fruit; stigma truncate. Nutlets broadly depressed-ovoid, convex dorsally, immarginate, 7–8 mm. long, 5–7 mm. wide, pale brown when ripe, more or less uniformly covered with thick, rigid, glochidiate spinules, the spaces between the bases of the spinules densely verruculose-papillose.

HAB.: Margins of fields, roadsides, vineyards, often in moist sites by streams or irrigation channels, sometimes on open grassy slopes or in Pine forest; sea-level to 4,500 ft. alt.; fl. Febr.–June.

DISTR.: Divisions 1–8. S. Europe and Mediterranean region eastwards to Central Asia.

1. Dhrousha, 1941, *Davis* 3295! Khrysokhou Bay, 1960, *N. Macdonald* 62! Ayios Minas near Smyies, 1962, *Meikle* 2171! Fontana Amorosa, 1978, *Holub* s.n.!
2. Prodhromos, 1859, *Kotschy* 1024, and, 1862, *Kotschy* 868! Same locality, 1961, *D. P. Young* 7308! Near Kambos, 1937, *Syngrassides* 1575! Platres, 1937, *Kennedy* 692! and, 1955, *Kennedy* 1863! Platania, 1947, *Mavromoustakis* s.n.! Cedars Valley, Tripylos, 1968, *Economides* in ARI 1247!
3. Mazotos, 1905, *Holmboe* 196; Paramali, 1964, *J. B. Suart* 209! Phasouri, 1970, *A. Genneou* in ARI 1390!
4. Cape Greco, 1862, *Kotschy* 123!

5. Kythrea, 1880, *Sintenis & Rigo*, and 1932, *Syngrassides* 258! Dhali, 1936, *Syngrassides* 1119! Athalassa, 1967, *Merton* in ARI 147!
6. Ayia Irini, 1941, *Davis* 2568A! Strovolos, 1901, *A. G. & M. E. Lascelles* s.n.!
7. Pentadaktylos, 1880, *Sintenis & Rigo*; Dhavlos, 1880, *Sintenis & Rigo*; Kyrenia, 1949, *Casey* 510! also, 1954, *Kennedy* 1808! 1955, *Miss Mapple* 19! 1956, *G. E. Atherton* 1000! 1129! and, 1973, *P. Laukkonen* 186!
8. Rizokarpaso, 1941, *Davis* 2386! Ayios Andronikos, 1948, *Mavromoustakis* s.n.! and, 1970, *A. Genneou* in ARI 1472!

[SOLENANTHUS APENNINUS (*L.*) Fisch. et Mey. (*Cynoglossum apenninum* L.), a native of Italy and Sicily, recorded by Sibthorp (Fl. Graec. Prodr., 1: 118; 1806) "In Graeciae campestribus, et in Zacyntho et Cypro", is certainly an error. Holmboe (Veg. Cypr., 147), following Halácsy (Consp. Fl. Graec., 2: 359) refers the record to *Cynoglossum creticum* Mill., which is probably correct, though difficult to prove. *Solenanthus apenninus* superficially resembles *Cynoglossum*, but is at once distinguished by its narrowly tubular flowers and shortly exserted stamens.]

3. ASPERUGO *L.*

Sp. Plant., ed. 1, 138 (1753).
Gen. Plant., ed. 5, 67 (1754).

Sprawling annual; stems branched at and above the base, thinly hispidulous; leaves alternate and petiolate towards base of plant, opposite and sessile above, thinly hispidulous; flowers solitary or few in the axils of opposite, foliaceous bracts; calyx at first shallowly cupular, 5-dentate, much accrescent in fruit and strongly compressed laterally to form an envelope (shaped like a cockscomb) around the mature nutlets; corolla infundibuliform-rotate, 5-lobed, with a short tube and small faucal scales; filaments short; anthers included; style very short; stigma capitate. Nutlets strongly compressed laterally, attached ventrally over a small area.

One species widely distributed in temperate Europe, Asia and the Mediterranean region.

1. **A. procumbens** *L.*, Sp. Plant., ed. 1, 138 (1753); Sibth. et Sm., Fl. Graec. Prodr., 1: 123 (1806), Fl. Graec., 2: 65, t. 177 (1813); Poech, Enum. Plant. Ins. Cypr., 24 (1842); Unger et Kotschy, Die Insel Cypern, 284 (1865); Boiss., Fl. Orient., 4: 275 (1875); H. Stuart Thompson in Journ. Bot., 44: 335 (1906); Holmboe, Veg. Cypr., 147 (1914); Post, Fl. Pal., ed. 2, 2: 253 (1933); Osorio-Tafall et Seraphim, List Vasc. Plants Cyprus, 86 (1973); Davis, Fl. Turkey, 6: 264 (1978).

TYPE: "*in* Europae *ruderatis pinguibus*".

Sprawling annual; stems 10–50 (–70) cm. long, commonly branched at the base and sparingly so above, angled, thinly retrorse-hispidulous; leaves petiolate towards base of plant, sessile or subsessile above, lanceolate, oblanceolate or narrowly obovate, 4–15 cm. long, 1–4 cm. wide, thinly hispidulous, acute or obtuse at apex, tapering to base; inflorescence a very lax, dichotomously branched, bracteate cyme; bracts foliaceous, 1–4 cm. long, 0·5–2 cm. wide; flowers usually solitary (occasionally 2 or more) in the axils of the bracts, subsessile or shortly pedicellate; calyx-lobes oblong, about 1·5 mm. long, 0·5 mm. wide, hispidulous externally, apex obtuse or subacute; calyx strongly accrescent and deflexed in fruit, semicircular, laterally compressed, usually more than 1 cm. diam., reticulate-veined, with erect acute lobes 4–5 mm. long alternating with bifid, recurved appendages in the sinuses; corolla violet-blue with a pallid tube about 1 mm. long, lobes erect or spreading, semicircular, glabrous, about 1 mm. diam.; faucal appendages broadly oblong, truncate, papillose or shortly pilose, about 0·2 mm. long, 0·4 mm. wide; stamens attached about 0·5 mm. above base of corolla-tube; filaments very short; anthers ovate-oblong, about 0·8 mm. long, 0·4 mm. wide; ovary glabrous, laterally compressed; style about 0·3 mm. long; stigma capitate; disk inconspicuous. Nutlets pale brown when ripe, strongly compressed laterally, ovate-pyriform, about 3 mm. long, 2 mm. wide, minutely and bluntly tuberculate, with a very narrow marginal wing; ventral scar very small.

HAB.: Waste and cultivated ground, roadsides, often in moist situations; sea-level to 3,500 ft. alt.; fl. Jan.–May.

DISTR.: Divisions 2–7; distribution that of the genus.

2. Near Statos, 1974, *Meikle* 4017!
3. Lefkara, 1941, *Davis* 2714!
4. Varosha, ? 1862, *Kotschy*; Larnaca, 1880, *Sintenis & Rigo* 830!
5. Kythrea, 1927, *Rev. A. Huddle* 79! and, 1967, *Merton* in ARI 335! By Larnaca road near Nicosia, 1941, *Davis* 2142!
6. Nicosia, 1905, *Holmboe* 279; near Ayia Marina, Skylloura, 1957, *Merton* 2885!
7. Near Kantara, 1880, *Sintenis & Rigo*.

4. BORAGO L.
Sp. Plant., ed. 1, 137 (1753).
Gen. Plant., ed. 5, 67 (1754).

Hispid annuals or perennials; flowers in lax, nodding, bracteate cymes; calyx 5-lobed almost to base, persistent; corolla blue, pink or white, rotate or subrotate, with a very short tube and 5 stellately spreading or suberect, acute lobes; faucal scales conspicuous, pallid, deltoid or oblong; stamens inserted at base of corolla-tube; filaments short, thick, with an apical appendage; anthers large, connivent, dehiscing introrsely from the apex; style lengthening with maturity; stigma capitate. Nutlets oblong-ovoid, shortly stipitate, and with a concave base surrounded by a well-marked, ribbed annulus.

Four species in S. Europe, the Mediterranean region and western Asia.

1. **B. officinalis** *L.*, Sp. Plant., ed. 1, 137 (1753); Sibth. et Sm., Fl. Graec. Prodr., 1: 122 (1806); Poech, Enum. Plant. Ins. Cypr., 25 (1842); Unger et Kotschy, Die Insel Cypern, 283 (1865) as *Borrago*; Boiss., Fl. Orient., 4: 150 (1875); Holmboe, Veg. Cypr., 147 (1914) as *Borrago*; Post, Fl. Pal., ed. 2, 2: 224 (1933); Osorio-Tafall et Seraphim, List Vasc. Plants Cyprus, 86 (1973); Davis, Fl. Turkey, 6: 434 (1978).
TYPE: "*hodie in Normannia ad Colbeck & alibi in Europa; venit olim ex Aleppo*".

Spreading annual 10–60 cm. high with a short thick taproot; stems and leaves thinly clothed with white, prickly bristles; stems rather fleshy, longitudinally striate; leaves ovate, 4–15 cm. long, 2·5–10 cm. wide, dull green, nervation rather prominent below, apex obtuse or subacute, base rounded or subcordate; petioles channelled, up to 9 cm. long in the lower leaves, 1·5–4 cm. long in the upper; inflorescence terminal, usually much branched; bracts foliaceous, sessile, narrowly ovate or lanceolate, acute, 0·8–2 cm. long, 0·3–0·8 cm. wide; pedicels bristly, 1·5–3 cm. long at anthesis, lengthening (sometimes to 6 cm. or more) and becoming pendulous in fruit; flowers nodding; calyx-lobes erect, lanceolate, acute, bristly externally, 8–10 mm. long, 2–2·5 mm. wide; corolla 2–3 cm. diam., stellate-rotate with patent lobes, tube shallowly concave, about 2 mm. long, lobes broadly obovate, acute, 1–1·3 cm. long, 0·8–1 cm. wide; faucal scales glabrous, deltoid or truncate-deltoid, about 2·5 mm. long, 2 mm. wide, apex emarginate; filaments flattened, oblong, about 2 mm. long, 1·5 mm. wide, with a deltoid-subulate apical appendage about 1·5 mm. long; anthers blackish, connivent into a cone, narrowly lanceolate, 5–6 mm. long, 1·3–1·5 mm. wide, with a short apical cusp; ovary glabrous; style about 6–7 mm.

long at anthesis; stigma capitate. Nutlets enveloped by persistent calyx, 4–5 mm. long, about 3 mm. wide, attached to a short, pallid, peg-like extension of the disk, sharply keeled ventrally, fuscous, verruculose, with prominent longitudinal striae, base concave, surrounded by a conspicuous, thickened, regularly costate annulus.

HAB.: Gardens, waste ground, hedgerows and roadsides, probably always a garden escape; sea-level to 500 ft. alt.; fl. March–April.

DISTR.: Divisions 1, 4, 7, 8. S. Europe and Mediterranean region, frequently cultivated and naturalized in many temperate regions of the world.

1. Khrysokhou, 1957, *Merton* 3028! Polis, 1962, *Meikle* 2238!
4. Larnaca, 1862, *Kotschy*.
7. Ayios Epiktitos, 1936, *Syngrassides* 1033! Lambousa, 1936, *Syngrassides* 1034! Kazaphani, 1937, *Syngrassides* 1451! Kyrenia, 1949, *Casey* 492! and, 1956, *G. E. Atherton* 1124! Lapithos, 1967, *Merton* in ARI 551!
8. Khelones, 1970, *A. Hansen* 202!

NOTES: The flowers and leaves were formerly used to impart a refreshing, cooling flavour to drinks, and the whole plant, infused, was considered valuable in fevers and pulmonary complaints. It seems likely that, in Cyprus, the plant is always a relic of former cultivation.

5. ANCHUSA L.

Sp. Plant., ed. 1, 133 (1753).
Gen. Plant., ed. 5, 64 (1754).
M. Guşuleac in Bul. Fac. Ştiinţe, Cernăuti, 1: 73–123 (1927); ibid., 1: 235–325 (1927) et in Fedde Repert., 26: 286–322 (1929).
Lycopsis L., Sp. Plant., ed. 1, 138 (1753), Gen. Plant., ed. 5, 68 (1754).

Biennial or perennial, or rarely annual herbs, generally with strigose or hispid stems and leaves, the latter entire or irregularly dentate; inflorescence terminal, generally bracteate, often lengthening considerably after anthesis; calyx 5-dentate or 5-partite with the lobes reaching almost to the base; corolla blue, purple, white or yellow, infundibuliform, hypocrateriform or subrotate, the tube straight or curved, short or long, the limb horizontal or oblique, 5-lobed; faucal scales 5, ovate or oblong, erect or recurved, pilose or papillose; stamens 5, inserted at or above the middle of the tube; filaments usually short; style included; stigma clavate or capitate. Nutlets 4, erect or oblique with a basal attachment scar, or transverse with a subventral scar, the scar generally surrounded by a swollen, crenate or dentate annulus.

About 30 species widely distributed in Europe, the Mediterranean region, western Asia, tropical Africa and South Africa, some of the more decorative species naturalized elsewhere as garden escapes.

Corolla pale yellow; flowers apparently solitary, axillary on an elongate, often unbranched inflorescence - - - - - - - - - **4. A. aegyptiaca**
Corolla blue, pink, violet, purple or whitish:
Leaves conspicuously undulate-dentate; lobes about ⅓–½ total length of calyx
1. A. undulata ssp. **hybrida**
Leaves not conspicuously undulate-dentate; lobes reaching to near base of calyx:
Robust biennials or perennials 20–150 (–200) cm. high; inflorescence lax, paniculate; corolla 7–20 mm. diam.:
Calyx-lobes linear-acuminate; corolla rich blue or purplish; leaves and stems green, rather densely, but not very pungently strigose-hispid - - - **2. A. azurea**
Calyx-lobes linear-oblong, obtuse or subacute; corolla pale blue or whitish-blue; leaves and stems glaucous, rather thinly clothed with stout, pungent bristles
3. A. strigosa
Low growing annuals (3–) 8–30 (–50) cm. high; inflorescence dense, congested or glomerulate, conspicuously white-setose; corolla 3–6 mm. diam. - **5. A. humilis**

SUBGENUS 1. **Anchusa** (including subgenus *Buglossum* (Gaertn.) Guşul.)
Biennial, perennial or annual herbs; corolla-tube straight, usually exceeding

the calyx, limb actinomorphic. Nutlets oblong or ovoid, erect or oblique, with a basal or sub-basal attachment scar.

1. **A. undulata** *L.*, Sp. Plant., ed. 1, 133 (1753); Boiss., Fl. Orient., 4: 152 (1875) quoad nomen; Guşuleac in Bul. Fac. Ştiinţe, Cernăuti, 1: 263 (1927), et in Fedde Repert., 26: 298 (1929); Osorio-Tafall et Seraphim, List Vasc. Plants Cyprus, 86 (1973); Davis, Fl. Turkey, 6: 392 (1978).

Erect or spreading biennial or perennial 10–50 cm. high; stems unbranched or branched at the base, angled, harshly pilose or strigose with a mixed indumentum of long, slender bristles, and shorter hairs; basal leaves forming a rosette, ± strigose, variable, narrowly to broadly oblanceolate or elliptic-obovate, apex obtuse or rounded, margins undulate, repand-dentate, base tapering to a short, inconspicuous petiole; cauline leaves similar to the basal, but sessile and often amplexicaul; inflorescence usually much branched, the cymes at first compact and dense, becoming lax in fruit and 10 cm. or more long; bracts foliaceous, hispid-strigose, lanceolate or narrowly ovate, 5–15 mm. long, 2–8 mm. wide, diminishing rapidly upwards; pedicels at first very short, about 1–2 mm. long, elongating to 5 mm. or more and becoming deflexed in fruit; calyx cylindrical or narrowly infundibuliform, rather densely strigose, 7–11 mm. long, 3–4 mm. wide at anthesis, divided to one-third or half its length into narrow, oblong, obtuse lobes about 1–1·3 mm. wide; corolla dark blue, violet or purple, infundibuliform, tube 1·2–1·8 cm. long, glabrous, generally much exceeding the calyx, lobes suberect, variable, narrowly and acutely deltoid, or almost semicircular, 3–5 mm. long, 1·5–4 mm. wide; faucal scales oblong, 1·3–2 mm. long, 0·8–1 mm. wide, white or brownish, shortly pilose, apex rounded or truncate; stamens inserted between, and level with the bases of the faucal scales (ssp. *undulata*) or rather more than 2 mm. lower down the corolla-tube (ssp. *hybrida* (Ten.) Béguinot); filaments about 0·4 mm. long; anthers dark blue before dehiscence, narrowly oblong or linear, 2–3 mm. long, 0·4–0·5 mm. wide, almost equalling the faucal scales in ssp. *undulata*, falling short of their bases in ssp. *hybrida*; ovary glabrous, small, barely 1 mm. diam.; disk very inconspicuous; style 9–11 mm. long; stigma broadly clavate or subcapitate. Nutlets enclosed in the persistent, accrescent, inflated calyx, obliquely oblong-ovoid with a blunt, slightly incurved apex, greyish when ripe, densely verruculose with numerous, prominent, longitudinal striae.

ssp. **hybrida** *(Ten.)* *Béguinot* in Nuov. Giorn. Bot. Ital., n.s. 17: 634 (1910); Osorio-Tafall et Seraphim, List Vasc. Plants Cyprus, 86 (1973); Davis, Fl. Turkey, 6: 392 (1978).
  A. hybrida Ten., Prodr. Fl. Nap., XIV, t. XI (1811–15), Fl. Nap., 1: 45 (1811–15); Unger et Kotschy, Die Insel Cypern, 280 (1865); Boiss., Fl. Orient., 4: 152 (1875); Holmboe, Veg. Cypr., 147 (1914); Guşuleac in Bul. Fac. Ştiinţe, Cernăuti, 1: 254 (1927) as "*A. hydrida*", Guşuleac in Fedde Repert., 26: 293 (1929); Post, Fl. Pal., ed. 2, 2: 224 (1933); Lindberg f., Iter Cypr., 27 (1946); Rechinger f. in Arkiv för Bot., ser. 2, 1: 429 (1950).
  TYPE: Italy; Naples, "E communissima questa *buglossa* per le strade di campagna" (NAP, K!).

Stamens inserted at least 2 mm. below the bases of the faucal scales, the anthers usually falling short of the bases of the scales. Otherwise indistinguishable in vegetative or floral characters from *A. undulata* L., ssp. *undulata*.

HAB.: Sandy coastal flats; dry hillsides amongst garigue; roadsides, waste and cultivated ground; on calcareous and igneous formations; sea-level to 4,500 ft. alt.; fl. March–May.

DISTR.: Divisions 1–8, often common. Widely distributed in the Mediterranean region, ssp. *hybrida* from Corsica and S. Italy eastwards to Turkey, Syria, Palestine and Egypt.

1. Lyso, 1913, *Haradjian* 854.
2. Prodhromos, 1862, *Kotschy* 770! and, same locality, 1961, *D. P. Young* 7392! Panayia, 1913, *Haradjian* 898! Kryos Potamos, 3,100 ft. alt., 1937, *Kennedy* 700! 701! 702! Platres, 1937,

Kennedy 703! Stavros tis Psokas, 1937, *Syngrassides* 1590! and, same locality, 1953, *Kennedy* 1797! Lemithou, 1949, *Casey* 720! Apliki, 1967, *Merton* in ARI 505! Galata, 1974, *A. Genneou* in ARI 1641! 1642!

3. Limassol, 1913, *Haradjian* 567! Arsos, 1969, *A. Genneou* in ARI 1352! Khalassa, 1970, *A. Genneou* in ARI 1494!

4. Larnaca, 1862, *Kotschy* 49, also, 1880, *Sintenis & Rigo* and, 1912, *Haradjian* 67! Mile 19, Nicosia-Larnaca road, 1950, *Chapman* 406! Cape Greco, 1958, *N. Macdonald* 12! Dhekelia, 1977, *A. Della* 1673!

5. Margo, 1936, *Syngrassides* 1056! Ayios Sozomenos, 1957, *Merton* 3098! Salamis, 1969, *A. Genneou* in ARI 1316!

6. Nicosia, 1930, *F. A. Rogers* 0660! 0723! Ayia Irini, 1937, *Syngrassides* 1414! and, 1941, *Davis* 2542! Nicosia airport, 1952, *F. M. Probyn* 31! Cape Kormakiti, 1956, *Merton* 2509! and, 1967, *Merton* in ARI 124!

7. St. Hilarion, 1880, *Sintenis & Rigo* 662!, also, 1939, *Lindberg f.* s.n.! Kyrenia, 1931, *J. A. Tracey* 77! Panagra Pass, 1936, *Mrs. Dray* 3! Dhavlos, 1880, *Sintenis & Rigo*, also, 1941, *Davis* 2450! Ayios Yeoryios, 1949, *Casey* 370! S.W. of Kyrenia, 1950, *Casey* 1031!

8. Ephtakomi, 1880, *Sintenis & Rigo* 110! Leonarisso, 1880, *Sintenis & Rigo*; Cape Andreas, 1880, *Sintenis & Rigo*; near Komi Kebir, 1912, *Haradjian* 289!

NOTES: The Cyprus plants vary greatly in the size of the flowers and in the shape of the corolla-lobes. All seem, however, to be uniformly distinct from typical *A. undulata* (from Spain and Portugal) in having the stamens inserted well below the faucal scales. The distinction seems to hold good elsewhere, though in the vicinity of Naples (the type locality of *A. undulata* ssp. *hybrida*) plants with the stamens of typical *A. undulata* and with those of ssp. *hybrida* are both to be found. Tenore's illustration of *A. hybrida* (Fl. Nap. Atlas t. XI) seems, in so far as the analysis can accurately be judged, to be of the plant generally so named, as are also specimens sent by Tenore to J. Gay (K!) in 1814.

2. A. azurea Mill., Gard. Dict., ed. 8, no. 9 (1768); Post, Fl. Pal., ed. 2, 2: 224 (1933); Osorio-Tafall et Seraphim, List Vasc. Plants Cyprus, 86 (1973); Davis, Fl. Turkey, 6: 393 (1978).

A. italica Retz., Obs. Bot., 1: 12 (1779); Unger et Kotschy, Die Insel Cypern, 280 (1865); Boiss., Fl. Orient., 4: 154 (1875); Holmboe, Veg. Cypr., 147 (1914); Guşuleac in Fedde Repert., 26: 306 (1929); Lindberg f., Iter Cypr., 27 (1946).

A. paniculata Aiton, Hort. Kew., ed. 1, 1: 177 (1789); Sibth. et Sm., Fl. Graec., 2: 53, t. 163 (1813).

TYPE: Not indicated, but description probably based on specimens cultivated in Chelsea Physic Garden.

Robust, erect, strigose-hispid perennial, 20–100 (–150) cm. high; stems branched or unbranched, subterete or obscurely angled, sparingly or densely clothed with long, spreading, white, bristly hairs; basal leaves forming a loose tuft, narrowly obovate-elliptical, 8–15 (–30) cm. long, 1·5–6 (–10) cm. wide, dull green, thinly or densely strigose, apex acute, margins entire, base tapering to a flattened, narrowly winged petiole 2–8 cm. long; cauline leaves sessile, linear-lanceolate or narrowly oblong-lanceolate, 6–20 cm. long, 0·8–5 cm. wide, acute or acuminate, often with obscurely erose-dentate margins; inflorescence terminal, lax, paniculate, consisting of numerous, branched cymes; bracts foliaceous, the lowermost ovate-acuminate, the upper much smaller, narrowly lanceolate-acuminate; pedicels usually less than 1 cm. long at anthesis, lengthening to 1·5–3 cm. or more in fruit, hispid, remaining erect after anthesis; calyx hispid, divided almost to base into 5 linear-acuminate lobes, 10–12 mm. long, 0·8–1 mm. wide at anthesis, accrescent to 1·5–2 cm. long and 0·3 cm. wide and persistent in fruit; corolla hypocrateriform, bright blue, tube usually a little longer than the calyx, glabrous, limb actinomorphic, 10–20 mm. diam., divided into 5 broadly oblong or suborbicular, glabrous lobes 5–8 mm. long and about as wide; faucal scales oblong, blunt, about 2 mm. long, 1·3 mm. wide, densely papillose-pilose; stamens inserted between, and a little below the bases of the faucal scales; filaments glabrous, about 1–1·5 mm. long; anthers narrowly oblong, about 2·5 mm. long, 0·8 mm. wide, protruding slightly from corolla-tube; ovary glabrous, about 2 mm. diam.; disk inconspicuous; style glabrous, 10–11 mm. long; stigma capitate. Nutlets oblong, erect, large, about 5–7 mm. long, 2–3 mm. wide, greyish-brown, minutely verruculose with prominent longitudinal ridges; basal annulus obscure, closely ribbed.

HAB.: Cultivated and waste ground, fallow fields, roadsides; occasionally amongst garigue on dry hillsides; sea-level to 4,500 ft. alt.; fl. March–May.

DISTR.: Divisions 1–8, locally common. S. Europe and Mediterranean region eastwards to Pakistan and Central Asia; Atlantic Islands.

1. Arodhes ["Arora"], 1862, *Kotschy*; Stroumbi, 1913, *Haradjian* 730! Dhrousha, 1941, *Davis* 3263! Near Kissonerga, 1967, *Economides* in ARI 969! Near Baths of Aphrodite, 1979, *Edmondson & McClintock* E 2742!
2. Prodhromos, 1862, *Kotschy* 846! Lefka, 1932, *Syngrassides* 240! Platres, 1938, *Kennedy* 1360! 1361! and, 1955, *Kennedy* 1864! Saïttas, 1950, *F. J. F. Barrington* 23!
3. Kolossi, 1939, *Lindbergf.* s.n.; Pyrgos, 1963, *J. B. Suart* 22! Malounda, 1967, *Merton* in ARI 496!
4. Larnaca, 1880, *Sintenis & Rigo.*
5. Athalassa, 1939, *Lindbergf.* s.n.; Kythrea, 1950, *Chapman* 83!
6. Lakatamia, 1932, *Syngrassides* 690! Morphou, 1972, *W. R. Price* 988! 1048!
7. Dhavlos, 1880, *Sintenis & Rigo*; Kyrenia, 1932, *Syngrassides* 688! also, 1949, *Casey* 234!, 1955, *Miss Mapple* 17!, 1956, *G. E. Atherton* 891! 980!, 1973, *P. Laukkonen* 73! Dhikomo, 1936, *Syngrassides* 1326! Ayios Amvrosios, 1937, *Syngrassides* 1455! St. Hilarion, 1939, *Lindbergf.* s.n.; Myrtou, 1955, *G. E. Atherton* 1102a! Vasilia, 1956, *G. E. Atherton* 1102!
8. Ephtakomi, 1880, *Sintenis & Rigo* 111! Near Apostolos Andreas Monastery, 1880, *Sintenis & Rigo*; Yialousa, 1970, *A. Genneou* in ARI 1456!

3. A. strigosa *Labill.*, Icon. Plant. Syr. Rar., 3: 7, t. 4 (1809); Unger et Kotschy, Die Insel Cypern, 280 (1865); Boiss., Fl. Orient., 4: 155 (1875); H. Stuart Thompson in Journ. Bot., 44: 335 (1906); Holmboe, Veg. Cypr., 147 (1914); Guşuleac in Fedde Repert., 26: 308 (1929); Post, Fl. Pal., ed. 2, 2: 225 (1933); Lindberg f., Iter Cypr., 27 (1946); Osorio-Tafall et Seraphim, List Vasc. Plants Cyprus, 86 (1973); Davis, Fl. Turkey, 6: 396 (1978).

TYPE: Lebanon; "Juxtà Tripolim Syriae", *Labillardière* (FI).

Stout, erect biennial or perennial, 40–150 (–200) cm. high; stems usually unbranched, sometimes branched at the base, angled, glaucous, thinly or rather densely clothed with rigid, prickly bristles; basal leaves forming a loose rosette, obovate-elliptical, 7–22 cm. long, 2·5–10 cm. wide, thinly or densely pustulose-hispid, glaucous, apex acute or obtuse, margins entire or subentire, base tapering to a flattened, narrowly winged petiole 2–8 cm. long; cauline leaves rather few, sessile or subsessile, narrowly elliptic or oblong-lanceolate, 5–10 cm. long, 1–3 cm. wide, acute or acuminate, entire; inflorescence terminal, lax, paniculate, consisting of numerous branched cymes; bracts foliaceous, inconspicuous, the lowermost ovate-acuminate, the upper linear or subulate; pedicels 3–5 mm. long at anthesis, lengthening to 1–2 (or more) cm., but remaining erect in fruit, hispid or setulose; calyx hispid, often purplish, deeply divided into 5 linear-oblong, obtuse or subacute lobes 5–10 mm. long, 1–3 mm. wide at anthesis, persistent and accrescent in fruit, the lobes remaining erect; corolla hypocrateriform, pale or bright sky-blue, sometimes milky-blue or pale purplish, tube usually exceeding, and sometimes conspicuously exceeding, the calyx, glabrous; limb actinomorphic, 7–16 mm. diam., divided into 5 suborbicular, glabrous lobes 3–6 mm. diam.; faucal scales oblong, blunt, 1·5–3 mm. long, 1–1·5 mm. wide, densely papillose-pilose; stamens inserted between, and a little below the bases of the faucal scales; filaments glabrous, very short, usually less than 0·5 mm. long; anthers narrowly oblong, about 3–3·5 mm. long, 1 mm. wide, protruding slightly from corolla-tube; ovary glabrous, 2–3 mm. diam.; disk rather conspicuous, cupular; style glabrous, 8–12 mm. long; stigma clavate or subcapitate. Nutlets commonly 2 by abortion, oblong, erect, large, 5–8 mm. long, 2·5–3 mm. wide, greyish-brown, densely minutely papillose, and more sparingly verruculose, with very prominent, sharp longitudinal ridges; basal annulus scarcely developed.

HAB.: Roadsides, fallow fields; amongst garigue on dry calcareous or non-calcareous mountainsides; 300–3,500 ft. alt.; fl. March–June (to Sept.).

DISTR.: Divisions 1–4, 6, 7. Eastern Mediterranean region from Rhodes and southern Turkey to Egypt, and eastwards to Iran.

1. Lyso ["Lisso"], 1913, *Haradjian* 824! Dhrousha, 1941, *Davis* 3258! Pelathousa, 1962, *Meikle* 2274! Near Meladhia, 1979, *Edmondson & McClintock* E 2784!
2. Lefka to Moutoullas, 1937, *Syngrassides* 1555! Between Ambelikou and Kambos, 1939, *Lindberg f.* s.n.! Near Kalopanayiotis, 1949, *Casey* 941! Asinou valley, 3,500 ft. alt., 1953, *Kennedy* 1795!
3. Lania, 1970, *A. Genneou* in ARI 1513!
4. Kouklia (Mesaoria), 1905, *Holmboe* 376.
6. N.W. of Pano Dheftera, 1979, *Edmondson & McClintock* E 2940!
7. Between Nicosia and Kyrenia near "Tricomo" [Dhikomo is almost certainly intended], 1862, *Kotschy* 453!

SUBGENUS 2. **Lycopsis** (*L.*) *Guşuleac* (including subgenus *Buglossoides* Guşul.) Annuals; corolla-tube straight or curved, limb zygomorphic. Nutlets obliquely ovoid with a basal attachment scar.

4. A. aegyptiaca (*L.*) *DC.*, Prodr., 10: 48 (1846); Unger et Kotschy, Die Insel Cypern, 280 (1865); Boiss., Fl. Orient., 4: 159 (1875); Holmboe, Veg. Cypr., 147 (1914); Guşuleac in Fedde Repert., 26: 312, t. LXXXIX A–D (1929); Post, Fl. Pal., ed. 2, 2: 227 (1933); Lindberg f., Iter Cypr., 27 (1946); Rechinger f. in Arkiv för Bot., ser. 2, 1: 429 (1950); Osorio-Tafall et Seraphim, List Vasc. Plants Cyprus, 86 (1973); Davis, Fl. Turkey, 6: 399 (1978).
  Lycopsis aegyptiaca L., Sp. Plant., ed. 1, 138 (1753).
  TYPE: "*in* Aegypto".

Prostrate or sprawling annual; stems usually branched, especially towards base, obscurely angled or ridged, 10–50 cm. long, thinly or rather densely prickly-hispid; basal leaves forming a loose tuft, obovate or oblanceolate, 4–15 cm. long, 0·8–3 cm. wide, bright green, thinly pustulose-hispid, apex acute or obtuse, margins irregularly erose-dentate, base tapering to a flattened or channelled petiole 1–3 cm. long; cauline leaves similar, sessile or subsessile, often somewhat amplexicaul, 4–10 cm. long, 1–4 cm. wide; inflorescences lax and elongate, commonly unbranched, the flowers apparently solitary and axillary, subtended by leaf-like bracts up to 9 cm. long and 3·5 cm. wide; pedicels prickly-hispid, generally less than 5 mm. long at anthesis, lengthening to 1 cm. or more in fruit and often becoming patent or deflexed; calyx rather densely hispid, divided almost to base into subacute, lanceolate lobes 4–5 mm. long, 1 mm. wide, persistent and somewhat accrescent in fruit and loosely embracing the nutlets; corolla sulphur-yellow, tube straight, 5–6 mm. long, glabrous, a little exceeding the calyx; limb inconspicuously zygomorphic, 5–6 mm. diam., lobes suborbicular, 1·5–2 mm. diam.; faucal scales brownish, densely papillose-pilose, oblong, about 1 mm. long, 0·5 mm. wide, apex sharply recurved or revolute; stamens inserted at 2 levels near base of corolla-tube; filaments glabrous, about 0·5 mm. long; anthers brown, suborbicular, about 0·8 mm. diam.; ovary glabrous, about 1·5 mm. diam.; disk inconspicuous; style straight, glabrous, 2–2·5 mm. long; stigma subcapitate. Nutlets obliquely ovoid, subacute, about 5 mm. long, 2·5–3·5 mm. wide, pale yellow-brown when ripe, densely papillose-verruculose, with prominent, interconnecting ridges.

HAB.: Cultivated and fallow fields; waste ground; dry hillsides amongst garigue; sandy seashores; sea-level to 1,500 ft. alt.; fl. Febr.–May.

DISTR.: Divisions 1, 3–8. Eastern Mediterranean region from Sicily to Cyrenaica and eastwards to Iran.

1. Ktima, 1913, *Haradjian* 675! Ayios Yeoryios (Akamas), 1962, *Meikle* 2109! Kato Paphos, 1978, *Holub* s.n.!
3. Amathus, 1964, *J. B. Suart* 149!
4. Larnaca, 1862, *Kotschy* 8, also, 1880, *Sintenis & Rigo* 112! and, 1897, *Post* 475! Famagusta Castle, 1936, *Syngrassides* 1069! Dhromolaxia, 1970 and 1977, *A. Genneou* (*A. Della*) in ARI 1411! 1681! Dhekelia, 1973, *P. Laukkonen* 129! Near Ayia Napa, 1979, *Chesterman* 109!

5. Salamis, 1949, *Casey* 482! Kythrea, 1950, *Chapman* 342! and, 1956, *G. E. Atherton* 1066!
6. Karpasha Forest, 1956, *Merton* 2576!
7. Above Kyrenia, 1880, *Sintenis & Rigo* 112A! Kyrenia, 1938, *C. H. Wyatt* 45! Kyrenia Pass, 1939, *Lindberg f.* s.n.; Antiphonitis Monastery, 1941, *Davis* 2153! Tjiklos, Kyrenia, 1949, *Casey* 427!
8. Near Komi Kebir, 1912, *Haradjian* 306! Koma tou Yialou, 1935, *Syngrassides* 708! Ayios Philon near Rizokarpaso, 1941, *Davis* 2274!

5. A. humilis (*Desf.*) *I. M. Johnst.*, Contrib. Gray Herb., n.s., 73: 55 (1924).
   Echium humile Desf., Fl. Atlant., 1: 166 (1798).
   Anchusa aggregata Lehm., Plant. Asperif., 219 (1818), Icon. Rar. Plant. Fam. Asperif., 27, t. 47 (1821); Unger et Kotschy, Die Insel Cypern, 281 (1865); Boiss., Fl. Orient., 4: 158 (1875); Holmboe, Veg. Cypr., 147 (1914); Post, Fl. Pal., ed. 2, 2: 226 (1933); Lindberg f., Iter Cypr., 27 (1946); Davis, Fl. Turkey, 6: 401 (1978).
   Hormuzakia aggregata (Lehm.) Guşuleac in Publ. Soc. Nat. Român., no. 6: 8, fig. 2, 15 (1923); Rechinger f. in Arkiv för Bot., ser. 2, 1: 429 (1950); Osorio-Tafall et Seraphim, List Vasc. Plants Cyprus, 86 (1973).
   [*Anchusa parviflora* (non Willd.) Sm. in Sibth. et Sm., Fl. Graec. Prodr., 1: 117 (1806), Fl. Graec., 2: 57, t. 167 (1813).]
   TYPE: Tunisia; Gofsa, "in arenis deserti prope Cafsam", *Desfontaines* (P).

Erect or spreading annual (3–) 8–30 (–50) cm. high; stems usually much branched from the base, obscurely ridged or angular, densely white-hispid with pungent bristles; basal leaves forming a loose rosette, oblanceolate, 7–15 cm. long, 0·7–1·5 cm. wide, adpressed-hispid, apex obtuse or subacute, margins irregularly erose-dentate or subentire, base tapering to a short petiole, cauline leaves similar to basal, the uppermost sublinear, less than 4 cm. long and 0·5 cm. wide; cymes terminal and subterminal, remaining dense, glomerulate or congested-elongate even in fruit; bracts foliaceous, similar to the uppermost leaves but usually less than 20 mm. long and 2 mm. wide, persistent and often regularly and alternately patent (forming a fishbone pattern) in the infructescences; flowers sessile or subsessile; calyx densely white-setose, divided almost to base into 5 linear lobes, about 5 mm. long, 0·8 mm. wide; corolla bright cobalt blue, glabrous, tube straight, about 4–7 mm. long, equalling or slightly longer than the calyx, limb actino-morphic, about 3–6 mm. diam., divided into 5 broadly ovate, rounded or shortly acute lobes about 1·5–2 mm. long and almost as wide; faucal scales ovate, subacute, about 0·5 mm. long, 0·4 mm. wide, densely white-pilose; stamens inserted at one level about 1 mm. below faucal scales; filaments very short, less than 0·5 mm. long; anthers narrowly oblong, about 1·5 mm. long, 0·5 mm. wide; ovary glabrous, about 1 mm. diam.; style 3–3·5 mm. long; stigma capitate. Nutlets ovoid, distinctly curved and crescent-like, attached ventrally with a plicate-denticulate, skirt-like annulus, surface grey or pale brown, minutely and densely papillose, with numerous slightly larger, conical protuberances and an open reticulation of low, blunt ridges, apex pointed, concave, resembling the visor of a helmet.

HAB.: Coastal sand dunes and seashores, occasionally on rocky ground by the sea, or as a weed on sandy, cultivated ground; near sea-level; fl. Febr.–May.

DISTR.: Divisions 3–8. Eastern Mediterranean region from Sicily and Algeria eastwards to Palestine and Egypt.

3. Limassol, 1787, *Sibthorp*; Amathus, 1862, *Kotschy* 580, and, 1947, *Mavromoustakis* s.n.! Curium, 1939, *Kennedy* 1364! Petra tou Romiou, 1950, *F. J. F. Barrington* 17; Episkopi, 1960, *C. E. H. Sparrow* 49!
4. Near Ayia Napa, 1862, *Kotschy* 94a; between Livadhia and Cape Pyla, 1935, *Syngrassides* 837! Famagusta, 1936, *Syngrassides* 1064! Perivolia, 1938, *Syngrassides* 1750! and, same locality, 1939, *Lindberg f.* s.n.; Ayios Memnon, 1948, *Mavromoustakis* s.n.! Cape Kiti, 1955, *Merton* 2022!
5. Trikomo, 1939, *Lindberg f.* s.n.! Salamis, 1939, *Lindberg f.* s.n.! also, 1957, *Merton* 3209!, 1959, *P. H. Oswald* 55!, 1971, *C. C. Townsend* 71/110!
6. Ayia Irini, 1936, *Syngrassides* 1208! also, 1941, *Davis* 2120! 2602!, 1952, *Merton* 712!, 1962, *Meikle* 2390! and, 1967, *Economides* in ARI 940!

7. Kyrenia, 1949, *Casey* 395! Vachyammos, 1955, *Merton* 2288!
8. Near Yialousa, 1880, *Sintenis & Rigo* 113! Near Rizokarpaso, 1912, *Haradjian* 249! Ayios Philon, 1941, *Davis* 2271! Apostolos Andreas, 1941, *Davis* 2311!

6. NONEA *Medik.*
Phil. Bot., 1: 31 (1789).

Annual, biennial or perennial, strigose or glandular-pilose herbs; leaves entire or erose-dentate, the basal attenuate or obscurely petiolate, the cauline sessile and often amplexicaul or decurrent at the base; inflorescence terminal, branched or unbranched, with conspicuous foliaceous bracts; calyx 5-lobed to less or more than half its length, accrescent, inflated and ventricose in fruit; corolla blue, purple, pink, red, white or yellow, infundibuliform or hypocrateriform, tube straight or twisted at apex, limb horizontal and actinomorphic or oblique, and somewhat zygomorphic; faucal scales very small, sometimes reduced to a zone of hairs; stamens inserted near apex of corolla-tube; filaments short; style included; stigma clavate or capitate. Nutlets 4, erect, oblique or transverse with a basal, sub-basal or ventral attachment scar, the latter often surrounded by a prominent, costate or crenate-dentate annulus.

About 30 species in the Mediterranean region and western Asia.

Nutlets obliquely ovoid with a well-developed, crenate-lobulate annulus surrounding the sub-basal attachment scar; inflorescence 2–3 (or more)-branched, at length corymbiform; corolla-limb 6–8 mm. wide, noticeably oblique and zygomorphic -   - **1. N. philistaea**
Nutlets reniform, attachment scar ventral, median, without an annulus; inflorescence simple or 2-branched, at length spiciform; corolla-limb 2·5–4 mm. wide, not markedly oblique or zygomorphic - - - - - - - - - - **2. N. ventricosa**

1. **N. philistaea** *Boiss.*, Diagn., 1, 11: 96 (1849), Fl. Orient., 4: 166 (1875); H. Stuart Thompson in Journ. Bot., 44: 335 (1906); Holmboe, Veg. Cypr., 147 (1914); Post, Fl. Pal., ed. 2, 2: 229 (1933); Osorio-Tafall et Seraphim, List Vasc. Plants Cyprus, 86 (1973).
   TYPE: Palestine; "circà *Gaza* ubi legi Apr. 1846" *Boissier* (G, K!).

Erect or sprawling annual 4–30 (–40) cm. high; stems simple or branched from near the base, angled, harshly pilose with a mixed indumentum of short hairs, and long, patent bristles; basal leaves oblanceolate 3–8 (–10) cm. long, 0·5–1·5 cm. wide, strigose or pustulose-strigose, apex acute or obtuse, margins entire or obscurely erose, base tapering to a slightly expanded sheath; cauline leaves usually larger than basal, up to 14 cm. long, narrowly oblong or oblanceolate, base slightly dilated and amplexicaul; inflorescence terminal, 2–3 (or more)-branched, at first dense, later subcorymbiform; bracts foliaceous, ovate, acute, persistent, 0·5–3·5 cm. long, 0·3–1·3 cm. wide; flowers at first subsessile or very shortly pedicellate, later nutant on short, down-curved pedicels; calyx 5–6 mm. long, 3–4 mm. wide, cylindrical or narrowly campanulate at anthesis, accrescent and becoming inflated and ventricose in fruit, rather densely strigose-pilose, lobes deltoid about 2mm. long, 1–1·5 mm. wide at base; corolla whitish, creamy or pale yellow, tube 4–5 mm. long, slightly expanded and curved towards apex, limb oblique 6–8 mm. wide with 5 semicircular lobes about 2 mm. long; faucal scales minute, clothed with long bristles; stamens inserted in one series just below throat; filaments very short; anthers oblong, brownish, about 1–1·5 mm. long, 0·5 mm. wide; ovary glabrous about 1 mm. diam.; style rather thick, about 2 mm. long; stigma clavate, of 2 connate lobes about 0·5 mm. long. Nutlets broadly ovoid, blunt, oblique, about 2·5–3 mm. long, 2–2·5 mm. wide, fuscous, thinly pilose, coarsely and irregularly areolate with conspicuous interconnecting ridges, surface minutely verruculose-papillose, annulus surrounding the sub-basal attachment scar well-developed, conspicuously crenate-lobulate; caruncle small, inconspicuous.

HAB.: Cultivated and waste ground, occasionally on sand-dunes; sea-level to 500 ft. alt.; fl. Jan.–April.

DISTR.: Divisions 3–7. Also Palestine and S. Syria.

3. Mazotos, 1905, *Holmboe* 194.
4. Near Larnaca, 4th March, 1880, *Sintenis & Rigo* 114!
5. Athalassa, 1933, *Syngrassides* 1372! Lysi, 1954, *Merton* 2029! Kythrea, 1950, *Chapman* 343! Cornfields along the Nicosia–Famagusta road S. of Kythrea, 1970, *A. Hansen* 507!
6. Nicosia, 1937, *Syngrassides* 1402! also, same area, 1941, *Davis* 2141! English School, Nicosia, 1963, B. R. *Trenbath* s.n.! Omorphita, 1973, *P. Laukkonen* 59!
7. "In montibus prope Kythrea", April 1880, *Sintenis & Rigo* 114a partly, but see Notes.

NOTES: The unlocalized A. G. & M. E. Lascelles specimen, cited by Stuart Thompson and Holmboe is no longer to be found under the name *Nonea* (or *Nonnea*) *philistaea* in herb. Kew. It may have been a misidentification. The Sintenis & Rigo record, from above Kythrea, is also suspect, the specimen is labelled *N. ventricosa*, but is in fact *N. philistaea*. It is clear that material of both species, from different localities, was distributed under the one label and number.

In Cyprus the flowers of *N. philistaea* appear to be creamy-white or pale yellow, not pale orange as described by Boissier.

2. N. ventricosa (Sm.) Griseb., Spic. Fl. Rum. et Bith., 2: 93 (1844); Unger et Kotschy, Die Insel Cypern, 280 (1865); Boiss., Fl. Orient., 4: 169 (1875); Holmboe, Veg. Cypr., 147 (1914); Post, Fl. Pal., ed. 2, 2: 229 (1933); Rechinger f. in Arkiv för Bot., ser. 2, 1: 429 (1950); Osorio-Tafall et Seraphim, List Vasc. Plants Cyprus, 86 (1973); Davis, Fl. Turkey, 6: 413 (1978).

Anchusa ventricosa Sm. in Sibth. et Sm., Fl. Graec. Prodr., 1: 117 (1806), Fl. Graec., 2: 58, t. 168 (1813); Poech, Enum. Plant. Ins. Cypr., 25 (1842).

[*Nonea echioides* (non (L.) Roemer et Schultes) J. Edmondson in Willdenowia, 8: 29 (1977).]

TYPE: "In Graeciae campestribus, et in insulâ Cypro" *Sibthorp* (OXF).

Sprawling annual, 4–30 (–40) cm. high; stems usually much branched from near the base, occasionally simple, angled, harshly pilose with a mixed indumentum of short hairs and long, patent bristles; basal leaves narrowly oblanceolate (2–) 5–7 cm. long, (0·3–) 0·7–0·8 cm. wide, strigose or pustulose-strigose, apex shortly acute or obtuse, margins entire or obscurely erose, base tapering to a slightly expanded sheath; cauline leaves often rather larger, regularly oblong or narrowly ovate-oblong, base scarcely dilated, semi-amplexicaul; inflorescence terminal, simple or commonly geminate, at first dense, but soon elongate and spiciform; bracts foliaceous, narrowly ovate or lanceolate, acuminate, persistent, 0·8–2 cm. long, 0·3–1 cm. wide; flowers sessile or very shortly pedicellate, remaining erect or patent after anthesis; calyx about 5 mm. long, 2·5 mm. wide, narrowly campanulate at anthesis, accrescent to 1 cm. or more long, and becoming ventricose and inflated in fruit, rather densely strigose-pilose, lobes deltoid, about 1·5 mm. long, 1 mm. wide at base; corolla whitish or pale yellow, tube 4–5 mm. long, straight, expanded above, limb horizontal, about 2·5–4 mm. wide, with 5 minute semicircular lobes about 0·5 mm. long; faucal scales almost obsolete, reduced to a zone of bristly hairs; stamens inserted in one series just below throat; filaments very short; anthers oblong, brownish, about 1·3 mm. long, 0·4 mm. wide; ovary glabrous, about 1 mm. diam.; style rather thick, about 1–1·2 mm. long; stigma clavate, of 2 connate lobes about 0·4 mm. long. Nutlets reniform, transverse, 3–3·5 mm. wide, 2–2·5 mm. long, fuscous, glabrous or very sparsely hairy, rather regularly dorsiventrally ridged, surface minutely verruculose-papillose, attachment scar ventral, median, without any surrounding annulus, but with a conspicuous, yellowish caruncle.

HAB.: Cultivated and waste ground, roadsides; occasionally on hillsides amongst garigue; sea-level to 3,500 ft. alt.; fl. Jan.–May.

DISTR.: Divisions 1–3, 5, 7, 8. Northern Mediterranean from Spain to Turkey and N. Syria [and ? Palestine], Crete, Rhodes, Khios.

1. Ktima, 1913, *Haradjian* 703; Stroumbi, 1913, *Haradjian* 732.
2. Prodhromos, 1862, *Kotschy* 869; Pano Platres, 1930, *Wall* s.n.; same locality, 1937–38, *Kennedy* 698! 1362! 1363! 1368! Saïttas, 1960, *N. Macdonald* 7!
3. Skarinou, 1941, *Davis* 2675! Zagala above Trimiklini, 1941, *Davis* 3130! Phasouri, 1966, *Ann Matthews* 10! Amathus, 1978, *Holub* s.n.!
5. Kythrea, 1932, *Syngrassides* 227!
7. Ayios Khrysostomos Monastery, 1941, *Davis* 2179! Kyrenia, 1949, *Casey* 376! and, 1956, *G. E. Atherton* 852!
8. Rizokarpaso, 1941, *Davis* 2380!

NOTES: In the absence of fruits it is sometimes difficult to distinguish *N. ventricosa* from *N. philistaea*, and the two species have been much confused.

If the lectotypification proposed by J. Edmondson (in Willdenowia, 8: 29; 1977) is accepted, then the correct name for this species is *Nonea echioides* (L.) Roemer et Schultes. I am convinced, however, that the protologue of *Lycopsis echioides* L., Sp. Plant., ed. 2, 1: 199 (1762) is based entirely upon the text and illustration in Buxbaum, Cent. 1 Plant., 1, t. 1 (1728). The specimens collected by Loefling in Spain and preserved in herb. LINN and herb. S are nomenclaturally irrelevant and certainly cannot be used as lectotypes for the name *Lycopsis echioides* L. In my view the names *Nonea ventricosa* and *Arnebia echioides* should both be retained in their long-established and traditional sense.

7. ALKANNA *Tausch*
in Flora, 7: 234 (1824) nom. cons.

Perennial, pilose or setose herbs; leaves generally entire; inflorescences terminal, often branched, with persistent foliaceous bracts; calyx deeply 5-lobed, somewhat accrescent in fruit; corolla blue, white or yellow, infundibuliform or hypocrateriform, throat with a ring of hairs, or sometimes with inconspicuous swellings or invaginations; stamens 5, inserted near the middle of the corolla-tube, sometimes at more than one level; filaments very short; anthers included; style included; stigma small, clavate or capitate. Nutlets generally 1–2 by abortion, sometimes up to 4, ± stipitate, the upper half commonly bent horizontally or downward, surface usually rugose-reticulate or tuberculate.

About 30 species in S. Europe, the Mediterranean region and western Asia.

1. **A. lehmanii** (*Tineo*) *A.DC.* in DC., Prodr., 10: 588 (1846).
   Lithospermum tinctorium L., Sp. Plant., ed. 1, 132 (1753).
   Anchusa tinctoria (L.) L., Sp. Plant., ed. 2, 192 (1762) quoad nomen; Sibth. et Sm., Fl. Graec. Prodr., 1: 116 (1806), Fl. Graec., 2: 57, t. 166 (1813); Poech, Enum. Plant. Ins. Cypr., 25 (1842).
   Anchusa tuberculata Forssk., Fl. Aegypt.-Arab., 41 (1775).
   Anchusa bracteolata Viv., Fl. Libyc. Spec., 10, t. 4 (1824).
   Alkanna matthioli Tausch in Flora, 7: 235 (1824) nom. illeg.
   Alkanna tinctoria Tausch in Flora, 7: 234 (1824); Unger et Kotschy, Die Insel Cypern, 281 (1865); Boiss., Fl. Orient., 4: 227 (1875); Holmboe, Veg. Cypr., 147 (1914); Post, Fl. Pal., ed. 2, 2: 243 (1933); Osorio-Tafall et Seraphim, List Vasc. Plants Cyprus, 86 (1973); Davis, Fl. Turkey, 6: 425 (1978) nom. illeg.
   Anchusa rhizochroa Viv., Fl. Aegypt. Dec., 16 (1830) nom. superfl. illeg.
   Lithospermum lehmani Tineo in Guss., Fl. Sic. Syn., 2: 791 (1845).
   Alkanna tuberculata (Forssk.) Meikle in Kew Bull., 34: 823 (1980) non *Alkanna tuberculata* Greuter in Exsicc. Genav., fasc. 3: 38 (1972) nom. illeg.
   TYPE: Sicily; "In collibus aridis calcareis; *Ragusa* (*Tin.*) *Mazzara* (*Alex.*)".

Tufted perennial (10–) 15–30 (–40) cm. high with a thick, woody taproot exuding a violet-purple dye; stems and leaves clothed with a mixed indumentum of short tomentellous hairs and long, white, harsh, spreading bristles; basal leaves crowded, narrowly obovate or oblong-oblanceolate, 2–8 cm. long, 0·3–1·3 cm. wide, apex acute or obtuse, margins entire or crisped, often sharply recurved, base tapering to a narrow sheath; cauline leaves rather short, 1–5 cm. long, oblong or ovate-deltoid, apex acute or obtuse, base amplexicaul; inflorescence branched, at first dense and subcorymbiform, becoming lax, with elongate branches in fruit; bracts conspicuous, spreading, persistent, subequal, resembling the uppermost cauline leaves, 0·8–2 cm. long, 0·2–0·8 cm. wide; flowers sessile or very shortly pedicellate, the pedicel lengthening and becoming patent or deflexed in fruit; calyx campanulate, not much accrescent in fruit, about 7 mm. long, 4 mm. wide, silvery-hirsute or strigose, divided almost to base into 5 lanceolate-acuminate, strongly costate lobes 5–6 mm. long, 1·5–2 mm. wide; corolla vivid blue (seldom pale or purplish), tube straight, 5–8 mm. long, glabrous, limb actinomorphic, 3·5–6 mm. diam., divided into 5 semicircular lobes 1·5–2 mm. long; faucal scales obscure or reduced to an inconspicuous zone of short hairs; stamens inserted at two levels, one series (usually of 3 stamens) just below the throat, the remaining 2 stamens about 1 mm. lower down tube; filaments very short; anthers yellow, ovate-oblong, about 1 mm. long, 0·7 mm. wide; ovary glabrous, about 0·5 mm. diam.; disk conspicuous, slightly sinuate; style slender, about 4–6 mm. long, tapering to a small, capitate apex. Nutlets 1–3 (–4), their size proportionate to their number, 2·5–3·5 mm. long, 1·5–2·5 mm. wide, pale grey-brown, glabrous and rather glossy, densely and irregularly rugose-reticulate and/or tuberculate, apex blunt, curved downwards, base with a conspicuous, narrow annulus attached to a peg-like extension of the receptacle.

HAB.: Dry hillsides amongst garigue; fallow fields, roadsides; sandy or rocky seashores; sea-level to 2,800 ft. alt.; fl. Febr.–April.

DISTR.: Divisions 2–8, locally common; south-central Europe and the Mediterranean region.

2. Near Philani, 1908, *Clement Reid* s.n.! Evrykhou, 1935, *Syngrassides* 820! Makheras road, 1949, *Casey* 302!

3. Akrotiri, 1939, *Mavromoustakis* 107! also, 1960, *N. Macdonald* 78! and, 1963, *J. B. Suart* 33! Mile 22, Nicosia-Limassol road, 1950, *Chapman* 549!

4. Famagusta, 1936, *Syngrassides* 876! Dhekelia, 1977, *A. Della* (*A. Genneou*) in ARI 1674!

5. Vatili, 1905, *Holmboe* 322; Pyroï, 1936, *Syngrassides* 1079! Athalassa, 1936, *Syngrassides* 1123! also, same locality, 1967, *Merton* in ARI 156! Kythrea, 1937, *Miss Godman* 20! Locally abundant in this division.

6. Locally common about Nicosia 1930–1973, *F. A. Rogers* 0723! *Syngrassides* 69! *Chapman* 570! *P. Laukkonen* 26!. Also, Ayia Irini, 1936, *Syngrassides* 1210! 1413! also, 1941, *Davis* 2126! 2591! and, 1950, *Mrs. Grove* in *Casey* 1102!

7. Phlamoudhi, 1880, *Sintenis & Rigo* 583! Kyrenia, 1931–1956, *J. A. Tracey* 76! *C. H. Wyatt* 41! *G. E. Atherton* 892! 908! 1193! 1309! Lapithos, 1937, *Roger-Smith* s.n.! Ayios Khrysostomos Monastery, 1939, *Kennedy* 1726! and, 1941, *Davis* 2185! Yaïla, 1941, *Davis* 2862! Akanthou, 1950, *Casey* 987! Kyparissovouno, 1970, *I. M. Hecker* 45! Panagra, 1972, *G. Joscht* 6702!

8. Ovgoros, 1941, *Davis* 2473!

NOTES: Specimens from Division 6, Ayia Irini, are, like other plants from this area, exceptionally compact, with a dense indumentum of silvery hairs. Otherwise *Alkanna lehmanii* is not very variable in Cyprus.

My note on the involved nomenclature of this species (Kew Bull., 34: 821–824; 1980) failed to take account of W. Greuter's contribution to the same matter in Exsicc. Genav., fasc. 3: 38 (1972), with the consequence that the combination *Alkanna tuberculata* (Forssk.) Meikle becomes an illegitimate later homonym of *A. tuberculata* Greuter. In a skilful attempt to safeguard the well-known name *Alkanna tinctoria*, Dr Greuter has used (effectively, as it proved) a blocking mechanism to prevent the transfer of *Anchusa tuberculata* Forssk. and *Anchusa bracteolata* Viv. to the genus *Alkanna*. I do not think, however, he has been wholly successful in attaining his ultimate objective: in my opinion the name *Alkanna tinctoria* Tausch was rendered from the outset illegitimate, by the joint operation of Articles 53.1. and 67.1. of ICBN, since Tausch, on dividing *Anchusa tinctoria* (L.) L. did not adopt the final epithet of the earliest legitimate name (*Lithospermum tinctorium* L.) for the taxon, though it was available when he transferred *Lithospermum tinctorium* to the genus *Alkanna*; on the contrary he adopted the epithet for that part of the subdivided *Anchusa tinctoria* for which it was not available.

8. MYOSOTIS *L.*

Sp. Plant., ed. 1, 131 (1753).
Gen. Plant., ed. 5, 63 (1754).

Annual, biennial or perennial herbs; leaves entire, generally softly hairy; inflorescences terminal, often paired, ebracteate, or sometimes bracteate towards base, rarely throughout their length, elongating in fruit; calyx 5-lobed to, or beyond, the middle, somewhat accrescent in fruit; corolla commonly blue, rotate, with a short tube and a flat or concave limb, 5-lobed, the lobes rounded; faucal swellings whitish or yellowish, papillose; stamens included; filaments inserted near middle of tube; anthers with a terminal lingulate appendage; style included; stigma capitate. Nutlets 4 (or fewer by abortion), erect, ovoid, ± compressed, shining, sometimes with a narrow marginal wing or rim; attachment scar small, sometimes with a spongy appendage.

About 80 species widely distributed in temperate regions of the world and on the mountains of Tropical Africa.

Corolla minute, less than 4 mm. diam.; plants generally less than 20 cm. high:
  Pedicels erect or patent after anthesis, or flowers subsessile, generally 2-ranked:
    Inflorescences bracteate throughout their length; calyces uniformly clothed with straight,
      adpressed hairs - - - - - - - - - - - **1. M. pusilla**
    Inflorescences ebracteate or bracteate only towards base; calyces clothed with a mixed
    indumentum of adpressed straight hairs and patent or deflexed, hooked hairs:
      Flowers subsessile or very shortly pedicellate; fruiting calyces erect, often adpressed to
        rhachis; rhachis with spreading hairs - - - - - **3. M. minutiflora**
      Flowers distinctly pedicellate, the pedicels patent or erecto-patent in fruit; rhachis
        adpressed-hispidulous - - - - - - - - **2. M. ramosissima**
  Pedicels sharply deflexed after anthesis; fruiting calyx pointing downward; inflorescence
    commonly secund and 1-ranked - - - - - - - - - **4. M. refracta**
Corolla conspicuous, 4–7 mm. diam.; plants usually 10–30 (–50) cm. high - **5. M. sylvatica**

1. M. pusilla *Loisel.* in Desv., Journ. de Bot., 2: 260, t. 8, fig. 2 (1809); Boiss., Fl. Orient., 4: 237 (1875); Holmboe, Veg. Cypr., 148 (1914).
    [*M. idaea* (non Boiss. et Heldr.) Unger et Kotschy, Die Insel Cypern, 281 (1865); H. Stuart Thompson in Journ. Bot., 44: 335 (1906).]
    [*M. cretica* (non Boiss. et Heldr.) Lindberg f., Iter Cypr., 28 (1946).]
    [*M. incrassata* (non Guss.) Osorio-Tafall et Seraphim, List Vasc. Plants Cyprus, 87 (1973).]
    TYPE: Corsica; "découverte dans les champs en Corse, par M. G. Robert." (AV).

Small erect, spreading or decumbent annual, generally less than 10 cm. high, sometimes scarcely exceeding 1 cm.; stems usually much branched from the base, thinly or densely clothed with straight, white bristly hairs, irregularly spreading or subadpressed towards the base of the stems, more closely adpressed above; basal leaves forming a rather dense tuft, lingulate-obovate, 0·5–1 cm. long, 0·3–0·5 cm. wide, thinly pustulose-hispidulous, with subadpressed, straight hairs above, subglabrous or thinly hispidulous below, apex rounded or obtuse, base tapering to an indistinct petiole; stem leaves similar, often distinctly smaller, grading into small foliaceous bracts; inflorescence usually short, seldom exceeding 4 cm. in length, lax, rather irregularly bracteate from base to apex; rhachis adpressed-hispidulous, sometimes irregularly so, with long and short bristles; flowers subsessile or with short, thick pedicels generally less than 2 mm. long, at first erect, becoming erecto-patent in fruit; calyx rather densely subadpressed white-hispidulous, about 2 mm. long, 1·5 mm. wide at anthesis, 5-lobed for more than half its length, lobes erect, narrowly oblong-linear, obtuse or subacute, hispidulous; corolla pale milky-blue, minute; corolla-tube glabrous, about 1·3 mm. long, lobes erecto-patent, broadly lingulate, rounded, about 0·5 mm. long, 0·4 mm. wide; stamens inserted about middle of corolla-tube; filaments very short; anthers narrowly oblong, about 0·5 mm. long, 0·2 mm. wide, with a short, membranous apical appendage and subsagittate base; ovary depressed-globose, glabrous, about 0·5 mm. diam.; style 0·5 mm. long; stigma capitate, obscurely lobed. Nutlets ovoid, about 1·3 mm. long, 1 mm. wide, strongly compressed, shining fuscous-brown when fully ripe, with an obscure marginal rim and a small, impressed attachment scar.

HAB.: Rocky mountainsides on serpentine, frequently by springs or streams; 3,850–5,600 ft. alt.; fl. April–May.

DISTR.: Division 2, very local, but often forming dense colonies. S. France, Corsica, Sardinia, Algeria, ? Libya.

2. Restricted to the upper slopes of Khionistra, Platres, Kryos Potamos, Livadhi tou Pasha, etc. *Kotschy* 716 (? partly); *Holmboe* 930! 977! *Kennedy* 711–713! 1366! *Poore* 53! *Meikle* 2880! *N. Macdonald* 140!

NOTES: Phytogeographical consideration would indicate *M. incrassata* Guss. as the more likely occurrence in Cyprus, since it generally replaces *M. pusilla* in the eastern Mediterranean region. The inflorescences of Cyprus plants are bracteate (but rather irregularly so) throughout their length, while those of *M. incrassata* are bracteate only at the base. *M. pusilla* is also less robust and less erect than *M. incrassata*.

Boissier may be correct in referring the material seen by him of *Kotschy* 716 to *M. pusilla* (Fl. Orient., 4: 237), but the only specimen I have seen of this number (*Kotschy* 716, W!) is *M. ramosissima* Rochel.

2. M. ramosissima *Rochel* in Schultes, Oesterr. Fl., ed. 2, 1: 366 (1814); Stroh in Not. Bot. Gart. Berlin, 12: 473 (1935); Osorio-Tafall et Seraphim, List Vasc. Plants Cyprus, 87 (1973); Davis, Fl. Turkey, 6: 269 (1978).

M. hispida Schlechtendal in Ges. Nat. Freunde Berlin Mag., 8: 230 (1817); Unger et Kotschy, Die Insel Cypern, 281 (1865); Boiss., Fl. Orient., 4: 239 (1875); Sintenis in Oesterr. Bot. Zeitschr., 32: 194 (1882).

[*M. stricta* (non Link ex Roemer et Schultes) Sintenis in Oesterr. Bot. Zeitschr., 32: 51, 53 (1882).]

[*M. collina* (non Hoffm.) Holmboe, Veg. Cypr., 148 (1914); Post, Fl. Pal., ed. 2, 2: 245 (1933); Lindberg f., Iter Cypr., 28 (1946).]

[*M. discolor* (non Pers.) Osorio-Tafall et Seraphim, List Vasc. Plants Cyprus, 87 (1973).]

TYPE: Czechoslovakia; Velké Rovné ["Rownye"], 1807, *Rochel* (W).

Erect or spreading annual 2–20 (–30) cm. high; stems usually much branched from near the base, thinly or rather densely clothed with straight white hairs, subadpressed or irregularly spreading towards base of plant, closely adpressed above; basal leaves forming a loose or dense tuft, narrowly obovate, 0·5–3·5 cm. long, 0·3–1·3 cm. wide, thinly or densely hispidulous with subadpressed, straight hairs above, subglabrous or more thinly hispidulous below, apex rounded or obtuse, base tapering to an indistinct petiole or leaves subsessile; stem leaves similar, becoming smaller and more acute upwards; inflorescence ebracteate, at first dense, becoming lax and elongating to 7–14 cm. or more after anthesis; rhachis adpressed-hispidulous; flowers distinctly pedicellate, the pedicels adpressed-hispidulous, elongating to 2–3 mm., and becoming patent or recurved in fruit; calyx campanulate in fruit, 2–4 mm. long, 1–3 mm. wide, thinly or densely clothed with short, adpressed hairs and longer, spreading or reflexed, hooked hairs, 5-lobed for slightly more than half its length, lobes erect or slightly spreading, but not connivent in fruit, narrowly oblong, subacute, 1–2 mm. long, 0·5–0·8 mm. wide at base; corolla blue, minute, tube glabrous, about 1·5 mm. long, lobes spreading, rounded, about 0·4 mm. diam.; stamens inserted about two-thirds way down corolla-tube; filaments very short; anthers oblong, about 0·4 mm. long, 0·3 mm. wide, with a conspicuous rounded apical appendage and a bluntly subsagittate base; ovary globose, glabrous, about 0·4 mm. diam.; style 0·6 mm. long; stigma capitate. Nutlets ovoid, strongly compressed, about 1·3 mm. long, 1 mm.

wide, dark shining brown when fully ripe, margin without a distinct rim; attachment scar very small, whitish.

HAB.: Screes and stony slopes, amongst garigue in open sites or in *Pinus* or *Cedrus* woodland, or by roadsides; 1,700–5,000 ft. alt.; fl. March–June.

DISTR.: Divisions 2, 3, 7, locally common. Widespread in Europe, the Mediterranean region and eastwards to Iran.

2. Common: Prodhromos, Kambos, Moutoullas, Platres, Kionia, Papoutsa, Tripylos, Platania, Makheras, Polystipos, Kannoures Springs, Troodhitissa, Kykko, Ayia, Saïttas, Spilia, etc.; *Kotschy* 716! 716a (partly)! *Holmboe* 900, 929!; *Syngrassides* 1557! *Kennedy* 705–707! 709! 710! 1367! *Davis* 2709! 3097! 3111! 3493! *Mavromoustakis* 52! *Casey* 680! *Chapman* 300! *Merton* 2996! 3006! 3119! *P. H. Oswald* 110! *Meikle* 2657! 2792! 2822! *Merton* in ARI 252! *Edmondson & McClintock* E 2814! etc.

3. Above Kellaki, 1956, *Merton* 2554!

7. Kantara, 1880, *Sintenis & Rigo* 107! Sina Oros, 1941, *Davis* 2469! Yaïla, 1941, *Davis* 2837! 3½ m. W. of Kantara, 1952, *Merton* 766! Phterykha, 1953, *Casey* 1294! Kyparissovouno, 1962, *Meikle* 2424A!

3. M. minutiflora *Boiss. et Reuter*, Pugill. Plant. Nov., 80 (1852); Davis, Fl. Turkey, 6: 271 (1978).

[*M. stricta* (non Link ex Roemer et Schultes) Kotschy in Unger et Kotschy, Die Insel Cypern, 281 (1865); H. Stuart Thompson in Journ. Bot., 44: 335 (1906); Holmboe, Veg. Cypr., 148 (1914).]

TYPE: Spain: "in arenosis regionis alpinae regni Granatensis, *Sierra Tajeda* cacumen (Boiss.), *Sierra Nevada, a la Cueva de Panderon* inter *Juniperos* (Reuter)" (G).

Erect annual (2–) 4–10 (–15) cm. high; stems unbranched or branched from near the base, thinly or rather densely clothed with spreading, crispate hairs; basal leaves forming a loose tuft, commonly brownish and partly withered at anthesis, narrowly obovate or spathulate, 5–15 (–20) mm. long, 3–8 mm. wide, thinly hairy, apex rounded or obtuse, base tapering, sessile; cauline leaves similar, but usually acute or subacute with a more rounded base, decreasing in size upwards; inflorescence ebracteate or bracteate only towards base, at first dense, becoming elongate and 4–8 cm. long in fruit; rhachis thinly white-hairy with spreading, straight or hooked hairs; flowers erect or erecto-patent, subsessile or very shortly pedicellate, the pedicels with short, spreading, white hairs; calyx narrowly campanulate, 2–3 mm. long, 1–1·5 mm. wide, rather densely clothed with adpressed, straight, white hairs and spreading or deflexed, hooked hairs, 5-lobed for more than half its length, lobes erect or slightly spreading in fruit, narrowly oblong, acute or subacute, 1·5–2 mm. long, 0·6 mm. wide at base; corolla pale blue, minute, tube about 1 mm. long, rather shorter than calyx, glabrous, lobes spreading, rounded or subtruncate, about 0·5 mm. long, 0·6 mm. wide; stamens inserted about two-thirds way down tube; filaments very short; anthers ovate, about 0·6 mm. long, 0·5 mm. wide, with a conspicuous, obtuse, lingulate apical appendage and a bluntly subsagittate base; ovary subglobose, glabrous, about 0·5 mm. diam.; style about 0·5 mm. long; stigma capitate. Nutlets ovoid, strongly compressed, about 1·5 mm. long, 1·2 mm. wide, dark shining brown when ripe, margin with a narrow rim, ventral face longitudinally keeled; attachment scar very small, obliquely triangular.

HAB.: Gravelly slopes and screes, sometimes under *Juniperus foetidissima*; 5,500–6,400 ft. alt.; fl. April–June.

DISTR.: Division 2, rare. Spain, Balkans, Turkey, Iraq, Iran, Afghanistan.

2. Khionistra, above 5,500 ft. alt. to summit. *Kotschy*; *Holmboe* 978! *Kennedy* 714! *Davis* 3168! 3513!

NOTES: Records for *Myosotis stricta* Link ex Roemer et Schultes from Buffavento (*Sintenis & Rigo* in Oesterr. Bot. Zeitschr., 32: 51) and Stavrovouni ["Sta. Croce"] (*Kotschy* 201 in Unger et Kotschy, Die Insel Cypern, 281) are undoubtedly misidentifications of *M. ramosissima* Rochel; *Myosotis minutiflora* is evidently confined to the topmost slopes of Khionistra. *Myosotis stricta* Link ex Roemer et Schultes, with which it has been confused, is apparently absent from Cyprus.

4. M. refracta *Boiss.*, Voy. Bot. Espagne, 2: 433, t. 125a (1839); Unger et Kotschy, Die Insel Cypern, 282 (1865); Boiss., Fl. Orient., 4: 240 (1875); Holmboe, Veg. Cypr., 148 (1914); Post, Fl. Pal., ed. 2, 2: 246 (1933); Osorio-Tafall et Seraphim, List Vasc. Plants Cyprus, 87 (1973); Davis, Fl. Turkey, 6: 272 (1978).

TYPE: Spain; "In arenosis regionis alpinae, in monte *Sierra de la Nieve en el Pilar de Tolox* legit amic. Prolongo. Alt. circ. 6500'. Fl. aestate". (G).

Erect or diffuse annual 5–20 (–25) cm. high; stems usually much branched from the base, rarely unbranched, shortly pilose with spreading crispate or hooked hairs; leaves forming a dense basal tuft or rosette, often partly withered at anthesis, narrowly obovate or subspathulate, sessile or obscurely petiolate, 10–15 (–25) mm. long, 3–10 mm. wide, hispidulous with spreading straight or hooked, white, bristly hairs; apex obtuse or rounded, base tapering; cauline leaves sessile, oblong, obtuse or subacute, decreasing upwards; inflorescence ebracteate, at first dense, elongating very conspicuously and becoming lax in fruit; rhachis shortly pubescent with crispate hairs; flowers mostly unilaterally secund, strongly deflexed after anthesis; pedicels very short, usually less than 1 mm. long at anthesis, rather thick, thinly clothed with spreading, straight or hooked hairs, sharply recurved after anthesis; calyx narrowly campanulate or subcylindrical, about 2–2·5 mm. long, 0·8 mm. wide at anthesis, accrescent to 4–5 mm. long in fruit, rather densely clothed with short adpressed, straight, white hairs and longer, patent or deflexed, hooked hairs, 5-lobed almost to the middle, lobes subulate, acuminate, about 0·5 mm. wide at base, erect or very slightly spreading, but not connivent, in fruit; corolla bright blue, minute, tube about 1·5 mm. long, glabrous, lobes spreading, rounded or subtruncate, about 1 mm. long and as wide or wider; stamens inserted about two-thirds way down corolla-tube; filaments very short; anthers oblong-elliptic, about 0·4 mm. long, 0·3 mm. wide, with a conspicuous lingulate appendage, base shortly sagittate; ovary glabrous, about 0·5 mm. diam.; style about 0·3 mm. long; stigma capitate. Nutlets ellipsoid, tapering to apex and base, about 1·5 mm. long, 1 mm. wide, dark shining brown when ripe, margin with a distinct, narrow rim, ventral face keeled; basal attachment scar minute.

HAB.: Gravelly screes and slopes on diabase or serpentine; 4,000–5,000 ft. alt.; fl. March–May.

DISTR.: Division 2, rare. Spain, Balkans, Crete, Palestine eastwards to Afghanistan.

2. Damaskinari ["Ta Maschinari"] above Prodhromos, 1862, *Kotschy* 716a (partly)! Selladhi tou Mavrou Dasous above Spilia, 1962, *Meikle* 2821! Papoutsa, 1979, *Edmondson & McClintock* E 2877!

NOTES: The Cyprus plant is typical *M. refracta*, matching closely with the type illustration.

5. M. sylvatica *Hoffm.*, Deutschl. Fl., 1: 85 (1791); Boiss., Fl. Orient., 4: 237 (1875); Post, Fl. Pal., ed. 2, 2: 245 (1933); Davis, Fl. Turkey, 6: 274 (1978).

TYPE: Germany; "*Hannoverae*" Ehrhart (Beiträge, 5: 176; 1790 sub *Myosotis scorpioides sylvatica* E.).

Erect biennial or perennial up to 50 cm. high; stems usually branched except in starved specimens, rather thinly crispate-hairy with spreading hairs in the lower part, adpressed-hairy above; basal leaves usually withered at anthesis, obovate-spathulate, lamina 1–4 cm. long, 0·5–2 cm. wide, white-hairy, with spreading hairs, apex rounded, base tapering to a petiole sometimes more than 6 cm. long; cauline leaves narrowly oblong, ovate or obovate, sometimes 5 cm. or more long, decreasing in size towards apex of stems, petiolate towards base of stem, sessile above, apex acute or subacute; inflorescence ebracteate, commonly bifurcate; rhachis densely adpressed-pilose; flowers bifarious on spreading, adpressed-pilose pedicels 1–3 mm. long, the pedicels lengthening to 6 mm. or more in fruit; calyx narrow-campanulate, 1·5–2 mm. long, 0·8–1 mm. wide at anthesis, accrescent to 4 mm. long (or more) in fruit, thinly or densely grey-pilose with adpressed straight hairs and spreading or deflexed, curved or hooked (often brownish-tipped) hairs, 5-lobed to half-way or more, the lobes subulate or narrowly oblong, acute or subacute, erect or spreading in fruit; corolla bright blue with a pallid "eye", tube about as long as calyx, glabrous, limb 4–7 mm. diam., flat, divided into 5 rounded lobes 1–1·5 mm. diam.; stamens inserted about two-thirds way down corolla-tube; filaments very short; anthers oblong, about 1 mm. long, 0·5 mm. wide, apex with a recurved lingulate appendage, base shortly sagittate; ovary about 0·6 mm. diam., style 1 mm. long; stigma capitate. Nutlets ovoid, strongly compressed, 1·2–2 mm. long, 1–1·5 mm. wide, shining dark brown with a distinct marginal rim, and a small, impressed attachment scar.

HAB.: Fallow land; 3,750 ft. alt.; fl. April.

DISTR.: Division 2, very rare. Widespread in Europe and western Asia.

2. Platres, 1938, *Kennedy* 1365!

NOTES: This sheet, the total representation of the species from Cyprus, consists of three very young, depauperate specimens, and the identification is published with diffidence, since the material cannot be regarded as adequate for confident determination. *Myosotis sylvatica* is, however, recorded from areas adjacent to Cyprus, and may be more widespread on the island than would appear to be the case. It is also possible that *Kennedy* 1365 may have been a garden escape though the collector does not suggest this.

9. LITHODORA *Griseb.*
Spic. Fl. Rum. et Bith., 2: 85 (1844).
I. M. Johnston in Journ. Arn. Arb., 34: 259–268 (1953).
Lithospermum L., Sp. Plant., ed. 1, 132 (1753) pro parte; Gen. Plant., ed. 5, 64 (1754) pro parte.

Shrubs and subshrubs; flowers sometimes heterostylous, in small, leafy, non-scorpioid, clusters or cymes; calyx 5-lobed almost to base, somewhat accrescent in fruit; corolla blue, red, pink, purple or white, infundibuliform, corolla-lobes rounded, imbricate, sometimes ill-defined; faucal appendages and basal annulus wanting; stamens at one or more levels near apex of corolla-tube; filaments short in long-styled flowers or long in short-styled flowers; anthers included or very slightly exserted, oblong, apex obtuse or retuse, base emarginate, thecae united for the greater part of their length or at least to the middle; style shorter than, or much exceeding, the calyx, filiform, simple or forked at the apex; stigmas small, terminal, included or shortly exserted from corolla-throat. Nutlets commonly 1 (or 2), straight or curved, ovoid, smooth, rugulose or tuberculate, with a ventral keel, circumscissile above the base with a sterile basal portion remaining attached to the torus as a cupular appendage; attachment-scar with a central or oblique, peg-like or pyramidal appendage.

About 7 species chiefly in the Mediterranean region.

1. **L. hispidula** *(Sm.) Griseb.*, Spic. Fl. Rum. et Bith., 2: 85 (1844); 531 (1846); I. M. Johnston in Journ. Arn. Arb., 34: 266 (1953); Davis, Fl. Turkey, 6: 315 (1978).
   Lithospermum hispidulum Sm. in Sibth. et Sm., Fl. Graec. Prodr., 1: 114 (1806), Fl. Graec., 2: 53, t. 162 (1813); Unger et Kotschy, Die Insel Cypern, 279 (1865); Boiss., Fl. Orient., 4: 220 (1875), Suppl., 352 (1888); Holmboe, Veg. Cypr., 148 (1914); Post, Fl. Pal., ed. 2, 2: 241 (1933); Lindberg f., Iter Cypr., 28 (1946); Chapman, Cyprus Trees and Shrubs, 63 (1949); Osorio-Tafall et Seraphim, List Vasc. Plants Cyprus, 87 (1973).

Erect or spreading, much-branched shrub to 1 m. high; twigs at first canescent with a dense indumentum of short, closely adpressed hairs, this outer layer ultimately flaking off to expose a dark, fuscous-brown underlayer; flowering branches persistent and spinescent when dead and leafless; leaves oblong, 5–15 (–25) mm. long, 2–4 (–7) mm. wide, midrib strongly impressed, nervation obscure, lower surface densely canescent with closely adpressed hairs, upper surface sometimes thinly canescent or becoming glabrous and dark green with a sparse covering of spreading or subadpressed bristles, apex obtuse, margins revolute, base tapering to a very short, obscure petiole; inflorescences terminal, forming small 1–6 (or more)-flowered clusters; bracts foliaceous, resembling the upper leaves but usually more acute and sometimes distinctly carinate; flowers sessile or subsessile; calyx divided almost to the base into 5 erect, linear-lanceolate, acute or subacute, somewhat carinate, densely adpressed-strigose lobes 6–8 mm. long, about 1 mm. wide; corolla blue, pink, reddish, white, mauve or purple, infundibuliform or trumpet-shaped with a slender tube and suddenly expanded apex, tube 8–15 mm. long, 5–8 mm. wide at apex, lobes erect or spreading, semicircular, about 3 mm. diam.; stamens inserted at one level near apex of corolla-tube; filaments very short in long-styled flowers, 1 mm. long or more in short-styled flowers; anthers included or exserted, narrowly oblong, 2–2·5 mm. long, 0·5–0·6 mm. wide; ovary glabrous, 4-lobed; style 5–10 mm. long, shorter than calyx (in short-styled plants) or almost twice its length (in long-styled plants), very shortly 2-lobed at apex; stigmas subcapitate. Nutlets ovoid, somewhat incurved, blunt, pallid or fuscous, densely scrobiculate-papillose; attachment scar suborbicular, with a conspicuous, pale, fleshy appendage.

ssp. **versicolor** *Meikle* in Ann. Musei Goulandris, 6: 90 (1983). See App. V, p. 1898.

TYPE: Cyprus; "Prope Klafdea [Klavdhia]. 29. Februario 1880". *Sintenis & Rigo* 102 (K!).

Leaves persistently adpressed-strigose above, with scattered, stouter, tuberculate-based, spreading bristles; inflorescence generally 3-flowered; corolla infundibuliform with erect lobes, pink or reddish or white, sometimes purplish, rarely (if ever) blue. Nutlets pallid.

HAB.: In garigue or under *Pinus brutia* on dry hillsides and rocky ground; sea-level to 3,400 ft. alt.; fl. Febr.–May.

DISTR.: Divisions 1–3, 7, 8, locally very common. Also Turkey and Syria.

1. Lyso, 1913, *Haradjian* 847! Seashore at Lachi (Polis), 1979, *Hewer* s.n.!
2. Khrysorroyiatissa Monastery, 1862, *Kotschy*; Moniatis R., 3,400 ft. alt., 1937, *Kennedy* 715! Platania, 1939, *Lindberg f.* s.n.! Platres, 1937, *Miss Godman* 21! Asinou, 1974, *Meikle*!
3. Stavrovouni, 1862, *Kotschy*, and, 1880, *Sintenis & Rigo*; Klavdhia, 1880, *Sintenis & Rigo* 102! Mazotos, 1905, *Holmboe* 162; Akrotiri, 1939, *Lindberg f.* s.n.; Mosphiloti, 1962, *Meikle*!
7. Pentadaktylos, 1862, *Kotschy*, also, 1880, *Sintenis & Rigo*; Ayios Khrysostomos Monastery, 1862, *Kotschy*; Kantara, 1880, *Sintenis* 102a! also, 1941, *Davis* 2434! Myrtou, 1932, *Syngrassides* 85! Kyrenia, 1933, *Syngrassides* 1363! also, 1938, *C. H. Wyatt* 38! 63! 1943, *Casey* 333! 1956, *G. E. Atherton* 805! 817! St. Hilarion, 1939, *Lindberg f.* s.n.! and, 1963, *Townsend* 63/61! Kythrea, 1937, *Miss Godman* 26! Akanthou, 1940, *Davis* 2026! Sisklipos, 1941, *Davis* 211! Paleosophos, 1968, *Economides* in ARI 1067!
8. Rizokarpaso, 1880, *Sintenis & Rigo* 102b (form with exceptionally narrow, acuminate corolla-lobes)! Koronia, 1933, *Chapman* 10! Yialousa, 1941, *Davis* 2403!

NOTES: Aggregate *Lithodora hispidula* is divisible into three well-marked subspecies.

1. *L. hispidula* (Sm.) Griseb. ssp. *hispidula*. Subshrub usually less than 35 cm. high; flowers mostly solitary at the tips of the branches; corolla blue, trumpet-shaped; leaves subglabrous above with scattered, spreading, tuberculate-based bristles. Crete, Karpathos, Samos, Tilos, Rhodes, W. Turkey.

2. *L. hispidula* (Sm.) Griseb. ssp. *versicolor* Meikle Subshrub to 1 m. high; flowers usually in threes at the tips of the branches; corolla infundibuliform, pink, red or white, sometimes purple or mauve; leaves thinly clothed above with persistent, short, adpressed hairs and longer, scattered, spreading, tuberculate-based bristles. Cyprus, S. Turkey, Amanus, N. Syria.

3. *L. hispidula* (Sm.) Griseb. ssp. *cyrenaica* (Pamp.) Brullo et Furnari in Webbia, 34: 167 (1979) (*Lithospermum hispidulum* Sm. var. *cyrenaicum* Pamp. in Archivio Bot., 12: 41; 1936). Robust subshrub to 1 m. or more; flowers 4–6 (or more) in terminal heads; corolla dark blue, trumpet-shaped; leaves larger than in ssp. *hispidula* or ssp. *versicolor*, subglabrous above with scattered, short adpressed hairs and longer, spreading, tuberculate-based bristles. Libya.

In the few fruiting specimens available, examination suggests that the nutlets of *L. hispidula* ssp. *hispidula* are fuscous; those of ssp. *versicolor* pallid, a distinction given further weight by the reference in Fl. Turkey, 6: 315 to a specimen from Samos "nr Despote Vrysis, *Gathorne-Hardy* 229 (form with black nutlets)". But the value of the distinction cannot be adjudged without seeing more material.

While the illustration in Flora Graeca, 2: t. 162, which may be regarded as the type of *L. hispidula*, is almost certainly the Rhodes plant, as stated, the Sibthorp duplicate at Kew, though fragmentary and unsatisfactory, corresponds much more closely to ssp. *versicolor*, and comes, I suspect, from Cyprus and not from Rhodes.

The relationships between all three subspecies are close, but ssp. *versicolor* would appear to have closer affinities with ssp. *cyrenaica* than with ssp. *hispidula*.

10. BUGLOSSOIDES *Moench*

Meth., 418 (1794).

I. M. Johnston in Journ. Arn. Arb., 35: 38–44 (1954) partim quoad sect. *Eubuglossoides*.

Erect or decumbent, strigillose or hispidulous annuals; roots containing a purple dye; leaves entire, obscurely veined; inflorescences terminal, usually unbranched, bracteate, elongating conspicuously and becoming racemose after anthesis; calyx divided to near base into 5 narrow lobes, accrescent and sometimes becoming oblique or asymmetric in fruit through the unilateral swelling of the pedicel; corolla blue, pink or milky-white, narrowly infundibuliform or hypocrateriform, distinctly 5-lobed with rounded or truncate lobes; faucal swellings absent or obscure, the throat with 5 longitudinal bands of hairs extending downwards from the base of the corolla-lobes to the tips of the anthers; stamens included; filaments equal, inserted at different levels towards base of corolla-tube; anthers with a short apical appendage; style short, included; stigmas 2, sometimes close together and apparently capitate, or terminating short style-branches. Nutlets 4 (or fewer by abortion), erect or divergent, smooth or tuberculate-rugose, ovoid or pyriform, sometimes with 2 lateral swellings; attachment scar usually conspicuous, flattish or depressed.

Two (or possibly three) species in Europe, the Mediterranean region and western Asia.

Corolla-limb minute, about 2 mm. diam., blue; inflorescence hairs yellowish; flowers crowded at anthesis, with bracts neatly arranged in two rows; nutlets with distinct lateral swellings

        1. B. tenuiflora

Corolla-limb 3–4 mm. (or more) diam., whitish or blue; inflorescence hairs white or greyish; flowers not crowded, nor the bracts neatly arranged; nutlets without obvious lateral swellings:

  Fruiting pedicels not noticeably incrassate; base of calyx truncate or shallowly convex; corolla-limb milky-white:

    Plants erect or suberect, 15–30 (–50) cm. high; calyx often strongly accrescent after anthesis; nutlets 3 mm. long - - - - - **2. B. arvensis** ssp. **arvensis**

    Plants prostrate or decumbent, usually less than 10 cm. high; calyx not much accrescent after anthesis; nutlets 2 mm. long - - - - **2. B. arvensis** ssp. **sibthorpiana**

  Fruiting pedicels much incrassate, merging with base of calyx to form an oblique, obconic hypanthium; corolla-limb usually blue - - - **2. B. arvensis** ssp. **gasparrinii**

1. B. tenuiflora (*L. f.*) *I. M. Johnston* in Journ. Arn. Arb., 35: 42 (1954); Davis, Fl. Turkey, 6: 316 (1978).

    Lithospermum tenuiflorum L. f., Suppl. Plant., 130 (1781); Sibth. in Walpole, Travels, 15 (1820); Unger et Kotschy, Die Insel Cypern, 279 (1865); Boiss., Fl. Orient., 4: 217 (1875); Sintenis in Oesterr. Bot. Zeitschr., 31: 193, 392 (1881), 32: 122 (1882); Holmboe, Veg. Cypr., 148 (1914); Post, Fl. Pal., ed. 2, 2: 240 (1933); Osorio-Tafall et Seraphim, List Vasc. Plants Cyprus, 37 (1973).

    [*L. incrassatum* (non Guss.) Lindberg f., Iter Cypr., 28 (1946).]

    TYPE: "in Aegypto. *La Tourette*"cult. Uppsala (LINN).

Erect annual 8–30 (–50) cm. high; stems commonly unbranched except towards apex, frequently tinged violet or purple, thinly or densely adpressed-strigillose; basal leaves 2–7 cm. long, 0·5–1 cm. wide, thinly adpressed-strigillose on both surfaces, apex rounded, base tapering to an indistinct petiole; cauline leaves sessile and often somewhat amplexicaul, narrowly oblong or oblanceolate, 1–3·5 cm. long, 0·3–0·7 cm. wide, apex obtuse or subacute; inflorescences simple, terminal, or several arising from the axils of the upper leaves, erecto-patent, at first very dense with the protruding bracts often arranged bifariously in a regular, herringbone pattern, becoming elongate and rather lax after anthesis; bracts foliaceous, oblong-lanceolate, subacute, clothed with adpressed, yellowish bristles; flowers subsessile or very shortly pedicellate; calyx campanulate, densely clothed with yellowish bristles, somewhat accrescent after anthesis, base shallow, cupular, lobes erect, blunt, linear, about 5 mm. long, 0·5 mm. wide; corolla-tube about 8 mm. long, narrowly cylindrical with a slighty expanded base, whitish above, blue towards base, thinly strigose externally, glabrous within except for 5 longitudinal pilose bands terminating apically in obscure faucal swellings, basal annulus well marked, lobulate; corolla-lobes suberect, blue, oblong, rounded or truncate, about 1 mm. long, 0·6 mm. wide; stamens inserted about 1 mm. above base of corolla-tube; filaments very short; anthers oblong, about 0·6 mm. long, 0·3 mm. wide, base shortly cordate, apex with a small lingulate appendage; ovary glabrous, 4-lobed (or with fewer lobes through abortion); style about 1·4 mm. long; stigma obscurely 2-lobed, quasi-capitate. Nutlets ovoid, about 2 mm. long, 1·5 mm. wide, whitish or dull brown, tuberculate, rather fragile, with distinct lateral swellings, and a blunt, slightly incurved, median beak, ventral face lightly keeled; attachment scar suborbicular or 3-angled, flat, about 0·6 mm. diam.

HAB.: Cultivated and fallow fields,.waste ground; sea-level to 2,500 ft. alt.; fl. Jan.–May.

DISTR.: Divisions 2–8. Greece and eastern Mediterranean region eastwards to Afghanistan; North Africa from Algeria to Egypt.

2. Gourri, 1950, *Chapman* 312 (partly)!
3. Stavrovouni, 1787, *Sibthorp*; S. of Lefkara, 1974, *Meikle* 5004!
4. Larnaca, 1880, *Sintenis*; Phaneromene near Larnaca, 1905, *Holmboe* 149.
5. Near Kythrea, 1880, *Sintenis & Rigo* 104! Near Ayios Varnavas Monastery, 1948, *Mavromoustakis* s.n.! Pera, 1949, *Casey* 304! Athalassa, 1957, *Merton* 2900! Between Mia Milea and Koutsovendis, 1974, *Meikle & Osorio-Tafall* in *Meikle* 4073! Ayia, 1977, *A. Della* in ARI 1684!
6. Nicosia, 1905, *Holmboe* 266! also, 1934, *Syngrassides* 1345! and, 1940, *Davis* 2137!
7. Kyrenia, 1950, *Casey* 1008! Ayios Epiktitos, 1951, *Casey* 1127! Templos, 1956, *G. E. Atherton* 1203! Kambyli, 1956, *Merton* 2568!
8. Ayios Theodhoros, 1947, *Mavromoustakis* s.n.!

2. B. arvensis (*L.*) *I. M. Johnston* in Journ. Arn. Arb., 35: 42 (1954); Davis, Fl. Turkey, 6: 316 (1978).

   Lithospermum arvense L., Sp. Plant., ed. 1, 132 (1753); Unger et Kotschy, Die Insel Cypern, 279 (1865); Boiss., Fl. Orient., 4: 216 (1875); H. Stuart Thompson in Journ. Bot., 44: 335 (1906); Holmboe, Veg. Cypr., 148 (1914); Post, Fl. Pal., ed. 2, 2: 240 (1933); Osorio-Tafall et Seraphim, List Vasc. Plants Cyprus, 87 (1973).

Erect or decumbent annual (2–) 6–30 (–50) cm. high; stems branched from near the base or unbranched, thinly or densely adpressed-strigillose, sometimes tinged purplish; basal leaves 1·5–7 cm. long, 0·5–1 cm. wide, thinly adpressed-strigillose on both surfaces, apex blunt or rounded, base tapering to an indistinct petiole; cauline leaves sessile or somewhat amplexicaul, narrowly oblong or lanceolate, up to 10 cm. long, 2 cm. wide, becoming smaller, narrower and more acute towards the tips of the stems; inflorescences simple, terminal, or several arising from the axils of the uppermost leaves, erecto-patent, becoming lax very soon after anthesis, the protruding bracts not regularly arranged; bracts foliaceous, oblong-lanceolate, subacute, clothed with adpressed, whitish bristles; flowers at first very shortly pedicellate, the pedicel lengthening, and sometimes becoming markedly and asymmetrically incrassate in fruit; calyx campanulate, densely clothed with pallid bristles, base shallow, cupular, lobes erect, blunt or acute, linear-subulate, at first about 4–6 mm. long, 0·5 mm. wide, accrescent to 8–15 mm. long, 1–2 mm. wide and becoming erecto-patent or spreading in fruit; corolla blue or white, occasionally pink, infundibuliform, or almost hypocrateriform; tube as long as calyx or a little longer, narrowly cylindrical with a slightly expanded base, thinly strigose externally, glabrous within except for 5 longitudinal pilose bands terminating apically in obscure faucal swellings, basal annulus well marked, lobulate; corolla-lobes suberect, oblong, rounded or subtruncate, 1·5–2·5 mm. long, 1–2 mm. wide; stamens inserted about 1 mm. above base of corolla-tube; filaments very short; anthers narrowly oblong, about 1 mm. long, 0·4 mm. wide, base shortly cordate, apex with a small lingulate appendage; ovary glabrous, usually 4-lobed; style about 1 mm. long; stigma capitate, sometimes with a short, conical, sterile, apical appendage. Nutlets ovoid, 2–3 mm. long, 1·5–2 mm. wide, whitish or pale brown, with a hard, flinty pericarp, minutely or coarsely rugulose-tuberculate, ventral face keeled, lateral swellings absent, beak short, rather blunt, slightly incurved; attachment scar basal or slightly tilted upwards towards dorsal face, suborbicular or 3-angled, flattish or slightly convex, about 1–1·5 mm. diam.

ssp. **arvensis**
TYPE: "*in Europae agris & arvis*" (LINN).
Erect or suberect annual usually 15–30 (–50) cm. high; calyx often strongly accrescent after anthesis, the lobes attaining 1–1·5 cm. in length; corolla milky-white, the base of the tube often livid-blue; calyx in fruit not noticeably asymmetrical, the pedicel not, or very slightly incrassate. Nutlets usually large, about 3 mm. long, coarsely tuberculate.

HAB.: Cultivated fields, waste ground, roadsides; near sea-level to 4,500 ft. alt.; fl. Jan.–May.

DISTR.: Divisions 1–8. Widespread as a weed in Europe and the Mediterranean region eastwards to China and India, also as a recent introduction in most temperate areas of the world.
1. Neokhorio, 1962, *Meikle* 2213!
2. Prodhromos, 1862, *Kotschy* 890; Platres, 1938, *Kennedy* 1371! 1372! Chakistra, 1962, *Meikle* 2726!
3. Phasouri, 1957, *Poore* 44!
4. Larnaca, 1880, *Sintenis*; Ormidhia, 1880, *Sintenis*; the Phaneromene, 1905, *Holmboe* 118; Paralimni, 1940, *Davis* 2053! Larnaca road near Nicosia, 1940, *Davis* 2136!
5. Syngrasis, 1936, *Syngrassides* 1003! Near Salamis, 1948, *Mavromoustakis* s.n.! Athalassa, 1957, *Merton* 2899! Ayia, 1977, *A. Della* in ARI 1683!
6. Nicosia, 1934, *Syngrassides* 709! 1346! Near Ayia Marina, 1957, *Merton* 2884! Near Dhiorios, 1962, *Meikle* 2337! Kato Kopia, 1970, *A. Genneou* in ARI 1382!
7. Kyrenia, 1949, *Casey* 319! Templos, 1956, *G. E. Atherton* 1110! Kambyli, 1956, *Merton* 2569!
8. Gastria, 1941, *Davis* 2415!

ssp. **sibthorpiana** (*Griseb.*) *R. Fernandes* in Bot. Journ. Linn. Soc., 64: 379 (1971) et in Tutin et al., Fl. Europ., 3: 87 (1972).
  Lithospermum splitgerberi Guss., Fl. Sic. Syn., 1: 217 (1842).
  L. sibthorpianum Griseb., Spic. Fl. Rum. et Bith., 2: 86 (1844); Boiss., Fl. Orient., 4: 216 (1875); Post, Fl. Pal., ed. 2, 2: 240 (1933); Rechinger f. in Arkiv för Bot., ser. 2, 1: 429 (1950); Osorio-Tafall et Seraphim, List Vasc. Plants Cyprus, 87 (1973).
  L. arvense L. var. *sibthorpianum* (Griseb.) Hal., Consp. Fl. Graec., 2: 349 (1902).
  L. arvense L. var. *splitgerberi* (Guss.) Fiori in Fiori et Paoletti, Fl. Anal. Ital., 2: 369 (1902).
  L. arvense L. ssp. *sibthorpianum* (Griseb.) Holmboe, Veg. Cypr., 148 (1914); Lindberg f., Iter Cypr., 28 (1946).
  [*L. tenuiflorum* (non L. f.) Sm. in Sibth. et Sm., Fl. Graec. Prodr., 1: 113 (1806), Fl. Graec., 2: 50, t. 159 (1813); Poech, Enum. Plant. Ins. Cypr., 25 (1842).]
  [*L. incrassatum* (non Guss.) Holmboe, Veg. Cypr., 148 (1914) pro parte quoad *Holmboe* 979.]

TYPE: "In insulâ Cypro", figured in *Fl. Graeca*, 2: t. 159 (lectotype!).

Prostrate or decumbent annual, rarely more than 10 cm. high; stems usually numerous, radiating from the apex of the rootstock; calyx not much accrescent after anthesis, the lobes generally less than 1 cm. long and spreading in fruit; corolla milky-white, the base of the tube often inky-blue; calyx in fruit not noticeably asymmetrical, the pedicel not incrassate. Nutlets small, about 2 mm. long, pitted, scrobiculate or minutely rugulose-tuberculate.

HAB.: Gravelly mountainsides and summits at high altitudes; (3,400–) 5,000–6,400 ft. alt.; fl. March–June.

DISTR.: Division 2, locally common and often forming dense colonies. Also? Spain, Corsica, Italy, Sicily, Greece, Aegean Islands, Turkey, Syria, Lebanon.

2. Common on the uppermost slopes and summit of Khionistra, becoming rare below 5,000 ft. alt., but recorded from Xerokolymbos, 4,300 ft. alt. (*Kennedy* 722!) and Platres, 3,400 ft. alt. (*Kennedy* 1370!). *Sintenis & Rigo* 105 (one sheet at Kmislabelled "In m. Pentadactylos" and identified as *Lithospermum incrassatum* Guss.)! *Haradjian* 460! *Kennedy* 717–723! 1369! 1370! *Lindberg f.* s.n.! *Davis* 1813! 3166! *Merton* 2374! *N. Macdonald* 193! *D. P. Young* 7435! *Meikle* 2777!

NOTES: Holmboe (Veg. Cypr., 148) is mistaken in supposing that the plant figured as *Lithospermum tenuiflorum* in *Flora Graeca*, 2: t. 159 is the same as that collected by Sibthorp on 13 April 1787, at the foot of Stavrovouni (Walpole, Travels, 15). The latter was almost certainly correctly identified by Sibthorp, while the plant figured in *Flora Graeca* was most probably found on 1 May, 1787, on the summit of Khionistra.

ssp. **gasparrinii** (*Heldr. ex Guss.*) *R. Fernandes* in Bot. Journ. Linn. Soc., 64: 379 (1971) et in Tutin et al., Fl. Europ., 3: 87 (1972).
 Lithospermum incrassatum Guss., Ind. Sem. Hort. Boccad., 1826: 6 (1826), Fl. Sic. Prodr., 1: 211 (1827); Unger et Kotschy, Die Insel Cypern, 279 (1865); Boiss., Fl. Orient., 4: 217 (1875); Holmboe, Veg. Cypr., 148 (1914) pro parte; Lindberg f., Iter Cypr., 28 (1946); Osorio-Tafall et Seraphim, List Vasc. Plants Cyprus, 87 (1973).
 L. arvense L. var *minus* [*minor*] Ten., Fl. Nap., 3: 175 (1824–29).
 L. gasparrinii Heldr. ex Guss., Fl. Sic. Syn., 1: 217 (1842).
 L. arvense L. var. *coerulescens* DC., Prodr., 10: 74 (1846).
 L. incrassatum Guss. var. *gasparrinii* (Heldr. ex Guss.) Cesati et al., Compend. Fl. Ital., 377 (1876).
 L. incrassatum Guss. ssp. *gasparrinii* (Heldr. ex Guss.) Nyman, Consp. Fl. Europ., 518 (1881).
 Buglossoides incrassata (Guss.) I. M. Johnston in Journ. Arn. Arb., 35: 43 (1954); Davis, Fl. Turkey, 6: 317 (1978).
TYPE: Sicily; "In aridis montosis; *Madonie* (Gasparr. Heldr.): *Busambra*" (NAP).

Erect or decumbent annual 6–30 cm. high; stems few or numerous, sometimes radiating from apex of rootstock; calyx not much accrescent after anthesis, the lobes usually less than 1 cm. long, erect or suberect in fruit; corolla usually blue or occasionally milky-white with a blue tube; calyx becoming asymmetrical in fruit, the pedicel strongly incrassate especially on the adaxial face, adpressed to the rhachis, and resembling an obconical inferior ovary. Nutlets small, about 2 mm. long, generally rugulose-tuberculate.

HAB.: Dry banks and rocky hillsides; 3,000–5,000 ft. alt.; fl. March–May.

DISTR.: Division 2, uncommon. Widespread in the Mediterranean region (usually at high altitudes) and eastwards to Afghanistan.

2. Between Prodhromos and Tris Elies, 1862, *Kotschy*; near Lemithou, 1862, *Kotschy* 878; below Tripylos near a group of Cedars, 1930, *C. Norman* 435! Askas, near Palekhori in Hazel groves, 1957, *Merton* 3137! Livadhi, near the Cedar forest, 1962, *Meikle* 2676! Near Polystipos, in a Hazel grove, 1974, *Meikle* 4044!

NOTES: All the specimens examined are immature and unsatisfactory for certain identification. The characteristic swelling of the pedicel becomes more conspicuous as the fruits mature and is not evident in juvenile plants. In general, plants with incrassate fruiting pedicels are also distinguished by having blue corollas, but this is not always so, and the two colour variants are here united. The existence of blue-flowered *Buglossoides arvensis* without thickened pedicels in the Baltic area (S. Sweden, Aland Islands, W. Finland) weakens the distinction between taxa which are frequently given specific rank. I have followed Dr. R. Fernandes in preferring to regard *B. incrassata* and *Lithospermum sibthorpianum* as subspecies of *B. arvensis*.

11. NEATOSTEMA *I. M. Johnston*
Journ. Arn. Arb., 34: 2, 5 (1953).

Erect hispid annual; flowers small, yellow, sometimes cleistogamous, in dense terminal, elongating, bracteate cymes; calyx 5-lobed, the lobes connivent in fruit; corolla narrowly infundibuliform, symmetrical, somewhat hairy externally, throat with a broken or continuous band of hairs, but without faucal scales, tube thinly hairy within, with a basal, 10-lobed, nectariferous annulus; stamens 5, inserted just above annulus; filaments very short; anthers oblong, apex obtuse, base obtuse or subcordate; style very short; stigma capitate, obscurely 2-lobed. Nutlets weakly incurved, tuberculate; attachment scar small, oblique.

One species in Mediterranean region, western Asia, Atlantic Islands.

1. **N. apulum** (*L.*) *I. M. Johnston* in Journ. Arn. Arb., 34: 6 (1953); Davis, Fl. Turkey, 6: 318 (1978).

    Myosotis apula L., Sp. Plant., ed. 1, 131 (1753).
    Lithospermum apulum (L.) Vahl, Symb. Bot., 2: 33 (1791); Sibth. et Sm., Fl. Graec. Prodr., 1: 113 (1806), Fl. Graec., 2: 49, t. 158 (1813); Unger et Kotschy, Die Insel Cypern, 279 (1865); Boiss., Fl. Orient., 4: 218 (1875); Holmboe, Veg. Cypr., 148 (1914); Post, Fl. Pal., ed. 2, 2: 241 (1933); Osorio-Tafall et Seraphim, List Vasc. Plants Cyprus, 87 (1973).
    TYPE: "*in* Italia, Hispania, Narbona" (LINN).

Erect hispid annual, (5–) 8–15 (–20) cm. high; stems unbranched or branched from the base, terete, adpressed-strigillose; basal leaves commonly persisting until anthesis, narrowly oblanceolate, 1·5–3 cm. long, 0·2–0·3 cm. wide, tuberculate-strigillose above, strigillose below, apex obtuse or subacute, base tapering to an indistinct, slightly expanded petiole, margins narrowly recurved, midrib prominent below; cauline leaves sessile, linear or very narrowly oblong, acute or subacute, 2–3·5 cm. long, 0·2–0·3 cm. wide; inflorescences terminal, usually much branched; bracts bifariously arranged in a regular herringbone pattern, lanceolate, subacute, foliaceous, about 7–10 mm. long, 2–3 mm. wide, conspicuously hispidulous; flowers subsessile or very shortly pedicellate; calyx densely hispidulous, narrowly campanulate, divided almost to base into 5 lanceolate lobes about 4 mm. long, 0·6 mm. wide; corolla bright yellow, tube straight, about 4 mm. long, 1 mm. wide, thinly hairy externally, lobes suberect, oblong, obtuse, about 1·5 mm. long, 1 mm. wide, minutely glandular-papillose; interior of tube thinly hairy, with 5 conspicuous tufts of hairs at the throat (the tufts sometimes coalescing into a band); basal annulus very prominent, forming 10, blunt, barbate lobes; stamens inserted just above basal annulus; filaments very short; anthers oblong, about 0·8 mm. long, 0·4 mm. wide, without any obvious apical appendage; ovary glabrous, 4-lobed, about 0·8 mm. wide; style straight, about 0·5 mm. long; stigma capitate, very obscurely 2-lobed. Nutlets pale brown, ovoid, about 2 mm. long, 1·5 mm. wide, lateral margins distinctly gibbous, median beak somewhat incurved, pericarp tuberculate, the tubercles arranged dorsally in 2 median, longitudinal rows; attachment scar oblique, horizontally subelliptic, about 1 mm. wide, flat.

    HAB.: Rocky and gravelly hillsides in garigue; roadsides, cultivated and fallow fields, on calcareous and igneous rocks; sea-level to 1,800 ft. alt.; fl. Febr.–May.

    DISTR.: Divisions 1, 3, 5–8. General distribution that of genus.

1. Ayia Nikola (Akamas), 1941, *Davis* 3329 !

3. Stavrovouni, 1862, *Kotschy* 201a; also, 1937, *Syngrassides* 1471! 1957, *Merton* 3154! 1974, *Meikle* 4060! Episkopi, 1960, *N. Macdonald* 38!
5. Between Mia Milea and Koutsovendis, 1974, *Osorio-Tafall & Meikle* in *Meikle* 4072!
6. Near Nicosia, 1905, *Holmboe* 295!
7. Pentadaktylos, 1880, *Sintenis & Rigo* 103! 6 m. E. of Kyrenia, 1949, *Casey* 380!
8. Ronnas Bay near Rizokarpaso, 1941, *Davis* 2387! Kantara Castle, 1941, *Davis* 2456! Apostolos Andreas Forest, 1957, *Merton* 2923!

NOTES: I. M. Johnston (*loc. cit.*) mentions the frequent occurrence of cleistogamous flowers in this species, with "tiny obconic corollas which eventually fall off unopened". All the Cyprus material examined appears to have normal, chasmogamous flowers.

Arnebia hispidissima (Sieber ex Lehm.) DC. (*Anchusa hispidissima* Sieber ex Lehm.) with densely white-hispid stems and leaves, and yellow, long-tubular flowers in elongating, bracteate cymes, superficially resembles *Neatostema apulum* (L.) I. M. Johnston, but differs, apart from indumentum, in the insertion of the stamens high up in the corolla-tube. It has a wide distribution in arid areas of Africa and western Asia, as far east as India, and is recorded from Cyprus by Osorio-Tafall et Seraphim, (List Vasc. Plants Cyprus, 87; 1973). In the absence of a specimen, or more detailed information, the record must, at least for the present, be set aside as unsatisfactory.

12. ECHIUM *L.*

Sp. Plant., ed. 1, 139 (1753).
Gen. Plant., ed. 5, 68 (1754).

G. Klotz in Wiss. Zeitschr. Martin-Luther-Univ. Halle-Witt., IX/3: 363–378 (1960); XI/2: 293–302 (1962); XI/5: 703–711 (1962); XI/9: 1087–1104 (1962); XII/2: 137–142 (1963).

Annual, biennial or perennial, setose or hispid herbs, often with purple-staining roots; basal leaves commonly forming a rosette, often withering before anthesis, cauline leaves numerous, frequently sessile and somewhat amplexicaul; flowers usually very numerous in compound, spicate, thyrsoid or paniculate, bracteate cymes; calyx divided almost to base into 5 narrow, somewhat accrescent lobes; corolla narrowly or widely infundibuliform, ± zygomorphic, blue, purple, red, pink, yellow or whitish, sometimes changing colour as it develops; faucal scales or swellings absent; basal annulus present, sometimes represented by tufts of hairs; stamens included or exserted, attached to the tube at varying levels; style exserted; stigma obscurely or conspicuously 2-lobed. Nutlets 4, ovoid or subglobose, sometimes angled, irregularly rugose, with a flattish attachment scar.

About 50 species in Europe, Mediterranean region, western Asia and Atlantic Islands.

Stamens exserted from corolla:
  Inflorescence narrowly cylindrical or spiciform, usually unbranched; corolla pink or whitish
                                                                    1. E. glomeratum
  Inflorescence pyramidal or irregularly branched, not narrowly cylindrical:
    Corolla 2–3 cm. long, 1–1·5 cm. wide at mouth, glabrous or hispid only along the nerves
      externally   -   -   -   -   -   -   -   -   -   -   -   **4. E. plantagineum**
    Corolla less than 1·5 cm. long and 1 cm. wide at mouth, pubescent or hispid all over:
      Plant biennial; whole inflorescence conical or pyramidal; flowers pale blue or whitish
                                                                    2. E. italicum
      Plant perennial with a woody rootstock; whole inflorescence irregularly branched;
        flowers red or purple   -   -   -   -   -   -   -   -   **3. E. angustifolium**
Stamens included in corolla; corolla usually less than 1 cm. long, violet-blue or mauve; stems
  prostrate or sprawling   -   -   -   -   -   -   -   -   -   **5. E. arenarium**

1. E. glomeratum *Poir.* in Lam., Encycl. Méth., 8: 670 (1808); Boiss., Fl. Orient., 4: 206 (1875); de Coincy in Morot, Journ. de Bot., 16: 110 (1902); Post, Fl. Pal., ed. 2, 2: 235 (1933); A. K. Jackson in Kew Bull., 1936: 16 (1936); Osorio-Tafall et Seraphim, List Vasc. Plants Cyprus, 87 (1973); Davis, Fl. Turkey, 6: 321 (1978).
    [*E. italicum* (non L.) Lindberg f., Iter Cypr., 28 (1946) pro parte]
    TYPE: "Cette plante a été recueillie en Syrie par M. de Labillardière" (P).

Erect, commonly unbranched biennial, up to 1 m. or a little more in height; stem shallowly sulcate, densely hispid with long, spreading, white bristles intermixed, at least towards the apex of the stem, with much shorter, subadpressed pubescence; basal leaves forming a loose rosette, persisting until anthesis, narrowly oblong-oblanceolate, 20–30 cm. long, 3–4 cm. wide, rather densely adpressed-strigose above and below, or sometimes pustulate-strigose, especially on the undersurface, apex acute or acuminate, base tapering, but scarcely petiolate; cauline leaves similar, but shorter, the uppermost becoming bract-like, sometimes with a dilated, semi-amplexicaul base; inflorescence rigid, narrowly cylindrical (or poker-like), sometimes more than 50 cm. long, 2–7 cm. diam.; partial inflorescences short, condensed, sessile in the axils of the bract-like leaves, commonly geminate, elongating a little after anthesis; flowers crowded, sessile or subsessile; bracts ovate-acuminate, imbricate, 5–7 mm. long, 3–4 mm. wide, densely white-hispid, especially along margins and midrib, with an additional indumentum of short pubescence; calyx narrowly campanulate, densely hispid, lobes narrowly oblong or linear, subacute, 5–6 mm. long, 1·2 mm. wide, erect, not much accrescent in fruit; corolla pink or whitish with mauve blotches or staining, obliquely infundibuliform, tube 5–7 mm. long, slightly curved, hairy externally, glabrous internally, lobes erect, oblong-suborbicular, blunt, about 3 mm. long, 2–2·5 mm. wide, hairy externally, basal annulus rather obscure, of 10, small, oblong, pilose lobes; stamens inserted at slightly different levels about half-way down corolla-tube; filaments reddish, filiform, curved, about 10–12 mm. long, much exserted from corolla, glabrous or sometimes thinly pilose; anthers oblong, about 0·6 mm. long, 0·5 mm. wide; ovary glabrous, deeply 4-lobed, about 1 mm. diam.; style about 15 mm. long, tapering to a 2-lobed apex, pilose in the lower half; stigmas short, terminating the stylar lobes. Nutlets ovoid, about 3 mm. long, 2·5 mm. wide, pale brown, tuberculate-rugose, with the sides gibbous and flanking a slightly incurved median beak which is distinctly keeled dorsally; attachment scar large, flat, about 2 mm. diam.

HAB.: Roadsides and rocky, calcareous hillsides; 500–1,000 ft. alt.; fl. May–June.

DISTR.: Division 7, very rare. S.E. Turkey, Syria, Lebanon and Palestine.

7. Lapithos, 1880, *Sintenis & Rigo* 623! Same locality, 1939, *Lindberg f.* s.n.!

NOTES: So closely related to *E. boissieri* Steud. (*E. pomponium* Boiss.) from Spain, Portugal and western N. Africa, that one suspects this latter should be regarded rather as a subspecies of *E. glomeratum*, differing chiefly in its larger flowers and distinctly pectinate-hispid bracts. De Coincy (*loc. cit.*) draws attention to hairs at the base of the filaments and on the interior of the corolla-tube in *E. boissieri*, remarking that in *E. glomeratum* the filaments and interior of the corolla-tube are glabrous. On comparing material of both species I am not satisfied that these distinctions, nor those based on the shape of the corolla-lobes, provide reliable distinctions. It is, however, true that the uppermost leaves in *E. boissieri* tend to be dilated and subamplexicaul at the base, though the distinction does not always hold good.

2. E. italicum L., Sp. Plant., ed. 1, 139 (1753); Boiss., Fl. Orient., 4: 205 (1875); H. Stuart Thompson in Journ. Bot., 44: 335 (1906); Holmboe, Veg. Cypr., 150 (1914); Post, Fl. Pal., ed. 2, 2: 235 (1933); Lindberg f., Iter Cypr., 28 (1946) pro parte; Osorio-Tafall et Seraphim, List Vasc. Plants Cyprus, 87 (1973); Davis, Fl. Turkey, 4: 321 (1978).

E. pyramidatum DC., Prodr., 10: 23 (1846) pro parte; Unger et Kotschy, Die Insel Cypern, 279 (1865).

TYPE: "*in* Anglia, Italia".

Robust, erect biennial, 40–70 (–90) cm. high; stem angled, usually simple, densely hispid with spreading white bristles; basal leaves usually persistent, at least until anthesis, forming an irregular rosette, oblanceolate, sessile or subsessile, 16–25 cm. long, 1·5–5 cm. wide, densely white-hispid, apex acute or subacute, base tapering, nervation obscure, cauline leaves linear-lanceolate, 5–17 cm. long, 0·5–2 cm. wide; inflorescences arising from the axils of all but the lowermost leaves, inflorescence-branches becoming progressively shorter upwards and giving the whole structure a pyramidal or conical appearance; bracts foliaceous, narrowly lanceolate or linear-subulate, 0·5–1·5 cm. long, 0·2–0·4 cm. wide, densely white-hispid or becoming yellowish with age (at least in the herbarium); flowers very numerous, subsessile; calyx narrowly campanulate, divided to the base into 5 linear, subacute, densely hispid lobes, the adaxial about 10 mm. long, 1 mm. wide, the abaxial about 8 mm. long; corolla pale blue, pinkish or whitish, narrowly infundibuliform, thinly hispid externally, glabrous within, tube 9–10 mm. long, lobes rounded, subequal, erect, about 2 mm. diam.; basal annulus of 10 small, oblong, pilose lobes; stamens long-exserted, declinate; filaments 13–17 mm. long, glabrous; anthers oblong, about 0·7 mm. long, 0·5 mm. wide; ovary about 1 mm. diam., glabrous, somewhat rugulose, deeply 3–4-lobed, style slender, pilose, about 17 mm. long, divided at apex into 2, filiform, pointed lobes about 2 mm. long; stigmas minute, apical. Nutlets ovoid, about 3 mm. long, 2·5 mm. wide, pale greyish-brown, coarsely tuberculate-rugose, with the sides gibbous and flanking a slightly incurved median beak which is distinctly keeled dorsally; attachment scar flat, about 1·5 mm. diam.

HAB.: Roadsides and vineyards on chalky ground; 800–3,000 ft. alt.; fl. April–June.

DISTR.: Divisions 1–3, locally common. Central and southern Europe, Mediterranean region, western Asia.

1. Near Arodhes ["Arora"], 1862, *Kotschy* 667; Stroumbi, 1913, *Haradjian* 719! and, 1934, *Syngrassides* 848! Ayios Neophytos, 1939, *Lindberg f.* s.n.
2. Mandria, 1955, *P. Paolides* in *Kennedy* s.n.! also, 1970, *A. Genneou* in ARI 1515!
3. Episkopi, 1905, *Holmboe* 692; Trimiklini, 1941, *Davis* 3135! Kannaviou, 1962, *Meikle* 2647!

E. VULGARE *L.* (Sp. Plant., ed. 1, 139; 1753), a robust biennial, with spiciform, simple or branched inflorescences of rather large, bright blue, strongly zygomorphic flowers, and with the corolla-tube thinly pilose externally, is recorded for Cyprus in Osorio-Tafall et Seraphim, List Vasc. Plants Cyprus, 88 (1973). *E. vulgare* is absent from areas adjacent to Cyprus, and its occurrence there must be regarded as doubtful. The record is most probably based on a misidentification of the widespread *E. angustifolium* Mill. (*E. sericeum* Vahl).

3. **E. angustifolium** *Mill.*, Gard. Dict., ed. 8, no. 6 (1768); Lacaita in Journ. Linn. Soc., 44: 437 (1919); Post, Fl. Pal., ed. 2, 2: 236 (1933); Osorio-Tafall et Seraphim, List Vasc. Plants Cyprus, 87 (1973); Davis, Fl. Turkey, 6: 323 (1978).
  E. hispidum Sm. in Sibth. et Sm., Fl. Graec. Prodr., 1: 125 (1806), Fl. Graec., 2: 68, t. 181 (1813) non Burm. f. nec Thunb. nom. illeg.
  E. elegans Lehm., Asper., 459 (1818).
  E. sibthorpii Roemer et Schultes, Syst. Veg., 4: 26 (1819).
  E. sericeum Vahl var. *hispidum* Boiss., Fl. Orient., 4: 207 (1875).
  E. sericeum Vahl ssp. *elegans* (Lehm.) Holmboe, Veg. Cypr., 150 (1914); Lindberg f., Iter Cypr., 28 (1946).
  E. sericeum Vahl ssp. *halacsyi* Holmboe, Veg. Cypr., 150 (1914).
  E. rubrum Forssk. ssp. *halacsyi* (Holmboe) Osorio-Tafall et Seraphim, List Vasc. Plants Cyprus, 87 (1973) nom. invalid.
  [*E. creticum* (non L.) Hume in Walpole, Mem. Europ. Asiatic Turkey, 254 (1817).]
  [*E. sericeum* Vahl var. *diffusum* Boiss., Fl. Orient., 4: 207 (1875) quoad plant. vix *E. diffusum* Sm. in Sibth. et Sm.]
  [*E. rauwolfii* (non Del.) Boiss., Fl. Orient., Suppl., 351 (1888) quoad plant. cypr.; Holmboe, Veg. Cypr., 150 (1914); Osorio-Tafall et Seraphim, List Vasc. Plants Cyprus, 87 (1973).]
  [*E. sericeum* (non Vahl) Chapman, Cyprus Trees and Shrubs, 64 (1949).]
  TYPE: "grows naturally in Crete" (probably described from plants grown in Chelsea Physic Garden, but Lacaita (Journ. Linn. Soc., 44: 438; 1919) says that specimen at BM is mislabelled and is not *E. angustifolium* Mill. The holotype of *E. hispidum* Sm. (OXF) has been selected as neotype of *E. angustifolium* Mill. by J. R. Edmondson (Fl. Turkey; 6: 323).

Erect or sprawling perennial (10–) 15–40 (–60) cm. high; rootstock woody, tough, usually much branched at apex; stems generally numerous, obscurely angled, densely clothed with a mixed indumentum of long, spreading, white, pungent bristles and short, adpressed hairs; basal leaves often densely tufted, persisting at least until anthesis, sessile, 3–10 (or more) cm. long, usually less than 1 cm. wide, linear or narrowly oblong, acute or obtuse, greyish, with long, often pustulate-based, bristles and short pubescence; cauline leaves similar, but often narrower and more acute, sometimes twisted with undulate margins; inflorescence at first thyrsoid, becoming lax and irregularly spreading with age; flowers numerous, sessile; bracts linear-lanceolate, 3–10 mm. long, 1–4 mm. wide, densely hispid; calyx narrowly campanulate, divided to the base into 5 subequal, linear-lanceolate, hispid lobes about 8–10 mm. long, 0·5–1·5 mm. wide; corolla red or purple, commonly changing from one colour to the other with age, narrowly infundibuliform, about 1·5 cm. long, about 0·5–0·8 cm. wide at apex, pubescent and thinly hispidulous all over externally, glabrous internally; basal annulus of 10 small, oblong-hemispherical, pilose lobes; mouth of corolla markedly oblique, the adaxial, rounded lobes 2·5–3 mm. long, erect or slightly spreading, the abaxial similar but generally about 1·5 mm. long; stamens long-exserted; filaments glabrous, 12–14 mm. long; anthers oblong, about 1 mm. long, 0·6 mm. wide; ovary glabrous, about 1 mm. diam., deeply 4-lobed; style about 20 mm. long, pilose, divided at apex into 2 subulate lobes about 1·5–2 mm. long; stigmas minute, apical. Nutlets broadly ovoid, about 2·5 mm. long, 2 mm. wide, greyish-brown, rugulose, keeled dorsally and ventrally, the sides obscurely gibbous, the median beak short and scarcely incurved.

HAB.: Rocky and sandy seashores; roadsides and waste ground; dry banks and hillsides; sea-level to 3,100 ft. alt.; fl. March–July.

DISTR.: Divisions 1–8, locally common. Eastern Mediterranean region, from Greece to Egypt and Libya.

1. Stroumbi, 1913, *Haradjian* 782! Kato Paphos, 1978, *Holub* s.n.! Near Lachi, 1979, *Edmondson & McClintock* E 2748! Near Lyso, 1979, *Edmondson & McClintock* E 2828!
2. Platres, 1937, *Kennedy* 1373! Kannaviou, 1939, *Lindberg f.* s.n.
3. Near Moni and Kolossi, 1862, *Kotschy*; Stavrovouni, 1939, *Lindberg f.* s.n.! Between Limassol and Amathus, 1941, *Davis* 3590! Kolossi, 1961, *Polunin* 6656! Curium, 1961, *Polunin* 6694! Limassol, 1966, *A. Matthews* 11! Amathus, 1978, *Holub* s.n.!
4. Common about Larnaca, Famagusta, Cape Greco, etc. *Hume*; *Kotschy* 66; *Syngrassides* 483! *Casey* 551! *Chapman* 422! *N. Macdonald* 13! *A. Genneou* in ARI 1592! *W. R. Price* 999! etc.
5. Athalassa, 1967, *Merton* in ARI 152!
6. Nicosia, 1930–1931, *F. A. Rogers* 0721! 7091! Ayia Irini, 1936, *Syngrassides* 1209! Nicosia, 1973, *P. Laukkonen* 113!
7. Common; Kythrea, Kyrenia, Myrtou, Ardhana, Larnaka tis Lapithou, Kalogrea, Akanthou, etc., *Sintenis & Rigo* 106! *Syngrassides* 21! 38! 689! *Chapman* 27! 266! *Davis* 2987! *Mrs. Grove* in *Casey* 1037! *Miss Mapple* 35! *G. E. Atherton* 66! 163! 208! *I. M. Hecker* 22! *A. Genneou* in ARI 1620! etc.
8. Near Leonarisso, 1880, *Sintenis & Rigo*; Cape Andreas, 1880, *Sintenis & Rigo*; near Komi Kebir 1912, *Haradjian* 310! Ayios Philon near Rizokarpaso, 1941, *Davis* 2228!

NOTES: Variable elsewhere but rather uniform in Cyprus, and for the most part closely resembling the plant figured in *Flora Graeca*, t. 181 as *E. hispidum*.

4. **E. plantagineum** L., Mantissa Altera, 202 (1771); Boiss., Fl. Orient., 4: 208 (1875); H. Stuart Thompson in Journ. Bot., 44: 335 (1906); Holmboe, Veg. Cypr., 150 (1914); Post, Fl. Pal., ed. 2, 2: 236 (1933); Rechinger f. in Arkiv för Bot., ser. 2, 1: 429 (1950); P. E. Gibbs in Lagascalia, 1: 57 (1971); Osorio-Tafall et Seraphim, List Vasc. Plants Cyprus, 88 (1973); Davis, Fl. Turkey, 6: 322 (1978).
[*E. lycopsis* (non L., Fl. Angl., 12; 1754) Dandy, List Brit. Vasc. Plants, 93 (1958); Klotz in Wiss. Zeitschr. Univ. Halle, 9 (3): 375 (1960).]
TYPE: "in Italia".

Erect, branched or unbranched annual or biennial; roots containing a purple dye; stems obscurely angled, rather densely clothed with spreading, soft bristles; basal leaves oblong-ovate, 5–10 cm. long, 2–4 cm. wide, rather densely greyish-strigose with conspicuous, ascending nervation, apex obtuse or acute, base narrowing abruptly to a flattened petiole 3–6 cm. long; cauline leaves sessile, somewhat amplexicaul, narrowly oblong, 2–10 cm. long, 0·5–2 cm. wide, densely strigose; inflorescences arising from the axils of the uppermost leaves, together forming a broadly pyramidal or dome-shaped mass in well-developed specimens; bracts lanceolate or deltoid-acuminate, strongly amplexicaul, 0·5–2 cm. long, 0·3–1 cm. wide; flowers subsessile or very shortly pedicellate; calyx narrowly campanulate, divided almost to base into 5 linear, hispid lobes, 8–10 mm. long, 1–1·5 mm. wide; corolla at first reddish, becoming violet-blue, purple or lilac with age, broadly and obliquely infundibuliform, 2–3 cm. long, 1–1·5 cm. wide at the mouth, glabrous or thinly hispid along the nerves externally, glabrous internally; corolla-lobes rather shallow, rounded, to about 5 mm. diam.; basal annulus of 10 small, oblong, pilose lobes about 0·4 mm. long; stamens (or some of them) distinctly exserted from corolla; filaments unequal, inserted at different levels in the basal half of the corolla-tube, declinate, purplish, thinly clothed with long, woolly hairs, 2–3 cm. long; anthers oblong, 0·8–1 mm. long, about 0·6 mm. wide, purplish, shortly lanuginose; ovary thinly hispid, about 1 mm. diam., deeply 4-lobed; style about 2·5 cm. long, densely hispidulous in the lower half, divided apically into 2 slender branches about 1 mm. long; stigmas minute, terminal. Nutlets ovoid, angular, about 3 mm. long, 2·5 mm. wide, greyish-brown, densely tuberculate, keeled dorsally and ventrally, with distinct lateral gibbosities and a short, blunt median beak; attachment-scar large, flat, about 1·5 mm. wide.

HAB.: Roadsides, fallow fields, waste ground, occasionally on sandy seashores; sea-level to 3,500 ft. alt., usually lowland; fl. Febr.–July.

DISTR.: Divisions 2, 7, 8, locally abundant. South Europe and Mediterranean region, extending north to France and Great Britain, and occurring as an introduced weed elsewhere.

2. Between Lefka and Moutoullas, 1937, *Syngrassides* 1549! Platres, 1955, *Kennedy* 1865! Yialia, 1962, *Meikle* 2259!
7. Kyrenia, 1880, *Sintenis & Rigo* 532! also *Syngrassides* 1322! *Casey* 597! *Miss Mapple* 13! *G. E. Atherton* 936! *S. Economides* in ARI 1057! Trimithi, 1941, *Davis* 2781! Vasilia, 1941, *Davis* 3045! Ayios Epiktitos, 1955, *G. E. Atherton* 238! and, 1967, *Merton* in ARI 326! Karakoumi, 1956, *G. E. Atherton* 1235! Lambousa, 1973, *P. Laukkonen* 269!
8. Ayios Andronikos, 1970, *A. Genneou* in ARI 1476!

NOTES: The first definite Cyprus record for *E. plantagineum* dates from 1880, and the species was evidently not seen by Holmboe in 1905. In the circumstances one suspects it may be a recent introduction, now abundant and actively spreading about Kyrenia and Lefka, and perhaps on its way to becoming a widespread weed on the island.

The earlier name *E. lycopsis* L. (1754) has been rejected by P. E. Gibbs (in Lagascalia, 1: 60; 1971) on the grounds that it is based on a plant listed by Ray in his *Synopsis*, ed. 3, 227 (1729), but not described, so that the name can be validated only by reference to an illustration in Dodoens, *Stirp. Hist. Pempt. Sex*, 630 (1616), which is suggested as the lectotype, and tentatively identified as *E. vulgare* L.

5. E. arenarium *Guss.*, Ind. Sem. Hort. Boccadifalco, 5 (1825); Plant. Rar., 88, t. 17 (1826); Boiss., Fl. Orient., 4: 210 (1875), Suppl., 351 (1888); Post, Fl. Pal., ed. 2, 2: 237 (1933); Seligman in Quart. Bull. Alp. Gard. Soc., 20: 228 (1952); Osorio-Tafall et Seraphim, List Vasc. Plants Cyprus, 88 (1973); Davis, Fl. Turkey, 6: 324 (1978).
[*E. diffusum* (non Sm.) Holmboe, Veg. Cypr., 150 (1914).]
TYPE: Italy; "In arenosis maritimis Calabriae orientalis, *Reggio, S. Leonardo, Capo Rizzuto*; Japygiae *Taranto, Pulsano, S. Cesarea*". (NAP ?).

Prostrate or sprawling biennial 6–20 (–25) cm. high with a purple-staining root; stems generally much branched from the base, adpressed-strigose, subterete or obscurely angled; basal leaves narrowly oblong-oblanceolate or subspathulate, 1·5–6 cm. long, 0·5–1·3 cm. wide, adpressed-strigose above and below, apex rounded or subacute, base tapering to a slender (sometimes obscure), flattened petiole 0·5– 2·5 cm. long; nervation obscure; cauline leaves smaller, oblong, usually less than 2 cm. long, 0·5 cm. wide; inflorescences terminal, usually simple, elongating greatly in fruit; bracts foliaceous, lanceolate or narrowly deltoid, subamplexicaul, up to 1 cm. long, 0·4 cm. wide; flowers subsessile; calyx narrowly campanulate, accrescent in fruit, densely hispid with spreading, pungent, yellowish bristles, divided almost to base into 5 narrow, lanceolate lobes 4–6 mm. long, 1–1·3 mm. wide, accrescent to 8 mm. or more in fruit; corolla violet-blue or mauve, narrowly infundibuliform, about 8 mm. long, 3·5 mm. diam. at mouth, adpressed-pubescent all over externally, glabrous internally, lobes oblique, rounded, the abaxial about 1·5 mm. long, the adaxial less than 1 mm. long; basal annulus of 10 very short, rounded, pilose lobes; stamens inserted at different levels near base of corolla-tube; filaments unequal, 3–5 mm. long, glabrous, not exserted; anthers oblong, yellow, about 0·6 mm. long, 0·4 mm. wide; ovary about 0·5 mm. diam., glabrous, deeply 4-lobed; style about 5 mm. long, densely subadpressed-pilose throughout its length, divided at apex into 2 slender branches about 0·5 mm. long; stigmas decurrent down the inner face of the style-branches. Nutlets broadly ovoid, about 2 mm. long, 1·5 mm. wide, pale grey-brown, minutely verruculose, conspicuously keeled dorsally and ventrally, with 2 distinct lateral gibbosities flanking an acute, slightly incurved median beak, each of the gibbosities terminating in a short, sharp nipple; attachment-scar about 1 mm. wide.

HAB.: On rocky ground, in sandy fields, or in *Pistacia-Juniperus* forest near the sea; sea-level or just above sea-level; fl. March–May.

DISTR.: Divisions 1, 4, 6, 8. Chiefly islands in the Mediterranean (Balearics, Sardinia, Sicily, Crete, Ionian and Aegean Islands, Lampedusa, Malta), also Italy, Greece, N. Africa.

1. Skali near Ayios Yeoryios (Akamas), 1962, *Meikle* 2081 !
4. Famagusta, 1936, *Syngrassides* 877 !
6. About Ayia Irini, 1949 or 1950, *Seligman*.
8. Locally common about Apostolos Andreas Forest and Cape Andreas, 1880–1962, *Sintenis & Rigo* 109 ! *Davis* 2281 ! *Mavromoustakis* s.n. ! *Meikle* 2484 !

13. ONOSMA *L.*

Sp. Plant., ed. 2, 196 (1762).
Gen. Plant., ed. 6, 76 (1764).
G. Stroh in Beih. Bot. Centralbl., 59B: 430–454 (1939).
H. Riedl in Oesterr. Bot. Zeitschr., 109: 213–249 (1962).

Perennial, annual or biennial herbs or subshrubs; leaves usually hispid or strigose, the bristles arising from glabrous or stellate-pilose tubercles; flowers in terminal, or terminal and lateral, bracteate cymes, occasionally solitary; calyx campanulate, divided almost or quite to base, lobes sometimes markedly accrescent after anthesis; corolla cylindrical, fusiform or narrowly infundibuliform, usually yellow or creamy-white, sometimes becoming pink, purple or blue with age, lobes usually very small, patent or deflexed; faucal scales wanting; basal, nectariferous annulus present, generally glabrous; stamens attached to middle or towards base of corolla; filaments flattened; anthers exserted or included, usually coherent at the base, mostly with a produced connective; style filiform, usually exserted; stigma capitate or minutely 2-lobed. Nutlets 4 or often 1 by abortion, straight, trigonous-ovoid, often shortly beaked, smooth or less often tuberculate-rugose; basal attachment-scar flattish.

About 150 species, chiefly in south-western Asia, but extending from the Mediterranean region eastwards to China.

Many authors have followed Linnaeus in regarding the name *Onosma* as feminine. Linnaeus, however, borrowed the name from Dioscorides (see Critica Botanica, 248; 1737), and *Onosma* is neuter in both Latin and Greek.

Anthers exserted from corolla-tube:
Indumentum simple, not stellate-pilose; flowers solitary or few at the apex of rigid, woody,
  subspinose branches - - - - - - - - - - **1. O. fruticosum**
Indumentum closely stellate-pilose; flowers numerous in unilateral, scorpioid cymes
  4. O. mite
Anthers included in corolla-tube:
Robust, erect biennial, more than 30 cm. high; indumentum stellate-pilose; inflorescence an
  elongate panicle of cymes - - - - - - **5. O. giganteum** ssp. **hispidum**
Suffruticose, tufted perennials less than 30 cm. high; indumentum simple, not stellate-pilose;
  inflorescence a terminal, corymbiform cyme:
Basal leaves crowded, rosulate, spathulate, 10–20 mm. long, 4–8 mm. wide **2. O. troodi**
Basal leaves not crowded or rosulate, oblanceolate or narrowly spathulate, 10–40 mm.
  long, 3–13 mm. wide - - - - - - - - **3. O. caespitosum**

1. **O. fruticosum** *Sm.* in Sibth. et Sm., Fl. Graec. Prodr., 1: 122 (1806), Fl. Graec., 2: 63, t. 174 (1813); Poech, Enum. Plant. Ins. Cypr., 25 (1842); Holmboe, Veg. Cypr., 148 (1914); Chapman, Cyprus Trees and Shrubs, 64 (1945); Lindberg f., Iter Cypr., 28 (1946); Osorio-Tafall et Seraphim, List Vasc. Plants Cyprus, 87 (1973); Elektra Megaw, Wild Flowers of Cyprus, 10, t. 15 (1973).
  O. fruticosum Labill., Icones Plant. Syr. Rar., 3: 10, t. 6 (1809); Kotschy in Unger et Kotschy, Die Insel Cypern, 282 (1865); Boiss., Fl. Orient., 4: 85 (1875); Stroh in Beih. Bot. Centralbl., 59B: 435 (1939).
  TYPE: "In insulâ Cypro", 1787, *Sibthorp* (OXF).

Erect, much-branched, canescent shrub 15–50 (–90) cm. high; bark greyish, peeling off to expose a dark brown underlayer; twigs rigid, subspinose, closely adpressed-pubescent with scattered bristles; leaves sessile, narrowly oblong or oblanceolate, 7–18 mm. long, 2–4 mm. wide, grey-green, with a mixed indumentum of short adpressed pubescence and long, subadpressed (or rarely spreading), scabridulous bristles arising from a glabrous, sulcate base, apex acute, obtuse or rounded, margins revolute, base cuneate; flowers nodding, usually solitary, terminal, occasionally 2–3 in a lax terminal cyme; bracts foliaceous, like the leaves but smaller and narrower; pedicel usually less than 3 mm. long or almost wanting; calyx narrowly campanulate, divided to the base into 5 linear lobes 7–8 mm. long, 1–1·5 mm. wide; corolla golden-yellow, turning orange or brownish with age, cylindrical or somewhat fusiform, 10–14 mm. long, 4–6 mm. diam., minutely papillose externally and internally; corolla-lobes small, deflexed, bluntly deltoid; basal annulus consisting of 10 short, rounded, glabrous lobes; stamens inserted about 4 mm. above base of corolla-tube; filaments flattened, 6–7 mm. long; anthers half-exserted, linear, 6–7 mm. long, about 1 mm. wide, base shortly and bluntly sagittate, apex produced into a subacute, membranous appendage; ovary glabrous, about 1 mm. diam., deeply 4-lobed; style 18–19 mm. long, tapering upwards from a stout base, slightly exceeding the anthers; stigma terminal, very shortly 2-lobed. Nutlets ovoid, about 3·5 mm. long, 2 mm. wide, smooth, rather glossy, marbled brown and creamy-white, apex shortly and bluntly beaked, ventral face obscurely keeled; basal attachment-scar about 1 mm. diam.

HAB.: On dry, stony hillsides in garigue; near sea-level to 3,600 ft. alt.; fl. March–May.

DISTR.: Divisions 1–8, locally common. Endemic.

1. Smyies, 1937, *Chapman* 288! Paphos, 1937, *Miss Godman* 24! Dhrousha, 1941, *Davis* 3260! Psindron near Ayios Yeoryios (Akamas), 1962, *Meikle* 2121! Near Baths of Aphrodite, 1979, *Hewer* 4559!
2. Aphamis, 1938–1939, *Kennedy* 1678! 1742!

3. Stavrovouni, 1787, *Labillardière*! Mazotos, 1862, *Kotschy* 553 and, 1905, *Holmboe* 201; Kophinou, 1905, *Holmboe* 571; Limassol, 1913, *Haradjian* 565! Oritaes Forest, 1937, *Chapman* 303! Episkopi, 1939, *C. E. H. Sparrow* 1! and, 1950, *F. J. F. Barrington* s.n.! Amathus, 1978, *Holub* s.n.!

4. Near Famagusta, 1862, *Kotschy*; Ormidhia, 1905, *Holmboe* 78; near Goshi, 1967, *Merton* in ARI 374!

5. Road to Makheras, 1937, *Chapman* 267! Athalassa, 1939, *Lindberg f.* s.n.! also, 1941, *Davis* 3194!

6. Ayia Irini, 1941, *Davis* 2574! Between Myrtou and Cape Kormakiti, 1970, *I. M. Hecker* 4!

7. Locally common; Kantara, Kyrenia, Pentadaktylos, Akanthou, Kornos, etc. *Sintenis & Rigo* 108! *Syngrassides* 20! *Mrs. Dray* 2! *C. H. Wyatt* 6! *Davis* 3034! *Casey* 359! *G. E. Atherton* 1028! *Polunin* 8541! etc.

8. Cape Andreas, 1880, *Sintenis & Rigo*; between Yialousa and Ayios Andronikos, 1941, *Davis* 2395! Koma tou Yialou, 1950, *Chapman* 256!

2. O. troodi *Kotschy* in Unger et Kotschy, Die Insel Cypern, 283 (1865); Boiss., Fl. Orient., 4: 189 (1875); Holmboe, Veg. Cypr., 149 (1914); Warr et Gresham in Journ. R.H.S., 59: 71 (1934); Lindberg f., Iter Cypr., 28 (1946); Chapman, Cyprus Trees and Shrubs, 64 (1949); Rechinger f. in Arkiv för Bot., ser. 2, 1: 429 (1950); Osorio-Tafall et Seraphim, List Vasc. Plants Cyprus, 87 (1973).

TYPE: Cyprus; "Auf der Spitze des Troodos in Felsenspalten häufig, hathe aber am 20. Mai nur noch wenige Blüthen entwickelt. n. 754" (W, K!).

Tufted perennial 15–20 cm. high; rootstock and base of stems woody, much branched, covered with dark fuscous-brown bark; basal leaves crowded, spathulate, silvery-grey, 10–20 mm. long, 4–8 mm. wide, densely clothed with long, adpressed, scabridulous bristles above and below, covering an underlayer of short, subadpressed hairs, apex acute or obtuse, base tapering to an obscure petiole; flowering stems herbaceous, annual, canescent with an indumentum of short hairs and long, spreading bristles; cauline leaves sessile, oblong, acute, 8–25 mm. long, 2–5 mm. wide, indumentum as in the basal leaves but the bristles usually sparser; inflorescence a dense, branched, corymbiform, many-flowered cyme; bracts foliaceous, narrowly oblong or lanceolate, up to 2 cm. long, usually less than 0·3 cm. wide, densely strigose with silvery bristles; flowers subsessile or very shortly pedicellate at anthesis, the pedicel sometimes lengthening to 5 mm. in fruit; calyx narrowly campanulate, densely sericeous-strigose, about 13 mm. long, less than 4 mm. wide, with a short basal tube, divided above into 5 linear, acute lobes accrescent to 1·5 cm. or more in fruit; corolla yellow, narrowly cylindrical or sub-fusiform, 15–17 mm. long, 3 mm. diam., glabrous externally except for a few short bristles along the principal nerves; lobes spreading, very short, rounded, about 0·8 mm. long, 1·5 mm. wide at base; basal annulus of 10 obscure, glabrous gibbosities; stamens inserted about 7 mm. above base of corolla; free part of filament about 4 mm. long, flattened, glabrous; anthers included, linear-lanceolate, about 7 mm. long, 1 mm. wide, apex acuminate, membranous, base very obscurely sagittate; ovary glabrous, about 1·5 mm. diam., deeply 4-lobed; style about 17 mm. long, not much tapering, distinctly exserted from corolla; stigma terminal, shortly 2-lobed. Nutlets broadly ovoid, dorsiventrally compressed, about 3·5 mm. long, 3 mm. wide, very pale brown, tinged fuscous, glossy, strongly keeled ventrally, with a short, conspicuous apical beak; basal attachment scar flat, triangular, about 1·5 mm. wide.

HAB.: Rocky slopes and screes on serpentine; 5,200–6,400 ft. alt., rarely on diabase or gabbro, 3,400–4,800 ft. alt.; fl. April–June.

DISTR.: Division 2. Endemic.

2. Summit of Khionistra; also by Kryos Potamos, 3,400 ft. alt., and about Platres. *Kotschy* 754! *Sintenis & Rigo* 827! *Holmboe* 876, 987; *Haradjian* 517! *C. B. Ussher* 45! *Kennedy* 724–730! *C. H. Wyatt* 11! *Lindberg f.* s.n.! *Davis* 1866! 3525! *Mavromoustakis* s.n.! *Casey* 830! *Merton* 2329! *D. P. Young* 7337! etc. Papoutsa, 1979, *Edmondson & McClintock* E 2890!

NOTES: Allied to *O. frutescens* Lam., which has a wide distribution in the eastern Mediterranean region, but differing in habit, indumentum and included anthers. Specimens of *O. frutescens* at Kew, labelled "Nicosia, Mar. 1931, *Rev. F. A. Rogers* 0643" are part of a collection made by Rogers on 24 March, 1930, at Iskenderum (Alexandretta). *O. frutescens* is not known to occur in Cyprus. Similar mislabellings are noted under *Fibigia eriocarpa* and *Cytisopsis dorycniifolia*.

3. O. caespitosum *Kotschy* in Unger et Kotschy, Die Insel Cypern, 282 (1865); Boiss., Fl. Orient., 4: 190 (1875), Suppl., 351 (1888); Holmboe, Veg. Cypr., 149 (1914); Lindberg f., Iter Cypr., 28 (1946); Chapman, Cyprus Trees and Shrubs, 64 (1949); Seligman in Bull. Alp. Gard. Soc., 20: 234 (1952); Osorio-Tafall et Seraphim, List Vasc. Plants Cyprus, 87 (1973).
[*O. frutescens* (non Lam.) H. Stuart Thompson in Journ. Bot., 44: 335 (1906).]
TYPE: Cyprus; "In Spalten der Felwände auf dem Buffavento unter dem Castello della Regina an unzugänglichen Stellen mit Brassica cretica [*B. hilarionis*]; an den citronengelben Blumen aus der Ferne zu erkennen. Am 15. April. n. 445" (W).

Loosely tufted perennial 10–30 cm. high, with numerous straggling branches, fuscous and woody at the base, herbaceous and rather densely strigillose above, with white, patent rigid hairs or fine bristles; leaves scattered, not distinctly rosulate, oblanceolate or narrowly spathulate, 1–4 cm. long, 0·3–1·3 cm. wide, rather densely adpressed-strigose on both surfaces, the scabridulous bristles arising from a glabrous, sulcate-tuberculate base, apex usually obtuse, base tapering to an indistinct petiole, margins narrowly revolute; inflorescence a dense, branched, corymbiform, many-flowered cyme; bracts foliaceous, strigose, lanceolate or narrowly ovate, 10–15 mm. long, 2–5 mm. wide, tapering from a broad, rounded base to a slender acumen; flowers distinctly pedicellate, with slender pedicels 2–7 mm. long, the pedicels sometimes lengthening to 10 mm. or more in fruit; calyx narrowly campanulate, densely sericeous-strigose with spreading hairs, about 10–13 mm. long, less than 4 mm. wide, with a short basal tube, divided above into 5 linear, acute lobes, persistent but not much accrescent in fruit; corolla pale or bright yellow, narrowly cylindrical or sub-fusiform, about 15 mm. long, glabrous externally; lobes spreading, very short, widely deltoid, about 0·8 mm. long, 1·5 mm. wide at base; basal annulus of 10, small, shortly lingulate, unequal, glabrous lobes; stamens inserted about 6 mm. above base of corolla-tube; free part of filaments about 3 mm. long, flattened, glabrous; anthers included, linear-lanceolate, about 7 mm. long, 1 mm. wide, apex acuminate, membranous, base very shortly sagittate; ovary glabrous, about 1·5 mm. diam., deeply 4-lobed; style about 16 mm. long, not much tapering, shortly but distinctly exserted from corolla; stigma terminal, shortly 2-lobed. Nutlets broadly ovoid, about 3·5 mm. long, 2·5–3 mm. wide, very pale brown, marbled darker brown, glossy, strongly keeled ventrally, with a short, conspicuous apical beak; basal attachment scar flat, triangular, about 1·5 mm. wide.

HAB.: Clefts of limestone rocks; 1,000–3,000 ft. alt.; fl. March–May.

DISTR.: Division 7. Endemic.

7. Widely, but locally, distributed on the Northern Range; Buffavento, Pentadaktylos, St. Hilarion, Lapithos, Larnaka tis Lapithou, Yaïla, Kornos, etc. *Kotschy* 445; *Sintenis & Rigo* 531! *Holmboe* 823; *A. G. & M. E. Lascelles* s.n.! *J. A. Tracey* 46! 47! *Syngrassides* 911! *Lindberg f.* s.n.! *Kennedy* 1561! *Davis* 2500! 2898! 2965! *Casey* 633! *G. E. Atherton* 733! *Polunin* 6601! etc.

NOTES: More closely related to *O. frutescens* Lam. than *O. troodi* Kotschy, but differing in its included anthers and shortly exserted style. *O. caespitosum* and *O. troodi* provide a good example of "species pairs", one confined to the limestones of the Northern Range, the other to the igneous rocks of Troödos.

4. O. mite *Boiss. et Heldr.* in Boiss., Diagn., 1, 11: 111 (1849); Unger et Kotschy, Die Insel Cypern, 282 (1865); Boiss., Fl. Orient., 4: 200 (1875); Holmboe, Veg. Cypr., 150 (1914); Chapman, Cyprus Trees and Shrubs, 64 (1949); Osorio-Tafall et Seraphim, List Vasc. Plants Cyprus, 87 (1973); Davis, Fl. Turkey, 6: 364 (1978).

TYPE: Turkey; "in pinetis apricis ad radices montis *Taktalu* Lyciae supra portum *Tcherali* (Heldr[e]ich)." (G).

Tufted subshrub 15–30 (–55) cm. high; lower parts of stems woody, covered with flaking, dark brown bark; branches usually numerous, densely clothed with adpressed, greyish, bristly hairs; leaves narrowly oblanceolate, 1–4·5 cm. long, 0·2–0·8 cm. wide, densely grey-tomentose with an adpressed indumentum of long, minutely scabridulous bristly hairs arising from a stellate-pilose, tuberculate base, apex acute or obtuse, margins revolute, base tapering to a very short, indistinct petiole or leaves sessile; inflorescence an elongating, 1-sided, scorpioid cyme, 4–6 cm. long at anthesis, lengthening to 15 cm. (or more) in fruit, borne at the apex of a well-developed, sparsely leafy flowering stem; bracts narrowly lanceolate, foliaceous, 8–15 mm. long, 2–5 mm. wide; flowers distinctly pedicellate, with a slender, villous pedicel up to 10 mm. long; calyx narrowly campanulate, white-villous, divided almost to the base into 5 linear, acute lobes about 10 mm. long, 1·5 mm. wide; corolla golden-yellow, narrowly cylindrical or sub-fusiform, 12–15 mm. long, 7 mm. wide, minutely glandular-papillose externally and glabrous except for a few bristles around the base of the lobes; corolla-lobes spreading, deltoid, about 1·5 mm. long, 1·5 mm. wide at base; basal annulus of 10 short, glabrous, horn-like, richly nectariferous lobes; stamens inserted about 5 mm. above base of corolla-tube; free part of filament about 6 mm. long, flattened, glabrous; anthers long-exserted; brownish, linear, about 7 mm. long, 1 mm. wide, base shortly sagittate, apex with a long, acuminate, membranous appendage; ovary about 1·5 mm. diam., glabrous; style about 20 mm. long, glabrous, tapering to apex; stigma terminal, minutely, obscurely 2-lobed. Nutlets ovoid, about 4·5 mm. long, 3 mm. wide, very pale brown with darker marbling, smooth and rather glossy, with a very short, blunt beak, ventral face shallowly keeled; basal attachment scar flat, suborbicular or somewhat 3-angled, about 1·5 mm. wide.

HAB.: On rocky igneous hillsides, often in *Pinus brutia* forest; 1,500–3,000 ft. alt.; fl. April–June.

DISTR.: Divisions 1, 2. Also in S. Turkey.

1. Near Lyso ["Lisso"], c. 2,000 ft. alt., 1913, *Haradjian* 884 ! N. of Melandra, 1949, *Casey* 746 !
2. "In declivibus raro provenit inter Prodromo et Pagum Fini die 17. Mai, 1862", *Kotschy* 705 partly (see Notes) ! Ayia valley, 1905, *Holmboe* 1042; also, same area, 1934, *Chapman* 148 ! and, 1962, *Meikle* 2651 ! Between Pano Panayia and Kykko, 1941, *Davis* 3383 ! and, 1979, *Hewer* 4607 ! Near Khrysorroyiatissa Monastery, 1952, *F. M. Probyn* 104 !

NOTES: Some doubt must attach to Kotschy's record from between Prodhromos and Phini. The specimen (*Kotschy* 705) at Kew consists of a mixture of *O. mite* and *O. troodi*, and while the latter is known from this area, *O. mite* is otherwise confined to the Paphos Forest and western extension of the Troödos Range. It is more likely that Kotschy found *O. mite* in the neighbourhood of Khrysorroyiatissa, and subsequently mixed it with material from the Prodhromos–Phini area.

5. O. giganteum *Lam.*, Tabl. Encycl., 1: 407 (1792), Encycl. Méth., 4: 584 (1798); Unger et Kotschy, Die Insel Cypern, 283 (1865); Boiss., Fl. Orient., 4: 202 (1875); Holmboe, Veg. Cypr., 150 (1914); Post, Fl. Pal., ed. 2, 2: 234 (1933); Osorio-Tafall et Seraphim, List Vasc. Plants Cyprus, 87 (1973); Davis, Fl. Turkey, 6: 375 (1978).

Erect biennial up to 75 cm. high; stems robust, usually unbranched, or occasionally branched at the base, obscurely sulcate, densely clothed with adpressed (var. *giganteum*) or spreading (var. *hispidum* Boiss.), scaberulous bristles arising from a stellately pilose, tuberculate base; basal leaves crowded into an irregular tuft or rosette, narrowly oblanceolate, 11–25 cm. long, 1–2·5 cm. wide, hispid with pungent bristles arising from a stellate-pilose base, apex acute, margins narrowly revolute, base tapering to a flattened, ill-defined petiole, nervation sometimes rather prominent and arcuate on the undersurface of the lamina; cauline leaves sessile, lanceolate-acuminate or narrowly oblong, becoming progressively smaller and markedly amplexicaul upwards; inflorescence a long, narrow panicle of congested, branched cymes, commonly 20–25 cm. long, 8–12 cm. wide; rhachis densely strigose or hispid; bracts foliaceous, lanceolate-acuminate, 5–12 mm. long, 2–5 mm. wide, densely hispid; flowers very numerous, pedicellate, with slender white-strigose pedicels commonly up to 1 cm. or more in length; calyx narrowly campanulate, densely white-strigose, divided almost to the base into linear, acute lobes 10–13 mm. long, 0·8–1·5 mm. wide; corolla golden-yellow, fusiform or sometimes narrowly infundibuliform, 1·5–2 cm. long, 0·6–0·8 cm. wide, glandular-papillose and shortly and closely strigillose externally; lobes broadly deltoid, spreading, about 1·5 mm. long, 2·5 mm. wide at base; basal annulus of 10 small, glabrous flaps, or sometimes with these united to form a continuous narrow rim; stamens inserted 7–10 mm. above base of corolla-tube; free part of filaments about 6 mm. long, glabrous, flattened; anthers included, linear, about 8 mm. long, 1–1·5 mm. wide, shortly sagittate at base, apex produced into a membranous, acuminate appendage; ovary glabrous, about 1·5 mm. diam., deeply 4-lobed; style about 2·4 cm. long, glabrous, not much tapering, exserted for 3–5 mm. from mouth of corolla; stigma terminal, shortly but distinctly 2-lobed. Nutlets very broadly ovoid, about 4 mm. long, 3·5 mm. wide, pale grey with fuscous marbling, smooth, glossy, beak very short and obscure, ventral face somewhat keeled; basal attachment-scar flattened-deltoid, about 1·5 mm. wide.

var. **hispidum** *Boiss.*, Fl. Orient., 4: 203 (1875); Lindberg f., Iter Cypr., 28 (1946); Rechinger f. in Arkiv för Bot., ser. 2, 1: 429 (1950).

TYPE: Turkey; "ad Bouloukli Ciliciae (Bal.)" ["Village de Bouloukli, près de Mersina (Cilicie)", 3 June, 1855, *Balansa* 545] (lectotype G, K!).

Stems and leaves hispid with spreading white bristles; otherwise as in var. *giganteum*.

HAB.: Dry chalk or limestone banks and roadsides; 500–2,500 ft. alt.; fl. April–July.

DISTR.: Divisions 1–3, 7, locally common. S. Turkey, Syria, Palestine; var. *hispidum* Boiss. perhaps confined to Cyprus and S. Turkey.

1. Near Tsadha ["Tsarda"], 1901, *A. G. & M. E. Lascelles* s.n.! Stroumbi, 1913, *Haradjian* 790! Also, same locality, 1939, *Lindberg f.* s.n.! 1941, *Davis* 3347! Dhrymou, 1935, *Chapman* 194! Above Letimbou, 1979, *Hewer* 4614!
2. Near Anadhiou ["Aeodiu"], 1862, *Kotschy* 680! Lasa, 1941, *Davis* 3352! Kannaviou, 1962, *Meikle* 2646!
3. Lemona, 1905, *Holmboe* 749.
7. Between Myrtou and Kyrenia, 1932, *Syngrassides* 173! Between Orga and Vasilia, 1940, *Kennedy* 1560! Panagra, 1948, *Casey* 189!

NOTES: Apparently confined to two widely separated areas of the island. All the Cyprus material is clearly referable to *O. giganteum* var. *hispidum*, and because of this uniformity, the variety, probably of minor taxonomic significance, has been given recognition. While the distinction holds good in Cyprus, some intergradation is evident in Turkish specimens.

O. ORIENTALE (*L.*) *L.*, Sp. Plant., ed. 2, 196 (1762); Sibth. et Sm., Fl. Graec. Prodr., 1: 120 (1806); Hume in Walpole, Mem. Europ. Asiatic Turkey, ed. 1, 253 (1817); H. Stuart Thompson in Journ. Bot., 44: 335 (1906); Holmboe, Veg. Cypr., 150 (1914).
    Cerinthe orientalis L., Centuria I Plant., 7 (1755).

A straggling subshrubby perennial, clothed with a mixed indumentum of short pubescence and long, simple bristles; leaves sessile, scattered along the stems, oblong-elliptic, 1–4 cm. long, 0·4–1·3 cm. wide; inflorescence terminal, at first dense, elongating after anthesis; corolla-tube cylindrical, about 1 cm. long, usually blue, with strongly reflexed, acute, yellow lobes.

Recorded by Sibthorp (Walpole, Travels, 15; 1820) from near the base of Stavrovouni (April 13, 1787), and by Hume, from Limassol (July 1801). In both instances *Onosma fruticosum* Sibth. et Sm. was most probably the plant seen, for *O. orientale* has not since been collected in Cyprus. However, as it has a wide distribution in areas adjacent to the island a short description is given, just in case it should yet be refound.

65. CONVOLVULACEAE

Erect, sprawling or climbing herbs or (less commonly) woody plants, with alternate, mostly exstipulate, simple leaves, or leafless parasites (*Cuscuta*); flowers solitary or in cymes, axillary or terminal, generally hermaphrodite; bracteoles often conspicuous and enveloping the calyx; sepals free or connate at the base, imbricate; corolla sympetalous, frequently funnel-shaped, with 4–5 distinct or obscure lobes, plicate, or contorted in bud or occasionally imbricate; stamens 4–5, usually inserted near the base of the corolla-tube; anthers 2-thecous; disk commonly present; ovary superior (1–) 2 (–5)-locular, each loculus containing 1–2 erect ovules; placentation axile; styles 1–2 (–3); stigmas various. Fruit usually a capsule, rarely fleshy or nut-like and indehiscent; seeds sometimes hairy; endosperm present; embryo curved; cotyledons folded or crumpled.

A cosmopolitan family, represented in Cyprus by 6 genera.

Parasitic plants without leaves; stems white, yellow or reddish - - - **5. Cuscuta**
Non-parasitic plants with green leaves:
  Ovary 2-lobed; prostrate plants, the stems rooting at the nodes; flowers minute
                                             DICHONDRA|(p. 1176)
  Ovary not lobed:
    Corolla funnel-shaped, not or only slightly lobed; stamens included:
      Stigma capitate, subglobose, undivided or at most shortly lobed - - **1. Ipomoea**
      Stigma divided to the base into 2 narrow lobes:
        Bracteoles large, enveloping the calyx - - - - - - **2. Calystegia**
        Bracteoles not enveloping the calyx - - - - - - **3. Convolvulus**
    Corolla deeply lobed; stamens exserted; flowers small, inconspicuous - - **4. Cressa**

1. IPOMOEA *L.*

Sp. Plant., ed. 1, 159 (1753).
Gen. Plant., ed. 5, 76 (1754).

Mostly climbing or trailing herbs; leaves entire or variously lobed or dissected; flowers solitary or in cymes; bracteoles wanting or inconspicuous; sepals equal or unequal, the exterior usually larger than the interior, frequently enclosing the fruit; corolla infundibuliform or salver-shaped; stamens included; stigma subglobose, capitate or shortly 2-lobed; ovary 2–4 (–5)-locular, usually with 2 ovules in each loculus. Fruit a 2–10-valved capsule. Pollen grains pantoporate, exine spinose.

A very extensive genus, of at least 500 species, often subdivided into several critical genera, and widely distributed throughout the tropics and subtropics, but rare in temperate regions. Three species occur in Cyprus, one of them an escape from gardens.

Leaves retuse, panduriform or digitately lobed - - - - - - **1. I. stolonifera**
Leaves acute, cordate or sagittate:
  Upper leaves narrowly sagittate with distinct lateral lobes; calyx-lobes obtuse, glabrous or
    subglabrous - - - - - - - - - - - - **2. I. sagittata**
  Upper leaves broadly cordate; calyx-lobes acute, distinctly hairy - - **3. I. purpurea**

1. I. stolonifera (*Cyr.*) *J. F. Gmel.* in L., Syst. Nat., ed. 13, 2 (1): 345 (1791); Post, Fl. Pal., ed. 2, 2: 212 (1933); A. K. Jackson et Turrill in Kew Bull., 1938: 466 (1938); Osorio-Tafall et Seraphim, List Vasc. Plants Cyprus, 85 (1973); Davis, Fl. Turkey, 6: 221 (1978).

    Convolvulus stolonifer[us] Cyr., Plant. Rar. Regn. Neap., fasc. prim., 14, t. 5 (1788).
    C. littoralis L., Syst. Nat., ed. 10, 924 (1759).
    Ipomoea littoralis (L.) Boiss., Fl. Orient., 4: 112 (1875); Lindberg f., Iter Cypr., 27 (1946) non *I. littoralis* Blume, Bijdr., 713 (1826) nom. illeg.

    TYPE: Italy; "*in arena Littoris ubi dicitur* Bagnoli", *Cyrillus* (Cirillo).

A fleshy perennial, with trailing stems often several feet long; leaves variable in shape, sometimes broadly oblong or panduriform with a retuse apex, about 2 cm. long and 1 cm. wide, or elsewhere deeply palmate-lobed, 3 cm. or more long and 2·5 cm. wide with 5 obtusely obovate lobes; flowers axillary, solitary; peduncle to about 3·5 cm. long; bracteoles linear-subulate, about 2–3 mm. long; sepals oblong, obtuse, mucronate, glabrous, about 13 mm. long and 6 mm. wide; corolla white or cream-coloured, infundibuliform, about 6 cm. wide at mouth; stamens about 1·2 cm. long, inserted near base of corolla-tube; filaments about 8 mm. long, glabrous above, distinctly hairy at base; anthers linear-sagittate; ovary conical, glabrous; style slender, about 1·5 cm. long; stigma capitate. Fruit globose, glabrous, about 1·3 cm. diam., partly enclosed by the persistent calyx-lobes; seeds broadly ovoid, about 7 mm. long and 6 mm. wide; testa fuscous, covered with pale brown woolly hairs.

    HAB.: In sand just above tide-level; fl. July–Nov.

    DISTR.: Divisions 4, 6. S. Italy, Crete, Turkey, Syria, Palestine, Arabia, Egypt and N. Africa; Azores. Widespread on sandy seashores in the tropics and subtropics.

    4. Famagusta, 1937, *Lady Loch* 12! also, 1939, *Lindberg f.*, s.n.! and, 1940, *Davis* 2049!
    6. Ayia Irini, 1952, *Lady Loch* 142!

2. I. sagittata *Poir.*, Voy. en Barbarie, 2: 122 (1789); Boiss., Fl. Orient., 4: 113 (1875); Post, Fl. Pal., ed. 2, 2: 212 (1933); Davis, Fl. Turkey, 6: 222 (1978).

    TYPE: Algeria-Tunisia frontier; "près du Bastion de France & sur les bords du lac infect qui en est peu distant, dans le voisinage du Souk", *Poiret & Desfontaines* (P).

A slender twining or trailing perennial; leaves and stems glabrous or almost so; lower leaves cordate-sagittate to about 5 cm. long and 2·5 cm. wide, upper leaves sagittate, the lamina and divergent lateral lobes often linear; flowers axillary, solitary; peduncle relatively stout, 1·5–2 cm. long, often furnished midway with 2 inconspicuous bracteoles; sepals oblong, rounded, mucronulate, glabrous, about 10 mm. long and 7 mm. wide, the outer rather shorter than the inner; corolla large, bright purple, about 5–6 cm. wide at the mouth; stamens about 2 cm. long; filaments 1·5 cm. long, glabrous above, hairy at the base, inserted near base of corolla; anthers narrowly oblong, 5–6 mm. long, 1·5 mm. wide; ovary conical, glabrous, about 2 mm. long; style 2·5 mm. long; stigma capitate, obscurely 2-lobed. Fruit globose, glabrous, about 1·3 cm. diam., partly concealed by the persistent calyx-lobes; seeds ovoid, rather angular, about 7 mm. long and 5 mm. wide; testa fuscous, thinly woolly-hairy especially at the angles.

    HAB.: In marshes and by the sides of ditches; near sea-level; fl. July–Aug.

    DISTR.: Division 3, rare. Spain, Balearics, Corsica, Italy, Greece, Turkey, Syria, Palestine, N. Africa, Florida to Texas, Jamaica, Bermuda, Bahamas, Cuba.

    3. Asomatos (near Limassol), 1939, *Mavromoustakis* 127!

3. I. purpurea (*L.*) *Roth*, Bot. Abh., 27 (1787); Post, Fl. Pal., ed. 2, 2: 212 (1933); Meeuse in Bothalia, 6: 734 (1957); Davis, Fl. Turkey, 6: 222 (1978).
  Convolvulus purpureus L., Sp. Plant., ed. 2, 219 (1762).
  TYPE: "*in* America".

A robust, twining, thinly strigillose annual; leaves broadly cordate, 5–8 cm. long, and about as wide, with entire margins and a shortly cuspidate apex; petiole slightly channelled, 2–5 cm. long; flowers axillary, usually in pairs or in few-flowered cymes; peduncle 2·5–4 cm. long; bracteoles linear, about 5 mm. long; pedicels about 7 mm. long; sepals narrowly ovate, about 1·2–1·5 cm. long and 0·4–0·5 cm. wide, distinctly hairy, with an acute apical appendage; corolla white, pink or purple, 4–5 cm. wide at mouth; stamens 2–2·5 cm. long; filaments 1·8–2·3 cm. long, glabrous above, hairy towards base; anthers oblong, about 2 mm. long; ovary conical, glabrous, about 2 mm. long; style glabrous, 2·5 mm. long; stigma capitate, obscurely lobed. Fruit globose, about 8 mm. diam., enclosed within the persistent calyx; seeds ovoid, about 5 mm. long and 3 mm. wide; testa fuscous, glabrous.

HAB.: Waste ground; introduced; near sea-level; fl. July–Aug.

DISTR.: Division 7. Native of N. & S. America from Virginia to the Argentine; common as a weed almost throughout the tropics and subtropics.

7. Kyrenia, 1955, *G. E. Atherton* 429!

2. CALYSTEGIA *R. Br.*

Prodr. Fl. Nov. Holl., 483 (1810) nom. cons.

Twining or trailing, perennial herbs, very similar to *Convolvulus* in appearance, but with two large bracteoles subtending and usually enveloping the 5-lobed calyx; corolla infundibuliform, 5-pleated; stigma divided into 2 ovate lobes; ovary and capsule imperfectly 2-locular, each loculus containing 2 ovules or seeds. Pollen grains pantoporate; exine more or less smooth.

About 14 species widely distributed in temperate regions.

1. C. sepium (*L.*) *R. Br.*, Prodr. Fl. Nov. Holl., 483 (1810); Kotschy, Die Insel Cypern, 285 (1865); Boiss., Fl. Orient., 4: 111 (1875); H. Stuart Thompson in Journ. Bot., 44: 334 (1906); Post, Fl. Pal., ed. 2, 2: 211 (1933); Osorio-Tafall et Seraphim, List Vasc. Plants Cyprus, 85 (1973); Davis, Fl. Turkey, 6: 220 (1978).
  Convolvulus sepium L., Sp. Plant., ed. 1, 153 (1753); Holmboe, Veg. Cypr., 145 (1914).
  TYPE: "*in* Europae *sepibus*".

Twining perennial herb; stems glabrous, distinctly ridged; petiole to 5 cm. long, leaves sagittate, 4–10 cm. or more long, 2·5–5 cm. wide, glabrous except at the point where the petiole joins the lamina, where there is a varying amount of hair; lateral lobes often irregularly toothed and lobed; flowers solitary, axillary, peduncle to 6 cm. long; bracteoles ovate-cordate, acute, 1·5–2 cm. long, about 0·8 cm. wide, not overlapping; sepals narrowly ovate, 1·3–1·5 cm. long, 0·5 cm. wide, acute, glabrous; corolla white, 3–7 cm. long, 4–6 cm. wide at mouth; stamens c. 1·8 cm. long; filaments c. 1·4 cm. long, inserted near base of corolla-tube, flattened, glabrous above, glandular-hairy towards base; anthers narrowly oblong, about 4 mm. long and 1 mm. wide; ovary conical, glabrous, about 3 mm. long; style glabrous, about 5 mm. long; stigma divided into 2 blunt, ovate lobes, each about 2 mm. long. Fruit globose, glabrous, apiculate, about 1·2 cm. diam.; seeds ovoid, angular, about 4 mm. long and 3 mm. wide; testa dark brown, glabrous.

HAB.: Amongst reeds and in thickets by streams and marshes; near sea-level; fl. June–Aug.

DISTR.: Divisions 3, 7, 8. Widespread in temperate regions of both hemispheres.

3. Zakaki marshes (Limassol), 1943, *Mavromoustakis* 7!
7. Near Ayios Khrysostomos, *Kotschy* s.n.; Akhiropiitos (Karavas), 1936, *Syngrassides* 1296!
8. Near Ayios Thyrsos, 1937, *Syngrassides* 1658! Rizokarpaso, Ronnas Valley, 1957, *Merton* 2933!

3. CONVOLVULUS *L.*

Sp. Plant., ed. 1, 153 (1753).
Gen. Plant., ed. 5, 76 (1754).
F. Sa'ad, The Convolvulus Species of the Canary Isles, the Mediterranean Region, etc., 288 pp., Rotterdam (1967).

Climbing, trailing, or erect herbs or subshrubs; leaves entire or variously toothed or lobed; flowers usually solitary, axillary, occasionally paired or in cymes; bracteoles wanting or inconspicuous; sepals 5; corolla usually infundibuliform, 5-pleated; stamens attached near base of corolla-tube; ovary 2-locular, with 2 ovules in each loculus; stigma deeply bilobed. Fruit a 2-valved or irregularly dehiscing capsule. Pollen grains tricolpate; exine not spinose.

More than 200 species, chiefly in temperate and subtropical regions, and particularly abundant in the Mediterranean region and the Orient.

Perennials:
 Erect or suberect herbs or subshrubs; rootstock usually woody:
  Stems woody, persistent:
   Sepals clothed with appressed hairs:
    Plant erect, 30–60 cm. high; inflorescence almost leafless with numerous, divaricate
     branches - - - - - - - - - **-1. C. dorycnium**
    Plant decumbent, less than 30 cm. high; inflorescence terminal on sparsely branched,
     leafy stems - - - - - - - - - **-4a. C. × cyprius**
   Sepals clothed with spreading hairs - - - - - - **2. C. oleifolius**
  Stems herbaceous, dying back in winter:
   Sepals clothed with spreading hairs - - - - - **[3. C. cantabrica]**
   Sepals clothed with appressed hairs - - - - - - **4. C. lineatus**
 Climbing or trailing herbs; rootstock usually not woody:
  Upper cauline leaves deeply palmatisect - - - - **6. C. althaeoides**
  Upper cauline leaves not palmatisect:
   Leaves oblong, oblanceolate or elliptic - - - - - **4. C. lineatus**
   Leaves cordate or hastate:
    Plants glabrous or very sparsely pubescent - - - - **7. C. arvensis**
    Plants hairy or distinctly pubescent - - - - - **5. C. betonicifolius**
Annuals:
 Flowers distinctly stalked; capsule glabrous:
  Corolla 1·5–2·8 cm. diam., pink or purplish; lower leaves broadly cordate or reniform,
   upper often lobed - - - - - - - - - **8. C. coelesyriacus**
  Corolla not more than 1·5 cm. diam., blue; upper leaves not lobed:
   Leaves oblong or narrowly cordate, distinctly petiolate - - - **9. C. siculus**
   Leaves narrowly obovate or oblanceolate, sessile or narrowed to an indistinct petiole
                     10. C. pentapetaloides
 Flowers sessile or subsessile; capsule hirsute - - - - - **11. C. humilis**

1. **C. dorycnium** *L.*, Sp. Plant., ed. 2, 224 (1762); Clarke, Travels, ed. 2, 2 (3): 718 (1816); Unger et Kotschy, Die Insel Cypern, 285 (1865); Boiss., Fl. Orient., 4: 91 (1875); Holmboe, Veg. Cypr., 145 (1914); Post, Fl. Pal., ed. 2, 2: 204 (1933); Lindberg f., Iter Cypr., 27 (1946); Osorio-Tafall et Seraphim, List Vasc. Plants Cyprus, 84 (1973); Davis, Fl. Turkey, 6: 204 (1978).
 C. dorycnium L. var. *oxysepalus* Boiss., Fl. Orient., 4: 92 (1875) pro parte quoad spec. cypr.
 C. dorycnium L. ssp. *oxysepalus* (Boiss.) Rechinger f. in Oesterr. Bot. Zeitschr., 94: 170 (1948) pro parte quoad spec. cypr.
 TYPE: "*in* Oriente".

Perennial subshrubs; stems erect, much-branched, terete, 30–60 cm. or more high, distinctly hairy towards the base, thinly hairy above, with closely adpressed hairs; leaves sparse, linear or linear-oblong, 10–30 mm. or more long, 3–8 mm. or more wide, apex acute or subacute; basal leaves closely hairy, upper thinly adpressed-hairy; petiole short or wanting in the upper leaves; inflorescence with numerous, divaricate, rigid branches; flowers terminal on the branches, solitary or less commonly in twos or threes; 2 outer sepals concave, oblong, shortly mucronate or acute, about 3 mm. long and 1·8 mm. wide, the outer surface adpressed-hairy; 3 inner sepals similar but rather broader with distinctly mucronate or cuspidate apices; corolla 2–2·5 cm. diam. at mouth, rose-pink with darker stripes, the stripes adpressed silky-hairy externally; stamens 8–9 mm. long; filaments slender, flattened, glabrous; anthers narrowly oblong, about 2 mm. long and 1 mm. wide; ovary glabrous, ovoid-conical, about 1 mm. long and 0·8 mm. wide; style about 2 mm. long, stigma-lobes about 5 mm. long. Fruit brown, globose, about 3–5 mm. diam. surmounted by the persistent style; seed blackish, angular, 2–4 mm. long, covered with very short whitish hairs.

HAB.: Dry stony fields and roadsides; sea-level to 800 ft. alt.; fl. May–June.

DISTR.: Divisions 1, 5, 6, locally common. Greece, Crete and Tunisia eastwards to Iran.

1. Near Paphos, between Kouklia and Yeroskipos, 1862, *Kotschy* 638.
5. Locally common; Lefkoniko, Trikomo, Tymbou, Athalassa, Nisou, etc.; *Holmboe* 842; *Syngrassides* 396! *Lindberg f.* s.n.! *Davis* 3553! *Economides* in ARI 1188! *A. Genneou* in ARI 1543–1546! etc.
6. Between Peristerona and Potami, Yerolakkos, Morphou, Nicosia, Orounda, etc.; *Sintenis & Rigo* 589! *Casey* 759! *Merton* 2417! etc.

NOTES: *C. dorycnium* L. var. *oxysepalus* Boiss. (Fl. Orient., 4: 92; 1875), said to have rather narrower sepals tapering to a longer mucro, does not occur in its extreme form (e.g. *Balansa* 698 from Turkey) in Cyprus. The specimen (*Kotschy* 638) cited by Boissier is best left under *C. dorycnium* var. *dorycnium*. Apart from the questionable taxonomic significance of the variety, it is clear that Boissier was prepared to allow a considerable range of forms to be associated under the varietal epithet.

2. C. oleifolius *Desr.* in Lam., Encycl. Méth., 3: 552 (1792) — as *C. oleaefolius*; Boiss., Fl. Orient., 4: 93 (1875); Holmboe, Veg. Cypr., 145 (1914); Post, Fl. Pal., ed. 2, 2: 204 (1933); Rechinger f. in Arkiv för Bot., ser. 2, 1: 429 (1950); Meikle in Kew Bull., 1956: 547 (1957); Davis, Fl. Turkey, 6: 208 (1978).
    [*C. cyprius* (non Boiss.) Lindberg f., Iter Cypr., 27 (1946).]

Erect, spreading or tufted, leafy subshrub (3–) 10–30 (–40) cm. high; stems terete, much branched from the base, clothed with dense, adpressed, silvery-silky indumentum; leaves oblanceolate, or linear, acute or obtuse, 1–5 cm. or more long, 0·2–1 cm. wide, the lower gradually narrowed to a distinct petiole, the upper sessile or almost so; lamina distinctly or obscurely nerved, greenish or covered with silvery adpressed hairs; flowers terminal in rather congested 2–5-flowered, sub-capitate cymes; pedicels shorter than calyx or almost wanting; bracts subulate; calyx 5-fid, 3 outside sepals lanceolate, 0·7–1·4 cm. long, 0·2–0·3 cm. wide with a slender apical cusp, concave, glabrous internally, often streaked black externally and ± densely clothed with long, spreading, silky hairs; inner sepals similar but rather shorter and wider with a distinct membranous margin; corolla 2–3 cm. diam., pink; stamens unequal, the shorter about 7 mm. long, the longer 10–11 mm. long, attached 2–3 mm. above base of corolla; filaments slender, glabrous; anthers narrowly oblong, about 3 mm. long, 1 mm. wide; ovary ovoid-conical, about 1·5 mm. long, 1 mm. wide, densely hairy; style 3–5 mm. long, hairy; stigma-lobes 2–3 mm. long. Capsule broadly ovoid, about 4 mm. long, 2–3 mm. wide, subglabrous towards base, hairy above; seeds about 3 mm. long and almost as wide; testa dark brown, minutely rugulose, subglabrous or shortly hairy.

var. **oleifolius**
    TYPE: "vient du Levant, & est cultivée au Jardin du Roi" (P).

Spreading subshrub 10–30 (40–) cm. high; leaves numerous, oblanceolate; bracts and calyx usually rather densely clothed with soft, spreading hairs. *Plate 69, figs. 1–3.*

HAB.: In garigue and on dry rocky slopes; from sea-level to 2,000 ft. alt.; fl. March–June.

DISTR.: Divisions 1, 3, 4, 6–8, locally common. Sicily, Malta, Greece, Crete, Aegean Islands, Palestine, Egypt, Libya.

1. Common; Akamas, Ktima, Polis, etc.; *Lindberg f.* s.n.! *Davis* 3286! *Kennedy* 1601! 1632! *Merton* 2125! 2168! etc.
3. Common; Episkopi, Curium, Pissouri, between Pano Arkhimandrita and Kouklia, Akrotiri, Cape Gata, etc. *Kotschy*; *Kennedy* 1561! 1804! *Mavromoustakis* 23! *Casey* 1653–1655! *Meikle* 2914! etc.
4. Cape Greco, 1958, *N. Macdonald* 4! and, 1962, *Meikle* 2600!
6. Nicosia, 1931, *J. A. Tracey* 40!
7. Between Kyrenia and Lapithos, 1862, *Kotschy* 480! 485; Dhavlos, 1933, *Chapman* 41! and, 1957, *H. Painter* 22! Lefkoniko Pass, 1955, *Merton* 2289! W. of Kyrenia, 1967, *Merton* in ARI 530!
8. Common; Yialousa, Ephtakomi, between Rizokarpaso and Ayios Philon, Apostolos Andreas Monastery, etc.; *Sintenis & Rigo* 58! *Holmboe* 519! *Davis* 2252! 2307! 2449! *Meikle* 2476! *A. Genneou* in ARI 1453! etc.

var. **deserti** *Pamp.* in Archivio Bot., 12: 40 (1936); Meikle in Kew Bull., 1956: 548 (1957).
   [*C. cyprius* (non Boiss.) Chapman, Cyprus Trees and Shrubs, 63 (1949).]
   TYPE: Cyrenaica, without precise locality, *Pampanini* (FI).

Rigid, broom-like, ± erect, much-branched shrub (10–) 15–40 cm. high, the branches often leafless towards the base; leaves narrowly linear, 1–1·5 cm. long, about 0·2 cm. wide, silvery-sericeous; inflorescence often rather lax and branched; bracteoles and calyx with rather short spreading hairs; outer sepals about 9 mm. long, 3 mm. wide, inner sepals 7 mm. long, 2·5 mm. wide. Seeds dull brown, shortly hairy, 3–3·5 mm. long, 2–2·5 mm. wide. *Plate 69, fig. 4.*

HAB.: On dry, rocky, calcareous hillside; c. 500 ft. alt.; fl. March–June.

DISTR.: Division 5, locally common. Also Libya.

5. Athalassa, 1932, *Syngrassides* 381! Also, same locality, *Davis* 3195! *Kennedy* 1721! 1806! *Chapman* 630! *Seligman* in *Casey* 1657!

NOTES: Very distinct and uniform, and apparently restricted to this locality.

var. **pumilus** *Pamp.* in Archivio Bot., 12: 41 (1936); Meikle in Kew Bull., 1956: 548 (1957); Turrill in Bot. Mag., 172: n.s. t. 324 (1958).
   TYPE: Cyrenaica, without precise locality, *Pampanini* (FI).

Dwarf, much-branched shrublet, 3–10 cm. high, forming dense, rounded tufts 5–10 cm. diam.; leaves small, oblanceolate, 1–2 cm. long, up to about 0·6 cm. wide, conspicuously silvery-sericeous; inflorescences usually few-flowered; bracteoles and calyx long-hairy; outer sepals about 7 mm. long, 2·5 mm. wide; inner sepals 7 mm. long, about 3 mm. wide. *Plate 69, fig. 5.*

HAB.: On sand-covered rocks and amongst loose sand close to seashore; near sea-level; fl. March–June.

DISTR.: Division 6. Also Libya.

6. Ayia Irini, 1933, *Kennedy* 1783! 1784! 1805! Also, same locality, *Seligman* in *Casey* 1656! *Barclay* 1075!

NOTES: A maritime ecotype of *C. oleifolius*, apparently retaining its characteristics in cultivation.

[**3. C. cantabrica** *L.*, Sp. Plant., ed. 1, 158 (1753); Kotschy in Oesterr. Bot. Zeitschr., 12: 277 (1862); Boiss., Fl. Orient., 4: 95 (1875); Post, Fl. Pal., ed. 2, 2: 205 (1933); Davis, Fl. Turkey, 6: 206 (1978).
   TYPE: "*in* Italia, Sicilia, Narbona, Verona".

Tufted perennial with a tough woody rootstock; stems erect, spreading or decumbent, up to 50 cm. or more long, terete, ± densely clothed with long

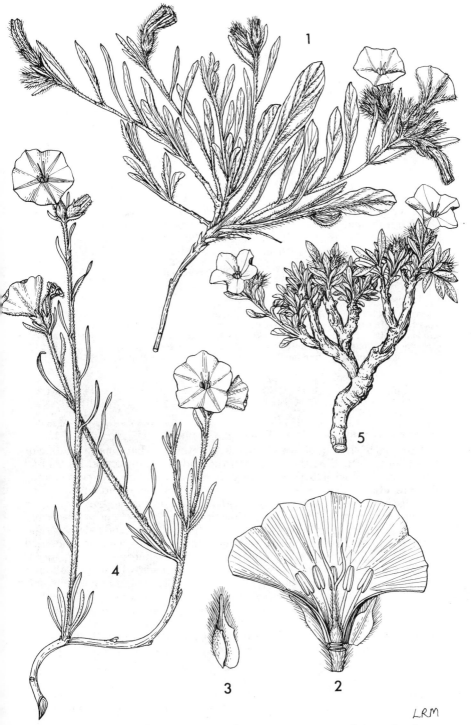

LRM

Plate 69. Figs. **1–3.** CONVOLVULUS OLEIFOLIUS Desr. var. OLEIFOLIUS **1,** habit, ×⅔; **2,** flower, ×2; **3,** inner sepal, ×2; fig. **4.** CONVOLVULUS OLEIFOLIUS Desr. var. DESERTI Pamp. **4,** habit, ×⅔; fig. **5.** CONVOLVULUS OLEIFOLIUS Desr. var. PUMILUS Pamp. **5,** habit, ×⅔. (**1–3** from *Kennedy* 1954; **4** from *Kennedy* 1721; **5** from *Seligman* 1656.)

adpressed or spreading hairs; basal leaves oblong or spathulate, 2–4 cm. long, 0·5–1 (–1·5) cm. wide, venation often rather distinctly impressed, apex obtuse or subacute, base gradually narrowed to a distinct petiole 2 cm. or more long, lamina green, thinly clothed with long, subadpressed hairs; upper leaves usually shorter and narrower, sometimes linear or linear-lanceolate, sessile or very shortly petiolate; inflorescences terminal on slender leafless branches often 7 cm. or more long, arising from the axils of the upper leaves; flowers usually in a loose cluster of about 3; bracteoles linear-subulate, 0·5–1 cm. or more long, ± densely villose; calyx hirsute, outer sepals narrowly ovate-acuminate about 7–8 mm. long and 4 mm. wide, inner sepals a little shorter, sometimes obtuse with a cuspidate apex; corolla pink, about 2·5 cm. diam.; stamens unequal, 8–10 mm. long, attached about 2 mm. above base of corolla-tube; filaments flattened; anthers narrow, sagittate, about 2 mm. long; ovary ovoid-conical, hairy, about 1·5 mm. long and 1 mm. wide; style about 5 mm. long, hairy; stigma-lobes 3–4 mm. long. Capsule subglobose, hairy, about 5–6 mm. diam.; seeds angular, dark grey-brown, about 2·5 mm. long, ± densely covered with short adpressed hairs.

HAB.: Dry, calcareous pastures; c. 100 ft. alt.; fl. March–June.

DISTR.: Division 1. Widespread in the Mediterranean region.

1. "Plains of Paphos", 1862, *Kotschy*. See Notes.

NOTES: This is the sole record for *C. cantabrica* from Cyprus and is not repeated in Unger & Kotschy's "Die Insel Cypern" (1865). In the circumstances it seems likely that it was a misidentification of *C. dorycnium* L., recorded by Kotschy from the area, but since *C. cantabrica* is common in almost every country adjacent to Cyprus, it is thought advisable to insert the record and description, in the hope that (like *C. humilis*) the plant may be refound.]

4. C. lineatus *L.*, Fl. Monspel., 11, (1756); Boiss., Fl. Orient., 4: 97 (1875); Post, Fl. Pal., ed. 2, 2: 205 (1933); Osorio-Tafall et Seraphim, List Vasc. Plants Cyprus, 84 (1973); Davis, Fl. Turkey, 6: 207 (1978).

TYPE: France; Montpellier, "in incultis sterilibus ultra *La Colombière* quà itur ad fontem *Picquet*, & circa Salenovam" *Magnol* (Bot. Monsp., 74; 1676).

Perennial with a tough woody rootstock; stems herbaceous, spreading or decumbent, clothed at the base with the scaly remains of old leaves and densely covered with adpressed silky hairs; basal leaves narrowly oblong, obovate, oblanceolate or elliptic, up to about 7 cm. long and 1·5 cm. wide, with distinctly impressed nervation, apex acute or obtuse, base gradually narrowed to a distinct flattened petiole, lamina ± densely adpressed-silky on both surfaces; upper leaves similar but usually shorter and narrower, and sessile or shortly petiolate; flowers axillary and terminal in lax or condensed 1–5-flowered, shortly stalked clusters; bracteoles linear-subulate, inconspicuous; calyx adpressed-silky, sepals subequal, narrowly oblong, about 6 mm. long and 2–2·5 mm. wide, apex rather broadly acuminate; corolla pale pink, 2–2·5 cm. diam.; stamens attached near base of corolla-tube, unequal, the longest about 6 mm. long; anthers narrowly oblong about 2·5 mm. long and 1 mm. wide; ovary narrowly ovoid-conical, about 2 mm. long, densely hairy; style about 5 mm. long, hairy; stigma-lobes about 5 mm. long. Capsule 6 mm. long and 5 mm. wide; seeds angular, dark brown, hairy, 4 mm. long and 3 mm. wide.

HAB.: Sandy banks; near sea-level; fl. April–June.

DISTR.: Division 3, rare. S. Europe and Mediterranean region eastwards to Central Asia.

3. Akrotiri, 1941, *Davis* 3573! Cape Gata, 1954, *Seligman & Scott-Moncrieff* in *Casey* 1654 (partly)! Same locality, 1962, *Meikle* 2912!

NOTES: Some of the material cited above approaches *C. oleifolius* Desr., and is mixed with specimens of the latter. It may be referable to *C.* × *cyprius* Boiss. rather than to "pure" *C. lineatus* (see Notes under 4a *C.* × *cyprius*).

4a C. × cyprius *Boiss.*, Fl. Orient., 4: 93 (1875) pro sp.; Holmboe, Veg. Cypr., 145 (1914); Osorio-Tafall et Seraphim, List Vasc. Plants Cyprus, 84 (1973).

C. lineatus L. var. *angustifolius* Kotschy ex Sa'ad, The Convolvulus Species, 130 (1967); Kotschy in Unger et Kotschy, Die Insel Cypern, 285 (1865) nom. nud.; H. Stuart Thompson in Journ. Bot., 44: 334 (1906) nom. nud.

C. lineatus L. × *C. oleifolius* Desr.

TYPE: Cyprus; "in fissuris rupium regionis inferioris Cypri prope Laminas [Lania] ad promontorium Capo Gatto [Cape Gata] (Ky. 627!)" (G, W!).

Similar to *C. lineatus* L., but with more persistent, subligneous stems, narrower, linear-lanceolate leaves, usually less than 5 mm. wide, and with the flowers commonly clustered towards the apex of the stems as in *C. oleifolius*, from which it differs in the calyx being clothed with adpressed hairs.

HAB.: On sandy banks, with *C. lineatus* and *C. oleifolius*; near sea-level; fl. April–June.

DISTR.: Division 3, locally frequent. ? Endemic.

3. Cape Gata, near Lania, 1862, *Kotschy* 627! Same locality, 1951, *Mavromoustakis* 23 (partly)! and, 1962, *Meikle* 2908!

NOTES: I have already suggested (Kew Bull. 1956: 548–549; 1957) that *C. cyprius* Boiss. may be a hybrid. Since then an opportunity to visit the type locality has convinced me that the suggestion is correct. Both putative parents grow together at Cape Gata, and *C. cyprius* is but one of a series of interconnecting plants. Indeed there is some evidence that introgression may be extinguishing "pure" *C. lineatus* at Cape Gata, which has here its sole station on the island.

As mentioned in the note cited above, two specimens of typical *C. cyprius* are mounted with *C. oleifolius* on a Kew sheet of *Kotschy* 480 ("Inter Cerinia et Lapithos ad margines fruticetorum die 17. Aprilis [1862]"). It seems likely that two separate collectings (*Kotschy* 627 — *C. cyprius*, and *Kotschy* 480 — *C. oleifolius*) were accidentally mixed when the duplicates were distributed.

5. C. betonicifolius *Mill.*, Gard. Dict., ed. 8, no. 20 (1768); Post, Fl. Pal., ed. 2, 2: 207 (1933); Osorio-Tafall et Seraphim, List Vasc. Plants Cyprus, 84 (1973); Davis, Fl. Turkey, 6: 216 (1978).

C. hirsutus M. Bieb., Fl. Taur.-Cauc., 1: 422 (1808); Boiss., Fl. Orient., 4: 105 (1875); Holmboe, Veg. Cypr., 146 (1914); Lindberg f., Iter Cypr., 27 (1946); Rechinger f. in Arkiv för Bot., ser. 2, 1: 429 (1950).

TYPE: Cultivated at Chelsea Physic Garden (BM).

Prostrate or climbing perennial with a slender creeping rootstock; stems distinctly angled, to 1 m. or more in length, ± densely and softly pilose with spreading hairs; leaves cordate-hastate or sagittate, about 3·5–8 cm. long, 2–4·5 cm. wide, apex acute or obtuse, basal lobes rounded, acute or sometimes irregularly toothed, lamina softly pubescent above and below; petiole 1–2·5 cm. or more long; inflorescence axillary, flowers shortly pedicellate, solitary or in loose, few-flowered cymes terminating a stout peduncle up to 12 cm. or more long; bracts linear-subulate to about 1 cm. long, bracteoles similar but smaller; sepals narrowly ovate-acuminate, the outer about 1·4 cm. long and 0·4–0·5 cm. wide, hairy externally with a distinct, slightly reflexed herbaceous apex, the inner rather smaller with a less conspicuous apex; corolla white above, pinkish or purplish towards base, with the stripes pinkish or buff externally, up to about 7 cm. diam., thinly hairy externally; stamens 1·2–1·3 cm. long; filaments papillose at base; anthers narrowly oblong, 4 mm. long and about 1·5 mm. wide; ovary conical, about 2 mm. long, densely covered with long hairs; style about 7 mm. long, glabrous above, hairy towards base; stigma-lobes about 1·5–2 mm. long. Fruit ovoid, about 10 mm. long and 8 mm. wide, hairy; seeds brown, angular, distinctly papillose, about 4·5 mm. long and 3·5 mm. wide.

HAB.: Cultivated fields and vineyards; sea-level to c. 2,000 ft. alt.; fl. April–July.

DISTR.: Divisions 1, 3, 5, 8. S.E. Europe and eastern Mediterranean region eastwards to Iraq.

1. Stroumbi, 1913, *Haradjian* 729! and, 1941, *Davis* 3353!

3. Akhelia, 1939, *Lindberg f.* s.n., and, 1956, *Poore* 23!

5. Syngrasis, 1957, *Merton* 2970!
8. Between Ephtakomi and Leonarisso, 1880, *Sintenis & Rigo* 56! Between Tavros and Koma tou Yialou, 1957, *Poore* 60! Monarga, 1962, *Meikle* 2626!

6. C. althaeoides *L.*, Sp. Plant., ed. 1, 156 (1753); Boiss., Fl. Orient., 4: 106 (1875); Holmboe, Veg. Cypr., 146 (1914); Post, Fl. Pal., ed. 2, 2: 208 (1933); Lindberg f., Iter Cypr., 27 (1946); Osorio-Tafall et Seraphim, List Vasc. Plants Cyprus, 84 (1973); Davis, Fl. Turkey, 6: 214 (1978).
TYPE: "*in* Europa *meridionali*".

Trailing or twining, perennial herb with a slender creeping rootstock; stems subterete, thinly pilose; lowermost leaves cordate or reniform, obtuse, entire or undulate, pilose, 2–7 cm. long, 1·5–6 cm. wide, upper stem-leaves progressively more deeply laciniate, the uppermost commonly divided almost to the base into 5–7 linear, palmately spreading lobes; petioles slender, 1–4 cm. long; peduncles up to 12 cm. long; bracts filiform, pilose; pedicels usually less than 2 cm. long; flowers solitary or often in pairs, rarely 3 or 4; sepals oblong-ovate, thinly pilose with distinctly membranous margins, about 1 cm. long and 0·5–0·7 cm. wide; corolla pink, up to about 6 cm. diam.; stamens about 1·5 cm. long with distinctly flattened filaments; anthers narrowly oblong-linear, about 4 mm. long and 1–1·5 mm. wide; ovary conical, thinly pilose, about 2 mm. long; style glabrous, 6–8 mm. long; stigma-lobes about 2 mm. long. Capsule glabrous; seeds broadly ovoid or subglobose, 4–5 mm. long, 3–4 mm. wide; testa fuscous, conspicuously rugose-verruculose, glabrous.

HAB.: Roadsides, waste ground and rocky slopes; sea-level to 3,000 ft. alt.; fl. March–July.

DISTR.: Divisions 3, 4, 6–8, common in Division 7. Widespread in S. Europe and the Mediterranean region.
3. Near Limassol, 1939, *Lindberg f.* s.n.! also, 1941, *Davis* 3580! and, 1950, *F. J. F. Barrington* 32!
4. Ormidhia, 1880, *Sintenis & Rigo* 59a! Ayia Napa, 1905, *Holmboe* 21; near Famagusta, 1948, *Mavromoustakis* s.n.! Cape Greco, 1958, *N. Macdonald* 8!
6. Near Morphou, 1927, *Rev. A. Huddle* 81! Nicosia, 1950, *Chapman* 465!
7. Common; *Kotschy* s.n.; *Sintenis & Rigo* 59! *Syngrassides* 225! *Davis* 2833! *Casey* 804! *Chapman* 178! *G. E. Atherton* 1213! etc.
8. Rizokarpaso, 1880, *Sintenis & Rigo* s.n.

7. C. arvensis *L.*, Sp. Plant., ed. 1, 153 (1753); Boiss., Fl. Orient., 4: 108 (1875); Holmboe, Veg. Cypr., 146 (1914); Post, Fl. Pal., ed. 2, 2: 209 (1933); Lindberg f., Iter Cypr., 27 (1946); Osorio-Tafall et Seraphim, List Vasc. Plants Cyprus, 84 (1973); Davis, Fl. Turkey, 6: 213 (1978).

Trailing or twining perennial with an extensively creeping rootstock; stems distinctly ribbed, glabrous or thinly pubescent; leaves sagittate or hastate, variable in size, up to about 5 cm. long and 3 cm. wide, glabrous or subglabrous, apex acute or obtuse, lateral lobes usually spreading and acute; petiole rather flattened, up to about 2·5 cm. long; flowers usually solitary, axillary, sometimes in pairs; peduncles to about 5 cm. long; bracts and bracteoles subulate, about 2·5–3 mm. long; pedicels 1–2 cm. long; sepals subequal, oblong, obtuse, truncate or mucronulate, about 4 mm. long and 2·5 mm. wide; corolla white or pink, glabrous, 2–3 cm. diam.; stamens unequal, 8–14 mm. long; filaments papillose towards base; anthers oblong, about 2 mm. long and 1·5 mm. wide; ovary conical, glabrous, about 1·5 mm. long; style about 1 cm. long; stigma-lobes about 3 mm. long. Capsule broadly ovoid or subglobose, about 6 mm. diam.; seeds angular, dark brown, about 3·5 mm. long, distinctly scabridulous.

var. **arvensis**
TYPE: "*in* Europae *agris*".

Leaves with an oblong or ovate-deltoid mid-lobe and short, often deflexed, lateral lobes.

HAB.: Cultivated places, roadsides, waste-land; sea-level to 5,600 ft. alt.; fl. April–Sept.

DISTR.: Divisions 1–8, common in Divisions 2, 4, 7. Widespread in temperate regions of both hemispheres.

1. Dhrousha, 1941, *Davis* 3276!
2. Common; Platres, Troödos, Kakopetria, Prodhromos, etc. *Kennedy* 1354! *Lindberg f.* s.n.; *N. Chiotellis* 461! *G. E. Atherton* 615! *Economides* 1223! etc.
3. Limassol, ? 1905, *Michaelides* s.n.; Amathus, 1963, *J. B. Suart* 40!
4. Common about Larnaca and Famagusta, etc. *Kotschy*; *Syngrassides* 480! *Mavromoustakis* s.n.! etc.
5. Near Apostolos Varnavas Monastery, 1948, *Mavromoustakis* s.n.!
6. Omorphita, 1965, *Huovila* 6! Nicosia, 1973, *P. Laukkonen* 199!
7. Common about Kyrenia; *Casey* 775! *G. E. Atherton* 177! 327! 412! 695! etc.
8. Near Yialousa, 1962, *Meikle* 2451!

var. **linearifolius** *Choisy* in DC., Prodr., 9: 407 (1845); Hayek, Prodr. Fl. Pen. Balc., 2: 39 (1931).
   C. cherleri Agardh in Roem. et Schult., Syst. Veg., 4: 261 (1819).
   C. arvensis L. var. *cherleri* (Agardh) Hal., Consp. Fl. Graec., 2: 307 (1902); Lindberg f., Iter Cypr., 27 (1946).
   TYPE: U.S.S.R.; by the shores of the Caspian, *Karelin* (LEN).

Leaves with a narrowly oblong or linear mid-lobe and spreading acute lateral lobes.

HAB.: Cultivated places, roadsides and waste-land; sea-level to 650 ft. alt.; fl. April–Sept.

DISTR.: Divisions 3, 6. Mediterranean region, western Asia.

3. Akhelia, 1939, *Lindberg f.* s.n.!
6. Kokkini Trimithia, 1955, *Merton* 2405!

NOTES: Not convincing even as a variety; perhaps a seasonal condition of the typical plant.

8. **C. coelesyriacus** *Boiss.*, Diagn., 1, 11: 85 (1849); Fl. Orient., 4: 109 (1875) as *C. caelesyriacus*; H. Stuart Thompson in Journ. Bot., 44: 334 (1906); Holmboe, Veg. Cypr., 146 (1914); Post, Fl. Pal., ed. 2, 2: 209 (1933); Osorio-Tafall et Seraphim, List Vasc. Plants Cyprus, 84 (1973).
   C. sintenisii Boiss., Fl. Orient., Suppl., 349 (1888); Holmboe, Veg. Cypr., 146 (1914); Rechinger f. in Arkiv för Bot., ser. 2, 1: 429 (1950).
   TYPE: Lebanon; "in cultis Coelesyriae inter *Hasbeya* et *Rascheya* et colui e seminibus relatis" *Boissier* (G).

A decumbent or ascending annual with thinly pubescent, angled or obscurely ribbed stems up to 30 cm. (or occasionally more) long; lower leaves rather distant, broadly cordate or reniform (1·5–) 2–4 (–5) cm. long, (1–) 1·5–3 (–4) cm. wide, thinly pubescent or subglabrous, margins entire or undulate; petiole up to 10 cm. long, thinly pubescent, canaliculate above; upper leaves similar, but smaller and more acute, with shorter petioles, the uppermost often irregularly dentate or divided palmately into 5 acute lobes; flowers axillary, solitary, peduncle slender at first, thickening and reflexing in fruit, 1–3 cm. long, thinly pubescent; bracts filiform, pubescent, 3–4 mm. long; pedicels 0·8–1·5 cm. long; sepals hairy, broadly oblong, 3–5 mm. long and almost as wide, with an obtuse, truncate or emarginate apex, the midrib frequently produced into a long, filiform, hairy cusp; corolla pink or rose-purple, 1·5–2 (–2·8) cm. diam., ribs hairy externally; stamens subequal, about 6 mm. long; filaments flattened, papillose towards base; anthers oblong about 1·5 mm. long and 1 mm. wide; ovary ovoid-conical about 1·2 mm. long, glabrous; style glabrous, about 4 mm. long; stigma-lobes about 2 mm. long. Capsule glabrous, globose, 6–8 mm. diam.; seeds dark brown, obscurely angular, 3–4 mm. long; testa distinctly verruculose.

HAB.: On rocky or stony slopes and banks, or in stony fallow fields; sea-level to 2,800 ft. alt.; fl. Febr.–April.

DISTR.: Divisions 1, 3, 7, 8, common in Division 7. Also Lebanon and Palestine.

1. Ayios Neophytos, 1901, *A. G. & M. E. Lascelles* s.n.! Akamas, Kako Skala, 1950, *Casey* 1015! Ayios Yeoryios (Paphos distr.), 1960, *N. Macdonald* 48! and, 1962, *Meikle* 2007!
3. Akhelia, 1956, *Merton* 2499! Kithasi, 1981, *Hewer* 4783!
7. Common; Kornos, Kyparissovouno, Kyrenia, Aghirda, Larnaka tis Lapithou, etc. *Sintenis & Rigo* 55! *Syngrassides* 1194! *Davis* 2486! 2912! 2979! 3033! *Casey* 130! *Kennedy* 1658! *G. E. Atherton* 1380! *Osorio-Tafall & Seraphim* 10235! *I. M. Hecker* 46! etc.
8. Rizokarpaso, 1912, *Haradjian* 152; Apostolos Andreas, 1941, *Davis* 2295! Platanisso (Karpas), 1950, *Chapman* 241a! Ronnas Valley, 1957, *Merton* 2944!

NOTES: Examination of isotype material of *Convolvulus sintenisii* Boiss. shows that, contrary to the original description, the upper leaves of some of the specimens are lobed, and that the capsules are glabrous. The other characters cited by Boissier are of trivial importance; in the absence of trustworthy alternative distinctions, I have been obliged to unite *C. sintenisii* with *C. coelesyriacus.*

9. C. siculus *L.*, Sp. Plant., ed. 1, 156 (1753); Kotschy, Die Insel Cypern, 286 (1865); Boiss., Fl. Orient., 4: 109 (1875); Holmboe, Veg. Cypr., 146 (1914); Post, Fl. Pal., ed. 2, 2: 210 (1933); Osorio-Tafall et Seraphim, List Vasc. Plants Cyprus, 84 (1973); Davis, Fl. Turkey, 6: 212 (1978).
TYPE: "*in* Sicilia" (LINN).

Trailing or spreading annual, stems subterete, distinctly pubescent with curling hairs; leaves variable in size, 1·5–6 cm. long, 1–4 cm. wide, oblong or cordate, thinly hairy or subglabrous, acute or acuminate, usually with distinct nervation; petiole short, channelled, rarely more than 1·5 cm. long, often very much shorter; flowers axillary, solitary; peduncle slender, usually much shorter than the subtending leaf, and seldom more than 2 cm. long, conspicuously reflexed in fruit; bracts subulate or linear-lanceolate, situated close to the flower, usually about 4–7 mm. long, but occasionally up to 20 mm. long and 4 mm. wide; pedicels rarely more than 3 mm. long; sepals hairy, ovate-acute, about 7 mm. long and 2·5–3 mm. wide, gradually narrowed to a long, cuneate, purple-stained base; corolla small, distinctly lobed, blue, about 0·8–1·3 cm. diam., glabrous or hairy; stamens unequal, the longest about 4 mm. long; filaments smooth throughout; anthers narrowly oblong, about 1 mm. long; ovary conical, glabrous, about 1·5 mm. long; style about 5 mm. long; stigma-lobes about 2 mm. long. Capsule globose, about 5 mm. diam., glabrous; seeds angular, about 2·5 mm. long; testa dark brown, obscurely scabridulous.

HAB.: On dry hillsides and in rocky places; sea-level to 3,000 ft. alt.; fl. Febr.–May.

DISTR.: Divisions 1, 3, 4, 6–8. Widespread in the Mediterranean region.

1. Below Ayia Nicola, 1941, *Davis* 2309! Near Ayios Yeoryios (Akamas), 1962, *Meikle* 2014!
3. Near Kouklia, 1960, *N. Macdonald* 21! Amathus, 1964, *J. B. Suart* 161! Paramali, 1964, *J. B. Suart* 210! Kithasi, 1981, *Hewer* 4785!
4. Cape Greco, 1862, *Kotschy* 135!
6. Near Orga, 1956, *Merton* 2581!
7. Common! Kornos, Kyrenia, Yaïla, Melounda, etc. *Sintenis & Rigo* 54! *Davis* 2506! 2854! 3032! *G. E. Atherton* 1019! *J. B. Suart* 77! etc.
8. Common; Koma tou Yialou, Apostolos Andreas, Ronnas Bay, Ayios Theodoros, etc. *Syngrassides* 706! Davis 2294! 2341! *Mavromoustakis* s.n.! etc.

10. C. pentapetaloides *L.*, Syst. Nat., ed. 12, 3: 229 (1768); Unger et Kotschy, Die Insel Cypern, 285 (1865); Boiss., Fl. Orient., 4: 110 (1875); Holmboe, Veg. Cypr., 146 (1914); Post, Fl. Pal., ed. 2, 2: 210 (1933); Lindberg f., Iter Cypr., 27 (1946); Osorio-Tafall et Seraphim, List Vasc. Plants Cyprus, 84 (1973); Davis, Fl. Turkey, 6: 212 (1978).
TYPE: Balearic Islands; "*in* Majorca" *Gerard* (LINN).

An erect or spreading annual, up to about 40 cm. high; stems somewhat angular, ± densely pubescent; leaves narrowly obovate or oblanceolate, up to about 5 cm. long and 1·5 cm. wide, but usually much smaller, thinly pubescent, the lowermost with the bases gradually narrowed to a short flattened petiole, the upper sessile, rounded at the base or sometimes almost amplexicaul; flowers axillary, solitary; peduncle about 2 cm. long, distinctly hairy, reflexed in fruit; bracts minute, filiform or subulate, about 1–2 mm. long; pedicels variable in length, up to about 1 cm. long; sepals oblong, acute, about 6 mm. long and 3 mm. wide, sparsely hairy; corolla distinctly 5-lobed, blue, about 1–1·5 cm. diam., hairy externally; stamens unequal, 3–4 mm. long; filaments flattened, glabrous; anthers oblong, about 0·8 mm. long; ovary subglobose, thinly hairy, about 1 mm. diam.; style about 1·8 mm. long; stigma-lobes each about as long as the style. Capsule globose, about 6 mm. diam., glabrous or thinly hairy, seeds angular, about 2 mm. long; testa dark brown, echinate-papillose.

HAB.: In cultivated and waste ground, on roadside banks, and on stony, well-drained pastures and slopes; sea-level to 3,000 ft. alt.; fl. March–May.

DISTR.: Divisions 1–4, 6–8, common in Division 7. Widespread in the Mediterranean region and eastwards to Iraq.

1. Avgas valley near Ayios Yeoryios (Akamas), 1962, *Meikle* 2006!
2. Between Khyrsorroyiatissa Monastery and Statos, 1941, *Davis* 3414!
3. Limassol, 1862, *Kotschy* 484; also, 1939, *Lindberg f.* s.n.! and, 1950, *Chapman* 560! Sotira, 1905, *Holmboe* 408; Kakomallis near Louvaras, 1962, *Meikle* 2900!
4. Cape Greco, 1862, *Kotschy* 135; near Athienou, 1862, *Kotschy* 174; Pergamos, 1955, *Merton* 1968! and, 1956, *Merton* 2630! Dhromolaxia, 1970, *A. Genneou* in ARI 1417!
6. Evrykhou, 1935, *Syngrassides* 829!
7. Common; Pentadaktylos, Kyrenia, Ayios Epiktitos, Larnaka tis Lapithou, Ayios Khrysostomos, St. Hilarion, etc. *Sintenis & Rigo* 60! *Syngrassides* 903! 1030! *Davis* 2505! 2819! 2917! 2973! *Casey* 346! *G. E. Atherton* 1266! *Merton* in ARI 317! 336! etc.
8. Cape Andreas, 1880, *Sintenis & Rigo*; Ronnas Valley, Rizokarpaso, 1957, *Merton* 2943!

11. **C. humilis** *Jacq.*, Collect., 4: 209, t. 22, fig. 2 (1791); Post, Fl. Pal., ed. 2, 2: 210 (1933); A. K. Jackson in Kew Bull., 1937: 344 (1937); Osorio-Tafall et Seraphim, List Vasc. Plants Cyprus, 84 (1973).
   C. undulatus Cav., Icones, 3: 39, t. 277 (1796); Kotschy, Die Insel Cypern, 285 (1865); Boiss., Fl. Orient., 4: 110 (1875).
   C. evolvuloides Desf., Fl. Atlant., 1: 176, t. 49 (1798); Sibth. et Sm., Fl. Graec. Prodr., 1: 134 (1806), Fl. Graec., 2: 81, t. 198 (1813).
   TYPE: "Patriam ignoro", cult. Vienna. *Jacquin*, Collect., 4: t. 22, fig. 2!

A decumbent or ascending, thinly pilose annual, to about 25 cm. high, usually with numerous branches arising from a central rosette; lower leaves spathulate, obtuse, tapering to a flattened indistinct petiole, 0·5–4 cm. long, 0·3–1 cm. wide, the upper often larger, oblong, obtuse, sessile or sometimes distinctly amplexicaul with rounded auricles; flowers sessile or subsessile in the axils of the uppermost leaves, by which they are more or less concealed; sepals glabrous or subglabrous, often streaked with black, ovate, acuminate, 2–2·5 mm. long, 1·5–2 mm. wide; corolla blue with a pale tube, 4 or 5 times as long as calyx, with 5 shallow triangular lobes, hairy on the ribs externally; stamens unequal, 3–4 mm. long; anthers oblong-sagittate, about 1 mm. long; ovary densely clothed with long erect hairs, subglobose, about 1·5 mm. diam.; style about 2 mm. long, glabrous; stigma-lobes linear, 2–2·5 mm. long. Capsule globose, thinly hairy, about 7–8 mm. diam.; seeds angular, about 3 mm. long; testa dark fuscous-brown, very strongly rugose-verruculose.

HAB.: In fields and by roadsides; sea-level to 750 ft. alt.; fl. March–June.

DISTR.: Divisions 5, 6, 7, rare. Also Portugal, Spain, Italy, Sicily, Morocco, Algeria, Libya, Palestine.

5. Salamis, among the tombs, 1971, *Townsend* s.n.!
6. South of Aghirda, 1954, *Casey* 1329!
7. Near Buffavento Castle, 1880, *Sintenis & Rigo* s.n.; near Ayios Yeoryios, 1880, *Sintenis & Rigo* 57! Dhikomo, 1936, *Syngrassides* 1195!

4. CRESSA *L.*

Sp. Plant., ed. 1, 223 (1753).
Gen. Plant., ed. 5, 104 (1754).

Small subshrubs with entire leaves, or occasionally leafless; flowers solitary in the axils of the upper leaves, sometimes appearing to form a dense capitulum; sepals free, imbricate; corolla small, deeply 5-lobed; stamens 5, alternating with the corolla lobes, exserted; ovary 2-locular, containing 4 ovules; styles 2; stigmas capitate. Capsule 2–4-valved, generally 1–2-seeded. Pollen grains tricolpate, exine not spinose.

A cosmopolitan genus, comprising at least 2, and possibly several species, as yet inadequately analysed. One species in Cyprus.

1. **C. cretica** *L.*, Sp. Plant., ed. 1, 223 (1753) [Amoen. Acad., 1: 395 (1747)]; Unger et Kotschy, Die Insel Cypern, 284 (1865); Boiss., Fl. Orient., 4: 114 (1875); Holmboe, Veg. Cypr., 146 (1914); Post, Fl. Pal., ed. 2, 2: 213 (1933); Lindberg f., Iter Cypr., 27 (1946); Osorio-Tafall et Seraphim, List Vasc. Plants Cypr., 85 (1973); Davis, Fl. Turkey, 6: 198 (1978).
   TYPE: "*in* Cretae *litoribus salsis*".

A slender erect or spreading subshrub, 5–25 cm. high, with a tough woody rootstock and numerous, hairy, obscurely angled branches; leaves sessile, ovate or lanceolate, 1·5–6 mm. long, 1–4 mm. wide, hairy on both surfaces, each of the lower leaves usually subtending an axillary tuft of smaller leaves; flowers scented, sessile or subsessile in the axils of the upper leaves, often crowded into a dense spike or cluster; bracts 1 or 2, linear-oblong, about 2·5 mm. long and 1 mm. wide; sepals 5 (occasionally 4), hairy externally, obovate, about 3·5 mm. long and 1–2 mm. wide; corolla small, whitish or tinged purple, rarely more than 4 mm. across, with a tube about as long as the calyx, and 5 narrowly triangular, spreading or recurved lobes about 1·5 mm. long; stamens 4–5 mm. long, with slender glabrous filaments; anthers shortly oblong, less than 1 mm. long; ovary ovoid about 1·8 mm. long, clothed with long erect hairs; styles glabrous, about 4–5 mm. long; stigmas rather large, capitate. Capsule ovoid, thinly hairy, about 4 mm. long; seed ovoid, sometimes with one side flattened, about 2 mm. long; testa brown, minutely asperate.

HAB.: Sandy ground; mud flats; lake shores, etc., usually under saline conditions; near sea-level; fl. June–Oct.

DISTR.: Divisions 1, 3, 4, 6, common in Divisions 3, 4. Mediterranean region and eastwards to India; tropical Africa; ? Australia ? N. America.

1. Near Paphos, 1939, *Lindberg f.* s.n.
3. Common about Limassol; *Kotschy* s.n.; *Lindberg f.* s.n.; *Mavromoustakis* 36! 44! 46! *Davis* 3574!
4. Common about Larnaca and Famagusta; *Sintenis & Rigo* 755! *Haradjian* 1017! *Lindberg f.* s.n.! *Merton* 2256! *Erik Julin* s.n.! *A. Genneou* in ARI 1599!
6. Syrianokhori, 1954, *Merton* 1960!

DICHONDRA MICRANTHA *Urban*, Symb. Antill., 9: 243 (1924) (*D. repens* auctt.), from E. Asia, a prostrate perennial, with slender stems rooting at the nodes, small, long-petiolate, reniform leaves, and minute whitish flowers with a deeply 5-lobed corolla and 2-lobed ovary, is sometimes planted as ground cover, and survives as a weed in lawns. It has been collected in the Municipal Gardens, Nicosia (1974, *Meikle* 4084!).

5. CUSCUTA *L.*

Sp. Plant., ed. 1, 124 (1753).
Gen. Plant., ed. 5, 60 (1754).
T. G. Yuncker in Memoirs Torrey Bot. Club, 18: 113–331 (1932).
N. Feinbrun in Israel Journ. Bot., 19: 16–29 (1970).

Parasitic herbs, without, or with only a trace of chlorophyll; stems filiform, twining, attached to the host plants by haustoria; leaves reduced to minute scales; flowers small, usually in cymose clusters; sepals ± united, 3–5-merous; corolla 3–5-merous, with small fringed or fimbriated scales attached internally opposite the stamens; stamens inserted in the throat of the corolla, alternating with the corolla-lobes; ovary 2-locular, each loculus containing 2 ovules. Fruit a capsule, dehiscing irregularly or by a line of circumscission near the base; embryo without cotyledons or with rudimentary cotyledons. Pollen grains tricolpate (or sometimes 4–6-colpate), exine pitted or more rarely reticulate.

About 150 species scattered almost throughout the world, but most abundantly represented in North and South America.

Styles united; stigma capitate - - - - - - - - - **1. C. monogyna**
Styles 2, free:
  Stigmas capitate - - - - - - - - - - **2. C. campestris**
  Stigmas elongate, cylindrical, not thicker than style:
    Glomerules of flowers about 10 mm. diam.; stems simple or sparingly branched (a parasite
      of cultivated Flax)- - - - - - - - C. EPILINUM (p. 1179)
    Glomerules of flowers 3–7 mm. diam.; stems much branched:
      Corolla-lobes 4, (rarely 3 or 5), connivent or suberect; anthers almost orbicular, scarcely
        longer than wide; flowers in small irregular clusters - - **3. C. palaestina**
      Corolla-lobes 5 (rarely 4), spreading; anthers oblong, distinctly longer than wide; flowers
        in dense, compact clusters - - - - - - - **4. C. planiflora**

1. C. monogyna *Vahl*, Symb. Bot., 2: 32 (1791); Boiss., Fl. Orient., 4: 121 (1875); Holmboe, Veg. Cypr., 146 (1914); Yuncker in Mem. Torrey Bot. Club, 18: 256 (1932); Post, Fl. Pal., ed. 2, 2: 215 (1933); Osorio-Tafall et Seraphim, List Vasc. Plants Cyprus, 85 (1973); Davis, Fl. Turkey, 6: 236 (1978).
    [*C. major* (non Choisy) P. Mouillefert, Report on the Vineyards of Cyprus, 3–13 (1893).]
    TYPE: "In Oriente" *Vahl* (C).

Stems rather robust (1–2 mm. diam.), tinged purple; flowers distinctly pedicellate, in irregular 2–8-flowered cymose clusters; bracts broadly oblong, obtuse, concave, 1–2 mm. long and almost as wide; calyx glabrous, divided almost to the base into 5 concave suborbicular lobes about 2 mm. diam.; corolla pale violet, 2·5–3 mm. long, narrowing to a distinct median zone above which it expands into 5 bluntly ovate or ovate-deltoid suberect lobes which are ± papillose internally; anthers sessile, tinged blue, oblong, about 0·8 mm. long and 0·5 mm. wide; ovary subglobose, 1–1·5 mm. diam., glabrous; styles united, about 0·5 mm. long; stigmas capitate. Capsule circumscissile, globose, 5–8 mm. diam., surmounted by the persistent style and stigma; seeds flattened, reniform, about 3 mm. wide; testa dark brown, minutely rugulose.

    HAB.: Parasitic on *Vitis, Pistacia, Cannabis, Citrus* and other plants; 500–4,500 ft. alt.; fl. June–Sept.

    DISTR.: Divisions 2, 3, 6. S. Europe and Mediterranean region eastwards to Afghanistan and Pakistan.

2. Prodhromos, 1893, *Mouillefert*; Omodhos, 1905, *Holmboe* 1151; Platres, 1937, *Kennedy* 314 ! Aphamis, 1981, *Meikle* !
3. Beyond Trimiklini towards Limassol, 1935, *Syngrassides* 713 !
6. Kato Kopia, 1970, *A. Genneou* in ARI 1555 !

2. C. campestris *Yuncker* in Mem. Torrey Bot. Club, 18: 138 (1932); Feinbrun in Israel Journ.
Bot., 19: 20 (1970); Davis, Fl. Turkey, 6: 225 (1978).
TYPE: U.S.A.; Texas, 1843, *F. Lindheimer* (MO, K !).

Stems slender, much branched, yellow; flowers shortly pedicellate, in
rather irregular clusters 6–10 mm. diam.; bracts ovate, acute, about 1·5 mm.
long, 1 mm. wide; calyx glabrous, not very fleshy, with 5 overlapping,
suborbicular lobes about 1·5 mm. long and almost as wide; corolla cup-
shaped, glabrous, slightly exceeding calyx, with 5 (4), triangular, acute,
erect or reflexed lobes, about 0·5 mm. long; filaments about 0·5 mm. long,
attached near apex of corolla-tube, tapering upwards; anthers exserted,
shortly oblong, about 0·3 mm. long, 0·2 mm. wide, connective not forming
an apiculus; corolla-scales blunt, long-fimbriate, extending to the base of the
stamen; ovary subglobose, about 1·5 mm. diam.; styles 2, free, about 0·5
mm. long; stigmas capitate. Capsule depressed-globose, about 4 mm. diam.,
circumscissile towards base; seeds broadly oblong or subglobose, about 1·5
mm. long and almost as wide; testa dark brown, minutely and obscurely
rugulose.

HAB.: Parasitic on a wide variety of herbs; ? sea-level to 500 ft. alt.; fl. April–Oct.

DISTR.: Division 8, an introduced weed. Probably of North American origin, now found in
almost every part of the world as a weed of cultivation.

8. Ayios Andronikos (Yialousa), on Alfalfa (*Medicago sativa* L.), 1970, *A. Genneou* in ARI
1588!

NOTES: Feinbrun (*loc. cit. supra*) notes that it "often causes serious damage to fields of
cultivated clover and alfalfa, its seeds being apparently sown together with those of the host".

3. C. palaestina *Boiss.*, Diagn., 1, 11: 86 (1849), Fl. Orient., 4: 116 (1875); Unger et Kotschy, Die
Insel Cypern, 286 (1865); Yuncker in Mem. Torrey Bot. Club, 18: 279 (1932); Post, Fl. Pal.,
ed. 2, 2: 214 (1933); Osorio-Tafall et Seraphim, List Vasc. Plants Cyprus, 85 (1973); Davis,
Fl. Turkey, 6: 229 (1978).
C. globularis Bert., Fl. Ital., 7: 625 (1850); Holmboe, Veg. Cypr., 146 (1914).
[*C. epithymum* (non Murr.) H. Stuart Thompson in Journ. Bot., 44: 334 (1906) pro parte;
Lindberg f., Iter Cypr., 27 (1946).]
TYPE: "in Palaestinâ in *Poterio spinoso* parasitica (Boiss.)" (G).

Stems very slender (less than 1 mm. diam.) reddish or yellow; flowers
sessile or shortly pedicellate, in small, rather irregular clusters; bracts ovate-
acute, concave, about 1 mm. long; calyx glabrous, rather fleshy, with 4
(rarely 3 or 5) shallow triangular lobes about 1·5 mm. long and 1·5 mm. wide;
corolla subglobose, glabrous, scarcely exceeding the calyx, with 3–4 (–5)
triangular-ovate, suberect or connivent lobes, the apex of each lobe
thickened and often distinctly cucullate; filaments 0·2 mm. long; anthers not
exserted, suborbicular, about 0·4 mm. diam., the connective produced into a
short apiculus; ovary subglobose, about 0·5 mm. diam.; styles 2, free, about
0·2 mm. long; stigmas cylindrical, not thicker than style and about as long.
Capsule circumscissile at base, about 1·3 mm. diam., membranous,
translucent, containing 2–4 seeds; seeds rather flattened, oblong, ovate or
slightly reniform, about 0·7 mm. long; testa rich brown, coarsely reticulate-
rugose.

HAB.: Parasitic on *Linum*, *Trifolium*, *Erigeron*, *Telephium*, *Medicago*, *Helianthemum*,
Pterocephalus, *Alyssum*, *Galium*, *Hypericum*, *Nicotiana*, *Salvia*, *Thymbra*, *Lithospermum*,
Ziziphus and a wide variety of herbs and shrubs; sea-level to 6,401 ft. alt.; fl. April–Oct.

DISTR.: All Divisions, common in Division 2. Widespread in the eastern Mediterranean
region, from Sicily eastwards.

1. Near Paphos, *Kotschy* s.n.; Paphos, 1936, *Chapman* 198!
2. Common; Yialia, Prodhromos, Perapedhi, Platres, Khionistra, etc. *Syngrassides* 628!
Kennedy 1355–1359! *Casey* 907! *F. M. Probyn* 133! *N. Macdonald* 148! *Merton* 2160! etc.
3. Near Limassol, *Kotschy* s.n.; Mazotos, 1905, *Holmboe* 200; Kouklia (distr. Paphos), 1935, *R.
M. Nattrass* s.n.!

4. Larnaca, 1905, *Holmboe* 90; Cape Greco, 1958, *N. Macdonald* 37!
5. Athalassa, 1932, *Syngrassides* 406! Mia Milea, 1950, *Chapman* 671!
6. Orga, 1956, *Poore* 2651!
7. Kythrea, 1938, *Syngrassides* 659! Boghaz, 1939, *Lindberg f.* s.n.! N.E. of Aghirda, 1940, *Kennedy* 1562!
8. Near Apostolos Andreas Monastery, 1880, *Sintenis & Rigo* 61!

4. **C. planiflora** *Ten.*, Fl. Nap., 3: 250 (1824–1829); Boiss., Fl. Orient., 4: 116 (1875); Holmboe, Veg. Cypr., 146 (1914); Yuncker in Mem. Torrey Bot. Club, 18: 292 (1932); Post, Fl. Pal., ed. 2, 2: 214 (1933); Osorio-Tafall et Seraphim, List Vasc. Plants Cyprus, 85 (1973); Davis, Fl. Turkey, 6: 234 (1978).

[*C. minor* (non DC.) Kotschy in Unger et Kotschy, Die Insel Cypern, 286 (1865).]
[*C. epithymum* (non Murr.) H. Stuart Thompson in Journ. Bot., 44: 334 (1906) pro parte; Holmboe, Veg. Cypr., 146 (1914); Osorio-Tafall et Seraphim, List Vasc. Plants Cyprus, 85 (1973).]

TYPE: Italy; Naples, on *Plantago lanceolata* (NAP).

Very similar to *C. palaestina* in general aspect, but with the flowers forming compact spherical glomerules up to about 7 mm. diam.; calyx glabrous and often rather fleshy, up to 2 mm. long and 2 mm. wide with 5 ovate-deltoid lobes divided to more than half the total length of the calyx; corolla-tube about as long as the calyx, and divided above into 5 spreading, ovate-deltoid, acute, corolla-lobes; filaments 0·4–0·5 mm. long; anthers distinctly exserted, oblong, about 0·5 mm. long and 0·3 mm. wide, the connective produced into a short apiculus; ovary subglobose, 0·5–0·6 mm. diam.; styles 2, free, about 0·2 mm. long; stigmas cylindrical, not thicker than style, but a little longer. Capsule circumscissile at base, about 1·5 mm. diam., membranous, translucent, containing 2–4 seeds; seeds ± globose, a little flattened, just under 1 mm. diam.; testa bright brown, distinctly reticulate-verruculose with swollen, hyaline cells.

HAB.: On *Cirsium, Urginea, Linum* and other herbs; about 2,000 ft. alt.; fl. March–July.

DISTR.: Divisions 2, 7. Widespread in the Mediterranean region.

2. Near Evrykhou, 1880, *Sintenis & Rigo* 61a!
7. Buffavento, 1862, *Kotschy* 421; Yaïla, 1941, *Davis* 2900! Kantara, 1948, *Mavromoustakis* s.n.!

NOTES: Some doubt attaches to the localization of the *Sintenis & Rigo* material, particularly as it is evident that these collectors put together material collected from different localities when sending out their duplicates. The other records suggest that *C. planiflora* may be more frequent on the Northern (Kyrenia) Range than elsewhere in Cyprus. It must, however, be admitted that the distinctions between *Cuscuta* species of the subsection *Planiflorae* Yuncker are anything but satisfactory, and further field-work in Cyprus may show either that *C. planiflora* has wider distribution or that it merges with *C. palaestina*. True *C. epithymum* Murr. does not seem to occur on the island.

C. EPILINUM Weihe (in Arch. Apothekerver. Nördl. Teutschl., 8: 50; 1824), a parasite of cultivated Flax, *Linum usitatissimum* L., is noted as having been collected at (3) Kouklia, near Paphos (*Syngrassides* 502 in herb. Kew.) on 28th May, 1935. *C. epilinum* has glomerules of flowers about 1 cm. diam., larger than in *C. palaestina* or *C. planiflora*; the individual flowers are 5-merous, and the stems are simple, or at most very sparingly branched. The original home of *C. epilinum* is uncertain; it is now widely distributed as a weed, and may still occur in Cyprus. Unfortunately the material available is unsatisfactory for accurate determination.

66. SOLANACEAE

Herbs, shrubs or small trees; leaves generally alternate, exstipulate, simple, pinnatisect or pinnate; flowers solitary, axillary or in axillary, extra-axillary or terminal cymes, actinomorphic or zygomorphic, hermaphrodite; calyx (3–) 5 (–6)-lobed or -dentate, often persistent and accrescent; corolla rotate, campanulate, infundibuliform or tubular, usually with 5 (or more) plicate, valvate, contorted or imbricate lobes; stamens (4–) 5 (–8), filaments adnate to the corolla-tube and alternating with the lobes; anthers sometimes connivent, generally 2-thecous, introrse, dehiscing longitudinally or by apical pores; hypogynous disk normally present; ovary superior, usually 2-locular with an oblique septum, sometimes apparently 4-locular, or 1-locular by the decay of the septum; style simple; stigma capitate or 2-lobed; placentation axile, commonly on extruded placentas. Fruit a berry or capsule; seeds often numerous; endosperm present, fleshy; embryo straight or curved.

About 90 genera with a wide distribution, but particularly well represented in tropical America. The family is important economically, and includes such valuable crop plants as the Potato (*Solanum tuberosum* L.), Tobacco (*Nicotiana tabacum* L.) and the Tomato (*Lycopersicon lycopersicum* (L.) Farw.) as well as such drug plants as Belladonna (*Atropa bella-donna* L.) and Henbane (*Hyoscyamus niger* L.)

Plant acaulescent; leaves in a basal rosette; fruit a large berry 2–2·5 cm.
diam. - - - - - - - - - - - - **4. Mandragora**
Plant with a distinct stem; leaves not in a basal rosette:
  Corolla with a very short tube and a widely spreading or recurved, often stellately lobed, limb:
    Anthers connivent into a central cone; filaments very short:
      Anthers dehiscing by apical pores, without sterile apical appendages - **1. Solanum**
      Anthers dehiscing by longitudinal slits, with slender, acuminate, sterile apical appendages - - - - - - - LYCOPERSICON (p. 1180)
    Anthers not connivent into a central cone - - - - - CAPSICUM (p. 1185)
  Corolla tubular, infundibuliform or campanulate:
    Fruit a berry:
      Spiny shrubs; calyx not much accrescent in fruit - - - - **3. Lycium**
      Unarmed herbs; calyx strongly accrescent and completely enveloping the fruit
        2. Withania
    Fruit a capsule:
      Capsule circumscissile; corolla oblique, often distinctly zygomorphic; inflorescences spicate or scorpioid - - - - - - - **6. Hyoscyamus**
      Capsule dehiscing longitudinally into valves; corolla actinomorphic or very obscurely zygomorphic:
        Flowers solitary in the bifurcations of the branches; capsule generally spinose
          5. Datura
        Flowers numerous in terminal panicles; capsule smooth - - - **7. Nicotiana**

LYCOPERSICON LYCOPERSICUM (*L.*) *Farw.* in Ann. Rep. Commissioners Parks & Boulevards Detroit, 11: 84 (1900) (*Solanum lycopersicum* L.; *Lycopersicon esculentum* Mill.), the Tomato, is an important and extensively cultivated crop in Cyprus, first recorded for the island by Hume in 1801 (Walpole, Mem. Europ. Asiatic Turkey, ed. 1, 254; 1817), but no doubt in cultivation long before this date. It is believed to be a native of western tropical South America, first brought into cultivation by the Amerindians, and imported into Europe during the first half of the 16th century. For almost three centuries it was valued chiefly as an ornamental or medicinal plant (the vulgar name Love Apple suggesting aphrodisiac properties), and it is only within the last hundred years that it has become really popular as a comestible. *Lycopersicon* is very closely related to *Solanum*, differing technically in the structure of the anthers, which dehisce by longitudinal slits, and not by pores, and which are produced apically into slender, acuminate, sterile appendages. The species is represented in cultivation by a very large range of named cultivars.

1. SOLANUM L.

Sp. Plant., ed. 1, 184 (1753).
Gen. Plant., ed. 5, 85 (1754).

Annual or perennial herbs, shrubs or small trees, occasionally scandent, usually ± pilose, sometimes prickly, often foetid-smelling when crushed; leaves alternate, entire, toothed or lobed, or imparipinnate; flowers in terminal or lateral cymes or rarely solitary; calyx campanulate or patent, usually 5-lobed, often persistent and accrescent; corolla white, purple or yellow, tube short, lobes spreading or recurved; stamens inserted in throat of corolla; filaments generally very short; anthers connivent into a central cone, dehiscing by apical pores which extend, with age, into introrsely longitudinal slits; disk inconspicuous; ovary usually 2-locular; style simple; stigma small. Fruit a globose (rarely pyriform) berry; seeds numerous, usually suborbicular or reniform; embryo curved.

About 1,700 species with a cosmopolitan distribution, but predominantly tropical.

Leaves pinnate or deeply pinnatisect:
  Tuberous herbs with thick, fleshy stems - - - - - S. TUBEROSUM (p. 1183)
  Woody climbers with slender stems - - - - - S. SEAFORTHIANUM (p. 1184)
Leaves simple, entire, or variously lobed:
  Shrubs with entire, lanceolate, grey or glaucous leaves:
    Leaves densely stellate-pilose - - - - - S. ELAEAGNIFOLIUM (p. 1184)
    Leaves glabrous, glaucous - - - - - - - S. GLAUCUM (p. 1185)
  Herbs with green, ovate or oblong, lobed leaves:
    Corolla 2–4 cm. diam.; leaves stellate-pilose; fruit very large, sometimes 30 cm. long
                          S. MELONGENA (p. 1184)
    Corolla less than 1·5 cm. diam.; leaves without stellate hairs; fruit a small berry 7–10 mm. diam.:
      Ripe fruit red or orange; upper parts of stems villose, with long, white, multicellular hairs - - - - - - - - - - - - **2. S. villosum**
      Ripe fruit black; upper parts of stems thinly clothed with short, incurved, subadpressed hairs - - - - - - - - - - - - **1. S. nigrum**

1. S. nigrum *L.*, Sp. Plant., ed. 1, 186 (1753); Hume in Walpole, Mem. Europ. Asiatic Turkey, 254 (1817); Gaudry, Recherches Sci. en Orient, 198 (1855); Unger et Kotschy, Die Insel Cypern, 287 (1865); Boiss., Fl. Orient., 4: 284 (1879); H. Stuart Thompson in Journ. Bot., 44: 335 (1906); Holmboe, Veg. Cypr., 163 (1914); Post, Fl. Pal., ed. 2, 2: 256 (1933); J. Wessely in Fedde Repert., 63: 306 (1960); Osorio-Tafall et Seraphim, List Vasc. Plants Cyprus, 93 (1973); R. J. F. Henderson in Contrib. Queensland Herb., no. 16: 19 (1974); J. M. Edmonds in Bot. Journ. Linn. Soc., 75: 141–178 (1977); Davis, Fl. Turkey, 6: 439 (1978); Feinbrun, Fl. Palaest., 3: 451 (1978).
  [*S. villosum* (non Mill.) H. Stuart Thompson in Journ. Bot., 44: 335 (1906) pro parte quoad *Sintenis & Rigo* 675.]
  TYPE: "*in* Orbis *totius cultis*".

An erect or straggling, usually much-branched annual or short-lived perennial up to 1 m. high; stems terete or obscurely angled or winged, greenish or stained purple, subglabrous or thinly pubescent with short, incurved eglandular hairs occasionally intermixed with shorter glandular hairs or remotely scaberulous along the angles of the stem; leaves dark green, ovate, 2–10 (–13) cm. long, 0·8–5 (–7·5) cm. wide, subglabrous or rather thinly pubescent above and below, apex obtuse, acute, or acuminate, margins entire, subentire or irregularly lobed, base tapering rather abruptly to a narrowly winged petiole 0·5–3 cm. long; inflorescences extra-axillary, umbellate or shortly racemose, 3–8 (–12)-flowered; peduncle erecto-patent, 1–3 cm. long, lengthening in fruit, thinly pubescent with curved subadpressed eglandular hairs, sometimes intermixed with shortly glandular hairs; pedicels 0·5–1·5 cm. long at anthesis, at first spreading, becoming deflexed in fruit, indumentum like that of the peduncles; calyx-lobes erect, oblong, obtuse, about 2 mm. long, 1 mm. wide; corolla white, rotate or shallowly infundibuliform, shortly puberulous externally, tube about 2 mm. long, lobes ovate, acute, about 5 mm. long, 3 mm. wide; filaments flattened dorsiventrally, about 1–1·5 mm. long, pilose internally; anthers yellow, narrowly oblong, about 2 mm. long, 0·5 mm. wide, shortly sagittate-lobed at base; ovary subglobose, about 1·5 mm. diam., glabrous; style 3 mm. long, shortly pilose towards base; stigma about 0·5 mm. diam., capitate or obscurely 2-lobed. Fruit a subglobose, dull or shining, purple-black berry 7–10 mm. diam.; seeds sub-reniform, strongly compressed, about 1·8 mm. long, 1·5 mm. wide; testa pale-brown, closely foveolate.

HAB.: Cultivated and waste ground; sea-level to 5,000 ft. alt.; fl. Jan.–Dec.

DISTR.: Divisions 1–6, 8. A cosmopolitan weed.

1. Prodhromi near Polis, 1937, *Syngrassides* 1714!
2. Troödhitissa Monastery, 1853–54, *Gaudry*; Palekhori, 1940, *Davis* 1989!
3. Limassol, 1801, *Hume*; between Kolossi and Paphos, 1862, *Kotschy* 613; Silikou, 1940, *Davis* 2080! Pyrgos, 1963, *Suart* 48! Khalassa, 1970, *A. Genneou* in ARI 1496!
4. Larnaca, 1801, *Hume*, and, 1880, *Sintenis & Rigo* 895!
5. Ayios Seryios, 1973, *P. Laukkonen* 18!
6. Kykko near Nicosia, 1930, *Druce* s.n.
8. Ayios Andronikos, 1880, *Sintenis & Rigo* 675!

NOTES: In the absence of notes on the colour of ripe fruits, three additional specimens cannot be identified with certainty; they are: (Div. 3) Near Amathus, 1947, *Mavromoustakis* s.n.! (Div. 8) Apostolos Andreas, 1930, *Druce* s.n.! Near Platanisso, 1962, *Meikle* 2508! All three resemble *S. nigrum*, but have coarsely serrate or lobed leaves. Two of them (*Mavromoustakis* s.n.; *Meikle* 2508) have been submitted to Dr. Jennifer M. Edmonds (Dept. of Botany, Cambridge), who reports (6 May, 1980) as follows:

"*Mavromoustakis* s.n. Superficially, this specimen resembles *S. villosum* Mill. subsp. *puniceum* (Kirschleger) Edmonds, through its short peduncles, prominently sinuate-dentate leaves and sparse eglandular-haired pubescence. However, both its pollen and stomatal aperture dimensions strongly indicate it being hexaploid rather than tetraploid. Unless there is a hexaploid endemic to Cyprus, or indeed to any of the nearby Arab countries, which I have not yet come across, this would suggest that the plant belongs to *S. nigrum* subsp. *nigrum*. The flowers, with anthers of *c.* 3 mm., are rather larger than expected in this species (or indeed even in *S. villosum*), where anthers are usually *c.* 2 mm. However, anthers in excess of 3 mm. have been reported by a number of recorders for *S. nigrum* in my current BSBI Black Nightshade Survey of the British Isles, so the floral dimensions do not entirely exclude it from belonging to this species.

Meikle 2508. Through its lax cymose inflorescences, long peduncles and eglandular-haired pubescence, this specimen resembles *S. nigrum* L. ssp. *nigrum*. Its flowers, with anthers of *c.* 2·7 mm., are again a little larger than is usual in this species, but, as explained above, this does not preclude the specimen from being conspecific with *S. nigrum*. The stomatal aperture dimensions point to the plant being more typical of tetraploid species in this *Solanum* group. However, hexaploid plants are occasionally found with small pollen grains. The plant, and especially the leaves, is covered with a denser pubescence than the *Mavromoustakis* specimen, but this may be due to its collection by side of spring in a bushy place."

Druce has added a pencil note to the third specimen from Apostolos Andreas, indicating that it had red berries, but the value of this evidence is lessened by the fact that Druce also collected *S. villosum* in Cyprus, and may possibly have attributed the fruit-colour of one species to the unripe fruits of another.

2. S. villosum *Mill.*, Gard. Dict., ed. 8, no. 2 (1768); Stebbins et Paddock in Madroño, 10: 74 (1949); R. J. F. Henderson in Contrib. Queensland Herb., no. 16: 54 (1974); J. M. Edmonds in Bot. Journ. Linn. Soc., 75: 141–178 (1977).
     S. nigrum L. var. *villosum* L., Sp. Plant., ed. 1, 186 (1753).

S. luteum Mill., Gard. Dict., ed. 8, no. 3 (1768); J. Wessely in Fedde Repert., 63: 314 (1960); Osorio-Tafall et Seraphim, List Vasc. Plants Cyprus, 93 (1973); Davis, Fl. Turkey, 6: 440 (1978); Feinbrun, Fl. Palaest., 3: 165, t. 273 (1978).

S. rubrum Mill., Gard. Dict., ed. 8, no. 4 (1768); Druce in Rep. B.E.C., 9: 470 (1931).

S. nigrum L. ssp. *villosum* (L.) Ehrh. in Hannov. Mag., 1780 (Stück 14): 218 (1780); Holmboe, Veg. Cypr., 163 (1914).

S. villosum (L.) Lam., Tabl. Encycl., 2: 18 (1794), Encycl. Méth., 4: 289 (1797); Boiss., Fl. Orient., 4: 285 (1879); H. Stuart Thompson in Journ. Bot., 44: 335 (1906) pro parte; Post, Fl. Pal., ed. 2, 2: 257 (1933).

[*S. alatum* (non Moench) Lindberg f., Iter Cypr., 31 (1946).]

TYPE: Based on material cultivated at Chelsea Physic Garden (BM).

Erect or sprawling annual or short-lived perennial, to about 50 cm. high, but usually smaller than *S. nigrum* L., commonly much branched; stems terete or less often obscurely angled, greenish, often densely clothed with coarse, spreading, white, multicellular hairs, occasionally mixed with shorter, gland-tipped hairs; leaves generally paler than in *S. nigrum*, broadly ovate, 2–6 cm. long, 1–5 cm. wide, usually pubescent above and below, apex acute or rarely obtuse, margins conspicuously but rather irregularly sinuate-dentate, base tapering abruptly to a slender, hairy, scarcely winged petiole 1–2 (–3) cm. long; inflorescences extra-axillary, umbellate or shortly racemose, 2–6 (–8)-flowered; peduncle erecto-patent, 5–15 (–20) mm. long at anthesis, lengthening a little in fruit, usually pilose with spreading hairs; pedicels slender, hairy, about 5–8 mm. long at anthesis, attaining 1 cm. or more, and becoming strongly deflexed in fruit; calyx-lobes erect, oblong, blunt, hairy, about 2 mm. long, 1 mm. wide, commonly joined by a membranous web for almost half their length; corolla white, rotate or shallowly infundibuliform, subglabrous or shortly pubescent especially towards the tips of the lobes externally, tube about 2·5 mm. long, lobes ovate, acute, about 5 mm. long, 3–4 mm. wide; filaments flattened dorsiventrally, about 1·5 mm. long, hairy internally; anthers yellow, narrowly oblong, about 2 mm. long, 1 mm. wide, shortly and bluntly sagittate-lobed at the base; ovary glabrous, subglobose, about 1·4 mm. diam.; style 3 mm. long, pilose towards base; stigma about 0·5 mm. diam., capitate or obscurely 2-lobed. Fruit subglobose, about 9 mm. diam., said to be sometimes slightly longer than wide, orange-yellow or red when ripe; seeds reniform, strongly compressed, about 2 mm. long, 1·5 mm. wide; testa pale brown, closely foveolate.

HAB.: Cultivated and waste ground; sea-level to 4,400 ft. alt.; fl. Jan.–Dec.

DISTR.: Divisions 1–3, 5–7. Widespread in central and southern Europe and the Mediterranean region eastwards to Iran; elsewhere as an introduced weed.

1. Yeroskipos, 1939, *Lindberg f.* s.n.! Kissonerga to Maa Beach, 1967, *Economides* in ARI 980!
2. Kaminaria, 1905, *Holmboe* 1072; Lefka, 1932, *Syngrassides* 244! Kambos, 1939, *Lindberg f.* s.n.! Platres, 1948, *S. G. Cowper* s.n.! Prodhromos, 1955, *G. E. Atherton* 669! Mavres Sykies, 1962, *Meikle* 2735!
3. Kalavasos, 1905, *Holmboe* 613!
5. Kythrea, 1928, *Druce* s.n.! and, 1939, *Lindberg f.* s.n.!
6. Nicosia, 1901, *A. G. & M. E. Lascelles* s.n.! Kykko, 1930, *Druce* s.n.! English School, Nicosia, 1958, *P. H. Oswald* 22! Kokkini Trimithia, 1966, *Merton* in ARI 8!
7. Kyrenia, 1939, *Lindberg f.* s.n.! also, 1948, *Casey* 24! and, 1955, *G. E. Atherton* 301! Near Thermia, 1955, *G. E. Atherton* 387! Bellapais, 1956, *G. E. Atherton* 861!

NOTES: Despite numerous records, there is no satisfactory evidence that *S. villosum* Mill. ssp. *alatum* (Moench) Edmonds (*S. miniatum* Bernh.) occurs on Cyprus, and all the specimens so named appear on examination to be minor variants of the variable *S. villosum* ssp. *villosum*.

S. TUBEROSUM *L.*, Sp. Plant., ed. 1, 185 (1753), the Potato, a stoloniferous, tuber-bearing perennial herb, with erect, fleshy, angled or narrowly winged stems, imparipinnate leaves with 3–4 pairs of leaflets, and lax terminal cymes of heavily scented flowers with white or purplish, rotate-pentagonal corollas, is generally believed to have originated in Chile, whence it became disseminated in cultivation throughout western South America, reaching Europe (Spain) in the early part of the 16th century. Bevan (Notes on Agriculture in Cyprus, 55; 1919) refers to "a native variety of Potato, believed to have been imported by Syrian Arabs in the sixteenth century . . . still grown on a small scale in the Marathassa valley", but even substituting "local" for the erroneous term "native", it is hard to believe (in the absence of precise records) that the Potato could have reached Cyprus so early. Kotschy (Die Insel Cypern, 287) is more likely to be correct in stating that it has been cultivated on the island since 1820, though again, I have not been able to trace his authority for this very exact date; it is clear, however, that the Potato was already well known when Gaudry visited Cyprus in 1853 (Recherches Sci. en Orient, 184; 1855). It has become an increasingly important item in the economy of Cyprus, many thousands of tons being exported annually and reaching western Europe before the earliest of local supplies of "new" Potatoes becomes available; exports in 1971 were valued at well over six million Cyprus pounds (C. V. Economides et al. in ARI Techn. Bull., 11: 3; 1973).

S. MELONGENA *L.*, Sp. Plant., ed. 1, 186 (1753) (*S. esculentum* Dunal, Hist. Solan., 208, t. 3E; 1813), the Egg-plant or Aubergine, an erect, branching annual or short-lived subshrubby perennial 0·5–1·5 m. high, with unarmed or sometimes spinose stems, and large, ovate or oblong, sinuate-lobed leaves clothed, especially on the lower surface, with a greyish or violet-tinged indumentum of stellate hairs; the flowers are either solitary or in small extra-axillary cymes, with a campanulate, deeply 5-lobed, accrescent, tomentose (and sometimes spinose) calyx, and a purplish, pentagonal-rotate corolla 2–4 cm. diam. The fruit is a large ovoid, obovoid or pyriform, lustrous, white, purple, yellowish or violet-black berry, up to 30 cm. long, containing numerous, small, pale brown, strongly compressed seeds.

Generally considered a native of tropical Asia, but now widely distributed as a cultivated plant in tropical and warm temperate regions of the world, and represented in gardens by numerous cultivars differing in the armature of stems and leaves, and in the size, shape and colour of their fruits. Kotschy (Die Insel Cypern, 288; 1865) records the Aubergine from gardens about Varosha and Nicosia. It is still widely grown on the island, but is of less economic importance than the Tomato.

S. SEAFORTHIANUM *Andrews*, Bot. Reposit., t. 54 (1808), a slender, glabrous climber; leaves long-petiolate, imparipinnate or imparipinnatisect, with entire, lanceolate or narrowly ovate-elliptic lobes; inflorescences leaf-opposed, extra-axillary or terminal, forming lax, many-flowered panicles of 5-lobed, lavender-blue flowers 1·5–2·5 cm. diam. Fruits globose, shining, scarlet.

A native of the West Indies and eastern tropical S. America, popular as a garden or greenhouse plant in other areas. *S. seaforthianum* has been collected in Kyrenia (1955, *G. E. Atherton* 86!), and is probably not uncommon in Cyprus gardens.

S. ELAEAGNIFOLIUM *Cav.*, Icones, 3(1): 22, t. 243 (1795), a slender, much-branched shrub, 0·5–1 m. high, sparsely armed with slender, tawny acicles, and densely clothed with a close, grey indumentum of stellate hairs; leaves lanceolate, entire, 4–7 cm. long, 0·8–2 cm. wide, tomentose; inflorescences 2–6-flowered, forming lax, extra-axillary cymes; calyx truncate-campanulate, with short, subulate lobes; corolla purple, shallowly infundibuliform, about 2·5 cm. diam., deeply divided into 5 narrowly ovate-acute lobes. Fruit a globose, shining, yellow berry.

A native of Central and South America, probably introduced as an ornamental, and now escaped, and locally naturalized on waste ground and roadsides about Nicosia, where it was first collected in 1958 by *P. H. Oswald* (no. 15!) and subsequently in 1959 (*C. E. H. Sparrow* 46!) and in 1973 (*P. Laukkonen* 333!). A. Yannitsoros and E. Economidou (in Candollea, 29: 114; 1974) note that *S. elaeagnifolium* has become locally abundant about Thessaloniki and occurs in other parts of Greece and Crete. It is also recorded from Sicily, Israel, Egypt, Ionian Islands, northern Sporades and from Lesvos (in Davis, Fl. Turkey, 6: 443; 1978).

S. GLAUCUM *Dunal* in DC., Prodr., 13 (1): 100 (1852), a native of eastern S. America is recorded from Famagusta by Druce (in Rep. B.E.C., 9: 470; 1931); it is a rhizomatous subshrub, 1·5–2 m. high, with glabrous, glaucous leaves, not unlike those of the Oleander (*Nerium oleander* L.) in size and outline, and branched panicles of blue or purple, star-shaped flowers, about 2 cm. diam., followed by black, globose berries about 2·5 cm. diam. The identification of the Famagusta specimen has not been checked, and may be erroneous; there are no other records of the species from Cyprus.

CAPSICUM ANNUUM *L.*, Sp. Plant., ed. 1, 188 (1753), Paprika, Sweet Pepper, an erect, branched annual up to 1 m. high; stems glabrous, distinctly ridged; leaves ovate or ovate-elliptic, 2–12 cm. long, 2–6 cm. wide, long-petiolate, glabrous, entire, acute or caudate-acuminate; flowers usually solitary in the axils of the upper leaves; peduncle patent or distinctly recurved; calyx glabrous, campanulate, obscurely lobed; corolla subrotate or shallowly infundibuliform, whitish, deeply 5-lobed; stamens erect but not connivent, with bluish anthers. Fruit very variable in shape, size and colour, commonly a hollow berry, 18 cm. or more in length, tapering to an acute apex, and becoming scarlet when ripe; flesh and seeds mild or pungent.

The species comprises a vast range of named cultivars, as yet imperfectly classified, nor are taxonomists in total agreement as to the number of natural ("wild") species included in the genus. *C. annuum* is a native of tropical America, now cultivated in warmer regions throughout the world. It is commonly grown in Cyprus, where it was first noted by Kotschy (Die Insel Cypern, 287; 1865).

CAPSICUM BACCATUM *L.*, Mantissa I, 47 (1767), Bird Pepper, also a native of tropical America, is recorded as planted in Salamis Plantation by Bovill (Rep. Plant. Work Cyprus, 14; 1915); it is a shrubby glabrous or pubescent perennial, often 2 m. high, with small, ovate-acuminate leaves, and solitary or paired flowers on long, erect, axillary or extra-axillary peduncles; the corolla is smaller than that of *C. annuum*, and the subglobose or ovoid fruits are commonly less than 2 cm. long, red and pungent when ripe.

2. WITHANIA *Pauquy*

De la Belladonne, 14 (1825) nom. cons.

Subshrubby perennials; leaves entire, petiolate; flowers small, in congested axillary clusters; calyx campanulate, 5-dentate, accrescent after anthesis and surrounding the fruit; corolla narrowly campanulate, with 5 (sometimes more or fewer) valvate lobes; stamens 5 (–7) inserted near base of corolla, filaments free; anthers not connivent, dehiscing longitudinally;

ovary 2-locular; style filiform; stigma shortly 2-lobed. Fruit a many-seeded berry; seeds compressed, with a strongly incurved or spiral embryo.

About 7 species in S. Europe, W. Asia, Atlantic Islands and throughout Africa.

1. **W. somnifera** (*L.*) *Dunal* in DC., Prodr., 13: 453 (1852); Unger et Kotschy, Die Insel Cypern, 288 (1865); Boiss., Fl. Orient., 4: 287 (1879); Holmboe, Veg. Cypr., 163 (1914); Post, Fl. Pal., ed. 2, 2: 260 (1933); Lindberg f., Iter Cypr., 31 (1946); Chapman, Cyprus Trees and Shrubs, 75 (1949); Rechinger f., in Arkiv för Bot., ser. 2, 1: 431 (1950); Osorio-Tafall et Seraphim, List Vasc. Plants Cyprus, 93 (1973); Davis, Fl. Turkey, 6: 445 (1978); Feinbrun, Fl. Palaest., 3: 164, t. 272 (1978).
    Physalis somnifera L., Sp. Plant., ed. 1, 182 (1753); Sibth. et Sm., Fl. Graec. Prodr., 1: 154 (1806), Fl. Graec., 3: 28, t. 233 (1819); Clarke, Travels, ed. 1, 2 (3): 722 (1816).
    TYPE: "*in* Mexico, Creta, Hispania".

An erect subshrub or woody-based perennial to about 1·5 m. high; stems not much branched, terete, covered with a close tomentum of whitish, stellate hairs; leaves broadly ovate, 4–10 cm. long, 3–7 cm. wide, at first thinly tomentellous with stellate hairs, becoming subglabrous with age, apex acute, margins entire, base broadly cuneate, narrowing to a petiole 1–3 cm. long; nervation obscure; flowers 4–16 in congested axillary clusters; pedicels usually less than 5 mm. long, densely tomentose; calyx (at anthesis) about 5 mm. long, 3·5 mm. wide, campanulate, densely tomentose, with 5, erect, subulate apical teeth about 2 mm. long; corolla greenish, tomentose externally, tube about 2 mm. long, lobes generally 5, ovate, about 2 mm. long, 1·5 mm. wide at base; filaments glabrous, about 1·5 mm. long; anthers yellow, oblong, about 1 mm. long, 0·8 mm. wide; ovary ovoid, glabrous, about 1·8 mm. long, 1·5 mm. diam.; style straight, about 1·4 mm. long; stigma capitate, about 0·4 mm. diam. Fruit a glossy, globose, scarlet berry about 8–10 mm. diam., enclosed within a much accrescent, papery calyx; seeds strongly compressed, reniform-suborbicular, about 2 mm. diam.; testa pale brown, rather regularly foveolate.

    HAB.: Roadsides, waste ground, gardens; sea-level to 500 ft. alt.; fl. throughout the year.

    DISTR.: Divisions 1, 4, 5, 8. External distribution that of the genus.

1. Paphos, 1787, *Sibthorp*, also, 1901, *A. G. & M. E. Lascelles* s.n.!, 1913, *Haradjian* 650! and, 1967, *Polunin* 8564!
4. Ayia Napa, 1862, *Kotschy* 113, and, 1939, *Lindberg f.* s.n.! Larnaca, 1880, *Sintenis & Rigo* 907! Famagusta, 1912, *Haradjian* 97!
5. Tymbou, 1932, *Syngrassides* 385! Trikomo, 1956, *Merton* 2469!
8. Komi Kebir, 1912, *Haradjian* 297! Rizokarpaso, 1933, *Chapman* 25! Koma tou Yialou, 1970, *A. Genneou* in ARI 1480!

3. LYCIUM *L.*

Sp. Plant., ed. 1, 191 (1753).
Gen. Plant., ed. 5, 88 (1754).
N. Feinbrun in Collectanea Botanica, 7 (1): 359–379 (1968).

Spiny shrubs; leaves entire, generally tapering basally to a short petiole, commonly in axillary clusters; flowers solitary or sometimes in clusters; calyx campanulate or cupuliform, truncate or irregularly toothed or lobed, sometimes 2-lipped; corolla tubular-infundibuliform, tube short or elongate, lobes usually 5, patent, imbricate in bud; stamens as many as corolla-lobes, inserted about the middle of the corolla-tube; filaments commonly unequal; anthers exserted or included, not connivent, dehiscing longitudinally; ovary 2-locular; style filiform; stigma capitate or obscurely 2-lobed. Fruit a globose or ovoid, few- or many-seeded berry; seeds compressed with a strongly curved or spiralling embryo.

About 80 species widely distributed in temperate and tropical regions.

Corolla-tube elongate, slender, much exceeding the calyx, the latter not much accrescent in
fruit - - - - - - - - - - - - - **1. L. schweinfurthii**
Corolla-tube scarcely exceeding the conspicuously accrescent calyx - **2. L. ferocissimum**

1. **L. schweinfurthii** *U. Dammer* in Engl., Bot. Jahrb., 48: 224 (1913); N. Feinbrun in Col-
lectanea Botanica, 7 (1): 367 (1968); Davis, Fl. Turkey, 6: 447 (1978); Feinbrun, Fl.
Palaest., 3: 160, t. 263 (1978).
  [*L. europaeum* (non L.) Sm. in Sibth. et Sm., Fl. Graec., 3: 30, t. 236 (1819); Sintenis in
Oesterr. Bot. Zeitschr., 31: 194 (1881); H. Stuart Thompson in Journ. Bot., 44: 335 (1906);
Holmboe, Veg. Cypr., 163 (1914); Lindberg f., Iter Cypr., 31 (1946); Chapman, Cyprus
Trees and Shrubs, 75 (1949); Osorio-Tafall et Seraphim, List Vasc. Plants Cyprus, 92
(1973).]
  [*L. vulgare* (non Dunal) Kotschy in Unger et Kotschy, Die Insel Cypern, 288 (1865).]
  TYPE: Egypt: [Mex] near Alexandria, 1868, *Schweinfurth* 67 (B, K!).

Deciduous, rigid shrub 3–5 m. high, intricately branched, many of the
lateral branches modified into stout, straight spines, shoots at first densely
pubescent, becoming glabrous and greyish with age; leaves often in clusters,
oblanceolate, 10–50 mm. long, 2–8 mm. wide, bluish-green, glabrous or very
sparsely puberulous, apex obtuse or acute, base tapering to a short, often
puberulous, petiole; flowers usually solitary, or sometimes up to 3 in a
cluster; pedicel slender, often very short, occasionally 3–4 mm. long,
puberulous with short, spreading glandular hairs; calyx cup-shaped, about
1·5 mm. long and almost as wide, not much accrescent in fruit, thinly
glandular-puberulous, lobes 5, small, blunt, ciliate, base of calyx angular or
narrowly winged, tapering to the peduncle; corolla mauve, trumpet-shaped,
with a slender tube 15–20 mm. long and 1·5 mm. wide at base, expanding
gradually in the upper half, apparently glabrous externally and internally,
lobes patent, suborbicular, 2·5–3 mm. diam.; stamens inserted at different
levels in the upper half of the corolla-tube, filaments slender, glabrous, 4–7
mm. long; anthers oblong, about 0·8 mm. long, 0·5 mm. wide, some of them
shortly exserted; ovary glabrous, ovoid, about 1 mm. long, 0·8 mm. wide;
style as long as corolla tube; stigma capitate, about 0·5 mm. diam. Fruit
globose, about 4 mm. diam., [black or red?]; seeds 5–6, irregularly
compressed-reniform, about 2·5–3 mm. long, 1·8–2·5 mm. wide; testa pale
brown, verruculose.

HAB.: Sandy seashores and sand-dunes, occasionally on sandy ground inland, but never far
from the sea; sea-level to 100 ft. alt.; fl. ? Nov.–July.

DISTR.: Divisions 1, 3–5, 7. Eastern Mediterranean region from Sicily to Egypt; North Africa
westwards to Algeria.

1. Paphos, 1901, *A. G. & M. E. Lascelles* s.n.! Skali near Ayios Yeoryios (Akamas), 1962,
Meikle 2083!
3. Near Amathus, 1862, *Kotschy*, and, 1947, *Mavromoustakis* 18!
4. Near Larnaca, 1880, *Sintenis & Rigo* 926! also, 1905, *Holmboe* 95, and, 1939, *Lindberg f.* s.n.!
Varosha, 1970, *A. Hansen* 725!
5. Salamis, 1934, *Chapman* 133!
7. East of Kyrenia, 1948, *Casey* 151!

NOTES: The above description is based upon Cyprus material, which differs in several respects
from typical *L. schweinfurthii* from the neighbourhood of Alexandria. The latter is altogether
more slender, with the shoots generally glabrous or subglabrous, the spines shorter and more
numerous, the leaves smaller and narrower, and the flowers rather smaller; it approaches the
Spanish *L. intricatum* Boiss., but is, I think, specifically distinct from this. The Cyprus plant
(closely resembling the plant depicted in *Fl. Graeca* t. 236) is generally robust, with densely
puberulous shoots, stout, scattered spines, long leaves and relatively large flowers; apart from
the slender, elongate corolla-tube, it much resembles *L. europaeum* L.
  Feinbrun notes that the fruit of *L. schweinfurthii* is black; Chapman (Cyprus Trees and
Shrubs, 75) says that the Cyprus plant has an orange berry, like that figured in *Fl. Graeca*. Only
one of the specimens seen has fruits, and these are of no discernible hue, nor have collectors
elsewhere noted fruit-colour for Cyprus specimens. In the circumstances, and in the absence of
adequate data, I prefer to leave the plant as *L. schweinfurthii*, but with some doubts. The same
plant is also to be found in Crete, the Cyclades and in Palestine.

2. L. ferocissimum *Miers* in Ann. Mag. Nat. Hist., ser. 2, **14**: 187 (1854), Illustr. S. Amer Plants, **2**: t. 70, fig. D (1857).

[*L. horridum* (non Thunb.) Hutchins, Rep. Cyprus Forests, 68 (1909); Bovill, Rep. Plant. Work Cyprus, 14 (1915).]

[? *L. chinense* (non Mill.) Bull. Imp. Inst. 26: 330 (1928).]

[? *L. halimifolium* (non Mill.) Osorio-Tafall et Seraphim, List Vasc. Plants Cyprus, 92 (1973).]

TYPE: South Africa; Uitenhage, *Zeyher* 105 (K!); erroneously recorded as "*Harvey* 105" by Miers.

Much-branched, spiny shrub or small tree to about 7 m. high; bark greyish, lightly fissured; branches divaricate, pallid, glabrous, commonly modified into rigid spines; leaves oblanceolate or narrowly obovate, 0·8–4·5 cm. long, 0·3–1·5 cm. wide, glabrous, apex rounded or obtuse, rarely shortly acute, base cuneate, tapering to a very short and obscure petiole; flowers usually solitary; peduncle slender, 2–5 mm. long, thinly glandular; calyx campanulate, 6–8 mm. long, 4–5 mm. wide, thinly glandular, conspicuously accrescent in fruit, lobes irregularly crenate-dentate, tomentose, ciliate, splitting irregularly down the calyx-tube in fruit; corolla shortly tubular with spreading or reflexed lobes, white with purple marks at the base of the lobes, tube infundibuliform, about as long as calyx, 1·5 mm. wide at base, lobes suborbicular, about 4·5 mm. diam.; stamens inserted, at more or less the same level, near base of corolla-tube, filaments slender, about 7 mm. long, densely lanuginose above point of insertion; anthers ovate-oblong, yellow, about 1·3 mm. long, 1 mm. wide, conspicuously exserted; ovary ovoid, glabrous, about 1·6 mm. long, 1·2 mm. wide; style 10 mm. long, glabrous; stigma capitate, about 1·3 mm. diam. Fruit globose, about 1·3 cm. diam., at first red, later purple-black; seeds very numerous, compressed-reniform, about 2·5 mm. long, 1·5 mm. wide, testa brown, rugulose-verruculose.

HAB.: Gardens, roadsides; seashores; sea-level to 600 ft. alt.; fl. Jan.–Dec.

DISTR.: Divisions 4–7, planted; native of S. Africa, naturalized in S. Australia and New Zealand.

4. Larnaca, 1909, *Hutchins*; Famagusta, 1909, *Hutchins*.
5. Salamis, 1915, *Bovill*, also, 1973, *P. Laukkonen* 42! Mia Milea, 1967, *Merton* in ARI 758! Ayios Varnavas Monastery, 1968, *A. Genneou* in ARI 1270!
6. Nicosia, 1934, *Chapman* 125! Yerolakkos, 1967, *Economides* in ARI 1029!
7. Kyrenia, 1955, *G. E. Atherton* 161!

NOTES: Apparently introduced with other exotics into Salamis Plantation about the beginning of the present century, and now widely planted as a hedge or windbreak, and becoming naturalized in some areas, though still uncommon. Most authorities state that the fruits are red, but G. E. Atherton notes: "fruit first red, then dark purple".

L. AFRUM *L.*, Sp. Plant., ed. 1, 191 (1753), also from S. Africa, is distinguished from *L. ferocissimum* Miers by its very narrow, linear leaves, less than 3 mm. wide, and narrow-infundibuliform or subcylindrical corollas, the tube much exceeding the calyx, and the lobes very short, erect or suberect. It is said to have been grown in Salamis Plantation (Bovill, Rep. Plant. Work Cyprus 14; 1915), but there are no recent records from Cyprus.

4. MANDRAGORA *L.*

Sp. Plant., ed. 1, 181 (1753).
Gen. Plant., ed. 5, 84 (1754).
Vierhapper in Oesterr. Bot. Zeitschr., **65**: 124–138 (1915).
E. Maugini in Nuov. Giorn. Bot. Ital., n.s. **66**: 34–60 (1959).

Perennial herbs with a large, elongate, fleshy vertical rhizome; stem generally very short; leaves often large, forming a spreading basal rosette;

flowers solitary, axillary, pedunculate, forming a cluster at the centre of the leaf-rosette; calyx campanulate, 5-lobed, accrescent after anthesis; corolla campanulate or infundibuliform, lobes 5, imbricate, the sinuses plicate; stamens 5, inserted below the middle of the corolla; filaments slender, villous near the base; anthers dorsifixed, usually exserted; ovary 2-locular, ovules numerous; style elongate; stigma large, capitate, often with reflexed margins. Fruit a globose or ovoid, yellow or orange berry; seeds strongly compressed; embryo subperipheral, strongly curved.

Probably not more than 3 species in S. Europe, the Mediterranean region, Central Asia.

1. **M. officinarum** *L.*, Sp. Plant., ed. 1, 181 (1753); Boiss., Fl. Orient., 4: 291 (1879); Holmboe, Veg. Cypr., 163 (1914); Post, Fl. Pal., ed. 2, 2: 261 (1933); Osorio-Tafall et Seraphim, List Vasc. Plants Cyprus, 93 (1973).

   M. vernalis Bert., Elench. Plant. Hort. Bot. Bonon., 6 (1820); Unger et Kotschy, Die Insel Cypern, 288 (1865); Sintenis in Oesterr. Bot. Zeitschr., 31: 191 (1881).

   M. autumnalis Bert., Elench. Plant. Hort. Bot. Bonon., 6 (1820); Davis, Fl. Turkey, 6: 450 (1978); Feinbrun, Fl. Palaest., 3: 167, t. 278 (1978).

   M. haussknechtii Heldr. in Mitt. Geogr. Ges. Jena, 4: 77 (1886).

   M. officinarum L. ssp. *haussknechtii* (Heldr.) Vierhapper in Oesterr. Bot. Zeitschr., 65 135 (1915).

   TYPE: "*in* Hispaniae, Italiae, Cretae, Cycladum *apricis*".

Acaulescent perennial, generally less than 10 cm. high with a long, fleshy rootstock (chiefly rhizomatous); leaves spreading on the ground, ovate-elliptic, 12–23 cm. long, 4–12 cm. wide when fully developed, glabrous or thinly hairy along the midrib and nerves, sometimes covered all over with a thin, scurfy indumentum, flat or more or less bullate, apex obtuse or subacute, margins entire or irregularly sinuate-lobulate, base tapering to a flattened petiole 2–8 cm. long; nervation conspicuous or obscure; flowers generally numerous, crowded at the centre of the leaf-rosette; peduncles slender, 2–8 cm. long, glabrous or pilose with spreading hairs; calyx generally clothed with long multicellular hairs, intermixed with sessile or shortly stipitate glands, lobes lanceolate-acuminate, 8–15 mm. long, 2–4 mm. wide at base, erect, accrescent and persistent in fruit; corolla whitish veined blue or purple or sometimes purplish all over, tube about 10 mm. long, lobes narrowly obovate, oblanceolate or almost oblong, 15–20 mm. long, 5–8 mm. wide, spreading, thinly glandular-pilose; filaments 8–15 mm. long, glabrous or sparsely glandular above, densely villous at point of insertion on corolla-tube; anthers bluish, oblong or subsagittate, about 2–2·5 mm. long, 1–1·5 mm. wide; ovary glabrous, subconical, about 3–4 mm. long, 2·5–3 mm. wide; disk consisting of 2 conspicuous, opposite tubercles or swellings adnate to base of ovary; style 1·5–2·5 cm. long; stigma capitate, obscurely 2-lobed, with reflexed margins, about 2 mm. diam. Fruit globose or broadly ovoid or ellipsoid, 2–2·5 cm. diam., mucronate with the persistent base of the style, yellow or orange when ripe; seeds broadly reniform, strongly compressed, about 5–6 mm. long, 5 mm. wide; testa pale brown, shallowly rugulose-foveolate.

HAB.: Roadsides, waste ground, Olive groves, occasionally on sandy seashores; sea-level to 2,000 ft. alt.; fl. Dec.–April.

DISTR.: Divisions 1, 3, 4, 6, 7, locally common. Widespread in the Mediterranean region.

1. Ayios Yeoryios (Akamas) 1962, *Meikle*! Near Polis, 1962, *Meikle*!
3. Near Limassol, 1859, *Kotschy*; Phasouri, 1970, *A. Genneou* in ARI 1391!
4. Larnaca, 1862, *Kotschy*, also, 1880, *Sintenis & Rigo* 8! and, 1881, *C. & W. Barbey*.
6. Common about Nicosia, 1962, *Meikle*! Kykko Camp near Nicosia, 1973, *P. Laukkonen* 55a!
   Also noted from Ayia Irini in Cyprus Journ. 2 (6): 84 (1905).
7. Common about Kyrenia, *Sintenis & Rigo* 8a! *Syngrassides* 1464! *Davis* 2108! *Casey* 221! *G. E. Atherton* 793! 812! 998! *Polunin* 6616! *Economides* in ARI 1055! *P. Laukkonen* 55! etc.

NOTES: I am not at all convinced that there are valid distinctions, at least at species level, between *Mandragora officinarum* L. (*M. vernalis* Bert.) and *M. autumnalis* Bert., unless the differences in the fruits (globose and much exceeding the persistent calyx in *M. officinarum*; ovoid (or ellipsoid) and smaller in *M. autumnalis*) are more reliable than the other characters quoted by authors to distinguish the two. Unfortunately fruiting specimens of *Mandragora* are rarely collected, and, from the scant evidence available, it is impossible to be certain that the fruit characters are not also illusory. All the Cyprus material appears to flower between December and the end of April and might, one supposes, be referred to *M. officinarum* sens. strict. (*M. vernalis*) rather than *M. autumnalis* on this count. But in colour, size and structure of corolla most (but not all) of the specimens accord better with *M. autumnalis*. They have been cited under *M. officinarum* ssp. *haussknechtii* by Vierhapper (loc. cit.). This infraspecies seems, however, to consist very largely of intermediates connecting *M. officinarum* sens. strict. and *M. autumnalis*, nor do I find any good reason for maintaining it as distinct at any rank. The leaf and indumentum characters given by Vierhapper and others in support of their taxonomy seem to be of little significance, and depend, to a degree, upon the time of year when the specimen is collected.

5. DATURA *L.*

Sp. Plant., ed. 1, 179 (1753).
Gen. Plant., ed. 5, 83 (1754).
A. G. Avery et al., Blakeslee: The Genus Datura, 289 pp. New York (1959).

Erect annuals; leaves alternate, subentire or coarsely sinuate-dentate; flowers solitary, axillary, shortly pedunculate, often large and infundibuliform; calyx elongate, tubular, 5-lobed or sometimes spathaceous, circumscissile about the base after anthesis; corolla with plicate aestivation, lobes 5, often long-acuminate; stamens 5, free, usually included, inserted near base of corolla-tube; filaments slender; anthers linear; ovary 2-locular, or 4-locellate with false septa; style elongate; stigma 2-lobed. Fruit a spinose (rarely smooth) 4-valved capsule; seeds usually numerous, compressed; embryo subperipheral, strongly curved.

Ten species widely distributed in warm-temperate and tropical regions.

Fruit erect; stems glabrous or subglabrous; leaves lacerate-lobed - - **1. D. stramonium**
Fruit nodding on a decurved peduncle; stems clammy-pubescent; leaves subentire or irregularly dentate - - - - - - - - - - - **2. D. innoxia**

1. **D. stramonium** *L.*, Sp. Plant., ed. 1, 179 (1753); Boiss., Fl. Orient., 4: 292 (1879); Holmboe, Veg. Cypr., 163 (1914); Post, Fl. Pal., ed. 2, 2: 262 (1933); Osorio-Tafall et Seraphim, List Vasc. Plants Cyprus, 93 (1973); Davis, Fl. Turkey, 6: 451 (1978); Feinbrun, Fl. Palaest., 3: 168, t. 279 (1978).
   TYPE: "*in* America, *nunc vulgaris per* Europam".

Glabrous or puberulous, erect, foetid-smelling annual 50–150 (–200) cm. high; stems stout, terete, usually much branched in well-grown specimens; leaves ovate-oblong, 5–12 (–17) cm. long, 2·5–8 (–14) cm. wide, thinly puberulous or glabrous, apex acute, margins sharply lacerate-lobed, base cuneate, narrowing rather abruptly to a petiole 1–6 cm. long; flowers solitary in the bifurcations of the branches, subsessile, the peduncle lengthening to 1 cm. in fruit; calyx-tube narrowly cylindrical, glabrous, 2–5 cm. long, 0·6–1 cm. wide, lobes 5, rather unequal, subulate, 4–5 mm. long, 1–1·5 mm. wide at base, margins minutely ciliate; corolla white or purplish (var. *tatula* (L.) Torr.), infundibuliform with an elongate tube 7–10 cm. long spreading rather abruptly near the apex and terminating in 5 small acuminate-caudate lobes 3–7 mm. long; stamens included; filaments about 2·5 cm. long, glabrous; anthers yellow, narrowly oblong, about 6–7 mm. long, 1·5–2 mm. wide; ovary conical, muricate, about 4–5 mm. long and about as wide at the base, surrounded by a fleshy lobulate disk; style glabrous, about 5 cm. long; stigma 2-lobed, about 4 mm. diam. Fruit an erect, ovoid, woody, echinate capsule 3–5 cm. long, 2–4 cm. wide, splitting almost to base into 4 lobes; seeds numerous, reniform, strongly compressed, about 3·5 mm. long, 2·5 mm. wide; testa blackish, closely foveolate.

HAB.: Waste ground, field margins, roadsides, sometimes on sandy spits by streams or rivers; sea-level to 4,000 ft. alt.; fl. May–Sept.

DISTR.: Divisions 2–4, 6. A cosmopolitan weed, possibly of American origin.

2. Platres, 1937, *Kennedy* 835 !
3. Kilani, 1905, *Holmboe* 1148; Khalassa, 1970, *A. Genneou* in ARI 1500 !
4. Famagusta, 1936, *Nattrass* 718.
6. Nicosia, 1905, *Holmboe*.

2. D. innoxia *Mill.*, Gard. Dict., ed. 8, no. 5 (1768) [as *D. inoxia*]; Osorio-Tafall et Seraphim, List Vasc. Plants Cyprus, 93 (1973); Davis, Fl. Turkey, 6: 452 (1978); Feinbrun, Fl. Palaest., 3: 168, t. 280 (1978).
[*D. metel* (non L.) Lindberg f., Iter Cypr., 31 (1946).]
TYPE: "grows naturally at La Vera Cruz, from whence I received the seeds" — cultivated in Chelsea Physic Garden (BM).

Erect annual up to 200 cm. high; stems and leaves rather densely clothed with soft, clammy hairs; stems usually repeatedly branched; leaves broadly ovate, 3–14 cm. long, 3–10 cm. wide, acute, margins subentire or irregularly dentate, base tapering abruptly to a petiole 1–7 cm. long, sometimes truncate or subcordate; flowers solitary in the bifurcations of the branches; peduncle stout, about 1 cm. long; calyx-tube narrowly cylindrical, 6–7 cm. long, 1–1·5 cm. wide, densely pubescent, lobes 5, subulate, about 5–6 mm. long, 3–4 mm. wide at base; corolla white, infundibuliform, with a glabrous tube 15–16 cm. long, spreading gradually towards the apex and terminating in 10, very small, subulate lobes; stamens included; filaments glabrous, about 6 cm. long; anthers yellow, narrowly oblong, about 10 mm. long, 2 mm. wide; ovary conical, about 7 mm. long, 5–6 mm. wide, densely spinose; style 10–12 cm. long, glabrous; stigma 2-lobed, about 4 mm. diam. Fruit a nodding, ovoid, irregularly dehiscing, softly spinose capsule, 4–5 cm. long, 3–4 cm. wide, the base surrounded by a frill formed from the much-accrescent base of the circumscissile calyx-tube; fruiting peduncle strongly arcuate-decurved; seeds numerous, reniform, strongly compressed, about 5 mm. long, 4 mm. wide; testa pale brown, minutely and closely foveolate with a number of larger, deeper pits along the dorsal surface.

HAB.: Gardens, roadsides, waste ground, sea-level to 500 ft. alt.; fl. June–Oct.

DISTR.: Divisions 3, 4, 6, 7. Native in Central America, now widely naturalized in tropical and subtropical regions.

3. Kouklia, 1939, *Lindberg f.* !
4. Larnaca, 1939, *Lindberg f.*
6. Nicosia, 1935, *Syngrassides* 663 ! and, 1976, *A. Della* in ARI 1691 !
7. Kyrenia, 1970, *A. Genneou* in ARI 1614 !

In addition to the above annual weeds, two woody representatives of *Brugmansia* Pers., a genus closely allied to *Datura*, and commonly united with it, are widely cultivated in the Mediterranean region, and are no doubt to be seen in Cyprus gardens. Apart from its arborescent habit, *Brugmansia* differs from *Datura* in its non-circumscissile calyx; unarmed, irregularly rupturing fruit, and large, corky seeds. The two species mentioned above are: BRUGMANSIA SUAVEOLENS (*Humb. et Bonpl. ex Willd.*) *Bercht. et J. Presl*, (Angels' Trumpet), from South America, a shrub or small tree up to 5 m. high, with ovate or elliptic, entire leaves, and very large, nodding, white, sweet-scented, trumpet-shaped flowers, sometimes as much as 30 cm. long. B. SANGUINEA (*Ruiz et Pavon*) *D. Don* is similar in habit, but with scentless orange-red flowers shading to greenish-yellow at the base of the corolla tube. It is also a native of South America, but from high elevations, and is consequently hardier than *B. suaveolens*. A third *Brugmansia*, B. × CANDIDA

Pers., a hybrid between *B. suaveolens* and *B. aurea* Lagerh., is also commonly cultivated. It much resembles *B. suaveolens*, but has a spathe-like (not lobed) calyx-apex, and a more widely flaring corolla-tube, not so obviously constricted above the mouth of the calyx as in *B. suaveolens*. The flowers of *B.* × *candida* are commonly creamy-white, yellowish or tinged pink.

6. HYOSCYAMUS *L.*

Sp. Plant., ed. 1, 179 (1753).
Gen. Plant., ed. 5, 84 (1754).

Annual, biennial or perennial herbs; leaves and stems commonly clammy-villous; leaves often coarsely dentate, incised or pinnatisect, rarely entire; flowers solitary, axillary, or the upper forming crowded, bracteate spikes or scorpioid cymes; calyx tubular-campanulate or urceolate, accrescent and enveloping the fruit; corolla somewhat zygmorphic, oblique, with 5 subequal or unequal lobes; stamens inserted about the middle or near the base of the corolla-tube; anthers generally exserted, the thecae dehiscing longitudinally; ovary 2-locular, style slender; stigma capitate. Fruit an operculate, circumscissile, many-seeded capsule.

About 20 species in the Mediterranean region, Atlantic Islands and western Europe eastwards to Central Asia.

Corolla greenish-white; base of calyx distinctly ventricose in fruit, narrowing abruptly to pedicel; leaves bluntly dentate-lobulate - - - - - - - - **1. H. albus**
Corolla golden-yellow with a dark purple throat; base of calyx tapering to pedicel, not ventricose; leaves sharply and irregularly incised-dentate - - - **2. H. aureus**

1. H. albus *L.*, Sp. Plant., ed. 1, 180 (1753); Sibth. et Sm., Fl. Graec. Prodr., 1: 153 (1806), Fl. Graec., 3: 24, t. 230 (1819); Unger et Kotschy, Die Insel Cypern, 287 (1865); Boiss., Fl. Orient., 4: 295 (1879); Holmboe, Veg. Cypr., 163 (1914); Post, Fl. Pal., ed. 2, 2: 264 (1933); Lindberg f., Iter Cypr., 31 (1946); Osorio-Tafall et Seraphim, List Vasc. Plants Cyprus, 93 (1973); Davis, Fl. Turkey, 6: 455 (1978); Feinbrun, Fl. Palaest., 3: 163, t. 269 (1978).
TYPE: *"in* Europa *australi"*.

Erect annual or biennial up to 1 m. high; stems usually much branched, somewhat viscid, with a mixed indumentum of long and short, spreading hairs; leaves broadly ovate, 5–10 cm. long, 3–8 cm. wide, thinly hairy or glabrescent, apex blunt or shortly acute, margins coarsely, but rather bluntly, dentate-lobulate, sometimes strongly undulate, base broadly cuneate, truncate or subcordate, narrowing abruptly to a stout, channelled, hairy petiole 2–8 cm. long; lowermost flowers often solitary in the axils of the branches, upper crowded into a scorpioid cyme; bracts foliaceous, sessile, narrowly oblong or oblanceolate, 1–5 cm. long, 0·5–1·5 cm. wide; pedicels very short or almost wanting, accrescent to 1 cm. in fruit; calyx campanulate, villous, about 13 mm. long and 8–10 mm. wide at anthesis, accrescent to 2 cm. or more long, and becoming urceolate and conspicuously costate in fruit; calyx-lobes 5, sharply deltoid, unequal, 1–3 mm. long, 1–2·5 mm. wide at base, sometimes with small secondary lobes in the sinuses; corolla pale greenish-white, broadly infundibuliform, strongly oblique, thinly hairy externally, tube 1·5–2·5 cm. long, lobes broad, rounded, the adaxial up to 8 mm. long, and about as wide, the abaxial distinctly smaller; stamens inserted near base of corolla-tube; filaments slender, glabrous or pilose, declinate, 1–2 cm. long; anthers shortly exserted, yellow, narrowly oblong, almost 4 mm. long, 1·5 mm. wide; ovary sessile on a small, deeply lobulate disk, ovoid, about 2·5–3 mm. long, 2–2·5 mm. wide; style glabrous or thinly pilose, 1·6–2·3 cm. long; stigma capitate. Fruit a bluntly ovoid, circumscissile capsule about 1 cm. long, 0·8 cm. diam., concealed within the accrescent calyx; seeds irregularly oblong, about 1·5 mm. long and almost as wide; testa pale grey, deeply and closely foveolate with sinuous ridges between the pits.

HAB.: Waste ground, roadsides, sandy river banks, sometimes growing out of old walls; sea-level to 3,000 ft. alt.; fl. Jan.–Sept.

DISTR.: Divisions 1–4, 6, 7. S. Europe and Mediterranean region eastwards to Iraq.

1. Stephani near Ayios Neophytos Monastery, and at the monastery, 1939, *Lindberg f.* s.n.; Paphos, 1960, *C. E. H. Sparrow* 48! and, 1973, *P. Laukkonen* 317!
2. Makheras Monastery, 1937, *Syngrassides* 1633! Kambos, 1939, *Lindberg f.* s.n.; near Mallia, 1950, *F. J. F. Barrington* 73! Chakistra, 1962, *Meikle* 2743!
3. Mazotos, 1862, *Kotschy* 552; Kolossi, 1961, *Polunin* 6655! Khalassa, 1970, *A. Genneou* in ARI 1492!
4. Larnaca, 1880, *Sintenis*, and, 1939, *Lindberg f.* s.n.!
6. Nicosia, 1905, *Holmboe* 303; Arkhangelos Monastery, 1935, *Syngrassides* 1331!
7. Buffavento, 1862, *Kotschy*; monastery near Karavas, 1930, *C. B. Ussher* 127! Akhiropiitos Monastery, 1936, *Syngrassides* 1297! Kyrenia, 1949, *Casey* 456! also, 1955, *G. E. Atherton* 24! 209! Monastery at Lambousa, 1955, *G. E. Atherton* 341!

NOTES: The apparent prevalence of this species in the neighbourhood of monasteries may be indicative of its former use as a narcotic and analgesic.

2. H. aureus L., Sp. Plant., ed. 1, 180 (1753); Sibth. et Sm., Fl. Graec. Prodr., 1: 153 (1806), Fl. Graec., 3: 25, t. 231 (1819); Unger et Kotschy, Die Insel Cypern, 287 (1865); Boiss., Fl. Orient., 4: 296 (1879); Holmboe, Veg. Cypr., 163 (1914); Post, Fl. Pal., ed. 2, 2: 264 (1933); Lindberg f., Iter Cypr., 31 (1946); Osorio-Tafall et Seraphim, List Vasc. Plants Cyprus, 93 (1973); Davis, Fl. Turkey, 6: 456 (1978); Feinbrun, Fl. Palaest., 3: 163, t. 271 (1978).
TYPE: "*in* Creta".

Woody-based perennial 30–60 (–100) cm. high; stems sprawling or pendulous, thinly or densely glandular-pilose with spreading hairs, generally much branched; leaves very broadly oblong or almost suborbicular in outline, 2–8 cm. long and about as wide, thinly clothed with spreading, viscid hairs, apex usually sharply acute, margins sharply and irregularly incised-dentate, base cordate or subcordate; petiole up to 5 cm. long, canaliculate above, pilose with spreading hairs; flowers in rather lax, elongate scorpioid cymes; bracts foliaceous, dentate, similar to the leaves, about 1·5–2·5 cm. long, 1–1·5 cm. wide; pedicels slender, pilose, about 2–3 mm. long, lengthening to 5–10 mm. in fruit; calyx campanulate, pilose, about 1·8 cm. long, 1 cm. wide at apex, accrescent to 2 cm. long in fruit and becoming strongly costate, but without the markedly ventricose base of *H. albus*, calyx-lobes deltoid-acuminate, about 2 mm. long, 1·5 mm. wide; corolla golden-yellow with a purple throat and elongate, purple-stained tube 3–3·5 cm. long, lobes rounded or subtruncate, spreading, the adaxial up to 8 mm. long and almost as wide, the abaxial distinctly smaller; stamens inserted near base of the corolla-tube; filaments slender, pilose, purplish, declinate, 2–2·5 cm. long; anthers long-exserted, purplish, narrowly oblong, about 3 mm. long, 1·3 mm. wide; ovary glabrous, ovoid, about 5 mm. long, 4 mm. wide, with a well-marked dome-like operculum; disk inconspicuous; style slender, glabrous, up to 4 cm. long, upcurved at apex; stigma capitate. Fruit a bluntly ovoid or obovoid, circumscissile capsule about 8 mm. long, 6 mm. wide, concealed within the calyx; seeds compressed, suborbicular or broadly reniform, about 1 mm. diam.; testa yellowish-brown, conspicuously rugulose-foveolate.

HAB.: Old walls; limestone cliffs and rocks; sea-level to 2,000 ft. alt.; fl. Febr.–July.

DISTR.: Divisions 1, 3, 4, 6, 7, locally common; Crete, Aegean Islands, Turkey, Syria, Lebanon, Palestine, Egypt, eastwards to Iraq.

1. Between Androlikou and Prodhromi, 1957, *Merton* 3031!
3. Stavrovouni Monastery, 1787, *Sibthorp*.
4. Famagusta, 1862, *Kotschy*; Larnaca, 1880, *Sintenis & Rigo* 9! Cape Greco, 1905, *Holmboe* 414! Between Paralimni and Ayia Napa, 1979, *Chesterman* 108!

6. W. of Orga, 1956, *Merton* 2603!
7. Locally common; Kyrenia, St. Hilarion, Panagra, Kantara, etc. *Syngrassides* 268! 1131! *Lindberg f.* s.n.! *Davis* 2433! *Casey* 297! *G. E. Atherton* 180! 210! 951! *C. E. H. Sparrow* 19! *Meikle* 2373! *P. Laukkonen* 90! etc.

7. NICOTIANA *L.*

Sp. Plant., ed. 1, 180 (1753).
Gen. Plant., ed. 5, 84 (1754).
T. H. Goodspeed, The Genus Nicotiana, in Chronica Botanica, 16 (1/6): I–XXII & 1–536 (1954).

Annual or perennial herbs or sometimes shrubs, often with a clammy or viscid-glandular indumentum; leaves simple, entire, alternate, sessile or petiolate; flowers in panicles or false racemes, occasionally in crowded heads, sometimes fragrant towards evening; calyx tubular, 5-lobed or irregularly cleft, persistent and often accrescent in fruit, always considerably smaller than the corolla, but sometimes exceeding the capsule; corolla actinomorphic or somewhat zygomorphic, tubular, infundibuliform or hypocrateriform, limb deeply or shallowly 5-lobed, the lobes spreading or recurved, contorted-plicate or rarely imbricate in aestivation; stamens 5, free; filaments equal or unequal, inserted below throat of corolla; anthers usually included, dehiscing longitudinally; ovary 2-locular; style terminal; stigma slightly grooved; hypogynous disk sometimes annular and nectariferous. Fruit a capsule, dehiscing septicidally and loculicidally in the upper part; seeds numerous, small; embryo straight or curved.

About 60 species native in South America, North America, Australia and the S. Pacific region, naturalized and cultivated throughout temperate and tropical regions of the world.

Shrub or small tree; leaves glabrous, glaucous; flowers greenish-yellow   -   -   **1. N. glauca**
Herbs; leaves green, viscid-puberulous:
  Leaves with a distinct, unwinged petiole; flowers greenish-yellow   -   N. RUSTICA (p. 1195)
  Leaves decurrent, the lamina extending down the margins of the indistinct petiole; flowers white, pink or reddish   -   -   -   -   -   -   -   N. TABACUM (p. 1195)

1. N. glauca *Graham* in Edinb. New Phil. Journ., 5: 175 (1828); Post, Fl. Pal., ed. 2, 2: 265 (1933); Rechinger f. in Arkiv. för Bot., ser. 2, 1: 431 (1950); Goodspeed, The Genus Nicotiana, 335, fig. 59 (1954); Osorio-Tafall et Seraphim, List Vasc. Plants Cyprus, 93 (1973); Davis, Fl. Turkey, 6: 457 (1978); Feinbrun, Fl. Palaest., 3: 169, t. 281 (1978).
TYPE: "raised in 1827 from seeds communicated ... to the Royal Botanic Gardens, Edinburgh ... from Buenos Ayres" (E).

Loosely branched, soft-wooded shrub or short-lived tree 3–6 (–10) m. high; stems terete, glabrous, glaucous, laxly branched; leaves ovate, acute, 3–8 (–16) cm. long, 1·5–6 (–12) cm. wide, glabrous, glaucous, base tapering abruptly to a slender petiole 3–5 cm. long; flowers in lax, subcorymbose terminal panicles; bracts caducous, minute, subulate, about 2 mm. long, less than 0·5 mm. wide, ciliate with viscid multicellular hairs; pedicels 2–8 mm. long, thinly viscid-pilose, elongating and curving downwards in fruit; calyx tubular, about 10 mm. long, 4 mm. wide, glabrous, with 5 short, subulate lobes, about 1·5–2 mm. long, 0·6 mm. wide at base; corolla greenish-yellow, narrowly tubular, 30–40 mm. long, 4 mm. wide, rather densely viscid-pilose externally in the upper part, markedly constricted and glabrous towards base, lobes very small, erect, semicircular, apiculate, about 2 mm. long, 3 mm. wide; stamens inserted near base of corolla-tube; filaments glabrous, slender, about 2·4 cm. long, anthers suborbicular, about 1·3 mm. diam.; disk nectariferous; ovary ovoid, glabrous, about 2·5 mm. long, 1·5 mm. wide; style about 3 cm. long; stigma capitate, obscurely 2-lobed, about 0·7 mm. diam. Fruit ovoid, about 10 mm. long, 7 mm. wide, slightly exceeding the persistent, pale brown, papery calyx and dehiscing by 4 short, deltoid apical teeth; seeds irregularly rounded-oblong, about 0·6 mm. long, 0·4 mm. wide; testa yellowish-brown, loosely reticulate with sinuous ridges; embryo straight.

HAB.: Gardens, waste ground, roadsides; sea-level to 500 ft. alt.; fl. Jan.–Dec.

DISTR.: Divisions 4, 6. Native of Argentina and possibly Bolivia, now naturalized in many warm-temperate and tropical countries.

4. Larnaca, 1913, *Haradjian* 1020! Famagusta, 1949, *Casey* 575!
6. Omorphita, Nicosia, 1973, *P. Laukkonen* 124!

NOTES: Little has been recorded concerning the spread of this species, now thoroughly naturalized not only in the Mediterranean region, but in Africa, Australia, southern U.S.A. and elsewhere. The earliest records of *N. glauca* as a naturalized plant in our area are very recent, but it is noted from Egypt in 1889, (Ascherson & Schweinfurth, Suppl., Illustr. Fl. Egypte, 770) and was probably well established even earlier. The leaves are said to be poisonous to stock (Steyn, Toxicology of Plants in South Africa, 365; 1934).

N. RUSTICA *L.*, Sp. Plant., ed. 1, 180 (1753), an annual herb, up to to 1·5 m. high, with shortly petiolate, ovate, elliptic or subcordate, viscid-puberulous leaves, 10–15 (–30) cm. long, and terminal panicles of numerous, tubular, yellow-green flowers 1–2 cm. long, 0·6–0·8 cm. wide, was formerly cultivated in Cyprus (Unger et Kotschy, Die Insel Cypern, 287; 1865) as a source of snuff and tobacco, and may possibly persist either in cultivation or as an escape from cultivation. It is probably of South American origin, but was grown in many parts of Central and North America in the pre-Columbian period, and spread to Europe soon after the discovery of the Americas; it is now very largely replaced in cultivation by the following.

N. TABACUM *L.*, Sp. Plant., ed. 1, 180 (1753), Tobacco, an erect annual or perennating herb up to 3 m. high, with very large, decurrent, ovate or elliptic, viscid-puberulous leaves, and much-branched, terminal panicles of tubular, white, pink or reddish flowers, 4–5 cm. long, 0·5–0·8 cm. wide, has long been cultivated in Cyprus, especially in the Karpas peninsula, also about Paphos and Omodhos (Unger et Kotschy, Die Insel Cypern, 286; 1865). Its origin is uncertain, though most probably South American, and its use would appear to have been restricted to Central and South America and the West Indies in the pre-Columbian period, being replaced by *N. rustica* L. in North America. Both species were brought to the Old World early in the 16th century, but *N. tabacum* has now very largely supplanted *N. rustica* as a smoking tobacco, though the latter is still locally grown as a source of snuff and nicotine.

[N. PUSILLA *L.*, Syst. Nat., ed. 10, 933 (1759), a slender annual with rugulose leaves, and narrow-tubular, yellow-green flowers, was recorded from Limassol in 1801, by Hume (in Walpole, Mem. Europ. Asiatic Turkey, ed. 1, 253; 1817). The record is almost certainly erroneous, but it is impossible, in the absence of a specimen, to identify the plant so named. It may have been *N. rustica* L.]

67. SCROPHULARIACEAE

R. v. Wettstein in Engl. et Prantl, Pflanzenfam., ed. 1, IV, 3b: 38–107
(1891–93).

Herbs or rarely shrubs or trees, sometimes parasitic (and blackening when dried); leaves alternate, opposite or rarely verticillate, simple, entire, toothed or variously lobed, occasionally very reduced or wanting; stipules absent; inflorescence a spike, raceme or branched cyme, or flowers solitary and axillary; bracts and bracteoles present or absent; flowers hermaphrodite; calyx often persistent, tubular or divided to, or almost to, the base into 5 (or 4) valvate, imbricate or open equal or unequal lobes; corolla usually zygomorphic, gamopetalous, sometimes spurred and 2-lipped, with 4–5 (–8) imbricate lobes, the adaxial lobe sometimes hooded and lobulate at the apex, •the abaxial frequently enlarged, 3-lobulate, or concave, occasionally saccate; stamens usually 4, didynamous, or 2, inserted at the base of the corolla-tube and alternate with the lobes, the fifth stamen absent, or reduced to a staminode, seldom perfect; anthers 2-thecous, or less often 1-thecous, introrse, dehiscing longitudinally, sometimes coherent in pairs; nectariferous disk present or absent; ovary superior, sessile, generally 2-locular; placentation generally axile, placentas adnate to the septum, often tumid; ovules commonly numerous; style terminal, simple or shortly lobed at the apex; stigma terminal, truncate or capitate, or sometimes bordering the stylar lobes. Fruit generally a septicidal or poricidal capsule, rarely baccate or indehiscent; seeds usually numerous; endosperm normally present; embryo straight or curved.

More than 200 genera and about 3,000 species with a cosmopolitan distribution. Species of *Verbascum, Calceolaria, Linaria, Antirrhinum, Penstemon, Mimulus, Veronica* and *Hebe* are (together with members of other genera) valued as ornamental garden plants. *Digitalis* is grown as an ornamental and as a drug plant, while the arborescent *Paulownia* is widely planted as a street tree in S. Europe and the Mediterranean region.

Trees with broad, cordate, petiolate leaves; flowers large, tubular, mauve-blue, borne on leafless
    branches in spring - - - - - - - - - PAULOWNIA (p. 1216)
Herbs or subshrubs:
  Aquatic herbs; leaves all basal; corolla minute, rotate, 5-lobed - - - **9. Limosella**
  Terrestrial herbs or subshrubs; leaves not all basal:
    Stamens 2; corolla small, subrotate, with a very short tube and 4 spreading lobes
                                        10. Veronica
    Stamens 4 or 5:
      Corolla subrotate or widely infundibuliform, lobes 5, subequal - - **1. Verbascum**
      Corolla with a distinct tube and bilabiate limb:
        Corolla with a basal spur or pouch:
          Sprawling or trailing herbs; leaves distinctly petiolate, cordate, hastate, sagittate
          or reniform:
            Leaves palmately nerved, long-petiolate; capsule dehiscing longitudinally or
              indehiscent - - - - - - - - **3. Cymbalaria**
            Leaves pinnately nerved, short-petiolate; capsule dehiscing by circular lids or
              opercula - - - - - - - - **4. Kickxia**
        Erect annuals or perennials; leaves tapering to base, often indistinctly petiolate,
          not cordate, hastate, sagittate or reniform:
          Base of corolla shortly and bluntly pouched or gibbous:
            Calyx-lobes unequal; corolla small, scarcely exceeding calyx; seeds with one
            face convex, keeled, the other concave with an incurved sinuate-dentate
            border - - - - - - - - **6. Misopates**
            Calyx-lobes subequal; corolla large, much exceeding calyx; seeds oblong,
            foveolate or reticulate - - - - - - **7. Antirrhinum**

Base of corolla with a distinct, narrow spur:
  Throat of corolla closed by a prominent gibbous palate - - - **2. Linaria**
  Throat of corolla open, palate not conspicuously gibbous **5. Chaenorhinum**
Corolla without a basal pouch or spur:
  Calyx-lobes 5, corolla-tube short, ventricose; corolla-lobes 5, the adaxial not
    connate or forming a hood - - - - - - - **8. Scrophularia**
  Calyx-lobes 4; corolla-tube long, cylindrical or narrowly infundibuliform; adaxial
    corolla-lobes connate forming a hood over the anthers:
    Leaves toothed or lobed; glandular-viscid annuals:
      Capsule narrow, fusiform, not much swollen; seeds with a smooth or obscurely
        rugulose testa - - - - - - - **11. Parentucellia**
      Capsule ovoid-globose, turgid; seeds with a longitudinally ridged, and
        transversely striate testa - - - - - - **12. Bellardia**
    Leaves entire, not glandular-viscid; suffruticose perennial - **13. Odontites**

TRIBE 1. **Verbasceae** *Benth.* Leaves alternate, or very rarely opposite; corolla subrotate with a short tube, or widely infundibuliform; lobes 5, subequal, the adaxial external in bud; anthers 1-thecous.

1. VERBASCUM L.

Sp. Plant., ed. 1, 178 (1753).
Gen. Plant., ed. 5, 83 (1754).
Celsia L., Sp. Plant., ed. 1, 621 (1753).

Annual, biennial or perennial herbs, or rarely subshrubs, commonly clothed with a dense lanuginose indumentum of simple or compound, glandular or eglandular hairs; leaves generally alternate, entire or variously toothed or lobed, the basal often forming a rosette; inflorescence a terminal spike, raceme or panicle; calyx deeply 5-lobed, lobes generally equal, imbricate; corolla subrotate or widely infundibuliform, almost actinomorphic, with 5 subequal lobes, or occasionally somewhat zygomorphic; stamens 5 or 4, sometimes 4 fertile and the fifth reduced to a staminode; filaments equal or unequal, commonly clothed with coloured hairs; anthers uniform, transversely medifixed, reniform, or dimorphic, the lower (or abaxial) elongate and decurrent along the filament, or occasionally shorter and obliquely inserted; style filiform or slightly swollen above; stigma capitate, obovate or spathulate. Capsule broadly ovoid or subglobose, septicidal, 2-valved; seeds numerous, small, usually angular.

About 400 species, chiefly in Europe, western Asia and North Africa, but very sparsely represented in Cyprus.

Leaves and stems shortly tomentose with stellate hairs; basal leaves conspicuously sinuate-
  lobed - - - - - - - - - - - - **4. V. sinuatum**
Leaves and stems subglabrous or thinly pilose with simple glandular or eglandular hairs:
  Flowers very shortly pedicellate or subsessile; annual with the cauline leaves pinnatisect or
    bipinnatisect into narrow lobes - - - - - - - **3. V. orientale**
  Flowers distinctly pedicillate; biennials or perennials; cauline leaves not as above:
    Basal leaves crenate or shortly sinuate-lobed; cauline alternate; stem sparsely glandular
      above, glabrous towards base; stamens 5 - - - - - **1. V. blattaria**
    Basal leaves lyrate-pinnatisect; cauline opposite; stems generally glandular-pilose
      throughout; stamens 4 - - - - - - - - **2. V. levanticum**

1. V. blattaria *L.*, Sp. Plant., ed. 1, 178 (1753); Boiss., Fl. Orient., 4: 308 (1879); Holmboe, Veg. Cypr., 164 (1914); Osorio-Tafall et Seraphim, List Vasc. Plants Cyprus, 93 (1973); Davis, Fl. Turkey, 6: 491 (1978).
TYPE: "*in* Europae *australioris locis argillaceis*".

Erect biennial 30–100 (–150) cm. high; stems subterete or obscurely striate, glabrous below, sparsely glandular above, usually unbranched; basal leaves forming a loose rosette, narrowly oblong or oblanceolate, 3–15 (–25) cm. long, 1–5 (–7) cm. wide, subsessile, rugulose, glabrous or very thinly hispidulous, apex obtuse or subacute, margins crenate or sinuate-lobed, base tapering; cauline leaves similar but with sessile, amplexicaul bases, becoming progressively smaller and with wider bases towards the apex of the stem; inflorescence a lax, slender, unbranched raceme often 30–50 cm. long; rhachis glandular; flowers numerous, solitary in each axil; bracts ovate, the lowermost up to 10 mm. long, 8 mm. wide, obscurely dentate, glandular, the upper progressively smaller, narrower and becoming entire; bracteoles absent; pedicels patent, glandular, up to 1 cm. long; calyx persistent, lobes narrowly oblong or lanceolate, entire, subequal, about 5 mm. long, 1·5 –2 mm. wide, glandular, apex subacute or obtuse; corolla yellow, about 3 cm. diam.; lobes subequal, almost orbicular, about 10–12 mm. diam.; stamens 5, the 3 adaxial about 5–7 mm. long, the filaments densely clothed with long purple hairs, the 2 abaxial 7–8 mm. long, glabrous towards apex; anthers of the adaxial stamens reniform-medifixed, about 1·5 mm. wide, those of the abaxial stamens shortly decurrent, about 2 mm. long; ovary about 1·5–2 mm. diam., densely glandular; style about 6 mm. long, filiform, glandular towards base; stigma capitate. Capsule subglobose, 6–7 mm. diam., thinly glandular; seeds oblong, about 0·8 mm. long, 0·5 mm. wide; testa dull brown, minutely reticulate with prominent, regular, transverse rugosities.

HAB.: Moist, grassy places, or as a weed in irrigated gardens and orchards; 4,000–5,000 ft. alt.; fl. July–Sept.

DISTR.: Division 2, rare. Widespread in Europe and eastwards to Afghanistan and Central Asia; North Africa, and introduced elsewhere.

2. Prodhromos, 1905, *Holmboe* 907, and, 1955, *G. E. Atherton* 645! Platres, 1905, *Holmboe*, and (at Kephalovryso), 1940, *Davis* 1976!

2. **V. levanticum** *I. K. Ferguson* in Bot. Journ. Linn. Soc., 64: 230 (1971); Tutin et al., Fl. Europaea, 3: 209 (1972); Davis, Fl. Turkey, 6: 487 (1978); Feinbrun, Fl. Palaest., 3: 182, t. 302 (1978).
    Celsia glandulosa Bouché in Verh. Ges. Naturf. Fr., Berlin, 1: 395 (1829); Boiss., Fl. Orient., 4: 350 (1879); Post, Fl. Pal., ed. 2, 2: 279 (1933).
    Verbascum glandulosum (Bouché) O. Kuntze, Rev. Gen., 2: 469 (1891) non Del. (1849) nom. illeg.
    [*Celsia arcturus* (non (L.) Jacq.) Sm. in Sibth. et Sm., Fl. Graec. Prodr., 1: 438 (1809); Poech, Enum. Plant. Ins. Cypr., 25 (1842); Unger et Kotschy, Die Insel Cypern, 289 (1865); Sintenis in Oesterr. Bot. Zeitschr., 32: 399 (1882); Osorio-Tafall et Seraphim, List Vasc. Plants Cyprus, 93 (1973).]
    [*C. horizontalis* (non Moench) Lindberg f., Iter Cypr., 31 (1946).]
    TYPE: Cyprus; "ad rupes montis Pentadactylos" May, 1880, *Sintenis & Rigo* 144 (K !).

Erect biennial or short-lived perennial 30–150 cm. high, often with a procumbent woody base clothed with the remnants of decayed leaves and petioles; stems generally branched only at the base, subterete or bluntly angled, thinly or rather densely clothed with long eglandular and shorter glandular, spreading hairs; basal leaves rather crowded, but not forming a distinct rosette; petiole 2–7 cm. long, channelled above, patently glandular-pilose; lamina oblong, 7–20 cm. long, 2–10 cm. wide, lyrate-pinnatisect with a much enlarged, ovate, rounded or obtuse, crenate-dentate, terminal lobe, and 2–5 irregularly disposed pairs of much smaller, oblong-elliptic, dentate lateral lobes, gradually increasing in size towards apex of rhachis; upper surface of leaflets (or leaf-lobes) subglabrous or thinly pilose, undersurface more densely pilose especially along midrib and nerves; cauline leaves opposite, shortly petiolate or sessile, broadly ovate, 3–10 cm. long, 3–8 cm. wide, often simple, occasionally pinnatisect with 1–2 pairs of lobes, margins often sharply serrate; inflorescence a lax, terminal unbranched or very sparingly branched raceme 20–30 (or more) cm. long; bracts mostly alternate, ovate, denticulate, acute or cuspidate, 5–15 mm. long, 5–8 mm.

wide; bracteoles absent; pedicels patent, slender, up to 2 cm. long at anthesis, shortly glandular; calyx-lobes oblong, acute, entire, 3–4 mm. long, 1·5–2 mm. wide, glabrous, glandular; corolla bright yellow, 2–3 cm. diam., lobes spreading, suborbicular, about 1 cm. diam.; stamens 4, the 2 adaxial filaments 5–6 mm. long, densely clothed throughout their length with long purplish, globe-tipped hairs; abaxial filaments up to 8 mm. long, hairy below, glabrous towards apex; anthers of adaxial stamens reniform-medifixed, about 0·8 mm. diam., those of abaxial stamens elongate-decurrent, about 1·5 mm. long; pollen orange-yellow; ovary ovoid, about 2 mm. long, 1·5 mm. wide, thinly glandular; style about 8 mm. long; stigma slightly swollen with a short conical apex. Capsule subglobose, about 5 mm. diam.; seeds oblong, dull brown, conspicuously rugulose.

HAB.: On old walls and clefts of limestone rocks; sea-level to 3,800 ft. alt.; fl. March–May.

DISTR.: Divisions 1, 2, 4, 6, 7. Also S. Turkey, Lebanon, Palestine, naturalized in Portugal.

1. Paphos, 1905, *Holmboe* 717, also, 1913, *Haradjian* 679! and, 1973, *P. Laukkonen* 311!
2. Kykko monastery, 1905, *Holmboe* 1026, and, 1941, *Davis* 3463!
4. Cape Greco, 1905, *Holmboe* 411, and, 1957, *Poore* 52!
6. Near Liveras, 1962, *Meikle* 2412!
7. Common; Ayios Khrysostomos monastery, Akheropiitos, Pentadaktylos, Lapithos, Bellapais, St. Hilarion, Ayios Epiktitos, near Akanthou, Yaïla, etc. *Kotschy* 389; *Sintenis & Rigo* 144! *C. B. Ussher* 111! *J. A. Tracey* 57! *Syngrassides* 1252! 1380! 1535! *Kennedy* 1070! 1663! *Davis* 1750! 2030! 2870! 3630! *Casey* 661! *G. E. Atherton* 401! *C. E. H. Sparrow* 26! *P. H. Oswald* 134! etc.

3. V. orientale (*L.*) *All.*, Fl. Ped., 1: 106 (1785); Davis, Fl. Turkey, 6: 480 (1978); Feinbrun, Fl. Palaest., 3: 181 (1978).

    Celsia orientalis L., Sp. Plant., ed. 1, 621 (1753); Boiss., Fl. Orient., 4: 360 (1879); Sintenis in Oesterr. Bot. Zeitschr. 31: 393, 395 (1881); Holmboe, Veg. Cypr., 164 (1905); Murbeck in Lunds Univ., Årsskr. N.F. Avd. 2, 22 (1): 95 (1925); Osorio-Tafall et Seraphim, List Vasc. Plants Cyprus, 93 (1973).

TYPE: "*in* Cappadocia, Armenia".

Erect annual 6–80 (–100) cm. high; stems terete, commonly unbranched, sometimes with numerous erect branches, subglabrous or puberulous below, thinly glandular above, often purplish; basal leaves forming a loose rosette, oblong or obovate, 2–5 cm. long, 1–2·5 cm. wide, usually pinnatisect with numerous, narrow, bluntly toothed or laciniate lobes, occasionally irregularly lobulate or subentire, thinly glandular-puberulous above and below; petiole slender, 1–3 cm. long, concave above; cauline leaves alternate, similar to the basal, but frequently bipinnatisect into numerous, narrow, lanceolate, acute lobes; more shortly petiolate or subsessile, especially towards apex of stem; inflorescence a simple or branched, spiciform raceme; flowers often rather remote; pedicels very short, commonly less than 3 mm. long at anthesis, elongating and thickening in fruit; bracts narrowly lanceolate or almost linear, usually entire, shortly puberulous, up to 1 cm. long, generally less than 2 mm. wide; bracteoles absent; calyx lobed almost to base, lobes lanceolate or narrowly oblong, entire, acute or obtuse, sparsely glandular-puberulous, 4–7 mm. long, 1–2·5 mm. wide; corolla pale yellow, 1·5–1·8 cm. diam., lobes subequal, rounded, 6–8 mm. long and almost as wide, glabrous or subglabrous; stamens 4; filaments subequal, 3·5–4 mm. long, densely clothed with golden, club-tipped, woolly hairs; anthers yellow, uniformly reniform-medifixed, about 0·8 mm. diam.; ovary broadly ovoid, glabrous, about 1·5 mm. long, 1·4 mm. wide; style about 3·5 mm. long; stigma subcapitate, fringed with clavate hairs. Capsule subglobose or broadly ovoid, 4–5 mm. long, 3·5–4 mm. wide, glabrous, obscurely reticulate, tardily dehiscent, often crowned with the persistent style-base; seeds irregularly angular, sometimes curved, about 0·8 mm. long, 0·6 mm. wide; testa fuscous-brown, coarsely rugulose.

HAB.: Rocky limestone slopes and cliffs; 1,000–1,700 ft. alt.; fl. Feb.–May.

DISTR.: Division 7, rare. Eastern Mediterranean region and eastwards to Crimea and Caucasus.

7. Pentadaktylos, 1880, *Sintenis & Rigo* 143! Kornos, 1941, *Davis* 2198! 2999! Kyrenia Pass, 1953, *Casey* 1317!

4. V. sinuatum *L.*, Sp. Plant., ed. 1, 178 (1753); Unger et Kotschy, Die Insel Cypern, 289 (1865); Boiss., Fl. Orient, 4: 322 (1879); H. Stuart Thompson in Journ. Bot., 44: 335 (1906); Holmboe, Veg. Cypr., 164 (1914); Post, Fl. Pal, ed. 2, 2: 273 (1933); Lindberg f., Iter Cypr., 31 (1946); Osorio-Tafall et Seraphim, List Vasc. Plants Cyprus, 93 (1973); Davis, Fl. Turkey, 6: 538 (1978); Feinbrun, Fl. Palaest., 3: 174 (1978).

TYPE: "Monspelii, Florentiae".

Erect biennial 50–150 cm. high; stems and leaves clothed with a short tomentum of greyish or yellowish, stellate hairs; stems commonly branched from the base, terete or obscurely angled; basal leaves oblong or oblong-obovate, 10–30 cm. long, 2·5–7 cm. wide, margins undulate-sinuate, crenate or irregularly pinnatilobed, apex blunt or subacute, base narrowing to a short petiole, generally less than 2 cm. long; cauline leaves diminishing rapidly upwards, sessile, amplexicaul-decurrent, margins subentire or irregularly toothed or lobed; inflorescence a lax panicle with numerous, spreading branches; bracts cordate-deltoid, tomentose, acute, the lowermost 2·5–3 cm. long, 1·5–2 cm. wide, becoming smaller upwards, the uppermost generally minute; flowers in clusters of 2–7, occasionally solitary; pedicels very short, usually less than 5 mm. long; bracteoles minute, ovate; calyx deeply lobed, the lobes ovate-lanceolate, 2·5–3 mm. long, 1–1·5 mm. wide, densely stellate-tomentose externally; corolla yellow, about 2·5 cm. diam., glabrous internally, thinly stellate-pubescent and copiously gland-dotted externally, lobes bluntly obovate or suborbicular, subequal, about 10 mm. long, 8–9 mm. wide; stamens 5, filaments about 8 mm. long, densely clothed to near apex with purple, woolly hairs; ovary ovoid, about 2 mm. long, 1·5 mm. wide, densely stellate-tomentose; style glabrous, about 6 mm. long, slightly swollen towards apex; stigma subcapitate. Capsule broadly ovoid, glabrescent, about 4 mm. long, 3·5 mm. wide, dark brown when mature, tardily dehiscent; seeds irregularly oblong, about 0·4 mm. long, 0·3 mm. wide; testa dull-brown or subfuscous, minutely reticulate, conspicuously and rather regularly rugulose.

HAB.: Dry fields and gardens, roadsides; sandy seashores; sea-level to 2,000 ft. alt.; fl. April–July.

DISTR.: Divisions 1–7. Mediterranean region and eastwards to Iran.

1. Ktima, 1913, *Haradjian* 699! Ayios Yeoryios (Akamas), 1962, *Meikle*! near Polis, 1962, *Meikle*! Lachi, 1979, *Hewer* 4560!
2. Near Evrykhou, 1862, *Kotschy* 916; Galata, 1880, *Sintenis & Rigo* 745! Kannaviou, 1939, *Lindberg f.* s.n.
3. Limassol, 1905, *Holmboe* 647, and, 1963, *J. B. Suart* 119!
4. Dhekelia, 1936, *Syngrassides* 1000!
5. Salamis, 1931, *J. A. Tracey* 56!
6. Nicosia, 1939, *Lindberg f.* s.n., and 1970, *A. Genneou* in ARI 1570!
7. Near Myrtou, 1862, *Kotschy*, and, 1932, *Syngrassides* 210! Kyrenia, 1948, *Casey* 30! also, 1955, *Miss Mapple* 82! and, 1955, *G. E. Atherton* 78! Lambousa beach, 1973, *P. Laukkonen* 444!

NOTES: Ripe seed of this species has been difficult to obtain, most of the capsules being found to contain a relatively large beetle, and nothing besides.

TRIBE 2. **Antirrhineae** *Chav.* Leaves, or some of them, frequently opposite, seldom all alternate; corolla tubular, commonly with a saccate or spurred base, limb often 2-lipped, sometimes with a gibbous palate; stamens 4; anthers 2-thecous, the thecae generally confluent apically with divergent bases; capsule dehiscing by pores or apical slits.

2. LINARIA *Mill.*

Gard. Dict. Abridg., ed. 4 (1754).

B. Valdés, Revisión de las Especies Europeas de Linaria con Semillas Aladas, in Publ. Univ. Sevilla, ser. Cienc., no. 7, 288 pp. (1970).

Annual or perennial herbs; leaves simple, sessile, entire, generally narrow, the lower opposite or verticillate, the upper alternate; flowers in bracteate racemes or spikes, rarely solitary, axillary, calyx deeply and unequally 5-lobed, the adaxial lobe generally the longest; corolla oblique, tubular, 2-lipped, usually glabrous except at the palate, base produced into a spur, upper lip 2-lobed, lower lip 3-lobed, with a gibbous basal palate closing the mouth of the tube; stamens 4, didynamous, included; ovary 2-locular; ovules numerous; style simple; stigma small, sometimes emarginate. Capsule ovoid or subglobose, with 2 subequal loculi, dehiscing by apical fissures forming irregular teeth; seeds numerous, angled or discoid and winged.

About 150 species, mostly in the Mediterranean region and Western Asia.

Calyx glabrous or with a few sessile glands:
  Calyx-lobes tapering, acuminate; corolla 10–16 mm. long:
    Corolla violet-purple; inflorescence short, compact - - - - **1. L. pelisseriana**
    Corolla creamy white; inflorescence lax, elongate - - - - **5. L. chalepensis**
  Calyx-lobes obtuse or subacute, not tapering; corolla very small, less than 10 mm. long
                             4. L. albifrons

Calyx distinctly glandular-pilose:
  Corolla yellow, exceeding the calyx; spur 2·5–3·5 mm. long - - - **2. L. simplex**
  Corolla blue or bluish, scarcely exceeding the calyx; spur less than 0·5 mm. long
                             3. L. micrantha

1. L. pelisseriana (*L.*) *Mill.*, Gard. Dict., ed. 8, no. 11 (1768); Boiss., Fl. Orient., 4: 375 (1879); Post, Fl. Pal., ed. 2, 2: 286 (1933); Valdés, Rev. Linaria Sem. Alad., 71 (1970); Osorio-Tafall et Seraphim, List Vasc. Plants Cypr., 94 (1973); Davis, Fl. Turkey, 6: 672 (1978); Feinbrun, Fl. Palaest., 3: 189, t. 313 (1978).
Antirrhinum pelisserianum L., Sp. Plant., ed. 1, 615 (1753).
TYPE: "*in* Gallia".

Erect annual 8–20 (–30) cm. high; stems slender, glabrous, obscurely ridged, usually unbranched except in luxuriant specimens; basal leaves ternate or opposite, narrowly obovate or oblanceolate, 10–20 mm. long, 3–6 mm. wide, glabrous, apex acute, base tapering to a short or indistinct petiole; cauline leaves remote, alternate, linear or almost filiform, 10–30 mm. long, seldom more than 2 mm. wide, glabrous, sessile; inflorescence a short, crowded raceme, usually less than 3 cm. long at anthesis, elongating to 10 cm. or more in fruit; bracts linear, about 2–3 mm. long, 0·5 mm. wide, glabrous or thinly sprinkled with pallid, sessile glands; pedicels 2–3 mm. long, thinly glandular; calyx-lobes subequal, linear-lanceolate, acuminate, 3–3·5 mm. long, thinly glandular, keeled, with narrow membranous margins; corolla violet-purple with a paler palate; tube about 10 mm. long, thinly glandular externally, produced into a slender, acuminate, downcurved spur 5–8 mm. long; upper lip narrowly oblong, about 5 mm. long, 2 mm. wide, deeply 2-lobed; lower lip recurved, about 3 mm. long, 3-lobed, with a prominent, sulcate palate; filaments of longer pair of stamens about 3·5 mm. long, those of the shorter pair about 2 mm. long; anthers yellow, with divaricate thecae, about 0·8 mm. wide; ovary compressed-subglobose, about 1 mm. diam., thinly glandular; style about 2·5 mm. long; stigma slightly swollen, not emarginate. Capsule subglobose, strongly emarginate or subdidymous, somewhat compressed, pale brown, about 4–5 mm. diam., dehiscing from the apex into irregular teeth; seeds discoid, about 0·8 mm.

diam.; testa fuscous, papillose, with a conspicuous fimbriate-laciniate marginal wing.

HAB.: In open garigue or dry grassland; 800–900 ft. alt.; fl. March–April.

DISTR.: Division 6, rare. Widespread in the Mediterranean region and S.W. Europe and eastwards to the Caucasus.

6. Kormakiti-Liveras area, 1937, *Syngrassides* 1419! Dhiorios, 1962, *Meikle* 2326!

2. L. simplex Desf., Tabl. École Bot., 65 (1804); DC. in Lam. et DC., Fl. Franç., 3: 588 (1805); Sintenis in Oesterr. Bot. Zeitschr., 31: 255, 393 (1881); Valdés, Rev. Linaria Sem. Alad., 89 (1970); Davis, Fl. Turkey, 6: 670 (1978); Feinbrun, Fl. Palaest., 3: 189, t. 314 (1978).
  Antirrhinum parviflorum Jacq., Collectanea, 4: 204 (1791), Icones Plant. Rar., 3: 7, t. 499 (1793).
  A. simplex Willd., Sp. Plant., 3: 243 (1800) nom. superfl. illeg.
  Linaria arvensis (L.) Desf. var. *flaviflora* Boiss., Fl. Orient., 4: 375 (1879); Holmboe, Veg. Cypr., 164 (1914).
  L. parviflora (Jacq.) Hal., Consp. Fl. Graec., 2: 413 (1902); Post, Fl. Pal., ed. 2, 2: 286 (1933) non *L. parviflora* Desf. (1798).
  L. arvensis (L.) Desf. ssp. *simplex* Fourn., Quatre Fl. France, 764 (1937).
  TYPE: Not localized; the plate in Jacquin, *Icones Plant. Rar.*, 3: t. 499!

Erect annual, 5–30 cm. high; stems simple or rather sparingly branched, terete or subterete, glabrous; basal leaves rather crowded, opposite or verticillate, linear-oblong, 10–15 mm. long, 1–2 mm. wide, sessile, obtuse or subacute; cauline leaves usually remote, the lower opposite or verticillate, the upper alternate, similar to the basal, but up to 3 cm. long, with narrowly recurved margins; inflorescence a short, congested terminal raceme, not much exceeding 1 cm. long at anthesis, elongating to 10 cm. or more in fruit; bracts foliaceous, similar to the basal leaves, but smaller and sparsely glandular-pilose; flowers subsessile, or with short glandular pedicels less than 2 mm. long at anthesis; calyx-lobes divided almost to base, oblong-lanceolate, shortly acute, thinly glandular-pilose, the adaxial about 3 mm. long, 1 mm. wide, the abaxial 2–2·5 mm. long; corolla yellow, often veined purple or violet, with a paler spur, tube about 6 mm. long (including the slender, straight or curved spur, 2·5–3·5 mm. long), adaxial lip deeply divided into 2 acute lobes about 3·5 mm. long, abaxial lip recurved, 3-lobed, with short, rounded lobes about 1 mm. long, palate prominent, sulcate, shortly papillose; filaments glabrous, 1·5–2 mm. long; anthers yellow, about 0·5 mm. wide, with diverging thecae; ovary glabrous, broadly ovoid or suborbicular, about 1 mm. wide; style glabrous, about 2 mm. long; stigma swollen, but scarcely emarginate. Capsule subglobose, about 5–6 mm. diam., pale brown, dehiscing from the apex into irregular teeth; seeds discoid, about 1·5 mm. diam., with a narrow erosulate wing; testa blackish, sparsely papillose with acute papillae.

HAB.: In open garigue on dry hillsides, or on gravelly roadsides; sea-level to 3,800 ft. alt.; fl. March–May.

DISTR.: Divisions 2, 4, 7, rare. Widespread in the Mediterranean region eastwards to Caucasus and Iran.

2. Kryos Potamos, 3000 ft. alt., 1937, *Kennedy* 837! Platres, 1937, *Kennedy* 838! also, 1960, *N. Macdonald* 79!
4. Cape Pyla, 1880, *Sintenis.*
7. Mountains above Kythrea, 1880, *Sintenis & Rigo* 135! Kambyli, 1956, *Merton* 2525!

3. L. micrantha (*Cav.*) *Hoffsgg. et Link*, Fl. Port., 1: 258 (1813); Boiss., Fl. Orient., 4: 375 (1879); Sintenis in Oesterr. Bot. Zeitschr., 31: 194 (1881); Holmboe, Veg. Cypr., 164 (1914); Post, Fl. Pal., ed. 2, 2: 286 (1933); Valdés, Rev. Linaria Sem. Alad., 96 (1970); Osorio-Tafall et Seraphim, List Vasc. Plants Cyprus, 94 (1973); Davis, Fl. Turkey, 6: 672 (1978); Feinbrun, Fl. Palaest., 3: 189, t. 315 (1978).
  Antirrhinum micranthum Cav., Icones, 1: 51, t. 69, fig. 3 (1791).
  TYPE: Spain; "in incultis *del Real Retiro*" (MA).

Erect or spreading annual, 5–30 cm. high; stems simple or branched, particularly from the base, glabrous, striate or subterete; leaves sessile, glaucous, the basal whorled, linear-lanceolate, 8–12 mm. long, 1–2·5 mm. wide, acute or obtuse, the upper generally larger, opposite or alternate, lanceolate or narrowly elliptic, 10–25 mm. long, 3–8 mm. wide, longitudinally 3-nerved, apex usually blunt or subacute, margins narrowly recurved; inflorescence congested, subcapitate at anthesis, about 1 cm. diam., lengthening to 5–10 cm. in fruit; flowers sessile or subsessile; bracts foliaceous, narrowly oblong, recurved, 2–8 mm. long, 1–2·5 mm. wide; calyx-lobes divided almost to base, narrowly oblanceolate, subacute, thinly glandular-pilose externally, the adaxial about 3 mm. long, 0·8 mm. wide at anthesis, distinctly accrescent in fruit, the abaxial a little shorter; corolla minute, blue, mauve or whitish, striate, fugacious, tube (including spur) about 2·5 mm. long, adaxial lip about 1·5 mm. long, divided into 2 acute lobes, abaxial lip recurved, about 1 mm. long, divided into 3 rounded lobes, palate prominent, sulcate, minutely papillose, spur shortly conical, less than 0·5 mm. long, set almost at a right angle to the tube; filaments glabrous, 1–1·2 mm. long; anthers bluish, about 0·4 mm. wide, with divaricate thecae; ovary compressed-globose, about 0·8 mm. diam., glabrous; style about 1·2 mm. long; stigma slightly thickened, but not emarginate, a little oblique. Capsule globose, about 5 mm. diam., crowned with the persistent style, dehiscing from apex almost to base into irregular teeth; seeds discoid, 1·5–1·8 mm. diam., with a broad, subentire marginal wing; testa fuscous, rather densely papillose with acute papillae.

HAB.: Cultivated and fallow fields; pastures; waste ground, or rocky places near the sea; sea-level to 2,400 ft. alt.; fl. (Dec.–) Febr.–May.

DISTR.: Divisions 3–7. Widespread in the Mediterranean region and eastwards to Iraq, but commoner in the west than in the east.

3. Lefkara, 1941, *Davis* 2747!
4. Phaneromene near Larnaca, 1880, *Sintenis*; also, 1905, *Holmboe* 121; Paralimni, 1940, *Davis* 2062! Athna Forest, 1952, *Merton* 774!
5. Near Lefkoniko, 1880, *Sintenis & Rigo* 138! Near Kythrea, 1880, *Sintenis & Rigo* 136! Mile 11 Nicosia–Famagusta road, 1951, *Merton* 231! Athalassa, 1970, *A. Genneou* in ARI 1410!
6. Nicosia, 1937, *Syngrassides* 1401!
7. Kyrenia, 1938, *C. H. Wyatt* 43! also, 1948, *Kennedy* 1659! and, 1949, *Casey* 254!

L. PELOPONNESIACA *Boiss. et Heldr.* in Boiss., Diagn., 2, 3: 163 (1856); *Antirrhinum strictum* Sm. in Sibth. et Sm., Fl. Graec. Prodr., 1: 433 (1809), Fl. Graec., 6: 75, t. 594 (1826); *Linaria stricta* (Sm.) Guss., Plant. Rar., 250 (1826) non Hornem. (1815); *L. aparinoides* Chav., Mon. Antirrh., 138 (1833) pro parte nec *Antirrhinum aparinoides* Willd., Sp. Plant., 3: 247 (1801); Kotschy in Unger et Kotschy, Die Insel Cypern, 290 (1865); *L. sibthorpiana* Boiss. et Heldr. ex Boiss. var. *peloponnesiaca* (Boiss. et Heldr.) Boiss., Fl. Orient., 4: 378 (1879); *L. sibthorpiana* Boiss. et Heldr. ex Boiss. var. *parnassica* (Boiss. et Heldr.) Boiss., Fl. Orient., 4: 378 (1879).

A robust perennial, up to 1 m. high; leaves numerous, linear, 25–60 mm. long; inflorescence a dense, spiciform raceme, sometimes 10 cm. long; flowers yellow; seeds angled (not discoid), rugose-papillose. Recorded by Sibthorp (Fl. Graec. Prodr., 1: 434; 1809) "in agro Cariensi, et insulâ Cypro; nec non in Siciliâ", but the Cyprus record is almost certainly erroneous, for the plant has not since been refound on the island, nor in adjacent areas. It has a limited distribution in southern and western parts of the Balkan Peninsula.

4. L. albifrons *(Sm.) Spreng.*, Syst. Veg., 2: 793 (1825); Boiss., Fl. Orient., 4: 382 (1879); H. Stuart Thompson in Journ. Bot., 44: 335 (1906); Holmboe, Veg. Cypr., 165 (1914); Post, Fl. Pal., ed. 2, 2: 289 (1933); Osorio-Tafall et Seraphim, List Vasc. Plants Cyprus, 94 (1973); Feinbrun, Fl. Palaest., 3: 192, t. 321 (1978).

Antirrhinum albifrons Sm. in Sibth. et Sm., Fl. Graec. Prodr., 1: 432 (1809), Fl. Graec., 6: 71, t. 588 (1826).

TYPE: "In insulâ Rhodo" (OXF).

Erect, glabrous, glaucous annual 4–25 cm. high; stems terete, branched or unbranched; leaves narrowly oblong-elliptic, 8–25 mm. long, 3–7 mm. wide, sessile, acute, longitudinally 3-nerved, the basal opposite or ternate, the upper generally alternate; inflorescence a congested terminal raceme, usually less than 1 cm. wide, elongating to 5 cm. (or sometimes more) after anthesis; flowers rather few, subsessile or very shortly pedicillate, with glabrous pedicels; bracts foliaceous, lanceolate, 4–8 mm. long, 1–2 mm. wide; reflexed; calyx-lobes subequal, divided almost to base, glabrous, narrowly oblong, subacute or obtuse, about 4 mm. long, 1–2 mm. wide; corolla whitish or blue with a yellowish palate, or sometimes wholly yellowish, tube (including spur) about 5 mm. long, adaxial lip deeply divided into 2 oblong, subacute lobes about 2·5 mm. long, abaxial lip strongly recurved, with 3, broad, blunt lobes about 0·5 mm. long, palate very prominent, sulcate, shortly papillose, spur slender, acuminate, 2–2·5 mm. long, almost at a right angle to the tube; filaments 2–2·5 mm. long, glabrous; anthers tinged blue, almost 0·4 mm. wide, thecae divergent; ovary compressed-globose, about 2 mm. diam., glabrous; style 1·5 mm. long; stigma somewhat thickened, not emarginate. Capsule about 5 mm. long, 4 mm. wide, compressed, pale brown with a distinct median furrow; seeds shaped like the segments of an orange, not winged, about 0·8 mm. long, 0·6 mm. wide; testa fuscous, strongly rugose, with verruculose rugosities.

HAB.: Cultivated ground and field margins; sea-level to c. 500 ft. alt.; fl. Febr.–April.

DISTR.: Divisions 4, 6, rare. Eastern Mediterranean region from Rhodes, Syria, and Palestine to Egypt, and eastwards to Transcaucasia and Iran.

4. Phaneromene near Larnaca, 1905, *Holmboe* 119.
6. Nicosia, a garden weed, 1901, *A. G. & M. E. Lascelles* s.n. ! Ayia Irini, 1941, *Davis* 2588 !

NOTES: Superficially similar to *L. micrantha* (Cav.) Hoffsgg. & Link, and sometimes growing with it, so perhaps overlooked and less rare than the records would indicate.

5. **L. chalepensis** (*L.*) *Mill.*, Gard. Dict., ed. 8, no. 12 (1768); Poech, Enum. Plant. Ins. Cypr., 26 (1842); Unger et Kotschy, Die Insel Cypern, 290 (1865); Boiss., Fl. Orient., 4: 381 (1879); Holmboe, Veg. Cypr., 165 (1914); Post, Fl. Pal., ed. 2, 2: 287 (1933); Osorio-Tafall et Seraphim, List Vasc. Plants Cyprus, 94 (1973); Davis, Fl. Turkey, 6: 664 (1978); Feinbrun, Fl. Palaest., 3: 190, t. 317 (1978).
　　Antirrhinum chalepense L., Sp. Plant., ed. 1, 617 (1753); Sibth. et Sm., Fl. Graec. Prodr., 1: 433 (1809), Fl. Graec., 6: 73, t. 592 (1826).

TYPE: "*in* Italia".

Erect annual, 10–30 (–50) cm. high; stems usually simple, sometimes with a few slender, sterile branches near the base, subterete or striate, glabrous; leaves sessile, glabrous, rather glaucous, those of the basal, sterile shoots narrowly lanceolate or oblanceolate, 15–25 mm. long, 2–5 mm. wide, obtuse or shortly acute, those of the flowering stems narrower, often linear, 20–50 mm. long, 1–4 mm. wide, tapering towards apex and base, nervation obscure; inflorescence a lax terminal raceme 3–10 cm. long at anthesis, elongating considerably in fruit; bracts linear-subulate, foliaceous, 3–15 mm. long, 0·5–2·5 mm. wide, glabrous, acuminate; pedicels very short, usually about 2 mm. long, elongating slightly in fruit; calyx-lobes divided almost to base, linear-subulate, glabrous, 5–10 mm. long, about 1 mm. wide, the adaxial sometimes, but not always, slightly longer than the abaxial, at first directed forwards, often spreading or slightly recurved after anthesis; corolla creamy white, glabrous, tube (including spur) 12–16 mm. long, adaxial lip divided almost to the base into 2 narrowly oblong, blunt or subacute lobes, 4–5 mm. long, 0·8–1 mm. wide, abaxial lip 6–7 mm. long, divided into 3 narrowly oblong, blunt lobes, about 3–4 mm. long, 0·8–1 mm. wide, the 2 lateral lobes commonly directed forwards, the median lobe deflexed like the labellum of an orchid; palate not very prominent; spur very slender, tapering, curved, 8–10 mm. long; filaments 1·5–2·5 mm. long, glabrous; anthers yellow, about 0·6 mm. wide, thecae divergent; ovary compressed-globose, glabrous, about 1 mm. diam.; style about 1 mm. long; stigma slightly thickened and deflexed. Capsule subglobose, somewhat compressed, about 4–5 mm. diam., dehiscing from the apex into irregular teeth, glabrous with a distinct median furrow, usually crowned with the persistent style; seeds irregularly angular, about 1 mm. long, 0·6 mm. wide; testa fuscous, irregularly rugose, the surface minutely scrobiculate.

HAB.: Cultivated and fallow fields, waste ground, occasionally in open parts of Pine forest; sea-level to 3,500 ft. alt.; fl. Jan.–May.

DISTR.: Divisions 1–5, 7, 8. Southern Europe and Mediterranean region eastwards to Iran and Arabia.

1. Ktima, 1913, *Haradjian* 643! Aspropotamos near Ayios Yeoryios (Akamas), 1962, *Meikle* 2087!
2. Kato Platres, 1937, *Kennedy* 839! Agros, 1941, *Davis* 3060! Kapoura Forest, 1952, *F. M. Probyn* 30! Palekhori, 1969, *Economides* in ARI 1320!
3. Polemidhia 1948, *Mavromoustakis* s.n.! Yerasa, 1948, *Mavromoustakis* s.n.! Morokambos near Akhelia, 1956, *Merton* 2500! 2m. W. of Vikla, 1959, *C. E. H. Sparrow* 9! Pyrgos, 1963, *J. B. Suart* 23! 2 m. W. of Limassol, 1963, *J. B. Suart* 28! Kolossi, 1971, *Townsend* 71/124.
4. Near Larnaca, 1880, *Sintenis & Rigo* 136! Ayia Napa, 1905, *Holmboe* 45; Phaneromene near Larnaca, 1905, *Holmboe* 120.
5. Apostolos Varnavas Monastery, 1948, *Mavromoustakis* s.n.! Athalassa, 1967, *Merton* in ARI 87!
7. Karavostasi, 1932, *Syngrassides* 254! Kyrenia, 1938, *C. H. Wyatt* 25! also, 1956, *G. E. Atherton* 967! Karmi, 1941, *Davis* 2773! Klepini, 1960, *N. Macdonald* 76! Tjiklos, 1973, *P. Laukkonen* 344!
8. Apostolos Andreas, 1934, *Syngrassides* 1395! Ayios Philon near Rizokarpaso, 1941, *Davis* 2249!

NOTES: All the material seen belongs to the typical variety with long, tapering calyx-lobes; the distinctive var. *brevicalyx* Davis (Fl. Turkey 6: 664–5) with obtuse or subobtuse calyx-lobes equalling or shorter than the capsule, occurs in S. Turkey near Antalya, and may yet be found in our area.

3. CYMBALARIA *Hill*

British Herbal, 113 (1756).
Cufodontis in Archivio Bot., 12: 54–81, 135–158, 233–254 (1936);
Bot. Notiser, 1947: 135–156 (1947).

Slender trailing perennials or annuals; leaves usually alternate, sometimes opposite, cordate, suborbicular or reniform, crenate-dentate or lobed, palmatinerved, generally long-petiolate; flowers small, solitary, axillary, long-pedunculate; calyx with unequal lobes; corolla shortly and bluntly spurred, sometimes ecalcarate. Capsule with equal loculi, dehiscing from the apex into irregular teeth, sometimes almost indehiscent; seeds winged or unwinged.

Ten species, mostly in S. Europe and the Mediterranean region.

1. **C. longipes** (*Boiss. et Heldr.*) *Cheval.* in Bull. Soc. Bot. Fr., 83: 641 (1937); Cufodontis in Bot. Notiser, 1947: 152 (1947); Osorio-Tafall et Seraphim, List Vasc. Plants Cyprus, 94 (1973); Davis, Fl. Turkey, 6: 674 (1978).
      Linaria longipes Boiss. et Heldr. in Boiss., Diagn., 1, 12: 40 (1853), Fl. Orient., 4: 365 (1879); Druce in Proc. Linn. Soc., sess. 141: 51 (1930), Rep. B.E.C., 9: 470 (1931); A. K. Jackson in Kew Bull., 1937: 344 (1937).
      [*Antirrhinum cymbalaria* (non L.) Sm. in Sibth. et Sm., Fl. Graec. Prodr., 1: 430 (1809) pro parte quoad spec. cypr.]
      [*Linaria cymbalaria* (non (L.) Mill.) Poech, Enum. Plant. Ins. Cypr., 26 (1842); Unger et

Kotschy, Die Insel Cypern, 289 (1875); Sintenis in Oesterr. Bot. Zeitschr., 31: 192 (1881); Holmboe, Veg. Cypr., 164 (1914); Rechinger f. in Arkiv för Bot., ser. 2, 1: 431 (1950).]
    [*Cymbalaria muralis* (non Gaertn.) Osorio-Tafall et Seraphim, List Vasc. Plants Cyprus, 94 (1973).]
    TYPE: Turkey; "ad muros et rupes circà *Adalia* Pamphyliae (Heldr.)". (Boiss. et Heldr. in Plant. Anat. exsicc. 1846) (G, ? K).

Slender trailing perennial; stems glabrous, terete, usually much branched, rooting at the nodes; leaves glabrous, reniform or broadly cordate, 5–20 mm. wide, subentire or bluntly 3–5-lobed; petioles commonly very long, sometimes exceeding 4 cm. and generally longer than the lamina; peduncles long, commonly 4 cm. or more, elongating greatly in fruit; calyx-lobes unequal, shortly oblong, obtuse, about 0·5 mm. long, 0·3 mm. wide, glabrous or thinly dotted over with sessile glands; corolla-tube (including spur) 7–10 mm. long, glabrous, lavender-blue, adaxial lip striate, 2-lobed, the lobes oblong, obtuse or subacute, about 2 mm. long, 1 mm. wide, abaxial lip strongly recurved, about 2·5 mm. long, with 3 short, rounded lobes about 1 mm. long, palate very prominent, orange-yellow, distinctly papillose, spur 2·5–4 mm. long, straight or slightly curved, acute; stamens 4, filaments 2–3 mm. long, glabrous; anthers yellow, about 0·8 mm. wide with divaricate thecae; ovary compressed-subglobose, about 0·5 mm. diam., glabrous; style about 2 mm. long; stigma distinctly swollen, shortly emarginate. Capsule irregularly dehiscent, subglobose, 5–8 mm. diam., pale brown and thinly glandular-pubescent or subglabrous when ripe; seeds coalescing into an irregularly rugose mass; testa mid-brown, minutely striate-reticulate.

HAB.: Shingle banks by the sea; near sea-level; fl. March–May.

DISTR.: Division 4, rare. Greece, Crete, Aegean Islands, Turkey and Syria.

4. Larnaca, 1880, *Sintenis & Rigo* 782! also, 1930, *Druce* s.n.! Miles 3–10, Larnaca–Famagusta road, on shingle banks by the sea, 1950, *Chapman* 703!

NOTES: Holmboe (Veg. Cypr., 164), Druce (Proc. Linn. Soc., *loc. cit. supra*) and Rechinger f. (Arkiv för Bot., ser. 2, 1: 431) have noted that Sibthorp's and Sintenis's records for *Cymbalaria muralis* are misidentifications of *C. longipes*.

4. KICKXIA *Dumort.*
Florula Belgica, 35 (1827).

Annual or perennial herbs, generally with trailing or sprawling stems; leaves mostly alternate, petiolate, generally cordate, deltoid or hastate; flowers usually solitary in the axils of reduced or normally developed leaves, frequently with long, filiform peduncles; calyx and corolla as in *Linaria* Mill. Capsule (in our species) compressed-globose or broadly ovoid, divided into 2 equal loculi, and dehiscing by 2 large, lateral, circular lids or opercula, occasionally indehiscent; seeds usually numerous, rugulose-alveolate or tuberculate.

About 25 species in Europe, Mediterranean region, S.W. Asia, Africa, and as introduced weeds elsewhere.

Peduncles glabrous; seeds acutely tuberculate; leaves usually hastate; corolla milky-white
    tinged mauve - - - - - - - **1. K. commutata** ssp. **graeca**
Peduncles hairy; seeds cerebriform-alveolate:
  Calyx-lobes lanceolate or deltoid, not noticeably accrescent or plicate at the base:
    Peduncles 0·8–2 (–3) cm. long, much longer than the subtending leaves; corolla yellow with
      violet adaxial lobes - - - - - **2. K. elatine** ssp. **sieberi**
    Peduncles very short (less than 1 cm. long) or flowers commonly subsessile; corolla usually
      white with blue or purplish adaxial lobes - - - **3. K. lanigera**
  Calyx-lobes broadly lanceolate, distinctly accrescent in fruit and plicate at the base; corolla
    yellow with blackish-maroon adaxial lobes- - - - **4. K. spuria**

1. **K. commutata** (*Bernh. ex Reichb.*) *Fritsch*, Excursionfl. Oesterr., 492 (1897);Osorio-Tafall et Seraphim, List Vasc. Plants Cyprus, 94 (1973); Davis, Fl. Turkey, 6: 675 (1978).

Linaria commutata Bernh. ex Reichb., Fl. Germ. Excurs., 1: 373 (1831); Holmboe, Veg. Cypr., 164 (1914); Lindberg f., Iter Cypr., 31 (1946); Rechinger f. in Arkiv för Bot., ser. 2, 1: 431 (1950).

TYPE: "Auf Aeckern im südl. Gebiete, Istrien, auf den Inseln, Oberitalien (Corsica.)" (? W).

Trailing annual; stems slender, usually much branched from the base, 15–50 cm. long, subterete, pilose with a mixed indumentum of long and short hairs; basal leaves broadly obovate, up to 3 cm. long, 1·5 cm. wide, subsessile, median leaves 1–2·5 cm. long, 0·5–2 cm. wide, subglabrous or thinly pilose on both surfaces, apex acute or obtuse; petiole slender, 2–5 mm. long; upper leaves smaller, 0·5–1·5 cm. long, with or without distinct, spreading basal lobes; flowers borne along most of the stem; peduncles patent, filiform, glabrous except at the very apex, 1·3–5 cm. long, sometimes much exceeding the leaves; calyx divided almost to base in 5 lanceolate-acuminate, pilose lobes, 2·5–3 mm. long, 0·8–1·2 mm. wide, lobes with a conspicuous or obscure membranous margin; corolla milky-white tinged mauve, with a bluish-mauve adaxial lip and spur, the palate yellowish minutely spotted purple, tube (including spur) 7–10 mm. long, minutely pilose or subglabrous, adaxial lip about 4 mm. long, divided half-way into 2 acute or subacute, deltoid or oblong lobes, abaxial lip recurved, bluntly 3-lobed, palate prominent, pilose, spur slender, acuminate, usually downcurved, 4–5 mm. long; filaments 2·5–3·5 mm. long, long-pilose; anthers violet, about 0·5 mm. wide, with diverging, barbellate thecae; ovary compressed-ovoid, about 0·8–1 mm. long, 0·7–0·8 mm. wide, shortly pilose; style rather thick, 1·5–2 mm. long; stigma deflexed, concave, often shortly emarginate at apex. Capsule globose, 3–3·5 mm. diam., pale brown, minutely pilose; seeds irregularly oblong, 6–7 mm. long, 5–6 mm. wide, obscurely angled; testa dull brown, densely and acutely tuberculate.

ssp. **graeca** (*Bory et Chaub.*) *R. Fernandes* in Bot. Journ. Linn. Soc., 64: 74 (1971); Davis, Fl. Turkey, 6: 675 (1978).

Antirrhinum graecum Bory et Chaub. in Bory, Exped. Sci. Morée, 175, t. 21 (1832).
    Linaria commutata Bernh. ex Reichb. var. *polygonoides* Hal., Consp. Fl. Graec., 2: 416 (1902); Lindberg f., Iter Cypr., 31 (1946).
TYPE: Greece; "E Scardamula et Chimova", *Bory* (P).

Stems often much branched; median and upper leaves hastate-acute, with prominent basal lobes; peduncles commonly 2–5 cm. long, much exceeding the subtending leaves; corolla-tube (excluding spur) 4–6 mm. long.

HAB.: Cultivated and waste ground, roadsides, bare ground on hillsides; sea-level to 1,500 ft. alt.; fl. April–July.

DISTR.: Divisions 1, 3, 5–8. Sicily, Greece, Crete, Aegean area, Turkey, Syria.

1. Letimbou, 1905, *Holmboe* 748; near Ayios Neophytos monastery, 1939, *Lindberg f.* s.n.! Stroumbi, 1941, *Davis* 3372! S.W. side Akamas, 1956, *Merton* 2780!
3. Pissouri, 1905, *Holmboe* 711.
5. Athalassa, 1939, *Lindberg f.* s.n.!
6. Between Evrykhou and Peristerona, 1880, *Sintenis & Rigo* 744!
7. Near Kyrenia, 1936, *Mrs Dray* 11!
8. Near Rizokarpaso, 1912, *Haradjian* 173.

2. K. elatine (*L.*) *Dumort.*, Florula Belgica, 35 (1827); Davis, Fl. Turkey, 6: 676 (1978).
    Antirrhinum elatine L., Sp. Plant., ed. 1, 612 (1753); Sibth. et Sm., Fl. Graec. Prodr., 1: 431 (1809).
    Linaria elatine (L.) Mill., Gard. Dict., ed. 8, no. 16 (1768); Boiss., Fl. Orient., 4: 367 (1879).

Trailing or sprawling annual; stems slender, terete, usually much branched at base, sparingly or copiously branched above, thinly or densely pilose with a mixed indumentum of long and short, glandular or eglandular hairs; basal leaves ovate or broadly elliptic, entire, obtuse, 2–3 cm. long, 1–2 cm. wide, with a petiole 0·5–1 cm. long, median leaves bluntly or acutely hastate, 0·5–3 cm. long, 0·3–2·5 cm. wide, thinly or rather densely lanuginose on both surfaces, usually entire or subentire, but sometimes toothed towards base, basal lobes usually distinct, but sometimes wanting; petiole very short, usually less than 5 mm. long, upper leaves progressively smaller, the uppermost often less than 3 mm. long; peduncles filiform, glabrous or hirsute, 0·8–2 (–3) cm. long; calyx-lobes divided almost to base, lanceolate or narrowly deltoid, 3·5–4 mm. long, 0·8–1·5 mm. wide, acute or acuminate, thinly or densely pilose, with or without distinct membranous margins; corolla pale yellow with violet adaxial lobes, tube (including spur) about 1 cm. long, shortly pilose, adaxial lobes oblong, blunt, about 2·5 mm. long, 1 mm. wide, abaxial lobes short, rounded, recurved, palate prominent, spur slender, acuminate, usually downcurved; filaments 2–2·5 mm. long, pilose; anthers violet, barbellate, about 0·5 mm. wide, with diverging thecae; ovary compressed, subglobose, 1–1·5 mm. diam., shortly pilose; style about 2 mm. long; stigma elongate, slightly thickened. Capsule globose, about 5 mm. diam., pale brown, glabrous or thinly pilose; seeds rather regularly and bluntly oblong, 1–1·2 mm. long, about 1 mm. wide; testa dark brown, conspicuously cerebriform-alveolate.

ssp. **sieberi** (*Arcang.*) *Hayek*, Prodr. Fl. Pen. Balc., 2: 144 (1929).

Linaria sieberi Reichb. [Fl. Germ. Excurs., 374 (1831) nomen] ex Heldr., Herb. Graec. Norm. 867a (1886) nom. illeg.; Post, Fl. Pal., ed. 2, 2: 284 (1933).

L. prestandreae Tineo in Gussone, Fl. Sic. Syn., 2 (2): 842 (1845).

L. bombycina Boiss. et Bl. in Boiss., Diagn., 2, 3: 161 (1856).

L. crinita Mabille, Recherches Plant. Corse, 31 (1867).

L. elatine (L.) Mill. var. *villosa* Boiss., Fl. Orient., 4: 367 (1879); H. Stuart Thompson in Journ. Bot., 44: 335 (1906).

L. sieberi Reichb. ssp. *crinita* (Mabille) Nyman, Consp. Fl. Europ., 542 (1881) nom. invalid.

L. elatine (L.) Mill. ssp. *sieberi* Arcang., Comp. Fl. Ital., ed. 2, 397 (1894); Holmboe, Veg. Cypr., 164 (1914).

L. sieberi Reichb. ex Heldr. var. *bombycina* (Boiss. et Bl.) Hal., Consp. Fl. Graec., 2: 415 (1902); Lindberg f., Iter Cypr., 31 (1946).

K. sieberi H. H. Allan, Handb. Naturalised Fl. N. Zealand, 300 (1940); Rechinger f., Fl. Aegaea, 476 (1943); Feinbrun, Fl. Palaest., 3: 186, t. 309 (1978).

K. elatine (L.) Dumort. ssp. *crinita* (Mabille) Greuter in Boissiera, 13: 108 (1967); Davis, Fl. Turkey, 6: 676 (1978).

[*Linaria elatine* (non (L.) Mill.) Poech, Enum. Plant. Ins. Cypr., 26 (1842); Unger et Kotschy, Die Insel Cypern, 290 (1865).]

TYPE: Crete, *Sieber* (? W).

Plant rather densely pilose; upper parts of stems generally much branched; leaves blunt or subacute, the basal lobes often small or obscure, sometimes adjoined by 2–3 small lateral teeth; peduncles hairy.

HAB.: Cultivated and waste ground, roadsides, earthy banks often in moist situations; sea-level to 4,000 ft. alt.; fl. May–Nov.

DISTR.: Divisions 1–4, 6–8. S. Europe and Mediterranean region.

1. Kissonerga, 1967, *Economides* in ARI 987 !
2. W. of Platres, 1900, *A. G. & M. E. Lascelles* s.n.! Agros, 1940, *Davis* 1883! Near Kyperounda, 1954, *Casey* 1665! Ayios Nikolaos near Kakopetria, 1962, *Meikle* 2852!
3. Below Arminou, 1941, *Davis* 3424!
4. Near Larnaca, 1939, *Lindberg f.* s.n.
6. Kykko metokhi near Nicosia, 1932, *Syngrassides* 399! and, 1952, *Probyn* 144! Near Nicosia, 1939, *Lindberg f.* s.n.! Morphou, 1955, *Chiotellis* 472!
7. Near Kyrenia, 1880, *Sintenis & Rigo* 607! also, 1936, *Kennedy* 836! 1948, *Casey* 43! 1949, *Casey* 866! 1955, *G. E. Atherton* 30! 185! 219! 714! Near Glykyotissa Island, 1958, *P. H. Oswald* 7! Myrtou, 1955, *G. E. Atherton* 421!
8. Ayios Andronikos, 1934, *Syngrassides* 964!

NOTES: All the material seen belongs to the hirsute, blunt-leaved ssp. *sieberi*. It may be assumed that the Cyprus part of Sibthorp and Smith's references to the distribution of *K.* (*Antirrhinum*) *elatine* likewise relates to the above subspecies.

3. K. lanigera (*Desf.*) *Hand.-Mazz.* in Ann. Nat. Mus. Wien, 27: 403 (1913); Davis, Fl. Turkey, 6: 676 (1978); Feinbrun, Fl. Palaest., 3: 187, t. 311 (1978).
  Linaria lanigera Desf., Fl. Atlant., 2: 38, t. 130 (1798); Boiss., Fl. Orient., 4: 366 (1879); Post, Fl. Pal., ed. 2, 2: 284 (1933).
  TYPE: Tunis; "in arvis cultis prope veterem Carthaginem" (P).

Sprawling, glandular-pilose annual; stems slender, terete, much branched from the base, and generally with numerous patent branches above, up to 50 cm. long; basal leaves broadly ovate, 2–3 cm. long, 1·5–2·5 cm. wide, apex shortly acute, base rounded, margins entire or irregularly toothed but not hastate, upper leaves 4–15 mm. long, 3–13 mm. wide, becoming progressively smaller upwards, broadly cordate, entire, subsessile or with slender petioles generally less than 2 mm. long; flowers scattered along the whole length of the lateral branchlets, subsessile or, less often, with slender glandular-pilose peduncles up to 1 cm. long; calyx-lobes lanceolate-acuminate, divided almost to base, 3–4 mm. long, 1–1·5 mm. wide, densely glandular-pilose; corolla white, cream or very pale yellow with a violet, lilac or purple adaxial lip, glandular-pubescent externally, tube (including spur) about 8 mm. long, adaxial lip about 3 mm. long, shortly and bluntly 2-lobed, abaxial lip recurved, about 5 mm. long, palate prominent, spur slender, acuminate, about 4–5 mm. long, usually downcurved; filaments pilose, about 2–2·5 mm. long; anthers violet, about 0·5 mm. wide, with diverging thecae; ovary compressed-subglobose, about 1 mm. diam., densely pubescent, apex distinctly emarginate; style about 2 mm. long; stigma deflexed, about 0·5 mm. long, tapering towards a truncate apex. Capsule small, compressed-subglobose, distinctly emarginate, pale brown, about 3 mm. diam., largely concealed by the persistent, adpressed calyx; seeds shortly oblong or subovoid, about 0·8 mm. long, 0·6 mm. wide; testa fuscous-brown, cerebriform-alveolate, but rather shallowly so.

HAB.: Cultivated and waste ground, roadsides; sea-level to 500 ft. alt.; fl. Jan.–Oct.

DISTR.: Divisions 3, 7, 8. Widespread (but disjunct) in the Mediterranean region.
3. Skarinou, 1940, *Davis* 1987! Cherkez, 1945, *Mavromoustakis* 35!
7. Kyrenia, 1948, *Casey* 14! also, 1955, *G. E. Atherton* 220! 745! and, 1956, *G. E. Atherton* 841! 875!
8. Koma tou Yialou, 1970, *A. Genneou* in ARI 1587!

4. K. spuria (*L.*) *Dumort.* Florula Belgica, 35 (1827); Osorio-Tafall et Seraphim, List Vasc. Plants Cyprus, 94 (1973); Davis, Fl. Turkey, 6: 677 (1978); Feinbrun, Fl. Palaest., 3: 187, t. 310 (1978).
  Antirrhinum spurium L., Sp. Plant., ed. 1, 613 (1753); Hume in Walpole, Mem. Europ. Asiatic Turkey, 254 (1817).
  Linaria spuria (L.) Mill., Gard. Dict., ed. 8, no. 15 (1768); Poech, Enum. Plant. Ins. Cypr., 26 (1842); Unger et Kotschy, Die Insel Cypern, 290 (1865); Boiss., Fl. Orient., 4: 366 (1879); Holmboe, Veg. Cypr., 164 (1914); Post, Fl. Pal., ed. 2, 2: 283 (1933).
  TYPE: "*in Germaniae, Angliae, Galliae, Italiae arvis*".

Sprawling, glandular-pilose annual; stems slender, terete, copiously branched at the base and throughout their length; basal leaves broadly ovate-oblong, 3–4·5 cm. long, 2–3·5 cm. wide, subsessile or with petioles up to 1 cm. long, apex obtuse or shortly acute, margins subentire or shortly and shallowly dentate; cauline leaves suborbicular or widely cordate, subsessile or very shortly petiolate, 8–20 mm. long, 8–15 mm. wide, apex shortly acute, often mucronate, margins usually entire; flowers borne along the length of the lateral branches; peduncles slender, thinly or densely glandular-pilose, 3–20 mm. long, generally much exceeding the leaves, commonly deflexed at apex in fruit; calyx-lobes broadly lanceolate-acuminate, about 4 mm. long, 2 mm. wide at anthesis, plicate at base, distinctly accrescent in fruit; corolla yellow with a dark maroon upper lip, subglabrous or glandular-pubescent externally, tube (including spur) about 7 mm. long, upper lip about 2·5 mm. long, divided into 2 deltoid, acute or subacute lobes, lower lip about 3·5 mm. long, reflexed, bluntly 3-lobed, palate not very prominent, spur slender, acuminate, usually downcurved, about 5 mm. long; filaments pilose, about 3 mm. long; anthers pallid, about 0·5 mm. wide, barbellate; ovary compressed-globose, about 2 mm. diam., pubescent, apex shortly emarginate; style about 2 mm. long; stigma subspathulate, about 0·5 mm. long, slightly deflexed. Capsule compressed, subglobose, slightly emarginate, pale brown, about 4 mm. wide; seeds bluntly oblong, about 1 mm. long, 0·8 mm. wide; testa fuscous, shallowly cerebriform-alveolate.

HAB.: Dry stream-beds; cultivated and waste ground; 1,500–2,000 ft. alt.; fl. ? May–Oct.

DISTR.: Division 2, very rare. Widespread in Europe and the Mediterranean region, and as a weed elsewhere.

2. Cotton fields at Evrykhou, 1840, *Kotschy* 5; between Platres and Perapedhi, 1937, *Syngrassides* 1706!

NOTES: Only one specimen has been seen, on the strength of which it is impossible to say whether the Cyprus plant should be referred to ssp. *spuria* or to the more densely pilose, smaller-flowered (and fruited) ssp. *integrifolia* (Brot.) R. Fernandes, but it most probably belongs to the latter. To at least some degree the differences seem to depend upon the season when the specimens are collected, those collected at the beginning of the flowering period being more glabrous and less branched than those collected in late summer or autumn.

5. CHAENORHINUM (*DC.*) *Reichb.*

Consp. Regn. Veg., 123 (1828) ["Chaenarrhinum"].

Linaria Mill. sect. ["groupe"] *Chaenorhinum* DC. in Lam. et DC., Fl. Franç., 5 [vol. 6]: 410 (1815).

Annual herbs (in Cyprus); leaves opposite and petiolate towards base, alternate and subsessile above, entire; flowers usually in lax, bracteate, often flexuous terminal racemes, rarely solitary and axillary; calyx deeply and rather unequally 5-lobed; corolla 2-lipped with a short, straight spur; palate not closing throat of corolla; stamens 4, didynamous, included. Capsule obliquely ovoid or subglobose, unequally 2-locular, opening by 2–3 apical pores; seeds numerous, oblong-ellipsoid or truncate-conical; testa smooth or variously sculptured with ridges and papillae.

About 20 species in Europe, the Mediterranean region and western Asia.

1. **C. rubrifolium** (*Robill. et Cast. ex DC.*) *Fourr.* in Ann. Soc. Linn. Lyon, n.s., 17: 127 (1869) [as "*rubriflorum*"]; Osorio-Tafall et Seraphim, List Vasc. Plants Cyprus, 94 (1973); Davis, Fl. Turkey, 6: 652 (1978).
Linaria rubrifolia Robill. et Cast. ex DC. in Lam. et DC., Fl. Franç., 5 [vol. 6]: 410 (1815); Boiss., Fl. Orient., 4: 383 (1879); Holmboe, Veg. Cypr., 165 (1914).
TYPE: France; "sur les collines rocailleuses des environs de Marseille" *Robillard & Castagne* (G).

Slender, erect annual 6–30 cm. high; stems terete, branched or unbranched, thinly glandular-pilose, often tinged purple; basal leaves broadly obovate or spathulate, 1–2 cm. long, 0·8–1·5 cm. wide, dark green above, purple below, subglabrous or thinly glandular-pilose, apex obtuse, margins flat, entire, base tapering to a petiole 0·5–1 cm. long, upper leaves progressively narrower and more elliptic in outline, petioles shorter or sometimes almost wanting; flowers in elongating, lax, flexuous terminal racemes; rhachis glandular-pilose; bracts foliaceous, lanceolate, 0·5–1·5 cm. long, 0·1–0·5 cm. wide, glandular-pilose; pedicels slender, erecto-patent, 10–15 mm. long, glandular-pubescent; calyx-lobes divided almost to base, narrowly oblanceolate, the adaxial rather larger than the abaxial, 4–5 mm. long, 1–1·5 mm. wide, glandular-pilose, apex obtuse or subacute; corolla white with a reddish upper lip, tube (including spur) about 5 mm. long, upper lip divided into 2 bluntly oblong lobes about 1 mm. long, 0·8 mm.

wide, lower lip a little longer, divided into 3 blunt, oblong lobes about 1 mm. long, 0·8 mm. wide; palate obscure; spur straight, acuminate, about 1·5 mm. long; filaments unequal, about 1 mm. and 1·8 mm. long, glabrous; anthers pale, about 0·5 mm. wide, thecae strongly divergent; ovary compressed-globose, about 0·7 mm. diam., densely glandular-pubescent; style about 1 mm. long, glandular-pilose towards base; stigma slightly expanded, about 0·4 mm. long. Capsule subglobose, pale brown, about 3 mm. diam.; seeds bluntly oblong, about 0·4 mm. long, 0·3 mm. wide; testa blackish, conspicuously alveolate with tuberculate-margined walls.

HAB.: Walls and limestone cliffs; c. 500 ft. alt.; fl. March–April.

DISTR.: Divisions 6, 7, rare. Chiefly S. Europe and western Mediterranean region, but sporadically eastwards to Iran.

6. Famagusta Gate, Nicosia, 1905, *Holmboe* 281.
7. Limestone cliffs near Panagra, 1962, *Meikle* 2369!

6. MISOPATES *Rafin.*

Autikon Botanikon, 155, 158 (1840).
W. Rothmaler in Fedde Repert., 52: 33 (1943).

Annual herbs with erect stems; leaves narrow, entire, shortly petiolate, the basal opposite, the upper usually alternate; inflorescence an elongate, bracteate terminal raceme; pedicels short; calyx deeply 5-lobed, the adaxial lobes conspicuously longer than the abaxial; corolla shorter than, or not much exceeding calyx, tube gibbous or shortly saccate at base; stamens 4. Capsule 2-locular, the adaxial loculus indehiscent, the abaxial dehiscing by 2 apical pores; seeds compressed, one face smooth and keeled, the other papillose with an incurved, sinuate-dentate border.

Three species in the Mediterranean region, *M. orontium* also widespread as a weed in central and southern Europe and in other parts of the world.

1. M. orontium (*L.*) *Rafin.*, Autikon Botanikon, 158 (1840); Davis, Fl. Turkey, 6: 649 (1978).
Antirrhinum orontium L., Sp. Plant., ed. 1, 617 (1753); Unger et Kotschy, Die Insel Cypern, 290 (1865); Boiss., Fl. Orient., 4: 385 (1879); Holmboe, Veg. Cypr., 165 (1914); Post, Fl. Pal., ed. 2, 2: 290 (1933); Lindberg f., Iter Cypr., 31 (1946); Osorio-Tafall et Seraphim, List Vasc. Plants Cyprus, 94 (1973); Feinbrun, Fl. Palaest., 3: 192, t. 322 (1978).
TYPE: "*in* Europae *agris & arvis*".

Erect annual 10–60 (–100) cm. high; stems terete, subglabrous or glandular-pubescent, simple or variously branched, with erect branches; leaves narrowly lanceolate or linear, 1–6 cm. long, 0·2–0·8 cm. wide, glabrous or subglabrous, the lower relatively broad with an obtuse or subacute apex, the upper tapering to an acute apex, margins distinctly recurved, nervation obscure, base tapering to a short, indistinct petiole; inflorescence elongating greatly after anthesis, at first rather congested, but lengthening to 20–30 cm. (or more) in fruit; rhachis rather rigid and straight, glandular-pubescent; bracts narrowly linear, foliaceous, to 3 cm. long, usually much exceeding the flowers; pedicels very short, rarely exceeding 3 mm., sometimes almost wanting; calyx-lobes unequal, linear, glandular-pubescent, the adaxial 8–15 mm. long, 1–2 mm. wide at anthesis, the abaxial 5–12 mm. long and about as wide as the adaxial; corolla pink, tube glabrous, about 5 mm. long, 2·5 mm. wide, slightly gibbous but not spurred at the base; upper lip spreading, about 5 mm. long, divided to about half-way into 2 blunt or rounded lobes about 3 mm. wide, lower lip slightly shorter than the upper, with a shallowly and bluntly 3-lobed, strongly recurved margin; palate prominent, sulcate, bearded; filaments glabrous, the adaxial about 4 mm. long, the abaxial a little longer; anthers yellow, about 0·8 mm. wide, thecae strongly divergent; ovary broadly ovoid, compressed, about 2·5 mm.

long, 2 mm. wide, densely covered with long straight hairs; style about 3·5 mm. long; stigma about 0·5 mm. long, truncate or shortly emarginate. Capsule ovoid, slightly oblique, about 8–10 mm. long, 5–7 mm. wide, dull brown, rather roughly hairy, with 2, roughly circular apical pores opening by 2-lobed lids attached along the upper margin of the pore; seeds scarab-shaped, about 0·8 mm. long, 0·6 mm. wide, one face flattish or slightly convex with a median, longitudinal keel and a narrow membranous margin, the other concave, with a strongly incurved sinuate-dentate marginal wing; testa fuscous, one surface minutely papillose, the other more coarsely and acutely papillose.

HAB.: Sandy fields, roadsides; dry hillsides and clearings in Pine forest; sea-level to 4,000 ft. alt.; fl. March–July.

DISTR.: Divisions 2–4, 6, 7. S. and W. Europe and the Mediterranean region, also as a weed almost throughout the world.

2. Prodhromos, 1862, *Kotschy* 931; Karavostasi-Lefka, 1935, *Syngrassides* 881! Makheras Monastery, 1937, *Syngrassides* 1632! Mandria, 1938, *Kennedy* 1409! Kakopetria, 1939, *Lindberg f.* s.n.! Agros, 1940, *Davis* 1884! Troödos area, 1950, *F. M. Probyn* s.n.! Kapoura Forest, 1952, *F. M. Probyn* 29! Between Vavatsinia and Vavatsinias, 1967, *Merton* in ARI 716! Palekhori, 1969, *Economides* in ARI 1336!
3. Near Stavrovouni, 1937, *Syngrassides* 1489! Polemidhia hills, 1948, *Mavromoustakis* s.n.!
4. Larnaca, 1862, *Kotschy* 76; also, 1905, *Holmboe* 86, and, S.W. of Larnaca, 1967, *Merton* in ARI 377!
6. Nicosia, 1905, *Holmboe* 280, and, 1927, *Rev. A. Huddle* 83!
7. Pentadaktylos, 1880, *Sintenis & Rigo*, also, 1901, *A. G. & M. E. Lascelles* s.n.! Kyrenia, 1902, *A. G. & M. E. Lascelles* s.n.! also, 1908, *Clement Reid* s.n.! Yaïla, 1941, *Davis* 2851! Melounda, 1963, *J. B. Suart* 78!

7. ANTIRRHINUM *L.*

Sp. Plant., ed. 1, 612 (1753).
Gen. Plant., ed. 5, 268 (1754).
W. Rothmaler in Fedde Repert., Beih., 136: 1–124 (1956).

Annuals, biennials, or perennials, sometimes suffruticose; leaves usually opposite towards base of plant, alternate above, rarely all opposite, simple, entire, shortly petiolate; flowers generally in terminal, bracteate racemes, calyx 5-lobed, the lobes much shorter than the corolla-tube; corolla large, gibbous or shortly saccate at the base, but not spurred, palate conspicuous, closing the mouth of the corolla; stamens 4, inserted near the base of the corolla-tube; staminode minute, squamiform. Capsule 2-locular, ovoid, oblique, abaxial loculus wide, short, dehiscing by 2 apical pores, adaxial loculus long and narrow, dehiscing by a single pore; seeds oblong, reticulate-foveolate.

About 40 species in S. Europe, Mediterranean region and N. America; *A. majus* L. widely grown for its ornamental flowers.

1. **A. majus** *L.*, Sp. Plant., ed. 1, 617 (1753); Sibth. et Sm., Fl. Graec. Prodr., 1: 434 (1809); Poech, Enum. Plant. Ins. Cypr., 26 (1842); Post, Fl. Pal., ed. 2, 2: 290 (1933); Rothmaler in Fedde Repert., Beih., 136: 92 (1956); Davis, Fl. Turkey, 6: 649 (1978).

Erect or sprawling perennial 30–100 (–120) cm. high; stems terete, usually much branched, glabrous or ± glandular-pilose; leaves narrowly ovate, lanceolate or almost linear, 10–70 mm. long, 0·5–15 (–20) mm. wide, usually glabrous, the lower opposite or ternate, the upper often alternate, apex acute or subacute, base tapering to a short petiole (generally less than 5 mm. long) or subsessile; flowers 10–30 in a lax or congested terminal raceme; rhachis glabrous or glandular-pubescent; bracts herbaceous, ovate-acute, (1–) 3–10 mm. long, exceeding or shorter than the pedicels; calyx-lobes subequal, broadly ovate, obtuse, 6–8 mm. long, 3–5 mm. wide, usually glandular-pubescent; corolla purple, pink or white, 2·5–4·5 cm. long, about 1 cm. wide, thinly glandular externally, upper lip porrect, about 1 cm. long, deeply divided into 2 broad, bluntly oblong lobes, lower lip strongly reflexed, bluntly 3-lobed, the 2 lateral lobes much larger than the median; palate conspicuous, yellow (or white), sulcate, hairy, base of corolla-tube very shortly saccate or gibbous; filaments 1·5–2·5 cm. long, glabrous or glandular-pubescent; anthers yellow, oblong, about 2·5 mm. long, 1·5 mm. wide, thinly barbellate; ovary broadly ovoid, about 5 mm. long, 4 mm. wide, densely glandular-pilose; style about 1·5 cm. long, glabrous or thinly glandular; stigma truncate or shortly emarginate. Capsule obliquely ovoid, 10–12 mm. long, 6–9 mm. wide, adaxial pore generally a little larger than the 2 abaxial pores; seeds oblong, about 0·7–1 mm. long, 0·5–0·8 mm. wide; testa fuscous, deeply foveolate.

var. **angustifolium** *Chav.*, Mon. Antirrh., 86 (1833); Boiss., Fl. Orient., 4: 385 (1879); Holmboe, Veg. Cypr., 165 (1914); Post, Fl. Pal., ed. 2, 2: 290 (1933).
    A. majus L. ssp. *angustifolium* (Chav.) Lindberg f., Iter Cypr., 31 (1946); Osorio-Tafall et Seraphim, List Vasc. Plants Cyprus, 94 (1973).
    A. majus L. ssp. *tortuosum* (Bosc. ex Lam.) Rouy f. *glandulosum* Rothm. in Fedde Repert., Beih., 136: 103 (1956).
    A. majus L. ssp. *majus* sec. Davis, Fl. Turkey, 6: 649 (1978).
    A. majus L. ssp. *tortuosum* (Bosc. ex Lam.) Rouy sec. Feinbrun, Fl. Palaest., 3: 193 (1978) pro parte quoad plantas glandulosas.
    [*A. siculum* (non Mill.) Kotschy in Unger et Kotschy, Die Insel Cypern, 290 (1865).]
    TYPE: Turkey; "in locis aridis et calidioribus; super muros Constantinopolis" *Olivier & Bruguière* (P).

Leaves opposite or ternate, rarely alternate, linear or narrowly lanceolate, generally less than 8 mm. wide; flowers commonly opposite or in threes; rhachis, calyx and corolla glandular-pilose. Capsule rather small, usually less than 12 mm. long, 9 mm. wide, glandular-pubescent; seeds small, about 0·7 mm. long, 0·5 mm. wide. Otherwise as in *A. majus* L. var. *majus*.

HAB.: Limestone cliffs and old walls; sea-level to 2,500 ft. alt.; fl. March–November.

DISTR.: Division 7, perhaps an old escape from cultivation. Widespread in the Mediterranean region.

7. Buffavento, Bellapais, St. Hilarion, Lapithos, Kyrenia Pass, Halevga, Kyrenia Castle, Kornos. *Kotschy* [1859], 496, [1862] 385; *Sintenis & Rigo* 535 ! *A. G. & M. E. Lascelles* s.n. ! *Holmboe* 826; *C. B. Ussher* 69 ! 94 ! *Syngrassides* 672 ! 1020 ! *Kennedy* 840 ! *Lindberg f.* s.n.; *Davis* 1751 ! 2045 ! 3594 ! *Casey* 808 ! *Miss Mapple* 44 ! *G. E. Atherton* 193 ! 194 ! 399 ! 400 ! *I. M. Hecker* 1 !

NOTES: Rothmaler (loc. cit., 105) records a form with glabrous stems, rhachises and calyx (= *A. majus* L. ssp. *tortuosum* (Bosc. ex Lam.) Rouy) from Bellapais ["Bella Paese"] collected [1853 or 1854] by Gaudry (P). In all the material I have seen, however, including several specimens from Bellapais, the inflorescence region is conspicuously glandular-pilose.
    To judge from collectors' notes, the corollas of the Cyprus plant vary in colour through shades of pink to white.

TRIBE 3. **Scrophularieae** Leaves all opposite, or the basal opposite; inflorescence cymose, usually many-flowered; corolla 2-lipped, tube not saccate or spurred at the base, adaxial lip exterior in bud; perfect stamens 4, the fifth generally present as a staminode; anthers with free or confluent thecae. Capsule 2- or 4-valved, or fruit rarely a berry.

8. SCROPHULARIA L.

Sp. Plant., ed. 1, 619 (1753).
Gen. Plant., ed. 5, 271 (1754).
H. Stiefelhagen in Engl., Bot. Jahrb., 44: 406–496 (1910).
A. Eig in Pal. Journ. Bot., J. series, 3: 79–93 (1944).

Annual, biennial or perennial herbs or subshrubs; leaves usually opposite, undivided or variously dissected; inflorescences terminal or lateral, generally compound, or flowers rarely solitary, axillary; flowers generally small, yellowish, greenish or brownish-red; calyx 5-lobed; corolla-tube short, ventricose, corolla-lobes 5, the adaxial usually rather longer than the abaxial; stamens didynamous, declinate, inserted on the corolla-tube, included or exserted; anther-thecae confluent, glabrous, or glandular; staminode often distinct, squamiform, sometimes absent. Fruit an ovoid or subglobose, septicidal capsule; seeds usually numerous, ovoid, foveolate.

About 150 species widely distributed in the northern hemisphere.

Leaves undivided, ovate-cordate; calyx-lobes ovate-deltoid, acuminate; plant annual
1. S. peregrina
Leaves 3-pinnatisect; calyx-lobes almost orbicular; stout perennial with a woody base
2. S. peyronii

1. **S. peregrina** *L.*, Sp. Plant., ed. 1, 621 (1753); Boiss., Fl. Orient., 4: 395 (1879); Sintenis in Oesterr. Bot. Zeitschr., 32: 193, 194 (1882); Stiefelh., in Engl., Bot. Jahrb., 44: 459 (1910); Druce in Rep. B. E. C., 9: 470 (1931); Post, Fl. Pal., ed. 2, 2: 291 (1933); A. K. Jackson in Kew Bull., 1934: 272 (1934); Lindberg f., Iter Cypr., 31 (1946); Osorio-Tafall et Seraphim, List Vasc. Plants Cyprus, 94 (1973); Davis, Fl. Turkey, 6: 614 (1978).
   TYPE: "*in* Italia".

Erect annual 20–70 (–100) cm. high; stems 4-angled, glabrous or glandular-pubescent, frequently purplish, branched or unbranched; leaves ovate-cordate, 3–7 cm. long, 2–5 cm. wide, mostly opposite, or the uppermost sometimes alternate; apex acute or obtuse, margins usually coarsely and sharply serrate, sometimes shortly crenate-serrate, lamina generally glabrous and bright green above, glandular along the midrib and nerves below; petiole up to 7 cm. long, thinly glandular, canaliculate above; inflorescence a narrow, elongate thyrse, frequently more than 30 cm. long, generally less than 7 cm. wide, lateral branches alternate, 2–8-flowered; bracts foliaceous, the lower resembling the leaves, the upper becoming progressively smaller and narrower, and usually sessile or subsessile; inflorescence-branches 1–3-times divided, angular, thinly glandular; bracteoles minute, herbaceous, linear-subulate, about 2 mm. long, usually less than 0·5 mm. wide; pedicels slender, 5–10 mm. long, glandular; calyx-lobes subequal, glabrous, rather narrowly ovate-deltoid, acuminate, about 1–2 mm. long, 0·5–1 mm. wide at anthesis, accrescent and persistent in fruit; corolla dark red-brown, about 2·5–3 mm. long, lobes rounded, the 2 adaxial 1·5–2 mm. long and about as wide, the 2 lateral distinctly shorter, the abaxial very small; filaments about 1·5 mm. long, included in corolla-tube; anthers pallid, about 0·5 mm. wide, thecae confluent; staminode fan-shaped, about 1 mm. wide; ovary subglobose, about 1 mm. diam., sessile in a conspicuous disk, glabrous; style about 1·5 mm. long; stigma shortly bifid. Capsule broadly ovoid-subglobose, acute, about 5–7 mm. long and almost as wide, glabrous, the persistent style-base forming a short apical beak; seeds oblong, about 1 mm. long, 0·6 mm. wide; testa fuscous, rugulose-reticulate, with rather regular, narrowly rectangular, minutely reticulate depressions.

HAB.: Cultivated ground; by pathsides; on rocks and old walls, or by gravelly streamsides and margins of irrigation channels; 300–2,000 ft. alt.; fl. March–June.

DISTR.: Divisions 2, 7. Widespread in the Mediterranean region and eastwards to the Caucasus, but becoming rather uncommon in the eastern Mediterranean.

2. Near Galata, 1880, *Sintenis & Rigo* 142a! Lefka, 1932, *Syngrassides* 246! Paliambela, 1962, *Meikle* 2249!

7. Near Kantara, 1880, *Sintenis & Rigo* 142! Lapithos, 1928 or 1930, *Druce* s.n., also, 1939, *Lindberg f.* s.n.; 1953, *Austin-Harrison* in *Casey* 1318! 1955, *Merton* 2229! and, 1967, *Merton* in ARI 535! Vasilia, 1941, *Davis* 3026!

2. S. peyronii *Post* in Bull. Herb. Boiss., 1: 28 (1893); Post, Fl. Pal., ed. 2, 2: 296 (1933); Eig in Pal. Journ. Bot., J. series, 3: 89 (1944); Davis, Fl. Turkey, 6: 645 (1978); Feinbrun, Fl. Palaest., 3: 198, t. 332 (1978).

[*S. canina* (non L.) Sm. in Sibth. et Sm., Fl. Graec. Prodr., 1: 436 (1809), Fl. Graec., 6: 78 (1826) pro parte quoad plant. cypr.; Poech, Enum. Plant. Ins. Cypr., 25 (1842); Unger et Kotschy, Die Insel Cypern, 289 (1865).]

[*S. sphaerocarpa* (non Boiss. et Reut.) Post in Mém. Herb. Boiss., 18: 97 (1900); H. Stuart Thompson in Journ. Bot., 44: 335 (1906); Holmboe, Veg. Cypr., 165 (1914); Lindberg f., Iter Cypr., 31 (1946); Osorio-Tafall et Seraphim, List Vasc. Plants Cyprus, 94 (1973).]

[*S. xanthoglossa* (non Boiss.) Stiefelh. in Engl., Bot. Jahrb., 44: 473 (1910) pro parte.]

TYPE: Lebanon; "in fissuris rupium prope Beirut" *Post* 113 (BEI, K!).

Erect, much-branched perennial from a woody base, 60–150 cm. high; stems angled, glabrous or minutely scabridulous-pubescent; branches spreading; leaves mostly alternate, the basal narrowly deltoid in outline, about 10 cm. long, 6 cm. wide, tripinnatisect, the ultimate divisions narrowly oblong, 4–10 mm. long, 1·5–4 mm. wide, grey-green, glabrous, margins narrowly recurved, remotely lobulate; petioles 4–5 cm. long, flattish above, dilated at base; upper stem leaves similar to basal, but more shortly petiolate; inflorescence broadly paniculate, generally with very numerous, spreading, bifurcating branches terminating in elongate 6–15 (–30)-flowered monochasia; bracts foliaceous, usually consisting of a small, undivided or lobed, sessile segment; bracteoles narrowly oblong, herbaceous, about 2 mm. long, 1 mm. wide; pedicels usually very short, seldom exceeding 2 mm., sparsely glandular; calyx-lobes subequal, almost orbicular, about 0·6–0·8 mm. diam., glabrous with a narrow membranous margin; flowers very small; adaxial lobes of corolla maroon-red, suborbicular, about 2 mm. diam., median and abaxial lobes whitish, sometimes speckled red, a little smaller than the adaxial; stamens included; filaments about 1 mm. long, minutely glandular; anthers pallid, about 0·3 mm. diam., thecae confluent; staminode minute, ovate, pallid, less than 0·5 mm. long, 0·3 mm. wide; ovary subglobose, glabrous, about 0·8 mm. diam., sessile in a conspicuous disk; style about 2·5 mm. long; stigma minute. Capsule brown and woody when ripe, globose or shortly mucronate, about 3–4 mm. diam.; seeds few, oblong, about 1·5 mm. long, 1 mm. wide; testa black, minutely reticulate, deeply rugulose-pitted, with regular, narrowly oblong pits.

HAB.: Dry fields and fallows, roadsides; on dry hillsides amongst garigue; sea-level to 4,500 ft. alt.; fl. April–July.

DISTR.: Divisions 1, 2, 6, 7. S. Turkey, Syria, Lebanon.

1. Paphos, 1913, *Haradjian* 629! and, 1940, *Davis* 1779! Ayios Neophytos, 1939, *Lindberg f.* s.n.

2. Near Galata, 1880, *Sintenis & Rigo* 141! Kato Platres, 1937, *Kennedy* 841! Trikoukkia, 1961, *D. P. Young* 7460! 2 miles S.W. of Palekhori, 1973, *P. Laukkonen* 431!

6. Yerolakkos, 1932, *Syngrassides* 106!

7. Dhavlos, 1905, *Holmboe* 543; Panagra, 1936, *Mrs. Dray* 6! Kyrenia, 1936, *Kennedy* 842! also 1948, *Casey* 70! and, 1955, *G. E. Atherton* 245! 709! St. Hilarion, 1939, *Lindberg f.* s.n.!

[*S. HYPERICIFOLIA* *Wydl.*, a woody perennial with small, often entire, acute leaves, was distributed by Sintenis & Rigo, as no. 1041, with a label headed *Iter Cyprium*. The specimens are not from Cyprus, however, but from Sarona or Hakirya, in Israel.]

Paulownia tomentosa *(Thunb.) Steud.*, Nomencl. Bot., ed. 2, 2: 278 (1841) (*Bignonia tomentosa* Thunb., Fl. Jap., 252; 1784; *Paulownia imperialis* Sieb. et Zucc., Fl. Jap., 1: 27, t. 10; 1835). A deciduous tree, flowering on the leafless branches in spring; twigs thick, pithy, forming a rounded crown; leaves opposite, long-petiolate, broadly ovate-cordate, usually with 3–5 shallow lobes, 15–30 (or more) cm. wide, softly pubescent; flowers in terminal panicles; corolla mauve-blue, 4–5 cm. long, tubular or narrowly funnel-shaped, with 5 conspicuous, spreading, rounded lobes. Fruit a large, woody, ovoid-acute capsule, about 5 cm. long, dehiscing septicidally, and containing numerous, large, winged seeds.

A native of China, long cultivated in Japan, and introduced into Europe in 1834. It is now a popular ornamental tree in many countries adjacent to the Mediterranean, but fails to flower satisfactorily further north, the developing inflorescences being frequently damaged during the winter. G. Frangos (Cypr. Agric. Journ., 18: 88; 1923) notes specimens in nursery gardens at Larnaca and Nicosia, and the tree, which grows quickly from seed, may now be well established on the island.

TRIBE 4. **Gratioleae** *Benth.* Leaves generally opposite, sometimes all basal; inflorescence racemose or flowers solitary, axillary; corolla sometimes 2-lipped, tube not saccate or spurred at the base; stamens 4 (or 2); anthers with distinct, or rarely with confluent thecae. Fruit a 2- or 4-valved capsule, or, rarely, indehiscent or irregularly dehiscent.

9. LIMOSELLA *L.*
Sp. Plant., ed. 1, 631 (1753).
Gen. Plant., ed. 5, 280 (1754).

Small annual aquatic herbs, commonly with rooting stolons; leaves basal, petiolate, with or without a distinct lamina; flowers solitary, axillary, often cleistogamous; calyx subequally 5 (rarely 4)-lobed; corolla rotate, almost actinomorphic; stamens 4; inserted on the corolla-tube; anthers confluent, apparently 1-thecous; ovary incompletely 2-locular; style subulate; stigma subcapitate. Capsule septicidally 2-valved, sometimes tardily dehiscent; seeds numerous, with a ribbed testa.

About 10 species with a cosmopolitan distribution.

1. **L. aquatica** *L.*, Sp. Plant., ed. 1, 631 (1753); Boiss., Fl. Orient., 4: 428 (1879); Post, Fl. Pal., ed. 2, 2: 298 (1933); Davis, Fl. Turkey, 6: 680 (1978).
    TYPE: "*in* Europae *septentrionalis inundatis*".

Loosely tufted annual 2–7 cm. high, with slender, glabrous stolons up to 7 cm. long, rooting at the apex; leaves all basal, the floating with slender petioles 2–7 cm. long and an oblong or bluntly elliptical, dark green (sometimes purple-margined), glabrous, obscurely nerved lamina 3–16 (–20) mm. long, 2–9 mm. wide, submerged leaves often subulate or almost filiform without a distinct lamina; flowers numerous, arising from the leaf-axils; peduncle 1–2 cm. long, patent or recurved; calyx campanulate, glabrous, tube about 1·5 mm. long, lobes 5, equal, ovate, about 1 mm. long, 0·8 mm. wide; corolla whitish or very pale purple, tube 1·5 mm. long, lobes 5, ovate, subequal, about 0·8 mm. long, 0·5 mm. wide; stamens 4, inserted high up the corolla-tube; filaments glabrous, 0·4 mm. long; anthers minute, pallid, apparently 1-thecous, about 0·1 mm. wide; ovary broadly ellipsoid, about 1·5 mm. long, 1 mm. wide, glabrous; style distinct, about 0·4 mm. long; stigma rather large, capitate; disk small, almost hidden under the ovary.

Capsule bluntly ellipsoid, pale brown, papery, about 2·5 mm. long, 2 mm. wide, protruding beyond the persistent calyx; seeds oblong, about 0·5 mm. long, 0·3 mm. wide; testa brown, conspicuously ribbed longitudinally, and finely reticulate overall.

HAB.: In rock pools and muddy puddles; sea-level to 900 ft. alt.; fl. March–April.

DISTR.: Divisions 1, 6, rare. Widespread in temperate Europe and Asia, but rare in the Mediterranean region, though recorded from the Nile Delta, and from 'Ajlun, 25 miles N.N.W. of Amman', Jordan. Perhaps overlooked elsewhere in the eastern Mediterranean.

1. Erimidhes near Ayios Yeoryios (Akamas), 1962, *Meikle* 2139! Between Kato Paphos and Yeroskipos, 1982, *A. A. Butcher* s.n.!
6. Mammari, 1954, *Merton* 1869! Prophitis Elias monastery near Ayia Marina, 1957, *Merton* 2887!

TRIBE 5. **Digitaleae** *Benth.* Calyx 4–5-lobed; corolla with 4–5 spreading lobes, the lateral, or one of them, external in bud; stamens usually 4, sometimes 2, the anther-thecae contiguous or divergent, sometimes confluent. Fruit generally a loculicidal or septicidal, 2-valved capsule.

10. VERONICA *L.*

Sp. Plant., ed. 1, 9 (1753).
Gen. Plant., ed. 5, 10 (1754).
H. Römpp in Fedde Repert., Beih., 50: 1–172 (1928).
G. Stroh in Beih. Bot. Centralbl., 61B: 384–451 (1942).

Annual or perennial herbs, occasionally subshrubs; leaves opposite, sometimes deeply lobed; inflorescence a terminal or lateral raceme or spike, or flowers sometimes apparently solitary, axillary, arising from the axils of alternate, leaf-like bracts; calyx divided almost to base into 4 (–5) unequal lobes, the abaxial generally larger than the adaxial; corolla almost actinomorphic or slightly zygomorphic, rotate, with a very short tube and 4 spreading lobes; stamens 2, exserted. Capsule compressed transversely to septum, 2-locular, generally emarginate at apex, dehiscing loculicidally or sometimes septicidally; seeds numerous (less often, few), commonly cup-shaped, with a distinct elevation on the concave, ventral face.

About 250 species throughout temperate regions of the world or on mountains in the tropics.

Floral leaves (or bracts) similar to opposite basal leaves; flowers apparently solitary, axillary:
  Leaves and bracts palmately divided into finger-like lobes - - - - **3. V. triphyllos**
  Leaves and bracts entire, toothed or shortly lobed:
    Calyx-lobes broadly ovate-deltoid with a wide base, shortly stipitate; corolla pallid; leaves
      reniform-suborbicular, 3–5-lobed - - - - - - **7. V. hederifolia**
    Calyx-lobes ovate, obovate or suborbicular, not deltoid:
      Lobes of capsule strongly divergent; pedicels up to 3 cm. long; corolla blue and white
        5. V. persica
      Lobes of capsule not markedly divergent:
        Leaves ovate, bluntly serrate; corolla bright blue; pedicels to about 10–12 mm. long
          4. V. polita
        Leaves suborbicular-flabellate or reniform, 7–9-lobed; corolla white; pedicels 8–35
          mm. long - - - - - - - - - **6. V. cymbalaria**
Floral leaves much reduced, distinct from opposite cauline leaves; flowers in lateral or terminal racemes:
  Racemes terminal; leaves small, 0·6–1·5 (–2·0) cm. long, dull, often pubescent:
    Pedicels very short, indistinct, rarely exceeding 2 mm.; capsule generally shorter than the
      calyx-lobes - - - - - - - - - - - **2. V. arvensis**
    Pedicels distinct, 3–5 (–8) mm. long, upcurved in fruit; capsule generally equalling or
      exceeding the persistent calyx-lobes - - - - - **1. V. ixodes**
  Racemes lateral; leaves large, 2–17 cm. long, glabrous, lustrous or sublustrous
      8. V. anagallis-aquatica

1. V. ixodes Boiss. et Bal., Diagn., 2, 3: 172 (1856), Fl. Orient., 4: 459 (1879); Holmboe, Veg.
Cypr., 165 (1914); Lindberg f., Iter Cypr., 31 (1946); Osorio-Tafall et Seraphim, List Vasc.
Plants Cyprus, 94 (1973).
   V. hispidula Boiss. et Huet ssp. ixodes (Boiss. et Bal.) M. Fischer in Davis, Fl. Turkey, 6:
709 (1978).
   [V. pusilla (non Kotschy, 1845) Kotschy in Oesterr. Bot. Zeitschr., 12: 279 (1862).]
   [V. acinifolia (non L.) Kotschy in Unger et Kotschy, Die Insel Cypern, 291 (1865);
Boiss., Fl. Orient., 4: 458 (1879) pro parte quoad spec. cypr.]
   [V. hispidula (non Boiss. et Huet) Römpp in Fedde Repert., Beih., 50: 64 (1928); Stroh in
Beih. Bot. Centralbl., 61B: 399 (1942) pro parte]
   TYPE: Turkey; "in regione alpinâ montis *Masmeneu Dagh* in viâ a Ciliciâ ad *Caesaream* siti cl.
Balansa" [1855] (G, K!).

Erect annual 4–12 (–17) cm. high; stems terete, purplish, clothed with
patent gland-tipped hairs throughout their length, generally much
branched, with opposite erecto-patent branches arising from near base of
plant; leaves rather few, oblong, ovate or obovate, 6–15 mm. long, 2·5–8
mm. wide, glandular-hispidulous on both faces, commonly bronze- or
purple-tinged, turning blackish when dried, apex obtuse, margins entire or
occasionally with a few shallow lobes, base tapering gradually or abruptly to
a short, indistinct petiole; inflorescence a rather crowded, 8–25 (–30)-
flowered, terminal raceme, lengthening greatly in fruit and forming most of
the length of the branches; lowermost bracts resembling the leaves, but
smaller and alternate, upper bracts narrowly oblong or almost linear, c. 3
mm. long, 1 mm. wide, usually rather shorter than the fruiting pedicels,
glandular-pubescent; pedicels slender, glandular-pubescent, about 2–3 mm.
long at anthesis, lengthening to 5 (–8) mm. and curving upwards in fruit;
calyx-lobes bluntly oblong-obovate, densely glandular-pilose, about 1·5
mm. long, 1 mm. wide at anthesis, persistent and accrescent in fruit; corolla
minute, white, fugacious, adaxial lobes broadly and bluntly oblong or
suborbicular, about 1 mm. long, almost 1 mm. wide, abaxial lobe narrowly
oblong; filaments about 5 mm. long; anthers violet, broadly oblong, about
0·4 mm. long and almost as wide; ovary compressed, suborbicular, about 0·5
mm. diam., thinly glandular-pubescent; style about 0·5 mm. long,
lengthening to c. 1 mm. in fruit; stigma capitate. Capsule brownish when
ripe, about 2–3 mm. long, 3–5 mm. wide, bluntly 2-lobed, with suberect,
thinly glandular-pubescent lobes, median sinus usually narrow, considerably longer than the persistent style; seeds strongly compressed, bluntly
oblong, 1–1·2 mm. long, 0·7–0·9 mm. wide; testa pale yellowish-brown,
minutely rugulose-verruculose.

HAB.: Rocky mountainsides on serpentine, often by the sides of streams or springs;
4,300–6,400 ft. alt.; fl. May–August.

DISTR.: Division 2, locally common. Also south-central Turkey.

2. On the uppermost slopes of Khionistra; Kannoures Springs, Xerokolymbos, Kryos
Potamos, etc. *Kotschy* 852; *Sintenis & Rigo* 743! *Holmboe* 973; *Kennedy* 848–855!
1410–1412! *Lindberg f.* s.n.! *Davis* 1860! 3514! *N. Macdonald* 185!

NOTES: Closely allied to *V. hispidula* Boiss. et Huet, and sometimes united with it, but
uniformly distinct in habit, foliage, flower-colour and shape of capsule, and here maintained as
a species. The type specimen is very robust, with 30–40 flowers in each raceme, whereas Cyprus
specimens are uniformly less luxuriant, with 12–20 flowers per inflorescence. The difference is
not of great moment, and it would be premature to accord it nomenclatural recognition.

2. V. arvensis L., Sp. Plant., ed. 1, 13 (1753); Sibth. et Sm., Fl., Graec. Prodr., 1: 9 (1806); Poech,
Enum. Plant. Ins. Cypr., 26 (1842); Unger et Kotschy, Die Insel Cypern, 291 (1865); Boiss.,
Fl. Orient., 4: 457 (1879); Holmboe, Veg. Cypr., 165 (1914); Römpp in Fedde Repert.,
Beih., 50: 75 (1928); Post, Fl. Pal., ed. 2, 2: 303 (1933); Stroh in Beih. Bot. Centralbl., 61B:
401 (1942); Osorio-Tafall et Seraphim, List Vasc. Plants Cyprus, 94 (1973); Davis, Fl.
Turkey, 6: 713 (1978); Feinbrun, Fl. Palaest., 3: 204, t. 344 (1978).
   TYPE: "*in* Europae *arvis, cultis*".

Erect annual 2–15 (–25) cm. high; stems slender, much branched (except in starved specimens), terete, generally clothed with 2 (or more) longitudinal bands of spreading, multicellular, eglandular hairs separated by glabrous or subglabrous bands; leaves broadly ovate-deltoid, 7–15 (–20) mm. long, 4–15 (–20) mm. wide, subglabrous above, thinly hispidulous-pilose on the nerves below, apex blunt or subacute, margins crenate or bluntly serrate, base tapering abruptly to a flattened petiole usually less than 5 mm. long; inflorescence spiciform, 10–25 (–30)-flowered, lengthening rapidly during anthesis, ultimately 10–15 (or more) cm. long, the developing fruits rather widely spaced; bracts herbaceous, thinly pubescent, the lowermost foliaceous, broadly or narrowly rhomboid, 6–15 mm. long, 3–10 mm. wide, crenate-serrate, the upper lanceolate or oblong-rhomboid, 5–8 mm. long, 1·5–6 mm. wide, acute, entire or remotely lobulate; pedicels very short, rarely exceeding 2 mm., pilose; calyx-lobes unequal, lanceolate, pubescent, the adaxial about 2·5 mm. long, 1–1·5 mm. wide, the abaxial 3–5 mm. long, 1·5–2 mm. wide, lobes persistent but not much accrescent in fruit; corolla blue or whitish, minute, fugacious, lobes broadly ovate or almost orbicular, subacute, 1–1·2 mm. long and almost as wide; filaments about 0·4 mm. long; anthers pale, suborbicular, about 0·4 mm. diam.; ovary strongly compressed, broadly obcordate, about 1 mm. long, 1·2 mm. wide, pubescent; style about 0·5 mm. long; stigma subcapitate. Capsule pale brown, widely obcordate, about 2·5–3 mm. wide, 2–2·5 mm. long, thinly glandular-ciliate, apex with a wide sinus and 2 suberect or slightly divergent, rounded lobes; seeds about 10 in each capsule, bluntly oblong, compressed, about 0·8 mm. long, 0·6 mm. wide; testa yellowish or pale brown, obscurely rugulose and minutely pitted.

HAB.: In fields; by roadsides, or amongst garigue on hillsides; 800–4,500 ft. alt.; fl. March–April.

DISTR.: Divisions 1–3, rare. Widespread in Europe and eastwards to Central Asia, western N. Africa; introduced elsewhere.

1. Ayios Minas near Smyies, 1962, *Meikle* 2179!
2. Prodhromos, 1905, *Holmboe* 931; Kryos Potamos, 3,300 ft. alt., 1937, *Kennedy* 1413!
3. Khoulou, 1905, *Holmboe* 752.

3. V. triphyllos *L.*, Sp. Plant., ed. 1, 14 (1753); Unger et Kotschy, Die Insel Cypern, 291 (1865); Boiss., Fl. Orient., 4: 463 (1879); Sintenis in Oesterr. Bot. Zeitschr., 31: 192, 257 (1881); H. Stuart Thompson in Journ. Bot., 44: 335 (1906); Holmboe, Veg. Cypr., 166 (1914); Römpp in Fedde Repert., Beih., 50: 70 (1928); Post, Fl. Pal., ed. 2, 2: 304 (1933); Stroh in Beih. Bot. Centralbl., 61B: 400 (1942); Osorio-Tafall et Seraphim, List Vasc. Plants Cyprus, 94 (1973); Davis, Fl. Turkey, 6: 716 (1978).
TYPE: *"in* Europae *agris"*.

Erect or sprawling annual 6–15 cm. high; stems usually much branched from near the base, terete, shortly pubescent, often purplish; basal leaves broadly ovate, 8–12 mm. long, 5–8 mm. wide, bluntly lobed, thinly hispidulous; petiole short, flattened, rarely exceeding 3 mm.; upper leaves similar to lower, but palmately-divided almost to base into 5, finger-like, blunt lobes, generally less than 2 mm. wide; inflorescence a terminal 8–20 (–30)-flowered raceme, at first congested, but elongating greatly at anthesis, attaining 8–15 (–20) cm. in fruit, the capsules becoming remote; rhachis with a mixed indumentum of short, eglandular and longer, glandular hairs; bracts foliaceous, 5–12 mm. long and almost as wide, palmately divided into 3–5 blunt, narrow lobes; pedicels slender, glandular-pubescent, equalling or somewhat exceeding bracts, patent or curved upwards in fruit; calyx-lobes subequal, narrowly oblong-obovate, about 5–6 mm. long, 2·5 mm. wide, accrescent and persistent in fruit, thinly glandular-pubescent, apex obtuse; corolla bright blue, lobes broadly and bluntly ovate, about 2–2·5 mm. long,

1·8–2·2 mm. wide; filaments slender, about 1 mm. long; anthers violet, broadly and bluntly ovoid, about 0·4 mm. long and almost as wide; ovary suborbicular, about 1 mm. diam., thinly glandular; style about 1 mm. long, elongating in fruit; stigma subcapitate. Capsule broadly obcordate or suborbicular, strongly compressed, 5–6 mm. long and almost as wide, pale brown when ripe and thinly glandular, apical sinus rather narrow, lobes erect, blunt, usually a little shorter than the persistent, accrescent style; seeds ear-shaped or almost orbicular, about 1·5 mm. long, 1·4 mm. wide, convex dorsally, concave ventrally with a distinct median ridge; testa dark brown, verruculose on the dorsal surface.

HAB.: Gravelly mountainsides, or as a weed in cultivated land; sea-level to 4,500 ft. alt.; fl. March–April.

DISTR.: Divisions 2, 4. Widespread in Europe and eastwards through Turkey and Syria to Iran; N.W. Africa.

2. Prodhromos, 1862, *Kotschy*; Platres, 1938, *Kennedy* 847! Gourri, 1950, *Chapman* 323! Slopes of Adelphi above Lagoudhera, 1974, *Meikle* 4050!

4. Phaneromene near Larnaca, 1880, *Sintenis*; Ormidhia, 1880, *Sintenis*.

NOTES: Almost certainly indigenous in mountain areas, but perhaps introduced with crops in the lowlands.

4. V. polita *Fries*, Novit. Fl. Suec., 63 (1819); E. Lehm. in Bull. Herb. Boiss., ser. 2, 7: 546 (1907); Schinz et Thellung in Vierteljahr. Nat. Ges. Zürich, 53: 561 (1909); Römpp in Fedde Repert. Beih., 50: 84 (1928); Lindberg f., Iter Cypr., 31 (1946); Osorio-Tafall et Seraphim, List Vasc. Plants Cyprus, 95 (1973); Davis, Fl. Turkey, 6: 720 (1978); Feinbrun, Fl. Palaest., 3: 206, t. 348 (1978).

? *V. didyma* Ten., Fl. Nap. Prodr., 6 (1811); Boiss., Fl. Orient., 4: 466 (1879); Druce in Rep. B.E.C., 9: 470 (1931); Post, Fl. Pal., ed. 2, 2: 305 (1933); Stroh in Beih. Bot. Centralbl., 61B: 403 (1942).

V. polita Fr. ssp. *thellungiana* E. Lehm. in Oesterr. Bot. Zeitschr., 50: 256 (1909).

V. thellungiana (E. Lehm.) Dalla Torre et Sarnth., Fl. Tirol., 6: 272 (1912).

V. didyma Ten. var. *thellungiana* (E. Lehm.) Druce in Rep. B.E.C., 9: 470 (1931).

TYPE: Sweden; "ubique in arvis Scaniae" *Fries* (UPS).

Prostrate or sprawling annual; stems slender, 4–20 cm. long, crispate-pubescent, usually much branched from base; leaves opposite towards base of plant, broadly ovate, obtuse or subacute, 5–10 mm. long, 4–9 mm. wide, or sometimes a little wider than long, subglabrous above, thinly pubescent along the nerves below, margins narrowly recurved, conspicuously, but bluntly, serrate, base very widely cuneate or truncate; petiole usually very short, to about 5 mm. long; inflorescence running almost the full length of the branches; bracts foliaceous, resembling the basal leaves but smaller and commonly alternate, the flowers appearing to be solitary and axillary; pedicels to about 10–12 mm. long, equalling or slightly exceeding the bracts, crispate-pubescent, usually deflected in fruit; calyx-lobes broadly ovate, about 3·5 mm. long, 2–2·5 mm. wide, overlapping at the base, strongly 5-nerved, ciliate along the margins and nerves, apex blunt; corolla blue, scarcely exceeding calyx-lobes, glabrous, the corolla-lobes rounded, about 3 mm. long, 2·5 mm. wide; filaments about 1·5 mm. long; anthers violet, broadly ovate-elliptic, about 0·5 mm. long, and almost as wide; ovary compressed-suborbicular, about 0·5 mm. diam., shortly puberulous; style about 1·5 mm. long; stigma capitate. Capsule about 3·5 mm. long, 4·5 mm. wide, slightly compressed, but not keeled dorsally, clothed with numerous short eglandular hairs interspersed with a few long glandular ones, lobes erect, rounded, median sinus rather wide, shorter than the persistent style; seeds cymbiform or almost cup-shaped, usually about 1·3 mm. long, 0·8–1 mm. wide; testa yellowish, rugose with parallel ridges on the convex surface, smooth on the concave surface, with a conspicuous median umbilicoid appendage.

HAB.: Cultivated fields and gardens; roadsides; damp grassland and stony hillsides; sea-level to 5,600 ft. alt.; fl. Dec.–June.

DISTR.: Divisions 1–7. Widespread in Europe and Asia; Mediterranean region; elsewhere as a weed of cultivation.

1. Paphos, 1928 or 1930, *Druce* s.n.
2. Kambos, 1928 or 1930, *Druce* s.n.; Troödos, 1928 or 1930, *Druce* s.n.; Platres, 1938, *Kennedy* 1414!
3. Amathus, 1964, *J. B. Suart* 160! Episkopi, 1964, *J. B. Suart* 198!
4. Paralimni, 1940, *Davis* 2064! Near Famagusta, 1969, *S. Genneou* in ARI 1375!
5. Kythrea, 1880, *Sintenis & Rigo* 133!
6. Mile 30, Nicosia–Troödos road, 1955, *Merton* 1990!
7. Lapithos, 1939, *Lindberg f.* s.n.; Kyrenia, 1949, *Casey* 223! also, 1950, *Casey* 1009! and, 1956, *G. E. Atherton* 828! 1037! Bellapais, 1956, *G. E. Atherton* 863!

NOTES: Some doubt remains as to the correct name for this species: Schinz & Thellung (*loc. cit.*, 1909) have argued that *V. didyma* Tenore (1811) is a *nomen ambiguum*, applied by Tenore, prior to 1824, to several species, and not exclusively to the plant described above. Their evidence for this is, however, not wholly convincing, though it would appear that, at least until this date, Tenore did not fully appreciate the differences between *V. didyma* and *V. agrestis* L. After 1824, he seems to have consistently applied the name to the plant more generally named *V. polita* Fr. (1819), and if it can be shown that his earlier usage, as demonstrated by specimens from which the lectotype of *V. didyma* might be chosen, is no different from this later application, then the name *V. didyma* must be restored for the species.

5. **V. persica** *Poir.* in Lam., Encycl. Méth., 8: 542 (1808); Post, Fl. Pal., ed. 2, 2: 305 (1933); Stroh in Beih. Bot. Centralbl., 61B: 404 (1942); Lindberg f., Iter Cypr., 31 (1946); Osorio-Tafall et Seraphim, List Vasc. Plants Cyprus, 95 (1973); Davis, Fl. Turkey, 6: 721 (1978); Feinbrun, Fl. Palaest., 3: 205, t. 347 (1978).

    V. tournefortii C. C. Gmel., Fl. Bad., 1: 39 (1805) pro parte non Vill. (1787); Römpp in Fedde Repert., Beih., 50: 85 (1928).

    V. buxbaumii Ten., Fl. Nap., 1: 7, t. 1 (1811) non F. W. Schmidt (1791); Boiss., Fl. Orient., 4: 465 (1879).

TYPE: "Croît dans la Perse. On le cultive au Jardin des Plantes de Paris" (P).

Prostrate or sprawling annual; stems slender, up to 30 cm. long (or occasionally more), often purplish and usually much divided from the base, crispate-pubescent, the indumentum sometimes concentrated in longitudinal bands; basal leaves opposite, broadly ovate, subcordate, 1–2 cm. long, 1–1·5 cm. wide, thinly pubescent above and below, apex shortly acute or obtuse, margins rather boldly and sharply serrate, base very broadly cuneate, truncate or subcordate; petiole flattened, pubescent, generally less than 8 mm. long; inflorescence running almost the full length of the branches; bracts usually alternate, resembling the leaves but often rather smaller and more shortly petiolate; flowers apparently solitary in the axils of the foliaceous bracts; pedicels slender, up to 3 cm. long, much exceeding the bracts, rather densely crispate-pubescent, not much curved or deflected in fruit; calyx-lobes strongly divergent in pairs, ovate, not much overlapping at base, 4–5 mm. long, 2–3 mm. wide, pilose on the margins and prominent nerves, apex obtuse or subacute; corolla-lobes broadly ovate, almost as long as the calyx-lobes, 2·5–3 mm. wide, the 3 adaxial blue with darker veins, the abaxial (basal) paler or white with pale blue veins; filaments about 1·5 mm. long, minutely papillose-verruculose; anthers violet, broadly oblong, about 0·8 mm. long and almost as wide; ovary compressed, broadly obovate or suborbicular, almost 0·5 mm. diam., thinly pubescent towards the truncate or shallowly emarginate apex; style 2–2·5 mm. long; stigma capitate. Capsule pale brown when ripe, compressed, 7–8 mm. wide, about 3–4 mm. long, distinctly veined, thinly pubescent, with glandular and eglandular hairs, lobes blunt, keeled, strongly divergent, sinus wide, exceeded by the persistent style; seeds about 15, bluntly oblong, cymbiform, about 1·5 mm. long, 1 mm. wide; testa pale yellowish-brown, rugose with parallel ridges on the convex surface, smooth on the concave surface with a conspicuous median, umbilicoid appendage.

HAB.: Gardens and orchards, almost certainly introduced; 2,000–4,500 ft. alt.; fl. March–Sept.

DISTR.: Division 2. Probably a native of the Caucasus area and S.W. Asia; now widespread as a weed in most temperate countries.

2. Pedhoulas, 1932, *Syngrassides* 46! Moutoullas, 1937, *Syngrassides* 1550! Prodhromos, 1939, *Lindberg f.* s.n.! also, 1949, *Casey* 741! and, 1955, *G. E. Atherton* 612! Phini, 1973, *P. Laukkonen* 87!

6. V. cymbalaria *Bodard*, Mém. Véronique Cymb., 3 (1798); Unger et Kotschy, Die Insel Cypern, 291 (1865); Boiss., Fl. Orient., 4: 467 (1879); Holmboe, Veg. Cypr., 166 (1914); Römpp in Fedde Repert., Beih., 50: 93 (1928); Post, Fl. Pal., ed. 2, 2: 306 (1933); Stroh in Beih. Bot. Centralbl., 61B: 406 (1942); Lindberg f., Iter Cypr., 31 (1946); Osorio-Tafall et Seraphim, List Vasc. Plants Cyprus, 94 (1973); M. Fischer in Plant Systematics and Evolution, 123: 97–105 (1975); M. Fischer et W. Greuter in Plant Systematics and Evolution, 125: 245–252 (1976); Davis, Fl. Turkey, 6: 723 (1978); Feinbrun, Fl. Palaest., 3: 206, t. 349 (1978).
TYPE: Italy; from the vicinity of Pisa.

Prostrate, trailing or ascending annual, sometimes conspicuously tinged purple; cotyledons usually persisting at least until anthesis, broadly oblong or suborbicular, sometimes shortly emarginate, long-stipitate, 0·5–1·5 cm. long, 0·3–1 cm. wide; stems 7–30 cm. long, usually much branched from base, subglabrous or thinly clothed with patent, eglandular (or rarely glandular), white, multicellular hairs 0·8–2 (–3) mm. long; leaves opposite towards base of plant, suborbicular-flabellate or reniform, 0·8–2·5 cm. long, 0·6–3 cm. wide, subglabrous or thinly pilose on both surfaces, apex rounded or subacute, margins 7–9-lobed, with short, obtuse or subacute teeth or lobes, base widely cuneate, truncate or shallowly cordate; petiole flattened, 1–2·5 cm. long; inflorescence running almost the whole length of the branches; bracts foliaceous, so closely resembling the basal leaves that the flowers appear to be solitary, axillary; pedicels 0·8–3·5 cm. long at anthesis, often much exceeding the leaves, not elongating greatly in fruit, subglabrous or ± densely pilose with patent, eglandular (and a few glandular) hairs; calyx-lobes broadly ovate or almost orbicular, obtuse or subacute, 3–4 mm. long, 2–3 mm. wide, ± densely pilose with long, eglandular hairs, persistent and somewhat accrescent in fruit; corolla white, glabrous, lobes bluntly oblong-elliptic or suborbicular, 2·5–3·5 mm. long, 2–3 mm. wide; filaments 1·5 mm. long; anthers oblong, violet, about 0·8 mm. long, 0·6–0·7 mm. wide; ovary subglobose, about 1·5 mm. diam., thinly or densely clothed with long, multicellular hairs, rarely glabrous; style 1·5 mm. long; stigma capitate. Capsule not much compressed, depressed-subglobose, usually pilose, about 3–4 mm. wide, 2·5–3 mm. long, 2–4-seeded, walls greenish or pale brown, thin and easily broken, apical lobes short and blunt, rounded dorsally; seeds deeply cymbiform or cup-shaped, about 2–2·5 mm. long, 1·8–2 mm. wide; testa brown or yellowish, almost smooth dorsally and internally, regularly fluted or rugulose about the rim; elaiosome conspicuous on the internal surface.

HAB.: Cultivated and fallow fields; roadsides; damp walls; stony mountainsides, usually in moist situations by springs or streams; occasionally on sand dunes; sea-level to 4,500 ft. alt.; fl. Dec.–June.

DISTR.: Divisions 1–8, locally common. Widespread in S. Europe and the Mediterranean region.

1. Peyia, 1962, *Meikle* 2028!
2. Makheras Monastery, 1937, *Syngrassides* 1624! Platres, 1938, *Kennedy* 1415! Prodhromos, 1939, *Lindberg f.* s.n.! Gourri, 1950, *Chapman* 301! Vouni, 1957, *Merton* 2866! Khorteri, 1962, *Meikle* 2712! Selladhi tou Mavrou Dasous, above Spilia, 1962, *Meikle* 2819!
3. Stavrovouni, 1880, *Sintenis*; Amathus, 1947, *Mavromoustakis* s.n.! Limassol, 1951, *Mavromoustakis* 37! Episkopi, 1964, *J. B. Suart* 165! Polymidhia, 1964, *J. B. Suart* 192!
4. Cape Greco, 1862, *Kotschy* 147; Phaneromene near Larnaca, 1880, *Sintenis*, and, 1905, *Holmboe* 147!

5. Kythrea, 1967, *Merton* in ARI 354!
6. Mile 29½ Nicosia–Troödos road, 1951, *Merton* 97! Kato Kopia, 1970, *A. Genneou* in ARI 1379! Nicosia, 1973, *P. Laukkonen* 83!
7. Pentadaktylos, 1880, *Sintenis & Rigo* 526! St. Hilarion, 1940, *Mavromoustakis* 26! Kyrenia, 1949, *Casey* 229! also, 1956, *G. E. Atherton* 831! 850! 1210!
8. Yialousa, 1941, *Davis* 2391!

NOTES: Very variable in habit and leaf-size. Most of the Cyprus material is wholly or largely eglandular, with a rather coarse indumentum of long, white multicellular hairs; in two specimens, however, (*Kennedy & Davis* 2698! and 2699! both collected on N. slopes of Kionia, c. 3,500 ft. alt., 18th March, 1941), the indumentum is very largely glandular, and suggests that the specimens might be referred to *V. trichadena* Jord. et Fourr. (Brev. Plant. Nov., 1: 42; 1866), a species closely allied to *V. cymbalaria*, but differing in its slender habit, small, subentire or bluntly crenate leaves, and dense indumentum of short, slender, glandular hairs. *V. trichadena* has a wide distribution in the Mediterranean region, ranging from the Balearic Islands eastwards to southern Turkey. The species might well occur in Cyprus, but the two specimens cited above (which agree closely with *Forsyth Major* 749 from Samos, cited in Fl. Turkey, 6: 724 under *V. trichadena*) are not satisfactory proofs of its occurrence, being intermediate between typical *V. trichadena* and *V. cymbalaria* in habit, leaf-shape and leaf-toothing, and in the relative coarseness of their glandular indumentum. Until good, characteristic specimens of *V. trichadena* are found on the island, I prefer to include these aberrant sheets within *V. cymbalaria* aggr.

7. **V. hederifolia** L., Sp. Pl., ed. 1, 13 (1753); Sibth. et Sm., Fl. Graec. Prodr., 1: 9 (1806); Poech, Enum. Plant. Ins. Cypr., 26 (1842); Unger et Kotschy, Die Insel Cypern, 291 (1865); Boiss., Fl. Orient., 4: 468 (1879); Holmboe, Veg. Cypr., 167 (1914); Römpp in Fedde Repert., Beih., 50: 92 (1928); Post, Fl. Pal., ed. 2, 2: 306 (1933); Stroh in Beih. Bot. Centralbl., 61 B: 406 (1942); M. Fischer in Oesterr. Bot. Zeitschr., 114: 189–233 (1967); Osorio-Tafall et Seraphim, List Vasc. Plants Cyprus, 95 (1973); Davis, Fl. Turkey, 6: 725 (1978); Feinbrun, Fl. Palaest., 3: 206, t. 350 (1978).

[*V. hederifolia* L. var. *triloba* (non Opiz) Lindberg f., Iter Cypr., 31 (1946).]

TYPE: "*in* Europae *ruderatis*".

Trailing annual; cotyledons usually persisting at least until anthesis, suborbicular, 1–1·8 cm. long and almost as wide, long-stipitate; stems 20–30 cm. or more long, usually much branched from base, thinly clothed with long, white, eglandular hairs 1–2 mm. long, forming rather ill-defined vertical bands; leaves opposite only at very base of plant, reniform-suborbicular, 0·5–2 cm. long and almost as wide or a little wider, subglabrous or very sparsely white-pilose above and below, apex shortly acute or obtuse, margins 3–5-lobed, with blunt or subacute, spreading lobes, base truncate or very broadly cuneate; petiole flattened 5–10 (–15) mm. long; inflorescence running almost the whole length of the branches; bracts foliaceous, so closely resembling the leaves that the flowers appear to be solitary and axillary; pedicels 5–15 mm. long at anthesis, usually rather shorter than the leaves, not elongating greatly in fruit, subglabrous or thinly eglandular-pilose; calyx-lobes shortly stipitate, broadly ovate-deltoid, acute or acuminate, 5–6 mm. long, 3–4 mm. wide at anthesis, distinctly accrescent and persistent in fruit, glabrous or subglabrous dorsally, conspicuously pectinate-ciliate along the margins with long white hairs; corolla white, veined pale blue, fugacious, lobes bluntly ovate or suborbicular, about 1·5–2 mm. long, 1–1·5 mm. wide; filaments about 0·8–1 mm. long; anthers yellow, oblong-ovate, about 0·5 mm. long and almost as wide; ovary compressed, suborbicular, about 0·8 mm. diam.; style about 0·8 mm. long; stigma capitate. Capsule not much compressed, strongly depressed-subglobose, about 4–5 mm. wide, 2·5–3 mm. long, c. 4-seeded, walls greenish or pale brown, glabrous, thin and easily broken; apical lobes very short and blunt, rounded dorsally; sinus wide and shallow; seeds deeply cymbiform or cup-shaped, 2–2·5 mm. long, about 2 mm. wide; testa dark or light brown, almost smooth dorsally and internally, regularly fluted or rugulose about the rim; elaiosome conspicuous on the internal surface.

HAB.: Cultivated ground, roadsides; walls; sea-level to 4,500 ft. alt.; fl. ? Dec.–June.

DISTR.: Divisions 2, 4, rather rare. Widespread in Europe and western Asia, and elsewhere as an introduced weed.

2. Platres, 1938, *Kennedy* 1415! Prodhromos, 1939, *Lindberg f.* s.n.! and, 1970, *A. Hansen* 761 (in lit.); Polystipos, 1957, *Merton* 3126!

4. Phaneromene near Larnaca, 1880, *Sintenis*; Cape Pyla, 1880, *Sintenis*.

NOTES: Apart from the Sintenis records (unsupported by specimens) *V. hederifolia* is a plant of high altitudes in Cyprus, sometimes growing in association with *V. cymbalaria* Bod.

The specimen cited by Lindberg f. as *V. hederifolia* L. var. *triloba* Opiz (*V. triloba* (Opiz) Kerner) was collected in mid-June, when it was already somewhat shrivelled; the dorsal surface of the calyx-lobes is glabrous, and not pubescent as is usual in *V. triloba*, and I am fairly confident that the plant is simply a late-season state of *V. hederifolia*.

8. **V. anagallis-aquatica** L., Sp. Plant., ed. 1, 12 (1753); Hume in Walpole, Mem. Europ. Asiatic Turkey, 254 (1817); Gaudry, Recherches Sci. en Orient., 190 (1855); Unger et Kotschy, Die Insel Cypern, 291 (1865); Boiss., Fl. Orient., 4: 437 (1879); Holmboe, Veg. Cypr., 165 (1914); Römpp in Fedde Repert., Beih., 50: 159 (1928); Post, Fl. Pal., ed. 2, 2: 300 (1933); G. Schlenker in Fedde Repert., Beih., 90: 4 (1936), Flora, 130: 310 (1936); Lindberg f., Iter Cypr., 31 (1946); Osorio-Tafall et Seraphim, List Vasc. Plants Cyprus, 94 (1973); Davis, Fl. Turkey, 6: 726 (1978).

[*V. beccabunga* (non L.) H. Stuart Thompson in Journ. Bot., 44: 335 (1906); Holmboe, Veg. Cypr., 165 (1914); Osorio-Tafall et Seraphim, List Vasc. Plants Cyprus, 94 (1973).]

Perennial or annual herb (5–) 20–100 (–120) cm. high, often blackening when dried; stems erect, ascending or decumbent, often 4-angled at least towards base, usually glabrous and purple-tinged, occasionally very sparsely glandular, sometimes unbranched, more frequently with several spreading branches arising at or near the base of the plant; leaves opposite, the basal commonly obovate or elliptic-oblong, 1–4 cm. long, 0·7–1·5 cm. wide, glabrous, lustrous, apex acute or obtuse, margins shallowly serrate, base narrowing to a distinct petiole 5–15 mm. long, upper leaves occasionally similar to basal, but generally distinct, narrowly oblong, obovate or oblanceolate, sessile and subamplexicaul, 1·5–17 cm. long, 0·8–3·5 cm. wide, apex acute or acuminate (rarely obtuse or rounded), margins regularly and shallowly serrate (rarely subentire); inflorescence a compact or lax, axillary raceme, elongating to 4–15 cm. long, 1–2·5 cm. wide in fruit; peduncles spreading, 2–4 cm. long; bracts linear, lanceolate or narrowly obovate, 2–10 (–15) mm. long, 0·5–4 mm. wide, sometimes exceeding the pedicels at anthesis, usually shorter than pedicels in fruit; rhachis angular, glabrous, subglabrous or ± densely glandular-pubescent, with multicellular, gland-tipped hairs; pedicels filiform, erecto-patent or spreading, at first very short, elongating to 10 mm. or more in fruit, glabrous or ± glandular-pubescent; calyx-lobes ovate-elliptic or oblong, sometimes almost linear-oblong, 2·5–4 mm. long, 1–1·5 mm. wide, glabrous or sparsely glandular, persisting and ± spreading in fruit, apex acute or subacute, margins minutely papillose-erosulate; corolla blue, lilac or mauve, with darker veining, abaxial lobe often paler than the others, lobes broadly and bluntly ovate, 2–3 mm. long, 1·5–2·5 mm. wide; filaments 1–1·5 mm. long; anthers violet, oblong, about 0·7 mm. long, 0·5 mm. wide; ovary glabrous or thinly glandular, suborbicular or broadly elliptic, about 1–1·5 mm. diam.; style 1·5–2·5 mm. long; stigma capitate or capitate-truncate. Capsule suborbicular or very broadly ellipsoid, about 3 mm. long, 2·5–3 mm. wide, distinctly compressed, apex widely but very shallowly emarginate; seeds bluntly oblong-elliptic, 0·4–0·6 mm. long, 0·3–0·5 mm. wide, strongly compressed or plano-convex, but not cymbiform or cup-shaped; testa yellow-brown, minutely rugulose, with a minute, dot-like chalaza.

f. **anagallis-aquatica**
TYPE: "*in* Europa *ad fossas*".

Plant robust, often 1 m. high; base of stem usually fistulose; cauline leaves sessile, semi-amplexicaul, narrowly oblong, acute or acuminate; inflorescence subglabrous or very sparsely glandular; pedicels often forward-pointing, even in fruit, exceeded by the narrow bracts at anthesis; calyx-lobes narrowly oblong-ovate, often exceeding the capsule.

HAB.: Marshy ground by the side of streams, springs and irrigation channels; sea-level to 5,000 ft. alt.; fl. Febr.–July.

DISTR.: Divisions 2–5, 7, 8. Widespread in Europe, Asia, Africa and temperate America.

2. Troödos, 1901, *Bovill* in *A. G. & M. E. Lascelles* s.n.! Lefka, 1932, *Syngrassides* 247! Platres, 1937, *Kennedy* 843! Kykko, 1939, *Lindberg f.* s.n.! Agros, 1940, *Davis* 1886! Near Polystipos, 1957, *Merton* 3124! Kalopanayiotis, 1968, *A. Genneou* in ARI 1283!
3. Limassol, 1801, *Hume*; Kalavasos, 1905, *Holmboe* 626; Pano Polemidhia, 1948, *Mavromoustakis* 1!
4. Larnaca, 1801, *Hume*.
5. Kythrea, 1880, *Sintenis & Rigo* 134! also, 1932, *Syngrassides* 224! and, 1967, *Merton* in ARI 357! Dhali, 1936, *Syngrassides* 1118!
7. Kyrenia, 1939, *Lindberg f.* s.n., also, 1956, *G. E. Atherton* 999! Lapithos, 1939, *Lindberg f.* s.n., also, 1951, *Merton* 300! and, 1955, *Merton* 2245! Bellapais, 1939, *Lindberg f.* s.n., also 1956, *G. E. Atherton* 1319!
8. Gastria, 1905, *Holmboe* 472; between Komi Kebir and Dhavlos, 1941, *Davis* 2443!

f. **anagallidiformis** *Boreau*, Fl. Centre France, ed. 3, 2: 489 (1857) — as "*anagalliformis*"; Lindberg f., Iter Cypr., 31 (1946).
    V. montioides Boiss., Diagn., 1, 7: 43 (1846).
    V. anagallis-aquatica L. var. *pseudo-anagalloides* Gren., Fl. Chaine Jurass., 579 (1865).
    V. anagallis-aquatica L. var. *montioides* (Boiss.) Boiss., Fl. Orient., 4: 437 (1879); Holmboe, Veg. Cypr., 165 (1914).
    V. anagallis-aquatica L. var. *anagallidiformis* (Boreau) Franchet, Fl. Loir-et-Cher, 434 (1885).
    V. anagallis-aquatica L. ssp. *divaricata* Krösche in Allg. Bot. Zeitschr., 18: 83 (1912); G. Schlenker in Fedde Repert., Beih., 90: 8 (1936) et in Flora, 130: 310 (1936); Stroh in Beih. Bot. Centralbl., 61B: 427 (1942); Feinbrun, Fl. Palaest., 3: 203 (1978).
    [*V. beccabunga* (non L.) Lindberg f., Iter Cypr., 31 (1946).]
TYPE: Locality not indicated (Boreau in P sec. Schlenker).

Plant usually less robust than in typical *V. anagallis-aquatica*; stems frequently decumbent, sometimes fistulose at the base, sometimes not; cauline leaves often narrowed towards base or even shortly petiolate, sometimes sessile and amplexicaul, acute or obtuse; inflorescence generally glandular-pubescent, sometimes sparsely so; pedicels patent, usually much longer than the short, narrowly ovate bracts; calyx-lobes narrowly ovate or lanceolate, equalling or not much exceeding the capsule.

HAB.: By streams and springs; sea-level to 5,000 ft. alt.; fl. March–Dec.

DISTR.: Divisions 2, 6, 7. Europe and Asia, but probably just as widespread as *V. anagallis-aquatica* f. *anagallis-aquatica*.

2. Prodhromos, 1905, *Holmboe* 901; also, 1939, *Lindberg f.* s.n.! and, 1955, *G. E. Atherton* 558! Pomos, 1934, *Syngrassides* 958! Between Platres and Phini, 1936, *Kennedy* 846! Xerokolymbos, 4,500 ft. alt., 1937, *Kennedy* 844! also, 1961, *D. P. Young* 7440! Kambos, 1939, *Lindberg f.* s.n.! Milikouri, 1939, *Lindberg f.* s.n.! Platania, 1939, *Lindberg f.* s.n.! Near Polystipos, 1957, *Merton* 3125! Above Klirou, 1957, *Merton* 3134! Prodhromos–Lemithou road, 1957, *Merton* 3186! Paliambela, 1962, *Meikle* 2255!
6. Nicosia–Troödos road, mile 29, 1951, *Merton* 422!
7. Kyrenia, 1936, *Kennedy* 845! also, 1955, *G. E. Atherton* 309! Lapithos, 1955, *G. E. Atherton* 684!

NOTES: I suspect a mere habitat-state of typical *V. anagallis-aquatica*, prevalent in areas which are temporarily moist, but beginning to dry out, or in sites which become dry before the plant is fully developed. Every possible intermediate can be found between f. *anagallidiformis* and the typical plant, and the two have been collected in close proximity in Cyprus.
All records for *V. beccabunga* L. belong here; the true plant has not been found in Cyprus.

[V. CAESPITOSA *Boiss.*, Diagn., 1, 4: 79 (1844), a suffruticose, cushion-forming perennial, with crowded, Thyme-like leaves with strongly revolute margins, and short subterminal racemes of large, clear blue or purplish flowers, is recorded by Kotschy (Die Insel Cypern, 291; 1865) "Auf der Spitze des Troödos an der Nordseite der Felstrümmer. 1859 lebend in den bot. Garten von Wien gesandt". It has not since been seen or collected on or near the summit of Troödos — a very well-worked area — nor in any other part of Cyprus, and one must suspect some error in labelling in the Botanic Garden at Vienna. *V. caespitosa* is widely distributed in Turkey, with a glabrous-leaved variant (var. *leiophylla* Boiss.) in Lebanon.]

TRIBE 6. **Euphrasieae** *Benth.* Leaves opposite or alternate; inflorescence usually a raceme; corolla 2-lipped, the adaxial lip erect, concave or galeate, entire, emarginate or bifid, abaxial lip often spreading, 3-lobed, exterior in bud; spur absent; stamens 4, didynamous, rarely 2; anthers 2-thecous, often muticous or mucronate basally. Capsule generally dehiscing loculicidally. Plants commonly hemiparasitic.

11. **PARENTUCELLIA** *Viv.*

Fl. Libyc. Spec., 31 (1824).

Hemiparasitic annual, turning blackish when dried; stems erect, glandular; leaves (except sometimes the uppermost) opposite, shortly serrate or dentate, glandular-pubescent; inflorescence a raceme; bracts foliaceous, alternate; pedicels very short; calyx tubular, ± equally 4-lobed, membranous between the nerves; corolla white, yellow or red-purple, adaxial lip galeate, bifid, abaxial lip longer than adaxial, 3-lobed; stamens 4, didynamous; anthers mucronate, glabrous or pilose. Capsule oblong; seeds numerous, minute; testa smooth or reticulate.

Three species in southern Europe, Mediterranean region and western Asia.

Corolla 10–12 mm. long, purple, white or pale yellow; inflorescence seldom more than 8 cm. long, even in fruit; plant (3–) 5–20 (–27) cm. high - - - - - **1. P. latifolia**
Corolla 15–25 mm. long, yellow; inflorescence to 25 cm. or more long in fruit; plant 16–50 cm. high - - - - - - - - - - - - **2. P. viscosa**

1. **P. latifolia** (*L.*) *Caruel* in Parl., Fl. Ital., 6: 482 (1885); Holmboe, Veg. Cypr., 167 (1914); Post, Fl. Pal., ed. 2, 2: 307 (1933); Osorio-Tafall et Seraphim, List Vasc. Plants Cyprus, 95 (1973); Davis, Fl. Turkey, 6: 677 (1978).
   Euphrasia latifolia L., Sp. Plant., ed. 1, 604 (1753); H. Stuart Thompson in Journ. Bot., 44: 336 (1906).
   Bartsia latifolia (L.) Sm. in Sibth. et Sm., Fl. Graec. Prodr., 1: 428 (1809), Fl. Graec., 6: 69, t. 586 (1826); Poech, Enum. Plant. Ins. Cypr., 26 (1842).
   Eufragia latifolia (L.) Griseb., Spic. Fl. Rum. et Bith., 2: 14 (1844); Unger et Kotschy, Die Insel Cypern, 292 (1865); Boiss., Fl. Orient., 6: 473 (1879); Sintenis in Oesterr. Bot. Zeitschr., 31: 393 (1881).
   Eufragia latifolia (L.) Griseb. var. *flaviflora* Boiss., Fl. Orient., 4: 473 (1879).
   Parentucellia latifolia (L.) Caruel var. *albiflora* Hal., Consp. Fl. Graec., 2: 437 (1902); Holmboe, Veg. Cypr., 167 (1914).
   P. latifolia (L.) Caruel ssp. *flaviflora* (Boiss.) Hand.-Mazz. in Ann. Nat. Hofmus. Wien, 27: 16 (1913); I. C. Hedge in Notes Roy. Bot. Gard. Edinb., 36: 11 (1978) et in Davis, Fl. Turkey, 6: 767 (1978).
   P. flaviflora (Boiss.) Nevski in Acta Inst. Bot. Acad. Sci. URSS, ser. 1, 4: 321 (1937); Feinbrun in Fl. Palaest., 3: 207, t. 351 (1978).
   TYPE: "*in* Apulia, Italia, Monspelii".

Erect annual (3–) 5–20 (–27) cm. high; stem usually unbranched, occasionally branched from near base, pale green or purplish, ± densely glandular-pilose and viscid; leaves opposite, sessile, remote, 0·3–2 cm. long, 0·3–8 cm. wide, broadly ovate or oblong-ovate, nervation impressed above, prominent and glandular-pilose below, apex acute or obtuse, margins strongly revolute, bluntly or sharply 2–6-toothed, strigillose-ciliate; inflorescence a dense terminal spiciform raceme 1–4 cm. long, 1–1·5 cm. wide at anthesis, elongating to 8 cm. or more in fruit; bracts foliaceous, broadly obcuneate 5–10 mm. long, 3–8 mm. wide, deeply 3–7-lobed, glandular-pubescent; flowers sessile or subsessile; calyx-tube cylindrical, 4-nerved, about 7–10 mm. long, 2–3 mm. wide at anthesis, densely glandular-pilose, accrescent and persistent in fruit; lobes narrowly deltoid, erect, about 2·5 mm. long, 1 mm. wide, subacute; corolla crimson, purple, yellow (ssp. *flaviflora* (Boiss.) Hand.-Mazz.) or entirely white (var. *albiflora* Hal.), tube narrowly cylindrical, 8–10 mm. long, 1–1·5 mm. wide, subglabrous or glandular-hairy, 2-lipped, adaxial lip galeate, entire, about 2 mm. long, 1·5 mm. wide, enclosing the anthers, abaxial lip about 2–2·5 mm. long, 3 mm. wide, conspicuously 3-lobed, with 2 distinct yellow gibbosities on the palate just above the median lobe; filaments adnate to tube for the greater part of their length, free for about 1–1·5 mm. from apex; anthers yellow, glabrous or subglabrous, about 0·8–1 mm. long and a little wider, suborbicular, the thecae-bases sharply pointed or mucronate; ovary fusiform, glabrous, about 3 mm. long, 1 mm. wide; style filiform, about 7–8 mm. long; stigma strongly compressed laterally, subglobose. Capsule fusiform, about 10 mm. long, 3 mm. wide, sublustrous dark brown; seeds oblong-ovoid, about 0·4 mm. long, 0·3 mm. wide; testa dull brown, minutely and irregularly rugulose.

HAB.: Dry grassland, pastures, tracksides, or in open garigue on calcareous or non-calcareous hillsides; sea-level to 4,400 ft. alt.; fl. March–May.

DISTR.: Divisions 1–8, locally abundant. Western Europe and Mediterranean region eastwards (ssp. *flaviflora*) to Afghanistan and Central Asia.

1. Fontana Amorosa, 1962, *Meikle* 2295! Polis, 1962, *Meikle*!
2. Platres, 1937, *Kennedy* 856! 1418! Mandria-Perapedhi, 1938, *Kennedy* 857! also, 1960, *N. Macdonald* 45! By Kryos Potamos, 1937, *Kennedy* 858! Kionia, 1941, *Davis* s.n.! and, 1967, *Merton* in ARI 512!
3. Ayia Varvara, Stavrovouni, 1862, *Kotschy* 206; Episkopi, 1959, *C. E. H. Sparrow* 2! 2 m. W. of Limassol, 1964, *J. B. Suart* 178!
4. Larnaca, 1862, *Kotschy*, and, 1950, *Chapman* 48! Ayia Napa, 1905, *Holmboe* 46; Xylophagou, 1905, *Holmboe* 75; Cape Greco, 1958, *N. Macdonald* 20! Dhekelia, 1977, *A. Della* in ARI 1680!
5. Kythrea, 1956, *G. E. Atherton* 1067! Athalassa, 1967, *Merton* in ARI 88! and *A. Genneou* in ARI 1419! Salamis, 1962, *Meikle*!
6. Nicosia, 1927, *Rev. A. Huddle* 84! Ayia Irini, 1941, *Davis* 2530! Dhiorios, 1962, *Meikle*!
7. Pentadaktylos, 1880, *Sintenis & Rigo* 140! Kyrenia, 1949, *Casey* 256! 257!, also, 1950, *Casey* 1005! and, 1956, *G. E. Atherton* 1126! 1171! 1196!
8. Ronnas Bay near Rizokarpaso, 1941, *Davis* 2376! Yialousa, 1941, *Davis* 2393! 2394! Akradhes, 1962, *Meikle*! Koma tou Yialou, 1968, *S. Economides* in ARI 1113!

NOTES: Cyprus is one of the few areas where typical purple- and/or white-flowered *P. latifolia* (ssp. *latifolia*) and the yellow-flowered, *P. latifolia* ssp. *flaviflora* are found growing together (see Hedge, *loc. cit.*, 10, fig. 2). However, it will, I think, be found that the two form distinct populations, even when occurring in close proximity. Unfortunately, in the absence of notes on flower-colour, it is difficult to discriminate between the two, or to indicate their separate distributions. Both occur in Divisions 1, 2 and 8, while, to date, all records from 3 are for ssp. *latifolia*, and all from 4, 5 and 6 for ssp. *flaviflora*. Both seem to occur in Division 7, but the evidence is not wholly satisfactory. Apart from flower-colour, the two subspecies appear to be indistinguishable.

2. **P. viscosa** (*L.*) *Caruel* in Parl., Fl. Ital., 6: 482 (1885); Holmboe, Veg. Cypr., 167 (1914); Post, Fl. Pal., ed. 2, 2: 308 (1933); Rechinger f. in Arkiv för Bot., ser. 2, 1: 431 (1950); Osorio-Tafall et Seraphim, List Vasc. Plants Cyprus, 95 (1953); Davis, Fl. Turkey, 6: 766 (1978); Feinbrun, Fl. Palaest., 3: 207, t. 352 (1978).
   Bartsia viscosa L., Sp. Plant., ed. 1, 602 (1753).
   Euphrasia viscosa Sieb., Herb. Cret. (1819) non L. (1767); H. Stuart Thompson in Journ. Bot., 44: 336 (1906) nom. illeg.
   Eufragia viscosa (L.) Benth. in DC., Prodr., 10: 543 (1846); Unger et Kotschy, Die Insel Cypern, 292 (1865); Boiss., Fl. Orient., 4: 474 (1879).
   TYPE: "*in* Angliae, Galliae, Italiae *paludibus ad rivulos*".

Erect annual, 16–50 cm. high; stem usually unbranched, occasionally branched from near base, densely glandular-pilose and viscid; leaves opposite or alternate, sessile, not very remote, 1–3·5 cm. long, 0·5–1·5 cm. wide, narrowly ovate or oblong, nervation impressed above, prominent below, glandular-pilose, apex subacute or obtuse, margins narrowly revolute, bluntly or shortly serrate, strigillose-ciliate; inflorescence a spiciform raceme, elongating to 25 cm. or more and becoming lax in fruit; bracts foliaceous, very similar to the upper cauline leaves, the uppermost commonly lanceolate or linear-oblong, subentire or remotely serrate; flowers sessile or subsessile; calyx-tube cylindrical, 8-nerved, 8–12 mm. long, 4–5 mm. wide, glandular-pilose, accrescent and persistent in fruit, lobes bluntly linear-subulate, 5–8 mm. long, 1–2 mm. wide; corolla yellow, tube narrowly infundibuliform, about 1·5 cm. long, less than 3 mm. wide at the middle, thinly glandular-pubescent, 2-lipped, adaxial lip galeate, entire, about 7 mm. long, 4 mm. wide, abaxial lip 7–8 mm. long, 8–9 mm. wide, conspicuously 3-lobed, the lobes subequal, bluntly oblong, about 2 mm. long and almost as wide, with 2 linear gibbosities on the palate just above the median lobe; filaments free for 4–5 mm. from apex; anthers yellow, pilose, oblong, about 1·8 mm. long, 1 mm. wide, the thecae bases sharply pointed or mucronate; ovary narrowly ovoid, 4–5 mm. long, 2 mm. wide, shortly pilose especially towards apex; style 1·6 cm. long, shortly pilose; stigma strongly compressed laterally, about 1 mm. long. Capsule fusiform, 8–9 mm. long, 2–3 mm. wide, fuscous, pilose; seeds oblong-ovoid, about 0·4 mm. long, 0·3 mm. wide; testa dull brown, minutely and obscurely rugulose.

HAB.: By springs, streams and irrigation channels; sea-level to 3,050 ft. alt.; fl. April–June.

DISTR.: Divisions 1–3, rather rare. Western and southern Europe, the Mediterranean region, Atlantic Islands and eastwards to Iran. An introduced weed elsewhere.

1. "By the aqueduct near Khrysokhou towards the village of Slewra [sic]", 1862, *Kotschy* 677 ! Stroumbi, 1913, *Haradjian* 720; between Evretou and Sarama, 1941, *Davis* 3350 !
2. Platres, 1938, *Kennedy* 1417 !
3. Lemona, 1905, *Holmboe* 747 !

12. BELLARDIA *All.*

Fl. Pedem., 1: 61 (1785).

Hemiparasitic annual, turning blackish when dried; stems erect, glandular; leaves opposite below, sometimes alternate or subalternate above, sessile, narrow, serrate or serratilobed, glandular; inflorescence a dense terminal spike or spiciform raceme; bracts foliaceous; calyx 2-labiate, each lip shortly and subequally lobed; corolla showy, 2-labiate, white, or white and pink, or occasionally yellow; adaxial lip bifid or subentire; abaxial lip 3-lobed, the lateral lobes larger than the median; stamens 4, didynamous; anthers mucronate. Capsule loculicidal, ovoid-globose, turgid; placentas thick; seeds numerous, small, longitudinally ridged.

One species in southern Europe, Mediterranean region east to Iran, Atlantic Islands, temperate and tropical Africa; an introduced weed in Australia, N. America, S. America, etc.

1. **B. trixago** (*L.*) *All.*, Fl. Pedem., 1: 61 (1785); Holmboe, Veg. Cypr., 167 (1914); Post, Fl. Pal., ed. 2, 2: 308 (1933); Osorio-Tafall et Seraphim, List Vasc. Plants Cyprus, 95 (1973); Davis, Fl. Turkey, 6: 768 (1978); Feinbrun, Fl. Palaest., 3: 208, t. 353 (1978).
   Bartsia trixago L., Sp. Plant., ed. 1, 602 (1753).
   Rhinanthus trixago (L.) L., Syst. Nat., ed. 10, 1107 (1759), Sp. Plant., ed. 2, 840 (1763), Syst. Nat., ed. 12, 405 (1767).
   Trixago apula Steven in Mém. Soc. Nat. Mosc., 6: 4 (1823); Unger et Kotschy, Die Insel Cypern, 292 (1865); Boiss., Fl. Orient., 4: 474 (1879).
   TYPE.: "*in* Italiae *maritimis, humentibus*".

Erect annual, (6–) 18–30 (–50) cm. high; stems usually unbranched, sometimes with several erect branches, terete, thinly hispidulous-pilose towards base, often more densely and retrorsely pilose above; leaves often rather dense, linear-oblong, 1·5–7 cm. long, 0·3–1·2 cm. wide, glandular-pubescent, apex acute or obtuse, margins narrowly revolute, bluntly serrate or serratilobed, nervation obscure; inflorescence a dense, terminal spike or spiciform raceme, 2–8 cm. long, 1·3–2 cm. wide, lengthening to 12 cm. or more in fruit; bracts ovate-acuminate, 1–2 cm. long, 0·8–1 cm. wide, densely glandular-pubescent, the upper usually entire,the lower remotely serrate or serratilobed; flowers sessile or subsessile; calyx laterally compressed, broadly ovate-oblong, the adaxial lobes with a short median sinus, the abaxial divided almost to base of calyx-tube, so that the calyx appears to have 2 lateral, unequally lobed lips, lobes broadly deltoid, about 2–3 mm. long, 3–5 mm. wide at base, densely glandular; corolla with the upper or lower lip white or pink, or both lips white or both occasionally yellow (var. *flaviflora* (Boiss.) Hayek; *Bartsia trixago* L. var. *flaviflora* Boiss., Fl. Orient., 4: 474; 1879), corolla-tube about 1 cm. long, 0·3 cm. wide, adaxial lip galeate, about 1 cm. long, 0·3 cm. wide, laterally compressed, glandular-pubescent; abaxial lip broadly flabellate, about 1·3 cm. long, 1·4–1·5 cm. wide, divided into 3 rounded lobes about 3 mm. long, palate with two linear swellings above the median lobe; filaments flattened, ribbon-like, the shorter about 10 mm. long, the longer 12 mm. long; anthers concealed by upper lip of corolla, bluntly oblong, about 2·5 mm. long, 1·8 mm. wide, densely pilose; ovary ovoid, about 5 mm. long, 3·5 mm. wide, pilose; style about 1·5 cm. long, shortly pilose; stigma decurved, clavate, laterally compressed, about 1·5 mm. long. Capsule ovoid, acuminate, pilose, about 10 mm. long, 6 mm. wide, valves rather thick and woody; seeds bluntly oblong, about 0·6 mm. long, 0·3 mm. wide; testa brown, with about 14 pale longitudinal ridges or very narrow wings, connected transversely by numerous, close, parallel striae.

HAB.: Cultivated and fallow fields, waste ground, roadsides; occasionally in grass steppe or on hillsides in garigue; sea-level to 3,800 ft. alt.; fl. March–May.

DISTR.: Divisions 1–8. External distribution that of the genus.

1. On the way from Ktima to Arodhes ["Arora"], 1862, *Kotschy* 676! Between Dhrousha and Kato Arodhes, 1941, *Davis* 3281!
2. Platres, 1938, *Kennedy* 1416! Ayios Merkourios, 1962, *Meikle* 2284!
3. Kophinou, 1905, *Holmboe* 593! Kolossi, 1959, *C. E. H. Sparrow* 32!, also, 1961, *Polunin* 6659! Near Souni, 1960, *N. Macdonald* 125! 2 m. W. of Limassol, 1963, *J. B. Suart* 30!
4. Larnaca aerodrome, 1950, *Chapman* 418! Larnaca Salt Lake Plantation, 1955, *Merton* 1997!
5. Kythrea, 1934, *Syngrassides* 481! Mia Milea, 1967, *Merton* in ARI 174!
6. Syrianokhori, 1941, *Davis* 2608!
7. Kyrenia, 1931, *J. A. Tracey* 50!, 1956, *G. E. Atherton* 1260! and, 1968, *Economides* in ARI 1143! Karakoumi, 1956, *G. E. Atherton* 1231! Ayios Yeoryios, 1959, *P. H. Oswald* 124! Lambousa Beach, 1973, *P. Laukkonen* 273!
8. Ayios Symeon, 1880, *Sintenis & Rigo* 139! Yioti, 1962, *Meikle* 2526! Ayios Andronikos, 1970, *A. Genneou* in ARI 1485!

NOTES: Most of the material from Cyprus is noted as having white and pink corollas; in one example, however (*Economides* in ARI 1143, from Kyrenia), the flowers are said to be yellow and dark red, approaching the uniformly yellow var. *flaviflora* (Boiss.) Hayek (Prodr. Fl. Pen. Balc., 2: 177 (1929); *Trixago apula* Steven var. *lutea* Willk. in Willk. et Lange, Prodr. Fl. Hisp., 2: 613 (1870) nom. illeg.). The two colour variants do not appear to have distinctive distributions, though the typical variant with white and pink flowers predominates in most areas. *B. trixago* (L.) All. var. *flaviflora* (Boiss.) Hayek is figured (as *Bartsia trixago*) in Sibth. et Sm., Fl. Graec., 6: t. 585 (1826).

13. ODONTITES *Ludwig*
Inst. Regn. Veg., ed. 2, 120 (1757).

Hemiparasitic annuals or suffruticose perennials, turning blackish when dried; stems usually erect, strigose or hairy; leaves opposite or sometimes alternate, simple, usually entire; inflorescence a terminal, secund, spiciform raceme; bracts foliaceous; calyx tubular or campanulate, subequally 4-toothed; corolla 2-labiate, adaxial lip entire or bifid, abaxial lip 3-lobed; stamens 4, didynamous; anthers mucronate at base. Capsule loculicidal, ovoid-ellipsoid; seeds rather few, longitudinally ridged or striate.

About 25 species, chiefly in Europe, the Mediterranean region and western Asia.

1. **O. cypria** *Boiss.*, Fl. Orient., 4: 477 (1879); Holmboe, Veg. Cypr., 167 (1914); Lindberg f., Iter Cypr., 31 (1946); Osorio-Tafall et Seraphim, List Vasc. Plants Cyprus, 95 (1973).

    [*O. frutescens* (non Sieber ex Hal.) Poech, Enum. Plant. Ins. Cypr., 26 (1842).]
    [*O. bocconei* (non (Guss.) Walpers) Kotschy in Unger et Kotschy, Die Insel Cypern, 292 (1865).]
    [*O. lutea* (non (L.) Clairv.) H. Stuart Thompson in Journ. Bot., 44: 335 (1906).]

    TYPE: Cyprus; "in insulâ Cypro prope Prodromo, Evrico, ad prom. Capo Gallo [sic] et ad radices montis Troodos", *Kotschy* (G, K !).

Erect, suffruticose perennial 15–50 cm. high; stems usually much branched at base and apex, with a long unbranched middle region, subglabrous towards base, thinly strigillose towards apex, often violet or purple; leaves commonly alternate on sterile shoots, opposite or subopposite on flowering shoots, sessile, linear or linear-lanceolate, 5–12 (–25) mm. long, 1·5–2 mm. wide, subglabrous, often purplish, nervation obscure, apex acute or subacute, margins flat, entire; inflorescence a congested, terminal, spiciform raceme, 1–5 cm. long, 1·5–2 cm. wide, flowers generally secund; bracts foliaceous, narrowly lanceolate, about 6–8 mm. long, 1·5 mm. wide; pedicel 0·5–1 mm. long; calyx cylindrical-campanulate, subglabrous, about 4 mm. long, 2 mm. wide, apex divided into 4 deltoid, glabrous or ciliate lobes about 1·2 mm. long, 0·8 mm. wide at base; corolla yellow, deflexed, tube cylindrical, 3·5–4 mm. long, 1 mm. wide, adaxial lip shallowly galeate, about 3–3·5 mm. long, 1·5 mm. wide, apex rounded or shortly emarginate, abaxial lip deeply divided into 3 suborbicular or shortly oblong lobes about 1–1·5 mm. diam.; palate with 2 linear, longitudinal gibbosities above the median corolla-lobe; filaments about 2 mm. long, glabrous; anthers oblong, pilose, about 0·8 mm. long, 0·5 mm. wide; ovary obovate, shortly pilose, about 1·5 mm. long, 1 mm. wide; style about 6 mm. long, shortly pilose; stigma capitate. Capsule oblong, about 5 mm. long, 3·5–4 mm. wide, glabrous or sparsely pilose, apex emarginate; seeds narrowly oblong, about 2 mm. long, 0·8 mm. wide, about 4–5 in each loculus of capsule; testa pale brown, 20–24-ribbed longitudinally, with numerous, close, parallel transverse striae. *Plate 70.*

    HAB.: On rocky slopes in garigue or under Pines or Cypresses; 100–5,500 ft. alt.; fl. July–Dec.
    DISTR.: Divisions 1–3, 5, 7, locally common. Endemic.

1. Near Ayios Neophytos monastery, 1939, *Lindberg f.* s.n !
2. Common; Prodhromos, Phini, Platres, Galata, Trikoukkia, Kalopanayiotis, etc. *Kotschy* 29, 858; *A. G. & M. E. Lascelles* ! *Chapman* 113 ! *Kennedy* 859 ! 860 ! 861 ! 862 ! *Syngrassides* 1695 ! *Davis* 1973 ! *Lindberg f.* s.n. ! *G. E. Atherton* 527 ! *P. H. Oswald* 165 ! *A. Genneou* in ARI 1287 ! etc.
3. Cape Gata ["Gatto"], 1862, *Kotschy* 602 ! Episkopi, 1963, *J. B. Suart* 135 !
5. Kythrea, 1940, *Davis* 1963 !
7. St. Hilarion, 1901, *A. G. & M. E. Lascelles* s.n. ! also, 1936, *Lady Loch* 2 ! 1940, *Davis* 2041 ! Kyrenia, 1936, *Kennedy* 863 ! also, 1955 and 1956, *G. E. Atherton* 719 ! 814 ! Yaïla, 1940, *Davis* 1966 ! Near Myrtou, 1940, *Davis* 1969 ! Vasilia, 1940, *Davis* 2014 ! Kyrenia Pass, 1948, *Casey* 127 !

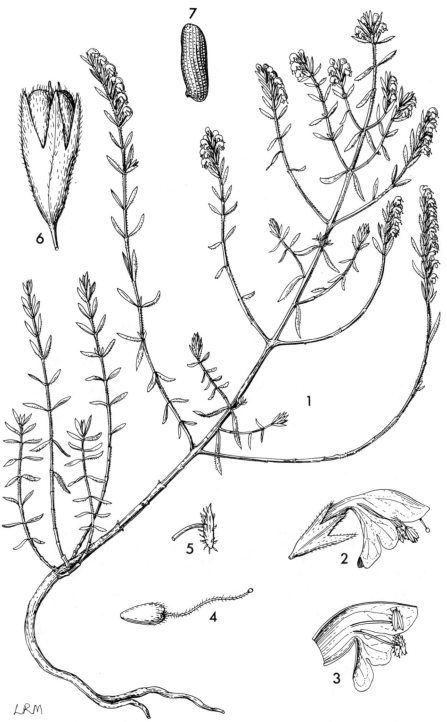

LRM

Plate 70. ODONTITES CYPRIA Boiss. **1,** habit, × ⅔; **2,** flower, × 4; **3,** flower, longitudinal section, × 4; **4,** gynoecium, × 4; **5,** stamen, × 6; **6,** fruit, × 6; **7,** seed, × 10. (**1–6** from *Kennedy* 863.)

[RHINANTHUS MINOR *L.*, Amoen. Acad., 3: 54 (1756) (*Alectorolophus minor* (L.) Wimm. et Grab.), an erect hemiparasitic annual with narrow crenate-serrate leaves, and terminal spiciform racemes of yellow, bilabiate flowers with conspicuous, persistent, inflated calyces, is recorded, without precise locality, by H. Stuart Thompson (in Journ. Bot., 44: 335; 1906), apparently on the strength of a specimen said to have been collected in Cyprus by Miss E. A. Samson in 1904. Although Miss Samson's specimens are generally to be found in the Kew Herbarium, this particular record is missing, though, perhaps significantly, there is an unlocalized Cyprus specimen of *Bellardia trixago* (L.) All. collected by Miss Samson in 1904, but not mentioned by H. Stuart Thompson in his notes on the Cyprus flora. It is just possible that the specimen was erroneously listed as *Rhinanthus minor*, though it is clearly labelled "*Trixago apula* Stev." in Thompson's own hand.

Holmboe's record for *Alectorolophus minor* is based upon a reference to Thompson's paper, and there is no independent evidence for the occurrence on Cyprus of a species which is generally absent from this part of the Mediterranean.]

68. OROBANCHACEAE

G. Beck-Mannagetta in Engl., Pflanzenr., 96 (IV.261): 1-348 (1930).

Parasitic herbs without chlorophyll; base of stem often swollen; leaves reduced to scales, alternate; flowers commonly in terminal, bracteate spikes or racemes, rarely in panicles or solitary, hermaphrodite; calyx gamosepalous, 4–5-toothed or -lobed, sometimes distinctly 2-lipped; corolla tubular, subequally 4–5-lobed, or often distinctly zygomorphic and 2-lipped; stamens 4, didynamous (the fifth reduced to a staminode or wanting); filaments inserted towards base of corolla-tube; anthers dorsifixed, connivent, 2-thecous, often mucronate at the base, generally included and often concealed under the adaxial lip of the corolla; ovary superior, 1-locular, with 2–4 parietal placentas; style simple, terminal; stigma capitate or peltate; or commonly 2-lobed. Fruit a loculicidal capsule, generally enveloped by the persistent calyx and corolla; seeds numerous, minute; testa rugose or reticulate-faveolate; endosperm oily; embryo undifferentiated.

About 13 genera, widely distributed, and particularly well represented in temperate regions of Europe, Asia and North America.

Flower almost actinomorphic with 5, subequal, rounded corolla-lobes; calyx 5-lobed
                                                                        1. Cistanche
Flower distinctly zygomorphic, 2-lipped, the lower lip 3-lobed; calyx 4-lobed or cleft into 2
    divisions   -   -   -   -   -   -   -   -   -   -   -   -   **2. Orobanche**

1. CISTANCHE *Hoffsgg. et Link*

Fl. Port., 1: 318 (1813).

G. Beck-Mannagetta in Engl., Pflanzenr., 96 (IV.261): 26 (1930).

Stem robust, simple; inflorescence showy, with bracts and bracteoles; calyx campanulate, 5 (or rarely 4)-lobed, lobes imbricate, rounded, sometimes dentate or lobulate; corolla tubular or infundibuliform, tube curved or straight, limb scarcely bilabiate, the corolla-lobes patent, subequal, rounded; glandular nectary-disk present; ovary ellipsoid; style inflexed at apex; stigma large, subentire; placentas 4 (rarely more or fewer). Seeds subglobose; testa reticulate-faveolate.

About 12 species widely distributed in southern Europe, Asia and Africa.

1. **C. phelypaea** (*L.*) *P. Cout.*, Fl. de Port., ed. 1, 571 (1913) – as *C. phelipaea*.
  Lathraea phelypaea L., Sp. Plant., ed. 1, 606 (1753).
  Orobanche tinctoria Forssk., Fl. Aegypt.-Arab., 112 (1775).
  Phelipaea lutea Desf., Fl. Atlant., 2: 61, t.146 (1798); Boiss., Fl. Orient., 4: 500 (1879).
  Cistanche lutea (Desf.) Hoffsgg. et Link, Fl. Port., 1: 319 (1813); Post, Fl. Pal., ed. 2, 2: 312 (1933).
  C. tinctoria (Forssk.) G. Beck-Mannagetta in Bull. Herb. Boiss., ser. 2, 4: 685 (1904) et in Engl., Pflanzenr., 96 (IV. 261): 30 (1930); Holmboe, Veg. Cypr., 167, 255, fig. 95 (1914).
  TYPE: Portugal; "*in* Lusitaniae *umbrosis*".

Erect (? perennial) parasitic herb 20–70 (–100) cm. high; stem unbranched, swollen, oblique and densely scaly at the base, more remotely scaly above, scales ovate, tapering to an obtuse apex, 1–2·5 cm. long, 0·8–1·2 cm. wide, subadpressed, glabrous, with paler, scarious, erose-fimbrillate margins; inflorescence a stiff, crowded, cylindrical spike (10–) 15–20 (–50) cm. long, 5–8 cm. wide; bracts similar to the stem-scales, sometimes drawn out into a long, blunt acumen; bracteoles oblong, 12–15 mm. long, 4–5 mm. wide, with scarious, erosulate margins; calyx campanulate, about 15 mm. long, 12 mm. wide, divided to less than half its length into 5 broadly oblong-suborbicular, scarious-margined lobes about 7 mm. long, 5–6 mm. wide; corolla bright yellow, broadly infundibuliform, tube 15–20 mm. long, 8 mm. wide, distinctly expanded above, straight or curved, lobes spreading, suborbicular, about 10–12 mm. long and almost as wide, margins subentire, undulate-crispate; stamens inserted near base of corolla-tube; filaments about 20 mm. long, 0·8 mm. wide, hairy at the base, glabrous above; connective produced as a short apiculus; anthers yellow, oblong, with a sharply apiculate base, about 5 mm. long, 4 mm. wide, densely and softly villose; ovary conical, about 5 mm. long, 4 mm. wide; style about 25 mm. long; stigma compressed-capitate, about 3 mm. wide, obscurely 2-lobed. Capsule (from Cretan specimens) partly concealed by persistent corolla, bracts and bracteoles, ovoid, about 2 cm. long, 1·2 cm. wide, acute, walls thick; seeds globose, about 1 mm. diam.; testa fuscous, deeply and regularly reticulate-faveolate.

HAB.: On *Suaeda vera* J. F. Gmel., *Atriplex halimus* L., and possibly other *Chenopodiaceae*, near the sea; about sea-level; fl. March–May.

DISTR.: Divisions 1, 3, apparently rare. Spain, Portugal, Crete, N. Africa, and perhaps more widely in western Asia if allowed a greater measure of variability.

1. Ayios Yeoryios Island (Akamas), 1962, *Meikle* 2164!
3. Near the lighthouse at Cape Gata, 1905, *Holmboe* 679.

2. OROBANCHE *L.*

Sp. Plant., ed. 1, 632 (1753).
Gen. Plant., ed. 5, 281 (1754).
G. Beck-Mannagetta in Engl., Pflanzenr., 96 (IV. 261): 44 (1930).

Annual, biennial or perennial parasites; stems usually erect, from a swollen base, branched or unbranched, variously coloured, often glandular-pilose; flowers in terminal spikes or racemes; bracts present, bracteoles commonly wanting, when present adnate to the calyx; calyx tubular or campanulate, 4(–5)-lobed or split almost to base into 2, entire or bifid, lateral segments; corolla tubular, often conspicuously 2-lipped, marcescent, upper lip entire, emarginate or 2-lobed, lower lip 3-lobed; stamens 4, didynamous, included; filaments inserted below the middle of the corolla-tube, often thickened towards the base; anthers commonly coherent below the stigma, glabrous or hairy; ovary 2-sulcate; placentas 4 in 2 pairs; style commonly decurved or incurved at apex; stigma rather large and thick, 2- or 4-lobed. Capsule often enveloped in the persistent floral organs, dehiscing dorsiventrally into 2 valves at maturity, the valves often cohering apically at the persistent style-base; seeds numerous, minute, subglobose.

About 100 species with a cosmopolitan distribution in temperate and tropical regions, but most abundantly represented in Europe and western Asia.

Bracteoles 2, inserted on calyx-tube; corolla blue, mauve, purple or violet; stem usually branched:
　　Bracts lanceolate or ovate, acuminate, 5–10 mm. long; calyx-lobes 4–5 mm. long; corolla-tube markedly constricted above ovary:
　　　Corolla 12–20 mm. long; tube 4–5 mm. wide at apex　-　-　-　- **1. O. ramosa**
　　　Corolla 20–30 mm. long; tube 6–10 mm. wide at apex　-　-　- **2. O. aegyptiaca**
　　Bracts caudate-acuminate, 15–20 mm. long, far overtopping buds at apex of inflorescence; calyx-lobes 6–7 mm. long; corolla-tube narrowly subcylindrical, not markedly constricted above ovary　-　-　-　-　-　-　-　-　- **3. O. orientalis**
Bracteoles absent; calyx 2-cleft; corolla whitish, yellowish, brownish, reddish or maroon, sometimes tinged violet or purple; stems normally unbranched:
　　Corolla-tube about 4 mm. wide at apex, maroon-red, corolla-lobes finely fimbriate-denticulate; inflorescence compact, crowded　-　-　-　-　-　**4. O. cypria**
　　Corolla-tube 5–10 mm. wide at apex; corolla not maroon-red:
　　　Calyx-divisions entire; corolla with 2 conspicuous yellow folds on lower lip　- **5. O. alba**
　　　Calyx-divisions deeply 2-lobed:
　　　　Corolla-tube indistinctly curved, 10–15 mm. long, corolla-lobes conspicuous, spreading, the adaxial 7–12 mm. wide　-　-　-　-　-　-　-　- **6. O. crenata**
　　　　Corolla-tube distinctly curved throughout its length, 8–10 mm. long; lobes inconspicuous, the adaxial porrect, the others spreading:
　　　　　Corolla-tube subglabrous or shortly glandular-pubescent dorsally
　　　　　　　　　　　　　　　　　　　　　　　　　　7. O. minor var. **minor**
　　　　　Corolla-tube with white, flattened, crispate hairs dorsally **7. O. minor** var. **pubescens**

SECT. 1. **Trionychon** *Wallr.* Bracteoles 2, adnate to the calyx; calyx subequally 4-lobed (rarely with a small additional lobe); corolla commonly blue or mauve, tube often markedly constricted above the ovary; stems frequently branched.

1. O. ramosa *L.*, Sp. Plant., ed. 1, 633 (1753); Beck-Mannagetta in Bibl. Bot., 4 (19): 87 (1890) et in Engl., Pflanzenr., 96 (IV. 261): 66 (1930); Druce in Rep. B.E.C., 9: 470 (1931); Post, Fl. Pal., ed. 2, 2: 314 (1933); Osorio-Tafall et Seraphim, List Vasc. Plants Cyprus, 95 (1973); Davis, Fl. Turkey, 7: 6 (1982).
　　Phelipaea ramosa (L.) C. A. Mey., Verz. Pflanz. Cauc., 104 (1831); Unger et Kotschy, Die Insel Cypern, 293 (1865); Boiss., Fl. Orient., 4: 498 (1879); Post in Mém. Herb. Boiss., 18: 97 (1900).
　　P. ramosa (L.) C. A. Mey. var. *simplex* Vis., Fl. Dalm., 2: 180 (1847).
　　P. mutelii (F. Schultz) Reuter var. *nana* Reuter in DC., Prodr., 11: 9 (1847).
　　P. oxyloba Reuter in DC., Prodr., 11: 9 (1847); Boiss., Fl. Orient., 4: 497 (1879).
　　P. ramosa (L.) C. A. Mey. var. *nana* (Reuter) Boiss., Fl. Orient., 4: 499 (1879).
　　Orobanche nana (Reuter) Noë ex Beck-Mannagetta in Bibl. Bot., 4 (19): 91 (1890).
　　O. oxyloba (Reuter) Beck-Mannagetta in L. Koch, Entwickl.-Gesch. Orob., 209 (1887), Bibl. Bot., 4 (19): 108 (1890), et in Engl., Pflanzenr., 96 (IV. 261): 88 (1930); Holmboe, Veg. Cypr., 168 (1914); Osorio-Tafall et Seraphim, List Vasc. Plants Cyprus, 96 (1973).
　　O. ramosa L. ssp. *nana* (Reuter) Coutinho, Fl. de Port., ed. 1, 566 (1913); Osorio-Tafall et Seraphim, List Vasc. Plants Cyprus, 95 (1973).

Stem branched or sometimes simple, 6–24 cm. high, brownish, thinly glandular-pilose with crispate, multicellular hairs; cauline scales usually sparse and remote; inflorescences rather lax or sometimes dense and crowded; bracts lanceolate-acuminate, 5–10 mm. long, 3–5 mm. wide, pilose, pale or dark brownish; bracteoles linear-subulate, 5–8 mm. long, 0·5–1·5 mm. wide; calyx campanulate, thinly pubescent, tube 4–5 mm. long,

5–6 mm. wide, lobes 4, linear-filiform, 4–5 mm. long, erect; corolla lavender-blue or purplish with white gibbous blotches on the palate, thinly glandular-pubescent externally, tube 12–20 mm. long, 4–5 mm. wide at apex, curved, pallid and ventricose at the base, distinctly constricted above the ovary, narrowly infundibuliform towards apex, lobes porrect, broadly ovate, 2–3 mm. long, 3–3·5 mm. wide at base, margins erosulate; stamens inserted about 3 mm. above base of corolla-tube; filaments glabrous, 8–9 mm. long; anthers pallid, glabrous or thinly pilose towards the rounded base, apex acuminate-apiculate; ovary glabrous, bluntly ovoid, about 3 mm. long, 2·5 mm. wide; style glabrous, about 10 mm. long, deflexed at apex; stigma pallid, capitate-reniform, shallowly 2-lobed. Capsule about 6 mm. long, 3·5 mm. wide; seeds numerous, subglobose, about 0·2 mm. diam.; testa dark brown, conspicuously reticulate.

var. **ramosa**
TYPE: "*in* Europae *siccis*".

Stems usually with 3–9 or more erect branches; cauline scales and bracts often dark brown; inflorescence usually rather lax and elongate; flowers frequently rather small, with a narrow corolla-tube 12–15 mm. long.

HAB.: On *Vicia, Melilotus, Trifolium* and other *Leguminosae*; also commonly on *Oxalis pescaprae*; in cultivated fields and waste ground; sea-level to 1,000 ft. alt.; fl. Febr.–April.

DISTR.: Divisions 1, 3, 4, 6–8. Southern Europe, western Asia and Africa; naturalized in many parts of the world.
1. Neokhorio (Akamas), 1956, *Merton* 678! Ayios Yeoryios (Akamas), 1962, *Meikle* 2107!
3. Episkopi, 1964, *J. B. Suart* 203!
4. Phaneromene, Larnaca, 1905, *Holmboe* 260; Larnaca Salt Lake Plantation, 1955, *Merton* 1996! Kouklia, 1967, *Merton* in ARI 453! Larnaca Salt Lake, 1977, *J. J. Wood* 7!
6. Dhiorios, 1952, *F. M. Probyn* 97! Near Orga, 1956, *Merton* 2584!
7. Common near Kyrenia, *J. A. Tracey* 60! Syngrassides 1024! *Casey* 1233! 1718! *Merton* 2096! *Merton* in ARI 575! *G. E. Atherton* 992! etc. Karmi, 1941, *Davis* 2766! Bellapais, 1963, *J. B. Suart* 95! and, 1970, *I. M. Hecker* 57! Lambousa Beach, 1973, *P. Laukkonen* 278!
8. Near Rizokarpaso, 1880, *Sintenis & Rigo* 38! Ronnas Bay, 1941, *Davis* 2351!

var. **brevispicata** (*Ledeb.*) *R. A. Graham* in Kew Bull., 1955: 467 (1955).
   O. mutelii F. Schultz in Mutel, Fl. Franç., 2: 353, t. 43, fig. 314 (1835); Beck-Mannagetta in Bibl. Bot., 4 (19): 95 (1890) et in Engl., Pflanzenr., 96 (IV. 261): 75 (1930); Holmboe, Veg. Cypr., 168 (1914); Rechinger f. in Arkiv för Bot., ser. 2, 1: 431 (1950); Feinbrun, Fl. Palaest., 3: 211, t. 356 (1978); Davis, Fl. Turkey, 7: 7 (1982).
   Phelipaea mutelii (F. Schultz) Reuter in DC., Prodr., 11: 8 (1847).
   P. ramosa (L.) C. A. Mey. var. *brevispicata* Ledeb., Fl. Ross., 3: 313 (1849).
   Orobanche ramosa L. var. *minor* Loret et Barr., Fl. Montpell., 495 (1876).
   Phelipaea ramosa (L.) C. A. Mey. var. *mutelii* (F. Schultz) Boiss., Fl. Orient., 4: 499 (1879); H. Stuart Thompson in Journ Bot., 44: 336 (1906).
   Orobanche ramosa L. ssp. *mutelii* (F. Schultz) Coutinho, Fl. de Port., ed. 1, 566 (1913); Osorio-Tafall et Seraphim, List Vasc. Plants Cyprus, 95 (1973).
   [*O. schultzii* (? Mutel) Beck-Mannagetta in Bibl. Bot., 4 (19): 113 (1890) et in Engl., Pflanzenr., 96 (IV. 261): 92 (1930) pro parte quoad plant. cypr.]
   TYPE: Algeria; "à Bone sur des coteaux escarpés au bord de la mer", *Mutel* (GRM).

Stems commonly unbranched or sparingly branched, often thick and robust; cauline scales and bracts frequently pale; inflorescence compact, rather congested; flowers usually larger than in var. *ramosa*; corolla-tube often more than 15 mm. long.

HAB.: Dry hillsides and sandy flats by the sea; on *Plantago, Anchusa, Umbelliferae* and *Compositae*; sea-level to 3,500 ft. alt.; fl. March–May.

DISTR.: Divisions 1–3, 6–8. Southern Europe, western Asia, Africa.
1. W. of Baths of Aphrodite, 1979, *Edmondson & McClintock* E 2773!
2. Perapedhi, 1894, *Post* sec. *A. Gilli*; Tripylos, 1941, *Davis* 3498! Platania, 1947, *Mavromoustakis* s.n.! Ayia, 1962, *Meikle* 2653!
3. Between Limassol and Moni, 1859, *Kotschy* 166! 451 (sec. Holmboe); Near Apsiou, 1979, *Edmondson & McClintock* E 2706!

6. Ayia Irini, 1941, *Davis* 2578!
7. Pentadaktylos, 1880, *Sintenis & Rigo* 632! also, 1902, *A. G. & M. E. Lascelles* s.n.! St. Hilarion, 1963, *Townsend* 63/60!
8. Seashore south of Galinoporni, 1962, *Meikle* 2504!

NOTES: The distinctions between *O. ramosa*, *O. nana*, *O. mutelii* and *O. schultzii* are, at least as regards Cyprus material, so unsatisfactory that all records have been referred to *O. ramosa*. It may be that field and experimental study of the group will show that it can be subdivided into meaningful, recognizable taxa. Existing classifications cannot be said to achieve this goal.

2. O. aegyptiaca Pers., Syn. Plant., 2: 181 (1807); Beck-Mannagetta in Bibl. Bot., 4 (19): 100 (1890); Holmboe, Veg. Cypr., 168 (1914); Beck-Mannagetta in Engl., Pflanzenr., 96 (IV. 261): 81 (1930); Post, Fl. Pal., ed. 2, 2: 314 (1933); Lindberg f., Iter Cypr., 31 (1946); Osorio-Tafall et Seraphim, List Vasc. Plants Cyprus, 95 (1973); Feinbrun, Fl. Palaest., 3: 211, t. 357 (1978); Davis, Fl. Turkey, 7: 8 (1982).
  Phelipaea aegyptiaca (Pers.) Walpers, Repert., 3: 463 (1844–45); Unger et Kotschy, Die Insel Cypern, 293 (1865); Boiss., Fl. Orient., 4: 499 (1879); Sintenis in Oesterr. Bot. Zeitschr., 32: 365, 397 (1882).
  [? *O. ramosa* (non L.) Sm. in Sibth. et Sm., Fl. Graec., 7: t. 608 (1830).]
  [*O. longiflora* (non Pers.) Poech, Enum. Plant. Ins. Cypr., 26 (1842).]
  [*O. lavandulacea* (non Reichb.) Post in Mém. Herb. Boiss., 18: 97 (1900); H. Stuart Thompson in Journ. Bot., 44: 336 (1906).]
  TYPE: "in Aegypto (Herb. Decand.)" (G).

Stem branched or sometimes simple, erect, 12–30 (–40) cm. high, thinly glandular-pilose, brownish; cauline scales few, remote, ovate-deltoid, acute, 5–10 mm. long, 3–8 mm. wide, brownish, sparsely pubescent; inflorescence usually lax and elongate, often exceeding the stems; bracts narrowly ovate-acuminate, up to 10 mm. long, 3–4 mm. wide; flowers frequently remote, subsessile or the lowermost sometimes distinctly pedicellate; bracteoles linear-subulate, thinly glandular-pubescent, about 7–8 mm. long, 0·5–0·6 mm. wide; calyx broadly campanulate, thinly glandular-pubescent, tube about 3–4 mm. long, 5 mm. wide at apex, lobes erect, deltoid-acuminate, produced into a slender acumen about 4–5 mm. long; corolla blue or mauve-purple, with white gibbous blotches on the palate, thinly glandular-pubescent externally, tube 20–25 (–30) mm. long, straightish and erecto-patent or ± distinctly downcurved, pallid and ventricose at the base, strongly constricted above the ovary, narrowly infundibuliform above and 7–10 mm. wide at apex, lobes porrect or spreading, ovate, acute, about 4 mm. long, 3–4 mm. wide at base, margins erosulate; stamens inserted about 0·5 mm. above base of corolla-tube; filaments shortly pubescent towards base, glabrous above, about 10 mm. long; anthers pallid, oblong, about 1·5 mm. long, 1·2 mm. wide, bearded towards the rounded base; style glabrous, about 15 mm. long, deflexed at apex; stigma pallid, shallowly 2-lobed, about 2 mm. wide. Capsule ovoid, acute, about 10 mm. long, 6 mm. wide; seeds numerous, minute, ovoid, about 0·4 mm. long, 0·3 mm. wide; testa dark brown, conspicuously reticulate.

HAB.: Cultivated fields and gardens, on *Zea*, *Cucurbita*, *Phaseolus*, *Lycopersicon*, etc., occasionally on *Smyrnium* and other native species; sea-level to 4,500 ft. alt.; fl. throughout the year.

DISTR.: Divisions 2–4, 6–8. S.E. Europe and N. Africa, eastwards to Central Asia.

2. Palekhori, 1905, *Holmboe* 1138; Platres, 1938, *Kennedy* 867! 1419! 1422! Prodhromos, 1949, *Casey* 956! and, 1955, *G. E. Atherton* 596! 666!
3. Limassol, 1840, *Kotschy* 49.
4. Perivolia, 1939, *Lindberg f.* s.n.!
6. Nicosia, 1894, *Post* s.n., also, 1901, *A. G. & M. E. Lascelles* s.n.! and, 1936, *Syngrassides* 868! Dheftera, 1936, *Syngrassides* 1289! Ayia Irini, 1962, *Meikle* 2392!
7. Kyrenia, 1955, *G. E. Atherton* 729!
8. Rizokarpaso, 1880, *Sintenis & Rigo*; Cape Andreas, 1880, *Sintenis & Rigo* s.n.

NOTES: Holmboe (Veg. Cypr., 168) notes that this is a troublesome weed, parasitizing and injuring potatoes and other crops, especially at higher altitudes. His suggestion that the large-flowered *Orobanche*, figured as *O. ramosa* L. in Fl. Graec., t. 608, is more probably *O. aegyptiaca*, seems to be correct, though the provenance of the plant figured is uncertain.

3. O. orientalis *Beck-Mannagetta* in Bibl. Bot., 4 (19): 110, t. 19 (1890) et in Engl., Pflanzenr., 96 (IV. 261): 90 (1930).

TYPE: W. Himalayas; "in regione temperata Himalayae boreali-occidentalis (prope Jamu et Banahal) 5–7000' s.m. leg. Hooker f. et Thomson" (W, K !).

Stems robust, erect, 18–30 (–40) cm. high, thinly glandular-pilose, frequently with numerous erect, almost fastigiate branches, but sometimes unbranched; cauline scales rather sparse and remote, ovate-acuminate, 5–15 (–20) mm. long, 3–5 mm. wide, subglabrous, thick, blackened (as if burnt) in many dried specimens; inflorescence elongate, tapering, spire-like, often rather lax, 13–30 cm. long, usually less than 3 cm. wide, rhachis, bracts and bracteoles densely glandular-pubescent, the hairs conspicuously gland-tipped; bracts narrowly ovate, caudate-acuminate, 15–20 mm. long, 3–5 mm. wide, produced far beyond the buds and forming a comose apex to the inflorescence; flowers subsessile or the lowermost with erect pedicels up to 5 mm. long; calyx campanulate, tube about 4–5 mm. long, 4–5 mm. wide, glandular externally, lobes linear-filiform, 6–7 mm. long, exceeding the tube, slightly recurved towards apex, conspicuously glandular; bracteoles linear-filiform, 7–8 mm. long, sometimes adnate to the calyx-tube and resembling an additional calyx-lobe, but often free except for a short adnate basal part, occasionally with a secondary apical lobe; corolla livid, purple, pink or mauve-blue, erect, shortly glandular-pubescent, tube slightly curved, about 18–19 mm. long, 4 mm. wide at apex, subcylindrical, not very conspicuously constricted above ovary nor widened above; lobes porrect, small, deltoid-acute, about 2–2·5 mm. long, 2·5 mm. wide at base; stamens inserted about 4 mm. above base of corolla-tube; filaments glabrous, about 12 mm. long; anthers pallid, about 1·5 mm. long, 1 mm. wide, thinly lanuginose towards base, apex with narrowly tapering lobes; ovary ovoid-pyriform, about 6 mm. long, 3 mm. wide, glabrous or minutely glandular towards apex; style about 10 mm. long, minutely glandular; stigma pallid, 2-lobed, about 1·5 mm. wide. Capsule blunt, about 7 mm. long, 4 mm. wide; seeds ovoid, about 0·5 mm. long, 0·3 mm. wide; testa greyish-fuscous, coarsely and conspicuously winged-reticulate.

HAB.: Hillsides, vineyards, pathsides, parasitic on *Prunus dulcis*, *Medicago*, *Astragalus lusitanicus*, etc.; 350–3,700 ft. alt.; fl. April–June.

DISTR.: Divisions 2–4, 6, 7. ? E. Turkey, ? Lebanon, ? Iran, Azerbaidjan, Afghanistan, Pakistan.

2. Platres, 1960, *N. Macdonald* 170! Mandria, 1970, *A. Genneou* in ARI 1540! Below Phterykoudhi, 1979, *Edmondson & McClintock* E 2925!
3. Kakomallis, 1962, *Meikle* 2929!
4. Pergamos, 1955, *Merton* 2213!
6. Vizakia, 1938, *Syngrassides* 1014!
7. Akanthou, 1967, *Merton* in ARI 794!

NOTES: Distinguished by its long, tapering inflorescences, caudate-acuminate bracts, linear-filiform, glandular bracteoles and calyx-lobes, and narrow-tubular corollas. It has been much confused with the distinct *O. lavandulacea* Reichb., which has proportionately shorter bracts and crowded inflorescences of larger flowers, and with *O. oxyloba* (Reuter) Beck-Mannagetta, which, to judge from type material (Turkey; Alaya, 1845, *Heldreich*!), is simply a form of *O. ramosa* L., indistinguishable from the plant often called *O. nana* (Reuter) Noë ex Beck-Mannagetta.

Because of confusion with other species it is difficult to ascertain the distribution of *O. orientalis*. It certainly occurs in Baluchistan, Afghanistan and western Pakistan, and is very probably also found in Iran and eastern Turkey (Amanus Range and Lake Van area).

SECT. 2. **Orobanche** Bracteoles absent; calyx split dorsiventrally, the 2 lateral divisions generally 2-lobed; corolla usually brownish, yellowish, whitish or maroon, tube rarely much constricted above ovary; stems mostly simple.

4. O. cypria *Reuter* in Unger et Kotschy, Die Insel Cypern, 294 (1865); Boiss., Fl. Orient., 4: 513 (1879); Beck-Mannagetta in Bibl. Bot., 4 (19): 173 (1890) et in Engl., Pflanzenr., 96 (IV. 261): 252 (1930) quoad plant. cypr.; Holmboe, Veg. Cypr., 168 (1914); Osorio-Tafall et Seraphim, List Vasc. Plants Cyprus, 96 (1973).
TYPE: Cyprus; "Um Prodromo im Mai" [1862], *Kotschy* 854 (W).

Stems simple, erect, reddish-purple, 10–22 cm. high, arising from a swollen scaly base, thinly or rather densely pubescent with viscid-glandular, multicellular hairs; cauline scales numerous, erect, ovate-acuminate, 8–10 mm. long, 4–5 mm. wide, reddish, glandular-pubescent dorsally; inflorescence compact, crowded, 2–8 cm. long at anthesis, 1·5–3 cm. wide; bracts resembling the cauline scales in shape and size, densely glandular-pubescent; flowers subsessile; calyx divisions narrowly oblong, 10–12 mm. long, 3–4 mm. wide, densely glandular-pubescent, apex entire, acuminate, or divided, sometimes to the middle of the division, into 2, unequal, linear-subulate lobes; corolla maroon-red, thinly glandular externally, tube slightly curved, 10–12 mm. long, about 4 mm. wide at apex, not markedly constricted above the ovary, lobes rounded, about 2·5 mm. diam., the 2 adaxial porrect, the 3 abaxial spreading or reflexed, with strongly plicate sinuses, margins conspicuously and finely fimbriate-denticulate; stamens inserted about 3 mm. above base of corolla-tube; filaments about 7 mm. long, thinly pilose near base, glabrous above, or thinly pilose throughout their length; anthers cohering, oblong, pallid, about 1·5 mm. long, 1 mm. wide, apices cuspidate-acuminate; ovary narrowly ellipsoid, about 6 mm. long, 2 mm. wide, glabrous or sparsely glandular; style 7–8 mm. long, glabrous or thinly pilose, decurved at apex; stigma bilobed, purple (or sometimes pallid), 1·5 mm. wide. Capsule compressed-ellipsoid, thick-walled, about 9 mm. long, 3–4 mm. wide; seeds irregularly oblong, about 0·4 mm. long, 0·3 mm. wide; testa fuscous; coarsely reticulate. *Plate 71.*

HAB.: Rocky hillsides on chalk or igneous rocks; in vineyards or *Pinus nigra* forest; parasitic on *Pterocephalus, Cistus, Salvia* and *Scutellaria*; (500–) 1,000–5,500 ft. alt.; fl. May–June.

DISTR.: Divisions 2, 3. Endemic (see Notes).

2. Prodhromos, 1862, *Kotschy* 854, also, 1961, *D. P. Young* 7265! 7373! and, 1968, *Economides* in ARI 1226! Platres, 1936, *Kennedy* 866! Mandria, 1938, *Kennedy* 1421! Khrysorroyiat-issa, 1941, *Davis* 3416! Ayios Nikolaos, 1958–60, *N. Macdonald* 164! Statos, 1979, *Hewer* 4616!

3. Kalavasos, 1905, *Holmboe* 611; Kakomallis, 1962, *Meikle* 2885!

NOTES: Uniform and distinct, with small maroon-red flowers. Records from Iraq and Iran are very doubtful; certainly an isotype of *Orobanche cypria* Reuter var. *pterocephali* Beck-Mannagetta from Kuh-Sefin (*Bornmüller* 1646) has no claim to be considered conspecific with the Cyprus plant. It closely resembles *O. ovata* Blakelock from the same area.

5. O. alba *Stephan ex Willd.*, Sp. Plant., 3: 350 (1800); Boiss., Fl. Orient., 4: 507 (1879); Beck-Mannagetta in Bibl. Bot., 4 (19): 208 (1890) et in Engl., Pflanzenr., 96 (IV. 261): 145 (1930); Holmboe, Veg. Cypr., 169 (1914); Osorio-Tafall et Seraphim, List Vasc. Plants Cyprus, 96 (1973); Davis, Fl. Turkey, 7: 14 (1982).
TYPE: "*in* Sibiria *versus* Mare Caspium" (LE).

Stems simple, erect, 8–20 cm. high, purplish, densely glandular-pubescent; cauline scales often numerous and approximate, ovate-acuminate, 10–15 mm. long, 5–6 mm. wide, glandular-pubescent; inflorescence often short, blunt and compact, 3–15 cm. long, 2·5–4 cm. wide; bracts ovate, long-acuminate, 12–15 mm. long, 4–6 mm. wide, densely glandular-pubescent; flowers sessile, usually rather crowded, occasionally remote; calyx-divisions narrowly ovate, long-acuminate, about 10 mm. long, 4 mm. wide, densely glandular-pubescent, apex (in all the Cyprus

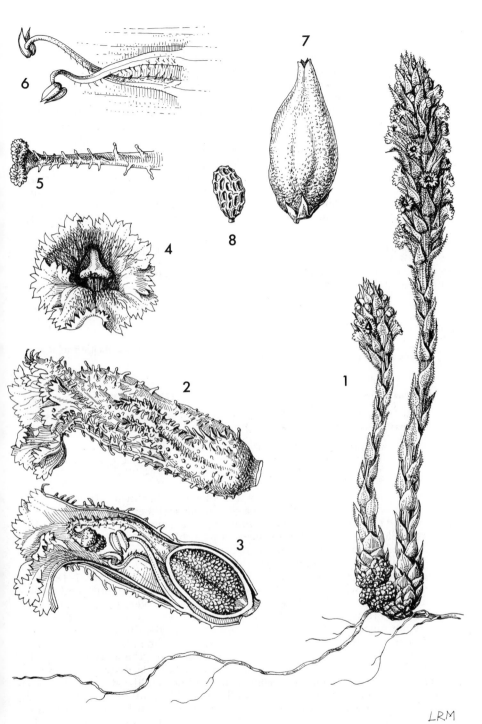

Plate 71. OROBANCHE CYPRIA Reuter **1,** habit, ×1; **2,** flower, side view, ×6; **3,** flower, longitudinal section, ×6; **4,** flower, front view, ×6; **5,** stigma, ×12; **6,** stamens, ×6; **7,** capsule, ×9; **8,** seed, ×84. (**1–6** from *Davis* 3416 & *Meikle* 2885; **7, 8** from *Meikle* 2885.)

specimens) entire, without lobes; corolla-tube 10–20 mm. long, 7–8 mm. wide at mouth, gently curved dorsally, not constricted above ovary nor much expanded towards apex, densely glandular-pubescent, reddish, purplish, yellowish or whitish, with 2 conspicuous orange-yellow folds on the lower lip; corolla-lobes spreading, undulate-crispate, the adaxial rounded or apiculate, about 10 mm. wide, the 3 abaxial rounded, subequal, about 4 mm. wide; stamens inserted about 2 mm. above base of corolla-tube; filaments stout, angled, about 12 mm. long, shortly pilose near base, subglabrous or thinly glandular above; anthers pale, about 1·5 mm. long, 1 mm. wide, shortly caudate; ovary narrowly oblong, about 7 mm. long, 3 mm. wide, glandular towards apex; style stout, about 10 mm. long, sparsely glandular; stigma usually pallid, bilobed, about 2·5 mm. wide. Capsule about 8 mm. long, 4 mm. wide, walls rather thick; seeds irregularly oblong or subglobose; very numerous, about 0·4 mm. long and almost as wide; testa blackish, coarsely reticulate.

HAB.: In garigue on rocky hillsides; in Pine forest; in pastures; parasitic on *Thymus*; sea-level to 4,000 ft. alt.; fl. March–May.

DISTR.: Divisions 1–4, 6, 7. Widespread in Europe and Asia, east to the Himalayas.

1. Fontana Amorosa, 1962, *Meikle* 2303 !
2. Platres, 1938, *Kennedy* 1420 !
3. Mazotos, 1905, *Holmboe* 199; Zagala, 1941, *Davis* 3132 ! Kato Lefkara, 1955, *Merton* 2056 ! Kakomallis, 1962, *Meikle* 2903 ! Pissouri, 1967, *Merton* in ARI 673 !
4. Between Paralimni and Ayia Napa, 1979, *Chesterman* 114 !
6. Dhiorios, 1952, *F. M. Probyn* 98 !
7. Near St. Hilarion, 1933, *Syngrassides* 50 ! Yaïla, 1941, *Davis* 2899 ! Kambyli, 1956, *Merton* 2570 ! Kyrenia, 1956, *G. E. Atherton* 1326 !

6. O. crenata *Forssk.*, Fl. Aegypt.-Arab., LXVIII et 113 (1775); Beck-Mannagetta in Bibl. Bot., 4 (19): 225 (1890) et in Engl., Pflanzenr., 96 (IV. 261): 136 (1930); Holmboe, Veg. Cypr., 168 (1914); Post, Fl. Pal., ed. 2, 2: 315 (1933); Osorio-Tafall et Seraphim, List Vasc. Plants Cyprus, 96 (1973); Feinbrun, Fl. Palaest., 3: 213, t. 362 (1978); Davis, Fl. Turkey, 7: 14 (1982).

O. speciosa DC. in Lam. et DC., Fl. Franç., 6: 393 (1815); Boiss., Fl. Orient., 4: 506 (1879).
O. pruinosa Lapeyr., Hist. Abrég. Plant. Pyr. Suppl., 87 (1818); Unger et Kotschy, Die Insel Cypern, 294 (1865); Sintenis in Oesterr. Bot. Zeitschr., 32: 121 (1882).

TYPE: Egypt; "*Káhirae*" (C).

Stem simple, erect, robust, 15–50 (–100) cm. high, sulcate, thinly pilose, arising from a swollen, bulb-like base; cauline scales crowded at the base, sparse and remote above, ovate-acuminate, 1·5–3 cm. long, 7–10 mm. wide, shortly pilose dorsally; inflorescence compact, blunt, 2·5–30 cm. long, 2·5–7 cm. wide, rhachis thick, shortly pilose; bracts narrowly ovate-caudate, 10–20 mm. long, 3–5 mm. wide, pilose dorsally, apex usually recurved; flowers crowded, sessile, showy, fragrant; calyx-divisions oblong, 7–18 mm. long, 3–6 mm. wide, deeply 2-lobed, the lobes unequal, subulate-filiform, 3–12 mm. long; corolla white or whitish with violet veining, tube slightly curved, 10–15 mm. long, 5–10 mm. wide at apex, not constricted above ovary nor much expanded towards apex, sparsely glandular and with a few flattened, arachnoid hairs externally, lobes spreading, elegantly crispate-denticulate, the adaxial emarginate, 7–12 mm. wide; the abaxial unequal, with the median up to 10 mm. wide, the 2 laterals distinctly smaller, separated from the median by conspicuous gibbous folds; stamens inserted about 1·5–2 mm. from base of corolla-tube; filaments shortly pilose or subglabrous towards base, usually ± glandular towards apex, 9–12 mm. long; anthers livid, oblong, 1·5–2·5 mm. long, 1–2 mm. wide, shortly cuspidate-caudate; ovary narrowly ellipsoid, 7–8 mm. long, 3–4 mm. wide, subglabrous or thinly glandular; style 7–10 mm. long, thinly glandular, decurved towards apex; stigma orange, livid or violet, 2-lobed, 2–3 mm.

wide; seeds very numerous, minute, irregularly oblong, 0·3–0·4 mm. long, 0·3 mm. wide; testa blackish, coarsely reticulate.

HAB.: Cultivated fields and gardens, occasionally on open ground away from cultivation; on *Vicia faba*, *Pisum sativum* and other *Leguminosae*, occasionally on other garden plants (*Antirrhinum*, *Tropaeolum*, etc.); sea-level to 500 ft. alt.; fl. March–May.

DISTR.: Divisions 1, 4–7. Widespread in the Mediterranean region and eastwards to Turkey and the Caucasus; Atlantic Islands.

1. Ktima, 1955, *Merton* 2162!
4. Kiti, 1862, *Kotschy*.
5. Kythrea, 1880, *Sintenis & Rigo*, also, 1932, *Syngrassides* 226! Lefkonico, 1950, *Chapman* 216!
6. Nicosia, 1934, *Syngrassides* 491! Dhiorios, 1952, *F. M. Probyn* 99! 101! Near Orga, 1956, *Merton* 2585! Morphou, 1972, *W. R. Price* 1087! 1088!
7. Kyrenia, 1956, *Casey* 1719! and, 1956, *G. E. Atherton* 1357!

NOTES: Two specimens, both from Div. 4, Cape Greco (1958, *N. Macdonald* 5! and, 1960, *Mrs N. Macdonald* 105!) may possibly be referable to *O. crenata*, but are abnormally small, 12–14 cm. high, with short, dense spikes of whitish or yellowish flowers veined pink. They were apparently parasitic on *Compositae*, which suggests *O. loricata* Reichb. (incl. *O. picridis* F. W. Schultz), but differ from the latter in stature, and in the proportionately short, very slightly curved corolla-tube, with widely flaring lobes. Further material is required for more exact determination.

O. crenata is locally a serious pest of leguminous crops.

7. O. minor *Sm.*, English Bot., 6: t. 422 (1797); Sutton in Trans. Linn. Soc., 4: 179 (1798); Boiss., Fl. Orient., 4: 512 (1879); Beck-Mannagetta in Bibl. Bot., 4 (19): 251 (1890) et in Engl., Pflanzenr., 96 (IV. 261): 205 (1930); Post, Fl. Pal., ed. 2, 2: 316 (1933); Davis, Fl. Turkey, 7: 17 (1982).

Stems simple, erect, rather slender, 8–30 (–50) cm. high, thinly or rather densely whitish-lanuginose, generally dull brownish-purple, sometimes yellowish; cauline scales linear-lanceolate, 2–2·5 cm. long, 0·3–0·6 cm. wide, acuminate, usually rather remote, sometimes lanuginose dorsally; inflorescence lax or dense, 2–18 cm. long, 2–3 cm. wide; bracts lanceolate-acuminate or shortly caudate, 1·5–2 cm. long, 0·3–0·6 cm. wide, glandular-pilose or thinly lanuginose dorsally; flowers sessile, dull brownish-purple, brownish-mauve or occasionally yellowish; corolla-tube distinctly curved throughout its length, about 8–10 mm. long, 5 mm. wide, not constricted above ovary nor expanded towards apex, subglabrous or thinly glandular-pubescent, sometimes clothed dorsally with matted, whitish hairs, adaxial lobe hemispherical, 5–7 mm. diam., porrect, denticulate-erosulate, rounded, apiculate, emarginate or sometimes 2-lobulate, abaxial lobes much smaller, spreading, the median about 3 mm. wide, the laterals smaller; stamens inserted 2–3 mm. above base of corolla-tube; filaments 7–8 mm. long, pilose at the base, glabrous or more thinly pilose above; anthers oblong, about 1 mm. long, 0·8 mm. wide, shortly cuspidate, livid or yellowish; ovary oblong-ellipsoid, 5 mm. long, 3 mm. wide, glabrous or sparsely glandular-pubescent; styles 9–10 mm. long, downcurved at apex, glabrous or sparsely glandular; stigma livid, purplish or yellowish, bilobed, about 2 mm. wide. Capsule 8–9 mm. long, 4 mm. wide; seeds minute, very numerous, about 0·3 mm. long, 0·2 mm. wide; testa blackish, coarsely but bluntly reticulate.

var. **minor**
TYPE: England; Norfolk, "near Sheringham", *Sutton* (Eng. Bot. t. 422!)

Dorsal surface of corolla-tube and adaxial lobe shortly glandular-pubescent with patent hairs, or subglabrous.

HAB.: On sandy seashores and stabilised sand-dunes, or in garigue on rocky ground inland; sea-level to 150 ft. alt.; parasitic on *Leguminosae*, *Echium angustifolium* Mill., etc.; fl. March–May.

DISTR.: Divisions 1, 3, 4, 6, 8. West, Central and southern Europe, Mediterranean region, Atlantic Islands; widespread in tropical Africa, where it is possibly a recent introduction. Introduced into North America and elsewhere.

1. Between Baths of Aphrodite and Fontana Amorosa, 1977, *J. J. Wood* 79!
3. Pyrgos, 1963, *J. B. Suart* 43!
4. Beach near Dhekelia, 1956, *Merton* 2663!
6. Ayia Irini, 1936, *Syngrassides* 1026!, also, 1941, *Davis* 2600! and, 1962, *Meikle* 2380! Near Syrianokhori, 1956, *Merton* 2628!
8. Aphendrika ("Epiotissa"), 1880, *Sintenis & Rigo* 37 (partly)! Shiramilia near Platanisso, 1962, *Meikle* 2510!

var. **pubescens** (*Urv.*) *Meikle* **stat. nov.**

    O. pubescens Urv. in Mém. Soc. Linn. Par., 1: 332 (1822); Boiss., Fl. Orient., 4: 507 (1879); Osorio-Tafall et Seraphim, List Vasc. Plants Cyprus, 97 (1973); Feinbrun, Fl. Palaest., 3: 214 (1978); Davis, Fl. Turkey, 7: 15 (1982).

    O. versicolor F. W. Schultz in Flora, 1: 129 (1843); Beck-Mannagetta in Bibl. Bot., 4 (19): 237 (1890), et in Engl., Pflanzenr., 96 (IV. 261): 183 (1930); Holmboe, Veg. Cypr., 168 (1914); Post, Fl. Pal., ed. 2, 2: 316 (1933); Rechinger f. in Arkiv för Bot., ser. 2, 1: 431 (1950).

TYPE: Milos; "in collibus siccis insulae Meli", *Dumont d'Urville* (P).

Dorsal surface of corolla-tube and adaxial lobe clothed with flattened, somewhat matted, crispate, whitish hairs.

HAB.: In garigue on hillsides; in pastures or on cultivated land; parasitic on a variety of herbs in *Compositae, Umbelliferae, Leguminosae*, etc.; sea-level to 2,000 ft. alt.; fl. March–May.

DISTR.: Divisions 1, 3, 4, 6–8. S.E. Europe and eastern Mediterranean region, perhaps a recent introduction west of Sicily.

1. Near Fontana Amorosa (with *O. alba* Stephan ex Willd.), 1962, *Meikle* 2303a!
3. Khalassa, 1960, *N. Macdonald* 92! Five miles W. of Malounda, 1963, *J. B. Suart* 56!
4. Kouklia, 1967, *Merton* in ARI 454!
6. Near Ayia Irini, 1962, *Meikle* 2393!
7. Pentadaktylos, 1902, *A. G. & M. E. Lascelles* s.n.! Kyrenia, 1931, *J. A. Tracey* 61! Lapithos, 1955, *Merton* 2233! Above Antiphonitis Monastery, 1955, *Merton* 2282! Trypa Vouno, 1956, *G. E. Atherton* 1385! Akanthou, 1967, *Merton* in ARI 795! Buffavento, 1974, *Meikle* 4098!
8. Aphendrika ("Epiotissa"), 1880, *Sintenis & Rigo* 37 (partly)!

NOTES: Apart from the differences in corolla-indumentum, and perhaps a tendency for var. *pubescens* to have rather larger flowers than var. *minor*, I can find no satisfactory distinctions between the two taxa. Typical *O. minor* L. said (by Beck-Mannagetta) to have the adaxial lobe of the corolla "bilobum vel plicato-emarginatum", while that of var. *pubescens* is said to be "integrum vel plicato-emarginatum rarius bilobum". But Smith (Eng. Bot., t. 422) says of *O. minor* "Upper lip of the corolla undivided, at least till it splits by age or violence". Differences in the insertion of the stamens (2–3 mm. above base of corolla in *minor*, 3–4 mm. in *pubescens*) do not furnish a useful distinction. Corolla-indumentum generally separates the two, though, in Cyprus at least, some transition is found, even in this character.

69. LENTIBULARIACEAE
by P. G. Taylor

    Annual or perennial herbs with specialized organs for the capture and digestion of small organisms; roots frequently absent; leaves rosulate or alternate, or verticillate on stolons, entire or divided, often dimorphic or polymorphic; flowers mostly small, solitary or in racemes, hermaphrodite, zygomorphic; calyx deeply 2–4- or 5-partite; sepals persistent and often accrescent; corolla gamopetalous, 2-lipped, usually spurred, rarely saccate, usually yellow or violet; tube very short; upper lip interior in bud, entire or 2- or rarely 4-lobed; lower lip entire or 2–5-lobed; stamens 2, anticous, inserted at the base of the corolla; filaments usually short; anthers 2-thecous, the thecae sometimes confluent, dehiscing longitudinally; ovary superior, 1-locular; carpels 2, median; placentation free-basal; ovules usually many, rarely fewer or 2, sessile, anatropous; style simple, usually short; stigma unequally 2-lipped, the upper lip smaller or sometimes obsolete. Fruit a 1-celled, 1–many-seeded capsule, dehiscing by longitudinal valves or slits, or by pores, or circumscissile, or rarely indehiscent; seeds small or very small, variously shaped; testa usually reticulate, thin or spongy or corky, rarely mucilaginous; endosperm absent; embryo undifferentiated.

Four genera distributed throughout the world, but most abundant in the tropics. *Pinguicula* alone is represented in Cyprus by a single species.

1. PINGUICULA *L.*

Sp. Plant., ed. 1, 17 (1753)—as *Pingvicula*.
Gen. Plant., ed. 5, 11 (1754).
S. J. Casper in Bibliotheca Bot., 127/128: 209 pp., 16 plates (1966).

Annual or perennial, somewhat fleshy herbs; roots numerous, filiform; leaves radical, entire, sessile or petiolate, fleshy, covered on the upper surface with sessile and stipitate glands which trap and digest small insects; flowers scapose; calyx 5-lobed, 2-labiate; corolla spurred, mostly in shades of violet or yellow; upper lip 2-lobed; lower lip 3-lobed; anther-thecae confluent; style very short and thick; ovules numerous. Fruit laterally 2-valved; seeds very small, conspicuously reticulate.

About 50 species in northern temperate regions, the Mediterranean and in North, Central and South America.

1. **P. crystallina** *Sm.* in Sibth. et Sm., Fl. Graec. Prodr., 1: 11 (1806), Fl. Graec., 1: 8, t. 11 (1806); Poech, Enum. Plant. Ins. Cypr., 27 (1842); Unger et Kotschy, Die Insel Cypern, 294 (1865); Boiss., Fl. Orient., 4: 2 (1875); Holmboe, Veg. Cypr., 167 (1914); Casper in Bibl. Bot., 127/128: 108, t. 32, fig. 6 (1966); Osorio-Tafall et Seraphim, List Vasc. Plants Cyprus, 96 (1973); Davis, Fl. Turkey, 6: 108 (1978).

TYPE: Cyprus; "In rivulis prope vicum Comandriae [? Kato Amiandos] in insulâ Cypro. *D. F. Bauer*" (OXF).

Rosette-forming, often gregarious, perennial herb, 4–15 cm. high, with a short erect or ascending rootstock and numerous filiform simple roots; leaves 6–10 or sometimes more, oblong-elliptic, obovate or ovate, 2–5 cm. long, 1–2·5 cm. wide, apex rounded or emarginate, base tapering gradually to a short rather indistinct flattened petiole, lateral margins slightly involute, upper surface densely beset with minute sessile and stipitate, viscid glands; flowers solitary, scapose; pedicels 2–6 or more, filiform, erect, lengthening somewhat in fruit, densely stipitate-glandular; calyx bilabiate, about 3 mm. long, externally densely stipitate-glandular; upper lip 3-lobed, the lobes equal, oblong or oblong-obovate, apex rounded, truncate or retuse; lower lip obovate, apex emarginate; corolla yellowish-white with black lines at the base of the upper lip, lilac lobes and yellow throat and spur, 12–22 mm. long including spur, externally stipitate-glandular; upper lip 2-lobed, lobes obovate, apex rounded; lower lip 3-lobed, the midlobe wider than the side lobes, lobes obovate, apex rounded, truncate or emarginate; tube cylindrico-conical; palate densely pilose, the hairs multicellular, irregularly capitate; spur subulate, apex acute or obtuse, about ⅓ of the total length of the corolla; stamens about 2 mm. long, filaments linear, curved; anthers globose, less than 1 mm. long, thecae confluent; ovary globose; stigma almost sessile, lower lip flabellate, about 2 mm. wide, the apical margin fimbriate, upper lip very much smaller, narrowly deltoid. Capsule globose, about 3 mm. diam., laterally 2-valved; seeds numerous, fusiform, about 1 mm. long; testa thin, brown, with conspicuous regular rows of very small isodiametric cells. *Plate 72, figs. 1–5.*

HAB.: Wet rocks and by streamsides; 2,500–6,400 ft. alt.; fl. March–Sept.

DISTR.: Division 2. Also S.W. Turkey.

2. "Comandria" [possibly Kato Amiandos], 1787, *Ferd. Bauer*; Evrykhou, 1787, *Sibthorp* teste *Kotschy*; Prodhromos and S.W. slopes of Khionistra, 1862, *Kotschy* 765! also, 1930, *Druce* s.n.! and, 1955, *G. E. Atherton* 596! Panayia, near Prodhromos, 1880, *Sintenis & Rigo* 741! Head of Ayios valley, 1900, *A. G. & M. E. Lascelles* s.n.! Between Troödos and Kyperounda, 1905, *Holmboe* 857; Troödos, 1912, *Haradjian* 399! Kryos Potamos, 3,700–4,000 ft. alt., 1937, *Kennedy* 864! also, 1941, *Davis* 3155! and, 1960, *N. Macdonald* 34! Kalogeros, S.W. of Khionistra, 1937, *Kennedy* 865! Khrysovrysi near Amiandos, 1940, *Davis* 1888! Between Platania and Kakopetria, 1941, *Davis* 3206! Near Trikoukkia, 1949, *Casey* 831! Amiandos, 1962, *Meikle* 2775!

NOTES: The plate in Flora Graeca, 1: t. 11 (1806) shows a plant with the median lobe of the lower corolla-lip deeply bifid, and is uncharacteristic of Cyprus material of this species. Sibthorp (in Walpole, Travels, 23: 1820) notes, somewhat testily: "My draughtsman [Bauer] stopping to sketch these plants [*Solenopsis minuta* and *Pinguicula crystallina*] was the cause of my losing my companion [John Hawkins], who slept at a neighbouring monastery." Bauer, normally the most accurate of draughtsmen, may have had insufficient time to complete his sketch from fresh material.

69a. BIGNONIACEAE
Syngrassides in Cyprus Agric. Journ., 34: 71 (1939).

Trees, shrubs and woody climbers, rarely herbs. Leaves opposite, rarely alternate or basal, generally pinnately compound, the terminal leaflet sometimes transformed into a branched tendril. Flowers showy, zygomor-phic; calyx campanulate, truncate or lobed; corolla tubular, campanulate or infundibuliform, lobes 5, often unequal (and corolla distinctly 2-lipped), imbricate; stamens 4 (2 fertile, 2 sterile), didynamous, the fifth reduced to a staminode or absent; anther-thecae usually one above the other, dehiscing longitudinally. Disk present. Ovary superior, 2 (or 1)-locular, with axile (or parietal) placentation, the placentas often paired; style simple, terminal; stigma 2-lobed; ovules numerous. Fruit normally a capsule, rarely indehiscent; seeds frequently winged; endosperm absent; embryo straight.

A large and relatively homogeneous family, including more than 100 genera, widely distributed in tropical and subtropical regions, but not represented by any indigenous species in Cyprus. One member of the family, *Jacaranda mimosifolia* D. Don is, however, frequently planted in gardens and promenades throughout the island. It is a tall tree, with elegant, dissected foliage, and panicles of mauve-blue flowers in late spring and early summer. There are also records for *Catalpa bignonioides* Walt., *Chilopsis linearis* (Cav.) Sweet and *Podranea ricasoliana* (Tanf.) Sprague (Bovill, Rep. Plant. Work Cyprus, 14–15; 1915), and other Bignoniaceae have been seen by the author in Cyprus gardens. The following key, while probably incomplete, may assist in identifying the cultivated species:

Leaves simple:
    Leaves broadly ovate or cordate; petioles long -   - CATALPA BIGNONIOIDES *Walt.* (U.S.A.)
    Leaves narrowly lanceolate or linear; petioles short
                  CHILOPSIS LINEARIS (*Cav.*) *Sweet* (South U.S.A., Mexico)
Leaves compound:
    Leaves bipinnate; leaflets numerous, small; corolla mauve-blue. Tall trees
                  JACARANDA MIMOSIFOLIA *D. Don* (S. Amer.)

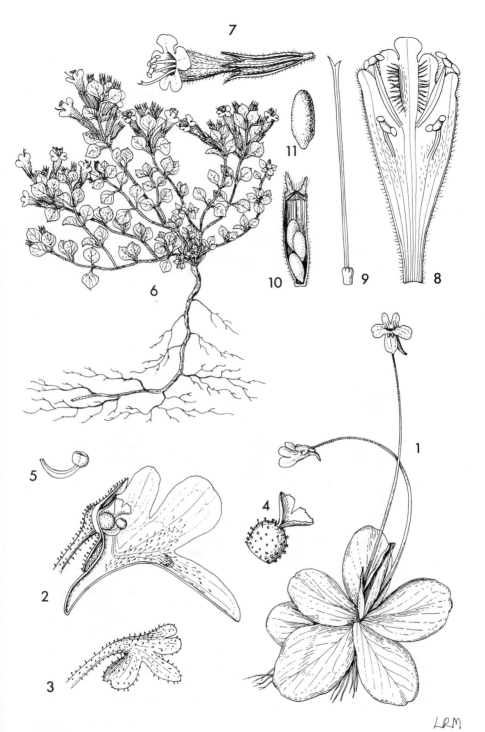

Plate 72. Figs. **1–5.** PINGUICULA CRYSTALLINA Sm. **1,** habit, ×⅔; **2,** flower, longitudinal section, ×4; **3,** calyx, ×4; **4,** gynoecium, ×8; **5,** stamen, ×8; figs. **6–11.** ACINOS TROODI (Post) Leblebici **6,** habit, ×⅔; **7,** flower, ×2; **8,** flower opened out, ×4; **9,** gynoecium, ×4; **10,** fruit opened to show seeds, ×4; **11,** seed, ×8. (**1–5** from *Kennedy* 864; **6–9** from *Kennedy* 745; **10, 11** from *Kennedy* 744.)

Leaves simply pinnate or 1–3-foliolate; shrubs and climbers:
  Leaflets entire:
    Leaves (or some of them) with tendrils:
      Corolla red, narrowly tubular; tendrils irregularly branched
                            PYROSTEGIA IGNEA (*Vell.*) *C. Presl* (S. Amer.)
      Corolla yellow, infundibuliform; tendrils regularly 3-cleft, claw-like
                            DOXANTHA UNGUIS-CATI (*L.*) *Miers* (Central & S. Amer. & W. Indies)
    Leaves without tendrils; corolla white or pink, with a purple throat
                            PANDOREA JASMINOIDES (*Lindl.*) *K. Schum.* (Australia)
  Leaflets toothed:
    Corolla flesh-pink; calyx inflated   PODRANEA RICASOLIANA (*Tanf.*) *Sprague* (S. Africa)
    Corolla yellow, orange or red; calyx not inflated:
      Corolla yellow   -   - TECOMA STANS (*L.*) *Kunth* (Central & S. Amer. & W. Indies)
      Corolla orange or red:
        Stamens exserted from corolla
                            TECOMARIA CAPENSIS (*Thunb.*) *Spach* (E. & S. Africa)
        Stamens included in corolla:
          Leaflets hairy beneath; corolla-limb 3–4·5 cm. wide, tube 6–7·5 cm. long
                            CAMPSIS RADICANS (*L.*) *Seem.* (U.S.A.)
          Leaflets glabrous beneath; corolla-limb 6–8·5 cm. wide; tube 3–5 cm. long
                            CAMPSIS GRANDIFLORA (*Thunb.*) *K. Schum.* (China & Japan)

69b. PEDALIACEAE

Annual or perennial herbs, or sometimes shrubs; leaves opposite or the upper alternate, simple; stipules wanting; flowers solitary, paired, or in cymes, with conspicuous glands (metamorphosed flowers) at the base of the stalks, hermaphrodite; calyx 5-lobed; corolla sympetalous, zygomorphic, lobes 5, imbricate; stamens 4, didynamous, with a postical staminode, or sometimes 2 stamens fertile and 2 reduced and infertile; anthers connivent in pairs, 2-thecous, opening longitudinally; disk hypogynous; ovary superior, (1–) 2–4-locular, the loculi sometimes subdivided by false septa; style simple, terminal, often long with 2 stigmas; ovules 1–many; placentation axile. Fruit a capsule or a nut, sometimes horned, hooked or spinose; endosperm scanty; embryo straight.

About 12 genera, mostly from the tropics and subtropics of the Old World. None of the species included in the family is indigenous in Cyprus, but Sesame, *Sesamum indicum* L., is grown on a limited scale, its seeds being used as a sweetmeat, or a condiment, or for the extraction of an oil. It is an erect, glabrous or pubescent annual, about 0·5–1 m. high, with a bluntly angular, sulcate stem, ovate or lanceolate, entire or irregularly toothed or lobed leaves, opposite towards the base of the plant, alternate above, with solitary or paired, shortly pedicellate or subsessile flowers in their axils. The corolla is white or tinged with pink, obliquely tubular or somewhat funnel-shaped, 1·5–2 cm. long, 0·8–1·5 cm. wide at the bluntly lobed mouth. The fruit is an oblong-quadrangular, shortly beaked, somewhat compressed, puberulous or pubescent capsule, 1·5–3 cm. long, 0·5–0·8 cm. wide, containing numerous, ovate-elliptic, compressed, pale brown seeds, 2·5–3 mm. long, 1·5 mm. wide.

Though probably of African origin, *Sesamum indicum* has long been cultivated in India, and indeed over much of warm-temperate and tropical Asia, from Turkey to China. It is noted by Kotschy, (Die Insel Cypern, 293; 1865) from Dhali, Lapithos and Soli, and has been collected at Morphou (*Syngrassides* 549 !) and Kyrenia (*G. E. Atherton* 345 !).

69c. ACANTHACEAE

Herbs, shrubs and climbers, rarely trees; leaves opposite, often decussate and usually entire, with cystoliths; stipules wanting; flowers commonly in dichasial cymes, sometimes in spikes or solitary, hermaphrodite, zygomorphic; bracts frequently conspicuous and coloured; calyx 5–4-lobed, or the lobes occasionally obsolete; corolla sympetalous, often distinctly 2-lipped, lobes imbricate or contorted; stamens (5–) 4, didynamous, or 2, inserted on the corolla-tube; staminodes often present; anthers 2-thecous, or 1-thecous through abortion, or with 1 theca much smaller than the other; disk usually present, nectariferous; ovary superior, sessile, 2-locular; ovules numerous or few, often in 2 rows; placentation axile; style simple, stigmas 2; unequal, or unequally 2-lobed. Fruit a 2-locular capsule, often dehiscing elastically from apex to base, the dispersal of the seeds being assisted by hook-like outgrowths of the funicle (*retinacula* or *jaculators*); seeds without, or with scanty endosperm; embryo large.

A very large and important family in tropical and subtropical regions, comprising more than 200 genera. A limited number of species extends into temperate regions, and the genus *Acanthus* L. is fairly well represented in the Mediterranean region and western Asia; it is distinguished chiefly by its compact terminal spikes of conspicuously bracteate flowers. The flowers are 1-lipped, the lip broadly 3-lobed, and the leaves, which are mostly basal, are commonly large and pinnatisect. One species, *Acanthus mollis* L., has been collected in Cyprus by A. G. & M. E. Lascelles! (H. Stuart Thompson in Journ. Bot., 44: 336; 1906; Holmboe, Veg. Cypr., 169; 1914), but without precise locality, and almost certainly from a garden. It is a robust perennial, with large, flat, pinnatisect basal leaves and spikes of 1-lipped white flowers with spinose purplish or greenish bracts. *Acanthus mollis* has a wide distribution in the western and central Mediterranean region, but is generally absent from the eastern Mediterranean except as a cultivated plant. Holmboe suggests that the Lascelles specimens may be referable to *A. spinosus* L. or *A. syriacus* Boiss., but examination proves it to have been correctly identified as *A. mollis*.

One other member of the family, *Justicia adhatoda* L. (*Adhatoda vasica* Nees), a native of India, is to be seen in gardens at Nicosia (*Meikle* 5002 !) and elsewhere on the island. It is a spreading shrub, up to 4·5 m. high, with dark green, opposite, entire, elliptic or elliptic-lanceolate leaves, and opposing pairs of stalked, spicate inflorescences of white, 2-lipped flowers, subtended by entire, ovate bracts.

70. VERBENACEAE

Briquet in Engl. et Prantl, Pflanzenfam., ed. 1, IV, 3a: 132–182 (1895).

Herbs, shrubs and trees; leaves exstipulate, generally opposite, simple or digitately compound, entire, toothed or variously lobed, commonly aromatic when crushed; inflorescences paniculate, cymose or spicate, bracteate; flowers 4–5-merous, usually hermaphrodite, generally somewhat zygomorphic and 2-lipped, occasionally almost actinomorphic; calyx tubular or campanulate, lobed or subentire, sometimes 2-lipped; corolla gamopetalous, generally with a distinct, narrow tube and spreading limb, lobes subequal, imbricate in bud; stamens 4, didynamous, rarely fewer or more, or of equal length; filaments free; anthers free or connivent, dorsifixed, introrse, 2-thecous, the thecae parallel or divergent; disk generally present, sometimes conspicuous; ovary superior, at first usually simple, 2-locular, with axile placentation and 2 ovules in each loculus; style terminal; stigma entire or lobed. Fruit a drupe, or sometimes dividing at maturity into 2 or 4 nutlets, rarely a capsule; seeds without endosperm; testa membranous.

About 70 genera widely distributed in tropical and temperate regions; many members of the family are valued as ornamentals; Teak, *Tectona grandis* L.f., provides a very valuable, durable timber.

Leaves compound, palmate, of 5–7 narrow leaflets - - - - - - **3. Vitex**
Leaves simple, sometimes deeply lobed or dissected, but without leaflets:
  Shrub with blackish, fleshy fruits - - - - - - - LANTANA (p. 1248)
  Herbs with dry fruits:
    Fruit comprising 2 nutlets; inflorescences axillary, bracts closely imbricate; leaves shortly
    toothed - - - - - - - - - - - - - **1. Phyla**
    Fruit comprising 4 nutlets; inflorescences terminal, bracts not closely imbricate; leaves
    irregularly lobed, or dissected - - - - - - - - **2. Verbena**

SUBFAMILY 1. **Verbenoideae** Mostly herbs or shrubs; leaves generally simple, opposite; inflorescences racemose or spicate. Fruit drupaceous or of 2 or 4 free nutlets.

LANTANA CAMARA *L.*, Sp. Plant., ed. 1, 627 (1753), a much-branched, scabridulous or thinly prickly shrub 1–2 m. high with 4-angled, pale brown twigs; leaves opposite, ovate, 2–5 cm. long, 0·8–3 cm. wide, rugulose-scabridulous with shortly recurved crenate-serrate margins; inflorescence a dense, bracteate, corymbose spike, less than 1 cm. long, terminal on a naked, angled peduncle 3–5 (or more) cm. long; calyx small, membranous; corolla yellow, orange, red, pink or white, with a slender tube 7–10 mm. long, and a subrotate, slightly zygomorphic limb, 4–5 mm. diam. Fruit a small, shining, blackish drupe, about 4 mm. diam., containing 2 nutlets.

A native of tropical America, now cultivated and naturalized almost throughout temperate and tropical regions. Specimens have been collected at Nicosia (1938, *Syngrassides* 1818!) and at Kyrenia (1955, *G. E. Atherton* 45!) but *Lantana camara* is common in Cyprus gardens, sometimes planted to make an informal, but not reliably hardy, hedge.

Lantana hybrida Neubert, a variety of *L. camara* (var. *hybrida* (Neub.) Mold.), is recorded as cultivated at Salamis Plantation by Bovill (Rep. Plant. Work Cyprus, 14; 1915).

1. PHYLA *Lour.*
Fl. Cochin., 66 (1790).
E. L. Greene in Pittonia, 4: 45–48 (1899).

Trailing perennials, rooting at the nodes; stems 4-angled; leaves opposite, simple, toothed or subentire, generally tapering basally to an obscure petiole, lamina sparsely or densely clothed with adpressed, medifixed hairs; inflorescence a short, dense, ovoid, bracteate spike, terminal on a conspicuous, naked, axillary peduncle; bracts closely imbricate; flowers sessile; calyx membranous, 2-lobed, strongly compressed; corolla with a short tube and spreading, somewhat zygomorphic, 4-lobed, white or purplish limb; stamens 4, included. Fruit consisting of 2 nutlets enveloped by the persistent calyx.

About 6 species, chiefly American; *P. nodiflora* is found almost throughout the warmer regions of the world.

Corolla-tube about 1·5 mm. long, scarcely exceeding the subtending obovate-, or oblate-cuspidate bract; corolla-limb small, about 1·5 mm. diam.; calyx lobed almost to base abaxially, thinly and shortly pubescent - - - - - - - - **1. P. nodiflora**
Corolla-tube about 2·5 mm. long, generally much exceeding the subtending, ovate-acuminate bract; corolla-limb 2·5–3 mm. diam.; calyx shortly lobed abaxially, densely long-pubescent along lateral margins - - - - - - - - **2. P. filiformis**

1. **P. nodiflora** (*L.*) *Greene* in Pittonia, 4: 46 (1899); Feinbrun, Fl. Palaest., 3: 94, t. 153 (1978); Davis, Fl. Turkey, 7: 32 (1982).
   Verbena nodiflora L., Sp. Plant., ed. 1, 20 (1753).
   Lippia nodiflora (L.) Michx., Fl. Bor.-Amer., 2: 15 (1803); Boiss., Fl. Orient., 4: 532 (1879); Post, Fl. Pal., ed. 2, 2: 231 (1933); Lindberg f., Iter Cypr., 28 (1946) pro parte; Osorio-Tafall et Seraphim, List Vasc. Plants Cyprus, 88 (1973).
   TYPE: "*in* Virginia".

Trailing perennial with a woody rootstock; stems rooting at the nodes, thinly adpressed-pubescent with medifixed hairs, often purplish; leaves oblanceolate or subspathulate, 0·5–4 cm. long, 0·2–1·8 cm. wide, sub-glabrous or thinly pubescent with adpressed, medifixed hairs, nervation obscure, apex sharply 5–9-toothed, base entire, tapering to a short, indistinct petiole; peduncle 1·3–7 cm. long, usually much exceeding the leaves; spike ovoid, 2–12 mm. long, 2–7 mm. wide; bracts broadly obovate- or oblate-cuspidate, 2–3 mm. long, 3–5 mm. wide, often purplish, thinly adpressed-pubescent; calyx about 1·4 mm. long, 1 mm. wide, membranous, sparsely and shortly pubescent, lobed almost to base abaxially, more shortly lobed adaxially, lobes acute or subacute; corolla whitish, lilac or purplish, tube about 1·5 mm. long, 0·5 mm. wide, scarcely exceeding bracts, limb spreading, lobes rounded, puberulous externally, the abaxial about 0·8 mm. diam., the 3 adaxial about 0·6 mm. diam., the median sometimes emarginate or toothed; stamens inserted at 2 levels near apex of corolla-tube; filaments very short; anthers suborbicular, about 0·3 mm. diam.; ovary ovoid, glabrous, about 1 mm. long, 0·5 mm. wide; style about 0·4 mm. long; stigma capitate. Nutlets pale or dark brown, about 1·2 mm. long, 1 mm. wide, smooth, strongly convex dorsally with flat commissural faces.

HAB.: Moist ground and marshes; near sea-level; fl. April–August.

DISTR.: Divisions 3, 6. Throughout the warmer regions of the world.

3. Near Limassol Salt Lake, 1939, *Lindberg f.* s.n.; also *Mavromoustakis* 62! 63! 68! and, 1941, *Davis* 3575! Asomatos, 1962, *Meikle* 2927!

6. Syrianokhori, 1935, *Syngrassides* 637! and, 1952, *Merton* 797!

2. **P. filiformis** (*Schrader*) *Meikle* **comb. nov.**
   Lippia filiformis Schrader, Ind. Sem. Hort. Gotting., (1834) et in Ann. Sci. Nat., ser. 2, 6: 99 (1836); Walpers, Repert., 4: 50 (1845).
   Zappania nodiflora (L.) Lam. var. *rosea* D. Don in Sweet, British Flower Gard., 6: t. 225 (1834).
   Phyla nodiflora (L.) Greene var. *rosea* (D. Don) Mold. in Phytologia, 2: 22 (1941).
   Lippia nodiflora (L.) Michx. var. *rosea* (D. Don) Macbride in Bot. Ser. Field Mus. Nat. Hist., 13 (v), No. 2: 651 (1960).
   [*Lippia canescens* (non Kunth) Schauer in DC., Prodr., 11: 585 (1847) pro parte, et auctt. mult.]
   [*Lippia nodiflora* (non (L.) Michx.) Lindberg f., Iter Cypr., 28 (1946) pro parte]
   [*Phyla canescens* (non (Kunth) Greene) Davis, Fl. Turkey, 7: 32 (1982).]
   TYPE: from Chile, cult. Göttingen Bot. Gard.

Trailing perennial with a woody rootstock; stems rooting at the nodes, thinly adpressed-pubescent with medifixed hairs; leaves narrowly obovate, 0·5–3 cm. long, 0·2–1 cm. wide, thinly adpressed-pubescent, nervation obscure, apex acute or rounded, sharply and rather regularly (7-) 9-toothed, base entire, tapering to an indistinct petiole; peduncle slender, up to 8 cm. long, usually much exceeding the leaves; spike ovoid, 0·5–1·8 cm. long, about 0·6–0·7 cm. wide; bracts ovate-acuminate, 2–3 mm. long, 1·5–2 mm. wide, rather densely adpressed-pubescent, often purple-margined; calyx about 1·5–2 mm. long, 1 mm. wide, membranous, rather conspicuously pilose especially along the lateral margins, lobed subequally on both sides to less than half the length of the tube, lobes acute or subacute; corolla pink or lilac-pink, tube about 2·5 mm. long, 1 mm. wide, generally much exceeding bracts, limb spreading, lobes rounded or subtruncate, the abaxial 1·5–2 mm. long, the 3 adaxial about 1 mm. diam., the median often emarginate or toothed; stamens inserted at 2 levels near apex of corolla tube, filaments very short; anthers ovate or suborbicular, about 0·4 mm. long and about as wide; ovary glabrous, about 0·4 mm. long, 0·3 mm. wide; style about 0·5 mm. long; stigma capitate, oblique. Nutlets light brown, about 1·5 mm. long, 1 mm. wide, smooth, glabrous, strongly convex dorsally, with flat commissural faces, enveloped in the pilose calyx even at maturity.

HAB.: Grassy paths, road verges, lawns and garden beds, occasionally in moist places by springs and streams; sea-level to 500 ft. alt.; fl. March–October.

DISTR.: Divisions 1, 4, 6, 7. A native of Chile, Argentina, Paraguay and S. Brazil, now widely cultivated as lawn-cover, and naturalised in many parts of the Mediterranean region, also in California, Kenya, Australia and elsewhere.

1. Maa Beach, 1967, *Economides* in ARI 990 !
4. Varosha, Famagusta, 1970, *A. Hansen* 735 ! Near Othello's Tower, Famagusta, 1971, *Townsend* 71/105 !
6. Nicosia, 1939, *Lindberg f.* s.n. ! also 1962, *Meikle* 2640 ! and, 1963, *Townsend* 63/66 ! Cape Kormakiti, 1968, *I. M. Hecker* 24 ! Morphou, 1972, *W. R. Price* 1049 !
7. Kyrenia 1955, *Miss Mapple* 79 ! also, 1955, *G. E. Atherton* 243 ! Bellapais, 1955, *G. E. Atherton* 397 !

NOTES: Commonly used as a grass substitute for lawns in Cyprus, whereas *P. nodiflora*, unquestionably indigenous in the island, does not appear to be cultivated. The distinctions between the two species, though slight, appear to be constant throughout their range. True *Lippia canescens* Kunth is much more closely allied to *P. nodiflora*, and is quite distinct from the present plant.

2. VERBENA *L.*

Sp. Plant., ed. 1, 18 (1753).
Gen. Plant., ed. 5, 12 (1754).

Annual or perennial herbs or subshrubs; leaves opposite or sometimes in whorls of 3, toothed or dissected; flowers usually in simple or compound, terminal, bracteate spikes; calyx tubular, 5-costate, 5-toothed; corolla hypocrateriform with a patent, somewhat 2-lipped limb divided into 5 rounded or emarginate lobes; stamens usually 4, inserted about the middle of the corolla-tube, included; anthers ovate, usually exappendiculate and with parallel thecae; ovary entire or 4-lobed, with a single ovule in each loculus; style short; stigma unequally 2-lobed. Fruit enveloped by persistent calyx, splitting into 4 nutlets at maturity.

More than 200 species, chiefly American, of which several (and their hybrids) are popular garden ornamentals. The "Lemon-scented Verbena", a shrub with whorls of lemon-scented, lanceolate leaves and lax panicles of small, lilac flowers, is generally referred nowadays to a distinct genus, as *Aloysia triphylla* (L'Hérit.) Britt. (*A. citriodora* L'Hérit.). It is a native of Chile, and is probably grown in Cyrpus gardens, as elsewhere in the Mediterranean region, though precise records are wanting.

Inflorescences lax, 5–30 cm. long; leaves coarsely and irregularly pinnatisect; nutlets prominently rugulose-reticulate dorsally - - - - - - - **1. V. officinalis**
Inflorescences compact, 1·5–7 cm. long; leaves finely pinnatisect; nutlets obscurely veined-reticulate or smooth dorsally- - - - - - - - **2. V. supina**

1. V. officinalis *L.*, Sp. Plant., ed. 1, 20 (1753); Unger et Kotschy, Die Insel Cypern, 277 (1865); Boiss., Fl. Orient., 4: 534 (1879); H. Stuart Thompson in Journ. Bot., 44: 336 (1906); Holmboe, Veg. Cypr., 150 (1914); Post, Fl. Pal., ed. 2, 2: 321 (1933); Lindberg f., Iter Cypr., 28 (1946); Osorio-Tafall et Seraphim, List Vasc. Plants Cyprus, 88 (1973); Feinbrun, Fl. Palaest., 3: 93, t. 151 (1978); Davis, Fl. Turkey, 7: 33 (1982).

TYPE: *"in Europae mediterraneae ruderatis"*.

Erect perennial 30–100 cm. high; stems 4-angled, sulcate, subglabrous or sparsely scabridulous; branches lax, spreading; leaves opposite, the lower ovate in outline, 2–7 cm. long, 0·8–4·5 cm. wide, pinnatisect or pinnatifid, with broad, acute, bluntly serrate lobes, nervation prominent, scabridulous below, upper surface of lamina adpressed-hispidulous, apex blunt or acute, base tapering to an indistinct, winged petiole, 1–2 cm. long, upper cauline leaves commonly narrower than the lower, often lanceolate or oblanceolate, coarsely but bluntly serrate, without, or with very few, lobes; inflorescence lax, of numerous, elongate spikes, 5–30 cm. long; flowers minute, sessile; rhachis angular, glandular-pubescent; bracts narrowly ovate-acuminate, glandular-pubescent, about 2 mm. long, 1 mm. wide, somewhat keeled dorsally; calyx about 2 mm. long, 0·8 mm. wide, glandular-pubescent externally, calyx-teeth minute, deltoid, subequal, about 0·4 mm. long; corolla pale lilac-purple, glabrous, tube about 3 mm. long, limb subequally 5-lobed, the abaxial lobe rounded, about 1 mm. diam., the lateral and adaxial a little smaller; stamens inserted at 2 levels about the middle of the corolla-tube; filaments very short; anthers ovate, about 0·4 mm. long, 0·3 mm. wide; ovary glabrous, ovoid, obscurely 4-lobed, about 0·8 mm. long, 0·5 mm. wide; style stout, about 0·4 mm. long; stigma minute. Nutlets narrowly oblong, about 2 mm. long, 0·6 mm. wide, bright brown, glabrous, longitudinally rugulose-reticulate dorsally, whitish and scabridulous-papillose ventrally.

HAB.: Roadsides, waste ground, ditches, sides of streams and irrigation channels or in marshes; generally near water; sea-level to 4,500 ft. alt.; fl. June–Oct.

DISTR.: Divisions 1–8; widely distributed, almost throughout temperate and tropical regions.
1. Kissonerga, 1967, *Economides* in ARI 986!
2. Platres, 1948, *S. G. Cowper* s.n.! Troödos area, 1950, *F. M. Probyn* s.n.! Prodhromos, 1955, *G. E. Atherton* 565! Cedars Valley, 1968, *Economides* in ARI 1236!
3. Episkopi, 1905, *Holmboe* 700; Zakaki marshes, 1939, *Mavromoustakis* 43!
4. Larnaca, 1862, *Kotschy.*
5. Tymbou, 1932, *Syngrassides* 390! Near Athalassa, 1959, *P. H. Oswald* 162!
6. Near Strovolos, 1901, *A. G. & M. E. Lascelles* s.n.! Nicosia, 1939, *Lindberg f.* s.n.! Morphou, 1955, *N. Hiotellis* in *G. E. Atherton* 470!
7. Bellapais, 1939, *Lindberg f.* s.n.; Akanthou, 1940, *Davis* 2024! Kyrenia, 1955, *G. E. Atherton* 51! 228! 312! Ayios Epiktitos, 1955, *G. E. Atherton* 230! Lapithos, 1955, *G. E. Atherton* 694! Kambyli, 1957, *Merton* 2973!
8. Rizokarpaso, Ronnas Valley, 1957, *Merton* 2935A!

2. V. supina *L.*, Sp. Plant., ed. 1, 21 (1753); Boiss., Fl. Orient., 4: 534 (1879); Post, Fl. Pal., ed. 2, 2: 322 (1933); Lindberg f., Iter Cypr., 28 (1946); Osorio-Tafall et Seraphim, List Vasc. Plants Cyprus, 88 (1973); Feinbrun, Fl. Palaest., 3: 93, t. 152 (1978); Davis, Fl. Turkey, 7: 34 (1982).

Straggling annual or perennial, sometimes with an elongate rhizome, rooting at the nodes; stems 5–30 (–40) cm. long, slender or robust, branched or unbranched, 4-angled, glabrous or ± densely adpressed-strigose; leaves opposite, broadly rhomboid, about 1–2 cm. long, 0·8–1·5 cm. wide, pinnatisect into numerous, narrow, blunt or subacute lobes, subglabrous or ± densely strigillose, tapering basally to a short, flattish, canaliculate petiole; nervation usually prominent below; inflorescence a simple or sparingly branched, rather crowded spike, 1·5–7 cm. long at anthesis, elongating to 10 cm. in fruit; rhachis densely glandular-strigillose; bracts narrowly oblong, about 1·5 mm. long, 0·4 mm. wide, thinly strigillose, prominently keeled; calyx about 2·5 mm. long, 1 mm. wide, adpressed-strigillose with 5 minute, purplish, deltoid apical teeth; corolla pale heliotrope, tube about 2·5 mm. long, thinly pubescent externally, abaxial lobe about 1·7 mm. long, 0·8 mm. wide, bluntly suborbicular-oblong, pilose basally around the throat; stamens inserted about half-way up tube; filaments very short; anthers oblong, about 0·6 mm. long, 0·4 mm. wide; ovary glabrous, obscurely 4-lobed, ovoid, about 0·7 mm. long, 0·4 mm. wide; style about 0·9 mm. long; stigma subclavate, obscurely 2-lobed. Nutlets narrowly oblong, about 2·5 mm. long, 1·2 mm. wide, rich brown and obscurely veined-reticulate dorsally (sometimes the venation reduced to a longitudinal median stripe), paler and sparingly papillose ventrally.

f. supina
TYPE: "*in* Hispania"

Stems robust, suberect; leaves thinly strigillose, divided into narrowly oblong, broad-based lobes; plant evidently perennial, with creeping rhizomes.

HAB.: Marshy ground and seasonally inundated meadows; sea-level to 200 ft. alt.; fl. April–June.

DISTR.: Division 4, rare. S. Europe and Mediterranean region eastwards to Iraq and Arabia.

4. Sandy ground near Larnaca, 1939, *Lindberg f.*, s.n.; Kouklia, on site of old reservoir, 1962, *Meikle* 2615!

f. petiolulata *Lindberg f.*, Iter Cypr., 28 (1946).
TYPE: Cyprus; "in arenosis juxta opp. Larnaca" (H, K!)

Stems slender, diffuse; leaves densely strigillose, divided into narrowly oblanceolate-petiolulate lobes, the terminal usually much elongated; plant apparently annual.

HAB.: Sandy ground by the sea; near sea-level; fl. April–June.

DISTR.: Division 4, rare. ? Endemic.

4. Larnaca, 1939, *Lindberg f.* s.n.!

NOTES: Perhaps no more than a starved and desiccated condition of *V. supina*, with which it was apparently growing, but remarkable for its slenderness, canescence, and narrow, petiolulate leaf-divisions.

Verbena supina is commonly said to be an annual, but specimens from Kouklia are clearly perennial, with a horizontal, rooting rhizome. Similar plants are found throughout the range of the species, and one may conclude that the duration of the species is tied to habitat, annual plants growing in dry, sandy places, perennials in seasonally inundated pastures and meadows.

V. ARISTIGERA *S. Moore* (in Trans. Linn. Soc., ser. 2, 4: 439 (1895); *Glandularia aristigera* (S. Moore) Troncoso; *Verbena tenuisecta* Briquet) from South America, is frequently cultivated as an ornamental in Cyprus. It is a trailing, rooting perennial, with dissected leaves like *V. supina*, and flat-topped spikes of relatively large, vivid purple flowers. Specimens have been collected in the Municipal Gardens, Nicosia, (*Meikle* 4083!). The genus *Glandularia* G. F. Gmelin is sometimes distinguished from *Verbena*, chiefly by its flat-topped inflorescences and gland-tipped anther-connectives.

DURANTA REPENS *L.* (Sp. Plant., ed. 1, 637; 1753) a subscandent, often thorny, shrub, with lax racemes of lavender-blue flowers, and small orange-yellow, fleshy fruits, has been seen in gardens at Kato Paphos, and may be more widely grown in Cyprus. It is a native of tropical America, and is cultivated almost throughout the tropics.

SUBFAMILY 2. **Viticoideae** Mostly trees and shrubs; leaves sometimes compound, palmate; inflorescence cymose. Fruit generally drupaceous.

3. VITEX *L.*

Sp. Plant., ed. 1, 638 (1753).
Gen. Plant., ed. 5, 285 (1754).

Trees and shrubs, generally with compound, palmate leaves; flowers in lax or congested (sometimes spiciform), bracteate cymes; calyx usually campanulate, 5-toothed; corolla 2-lipped, the adaxial lip 2-lobed, the abaxial lip 3-lobed; stamens 4, inserted on the corolla-tube, didynamous, exserted; ovary usually 4-locular at maturity, with a solitary, laterally-attached ovule in each loculus; stigma shortly 2-fid. Fruit a drupe, surrounded by the persistent calyx; seeds without endosperm.

About 250 species distributed almost throughout tropical regions, with a few species extending into colder climates.

1. **V. agnus-castus** *L.*, Sp. Plant., 638 (1753); Gaudry, Recherches Sci. en Orient, 189, 197 (1855); Unger et Kotschy, Die Insel Cypern, 278 (1865); Boiss., Fl. Orient., 4: 535 (1879); Holmboe, Veg. Cypr., 150 (1914); Post, Fl. Pal., ed. 2, 2: 322 (1933); Lindberg f., Iter Cypr., 28 (1946); Chapman, Cyprus Trees and Shrubs, 65 (1949); Osorio-Tafall et Seraphim, List Vasc. Plants Cyprus, 88 (1973); Feinbrun, Fl. Palaest., 3: 95, t. 154 (1978); Davis, Fl. Turkey, 7: 35 (1982).

TYPE: *"in* Siciliae *et* Neapolis *paludosis"*

Deciduous shrub 1–3 m. high; branches at first somewhat 4-angled and tomentellous, becoming subterete, sulcate, puberulous, dull brown with age; internodes usually long; leaves opposite; petioles 2–7 cm. long, rather slender, tomentellous, canaliculate above; lamina palmately compound, divided into 5 or 7 linear-lanceolate, entire, petiolulate leaflets; petiolules up to 1·5 cm. long; leaflets 2–12 cm. long, 0·3–1·8 cm. wide, green and minutely puberulous above, whitish-tomentellous below, apex acuminate; margins narrowly recurved; nervation obscure; inflorescence terminal, much-branched, consisting of 10–20 or more pairs of congested cymes superimposed upon a common axis 7–20 cm. or more long, with 6–10 (or more) flowers in each cyme; pedicels very short, less than 1 mm. long, tomentellous; calyx campanulate, 2·5 mm. long, 2 mm. wide at apex, tomentellous, very shortly 5-dentate; corolla whitish, mauve or lavender-blue, thinly tomentellous externally, tube narrowly infundibuliform, about 5 mm. long, 3 mm. wide at apex, glabrous internally save for a conspicuous lanuginose zone around the point of insertion of the filaments, corolla-lobes subequal, oblong, rounded, about 1·5–2 mm. long, 1–1·5 mm. wide; stamens conspicuously exserted; filaments glabrous, about 6 mm. long, attached near apex of corolla-tube; anthers ovate-oblong, about 1 mm. long, 0·8 mm. wide, thecae free and tapering towards the base; ovary subglobose, about 0·8 mm. diam., glabrous; style straight, about 6 mm. long, glabrous; stigma 2-fid, the lobes narrow, acute, about 0·7 mm. long. Fruit globose, about 3 mm. diam., like a peppercorn; seeds usually 3–4; testa brown, closely adherent to the hard, bony endocarp.

HAB.: River-beds, sides of streams and irrigation channels, roadsides and sandy or gravelly seashores, usually in moist situations or near underground water; sea-level to 2,000 ft. alt., but usually lowland; fl. June–Dec.

DISTR.: Divisions 1–3, 7, 8, locally common. Widely distributed in S. Europe and the Mediterranean region.

1. Polis, 1936, *Lady Loch* 5! Prodhromi near Polis, 1937, *Syngrassides* 1711! Kissonerga to Maa Beach, 1967, *Economides* in ARI 971! 972!
2. Near Evrykhou, 1880, *Sintenis & Rigo* 683! Near Kalopanayiotis, 1900, *A. G. & M. E. Lascelles* s.n.! Near Pomos, 1934, *Syngrassides* 959! 960! also, 1939, *Lindberg f.* s.n.!
3. Near Stavrovouni, 1880, *Sintenis & Rigo*; R. Dhiarizos near Souskiou, 1959, *P. H. Oswald* 170!

7. Near Kyrenia, 1922, *Mrs. Houston* s.n.! also, 1955, *G. E. Atherton* 8! and, 1968, *I. M. Hecker* 28! Lapithos, 1938, *Syngrassides* 397! Lambousa Beach, 1973, *P. Laukkonen* 447!
8. Cape Andreas, 1937, *Syngrassides* 1662!

71. LABIATAE

J. Briquet in Engl. et Prantl, Pflanzenfam., ed. 1, IV, 3a: 183–380 (1895–97).
S. Junell in Symb. Bot. Upsal., 4: 1–219 (1934).

Shrubs, subshrubs, annual and perennial herbs, commonly glandular and aromatic; stems often tetragonous; leaves generally opposite, decussate, simple, exstipulate; flowers mostly hermaphrodite, often in more or less condensed cymes or verticillasters, sometimes in racemes or spikes; bracts foliaceous or reduced; bracteoles small or wanting; calyx often tubular or infundibuliform, persistent, commonly with prominent nervation, 4–5-toothed or lobed, occasionally 2-lipped with emarginate or toothed lips, or subentire; corolla tubular, sympetalous, the limb often 2-lipped, with the adaxial lip frequently emarginate, the abaxial 3-lobed; stamens usually 4, didynamous, sometimes 2, inserted on the corolla-tube; anthers 1–2-thecous, introrse; ovary superior, generally seated on a nectariferous disk, 2-carpellate, but ultimately divided almost to base into 4 divisions; style generally gynobasic, arising from the base of the ovary-divisions; stigma commonly 2-lobed. Fruit usually consisting of 4, 1-seeded nutlets enveloped by the persistent calyx, rarely drupaceous; seed without, or with very scanty endosperm; embryo straight or curved, the radicle pointing downwards.

About 180 genera and more than 3,000 species with a cosmopolitan distribution, but exceptionally well represented in the Mediterranean region. Many genera (*Salvia, Thymus, Origanum, Ocimum, Mentha,* etc.) furnish aromatic potherbs and are widely cultivated; others are valued for their fragrant oils.

Sexual dimorphism and cleistogamy are found in the flowers of many *Labiatae*, and may account for misleading differences within a single species.

Stamens 2:
  Leaves coriaceous, linear; flowers in short axillary racemes; erect or prostrate, aromatic
    shrubs - - - - - - - - - - **13. Rosmarinus**
  Leaves and inflorescences not as above:
    Calyx-tube narrowly cylindrical, about 1·5 mm. diam.; adaxial (upper) lip of corolla small,
      flattish, emarginate - - - - - - - - - **11. Ziziphora**
    Calyx-tube campanulate, more than 1·5 mm. diam.; adaxial (upper) lip of corolla large,
      concave, often falcate - - - - - - - - - - **12. Salvia**
Stamens 4:
  Corolla subequally 4-lobed; stamens commonly long-exserted; aromatic herbs of moist
    situations - - - - - - - - - - - **2. Mentha**
  Corolla distinctly 1- or 2-lipped:
    Corolla apparently 1-lipped, the adaxial (upper) lip much reduced or wanting; style not
      gynobasic:
      Abaxial (lower) lip of corolla apparently 5-lobed, the 2 basal lobes replacing the adaxial
        (upper) lip; median lobe concave, usually entire or denticulate **25. Teucrium**
      Abaxial (lower) lip of corolla 3-lobed, the median lobe emarginate or bifid - **26. Ajuga**
    Corolla 2-lipped; style gynobasic:
      Calyx with a dorsal flap (or *scutellum*), calyx-lips entire, mouth of calyx closed in fruit
        24. Scutellaria
      Calyx without a dorsal flap:
        Mouth of calyx closed in fruit; bracts unlike cauline leaves - - **22. Prunella**
        Mouth of calyx open in fruit:
          Nutlets fleshy, black, shining, like drupelets - - - - **23. Prasium**

Nutlets not fleshy or black:
  Bracteoles forming rigid, curved spines - - - - - **20 Ballota**
  Bracteoles not spinose:
    Stamens declinate, lying along the abaxial (lower) lip of corolla
                         OCIMUM (p. 1256)
    Stamens ascending, lying under the adaxial (upper) lip of corolla:
      Bracts distinct from cauline leaves, not leaf-like:
        Inflorescence a dense, oblong, 4-ranked, pedunculate spike **1. Lavandula**
        Inflorescence not as above:
          Corolla yellow or yellow and brown; inflorescence spiciform **15. Sideritis**
          Corolla pink, purple or white, inflorescence lax, branched, not spiciform
                                   3. Origanum
      Bracts foliaceous, resembling the cauline leaves but often smaller:
        Calyx 1·5–3 cm. wide, broadly infundibuliform - - **19. Moluccella**
        Calyx much less than 1·5 cm. wide:
          Calyx 10-toothed; leaves broadly obovate, reniform or suborbicular,
            usually ± tomentose - - - - - - **16. Marrubium**
          Calyx 5 (4)-toothed:
            Corolla bright yellow, the median lobe of the abaxial (lower) lip more
              than 1 cm. wide - - - - - - **21. Phlomis**
            Corolla not as above:
              Calyx-tube curved, gibbous or saccate at base; pedicels dorsivent-
                rally flattened or compressed:
                Calyx-teeth recurved, aristate; inflorescence elongate; corolla
                  whitish or sometimes with a pale mauve adaxial (upper) lip
                                15. Sideritis
                Calyx-teeth not recurved or aristate; inflorescence compact;
                  corolla pink or purple - - - - - **8. Acinos**
              Calyx-tube not curved, gibbous or saccate at base; pedicels terete
               or subterete:
                Adaxial (upper) lip of corolla deeply bifid - - - **14. Nepeta**
                Adaxial (upper) lip of corolla not bifid:
                  Adaxial (upper) lip of corolla strongly concave, hooded (or
                  galericulate):
                  Lateral lobes of abaxial (lower) lip of corolla scarcely
                    developed - - - - - - **18. Lamium**
                  Lateral lobes of abaxial (lower) lip of corolla well developed:
                    Flowers sessile or subsessile forming dense, compact
                    verticels - - - - - **17. Stachys**
                    Flowers pedicellate or in pedunculate cymes, not forming
                    compact verticels - - - - **20. Ballota**
                Adaxial (upper) lip of corolla flattish or shallowly concave, but
                not hooded:
                Inflorescence terminal, capitate or subcapitate; leaves
                  linear or narrowly lanceolate, less than 2 mm. wide,
                  gland-dotted - - - - - **4. Thymus**
                Inflorescence spiciform, consisting of several superimposed
                whorls or verticillasters:
                  Flowers in compact, crowded whorls; pedicels and
                  peduncles very short:
                  Bracteoles inconspicuous, up to 6 mm. long; leaves
                  small, rigid, acute:
                  Leaves carinate, thick, conspicuously gland-dotted
                                5. Satureja
                  Leaves not carinate, not thick nor conspicuously
                  gland-dotted - - - - **6. Micromeria**
                Bracteoles conspicuous, filiform, setose, up to 10 mm.
                  long; leaves not small or rigid - **9. Clinopodium**
                Flowers in cymes with distinct peduncles and pedicels:
                  Bracteoles minute, subulate, to 2 mm. long:
                  Leaves small, rigid, acute; corolla 5–6 mm. long
                              6. Micromeria
                  Leaves broadly and bluntly ovate-cordate or
                  suborbicular, papery; corolla about 10 mm. long
                              7. Calamintha
                Bracteoles foliaceous, lanceolate, about 5 mm. long,
                2 mm. wide - - - - **10. Melissa**

TRIBE 1. **Ocimeae** *Dumort.* Stamens 4, declinate, lying on the lower lip of the corolla, or sometimes enveloped by it; style gynobasic.

OCIMUM BASILICUM *L.*, Sp. Plant., ed. 1, 597 (1753), Basil, an erect, thinly pubescent or subglabrous annual, 15–45 cm. high, with entire or remotely serrulate, pleasantly aromatic, often purplish, ovate leaves, 1–5 cm. long, and whitish flowers in closely superimposed whorls forming an elongate, terminal, inconspicuously bracteate verticillaster, is recorded as cultivated in Cyprus by Kotschy (Die Insel Cypern, 265; 1865) and is represented by a specimen (*G. E. Atherton* 70!) said to have been collected, in 1955, in an old garden in Kyrenia. Basil, generally believed to be a native of tropical Asia, is widely cultivated as a culinary herb in warm-temperate and tropical regions, and is probably common in Cyprus.

An unlocalized specimen, collected by A. G. & M. E. Lascelles in 1903, has been identified as the closely allied *O. canum* Sims (= *O. americanum* L.), which differs from *O. basilicum* in its smaller flowers, denser indumentum and chromosome number (2n = 24 in *O. canum*, 2n = 48 in *O. basilicum*). It may be doubted, however, if the Lascelles's plant is more than an impoverished condition of *O. basilicum*.

TRIBE 2. **Lavanduleae** *Boiss.* Stamens 4, included; anthers 1-thecous at tip; thecae divergent; calyx 13–15-nerved; lobes of ovary opposite disk-lobes; style gynobasic; nutlet with dorsal-basal attachment; embryo straight.

1. LAVANDULA *L.*

Sp. Plant., ed. 1, 572 (1753).
Gen. Plant., ed. 5, 249 (1754).
D. A. Chaytor in Journ. Linn. Soc. (Bot.), 51: 153–204 (1937).

Perennial herbs or shrubs; stems angled or terete; leaves simple, entire, dentate or pinnatisect; inflorescence simple or branched, spicate, generally borne on a long, leafless peduncle; bracts opposite, alternate or spirally arranged, subtending 1–7 flowers, membranous or herbaceous, the uppermost sometimes enlarged and coloured; bracteoles present or absent; flowers sessile or subsessile; calyx cylindrical or urceolate, often accrescent, 5-toothed, the adaxial tooth sometimes enlarged and appendiculate; corolla usually bluish or purplish, tube equalling or exceeding calyx, lobes 5, subequal, spreading, or ± distinctly bilabiate; stamens 4, included; anthers reniform or suborbicular; style about as long as corolla-tube, gynobasic; stigma subglobose, the 2 lobes connivent save when receptive. Nutlets 4, ellipsoid or oblong, attached dorsally or basally to the disk, exocarp sometimes mucilaginous when wetted.

About 30 species, chiefly in the Mediterranean region and eastwards through Arabia to India; also Atlantic Islands and northern temperate and tropical Africa.

1. **L. stoechas** *L.*, Sp. Plant., ed. 1, 573 (1753); Unger et Kotschy, Die Insel Cypern, 265 (1865); Boiss., Fl. Orient., 4: 540 (1879); Holmboe, Veg. Cypr., 153 (1914); Post, Fl. Pal., ed. 2, 2: 328 (1933); Chaytor in Journ. Linn. Soc. (Bot.), 51: 163 (1937); Lindberg f., Iter Cypr., 28 (1946); Chapman, Cyprus Trees and Shrubs, 68 (1949); Rozeira in Brotéria, 18: 42 (1949); Osorio-Tafall et Seraphim, List Vasc. Plants Cyprus, 92 (1973); Feinbrun, Fl. Palaest., 3: 109, t. 175 (1978); Davis, Fl. Turkey, 7: 76 (1982).
TYPE: "*in* Europa *australi*".

Much-branched shrub 40–120 cm. high; bark grey-brown, flaking in narrow strips on old branches; young growths tetragonous, closely crispate-pubescent with greyish stellate hairs; leaves sessile, linear or linear-lanceolate, 1–4 cm. long, 0·2–0·5 cm. wide, usually with condensed clusters of smaller leaves in their axils, lamina obtuse or subacute, greyish, slightly viscid, closely stellate-pubescent on both faces, margins entire, recurved or revolute, nervation obscure; inflorescence a dense, oblong spike, 1·3–4 cm. long, 0·7–1·2 cm. wide; peduncle generally short, less than 2·5 cm. long, sometimes almost wanting; flowers in 4 vertical ranks; fertile bracts closely imbricate, broadly obovate, cuspidate or apiculate, 2–6 mm. long, 2–7 mm. wide, purplish, thinly lanuginose, reticulate-veined, each subtending a condensed cyme of 2–7 flowers; spike usually terminating in several, conspicuous, much-enlarged, violet-blue, erose, sterile bracts 1–2 cm. long, 0·5–1 cm. wide; calyx cylindrical, about 5 mm. long, 1·5 mm. wide, shortly lanuginose externally, 13-nerved, calyx-teeth short, broadly deltoid, less than 0·4 mm. long, the adaxial much enlarged, flabellate, about 1 mm. diam.; corolla dark violet-purple, tube 5–6 mm. long, 1 mm. wide, glabrous except at apex, bearded internally in a wide zone around the insertion of the stamens, lobes suborbicular, spreading, about 1 mm. diam.; stamens inserted almost 1·5 mm. below throat; filaments glabrous, almost 1 mm. long; anthers pale, broadly reniform or suborbicular, about 0·5 mm. diam.; ovary glabrous, about 0·4 mm. diam., lobed almost to base and attached to the apex of a stipitate 4-grooved disk; style about 3 mm. long, sparsely pilose above; stigma subglobose or capitate, about 0·5 mm. diam. Nutlets oblong, compressed, about 1·5 mm. long, 1·2 mm. wide, rich brown, glandular-verruculose, not mucilaginous when wetted.

HAB.: Dry rocky hillsides, usually on pillow lava; 500–1,500 (–2,400) ft. alt.; fl. March–May.

DISTR.: Divisions 1–3, 5, locally abundant. Mediterranean region east to Palestine; Atlantic Islands.

1. N.E. of Lyso, 1979, *Edmondson & McClintock* E 2792!
2. Between Lefkara and Makheras Monastery, 1862, *Kotschy*; Makheras Forest, 1937, *Chapman* 268! also, 1952, *F. M. Probyn* 25! near Ayios Merkourios, 1962, *Meikle* 2282!
3. Stavrovouni, 1862, *Kotschy* 193, also, 1937, *Syngrassides* 1487! 1939, *Lindberg f.* s.n.! and, 1959, *C. E. H. Sparrow* 3! Cape Gata, 1862, *Kotschy* 603; Alethriko, 1905, *Holmboe* 232; Lefkara, 1931, *J. A. Tracey* 55! and, 1934, *Syngrassides* 1388! Skarinou (on chalk), 1941, *Davis* 2789! Mile 19, Nicosia–Limassol road, 1950, *Chapman* 546! Between Vavla and Khirokitia, 1956, *Merton* 2551!
5. Pera, 1908, *Clement Reid* s.n.! and, 1949, *Casey* 706!

NOTES: All the Cyprus material seen to date is uniformly short-pedunculate, and referable to ssp. *stoechas*.

Kotschy's record from Cape Gata requires confirmation, I am not aware that any other collector has recorded the species (which has a limited range in Cyprus) from this locality, but Kotschy may be correct.

Lavandula stoechas, apart from its beauty, is a valuable bee-plant, and provides excellent coverage for bare, unstable slopes, especially on lava. Its spread might well be encouraged in suitable localities.

L. AUGUSTIFOLIA *Mill.*, Gard. Dict., ed. 8, no. 2 (1768) (*L. spica* L. pro parte, *L. officinalis* Chaix, *L. vera* DC.) the Lavender of commerce, with long, interrupted spikes, and without the coma of coloured, sterile bracts characteristic of *L. stoechas*, is recorded from Lapithos, 1939, by Lindberg f. (Iter Cypr., 28; 1946) and from Tjiklos, 1973, by P. Laukkonen 354! In both instances the records are based on cultivated plants; *L. augustifolia* is nowhere native in the eastern Mediterranean region. The reference in Osorio-Tafall & Seraphim (List Vasc. Plants Cyprus, 92; 1973) is almost certainly based on the Lindberg f. record.

TRIBE 3. **Saturejeae** *Benth.* Stamens generally 4 (2 in *Ziziphora*), divergent or ascending, calyx campanulate or tubular, usually (10-) 13-nerved; corolla-lobes subequal, flattish or slightly concave, truncate or emarginate; lobes of ovary alternating with disk-lobes; style gynobasic. Nutlets dry, attached basally; embryo straight.

2. MENTHA *L.**

Sp. Plant., ed. 1, 576 (1753).
Gen. Plant., ed. 5, 250 (1754).
J. Briquet in Bull. Soc. Bot. Genève, 5: 20–122 (1889).
R. Harley & C. A. Brighton in Journ. Linn. Soc. (Bot.), 74: 71–96 (1977).

Perennial (rarely annual) herbs of damp places, with creeping rhizomes rooting at the nodes; leaves with characteristically scented epidermal glands; flowers hermaphrodite or female on the same (gynomonoecious) or separate (gynodioecious) plants, usually in dense many-flowered bracteate verticillasters; bracts leafy, or if reduced, the verticillasters condensed to form a long spike-like or capitate terminal inflorescence; calyx actinomorphic or weakly 2-lipped, tubular or campanulate, 10- to 13-veined, with 5 (4 in one W. European species) subequal or unequal teeth; corolla weakly 2-lipped, with 4 subequal lobes, the upper lobe wider and usually emarginate; tube shorter than the calyx; stamens about equal, divergent or ascending under the upper lip of the corolla, exserted (except in female flowers and most hybrids, where the stamens are reduced or absent); anthers 2-locular with equal and parallel loculi; nutlets smooth, foveolate, reticulate or rugulose.

About 20 species chiefly in temperate and subtropical regions of both hemispheres; introduced in South America.

All Mints are extremely variable, and their morphological features are easily altered by such environmental factors as humidity and light. Hybrids may occur when one or more species of Section *Mentha* grow together. Both the above factors, together with the complexity of cultivated forms of *M. spicata* and its hybrids, which may become naturalized, combine to make this one of the most difficult taxonomic groups.

Calyx hairy in throat, with unequal teeth. Flowers borne in whorls subtended by leaf-like
  bracts - - - - - - - - - - - - **1. M. pulegium**
Calyx glabrous in throat with ± equal teeth. Flowers borne in a terminal head or spike:
  Leaves with petioles 10 mm. or more. Flowers borne in a terminal head - **2. M. aquatica**
  Leaves with very short petioles usually less than 5 mm., or sessile; flowers borne in a
    congested or interrupted terminal spike:
    Cauline leaves broadest at or above the middle, grey-green and finely but densely
      puberulous on both surfaces, with simple curved hairs with a felted appearance when
      dry - - - - - - - - - - - - - **3. M. longifolia**
    Cauline leaves usually broadest below the middle, green and glabrous to coarsely grey-
      green or grey-villous beneath, with simple hairs, branched hairs often also present
                                                              4. M. spicata

SECT. 1. **Pulegium** (*Mill.*) *DC.* Bracts similar to the leaves; calyx tubular, weakly two-lipped with unequal teeth, throat hairy within; corolla-tube gibbous.

1. **M. pulegium** *L.*, Sp. Plant., ed. 1, 577 (1753); Gaudry, Recherches Sci. en Orient, 190 (1855); Unger et Kotschy, Die Insel Cypern, 266 (1865); Boiss., Fl. Orient., 4: 545 (1879) as *M. palegium*; H. Stuart Thompson in Journ. Bot., 44: 336 (1906); Holmboe, Veg. Cypr., 163 (1914); Post, Fl. Pal., ed. 2, 2: 331 (1933); Lindberg f., Iter Cypr., 29 (1946); Osorio-Tafall et Seraphim, List Vasc. Plants Cyprus, 92 (1973); Davis, Fl. Turkey, 7: 385 (1982).
  Pulegium vulgare Mill., Gard. Dict., ed. 8, no. 1 (1768).
  TYPE: "*in* Angliae, Galliae, Helvetiae *inundatis*" (LINN 730, 19).

* by R. M. Harley

Subglabrous to densely villous, usually red-tinged perennial herb with a pungent odour; stems 10–40 cm. high, procumbent to ascending or erect; rhizomes mainly epigeal; leaves 8–30 mm. long, 4–12 mm. wide, narrowly elliptic with an attenuate base to rarely suborbicular, shortly petiolate, with margin entire or with up to six often obscure teeth on each side, hairy beneath; flowering stems ascending or erect, flexuous, simple or much branched; verticillasters many flowered with bracts similar to the leaves but usually smaller; calyx (2–) 2·5–3 mm. long, tubular, weakly two-lipped, throat hairy within, calyx-teeth ciliate, the lower subulate, the upper shorter and wider; corolla (4–) 4·5–6 mm. long, with tube gibbous beneath, lilac; stamens exserted or rarely included; fertile anthers 0·4 mm. long. Nutlets 0·75 mm. long, pale brown.

HAB.: Streamsides and marshy ground; sea-level to 1,000 ft. alt.; fl. June–Aug.

DISTR.: Divisions 1–7, locally common. Widespread in Europe, except in N. Mediterranean area, E. to Iran.

1. Dhrousha, 1900, *A. G. & M. E. Lascelles* s.n. ! Cape Arnauti, 1913, *Haradjian* 796 !
2. Kilani, 1905, *Holmboe* 1150.
3. Limassol, ? 1905, *Michaelides* s.n. sec. *Holmboe*; Kannaviou, 1939, *Lindberg f.* s.n.
4. Larnaca, 1862, *Kotschy*.
5. Kythrea, 1905, *Holmboe* 844; also, 1939, *Lindberg f.* s.n. !
6. Near Nicosia, 1932, *Syngrassides* 400 ! and, 1939, *Lindberg f.* s.n.
7. Akhiropiitos, 1862, *Kotschy* 970; Karavas–Paleosophos road, 1968, *Economides* in ARI 1141 ! Kannoures (Asomatos), 1970, *A. Genneou* in ARI 1576a !

SECT. 2. **Mentha** Bracts variable, often very reduced. Calyx tubular or campanulate with ± equal teeth, throat glabrous within. Corolla-tube straight.

2. M. aquatica *L.*, Sp. Plant., ed. 1, 576 (1753); Boiss., Fl. Orient., 4: 544 (1879); Post, Fl. Pal., ed. 2, 2: 330 (1933); Davis, Fl. Turkey, 7: 387 (1982).
   M. hirsuta Huds., Fl. Angl., ed. 1, 223 (1762).
   TYPE: "*in* Europa *ad aquas*", Herb. Cliff. (BM).

Extremely variable subglabrous to hirsute perennial herb, often purple-tinged and with characteristic pungent odour; rhizomes usually epigeal, green or purple-tinged, not brittle and with small usually reniform leaves; flowering stems (10–) 20–90 cm. high, ascending to erect, usually much branched below; leaves (15–) 30–90 mm. long, (10–) 15–40 mm. wide, ovate to ovate-lanceolate with an acute to often blunt apex and usually truncate but occasionally cuneate or weakly cordate base; petioles 9–14 mm. long, sometimes more; leaf margin variably serrate; inflorescence of 2–3 congested verticillasters with inconspicuous bracts, forming a terminal head to 20 mm. diam., and occasionally with 1–3 distant verticillasters below in the axils of leaf-like bracts, lower verticillasters sometimes shortly pedunculate; calyx (2·5–) 3–4 mm. long, tubular, distinctly veined, and with subulate or narrowly triangular teeth; pedicels hairy; corolla lilac. Nutlets pale brown, smooth or sometimes weakly foveolate.

HAB.: Streamsides and marshy ground, near sea-level; fl. June–Aug.

DISTR.: Divisions 3, 8, rare. Widespread in Europe and N. Africa, occurring sporadically southwards to South Africa, and eastwards to Caucasia and Iran.

3. Asomatos near Limassol, 1962, *Meikle* 2917 !
8. Ronnas R. near Rizokarpaso, 1962, *Meikle* !

3. M. longifolia (*L.*) *L.*, Fl. Monspel., 19 (1756); Post, Fl. Pal., ed. 2, 2: 330 (1933); Davis, Fl. Turkey, 7: 388 (1982).
   M. spicata L. var. *longifolia* L., Sp. Plant., ed. 1, 576 (1753).
   M. sylvestris L., Sp. Plant., ed. 2, 804 (1763); Boiss., Fl. Orient., 4: 543 (1879).

Very variable hairy perennial with musty or pungent odour; rhizomes mainly hypogeal, glabrous with squamiform leaves; young shoots glabrous when first appearing above ground, soon becoming hairy; flowering stems 40–120 cm. high, usually much branched; leaves (20–) 30–90 mm. long, (5–) 10–32 (–40) mm. wide, oblong-elliptic, oblong-lanceolate or lanceolate, widest at or above the middle, with a more or less acute apex and cordate to subcordate base, sessile or subsessile, rarely petiolate; lamina smooth or very weakly rugose, green- to grey-tomentose on upper surface, green- to white-tomentose beneath (discolorous in ssp. *longifolia*), sometimes minutely puberulous on both surfaces; leaf-hairs unbranched, finely matted beneath when dry; basal cell 18–36 (–41) μm. diam.; leaf margin sharply serrate with many irregular and often spreading teeth; inflorescence of many usually congested verticillasters forming a terminal often much-branched spike (30–) 40–100 mm. long, (6–) 8–15 mm. diam.; calyx 1–3 mm. long, narrowly campanulate with subulate teeth; corolla lilac or white. Nutlets castaneous, strongly reticulate.

ssp. **cyprica** (*H. Braun*) *R. Harley* stat. nov.

　　M. cyprica H. Braun in Verhl. Zool.-Bot. Ges. Wien, 39: 217 (1889); Druce in Rep. B.E.C., 9: 470 (1931) as *M. cypria*.

　　M. cyprica H. Braun var. *galatae* H. Braun in Verhl. Zool.-Bot. Ges. Wien, 39: 217 (1889).

　　M. longifolia L. var. *cyprica* (H. Braun) Briq. in Bull. Herb. Boiss., 2: 696 (1894).

　　M. longifolia L. var. *galatae* (H. Braun) Briq. in Bull. Herb. Boiss., 2: 696 (1894).

　　[*M. sylvestris* L. var. *nemorosa* (non Reichb.) Kotschy in Unger et Kotschy, Die Insel Cypern, 266 (1865).]

　　[*M. tomentosa* (non Urv.) Sintenis in Sintenis et Rigo, Iter Cypr., exsicc. no. 732 (1880).]

　　[*M. longifolia* (non (L.) L.) Holmboe, Veg. Cypr., 163 (1914); Lindberg f., Iter Cypr., 29 (1946).]

　　[*M. microphylla* (non C. Koch) Rechinger f. in Arkiv för Bot., ser. 2, 1: 430 (1950); Osorio-Tafall et Seraphim, List Vasc. Plants Cyprus, 92 (1973).]

　　TYPE: Cyprus; "In den Schluchten um das Kloster Trooditissa. 10 Oct. 1840." *Kotschy* 23 (W).

A slender plant with narrowly lanceolate leaves 20–50 (–70) mm. long, 5–10 (–14) mm. wide, with shallowly serrate teeth; apex acute and the base truncate to weakly cordate with a distinct petiole 1–3 mm. long or rarely more; lamina greyish-green above and below, densely puberulous with sessile glands, visible with a hand lens, beneath; spikes slender, (6–) 8 (–10) mm. diam. at flowering, interrupted below, unbranched or with few, diffuse, spreading branches. 2n = 24.

HAB.: Moist ground by streams and springs; 200–5,400 ft. alt., generally above 2,000 ft. alt.; fl. June–Nov.

DISTR.: Divisions 1–3, 5–7, locally common. Endemic.

1. Yeroskipos, 1939, *Lindberg f.* s.n. ! Skoulli, 1939, *Lindberg f.* s.n. !
2. Common; Troödhitissa Monastery, Troödos, Galata, Platania, Prodhromos, Trikoukkia, Platres, Kykko Monastery, Milikouri, Mesapotamos, Ambelikou, Kambos, Tripylos, etc. *Kotschy* 23 ! *Sintenis & Rigo* 732 ! *A. G. & M. E. Lascelles* s.n. ! *Haradjian* 1006 ! *Syngrassides* 741 ! 1687 ! *Kennedy* 732–734 ! *Lindberg f.* s.n. ! *Merton* 1599 ! *G. E. Atherton* 670 ! *Economides* in ARI 1243 ! etc.
3. Stavrovouni, 1939, *Lindberg f.* s.n.; Kannaviou, 1939, *Lindberg f.* s.n.
5. Kythrea, 1934, *Syngrassides* 535 ! Trikomo, 1939, *Lindberg f.* s.n.
6. Morphou, 1955, *N. Chiotellis* 480 !
7. Upper Thermia, 1955, *G. E. Atherton* 485 ! Lapithos, 1966, *Merton* in ARI 32 !

NOTES: This characteristic plant with its slender stems and narrow leaves is the commonest Mint in Cyprus. Robust plants, which sometimes occur, are probably due to environmental conditions. *M. longifolia* var. *galatae* differs from typical var. *cyprica* in its rather broader, more lanceolate, and often more sharply toothed leaves. Some specimens, notably the unlocalized *Aucher-Eloy* 1803 (possibly from the Troödos area), and some collections from Israel and neighbouring territories, recall *M. longifolia* (L.) L. ssp. *typhoides* (Briq.) R. Harley from Turkey, but except for the Aucher-Eloy specimen, differ from these in the shortly petiolate leaves, characteristic of ssp. *cyprica*.

Two collections referable to *M. longifolia* sensu lato require further study: these are *G. E. Atherton* 679 and *Merton* 32, both from (7) Lapithos. These have rather broad spikes, ca. 11 mm. diam., and stalked leaves, which in the Merton collection bear petioles up to 8 mm. long. Petiolate-leaved *M. longifolia* occurs sporadically through its eastern range, most notably in *M. longifolia* ssp. *noeana* (Boiss. ex Briq.) Briq. from Iraq and S.E. Anatolia, and in *M. longifolia* ssp. *hymalaiensis* Briq. (including ssp. *royleana* (Benth.) Briq.) from Afghanistan and the Himalayas. Indeed Briquet placed "var. *cyprica*" under the latter subspecies. Similar leaf-characters also frequently occur in forms of *M. spicata*.

The following sterile specimens belong to aggregate *M. longifolia*, but cannot, in the absence of inflorescences, be identified at infraspecific level (1) Phrangos near Neokhorio, 1962, *Meikle* 2207! (2) Khionistra, 1941, *Davis* 3505! Kakopetria, 1966, *Y. Ioannou* s.n.! Near Klirou, 1967, *Merton* in ARI 288! (4) Ayia Anna, 1957, *Merton* 2879! (7) Stream below Bellapais, 1948, *Casey* 104!

4. M. spicata L., Sp. Plant., ed. 1, 576 (1753); R. M. Nattrass, First List Cyprus Fungi, 18 (1937); Davis, Fl. Turkey, 7: 391 (1982).
    M. viridis (L.) L., Sp. Plant., ed. 2, 804 (1763); Post, Fl. Pal., ed. 2, 2: 331 (1933).
    M. sylvestris L. var. *viridis* (L.) Boiss., Fl. Orient., 4: 544 (1879).
    M. sylvestris L. var. *glabra* W. Koch, Syn. Fl. Germ. et Helv., 550 (1837).
    M. crispa L., Sp. Plant., ed. 1, 576 (1753), non *M. crispa* L., Sp. Plant., ed. 2, 805 (1763).

Extremely variable, glabrous to coarsely grey-villous perennial with musty, pungent or, in most cultivated forms, a sweet odour; rhizomes usually hypogeal; flowering stems 30–100 cm., leaves (18–) 30–75 (–90) mm. long, 8–25 (–32) mm. wide, sessile or rarely shortly petiolate, oblong-ovate to lanceolate, usually wider below the middle in middle cauline leaves, apex obtuse or acute, base round to cordate, margin serrate; lamina smooth or rugose, glabrous to coarsely grey-villous; hairs, when present on leaf undersurface, simple or branched (often mixed), basal cell 35–49 µm. diam.; verticillasters many, forming a terminal spike (30–) 40–110 (–140) mm. long, (6–) 8–12 (–14) mm. diam. usually interrupted at least below and often unbranched; corolla white, pink or lilac. Nutlets chestnut to dark brown, reticulate in hairy plants and smooth in glabrous plants.

ssp. **spicata**
    TYPE: "Crescit in Anglia, Gallia, Germania *parcius*". Herb. Cliff. (BM!).

Plant glabrous or hairy, cauline leaves 35–75 mm. long and usually over 12 mm. wide, with a flat margin and often simple hairs beneath; spike 40–80 (–110) mm., interrupted below.

HAB.: Waste ground, gardens; sea-level to 3,200 ft. alt.; fl. July–Nov.

DISTR.: Divisions 2–4, 7. An escape from cultivation. Widespread in Europe, W. Asia and N. Africa; cultivated almost throughout the world.

2. Kakopetria, 1955, *N. Chiotellis* 456! Pano Platres, 1981, *Meikle*!
3. Sanctuary of Apollo, Curium, 1982, *A. Fell* s.n.!
4. Pergamos, 1935, *Nattrass* 547.
7. Kyrenia, 1955, *G. E. Atherton* 69!

NOTES: The glabrous form is frequently cultivated and may perhaps be found elsewhere as a garden escape. Seedlings do not necessarily breed true, and in the wild a range of glabrous and hairy plants, variable in morphology and scent, can be expected.

ssp. **tomentosa** (*Briq.*) R. Harley in Notes Roy. Bot. Gard. Edinb., 38, 1: 38 (1980); Davis, Fl. Turkey, 7: 392 (1982).
    M. tomentosa d'Urv., nom. illeg. ssp. *tomentosa* Briq. in Bull. Trav. Soc. Bot. Genève, 5: 97 (1889).
    M. microphylla sensu Hayek, Prodr. Fl. Pen. Balc., 2: 397 (1930); F. Petrak in Rechinger f., Fl. Aegaea, 542 (1943); R. Harley in Tutin et al., Fl. Europaea, 3: 186 (1972); ? an C. Koch in Linnaea, 21: 648 (1849).
    M. sieberi C. Koch in Linnaea, 21: 649 (1849).
    M. sylvestris L. var. *stenostachya* Boiss., Fl. Orient., 4: 543 (1879).
    TYPE: Greece; "ad fontes insulae Scyri" *Dumont d'Urville* 618 (G–DC).

Plant always hairy, leaves 30–38 mm. long and usually under 12 cm. wide, often with undulate margins and with numerous branched hairs beneath;

spike (35–) 70–110 (–140) mm. long, usually much interrupted, especially in fruit.

HAB.: By streams, ditches and irrigation channels; sea-level to 3,900 ft. alt.; fl. July–Nov.

DISTR.: Divisions 2, ? 3, 7. Also S. Italy, Balkans, W. Turkey.

2. Platres, 1937, *Kennedy* 731 ! Kalopanayiotis, 1968, *A. Genneou* in ARI 1289 !
?3. "Ashiela (? Akhelia) to the sea", 1967, *Economides* in ARI 1007 !
7. Kyrenia, 1936, *Kennedy* 735 ! also, 1948, *Casey* 19 ! and, 1955, *G. E. Atherton* 300 !

NOTES: The material from Cyprus is extremely variable, and while this is a feature of *M. spicata* as a whole, it is possible that some introgression with ssp. *spicata* has occurred.

3. ORIGANUM *L.*

Sp. Plant., ed. 1, 588 (1753).
Gen. Plant., ed. 5, 256 (1754).
J. H. Ietswaart, A Taxonomic Revision of the Genus Origanum (Leiden Botanical Series, Vol. 4), 153 pp. (1980).

Perennial herbs and subshrubs; leaves generally broadly elliptic, ovate or suborbicular, entire or serrulate; flowers usually in crowded or lax, erect or pendulous spikes, solitary at the tips of the branches or forming lax, branched corymbs or panicles; bracts sometimes large and coloured, or small and green, but not foliaceous; calyx 2-labiate or subequally 5-toothed, or sometimes subentire with a deep abaxial slit; corolla purplish, pinkish or whitish, the tube equalling or much exceeding calyx, limb 2-lipped, the upper lip usually emarginate or 2-lobed, the lower lip 3-lobed; stamens 4, ascending or spreading, included or exserted; anthers 2-thecous, the thecae divergent; stigma with 2 acute lobes. Nutlets smooth, ovoid or oblong.

About 35 species widely distributed in Europe and the Mediterranean region eastwards to Central Asia. Many are pleasantly aromatic, and are valued as culinary herbs.

Bracts large, showy, pink or purplish, suborbicular, imbricate, 0·8–2 cm. diam.; leaves broadly ovate-cordate, sessile, glabrous - - - - - - **1. O. cordifolium**
Bracts inconspicuous, small, less than 0·8 cm. long; leaves not as above:
  Bracts loosely imbricate, oblong or lanceolate, distinctly longer than wide, glabrous or sparsely strigillose; calyx tubular:
    Corolla-tube about 12 mm. long; flower purple; leaves glabrous - **2. O. laevigatum**
    Corolla-tube short, about 4 mm. long; flowers white; leaves thinly hairy on both surfaces **3. O. vulgare** ssp. **hirtum**
  Bracts closely imbricate, broadly ovate, obovate or suborbicular, densely pubescent or tomentose; calyx sheath-like, slit almost to base abaxially:
    Underside of leaf prominently and conspicuously reticulate-veined- - **4. O. syriacum**
    Underside of leaf inconspicuously veined:
      Inflorescence thyrsoid or subcorymbose, compact and not extending far down the stem **5. O. dubium**
      Inflorescence elongate, narrow, lax, extending far down the stem - **6. O. majorana**

SECT. 1. **Amaracus** *Benth.* Inflorescence commonly pendulous, with conspicuous, coloured bracts; calyx 2-lipped; stamens ascending.

1. **O. cordifolium** (*Aucher-Eloy et Montbret ex Benth.*) *Vogel* in Linnaea, 15: 76 (1841); Benth. in DC., Prodr., 12: 191 (1848); Boiss., Fl. Orient., 4: 548 (1879); Post in Mém. Herb. Boiss., 18: 97 (1900); H. Stuart Thompson in Journ. Bot., 44: 336 (1906); Holmboe, Veg. Cypr., 161 (1914); Rechinger f. in Arkiv. för Bot., ser. 2, 1: 430 (1950); Ietswaart, Taxon. Rev. Gen. Origanum, 41 (1980).
Amaracus cordifolius Aucher-Eloy et Montbret ex Benth. in Ann. Sci. Nat., ser. 2, 6: 43 (1836); Briquet in Engl. et Prantl, Pflanzenfam., ed. 1, IV, 3a: 304 (1896); Chapman, Cyprus Trees and Shrubs, 74 (1949); Osorio-Tafall et Seraphim, List Vasc. Plants Cyprus, 92 (1973).
TYPE: "in insulâ Cypro", *Aucher-Eloy* (K !). See Notes.

Perennial subshrub to about 60 cm. high; stems much-branched from base, unbranched or sparingly branched above, terete, glabrous, often

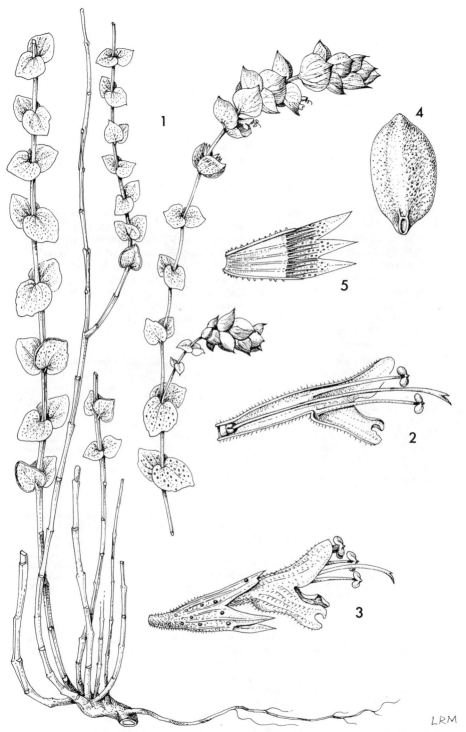

Plate 73. ORIGANUM CORDIFOLIUM (Aucher-Eloy et Montbret ex Benth.) Vogel **1,** habit, × 1; **2,** flower, longitudinal section, × 6; **3,** flower, × 6; **4,** nutlet, × 36; **5,** inside of calyx, × 9. (**1, 2, 3, 5** from *Kennedy* 1633; **4** from *Probyn* 137.)

purplish; leaves sessile, rather regularly spaced in pairs at intervals of about 2 cm., broadly ovate-cordate, 1–1·5 cm. long, 0·7–1·8 cm. wide, glabrous, glaucous, gland-dotted, apex acute or subacute, margins entire or irregularly undulate-crenate, distinctly but narrowly recurved; inflorescence a solitary, pendulous spike or loose panicle of spikes, each 4–7 cm. long, 1·3–2·5 cm. wide; bracts subtending 1–4 flowers, suborbicular, apiculate, entire, loosely imbricate, 0·8–2 cm. diam., glabrous, purple or pinkish, obscurely reticulate-veined, sparingly gland-dotted on the adaxial surface; calyx narrowly infundibuliform or almost tubular, tube about 4–5 mm. long, 1·5 mm. wide at apex, 13-nerved, glabrous externally and internally, except for a bearded zone around the base of the lips, adaxial lip divided into 2 subulate lobes 1–2 mm. long, abaxial lip 3-lobed, the lobes shortly deltoid, subequal, about 0·6 mm. long; corolla pink, tube about 5 mm. long, distinctly exserted from calyx, thinly puberulous externally, distinctly gibbous abaxially for about 2 mm. from apex, adaxial lip about 3 mm. long, 2 mm. wide, a little concave, shortly emarginate, puberulous externally, abaxial lip 3-lobed, the lobes subequal, shortly oblong, rounded, each about 1·4 mm. long, 1 mm. wide; stamens inserted at base of gibbous zone of corolla-tube; filaments 6–7 mm. long, conspicuously exserted, glabrous; anthers pallid with strongly divaricate, ovoid thecae about 0·4 mm. long, 0·2 mm. wide; ovary glabrous, deeply 4-lobed, about 0·5 mm. diam.; style glabrous, about 10 mm. long; stigma shortly and acutely 2-lobed. Nutlet broadly ellipsoid, compressed, about 1·5 mm. long, 1 mm. wide, dark brown, minutely foveolate. *Plate 73.*

HAB.: Moist rocky gorges by mountain streams on igneous formations; 1,500–3,000 ft. alt.; fl. June–Aug.

DISTR.: Division 2, in one area only. Endemic.

2. Roudhkias Valley, between Pano Panayia and Kykko Monastery, locally plentiful along the Palekhori, Stenous and Alonoudhi valleys, 1831–1954, *Aucher-Eloy* 1656! *Post* s.n.! *Haradjian* 954! *Chapman* 378! *Davis* 3388! *Kennedy* 1633! *F. M. Probyn* 137! *Merton* 1944!

NOTES: Following Boissier's lead, authors still occasionally cite Aucher-Eloy specimens as coming from Syria, though as Holmboe (Veg. Cypr., 161–162) observes, and as subsequent records have confirmed, this is certainly a mistake: the type material of *O. cordifolium* and all other specimens found to date come from a very restricted area of the Troödos Range, where the species is endemic. Without question Aucher-Eloy must have visited this area in August 1831 and collected the plant in full flower.

SECT. 2. **Prolaticorolla** *Ietswaart* Inflorescence erect; bracts foliaceous, small, greenish or coloured; calyx subequally 5-dentate; corolla 2-lipped, tube about 3 times longer than calyx.

2. **O. laevigatum** *Boiss.* in Ann. Sci. Nat., ser. 4, 2: 252 (1854); Fl. Orient., 4: 550 (1879); Rechinger f. in Arkiv för Bot., ser. 2, 1: 430 (1950); Osorio-Tafall et Seraphim, List Vasc. Plants Cyprus, 92 (1973); Ietswaart, Taxon. Rev. Gen. Origanum, 126 (1980); Davis, Fl. Turkey, 7: 310 (1982).
TYPE: Turkey; "in *Cataonia meridionali* et *Anti Tauro*. Cl. P. a Tchihatcheff" (G).

Perennial subshrub to about 50 cm. high; stems much-branched at base, unbranched or sparingly branched above, bluntly tetragonal, glabrous, purplish; leaves subsessile or very shortly petiolate, generally with condensed, small-leaved shoots in their axils, ovate, 5–13 mm. long, 2–8 mm. wide, glabrous, gland-dotted, apex acute or obtuse, margins minutely scabridulous with curved bristles, flat; nervation very obscure; inflorescence terminal, lax, subcorymbose, repeatedly branched; bracts foliaceous, lanceolate, 4–5 mm. long, 1·5–2 mm. wide, acute, scabridulous-serrulate towards apex, each subtending a single, subsessile flower; calyx tubular, about 4·5 mm. long, 1 mm. wide at apex, glabrous externally, densely lanuginose internally at the throat, conspicuously 10-nerved, teeth subequal, narrowly deltoid, about 1 mm. long, 0·5 mm. wide at base; corolla (from Turkish specimens) purple, tube narrowly cylindrical-infundibuliform, about 12 mm. long, 3 mm. wide at apex, thinly puberulous externally, limb obliquely 5-lobed, the 2 adaxial lobes rounded, about 1·3 mm. diam., the 3 abaxial lobes broadly oblong, about 1·5 mm. long, 1·3 mm. wide with rounded apices; stamens inserted about 2 mm. below base of corolla-lobes; filaments glabrous, the abaxial about 3·5 mm. long, slightly exceeding corolla-tube, the adaxial about 2 mm. long, scarcely exserted; anthers pallid with divaricate thecae about 0·8 mm. wide; ovary glabrous, about 0·4 mm. diam. with a conspicuous disk and 4 prominent, discrete lobes; style glabrous, about 12 mm. long; stigmatic lobes short, subulate, subequal. Nutlets oblong, angled on the ventral face, about 1 mm. long, 0·6 mm. wide, light brown, minutely granulose-rugulose.

HAB.: Not recorded, usually grows in garigue or open forest on hillsides, at 1,200–4,000 ft. alt.; fl. July–Sept.

DISTR.: Division 4, very rare. Also S.E. Turkey and N. Syria.

4. "Environs de Larnaka", August 1913, *Haradjian* 1013!

NOTES: The localization is somewhat vague, and no other collectors have rediscovered *O. laevigatum* in the vicinity of Larnaca. One suspects that the plant may actually have been found on igneous rocks at Stavrovouni, or on the lower slopes of the Troödos Range, possibly near Makheras Monastery. Outside Cyprus it is most frequent on the largely igneous Amanus Range.

SECT. 3. **Origanum** Inflorescence erect; bracts small, herbaceous, greenish or coloured; calyx subequally 5-dentate; corolla 2-lipped, tube narrowly infundibuliform, not more than twice as long as calyx.

3. O. vulgare L., Sp. Plant., ed. 1, 590 (1753); Boiss., Fl. Orient., 4: 551 (1879); Post, Fl. Pal., ed. 2, 2: 334 (1933); Ietswaart, Taxon. Rev. Gen. Origanum, 106 (1980); Davis, Fl. Turkey, 7: 309 (1982).

Erect, rhizomatous perennial to 80 cm. high; stems branched and often woody at base, unbranched or sparingly branched above, subterete or obscurely tetragonal, thinly pilose with adpressed or spreading hairs or subglabrous, sometimes purplish; leaves generally with condensed, small-leaved shoots in their axils, ovate, 5–25 (–40) mm. long, 3–15 (–30) mm. wide, thinly hairy on both surfaces or subglabrous, conspicuously or inconspicuously gland-dotted, apex obtuse or shortly acute, margins flat, entire or remotely serrulate, minutely scabridulous with curved bristles; nervation rather distinct; petiole slender, to 1 cm. long; inflorescence terminal, generally rather elongate and narrow, 15–30 cm. long, 2·5–8 (–11) cm. wide, repeatedly branched; flowers subsessile, solitary in the axils of the bracts, forming small, oblong spikes 5–15 mm. long, 4–5 mm. wide; bracts oblong-obovate (2–) 4–5 (–11) mm. long, (1–) 2–4 (–7) mm. wide, loosely imbricate, purplish or greenish, sparsely strigillose externally with subincurved, adpressed bristles or subglabrous, conspicuously or inconspicuously gland-dotted internally, apex obtuse or subacute, margins entire; calyx tubular, about 2–2·5 mm. long, 1–1·3 mm. wide at apex, strigillose or glabrous and densely or sparsely gland-dotted externally, glabrous internally, teeth erect, subequal, deltoid, strigillose, about 0·8 mm. long, 0·5 mm. wide at base; corolla purplish-pink or white, puberulous and gland-dotted externally, tube narrowly cylindrical-infundibuliform, about 4 mm. long, 1·5 mm. wide at apex, limb oblique, 5-lobed, lobes subequal, oblong, about 1 mm. long, 0·8 mm. wide at base, apex rounded; stamens inserted about 1 mm. above base of corolla-tube; filaments glabrous, subequal, about 2–5 mm. long, slightly exceeding corolla-tube; anthers pallid, with divergent thecae, about 0·6 mm. wide; ovary about 2 mm. wide, glabrous; disk very small, obscurely lobed; style 6–7 mm. long; stigmatic-lobes subequal, subulate, about 0·4 mm. long. Nutlets oblong, terete, about 0·8 mm. long, 0·4 mm. wide, dark brown, minutely granulose; attachment area small, with 2 shallow, collateral depressions.

ssp. **hirtum** (*Link*) *Ietswaart* in Taxon. Rev. Gen. Origanum, 112 (1980); Davis, Fl. Turkey, 7: 310 (1982).

    O. hirtum Link, Enum. Plant. Hort. Berol., 2: 114 (1822); Boiss., Fl. Orient., 4: 552 (1879); Holmboe in Cypr. Journ., no. 4: 93 (1907).

    [*O. vulgare* L. ssp. *heracleoticum* (non *O. heracleoticum* L.) Holmboe, Veg. Cypr., 162 (1914).]

    [*O. heracleoticum* (non L.) Osorio-Tafall et Seraphim, List Vasc. Plants Cyprus, 92 (1973).]

    TYPE: Greece; Hagion Oros peninsula, Kerasia, 1891, *Sintenis & Bornmüller* 848 (neotype, G, W, K !).

Stems and leaves thinly or rather densely pilose; leaves densely and conspicuously gland-dotted below; bracts generally greenish, 2–5 mm. long; corollas rather small, generally white.

HAB.: Stony ground; near sea-level; fl. July–Sept.

DISTR.: Division 1, very rare. East Mediterranean region, from Ischia to Turkey.

1. Paphos, "endroits pierreuses. 27.8.01" *M. G. Michaelides* s.n. (O !)

NOTES: Evidently very rare, but perhaps overlooked because of its late flowering.

SECT. 4. **Majorana** (*Mill.*) *DC.* Inflorescence branched, consisting of clustered, condensed, spiciform verticillasters; bracts not large or foliaceous, generally greenish; calyx turbinate, usually with a deep abaxial slit.

4. O. syriacum *L.*, Sp. Plant., ed. 1, 590 (1753); Ietswaart, Taxon. Rev. Gen. Origanum, 87 (1980); Davis, Fl. Turkey, 7: 309 (1982).

    O. maru L., Sp. Plant., ed. 2, 825 (1763); Boiss., Fl. Orient., 4: 553 (1879); H. Stuart Thompson in Journ. Bot., 44: 336 (1906) pro parte; Post, Fl. Pal., ed. 2, 2: 334 (1933).

    Majorana syriaca (L.) Rafin., Autikon Bot., 119 (1840); Feinbrun, Fl. Palaest., 3: 153, t. 253 (1978).

    O. maru L. var. *viridulum* H. Stuart Thompson in Journ. Bot., 44: 336 (1906) nomen.

    O. pseudo-onites Lindberg f., Iter Cypr., 29 (1946).

Erect shrub or subshrub, 30–130 cm. high; stems much branched, the older ones woody, subterete, covered with flaking, dull brown bark, the younger ones bluntly tetragonal, glandular, rather densely clothed with a mixed indumentum of short and long spreading hairs; leaves subsessile or shortly petiolate, with petioles up to 5 mm. long, generally subtending condensed, small-leaved axillary shoots, lamina ovate, 0·8–3 cm. long, 0·5–2 cm. wide, sparsely or densely canescent-tomentellous, margins entire or subentire, narrowly recurved; nervation distinctly impressed above, often conspicuously and prominently reticulate below; inflorescence not extending far down the stem, subcorymbose or thyrsoid, consisting of 2–5 (or more) lateral pairs of clustered partial inflorescences topped by a terminal cluster, branches of lateral inflorescences rather rigid, divaricate, often very short, but occasionally up to 3 cm. long; flowers in dense ovoid, oblong or subglobose, 4-ranked spikes lengthening to 2 cm. (or more) in fruit; bracts imbricate, broadly ovate or obovate, subacute, obtuse or sometimes emarginate, about 2·5 mm. long, 2 mm. wide, concave, gland-dotted and densely grey-pubescent externally and ribbed with prominent nerves, smooth and subglabrous internally; calyx sheath-like, slit almost to the base abaxially, about 1·8 mm. long, 1·5 mm. wide, gland-dotted externally, apex entire, rounded, subtruncate or slightly emarginate, margins grey-ciliate; corolla white, tube infundibuliform, about 4 mm. long, 2 mm. wide at apex, gland-dotted and puberulous externally, lobes about 1 mm. long, broadly and bluntly deltoid, the adaxial shortly emarginate, the 3 abaxial subequal; stamens inserted in a thinly lanuginose zone about 1 mm. from apex of corolla-tube; filaments glabrous, about 3 mm. long, conspicuously exserted; anthers lilac-purple, reniform, about 0·5 mm. wide; thecae divergent; ovary glabrous, about 0·5 mm. diam., the 4 lobes seated on a conspicuous, shortly stipitate disk; style glabrous, about 3 mm. long in androdynamous flowers, 6 mm. long, and exserted in gynodynamous flowers; stigma shortly and acutely 2-lobed. Nutlets broadly oblong or suborbicular, about 0·8 mm. long, 0·7 mm. wide, distinctly compressed, rich brown, minutely granulose.

var. **syriacum**
TYPE: Probably the illustration in Lobel, *Plantarum seu Stirpium Icones*, t. 499 (1591), said by Lobel (*Plant. Hist.*, 265, *Advers.*, 213; 1576) to be based on specimens collected in Syria: "Venetiis Alepo Syrię habuimus".

Leaves and inflorescences rather densely canescent-tomentellous, the indumentum obscuring the leaf-nervation and sessile glands.

HAB.: Steep rocky banks on limestone, sometimes on limestone walls, 500–700 ft. alt.; fl. June–Nov.
DISTR.: Division 7, rare. S.E. Turkey (chiefly Amanus Range), Syria, Lebanon, Palestine, ? Egypt.
7. Lapithos, 1880–1951, *Sintenis & Rigo* 612! *Lindberg f.* s.n.! *Kennedy* 1639! *Casey* 964! *Merton* 604! St. Hilarion, 1900, *A. G. & M. E. Lascelles* s.n.!

var. **bevanii** (*Holmes*) *Ietswaart* in Taxon. Rev. Gen. Origanum, 88 (1980); Davis, Fl. Turkey, 7: 309 (1982).
O. bevanii Holmes in the Perfumery and Essential Oil Record, 6: 19 (Jan. 1915); B. L. Burtt in Kew Bull., 1949: 40 (1949); Osorio-Tafall et Seraphim, List Vasc. Plants Cyprus, 92 (1973).
Majorana bevanii (Holmes) A. Cheval. in Rev. Bot. Appliq., 18: 596 (1938).
TYPE: Cyprus; "Prope Lapitham [Lapithos], Cyprus, Junio 10, 1913, W. Bevan. Nom. vernac. Rikhanon." (K!)

Leaves and inflorescences thinly and shortly canescent-tomentellous; undersides of leaves conspicuously and closely reticulate-veined and gland-dotted.

HAB.: Rocky slopes on limestone; c. 700 ft. alt.; fl. June–Nov.
DISTR.: Division 7, rare. Apparently endemic, but see Notes.
7. Near Lapithos, 1913, *W. Bevan* s.n.!
NOTES: The distinctions between typical *O. syriacum* and var. *bevanii* are scarcely worth maintaining, the latter being little more than a calvescent state of the former. Cyprus material agrees in all respects with specimens from S.E. Turkey, Syria and Lebanon. Palestine specimens of *O. syriacum* depart from the type in having densely tomentose inflorescences and small, velutinous leaves. Although in the past sometimes distinguished as a species (*Origanum maru* L., Sp. Plant., ed. 2, 825 (1763); *O. vestitum* Clarke, Travels, 2 (1): 451 (1812); *Majorana crassifolia* Benth., Labiat. Gen. et Sp., 339; 1834), the Palestine plant is scarcely more than a third variety of *O. syriacum*.
The oil-content of Cyprian *O. syriacum* has been investigated (Bull. Imp. Inst., 15: 305–307; 1917): the total yield of oil has been found to be only about half that obtained from *O. dubium* Boiss., and the phenols present include a mixture of about 41 parts of carvacrol and 34 of thymol, whereas *O. dubium* contains carvacrol alone. For several reasons *O. syriacum* is considered less valuable economically than *O. dubium*.

5. O. dubium *Boiss.*, Fl. Orient., 4: 553 (1879); A. H. Unwin, Goat grazing and Forestry in Cyprus, 102–103 (1928); Osorio-Tafall et Seraphim, List Vasc. Plants Cyprus, 92 (1973).
Majorana dubia (Boiss.) Briq. in Engl. et Prantl, Pflanzenfam., IV, 3a: 307 (1896).
O. syriacum L. ssp. *dubium* (Boiss.) Holmboe, Veg. Cypr., 162 (1914) pro parte et quoad nomen; Lindberg f., Iter Cypr., 29 (1946) pro parte
[*O. majoranoides* (non Willd.) Holmes in The Perfumery and Essential Oil Record, 4: 41 (1913).]
[*O. syriacum* (non L.) Chapman, Cyprus Trees and Shrubs, 74 (1949).]
[*O. majorana* (non L.) Ietswaart, Taxon. Rev. Gen. Origanum, 83 (1980) pro parte; Davis, Fl. Turkey, 7: 308 (1982).]

TYPE: To be selected from the material cited by Boissier in Fl. Orient., 4: 553 (G).

Erect shrub or subshrub, 30–70 (–100) cm. high; stems much branched, the older ones woody, subterete, covered with flaking, dull brown bark, the younger ones sharply tetragonal, glandular, rather densely pubescent with short crispate hairs sometimes intermixed with a few longer, straighter hairs; leaves shortly petiolate, with petioles up to 5 mm. long, generally subtending condensed, small-leaved axillary shoots, lamina broadly and bluntly ovate or elliptic, 5–25 mm. long, 3–20 mm. wide, gland-dotted and canescent-furfurascent with short crispate hairs, margins entire; nervation fairly prominent but not reticulate below; inflorescence not extending far down the stem, thyrsoid or subcorymbose, consisting of 2–5 (–7) lateral pairs of clustered partial inflorescences topped by a terminal cluster, branches of lateral inflorescences rather rigid, divaricate, sometimes short and crowded, but up to 3·5 cm. long; flowers in subglobose or oblong, 4-ranked spikes, generally less than 1 cm. long in fruit; bracts imbricate, suborbicular or broadly obovate, rounded, concave, gland-dotted and densely pubescent externally, subglabrous and gland-dotted internally, inconspicuously nerved; calyx sheath-like, slit almost to base abaxially, gland-dotted externally with ciliate margins, about 1·5 mm. long, 1·3 mm. wide; corolla white, tube infundibuliform, about 2·5 mm. long, 1 mm. wide at apex, gland-dotted and minutely puberulous externally, lobes about 1 mm. long, deltoid, the adaxial shortly emarginate, the abaxial subequal, acute with recurved margins; stamens inserted in a thinly lanuginose zone about 0·5 mm. below base of corolla-lobes; filaments glabrous, exserted, about 2 mm. long; anthers (?) pallid, reniform, about 0·3 mm. wide with divergent thecae; ovary glabrous, about 0·3 mm. diam., the 4 lobes seated on a conspicuous, shortly stipitate disk; style glabrous, about 4 mm. long, exserted; stigma shortly and acutely 2-lobed. Nutlets bluntly oblong, about 0·8 mm. long, 0·6 mm. wide, distinctly compressed, rich brown, minutely granulose.

HAB.: Amongst garigue on dry, rocky hillsides, on igneous rocks; 2,000–4,000 ft. alt.; fl. June–Aug.

DISTR.: Divisions 1–3, [7]. Also [? Naxos], S. Turkey (chiefly Amanus Range).

1. Lyso, 1913, *Haradjian* 886!
2. Locally common; Yialia, 1905, *Holmboe* 798; Stavros tis Psokas, 1905–1948, *Holmboe* 1040; *Chapman* 170! 171! *Syngrassides* 1605! *Lindberg f.* s.n.! *Davis* 1774! *Kennedy* s.n.! Platres, 1937, *Kennedy* 740! 741! Kambos, 1948, *Kennedy* s.n.! and, 1968, *Economides* in ARI 1202! 1973, *P. Laukkonen* 373!
3. Limassol, 1905, *Michaelides* teste *Holmboe*.
[7.] Pentadaktylos, near Armenian Monastery, 1939, *Lindberg f.* s.n.! See Notes.

NOTES: This is the *Rigani* or *Riganis* of the Cypriots, and has long been valued as the source of a valuable aromatic oil. It occurs chiefly in the Paphos Forest, where it is locally abundant, and seems to be confined to igneous rocks, so that one must question the accuracy of Lindberg's record from Pentadaktylos, though the specimen (K!) is correctly identified. It is just possible that the plant may occur on the Northern Range, but some confusion is more likely, since Lindberg did not distinguish the *Rigani* of the south-western forests from the *Sapsishia* (*O. majorana* L.) which is so common on the Northern Range.

Investigations into the oil content of *O. dubium* (Bull. Imp. Inst., 11: 50 (1913); Holmes in The Perfumery and Essential Oil Record, 4: 41; 1913) have shown that it contains over 80 per cent of carvacrol, while *O. majorana* yields only terpineol and terpinene.

6. O. majorana *L.*, Sp. Plant., ed. 1, 590 (1753); Ietswaart, Taxon. Rev. Gen. Origanum, 83 (1980) pro parte

Majorana hortensis Moench, Method., 406 (1794); Briquet in Engl. et Prantl, Pflanzenfam., IV, 3a: 307 (1896).

Much-branched, bushy shrub 50–100 cm. high; old branches woody, subterete, with flaking greyish-brown bark, young stems sharply tetragonal, grey-pubescent with crispate hairs; leaves oblong, obovate, suborbicular or subspathulate, 2–25 mm. long, 2–15 mm. wide, distinctly petiolate, petiole slender, generally less than 8 mm. long, lamina thinly or densely grey-pubescent with short, adpressed, crispate hairs, rarely greenish and subglabrous, apex rounded or subacute, base broadly or narrowly cuneate, or rounded, occasionally subcordate, margins entire; nervation very obscure, not prominent or reticulate below; inflorescence narrow, consisting of 5–7 (or more) pairs of clustered partial inflorescences extending far down the flowering stems, branches of lateral inflorescences divaricate, subequal, usually less than 2 cm. long, commonly very short; flowers in compact, subglobose or very shortly oblong, 4-ranked spikes, generally less than 1 cm. long in fruit; bracts broadly obovate or spathulate, 3–3·5 mm. long, 2–2·5 mm. wide, gland-dotted and tomentellous externally, glabrous or subglabrous and gland-dotted internally, apex rounded or obtuse; calyx sheath-like, split almost to the base abaxially, about 2 mm. long, 1·5 mm. wide, conspicuously gland-dotted and tomentellous externally; corolla white, tube infundibuliform, about 2 mm. long, 1·5 mm. wide at apex, pubescent and thinly gland-dotted externally, lobes about 1 mm. long, deltoid, the adaxial shortly emarginate, the abaxial subequal, acute with recurved margins; stamens inserted in a thinly lanuginous zone about 0·5 mm. below base of corolla-lobes; filaments glabrous, exserted, about 2 mm. long; anthers pallid, reniform, about 0·3 mm. wide with divergent thecae; ovary glabrous, about 0·3 mm. diam., the 4 lobes seated on a conspicuous, shortly stipitate disk; style glabrous, 3–4 mm. long, exserted; stigma shortly and acutely 2-lobed. Nutlets suborbicular, about 1 mm. diam., distinctly compressed, rich brown, minutely granulose.

var. **tenuifolium** *Weston*, Universal Botanist, 3: 530 (1772).
    O. majoranoides Willd., Sp. Plant., 3: 137 (1800); Stapf in Bull. Imp. Inst., 11: 50 (1913).
    [*O. majorana* (? L.) Kotschy in Unger et Kotschy, Die Insel Cypern, 269 (1865).]
    [*O. microphyllum* (non (Benth.) Boiss.) Sintenis in Oesterr. Bot. Zeitschr., 32: 192, 291 (1882); Osorio-Tafall et Seraphim, List Vasc. Plants Cyprus, 92 (1973).]
    [*O. maru* (non L.) Holmes in The Essential Oil and Perfumery Record, 4: 69 (1913).]
    TYPE: Not indicated, but based on *Majorana tenuifolia* C. Bauhin, *Pinax*, 224 (1671).

Plant woody, suffruticose; leaves small, often broadly obovate or suborbicular, densely grey-pubescent; the whole inflorescence long, narrow, extending far down the flowering stems, with short lateral branches; otherwise as in *O. majorana* var. *majorana*, the cultivated Sweet or Knotted Marjoram.

    HAB.: In garigue on dry limestone hillsides, or in open spaces in *Pinus brutia* forest; sea-level to 3,000 ft. alt.; fl. May–Oct.

    DISTR.: Divisions 1–3, 7, 8, locally abundant. Endemic, but occurs as a cultivated (and naturalized) plant in many parts of the world.
1. Lyso, 1913, *Haradjian* 836! Between Polis and Stroumbi, 1934, *Syngrassides* 845! Akamas, 1948, *Kennedy* 1640! Smyies, 1948, *H. Michaelides* 2! Kakoskala, 1957, *Merton* 3050!
2. Aphamis, 1937, *Kennedy* 742!
3. Near Trimiklini, 1954, *Mavromoustakis* 17!
7. Common; Myrtou, Pentadaktylos, St. Hilarion, Ayios Epiktitos, Ayios Amvrosios, Koronia, Ardhana, Buffavento, Halevga, Yaïla, Mandres, Mersinniki, Lapithos, Kyrenia, etc. 1862–1955, *Kotschy* 937, *Sintenis & Rigo* 678! *A. G. & M. E. Lascelles* s.n.! *Syngrassides* 189! 1294! 1736! *Chapman* 243! 241! 247! 265! *Kennedy* 738! 1638! *Davis* 1730! 3605! *D. F. Davidson* 1! *C. Koutsoftas* 10! *V. Yakoumi* 11! *Miss Mapple* 110! *G. E. Atherton* 49! 240! 259! etc.
8. Common, 1880–1968; Apostolos Andreas, Ambeli, Aphendrika, Kantara Castle, Yioti, Gonies, Peristerges, Yialousa, etc. *Davis* 2313! *J. Voskarides* 4! *J. Spyrou* 13! *N. Faik* 5! *K. Payatsos* 9! *Economides* in ARI 1190!

    NOTES: Although regarded as a distinct species by Stapf and others, there can be little doubt that var. *tenuifolium* Weston (*O. majoranoides* Willd.) is a habitat variant, of very minor taxonomic significance, of the well-known and widely cultivated Sweet Marjoram, *O. majorana* L. The latter is thinly canescent or greenish, with larger, subspathulate leaves, and is generally grown as an annual, flowering in its first year, though it will perennate in warm climates, and then becomes indistinguishable from var. *tenuifolium*. This latter is the Marjoram Gentle of the old herbals, but, as Bauhin (1671) remarks "Haec apud nos in Majoranam vulgarem [*Origanum majorana* L.] degenerat". J. R. T. Vogel (in Linnaea, 15: 78) concludes that *O. majoranoides* Willd. consists of specimens of *O. majorana* kept alive in a cool greenhouse. The species was never reliably hardy in N. Europe, and rarely ripened fruits there, being grown as an annual from seed collected in the Mediterranean area.

If it is correct that *O. majorana* var. *majorana* and *O. majorana* var. *tenuifolium* (*O. majoranoides*) represent cultivated and wild variants (or states) of the same species, then we may also conclude that Sweet Marjoram originated in, and is a native of Cyprus, where it is unquestionably wild, and apparently endemic. This solves a problem which has long vexed botanists. Furthermore a fresh significance is added to Pliny's comment (Nat. Hist., xxi, xciii) that "the most valued and most fragrant samsuchum or amaracum comes from Cyprus". It also adds relevance to Virgil's charming verses on the removal of the child Ascanius to Cyprus (Idalium), "ubi mollis amaracus illum floribus et dulci adspirans complectitur umbra" (Aeneid, 1: 692–694), and to the fable (see Poiret, Hist. Plantes Europ., 4: 439; 1827) that Amaracus was perfumer (or possibly son) to Cinyras, King of Cyprus, metamorphosed into Marjoram after his untimely death. *Amaracus* (or *amaracum*) and *samsuchum* (the basis of the modern "Sapsishia" or "Sampsishia" of the Cypriots) are the most generally used names for the plant and its allies in ancient literature.

4. THYMUS *L.*

Sp. Plant., ed. 1, 590 (1753).
Gen. Plant., ed. 5, 257 (1754).

Aromatic shrubs and subshrubs; leaves small, entire, usually gland-dotted; flowers often gynodioecious, borne in spikes, heads or verticillasters; bracts foliaceous or sometimes enlarged and coloured; bracteoles minute; calyx ovoid, campanulate or tubular, pilose internally around the throat, 10–13-nerved, 2-lipped, upper lip 3-dentate, lower lip with 2 longer, narrower teeth; corolla-tube included or exserted, limb 2-lipped, the upper (adaxial) lip entire or emarginate, the lower (abaxial) lip equally or unequally 3-lobed; stamens 4, included or exserted, subequal or didynamous; anthers with parallel or divergent thecae; disk entire; style divided apically into 2 short, equal or unequal stigmatic lobes. Nutlets ovoid or oblong, smooth.

About 350 species, chiefly in Europe, the Mediterranean region and western Asia.

Corolla-tube 10–15 mm. long; leaf-margins revolute; sprawling, lax subshrub -　　**1. T. integer**
Corolla-tube about 6 mm. long; leaf-margins flat, not revolute; rigid, erect, small bushy shrubs
　　　　　　　　　　　　　　　　　　　　　　　　　　　　　　2. T. capitatus

1. T. integer *Griseb.*, Spic. Fl. Rum. et Bith., 2: 116 (1844); Holmboe, Veg. Cypr., 162 (1914); Lindberg f., Iter Cypr., 31 (1946); Chapman, Cyprus Trees and Shrubs, 74 (1949); Osorio-Tafall et Seraphim, List Vasc. Plants Cyprus, 92 (1973).

　　T. billardieri Boiss., Diagn., 2, 4: 8 (1859), Fl. Orient., 4: 560 (1879); Unger et Kotschy, Die Insel Cypern, 269 (1865).

　　[*T. villosus* (non L.) Sm. in Sibth. et Sm., Fl. Graec. Prodr., 1: 422 (1809), Fl. Graec., 6: 62, t. 578 (1826).]

TYPE: "In Archipelagi insulis" (OXF), but see Notes.

Sprawling, gnarled subshrub generally less than 10 cm. high; branches numerous, erect or prostrate, sometimes rooting, the older covered with fissured greyish-brown bark, young shoots obscurely tetragonal, reddish-purple, thinly clothed with short, retrorse, white bristly hairs; leaves sessile, crowded on short axillary branchlets, sessile, linear-lanceolate, acute, 3–10 mm. long, 1–2 mm. wide, gland-dotted, with a mixed indumentum of very short pubescence and long, white multicellular hairs, the latter frequently confined to the strongly revolute margins; nervation obscure; inflorescence terminal, subcapitate, consisting of 5–12 (or more) shortly pedicellate flowers; bracts lanceolate, 5–7 mm. long, 1–2 mm. wide, purplish, ciliate with white bristles, apex acute, subentire or obscurely serrulate, margins recurved; pedicels about 1 mm. long, densely pubescent; calyx purple, tube narrowly campanulate, about 3 mm. long, 1·5 mm. wide, distinctly 10-nerved, pubescent externally, bearded internally at apex, adaxial lip 3 mm. long, 2·5 mm. wide at base, with 3 short, subulate apical teeth, abaxial lip divided into 2 subulate, pectinate-ciliate teeth about 3 mm. long, 0·5 mm. wide at base; corolla varying in colour from white to dark rosy-purple, tube narrowly cylindrical-infundibuliform, 10–15 mm. long, about 3 mm. wide at apex, shortly pubescent externally, adaxial lip semicircular, very shortly emarginate, about 2 mm. diam., abaxial lip deeply 3-lobed, the lobes suborbicular, reflexed, the median about 2 mm. diam., the laterals slightly smaller; stamens strongly didynamous, the 2 adaxial included or shortly exserted with glabrous filaments about 3 mm. long, the 2 abaxial exserted, with filaments about 5 mm. long; disk and ovary glabrous, about 0·5 mm. diam.; style glabrous, 15–17 mm. long; stigma-lobes acute, about 0·3 mm. long. Nutlets subglobose, slightly compressed, about 0·6 mm. diam., dark reddish-brown, minutely pitted.

HAB.: Dry, rocky, igneous hillsides; 300–5,600 ft. alt.; fl. March–June.

DISTR.: Divisions 1–3 [7]. Endemic.

1. Akamas, 1941, *Davis* 3316! Karavopetres near Ayios Yeoryios, 1962, *Meikle* 2061! Melanos near Ayios Yeoryios, 1962, *Meikle* 2151! Between Inia and Smyies, 1979, *Edmondson & McClintock* E 2756!
2. Evrykhou, 1859, *Kotschy* s.n.; Khrysorroyiatissa Monastery, 1862, *Kotschy* 691; Prodhromos, 1862, *Kotschy* 832! also, 1961, *D. P. Young* 7280! Troödhitissa Monastery, 1880, *Sintenis & Rigo* 729! Stavros tis Psokas, 1933, *A. Foggie* 150! Platania, 1933, *Chapman* 57! and, 1939, *Lindberg f.* s.n.! Platres, 1937, *Kennedy* 737! also, 1941, *Davis* 3151! and, 1958, *N. Macdonald* 68! 1973, *P. Laukkonen* 112! Vavatsinia, 1941, *Davis* 2694! Makheras road, 1949, *Casey* 277! and, 1950, *Chapman* 329! Phileyia sawmills above Pyrgos, 1957, *Merton* 2868! Between Prodhromos and Ayios Nikolaos, 1962, *Meikle* 2841! Asinou, 1967, *Polunin* 8537! Khionistra, 1970, *I. M. Hecker* 43! Pedhoulas, 1976, *Holubová* s.n.!
3. Stavrovouni, 13 April, 1787, *Sibthorp*; same locality, 1862, *Kotschy* 195; and, 1934–1959, *Syngrassides* 1389! 1480! *C. E. H. Sparrow* 4! Kophinou, 1901, *A. G. & M. E. Lascelles* s.n.! Alethriko, 1905, *Holmboe* 233! Near Mosphiloti, 1956, *Merton* 2546! Near Apsiou, 1979, *Edmondson & McClintock* E 2694!

[7.] "Am Fusse des Buffavento auf Wiener Sandstein"? 1862, *Kotschy*.

NOTES: Almost certainly a Cyprus endemic. Sibthorp and Smith's record from the Aegean Islands is doubtless an error, and Boissier (Fl. Orient., 4: 560) acknowledges that his earlier reference to a Labillardière specimen from Lebanon was a mistake. Most probably Labillardière material, like Sibthorp's, came from Stavrovouni. Apart from a few outlying stations in the Akamas, *T. integer* is confined to the igneous rocks of the Troödos massif, and Kotschy's record from Buffavento must be dismissed as highly questionable, especially as the species has not since been found in this area.

2. T. capitatus (*L.*) *Hoffsgg. et Link*, Fl. Port., 1: 123 (1809); Unger et Kotschy, Die Insel Cypern, 269 (1865); Boiss., Fl. Orient., 4: 560 (1879); Holmboe, Veg. Cypr., 162 (1914); Post, Fl. Pal., ed. 2, 2: 337 (1933); Lindberg f., Iter Cypr., 31 (1946); Chapman, Cyprus Trees and Shrubs, 75 (1949); Osorio-Tafall et Seraphim, List Vasc. Plants Cyprus, 92 (1973).

Satureja capitata L., Sp. Plant., ed. 1, 568 (1753); Sibth. et Sm., Fl. Graec., 6: 36, t. 544 (1826).

Coridothymus capitatus (L.) Reichb. f. in Oesterr. Bot. Wochenbl., 7: 161 (1857), Icones Fl. Germ. et Helv., 18: 40, t. MCCLXXI, fig. II (1858); Feinbrun, Fl. Palaest., 3: 155, t. 256 (1978); Davis, Fl. Turkey, 7: 382 (1982).

[*Thymus tragoriganum* (non L.) Sm. in Sibth. et Sm., Fl. Graec. Prodr., 1: 421 (1809) pro parte quoad plant. cypr.; Sibth. in Walpole, Travels, 15 (1820).]

[*Satureia spinosa* (non L.) Gaudry, Recherches Sci. en Orient, 195 (1855); Unger et Kotschy, Die Insel Cypern, 270 (1865); H. Stuart Thompson in Journ. Bot., 44: 336 (1906).]

[*Satureia thymbra* (non L.) Kotschy in Unger et Kotschy, Die Insel Cypern, 270 (1865); Holmboe, Veg. Cypr., 159 (1914).]

[*Thymbra spicata* (non L.) H. Stuart Thompson in Journ. Bot., 44: 336 (1906).]

TYPE: "*in* Creta, Baetica, Hispali, Graecia".

Erect, rigid, much-branched, aromatic shrub, forming low, dome-shaped bushes 5–25 cm. high, old wood covered with pale brown, fissured bark, twigs obscurely tetragonal, whitish or pale grey, closely pubescent with curved hairs; leaves sessile, usually subtending condensed shoots covered with tufts of reduced leaves, lamina 2–10 mm. long, 1–1·5 mm. wide, linear, subacute, thickish and shallowly carinate, conspicuously gland-dotted, clothed with a mixed indumentum of short greyish pubescence with a few long white, multicellular hairs usually confined to the margins, the latter flat and entire, flowers sessile in dense terminal heads about 10–15 mm. long, 8–10 mm. wide; bracts imbricate, foliaceous, linear-lanceolate or narrowly ovate, about 5 mm. long, 1·5–2 mm. wide, strongly keeled, conspicuously gland-dotted, with flat, pectinate-ciliate margins; calyx flattened dorsally, acutely angled or narrowly winged along the junction of the upper (adaxial) and convex lower lip, tube about 3 mm. long, 2 mm. wide, subglabrous externally except along the angled sutures, conspicuously gland-dotted, bearded internally at base of calyx-teeth, upper lip 3-toothed, the teeth narrowly deltoid, subacute, about 1 mm. long, 0·5 mm. wide at base, ciliate, the lateral teeth keeled in continuation of the angled sutures, lower lip with 2 subulate, ciliate teeth, about 2 mm. long, 0·3 mm. wide at base; corolla pinkish-purple (sometimes white), minutely pubescent externally, tube exserted, infundibuliform, about 6 mm. long, 2–3 mm. wide at apex, upper lip about 1 mm. long, 2 mm. wide, deeply emarginate, lower lip subequally 3-lobed, the lobes oblong, rounded, about 1·5 mm. long, 1 mm. wide at base; palate thinly pilose; stamens subequal, exserted; filaments glabrous, about 3 mm. long, inserted about 1 mm. below corolla-lobes; anthers purplish, reniform, about 0·5 mm. diam., thecae almost parallel, connective discoloured (? glandular) at apex; disk and ovary glabrous, about 0·5 mm. diam.; style about 4–5 mm. long; stigmatic lobes short, acute, apparently connivent until at the receptive stage. Nutlets subglobose, slightly compressed, about 0·8 mm. diam., bright golden-brown, minutely granulose.

HAB.: In garigue on dry rocky slopes and waste ground, occasionally on sand-dunes or on rocks by the sea; sea-level to 2,900 ft. alt.; fl. end of May–Oct.

DISTR.: Divisions 1–7, common in the lowlands. Widespread in the Mediterranean region.

1. Ayios Neophytos, 1939, *Lindberg f.* s.n.; Maa beach, 1967, *Economides* in ARI 274!
2. Evrykhou, ? 1950, *Chapman* 82; Platres, 1937, *Kennedy* 743!
3. Lefkara, 1905, *Holmboe* 1100; Yermasoyia River, 1905, *Michaelides* s.n.; Akrotiri Forest, 1939, *Mavromoustakis* 95! Stavrovouni, 1939, *Lindberg f.* s.n.; Ayios Therapon, 1957, *H. Painter* 27!
4. Larnaca, 1880, *Sintenis* s.n.; Cape Greco, 1958, *N. Macdonald* 83!
5. Between Xeri and Laxia, 1900, *A. G. & M. E. Lascelles* s.n.! Vatili, 1905, *Holmboe* 333; Athalassa, 1932, *Syngrassides* 409! and, 1939, *Lindberg f.* s.n.; Mia Milea-Mandres, 1933, *Syngrassides* 330! Trikomo, 1939, *Lindberg f.* s.n.
6. Nicosia, 1914, *Feilden* s.n.!
7. Common; Kyrenia Pass, Kazaphani, Kyrenia, Platani, etc. *Lindberg f.* s.n.! *G. E. Atherton* 82! 117! 130! 290! 721! *I. M. Hecker* 21! *Barclay* 1067! *P. Laukkonen* 448! etc.

5. SATUREJA *L.*

Sp. Plant., ed. 1, 567 (1753).
Gen. Plant., ed. 5, 247 (1754).

Annual or perennial herbs or subshrubs; leaves usually small, gland-dotted, entire; flowers in verticillasters or cymes; bracts inconspicuous, foliaceous; calyx tubular or campanulate, rarely gibbous, 10 (–13)-nerved, glabrous or pilose at the throat, teeth 5, generally subequal, sometimes obscurely 2-labiate and unequal; corolla with a straight tube and 2-labiate limb, adaxial lip flattish, ± emarginate, abaxial lip 3-lobed; stamens 4, generally included and divergent, didynamous, the adaxial pairs shorter than the abaxial; stigmatic-lobes nearly equal, subulate. Nutlets ovoid or oblong, nearly smooth.

About 30 species (in the restricted sense) widely distributed but particularly well represented in the Mediterranean region. Briquet (in Engl. et Prantl, Pflanzenfam., ed. 1, IV, 3a: 296) enlarges the genus to include *Calamintha, Acinos, Clinopodium* and *Micromeria*, here regarded as separate genera.

1. S. thymbra *L.*, Sp. Plant., ed. 1, 567 (1753); Boiss., Fl. Orient., 4: 567 (1879) pro parte excl. plant. cypr.; Post, Fl. Pal., ed. 2, 2: 339 (1933); Osorio-Tafall et Seraphim, List Vasc. Plants Cyprus, 91 (1973); Feinbrun, Fl. Palaest., 3: 147, t. 243 (1978); Davis, Fl. Turkey, 7: 315 (1982).

Thymus tragoriganum L., Mantissa 1, 84 (1767).
TYPE: "*in* Creta".

Erect, much-branched, aromatic shrub 10–30 (–50) cm. high; old branches covered with flaking, fissured, fuscous-brown bark, young shoots rather indistinctly tetragonal, gland-dotted, pubescent with short, downcurved white hairs; leaves sessile, generally subtending condensed growths with clustered, reduced leaves; nervation obscure, lamina narrowly obovate or elliptic, acute, rather thick, 7–18 mm. long, 2–7 mm. wide, carinate, shortly strigillose with white bristles, conspicuously dotted with numerous, brown, protuberant glands; inflorescence an elongate, terminal verticillaster, consisting of 3–6 (or more) equally spaced, dense, compact whorls; bracts very similar to cauline leaves in size, shape and indumentum; bracteoles lanceolate, acuminate, about 5 mm. long, 2 mm. wide; flowers sessile or subsessile; calyx narrowly campanulate, thinly hispidulous-pubescent and densely gland-dotted externally, glabrous internally, tube about 3·5 mm. long, 2·5 mm. wide at apex, teeth 5, subequal, erect, narrowly deltoid-subulate, about 3·5 mm. long, 1 mm. wide at base; corolla rosy-purple, tube straight, slightly gibbous at apex, thinly puberulous or subglabrous externally, glandular-puberulous internally, about 7 mm. long, adaxial lip oblong, shortly emarginate, flattish, about 3 mm. long, 2 mm. wide, abaxial lip with 3, reflexed, rounded lobes, the median about 2·5 mm. diam., the laterals distinctly smaller, palate concave, thinly pilose; stamens downcurved, strongly didynamous, exserted; filaments glabrous, the adaxial about 2·5 mm. long, inserted about 1 mm. below apex of tube, the abaxial about 5 mm. long, inserted about 2 mm. below apex of tube; anthers reniform, about 1 mm. wide, thecae strongly divergent, connective brownish, almost semicircular; ovary shortly papillose, about 0·5 mm. diam.; disk small; style downcurved at apex, about 10 mm. long, stigmatic lobes subequal, subulate. Nutlets oblong, about 1·8 mm. long, 0·8 mm. wide, fuscous, minutely granulose.

HAB.: Rocky limestone gully, in garigue; c. 600 ft. alt.; fl. April–July.

DISTR.: Division 7, very rare; Sardinia, Greece, Crete, and eastern Mediterranean region from Turkey to Palestine.

7. Mavriskala gully above Akanthou, ? 600 ft. alt., 1956, *D. Davidson in Merton* 2632!

NOTES: Only once collected in Cyprus, in January 1956, without flowers, but with a few persistent fruits. All other records for *Satureja thymbra* trace back to Sibthorp or Kotschy, and are misidentifications of the common *Thymus capitatus* (L.) Hoffsgg. et Link.

[HYSSOPUS OFFICINALIS *L.*, Sp. Plant., ed. 1, 569 (1753), Hyssop, an aromatic subshrub 20–60 cm. high, with rather crowded, lanceolate or narrowly oblong leaves, and spiciform, terminal inflorescences of rather large violet blue flowers with exserted stamens, is represented in herb. Kew by a specimen purporting to come from Cyprus, and labelled "Sintenis &

Rigo 563. *Micromeria juliana* Benth. In vineis pr. Galata 16/6 [1880]". No doubt the plant originally so labelled was *Micromeria*, and at some stage in the process of mounting was replaced by *Hyssopus*. The specimen came to Kew (1906) as part of Herbarium Churchillianum, and there is evidence (see Fl. Cypr., 1: 476) of other, similar misplaced labelling in this collection. Hyssop may, however, be cultivated as an ornamental plant in Cyprus gardens.]

6. MICROMERIA *Benth.*

in Bot. Register, 15: t. 1282 (1829) nom. cons.

Perennial herbs and subshrubs; leaves small, entire or toothed; inflorescences usually spiciform, consisting of superimposed, few- or many-flowered verticillasters; flowers small; calyx tubular, 13 (or 15)-nerved, teeth 5, subequal or slightly unequal and bilabiate, glabrous or pilose at the throat; corolla generally pink or purple, tube straight, limb bilabiate, the adaxial lip flattish, entire or emarginate, abaxial lip 3-lobed with the median lobe often larger than the lateral; stamens 4, didynamous, the abaxial longer than the adaxial; anthers connivent under the adaxial lip, thecae sub-parallel or divergent; stigmatic lobes subequal, subulate. Nutlets ovoid or oblong, almost smooth.

About 100 species widely distributed in warm-temperate and tropical regions; the genus is often united, wholly or partly, with *Satureja* L. and *Calamintha* Mill.

Stems subadpressed-pubescent with retrorsely curved, white hairs:
Flowers distinctly pedicellate; bracteoles about as long as pedicels; calyx villous with long white hairs - - - - - - - - - - - **1. M. nervosa**
Flowers subsessile, forming compact, crowded verticillasters; bracteoles much longer than pedicels; calyx shortly pubescent - - - - - - - **2. M. myrtifolia**
Stems clothed with very short spreading hairs, generally intermixed with much longer, spreading hairs:
Flowers distinctly pedicellate; bracteoles about as long as pedicels; nutlets ovoid-oblong, about 0·7 mm. long, 0·4 mm. wide - - - - - - **3. M. microphylla**
Flowers subsessile; bracteoles much exceeding the very short pedicels; nutlets oblong, about 1 mm. long, 0·4 mm. wide - - - - - - - - **4. M. chionistrae**

1. **M. nervosa** *(Desf.) Benth.*, Labiat. Gen. et Sp., 376 (1834); Unger et Kotschy, Die Insel Cypern, 271 (1865); Boiss., Fl. Orient., 4: 569 (1879); C. et W. Barbey, Herborisations au Levant, 99 (1880); Post, Fl. Pal., ed. 2, 2: 339 (1933); Chapman, Cyprus Trees and Shrubs 73 (1949); Osorio-Tafall et Seraphim, List Vasc. Plants Cyprus, 91 (1973); Feinbrun, Fl. Palaest., 3: 149, t. 245 (1978); Davis, Fl. Turkey, 7: 340 (1982).
Satureia nervosa Desf., Fl. Atlant., 2: 9, t. 121, fig. 2 (1978); Holmboe, Veg. Cypr., 159 (1914).
TYPE: "in fissuris rupium Atlantis" (P).

Erect, much-branched subshrub 20–50 cm. high; old branches with flaking, pale, grey-brown bark, young shoots sharply tetragonal, thinly glandular or eglandular, rather densely clothed with a subadpressed indumentum of strongly retrorse, curved, white hairs; leaves subsessile, generally without axillary tufts of reduced leaves, lamina ovate, acute, 5–12 mm. long, 2–8 mm. wide, subglabrous or thinly white-pilose, gland-dotted below, margins narrowly recurved, nervation subimpressed above, usually prominent below; inflorescence spiciform, 6–18 (or more) cm. long, 1–1·5 cm. wide, consisting of 6–14 (or more) spaced or approximate, rather loosely clustered, many-flowered verticillasters; rhachis clothed with numerous, retrorse-subadpressed hairs; bracts foliaceous, closely resembling leaves in size and shape; bracteoles filiform, 1·5–2 mm. long, hispidulous, about as long as the hispidulous pedicels; calyx-tube narrowly cylindrical, about 2·5 mm. long, 0·8 mm. wide, villous with long white, spreading hairs externally, bearded internally at throat, 13-nerved, commonly purplish, inconspicuously bilabiate, the 3 adaxial teeth about 0·8 mm. long, subulate, villous, reflexed, the 2 abaxial similar, but about 1·2 mm. long; corolla purple-pink, minutely puberulous externally, tube about 5 mm. long, 2 mm. wide at infundibuliform apex, minutely retrorse-strigillose internally, adaxial lip oblong, slightly concave, about 1·2 mm. long, 1 mm. wide, very shortly emarginate, abaxial lip subequally 3-lobed, the lobes rounded, about 1 mm. diam.; stamens inserted about 1 mm. below throat of corolla; filaments glabrous, those of the adaxial stamens about 0·5 mm. long, the abaxial about 1 mm. long; anthers about 0·3 mm. wide with strongly divergent thecae and an expanded connective; ovary glabrous, about 0·4 mm. diam.; disk with distinct vertical furrows; style glabrous, about 3 mm. long; stigmatic lobes subequal, subulate, about 0·4 mm. long. Nutlets narrowly and bluntly oblong, slightly compressed, about 1 mm. long, 0·5 mm. wide, mid-brown, minutely granulose.

HAB.: On dry rocky ground in garigue and batha; sea-level to 2,000 ft. alt.; fl. Febr.–April.

DISTR.: Divisions 1, 3, 4, 6–8, often common. Mediterranean region from Balearics, Italy and Sicily to Palestine, Egypt, Libya, Tunisia and Algeria.

1. Peyia, 1937, *Chapman* 278! Smyies, 1948, *Kennedy* 1608! Near Baths of Aphrodite, 1962, *Meikle* 2220! and, 1979, *Edmondson & McClintock* E 2724! Fontana Amorosa, 1978, *J. Holub* s.n.!
3. Mazotos, 1862, *Kotschy*; Limassol, 1903, *A. G. & M. E. Lascelles* s.n.! Alethriko, 1905, *Holmboe* 229; between Kophinou and Stavrovouni, 1937, *Syngrassides* 1502! Yerasa, 1941, *Davis* 3080! Episkopi, 1954 *Mavromoustakis* 1! Amathus, 1960, *N. Macdonald* 16! and 1978, *J. Holub* s.n.! Kolossi, 1961, *Polunin* 6678! Lefkara, 1967, *Merton* in ARI 724!
4. Larnaca, 1862, *Kotschy* 247! Cape Greco, 1862, *Kotschy*; Psevdhas, 1908, *Clement Reid* s.n.! Near Larnaca, 1912, *Haradjian* 52! E. of Paralimni, 1958, *N. Macdonald* 52! Athna Forest, 1967, *Merton* in ARI 429! Dhekelia, 1972, *P. Laukkonen* 222! Near Hala Sultan Tekké, Larnaca, 1979, *Edmondson & McClintock* E 2951!
6. Nicosia, 1927, *Rev. A. Huddle* 86! Ayia Irini, 1941, *Davis* 2521! Near Orga, 1956, *Poore* 2640! 2659! and, 1967, *Merton* in ARI 119!
7. Kyrenia, 1932–1956, *Syngrassides* 277! *Kennedy* 766! *C. H. Wyatt* 34! *Casey* 424! *G. E. Atherton* 957! Ayios Amvrosios, 1937, *Miss Godman* 31! St. Hilarion, 1933, *Syngrassides* 60! Koronia, 1933, *Chapman* 48! 230! Halevga, 1952, *F. M. Probyn* 43! Kharcha, 1956, *G. E. Atherton*, 1055!
8. Near Rizokarpaso, 1880, *Sintenis & Rigo* 128! Valia Forest, 1938, *Chapman* 334! Between Ayios Philon and Rizokarpaso, 1941, *Davis* 2250! Apostolos Andreas Forest, 1956, *Merton* 2474! Yialousa, 1970, *A. Genneou* in ARI 1448!

2. M. myrtifolia *Boiss. et Hoh.* in Boiss., Diagn., 1, 5: 19 (1844); Lindberg f., Iter Cypr., 29 (1946); Rechinger f. in Arkiv för Bot., ser. 2, 1: 430 (1950); Osorio-Tafall et Seraphim, List Vasc. Plants Cyprus, 91 (1973); Feinbrun, Fl. Palaest., 3: 149, t. 246 (1978); Davis, Fl. Turkey, 7: 341 (1982).

　　M. juliana (L.) Benth. var. *myrtifolia* (Boiss. et Hoh.) Boiss., Fl. Orient., 4: 570 (1879); Post, Fl. Pal., ed. 2, 2: 340 (1933).

　　[*M. juliana* (non L.) Benth.) Sintenis in Oesterr. Bot. Zeitschr., 32: 124, 192 (1882); A. K. Jackson in Kew Bull., 1934: 272 (1934); Chapman, Cyprus Trees and Shrubs, 73 (1949).]

　　[*M. graeca* (non L.) Benth.) Chapman, Cyprus Trees and Shrubs, 173 (1949).]

TYPE: Iraq; "in rupestribus *Gara* [Gara Dagh] *Kurdistaniae* Kotschy" (K!).

Erect, much-branched subshrub 20–50 cm. high; old branches with flaking, pale, grey-brown bark, young shoots sharply tetragonal, thinly glandular or eglandular, pubescent with a subadpressed (rarely somewhat spreading) indumentum of strongly retrorse, short, curved, white hairs; leaves subsessile or shortly petiolate, generally without axillary tufts of reduced leaves, lamina ovate (rarely lanceolate), acute, 5–12 mm. long, 1–6 mm. wide, subglabrous or thinly white-pilose, gland-dotted below, margins narrowly recurved; nervation indistinct above, usually prominent below; inflorescence rigidly spiciform, up to 30 cm. long, usually less than 1·2 cm. wide, consisting of 10–23 rather regularly spaced, compact, dense, hemispherical, many-flowered verticillasters; rhachis clothed with numerous, retrorse-subadpressed hairs; bracts similar to cauline leaves, but often narrower; bracteoles linear-subulate or narrowly lanceolate, up to 6 mm. long, about 0·5 mm. wide, thinly or densely and shortly strigillose with a prominent midrib and strongly recurved margins; pedicels generally very short, or flowers subsessile; calyx-tube narrowly cylindrical, 2·5–3 mm. long, 0·8 mm. wide, shortly pubescent with spreading hairs externally, bearded at the throat internally, strongly 13-nerved, teeth subequal, erect or slightly spreading, subulate or very narrowly deltoid, about 0·5 mm. long, 0·2 mm. wide at base, shortly pubescent; corolla pink-purple, occasionally white (f. *albiflora* Lindberg f.), often very much reduced (? flowers cleistogamous), when fully developed minutely puberulous externally, tube about 3·5 mm. long, 1 mm. wide at apex, minutely retrorse-strigillose internally, adaxial lip oblong, flattish, about 1·2 mm. long, 0·8 mm. wide, distinctly emarginate, abaxial lip subequally 3-lobed, the lobes rounded-oblong, about 1 mm. long, 0·8 mm. wide; stamens inserted about 0·5 mm. below throat of corolla, filaments glabrous, those of the adaxial stamens about 0·5 mm. long, the abaxial about 1 mm. long; anthers about 0·5 mm. wide with strongly divergent thecae and an expanded connective; ovary glabrous, about 0·3 mm. diam.; disk slightly gibbous below the immature nutlets but not markedly sulcate; style glabrous, about 3·5 mm. long; stigmatic lobes subequal, subulate, about 0·4 mm. long. Nutlets oblong, bluntly angled, about 0·8 mm. long, 0·4 mm. wide, apex often subconical, base mucronate, pericarp mid-brown, minutely granulose.

HAB.: Rocky limestone and igneous hillsides, sometimes in open Pine forest; sea-level to 4,000 ft. alt.; fl. May–July.

DISTR.: Divisions 1–3, 6–8. Eastern Mediterranean region from Greece and Crete to Palestine, and eastwards to Iraq and Iran.

1. Cape Arnauti, 1913, *Haradjian* 809; Ayios Neophytos, 1939, *Lindberg f.* s.n.
2. Near Galata, 1880, *Sintenis & Rigo* 563! Kykko Monastery, 1913, *Haradjian* 923 pro parte; Ayia, 1934, *Chapman* 174! 175! Platres, 1937, *Kennedy* 767! Kryos Potamos, 2,900 ft. alt., 1937, *Kennedy* 764! Mesopotamos, 1939, *Lindberg f.* s.n.; between Ambelikou and Kambos, 1939, *Lindberg f.* s.n.
3. Kolossi, 1939, *Lindberg f.* s.n.; Stavrovouni, 1939, *Lindberg f.* s.n.; Akrotiri Forest, 1941, *Mavromoustakis* s.n.!
6. Near Ayios Yeoryios, Orga, 1956, *Merton* 2637!
7. Ayios Epiktitos, 1932, *Syngrassides* 280! Kyrenia, 1937, *Kennedy* 768! also, 1955, *G. E. Atherton* 258! 720! St. Hilarion, 1939, *Kennedy* 1491! also, 1939, *Lindberg f.* s.n.! 1941, *Davis* 3617! Bellapais, 1939, *Lindberg f.* s.n.; Lapithos, 1939, *Lindberg f.* s.n.; Pentadaktylos, 1939, *Lindberg f.* s.n., and *Kennedy* in K 2280! Buffavento, 1939, *Kennedy* 1493!
8. Kantara, 1880, *Sintenis & Rigo*.

NOTES: For the most part Cyprus plants appear to be cleistogamous with very small closed corollas, minute stamens, but well-developed ovaries and nutlets; chasmogamous flowers have been seen, however, but almost always on what appear to be weak, etiolate, grazed or starved plants.

3. M. microphylla (*Urv.*) *Benth.*, Labiat. Gen. et. Sp., 377 (1834); Boiss., Fl. Orient., 4: 572 (1879).

 Thymus microphyllus Urv. in Mém. Soc. Linn. Par., 1: 327 (1822).

 Micromeria cypria Kotschy in Unger et Kotschy, Die Insel Cypern, 270 (1865); Lindberg f., Iter Cypr., 29 (1946); Osorio-Tafall et Seraphim, List Vasc. Plants Cyprus, 91 (1973).

 Satureia graeca L. ssp. *cypria* (Kotschy) Holmboe, Veg. Cypr., 159 (1914).

 Micromeria cypria Kotschy var. *villosissima* Lindberg f., Iter Cypr., 29 (1946).

 M. graeca (L.) Benth. ssp. *cypria* (Kotschy) Chapman, Cyprus Trees and Shrubs, 73 (1949); Rechinger f. in Arkiv för Bot., ser. 2, 430 (1950).

 [*M. graeca* (L.) Benth. var. *latifolia* (non Benth.) Kotschy in Unger et Kotschy, Die Insel Cypern, 270 (1865).]

TYPE: Malta; "In collibus aridis insulae Melitae copiosissimè" *Dumont d'Urville* (P).

Prostrate or sprawling subshrub, forming loose tufts usually less than 15 cm. high; stems slender, much branched, the oldest covered with grey-brown, lightly fissured bark, the young shoots sharply tetragonal, clothed with a mixed indumentum of numerous short glandular hairs interspersed with a varying number of long, white, spreading or recurved eglandular hairs, occasionally almost without long hairs or densely white-villose; leaves shortly petiolate, generally without axillary tufts of reduced leaves, lamina ovate, acute, 3–9 mm. long, 1·5–5 mm. wide, puberulous with interspersed sparse, long white hairs, gland-dotted below, margins narrowly recurved, nervation indistinct above, usually prominent below; inflorescence lax, apparently consisting of numerous, loose, few (1–7)-flowered, axillary cymes or verticillasters, spaced or rather crowded along the apices of the slender flowering stems; rhachis densely glandular-pubescent, sometimes villose; bracts very similar to the cauline leaves; bracteoles linear-lanceolate, about 1–1·5 mm. long, 0·3–0·4 mm. wide, glandular-pubescent, about as long as, or somewhat shorter than the glandular-pubescent pedicels; calyx-tube cylindrical, sometimes slightly ventricose, 2–2·5 mm. long, about 0·8 mm. wide, glandular-pubescent externally, bearded at the throat internally, strongly 13-nerved, inconspicuously 2-lipped, the 3 adaxial teeth deltoid-subulate, about 0·5 mm. long, the 2 abaxial teeth subulate, about 0·8 mm. long, thinly villose; corolla pink-purple with white blotches on the palate, pubescent externally, tube about 4 mm. long, 1·5 mm. wide at apex, minutely retrorse-strigillose internally, adaxial lip broadly oblong, flattish, about 1–1·5 mm. long, 1–1·2 mm. wide, distinctly emarginate, abaxial lip unequally 3-lobed, lobes rounded-oblong, the median about 1·5 mm. diam., the laterals about 1·2 mm. diam.; stamens inserted about 1 mm. below throat of corolla; filaments glabrous, those of the adaxial stamens about 0·8 mm. long, the abaxial about 2 mm. long; anthers purplish, about 0·6 mm. wide, with strongly divergent thecae and an expanded, brown connective; ovary glabrous, about 0·3 mm. diam.; disk not noticeably gibbous or sulcate; style glabrous, about 5 mm. long; stigmatic lobes subequal, subulate, about 0·8 mm. long. Nutlets very small, ovoid-oblong with a rounded apex, about 0·7 mm. long, 0·4 mm. wide, minutely furfuraceous-granulose.

HAB.: Crevices of limestone rocks; 1,000–3,100 ft. alt.; fl. Jan.–Nov.

DISTR.: Division 7. Also Balearics, S. Italy, Sicily, Malta, ? Crete.

7. Pentadaktylos, 1862, *Kotschy* 338! also, 1880, *Sintenis & Rigo* 121! 1938, *Kennedy* 1226! Buffavento, 1862, *Kotschy* 390! 396! also, 1889, *Pichler*! 1937, *Syngrassides* 1737! 1939, *Kennedy* 1492! and, 1974, *Meikle* 4096! Antiphonitis Monastery, 1862, *Kotschy* 508! St. Hilarion, 1931, *J. A. Tracey* 70! 72! also, 1934, *Syngrassides* 670! 1939, *Kennedy* 1489! 1490! 1949, *Casey* 729! and, 1950, *Casey* 1030! Dhikomo, 1936, *Syngrassides* 992! Kyrenia 1936, *Kennedy* 765! also, 1955, *Miss Mapple* 70! and, 1973, *P. Laukkonen* 296! Yaïla 1938, *Kennedy* 1227! also, 1941, *Davis* 2855! 3636! Larnaka tis Lapithou, 1941, *Davis* 2096! Kornos, 1941, *Davis* 2995! Above Kythrea, 1966, *O. Huovila* 65!

NOTES: Examination of a range of specimens from Cyprus, Malta, Sicily and Italy, satisfies me that no clear distinction can be drawn between *Micromeria cypria* and the variable *M. microphylla*. Earlier authors were so intent on distinguishing the plant from *M. graeca*, that they seem to have overlooked this more obvious affinity. Cyprus plants tend to be rather more robust, villose and larger leaved than in other areas of the species' distribution, but there are many exceptions to this, too many to permit distinction at any rank.

4. M. chionistrae *Meikle* in Ann. Musei Goulandris, 6: 92 (1983). See App. V, p. 1898.

TYPE: Cyprus; Phini, 3,200 ft. alt., in cracks of bare rock; 6 June, 1939, *Kennedy* 1495 (K!).

Suberect or sprawling subshrub, forming rather dense, many-branched cushions up to 30 cm. high; stems slender, the oldest covered with dark, grey-brown bark, the young shoots sharply tetragonal, generally clothed with a dense, mixed indumentum of very short glandular pubescence and longer white, eglandular, patent hairs, sometimes without the longer hairs; leaves shortly but distinctly petiolate with a glandular-pilose petiole, generally without axillary tufts of reduced leaves, the basal leaves generally crowded, the upper leaves widely spaced, lamina ovate, acute, 3–5·5 mm. long, 2–2·5 mm. wide, glandular-puberulous, with intermixed, longer, eglandular hairs, or sometimes without the latter, not or very obscurely gland-dotted below, margins narrowly recurved; nervation obscure above, generally rather prominent below; inflorescence rather rigid, spiciform, 5–15 cm. (or more) long, about 0·8 cm. wide, consisting of 5–15 (or more) superimposed, equidistant, dense, 3–8-flowered verticillasters; rhachis glandular-puberulous, often pilose with spreading eglandular hairs; bracts very similar to cauline leaves; peduncles generally less than 1·5 mm. long, often very short, but usually distinct, glandular-puberulous; bracteoles linear-subulate, 2–2·5 mm. long, generally less than 0·3 mm. wide, glandular-puberulous, much exceeding the very short (almost non-existent) pedicels, and frequently extending beyond half the length of the calyx; calyx-tube cylindrical, 2·8–3 mm. long, 0·8 mm. wide, glandular-pubescent externally, bearded at the throat internally, strongly 13–nerved, scarcely 2-lipped, calyx-teeth subequal, about 1 mm. long, 0·3 mm. wide at base, narrowly deltoid-subulate, pubescent, often reddish; corolla pink-purple with a white throat, pubescent externally, tube about 5 mm. long, 1·5 mm. wide at apex, retrorse-strigillose internally, adaxial lip oblong, flattish, about 1·5 mm. long, 1–1·2 mm. wide, distinctly emarginate, abaxial lip subequally 3-lobed, lobes rounded–oblong, the median about 1·5 mm. long, 1 mm. wide, the laterals a little smaller; stamens inserted about 1 mm. below throat of corolla; filaments glabrous, those of the adaxial stamens about 0·8 mm. long, the abaxial about 1·8 mm. long; anthers about 0·6 mm. wide, with strongly divergent thecae; ovary glabrous, about 0·3 mm. diam.; style glabrous, 4–5 mm. long; stigmatic lobes subulate, subequal, about 0·5 mm. long. Nutlets oblong, about 1 mm. long, 0·4 mm. wide, somewhat angled, apex subconical, pericarp rich brown, minutely granulose.

HAB.: Crevices of igneous rocks; 2,800–5,000 ft. alt.; fl. June–Nov.

DISTR.: Division 2. Endemic.

2. Near Kykko Monastery, 1880, *Sintenis & Rigo* 121b! Kryos Potamos, 2,800–2,900 ft. alt., 1937 and 1939, *Kennedy* 763! 1497! Xerokolymbos, 4,300 ft. alt., 1938, *Kennedy* 1256! Khionistra, S.W. slopes, 4,800 ft. alt., 1938, *Kennedy* 1257! and, 1939, *Kennedy* 1494! Phini, 1939, *Kennedy* 1495! 1496! Above Stavros tis Psokas, 4,000 ft. alt., 1940, *Davis* 1788! Above Prodhromos, 1949, *Casey* 911! Above Troödhitissa Monastery, 1951, *Chapman* s.n.!

NOTES: Superficially very similar to *M. microphylla* (Urv.) Benth., but consistently distinguished by its subsessile flowers forming compact verticillasters, with very short pedicels much exceeded by the bracteoles. The nutlets are very like those of *M. myrtifolia* Boiss. et Hoh., small forms of which might be confused with *M. chionistrae*, but will be found to differ, *inter alia*, in the stem-indumentum of short, retrorse, curved hairs, that of *M. chionistrae* consisting of short, spreading glandular hairs, generally overtopped by an even layer of longer, eglandular hairs, which are usually much more numerous, and much less villous than the long hairs found in *M. microphylla* (Urv.) Benth.

[THYMBRA SPICATA *L.*, Sp. Plant., ed. 1, 569 (1753), an erect subshrub up to 55 cm. high, with linear or linear-lanceolate, sessile, gland-dotted, entire, leaves 1·5–2·5 cm. long; dense, terminal spiciform inflorescences with erect, entire, purplish, ciliate bracts, dorsally flattened, 13-nerved calyces, and pink-purple corollas with a straight tube and long, porrect upper lip, has been often recorded from Cyprus (Clarke, Travels, 2 (3): 724 (1816); Hume in Walpole, Mem. Europ. Asiatic Turkey, 254 (1817); Kotschy in Unger et Kotschy, Die Insel Cypern, 272 (1865); H. Stuart Thompson in Journ. Bot., 44: 336 (1906); Holmboe, Veg. Cypr., 161 (1914), etc.), but without satisfactory evidence of its occurrence on the island. In most, if not all, instances, the plant actually seen, but misidentified, was the common *Thymus (Coridothymus) capitatus* (L.) Hoffsgg. et Link. *Thymbra spicata* has a wide distribution in the eastern Mediterranean area, from Greece to

Palestine, and might be expected to occur in Cyprus. It is normally a plant of batha and garigue on dry, rocky, calcareous slopes, generally at low altitudes.]

7. CALAMINTHA *Mill.*

Gard. Dict. Abridg., ed. 4 (1754).

Aromatic perennial herbs, sometimes woody at the base; leaves frequently petiolate, ovate, subcordate or suborbicular, often crenate or crenate-serrate and prominently nerved; flowers in lax, few-flowered axillary cymes, forming lax verticillasters; bracts foliaceous; bracteoles filiform or subulate, usually shorter than the pedicels; calyx tubular, hairy internally at the throat, not noticeably gibbous, 13-nerved, 2-lipped, the adaxial lip with 3 short porrect teeth, the 2 abaxial teeth longer, spreading or recurved; corolla usually pink or purplish, 2-lipped, the upper lip truncate or emarginate, shallowly concave, the lower 3-lobed, the median lobe larger than the lateral; stamens 4, included or shortly exserted; anthers connivent, thecae divergent; style exserted; stigmatic lobes unequal, the adaxial subulate, the abaxial longer and wider. Nutlets sub-globose, ovoid or oblong, almost smooth.

About 7 species in Europe, the Mediterranean region and western Asia.

Leaves densely grey-pubescent or tomentellous, normally entire; adaxial calyx-teeth deltoid, less than 1 mm. long - - - - - - - - - - - **1. C. incana**
Leaves thinly pubescent, greenish, normally crenate or crenate-serrate; adaxial calyx-teeth subulate, commonly more than 1 mm. long - - - - - C. NEPETA (p. 1279)

C. NEPETA (*L.*) *Savi*, Fl. Pisana, 2: 63 (1798) (*Melissa nepeta* L., Sp. Plant., ed. 1, 593 (1753); *Satureja nepeta* (L.) Scheele in Flora, 26: 577; 1843), recorded by Post (in Mém. Herb. Boiss., 18: 97; 1900) "Près de Chrysorogiatiza . . . août [1894]", but the specimen has not been examined, nor are there any other records for the species from Khrysorroyiatissa, or from any other locality on the island. Since Post omits to mention the common *C. incana* (Sibth. et Sm.) Boiss., one suspects that the plant recorded was really this, though the identification may have been correct. *Calamintha nepeta* much resembles *C. incana*, but has green, pubescent (not grey-tomentose) leaves, and longer adaxial calyx-teeth; it has a wide distribution in southern and western Europe, but becomes rare in the eastern Mediterranean area, though recorded from Crete and S. Turkey. All records for *Calamintha* (or *Satureja*) *nepeta*, from Cyprus (H. Stuart Thompson in Journ Bot., 44: 336 (1906); Holmboe, Veg. Cypr. 160 (1914); Osorio-Tafall et Seraphim, List Vasc. Plants Cyprus, 91; 1973) are based upon the reference in Mém. Herb. Boiss. cited above.

1. C. incana (*Sm.*) *Boiss. ex Benth.* in Heldr., Plant. Atticae, 1844 (1845) et in DC., Prodr., 12: 226 (1848); Unger et Kotschy, Die Insel Cypern, 271 (1865); Boiss., Fl. Orient., 4: 578 (1879); Post, Fl. Pal., ed. 2, 2: 343 (1933); Osorio-Tafall et Seraphim, List Vasc. Plants Cyprus, 91 (1973); Feinbrun, Fl. Palaest., 3: 150, t. 249 (1978); Davis, Fl. Turkey, 7: 328 (1982).

Thymus incanus Sm. in Sibth. et Sm., Fl. Graec. Prodr., 1: 421 (1809), Fl. Graec., 6: 62, t. 577 (1827).

Satureia incana (L.) Briq., Labiat. Alp. Marit., 432 (1895); Holmboe, Veg. Cypr., 160 (1914); Lindberg f., Iter Cypr., 30 (1946); Chapman, Cyprus Trees and Shrubs, 73 (1949); Rechinger f. in Arkiv för Bot., ser. 2, 1: 430 (1950).

[*C. cretica* (non (L.) Lam.) Unger et Kotschy, Die Insel Cypern, 271 (1865); H. Stuart Thompson in Journ. Bot., 44: 336 (1906).]

TYPE: "In Archipelagi insulis frequens, et circa Athenas" *Sibthorp* (OXF).

Prostrate, sprawling or rarely erect perennial, up to 60 cm. high; old stems usually woody with flaking, brown bark, young stems rather sharply tetragonal, densely clothed with spreading, white hairs; leaves suborbicular, densely grey-pubescent with subadpressed, crispate pubescence, very variable in size, 4–10 (–16) mm. long, and almost as wide, aromatic with a strong, heavy smell ("like ether" *Kennedy*), apex rounded or obtuse, base truncate or rounded, margins entire or subentire, nervation usually fairly distinct and prominent on the undersurface; petiole very short or almost wanting, occasionally up to 6 mm. long; inflorescence lax, the 2–6-flowered cymes, occupying almost the whole length of the branches; rhachis either shortly glandular-puberulous, or glandular and with an additional indumentum of longer, white, eglandular hairs; bracts foliaceous, the lower much resembling the leaves, the upper generally much reduced; bracteoles lanceolate-subulate, very small, about 0·8–1 mm. long, 0·2–0·4 mm. wide, shortly glandular-puberulous; pedicels slender, 3–6 mm. long, glandular, or glandular-pubescent; calyx tubular, 2–2·5 mm. long, 1·5 mm. wide, puberulous or pubescent, gland-dotted, conspicuously nerved, 3 adaxial teeth deltoid, less than 1 mm. long, 2 abaxial teeth slender, subulate, about 1·5 mm. long, ciliate with white, bristly hairs, throat distinctly bearded internally; corolla pale mauve or lavender, often whitish towards the centre, with darker spots on the lower lip, tube narrowly infundibuliform, 7–8 mm. long, pubescent externally, thinly puberulous internally with retrorse hairs, upper lip broadly oblong, about 3 mm. long and almost as wide, shortly emarginate, lower lip 3-lobed, the lobes oblong-suborbicular, unequal, the median about 2·5 mm. long, 2 mm. wide, the laterals about 1·5 mm. long and almost as wide; stamens inserted about 5–6 mm. above base of corolla-tube; filaments curved, connivent, glabrous, the adaxial about 2 mm. long, the abaxial 4·5 mm. long, distinctly exserted; anthers connivent, often adherent in pairs, about 1 mm. wide, the thecae strongly divergent, connectives strongly expanded, almost pulviniform; ovary glabrous, about 0·5 mm. diam., seated on a short, sulcate, orange-brown disk; style 8–9 mm. long, glabrous, much exserted; stigmatic lobes very unequal, the adaxial slender, subulate, about 0·3 mm. long, the abaxial thicker, up to 0·6 mm. long. Nutlets suborbicular or lens-shaped, strongly compressed, about 0·8 mm. diam., dark brown, furfuraceous, minutely granulose.

HAB.: Dry rocky ground, fallow fields, roadsides, occasionally on sand-dunes or by dried up streams and irrigation channels; sea-level to 4,900 ft. alt.; fl. June–Dec.

DISTR.: Divisions 1–7, common. Eastern Mediterranean region from Greece to Palestine.

1. Ayios Neophytos Monastery, 1939, *Lindberg f.* s.n.; Ktima, 1940, *Davis* 1775! Lyso, 1970, *A. Genneou* in ARI 1631!
2. Near Troödhitissa Monastery, 1862, *Kotschy* 734a! Evrykhou, 1880, *Sintenis & Rigo* 727! Galata, 1913, *Haradjian* 1005! Mandria valley, 1937, *Kennedy* 759! Moniatis River, 3,900 ft. alt., 1937, *Kennedy* 1498! Below Platania, 1966, *Merton* in ARI 74!
3. Sand-dunes near Amathus, 1947, *Mavromoustakis* s.n.!
4. Larnaca, 1880, *Sintenis & Rigo* 727a! Between Prastio and Kouklia, 1939, *Lindberg f.* s.n.; Paralimni, 1939, *Lindberg f.* s.n.
5. Athalassa, 1932, *Syngrassides* 408!
6. Near Nicosia, 1862, *Kotschy* 980, and 1939, *Lindberg f.* s.n., also *Chapman* 182, and, 1940, *Davis* 2071! Myrtou, 1940, *Davis* 1970! and, 1955, *G. E. Atherton* 405! English School, Nicosia, 1959, *P. H. Oswald* 18! Nicosia–Morphou road, 1968, *A. Genneou* in ARI 1257!
7. Coast near Myrtou, 1862, *Kotschy*; Kyrenia, 1936, *Kennedy* 762! also, 1955, *G. E. Atherton* 424! 759! Yaila, 1940, *Davis* 1929! Vasilia, 1940, *Davis* 2002! Kazaphani, 1948, *Casey* 114! also, 1955, *G. E. Atherton* 111! 770! 771! Ayios Epiktitos, 1955, *G. E. Atherton* 239! Lapithos, 1955, *G. E. Atherton* 680! Kambyli, 1957, *Merton* 2990!

8. ACINOS *Mill.*

Gard. Dict. Abridg., ed. 4 (1754).

Annual or perennial herbs similar to *Calamintha*; leaves aromatic, petiolate or subsessile, crenate-serrate or entire, often conspicuously nerved; flowers in loose or compact, axillary cymes or occasionally solitary, forming lax or condensed verticillasters; bracts foliaceous; bracteoles small, subulate, usually shorter than the relatively robust, dorsiventrally flattened pedicels; calyx tubular, obscurely 2-lipped, gibbous towards base, somewhat constricted above, bearded internally at throat, nerves 13, prominent, often hispidulous; calyx-teeth 5, subequal, subulate, erect in flower and fruit; corolla narrowly infundibuliform, 2-lipped, pubescent externally, minutely glandular internally, adaxial lip truncate or emarginate, shallowly concave, abaxial (lower) lip 3-lobed, the median lobe larger than the laterals; stamens 4, the adaxial included, the abaxial shortly exserted; anthers with strongly divergent thecae, connective dilated; style exserted; stigmatic lobes unequal, the adaxial subulate, very short, the abaxial longer and wider. Nutlets subglobose, compressed or oblong, almost smooth.

About 10 species chiefly in southern and western Europe, the Mediterranean region and western Asia.

Annual; calyx hispid; corolla small, tube 6–7 mm. long, scarcely exceeding calyx
1. A. exiguus
Perennial; calyx shortly puberulous; corolla conspicuous, tube 9–10 mm. long, much exceeding calyx - - - - - - - - - - - - - - - **2. A. troodi**

1. **A. exiguus** (*Sm.*) *Meikle* **comb. nov.**
   Thymus exiguus Sm. in Sibth. et Sm., Fl. Graec. Prodr., 1: 421 (1809), Fl. Graec., 6: 61, t. 575 (1826).
   Calamintha exigua (Sm.) Hal., Consp. Fl. Graec., 2: 546 (1902).
   Satureia exigua (Sm.) Holmboe var. *integrifolia* Holmboe, Veg. Cypr., 161 (1914) quoad plant. cypr. nec *Acinos graveolens* M. Bieb. var. *integrifolius* Raulin, Descr. Phys. Ile Crète, 828 (1858) nom. nud.
   Satureia crassinervis Lindberg f. in Soc. Sci. Fenn. Årsbok, 20B, no. 7: 6 (1942), Iter Cypr., 30 (1946).
   Calamintha crassinervis (Lindberg f.) Osorio-Tafall et Seraphim, List Vasc. Plants Cyprus, 91 (1973) nom. invalid.
   [*Melissa graveolens* (non *Thymus graveolens* M. Bieb.) Benth., Labiat. Gen. et Sp., 390 (1834) pro parte quoad spec. cypr.; Poech, Enum. Plant. Ins. Cypr., 23 (1842).]
   [*Calamintha graveolens* (non (M. Bieb.) Benth.) Kotschy in Unger et Kotschy, Die Insel Cypern, 271 (1865); Boiss., Fl. Orient., 4: 583 (1879).]
   TYPE: "In insulae Cypri montosis" *Sibthorp* (OXF).

Diminutive annual, 8–40 mm. high; stems erect or spreading, simple or branched, minutely glandular and generally rather densely clothed with short, retrorse, white bristly hairs, often purplish; leaves broadly and bluntly ovate, 2–8 mm. long, 2–7 mm. wide, shortly puberulous, often purplish, margins entire or occasionally with a very few obscure, blunt teeth, narrowly recurved, base rounded or broadly cuneate, nervation rather obscure, the lowermost pair of lateral nerves not quite extending to leaf-margin; petiole slender, pubescent, generally less than 5 mm. long; inflorescence a compact, crowded verticillaster, consisting of closely superimposed pairs of axillary flowers, or rarely of 2 (–3)-flowered axillary cymes; bracts foliaceous, similar to leaves in shape and size, but often with the nervation conspicuously thickened and prominent on the undersurface; bracteoles filiform-subulate, pubescent, less than 0·8 mm. long; pedicels 1–2 mm. long, densely pubescent; calyx-tube 4–5 mm. long, 1·4 mm. wide, somewhat curved, distinctly gibbous towards base, prominently 13-nerved with hispid nerves, apex obscurely 2-lipped, the calyx-teeth subequal, subulate, the adaxial about 1·5 mm. long, the abaxial about 2 mm. long, hispidulous; corolla lavender, mauve or pinkish-purple, tube narrowly infundibuliform, 6–7 mm. long, 2 mm. wide at apex, puberulous externally, minutely and densely glandular within, adaxial lip porrect, broadly oblong, weakly concave, shallowly emarginate, about 0·8 mm. long, and about as wide; abaxial lip reflexed, divided into 3 rounded lobes about 0·7 mm. long and about as wide; stamens inserted about 1 mm. down corolla-tube; filaments glabrous, those of the adaxial stamens about 0·8 mm. long, included, those of the abaxial stamens about 1 mm. long, very shortly exserted; anthers purple, about 0·5 mm. wide, with strongly divergent thecae and a conspicuously expanded, glandular connective; ovary glabrous, about 0·5 mm. diam., disk distinctly sulcate; style 6–7 mm. long, stigmatic lobes unequal, the adaxial minute, subulate, the abaxial a little wider, about 0·5 mm. long. Fruiting pedicels erect, adpressed to rhachis, strongly flattened; nutlets oblong, about 1·7 mm. long, 1 mm. wide, subacute and shallowly excavate at the base, dark brown, furfuraceous, minutely granulose.

HAB.: Stony mountain summits, screes and gravelly streamsides or stream-beds, generally on igneous (mostly serpentine) formations; 2,500–6,400 ft. alt.; fl. April–July.

DISTR.: Divisions 2, 3. Endemic.

2. Near Phini, 1862, *Kotschy* 762! Summit of Khionistra, 1862, *Kotschy* 774, also, 1880, *Sintenis & Rigo* 737! and, 1937, *Kennedy* 754!, 1939, *Lindberg f.* s.n.! 1941, *Davis* 3515! Between Platres and Troödos, 1937, *Kennedy* 750! 751! Xerokolymbos, 4,400–5,800 ft. alt., 1937, *Kennedy* 752! 755! Kryos Potamos, 5,600 ft. alt., 1937, *Kennedy* 753! 758! E. of Khionistra, 5,500 ft. alt., 1937, *Kennedy* 756! Livadhi tou Pasha, 1937, *Kennedy* 757! Troödos, 1940, *Davis* 1796! and, 1960, *N. Macdonald* 192! Near Prodhromos, 1961, *D. P. Young* 7339! 7365! Tripylos, 1962, *Meikle* 2682! Near Kakopetria, 1962, *Meikle* 2846!
3. Hill behind Kakomallis above Louvaras, 1962, *Meikle* 2901!

NOTES: Closely allied to *Acinos rotundifolius* Pers., but uniformly minute, with compact inflorescences of small, pale flowers. A similar dwarf species replaces *A. rotundifolius* in Crete (and, possibly, Samos) and has been named *A. graveolens* M. Bieb. var. *integrifolius* by Raulin (op. cit. supra), but without description. In this the leaves and bracts are constantly bordered by a conspicuous, thickened lateral nerve; in Cyprus material this nerve is sited some distance within the margin, and the leaves are not distinctly bordered.

Sintenis & Rigo 570 (K!), labelled "In mont. supra Melanissiko 26/5/" [1880], appears to be fruiting material of *A. exiguus*, but I have not been able to find the locality in any gazetteer or atlas. The two collectors appear to have been in the neighbourhood of Kyrenia on 26 May 1880, but no other collector has since found *A. exiguus* or any other *Acinos* species in this area.

2. A. troodi (*Post*) *Leblebici* in Bitki, 1: 405 (1974).
    Calamintha troodi Post in Mém. Herb. Boiss., 18: 97 (1900) [as *C. troodii*]; H. Stuart Thompson in Journ. Bot., 44: 336 (1906); Osorio-Tafall et Seraphim, List Vasc. Plants Cyprus, 91 (1973).
    Satureia troodi (Post) Holmboe, Veg. Cypr., 161 (1914); Warr et Gresham in Journ. Roy. Hort. Soc., 59: 71 (1934); Lindberg f., Iter Cypr., 30 (1946).
    TYPE: "in Monte Troodi Cypri" (BEI).

Creeping or sprawling perennial, generally less than 5 cm. high; stems woody at the base, forming loose, intricately branched mats, glandular-pubescent above, with short, spreading hairs, commonly purplish and subterete or obscurely tetragonal; leaves broadly ovate or suborbicular, 2–7 mm. long, and as wide or a little wider, puberulous on both surfaces, apex apiculate, margins subentire or very obscurely crenulate, narrowly recurved, base subtruncate, narrowing abruptly to a petiole 1–4 mm. long; nervation obscure above, prominent below; inflorescence a short, rather congested verticillaster, consisting of approximate, 1–3-flowered, rather congested, axillary cymes; bracts similar to leaves, but often with more strongly defined nervation; bracteoles minute, subulate, puberulous, less than 1 mm. long; calyx-tube 5–6 mm. long, about 2 mm. wide, shortly and sparsely puberulous externally, bearded internally at throat, slightly curved, with an inconspicuously gibbous base and distinct longitudinal nervation, apex not distinctly 2-lipped, the adaxial calyx-teeth about 1·4 mm. long, subulate, hispid-ciliate, the abaxial longer and more slender, about 2 mm. long; corolla pink or purplish, or sometimes tube buff-coloured and lobes pink or purple, proportionately large, tube narrowly infundibuliform, puberulous externally, about 9–10 mm. long, 3 mm. wide at apex, upper (adaxial) lip broadly oblong, about 2 mm. long and almost as wide, shallowly concave, emarginate, lower (abaxial) lip about 3 mm. long, 4 mm. wide, divided into 3 rounded lobes 1–1·5 mm. diam., the sutures between the median and lateral lobes conspicuously pectinate-strigillose with retrorse bristles; stamens inserted 2–4 mm. down corolla-tube; filaments flattened, glabrous, the adaxial 1 mm. long, included, the abaxial 3 mm. long, exserted; anthers about 1·5 mm. wide, with patently divergent thecae and a conspicuously expanded connective; ovary about 0·5 mm. diam., glabrous; disk sulcate; style glabrous, about 12–13 mm. long, exserted; stigmatic lobes unequal, the adaxial minute, about 0·2 mm. long, the abaxial subulate, about 0·5 mm. long. Fruiting pedicels erect, subadpressed, strongly flattened; nutlets oblong, about 1·5 mm. long, 0·8 mm. wide, pointed and excavate at the base, rich brown, minutely granulose. *Plate 72, figs. 6–11.*

HAB.: Rocky slopes and screes at high elevations on serpentine; (5,000–) 5,500–6,400 ft. alt.; fl. June-Aug.

DISTR.: Division 2. Endemic.

2. Troödos, 1880, *Sintenis & Rigo* 736! also, 1894, *Post* s.n., and, 1912, *Haradjian* 478! Near Government House, Troödos, 1928, *Forestry Dept.* comm. *Druce*! E. side Khionistra, 1937, *Wyatt* 12! Summit of Khionistra, 1937, *Kennedy* 744–749! also, 1939, *Lindberg f.* s.n.! and, 1940, *Davis* 1801! 1856! Rocky slope of Moni tou Pellou, 5,000 ft. alt., 1949, *Casey* 917! Also Troödos area, 1952, *F. M. Probyn* 123! and, 1963, *D. P. Young* 7801! 7803!

9. CLINOPODIUM *L.*

Sp. Plant., ed. 1, 587 (1753).
Gen. Plant., ed. 5, 256 (1754).
R. von Bothmer in Bot. Notiser, 120: 202–208 (1967).

Perennial herbs resembling *Calamintha* Mill.; stems usually robust, tetragonal, retrorsely strigose; leaves generally petiolate, crenate-serrate, or subentire; flowers in dense axillary cymes, forming remote verticillasters; bracts foliaceous; bracteoles filiform, usually much longer than the short, terete pedicels; calyx tubular, slightly curved, not, or indistinctly gibbous at base, 13-nerved, rather indistinctly 2-lipped, the teeth of the adaxial (upper) lip a little shorter than those of the abaxial (lower) lip; corolla 2-lipped, the adaxial lip shallowly concave, subentire or emarginate, abaxial lip 3-lobed, the median lobe rather larger than the laterals; stamens 4, the adaxial included, the abaxial shortly exserted; anthers with strongly divergent thecae and a dilated connective; style exserted; stigmatic lobes very unequal. Nutlets oblong or ovoid, almost smooth.

About 3 or 4 species in Europe, Asia and North America. *C. vulgare* introduced elsewhere.

1. **C. vulgare** *L.*, Sp. Plant., ed. 1, 587 (1753); Osorio-Tafall et Seraphim, List Vasc. Plants Cyprus, 91 (1973); Feinbrun, Fl. Palaest., 3: 151, t. 250 (1978); Davis, Fl. Turkey, 7: 329 (1982).
    Melissa clinopodium Benth., Labiat. Gen. et Sp., 392 (1834); Poech, Enum. Plant. Ins. Cypr., 24 (1842).
    Calamintha clinopodium Spenner, Handb., 2: 429 (1835); Benth. in DC., Prodr., 12: 233 (1848); Unger et Kotschy, Die Insel Cypern, 271 (1865); Boiss., Fl. Orient., 4: 579 (1879).
    Calamintha vulgaris (L.) Karsten, Deutsche Fl., 1002 (1882); Druce in Ann. Scott. Nat. Hist., 1906: 224 (1906); Post, Fl. Pal., ed. 2, 2: 343 (1933) nom. illeg.
    Satureja vulgaris (L.) Fritsch, Excursionfl. Oesterr., 632 (1897); Holmboe, Veg. Cypr., 161 (1914); Lindberg f., Iter Cypr., 30 (1946); Rechinger f., in Arkiv för Bot., ser. 2, 1: 430 (1950).
    TYPE: "*in rupestribus* Europae, Canadae".

Erect perennial 20–80 cm. high; stems numerous, arising from slender, much-branched, subligneous rhizomes, usually rather sparingly branched or simple above, sharply tetragonal, clothed with numerous retrorse, white hairs; leaves ovate, 1–4·5 cm. long, 1–2·5 cm. wide, shortly hairy above and below, apex subacute or obtuse, sometimes apiculate, margins obscurely crenate-serrate or subentire, narrowly recurved, base rounded or broadly cuneate; petiole 2–7 mm. long, channelled above, hirsute; inflorescences usually of 2–3 remote, compact, many-flowered whorls, sometimes consisting of 1 subglobose terminal whorl; bracts closely resembling the leaves, but usually rather smaller; bracteoles filiform, up to 10 mm. long, setose with spreading bristles; cymes dense, many-flowered, repeatedly branched; pedicels short, 1–1·5 mm. long, terete, densely pubescent; calyx-tube 6–8 mm. long, 2·8–3 mm. wide at apex, with a mixed indumentum of short, glandular, and long, eglandular, spreading hairs, prominently 13-nerved, slightly curved but not noticeably gibbous at the base, calyx-limb obscurely 2-lipped, the adaxial teeth subulate, setulose, 2–3·5 (–4) mm. long, the abaxial 3·5–5·5 mm. long; corolla bright pink, puberulous externally, tube narrowly infundibuliform, 11–16 mm. long, 4–5 mm. wide at apex, upper lip slightly concave, oblong, about 4–5 mm. long, almost as wide, distinctly emarginate, lower lip 3-lobed, median lobe suborbicular, emarginate, 3·5–4·5 mm. diam., concave, lateral lobes 2·5–3 mm. diam., convex, palate bearded with thick, white hairs; stamens inserted about 2·5 mm. down corolla-tube; filaments flattened, glabrous, the adaxial about 2·5–3 mm. long, the abaxial 4–5 mm. long; anthers purplish, about 1 mm. wide, thecae strongly divergent, with a conspicuously expanded connective; ovary glabrous, about 0·8 mm. diam.; disk obscurely sulcate; style 11–15 mm. long, sparsely scabridulous towards apex; stigmatic lobes very unequal, the adaxial narrow, subulate, about 0·8 mm. long, the abaxial flattened, linear-lanceolate, 1–1·5 mm. long. Nutlets broadly oblong, about 1 mm. long, 0·8 mm. wide, bluntly trigonous, rich brown, granulose, with small, shallow, whitish, basal pits.

HAB.: On rocky hillsides, usually in the shade of bushes or near streams or springs, sometimes in open Pine forest, or by roadsides; 2,000–6,300 ft. alt.; fl. May–August.

DISTR.: Divisions 1, 2. Widespread in Europe and eastwards to Central Asia and India; Canada and U.S.A. Naturalized elsewhere.

1. Lyso, 1913, *Haradjian* 860!
2. Troödos area, 1862, *Kotschy*; 1900, *A. G. & M. E. Lascelles* s.n.! 1912, *Haradjian* 446! 1952, *F. M. Probyn* 127! 1966, *Merton* in ARI 55! 1973, *P. Laukkonen* 425! Yialia, 1905, *Holmboe* 803; Platania, 1935–36, *Syngrassides* 733! 1105! and, 1939, *Lindberg f.* s.n.! Xerokolymbos, 4,300 ft. alt., 1937, *Kennedy* 769! Kryos Potamos, 1937, *Kennedy* 770! Platres, 1937, *Kennedy* 772! Kykko, 1939, *Lindberg f.* s.n.; Stavros tis Psokas, 1940, *Davis* 1761! Khrysorroyiatissa, 1941, *Davis* 3366! Prodhromos, 1955, *G. E. Atherton* 554! and, 1961, *D. P. Young* 7291! also, 1968, *Economides* in ARI 1219! Cedar Valley, 1968, *Economides* in ARI 1240! Nicosia–Troödos road, at junction with Kyperounda road, 1973, *P. Laukkonen* 425!

NOTES: It is probable that all the above fall within the scope of *C. vulgare* L. ssp. *orientale* v. Bothmer (in Bot. Notiser, 120: 206; 1967). This has, on the whole, a longer calyx than typical *C. vulgare*, with the adaxial teeth less noticeably shorter than the abaxial teeth. Rather cursory examination also suggests that the median lobe of the lower corolla-lip tends to be larger, wider and more distinctly emarginate in western European material than in Cyprus specimens, but it is not easy to see a clear distinction between typical *C. vulgare* and ssp. *orientale*.

10. MELISSA *L.*

Sp. Plant., ed. 1, 592 (1753).
Gen. Plant., ed. 5, 257 (1754).

Aromatic perennial herbs; stems robust, tetragonal, glabrous or pilose; leaves petiolate, cordate, crenate-serrate; inflorescence lax, consisting of numerous, remote, superimposed verticillasters, the individual flowers often secund; bracts and bracteoles foliaceous; calyx tubular-campanulate, tube

13-nerved, not bearded at the throat internally, distinctly 2-lipped, upper
lip somewhat recurved, shortly 3-toothed, lower lip porrect, with 2 subulate
teeth; corolla 2-lipped, tube exserted, curved, without an internal ring of
hairs; upper lip erect, flattish or concave, emarginate, lower lip 3-lobed;
stamens 4, didynamous, included or very shortly exserted; anthers with
divergent thecae; stigmatic lobes subulate, subequal. Nutlets ovoid, almost
smooth.

Three species in Europe and Asia; *M. officinalis* L. cultivated and
sometimes naturalized outside this area.

1. **M. officinalis** *L.*, Sp. Plant., ed. 1, 592 (1753); Unger et Kotschy, Die Insel Cypern, 272 (1865);
Boiss., Fl. Orient., 4: 584 (1879); H. Stuart Thompson in Journ. Bot., 44: 336 (1906);
Holmboe, Veg. Cypr., 159 (1914); Post, Fl. Pal., ed. 2, 2: 345 (1933); Chapman, Cyprus
Trees and Shrubs, 72 (1949); Rechinger f. in Arkiv för Bot., ser. 2, 1: 430 (1950); Osorio-
Tafall et Seraphim, List Vasc. Plants Cyprus, 92 (1973); Feinbrun, Fl. Palaest., 3: 146, t.
242 (1978); Davis, Fl. Turkey, 7: 262 (1982).

    M. romana Mill., Gard. Dict., ed. 8, no. 2 (1768).

    M. officinalis L. var. *romana* (Mill.) Woodv., Med. Bot., 3: 398 (1792).

    M. altissima Sm. in Sibth. et Sm., Fl. Graec. Prodr., 1: 423 (1809), Fl. Graec., 6: 63, t. 579
(1826).

    M. officinalis L. var. *villosa* Benth., Labiat. Gen. et. Sp., 393 (1834); Lindberg f., Iter
Cypr., 29 (1946).

    TYPE: "*in montibus* Genevensibus, Allobrogicis, Italicis".

Robust erect perennial to 80 (–120) cm. high; stems arising from a
branched, creeping rhizome, generally unbranched except in the region of
the inflorescence, sharply tetragonal, subglabrous or ± hirsute with
spreading hairs; leaves pleasantly scented of lemons and thinly pubescent
(var. *officinalis*) or somewhat foetid and rather densely pubescent (var.
romana (Mill.) Woodv.), the lower cauline cordate, acute or subacute, 4–10
cm. long, 1·5–7 cm. wide, margins crenate or bluntly serrate, nervation
distinct but not very prominent; petiole slender, 2–6 cm. long, channelled
above, pubescent or hirsute; upper cauline leaves similar, but smaller and
often truncate or broadly cuneate at the base; inflorescence a large, lax
panicle, often 30–50 cm. long, with erecto-patent branches clothed with a
mixed indumentum of short glandular, and longer eglandular hairs,
verticillasters remote, 4–12-flowered, lax or rather congested, the individual
flowers generally secund; bracts foliaceous, the lowermost resembling the
cauline leaves, the upper much smaller, often under 2 cm. long, 1·5 cm. wide,
ovate or oblong with a truncate or cuneate base, shortly petiolate or
subsessile; bracteoles foliaceous, lanceolate, entire, about 5 mm. long, 2 mm.
wide, glandular-pubescent; pedicels slender, up to 6 mm. long, glandular;
calyx-tube tubular-campanulate, flaring at apex, 5–6 mm. long, 3–3·5 mm.
wide at mouth, glandular and sparsely or densely hirsute, distinctly nerved,
adaxial teeth recurved, connate almost throughout their length with 3 short
deltoid-spinulose apices, abaxial teeth subulate, to 3 mm. long; corolla
whitish tinged lilac, tube glabrous or subglabrous externally, pilose
internally, infundibuliform, curved, 8–11 mm. long, about 5 mm. wide at
apex, adaxial lip concave, puberulous, emarginate, about 2 mm. long, 2·5
mm. wide, abaxial lip 3-lobed, the lateral lobes suborbicular, about 2 mm.
diam., the median about 2 mm. long, 3–4 mm. wide; palate thinly pilose or
subglabrous; stamens inserted about 2 mm. down corolla-tube; filaments
glabrous, flattened, the adaxial about 2 mm. long, the abaxial about 3·5 mm.
long; anthers about 1·2 mm. wide, thecae strongly divergent, connective
inconspicuous; ovary glabrous, about 0·8 mm. diam.; disk distinctly gibbous
between the ovules; style glabrous, about 10 mm. long; stigmatic lobes
subequal, subulate, about 1 mm. long, recurved at maturity. Nutlets
narrowly ovoid, about 1·8–2 mm. long, 0·9–1 mm. wide, obscurely

trigonous, fuscous, granulose, tapering to a whitish base marked with a strongly curved, almost circular, scar.

HAB.: Damp thickets and streamsides, occasionally in open *Cistus* garigue, by roadsides or in cultivated ground; sea-level to 4,500 ft. alt.; fl. May–Sept.

DISTR.: Divisions 1–3, 5, 7, 8. S. Europe and Mediterranean region eastwards to Caucasus and Iran.

1. Paphos, 1901, *A. G. & M. E. Lascelles* s.n.! also, 1913, *Haradjian* 706! Lyso, 1913, *Haradjian* 853! Baths of Aphrodite, 1949, *Casey* 755!
2. Prodhromos, 1862, *Kotschy* 877! and, 1955, *G. E. Atherton* 624! Between Galata and Spilia, 1880, *Sintenis & Rigo* 942! Kykko, 1913, *Haradjian* 999; Platres, 1937, *Kennedy* 806! Near Phini 1939, *Kennedy* 1499! Aphamis, 1939, *Kennedy* 1500! Milikouri, 1939, *Lindberg f.* s.n.! Below Galata, 1956, *Merton* 2806! Stavros tis Psokas, without date, *Chapman* 176!
3. Kolossi Castle, 1939, *Lindberg f.* s.n.; and, 1941, *Davis* 3536!
5. Kythrea, 1880, *Sintenis & Rigo* 680! also, 1934, *Syngrassides* 536! and, 1939, *Lindberg f.* s.n.
7. Armenian Monastery, 1912, *Haradjian* 347! St. Hilarion, 1934, *Syngrassides* 678! Kyrenia 1936–1973, *M. E. Dray* 1a! *Syngrassides* 1332! *Casey* 777! *G. E. Atherton* 77! 187! 303! *P. Laukkonen* 445! Kyrenia Pass, 1939, *Lindberg f.* s.n.! Lapithos Forest, 1939, *Lindberg f.* s.n., and, 1948, *Kennedy* 1637!
8. Ayios Andronikos, 1937, *Syngrassides* 1657! Rizokarpaso, 1957, *Merton* 2934!

NOTES: Such is the variation in Cyprus populations of *Melissa officinalis* that subdivision at any rank is impracticable; it is possible that the thinly pubescent, lemon-scented var. *officinalis* has been selected out of populations of the commoner var. *romana* (Mill.) Woodv. and perpetuated in cultivation. To judge from herbarium labels, at least some of the Cyprus specimens are from cultivation, or escapes from cultivation, though others are unquestionably from spontaneous and indigenous plants. I can see no useful taxonomic distinctions in the adaxial calyx-teeth of var. *officinalis* and var. *romana* (*M. altissima* Sm.).

11. ZIZIPHORA *L.*

Sp. Plant., ed. 1, 21 (1753).
Gen. Plant., ed. 5, 13 (1754).

Annual or rarely perennial herbs; leaves entire or obscurely toothed, aromatic; inflorescence capitate or spiciform, consisting of remote, or crowded, few- or many-flowered verticillasters; bracts foliaceous; bracteoles minute; calyx-tube narrowly cylindrical, 13-nerved, obscurely 2-lipped, hairy internally at the throat, calyx-teeth often connivent; corolla pink or purple, 2-lipped, tube slender, without an internal hairy zone, adaxial lip erect, entire or emarginate, abaxial lip 3-lobed, lobes suborbicular, the median larger than the laterals; 2 adaxial stamens rudimentary or absent, 2 abaxial stamens exserted, connivent under the adaxial corolla-lip; anthers fused along their internal margins, one theca fertile, the other sterile, sometimes reduced to a small appendage; style included or shortly exserted; stigmatic lobes very unequal, the abaxial subulate, the adaxial minute. Nutlets ovoid or oblong, almost smooth.

About 25 species in S. Europe, the Mediterranean region, and eastwards to Central Asia.

1. **Z. capitata** *L.*, Sp. Plant., ed. 1, 21 (1753); Sibth. et Sm., Fl. Graec. Prodr., 1: 12 (1806); Fl. Graec., 1: 10, t. 13 (1806); Poech, Enum. Plant. Ins. Cypr., 23 (1842); Unger et Kotschy, Die Insel Cypern, 268 (1865); Boiss., Fl. Orient., 4: 586 (1879); Holmboe, Veg. Cypr., 159 (1914); Post, Fl. Pal., ed. 2, 2: 346 (1933); Osorio-Tafall et Seraphim, List Vasc. Plants Cyprus, 92 (1973); Feinbrun, Fl. Palaest., 3: 146, t. 240 (1978); Davis, Fl. Turkey, 7: 397 (1982).
TYPE: "*in* Syria".

Erect annual 3–14 cm. high; stems simple (in starved specimens) or much branched, bluntly tetragonal, densely pubescent with retrorse, crispate hairs; cauline leaves lanceolate or narrowly ovate, acute, 0·8–4 cm. long, 0·3–1·2 cm. wide, mid-green, subglabrous or thinly scabridulous above, pale green or purplish, sparsely puberulous and thinly gland-dotted below, margins entire or remotely and bluntly serrulate, nervation generally obscure; petiole to about 1 cm. long; inflorescence terminal, capitate, 1·5–3 cm. diam.; bracts foliaceous, broadly ovate, acute or cuspidate, shortly petiolate or subsessile, 10–25 mm. long, 5–15 mm. wide, nervation prominent, sub-parallel below, margins narrowly recurved, entire, ciliate partly or all around, with conspicuous white bristles; bracteoles minute, oblong-acute, about 1 mm. long; flowers numerous, crowded; pedicels very short, usually less than 2 mm. long, pubescent; calyx-tube narrow, 7–9 mm. long, about 1·5 mm. wide, straight or slightly curved, scarcely gibbous, minutely puberulous externally with few or numerous long, eglandular hairs, densely bearded at the throat internally, teeth subequal, erect or usually connivent, narrowly subulate, about 2 mm. long; corolla pale or bright mauve, pink or purple, tube very slender, about 9 mm. long, 0·8 mm. wide at apex, glabrous towards base, glandular-puberulous towards apex externally, glabrous internally, adaxial lip about 1·5 mm. long, 1 mm. wide, suborbicular or bluntly oblong, emarginate, abaxial lip 3-lobed, about 2 mm. long, lateral lobes suborbicular, about 1 mm. diam., entire, median lobe bluntly oblong, about 1·5 mm. long, emarginate; palate glabrous; adaxial stamens wanting, abaxial stamens inserted about 1 mm. down corolla-tube; filaments about 1·8 mm. long, glabrous, purplish; anthers fused along their inner margins, narrowly oblong, about 0·7 mm. long, 0·2 mm. wide, purple; ovary and disk glabrous, about 0·5 mm. diam.; style glabrous, included, about 8 mm. long; stigmatic lobes very unequal, the abaxial narrowly subulate, about 0·4 mm. long, the adaxial minute. Nutlets oblong, about 1·5 mm. long, 0·8 mm. wide, mid-brown, obscurely trigonous, tapering to a subacute, whitish, shallowly excavate base, minutely granulose.

HAB.: In garigue on calcareous or igneous hillsides; occasionally in shingle of dried-up stream-beds or by margins of cultivated fields; 200–3,800 ft. alt.; fl. March–May.

DISTR.: 1–3, 7, 8. S.E. Europe and eastern Mediterranean region eastwards to Caucasus and Iran.

1. Stroumbi, 1913, *Haradjian* 749!
2. Platres, 1938, *Kennedy* 1339!
3. "In monte Crucis insulae Cypri. [1787] D. F. Bauer", same locality [Stavrovouni], 1974, *Meikle* 4061! Kakoradjia, 1937, *Syngrassides* 1509! Akrotiri Forest, 1939, *Mavromoustakis* 123! Skarinou, 1941, *Davis* 2629! Mile 25, Nicosia–Limassol road, 1950, *Chapman* 607! Near Souni, 1960, *N. Macdonald* 123! Near Apsiou, 1962, *Meikle* 2888!
7. Near Ayios Khrysostomos Monastery, 1862, *Kotschy* 433! Kalogrea ["Kalorgha"], 1880, *Sintenis & Rigo* 124! Yaïla, 1941, *Davis* 2829! Larnaka tis Lapithou, 1941, *Davis*, s.n.! Above Kyrenia Pass, 1953, *Casey* 1304!
8. Near Kantara, 1880, *Sintenis & Rigo* 124a! Yioti, 1905, *Holmboe* 532; Matakos near Galinoporni, 1962, *Meikle* 2446! Mutsefla, Khalosta Forest, 1962, *Meikle* 2454!

TRIBE 4. **Salvieae** *Dumort.* Corolla strongly 2-lipped, the adaxial lip commonly curved with inflexed margins; stamens 2; filaments articulated; anthers with narrow thecae and a linear-filiform connective; style gynobasic. Nutlets dry; embryo straight.

12. SALVIA *L.**
Sp. Plant., ed. 1, 23 (1753).
Gen. Plant., ed. 5, 15 (1754).

Shrubs, perennial, biennial or rarely annual herbs, usually strongly and pleasantly aromatic; leaves undivided, lobed or pinnatisect; inflorescence of variously arranged cymes, often in false whorls (verticillasters); verticillasters 2–12 (–40)-flowered, remote or approximating; bracts always present, bracteoles sometimes so; calyx bilabiate, tubular, urceolate to infundibuliform, slightly or clearly expanding in fruit, with 10–15 veins,

* by I. C. Hedge

prominent or not; upper lip with three teeth, median usually smaller than laterals; lower lip with two equal teeth, longer than or equal to those of upper lip; corolla 2-lipped, white, cream to rose or violet; upper lip (hood) straight, curved or clearly falcate, emarginate or bifid, lower lip (labellum) 3-lobed with median lobe clearly larger than laterals and usually concave, often reflexed, tube included in or exserted beyond the calyx teeth, straight, curved or ventricose, annulate within or not, invaginated and with a small internal scale (squamulate) or not (esquamulate); stamens 2, each consisting of a large fertile theca at the upper end, and at the lower either a subfertile smaller theca or variously shaped often dolabriform sterile tissue, connectives clearly separating the thecae, short or much elongated, filaments short, stamens usually articulating at the point of attachment of the filament and connective; staminodes (posterior pair of stamens) always present, very small; ovary borne on a short fleshy disc, style elongated; stigmatic lobes subulate or flattened, unequal. Nutlets usually 4, glabrous, oblong to spherical, often trigonous, smooth, veined or not, frequently producing mucilage on wetting.

About 800 species throughout most warmer parts of the New and Old World. South-west Asia is a major centre for the genus.

Leaves pinnate or pinnatisect:
  Calyx 5–8 mm. long, densely covered with long white eglandular hairs; leaf-lobes ± linear
10. S. lanigera
  Calyx 8–14 mm. long, ± densely covered with long glandular hairs; leaf lobes oblong to ovate
3. S. pinnata
Leaves simple (not divided), entire, lobed or trilobed with a large terminal leaflet and 1 pair of small basal lobules:
  Annuals, with or without a coma of coloured sterile bracts topping inflorescence; fruiting calyces strongly deflexed   -   -   -   -   -   -   -   -   -   **4. S. viridis**
  Perennials or biennials, without a coma; fruiting calyces not or somewhat deflexed:
    Leaf-margins subentire to finely crenulate; leaves simple or some trilobed, rugulose; lower thecae fertile:
      Shrub; calyx 6–10 mm. long; corolla lavender, pink or white   -   -   **2. S. fruticosa**
      Suffruticose herb; calyx 10–14 mm. long, corolla white   -   -   -   **1. S. willeana**
    Leaf-margins irregularly crenulate, erose or lobed; leaves all simple, rugose to ± flat; lower thecae sterile:
      Leaves white-lanate/pannose, at least when young:
        Inflorescence widely candelabriform; leaves ovate to oblong; flowers white
5. S. aethiopis
        Inflorescence not candelabriform; leaves very broadly ovate to suborbicular; flowers lilac   -   -   -   -   -   -   -   -   -   -   **7. S. veneris**
      Leaves green with a pilose to villous indumentum:
        Leaves crenate, serrate, erose or lobed; corolla c. 12 mm.   -   -   -   **9. S. verbenaca**
        Leaves irregularly crenulate or serrulate or sub-erose; corolla 15–25 (–30) mm. long:
          Corolla white to cream; calyx with a dense indumentum of long white eglandular hairs   -   -   -   -   -   -   -   -   -   -   **6. S. dominica**
          Corolla dark pink to purple; calyx with a capitate-glandular indumentum
8. S. hierosolymitana

1. **S. willeana** (*Holmboe*) *Hedge* in Notes Roy. Bot. Gard. Edinb., 23: 47 (1959).
    S. grandiflora Etl. ssp. *willeana* Holmboe, Veg. Cypr., 157, fig. 53 (1914); Lindberg f., Iter Cypr., 29 (1946); Chapman, Cyprus Trees and Shrubs, 71 (1949); Osorio-Tafall et Seraphim, List Vasc. Plants Cyprus, 90 (1973).
    S. grandiflora Etl. f. *albiflora* Lindberg f., Iter Cypr., 29 (1946).
    [*S. grandiflora* (non Etl.) Unger et Kotschy, Die Insel Cypern, 267 (1865); Boiss., Fl. Orient., 4: 593 (1879); Post in Mém. Herb. Boiss., 18: 97 (1900).]

    TYPE: (lectotype selected here) "Cyprus, circa Prodromo in monte Troodos, 30 Junio 1880, *Sintenis et Rigo* 726" (K!) — distributed as *S. grandiflora* Etl. var. *brachyodonta* Boiss. nom. nud.

Low-growing, strongly aromatic suffruticose herb, sometimes carpeting the ground; stems ascending to erect, c. 25–55 cm. long, tetragonal, below with mostly pilose to villous eglandular hairs, above with a dense indumentum of short capitate glandular hairs of varied length, sessile glands and sometimes also with eglandular hairs; leaves distributed over stem, ovate to elliptic, 1·5–6 cm. long, 0·8–3·2 cm. wide, simple or with a pair of small basal lobules, thickly pubescent to velutinous above, below with a dense indumentum of eglandular and small capitate glandular hairs and numerous sessile glands, apex ± obtuse, margins subentire to finely and regularly crenulate, base rounded or cordate, clearly rugulose; petiole 0·8–4·2 cm. long below, decreasing in length upwards; inflorescence shoots usually with lateral branches, verticillasters c. 5–8, 6–10-flowered, distant below, approximating above, viscid; lower bracts leaf-like, median ovate, c. 1–1·5 cm. long, 0·5–1 cm. wide, sometimes purplish; bracteoles present; pedicels 2–8 mm. long, erect-spreading; calyx ± tubular, 10–14 mm. long, scarcely elongating in fruit, green or veined purple, 15-veined, densely covered with capitate glandular hairs and few villous eglandular hairs; upper lip with three subequal triangular teeth, lower lip with two c. 4 mm. teeth, equal to upper; corolla white or tinged mauve, (15–) 20–25 mm. long; tube c. 15 mm. long, gradually widening towards throat, annulate c. 7 mm. from base; upper lip straight, scarcely emarginate, glandular-pilose externally; lower lip ± equalling upper, with a suborbicular not deflexed median lobe and short very broadly ovate lateral lobes; staminal connectives c. 6 mm. long; lower thecae fertile, clearly smaller than upper; staminal filaments c. 7 mm. long; style glabrous with unequally bifid stigmatic lobes. Nutlets c. 3 mm. long and 2·5 mm. wide, ± ovoid, rounded-trigonous, dark brown, without venation, not (?) mucilaginous on wetting.

HAB.: Moist, rocky mountainsides under Pines, Junipers and *Quercus alnifolia*; 3,450–6,400 ft. alt.; fl. May–Oct.

DISTR.: Division 2. Endemic.

2. Common on the upper slopes of Khionistra; Troödhitissa Monastery, Prodhromos, Pedhoulas, Platres, Troödos, Amiandos, etc. *Kotschy* 18; *Sintenis & Rigo* 118! *A. G. & M. E. Lascelles* s.n.! *Holmboe* 1086; *Haradjian* 480! *Chapman* 103! *Syngrassides* 725! *Kennedy* 776–789! *C. H. Wyatt* 7! *Lindberg f.* s.n.! *Davis* 1824! *H. R. P. Dickson* 03! *G. E. Atherton* 514! *D. P. Young* 7374! 7806! 7807! *Merton* in ARI 56! *Economides* in ARI 1221! 1244! *P. Laukkonen* 395! etc.

NOTES: Although clearly allied to the common and widespread, mainly E. Mediterranean, *S. tomentosa* Miller (syn. *S. grandiflora* Etl.), the Cyprian plant can be distinguished by the lower stature, broader leaves with obtuse apices, glandular inflorescence axes, the smaller green calyces with shorter teeth, and the generally white corollas. Its closest ally in Cyprus is the very widespread *S. fruticosa*, from which it can usually be distinguished, in addition to differences given in the key, by its paler yellowish-green (not grey) leaves and calyces. *S. willeana* is restricted to the Troödos range, where it flowers relatively late in the season. See remarks under the following species.

2. S. fruticosa *Mill.*, Gard. Dict., ed. 8, no. 5 (1768); Hedge in Notes Roy. Bot. Gard. Edinb., 33: 23, fig. 4, map (1974); Feinbrun, Fl. Palaest., 3: 135, t. 219 (1979); Davis, Fl. Turkey, 7: 413 (1982).
    S. triloba L. f., Suppl. Plant., 88 (1781); Sibth. et Sm., Fl. Graec. Prodr., 1: 14 (1806), Fl. Graec., 1: 13, t. 17 (1806); Poech, Enum. Plant. Ins. Cypr., 23 (1842); Boiss., Fl. Orient., 4: 595 (1879); Post, Fl. Pal., ed. 2, 2: 349 (1933); Rechinger f. in Arkiv för Bot., ser. 2, 1: 430 (1951); Hedge in Notes Roy. Bot. Gard. Edinb., 23: 47 (1959).
    S. libanotica Boiss. et Gaill. in Boiss., Diagn., 2, 4: 16 (1859); Unger et Kotschy, Die Insel Cypern, 266 (1865); Boiss., Fl. Orient., 4: 594 (1879); Post, Fl. Pal., ed. 2, 2: 349 (1933).
    S. cypria Kotschy in Unger et Kotschy, Die Insel Cypern, 266 (1865); Boiss., Fl. Orient., 4: 596 (1879); F. Tracey in Journ. Roy. Hort. Soc., 58: 304 (1933); Chapman, Cyprus Trees and Shrubs, 71 (1949); Hedge in Notes Roy. Bot. Gard. Edinb., 23: 49 (1959); Osorio-Tafall et Seraphim, List Vasc. Plants Cyprus, 90 (1973).
    S. triloba L. f. ssp. *cypria* (Kotschy) Holmboe, Veg. Cypr., 158 (1914).
    S. triloba L. f. ssp. *libanotica* (Boiss. et Gaill.) Holmboe, Veg. Cypr., 158 (1914); Lindberg f., Iter Cypr., 29 (1946); Osorio-Tafall et Seraphim, List Vasc. Plants Cyprus, 90 (1973).
    S. triloba L. f. var. *cypria* (Kotschy) Lindberg f., Iter Cypr., 30 (1946).
TYPE: BM! (a cultivated specimen labelled 'Hort. Miller' — possibly raised from seed originally collected at Smyrna [Izmir] in Turkey).

Shrub, strongly aromatic; stems erect up to 1 m. high, sometimes with galls, tetragonal only on younger growth, often purplish, with a variable indumentum, eglandular-pubescent to lanate or glandular below, above usually with a dense indumentum of eglandular and short or longer capitate glandular hairs, but sometimes glabrous; leaves distributed over stems, grey-green, elliptic to ovate-oblong, variable in size, 1·2–4·5 cm. long and 0·7–2·5 cm. wide, simple or trilobed with a large terminal leaflet and a pair of much smaller basal lobes, below densely tomentose or pubescent with numerous sessile glands, apex acute or rounded, margins finely crenulate, base cordate or cuneate, rugulose; petiole 0·5–1·5 (–3·5) cm. long; inflorescence shoots with usually one pair of lateral branches, verticillasters 4–10, (2–) 6–10-flowered, approximating at least above; lower bracts leaf-like, median broadly ovate, c. 7 mm. long and 6 mm. wide, caducous, bracteoles present; pedicels 2–3·5 mm. long, spreading-erect, up to 5 mm. in fruit; calyx tubular to tubular-campanulate, not markedly bilabiate, 6–10 mm. long in flower, scarcely elongating in fruit, usually flushed purple at tips, 14–15-veined, with a dense indumentum of short and long capitate glandular hairs and some sessile glands, upper lip with three equal, broadly triangular, acuminate, 1·5 mm. long teeth, lower lip equal to upper and with similar teeth; corolla lavender to violet-blue, pink or white, 16–24 mm. long; tube 11–14 mm. long, gradually widening towards throat, clearly annulate c. 6 mm. from base; upper lip straight, bifid or scarcely so, glandular-pilose externally; lower lip equalling upper, with a suborbicular median lobe, not deflexed, and broadly ovate lateral lobes; staminal connectives c. 4·5 mm. long; lower thecae fertile clearly smaller than upper; staminal filaments c. 5 mm. long; style glabrous with unequal bifid stigmatic lobes. Nutlets c. 3 mm. long and 2 mm. wide, ± ovoid, rounded-trigonous, dark brown without venation, mucilaginous on wetting.

HAB.: Dry, rocky limestone slopes, edges of or clearings in Pine forest; river-beds; in garigue by the sea or on sand-dunes; roadsides; sea-level to 5,100 ft. alt.; fl. Febr.–July (to Dec.).

DISTR.: Divisions 1–3, 5–8. Eastern Mediterranean region, from S. Italy to Palestine.

1. Lyso, 1913, *Haradjian* 830! Smyies, 1937, *Chapman* 276! 290! Limni plantation, 1938, *Chapman* 339! Near Ayios Yeoryios (Akamas), 1962, *Meikle* 2018! 2022! 2125! Near Neokhorio, 1962, *Meikle* 2192!
2. Between Prodhromos and Lemithou, 1862, *Kotschy* 724! Troödos area, 1912, *Haradjian* 491! Saïttas, 1931, *J. A. Tracey* 49! Platania, 1936, *Syngrassides* 788! and, 1939, *Lindberg f.* s.n.! Platres, 1937, *Kennedy* 785! 786! 787! also, 1939, *Lindberg f.* s.n.! and, 1941, *Davis* 3530! Prodhromos, 1940, *Davis* 1828! and, 1961, *D. P. Young* 7350! also, 1968, *Economides* in ARI 1220! Psilondendron, 1962, *Meikle* 2787! Mallia, 1950, *Barrington* 62! E. of Polystipos, 1957, *Merton* 2995!
3. Alethriko, 1905, *Holmboe* 207; between Ayia Varvara Monastery and Stavrovouni, 1937, *Syngrassides* 1477! 1488! also, 1939, *Lindberg f.* s.n.! Skarinou, 1941, *Davis* 2788! Near Limnatis, 1941, *Davis* 3088!
5. Apostolos Varnavas Monastery near Famagusta, 1947, *Mavromoustakis* 36! Lakovounara, 1950, *Chapman* 347!
6. Nicosia, 1927, *Rev. A. Huddle* 85! Ayia Irini, 1937, *Syngrassides* 1412! 1470!
7. Ayios Khrysostomos Monastery, 1859, *Kotschy* 483; Pentadaktylos, 1862, *Kotschy* 369; also 1901, *A. G. & M. E. Lascelles* s.n.! Near Lapithos, 1862, *Kotschy* 483; Kantara, 1880, *Sintenis & Rigo* 118! also, 1948, *Mavromoustakis* s.n.! Mylous Forest Station, 1932, *Syngrassides* 30! 213! also, 1937, *Syngrassides* 1434! and, 1935, *Chapman* 191! Koronia Forest Station, 1933, *Chapman* 11! 13! 30! and, 1937, *Chapman* 225! 228! 263! Platani, 1933, *Chapman* 52! 56! 74! Kyrenia, 1937, *Mrs. Stagg* s.n.! also, 1937, *Syngrassides* 1437! and 1938–1956, *C. H. Wyatt* 35! *G. E. Atherton* 119! 176! 268! 271! 809! *Lindberg f.* s.n.! Kazaphani, 1955, *G. E. Atherton* 116! Bellapais, 1963, *J. B. Suart* 96! Karakoumi, 1968, *Economides* in ARI 1047! 1172!
8. Rizokarpaso, 1941, *Davis* 2384! Peristerges, Kilanemos, 1962, *Meikle* 2435!

NOTES: It is in Cyprus, where it grows practically all over the island, that the strictly E. Mediterranean *S. fruticosa* shows its greatest range of variation. The most extreme form is that previously called *S. cypria* which has the lowest ranges of leaf measurements given above and a compacted inflorescence; such forms are usually found at higher altitudes. There is also throughout the island a wide range of stem indumentum types; the inflorescence axis may be densely covered with capitate glandular hairs, or eglandular white-lanate, or completely glabrous.

Although *S. fruticosa* and *S. willeana* can normally be readily distinguished on the characters given in the key, they are related species and where they grow together they may hybridise. No such hybrids have been recorded, but they should be looked for.

The leaves of *S. fruticosa* are sometimes used in Cyprus, and elsewhere, for making a tea-like drink. In Palestine it is noted as being an excellent honey-plant.

3. S. pinnata *L.*, Sp. Plant., ed. 1, 27 (1753); Unger et Kotschy, Die Insel Cypern, 267 (1865); Boiss., Fl. Orient. 4: 601 (1879); H. Stuart Thompson in Journ. Bot., 44: 336 (1906); Holmboe, Veg. Cypr., 158 (1914); Post, Fl. Pal., ed. 2, 2: 351 (1933); Osorio-Tafall et Seraphim, List Vasc. Plants Cyprus, 90 (1973); Feinbrun, Fl. Palaest., 3: 135, t. 220 (1979); Davis, Fl. Turkey, 7: 417 (1982).
TYPE: "*In* Oriente & Arabia" (LINN 42/51 !).

Perennial herb; stems 18–60 cm. long, procumbent or ascending, rarely erect, tetragonal, often purplish at angles, densely glandular-villous throughout with numerous long spreading capitate glandular hairs, some longer eglandular hairs and sessile glands; leaves pinnatisect with 2–3 pairs of sessile or shortly petiolulate, oblong lateral segments and a larger, c. 4 cm. long and 1·5 cm. wide (to 7 × 4 cm.), oblong to oblong-ovate terminal segment, apex acute to rounded, margins crenate to serrulate, base cuneate to subcordate, with an indumentum of capitate glandular and eglandular hairs, especially on veins beneath, and sessile glands; petiole channelled 3–4 (–12) cm. long, glandular-villous; inflorescence unbranched or with few short lateral branches; verticillasters c. 6–8, 4–6-flowered, clearly distant; bracts oblong-ovate, c. 5 mm. long, soon deciduous; pedicels 5–15 mm. long, spreading-erect, glandular-villous; calyx urceolate, 8–12 (–14) mm. long, and up to 10 mm. wide at throat, often tinged purple, slightly expanding in fruit, densely glandular-villous with numerous very long capitate glandular and eglandular hairs and sessile glands, internally with numerous short glands; upper lip truncate, obscurely tridentate; lower lip with two very short, broad mucronulate teeth, equalling upper lip; corolla purplish to pink, c. 20–25 mm. long; tube exserted beyond calyx, c. 15 mm. long, widening towards throat, tufted pilose inside towards base of tube; upper lip ± straight, deeply bifid, lower lip somewhat longer than upper; staminal connectives c. 5 mm. long, lower thecae subfertile, much smaller than upper; staminal filaments c. 5 mm. long; style glabrous, stigmatic lobes unequal, short, recurved. Nutlets c. 2·5 mm. long and broad, almost spherical, not trigonous, dark brown, without veins, not mucilaginous on wetting.

HAB.: In garigue on calcareous rocks; near sea-level to 1,500 ft. alt.; fl. April–May.

DISTR.: Division 7, rare. Also W. & S. Turkey, Lebanon, Palestine.

7. Melandrina (near Ayios Amvrosios), 1862, *Kotschy*; Antiphonitis, 1862, *Kotschy* 528; Kyrenia, 1955, *Mrs. Scott-Moncrieff* s.n. !

NOTES: A rare plant in Cyprus, where, as elsewhere in its range, it is a segetal species. Little adequate Cyprian material has been available for examination and the description is partly based on Turkish and Palestinian specimens.

4. S. viridis *L.*, Sp. Plant., ed. 1, 24 (1753); Sibth. et Sm., Fl. Graec. Prodr., 1: 14 (1806), Fl. Graec., 1: 15, t. 19 (1806); Poech, Enum. Plant. Ins. Cypr., 23 (1842); Unger et Kotschy, Die Insel Cypern, 267 (1865); Boiss., Fl. Orient., 4: 630 (1879); H. Stuart Thompson in Journ. Bot., 44: 336 (1906); Holmboe, Veg. Cypr., 159 (1914); Post, Fl. Pal., ed. 2, 2: 360 (1933); Osorio-Tafall et Seraphim, List Vasc. Plants Cyprus, 91 (1973); Hedge in Notes Roy. Bot. Gard. Edinb., 33: 94 (1974); Feinbrun, Fl. Palaest., 3: 143, t. 236 (1979); Davis, Fl. Turkey, 7: 434 (1982).
S. horminum L., Sp. Plant., ed. 1, 24 (1753); Sibth. et Sm., Fl. Graec. Prodr., 1: 14 (1806), Fl. Graec., 1: 15, t. 20 (1806); Unger et Kotschy, Die Insel Cypern, 267 (1865); Boiss., Fl. Orient., 4: 631 (1879); H. Stuart Thompson in Journ. Bot., 44: 336 (1906); Post, Fl. Pal., ed. 2, 2: 360 (1933); Feinbrun, Fl. Palaest., 3: 142, t. 235 (1979).
S. viridis ssp. *horminum* (L.) Holmboe, Veg. Cypr., 159 (1914).
TYPE: (LINN 42/11 !) (No locality or provenance indicated).

Annual of neat habit, scarcely aromatic; stems 10–45 cm. high, erect, branched from base or with lateral stem branches, tetragonal, sometimes purplish at base, with many white tangled eglandular flattened multicellular hairs below, similar above and with some sessile glands and glandular hairs; leaves distributed over stem, elliptic to broadly oblong, (2–) 4 (–6) cm. long, (1–) 2–3 (–5) cm. wide, with many eglandular hairs below, mostly on veins, and some sessile glands, above with pilose eglandular hairs usually in clusters, apex rounded, margins regularly crenulate to serrulate, base rounded, flat; petiole 1–5·5 cm. long; inflorescence showy, verticillasters (5–) 10–15, 4–6-flowered, clearly distant to conferted, elongating in fruit or not; bracts usually prominent, deep violet or green, sometimes forming a sterile coma topping flowering shoots, c. 9 mm. long, 8 mm. wide, broadly obovate, acuminate; pedicels 2–3 (–4) mm. long, erect–spreading in flower, strongly deflexed in fruit, flattened; calyx tubular, 8–10 mm. long in flower, c. 12 mm. long in fruit, 13-ribbed, with a sparse to fairly dense indumentum of pilose to villous eglandular hairs, capitate glandular hairs and sessile glands; upper lip truncate with two c. 0·5–1 mm. long lateral teeth and a very reduced median tooth, lower lip with two acuminate 2–3 mm. long teeth, ± equalling upper lip; corolla with a lilac to violet hood and a paler labellum, sometimes all white, (12–) 14–16 mm. long; tube straight, 9–12 mm. long, included within calyx, slightly wider at throat, glabrous within; upper lip falcate, entire, eglandular-pilose externally; lower lip shorter than upper, with a cucullate reflexed median lobe and distinct oblong lateral lobes; staminal connectives c. 5 mm. long, pilose near base; lower thecae of dolabriform sterile tissue, not adhering; staminal filaments c. 1·5 mm. long; style glabrous with markedly unequal stigmatic lobes. Nutlets oblong-elliptic, 2·5–3 mm. long, 1·5 mm. wide, pale brown, rounded-trigonous, without venation, strongly mucilaginous on wetting.

HAB.: Grassy banks, dry limestone slopes, rock ledges, fields, roadsides, waste places; sea-level to 800 ft. alt.; fl. Febr.–April.

DISTR.: Divisions 1–8, locally common. S. Europe and Mediterranean region east to Iran.

1. Skali near Ayios Yeoryios (Akamas), 1962, *Meikle* 2080!
2. Near Vouni, 1948, *Kennedy* 1703!
3. Near Phasouri, 1862, *Kotschy* s.n.; by Yermasoyia River, 1948, *Mavromoustakis* 3! Cherkez, 1948, *Mavromoustakis* s.n.! Kouklia, 1960, *N. Macdonald* 58! Amathus, 1964, *J. B. Suart* 170! Mathicoloni, 1969, *Economides* in ARI 1337!
4. Larnaca, 1862, *Kotschy* 38! Near Cape Pyla, 1862, *Kotschy* 175a! also, 1880, *Sintenis & Rigo*; Athienou, 1880, *Sintenis & Rigo*; Akhyritou, 1901, *A. G. & M. E. Lascelles* s.n.! and, 1949, *Casey* 568! Between Famagusta and Salamis, 1934, *Syngrassides* 484! Pergamos, 1955, *Merton* 1966! Kouklia reservoir, 1963, *Merton* in ARI 441! Dhekelia, 1977, *A. Della* in ARI 1672!
5. Kythrea, 1880, *Sintenis & Rigo*; near Lefkoniko, 1880, *Sintenis & Rigo*; Vatili, 1905, *Holmboe* 321! Mia Milea, 1933, *Syngrassides* 308! also, 1952, *F. M. Probyn* 23! and, 1967, *Merton* in ARI 176! Near Eylenja, 1933, *Syngrassides* 2! Lakovounara, 1950, *Chapman* 51! 509! Between Ardhana and Trikomo, 1880, *Sintenis & Rigo* 119! also, 1948, *Mrs. Marshall* in *Kennedy* 1675! Apostolos Varnavas Monastery, 1948, *Mavromoustakis* s.n.! S. of Aghirda, 1949, *Casey* 284! Petra tou Dhiyeni, 1951, *Merton* 207! Aphania, 1955, *Merton* 1965! 2 m. S. of Koutsovendis, 1959, *C. E. H. Sparrow* 6! Athalassa, 1962, *Meikle* 2973! and, 1966, *Y. Ioannou* in ARI s.n.!
6. Near Nicosia, 1930, *F. A. Rogers* 0664!
7. Kyrenia, 1902, *A. G. & M. E. Lascelles* s.n.! also, 1938–1956, *C. H. Wyatt* 20! *Casey* 381! *G. E. Atherton* 1202! Lefkoniko Pass, 1949, *Casey* 479! Dhavlos, 1971, *Guichard* KG/CYP 10!
8. Yialousa, 1880, *Sintenis & Rigo* 119b! Rizokarpaso, 1953, *Kennedy* 1785! Ayios Philon near Rizokarpaso, 1961, *Davis* 2267! Between Apostolos Andreas and Rizokarpaso, 1941, *Davis* 2287! also, Karpas peninsula, 1950, *Chapman* 206!

NOTES: The field note of *Merton* 207 comments: "all large populations in Cyprus show wide variation in flower and bract colour, development of sterile bracts, etc." In other parts of its, mostly Mediterranean, range the situation is different, particularly with regard to the development of the sterile bracts. In N.W. Africa, for instance, plants without a showy coma are much more frequent than those with, whereas in Turkey and Iran, the violet-topped form is much more common than that without a coma.

S. viridis is a strikingly hygrochastic species with the strongly deflexed calyces soon raised to a horizontal position on wetting.

S. viridis and *S. horminum* were both described by Linnaeus in the *Species Plantarum*, 1753, and although they are still occasionally (as in *Flora Palaestina*, 1979) regarded as independent species, they are generally considered conspecific. J. E. Smith in 1806 appears to have been the first author to have united them, under the name *S. viridis*, at least in intent. Although both species are illustrated and described in *Flora Graeca*, there is a discussion by Smith at the end of the description of *S. horminum* (p. 16) in which he points out that there are very few differences between the two species; he states "...caeterum nullo discrimine a *S. viridis* dignoscenda, cujus, me judice, varietas est." He clearly thought that they were conspecific.

5. S. aethiopis L., Sp. Plant., ed. 1, 27 (1753); Boiss., Fl. Orient., 4: 616 (1879); Post, Fl. Pal., ed. 2, 2: 356 (1933); Davis, Fl. Turkey, 7: 440 (1982).

TYPE: "in Illyria [coastal Jugoslavia], Graecia, Africa" (LINN 42/48!)

Biennial, sometimes perennial herb; stems stiffly erect, c. 30–50 cm. high, sturdy, c. 6–8 mm. thick at base, tetragonal, densely white-lanate at base with long eglandular hairs and with a similar but thinner indumentum above with some sessile glands; leaves mostly near base of stem, sometimes rosette-forming, ovate to oblong in outline, c. 10–20 cm. long, 4–10 (–18) cm. wide, rugose, softly white-lanate on younger leaves to arachnoid on older, apex acute, margins serrate, erose or lobed, base cuneate to subcordate, nervation prominent at least below; petiole 4–13 cm. long, lanate; inflorescence showy, widely candelabriform, up to 25 cm. long and 25 cm. wide; verticillasters numerous, 4–6-flowered, mostly approximating; bracts broadly obovate to almost circular, cuspidate, c. 12 mm. long and 15 mm. wide, with an indumentum of eglandular hairs and sessile glands, green or tinged purple; pedicels c. 2 mm. long, spreading–erect; calyx tubular to tubular-ovate, 8–10 mm. long in flower enlarging to c. 12–15 mm. long in fruit, with a dense lanate indumentum of long white eglandular hairs and numerous sessile glands, almost glabrous within; upper lip tridentate, median shorter than laterals, teeth cuspidate, somewhat recurved; lower lip with 2 lanceolate cuspidate teeth, ± as long as upper lip; corolla white, sometimes with a yellowish lip, c. 14–15 mm. long; tube included within calyx, c. 5–6 mm. long, abruptly ventricose, squamulate with a fringed scale; upper lip strongly falcate, bifid, with long eglandular hairs and sessile glands externally, lower lip with a broad suborbicular, cucullate middle lobe and two oblong lateral lobes, shorter than upper lip; staminal connectives c. 7 mm. long; lower thecae represented by a plate of sterile dolabriform tissue; staminal filaments c. 2 mm. long; style finely pilose towards apex; stigmatic lobes unequal, flattened, recurved. Nutlets c. 2 mm. long and c. 1·5 mm. broad, obovoid, brown with darker venation, trigonous, mucilaginous on wetting.

HAB.: In *Pinus nigra* forest; 5,600–5,700 ft. alt.; fl. June–July.

DISTR.: Division 2; rare and perhaps introduced. Central and South Europe, Turkey, Caucasia, Crimea, Iran.

2. Troödos; "In the Pine forest . . . by the old Government Dispensary. Many in a ditch between the building and the mountain slope, and some close above", 1956, *Kennedy* 1872–1875! Also, probably from the same locality, 1956, *Merton* 2784!

NOTES: The first (and only) collections of this unmistakable species were made in 1956, when Mrs. Kennedy noted: "an innumerable colony". No recent gatherings have been seen, and since *S. aethiopis* is not an easily overlooked plant, there is a possibility that it was originally introduced, and did not persist.

6. S. dominica L., Sp. Plant., ed. 1, 25 (1753); Post, Fl. Pal., ed. 2, 2: 356 (1933); Hedge in Notes Roy. Bot. Gard. Edinb., 23: 55 (1959); Osorio-Tafall et Seraphim, List Vasc. Plants Cyprus, 91 (1973); Hedge in Notes Roy. Bot. Gard. Edinb., 33: 53 (1974); Feinbrun, Fl. Palaest., 3: 138, t. 227 (1979).

S. graveolens Vahl, Enum., 1: 273 (1804); Boiss., Fl. Orient., 4: 615 (1879); A. K. Jackson et Turrill in Kew Bull., 1939: 478 (1939); Chapman, Cyprus Trees and Shrubs, 72 (1949).

TYPE: "in Domingo" (LINN 42/19 !) [the species is restricted to the E. Mediterranean; it does not grow in the W. Indies].

Shrub or subshrub with herbaceous branches, strongly aromatic; stems ± erect, tetragonal, with a dense grey indumentum of short capitate glandular hairs, few or many longer eglandular hairs and some sessile glands towards base of stem; leaves spread over stem, ± ovate to ovate-oblong, 2·8–6·5 (–7·5) cm. long and 1·8–3 (–5) cm. wide, with numerous white eglandular hairs, sessile glands and very small capitate glandular hairs, rugose, apex ± acute, margins undulate, irregularly serrulate to sub-erose, base cordate to truncate; petiole on lower leaves 5–16 mm. long, absent above, with a similar indumentum to the leaves; inflorescence with rather slender main and lateral branches; verticillasters c. 8–15, 2–6-flowered, mostly approximating; bracts broadly ovate 3·5–17 mm. long and 3–7 mm. wide, decreasing above, acuminate, glandular-pilose with some longer eglandular hairs; pedicels c. 2 mm. long, spreading-erect; calyx campanulate-obtriangular, 7–10 mm. long, enlarging in fruit to c. 13 mm. long and with widely diverging lips, nerves prominent c. 10, with a very dense indumentum of long, flattened, villous eglandular hairs, very small capitate glands and sessile glands, almost glabrous within; upper lip with three subequal, small, connivent teeth; lower lip ± equalling upper; corolla white to cream c. 15–18 mm. long; tube included within calyx, c. 8 mm. long, abruptly ventricose, squamulate with a prominent fringed scale; upper lip clearly falcate, scarcely bifid, with many villous hairs and some sessile glands externally; lower lip with a cucullate median lobe and two long oblong lateral lobes, deflexing, ± equal to upper lip; staminal connectives c. 11 mm. long; lower thecae represented by sterile, dolabriform tissue, adhering; staminal filaments c. 3 mm.; style glabrous; stigmatic lobes unequal, flattened, recurved. Nutlets c. 3 mm. long and 2·2 mm. wide, broadly oblong, pale brown with darker venation, rounded-trigonous, mucilaginous on wetting.

HAB.: Rocky hillsides; 200–500 ft. alt.; fl. April–May.

DISTR.: Divisions 2, 3, rare. Palestine, Syria, Egypt.

2. Between Ambelikou and Karavostasi, by the roadside, 1932, *Syngrassides* 1613 ! and, same locality, 1938, *Chapman* 364 !
3. Curium, 1961, *Polunin* 6697 ! and, 1981, *G. J. van Wieringen* 27 !

NOTES: Apparently a rare plant in Cyprus; *S. dominica* is not uncommon in Palestine, and is also known from Jordan, Syria and the Isthmic desert of Egypt. Characteristic features of the species are its habit, calyx indumentum, and the widely divergent lips of the fruiting calyces. Although it has no close allies among the S.W. Asiatic Salvias, it shows several points of similarity with the geographically far distant *S. garipensis* E. Mey. from S.W. Africa [Namibia].

7. S. veneris *Hedge*, nom. nov.

S. crassifolia Sm. in Sibth. et Sm., Fl. Graec. Prodr., 1: 17 (1806), Fl. Graec., 1: 19, t. 26 (1806 probably Nov.); non Jacq., Fragm., 47, t. 60 (1800–09); Boiss., Fl. Orient., 4: 622 (1879) pro parte; Holmboe, Veg. Cypr., 159 (1949); Osorio-Tafall et Seraphim, List Vasc. Plants Cyprus, 90 (1973).

S. candidissima (non Vahl) Poech, Enum. Plant. Ins. Cypr., 23 (1842); Unger et Kotschy, Die Insel Cypern, 267 (1865).

TYPE: "In insula Cypro, at rarissime, D. Ferd. Bauer" (holo. OXF, photo !)

Perennial tufted herb; stems c. 20–40 cm. high from a woody rootstock, erect, tetragonal, lanate at base of stem, above with a dense indumentum of capitate glandular and eglandular multicellular scabridulous hairs; leaves almost confined to basal rosettes, very broadly ovate to suborbicular, 2·5–8 cm. long and 1·7–8 cm. wide, with a dense white soft lanate or pannose indumentum on both surfaces, somewhat thinner above, apex rounded to acute, margins sub-entire to serrate, base truncate to subcordate; petiole lanate to pannose, 1·5–5·2 cm. long; inflorescence with a main shoot and up to 4 smaller lateral branches; verticillasters 6–12, 2–6-flowered, mostly distant; bracts broadly ovate to suborbicular, lanate on lower surface, c. 7 mm. long and 6 mm. broad, bracteoles present; pedicels 2·5–3·5 mm. long, spreading-erect; calyx tubular campanulate, 7–10 mm. long in flower to c. 12 mm. long in fruit and broadening with a bisulcate upper lip, 13–14-nerved with a ± dense indumentum of eglandular multicellular scabridulous hairs, short capitate glandular hairs and numerous sessile glands; upper lip with three short ± connivent mucronulate teeth; lower lip slightly longer than upper with two c. 4 mm. long, mucronate teeth; corolla with a pale lilac hood and a yellowish labellum, 17–24 mm. long; tube c. 9 mm. long, ± included in calyx, ventricose, squamulate; upper lip falcate, bifid, with eglandular pilose and short capitate glandular hairs externally; lower lip shorter than upper, with a deflexed cucullate orbicular median lobe and two long linear-oblong lateral lobes; staminal connectives c. 16 mm. long; lower thecae represented by sterile dolabriform adhering tissue; staminal filaments c. 3 mm. long, style glabrous with unequally bifid recurved stigmatic lobes. Nutlets ± ovoid, c. 2·5 mm. long and 2 mm. wide, pale brown with darker venation, rounded-trigonous, mucilaginous on wetting.

HAB.: Eroded sandstone hills; steep lava slopes; 500–1,000 ft. alt.; fl. March–April.

DISTR.: Divisions 5, 7, rare. Endemic.

5. On marl and sandstone hummocks near Kythrea, 1862–1974, *Kotschy* 333! *Sintenis & Rigo* 117! *Davis* 2292! *Chapman* 473! *Osorio-Tafall & Meikle* in *Meikle* 4042!
7. On lava intrusions above Kephalovryso, Kythrea, 1962, *Meikle* 2551!

NOTES: Endemic to Cyprus, where it is restricted to the Kythrea area. Easily distinguished from all other Cyprus species on account of the rosette-forming pannose/lanate leaves and the lilac/yellow corollas. Its closest allies are the Turkish endemics *S. cilicica* Boiss. et Ky., *S. cassia* Rechinger f. and *S. cyanescens* Boiss. et Bal., and the mainly Turkish *S. candidissima* Vahl. The Cyprus plant differs from its mainland allies in the arrangement, shape and indumentum of the leaves and in the dimensions and colour of the flowers.

It is unfortunately necessary to change the name by which it has always been known both on account of Jacquin's earlier use of the same epithet for a different species and the citation by Smith of *S. candidissima* Vahl as a synonym of *S. crassifolia*, thus rendering their new species technically a *nomen superfluum*. It is not possible to establish with certainty the precise publication date of *S. crassifolia* Jacq. in *Fragmenta Botanica*. However, the work appeared in six fascicles for which there are publication dates, but no corresponding details of pagination; because the *Salvia* is t. 60 out of a total of 138 plates it is likely that it appeared in fascicle 4 which is dated 1804, and most unlikely that it appeared after mid-1806 (see Stafleu, *Taxonomic Literature* ed. 2, 2: 413, 1979).

Holmboe (Veg. Cypr., 159) is mistaken in supposing that Sibthorp (or, rather, Bauer) first collected *S. veneris* (*S. crassifolia*) at Nicosia. The plant collected there, as "*Salvia Argentea*", on 22 April 1787, was almost certainly *S. lanigera* Poir. So far as is known, *S. veneris* does not occur anywhere near Nicosia.

8. S. hierosolymitana *Boiss.*, Diagn., 1, 12: 61 (1853); Boiss., Fl. Orient., 4: 627 (1879); Post, Fl. Pal., ed. 2, 2: 358 (1933); Druce in Rep. B.E.C., 9: 470 (1931); A. K. Jackson in Kew Bull., 1936: 16 (1936); Osorio-Tafall et Seraphim, List Vasc. Plants Cyprus, 91 (1973); Feinbrun, Fl. Palaest., 3: 140, t. 231 (1979).

TYPE: Israel; "in rupestribus frigidis circa Hierosolymam frequenter, *Boissier*" (G photo!).

Sturdy perennial herb; stems erect, c. 40–100 cm. high, up to c. 8 mm. thick at base, clearly tetragonal, scabridulous at margins, glabrous below, with few to numerous eglandular coarse white hairs above and densely glandular-pilose in inflorescence region; leaves coarse, 10–24 cm. long and 3·5–13·5 cm. wide, thin-textured, with many coarse ± scabrid hairs below confined to the veins, apex acute to obtuse, margins often lobed, wavy, crenulate, base cordate; petiole of lower leaves 3–13 cm. long, with ± scabrid white eglandular hairs, cauline leaves few, sessile; inflorescence a loose panicle; verticillasters up to c. 12, 2–6-flowered, clearly distant, stems often purplish at edges; bracts broad ovate to suborbicular, c. 6–14 mm. long, 6–12 mm. wide, acuminate, usually purplish towards apex, with many glandular capitate hairs; pedicels c. 3–4 mm. long, spreading-erect; calyx ± tubular, 10–14 (–15) mm. long, enlarging in fruit to c. 15–16 mm. long, with a dense indumentum of short and longer glandular hairs and some eglandular villous hairs, inside with numerous very small glandular hairs; upper lip with three sub-equal 0·8–1·5 mm. long, mucronate teeth; lower lip with two somewhat recurved 3–4 mm. long teeth, ± equalling upper lip; corolla dark pink to deep rosy purple, (16–) 20–25 (–30) mm. long; tube c. 15 mm., exserted from calyx, widening towards throat, esquamulate and ± glabrous within; upper lip ± falcate, compressed, slightly emarginate, externally with numerous small capitate glandular hairs; lower lip shorter than upper, with a large suborbicular ± flat median lobe and two oblong lateral lobes; staminal connectives c. 13 mm. long; lower thecae represented by sterile dolabriform tissue, adhering; staminal filaments c. 2–2·5 mm. long; style glabrous; stigmatic lobes subequal, recurved. Nutlets c. 2·7 mm. long and broad, ± orbicular, rounded-trigonous, dark brown without obvious venation, somewhat mucilaginous on wetting.

HAB.: Banks, meadows, steep north-facing slopes with *Styrax*; near sea-level to 300 ft. alt.; fl. April–May.

DISTR.: Divisions 7, 8, rather rare. Israel, Lebanon, Syria.

7. Above Bellapais, 1880, *Sintenis & Rigo* 116! Zigaret, 4 miles E. of Kyrenia, 1936, *Mrs. Dray* 1! Dramia, Ayios Epiktitos, 1953, *Kennedy* 1787!
8. Ayios Andronikos, 1880, *Sintenis & Rigo*; Yialousa, 1880, *Sintenis & Rigo*; also, 1928 or 1930, *Druce* s.n.; Eleousa, 1880, *Sintenis & Rigo* s.n.; Platanisso, 1962, *Meikle* 2506!

NOTES: Distinctive on account of the rather coarse, yet thin-textured, leaves and the long dark pink to purple corollas. It would appear to be confined in Cyprus to the eastern part of the Northern Range and to the Karpas Peninsula.

9. S. verbenaca *L.*, Sp. Plant., ed. 1, 25 (1753); Unger et Kotschy, Die Insel Cypern, 268 (1865); Boiss., Fl. Orient., 4: 629 (1879); Holmboe, Veg. Cypr., 159 (1914); Post, Fl. Pal., ed. 2, 2: 359 (1933); Lindberg f., Iter Cypr., 30 (1946); Osorio-Tafall et Seraphim, List Vasc. Plants Cyprus, 90 (1973); Hedge in Notes Roy. Bot. Gard. Edinb., 33: 95 (1974); Feinbrun, Fl. Palaest., 3: 141, t. 233 (1979); Davis, Fl. Turkey, 7: 458 (1982).

    S. clandestina L., Sp. Plant., ed. 2, 36 (1762).
    S. multifida Sm. in Sibth. et Sm., Fl. Graec. Prodr., 1: 16 (1806); Fl. Graec., 1: 17, t. 23 (1806).
    S. controversa Ten., Fl. Nap., 18 (1831) non auctt.
    S. verbenaca L. var. *serotina* Boiss., Voy. Bot. Espagne, 484 (1841); Post, Fl. Pal., ed. 2, 2: 359 (1933).
    S. verbenaca L. var. *vernalis* Boiss., Voy. Bot. Espagne, 484 (1841); Druce in Rep. B.E.C., 9: 470 (1931).
    S. clandestina var. *multifida* (Sm.) DC., Prodr., 12: 295 (1848); Unger et Kotschy, Die Insel Cypern, 268 (1865).
    S. verbenaca L. subsp. *clandestina* (L.) Briq., Labiat. Alpes Marit., 518 (1891); Osorio-Tafall et Seraphim, List Vasc. Plants Cyprus, 90 (1973).
    S. verbenaca L. var. *clandestina* (L.) Halácsy, Consp. Fl. Graec. 2: 490 (1902); Holmboe, Veg. Cypr., 159 (1914); Post, Fl. Pal., ed. 2, 2: 359 (1933).
    TYPE: "*in* Europae *pascuis*" (LINN 42/20!).

Perennial herb; stems c. 10–70 (–90) cm. high, ± erect from a woody rootstock, often purplish below, usually branched above, tetragonal, with short eglandular often retrorse hairs below, above with numerous capitate glandular hairs and shorter eglandular ones, occasionally without glandular hairs; leaves often mostly basal, variable in shape, 2·7–14 cm. long, 2–7 cm. wide, oblong to ovate, usually with a sparse indumentum below of short eglandular hairs, mainly on veins, and some sessile glands, apex acute, margins crenate, serrate, lobed or erose, base cordate or truncate, ± flat; petiole of lowermost leaves 2–14 cm. long, upper leaves sessile; inflorescence branches with one pair of short lateral shoots; verticillasters up to 12, 4–6-

flowered, distant or approximating; bracts broad ovate, c. 6 mm. long and 6 mm. wide, acuminate, pedicels 1–2 mm. long, spreading-erect, recurving in fruit and to 4 mm. long; calyx clearly bilabiate, ± campanulate, 5–7 mm. long, enlarging in fruit to 7–10 mm., with many long spreading or forward-pointing eglandular hairs, shorter capitate glandular hairs, sometimes absent, and some sessile glands, internally with long eglandular hairs at throat; upper lip green or violet-blue, with three connivent subequal teeth 0·5 mm. long, broadening in fruit and bisulcate-concave; lower lip with two c. 4 mm. long, shortly mucronate teeth; corolla deep lavender blue, pink to mauve, c. 12 mm. long; tube c. 6 mm. long, relatively broad, ± included within calyx, ventricose, esquamulate, glabrous within; upper lip hooded scarcely falcate, emarginate, with few eglandular pilose hairs; lower lip ± equal to upper, with a cucullate median lobe, deflexed; staminal connectives c. 5 mm. long; lower thecae represented by sterile dolabriform tissue, adhering; staminal filaments c. 1·5 mm. long; style glabrous, deeply cleft at apex with flattened unequal stigmatic lobes. Nutlets oblong-ovoid, 2–2·5 mm. long, 1·5–2 mm. broad, rounded-trigonous, ± smooth-textured, very dark brown without obvious venation, strongly mucilaginous on wetting.

HAB.: In *Pinus* forest, Olive groves, fields, roadsides, damp places in garigue, seashores; near sea-level to 500 (–5,750) ft. alt.; fl. Jan.–April (and Nov.).

DISTR.: Divisions 1–8. S. & W. Europe, Mediterranean region east to Caucasus; naturalized in many other parts of the world.

1. Baths of Aphrodite, 1962, *Meikle* 2233!
2. Prodhromos, 1862, *Kotschy* 694; Phiti, 1905, *Holmboe* 773! Mesapotamos, 1939, *Lindberg f.* s.n.! Troödos, 1956, *Kennedy* 1876!
3. Lania, 1928 or 1930, *Druce* s.n.; Limassol, 1947, *Mavromoustakis* 21! Episkopi, 1964, *J. B. Suart* 146!
4. Larnaca, 1859, *Kotschy* 58; also, 1880, *Sintenis & Rigo*; Ayia Napa, 1905, *Holmboe* 52; Psevdhas, 1908, *Clement Reid* s.n.!
5. Gypsos, 1969, *A. Genneou* in ARI 1315!
6. Near Nicosia, 1930–1973, *F. A. Rogers* 0664! *Davis* 2144! *P. Laukkonen* 3! Kykko metokhi, 1952, *F. M. Probyn* 41! English School, Nicosia, 1958, *P. H. Oswald* 25! Myrtou, 1968, *Economides* in ARI 1179!
7. Near Kantara, 1880, *Sintenis & Rigo* 120! Kyrenia, 1932–1956, *Syngrassides* 358! *Casey* 252! 560! *J. E. Atherton* 764! 821! 912! Kazaphani, 1956, *G. E. Atherton* 773! 1149!
8. Rizokarpaso, 1941, *Davis* 2357!

NOTES: Variable in stature, leaf-size and lobing, indumentum and form of inflorescence. Sex-forms with male-sterile flowers are not infrequent; cleistogamous forms may also occur. A common species in Cyprus and throughout the Mediterranean area.

10. S. lanigera *Poir.* in Lam., Encycl. Méth. Suppl., 5: 49 (1817); Post, Fl. Pal., ed. 2, 2: 359 (1933); Hedge in Notes Roy. Bot. Gard. Edinb., 33: 99 (1974); Feinbrun, Fl. Palaest., 3: 142, t. 234 (1979).
   [*S. clandestina* (non L.) Sm. in Sibth. et Sm., Fl. Graec. Prodr., 1: 16 (1806) et Fl. Graec., 1: 18, t. 24 (1806), 7: 43 (1830) — sub *Brassica eruca* t. 647; prob. Poech, Enum. Plant. Ins. Cypr., 23 (1842).]
   [*S. controversa* (non Ten.) Unger et Kotschy, Die Insel Cypern, 268 (1865); Boiss., Fl. Orient., 4: 630 (1879); Holmboe, Veg. Cypr., 159 (1914); Lindberg f., Iter Cypr., 29 (1946); Chapman, Cyprus Trees and Shrubs, 72 (1949); Osorio-Tafall et Seraphim, List Vasc. Plants Cyprus, 91 (1973).]
   TYPE: "Cette plante croît en Perse et dans l'Egypte" (no type specimen has been found).

Perennial herb, strong-smelling; stems 8–40 cm. long, ± erect from a woody rootstock, often branched from the base, somewhat purplish below, tetragonal, with tangled eglandular villous hairs below and a mixture of short pilose glandular and eglandular hairs above, together with long spreading white eglandular hairs; leaves deeply pinnatisect, distributed over stems, narrow oblong to oblong-ovate in outline, 2–5·3 (–10·5) cm. long, lobes ± linear and lobed, to c. 12 mm. long, beneath with many sessile glands, small eglandular pilose and long white ± lanate hairs, revolute at margins, rugose; petiole of lower leaves up to c. 2·5 cm. long, upper leaves sessile; inflorescence white-woolly, stems branched or not; verticillasters 8–12, c. 6-flowered, ± approximating; bracts broad ovate to suborbicular, c. 2·5–4 mm. long, 3–4 mm. broad; pedicels 1–2 mm. long, spreading-erect, white-villous; calyx clearly bilabiate, tubular-campanulate to infundibuliform, 5–7 mm. long in flower and c. 7–8 mm. long in fruit, very densely covered with long white eglandular hairs, sessile glands and few to many short glandular hairs, internally eglandular villous at throat; upper lip often purplish, with three subequal 0·8 mm. long teeth, broadening in fruit and ± bisulcate; lower lip ± equal to upper, with two c. 3 mm. long, scarcely mucronate teeth; corolla deep clear blue to dark violet, 13–15 mm. long; tube 8–9 mm. long, exserted from calyx, ventricose, esquamulate, glabrous within; upper lip hooded to ± falcate, bifid, eglandular villous with some sessile glands externally; lower lip ± equal to upper, median lobe cucullate, deflexed; staminal connectives c. 9 mm. long; lower thecae represented by sterile dolabriform tissue, adhering; staminal filaments c. 1·5 mm. long; style glabrous, stigmatic lobes unequal, flattened, to c. 2·3 mm. long. Nutlets obovoid, c. 2·4 mm. long and 1·4 mm. broad, narrowed towards attachment scar, ± smooth-textured, blackish without visible venation, strongly mucilaginous on wetting.

HAB.: Dry rocky slopes, garigue, sandy flats and dunes; near sea-level to c. 600 ft. alt.; fl. Jan.–April.

DISTR.: Divisions 4–6. N. Africa, Israel, Jordan, Iraq, Iran, Arabia.

4. Between Stavrovouni and Larnaca, 1787, *Sibthorp* s.n.! Larnaca, 1862, *Kotschy* 256! also, 1900, *A. G. & M. E. Lascelles* s.n.! and, 1912, *Haradjian* 35! Near Ormidhia, 1880, *Sintenis & Rigo* 567!
5. Margi, 1935, *Syngrassides* 665! Athalassa, 1939–1967, *Lindberg f.* s.n.! *Davis* 3196! *Chapman* 632! *Merton* in ARI 145! Ayios Sozomenos, 1957, *Merton* 3099! Salamis, 1962, *Meikle* 2593!
6. Nicosia, 1880, *Sintenis & Rigo* 567b! also, 1941, *Davis* 2143! Ayia Irini, 1941, *Davis* 2119! 2567!

NOTES: Related to *S. verbenaca*, but easily separated by the pinnatisect leaves, white-villous calyces and exserted corolla-tubes. It is a conspicuous and characteristic plant of the Mesaoria, scarcely extending beyond this region of the island.

In addition to the species cited above, the following are doubtfully or wrongly recorded from Cyprus:

S. VIRGATA *Jacq.*, Hort. Vindob., 1: 14, t. 37 (1770); Holmboe, Veg. Cypr., 159 (1914); Osorio-Tafall et Seraphim, List Vasc. Plants Cyprus, 90 (1973).

S. sibthorpii Sm. in Sibth. et Sm., Fl. Graec. Prodr., 1: 15 (1806), Fl. Graec., 1: 17, t. 22 (1806); Unger et Kotschy, Die Insel Cypern, 268 (1865). ? *S. cerignensis* Sibth. in Walpole, Travels, 18 (1820) nom. nud.

A rather coarse perennial herb with simple, ovate leaves, a widely branched panicle, a fruiting calyx with a recurved, bisulcate upper lip and violet-blue (rarely white) corollas. It is doubtful if this species really grows in Cyprus. It is recorded from Larnaca by Unger et Kotschy ("In Larnaca gegen Wlachy [Vlakhos] n. 88a") as *S. sibthorpii*, and the record is repeated — under *S. virgata* — by Holmboe and Osorio-Tafall & Seraphim. *S. virgata* is a widespread species in S.W. Asia, also growing in Italy and the Balkans, so its occurrence in Cyprus would not be unexpected. However in the absence of any confirmatory herbarium specimen, the record must be viewed with much doubt; it may be no more than a misidentification of *S. verbenaca*. No less doubtful is the identity of *Salvia cerignensis* Sibth. (a *nomen nudum*), regarded as synonymous with *S. sibthorpii* by Kotschy. Sibthorp found it near Kyrenia on 20 April 1787 and refers to it as "a beautiful species". Locality and comment suggest that it may have been *S. hierosolymitana*.

S. MULTICAULIS *Vahl*, Enum., 1; 225 (1804).

There is a specimen of this at Kew (labelled "*S. acetabulosa*") collected by F. A. Rogers and annotated "probably road from Famagusta to Nicosia". It is very unlikely that it was in fact collected anywhere in Cyprus. *S. multicaulis* has membranous-reticulate fruiting calyces, and the indumentum on the stems and leaves consists of at least some branched hairs.

TRIBE 5. **Rosmarineae** *Briq.* Calyx 2-lipped, 11-nerved; corolla 2-lipped, the adaxial lip bifid, the abaxial lip unequally 3-lobed; stamens 2, exserted; filaments arching, often with a small, tooth-like, lateral appendage; anthers connivent or fused, 1-thecous, with linear thecae.

13. ROSMARINUS *L.*

Sp. Plant., ed. 1, 23 (1753).
Gen. Plant., ed. 5, 14 (1754).

Evergreen shrubs; leaves coriaceous, aromatic; flowers in short axillary racemes; calyx campanulate, 2-lipped with a naked throat; corolla shortly exserted, tube dilated at the throat, without an internal hairy zone; median lobe of abaxial (lower) lip enlarged, concave, cochleariform, pendulous; stamens distinctly exserted; stigmatic lobes very unequal, the adaxial minute, the abaxial narrowly deltoid, concave. Nutlets ovoid, almost smooth.

Three species in the Mediterranean region.

1. **R. officinalis** *L.*, Sp. Plant., ed. 1, 23 (1753); Hume in Walpole, Mem. Europ. Asiatic Turkey, ed. 1, 254 (1817); Boiss., Fl. Orient., 4: 636 (1879); Holmboe, Veg. Cypr., 152 (1914); Post, Fl. Pal., ed. 2, 2: 362 (1933); Chapman, Cyprus Trees and Shrubs, 68 (1949); Rechinger f. in Arkiv för Bot., ser. 2, 1: 429 (1950); Osorio-Tafall et Seraphim, List Vasc. Plants Cyprus, 89 (1973); Davis, Fl. Turkey, 7: 76 (1982).
TYPE: "*in* Hispania, G. narbonensi, Galilaea".

Erect or prostrate shrub, up to 2 m. high; bark grey-brown, longitudinally fissured; leaves sessile, crowded, coriaceous, aromatic (smelling of ginger), linear, 10–20 mm. long, 2–3 mm. wide, dark green, glabrous and glandular-punctate above, canescent with stomata in hairy grooves below, apex obtuse, margins entire, strongly revolute; inflorescence a shortly pedunculate, condensed, axillary 10–20-flowered raceme, 1–4 cm. long; rhachis 4-angled, stellate-tomentose; bracts ovate, acute, persistent, about 2 mm. long, 1–2 mm. wide, stellate-tomentose externally; bracteoles minute, subulate; pedicels 2–3 mm. long, erecto-patent, persistent, articulated at apex, stellate-tomentose; calyx campanulate, 4–5 mm. long, 4–5 mm. wide at the conspicuously flared apex, gland-dotted and thinly tomentellous externally, glabrous internally, nervation obscure, adaxial lip about 2 mm. long, 3·5 mm. wide, deltoid, minutely 3-toothed at apex, abaxial lip divided into 2 deltoid lobes about 2 mm. long, 2 mm. wide; corolla whitish or blue, subglabrous or thinly puberulous externally, tube infundibuliform, gibbous abaxially, about 4·5 mm. long, 2·5 mm. wide at apex, adaxial lip oblong, 5–6 mm. long, 2·5–3 mm. wide, porrect, deeply 2-lobed, with rounded lobes at apex, abaxial lip 3-lobed, the lateral lobes oblong, obtuse, about 3·5 mm. long, 2 mm. wide, median lobe concave-cochleariform, about 3·5 mm. long, 5–6 mm. wide, shortly stipitate at base and with erose-denticulate margins; adaxial stamens wanting, abaxial stamens with upwards-arching, glabrous filaments about 8 mm. long, inserted at the mouth of the corolla-tube near the base of the lateral lobes of the lower corolla-lip; anthers dorsifixed, linear, about 1·8 mm. long, 0·4 mm. wide; connective linear, running almost the full length of the theca; ovary glabrous, about 1 mm. diam.; disk slightly sulcate longitudinally and gibbous between the ovules; style about 16 mm. long, glabrous, long-exserted; stigmatic lobes unequal, the adaxial minute, the abaxial about 0·8 mm. long. Nutlets ovoid-oblong, about 2·5 mm. long, 1·4 mm. wide, dark brown, minutely granulose, with an oblique, white, basal attachment area, and a conspicuous circular scar.

HAB.: Maritime rocks and fixed sand-dunes; near sea-level; fl. Jan.–March.

DISTR.: Division 8, in one area only; cultivated and an occasional escape in other parts of the island. N. Mediterranean region to Lebanon.

8. On rocks and sand along the coast between Yialousa and Ronnas Bay, apparently indigenous, 1930–1962, *Chapman* 26! *Davis* 2383!, etc.

NOTES: Extensively cultivated in Cyprus, and recorded as a cultivated plant from Larnaca, Stavrovouni, Platres and elsewhere (*Hume*; *Sintenis & Rigo* 937! *Haradjian* 23, etc.) occasionally escaping to wild ground, as on cliffs at Bellapais (1968, *I. M. Hecker* 4!). It is locally abundant, and apparently quite spontaneous, on a limited stretch of coast near Yialousa, and may be accepted as indigenous there, though the possibility remains that it may have been accidentally introduced at some remote period. In the one specimen (*Davis* 2383) available for examination the filaments appear to lack the small, tooth-like appendage so frequently found in the species.

TRIBE 6. **Nepeteae** *Dumort.* Calyx 5-toothed, obscurely or distinctly 2-lipped; corolla 2-lipped, tube usually exserted; stamens 4, the adaxial longer than the abaxial; style gynobasic. Nutlets dry.

14. NEPETA *L.*

Sp. Plant., ed. 1, 570 (1753).
Gen. Plant., ed. 5, 249 (1754).

Perennial, or rarely annual, herbs, sometimes woody at the base; leaves generally dentate or crenate, often strongly (and rather unpleasantly) aromatic; inflorescences usually spiciform, consisting of few- or many-flowered, remote or crowded, lax or dense verticillasters; bracts foliaceous or coloured; bracteoles commonly narrowly ovate or linear, entire; calyx tubular, 15-nerved, straight or curved, teeth 5, subequal; corolla 2-lipped, tube included in calyx or exserted, without an internal hairy zone, adaxial lip flattish or concave, emarginate or bifid, abaxial lip 3-lobed with an enlarged, concave median lobe; stamens 4, parallel, the adaxial longer than the abaxial; anthers connivent under the adaxial lip of corolla; anther-thecae divergent, opening by a common slit; stigmatic lobes subulate, subequal. Nutlets ovoid or oblong, smooth, tuberculate or variously sculptured.

About 250 species in temperate regions of the Old World or on mountains in tropical Africa.

Leaves coarsely serrate, 2–7 cm. long, 1·5–5 cm. wide; inflorescences crowded, spiciform
N. CATARIA (p. 1300)
Leaves shortly crenate-serrate, 1–3 cm. long, 0·6–2 cm. wide; inflorescence lax, verticillasters
generally remote -   -   -   -   -   -   -   -   -   -   **1. N. troodi**

N. CATARIA *L.*, Sp. Plant., ed. 1, 570 (1753), Catmint, a heavily aromatic, robust perennial 50–100 cm. high, with deltoid-cordate, pubescent, coarsely serrate leaves 2–7 cm. long, 1·5–5 cm. wide, and small white flowers in crowded, branched, spiciform inflorescences, is recorded by Kotschy (Die Insel Cypern, 273; 1865) from Paphos and S. of Prodhromos. His records have been repeated by H. Stuart Thompson (Journ. Bot., 44: 336; 1906) and

Holmboe (Veg. Cypr., 155; 1914) but, so far as I am aware, the plant has not since been collected in Cyprus. If not a misidentification, then *N. cataria* was most probably a cultivated alien, and may not have persisted. It cannot be considered an established member of the Cyprus flora.

Catmint is widely distributed in Europe and western Asia, but is often an obvious introduction where it occurs, being formerly valued medicinally as a tonic, carminative and diaphoretic. In common with other members of the genus, the odour of the crushed leaves is peculiarly attractive to cats.

1. **N. troodi** *Holmboe*, Veg. Cypr., 155 (1914); Lindberg f., Iter Cypr., 29 (1945); Chapman, Cyprus Trees and Shrubs, 69 (1949); Rechinger f. in Arkiv för Bot., ser. 2, 1: 429 (1950); Osorio-Tafall et Seraphim, List Vasc. Plants Cyprus, 90 (1973).

   [*N. mussinii* (non Spreng. ex Henck.) Kotschy in Unger et Kotschy, Die Insel Cypern, 273 (1865); H. Stuart Thompson in Journ. Bot., 44: 336 (1906).]

   [*N. orientalis* (non Mill.) Post in Mém. Herb. Boiss., 18: 98 (1900); H. Stuart Thompson in Journ. Bot., 44: 336 (1906).]

   [*N. sibthorpii* (non Benth.) H. Stuart Thompson in Journ. Bot., 44: 336 (1906).]

   TYPE: Not precisely indicated. *Sintenis & Rigo* 723 from near Prodhromos, July 1880, would be a suitable lectotype (K !).

Erect or sprawling perennial 20–50 cm. high; stems much branched at the woody base, sparingly branched above, sharply tetragonal with a rather dense indumentum of short glandular and long, spreading, white, multicellular hairs; leaves cordate-deltoid, 1–3 cm. long, 0·6–2 cm. wide, rugulose, softly greyish-pubescent, apex subacute or obtuse, margins crenate-serrate, base shortly cordate or subtruncate, nervation distinct but not very prominent; petiole slender, pubescent, channelled above, 3–10(–20) mm. long; inflorescence lax, spiciform, 10–24 cm. long, consisting of 5–8, remote, congested, many-flowered verticillasters; bracts foliaceous, the lowermost similar to the cauline leaves, the upper lanceolate-acuminate, entire, often very small; bracteoles narrowly linear-lanceolate, 3–10 (–12) mm. long, 0·4–1 mm. wide, apex finely setaceous, densely glandular-pubescent with a sparser indumentum of long, spreading, eglandular hairs; flowers subsessile or very shortly pedicellate; calyx-tube 3–4 mm. long, 1·5 mm. wide, straight or slightly curved, softly pubescent with short, glandular and longer, eglandular hairs externally, glabrous within, obscurely 2-lipped, teeth lanceolate-subulate, about 3 mm. long, the adaxial a little shorter than the abaxial; corolla whitish or very pale lilac, the abaxial lip dotted purple, tube about 8 mm. long, 3 mm. wide at the abruptly dilated apex, pubescent externally, strongly curved, adaxial lip broadly oblong, about 2·5 mm. long, 3·5 mm. wide, hairy externally, deeply bifid with rounded lobes and inflexed margins, abaxial lip 3-lobed, the lateral lobes about 1 mm. long, almost semicircular, the median shortly stipitate, suborbicular, about 2 mm. diam., margin irregularly toothed or lacerate; stamens inserted a little way down the corolla-tube, the adaxial pair about 0·8 mm. above the abaxial; filaments subequal, about 3·5 mm. long, flattened, thinly hairy; anthers dark purple, about 0·8 mm. diam., thecae strongly divergent; connective very small; ovary about 1 mm. diam., minutely glandular at apex; disk small, longitudinally sulcate; style glabrous, about 12 mm. long; stigmatic lobes equal, subulate, about 0·7 mm. long. Nutlets oblong, about 1·8 mm. long, 0·9 mm. wide, distinctly trigonous, dark brown, verruculose, area of attachment white, bluntly sublunate, scar minute. *Plate 74.*

HAB.: Rocky mountainsides under *Pinus nigra* ssp. *pallasiana*, *Juniperus foetidissima* and *Quercus alnifolia*; 3,600–6,400 ft. alt.; fl. June–Oct.

DISTR.: Division 2, locally common. Endemic.

2. Summit of Khionistra, 1862–1970, *Kotschy* 773; *A. G. & M. E. Lascelles* s.n. ! *Holmboe* 989; *Haradjian* 458 ! *Syngrassides* 726 ! *Kennedy* 790 ! 791 ! *Lindberg f.* s.n. ! *Davis* 1793 ! 1868 !

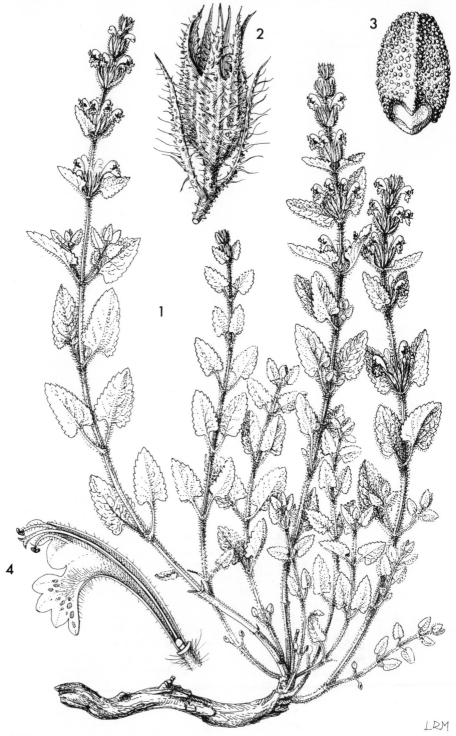

Plate 74. NEPETA TROODI Holmboe **1**, habit, × 1; **2**, fruiting calyx, × 12; **3**, nutlet, × 24; **4**, corolla, longitudinal section, × 6. (**1, 4** from *D. P. Young* 7436; **2, 3** from *Syngrassides* 726.)

Mavromoustakis s.n.! *Casey* 844! *A. Genneou* in ARI 1629! Prodhromos, 1880, *Sintenis & Rigo* 723!, and, 1949, *Casey* 909! Kryos Potamos, 5,000 ft. alt., 1937, *Kennedy* 792! Xerokolymbos, 4,950 ft. alt., 1937, *Kennedy* 792A! Moniatis R., 3,800 ft. alt., 1937, *Kennedy* 793! Kykko, 1913, *Haradjian* 969! and, 1940, *Davis* 1821! Madhari, 1940, *Davis* 1879! and, 1955, *Merton* 2451! Papoutsa, 1940, *Davis*; S. of Trikoukkia, 1961, *D. P. Young* 7436!

TRIBE 7. **Lamieae** Calyx 5-toothed, obscurely or distinctly 2-lipped; throat generally open in fruit (except in *Prunella*); corolla 2-lipped, the adaxial lip often strongly concave or galericulate; stamens 4, didynamous, the adaxial filaments shorter than the abaxial; anthers 2-thecous; style gynobasic. Nutlets dry, glabrous or hairy.

15. SIDERITIS *L.*
Sp. Plant., ed. 1, 574 (1753).
Gen. Plant., ed. 5, 250 (1754).

Annual or perennial herbs or subshrubs; leaves entire or toothed, often pilose or tomentose; inflorescence generally spiciform, consisting of several, superimposed, 4–many-flowered verticillasters; bracts foliaceous, or often differing markedly from the leaves; calyx tubular-campanulate, 5–10-nerved, sometimes obscurely 2-lipped, with 5 erect or recurved, often spinous-tipped teeth; corolla yellowish or whitish, sometimes pink or purple, 2-lipped, adaxial lip flattish, entire, emarginate or 2-lobed, abaxial lip 3-lobed, the median lobe generally emarginate, larger than the lateral lobes; stamens included, the adaxial much shorter than the abaxial, with divergent 2-thecous anthers, the abaxial anthers often much reduced or deformed; style included; stigmatic lobes unequal, the abaxial short, obtuse, the adaxial longer, often sheathed at the base by the abaxial lobe. Nutlets ovoid or oblong, rounded at apex, almost smooth or distinctly tuberculate.

About 100 species in S. Europe, Mediterranean region and temperate Asia.

Sprawling annual 5–25 (–40) cm. high; calyx-teeth recurved, aristate, the adaxial much larger than the abaxial; corolla white or tinged mauve - - - - - - **1. S. curvidens**
Erect perennial 40–100 cm. high from a woody base; calyx-teeth erect, subequal; corolla yellow and brown:
  Cauline leaves densely white-tomentose; calyx-teeth oblong, subacute; corolla yellow, streaked and margined brown - - - - - - - - **2. S. cypria**
  Cauline leaves thinly hispidulous; calyx-teeth deltoid-acute; corolla uniformly pale yellow **3. S. perfoliata**

1. S. curvidens *Stapf* in Denkschr. Akad. Wiss. Wien, 50 (2): 100 (1885); Rechinger f. in Arkiv för Bot., ser. 2, 1: 429 (1950); Feinbrun, Fl. Palaest., 3: 113, t. 182 (1978); Davis, Fl. Turkey, 7: 182 (1982).
  S. romana L. ssp. *curvidens* (Stapf) Holmboe, Veg. Cypr., 153 (1914); Osorio-Tafall et Seraphim, List Vasc. Plants Cyprus, 89 (1973).
  [*S. romana* (non L.) Kotschy in Unger et Kotschy, Die Insel Cypern, 274 (1865); Boiss., Fl. Orient., 4: 706 (1879); Sintenis in Oesterr. Bot. Zeitschr., 31: 393 (1881); 32: 21, 124, 397 (1882); Post, Fl. Pal., ed. 2, 2: 373 (1933).]
  TYPE: Turkey; "In Acropoli urbis Xanthos", 1882, *F. Luschan* (WU).

Sprawling annual; stems 5–25 (–40) cm. long, usually much branched at base, unbranched or sparingly branched above, sharply tetragonal, clothed with a mixed indumentum of long and short, eglandular, spreading hairs; basal leaves obovate or subspathulate, 1–5 cm. long, 0·8–2 cm. wide, thinly villose on both surfaces, apex acute or obtuse, margins remotely serrulate, base subsessile or tapering to a flattened petiole up to 2 cm. long; stem leaves similar to basal, but generally sessile or subsessile and smaller; inflorescence elongate, narrow, consisting of 6 or more, remote or crowded, 4–6-flowered verticels; bracts narrowly ovate or lanceolate, foliaceous, 5–10 mm. long,

2·5–5 mm. wide, acute, entire, villous, generally recurved; bracteoles wanting; pedicels very short, usually less than 2·5 mm. long, dorsiventrally compressed, densely villous; calyx campanulate, tube 4–5 mm. long, 2–3 mm. wide, 10-nerved, conspicuously saccate at the base, villous externally, densely bearded at the throat, otherwise glabrous internally, teeth recurved, terminating in a pilose awn about 2 mm. long, the adaxial ovate, about 6 mm. long, 2·5–4 mm. wide at base, the 4 abaxial much smaller and narrower, about 4 mm. long, 0·8 mm. wide at base; corolla white or with a pale mauve adaxial lip, tube 4–5 mm. long, included in calyx, pubescent externally and internally; adaxial lip oblong, truncate, flattish, about 1 mm. long, 0·8 mm. wide, abaxial lip 3-lobed, the median lobe suborbicular, about 3 mm. diam., the lateral lobes about 1 mm. long, connate with the median lobe along the greater part of their inner margin; stamens inserted about half-way down tube, included; filaments of adaxial stamens very short, about 0·2 mm. long, pilose at base; anthers perfect, thecae strongly divaricate, about 0·8 mm. wide; filaments of abaxial stamens about 1 mm. long, pilose at base, anthers deformed, only one of the thecae properly developed; ovary glabrous, about 0·5 mm. diam., seated on a short glabrous disk; style glabrous, 1·5–1·8 mm. long; stigmatic lobes very short, blunt, concave. Nutlets broadly obovoid, about 1·5 mm. long, 1·2 mm. wide, very obscurely trigonous, brown, tuberculate-verruculose, attachment area very small and inconspicuous.

HAB.: Dry stony slopes; waste ground; cultivated and fallow fields; sea-level to 2,700 ft. alt.; fl. March–May.

DISTR.: Divisions 1, 3–8. Widespread in the eastern Mediterranean region from Greece and Crete to Palestine.

1. Ktima, 1913, *Haradjian* 664! Paphos, 1930, *E. Wall* s.n.; Peyia Forest, Kalifes, 1962, *Meikle* 2047! Polis, 1962, *Meikle*!
3. Malounda, 1963, *J. B. Suart* 10! 58!
4. Larnaca, 1862, *Kotschy* 39; Cape Greco, 1862, *Kotschy* 134; same locality, 1905, *Holmboe* 416, also, 1957, *Poore* in *Merton* 3165! and, 1960, *N. Macdonald* 99!
5. Mile 3, Nicosia-Famagusta road, 1950, *Chapman* 398!
6. Kormakiti, 1936, *Syngrassides* 1222! Dhiorios, 1962, *Meikle*!
7. Buffavento, 1862, *Kotschy* 442; also, 1880, *Sintenis & Rigo*; Pentadaktylos, 1880, *Sintenis & Rigo* 127! Larnaka tis Lapithou, 1936, *Syngrassides* 912! Kyrenia, 1936, *Mrs. Dray* 12! also, 1949, *Casey* 438!, 1956, *G. E. Atherton* 1388! Aghirda, 1941, *Davis* 2488! Yaïla, 1941, *Davis* 2847! Ayios Epiktitos, 1951, *Casey* 1124!
8. Cape Andreas, 1880, *Sintenis & Rigo*; near Rizokarpaso, 1912, *Haradjian* 141; Apostolos Andreas, 1948, *Mavromoustakis* s.n.!

2. **S. cypria** *Post* in Mém. Herb. Boiss., 18: 98 (1900); H. Stuart Thompson in Journ. Bot., 44: 336 (1906); Holmboe, Veg. Cypr., 153 (1914); Chapman, Cyprus Trees and Shrubs, 69 (1949); Rechinger f. in Arkiv för Bot., ser. 2, 1: 429 (1950); Seligman in Quart. Bull. Alpine Gard. Soc., 20: 234 (1952); Osorio-Tafall et Seraphim, List Vasc. Plants Cyprus, 89 (1973). *S. cilicica* Boiss. var. *cypria* (Post) Lindberg f., Iter Cypr., 30 (1946). [*S. pullulans* (non Vent.) Kotschy in Unger et Kotschy, Die Insel Cypern, 274 (1865); H. Stuart Thompson in Journ. Bot., 44: 336 (1906).] TYPE: Cyprus; "ad castellum Sancti Hilarionis Cypri", 1894, *Post* 915 (BEI).

Erect, suffruticose perennial up to 60 cm. high, base of stems woody, branched, densely white-tomentose, flowering shoots unbranched, tetragonal, shortly glandular, puberulous with a few, scattered, longer, eglandular hairs, commonly reddish; lower cauline leaves oblanceolate or obovate, 3–12 cm. long, 1–5 cm. wide, densely tomentose, apex obtuse, base tapering gradually to an indistinct petiole, margins finely serrulate (the serrulations, like the close reticulate venation, concealed by the dense tomentum), upper cauline leaves very similar to the lower, but usually a little smaller, the uppermost pair sometimes calvescent, greenish and distinctly reticulate-veined towards the base; inflorescences generally

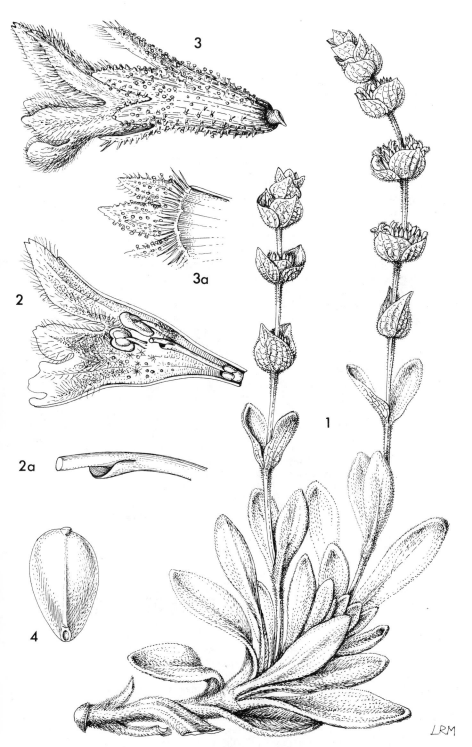

Plate 75. SIDERITIS CYPRIA Post **1,** habit, × 1; **2,** corolla, longitudinal section, × 6; **2a,** apex of style, × 24; **3,** flower, × 6; **3a,** calyx interior, × 6; **4,** nutlet, × 15. (**1** from *Lindberg f.* s.n.; **2, 2a** from *Chapman* 404; **3, 3a** from *Davis* 2040; **4** from *Davis* 1957.)

unbranched or sometimes with 2 elongate basal branches, 12–35 cm. long, spiciform, consisting of 8–15 or more dense verticillasters regularly superimposed along a glandular rhachis; bracts connivent into a cup, very broadly ovate or suborbicular, 0·8–1·5 cm. long, 1–2 cm. wide, shortly glandular, conspicuously reticulate, apex rounded, mucronate-apiculate or subacute, very distinct from the cauline leaves; flowers crowded, subsessile, smelling of hops or vinegar when bruised; calyx tubular-campanulate, densely glandular externally, glabrous internally except for the bearded throat, tube about 7 mm. long, 3·5 mm. wide, teeth erect, oblong, subacute, ciliate with long, eglandular hairs, the adaxial about 4 mm. long, 1·3 mm. wide, the abaxial 4 a little shorter, about 3 mm. long; corolla bright yellow, boldly striped and bordered with brown, limb shortly adpressed-pilose externally and internally on the palate, tube about 11 mm. long, 2·5 mm. wide, slightly curved, adaxial lip oblong, about 5 mm. long, 3 mm. wide, deeply bifid at apex, abaxial lip 3-lobed, the lateral lobes deltoid, about 2 mm. long and 2 mm. wide at base, the median semicircular, crenate-undulate, about 2·5 mm. long, 4 mm. wide; stamens inserted about half-way down corolla-tube, included; filaments subequal, thick, about 1·5 mm. long, pilose at base; anthers of adaxial stamens perfect, the thecae divaricate, about 2 mm. wide, thecae of abaxial stamens deformed, 1-thecous; ovary glabrous, about 1·5 mm. diam., seated on a short, shallowly lobed, glabrous disk; style fragile at base, thickish, about 4 mm. long; stigmatic lobes blunt, less than 1·5 mm. long, subtruncate, connivent, the abaxial sheathing the base of the adaxial. Nutlets obovoid, about 2·8 mm. long, 1·8 mm. wide, fuscous-brown, verruculose, attachment area very small. *Plate 75*.

HAB.: Fissures of dry, south-facing, limestone cliffs; 1,000–3,000 ft. alt.; fl. June–Aug.

DISTR.: Division 7. Endemic.

7. Buffavento, Pentadaktylos, St. Hilarion, Halevga, Bellapais, Yaïla, above Kythrea, above Sisklipos, etc., 1862–1955, *Kotschy* 391; *Sintenis & Rigo* 571! *Post* 915! *Haradjian* 371! *Holmboe* 822; *C. B. Ussher* 67! *J. A. Tracey* 48! *M. E. Dray* 5! *Roger-Smith* s.n.! *Syngrassides* 1540! 1734! *Kennedy* 1077–1079! *Chapman* 404! *Lindberg f.* s.n.! *Davis* 1926! 1932! 2040! 2112! 3629! 3638! *F. M. Probyn* 145! *Miss Mapple* 10! *G. E. Atherton* 148! 732!

NOTES: Although the affinity with *S. cilicica* Boiss. et Bal., from S. Turkey, is close, differences in habit, indumentum and corolla are, as pointed out by Holmboe (Veg. Cypr., 153), sufficient to distinguish *S. cypria* as a species, though I can find no significant distinction in corolla/calyx length. *S. cilicica* appears to be a plant of mountain slopes, growing amongst other herbage, whereas *S. cypria* is strictly a plant of fissures in sun-baked limestone cliffs. In *S. cypria* the cauline and basal leaves persist throughout anthesis, and are uniformly densely tomentose. In *S. cilicica* the lower cauline and basal leaves are marcescent, and appear to wither before anthesis, while the upper cauline leaves are green, calvescent, and conspicuously reticulate-veined.

3. S. perfoliata *L.*, Sp. Plant., ed. 1, 575 (1753); Boiss., Fl. Orient., 4: 714 (1879); Post, Fl. Pal., ed. 2, 2: 375 (1933); A. K. Jackson et Turrill in Kew Bull., 1938: 466 (1938); Osorio-Tafall et Seraphim, List Vasc. Plants Cyprus, 89 (1973); Feinbrun, Fl. Palaest., 3: 114, t. 184 (1978); Davis, Fl. Turkey, 7: 199 (1982).
TYPE: "*in* Oriente?".

Erect perennial with a woody rootstock, smelling of Poppies when bruised; stems to 1 m. high, sharply tetragonal, pilose with a mixed indumentum of short, glandular, and long, eglandular hairs or bristles; branches mostly basal, or occasionally produced in the region of the inflorescence; leaves of two kinds, the basal narrowly obovate or lanceolate, 1–5 cm. long, 0·5–1·2 cm. wide, densely or thinly lanuginose, acute or subacute, margins subentire or crenate, base tapering to an indistinct, or occasionally distinct, flattened petiole, cauline leaves sessile, amplexicaul, oblong-lanceolate, 4–7 cm. long, 1–2 cm. wide, thinly hispidulous with very conspicuous, prominent, coarsely reticulate venation, apex acute or acuminate, margins subentire or crenate; inflorescence spiciform, 8–20 (–30) cm. long, branched or unbranched, consisting of 8–17 closely superimposed verticels; bracts broadly ovate or suborbicular, cuspidate-acuminate, about 1·5–2 cm. long and wide, hispidulous-ciliate, prominently reticulate-veined, forming a cup around the rhachis; flowers crowded, subsessile; calyx tubular, glandular-pubescent with scattered, long, eglandular hairs, tube 8 mm. long, 2·5 mm. wide at apex, obscurely 10-nerved, shortly pubescent internally, teeth subequal, erect, narrowly deltoid, about 2 mm. long, 1·2 mm. wide at base, long-ciliate; corolla pale yellow, tube narrowly infundibuliform, about 8 mm. long, 3·5 mm. wide at apex, glabrous externally towards base, densely stellate-tomentellous towards apex and on limb, pubescent at the throat internally, adaxial lip oblong, about 2 mm. long, 2·5 mm. wide, emarginate with rounded lobes, abaxial lip 3-lobed, lateral lobes semicircular, about 0·8 mm. long, median lobe concave, undulate-crenulate, rounded, about 2 mm. long, 4 mm. wide; adaxial stamens inserted about 4 mm. down corolla-tube, abaxial stamens inserted about 2·5 mm. down corolla-tube; filaments less than 1·5 mm. long, pilose at base; anthers all perfect, about 1·5 mm. wide, with divaricate thecae; ovary glabrous, about 1 mm. diam., seated in a cupuliform, glabrous, shallowly lobed, longitudinally sulcate disk; style thick, glabrous, about 1·5 mm. long; stigmatic lobes unequal, truncate, the adaxial about 1 mm. long, the abaxial about 0·8 mm. long, concave, sheathing the adaxial at its base. Nutlets broadly obovate, about 2 mm. long, 1·5 mm. wide, distinctly trigonous, pale brown, minutely verruculose, with a very small attachment area.

HAB.: In vineyards or amongst garigue on open chalk hillsides; 2,500–3,600 ft. alt.; fl. May–July.

DISTR.: Divisions 2, 3, rare. Eastern Mediterranean region from Greece to Palestine.

2. Aphamis, 1937–38, *Kennedy* 794! 1076! Above Arminou, 1941, *Davis* 3400! Above Ayios Mamas, 1956, *Merton* 2814!
3. Zagala above Trimiklini, 1941, *Davis* 3148!

NOTES: Recorded only from chalk slopes around Perapedhi, but not unlikely to occur on similar ground about Khrysorroyiatissa, Lefkara, etc.

[S. INCANA *L.*, Sp. Plant., ed. 2, 802 (1763), recorded from "Larnica and Limassol" by Hume (in Walpole, Mem. Europ. Asiatic Turkey, ed. 1, 254; 1817), but undoubtedly an error. *S. incana* is a shrubby perennial, not unlike Lavender in general habit, with similar long-stalked, spiciform inflorescences, but of sulphur-yellow or pink flowers. It is widely distributed in Spain, but occurs nowhere in the eastern Mediterranean. It is impossible, in the absence of a specimen, to say what Hume intended.]

16. MARRUBIUM *L.*

Sp. Plant., ed. 1, 582 (1753).
Gen. Plant., ed. 5, 254 (1754).

Perennial, or rarely annual, herbs, often tomentose or canescent; leaves generally toothed or lobed, often rugose or with prominent nervation; inflorescences usually spiciform, consisting of remote, crowded verticillasters; bracts foliaceous, similar to cauline leaves; bracteoles filiform; calyx tubular-campanulate, 5- or 10-nerved, generally barbate at throat, calyx-teeth 5 or 10, commonly spreading or reflexed, spinescent, sometimes uncinate; corolla-tube included in calyx, naked or with a hairy zone internally, limb 2-lipped, the adaxial lip flattish, emarginate or bifid, the abaxial lip 3-lobed, with an enlarged median lobe; stamens 4, didynamous, included, the abaxial with longer filaments than the adaxial; anthers with divergent thecae; style included; stigmatic lobes short, blunt. Nutlets ovoid, rounded or truncate at apex, almost smooth.

About 40 species in Europe, the Mediterranean region and temperate Asia; introduced elsewhere.

1. **M. vulgare** *L.*, Sp. Plant., ed. 1, 583 (1753); Boiss., Fl. Orient., 4: 703 (1879); Post in Mém. Herb. Boiss., 18: 98 (1900); H. Stuart Thompson in Journ. Bot., 44: 336 (1906); Post, Fl. Pal., ed. 2, 2: 372 (1933); Feinbrun, Fl. Palaest., 3: 122, t. 180 (1978); Davis, Fl. Turkey, 7: 168 (1982).

    M. apulum Ten., Prodr. Fl. Nap., 34 (1811), Fl. Nap., 5: 16, t. 154 (1835).

    M. vulgare L. var. *lanatum* Benth., Labiat. Gen. et Sp., 591 (1834); Unger et Kotschy, Die Insel Cypern, 274 (1865).

    M. vulgare L. var. *apulum* (Ten.) Heldr. in Bull. Herb. Boiss., 6: 385 (1898); Holmboe, Veg. Cypr., 153 (1914); Lindberg f., Iter Cypr., 28 (1946).

    M. vulgare L. ssp. *apulum* (Ten.) Osorio-Tafall et Seraphim, List Vasc. Plants Cyprus, 89 (1973) nom. invalid.

TYPE: "*in* Europae *borealioris ruderatis*".

Erect perennial from a woody rootstock, 30–60 cm. high; stems branched at the base, branched or unbranched above, tetragonal, generally densely white-tomentose; leaves broadly obovate, reniform or suborbicular, 1–4 cm. long, 1–5 cm. wide, conspicuously rugose, grey-green and thinly tomentose above, often rather densely tomentose below, apex rounded or blunt, margins undulate-crenulate, base cordate, truncate or broadly cuneate; petiole distinct, 0·5–3 cm. long, channelled above, lanuginose-tomentose; inflorescence generally branched, spiciform, 8–20 cm. long, consisting of 6–10 (or more), evenly spaced, crowded, many-flowered verticillasters; bracts foliaceous, similar to the leaves but smaller; bracteoles filiform, spinulose, pilose, 3–6 mm. long; flowers subsessile; calyx tubular, 5–7 mm. long, 2 mm. wide, gland-dotted and rather densely stellate-pubescent, tube 10-nerved, slightly constricted below the apex, bearded internally at the throat, teeth 10, spreading, subequal or alternately longer and shorter, 1–1·5 mm. long, spinulose or uncinate, thinly pilose towards the base which forms a continuous cartilaginous rim to the mouth of the calyx-tube; corolla white, tube included in calyx, 5–6 mm. long, slightly curved, glabrous externally, with a bearded zone internally about 2 mm. from apex, limb stellate-pilose externally, 2-lipped, the adaxial lip narrowly oblong, about 2 mm. long, 1 mm. wide, deeply bifid, abaxial lip very unequally 3-lobed, the lateral lobes oblong, about 1 mm. long, 0·5 mm. wide, the median flabelliform, shortly stipitate at base, about 2 mm. long, 2·5 mm. wide, margin crispate-erosulose; stamens inserted about 2 mm. down corolla-tube, abaxial filaments about 0·8 mm. long, adaxial filaments about 0·5 mm. long; anthers about 0·5 mm. wide, with a conspicuous, dark, dilated connective; ovary about 1·5 mm. diam., glabrous, seated on a very short, lobed, glabrous disk; style about 3·5 mm. long, glabrous; stigmatic lobes ovate, concave, subacute, the adaxial about 0·1 mm. long, the abaxial about 0·2 mm. long, the two separated by a membranous, median, tongue-like appendage. Nutlets narrowly oblong, trigonous, about 2 mm. long, 1·3 mm. wide, fuscous-brown, minutely verruculose, attachment area very small.

HAB.: Roadsides and waste ground, occasionally on rocky, calcareous hillsides; sea-level to 4,500 ft. alt.; fl. Febr.–Oct.

DISTR.: Divisions 1–7. Widespread in Europe, the Mediterranean region and western Asia.

1. Kato Paphos, 1973, *P. Laukkonen* 310!
2. Prodhromos, 1862, *Kotschy*; Kykko Monastery, 1880, *Sintenis & Rigo* 568b! also, 1913, *Haradjian* 962! 1939, *Lindberg f.* s.n.!. 1973, *P. Laukkonen* 371! Lefka, 1932, *Syngrassides* 87! Makheras Monastery, 1937, *Syngrassides* 1630! Platres, 1937–1955, *Kennedy* 795! 1862! Aphamis, 1937, *Kennedy* 796! Kissousa, 1959, *C. E. H. Sparrow* 34!
3. Pissouri, 1862, *Kotschy* 628; Limassol, 1905, *Michaelides*.

4. Larnaca, 1860, *Hooker & Hanbury* s.n.! Athienou, 1862, *Kotschy* 972! *Larnaca*, 1912, *Haradjian* 57! Kouklia, 1967, *Merton* 443! Lysi, 1970, *A. Genneou* in AR1 1632!
5. Salamis, 1962, *Meikle* 2573!
6. Nicosia, 1898, *Post* s.n.; also, 1950, *Chapman* 685! Myrtou, 1932, *Syngrassides* 456!
7. Armenian Monastery, 1912, *Haradjian* 365! Kyrenia, 1949, *Casey* 815! also, 1955, *Miss Mapple* 105! *G. E. Atherton* 55!

17. STACHYS *L.*

Sp. Plant., ed. 1, 580 (1753).
Gen. Plant., ed. 5, 253 (1754).

Annual or perennial herbs or small shrubs; leaves entire or variously toothed; inflorescence generally spiciform, consisting of several few- or many-flowered verticillasters, occasionally subcapitate; bracts foliaceous, resembling or distinct from the leaves; bracteoles present or absent; calyx tubular or campanulate, 5- or 10-nerved, sometimes obscurely 2-lipped, calyx-teeth acute, subequal; corolla purple, pink, white or yellow, 2-lipped, adaxial lip generally concave or galericulate, sometimes flattish; abaxial lip 3-lobed with a large median lobe, tube usually with a hairy zone internally, sometimes glabrous; stamens 4, didynamous, generally exserted; anthers 2-thecous, with parallel or divergent thecae; style included or exserted; stigmatic lobes short, subequal. Nutlets obovoid or oblong with a rounded apex.

About 300 species widely distributed in temperate regions, and on mountains in the tropics.

1. S. cretica *L.*, Sp. Plant., ed. 1: 581 (1753); Boiss., Fl. Orient., 4: 719 (1879); Holmboe, Veg. Cypr., 157 (1914); Post, Fl. Pal., ed. 2, 2: 377 (1933); Lindberg f., Iter Cypr., 30 (1946); Rechinger f. in Arkiv för Bot., ser. 2, 1: 430 (1950); Osorio-Tafall et Seraphim, List Vasc. Plants Cyprus, 89 (1973); Feinbrun, Fl. Palaest., 3: 129, t. 208 (1978); Davis, Fl. Turkey, 7: 216 (1982).

S. italica Mill., Gard. Dict., ed. 8, no. 3 (1768); Gaudry, Recherches Sci. en Orient, 195 (1855); Unger et Kotschy, Die Insel Cypern, 273 (1865).

TYPE: "*in* Creta".

Erect perennial 60–80 (–100) cm. high; rhizome shortly creeping; stems robust, tetragonal, densely white-tomentose, generally unbranched; lower cauline leaves narrowly oblong, 4–8 (–10) cm. long, 1–2 (–3) cm. wide, rugulose, thinly or densely white-tomentose, apex obtuse or subacute, margins crenulate, base rounded or narrowing abruptly to a flattened petiole 2–4 cm. long, upper cauline leaves similar, but shortly petiolate or subsessile; inflorescence spiciform, 15–40 cm. long, consisting of 5–10 remote, crowded, many-flowered verticillasters; bracts foliaceous, resembling the leaves, but smaller and usually more acuminate; flowers subsessile; bracteoles linear or lanceolate, 6–8 mm. long, 0·8–1·5 mm. wide, densely tomentose; calyx campanulate, densely tomentose, tube about 7 mm. long, 4–5 mm. wide at apex, thinly bearded internally at throat, otherwise glabrous, teeth erect, subequal, deltoid, 1·8–2 mm. long, 2–2·5 mm. wide at base, tomentose externally, shortly spinulose at apex; corolla pink or purplish, tube about 9 mm. long, 4–5 mm. wide at throat, slightly bent, puberulous externally, glabrous internally except for a narrow bearded zone about 2 mm. above base, adaxial lip pilose externally, hooded, about 8 mm. long, 4 mm. wide, apex shortly emarginate, margins undulate-erosulose, abaxial lip 3-lobed, lateral lobes oblong, about 3 mm. long, 2 mm. wide, median lobe shortly stipitate, flabelliform, about 5 mm. long, 8 mm. wide, undulate-erosulose; stamens inserted about 2 and 3 mm. down corolla-tube; filaments robust, about 5 mm. long, shortly pilose; anthers about 2 mm. wide, with strongly divergent thecae; ovary glabrous, about 2 mm.

diam., seated in a shallow, cupuliform, shortly lobed disk; style glabrous, about 10 mm. long, exserted; stigmatic lobes acute, concave, about 0·5 mm. long. Nutlets broadly obovate, about 2·8 mm. long, 2 mm. wide, distinctly trigonous, pale brown, minutely verruculose, attachment area very small.

HAB.: Roadsides; vineyards; dry grassy banks or in garigue on rocky hillsides; sea-level to 5,700 ft. alt.; fl. April–July.

DISTR.: Divisions 1–6. Widely distributed in the Mediterranean region.

1. Between Khrysokhou and Arodhes ["Arora"], 1862, *Kotschy* 906! Ktima, 1913, *Haradjian* 694! Lyso, 1913, *Haradjian* 848! Between Istinjo and Philousa, 1962, *Meikle* 2765! Near Stroumbi, 1979, *Hewer* 4554A!
2. Frequent; Prodhromos, Panayia, Kalopanayiotis, Kryos Potamos, Platres, Aphamis, Perapedhi, Statos, Khrysorroyiatissa, Troödos, Stavros tis Psokas, Kambos, etc. 1862–1973, *Kotschy*; *Sintenis & Rigo* 734! *G. E. Atherton* 586! *D. P. Young* 7349! *Haradjian* 897! *Syngrassides* 1552! *Kennedy* 801–804! *Davis* 1832! 3398! 3412! *F. M. Probyn* s.n.! *Kennedy* 1877! *Meikle* 2694! *Economides* in ARI 1205! *P. Laukkonen* 483!
3. Kannaviou, 1939, *Lindberg f.* s.n.; Evdhimou, 1950, *F. J. F. Barrington* 13! Cherkez, 1952, *Mavromoustakis* 33! Kolossi, 1959, *C. E. H. Sparrow* 29! 29A! S. of Kalokhorio, 1959, *P. H. Oswald* 145! Akrotiri, 1967, *Merton* in ARI 665! Amathus, 1978, *J. Holub* s.n.!
4. Between Larnaca and Alethriko, 1905, *Holmboe* 568.
5. Nisou, 1901, *A. G. & M. E. Lascelles* s.n.! Athalassa, 1939, *Lindberg f.* s.n.! Pano Dheftera, 1979, *Edmondson & McClintock* E 2931!
6. Myrtou, 1932, *Syngrassides* 454!

NOTES: It is noteworthy that this conspicuous plant has yet to be collected in Divisions 7 or 8.

18. LAMIUM *L.*

Sp. Plant., ed. 1, 579 (1753).
Gen. Plant., ed. 5, 252 (1754).

Annual or perennial herbs; leaves crenate, dentate or lobed, the basal generally petiolate; inflorescences lax or crowded, usually consisting of dense, many-flowered, superimposed verticillasters; bracts foliaceous, sometimes coloured; bracteoles small, subulate or lanceolate, or absent; calyx tubular or campanulate, tube 5-nerved, limb subequally 5-toothed; corolla 2-lipped, tube usually exserted, markedly dilated at apex, glabrous or with a pilose zone internally, adaxial lip concave or galericulate, entire, toothed or 2-fid, abaxial lip with an enlarged, shortly stipitate, emarginate or 2-fid median lobe, lateral lobes short or wanting; stamens 4, didynamous, exserted, the abaxial pair with longer filaments than the adaxial; anthers with divergent, sometimes hairy, thecae; style exserted; stigmatic lobes subulate, subequal. Nutlets trigonous, truncate or rounded at apex, smooth or verruculose.

About 40 species in temperate regions of the northern hemisphere. Some of the species may have small, cleistogamous flowers with closed corollas.

Corolla-tube long, slender, more than 10 mm. long, exceeding the corolla-limb; bracts not blotched:
    Bracts petiolate; adaxial lip of corolla bifid; flowers striped with darker venation; plants perennial - - - - - - - - - **1. L. garganicum** ssp. **striatum**
    Bracts sessile, amplexicaul; flowers not striped; adaxial lip of corolla entire or obscurely emarginate; plant annual - - - - - - - - **2. L. amplexicaule**
Corolla-tube short, less than 10 mm. long, shorter than corolla-limb; bracts often blotched white or pink; adaxial lip of corolla entire or subentire, not bifid; plant annual
                            3. L. moschatum

1. **L. garganicum** *L.*, Sp. Plant., ed. 2, 808 (1763); Osorio-Tafall et Seraphim, List Vasc. Plants Cyprus, 90 (1973); Feinbrun, Fl. Palaest., 3: 122, t. 198 (1978); Davis, Fl. Turkey, 7: 131 (1982).

Suberect or sprawling perennial from a woody rootstock; stems 5–50 cm. long, distinctly tetragonal, densely or sparsely pubescent or subglabrous, often purplish, sometimes glaucous, usually much branched at the base, unbranched or sparingly branched above; leaves ovate-deltoid, 0·7–5 cm. long, 0·5–5 cm. wide, densely or sparsely pubescent on both surfaces, apex acute or blunt, margins crenate-dentate, bluntly serrate, or rather irregularly lobed, base cordate or truncate; petiole distinct, 0·5–5 cm. long; inflorescence a condensed or loose spike, consisting of 2–7 (or more) crowded or remote, 2–10-flowered verticillasters; bracts very similar to the cauline leaves, but generally more shortly petiolate; bracteoles linear-lanceolate, 5–8 mm. long 0·5–1 mm. wide, shortly pubescent or subglabrous; flowers sessile or subsessile; calyx narrowly campanulate, tube 8–9 mm. long, 4–5 mm. wide at throat, straight or slightly curved, subglabrous or pubescent externally, glabrous or sparsely glandular internally, teeth suberect or spreading, subequal, narrowly deltoid-acuminate, 5–6 mm. long, 2–2·5 mm. wide at base, subglabrous or thinly pubescent externally; corolla white, pink or purple, generally striate with darker veins, tube 2–2·5 cm. long, 0·5–1·3 cm. wide at the markedly dilated apex, subglabrous or thinly pubescent externally, glabrous internally; adaxial lip concave, subentire, irregularly denticulate or shortly bifid with entire or denticulate lobes, densely pilose externally, abaxial lip with broadly depressed-deltoid, inconspicuous lateral lobes, and a shortly stipitate, reflexed, emarginate or bifid, median lobe, 5–7 mm. long, 7–12 mm. wide; stamens exserted; filaments inserted about 2 or 4 mm. down corolla-tube, downcurved, glabrous, the adaxial about 10 mm. long, the abaxial about 14 mm. long; anthers about 2 mm. wide, thecae strongly divaricate; ovary glabrous, about 2 mm. long, 2·5 mm. wide, lobes with a membranous apical wing; style glabrous, 2·5–3 cm. long; stigmatic lobes subequal, subulate, 1·5–2 mm. long. Nutlets 2–3 mm. long, 1·5–2 mm. wide, sharply trigonous, olive-brown, minutely verruculose with pallid warts.

ssp. **striatum** (*Sm.*) *Hayek*, Prodr. Fl. Pen. Balc., 2: 275 (1929); Davis, Fl. Turkey, 7: 132 (1982).
    L. striatum Sm. in Sibth. et Sm., Fl. Graec. Prodr., 1: 405 (1809), Fl. Graec., 6; 46, t. 557 (1826); Boiss., Fl. Orient., 4: 757 (1879); Post, Fl. Pal., ed. 2, 2: 385 (1933).
    [*L. maculatum* (non L.) A. K. Jackson in Kew Bull., 1934: 272 (1934); Osorio-Tafall et Seraphim, List Vasc. Plants Cyprus, 90 (1973).]
    [*L. garganicum* ssp. *garganicum* (non *L. garganicum* L.) Osorio-Tafall et Seraphim, List Vasc. Plants Cyprus, 90 (1973).]
    TYPE: "in ruderatis Graeciae et Archipelagi copiosè", *Sibthorp* (OXF).

Stems and leaves thinly pubescent or subglabrous; corolla proportionately long and slender, tube about 2 cm. long, 0·5 cm. wide at apex, adaxial lip small, narrow, deeply bifid, generally with denticulate, spreading lobes.

HAB.: In clefts of limestone rocks, on damp walls and rocky banks; 2,000–3,300 ft. alt.; fl. Febr.–April.

DISTR.: Divisions 2, 7, rare. Greece, Crete, Aegean Islands, ? W. & S. Turkey, ? Syria, ? Lebanon, ? Palestine.

2. Pedhoulas, 1932, *Syngrassides* 51 ! and, 1955, *Merton* 2363 ! Moutoullas, 1937, *Syngrassides* 1551 !
7. Kornos, 1951, *Merton* 219 ! Kyparissovouno, 1954, *Merton* 1852 ! and, 1962, *Meikle* 2427 ! Kourtella Peak above Karmi, 1955, *Casey* 1674 !

NOTES: Readily distinguished from typical *L. garganicum* by its slender corolla-tube and narrow, bifid, adaxial corolla-lip, but connected to the type by a puzzling range of variants in Greece and Turkey. In Cyprus, the plants from around Pedhoulas differ from the Northern Range populations in having the lobes of the adaxial corolla-lip subentire, and not conspicuously and irregularly denticulate.

2. L. amplexicaule *L.*, Sp. Plant., ed. 1, 579 (1753); Unger et Kotschy, Die Insel Cypern, 273 (1865); Boiss., Fl. Orient., 4: 760 (1879); H. Stuart Thompson in Journ. Bot., 44: 336 (1906); Holmboe, Veg. Cypr., 156 (1914); Post, Fl. Pal., ed. 2, 2: 386 (1933); Osorio-Tafall et

Seraphim, List Vasc. Plants Cyprus, 90 (1973); Feinbrun, Fl. Palaest., 3: 123, t. 199 (1978); Davis, Fl. Turkey, 7: 138 (1982).
[*L. purpureum* (non L.) Sintenis in Oesterr. Bot. Zeitschr., 31: 192, 229, 231 (1881); Holmboe, Veg. Cypr., 157 (1914); Osorio-Tafall et Seraphim, List Vasc. Plants Cyprus, 90 (1973).]

TYPE: "*in* Europae *cultis*".

Erect or sprawling annual 4–30 cm. high; stems tetragonal, glabrous or thinly retrorse-strigose towards apex, often stained purple, unbranched in starved specimens, much branched near the base when luxuriant; leaves remote, separated by long, bare internodes, broadly and bluntly ovate, 0·7–3 cm. long, 0·5–3 cm. wide, dull green, thinly pubescent on both surfaces, margins crenate, or coarsely and bluntly dentate, occasionally lobed, base shallowly cordate, truncate or rounded; petiole 0·5–4 cm. long; inflorescences commonly crowded, or lax in luxuriant plants, consisting of 2–4 (or more) compact verticillasters; flowers subsessile, sometimes cleistogamous, but apparently always chasmogamous in our area; bracts foliaceous, similar to the cauline leaves, but often larger, more coarsely toothed or lobed, sessile and amplexicaul, forming a shallow cup around the flowers, frequently purplish; bracteoles wanting or very much reduced; calyx narrowly campanulate, pubescent or villose, tube 3–4 mm. long, 2–2·5 mm. wide at mouth, glabrous and thinly glandular within, teeth erect, subequal, narrowly deltoid-acuminate, about 2·5–3 mm. long, 1 mm. wide at base, pilose externally; corolla bright purple with darker spots on the palate, tube slender, 15–20 mm. long, about 5 mm. wide at the abruptly dilated apex, pubescent externally, glabrous internally, adaxial lip galericulate, 5–6 mm. long, 4–5 mm. wide, entire or obscurely emarginate, densely pilose externally, abaxial lip obscurely 3-lobed, the lateral lobes so flattened as to be almost obsolete, about 3–4 mm. wide, less than 1 mm. long, median lobe reflexed, stipitate, suborbicular, about 3 mm. diam.; stamens inserted 1 or 2 mm. down corolla-tube, filaments thinly pilose, the adaxial about 4 mm. long, the abaxial about 5·5 mm. long; anthers about 1 mm. wide, thecae strongly divergent, pilose; ovary glabrous, about 1 mm. wide, disk very reduced; style slender, glabrous, 2·4–3 cm. long; stigmatic lobes subequal, subulate, about 1 mm. long. Nutlets narrowly oblong, 2–3 mm. long, 1–1·3 mm. wide, sharply trigonous; dark brown, densely verruculose with whitish warts, attachment area very small.

HAB.: Cultivated and fallow fields, vineyards, gardens; occasionally on seashores or open mountainsides; sea-level to 5,000 ft. alt.; fl. Dec.–May.

DISTR.: Divisions 1–7, common. Widespread in Europe and western Asia eastwards to Siberia and the Himalayas.

1. Near Pelathousa, Polis, 1962, *Meikle* 2275! Ayios Yeoryios (Akamas), 1962, *Meikle*!
2. Prodhromos, 1862, *Kotschy* 895; Evrykhou, 1935, *Syngrassides* 832! and, 1951, *Merton* 816! Saïttas, 1938, *Kennedy* 1400! Kryos Potamos, 2,900 ft. alt., 1938, *Kennedy* 1401! Platres, 1939, *Kennedy* 1402! Phini, 1939, *Kennedy* 1403! Mandria 1938, *Kennedy* 1404! Platres, 1938, *Kennedy* 1405! Gourri 1950, *Chapman* 327! Kykko Monastery, 1978, *J. Holub* s.n.! Phiti, 1978, *J. Holub* s.n.! Between Troödos and Pano Amiandos, 1979, *Edmondson & McClintock* E 2857!
3. Stavrovouni, 1880, *Sintenis & Rigo*; Amathus, 1964, *J. B. Suart* 171!
4. Ayia Napa, 1862, *Kotschy* 112a; Larnaca, 1880, *Sintenis & Rigo* 945! Paralimni, 1940, *Davis* 2066! Famagusta, 1957, *Merton* 2889!
5. Kythrea, 1932, *Syngrassides* 53! Athalassa, 1936, *Syngrassides* 1175!
6. Dhiorios, 1962, *Meikle*! Myrtou, 1968, *Economides* in ARI 1153! Nicosia, Omorphita, 1973, *P. Laukkonen* 136!
7. Kyrenia, 1949, *Casey* 219! and, 1956, *G. E. Atherton* 827! Pentadaktylos, 1950, *Casey* 983!

3. **L. moschatum** *Mill.*, Gard. Dict., ed. 8, no. 4 (1768); Unger et Kotschy, Die Insel Cypern, 273 (1865); Boiss., Fl. Orient., 4: 765 (1879); Holmboe, Veg. Cypr., 159 (1914); Post, Fl. Pal., ed. 2, 2: 387 (1933); Rechinger f. in Arkiv för Bot., ser. 2, 1: 430 (1950); Osorio-Tafall et

Seraphim, List Vasc. Plants Cyprus, 90 (1973); Feinbrun, Fl. Palaest., 3: 123, t. 200 (1978); Davis, Fl. Turkey, 7: 146 (1982).

L. moschatum Mill. var. *micranthum* Boiss., Fl. Orient., 4: 765 (1879); Post, Fl. Pal., ed. 2, 2: 387 (1933).

TYPE: Cultivated Chelsea Physic Garden; "grows naturally in the Archipelago" (BM).

Erect annual up to 60 cm. high; stems rather fleshy, glabrous, glaucous, often purplish, branched or unbranched; leaves ovate or cordate, 2–10 cm. long, 1·5–6 cm. wide, thinly or densely adpressed-pubescent on both surfaces, apex acute or obtuse, margins crenate or coarsely and irregularly and bluntly serrate, base generally cordate, nervation distinct; petiole slender, thinly pubescent, up to 7 cm. long; inflorescence usually lax, consisting of 5–8 (or more) remote, compact, many-flowered verticillasters; flowers sessile or subsessile; bracts foliaceous, very similar to the leaves but usually more acuminate and more sharply serrate, often with a conspicuous white or pink basal blotch, and subsessile or with a short, flattened petiole; bracteoles linear-subulate, strigose, about 3 mm. long, 0·8 mm. wide; calyx campanulate-infundibuliform, tube about 4 mm. long, 3·5 mm. wide at apex, glabrous or subglabrous externally, glabrous internally, teeth spreading, subequal, subulate from a wide base, shortly strigillose, 3–4 mm. long, 1·5–2 mm. wide at base; corolla white, tube short, 5–7 mm. long, 1·5–2·5 mm. wide, glabrous externally, with a bearded zone internally near the throat, adaxial lip galericulate, 7–20 mm. long, 5–14 mm. wide, densely pilose externally, abaxial lip 3-lobed, lateral lobes sharply deltoid, 2–3 mm. long, 1·5–2 mm. wide, median lobe suborbicular 4–9 mm. diam.; stamens inserted near apex of corolla-tube, filaments 11–15 mm. long, pubescent; anthers 2–4 mm. wide, thecae pilose, strongly divaricate; ovary glabrous, 1·5–2 mm. diam.; style glabrous, 1·5–2 cm. long; stigmatic lobes subulate, the adaxial 0·5–1 mm. long, the abaxial up to 2 mm. long. Nutlets oblong-obovate, sharply trigonous, about 2 mm. long, 1 mm. wide, fuscous-brown, pallid-verruculose, base with a well-developed elaiosome, area of attachment very small.

HAB.: Waste ground, field margins, sides of irrigation channels, occasionally on walls or limestone screes; sea-level to 2,000 ft. alt.; fl. Febr.–May.

DISTR.: Divisions 1–7. Eastern Mediterranean region from Greece to Palestine.

1. Lyso, 1913, *Haradjian* 855! Ktima, 1913, *Haradjian* 668! Near Meladhia, 1979, *Edmondson & McClintock* E 2794!
2. Lefka, 1932, *Syngrassides* 42! Phiti, 1978, *J. Holub* s.n.!
3. Episkopi, 1905, *Holmboe* 704; Kolossi Castle, 1955, *Merton* 2064!
4. Larnaca, 1862, *Kotschy* 295.
5. Kythrea, 1967, *Merton* in ARI 215!
6. W. of Orga, 1956, *Merton* 2592! Ayios Yeoryios near Liveras, 1962, *Meikle* 2401!
7. Pentadaktylos, 1880, *Sintenis & Rigo* 125! also, 1959, *P. H. Oswald* 49! Kyrenia, 1901, *A. G. & M. E. Lascelles* s.n.! also, 1949, *Casey* 263! 666! 1956, *G. E. Atherton* 869! 1003! and, 1967, *Merton* in ARI 579! Armenian Monastery, 1912, *Haradjian* 359! St. Hilarion, 1931, *J. A. Tracey* 10! and, 1963, *J. B. Suart* 97! Above Vasilia, 1941, *Davis* 2204! Buffavento, 1941, *Davis* 2800! Karavas, 1968, *Economides* in ARI 1093!

NOTES: All the Cyprus specimens have rather small flowers, with corollas rarely exceeding 15 mm. in length, whereas some specimens from Greece have corollas as much as 2 cm. long or even slightly longer. In the circumstances one might refer the Cyprus population to *L. moschatum* Mill. var. *micranthum* Boiss., but I am by no means satisfied that a clear distinction can be drawn between large- and small-flowered variants, nor that typical *L. moschatum* is the large-flowered plant.

19. MOLUCCELLA *L.*

Sp. Plant., ed. 1, 587 (1753).
Gen. Plant., ed. 5, 255 (1754).

Glabrous or subglabrous annuals; leaves petiolate, broad, crenate, toothed or lobed; inflorescence loose or dense, consisting of several, remote or closely superimposed, many-flowered verticillasters; bracts foliaceous; bracteoles subulate, spinescent, commonly reflexed; flowers sessile or subsessile; calyx very large, infundibuliform, subactinomorphic or somewhat 2-lipped, with a papyraceous or subcoriaceous, reticulate limb, 5- or 10-nerved; corolla proportionately small, tube glabrous or with an oblique bearded zone internally, limb 2-lipped, the adaxial lip galericulate, the abaxial 3-lobed, the median lobe much larger than the laterals; stamens 4, didynamous, ascending under the adaxial corolla-lip, filaments of the abaxial stamens longer than those of the adaxial; stigmatic lobes subulate, subequal. Nutlets sharply trigonous, smooth or verruculose, apex truncate.

Two species in the Mediterranean region and western Asia.

Calyx infundibuliform, almost actinomorphic, margin wth 5 minute apiculi -   **1. M. laevis**
Calyx distinctly 2-lipped, the adaxial lip of one broad, porrect spine, the abaxial lip bordered
      with 7 slender, spreading spines -   -   -   -   -   -   -   **2. M. spinosa**

1. M. laevis *L.*, Sp. Plant., ed. 1, 587 (1753); Unger et Kotschy, Die Insel Cypern, 275 (1865); Boiss., Fl. Orient., 4: 768 (1879); H. Stuart Thompson in Journ. Bot., 44: 336 (1906); Holmboe, Veg. Cypr., 157 (1914); Post, Fl. Pal., ed. 2, 2: 389 (1933); Osorio-Tafall et Seraphim, List Vasc. Plants Cyprus, 90 (1973); Feinbrun, Fl. Palaest., 3: 125, t. 202 (1978); Davis, Fl. Turkey, 7: 155 (1982).
TYPE: "*in* Syria".

Erect annual 30–80 cm. high; stems bluntly tetragonal, at first thinly lanuginose, soon glabrous, branched at the base or unbranched; leaves very broadly ovate or suborbicular, 1·5–5 (–7) cm. long and about as wide or a little wider, glabrous or thinly pubescent on the nerves below, apex obtuse or rounded, margins coarsely and bluntly dentate, base rounded or broadly cuneate, nervation rather distinct; petiole slender, thinly hairy or glabrous, 2–4 cm. long; inflorescence usually rather dense, cylindrical, consisting of numerous superimposed verticillasters; bracts foliaceous, almost indistinguishable from the cauline leaves; bracteoles subulate, spinulose, about 5–10 mm. long, spreading or reflexed; calyx somewhat oblique, very broadly infundibuliform, glabrous, reticulate, papyraceous, about 2–2·5 cm. long, 2·5–3 cm. wide at mouth, 5-nerved, the nerves produced beyond the calyx-margin as short spines or apiculi; corolla white with a purplish adaxial lip, included in calyx, tube 5–6 mm. long, glabrous, scattered over with sessile glands externally, adaxial lip galericulate, 10–12 mm. long, 7–8 mm. wide, shortly pilose externally, apex entire, abaxial lip 3-lobed, the lateral lobes reduced to broad, rounded wings, the median lobe 5–6 mm. long, 10–12 mm. wide, deeply emarginate; stamens inserted at, and about 2 mm. below, apex of corolla-tube, filaments thinly pilose, the adaxial about 7 mm. long, the abaxial about 10 mm. long; anthers 2·5 mm. wide, thecae glabrous, strongly divaricate; ovary about 2 mm. diam.; disk well developed, strongly lobed between the ovary-lobes, the latter glabrous with a gland-covered apex; style glabrous, about 15 mm. long; stigmatic arms subulate, about 1–1·5 mm. long, subequal. Nutlets wedge-shaped, sharply triquetrous, about 3·5 mm. long, 2·5 mm. wide, rich brown, smooth or thinly glandular on the truncate apex; attachment area small.

HAB.: Cultivated fields, fallows and waste ground; 500–600 ft. alt.; fl. May–June.

DISTR.: Divisions 4–6, uncommon. Eastern Mediterranean region from Turkey to Palestine eastwards to Iraq.

4. Between Athienou and Larnaca, 1862, *Kotschy* 975.
5. Between Alambra and Mosphiloti, 1900, *A. G. & M. E. Lascelles* s.n.!
6. Between Peristerona and Potami, 1880, *Sintenis & Rigo* 728! Angolemi near Lefka, 1905, *Holmboe* 812; Peristerona, 1934, *Syngrassides* 447! Between Koutraphas and Astromeritis, 1955, *Merton* 2326!

2. M. spinosa *L.*, Sp. Plant., ed. 1, 587 (1753); Boiss., Fl. Orient., 4: 768 (1879); H. Stuart Thompson in Journ. Bot., 44: 336 (1906); Holmboe, Veg. Cypr., 157 (1914); Post, Fl. Pal., ed. 2, 2: 389 (1933); Osorio-Tafall et Seraphim, List Vasc. Plants Cyprus, 90 (1973); Feinbrun, Fl. Palaest., 3: 125, t. 203 (1978); Davis, Fl. Turkey, 7: 155 (1982).

TYPE: "*in* Moluccis".

Erect annual 40–100 cm. high; stem bluntly tetragonal, glabrous, often purplish, generally unbranched; leaves broadly ovate or suborbicular, 4–7 cm. long and about as wide, glabrous, palmately 5-lobed, the lobes coarsely dentate, base cordate or truncate, nervation rather distinct; petiole slender, 3–4 cm. long; inflorescence lax, spiciform, consisting of 7 or more, remote, regularly spaced verticillasters; flowers subsessile, bracts foliaceous, very similar to the cauline leaves, diminishing in size upwards; bracteoles as in *M. laevis*, subulate, spinulose, spreading or reflexed; calyx coriaceous, obliquely and obscurely 2-lipped, tube infundibuliform, about 1·5 cm. long, 1·5 cm. wide at apex, glabrous or scattered over with sessile glands, conspicuously reticulate, with 10 primary nerves reaching the rim of the calyx-tube, separated by 10 secondary nerves arising from the base of the calyx-tube and reaching about half its length, calyx-teeth 8, spinose, unequal, the adaxial deltoid, 10–12 mm. long, 5–7 mm. wide at base, porrect, the 7 abaxial subulate-spinose, 7–10 mm. long, spreading; corolla white, tube 14–15 mm. long, 4–5 mm. wide at apex, thinly glandular with sessile glands externally, with an oblique bearded zone internally, adaxial lip galericulate, about 20 mm. long, 9–10 mm. wide, shortly pilose externally, apex entire or shallowly emarginate, abaxial lip 3-lobed, the lateral lobes bluntly oblong, about 3 mm. long, 2 mm. wide, the median lobe oblate-suborbicular, about 8 mm. diam., shallowly emarginate; stamens inserted at, and about 2 mm. below apex of corolla-tube, filaments thinly pilose, the adaxial about 9 mm. long, the abaxial about 13 mm. long; anthers about 3 mm. wide, thecae glabrous, strongly divaricate; ovary about 2 mm. diam., glabrous; disk well-developed, strongly lobed between the ovules; style about 25 mm. long, glabrous; stigmatic lobes subulate, about 1 mm. long, subequal. Nutlets wedge-shaped, sharply triquetrous, about 4 mm. long, 3 mm. wide, fuscous, minutely verruculose, apex truncate; attachment area small.

HAB.: Cultivated fields and waste ground; sea-level to 300 ft. alt.; fl. May–June.

DISTR.: Divisions 1, 7, rare. Widespread, but very local, in the Mediterranean region.

1. Paphos, 1901, *A. G. & M. E. Lascelles* s.n.! also, 1905, *Holmboe* 721, and, 1913, *Haradjian* 633!

7. Lapithos, 1880, *Sintenis & Rigo* 666! also, 1940, *Davis* 1738!

20. BALLOTA *L.*

Sp. Plant., ed. 1, 582 (1753).
Gen. Plant., ed. 5, 253 (1754).
A. Patzak in Ann. Naturhist. Mus. Wien, 62: 57–86 (1958);
63: 33–81 (1959); 64: 42–56 (1960).

Perennial herbs or shrubs; stems bluntly or sharply tetragonal; leaves ovate, oblong or cordate, entire, crenate or dentate, usually petiolate; inflorescences lax or dense, consisting of remote or congested, few- or many-flowered, superimposed verticillasters; bracts foliaceous; bracteoles subulate-filiform or sometimes rigid, spinose; calyx tubular-infundibuliform, 10-nerved, often sulcate longitudinally with an expanded, 5–10-dentate, mucronate or crenate limb; corolla usually pink, purple or white, 2-lipped, sometimes cleistogamous, tube shorter than, or equalling calyx, with an internal pilose zone, adaxial lip concave or galericulate, often pilose, abaxial lip 3-lobed, with a broad, retuse or emarginate median lobe;

stamens 4, didynamous, ascending under the adaxial corolla-lip, the abaxial pair longer than the adaxial; anther-thecae divergent; stigmatic lobes subulate, subequal. Nutlets ovoid or oblong, rounded at apex, smooth.

About 35 species in Europe, the Mediterranean region, Africa and western Asia.

Erect, unarmed herb; verticillasters many-flowered -   -   -   -   - **1. B. nigra**
Sprawling, spiny shrub; flowers in pairs at each node of the inflorescence   **2. B. integrifolia**

SECT. 1. **Ballota** Perennial herbs; bracteoles subulate-filiform, herbaceous; calyx 5-toothed, sometimes with minute intercalary teeth.

1. **B. nigra** L., Sp. Plant., ed. 1, 582 (1753); Poech, Enum. Plant. Ins. Cypr., 24 (1842); Unger et Kotschy, Die Insel Cypern, 274 (1865); Boiss., Fl. Orient., 4: 775 (1879); H. Stuart Thompson in Journ. Bot., 44: 337 (1906); Holmboe, Veg. Cypr., 157 (1914); Post, Fl. Pal., ed. 2, 2: 391 (1933); Rechinger f. in Arkiv för Bot., ser. 2, 1: 430 (1950); Osorio-Tafall et Seraphim, List Vasc. Plants Cyprus, 89 (1973); Davis, Fl. Turkey, 7: 163 (1982).

Erect perennial up to 100 cm. high, with a shortly creeping, rhizomatous rootstock; stems tetragonal, retrorse-strigillose, often purplish, branched or unbranched; leaves ovate, 2·5–8 cm. long, 1·5–7 cm. wide, dull green, foetid when bruised, thinly or densely strigillose above and below, sometimes with scattered, sessile, shining glands on the underside, apex acute or obtuse, base truncate, broadly cuneate or shallowly cordate, margins bluntly or sharply, regularly or irregularly crenate-serrate, sometimes obscurely lobulate; petiole 1–3 cm. long, channelled above, pilose or strigose; inflorescence branched or unbranched, pyramidal or spiciform, usually rather lax, consisting of 4–10 (or more) remote, lax or crowded, superimposed verticillasters; bracts foliaceous, closely resembling the leaves, becoming progressively smaller upwards; bracteoles subulate-filiform, 5–9 mm. long, pilose; flowers sessile or subsessile in dense, shortly pedunculate clusters of 5–8 or more; calyx tubular-infundibuliform, 10-nerved, pilose with scattered, sessile glands, tube 5–11 mm. long, 4–6 mm. wide, glabrous within, limb with 5 erect or spreading, shortly spinescent-acuminate or acuminate-aristate teeth 1·5–3 mm. long; corolla purple, pink or white, tube straight, 8–11 mm. long, equalling or exceeding calyx, papillose and with an oblique bearded zone internally about 2–4 mm. above base, adaxial lip galericulate, about 5 mm. long, 3 mm. wide, pilose externally, abaxial lip 3-lobed, lateral lobes bluntly and obliquely oblong, about 2 mm. long, median lobe suborbicular, emarginate, 4–5 mm. diam.; stamens inserted about 1–2 mm. down corolla-tube, filaments thinly pubescent, 4–6 mm. long; anthers about 1·5 mm. wide, thecae strongly divaricate; ovary about 1·2 mm. diam., lobes sometimes with an apical covering of sessile glands; disk small, lobed; style glabrous, 11–12 mm. long; stigmatic lobes subequal, subulate, about 1 mm. long. Nutlets oblong-obovate, trigonous, about 2–2·3 mm. long, 1–1·5 mm. wide, dark brown, shining, smooth or minutely granulose, apex rounded, sometimes with a few sessile glands, attachment area small.

ssp. **foetida** Hayek, Prodr. Fl. Pen. Balc., 2: 278 (1929); Davis in Notes Roy. Bot. Gard. Edinb., 21: 61–63 (1952); Patzak in Ann. Naturhist. Mus. Wien, 62: 71 (1958); Davis, Fl. Turkey, 7: 164 (1982).
   B. foetida Lam., Fl. Franç., 2: 381 (1778) pro parte; nom. illeg.
   TYPE: not indicated.

Calyx-tube 7–11 mm. long, teeth erecto-patent, abruptly narrowed to a short, spinose mucro, much less than half the length of the tube; corolla pinkish-purple or often white.

HAB.: Roadsides, waste and cultivated ground; 3,300–4,500 ft. alt.; fl. June–Dec.

DISTR.: Division 2. Widespread in western and central Europe, becoming rare in the Mediterranean area.

2. Kykko, 1913, *Haradjian* 975! Platres, 1936, *Kennedy* 805!

NOTES: Very closely allied to the following (*B. nigra* L. ssp. *uncinata* (Fiori et Bég.) Patzak), but distinguished, over very large areas of its distribution, by its erecto-patent, not horizontal-recurved, calyx-teeth. Additional material from high altitudes on Troödos would be required to show if this distinction between the two subspecies holds good there.

ssp. **uncinata** (*Fiori et Bég.*) *Patzak* in Ann. Naturhist. Mus. Wien, 62: 64 (1958); Feinbrun, Fl. Palaest., 3: 128, t. 207 (1978); Davis, Fl. Turkey, 7: 164 (1982).
   B. nigra L. var. *meridionalis* Bég. f. *uncinata* Bég. in Fiori et Paoletti, Fl. Anal. Ital., 3: 39 (1903).
   [*B. nigra* L. ssp. *foetida* (non Hayek) Lindberg f., Iter Cypr., 28 (1946).]
   TYPE: Sicily; Palermo, *Todaro* "ex spec. herb. Rom." (RO).

Calyx-tube 7–11 mm. long, often rather densely covered with sessile glands, teeth patent or recurved, abruptly narrowed to a short, spinose, often somewhat hooked mucro, much less than half the length of the tube; corolla generally pinkish-purple.

HAB.: Roadsides, waste and cultivated ground, commonly near water; sea-level to 4,500 ft. alt.; fl. May–Sept.

DISTR.: Divisions 2, 3, 5–7. Widespread in the Mediterranean region; Atlantic Islands.

2. Evrykhou, 1840, *Kotschy* 8; Troödhitissa, 1903, *A. G. & M. E. Lascelles* s.n.! Yialia, 1905, *Holmboe* 799; Milikouri, 1939, *Lindberg f.* s.n.; Kambos, 1939, *Lindberg f.* s.n.; Prodhromos, 1955, *G. E. Atherton* 625!
3. Kolossi, 1939, *Lindberg f.* s.n.
5. Kythrea, 1932, *Syngrassides* 321! and, 1939, *Lindberg f.* s.n.!
6. Syrianokhori, 1936, *Syngrassides* 1247!
7. Armenian Monastery, 1912, *Haradjian* 376! Bellapais, 1955, *G. E. Atherton* 392!

SECT. 2. **Acanthoprasium** *Benth.* Small shrubs; bracteoles hard, spinose, often recurved; calyx 5–10-toothed.

2. **B. integrifolia** *Benth.*, Labiat. Gen. et Sp., 598 (1834); Poech, Enum. Plant. Ins. Cypr., 24 (1842); Unger et Kotschy, Die Insel Cypern, 274 (1865); Boiss., Fl. Orient., 4: 776 (1879); Holmboe, Veg. Cypr., 157 (1914); Lindberg f., Iter Cypr., 28 (1946); Chapman, Cyprus Trees and Shrubs, 71 (1949); Patzak in Ann. Naturhist. Mus. Wien, 63: 42 (1959); Osorio-Tafall et Seraphim, List Vasc. Plants Cyprus, 89 (1973).
   B. wettsteinii Rechinger in Oesterr. Bot. Zeitschr., 40: 153 (1890); Holmboe, Veg. Cypr., 157 (1914); Rechinger f. in Arkiv för Bot., ser. 2, 1: 430 (1950); Patzak in Ann. Naturhist. Mus. Wien, 63: 43 (1959); Osorio-Tafall et Seraphim, List Vasc. Plants Cyprus, 89 (1973).
   [*Moluccella frutescens* (non L.) Sm. in Sibth. et Sm., Fl. Graec. Prodr., 1: 415 (1809), Fl. Graec., 6: 55, t. 568 (1827).]
   TYPE: Cyprus; "in ins. Cypro *Sibthorp*! (*v.s. sp. in herb. Sibthorp et Banks.*)" (OXF, BM).

Slender sprawling shrub to 100 cm. high; old branches dull brown, young twigs bluntly tetragonal, densely strigillose with retrorsely adpressed, bristly hairs, sometimes thinly lanuginose with a mixed indumentum of short, glandular and long, eglandular hairs; leaves dull or lustrous green, 0·7–2 (–3) cm. long, 0·5–1·5 (–2·5) cm. wide, oblong, obovate, suborbicular or occasionally ovate, thinly adpressed-strigillose above, subglabrous below, or occasionally thinly lanuginose on both surfaces, apex obtuse or subacute, sometimes shortly apiculate, margins entire or sometimes deeply but bluntly serrate or irregularly lobed, base cuneate or rounded; nervation obscure; petioles distinct, slender, up to 1·5 cm. long, adpressed-strigillose or lanuginose; inflorescence very lax, running almost the full length of the branches; flowers solitary in each axil; bracts foliaceous; bracteoles in pairs at the base of the pedicel, hard, spinose, brownish, 8–15 mm. long, straight or recurved; pedicel short but distinct, generally less than 3 mm. long, strigillose; calyx infundibuliform, 11–12 mm. long, 7–9 mm. wide at anthesis, accrescent and persistent in fruit, tube glabrous or pilose

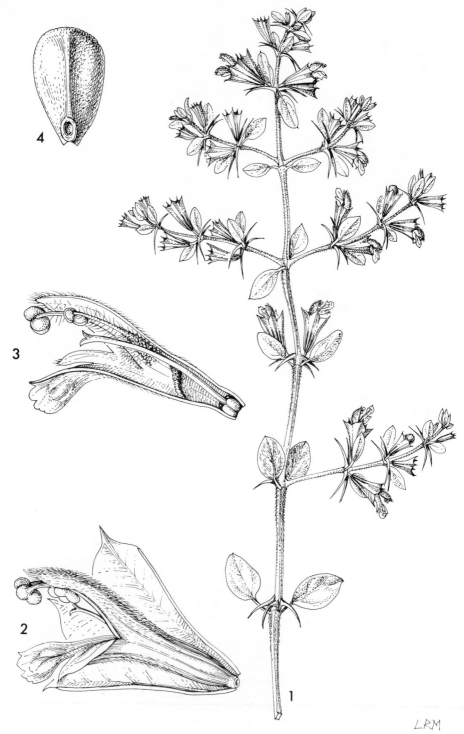

LRM

Plate 76. BALLOTA INTEGRIFOLIA Benth. **1,** habit, × 1; **2,** flower, × 6; **3,** corolla, longitudinal section, × 6; **4,** nutlet, × 12. (**1** from *Lindberg f.* s.n. & *Chapman* 72; **2, 3** from *Lindberg f.* s.n.; **4** from *Syngrassides* 1367.)

externally, glabrous internally, prominently 10-nerved, somewhat reticulate, teeth 5, subequal, broadly and bluntly deltoid, about 2–3 mm. long, 4–5 mm. wide, shortly apiculate at the apex, and sometimes in the sinuses; corolla white, with purple or red stripes, tube about 9 mm. long, 3·5 mm. wide at apex, pubescent externally, glabrous internally with a bearded zone about 5 mm. from the base, adaxial lip galericulate, about 7 mm. long, 3·5 mm. wide, pilose externally, abaxial lip 3-lobed, the lateral lobes oblong, about 2 mm. long, 2 mm. wide, the median lobe wide, about 3·5 mm. long, 6–7 mm. wide, deeply emarginate; stamens inserted at about the same level, 2 mm. down the corolla-tube, filaments thinly hairy, the adaxial about 6 mm. long, the abaxial about 8 mm. long; anthers about 2·3 mm. wide, the thecae strongly divaricate; ovary glabrous, about 1·8 mm. diam.; disk small, distinctly lobed; style about 12 mm. long, rather thick; stigmatic lobes narrowly deltoid, the adaxial less than 0·5 mm. long, the abaxial a little longer. Nutlets obovoid, trigonous, about 3·5 mm. long, 2·5 mm. wide, dark brown, smooth or slightly granulose, attachment area small. *Plate 76*.

HAB.: Rocky hillsides or in clefts of limestone or sandstone rocks, occasionally on walls; sea-level to 2,500 ft. alt.; fl. April–June.

DISTR.: Divisions 1–3, 7. Endemic.

1. Paphos, 1905, *Holmboe* 719; Ktima, 1940, *Davis* 1778! and, 1974, *Meikle* 4001! Lyso, 1913, *Haradjian* 856; Rigopoulos, 1957, *Merton* 3063! Cliffs above Peyia, 1962, *Meikle* 2041!
2. Vouni Palace, 1939, *Chapman* 398!
3. Lania, 1933, *Chapman* 72! Between Limassol and Trimiklini, 1934, *Syngrassides* 1367! Arminou, 1941, *Davis* 3405! Between Akhelia and Pissouri, 1955, *Merton* 2132!
7. Kyrenia Pass, 1787, *Sibthorp*, also, 1939, *Lindberg f.* s.n.! and, 1955, *Merton* 2098! Between Buffavento Castle and Ayios Khrysostomos Monastery, 1862, *Kotschy* 390; near Pentadaktylos, 1880, *Sintenis & Rigo* 564! Near Ayios Khrysostomos Monastery, 1889, *Pichler*; Panayia Plataniotissa, 1901, *A. G. & M. E. Lascelles* s.n.! S. side Buffavento, 1955, *Merton* 2032!

NOTES: *Ballota integrifolia* is variable in leaf-shape and indumentum, being generally glabrescent and entire-leaved, but sometimes, especially in the earlier part of its growing season, thinly lanuginose and with lobed leaves; other supposed differences between it and *B. wettsteinii* will not stand up to examination, nor should it be overlooked that *B. integrifolia* has been collected more than once in the type locality of *B. wettsteinii*. Merton notes (on his spec. no. 2032): "The young spring shoots are *'wettsteinii'*, the older ones *'integrifolia'* ". To judge from the fact that the flowers were poorly developed in Pichler's material of *B. wettsteinii*, one suspects it was simply an immature state of *B. integrifolia*.

21. PHLOMIS *L.*

Sp. Plant., ed. 1, 584 (1753).
Gen. Plant., ed. 5, 255 (1754).
Vierhapper in Oesterr. Bot. Zeitschr., 65: 205–236; 252–257 (1915).

Shrubs and perennial herbs, commonly with tomentose stems and leaves, and showy inflorescences; leaves simple, rugose or conspicuously reticulate-veined; inflorescences lax or crowded, few- or many-flowered; flowers sessile or subsessile; bracts foliaceous; bracteoles ovate, obovate, lanceolate or linear-subulate, occasionally wanting; calyx tubular or campanulate, 5–10-veined, with 5, subequal, acute, obtuse or sometimes retuse teeth; corolla often yellow, sometimes purple or white, 2-lipped, tube generally with a bearded zone internally, adaxial lip galericulate, often laterally compressed and falcate, abaxial lip patent, 3-lobed; stamens 4, didynamous, ascending under the adaxial corolla-lip, the filaments of the abaxial stamens longer than the adaxial, the latter often appendiculate at the base; anthers with divergent thecae; stigmatic lobes unequal, the abaxial longer than the adaxial. Nutlets oblong or obovoid, trigonous, glabrous or sometimes pilose.

About 100 species chiefly in the Mediterranean region and western Asia.

All the Cyprian species belong to *Phlomis* L. sect. *Phlomis* subsect. *Phlomis* comprising shrubs and subshrubs with large flowers, the adaxial lip of the corolla arcuate or falcate and laterally compressed, the abaxial lip with a very large, broad, emarginate or bifid median lobe.

Bracteoles very small, 2–5 mm. long, less than half the length of the calyx
        3. P. brevibracteata
Bracteoles more than 5 mm. long, usually almost as long as the calyx or a little longer:
  Bracteoles linear-subulate, 10–15 mm. long, 1–3 mm. wide-   **5. P. longifolia** var. **bailanica**
  Bracteoles oblong, oblong-elliptic, oblong-obovate or rhomboid, 7–18 mm. long, 3–7 mm. wide:
    Bracteoles thinly pilose, with hispid-ciliate margins; apex of calyx bearing 5, spreading, spinous awns 2·5–4 mm. long - - - - - - **4. P. lunariifolia**
    Bracteoles rather densely stellate-pubescent externally; apex of calyx with very short spines or cusps:
      Calyx-tube about 20 mm. long, apical cusps slender, inconspicuous; nutlets often pilose at apex - - - - - - - - - - **1. P. fruticosa**
      Calyx-tube 7–12 mm. long, apical cusps rather rigid, spinous; nutlets glabrous or sparsely glandular-pilose at apex - - - - - - **2. P. cypria**

1. P. fruticosa *L.*, Sp. Plant., ed. 1, 584 (1753); Boiss., Fl. Orient., 4: 784 (1879); Post, Fl. Pal., ed. 2, 2: 396 (1933); Chapman, Cyprus Trees and Shrubs, 70 (1949); Osorio-Tafall et Seraphim, List Vasc. Plants Cyprus, 90 (1973); Davis, Fl. Turkey, 7: 112 (1982).
   TYPE: "*in* Sicilia, Hispania".

Much-branched shrub 1–1·5 m. high, old wood covered with pale brown, lightly fissured bark, young shoots greyish stellate-tomentose, tetragonal; leaves ovate-oblong, 4–10 cm. long, 1·5–6 cm. wide, grey-green and densely stellate-tomentose on both surfaces, rugulose-bullate with finely reticulate venation impressed above and prominent below, apex blunt or rounded, margins minutely crenulate, base rounded, truncate or shallowly cordate; petiole stout, 1–3 cm. long, tomentose, channelled above; inflorescence consisting of 1 or 2, dense, many-flowered heads or verticillasters 3–3·5 cm. diam.; bracts foliaceous, similar to the leaves but generally subsessile, narrower and more acuminate; bracteoles adpressed, oblong-elliptic or oblong-lanceolate, 10–18 mm. long, 4–7 mm. wide, rather densely stellate-pubescent with a marginal fringe of longer hairs; calyx tubular, about 2 cm. long, 0·6–1 cm. wide at apex, densely stellate-pubescent with scattered longer hairs externally, pilose along the nerves and at the mouth internally, nerves 10, the primary very prominent, the secondary less so, teeth 5, slender, spreading, usually less than 2 mm. long, arising abruptly from the truncate apex of the calyx; corolla yellow, tube 13–14 mm. long, 7–8 mm. wide at the markedly dilated apex, thinly pubescent externally, glabrous internally with a conspicuous, oblique bearded zone about 6–8 mm. from the base, adaxial lip falcate, strongly compressed laterally, c. 18 mm. long, 26 mm. wide, densely pilose externally, abaxial lip about 25 mm. long, pilose externally, lateral lobes small, oblong, blunt, with caudate appendages about 4 mm. long, 1·5 mm. wide, median lobe abruptly expanded, about 25–28 mm. wide at apex, shallowly emarginate; stamens inserted near mouth of corolla-tube, filaments arcuate, thinly pilose, the adaxial about 18 mm. long, the abaxial about 24 mm. long; anthers about 2·5 mm. wide, with short, divergent thecae; ovary about 3 mm. diam., the lobes shortly and densely pilose; disk glabrous, obscurely lobed; style glabrous, about 40 mm. long; stigmatic lobes very unequal, the adaxial scarcely 0·4 mm. long, the abaxial subulate, about 2–2·5 mm. long. Nutlets narrowly oblong, 6–7 mm. long, 2 mm. wide, sharply triquetrous, dark brown, smooth with a densely pilose, truncate apex; attachment area small.

  HAB.: Dry, rocky, limestone hillsides; c. 1,000–2,000 ft. alt.; fl. April–June.

  DISTR.: Division 7, rare. Mediterranean region from Sardinia to Cyprus.

7. Mylous Forest Station, Kyrenia, 1932, *Syngrassides* 420! St. Hilarion, *Chapman* 430.

NOTES: In the past much confused with *P. lunariifolia* and *P. cypria*, so that, in the absence of specimens, identifications must be regarded as dubious. The Syngrassides specimen cited above is the only genuine material of *P. fruticosa* I have seen from the island; it was originally named *P. lunariifolia*, this identification being subsequently corrected to *P. fruticosa* by P. H. Davis.

2. P. cypria *Post* in Mém. Herb. Boiss., 18: 99 (1900); H. Stuart Thompson in Journ. Bot., 44: 337 (1906); Holmboe, Veg. Cypr., 155 (1914); Chapman, Cyprus Trees and Shrubs, 70 (1949); Rechinger f. in Arkiv för Bot., ser. 2, 1: 430 (1950); Osorio-Tafall et Seraphim, List Vasc. Plants Cyprus, 90 (1973).

    P. fruticosa L. ssp. *cypria* (Post) Lindberg f., Iter Cypr., 29 (1946).

Much-branched shrub 50–150 cm. high, old wood covered with fissured or flaking grey-brown bark, young shoots closely stellate-tomentose, with greyish or yellowish hairs, rather slender and sharply tetragonal; leaves ovate-oblong or oblong-lanceolate, 2–7 cm. long, 0·5–3 cm. wide, dull green above, closely greyish or yellowish, stellate-tomentose below, rugulose-bullate with finely reticulate venation impressed above and prominent below, apex subacute, obtuse or rounded, base broadly cuneate or subcordate, margins subentire or minutely and obscurely crenulate; petiole stout, tomentose, 0·5–2·5 cm. long, channelled above; inflorescence consisting of 1 or 2, dense, many-flowered heads or verticillasters 1·5–3·5 cm. diam.; bracts foliaceous, similar to the leaves but usually subsessile and smaller; bracteoles adpressed, obovate or rhomboid-obovate, 7–12 mm. long, 3–7 mm. wide, rather densely stellate-pubescent with a marginal fringe of long hairs; calyx tubular, 13–15 mm. long, 6–8 mm. wide at apex, densely or sparsely stellate-pubescent, mixed with a varying number of longer hairs externally, conspicuously plicate towards apex, glabrous internally except for a dense zone of yellowish or whitish hairs at the mouth; nerves 10, alternately prominent and inconspicuous, the prominent nerves exserted beyond the truncate apex of calyx-tube as rather rigid, pilose cusps 2–2·5 mm. long; corolla pale yellow, the upper lip often paler than the lower, tube about 11 mm. long, 6 mm. wide at the markedly dilated apex, subglabrous or thinly stellate-pubescent externally, glabrous internally with a conspicuous oblique, bearded zone 4–5 mm. from base, adaxial lip falcate, strongly compressed laterally, about 16 mm. long, 16 mm. wide, densely stellate-pilose externally, with distinct lateral lobes and an emarginate apex, abaxial lip about 15 mm. long, pilose externally, lateral lobes narrowly winged with a caudate appendage 2–2·5 mm. long, median lobe about 8 mm. long, 12 mm. wide, shallowly emarginate, stamens inserted near mouth of corolla-tube, filaments arcuate, very thinly pilose, the adaxial about 18 mm. long, the abaxial about 20 mm. long; anthers about 2 mm. wide, with short, divergent thecae; ovary about 2 mm. diam., the lobes shortly glandular-papillose at apex; disk glabrous, shallowly but distinctly lobed; style glabrous, about 33 mm. long; stigmatic lobes very unequal, the adaxial about 0·5 mm. long, the abaxial 1–1·5 mm. long, subulate. Nutlets narrowly oblong, about 5 mm. long, 2–2·5 mm. wide, sharply triquetrous, dark brown or greyish-brown, smooth with a shortly and inconspicuously glandular-papillose apex; attachment area small.

var. **cypria**
    [*P. fruticosa* (non L.) H. Stuart Thompson in Journ. Bot., 44: 337 (1906).]
    TYPE: Cyprus; "Habitat ad castellum Sancti Hilarionis Cypri; floret junio" [1898], *Post* 917 (BEI, K!).

Leaves shortly and broadly oblong, about 3–4 cm. long, 1·5–2 cm. wide, whitish- or greyish-tomentose, with a rounded or very obtuse, broad apex. *Plate 77, figs 1–2a.*

Plate 77. Figs. **1–2a.** PHLOMIS CYPRIA Post var. CYPRIA **1,** habit, ×1; **2,** corolla, longitudinal section, ×3; **2a,** calyx, longitudinal section, ×3; figs. **3–5a.** PHLOMIS CYPRIA Post var. OCCIDENTALIS Meikle **3,** habit, ×1; **4,** leaf, ×1; **5,** corolla, longitudinal section, ×3; **5a,** calyx, longitudinal section, ×3; figs. **6–7a.** PHLOMIS BREVIBRACTEATA Turrill **6,** whorl of flowers, ×1; **7,** corolla, longitudinal section, ×3; **7a,** calyx, longitudinal section, ×3. (**1–2a** from *G. E. Atherton* 1372; **3–5a** from *Meikle* 2770; **6–7a** from *Davis* 3444 & *Merton* 2776.)

HAB.: In garigue on dry, sunny, limestone hillsides; 500–2,300 ft. alt.; fl. April–June.

DISTR.: Division 7. Endemic.

7. St. Hilarion, 1880, *Sintenis & Rigo* 572! also, 1898, *Post* 917! 1905, *Holmboe* 821; 1931, *J. A. Tracey* 53! 1937, *Syngrassides* 1538! 1939, *Lindberg f.* s.n.! 1940, *Kennedy* 1558! and, 1940, *Davis* 2043! Kornos, 1940 and 1941, *Davis* 1995! 2959! 3046! Karavas, 1949, *Casey* 969! and, 1956, *G. E. Atherton* 1372!

var. **occidentalis** *Meikle* in Ann. Musei Goulandris, 6: 93 (1983). See App. 5, p. 1898.

TYPE: Cyprus; between Lyso and Abdoulinas Junction, 1,700 ft. alt.; 1962, *Meikle* 2770 (K!).

Leaves narrowly oblong, subacuminate, 3–7 cm. long, 1–2·3 cm. wide, often yellowish-tomentose, tapering to a narrow, obtuse or subacute apex. *Plate 77, figs. 3–5a.*

HAB.: In garigue on stony, igneous hillsides; 500–2,500 ft. alt.; fl. April–June.

DISTR.: Divisions, 1, 2. Endemic.

1. Mavrohomata (Akamas), 1937, *Chapman* 287! near Smyies, 1941, *Davis* 3299! Between Lyso and Abdoulinas Junction, 1962, *Meikle* 2770!
2. Atratsa, Marathasa Valley, 1937, *S. Tsiakouris* in *Chapman* 206! Stavros tis Psokas valley, 1949, *Casey* 762! Near Alonoudhi Junction, 1954, *Merton* 1941! Abundant about 4 miles from Stavros tis Psokas on the Lyso road, 1981, *Meikle*!

NOTES: This species and the following (*P. brevibracteata* Turrill) have similar disjunct distributions in Cyprus, occurring on the Northern Range and on the lower flanks of the Troödos Range.

In *P. brevibracteata* no distinction can be found between the northern and southern populations, but in *P. cypria* there is evidence of incipient speciation, and one can recognize certain trivial differences apparently peculiar to the plants from either area. It is also noteworthy that while *P. cypria* var. *cypria* grows on the Northern Range limestones, *P. cypria* var. *occidentalis* appears to be restricted to serpentine, diabase and pillow lava.

3. P. brevibracteata *Turrill* in Gard. Chron., ser. 3, 117: 48 (1945); Chapman, Cyprus Trees and Shrubs, 70 (1949); Osorio-Tafall et Seraphim, List Vasc. Plants Cyprus, 90 (1973).

TYPE: Cyprus; "Kakoratyia [Kakoradjia], 152 m., on dry rocky forest soil, together with *Cistus*, etc., 10.10.37, *A. Syngrassides* 1710" (K!).

Lax, spreading, much-branched shrub to 1·5 m. high, old wood covered with irregularly fissured or flaking grey-brown bark, young shoots closely stellate-tomentellous or subglabrous, often reddish, rather slender and sharply tetragonal; leaves oblong, 2–5 cm. long, 0·7–1·5 cm. wide, bright green above, closely white-tomentose below, inconspicuously rugulose above, distinctly and prominently reticulate-veined below, apex obtuse or rounded, base rounded or broadly cuneate, margins entire or somewhat undulate; petiole short, channelled above, 0·4–0·8 cm. long; inflorescence lax, often branched, consisting of (1–) 2–3, dense, 4–12-flowered verticillasters 3–6 cm. diam.; bracts foliaceous, very similar to the leaves; bracteoles very small, linear-subulate, 2–5 mm. long, about 1 mm. wide, densely greyish-tomentellous; calyx tubular, 10–12 mm. long, 7–8 mm. wide at apex, densely and closely stellate-tomentellous, distinctly plicate towards apex, glabrous internally except for a zone of short, sparse hairs at the mouth, nerves 10, almost equally prominent, calyx-lobes very obscure, reduced to crenate, sometimes shortly muticous prominences; corolla rich yellow; tube about 13 mm. long, 8 mm. wide at the markedly dilated apex, subglabrous externally, glabrous internally with a conspicuous oblique, bearded zone 6–8 mm. from base, adaxial lip falcate, strongly compressed laterally, about 2–2·5 cm. long, 2 cm. wide, densely stellate-pilose externally, with indistinct lateral lobes and an emarginate apex, abaxial lip about 2–2·5 cm. long, pilose externally, lateral lobes narrowly winged with a caudate-emarginate appendage about 4 mm. long, 2 mm. wide, median lobe about 14–15 mm. long, 14–16 mm. wide, with a sulcate median ridge and strongly reflexed, undulate margins; stamens inserted near mouth of corolla-tube, filaments arcuate, softly pilose, the adaxial about 20 mm. long, the abaxial a little longer; anthers about 2 mm. wide, with short, divergent thecae; ovary about 2·5 mm. diam., the lobes very broad and blunt, glabrous or sparsely glandular at the apex; disk glabrous, shallowly lobed; style glabrous, about 4 cm. long; stigmatic lobes narrowly subulate, unequal, the adaxial about 1 mm. long, the abaxial about 3 mm. long. Nutlets narrowly oblong, about 4 mm. long, 2 mm. wide at truncate apex, tapering to base, sharply triquetrous, dark brown, glabrous, minutely granulose; attachment area very small. *Plate 77, figs. 6–7a.*

HAB.: Amongst garigue on sunny chalk, limestone or rarely igneous hillsides; 300–3,000 ft. alt.; fl. May–June.

DISTR.: Divisions 3, 7. Endemic.

3. Kakoradjia (Nicosia-Limassol road), 1937–1938, *Syngrassides* 1710! 1810! Trimiklini-Lania, 1940, *Kennedy* 1557! Lefkara, 1940–1941, *Davis* 1901! 2736! 3589! Zagala above Trimiklini, 1941, *Davis* 3134! 3444! Trimiklini, 1948, *Mavromoustakis* s.n.! and, 1954, *Mavromoustakis* 16! Oritaes Forest, 1956, *Merton* 2776!
7. Kantara–Dhavlos road, 1937, *Chapman* 294! Near Kantara road, 1950, *Casey* 1047! Panagra gorge, 1954, *Casey* 1568! Ardhana, 1962, *Meikle* 2952! Akanthou, 1967, *Merton* in ARI 764!

4. **P. lunariifolia** *Sm.* in Sibth. et Sm., Fl. Graec. Prodr., 1: 414 (1809) [*"lunarifolia"*]; Unger et Kotschy, Die Insel Cypern, 275 (1865); Boiss., Fl. Orient., 4: 785 (1879) pro parte; H. Stuart Thompson in Annals of Bot., 19: 439 (1905) et in Journ. Bot., 44: 337 (1906); Holmboe, Veg. Cypr., 156 (1914); Lindberg f., Iter Cypr., 29 (1946); Chapman, Cyprus Trees and Shrubs, 70 (1949); Osorio-Tafall et Seraphim, List Vasc. Plants Cyprus, 90 (1973); Elektra Megaw, Wild Flowers of Cyprus, 11, t. 16 (1973); Davis, Fl. Turkey, 7: 112 (1982).
TYPE: "In variis Peloponnesi locis; etiam in monte Athô". An error; the type material almost certainly came from Cyprus, probably from somewhere between Paphos and Polis (OXF). See note by H. Stuart Thompson in Annals of Bot., 19: 439 (1905).

Robust, erect, much-branched shrub 1–2 m. high, old wood with irregularly fissured brown bark, young shoots at first stellate-tomentose, soon becoming subglabrous and reddish, stout and sharply tetragonal; leaves oblong or narrowly ovate, occasionally subhastate, 3–10 cm. long, 1–6 cm. wide, dark green above, densely grey-tomentose below, rugulose with impressed, reticulate venation, apex subacute or rounded, base truncate, cuneate or subcordate, margins bluntly serrulate; petiole up to 3 cm. long, tomentose, channelled above, or the upper leaves often subsessile; inflorescence lax, generally unbranched, consisting of 1–2, remote, dense, many-flowered verticillasters, 3·5–4·5 (–6) cm. diam.; bracts foliaceous, very similar to the leaves but usually sessile, longer and proportionately narrower; bracteoles oblanceolate or narrowly elliptic, 10–18 mm. long, 3–7 mm. wide, acuminate-spinose, obscurely penninerved, thinly stellate-pubescent dorsally, the margins conspicuously hispid-ciliate with long, yellowish, multicellular bristles; calyx tubular, about 12 mm. long, 6 mm. wide at the truncate apex, glabrous except for stellate-pubescence along the nerves, glabrous internally except for a short hispid zone at the mouth, nerves 10, alternately produced into a spreading, spinous awn 2·5–4 mm. long; corolla bright yellow; tube almost 16 mm. long, 8 mm. wide at the not very markedly dilated apex, glabrous externally and internally save for an oblique bearded internal zone 9–10 mm. from base of corolla-tube, adaxial lip falcate, about 1·8 cm. long, 2 cm. wide, strongly compressed laterally, densely stellate-pilose externally, with indistinct lateral lobes and a shortly emarginate apex, abaxial lip 3-lobed, about 18 mm. long, pilose externally, lateral lobes narrowly winged with an oblong appendage 3–4 mm. long, 2–3 mm. wide at base, median lobe suborbicular, about 10 mm. diam., with a sulcate median ridge and reflexed sides; stamens inserted near mouth of corolla-tube; filaments arcuate, softly pilose, the adaxial about 16 mm. long, the abaxial about 18 mm. long; anthers about 2·5 mm. wide, with short, divergent thecae; ovary about 2·5 mm. diam., the lobes sparingly glandular-papillose at the apex; disk glabrous, shallowly lobed; style glabrous, about 3·5 cm. long; stigmatic lobes very unequal, subulate, the abaxial about 2 mm. long, the adaxial barely 0·5 mm. long. Nutlets narrowly oblong, about 5 mm. long, 2 mm. wide at truncate apex, tapering to base, triquetrous, mottled grey-brown, glabrous, minutely granulose; attachment area very small.

HAB.: Amongst garigue, usually in rocky gullies by sides of streams; sea-level to 600 ft. alt.; fl. March–May.

DISTR.: Divisions 1, 3. Also S. Turkey.

1. Between Khrysokhou and "Slewra" on the way to Khrysorroyiatissa Monastery, 1862, *Kotschy* 678! Tsadha, 1905, *Holmboe* 737; Ktima, 1913, *Haradjian* 654! Evretou, 1934, *Chapman* 146! Stroumbi, 1934, *Syngrassides* 846! Between Evretou and Sarama, 1937, *Syngrassides* 1718! 4 miles from Polis on the way to Ktima, 1938, *Syngrassides* 1758! Yiolou near Stroumbi, 1938, *Syngrassides* 1846! Near Ayios Neophytos Monastery, 1939, *Lindberg f. s.n.*! Stroumbi, 1941, *Davis* 3342! Lara near Ayios Yeoryios (Akamas), 1962, *Meikle* 2130! Kissonerga to Maa Beach, 1967, *Economides* in ARI 973! N.E. of Ktima, 1971, *G. Joscht* 6520! Between Lachi and Baths of Aphrodite, 1979, *Edmondson & McClintock* E 2752!
3. Kouklia, 1940, *Kennedy* 1556! Also noted from Pissouri, 1934, *Syngrassides*.

NOTES: Also recorded from Division 2. "Stavros [tis Psokas]", 2,500 ft. alt. 1956, *G. E. Atherton* 1373!, but this is right outside its normal distribution on the island, and some doubt must attach to the localization, especially as, from other data on the label, it would appear that the locality was noted first as "Kyrenia", which was deleted and "Stavros" added, apparently some time after the plant was collected.

5. P. longifolia *Boiss. et Bl.* in Boiss., Diagn., 2, 4: 47 (1859), Vierhapper in Oesterr. Bot. Zeitschr., 65: 220 (1915); Post, Fl. Pal., ed. 2, 2: 397 (1933); Davis, Fl. Turkey, 7: 115 (1982).
   P. viscosa Poir. var. *angustifolia* Boiss., Fl. Orient., 4: 788 (1879).

Slender shrub or subshrub to 1 m. high; stems usually much-branched, the older covered with greyish, fissured bark, the younger sharply tetragonal, subglabrous or thinly stellate-tomentose; leaves narrowly oblong-lanceolate or oblong-ovate, 4–10 cm. long, 0·8–4 cm. wide, dull green and distinctly rugulose above, thinly greyish, stellate-tomentose below, apex obtuse or subacute, base subcordate, truncate or broadly cuneate, margins distinctly crenulate-serrulate; petioles rather slender, to 5 cm. long, stellate-tomentose or subglabrous, channelled above; inflorescences lax, consisting of 1–3, remote, compact, many-flowered verticillasters 2–6 cm. diam.; bracts foliaceous, resembling the leaves but more shortly petiolate or subsessile; bracteoles linear-subulate, 10–15 mm. long, 1–3 mm. wide, thinly or rather densely stellate-hispidulous, the hair-clusters with 1 or more elongate rays; calyx tubular, about 14 mm. long, 7–10 mm. wide at the truncate apex, shortly stellate-pubescent externally, glabrous internally except for a bearded zone at the apex, nerves 10, alternately produced into a spinose, pubescent awn 2–5·5 mm. long; corolla bright yellow, tube about 16 mm. long, 6–7 mm. wide at the dilated apex, glabrous externally and internally save for an oblique, internal bearded zone about 9 mm. from the base, adaxial lip falcate, 16–17 mm. long, 16–18 mm. wide, strongly compressed laterally, densely stellate-pilose externally, apex shallowly emarginate, lateral lobes obscure, abaxial lip about 16 mm. long, pilose externally, median lobe suborbicular, about 12 mm. diam., with a sulcate central ridge and reflexed, sinuate-undulate sides, lateral lobes narrowly deltoid, about 4·5 mm. long, 2 mm. wide at base; stamens inserted at mouth of corolla-tube; filaments arcuate, 18 and 20 mm. long, thinly and softly pilose; anthers about 2·5 mm. wide, with short divergent thecae; ovary glabrous, about 2·5 mm. diam.; disk glabrous, bluntly lobed; style glabrous, about 4 cm. long;

stigmatic lobes subulate, very unequal, the adaxial about 0·5 mm. long, the abaxial about 2 mm. long. Nutlets not seen.

var. **bailanica** (*Vierh.*) *Huber-Morath* in Bauhinia, 1: 112, 121 (1958); Davis, Fl. Turkey, 7: 115 (1982).
    P. bailanica Vierh. in Oesterr. Bot. Zeitschr., 65: 219 (1915) species vel forma nova?
    [*P. viscosa* (non Poir.) Kotschy in Unger et Kotschy, Die Insel Cypern, 275 (1865); Boiss., Fl. Orient., 4: 788 (1879); Holmboe, Veg. Cypr., 156 (1914); Vierh. in Oesterr. Bot. Zeitschr., 65: 217 (1915); Osorio-Tafall et Seraphim, List Vasc. Plants Cyprus, 90 (1973) pro parte quoad plant. cypr.]
    TYPE: Two syntypes cited by Vierhapper, both from S.E. Turkey: "1. Syria septentrionalis. Prope Alexandrette [Iskenderun]. Orient. herb. Montbret (M; K!]); 2. Frequens ad. aquaeductum Bailanensem in calcariis devexis. Th. Kotschy, Pl. Syriae bor. ex Amano prope Bailan 1862, Nr. 38 (M[K!])".

Leaves shorter and proportionately broader than in *P. longifolia* Poir. var. *longifolia*; calyx-spines longer than in the type, up to 5·5 mm. long.

HAB.: Calcareous hillsides; c. 1,000–2,000 ft. alt.; fl. ? April–May.

DISTR.: ? Division 3, very rare. Also in S.E. Turkey, Syria, Lebanon.
? 3. "Zwischen Limasol et Omodos [between Limassol and Omodhos] 1859. [*Kotschy*] n. Suppl. 464".

NOTES: Not seen since, and evidently very rare. Kotschy must have collected it on 8th April, 1859, when he was travelling from Omodhos *via* Erimi to Limassol. It should be looked for in this area.
    Several authors, following Kotschy, have identified the plant as *P. viscosa* Poir., which differs most obviously in its glandular indumentum. The error is corrected by Huber-Morath (Bauhinia 1: 112, 121; 1958). I have not examined Kotschy's specimens.

[*P.* BERTRAMII *Post* in Mém. Herb. Boiss., 44: 337 (1906), as already pointed out by Holmboe (Veg. Cypr., 156), this was recorded from Cyprus in error by H. Stuart Thompson (Journ. Bot., 44: 337; 1906). The type is from Lebanon, and not from Cyprus, as Stuart Thompson supposed.]

22. PRUNELLA L.

Sp. Plant., ed. 1, 600 (1753).
Gen. Plant., ed. 5, 261 (1754).

Perennial herbs; leaves simple, entire, toothed, pinnatifid or lobed, generally petiolate; inflorescence spiciform, consisting of several, closely superimposed, few-flowered verticillasters; bracts distinct from leaves, very broadly ovate or suborbicular, papyraceous, often cuspidate; bracteoles much reduced or absent; calyx tubular-campanulate, 2-lipped, closed in fruit; adaxial lip usually 3-toothed, abaxial lip with 2 larger teeth; corolla purple, pink, white or yellowish, tube somewhat exserted, with a hairy zone internally, limb 2-lipped, adaxial lip galericulate, abaxial lip 3-lobed, median lobe often denticulate; stamens 4, unequal, the abaxial longer than the adaxial; filaments with a subapical appendage; anther-thecae divergent; stigmatic lobes subequal. Nutlets oblong or obovoid, smooth or almost so.

Seven species widely distributed in the northern hemisphere.

1. **P. vulgaris** *L.*, Sp. Plant., ed. 1, 600 (1753); Gaudry, Recherches Sci. en Orient, 198 (1855); Unger et Kotschy, Die Insel Cypern, 272 (1865); Boiss., Fl. Orient., 4: 691 (1879); Holmboe, Veg. Cypr., 155 (1914); Post, Fl. Pal., ed. 2, 2: 371 (1933); Lindberg f. Iter Cypr., 28 (1946); Osorio-Tafall et Seraphim, List Vasc. Plants Cyprus, 90 (1973); Davis, Fl. Turkey, 7: 295 (1982).
    TYPE: "*in* Europae *pascuis*".

Rhizomatous perennial; stems branched or unbranched, erect or decumbent-ascending, 7–30 (–50) cm. long, tetragonal, at first often thinly pilose with white, multicellular hairs, soon glabrous or subglabrous and often purplish; leaves oblong-ovate, 1·5–7 cm. long, 0·7–3·5 cm. wide, glabrous or thinly pilose along the nerves, apex obtuse, base broadly cuneate, margins entire, subentire or obscurely or irregularly toothed or lobed; nervation generally distinct; petiole slender, 1·5–3 cm. long; inflorescence subsessile or shortly pedunculate, dense, cylindrical, 2–4 cm. long, 1–1·5 cm. diam.; bracts suborbicular-cuspidate, papery, 0·5–1 cm. long, 0·8–1·5 cm. wide, conspicuously reticulate-veined, ciliate with long, white, multicellular hairs; flowers subsessile; calyx tubular, 6–9 mm. long, 3–4 mm. wide at apex, 5-nerved, with prominent anastomosing venation, pilose with multicellular hairs externally, glabrous within, usually bronze-purple, adaxial lip almost truncate with 3 very small teeth, abaxial lip divided into 2 narrow-deltoid, acuminate lobes, 2–2·5 mm. long, 0·8–1 mm. wide at base; corolla violet (sometimes white), tube shortly exserted, about 11 mm. long, 4 mm. wide at apex, glabrous externally, with an internal hairy zone about 4 mm. from base, adaxial lip oblong, 5 mm. long, 4 mm. wide, shallowly galericulate, abaxial lip 3-lobed, lateral lobes oblong-suborbicular, about 1·5 mm. long and almost as wide, median lobe suborbicular, deflexed, about 3 mm. diam., margins denticulate; adaxial stamens inserted at mouth of corolla-tube, abaxial stamens about 5 mm. down tube; filaments glabrous, the adaxial about 2 mm. long, the abaxial about 9 mm. long; anthers small, about 1 mm. wide, thecae strongly divergent; ovary glabrous, about 1 mm. diam.; disk glabrous, distinctly lobed; style glabrous, 13–14 mm. long; stigmatic lobes subequal, narrow, subulate, 1–1·5 mm. long. Nutlets obovoid, about 2 mm. long, 1·3 mm. wide, obscurely trigonous, minutely granulose, yellow-brown, with 2 fine, parallel stripes running longitudinally down the 2 faces and sides; area of attachment very small, with a minute, conical elaiosome. The persistent calyx is closed at the mouth, retaining the nutlets, in dry weather, and opens in wet weather to release them.

HAB.: Moist, shady ground by streams and springs at high altitudes; 3,900–5,100 ft. alt.; fl. June–July.

DISTR.: Division 2 only. Widespread in Europe, temperate Asia, N. Africa and N. America. Probably not indigenous in Australasia and elsewhere.

2. Troödos, 1900, *A. G. & M. E. Lascelles* s.n.! also, 1912, *Haradjian* 508! Paleomylos, 1905, *Holmboe* 914; Kykko, 1913, *Haradjian* 967! Trikoukkia, 1935, *Syngrassides* 745! and, 1936, *Kennedy* 799! Kryos Potamos, 1937, *Kennedy* 797! 798! Platania, 1939, *Lindberg f.* s.n.! and, 1940, *Davis* 1846! Kakopetria, 1949, *Casey* 825! Prodhromos, 1955, *G. E. Atherton* 555! Cedar Valley, Tripylos, 1968, *Economides* in ARI 1198!

TRIBE 8. **Prasieae** *Benth.* Shrubs or herbs; calyx 5–10-nerved, 5-dentate; corolla 2-lipped; stamens 4, didynamous; style gynobasic. Nutlets attached basally, drupe-like, with a fleshy exocarp and hard endocarp.

23. PRASIUM *L.*
Sp. Plant., ed. 1, 601 (1753).
Gen. Plant., ed. 5, 261 (1754).

Erect or scrambling shrubs; leaves petiolate; inflorescences terminal, lax, racemose, the flowers apparently solitary (or rarely paired) in the axils of leaf-like bracts; bracteoles linear-subulate; calyx campanulate, persistent and somewhat accrescent, subequally 5-dentate; corolla conspicuously 2-lipped, white or tinged mauve; stamens concealed under adaxial corolla-lip or shortly exserted, filaments of the abaxial pair scarcely longer than the adaxial; style exserted; stigmatic lobes subequal, subulate. Nutlets broadly oblong, fleshy, black, shining.

One species widely distributed in the Mediterranean area and Atlantic Islands.

1. **P. majus** *L.*, Sp. Plant., ed. 1, 601 (1753); Unger et Kotschy, Die Insel Cypern, 275 (1865); Boiss., Fl. Orient., 4: 798 (1879); Post in Mém. Herb. Boiss., 18: 99 (1900); H. Stuart Thompson in Journ. Bot., 44: 337 (1906); Holmboe, Veg. Cypr., 152 (1914); Post, Fl. Pal., ed. 2, 2: 400 (1933); Lindberg f., Iter Cypr., 29 (1946); Chapman, Cyprus Trees and Shrubs, 68 (1949); Osorio-Tafall et Seraphim, List Vasc. Plants Cyprus, 89 (1973); Feinbrun, Fl. Palaest., 3: 107, t. 171 (1978); Davis, Fl. Turkey, 7: 78 (1982).
TYPE: "*in* Sicilia, Rómae & *in agro* Tingitano".

Erect or scrambling much-branched shrub, 1–4 m. in height; branches tetragonal, the old ones covered with pale grey-brown bark, the young shoots glabrous or thinly hispidulous at the nodes; leaves ovate-cordate, 1–4 cm. long, 0·5–2 cm. wide, glabrous or thinly hispidulous along the nerves below, apex acute, margins sharply serrate, base cordate, truncate or widely cuneate; nervation distinct; petioles slender, 0·5–1·5 cm. long, flattened above, often thinly hispidulous, the bases commonly connate and forming a ring around the stem; inflorescence short or elongate, usually rather lax; bracts foliaceous, indistinguishable from the cauline leaves, becoming progressively smaller and more shortly petiolate upwards; flowers usually solitary (sometimes paired) in the axil of the bract; bracteoles linear-subulate, 3–6 mm. long, 1–1·5 mm. wide at base, margins thinly hispidulous; pedicels 2–5 mm. long, sparsely glandular; calyx widely campanulate, 10-nerved, sparsely glandular-puberulous externally, tube about 5 mm. long, 5 mm. wide at apex, teeth 5, subequal, ovate, spreading, terminating in a short, spinose cusp, about 7 mm. long, 4 mm. wide at base; corolla white (occasionally tinged mauve), tube straight, infundibuliform, about 14 mm. long, 6–7 mm. wide at apex, thinly glandular-puberulous externally, with an internal bearded zone about 5 mm. from the base; adaxial lip hooded, 8–10 mm. long, 6–7 mm. wide, apex retuse, abaxial lip about 12–14 mm. long, 3-lobed, lateral lobes bluntly oblong, about 3 mm. long, 4 mm. wide at base, median lobe suborbicular, concave, about 6 mm. diam., palate furrowed with several, well-marked transverse corrugations; stamens subequal, inserted about 5 mm. down corolla-tube; filaments flattened, thinly lanuginose; anthers about 2·5 mm. wide, with a dilated connective and strongly divergent thecae; ovary glabrous, about 1·8 mm. diam.; disk conspicuous, slightly undulate; style straight, glabrous, about 15 mm. long; stigmatic lobes subulate, subequal, about 1 mm. long. Nutlets broadly obovoid, about 4·5 mm. long, 3 mm. wide, shining black, exocarp fleshy; attachment area basal, very small.

HAB.: Amongst bushes in garigue or maquis, by roadsides, or occasionally in rock-crevices; sea-level to 1,500 ft. alt.; fl. Jan.–May.

DISTR.: Divisions 1, 3–8. Distribution that of the genus.

1. Paphos, 1937, *Miss Godman* 29! Mirillis River near Ayios Yeoryios (Akamas), 1962, *Meikle* 2071! Polis, 1962, *Meikle*!
3. Near Cape Gata, 1862, *Kotschy* 606; Mazotos, 1905, *Holmboe* 178; Limassol, 1913, *Haradjian* 589! Kolossi, 1939, *Lindberg f.* s.n.! Yerasa, 1947, *Mavromoustakis* s.n.! Paramali, 1960, *N. Macdonald* 11! Pyrgos, 1963, *J. B. Suart* 47! 1 km. N.E. of Apsiou, 1979, *Edmondson & McClintock* E 2715!
4. Ayia Napa, 1862, *Kotschy* 133! Cape Greco, 1958, *N. Macdonald* 2! 16! Athna Forest, 1967, *Merton* in ARI 422! Dhekelia, 1977, *A. Della* in ARI 1671!
5. Near Kythrea, 1880, *Sintenis & Rigo* 126! and, 1941, *Davis* 2941! Between Sykhari and Dhikomo, 1901, *A. G. & M. E. Lascelles* s.n.! Lakovounara Forest, 1950, *Chapman* 350! Salamis, 1962, *Meikle*!
6. Dhiorios, 1962, *Meikle*!
7. Ayios Khrysostomos Monastery, 1862, *Kotschy* 411; Koronia, 1935, *Chapman* 9; Ayios Epiktitos, 1937, *Roger-Smith* 7! Larnaka tis Lapithou, 1941, *Davis* 2986! Kyrenia, 1932–1956, *Syngrassides* 293! *Casey* 334! 472! *G. E. Atherton* 806! 898! Kyrenia Pass, 1949,

Casey 515 ! Thermia, 1956, *G. E. Atherton* 879 ! Paleosophos, 1968, *Economides* in ARI 1064 !
Bellapais, 1972, *W. R. Price* 974 ! Tjiklos, 1973, *P. Laukkonen* 91 !
8. Rizokarpaso, 1912, *Haradjian* 160 ! Apostolos Andreas, 1941, *Davis* 2322 ! Akradhes, 1962,
   Meikle !

TRIBE 9. **Scutellarieae** *Benth.* Calyx 2-lipped, often with a dorsal
appendage; lips generally entire, throat closed in fruit; corolla 2-lipped,
adaxial (upper) lip galericulate; stamens 4, ascending under the upper
corolla-lip, abaxial filaments longer than adaxial, adaxial anthers with 2
divergent thecae, abaxial anthers 1-thecous by abortion; style gynobasic.
Nutlets 4, commonly tuberculate or verruculose; seed transverse, embryo
with a curved radicle lying against one of the cotyledons.

24. SCUTELLARIA *L.*

Sp. Plant., ed. 1, 598 (1753).
Gen. Plant., ed. 5, 260 (1754).

Annual or perennial herbs, rarely shrubs or subshrubs; leaves usually
petiolate, entire or variously toothed or lobed, not aromatic; inflorescence
racemose or spiciform, flowers solitary at each bract, often secund; bracts
foliaceous or sometimes coloured; bracteoles minute or absent; calyx
campanulate, dorsiventrally compressed, conspicuously 2-lipped, accrescent and closed at the throat in fruit, splitting longitudinally at maturity, the
lower (abaxial) half persistent, adaxial (upper) half usually with a
conspicuous dorsal, flap-like appendage or *scutellum*, calyx-lips generally
entire; corolla 2-lipped, tube generally long, much exserted, curved
upwards, naked within and markedly dilated towards apex, adaxial (upper)
lip galericulate, entire or emarginate, abaxial (lower) lip apparently 1-lobed,
the lateral lobes connivent with or partly fused with the adaxial lip; stamens
4, parallel, ascending under the adaxial lip of the corolla; anthers
approximate, ciliate, the adaxial 2-thecous with divergent thecae, the
abaxial 1-thecous by abortion; disk with a conical or rostrate apex;
stigmatic lobes unequal, the adaxial much shorter than the abaxial. Nutlets
subglobose or ovoid, commonly tuberculate and pubescent.

About 300 species, widely distributed in temperate and tropical regions,
but rare in Africa, and absent from S. Africa.

Stems and leaves sparsely pubescent or strigillose, or subglabrous; leaves generally with
    markedly cordate bases - - - - - - - - - **1. S. sibthorpii**
Stems and leaves densely pubescent or villose; leaves generally ovate with a truncate or broadly
    cuneate base:
    Stems diffuse, villose with long, patent hairs; leaves small, less than 2 cm. long, villose
        2. S. cypria var. **cypria**
    Stems erect or ascending, pubescent with recurved, crispate hairs; leaves 2–3·5 cm. long,
    densely pubescent - - - - - - - - **2. S. cypria** var. **elatior**

1. **S. sibthorpii** (*Benth.*) *Hal.* [in Oesterr. Bot. Zeitschr., 42: 375, 420 (1892) ? nom. invalid.],
   Consp. Fl. Graec., 2: 494 (1902); Rechinger f. in Bot. Archiv., 43: 50 (1941).
      S. columnae All. var. *sibthorpii* Benth. in DC., Prodr., 12: 419 (1848); Unger et Kotschy,
   Die Insel Cypern, 272 (1865) pro parte.
      S. peregrina L. var. *sibthorpii* (Benth.) Boiss. et Reut. in Boiss., Diagn., 2, 4: 28 (1859),
   quoad nomen, Fl. Orient., 4: 688 (1879) quoad nomen; Sintenis in Oesterr. Bot. Zeitschr.,
   32: 400 (1882) [*"sibthorpiana"*]
      S. peregrina L. ssp. *sibthorpii* (Benth.) Holmboe, Veg. Cypr., 153 (1914); Lindberg f.,
   Iter Cypr., 30 (1946).
      S. rubicunda Hornem. ssp. *sibthorpii* (Benth.) Osorio-Tafall et Seraphim, List Vasc.
   Plants Cyprus, 89 (1973) nom. invalid.
      [*S. peregrina* (non L.) Sm. in Sibth. et Sm., Fl. Graec. Prodr., 1: 424 (1809), Fl. Graec., 6:
   66, t. 582 (1826); Post in Mém. Herb. Boiss., 18: 98 (1900); Lindberg f., Iter Cypr., 30 (1946)
   pro parte.]

[*S. columnae* (non All.) Poech, Enum. Plant. Ins. Cypr., 24 (1842).]
[*S. utriculata* (non Labill.) H. Stuart Thompson in Journ. Bot., 44: 336 (1906).]
TYPE: The plate in Flora Graeca, 6: t. 582!

Erect or spreading perennial 12–50 (–60) cm. high, arising from a woody base; stems purplish, markedly tetragonal, thinly and retrorsely crispate-pubescent or subglabrous, generally branched only at the base and in the region of the inflorescence; leaves rather remote, ovate-cordate, 2–4·5 cm. long, 1·5–3·5 cm. wide, dull green and thinly adpressed-strigillose above, purplish and more densely strigillose below, apex acute or obtuse, base cordate, margins conspicuously crenate-serrate; nervation distinct but not very prominent; petiole slender, pilose, 0·5–3 cm. long, inflorescence a lax raceme 10–30 cm. long; flowers usually secund; bracts lanceolate or narrowly ovate, 6–20 mm. long, 3–8 mm. wide, thinly pubescent, acute, entire or the lowermost sometimes remotely serrate; pedicels very short, usually less than 5 mm. long, densely pilose; calyx campanulate, 4–5 mm. long, 3·5 mm. wide at apex, pilose and thinly glandular externally, glabrous within, lips shallow, entire or the upper slightly emarginate, dorsal flap (or scutellum) conspicuous, about 2 mm. long; corolla dark crimson with a pale blotch at base of abaxial lip, tube curved upwards near base, about 15 mm. long, 5 mm. wide at apex, thinly pilose externally, with a pilose zone internally near base; adaxial lip galericulate, about 3·5 mm. long, 2·5 mm. wide, abaxial lip with a reflexed, suborbicular, undulate, median lobe about 7 mm. wide, lateral lobes broadly deltoid, acute, about 2 mm. long, partly fused with base of adaxial lip; adaxial stamens inserted about 2·5 mm. from apex of tube, filaments about 5 mm. long, flattened, thinly puberulous owards base; abaxial stamens inserted about 5 mm. from apex of tube, filaments about 12 mm. long, puberulous towards base; anthers of adaxial stamens about 1 mm. wide, with 2 divergent, ciliate thecae and a dilated connective, anthers of abaxial stamens with one of the thecae abortive; ovary glabrous, about 0·5 mm. diam., seated on a conical protuberance arising near the edge of a glabrous disk 1 mm. diam.; style filiform, glabrous, 19–20 mm. long; abaxial stigmatic lobe narrowly subulate, about 0·8 mm. long, adaxial lobe minute. Nutlets oblately ovoid, about 1·8 mm. wide, 1·3 mm. long, fuscous, rugulose-verruculose, shortly hairy.

HAB.: Rocky calcareous hillsides, sometimes in coastal garigue; sea-level to 2,000 ft. alt.; fl. April–May.

DISTR.: Divisions 7, 8. Endemic.

7. Locally common: Antiphonitis Monastery, Pentadaktylos, Ayios Khrysostomos Monastery, Kyrenia, Lapithos, Kantara, St. Hilarion, Koronia, above Kythrea, Halevga, Karavas, Kornos, Larnaka tis Lapithou, etc., 1787–1973; *Sibthorp*; *Kotschy* 353; 457; 489! *Sintenis & Rigo* 574! *A. G. & M. E. Lascelles* s.n.! *C. B. Ussher* 118! *J. A. Tracey* 51! *Syngrassides* 675! *Chapman* 248! *C. H. Wyatt* 42! *Lindberg f.* s.n.! *Davis* 1746! 2982! 3031! *Mavromoustakis* s.n.! *Casey* 732! *Merton* 2190! 2276! *Miss Mapple* 59! *G. E. Atherton* 1303! *C. E. H. Sparrow* 33! *J. B. Suart* 7! *Townsend* 63/63! *G. Joscht* 6340! *P. Laukkonen* 300! etc.
8. Livadhia, 1905, *Holmboe* 514; Komi Kebir, 1912, *Haradjian* 291! Ayios Theodhoros, 1939, *Lindberg f.* s.n.; Yioti, 1962, *Meikle* 2520!

2. **S. cypria** *Rechinger f.* in Bot. Archiv., 43: 48 (1941); Arkiv för Bot., ser. 2, 1: 429 (1950); Die Kulturpflanze, Beih. 3: 56 (1962); Osorio-Tafall et Seraphim, List Vasc. Plants Cyprus, 89 (1973).
  Scutellaria fl. rubro nova Kotschy in Oesterr. Bot. Zeitschr., 12: 279 (1862) nomen.
  [*S. hirta* (non Sm.) Kotschy in Unger et Kotschy, Die Insel Cypern, 273 (1865); Boiss., Fl. Orient., 4: 690 (1879) pro parte quoad plant. cypr.; Holmboe, Veg. Cypr., 153 (1914); Lindberg f., Iter Cypr., 30 (1946).]
  [*S. peregrina* (non L.) Kotschy in Unger et Kotschy, Die Insel Cypern, 273 (1865); Post in Mém. Herb. Boiss., 18: 98 (1900) pro parte; Lindberg f., Iter Cypr., 30 (1946) pro parte.]
  [*S. columnae* All. var. *sibthorpii* (non Benth.) Kotschy in Unger et Kotschy, Die Insel Cypern, 272 (1865) pro parte.]

Erect or spreading perennial 7–25 (–50) cm. high; stems purplish, distinctly tetragonal, often densely pubescent with a mixed indumentum of short glandular hairs and long, patent eglandular ones, sometimes clothed solely with spreading or subadpressed, straight or curved, eglandular hairs, much-branched at the woody base, sparingly branched above except in the region of the inflorescence; leaves often small, ovate or obscurely cordate, 0·8–2 (–3·5) cm. long, 0·5–1·5 (–2·5) cm. wide, grey-pilose with spreading or subadpressed hairs; apex acute or obtuse, margins bluntly serrate or crenate-serrate, base truncate or broadly cuneate, occasionally subcordate; nervation obscure; petiole slender, 0·5–1·5 cm. long, generally glandular-pilose; inflorescence very similar to but usually shorter and more crowded than in *S. sibthorpii*; bracts often a little shorter and blunter than in *S. sibthorpii*; bracteoles minute or wanting; pedicels generally less than 5 mm. long, densely glandular-hairy; calyx campanulate, about 3·5 mm. long, 2·5 mm. wide at apex, rather densely pilose with spreading hairs externally, glabrous within, lips shallow, entire, bordered crimson; dorsal flap (or scutellum) conspicuous, about 2 mm. long; corolla crimson with a yellow or greenish blotch on the abaxial lip, or uniformly sulphur-yellow; tube curved upwards near base, about 15 mm. long, 3·5–5 mm. wide at apex, distinctly pilose with spreading glandular and eglandular hairs externally, glabrous internally except for a glandular-pilose zone near base; adaxial lip galericulate, about 3·5 mm. long, 2·5 mm. wide, abaxial lip with a reflexed, suborbicular, undulate, median lobe about 7 mm. wide, lateral lobes broadly deltoid, acute, about 2 mm. long, partly fused with base of adaxial lip; androecium and gynoecium as in *S. sibthorpii*. Nutlets depressed-subglobose, about 2 mm. long and almost as wide, fuscous, bluntly verruculose and shortly puberulous; attachment scar very small.

var. **cypria**

TYPE: Cyprus; "In montibus Troodos dispersa alt. 3000–5000′ diebus Maii floret", 1862, *Kotschy* 699 (JE, W, K !).

Stems diffuse, 7–15 (–20) cm. long, densely clothed with long, spreading, white, eglandular hairs; leaves small, 0·8–2 cm. long, 0·5–1·5 cm. wide, rather densely clothed with long, eglandular hairs; inflorescences generally less than 12 cm. long, usually rather crowded.

HAB.: Rocky or gravelly slopes on open igneous mountainsides, sometimes in open Pine forest; 1,700–6,400 ft. alt.; fl. May–July.

DISTR.: Division 2, common. Endemic.

2. Common; Khionistra, Platania, Platres, Xerokolymbos, Kryos Potamos, Phini, Kykko, Stavros tis Psokas, Lemithou, Amiandos, Prodhromos, Ayia, above Spilia, Kakopetria, etc., 1862–1973; *Kotschy* 699! *Sintenis & Rigo* 731! *Haradjian* 422! 995! *Feilden* s.n.! *C. B. Ussher* 11! *Syngrassides* 743! 1584! *Kennedy* 807! 808! 810! 812–815! *Wyatt* 1! *Lindberg f.* s.n.! *Davis* 1766! 1806! *Mavromoustakis* s.n.! *Casey* 721! 838! *Probyn* 105! *D. P. Young* 7281! 7282! 7800! *Meikle* 2652! 2823B! *Economides* in ARI 945! 1212! 1213! *P. Laukkonen* 384! 397! 410! 419! etc.

var. **elatior** *Meikle* in Ann. Musei Goulandris, 6: 93 (1983). See App. V, p. 1898.

[*S. columnae* All. var. *sibthorpii* (non Benth.) Kotschy in Unger et Kotschy, Die Insel Cypern, 272 (1865) pro parte.]

[*S. peregrina* L. ssp. *sibthorpii* (non (Benth.) Holmboe) Holmboe, Veg. Cypr., 153 (1914) pro parte.]

[*S. albida* (non L.) H. Stuart Thompson in Journ. Bot., 44: 336 (1906); Holmboe, Veg. Cypr., 153 (1914); Osorio-Tafall et Seraphim, List Vasc. Plants Cyprus, 89 (1973).]

TYPE: Cyprus; 1 km. N.E. of Apsiou, 15 km. N. of Limassol. Macchie on loose stony serpentine slopes; 450 m. alt.; 8 April, 1979, *Edmondson & McClintock* E 2692 (K !).

Stems erect, 20–50 cm. high, densely clothed with recurved, crispate, white, eglandular hairs, sometimes mixed with long, spreading hairs especially towards base of stems; leaves 2–3·5 cm. long, 1–2·5 cm. wide, densely crispate-pubescent on both surfaces; inflorescences elongate, rather lax, up to 25 cm. long.

HAB.: In garigue on calcareous or igneous hillsides, occasionally by roadsides or in vineyards; 1,500–3,800 ft. alt.; fl. April–June.

DISTR.: Divisions 1–3. Endemic.

1. Tsadha, 1905, *Holmboe* 736; Stroumbi, 1913, *Haradjian* 780! also, 1941, *Davis* 3354! Dhrousha, 1941, *Davis* 3212!
2. Khrysorroyiatissa Monastery, 1862, *Kotschy* 699a, also, 1941, *Davis* 3362! 3404! Platres, 1937, *Kennedy* 809! Kykko Monastery, 1940, *Davis* 1820! Zoopiyi, 1941, *Davis* 3082! Perapedhi, 1970, *A. Genneou* in ARI 1520!
3. Phasoula, 1950, *F. J. F. Barrington* 74! Lania, 1970, *A. Genneou* in ARI 1534! Near Apsiou, 1979, *Edmondson & McClintock* E 2692! Lefkara, 1979, *Hewer* 4598 B!

NOTES: The relationship between *S. sibthorpii* and *S. cypria* is a very close one, but, by analogy with such pairs of species as *Arabis cypria*/*A. purpurea*, *Origanum dubium*/*O. majorana*, it seems best to regard them as distinct species since, despite the apparent triviality of the differences, they are apparently always distinct, and readily distinguishable. The situation with *S. cypria* var. *cypria* and *S. cypria* var. *elatior* is rather different; here the extremes, that is, the dwarf, villose high-altitude plant and the taller, erect, close-pubescent low-altitude one, are readily identifiable, but at or around 3,000 ft., in areas about Kykko and Platres, the two begin to merge, as do also certain specimens from Stroumbi, Dhrousha and Lyso in the south-west of the island. Here the situation is analogous to *Genista sphacelata* var. *sphacelata*/var. *crudelis*, *Astragalus echinus* var. *echinus*/var. *chionistrae*, and, such being the case, variety seems the proper rank for the two taxa.
It should be noted that whereas the flowers of *S. sibthorpii* appear to be uniformly crimson, those of *S. cypria* can be crimson or yellow even within the same population.

TRIBE 10. **Teucrieae** *Dumort.* Herbs or shrubs; corolla 2-lipped or 1-lipped, with the adaxial lip much reduced or wanting; stamens 4 or 2; style not gynobasic. Nutlets dry, attachment lateral-ventral; attachment area usually large.

25. TEUCRIUM *L.*

Sp. Plant., ed. 1, 562 (1753).
Gen. Plant., ed. 5, 247 (1754).

Annual or perennial herbs or shrubs; leaves variable, simple or deeply dissected, entire or toothed, petiolate or subsessile; inflorescences lax or dense, sometimes spiciform or capitate, consisting of remote or crowded, superimposed, few- or many-flowered verticillasters; bracts resembling leaves or distinct from them; flowers commonly pedicellate; calyx tubular or campanulate, 10-nerved, teeth subequal or the adaxial larger than the abaxial; corolla-tube glabrous internally, limb apparently 1-lipped, the adaxial lip being reduced to 2 lobes at the base of the 5-lobed abaxial lip, median lobe of abaxial lip enlarged, often concave; stamens 4, didynamous, ascending-arcuate, usually exserted, the abaxial filaments longer than the adaxial; style not gynobasic; stigmatic lobes subequal, acute. Nutlets obovoid, smooth, rugose or reticulate, with a large, oblique attachment area.

About 300 species found throughout the world, but chiefly in Europe, the Mediterranean region and western Asia.

Shrub 1–2 m. high; leaves linear with revolute margins; flowers mauve or pink in a narrow, elongate, terminal raceme - - - - - - - - - **1. T. creticum**
Subshrubs less than 50 cm. high, or perennial herbs:
  Inflorescences capitate; stems terete or subterete; low much-branched subshrubs:
    Stems clothed with a mixed indumentum of short, simple glandular hairs and long, spreading, simple eglandular hairs - - - - - - **5. T. cyprium**
    Stems densely white-tomentose with an indumentum of adpressed (rarely spreading) compound hairs - - - - - - - - **6. T. micropodioides**

Inflorescences elongate, racemose, or flowers apparently axillary in the axils of the upper
    leaves:
    Leaves sessile, often amplexicaul; stoloniferous, perennial herbs of marshy ground
                                               3. T. scordium
    Leaves petiolate:
        Flowers greenish-yellow; calyx 2-lipped, teeth very unequal; robust, rhizomatous
           perennial herb - - - - - - - - - **2. T. kotschyanum**
        Flowers pink, purple or reddish; calyx not distinctly 2-lipped, teeth subequal; much-
           branched subshrub - - - - - - - - **4. T. divaricatum**

SECT. 1. **Teucrium** Perennial herbs and shrubs; inflorescence lax,
racemose or paniculate, flowers not conspicuously secund; calyx campanulate, subequally 5-dentate, not gibbous at base.

1. **T. creticum** L., Sp. Plant., ed. 1, 563 (1753); Sibth. et Sm., Fl. Graec. Prodr., 1: 391 (1809), Fl.
    Graec., 6: 25, t. 529 (1826); Poech, Enum. Plant. Ins. Cypr., 24 (1842); Unger et Kotschy,
    Die Insel Cypern, 275 (1865); Holmboe, Veg. Cypr., 151 (1914); Post, Fl. Pal., ed. 2, 2: 403
    (1933); Lindberg f., Iter Cypr., 30 (1946); Chapman, Cyprus Trees and Shrubs, 66 (1949);
    Osorio-Tafall et Seraphim, List Vasc. Plants Cyprus, 88 (1973); Feinbrun, Fl. Palaest., 3:
    102, t. 160 (1978); Davis, Fl. Turkey, 7: 56 (1982).
    T. rosmarinifolium Lam., Encycl. Méth., 2: 693 (1788); Boiss., Fl. Orient., 4: 806 (1879).
    TYPE: *"in Creta"*.

Slender, erect, much-branched shrub 1–2 m. high, old wood covered with
pale brown, lightly fissured bark, young shoots sharply tetragonous, shortly
white-tomentose or glabrescent; leaves linear, 7–40 mm. long, 1–4 mm.
wide, much resembling those of *Rosmarinus*, the upper surface lustrous dark
green, the lower surface white-tomentose, apex obtuse or subacute, base
tapering to an indistinct petiole, margins entire, revolute; nervation
obscure; inflorescence terminating an elongate, sparsely leaved flowering
shoot, narrowly racemose or almost spiciform, 10–30 (or more) cm. long,
usually less than 2 cm. wide, consisting of numerous, few (usually 2–8)-
flowered verticillasters, the lower remote, the upper often rather crowded;
flowers shortly pedicellate, with tomentose pedicels usually less than 5 mm.
long; bracts foliaceous, resembling the leaves, becoming progressively
smaller upwards; bracteoles subulate, less than 4 mm. long, caducous or
absent; calyx campanulate, tomentose, tube about 5 mm. long, 3 mm. wide
at apex, teeth subequal, deltoid, about 3 mm. long, 2 mm. wide at base,
spinulose-mucronate, the prominent midrib excurrent as a short point;
corolla mauve-pink, shortly glandular externally, tube campanulate, 4–5
mm. long, 5–6 mm. wide at apex, lip about 14 mm. long, 8 mm. wide, 5-
lobed, bearded at base, basal lobes broadly oblong, about 3·5 mm. long, 3
mm. wide, lateral lobes oblong, about 3·5 mm. long, 2 mm. wide, median
lobe cochleariform, about 8 mm. long, 5 mm. wide, subentire or obscurely
lobed; stamens inserted at mouth of corolla-tube; filaments shortly
glandular-pubescent, the adaxial about 8 mm. long, the abaxial about 9·5
mm. long; ovary subglobose, about 1·2 mm. diam., obscurely 4-sulcate
longitudinally, tomentose at apex; disk inconspicuous, obscurely lobed;
style glabrous, about 9 mm. long, directed upwards above the exserted
stamens; stigmatic lobes narrowly deltoid, subequal, about 0·5 mm. long.
Nutlets oblong-obovoid, 4–5 mm. long, 2–2·5 mm. wide, trigonous, pale
brown with a tomentose apex, dorsal surface coarsely reticulate-rugulose;
attachment area ventral-basal, large, somewhat excavate.

HAB.: Dry rocky hillsides on calcareous or igneous rocks; near sea-level to 3,000 ft. alt., but
usually lowland; fl. March–June.

DISTR.: Divisions 1–4, 7, 8. Eastern Mediterranean region from Turkey to Palestine.

1. "an der Südküste bei Papho nicht selten", 1862, *Kotschy*.
2. Pano Panayia, 1905, *Holmboe* 1063; E. of Perapedhi, 1938, *Kennedy* 1406!

3. Cape Gata, 1905, *Holmboe* 677; Ayios Nikolaos, 1937, *Chapman* 309! Between Limassol and Amathus, 1939, *Lindberg f.* s.n.! Lefkara, 1940, *Davis* 1912! Zyyi, 1949, *C. E. H. Sparrow* 38! Akrotiri, 1963, *J. B. Suart* 116! Lania, 1970, *A. Genneou* in ARI 1533!

4. Cape Kiti, near Larnaca, 1979, *Edmondson & McClintock* E 2967!

7. Common; Kyrenia, near Myrtou, between Antiphonitis Monastery and Bellapais, between Kyrenia and Lapithos, Dhavlos, St. Hilarion, near the Armenian Monastery, between Ayios Amvrosios and Ayios Epiktitos, near Panagra, Kyrenia Pass, etc. *Sibthorp*; *Kotschy* 334, 934! *Sintenis & Rigo* 565! *A. G. & M. E. Lascelles* s.n.! *Haradjian* 346! *J. A. Tracey* 54! *Syngrassides* 421! *Chapman* 151! 309! *Kennedy* 1407! *C. H. Wyatt* 26! *Lindberg f.* s.n.! *Davis* 1729! *Casey* 796! *Merton* 2280! *Miss Mapple* 5! *G. E. Atherton* 95! 191! *Meikle* 2362! *A. Genneou* in ARI 1550! *P. Laukkonen* 329!

8. Near Rizokarpaso, 1912, *Haradjian* 161! Ayios Theodhoros, 1939, *Lindberg f.* s.n.; Yialousa, 1940, *Davis* 2401!

NOTES: Several collectors remark on the attractiveness of *T. creticum* to bees, a fact to be borne in mind by Cyprus bee-keepers.

SECT. 2. **Scorodonia** *Schreber* Perennial herbs or subshrubs; inflorescence racemose or paniculate; flowers secund; calyx campanulate, 2-lipped, tube curved, gibbous at base, adaxial tooth much larger and broader than the others.

2. **T. kotschyanum** *Poech*, Enum. Plant. Ins. Cypr., 24 (1842), et in Flora, 27: 454 (1844); Benth. in DC., Prodr., 12: 585 (1848); Boiss., Fl. Orient., 4: 812 (1879); Holmboe, Veg. Cypr., 151 (1914); A. K. Jackson in Hooker's Icones Plant., 34: t. 3327 (1937); Lindberg f., Iter Cypr., 24 (1946); Rechinger f. in Arkiv för Bot., ser. 2, 1: 429 (1950); Osorio-Tafall et Seraphim, List Vasc. Plants Cyprus, 88 (1973); Davis, Fl. Turkey, 7: 74 (1982).

T. smyrnaeum Boiss., Diagn., 1, 5: 43 (1844); Unger et Kotschy, Die Insel Cypern, 276 (1865).

TYPE: "in ins. Cypro. (Kotschy *pl. Cypr.*!)".

Perennial, rhizomatous herb to 80 cm. high; stems branched at base, usually unbranched and erect above, tetragonal, pilose with recurved, white hairs and sparsely glandular with scattered, shining, sessile glands, sometimes purplish; leaves rather evenly spaced, oblong-lanceolate, 2–5 cm. long, 0·6–2 cm. wide, pale green, markedly reticulate-rugose above and below, softly glandular-pilose, apex obtuse or rounded, base broadly cuneate, margins finely crenulate; petiole 3–10 mm. long, channelled above; inflorescence a narrow, elongate, erect, terminal raceme 10–30 cm. long, about 1·5 cm. wide, sometimes with 2 basal branches; bracts herbaceous, lanceolate, becoming progressively smaller upwards, 4–10 mm. long, 2–5 mm. wide; bracteoles wanting; verticillasters rather closely superimposed, generally 2-flowered; pedicels 1–4 mm. long, pilose; flowers subsecund; calyx campanulate, 2-lipped, tube about 5 mm. long, 4 mm. wide at apex, shortly pilose externally, distinctly gibbous abaxially at base, bearded internally just below the abaxial lip, otherwise glabrous, adaxial lip consisting of a single, broadly ovate or semicircular, mucronate lobe about 3 mm. long, 4·5 mm. wide, porrect in flower, becoming erect in fruit, abaxial lip of 4 short, subequal, ovate-acuminate, spine-tipped teeth about 2 mm. long, 1·5 mm. wide at base; corolla pale greenish-yellow, tube slightly curved, about 8 mm. long, 2·5 mm. wide at apex, thinly pubescent externally, glabrous internally, lip deflexed, 5-lobed, glandular-puberulous externally, 2 basal lobes very broadly depressed-deltoid, about 1 mm. long, lateral lobes oblong-acute, about 1·5 mm. long, 1 mm. wide at base, median lobe cochleariform, about 4·5 mm. diam.; stamens inserted at mouth of corolla-tube; filaments glabrous below, shortly glandular-puberulous above, the adaxial about 5 mm. long, the abaxial about 6 mm. long; anthers about 1 mm. wide, thecae united, dehiscing by one apical slit; ovary about 2 mm. diam., lobes minutely papillose-hispidulous at apex; style attached somewhat unilaterally, glabrous, about 10 mm. long; stigmatic lobes acute, about 0·5 mm. long. Nutlets subglobose, about 1·8 mm. long, 1·5 mm. wide, rich brown, minutely and sparsely papillose, attachment area large, orbicular.

HAB.: On rocky, igneous mountainsides; 1,800–5,100 ft. alt.; fl. May–July.

DISTR.: Divisions 1–3. Also E. Aegean Islands and S. Turkey.

1. Lyso, 1913, *Haradjian* 861!
2. Common; Galata, Prodhromos, Kambos, Platania, Platres, Khionistra, Kykko, Stavros tis Psokas, Mesapotamos, Moniatis, etc. 1840–1973, *Kotschy* 13, 921! *Sintenis & Rigo* 725! *Post* s.n.! *A. G. & M. E. Lascelles* s.n.! *Holmboe* 1008; *Haradjian* 413! *Syngrassides* 737! 1327! *Kennedy* 832! 833! 1080! *Lindberg f.* s.n.! *Davis* 1764! 3449! *Casey* 747! *F. M. Probyn* 111! *Mavromoustakis* 63! *G. E. Atherton* 587! *Economides* in ARI 1207! *P. Laukkonen* 440!
3. Between Yerasa and Kalokhorio, 1962, *Meikle* 2895!

SECT. 3. **Scordium** *Reichb.* Annual or perennial herbs; flowers not secund; calyx tubular, curved, gibbous at base, teeth subequal or the adaxial slightly enlarged.

3. T. scordium *L.*, Sp. Plant., ed. 1, 565 (1753); Boiss., Fl. Orient., 4: 813 (1879); Post, Fl. Pal., ed. 2, 2: 405 (1933); Osorio-Tafall et Seraphim, List Vasc. Plants Cyprus, 88 (1973); Davis, Fl. Turkey, 7: 61 (1982).

Erect or sprawling, rhizomatous or stoloniferous perennial; stems slender or robust, much branched or sparingly branched, 8–60 cm. high, tetragonal, pubescent or densely villose, with a mixed indumentum of spreading hairs and stalked glands, sometimes purplish; leaves with a slightly unpleasant smell when crushed, sessile, sometimes amplexicaul, or shortly petiolate, oblong, 1–4 cm. long, 0·5–2 cm. wide, pubescent or villose, often strongly rugose, apex blunt or acute, base broadly cuneate, rounded or subcordate, margins crenate or serrate, narrowly recurved; inflorescence lax, occupying the greater part of the length of the stems and branches and consisting of remote, 4-flowered verticillasters (i.e. paired flowers in the axil of each of the opposite bracts); flowers not secund; bracts indistinguishable from cauline leaves, diminishing gradually upwards; bracteoles absent; pedicels slender, pubescent or pilose, 2–5 mm. long; calyx campanulate, tube about 3 mm. long, 2·5 mm. wide at apex, thinly or densely villose externally, glabrous internally except for a narrow zone of hairs at apex, base gibbous, teeth 5, erect, acutely deltoid, about 1 mm. long, 0·8 mm. wide at base; corolla purple, pink, lilac or whitish, 3–3·5 mm. long, 2·5 mm. wide at apex, shortly puberulous externally, lip about 7 mm. long, 5-lobed, basal lobes acutely deltoid, about 1 mm. long, 0·8 mm. wide at base, lateral lobes slightly shorter and blunter, median lobe cochleariform, about 3 mm. long, 2·5 mm. wide, pilose on the underside; stamens inserted at mouth of corolla-tube; filaments thinly pilose, the adaxial about 4 mm. long, the abaxial about 4·5 mm. long; anthers about 0·5 mm. wide, thecae divergent, opening by a single, common slit; ovary about 0·5 mm. diam., lobes glabrous, with an apical covering of sessile glands; style about 5 mm. long, thinly hairy towards base; stigmatic lobes subulate, subequal, about 0·5 mm. long. Nutlets subglobose, about 1 mm. diam., rich brown, apex thinly glandular-verruculose, attachment area large, orbicular.

ssp. **scordium**
TYPE: "*in* Europae *paludosis*".

Stems slender, diffuse, generally irregularly and often sparsely branched, thinly pubescent; leaves often smallish, to 2·5 cm. long, 1·2 cm. wide, thinly hairy, not very rugose, apex generally acute or subacute, base cuneate or rounded, margins often rather sharply serrate or serrate-crenate; calyx thinly hairy; corolla bright purple-pink.

HAB.: Damp ground by mountain streams, rare at lower altitudes; 300–4,500 ft. alt.; fl. June–Sept.

DISTR.: Divisions 2, 3, 7. Widespread in Europe, becoming rare in the Mediterranean region.

2. Near Evrykhou, 1880, *Sintenis & Rigo* 735! Near Galata, 1913, *Haradjian* 1007! Platanassa N. of Khionistra, 1937, *Kennedy* 816! Kryos Potamos, 3,900 ft. alt., 1937, *Kennedy* 817! Xerokolymbos, 4,500 ft. alt., 1937, *Kennedy* 818! Platres, Kephalovryso, 1937, *Kennedy* 819! Amiandos, 1938, *Kennedy* 1408! Mesapotamos Monastery, 1952, *F. M. Probyn* 138! S. side of Tripylos, c. 3,500 ft. alt., 1956, *Merton* 2827! Platres, 1960, *N. Macdonald* 203!
3. Near Limassol, 1947, *Mavromoustakis* 62!
7. Vasilia, 1970, *A. Genneou* in ARI 1578!

ssp. **scordioides** (*Schreber*) *Arcang.*, Comp. Fl. Ital., ed. 1, 559 (1882); Holmboe, Veg. Cypr., 151 (1914); Osorio-Tafall et Seraphim, List Vasc. Plants Cyprus, 88 (1973); Davis, Fl. Turkey, 7: 62 (1982).
    T. scordioides Schreber, Plant. Verticill. Unilab. XXXVII (1773); Unger et Kotschy, Die Insel Cypern, 276 (1865); Boiss., Fl. Orient., 4: 813 (1879); H. Stuart Thompson in Journ. Bot., 44: 337 (1906); Post, Fl. Pal., ed. 2, 2: 405 (1933); Lindberg f., Iter Cypr., 31 (1946); Feinbrun, Fl. Palaest., 3: 104, t. 165 (1978).
    T. scordium L. var. *scordioides* (Schreber) Sm. in Sibth. et Sm., Fl. Graec. Prodr., 1: 393 (1809); Camus in Bull. Soc. Bot. Fr., 34: 54 (1887).
    TYPE: "in Creta" (? M).

Stems rather robust, erect, branched at almost every node, and often forming a neat conical bush, with the lateral branches diminishing in length upwards, rather densely white-pilose; leaves usually large, up to 4 cm. long, 2 cm. wide, softly lanuginose, conspicuously rugose, apex blunt or rounded, base amplexicaul-subcordate, margins bluntly toothed or crenate; calyx lanuginose; corolla often pale pink, lilac or whitish.

HAB.: Marshy ground with *Juncus, Carex*, etc.; sea-level to 1,500 ft. alt.; fl. June–Sept.

DISTR.: Divisions 2–4, 6, 7. S. Europe and Mediterranean region east to Iraq and Iran.

2. Vavatsinia, 1905, *Holmboe* 1103.
3. Zakaki marshes, 1939 & 1943, *Mavromoustakis* 55! 84! Salt marshes near Limassol, 1939, *Lindberg f.* s.n.; Asomatos marshes, 1940, *Mavromoustakis* 78! also, 1962, *Meikle* 2918! and 1970, *A. Genneou* in ARI 1576a!
4. Between Prastio and Kouklia, 1939, *Lindberg f.* s.n.!
6. Dhiorios Forest, 1954, *Merton* 1917!
7. Moist places near Ayios Khrysostomos Monastery, 1862, *Kotschy*; near Panteleimon Monastery, Myrtou, 1862, *Kotschy* 949.

NOTES: While there is no denying that Cyprus populations of *T. scordium* are, on the whole, readily separable into the slender, glabrescent *T. scordium* ssp. *scordium*, mostly found at high altitudes on the Troödos Range, and the more robust, lanuginose *T. scordium* ssp. *scordioides* from the lowlands, it is not easy to find reliable characters for expressing these distinctions, nor is it always possible to draw a sharp line between the two populations. In the circumstances, I cannot accept *T. scordioides* as specifically distinct from *T. scordium*, though I think the differences between the two — ecological as well as morphological — are worth more than varietal rank, as recommended by Camus (loc. cit. supra).

Holmboe's record for ssp. *scordioides* from Vavatsinia, and Lindberg's record for the same subspecies from Mesapotamos Monastery, may both, by my reckoning, be referable rather to ssp. *scordium*, while *Sintenis & Rigo* 735 referred to ssp. *scordium* shows some transition to ssp. *scordioides*.

SECT. 4. **Chamaedrys** *Schreber* Subshrubs; inflorescences loosely or densely racemose; flowers often somewhat secund; calyx campanulate, inconspicuously curved and gibbous, teeth subequal.

4. **T. divaricatum** [*Sieber ex*] *Heldreich*, Herb. Graec. Norm. no. 290 (1856); Boiss., Fl. Orient., 4: 816 (1879); Post, Fl. Pal., ed. 2, 2: 407 (1933); Chapman, Cyprus Trees and Shrubs, 67 (1949); A. Strid in Bot. Notiser, 118: 116 (1965); Greuter in Boissiera, 13: 112 (1967); Osorio-Tafall et Seraphim, List Vasc. Plants Cyprus, 88 (1973); Feinbrun, Fl. Palaest., 3: 105, t. 168 (1978); Davis, Fl. Turkey, 7: 66 (1982).
    T. flavum L. var. *purpureum* Benth., Labiat. Gen. et Sp., 682 (1835).
    T. flavum L. var. *purpureum* Benth. f. *divaricatum* Kotschy in Unger et Kotschy, Die Insel Cypern, 276 (1865).
    T. sieberi Čelak. in Bot. Centralbl., 14: 187 (1883).

T. divaricatum Heldr. ssp. *sieberi* (Čelak.) Holmboe, Veg. Cypr., 151 (1914); Rechinger f. in Bot. Archiv., 42: 391 (1941); Lindberg f., Iter Cypr., 30 (1946).

[*T. lucidum* (non L.) Sm. in Sibth. et Sm., Fl. Graec. Prodr., 1: 393 (1809), Fl. Graec., 6: 27, t. 532 (1826); Poech, Enum. Plant. Ins. Cypr., 24 (1842).]

Erect or sprawling, much-branched subshrub, 10–30 (–50) cm. high; old branches gnarled, covered with dark, greyish, irregularly fissured bark, young shoots tetragonal, subglabrous, pubescent with deflexed hairs, or densely tomentellous with a short white indumentum; leaves numerous, rather close, obovate-spathulate, 5–25 mm. long; 3–15 mm. wide, green and subglabrous or thinly canescent above, subglabrous and gland-dotted below, or rather densely greyish or whitish tomentellous, apex blunt or acute, base tapering to a short, but distinct petiole, margins recurved, closely or remotely, and often rather irregularly crenate-dentate; nervation frequently impressed above and prominent below, sometimes obscure; inflorescence lax or rather dense, 2–20 (or more) cm. long, consisting of few or many, remote or rather crowded, superimposed, (2–) 4–6-flowered verticillasters; bracts foliaceous, resembling the leaves, the uppermost usually entire or almost entire; bracteoles wanting or very reduced; pedicels distinct, slender, 3–8 mm. long, subglabrous, pubescent (sometimes glandular-pubescent) or white-villose; calyx campanulate, tube 4–5 mm. long, 4 mm. wide at apex, often purplish, shortly puberulous externally, or sometimes with a mixed indumentum of long and short hairs, occasionally ± densely glandular-hairy, glabrous internally except at the throat; teeth subequal, erect, sharply deltoid, 2–3 mm. long, 1·5 mm. wide at base; corolla pink, purple or reddish, tube distinctly curved, about 7 mm. long, 3–4 mm. wide at apex, glabrous in the lower part externally, thinly glandular or pubescent above, shortly retrorse-strigillose within, lip 11–12 mm. long, pilose externally, basal lobes oblong-deltoid, subacute, about 4 mm. long, 2 mm. wide at base, lateral lobes bluntly oblong or deltoid, 1·5–2 mm. long, 1·5–2 mm. wide at base, median lobe cochleariform, about 4–5 mm. diam., strongly concave, with 2 bearded ridges at its base; stamens inserted at apex of corolla-tube, filaments arcuate, 8–10 mm. long, thinly pilose, distinctly glandular at apex; anthers 1–1·3 mm. wide, thecae dehiscing by a common, apical slit; ovary 1–1·3 mm. diam., glabrous, the lobes covered apically with sessile glands; style about 14 mm. long; stigmatic lobes subequal, subulate, about 0·8 mm. long. Nutlets subglobose, about 2 mm. long, 1·6 mm. wide, dark blackish-brown, minutely granulose and obscurely and coarsely reticulate-rugulose, apex with scattered hyaline glands, attachment area large, suborbicular.

ssp. **canescens** (*Čelak.*) *Holmboe*, Veg. Cypr., 151 (1914); Bornm. in Mitt. Thüring. Bot. Ver. n.f., 38: 29 (1929); Rechinger f. in Bot. Archiv., 42: 392 (1941); Lindberg f., Iter Cypr., 30 (1946); Osorio-Tafall et Seraphim, List Vasc. Plants Cyprus, 88 (1973).

    T. sieberi Čelak. var. *canescens* Čelak. in Bot. Centralbl., 14: 190 (1883).

    T. divaricatum Boiss. ssp. *canescens* (Čelak.) Holmboe var. *kotschyanum* Bornm. in Mitt. Thüring. Bot. Ver. n.f., 38: 29 (1929).

    T. divaricatum Boiss. ssp. *canescens* (Čelak.) Holmboe f. *kotschyanum* (Bornm.) Rechinger f. in Bot. Archiv., 42: 392 (1941).

    ? *T. pseudo-chamaedrys* Sibth. in Walpole, Travels, 25 (1820) nom. nud.

    TYPE: Cyprus; probably *Sintenis & Rigo* 573 from between Bellapais and Kyrenia, 28 May, 1880 (PR, W, K!).

Branches slender, densely white-tomentellous when young; leaves generally small, obovate, about 1–1·5 cm. long, 0·8–1 cm. wide, green and thinly strigillose above, conspicuously white-tomentellous below with a dense indumentum of short subadpressed hairs; calyx and pedicels shortly crispate-canescent (f. *canescens*) or ± pilose with an intermixture of long, eglandular hairs (f. *kotschyanum* (Bornm.) Rechinger f.).

HAB.: In garigue on dry, rocky ground, on limestone or igneous formations; sea-level to 5,500 ft. alt.; fl. May–July.

DISTR.: Divisions 2–4, 6–8, locally common. Endemic. Typical *T. divaricatum* (and a multiplicity of variants) is widely distributed in the eastern Mediterranean region from Greece to Palestine and eastwards to Iraq and Iran.

2. Near Galata, 1880, *Sintenis & Rigo* 731! Platres, 1937, *Kennedy* 827! 829! Mandria, 1937, *Kennedy* 826! Kryos Potamos, 2,900 ft. alt., 1937, *Kennedy* 828! Perapedhi, 1940, *Davis* 1852! Khionistra, 1937, *Kennedy* 830! Aphamis, 1937, *Kennedy* 831!
3. Cape Gata, 1862, *Kotschy* 596! also, 1962, *Meikle* 2650! Amathus, 1905, *Holmboe* 642, and, 1978, *J. Holub* s.n.! Oritaes Forest, 1937, *Chapman* 306! Akrotiri Forest, 1939, *Lindberg f.* s.n., also, 1947, *Mavromoustakis* s.n.! and, 1963, *J. B. Suart* 117! Kellaki, 1948, *Kennedy* 1613! Near Kouklia, 1960, *N. Macdonald* 167!
4. Between Cape Greco and Ayia Napa, 1958, *N. Macdonald* 74!
6. Ayia Irini, 1967, *Economides* in ARI 928!
7. Common; between Bellapais and Kyrenia, 1880, *Sintenis & Rigo* 573! Kyrenia, 1901, *A. G. & M. E. Lascelles* s.n.! Armenian Monastery, 1912, *Haradjian* 367! Halevga, 1939, *Lindberg f.* s.n.! and, 1940, *Davis* 1733! Kyrenia Pass, 1939, *Lindberg f.* s.n.; St. Hilarion, 1939, *Lindberg f.* s.n.; below N. cliffs, St. Hilarion, 1951, *Merton* 1296!
8. Near Komi Kebir, 1912, *Haradjian* 285!

The following are not assignable with certainty to ssp. *canescens* (Čel.) Holmboe, or to any other named subspecies:

1. Near Paphos, 1787, *Sibthorp*; Lyso, 1913, *Haradjian* 839! Akamas, 1941, *Davis* 3319! Kakoskala near Baths of Aphrodite, 1949, *Casey* 756!
2. Platres, 1914, *Feilden* 13!
6. Kormakiti, 1862, *Kotschy*.
7. Between Lefkoniko and Kalogrea, 1932, *Syngrassides* 422! Lefkoniko Pass, 1934, *Syngrassides* 441! Kantara, 1948, *Mavromoustakis* s.n.! Lapithos, 1955, *Merton* 2239! Kyrenia, 1955, *Miss Mapple* 21! 104! *G. E. Atherton* 41!
8. Near Ayios Theodhoros, by the sea, 1939, *Lindberg f.* s.n.

NOTES: Authorship of the name *Teucrium divaricatum* is generally attributed to "Sieber" or "Sieber ex Boiss.", but Sieber never published a description or diagnosis, and Boissier's publication of the name (Fl. Orient., 4: 816, 1879) is antedated by Heldreich, who, as pointed out by Greuter (Boissiera, 13: 112, 1967) had, more than twenty years earlier, validated the name by issuing mechanically reproduced autograph labels bearing not only the name, but references to Flora Graeca and de Candolle's Prodromus. Čelakovsky (Bot. Centralbl., 14: 187–190; 1883) has, however, shown that unbelievable confusion surrounds the distribution of Sieber's specimens and labels, at least two species (and possibly more) having been sent out with the names *T. divaricatum* and *T. saxatile* indiscriminately attached to them. Čelakovsky argues that the name *Teucrium saxatile* Sieber (a later homonym) was intended by Sieber to be attached to the plant which we now call *T. divaricatum*, while *T. divaricatum* Sieber was intended for the plant now called *T. alpestre* Sm. Heldreich (and Boissier) inadvertently and excusably attached the name *T. divaricatum* to the wrong species. Accepting Čelakovsky's argument, I do not think, however, that he was justified in replacing the name *T. divaricatum* with *T. sieberi* Čelak., though I think it is historically incorrect and possibly misleading to attribute the name *T. divaricatum* to "Sieber ex Heldreich" or "Sieber ex Boiss.", even if the validating authors attributed the name to Sieber.

The species is variable in Cyprus, as elsewhere, and while most of the specimens examined are distinguishable from *T. divaricatum* from Crete and Greece by their slender habit, white canescent shoots, and small leaves conspicuously white-canescent on the undersurface, a small residue is left (especially of specimens from the south-west and north-east of the island) which, though obviously close to ssp. *canescens*, do not fall very neatly into the subspecies. I do not regard the pilosity of the calyx, pedicels and bracts of much significance, and have not maintained the distinction between f. *canescens* (with a short indumentum) and f. *kotschyanum* (with scattered long hairs).

SECT. 5. **Polium** *Schreber* Perennial herbs and small subshrubs, sometimes with an indumentum of compound hairs; inflorescences condensed, capitate or subcapitate; calyx tubular, not noticeably gibbous at base, teeth subequal.

5. **T. cyprium** *Boiss.*, Diagn., 1, 5: 43 (1844); Unger et Kotschy, Die Insel Cypern, 276 (1862); Boiss., Fl. Orient., 4: 820 (1879); Post in Bull. Herb. Boiss., 7: 161 (1899); Holmboe, Veg. Cypr., 151 (1914); A. K. Jackson in Hooker's Icones Plant., 34: t. 3328 (1937); Lindberg f., Iter Cypr., 30 (1946); Chapman, Cyprus Trees and Shrubs, 67 (1949); Osorio-Tafall et Seraphim, List Vasc. Plants Cyprus, 89 (1973).

T. cypricum Post in Bull. Herb. Boiss., 5: 759 (1897).

Dwarf, much-branched subshrub, usually less than 10 cm. high, forming dense cushions or mats; old wood covered with flaking, dull brown bark, young shoots terete or subterete, slender, diffuse, clothed with a mixed indumentum of short, glandular and long, white, spreading, eglandular hairs; leaves usually crowded, oblong or obcuneate, 5–15 mm. long, 3·5–8 mm. wide, rather densely villous on both surfaces, and usually besprinkled with sessile, shining glands, apex blunt or rounded, margins shortly toothed in the upper half, generally entire below and tapering to a subsessile base, frequently recurved or revolute; inflorescence terminal, capitate, 1·3–2·5 cm. diam., or flowers occasionally geminate in the uppermost pairs of leaves (or bracts) in shade-grown specimens; bracts foliaceous, oblong, 5–7 mm. long, 2–3·5 mm. wide, softly glandular-villous, entire or with a few small apical teeth; bracteoles wanting; flowers subsessile; calyx tubular, 5–6 mm. long, 3 mm. wide at apex, gland-dotted and villous with spreading hairs externally, glabrous internally except for a zone of hairs just below the throat, obscurely 5-nerved, teeth erect, acutely deltoid, the 3 adaxial about 1·5 mm. long, 1 mm. wide at base, the 2 abaxial longer and more acuminate, about 2 mm. long; corolla pink, yellowish-purple, yellowish or white, glandular and thinly pubescent externally, tube slightly curved, about 5–6 mm. long, 1·5 mm. wide at apex, lip 5–6 mm. long, basal lobes oblong, subacute or rounded, 3 mm. long, 1·5 mm. wide at base, lateral lobes concave, 1·5 mm. long, 0·8 mm. wide, usually pressed against the cochleariform, subacute, 2·5–3 mm. long, 2 mm. wide median lobe; stamens inserted at mouth of corolla-tube, filaments thinly pilose, the adaxial about 3 mm. long, the abaxial about 4 mm. long; anthers about 1 mm. wide, thecae divergent, with a single, common apical slit; ovary glabrous, about 0·8 mm. diam., disk minute; style glabrous, about 7 mm. long; stigmatic lobes subulate, subequal, about 0·4 mm. long. Nutlets broadly obovoid, about 2 mm. long, 1·5 mm. wide, dark brown, granulose, distinctly foveolate, attachment area large, suborbicular, about 1 mm. diam.

ssp. **cyprium**
TYPE: "in insulâ *Cypro* Aucher [Eloy] No 1595" (G, K !).

Stems slender, sprawling; leaves small, 5–10 mm. long, 3–8 mm. wide; capitula usually less than 1·5 cm. diam.; corolla mostly pink or purplish, rarely pale yellow or white. *Plate 78, figs. 4–7.*

HAB.: Dry igneous screes and stony mountainsides, often in open Pine forest; 1,400–6,300 ft. alt.; fl. June–July.

DISTR.: Division 2. Endemic.

2. Khionistra, Platres, Kryos Potamos, Kykko, between Ambelikou and Kambos, Madhari, Alonoudhi Junction, Ayia valley, Troödos, Prodhromos, etc., 1831–1968, *Aucher-Eloy* 1595! *Kotschy* 783; *Sintenis & Rigo* 727! *Post* 26; *A. G. & M. E. Lascelles* s.n.! *Holmboe* 936; *Haradjian* 985! *C. B. Ussher* 28! *Kennedy* 820–822! 825! *Davis* 1811! 1822! 1876! *Lindberg f.* s.n.! *Merton* 1945! 2824! *D. P. Young* 7805! *Barclay* 1104!

ssp. **kyreniae** *P. H. Davis* in Kew Bull., 4: 111 (1949); Osorio-Tafall et Seraphim, List Vasc. Plants Cyprus, 89 (1973).
Teucrium nov. sp. Chapman, Cyprus Trees and Shrubs, 67 (1949).
TYPE: Cyprus; "N. side of Yaïla near Halefka [Halevga]; 600–750 m.; 4 Aug. 1940, *Davis* 1921" (K !).

More robust than ssp. *cyprium*; stems frequently suberect; leaves larger, commonly 8–15 mm. long, 5–8 (occasionally more) cm. wide; capitula 1·5–2·5 cm. diam.; corolla white or pale yellow. *Plate 78, figs. 1–3.*

HAB.: Crevices of limestone rocks, or on limestone screes; 1,000–3,000 ft. alt.; fl. May–July.

DISTR.: Division 7. Endemic.

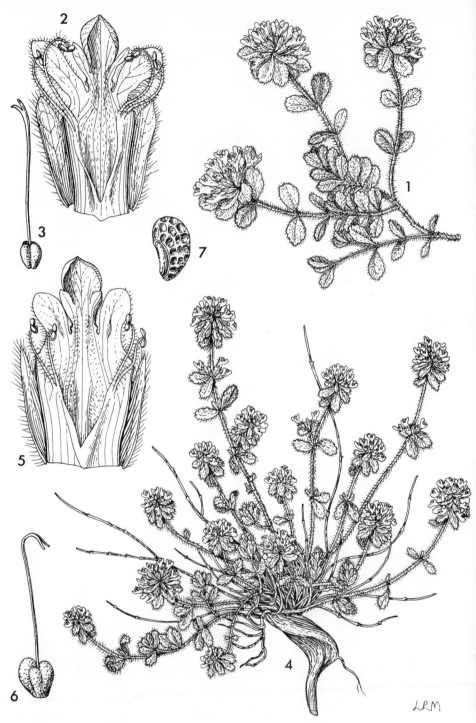

Plate 78. Figs. **1–3.** TEUCRIUM CYPRIUM Boiss. ssp. KYRENIAE P. H. Davis **1,** flowering branch, ×⅔; **2,** corolla opened out, ×4; **3,** gynoecium, ×4; figs. **4–7.** TEUCRIUM CYPRIUM Boiss. ssp. CYPRIUM **4,** habit, ×⅔; **5,** corolla opened out, ×6; **6,** gynoecium, ×6; **7,** nutlet, ×8. (**1–3** from *Davis* 3608; **4–6** from *Kennedy* 820; **7** from *Kennedy* 825.)

7. Buffavento, 1938, *Kennedy* 1502! N. side Yaïla, 1940, *Davis* 1921! and, 1941, *Davis* 3608! Larnaka tis Lapithou, 1941, *Davis* 2084! St. Hilarion, 1940, *Davis* 2044! and, 1949, *Casey* 215! End of Forest road on Kyrenia Range, 1955, *G. E. Atherton* 124!

NOTES: The distinctions between *T. cyprium* Boiss. ssp. *cyprium* and *T. cyprium* ssp. *kyreniae* are perhaps more obvious in the field than in the herbarium. I do not find any useful difference in the size or shape of the calyx-teeth, and the difference in flower-colour is unfortunately made less useful by the occurrence of pallid-flowered populations of ssp. *cyprium* in the Troödos area.

6. T. micropodioides *Rouy* in Le Naturaliste, 2: 16 (1882); Čelak. in Bot. Centralbl., 14: 154 (1883); Boiss., Fl. Orient., Suppl., 364 (1888); Rechinger f. in Arkiv för Bot., ser. 2, 1: 429 (1949).

T. pseudo-polium Sibth. in Walpole, Travels, 26 (1820) nom. nud.

T. polium L. ssp. *micropodioides* (Rouy) Holmboe, Veg. Cypr., 152 (1914); Lindberg f., Iter Cypr., 31 (1946); Osorio-Tafall et Seraphim, List Vasc. Plants Cyprus, 89 (1973).

[*T. polium* (non L.) Clarke, Travels, ed. 1, 2: 724 (1816); Kotschy in Unger et Kotschy, Die Insel Cypern, 277 (1865); Chapman, Cyprus Trees and Shrubs, 67 (1949).]

[*T. polium* L. var. *purpurascens* (non Benth.) Kotschy in Unger et Kotschy, Die Insel Cypern, 277 (1865).]

[*T. polium* L. var. *roseum* (non *T. capitatum* Fl. Graec., t. 536) Boiss., Fl. Orient., 4: 821 (1879) pro parte quoad plant. cypr.]

TYPE: Cyprus; "in vineis pr. *Galata* — 16 Juin 1880 (Sintenis et Rigo)" — an error, *Sintenis & Rigo* 566 (the number quoted by Rouy) is labelled "In montibus circa Kythraea Majo [or 7 June] 1880". See Notes.

Small much-branched subshrub, forming low, intricate, diffuse or domed bushes 3–20 (–30) cm. high; old wood greyish, coarsely fissured, young shoots white-tomentose with a dense indumentum of closely adpressed, compound hairs, terete or subterete, persistent and becoming leafless, rigid or spinescent, but remaining white-tomentose, after their first year; leaves oblong, 5–10 (–12) mm. long, 1–3·5 mm. wide, subsessile, closely grey-tomentellous above and below, apex blunt, margins strongly recurved or revolute, apparently entire, but often with a few, obscure teeth near the apex; inflorescences capitate, almost 1–1·3 cm. diam., sometimes solitary and terminal, but more commonly arising on short lateral shoots from the axils of cauline leaves, and forming a narrow, spiciform inflorescence running the greater length of the branches; flowers sessile; bracts foliaceous, similar to the leaves but rather shorter and narrower, about equalling the flowers; bracteoles absent; calyx tubular, about 3 mm. long, 2 mm. wide at apex, densely lanuginose externally, glabrous internally, teeth unequal, the 4 abaxial deltoid, about 1 mm. long, 0·8 mm. wide at base, the adaxial much broader, about 1 mm. long, nearly 2 mm. wide at base; corolla maroon, purple or brownish-pink, tube 3–4 mm. long, 1·5 mm. wide at apex, slightly curved, subglabrous or very thinly glandular and pubescent externally, glabrous internally except at the throat, limb gland-dotted and shortly pilose externally, 4–5 mm. long, basal lobes bluntly oblong, about 1 mm. long, 1 mm. wide at base, lateral lobes adpressed to median lobe, similar to basal, but narrow, about 0·8 mm. wide at base, median lobe cochleariform, about 2·5 mm. diam.; stamens inserted at mouth of corolla-tube; filaments purple, thinly pilose, the adaxial about 3 mm. long, the abaxial about 4 mm. long; anthers purple, about 0·8 mm. wide, thecae strongly divergent, opening by a common, apical slit; ovary glabrous, about 0·8 mm. diam.; disk very small; style glabrous, about 5 mm. long; stigmatic lobes subulate, subequal, about 0·3 mm. long. Nutlets broadly obovate, about 1·8 mm. long, 1·5 mm. wide, dark brown, granulose, distinctly foveolate, attachment area large, about 0·8 mm. diam.

HAB.: In garigue on dry, rocky ground on calcareous or igneous formations; sometimes in open Pine forest or *Juniperus* maquis, or on sand-covered rocks by the sea; sea-level to 2,800 ft. alt.; fl. April–July.

DISTR.: Division 1–8, locally abundant. Endemic.

1. Paphos, 1787, *Sibthorp*, also, 1913, *Haradjian* 625! 626! Lyso, 1913, *Haradjian* 822! Akamas, 1941, *Davis* 3314! and, 1948, *Kennedy* in *Casey* 162!
2. Mandria, 1939, *Kennedy* 1501! Adelphi Forest, 1933, *Chapman* 85!
3. Cape Zevgari, 1905, *Holmboe* 670! Oritaes Forest, 1937, *Chapman* 304! Near Limassol, 1939, *Lindberg f.* s.n.; near Akrotiri, 1909, *Lindberg f.* s.n., also, 1947, *Mavromoustakis* s.n.! Cape Gata, 1941, *Davis* 3549! Lefkara, 1940, *Davis* 1911!
4. Larnaca, 1939, *Lindberg f.* s.n.; Cape Greco, 1973, *E. Chicken* 472!
5. Athalassa, 1932, *Syngrassides* 380! and, 1939, *Lindberg f.* s.n.; road from Nicosia to Famagusta, 1972, *W. R. Price* 1001! 1015!
6. Sandy ground between Morphou and Kormakiti, 1862, *Kotschy* 925! Nicosia, 1950, *Chapman* 621! also, 1972, *W. R. Price* 1025! and, 1973, *P. Laukkonen* 263! Ayia Irini, 1967, *Economides* in ARI 927!
7. Mountains above Kythrea, 1880, *Sintenis & Rigo* 566! Shore between Akanthou and Kyrenia, 1934, *Syngrassides* 445! 446! Pentadaktylos, 1939, *Lindberg f.* s.n.! Kyrenia, 1939, *Lindberg f.* s.n.; also, 1955–1968, *Miss Mapple* 59! *G. E. Atherton* 43! 123! 132! *Barclay* 1056! Halevga, 1940, *Davis* 1754! Upper Thermia, 1955, *G. E. Atherton* 488!
8. Near Komi Kebir, 1912, *Haradjian* 281! Ayios Andronikos, 1934, *Syngrassides* 541! Near Ayios Elias, 1962, *Meikle* 2628!

NOTES: For the most part very uniform and distinct, though a few plants from the Akamas (*Davis* 3314!), Div. 2 (*Kennedy* 1501! *Chapman* 85!) and Div. 3 (*Chapman* 304!) are exceptional in having a longer, more lanuginose indumentum, and might possibly merit recognition as varieties. A few specimens, collected without inflorescences, or with abnormal inflorescences on secondary growths (Div. 7, *Casey* 146! *G. E. Atherton* 232! Div. 8, *Davis* 2253! *Merton* 3065!) cannot be assigned with certainty to any species. *T. micropodioides* has yet to be discovered outside Cyprus; its nearest relatives are to be found amongst segregates of *T. polium* L. occurring in the Aegean Islands.

Rouy (Le Naturaliste, 2: 16; 1882) cites the type of *T. micropodioides* (*Sintenis & Rigo* 566) as coming from vineyards near Galata, but all the material I have seen of *Sintenis & Rigo* 566 is labelled "In montibus circa Kythraea [sic]", nor is there any evidence that the species is found about Galata, though, significantly, *T. divaricatum* Boiss. ssp. *canescens* (Čelak.) Holmboe was recorded (*Sintenis & Rigo* 731!) from precisely this locality. I suspect some confusion in labelling, not infrequent with *Sintenis & Rigo* specimens.

[T. BREVIFOLIUM *Schreber*, Plant. Verticill. Unilab., 27 (1773), a low shrub with large, blue flowers, much resembling the western Mediterranean *T. fruticans* L., has been recorded from Kyrenia by C. E. Gresham (New Flora & Sylva, 5: 271; 1933) but unquestionably in error, for though *T. brevifolium* is found in Greece, Crete, Aegean Islands and Turkey, it has not since been seen at Kyrenia, or anywhere else in Cyprus.]

26. AJUGA *L.*

Sp. Plant., ed. 1, 561 (1753).
Gen. Plant., ed. 5, 246 (1754).

Annual or perennial herbs, sometimes stoloniferous; leaves variable, entire, toothed or lobed, glabrous or pilose; inflorescences lax or condensed, sometimes spiciform; bracts foliaceous, resembling the cauline leaves or distinct, sometimes coloured; bracteoles very small or absent; verticillasters few- to many-flowered; flowers sessile; calyx tubular-campanulate, commonly 10-nerved, subequally toothed or lobed, glabrous at the throat; corolla-tube straight or curved, often with a bearded zone internally, limb apparently 1-lipped, the adaxial lip commonly much reduced or obsolete, the abaxial lip 3-lobed, the median lobe often bilobed and much larger than the 2 laterals; stamens 4, didynamous, abaxial filaments longer than the adaxial; style not gynobasic; stigmatic lobes subequal. Nutlets obovoid, reticulate, rugose or foveolate.

About 40 species widely distributed in temperate regions of the Old World.

Basal leaves oblong, 6–15 cm. long, 1·5–3·5 cm. wide, forming a loose rosette; inflorescence spiciform, flowers resupinate, with a long slender corolla-tube, included stamens, and a violet-blue or claret-red corolla limb - - - - - - - - **1. A. orientalis**

Basal leaves narrowly oblong, oblanceolate, narrowly spathulate or linear, generally less than 7 cm. long and 1 cm. wide; inflorescence elongate, flowers not resupinate, corolla-tube narrowly infundibuliform; stamens exserted; corolla-limb yellow (sometimes pink when dried):

Cauline leaves and bracts almost linear, entire or very obscurely 3-lobed at apex; nutlets foveolate dorsally with rounded pits - - - - - - - - **2. A. iva**

Cauline leaves and bracts cuneate, usually 3-lobed apically; nutlets transversely rugose dorsally or with oblong reticulation:

Cauline leaves and bracts distinctly 3-lobed; corolla up to 20 mm. long, 9 mm. wide; nutlets closely transverse-rugose - - - - **3. A. chamaepitys** ssp. **palaestina**

Cauline leaves and bracts very shortly and obtusely 3-lobed or subentire; corolla 8–11 mm. long, less than 5 mm. wide; nutlets rather openly transverse-rugose

3. A. chamaepitys ssp. **cypria**

1. **A. orientalis** *L.*, Sp. Plant., ed. 1, 561 (1753); Unger et Kotschy, Die Insel Cypern, 277 (1865); Boiss., Fl. Orient., 4: 800 (1879); Holmboe, Veg. Cypr., 151 (1914); Post, Fl. Pal., ed. 2, 2: 400 (1933); Lindberg f., Iter Cypr., 28 (1946); Osorio-Tafall et Seraphim, List Vasc. Plants Cyprus, 88 (1973); Feinbrun, Fl. Palaest., 3: 99, t. 157 (1978); Davis, Fl. Turkey, 7: 43 (1982).

TYPE: "*in* Oriente".

Spreading or ascending, shortly rhizomatous perennial 9–25 cm. high; stems angled, rather densely and softly lanuginose with crisped, multicellular hairs; basal leaves forming a loose rosette, 6–15 cm. long, 1·5–3·5 cm. wide, oblong, thinly lanuginose-pilose, apex obtuse or rounded, margins entire, subentire, or bluntly undulate-dentate, base tapering to a long, flattish petiole 3–6 cm. or more in length; cauline leaves similar to basal but smaller, more shortly petiolate, and often more conspicuously dentate; inflorescence spiciform, unbranched, terminal, consisting of 6–17 (or more) closely superimposed verticillasters; flowers 4–6 (sometimes more) in each verticillaster, sessile or subsessile, resupinate; bracts foliaceous, the lowermost resembling the cauline leaves, the upper broadly oblong or obovate, 1·5–2 cm. long, 1–1·5 cm. wide, sessile, thinly lanuginose, often tinged purple or violet, margins dentate or shortly lobed; calyx tubular-campanulate, about 7 mm. long, 5 mm. wide, violet-purple, densely lanuginose externally, divided more than half its length into 5 narrow, deltoid, acute teeth about 4 mm. long, 2 mm. wide at base; corolla-tube slender, twisted, slightly curved, pallid, about 10 mm. long, 1·8 mm. wide, subglabrous or thinly pilose externally, glabrous internally except for a shortly bearded zone just below insertion of stamens, limb violet-blue rarely claret-red, abaxial lobe about 3 mm. long, 2 mm. wide, pilose externally, apex emarginate, 4 adaxial lobes oblong, obtuse, about 2 mm. long, 1 mm. wide at base; stamens inserted about half-way down corolla-tube; filaments glabrous, subequal, about 1·2 mm. long; anthers coherent, thecae about 0·5 mm. wide, divergent; ovary about 1 mm. diam., lobes subglabrous or very thinly pilose at apex; disk inconspicuous, style glabrous, about 4 mm. long; stigmatic lobes subequal, blunt, about 0·5 mm. long. Nutlets oblong, about 3 mm. long, 1·5 mm. wide, dark brown, irregularly reticulate, glabrous or with a very few hairs along the margins of the very large attachment area, the latter divided down the middle by a conspicuous longitudinal ridge.

HAB.: On igneous mountainsides, in Pine forest, by roadsides, or sometimes in fallow fields at high altitudes, usually in damp places, or near water; 3,000–5,000 ft. alt.; fl. March–July.

DISTR.: Division 2 only; eastern Mediterranean region from Italy and Sicily to Palestine, and eastwards to Transcaucasia, Iraq and Iran.

2. Troödos above Prodhromos, 1862, *Kotschy* 752! also, 1880–1957, *Sintenis & Rigo* 733! *A. G. & M. E. Lascelles* s.n.! *Holmboe* 862; *Lindberg f.* s.n.! *F. M. Probyn* 115! *G. E. Atherton* 518! *Merton* 2789! 3171! Platania, 1936, *Syngrassides* 1092! and, 1952, *F. M. Probyn* 96! Platres, 1937–1960, *Kennedy* 704! 704B! 704E! *Davis* 3176! *N. Macdonald* 84! Khionistra, 1937, *Kennedy* 704A! 704C! Kryos Potamos, 5,000 ft. alt., 1937, *Kennedy* 704D! Kionia, 1937,

Syngrassides 1622! also, 1941, *Davis* 2695! and, 1967, *Merton* in ARI 503! Agros, 1941,
Davis 3065! Lagoudhera, 1957, *Merton* 3144! Tripylos, swamp E. of Cedar valley, 1962,
Meikle 2671! 2 miles S.W. of Palekhori, 1973, *P. Laukkonen* 435! Phterykoudhi, 1979,
Edmondson & McClintock E 2929!

2. A. iva (*L.*) *Schreber*, Plant. Verticill. Unilab., xxv (1773); Sibth. et Sm., Fl. Graec. Prodr., 1:
389 (1806), Fl. Graec., 6: 22, t. 525 (1826); Boiss., Fl. Orient., 4: 802 (1879); Post, Fl. Pal.,
ed. 2, 2: 401 (1933); Feinbrun, Fl. Palaest., 3: 100, t. 158 (1978); Davis, Fl. Turkey, 7: 47
(1982).

   Teucrium iva L., Sp. Plant., ed. 1, 563 (1753).
   Ajuga pseudo-iva Robill. et Cast. ex DC., Fl. Franç., 6: 395 (1815).
   A. iva (L.) Schreber var. *spatulifolia* [*"spatulaefolia"*] Mutel, Fl. Franç., 3: 55 (1836).
   A. iva (L.) Schreber var. *pseudo-iva* (Robill. et Cast. ex DC.) Steud., Nomencl. Bot., ed. 2,
1: 46 (1840); Benth. in DC., Prodr., 12: 600 (1848); Unger et Kotschy, Die Insel Cypern, 277
(1865).
   A. iva (L.) Schreber ssp. *pseudo-iva* (Robill. et Cast. ex DC.) Holmboe, Veg. Cypr., 151
(1914). Lindberg f., Iter Cypr., 28 (1946); Osorio-Tafall et Seraphim, List Vasc. Plants
Cyprus, 88 (1973).
   TYPE: "*in* Lusitania, G. Narbonensi, Monspelii".

Sprawling perennial 7–20 cm. high, stems radiating from a woody
rootstock, angled, softly white-lanuginose, generally unbranched; leaves
narrowly oblong or almost linear, usually crowded and overlapping, 1–4 cm.
long, 0·3–0·6 cm. wide, strigillose or thinly lanuginose on both surfaces, apex
obtuse or subacute, margins entire or bluntly 1–3-lobulate, often strongly
revolute, base sessile, subamplexicaul; inflorescences elongate, leafy,
consisting of numerous, closely superimposed, 2–4-flowered verticillasters;
flowers sessile or subsessile; bracts foliaceous, indistinguishable from the
cauline leaves; calyx campanulate, about 4 mm. long, 3 mm. wide, densely
lanuginose externally, glabrous internally, teeth 5, erect, subequal, deltoid,
subacute, about 2 mm. long, 1·5 mm. wide at base; corolla yellow in our area,
purple or pink elsewhere, tube narrowly infundibuliform, about 6 mm. long,
3 mm. wide at apex, thinly lanuginose externally, bearded internally about
3 mm. from base, limb pilose externally, abaxial lip about 8 mm. long,
lateral lobes bluntly deltoid, about 1·5 mm. long, 1·5 mm. wide at base,
median lobe about 5 mm. long, 4 mm. wide, apex distinctly emarginate;
stamens exserted, filaments inserted just below throat of corolla-tube,
subequal, thinly pilose, about 4·5 mm. long; anthers reniform, about 0·5
mm. wide, thecae dehiscing by a common, apical slit; ovary glabrous, about
1 mm. diam.; disk very small; style about 9 mm. long, glabrous; stigmatic
lobes subequal, blunt, flabelliform, about 0·5 mm. long. Nutlets oblong,
about 3 mm. long, 1·8 mm. wide, dark fuscous-brown, alveolate or foveolate
dorsally with rounded pits, the lateral sculpturing more rectangular;
attachment area large, ovate-oblong, with a well-developed inner peripheral
ridge.

   HAB.: Waste ground, roadsides, open areas in garigue; sea-level to ? 2,500 ft. alt.; fl.
April–August.

   DISTR.: Divisions [2]3–8. Throughout the Mediterranean region.

2. "In Prodromo [Prodhromos]", *Kotschy*. See Notes.
3. Akrotiri Forest, 1939, *Mavromoustakis* s.n. ! Mile 20, Nicosia–Limassol road, 1950, *Chapman*
552! Amathus, 1978, *J. Holub* s.n. !
4. Near Larnaca, 1905, *Holmboe* 257; also, 1939, *Lindberg f.* s.n.; near Famagusta, 1939,
Lindberg f. s.n. !
5. Between Athalassa and Nicosia, 1939, *Lindberg f.* s.n.; Nicosia–Famagusta road, 1972, *W.
R. Price* 1000 !
6. Nicosia, 1950, *Chapman* 493 (see Notes) ! Nicosia, Kykko Camp, 1973, *P. Laukkonen* 104.
7. Ayios Khrysostomos, 1865, *Kotschy*; and, 1880, *Sintenis & Rigo* 122 ! Pentadaktylos, 1939,
Lindberg f. s.n.; Halevga, 1939, *Lindberg f.* s.n.; Myrtou, 1955, *G. E. Atherton* 408 ! Tjiklos,
1973, *P. Laukkonen* 75 !
8. Near Galinoporni, 1880, *Sintenis & Rigo*.

NOTES: Kotschy's record from Prodhromos is very questionable; neither *A. iva* nor *A. chamaepitys* is normally to be found at such an altitude on Troödos.

Chapman 493 from Nicosia is exceptionally slender and etiolate, with small leaves, scarcely 1 cm. long; flowers and fruits are, however, those of *A. iva*.

3. A. chamaepitys (*L.*) *Schreber*, Plant. Verticill. Unilab., xxiv (1773); Boiss., Fl. Orient., 4: 802 (1879); Post, Fl. Pal., ed. 2, 2: 401 (1933); Turrill in New Phytologist, 33: 218–230 (1934); Davis in Notes Roy. Bot. Gard. Edinb., 38: 24 (1980), Fl. Turkey, 7: 47 (1982).

Teucrium chamaepitys L., Sp. Plant., ed. 1, 562 (1753).

Sprawling annual, biennial or perennial herb, rootstock often woody; stems generally unbranched or sparingly branched except at the base, angled, subglabrous or ± densely villose or lanuginose, sometimes purple; basal leaves oblanceolate or narrowly spathulate, 2–7 cm. long, 0·3–1 cm. wide, subglabrous, hispidulous or lanuginose, apex acute, obtuse or rounded, margins entire, subentire or variously lobed, base tapering to an indistinct petiole; cauline leaves very variable, cuneate, typically divided almost to base into 3 narrow linear or oblanceolate, obtuse lobes, commonly 20 mm. long and less than 2 mm. wide, but lobes often much shorter and broader, or sometimes obsolescent, not infrequently irregular and extending down both margins of the leaf; inflorescence extending along the greater part of the stem, consisting of numerous usually rather closely superimposed 2-flowered verticillasters; flowers sessile; bracts foliaceous, as variable as the cauline leaves, but generally ± 3-lobed, glabrous, hispidulous or loosely lanuginose; calyx campanulate, 4–6 mm. long, 2·5–3·5 mm. wide, pilose externally, indistinctly nerved, divided apically into 5 subequal, obtuse or subacute, erect teeth 2–2·5 mm. long, 0·5–1·5 mm. wide at base; corolla generally yellow (often drying pink), the lip spotted or streaked red, tube narrowly infundibuliform, distinctly constricted at the middle, 5–8 mm. long, 2·5–3·5 mm. wide at apex, thinly pilose externally with a bearded zone internally just below insertion of stamens, abaxial lip 7–14 mm. long, with 2 parallel tomentose stripes running from the base of the median lobe to the mouth of the tube, lateral lobes oblong, rounded, 0·8–2 mm. long, 0·5–1·5 mm. wide at base, median lobe 3·5–8 mm. long, 3–9 mm. wide, flabelliform, distinctly emarginate, adaxial lip minute or almost obsolete, 2-lobed; stamens exserted; filaments 3–8 mm. long, the abaxial distinctly longer than the adaxial, pilose at least in the upper half; anthers ovate-reniform, 0·5–0·8 mm. long, 0·3–0·5 mm. wide, with a tongue-like, basal flap; ovary glabrous 0·8–1·2 mm. diam.; disk minute; style 7–12 mm. long, glabrous or pilose at apex; stigmatic lobes 0·5–1 mm. long, the adaxial usually longer, narrower and more acute than the abaxial. Nutlets oblong, 2·5–3·5 mm. long, 1·2–1·8 mm. wide, dark brown, loosely rectangular-reticulate or closely transversely rugose; attachment area oblong, 1·5–2 mm. long, 0·5–0·8 mm. wide with a marginal ridge or narrow wing and a longitudinal median ridge.

ssp. **palaestina** (*Boiss.*) *Bornm.* in Beih. Bot. Centralbl., 31: 254 (1914); Davis in Notes Roy. Bot. Gard. Edinb., 38: 27 (1980), Fl. Turkey, 7: 50 (1982).

    A. palaestina Boiss., Diagn., 1, 12: 92 (1853).

    [*A. tridactylites* (non Gingins ex Benth.) Kotschy in Unger et Kotschy, Die Insel Cypern, 277 (1865); Sintenis in Oesterr. Bot. Zeitschr., 35: 393 (1881); 32: 52, 193 (1882) pro parte; Post, Fl. Pal., ed. 2, 2: 401 (1933) pro parte.]

    [*A. chia* Schreber var. *tridactylites* (non *A. tridactylites* Gingins ex Benth.) Boiss., Fl. Orient., 4: 803 (1879) pro parte quoad plant. cypr.]

    [*A. chia* (non Schreber) H. Stuart Thompson in Journ. Bot., 44: 337 (1906).]

    [*A. chia* Schreber ssp. *tridactylites* (non *A. tridactylites* Gingins ex Benth.) Holmboe, Veg. Cypr., 151 (1914).]

    [*A. chamaepitys* (L.) Schreber ssp. *chia* (non *A. chia* Schreber) Osorio-Tafall et Seraphim, List Vasc. Plants Cyprus, 88 (1973).]

TYPE: "in collibus circâ *Hierosolymam* et *Damascum*", [1846] *Boissier* (G).

Perennial or perennating herb, usually with a woody rootstock; stems sprawling or prostrate, subglabrous or thinly lanuginose; cauline leaves and bracts generally less than 2 cm. long, 3-lobed, the lobes rarely more than half the length of the leaf or bract, oblong, blunt, subequal or the median much longer than the lateral; inflorescence usually rather dense, or becoming lax with age; corolla yellow with crimson spots on the lip, up to 20 mm. long, 9 mm. wide. Nutlets closely transverse-rugose.

HAB.: Open, rocky hillsides; roadsides, fallow fields; sea-level to 3,000 ft. alt.; fl. Jan.–May.

DISTR.: Divisions 1–4, 6, 7. W. & S. Turkey, E. Aegean Islands, Lebanon and Palestine.

1. Androlikou, 1957, *Merton* 3030! Avgas near Ayios Yeoryios, 1962, *Meikle* 2005!
2. Near Omodhos, 1960, *N. Macdonald* 36!
3. Alethriko, 1905, *Holmboe* 218; Trimiklini, 1941, *Davis* 3126! Amathus, 1964, *J. B. Suart* 155!
4. Ayia Napa, 1905, *Holmboe* 41!
6. Ayia Irini, 1941, *Davis* 2570!
7. Rather common; Ayios Khrysostomos Monastery, 1859, *Kotschy* 426; Kantara, 1880, *Sintenis & Rigo* 123 partly! Pentadaktylos, 1901, *A. G. & M. E. Lascelles* s.n.! Kyrenia, 1932, *Syngrassides* 682! also, 1937, *Mrs Stagg* s.n.! 1955, *Miss Mapple* 100! 1956, *G. E. Atherton* 1084! Myrtou, 1932, *Syngrassides* 34! Larnaka tis Lapithou, 1936, *Syngrassides* 915! St. Hilarion, 1937, *H. Roger-Smith* s.n.! also, 1937, *Miss Godman* 30! 1941, *Davis* 2520! 1949, *Casey* 362! 1952, *F. M. Probyn* 46! Above Sisklipos, 1941, *Davis* 2110! Kornos, 1941, *Davis* 2197! Kantara, 1941, *Davis* 2437! and, 1955, *J. Hughes* s.n.! Buffavento, 1941, *Davis* 2184! Ayios Epiktitos, 1968, *Economides* in ARI 1078!

ssp. **cypria** *P. H. Davis* in Notes Roy. Bot. Gard. Edinb., 38: 30 (1980); Fl. Turkey, 7: 51 (1982). *A. tridactylites* Gingins ex Benth. var. *integrifolia* Sintenis in Oesterr. Bot. Zeitschr., 32: 52 (1882) nomen.

TYPE: Cyprus; Near Dhiorios, 1962, *Meikle* 2345 (K!).

Biennial or perennial from a woody taproot; stems 3–8 (–16) cm. long, prostrate or ascending, softly white-lanuginose; cauline leaves and bracts narrowly oblong-cuneate, 8–14 (–18) mm. long, usually less than 5 mm. wide, apex subentire or very shortly and obtusely 3 (–5) -dentate or lobulate; corolla small, yellow, 8–11 mm. long. Nutlets reticulate-foveolate at either extremity, closely or rather openly transverse-rugose dorsally.

HAB.: Rocky calcareous ground; sea-level to 3,000 ft. alt.; fl. Jan.–June.

DISTR.: Divisions 4–7. Also S. Turkey.

4. Cape Greco, 1958, *N. Macdonald* 82!
5. Athalassa, 1936, *Syngrassides* 1167! 1328!
6. Near Dhiorios, 1962, *Meikle* 2345!
7. Pentadaktylos, 1880, *Sintenis & Rigo* 123 partly! Kantara, 1880, *Sintenis & Rigo* 123 partly! Larnaka tis Lapithou, 1941, *Davis* 2094! Buffavento, 1941, *Davis* 2186! 2798! Yaïla, 1941, *Davis* 2865!

NOTES: Superficially resembles *A. iva* (L.) Schreber more than *A. chamaepitys* (L.) Schreber, but connected to the latter by a range of subspecies. The nutlets are rugose with transverse ridges dorsally, not foveolate with rounded pits as in *A. iva*. *A. chamaepitys* ssp. *cypria* and *A. chamaepitys* ssp. *palaestina* sometimes grow in mixed populations in Cyprus, but the former has, on the whole, a much more limited distribution than the latter.

72. PLANTAGINACEAE

R. Pilger in Engl., Pflanzenr., 102 (IV. 269): 1–466 (1937).

Annual or perennial herbs, or sometimes woody subshrubs; leaves usually in basal rosettes, alternate or sometimes opposite, exstipulate; flowers in stalked, axillary heads or spikes, actinomorphic, mostly hermaphrodite; sepals 4, generally imbricate, 1-nerved, persistent; corolla scarious, with 4 spreading or reflexed lobes, persistent; stamens 4 (rarely fewer), alternating with corolla-lobes; filaments slender, anthers usually long-exserted, versatile, introrse, 2-thecous, dehiscing longitudinally; ovary superior, usually 2-locular, ovules 1–many in each loculus, axile or basal; style 1; stigma simple, elongate.

Fruit a circumscissile capsule, or rarely an indehiscent, 1-seeded nutlet; seeds peltately attached; testa often mucilaginous; endosperm fleshy; embryo straight or rarely curved.

One large cosmopolitan genus (*Plantago*) and two very small ones (*Littorella, Bougueria*). *Plantago* alone occurs in Cyprus.

1. PLANTAGO *L.*

Sp. Plant., ed. 1, 112 (1753).
Gen. Plant., ed. 5, 52 (1754).

Fruit a circumscissile, 2–many-seeded capsule. Flowers usually hermaphrodite and anemophilous. Characters otherwise those of the family.

About 250 species with a cosmopolitan distribution.

Stem developed; leaves not in a basal rosette:
  Leaves alternate:
    Leaves sparsely pilose or glabrescent; plant annual   -   -   -   **4. P. amplexicaulis**
    Leaves silky-hairy, silvery; plant perennial with a tough, woody rootstock **7. P. albicans**
  Leaves opposite:
    Lowermost bracts similar to upper; whole plant viscid-glandular; leaves not fleshy
                                                                    14. P. afra
    Lowermost bracts foliaceous, much larger and more conspicuous than the upper; whole plant not viscid-glandular; leaves fleshy -   -   -   -   -   **13. P. squarrosa**
Stem not developed, or very short; leaves in a basal rosette:
  Bracts and sepals conspicuously hairy or lanuginose:
    Scapes strongly ribbed, usually much longer than leaves; inflorescences conspicuously silky-hairy   -   -   -   -   -   -   -   -   -   -   **6. P. lagopus**
    Scapes obscurely ribbed or terete, often shorter than leaves, or not much exceeding them; inflorescences hairy or woolly, but not conspicuously silky:
      Bracts and corolla-lobes broadly and bluntly ovate-orbicular; leaves tapering to a slender apex, commonly with a few, conspicuous, upward-pointing marginal teeth
                                                                    10. P. notata
      Bracts and corolla-lobes ovate-acute or ovate-acuminate; leaves entire or subentire, acute but not long-acuminate:
        Corolla-lobes broadly ovate, shortly acuminate; inflorescence subglobose or semi-globose, woolly; scapes thickening and becoming rigidly revolute in fruit
                                                                    11. P. cretica
        Corolla-lobes ovate-acuminate, acumen often long and slender; inflorescences ovoid or cylindrical, hairy but scarcely woolly; scapes not much thickened nor rigidly revolute in fruit -   -   -   -   -   -   -   -   -   **12. P. bellardii**
  Bracts and sepals sometimes ciliate, but not conspicuously hairy or lanuginose:
    Scapes conspicuously ribbed; bracts usually long-acuminate; abaxial sepals connate almost to apex   -   -   -   -   -   -   -   -   -   **5. P. lanceolata**
    Scapes terete or obscurely striate; bracts not long-acuminate; abaxial sepals free:
      Inflorescence a subglobose or shortly oblong spike:
        Corolla-lobes broadly ovate or almost orbicular, about 2 mm. wide; median nerve ending in apex of bract -   -   -   -   -   -   -   -   **8. P. ovata**

Corolla-lobes ovate-acute, about 0·8 mm. wide; median nerve of bract shortly
    excurrent - - - - - - - - - - **9. P. loeflingii**
Inflorescence a narrow, cylindrical or caudate spike:
    Leaves broadly ovate or elliptical - - - - - - - **1. P. major**
    Leaves not as above:
        Leaves elongate, narrowly linear, entire or subentire, fleshy; bracts and sepals not
            conspicuously ciliate - - - - - **3. P. maritima** ssp. **crassifolia**
        Leaves usually pinnatisect or lobed, not fleshy; bracts and sepals usually
            conspicuously ciliate - - - - **2. P. coronopus** ssp. **commutata**

SUBGENUS 1. **Plantago** Leaves alternate, usually forming basal rosettes.

1. P. major *L.*, Sp. Plant., ed. 1, 112 (1753); Boiss., Fl. Orient., 4: 878 (1879); H. Stuart
Thompson in Journ. Bot., 44: 337 (1906); Holmboe, Veg. Cypr., 169 (1914); Post, Fl. Pal.,
ed. 2, 2: 417 (1933); Pilger in Engl., Pflanzenr., 102 (IV. 269): 41 (1937); Lindberg f., Iter
Cypr., 32 (1946); Osorio-Tafall et Seraphim, List Vasc. Plants Cyprus, 96 (1973); Davis, Fl.
Turkey, 7: 507 (1982).
TYPE: *"in* Europa *ad vias"*.

Acaulescent perennial 8–50 cm. high; leaves ovate or broadly elliptic,
3–15 (–20) cm. long, 2·5–10 (–15) cm. wide, glabrous or ± pubescent with
white multicellular hairs, longitudinally 5–7 (–9)-nerved, apex obtuse or
acute, margins entire or remotely and irregularly undulate-dentate, base
tapering abruptly or gradually to a flattened petiole 1–12 cm. long; scape
equalling or exceeding leaves, terete or obscurely striate, glabrous or thinly
pubescent; flowers in a narrow, caudate spike 4–30 cm. long, 0·3–0·7 cm.
wide, the lowermost sometimes remote, the upper crowded; bracts ovate,
1·5–3 mm. long, 1·5–2·5 mm. wide, keeled, with a thick median nerve, and
membranous, glabrous or ciliate margins; sepals concave, ovate-oblong,
obtuse or subacute, about 2 mm. long, 1·5 mm. wide, with broadly
membranous margins; corolla-tube about 1·5 mm. long, lobes ovate, acute,
strongly reflexed, about 0·8 mm. long, 0·5 mm. wide; filaments about 2 mm.
long; anthers oblong, yellow, about 1 mm. long, 0·8 mm. wide, with a
conspicuous apiculus; ovary glabrous, broadly ovoid, about 0·8 mm. long
and almost as wide; style about 0·5 mm. long; stigma filiform, up to 2 mm.
long. Capsule ovoid, about 3 mm. long, 2·5 mm. wide; seeds 8–10, oblong,
somewhat flattened, about 1·2 mm. long, 0·8 mm. wide; testa dark brown,
cerebriform-rugose.

HAB.: Moist situations by roadsides, field-borders, springs or irrigation channels; sea-level to
4,500 ft. alt.; fl. March–Oct.

DISTR.: Divisions 2–4, 6, 7. Cosmopolitan.

2. Platres, 1938, *Kennedy* 1425! Kykko, 1939, *Lindberg f.* s.n.! Kakopetria, 1955, *N. Chiotellis*
448! Prodhromos, 1955, *G. E. Atherton* 569! 578! Kalopanayiotis, 1968, *A. Genneou* in ARI
1281!
3. Limassol, ? 1905, *Michaelides* teste *Holmboe*.
4. Larnaca, 1902, *A. G. & M. E. Lascelles* s.n.; K
ouklia, 1905, *Holmboe* 388.
6. Syrianokhori, 1936, *Syngrassides* 1248!
7. Lapithos, 1880, *Sintenis & Rigo* 621! also, 1955, *Miss Mapple* 91! and, 1966, *Merton* in ARI
30! Near Bellapais, 1948, *Casey* 112! Ayios Epiktitos, 1955, *G. E. Atherton* 234! Kyrenia,
Tchingen, 1955, *G. E. Atherton* 316!

2. P. coronopus *L.*, Sp. Plant., ed. 1, 115 (1753); Unger et Kotschy, Die Insel Cypern, 228
(1865); Boiss., Fl. Orient., 4: 888 (1879); Holmboe, Veg. Cypr., 170 (1914); Post, Fl. Pal.,
ed. 2, 2: 421 (1933); Pilger in Engl., Pflanzenr., 102 (IV. 269): 126 (1937); Lindberg f., Iter
Cypr., 32 (1946); Osorio-Tafall et Seraphim, List Vasc. Plants Cyprus, 96 (1973); Davis, Fl.
Turkey, 7: 508 (1982).

Acaulescent annual (or biennial) herb usually with a single rosette; leaves
very variable (1–) 4–8 (–15) cm. long, (0·3–) 0·5–1·5 (–3) cm. wide,
oblanceolate in outline, ± densely pilose or subglabrous, usually deeply 1-
or 2-pinnatisect with numerous narrow, acute lobes, sometimes (especially

in depauperate specimens) subentire or sparingly lobed, base tapering to a flattened, winged petiole up to 5 cm. long; scapes usually numerous, shorter than, or exceeding leaves, terete or obscurely sulcate, generally clothed with numerous adpressed hairs; flowers in a narrow, crowded, caudate spike, 4–10 cm. long, 0·3–0·7 cm. wide, or inflorescence sometimes shorter, oblong or even ovoid; bracts ovate, 2–2·5 mm. long, 1–1·5 mm. wide, convex, with a thick median nerve, apex acute or cuspidate, margins membranous, hyaline, usually ciliate; sepals in unequal pairs, the abaxial concave, 2–2·5 mm. long, 1·5–2 mm. wide, acute or subacute with broad membranous, ciliate margins, but without a dorsal wing, the adaxial boat-shaped, rather broader than the abaxial, with a narrow or broad, ciliate, dorsal wing; corolla-tube ± pilose, about 2 mm. long, lobes ovate, acute, spreading, about 1 mm. long, 0·8 mm. wide; filaments about 0·5 mm. long, glabrous; anthers narrowly oblong, yellow, about 0·8 mm. long, 0·4 mm. wide, base subsagittate, apex with a distinct membranous appendage; ovary glabrous or pilose near apex, ovoid, about 1·5 mm. long, 0·8 mm. wide; style very short; stigma filiform, about as wide as style, tapering slightly upwards, about 0·8–1·5 mm. long, or exceptionally much longer. Capsule ovoid, about 2 mm. long, 1·8 mm. wide; seeds usually 2, oblong-elliptic, rather compressed, about 1–1·5 mm. long, 0·8 mm. wide; testa dull brown, smooth or minutely whitish-pustulate and mucilaginous when wet.

ssp. **commutata** (*Guss.*) *Pilger* in Feddes Repert., 28: 287 (1930), Engl., Pflanzenr., 102 (IV. 269): 142, fig. 19 (1937); Rechinger f., Fl. Iran., n. 15: 6 (1965); Runemark in Bot. Notiser, 120: 10 (1967); Davis, Fl. Turkey, 7: 508 (1982).

    P. commutata Guss., Suppl. Fl. Sic. Prodr., 46 (1832).

    P. coronopus L. var. *simplex* Decne. in DC., Prodr., 13 (1): 732 (1852); Unger et Kotschy, Die Insel Cypern, 229 (1865).

    TYPE: "In arenosis maritimis; *Ustica, Lampedusa*" (? NAP).

Inflorescence dense, the flowers closely imbricate; adaxial sepals strongly ciliate-winged dorsally; scapes often becoming conspicuously thickened in fruiting specimens; flowers and seeds often larger than in ssp. *coronopus*.

HAB.: Dry hillsides, rocky banks; fallow fields; roadsides, waste land; seashores, salt-marshes; sea-level to 4,000 ft. alt.; usually lowland; fl. Febr.–Oct.

DISTR.: Divisions 1–8, locally abundant. Widespread in the Mediterranean region, especially in the eastern Mediterranean, and eastwards to Iran and Afghanistan.

1. Ayios Yeoryios (Akamas), 1962, *Meikle*! Polis, 1962, *Meikle*!
2. Evrykhou, 1935, *Syngrassides* 821! Kannaviou, 1939, *Lindberg f.* s.n.; Troödhitissa Monastery, 1939, *Lindberg f.* s.n.
3. Cape Gata, 1862, *Kotschy* 603a; Akrotiri, 1905, *Holmboe* 685; 2 miles S. of Pyrgos, 1963, *J. B. Suart* 87! Khalassa, 1970, *A. Genneou* in ARI 1497!
4. Larnaca, 1862, *Kotschy* 324; also, 1881, *C. & W. Barbey* s.n.; Cape Pyla, 1880, *Sintenis & Rigo*; Ayia Napa, 1905, *Holmboe* 67; Larnaca aerodrome, 1950, *Chapman* 18!
5. Lefkoniko, 1880, *Sintenis & Rigo* 66! Laxia, 1932, *Syngrassides* 78! Mia Milea–Mandres, 1933, *Syngrassides* 327! Between Nisou and Stavrovouni, 1933, *Syngrassides* 29! Piyi, 1936, *Syngrassides* 938! Kythrea, 1941, *Davis* 2949! Mia Milea, 1967, *Merton* in ARI 178!
6. Strovolos, 1902, *A. G. & M. E. Lascelles* s.n.; Nicosia, 1905, *Holmboe* 300b; also, 1973, *P. Laukkonen* 267! Kokkini Trimithia, 1935, *Syngrassides* 783!
7. Kyrenia, 1939, *Lindberg f.* s.n.! also, 1949, *Casey* 384! 621! 622! Near Thermia, 1948, *Casey* 84! Lakovounara Forest, 1950, *Chapman* 65! 139! Karakoumi, 1956, *G. E. Atherton* 1230! Lambousa Beach, 1973, *P. Laukkonen* 277!
8. Near Leonarisso, 1880, *Sintenis & Rigo*; Boghaz, 1941, *Davis* 2409! Cape Andreas, 1948, *Mavromoustakis* s.n.! Akradhes, 1962, *Meikle*!

NOTES: Very variable in size and shape of leaves, and in inflorescence-length, but uniformly distinct from western European *P. coronopus* in its close, smooth, funiform fruiting spikes and broadly winged adaxial sepals; it is, however, connected to the type by a series of intermediates occurring in the western Mediterranean, and in southern Europe.

3. P. maritima L., Sp. Plant., ed. 1, 114 (1753); Unger et Kotschy, Die Insel Cypern, 229 (1865); Boiss., Fl. Orient., 4: 889 (1879); Pilger in Engl., Pflanzenr., 102 (IV. 269): 169 (1937).

Acaulescent perennial herb with a tough, branched rootstock, and usually several rosettes; apex of rootstocks often lanuginose or hairy; leaves linear, (5–) 10–20 (–30) cm. long, 0·2–1 cm. wide, glabrous, rather fleshy, entire or remotely toothed, acute or obtuse; petiole indistinct or wanting; scapes few or numerous, erect, usually exceeding the leaves, terete or almost so, generally clothed with numerous, adpressed, multicellular hairs; flowers in a narrow, lax or crowded, caudate spike, (2–) 4–8 (–10) cm. long, 0·4–0·5 cm. wide; bracts ovate, concave or obscurely keeled with an incrassate nerve, 2–3 mm. long, 1·5–2 mm. wide, apex acute or obtuse, margins membranous, glabrous or sparsely ciliate; sepals in unequal pairs, the abaxial concave, 2–2·5 mm. long, 1·5 mm. wide, obtuse, with a wide membranous or ciliate margin, but without a dorsal wing, the adaxial boat-shaped, distinctly broader than the abaxial, with a narrow or broad, ciliate, dorsal wing; corolla-tube ± pilose, about 2 mm. long, lobes ovate, acute, spreading or reflexed, about 1·5 mm. long, 1 mm. wide; filaments very short, or up to 3 mm. long, glabrous; anthers yellow, oblong, about 1·5 mm. long, 0·8–1 mm. wide, base subsagittate, apex with a distinct membranous appendage; ovary glabrous, ovoid, about 1 mm. long, 0·8 mm. wide; style very short; stigma filiform, about as wide as style, tapering slightly upwards, about 3–6 mm. long. Capsule ovoid, 3–4 mm. long, 2 mm. wide; seeds usually 2, oblong-elliptic, convex dorsally, 1·5–2·5 mm. long, 0·8–1·2 mm. wide; testa dull brown, smooth, mucilaginous when wet.

ssp. crassifolia (*Forssk.*) [*Batt. et Trabut*, Fl. de l'Algerie, 743 (1890) absque gradu tax.] *Holmboe*, Veg. Cypr., 170 (1914); Lindberg f., Iter Cypr., 32 (1946).
   P. crassifolia Forssk., Fl. Aegypt.-Arab., 31 (1775); Post, Fl. Pal., ed. 2, 2: 421 (1933); Pilger in Engl., Pflanzenr., 102 (IV. 269): 160, fig. 21 (1937); Osorio-Tafall et Seraphim, List Vasc. Plants Cyprus, 96 (1973); Davis, Fl. Turkey, 7: 509 (1982).
   [*P. maritima* (non L.) Sibth. et Sm., Fl. Graec. Prodr., 1: 101 (1806), Fl. Graec., 2: 37, t. 148 (1813); Hume in Walpole, Mem. Europ. Asiatic Turkey, ed. 1, 254 (1817); Poech, Enum. Plant. Ins. Cypr., 15 (1842); Nattrass, First List Cyprus Fungi, 45 (1937).]
   TYPE: Egypt; "*Alexandriae*", *Forsskål* (C).

Apex of rootstock generally densely lanuginose; inflorescence usually dense, flowers often becoming imbricate after anthesis; bracts short and widely ovate, generally about half as long as the sepals; adaxial sepals broadly ciliate-winged dorsally. Seed usually about 1·5 mm. long, considerably smaller than in *P. maritima* L. ssp. *maritima*.

   HAB.: Salt-marshes and sandy fields near the sea, or on saline ground inland; sea-level to c. 500 ft. alt.; fl. March–Oct.

   DISTR.: Divisions 3, 4, 6. Widespread in the Mediterranean region.
   3. Akrotiri Bay, 1942, *Mavromoustakis* s.n. !
   4. Larnaca Salt Lake and nearby, 1862, *Kotschy* 310a; also, same area, *Hume*; *Sintenis & Rigo* 706! *A. G. & M. E. Lascelles*! *Holmboe* 106; *Chapman* 415! 439! *Nattrass* 476! *Merton* in ARI 371 (starved form)! etc.
   6. Makhedonitissa Monastery, 1936, *Syngrassides* 1095! Syrianokhori, 1941, *Davis* 2610!

   NOTES: Usually regarded as a distinct species, but exactly analogous to *Plantago coronopus* L. ssp. *commutata* (Guss.) Pilger, and, like the latter, so closely allied to the western European species, that subspecies seems the more appropriate rank.

4. P. amplexicaulis *Cav.*, Icones, 2: 22, t. 125 (1793); Boiss., Fl. Orient., 4: 883 (1879), Suppl., 366 (1888); Sintenis in Oesterr. Bot. Zeitschr., 31: 326, 393 (1881); 32: 259 (1882); Holmboe, Veg. Cypr., 169 (1914); Post, Fl. Pal., ed. 2, 2: 418 (1933); Pilger in Engl., Pflanzenr., 102 (IV. 269): 310, fig. 31 (1937); Rechinger f. in Arkiv för Bot., ser. 2, 1: 431 (1950); Osorio-Tafall et Seraphim, List Vasc. Plants Cyprus, 97 (1973).
   TYPE: Spain; "in Saguntinae [Sagunto] arcis vetustissimo muro" *Cavanilles* (MA).

Erect or decumbent annual; stems very short, or elongate, up to 30 cm. long, branched at the base or unbranched, striatulate, glabrous or thinly pilose; leaves alternate, oblanceolate or narrowly elliptic, 2·5–10 cm. long, 0·5–1·5 cm. wide, indistinctly 3–5-nerved, sparsely long-pilose, apex acuminate, base tapering gradually to a conspicuous, amplexicaul sheath, margins entire or subentire; scapes axillary, usually numerous, much exceeding the leaves, up to 27 cm. long, glabrous, striatulate; flowers in a dense ovoid or oblong spike, 1–3·5 cm. long, 0·8–1 cm. wide; bracts broadly ovate or oblate-orbicular, 3–5 (–6) mm. long, 4–6 (–7) mm. wide, shortly acute or mucronate, glabrous, with an incrassate nerve, and wide membranous margins; sepals in very unequal pairs, the abaxial concave, broadly ovate, about 3–3·5 mm. long, 2·5 mm. wide, almost wholly hyaline-membranous, glabrous, the adaxial boat-shaped, about 4–4·5 mm. long, 3–3·5 mm. wide, with a conspicuous nerve and a narrow, long-pilose dorsal wing; corolla-tube very short, about 1·5 mm. long, glabrous, lobes ovate, acuminate, spreading, about 3–3·5 mm. long, 2–2·5 mm. wide; filaments short, about 2·5 mm. long, glabrous; anthers yellow, ovate-oblong, about 1·5 mm. long, 1·2 mm. wide, apex with a very small appendage, base subsagittate; ovary broadly ovoid, glabrous, about 1–1·5 mm. long and almost as wide; style very short; stigma filiform, about as wide as style, tapering slightly upwards, about 4–5 mm. long, Capsule ovoid, about 5 mm. long, 3 mm. wide; seeds usually 2, oblong, about 5 mm. long, 2·5 mm. wide, convex dorsally, concave ventrally with an elongate, linear hilum; testa pale or mid-brown, shining, minutely rugulose, mucilaginous when wet. *Plate 79, figs. 7–11.*

HAB.: Rocky slopes on limestone or sandstone; roadsides; fallow fields, usually on the Kythrean formation; sea-level to 1,600 ft. alt.; fl. Febr.–May.

DISTR.: Divisions 5, 7, 8. Spain, North Africa and eastern Mediterranean from Greece and Crete eastwards to Pakistan (var. *bauphula* (Edgew.) Pilger); Atlantic Islands; Socotra.

5. Mia Milea, 1952, *Merton* s.n.!
7. "In montibus pr. Kythream", 1880, *Sintenis & Rigo* 67! (Pentadaktylos sec. *Pilger*); Ayios Amvrosios, 1932, *Syngrassides* 448! Ayios Dhimitrianos and Lakovounara, 1950, *Chapman* 114! 507! N.W. of Bellapais, 1952, *Casey* 1268! Above Kyrenia, 1953, *Casey* 1316!
8. Rizokarpaso, 1912, *Haradjian* 208! Between Apostolos Andreas and Khelones, 1941, *Davis* 2285! Mutsefla, Khalosta Forest, 1962, *Meikle* 2456!

NOTES: All the Cyprus material has adaxial sepals with a pilose dorsal wing, as in typical *P. amplexicaulis*, which grows in S. Spain, Greece, Crete and Aegean Islands, North Africa (Morocco to Tunis), Canary Islands. This seems to reach its eastern limit in Cyprus, to be replaced eastwards by the glabrous-sepalled var. *bauphula* (Edgew.) Pilger.

5. P. lanceolata *L.*, Sp. Plant., ed. 1, 113 (1753); Boiss., Fl. Orient., 4: 881 (1879); H. Stuart Thompson in Journ. Bot., 44: 337 (1906); Holmboe, Veg. Cypr., 169 (1914); Post, Fl. Pal., ed. 2, 2: 417 (1933); Pilger in Engl., Pflanzenr., 102 (IV. 269): 313 (1937); Lindberg f., Iter Cypr., 32 (1946); Osorio-Tafall et Seraphim, List Vasc. Plants Cyprus, 96 (1973); Davis, Fl. Turkey, 7: 513 (1982).

Acaulescent perennial with a thick, tough rootstock, often densely lanuginose at the apex, and usually divided above into several rosettes; leaves lanceolate or elliptic, 2–30 cm. long, 0·5–5 cm. wide, glabrous or ± densely hirsute, prominently 3–5 (–7)-nerved, apex acute or acuminate, base tapering to a flattened or canaliculate petiole 0·5–12 cm. long, margins entire or remotely serrulate; scapes strongly ribbed or angled, 10–30 (–60) cm. long, adpressed-pilose; flowers in a dense ovoid, oblong, cylindrical or bluntly conical spike 0·7–4 (–5) cm. long, lengthening to 7 cm. or more (exceptionally to 12 cm.) in fruit; bracts obovate, 3–5 mm. long, 1·5–3 mm. wide, membranous, acuminate or with a long-attenuate cusp, pale brown or fuscous towards apex, glabrous, faintly nerved; sepals membranous, the abaxial connate almost to apex, 2·5–3 mm. long, 2–2·5 mm. wide (jointly), glabrous or thinly pilose dorsally towards apex, stramineous or fuscous above, adaxial sepals deeply concave or boat-shaped, about 3 mm. long, 2

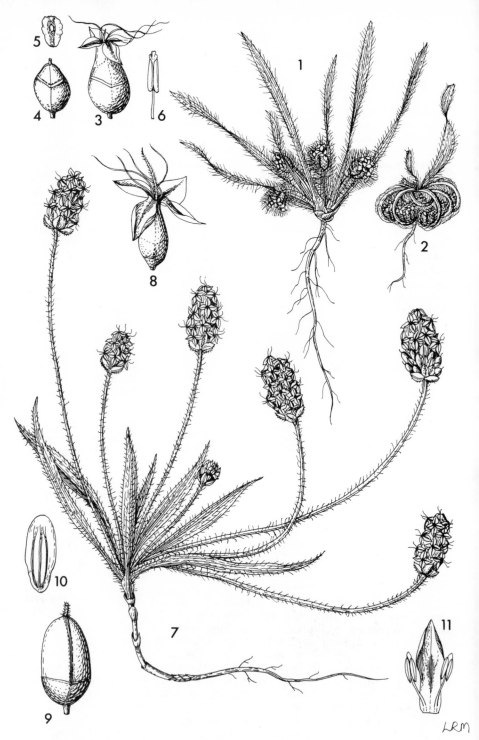

Plate 79. Figs. **1–6.** PLANTAGO CRETICA L. **1,** habit, ×⅔; **2,** fruiting plant, habit, ×⅔; **3,** flower, ×4; **4,** fruit, ×4; **5,** seed, ×4; **6,** stamen, ×4; figs. **7–11.** PLANTAGO AMPLEXICAULIS Cav. **7,** habit, ×⅔; **8,** flower, ×4; **9,** fruit, ×4; **10,** seed, ×4; **11,** stamens and part of corolla, ×4. (**1, 6** from *Casey* 391; **2, 4, 5** from *Casey* 782; **3** from *G. E. Atherton* 1390; **7, 9, 10** from *Chapman* 114; **8, 11** from *Davis* 2285.)

mm. wide, distinctly nerved, the nerve sometimes narrowly winged dorsally, apex acute or obtuse, sometimes thinly pilose, pale or fuscescent; corolla-tube 1·5–3 mm. long, often shorter than the sepals, glabrous, lobes ovate, acuminate, spreading or reflexed, 1·5–2 mm. long, about 1–1·5 mm. wide, with an obscure or distinct and fuscescent nerve; filaments slender, up to 5 mm. long, glabrous; anthers yellowish, oblong, about 2·5 mm. long, 0·8 mm. wide, apex with a small acute appendage, base very shortly lobed; ovary ovoid, almost 0·5 mm. long and almost as wide, glabrous; style about 2 mm. long; stigma filiform, 2–3 mm. long, about as wide as style at base, tapering gradually to apex. Capsule subglobose to ovoid, 2–2·5 mm. long, about 2 mm. wide; seeds 2, oblong, 2–2·5 mm. long, 1 mm. wide, convex dorsally, concave ventrally with a short, oblong hilum; testa bright or dark shining brown, minutely rugulose, mucilaginous when wet, the testa becoming papillose with projecting cell-walls.

var. **lanceolata**
TYPE: "*in* Europae *campis sterilibus*".

Plants not very robust; scapes usually less than 30 cm. long; leaves generally with short petioles, or with the lamina gradually tapering to the base; lamina lanceolate, usually 0·5–3 cm. wide, glabrous or thinly hirsute, margins entire or remotely serrulate; spikes subglobose, ovoid or shortly oblong, mostly less than 3 cm. long at anthesis, not much lengthening in fruit, brown or fuscescent; bracts acuminate, but without a long, attenuate, recurved apical cusp; sepals usually brownish; corolla-lobes generally shorter than tube. Capsule ovoid, often protruding rather conspicuously from the ripe fruiting spike.

HAB.: Mountainsides, roadsides, waste ground; c. 4,500 ft. alt.; fl. ? March–Oct.

DISTR.: Division 2. Widespread in Europe, and found as a weed almost throughout the world.

2. Prodhromos, 1949, *Casey* 863 !

NOTES: The specimen was collected in July, and may be an abnormal form of the following variety (var. *bakeri* C. E. Salmon), none the less it is the only Cyprus specimen seen which agrees with typical *P. lanceolata*. It may be that further collecting, especially at high altitudes, would show that *P. lanceolata* var. *lanceolata* is not so rare as would appear. It is also possible that some of the specimens listed from Division 2 under var. *bakeri* should be referred to the type variety.

var. **bakeri** C. E. *Salmon*, Flora of Surrey, 540 (1931).
    P. lanceolata L. var. *mediterranea* Pilger in Engl., Pflanzenr., 102 (IV. 269): 320 (1937).
    [*P. lanceolata* L. var. *timbali* (non Jord.) Syme in Sm., English Bot., ed. 3, 7: 171, t. MCLXV (1867).]
    TYPE: England; Essex, Leigh-on-Sea, 1860, *J. T. Syme* (K, in herb. H. C. Watson !), "in fields of clover, sainfoin, and lucerne, but apparently not indigenous".

Plants robust; scapes often 40–60 cm. long; leaves long-petiolate, broadly lanceolate or narrowly elliptic, 2·5–5 cm. wide, usually glabrous, margins commonly serrulate; spikes elongate, narrowly and bluntly conical, frequently 3–5 cm. at anthesis, lengthening to 7 (or exceptionally 12) cm. and becoming cylindrical in fruit, pale brown or silvery; bracts generally with a long, attenuate, often recurved, apical cusp; sepals pale, membranous; corolla-lobes often almost as long as corolla-tube. Capsule often rather short and subglobose, not much protruding in fruiting spike; seeds a little smaller than is usual in typical *P. lanceolata*.

HAB.: Roadsides, fallow fields, waste ground, or mountainsides under Pines, often in moist ground by springs, irrigation channels or ditches; sea-level to 4,500 ft. alt.; fl. March–Sept.

DISTR.: Divisions 1–3, 6, 7, not common. Mediterranean region and eastwards to Pakistan, occurring as an introduction in W. Europe, Australia, N. America, etc.

1. Lyso ["Lisso"], 1913, *Haradjian* 883 !
2. Prodhromos, 1905, *Holmboe* 902, also, 1939, *Lindberg f.* s.n., and, *G. E. Atherton* 531 ! Platres, 1938, *Kennedy* 1426 ! Mandria, 1938, *Kennedy* 1424 ! Agros, 1941, *Davis* 3072 ! Kakopetria, 1955, *Merton* 2344 ! and, 1955, *N. Chiotellis* 441 ! Saïttas, 1962, *Meikle* 2793 !

3. Kato Lefkara, 1937, *Syngrassides* 1510! Kolossi, 1939, *Lindberg f.* s.n.! Temple of Apollo near Episkopi, 1950, *Chapman* 564!
6. Morphou, 1955, *N. Chiotellis* 477!
7. St. Hilarion, 1880, *Sintenis & Rigo* 585! Lapithos, 1939, *Lindberg f.* s.n.; Kyrenia, 1949, *Casey* 459! Kyrenia, Tchingen, 1955, *G. E. Atherton* 294!

NOTES: Perhaps because of the protean variability of *P. lanceolata*, most botanists have ignored var. *bakeri* C. E. Salmon (var. *mediterranea* Pilger), though it is distinctive, and the prevailing variant of the species over large areas of the eastern Mediterranean and western Asia, and not uncommon as an alien elsewhere. Syme (English Bot., *loc. cit.*) hints that it might merit subspecific status, and one would be inclined to agree, were it not for the fact that it is evidently linked to typical *P. lanceolata* in some areas by a range of intermediates, and also because there is no one diagnostic character which can be relied upon to identify starved or imperfect specimens, such as the material (*Sintenis & Rigo* 973!) collected on July 15, 1880 "in agris pr. Larnaka vecchia" (Div. 4). Although Holmboe (Veg. Cypr., 169) notes that *P. lanceolata* is "not rare" in Cyprus, it cannot be regarded as a common plant, unless collectors, conscious of its abundance elsewhere, have thought it not worth collecting.

var. **dubia** (*L.*) *Liljeblad*, Utk. Svensk Fl., ed. 3, 93 (1816); Post, Fl. Pal., ed. 2, 2: 417 (1933); Pilger in Engl., Pflanzenr., 102 (IV. 269): 322 (1937).
   P. dubia L., Fl. Suec., ed. 2, XVI, 46 (1755).

Leaves conspicuously villose; apex of rootstock densely lanuginose.

subvar. **euryphylla** *Pilger* in Engl., Pflanzenr., 102 (IV. 269): 324 (1937).
   TYPE: Cyprus; "In montibus supra Phlamudi [Phlamoudhi]", 4 May, 1880, *Sintenis & Rigo* 584 (lectotype, K!).

Plants not very robust; scapes 12–30 cm. long; leaves subsessile or very shortly petiolate, broadly elliptic, acute or subacute, 3–12 cm. long, 1–3·5 cm. wide, clothed with a conspicuous indumentum of long, soft hairs, margins entire or subentire; spikes short, broadly and bluntly ovoid, oblong or conical, 1·3–2 cm. long, 1–1·3 cm. wide at anthesis, rich brown; bracts acuminate, but without a long, attenuate cusp; corolla-lobes distinctly shorter than corolla-tube. Fruits not seen.

HAB.: Mountainsides, on limestone; 1,000–2,000 ft. alt.; fl. ? April–May.
DISTR.: Division 7. Also ? Balkan Peninsula and Aegean Islands.
7. Mountains above Phlamoudhi, 1880, *Sintenis & Rigo* 584!

NOTES: an easily recognised variant, with short, broad, softly hairy leaves. The indumentum is said by Pilger to be fulvous-canescent or yellowish, but this is true only of old herbarium specimens; it is probably white or greyish in fresh specimens. No one has collected the plant in Cyprus since 1880, and considering how much work has been done on the Northern Range, one must conclude that it is uncommon. Pilger records subvar. *euryphylla* from the island of Zante (*Bornmüller* 1339); there is a very similar plant in herb. Kew from the island of Hydra (*Heldreich & Halácsy* s.n., coll. 1 May, 1889) which (Pflanzenr. *loc. cit.*, 325) Pilger has named *P. lanceolata* L. var. *macrocephala*, and another from Yugoslavia (Spalato, May 1868, *Pichler* s.n.) which is not cited in the monograph.

6. **P. lagopus** *L.*, Sp. Plant., ed. 1, 114 (1753); Unger et Kotschy, Die Insel Cypern, 229 (1865); Boiss., Fl. Orient., 4: 886 (1879); H. Stuart Thompson in Journ. Bot., 44; 337 (1906); Holmboe, Veg. Cypr., 169 (1914); Post, Fl. Pal., ed. 2, 2: 420 (1933); Pilger in Engl., Pflanzenr., 102 (IV. 269): 332, fig. 33 (1937); Lindberg f., Iter Cypr., 32 (1946); Osorio-Tafall et Seraphim, List Vasc. Plants Cyprus, 96 (1973); Davis, Fl. Turkey, 7: 514 (1982).
   P. lagopus L. var. *minor* Ten., Syll. Fl. Neap., 72 (1831) pro parte.
   P. lagopus L. f. *minor* (Ten.) Pilger in Engl., Pflanzenr., 102 (IV. 269): 335 (1937).
   TYPE: "*in* G. Narbonensi, Hispania, Lusitania".

Acaulescent annual; rootstock slender, generally lanuginose at apex; leaves usually forming a single rosette, lanceolate or oblanceolate, 2–15 cm. long, 0·2–4 cm. wide, glabrous or ± hirsute, 3–5-nerved, apex acute or obtuse, usually mucronate, base tapering to a flattened petiole up to 12 cm. long, margins entire or remotely denticulate; scapes 3–35 cm. long, generally much exceeding leaves, strongly ribbed, glabrous or adpressed-pilose; flowers in a dense globose, ovoid or shortly and bluntly cylindrical, sericeous-hirsute spike, 0·8–2·5 (–5) cm. long, 0·6–1·3 cm. wide; bracts narrowly or broadly obovate, slightly keeled, acute, 2·5–3 mm. long, 1–2 mm. wide, membranous, apex fuscescent, densely clothed with long hairs; sepals membranous, the abaxial pair connate almost to apex, about 3 mm. long, 2 mm. wide (jointly), pallid, densely hirsute; adaxial sepals narrowly boat-shaped, acuminate, about 3 mm. long, 1 mm. wide, hirsute above, with an incrassate greenish nerve forming a slight dorsal keel; corolla-tube about 2 mm. long, glabrous, lobes spreading, ovate-acuminate, somewhat keeled, fuscescent, about 2 mm. long, 0·8 mm. wide, pilose dorsally; filaments slender, glabrous, 4 mm. or more long; anthers yellowish, broadly oblong, about 1·5 mm. long and almost as wide, with a small, but distinct, apical appendage; ovary glabrous, obovoid, about 1·3 mm. long, 0·8 mm. wide; style very short; stigma filiform, 5 mm. or more long. Capsule subglobose, about 2 mm. diam.; seeds 2, narrowly oblong, about 1·8 mm. long, 0·8 mm. wide, convex dorsally, deeply concave ventrally with a small hilum; testa dark brown, shining, mucilaginous and minutely papillose when wet.

HAB.: Hillsides, grass steppe, fallow fields and sandy flats by the sea; sea-level to 3,800 ft. alt.; fl. Febr.–June.

DISTR.: Divisions 1–8, common in the lowlands. Mediterranean region eastwards to Iran; Atlantic Islands.

1. Ayios Yeoryios (Akamas), 1962, *Meikle*! Polis, 1962, *Meikle*!
2. Pano Panayia, 1905, *Holmboe* 1059; Platres, 1938, *Kennedy* 1427!
3. Episkopi, 1862, *Kotschy* 654; Akhelia, 1939, *Lindberg f.* s.n.; Yermasoyia, 1950, *F. J. F. Barrington* s.n.! Kolossi, 1961, *Polunin* 6683! Cape Gata, 1962, *Meikle*!
4. Common; Larnaca, Ayia Napa, Aradhippou, etc. *Kotschy* 39, 148, 323; *Sintenis & Rigo*; *A. G. & M. E. Lascelles*; *Holmboe* 97; *Lindberg f.* s.n.; *Chapman* 15! 694! *Merton* in ARI 414! *P. Laukkonen* 33! etc.
5. Salamis, 1905, *Holmboe* 444; Laxia, 1932, *Syngrassides* 82! Between Nisou and Stavrovouni, 1933, *Syngrassides* 30! Mia Milea–Mandres road, 1933, *Syngrassides* 312! Mile 5, Nicosia–Famagusta road, 1950, *Chapman* 43! Mia Milea, 1967, *Merton* in ARI 197!
6. Ayia Irini, 1941, *Davis* 2583! Dhiorios, 1962, *Meikle*! Nicosia, 1923, *P. Laukkonen* 41!
7. Common; Kyrenia, Kantara, St. Hilarion, Larnaca tis Lapithou, Lakovounara, Kazaphani, etc. *Sintenis & Rigo*; *Syngrassides* 267! *Miss Godman* 36! *Lindberg f.* s.n.; *Davis* 3019! *Casey* 288! 311! *Chapman* 138! 196! *G. E. Atherton* 920! *Miss Mapple* 38! etc.
8. Cape Andreas, 1880, *Sintenis & Rigo*; Ayios Philon near Rizokarpaso, 1941, *Davis* 2226! Akradhes, 1962, *Meikle*!

NOTES: Specimens of *P. lagopus* communicated by the Director of Agriculture, Cyprus (letter dated 19th June, 1931) are said to have been cultivated on the island as a source of "Psyllium" seed (see F. N. Howes in Kew Bull., 1931: 62; 1931).

7. P. albicans L., Sp. Plant., ed. 1, 114 (1753); Unger et Kotschy, Die Insel Cypern, 228 (1865); Boiss., Fl. Orient., 4: 882 (1879); Holmboe, Veg. Cypr., 169 (1914); Post, Fl. Pal., ed. 2, 2: 418 (1933); Pilger in Engl., Pflanzenr., 102 (IV. 269): 341 (1937); Lindberg f., Iter Cypr., 32 (1946); Osorio-Tafall et Seraphim, List Vasc. Plants Cyprus, 96 (1973); Davis, Fl. Turkey, 7: 515 (1982).

TYPE: "*in Hispaniae & Narbonae aridis*".

Loosely tufted perennial with a tough, woody, branched rootstock; leaves linear or narrowly lanceolate, 4–20 cm. long, 0·2–1·5 cm. wide, obscurely 1–5-nerved, thinly or densely clothed with long, adpressed or spreading, whitish or silvery, silky hairs, apex tapering to a slender point, base decurrent into an obscure, flattened petiole terminating in a dilated, ribbed, villose basal sheath; scapes usually solitary from each rosette, 4–20 (–30) cm. long, terete or obscurely striatulate, generally clothed with a dense indumentum of pale woolly hairs; flowers in a dense or lax, ovoid, oblong or elongate-cylindrical spike 1–6 (–10) cm. long, 0·8–1 cm. wide; bracts broadly ovate, concave, 4–4·5 mm. long, 3·5–4 mm. wide, membranous with a hirsute apex and a very prominent incrassate, greenish or purplish nerve excurrent into a short, blunt mucro; sepals unequal, the abaxial ovate-oblong, concave shortly acute, about 4 mm. long, 1·5 mm. wide, membranous, the very thick, prominent nerve hirsute dorsally; adaxial sepals boat-shaped, acute, about 4 mm. long, 1–1·5 mm. wide, membranous, with a villose apex, and a slender, slightly thickened, median nerve; corolla-tube glabrous, about 4 mm. long, 1·5 mm. wide, lobes spreading or reflexed, broadly ovate-acuminate, about 2 mm. long and almost as wide, glabrous; filaments slender, glabrous, about 5 mm. long; anthers yellowish, broadly oblong or ovate, about 3 mm. long, 2–2·5 mm. wide, with a small apical appendage; ovary glabrous, broadly ovoid or subglobose, about 0·8 mm. long, 0·7 mm. wide; style very short; stigma filiform about 6 mm.ʼ long (or longer), angled. Capsule ovoid-conic, about 3 mm. long, 2·5 mm. wide; seeds 2, oblong, about 2·5 mm. long, 1·5 mm. wide, convex dorsally, concave ventrally, with a small hilum; testa dark brown, minutely shagreened, mucilaginous when wet.

HAB.: Dry banks inland, and sandy ground by the sea; sea-level to c. 500 ft. alt.; fl. March–May.

DISTR.: Divisions 3–6, local. Widespread in the Mediterranean region; Ethiopia.

3.   Mile 43, Nicosia–Limassol road, 1950, *Chapman* 554! Phasoula, 1950, *F. J. F. Barrington* s.n.!

4.   Larnaca, 1862, *Kotschy* 265! also, 1880, *Sintenis & Rigo! C. & W. Barbey*; *Haradjian* 7! *Merton* in ARI 415! Cape Pyla, 1905, *Holmboe* 566; Dhekelia, 1972, *P. Laukkonen* 221!

5.   Athalassa, 1932, *Syngrassides* 220! 1939, *Lindberg f.*; and, 1949, *Casey* 570! 707! Yeri, near Athalassa, 1955, *Merton* 2074!

6.   Nicosia, 1905, *Holmboe* 306; Ayia Irini, 1936, *Syngrassides* 1040! 1941, *Davis* 2584! and, 1962, *Meikle* 2391!

8. P. ovata *Forssk.*, Fl. Aegypt.-Arab., 31 (1775); Boiss., Fl. Orient., 4: 885 (1879); Suppl., 366 (1888); Holmboe, Veg. Cypr., 169 (1914); Post, Fl. Pal., ed. 2, 2: 419 (1933); Pilger in Engl., Pflanzenr., 102 (IV. 269): 347 (1937); Rechinger f. in Arkiv för Bot., ser.2, 1: 431 (1950); Osorio-Tafall et Seraphim, List Vasc. Plants Cyprus, 96 (1973).

TYPE: Egypt; "*Alexandriae*", *Forsskål* (C).

Low-growing, acaulescent annual (or sometimes perennating) with a slender tap-root; leaves variable, linear or almost filiform, 2·5–8 (–12) cm. long, 0·1–0·5 (–1) cm. wide, subglabrous or lanuginose, tapering to a slender, pointed apex, base dilated and forming a rather conspicuous sheath, margins entire or remotely denticulate, nervation obscure; scapes usually numerous, often forming a low dome, slender, 2·5–5 (–10) cm. long, generally upcurved from a spreading base, thinly or rather densely hirsute, terete; flowers in a dense subglobose or shortly oblong spike, 0·6–1·5 (–2·5) cm. long, 0·6–1 cm. wide; bracts suborbicular or very broadly ovate, about 3 mm. long, 2·5–3 mm. wide, hyaline-membranous or fuscescent, glabrous or ciliate with an incrassate, conspicuous, median nerve; sepals subequal, broadly obovate, 2–2·5 mm. long, 1–1·5 mm. wide, concave, obtuse or rounded at the apex, glabrous or thinly ciliate, with a distinct, incrassate nerve; corolla-tube 2–2·5 mm. long, lobes sharply reflexed, broadly ovate, or almost orbicular, about 2 mm. long and almost as wide, with a shortly acute apex and a slender, fuscous-stained median nerve, glabrous; filaments slender, glabrous, 2–3 mm. long; anthers yellow, broadly oblong, about 1·2–1·5 mm. long, 0·8–1 mm. wide, with a small apical appendage; ovary glabrous, ovoid, about 0·8 mm. long, 0·6 mm. wide; style very short; stigma filiform, 2–5 mm. long. Capsule broadly ovoid, about 2·5 mm. long and almost as wide; seeds oblong-elliptic, about 2·5 mm. long, 1·5 mm. wide, convex dorsally, concave ventrally with a small hilum; testa rich brown, smooth, thickly mucilaginous when wet.

HAB.: Dry sandy fields and hillsides, often on Kythrean marls; sea-level to c. 550 ft. alt.; fl. March–May.

DISTR.: Divisions 5–7, rare. Mediterranean region and eastwards to Pakistan and India.

5. Mia Milea to Mandres, 1933, *Syngrassides* 313! Mia Milea, 1950, *Chapman* 469!
6. Nicosia, 1905, *Holmboe* 300 a.
7. Lakovounara Forest, 1950, *Chapman* 72! 137!

9. P. loeflingii *L.*, Sp. Plant., ed. 1, 115 (1753); Boiss., Fl. Orient, 4: 883 (1879); Post, Fl. Pal., ed. 2, 2: 419 (1933); Pilger in Engl., Pflanzenr., 102 (IV. 269): 352 (1937); Davis, Fl. Turkey, 7: 516 (1982).

TYPE: Spain; "*in* Hispaniae *collibus agrorumque marginibus*" (LINN).

Acaulescent annual with a slender taproot; leaves forming a loose rosette, narrowly linear-lanceolate, 30–70 mm. long, 2–4 mm. wide, thinly long-pilose above and below, apex tapering to a slender acumen, base slightly dilated, densely white-lanuginose, margins remotely lobulate, the slender, blunt, forward-pointing lobules subtending indistinct tufts of hairs, nervation obscure; scapes usually numerous, spreading or erect, 3·5–9 cm. long, slender, terete, thinly adpressed-pilose; flowers in a dense, blunt, oblong spike, 7–20 mm. long, 5–7 mm. wide; bracts concave, suborbicular or very broadly obovate, about 3 mm. long, 3·5 mm. wide, hyaline-membranous, glabrous or very thinly long-pilose at base, median nerve conspicuous, incrassate, green, protruding shortly from shallowly emarginate apex of bract; sepals subequal, broadly obovate, about 2·5 mm. long, 2 mm. wide, concave, hyaline-membranous, glabrous, apex obtuse or subacute, median nerve incrassate, green, scarcely extending beyond a quarter the length of the sepal; corolla-tube about 2 mm. long, lobes sharply reflexed, ovate, acute, concave, glabrous, about 1·5 mm. long, 0·8 mm. wide, with an obscure, brownish nerve; filaments slender, glabrous, about 1·5 mm. long; anthers yellow, broadly oblong, about 1·2 mm. long, 0·8 mm. wide, with a small apical appendage; ovary glabrous, ovoid, about 1 mm. long, 0·8 mm. wide; style very short; stigma filiform, about 2 mm. long. Capsule ovoid, about 3 mm. long, 2·5 mm. wide; seeds narrowly oblong, about 2 mm. long, 0·8 mm. wide, convex dorsally, concave ventrally, with a small hilum; testa olive-brown, smooth, thickly mucilaginous when wet.

HAB.: Dry, rocky calcareous hillsides, c. 1,500 ft. alt.; fl. March–May.

DISTR.: Division 1, rare. Portugal, Spain, N. Africa, Palestine, Syria, S. Turkey eastwards to Iran; Atlantic Islands.

1. Near Ayios Neophytos Monastery, 1978, *J. Holub* s.n.!

10. P. notata *Lag.*, Gen. et Sp. Nov., 7 (1816); Boiss., Fl. Orient., 4: 885 (1879); J. Freyn in Bull. Herb. Boiss., 5: 798 (1897); Holmboe, Veg. Cypr., 169 (1914); Post, Fl. Pal., ed. 2, 2: 420 (1933); Pilger in Engl., Pflanzenr., 102 (IV. 269): 354 (1937); Osorio-Tafall et Seraphim, List Vasc. Plants Cyprus, 96 (1973).
[*P. ovata* (non Forssk.) Sintenis in Oesterr. Bot. Zeitschr., 32: 19, 122, 259, 397 (1882).]

TYPE: Spain, Murcia; "venit locis ruderatis et juxta vias circa *Pulpi* Pagum, à *Cuevas overa* oppido ad Heliocratam [Cliocratra–Lorca] urbem eundo" (G, *teste* Holmboe).

Low-growing, acaulescent annual with a slender taproot; leaves forming a single rosette, linear or narrowly lanceolate, or sometimes almost filiform, 2·5–10 cm. long, (1–) 3–10 mm. wide, thinly or rather densely pilose, 1–3-nerved, tapering to a slender acuminate apex, base dilated and forming a sheath, margins sharply laciniate-dentate, with long arcuate, upward-pointing teeth, and commonly with a conspicuous tuft of hairs at the base of the sinus, or leaves sometimes entire in depauperate, narrow-leaved forms; scapes usually numerous, slender, spreading or arcuate-ascending, 2–15 cm. long, as long as or a little longer than the leaves, thinly or densely adpressed-pilose, subterete or obscurely ribbed; flowers in a dense, subglobose or shortly oblong, pallid, lanuginose spike 0·5–2·5 (–4) cm. long, 0·5–0·8 cm. wide; bracts broadly ovate-orbicular, 2·5–3 mm. long and almost as wide, blunt, concave, with a prominent, incrassate, greenish median nerve, densely lanuginose externally; sepals subequal, broadly and bluntly ovate, about 1·5 mm. long, 1·3 mm. wide, hyaline-membranous, lanuginose at apex and base, with a short, greenish nerve stopping far below the apex; corolla-tube glabrous, about 2 mm. long, 0·8 mm. wide; lobes stained brown, broadly ovate-orbicular, obtuse, concave, spreading or reflexed, about 1·5 mm. long and almost as wide; filaments slender, glabrous, about 2·5 mm. long; anthers yellowish, broadly ovate, about 1 mm. long, 0·8 mm. wide, apex with a small appendage; ovary subglobose, about 0·8 mm. diam.; style very short; stigma filiform, 3·5–4 mm. long, tapering very slightly towards apex, and protruding conspicuously in the flowering spikes. Capsule broadly ovoid, 2·5–3 mm. long, 2–2·5 mm. wide; seeds 2, oblong, about 1·8–2 mm. long, 0·8–1 mm. wide, convex dorsally, narrowly concave-channelled ventrally, with a small hilum; testa pale brown, smooth and shining when dry, thickly mucilaginous when wet.

HAB.: Roadsides, walls, waste ground; sea-level to c. 1,000 ft. alt.; fl. March–May.

DISTR.: Divisions 4–7. S. Spain, N. Africa, Palestine and eastwards to Iraq, S. Iran and Transcaucasia.

4. Larnaca, 1893, *Deschamps* 424a-d sec. *Freyn*.
5. Lefkoniko, 1880, *Sintenis & Rigo* 63a!
6. Nicosia, 1905, *Holmboe* 300c.
7. Above Kythrea, 1880, *Sintenis & Rigo* 63b! Kyrenia, 1948, *Casey* 402!

NOTES: The original description of *P. notata* is so short that some authors (including Pilger) have questioned if the plant so named has been correctly identified, a doubt amplified by the fact that no one else seems to have collected *P. notata* in Spain. Boissier, however, was given a specimen of *P. notata* by Lagasca (Voy. Bot. Esp., 2: 535–536; 1841), and his more detailed description agrees exactly with that of the plant described above. Holmboe (Veg. Cypr., 169) notes that he has seen the specimen, and confirms the identification.

11. **P. cretica** *L.*, Sp. Plant., ed. 1, 114 (1753); Sibth. et Sm., Fl. Graec. Prodr., 1: 100 (1806), Fl. Graec., 2: 37, t. 147 (1813); Poech, Enum. Plant. Ins. Cypr., 15 (1842); Unger et Kotschy, Die Insel Cypern, 229 (1865); Boiss., Fl. Orient., 4: 834 (1879); Holmboe, Veg. Cypr., 169 (1914); Post, Fl. Pal., ed. 2, 2: 419 (1933); Pilger in Engl., Pflanzenr., 102 (IV. 269): 410 (1937); Osorio-Tafall et Seraphim, List Vasc. Plants Cyprus, 96 (1973); Davis, Fl. Turkey, 7: 516 (1982).
TYPE: "*in* Creta".

Low-growing, gregarious, acaulescent annual with a slender taproot; leaves forming a single rosette, linear or narrowly oblanceolate, 1·5–10 cm. long, 0·1–0·6 cm. wide, thinly or rather densely hirsute, 1–3-nerved, tapering to an acute, (but not very slender) apex, dilated and sheathing at the base, margins generally entire, occasionally with a few minute, remote teeth; scapes numerous, at first erect or with the spike slightly inclined, 0·5–3 cm. long, usually much shorter than the leaves, angled or subterete, densely lanuginose, becoming thickened, revolute and rigid in fruit, so that the whole plant becomes a detached spherical tumble-weed, the revolute scapes straightening when wet; spikes short, subglobose or semi-globose, 0·6–1 cm. diam., densely lanuginose; bracts ovate-acuminate, 4–5 mm. long, 3–4 mm. wide, concave, lanuginose externally, obscurely nerved; sepals ovate-oblong, 3–3·5 mm. long, 2 mm. wide, subacute or obtuse, lanuginose externally, nerve ill defined; corolla-tube glabrous, about 2 mm. long, 1 mm. wide, lobes spreading, broadly ovate, about 2 mm. long, 1·5 mm. wide, shortly acuminate, glabrous and fuscous at the base; filaments slender, glabrous, about 3 mm. long; anthers yellow, oblong, about 1·3 mm. long, 0·8 mm. wide, apex with a very large and conspicuous, membranous, appendage; ovary subglobose, glabrous, 1–1·5 mm. diam.; style very short; stigma filiform, about 3·5 mm. long, scarcely tapering to apex. Capsule ovoid-conical, 2·5–3 mm. long, 1·3 mm. wide; seeds 2, enclosed at maturity in an envelope formed by the erect indurated sepals, this envelope opening when wet, shortly oblong, about 2 mm. long, 1·3 mm. wide, convex dorsally with a distinct, transverse, median groove, somewhat concave ventrally, with a small hilum; testa dull brown, regularly faveolate when dry, thickly mucilaginous when wet. *Plate 79, figs. 1–6.*

HAB.: Dry sandy or gravelly ground, often very abundant on pillow lava and Kythrean marls; sea-level to 3,350 ft. alt.; fl. Febr.–May.

DISTR.: Divisions 2–8, locally common. Eastern Mediterranean region from Greece to Palestine.

2. Platres, 1939, *Kennedy* 1423! Ayios Merkourios, 1962, *Meikle* 2286! Soli, 1963, *Townsend* 63/102! Near Klirou, 1967, *Merton* in ARI 271!
3. Skarinou, 1941, *Davis* 2682! Below Lania, 1960, *N. Macdonald* 91! Kolossi, 1961, *Polunin* 6682! 4 miles W. of Kandou, 1963, *J. B. Suart* 51!
4. Larnaca, 1862, *Kotschy* 322; also, 1880, *Sintenis & Rigo*, and, 1950, *Chapman* 23! Dhekelia, 1880, *Sintenis & Rigo* 62! Ayia Napa, 1905, *Holmboe* 26; Cape Greco, 1958, *N. Macdonald* 17! 26!
5. Near Athalassa, 1933, *Syngrassides* 10! Mia Milea–Mandres, 1933, *Syngrassides* 311! Between Nisou and Stavrovouni, 1933, *Syngrassides* 19! Lakovounara Forest, 1950, *Chapman* 64!
6. Ayia Irini, 1941, *Davis* 2587! Cape Kormakiti, 1956, *Merton* 2589! Dhiorios, 1962, *Meikle* 2332! Nicosia, 1973, *P. Laukkonen* 122!
7. Ayios Khrysostomos Monastery, 1862, *Kotschy* 440! Near Ardhana, 1880, *Sintenis & Rigo* 62a! Near Kythrea, 1880, *Sintenis & Rigo* 62b! Kyrenia, 1932, *Syngrassides* 17! also, 1937, *Miss Godman* 35! and *Casey* 293! 391! 713! 782! *G. E. Atherton* 1220! 1390! *I. M. Hecker* 56!
8. Near Rizokarpaso, 1912, *Haradjian* 186! Gastria, 1941, *Davis* 2412! Ovgoros, 1941, *Davis* 2476!

12. **P. bellardii** *All.*, Fl. Pedem., 1: 82, t. 85, fig. 3 (1785); Boiss., Fl. Orient., 4: 834 (1879); Post, Fl. Pal., ed. 2, 2: 419 (1933); A. K. Jackson in Kew Bull., 1934: 272 (1934); Pilger in Engl., Pflanzenr., 102 (IV. 269): 411 (1937); Osorio-Tafall et Seraphim, List Vasc. Plants Cyprus, 96 (1973); Davis, Fl. Turkey, 7: 517 (1982).
TYPE: France; "invenit *Cl.* BELLARDI locis sterilibus *Villaefrancae Nicaeensis* prope *Lanternam*" (TO).

Low-growing, acaulescent annual with a slender taproot; leaves forming a single rosette, narrowly oblanceolate, 1·5–10 cm. long, 0·15–0·6 cm. wide, thinly or rather densely hirsute, 1–3-nerved, tapering to an acute (but not very slender) apex, dilated and sheathing at the base, margins generally entire or occasionally with a few, minute, remote teeth; scapes rather few (1–6 per rosette), occasionally numerous, at first erect or ascending, sometimes recurved or revolute in fruiting specimens, but never so strikingly thickened and rigid as in *P. cretica* L., 1·5–10 cm. long, commonly a little shorter than the leaves, rather densely hirsute with spreading hairs; spikes ovoid or cylindrical, 0·7–3 (–6) cm. long, 0·6–1 cm. wide, hirsute with spreading hairs, but not woolly; bracts narrowly ovate, 5–7 mm. long, 1–1·5 mm. wide, hirsute externally, somewhat concave, with a long-acuminate or caudate apex; sepals ovate-acuminate, subequal or the abaxial pair rather larger and less membranous than the adaxial, 4–5 mm. long, 1–1·5 mm. wide, nerve obscure, and ill-defined, appearing to occupy most of the abaxial sepals in some specimens; corolla-tube glabrous, about 3 mm. long, 1 mm. wide, lobes spreading or reflexed, ovate-acuminate, 1·5–2 mm. long, 0·6–1 mm. wide, acumen often long and slender; filaments slender, glabrous, about 3 mm. long; anthers yellow, very similar to those of *P. cretica*, oblong, about 1·3 mm. long, 0·8 mm. wide, with a very large apical appendage; ovary glabrous, subglobose, about 1·2 mm. diam.; style very short; stigma filiform, about 4 mm. long. Capsule ovoid, about 3 mm. long, 2·5 mm. wide; seeds 2, apparently enclosed in an indurated envelope as in *P. cretica*, shortly oblong, about 1·5 mm. long, 1 mm. wide, convex dorsally with a transverse groove, slightly concave ventrally, with a small hilum; testa dull brown, regularly faveolate when dry, thickly mucilaginous when wet.

HAB.: Dry rocky fields and on hillsides in open garigue; sea-level to 500 ft. alt.; fl. Febr.–May.

DISTR.: Divisions 5, 8, apparently rare. Widespread in the Mediterranean region and eastwards to Iran.

5. Near Athalassa, 1933, *Syngrassides* 13!
8. Gastria, 1941, *Davis* 2419! Boghaz, 1950, *Chapman* 210!

NOTES: None of the Cyprus material is typical, and deviates in the direction of *P. cretica* L., having rather small, villose inflorescences and relatively short bracts. It may well be *P. bellardii* All. var. *deflexa* Pilger (in Fedde Repert., 18: 474; 1922), which occurs in Greece, Crete and the Aegean Islands, and which has strongly deflexed fruiting scapes. Unfortunately none of the Cyprus material is sufficiently mature to show this characteristic, and must, for the present, be referred to the aggregate species.

SUBGENUS 2. **Psyllium** *Harms* Leaves opposite; stem elongate, often branched.

13. P. squarrosa *Murr.* in Comment. Soc. Reg. Sci. Gotting., 4: 38, t. III (1782); Boiss., Fl. Orient., 4: 892 (1879); Post, Fl. Pal., ed. 2, 2: 423 (1933); Pilger in Engl., Pflanzenr., 102 (IV. 269): 416 (1937); Osorio-Tafall et Seraphim, List Vasc. Plants Cyprus, 96 (1973); Davis, Fl. Turkey, 7: 518 (1982).

P. squarrosa Murr. var. *brachystachys* Boiss., Fl. Orient., 4: 893 (1879), Suppl., 366 (1888); Sintenis in Oesterr. Bot. Zeitschr., 31: 194 (1881); Holmboe, Veg. Cypr., 170 (1914).

TYPE: Cultivated at Göttingen from seed sent to J. A. Murray, as *P. aegyptiaca*, by Giovanni Scopoli.

Erect or decumbent annual; stems 4–10 cm. long (in our area, up to 30 cm. long elsewhere), branched or unbranched, shortly pubescent, often purple; leaves spreading or recurved, fleshy, dull green, linear or narrowly oblanceolate, 1–4 cm. long, 0·1–0·3 cm. wide, glabrous or subglabrous above, ciliate at the narrowly sheathing base, apex acute or obtuse, margins minutely scabridulous; scapes usually numerous, commonly very short, occasionally up to 3 cm. long, ± densely viscid-glandular; spikes subglobose or shortly ovoid, 7–14 mm. long, 5–10 mm. wide; lowermost bracts often foliaceous, with an elongate, caudate apex, 6–10 mm. long, 2–3 mm. wide at the dilated, concave, membranous-margined base; upper bracts about 3–5 mm. long, 1·5–3 mm. wide, concave, glandular, with a short acute acumen; sepals very unequal, the abaxial pair obovate, subacute, about 3 mm. long, 1·5 mm. wide, shallowly concave, gibbous towards the apex and glandular dorsally; adaxial pair boat-shaped, acuminate, membranous, about 2·5 mm. long, 1 mm. wide, with a conspicuous, glandular dorsal keel; corolla-tube rugose, about 2·5–3 mm. long, 1 mm. wide, lobes ovate-acuminate, spreading, about 2–2·5 mm. long, 1 mm. wide; filaments slender, glabrous, 3–4 mm. long; anthers yellow, oblong, about 2 mm. long, 1 mm. wide, with a very small apical appendage; ovary subglobose, about 0·7 mm. diam., glabrous; style about 1 mm. long; stigma filiform 1·5–2 mm. long, not much tapering to apex. Capsule subglobose, about 2–2·5 mm. diam.; seeds shortly oblong, 1·5–2 mm. long, 1·3 mm. wide, strongly convex dorsally, deeply concave ventrally with a small hilum; testa rich brown, minutely rugulose or pitted when dry, thickly mucilaginous when wet.

HAB.: Sand-dunes, and sandy ground near the sea; near sea-level; fl. Febr.–May.

DISTR.: Divisions 1, 3–6. Eastern Mediterranean region from Greece and Crete to Egypt.

1. Cape Arnauti, 1962, *Meikle* 2304!
3. Akrotiri Forest, 1939, *Mavromoustakis* 122!
4. Near Dhekelia [Redgelia], 1880, *Sintenis & Rigo* 705! Perivolia, 1938, *Syngrassides* 1752! Cape Kiti, 1955, *Merton* 2019!
5. Salamis, 1905, *Holmboe* 443; also, *Casey* 481! *Merton* 3210! *A. Genneou* in ARI 1311! *E. Chicken* 284!
6. Ayia Irini, 1936, *Syngrassides* 1037! Same locality, 1941, *Davis* 2549! Syrianokhori, 1956, *Poore* in *Merton* 2622!

NOTES: In the key to sect. *Psyllium*, Pilger (Engl., Pflanzenr., 102: 415; 1937) notes against *P. squarrosa*, "corollae tubus haud rugulosus". So far as concerns Cyprus material, this is not correct. In all specimens examined, the corolla-tube is very distinctly rugose.

14. P. afra L., Sp. Plant., ed. 2, 168 (1762); Grande in Nuov. Giorn. Bot. Ital., n.s., 32: 76 (1925); Verdcourt in Kew Bull., 23: 509 (1969); Davis, Fl. Turkey, 7: 519 (1982).

    P. stricta (? Schousb., Iagtt. Vextr. Marokko, 69; 1800) Sintenis in Oesterr. Bot. Zeitschr., 31: 229 (1881).

    P. psyllium (non L., Sp. Plant., ed. 1, 115; 1753) L., Sp. Plant., ed. 2, 167 (1762); Unger et Kotschy, Die Insel Cypern, 229 (1865); Boiss., Fl. Orient., 4: 891 (1879); H. Stuart Thompson in Journ. Bot., 44: 337 (1906); Holmboe, Veg. Cypr., 170 (1914); Post, Fl. Pal., ed. 2, 2: 422 (1933); Pilger in Engl., Pflanzenr., 102 (IV. 269): 422 (1937); Lindberg f., Iter Cypr., 32 (1946).

    [*P. indica* (non L.) Osorio-Tafall et Seraphim, List Vasc. Plants Cyprus, 96 (1973).]

TYPE: "*in* Sicilia, Barbaria".

Erect, or occasionally decumbent annual 2–20 (–30) cm. high, commonly unbranched, but sometimes branched from near the base; stems subterete or obscurely angled, generally rather densely glandular-pubescent and somewhat viscid, often tinged purple; leaves sessile, spreading, linear or narrowly lanceolate, 10–50 mm. long, 1–3 mm. wide, tapering to apex and base, entire or remotely serrulate, thinly glandular-pubescent, and sometimes clothed with long, white, multicellular hairs near the base; scapes usually numerous, rather crowded near the apex of the stems, slender, 1–4 cm. long, generally viscid-glandular; spikes subglobose or shortly ovoid, 5–10 mm. long, and about as wide, glandular; bracts all ± similar, about 5 mm. long, with a concave, ovate base about 1·5 mm. wide and tapering, caudate-acuminate apex; median nerve distinct; sepals subequal, or the abaxial a little longer and narrower than the adaxial, about 4 mm. long, 1·3 mm. wide, ovate-acuminate, membranous, with a distinct, glandular, median nerve; corolla-tube about 3·5 mm. long, 1·5 mm. wide, glabrous, conspicuously rugose; lobes spreading or strongly reflexed, ovate-acuminate, about 2–2·5 mm. long, 1 mm. wide; filaments slender, glabrous, about 3 mm. long; anthers yellowish, ovate, about 1·5 mm. long, 1 mm. wide, with a small apical appendage; ovary ovoid, glabrous, about 1·5 mm. long, 1 mm. wide; style very short; stigma filiform, about 4–5 mm. long. Capsule ovoid, about 2 mm. long, 1·5 mm. wide; seeds 2, narrowly oblong, about 2·2 mm. long, 0·8 mm. wide, convex dorsally, but without a transverse groove, concave ventrally with a small hilum; testa shining brown, almost smooth, thickly mucilaginous.

HAB.: Roadsides, fallows, edges of cultivated fields, grass steppe, and amongst garigue on rocky hillsides; sea-level to 2,500 ft. alt., usually lowland; fl. Febr.–June.

DISTR.: Divisions 2–8, locally common; Mediterranean region eastwards to Pakistan; Atlantic Islands.

2. Near Vouni, 1902, *A. G. & M. E. Lascelles* s.n. ! Gourri, 1950, *Chapman* 337 !
3. Yerasa, 1948, *Mavromoustakis* s.n. ! Parekklisha, 1948, *Mavromoustakis* 25 ! Limassol, 1949, *Mavromoustakis* 27 ! Below Kouklia, 1960, *N. Macdonald* 23 ! 4 miles N. of Kandou, 1963, *J. B. Suart* 65 !
4. Near Larnaca, 1862, *Kotschy* 40; also, from the same area, 1905, *Holmboe* 134; 1912, *Haradjian* 61 ! 1950, *Chapman* 7 !
5. Vatili, 1905, *Holmboe* 335; Mia Milea–Mandres, 1933, *Syngrassides* 314 ! Between Nisou and Stavrovouni, 1933, *Syngrassides* 16 ! Mile 7, Nicosia–Famagusta road, 1950, *Chapman* 369 !
6. Nicosia, 1973, *P. Laukkonen* 120 !
7. Ayios Khrysostomos, 1862, *Kotschy* 398; above Kythrea, 1880, *Sintenis & Rigo* 65 ! Near Kyrenia, 1908, *Clement Reid* s.n. ! also, same area, *Syngrassides* 360 ! *Casey* 243 ! *Chapman* 626 ! *Casey* 1110 ! *G. E. Atherton* 1122 ! St. Hilarion, 1939, *Lindberg f.*; and, same locality, 1940, *Mavromoustakis* 16 !
8. Apostolos Andreas, 1941, *Davis* 2309 !

NOTES: The distinctions between *P. afra* L. (*P. psyllium* auctt.) and *P. stricta* Schousboe are difficult to maintain, even at varietal level. Some of the Cyprus specimens have very narrow, entire, linear leaves, but in other respects they are indistinguishable from forms with broader, serrulate leaves; if a distinction can be made elsewhere, it certainly cannot be made in our area.

72a. NYCTAGINACEAE

Trees, shrubs, climbers and herbs; stems sometimes armed; leaves simple, opposite (often unequal), exstipulate; flowers in cymes, umbels or verticels, hermaphrodite or unisexual, actinomorphic; bracts and bracteoles generally present, often conspicuous and sometimes simulating a calyx; perianth tubular, campanulate or infundibuliform, valvate or plicate in bud, the lower part frequently persistent, enveloping the fruit to form an anthocarp; stamens 1 to many, often unequal, hypogynous, the filaments free or often connate at the base, involute in bud; anthers 2-thecous, dehiscing longitudinally; ovary superior, 1-locular; ovule basal, erect, anatropous or campylotropous; style simple, terminal; stigma capitate or shortly lobed. Fruit indehiscent, generally enclosed in the persistent base of the perianth; seed with a straight or curved embryo; endosperm copious or scanty, mealy or gelatinous.

About 30 genera and 300 species, chiefly in the tropics of the New World. No indigenous representatives in Cyprus, but *Bougainvillea* Commerson ex Juss. (woody climbers) and *Mirabilis* L. (tuberous herbs) are frequently planted. Of the former, two species: *B. glabra* Choisy (leaves subglabrous or sparsely puberulous; perianth shortly pilose with curved hairs; bracts purple) and *B. spectabilis* Willd. (leaves softly pubescent below; perianth densely pilose with ± straight hairs; bracts reddish-purple), both from S. America, are commonly cultivated. The hybrid *B.* × *buttiana* Holttum et Standl. (*B. glabra* Choisy × *B. peruviana* Kunth) with crimson or orange bracts, sparsely puberulous perianths, and broad, densely puberulous leaves, is a more recent introduction, and not so generally common in gardens as the other two. It will be noted that it is the bracts, which envelop the relatively insignificant flowers, that lend beauty to the genus. In *Mirabilis* the bracts are green and calyx-like, while the long-tubular, trumpet-shaped perianths are white, crimson, yellow or parti-coloured and showy. Only one species, *M. jalapa* L., the Marvel of Peru, or Four O'clock Flower, is in general cultivation. It is a robust, bushy, rather fleshy herb, with terminal clusters of flowers which expand in the late afternoon and close in the morning. Linnaeus (Hort. Cliff., 54; 1737) gives five reasons for naming the genus *Mirabilis* ("Wonderful") concluding, sarcastically, "Mirabilis est planta quae primariis Botanicis miro modo imposuit, ut fines multiplicationis unius speciei vix attingere potuerint".* Because of its variability, no fewer than thirty-three names are cited as synonyms in the *Hortus Cliffortianus*.

* "Wonderful is the plant which so marvellously fooled the early botanists that they could hardly draw a limit to their multiplication of a single species."

73. AMARANTHACEAE
H. Schinz in Engler et Prantl, Pflanzenfam., ed. 2, 16c: 7–85 (1934).

by C. C. Townsend

Annual or perennial herbs or subshrubs, rarely lianes; leaves simple, alternate or opposite, exstipulate, entire or almost so; inflorescence a dense head, loose or spike-like thyrse, spike, raceme or panicle, basically cymose, bracteate; bracts hyaline to white or coloured, subtending one or more flowers; flowers hermaphrodite or unisexual (plants dioecious or monoecious), actinomorphic, usually bibracteolate, frequently in ultimate 3-flowered cymules (triads), lateral flowers of such cymules sometimes (not in Cyprus) modified into scales, spines or hooks; perianth uniseriate, membranous to firm and often finally ± indurate, usually falling with the ripe fruit included, tepals free or somewhat fused below, frequently ± pilose or lanate, green to white or variously coloured; stamens as many as and opposite to the tepals, rarely fewer; filaments free or fused into a cup below, sometimes almost completely fused and 5-toothed at the apex with entire or deeply lobed teeth, some occasionally ananantherous, alternating with variously shaped pseudo-staminodes or not; anthers 1- or 2-thecous; ovary superior, unilocular; ovules 1–many, erect to pendulous, placentation basal; style very short or long and slender; stigmas capitate to long and filiform. Fruit an irregularly rupturing, circumscissile or indehiscent capsule, rarely a berry or crustaceous, usually with thin, membranous walls; seeds round to lenticular or ovoid; embryo curved or circular, surrounding the ± copious endosperm.

A large and predominantly tropical family of some 65 genera and over 1,000 species, including some cosmopolitan weeds and a large number of xerophytic plants.

Shrubs, flowers subtended by 3–4 bracteoles - - - - - - - **1. Bosea**
Herbs with bibracteolate flowers - - - - - - - - **2. Amaranthus**

1. BOSEA L.
Sp. Plant., ed. 1, 225 (1753).
Gen. Plant., ed. 5, 105 (1754).

Erect or scrambling shrubs, usually intricately much branched, branches alternate, commonly drooping, generally elongate; leaves entire, alternate; flowers small, greenish-brown or greenish, 2–4-bracteolate, hermaphrodite or unisexual (polygamous), in terminal and axillary bracteate racemes or spikes, the uppermost with much reduced leaves and often forming a well-developed panicle; perianth-segments 5, free, equal; stamens 5, filiform, fused basally and alternating with 5 short, thick lobes, having the appearance of an interrupted hypogynous disk; stigmas 2–3; ovary with a single suberect ovule. Fruit a globose berry; seed large, with abundant endosperm.

Three species, all narrow endemics, one in Macaronesia, one in the N.W. Himalayas and one in Cyprus. The three species may be readily separated on the basis of seed characters alone (Townsend in Kew Bull., 28: 144–5; 1973).

1. **B. cypria** *Boiss. ex* [Hook. f., Fl. Brit. India, 4: 716 (1885), nomen subnudum; Sintenis in Oesterr. Bot. Zeitschr., 32: 291, 293, 365 (1882), nomen] *Schinz et Autran* in Bull. Herb.

Boiss., 1: 12 (1893); Post in Mém. Herb. Boiss., 18: 99 (1900); Holmboe, Veg. Cypr., 66 (1914); Lindberg f., Iter Cypr., 13 (1946); Chapman, Cyprus Trees and Shrubs, 37 (1949); Osorio-Tafall et Seraphim, List Vasc. Plants Cyprus, 33 (1973).

[*B. yervamora* (non L.) Kotschy in Unger et Kotschy, Die Insel Cypern, 224 (1865); Sintenis in Oesterr. Bot. Zeitschr., 31: 192 (1881).]

TYPE: Cyprus; Larnaca, without collector, almost certainly *Sintenis & Rigo* 13 [a] (G).

Intricately branched shrub, about 0·9–1·5 m. high, bushy and erect, or pendulous, hanging over walls or cliffs, with the habit of a *Lycium*; stem and branches glabrous, pale to dark green, the young shoots ± quadrangular, the older parts terete, striate and frequently ridged; leaves glabrous, broadly lanceolate or lanceolate-ovate to elliptic, those of the mature parts of the stems and branches 2·5–6 cm. long, 1·1–3 cm. wide, (including the 2·5–5 mm. long petiole), cuneate at the base, apex acute to obtuse, mucronulate, the midrib pale and prominent on the lower surface; inflorescence glabrous, consisting of terminal and axillary spikes, the upper leafless and forming a variably developed terminal panicle; lower axillary spikes at first patent-ascending, finally elongating, patent or recurved, peduncle to about 1·5 cm. long; bracts in lower part of terminal panicle lanceolate-ovate, about 1·5 mm. long, elsewhere broadly deltoid-ovate, about 1 mm. long, scarious-margined; flowers sessile, hermaphrodite, the base surrounded by (3–) 4 imbricate, suborbicular, broadly scarious-margined bracteoles about 1 mm. long; perianth-segments also broadly scarious-margined, broadly oblong, about 2·5 mm. long, 1·5 mm. wide, greenish-brown, with darker central streaks, deeply concave; filaments exserted, 3–3·5 mm. long, filiform, the short intermediate lobes rounded or retuse; ovary narrowly pyriform, smooth; style very short, with 2 divergent stigmatic branches. Berry globose, 4–5 mm. diam., red; seed subglobose, somewhat compressed laterally, about 4 mm. diam., indented about the hilum, with a large shallow areole to one side of it; testa black, shining, minutely reticulate. *Frontispiece.*

HAB.: Dry rocky ground, cliffs, old walls; sea-level to 2,000 ft. alt.; fl. April–July.

DISTR.: Divisions 1–4, 6–8, locally abundant. Endemic.

1. Paphos, 1894, *Post* 918; Dhrousha, 1941, *Davis* 3241! Peyia, 1962, *Meikle*!
2. Lefka, 1940, *Davis* 1983!
3. Kalavasos, 1905, *Holmboe* 624.
4. Larnaca, 1860, *J. D. Hooker* et *D. Hanbury* s.n.! also, 1862, *Kotschy*; and, 1880, *Sintenis & Rigo* 13a; near Famagusta, 1934, *Syngrassides* 860A! Sotira, Ayios Antonios, 1950, *Chapman* 587!
6. Nicosia, *Chapman* 409; Ayia Irini, 1962, *Meikle* 2395! Dhiorios, 1962, *Meikle*!
7. Common; Lapithos, Akhiropiitos, Kyrenia, Ayios Epiktitos, Ayios Amvrosios, Akanthou, St. Hilarion, Halevga, Vasilia, Kazaphani, etc. *Kotschy* 492! *Post*; *Holmboe* 833; *Kennedy* 1343! *Syngrassides* 1730! 1731! *Lindberg f.* s.n.! *Davis* 1928! 2958! 3595! *Casey* 113! 655! *G. E. Atherton* 707! 1354! *P. H Oswald* 133!
8. Ephtakomi, 1880, *Sintenis et Rigo* 13! Yialousa, 1880, *Sintenis & Rigo*; Rizokarpaso, 1880, *Sintenis & Rigo*; Akradhes, 1962, *Meikle*!

NOTES: Following their usual practice, it would appear that all the specimens of *Bosea cypria* collected by Sintenis & Rigo, whatever their origin, have been equally numbered "13", hence the apparent conflict between the numbering of specimens collected at Ephtakomi in April, 1880, and at Larnaca in Sept. 1880; their latter specimen (the type of the species) should be re-numbered "13a" to avoid the ambiguity.

The abundance of this plant in Cyprus, and the apparent attractiveness to birds of its copious, translucent, crimson fruits, make it all the more difficult to explain why it should be absent from adjacent regions of Turkey and Syria. It would be interesting to know if, in fact, its fruits are eaten by birds, and, if so, by what sort of birds. So far as I am aware, nothing has been published on the biology of this remarkable endemic, nor about its allies in the Canaries and Himalayas.

Bosea cypria is most probably an ancient element in the Cyprus flora, forming part of the distinctive "Tethyan-Tertiary" residue in the Mediterranean flora, in company with *Laurus*, *Myrtus*, *Nerium*, *Ceratonia*, *Ficus* and probably *Olea*. [R.D.M.]

2. AMARANTHUS *L.*

Sp. Plant., ed. 1, 989 (1753).
Gen. Plant., ed. 5, 427 (1754).
P. Aellen in Hegi, Illustr. Fl. Mitteleuropa, ed. 2, Bd. 3/2, Lief. 1: 465–516 (1959).
J. Sauer, "The grain amaranths and their relatives: a revised taxonomic and geographic study" in Ann. Miss. Bot. Gard., 54: 103–137 (1967).

Annual or more rarely perennial herbs, glabrous or furnished with short and gland-like or multicellular hairs; leaves alternate, long-petiolate, simple and entire or sinuate; inflorescence basically cymose, bracteate, consisting entirely of dense to lax axillary clusters, or the upper clusters leafless and ± approximate, to form a lax or dense "spike" or panicle; flowers dioecious (not in Cyprus) or monoecious, bibracteolate; perianth-segments (2–) 3–5, free or connate at the base, membranous, those of the female flowers slightly accrescent in fruit; stamens free, usually similar in number to the perianth-segments; anthers 2-thecous; stigmas 2–3; ovule solitary, erect. Fruit a dry capsule, indehiscent, irregularly rupturing or commonly dehiscing by a circumscissile lid; seeds usually black and shining, testa thin; embryo annular, endosperm present.

About 50 species, chiefly in the warmer temperate and subtropical regions of the world, about a dozen species being cosmopolitan or very widely distributed weeds in the tropics also. A genus of considerable difficulty to the taxonomist, especially the "grain amaranths" such as *A. caudatus* and *A. hybridus*, which have been cultivated from ancient times.

Inflorescence not leafy to the top, a terminal leafless spike or panicle present:
  Capsule indehiscent, irregularly rupturing, very strongly muricate; seeds with low, scurfy verrucae with the same shape as the reticulum of the testa - - - **3. A. viridis**
  Capsule dehiscent by a circumscissile lid, sometimes wrinkled but never strongly muricate; seeds without low papillae:
    Female perianth-segments spathulate to narrowly oblong-spathulate, obtuse or emarginate, longer than the fruit - - - - - - **2. A. retroflexus**
    Female perianth-segments lanceolate to oblong, acute, or if obtuse then not longer than the fruit - - - - - - - - - - - **1. A. hybridus**
Inflorescence leafy to the top, consisting entirely of axillary cymose clusters:
  Bracteoles strongly spinous-acuminate, about twice as long as the perianth **5. A. albus**
  Bracteoles not spinous-acuminate, subequalling or shorter than the perianth
                                          4. A. graecizans

1. A. hybridus *L.*, Sp. Plant., ed. 1, 990 (1753); Zohary, Fl. Palaest., 1: 181, 264 (1966); Osorio-Tafall et Seraphim, List Vasc. Plants Cyprus, 33 (1973).

Annual herb, erect or less commonly ascending, up to about 2 (–3) m. in height in cultivated forms but much less in spontaneous plants, sometimes reddish-tinted throughout; stems stout, branched, angular, glabrous or thinly to moderately furnished with short or long multicellular hairs (increasingly so above, especially in the inflorescence); leaves glabrous, or thinly pilose on the lower margins and underside of the primary nervation, long-petiolate (petioles up to 15 cm. long but even then scarcely exceeding the lamina), lamina broadly lanceolate to rhomboid or ovate, 3–19 (–30) cm. long, 1·5–8 (–12) cm. wide, gradually narrowed to the blunt or subacute, mucronulate tip, base attenuate or shortly cuneate, running into the petiole; flowers in yellowish, green, reddish or purple, axillary and terminal spikes, formed of cymose clusters which are increasingly approximate upwards, the terminal inflorescence varying from a single spike to a broad, much-branched, panicle up to about 45 cm. long, 25 cm. wide, the ultimate spike not infrequently nodding; male and female flowers intermixed throughout the spikes; bracts and bracteoles deltoid-ovate to deltoid-lanceolate, pale-membranous, acuminate and with a long, pale to reddish-tipped, erect arista formed by the stout, excurrent, yellow or greenish midrib, subequalling to much exceeding the perianth; perianth-segments 5, lanceolate or oblong, 3·5–5·5 mm. long, acute-aristate or the inner sometimes blunt in the female flowers, only the midrib at most greenish; stigmas (2–) 3, erect, flexuose or recurved, about 0·75–1·25 mm. long. Capsule subglobose to ovoid or ovoid-urceolate, 2·5 mm. diam., circumscissile, with a moderately distinct to obsolete "neck", lid smooth, longitudinally sulcate, or sometimes rugulose below the neck; seed black and shining or pale, compressed, 0·75–1·25 mm. diam., testa almost smooth centrally, faintly reticulate around the margins.

ssp. **hybridus**

    A. chlorostachys Willd., Hist. Amaranth., 34, t. 10, fig. 19 (1790); Boiss., Fl. Orient., 4: 988 (1879); Holmboe, Veg. Cypr., 66 (1914); Lindberg f., Iter Cypr., 13 (1946); Davis, Fl. Turkey, 2: 341 (1967).

    A. patulus Bertol., Comment. It. Neap., 19, t. 2 (1837); Boiss., Fl. Orient., 4: 989 (1879); Holmboe, Veg. Cypr., 66 (1914).

    [*A. caudatus* (non L.) Post in Mém. Herb. Boiss., 18: 99 (1900); Holmboe, Veg. Cypr., 66 (1914); Osorio-Tafall et Seraphim, List Vasc. Plants Cyprus, 33 (1973).]

    TYPE: "*in* Virginia"; Linnaean specimen 1117/19 (LINN, lecto.).

Style-branch bases and upper part of the lid of the fruit swollen, so that the ripe utricle has a distinct inflated "beak"; inner perianth-segments of female flowers commonly acute; longer bracteoles of the female flowers mostly about twice as long as the perianth. Occurs in forms with both green (var. *hybridus*) and red (var. *erythrostachys* Moq.) inflorescences.

HAB.: Cultivated and waste ground, occasionally in garigue or on sandy seashores; sea-level to 4,500 ft. alt.; fl. June–Oct.

DISTR.: Divisions 1–3, 6, 7. Practically throughout the warmer regions of the world as a spontaneous weed, a native of the Americas; also as a casual in temperate regions. Grown in many regions as a grain crop, and coloured forms widespread as decoratives under the name "Prince of Wales' Feathers".

1. Kissonerga to Maa Beach, 1967, *Merton* in ARI 995!
2. Between Galata and Evrykhou, 1880, *Sintenis & Rigo* 747 partly! Agros, 1894, *Post*! Kaminaria, 1905, *Holmboe* 1073; Kyperounda, 1937, *Syngrassides* 1703! 1704! Milikouri, 1939, *Lindberg f.* s.n.; Kambos, 1939, *Lindberg f.* s.n.; Kakopetria, 1939, *Lindberg f.* s.n.; Prodhromos, 1955, *G. E. Atherton* 521!
3. Limassol, 1905, *Michaelides*.
6. Morphou, 1955, *N. Chiotellis* 469! 484! Kokkini Trimithia, 1966, *Merton* in ARI 11!
7. Kyrenia, 1948, *Casey* 38! also, 1955, *G. E. Atherton* 274! 1956, *Casey* 1734! 1958, *P. H. Oswald* 10! Upper Thermia, 1955, *G. E. Atherton* 671!

ssp. **cruentus** (*L.*) *Thell.* in Fl. Adv. Montpellier, 205 (1912).

    A. cruentus L., Syst. Nat., ed. 10, 2: 1269 (1759); Osorio-Tafall et Seraphim, List Vasc. Plants Cyprus, 33 (1973).

    A. paniculatus L., Sp. Plant., ed. 2, 1406 (1763); Boiss., Fl. Orient., 4: 989 (1879).

    TYPE: Locality not indicated; Linnaean specimen 1117/25 (LINN, lecto.).

Style-branch bases and upper part of lid of fruit not swollen, so that the fruit has a smooth, narrow "beak"; inner perianth-segments of female flowers commonly obtuse; longer bracteoles of the female flowers 1–1·5 times as long as the perianth. As with ssp. *hybridus*, green and red forms occur.

HAB.: Cultivated ground; near sea-level; fl. ? April–Oct.

DISTR.: Division 7 (see Notes). Probably of Central American origin, now found throughout the warmer regions of the world as a garden ornamental and in some areas as a grain crop or potherb. As a spontaneous weed, found principally in Asia eastwards from Malaya and in tropical Africa.

7. Kyrenia, in an Olive grove, 19 April, 1952, *Casey* 1249!

NOTES: The specimen upon which this record is based, has been referred to ssp. *cruentus* by an authority on the genus. It is, however, very young, with the ovary scarcely commencing to develop, and in my opinion could equally well be ssp. *hybridus* (var. *erythrostachys*).

2. A. retroflexus *L.*, Sp. Plant., ed. 1, 991 (1753); Boiss., Fl. Orient., 4: 989 (1879); Zohary, Fl. Palaest., 1: 182, t. 265 (1966); Davis, Fl. Turkey, 2: 340 (1967).

A. delilei Richter et Loret in Bull. Soc. Bot. France, 13: 316 (1866).

A. retroflexus L. var. *delilei* (Richter et Loret) Thell. in Vierteljahrsschr. Nat. Ges. Zürich, 52: 442 (1907); Zohary, Fl. Palaest., 1: 182 (1966).

TYPE: Cultivated material from Uppsala Botanic Garden, Linnaean specimen 1117/22 (LINN, lecto.).

Annual herb, erect or with ascending branches, (6–) 15–80 (–100) cm. high, simple or branched (especially from the base to about the middle of the stem); stem stout, subterete to angled, densely furnished with multicellular hairs; leaves hairy along the lower surface of the primary venation and often the basal margins, long-petiolate (petioles to c. 6 cm. long and in large plants not rarely equalling the lamina), lamina ovate to rhomboid or oblong-ovate, (1–) 5–11 cm. long, (0·6–) 3–6 cm. wide, obtuse to subacute at the mucronulate tip, base shortly cuneate or attenuate into the petiole; flowers in greenish or rarely somewhat pinkish-suffused, stout, axillary and terminal spikes, which are usually shortly branched to give a lobed appearance, more rarely with longer branches; terminal inflorescence paniculate, very variable in size, male and female flowers intermixed, the latter generally much more plentiful except sometimes at the tips of the spikes; bracts and bracteoles lanceolate-subulate, pale, membranous, with a prominent green midrib excurrent into a stiff, colourless arista, longer bracteoles subequalling to twice as long as the perianth; perianth-segments 5, those of the male flowers 1·75–2·25 mm. long, lanceolate-oblong, blunt to subacute, those of the female flowers 2–3 mm. long, narrowly oblong-spathulate to spathulate, obtuse or emarginate, $\pm$ green-vittate along the midrib, which ceases below the apex or is excurrent in a short mucro; stigmas 2–3, patent-flexuous or erect, c. 1 mm. long. Capsule subglobose, c. 2 mm. diam., usually shorter than the perianth, circumscissile, rugose below the lid, beak indistinct; seed black and shining, compressed, c. 1 mm. diam., almost smooth centrally, faintly reticulate around the margins.

HAB.: Cultivated and waste ground, roadsides; sea-level to 4,500 ft. alt.; fl. June–Oct.

DISTR.: Divisions 2, 5. A native of N. America introduced into the Old World as a weed, less widely distributed than many of its allies, and chiefly in warmer temperate Eurasia, also adventive in Australia, N. & S. America and probably elsewhere.

2. Between Galata and Evrykhou, 1880, *Sintenis & Rigo* 747 partly! Prodhromos, 1955, *G. E. Atherton* 519! Trikoukkia, 1970, *A. Genneou* in ARI 1568!

5. Kythrea, 1932, *Syngrassides* 693!

NOTES: The var. *delilei* seems too ill marked to be worth formal taxonomic recognition.

3. A. viridis *L.*, Sp. Plant., ed. 2: 1405 (1763); Lindberg f., Iter Cypr., 13 (1946); Davis, Fl. Turkey, 2: 341 (1967); Osorio-Tafall et Seraphim, List Vasc. Plants Cyprus, 33 (1973).

A. gracilis Desf. [Tabl. Ecole Bot., 43 (1804) nomen] ex Poir. in Lam., Encycl. Méth. Suppl., 1: 312 (1810); Zohary, Fl. Palaest., 1: 186, t. 272 (1966).

TYPE: "*in* Europa, Brasilia"; Linnaean specimen 1117/15 (LINN, lecto.).

Annual herb, erect or more rarely ascending, 10–75 (–100) cm. high; stem rather slender, sparingly or considerably branched, angular, glabrous or more frequently increasingly hairy upwards (especially in the inflorescence) with short or longer, and rather floccose, multicellular hairs; leaves glabrous, or with short or fairly long hairs on the lower surface of the primary nerves and most of the venation, long-petiolate (petioles to about 10 cm. long and the longest commonly longer than the lamina), lamina deltoid-ovate to rhomboid-oblong, 2–7 cm. long, 1·5–5·5 cm. wide, the margins occasionally obviously sinuate, base shortly cuneate to sub-truncate, apex obtuse and narrowly to clearly emarginate and minutely mucronate; flowers green, in slender, axillary or terminal, often paniculate spikes about 2·5–12 cm. long and 2–5 mm. wide, or in the lower part of the stem in dense axillary clusters to about 7 mm. diam.; male and female flowers intermixed but the latter more numerous; bracts and bracteoles deltoid-ovate to lanceolate-ovate, whitish-membranous with a very short, pale or reddish awn formed by the excurrent green midrib, bracteoles shorter than the perianth (about 1 mm. long); perianth-segments 3, very rarely 4, those of the male flowers oblong-oval, acute, concave, about 1·5 mm. long, shortly mucronate; those of the female flowers narrowly oblong to narrowly spathulate, finally 1·25–1·75 mm. long, the borders white-membranous, minutely mucronate or without a mucro, midrib green and often thickened above; stigmas 2–3, short, erect or almost so. Capsule subglobose, 1·25–1·5 mm. diam., not or slightly exceeding the perianth, indehiscent or rupturing irregularly, very strongly rugose throughout; seed about 1–1·25 mm. diam., round, only slightly compressed, dark brown to black, often with a paler thick border, ± shining; testa reticulate, and with shallow scurfy verrucae on the reticulum, the verrucae with the shape of the areolae.

HAB.: Cultivated and waste ground, roadsides; sea-level to 450 ft. alt.; fl. [March] July–Nov.

DISTR.: Divisions 4, 8. Probably the most cosmopolitan species of the entire genus; found throughout the tropics and subtropics, and more tolerant of temperate regions than most other weedy species of the genus.

4. Famagusta, 1939, *Lindberg f.* s.n.! also, 1969, *A. Genneou* in ARI 1358! Ayios Memnon, 1970, *A. Genneou* in ARI 1635!
8. Rizokarpaso, 1970, *A. Hansen* 206!

4. A. graecizans *L.*, Sp. Plant., ed. 1: 990 (1753); Zohary, Fl. Palaest., 1: 185, t. 271 (1966); Davis, Fl. Turkey, 2: 343 (1967).

Annual herb, branched from the base and usually also above, erect, decumbent or prostrate, mostly up to 45 cm. long (more rarely to 70 cm. long); stem slender to stout, angular, glabrous or thinly to moderately furnished with frequently crisped multicellular hairs which become denser above, especially in the inflorescence; leaves glabrous or sometimes sparingly furnished on the under surface of the primary venation with very short gland-like hairs, long-petiolate (petiole from 3–45 mm. long, sometimes longer than the lamina), lamina broadly ovate or rhomboid-ovate to narrowly linear-lanceolate, 4–55 mm. long, 2–30 mm. wide, acute to obtuse or obscurely retuse at the mucronulate tip, cuneate to long-attenuate at the base; flowers all in axillary cymose clusters, male and female intermixed, males commonest in the upper cymes; bracts and bracteoles narrowly lanceolate-oblong, pale, membranous, acuminate and with a pale or reddish arista formed by the excurrent green midrib, bracteoles subequalling or usually shorter than the perianth; perianth-segments 3, all 1·5–2 mm. long; those of the male flowers lanceolate-oblong, cuspidate, pale membranous with a narrow green midrib excurrent in a short, pale arista; those of the female flowers lanceolate-oblong to linear-oblong, gradually to abruptly narrowed to the mucro, the midrib often bordered by a green vitta above and apparently thickened, the margins pale, whitish to greenish; stigmas 3, slender, usually pale, flexuous, about 0·5 mm. long. Capsule subglobose to shortly ovoid, 2–2·25 mm. diam., usually strongly wrinkled throughout with a very short, smooth beak, exceeding the perianth, usually circumscissile; seeds shining, compressed, black, 1–1·25 mm. diam.; testa faintly reticulate.

ssp. **graecizans**
    A. angustifolius Lam., Encycl. Méth., 1: 115 (1783), nom. illeg.
    A. blitum L. var. *graecizans* (L.) Moq. in DC., Prodr., 13 (2): 363 (1849).

A. silvestris Vill. var. *graecizans* (L.) Boiss., Fl. Orient., 4: 990 (1879).
A. blitum L. ssp. *graecizans* (L.) Lindberg f., Iter Cypr., 13 (1946).
TYPE: "*in* Virginia"; material from Uppsala Botanic Garden, Linnaean specimen 1117/3 (LINN, lecto.).

Leaf-blade (particularly the larger leaves of the main stems) at least 2·5 times as long as broad, narrowly oblong to linear-lanceolate; perianth-segments shortly (to c. 0·25 mm. long) mucronate.

HAB.: Cultivated and waste ground; near sea-level; fl. June–Aug.

DISTR.: Divisions 4, 7. The characteristic subspecies thoughout S.W. Asia (Jordan, Iraq, Arabia, S. Iran, etc.); also occurs in N.W. India and most of Africa, and in other regions (e.g. Northern Europe) chiefly as a casual.

4. Larnaca, 1880, *Sintenis* 963! and, 1939, *Lindberg f.* s.n.!
7. Loukkos near Thermia, 1955, *G. E. Atherton* 365!

ssp. **silvestris** (*Vill.*) *Brenan* in Watsonia, 4: 273 (1961).
A. silvestris Vill., Cat. Plant. Jard. Strasb., 111 (1807); Boiss., Fl. Orient., 4: 990 (1879).
A. graecizans var. *silvestris* (Vill.) Aschers. in Schweinf., Beitr. Fl. Aethiop., 176 (1867); Zohary, Fl. Palaest., 1: 185, t. 271 (1966); Davis, Fl. Turkey, 2: 343 (1967).
[*A. blitum* (non L.) Unger et Kotschy, Die Insel Cypern, 224 (1865); Holmboe, Veg. Cypr., 66 (1914); Lindberg f., Iter Cypr., 13 (1946).]
TYPE: France; Paris, "presque dans tous les jardins, dans les cours des maisons, sur les ramparts & sur les quais de la ville"; Tournefort Herbarium, specimen No. 1849 (P).

Leaf-blade (particularly the large leaves of the main stems) broadly to rhomboid-ovate or elliptic-ovate, less than 2·5 times as long as broad; perianth-segments extremely shortly (sometimes scarcely at all) mucronate.

HAB.: Cultivated and waste ground; sea-level to 720 ft. alt.; fl. June–Sept.

DISTR.: Divisions 1, 3–7. Mostly in more temperate regions of the Old World than ssp. *graecizans*, occurring from the warmer parts of Europe to the cooler regions of western Asia (Caucasus, N. Iran, etc.) and N.W. India; also in N. and tropical (chiefly eastern) Africa.

1. Kissonerga to Maa beach, 1967, *Merton* in ARI 996!
3. Skarinou, 1935, *Syngrassides* 714! Kouklia, 1939, *Lindberg f.* s.n.
4. Larnaca, 1865, *Kotschy* 257; Ayia Napa, 1939, *Lindberg f.* s.n.
5. Kythrea, 1939, *Lindberg f.* s.n.
6. Kokkini Trimithia, 1966, *Merton* in ARI 1!
7. Kyrenia, 1948, *Casey* 53! also, 1955, *G. E. Atherton* 22! and, 1956, *Casey* 1733!

5. A. albus L., Syst. Plant., ed. 10, 2: 1268 (1759); Boiss., Fl. Orient., 4: 990 (1879); Holmboe, Veg. Cypr., 66 (1914); Lindberg f., Iter Cypr., 13 (1946); Zohary, Fl. Palaest., 1: 185, t. 270 (1966); Davis, Fl. Turkey, 2: 343 (1967); Osorio-Tafall et Seraphim, List Vasc. Plants Cyprus, 33 (1973).
TYPE: Locality not indicated; Linnaean specimens exist in both LINN and S.

Annual herb, (10–) 20–45 (–55) cm. high, erect to prostrate, much-branched from the base and above; stems tough and wiry, whitish to pale green, angular, glabrous or frequently ± whitish-floccose with multicellular hairs at least in the youngest parts; leaves glabrous, oblong-obovate to spathulate, rounded to truncate or slightly emarginate at the pale-mucronate tip (mucro 0·5–1 mm. long), long-attenuate below, those of the main stem and branches 1–5 (–8·5) cm. long, 0·4–1·5 (–2·25) cm. wide (including the 0·3–1·5 (–4) cm. long petiole), the margins frequently minutely crispate, upper leaves rapidly diminishing in size; flowers in congested, axillary, rather prickly cymes to about 2 cm. diam., the cyme-branches spike-like; male flowers scattered, or some solitary in the axils of the uppermost leaves; bracteoles much exceeding the perianth, subulate or lanceolate-subulate, 2–3 mm. long, pale-margined with the margin widest below, sharply pungent with a stoutly aristate, excurrent midrib; perianth-segments 3, all 0·75–1·2 mm. long, those of the male flowers lanceolate-oblong, pale-membranous with a narrow green midrib excurrent in a short, pale mucro, those of the female flowers narrowly oblong to narrowly spathulate, acute or more commonly bluntish with a distinct sharp mucro, the narrow green vitta at the midrib often widened above; stigmas 3, slender, pale, recurving. Capsule 1·5–2 mm. long, pyriform, circumscissile, greenish, and wrinkled except for the swollen, whitish beak, obviously exceeding the perianth; seeds somewhat compressed, about 1–1·2 mm. diam.; testa shining black, faintly reticulate.

HAB.: Cultivated and waste ground; sea-level to 700 ft. alt.; fl. July–Dec.

DISTR.: Divisions 1, 3, 4, 6, 7. A native of N. America, now dispersed over many of the warmer N. temperate regions of the world as an introduced weed.

1. Between Paphos and Yeroskipos, 1981, *Meikle* 5052!
3. Alethriko, 1905, *Holmboe* 1098; Kouklia, 1939, *Lindberg f.* s.n.
4. Near Famagusta, 1969, *A. Genneou* in ARI 1376!
6. Kokkini Trimithia, 1966, *Merton* in ARI 2!
7. Kyrenia, 1948, *Casey* 33! also, 1955, *G. E. Atherton* 382! 385! and, 1956, *Casey* 1732!

74. CHENOPODIACEAE

E. Uhlbrich in Engler et Prantl, Pflanzenfam., ed. 2, 16c: 379–584 (1934).

Herbs or shrubs, sometimes small trees; stems continuous or articulated; terete or angular, sometimes fleshy; leaves alternate or nearly opposite, exstipulate, simple, dorsiventrally flattened or terete, commonly entire, often glaucous, pruinose or pilose, sometimes reduced to scales or wanting; flowers in terminal or lateral cymes or clusters, sometimes sunk in the fleshy rhachis, generally small and greenish, hermaphrodite, polygamous or unisexual, commonly bracteate and bracteolate; perianth actinomorphic, sepaloid, persistent or caducous, (3–) 5-lobed, with imbricate or subvalvate lobes; stamens usually as many as, and opposite the perianth-lobes, sometimes fewer or more, or reduced to staminodes, hypogynous or inserted on a disk; anthers dorsifixed, rarely basifixed, oblong or sagittate, 2-thecous, dehiscing longitudinally, connective sometimes produced into an appendage; ovary usually free, sometimes sunk in base of perianth, superior, ovoid or subglobose, occasionally elongate, 1-ovulate, ovule basal; style short or elongate, simple or divided; stigmas 2–3 (–5). Fruit usually a membranous, coriaceous or fleshy achene or nutlet, generally enclosed in the persistent perianth; seed horizontal or vertical, compressed or subglobose; testa various; endosperm generally copious, sometimes scanty; embryo usually peripheral, curved, sometimes coiled into a spiral.

About 102 genera, widely distributed, but most abundantly represented in salt-deserts, seashores and similar saline habitats.

Thorny shrubs or subshrubs; fruiting perianth pink, winged - - - - **12. Noaea**
Unarmed shrubs or herbs (sometimes with spine-tipped leaves):
  Inflorescences cylindrical-spicate, fleshy; flowers embedded in cavities in the fleshy stems or largely concealed by bracts:
    Bracts spirally arranged in inflorescence; leaves fleshy, subglobose or ovoid, generally shed before anthesis - - - - - - - - - - **6. Halopeplis**
    Bracts opposite-decussate in inflorescence; leaves all reduced to narrow sheaths or scales or apparently wanting:
      Bracts free; lateral perianth-segments gibbous-galericulate; flowers in shortly cylindrical or subglobose-spikes - - - - - **7. Halocnemum**
      Bracts connate in pairs; perianth-segments not gibbous-galericulate; inflorescence elongate-cylindrical:
        Flowers rather loosely arranged in groups of 3, protruding a little beyond bracts **8. Arthrocnemum**
        Flowers pressed tightly together or connate, not protruding beyond subtending bracts - - - - - - - - - - - - **9. Salicornia**

Inflorescences not cylindrical-spicate; flowers not embedded in cavities in fleshy stems nor
concealed by bracts:
    Leaves flat, petiolate, usually broad, toothed or lobed:
        Bracteoles absent or inconspicuous in fruit; flowers generally hermaphrodite, and with a
        perianth:
            Ovary semi-inferior, immersed in perianth-base; perianth persistent, accrescent and
            indurated in fruit -   -   -   -   -   -   -   -   -   -   **2. Beta**
            Ovary superior; perianth either unchanged in fruit, or becoming fleshy and coloured
                                                                            3. Chenopodium
        Bracteoles conspicuous, accrescent in fruit; flowers unisexual; female flowers without a
        perianth:
            Male and female flowers on separate (dioecious) plants -   -   SPINACIA (p. 1383)
            Male and female flowers on the same (monoecious) plant:
                Fruiting bracteoles deltoid, almost free or free to below middle; leaves usually
                toothed or lobed -   -   -   -   -   -   -   -   -   **4. Atriplex**
                Fruiting bracteoles obdeltoid, connate to above middle; leaves entire
                                                                            5. Halimione
    Leaves narrow, linear, subulate or filiform, sessile or subsessile, often dilated or slightly
    amplexicaul at base:
        Leaves subulate, keeled, with a prominent midrib and shortly spinous apex
                                                                            1. Polycnemum
        Leaves oblong or linear, not keeled nor with a prominent midrib:
            Perianth-segments with a conspicuous dorsal wing in fruit; the (apparent) fruit
            discoid-winged and conspicuous -   -   -   -   -   -   -   **11. Salsola**
            Perianth-segments not winged in fruit:
                Bracteoles very small, membranous -   -   -   -   -   -   **10. Suaeda**
                Bracteoles 2–8 mm. long, fleshy-herbaceous, distinctly keeled dorsally **11. Salsola**

TRIBE 1. **Polycnemeae** (*Dumort.*) *Moq.* Leaves linear or subulate; flowers
hermaphrodite, solitary, bracteolate; stamens 2–5, united toward base;
embryo curved, peripheral, enclosing endosperm.

1. POLYCNEMUM *L.*

Sp. Plant., ed. 1, 35 (1753).
Gen. Plant., ed. 5, 22 (1754).

Annual herbs; leaves alternate, subulate, spine-tipped; flowers hermaphrodite, solitary, sessile, axillary in the axils of foliaceous bracts;
bracteoles 2; perianth-segments 5, equal or subequal; stamens (1–) 3 (–5),
shortly united at the base; filaments slender; anthers 2-thecous; style short,
filiform; stigmas 2, spreading. Fruit a 1-seeded, indehiscent utricle or nutlet
enclosed in the persistent perianth; seed vertical, reniform-lenticular.

Six or seven species in central and southern Europe, the Mediterranean
region and eastwards to Central Asia.

1. **P. arvense** *L.*, Sp. Plant., ed. 1, 35 (1753); Boiss., Fl. Orient., 4: 995 (1879); Davis, Fl. Turkey,
    2: 296 (1967).
    TYPE: "*in* Galliae, Italiae, Germaniae *arvis*".

Sprawling annual; stems 2–17 cm. long, much branched, angular, thinly
lanuginose; leaves erecto-patent, subulate, keeled, 5–14 mm. long, about 1
mm. wide at base, glabrous or thinly scabridulous, midrib prominent below,
apex spine-tipped, margins entire, base amplexicaul, somewhat sheathing;
bracts very similar to leaves; bracteoles ovate, about 1·5 mm. long, 0·8 mm.
wide, apex long-acuminate, base membranous, midrib prominent; perianth
divided almost to base into 5 erect, ovate-acuminate segments 1·5 mm. long,
0·6 mm. wide; filaments slender, about 0·6 mm. long; anthers yellow,
suborbicular, about 0·4 mm. diam.; ovary broadly ovoid, about 0·5 mm.
long, 0·4 mm. wide, glabrous; style short, thick, about 0·2 mm. long; stigmas
blunt, about 0·1 mm. long. Fruit compressed, suborbicular, membranous,
about 1·2 mm. long, 1 mm. wide, with a distinct peripheral ridge near the

apex, forming a small cap or lid; seed filling utricle, strongly compressed; testa fuscous-brown, minutely and bluntly papillose-tuberculate.

HAB.: Sun-baked, muddy hollow in open part of Pine forest; 5,200 ft. alt.; fl. June.

DISTR.: Division 2 only, very rare. Central and southern Europe, extending east to central Russia and Turkey.

2. Prodhromos, towards Khionistra, 1961, *D. P. Young* 7455!

NOTES: One, minute specimen, but unquestionably this. From the collector's notes it would appear to occupy the same habitat as *Myosurus minimus* and *Herniaria micrantha*.

TRIBE 2. **Beteae** *Moq.* Leaves broad, fleshy; flowers solitary or united in clusters; perianth-segments 5; stamens 5; ovary semi-inferior; stigmas 2–3, sessile. Fruit buried in the hardened perianth-base; seeds horizontal; embryo curved, peripheral, encircling endosperm.

2. BETA *L.*

Sp. Plant., ed. 1, 222 (1753).
Gen. Plant., ed. 5, 103 (1754).

Annual or perennial, fleshy herbs, sometimes with swollen, sugar-rich roots; leaves alternate, petiolate; inflorescences elongate, spiciform, bracteate; bracteoles 3, much reduced, or wanting; flowers hermaphrodite, sessile, solitary or in clusters in the axils of the bracts; perianth deeply 5-lobed, the base persistent, becoming hard and surrounding the fruit; stamens 5, perigynous, opposite the perianth-segments, inserted on a fleshy basal disk or annulus; anthers 2-thecous; ovary semi-inferior, immersed in the perianth-base; stigmas 2–3, sessile. Utricle subglobose, buried in the fruiting perianth-base; seed smooth, horizontal.

About 12 species in Europe, the Mediterranean region, Atlantic Islands and western Asia. Sugar Beet, Mangold, Spinach Beet and Beetroot, all derived from *B. vulgaris* L., are important farm and garden crops in many areas of northern and western Europe.

1. **B. vulgaris** *L.*, Sp. Plant., ed. 1, 222 (1753); Unger et Kotschy, Die Insel Cypern, 223 (1865); Boiss., Fl. Orient., 4: 898 (1879); Holmboe, Veg. Cypr., 65 (1914); Post, Fl. Pal., ed. 2, 2: 429 (1933); Zohary, Fl. Palaest., 1: 139, t. 196 (1966); Davis, Fl. Turkey, 2: 299 (1967).

Erect or decumbent annual or perennial, 12–100 (–200) cm. high, sometimes with a thick, fleshy taproot; stems conspicuously ribbed, often stained crimson, glabrous or subglabrous; leaves ovate or rhomboid, 2·5–30 (or more) cm. long, 1·5–15 (or more) cm. wide, rather fleshy, glabrous or pilose, glossy or dull, apex acute or obtuse, often mucronate, margins flat or crisped; petiole channelled, 1·5–20 (or more) cm. long, sometimes thick, succulent and coloured; inflorescence paniculate, usually divided into numerous, elongate, spiciform branches; bracts foliaceous, often large and resembling the cauline leaves towards the base of the flower-spikes, diminishing upwards and obsolete at the tips of the spikes; bracteoles obsolete; flowers sessile, sometimes solitary, but usually in groups of 2 or more, perianth-segments oblong, cucullate, fleshy, carinate dorsally, 1–3 mm. long, 1–1·5 mm. wide; filaments 0·5 mm. long; anthers yellow, broadly oblong, about 0·8 mm. long, 0·5 mm. wide; apex of immersed ovary conical; stigmas (2–) 3, less than 0·5 mm. long, papillose internally. Fruit enclosed with the persistent, indurated perianth or perianth-base (the flower-clusters coalescent in fruit); seed horizontal, suborbicular, 2–2·5 mm. diam.; testa rich, shining brown, minutely verruculose.

ssp. **maritima** (*L.*) *Arcang.*, Comp. Fl. Ital., 593 (1882); Holmboe, Veg. Cypr., 65 (1914); Zohary, Fl. Palaest., 1: 139 (1966).
    B. maritima L., Sp. Plant., ed. 2, 322 (1762); Sintenis in Oesterr. Bot. Zeitschr., 32: 292, 397 (1882); Davis, Fl. Turkey, 2: 298 (1967); Osorio-Tafall et Seraphim, List Vasc. Plants Cyprus, 31 (1973).
    B. vulgaris L. var. *maritima* (L.) Moq. in DC., Prodr., 13, 2: 56 (1849); Boiss., Fl. Orient., 4: 899 (1879).
    TYPE: "*in* Angliae, Belgii *littoribus maris*".

Annual or perennial; roots not much thickened; stems erect or decumbent; leaves usually small, generally less than 10 cm. long, 7 cm. wide, glabrous or thinly hairy (var. *pilosa* Del.), margins generally flat; flowers solitary or 2–3 together, perianth-segments usually less than 2 mm. long; bracts small, often linear, diminishing upwards along the flower-spikes, or becoming almost obsolete.

HAB.: Rock fissures and walls by the sea; salt-marshes, occasionally as a weed of cultivated or fallow fields inland; sea-level to 500 ft. alt.; fl. Febr.–May.

DISTR.: Divisions 1, 3–8. Widespread in western Europe, the Mediterranean region, Atlantic Islands and western Asia.

1. Ayios Yeoryios (Akamas), 1962, *Meikle* 2111! Kato Paphos, 1978, *J. Holub* s.n.!
3. Curium, 1961, *Polunin* 6692!
4. Larnaca, 1862, *Kotschy* 306 (var. *pilosa* Del. teste Holmboe), also 1935, *Nattrass* 311, 470; Kondea, 1905, *Holmboe*; Paralimni, 1905, *Holmboe* 428; Famagusta, 1912, *Haradjian* 102; Larnaca–Famagusta road, Mile 3, 1950, *Chapman* 438! Xylophagou, 1969, *A. Genneou* in ARI 1363! Dromolaxia, 1970, *A. Genneou* in ARI 1415! 1416!
5. Prastio, 1935, *Nattrass* 526; near Pyroi, 1936, *Syngrassides* 1050! Ayios Seryios, 1936, *Syngrassides* 1065!
6. Between Syrianokhori and Ayia Irini, 1941, *Davis* 2601! Mile 3, Nicosia–Myrtou road, 1950, *Chapman* 579!
7. Kyrenia, 1932, *Syngrassides* 655! also, 1956, *G. E. Atherton* 1400! Myrtou, 1932, *Syngrassides* 35! Kharcha, 1950, *Chapman* 285! Ayios Epiktitos, 1951, *Casey* 1115! Karakoumi, 1956, *G. E. Atherton* 1229!
8. Cape Andreas, 1880, *Sintenis & Rigo* 595! Boghazi, 1941, *Davis* 2408!

NOTES: A polymorphic plant in Cyprus, sometimes annual, sometimes perennial, with erect or decumbent stems, glabrous or thinly hairy leaves and congested or lax infructescences. I can find no satisfactory distinctions between these variants and ecotypes, and have left the subspecies intact.
    Nattrass (First List Cypr. Fungi, 3; 1937) records Spinach Beet (as *B. vulgaris* L., var. *cicla* L.) and Beetroot (as *B. vulgaris* L. var. *rubra* (L.) Moq.) as cultivated at Nicosia.

TRIBE 3. **Chenopodieae** Leaves flat, often mealy; flowers in cymes or clusters; perianth-segments usually 5, generally divided to beyond the middle and seldom noticeably accrescent or indurated in fruit; stamens 1–5, free or united at base; ovary superior; stigmas 2 (–5). Seed generally horizontal; embryo curved, peripheral, encircling the endosperm.

3. CHENOPODIUM *L.*

Sp. Plant., ed. 1, 218 (1753).
Gen. Plant., ed. 5, 103 (1754).

Annual or rarely perennial herbs, occasionally suffruticose; stems often prominently sulcate or striate; leaves alternate, mostly petiolate, lamina flat, often scurfy or mealy, sometimes glandular and aromatic or malodorous, entire or variously toothed or lobed; inflorescence commonly branched; flowers hermaphrodite or female in loose or compact cymose clusters; perianth-segments 5 (rarely 2–3), persistent, but generally not much altered or accrescent in fruit, free or partly united, stamens mostly 5 (sometimes fewer), free or occasionally united at base; stigmas 2 (–5), sometimes lobed. Fruit a membranous utricle; seed lenticular, generally horizontal.

About 250 species with a cosmopolitan distribution, often as weeds or adventives.

Plant glandular, pleasantly aromatic; inflorescence elongate, thyrsoid, lax, many-flowered
                  1. C. botrys
Plant glabrous or mealy, but not glandular or pleasantly aromatic:
 Inflorescence narrow, spiciform, usually unbranched; flower-clusters becoming fleshy and
  crimson in fruit -   -   -   -   -   -   -   -   -   -   **2. C. foliosum**
 Inflorescence not as above, usually branched; flower-clusters not becoming fleshy and
  crimson in fruit:
  Plants stinking of rotting fish:
   Leaves entire or subentire -   -   -   -   -   -   -   **3. C. vulvaria**
   Leaves distinctly toothed or lobed   -   -   -   -   - C. HIRCINUM (p. 1377)
  Plants odourless or almost so, not stinking:
   Inflorescence greenish, not or indistinctly mealy; leaves coarsely and deeply incised-
    dentate; testa of seeds minutely pitted   -   -   -   -   **4. C. murale**
   Inflorescence mealy; leaves irregularly and often shortly toothed or lobed; testa of seeds
    not pitted:
    Leaves about as long as wide, often with distinct, subhastate lobes towards the base of
     the lamina; inflorescence often lax, with spreading branches   **5. C. opulifolium**
    Leaves generally twice as long as wide, not often distinctly lobed towards the base of
     the lamina; inflorescence usually rather congested and narrow, with erect
     branches   -   -   -   -   -   -   -   -   -   **6. C. album**

1. C. botrys *L.*, Sp. Plant., ed. 1, 219 (1753); Boiss., Fl. Orient., 4: 903 (1879); Holmboe, Veg. Cypr., 65 (1914); Post, Fl. Pal., ed. 2, 2: 431 (1933); Lindberg f., Iter Cypr., 13 (1946); Davis, Fl. Turkey, 2: 300 (1967); Osorio-Tafall et Seraphim, List Vasc. Plants Cyprus, 31 (1973).
TYPE: "*in* Europae *australis arenosis*".

Erect, glandular-aromatic annual (4–) 10–30 (–70) cm. high; stems angled, usually much branched; leaves often withered before fruits are ripe, oblong, 1–5 cm. long, 0·5–2·5 cm. wide, obtuse, coarsely and irregularly lacerate-pinnatifid; petiole slender 1–4 cm. long; inflorescence narrow, elongate, thyrsoid, many-flowered; bracts foliaceous, often much reduced or almost wanting; flowers in dense cymes or glomerules; perianth-segments 5, free almost to base, ovate, concave but not keeled dorsally, about 0·7 mm. long, 0·4 mm. wide, densely glandular externally with an acute apex and membranous margins; stamens 1 (–2), sometimes wanting; filaments 0·1–0·2 mm. long, glabrous; anthers yellow, suborbicular, about 0·3 mm. diam.; ovary subglobose, glabrous, about 0·2 mm. diam.; stigmas 2, filiform, about 0·6 mm. long. Seed horizontal or vertical, lenticular, 0·6–0·8 mm. diam.; testa glossy, fuscous, indistinctly marked with minute curved or spiral ridges.

HAB.: Cultivated and fallow fields, vineyards, roadsides, occasionally on stony mountainsides or dried-up river-beds; 500–5,000 ft. alt.; fl. May–July.

DISTR.: Divisions 2, 3, 6, 7. Central and eastern Europe, Mediterranean region, Asia.

2. Galata, 1880, *Sintenis & Rigo* 924! also, 1913, *Haradjian* 1002! Vavatsinia, 1905, *Holmboe* 1110! Mt. Troödos, 1912, *Haradjian* 509! Platres, 1937, *Kennedy* 931! and, 1941, *Davis* 3478! also, 1960, *N. Macdonald* 180! Kannaviou, 1939, *Lindberg f.* s.n.; Kakopetria, 1939, *Lindberg f.* s.n.; Prodhromos, 1961, *D. P. Young* 7418! Spilia, 1966, *Economides* s.n.! S.W. of Palekhori, 1973, *P. Laukkonen* 437!
3. Kalavasos, 1905, *Holmboe* 616.
6. Pedieos R., Nicosia, 1935, *Syngrassides* 653! Near Pano Dheftera, 1979, *Edmondson & McClintock* E2937!
7. Village E. of Kyrenia, 1972, *Breyntraj* per *A. Hansen* s.n.!

2. C. foliosum (*Moench*) *Aschers.*, Fl. Prov. Brandenb., 1: 572 (1864); Holmboe, Veg. Cypr., 64 (1914); Post, Fl. Pal., ed. 2, 2: 431 (1933); Lindberg f., Iter Cyper., 13 (1946); Davis, Fl. Turkey, 2: 301 (1967); Osorio-Tafall et Seraphim, List Vasc. Plants Cyprus, 31 (1973).
Blitum virgatum L., Sp. Plant., ed. 1, 4 (1753); Boiss., Fl. Orient., 4: 905 (1879); Post in Mém. Herb. Boiss., 18: 99 (1900); H. Stuart Thompson in Journ. Bot., 44: 337 (1906).
Morocarpus foliosus Moench, Meth., 342 (1794).

Chenopodium virgatum (L.) Ambrosi, Fl. Tirol. Austral., 2, 1: 179 (1857); Osorio-Tafall et Seraphim, List Vasc. Plants Cyprus, 32 (1973) non Thunb. (1815).

TYPE: "*in* Tataria, Hispania".

Erect or spreading, glabrous or subglabrous, much-branched annual, (2–) 10–40 (–100) cm. high; stems bluntly angled or sulcate; leaves narrowly deltoid, (1·5–) 2–5 (–13) cm. long, (0·8–) 1–2 (–9) cm. wide, persistent and often turning crimson in fruiting specimens, apex acute or acuminate, base sometimes hastate, margins coarsely lacerate-dentate; petioles slender, 0·5–4 (–14) cm. long; inflorescence narrow, spiciform, generally unbranched, the flowers numerous, forming dense, sessile, subglobose clusters in the axils of foliaceous bracts, the latter diminishing upwards or almost obsolete at the tips of the inflorescence-spikes; perianth-segments 3–5, broadly deltoid, subacute, united almost to the middle, about 0·6 mm. long, 0·8 mm. wide at the base of the free part, at first membranous, generally accrescent and becoming fleshy and crimson in fruit; stamen often solitary, or the flowers wholly female; filaments slender, about 0·8 mm. long; anthers yellow, subreniform, about 0·3 mm. long, 0·4 mm. wide; ovary laterally compressed, suborbicular, about 1 mm. diam.; style very short; stigmas 2, spreading, about 0·4 mm. long. Fruit crimson, about 1·4 mm. diam.; seed vertical; testa fuscous, minutely punctate.

HAB.: Rocky mountainsides, clearings in Pine Forest, roadsides, vineyards and waste ground; 4,000–6,300 ft. alt.; fl. June–Aug.

DISTR.: Division 2 only; common locally. Mountainous regions of Europe, Asia and N. Africa.

2. Common on the summit of Khionistra, above Amiandos, Prodhromos and Troödos, 1880–1973, *Sintenis & Rigo* 753! *A. G. & M. E. Lascelles* s.n.! *Feilden* s.n.! *Syngrassides* 716! *Kennedy* 318–320! *C. Wyatt* 24! *Lindberg f.* s.n.! *Davis* 1818! *Mavromoustakis* 47! *Mrs H. R. P. Dickson* 05! *F. M. Probyn* 120! *G. E. Atherton* 508! *Merton* 53! *Economides* in ARI 1216!

[C. RUBRUM *L.*, Sp. Plant., ed. 1, 218 (1753) is recorded for Cyprus, without precise locality ("Cyprus, *Samson!*") by H. Stuart Thompson in Journ. Bot., 44: 337 (1906). The record is mentioned in Holmboe, Veg. Cypr., 64 (1914), but the specimen upon which it is based, which should be in the Kew collection, cannot be traced.

Chenopodium rubrum is a rare plant in the Mediterranean area, and absent from areas adjacent to Cyprus. In the circumstances one must suspect that the identification was erroneous.]

3. **C. vulvaria** *L.*, Sp. Plant., ed. 1, 220 (1753); Boiss., Fl. Orient., 4: 901 (1879); Holmboe, Veg. Cypr., 65 (1914); Post, Fl. Pal., ed. 2, 2: 430 (1933); Lindberg f., Iter Cypr., 13 (1946); Rechinger f. in Arkiv för Bot., ser. 2, 1: 420 (1950); Zohary, Fl. Palaest., 1: 141, t. 199 (1966); Davis, Fl. Turkey, 2: 303 (1967); Osorio-Tafall et Seraphim, List Vasc. Plants Cyprus, 32 (1973).

TYPE: "*in* Europae *cultis oleraceis*".

Erect or decumbent, malodorous annual (4–) 8–30 (–60) cm. high; stems ridged or angled, subglabrous or grey-mealy, usually much branched at the base; leaves ovate or ovate-rhomboid, 0·5–2·5 cm. long, 0·4–2 cm. wide, thinly grey-mealy, stinking of rotten fish, apex acute or obtuse, base broadly cuneate, margins entire, or obscurely toothed near the base; petioles 0·4–1·5 cm. long; inflorescences terminal and axillary, consisting of small, loose or rather congested cymes; bracts foliaceous, resembling the leaves towards base of partial inflorescences, much reduced or wanting towards apex; perianth-segments 5, densely mealy externally, subacute, about 0·8 mm. long, 0·4 mm. wide, united to beyond the middle, rounded dorsally; stamens 1–5, sometimes wanting in female flowers; filaments glabrous, about as long as perianth; anthers yellow, about 0·4 mm. long, 0·3 mm. wide; ovary subglobose, glabrous, about 0·3 mm. diam.; style short but distinct; stigmas spreading, about 0·3 mm. long. Seeds horizontal, depressed-globose or lenticular, slightly more than 1 mm. diam.; testa shining, blackish, minutely and openly ridged.

HAB.: Gardens, plantations, waste ground, roadsides; occasionally on moist slopes in Pine forest at high altitudes; sea-level to 6,000 ft. alt.; fl. June–Oct.

DISTR.: Divisions 2, 4–7. Widespread in Europe, Asia and North Africa.

2. Summit of Khionistra, 1905, *Holmboe* 966! also, 1939, *Lindberg f.* s.n.; near Galata, 1913, *Haradjian* 1003! Livadhi tou Pasha, 1937, *Kennedy* 317! Prodhromos, 1949, *Casey* 914!
4. Ayia Napa, 1939, *Lindberg f.* s.n.
5. Kythrea, 1932, *Syngrassides* 692! and, 1939, *Lindberg f.* s.n.! Agricultural Research Institute, Athalassa, 1966, *Merton* in ARI 6! North of Salamis, 1970, *A. Hansen* 721!
6. English School, Nicosia, 1958, *P. H. Oswald* 19!
7. Kazaphani, 1949, *Casey* 120!

4. C. murale L., Sp. Plant., ed. 1, 219 (1753); Boiss., Fl. Orient., 4: 902 (1879); Druce in Rep. B.E.C., 9: 470 (1931); Post, Fl. Pal., ed. 2, 2: 430 (1933); A. K. Jackson in Kew Bull., 1934: 272 (1934). Nattrass, First List Cyprus Fungi, 46 (1937); Lindberg f., Iter Cypr., 13 (1946); Rechinger f. in Arkiv för Bot., ser. 2, 1: 420 (1950); Zohary, Fl. Palaest., 1: 142, t. 202 (1966); Davis, Fl. Turkey, 4: 302 (1967); Osorio-Tafall et Seraphim, List Vasc. Plants Cyprus 32 (1973).
TYPE: "*in* Europae *muris aggeribusque*".

Erect annual 14–60 (–90) cm. high; stems subglabrous or thinly mealy, strongly angled, usually with erect or spreading branches; leaves ovate-rhomboid, 0·7– 8 cm. long, 0·5–6 cm. wide, dull green or rather mealy, apex usually acute, base broadly or narrowly cuneate, margins coarsely and irregularly incised-serrate; petiole slender, up to 3 cm. long; inflorescences terminal and axillary, with erecto-patent branches; flowers numerous in loose or rather dense cymes; bracts foliaceous, the lower resembling the leaves, the upper much reduced or wanting; perianth-segments 5, broadly obovate, about 1 mm. long, 1·2 mm. wide, free for about two thirds their length, cucullate, obtuse or subacute, bluntly keeled dorsally towards apex; stamens usually 5; filaments about 1·3 mm. long; anthers yellow, about 0·4 mm. long and almost as wide; ovary glabrous, subglobose, about 0·4 mm. diam.; style very short; stigmas about 0·2 mm. long. Seeds horizontal, lenticular, about 1·3 mm. diam.; testa black, minutely pitted.

HAB.: Gardens, cultivated fields and waste ground, occasionally on sandy seashores; sea-level to 4,700 ft. alt.; fl. Febr.–Sept.

DISTR.: Divisions 1–8, locally common. Widely distributed in Europe, Asia, N. Africa and Atlantic Islands.

1. Lyso, 1913, *Haradjian* 834; Ayios Neophytos, 1939, *Lindberg f.* s.n.
2. Kykko, 1913, *Haradjian* 961! Prodhromos, Damaskinari, 1966, *Merton* in ARI 54!
3. Limassol, 1928 or 1930, *Druce* s.n.; Amathus, 1978, *J. Holub* s.n.!
4. Near Famagusta, 1936, *Syngrassides* 1071! Larnaca, 1939, *Lindberg f.* s.n.; also, 1967, *Merton* in ARI 362!
5. Kythrea, 1880, *Sintenis & Rigo* 587! and, 1932, *Syngrassides* 375! Nisou, 1935, *Nattrass* 707; Salamis, 1962, *Meikle* 2565! Ayios Seryios, 1973, *P. Laukkonen* 17!
6. Nicosia, 1950, *Chapman* 458! and, 1958, *P. H. Oswald* 5!
7. St. Hilarion, 1939, *Lindberg f.* s.n.! Kyrenia, 1948, *Casey* 52! also, 1955–56, *G. E. Atherton* 23! 1256!
8. Komi Kebir, 1912, *Haradjian* 271!

5. C. opulifolium *Schrader ex Koch et Ziz*, Cat. Plant. Palat., 6 (1814); Boiss., Fl. Orient., 4: 901 (1879); Druce in Rep. B.E.C., 9: 470 (1931); Post, Fl. Pal., ed. 2, 2: 430 (1933); Lindberg f., Iter Cypr., 13 (1946); Zohary, Fl. Palaest., 1: 142, t. 201 (1966); Davis, Fl. Turkey, 2: 302 (1967); Osorio-Tafall et Seraphim, List Vasc. Plants Cyprus, 32 (1973).
TYPE: Germany; "Bey Arheilgen an Zäuner und auf Schutthaufen".

Erect, much-branched annual (10–) 30–100 (–150) cm. high; stems glabrous or thinly mealy, distinctly angled; leaves generally grey-green, thinly mealy, broadly ovate-rhomboid, 1–5 cm. long, 0·5–4·5 cm. wide, apex acute or obtuse, margins subentire with conspicuous subhastate lateral lobes, or sometimes irregularly dentate; petiole slender, 1–4 cm. long; inflorescences terminal and axillary with numerous divaricate branches; flowers densely grey-mealy, usually in compact clusters; bracts foliaceous, those at the base of the inflorescences similar to the leaves, the upper much reduced or wanting; perianth-segments broadly obovate, about 1 mm. long and almost as wide, free for about two thirds their length, cucullate, subacute, keeled dorsally; stamens usually 5; filaments about 1 mm. long, glabrous; anthers yellow, about 0·4 mm. long, 0·3 mm. wide; ovary subglobose, glabrous, about 0·6 mm. diam.; style very short; stigmas spreading, about 0·2 mm. long. Seeds horizontal, depressed-globose or lenticular, about 1·4 mm. diam.; testa shining black, minutely rugulose radially, but not pitted.

HAB.: Cultivated and waste ground; sea-level to 6,400 ft. alt.; fl. June–Sept.

DISTR.: Divisions 1, 2, 4–7. Widely distributed in Europe, Asia and Mediterranean region.

1. Ayios Neophytos, 1939, *Lindberg f.* s.n.
2. Milikouri, 1939, *Lindberg f.* s.n.! "Olympus Camp Hotel", Troödos, 1939, *Lindberg f.* s.n.; summit of Khionistra, 1955, *G. E. Atherton* 571!
4. Larnaca, 1939, *Lindberg f.* s.n.
5. Salamis, 1928 or 1930, *Druce* s.n.; Kythrea, 1939, *Lindberg f.* s.n.; also, 1941, *Davis* 3631! Trikomo, 1939, *Lindberg f.* s.n. (as *C. album* L.)!
6. Nicosia, 1939, *Lindberg f.* s.n.; Morphou, 1955, *N. Chiotellis* 473! 478!
7. Kyrenia, 1955, *G. E. Atherton* 44! 360! also, 1956, *Casey* 1735!

C. HIRCINUM Schräder, Ind. Sem. Hort. Gotting., 2 (1833), a native of S. America, frequently occurring as a casual in Great Britain and elsewhere in Europe, resembles *C. opulifolium*, but has a strong smell of decaying fish (trimethylamine), and numerous narrow, oblong-rhomboid mealy leaves usually with well-marked subhastate basal lobes and irregular, sharp marginal teeth; the inflorescences are densely congested, forming short, crowded, leafy lateral and terminal spikes; the perianth-segments are very mealy, and keeled dorsally, and the seeds have numerous radial furrows with pits between them.

C. hircinum has been collected Varosha, Famagusta, in 1970, by Dr. A. Hansen (no. 734!). The specimen (without flowers or fruits) is said to have had the characteristic smell of *C. hircinum* when fresh. Although omitted from most floras of the Mediterranean, it is not improbable that *C. hircinum* occurs on waste ground about ports and harbours.

6. C. album L., Sp. Plant., ed. 1, 219 (1753); Hume in Walpole, Mem. Europ. Asiatic Turkey, 254 (1817); Boiss., Fl. Orient., 4: 901 (1879); Druce in Rep. B.E.C., 9: 470 (1931); Post, Fl. Pal., ed. 2, 2: 430 (1933); Lindberg f., Iter Cypr., 13 (1946); Zohary, Fl. Palaest., 1: 142, t. 200 (1966); Davis, Fl. Turkey, 2: 304 (1967); Osorio-Tafall et Seraphim, List Vasc. Plants Cyprus, 32 (1973).

TYPE: "*in agris* Europae".

Erect, branched or unbranched annual (15–) 30–100 (–150) cm. high; stems subglabrous, sometimes reddish or sparsely mealy, distinctly angled; leaves ovate, rhomboid or sometimes almost lanceolate, 1–5 (–8) cm. long, 0·5-3·5 cm. wide, green or grey-mealy, apex blunt or acute, base broadly or narrowly cuneate, margins often irregularly lacerate-dentate, sometimes almost entire, generally (in Cyprus) without strongly developed lateral lobes; petiole slender, 1–3 cm. long; inflorescences terminal and axillary, forming narrow, subspiciform panicles, or sometimes lax and spreading; flowers grey-mealy, frequently rather crowded; perianth-segments 5, broadly obovate, about 1·2 mm. long, 1 mm. wide, divided almost to base, densely mealy externally with membranous margins, strongly cucullate, subacute or obtuse, gibbous or bluntly keeled dorsally; stamens usually 5;

filaments glabrous, about 1 mm. long; anthers yellow, about 0·4 mm. long, 0·3 mm. wide; ovary subglobose; about 0·4 mm. diam., glabrous; style very short; stigmas about 0·6 mm. long. Seeds lenticular, about 1·5 mm. diam.; testa black, shining, minutely striate radially.

HAB.: Cultivated and waste ground, roadsides; occasionally in Pine forest or on open mountainsides at high altitudes; sea-level to 6,400 ft. alt.; fl. June–Sept.

DISTR.: Divisions 1–5. Widespread in Europe, Asia and North Africa, and as an adventive almost throughout temperate regions of the world.

1. Between Kissonerga and Maa Beach, 1967, *Economides* in ARI 997!
2. Galata, 1880, *Sintenis & Rigo* 773 (as *C. opulifolium* Schrad.)! also, 1936, *Syngrassides* 1272! Livadhi tou Pasha, 1937, *Kennedy* 317A! Khionistra, 1940, *Davis* 1944! Prodhromos, 1949, *Casey* 889! also, 1955, *G. E. Atherton* 661! and, 1966, *Merton* in ARI 51! Trikoukkia, 1955, *Merton* 2456! Kokkinotrimithia, 1966, *Merton* in ARI 7!
3. Limassol, 1801, *Hume* s.n., and, 1941, *Davis* 3546!
4. Larnaca, 1801, *Hume* s.n., and, 1939, *Lindberg f.* s.n.; Famagusta, 1928 or 1930, *Druce* s.n.; Perivolia, 1939, *Lindberg f.* s.n.
5. Salamis, 1928 or 1930, *Druce* s.n., also, 1939, *Lindberg f.* s.n.; Kythrea, 1939, *Lindberg f.* s.n.

NOTES: Some of the above records may be misidentifications of *C. opulifolium*, with which *C. album* is frequently confused.

TRIBE 4. **Atripliceae** *C. A. Meyer* Leaves flat, often mealy; flowers generally unisexual, in clusters; male flowers ebracteolate, with a perianth; female flowers 2-bracteolate, without a perianth. Fruit enveloped by persistent bracteoles; seed often vertical; embryo peripheral.

4. ATRIPLEX *L.*

Sp. Plant., ed. 1, 1052 (1753).
Gen. Plant., ed. 5, 472 (1754).
P. Aellen in Engl., Bot. Jahrb., 70: 1–66 (1939).

Annual or perennial herbs or shrubs; leaves alternate or opposite, petiolate or sessile, glabrescent or mealy, often toothed or lobed; inflorescence spiciform or paniculate; flowers unisexual in terminal or axillary clusters, or occasionally solitary, the males with 5 (3) perianth-segments and 5 (3) stamens, the females without a perianth, but enveloped by 2 persistent, often accrescent, free or partly connate bracteoles; stigmas 2. Seeds generally vertical, sometimes dimorphic; testa smooth or pitted.

About 200 species widely distributed in temperate and subtropical regions.

Shrubs and woody subshrubs:
　Erect shrub; inflorescence leafless, terminal, paniculate; leaves broadly ovate, deltoid or
　　subhastate, 0·5–3 (–5) cm. wide -　-　-　-　-　-　-　-　**1. A. halimus**
　Prostrate or sprawling subshrub; inflorescence leafy; flowers in small axillary clusters; leaves
　　ovate or narrowly oblong, 0·5–0·8 cm. wide　-　-　-　-　**5. A. semibaccata**
Herbaceous annuals:
　Leaves densely grey-mealy or lepidote, margins erose-dentate or lobulate all round;
　　bracteoles becoming hard in fruit -　-　-　-　-　-　-　**2. A. rosea**
　Leaves green or thinly mealy, not erose-dentate or lobulate all round; bracteoles not
　　becoming hard in fruit:
　　Leaves (even the upper) usually hastate; inflorescence usually leafy - **3. A. prostrata**
　　Leaves (especially the upper) not hastate, without distinct basal lobes; inflorescence
　　　usually leafless -　-　-　-　-　-　-　-　-　-　-　-　**4. A. patula**

1. **A. halimus** *L.*, Sp. Plant., ed. 1, 1052 (1753); Sibth. et Sm., Fl. Graec. Prodr., 2: 266 (1816), Fl. Graec., 10: 52, t. 962 (1840); Poech, Enum. Plant. Ins. Cypr., 13 (1842); Unger et Kotschy, Die Insel Cypern, 222 (1865); Boiss., Fl. Orient., 4: 916 (1879); H. Stuart Thompson in Journ. Bot., 44: 337 (1906); Holmboe, Veg. Cypr., 65 (1914); Post, Fl. Pal., ed. 2, 2: 436 (1933); Lindberg f., Iter Cypr., 12 (1946); Chapman, Cyprus Trees and Shrubs, 35 (1949); Zohary, Fl. Palaest., 1: 145, t. 204 (1966); Davis, Fl. Turkey, 2: 306 (1967); Osorio-Tafall et Seraphim, List Vasc. Plants Cyprus, 32 (1973).
TYPE: "*in Hispaniae, Lusitaniae, Virginiae sepibus maritimis*".

Erect shrub to 2 m. high; older branches covered with pale grey bark peeling off in narrow longitudinal strips, young shoots pallid, mealy, obtusely angled; leaves alternate, persistent, ovate, deltoid or subhastate; silvery-grey, 1–4 (–6) cm. long, 0·5–3 (–5) cm. wide, apex obtuse or mucronate, base broadly cuneate, margins entire, dentate or lobed; petiole generally less than 1 cm. long, often very short; inflorescences terminal, paniculate; flowers in dense, congested or remote, lateral and terminal clusters; bracts much reduced or wanting; male flowers often at the centre of the flower-clusters, perianth-segments 5, oblong-obovate, subacute, incurved, membranous, lanuginose externally, about 0·8 mm. long, 0·6 mm. wide; stamens 5, filaments glabrous, up to 1 mm. long; anthers yellow, about 0·5 mm. long and almost as wide; female flowers with 2, free, ovate or deltoid, sessile, lanuginose bracteoles, about 1 mm. long, 1·3 mm. wide at anthesis, accrescent to 4–6 mm. in fruit; ovary glabrous, laterally compressed, suborbicular, about 0·2 mm. diam.; style very short; stigmas 2, about 0·8 mm. long, often unequal. Seed strongly compressed, suborbicular, about 1·5 mm. diam.; testa dark brown, conspicuously pitted.

HAB.: Sandy and rocky seashores, sometimes in dry exposed situations inland; sea-level to 500 ft. alt.; fl. July–Oct.

DISTR.: Divisions 3–8. Widespread in the Mediterranean region.

3. Cape Gata near Akrotiri, 1862, *Kotschy*; Episkopi, 1967, *Merton* in ARI 666!
4. Larnaca, 1860, *Hanbury & Hooker* s.n.! also, 1862, *Kotschy*, and, [12 Sept.] 1880, *Sintenis* 936! Perivolia, 1939, *Lindberg f.* s.n.; Larnaca Salt Lake, 1967, *Merton* in ARI 393!
5. Salamis, 1939, *Lindberg f.* s.n.! also, 1973, *P. Laukkonen* 389! Trikomo, 1939, *Lindberg f.* s.n.
6. Yerolakkos, 1967, *Economides* in ARI 1022!
7. Glykyotissa Island near Kyrenia, 1955, *G. E. Atherton* 337! Lefkoniko–Akanthou road, 1970, *A. Genneou* in ARI 1623!
8. Klidhes Islands [? date], *Chapman* 20!

A. NUMMULARIA *Lindl.* (in Mitchell, Journ. Trop. Austral., 64; 1848), a shrub 1–3 m. high, superficially very similar to *A. halimus* L., but dioecious, the male plants with conspicuous, much-branched terminal and subterminal panicles, the females with large, orbicular or broadly rhomboid bracteoles, free or almost free to the shortly stipitate base, is reported as planted at Nicosia (Bovill, Rep. Plant. Work Cyprus, 8; 1915) and may have persisted. It is a native of Australia.

[A. LEUCOCLADA *Boiss.* (Diagn., 1, 12: 95; 1853) a perennial herb 30–100 cm. high with a woody base and grey-mealy leaves similar to those of *A. halimus* L., is reported from Nicosia by Post (in Mém. Herb. Boiss., 18: 99; 1900), but some error is likely, since the species, normally confined to desert regions of western Asia, has not been found by any other collector. *A. leucoclada* occurs from Syria, Palestine and Egypt eastwards to Central Asia.]

2. **A. rosea** L., Sp. Plant., ed. 2, 1493 (1763); Boiss., Fl. Orient., 4: 911 (1879); Post, Fl. Pal., ed. 2, 2: 434 (1933); Lindberg f., Iter Cypr., 12 (1946); Rechinger f. in Arkiv för Bot., ser. 2, 1: 420 (1950); Zohary, Fl. Palaest., 1: 148, t. 211 (1966); Davis, Fl. Turkey, 2: 310 (1967); Osorio-Tafall et Seraphim, List Vasc. Plants Cyprus, 32 (1973).
    [*A. laciniata* (non L.) Kotschy in Unger et Kotschy, Die Insel Cypern, 222 (1865).]
    [*A. tatarica* (non L.) Druce in Rep. B.E.C., 9: 470 (1931).]
    TYPE: "*in* Europa *australiore*".

Erect, divaricately branched, grey-lepidote annual 30–80 cm. high; stems smooth, subterete or bluntly angled, leafy; leaves ovate- or deltoid-rhomboid, 1–5 cm. long, 0·5–4 cm. wide, apex acute, margins irregularly erose-dentate or lobulate, base cuneate; nervation rather obscure; petiole occasionally up to 2 cm. long, but generally short or sometimes wanting; inflorescence commonly much branched, pyramidal, the flowers clustered in the axils of small, foliaceous bracts, or forming ebracteate spikes at the tips of the partial inflorescences; male flowers with 5 oblong, subacute perianth-segments about 0·6 mm. long, 0·4 mm. wide, segments membranous, lanuginose-papillose externally; stamens usually 5; filaments glabrous, about 0·8 mm. long; anthers yellow, suborbicular, about 0·4 mm. diam.; female flowers with 2 broadly deltoid-rhomboid bracteoles about 2 mm. long and a little wider, with denticulate margins and a connate base; ovary strongly compressed, suborbicular, glabrous, about 1·3 mm. wide, with conspicuous, membranous lateral wings; style very short; stigmas about 0·3 mm. long. Fruiting bracteoles indurated, accrescent, 5–8 mm. wide, rugose or tuberculate dorsally; seed suborbicular, strongly compressed, about 2 mm. diam.; testa brown, pitted.

HAB.: Cultivated and waste ground, roadsides; sea-level to 5,650 ft. alt.; fl. June–Oct.

DISTR.: Divisions 2–4, 6–8. Mediterranean region eastwards to Central Asia.

2. Kykko, 1913, *Haradjian* 960! also, 1939, *Lindberg f.* s.n.! Amiandos, 1937, *Syngrassides* 1689! and, 1956, *Merton* 2809! Troödos, 1938, *Kennedy* 1730! and, 1950, *F. M. Probyn* s.n.!
3. Near Limassol, 1859, *Kotschy*.
4. Umber pits near Larnaca, 1862, *Kotschy* 35; also near Larnaca, 1880, *Sintenis* 754! and, 1928 or 1930, *Druce* s.n.; and, 1939, *Lindberg f.* s.n.!
6. Khrysiliou, 1936, *Syngrassides* 6315! Syrianokhori, 1952, *Merton* 787!
7. Ayios Andronikos above Kythrea, 1880, *Sintenis & Rigo* 754 a!
8. E. of Khelones (Karpas), 1970, *A. Hansen* 203!

3. A. prostrata *Boucher ex DC.*, Fl. Franç., 3: 387 (1805); Gustafsson in Opera Bot., 39: 21–27 (1976).
    A. triangularis Willd., Sp. Plant., 4: 963 (1806); Taschereau in Canad. Journ. Bot., 50: 1583 (1972).
    A. hastata L. var. *salina* (? (Wallr.) Gren. et Godr.) Lindberg f., Iter Cypr., 12 (1946).
    [*A. hastata* (non L.) Boiss., Fl. Orient., 4: 909 (1879); Druce in Rep. B.E.C., 9: 470 (1931); Post, Fl. Pal., ed. 2, 2: 433 (1933); Zohary, Fl. Palaest., 1: 149, t. 214 (1966); Davis, Fl. Turkey, 2: 311 (1967); Osorio-Tafall et Seraphim, List Vasc. Plants Cyprus, 32 (1973).]
    TYPE: France; "Elle a été découverte par M. Boucher le long du canal de Saint-Valery".

Erect or decumbent annual, up to 100 cm. high; stems usually much branched, bluntly angular, subglabrous or thinly mealy; leaves opposite or the upper alternate, green or grey-mealy, 1·5–8 cm. long, 0·4–5 cm. wide, broadly or narrowly deltoid, with an acute or subacute apex and distinctly hastate-lobed base, margins entire or irregularly dentate; nervation obscure; petiole usually distinct, 1–4 cm. long; inflorescence paniculate, generally consisting of flower-clusters distributed along lax or congested, spiciform, partial inflorescences; bracts foliaceous, the lower similar to the leaves, the upper much reduced or wanting; male flowers intermixed with female, perianth-segments 5, bluntly oblong, about 0·5 mm. long, 0·3 mm. wide, membranous, thinly lanuginose externally; stamens 5; filaments very short, glabrous; anthers suborbicular, about 0·4 mm. diam.; bracteoles of female flowers deltoid-rhomboid, acute, mealy and often roughly tuber-culate or rugose externally, about 2 mm. long, 1·5 mm. wide, connate along basal margins; ovary glabrous, strongly compressed, suborbicular, about 1 mm. diam.; style almost obsolete; stigmas 0·4 mm. long. Seeds dimorphic, the smaller bluntly lenticular, about 1·5 mm. diam., with a blackish, minutely striatulate testa, and usually enveloped by small bracteoles; the larger lenticular with a distinct submarginal rim, about 2 mm. diam., with a dark, reddish-brown, minutely rugulose-punctate testa, and bracteoles up to 5 mm. long, 4 mm. wide.

HAB.: Sandy seashores; near sea-level; fl. July–Oct.

DISTR.: Divisions 1, 3–5. Widespread in Europe and the Mediterranean region east to Central Asia.

1. Paphos, 1928 or 1930, *Druce* s.n.; also, 1939, *Lindberg f.* s.n.

3. Kouklia, 1928 or 1930, *Druce* s.n.
4. Larnaca, 1928 or 1930, *Druce* s.n.; Ayia Napa, 1939, *Lindberg f.* s.n.
5. Salamis, 1939, *Lindberg f.* s.n.!

NOTES.: The application of the well-known and long-established name *A. hastata* to the above species has been challenged since 1800 (cf. Smith in Eng. Bot., 13: t. 936), and, on the strength of a specimen in herb. Linn. (1221/17) has been referred to the plant previously named *A. calotheca* (Rafn) Fr., which is confined to a limited area of Scandinavia and the Baltic.

It may be questioned, however, if *A. hastata*, described by Linnaeus as "valvulis femineis magnis deltoidibus sinuatis" can be lectotypified by LINN 1221/17, which has the conspicuously lacerate bracteoles of *A. calotheca*, and which is more likely to have been the basis of the footnote added by Linnaeus to the diagnosis of *A. laciniata* L. in Fl. Suecica, ed. 2, 364 (1755).

4. A. patula *L.*, Sp. Plant., ed. 1, 1053 (1753); Boiss., Fl. Orient., 4: 909 (1879); Post, Fl. Pal., ed. 2, 2: 433 (1933); A. K. Jackson in Kew Bull., 1937: 345 (1937); Lindberg f., Iter Cypr., 12 (1946); Davis, Fl. Turkey, 2: 311 (1967); Osorio-Tafall et Seraphim, List Vasc. Plants Cyprus, 32 (1973).

TYPE: "*in* Europae *cultis, ruderatis*".

Erect or sprawling, much-branched, glabrous or thinly mealy annual to 1 m. high; stems angular, usually with long internodes; leaves alternate or opposite, the lower oblong-hastate or ovate-oblong without lateral lobes, 4–7 cm. long, 1·5–3 cm. wide, green or slightly glaucous, obtuse or subacute with entire or irregularly toothed margins and a cuneate base; the upper generally narrowly oblong, lanceolate, or linear-lanceolate, 2–7 cm. long, 0·3–2 cm. wide, entire or subentire, acute, with a tapering, cuneate base; nervation obscure; petioles usually distinct, up to 3 cm. long; inflorescence lax, paniculate, the flowers solitary or clustered in elongate, largely ebracteate, spiciform partial inflorescences; bracts (where developed) foliaceous, similar to the leaves; male flowers intermixed with females; perianth-segments 5, united almost half-way up, oblong, obtuse, cucullate, mealy or thinly lanuginose externally, about 0·8 mm. long, 0·4 mm. wide; stamens 5; filaments glabrous, 0·8 mm. long; anthers yellow, suborbicular, about 0·4 mm. diam.; bracteoles of female flowers connate along basal margin, broadly deltoid-rhomboid, about 1·2 mm. long, 1 mm. wide, thinly lanuginose, with a prominent acute or subacute apex, dentate margins, and a dentate-tuberculate dorsal surface; ovary compressed-subglobose, about 0·5 mm. diam., glabrous; style obsolete; stigmas spreading, about 0·3 mm. long. Seeds apparently uniform, lenticular, about 2 mm. diam.; testa dark shining brown, almost smooth or very minutely punctulate.

HAB.: Gardens, cultivated fields, roadsides, waste ground; 3,500–5,000 ft. alt.; fl. June–Oct.

DISTR.: Division 2. Widespread in Europe and Asia, but evidently rare in the eastern Mediterranean region.

2. Platres, 1934, *Syngrassides* 522! Prodhromos, 1939, *Lindberg f.* s.n.! also, 1949, *Casey* 895! 962! and, 1955, *G. E. Atherton* 618! Troödos, 1950, *F. M. Probyn* s.n.! Polystipos, 1955, *Merton* 2429!

5. A. semibaccata R.Br., Prodr. Fl. Nov. Holl., 1: 406 (1810); Post, Fl. Pal., ed. 2, 2: 434 (1933); Zohary, Fl. Palaest., 1: 150, t. 215 (1966).

TYPE: Australia; New South Wales, near Port Jackson, 1802–1805, *R. Brown* 3022 (K!).

Prostrate or sprawling subshrub; stems subglabrous, pallid, bluntly angled, up to 50 cm. long, usually much branched; leaves ovate or narrowly oblong, 0·6–3 cm. long, 0·5–0·8 cm. wide, subglabrous or thinly mealy on both surfaces, apex acute or obtuse, margins entire or serrulate, base tapering to a short, indistinct petiole; nervation obscure, minutely reticulate-anastomosing; flowers solitary or clustered in the axils of leaf-like bracts towards the tips of the branches; male flowers usually several in the centre of the flower-clusters; perianth-segments 5, connate almost to apex, about 0·8 mm. long, 0·4 mm. wide, thinly lanuginose externally, incurved or cucullate; stamens 5; filaments flattened, about 0·7 mm. long; anthers yellow, suborbicular, about 0·4 mm. diam.; bracteoles of female flowers broadly rhomboid, connate along their basal margins, shortly stipitate, about 2–2·5 mm. long, 2 mm. wide at anthesis, accrescent to 4–5 mm. and indurating or becoming red and fleshy in fruit, surface rugose-reticulate, apex acute, margins entire or sharply serrulate; ovary glabrous, compressed-globose, about 0·7 mm. diam.; style almost absent; stigmas 4–5 mm. long. Seeds lenticular, about 2 mm. diam.; testa dark, shining brown, minutely punctulate.

HAB.: Brackish pastures and garigue; 100–500 ft. alt.; fl. Jan.–Dec.

DISTR.: Divisions 4, 6. A native of Australia, naturalized in Palestine, New Mexico, Arizona and California.

4. Kouklia reservoir, 1968, *Economides* in ARI 1186!
6. Nicosia, near the English School, 1958, *P. H. Oswald* 26!

NOTES: Creeping Saltbush. Said to be a good fodder-plant for sheep, and able to survive in arid, saline habitats. Probably introduced as a pasture plant.

5. HALIMIONE *Aellen*

Verh. Naturf. Ges. Basel, 49: 121 (1938).

Annual or perennial dioecious or monoecious herbs or subshrubs; stems usually angular, leaves opposite or alternate, sessile or petiolate, ovate-spathulate or oblong-elliptic; entire or subentire; flowers unisexual, in terminal spikes or panicles; male flowers with a 4–5-lobed perianth, but without bracteoles; stamens 4–5; female flowers with 2, obdeltoid bracteoles united almost to apex; style very short; stigmas 2. Seeds vertical; testa almost smooth.

Three species in Europe, the Mediterranean region and eastwards to Central Asia; closely allied to *Atriplex*, from which it differs chiefly in the shape and structure of the bracteoles of the female flowers.

1. **H. portulacoides** (*L.*) *Aellen* in Verh. Naturf. Ges. Basel, 49: 126 (1938); Zohary, Fl. Palaest., 1: 151, t. 216 (1966); Davis, Fl. Turkey, 2: 312 (1967); Osorio-Tafall et Seraphim, List Vasc. Plants Cyprus, 32 (1973).
    Atriplex portulacoides L., Sp. Plant., ed. 1, 1053 (1753); Boiss., Fl. Orient., 4: 913 (1879); Post in Mém. Herb. Boiss., 18: 89 (1900); H. Stuart Thompson in Journ. Bot., 44: 337 (1906); Holmboe, Veg. Cypr., 65 (1914); Post, Fl. Pal., ed. 2, 2: 434 (1933); Lindberg f., Iter Cypr., 12 (1946); Chapman, Cyprus Trees and Shrubs, 36 (1949).
    Obione portulacoides (L.) Moq., Chenop. Monogr. Enum., 75 (1840); Unger et Kotschy, Die Insel Cypern, 222 (1865); Sintenis in Oesterr. Bot. Zeitschr., 31: 191, 194, (1881).
    TYPE: "*in* Oceani: Europae *septentrionalis littoribus*".

Sprawling subshrub with much-branched, angular, pallid stems up to 1 (–1·5) m. long; leaves persistent, opposite, subsessile or obscurely petiolate, narrowly oblong-elliptic, 2–5 (–7) cm. long, 0·3–1·5 cm. wide, densely grey-lepidote, apex obtuse or subacute, margins entire, base gradually tapering; nervation obscure; inflorescence paniculate; flowers in discrete, nodulose clusters; the male flowers intermixed with female or on separate plants, perianth-segments 5, oblong, obtuse, incurved, about 0·8 mm. long, 0·4 mm. wide, united towards base, lanuginose externally; stamens 5; filaments glabrous, about 0·8 mm. long; anthers yellow, about 0·4 mm. long, 0·3 mm. wide; female flowers with 2, broadly obdeltoid, 3-lobed, rather fleshy bracteoles about 3 mm. long, 3·8 mm. wide, median lobe of bracteole generally longer than the 2 laterals, and external surface of bracteole furnished with several, finger-like appendages; ovary closely invested by bracteoles, compressed-subglobose, about 1·3 mm. diam.; style about 1 mm. long, long-papillose, stigmas spreading, about 1 mm. long. Seed lenticular, about 2 mm. diam.; testa black, shining, minutely, densely pitted.

HAB.: Edges of salt-marshes; near sea-level; fl. June–Oct.

DISTR.: Divisions 3–5. W. Europe and the Mediterranean region.

3. Between Kolossi and Akrotiri, 1862, *Kotschy* s.n.
4. Larnaca, 1880–1970, *Sintenis* 976! *Syngrassides* 493! *Chapman* 410! *Merton* in ARI 413! *A. Genneou* in ARI 1590! etc. Famagusta, 1938, *Chapman* 331! and, 1929, *Lindberg f.* s.n.! Near Paralimni, 1948, *Mavromoustakis* s.n.!
5. Salamis, 1968, *A. Genneou* in ARI 1263!

NOTES: In Cyprus, as elsewhere, *H. portulacoides* appears to be more often dioecious than most authors indicate; female flowers are not available from Cyprus material, and have been described from British specimens.

SPINACIA OLERACEA *L.*, Sp. Plant., ed. 1, 1027 (1753), Spinach, a succulent-leaved, dioecious annual, resembles *Atriplex*, but has 4–5 stigmas, and fruits enveloped by (2–) 4–5, more or less connate, spinous-tipped bracteoles. It is recorded by Kotschy (Die Insel Cypern, 223; 1865) as "semi-wild and cultivated at Livadhia near Larnaca, Nicosia", and is no doubt still grown as a vegetable, though it cannot be regarded as a naturalized plant on the island.

TRIBE 5. **Camphorosmeae** *Moq.* Herbs or shrubs with sessile, linear or subulate, entire, hairy leaves; flowers in cymes or solitary and axillary, hermaphrodite or unisexual, the latter often with rudimentary stamens or pistils; bracteoles wanting; perianth-segments 4–5, more or less connate; stamens 4–5; stigmas usually 2. Seeds with a peripheral, curved embryo.

[CAMPHOROSMA MONSPELIACA *L.*, Sp. Plant., ed. 1, 112 (1753) is recorded from Cyprus by Boissier (Fl. Orient., 4: 920; 1879), who attributes the record to Sibthorp. The latter, however, mentions only *Camphorosma pteranthus* L. (= *Pteranthus dichotomus* Forssk.), nor, so far as I am aware, has any other collector added *C. monspeliaca* to the flora of the island.]

KOCHIA SCOPARIA (*L.*) *Schrader* in Neues Journ., 3: 85 (1809) (*Chenopodium scoparia* L., Sp. Plant., ed. 1, 221; 1753) is represented in Cyprus gardens by the cultivars 'Trichophylla' and 'Childsii'. These are neat, bushy annuals 50–100 cm. high, with narrow-linear or strap-shaped leaves, and inconspicuous, greenish flowers; they are popularly (and aptly) named "Summer Cypress", and are mostly employed to relieve the flatness of summer flower-beds. Specimens have been collected in the grounds of the Department of Agriculture (1937, *Syngrassides* 1724!).

TRIBE 6. **Salicornieae** (*Dumort.*) *Moq.* Herbs and shrubs, often with fleshy, jointed stems; leaves and bracts (or their vestiges) opposite or alternate; flowers often embedded in the stems, usually hermaphrodite, commonly in threes; perianth-segments generally 3–4, usually connate; stamens 1–2. Seed with a curved embryo; endosperm sometimes sparse or wanting.

6. **HALOPEPLIS** *Bunge ex Ungern-Sternb.*
Vers. Syst. Salicorn., 102 (1866).

Annual herbs and subshrubs; stems swollen at the nodes or moniliform but not jointed; leaves mostly alternate, amplexicaul, fleshy, subglobose or ovoid, frequently shed or shrivelled before plants are fully mature; inflorescences densely spicate, fleshy; bracts spirally arranged; flowers hermaphrodite or female, usually in groups of 3, sometimes connate and adnate to the subtending bract; perianth-segments 3, or 3-lobed, small, not winged; stamens 1–2; ovary laterally compressed; stigmas 2, subulate. Utricle included, membranous; seeds with a hooked embryo encircling copious endosperm.

Three species in the Mediterranean region, western Asia, South Africa.

1. **H. amplexicaulis** (*Vahl*) *Ungern-Sternb. ex Cesati* in Cesati, Passer. et Gibelli, Comp. Fl. Ital., 271 (1874); Boiss., Fl. Orient., 4: 934 (1879); Post, Fl. Pal., ed. 2, 2: 441 (1933); Lindberg f., Iter Cypr., 13 (1946); Zohary, Fl. Palaest., 1: 154, t. 222 (1966); Davis, Fl. Turkey, 2: 319 (1967); Osorio-Tafall et Seraphim, List Vasc. Plants Cyprus, 32 (1973).

   Salicornia amplexicaulis Vahl, Symb. Bot., 2: 1 (1791).

   TYPE: Tunis; "Habitat ad littora lacus prope Bardo" (C).

Erect, much-branched, fleshy annual 6–25 cm. high; stems terete, slightly swollen at the nodes, glabrous, glaucous, often rather woody at the base; leaves globose or depressed-globose, 1·5–5 mm. diam., alternate, amplexicaul, green, glaucous or pinkish; flower-spikes numerous, terminal and lateral, cylindrical with a blunt or subacute apex, 5–20 mm. long, 2·5–4 mm. wide; bracts fleshy, closely imbricate, spirally arranged, about 1–1·5 mm. diam., obtuse or subacute; flowers in groups of 3, perianth-segments connate, membranous; stamen 1; filament glabrous about 1 mm. long; anther yellow, oblong, about 1 mm. long, 0·5 mm. wide; ovary minute, compressed, about 0·2 mm. long; style about 0·2 mm. long; stigmas about 0·2 mm. long. Seed lenticular, or somewhat oblong, about 1 mm. diam.; testa brown, conspicuously hyaline-papillose.

HAB.: Margins of salt-lakes; near sea-level; fl. July–Sept.

DISTR.: Division 4. Widely distributed in the Mediterranean region.

4. Larnaca Salt Lake, 1930, *C. B. Ussher* 95! also, 1939, *Lindberg f.* s.n.! and, 1978, *J. Holub* s.n.! Salt-marshes near Famagusta, 1939, *Lindberg f.* s.n.!

7. HALOCNEMUM *M. Bieb.*
Fl. Taur.-Cauc., 3: 3 (1819).

Much-branched subshrubs; stems at first fleshy and apparently jointed, becoming woody with age; leaves reduced to opposite, connate scales; inflorescences spicate, terminal and lateral, opposite; flowers hermaphrodite, in groups of 3, not connate; perianth-segments membranous, free to near the base, concave-galericulate; stamen 1; ovary ovoid; stigmas 2, subulate. Utricles membranous, compressed; seed pendulous from a curved funicle; embryo curved, peripheral; endosperm present.

One species in the Mediterranean region eastwards to Central Asia.

1. **H. strobilaceum** (*Pall.*) *M. Bieb.*, Fl. Taur.-Cauc., 3: 3 (1819); Boiss., Fl. Orient., 4: 936 (1879); Post, Fl. Pal., ed. 2, 2: 441 (1933); Lindberg f., Iter Cypr., 13 (1946); Zohary, Fl. Palaest., 1: 155, t. 223 (1966); Davis, Fl. Turkey, 2: 320 (1967); Osorio-Tafall et Seraphim, List Vasc. Plants Cyprus, 32 (1973).

   Salicornia strobilacea Pall. [Reise Prov. Russ. Reichs, 1: 481, t. B, figs. 1, 2 (1771)] Illustr. Plant., 9, t. 4 (1803).

   TYPE: Pallas cites numerous localities in Central Asia; species probably best typified by plate in Pallas, Illustr. Plant., t. 4 (K!).

Erect subshrub 15–50 cm. high, stems much branched, green or glaucous, at first fleshy, becoming woody and pale grey-brown with age; leaves reduced to small, connate, papyraceous scales forming shallow cups along the stems; internodes usually very short; inflorescences on short, subglobose or cylindrical, opposite or whorled branches; bracts reniform or suborbicular, 2–3 mm. wide; perianth-segments membranous, the 2 lateral distinctly gibbous-galericulate, about 1·2 mm. long, 0·8 mm. wide, the median somewhat smaller and less concave; stamen 1; filament about 1·5 mm. long, glabrous; anther oblong, yellow, about 1·2 mm. long, almost 1 mm. wide; ovary flask-shaped, about 1 mm. long; style about 0·3 mm. long; stigmas subulate, about 0·5 mm. long. Seed broadly oblong, compressed laterally, about 1 mm. long, 0·8 mm. wide; testa brown, minutely and densely papillose.

HAB.: Salt-marshes and margins of salt-lakes; near sea-level; fl. Aug.–Oct.

DISTR.: Divisions 3–5. Mediterranean region and eastwards to Central Asia.

3. Limassol Salt Lake, 1939, *Mavromoustakis* 93 !
4. Famagusta, 1939, *Lindberg f.* s.n.! Larnaca Salt Lake, 1967, *Merton* in ARI 395! and, 1978, *J. Holub* s.n.!
5. Near Salamis, 1956, *Poore* 37 !

8. ARTHROCNEMUM *Moq.*

Chenop. Monogr. Enum., 111 (1840).

Perennial subshrub; stems erect or spreading, usually much branched, at first fleshy, becoming woody later; leaves reduced to scales, opposite, connate, forming a ring or shallow cup around the stem; inflorescences spicate, cylindrical, fleshy, terminal and lateral; bracts suborbicular, persistent; flowers in threes in cavities in the rhachis, protruding a little beyond subtending bract; perianth tubular, the segments united almost to apex; stamens 2; ovary ovoid, compressed; style usually long; stigmas 2, subulate. Utricles membranous; seeds ovoid, oblong, or lenticular; testa smooth or papillose; embryo curved; endosperm copious.

About 10 species, found in saline habitats almost throughout temperate and tropical regions of the world.

1. **A. macrostachyum** (*Moric.*) *Moris et Delponte* in Ind. Sem. Hort. Reg. Taur., 35, t. 2 (1854); Zohary, Fl. Palaest., 1: 156, t. 226 (1966).
    Salicornia glauca Del., Descr. Égypte Hist. Nat., 2: 49 (1813) non *S. glauca* Stokes (1812) nom. illeg.
    S. macrostachya Moric., Fl. Venet., 1: 2 (1820).
    Arthrocnemum glaucum Ungern-Sternb. in Atti Congr. Bot. Firenze, 283 (1876); Boiss., Fl. Orient., 4: 932 (1879); Post, Fl. Pal., ed. 2, 2: 440 (1933); Lindberg f., Iter Cypr., 12 (1946); Davis, Fl. Turkey, 2: 320 (1967); Osorio-Tafall et Seraphim, List Vasc. Plants Cyprus, 32 (1973).
    TYPE: Italy; Venice, "secus viam quae ad portum *Malamocco* ducit".

Erect, much-branched subshrub 20–50 cm. high; young stems succulent, glaucous, older stems woody, pale grey or ochraceous; reduced leaves forming shallow, papyraceous cups at intervals of 7–10 mm. along the sterile portions of the stem; inflorescence spicate, cylindrical-caudate, often 7–8 cm. long, tapering slightly to apex, the basal bracts interrupted, the upper closely overlapping; bracts papery, connate, forming shallow cups about 5–6 mm. diam; flowers in groups of three, subequal, free, ranged horizontally and slightly exceeding the subtending bract; perianth tubular, narrowly obovate, bluntly angled, about 2 mm. long, 1 mm. wide, tapering to base, apex with 4 short, obscure, incurved, subacute lobes; stamens 2; filaments glabrous, about 1·8 mm. long; anthers yellow, oblong, about 1·2 mm. long, 0·8 mm. wide; ovary flask-shaped, about 1·8 mm. long, 0·8 mm. wide; style 1·3 mm. long; stigmas spreading, about 1 mm. long. Fruit enveloped by a persistent, rather spongy perianth; pericarp membranous; seed broadly oblong, strongly compressed, about 1 mm. long, 0·7 mm. wide; testa rich brown, minutely papillose-rugulose; embryo peripheral, somewhat curved, situated to the side of the mass of endosperm.

HAB.: Salt-marshes and salt-lakes; near sea-level; fl. May–July.

DISTR.: Divisions 3, 4. Mediterranean region; Atlantic Islands and south to Senegal; Red Sea.

3. Limassol Salt Lake, 1939, *Lindberg f.* s.n.! and, 1941, *Davis* 3548!
4. Larnaca, 1880, *Sintenis & Rigo* s.n.! also, 1939, *Lindberg f.* s.n.! Famagusta, 1939, *Lindberg f.* s.n.!

NOTES: Davis notes that at Limassol Salt Lake this species is locally dominant, "going nearer the water than any other perennial". Unlike most halophytic *Chenopodiaceae*, it seems to flower in early summer and make its vegetative growth later in the year.

9. SALICORNIA *L.*

Sp. Plant., ed. 1, 3 (1753).
Gen. Plant., ed. 5, 4 (1754).

Annual or perennial herbs or subshrubs; stems usually branched, jointed, at first fleshy, sometimes woody later; leaves reduced to opposite, connate scales, forming shallow cups or narrow rings about the nodes; inflorescences terminal and lateral, spicate, cylindrical, fleshy; bracts decussate, scale-like, persistent; flowers sessile in groups of three, tightly pressed together, embedded in separated cavities in the lower abaxial surface of the superimposed bracts and not protruding; perianth thick, spongy, persistent, often angular, completely enveloping the ovary, with the stigmas and anthers alone exposed at anthesis; stamen usually 1; ovary flask-shaped; style short; stigmas subulate. Utricle membranous, immersed in the persistent, spongy perianth; seed compressed, ovoid, oblong or lenticular; testa covered with curved or hooked hairs or tuberculate; embryo curved or conduplicate; endosperm wanting.

About 30 species with a cosmopolitan distribution in saline habitats.

Plants suffruticose, with a hard, woody base; inflorescences usually compact, short, seldom exceeding 2 cm. in length - - - - - - - - - - - - **1. S. fruticosa**
Plants annual, without a woody base; inflorescences often elongate, commonly more than 5 cm. long - - - - - - - - - - - - - - **2. S. europaea**

1. **S. fruticosa** (*L.*) *L.*, Sp. Plant., ed. 2, 5 (1762); Boiss., Fl. Orient., 4: 932 (1879); Holmboe, Veg. Cypr., 65 (1914); Post, Fl. Pal., ed. 2, 2: 440 (1933); Uhlbrich in Engl. et Prantl, Pflanzenfam., ed. 2, 16c: 548 (1934); Lindberg f., Iter Cypr., 13 (1946); Chapman, Cyprus Trees and Shrubs, 36 (1949).
    S. europaea L. var. *fruticosa* L., Sp. Plant., ed. 1, 3 (1753).
    Arthrocnemum fruticosum (L.) Moq., Chenop. Monogr. Enum., 111 (1840) pro parte quoad nomen; Zohary, Fl. Palaest., 1: 156, t. 224 (1966); Davis, Fl. Turkey, 2: 320 (1967); Osorio-Tafall et Seraphim, List Vasc. Plants Cyprus, 32 (1973).
    TYPE: "*in* Europae *maritimis*".

Erect, much-branched subshrub 20–50 cm. high, young stems succulent, older stems woody, pale greyish brown; reduced leaves forming shallow, papyraceous cups at intervals of 5–12 mm. along the stems; inflorescences spicate, terminal and lateral, cylindrical, seldom more than 2 cm. long, scarcely tapering to the blunt apex; bracts all closely overlapping, papery at apex, about 2·5 mm. diam., decussate, with subacute apices; flowers in groups of 3, closely pressed together, the median slightly raised above the 2 lateral; perianth about 1 mm. long, 0·8 mm. wide, thick and spongy, angled, apex without distinct lobes; stamen 1; filament glabrous about 1 mm. long; anther yellow, oblong, 1 mm. long, 0·5 mm. wide; ovary glabrous, flask-shaped, about 0·8 mm. long, 0·4 mm. wide; style about 0·2 mm. long; stigmas 2, about 0·5 mm. long. Utricle enclosed in the persistent spongy perianth; seed broadly oblong, about 1·2 mm. long, 0·8 mm. wide; testa dark brown, minutely papillose-tuberculate with short, hyaline papillae; embryo conduplicate.

HAB.: Salt-marshes and salt-lakes; near sea-level; fl. Sept.–Nov.

DISTR.: Divisions 3, 4. Widespread in the Mediterranean region; Atlantic Islands.

3. Akrotiri, 1905, *Holmboe* 690; also, 1939, *Mavromoustakis* 94! Limassol Salt Lake, 1939, *Mavromoustakis* 51! 64! also, 1939, *Lindberg f.* s.n.
4. Famagusta, 1929, *C. B. Ussher* 79! Larnaca Salt Lake, 1905, *Holmboe* 93; also, 1970, *A. Genneou* in ARI 1598! Ayia Napa, 1939, *Lindberg f.* s.n.!

NOTES: Without flowers or fruits this species is difficult to distinguish from *Arthrocnemum macrostachyum* (Moric.) Moris et Delponte. Several sterile specimens have, for this reason, been deliberately omitted from the above distribution. Unlike *Arthrocnemum macrostachyum*, the inflorescences of *Salicornia fruticosa* do not seem to develop until late August or early September. It is probably common in suitable habitats on the island.

2. S. europaea *L.*, Sp. Plant., ed. 1, 3 (1753); Zohary, Fl. Palaest., 1: 157, t. 227 (1966); Davis, Fl. Turkey, 2: 321 (1967); Osorio-Tafall et Seraphim, List Vasc. Plants Cyprus, 33 (1973).
S. herbacea L., Sp. Plant., ed. 2, 5 (1762); Unger et Kotschy, Die Insel Cypern, 222 (1965); Boiss., Fl. Orient., 4: 933 (1879); Holmboe, Veg. Cypr., 65 (1914); Post, Fl. Pal., ed. 2, 2: 440 (1933); Lindberg f., Iter Cypr., 13 (1946).
TYPE: "*in* Europae litoribus maritimis".

Erect, spreading or prostrate annual, 8–30 (–60) cm. high; stems fleshy, subterete, sparsely or richly branched, the branches frequently as long as the main stem and sometimes subdivided; leaves reduced to shallow cups or narrow sheaths as in *S. fruticosa*, frequently more than 1·5 cm. apart on the main stems; inflorescences terminal and lateral, spicate, cylindrical, subterete, commonly more than 5 cm. long, not tapering much to apex; bracts adpressed to stem, overlapping, shortly subacute, decussate, about 2 mm. wide; flowers in groups of 3 immersed in cavities in the stem, the median raised higher than the 2 laterals; median flower generally hermaphrodite, the laterals often staminate only; perianths connate, spongy, angular, about 1 mm. long and almost as wide at the truncate apex, quite enclosing ovary, filaments and style; stamens usually 1 (2), filaments about 1 mm. long; anthers yellow, oblong, about 1 mm. long, 0·8 mm. wide; ovary flask-shaped, membranous, about 1·5 mm. long, 1 mm. wide; style short and rather thick, about 0·5 mm. long; stigmas about 0·8 mm. long. Utricle immersed in persistent, spongy perianth; seeds oblong, about 1 mm. long, 0·8 mm. wide; testa rich brown, rather thinly covered with short curved or hooked hairs; embryo conduplicate; endosperm wanting or very sparse.

HAB.: Moist sandy shores, salt-marshes, salt-lakes, near sea-level; fl. May–Oct.
DISTR.: Divisions 1, 3–5. Europe, Mediterranean region and eastwards to Central Asia.
1. Kato Paphos, 1967, *Economides* in ARI 1011 !
3. Cape Gata, 1862, *Kotschy*.
4. Larnaca, 1862, *Kotschy*; also, 1877, *herb. Post* s.n. ! 1936, *Syngrassides* 978 ! 1939, *Lindberg f.* s.n.; and 1951–1967, *Merton* 447 ! and in ARI 412 ! Ayia Napa, 1939, *Lindberg f.* s.n.
5. Trikomo, 1939, *Lindberg f.* s.n. ! Salamis, 1939, *Lindberg f.* s.n.; also, 1957, *Merton* 3213 !

NOTES: An aggregate consisting of numerous poorly differentiated and often highly critical microspecies. Since only one of the Cyprus specimens seen has flowers, and all are without seeds (the description of the seeds being taken from a Syrian specimen), more detailed analysis of the Cyprus plants must await future investigation.

TRIBE 7. **Suaedeae** *Moq.* Herbs and shrubs; leaves alternate, often rather fleshy; flowers hermaphrodite or occasionally unisexual; bracteoles membranous; perianth-segments 5; stamens 5. Fruit free or adnate to perianth; seeds vertical or horizontal; endosperm sparse or wanting; embryo coiled spirally.

10. SUAEDA *Forssk. ex Scop.*
Introd. Hist. Nat., 333 (1777) nom. cons.

Annual or perennial herbs or shrubs; leaves alternate, simple, linear, lanceolate or sometimes subglobose, rather fleshy, terete or semiterete, usually glabrous, but sometimes with a terminal bristle; flowers bracteolate, solitary or clustered in the axils, or on the petiole, of the leaf-like bract, hermaphrodite or occasionally unisexual; perianth divided into 5 subequal segments which are persistent and sometimes accrescent in fruit; stamens 5,

inserted on the perianth-segments; ovary sessile, free or adnate to perianth; stigmas 2–5. Utricle compressed or depressed; seeds commonly dimorphic, those produced in late summer and early autumn differing in colour and sculpturing of testa from those produced towards the end of the fruiting season; endosperm sparse or absent; embryo coiled spirally.
About 100 species with a cosmopolitan distribution.

Annual herbs; fruit enveloped by, but free from the fruiting perianth   -   - **3. S. maritima**
Perennial shrubs and subshrubs:
 Fruit enveloped by, but free from the fruiting perianth   -   -   -   -   - **1. S. vera**
 Fruit immersed in, and fused with, the spongy, accrescent fruiting perianth
                                                                      2. S. aegyptiaca

1. **S. vera** *Forssk.* [Fl. Aegypt.-Arab., 69 (1775) nom invalid.] *ex J. F. Gmelin*, Syst. Nat., ed. 13, 2: 503 (1791); Boiss., Fl. Orient., 4: 939 (1879); H. Stuart Thompson in Journ. Bot., 44: 337 (1906); Post, Fl. Pal., ed. 2, 2: 442 (1933); Zohary, Fl. Palaest., 1:159, t. 229 (1966); Osorio-Tafall et Seraphim, List Vasc. Plants Cyprus, 33 (1973).
 Chenopodium fruticosum L., Sp. Plant., ed. 1, 221 (1753).
 Salsola fruticosa (L.) L., Sp. Plant., ed. 2, 324 (1762).
 Suaeda fruticosa (L.) Dum., Florula Belg., 22 (1827): Unger et Kotschy, Die Insel Cypern, 223 (1865); Boiss., Fl. Orient., 4: 939 (1879); Post, Fl. Veg. Cypr., 65 (1914); Post, Fl. Pal., ed. 2, 2: 442 (1933); Lindberg f., Iter Cypr., 13 (1946); Chapman, Cyprus Trees and Shrubs, 36 (1949); Rechinger f. in Arkiv för Bot., ser. 2, 1: 420 (1950) et auctt. mult. non *Suaeda fruticosa* Forssk. (1775) nom. illeg.
 TYPE: Locality not indicated. *Forsskål* (C).

Erect, much-branched, evergreen shrub, 20–80 cm. high; young shoots green, bluntly angular or subterete, old branches covered with pale grey-brown, fissured and flaking bark; leaves numerous, small, spreading, subsessile, generally linear-oblong, 5–10 mm. long, 1–1·5 mm. wide, glabrous, rather fleshy, semiterete, apex obtuse or rounded, base rather abruptly narrowed to a very short petiole, margins entire; inflorescences extending along the tips of the branches, frequently for 15 cm. or more; flowers solitary or in dense clusters in the axils of leaf-like bracts; bracteoles membranous, broadly ovate or suborbicular, about 1 mm. long and almost as wide, apex often erose-dentate; perianth-segments bluntly oblong, about 1 mm. long, 0·8 mm. wide, apex obtuse or rounded, margins entire, narrowly membranous; filaments glabrous, scarcely 0·4 mm. long; anthers relatively large, yellow, broadly oblong, about 1 mm. long and almost as wide; ovary pyriform, glabrous, about 1·2 mm. long, 1 mm. wide at base; stigmas usually 5, often small and indistinct, forming an irregular crown on the top of the ovary. Utricle brownish, scarious, enveloped by the persistent incurved perianth; seed vertical, compressed-subglobose, about 1·2 mm. diam.; testa shining black or dark brown, minutely punctulate.

HAB.: Sandy seashores, salt-marshes, hillsides and rocky slopes, usually near the sea; sea-level to 100 ft. alt. (see Notes); fl. Oct.–May.

DISTR.: Divisions 1, 3–8. Widespread in S. and W. Europe, the Mediterranean region, Atlantic Islands.

1. Ayios Yeoryios Island (Akamas), 1962, *Meikle* 2154!
3. Zakaki marshes, 1939, *Mavromoustakis* 35! and, 1940, *Mavromoustakis* 65!
4. Larnaca Salt Lake and near Larnaca, 1860, *Hooker & Hanbury* s.n.! also, 1862, *Kotschy* 248! and, 1903, *A. G. & M. E. Lascelles* s.n.! and, 1905, *Holmboe* 104; and, 1967, *Merton* in ARI 369! Larnaca Aerodrome, 1950, *Chapman* 411! Famagusta, 1938, *Chapman* 329! Perivolia, 1938, *Syngrassides* 1956! Kouklia reservoir, 1967, *Merton* in ARI 452! also, 1968, *Economides* in ARI s.n.!
5. Trikomo, 1939, *Lindberg f.* s.n.! Salamis, 1939, *Lindberg f.* s.n.! also, 1957, *Merton* 3218!
6. Syrianokhori, 1934, *Nattrass* 427.
7. Mountains above Kythrea, 1880, *Sintenis & Rigo* 752! (see Notes).
8. Klidhes Islands, 1933, *Chapman* 20!

NOTES: The Sintenis and Rigo record (no. 752 "In montibus circa Kythraea") is a little suspect, since *S. vera* is otherwise strictly confined to saline habitats in the coastal belt. It is just possible, however, that their specimens were collected in brackish flushes on the Kythrean marls near Kythrea, and probably in Div. 5 rather than Div. 7. These marls lie at altitudes between 500 and 1,000 feet.

2. S. aegyptiaca (*Hasselq.*) *Zohary* in Journ. Linn. Soc. Bot., 55: 635 (1957); Zohary, Fl. Palaest., 1: 161, t. 235 (1966).
    Chenopodium aegyptiacum Hasselq., Iter Palaest., 460 (1757).
    Suaeda baccata Forssk. ex J. F. Gmelin, Syst. Nat., ed. 13, 2: 503 (1791); Uhlbrich in Engl. et Prantl, Pflanzenfam., ed. 2, 16c: 558 (1934).
    Schanginia baccata (Forssk.) Moq., Chenop. Monogr. Enum., 119 (1840); Boiss., Fl. Orient., 4: 944 (1879); Post, Fl. Pal., ed. 2, 2: 444 (1933).
    TYPE: Egypt; Alexandria; *Hasselquist*.

Subshrub to about 40 cm. high; stems much branched, obscurely angular or subterete; leaves subsessile, fleshy, narrowly oblong, incurved, semi-terete, 1–2 cm. long, 0·2–0·3 cm. wide, inflorescences forming inter-rupted spikes towards the tips of the branches; bracts similar to the leaves, but generally less than 1 cm. long; flowers in dense axillary clusters, sessile or very shortly pedicellate; bracteoles membranous, erose at apex, broadly ovate, about 0·5 mm. long and almost as wide; perianth-segments spongy, incurved-cucullate, oblong, shortly acute, about 2 mm. long, 1·3 mm. wide at anthesis, accrescent to 3 mm. long, 2 mm. wide in fruit; stamens 5, filaments slender, glabrous, about 1 mm. long; anthers yellow, sub-orbicular, about 0·8 mm. diam.; ovary glabrous, pyriform, about 1·5 mm. long, 1 mm. wide at base; stigmas 3, about 0·4 mm. long. Base of utricle immersed in, and fused with, spongy base of fruiting perianth; seed vertical; compressed-globose, about 1 mm. diam.; testa shining black, minutely and obscurely rugulose.

HAB.: Salt-marshes and margins of salt-lakes; near sea-level; fl. Aug.–Oct.

DISTR.: Division 4, rare. Also Palestine, Egypt, Arabia and eastwards to Iran.

4. Larnaca, 1970, *A. Genneou* in ARI 1596!

3. S. maritima (*L.*) *Dumort.*, Florula Belg., 22 (1827); Boiss., Fl. Orient., 4: 941 (1879); Post, Fl. Pal., ed. 2, 2: 443 (1933); Lindberg f., Iter Cypr., 13 (1946).
    Chenopodium maritimum L., Sp. Plant., ed. 1, 221 (1753).
    TYPE: "*in Europae maritimis*".

Erect, sprawling or prostrate, glaucous annual; stems 20–50 (or more) cm. long, usually much branched, especially at the base, subterete or bluntly angular; leaves rather crowded, strap-shaped or narrowly linear, 0·5–5 cm. long, 0·1–0·2 cm. wide, sessile or subsessile, apex obtuse, shortly acute or mucronate, margins entire; inflorescences loosely spicate, extending along the tips of the branches for 10 cm. or more; flowers subsessile, solitary or in clusters in the axils of leaf-like bracts; bracteoles suborbicular, obtuse, membranous, about 1 mm. diam., with an erosulose apex; perianth-segments incurved, concave, suborbicular, about 0·8 mm. diam. with a narrow membranous margin; stamens 5; filaments slender, about 0·8 mm. long; anthers yellow, suborbicular, about 0·4 mm. diam.; ovary glabrous, pyriform, about 0·6 mm. long, 0·2 mm. wide at base; stigmas usually 2, about 0·4 mm. long. Utricle depressed-globose, enveloped by (but not adnate to), persistent perianth-segments; seeds horizontal, compressed-globose, about 1·5 mm. diam.; testa shining black, shagreened with minute flattened protuberances.

HAB.: Salt-marshes, salt-lakes, and moist sandy ground by the sea; near sea-level; fl. Aug.–Oct.

DISTR.: Divisions 1, 3. Widely distributed in Europe, the Mediterranean region, western Asia, N.E. U.S.A. and Canada.

1. Kato Paphos, 1939, *Lindberg f.* s.n.; and, 1967, *Economides* in ARI 1012! 1013!
3. Near Limassol, 1939, *Lindberg f.* s.n.! Zakaki, 1939, *Mavromoustakis* 48! Cherkez, 1939,

Mavromoustakis 42 ! 49 ! 50 ! Asomatos, 1939, *Mavromoustakis* 2 ! 41 ! Limassol, 1941, *Davis* 3535 ! Akrotiri Bay, 1949, *Mavromoustakis* 57 !

NOTES: The description of the seeds is taken from British specimens, and may not be correct for Cyprus material. All but one of the specimens listed above are sterile, without flowers or fruits; in the circumstances they can be referred only to *Suaeda maritima* in the aggregate sense.

TRIBE 8.   **Salsoleae** *Dumort.* Herbs and shrubs; stems often jointed; leaves generally small, sometimes apparently wanting; flowers with bracts and bracteoles; perianth usually 5-lobed; stamens usually 4–5. Fruiting perianth commonly winged or horned; utricle generally immersed in fruiting perianth; seed horizontal or vertical; embryo coiled spirally, endosperm absent or very sparse.

11. SALSOLA *L.*

Sp. Plant., ed. 1, 222 (1753).
Gen. Plant., ed. 5, 104 (1754).

Herbs or shrubs; leaves small, alternate or sometimes opposite, sessile, frequently pilose, occasionally reduced to scales; flowers sessile, axillary, solitary or sometimes in groups, hermaphrodite or rarely unisexual; bracts often conspicuous; bracteoles 2, perianth 5-lobed, the segments oblong or lanceolate, accrescent in fruit, and generally with a conspicuous, scarious, transverse dorsal wing; stamens 5 (sometimes fewer), inserted on an inconspicuous, hypogynous disk; filaments commonly exserted; anthers often with the connective produced into a short appendage; ovary ovoid or subglobose; style short or long; stigmas 2. Fruit generally a membranous utricle immersed in the fruiting perianth, rarely fleshy; seed horizontal, or rarely vertical; testa membranous; embryo often green, coiled spirally; endosperm absent.

About 150 species, found throughout the world, generally in deserts and in saline habitats.

Succulent, thick-stemmed, sparsely hispidulous or glabrous annuals:
Leaves and bracts spine-tipped, pungent -   -   -   -   -   -   -   **1. S. kali**
Leaves and bracts blunt or shortly mucronate, not pungent   -   -   -   **2. S. soda**
Slender, wiry annuals; stems, leaves and bracts densely grey pulverulent-hirsute
3. S. inermis

1. **S. kali** *L.*, Sp. Plant., ed. 1, 222 (1753); Boiss., Fl. Orient., 4: 954 (1879); Holmboe, Veg. Cypr., 65 (1914); Post, Fl. Pal., ed. 2, 2: 447 (1933); Lindberg f., Iter Cypr., 13 (1946); Rechinger f. in Arkiv för Bot., ser. 2, 1: 420 (1950); Zohary, Fl. Palaest., 1: 170, t. 246 (1966); Davis, Fl. Turkey, 2: 329 (1967); Osorio-Tafall et Seraphim, List Vasc. Plants Cyprus, 33 (1973).
TYPE: "*in* Europae *litoribus maris*".

Erect or sprawling, succulent annual 20–60 cm. high; stems usually much branched, bluntly angular, subglabrous or more or less densely scabridulous; leaves commonly shed before anthesis, the lowermost opposite, the upper alternate, linear, semi-terete, 1·5–4 (–8) cm. long, 0·1–0·2 cm. wide, sparsely hispidulous, apex shortly spine-tipped, margins entire; flowers in loose terminal and lateral spikes, mostly solitary or in groups of 2 or 3; bracts conspicuous, patent, subulate, 8–12 mm. long, 2–2·5 mm. wide at the somewhat amplexicaul base, generally with a slender, pungent apical spine; bracteoles similar but smaller, about 4–5 mm. long, 1 mm. wide at base; perianth-segments 5, membranous (becoming scarious in fruit), oblong, obtuse or cuspidate, about 2 mm. long, 1 mm. wide, entire or a little erose; stamens 5; filaments glabrous, about 1·5 mm. long; anthers broadly oblong, about 1·5 mm. long, 1 mm. wide, yellow, the 2 thecae free almost to the middle, and the free part sharply inflexed; connective not produced; ovary broadly ovoid or subglobose, about 1 mm. long and almost as wide; style short, about 0·5 mm. long; stigmas filiform, brownish, about 0·5–0·8 mm. long. Fruit enveloped by the indurated, winged perianth, the seed-bearing portion bluntly obconical, obscurely ribbed, about 3 mm. long and almost as wide at apex, wing discoid, membranous, veiny, fragile, about 12 mm. diam.; seed horizontal, depressed-globose or lenticular 2–2·5 mm. diam.; testa dull brown, minutely asperulous, membranous; embryo bright green, neatly coiled.

HAB.: Sandy or stony seashores, sand-dunes, or in sandy fields or waste ground, usually near the sea; near sea-level; fl. May–Sept.

DISTR.: Divisions 1 [2], 3–8. Widespread in Europe, Asia and the Mediterranean region.

1. Maa Beach, 1967, *Economides* in ARI 976!
[2.] Prodhromos, Damaskinari Dam (probably introduced with sand and shingle used in building the dam), 1966, *Merton* in ARI 59!
3. Limassol, 1905, *Holmboe* 645; also, 1941, *Davis* 3557! and, 1947, *Mavromoustakis* 9! Amathus, 1978, *Holub* s.n.!
4. Larnaca, 1880, *Sintenis* 918! also, 1913, *Haradjian* 1014, and, 1939, *Lindberg f.* s.n.! Between Larnaca and Cape Pyla, 1935, *Syngrassides* 834!
5. Trikomo, 1939, *Lindberg f.* s.n.; Salamis, 1957, *Merton* 3215! also, 1968, *A. Genneou* in ARI 1259!
6. Yerolakkos, 1967, *Economides* in ARI 1031!
7. Kyrenia, 1955, *G. E. Atherton* 80! 107! 212! 226!
8. Near Rizokarpaso, 1912, *Haradjian* 245; Ayios Thyrsos ("Therissos"), 1934, *Syngrassides* 956! Boghaz, 1968, *Barclay* 1059!

2. S. soda *L.*, Sp. Plant., ed. 1, 223 (1753); Boiss., Fl. Orient., 4: 953 (1879); Holmboe, Veg. Cypr., 66 (1914); Post, Fl. Pal., ed. 2, 2: 447 (1933); Zohary, Fl. Palaest., 1: 170, t. 247 (1966); Davis, Fl. Turkey, 2: 330 (1967); Osorio-Tafall et Seraphim, List Vasc. Plants Cyprus, 33 (1973).

TYPE: "*in* Europae *australis salsis*".

Erect or spreading, succulent annual 20–60 cm. high; stems glabrous, subterete or bluntly angled, often reddish, branches frequently opposite or subopposite; leaves opposite (or occasionally alternate), fleshy, linear-semiterete, 1–4 (or more) cm. long, 0·2–0·4 mm. wide, glabrous, apex blunt, shortly mucronulate, base dilated, amplexicaul, margins entire, flowers remote, solitary or paired, axillary, forming lax, lateral and terminal, spicate inflorescences; bracts similar to leaves, but usually alternate and shorter, 0·5–2·5 cm. long, patent or recurved; bracteoles 2–8 mm. long, 2–4 mm. wide, sharply keeled, acute, shortly cuspidate; perianth-segments 5, membranous, obovate-cuspidate, about 3·5 mm. long, 2·5 mm. wide, the 2 inner distinctly narrower than the 3 outer; stamens usually 5; filaments linear, glabrous, about 3 mm. long; anthers yellow, oblong, about 1·3 mm. long, 0·5 mm. wide, connective produced into a short, blunt mucro; ovary broadly ovoid or subglobose, about 1·5 mm. long, 1·4 mm. wide; style 1 mm. long; stigmas filiform, tapering, about 2·5 mm. long. Fruiting perianth shortly cylindrical, truncate, about 3·5 mm. long, 4 mm. wide, indurated, wings reduced to small apical gibbosities; seed strongly depressed or discoid, about 3 mm. diam.; testa dark brown, minutely rugulose; embryo neatly coiled, not conspicuously green.

HAB.: Sandy seashores; near sea-level; fl. May–Sept.

DISTR.: Division 3, rare. S. Europe and Mediterranean region eastwards to China and Japan.

3. Limassol, with *S. kali*, 1905, *Holmboe* 646.

NOTES: Apparently rare, and only once collected on the island, but perhaps ignored by collectors, who tend to pass by succulent, late-flowering, dowdy *Chenopodiaceae*.

3. S. inermis *Forssk.*, Fl. Aegypt.-Arab., 57 (1755); Boiss., Fl. Orient., 4: 955 (1879); Post in Mém. Herb. Boiss., 18: 99 (1900); Holmboe, Veg. Cypr., 66 (1914); Post, Fl. Pal., ed. 2, 2:

447 (1933); Zohary, Fl. Palaest., 1: 171, t. 249 (1966); Davis, Fl. Turkey, 2: 330 (1967); Osorio-Tafall et Seraphim, List Vasc. Plants Cyprus, 33 (1973).

Bassia pulverulenta Lindberg f., Iter Cypr., 12 (1946).

[*Salsola hirsuta* (non L.) Sm. in Sibth. et Sm., Fl. Graec. Prodr., 1: 170 (1806).]

[*S. laniflora* (non S.G. Gmel.) Hume in Walpole, Mem. Europ. Asiatic Turkey, 254 (1817).]

[*Echinopsilon hirsutus* (non (L.) Moq.) Poech, Enum. Plant. Ins. Cypr., 13 (1842); Unger et Kotschy, Die Insel Cypern, 223 (1865); H. Stuart Thompson in Journ. Bot., 44: 337 (1906).]

[*Kochia hirsuta* (non (L.) Nolte) Holmboe, Veg. Cypr., 65 (1914).]

[*Bassia hirsuta* (non (L.) Aschers.) Osorio-Tafall et Seraphim, List Vasc. Plants Cyprus, 32 (1973).]

TYPE: Egypt; "*Alexandriae*" (C).

Sprawling annual with a tough woody rootstock; stems rather slender and wiry, 10–20 (–40) cm. long, obscurely angled, much branched, densely pulverulent; leaves small, linear, the lowermost opposite, soon deciduous, up to 5 mm. long, 1·5 mm. wide, thinly hirsute, apex blunt, margins entire, base slightly expanded and amplexicaul, upper leaves alternate, similar to lower, but seldom half as long, sometimes almost scale-like, overlapping, grey-pulverulent, hirsute; flowers solitary, sessile, forming slender, loose or dense, caudate spikes; bracts ovate-acute, somewhat keeled dorsally, densely pulverulent, about 2–3 mm. long and about as wide at the amplexicaul base; bracteoles similar to and almost as large as bracts; perianth segments 5, free almost to base, ovate, about 2 mm. long, 1 mm. wide, grey-pulverulent with a membranous margin, apex acute; stamens 5; filaments glabrous, about 1·5 mm. long, dilated towards base; anthers yellow, broadly ovate with a cordate base, about 1 mm. long and almost as wide, connective produced as a short blunt mucro; ovary ovoid, glabrous, about 1·5 mm. long, 1 mm. wide; style obsolete; stigmas 2, very short, about 0·3 mm. long, brownish. Fruiting perianth discoid, 6–7 mm. diam., with brownish, readily disintegrating wings; seed discoid, about 1·5 mm. diam.; testa membranous, hyaline, smooth; embryo neatly coiled, not very green.

HAB.: Sandy seashores and margins of salt-lakes; occasionally on saline ground inland; sea-level to 500 ft. alt.; fl. Sept.–Nov.

DISTR.: Divisions 4, 6, 8. Also Turkey, Syria, Palestine, Egypt, Libya.

4. Larnaca Salt Lake, 1880–1978, *Sintenis* 602! *Post* s.n.! *Lindberg f.* s.n.! *J. Holub* s.n.! Larnaca aerodrome, 1952, *Merton* s.n.! Famagusta, 1939, *Lindberg f.* s.n.

6. Mile 3, Nicosia–Myrtou road, 1950, *Chapman* 578!

8. Boghaz, 1937, *Syngrassides* 1647! also, 1968, *Barclay* 1061!

NOTES: Only one specimen (*Barclay* 1061) collected on 22 Sept. 1968, is sufficiently mature to have flowers; all the rest are sterile, with a juvenile facies, often misleadingly different from that of the mature plant, and no doubt the basis of misidentifications whereby the plant has been variously referred to *Kochia*, *Bassia* and *Salsola*. Post, who was probably familiar with the species in Palestine, is the only author to recognize it as *Salsola inermis*. Apart from the fact that Cyprus specimens tend to be rather more slender and prostrate than plants from Palestine, again almost certainly an aspect of their immaturity, the resemblances are very close. In Cyprus, it is questionable if *S. inermis* ripens fruit before mid-October, and the description of fruit and seeds is taken from Palestine material.

12. NOAEA *Moq.*

in DC., Prodr., 13, 2: 207 (1849).

Annual herbs or small shrubs, sometimes with thorny branches; leaves sessile; flowers hermaphrodite, solitary, sessile in the axils of the bracts; bracteoles 2; perianth deeply 5-lobed, the segments winged dorsally in fruit; stamens 5, inserted on a small hypogynous disk; anthers almost linear, the connective produced into a short apical appendage; ovary ovoid; style often elongate; stigmas 2. Fruiting perianth scarious; winged, often coloured; utricle membranous, immersed in perianth, but not adnate to it; seeds vertical, compressed, orbicular; embryo curved spirally, green; endosperm wanting.

Seven species in the Mediterranean region and eastwards to Central Asia.

1. **N. mucronata** (*Forssk.*) *Aschers. et Schweinf.*, Ill. Fl. Égypte, 131 (1887); Holmboe, Veg. Cypr., 65 (1914); Post, Fl. Pal., ed. 2, 2: 451 (1933); Lindberg f., Iter Cypr., 13 (1946); Chapman, Cyprus Trees and Shrubs, 36 (1949); Zohary, Fl. Palaest., 1: 175, t. 257 (1966); Davis, Fl. Turkey, 2: 336 (1967); Osorio-Tafall et Seraphim, List Vasc. Plants Cyprus, 33 (1973).

Salsola mucronata Forssk., Fl. Aegypt.-Arab., 56 (1775).

S. echinus Labill., Icon. Plant. Syr. Rar., dec. 2: 10, t. 5 (1791); Gaudry, Recherches Sci. en Orient, 187 (1855).

Noaea spinosissima Moq. in DC., Prodr., 13, 2: 209 (1849); Unger et Kotschy, Die Insel Cypern, 224 (1865); Boiss., Fl. Orient., 4: 965 (1879); Post in Mém. Herb. Boiss., 18: 99 (1900); H. Stuart Thompson in Journ. Bot., 44: 337 (1906).

TYPE: Egypt; "*Alexandriae* ad Catacombas" (C).

Erect or sprawling, rigid, much-branched shrub 10–60 cm. high; old wood covered with pale grey, fissured bark, young twigs ochraceous, glabrous, bluntly angled, terminating in a sharp spine, leaves caducous, alternate, narrowly linear, 3–30 mm. long, 1–1·5 mm. wide, semiterete, glabrous, apex acute, margins entire, base inconspicuously decurrent; flowers solitary in the axils of bracts, forming lax or crowded spikes along the ultimate, divaricate, branchlets; bracts rigid, narrowly ovate-deltoid, apiculate, green, glabrous with a narrow membranous margin, 3–4 mm. long, 2 mm. wide at base; bracteoles very similar to bracts and almost as long; perianth-segments free almost to base, erect, narrowly oblong-acuminate, about 4 mm. long, 1 mm. wide, membranous with a distinctly incrassate base and dorsal flap (ultimately developing into a transverse wing); stamens 5; filaments ligulate, about 3 mm. long, 0·3 mm. wide, glabrous; anthers yellow, linear, about 2 mm. long, 0·5 mm. wide, apex shortly apiculate, base shortly sagittate; ovary flask-shaped, glabrous, about 1·5 mm. long, 1·2 mm. wide; style 1·5 mm. long; stigmas 0·8 mm. long. Fruiting perianth with broadly and bluntly ovate, papery, pink, rather irregularly spreading wings, about 5–6 mm. long, 4–5 mm. wide; seed discoid, 2–2·5 mm. diam.; testa membranous, smooth; embryo neatly coiled, green.

HAB.: Dry pastures, coastal garigue, kafkalla land, sand-dunes, rocky hillsides; sea-level to 1,000 ft. alt.; fl. July–Oct.

DISTR.: Divisions 2, 4–7. Eastern Mediterranean region and eastwards to Central Asia; N. Africa.

2. Between Evrykhou and Galata, 1880, *Sintenis & Rigo* 689! also, 1940, *Davis* 1951! and, *Chapman* 107! Beween Ambelikou and Kambos, 1939, *Lindberg f.* s.n.

4. Larnaca, 1860, *Hooker & Hanbury* s.n.! also, 1939, *Lindberg f.* s.n.; Famagusta, 1939, *Lindberg f.* s.n.; Paralimni, 1905, *Holmboe* 1164; Athna Forest, 1957, *Merton* 601! Athienou, 1967, *Economides* in ARI 1038!

5. Between Mia Milea and Mandres, 1933, *Syngrassides* 329! Athalassa, 1939, *Lindberg f.* s.n.! and, 1970, *A. Genneou* in ARI 1602!

6. Morphou, 1933, *Chapman* 121! Nicosia, 1935, *Nattrass* 486; Ayia Irini, 1968, *Barclay* 1074!

7. Kyrenia, 1932, *Nattrass* 189; also, 1955, *G. E. Atherton* 168! 704! and, 1968, *I. M. Hecker* 17A!

NOTES: Also recorded by Post from Troödos (Mém. Herb. Boiss., 18: 99, 1900), but the locality probably refers to the Troödos Range in general, rather than to the village of Troödos, which, at 5,600 ft. alt., seems well above the normal vertical limit of distribution of the species.

74a. BASELLACEAE

Rhizomatous or tuberous perennials with annual climbing shoots; leaves alternate, entire, often fleshy. Flowers hermaphrodite or unisexual, actinomorphic, small, in axillary spikes, racemes or panicles; bracts and bracteoles present; sepals 5, free or somewhat connate, imbricate, persistent, often coloured; petals absent. Stamens 5, inserted opposite to, and at the base of the sepals; filaments free; anthers 2-thecous, versatile, opening longitudinally or by slits or pores. Ovary superior, 1-locular, with 1 basal, campylotropous ovule; style terminal with 3 stigmas, or rarely 1 stigma. Fruit a fleshy berry or drupe; seed almost spherical; endosperm usually copious; embryo annular or strongly curved.

Four genera, represented throughout the tropics, with a few species in cultivation elsewhere; of these the best known is the Madeira or Mignonette Vine, *Anredera cordifolia* (Ten.) Steenis (*Boussingaultia baselloides* hort.) a vigorous climber, with a tuberous rootstock, fleshy, glossy green, cordate leaves and elongate spikes of small, white, fragrant flowers. According to Kotschy (Die Insel Cypern, 223; 1865), *A. cordifolia* is frequent in gardens at Larnaca, Limassol and Nicosia, but there are no recent records, though the plant is still fairly frequent in cultivation elsewhere. It is a native of S. America.

75. PHYTOLACCACEAE
H. Walter in Engl., Pflanzenr., 39 (IV. 83): 1–154 (1909).

Herbs, shrubs, trees or rarely climbers, generally glabrous; leaves alternate, simple, entire, petiolate or sessile; stipules absent or very small; flowers hermaphrodite or unisexual in terminal, axillary or leaf-opposed simple or compound racemes or cymes; bracts and bracteoles generally present; perianth green or coloured, 4–5-partite, the segments equal or unequal, imbricate in aestivation; stamens 4–many, generally inserted on a fleshy hypogynous disk; filaments free or connate at base; anthers 2-thecous, introrse, dorsifixed or basifixed, dehiscing longitudinally; staminodes sometimes present; ovary superior, consisting of 2 or more free, or partly, or completely connate carpels, ovule 1 in each carpel; styles as many as carpels or wanting; stigmas usually linear. Fruit a berry, drupe, schizocarp, nut or achene; seed usually reniform or lenticular, sometimes subglobose, arillate or exarillate; endosperm copious, mealy or fleshy; embryo curved, peripheral.

Fourteen genera chiefly in tropical and subtropical regions, particularly well represented in South America.

1. PHYTOLACCA *L.*
Sp. Plant., ed. 1, 441 (1753).
Gen. Plant., ed. 5, 200 (1754).

Herbs, shrubs, trees or rarely climbers, sometimes with the rhachis and younger growths pubescent; leaves exstipulate, alternate, simple, entire, usually petiolate; inflorescence racemose, paniculate or spiciform, terminal, lateral or leaf-opposed; flowers pedicellate or subsessile, often numerous, hermaphrodite or unisexual, actinomorphic; bracts and bracteoles present; perianth commonly coloured, 4–5-partite, often small, segments generally subequal; stamens 5–30; filaments free or partly united, inserted on a fleshy hypogynous disk; ovary depressed-globose, of 5–16 free or connate carpels; styles as many as carpels, subulate, erect or recurved, stigmatose along the inner surface. Fruit a fleshy berry or schizocarp; seeds compressed, reniform; testa shining black; cotyledons semiterete.

About 35 species, chiefly in the tropics and subtropics.

Trees; racemes drooping; flowers unisexual - - - - - - - P. DIOICA (p. 1395)
Herbs; racemes erect or spreading:
  Flowers hermaphrodite; peduncles generally long and rather slender; stamens usually 10
                                              1. P. americana
  Flowers unisexual; peduncles generally short and stout; stamens usually about 20
                                            2. P. pruinosa

P. DIOICA *L.*, Sp. Plant., ed. 2, 632 (1762); H. Walter in Engl., Pflanzenr., 39 (IV. 83): 47 (1909); Post, Fl. Pal., ed. 2, 2: 425 (1933); Osorio-Tafall et Seraphim, List Vasc. Plants Cyprus, 34 (1973).

Pircunia dioica (L.) Moq. in DC., Prodr., 13, 2: 30 (1849); G. Frangos in Cyprus Agric. Journ., 18: 87 (1923).

A robust, spreading evergreen tree to 15 m. or more; bark pale; branches stout, pithy, terete; leaves ovate, 8–12 cm. or more long, 3–8 cm. or more wide, glabrous, rather glaucous, apex subacute, shortly apiculate, base rounded, nervation distinct on the undersurface, of 8–10 pairs of prominent spreading nerves; petiole semiterete, 1·5–8 cm. long, inflorescences male or female on separate trees (dioecious), drooping, racemose, 10–20 cm. long; peduncle generally 4–6 cm. long; rhachis angled, glandular-pubescent; bracts and bracteoles narrowly ovate, about 2 mm. long, 1 mm. wide, pubescent; male flowers with spreading pedicels about 5 mm. long; tepals 5, oblong-elliptic, concave, greenish, about 3 mm. long, 2 mm. wide, reflexed at anthesis, stamens 20–30, filaments free, slender, whitish, glabrous, about 8 mm. long, slightly dilated at base; anthers yellow, narrowly oblong, about 1·3 mm. long, 0·8 mm. wide, notched at apex and base; ovary rudimentary; female flowers rather more shortly pedicellate than the males, and usually in shorter racemes; tepals 5, rather broadly ovate-elliptic; staminodes usually about 10, sometimes bearing rudimentary anthers, ovary depressed-globose, about 4 mm. diam., 7–10-carpellate, the carpels partly connate; styles subulate, less than 1 mm. long, recurved towards apex. Fruit juicy, black, shining, depressed-globose, about 1 cm. diam.; seed reniform, about 3 mm. long, 2·3 mm. wide; testa shining black, minutely foveolate.

A native of South America, popular as a quick-growing shade tree in the Mediterranean region, and not uncommon in Cyprus, where it is recorded from 4. Larnaca (1923, *G. Frangos*); 5. Salamis (1968, *A. Genneou* in ARI 1260!) and 7. Kyrenia (1955, *G. E. Atherton* 72!). Both male and female trees are reported. In S. America the tree is generally called "Ombú", in the Mediterranean region "Bellasombra".

1. P. americana *L.*, Sp. Plant., ed. 1, 441 (1753); H. Walter in Engl., Pflanzenr., 39 (IV. 83): 52 (1909); Zohary, Fl. Palaest., 1: 69, t. 83 (1966); Davis, Fl. Turkey, 2: 347 (1967).
  P. decandra L., Sp. Plant., ed. 2, 631 (1762); Boiss., Fl. Orient., 4: 895 (1879); Post, Fl. Pal., ed. 2, 2: 425 (1933).
  TYPE: "*in* Virginia, Mexico".

Stout, erect or spreading, glabrous, rather foetid perennial 80–150 cm. high, with a thick, fleshy rootstock; stems rather succulent, much-branched, finely striate, often purplish; leaves ovate-elliptic, 6–16 (–25) cm. long, 3–10 cm. wide, glabrous (the lamina minutely scabridulous), apex acuminate, base broadly cuneate; nervation usually distinct; petiole short, generally less than 3 cm. long; inflorescence a rather lax, spreading, cylindrical, leaf-opposed raceme 5–10 (–20) cm. long; peduncle 2 cm. long or more; flowers hermaphrodite; bracts subulate, 2–3 mm. long, 0·5–0·8 mm. wide at base, bracteoles similar but much smaller; pedicels slender, about 3–4 mm. long at anthesis, lengthening to 8–10 mm. in fruit; perianth-segments generally 5 (6), broadly oblong-elliptic, concave, about 2·8 mm. long, 1·8 mm. wide, glabrous, greenish, submembranous, speckled with small, pale rhaphides; stamens 10; filaments slender, filiform, about 2 mm. long; anthers oblong, about 0·8 mm. long, 0·4 mm. wide; ovary 10-carpellate, depressed globose, glabrous, about 1·8 mm. diam.; styles 10, subulate, erect, about 0·8 mm. long. Fruit a fleshy, black, depressed-globose, 10-ribbed berry, 8–10 mm. diam.; seeds lenticular, about 3 mm. diam.; testa shining black, minutely foveolate.

HAB.: Roadsides, waste ground; sea-level to 1,000 ft. alt.; fl. July–Oct.

DISTR.: Divisions 4, 7. Native of N. America, widely distributed in other temperate regions.

4. Larnaca, 1892–94, *Deschamps* teste *H. Walter.*
7. Bellapais, by spring below monastery, 1948, *Casey* 122 !

NOTES: Thoroughly established in N. Anatolia (Fl. Turkey, 2: 347), but scarcely more than an adventive in our area.

2. P. pruinosa *Fenzl*, Del. Sem. Hort. Univ. Vindob., 6 (1855); Unger et Kotschy, Die Insel Cypern, 351 (1865); Boiss., Fl. Orient., 4: 895 (1879); Post in Mém. Herb. Boiss., 18: 99 (1900); H. Walter in Engl., Pflanzenr., 39 (IV. 83): 61 (1909); Holmboe, Veg. Cypr., 66 (1914); Post, Fl. Pal., ed. 2, 2: 425 (1933); Lindberg f., Iter Cypr., 13 (1946); Davis, Fl. Turkey, 2: 347 (1966); Osorio-Tafall et Seraphim, List Vasc. Plants Cyprus, 34 (1973).
    [*P. stricta* (non Hoffm.) Poech, Enum. Plant. Ins. Cypr., 35 (1842).]
TYPE: Turkey; Bulgar Dağ, "in devexis silvarum prope Gullek Boghas alt. 4000 ped. Dieb. Aug. 1853" *Kotschy* 271 (W, K !).

Robust, erect perennial with a woody rootstock, 1–1·5 m. high; stems not much branched, pithy, glabrous, terete, longitudinally striatulate, some-times reddish; leaves glabrous, sometimes glaucous, ovate-elliptic, 6–15 (–30) cm. long, 1·5–10 cm. wide, apex acute, subacute or obtuse, base narrowed gradually or abruptly to a petiole generally less than 1 cm. long; nervation obscure or sometimes fairly distinct on the undersurface of the leaf; inflorescences terminal and leaf-opposed, rather dense and many flowered, (2·5–) 6–14 (–20) cm. long; peduncle usually very short or sometimes almost wanting, seldom exceeding 3 cm.; flowers male and female on separate plants (dioecious); pedicels patent, angular, glabrous or a little scabridulous, 3–8 mm. long; bracts subulate 2–3 mm. long, 0·8–1 mm. wide at base; bracteoles similar, usually less than 1·5 mm. long and 0·5 mm. wide at base; male inflorescences generally larger and more lax than female, the flowers larger and with longer pedicels; perianth at first greenish, later often becoming bright ruby-red, segments usually 5, ovate-elliptic, 3–4 mm. long, 2–2·5 mm. wide, concave, persistent and accrescent after anthesis; stamens about 20; filaments subulate, glabrous, about 2 mm. long, tapering from a wide base; anthers pallid, oblong, about 1·8 mm. long, 0·8 mm. wide; ovary rudimentary; female flowers smaller; perianth-segments very broadly ovate-elliptic, almost suborbicular, strongly concave, about 3 mm. long, 2·5 mm. wide; ovary depressed-globose, glabrous, about 3 mm. diam., of 7–9 connate carpels surrounded basally by about 15 linear staminodes, about 0·8 mm. long, terminating in minute atrophied anthers; styles strongly curved, spreading, about 0·8 mm. long, stigmatose for most of their length. Fruit a black, fleshy berry, 6–10 mm. diam., bluntly 5–7 lobed and borne on a reddish infructescence, subtended by persistent, accrescent, often reddish, incrassate perianth-lobes; seeds lenticellate, about 3–3·5 mm. diam., testa black, rather lustrous, minutely and regularly rugulose with numerous concentric ridges. *Plate 80.*

Plate 80. PHYTOLACCA PRUINOSA Fenzl **1,** habit (female plant), ×1; **2,** female flower, ×9; **3,** fruit, ×6; **4,** seed, ×9; **5,** male inflorescence, ×1; **6,** male flower, ×9. (**1–4** from *P. H. Oswald* 147; **5, 6** from *Lindberg f.* s.n.)

HAB.: Roadsides, dry rocky hillsides and forest clearings; 3,000–5,600 ft. alt.; fl. April–July.

DISTR.: Division 2. Also S. Turkey, Syria, Lebanon.

2. Evrykhou, 1840, *Kotschy* 41; between Prodhromos and Galata, 1862, *Kotschy*; near Galata, 1880, *Sintenis & Rigo* 684! Hills N. of Phini, 1900, *A. G. & M. E. Lascelles* s.n.! Troödos, 1898, *Post* s.n.; also, 1905, *Holmboe* 854; common, Marathasa valley and Pitsilia, 1905, *Holmboe*; Troödos-Amiandos area, 1912–1970, *Feilden* s.n.! *Haradjian* 487! *Syngrassides* 516! 724! *Lindberg f.* s.n.! *Casey* 933! *F. M. Probyn* s.n.! *P. H. Oswald* 147! *D. P. Young* 7808! 7858! *A. Genneou* in ARI 1561! Prodhromos, 1929, *C. B. Ussher* 38! also, 1955, *G. E. Atherton* 644! Phini, 1936, *Kennedy* 322! Platres, 1937, *Kennedy* 321 and, 1955, *Kennedy* 1867! Kykko, 1937, *Syngrassides* 1576! and, 1939, *Lindberg f.* s.n., also, 1941, *Davis* 1790!, 1968, *Economides* in ARI 1238! Near Kakopetria, 1939, *Lindberg f.* s.n., and, 1962, *Meikle* 2856! Below Saïttas, 1956, *Merton* 2805! Spilia, 1966, *Y. Ioannou & Economides* in ARI s.n.!

76. POLYGONACEAE
H. Gross in Engl., Bot. Jahrb., 49: 234–339 (1913).
K. Haraldson in Symb. Bot. Upsal., 22 (2): 95 pp. (1978).

Herbs, shrubs and climbers, rarely trees; leaves usually alternate, simple; stipules often united into a membranous sheath or *ochrea* around the stem; inflorescence paniculate, spicate or capitate, or flowers solitary or clustered in the leaf-axils; pedicels commonly articulated; flowers hermaphrodite or unisexual, actinomorphic; perianth sepaloid or petaloid, persistent and frequently altered and accrescent in fruit; perianth-segments 3–6, free or partly united; stamens (3–) 6–9 (occasionally more); anthers 2-thecous; ovary superior, 1-locular; ovule 1, basal; styles 2–4, free or connate. Fruit a trigonous, compressed or winged achene, frequently enveloped by the persistent perianth; embryo excentric or lateral, straight or curved; endosperm copious, mealy, sometimes ruminate.

About 40 genera, widely distributed, but chiefly in temperate regions of the northern hemisphere.

Perianth-segments 5, persistent but not much accrescent in fruit; stigmas capitate
    1. Polygonum
Perianth-segments 6, all or some of them accrescent in fruit; stigmas penicillate or plumose:
    Perianth-segments free or almost free, the inner strongly accrescent in fruit   **2. Rumex**
    Perianth-segments fused, the outer with spreading, spinous tips-   -   -   -   **3. Emex**

1. POLYGONUM *L.*
Sp. Plant., ed. 1, 359 (1753).
Gen. Plant., ed. 5, 170 (1754).

Annual or perennial herbs or subshrubs; stems erect, prostrate or climbing; leaves alternate; simple, subsessile or petiolate; ochreae membranous, truncate or lobed, sometimes ciliate, often lacerate with age; flowers in clusters, spikes or panicles; bracts foliaceous, sometimes reduced or obsolete; flowers generally hermaphrodite; perianth greenish or coloured, persistent, segments usually 5, subequal or occasionally unequal, often erect and enveloping fruit; stamens (5–) 8; anthers 2-thecous; ovary superior, 1-locular, often triquetrous; styles 2–3, sometimes almost obsolete; stigmas capitate. Achenes triquetrous and lenticular; embryo lateral.

About 300 species with a cosmopolitan distribution.

The genus is frequently subdivided into 8 or more segregate genera distinguished chiefly by anatomical characters and by pollen or trichome characters. The resemblances between such genera seem, to me, more striking than their differences, and they are here regarded as sections of a large, polymorphic genus.

Leaves sagittate-cordate, long-petiolate; stems climbing or scrambling; flowers in elongate, slender, interrupted spikes - - - - - - - - - **7. P. convolvulus**
Leaves not sagittate-cordate; stems not climbing or scrambling:
Perennial plants:
Inflorescences leafy, the flowers in small axillary clusters; stems prostrate or sprawling, often almost concealed by the conspicuous, silvery ochreae; leaves coriaceous
3. P. maritimum
Inflorescences spiciform, terminal, slender, not leafy, the bracts much reduced; ochreae not covering the stems; leaves not coriaceous:
Rootstock woody; leaves 1–4·5 cm. long; ochreae lacerate - - **5. P. equisetiforme**
Rootstock rhizomatous, not woody; leaves 6–15 cm. long; ochreae truncate, ciliate
1. P. salicifolium
Annual plants:
Inflorescences leafy; flowers solitary, axillary, or in small axillary clusters
4. P. aviculare
Inflorescences not leafy, spiciform:
Leaves glandular below, 4–22 cm. long; flowers approximate, forming dense pedunculate spikes - - - - - **2. P. lapathifolium** ssp. **maculatum**
Leaves eglandular below, 1·3–4 cm. long; flowers in slender, interrupted, non-pedunculate terminal spikes - - - - - - - **6. P. patulum**

SECT. 1. **Persicaria** (*L.*) *DC.* Annuals or perennials; ochreae often brownish, truncate, ciliate or eciliate; inflorescence dense or lax, ovoid or spiciform, not leafy; perianth often coloured; styles 2–3, commonly united towards base. Fruit commonly compressed-ovoid or lenticular, sometimes trigonous.

1. P. salicifolium *Brouss. ex Willd.*, Enum. Plant. Hort. Reg. Berol., 428 (1809); Zohary, Fl. Palaest., 1: 56, t. 61 (1966); Davis, Fl. Turkey, 2: 273 (1967).
P. serrulatum Lag., Gen. et Sp. Nov., 14 (1816); Boiss., Fl. Orient., 4: 1028 (1879).
P. serrulatum Lag. var. *salicifolium* Boiss., Fl. Orient., 4: 1028 (1879).
P. scabrum Poir. var. *salicifolium* (Boiss.) Dinsmore in Post, Fl. Pal., ed. 2, 2: 471 (1933).
Persicaria salicifolia (Brouss. ex Willd.) Assenov in Fl. Reipubl. Popul. Bulgar., 3: 243 (1966); K. Haraldson in Symb. Bot. Upsal., 22 (2): 1; 86 (1978) non *Persicaria salicifolia* S. F. Gray (1821), nom. illeg.
[*Polygonum lapathifolium* (non L.) A. K. Jackson in Kew Bull., 1937: 345 (1937) pro parte.]
TYPE: Canary Islands; "in Teneriffa" (B).

Erect or ascending perennial 50–100 (–120) cm. high, rooting at the lower nodes; stems generally branched, rather slender, subterete, glabrous; leaves linear-lanceolate, 6–15 cm. long, 0·8–1·5 cm. wide, acuminate, glabrous except for the adpressed-scabridulous or ciliate margins; petiole slender, generally less than 8 mm. long; ochreae membranous, brownish, 1–2·5 (or more) cm. long, sparsely adpressed-scabridulous externally, apex truncate, fringed with bristly cilia sometimes 5–8 mm. long or longer; inflorescences terminal, narrowly spiciform, 4–8 cm. long, sparingly branched; peduncles glabrous; flowers 2–6 in the axils of subtruncate or unilaterally subacute, shortly tubular, sparingly ciliate or subglabrous ochreoles 1·5–3 mm. long; pedicels glabrous, 3–4 mm. long, articulated just below perianth; perianth pink or whitish, segments scarious, concave, oblong, subacute, glabrous, about 2·5–3 mm. long, 1–1·5 mm. wide, indistinctly nerved; stamens 5–8; filaments very slender, glabrous, 1·5–2 mm. long; anthers oblong, about 0·5 mm. long, 0·4 mm. wide; ovary trigonous, glabrous, subellipsoid, about 2–2·5 mm. long, 1·5 mm. wide; styles 3, filiform, spreading, about 0·5 mm. long, shortly united at base; stigmas capitate. Achene 2–3 mm. long, completely enveloped by persistent fruiting perianth, dark brown or blackish, lustrous.

HAB.: By margins of water channels; about 50 ft. alt.; fl. May–July.

DISTR.: Division 6, rare. Widespread in the Mediterranean region, Atlantic Islands, and widespread in tropical Africa, also in tropical Asia, America and Australia.

6. Syrianokhori, 1936, *Syngrassides* 977!

2. P. lapathifolium *L.*, Sp. Plant., ed. 1, 360 (1753); Boiss., Fl. Orient., 4: 1030 (1879); Post, Fl. Pal., ed. 2, 2: 471 (1933); A. K. Jackson in Kew Bull., 1937: 345 (1937) pro parte; Lindberg f., Iter Cypr., 12 (1946); J. Timson in Watsonia, 5: 386 (1963); Zohary, Fl. Palaest., 1: 56, t. 60 (1966); Davis, Fl. Turkey, 2: 273 (1967); Osorio-Tafall et Seraphim, List Vasc. Plants Cyprus, 30 (1973).

Erect or ascending, branched annual, 50–150 cm. high; stems succulent, swollen above the nodes, terete, glabrous, green, reddish or speckled red; leaves lanceolate or ovate-elliptic, 4–22 cm. long, 1–6 cm. wide, uniformly green or with a dark median blotch, sometimes grey-tomentose below, apex acute or acuminate, base cuneate, margins entire, minutely adpressed-scabridulous, undersurface usually covered with numerous, translucent, sessile glands; petiole slender 0·8–1·5 cm. long; ochreae membranous, tubular, sometimes split laterally, 0·8–1·5 cm. long, apex truncate, shortly ciliate or without cilia, thinly adpressed-strigose externally, nervation generally rather obscure; inflorescences lateral and terminal, densely or loosely ovoid, cylindrical or spiciform, leafless; peduncles to 7 cm. long, often densely covered with sessile glands; ochreoles tubular, to 1 cm. long, thinly adpressed-strigose externally, apex truncate, often shortly ciliate; pedicels very short, scarcely exceeding ochreoles; flowers pale green, pink or reddish; perianth-segments 3–5, concave, oblong-obovate, about 2–2·5 mm. long, 1–1·5 mm. wide, apex blunt or subacute; stamens 5–6; filaments slender, about 1·5 mm. long; anthers broadly oblong, about 0·4 mm. long and almost as wide; ovary broadly ovoid, strongly compressed, glabrous, about 1 mm. long, 0·8 mm. wide; styles 2, spreading, about 0·8 mm. long, shortly united towards base; stigmas capitate. Achenes enveloped by persistent perianth, broadly ovate-lenticular with a distinct median depression and shortly subacute apex, about 2–3 mm. long, 1·5–2·5 mm. wide, rich, lustrous brown, minutely granulate.

ssp. **maculatum** (*S. F. Gray*) *Dyer et Trimen* in Journ. Bot., 9: 36 (1871).
　Polygonum nodosum Pers., Syn. Plant., 1: 440 (1805).
　Persicaria maculata S. F. Gray, Nat. Arrang. Brit. Plants, 2: 270 (1821).
　Polygonum tenuiflorum J. et C. Presl, Deliciae Pragenses, 1: 67 (1822).
　TYPE: The illustration of *Polygonum pensylvanicum* var. *caule maculata* in Curtis, Fl. Londin., 1: t. 74 (1777)!

Stems generally stained or speckled red, often conspicuously swollen at the nodes; leaves with or without a dark median blotch, sometimes tomentose below; inflorescences commonly elongate, lax, sometimes very slender and spiciform, attaining 5–6 cm. in length; perianth pink or reddish, rather small, segments frequently about 2 mm. long, 1 mm. wide. Achenes 2–2·5 mm. long, 1·5–2 mm. wide, often distinctly ovate and acute.

HAB.: Sides of streams and water channels; sea-level to 4,500 ft. alt.; fl. June–Oct.

DISTR.: Divisions 2, 3. Widespread in Europe and Asia.

2. Milikouri, 1939, *Lindberg f.* s.n.! Prodhromos, 1955, *G. E. Atherton* 616! Kalopanayiotis, 1967, *A. Genneou* in ARI 1286!
3. Yermasoyia, 1935, *Syngrassides* 754!

NOTES: Many authors have commented upon the taxonomic difficulties presented by the *P. lapathifolium* complex. Cyprus material is uniformly distinct from *P. lapathifolium* L. as interpreted by the majority of western European botanists, and, in common with most of the specimens from the eastern Mediterranean region, agrees much more closely with the critical *P. nodosum*, particularly with the variant sometimes named *P. tenuiflorum* J. et C. Presl, which is distinguished by its exceptionally long, slender, raceme-like or spiciform inflorescences. While

P. nodosum and *P. tenuiflorum* are inseparably connected by a series of intermediates, I feel that *P. nodosum* and *P. lapathifolium* should be recognized as distinct, at least as subspecies, and have accorded them this rank. Material of *P. nodosum* in herb. Persoon (L) includes one unlocalized sheet of what is generally regarded as typical *P. nodosum*, a second sheet sent by Gussone, and marked by him "In humidis", which is *P. tenuiflorum* J. et C. Presl, and a third, of an intermediate variant, labelled "Lectum prope Nancy".

SECT. 2. **Polygonum** Annuals or perennials; ochreae membranous, hyaline, silvery; inflorescence consisting of axillary clusters or interrupted, generally leafy, spikes; perianth often greenish; styles 2–3. Fruit lenticular or trigonous.

3. P. maritimum *L.*, Sp. Plant., ed. 1, 361 (1753); Sibth. et Sm., Fl. Graec. Prodr., 1: 266 (1809), Fl. Graec., 4: 55, t. 363 (1823); Poech, Enum. Plant. Ins. Cypr., 14 (1842); Unger et Kotschy, Die Insel Cypern, 225 (1865); Boiss., Fl. Orient., 4: 1037 (1879); Holmboe, Veg. Cypr., 64 (1914); Post, Fl. Pal., ed. 2, 2: 473 (1933); Lindberg f., Iter Cypr., 12 (1946); Zohary, Fl. Palaest., 1: 54, t. 57 (1966); Davis, Fl. Turkey, 2: 276 (1967); Osorio-Tafall et Seraphim, List Vasc. Plants Cyprus, 30 (1973).
TYPE: "Monspelii, *in* Italia, Virginia".

Prostrate or sprawling perennial with a thick, woody rootstock; stems very numerous, 5–50 cm. long, much branched, longitudinally ribbed, glabrous, sometimes almost covered by the conspicuous silvery ochreae; leaves lanceolate or oblanceolate, coriaceous, glaucous, 5–30 mm. long, 3–8 mm. wide, apex blunt or shortly acute, margins entire, strongly revolute, base tapering to a short, indistinct petiole; nervation generally obscure; ochreae up to 1 cm. long, at first deeply 2-lobed, soon irregularly lacerate, membranous, hyaline-silvery above, usually brownish towards base, glabrous, conspicuously nerved; flowers solitary or in 2–4-flowered clusters in the axils of leaf-like bracts; ochreoles conspicuous, silvery, glabrous; pedicels slender, 2–3 mm. long, glabrous, articulated at apex, often shorter than the ochreoles; perianth whitish above, greenish below; perianth-segments 5, broadly ovate, obtuse, about 3 mm. long, 2·5 mm. wide, obscurely nerved; stamens 6–8; filaments 0·8 mm. long, glabrous, suddenly and conspicuously dilated towards base; anthers suborbicular, about 0·4 mm. diam.; ovary ellipsoid, trigonous, glabrous, about 1·5 mm. long, 0·8 mm. wide; styles 3 (4), very short (about 0·2 mm. long), free; stigmas small, capitate. Achenes sharply trigonous, tapering to an acute apex, about 3 mm. long, 2 mm. wide, shining dark brown, minutely granulate, loosely enveloped by the persistent perianth.

HAB.: Sandy and shingly seashores; sand-dunes; about sea-level; fl. March–Aug.

DISTR.: Divisions 1–3, 6. Coasts of Europe, Mediterranean region, also N. America and Asia.

1. Ayios Yeoryios (Akamas), 1962, *Meikle* 2115! Lachi, 1979, *Hewer* 4562B!
2. Seashore at Yialia, 1962, *Meikle*!
3. Limassol, 1905, *Holmboe* 648; also, 1913, *Haradjian* 580! 1939, *Lindberg f.* s.n.; 1941, *Davis* 3540! Akrotiri Bay, 1939, *Mavromoustakis* 71! Cape Gata, 1941, *Davis* 3571! S. of Mari, 1956, *Poore* 27! Amathus, 1978, *J. Holub* s.n.! Ladys' Mile Beach, Limassol, 1977, *J. J. Wood* 12!
6. Ayia Irini, 1953, *Casey* 1281!

4. P. aviculare *L.*, Sp. Plant., ed. 1, 362 (1753); Boiss., Fl. Orient., 4: 1036 (1879); Holmboe, Veg. Cypr., 64 (1914); Post, Fl. Pal., ed. 2, 2: 473 (1933); Lindberg f., Iter Cypr., 12 (1946); B. T. Styles in Watsonia, 5: 204 (1962); Davis, Fl. Turkey, 2: 277 (1967); Osorio-Tafall et Seraphim, List Vasc. Plants Cyprus, 30 (1973).
P. heterophyllum Lindm. in Svensk. Bot. Tidskr., 6: 690 (1912) nom. illeg.
P. aviculare L. var. *heterophyllum* Druce in Rep. B.E.C., 9: 471 (1931).
[*P. arenastrum* (non Bor.) Zohary, Fl. Palaest., 1: t. 59 (1966).]
TYPE: "*in* Europae *cultis ruderatis*".

Erect, decumbent or prostrate annual; stems glabrous, longitudinally ribbed, much branched, 5–70 (–150) cm. long; leaves numerous, ovate-lanceolate or lanceolate, those of the main stems often noticeably larger

than those of the branches, 3–35 (–50) mm. long, 2–10 (–15) mm. wide, green,
glabrous, obscurely nerved, apex acute or subacute, margins entire,
sometimes minutely scabridulous, base narrowing abruptly or gradually to
a slender petiole 2–6 mm. long, or leaf subsessile; ochreae 6–8 mm. long, at
first 2-lobed, soon irregularly lacerate, membranous, silvery towards apex,
brownish towards base; flowers solitary or in groups of 2–6 in the axils of
leaf-like bracts spaced remotely, or rarely crowded towards the apices of the
ultimate branches; pedicels slender, usually less than 2 mm. long, shorter
than the membranous ochreoles, articulate at apex; perianth whitish or pink
with a greenish base, segments 5, erect, oblong, obtuse or rounded at apex,
obscurely nerved, about 2–2·5 mm. long, 1 mm. wide, free to below the
middle, concave, persistent in fruit; stamens usually 8; filaments glabrous,
0·5 mm. long, abruptly dilated at base; anthers suborbicular, about 0·3 mm.
diam.; ovary usually triquetrous, about 2 mm. long, 1 mm. wide, glabrous;
styles 3 (rarely 2) very short, about 0·2 mm. long; stigmas capitate. Achene
broadly ovate-triquetrous, about 2·3 mm. long, 1·8–2 mm. wide, the 3 sides
subequal and slightly concave, the apex shortly acute, pericarp dark dull
brown, minutely granulate.

HAB.: Cultivated and waste ground; roadsides; occasionally in Hazel copses or on open
ground in Pine forest; sea-level to 5,200 ft. alt.; fl. April–Sept.

DISTR.: Divisions 1, 2, 4, 6, 7. Throughout temperate regions of the world as a weed.

1. Yeroskipos, 1934, *Syngrassides* 503 ! 506 !
2. Prodhromos, 1905, *Holmboe* 883; also, 1955, *G. E. Atherton* 523 ! and, 1961, *D. P. Young*
   7449 ! 7450 ! Kykko Monastery, 1939, *Lindberg f.* s.n. ! Polystipos, 1955, *Merton* 2430 !
   Kakopetria, 1955, *G. E. Atherton* 432 ! Trikoukkia, 1955, *Merton* 2458 !
4. Near Larnaca, 1939, *Lindberg f.* s.n.
6. Nicosia, 1928 or 1930, *Druce* s.n.; also, 1950, *Chapman* 459 ! and, near Paphos Gate, Nicosia,
   1958, *P. H. Oswald* 16 !
7. Kyrenia, 1948, *Casey* 12 ! also, 1955, *G. E. Atherton* 19 !

NOTES: Some of the small-leaved specimens (e.g. *P. H. Oswald* 16 !) from roadsides, paths and
trampled ground, are deceptively similar to *P. arenastrum* Bor., but have the larger flowers and
subequally concavo-triquetrous fruits of *P. aviculare*, of which they are, almost certainly, mere
habitat-states.

5. P. equisetiforme *Sm.* in Sibth. et Sm., Fl. Graec. Prodr., 1: 266 (1809), Fl. Graec., 4: 56, t. 364
(1823); Unger et Kotschy, Die Insel Cypern, 225 (1865); Boiss., Fl. Orient., 4: 1036 (1879);
H. Stuart Thompson in Journ. Bot., 44: 337 (1906); Holmboe, Veg. Cypr., 64 (1914); Post,
Fl. Pal., ed. 2, 2: 473 (1933); Lindberg f., Iter Cypr., 12 (1946); Chapman, Cyprus Trees and
Shrubs, 35 (1949); Zohary, Fl. Palaest., 1: 53 (1966); Davis, Fl. Turkey, 2: 278 (1967);
Osorio-Tafall et Seraphim, List Vasc. Plants Cyprus, 30 (1973).
TYPE: "In sepibus insulae Cretae" (OXF).

Sprawling or decumbent (rarely erect) perennial with a woody base, or
sometimes a subshrub; stems much branched, 50–100 cm. long, glabrous,
longitudinally ribbed; internodes generally long; leaves remote, often
largely shed before anthesis, linear-lanceolate, oblanceolate or narrowly
oblong, 1–4·5 cm. long, 0·3–1·8 cm. wide, subsessile or very shortly petiolate,
glabrous, apex acute or obtuse, margins narrowly recurved, cartilaginous,
often undulate, entire, base broadly or narrowly cuneate; ochreae
membranous, silvery above, brownish towards base, strongly nerved, at
first bifid, soon irregularly lacerate; inflorescences terminal, slender,
spiciform, consisting of numerous closely superimposed flowers; bracts
much reduced, resembling the leaves but seldom more than 5 mm. long, 2
mm. wide, sometimes obsolete; flowers solitary or 2–4 together, axillary;
pedicels about 3 mm. long, angular, glabrous, articulated at apex, a little
longer than the membranous ochreoles; perianth-segments oblong, free for
about two-thirds their length, 2·5–3 mm. long, 1·8 mm. wide, glabrous,
white with a reddish midrib, concave, persistent in fruit, apex rounded or
subacute; stamens usually 8, unequal; filaments glabrous 1–1·5 mm. long,

conspicuously dilated towards base; anthers yellow, suborbicular, about 0·8 mm. diam.; ovary triquetrous, glabrous, ellipsoid, about 1 mm. long, 0·7 mm. wide; styles 3, thick, about 0·5 mm. long; stigmas capitate. Achenes concealed by persistent perianth, or shortly protruding, triquetrous, with subequal concave sides, about 3 mm. long, 1·5 mm. wide, rich dark brown, sublustrous, minutely granulate.

HAB.: Roadsides; field margins; sandy seashores and sand-dunes; occasionally on dry, stony hillsides; sea-level to 3,000 ft. alt.; fl. April–Oct.

DISTR.: Divisions 1–8. Mediterranean region and eastwards to Iran.

1. Stroumbi, 1913, *Haradjian* 718! Paphos, 1978, *J. Holub* s.n.! Near Polis, 1979, *Edmondson & McClintock* E 2780!
2. Kykko, 1913, *Haradjian* 965! Kakopetria, 1933, *Chapman* 120! also, 1939, *Lindberg f.* s.n., and, 1955, *Chiotellis* 431! 438! Kato Platres, 1937, *Kennedy* 316! Near Makheras Monastery, 1940, *Davis* 1897!
3. Kalavasos, 1905, *Holmboe* 625; Limassol, 1901, *Michaelides* in *Holmboe* s.n.! Akrotiri Bay, 1939, *Mavromoustakis* 59! Amathus, 1948, *Mavromoustakis* s.n.!
4. Larnaca, 1860, *Hooker & Hanbury* s.n.! also, 1880, *Sintenis & Rigo* 979! Pyla, 1934, *Syngrassides* 430! Near Athna, 1934, *Syngrassides* 511! Famagusta, 1940, *Davis* 2050! and, 1948, *Mavromoustakis* s.n.!
5. Kythrea, 1880, *Sintenis & Rigo* 575! 667! also, 1932, *Syngrassides* 282! 379! and, 1934, *Syngrassides* 534! and, 1940, *Davis* 1958! Salamis, 1905, *Holmboe* 450; also, 1939, *Lindberg f.* s.n.; Trikomo, 1939, *Lindberg f.* s.n.!
6. Nicosia, 1927, *Rev. A. Huddle* 89! also, 1957, *Merton* 2908! and, 1966, *Merton* in ARI 21! Kato Lakatamia, 1937, *Chapman* 273! Dhiorios, 1962, *Meikle*! Morphou, 1972, *W. R. Price* 1043! N.W. of Pano Dheftera, 1979, *Edmondson & McClintock* E 2942!
7. Kyrenia, 1948, *Kennedy* 55! also, 1948, *Casey* 94! and, 1955, *G. E. Atherton* 27! 206! 361! 701! and, 1973, *P. Laukkonen* 324! Thermia, 1948, *Casey* 80! and, 1955, *G. E. Atherton* 490! Ayios Epiktitos, 1955, *G. E. Atherton* 233!
8. Akradhes area, 1962, *Meikle*!

6. P. patulum *M. Bieb.*, Fl. Taur.-Cauc., 1: 304 (1808); Rouy, Fl. de France, 12: 108 (1910); Zohary, Fl. Palaest., 1: 55, t. 58 (1966).

[*P. bellardii* (non All.) Boiss., Fl. Orient., 4: 1034 (1879); Post in Mém. Herb. Boiss., 18: 99 (1900); H. Stuart Thompson in Journ. Bot., 44: 337 (1906); Holmboe, Veg. Cypr., 64 (1914); Post, Fl. Pal., ed. 2, 2: 472 (1933); Davis, Fl. Turkey, 2: 279 (1967); Osorio-Tafall et Seraphim, List Vasc. Plants Cyprus, 30 (1973).]

TYPE: USSR: Crimea and Caucasus; "occurrit locis incultis cum P. aviculari rarius".

Erect or suberect (or rarely prostrate) annual 30–60 (–100) cm. high; stems much branched, terete, distinctly ribbed longitudinally, glabrous, sometimes minutely scabridulous, internodes long; leaves lanceolate, 1·3–4 cm. long, 0·3–1 cm. wide, green, glabrous, apex acute, margins entire, sometimes papillose-scabridulous towards apex, base tapering to a short, indistinct petiole; ochreae membranous, silvery, about 8 mm. long, at first bifid, soon becoming irregularly lacerate; flowers in very lax, elongate, slender terminal spikes, solitary or 2–3 in the axils of much reduced, lanceolate-subulate bracts up to 1 cm. long towards base of inflorescence, but diminishing upwards, the upper usually shorter than the flowers, the uppermost almost obsolete; pedicels very short, rarely exceeding 3·5 mm. in length, articulated at apex; perianth pinkish, glabrous, segments erect, concave, oblong, obtuse, about 3·5 mm. long, 1·2 mm. wide; stamens usually 8; filaments glabrous, about 0·8 mm. long, 0·5 mm. wide at the conspicuously dilated base; anthers yellow, suborbicular, about 0·4 mm. diam.; ovary glabrous, ellipsoid, 2–3 mm. long, 1–1·5 mm. wide, conspicuously triquetrous; styles almost obsolete; stigmas capitate. Achene triquetrous, ellipsoid, about 3·5 mm. long, 1·5 mm. wide, enveloped by the fruiting perianth, rich dark, lustrous brown, minutely granulate.

HAB.: Cultivated ground; mountainsides, 3,300–6,300 ft. alt; fl. May–August.

DISTR.: Division 2. Widely distributed in central and southern Europe, Mediterranean area and eastwards to Central Asia.

2. Prodhromos, 1880, *Sintenis & Rigo* 772! Khionistra, 6,300 ft. alt.; 1938, *Kennedy* 1153! Troödos, 5,600 ft. alt., 1940, *Davis* 1869! Ayia Moni above Khrysorroyiatissa, 3,300 ft. alt., 1941, *Davis* 3435!

NOTES: Post's record, "Plaines de Chypre, printemps" (Mém. Herb. Boiss., 18: 99) is very questionable, since *P. patulum* (*P. bellardii* auct.) is a plant of high altitudes in Cyprus. Perhaps the plant seen was the relatively common and widespread *P. equisetiforme* Sm., which Post does not list.

SECT. 3. **Tiniaria** *Meissner* Annual or perennial climbers and twiners; leaves petiolate, deltoid or sagittate-cordate; flowers in lax, terminal or lateral, often clustered, spiciform inflorescences; perianth segments 5–6, the exterior often distinctly winged; stamens 8; styles 3, very short, or capitate stigmas subsessile. Achene triquetrous, enveloped by the persistent perianth.

7. **P. convolvulus** *L.*, Sp. Plant., ed. 1, 364 (1753); Boiss., Fl. Orient., 4: 1032 (1879); Holmboe, Veg. Cypr., 64 (1914); Post, Fl. Pal., ed. 2, 2: 472 (1933); Lindberg f., Iter Cypr., 12 (1946); Davis, Fl. Turkey, 2: 280 (1967).
    Bilderdykia convolvulus (L.) Dumort., Florula Belg., 18 (1827); Osorio-Tafall et Seraphim, List Vasc. Plants Cyprus, 31 (1973).
    TYPE: "*in* Europae *agris*".

Twining or scrambling annual up to 1 m. high; stems usually much branched, terete, glabrous or thinly scabridulous, conspicuously longitudinally ribbed; ochreae membranous, brownish, glabrous, to about 1 cm. long, inconspicuously nerved, apex subtruncate; leaves sagittate-cordate, acute or acuminate, glabrous or thinly scabridulous, 1·5–8 cm. long, 1–4 cm. wide, margins entire, nervation inconspicuous; petioles slender, scabridulous, up to 5 cm. long; inflorescences generally solitary, axillary, elongate-spiciform, to 13 cm. or more; peduncles to 8 cm. long, glabrous, ribbed; flowers solitary or 2–4 in the axils of much-reduced, acuminate, foliaceous bracts 2–5 mm. long; pedicels slender, 1–3 mm. long; perianth greenish-white, segments erect, oblong, obtuse, concave, divided almost to base, scabridulous externally, 2–3 mm. long, 1·5–2 mm. wide, the 3 outer distinctly larger than the 2 inner, all persistent and accrescent in fruit, the outer becoming sharply keeled or narrowly winged dorsally; stamens usually 8; filaments glabrous, slender, about 0·8 mm. long; anthers yellow, suborbicular, about 0·2 mm. diam.; ovary ellipsoid, sharply triquetrous, 1·8–2 mm. long, 1 mm. wide, glabrous; styles 3, almost obsolete; stigmas capitate. Achene about 3 mm. long, 2 mm. wide, sharply triquetrous, blackish, minutely granulate, lustrous along the angles, enveloped by the persistent, accrescent perianth.

HAB.: Gardens, vineyards, cultivated and waste ground; 3,000–5,000 ft. alt; fl. June–Sept.

DISTR.: Division 2. Widespread in Europe, Asia and North Africa, and as a weed elsewhere.

2. Prodhromos, 1880, *Sintenis & Rigo* 748! also, 1905–1970, *Holmboe* 912; *Lindberg f.* s.n.; *Casey* 881! *Merton* 2457! *G. E. Atherton* 583! *A. Genneou* in ARI 1567! Kambos, 1939, *Lindberg f.* s.n.; Milikouri, 1939, *Lindberg f.* s.n.

2. RUMEX *L.*

Sp. Plant., ed. 1, 333 (1753).
Gen. Plant., ed. 5, 156 (1754).
K. H. Rechinger f. in Candollea, 12: 9–152 (1949).

Annual, biennial or perennial herbs; leaves alternate; ochreae tubular; inflorescences racemose or paniculate, the flowers generally in whorls subtended by ochreoles; pedicels articulated; flowers unisexual, monoecious, or dioecious, or sometimes hermaphrodite; perianth usually greenish, segments 6 in 2 whorls, the 3 outer commonly spreading or reflexed in fruit, the 3 inner accrescent and enveloping the achene, and commonly developing a conspicuous tubercle dorsally on the midrib; stamens 6; anthers basi-fixed; styles 3, patent or recurved; stigmas fimbriate or penicillate. Achene triquetrous, enveloped by the persistent perianth; embryo lateral.

About 200 species with a cosmopolitan distribution. The mature (fruiting) inner perianth-segments are here termed *valves*.

Leaves hastate or sagittate:
  Annual; inflorescence racemose, unbranched or sparingly branched; valves 10–20 mm. long
                                              1. R. cyprius
  Perennial with tuberous roots; inflorescence paniculate, much-branched; valves less than 10
    mm. long - - - - - - - - - - - R. TUBEROSUS (p. 1406)
Leaves not hastate or sagittate:
  Valves equally tuberculate, tubercles or calluses conspicuous, occupying the greater part of
    the surface of the valve:
    Valves entire, about 2 mm. wide; plant perennial - - - - **3. R. conglomeratus**
    Valves spinulose-denticulate, 2·5–3 mm. wide; plant annual or biennial
                                 5. R. dentatus ssp. **mesopotamicus**
  Valves unequally tuberculate, or tubercle very small and inconspicuous:
    Plants annual, to about 30 cm. high (usually much smaller); valves 2–4 mm. long; fruiting
      pedicel often flattened or dilated - - - - - **6. R. bucephalophorus**
    Plants perennial, 10–150 cm. high; valves 4–8 mm. long:
      Valves shortly erose-denticulate; infructescence narrowly paniculate, crowded, with
        erect branches - - - - - - - - - - - **2. R. cristatus**
      Valves with pectinate-spinulose margins; infructescence unbranched or with lax,
        spreading branches - - - - - - - - - **4. R. pulcher**

SUBGENUS 1. **Acetosa** (*Mill.*) *Rechinger f.* Plants dioecious or polygamous; leaves hastate or sagittate; valves exceeding achene, ecallose or with a small, basal callus or protuberance.

1. R. cyprius *Murb.* in Lunds Univers. Årsskr., N. F., Afd. 2, 2 (14): 20 (1907); Samuelsson in Bot. Notiser, 1939; 509 (1939); Rechinger f. in Candollea, 12: 34 (1949) et in Arkiv för Bot., ser. 2, 1: 419 (1950); Zohary, Fl. Palaest., 1: 61, t. 69 (1966); Osorio-Tafall et Seraphim, List Vasc. Plants Cyprus, 31 (1973).
  R. vesicarius L. ssp. *cyprius* (Murb.) Holmboe, Veg. Cypr., 64 (1914).
  R. cyprius Murb. ssp. *disciformis* Samuelsson in Bot. Notiser, 1939: 512 (1939); Lindberg f., Iter Cypr., 12 (1946); Rechinger f. in Candollea, 12: 34 (1949); Osorio-Tafall et Seraphim, List Vasc. Plants Cyprus, 31 (1973).
  R. cyprius Murb. ssp. *eucyprius* Samuelsson in Bot. Notiser, 1939: 514 (1939); Rechinger f. in Candollea, 12: 35 (1949).
  [*R. roseus* (L. 1759 non 1753) Sm. in Sibth. et Sm., Fl. Graec. Prodr., 1: 247 (1809), Fl. Graec., 4: 40, t. 346 (1823); Campderá, Mon. Rumex, 128 (1819) pro parte; Unger et Kotschy, Die Insel Cypern, 226 (1865); Boiss., Fl. Orient., 4: 1018 (1879) pro parte; Sintenis in Oesterr. Bot. Zeitschr., 31: 326, 393 (1881); Post, Fl. Pal., ed. 2, 2: 467 (1933).]
  [*R. tingitanus* (non L.) Sm. in Sibth. et Sm., Fl. Graec. Prodr., 1: 247 (1809); Poech, Enum. Plant. Ins. Cypr., 14 (1842); Unger et Kotschy, Die Insel Cypern, 226 (1865).]
  [*R. vesicarius* L. ssp. *roseus* (non *R. roseus* L. 1753) Holmboe, Veg. Cypr., 64 (1914).]
  TYPE: Not indicated; to be selected from the Cyprus material cited by Murbeck, op. cit., 21 (1907).

Erect or sprawling, glabrous annual 5–30 cm. high; stems fleshy, usually much branched towards base, distinctly ribbed or angled, often reddish; leaves rather fleshy, subglaucous, often minutely papillose, trullate or subhastate, 1–6 cm. long, 0·3–3 cm. wide, apex obtuse, acute or acuminate, base narrowing rather abruptly to a flattened petiole 1–6 cm. long; nervation obscure; ochreae whitish, membranous, splitting with age; inflorescences terminal, lax, racemose, 4–10 cm. long; peduncle stout, 1·5–4 cm. long; flowers 1–3 from each ochreole, hermaphrodite or unisexual; bracts absent; ochreoles membranous, 2–4 mm. long; pedicel slender, 2–3 mm. long, sometimes elongating to 10 mm. in fruit, articulate about one-third way up, downcurved in fruit; perianth-segments free almost to base, the 3 outer membranous, oblong, subacute, about 2·5 mm. long, 1·3 mm. wide, the 3 inner suborbicular, rather thicker than the outer, about 1 mm.

diam. at anthesis, but very strongly accrescent in fruit; stamens usually 6; filaments glabrous, slender, very short, seldom exceeding 0·8 mm. long; anthers oblong, about 1·8 mm. long, 0·8 mm. wide; ovary triquetrous with narrowly winged angles, glabrous, about 0·6 mm. long, 0·5 mm. wide; styles 3, spreading, about 0·5 mm. long; stigmas sparsely penicillate. Fruiting valves suborbicular, lobed at apex and base, 1–2 cm. diam., membranous, with conspicuous, anastomosing, reticulate, reddish-pink veins, margins shortly spinulose or rarely subentire, with a distinct, thickened peripheral vein; achene sharply triquetrous, about 4 mm. long, 2–2·5 mm. wide, fuscous-brown, minutely rugulose, angles pale, very narrowly winged. *Plate 81.*

HAB.: Dry stony slopes on calcareous or non-calcareous rocks; stony cultivated fields, occasionally on walls or on edges of salt-marshes; sea-level to 2,000 ft. alt.; fl. March–May.

DISTR.: Divisions 1–8. Syria, Palestine, Cyprus and Egypt eastwards to Iran and Arabia.

1. Khrysokhou, 1862, *Kotschy*; Paphos, 1980, *J. E. De Langhe* 107/80!
2. Pomos Point, 1960, *N. Macdonald* 60!
3. Between Skarinou and Limassol, 1934, *Syngrassides* 486! Episkopi Bay, 1950, *F. J. F. Barrington* 35! Curium, 1961, *Polunin* 6686! Limassol Salt Marsh, 1967, *Polunin* 8546! Amathus, 1978, *J. Holub* s.n.!
4. Larnaca, 1862, *Kotschy*; Troulli, 1981, *Hewer* 4739!
5. Lakovounara Forest, 1950, *Chapman* 53! Between Ovgoros and Ayios Elias, 1941, *Davis* 2472!
6. Walls of Nicosia, near Famagusta Gate, 1905, *Holmboe* 278! Orta Keuy, 1936, *Syngrassides* 1164!
7. Mountains above Kythrea, 1880, *Sintenis & Rigo* 47! Near Armenian Monastery, 1912, *Haradjian* 373! By Nicosia–Kyrenia road, 1937, *Syngrassides* 1435! St. Hilarion, 1939, *Lindberg f.* s.n.! Kyrenia, 1956, *G. E. Atherton* 1333! Near Panagra, 1962, *Meikle* 2358! and, 1967, *Merton* in ARI 105!
8. Gastria, 1862, *Kotschy* 15; 282; Platanisso, 1950, *Chapman* 251!

R. TUBEROSUS *L.*, Sp. Plant., ed. 2, 481 (1762), an erect perennial 10–80 cm. high, with clustered, fleshy, fusiform roots, hastate-sagittate leaves up to 11 cm. long, much-branched panicles, and greenish, pinkish or reddish, suborbicular, reticulate-veined fruiting valves, 3–8 mm. diam., is recorded by Sibthorp (Fl. Graec. Prodr., 1: 248; 1809) "In Lemno et Cypro insulis, et in Asiâ minori", but has not since been collected on the island, and is probably an error. All subsequent records (Poech, Kotschy, Boissier, Holmboe, etc.) are based on the statement in *Florae Graecae Prodromus*. *Rumex tuberosus* (and its subspecies) is widely distributed in the eastern Mediterranean region, and eastwards to Central Asia, and might very well be found in Cyprus.

SUBGENUS 2. **Rumex** Flowers generally hermaphrodite; leaves not hastate or sagittate; fruiting valves exceeding achene, generally with 1–3 prominent median calluses or protuberances.

[R. AQUATICUS *L.*, Sp. Plant., ed. 1, 336 (1753), a robust perennial, commonly 1–2 m. high, with large, broadly ovate-deltoid, deeply cordate basal leaves, frequently 30–50 cm. long, 20–30 cm. wide, and dense, erect-branched infructescences, is recorded from Cyprus by Sibthorp (Fl. Graec. Prodr., 1: 246; 1809) and, more precisely, from near Larnaca, by Kotschy (in Unger et Kotschy, Die Insel Cypern, 225 (1865) no. 146a). Subsequent references to the species by Boissier (Fl. Orient., 4: 1008; 1879), Holmboe (Veg. Cypr., 63; 1914), Osorio-Tafall et Seraphim (List Vasc. Plants Cyprus, 31; 1973) are based on one or both of these records. Since *R. aquaticus* has a more northerly distribution in Europe and Asia, and is absent from all areas adjacent to Cyprus, one must conclude misidentification. Sibthorp's plant was probably *R. cristatus* DC. (infra); Kotschy's is less easily guessed at, but

LRM

Plate 81. RUMEX CYPRIUS Murb. **1,** habit, × $\frac{2}{3}$; **2,** flower, × 6; **3,** outer perianth-segment, × 10; **4,** inner perianth-segment, × 10; **5,** anther, × 10; **6,** gynoecium, × 10; **7, 10,** infructescences, × $\frac{2}{3}$; **8,** fruits, × 2; **9,** achene, × 2. (**1–6** from *Sintenis & Rigo* 37; **7** from *Syngrassides* 1435; **8, 9** from *Davis* 2472; **10, 11** from *Lindberg f.* s.n.)

may have been a luxuriant form of *R. pulcher* L. (infra), the only large species common about Larnaca.]

2. **R. cristatus** *DC.*, Cat. Plant. Hort. Bot. Monsp., 56, 139 (1813); Rechinger f. in Candollea, 12: 77 (1949); Davis, Fl. Turkey, 2: 288 (1967); Osorio-Tafall et Seraphim, List Vasc. Plants Cyprus, 31 (1973).
　　　R. graecus Boiss. et Heldr. in Boiss., Diagn., 2, 4: 80 (1859).
　　　R. orientalis Bernh. var. *graecus* (Boiss. et Heldr.) Boiss., Fl. Orient., 4: 1009 (1879).
　　　R. patientia L. ssp. *graecus* (Boiss. et Heldr.) Holmboe, Veg. Cypr., 63 (1914); Lindberg f., Iter Cypr., 12 (1946).
　　　[*R. patientia* (non L.) Kotschy in Unger et Kotschy, Die Insel Cypern, 225 (1865); H. Stuart Thompson in Journ. Bot., 44: 337 (1906); R. M. Nattrass, First List Cyprus Fungi, 4: 9, 24 (1937).]
　　　[*R. orientalis* (non Bernh.) Boiss., Fl. Orient., 4: 1009 (1879) pro parte quoad plant. cypr.]
　　TYPE: Cultivated at Montpellier, of unknown origin (G).

Robust, erect perennial 0·5–1·5 m. high; rootstock cylindrical, thick, woody; stems glabrous, reddish, conspicuously sulcate, generally unbranched, at least in the lower part; basal leaves ovate-lanceolate or narrowly oblong, 20–30 cm. long, 5–15 cm. wide, subglaucous, minutely scabridulous-papillose on the under-surface, apex tapering, acute, base rounded or abruptly cuneate, margins crispate-undulate; nervation prominent below, coarsely reticulate with wide-spreading primary nerves; petiole canaliculate above, sulcate below, 10–15 cm. long; ochreae thinly papery, truncate, 1–3 cm. long; upper cauline leaves similar to lower, but smaller, narrower and more shortly petiolate, less than 12 cm. long, 2 cm. wide, with petioles less than 2 cm. long; inflorescence narrowly paniculate, 30 cm. long or more, at first rather lax, with distant, erect branches 4–10 cm. long, the branches accrescent to 18 cm. or more in fruit and becoming rather crowded; flowers numerous in rather dense whorls; pedicels very slender, 2–3 mm. long at anthesis, articulated about one quarter way up from base, lengthening to 1 cm. or more in fruit; outer perianth-segments oblong, subacute, about 1·5 mm. long, 0·8 mm. wide at anthesis, a little accrescent and strongly reflexed in fruit, inner perianth-segments broadly ovate-oblong, concave, subacute, about 2 mm. long, 1·3 mm. wide; filaments very narrowly deltoid-subulate, glabrous, about 0·5 mm. long; anthers oblong, yellow, about 1·2 mm. long, 0·8 mm. wide; ovary obovoid, deeply triquetrous, about 1 mm. long, 0·8 mm. wide; styles slender, reflexed, about 0·3 mm. long; stigmas penicillate-plumose. Fruiting valves subequal, broadly cordate-deltoid, 5–8 mm. long, 5–7 mm. wide, brownish, reticulate-veined, margins irregularly erose-denticulate, calluses generally 3, one very prominent, often 2–3 mm. long, 1·5 mm. wide, the other 2 much smaller; achene ovoid, acutely triquetrous, about 3 mm. long, 2 mm. wide, dark brown, minutely granulate.

　　HAB.: Rocky hillsides, usually at high altitudes; roadsides, ditches; cultivated and waste ground [1,500–] 3,000–6,400 ft. alt.; fl. May–July.

　　DISTR.: Divisions 2, 3. S. Italy, Sicily, Balkan Peninsula, Aegean area, W. Turkey.

　　2. Troödhitissa Monastery, 1862, *Kotschy* 795; near Prodhromos, 1880, *Sintenis & Rigo* 749! Kykko, 1905, *Holmboe* 1101, and, 1937, *Syngrassides* 1579! Khionistra, 1905, *Holmboe* 1084, also, 1912, *Haradjian* 486! and, 1937–1973, *Kennedy* 313! *Lindberg f.* s.n.! *Davis* 3507! *Casey* 935! *Economides* in ARI 1228! P. *Laukkonen* 453! Pano Panayia, 1913, *Haradjian* 906! Platres, 1929, *C. B. Ussher* 43! also, 1931, *Nattrass* 78, 82; Mesapotamos Monastery, 1939, *Lindberg f.* s.n.; Kambos, 1968, *Economides* in ARI 1211!
　　3. Khoulou, 1905, *Holmboe* 756!

　　NOTES; One other specimen, collected F. M. Probyn s.n. in 1950, at Kannoures Springs near the summit of Khionistra, is probably *R. cristatus* DC., but has unusually small, obscurely denticulate fruiting valves, barely 4 mm. wide.

3. R. conglomeratus *Murr.*, Prodr. Designat. Stirp. Gotting., sect. 2: 52 (1770); Boiss., Fl. Orient., 4: 1010 (1879); Post, Fl. Pal., ed: 2, 2: 464 (1933); Lindberg f., Iter Cypr., 12 (1946); Rechinger f. in Candollea, 12: 96 (1949); Zohary, Fl. Palaest., 1: 63, t. 74 (1966); Davis, Fl. Turkey, 2: 289 (1967); Osorio-Tafall et Seraphim, List Vasc. Plants Cyprus, 31 (1973).

TYPE: Switzerland; "Ad *Birsam & Wiesam*", *Haller* (P).

Erect, often rather slender perennial 0·3–1 (–1·2) m. high; rootstock thick, fleshy; stems glabrous, reddish, angled, shallowly sulcate, frequently branched, with slender, spreading branches; basal leaves narrowly oblong, 12–16 cm. long, 2·5–5 cm. wide, green, papery, glabrous or minutely papillose on the undersurface, apex tapering, acute, base rounded or broadly cuneate, margins flat, entire; nervation not very prominent below, primary nerves wide-spreading; petiole rather slender, sulcate, 6–10 cm. long; ochreae rather loose, membranous, brownish, to 3 cm. long, coarsely and irregularly fimbriate-lacerate; upper cauline leaves similar to lower, but usually smaller and more shortly petiolate, the uppermost reduced to lanceolate bracts 1–3 cm. long, 0·3–0·8 cm. wide; inflorescence lax, paniculate, with long spreading branches, 30–60 cm. long and often almost as wide; flowers numerous, the whorls generally lax and evenly spaced; pedicels short, slender, 1–2 mm. long at anthesis, articulated near the base, lengthening to 8 mm. in fruit; outer perianth-segments oblong-lingulate, glabrous, about 1 mm. long, 0·5 mm. wide, apex obtuse or rounded, remaining erect and not much accrescent in fruit; inner perianth-segments broadly oblong, concave, obtuse, about 1–7 mm. long, 1 mm. wide at anthesis; filaments glabrous, subulate, about 0·3 mm. long; anthers rather narrowly oblong, about 1·2 mm. long, 0·4 mm. wide; ovary subglobose, triquetrous, glabrous, about 0·5 mm. long, 0·4 mm. wide; styles reflexed, about 0·2 mm. long; stigmas penicillate-plumose. Fruiting valves subequal, bluntly deltoid, 2·5–3 mm. long, 2 mm. wide, brownish, entire, indistinctly veined, calluses 3, subequal, oblong, about 1·8 mm. long, 0·8 mm. wide, very conspicuous and occupying most of the face of each valve; achene broadly ovoid, sharply triquetrous, about 1·8–2 mm. long, 1·3 mm. wide, dark brown, shining, minutely granulate.

HAB.: In ditches; by streams; occasionally in moist pastures or cultivated land; 600–4,500 ft. alt.; fl. May–Sept.

DISTR.: Divisions 2, 5, 7. Widespread in temperate Europe and Asia; introduced in E. Asia, S. Africa, N. America, Australasia.

2. Mesapotamos Monastery, 1939, *Lindberg f.* s.n.; Kykko, 1939, *Lindberg f.* s.n.; Prodhromos, 1951, *Casey* 1210! and, 1955, *G. E. Atherton* 614! Kyperounda, 1954, *Merton* 1934!

5. Kythrea, 1880, *Sintenis & Rigo* 677! also, 1932, *Syngrassides* 219! and, 1939, *Lindberg f.* s.n.!

7. Lapithos, 1939, *Lindberg f.* s.n.

4. R. pulcher *L.*, Sp. Plant., ed. 1, 336 (1753); Unger et Kotschy, Die Insel Cypern, 226 (1865); Boiss., Fl. Orient., 4: 1012 (1879); H. Stuart Thompson in Journ. Bot., 44: 337 (1906); Holmboe, Veg. Cypr., 63 (1914); Post, Fl. Pal., ed. 2, 2: 465 (1933); Rechinger f. in Candollea, 12: 102 (1949) et in Arkiv. för Bot., ser. 2, 1: 419 (1950); Zohary, Fl. Palaest., 1: 64, t. 75 (1966); Davis, Fl. Turkey, 2: 291 (1967); Osorio-Tafall et Seraphim, List Vasc. Plants Cyprus, 31 (1973).

Erect or spreading perennial; rootstock thick, fleshy, sometimes much branched at apex; stems 10–70 cm. high, reddish, angled, conspicuously sulcate, generally much branched, with spreading or suberect, flexuous branches; basal leaves oblong, 2–16 cm. long, 1·5–7 cm. wide, bright or pale green, glabrous or strigose-hispidulous on the undersurface, apex rounded or shortly acute, base distinctly cordate, margins crisped, often with a sinus on either side, forming a fiddle-shaped (*panduriform*) lamina; nervation prominent or obscure, the primary nerves wide-spreading, sometimes strigose; petiole often long, sometimes attainig 15 cm., rather slender, shallowly canaliculate above, sometimes densely strigose; ochreae membranous, close-fitting, truncate, 1–2 cm. long, glabrous or strigose; upper cauline leaves similar to lower, but smaller, usually more acute, with shorter petioles, floral leaves (or bracts) often linear, 1–3 cm. long, less than 0·5 cm. wide; inflorescence a lax panicle, frequently more than 30 cm. long and almost as wide, with long, spreading branches; flowers numerous, the whorls lax or compact and rather evenly spaced; pedicels 1–5 mm. long, conspicuously articulated near the base, often reflexed, but not lengthening much in fruit; outer perianth-segments oblong or lingulate, about 1 mm. long, 0·5–0·8 mm. wide, apex acute or rounded, a little accrescent in fruit but seldom becoming reflexed; inner perianth-segments ovate-oblong, concave, obtuse or subacute, about 2–2·5 mm. long, 1 mm. wide at anthesis; filaments very short, glabrous, subulate; anthers narrowly oblong, about 1·7 mm. long, 0·4 mm. wide; ovary glabrous, subglobose, triquetrous, about 0·5 mm. long, 0·4 mm. wide; styles glabrous, about 0·2 mm. long, reflexed; stigmas penicillate-plumose. Fruiting valves subequal, broadly or rather narrowly deltoid, 4–7 mm. long, 3–7 mm. wide, brownish, conspicuously reticulate, with pectinate-spinulose margins, or occasionally subentire; calluses 3, subequal, or very unequal, or occasionally 1 almost obsolete, the largest oblong, about 2 mm. long, 1 mm. wide; achene broadly ovoid, sharply triquetrous, 2–2·5 mm. long, 1·8–2 mm. wide, dark, sublustrous brown, minutely granulate.

ssp. **cassius** (*Boiss.*) Rechinger f. in Beih. Bot. Centralbl., 49, 2: 38 (1932) et in Candollea, 12: 104 (1949).

    R. cassius Boiss., Fl. Orient., 4: 1013 (1879); Zohary, Fl. Palaest., 1: 64 (1966).

TYPE: Turkey; "in Syriae monte Cassio", *Boissier* (G).

Fruiting valves oblong-deltoid, 4–6 mm. long, 3–5 mm. wide, teeth generally 4 on either margin of the valve, often rather long (1–2 mm. or more) and curved, and generally confined to the basal half of the valve, the apical half entire or subentire and lingulate; inflorescence unbranched or with a few, elongate branches; petioles, ochreae and undersides of leaves glabrous or very sparsely papillose-scabridulous; leaves not panduriform, rather obscurely veined.

HAB.: On igneous mountainsides by streams, or in moist situations in Pine forest; 1,800–5,100 ft. alt.; fl. April–June.

DISTR.: Division 2. Also S.E. Turkey, Syria, Palestine.

2. Xerokolymbos, 4,300 ft. alt., 1937, *Kennedy* 1155! Kryos Potamos, 5,100 ft. alt., 1938, *Kennedy* 1154! Stavros tis Psokas, 1937, *Syngrassides* 1591! Mavres Sykies, 1962, *Meikle* 2736!

NOTES: *R. pulcher* L. ssp. *pulcher*, with numerous, divaricate inflorescence-branches and panduriform basal leaves, does not seem to occur in Cyprus, being replaced at low altitudes by ssp. *divaricatus* (L.) Arcang., and at high altitudes by the above, which is rather tentatively named ssp. *cassius* (Boiss.) Rechinger f., since it differs from typical material of the latter in having rather shorter, less conspicuous marginal teeth on the fruiting valves. Further investigation may show that it is deserving of recognition as an independent, endemic taxon, but the evidence at present available is insufficient to justify such recognition. I cannot accept that ssp. *cassius* merits species status, though (like the Aegean *R. pulcher* ssp. *raulinii*) it has good claims to be considered distinct from typical *R. pulcher*.

ssp. **divaricatus** (*L.*) *Arcang.*, Comp. Fl. Ital., 585 (1882); Lindberg f., Iter Cypr., 12 (1946); Rechinger f. in Candollea, 12: 105 (1949); Zohary, Fl. Palaest., 1: 64 (1966); Osorio-Tafall et Seraphim, List Vasc. Plants Cyprus, 31 (1973).

    R. divaricatus L., Sp. Plant., ed. 2, 478 (1762).

    [*R. pulcher* L. ssp. *anodontus* (non (Hausskn.) Rechinger f.) Lindberg f., Iter Cypr., 12 (1946).]

TYPE: "*in* Italia".

Fruiting valves broadly deltoid or suborbicular, 6–7 mm. long, 6–7 mm. wide, marginal teeth numerous, short, unequal, the valves pectinate-spinulose almost to apex; leaves, petioles and ochreae generally strigose; leaves rarely panduriform, often conspicuously and prominently veined on the undersurface. Plants often much more robust than in *R. pulcher* ssp. *pulcher*.

HAB.: Moist roadsides, marshes and lake shores; sea-level to 3,800 ft. alt.; fl. April–July.

DISTR.: Divisions 1–8. Mediterranean region and eastwards to Iran; an introduced weed or casual elsewhere.

1. Lyso, 1913, *Haradjian* 842!
2. Phini, 1880, *Sintenis & Rigo* 750! Mesapotamos Monastery, 1939, *Lindberg f.* s.n.; Lagoudhera, 1955, *Merton* 2428! Platania, 1961, *D. P. Young* 7409!
3. Episkopi, 1905, *Holmboe* 697; Kolossi Castle, 1971, *Townsend* 71/118!
4. Larnaca, 1862, *Kotschy* 31a.
5. Kythrea, 1880, *Sintenis & Rigo* 540! also, 1932, *Syngrassides* 292! and, 1939, *Lindberg f.* s.n.! Vatili, 1905, *Holmboe* 353; Strongylos, 1967, *Merton* in ARI 467!
6. Kormakiti, 1936, *Syngrassides* 1237! 2 m. S. of Kato Moni, 1956, *Merton* 2765!
7. Kyrenia, 1949, *Casey* 627! also, 1956, *Merton* 2768! Kambyli, 1956, *Merton* 2750!
8. Galatia, 1962, *Meikle* 2538! Ayios Andronikos, 1970, *A. Genneou* in ARI 1453!

5. R. dentatus *L.*, Mantissa Altera, 226 (1771); Boiss., Fl. Orient., 4: 1013 (1879); Post, Fl. Pal., ed. 2, 2: 465 (1933); Rechinger f. in Candollea, 12: 116 (1949); Zohary, Fl. Palaest., 1: 65, t. 77 (1966); Davis, Fl. Turkey, 2: 292 (1967); Osorio-Tafall et Seraphim, List Vasc. Plants Cyprus, 31 (1973).
[*R. pulcher* L. ssp. *divaricatus* (non (L.) Arcang.) Lindberg f., Iter Cypr., 12 (1946) pro parte]

Annual (or ? biennial) with a slender taproot; stems 20–70 (–90) cm. high, erect, often unbranched or with erect branches, glabrous, reddish, obscurely angled and shallowly sulcate; basal leaves usually withered before anthesis, narrowly oblong or oblong-lanceolate, 8–12 cm. long, 1–4 cm. wide, subglabrous or minutely papillose-scabridulous on the underside, bright green, apex obtuse or shortly acute, margins crispate, sometimes with lateral sinuses (and lamina panduriform), base rounded, truncate, broadly cuneate or shallowly cordate; nervation obscure, ascending; petiole slender, glabrous or sparsely papillose, sometimes as long as the lamina; ochreae membranous, glabrous, to 1·8 cm. long, irregularly lacerate at apex; upper leaves usually sparse and small, the uppermost shortly petiolate and grading into small (1–2 cm. long), lanceolate, subsessile bracts; inflorescence 10–25 (or more) cm. long, simple and spiciform, or with a varying number of lax, erect or ascending branches; flowers numerous, in rather lax, evenly spaced whorls; pedicels slender, 2–5 mm. long at anthesis, elongating a little and becoming recurved in fruit, conspicuously articulated near the base; outer perianth-segments narrowly lingulate, obtuse, about 1·5 mm. long, 0·7 mm. wide, obtuse, not reflexing in fruit; inner perianth-segments oblong, about 2 mm. long, 0·8–1 mm. wide, subacute or obtuse, with a conspicuous median callus (even at anthesis), and distinct marginal teeth; filaments subulate, about 0·3 mm. long, glabrous; anthers oblong, almost 1 mm. long, 0·4 mm. wide; ovary subglobose, strongly triquetrous, about 0·8 mm. long, 0·6 mm. wide, glabrous; styles glabrous, reflexed, about 0·3 mm. long; stigmas penicillate-plumose. Fruiting valves ovate-deltoid, 3–5 mm. long, 2·5–3 mm. wide, reticulate-foveolate, apex often slightly recurved, margins generally with 4 (or more) spinulose teeth 1–4 mm. long, rarely subentire or very sparsely and shortly toothed, median calluses 3, subequal, very large and conspicuous, 2–3 mm. long, 1–1·5 mm. wide, sometimes (ssp. *dentatus*) almost occupying the whole valve; achene broadly ovate, sharply triquetrous, about 2·5 mm. long, 1·8 mm. wide, dark brown, lustrous, minutely granulate.

ssp. **mesopotamicus** *Rechinger f.* in Beih. Bot. Centralbl., 49, 2: 15 (1932) et in Candollea, 12: 121 (1949); Zohary, Fl. Palaest., 1: 65 (1966).

TYPE: Not indicated; to be selected from amongst the exsiccata cited by Rechinger f. in Beih. Bot. Centralbl., 49, 2: 15 (1932).

Fruiting valves 4–5 mm. long, 3–4 mm. wide, each with a conspicuous median callus, a narrow, reticulate-foveolate limb, and a border of slender marginal teeth, as long as, or longer than the width of the valve; plant often unbranched, with sparsely leafy or almost leafless inflorescences; leaves sometimes panduriform.

HAB.: Moist, brackish ground; sea-level to 100 ft. alt.; fl. April–July.

DISTR.: Division 5, rather rare. Eastern Mediterranean region and eastwards throughout warm-temperate and tropical Asia.

5. Near Famagusta, 1939, *Lindberg f.* s.n.! Mouth of Pedieos R., 1955, *Merton* 2261! Kouklia reservoir, 1962, *Meikle* 2617!

NOTES: Apparently rare, but easily overlooked as a form of *R. pulcher.*

SUBGENUS 3. **Platypodium** (*Willk.*) *Rechinger f.* Flowers hermaphrodite, often dimorphic; plants annual; leaves not hastate or sagittate; fruiting pedicels recurved, generally flattened or dilated; callus small, inconspicuous.

6. R. bucephalophorus L., Sp. Plant., ed. 1, 336 (1753); Sibth. et Sm., Fl. Graec. Prodr., 1: 246 (1809), Fl. Graec., 4: 39, t. 345 (1823); Poech, Enum. Plant. Ins. Cypr., 14 (1842); Unger et Kotschy, Die Insel Cypern, 225 (1865); Boiss., Fl. Orient., 4: 1014 (1879); H. Stuart Thompson in Journ. Bot., 44: 337 (1906); Holmboe, Veg. Cypr., 63 (1914); Post, Fl. Pal., ed. 2, 2: 466 (1933); Rechinger f. in Bot. Notiser, 1939: 485 (1939) et in Candollea, 12: 140 (1949); Zohary, Fl. Palaest., 1: 66, t. 79 (1966); Davis, Fl. Turkey, 2: 292 (1967); Osorio-Tafall et Seraphim, List Vasc. Plants Cyprus, 31 (1973).

Annual herb up to 30 cm. high; stems erect, simple or branched from the base, obscurely sulcate, generally glabrous; leaves small, generally less than 4 cm. long, 2 cm. wide, lanceolate, elliptic, spathulate or suborbicular; nervation obscure; petiole generally distinct, often much longer than the lamina; ochreae membranous, tubular, to about 1 cm. long; inflorescence racemose or spiciform, leafless or sparsely leafy; flowers very small, 1–4 from each ochreole, sometimes dimorphic, pedicels slender, filiform or becoming flattened or dilated after anthesis, recurved or incurved and up to 8 mm. long in fruit, glabrous or glandular-pubescent; outer perianth-segments lingulate, subacute, 1–1·3 mm. long, 0·3–0·5 mm. wide, strongly reflexed after anthesis; inner perianth-segments up to 2 mm. long, 1 mm. wide, acute or subacute, distinctly spinulose-toothed towards base; filaments very short, glabrous, narrowly conical; anthers oblong, 1–1·3 mm. long, 0·8–1 mm. wide; ovary ellipsoid, strongly triquetrous, about 0·8 mm. long, 0·4 mm. wide; styles about 0·5 mm. long, strongly reflexed; stigmas penicillate-plumose. Fruiting valves 2–5 mm. long, 2–4 mm. wide, deltoid-oblong, apex shortly acute, margins with 2–4, short or elongate, slender or rather broad, straight, curved or hooked spinulose teeth; callus sub-basal, oblong, truncate, inconspicuous; achene ovoid, sharply triquetrous, about 2 mm. long, 1·2 mm. wide, rich brown; lustrous, minutely granulate.

ssp. **graecus** (*Steinh.*) *Rechinger f.* in Bot. Notiser, 1939: 492 (1939), et in Candollea, 12: 142 (1949).

R. *bucephalophorus* L. var. *graecus* Steinh. in Ann. Sci. Nat., ser. 2, 9: 201 (1838).

[R. *bucephalophorus* L. ssp. *gallicus* (non (Steinh.) Rechinger f.) B. L. Burtt in Kew Bull., 9: 70 (1954).]

[R. *bucephalophorus* L. ssp. *bucephalophorus* Osorio-Tafall et Seraphim, List Vasc. Plant. Cypr., 31 (1973).]

TYPE: "In Graecia; vid. sicc. in herb. Lessert" (G).

Plants simple or branched from the base, 5–30 cm. high, without modified flowers arising from the apex of the rootstock; leaves spathulate, the basal 1–3 cm. long, 1–2 cm. wide, often withered before anthesis. Fruiting valves deltoid, 3–4 mm. long and almost as wide, with 3–4, spreading or slightly recurved marginal spine-teeth 1–2 mm. long on either side of the valve.

HAB.: Moist fields, river-shingle; sea-level to 600 ft. alt.; fl. March–April.

DISTR.: Divisions 2, 4. Mediterranean region from S. France to W. Turkey.

2. Valley above Pyrgos, 1957, *Merton* 2859! Pyrgos Valley below Phlevas sawmill, 1959, *P. H. Oswald* 98! Pomos Point, 1962, *Meikle* 2240!

4. Larnaca, 1880, *Sintenis & Rigo* 1008!

ssp. **aegaeus** *Rechinger f.* in Bot. Notiser, 1939: 495 (1939) et in Candollea, 12: 142 (1949); B. L. Burtt in Kew Bull., 9: 70 (1954); Osorio-Tafall et Seraphim, List Vasc. Plants Cyprus, 31 (1973).

TYPE: Not indicated; to be selected from amongst the specimens cited by Rechinger f. in Bot. Notiser, 9: 496 (1939).

Plants small, usually branched from the base, 2–10 (–20) cm. high, with modified (? cleistogamous) flowers arising on short, thick, non-articulated pedicels from the apex of the rootstocks, as well as normal inflorescences; leaves spathulate or suborbicular, 0·5–1·5 cm. long and almost as wide, usually persisting until anthesis; inflorescences slender, reddish, flowers small. Fruiting valves narrowly oblong-deltoid, about 2 mm. long, 1–1·5 mm. wide, with 3–4, spreading or slightly recurved marginal spine-teeth 1–2 mm. long on either side of the valve.

HAB.: Rocky land by the sea; 50 ft. alt.; fl. March–April.

DISTR.: Division 7. Aegean area, W. Turkey, Libya.

7. Ayios Epiktitos, 1951, *Mercy Casey* in *Casey* 1113!

NOTES: In addition to the above, *R. bucephalophorus* (not a common plant in Cyprus) is recorded by Kotschy (Die Insel Cypern, 225) as frequent in the neighbourhood of Larnaca; in the absence of specimens cited (*Kotschy* 16, 306) infraspecific identification is impossible, though they are most probably referable to ssp. *graecus* (Steinh.) Rechinger f., which has been collected in this locality.

"R. HYMENOCEPHALUS" recorded by Bovill (Rep. Plant. Work, Cyprus, 14; 1915) as "tried, without any marked success" in (5) Salamis Plantation, cannot be identified with certainty, but is probably *R. hymenosepalus* Torrey (in Rep. Mexic. Bound. Survey, 2 (1): 177; 1858) a native of the southernmost states of the U.S.A. and northern Mexico, recalling *R. cyprius* in its very large, membranous, pink fruiting-valves, but altogether more robust, with elliptic-acuminate leaves and many-branched, crowded infructescences. It is unlikely to have persisted at Salamis.

3. EMEX *Campderá*
Monogr. Rumex, 56 (1819) nom. cons.

Annual herbs; leaves alternate, simple, petiolate; ochreae membranous, soon lacerate; flowers monoecious, in clusters, forming axillary, racemose inflorescences; male flowers terminal and sub-terminal, perianth-segments 6, subequal, almost free; stamens 4–6; female flowers with a persistent, accrescent, indurating perianth, segments 6, united into an urceolate tube, the 3 outer with the tips spreading and spinous in fruit; styles 3; stigmas penicillate-plumose. Achene triquetrous, enclosed within, but free from the fruiting perianth.

Two species, one native in the Mediterranean region, the other in South Africa, both widespread as weeds in warm-temperate and tropical countries.

1. E. spinosa (*L.*) *Campderá*, Monogr. Rumex, 58, t. 1 (1819); Unger et Kotschy, Die Insel Cypern, 225 (1865); Boiss., Fl. Orient., 4: 1005 (1879); H. Stuart Thompson in Journ. Bot., 44: 337 (1906); Holmboe, Veg. Cypr., 63 (1914); Post, Fl. Pal., ed. 2, 2: 462 (1933); Zohary, Fl. Palaest., 1: 67, t. 80 (1966); Davis, Fl. Turkey, 2: 293 (1967); Osorio-Tafall et Seraphim, List Vasc. Plants Cyprus, 31 (1973).

Rumex spinosus L., Sp. Plant., ed. 1, 337 (1753).

TYPE: "*in* Creta".

Erect, sprawling or prostrate, glabrous, rather fleshy annual 5–50 cm. high; root thick, fleshy; stems branched or unbranched, bluntly angular, sulcate, often reddish; leaves oblong-hastate, oblong-ovate or oblong-cordate, 3–17 cm. long, 2–10 cm. wide, apex blunt or apiculate, margins entire or somewhat undulate; petiole 2–18 cm. long, flattish above; ochreae membranous, about 1 cm. long, soon laciniate, almost obsolete at the upper nodes; inflorescences axillary; rhachis to 10 cm. or more long, erect, terminating in 1–2 clusters of pedicellate male flowers, the sessile female flowers in remote clusters at the middle and base of the rhachis; bracts foliaceous, resembling the leaves, 1–2 cm. long; male flowers with 6, spreading, oblong, obtuse perianth-segments about 1·2 mm. long, 0·6 mm. wide; pedicels slender, 2–3 mm. long, articulated towards the base; filaments very short, thick, glabrous; anthers yellow, oblong, about 1·2 mm. long, 0·8 mm. wide; female flowers subsessile, about 5 mm. long, 2·5 mm. wide, segments united to form an angular, irregularly rugose-foveate base, apices free, those of the outer 3 segments spreading, spinose, almost 1·5 mm. long, those of the inner 3 erect, ovate, about 1·2 mm. long, 0·8 mm. wide; ovary ovoid, triquetrous, glabrous, about 3·5 mm. long, 1·8 mm. wide; styles very short; stigmas reddish, plumose, about 0·8 mm. long. Fruiting perianth woody, angular, rugose-foveate, often reddish, 8–10 mm. long, 6–8 mm. wide, apical spines rigid, spreading; achene free, ovoid-triquetrous, about 5·5 mm. long, 3·5 mm. wide, bright brown, minutely granulate.

HAB.: Cultivated fields and sand-dunes; sea-level to 500 ft. alt.; fl. Jan.–April.

DISTR.: Divisions 1–8. Widespread in the Mediterranean region, and in many other parts of the world as an introduced weed.

1. Paphos, 1969, *A. Genneou* in ARI 1351 ! Also Ayios Yeoryios (Akamas), 1962, *Meikle* ! and Polis, 1962, *Meikle* !
2. Mavrovouni near Lefka, 1932, *Syngrassides* 89 !
3. Akrotiri, 1862, *Kotschy* 626a; Kalavasos, 1905, *Holmboe* 630; Amathus, 1947, *Mavromoustakis* s.n. ! Episkopi Bay, 1950, *F. J. F. Barrington* 33 ! Kouklia, 1960, *N. Macdonald* 27 ! Phasouri, 1970, *A. Genneou* in ARI 1383 !
4. Larnaca, 1880, *Sintenis*; Famagusta, 1936, *Syngrassides* 1070 ! and, 1938, *Nattrass* 918; Cape Greco, 1958, *N. Macdonald* 40 !
5. Salamis, 1962, *Meikle* !
6. Syrianokhori, 1935, *Syngrassides* 758 ! also, 1956, *Merton* 2617 ! Nicosia, 1938, *Nattrass* 912 ! Morphou, 1969, *A. Genneou* in ARI 1349 ! Dhiorios, 1962, *Meikle* !
7. Ayios Epiktitos, 1951, *Casey* 1123 !
8. Ayios Philon near Rizokarpaso, 1941, *Davis* 2229 ! Akradhes, 1962, *Meikle* !

77. RAFFLESIACEAE

Fleshy root or stem parasites, the vegetative organs reduced to mycelium-like threads; flowers unisexual (dioecious or monoecious) sometimes polygamous, rarely hermaphrodite, solitary or in spikes, sessile or stipitate, the stipe often covered with coloured scales; perianth fleshy, coloured, 4–10-partite, with imbricate or valvate lobes or segments; stamens 8 or more, sometimes numerous, anthers sessile, 2-thecous, dehiscing longitudinally or by pores, generally inserted in several series at the apex of a column; ovary inferior or semi-inferior, 4–8-carpellate, 1-locular; ovules numerous; placentation parietal; style undivided or absent; stigma discoid, capitate or lobed. Fruit fleshy, indehiscent or irregularly dehiscent; seeds numerous, minute; endosperm present; embryo minute.

Nine genera, chiefly in warm-temperate or tropical regions of the Old World.

1. CYTINUS L.

Gen. Plant., ed. 6, 576 (sphalm. "566") (1764) nom. cons.

Fleshy dioecious or monoecious parasites; stems simple, thick, usually short, clothed with coloured, imbricate scales; flowers sessile or shortly pedicellate, often with 2 conspicuous bracteoles; perianth tubular, 4–8-lobed, with imbricate lobes; anthers 8–10, sessile in a single series at the apex of a staminal column, dehiscence extrorse by longitudinal slits; ovary inferior; style cylindrical; stigma capitate. Fruit a pulpy berry containing many, minute seeds.

Six species in the Mediterranean region, South Africa and Madagascar.

1. **C. hypocistis** (*L.*) *L.*, Syst. Nat., ed. 12, 2: 602 (1767); Boiss., Fl. Orient., 4: 1071 (1879); Holmboe, Veg. Cypr., 63 (1914); Post, Fl. Pal., ed. 2, 2: 488 (1933); Zohary, Fl. Palaest., 1: 50, t. 53 (1966); Osorio-Tafall et Seraphim, List Vasc. Plants Cyprus, 30 (1973); Elèktra Megaw, Wild Flowers of Cyprus, 8, t. 7 (1973); Davis, Fl. Turkey, 7: 549 (1982).

Asarum hypocistis L., Sp. Plant., ed. 1, 442 (1753).

C. hypocistis (L.) L. var. *kermesinus* Guss., Fl. Sic. Syn., 2: 619 (1844); Kotschy in Unger et Kotschy, Die Insel Cypern, 176 (1865) as "var. *cermesina*".

Hypocistis rubra Fourr. in Ann. Soc. Linn. Lyon, n.s., 17: 148 (1869).

Cytinus hypocistis (L.) L. ssp. *clusii* Nyman, Consp. Fl. Europ., 645 (1881).

C. hypocistis (L.) L. ssp. *kermesinus* (Guss.) Arcang., Comp. Fl. Ital., 612 (1882); Wettst. in Berichte Deutsch Bot. Gesellsch., 35: 91 (1917).

C. hypocistis (L.) L. ssp. *orientalis* Wettst. in Berichte Deutsch Bot. Gesellsch., 35: 91 (1917); Osorio-Tafall et Seraphim, List Vasc. Plants Cyprus, 30 (1973).

C. ruber (Fourr.) Komarov, Fl. U.R.S.S., 5: 442 (1936); Tutin et al., Fl. Europ., 1: 75 (1964).

TYPE: "*in* Hispania, Lusitania, *parasitica cisti*".

Erect, fleshy parasite 3–13 cm. high, growing from the roots of *Cistus* species; stem thick, unbranched, densely or rather loosely clothed with oblong, obtuse or subacute, crimson, carmine, scarlet or orange, glabrous scales 6–20 mm. long, 3–8 mm. wide, the scale-margins often minutely erose or ciliate; flowers 5–10, subsessile or very shortly pedicellate, forming dense terminal heads, unisexual (monoecious), the pistillate flowers usually marginal, the staminate central; bracteoles 2, narrowly oblong, obtuse, 8–13 mm. long, 2–4 mm. wide, exceeding, equalling, or falling short of the perianth, glandular-papillose externally; perianth white or yellow, papillose externally, about 10–13 mm. long, 6–8 mm. wide, slightly constricted at the middle, divided above into 4 erect, ovate, obtuse or subacute, concave lobes about 6–7 mm. long, 4–6 mm. wide; male flowers with a stout staminal column, about 4–5 mm. long; anthers 8, linear-oblong, sessile, pallid, about 3 mm. long, 0·8 mm. wide, each produced into a short, fleshy, deltoid connective; female flowers with a broadly ellipsoid, longitudinally deeply 8-sulcate ovary, 5–6 mm. long, 4–5 mm. wide; style stout, about 3 mm. long; stigma capitate, about 3 mm. diam., deeply 8-sulcate, densely papillose. Fruit subglobose, about 1 cm. diam.; seeds bluntly ellipsoid, minute, about 0·1 mm. long, 0·08 mm. wide; testa brown, minutely reticulate.

HAB.: In garigue at the base of *Cistus* bushes; sea-level to 4,500 ft. alt.; fl. Febr.–May.

DISTR.: Divisions 1–3, 7, 8. Widespread in the Mediterranean region.

1. Karavopetres near Ayios Yeoryios (Akamas), 1962, *Meikle*! W. end Khrysokhou Bay, 1979, *Edmondson & McClintock* E 2734!
2. Near Prodhromos, 1862, *Kotschy* 751! Saïttas, 1931, *Dawe* s.n., and, *J. A. Tracey* 69! Platres, 1938, *Kennedy* 1152!
3. N. of Kandou, 1963, *J. B. Suart* 106!
7. Kantara, 1880, *Sintenis & Rigo* 508! St. Hilarion, 1932, *Syngrassides* 55! also, 1940, *Mavromoustakis* 27! Kharcha, 1956, *G. E. Atherton* 1053! Blackalls Valley, 1956, *G. E. Atherton* 1252! Kyrenia, 1956, *G. E. Atherton* 1275! Melounda, 1963, *J. B. Suart* 8! Akanthou, 1967, *Merton* in ARI 796!
8. Yioti, 1905, *Holmboe* 538.

NOTES: Material available is totally inadequate to test the validity of the characters used to distinguish *C. hypocistis* from *C. ruber*, or *C. hypocistis* ssp. *hypocistis* from *C. hypocistis* ssp. *orientalis*. Plants vary greatly in overall size, and size of perianth, and plants with yellow perianths and scarlet bracteoles (*C. hypocistis*) are evidently found on *Cistus creticus*, as well as others with white perianths and crimson or carmine bracteoles (*C. ruber*). Merton (in ARI 796 from Akanthou) remarks: "When growing on *C.* [*Cistus*] *parviflorus* the whole plant is yellow, the bracts orange, but when growing on *C. creticus* the plant is white, with red bracts". Only by close study in the field and experimental cultivation will it be possible to judge the taxonomic value of the supposed differences.

78. ARISTOLOCHIACEAE

Herbs and shrubs, often with climbing or twining stems; leaves alternate, simple, entire, usually cordate, exstipulate, petiolate; flowers hermaphrodite, solitary or racemose, axillary, terminal or in clusters on the older wood, zygomorphic or actinomorphic, commonly 3-merous, often malodorous; perianth generally tubular, the limb expanded, symmetrically 3-lobed or entire and unilateral, frequently conspicuous and luridly coloured; stamens 6–36, inserted at the apex of the ovary, free or adnate to a stylar column; filaments short, thick; anthers 2-thecous, extrorse, dehiscing longitudinally; ovary inferior or semi-inferior, generally 4–6-locular; ovules numerous, anatropous; placentation axile. Fruit usually a capsule, rarely indehiscent; seeds numerous; endosperm copious, fleshy; embryo very small.

About 10 genera widely distributed, but chiefly in warm-temperate and tropical regions.

1. ARISTOLOCHIA *L.*

Sp. Plant., ed. 1, 960 (1753).

Gen. Plant., ed. 5, 410 (1754).

P. H. Davis et M. S. Khan in Notes Roy. Bot. Gard. Edinb., 23: 515–546 (1961).

Rhizomatous herbs or woody climbers and twiners; leaves generally petiolate, cordate or hastate, sometimes with an amplexicaul, stipuloid modified leaf at the base of the petiole; flowers axillary, pedicellate; perianth tubular, zygomorphic, swollen basally around the anthers and styles, with a small or conspicuous, unilateral, sometimes lobed or appendiculate limb; anthers 6, subsessile, clustered around a stylar column; stigmas (4–) 6. Fruit a septicidal capsule; seeds compressed, numerous.

About 350 species in temperate and tropical regions; the flowers are

generally foetid, and are pollinated by flies, which are temporarily trapped
in the swollen base of the perianth-tube.

Leaves evergreen, coriaceous, 3–7 cm. long, 1·8–5 cm. wide, perianth-limb 8–20 mm. long, 8–15
mm. wide, ovate; climbers; stems woody towards base   -   -   -   **1. A. sempervirens**
Leaves deciduous, not coriaceous, 1–3 cm. long, 0·8–2·5 cm. wide; perianth-limb 2–4 cm. long,
0·4–0·6 cm. wide, strap-shaped; sprawling or ascending herb; stems slender 10–40 cm. long
2. A. parvifolia

1. A. sempervirens *L.*, Sp. Plant., ed. 1, 961 (1753); Sibth. et Sm., Fl. Graec., 10: 26, t. 934
(1840); Unger et Kotschy, Die Insel Cypern, 228 (1865); Boiss., Fl. Orient., 4: 1075 (1879);
H. Stuart Thompson in Journ. Bot., 44: 338 (1906); Davis et Khan in Notes Roy. Bot.
Gard. Edinb., 23: 520 (1961); Zohary, Fl. Palaest., 1: 48, t. 48 (1966); Osorio-Tafall et
Seraphim, List Vasc. Plants Cyprus, 30 (1973); Davis, Fl. Turkey, 7: 554 (1982).
A. altissima Desf., Fl. Atlant., 2: 324, t. 249 (1799); Boiss., Fl. Orient., 4: 1075 (1879);
Holmboe, Veg. Cypr., 62 (1914); Post, Fl. Pal., ed. 2, 2: 489 (1933); Lindberg f., Iter Cypr.,
12 (1946).
TYPE: "*in* Creta".

Climbing evergreen subshrub, commonly 1 m. high, occasionally
scrambling to 3 m. or more; rootstock thick, cylindrical, branched; stems
glabrous, angular, twining in an anticlockwise direction, generally
branched; leaves coriaceous, dull yellowish-green, 3–7 cm. long, 1·8–5 cm.
wide, glabrous, apex acute, acuminate or blunt, base deeply cordate with
prominent rounded auricles, margins entire or minutely erose; nervation
rather inconspicuously reticulate; petiole usually less than 1·5 cm. long;
flowers solitary, axillary; pedicels slender, glabrous or shortly strigillose,
1–2·5 cm. long; perianth strongly curved, 2–4·5 cm. long, glabrous or
sparsely strigillose externally with a conspicuous basal swelling (or utricle)
3–8 mm. diam.; tube about 2 mm. wide above utricle, gradually dilating
above, and 5–11 mm. wide at apex, reddish-green with darker, longitudinal
stripes, limb oblique, ovate, 8–20 mm. long, 8–15 mm. wide, acute or obtuse,
yellowish internally, striped chocolate-brown, with a conspicuous dark
marginal zone, thinly and shortly hairy at the throat; anthers 6, adnate to
stylar column just below its expanded apical rim, broadly oblong, about 0·6
mm. long, 1 mm. wide, the thecae discrete; ovary evident only as a slight,
furfuraceous swelling about 4–5 mm. long at apex of pedicel; stigmas 6,
stout, oblong, about 0·8 mm. long, 0·6 mm. wide. Fruit an ellipsoid capsule
2–3·5 cm. long, 1·5–2 cm. wide, splitting incompletely from the base upward
into 6, spreading segments; seeds strongly compressed or flattened, deltoid,
about 6 mm. long and almost as wide, the dorsal face flat, the ventral
shallowly excavate with a prominent median ridge; testa dull, pale brown,
sparsely verruculose. *Plate 82.*

HAB.: Amongst *Quercus, Acer, Pistacia, Juniperus* or *Rubus* scrub, usually in moist situations
by streams; sea-level to 3,500 ft. alt.; fl. Febr.–July.

DISTR.: Divisions 1–3, 6–8. Sicily and Algeria eastwards to Greece, Crete, Turkey, Lebanon,
Palestine.

1. Neokhorio, 1957, *Merton* 3059!
2. Phini, 1862, *Kotschy* 736! Kyklo Monastery, 1880, *Sintenis & Rigo* 698! Stavros tis Psokas,
1934, *G. W. Chapman* in *Chapman* 140! also, 1962, *Meikle* 2688! Kalopanayiotis, 1937,
Syngrassides 1553! Platres, 1937, *Kennedy* 308! Mandria, 1937, *Kennedy* 309! 310!
Perapedhi, 1937, *Kennedy* 311! By waterfall at Mesapotamos, 1939, *Lindberg f.* s.n.; below
Madhari, 1941, *Davis* 3428! Adelphi Forest, 1957, *Merton* 3138! Mandria–Omodhos road,
1970, *A. Genneou* in ARI 1539!
3. Near Khirokitia, 1903, *A. G. & M. E. Lascelles* s.n.! Kalavasos, 1905, *Holmboe* 605; Kato
Lefkara, 1937, *Syngrassides* 1526! Evdhimou, 1959, *C. E. H. Sparrow* 39! Pyrgos, 1963,
J. B. Suart 45! 49! Between Akapnou and Vikla, 1979, *Edmondson & McClintock* E 2984!
6. Orga-Liveras road, 1957, *Merton* 2853!
7. Mylous Forest Station near Kyrenia, 1934, *Syngrassides* 1357! and, 1937, *Syngrassides*
1465! South-west of Kyrenia, 1949, *Casey* 650! Below Phterykha, 1956, *Merton* 2691!
Dhavlos, 1971, *Guichard* KG/Cyp/8!
8. Valia Forest, 1956, *Poore* 35!

Plate 82. ARISTOLOCHIA SEMPERVIRENS L.   **1**, habit, ×⅔; **2**, flower, longitudinal section, ×2; **3**, fruit, ×2; **4**, androecium and apex of gynoecium, ×6. (**1, 2, 4** from *Merton* 3138; **3** from *Sintenis & Rigo* 698.)

[A. BAETICA *L.*, Sp. Plant., ed. 1, 961 (1753) is recorded, "In Cypro, Cretâ, et Graeciâ orientali, haud infrequens" by Sibthorp (Fl. Graec. Prodr., 2: 222; 1816), and the record is repeated by Poech (Enum. Plant. Ins. Cypr., 14; 1842) and by Kotschy (in Unger et Kotschy, Die Insel Cypern, 228; 1865). *A. baetica*, a native of Spain, Morocco and Algeria, is superficially very similar to *A. sempervirens*, but with smaller, blunter, glaucescent leaves, glabrous ovaries and uniformly dull brown or yellow-tinged perianths. It has not been found since in Cyprus or in any other part of the eastern Mediterranean region, and one may safely assume that Sibthorp (or Smith) confused it with *A. sempervirens*.]

[A. HIRTA *L.*, Sp. Plant., ed. 1, 961 (1753); Sibth. et Sm., Fl. Graec. Prodr., 2: 224 (1816), Fl. Graec., 10: 28, t. 937 (1840); Poech, Enum. Plant. Ins. Cypr., 15 (1842); Unger et Kotschy, Die Insel Cypern, 228 (1865); Boiss., Fl. Orient., 4: 1079 (1879); Holmboe, Veg. Cypr., 63 (1914); Osorio-Tafall et Seraphim, List Vasc. Plants Cyprus, 30 (1973). An erect perennial, to 50 cm. high, with shortly hairy stems and leaves, and conspicuously large, curved perianths 3·5–8 (–15) cm. long, with a broad, ovate, dark maroon-brown limb 1·5–4·5 (–8) cm. long, is reported from Cyprus by Sibthorp, whose record is the basis of the later references cited above. Although *A. hirta* is widely distributed in the Aegean region and in Turkey, no one has since seen it in Cyprus, nor, in view of its bizarre conspicuousness, is it likely that it would have been overlooked. One must assume that the Cyprus record was an error on the part of Sibthorp or his editors.]

2. **A. parvifolia** Sm. in Sibth. et Sm., Fl. Graec. Prodr., 2: 222 (1816), Fl. Graec., 10: 27, t. 935 (1840); Unger et Kotschy, Die Insel Cypern, 228 (1865); Boiss., Fl. Orient., 4: 1076 (1879); Holmboe, Vég. Cypr., 63 (1914); Post, Fl. Pal., ed. 2, 2: 489 (1933); Rechinger f. in Arkiv för Bot., ser. 2, 1: 419 (1950); Davis et Khan in Notes Roy. Bot. Gard. Edinb., 23: 543 (1961); Zohary, Fl. Palaest., 1: 48, t. 49 (1966); Osorio-Tafall et Seraphim, List Vasc. Plants Cyprus, 30 (1973); Davis, Fl. Turkey, 7: 556 (1982).

TYPE: Greece, "Circa Athenas vulgaris; nec infrequens per totam Graecam" (OXF) — an erroneous localisation; Davis (*loc. cit.*, 544) suggests that Sibthorp may have collected the type material in Cyprus.

Sprawling or ascending perennial with a thick, woody, oblong or fusiform rootstock; stems slender, angular, glabrous, 10–40 cm. long, branched at the base, unbranched or sparingly branched above; leaves glabrous, slightly glaucous, bluntly cordate or subreniform, 1–3 cm. long, 0·8–2·5 cm. wide, apex obtuse or rounded, sometimes slightly emarginate, base deeply indented, with 2 prominent blunt or rounded auricles, margins entire, narrowly revolute; nervation rather prominently reticulate; petiole short, generally less than 1 cm. long; flowers solitary, axillary; pedicel generally very short, often almost wanting, glabrous or shortly pubescent; perianth narrowly tubular, 3–6 cm. long, straight or slightly curved, greenish-brown, sometimes blotched chocolate-brown at the throat, glabrous externally, utricle ovoid or subglobose, 3–6 mm. diam., tube 2–3 mm. wide at base, dilating to about 5–6 mm. at apex, limb oblique, strap-shaped, blunt, about 2–4 cm. long, 0·5 cm. wide, shortly and sparsely hairy internally at throat; anthers 6, adnate to stylar column, broadly oblong, about 0·3–0·4 mm. long, 0·8 mm. wide, the thecae discrete; ovary oblong, about 4–5 mm. long, 2·5 mm. wide, longitudinally costate, shortly pubescent with spreading hairs; stylar column very short with a crenulate rim; stigmas 6, very short and thick, less than 0·3 mm. long. Fruit spherical, dull brown, 10–15 mm. diam., splitting irregularly from apex; seeds heart-shaped, not much compressed, about 4 mm. long, 3·5 mm. wide, convex dorsally, excavate ventrally; testa fuscous, conspicuously verrucose.

HAB.: In crevices of calcareous rocks, or in crevices of old walls, occasionally on banks or open ground; sea-level to 2,500 ft. alt.; fl. Dec.–June.

DISTR.: Divisions 1, 3, 4, 7, 8. Also Aegean Islands, Turkey, Syria, Lebanon and Palestine.

1. Paphos, 1901, *A. G. & M. E. Lascelles* s.n. ! Ayios Epiphanios (Akamas), 1905, *Holmboe* 785; Paphos, 1913, *Haradjian* 636; Toxeftera near Ayios Yeoryios (Akamas), 1962; *Meikle* 2093 ! Erimidhes near Ayios Yeoryios (Akamas), 1962, *Meikle* 2143 !
3. Kithasi, 1981, *Hewer* 4782 !
4. Cape Greco, 1862, *Kotschy* 118 ! also, 1905, *Holmboe* 412, and, 1962, *Meikle* 2594 ! Ayios Antonios, Sotira, 1950, *Chapman* 588 ! Famagusta, Ayios Memnon, 1969, *A. Genneou* in ARI 1372 !
7. Pentadaktylos; 1880, *Sintenis & Rigo* 511 ! Halevga, 1941, *Davis* 2823 !
8. Ayios Philon near Rizokarpaso, 1941, *Davis* 2277 !

79. LAURACEAE

Trees and shrubs, or rarely parasitic climbers, mostly with aromatic oil-glands; leaves usually alternate, evergreen, simple, entire, without stipules; inflorescences axillary or terminal, generally cymose or racemose; flowers actinomorphic hermaphrodite or sometimes unisexual (the plants dioecious); perianth 3-merous, small, usually greenish or yellowish, lobes mostly 6, in 2, imbricate whorls; stamens inserted on perianth, commonly in 3–4 whorls, the innermost whorl suppressed or reduced to staminodes; filaments free, sometimes glandular at the base; anthers continuous with filament, 2- or 4-thecous, introrse, (or occasionally one whorl extrorse), dehiscing from the base upward by 2 or 4 flaps; ovary superior, sometimes partly immersed in a concave hypanthium, 1-carpellate, 1-locular; ovule solitary, pendulous; style terminal, simple; stigma small, entire or sometimes lobed. Fruit a berry or drupe; seed with a membranous testa; endosperm absent; embryo large, straight with fleshy cotyledons.

A very large family of more than 30 genera and 2,000–2,500 species widely distributed in the tropics (especially of Asia and America) but less common in temperate regions. Apart from Bay (*Laurus nobilis* L. infra), the family contains such important economic plants as Cinnamon (*Cinnamomum zeylanicum* Blume), Camphor (*Cinnamomum camphora* (L.) Sieb.), Avocado Pear (*Persea americana* Mill.) and Greenheart (*Nectandra rodiaei* Hook.).

1. LAURUS *L.*
Sp. Plant., ed. 1, 369 (1753).
Gen. Plant., ed. 5, 173 (1754).

Evergreen trees or shrubs; leaves alternate, coriaceous, pinnately nerved, aromatic when crushed; inflorescences axillary, fasciculate, umbellate or shortly racemose or cymose; flowers hermaphrodite or unisexual (dioecious), perianth segments 4, subequal, free almost to base, yellowish; stamens (in male flowers) (8) 12 (14); all or some of them with stipitate nectary glands on either side of the filament at or near its middle; anthers introrsely 2-thecous, dehiscing by flaps; female flowers often with 4 staminodes; ovary superior with a short style and 3-lobed stigma. Fruit an ovoid, 1-seeded berry, seated on the persistent base of the perianth.

Two species, one widespread in the Mediterranean region, the other (*L. azorica* (Seub.) Franco) in the Azores, Madeira and the Canaries.

1. **L. nobilis** *L.*, Sp. Plant., ed. 1, 369 (1753); Poech, Enum. Plant. Ins. Cypr., 14 (1842); Unger et Kotschy, Die Insel Cypern, 226 (1865); Boiss., Fl. Orient., 4: 1057 (1879); Holmboe, Veg. Cypr., 82 (1914); Post, Fl. Pal., ed. 2, 2: 482 (1933); Lindberg f., Iter Cypr., 16 (1946); Giacomini et Zaniboni in Archivio Bot., 23: 1–16 (1946); R. R. Waterer, Island of Cyprus, 153 (1947); Chapman, Cyprus Trees and Shrubs, 38 (1949); Zohary, Fl. Palaest., 1: 190. t. 278 (1966); Osorio-Tafall et Seraphim, List Vasc. Plants Cyprus, 43 (1973); Davis, Fl. Turkey, 7: 535 (1982).

TYPE: "*in* Italia, Graecia".

Erect bush or small tree to 7 m. high; branches terete, dull grey-brown, young shoots green, glabrous, subterete or bluntly angular; leaves coriaceous, broadly lanceolate, elliptic, ovate or oblong-obovate, 4–10 cm. long, 1·5–6 cm. wide, dark green above, slightly paler below, pleasantly aromatic when crushed, apex acute or obtuse, base rounded or cuneate, margins entire, flat or closely undulate-crispate; nervation prominent below with stout primary nerves inter-connected with close reticulate venation; petiole short, stout, generally less than 1 cm. long; flowers unisexual (dioecious), in dense sub-umbellate clusters enclosed until maturity in a globose, bud-like involucre of concave, suborbicular, pubescent bracts 5–8 mm. diam.; peduncle pubescent, 5–12 mm. long; pedicels very short, usually less than 2 mm. long, densely pilose; perianth-segments free to base, broadly obovate or suborbicular, concave, glabrous, whitish, 4–6 mm. long, 3·5–4·5 mm. wide; male flowers with 8–12 stamens; filaments thick, glabrous, about 2 mm. long, each with 2 submedian, lateral, very conspicuous, stipitate glands, the stipe about 0·4–0·8 mm. long, the pileus-like apex about 0·8–1 mm. wide; anthers ovate, about 2 mm. long, 1·5 mm. wide, flaps narrowly lanceolate with a rounded apex, almost as long as the theca; female flowers with 4 conspicuous staminodes about 1·5 mm. long, the atrophied anthers flanked by 2 large, reniform, shortly stipitate glands; ovary ellipsoid, glabrous, about 2·5 mm. long, 1·8 mm. wide; style thick, about 1 mm. long; stigma large, pileate, about 1·8 mm. diam., 3-lobed. Fruit a black, shining, ovoid-ellipsoid berry, 1–1·8 cm. long, 0·8–1 cm. wide; seed almost as large as fruit; cotyledons thick, oily, aromatic; plumule small.

HAB.: Moist, rocky ground by streams and springs; 100–4,000 ft. alt.; fl. Febr.–April.

DISTR.: Divisions 1–3, 7. Widespread in the Mediterranean region, both as a wild plant and in cultivation; also cultivated in other warm-temperate and temperate areas.

1. Erimidhes near Ayios Yeoryios (Akamas), 1962, *Meikle* 2131!
2. Melina, ? 1905, *Michaelides* sec. *Holmboe*; Kykko, 1905, *Holmboe* 1002; also, 1939, *Lindberg f.* s.n.; Makheras, 1905, *Holmboe* 1118! Saïttas, 1931, *Mrs. F. A. Tracey*; between Kato Platres and Kokkinovouno, 1937, *Kennedy* 447! Between Pano and Kato Platres, 1937, *Kennedy* 448! Moutoullas, 1937, *Syngrassides* 1560! Kryos Potamos, 3,900 ft. alt., 1937, *Kennedy* 449! Ayia Valley, 1947, *Chapman* 178.
3. Limassol Forest, 1956, *Poore* 25!
7. N. side Pentadaktylos, 1862, *Kotschy*; Buffavento, 1880, *Sintenis & Rigo*; Lapithos, 1905, *Holmboe* 835; St. Hilarion, 1937, *Syngrassides* 1739! Kyrenia Pass, 1939, *Lindberg f.* s.n.! Foot of cliffs near St. Hilarion, 1955, *G. E. Atherton* 731!

NOTES: A Cyprian variety of *Laurus nobilis*, first noted by Marcus Porcius Cato (234–149 B.C.) is briefly described by Pliny the Elder (Nat. Hist. 15, 39), quoting Pompeius Lenaeus: "Cypriam esse folio brevi, nigro, per margines imbricato crispam". Plants with unusually dark green leaves and crisped leaf-margins are still to be seen in cultivation, but are not representative of present-day populations of the species in Cyprus.

Kennedy 447 and 448, from near Platres, have exceptionally large and abnormal cymose inflorescences, consisting of numerous branches, each terminating in a globose involucre enclosing further partial inflorescences. The condition may well be pathological.

One other member of the family, namely the Avocado or Alligator Pear, PERSEA AMERICANA *Mill.*, Gard. Dict., ed. 8 (1768), *Laurus persea* L., Sp. Plant., ed. 1, 370 (1753), a native of Central America, is grown on a limited scale in Cyprus, and recorded from Ktima, 1931, *Nattrass* 121. It is an evergreen tree, up to 20 m. high, with alternate, long-petiolate, elliptic, entire leaves, 5–30 cm. long, 3–15 cm. wide, and axillary, many-flowered panicles of yellowish-green, pubescent, hermaphrodite flowers; perianth consists of 6 lanceolate, acute segments 4–5 mm. long; there are 9 stamens, the innermost whorl with glanduliferous filaments; 3 staminodes; anthers 4-thecous, dehiscing by 4 flaps. The fruit is large and pear-shaped, 7–20 cm. long, with a leathery green or purple-maroon exocarp, a yellowish, buttery mesocarp, and a very large, subglobose or broadly ovoid seed. The mesocarp is pleasantly flavoured, and very nutritious, and the Avocado Pear, virtually unknown in Europe at the beginning of the present century, is now widely marketed. Several named cultivars, differing in size of fruit, and relative hardiness of the plant, are currently listed in horticultural literature.

79a. PROTEACEAE

Trees, shrubs, subshrubs, rarely herbaceous perennials; leaves alternate or scattered, rarely opposite or verticillate, simple, entire or sometimes dissected, often coriaceous; stipules absent; inflorescence racemose, spiciform or capitate, often large and showy; flowers mostly hermaphrodite, actinomorphic or zygomorphic; perianth globular or elongate, divided into 4 valvate segments commonly reflexed or revolute at anthesis; stamens 4, usually inserted on the perianth-segments; anthers with 2 thecae dehiscing longitudinally; ovary superior, 1-locular; style terminal, simple, sometimes bent inwards; ovules 1–many, collaterally attached, or imbricate in 2 contiguous rows. Fruit a follicle, drupe or nut, rarely capsular; seeds sometimes winged; endosperm absent; embryo straight, with fleshy cotyledons.

About 60 genera and more than 1,000 species found in Asia, Africa, Australasia and S. America, but particularly characteristic of the floras of South Africa and Australia. There are no indigenous representatives in Cyprus, but Bovill (Rep. Plant. Work Cyprus, 14) records an unidentified *Protea* as having been planted at (5) Salamis Plantation, together with *Macadamia ternifolia* F. Muell. (probably a misidentification of *M. integrifolia* Maiden et Betcke, the Queensland Nut), an evergreen tree with whorled, narrow-oblong leaves, dense racemes of small greenish-creamy flowers, and spherical fruits, resembling a walnut, with a leathery exocarp, and a hard bony endocarp; the subglobose seeds are well-flavoured and nutritious. *M. integrifolia* is a native of Queensland, cultivated for its nuts elsewhere, and possibly still surviving as a cultivated plant in Cyprus. Bovill (*ibid.*) also records two Australian shrubs, *Hakea gibbosa* (Sm.) Cav. from eastern Australia, and *H. suaveolens* R. Br., from western Australia; both have terete, rigid, needle-like leaves or leaf-segments, simple in *H. gibbosa*, pinnately compound in *H. suaveolens*, and clusters or short racemes of small, yellow flowers. It is not known if either has survived at Salamis, probably not, and perhaps just as well, since both are noted as aggressive weeds in the Cape area of S. Africa, where they have likewise been introduced.

A fifth member of the family, *Grevillea robusta* A. Cunn., the Silky Oak, popular as a greenhouse ornamental in Britain and western Europe, is noted by G. Frangos (Cypr. Agr. Journ., 18: 87; 1923) from the Public Gardens at (6) Nicosia. If allowed to develop, it will become a tree 30 m. high, with large, alternate, fern-like leaves, pinnately dissected into numerous linear-lanceolate, acuminate segments. While generally cultivated for its foliage, the orange flowers, in dense, one-sided racemes, are not without ornamental value; the timber is also handsome and durable. *Grevillea robusta* is a native of Queensland and northernmost New South Wales.

80. THYMELAEACEAE
W. Domke in Bibl. Bot., 111: 1–148 (1934).

Trees, shrubs or perennial herbs with a woody base, rarely climbers or annual herbs; bark tough, fibrous, often reticulated; leaves alternate or sometimes opposite, simple, entire, exstipulate; flowers hermaphrodite or unisexual, actinomorphic, in racemes, spikes, fascicles, or terminal capitula surrounded by an involucre of bracts; receptacle or hypanthium tubular, surrounding the ovary; calyx often petaloid, divided into 4–5, imbricate or subvalvate lobes; petals often small and scale-like, inserted at the apex, middle or near the base of the tubular hypanthium, sometimes absent; stamens equal in number to the calyx-lobes or twice as many; filaments free, inserted in, or at the apex of, the hypanthium; anthers 2-thecous, introrse, dehiscing longitudinally by slits; disk annular, cupular or lobed, sometimes absent; ovary superior, generally 1-locular, rarely more; ovules solitary, generally pendulous from near the apex of the ovary, anatropous; style simple, often excentric; stigma capitate or subdiscoid. Fruit a nut or drupe; seed solitary, with or without endosperm; embryo straight; cotyledons often thick and fleshy.

About 40 genera and 800 species, mostly in temperate regions, especially in S. Africa, Australia and the Mediterranean area.

Fruit a fleshy drupe; flowers usually conspicuous, fragrant -   -   -   DAPHNE (p. 1423)
Fruit not fleshy; flowers inconspicuous, not fragrant   -   -   -   -   -   **1. Thymelaea**

DAPHNE *L.*

Sp. Plant., ed. 1, 356 (1753).
Gen. Plant., ed. 5, 167 (1754).
C. D. Brickell et B. Mathew, Daphne, 194 pp. (1976).

Erect, spreading or prostrate shrubs; leaves evergreen or deciduous, generally alternate and entire, glabrous or hairy, sessile or shortly petiolate; flowers shortly racemose or in axillary or terminal clusters, generally hermaphrodite; calyx-lobes 4, spreading, greenish or variously coloured; petals absent; stamens 8, inserted in, and usually included in, the hypanthium; filaments very short or anthers subsessile; disk absent; style short or almost absent; stigma large, capitate. Fruit a fleshy drupe, sometimes enveloped by the persistent hypanthium.

About 70 species, chiefly in Europe, the Mediterranean region, and temperate Asia.

D. OLEOIDES *Schreber*, Icon. Descr. Plant., 1: 13, t. 7 (1766); Unger et Kotschy, Die Insel Cypern, 227 (1865); Boiss., Fl. Orient., 4: 1047 (1879); Holmboe, Veg. Cypr., 133 (1914); Post, Fl. Pal., ed. 2, 2: 478 (1933); Chapman, Cyprus Trees and Shrubs, 58 (1949); Osorio-Tafall et Seraphim, List Vasc. Plants Cyprus, 74 (1973); Davis, Fl. Turkey, 7: 524 (1982).
TYPE: "*in* Cretae *montosis*".

A much-branched, dwarf, evergreen shrub rarely more than 50 cm. high; young branches pubescent, becoming glabrous with age; leaves coriaceous, obovate or elliptic, 1–4·5 cm. long, 0·3–1·2 cm. wide, greyish, glabrescent or pubescent; flowers 2–8 together in ebracteate terminal clusters, whitish or tinged pink, fragrant, subsessile, hypanthium 6–8 (–15) mm. long, pubescent externally, lobes narrowly deltoid-acuminate, sometimes recurved at apex, 5–7 mm. long. Fruit red, pubescent, included in hypanthium until mature.

HAB.: Igneous mountain slopes; ? 1,500–3,000 ft. alt.; fl. March–June.

DISTR.: Division 2. Widespread in the Mediterranean region, from Spain to Turkey, and eastwards to the Caucasus and western Himalayas.

2. "Bei Evrico [Evrykhou] gesammelt 1859, auch über Galata", *Kotschy*.

NOTES: All records for *D. oleoides* L. from Cyprus are traceable to this reference in *Die Insel Cypern*. The plant has not since been seen on the island, nor is it clear if any Cyprus material was collected by Kotschy, or if the record was based solely on notes or recollections. Keissler (in Engl., Bot. Jahrb., 25: 88; 1898) evidently had not seen any Kotschy specimens when he suggested that the Cyprus plant might be referable to *D. oleoides* L. var. *glandulosa* (Bertol.) Keissler or to var. *brachyloba* Meissner, rather than to typical *D. oleoides*.

Kotschy may have been correct, but it is strange that such an attractive plant should not have been observed by others. Pending its re-discovery, *D. oleoides* must be regarded as very doubtfully present on the island.

1. THYMELAEA *Mill.*

Gard. Dict. Abridg., ed. 4 (1754) nom. cons.
K. Tan in Notes Roy. Bot. Gard. Edinb., 38: 189–246 (1980).

Much-branched, low shrubs, suffruticose perennials and annuals; twigs often with prominent scars; leaves alternate, sessile or subsessile, entire, coriaceous or herbaceous, often adpressed-imbricate when young; flowers hermaphrodite or unisexual, axillary or terminal, solitary or in clusters, actinomorphic; bracts present or absent; hypanthium tubular, urceolate or dilated towards apex, lobes 4, petaloid, persistent; petals absent; stamens 8, in 2 series, filaments adnate to hypanthium; anthers included or the upper whorl shortly exserted; disk minute or absent; ovary rudimentary in staminate flowers; style short, terminal or subterminal; stigma capitate or subdiscoid. Fruit indehiscent, dry, 1-seeded.

Thirty species in S. and central Europe, the Mediterranean region and eastwards to Central Asia.

Shrubs:
Leaves small, thick, ovate-amplexicaul, scale-like, overlapping, about 2–5 (–8) mm. long, 1·5–3 (–5) mm. wide; flowers in terminal clusters   -   -   -   -   **1. T. hirsuta**
Leaves linear-oblanceolate, about 20 mm. long, 1·5–2 mm. wide, not thick, overlapping or scale-like; flowers axillary, solitary or in small clusters
                                                     2. T. tartonraira var. **linearifolia**
Annual herbs with slender stems and linear leaves   -   -   **3. T. passerina** ssp. **pubescens**

SECT. 1. **Piptochlamys** (*C. A. Mey.*) *Endl.* Dioecious or monoecious shrubs; flowers in terminal, bracteate or ebracteate clusters; perianth persistent or deciduous.

1. **T. hirsuta** (*L.*) *Endl.*, Gen. Plant. Suppl., 4 (2): 65 (1848); Unger et Kotschy, Die Insel Cypern, 227 (1865); Boiss., Fl. Orient., 4: 1054 (1879); H. Stuart Thompson in Journ. Bot., 44: 337 (1906); Holmboe, Veg. Cypr., 133, t. 113 (1914); Post, Fl. Pal., ed. 2, 2: 480 (1933); Lindberg f., Iter Cypr., 24 (1946); Chapman, Cyprus Trees and Shrubs, 57 (1949); Zohary, Fl. Palaest., 2: 330, t. 487 (1972); Osorio-Tafall et Seraphim, List Vasc. Plants Cyprus, 74 (1973); K. Tan in Notes Roy. Bot. Gard. Edinb., 38: 209 (1980); Davis, Fl. Turkey, 7: 528 (1982).
    Passerina hirsuta L., Sp. Plant., ed. 1, 559 (1753); Sibth. et Sm., Fl. Graec. Prodr., 1: 262 (1809), Fl. Graec., 4: 52, t. 360 (1823); Poech, Enum. Plant. Ins. Cypr., 14 (1842).
    TYPE: "*in* Hispania".

Erect or sprawling, functionally dioecious (trimonoecious, cf. K. Tan, *op. cit.*, 197) evergreen shrub up to 150 cm. high, old wood with greyish-brown, flaking bark; branches very numerous, densely white-tomentose, divergent or sometimes pendulous; leaves numerous, overlapping, sessile, coriaceous, a little concave and semi-amplexicaul, ovate, dark green 2–5 (–8) mm. long, 1·5–3 (–5) mm. wide, glabrous dorsally, white-tomentose ventrally, apex acute or obtuse; flowers in small (2–8-flowered) clusters at the tips of short, leafy, lateral branches, sessile; bracts absent; hypanthium infundibuliform, tomentose externally, about 2·5 mm. long and almost as wide; lobes yellow, spreading, bluntly ovate or suborbicular, about 1·5 mm. long and almost as wide; male flowers with 8 stamens inserted at the apex and middle of the hypanthium; free part of filaments very short; anthers broadly oblong, about 0·8 mm. long, 0·5 mm. wide; rudimentary ovary almost 0·2 mm. long; female flowers similar in general aspect; ovary ellipsoid, about 1·5 mm. long, 1 mm. wide, subglabrous or thinly hairy; style about 0·2 mm. long; stigma relatively large, capitate; stamens wanting (or present in occasional hermaphrodite flowers). Fruit ellipsoid, glabrous or sparsely pubescent, about 4·5 mm. long, 2·3 mm. wide, pericarp membranous, minutely and closely verruculose-papillose; seed large, not much smaller than fruit; testa blackish with longitudinal rows of minute tubercles or papillae.

HAB.: Dry sandy banks and grass-steppes, usually near the sea; sea-level to 100 ft. alt.; fl. Jan.–May.

DISTR.: Divisions 3, 4, [6] ? 8. Widely distributed in the Mediterranean region.

3. Cape Gata near Ayios Nikolaos Monastery, 1862, *Kotschy*; also, 1905, *Holmboe* 674, and, 1962, *Meikle* 2913! Curium, 1937, *C. B. Ussher* s.n.! Mile 3, Limassol–Paphos road, 1938, *Chapman* 345! 346! Episkopi, 1939, *Lindberg f.* s.n.; Limassol Salt Lake, 1941, *Davis* 3545! Near Mazotos, 1957, *Merton* 2854! Kolossi, 1961, *Polunin* 6676! Akrotiri, 1966, *Ann Matthews* 9!

4. By Larnaca Salt Lake, 1862, *Kotschy*; also 1880–1974, *Sintenis & Rigo* 686! *A. G. & M. E. Lascelles* s.n.! *Holmboe* 94! *Lindberg f.* s.n.! *N. Macdonald* 13! *P. Laukkonen* 58! Between Larnaca and Perivolia, 1934, 1938, *Syngrassides* 495! 1749! Larnaca aerodrome, 1950, *Chapman* 420! S.W. of Larnaca, 1967, *Merton* in ARI 361! Cape Kiti, 1981, *Hewer* 4731!

[6.] Nicosia, 1931, *F. A. Rogers* 5458! See Notes.

?8. Karpas, May 1937, *H. Roger-Smith* s.n.!

NOTES: Apart from two records, apparently restricted to the neighbourhoods of Larnaca and Limassol, where it is locally plentiful. The *F. A. Rogers* record from Nicosia is almost certainly an error, probably due to some misunderstanding when the specimen was mounted at Kew. The Karpas record (*Roger-Smith* s.n.!) may be correct, but is rather vague, nor has *T. hirsuta* been found by any other collector in this area.

SECT. 2. **Chlamydanthus** (*C. A. Mey.*) *Endl.* Dioecious or functionally dioecious (trimonoecious) shrubs; flowers axillary, solitary or in clusters; bracts present or absent; perianth persistent.

2. **T. tartonraira** (*L.*) *All.*, Fl. Pedem., 1: 133 (1785) as "*T. tarton-raira*"; Unger et Kotschy, Die Insel Cypern, 227 (1865); Boiss., Fl. Orient., 4: 1053 (1879); Holmboe, Veg. Cypr., 133 (1914); Chapman, Cyprus Trees and Shrubs, 57 (1949); Osorio-Tafall et Seraphim, List Vasc. Plants Cyprus, 74 (1973); K. Tan in Notes Roy. Bot. Gard. Edinb., 38: 215 (1980).
   Daphne tartonraira L., Sp. Plant., ed. 1, 356 (1753).

Much-branched, bushy, functionally dioecious shrub to 1 m. high; old wood with greyish-brown or ferrugineous, flaking bark and prominent, tuberculate leaf-scars; young shoots densely leafy, pubescent-sericeous or glabrescent; leaves sessile, spreading, subcoriaceous (but not very thick), ovate-spathulate in the western Mediterranean region, becoming narrower further east, and linear-oblanceolate in our area, 7–15 (–20) mm. long; 2–6 (–9) mm. wide, glabrous, pubescent or sericeous; flowers 2–5, sessile, in dense, bracteate, axillary clusters; bracts 5–15, ovate, acute or obtuse; perianth pale yellow, pubescent or sericeous externally; hypanthium narrowly infundibuliform, 4–5 mm. long, almost 2·5 mm. wide at apex, lobes spreading, 1–2·5 mm. long, rounded or subacute, alternately 2–2·5 mm. and

1–1·5 mm. wide at base, pubescent or sericeous externally; male flowers with 8 stamens inserted in 2 series towards the apex of the hypanthium; free part of filaments about 0·5 mm. long; anthers broadly oblong or suborbicular, about 0·4 mm. diam.; ovary rudimentary, about 0·5 mm. long; female flowers with a pubescent, narrowly oblong-ellipsoid ovary, about 2 mm. long, 0·8 mm. wide; style thick, subterminal, about 0·4 mm. long; stigma capitate. Fruit oblong-ellipsoid, about 4·5 mm. long, 2·5 mm. wide, partly enveloped by the longitudinally split, persistent perianth, pericarp translucent, slightly fleshy; seed ovoid-acuminate, about 2·8 mm. long, 1·5 mm. wide; testa black, regularly and closely punctate with longitudinal rows of small pits.

ssp. **argentea** (*Sm.*) *Holmboe*, Veg. Cypr., 133 (1914) quoad nomen; Lindberg f., Iter Cypr., 24 (1946) quoad nomen; Osorio-Tafall et Seraphim, List Vasc. Plants Cyprus, 74 (1973) quoad nomen.
    Daphne argentea Sm. in Sibth. et Sm., Fl. Graec. Prodr., 1: 258 (1809), Fl. Graec., 4: 48, t. 355 (1824); Poech, Enum. Plant. Ins. Cypr., 14 (1842).
    Thymelaea argentea (Sm.) Endl., Gen. Plant. Suppl., 4, 2: 65 (1848).

Leaves sericeous or adpressed-pubescent, (5–) 10–25 mm. long, 1·5–4 mm. wide; bracts obtuse or subacute.

var. **linearifolia** *K. Tan* in Notes Roy. Bot. Gard. Edinb., 38: 218 (1980); Davis, Fl. Turkey, 7: 527 (1982).
    T. tartonraira (L.) All. ssp. *argentea* (Sm.) Holmboe, Veg. Cypr., 133 (1914) quoad plant.; Lindberg f., Iter Cypr., 24 (1946) quoad plant.; Osorio-Tafall et Seraphim, List Vasc. Plants Cyprus, 74 (1973) quoad plant.
    TYPE: Cyprus: Mazotos, 1905, *Holmboe* 162 (O).

Leaves narrowly linear-oblanceolate, commonly about 20 mm. long and 1·5–2 mm. wide, occasionally up to 3 mm. wide; bracts broadly ovate, obtuse (in Cyprus) or subacute, 1–1·8 mm. long, 0·8–1 mm. wide at base.

HAB.: In garigue on limestone or igneous hillsides; 100–2,500 ft. alt.; fl. Febr.–April.

DISTR.: Divisions 2, 3, 7, 8. Also S.W. Turkey, Rhodes, Crete.
2. Between Lefkara and Vavatsinia, 1941, *Davis* 2734!
3. Pissouri, 1840, *Kotschy* 65, also, 1962, *Meikle* 2648! Mazotos, 1905, *Holmboe* 163, and, 1957, *Merton* 2882! Stavrovouni, 1934, *Syngrassides* 496! also, 1937, *Syngrassides* 1479! and, 1939, *Lindberg f.* s.n.! Skarinou, 1940, *Davis* 1986! Pyrgos, 1939, *Mavromoustakis* 12! Mile 22, Nicosia–Limassol road, 1938, *Chapman* 342!
7. Near Kantara, 1880, *Sintenis & Rigo* 10! St. Hilarion, 1940, *Mavromoustakis* 18! Akanthou, 1940, *Davis* 2029!
8. Between Komi Kebir and Dhavlos, 1941, *Davis* 2452!

SECT. 3. **Ligia** (*Fasano*) *Meissner* Annuals or suffrutescent perennials; stems simple or branched; flowers unisexual (monoecious) or hermaphrodite, axillary, solitary or in clusters of 2–4 (–7); bracts 2, green.

3. T. passerina (*L.*) *Coss. et Germ.*, Syn. Fl. Env. Paris, ed. 2, 360 (1861); Zohary, Fl. Palaest., 2: 331, t. 488 (1972); Osorio-Tafall et Seraphim, List Vasc. Plants Cyprus, 74 (1973); K. Tan in Notes Roy. Bot. Gard. Edinb., 38: 236 (1980); Davis, Fl. Turkey, 7: 530 (1982).
    Stellera passerina L., Sp. Plant., ed. 1, 559 (1753).
    Ligia passerina (L.) Fasano in Atti Accad. Nap., 1787: 245 (1788); Boiss., Fl. Orient., 4: 1052 (1879), as "*Lygia*"*passerina*; Post, Fl. Pal., ed. 2, 2: 479 (1933), as "*Lygia*"*passerina*.

Erect annual, 10–60 cm. high; stems slender, subterete or longitudinally ridged, usually unbranched for some distance above base, sparingly or profusely branched above with long erect or spreading branches, glabrous throughout or varyingly pilose with adpressed hairs towards apex, glaucous-green or purplish; leaves numerous, scattered, erect, linear, 5–15 (–20) mm. long, 1–2·5 mm. wide, glabrous or sometimes thinly pilose on the undersurface, apex acute, margins entire or very obscurely erose; flowers solitary or in small (2–5-flowered) clusters in the axils of the upper cauline leaves, forming long, slender, wand-like inflorescences, sessile or subsessile, usually hermaphrodite; bracts 2, lingulate or ovate, obtuse or subacute, densely pubescent, 1·5–2 mm. long, 0·8–1 mm. wide; perianth greenish-yellow, adpressed-pubescent externally, hypanthium (or perianth-tube) narrowly cylindrical, 2·5–3 mm. long, 0·8 mm. wide, densely white-strigose at base; lobes ovate-deltoid, about 0·8–1 mm. long, 0·5 mm. wide at base; stamens inserted about the middle and apex of the hypanthium; filaments very short; anthers yellow, oblong, about 0·4 mm. long, 0·25 mm. wide; ovary glabrous, narrowly oblong, shortly stipitate, about 0·8 mm. long, 0·3 mm. wide; style distinct, about 0·4 mm. long; stigma capitate. Fruit pyriform, enclosed within the persistent perianth, about 2·5 mm. long, 1·5 mm. wide; pericarp thin, translucent; seed almost as large as fruit, pyriform-acuminate; testa blackish, minutely and closely, longitudinally punctate-foveolate.

ssp. **pubescens** (*Guss.*) *Meikle* **comb. nov.**
 Stellera pubescens Guss., Fl. Sic. Prodr., 1: 466 (1827).
 Lygia pubescens (Guss.) C. A. Mey. in Bull. Phys.-Math. Acad. Imp. Petersb., 1: 358 (1843); Boiss., Fl. Orient., 4: 1052 (1879).
 Thymelaea gussonei Boreau in Mém. Soc. Acad. Angers, 4: 121 (1858); K. Tan in Notes Roy. Bot. Gard. Edinb., 38: 237 (1980); Davis, Fl. Turkey, 7: 530 (1982).
 T. arvensis Lam. ssp. *pubescens* (Guss.) Arcang., Comp. Fl. Ital., ed. 1, 604 (1882).
 Lygia passerina (L.) Fasano ssp. *pubescens* (Guss.) Lindberg f., Iter Cypr., 24 (1946).
 TYPE: Sicily; "In argillosis, inter segetes, et post messem in arvis; *Palermo, Altavilla, Alcamo, Trapani, Vicari, Caltanissetta, Alimena, Polizzi, Catania, Lentini, Palagonia*, ec.".

Stems unbranched or with a few (2–5), long, spreading branches, glabrous below, thinly adpressed-pilose in the region of the inflorescence, often purplish (especially in fruiting specimens), generally rather more robust than in ssp. *passerina*.

HAB.: Cultivated and fallow fields; grass steppe; 150–600 ft. alt.; fl. April–July.

DISTR.: Divisions 5, 7, rather rare. Widespread in the Mediterranean region and eastwards to Iraq.
5. Between Athalassa and Nicosia, 1939, *Lindberg f.* s.n.! Mia Milea, 1950, *Chapman 676*!
7. Kyrenia, 1948, *Mercy Casey* in *Casey 140*! also, 1949, *Casey 960*!

NOTES: All the Cyprus material is pubescent to some degree, and is referable to ssp. *pubescens*, rather than to typical ssp. *passerina*. The distinctions between the two are scarcely worth specific rank, though typical *T. passerina* predominates in northern and western parts of the aggregate distribution, and, indumentum apart, is often recognizable by its numerous, slender branches. The purple colour of the stems, evident in some specimens of ssp. *pubescens*, seems to me of seasonal rather than taxonomic significance.
 The plant figured as *Thymelaea pubescens* (non (L.) Meissner) in Flora Palaestina, 2: t. 489 (1972) looks more like the distinct *T. mesopotamica* (Jeffrey) Peterson.

81. ELAEAGNACEAE
C. Servettaz in Beih. Bot. Centralbl., 25 (1): 1–420 (1909).

Trees or shrubs, usually clothed with silvery, scaly (or *lepidote*) hairs; leaves entire, alternate or sometimes opposite; stipules absent. Flowers often unisexual and dioecious, sometimes hermaphrodite, in spikes, racemes or fascicles; perianth tubular, hypogynous, sometimes adnate to ovary, apex with 2 or 4, valvate lobes or sometimes truncate. Stamens inserted on, or at the base of the perianth, as many as, or twice as many as the perianth lobes. Ovary 1-locular; style simple, terminal; stigma usually lateral; ovule solitary, erect, anatropous. Fruit a pseudo-drupe, enclosed by the persistent, incrassate perianth; seed erect, with little or no endosperm; embryo straight, with thick cotyledons.

Three genera (*Hippophaë, Elaeagnus, Shepherdia*) widely distributed in the northern hemisphere, also in N.E. Australia. Only one genus in Cyprus.

1. ELAEAGNUS *L.*

Sp. Plant., ed. 1, 121 (1753).
Gen. Plant., ed. 5, 57 (1754).

Shrubs with silvery, lepidote indumentum; leaves alternate; flowers in axillary clusters, hermaphrodite or polygamous, pedicellate; perianth campanulate or tubular, 4-lobed; stamens 4. Fruit a fleshy, lepidote pseudo-drupe.

About 45 species in Europe, Asia and N. America.

1. **E. angustifolia** *L.*, Sp. Plant., ed. 1, 121 (1753); H. Stuart Thompson in Journ. Bot., 44: 337 (1906); Holmboe, Veg. Cypr., 133 (1914); Post, Fl. Pal., ed. 2, 2: 481 (1933); R. M. Nattrass, First List Cyprus Fungi, 45 (1937); Lindberg f., Iter Cypr., 24 (1946); Chapman, Cyprus Trees and Shrubs, 58 (1949); Zohary, Fl. Palaest., 2: 332, t. 490 (1972); Osorio-Tafall et Seraphim, List Vasc. Plants Cyprus, 74 (1973); Davis, Fl. Turkey, 7: 534 (1982).

   E. orientalis L., Mantissa 1, 41 (1767).
   E. hortensis M. Bieb., Fl. Taur.-Cauc., 1: 112 (1808); Boiss., Fl. Orient., 4: 1056 (1879).
   E. angustifolia L. var. *orientalis* (L.). O. Kuntze in Act. Hort. Petrop., 10: 235 (1887).
   TYPE: "*in* Bohemia, Hispania, Syria, Cappadocia".

A deciduous shrub or small tree up to 8 m. high, with lax, often somewhat pendulous, spinose or unarmed branches, bark at first silvery-lepidote, becoming glabrous and shining brown with age; leaves elliptic-lanceolate or ovate-elliptic, 2–7 cm. long, 1–2·5 cm. wide, apex acute, acuminate or obtuse, base tapering to a petiole 0·5–1 cm. long, margins entire, upper surface of lamina at first densely lepidote, becoming green with age; lower surface persistently silvery-lepidote, except in sterile, juvenile or sucker shoots, where it is commonly dull, stellate-pubescent; flowers sweet-scented, solitary or in clusters of 2–3 in the leaf-axils; pedicels about 2–5 mm. long; perianth narrowly tubular at base, expanded, and campanulate above, 5–7 mm. long, yellow internally and ± glabrous, densely lepidote externally; lobes deltoid-acute, spreading, about 3–4 mm. long and almost as wide at the base; stamens inserted on short filaments just below perianth-sinuses; anthers oblong, about 2–2·5 mm. long, 1·2 mm. wide, brown; ovary closely enveloped in base of perianth-tube; style 5–7 mm. long, rather thick, sharply curved at apex, and surrounded near base by a conical perigonial disk or collar; stigma small, lateral. Fruit ovoid-ellipsoid, like an olive, yellowish-brown, shining or thinly lepidote, about 2–2·5 cm. long, 1·5–1·8 cm. wide, fleshy, edible but insipid; stone narrowly ellipsoid, 1·3–1·8 cm. long, 0·4–0·5 cm. wide, pale brown with 8, darker, longitudinal stripes.

HAB.: Gardens, roadsides, or sometimes on uncultivated hillsides, but probably always an escape from cultivation; sea-level to 4,500 ft. alt.; fl. May–June.

DISTR.: Divisions 2, 6. Mediterranean region, western Asia eastwards to N. China.

2. Madhari, near Spilia, 1880, *Sintenis & Rigo* 688! Near Kakopetria, 1900, *A. G. & M. E. Lascelles* s.n.! and, 1905, *Holmboe*, 851; and, *Chapman* 90! and, *Syngrassides* 1673! Galatà, 1939, *Lindberg f.* s.n.! Prodhromos, 1955, *G. E. Atherton* 660!

6. Dheftera, 1931, *Nattrass* 150! Skylloura, 1932, *Syngrassides* 455! Nicosia, 1974, *Meikle* 5000!

NOTES: Doubtfully native in any part of the Mediterranean region, but frequently planted as an ornamental, and sometimes naturalized. The fragrant flowers have been used in perfumery, and are said to be a good source of honey, while the fruits, edible but not very interesting in their fresh condition, can be fermented to make alcohol or a spirituous liquor "with the headiness of vodka".

The supposed differences between *E. angustifolia* L. and *E. orientalis* L. (Mantissa 1, 41; 1767), are unsatisfactory. It is true that a narrow-leaved variant is prevalent in the western Mediterranean region, from Spain to Italy and Yugoslavia, but it intergrades with broader-leaved forms from the eastern Mediterranean and Asia. The specimen of *E. angustifolia* in herb. Linn. (IDC microfiche 160/1) is one of these intermediates, and matches closely with specimens from Cyprus. *E. orientalis* of Linnaeus (IDC microfiche 160/2) is based on a sterile, juvenile or sucker shoot, and has the dull, soft, stellate-hairy indumentum characteristic of such growths. I have little doubt that mature, flowering branches from the same source would have proved indistinguishable from *E. angustifolia*. The very narrow-leaved variant found in the western Mediterranean area is most probably *E. angustifolia* L. var. *inermis* (Mill.) Lam. et DC., Fl. Franç., 3: 354 (1805), though the varietal epithet is inappropriate since thorns are often to be found in this variety; it is likely that the fruits might furnish better distinctions between var. *inermis* and var. *angustifolia*.

82. SANTALACEAE

R. Pilger in Engl. et Prantl, Pflanzenfam., ed. 2, 16b: 52–91 (1935).

Hemiparasitic herbs, shrubs and trees; leaves alternate or opposite, exstipulate, sometimes reduced to scales; flowers often in axillary cymes, forming paniculate, spiciform or capitate inflorescences, hermaphrodite or unisexual; perianth simple, sepaloid or petaloid, often fleshy, hypanthium adnate to ovary; lobes 3–6, valvate; stamens as many as perianth-lobes, and opposite to them; anthers 2-thecous, dehiscing longitudinally; disk epigynous or perigynous; ovary inferior or semi-inferior, 1-locular; ovules 1–3 (–4) pendulous from a central placenta; style simple; stigma truncate, capitate or lobed. Fruit a nut or drupe; seed without a testa; endosperm copious; embryo straight, often oblique; cotyledons often terete.

About 30 genera and 400 species widely distributed in temperate and tropical regions. The fragrant heartwood of *Santalum album* L. and related Sandalwoods is highly valued, especially in tropical Asia.

Fruit a fleshy red drupe; flowers normally unisexual; a stout rhizomatous shrub - **1. Osyris**
Fruit a dry, greenish nut; flowers normally hermaphrodite; slender annual (in our area) or perennial herbs - - - - - - - - - - - **2. Thesium**

1. OSYRIS *L.*

Sp. Plant., ed. 1, 1022 (1753).
Gen. Plant., ed. 5, 448 (1754).

Evergreen shrubs or trees; stems angled; leaves alternate, linear, lanceolate or elliptic, coriaceous; flowers unisexual (dioecious or polygamous), axillary, the males in short racemes, the female and hermaphrodite solitary or few; bracts and bracteoles caducous; perianth 3–4 (–5)-lobed; male flowers with a very short hypanthium (or perianth-tube); stamens 3–4; filaments short, often arising from a cluster of hairs; anthers 2-thecous; female (and hermaphrodite) flowers with an elongate-conical hypanthium; ovary inferior, ovules 3–4; style short; stigma 3–4-lobed. Fruit a globose or ovoid drupe, crowned with the persistent remains of the perianth-lobes; exocarp thinly fleshy; endocarp crustaceous; seed large, occupying most of the drupe; endosperm copious; embryo small.

Six or seven species in the Mediterranean region, Africa and Asia.

1. **O. alba** *L.*, Sp. Plant., ed. 1, 1022 (1753); Unger et Kotschy, Die Insel Cypern, 227 (1865); Boiss., Fl. Orient., 4: 1058 (1879); H. Stuart Thompson in Journ. Bot., 44: 338 (1906); Holmboe, Veg. Cypr., 62 (1914); Post, Fl. Pal., ed. 2, 2: 483 (1933); Zohary, Fl. Palaest., 1:

44, t. 43 (1966); Osorio-Tafall et Seraphim, List Vasc. Plants Cyprus, 30 (1973); Davis, Fl. Turkey, 7: 545 (1982).

TYPE: "*in* Italia, Hispania, Monspelii".

Scoparioid shrub, 40–80 cm. high, with tough, creeping rhizomes and much-branched, erect, angular, glabrous, green stems, forming small thickets under *Pistacia*, *Ceratonia* and other, taller shrubs; leaves rather sparse and remote, linear, entire, sessile or subsessile, 8–20 mm. long, 1–3 mm. wide; male flowers in short, lax, axillary, 4–10-flowered racemes 1–2 cm. long; bracts lanceolate, glabrous, recurved, 3–4 mm. long, 1–2 mm. wide; pedicels glabrous, about 6 mm. long, slightly swollen at apex; perianth shallowly cup-shaped, hypanthium glabrous, about 2 mm. long, 3 mm. diam., lobes usually 3, spreading or recurved, broadly deltoid, about 1·5 mm. long, 3 mm. wide at base, glabrous except for a conspicuous tuft of hairs internally at the point of insertion of the stamens; filaments glabrous, about 0·5 mm. long; anthers broadly oblong, about 0·5 mm. long and as wide or wider, thecae discrete; disk adnate to hypanthium, 3-lobed, the lobes alternating with the perianth-lobes; female (and hermaphrodite) flowers subsessile, solitary at the tips of short branchlets, subtended by 2–3, lanceolate bracts; hypanthium glabrous, conical, about 6 mm. long, 3 mm. wide at apex; perianth-lobes and disk as in male flowers, but the lobes glabrous on both faces; stamens reduced to small, non-functional staminodes in female flowers; ovary immersed in hypanthium; style stout, conical, 2–3 mm. long; stigma deeply 3-lobed, the lobes about 0·8 mm. long, recurved or revolute. Fruit globose, orange or red when ripe, 6–8 mm. diam., exocarp thin, fleshy, endocarp pale brown with shallow, longitudinal furrows; seed rough externally with small, blunt, brownish projections.

HAB.: In garigue or maquis on rocky hillsides; 800–1,400 ft. alt.; fl. April–June.

DISTR.: Divisions 1, 3, 7. Widespread in S. Europe and the Mediterranean region.

1. Ayios Minas near Smyies (Akamas), 1962, *Meikle* 2177 !
3. Near Limassol, 1859, *Kotschy* 985.
7. Kyrenia Pass, 1901, *A. G. & M. E. Lascelles* s.n. ! also, 1951 and 1956, *Casey* 1187 ! 1721 ! Above Krini, 1962, *Casey* 1739 !

2. THESIUM *L.*

Sp. Plant., ed. 1, 207 (1753).
Gen. Plant., ed. 5, 97 (1754).
R. Hendrych in Acta Univ. Carol. Biol., 1970: 293–358 (1972).

Annual or perennial herbs or subshrubs; leaves alternate, linear or scale-like; flowers hermaphrodite, solitary or in cymes, generally with bracts and bracteoles; perianth 4–5-merous, with a cup-shaped or elongate hypan-thium; stamens 4–5, sometimes with a tuft of hairs at the point of insertion; ovary inferior, 1-locular; ovules 2–4; style simple, sometimes elongate; stigma capitate or somewhat lobed. Fruit a dry, globose or ovoid-ellipsoid nut, crowned with the persistent perianth-lobes; seed 1; embryo small, sited near the apex, or at the middle of the copious endosperm.

About 220 species, widely distributed in Europe, the Mediterranean region, and almost throughout temperate and tropical Africa.

1. **T. humile** *Vahl*, Symb. Bot., 3: 43 (1794); Boiss., Fl. Orient., 4: 1064 (1879); Sintenis in Oesterr. Bot. Zeitschr., 31: 259 (1881), 32: 290 (1882); Holmboe, Veg. Cypr., 62 (1914); Post, Fl. Pal., ed. 2, 2: 484 (1933); Zohary, Fl. Palaest., 1: 45, t. 45 (1966); Osorio-Tafall et Seraphim, List Vasc. Plants Cyprus, 30 (1973); Davis, Fl. Turkey, 7: 538 (1982).
TYPE: Tunis; "in cultis Tuneti" (C).

Suberect or spreading, much-branched, glabrous, glaucous annual, 9–30 (–40) cm. high; taproot slender; stems angled or sulcate; leaves very narrowly linear, 10–40 mm. long, 1–1·5 mm. wide, 1-nerved, a little fleshy, apex acute, margins generally scabridulous; flowers very numerous, subsessile, generally solitary, forming a spiciform or subracemose inflorescence; bracts foliaceous, 6–15 (or more) mm. long; bracteoles 2, linear-subulate, 3–4 mm. long, 0·5 mm. wide; hypanthium barrel-shaped, about 2 mm. long, 1 mm. wide, glabrous; perianth-lobes 5, whitish or greenish, ovate-acute, incurved-cucullate, about 0·8 mm. long, 0·4 mm. wide at base, glabrous except for a small tuft of hairs at the point of insertion of the stamens; filaments 0·2 mm. long; anthers broadly oblong, about 0·2 mm. long, and almost as wide; style 0·4 mm. long; stigma capitate. Fruit barrel-shaped, 3·5–4 mm. long, 2–2·5 mm. wide, greenish, obscurely reticulate.

HAB.: Cultivated and fallow fields; in garigue on rocky hillsides; in open Pine forest; in rock-crevices by the sea, or on fixed sand-dunes; sea-level to 1,500 ft. alt.; fl. March–May.

DISTR.: Divisions 1, 3–8. Mediterranean region.

1. Neokhorio (Akamas), 1948, *Kennedy* 1598! Yioulou near Baths of Aphrodite, 1962, *Meikle* 2224!
3. Kellaki, 1948, *Kennedy* 1676! Amathus, 1978, *Holub* s.n.!
4. Cape Pyla, 1880, *Sintenis*; Athna Forest, 1952, *Merton* 648! Cape Greco, 1962, *Meikle* 2605!
5. Vatili, 1905, *Holmboe* 315; between Nisou and Stavrovouni, 1933, *Syngrassides* 22! Near Apostolos Varnavas Monastery, 1948, *Mavromoustakis* s.n.! Athalassa, 1950, *Chapman* 634!
6. Near Dhiorios, 1962, *Meikle* 2341!
7. Bellapais, 1952, *Casey* 1247! Above Kephalovryso, Kythrea, 1967, *Merton* in ARI 219! (see Notes).
8. Near Rizokarpaso, 1880, *Sintenis & Rigo* 7! Apostolos Andreas Monastery, 1938, *Syngrassides* 1771! Ayios Philon near Rizokarpaso, 1941, *Davis* 2225! Near Apostolos Andreas, 1941, *Davis* 2323!

NOTES: Most of the Cyprus material is fairly uniform morphologically, but two specimens, both apparently from the hills above (5) Kythrea, show some tendency to vary towards the related *T. bergeri* Zucc. (in Abhandl. Bayer. Akad. Wiss., 2: 324; 1837), which is tentatively recorded for Cyprus in Osorio-Tafall et Seraphim, List Vasc. Plants Cyprus, 30 (1973). One of these specimens, *Merton* in ARI 219, is clearly annual (whereas *T. bergeri* Zucc. is perennial with a woody rootstock), but has shortly pedunculate inflorescences of minute flowers. It is obviously etiolate, and may have been struggling to the light from amongst competing vegetation. It is almost certainly just a form of *T. humile* Vahl. The other specimen, *Sintenis* 7 [a] from "montibus pr. Kythraea", coll. May 1880, is evidently perennial, since, in two examples, it bears the withered remains of the previous year's growth, it does not, however, have the markedly woody base of typical *T. bergeri*, and while the partial inflorescences are distinctly pedunculate (as in *T. bergeri*), the individual flowers are relatively large, whereas those of *T. bergeri* are normally minute. Pending further investigation in the field, it must remain unidentified. There are four sheets of *Sintenis & Rigo* 7 in the Kew collections, three labelled *Thesium humile* Vahl "Im campis prope Rhizo Carpaso" (or "Felden bei Rhizo Carpaso"), but only one, also named *T. humile*, labelled "In mont. pr. Kythraea"; it is this one which is puzzling, and which is presumably correctly localized, though it is also represented, as a single individual, with a specimen of typical *T. humile*, on one of the sheets labelled "In campis prope Rhizo Carpaso". This latter localization is almost certainly erroneous, and the consequence of these collectors' misguided policy of re-sorting and numbering their collections in systematic sequence in the aftermath of the expedition. To avoid confusion I have re-numbered the puzzling specimens *Sintenis & Rigo* 7a.

T. BERGERI *Zucc.* in Abhandl. Bayer. Akad. Wiss., 2: 324 (1837); Zohary, Fl. Palaest., 1: 45, t. 44 (1966); Osorio-Tafall et Seraphim, List Vasc. Plants Cyprus, 30 (1973). Superficially resembles *T. humile* Vahl, but a perennial, with a tough, woody root-system; partial inflorescences distinctly pedunculate, the peduncles 5–10 (or more) mm. long; flowers 1–3, subtended by long bracts and bracteoles similar to those of *T. humile.* Fruit barrel-shaped, rather obscurely reticulate.

As noted above, listed tentatively by Osorio-Tafall & Seraphim, but no acceptable evidence is available to add this species to our flora; the plant recorded by Osorio-Tafall & Seraphim may have been the puzzling *Sintenis & Rigo* 7a mentioned in the notes under *T. humile* (*supra*).

T. DIVARICATUM *Jan ex Mert. et Koch* in Röhling, Deutschl. Fl., ed. 3, 2: 285 (1826); Unger et Kotschy, Die Insel Cypern, 226 (1865); Boiss., Fl. Orient., 4: 1061 (1879); H. Stuart Thompson in Journ. Bot., 44: 338 (1906); Holmboe, Veg. Cypr., 62 (1914); Post, Fl. Pal., ed. 2, 2: 484 (1933); Osorio-Tafall et Seraphim, List Vasc. Plants Cyprus, 30 (1973). An ascending or erect, robust perennial, up to 35 cm. high with a woody base; leaves narrowly linear, 1-nerved, as in *T. humile* Vahl; inflorescence a lax, many-flowered narrowly pyramidal panicle, the partial inflorescences distinctly pedunculate, with the bracts and bracteoles often rather shorter than the fruit; perianth 5-lobed, broadly campanulate. Fruit obscurely reticulate.

All the Cyprus records are based on the entry in Die Insel Cypern, 227: "Bei Prodromo unter Föhren in Aufsteigen des Troodos selten" [Near Prodhromos under Pines on the ascent of Troödos rare]. No specimen is cited by Kotschy, nor has *T. divaricatum* been collected since in this particularly well-worked area. It is a fairly conspicuous plant, and, while it may yet be re-found, one must, until such confirmatory evidence comes to hand, suspect an error. *T. divaricatum* has a wide distribution in south and south-central Europe, and ranges as far east as Turkey and Lebanon. Its occurrence in Cyprus is quite possible.

83. EUPHORBIACEAE

Boissier in DC., Prodr., 15 (2): 1–188 (1862).
Muell. Arg. in DC., Prodr., 15 (2): 189–1286 (1866).
Pax et Hoffm. in Engler et Prantl, Pflanzenfam., 19c: 11–233 (1931).

by A. Radcliffe-Smith

Dioecious or monoecious annual, biennial or perennial herbs, shrubs, climbers, trees or succulents sometimes with milky latex; indumentum of simple or stellate hairs or peltate scales; leaves usually alternate, sometimes opposite, occasionally alternate, whorled and opposite, usually simple; stipules present or not, simple, lobed or multifid, sometimes glandular; inflorescences fasciculate, spicate, racemose, paniculate, cymose or cyathial, or flowers solitary; flowers bracteate or not, pedicellate or not, unisexual, usually actinomorphic; male flowers: sepals 0–6, free or united, valvate or imbricate; petals 0–6; disk present or absent; stamens 1–∞, with the filaments free or variously connate; anthers 2–4-thecous, erect or inflexed in bud, dorsifixed or basifixed, introrse, dehiscing longitudinally, rarely by pores; pistillode present or absent; female flowers: sepals, petals and disk usually as in the male flowers; staminodes present or absent; ovary superior, usually 2–3-locular, with axile placentation and 1–2 pendulous ovules per loculus; styles 2–3, free or united at the base. Fruit an explosively dehiscent regma, less often a drupe; pericarp dry or fleshy; endocarp usually woody or crustaceous; seeds carunculate or not; endosperm mostly copious, fleshy; embryo straight; cotyledons broad or narrow.

A predominantly tropical family, less well represented in temperate regions. Approximately 300 genera and 5,000 species.

Flowers consisting of single achlamydeous stamens and ovaries, several male and 1 female being enclosed in each glanduliferous involucre; latex present -   -   -   **1. Euphorbia**
Flowers with perianths, not involucrate; latex absent:
Leaves opposite; regma 2-locular   -   -   -   -   -   -   -   **4. Mercurialis**

Leaves alternate; regma 3-locular:
  Leaves entire; flowers solitary or fasciculate; stamens ± free; regma smooth; seeds
    trigonous, 2 per loculus - - - · - - - - - - **2. Andrachne**
  Leaves dentate or lobate; inflorescence paniculate; stamens connate; regma tuberculate or
    spiny; seeds ± ovoid, 1 per loculus:
    Plant with stellate hairs; leaves scarcely lobate; stamens 3–15 - - **3. Chrozophora**
    Plant glabrous; leaves palmate; stamens ∞ - - - - - **5. Ricinus**

1. EUPHORBIA *L.*

Sp. Plant., ed. 1, 450 (1753).
Gen. Plant., ed. 5, 208 (1754).

Monoecious annual, biennial or perennial herbs, suffrutices, shrubs, trees or succulents, with a milky latex; indumentum (when present) simple; cauline leaves usually alternate, rarely opposite, decussate or whorled, sessile, rarely petiolate; ray or pseudumbel leaves whorled; raylet or cyathial leaves whorled or opposite, free or united; stipules present or not; pseudumbel rays usually dichotomous; inflorescence cyathial, with 1 female flower and several male flowers enclosed within a cupular glanduliferous involucre; cyathia often pseudodichasially arranged on the rays of a terminal pseudumbel or pseudopleiochasium, below which axillary rays may or may not occur; cyathial involucre commonly 5-lobed, with (1–) 4–5 variously-shaped glands alternating with the lobes; male flowers each consisting of a single naked stamen on its own usually bracteate pedicel; female flower consisting of a trilocular naked ovary on a pedicel which commonly elongates in fruit; ovules 1 per loculus; styles 3, united at the base, the stigmas often bifid; fruit a 3-locular, smooth or tuberculate septicidally dehiscent 3-seeded regma; seeds commonly carunculate.

About 2,000 species, cosmopolitan, but chiefly subtropical and warm temperate.

Leaves oblique at the base; stipules membranous; glands appendiculate:
  Ripe seeds smooth; plant fleshy - - - - - - - - **1. E. peplis**
  Ripe seeds rugulose; plant not fleshy:
    Plant prostrate; leaves not longer than 1·1 cm. - - - - **2. E. chamaesyce**
    Plant ascending or erect; leaves up to 3·6 cm. long - - - - **3. E. nutans**
Leaves symmetrical at the base; stipules membranous, glandular or not; glands ex-
    appendiculate:
  Stipules membranous, minute; leaves spinulose-denticulate, densely woolly, petiolate;
    glands pectinate - - - - - - - - - - **4. E. petiolata**
  Stipules glandular or not; leaves entire or serrate, variously indumented, usually sessile;
    glands entire or 2-horned:
    Stipules glandular; cyathia with 1–2 (–3) glands - - - - **5. E. heterophylla**
    Stipules O; cyathia with 4–5 glands:
      Glands rounded on the outer edge:
        Perennial herbs or shrubs:
          Plants with a napiform root-tuber; autumn-flowering - **7. E. dimorphocaulon**
          Plants without a root-tuber; spring or summer-flowering:
            Seeds minutely tuberculate - - - - - **9. E. pubescens**
            Seeds smooth or shallowly pitted:
              A shrub; regmata verrucose - - - - - **6. E. hierosolymitana**
              Perennial herbs; regmata smooth or granulate:
                Plant up to 2 m., usually pilose - - - **8. E. altissima**
                Plants not exceeding 50 cm., glabrous:
                  Seeds shallowly pitted - - - - - **10. E. cassia**
                  Seeds smooth - - - - - - - [11. E. CYPRIA]
      Annuals:
        Seeds smooth:
          Leaves finely serrulate; regmata beset with subulate processes
                                                                12. E. valerianifolia
          Leaves coarsely serrate; regmata smooth - - - - **15. E. arguta**
        Seeds variously ornamented or sculptured:
          Seeds minutely tuberculate - - - - - **9. E. pubescens**

Seeds reticulate:
    Leaves serrulate; seeds brown, with a flat caruncle   -   **13. E. helioscopia**
    Leaves coarsely serrate; seeds black, with a protuberant caruncle
              14. E. sintenisii
Glands truncate, lunate or 2-horned:
  Annuals; seeds variously ornamented or sculptured:
    Seeds tuberculate or rugulose:
      Lower cauline leaves densely imbricate, linear-setaceous -   - **16. E. aleppica**
      Lower cauline leaves laxly arranged, linear to linear-oblanceolate **17. E. exigua**
    Seeds pitted or grooved or both:
      Cauline leaves petiolate; regma-valves 2-ridged   -   -   - **18. E. peplus**
      Cauline leaves sessile or subsessile; regma-valves smooth:
        Seeds longitudinally grooved and pitted   -   -   **19. E. chamaepeplus**
        Seeds transversely grooved - - - - - - **20. E. falcata**
  Perennials; seeds smooth (except 21):
    Plant prostrate; leaves suborbicular; regma-valves 2-winged; seeds cylindric, pitted
              21. E. herniariifolia
    Plant decumbent-ascending or erect; leaves not suborbicular; regma-valves not
      winged; seeds smooth:
      Regma triquetrous, the valves angled   -   -   -   - **22. E. veneris**
      Regma trilobate, the valves rounded:
        Plant very fleshy; leaves adaxially concave; caruncle minute
              23. E. paralias
        Plant not fleshy; leaves flat; caruncle prominent:
          Lower cauline leaves ± equalling the upper; raylet leaves free
              24. E. terracina
          Lower cauline leaves much larger than the upper; raylet leaves connate
              25. E. thompsonii

1. **E. peplis** *L.*, Sp. Plant., ed. 1, 455 (1753); Boiss. in DC., Prodr., 15 (2): 27 (1862), Fl. Orient., 4: 1086 (1879); Post in Mém. Herb. Boiss., 18: 100 (1900); H. Stuart Thompson in Journ. Bot., 44: 338 (1906); Holmboe, Veg. Cypr., 119 (1914); Post, Fl. Pal., ed. 2, 2: 494 (1933); Lindberg f., Iter Cypr., 22 (1946); Zohary, Fl. Palaest., 2: 272, t. 390 (1972); Osorio-Tafall et Seraphim, List Vasc. Plants Cyprus, 67 (1973); Davis, Fl. Turkey, 7: 579 (1982).
    TYPE: *"in Narbonae, Hispaniae maritimis"*. (LINN!).

Prostrate, fleshy, glaucous, glabrous annual, commonly 4-branched from the base; branches up to 40 cm. long, often purplish; leaves falcate-oblong, 5–15 mm. long, 2·5–10 mm. wide, obtuse or emarginate, obliquely truncate at the base, ± entire; petioles 2–3 mm. long; stipules filiform, 1–1·5 mm. long; cyathial glands reddish-brown, with narrow white or pinkish appendages. Regma 4–5 mm. diam., often purplish; seeds ovoid-pyriform, 2·5–3 mm. long, smooth, pale grey, ecarunculate.

    HAB.: Sandy, gravelly and pebbly seashores, often with *Salsola kali* L., and *Verbascum sinuatum* L.; marshes; sea-level; fl. May–July (? Nov.).

    DISTR.: Divisions 2, 3, 5–7. S.W. Europe, N.W. Africa, Mediterranean Basin, Black Sea, S. Caspian, ? N. Persian Gulf; rarely on lakeshores (e.g. Tuz Gölü in Turkey).
    2. Near Yialia, 1959, *P. H. Oswald* 171!
    3. Limassol, 1905, *Holmboe* 694, and, 1939, *Lindberg f.* s.n.; Akhelia, 1939, *Lindberg f.* s.n.; Zakaki, 1939, *Mavromoustakis* 102! Cape Gata, 1941, *Davis* 3649!
    5. Salamis, 1939, *Lindberg f.* s.n.!
    6. Xeros, 1949, *Casey* 869!
    7. Kyrenia, 1955, *G. E. Atherton* 272!

    NOTES: First recorded from Cyprus in 1894 by Post (in Mém. Herb. Boiss., 18: 100; 1900), but without precise locality.

2. **E. chamaesyce** *L.*, Sp. Plant., ed. 1, 455 (1753); Boiss. in DC., Prodr., 15 (2): 34 (1862), Fl. Orient., 4: 1088 (1879); Holmboe, Veg. Cypr., 119 (1914); Post, Fl. Pal., ed. 2, 2: 494 (1933); Lindberg f., Iter Cypr., 21 (1946); Rechinger f. in Arkiv för Bot., ser. 2, 1: 426 (1950); Zohary, Fl. Palaest., 2: 273, t. 392 (1972); Osorio-Tafall et Seraphim, List Vasc. Plants Cyprus, 67 (1973); Davis, Fl. Turkey, 7: 580 (1982).
    E. canescens L., Sp. Plant., ed. 2, 652 (1762).
    E. massiliensis DC. in Lam. et DC., Fl. Franç., 6: 357 (1815).
    E. chamaesyce L. var. *canescens* (L.) Boiss. in DC., Prodr., 15 (2): 35 (1862).
    E. chamaesyce L. ssp. *canescens* (L.) Holmboe, Veg. Cypr., 119 (1914).

E. chamaesyce L. ssp. *massiliensis* (DC.) Thellung in Aschers. et Graebn., Syn. Mitteleurop. Fl., 7: 457 (1914).

TYPE: "*in* Europa *australi* Sibiria" (*Loefling*, sheet no. 15 in LINN!).

Prostrate glabrous or pubescent, somewhat glaucous annual, usually many-branched from the base; branches up to 30 cm. long; leaves asymmetrically suborbicular-ovate to ovate-oblong, 1–11 mm. long, 1–6 mm. wide, obtuse or emarginate, oblique at the base, ± entire or obscurely serrulate at least in the upper half; petioles 1 mm. long; stipules linear-filiform, 1 mm. long; cyathial gland suborbicular to transversely ovate, reddish-brown, with white or pinkish petaloid appendages. Regma 2 mm. diam.; seeds ovoid-quadrangular, 1·2 mm. long, irregularly tuberculate-rugulose, pale grey, ecarunculate.

HAB.: Dry, stony, calcareous or igneous hillsides in garigue; sandy river beds, fields, irrigated ground; sea-level to 700 ft. alt.; fl. May–Nov.

DISTR.: Divisions 1–8. Common in Division 7. S.W. Europe, N.W. Africa, Mediterranean region, S. U.S.S.R. (Russia), S.W. Asia (excluding Arabia) eastwards to Pakistan.

1. Cape Arnauti, 1913, *Haradjian* 813; Yeroskipos, by Ezuza R., 1939, *Lindberg f.* s.n.; Moa, near Kissonerga, 1967, *Economides* 988!
2. Saïttas, 1970, *A. Genneou* in ARI 1609!
3. By Yermasoyia R., 1940, *Mavromoustakis* 37!
4. Larnaca, 1880, *Sintenis & Rigo* 599A! and, 1939, *Lindberg f.* s.n.; Ayia Napa, 1939, *Lindberg f.* s.n.; Perivolia, 1939, *Lindberg f.* s.n.!
5. Between Athalassa and Nicosia, 1939, *Lindberg f.* s.n.
6. Koutraphas, 1905, *Holmboe* 846.
7. Common; Kyrenia, Bellapais, Kythrea, Ayios Epiktitos, etc. *Sintenis & Rigo* 599! *Kennedy* 422! 611! *Casey* 73! *G. E. Atherton* 235! 307! 380! 381! 425! 426! *P. H. Oswald* 11! etc.
8. Ayios Andronikos, 1934, *Syngrassides* 540!

3. E. nutans *Lag.*, Gen. et Sp. Plant., 17 (1816); Lindberg f., Iter Cypr., 21 (1946); Zohary, Fl. Palaest., 2: 275, t. 396 (1972); Osorio-Tafall et Seraphim, List Vasc. Plants Cyprus, 67 (1973).

E. preslii Guss., Fl. Sic. Prodr., 1: 539 (1827); Boiss. in DC., Prodr., 15 (2): 23 (1862).

[*E. maculata* (non L.) Wheeler in Rhodora, 43: 143 (1941).]

TYPE: Mexico; "in N[ova] H[ispania]" (MA).

Procumbent–ascending to erect annual up to 60 cm. high; stems pubescent above when young, otherwise subglabrous; leaves elliptic-oblong, 1–3·6 cm. long, 0·5–1·4 cm. wide, obtuse or subacute, obliquely subcordate at the base, serrate, strongly triplinerved, often with a median purple blotch above; petioles 1–2 mm. long; stipules deltoid, 0·5–1 mm. long, connate or free; cyathial glands transversely ovate, yellow, with small pale pink appendages. Regma 2 mm. diam.; seeds ovoid-quadrangular, 1–1·5 mm. long, irregularly transversely rugulose, dark grey or blackish, ecarunculate.

HAB.: Weed of arable land with summer crops, in gardens and orchards; sea-level to 600 ft. alt.; fl. Aug.–Sept.

DISTR.: Divisions 4, 5, 7. Native of N., C. & S. America; introduced in the Old World, particularly the Mediterranean region.

4. Kouklia, 1939, *Lindberg f.* s.n.!
5. Potamia, 1935 and 1937, *Syngrassides* 794! 1743!
7. Lapithos, 1970, *A. Genneou* in ARI 1586!

`4. E. petiolata *Banks et Sol.* in Russell, Nat. Hist. Aleppo, ed. 2, 2: 253 (1794); Zohary, Fl. Palaest., 2: 275, t. 398 (1972); Osorio-Tafall et Seraphim, List Vasc. Plants Cyprus, 67 (1973); Davis, Fl. Turkey, 7: 581 (1982).

E. malacophylla Clarke, Travels, 2 (1): 354 (1812); Poech, Enum. Plant. Ins. Cypr., 36 (1842); Holmboe, Veg. Cypr., 119 (1914); Nattrass, First List Cyprus Fungi, 24 (1937); Lindberg f., Iter Cypr., 21 (1946).

E. lanata Sieb. ex Spreng., Syst., 3: 792 (1826); Unger et Kotschy, Die Insel Cypern, 358 (1865); Boiss. in DC., Prodr., 15 (2): 101 (1862), Fl. Orient., 4: 1092 (1879); H. Stuart Thompson in Journ. Bot., 44: 338 (1906); Post, Fl. Pal., ed. 2, 2: 495 (1933).

E. syriaca Spreng., Syst., 3: 792 (1826).

TYPE: Syria; Aleppo, *Russell* s.n. (BM!).

Densely villous, much-branched annual up to 60 cm. high; leaves ovate to elliptic-ovate, 1–3·5 cm. long, 0·5–2 cm. wide, obtuse, rounded at the base, spinulose-denticulate; petioles 3–8 mm. long; stipules minute, scarcely visible; rays 3; cyathial glands stipitate, pectinate. Regma 5 mm. diam., smooth; seeds quadrangular, truncate, 3·5–4 mm. long, shallowly pustulate, greyish-brown; caruncle patelliform.

HAB.: Fallow fields, arable land, gardens, plantations; 200–600 ft. alt.; fl. May–Sept.

DISTR.: Divisions 3–6. Algeria; S. Turkey, Syria, Lebanon and Israel eastwards to Turkmen S.S.R. and Pakistan.

3. Near Kophinou, 1900, *Bovill* in *A. G. & M. E. Lascelles* s.n.!
4. Near Larnaca, 1880, *Sintenis & Rigo* 894!
5. Near Athalassa, 1966, *Merton* in ARI 10!
6. Near Nicosia, 1935, *Syngrassides* 654! *Lindberg f.* s.n.! Between Astromeritis and Koutraphas, 1955, *Merton* 2325! 2403! English School, Nicosia, 1958, *P. H. Oswald* 3!

5. E. heterophylla L., Sp. Plant., ed. 1, 453 (1753); Boiss. in DC., Prodr., 15 (2): 72 (1862); Davis, Fl. Turkey, 7: 581 (1982).

E. geniculata Ortega, Hort. Matr. Dec., 18 (1797); Zohary, Fl. Palaest., 2: 287, t. 423 (1972); Osorio-Tafall et Seraphim, List Vasc. Plants Cyprus, 67 (1973).

E. prunifolia Jacq., Hort. Schoenbr., 3: 15, t. 277 (1798).

Poinsettia heterophylla (L.) Klotzsch et Garcke, Mon. Akad. Berl., 1859: 253 (1859); Dressler in Ann. Mo. Bot. Gard., 48: 339 (1961).

TYPE: "*in* America *calidiore*" (LINN).

A coarse, erect–ascending, glabrous or sparingly pubescent annual, sometimes subperennial, up to 70 cm. high; leaves alternate below, usually opposite above, 3–9 cm. long, 1·5–5 cm. wide, elliptic-obovate or panduriform, obtuse or acute, entire or dentate, the upper ones sometimes with a purplish basal blotch; petioles 0·5–2 cm. long; stipules sessile, glandular; cyathia in dense terminal clusters; glands 1 (–3), stipitate, orbicular, deeply concave. Regma 4·5–5·5 mm. diam., smooth; seeds truncately subquadrangular, 2·5 mm. long, shallowly tuberculate, greyish-brown, ecarunculate.

HAB.: Orange groves; introduced; near sea-level; fl. July–Oct.

DISTR.: Division 4. Native of Tropical America and the S. U.S.A. Widely introduced in the Old World tropics, subtropics and warm temperate regions.

4. Larnaca, 1930, *C. B. Ussher* 121!

6. E. hierosolymitana *Boiss.*, Diagn., 1, 12: 110 (1853); Zohary, Fl. Palaest., 2: 277, t. 402 (1972); Davis, Fl. Turkey, 7: 583 (1982).

E. dumosa Boiss., Diagn., 1, 12: 110 (1853) nec A, Rich. (1850).

E. thamnoides Boiss., Cent. Euph., 33 (1860), in DC., Prodr., 15 (2): 131 (1862), Fl. Orient., 4: 1104 (1879); Post, Fl. Pal., ed. 2, 2: 497 (1933); Osorio-Tafall et Seraphim, List Vasc. Plants Cyprus, 67 (1973).

E. thamnoides Boiss. var. *hierosolymitana* (Boiss.) Boiss. in DC., Prodr., 15 (2): 131 (1862), Fl. Orient., 4: 1104 (1879); Post, Fl. Pal., ed. 2, 2: 498 (1933).

TYPE: Palestine; "in Judeae circa Hierosolymam praesertim circâ tumulos antiquos *Regum et Judicum* dictos (Boiss.) in *Palaestina* quoque Aucher 2035" (G! P! K!).

Glabrous, sparingly-branched, unarmed shrub up to 3 m. high; cauline and ray-leaves elliptic to elliptic-obovate, 1–4·5 cm. long, 0·5–1·7 cm. wide, obtuse or subacute, entire; raylet-leaves elliptic-obovate or obovate, 0·5–1·5 cm. long, 0·5–1 cm. wide, yellowish; rays 5, up to 3 times di- or trichotomous;` axillary rays 0–1 (–3). Regma 5 mm. diam., verrucose; seeds ovoid, 2 mm. long, smooth, brown, shiny; caruncle hemispherical.

HAB.: On limestone; sea-level to 500 ft. alt.; fl. Jan.–April (data based on Turkish material).

DISTR.: Not precisely localized. S.W. Turkey, Lebanon, Syria, Israel, Jordan.

NOTES: Recorded from Cyprus by Labillardière (sec. Boiss., Fl. Orient., 4: 1104 sub *E. thamnoide*), but, as with some other Labillardière records, questionably correct, and possibly collected in Syria. *E. hierosolymitana* has not since been observed in Cyprus; there is, however, no reason why it should not grow there.

7. E. dimorphocaulon *P. H. Davis* in Phyton, 1: 196 (1949); Osorio-Tafall et Seraphim, List Vasc. Plants Cyprus, 67 (1973); Davis, Fl. Turkey, 7: 586 (1982).
[*E. apios* (non L.) A. K. Jackson et Turrill in Kew Bull., 1938: 467 (1938).]
TYPE: Crete; Hierapetra/Sitia, Mt. Aphendi Kavusi, 0–800 m., 3 Dec. 1939, *Davis* 1065 (K !).

Perennial with 2 types of stem arising from a napiform root-tuber; prostrate or decumbent, sparingly pilose, leafy, vegetative stems are produced in spring, and erect, glabrous, subaphyllous or scale-leaved flowering stems up to 30 cm. high are produced in autumn; the prostrate stems arise from near the base of the old flowering stems; leaves of the vegetative stems elliptic-ovate to suborbicular, 4–10 mm. long, 2–7 mm. wide, obtuse, entire or subentire, shortly petiolate; the petiole less than 1 mm. long; leaves of the flowering stems (when present) and ray-leaves oblong or elliptic-oblong, 2–3 mm. long, 0·5–1 mm. wide, obtuse, entire or subentire, subsessile; raylet-leaves obtrullate, 2–2·5 mm. long, 1·5–2 mm. wide, obtuse, cuneate, entire, often purplish; rays 3–5, not or once dichotomous; axillary rays (0–) 1–5. Regma 4 mm. diam., tuberculate, greenish, the tubercles purple; seeds ovoid to compressed-subglobose, 2 mm. long, smooth, dark grey or brownish, shiny; caruncle hemispherical. *Plate 83*.

HAB.: In garigue on limestone, sandstone, or marl hillsides; often on anthills; in open Pine woods; occasionally on stabilised sand-dunes; sea-level to 2,500 ft. alt.; fl. Sept.–Nov.

DISTR.: Divisions 1, 2, 6, 7. Also S. Turkey (Isauria), Crete.

1. Aprilokambos near Ayios Yeoryios (Akamas), 1962, *Meikle* 2013 ! By the shore between Paphos and Yeroskipos, 1981, *Meikle* 5051 !
2. Xerovouno near Pyrgos, 1937, *Syngrassides* 1721 !
6. Liveras, 1954, *Lady Loch* 1 ! Beyond Kormakiti, 1957, *Merton* 2850 !
7. Common; Kyrenia, Halevga, Kythrea, Myrtou, Vasilia, etc. *Kennedy* 610 ! *Davis* 1955 ! 1971 ! 1991 ! 1993 ! *Merton* 603 ! etc.

8. E. altissima *Boiss.*, Diagn., 1, 5: 52 (1844), in DC., Prodr., 15 (2): 116 (1862), Fl. Orient., 4: 1096 (1879); Sintenis in Oesterr. Bot. Zeitschr., 31: 392 (1881), 32: 121 (1882); Holmboe, Veg. Cypr., 119 (1914); Post, Fl. Pal., ed. 2, 2: 496 (1933); Osorio-Tafall et Seraphim, List Vasc. Plants Cyprus, 67 (1973); Davis, Fl. Turkey, 7: 590 (1982).
TYPE: Turkey; "ad rivulos circà *Denisleh* et *Laodiceam* in *Phrygia* australi, legi flor. Junio 1842", *Boissier* (G ! K !).

Glabrous or pilose caespitose perennial up to 2 m. high; cauline leaves linear-lanceolate, 4-12 cm. long, 0·5–1·5 cm. wide, acute or subacute, finely serrulate; ray-leaves ovate-rhombic to elliptic-lanceolate; raylet-leaves ovate-rhombic to suborbicular, 0·5–1·5 cm. long and wide, subacute or obtuse, cuneate or rounded, finely serrulate or subentire; rays 5–8, tri- or tetrachotomous then once or twice dichotomous; axillary rays 1–15 (–20). Regma 5 mm. diam., granulate-verruculose; seeds ovoid-subglobose, 2–2·5 mm. long, smooth, brownish, faintly mottled, with a small caruncle.

HAB.: Fields (Sintenis); alt. in Cyprus not known (in Turkey 700–1,400 m.); fl. May–July.

DISTR.: Division 5 or 7 (Kythrea). S. Turkey, Syria, Lebanon, N. Iraq.

5 or 7. Kythrea, 1880, *Sintenis*.

NOTES: No Cyprus material has been seen, and the above description is based on material from Turkey and Iraq. It is possible that the Sintenis record was based on a misidentification of the following, *E. pubescens* Vahl.

9. E. pubescens *Vahl*, Symb. Bot., 2: 55 (1791); Boiss. in DC., Prodr., 15 (2): 134 (1862), Fl. Orient., 4: 1106 (1879); Holmboe, Veg. Cypr., 120 (1914); Post, Fl. Pal., ed. 2., 2: 498 (1933); Lindberg f., Iter Cypr., 22 (1946); Rechinger f. in Arkiv för Bot., ser. 2, 1: 426 (1950); Davis, Fl. Turkey, 7: 592 (1982).
E. verrucosa L., Sp. Plant., 459 (1753), nom. ambig. rejic.; Zohary, Fl. Palaest., 2: 278, t. 403 (1972); Osorio-Tafall et Seraphim, List Vasc. Plants Cyprus, 67 (1973).
TYPE: "in Gallia, Helvetia" (LINN).

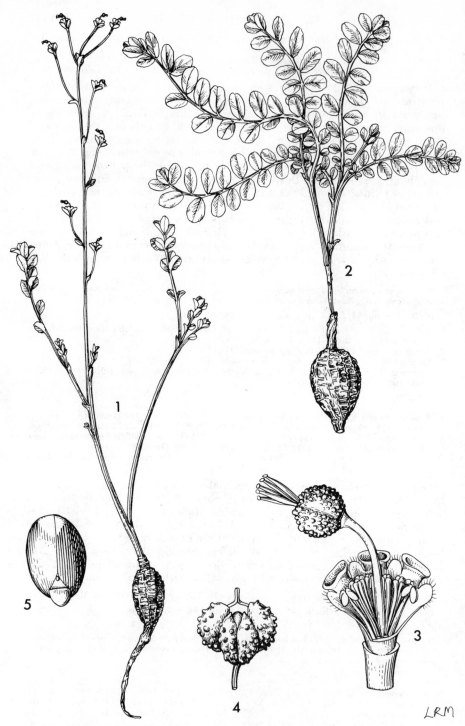

Plate 83. EUPHORBIA DIMORPHOCAULON P. H. Davis   **1,** flowering plant, habit,   × ⅔; **2,** foliage plant, habit,   × ⅔; **3,** inflorescence,   × 8; **4,** fruit,   × 2; **5,** seed,   × 8. (**1, 3** from *Davis* 1993; **2** from *Merton* 2850; **4, 5** from *Merton* 603.)

Sparingly pilose to densely villous annual, or caespitose perennial up to 80 cm. high; cauline and ray-leaves oblong-lanceolate to ovate-lanceolate, (1–) 2–5 (–6) cm. long, 0·3–1·5 cm. wide, acute, subacute or obtuse, cordate-auriculate, minutely serrulate, sometimes also dentate near the apex; raylet-leaves broadly ovate to ovate-deltoid, 0·7–2 cm. long, 0·7–1·8 cm. wide, subacute, subcordate, serrulate, denticulate or subentire; rays 5, at first trichotomous, then once or twice dichotomous; axillary rays 0–10. Regma 3·5–4 mm. diam., evenly verrucose; seeds ovoid, 2 mm. long, shallowly and minutely tuberculate or tuberculate-rugulose, dark brown; caruncle reniform.

HAB.: Marshland, streamsides, irrigation channels; sea-level to 100 ft. alt.; fl. Mar.–July.

DISTR.: Divisions 3, 6, 7. Mediterranean region, Black Sea, N. Iraq; Atlantic Islands.

3. Akhelia, 1939, *Lindberg f.* s.n.! Asomatos, 1939, *Mavromoustakis* 128!
6. Syrianokhori, 1935, *Syngrassides* 666! and 1941, *Davis* 2598!
7. Ayios Andronikos above Kythrea, 1880, *Sintenis & Rigo* 68 (as *E. altissima* Boiss. var.)! Kyrenia, 1949, *Casey* 810!

NOTES: Davis notes: "Vernacular name 'Tzoona'. Used by villagers for catching fish. They chop up the stems and throw them in the sea. The juice stupefies the fish and they float to the surface". The Irish Spurge, *Euphorbia hyberna* L., was used in the same way, in Ireland, until the beginning of the present century (Mackay, Flora Hibernica, 236; 1836. Scully, Flora of Kerry, 253; 1916).

10. E. cassia *Boiss.*, Diagn., 2, 12: 108 (1853), in DC., Prodr., 15 (2): 137 (1862), Fl. Orient., 4: 1108 (1879); Unger et Kotschy, Die Insel Cypern, 359 (1865); Sintenis in Oesterr. Bot. Zeitschr., 32: 19 (1882); Holmboe, Veg. Cypr., 120 (1914); Druce in Rep. B.E.C., 9: 471 (1931); Post, Fl. Pal., ed. 2, 2: 499 (1933); Lindberg f., Iter Cypr., 21 (1946); Osorio-Tafall et Seraphim, List Vasc. Plants Cyprus, 67 (1973); Davis, Fl. Turkey, 7: 592 (1982).

Glabrous, somewhat glaucous, perennial with several stiffly erect, decumbent or prostrate stems, commonly up to 50 cm. long, rarely up to 90 cm., arising from caudiculi borne on a small woody stock; cauline- and ray-leaves linear, linear-lanceolate, oblanceolate or obovate, 0·5–2 (–2·5) cm. long, 1–4 mm. wide, acute or obtuse, subentire; raylet-leaves ovate to rhombic, 2–7 (–9) mm. long, 1–6 mm. wide, acute or obtuse, rounded or cuneate at the base, entire or subentire, often purplish; rays 2–5, not, or once or twice, dichotomous, or cyathium terminal, solitary; axillary rays 0 (–6). Regma 2·5 mm. diam., smooth; seeds ovoid, subquadrangular, 2 mm. long, malleate or irregularly and minutely pitted, brownish or pale grey, ecarunculate.

ssp. **cassia**

TYPE: Turkey; "in jugis umbrosis *Cassii* in regione inferiori secus viam qua a *Laodiceâ* ad *Suadieh* iter circâ pagum *Cassab* frequenter sed ubique sparsa (Boiss.)" (G! K!).

Stems stiffly erect; cauline- and ray-leaves linear, linear-lanceolate or oblanceolate, 1–2 mm. wide; axillary rays 0 (–1); seeds irregularly and minutely pitted, pale grey.

HAB.: Sand dunes and sandy flats; on hillsides in garigue; in vineyards and fallow fields; sea-level to 3,000 ft. alt.; fl. May–Oct.

DISTR.: Divisions [1], 3–7. Turkey, Syria, Lebanon, Palestine.

[1.] Kathikas ["Catiga"], 1853, *Gaudry* s.n.! (see Notes under 11. *E. cypria* Boiss.).
3. Trimiklini, 1935, *Syngrassides* 755! Cherkez, 1939, *Mavromoustakis* 75! Akrotiri, 1939, *Mavromoustakis* 113! and, 1941, *Davis* 3582! Limassol, 1939, *Lindberg f.* s.n.
4. Larnaca, 1939, *Lindberg f.* s.n.! and, 1952, *Merton* 831! Perivolia, 1939, *Lindberg f.* s.n.; Famagusta, 1939, *Lindberg f.* s.n.
5. Athalassa, 1939, *Lindberg f.* s.n.! Salamis, 1928 or 1930, *Druce*, also 1939, *Lindberg f.* s.n., and, 1968, *A. Genneou* in ARI 1265!
6. Between Morphou and Panteleimon Mon., 1862, *Kotschy* 923! Syrianokhori–Ayia Irini, 1941, *Davis* 2611!

7. Near Kythrea, 1880, *Sintenis & Rigo* 600! also, 1940, *Davis* 1922! 1956! and, 1941, *Davis* 2936! 3634! and, 1962, *Meikle* 2560!; Buffavento, 1898, *Post*!; Myrtou, 1940, *Davis* 1968! and, 1967, *Economides* in ARI 1027!; Lakovounara, 1950, *Chapman* 661!

NOTES: Although included as a distinct species on p. 1440, in all probability *E. cypria* is conspecific with *E. cassia*.

ssp. **rigoi** (*Boiss. ex Freyn*) Holmboe, Veg. Cypr., 120 (1914); Lindberg f., Iter Cypr., 21 (1946); Rechinger f. in Arkiv för Bot., ser. 2, 1: 426 (1950); Osorio-Tafall et Seraphim, List Vasc. Plants Cyprus, 67 (1973).

E. *rigoi* Boiss. [in Sintenis et Rigo, Iter Cypr., 1880 no. 696, nomen] ex Freyn in Bull. Herb. Boiss., 6: 987 (1898).

E. *troodii* Post in Mém. Herb. Boiss., 18: 100 (1900); H. Stuart Thompson in Journ. Bot., 44: 338 (1906).

[*E. kotschyana* (non Fenzl) Kotschy in Unger et Kotschy, Die Insel Cypern, 359 (1865); H. Stuart Thompson in Journ. Bot., 44: 338 (1906).]

TYPE: Cyprus; "In monte Troodos. 19 Junio 1880". *Sintenis & Rigo* 696 (G! W, K!).

Stems prostrate to decumbent-ascending; cauline- and ray-leaves obovate to oblanceolate, 2–4 mm. wide, usually obtuse; rays 2–5, generally longer and better developed than in ssp. *cassia*; axillary rays 0–6. Seeds malleate, brownish. Otherwise as in ssp. *cassia*.

HAB.: On rocky, igneous slopes in garigue or Pine forest; 4,500–6,400 ft. alt.; fl. May–Sept.

DISTR.: Division 2. Endemic.

2. Common at high altitudes on Khionistra, above Platres, Prodhromos, Amiandos, etc. *Kotschy* 899; *Sintenis & Rigo* 696! *Post*! *A. G. & M. E. Lascelles*! *Haradjian* 516! *Syngrassides* 729! *Kennedy* 600–603! 1696–1698! *C. Wyatt* 23! *Lindberg f.* s.n.! *Davis* 1825! 1935! 3502! *H. R. P. Dickson* 11! *Merton* 503! 2787! *G. E. Atherton* 546! *N. Macdonald* 194! *D. P. Young* 7283! *Economides* in ARI 1224! 1245! *Barclay* 1108! *P. Laukkonen* 364!

11. E. CYPRIA *Boiss.* in DC., Prodr., 15 (2): 115 (1862), Fl. Orient., 4: 1096 (1879); Holmboe, Veg. Cypr., 119 (1914); Osorio-Tafall et Seraphim, List Vasc. Plants Cyprus, 67 (1973).

TYPE: Cyprus; "In cultis derelictis prope Catiga [Kathikas] insulae Cypri (Gaudry!)" (P!).

Glabrous much-branched subshrub up to 50 cm. high; cauline leaves few, linear, 4–5 mm. long, 1–2 mm. wide, subentire, coriaceous; ray-leaves ovate-oblong, mucronate, recurved at the apex, slightly longer than the cauline leaves; rays and raylet-leaves 0; cyathia terminal, solitary. Regma (immature) smooth; seeds unknown.

HAB.: Fallow fields, abandoned cultivated land; circa 2,000 ft. alt.; fl. ? April–June.

DISTR.: Division 1. Endemic.

1. Kathikas ["Catiga"], July 1854, *Gaudry* 144 (P, holotype!).

NOTES: Having examined the holotype of *E. cypria* Boiss. I am as certain as one can be in the absence of fruits and seeds, that it is indistinguishable from *E. cassia* Boiss. ssp. *cassia*. Boissier (in DC., Prodr., *loc. cit.*) remarks "Habitu *E. spinosam* vel *erinaceam* referens, sed leiocarpa. Ob semen ignotum locus subdubius". Nonetheless it is included in the group of species said to have smooth seeds. For this reason Boissier may have failed to realize its very close affinity with *E. cassia*, which was not known to be a Cyprus plant when he described *E. cypria* (1862), though Kotschy's Cyprus record of the latter was included in the Addenda to DC., Prodr., 15 (2): 1267 (1866).

12. E. valerianifolia *Lam.*, Encycl. Méth., 2: 435 (1788); Davis, Fl. Turkey, 7: 594 (1982).

E. *cybirensis* Boiss., Diagn., 1, 7: 89 (1846), 1, 12: 109 (1853), in DC., Prodr., 15 (2): 118 (1862), and Fl. Orient., 4: 1098 (1879); Post, Fl. Pal., ed. 2. 2: 497 (1933); Rechinger f. in Arkiv för Bot., ser. 2, 1: 426 (1950); Zohary, Fl. Palaest., 2: 281, t. 411 (1972); Osorio-Tafall et Seraphim, List Vasc. Plants Cyprus, 67 (1973).

E. *cybirensis* Boiss. var. *acutifolia* Boiss., Diagn., 1, 12: 109 (1853); Holmboe, Veg. Cypr., 120 (1914).

E. *cybirensis* Boiss. var. *dehiscens* Boiss. in DC., Prodr., 15 (2): 119 (1862), Fl. Orient., 4: 1099 (1879).

E. *zahnii* Heldr. ex Hal., Consp. Fl. Graec., 3: 100 (1904).

TYPE: Khios; "Tournefort a trouvé cet *Euphorbe* dans l'Isle de Chio, & en a fait faire un dessin que nous avons vu chez M. de Jussieu" (P).

Glabrous annual up to 40 cm. high, with simple stems; cauline leaves elliptic-obovate, elliptic or elliptic-oblanceolate, (1–) 2–5·5 cm. long, 1–2 cm. wide, acute, subacute or obtuse, finely serrulate; ray-leaves elliptic-lanceolate to ovate-lanceolate; raylet-leaves broadly obliquely ovate, 1–2·5 cm. long, 0·7–2 cm. wide, cuneate or rounded at the base; rays 5, up to 3 times dichotomous; axillary rays usually 0. Regma globose, 3–3·5 mm. diam., sparingly beset with filiform hair-tipped processes; seeds ovoid, 2 mm. long, smooth, dark brown; caruncle hemispherical.

HAB.: Cultivated and fallow fields; sea-level to 800 ft. alt.; fl. Febr.–May.

DISTR.: Divisions 1, 7, 8. Crete, Aegean Is., W. & S. Turkey, Syria, Lebanon, Palestine, Iraq.
1. Ayios Minas near Smyies, 1962, *Meikle* 2165!
7. Kephalovryso, Kythrea, 1880, *Sintenis & Rigo* 662! Vasilia, 1941, *Davis* 2194! Ayios Yeoryios, Kyrenia, 1941, *Davis* 2782! Thermia, 1949, *Casey* 581! Ayios Epiktitos, 1951, *Casey* 1129! Karavas, 1955, *Merton* 2177! Kyrenia, 1956, *G. E. Atherton* 1216!
8. Near Rizokarpaso, 1912, *Haradjian* 217! and, 1941, *Davis* 2372!

13. E. helioscopia *L.*, Sp. Plant., ed. 1, 459 (1753); Boiss. in DC., Prodr., 15 (2): 136 (1862), Fl. Orient., 4: 1107 (1879); H. Stuart Thompson in Journ. Bot., 44: 338 (1906); Holmboe, Veg. Cypr., 120 (1914); Post, Fl. Pal., ed. 2, 2: 498 (1933); Nattrass, First List Cyprus Fungi, 13 (1937); Zohary, Fl. Palaest., 2: 278, t. 404 (1972); Osorio-Tafall et Seraphim, List Vasc. Plants Cyprus, 67 (1973); Davis, Fl. Turkey, 7: 597 (1982).
TYPE: "*in* Europae *cultis*" (LINN).

Glabrous or sparingly pilose annual up to 40 cm. high, with stems simple or 2-branched from the base; cauline leaves obovate-spathulate, 1–4·5 cm. long, 0·3–2·5 cm. wide, obtuse or retuse, attenuate at the base, serrulate in the upper half; ray-leaves obovate; raylet-leaves obliquely obovate, 0·5–3 cm. long, 0·4–2 cm. wide, asymmetrically cuneate at the base; pseudumbel contracted, flattish; rays 5, trichotomous then once or twice dichotomous; axillary rays 0. Regma trilobate, 3·5 mm. diam., smooth; seeds ovoid, 2 mm. long, favose-reticulate, dark brown; caruncle applanate, transversely ovate, refringent, yellowish.

HAB.: Fields, waste ground, roadside ditches; sea-level to 3,700 ft. alt.; fl. Jan.–June.

DISTR.: Divisions 1, 2, 5–8. Throughout Europe, N. Africa and W. Asia; introduced elsewhere.
1. Ayios Minas near Smyies, 1962, *Meikle* 2166! Neokhorio (Akamas), 1962, *Meikle* 2212!
2. Platres, 1938, *Kennedy* 1324!
5. Athalassa, 1970, *A. Genneou* in ARI 1384! 1481!
6. Syrianokhori, 1941, *Davis* 2597! Nicosia, 1957, *Merton* 3076! and, 1973, *P. Laukkonen* 50!
7. Near Kythrea, 1880, *Sintenis & Rigo* 71! and, 1932, *Syngrassides* 61! Vasilia, 1941, *Davis* 2195! Near Dhikomo, 1949, *Casey* 212! Thermia, 1956, *G. E. Atherton* 876! Kyrenia, 1956, *G. E. Atherton* 1035!
8. Rizokarpaso, 1941, *Davis* 2373!

NOTES: *Meikle* 2344 from (6) Dhiorios is a dwarfish, prostrate, many-stemmed form of *helioscopia*, with seeds only 1 mm. long. It approximates closely to *E. helioscopioides* Loscos et Pardo, described from Spain, which is probably best treated as a var. of *E. helioscopia*; G. E. Atherton 1175 from (7) Templos also belongs here.

14. E. sintenisii *Boiss. ex Freyn* in Bull. Herb. Boiss., 6: 986 (1898); Holmboe, Veg. Cypr., 120 (1914); Osorio-Tafall et Seraphim, List Vasc. Plants Cyprus, 67 (1973); Davis, Fl. Turkey, 7: 598 (1982).
[*E. helioscopia* (non L.) Sintenis in Oesterr. Bot. Zeitschr., 31: 192, 194 (1881), 32: 193 (1882).]
TYPE: Cyprus; "In campis prope Kythraea die 21., 25. et 27. majo fructiferam (exs. 69) et in ageribus prope Larnaka die 25 februario 1880 juvenilem (exs. 69) leg. *Sintenis*! (G! K! W).

Glabrous or sparingly pilose annual up to 40 cm. high, with stems simple or 2-branched from the base; cauline leaves elliptic-oblanceolate, 1–5 cm. long, 0·5–1·5 cm. wide, obtuse, subacute or acute, attenuate at the base, coarsely serrate; ray-leaves ovate-trullate, ovate, rhombic or deltoid; raylet-leaves ovate to ovate-deltoid, 0·5–2 cm. long, 0·5–1·3 cm. wide, truncate or

cuneate at the base; rays 3–5, up to 3 times dichotomous; axillary rays 0 (–5).
Regma trilobate, 4–5 mm. diam., smooth; seeds ovoid, 2–2·5 mm. long,
shallowly and irregularly reticulate-rugulose, blackish; caruncle protuberant, hemispherical-conical, white.

HAB.: Marshes, damp meadows, cultivated and fallow fields; sea-level to 4,000 ft. alt.; fl.
Febr.–June.

DISTR.: Divisions 1, 4–8. Libya, Egypt, Palestine, Syria, Turkish Mesopotamia.

1. Between Sarama and Evretou, 1934, *Syngrassides* 841 ! Between Neokhorio and Polis, 1962,
    Meikle 2237 !
4. Larnaca, 1953, *Mrs. Scott-Moncrieff* in *Casey* 1286 !
5. Mia Milea, 1957, *Merton* 2968 ! Strongylos, 1967, *Merton* in ARI 438 !
6. Ayia Irini, 1941, *Davis* 2569 ! Mile 7, Nicosia–Troödos road, 1952, *Merton* 664 !
7. Near Kythrea, 1880, *Sintenis & Rigo* 69 ! Larnaka tis Lapithou, 1941, *Davis* 2978 ! Kyrenia,
    1949, *Casey* 803 !
8. Koma tou Yialou, 1970, *A. Genneou* in ARI 1466 !

15. E. arguta *Banks et Sol.* in Russell, Nat. Hist. Aleppo, ed. 2, 2: 253 (1794); Sibth. et Sm., Fl.
        Graec. Prodr., 329 (1809); Poech, Enum. Plant. Ins. Cypr., 36 (1842); Boiss. in DC., Prodr.,
        15 (2): 117 (1862), Fl. Orient., 4: 1097 (1879); Unger et Kotschy, Die Insel Cypern, 358
        (1865); Holmboe, Veg. Cypr., 119 (1914); Post, Fl. Pal., ed. 2, 2: 496 (1933); Nattrass et
        Papaioannou in Cypr. Agric. Journ., 34, 1: 26 (1939); Zohary, Fl. Palaest., 2: 280, t. 408
        (1972); Osorio-Tafall et Seraphim, List Vasc. Plants Cyprus, 67 (1973); Davis, Fl. Turkey,
        7: 599 (1982).
        E. calendulifolia Del., Fl. d'Égypte, 89, t. 30, fig. 1 (1813).
        TYPE: Syria, Aleppo, *Russell* (BM).

Very like *E. sintenisii*, but differing in having the cauline leaves auriculate
at the base, the rays commonly first trichotomous then dichotomous, and
the seeds smooth, with a patelliform caruncle.

HAB.: Cornfields, fallow fields; sea-level to 1,100 ft. alt.; fl. Febr.–April.

DISTR.: Divisions 2, 5, 7, 8. Greece, Aegean, S. Turkey, Lebanon, Palestine, Egypt.

2. Kanopetra Forest above Klirou, 1956, *Merton* 2594 !
5. Lakatamia, 1932, *Syngrassides* 364 !
7. Ayios Demetrianos, Kythrea, 1950, *Chapman* 169 ! Ayios Epiktitos, 1956, *Casey* 1702 !
8. Gastria, 1936, *Syngrassides* 929 !

NOTES: All these identifications are tentative: the plants are immature and it is not possible to
see the seeds, which provide the most important feature for distinguishing *E. arguta* from *E.
sintenisii*. In other respects the specimens are, to some degree, intermediate between these two
species.

16. E. aleppica *L.*, Sp. Plant., ed. 1, 458 (1753); Boiss. in DC., Prodr., 15 (2): 138 (1862), Fl.
        Orient., 4: 1109 (1879); Post, Fl. Pal., ed. 2, 2: 500 (1933); A. K. Jackson et Turrill in Kew
        Bull., 1938: 466 (1938); Zohary, Fl. Palaest., 2: 281, t. 412 (1972); Osorio-Tafall et
        Seraphim, List Vasc. Plants Cyprus, 67 (1973); Davis, Fl. Turkey, 7: 600 (1982).
        TYPE: "*in* Creta, Aleppo".

Minutely papillose, glaucous annual up to 40 cm. high, more commonly 20
cm. with the stems often 2- or more-branched from the base; lower cauline
leaves densely imbricate, linear-setaceous, 0·5–4·5 cm. long, 0·2–1 mm.
wide, acute, entire; upper cauline and ray-leaves linear-oblanceolate, 2–5
cm. long, 1–8 mm. wide; raylet leaves falcate-lanceolate to trullate or
ovate-rhombic, 0·5–3 cm. long, 0·3–1·3 cm. wide, acutely acuminate or
mucronate-aristate, entire or irregularly toothed, rounded or cuneate; rays
(2–) 3–5, up to 8 times dichotomous; axillary rays 0–15; glands long-horned.
Regma trilobate, 2·5 mm. diam., finely granulate; seeds ovoid-subglobose
and subquadrangular, 1·3 mm. long, minutely tuberculate, pale grey, with a
minute, readily deciduous caruncle.

HAB.: Damp fields, waste land, or in garigue on limestone; 30–200 ft. alt.; fl. June–Oct.

DISTR.: Division 7, rare. From Italy, Sicily and Tunisia eastwards to Crimea, Caucasus, W.
Iran, N. Iraq and Palestine.

7. Kyrenia, 1936, *Kennedy* 612! Also, 1949, *Casey* 789! and, 1952, *Casey* 1275!

17. E. exigua *L.*, Sp. Plant., ed. 1, 456 (1753); Boiss. in DC., Prodr., 15 (2): 139 (1862), Fl. Orient., 4: 1110 (1879); Unger et Kotschy, Die Insel Cypern, 358 (1865); H. Stuart Thompson in Journ. Bot., 44: 338 (1906); Holmboe, Veg. Cypr., 120 (1914); Post, Fl. Pal., ed. 2, 2: 500 (1933); Zohary, Fl. Palaest., 2: 282, t. 413 (1972); Osorio-Tafall et Seraphim, List Vasc. Plants Cyprus, 67 (1973); Davis, Fl. Turkey, 7: 602 (1982).

Glabrous somewhat glaucous annual up to 30 cm. high, more commonly 5–20 cm., with the stems simple or 2- or more-branched from the base; cauline and ray-leaves linear, linear-oblong or linear-lanceolate, 0·2–2 (–2·5) cm. long, 0·3–3 mm. wide, acute or subacute, or occasionally truncate, retuse or tricuspidate, entire; raylet-leaves linear-lanceolate to triangular-ovate, 0·2–1 (–1·5) cm. long, 1–4 mm. wide, acute or subacute, rounded to shallowly cordate at the base; rays 3–5, up to 8 times dichotomous; axillary rays 0–1 (–6); glands with 2 short or medium horns (i.e. shorter than or equalling the width of the gland), golden-brown. Regma trilobate, 1·5 mm. diam., ± smooth; seeds ovoid-quadrangular, 1·2 mm. long, tuberculate-rugulose, pale grey, with a bilobate white caruncle.

var. **exigua**
   E. exigua L. var. *acuta* L., Sp. Plant., ed. 1, 456 (1753); Post, Fl. Pal., ed. 2, 2: 500 (1933).
   TYPE: "*in* Lusatia, Gallia, Helvetia, Hispania, *inter segetes*" (LINN!).

All the leaves acute or subacute.

HAB.: On calcareous or igneous hillsides in garigue; in cultivated fields or by roadside ditches; sea-level to 800 ft. alt.; fl. Febr.–April.

DISTR.: Divisions 5–8. W. Europe, N. Africa, Atlantic Islands, east to Palestine and N. Iran.
5. Near Lefkoniko, 1880, *Sintenis & Rigo* 73 (in part)! Between Mia Milea and Piyi, 1936, *Syngrassides* 940! Perakhorio, 1955, *Merton* 2043!
6. Mile 3, Nicosia–Myrtou Road, 1950, *Chapman* 710!
7. Near Ayios Khrysostomos monastery, 1862, *Kotschy* 400; near Panagra, 1962, *Meikle* 2363 (in part)!
8. Near Apostolos Andreas, 1941, *Davis* 2330! and, 1962, *Meikle* 2459! Ronnas Bay near Rizokarpaso, 1941, *Davis* 2377! Koma tou Yialou, 1950, *Chapman* 202!

var. **retusa** *L.*, Sp. Plant., ed. 1, 456 (1753); Boiss. in DC., Prodr., 15 (2): 139 (1862), Fl. Orient., 4: 1110 (1879); Post, Fl. Pal., ed. 2, 2: 500 (1933); Zohary, Fl. Palaest., 2: 282 (1972); Davis, Fl. Turkey, 7: 603 (1982).
   E. tricuspidata La Peyr., Hist. Abr. Pyr., 271 (1813).
   E. exigua L. var. *tricuspidata* (La Peyr.) Koch, Syn. Fl. Germ. Helv., 1: 731 (1843); Boiss. in DC., Prodr., 15 (2): 139 (1862), Fl. Orient., 4: 1110 (1879); Post, Fl. Pal., ed. 2, 2: 500 (1933).
   E. exigua L. var. *truncata* Koch, Syn. Fl. Germ. Helv., 1: 731 (1843).
   [*E. rubra* (non Cav.) DC. in Lam. et DC., Fl. Franç., 5: 359 (1815).]
   TYPE: France; "Monspelii *in saxosis*" (BM! LINN!).

At least the lower cauline leaves truncate, retuse or tricuspidate at the apex, otherwise as in var. *exigua*.

HAB.: As for *E. exigua* L. var. *exigua*; sea-level to 800 ft. alt.; fl. Febr.–April.

DISTR.: Divisions 5, 7, 8. General distribution as for *E. exigua* L. var. *exigua*.
5. Near Lefkoniko, 1880, *Sintenis & Rigo* 73 (in part)!
7. Dhikomo, 1952, *Casey* 1221! Bellapais, 1952, *Casey* 1248! near Panagra, 1962, *Meikle* 2363 (in part)!
8. Near Boghaz, 1962, *Meikle* 2630!

NOTES: Frequently grows intermixed with var. *exigua*, and evidently of little taxonomic significance.

18. E. peplus *L.*, Sp. Plant., ed. 1, 456 (1753); Boiss. in DC., Prodr., 15 (2): 141 (1862), Fl. Orient., 4: 1112 (1879); Unger et Kotschy, Die Insel Cypern, 359 (1865); Sintenis in Oesterr. Bot. Zeitschr., 31: 192, 194, 257 (1881); H. Stuart Thompson in Journ. Bot., 44: 338 (1906); Holmboe, Veg. Cypr., 120 (1914); Post, Fl. Pal., ed. 2, 2: 501 (1933); R. M. Nattrass, First

List Cyprus Fungi, 13 (1937); Lindberg f., Iter Cypr., 22 (1946); Zohary, Fl. Palaest., 2: 283, t. 416 (1972); Osorio-Tafall et Seraphim, List Vasc. Plants Cyprus, 67 (1973); Davis, Fl. Turkey, 7: 605 (1982).

Glabrous annual up to 40 cm. high; stems simple or 2- or more-branched from the base; cauline leaves petiolate, the petioles up to 1 cm. long; lamina ovate, suborbicular or obovate, 0·2–1·5 (–2) cm. long, 0·2–1 (–1·2) cm. wide, obtuse or emarginate, entire, attenuate into the petiole; ray-leaves resembling the cauline, but more shortly petiolate; raylet-leaves sessile, somewhat obliquely ovate-rhombic, 0·3–1·5 cm. long, 0·2–1 cm. wide, rounded at the apex, rounded or cuneate at the base; rays 3 (–4), up to 6 times dichotomous; axillary rays (0–) 1–4; glands 2-horned, the horns shorter than, as long as, or longer than the width of the gland. Regma trilobate, with each valve having 2 dorsal ridges parallel to the keel, 2 mm. diam., smooth; seeds ovoid-hexagonal, 1–1·6 mm. long, the 2 ventral facets each with a longitudinal groove, the lateral and dorsal facets each with 2–4 pits, pale grey, but darker grey in the grooves and pits; caruncle conical, white.

var. **peplus**
    [*E. peploides* (non Gouan) Lindberg f., Iter Cypr., 21 (1946).]
    TYPE: "*in* Europae *cultis oleraceis*" (BM! LINN!).

Stems simple or sparsely branched; cauline leaves ovate or obovate; seeds 1·3–1·6 mm. long.

HAB.: *Pistacia–Juniperus* forest, shady valleys, Olive orchards, fields, gardens, roadside ditches, irrigation channels; sea-level to 2,000 ft. alt.; fl. Jan.–Nov.

DISTR.: Divisions 1–4, 6–8. Europe; N. Africa; Atlantic Islands; S.W. Asia eastwards to N. Iran. Introduced into E. Asia, Australia and N. America.

1. Ayios Neophytos, 1939, *Lindberg f.* s.n.
2. Evrykhou, 1935, *Syngrassides* 827! Galata, 1951, *Merton* 83!
3. Alethriko, 1905, *Holmboe* 246; Episkopi, 1960, *N. Macdonald* 18!
4. Larnaca, 1862, *Kotschy* 50, 300; also, 1880, *Sintenis & Rigo*, and, 1939, *Lindberg f.*; Cape Pyla, 1880, *Sintenis & Rigo*.
6. Morphou, 1935, *Nattrass*; Nicosia, 1939, *Lindberg f.* s.n.! Kormakiti, 1941, *Davis* 2104! Nicosia, 1950, *Chapman* 534! and, 1954, *Merton* 1874! and, 1973, *P. Laukkonen* 51! 128!
7. Kythrea, 1880, *Sintenis & Rigo* 70! and, 1932, *Syngrassides* 52! Common about Kyrenia, *Casey* 129! *G. E. Atherton* 308! 308A! 756! 824! *W. R. Price* 992! Lapithos, 1939, *Lindberg f.* s.n.! Larnaka tis Lapithou, 1941, *Davis* 2085! Thermia, 1956, *G. E. Atherton* 877!
8. Ayios Philon near Rizokarpaso, 1941, *Davis* 2254! Apostolos Andreas Mon., 1948, *Mavromoustakis* s.n.! Koma tou Yialou, 1950, *Chapman* 226! Ronnas Valley, 1956/57, *Merton* 3064! Matakos near Galinoporni, 1962, *Meikle* 2447!

var. **minima** *DC.* in Lam. et DC., Fl. Franç., 3: 331 (1805); Davis, Fl. Turkey, 7: 605 (1982).
    E. peploides Gouan, Fl. Monspel., 174 (1764); Boiss. in DC., Prodr., 15 (2): 141 (1862), Fl. Orient., 4: 1112 (1879); Post, Fl. Pal., ed. 2, 2: 501 (1933).
    E. peplus L. var. *peploides* (Gouan) Vis., Fl. Dalm., 3: 229 (1852).
    E. peplus L. var. *maritima* Boiss. in DC., Prodr., 15 (2): 141 (1862), Fl. Orient., 4: 1112 (1879); Holmboe, Veg. Cypr., 120 (1914).
    E. peplus L. ssp. *peploides* (Gouan) Rouy, Fl. France, 12: 175 (1910); Holmboe, Veg. Cypr., 120 (1914); Osorio-Tafall et Seraphim, List Vasc. Plants Cyprus, 67 (1973).
    E. peplus L. f. *peploides* (Gouan) Knoche, Fl. Balear., 2: 157 (1922).
    TYPE: France; specimen no. 9327β in herb. Willd. (B-WILLD).

Stems decumbent, often much branched; cauline leaves ± orbicular; seeds 1–1·4 mm. long; otherwise as in var. *peplus.*

HAB.: Rocks by seashore or waste land near the sea; sea-level to 50 ft. alt.; fl. April–July.

DISTR.: Divisions 3, 4, 7. Mediterranean region.

3. Mazotos, 1905, *Holmboe* 203.
4. Ayia Napa, 1905, *Holmboe* 19, 28.
7. Near Kharcha, by the beach, 1950, *Chapman* 274!

NOTES: Also recorded from "Ayios Therapon, seashore rocks", July 1957, *H. Painter in herb. C. West*, but this locality cannot be traced; it is clearly not Ayios Therapon (Div. 3) S. of Kilani.

19. E. chamaepeplus *Boiss. et Gaill.* in Boiss., Diagn., 2, 4: 88 (1859); Boiss. in DC., Prodr., 15 (2): 141 (1862), Fl. Orient., 4: 1113 (1879); Post, Fl. Pal., ed. 2, 2: 502 (1933); Zohary, Fl. Palaest., 2: 284, t. 417 (1972); Osorio-Tafall et Seraphim, List Vasc. Plants Cyprus, 67 (1973).

TYPE: "circâ *Hierosolymam* Boiss.! *Damascum* Gaillardot!" (G! P!).

Glabrous somewhat fleshy annual up to 10 cm. high; stems simple or 2- or more-branched from the base, often reddish-tinged; cauline and ray-leaves sessile, obtriangular to linear-oblanceolate, 2-9 mm. long, 0·5–2 mm. wide, retuse or emarginate, entire, attenuate at the base; raylet-leaves ovate-oblong to ovate or ovate-rhombic, 2–7 mm. long, 1·5–3·5 mm. wide, obtuse or truncate, mucronulate, erose or subentire, rounded or cuneate at the base; rays 2–3, up to 4 times dichotomous; axillary rays 0–2; glands long-horned. Regma subtrilobate-ellipsoid, 1·5 mm. diam., smooth, without dorsal ridges; seeds as in *E. peplus*, but with the grooves and pits each having dark median slits and pin-holes respectively; caruncle conical, stramineous.

HAB.: In *Pinus brutia* woodland; 900 ft. alt.; fl. March.

DISTR.: Division 6, rare. Also Syria, Iraq, Israel, Jordan, Egypt, Arabia.

6. Near Dhiorios, 1962, *Meikle* 2343!

20. E. falcata L., Sp. Plant., ed. 1, 456 (1753); Clarke, Travels, 2 (2): 719 (1816); Boiss. in DC., Prodr., 15 (2): 140 (1862), Fl. Orient., 4: 1111 (1879); Unger et Kotschy, Die Insel Cypern, 359 (1865); Sintenis in Oesterr. Bot. Zeitschr., 31: 257 (1881); H. Stuart Thompson in Journ. Bot., 44: 338 (1906); Holmboe, Veg. Cypr., 120 (1914); Post, Fl. Pal., ed. 2, 2: 500 (1933); Lindberg f., Iter Cypr., 21 (1946); Zohary, Fl. Palaest., 2: 282, t. 414 (1972); Osorio-Tafall et Seraphim, List Vasc. Plants Cyprus, 67 (1973); Davis, Fl. Turkey, 7: 607 (1982).

E. rubra Cav., Icones, 1, t. 34, fig. 1 (1791).

E. falcata L. var. *rubra* (Cav.) Boiss. in DC., Prodr., 15 (2): 140 (1862), Fl. Orient., 4: 1111 (1879).

TYPE: "*in* Europa *australi*" (LINN).

Glabrous, subglabrous or minutely puberulous, occasionally slightly pruinose-glaucous annual up to 30 cm. high; stems simple or 2- or more-branched from the base; cauline and ray-leaves linear-oblanceolate to oblanceolate or spathulate, 0·5–3 cm. long, 1–7 mm. wide, acute or subacute and mucronate, sometimes obtuse or truncate, entire or subentire, attenuate at the base; raylet-leaves falcate-elliptic to obliquely rhombic-ovate, 0·2–2 cm. long, 1–8 mm. wide, acute or obtuse, mucronate, subentire, cuneate or rounded at the base; rays 3–5, usually up to 5 times dichotomous, sometimes more; axillary rays 0–10 (–15); glands short-horned, or almost hornless and truncate. Regma ellipsoid, shallowly trisulcate, 1·5 mm. diam., smooth; seeds dorsiventrally compressed-ellipsoid, subquadrangular, 1·5 mm. long, with 5–7 transverse grooves per facet, pale grey when mature; caruncle prominent, terminal, conical, white, soon caducous.

HAB.: Dry slopes, shingly river-beds, marshes, sandy ground, cultivated and fallow fields, gardens, roadsides; sea-level to 4,400 ft. alt.; fl. March–Sept.

DISTR.: Divisions 1–8. Widespread from W. Europe and N. Africa eastwards to Pakistan.

1. Yeroskipos, 1939, *Lindberg f.* s.n.
2. Platres, 1938, *Kennedy* 1323! Kannaviou, 1939, *Lindberg f.* s.n.; Prodhromos, 1955, *G. E. Atherton* 638!
3. Akhelia, 1939, *Lindberg f.* s.n.; Limassol, 1939, *Lindberg f.* s.n.; Asomatos, 1939, *Mavromoustakis* 137! Skarinou, 1941, *Davis* 2674! Kolossi, 1971, *Townsend* 71/122!
4. Kophinou–Larnaca road, near Larnaca, 1970, *A. Genneou* in ARI 1593!
5. Athalassa, 1939, *Lindberg f.* s.n.; Mia Milea, 1950, *Chapman* 677! Mile 10, Nicosia–Famagusta road, 1952, *Merton* 836! Salamis, 1962, *Meikle* 2569! 2581! 2638!
6. "prope Solia versus Morphu", 1862, *Kotschy* 950! Nicosia, 1939, *Lindberg f.* s.n.! Between Koutraphas and Astromeritis, 1955, *Merton* 2324!
7. Near Kythrea, 1880, *Sintenis & Rigo* 72! Larnaka tis Lapithou, 1941, *Davis* 2980! Kythrea, 1950, *Chapman* 346! Kyrenia, 1949, *Casey* 788! 920! also, 1952, *Casey* 1256! and, 1955, *G. E. Atherton* 18! 417! 423! Ayios Epiktitos, 1967, *Merton* in ARI 314!
8. Koma tou Yialou, 1970, *A. Genneou* in ARI 1465!

21. E. herniariifolia *Willd.*, Sp. Plant., 2 (2): 902 (1800); Boiss. in DC., Prodr., 15 (2): 155 (1862), Fl. Orient., 4: 1123 (1879); Unger et Kotschy, Die Insel Cypern, 359 (1865); H. Stuart Thompson in Journ. Bot., 44: 338 (1906); Holmboe, Veg. Cypr., 121 (1914); Warr et Gresham in Journ. Roy. Hort. Soc., 59: 72 (1934); Osorio-Tafall et Seraphim, List Vasc. Plants Cyprus, 67 (1973); Davis, Fl. Turkey, 7: 609 (1982).

E. pumila Sm. in Sibth. et Sm., Fl. Graec. Prodr., 1: 324 (1809), Fl. Graec., 5: 47, t. 460 (1825); Kotschy in Oesterr. Bot. Zeitschr., 12: 279 (1862).

TYPE: "*in* Creta", specimen no. 9328 in herb. Willd. (B-WILLD).

Minutely puberulous or sometimes glabrous, glaucous, prostrate perennial with numerous much-branched stems arising from caudiculi borne on a woody stock; cauline, ray and raylet-leaves shortly petiolate, suborbicular, obovate or ovate, 0·2–1 cm. long, 2–8 mm. wide, obtuse or subacute, entire; rays (0–) 2–4, up to 3 times dichotomous; axillary rays 0–1; glands medium to long-horned. Regma trilobate, with 2 parallel wings on the back of each valve, 3–4 mm. diam., smooth; seeds ovoid-cylindric, 2·5–3 mm. long, shallowly and irregularly grooved and pitted, pale grey, sometimes darker in the depressions; caruncle prominent, terminal, conical, whitish.

HAB.: Rocky mountain slopes; circa 6,000 ft. alt.; fl. April–Sept.

DISTR.: Division 2, very rare. Albania, Greece, Turkey, Syria.

2. "Um die Spitze des Troodos auf der Nordost seite", 1862, *Kotschy*.

NOTES: No Cyprus material has been seen, nor does Kotschy cite a specimen. The above description is based on Turkish material.

22. E. veneris *M. S. Khan* in Kew Bull., 16: 447 (1963); Osorio-Tafall et Seraphim, List Vasc. Plants Cyprus, 67 (1973).

[*E. myrsinites* (non L.) Sm. in Sibth. et Sm., Fl. Graec. Prodr., 1: 331 (1809), Fl. Graec., 5: 55, t. 471 (1825); Poech, Enum. Plant. Ins. Cypr., 36 (1842); Boiss. in DC., Prodr., 15 (2): 173 (1862), Fl. Orient., 4: 1134 (1879); Unger et Kotschy, Die Insel Cypern, 359 (1865) ["*E. myrsinitis*"]; Holmboe, Veg. Cypr., 122 (1914).]

[*E. biglandulosa* (non Desf.) A. K. Jackson in Kew Bull., 1934: 273 (1934); A. K. Jackson et Turrill in Kew Bull., 1938: 467 (1938); Lindberg f., Iter Cypr., 21 (1946).]

TYPE: Cyprus; Kryos Potamos, near Platres, 870 m., 4 May 1937 *Kennedy* 607 (G ! K !).

Minutely pruinose-papillose, glaucous or purple-tinged, ascending or decumbent perennial, with several simple stems up to 35 cm. long arising from woody stock; cauline leaves elliptic-lanceolate, elliptic-oblanceolate or elliptic, 1–2·5 (–3) cm. long, 4–9 (–11) mm. wide, shortly acuminate, acute or subacute and mucronate, entire, coriaceous; ray-leaves elliptic-oblong to obovate; raylet leaves broadly ovate to ovate-suborbicular, 0·5–1·4 cm. long, 0·4–1·6 cm. wide, rounded at the base; rays (2–) 3–7 (–10), once or twice dichotomous; axillary rays 0 (–2); glands short- to medium-horned, with the horns dilated at the apex, purplish; male pedicels ebracteate; Regma triquetrous, 6–7 mm. diam., minutely granulate; seeds subquadrangular-cylindric, 4 mm. long, smooth, pale grey; caruncle prominent, terminal, truncate-conic, 1 mm. long.

HAB.: Rocky mountain slopes and screes, up to the snowline, in garigue or *Pinus nigra* woodland; by streamsides, or occasionally on walls; 2,900–5,600 ft. alt.; fl. Febr.–June.

DISTR.: Division 2. Endemic.

2. Locally common about Troödos, Prodhromos, Platres, Phini and Saïttas. *Sibthorp*; *Kotschy* 791; *Sintenis & Rigo* 695 ! *A. G. & M. E. Lascelles* ! *Holmboe* 942 ! *Feilden* 4 ! *C. B. Ussher* 70 ! *Syngrassides* 348 ! 948 ! *Kennedy* 604–609 ! *Wyatt* 22 ! *Lindberg f.* ! *Davis* 1780 ! 3157 ! *Mavromoustakis* ! *G. E. Atherton* 560 ! *Meikle* 2772 ! *J. B. Suart* 109 ! 114 ! *A. Genneou* in ARI 1297 ! 1573 !

NOTES: This species approaches *E. rigida* M. Bieb. very closely. The latter ranges from Portugal and Morocco eastwards to the Crimea, W. Caucasus, Syria and N.E. Iran, but *E. veneris* differs from it in its much less robust habit, smaller leaves, and perfectly smooth seeds.

23. E. paralias *L.*, Sp. Plant., ed. 1, 458 (1753); Boiss. in DC., Prodr., 15 (2): 167 (1862), Fl. Orient., 4: 1130 (1879); Holmboe, Veg. Cypr., 121 (1914); Post, Fl. Pal., ed. 2, 2: 504 (1933); Lindberg f., Iter Cypr., 21 (1946); Rechinger f. in Arkiv för Bot., ser. 2, 1: 426 (1950);

Zohary, Fl. Palaest., 2: 286, t. 422 (1972); Osorio-Tafall et Seraphim, List Vasc. Plants Cyprus, 67 (1973); Davis, Fl. Turkey, 7: 614 (1982).

TYPE: "*in* Europae *arena maritima*" (LINN).

Glabrous, glaucous, fleshy, caespitose perennial with numerous, ± simple, densely leafy stems up to 70 cm. high, arising from a woody stock; cauline leaves imbricate, linear-oblong, oblong or oblong-lanceolate, the uppermost ones usually ovate, 0·5–2·5 (–3) cm. long, 0·2–1 (–1·5) cm. wide, subacute or obtuse, entire, adaxially concave; ray-leaves ovate-lanceolate to broadly ovate; raylet-leaves suborbicular-rhombic to reniform, 0·5–1·5 cm. long, 0·7–1·7 cm. wide, cuneate to truncate at the base, strongly adaxially concave; rays 3–6, up to 3 times dichotomous; axillary rays 0–10; glands short-horned, orange. Regma strongly trilobate, 5–6 mm. diam., granulate-rugulose; seeds broadly ovoid, 3 mm. long, smooth, pale grey or whitish; caruncle minute.

HAB.: Sand-dunes, sandy or rocky seashores; about sea-level; fl. Febr.–Dec.

DISTR.: Divisions 1, 4, 5, 7. W. Europe, Mediterranean and Black Sea coasts; Atlantic Islands.

1. Cape Arnauti, 1962, *Meikle* 2301!
4. Famagusta, 1912, *Haradjian* 125! also, 1936, *Syngrassides* 1153! and, 1959, *P. H. Oswald* 172!
5. Salamis, 1905, *Holmboe* 458, 460; also, 1931–1957, *J. A. Tracey* 34! *Lindberg f.* s.n.! *Kennedy* in *Casey* 1038! *Poore* 36! *Merton* 3220! Trikomo, 1939, *Lindberg f.* s.n.
7. Kyrenia, 1949, *Mrs. Grove* in *Casey* 861! also, 1955, *G. E. Atherton* 76! 703!

24. E. terracina *L.*, Sp. Plant., ed. 2, 1: 654 (1762); Boiss. in DC., Prodr., 15 (2): 157 (1862), Fl. Orient., 4: 1123 (1879); Holmboe, Veg. Cypr., 120 (1914); Post, Fl. Pal., ed. 2, 2: 503 (1933); Lindberg f., Iter Cypr., 22 (1946); Zohary, Fl. Palaest., 2: 285, t. 420 (1972); Osorio-Tafall et Seraphim, List Vasc. Plants Cyprus, 67 (1973); Davis, Fl. Turkey, 7: 621 (1982).
　E. obliquata Forssk., Fl. Aegypt.–Arab., 93 (1775).
　E. obtusifolia Lam., Encycl. Méth., 2: 430 (1786).
　E. seticornis Poir., Voy. en Barbarie, 2: 173 (1789).
　E. provincialis Willd., Sp. Plant., 2 (2): 914 (1800).
　E. diversifolia Poir. in Lam., Encycl. Méth. Suppl., 2: 618 (1812).
　E. alexandrina Del., Fl. Égypte, 90, t. 30, fig. 2 (1813–14).
　E. leiosperma Sm. in Sibth. et Sm., Fl. Graec., 5: 51, t. 465 (1825); Poech, Enum. Plant. Ins. Cypr., 36 (1842).
　[*E. portlandica* (non L.) Sm. in Sibth. et Sm., Fl. Graec. Prodr., 1: 327 (1809).]
TYPE: Spain; "*in* Hispania, *Alstroemer*".

Glabrous, procumbent, ascending or erect perennial with several stems up to 70 cm. long, arising from the stock; cauline leaves oblanceolate, linear or elliptic-lanceolate, 0·5–2·5 (–3·5) cm. long, 2–7 mm. wide, acute, obtuse, retuse or tricuspidate, entire, minutely serrulate or sparingly dentate; ray-leaves lanceolate to ovate-lanceolate; raylet leaves ovate-rhombic, sometimes obliquely so, or occasionally ovate-deltoid or reniform, 0·5–1·5 cm. long, 0·5–1·5 cm. wide, acute or obtuse, cuneate or truncate, subentire, serrulate or dentate; rays 3–5, up to 7 times dichotomous; axillary rays 0–2 (–9); glands long-horned, the horns setaceous. Regma strongly trilobate, 4 mm. diam., smooth; seeds ± ovoid, obliquely truncate, 2·5 mm. long, smooth, pale grey; caruncle boat-shaped, 1·5 mm. long.

HAB.: Sand dunes, sandy flats, stony ground among ruins, by pathsides; sea-level to 100 ft. alt.; fl. (Dec.–) Febr.–June.

DISTR.: Divisions 4–8, locally common. Mediterranean region; Red Sea coasts; Georgia; Atlantic Islands.

4. Near Larnaca, 1880, *Sintenis & Rigo* 697! Famagusta, 1938, *Syngrassides* 1829! also, 1939, *Lindberg f.* s.n.! and, 1956, *Merton* 2840!
5. Trikomo, 1939, *Lindberg f.* s.n.! Common about Salamis, 1957–71, *Merton* 3204! *Meikle* 2562! *Townsend* 71/107!
6. Between Syrianokhori and Aloupos River, 1941, *Davis* 2613!
7. Between Ayios Epiktitos and Ayios Amvrosios, 1941, *Davis* 3616!
8. Ayios Philon near Rizokarpaso, 1941, *Davis* 2230!

25. E. thompsonii Holmboe, Veg. Cypr., 121, fig. 36 (1914); Osorio-Tafall et Seraphim, List Vasc. Plants Cyprus, 67 (1973); Davis, Fl. Turkey, 7: 629 (1982).

[*E. sylvatica* (non L.) Sm. in Sibth. et Sm., Fl. Graec. Prodr., 1: 332 (1809); Poech, Enum. Plant. Ins. Cypr., 36 (1842).]

[*E. amygdaloides* (non L.) Unger et Kotschy, Die Insel Cypern, 359 (1865); H. Stuart Thompson in Journ. Bot., 44: 338 (1906).]

[*E. kotschyana* (non Fenzl) Kotschy in Unger et Kotschy, Die Insel Cypern, 359 (1865); H. Stuart Thompson in Journ. Bot., 44: 338 (1906).]

[*E. characias* (non L.) H. Stuart Thompson in Journ. Bot., 44: 338 (1906).]

TYPE: Cyprus; Prodhromos, 1862, *Kotschy* 899 (W).

Robust tomentose, erect, caespitose perennial with several biennial flowering stems up to 1 m. high, woody and cicatricose towards the base; cauline leaves from the middle of the stem oblong-oblanceolate to linear-oblanceolate, 3–8·3 cm. long, 0·4–1·2 cm. wide, obtuse, rounded or emarginate, attenuate at the base, entire; those from the upper part of the stem, and the ray-leaves, elliptic–obovate, 1·5–3 cm. long, 0·8–1·5 cm. wide; raylet-leaves connate in pairs, forming ± suborbicular shallow cups 1–2 cm. across; rays 7–10, once or twice dichotomous; axillary rays 5–8; glands semilunar, short-horned. Regma trilobate, 5 mm. diam., smooth, tomentose; seeds oblong, 2·5 mm. long, smooth, grey; caruncle conical.

HAB.: Dry slopes in garigue and Pine-forest; 500–5,500 ft. alt.; fl. Febr.–May.

DISTR.: Divisions 2, 3, rare. S. Turkey (Cilician Taurus between 1,000 & 1,200 m. alt.).

2. Prodhromos, in *Pinus nigra* (ssp. *pallasiana*) forest, 1862, *Kotschy* 899.
3. Pissouri, 1902, *A. G. & M. E. Lascelles* s.n.! also, 1905, *Holmboe* 712.

NOTES: The combination of characters belonging to *E. characias* L. on the one hand and *E. amygdaloides* L. and *E. macrostegia* Boiss. on the other, might seem to suggest a hybrid origin for this species.

2. ANDRACHNE *L.*

Sp. Plant., ed. 1, 1014 (1753).
Gen. Plant., ed. 5, 444 (1754).

Monoecious perennial herbs or shrubs without a milky latex; indumentum (when present) simple, sometimes glandular; leaves alternate, simple, entire, petiolate; stipules green or chaffy; flowers unisexual, small, axillary, solitary or fascicled; sepals 5–6, free or nearly free, those of the female flowers larger than those of the male flowers; petals 5–6, smaller than the sepals, and those of the female flowers smaller than those of the male flowers; disk-glands of the male flowers 5–6, free, disk of the female flowers 5–6-lobed; stamens 5–6, ± free or connate at the base; pistillode present; ovary trilocular, with 2 ovules per loculus; styles 3, ± free, bipartite. Fruit a depressed-subglobose, smooth, 3-celled, septicidally dehiscent 6-seeded regma; seeds trigonous, with 2 flat lateral surfaces and 1 convex dorsal surface, rather like the segment of an orange, ecarunculate.

A widespread genus of some 25 species; only 1 species in Cyprus.

1. **A. telephioides** *L.*, Sp. Plant., ed. 1, 1014 (1753); Muell. Arg. in DC., Prodr., 15 (2): 235 (1866); Boiss., Fl. Orient., 4: 1138 (1879); Sintenis in Oesterr. Bot. Zeitschr., 31: 255, 259, 326 (1881); Post in Mém. Herb. Boiss., 18: 100 (1900); H. Stuart Thompson in Journ. Bot., 44: 338 (1906); Holmboe, Veg. Cypr., 119 (1914); Post, Fl. Pal., ed. 2, 2: 506 (1933); Lindberg f., Iter Cypr., 21 (1946); Rechinger f. in Arkiv för Bot., ser. 2, 1: 426 (1950); Zohary, Fl. Palaest., 2: 265, t. 382 (1972); Osorio-Tafall et Seraphim, List Vasc. Plants Cyprus, 66 (1973); Davis, Fl. Turkey, 7: 567 (1982).

A. rotundifolia C. A. Mey., in Eichw., Pl. Nov. It. Casp.–Cauc., 18 (1831).

A. telephioides L. var. *rotundifolia* (C. A. Mey.) Muell. Arg. in DC., Prodr., 15 (2): 236 (1866).

A. nummulariifolia Stapf in Denkschr. Akad. Wiss. Wien, Math.-Nat. Kl., 51: 314 (1886).

A. virescens Stapf in Denkschr. Akad. Wiss. Wien, Math.-Nat. Kl., 51: 314 (1886).

TYPE: "*in* Italia, Graecia, Media" (LINN!).

Glabrous, glaucous, prostrate perennial herb with numerous simple or sparingly branched stems up to 50 cm. long arising from a woody stock; leaves ovate or obovate to suborbicular, 2–12 mm. long, 1·5–8 mm. wide, subacute or obtuse, rounded or cuneate; petioles 0·5–3 (–6) mm. long; stipules subpeltate, ovate-lanceolate, 1 mm. long, fimbriate or entire, chaffy, white with a purple basal blotch; flowers solitary or fascicled, commonly with 2 male flowers and 1 female flower per fascicle; male flower 2 mm. across, yellowish; pedicels 1 mm. long; sepals ovate-rhomboid, 1·3 mm. long, yellow-green with a hyaline margin, entire; petals lanceolate, 1 mm. long, whitish; disk-glands smooth, bilobed; anthers 4-lobed, yellow; female flowers 3 mm. across, 5 mm. across in fruit; pedicels 2–5 mm. long; sepals ovate-rhomboid, 2 mm. long, dark green with a narrow whitish margin, entire; petals minute; disk-lobes emarginate-retuse; ovary globose, smooth. Regma 3–4 mm. diam.; seeds 1·5–2 mm. long, ± smooth, dull brown.

HAB.: Dry rocky limestone, sandstone, or igneous hillsides; sand-dunes; moist roadside banks; streamsides; vineyards and gardens; sea-level to 4,300 ft. alt.; Febr.–Oct.

DISTR.: Divisions 1–8, common in Divisions 2, 7; Mediterranean Europe, S.W. Asia E. to the Pamir-Alai and N.W. India, N. Africa S. to the Cape Verde Islands, Somalia and Abd al Kuri Island.

1. Lyso, 1913, *Haradjian* 840; Ayios Neophytos, 1939, *Lindberg f.* s.n.; near Ktima, 1940, *Davis* 1776!
2. Common; Galata, Troödos, Pharmakas, Milikouri, Kykko, Gourri, Vavatsinia, Mavres Sykies, Palekhori, etc., *Sintenis & Rigo* 75! *Post*; *Lindberg f.*; *Holmboe* 1127! *Haradjian* 947! *Kennedy* 315! *Davis* 1898! 1906! *Chapman* 322! *Meikle* 2740! *P. Laukkonen* 437a! etc.
3. Zagala above Trimiklini, 1941, *Davis* 3141!
4. Larnaca, 1939, *Lindberg f.* s.n.; Paralimni, 1939, *Lindberg f.* s.n.; Sotira, Ayios Antonios, 1950, *Chapman* 595!
5. Alambra, 1932, *Syngrassides* 388! Between Nisou and Stavrovouni, 1933, *Syngrassides* 23! Between Nicosia and Vouno, 1941, *Davis* 2821! Near Klirou, 1959, *P. H. Oswald* 54!
6. Sand-dunes at Liveras, 1962, *Meikle* 2415! ·
7. Common; above Kythrea, St Hilarion, Kyrenia, Bellapais, Yaïla, Ayios Epiktitos, etc. *Sintenis & Rigo* 75a! *C. H. Wyatt* 22! *Lindberg f.*; *Davis* 2907! 2943! *Casey* 111! 1138! *G. E. Atherton* 418! etc.
8. Cape Elea, 1905, *Holmboe* 509; Ephtakomi, 1905, *Holmboe* 521; Ayios Theodhoros, 1939, *Lindberg f.* s.n.! Koma tou Yialou, 1970, *A. Genneou* in ARI 1483!

3. CHROZOPHORA *A. H. L. Juss.*

Euph. Gen. Tent., 27 (1824) [as *Crozophora*], nom. cons.

Monoecious annual or perennial herbs or subshrubs without a milky latex; indumentum stellate; leaves alternate, simple, entire, dentate or lobed, with 2 basal glands beneath, long-petiolate; stipules filiform, chaffy; inflorescences congested, paniculate-subracemose, axillary, with male flowers above and female flowers below; male flowers subsessile; sepals 5, valvate; petals 5, equalling or shorter than the sepals, connate; disk 5-lobed, minute; stamens (3–) 5–12 (–15), the filaments united into a column, 1–3-seriate; anthers oblique, longitudinally dehiscent; pistillode 0; female flowers pedicellate, the pedicels becoming elongated and reflexed in fruit; sepals 5, narrow; petals 5, free, sepaloid, or occasionally 0; disk 5-lobed, the lobes alternating with the petals; staminodes 5 or 0; ovary trilocular, with 1 ovule per loculus; styles 3, ± free, bifid. Fruit a trilobate, stellate or lepidote, often tuberculate, 3-celled, septicidally dehiscent 3-seeded regma; columella persistent, prominent; seeds ovoid, somewhat angular, smooth or tuberculate, ecarunculate.

A genus of some 7 species from the Mediterranean region, tropical Africa and S.W. Asia; only 1 species in Cyprus.

1. **Chr. tinctoria** (*L.*) *Raf.*, Chlor. Aetn., 4 (1813); Muell. Arg. in DC., Prodr., 15 (2): 748 (1866); Boiss., Fl. Orient., 4: 1140 (1879); Post, Fl. Pal., ed. 2, 2: 507 (1933); Lindberg f., Iter Cypr., 21 (1946); Zohary, Fl. Palaest., 2: 267, t. 385 (1972); Osorio-Tafall et Seraphim, List Vasc. Plants Cyprus, 66 (1973); Davis, Fl. Turkey, 7: 568 (1982).

　　Croton tinctorius L., Sp. Plant., ed. 1, 1004 (1753).

　　C. obliquus Vahl, Symb. Bot., 1: 78 (1790).

　　C. verbascifolius Willd., Sp. Plant., 4: 539 (1805).

　　C. villosus Sm. in Sibth. et Sm., Fl. Graec. Prodr., 2: 249 (1816), Fl. Graec., 10: t. 951 (1840).

　　Chrozophora hierosolymitana Spreng., Syst. Veg., 3: 850 (1826).

　　Chr. obliqua (Vahl) A. H. L. Juss. ex Spreng., Syst. Veg., 3: 850 (1826); Muell. Arg. in DC., Prodr., 15 (2): 749 (1866) quoad nomen; Boiss., Fl. Orient., 4: 1141 (1879) quoad nomen; Post, Fl. Pal., ed. 2, 2: 507 (1933) quoad nomen; Zohary, Fl. Palaest., 2: 267, t. 386 (1972); Osorio-Tafall et Seraphim, List Vasc. Plants Cyprus, 66 (1973).

　　Chr. verbascifolia (Willd.) A. H. L. Juss. ex Spreng., Syst. Veg., 3: 851 (1826); Boiss., Fl. Orient., 4: 1141 (1879); Holmboe, Veg. Cypr., 119 (1914); Post, Fl. Pal., ed. 2, 2: 507 (1933); R. M. Nattrass, First List Cyprus Fungi, 5 (1937); Lindberg f., Iter Cypr., 21 (1946).

　　Chr. tinctoria (L.) Raf., var. *verbascifolia* (Willd.) Muell. Arg., in DC., Prodr., 15 (2): 748 (1866).

　　Chr. tinctoria (L.) Raf. var. *hierosolymitana* (Spreng.) Muell. Arg., in DC., Prodr., 15 (2): 749 (1866).

TYPE: France; "Monspelii" (LINN).

Sparingly stellate-pubescent to felty-tomentose, erect, ascending or procumbent, annual herb (0·1–) 0·3–1 m. high, with divaricately-branched stems often becoming woody at the base; leaves ovate-rhombic to ovate-lanceolate, occasionally indistinctly 3-lobed, 2–9 cm. long, 1–7 cm. wide, acute or obtuse, cuneate to subcordate, shallowly repand-dentate; petioles 1–9 cm. long; stipules 2–5 mm. long; inflorescences (1–) 2–4 cm. long; male flowers: sepals linear-lanceolate, 4 mm. long, stellate-pubescent without, glabrous within; petals elliptic-lanceolate, 4 mm. long, lepidote without, simply pubescent within, yellowish-green; stamens 3–12, biseriate; staminal column 3·5 mm. high, minutely papillose; anthers 2-lobed, 1·5 mm. long; female flowers: pedicels 3–5 mm. long, extending to 3–4 cm. in fruit; sepals and petals both resembling the male sepals; ovary densely lepidote; styles stellate-pubescent without, papillose within. Regma 5–8 mm. diam., purple; seeds 4 mm. long, grey.

HAB.: Gravelly soil; damp, irrigated land, in cultivated or fallow fields, or by roadsides; sea-level to 500 ft. alt.; fl. May–Sept.

DISTR.: Divisions 1, 2, 4–7. Mediterranean region, S.W. Asia E. to the Tien Shan and Pakistan and S. to S. Yemen and Socotra.

1. Kissonerga, 1967, *Economides* in ARI 982!
2. Near Omodhos, 1880, *Sintenis & Rigo* 627!
4. Larnaca, 1894, *Deschamps* 457; and, 1933, *Nattrass* 366; between Larnaca and Kiti, 1905, *Holmboe* 1093; near Salt Lake, Larnaca, 1939, *Lindberg f.* s.n.! Perivolia, 1939, *Lindberg f.* s.n.! Famagusta, 1939, *Lindberg f.* s.n.!
5. Athalassa, 1966, *Merton* in ARI 9!
6. Near Peristerona, 1880, *Sintenis & Rigo* 811! Nicosia, 1935, *Syngrassides* 651! Morphou, 1955, *N. Chiotellis* 475! 479! English School, Nicosia, 1958, *P. H. Oswald* 2!
7. Near Kyrenia, 1880, *Sintenis & Rigo* 624! Kyrenia, 1948, *Casey* 31! *G. E. Atherton* 221! coast W. of Glykyotissa Island, 1958, *P. H. Oswald* 9! 6 m. W. of Kyrenia, 1968, *Barclay* 1087!

NOTES: Very variable as regards leaf-shape, indumentum and stamen-number, in consequence several species have been recognized in the past. Since, however, there is a manifest continuum with regard to each of these characters, at least somewhere in the geographical range of the complex, it is best to regard the complex as representing one polymorphic species.

4. MERCURIALIS *L.*

Sp. Plant., ed. 1, 1035 (1753).

Gen. Plant., ed. 5, 457 (1754).

Dioecious or more rarely monoecious annual or perennial herbs without a milky latex; indumentum (when present) simple; leaves opposite, simple, crenate-serrate, often with 2 basal glands beneath, petiolate or subsessile; stipules chartaceous; male flowers subsessile, usually in clusters on long-peduncled interrupted axillary spikes; sepals 3–4, free, valvate; petals 0; disk 0; stamens 6–20, filaments free, anther-thecae longitudinally dehiscent, joined by a globular connective; pistillode 0; female flowers solitary or in few-flowered axillary fascicles, subsessile clusters or pedunculate spikes; sepals 3, free or connate at the base, imbricate; disk of 2 (–3) filiform glands alternating with the ovary lobes; ovary bilocular, rarely trilocular, with 1 ovule per loculus; styles 2 (–3), free or almost free, erect or divaricate, short, undivided, prominently papillose above; fruit a didymous or more rarely trilobate smooth or tuberculate 2 (–3)-celled septicidally dehiscent 2 (–3)-seeded regma; seeds ovoid or globose, smooth, rugulose or foveolate, carunculate.

A genus of 8 species, predominantly Mediterranean, but also in N. & C. Europe, Atlantic Islands, S.W. & E. Asia; only 1 species in Cyprus. The record by Hume (in Walpole, Mem. Europ. Asiatic Turkey, ed. 1, 254; 1817) of *M. tomentosa* L. for Cyprus is undoubtedly erroneous, since this species is confined to Spain, Portugal, S. France and the Balearic Is.

1. **M. annua** *L.*, Sp. Plant., ed. 1, 1035 (1753); Sibth. et Sm., Fl. Graec. Prodr., 2: 261 (1816); Poech, Enum. Plant. Ins. Cypr., 36 (1842); Unger et Kotschy, Die Insel Cypern., 360 (1865); Muell. Arg. in DC., Prodr., 15 (2): 797 (1866); Boiss., Fl. Orient., 4: 1142 (1879); Sintenis in Oesterr. Bot. Zeitschr., 192, 257, 325 (1881); H. Stuart Thompson in Journ. Bot., 44: 338 (1906); Holmboe, Veg. Cypr., 119 (1914); Post, Fl. Pal., ed. 2, 2: 508 (1933); Lindberg f., Iter Cypr., 22 (1946); Zohary, Fl. Palaest., 2: 268, t. 388 (1972); Osorio-Tafall et Seraphim, List Vasc. Plants Cyprus, 66 (1973); Davis, Fl. Turkey, 7: 569 (1982).

TYPE: "*in* Europae *temperatae umbrosis*" (LINN!).

Glabrous or subglabrous, erect, much-branched annual herb 5–50 cm. high, with the stems somewhat thickened at the nodes; leaves ovate to elliptic-lanceolate, 1–7·5 cm. long, 0·3–3·5 cm. wide, obtuse or obtusely acuminate, cuneate, rounded or shallowly cordate, crenate-serrate, the teeth usually sparingly ciliate, bright green; basal glands minute, ochraceous; petioles 0·1–4 cm. long; stipules ovate-deltoid to lanceolate, 1–2 mm. long, whitish; male spikes 2–10 cm. long; male flowers: sepals broadly ovate, 1·5–2 mm. long, acute, glabrous, pale yellowish-green, translucent; stamens 8–12; female flowers solitary or fascicled, 1–4 per axil, shortly pedicellate, rarely on a short spiciform cymule; sepals ovate, 1–2 mm. long, acute or subacute, glabrous, pale green or whitish; disk-glands 1 mm. long; ovary bilocular, sparingly to evenly tuberculate, with each tubercle bristle-tipped; styles less than 1 mm. long. Regma 3–4 mm. wide, often drying bluish-green; seeds ovoid-subglobose, 2 mm. long, tuberculate-rugulose, greyish-brown, shiny, with a small, white keel-shaped caruncle.

HAB.: Rock crevices, grassy hillsides; fields, irrigated orchards, Orange groves and vegetable gardens; waste places; pathsides; sea-level to 800 ft. alt.; fl. Oct.–Apr.

DISTR: Divisions 1, 2, 4, 5, 7, 8. Europe, S.W. Asia, N. Africa, Atlantic Islands; naturalized in S. Africa, N. America and the West Indies.

1. Above Peyia, 1962, *Meikle* 2053!
2. Lefka, 1932, *Syngrassides* 43!
4. Larnaca, 1862, *Kotschy;* same locality, 1939, *Lindberg f.* s.n.; Cape Pyla, 1880, *Sintenis & Rigo;* Sotira, Ayios Antonios, 1950, *Chapman* 598! Kouklia, 1967, *Merton* in ARI 444! Dhekelia, 1973, *P. Laukkonen* 43!
5. Nisou, 1936, *Syngrassides* 1133! 1135!
7. Kythrea, 1880, *Sintenis & Rigo;* Kyrenia, 1902, *A. G. & M. E. Lascelles* s.n.! also, 1949, *Casey* 211! and 1955–56, *G. E. Atherton* 725! 788! 820! 1123!
8. Ayios Philon near Rizokarpaso, 1941, *Davis* 2210!

5. RICINUS L.

Sp. Plant., ed. 1, 1007 (1753).
Gen. Plant., ed. 5, 437 (1754).

Large monoecious annual or perennial herb without a milky latex; indumentum 0; leaves alternate, peltate, palmately lobed, the lobes glandular-serrate, long-petiolate; stipules united to form a firmly chartaceous, readily caducous sheath; inflorescences interrupted-subpaniculate, leaf-opposed or subterminal, with male flowers in the lower part and female in the upper; bracts chaffy; male flowers pedicellate; buds subglobose, slightly apiculate; calyx 3–5-lobed, the lobes valvate; petals 0; disk 0; stamens numerous, sometimes up to 1,000, with the filaments variously and haphazardly united, anther-thecae 2, distinct, divaricate, subglobose, longitudinally dehiscent; pistillode 0; female flowers pedicellate; sepals 5, valvate, soon caducous; disk 0; ovary echinate, rarely smooth, trilocular, with 1 ovule per loculus; styles 3, ± free or shortly connate at the base, bipartite, plumose-papillose, dark red; fruit a trilobate, echinate or rarely smooth, 3-celled, septicidally dehiscent 3-seeded regma; spine-tipped tubercles accrescent; columella persistent, prominent; seeds compressed-ovoid, smooth, marbled and mottled, carunculate; endosperm copious, oleiferous.

A monotypic genus probably native to N.E. Tropical Africa, but now widely cultivated throughout the tropics, subtropics and warm temperate regions, and often becoming naturalized.

1. **R. communis** L., Sp. Plant., ed. 1, 1007 (1753); Sibth. et Sm., Fl. Graec. Prodr., 2: 249 (1816); Sibth. et Sm., Fl. Graec., 10: 42 (1840); Hume in Walpole, Mem. Europ. Asiatic Turkey, ed. 1, 254 (1817); Poech, Enum. Plant. Ins. Cypr., 36 (1842); Unger et Kotschy, Die Insel Cypern, 360 (1865); Muell. Arg. in DC., Prodr., 15 (2): 1017 (1866); Boiss., Fl. Orient., 4: 1143 (1879); H. Stuart Thompson in Journ. Bot., 44: 338 (1906); Holmboe, Veg. Cypr., 119 (1914); Post, Fl. Pal., ed. 2, 2: 509 (1933); Nattrass, First List Cyprus Fungi, 13, 34 (1937); Chapman, Cyprus Trees & Shrubs, 47 (1949); Zohary, Fl. Palaest., 2: 269, t. 389 (1972); Osorio-Tafall & Seraphim, List Vasc. Plants Cyprus, 66 (1973); Davis, Fl. Turkey, 7: 571 (1982).

TYPE: "*in* India *utraque*, Africa, Europa *australi*" (LINN).

Glabrous, somewhat glaucous, erect herb commonly up to 5 m. high, rarely up to 13 m., with thick, hollow stems; leaves deeply 5–11-lobed, the lobes lanceolate or ovate-lanceolate, acutely acuminate, coarsely serrate or biserrate, the teeth gland-tipped; lamina (6–) 10–50 (–100) cm. long and wide; petiole equalling or somewhat exceeding the lamina in length, with one or more stipitate, subsessile or sessile gland(s) on the adaxial surface; stipular sheath ovate, up to 2 cm. long; inflorescences up to 30 cm. long, floriferous almost from the base; bracts broadly triangular-lanceolate, 5–8 mm. long, membranous, soon shrivelling; male flowers up to 2 cm. across, long-pedicellate, the pedicels up to 2 cm. long, articulate; sepals ovate-lanceolate to triangular-ovate, 0·7–1 cm. long, acute, greenish-yellow; anthers in dense masses, thecae 0·3 mm. long, yellow; female flowers pedicellate, the pedicels 0·5–1 (–2) cm. long; sepals resembling those of the male flowers; ovary subglobose- to ovoid-trilobate, densely beset with cylindric-filiform protuberances, each tipped with an indurated, hyaline, claw-like spine; styles 0·5–1 cm. long. Regma 1–2 (–2·4) cm. long and 1–1·5 cm. diam.; seeds 0·8–1·5 (–2) cm. long, shiny, greyish and mottled dark brown, or concolorous; caruncle bilobate.

HAB.: Waste land, banks of dried-up streams, abandoned gardens; naturalized; sea-level to 500 ft. alt.; fl. June.

DISTR.: Divisions 1–7. Probably native to N.E. Tropical Africa, but widely cultivated in the warmer parts of the globe; the Castor Oil Plant often becomes naturalized.
1. Polis, 1934, *Syngrassides* 851! also *Chapman* 144.
2. Lefka, 1932, *Nattrass* 180.
3. Limassol, 1801, *Hume*.
4. Larnaca, 1801, *Hume*; Famagusta, 1931, *Nattrass* 92!
5. Syngrasis [Synkrasi], 1862, *Kotschy*.
6. Nicosia, *Meikle*!
7. Kyrenia, 1955, *G. E. Atherton* 40!

NOTES: The Castor Oil Plant exhibits much variation as regards the size and ornamentation of the regma and the size, colour and patterning of the seeds, and on these bases, nearly 20 varieties and some 20 formae have been recognized. The Cyprus material would appear to be referable to the var. *microcarpus* Muell. Arg. forma *viridis* (Willd.) Muell. Arg.

SAPIUM VERUM *Hemsley* in Hooker's Icones Plant., 27: t. 2647 (1900) (syn. *S. thomsonii* Godefroy-Lebeuf ex Jumelle) from S. America, an evergreen tree with oblong, coriaceous leaves, is reported by Bovill, Rep. Plant. Work Cyprus, 15 (1915) as being or having been cultivated as a rubber-tree in Salamis Plantation.

84. URTICACEAE

Annual or perennial herbs, shrubs, rarely trees or woody climbers, frequently armed with stinging hairs; leaves alternate or opposite, simple, generally stipulate, and often with conspicuous cystoliths; inflorescences cymose, sometimes crowded into heads; flowers mostly unisexual (monoecious, dioecious or polygamous) rarely hermaphrodite, small; perianth sepaloid, the lobes free or partly united, valvate or imbricate in bud; stamens generally as many as, and opposite to the perianth-lobes; filaments commonly inflexed, bursting outwards explosively at anthesis; anthers 2-thecous, dehiscing longitudinally; rudimentary ovary often present in male flowers; female flowers similar to male, but frequently accrescent in fruit; stamens reduced to small, scale-like staminodes; ovary adnate to perianth, superior, 1-locular, 1-ovulate; style simple, sometimes wanting; stigmas linear or capitate, often penicillate. Fruit an achene or drupe; seed usually with oily endosperm; embryo small, straight.

About 50 genera and more than 500 species with a world-wide distribution.

Leaves boldly toothed or serrate, opposite, usually armed with stinging hairs- - **1. Urtica**
Leaves entire, alternate, without stinging hairs - - - - - - **2. Parietaria**

1. URTICA *L*.
Sp. Plant. ed. 1, 983 (1753).
Sp. Plant. ed. 5, 423 (1754).

Annual or perennial herbs, usually clothed with stinging hairs; leaves opposite, petiolate, serrate or incised; stipules free or connate between the petiole-bases in pairs; inflorescence axillary, basically cymose, but often cincinnate, or sometimes condensed into a globose head; flowers unisexual (monoecious or dioecious) sometimes polygamous, ebracteate; perianth 4-lobed, lobes imbricate, unequal in female flowers; ovary erect; stigmas sessile or subsessile, capitate-penicillate. Achene compressed, generally enveloped by persistent perianth.

About 50 species, chiefly in temperate regions of the northern hemisphere.

Perennial plants with tough, creeping rhizomes - - - - - - **2. U. dioica**
Annual plants:
  Female inflorescence a pedunculate, dense, globose capitulum - - - **3. U. pilulifera**
  Female inflorescence not dense or globose:
    Male inflorescence a condensed axillary spike or cluster, much shorter than the subtending
    leaf, rhachis inconspicuous - - - - - - - - - **1. U. urens**
    Male inflorescence elongate, caudate, much longer than the subtending leaf, with a
    conspicuous, membranous, flattened rhachis - - - - **4. U. membranacea**

1. **U. urens** *L.*, Sp. Plant., ed. 1, 984 (1753); Boiss., Fl. Orient., 4: 1146 (1879); Druce in Rep.
B.E.C., 9: 471 (1931); Post, Fl. Pal., ed. 2, 2: 510 (1933); A. K. Jackson in Kew Bull., 1934:
273 (1934): Zohary, Fl. Palaest., 1: 40, t. 36 (1966); Osorio-Tafall et Seraphim, List Vasc.
Plants Cyprus, 29 (1973); Davis, Fl. Turkey, 7: 635 (1982).
    TYPE: *"in* Europae *cultis"*.

Erect, branched or unbranched annual, 10–60 cm. high; stems sharply 4-angled, thinly setose; leaves numerous, broadly ovate, 1·3–4 cm. long, 0·8–3·5 cm. wide, dark green, thinly clothed with stinging hairs, apex blunt or subacute, margins deeply and regularly, subacutely, lobulate-serrate, base broadly cuneate, rounded or occasionally subcordate; petiole slender, 0·5–3·5 cm. long; stipules 4, free, green, bluntly ovate, 2–3 mm. long, 1·5–2·5 mm. wide; inflorescences monoecious, spiciform, spreading, rarely more than 1·5 cm. long, often very short and condensed; flowers very shortly pedicellate, the males with 4, broadly ovate or suborbicular, green, membranous, subglabrous perianth-lobes about 0·8–1 mm. long and almost as broad; filaments about 1–1·2 mm. long; anthers pallid, broadly oblong or suborbicular, about 1 mm. diam.; female flowers compressed, 2 of the perianth-lobes ovate, about 1 mm. long (accrescent to 2·5 mm. in fruit), 0·8 mm. wide, the other 2 broadly ovate-suborbicular, scarcely 0·3 mm. long, and almost as wide; ovary sessile, compressed, ovate, about 1 mm. long, 0·8 mm. wide; style absent; stigmas minute, penicillate. Fruit bluntly ovate, strongly compressed, about 1·8–2 mm. long, 0·8–1 mm. wide, pale brown, minutely papillose-granulose, enveloped by the persistent perianth.

HAB.: Waste and cultivated ground, roadsides; sea-level to 800 ft. alt.; fl. Dec.–April.

DISTR.: Divisions 1, 3–7. Widespread in Europe, also Mediterranean region, Western Asia, and as an adventive elsewhere.

1. Tsadha, 1928 or 1930, *Druce*.
3. Limassol, 1928 or 1930, *Druce*; Ayia Varvara, Stavrovouni, 1937, *Syngrassides* 1492 !
4. Famagusta, 1928 or 1930, *Druce*; Ayia Napa, 1928 or 1930, *Druce*; Dherinia, 1969, *A. Genneou* in ARI 1367 !
5. Knodhara, E. of Kythrea, 1932, *Syngrassides* 374 ! Ayios Seryios 1973, *P. Laukkonen* 23 !
6. Nicosia, 1928 or 1930, *Druce*; also, 1950, *Chapman* 461 !
7. Kyrenia Castle moat, 1944, *Casey* 267 ! Kyrenia, 1956, *G. E. Atherton* 823 !

NOTES: Some of the collectors remark that the plant is stingless, which is certainly not true in western Europe, where *U. urens* stings very sharply, though the effect seems to wear off more rapidly than with *U. dioica*.

2. **U. dioica** *L.*, Sp. Plant., ed. 1, 984 (1753); Unger et Kotschy, Die Insel Cypern, 219 (1865);
Boiss., Fl. Orient., 4: 1146 (1879); Holmboe, Veg. Cypr., 62 (1914); Post, Fl. Pal., ed. 2, 2:
510 (1933); Osorio-Tafall et Seraphim, List Vasc. Plants Cyprus, 29 (1973); Davis, Fl.
Turkey, 7: 635 (1982).

Rhizomatous perennial with tough, yellowish rhizomes; stems erect, 4-angled, subglabrous to densely hispid-pilose, usually armed with stinging bristles, 30–100 cm. high, generally unbranched; leaves ovate or broadly lanceolate, (1–) 3–15 cm. long, 0·8–6 (–8) cm. wide, dull green, sometimes rugose, subglabrous to densely hispid-pubescent, generally with stinging hairs, apex acute, acuminate or shortly caudate, margins coarsely serrate with blunt, acute or acuminate serrations, base broadly cuneate, rounded or cordate; petiole slender, 0·5–6 (–8) cm. long, generally hispid; stipules 4, free, herbaceous, narrowly oblong, 3–10 mm. long, 1·5–3 mm. wide; inflorescences dioecious or sometimes monoecious, the males lax, much branched, up to 10 cm. long, often exceeding the subtending petioles, the females generally more compact and dense; male flowers shortly pedicellate or subsessile; perianth-lobes 4, ovate, about 1 mm. long, 0·7 mm. wide, hispid or glabrous; anthers 4, shortly oblong, about 0·8 mm. long, 0·6 mm. wide; female flowers subsessile; perianth-lobes 4, hispid or glabrous, very unequal, internal pair broadly ovate, about 0·8 mm. long, 0·6 mm. wide, the external pair about 0·6 mm. long, 0·3 mm. wide; ovary strongly compressed, ovate, glabrous, 1–1·2 mm. long, 0·8–1 mm. wide; style absent; stigmas penicillate. Fruit ovate, compressed, pale brown, minutely granulose-verruculose, about 1·2–1·5 mm. long, 0·8–1·2 mm. wide.

ssp. **dioica**
   TYPE: "*in* Europae *ruderatis*".

Stems and leaves generally pubescent-hispidulous; leaves commonly with a conspicuously cordate base; inflorescences normally dioecious.

   HAB.: Cultivated and waste ground; sea-level to 2,000 ft. alt.; March–May.

   DISTR.: Divisions 4, 7, evidently rare. Widespread in Europe and Western Asia, also as an introduced weed almost throughout temperate regions of the world.

4. In gardens at Larnaca, *Kotschy*.
7. Pentadaktylos, 1880, *Sintenis & Rigo.*

   NOTES: In all probability the above records are referable to the following (*U. dioica* L. ssp. *cypria* Lindberg f.), but this cannot be ascertained in the absence of specimens.

ssp. **cypria** *Lindberg f.*, Iter Cypr., 12 (1946); Osorio-Tafall et Seraphim, List Vasc. Plants Cyprus, 29 (1973).
   TYPE: "M. Troodos, Milikouri prope monasterium Kykko, in umbrosis in cultis. Kambos, in ruderatis". (H, K!).

Stems subglabrous, with a few, scattered, stinging bristles; leaves subglabrous or with a few stinging bristles on the nerves of the undersurface, base of lamina rounded or shallowly cordate; inflorescences monoecious, the male flowers usually in the lower half of the inflorescence, the females above; perianth-lobes glabrous or subglabrous.

   HAB.: Waste ground or moist walls and field-terraces; 1,900–4,000 ft. alt.; fl. May–July.

   DISTR.: Divisions 1, 2, 6, rare. ? Endemic (see Notes).

1. Tsadha, 1905, *Holmboe* 738!
2. Moutoullas, 1937, *Syngrassides* 1554! Milikouri, 1939, *Lindberg f.* s.n.! Kambos, 1939, *Lindberg f.* s.n.; Pedhoulas, 1955, *Merton* 2364!
6. Koutraphas, 1981, *N. Arnold* s.n.!

   NOTES: Differs from typical *U. dioica* in its overall glabrousness and monoecious inflorescences, in which respects it answers to *U. dioica* L. ssp. *gracilis* (Aiton) Selander in Svensk. Bot. Tidskr., 41: 271 (1947) (*U. gracilis* Aiton, Hort. Kew., ed. 1, 3: 341; 1789) the American Stinging Nettle, which is widely distributed and indigenous in North America, where *U. dioica* ssp. *dioica* is evidently a recent introduction. The only apparent distinction between ssp. *cypria* (1946) and ssp. *gracilis* (1947) lies in the marginal toothing of the leaves, which tends to be bolder, sharper, more deltoid and more upcurved in ssp. *gracilis* than in ssp. *cypria*. But Nettles are variable in this respect, and it may be a mistake to attach too much significance to the distinction. It will be noted that ssp. *cypria* has nomenclatural priority over ssp. *gracilis*.

   Why a Nettle with New World affinities should grow in Cyprus is hard to say, though it is just possible that it may have been imported directly or indirectly into the island at some recent date. Clearly it is a rare plant in Cyprus, nor does it fit into any of the natural plant communities on the island, but it may, none the less, be indigenous.

   For a detailed consideration of the American Nettles, see I. J. Basset, C. W. Crompton and D. W. Woodland in Canad. Journ. Bot., 52: 503–516 (1974) and in Canad. Journ. Plant. Sci., 57: 491–498 (1977).

3. U. pilulifera *L.*, Sp. Plant., ed. 1, 983 (1753); Unger et Kotschy, Die Insel Cypern, 219 (1865); Boiss., Fl. Orient., 4: 1147 (1879); Holmboe, Veg. Cypr., 62 (1914); Post, Fl. Pal., ed. 2, 2: 510 (1933); Lindberg f., Iter Cypr., 12 (1946); Zohary, Fl. Palaest., 1: 40, t. 38 (1966); Osorio-Tafall et Seraphim, List Vasc. Plants Cyprus, 29 (1973); Davis, Fl. Turkey, 7: 634 (1982).
TYPE: "*in* Europa *australi*".

Erect annual 30–130 cm. high; stems branched or unbranched, angular, sulcate, subglabrous or thinly setose with stinging bristles; leaves broadly ovate, 3–10 cm. long, 1–8 cm. wide, dark green, papery, subglabrous or thinly setose on both surfaces, apex acute, often shortly caudate, base rounded, truncate or shallowly cordate, margins very coarsely, subacutely serrate with bold, narrowly deltoid, upcurved teeth; petiole slender, setose, 1–8 cm. long; stipules 4, free, bluntly oblong, membranous, about 4–5 mm. long, 2–4 mm. wide; inflorescence monoecious, consisting of several, laxly superimposed mixed male and female partial inflorescences; male flowers in slender branched panicles, generally longer than the subtending petiole, pedicels 0·5–1 mm. long, thinly setose; perianth-lobes 4, membranous, ovate-oblong, about 2 mm. long, 1·4 mm. wide, thinly setose, subacute or obtuse; filaments slender, glabrous, about 1·5 mm. long; anthers broadly oblong, about 1·2 mm. long, 1 mm. wide; female flowers forming a dense globose capitulum 0·7–1 cm. diam. at the apex of a slender peduncle 1–3 cm. long; perianth-lobes very unequal, the outer minute, green, subglabrous, less than 0·5 mm. long, 0·8 mm. wide, the inner cucullate, about 2 mm. long, 1·5 mm. wide, membranous, densely setulose towards apex, connate in pairs to form an inflated utricle about the fruit; stigma sessile, penicillate. Fruit ellipsoid, strongly compressed, about 3 mm. long, 2 mm. wide, mid-brown, distinctly verruculose.

HAB.: Cultivated fields and waste ground, roadsides, dry-stone terraces; sea-level to 2,000 ft. alt.; fl. Febr.–April.

DISTR.: Divisions 1, 3–7, often common; widely distributed in the Mediterranean region and Western Asia.

1. Ayios Yeoryios (Akamas), 1962, *Meikle*! Polis, 1962, *Meikle*!
3. Ayia Varvara, Stavrovouni, 1937, *Syngrassides* 1491! Kolossi, 1939, *Lindberg f.* s.n.; Paramali, 1964, *J. B. Suart* 208!
4. Larnaca, 1880, *Sintenis*; Cape Pyla, 1880, *Sintenis*.
5. Kythrea, 1880, *Sintenis & Rigo* 51! Syngrasis, 1962, *A. Genneou* in ARI 1312!
6. Dhiorios, 1962, *Meikle*! Nicosia, 1962, *Meikle*!
7. Pentadaktylos, 1880, *Sintenis & Rigo*; Buffavento, 1880, *Sintenis & Rigo*; Kantara, 1880, *Sintenis & Rigo*; near the Armenian Monastery, 1939, *Lindberg f.* s.n.; St. Hilarion, 1939, *Lindberg f.* s.n.; Kyrenia, 1949, *Casey* 347! Temblos, 1956, *G. E. Atherton* 1108! Tjiklos, 1973, *P. Laukkonen* 126!

NOTES: Probably common throughout the lowlands, but Nettles tend, for obvious reasons, to remain uncollected.

4. U. membranacea *Poir.* in Lam., Encycl. Méth., 4: 638 (1978); Boiss., Fl. Orient., 4: 1147 (1879); Holmboe, Veg. Cypr., 62 (1914); Davis, Fl. Turkey, 7: 634 (1982).
   U. dubia Forssk., Fl. Aegypt.-Arab., cxxi (1775); Zohary, Fl. Palaest., 1: 40, t. 37 (1966); Osorio-Tafall et Seraphim, List Vasc. Plants Cyprus, 29 (1973) nom. invalid.
   U. caudata Vahl, Symb. Bot., 2: 96 (1791); Post, Fl. Pal., ed. 2, 2: 511 (1933) non Burm. f. (1768) nom. illeg.
   TYPE: Algeria-Tunisia, "Je l'ai trouvée en Barbarie . . . dans le pays habité par les Merdass", *Poiret* (P).

Erect annual 30–50 cm. high; stems branched or unbranched, angular, glabrous or sparingly setose; leaves bright green, ovate, glabrous or thinly setose, 2·5–7 cm. long, 1–5 cm. wide, apex acute or shortly caudate, base rounded, margins coarsely but regularly serrate with upcurved teeth; petioles slender, 2–5 cm. long, glabrous or thinly hispid; stipules connate into 1 pair, ovate, membranous, 3–6 mm. long, 2–5 mm. wide; inflorescences monoecious, terminating in male flowers, or sometimes dioecious; male flowers arranged unilaterally along caudate, flattened, straight or scorpioid-curved membranous rhachises 8–12 (or more) cm. long, generally much exceeding the subtending leaves; pedicels very short; perianth-lobes 4, free almost to base, oblong, concave-cucullate, about 1 mm. long, 0·8 mm. wide, obtuse and blackish at apex; stamens 4; filaments slender, glabrous, about 2 mm. long; anthers broadly oblong, about 0·8 mm. long, and as wide or wider; female flowers in short, dense spikes or racemes 1–2·5 cm. long; peduncle slender, 0·5–1·5 cm. long; pedicels very short; perianth-lobes 4, very unequal, shortly strigillose, the outer ovate, subacute, about 0·8 mm. long, 0·5 mm. wide, the inner oblong, obtuse, about 2 mm. long, 1·5 mm. wide; ovary ovoid, compressed, about 0·8 mm. long, 0·5–0·6 mm. wide; style absent; stigmas minute, penicillate. Fruit ovate, strongly compressed, blunt, about 1 mm. long, 0·8 mm. wide, pale brown, minutely and densely verruculose.

HAB.: Waste ground and clefts of limestone scarps near the sea; sea-level to 50 ft. alt.; fl. March–April.

DISTR.: Divisions 4, 6, rare. Mediterranean region, and eastwards to Iraq; also Portugal, Azores.

4. Famagusta, among ruins and in the moat, 1905 and 1957, *Holmboe* 431; *Merton* 2892!
6. Near Liveras, in clefts of scarps by the sea, 1962, *Meikle* 2411!

2. PARIETARIA *L.*

Sp. Plant., ed. 1, 1052 (1753).
Gen. Plant., ed. 5, 471 (1754).

Annual or perennial herbs, without stinging hairs; leaves alternate, simple, entire, petiolate, generally without stipules; inflorescences axillary, cymose, clustered, bracteate, mostly polygamous; flowers small, greenish; perianth 4-lobed, campanulate or tubular, that of the male and her-maphrodite flowers sometimes elongating and becoming conspicuous after anthesis; stamens 4, inserted at the base of the perianth; filaments springing out elastically at maturity; ovary 1-locular, ovoid; style very short or absent; stigma penicillate. Achene enveloped by perianth (and sometimes by bracts), compressed, ovate, often shining.

About 10 species with a cosmopolitan distribution.

Plants perennial; leaves 1–7 cm. long - - - - - - - - **1. P. judaica**
Plants annual; leaves 0·3–2·5 cm. long:
  Bracts conspicuous, accrescent, reddish-brown and indurated in fruit, enveloping and
    exceeding the fruiting-perianth - - - - - - - - **2. P. cretica**
  Bracts inconspicuous, not accrescent, scarcely exceeding the fruiting perianth
                                   3. P. lusitanica

1. **P. judaica** *L.*, Fl. Palaest., 32 (1756), Sp. Plant., ed. 2, 1492 (1763); Boiss., Fl. Orient., 4: 1149 (1879); Post, Fl. Pal., ed. 2, 2: 511 (1933); Chapman, Cyprus Trees and Shrubs, 35 (1949); C. C. Townsend in Watsonia, 6: 368 (1968); Davis, Fl. Turkey, 7: 636 (1982).
  P. vulgaris Hill, British Herbal, 491 (1757); Lindberg f., Iter Cypr., 12 (1946) nom. invalid.
  P. diffusa Mert. et Koch in Röhlings Deutschl. Fl., ed. 3, 1: 827 (1823); Zohary, Fl. Palaest., 1: 41, t. 39 (1966); Osorio-Tafall et Seraphim, List Vasc. Plants Cyprus, 24 (1973).
  P. officinalis L. ssp. *diffusa* (Mert. et Koch) Arcang., Comp. Fl. Ital., ed. 1, 600 (1882).
  P. officinalis L. ssp. *judaica* (L.) Bég. in Nuov. Giorn. Bot. Ital., 15: 342 (1908), Fl. Padov., 249 (1910); Holmboe, Veg. Cypr., 62 (1914).
  [*P. erecta* (non Mert. et Koch) Sintenis in Oesterr. Bot. Zeitschr., 31: 194, 392 (1881).]
  [*P. officinalis* (non L.) Sintenis in Oesterr. Bot. Zeitschr., 32: 397 (1882).]
  [*P. officinalis* L. ssp. *erecta* (non P. *erecta* Mert. et Koch) Holmboe, Veg. Cypr., 62 (1914).]
TYPE: "Palaestina" (LINN).

Sprawling, pubescent perennial; stems 20–60 cm. long, somewhat angled, usually woody and much branched towards the base; leaves broadly ovate-elliptic, 1–7 cm. long, 0·6–3 cm. wide, dull green, apex acute or acuminate, base cuneate; nervation rather obscure; petiole very slender 0·5–2·5 cm. long; flowers in small, subsessile, axillary clusters, polygamous; bracts crowded, unequal, foliaceous, shortly connate at base, ovate or lanceolate, 3–5 (–8) mm. long, 1·5–2·5 mm. wide; male flowers with a submembranous, deeply 4-lobed, pubescent perianth, lobes ovate, about 2 mm. long, 1·3 mm. wide; stamens 4; filaments stout, transversely corrugated dorsally, about 1·5 mm. long; anthers pallid, about 1·3 mm. long and wide; female perianth shortly tubular, about 2 mm. long, 1·5 mm. wide, lobes free for a short distance, caudate, pilose; hermaphrodite perianth tubular, about 3 mm. long, 1·5 mm. wide, accrescent to 4 mm. long in fruit, lobes pilose, broadly deltoid, about 1 mm. long; ovary ovoid, compressed, about 0·8 mm. long, 0·5 mm. wide; stigma sessile, shortly penicillate. Fruit ovoid, compressed, shining black, about 1–1·2 mm. long, 1 mm. wide.

HAB.: Damp slopes and walls; in caves; by rocky springs and rivulets; occasionally on marshy ground amongst *Juncus*; sea-level to 3,500 ft. alt.; fl. Jan.–Dec.

DISTR.: Divisions 1–8. S. and W. Europe, Mediterranean region eastwards to Central Asia and Pakistan; Atlantic Islands.

1. Ktima, 1905, *Holmboe* 723, also, 1939, *Chapman* 389! Baths of Aphrodite, 1962, *Meikle* 2217!
2. Lefka, 1932, *Syngrassides* 248! Kakopetria, 1955, *Chiotellis* 442! Pedhoulas, 1966, *Merton* in ARI 52! Above Askas, 1979, *Edmondson & McClintock* E2912!
3. Limassol, 1956, *Merton* 528!
4. Larnaca, 1880, *Sintenis*, and, 1905, *Holmboe* 250.
5. Kythrea, 1880, *Sintenis & Rigo* 50!
6. Nicosia, 1939, *Lindberg f.* s.n., and, 1978, *Holub* s.n.!
7. Pentadaktylos, 1880, *Sintenis & Rigo*; Lapithos, 1939, *Lindberg f.* s.n., and, 1955, *Merton* 2250! also, 1966, *Merton* in ARI 31! Kyrenia, 1948–1973, *Casey* 91! 266! *G. E. Atherton* 106! *P. Laukkonen* 108! Lambousa, 1955, *G. E. Atherton* 344! Kambyli, 1957, *Merton* 2975!
8. Cape Andreas, 1880, *Sintenis & Rigo!*

2. P. cretica *L.*, Sp. Plant., ed. 1, 1052 (1753); Unger et Kotschy, Die Insel Cypern, 219 (1865); Boiss., Fl. Orient., 4: 1150 (1879); H. Stuart Thompson in Journ. Bot., 44: 338 (1906); Holmboe, Veg. Cypr., 62 (1914); Osorio-Tafall et Seraphim, List Vasc. Plants Cyprus, 29 (1973); Davis, Fl. Turkey, 7: 638 (1982).
TYPE: "*in* Creta".

Slender, sprawling annual; stems angular, reddish, pubescent with spreading hairs, usually much branched especially towards the base; leaves broadly ovate-elliptic, small, 3–15 mm. long, 2·5–12 mm. wide, pale green, sparingly pubescent above and below, apex obtuse, base cuneate; nervation obscure; petiole very slender 2–12 mm. long; flowers polygamous, subsessile, in very small, axillary clusters; bracts connate basally, unequally 5-lobed, accrescent and indurated in fruit, dorsal lobe linear, 2·5–3 mm. long, 0·3 mm. wide, 2 median lobes oblanceolate-spathulate, 3–4 mm. long, 1–5 mm. wide at anthesis, 4–5 mm. long, 1·5–2 mm. wide in fruit, the 2 inner lobes bluntly acuminate-deltoid, 1·5 mm. long, 0·8 mm. wide at anthesis, 2 mm. long, 1 mm. wide in fruit; perianth of male flowers cup-shaped, membranous, pubescent, deeply 4-lobed, the lobes ovate-acuminate, about 1·3 mm. long, 0·6 mm. wide; filaments glabrous, stout, 1·5 mm. long; anthers pallid, about 0·6 mm. long, 0·8 mm. wide; female flowers with 4, ovate-caudate, erect, connivent or connate lobes 1·8 mm. long, 0·6 mm. wide; ovary compressed-ellipsoid, about 1 mm. long, 0·6 mm. wide; stigmas sessile, penicillate; hermaphrodite flowers with 4, ovate-acuminate, erect, connivent and partly connate lobes 2 mm. long, 0·8 mm. wide, accrescent to 3 (or more) mm. long after anthesis. Fruit ellipsoid, compressed, about 1·2 mm. long, 0·8 mm. wide, shiny dark brown, enveloped until maturity in the persistent, indurated bracts.

HAB.: In rock clefts, and in sandy ground at the foot of rocks and cliffs; sea-level–?1,000 ft. alt.; fl. Febr.–April.

DISTR.: Divisions 1, 4, 7, rather rare. Sicily, Crete, Greece, Aegean Islands, Libya.

1. Skali, near Ayios Yeoryios (Akamas), 1962, *Meikle* 2101 !
4. Larnaca, 1880, *Sintenis*.
7. Kyrenia area ("north side of northern range"), 1862, *Kotschy* 443; also, Kyrenia, 1938, *C. H. Wyatt* 17 !

3. P. lusitanica *L.*, Sp. Plant., ed. 1, 1052 (1753); Boiss., Fl. Orient., 4: 1150 (1879); Post, Fl. Pal., ed. 2, 2: 512 (1933); Zohary, Fl. Palaest., 1: 42, t. 40 (1966); Davis, Fl. Turkey, 7: 638 (1982).
TYPE: "*in* Lusitania, Hispania".

Slender, ascending or sprawling, pubescent annual 8–25 (–30) cm. high; stems obscurely angular, sometimes reddish, often very much branched, especially towards the base; leaves broadly ovate-elliptic, very similar to those of *P. cretica* L., 3–25 mm. long, 2·5–15 mm. wide, subglabrous or thinly pubescent on both surfaces, apex obtuse, base cuneate; nervation obscure; petiole slender, 2–10 mm. long; flowers polygamous, subsessile in dense axillary clusters; bracts unequally 5-lobed as in *P. cretica*, but the lobes not accrescent or indurated after anthesis, the largest lobe oblanceolate, about 2 mm. long, 0·7 mm. wide, the others lanceolate or linear, 0·5–1·5 mm. long, 0·3–0·5 mm. wide; male flowers with a cup-shaped perianth, divided almost to the base into 4 ovate, pubescent lobes about 1 mm. long, 0·4 mm. wide; stamens 4; filaments glabrous, about 0·8 mm. long; anthers pallid, about 0·3 mm. long, 0·4 mm. wide; female flowers with 4 oblong subacute, brownish, erect or connivent perianth-lobes about 1 mm. long, 0·4 mm. wide; ovary broadly ovate-ellipsoid, compressed, about 0·8 mm. long, 0·7 mm. wide; stigmas sessile, shortly penicillate; hermaphrodite flowers scarce, similar to the female, but a little longer. Fruit broadly ovoid-ellipsoid, not much compressed, shining greenish-brown, about 1 mm. long, 0·8 mm. wide, enveloped by the persistent, connivent perianth-lobes.

HAB.: Walls and rock-crevices; 400–2,400 ft. alt.; fl. March–April.

DISTR.: Divisions 3, 4, 6, rare. S.W. Europe, Mediterranean region eastwards to Iran.

3. Pano Lefkara, 1978, *J. Holub* s.n.!
4. Ayios Antonios, Sotira, 1951, *Merton* 181 !
6. Nicosia, walls of Market Hall, 1978, *J. Holub* s.n.!

85. ULMACEAE

Trees and shrubs; leaves simple, generally alternate, often asymmetric; stipules scarious, caducous; flowers usually in cymose clusters, often appearing in advance of the leaves, hermaphrodite or unisexual; perianth sepaloid, campanulate, divided into 4–8 free, or partly united, imbricate, persistent lobes; stamens as many as perianth-lobes and opposite to them, or sometimes more numerous, inserted at the base of the perianth; filaments free, often exserted; anthers 2-thecous, dehiscing longitudinally; ovary 1 (–2)-locular (one of the loculi often abortive); styles 2, divergent, stigmatic on their inner face; ovule solitary, pendulous. Fruit a nut, samara or drupe; seed generally without endosperm; embryo straight or curved.

About 15 genera and 200 species, mostly in temperate regions of the northern hemisphere; several species of *Ulmus* and *Celtis* are valued for their timber.

Leaves serrate; perianth divided almost to base into 4–5 (or more) lobes or segments:
Fruit a samara; flowers all hermaphrodite, produced in clusters on year-old twigs
1. Ulmus
Fruit a drupe; flowers male and hermaphrodite (or female) solitary or in clusters on current year's growths - - - - - - - - - - - - - **2. Celtis**
Leaves sinuate-lobulate; perianth-segments united to about the middle; fruit a rugose drupe
3. Zelkova

1. ULMUS *L.*

Sp. Plant., ed. 1, 225 (1753).
Gen. Plant., ed. 5, 106 (1754).

Trees; leaves alternate, asymmetric and unequal at the base, commonly scabridulous; stipules membranous, caducous; flowers generally hermaphrodite, usually in clusters on the leafless year-old twigs; perianth campanulate, (3–) 5 (–9)-lobed; stamens as many as lobes and opposite to them; anthers purplish-red; ovary compressed, usually 1-locular; styles 2. Fruit a samara, winged all round, with an emarginate apex.

About 45 species, mostly in temperate regions of the northern hemisphere.

1. **U. canescens** *Melville* in Kew Bull., 1957: 499 (1957); Zohary, Fl. Palaest., 1: 36, t. 33 (1966). *U. minor* Mill. ssp. *canescens* (Melville) Browicz et Zieliński in Fragm. Fl. Geobot., 23: 145 (1977); Davis, Fl. Turkey, 7: 647 (1982).
[*U. campestris* (non L.) Holmboe, Veg. Cypr., 61 (1914); Post, Fl. Pal., ed. 2, 2: 517 (1933); Lindberg f., Iter Cypr., 11 (1946);? Chapman, Cyprus Trees and Shrubs, 34 (1949).]
[*U. minor* (non Mill.) Osorio-Tafall et Seraphim, List Vasc. Plants Cyprus, 28 (1973).]
TYPE: Thrace; Karakeuy, 1932, *H. G. Tedd* 806 (K!).

Deciduous tree to 10 m. high with fissured bark; crown broad, spreading; twigs terete, grey-brown, sometimes with corky excrescences, at first thinly pilose, becoming glabrous at the end of their first year; buds broadly ovoid, 3–5 mm. long, with dark brown, hairy, obtuse or rounded scales; leaves ovate-elliptic, with a markedly unequal, semi-cordate base, bright green, at first softly pubescent, becoming subglabrous with age, 4·5–9 cm. long, 2·5–5 cm. wide, apex shortly acute or subacute, margin rather bluntly and closely undulate-serrate; nervation prominent below, with 12–16 (–18) pairs of straight lateral nerves, generally with small tufts of greyish hairs in their axils with the midrib; petioles short, pubescent, usually less than 7 mm. long; stipules membranous, brownish, lingulate, about 6–7 mm. long, 2 mm. wide; flowers in tight, subglobose clusters, 5–6 mm. diam., on the leafless year-old twigs; perianth campanulate-infundibuliform, subsessile, 2·5–3 mm. long, 3–3·5 mm. wide at apex, lobes 5, brownish-red, obtuse, about 1 mm. long, tomentellous-ciliate; stamens 5; filaments glabrous, 3·5–4 mm. long; anthers oblong, purplish, about 1·7 mm. long, 1·5 mm. wide; ovary suborbicular, about 1 mm. diam., strongly compressed, glabrous; styles stout, divergent, about 1·3 mm. long, stigmatose internally. Fruit a thin, papery, obovate samara, about 2 cm. long, 1·2 cm. wide, with a pallid wing and a crimson central zone, apex deeply emarginate, with rounded, overlapping lobes.

HAB.: Moist rocky slopes and ravines; sometimes planted by roadsides; sea-level to 1,500 ft. alt.; fl. Febr.–March.

DISTR.: Divisions 1–3, 7. Mediterranean region from Italy to Palestine.
1. By seashore at Baths of Aphrodite, 1962, *Meikle* 2234!
2. Lythrodhonda, 1947, *W. A. Gordon* comm. *R. R. Waterer* s.n.!
3. Khoulou, 1905, *Holmboe* 753.
7. Lapithos, 1905, *Holmboe*, also 1939, *Lindberg f.* s.n.! 1948, *D. F. Davidson* comm. *R. R. Waterer* s.n.! Kambyli, 1954, *Merton* 1862!

U. GLABRA *Huds.*, Fl. Angl., ed. 1, 95 (1762) (*U. montana* With., Bot. Arr., ed. 2, 1: 259; 1787), a spreading tree, similar to *U. canescens* but with rather larger, scabridulous leaves, with the base of the longer side of the lamina overlapping the short petiole, sharply and closely serrate leaf-margins, and larger, more elliptical samaras, often more than 2 cm. long, is reported as experimentally planted high up on Troödos, "on the Hospital Hill and near the Army Service Corps Camp" by Bovill (Rep. Plant. Work Cyprus, 30; 1915). *U. glabra* has a wide distribution in Europe. It has not been collected recently in Cyprus, but one specimen, from Amiandos, collected in 1936 (*Syngrassides* 1303!) provisionally labelled "*U. canescens* × *?glabra*" by R. Melville, agrees with *U. glabra*, and may be a survivor of the experimental planting, as may also be *Chapman* 94 from Livadhi tou Pasha, cited under *U. campestris* in Cyprus Trees and Shrubs. This last-named specimen has not been examined.

2. CELTIS *L.*

Sp. Plant., ed. 1, 1044 (1753).
Gen. Plant., ed. 5, 467 (1754).

Trees or shrubs; leaves alternate, asymmetric with an oblique base, pinnately nerved or sometimes 3-nerved from the base; stipules free, membranous, caducous; flowers hermaphrodite and/or male, rarely female, solitary or clustered on the young, leafy shoots; perianth divided to the base into 5 (or 4) segments; stamens usually as many as perianth-segments, ultimately exserted; anthers yellow; receptacle often pilose; ovary sessile, compressed-ovoid or subglobose; styles 2. Fruit an ovoid or subglobose drupe, exocarp thinly fleshy, endocarp bony; seed without endosperm, or with very scanty endosperm; embryo curved.

About 80 species widely distributed in the northern hemisphere, and in S. Africa.

Leaves caudate-acuminate; softly pubescent below   -   -   -   -   - **1. C. australis**
Leaves shortly acute; glabrous below   -   -   -   -   -   -   - **2. C. tournefortii**

1. C. australis *L.*, Sp. Plant., ed. 1, 1043 (1753); Boiss., Fl. Orient., 4: 1156 (1879); Holmboe, Veg. Cypr., 61 (1914); Post, Fl. Pal., ed. 2, 2: 516 (1933); Lindberg f., Iter Cypr., 11 (1946); Chapman, Cyprus Trees and Shrubs, 34 (1949); Zohary, Fl. Palaest., 1: 36 (1966); Osorio-Tafall et Seraphim, List Vasc. Plants Cyprus, 29 (1973); Davis, Fl. Turkey, 7: 650 (1982). TYPE: "*in* Europa *australi & Africa citeriore*".

Tree to 20 m. high with a broad, bushy crown; bark smooth, greyish; twigs slender, at first densely pubescent or pilose, becoming brown and subglabrous, with small, pallid lenticels; leaves narrowly and obliquely ovate, 4–8 (–13) cm. long, 1·5–3 (–5) cm. wide, thinly adpressed-pilose and scabridulous above, softly pubescent below, apex caudate-acuminate, base unequally cuneate or rounded, margins sharply and rather irregularly upcurved-serrate; nervation depressed above, very prominent below, with 3–5 ascending primary nerves; petiole short, slender, usually about 5–6 mm. long; stipules absent except at the tips of the strongest shoots, linear, 10–13 mm. long, 1–1·5 mm. wide, membranous, thinly pilose, soon deciduous; flowers appearing with the leaves on young growths, the hermaphrodite usually solitary, axillary, the male sometimes in small (4–6-flowered) fascicles; pedicels slender, pubescent, 8–15 mm. long at anthesis, elongating in fruit; receptacle densely pilose; perianth-segments usually 5, oblong, 3–4 mm. long, 1·5–2 mm. wide, obtuse, thinly tomentellous with fimbriate or lacerate margins, deciduous after anthesis; filaments glabrous, 2–2·5 mm. long; anthers oblong, about 1·8 mm. long, 1–1·3 mm. wide; ovary compressed-ovoid, about 5–6 mm. long, 3–4 mm. wide, glabrous; styles stout, tomentose, about 4–5 mm. long, divergent. Fruit globose, with a subacute apex, thinly fleshy, black when ripe, about 8–10 mm. diam.

HAB.: By roadsides, often near old churches or monasteries; generally planted, but reproducing itself naturally; 1,000–4,000 ft. alt.; fl. March–April.

DISTR.: Divisions 1, 2. Southern Europe and Mediterranean region, but often planted.

1. Ayios Neophytos Monastery, 1939, *Lindberg f.* s.n.; and, 1967, *Economides* in ARI 1018!
2. Near Kykko Monastery, 1905, *Holmboe* 1027; also, 1937, *Syngrassides* 1600! and 1939, *Lindberg f.* s.n.! Milikouri, 1939, *Lindberg f.* s.n.; below Pedhoulas, 1955, *Merton* 2443!

2. C. tournefortii *Lam.*, Encycl. Méth., 4: 138 (1797); Boiss., Fl. Orient., 4: 1157 (1879) pro parte; Holmboe, Veg. Cypr., 61 (1914); Osorio-Tafall et Seraphim, List Vasc. Plants Cyprus, 29 (1973); Davis, Fl. Turkey, 7: 651 (1982).
　　C. tournefortii Lam. var. *glabrata* Boiss., Fl. Orient., 4: 1157 (1879) pro parte quoad *Bourgeau* 250 et *Kotschy* 393 nec *C. glabrata* Steven ex Planchon (non Sprengel; 1828) in Ann. Sci. Nat., ser. 3, 10: 285 (1848) nom. illeg.
TYPE: Cultivated from seeds collected by Tournefort in E. Turkey or Iraq (P).

Tree or shrub 2–8 m. high; twigs dark brown, glabrous, rather lustrous, obscurely lenticellate; leaves broadly and obliquely ovate, 3–8 cm. long, 2–5·5 cm. wide, generally rather thick and rigid, and somewhat glaucescent, glabrous above and below, save for a few hairs sometimes present in the axils of the primary nerves on the undersurface, apex shortly and abruptly acute, base rounded or subtruncate, the two halves of the lamina subequal or unequal, either joining the apex of the petiole at the same point or distinctly unequal, margins serrate with broad-based, rather flattened irregular teeth, sometimes very obscurely and irregularly serrate; nervation scarcely impressed above, obscurely reticulate below with 3–5 pairs of prominent lateral nerves; petiole short, stout, glabrous or almost glabrous, canaliculate above, 3–12 (–15) mm. long; stipules not seen, evidently very caducous; flowers not seen, probably very similar to those of *C. australis* L. Fruit globose, like a Pea, about 8–10 mm. diam., glabrous, without a subacute apex; pericarp at first yellow, later orange-brown; endocarp ivory-coloured, hard, almost smooth, with 2 (or more) obscure, longitudinal ribs.

HAB.: Dry calcareous hillsides; 2,000–2,500 ft. alt.; fl. March-April.

DISTR.: Division 2, very rare. Sicily, ?Crete, ?Greece, Turkey, Iraq, Iran.

2. Near the old church of Ayios Arkhangelos between Lefkara and Vavatsinia, July, 1905, *Holmboe* 1101! [Panayia, 18–22 June, 1913, *Haradjian* 899! See Notes].

NOTES: The taxonomy of *Celtis tournefortii* aggr. stands in need of revision, largely in consequence of Boissier's failure to realize that *C. glabrata* Steven ex Planchon (non *C. glabrata* Sprengel) is specifically distinct from *C. tournefortii* Lam. and much more closely allied to *C. caucasica* Willd., so that his var. *glabrata* is a mixture, partly of *C. glabrata* ("Tauriâ (Stev.) Iberiâ Caucasicâ (Hoh. sub *C. caucasicâ*)" and partly of typical *C. tournefortii* ("in Lyciâ (Bourg. 250!)" and "monte Gara Kurdistaniae (Ky. 393!)").
Within the *C. tournefortii* complex it is possible to distinguish at least two recognizable segregates: (1) leaves glabrous or pilose only at the base of the primary nerves on the underside of the lamina; petiole glabrous; leaf margins with rather few, broad-based, flattened serrations. (2) leaves pubescent at least along the primary nerves below; petiole pubescent; leaf margins sharply serrate with numerous, acute, forward-pointing teeth. Variant (1) is *C. tournefortii* Lam. var. *tournefortii* (*C. australis* L. var. *lutescens* Guss., *C. tournefortii* Lam. var. *laevis* Spach, *C. kotschyana* Steven, *C. tournefortii* Lam. var. *aetnensis* Torn.). Variant (2) is *C. aspera* (Audib. ex Spach) Steven (*C. tournefortii* Lam. var. *aspera* Audib. ex Spach, *C. asperrima* Lojac.).
Holmboe 1101 is a robust and rather abnormal, sterile growth of typical *C. tournefortii* (variant 1); *Haradjian* 899, also robust, sterile and abnormal, is glabrous (like variant 1), but with acute, forward-pointing marginal teeth, and very conspicuous and finely reticulate venation. It matches *Haradjian* 3018 from Akra Dağ (Mons Cassius), and may be simply an abnormal coppiced growth of typical *C. tournefortii*, but it cannot be accurately identified from available material.

3. ZELKOVA *Spach*

Ann. Sci. Nat., ser. 2, 15: 356 (1841) nom. cons.
Abelicea Reichb., Consp. Regn. Veg., 84 (1828);
Baillon, Hist. Plant., 6: 185 (1875).

Trees or large shrubs with smooth bark; leaves deciduous, shortly petiolate or subsessile, serrate or lobulate, penninerved; stipules narrow, submembranous, caducous; flowers hermaphrodite or male, sometimes female, appearing with the leaves, the males usually at the lower leaf-axils, sometimes in clusters, the hermaphrodite and female solitary in the upper axils; perianth campanulate, shortly 5 (4)-lobed; male flowers with 5 (4) stamens; filaments short; anthers yellow; hermaphrodite flowers with all or some of the stamens perfect, or occasionally all reduced to staminodes in the female flowers; ovary sessile; styles 2, stigmatose internally. Fruit a small, dry drupe, with a short apical beak; endocarp rather hard; seed compressed; embryo curved; cotyledons emarginate.

Six or seven species in the Caucasus, Mediterranean region and eastern Asia.

1. **Z. abelicea** (*Lam.*) *Boiss.*, Fl. Orient., 4: 1159 (1879); Osorio-Tafall et Seraphim, List Vasc. Plants Cyprus, 29 (1973).
    Quercus abelicea Lam., Encycl. Méth., 1: 725 (1785).
    Ulmus? abelicea Sm. in Sibth. et Sm., Fl. Graec. Prodr., 1: 172 (1806).
    Planera abelicea (Lam.) J. A. Schultes, Syst. Veg., 6: 304 (1820).
    Zelkova cretica Spach, Hist. Nat. Veg., 21: 121 (1842).
    Planera cretica (Spach) Kotschy in Unger et Kotschy, Die Insel Cypern, 217 (1865).
    Abelicea cretica (Spach) Holmboe, Veg. Cypr., 61 (1914).
    TYPE: Crete, "Cet arbre croît naturellement dans l'Isle de Candie" *Tournefort* (P).

Tree up to 15 m. high with a thick trunk, or often a stunted, gnarled bush; bark greyish; young shoots rather densely lanuginose; leaves oblong-ovate, 1·2–4·5 cm. long, 0·6–3·5 cm. wide, dark sublustrous green and glabrous or thinly adpressed-strigose above, paler (often drying greyish) and sub-glabrous or softly pubescent below, apex obtuse, margins bluntly, but often rather deeply and regularly pinnatilobed, with 3–6 pairs of lateral lobes; nervation slightly impressed above, prominent or obscure below; petiole short, usually less than 3 mm. long; stipules soon deciduous, linear, obtuse, 2–3 mm. long, 0·8 mm. wide; male flowers in dense clusters, subsessile or shortly pedicellate; perianth campanulate, 4-lobed to about half-way, 2–2·5 mm. long and about as wide, lobes about 1 mm. long and wide, membranous, subacute or obtuse, thinly lanuginose-ciliate; stamens 4; filaments 1·5–2 mm. long, glabrous; anthers shortly oblong, about 1 mm. long and almost as wide; hermaphrodite flowers solitary, subsessile; perianth scarcely 1·5 mm. long, divided to about half-way into 4 broad blunt lobes; stamens 4; anthers as in male flowers, but filaments scarcely 0·5 mm. long; ovary sessile, subglobose, compressed, about 2·3 mm. diam., glabrous; styles stout, about 0·8 mm. long. Fruit subglobose, about 5–6 mm. diam., irregularly and deeply rugose, with bulges and longitudinal furrows (? derived from the persistent, incrassate perianth); endocarp pale brown, minutely alveolate.

HAB.: Calcareous mountainsides; ?1,000–2,000 ft. alt.; fl. April–May.

DISTR.: ?Division 7. Crete.

7. "Auf der Nordseite der Gebirge gegen Melandrina [about 3 miles E. of Ayios Amvrosios]", 19 April, 1862, *Kotschy* 503 (see Notes).

NOTES: No one has since collected *Zelkova abelicea* in Cyprus, nor have I been able to trace *Kotschy* 503. In the circumstances one must suspect a misidentification, with the cited specimens now filed away under a different name. There is, however, still a faint possibility that the record may be proved correct. It appears that Kotschy was travelling from Melandrina Monastery to Antiphonitis Church when he found the plant.

85a. CANNABACEAE

Erect or climbing annuals or perennials, without latex; leaves alternate or opposite, simple or palmately compound; stipules free; flowers unisexual (dioecious or sometimes monoecious), axillary, the males generally in panicles, pedicellate, perianth 5-partite, segments imbricate; stamens 5, filaments short, anthers erect in bud, longer than filaments, 2-thecous, dehiscing by longitudinal slits; rudimentary ovary absent; female flowers sessile, clustered or strobilate, often with large, conspicuous, imbricate bracts; perianth membranous, entire, enveloping the 1-locular ovary; style 2-partite, branches filiform, stigmatose internally, often caducous; ovule solitary, pendulous. Fruit an achene enveloped by the persistent perianth; seed with fleshy endosperm, and a curved or spirally twisted embryo.

Two genera widely distributed in North Temperate regions; the infructescences of the Hop (*Humulus* L.) are used to flavour beer, while Hemp (*Cannabis* L.) is well-known as a fibre-plant and as the source of an hypnotic. *Cannabis sativa* L., Sp. Plant., ed. 1, 1027 (1753); Gaudry, Recherches Sci. en Orient, 168 (1855); Unger et Kotschy, Die Insel Cypern, 219 (1865); Small & Cronquist in Taxon, 25: 405–435 (1976), an erect annual, 2–4 (–5) m. high, with strong-smelling, resinous shoots, and resinous, alternate leaves palmately divided into 5–7, narrowly lanceolate-acuminate, sharply serrate leaflets, is occasionally cultivated in Cyprus (Govt. Experimental Farm, 1935, *Syngrassides* 544!), but is of minimal importance in the economy of the island. It is a common adventive in many parts of Europe, and may occur here and there in Cyprus as a weed. The relatively large (3–5 mm. long), compressed-ovoid, pale brown seeds are a common ingredient of seed mixtures sold for cage-birds; they are also a source of oil, used as a substitute for, or additive to, linseed-oil, and in the manufacture of soap.

86. MORACEAE

Trees and shrubs, rarely herbs, generally with latex; leaves alternate or rarely opposite, pinnately or palmately nerved, simple; stipules 2, sometimes connate, often caducous, leaving an amplexicaul scar; inflorescences dioecious or monoecious, cymose, often forming spikes or heads, or flowers sometimes enclosed within a fleshy, accrescent receptacle (*syconium*); flowers small, actinomorphic; perianth 4 (3)-partite, with valvate or imbricate segments, sometimes wanting or fused with receptacle; male flowers with 4 or 2 stamens opposite the perianth-segments; filaments inflexed or straight in bud; anthers 2-thecous, dehiscing longitudinally; rudimentary ovary present or absent; female flowers with a 1 (–2)-locular ovary; ovule solitary, generally pendulous from apex of ovary; styles 2, usually slender. Fruit an achene or nut, sometimes surrounded by the fleshy, accrescent perianth, sometimes compound, consisting of clustered drupe-lets, or individual fruits united into a syncarp or syconium.

About 70 genera and 1,500 species widely distributed in tropical and subtropical regions, but rare in temperate regions.

Flowers enclosed in a hollow receptacle (*syconium*); stipules connate, caducous, leaving a
  conspicuous, amplexicaul or circular scar - - - - - - - - **1. Ficus**
Flowers in spikes, racemes or heads; stipules free, not leaving an amplexicaul or circular scar:
  Leaves entire; twigs often spinose; fruit 10–14 cm. diam., resembling an Orange when ripe
    MACLURA (p. 1466)
  Leaves toothed or lobed; fruit less than 4 cm. diam.:
    Fruit ovoid or shortly cylindrical, juicy, white, pink, purple or black when ripe, resembling
      a Blackberry (*Rubus*) - - - - - - - - MORUS (p. 1465)
    Fruit globose, orange or reddish, rather dry - - - - BROUSSONETIA (p. 1465)

BROUSSONETIA PAPYRIFERA (*L.*) *Vent.*, Tabl. Regne Végétal, 3: 548 (1799)
(*Morus papyrifera* L., Sp. Plant., ed. 1, 986; 1753), Paper Mulberry, a large
shrub or round-headed tree up to 16 m. high; twigs grey-brown, pubescent,
pithy; buds with 2–3 outer scales; leaves ovate or cordate, sometimes 3-
lobed, 9–20 cm. long, 6–16 cm. wide, softly pubescent on both surfaces, apex
acute, margins shortly dentate; petiole stout, 2–4 (or more) cm. long;
stipules conspicuous, oblong-acute; inflorescences dioecious, the male
flowers in dense tail-like catkins, commonly 10 cm. long, 1 cm. wide, the
females in dense, globose, shortly pedunculate clusters, 0·8–1·2 cm. diam.,
with long, purple filiform styles. Fruit a reddish globose syncarp 2–2·5 cm.
diam., consisting of numerous fused drupelets, each with a subglobose stone
about 2 mm. diam.
  Similar to the Mulberry in general appearance. The fruit is stony and
inedible,but the bark is (or was) widely used in China and Japan for making paper, and in the Pacific islands for making cloth. The tree is not
uncommonly planted as an ornamental in Cyprus, and is recorded from
Larnaca by Frangos (Cypr. Agric. Journ., 18: 86; 1923) and from the
Municipal Gardens, Nicosia (*Meikle* 4087 !). *Broussonetia papyrifera* is a
native of China and Japan.

MORUS *L.*
Sp. Plant., ed. 1, 986 (1753).
Gen. Plant., ed. 5, 424 (1754).

Deciduous trees or shrubs; buds with 3–6 imbricate scales; leaves
undivided or lobed, 3–5-nerved from base, margins serrate or dentate;
petiole generally long; stipules small, lanceolate, deciduous; inflorescences
monoecious or dioecious, spicate, axillary, solitary, the males long,
cylindric, catkin-like, the females often shorter; perianth 4-partite; stamens
4, filaments inflexed in bud; rudimentary ovary turbinate; ovary ovoid or
subglobose; style central, divided into 2 slender branches. Fruit an ovoid,
compressed achene, enveloped in the succulent, accrescent perianth, the
individual fruits connate, forming a fleshy ovoid or cylindrical syncarp.
  About 12 species, chiefly in temperate regions of the northern hemisphere.
Two species are cultivated in Cyprus.

MORUS ALBA *L.*, Sp. Plant., ed. 1, 986 (1753); Hume in Walpole, Mem.
Europ. Asiatic Turkey, 253 (1816); Unger et Kotschy, Die Insel Cypern, 218
(1865); Osorio-Tafall et Seraphim, List Vasc. Plants Cyprus, 29 (1973);
Davis, Fl. Turkey, 7: 641 (1982).
  A deciduous tree up to 15 m. high, with undivided or 3–5-lobed leaves,
6–16 cm. long, 4–12 cm. wide, base shallowly cordate, margins bluntly
serrate; undersurface subglabrous or very sparsely pilose at the base of the
nerves; male inflorescences rather lax, 1–2 cm. long, female inflorescences
oblong, about 1 cm. long, with slender peduncles 0·8–1 cm. long. Fruit
fleshy, insipid, whitish, pink or purple when ripe.
  The White Mulberry, formerly cultivated all over the island as a source of

food for silkworms, and still fairly frequent, though the silk industry is dead. Recorded from (3) Limassol, 1801, *Hume*; (4) Larnaca, 1932, *Nattrass* 805; Famagusta, 1933, *Nattrass* 286; (5) Nisou, 1931, *Nattrass* 696; Syngrasis, 1936, *Syngrassides* 996! (6) Nicosia, 1931, *Nattrass* 53, 715, also, 1934, *Syngrassides* 1304! 1305! 1306! Myrtou, 1934, *Syngrassides* 1351! 1352! 1353! (7) Lapithos, 1931, *Nattrass* 270; Kyrenia, 1931, *Nattrass* 132; Kazaphani, 1937, *Syngrassides* 1456!

Morus alba is a native of China.

MORUS NIGRA *L.*, Sp. Plant., ed. 1, 986 (1753); Unger et Kotschy, Die Insel Cypern, 218 (1865); R. M. Nattrass, First List Cyprus Fungi, 39 (1937); Osorio-Tafall et Seraphim, List Vasc. Plants Cyprus, 29 (1973); Davis, Fl. Turkey, 7: 642 (1982).

? *M. rubra* (non L.) Hume in Walpole, Mem. Europ. Asiatic Turkey, 253 (1817); Osorio-Tafall et Seraphim, List Vasc. Plants Cyprus, 29 (1973).

A rugged, spreading tree to about 10 m. high; leaves cordate, 5–15 cm. long, 4·5–13 cm. wide, scabridulous above, thinly pilose below, apex shortly acute, base deeply cordate, margins rather bluntly serrate or dentate; petiole shorter and stouter than in *M. alba*; male and female inflorescences subsessile or very shortly pedunculate, oblong, usually less than 2 cm. long, 0·8–1 cm. wide. Fruit very fleshy and juicy, refreshingly acid, dark blackish-crimson when fully ripe.

The Common Mulberry, grown partly for its luscious fruits and partly for its picturesque, rugged mode of growth; the leaves can be used to feed silkworms, but are said to result in a deterioration in the quality of the silk produced. It is less frequently seen in Cyprus than *M. alba*, but Syngrassides notes that it has been seen in several villages. Specimens have been collected at Nicosia (6) in 1935 (*Nattrass* 715) and 1936 (*Syngrassides* 1340!). Hume's record for *M. rubra* L., a N. American species, from Limassol (3) is almost certainly an error. His plant was most probably *M. nigra* which he does not mention. The record for *M. rubra* is repeated in Osorio-Tafall et Seraphim, List Vasc. Plants Cyprus.

MACLURA POMIFERA (*Rafin.*) *C. K. Schneider*, Illustr. Handb. Laubholzk., 1: 806 (1906); *Ioxylon pomiferum* Rafin. in Amer. Monthly Mag., 2: 118 (1817).

Maclura aurantiaca Nuttall, Gen. N. Amer. Plants, 2: 233 (1818); Bovill, Rep. Plant. Work Cyprus, 14 (1915); Frangos in Cypr. Agric. Journ., 18: 88 (1923).

A tree to 7 m. high, with deeply sulcate bark; twigs pale brown, glabrous, often armed with stout axillary spines; leaves ovate, 5–12 cm. long, 1·5–6 cm. wide, entire, acute or bluntly acuminate, rather lustrous green above, subglabrous or thinly pilose below; petioles 3–5 cm. long; inflorescences dioecious; male flowers in condensed clusters or racemes; peduncles slender, tomentellous, 1–2 cm. long; pedicels 1–3 mm. long; perianth thinly hairy, 4-lobed; stamens 4, exserted; female flowers sessile in dense, globose, shortly pedunculate heads; styles very long, commonly exceeding 1 cm., filiform, conspicuous. Fruit a large, subglobose, verruculose, orange syncarp, 10–14 cm. diam.; seeds numerous, compressed-ovoid, about 8 mm. long, 5 mm. wide.

The Osage Orange, a native of the southern states of the U.S.A., sometimes planted as an ornamental in the Mediterranean region, or used for hedging. The latex-bearing fruits are attractive but inedible. There are records for *Maclura pomifera* from (2) Linou (*Frangos*), (5) Salamis plantation (*Bovill*) and from (6) Nicosia (*Frangos*).

1. FICUS *L.*

Sp. Plant., ed. 1, 1059 (1753).
Gen. Plant., ed. 5, 482 (1754).
E. J. H. Corner in Gardens Bull. Singapore, 21: 186 pp. (1965).

Trees, shrubs and climbers, sometimes epiphytic, lactiferous; leaves generally alternate, simple, mostly entire, sometimes toothed or lobed, often with the undersurface conspicuously and elegantly reticulate-veined; stipules connate, enveloping the bud, caducous, leaving a conspicuous, amplexicaul scar; inflorescences monoecious or rarely dioecious, consisting of numerous minute flowers enclosed within a hollow receptacle with a small apical aperture (ostiole) furnished with spreading or inflexed, scale-like bracts; male and female flowers intermixed or in distinct zones, bracteoles generally present, sometimes longer than the flowers; male flowers with a 2–6-lobed perianth (rarely reduced to one scale), lobes imbricate; stamens 1–2, rarely 3–6; filaments straight; rudimentary ovary absent; female flowers usually with smaller perianth-lobes, or the perianth obsolete or absent; style excentric after fertilization; stigma peltate or filiform. Fruit a fleshy *syconium*; seeds numerous, small; endosperm often scanty; embryo curved; cotyledons equal or unequal.

About 800 species widely distributed in tropical and subtropical regions.

1. **F. carica** *L.*, Sp. Plant., ed. 1, 1059 (1753); Unger et Kotschy, Die Insel Cypern, 218 (1865); Boiss., Fl. Orient., 4: 1154 (1879); Holmboe, Veg. Cypr., 62 (1914); Post, Fl. Pal., ed. 2, 2: 515 (1933); Chapman, Cyprus Trees and Shrubs, 35 (1949); Zohary, Fl. Palaest., 1: 37 (1966); Osorio-Tafall et Seraphim, List Vasc. Plants Cyprus, 29 (1973); Davis, Fl. Turkey, 7: 643 (1982).

F. carica L. f. *genuina* Boiss., Fl. Orient., 4: 1154 (1879); Holmboe, Veg. Cypr., 62 (1914).
TYPE: "*in* Europa *australi*, Asia".

Large spreading shrub, or small tree; bark smooth, ashy-grey; twigs stout, terete, at first puberulous, soon glabrous, with prominent leaf- and stipule-scars, exuding copious white latex when wounded; leaves deciduous, very variable in size and shape, 6–20 (–30) cm. long, 5–18 cm. wide, or sometimes wider than long, usually broadly ovate or suborbicular and deeply 3–5-lobed, greyish-green, scabridulous above, pubescent below, base shallowly cordate, lobes palmate, broadly or narrowly oblong, obtuse with crenate or bluntly serrate margins; petiole stout, 2–7 cm. long, subglabrous or pubescent; stipules usually glabrous, caducous; inflorescence monoecious (or female only in some cultivated Figs), enclosed in a solitary, turbinate or subglobose receptacle, arising in the axils of the leaves towards the tips of the shoots, and shortly stipitate or subsessile; syconium 2–3 cm. long, 1·5–2·5 cm. wide in wild plants at anthesis, often strongly accrescent in fruit, glabrous or puberulous, fleshy; stipe stout, 4–10 mm. long, with 3, overlapping, semiorbicular bracts 2–4 mm. diam. at its apex, ostiole rather large, with numerous, imbricate, suborbicular, peripheral and internal, scale-bracts, about 1–1·5 mm. diam.; bracteoles membranous, narrowly oblong-acuminate, about 2–3 mm. long, 0·8–1 mm. wide; flowers minute, translucent, the males just below the ostiole, the females lower down the syconium; pedicels 1–1·5 mm. long; male flowers with 3–5, lanceolate-acuminate perianth-segments about 1·5 mm. long, 0·6 mm. wide; stamens (1–) 3 (–6); anthers yellow, shortly exserted; female flowers usually with 5 narrowly lanceolate-acuminate perianth-segments about 1 mm. long, 0·4 mm. wide; ovary strongly compressed, about 1 mm. long, 0·8 mm. wide; style tapering upwards, excentric, about 2 mm. long. Fruit soft and fleshy, often purplish, containing numerous pale brown, subglobose achenes about 1·5 mm. diam.

HAB.: In rock-fissures by springs and streams; sea-level to 1,000 ft. alt.; fl. March–April.

DISTR.: Divisions 1, 2, 5; apparently native in the eastern Mediterranean region and eastwards to Iran, but often cultivated.

1. Baths of Aphrodite, 1905, *Holmboe* (see Veg. Cypr., 214) and, 1962, *Meikle*!
2. Near Makheras Monastery, 1905, *Holmboe* 1119.
5. Athalassa, 1932, *Syngrassides* 412! Syngrasis, 1937, *Syngrassides* 1644!

NOTES: Rarely collected; Chapman (op. cit., 34) says "frequently found wild up to 900 m.". As with the Carob and the Olive, a study of wild and naturalized Figs would be of considerable scientific interest. The occurrence of *Ficus carica* in the tufa deposits of the Kyrenia Pass (see Holmboe, Veg. Cypr., 199, fig. 65) indicates that the species is truly indigenous, and not simply naturalized as some have suspected.

Ficus carica L. is the only Fig which can be regarded as part of the Cyprus flora. Numerous exotic Figs are to be seen in gardens and parks in many parts of Cyprus; few have been collected or identified, but the following can be said, with some certainty, to occur on the island:

F. SYCOMORUS *L.*, Sp. Plant., ed. 1, 1059 (1753); Druce in Rep. B.E.C., 9: 471 (1931); Lindberg f., Iter Cypr., 11 (1946); Zohary, Fl. Palaest., 1: 38 (1966) (*Sycomorus antiquorum* Gasparrini ex Miq. in Hooker's London Journ. Bot., 7: 109 (1848); Unger et Kotschy, Die Insel Cypern, 218; 1865). A tree 8–20 m. high, with a spreading crown, thinly pilose shoots and obtuse, oblong, entire leaves 3·5–20 cm. long, 2–15 cm. wide. The small Figs are borne in short branched clusters (like Grapes) on the trunks and older branches; they are individually broadly ovoid or depressed-globose, about 2 cm. diam., and yellowish when ripe.

A native of Ethiopia and tropical East Africa, long established as a cultivated tree in the eastern Mediterranean region and, in Cyprus, recorded from gardens at Larnaca (*Kotschy, Sintenis & Rigo* 983!), Nicosia (*Kotschy*), Famagusta (*Druce*) and Limassol (*Kotschy, Lindberg f.*). The fruits are edible but insipid.

For a detailed note on the distribution and biology of *F. sycomorus* see Galil, Stein & Horovitz in Garden's Bull. Singapore, 29: 191–205 (1977).

F. MICROCARPA *L.f.*, Suppl. Plant., 442 (1781); Corner in Garden's Bull. Singapore, 21: 22 (1965) (*F. nitida* auctt. non Thunb.; *F. benjamina* auctt. non L.; *F. retusa* auctt. non L.). An erect shrub or small tree, said to attain 12 m. or more in tropical Asia; branches greyish, rather slender; buds glabrous, narrowly rostrate-acuminate; leaves oblong-elliptic, 3–10 cm. long, 2–5 cm. wide, glabrous, shining dark green above, paler below; apex rounded or shortly and bluntly cuspidate; petiole short, usually less than 8 mm. long; nervation rather obscure, the lateral nerves running into a distinct, intramarginal nerve. Fruits small, about 8 mm. diam., globose or depressed-globose, glabrous, yellowish or brownish when ripe.

A native of tropical Asia, ranging from India eastwards to Queensland, now popular as a house plant in Europe and North America, and frequently planted outdoors in Cyprus. Specimens have been collected in Nicosia (*Meikle* s.n.).

F. DRUPACEA *Thunb.*, Diss. Ficus Gen., 11 (1786); Corner in Garden's Bull. Singapore, 21: 13 (1965) (*F. vidaliana* Warb. in Perkins, Fragm. Fl. Philipp., 197 (1905); ?*F. vegalensis* Frangos in Cypr. Agric. Journ., 18: 87; 1923).

A robust tree to 20 m. or more; branches stout, pale grey-brown; buds densely covered with rufous hairs; leaves broadly ovate-elliptic, coriaceous, 18–24 cm. long, 9–14 cm. wide, glabrous or pubescent below, apex shortly cuspidate, margins narrowly recurved; nervation very prominent below, with 14–16 lateral nerves interconnected by anastomosing secondary nervation, and joined apically by a series of well-marked intramarginal loops; petiole stout, 3–5 cm. long. Fruits oblong (roughly Plum-shaped) 2–4 cm. long, 1·5–2·5 cm. wide, glabrous (var. *drupacea*) or pubescent (var. *pubescens* (Roth) Corner), orange or reddish when ripe.

A native of tropical S.E. Asia, from Burma through the Philippines and New Guinea to Queensland, seen as a fine, lofty tree in a garden near the centre of Nicosia, and probably planted elsewhere. The *"Ficus vegalensis"* of Frangos, noted as growing at the Commercial Club, Larnaca, may well be *F. drupacea*.

F. MACROPHYLLA *Desf. ex Pers.* (Syn. Plant., 2: 609; 1807), a robust tree with oblong-ovate, obtuse or shortly acuminate, coriaceous leaves 10–23 cm. long, 7–12 cm. wide; petioles 10–15 cm. long, and subglobose, long-stipitate, yellowish-flecked fruits 1–2 cm. diam., is commonly grown as a shade tree in the Mediterranean region, and may occur in Cyprus, but has not been collected or recorded. It is a native of Australia and Lord Howe Island, and is popularly named the Moreton Bay Fig.

F. ELASTICA *Roxb. ex Hornem.* (Suppl. Hort. Bot. Hafn., 7; 1819), the India Rubber Tree, is usually seen as a pot-plant, but is said to reach a height of 60 m. in its native S.E. Asia; it has glossy, coriaceous, elliptical leaves, often 25 cm. long and 15 cm. wide, a rather obscure nervation of numerous, closely parallel lateral nerves, and small, sessile, oblong fruits. It is said to be more tender than *F. macrophylla*, but may be grown outdoors in Cyprus, as in Egypt and elsewhere in the eastern Mediterranean.

87. PLATANACEAE

L. E. Boothroyd in Amer. Journ. Bot., 17: 678–692 (1930).

Deciduous, anemophilous trees; bark generally exfoliating in large flakes or plates; young shoots and leaves often clothed with a dense indumentum of branched or stellate hairs; buds covered by the persistent, hollow, dilated petiole-bases; leaves alternate, simple, palmately 3–7-lobed, the lobes often irregularly serrate or dentate; petiole usually long; stipules large, foliaceous, encircling the stems; inflorescences monoecious, the males and females usually on distinct peduncles, commonly on separate branches or separate parts of branches, both forming dense, many-flowered, globose, sessile heads along or at the tip of a pendulous peduncle; bracts and bracteoles present; male flowers with (or rarely without) a minute, cupuliform, shallowly lobed, pilose calyx; petals minute, glabrous, strap-shaped, subglobose, truncate or irregularly lobed, alternating with the stamens, sometimes fused at the base; stamens 3–5, filaments very short; anthers long, 2-thecous, dehiscing longitudinally, connectives dilated, forming a peltate roof over the unopened anthers; staminodes generally present; rudimentary ovary present or absent; female flowers apocarpous, with 5–9 pistils arranged in 2–3 whorls; calyx as in male flowers; petals generally absent, occasionally strap-shaped; staminodes often in a whorl at the base of the pistils; ovary superior, 1-carpellate, 1-locular; ovules 1–2, pendulous; style long and slender, persistent, stigmatiferous for the greater part of its inner surface. Fruit a 1-seeded, fusiform, 4-angled achene (or follicle) surrounded with bristly hairs; seed with scanty endosperm and a straight embryo; cotyledons linear, often unequal.

One genus, comprising about 8 species (the figure is disputed), widespread in north and central America, also in S.E. Europe and western Asia as far east as India.

1. PLATANUS L.

Sp. Plant., ed. 1, 999 (1753).
Gen. Plant., ed. 5, 433 (1754).

Description that of the family.

1. **P. orientalis** L., Sp. Plant., ed. 1, 999 (1753); Poech, Enum. Plant. Ins. Cypr., 12 (1842); Kotschy in Oesterr. Bot. Zeitschr., 12: 278 (1862); Unger et Kotschy, Die Insel Cypern, 220 (1865); Boiss., Fl. Orient., 4: 1161 (1879); Holmboe, Veg. Cypr., 99 (1914); Post, Fl. Pal., ed. 2, 2: 518 (1933); Lindberg f., Iter Cypr., 12 (1946); Chapman, Cyprus Trees and Shrubs, 40 (1949); Zohary, Fl. Palaest., 2: 1, t. 1 (1972); Osorio-Tafall et Seraphim, List Vasc. Plants Cyprus, 51 (1973); Davis, Fl. Turkey, 7: 656 (1982).

P. orientalis L. var. *insularis* A. DC. in DC., Prodr., 16(2): 159 (1864).

TYPE: "*in* Asia, Tauro, Macedonia, Atho, Lemno, Creta, *locis riguis*".

Robust tree to 20 m. high, with a massive trunk and spreading crown; bark exfoliating in large flakes to expose a pallid undersurface; shoots and leaves at first covered with long silky hairs, soon becoming glabrous; twigs mid-brown, terete, minutely lenticellate; leaves coriaceous, 9–20 cm. long, and about as wide, palmately lobed for two-thirds their length into (3–) 5 (–7) narrowly oblong, acuminate, sub-entire or irregularly and coarsely serrate or sharply lobulate lobes; nervation obscure above, rather prominent and finely reticulate below; petiole 2–6 cm. long, terete, sometimes densely stellate-tomentose when young; stipules foliaceous, deeply dentate or lobulate, about 1·5–2·5 cm. diam., encircling the petiole-bases of young, leading shoots, soon deciduous; male and female inflorescences forming 3–6 globose, sessile or stipitate heads ranged along a peduncle 3–8 cm. long at anthesis, the female peduncle elongating to 14 cm. or more, and becoming pendulous in fruit; peduncle at first stellate-pilose, becoming glabrous with age; flower-heads 5–10 mm. diam. at anthesis, the female accrescent to 20–25 mm. diam. in fruit; male flowers with an obsolete, pilose calyx and 4, blunt, fleshy, scale-like fused petals about 0·5 mm. long; stamens usually united in groups of 4; filaments less than 0·2 mm. long; anthers yellow, about 1 mm. long, 0·5 mm. wide, crowned with a flat, shortly pilose cap of fused connectives; female flowers in groups of 5–6; petals absent; staminodes broadly clavate-stipitate, about 0·8 mm. long; pistils sessile, ovary narrowly ovoid, about 1 mm. long, 0·4 mm. wide, shortly pilose; style 3·5 mm. long, stigmatiferous for most of its length. Achene fusiform, about 10 mm. long, 2 mm. wide near the abruptly domed, pubescent apex, tapering to a basal coma of pale brown, multicellular bristles 8–11 mm. long, sides of achene glabrous or pilose, longitudinally sulcate, apex crowned with the persistent style.

HAB.: Usually by streams or springs; sea-level to 5,000 ft. alt., often planted; fl. April–May.

DISTR.: Divisions 1–3, 6, 7. S.E. Europe and western Asia, east to Kashmir, but planted in many areas, both within and outside its natural range.

1. Khrysokhou, 1862, *Kotschy.*
2. Prodhromos, 1862, *Kotschy* 757, also, 1955, *G. E. Atherton* 545! Vavatsinia, 1905, *Holmboe* 1109; below Troödos, 1908, *Clement Reid* s.n.! Stavros tis Psokas, 1936, *Nattrass* 719, and, 1939, *Lindberg f.* s.n.; Platania, 1936, *Syngrassides* 1145! Kryos Potamos, 3,600 ft. alt., *Kennedy* 532! Pelendria, 1973, *P. Laukkonen* 390!
3. Episkopi, 1862, *Kotschy*; Limassol, ? 1905, *Michaelides.*
6. Nicosia, 1931, *Nattrass* 103.

7. Kyrenia Pass, 1939, *Lindberg f.* s.n. ! Kephalovrysi, Kythrea, 1954, *M. Pallis* s.n. ! Karavas, 1955, *Merton* 2184 ! Lapithos, 1955, *G. E. Atherton* 681 !

NOTES: Several authors refer Cretan and Cyprian specimens to *P. orientalis* L. var. *insularis* A. DC., distinguishing this variety from typical *P. orientalis* by its deeply lobed leaves, the lobes extending beyond half the total length of the leaf, with the mid-lobe commonly less than 2·5 cm. wide at its base. This distinction holds good to some degree if one compares the majority of Cyprus specimens, with the majority of specimens from the Greek mainland, but exceptions would not be difficult to find, and might easily be multiplied if *P. orientalis* were more frequently collected. Nor am I satisfied that the Cyprus tree is distinct from Linnaean type material of *P. orientalis*. The infraspecific analysis proposed by A. de Candolle is to some extent vitiated by his failure to recognize *P.* × *acerifolia* (Ait.) Willd. (*P. occidentalis* L. × *P. orientalis* L., ? *P. hispanica* Muenchh., ? *P. hybrida* Brot.) as distinct. This latter, the London Plane, is distinguished from *P. orientalis* by its more broadly lobed leaves, fewer (2–3 rarely 4) fruiting heads and blunter achenes with a glabrous apex and densely pilose sides. It is a popular street tree in London, and thrives under conditions of atmospheric pollution which would kill or seriously injure many trees. *P.* × *acerifolia* may have been planted in Cyprus, but there are no records of its occurrence, nor of its other putative parent, the North American *P. occidentalis* L., which has shallow leaf-lobes, and a single, terminal fruiting head.

88. JUGLANDACEAE

Deciduous, anemophilous trees; leaves generally alternate, imparipinnate, exstipulate; flowers monoecious, the males forming pendulous catkins on the twigs of the previous year, with a perianth of 4 (or fewer) tepals and 23–40 stamens, the females few and sessile on the growths of the current year or numerous on long, stalked, pendulous spikes; bracts, bracteoles and perianth often adnate to ovary, and shortly toothed or lobed above; ovary inferior, 1-locular; style short; stigmas often large and plumose; ovule 1, erect, orthotropous. Fruit a drupe or a nut, sometimes winged; seed without endosperm; testa thin; cotyledons often conspicuously rugose.

Seven genera and about 40 species, chiefly in temperate and tropical regions of the northern hemisphere. The family has no indigenous representatives in Cyprus, but the Walnut, *Juglans regia* L. (Sp. Plant., ed. 1, 997; 1753), equally valued for its nuts and its timber, is not uncommon in cultivation.

Juglans regia L. is a robust tree, up to 30 m. high, with pale greyish, lightly fissured bark, twigs with lamellate pith, and aromatic leaves with 5–7, large (up to 15 cm. long, 10 cm. wide), ovate or obovate, entire, or almost entire, leaflets, which are glabrescent except along the midrib and in the axils of midrib and primary nerves. The male catkins are solitary and many-flowered; the fruit is a globose, green drupe, 4–6 cm. diam., containing a pale brown, wrinkled stone, which splits along the midribs to expose a very large, lobed and wrinkled seed with a thin brown testa.

The Walnut is recorded as a cultivated tree from Ktima (Div. 1) and from numerous localities on the Troödos Range (2) viz: Makheras Monastery (*Kotschy; Syngrassides* 1629 !), Troödhitissa Monastery (*Baker*), Palekhori (*Bevan*), Perapedhi (*Nattrass* 54), Platres (*Nattrass* 63), Kambos P. (*Laukkonen* 381 !). It is generally assumed to be native in Turkey, N. Iran, Transcaucasia and Central Asia.

88a. CASUARINACEAE

Monoecious or dioecious trees and shrubs, with whorls of slender, jointed, longitudinally sulcate or subterete, *Equisetum*-like branchlets, the leaves reduced to toothed sheaths around the branchlet-nodes; male flowers in superimposed whorls arising from sheaths at the tips of the branchlets; bracteoles 4, obovate, serrulate; calyx and petals absent; stamen 1, central; anther 4-thecous, dehiscing by longitudinal slits, basifixed; female flowers in spikes or capitula, borne laterally at the tips of short modified branches, each flower subtended by a bract and 2, persistent, woody bracteoles; calyx and petals absent; ovary superior, 1-locular; ovules 2 (or 1 by abortion), placentation parietal; styles shortly united at base, divided above into 2 long, linear stigmatiferous arms. Fruit a small, 1-seeded samara, enveloped by the woody bracteoles, the whole infructescence forming a subglobose or barrel-shaped cone; seed without endosperm; embryo straight, with a short radicle and broad cotyledons.

A small family of disputed affinity, occurring naturally in S.E. Asia, Australia and Polynesia, but extensively planted in tropical and subtropical countries, chiefly for ornament, but sometimes for firewood, or as a means of draining marshy ground. Until recently thought to comprise only one genus, *Casuarina* L., but currently in the process of re-classification and subdivision.

The 4 species noted as cultivated in Cyprus can be keyed out as follows:

Sheaths with (6–) 7 (–8) teeth; branchlets very numerous, lax, slender, almost hair-like:
 Male inflorescences dense, commonly 10–18 mm. long; bract-sheaths closely overlapping
  with regularly imbricate teeth; fruit-cones 13–15 mm. long, 10–12 mm. wide
                      C. EQUISETIFOLIA *L.*
 Male inflorescences vermiform, commonly 20–30 mm. long; bract-sheaths rather distant, and
  separated by distinct constrictions; fruit-cones 10–12 mm. long, 7–8 mm. wide
                  C. CUNNINGHAMIANA *Miq.*
Sheaths with (9-) 10–12 (–16) teeth; branchlets rather sparse, not slender or hair-like:
 Branchlets distinctly ribbed, dark green; fruit-cones 30–40 mm. long, 25–30 mm. wide
                    C. VERTICILLATA *Lam.*
 Branchlets almost terete, glaucous; fruit-cones 10–15 mm. long, 10–12 mm. wide
                C. GLAUCA *Sieb. ex Sprengel*

C. EQUISETIFOLIA *L.* (Amoen. Acad., 3: 143; 1759), a commonly cultivated species, comes from S.E. Asia and northern Australia; it is noted as growing in the Municipal Gardens, Nicosia and in the Hospital grounds, Nicosia (*Bovill*), also near the Forest Office, Nicosia (*Frangos*; *Nattrass*), but at least some of these records are probably referable to *C. cunninghamiana*.

C. CUNNINGHAMIANA *Miq.* (Rev. Crit. Casuar., 56, t. 6A; 1848), from Australia, is noted as growing in Nicosia (Hutchins, Rep. Cypr. Forestry, 63; 1909), and has been collected at Dhekelia, 1973, *P. Laukkonen* 46 !

C. VERTICILLATA *Lam.* (Encycl. Méth., 2: 501; 1788) is recorded (as *C. stricta* Ait. or *C. quadrivalvis* Labill.) from Nicosia Municipal Gardens (Frangos in Cypr. Agric. Journ., 18: 88; 1923). It is a native of Australia.

C. GLAUCA *Sieber ex Sprengel* (Syst. Veg., 3: 803; 1826), also Australian, is said to have been planted at Salamis (Bovill, Rep. Plant. Work Cyprus, 15; 1915).

89. BETULACEAE

Deciduous, anemophilous trees or shrubs; leaves alternate, simple, pinnately nerved; stipules caducous; inflorescences monoecious; male flowers forming pendulous, caudate catkins, usually in groups of 3, subtended by a catkin-scale consisting of a bract and fused bracteoles; perianth 4-lobed; stamens 4(2); filaments short, bifid or undivided; anthers glabrous, thecae separate or united; female flowers in pendulous or erect spikes, in groups of 2 or 3 subtended by a fused bract and bracteoles; perianth absent; ovary 2-locular, ovules 2 (or, by abortion, 1) in each loculus; style divided to base into 2 linear-subulate, stigmatiferous arms. Fruit a nut or samara; seed solitary, without endosperm; embryo straight; cotyledons flat.

Two genera widely distributed in the northern hemisphere and extending south along the Andes. Only one genus in Cyprus.

1. ALNUS *Mill.*
Gard. Dict. Abridg., ed. 4 (1754).

Deciduous trees or shrubs; winter-buds commonly constricted towards the base or shortly stipitate, with a large outer bud-scale; leaves broadly ovate, obovate or suborbicular, often shortly acute, truncate or emarginate; male catkins elongate, caudate, pendulous, each catkin-scale subtending 3 flowers; perianth 4-lobed; stamens 4, opposite the perianth-lobes; anthers undivided; female catkins shorter, suberect or spreading, flowers 2 in the axil of each scale; bracteoles 4, perianth absent; styles 2, short. Fruiting catkins ovoid, ellipsoid or subglobose, woody, persistent, cone-like, with 5-lobed scales (derived from the accrescent, incrassate bract and bracteoles); fruit a small compressed nut or samara; seed with a membranous testa and flat cotyledons.

About 35 species widely distributed in the northern hemisphere, and in the Andes.

1. **A. orientalis** *Decne.* in Ann. Sci. Nat., ser. 2, 4: 348 (1835); Unger et Kotschy, Die Insel Cypern, 215 (1865); Boiss., Fl. Orient., 4: 1179 (1879); Post in Mém. Herb. Boiss., 18: 100 (1900); H. Winkler in Engl., Pflanzenr., 19 (IV. 61): 113 (1904); Holmboe, Veg. Cypr., 59 (1914); Post, Fl. Pal., ed. 2, 2: 527 (1933); Lindberg f., Iter Cypr., 11 (1946); Chapman, Cyprus Trees and Shrubs, 31 (1949); Osorio-Tafall et Seraphim, List Vasc. Plants Cyprus, 28 (1973); Davis, Fl. Turkey, 7: 693 (1982).
   A. orientalis Decne. var. *longifolia* H. Winkler in Engl., Pflanzenr., 19 (IV. 61): 113 (1904); Holmboe, Veg. Cypr., 59 (1914); Chapman, Cyprus Trees and Shrubs, 31 (1949).
   A. orientalis Decne. var. *weissii* H. Winkler in Engl., Pflanzenr., 19 (IV. 61): 113 (1904); Holmboe, Veg. Cypr., 59 (1914); Chapman, Cyprus Trees and Shrubs, 31 (1949).
   A. orientalis Decne. var. *ovalifolia* H. Winkler in Engl., Pflanzenr., 19 (IV. 61): 114 (1904); Holmboe, Veg. Cypr., 59 (1914); Chapman, Cyprus Trees and Shrubs, 31 (1949).
   A. orientalis Decne. var. *weissii* H. Winkler f. *winkleri* Callier in Mitt. Deutsch. Dendr. Ges., 27: 66 (1918).
   A. orientalis Decne.var. *weissii* H. Winkler f. *puberula* Callier in Mitt. Deutsch. Dendr. Ges., 27: 66 (1918).
   [*A. oblongata* (non (Ait.) Willd.) Poech, Enum. Plant. Ins. Cypr., 12 (1842); Gaudry, Recherches Sci. en Orient, 197 (1854).]
   TYPE: Lebanon; "les bords du fleuve à Bairout" (*A. longifolia* Bové n. 496. P, K!).

Slender tree to 20 m. high with a narrow, pointed or pyramidal crown; bark grey, vertically fissured with age; twigs rather slender, glabrous; winter-buds oblong, blunt, glutinous, distinctly stipitate, about 4–7 mm.

long, 3–4 mm. wide; leaves ovate or oblong, mid-green, glabrous above, generally glabrous or glabrescent below with persistent tufts of hairs in the axils of midrib and nerves, 3–12 cm. long, 1·3–6 cm. wide, apex obtuse, acute or shortly acuminate, base rounded, truncate, broadly cuneate or shallowly cordate, sometimes the 2 halves unequal, margins subentire or irregularly erose or serrate; petioles rather slender, glabrous or thinly tomentellous, 1–3 cm. long; stipules oblong, membranous, brownish, caducous, 4–7 mm. long, 3–4 mm. wide; male catkins numerous, conspicuous, in lax, branched clusters of 3–6 at the tips of the branches, expanding in early spring before the development of the leaves, 4–12 cm. long, 0·6–0·8 cm. wide, pendulous; rhachis glabrous; catkin-scale (fused bract and bracteoles) semicircular, glabrous, about 2·5–3 mm. wide, 1·5 mm. long, apex shallowly emarginate; perianth submembranous, lobes bluntly oblong, erose, about 0·6 mm. long, 0·4 mm. wide; filaments 0·7 mm. long, glabrous; anthers oblong, about 1·5 mm. long, 1·2 mm. wide, at first reddish, later becoming pallid; female catkins few (1–3), at the tips of short lateral branches, usually developing with the leaves; catkin-scales thick, fleshy, broadly ovate, about 1·5 mm. wide, 1·2 mm. long, reddish, glabrous, apex shortly cuspidate or mucronate; ovary strongly compressed, glabrous, about 0·3 mm. long, 0·2 mm. wide; styles reddish, about 0·5–1 mm. long. Fruit-cone ovoid, glutinous, about 2 cm. long, 1·5 cm. wide; fruits flattened, suborbicular, about 3·5 mm. diam., sometimes narrowly winged, glabrous, pale brown. *Plate 84.*

HAB.: By streams in mountainous areas; 400–4,500 ft. alt.; fl. Jan.–April.

DISTR.: Divisions 1–3, 7. Also S.E. Turkey, Syria, Lebanon.

1. Evretou, 1937, *Syngrassides* 1716! also, 1938, *Chapman* 312!
2. Between Omodhos and Troödhitissa Mon., 1787, *Sibthorp*; "Slewra" (near Khrysorroyiatissa Mon.), 1840, *Kotschy* 12, and, 1862, *Kotschy* 679! Troödos, 1862, *Kotschy*; near Galata, 1880, *Sintenis & Rigo* 685! Yialia, 1905, *Holmboe* 797; Saïttas, 1933, *Nattrass* 354! Marathasa, 1936, *Nattrass* 745! Kryos Potamos, 2,800–3,400 ft. alt., 1936, *Kennedy* 288! 289! Mesapotamos Mon., 1936, *Kennedy* 290! Platania, 1938, *Chapman* 54; Stavros tis Psokas 1939, *Lindberg f.* s.n.; Prodhromos, 1939, *Lindberg f.* s.n.! Milikouri, 1939, *Lindberg f.* s.n.; Phlevas near Pyrgos, 1962, *Meikle* 2717! Kalopanayiotis, 1968, *A. Genneou* in ARI 1275! Pano Platres, 1973, *P. Laukkonen* 414! etc.; locally common.
3. Episkopi, 1862, *Kotschy* 618! 1 km. N.E. of Apsiou, 1979, *Edmondson & McClintock* E 2714! Kalokhorio to Athrakos, 1979, *Edmondson & McClintock* E 2979!
7. Kyrenia, watermill stream, 1955, *G. E. Atherton* 726!

NOTES: Perhaps planted in Division 7, but this is not indicated by collector.
Alnus orientalis is exceedingly variable in the size and shape of its leaves; some Cyprus specimens agree closely with the type, but many appear to have smaller, narrower, less coarsely erose leaves than are usual elsewhere, and correspond to *A. orientalis* var. *weissii* H. Winkler and *A. orientalis* var. *ovalifolia* H. Winkler. I cannot, however, regard these as satisfactorily distinguished from var. *orientalis* (var. *longifolia* H. Winkler).

90. CORYLACEAE

Deciduous, anemophilous trees and shrubs; buds with imbricate scales; leaves simple, alternate, pinnately nerved; stipules caducous; inflorescence monoecious; male flowers forming pendulous, caudate catkins, each solitary in the axil of a catkin-scale (a bract, or a bract united with 2 collateral bracteoles); perianth absent; stamens 3 or more, filaments bifid at apex or divided almost to base; anther-thecae separate, pilose at apex; female flowers in erect or pendulous spikes or in clusters enveloped by scaly buds,

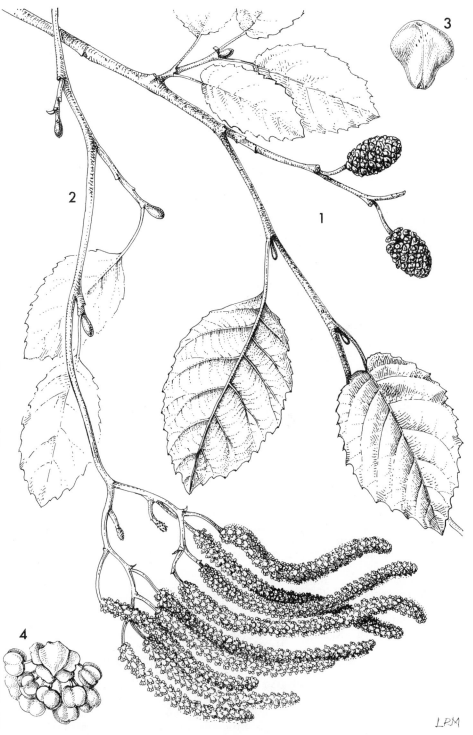

LPM

Plate 84. ALNUS ORIENTALIS Decne. **1,** branch with fruits, × 1; **2,** branch with catkins, × 1; **3,** seed, × 9; **4,** male flower, × 9. (**1** from *Kennedy* 289; **2–4** from *Kennedy* 288.)

the fused bract and bracteoles each subtending 2 flowers; perianth minute, adnate to ovary, 4–8-denticulate at apex; ovary minute, 2-locular; ovules usually 1 in each loculus, pendulous; style divided to base into 2 linear-subulate, stigmatiferous arms. Fruit a 1-seeded nut, subtended or enveloped by a lobed, foliaceous involucre developing from the fused bract and bracteoles; pericarp woody or coriaceous, smooth or longitudinally ribbed; seed without endosperm; testa membranous; cotyledons thick, fleshy, emergent or remaining in the nut at germination.

Four genera widely distributed in the northern hemisphere.

1. CORYLUS *L.*

Sp. Plant., ed. 1, 998 (1753).
Gen. Plant., ed. 5, 443 (1754).

Deciduous shrubs or small trees; inflorescences appearing on the naked branches well in advance of the leaves; catkin-scales of male catkins consisting of a bract united with 2 collateral bracteoles; stamens 4–8; female flowers clustered within a scaly bud, only the red style-arms protruding. Fruit a globose or ovoid nut enveloped by a cupular or tubular, foliaceous involucre, developed from the fused bract and bracteoles; pericarp woody, smooth; seed large, the thick, fleshy cotyledons remaining in the nut at germination.

About 15 species widely distributed in the northern hemisphere.

Involucre scarcely exceeding nut, campanulate, not constricted   -   -   -   **1. C. avellana**
Involucre much exceeding nut, and distinctly constricted about the middle
C. MAXIMA (p. 1477)

1. C. avellana *L.*, Sp. Plant., ed. 1, 998 (1753); Boiss., Fl. Orient., 4: 1176 (1879); H. Winkler in Engl., Pflanzenr., 19 IV 61): 46 (1904); Holmboe, Veg. Cypr., 59 (1914); Post, Fl. Pal., ed. 2, 2: 525 (1933); Chapman, Cyprus Trees and Shrubs, 31 (1949); Osorio-Tafall et Seraphim, List Vasc. Plants Cyprus, 28 (1973); Davis, Fl. Turkey, 7: 687 (1982).
TYPE: "*in* Europae *sepibus*".

Erect or spreading shrub or small, suckering tree to 10 m. high; bark rough, greyish; twigs terete, greyish-brown; young shoots pubescent-glandular, with numerous, long-stalked, reddish glands; buds blunt, about 3–4 mm. long, 3 mm. wide, with broad, rounded, ciliate scales; leaves broadly obovate or suborbicular, 5–10 cm. diam., dull green and thinly pilose along the nerves above, slightly paler below, apex acute or shortly cuspidate, base shallowly cordate, margins shortly, sharply and closely dentate, generally irregularly lobulate; nervation prominent below, with 7–8 pairs of spreading, pilose, lateral nerves; petiole 1–1·5 cm. long, glandular-pilose; stipules ovate-oblong, rounded or subacute, about 6 mm. long, 4 mm. wide, pubescent, submembranous, caducous; male catkins generally numerous, on short lateral branchlets along the year-old twigs, lax, caudate-cylindrical, 4–10 cm. long, 0·5–0·6 cm. diam., rhachis pilose; catkin-scales spathulate-flabellate, about 4 mm. long, 2·5 mm. wide, pubescent externally, with a wide median lobe (the bract) and 2 small lateral lobes (the bracteoles); stamens apparently 8 (sometimes fewer or more); filaments very short, slender; anthers apparently monothecous, oblong, about 1·2 mm. long, 0·8 mm. wide, yellow; pollen very copious and powdery; female flowers in clusters of about 8, concealed within a bud only a little larger than the vegetative buds; bracts ovate, concave, densely pubescent, about 2·5 mm. long, 2 mm. wide; perianth densely pilose, less than 0·4 mm. long, style-arms red, 4–5 mm. long. Nuts usually 2–4 together at the apex of a short lateral, leafy branch; involucre green, pubescent, with numerous, close, parallel nerves and a deeply, irregularly laciniate apex, about 1·5–2 cm. long, generally split to the base down one side, and wide at the apex with part of the nut exposed; nut ovoid, 1·5–2 cm. long, 1–1·5 cm. wide, brown when ripe, lustrous or shortly puberulous, with a large basal attachment scar.

HAB.: Moist ground by streams at high altitudes; 3,000–4,000 ft. alt.; fl. Febr.–March.

DISTR.: Division 2, evidently rare; perhaps planted. Widespread in Europe, also W. Asia as far south as Syria and east to Iran.

2. Kykko, 1905, *Holmboe* 1009; Platania, [? date], *Chapman* 64.

NOTES: Most authors have failed to distinguish the Hazel, *C. avellana*, from the Filbert, *C. maxima* Mill., and, as Holmboe notes, their records for cultivated Hazels refer to the latter rather than the former, which, according to Holmboe and Chapman, is rare on the island. True *C. avellana* is reported from the mainland of Asia adjacent to Cyprus, and may be indigenous, though so much has been experimentally planted on Troödos by the Forestry Department, that its status is bound to be rather dubious.

C. MAXIMA *Mill.*, Gard. Dict., ed. 8, no. 2 (1768); Holmboe, Veg. Cypr., 59 (1914); Osorio-Tafall et Seraphim, List Vasc. Plants Cyprus, 28 (1973); Davis, Fl. Turkey, 7: 688 (1982).
    C. tubulosa Willd., Sp. Plant., 4: 470 (1805); Boiss., Fl. Orient., 4: 1176 (1879).
    [*C. avellana* (vix L.) Kotschy in Unger et Kotschy, Die Insel Cypern, 215 (1865); Nattrass, First List Cyprus Fungi, 38 (1937); Chapman, Cyprus Trees and Shrubs, 31 (1949) pro parte]

Very similar to *C. avellana* L., but young shoots and petioles often densely hispid-glandular with conspicuous reddish stalked glands; leaves often rather larger, frequently 10 cm. long and wide. Fruiting involucre tubular, commonly 6 cm. long, 1·5 cm. wide, much exceeding the nut, with a distinct median constriction and serrate-laciniate, foliaceous lobes. Nut oblong-ovoid, often 2·5–3 cm. long, 1·5 cm. wide, base shining brown when ripe, apex dull and puberulous.

Extensively cultivated in the moist valleys of Pitsilia, at altitudes between 3,000 and 4,000 feet; also found in gardens here and there all around the Troödos Range; Kotschy notes it from Troödhitissa Monastery, and Lindberg f. (spec. K!) from Milikouri "pr. monast Kykko, in cultis". Hybrids, with deeply laciniate-lobed involucres shortly exceeding the nuts, are not infrequent in cultivation, and obscure the rather slender distinctions between *C. avellana* and *C. maxima*.

91. FAGACEAE

Deciduous or evergreen trees; leaves alternate, simple, pinnately nerved; stipules present, caducous or persistent; inflorescences monoecious; male flowers commonly in erect or pendulous spikes; perianth 4–6 (–7)-lobed, lobes imbricate; stamens few or numerous; filaments slender, free; anthers 2-thecous, dehiscing longitudinally; rudimentary ovary present or absent; female flower solitary in an involucre often covered with scales or spines; perianth 4–6-lobed, adnate to ovary, or absent; ovary 3–6-locular; ovules 2 in each loculus; styles as many as loculi; staminodes present or absent. Fruit a nut, seated in a woody, scaly or spiny cupule, or surrounded and sometimes adnate to a woody or spiny fruiting involucre; seed generally solitary (by abortion), without endosperm; cotyledons often thick and fleshy.

Eight genera, and about 400 species, widely distributed in temperate and

tropical regions, but without indigenous representatives in Africa south of
the Sahara.

Flowers in erect, spicate catkins, male in the upper part, female toward the base; involucre
   spinose, covering the nuts and splitting into lobes or valves at maturity
                                           CASTANEA (p. 1478)
Flowers in separate inflorescences, the males forming pendulous catkins; the females few, sessile
   or in pedunculate spikes; involucre (cupule) cup-shaped, not covering the nut (acorn) nor
   splitting at maturity   -   -   -   -   -   -   -   -   -   -   -   **1. Quercus**

CASTANEA SATIVA *Mill.*, Gard. Dict., ed. 8, no. 1 (1768); Osorio-Tafall et
Seraphim, List Vasc. Plants Cyprus, 28 (1973); *C. vulgaris* Lam., Encycl.
Méth., 1: 708 (1785); *C. vesca* Gaertn., Fruct. Sem. Plant., 1: 181, t. 37 (1788);
C. castanea Karsten, Deutsche Fl. Pharm.-Med. Bot., 495 (1883); Druce in
Rep. B.E.C., 9: 471 (1931).

A robust tree attaining 30 m. or more; bark deeply furrowed, the furrows
often twisting spirally round the trunk; twigs brown, glabrous or almost so;
buds ovoid, obtuse, 3–4 mm. long; leaves oblong-lanceolate, 10–30 cm. long,
4–8 cm. wide, bright green and glabrous above, glabrescent below, with an
acute apex, and numerous, prominent lateral nerves running into the
conspicuous, acuminate-aristate marginal serrations; petiole rather short,
about 2 cm. long; inflorescences terminal, in clusters, forming erect,
spiciform catkins, up to 20 cm. long, the greater part of the catkin consisting
of yellowish, 10–20-stamened male flowers, the female flowers few, confined
to the base of the catkin, generally in groups of 3 within each involucre. Nuts
1–3, compressed-ovoid, dark brown, 2–3·5 cm. diam., covered until ripe by
an involucre clothed with rigid, slender, branched spines.

The Sweet, or Spanish Chestnut, a native of S. Europe, N. Africa and
western Asia east to Iran. Not native in Cyprus, but Bevan (Notes on
Agriculture in Cyprus, 50; 1919) says, "some years ago good numbers of the
edible chestnut were raised at Pedoulas by the Agricultural Department and
distributed to villagers for growing in the hills. It is feared that the greater
part of these trees, through want of attention, unsuitability of soil or
climate, lack of moisture, and especially damage by goats, have been lost,
but some remain and well-grown trees may be found in certain localities and
in moderate numbers among the mountains."

How many of these planted trees still survive is hard to say, but *Castanea
sativa* has been noted at Platres, 1928 or 1930, by Druce, and collected
between Troödos and Pano Platres, 1973 (*P. Laukkonen* 405!)

1. QUERCUS *L.*

Sp. Plant., ed. 1, 994 (1753).
Gen. Plant., ed. 5, 431 (1754).
Kotschy, Die Eichen Europa's und des Orient's, 40 pls. (1862).
O. Schwarz in Fedde Repert., Sonderbeiheft D., 200 pp., 64 pls. (1936–37).
A. Camus, Les Chênes, 3 vols., atlas (1934–1954).
M. Zohary in Bull. Res. Council Israel, 9D, Bot.: 161–186 (1961).

Evergreen or deciduous, anemophilous trees or shrubs; leaves generally
toothed or lobed, or nearly entire; stipules often caducous; inflorescences
monoecious; male flowers solitary or in groups scattered along a slender
rhachis and forming drooping spikes or catkins; perianth 4–7-lobed; stamens
4–12, often 6; anthers usually rather large; rudimentary ovary present or
absent; female flowers subsessile or in solitary, erect (rarely pendulous)
spikes, involucre scaly; ovary 3-locular, with 2 ovules in each loculus; styles
usually 3, stigmatiferous along their inner surfaces. Fruit an ovoid, globose

or turbinate nut (acorn) seated in a saucer-shaped, scaly cupule (or acorn-cup); pericarp tipped with an umbo formed from the style-bases; seed large, solitary, with fleshy cotyledons; germination hypogeal.

About 450 species in Europe, North Africa, temperate and sub-tropical Asia, North America and western South America.

Many of the species provide valuable timber, also bark and galls used in tanning hides.

Leaves rigid, coriaceous; evergreen trees or shrubs:
 Leaf-margins spinose-dentate; lamina bright green, glabrous or subglabrous on both surfaces
  3. Q. coccifera ssp. **calliprinos**
 Leaf-margins not spinose-dentate; lamina dark shining green above, golden- or brown-tomentose below - - - - - - - - - - - **2. Q. alnifolia**
Leaves not rigid, thinly coriaceous or papyraceous, deciduous with the development of the young growth in spring; semi-evergreen trees:
 Cupule with short, adpressed scales; leaves subglabrous on both surfaces; lateral nerves not excurrent - - - - - - - - - **1. Q. infectoria** ssp. **veneris**
 Cupule with long, reflexed scales; leaves densely pubescent or tomentose below; lateral nerves excurrent from the marginal lobes as a short arista - Q. AEGILOPS (p. 1482)

1. Q. infectoria *Olivier*, Voy. Emp. Oth., 1: 252, tt. 14, 15 (1801); Post, Fl. Pal., ed. 2, 2: 520 (1933); A. Camus, Les Chênes, 2: 183, tt. 108–111 (figs. 1–8) (1939); Osorio-Tafall et Seraphim, List Vasc. Plants Cyprus, 28 (1973); Davis, Fl. Turkey, 7: 670 (1982).
 Q. lusitanica Lam. ssp. *orientalis* A.DC. in DC. Prodr., 16(2): 18 (1864).
 [*Q. lusitanica* (non Lam.) Boiss., Fl. Orient., 4: 1167 (1879).]

Tree to 10 m. or more with a massive trunk and thick grey, vertically fissured bark and a broad spreading crown, or sometimes a bush 3–5 m. high; twigs glabrous or more or less tomentose; buds bluntly ovoid, 4–5 mm. long, 3–4 mm. wide, with numerous, closely imbricate, broad, brownish, ciliate scales; leaves thinly coriaceous, persistent until the new foliage begins to develop in late winter or spring, very variable, oblong or oblong-obovate, 2·5–15 cm. long, 1·2–7 cm. wide, glabrous and rather lustrous above, paler and glabrescent below or thinly stellate-pubescent along the prominent midrib and nerves; apex rounded or shortly acute, base rounded, truncate or shallowly cordate, margins sinuate, or sinuate-serrate with shortly mucronate lobes, sometimes sharply serrate-lobulate, rarely entire or subentire; lateral nerves spreading or suberect, usually in 6–8 (–10) pairs, running directly into the marginal lobes; venation rather obscure, finely reticulate; petioles 5–20 (–30) mm. long, or leaves (especially those of sterile shoots) rarely subsessile; stipules linear, membranous, silky-pilose, caducous; male inflorescences generally crowded towards the tips of the branches, 3–5 cm. long; flowers solitary or clustered, sessile along a stellate-tomentose or subglabrous rhachis; perianth shallowly cup-shaped, about 4 mm. diam., stellate-pubescent, membranous, with 4–6 subacute lobes about 0·8 mm. long; stamens 4–6, filaments about 1 mm. long; anthers about 1·2 mm. long, 0·8 mm. wide; female inflorescences solitary, axillary, erect, usually 1–3 cm. long at anthesis, bearing 2–4 remote, sessile flowers, or flowers sometimes solitary and subsessile, peduncle and rhachis softly tomentose, involucre about 4 mm. diam., thinly tomentose; styles 3, stout, about 1 mm. long, dorsiventrally compressed and reflexed at the slightly dilated, subtruncate stigmatiferous apex. Fruit variable, the cupule 1·5–2·5 cm. diam., closely adpressed-scaly, thinly tomentose, with shortly and rather bluntly rostrate scales; acorn pale brown, narrowly ovoid, obovoid or subcylindrical 1·3–5 cm. long, 1–2 cm. wide, 3–5 times as long as the cupule, apex shortly mammillate; endocarp glabrous.

ssp. **veneris** (*A. Kerner*) *Meikle* **comb. nov.**
 Q. boissieri Reuter in Boiss., Diagn., 1, 12: 119 (1853); O. Schwarz in Fedde Repert., Sonderbeiheft D, 185 (1936–37); Zohary in Bull. Res. Council Israel 9D., Bot.: 165 (1961), Fl. Palaest., 1: 31, t. 30 (1966).

Q. petiolaris Boiss. et Heldr. in Boiss., Diagn., 1, 12: 120 (1853).
Q. syriaca Kotschy, Die Eichen, t. 1 (1858).
Q. tauricola Kotschy, Die Eichen, t. 10 (1859).
Q. pfaeffingeri Kotschy, Die Eichen, t. 23 (1860).
Q. cypria Hochst. et Kotschy in Oesterr. Bot. Zeitschr., 12: 278 (1862) nomen.
Q. lusitanica Lam. ssp. *orientalis* A.DC. var. *petiolaris* (Boiss. et Heldr.) A.DC. in DC.,
Prodr., 16(2): 19 (1864); Boiss., Fl. Orient., 4: 1167 (1879).
Q. lusitanica Lam. ssp. *orientalis* A.DC. var. *syriaca* (Kotschy) A.DC. in DC., Prodr.,
16(2): 19 (1864).
Q. pfaeffingeri Kotschy var. *cypria* Kotschy in Unger et Kotschy, Die Insel Cypern, 216
(1865).
Q. inermis Ehrenberg ex Kotschy in Unger et Kotschy, Die Insel Cypern, 216 (1865); O.
Schwarz in Fedde Repert., Sonderbeiheft D, 182 (1936–37).
Q. lusitanica Lam. var. *latifolia* Boiss., Fl. Orient., 4: 1167 (1879).
Q. veneris A. Kerner in Schneider, Ill. Handb. Laubholzk., 1: 191 (1904).
Q. infectoria Olivier var. *petiolaris* (Boiss. et Heldr.) Hand.–Mazz. in Ann. Naturh. Mus.
Wien, 26: 128 (1912); Post, Fl. Pal., ed. 2, 2: 521 (1933).
Q. lusitanica Lam. ssp. *veneris* (A. Kerner) Holmboe, Veg. Cypr., 60 (1914).
Q. infectoria Olivier ssp. *boissieri* (Reuter) O. Schwarz in Fedde Repert., 33: 336 (1934);
A. Camus, Les Chênes, 2: 187 (1939); Osorio-Tafall et Seraphim, List Vasc. Plants Cyprus,
28 (1973); Davis, Fl. Turkey, 7: 671 (1982).
Q. infectoria Olivier ssp. *petiolaris* (Boiss. et Heldr.) O. Schwarz in Fedde Repert., 33: 367
(1934).
Q. boissieri Reuter ssp. *latifolia* (Boiss.) O. Schwarz in Notizbl. Bot. Gart. Berlin, 13: 17
(1936), Fedde Repert., Sonderbeiheft D, 187 (1936–37).
Q. boissieri Reuter ssp. *tauricola* (Kotschy) O. Schwarz in Notizbl. Bot Gart. Berlin, 13:
17 (1936), Fedde Repert., Sonderbeiheft D, 188 (1936–37).
Q. boissieri Reuter ssp. *petiolaris* (Boiss.) O. Schwarz in Notizbl. Bot. Gart. Berlin, 13: 17
(1936), Fedde Repert., Sonderbeiheft D, 190 (1936–37).
Q. infectoria Olivier ssp. *boissieri* (Reuter) O. Schwarz var. *insularis* A. Camus, Les
Chênes, 2: 190 (1939).
Q. infectoria Olivier var. *veneris* (A. Kerner) Lindberg f., Iter Cypr., 11 (1946).
Q. boissieri Reuter var. *petiolaris* (Boiss. et Heldr.) Zohary in Bull. Res. Council Israel,
9D, Bot.: 165 (1961), Fl. Palaest., 1: 31 (1966).
[*Q. aegilops* (non L.) Hume in Walpole, Mem. Europ. Asiatic Turkey, 83 (1820).]
[*Q. infectoria* (non Olivier) Poech, Enum. Plant. Ins. Cypr., 12 (1842); Kotschy in Unger
et Kotschy, Die Insel Cypern, 215 (1865); Lindberg f., Iter Cypr., 17 (1946).]
[*Q. lusitanica* Lam. var. *genuina* Boiss., Fl. Orient., 4: 1167 (1879) pro parte quoad plant.
Cypr.]
[*Q. lusitanica* Lam. ssp. *infectoria* (non *Q. infectoria* Olivier) Holmboe, Veg. Cypr., 60
(1914).]
[*Q. lusitanica* (non Lam.) A. H. Unwin, Forests of Cyprus, 7 (1927); Druce in Proc. Linn.
Soc., 141 st. Session: 52 (1930); R. R. Waterer, The Island of Cyprus, 152 (1947); Chapman,
Cyprus Trees and Shrubs, 32 (1949).]

TYPE: Cyprus; "in valle Evrico versus Galata in hortis und ad pagum Trisedils [Trisedies]
prope Prodomo [Prodhromos]," 1862, *Kotschy* 720a (B, K !).

Generally robust trees with fissured bark; twigs stellate-pubescent or
stellate-tomentose, sometimes glabrescent with age; leaves usually rather
large, 4–15 cm. long, 2–7 cm. wide, thinly tomentose below at first, soon
becoming glabrous or thinly stellate-pubescent along midrib and nerves,
very variable in shape, margins sinuate, sinuate-serrate, bluntly or acutely
lobed, subentire or occasionally entire; petiole well developed, often 1·5–2
(–3) cm. long; female inflorescences elongate, spicate, generally with a well-
developed tomentose peduncle and 2–3 (or more) sessile flowers spaced along
the rhachis. Cupule usually large, often 2–2·5 cm. diam.; acorn narrowly
elongate-obovoid or subcylindrical, commonly 3–5 cm. long. Fruits ripening
(*teste* Chapman) in one season.

HAB.: Generally in mountain valleys on igneous rocks, but sometimes in lowland areas or on
limestone; sea-level to 4,500 ft. alt.: fl. March–April.

DISTR.: Divisions 1–3, [6], 7. S. Turkey, Syria, Palestine east to Iraq and S.W. Iran.

1. Kourtaka [Kurdaka], 1905, *Holmboe* 744; Ktima, 1913, *Haradjian* 709 ! Stroumbi, 1934,
   Syngrassides 847 ! Evretou, 1937, *Syngrassides* 1717 ! Stephani near Ayios Neophytos
   Monastery, 1939, *Lindberg f.* s.n.; Miliou, 1980, *De Langhe* 90/80 !
2. Common; near Evrykhou, 1859, 1862, *Kotschy* 259 ! 260 ! 371; 761; 962; 963; between

Evrykhou and Galata, 1937, *Syngrassides* 1676! Galata, 1939, *Lindberg f.* s.n.! Prodhromos, 1862, *Kotschy* 761a! (cited as "southern slopes near Kaminaria ["Chaminarga"]" in *Die Insel Cypern*); Prodhromos, 1955, G. E. *Atherton* 631! Kalopanayiotis, 1905, *Holmboe* 1001; and, 1968, *A. Genneou* in ARI 1296! Stavros tis Psokas, 1905, *Holmboe* 1039; also, 1937, *Chapman* 203! 221! 222! and, 1939, *Lindberg f.* s.n.! Pano Panayia, 1905, *Holmboe* 1051; Pharmakas, 1905, *Holmboe* 1130; Agros, 1905, *Holmboe* 1145; Omodhos, 1905, *Holmboe* 1155; Perapedhi, 1905, *Holmboe* 1158; Kato Yialia, 1934, *Syngrassides* 852! Platres, 1936–37 *Kennedy* 298! 300! 301! also, 1941, *Davis* 3147! Kambos, 1937, *Chapman* 310! and, 1939, *Chapman* 386! Mandria, 1937, *Kennedy* 299! Kannaviou, 1939, *Lindberg f.* s.n.! Kakopetria, 1955, *Chiotellis* 446! Spilia, 1966, *Economides* in ARI s.n.! Ora, 1967, *Merton* in ARI 722!
3. Episkopi, 1905, *Holmboe* 703; Kato Lefkara, 1967, *Merton* in ARI 655! Between Kalokhorio and Athrakos, 1977, *Edmondson & McClintock* E 2977!
[6. Roadside near Department of Agriculture, Nicosia, 1974, *Meikle* s.n.!, planted.]
7. Kantara, 1880, *Sintenis & Rigo* 253! and, 1905, *Holmboe* 553; near Bellapais, 1880, *Sintenis & Rigo* 518! Lapithos, 1898, *Post* s.n.!

NOTES: *Quercus* L. sect. *Gallifera* Spach, to which *Q. infectoria* and its allies *Q. canariensis* Willd., *Q. fruticosa* Brot. and *Q. faginea* Lam. belong, is notoriously difficult taxonomically, as is indicated by the inflated synonymy of *Q. infectoria* ssp. *veneris*. Those who take a broad view of species might arguably include all four cited above (and others) under a single binomial. At the other extreme, some taxonomists have favoured binomials for every conceivable variant. The taxonomy employed here is a compromise: *Q. faginea* (W. Medit.) and *Q. fruticosa* (W. Medit.) seem tolerably distinct in having the lower surface of the leaves persistently tomentose, whereas adult leaves of *Q. infectoria* (E. Medit.) are generally almost glabrous below, or at most very thinly pubescent. Within the *Q. infectoria* complex it is possible to distinguish ssp. *infectoria*, [a shrub] with small leaves, short petioles and smallish acorns, from ssp. *veneris*, [a tree] with larger, distinctly petiolate leaves and large acorns; there are, however, numerous connecting links between the two both in Cyprus and elsewhere, and *subspecies* seems the appropriate rank for them. At a lower level the polymorphism is inscrutable, nor can I see that any useful purpose is served by giving names to some of the vast array of variants, distinguished almost solely by the size of the leaf and the nature of the marginal lobing.

2. **Q. alnifolia** *Poech*, Enum. Plant. Ins. Cypr., 12 (1842); Kotschy, Die Eichen, t. 6 (1859); A.DC. in DC., Prodr., 16(2): 40 (1864); Unger et Kotschy, Die Insel Cypern, 217 (1865); Boiss., Fl. Orient. 4: 1169 (1879); Holmboe, Veg. Cypr., 61 (1914); A. Camus, Les Chênes, 1: 430 (1938); Lindberg f., Iter Cypr., 11 (1946); Chapman, Cyprus Trees and Shrubs, 32 (1949); Osorio-Tafall et Seraphim, List Vasc. Plants Cyprus, 28 (1973); Elektra Megaw, Wild Flowers of Cyprus, 11, t. 17 (1973).
    Q. cypria Jaub. et Spach, Illustr. Plant. Orient., 1: t. 56 (1843); Gaudry, Recherches Sci. en Orient, 197 (1855).
    [*Q. ilex* (non L.) Sibthorp in Walpole, Travels, 22 (1820).]
    [*Q. infectoria* (non Olivier) Gaudry, Recherches Sci. en Orient, 197 (1855).]
    TYPE: Cyprus; "in decliviis montis 'Olympi' ins. Cypri (Kotschy *pl. Cypr.*)" (W).

Evergreen shrub or small tree, exceptionally up to 10 m. high, usually much branched with a wide crown; bark grey, vertically fissured; twigs at first densely, greyish, stellate-tomentose, the tomentum sometimes persisting into the second year; buds minute, bluntly ovoid, tomentose; leaves thick, rigid, coriaceous, ovate, oblong, obovate or suborbicular, 1·5–6 (–10) cm. long, 1–5 (–8) cm. wide, dark shining green above, densely golden- or brownish-tomentose below, apex rounded or shortly acute, base rounded or broadly cuneate, margins conspicuously or obscurely serrate, or sometimes entire or almost entire; nervation obscure above, the midrib and lateral nerves rather prominent below; petiole stout, 6–10 (–12) mm. long, stellate-tomentose; stipules linear-oblanceolate, membranous, caducous, pilose externally, about 6 mm. long, 1·5 mm. wide, male inflorescences in dense clusters at the tips of the branches, individual catkins slender, spreading or pendulous, 5–8 cm. long, with a very short stellate-tomentose peduncle and a stellate-tomentose rhachis; flowers approximate or rather remote, solitary or in groups of 2 or 3, sessile; perianth cup-shaped, about 2 mm. long, 2·5–3 mm. diam., thinly tomentellous externally, lobes (5–) 6 (–8), shortly oblong, subacute, about 0·5 mm. long; stamens usually 6; filaments about 2·5 mm. long, glabrous; anthers yellow, oblong, about 1·8 mm. long, 1 mm. wide, thinly pilose with the connective produced into a distinct mucro;

female flowers solitary or in groups of 2–3, sessile, sub-sessile or shortly stipitate in the leaf-axils; involucre 3–3·5 mm. diam., thinly tomentose; ovary shortly exserted; styles 3, recurved, about 1·5 mm. long, thick, sulcate above, with a truncate or shortly emarginate apex. Fruits solitary or in clusters; cupule 8–15 mm. diam., shortly grey-tomentellous, closely covered with linear, strongly recurved scales; acorn narrowly obovate or sub-cylindrical, usually tapering towards base, 2–2·5 cm. long, 0·8–1·2 cm. wide, with a conspicuously apiculate apex; endocarp woolly. *Plate 85.*

HAB.: Igneous mountainsides, locally sub-dominant with *Pinus brutia*; 2,500-5,000 ft. alt.; fl. April–May.

DISTR.: Divisions 2[7]. Endemic.

2. Locally abundant above 2,500 ft. alt.: Troödos, Platres, Kakopetria, Vavatsinia, Makheras, Pharmakas, Prodhromos, Stavros tis Psokas, Galata, Kykko, Amiandos, Platania, Trikoukkia, Sina Oros, etc., 1745–1973, *Drummond, Sibthorp, Aucher-Eloy* 2881 ! *Kotschy* 409 ! 756 ! *Mrs F. Croker* s.n. ! *Sintenis & Rigo* 687 ! *Post* s.n. ! *Haradjian* 934 ! *Syngrassides* 519 ! 749 ! 1683 ! *Lindberg f.* s.n. ! *Kennedy* 293–297 ! *Davis* 1757 ! 3528 ! *Merton* 2379 ! 2811 ! ARI 8 ! *Polunin* 8539 ! *I. M. Hecker* 19 ! *W. R. Price* 1084 ! etc.

[7. "One stunted specimen was recently found in the northern range forests geographically far removed from the main golden oak forests." (Chapman, Cyprus Trees and Shrubs, 33; 1949).]

NOTES: Apart from this one, rather imprecise record from Div. 7, the Golden Oak is confined to the igneous rocks of the Troödos Range.

A. Camus (Les Chênes, 1: 374; 1938) puts *Q. alnifolia* in the monotypic subsection *Cypriotes* J. Gay of *Quercus* L. section *Cerris* Spach.

Q. AEGILOPS *L.*, Sp. Plant., ed. 1: 996 (1753), emend. Mill., Gard. Dict., ed. 8, no. 7 (1768); Cyprus Journal (Quarterly), 1: 10 (1906); Bovill, Rep. Plant. Work Cyprus, 8, 14, 30 (1914); Bevan, Notes Agric. Cyprus, 98 (1919); Frangos in Cypr. Agric. Journ., 18: 88 (1923).

Q. macrolepis Kotschy, Die Eichen, t. 16 (1859–60).

A tree up to 20 m. high, with grey, fissured bark, and a broad, spreading crown; leaves persistent, coriaceous, oblong or oblong-obovate, 5–14 cm. long, 2·5–8 cm. wide, glabrous and rather lustrous above, thinly but persistently tomentose below, base often shallowly cordate, margins coarsely serrate, with acute lobes, the midrib and lateral nerves often shortly excurrent. Fruits conspicuously large, the acorn-cup often 2·5 cm. diam., shortly tomentellous, with numerous long, narrow, reflexed scales; acorn up to 3 cm. long, 2·5 cm. wide.

Not native in Cyprus, but reported as planted at Troödos, Makheras, Pedhoulas, Prodhromos, Larnaca, Salamis and Nicosia, also in the Paphos area, and collected, in 1956, in Blackalls Valley, near Kyrenia (*G. E. Atherton* 819 !) where it was no doubt also planted. Bevan indicates that it did not survive at Salamis or Makheras, and that, after six years, it was only a foot high at Troödos, but the Kyrenia specimen is said to have been a tree 15–20 feet high.

The acorn-cups of *Q. aegilops*, the Valonia Oak, were formerly valued as a source of tannin, but its importance has declined, nor, to judge from reports, does the tree thrive in our area, though it is frequent, as a native, from Crete east to Iran.

Many authors reject the name *Q. aegilops* L. as of uncertain application, and prefer the name *Q. macrolepis* Kotschy for the eastern Mediterranean segregate, to which the Cyprus trees most probably belong. I have, however, followed Fl. Iraq, 4(1): 50 (1981) in retaining the name *Q. aegilops* for the Valonia Oak in the wide, aggregate sense.

Q. ILEX *L.*, Sp. Plant., ed. 1, 995 (1753); Bovill, Rep. Plant. Work Cyprus, 18, 28 (1915); Frangos in Cypr. Agric. Journ., 18: 88 (1923), the Evergreen or

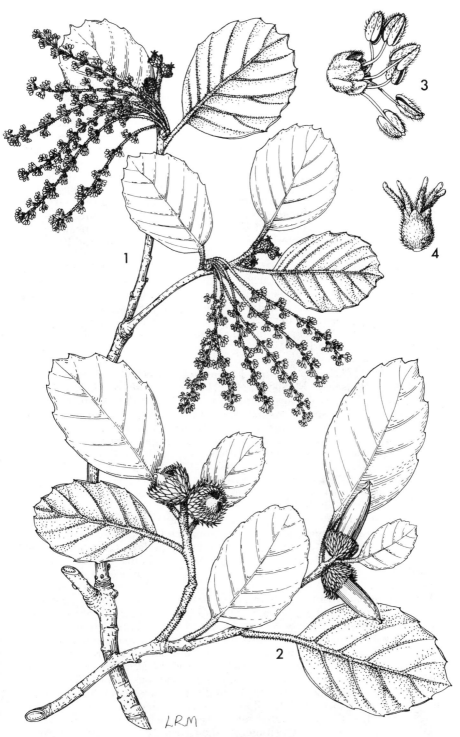

Plate 85. QUERCUS ALNIFOLIA Poech **1,** flowering branch, × ⅔; **2,** fruiting branch, × ⅔; **3,** male flower, × 6; **4,** female flower, × 6. (**1, 3, 4** from *Merton* 2379; **2** from *Merton* 2811 & *Kennedy* 293.)

Holm Oak, a handsome evergreen tree up to 25 m. high, with dark grey, finely fissured bark and rather small, coriaceous leaves, 3–5 (–7) cm. long, ovate-oblong or lanceolate in outline, dark green above, whitish-tomentose below, with entire or remotely serrate margins, has been experimentally planted at Koronia and Makheras, and is noted by Frangos as cultivated in the Public Gardens, Nicosia. It is a native of the Mediterranean region, though widely cultivated in western and southern Europe and elsewhere, chiefly for ornament or shelter. The acorns are smaller than those of *Q. aegilops*, and the greyish-tomentellous acorn-cups are covered with short, adpressed scales. QUERCUS SUBER *L.*, Sp. Plant., ed. 1, 995 (1753), superficially similar to *Q. ilex*, but at once distinguished by its thick, soft corky bark, which provides the familiar "cork" of commerce, is reported as having been grown at Larnaca Nursery (Frangos in Cypr. Agric. Journ., 18: 88; 1923) but is evidently not a common tree, nor planted commercially, in Cyprus. It is a native of S. Europe and the western Mediterranean region, and is cultivated, for commercial cork, in many parts of Spain and Portugal.

Q. MACROCARPA *Michaux*, Hist. Chênes Amér., 2: tt. 2, 3 (1801); Frangos in Cypr. Agric. Journ., 18: 88 (1923), the Burr Oak, a deciduous tree often 30–50 m. in height, with very large, deeply 5–7-lobed leaves, sometimes 20–30 cm. long, and with a distinctive acorn-cup, fringed apically with a border of long, erect, slender scales, is noted by Frangos as grown in the Public Gardens at Nicosia, and may occur elsewhere in gardens. It is a native of eastern North America.

[*Q.* PETRAEA (*Mattuschka*) *Liebl.*, Fl. Fuld., 403 (1784) (*Q. robur* L. Spielart *Q. petraea* Mattuschka, Fl. Siles., 2: 375 (1777); *Q. sessiliflora* Salisb., Prodr. Stirp., 392 (1796); Holmboe, Veg. Cypr., 60; 1914), the Durmast Oak, a deciduous tree with sessile or subsessile fruits, is erroneously recorded from Cyprus by E. Hartman (in Mitt. Deutsch. Dendrol. Gesellsch., 170; 1905). *Q. petraea*, which has a very wide distribution in Europe, does not occur in our area, nor is there any evidence that it has been planted in Cyprus.]

3. Q. coccifera *L.*, Sp. Plant., ed. 1, 995 (1753); Boiss., Fl. Orient., 4: 1169 (1879); Holmboe, Veg. Cypr., 61 (1914); Post, Fl. Pal., ed. 2, 2: 521 (1933); Chapman, Cyprus Trees and Shrubs, 33 (1949); Davis, Fl. Turkey, 7: 681 (1982).

Evergreen shrub or small tree 1–6 (–10) m. high; bark greyish, smooth or scaly with age; young shoots subglabrous, stellate-pubescent or thinly tomentose, generally becoming glabrous after the first year; buds ovoid, blunt, 3–4 mm. long, 2–3 mm. wide, with numerous, imbricate, broad, blunt, shortly pubescent, brownish scales; leaves variable, coriaceous, oblong, 1–5 cm. long, 0·5–2·8 cm. wide, generally glabrous and shining, bright green above, glabrous or thinly stellate-pubescent, or nearly tomentose below at first, usually becoming glabrous with age, apex acute or obtuse, base rounded or often shallowly cordate, margins entire, subentire or spinose-dentate, with spreading or upwards-directed short or long, pungent teeth; petioles short, rarely exceeding 5 mm., stellate-pubescent or subglabrous; nervation generally obscure; stipules linear, about 5 mm. long, 1 mm. wide, brownish, membranous, thinly pilose, soon deciduous; male inflorescence usually solitary, axillary, scattered along the young shoots, 3–5 cm. long, with a short, slender, stellate-pilose peduncle, and stellate-pilose rhachis; flowers mostly solitary, rather remote; bracts caducous, linear, about 3 mm. long, 0·8 mm. wide, brownish-membranous, thinly pilose; perianth cup-shaped, about 1·5 mm. long, 2–2·5 mm. diam., pubescent externally, irregularly lobed, with 2–3 (–5) short, subacute lobes; stamens 4–5 (–8);

filaments glabrous, about 2–2·5 mm. long; anthers oblong, about 1·2 mm. long, 1 mm. wide, glabrous, with a shortly mucronate apex; female flowers solitary or sometimes paired, axillary, with a short, stout stipe usually less than 5 mm. long; involucre 4–6 mm. diam., glabrous or pubescent, closely scaly; ovary scarcely exserted; styles 3–4 (–6), recurved, 1–2 mm. long, shallowly channelled above with a subtruncate apex. Fruits solitary or in pairs, stipe 8–12 mm. long, stout and rigid; cupule 1–3 cm. diam.; 1–2·5 cm. long, usually covering more than half the length of the acorn, glabrous or pubescent, with short, rigid, erect, prickly scales or longer linear-subulate or ligulate, loosely erect or distinctly recurved scales; acorn 1·5–3 (–3·5) cm. long; 0·8–1·5 cm. diam., dull light brown, attenuate or rounded at apex, with a short, but very distinct, dome-shaped apiculus; endocarp densely tomentose.

ssp. **calliprinos** (*Webb*) *Holmboe*, Veg. Cypr., 61 (1914); Lindberg f., Iter Cypr., 11 (1946); Osorio-Tafall et Seraphim, List Vasc. Plants Cyprus, 28 (1973).

    Q. calliprinos Webb, Iter Hispaniense, 15 (1838); Jaubert et Spach, Ill. Plant. Orient., 1: 57B (1842–43); A.DC. in DC., Prodr., 16(2): 54 (1864); Unger et Kotschy, Die Insel Cypern, 217 (1865); Sintenis in Oesterr. Bot. Zeitschr., 31: 229, 394 (1881), 32: 123, 291 (1882); A. Camus, Les Chênes, 1: 451 (1938), t. 50, 51 (1934); Zohary in Bull. Res. Council Israel, 9D, Bot.: 181 (1961), Fl. Palaest., 1: 33, t. 32 (1966).

    Q. palaestina Kotschy, Die Eichen, t. 19 (1859–60); Post, Fl. Pal., ed. 2, 2: 522 (1933).

    Q. calliprinos Webb var. *eucalliprinos* A.DC. in DC., Prodr., 16 (2): 55 (1864); Zohary in Bull. Res. Council Israel, 9D, Bot.: 183 (1961).

    Q. calliprinos Webb var. *arcuata* Kotschy ex A.DC. in DC., Prodr., 16 (2): 56 (1864).

    Q. coccifera L. var. *calliprinos* (Webb) Boiss., Fl. Orient., 4: 1169 (1879); Post, Fl. Pal., ed. 2, 2: 522 (1933).

    Q. coccifera L. var. *palaestina* (Kotschy) Boiss., Fl. Orient., 4: 1170 (1879).

    Q. coccifera L. ssp. *palaestina* (Kotschy) Holmboe, Veg. Cypr., 61 (1914).

    Q. calliprinos Webb var. *palaestina* (Kotschy) Zohary in Bull. Res. Council Israel, 9D, Bot.: 183 (1961), Fl. Palaest., 1: 35 (1966).

    [*Q. pseudo-coccifera* (non Desf.) Labill., Icones Plant. Syr. Rar., 5: 9, t. 6, fig. 1 (1812); Hooker f. in Trans. Linn. Soc., 23: 381 (1862).]

    [*Q. coccifera* (non L.) Sm. in Sibth. et Sm., Fl. Graec. Prodr., 2: 239 (1816), Fl. Graec., 10: 35, t. 944 (1840).]

    [*Q. coccifera* L. var. *pseudococcifera* (non *Q. pseudo-coccifera* Desf.) Boiss., Fl. Orient., 4: 1169 (1879); Post, Fl. Pal., ed. 2, 2: 522 (1933); Lindberg f., Iter Cypr., 11 (1946).]

    [*Q. coccifera* L. ssp. *pseudococcifera* (non *Q. pseudo-coccifera* Desf.) Holmboe, Veg. Cypr., 61 (1914).]

TYPE: Lebanon; "in Libano", *Labillardière* (FI).

Generally more robust than ssp. *coccifera*, not infrequently a small tree 6–10 m. high; leaves of mature growths commonly 4–5 cm. long, 2–2·5 cm. wide, often with rather few, upwards-directed marginal teeth (sucker or sterile, coppiced growth usually with small, markedly spinose leaves); cupules generally large, often 2–3 cm. diam., with linear-subulate or strap-shaped, loosely adpressed (var. *calliprinos*) or distinctly recurved (var. *palaestina* (Kotschy) Boiss. = *Q. calliprinos* Webb var. *arcuata* Kotschy ex A.DC.), not rigid or pungent scales; acorn generally with a rounded (not attenuate-acute) apex.

HAB.: In maquis or amongst garigue, commonly on dry hillsides, sometimes in Pine forest or by roadsides; sea level to 4,500 ft. alt.; fl. March–May.

DISTR.: Divisions 1–3, 6–8, locally common; S. Turkey, Syria, Palestine.

1. Lyso, 1913, *Haradjian* 885!
2. Near Prodhromos, 1880, *Sintenis & Rigo* 519! also, 1905, *Holmboe* 913; near Kambos, 1898, *Post* s.n.! also, 1973, *P. Laukkonen* 376! Lemithou, 1905, *Holmboe* 937; Kalopanayiotis, 1905, *Holmboe* 1000; also, 1968, *A. Genneou* in ARI 1294! 1295! and 1968, *Economides* in ARI 1299! Between Pano Panayia and Kaminaria, 1905, *Holmboe* 1067; Vavatsinia, 1905, *Holmboe* 1107; between Perapedhi and Platres, 1934, *Syngrassides* 500! also, 1937, *Syngrassides* 1696! Stavros tis Psokas valley, 1937, *Chapman* 209! 223! 293A! Platres, 1937, *Kennedy* 302–307! also, 1957, *Merton* 3187! Kannaviou, 1939, *Chapman* 397! also 1939, *Lindberg f.* s.n.

3. Between Pissouri and Kouklia, 1862, *Kotschy*; Stavrovouni, 1880, *Sintenis*; Panayia Glossa above Kellaki, 1956, *Merton* 2542! 2543!
6. S. of Vizakia, 1956, *Merton* 12! 13! 14!
7. St. Hilarion (near "Fungi") ?1862, *Carletti* teste *Kotschy*; Pentadaktylos, 1880, *Sintenis & Rigo* 519a! Kyrenia, 1928, *H. Campbell* s.n.! also 1955, *G. E. Atherton* 311! and 1956, *G. E. Atherton* 1325! Lapithos, 1955, *Merton* 2236! and, 1969, *A. Genneou* in ARI 1300!
8. Near Cape Andreas, 1880, *Sintenis & Rigo*; Valia, 1905; *Holmboe* 576; also 1956, *Merton* 2467! and *Poore* 19! Ronnas valley, 1956, *Merton* 3066! Yialousa, 1970, *A. Genneou* in ARI 1484!

NOTES: The differences between *Q. coccifera* and *Q. calliprinos* are no greater than those between *Q. infectoria* and *Q. infectoria* ssp. *veneris*, nor do I think they warrant more than subspecific rank. The species exhibits a clinal pattern of variation, the typical, low, shrubby, small-fruited race being characteristic of France, Spain and the western Mediterranean, while the robust, large-fruited ssp. *calliprinos* is evidently restricted to the eastern Mediterranean, though numerous transitional forms are to be found in Italy, Greece, N. Africa and elsewhere. As remarked by Zohary (Bull. Res. Council Israel 9D: 182; 1961) the "distinguishing characters are not strictly exclusive but are in ensemble quite sufficient to differentiate the bulk of individuals ... from one another". The Cyprus populations are not very variable, and conform generally with var. *palaestina* (Kotschy) Boiss., though I doubt if analysis at varietal level is either practicable or rewarding.

Q. coccifera ssp. *coccifera* was once famed for the crimson dye obtained from the scale insect or Kermes (*Kermes vermilio*) infesting the twigs. The procedure followed in obtaining the dye is given by Loudon, Arb. et Frut. Brit., 3: 1909 (1844), but the importation of Cochineal, and the subsequent development of aniline dyes, killed the Kermes trade in the 19th century. Since Cilicia was anciently one of the sources of Kermes, it is likely that the insect occurs on *Quercus coccifera* ssp. *calliprinos* as well as on *Q. coccifera* ssp. *coccifera*, but this is not wholly certain, nor am I aware that the dye was ever manufactured in Cyprus.

92. SALICACEAE

Dioecious, deciduous (or rarely evergreen) trees and shrubs with soft, light wood; leaves simple, usually alternate, stipulate, entire or variously toothed or lobed; flowers in spicate or racemose catkins, each flower subtended by a membranous bract or catkin-scale; perianth cupuliform, cyathiform, spathaceous or reduced to 1 or more small nectaries, rarely absent; stamens 2 to many; filaments free or sometimes connate; anthers 2-thecous, dehiscing laterally; ovary unilocular, 2–4-valved; placentation basal or parietal; ovules usually numerous. Fruit a capsule; seeds usually numerous, small, surrounded by a tuft of long, silky hairs arising from the funicle; endosperm wanting or very sparse; embryo straight; cotyledons flat.

Genera 2: *Salix* L. (including the closely related *Chosenia* Nakai and *Toisusu* Kimura) and *Populus* L., both genera widely distributed in temperate regions of the northern hemisphere, a few *Salix* species extending into the tropics and temperate regions of the southern hemisphere, while *Populus* is represented in E. tropical Africa.

Buds with one outer scale; catkin-scales entire; perianth absent or reduced to one or more small nectaries - - - - - - - - - - - - - - **1. Salix**
Buds with several imbricate scales; catkin-scales lobed, laciniate or dentate, not entire; perianth cupular or cyathiform - - - - - - - - - **2. Populus**

1. SALIX L.

Sp. Plant., ed. 1, 1015 (1753).
Gen. Plant., ed. 5, 447 (1754).

Trees, shrubs and subshrubs with sympodial branching; buds with one scale (very rarely more), seldom viscid or balsamic; leaves usually longer than wide, linear-lanceolate, oblong or obovate; petioles short, not mark-edly compressed laterally; stipules often conspicuous and persistent; inflorescences erect or spreading, occasionally pendulous, sessile or ped-unculate, the peduncle often with well-developed foliaceous bracts; catkin-scales entire, persistent; flowers usually insect-pollinated, rarely wholly or partly wind-pollinated; perianth reduced to 1–2 (or more) small nectaries, rarely absent; stamens usually 2 (sometimes up to 12) with long, free or rarely united filaments; ovary 2-carpellate, 1-locular; style frequently well developed; stigmas generally narrow, spreading or recurved, frequently bifid. Fruit a small 2-valved capsule, the valves mostly recurving from the apex at maturity.

About 300 species, chiefly in temperate regions of the northern hemisphere, with a few species extending to S. America, tropical and S. Africa. The frequent occurrence of hybrids gives rise to taxonomic problems, and estimates of the number of species differ widely.

1. S. alba *L.*, Sp. Plant., ed. 1, 1021 (1753); Sibth. in Walpole, Travels, 22 (1820); Sibth. et Sm.,
Fl. Graec. Prodr., 2: 254 (1816); Poech, Enum. Plant. Ins. Cypr., 13 (1842); Unger et
Kotschy, Die Insel Cypern, 221 (1865); Boiss., Fl. Orient., 4: 1185 (1879); Holmboe, Veg.
Cypr., 58 (1914); Post, Fl. Pal., ed. 2, 2: 530 (1933); Lindberg f., Iter Cypr., 11 (1946);
Chapman, Cyprus Trees and Shrubs, 31 (1949); Rechinger f. in Arkiv för Bot., ser. 2, 1: 419
(1950); Zohary, Fl. Palaest., 1: 26, t. 25 (1966); Osorio-Tafall et Seraphim, List Vasc.
Plants Cyprus, 27 (1973); Davis, Fl. Turkey, 7: 704 (1982).

[*S. australior* (non Anderss.) R. Görz in Fedde Repert., 28: 114 (1930) pro parte quoad
plant. cypr.]

[*S. fragilis* (non L.) Druce in Rep. B.E.C., 9: 471 (1931); Osorio-Tafall et Seraphim, List
Vasc. Plants Cypr., 27 (1973).]

TYPE: "*ad pagos & urbes* Europae".

Tree up to 30 m. high, but often less than 10 m. high in our area; bark grey-brown, fissured; crown narrow, pyramidal or broad and spreading; twigs slender, sometimes slightly pendulous, at first thinly sericeous, soon glabrous or subglabrous and lustrous brown; leaves broadly or narrowly lanceolate, 6–10 (–15) cm. long, 1–2 (–2·5)) cm. wide, at first densely or sparsely adpressed-sericeous on both surfaces, sometimes remaining sericeous, but in our area soon glabrescent, often quite glabrous and shining green above, paler or glaucous below with a few, persistent hairs especially along the midrib, apex tapering to a slender acumen, base narrowly (or rarely rather broadly) cuneate, margins subentire or minutely and regularly serrulate with numerous small, glandular teeth; petiole sulcate above, usually less than 6 mm. long, and commonly adpressed-hairy with 2 small, apical glands; stipules narrowly lanceolate or subulate, less than 3 mm. long, caducous and wanting on mature shoots; catkins appearing with the leaves, narrowly cylindrical (1·5–) 3–5 (–7) cm. long, 0·5–0·8 (–1·2) cm. wide, suberect or spreading; peduncle short but distinct, generally less than 1·5 cm. long, rather densely pilose, bearing 3–4 entire, leaf-like bracts 1–2 cm. long, 0·2–0·7 cm. wide; rhachis densely pubescent; male flowers with 2 (rarely more) stamens; catkin-scales oblong, subacute or obtuse, concave, about 1·5–2·5 mm. long, 1 mm. wide, uniformly pale tawny yellow (sometimes becoming reddish in dried specimens), hairy especially along the margins and near the base; filaments up to 4 mm. long, densely pubescent towards base; anthers yellow, oblong, about 0·7 mm. long, 0·5 mm. wide; nectaries 2, the adaxial oblong, about 0·5 mm. long, 0·3 mm. wide, the abaxial narrow, almost linear, usually less than 0·2 mm. wide; female flowers often with slightly longer and more acute catkin-scales; ovary flask-shaped, glabrous, subsessile or very shortly pedicellate, about 3 mm. long, 1·3 mm. wide; style about 0·4 mm. long; stigmas short, spreading, slightly notched;

adaxial nectary as in male flower, abaxial often wanting. Ripe capsule up to 7 mm. long, 3·5 mm. wide.

HAB.: Streamsides and moist places; sea-level to 4,500 ft. alt.; fl. Febr.–April.

DISTR.: Divisions 1–5, 7, 8, perhaps always planted, but status uncertain. Widespread in Europe, the Mediterranean region and eastwards to Central Asia, but often appearing to be planted or subspontaneous.

1. S. of Polis, ?1862, *Kotschy*; Pelathousa, 1938, *Chapman* 341! Near Ayios Neophytos Monastery, 1939, *Lindberg f.* s.n.! Near Tsadha, 1955, *Merton* 2128! Lyso, 1962, *Meikle* 2281!
2. Near Troödhitissa monastery, 1787, *Sibthorp*; Kambos, 1928 or 1930, *Druce*; Platres, 1930, *E. Wall*; also, 1937, *Kennedy* 286! 287! Perapedhi, 1952, *Merton* 690! Kalopanayiotis, 1968, *A. Genneou* in ARI 1272!
3. Episkopi, 1905, *Holmboe* 701; Silikou, 1930, *C. B. Ussher* 100! Mamonia, 1931, *Nattrass* 38, 51, 812; Kato Lefkara, 1937, *Syngrassides* 1521! Evdhimou, 1956, *Merton* 2501!
4. Larnaca, 1877, *J. Ball ex herb. Post* 584!
5. Kythrea, 1880, *Sintenis & Rigo* 15! also, 1931–35, *Nattrass* 690, 807; and, 1939, *Lindberg f.* s.n.!
7. Kornos, 1934, *Syngrassides* 435!
8. Ronnas valley, 1957, *Merton* 2940!

NOTES: All the Cyprus specimens belong to a glabrescent or subglabrous variant, showing some approach to *S. excelsa* S. G. Gmelin (*S. australior* Anderss.), and, at first sight, rather distinct from the typical sericeous-leaved *S. alba* of western Europe. The Swedish salicologist B. Floderus has determined two of the collectings (*C. B. Ussher* 100 and *Syngrassides* 435) as "*S. micans* Anderss. f. *kassanogluensis* R. Görz*", but Görz (Fedde Repert., 28: 116; 1930) regarded *S. micans* as a variety of *S. alba*, nor am I satisfied that the name *S. alba* f. *kassanogluensis* can be applied to any of the Cyprus material. A representative selection of specimens could show every transition between sericeous-leaved *S. alba* at one extreme, and glabrous *S. fragilis* at the other, with *S. excelsa* falling about mid-way between the two. The Cyprus plants, which are not quite homogeneous, come between *S. alba* and *S. excelsa*, but rather closer to the former than the latter. Similar glabrescent forms are widely distributed in the eastern Mediterranean and eastwards to Iraq and Iran; they are questionably native in any of these areas, and may possibly be of hybrid origin. Pending detailed examination of the *S. alba-fragilis* complex, it seems best to leave them under *S. alba*.

S. BABYLONICA *L.*, Sp. Plant., ed. 1, 1017 (1753), the Weeping Willow, is similar to *S. alba* L., but with slender, pendulous branches, narrow, long-acuminate leaves and small, slender catkins, the males with filaments rarely more than 2 mm. long, the female with glabrous ovaries about 2 mm. long, 1 mm. wide.

Recorded, without precise locality, by Sibthorp (Fl. Graec. Prodr., 2: 252; 1816), and subsequently, on the sole basis of this record, by Poech (Enum. Plant. Ins. Cypr., 13; 1842) and Kotschy (in Unger et Kotschy, Die Insel Cypern, 221; 1865). Holmboe (Veg. Cypr., 58; 1914) dismisses Sibthorp's record as a probable error for *S. alba* L., but it may have been correct, though based upon introduced and cultivated specimens. Frangos (Cypr. Agric. Journ., 18: 89; 1923) remarks: "Some trees are to be found near the spring of Filia [Philia] village, 2 at Lapithos and 3 or 4 in the properties of Mr Paschali Constantinides at Onishia and Mr Neoptolemos Paschalis at Fountji." The first two localities are in Divisions 6 and 7 respectively; the other two have not been located.

S. babylonica, generally supposed to have come from China, is rarely seen in cultivation, most of the "Weeping Willows" now grown being either *S. alba* × *S. babylonica* (*S. sepulcralis* Simonk.) or *S. babylonica* × *S. fragilis* (*S. × pendulina* Wend.). In the absence of specimens it is impossible to ascertain whether the Cyprus plants are true *S. babylonica* or *babylonica* hybrids, or perhaps a mixture of the two.

2. POPULUS L.

Sp. Plant., ed. 1, 1034 (1753).
Gen. Plant., ed. 4, 456 (1754).

Trees, mostly with monopodial branching; buds with several imbricate scales, often viscid and balsamic; leaves usually broad, commonly deltoid-ovate or rhombic-ovate; petioles long, often strongly compressed laterally; stipules generally inconspicuous and caducous; inflorescences mostly without foliaceous bracts, the males (and sometimes the females) lax and pendulous; catkin-scales generally lobed or laciniate, not entire; flowers wind-pollinated; perianth cupular or cyathiform; stamens numerous, with short filaments; ovary 2–4-carpellate; stigmas sessile or subsessile, frequently dilated apically. Fruit 2–4-valved, the valves sometimes suberect or indistinctly recurved at maturity.

About 35–40 species chiefly in temperate regions of the northern hemisphere.

Leaves deltoid-ovate or rhomboid-ovate, glabrous or glabrescent, margins shortly serrate-dentate or subentire; petiole strongly compressed laterally   -   -   -   **1. P. nigra**
Leaves palmate-lobed or sinuate-lobulate, at first densely white-tomentose, and often remaining so on the undersurface; petioles not strongly compressed laterally
P. ALBA (p. 1490)

1. P. nigra L., Sp. Plant., ed. 1, 1034 (1753); Sibth. et Sm., Fl. Graec. Prodr., 2: 260 (1816); Sibthorp in Walpole, Travels, 22 (1820); Unger et Kotschy, Die Insel Cypern, 221 (1865); Boiss., Fl. Orient., 4: 1194 (1879); H. Stuart Thompson in Journ. Bot., 44: 338 (1906); Holmboe, Veg. Cypr., 58 (1914); W. Bugaĺa in Arboretum Kórnickie, 12: 45–219 (1967); Osorio-Tafall et Seraphim, List Vasc. Plants Cyprus, 28 (1973); Davis, Fl. Turkey, 7: 719 (1982).

Tree up to 30 m. high; trunk deeply fissured, usually dark grey, sometimes pallid, frequently with conspicuous nodular swellings or burrs; crown wide and spreading, often with massive, downturned branches, or narrow with fastigiate branching; twigs terete or angled when young, becoming terete later, shining ochre-brown; buds elongate-ovoid, often slightly curved outwards, 1–2·5 cm. long, 0·3–0·5 cm. wide, scales viscid, aromatic, lustrous brown; leaves deltoid-ovate or rhomboid-ovate, 4–8 (–10) cm. long, 2·5–8 (–10) cm. wide, or sometimes larger on vigorous sucker or coppice shoots, at first glabrous, or pubescent, later glabrescent, dark green above, slightly paler below, apex acute or cuspidate-acuminate, base broadly cuneate or truncate, margins usually distinctly but bluntly serrate-dentate, occasion-ally subentire; petiole up to 5 cm. long, glabrous or at first thinly pubescent, strongly compressed laterally, eglandular or obscurely glandular at junction with lamina; stipules deltoid-ovate, acuminate, 5–7 mm. long, 3–5 mm. wide, usually glabrous, soon caducous; catkins appearing in advance of the leaves in early spring, narrowly cylindrical, rather dense, 3–6 cm. long, 0·5–0·7 cm. wide, spreading or pendulous, shortly pedunculate or subsessile; bracts wanting; rhachis glabrous or subglabrous, catkin-scales broadly obcuneate or flabelliform, about 3 mm. long, 5–6 mm. wide, membranous, brown or pallid, glabrous, apex deeply and irregularly fimbriate-laciniate; male flowers subsessile; perianth broadly cupular-spathaceous, about 1·5 mm. long, 2 mm. wide, entire, glabrous; stamens 12–15; filaments usually less than 1 mm. long; anthers oblong, about 1·3 mm. long, 0·4 mm. wide, crimson; female flowers shortly pedicellate; perianth cupuliform, about 2·5 mm. diameter, glabrous; ovary very broadly flask-shaped or subglobose, about 2 mm. diameter, glabrous; stigmas 2, almost sessile, suborbicular, about 1·5 mm. diam., strongly deflexed at maturity, deeply emarginate at apex and base. Fruits distinctly pedicellate, with slender pedicels to 5 mm.

long, broadly ovoid, 6–7 mm. long, 5–6 mm. wide, papillose-verruculose; dehiscing by 2 valves when ripe.

var. **afghanica** *Aitch.* et *Hemsley* in Journ. Linn. Soc., 18: 96 (1880).
 P. thevestina Dode in Bull. Soc. Amis des Arbres, ? page (1903), Extr. Mon. Inéd. Genre Populus, 52 (1905).
 P. nigra L. var. *thevestina* (Dode) Bean, Trees & Shrubs, ed. 1, 2: 217 (1914).
 P. usbekistanica Kom. cv. *Afghanica* W. Bugała in Arboretum Kórnickie, 12: 164 (1967).
 [*P. pyramidalis* (non Salisb.) Boiss., Fl. Orient., 4: 1194 (1879) pro parte]
 [*P. dilatata* (non Ait.) Kotschy in Unger et Kotschy, Die Insel Cypern, 221 (1865).]
 [*P. nigra* L. var. *italica* (non Muenchh.) Holmboe, Veg. Cypr. 58 (1914); Chapman, Cyprus Trees and Shrubs, 31 (1949).]
 [*P. nigra* L. [sphalm. "*alba*"] var. *pyramidalis* (non Spach) A. H. Unwin, Forests of Cyprus, 11 (1925).]
 TYPE: Afghanistan; "in the vicinity of Shalizán", 1879, *Aitchison* 161 (K!).

Trees columnar, with narrow, fastigiate branching; trunk and main branches covered with conspicuous, white bark (see Notes); twigs slender, shining ochre-brown; leaves generally rhomboid-ovate, with a rounded or cuneate base, rounded lateral angles and shallowly serrate or almost entire margins; inflorescences apparently always female (male in the superficially similar *P. nigra* L. var. *italica* Muenchh.).

HAB.: Planted in gardens, and by the margins of streams and springs, especially in mountain valleys; sea-level to 4,500 ft. alt.; fl. Febr.–March.

DISTR.: Divisions 1–2, 5–7. Balkan peninsula; North Africa, eastern Mediterranean region eastwards to Central Asia; planted in our area.

1. Khrysokhou, 1862, *Kotschy*.
2. Below Troödhitissa Mon., 1787, *Sibthorp*; near Prophitis Elias Mon. ["Elias"], 1862, *Kotschy*; Prodhromos, 1862, *Kotschy*; "Selia" [? "Xylias"] not far from Stavrovouni ["Sta. Croce"], 1862, *Kotschy*; Saïttas, 1932, *Nattrass* 216; Kalopanayiotis, 1968, *A. Genneou* in ARI 1276!
5. Kythrea, 1862, *Kotschy*; and, 1931, *Nattrass* 137; Dheftera, 1931, *Nattrass* 633.
6. Nicosia, 1931 and 1935, *Nattrass* 115; 267; 731.
7. Lapithos, 1929, *C. B. Ussher* 60!

NOTES: Seldom collected, but all the specimens seen by the author seem to belong to *P. nigra* var. *afghanica* and not to typical *P. nigra*, nor is there any certain evidence that the latter is found in Cyprus.

P. nigra var. *afghanica* is locally frequent as a planted tree in the valleys of the Troödos and Northern Ranges. It much resembles the Lombardy Poplar (*P. nigra* L. var. *italica* Muenchh.) and is most readily distinguished by its ghostly white trunk and main branches. Bugała (Arboretum Kórnickie, 12: 158–182, 213; 1967) regards var. *afghanica* as a cultivar of the Central Asian *P. usbekistanica* Komarov, distinguishing this latter from *P. nigra* largely by its thick glabrous leaves with subentire margins and rounded basal angles; he adds that female flowers and fruits are almost sessile in *usbekistanica*, but stipitate in *nigra*. In none of these characters do I find reliable specific distinctions, a view shared with D. L. Clarke in Bean, Trees & Shrubs, ed. 8, 3: 318 (1976).

In a very interesting article on "Leucism in the Bark of Trees in Sunny Climates" (Gardeners' Chronicle, ser. 3, 123–164, figs. 80–82; 1948) Collingwood Ingram suggests that the white bark of *P. nigra* var. *afghanica* (*P. thevestina* Dode) is a physiological response to high light intensity, also to be seen in Cyprus examples of *Prunus avium*, *Platanus orientalis* and *Ficus carica*. Bean (Trees & Shrubs, ed. 1, 217; 1914) remarks that *P. thevestina* fails to develop this white bark if grown in the British Isles. In the circumstances one may question if *P. nigra* var. *afghanica* is distinct, even at varietal level, from the Lombardy Poplar, *P. nigra* var. *italica*, so widely cultivated in western Europe. The latter, or, rather, genuine specimens of the latter are, however, said always to be male, whereas var. *afghanica* is said always to be female. This provides one reason (though not a very good one) for maintaining the two as distinct. The sex of the Cyprus trees has yet to be ascertained. The leaves of var. *afghanica* are, on the whole more ovate-rhomboid and less deltoid than those of var. *italica*.

P. ALBA *L.*, Sp. Plant., ed. 1, 1034 (1753); Boiss., Fl. Orient., 4: 1193 (1879); Post, Fl. Pal., ed. 2, 2: 533 (1933); W. Bugała in Nasze Drzewa Lésne (Monografie Popularnonaukowe), 12: 19 (1973), trans. "The Poplars", by the Foreign Scientific Publ. Dept. Nat. Center Sci. Techn. Econ. Inf. Warsaw, 15 (1976); Osorio-Tafall et Seraphim, List Vasc. Plants Cyprus, 28 (1973); Davis, Fl. Turkey, 7: 717 (1982).

A tree 10–20 (–30) m. high; trunk smooth, grey, often marked with horizontal rows of conspicuous, dark lenticels; twigs at first densely covered with white, floccose tomentum, becoming dull grey-brown with age; buds ovoid, blunt, densely white-tomentose; leaves of two kinds: those of the leading and sucker shoots often 10 cm. long and about as wide, with 3–7 deep, palmate, coarsely serrate lobes, at first densely white-tomentose on both surfaces, becoming dark green and rather glossy above, remaining persistently tomentose below, petiole subterete, 3–5 cm. long, densely tomentose at first; leaves of short lateral shoots broadly ovate, 4–10 cm. long, 3·5–9 cm. wide, deeply sinuate-lobulate, at first thickly white-tomentose on both surfaces, becoming glabrous or subglabrous with age, or remaining thinly tomentose on the lower surface; male catkins 5–8 cm. long, with a densely tomentose rhachis, catkin-scales obovate-cuneate with an irregularly dentate, long-ciliate apex; stamens 5–10; anthers purple; female catkins 3–5 cm. long; ovary ovoid, glabrous; stigmas 2, greenish, cleft almost to base into 4 linear lobes. Capsule about 3 mm. long, 2 mm. wide, dehiscing by 2 valves.

The White Poplar is widely distributed in western and central Europe, the Mediterranean region and eastwards to Central Asia, but is often an obvious introduction, though possibly indigenous in parts of central and south-east Europe and eastwards to Central Asia. It is not native in Cyprus, and uncommon even as an introduction, though specimens have been collected at (1) Maa Beach, 1967, *Economides* in ARI 975! and at (2) Lapithiou near Pano Panayia, 1957, *Merton* 475! and, between Pedhoulas and Moutoullas, 1968, *A. Genneou* in ARI 1279! The tree is sometimes grown in gardens as an ornamental, but is objectionable because of its suckering habit, though the tomentose, Maple-like leaves of the leading shoots are attractive; it makes a good wind-break in exposed maritime areas.

MONOCOTYLEDONES

93. ORCHIDACEAE

by J. J. Wood.

R. Schlechter in Notizbl. Bot. Gart. Berlin, 9: 582–591 (1926).
J. E. De Langhe et R. D'Hose in Bull. Soc. Roy. Bot. Belg., 115: 297–311 (1982).

Perennial herbs displaying an extremely variable range of growth-forms, developmentally sympodial (in our area), i.e., with determinate annual apical growth, the following year's growth commencing from a lateral bud, terrestrial (in our area), autotrophic (autophytic), sometimes saprophytic, never parasitic. Rootstock variously tuberous or rhizomatous (in our area). Stem rounded or angular, solid or fistular. Leaves entire (in our area), spirally arranged or distichous, rarely subopposite, sometimes reduced to sheaths only, particularly in saprophytic species, either grouped into a basal rosette or cauline, sometimes spotted. Inflorescence lateral or terminal, a one- to many-flowered spike or raceme. Flowers gynandrous, zygomorphic, usually hermaphrodite, epigynous; perianth segments 6, in 2 whorls, consisting of 3 outer sepals and 3 inner petals, each free from, or connate with, or adnate to, one another, the ventral median petal or lip (*labellum*) often being highly modified in structure as an adaptation to insect pollination, often different in colour and frequently provided with a basal spur, sometimes bipartite, divided into a basal portion (*hypochile*) and an apical portion (*epichile*); anthers and stigmas borne on a gynandrium (*column*), formed from the fusing of the filaments and style, which is fleshy, cylindrical, erect or arcuate, sessile or extended into a foot; fertile anthers 1, rarely 2, sessile or subsessile, bilocular, borne dorsally on or at the apex of the column; pollen grains solitary or in tetrads, bound by elastic filaments (*elaters*), farinose, granular or waxy, formed into 2–8 rounded or clavate bodies (*pollinia*), which may be attenuated into a sterile stalk (*caudicle*); stigmas 3, 2 fertile, the median sterile and usually forming a beak-like structure (*rostellum*) between the anther and fertile stigmas; rostellum often forming 1 or 2 viscid discs (*viscidia*) to which the pollinia are attached; viscidia sometimes enclosed in membranous pouches (*bursicles*) derived from the rostellum. Ovary inferior, unilocular, with parietal placentation, rarely trilocular, usually pedicellate, often twisted through 180° so that the lip is positioned lowermost (*resupinate*). Mature fruit a dry capsule or fleshy pod dehiscing by 1–6 longitudinal sutures. Seeds very numerous, usually minute, dust-like, without endosperm, the embryo undifferentiated and developing by way of a mycorrhizal fungus symbiosis (*mycotrophy*) into a protocorm which has one growing point (*meristem*) and no radicle.

The family comprises between 750 and 800 genera and between 25,000 and 35,000 species, according to differing opinion, and is found throughout the world except in the extreme Arctic and Antarctic. The greatest number of genera and species occur in the humid tropics of both Old and New Worlds.

The tubers of several European and Middle Eastern orchids, particularly species of *Orchis*, contain a highly nutritious starch-like substance called Bassorin. This substance has for many centuries been extracted by drying and powdering the tubers. Collection of tubers for this purpose is practised in central and southern Europe, the Middle East and parts of Asia. The powder is exported under the Arabian name "Sahlep", corrupted into English as "Saloop" or "Salep". A nutritious drink of the same name is prepared from this powder, which some would claim to have aphrodisiac properties.

Plants saprophytic, without green leaves; stems clothed only in violet, violet-red or bluish-
    green sheaths; lip entire   -   -   -   -   -   -   -   -   - **1. Limodorum**
Plants autophytic, with green leaves; stems with usually only a few sheaths; lip 3-lobed or
    entire:
  Pollinia without caudicles at the base; bursicles always absent:
    Inflorescence rhachis never spirally twisted; plants rhizomatous; lip bipartite, divided into
      a basal hypochile and an apical epichile, without basal nectaries; viscidium absent:
      Epichile of lip bearing 3–10 longitudinal keels; ovary not twisted; column 7–10 mm.
        long; rostellum absent; pollinia crescent-shaped   -   -   - **2. Cephalanthera**
      Epichile of lip without longitudinal keels; ovary twisted; column usually shorter, 2–3·5
        mm. long; rostellum present, large; pollinia clavate   -   -   - **3. Epipactis**
    Inflorescence rhachis spirally twisted; plants ± tuberous at the base; lip not bipartite,
      either entire or lobed, connivent with the dorsal sepal and petals to form a tube, with 2
      glandular, papillose to hairy nectaries at the base; viscidium present   **4. Spiranthes**
  Pollinia with caudicles at the base; bursicles normally present:
    Lip bipartite, divided into a basal hypochile bearing 2 (in our area) rounded, ridge-like calli
      at the base and a long, deflexed tongue-like epichile; spur absent   -   **5. Serapias**
    Lip not bipartite, either 3-lobed or entire; spur present or absent:
      Spur present:
        Lip 3-lobed or, if entire, then ovate to orbicular with a crenulate margin, as long as
          broad; viscidium or viscidia enclosed in a bursicle:
          Both pollinia attached to a single viscidium:
            Spur filiform, 12–15 mm. long; inflorescence conical to shortly cylindrical, 2–10

(–12) cm. long, composed of small pink, rarely blood-red or white, flowers; lip 6–9 mm. long, with 2 short, narrow, longitudinal calli at the base; stigmas 2, clearly separated and placed one on each side of the anther   **8. Anacamptis**
Spur conical, saccate, 4–6 mm. long; inflorescence elongate, 6–23 cm. long, composed of large, olive-green to purple-violet flowers; lip 15–20 mm. long, with 2 parallel median ridges running from the mouth of the spur to the middle of the mid-lobe; stigma 1, centrally placed on the front of the column
9. Barlia
Each pollinium attached to a separate viscidium:
   Lip variously-shaped, 3-lobed or entire, never linear-oblong, ligulate; viscidia enclosed in a bursicle:
      Spur conical to filiform, always more than 2 mm. long; stigma 1, centrally placed on the front of the column, forming a roof to the spur entrance:
         Floral bracts membranous; tubers entire -   -   -   -   **11. Orchis**
         Floral bracts leafy, often rather fleshy; tubers normally palmately-lobed (in our area either entire or only shortly lobed)   -   **12. Dactylorhiza**
      Spur broadly conical, very short, 1–2 mm. long; stigmas 2, placed one on each side of the column at the spur entrance -   -   -   **13. Neotinea**
   Lip linear-oblong, ligulate, always entire; viscidia naked, not enclosed in a bursicle -   -   -   -   -   -   -   -   -   **6. Platanthera**
Spur absent:
   Lip velvety, resembling an insect, with an often shiny, brightly coloured, variously-shaped central marking (*speculum*); each viscidium enclosed in a separate bursicle -   -   -   -   -   -   -   -   -   -   **7. Ophrys**
   Lip resembling a man in outline, without a speculum; both viscidia enclosed in a single bursicle -   -   -   -   -   -   -   -   -   **10. Aceras**

1. LIMODORUM *Boehm.*

Ludwig, Defin. Gen. Plant., ed. Boehmer, 358 (1760) nom. cons.

Saprophytic, with very little or no chlorophyll; rhizome short and thick; stem erect, clothed with violet scale-like sheaths; green leaves absent; peduncle glabrous; inflorescence an erect, spike-like raceme; perianth segments free, the dorsal sepal erect or curving forward, convex, the lateral sepals spreading, ± flat, the petals spreading; labellum spurred, entire or obscurely bilobed at the apex, horizontal or deflexed, margin undulate, without calli, glabrous; column long and slender; viscidium solitary; bursicles absent.

Two species distributed in central and southern Europe, south-west Asia and North Africa.

1. **L. abortivum** (*L.*) *Swartz* in Nov. Act. Reg. Soc. Sci. Upsal., 6: 80 (1799); Kotschy in Oesterr. Bot. Zeitschr., 12: 279 (1862); Unger et Kotschy, Die Insel Cypern, 209 (1865); Boiss., Fl. Orient., 5: 89 (1884); Holmboe, Veg. Cypr., 56 (1914); Soó in Botanisch. Archiv, 23: 107, 184 (1928); Post, Fl. Pal., ed. 2, 2: 581 (1933); Warr et Gresham in Journ. R.H.S., 59: 72, 73 (1934); Lindberg f., Iter Cypr., 11 (1946); R. S. Davis in Orch. Journ., 3: 164 (1954); Elektra Megaw, Wild Flowers of Cyprus, 11, t. 18 (1973); Osorio-Tafall et Seraphim, List Vasc. Plants Cyprus, 25 (1973); Sundermann, Europ. Medit. Orch., ed. 2, 199 (1975), ed. 3, 225 (1980); Landwehr, Wilde Orch. Europa, 2: 524 (1977); J. et P. Davies in Quart. Bull. Alpine Gard. Soc., 48: 227 (1980); Robatsch in Die Orchidee, 31: 195–200 (1980); Davies et Huxley, Wild Orch. Brit. Europe, 69 (1983).
   Orchis abortiva L., Sp. Plant., ed. 1: 943 (1753).
   TYPE: "*in* Gallia, Helvetia, Anglia, Italia".

Plant (20–) 40–75 (–80) cm. high; rhizome short, stout, up to 1 cm. thick, with numerous thick, fleshy roots; stem violet, violet-red or bluish-green, erect, robust, thick, rigid, bearing numerous violet, violet-red or bluish-violet, ovate, obtuse, rarely acute sheaths possessing very little or no chlorophyll; green leaves absent; inflorescence 4–25-flowered, an elongate spike-like raceme up to 33 cm. long, lax to ± dense; floral bracts bluish-green, usually flushed violet, oblong-elliptic, acute or acuminate, concave, membranous, up to 4 cm. long, the lowermost exceeding the ovary, the uppermost a little shorter than the ovary; sepals pale to dark violet, flushed greyish-violet on the exterior, rarely rose-pink, erect or slightly spreading; dorsal sepal ovate to obovate, obtuse, strongly concave, sparsely papillose on the undersurface, erect or curved forward, forming a hood over the column, 20–25 mm. long, 7–11 mm. wide; lateral sepals obliquely ovate to oblanceolate, acute or subacuminate, flat or slightly concave, glabrous or only very sparsely papillose on the undersurface, 16–20 mm. long, 5–7 mm. wide; petals pale violet towards the apex, whitish towards the base, narrowly elliptic to linear, acute, glabrous, spreading, 16–19 mm. long, 3–4 mm. wide; labellum violet, fading to dingy-yellow with age, rarely rose-pink, adnate to the base of the column, distinctly constricted between the hypochile and epichile, 10–15 (–17) mm. long, 7–12 mm. wide, hypochile whitish or pale violet, deltoid, porrect, parallel to the column, epichile bright violet with darker violet longitudinal nerves and often a yellow flush at the centre, entire, triangular-ovate to oblong, obtuse, sometimes obscurely bilobed at the apex, curved, margin slightly crenulate-undulate, turned upwards, particularly towards the apex, glabrous, spur pale violet or whitish, cylindrical, slender, obtuse, downturned, c. 15 (–17) mm. long, seldom shorter, rarely absent; column erect, slender, 13–17 mm. long; anther oblong, 5 mm. long, 3 mm. wide; ovary shortly pedicellate, subclavate-cylindrical, not twisted, glabrous, 1·5 cm. long, pedicel 1 cm. long, much enlarged after anthesis.

HAB.: In *Pinus brutia* forest, or in dry grassy places in clearings or woodland margins, mostly on calcareous soils, but also on some igneous formations; 2,500–6,000 ft. alt.; fl. April–June (–July).

DISTR.: Divisions 2, 7; central, southern and south-east Europe, northwestwards to Belgium, eastwards to Turkey, Syria, Lebanon, Crimea, Caucasus, Iraq and western and northern Iran; North Africa.

2. Common locally on the Troödos Range; Prodhromos, Troödos, Platania, Kykko Monastery, Khrysorroyiatissa Monastery, Stavros tis Psokas, Cedar Valley, Platres, etc. *Kotschy* 744 ! *Sintenis & Rigo* 869 ! *Haradjian* 436 ! 997 ! *Feilden* 3 ! *C. B. Ussher* 64 ! *Syngrassides* 1102 ! 1594 ! *Kennedy* 1518 ! *Davis* 3419 ! 3510 ! *Casey* 1188 ! *Merton* 2782 ! *D. P. Young* 7246 ! 7257 ! 7372 ! *Suart* 127 ! *Merton* in ARI 757 ! *Economides* in ARI 962 ! 1215 ! *Hewer* 4619 ! etc.

7. N. of Kantara Castle, 1905, *Holmboe* 545.

NOTES: This distinctive plant can often be found growing in colonies and is exclusively autogamous. The emergent inflorescences remind one of asparagus shoots. Forms from Cyprus and S. Turkey with rose-pink flowers have been named var. *rubrum* Ruckbrodt.

2. **CEPHALANTHERA** *L. C. M. Rich.*

Mém. Mus. Paris, 4: 51 (1818).

Plants erect, sometimes slender; rhizome short, creeping, horizontal or perpendicular, bearing numerous, clustered, filiform roots; stem straight or flexuous, leafy; leaves cauline, flat to plicate, unspotted; inflorescence lax or subdense, usually few-flowered; floral bracts leafy towards base of inflorescence; flowers medium to large, showy; sepals subequal, ± connivent, concealing the labellum, glabrous or puberulous; petals slightly shorter than sepals; labellum bipartite, constricted and divided into a suberect, saccate, basal hypochile having lateral lobes which clasp the column, and a reniform or ovate-elliptic, obtuse or acute, distal epichile, which bears 3–10 longitudinal keels on the inner surface and has a recurved apex; spur absent in Cyprus species; column long, slender, erect; rostellum absent; stigma rounded, slightly concave; viscidia absent; bursicles absent; pollinia crescent-shaped; ovary sessile or subsessile, not twisted.

About twelve species distributed in Europe, North Africa and western Asia, extending through the Himalayas, eastwards to China, Indo-China and Japan, with one saprophytic species native to the north-western U.S.A.

Flowers pure white or creamy-white; upper part of stem, sepals and ovaries glabrous; epichile of
    lip obtuse:
    Leaves 3–5, abbreviated, mostly ovate-elliptic to oblong-ovate, not distichous, remote;
    sepals obtuse - - - - - - - - - - **1. C. damasonium**
    Leaves numerous, elongated, mostly linear-lanceolate, distichous, ± crowded; sepals acute
                                           2. C. longifolia
Flowers bright rose to purplish-pink; upper part of stem, sepals and ovaries glandular-
    pubescent; epichile of lip acute to acuminate - - - - - - **3. C. rubra**

1. C. damasonium (*Mill.*) Druce in Ann. Scot. Nat. Hist., no. 60: 225 (1906); Sundermann,
    Europ. Medit. Orch., ed. 1, 189 (1970), ed. 2, 201 (1975), ed. 3, 227 (1980); Osorio-Tafall et
    Seraphim, List Vasc. Plants Cyprus, 24 (1973); Landwehr, Wilde Orch. Europa, 2: 512
    (1977); Davies et Huxley, Wild Orch. Brit. Europe, 66 (1983).
    Serapias damasonium Mill., Gard. Dict., ed. 8, no. 2 (1768).
    S. latifolia Mill., Gard. Dict., ed. 8, no. 4 (1768) [*"S. latifolium"*] non Huds. (1762) nom.
    illeg.
    Epipactis alba Crantz, Stirp. Austr., ed. 2, 6: 460 (1769).
    Serapias pallens Jundz., Fl. Lithuan., 268 (1791).
    Cephalanthera pallens (Jundz.) L. C. M. Rich., Orch. Eur. Annot., 38 (1817), pre-print of
    Mém. Mus. Hist. Nat. Paris, 4: 60 (1818); Boiss., Fl. Orient., 5: 85 (1884).
    C. grandiflora S. F. Gray, Nat. Arr. Brit. Plants, 210 (1821); Unger et Kotschy, Die Insel
    Cypern, 209 (1865), [as *"C. grandiflora* Babingt., Brit. Bot., 296"]; Boiss., Fl. Orient., 5: 86
    (1884); H. Stuart Thompson in Journ. Bot., 44: 339 (1906), [as *"C. grandiflora* Bab."].
    Post, Fl. Pal., ed. 2, 2: 580 (1933) non *Serapias grandiflora* L. (1767).
    C. alba (Crantz) Simonk., Enum. Fl. Transs., 504 (1886); Holmboe, Veg. Cypr., 56
    (1914).
    C. latifolia Janch. in Mitt. Nat. Ver. Univ. Wien, 5: 111 (1907); Bull. Herb. Boiss., Ser. 2,
    7: 560 (1907); Soó in Botanisch. Archiv, 23: 113, 188 (1928).
    TYPE: Not indicated.

Plant 15–60 (–70) cm. high; rhizome short, rather thick, woody, with
numerous tough roots; stem erect or sometimes flexuous, rigid, angled,
glabrous, bearing 2–3 brownish, ± obtuse sheaths below, and leaves above;
leaves 3–5, cauline, remote, the lowermost short, ovate-elliptic, the middle
oblong-ovate, the uppermost narrowly elliptic, grading into bracts, ±
obtuse to acuminate, 4–8 (–10) cm. long, 2–3·5 cm. wide; inflorescence 3–12
(–16)-flowered, 4–12 (–15) cm. long, lax; floral bracts ± obtuse to
acuminate, the lowermost leaf-like, ovate-elliptic to narrowly elliptic, up to
6 cm. long, 2 cm. wide, exceeding the ovary, the uppermost short, narrowly
elliptic to linear, equalling the ovary; flowers creamy-white or white, tinged
with orange-yellow on the labellum, erect, not opening widely; sepals oblong
to oblong-elliptic, ± obtuse to subacute, glabrous, ± connivent, (14–)
17–20 (–25) mm. long, 5–8 (–10) mm. wide; lateral sepals oblique, concave;
petals ovate-oblong, obtuse, (12–) 14–17 (–20) mm. long, 5–8 mm. wide;
labellum shorter than the sepals and petals, erect, c. 12–13 mm. long,
hypochile with a median orange-yellow crescentic patch on the inner
surface, subsaccate, with several prominent nerves, lateral lobes obliquely
triangular-ovate, clasping the column, 10–13 mm. wide between lobes,
epichile short, reniform, cordate at the base, broader than long, 10–12 mm.
long, 6–8 mm. wide, obtuse, with 3–5 orange-yellow, longitudinal keels on
the inner surface, margin slightly crenulate-undulate; spur absent; column
slender, ridged, 10 mm. long; ovary sessile, fusiform, glabrous. Capsule
erect, elongate when ripe, persistent until autumn.

    HAB.: In woodland and other shady places, usually calcicole; c. 5,000 ft. alt.; fl. late
May–June.

    DISTR.: Division 2; widespread in western, central and southern Europe, northwards to
England and south-east Sweden, eastwards to the Caucasus and northern Iran; North Africa.

    2. "Selten in Wäldern der Schwarzfähren 22. Mai in Blättern bei Prodromo" *Kotschy* 758a.

    NOTES: A doubtful record in view of the fact that Kotschy seems only to have seen sterile
material. *C. damasonium* has not since been collected on the island.

2. C. longifolia (*L.*) *Fritsch* in Oesterr. Bot. Zeitschr., 38: 81 (1888); Sundermann, Europ. Medit. Orch., ed. 1, 189 (1970), ed. 2, 203 (1975), ed. 3, 229 (1980); Landwehr, Wilde Orch. Europa, 2: 514 (1977); Davies et Huxley, Wild Orch. Brit. Europe, 67 (1983).

Serapias helleborine L. var. *longifolia* L., Sp. Plant., ed. 1, 950 (1753).

S. longifolia Huds., Fl. Angl., ed. 1: 341 (1762).

S. ensifolia Murray in L., Syst. Veg., ed. 14: 815 (1784).

Cephalanthera ensifolia (Murray) L. C. M. Rich., Orch. Eur. Annot., 38 (1817), preprint of Mém. Mus. Hist. Nat. Paris, 4: 60 (1818); Boiss., Fl. Orient., 5: 85 (1884); Post, Fl. Pal., ed. 2, 2: 579 (1933).

TYPE: "*in* Europae *asperis*."

Plant 10–50 (–60) cm. high; rhizome short, fleshy, with numerous wiry roots; stem erect or sometimes flexuous, striate above, glabrous, with 2–4 whitish or greenish-white, obtuse sheaths below and often densely leafy above; leaves 4–8 (–12), cauline, alternate, ± distichous, mostly linear-lanceolate, the lowermost often oblong-elliptic to narrowly elliptic, acute to acuminate, (4–) 7–15 (–20) cm. long, 0·5–3 (–4) cm. wide, spreading, often recurved, mostly conduplicate; inflorescence few- to many-flowered, 5–15 (–20) cm. long, 3–4 cm. wide, lax, sometimes dense; floral bracts unequal, acuminate, the lowermost leafy, linear to linear-lanceolate, reaching 10 cm. long, the uppermost very small, ovate or triangular-ovate, 0·2–0·5 cm. long; flowers pure white with orange-yellow on the labellum, opening wider than those of *C. damasonium*; sepals narrowly-elliptic to lanceolate, acute, glabrous, slightly concave, (10–) 14–18 mm. long; dorsal sepal 4–5 mm. wide, lateral sepals oblique, 5–6 mm. wide; petals ovate, obtuse, 10–12 mm. long, 5 mm. wide, margins minutely papillose; labellum shorter than the sepals and petals, erect, 7–10 mm. long, hypochile with an orange-yellow patch at the base, slightly concave to ± saccate, lateral lobes obliquely oblong or triangular, clasping the column, 8–9 mm. wide, epichile orange-yellow at the apex, broader than long, reniform or transversely ovate, obtuse, margin somewhat erose, with 4–6 orange-yellow, longitudinal, parallel keels, apex densely papillose; spur absent; column slender, ridged, 7–8 mm. long; ovary sessile, narrowly cylindrical, glabrous. Capsule erect, elongate when ripe, persistent until autumn.

HAB.: In woodland and other shady places; ? alt.; fl. late April–early June.

DISTR.: Division unknown; a widespread species occurring throughout most of Europe, except the extreme north and much of the north-east, extending eastwards through western Asia and Pakistan to south-west China, Korea and Japan; North Africa.

NOTES: Recorded by Sundermann (op. cit. supra) but without exact locality.

3. C. rubra (*L.*) *L. C. M. Rich.*, Orch. Eur. Annot., 38 (1817), preprint of Mém. Mus. Hist. Nat. Paris, 4: 60 (1818); Boiss., Fl. Orient., 5; 84 (1884); Holmboe, Veg. Cypr., 56 (1914); Mrs Frank Tracey in Journ. Roy. Hort. Soc., 58: 304 (1933); Lindberg f., Iter Cypr., 11 (1946); R. S. Davis in Orch. Journ., 3: 164 (1954); Sundermann, Europ. Medit. Orch., ed. 1, 191 (1970), ed. 2, 205 (1975), ed. 3, 231 (1980); Osorio-Tafall et Seraphim, List Vasc. Plants Cyprus, 24 (1973); J. et P. Davies in Quart. Bull. Alpine Gard. Soc., 48: 228 (1980); Robatsch in Die Orchidee, 31: 195–200 (1980); Davies et Huxley, Wild Orch. Brit. Europe, 68 (1983).

Serapias rubra L., Syst. Nat., ed. 12, 2: 594 (1767).

TYPE: Not indicated.

Plant 10–60 cm. high; rhizome slender, with numerous filiform roots; stem erect, often flexuous, striate, glandular-pubescent above, with 3–5 brown to brownish-green sheaths below, leafy above; leaves (2–) 3–6 (–8), cauline, plicate, remote, the lowermost short, oblong-elliptic to narrowly elliptic, the uppermost grading to linear-lanceolate, obtuse, particularly the lowermost leaves, acute or acuminate, (3–) 5–10 (–14) cm. long, 1–2 (–3) cm. wide; inflorescence few to many-flowered, rarely 1-flowered, 2–12 (–20) cm. long, lax; floral bracts narrowly elliptic to linear-lanceolate, acute to acuminate, the lowermost leafy, often exceeding the flowers, to 6 cm. long, the uppermost exceeding the ovary; flowers bright rose-pink to purplish-pink, the labellum whitish with a violet-rose apex and margin and brownish-yellow keels; sepals oblong-elliptic or narrowly elliptic, obtuse to subacute, glandular-pubescent on the exterior, (15–) 20–25 mm. long, 5–8 mm. wide, spreading; dorsal sepal concave; lateral sepals slightly oblique; petals ovate-elliptic, acute, glabrous, 15–18 (–20) mm. long, 7–9 mm. wide; labellum glabrous, usually equalling the sepals, 18–23 mm. long, hypochile erect, saccate, 6–8 mm. long, with several keels on the inner surface, and with oblique-oblong lateral lobes which clasp the column, epichile ovate-elliptic, acute to acuminate, 12–15 mm. long, with 7–10 longitudinal keels on the inner surface; spur absent; column violet-rose, slender, ridged, slightly curved, c. 10 mm. long; ovary subsessile, slender, cylindrical, glandular-pubescent, enlarged after flowering.

HAB.: In woodland and woodland clearings, on igneous formations [usually calcicole elsewhere]; 4,000–6,000 ft. alt.; fl. May–June (–July).

DISTR.: Division 2; widespread in Europe, northwards to southern England and southern Finland, eastwards to the Caucasus and northern Iran; North Africa.

2. Locally common on the higher slopes of Khionistra about Troödos. *Holmboe* 1088; *Haradjian* s.n.! 435! *Feilden* 13! *Kennedy* 284! 285! 1517! *Lindberg* f. s.n.! *Davis* 1947! *Casey* 824! *F. M. Probyn* 116! 117! *Merton* 2783! *D. P. Young* 7835!

NOTES: *C. rubra* is generally found under *Pinus nigra*, particularly in the Troödos Range. It often grows in association with *Limodorum abortivum* and *Epipactis* spp.

3. EPIPACTIS *Zinn*

Cat. Plant. Gotting., 85 (1757) nom. cons.

Plants erect, robust or slender, rhizome short, thickened, creeping, horizontal or vertical, bearing numerous fleshy roots; stem straight, or often flexuous, glabrous or puberulous, leafy; leaves cauline, flat to plicate, unspotted; inflorescence often secund, lax to dense, few- to many-flowered; floral bracts leafy; flowers spreading or pendent, often partly or totally autogamous; sepals and petals spreading or connivent, arranged into a campanulate perianth; sepals subequal, glabrous, puberulous or pubescent; petals slightly shorter than sepals; labellum bipartite, constricted into a narrow fold or an elastic hinge in the middle, divided into a concave and either cymbiform, acetabuliform or cupuliform basal hypochile and variously-shaped, but usually cordate to ligulate, distal epichile bearing basal tuberculate calli or keels; spur absent; column short, rarely medium-sized; rostellum prominent, globose, sometimes absent; stigma rounded or subquadrate, sometimes bilobed; viscidia absent; bursicles absent; pollinia clavate or pyriform, sulcate; ovary pedicellate, clavate, glabrous or minutely papillose, lightly twisted.

About 30 species distributed in Europe, southwards to Ethiopia, through temperate Asia as far east as Japan, and with 1 species native to western North America.

Flowers large, 20–40 mm. diam.; sepals and petals with reddish-brown or dark purple banding; sepals 10–22 mm. long; plants robust, often 120–150 cm. high; leaves 10–25 (–28) cm. long; lip (9–) 19–23 mm. long; hypochile and epichile of lip with lateral lobes; column 4–10 mm. long - - - - - - - - - - - - - - **1. E. veratrifolia**
Flowers small, 10–20 mm. diam.; sepals and petals not banded as above; sepals 6–13 mm. long; plants slender or robust, 15–80 (–100) cm. high; leaves (1·5–) 3–12 (–17) cm. long; lip 6–11 mm. long; hypochile and epichile of lip without lateral lobes; column 2–3·5 mm. long:
  Leaves reduced, relatively small, placed remotely along the stem, often shorter than or equalling the internodes, (1·5–) 2–5·5 (–9·5) cm. long, 0·7–3·5 (–4·5) cm. wide; upper part of stem and inflorescence-rhachis densely pubescent or puberulous; lip 6–9 mm. long:
    Inflorescence few- to several-flowered, lax; leaves dark green, stained or suffused purple-violet or purple-red; epichile of lip with 2 small, rugose or rugulose basal calli:

Sepals glabrous, 10–12 mm. long, 5 mm. wide, yellowish- to dull olive-green; ovary
short-pedicelled, clavate, puberulous - - - - - - - **2. E. troodi**
Sepals pubescent, 6–8 mm. long, 3 mm. wide, green, tinged purple-red outside, whitish-
green inside; ovary long-pedicelled, fusiform, tomentose - **3. E. microphylla**
Inflorescence many-flowered, dense; leaves brownish- to yellowish-green, sometimes
suffused reddish-purple on the undersurface; epichile of lip with a single v-shaped
basal callus; sepals puberulous, 11 mm. long, 7 mm. wide, green **4. E. condensata**
Leaves well-developed, relatively large, placed in close succession along the stem, longer than
the internodes, 6–12 (–17) cm. long, 3·5–7 (–10) cm. wide; upper part of stem minutely
pubescent, inflorescence-rhachis puberulous; inflorescence many-flowered, lax to dense;
sepals glabrous; lip 9–11 mm. long; ovary long-pedicelled, clavate, glabrous or
puberulous - - - - - - - - - - - **5. E. helleborine**

1. **E. veratrifolia** *Boiss. et Hohen.* in Boiss., Diagn., 1, 13: 11 (1853); Boiss., Fl. Orient., 5: 87
(1884); H. Stuart Thompson in Journ. Bot., 44: 339 (1906); Holmboe, Veg. Cypr., 56
(1914); Post, Fl. Pal., ed. 2, 2: 580 (1933); R. S. Davis in Orch. Journ., 3: 164 (1954);
Sundermann, Europ. Medit. Orch., ed. 1, 203 (1970); Taubenheim et Sundermann in Die
Orchidee, 25: 8 (1947); Sundermann, Europ. Medit. Orch., ed. 2, 215 (1975), ed. 3, 245
(1980); Landwehr, Wilde Orch. Europa, 2: 508 (1977); J. et P. Davies in Quart. Bull. Alpine
Gard. Soc., 48: 228 (1980); Robatsch in Die Orchidee, 31; 195–200 (1980); Davies et
Huxley, Wild Orch. Brit. Europe, 56 (1983).
E. consimilis Wall. ex Hook. f., Fl. Brit. Ind., 6: 126 (1891); Osorio-Tafall et Seraphim,
List Vasc. Plants Cyprus, 25 (1973).
TYPE: U.S.S.R.; "in faucibus *March Mahal* montis *Elbrus* propè *Derbend* Kotschy No. 401 et
632".

Plant erect or pendent, robust, 20–120 (–150) cm. high; rhizome short,
0·3–0·4 cm. diam., often woody, creeping; stem straight or flexuous, densely
leafy throughout, glabrous below, minutely pubescent to tomentose above;
leaves 8–20, ovate-elliptic, ovate-lanceolate to linear-lanceolate, plicate,
acute to acuminate, 10–25 (–28) cm. long, 1–3 (–6) cm. wide, sheathing at the
base, papery, prominently nerved on the lower surface, the uppermost
grading into bracts; inflorescence few- to many-flowered, erect or slightly
curved, 9–24 cm. long, lax or subdense; floral bracts ovate, narrowly elliptic
to linear-lanceolate, leaf-like, 1–5 (–15) cm. long, to 1·5 cm. wide, the
lowermost often up to 25 cm. long, the uppermost ± as long as the ovary,
acute to acuminate, minutely pubescent; flowers large, 20–40 mm. diam.,
becoming pendulous with age; sepals spreading, green, buff or yellow, with
broad reddish-brown or dark purple areas, particularly on the minutely
pubescent outer surface, prominently nerved, the median nerve often
forming a low keel; dorsal sepal ovate-elliptic or narrowly elliptic, 10–21
mm. long, 4–7 mm. wide, obtuse to acute; lateral sepals obliquely ovate,
10–22 mm. long, (4–) 7–10 mm. wide, acute; petals pale green, with reddish-
brown or dark purple marginal bands, broadly ovate to ovate-elliptic, 8–18
mm. long, 4–8 mm. wide, acute, minutely pubescent or tomentose on the
outside of the prominent median nerve, otherwise glabrous; labellum
sometimes slightly shorter than the sepals, curved, hypochile oblong-
cymbiform, saccate, 10–12 mm. long, 5–7 mm. wide, white or buff, with two
basal, erect, auriculate, obliquely-triangular lateral lobes, the interior with
dark purple granular bosses and two parallel pale reddish-brown calli in the
middle, epichile ovate-elliptic, 9–11 mm. long, 4–5 mm. wide, reddish-brown
with a purple band and a white apex, fleshy, with two small, suberect, semi-
ovate lateral lobes and a raised median nerve, acute; column 4–10 mm. long,
suberect, with two porrect stelidia on the ventral margins towards the apex;
anther dark green, large, narrowly subclavate or semi-ellipsoidal, 4–5 mm.
long, very shortly stalked; rostellum subtriangular, porrect; ovary narrowly
cylindrical, indistinctly pedicellate, minutely pubescent. Ripe capsule
accrescent, distinctly pedicellate, pendulous.

HAB.: On stream banks and in damp, shady ground around springs; 2,900–6,400 ft. alt.; fl.
May–July (–Aug.).

DISTR.: Division 2. From Turkey (Anatolia), Syria, Lebanon, Israel, Palestine, Iraq, Iran, Yemen Arab Republic and Oman, eastwards through Afghanistan and Baluchistan to the Himalayas, southwards to Somalia and Ethiopia; North Africa (Algeria, Egypt).

2. Locally frequent on the higher slopes of central Troödos; Kakopetria, Khrysovrysi near Prodhromos, Amiandos, Platania, Troödos, etc. *A. G. & M. E. Lascelles* s.n.! *Holmboe* 1161! *Feilden* 4! *Haradjian* 411! *Syngrassides* 1267! *Davis* 1890! *Casey* 853! *F. M. Probyn* 126! *D. P. Young* 7410!

NOTES: This very widespread species is variable throughout its range, particularly in robustness and flower-size. Holmboe (1914) states that in the Cyprian plant most of the floral bracts are shorter than the fruits.

2. E. troodi *Lindberg f.* in Soc. Sci. Fenn., Årsbok, 20 B, no. 7: 5 (1942), Iter Cypr., 11 (1946); Sundermann, Europ. Medit. Orch., ed. 1, 197 (1970), ed. 2, 209 (1975); D. P. Young in Jahresb. Naturwiss. Ver. Wuppertal, 23: 107, 108 (1970); Osorio-Tafall et Seraphim, List Vasc. Plants Cyprus, 25 (1973); Hautzinger in Verhandl. Zool. Bot. Ges. Wien, 115: 42–44 (1976); Williams et al., Field Guide, 150 (1978); J. et P. Davies in Quart. Bull. Alpine Gard. Soc., 48: 228 (1980); Robatsch in Die Orchidee, 31: 195–200 (1980); Davies et Huxley, Wild Orch. Brit. Europe, 62–63 (1983).

E. persica (Soó) Nannfeldt ssp. *troodi* (Lindberg f.) Landwehr, Wilde Orch. Europa, 2: 510, 511 (1977), comb. non rite publ.

E. helleborine (L.) Crantz ssp. *troodi* (Lindberg f.) Sundermann, Europ. Medit. Orch., ed. 3, 41, 235, t. 259 (1980).

[*E. latifolia* (L.) All. var. *parvifolia* (non (Pers.) Richter) Holmboe, Veg. Cypr., 56 (1914).]

TYPE: Cyprus; "m. Troodos, Platania, in margine silvae juxta viam publicam, 18.6.1939, H. Lindberg" (H, K !).

Plant erect, slender, (15–) 25–45 cm. high; rhizome vertical or horizontal, branched, bearing numerous thick roots; stems solitary or several together, straight, leafy, with 3–5 remote, amplexicaul, brownish-green or green, flushed purple-violet sheaths towards the base, glabrous below, puberulous above, dark green to purplish-grey; leaves 3–4 (–5), shorter than or equalling the internodes, ovate to ovate-elliptic, sometimes narrowly elliptic, (1·5–) 3·5–6 (–7) cm. long, 0·7–2·8 cm. wide, acute to acuminate, glabrous or sometimes puberulous along the nerves, margin scabridulous, dark green, stained deep purple-violet, particularly on the lower surface; inflorescence normally few-flowered, rarely with up to 30 flowers, lax, the rhachis puberulous; floral bracts narrowly elliptic, acute to acuminate, 1·5–3·5 (–5) cm. long, 0·4–1·5 (–2) cm. wide, the lowermost as long as or exceeding the flowers, glabrous or puberulous along the nerves, green, stained purple-violet; flowers pendulous; sepals yellowish-green to dull olive-green; dorsal sepal ovate-elliptic, 10–12 mm. long, 5–6 mm. wide, concave, acute, glabrous; lateral sepals obliquely ovate-elliptic, 10–12 mm. long, 4–6 mm. wide, slightly to strongly concave, acute, glabrous; petals ovate to ovate-elliptic, 8–10 (–12) mm. long, 5 (–7) mm. wide, acute, whitish-green, the margins often suffused purple; labellum 6–8 mm. long, hypochile semi-globose, saccate, cymbiform, 3–4 mm. long, up to 4 mm. wide, olive-green, edged purple within or completely reddish-purple, epichile triangular-ovate, 4–5 (–6) mm. long, subobtuse, the margin slightly crenulate, with 2 small, rugulose calli at the base, reddish-purple, green at the centre; column 2–3 mm. long; rostellum effective in young flowers; ovary clavate, puberulous, with a short, slender pedicel. Ripe seed-capsule accrescent, to 16 mm. long. *Plate 86.*

HAB.: Frequent in cool, shady *Pinus brutia* and *P. nigra* ssp. *pallasiana* forest, often beside streams under *Platanus orientalis*, amongst bracken and sometimes associated with *Cephalanthera rubra*, *Limodorum abortivum* and *Epipactis condensata*; 2,500–6,200 ft. alt.; fl. (late May–) mid-June–mid-July.

DISTR.: Division 2. Also Turkey (Amanus Mountains, Hatay Province).

2. Locally common on the higher slopes of central Troödos: Platania, Kakopetria, Xerokolymbos, Kryos Potamos, Khionistra, Troödos, Prodhromos, etc. *Syngrassides* 1268!

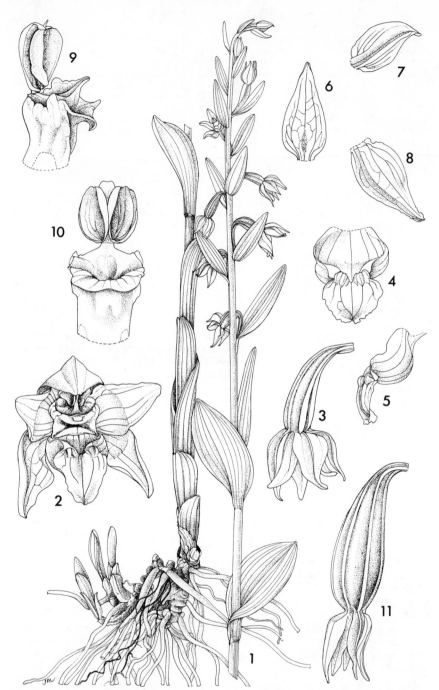

Plate 86. EPIPACTIS TROODI Lindberg f. **1**, habit, $\times \frac{2}{3}$; **2**, flower, front view, $\times 6.4$; **3**, flower, side view, $\times 3$; **4**, lip, front view, $\times 4$; **5**, lip, side view, $\times 4$; **6**, dorsal sepal, $\times 5$; **7**, petal, $\times 5$; **8**, lateral sepal, $\times 4$, **9**, column, side view with anther-cap lifted, $\times 4.6$; **10**, column, front view with anther-cap lifted, $\times 4.6$; **11**, capsule, $\times 2.2$. (**1** from *Lindberg f.* s.n. isotype; **2–10** from *Kennedy* 1509; **11** from *Davis* 1948.)

Kennedy 278! 279! 280a! 281! 282! 283! 1148! 1151! 1508! 1509! *Lindberg f.* s.n.! *Davis* 1781! 1795! 1937! 1948! *Casey* 955! *F. M. Probyn* 112a! *Holmboe* 934! *D. P. Young* 7837! 7852!

NOTES: This plant is to be found in full flower during June under the pines of the Troödos forests where it is often locally abundant. D. P. Young, in an unpublished note, points out that it is quite local and may possibly require a base-rich substrate, possibly of serpentine. All the collections I have examined have been from the igneous rocks of the Troödos massif.

The stems of the isotype are unusually pubescent. Lindberg f. in his amended description, published in 1946, describes the stem as follows: "Caulis 30–45 cm. altus, in suprema parte parce puberulus". All herbarium specimens that I have examined, apart from the isotype, agree well with this description.

Young (1970) describes the plant as varying little and appearing well-defined. The flowers are, from Young's field observations, probably facultatively autogamous and possibly self-fertile. It seems probable that this plant is closely allied to *E. microphylla*, a species having smaller, autogamous flowers. The larger flowers of *E. troodi* are likely to be more attractive to insects and cross-pollination by bees or wasps may take place. *E. troodi* possibly represents a transitional stage between cross-pollination and total autogamy.

Hautzinger (1976) treats *E. troodi* as being conspecific with *E. microphylla* (Ehrh.) Swartz var. *firmior* Schur, described from the Transylvania region of Romania in 1866. He admits, however, that he was unable to examine the type specimen of var. *firmior* owing to its inaccessibility. The variety is described as having oval leaves between 1¼ and 2 poll., i.e. to about 5 cm., long, which falls within the range found in *E. troodi*. According to Hautzinger, var. *firmior* occurs in Austria (southern Vienna woods), S.E. Europe, the larger Mediterranean islands, e.g. Sicily and Cyprus, and in Turkey. I am reluctant, however, to follow this treatment in this account as present knowledge of var. *firmior* remains obscure.

Taubenheim, in a personal communication to the author, points out that certain variants of *E. persica* (Soó) Nannfeldt from Iran and northern Turkey are morphologically very similar to the type material of *E. troodi*. *E. persica* is a variable species and further research is necessary to clarify the status of both these species.

3. E. microphylla (*Ehrh.*) *Swartz* in Kongl. Vetensk. Acad. Handl., 21: 232 (1800); Boiss., Fl. Orient., 5: 88 (1884); Robatsch in Die Orchidee, 31: 195–200 (1980); Davies et Huxley, Wild Orch. Brit. Europe, 62 (1988).
    Serapias microphylla Ehrh., Beitr. Naturk., 4: 42 (1789).
    Epipactis latifolia (L.) All. var. *microphylla* (Ehrh.) DC. in Lam. et DC., Fl. Franç., 6: 334 (1815); Boiss., Fl. Orient., 5: 88 (1884).
    TYPE: West Germany; "in Sylvis montanis Electoratus Brunsvico-Luneburgici".

Plant erect, slender, 15–40 (–50) cm. high; rhizome short, horizontal to vertical, bearing numerous partly elongated, wiry roots; stems solitary or grouped together, flexuous or straight, leafy, with 3–4 remote, amplexicaul, brownish-green sheaths towards the base, glabrous below, pubescent, sometimes greyish-tomentose above; leaves 2–4 (–6), shorter than or equalling the internodes, the lower and middle ovate to ovate-elliptic, 2–3 (–4) cm. long, (0·5–) 1–2 cm. wide, the uppermost bract-like, narrowly elliptic to linear-lanceolate, up to 3·5 cm. long, 0·3–0·8 cm. wide, acute to acuminate, glabrous, margin scabrid, often suffused purple-red; inflorescence few- to several-flowered, usually secund, lax, the rhachis densely pubescent; floral bracts narrowly elliptic to linear-lanceolate, acuminate, the lowermost as long as or exceeding the flowers, the uppermost gradually diminishing in length, glabrous, margin scabridulous; sepals and petals connivent, forming a campanulate perianth; sepals green, often tinged purple-red outside, whitish-green inside, oblong to ovate-elliptic, 6–8 mm. long, 3 mm. wide, slightly concave, acute, pubescent on the outer surface, the laterals oblique; petals whitish-green or greenish-rose, oblique or triangular-ovate, 6–7 mm. long, 3–4 mm. wide, acute; labellum 6 mm. long, subporrect, hypochile greenish-brown, semiglobose, saccate, acetabuliform, 2·5–3 mm. long and wide, abruptly contracted at the front, epichile whitish-green to pale pink, suborbicular-cordate to cordate-ligulate, 3–4 mm. long, subobtuse or acute, margin crenulate to slightly undulate, with two rugose calli at the base; column 3–3·5 mm. long; rostellum persistent; ovary fusiform, tomentose, with a long slender, tomentose pedicel. Ripe capsule accrescent, pendent.

HAB.: Shady woodland, often under hazel; vineyards, calcicole; 2,600–3,800 ft. alt.; fl. May–June (–early July).

DISTR.: Division 2. Also in eastern Spain, Balearic Islands, France, Benelux countries, Germany, Switzerland, Italy, Corsica, Sardinia, Sicily, Austria, Poland, Czechoslovakia, Hungary, Romania, Jugoslavia, Albania, Bulgaria, Greece, Crete, Turkey, Crimea, Caucasus and northern Iran, often local and sporadic.

2. Rare; Perapedhi, 1941, *Davis* 3454! Lagoudhera, 1955, *Merton* 2412!

4. E. condensata *Boiss. ex D. P. Young* in Jahresb. Naturwiss. Ver. Wuppertal, 23: 106 (1970); Sundermann, Europ. Medit. Orch., ed. 1; 197 (1970); Landwehr, Wilde Orch. Europa, 2: 504, 505 (1977); Williams et al., Field Guide, 152 (1978); J. et P. Davies in Quart. Bull. Alpine Gard. Soc., 48: 228 (1980); Robatsch in Die Orchidee, 31: 195–200 (1980); Davies et Huxley, Wild Orch. Brit. Europe, 60 (1983).
    E. condensata Boiss., Diagn., 2, 4: 91 (1859) pro syn.
    E. microphylla (Ehrh.) Swartz (var.) *β? congesta* Boiss., Fl. Orient., 5: 89 (1884).
    E. helleborine (L.) Crantz ssp. *condensata* (Boiss. ex D. P. Young) Sundermann, Europ. Medit. Orch., ed. 3, 41, 235, t. 257, 258 (1980).
    TYPE: Turkey; "Bounarbachi propè Smyrnam", *Balansa* 779 (G).

Plant erect, robust, 30–55 (–75) cm. high; rhizome thick, robust, vertical, bearing numerous, thick wiry roots; stems several, grouped together, straight, leafy, with progressively smaller, brownish-green amplexicaul sheaths towards and at the base, clothed with a brownish-yellow pubescent indumentum above, glabrous below; leaves 4–8, the lower and middle ovate to ovate-elliptic, 3·5–5·5 (–9·5) cm. long, 2·2–3·5 (–4·5) cm. wide, plicate, obtuse and mucronate to acute, glabrous, the uppermost bract-like, narrowly elliptic, 2·5–4 cm. long, up to 1·3 cm. wide, acute, puberulous on the undersurface, particularly towards the base, brownish to yellowish-green, sometimes suffused reddish-purple on the undersurface; inflorescence many-flowered, dense; floral bracts leafy, narrowly elliptic, acute, the lower and middle exceeding the flowers, the uppermost a little longer than the ovary, puberulous, particularly on the undersurface, flowers spreading to pendent; sepals green, ovate, 8–11 mm. long, 6–7 mm. wide, concave, acute, puberulous on the outer surface, the laterals oblique; petals delicate pink, flushed whitish in the centre and yellow at the apex, ovate, 9–10 mm. long, 5–6 mm. wide, acute, glabrous; labellum slightly shorter than the sepals, hypochile dark brown to purple-black inside, greenish, with a purple margin outside, hemispherical, cymbiform, saccate, 5 mm. long, 6 mm. wide, smooth on the outer surface, rather rugulose on the inner surface, containing nectar, epichile whitish, broadly deltoid to broadly cordate, 4 mm. long, 6 mm. wide, obtuse, margin rather crisped towards the apex, with a prominent reddish-violet, rugulose, shallowly V-shaped callus at the base; column 2–3 mm. long, up to 4 mm. wide, suberect or horizontal, with two small stelidia; anther large, ovate, 2·5 mm. long, 3·5 mm. wide, very shortly stalked; rostellum effective in young flowers, soon withering; ovary pyriform, papillose and usually clothed in a brownish-yellow indumentum.

HAB.: Amongst Bracken beneath *Pinus nigra* ssp. *pallasiana* in rocky places or on screes, often in rather exposed positions, on igneous formations; 5,000–6,000 ft. alt.; fl. July–Aug.

DISTR.: Division 2. Also in Turkey, Syria, Lebanon, recently found in the Caucasus (Georgia).

2. Khionistra, near Troödos and Kryos Potamos. *Kennedy* 280! 1149! 1150! 1510! *F. M. Probyn* 112! *D. P. Young* 7831! s.n.!

NOTES: This is a rare species in Cyprus and is apparently restricted to the *Pinus nigra* ssp. *pallasiana* forest zone in the Troödos Range, particularly around Khionistra (Mt. Olympus). It is a well-defined and easily recognisable species, most closely resembling the European *E. purpurata* Sm. but differing in its shorter, broader, yellowish-green leaves, brownish-yellow indumentum and smaller, differently coloured flowers. On first sight it may also resemble, perhaps, an odd form of *E. helleborine* (L.) Crantz, but again differs in its smaller leaves, sometimes shorter than the internodes, and in the dense-flowered inflorescence, bearing distinctive leafy bracts.

D. P. Young (1970) describes having seen, in Cyprus, young plants with the pollinia removed by insects. However, the rostellum soon withers and the flowers are probably autogamous in most instances.

5. E. helleborine (*L.*) *Crantz*, Stirp. Austr., ed. 2: 467 (1769); Sundermann, Europ. Medit. Orch., ed. 2, 207 (1975), ed. 3, 233 (1980); Landwehr, Wilde Orch. Europa, 2: 488 (1977); Davies et Huxley, Wild Orch. Brit. Europe, 57 (1983).

Serapias helleborine L., Sp. Plant., ed. 1, 949 (1753).

Epipactis latifolia (L.) All., Fl. Pedem., 2: 152 (1785); R. S. Davis in Orch. Journ., 3: 164 (1954).

TYPE: "*in* Europae *asperis*".

Plant erect, slender or robust, (15–) 35–80 (–100) cm. high; rhizome very short, thick, woody, oblique or horizontal, bearing numerous narrow roots; stem straight or slightly flexuous, leafy, with 2–4 remote, amplexicaul, brownish-green sheaths towards the base, glabrous and often purplish below, minutely pubescent above; leaves (3–) 5–9 (–12), evenly distributed along the stem, often large, the lowermost rather short and broad, ovate to ovate-elliptic, 3–8 cm. long, 2–6 cm. wide, amplexicaul, obtuse and mucronate to acute, the median the largest, ovate-elliptic to suborbicular, 6–12 (–17) cm. long, 3·5–7 (–10) cm. wide, amplexicaul, acute to shortly acuminate, the uppermost bract-like, narrowly elliptic, 3–10 cm. long, 0·7–3 cm. wide, acute to shortly acuminate, glabrous, margin and nerves on the abaxial and adaxial surfaces minutely papillose, scabrid; inflorescence many-flowered, sometimes secund, lax to dense, up to 25 (–35) cm. long, the rhachis puberulous; floral bracts ovate-elliptic to narrowly elliptic, acute to shortly acuminate, the lowermost exceeding the flowers, to 7 cm. long, the uppermost about as long as the ovary, glabrous, spreading; flowers spreading to pendent; sepals and petals connivent at the base, forming an open campanulate perianth; sepals green or olive-green, sometimes with brownish nerves, ovate-elliptic, (8–) 10–13 mm. long, 4–6 mm. wide, slightly concave, acute, glabrous, laterals oblique; petals pale green, suffused rose-violet, ovate or oblong, 8–11 mm. long, 4–6 mm. wide, acute or ± obtuse; labellum 9–11 mm. long, porrect, hypochile green outside, dark olive-brown, glossy inside, semi-globose, saccate, cupuliform, 4–6 mm. long and wide, containing nectar, epichile greenish-white, pink or violet, subreniform or cordate, 3·5–5 mm. long, 5–6 mm. wide, apex acute and deflexed, margin slightly crenulate or undulate, with two, sometimes three, smooth or rugulose calli at the base; column short, 2–3 mm. long, rather thick; rostellum with a white, round, persistent viscidium; anther ovate, to 3·5 mm. long; ovary clavate, glabrous or puberulous, with a long pedicel. Ripe capsule accrescent, ± spreading.

HAB.: In woodland, amongst scrub; 5,000–6,000 ft. alt.; fl. late June–early August.

DISTR.: Division 2, very rare. A very widespread species throughout most of Europe, east to Iran, Afghanistan and the Himalayas; North Africa; introduced into North America.

2. Kryos Potamos, 5,600 ft., 1938, *Kennedy* 1149a!

NOTES: There is also one unlocalized collection from Cyprus in the Kew herbarium. *E. helleborine* is most likely to be re-found in the forests of the Troödos massif.

E. PALUSTRIS (*L.*) *Crantz*, Stirp. Austr., 462 (1769); Post in Mém. Herb. Boiss., 18: 100 (1900); H. Stuart Thompson in Journ. Bot., 44: 339 (1906); Holmboe, Veg Cypr., 56 (1914).

Helleborine palustris Hill, Brit. Herb., 478 (1756); Soó in Botanisch. Archiv, 23: 117 (1928).

Rhizome stoloniferous; epichile of lip articulate, i.e. separated from the hypochile by a hinge. This species is recorded from northern Turkey (Anatolia), the Caucasus and Iran. Although a widespread plant in Europe and the Mediterranean Basin, it appears to be absent from Cyprus. Post's record of 1894 ("Eaux de Chypre, juillet") is most probably an erroneous one.

4. SPIRANTHES *L. C. M. Rich.*

De Orch. Eur. Annot., 20 (1817), preprint from
Mém. Mus. Hist. Nat. Paris, 4: 50 (1818) nom. cons.

Roots fleshy, fasciculate, ± tuberous, napiform to fusiform; stem slender, often glandular-pubescent above, leafy; leaves basally rosulate and cauline, conduplicate, fleshy, often glaucous, sometimes fugacious, unspotted; inflorescence rhachis usually spirally twisted; inflorescence dense; floral bracts small, membranous; flowers small, geniculate or arcuate, borne horizontally, sweetly scented; perianth segments subequal, the dorsal sepal erect and connivent with the petals to form a galea, the lateral sepals ± spreading; labellum entire or lobed, sessile or with a short claw at the base of which are usually 2 globular nectaries, margin undulate with an expanded, undulate or crenulate apex, forming a tube together with the dorsal sepal and petals; spur absent; column short, horizontal; clinandrium with prominent margins, often adnate to the rostellum; rostellum prominent, bifid, between the lobes of which is a single viscidium; stigmata 2, one on each side under the rostellum; anther erect, dorsal; caudicles of the pollinarium short.

Between 30 and 50 species distributed in the mostly temperate zones of Eurasia, east to Japan, south-east to Australia and New Zealand, North America, with a few Old and New World tropical species.

1. **S. spiralis** (*L.*) *Chevall.*, Fl. Gen. Env. Paris, 2: 330 (1827); Post, Fl. Pal., ed. 2, 2: 582 (1933); Summerhayes in Kew Bull., 4: 115 (1949); R. S. Davis in Orch. Journ., 3: 164 (1954); Osorio-Tafall et Seraphim, List Vasc. Plants Cyprus, 25 (1973); Sundermann, Europ. Medit. Orch., ed. 2, 191 (1975), ed. 3, 217 (1980); J. et P. Davies in Quart. Bull. Alpine Gard. Soc., 48: 228 (1980); Davies et Huxley, Wild Orch. Brit. Europe, 76 (1983).
 Ophrys spiralis L., Sp. Plant., ed. 1: 945 (1753).
 Spiranthes autumnalis L. C. M. Rich., Orch. Eur. Annot., 37 (1817) preprint from Mém. Mus. Hist. Nat. Paris, 4: 59 (1818).
 TYPE: "*in* Italiae, Galliae, Angliae *graminosis*".

Plant (6–) 10–25 (–35) cm. high; roots tuberous, 2–3, oblong-ellipsoid or napiform; stem slender, straight or flexuous, glandular-pubescent above, leafy at the base, with a few bract-like leaves above; radical leaves proteranthous, 2–10, rosulate, ovate to ovate-elliptic, attenuate towards the sheathing base, acute, 2–6 cm. long, 0·6–1·7 cm. wide, glabrous, glaucous, borne on the side of the flowering stem, erect to spreading, cauline leaves 3–7, lanceolate, acuminate, closely appressed, sheathing; inflorescence many-flowered, spirally twisted or with all of the flowers in a single row, 3–14 cm. long, dense; floral bracts ovate-elliptic, acuminate, slightly longer than or equal to the ovary; flowers scented during the daytime, white or greenish-white, the labellum often yellowish-green; sepals oblong or ovate-oblong, obtuse, sparsely glandular-pubescent, 4–5 (–7) mm. long, 1·5–2 mm. wide; dorsal sepal connivent with the petals; lateral sepals oblique, spreading; petals ligulate or oblong, obtuse, equal to or a little shorter than the sepals, 1 mm. wide; labellum, together with the dorsal sepal and petals, forming a trumpet-like tube, entire, ovate-oblong, minutely papillose, with 2 small nectaries at the base, margin crenulate towards the apex, 4–5 (–6) mm. long, 3·5–5 mm. wide; column short, 1·5–2 mm. long; rostellum triangular, apex shallowly bifid; ovary cylindrical, fusiform, curved at the apex, sparsely glandular-pubescent.

HAB.: In short turf and on grassy banks in garigue and *Pinus brutia* forest, preferring calcareous soils; sea-level–900 ft. alt.; fl. (Aug.–) Sept.–Nov.

DISTR.: Divisions 3, 6, 7. Widespread in western, central and southern Europe, east to Turkey, Caucasus area, Lebanon, Syria and northern Iran; North Africa.

3. Akrotiri, 1964, *P. H. Mason* s.n.!
6. Dhiorios, 1948, *Casey* 125!
7. Vasilia, 1940, *Davis* 1994!

5. SERAPIAS *L.*

Sp. Plant., ed. 1, 949 (1753) nom. cons.

Gen. Plant., ed. 5, 406 (1754).

Tubers 2–3 (–5), ovoid to globose, entire, sessile or stipitate; stem glabrous; leaves narrowly linear to broadly lanceolate, canaliculate, cauline, with spotted or unspotted sheathing bases; inflorescence lax to dense, 2–10 (–20)-flowered; floral bracts leafy, with numerous longitudinal nerves; perianth-segments connivent to form an acute to acuminate galea, the 3 outer sepals narrowly elliptic to ovate-elliptic, free at the apex only, the 2 inner petals ovate-elliptic, ovate-subulate to acuminate, free at the base only, glabrous; labellum without a spur, 3-lobed, usually broader and longer than the sepals, consisting of a basal hypochile, bearing 2 ± semicircular lateral lobes with (in Cyprus) 2 ridge-like calli arising laterally from the lower edges of the stigmatic cavity, extending on to a hirsute disc, and a usually prominent, narrowly elliptic to broadly ovate or cordate, 3-nerved, pendent epichile; column elongate, directed obliquely forwards; stigmatic cavity elongate or circular; rostellum small; anther-connective prominent, 7 (–10) mm. long; pollinia with caudicles; viscidium solitary, placed in a simple bursicle.

Between 5 and 10 species, according to authority, distributed in southern Europe and North Africa, extending west to the Canary Islands and Azores, east to Turkey and with the centre of distribution in the Mediterranean basin.

Lip (22–) 28–42 mm. long; galea 20–30 mm. long:
  Plants tall, (15–) 20–40 (–60) cm. high; inflorescence elongate, lax, rarely subdense; floral bracts noticeably exceeding the galea, the lowermost up to 7 cm. long; galea erect, rarely porrect; hypochile of lip concealed within or only slightly protruding from the galea; anther connective 5–7 mm. long- - - - - **1. S. vomeracea** ssp. **vomeracea**
  Plants normally short, (6–) 14–25 (–30) cm. high; inflorescence abbreviated, semi-lax to dense; floral bracts shorter than or only slightly exceeding the galea; galea porrect, becoming ± horizontal, particularly in the lowermost flowers; hypochile of lip ± protruding from the galea; anther connective 4–6 mm. long
              1. S. vomeracea ssp. **orientalis**
Lip (15–) 18–25 (–26) mm. long; galea 15–19 (–22) mm. long; inflorescence elongated, lax; floral bracts equal to or only slightly exceeding the galea; anther connective 3–4 mm. long
              1. S. vomeracea ssp. **laxiflora**

1. S. vomeracea *(Burm. f.) Briq.*, Prodr. Fl. Corse, 1: 378 (1910); Soó in Botanisch. Archiv, 23: 86, 167 (1928); Renz in Fedde Repert., 27: 218 (1929); Post, Fl. Pal., ed. 2, 2: 561 (1933); Rechinger f. in Arkiv för Bot., ser. 2, 1: 419 (1950); Sintenis in Die Orchidee, 20: 79 (1969); Sundermann, Europ. Medit. Orch., ed. 1, 97 (1970); Landwehr, Wilde Orch. Europa, 2: 356 (1977).

  Orchis vomeracea Burm. f. in Nov. Act. Phys.-med. Acad. Nat. Cur., 237 (1770).

  Helleborine longipetala Ten., Prodr. Fl. Nap., 53 (1810 or 1811).

  Helleborine pseudo-cordigera Sebast., Rom. Plant., fasc. 1: 14 (1813).

  Serapias pseudo-cordigera (Sebast.) Moric., Pl. Venet., 1: 374 (1820); Unger et Kotschy, Die Insel Cypern, 206 (1865); Sintenis in Oesterr. Bot. Zeitschr., 32: 262, 291, 398 (1882); C. E. Gresham in New Flora & Silva, 5: 273 (1933).

  Serapias longipetala (Ten.) Poll., Fl. Veron., 3: 30 (1824); Unger et Kotschy, Die Insel Cypern, 206 (1865); Holmboe, Veg. Cypr., 56 (1914); Post, Fl. Pal., ed. 2, 2: 561 (1933); R. S. Davis in Orch. Journ., 3: 164 (1954).

S. vomeracea (Burm. f.) Briq. l. *heldreichii* Soó in Fedde Repert., 24: 33 (1927); Renz in Fedde Repert., 27: 218 (1929).
   S. cordigera L. ssp. *vomeracea* (Burm. f.) Sundermann, Europ. Medit. Orch., ed. 3, 39, 125 (1980).

Plant 6–60 cm. high; tubers 2 (–3), globose to ovoid, sessile, the youngest sometimes shortly stipitate; stem erect, smooth, green, suffused red, reddish-violet, dull purple or dirty brown above, with 2 or several spotted, flecked or unspotted, membranous sheaths at the base; leaves 3–9, cauline, narrowly to broadly lanceolate, acute, sometimes obtuse and mucronate, with a speckled, flecked or unmarked sheathing base, the lowermost recurved, sometimes spreading, the uppermost erect or porrect, grading into narrowly-elliptic to ovate-elliptic, acute bracts, often reaching to or extending beyond the base of the inflorescence, the lowermost bluish- or yellowish-green, the uppermost often suffused with or entirely reddish-violet, purplish-red or dirty brown; inflorescence cylindrical, short or elongate, lax to dense; floral bracts narrowly elliptic to ovate-elliptic or ovate, acute to acuminate, shorter than or exceeding the galea, dull reddish- or bluish-violet or green with darker nerves; sepals and petals connivent, forming a horizontal, porrect or erect galea, the sepals often becoming free to various lengths towards the apex; sepals narrowly elliptic to ovate-elliptic, the laterals slightly oblique, acute to acuminate, 15–30 mm. long, dull greyish-, bluish- or purple-violet, sometimes dull pink or greenish-white on the exterior, darker purple, particularly towards the base, on the interior, nerves dark purple, dull purple-brown or greenish; petals broadly or triangular-ovate to circular at the base, attenuated into a subulate-acuminate apex, margin undulate-crenate to weakly notched or grooved, particularly towards the base, a little shorter than or equalling the sepals, blackish-purple or deep purple-red at the base, becoming paler, greenish-coloured distally; labellum longer than the sepals, 15–42 mm. long, hypochile concealed within or protruding from the galea, broad, erect or slightly spreading, pale purple-red to dark purplish-black, lobes rounded, basal calli 2, prominent or obscure, close or distant, parallel or slightly divergent, pinkish-violet, dull yellowish-red or brownish-green, sparsely or densely clothed above with long or short, white hairs; epichile narrowly lanceolate to narrowly cordate, 4–15 (–23) mm. wide, bright reddish-brown, dull ochre, purple-red, brownish-violet or pale yellowish-green, rarely white, nerves yellowish-green to brownish-purple, glabrous or with white hairs at the base, often strongly reflexed, constricted at the base, apex obtuse and mucronate or acute, margins flat or undulate; stigmatic cavity purple-red or pale purple-violet, short, rather broad; staminodes long or short; rostellum elongate; anther pale purple-violet or green, the connective 3–6 (–7) mm. long, purple-violet at the edge, pale green towards the centre; bursicle dark purple-red with a whitish margin, broad, elliptic to almost circular; caudicles of the pollinarium yellowish; pollinia greenish-yellow to dark bluish-green.

ssp. **vomeracea**
   TYPE: Corsica; not indicated.

Plant (15–) 20–40 (–60) cm. high; basal sheaths unspotted; leaves 4–6 (–7), narrowly linear-lanceolate to broadly lanceolate, with a green, unmarked sheathing base, (5–) 8–15 (–19) cm. long, 0·8–1 (–1·5) cm. wide; inflorescence elongate, (3–) 4–8 (–10)-flowered, lax, occasionally shorter and denser; floral bracts distinctly longer than the galea, the lowermost up to 7 cm. long; galea erect, rarely porrect; sepals dull grey to bluish-violet, rarely greenish-white externally, (20–) 23–28 (–30) mm. long; petals broadly ovate to triangular-ovate at the base, a little shorter than the sepals; labellum 30–40 (–42) mm. long; hypochile to 25 mm. wide when flattened, mostly concealed within or only slightly protruding from the galea, with broad, erect or slightly spreading, rounded lobes, basal calli 2, prominent, pinkish-violet, close or distant, with dense, long, white hairs in the upper part; epichile 9–12 mm. wide, bright rusty reddish-brown, brownish-violet, dull ochre, rarely pale yellowish-green or white, nerves brownish-purple, narrowly triangular-lanceolate, apex acute and often upturned, margins slightly undulate and sometimes erect, sparsely hairy at the base; staminodes long; anther-connective 5–6 (–7) mm. long.

HAB.: In damp meadows, garigue, Olive groves, *Pinus brutia* forest, or by roadsides; on calcareous or slightly acid soils; sea-level–1,200 ft. alt.; fl. late March–May.

DISTR.: Divisions 1–3, 7. Also Portugal, Spain, Balearic Islands, France, Switzerland, Italy, Corsica, Sardinia, Sicily, Jugoslavia, Albania, Bulgaria, Greece, Crete, Rhodes, Turkey, Caucasus area, Lebanon, Palestine; North Africa.

1. Tala, 1974, *Meikle* 4013! Between Khrysokhou and Polis, 1977, *J. J. Wood* 73!
2. Limnitis valley, 1954, *Mrs Grove in Casey* 1341!
3. Kakoradjia, 1938 *Syngrassides* 1808! Foot of Stavrovouni, 1959, *C. E. H. Sparrow* 31! Near Souni, 1960, *N. Macdonald* 127!
7. Kyrenia, 1948, *Duke-Woolley* 67!

NOTES: All the material cited intergrades with *S. vomeracea* ssp. *laxiflora*, as noted under the latter.

ssp. **orientalis** *W. Greuter*, Florist. Rep. Cretan Area, 19 (1972); Davies et Huxley, Wild Orch. Brit. Europe, 149 (1983).
  S. vomeracea (Burm. f.) Briq. Rassenkreis *Serapias orientalis* Erich Nelson, Mon. et Ikon. Serapias, etc., 16 (1968) nom. invalid.; Landwehr, Wilde Orch. Europa, 2: 366 (1977); J. et P. Davies in Quart. Bull. Alpine Gard. Soc., 48: 124, 229 (1980).
  S. vomeracea (Burm. f.) Briq. Rassenkreis *S. orientalis* Erich Nelson ssp. *orientalis* Erich Nelson, Mon. et Ikon. Serapias, etc., 17 (1968) nom. invalid.; Osorio-Tafall et Seraphim, List Vasc. Plants Cyprus, 26 (1973) nom. invalid.
  S. cordigera L. ssp. *orientalis* (W. Greuter) Sundermann, Europ. Medit. Orch., ed. 3, 39, 126 (1980).
  TYPE: Crete; "Kreta (Katochorio), in herbario Nelson Blütenanalyse Nr. 27, 20.4.1957".

Plant (6–) 14–25 (–30) cm. high; basal sheaths green or lightly speckled and flecked purple-red; leaves 4–6 (–9), broadly lanceolate, bluish- to yellowish-green with a green or lightly marked sheathing base, 6–13 (–16) cm. long, 1–2 (–2·5) cm. wide; inflorescence usually rather abbreviated, sometimes elongated, (2–) 3–6 (–10)-flowered, rather lax to dense; floral bracts shorter than to slightly longer than the galea, the lowermost sometimes noticeably longer in spikes that have only partially expanded; galea horizontal or porrect; sepals pale dull grey to dull purple-violet, sometimes yellowish-green on the exterior, (15–) 20–25 (–27) mm. long; petals ovate to circular at the base, a little shorter than or equalling the sepals; labellum (22–) 28–34 (–40) mm. long; hypochile 11–18 (–23) mm. wide when flattened, ± protruding from the galea, with 2 prominent, dull yellowish-red, close or distant calli at the base, clothed with rather long hairs above, lobes broad, erect or slightly spreading, rounded; epichile lanceolate, triangular-lanceolate or narrowly cordate, 10–15 (–23) mm. wide, dull ochre, yellowish-green, brownish to purple-red, nerves darker yellowish-green, brownish or purple-red, slightly constricted and long-pilose at the base, apex obtuse and mucronate or acute, margins slightly undulate; staminodes fairly long; anther-connective 4–6 mm. long.

HAB.: In marshy meadows, damp grassy places near streams, *Pinus brutia* forest, garigue, Olive groves, waste ground or by roadsides, primarily on calcareous soils; sea-level–2,000 ft. alt.; fl. (late Febr.–) March–April (–May).

DISTR.: Divisions 1–4, 6–8. Also southern Italy (Apulia, Calabria), Sicily, Jugoslavia, Greece, Crete, Aegean Islands, Rhodes, Turkey.

1. Near Ayios Yeoryios (Akamas), 1962, *Meikle* 2148! Between Baths of Aphrodite and Fontana Amorosa, 1977, *J. J. Wood* 78! Near Kathikas, 1977, *J. J. Wood* 116! Between Inia and Smyies, 1979, *Edmondson and McClintock* E 2759!
2. Between Ayios Amvrosios and Kissousa near Mallia, 1977, *J. J. Wood* 30!
3. Xylophaghou, 1862, *Kotschy* 178! Akrotiri Forest, 1939, *Mavromoustakis* 109!
4. Larnaca Salt Lake, 1977, *J. J. Wood* 3!
6. Kremama Kamilou, 1929, *Renz* 1772! Kormakiti, 1936, *Syngrassides* 1211! and, 1937, *Syngrassides* 1422! Between Kormakiti and Ayia Irini, 1941, *Davis* 2522! Dhiorios, 1952, *F. M. Probyn* 84! Cape Kormakiti, 1960, *N. Macdonald* 77!
7. Between Kyrenia and Lapithos, 1862, *Kotschy*; Near Dhavlos, 1880, *Sintenis & Rigo* 161! Kyrenia, 1900, *A. J. & M. E. Lascelles* s.n.! also, 1927, *Rev. A. Huddle* 92! 1929, *Renz* 1771! and, 1956, *G. E. Atherton* 1088! Halevga, 1941, *Davis* 2768! Lefkoniko Pass, 1949, *Casey* 654! Ayios Epiktitos, 1968, *Economides* in ARI 1142! Klepini, 1929, *Renz* 1770! Phlamoudhi and Ardhana, 1929, *Renz* 1774!
8. Valia, Ayios Theodhoros, 1905, *Holmboe* 497! Komi Kebir, 1929, *Renz* 1773! also, 1934, *Syngrassides* 1394! Ayios Andronikos, 1938, *Syngrassides* 1789! Rizokarpaso, 1941, *Davis* 2389! Leonarisso, 1968, *Economides* in ARI 1114!

NOTES: *S. vomeracea* ssp. *orientalis* superficially resembles *S. neglecta* De Not. and *S. cordigera* L., which are both western and central Mediterranean species. It intergrades with ssp. *vomeracea* in southern Italy (Apulia), Sicily, Crete and Cyprus, where the distributions of the two subspecies overlap (Nelson, 1968). Distinguishing characters such as dwarf stature, shorter inflorescence and floral bracts, usually broader epichile and a horizontal galea do, however, remain constant in many populations in Greece, Cyprus and Turkey, and are generally good guides in identification.

ssp. **laxiflora** (*Soó*) *Gölz et Reinhard* in Die Orchidee, 28: 114 (1977).
   S. laxiflora Chaub. in Bory et Chaub., Nouv. Fl. Pelop., 62 (1838); Unger et Kotschy, Die Insel Cypern, 206 (1865); Boiss., Fl. Orient., 5: 53 (1884); R. Stuart Thompson in Journ. Bot., 44: 338 (1906); Holmboe, Veg. Cypr., 56 (1914); Soó in Botanisch. Archiv, 23: 86, 167 (1928); Renz in Fedde Repert., 27: 210, 211, 218 (1929); R. S. Davis in Orch. Journ., 3: 164 (1954); Sundermann, Europ. Medit. Orch., ed. 1, 97 (1970); Davies et Huxley, Wild Orch. Brit. Europe, 149 (1983) nom. illeg.
   S. laxiflora Chaub. var. *columnae* Reichb. f., Icones Fl. Germ., 13 and 14: 13 (1851) nom. illeg.
   S. parviflora Parl. B (= var.) *columnae* Aschers. et Graebn., Syn. Mitteleur. Fl., 3: 779 (1907).
   S. columnae (Aschers. et Graebn.) Fleischm. in Oesterr. Bot. Zeitschr., 74: 190 (1925).
   S. parviflora Parl. ssp. *laxiflora* Soó in Fedde Repert., 24: 33 (1927); Soó in Botanisch. Archiv, 23: 86 (1928); Erich Nelson, Mon. et Ikon. Serapias, etc., 25 (1968); Osorio-Tafall et Seraphim, List Vasc. Plants Cyprus, 26 (1973); J. et P. Davies in Quart. Bull. Alpine Gard. Soc., 48: 124, 229 (1980).
   S. cordigera L. ssp. *laxiflora* (Soó) Sundermann, Europ. Medit. Orch., ed. 3, 125 (1980).
   TYPE: Greece: Peloponnese; "Sur le Plateau de Koubeh" (P, PC).

Plant (10–) 20–30 (–40) cm. high; basal sheaths frequently spotted and flecked purple-red; leaves (3–) 6 (–9), narrowly to broadly lanceolate, base sheathing, green or frequently heavily spotted and flecked purple-red, (6–) 10 (–18) cm. long, 0·5–1 (–1·5) cm. wide; inflorescence elongate, lax, peduncle occasionally slightly flexuous, (2–) 5–12 (or more)-flowered, the lower flowers often more remote than the upper; floral bracts equal to or slightly longer than the galea, the lowermost sometimes considerably longer; galea porrect; sepals pale green, dull pink or purple-violet externally, 15–19 (–22) mm. long; petals ovate at the base, a little shorter than the sepals; labellum (15–) 18–25 (–26) mm. long; hypochile 9–12 mm. wide when flattened, mostly concealed within or only slightly protruding from the galea, with 2 insignificant, brownish-green, close calli at the base, sparsely clothed with short hairs above, lobes erect, rounded; epichile narrowly lanceolate, 4–6 mm. wide, ± glabrous or with sparse, short hairs at the base, pale dull brownish to reddish-brown, sometimes a dirty yellowish-green, nerves greenish-brown, apex acute, margins flat; staminodes insignificant; anther-connective 3–4 mm. long.

HAB.: On dry, stony hillsides, dry meadows, garigue, roadsides, or sometimes in damper situations, frequently near the coast; on calcareous or slightly acid soils; sea-level–1,000 ft. alt.; fl. late March–early May.

DISTR.: Divisions 1, 3, 5. Also north west and southern Italy (Apulia, Calabria), Sicily, Yugoslavia, Albania, Greece, Crete, Aegean Islands, Rhodes, southern Turkey.

1. Between Lachi and Neokhorio (Akamas), 1977, *J. J. Wood* 104!
3. Lefkara, 1865, *Kotschy* 234; between Limassol and Omodhos, 1862, *Kotschy* 413; between Anglisidhes and Skarinou, 1929, *Renz* 1714! 1741! Between Lefkara and Kornos, 1929, *Renz*; Kakoradjia, 1937, *Syngrassides* 1528! Episkopi, 1959, *C. E. H. Sparrow* 31a! 1963, *J. B. Suart* A4! Akrotiri, 1963, *J. B. Suart* 98! 11 m. S.W. of Larnaca, 1978, *Chesterman* 38! Menoyia, *Chesterman* 113!
5. Delikipo, 1929, *Renz* 1742!

NOTES: Intermediates between ssp. *laxiflora* and ssp. *vomeracea* are frequent in many populations. A clear-cut line of demarcation cannot always be drawn between the two in those areas where their distributions overlap. *S. vomeracea* ssp. *laxiflora* is regarded by many authors as a subspecies of *S. parviflora* Parl., which is, however, a distinct and uniform small-flowered plant distributed from Portugal to western Turkey and Crete, but absent from Cyprus.

The presence or absence of purple markings on the basal and leaf sheaths seems to be a rather variable character. Nelson (1968) describes the sheaths as being frequently spotted, but many populations observed by the author in both Cyprus and eastern Crete had unspotted, green sheaths.

6. PLATANTHERA *L. C. M. Rich.*

De Orch. Eur. Annot., 20, 26 (1817), reprint from
Mém Mus. Hist. Nat. Paris, 4: 42, 48 (1818) nom. cons.

Tubers entire, napiform or oblong, attenuate; stem erect, slender, glabrous; basal leaves 1–2 (–3) (in sect. *Euplatanthera*), broad, unspotted, cauline leaves 2–4 (–5), bract-like; inflorescence lax to rather dense; floral bracts leafy; flowers white, white and green, yellowish-green or entirely green, often sweetly scented; dorsal sepal connivent with the slightly shorter petals to form a galea; lateral sepals spreading; labellum entire, linear-oblong, ligulate, pendent; spur long, filiform or narrowly cylindrical, sometimes clavate above the middle; column short, truncate; anther united with the column, the loculi parallel or diverging downwards; staminodes distinct, small and rounded; rostellum flattened, triangular, with a rudimentary median lobe; viscidia rounded or oval, without bursicles and naked; stigma concave; ovary sessile, twisted, glabrous.

Between 80 and 100 species distributed over Europe and North Africa, extending eastwards across Asia to New Guinea; North and Central America.

Flowers yellowish-green and white; dorsal sepal 5–10 mm. long; lateral sepals 9–12 mm. long; lip 11–15 (–20) mm. long; spur of lip slightly dilated to clavate above the middle, 2–3 (–4) cm. long - - - - - - - - **1. P. chlorantha** ssp. **chlorantha**
Flowers entirely green; dorsal sepal 5–7 mm. long; lateral sepals 7–10 mm. long; lip 7–10 mm. long; spur of lip narrow, attenuate, 1·7–2·1 cm. long - **1. P. chlorantha** ssp. **holmboei**

1. P. chlorantha *(Custer) Reichb.* in Moessler, Handb., ed. 2, 2: 1565 (1828); Briquet, Prodr. Fl. Corse, 1: 383 (1910); Soó in Botanisch. Archiv, 23: 103, 178 (1928); Sundermann, Europ. Medit. Orch., ed. 1: 173 (1970); Osorio-Tafall et Seraphim, List Vasc. Plants Cyprus, 25 (1973); Sundermann, Europ. Medit. Orch., ed. 2, 183 (1975); ed. 3, 205 (1980); Davies et Huxley, Wild Orch. Brit. Europe, 84 (1983).
Orchis chlorantha Custer in Neue Alpina, 2: 401 (1827).
[*Platanthera montana* Reichb. f., Icon. Fl. Germ. et Helv., 13: 123 (1851); Unger et Kotschy, Die Insel Cypern, 209 (1865); Boiss., Fl. Orient., 5: 83 (1884) non *Orchis montana* F. W. Schmidt (= *P. bifolia* (L.) L. C. M. Rich.).]

Plant erect, robust, (18–) 20–80 cm. high; tubers 2, ovoid or conical, abruptly attenuated into an elongate apical segment; stem straight, stout, bearing 2–5 bract-like, narrowly elliptic, acute cauline leaves; basal foliage leaves 2 (–3), sub-opposite, erect to spreading, broadly oblong to oblong-elliptic or obovate and attenuate towards the base, obtuse, 6–15 (–20) cm. long, (1·8–) 2·5–5 (–8) cm. wide; inflorescence few- to many-flowered, cylindrical, sometimes elongate, to 20 cm. long, lax to dense; floral bracts narrowly elliptic, the lowermost longer than the ovary, acute; flowers yellowish-green and white or wholly pale green or yellowish-green, sweetly scented during the evening and night; dorsal sepal broadly cordate-ovate, 5–10 mm. long, 7–9 mm. wide at the base, obtuse, forming a galea over the column together with the petals; lateral sepals obliquely ovate-elliptic, 7–12 mm. long, 3–6 mm. wide towards the base, obtuse, spreading, becoming slightly reflexed; petals linear-lanceolate, somewhat falcate, 6–9 mm. long, 2–3 mm. wide at the base, obtuse, apices touching; labellum linear-oblong, ligulate, longer than the sepals, 7–15 (–20) mm. long, 3–4 mm. wide, pendent, straight or curved, distinctly yellowish-green with a darker green apex or white with a green apex; spur filiform, either slightly dilated to clavate above the middle, or attenuate and rarely slightly dilated above the middle, 1·8–3 (–4) cm. long, horizontal, decurved, rarely ascending, being sometimes found in all positions on the same spike, yellowish-green, darker green towards the apex; column short, broad; stigma concave, scallop-shaped, green edged; anther loculi diverging towards the base; anther-connective broad, truncate or slightly emarginate above; viscidia large, circular, one on each side of the stigma; bursicles absent; pollinarium 4 mm. long, the caudicles longer than the pollinia, transparent bright yellow; ovary sessile, cylindrical, twisted.

ssp. **chlorantha**
TYPE: Switzerland; "E montibus Abbatiscellanis [Appenzell] in Helvetia". *Stein.* (? Steinmueller).

Plant 20–80 cm. high; leaves broadly oblong to oblong-elliptic or obovate and attenuate towards the base, 6–15 (–20) cm. long, 2·5–5 (–8) cm. wide; inflorescence lax to dense, many-flowered; flowers yellowish-green and white; sepals usually more oval than in ssp. *holmboei*; dorsal sepal 5–10 mm. long; lateral sepals 9–12 mm. long, 5–6 mm. wide at the base; labellum 11–15 (–20) mm. long, 3–4 mm. wide, straight; spur filiform, slightly dilated to clavate above the middle, 2–3 (–4) cm. long.

HAB.: In *Pinus brutia* forest, in clearings, in damp grassy places; 2,500–2,900 ft. alt.; fl. June–July.

DISTR.: Division 2. Distributed over much of Europe, from Iceland eastwards to Russia, the Caucasus, Turkey and northern Iran, extending into Siberia, northern China and Japan; North Africa.

2. Kryos Potamos, 2,900 ft. alt., 1937, *Kennedy* 272! Stavros tis Psokas, 1962, *Meikle* 2700a!

ssp. **holmboei** (*Lindberg f.*) *J. J. Wood* in Orch. Rev., 88: 122 (1980); Davies et Huxley, Wild Orch. Brit. Europe, 84 (1983).
P. *holmboei* Lindberg f. in Soc. Sci. Fenn. Arsbok, 20 B, 7: 5 (1942), Iter Cypr., 11 (1946); Summerhayes in Kew Bull., 4: 114 (1949); R. S. Davis in Orch. Journ., 3: 164 (1954) [as "*P. holombii*"]; N. Feinbrun in Bull. Research Council Israel, 8 D: 172 (1960); Dafni in Mitteilungsbl. Arbeitskr. Heim. Orch. Baden-Würt., 11: 208 (1979); J. et P. Davies in Quart. Bull. Alpine Gard. Soc., 48: 227 (1980); Robatsch in Die Orchidee, 31: 195–200 (1980); Sundermann, Europ. Medit. Orch., ed. 3, 205 (1980).
[P. *bifolia* (L.) L. C. M. Rich. ssp. *montana* Holmboe, Veg. Cypr., 58 (1914) non *Orchis montana* F. W. Schmidt]
[P. *chlorantha* (non (Custer) Reichb.) Sundermann in Die Orchidee, 20: 79 (1969).]
TYPE: Cyprus; Khionistra, "juxta «Military Camp»" (H !).

Plant (15–) 20–30 (–40) cm. high; stem somewhat square in cross-section; leaves 1–3 or 4, obovate or broadly oblong-elliptic, 6–12 (–18) cm. long, 1·8–4·5 cm. wide; inflorescence usually lax, few-flowered; flowers wholly pale green or yellowish-green with a whitish area around spur-mouth; sepals usually more deltoid than in ssp. *chlorantha*, slightly twisted at apex; dorsal sepal 5–7 mm. long; lateral sepals 7–10 mm. long, 3–4 mm. wide at the base;

labellum 7–10 mm. long, 1·5–3 mm. wide, curved; spur filiform, attenuate, rarely slightly dilated above the middle, 1·7–3 cm. long. *Plate 87.*

HAB.: On shaded slopes under *Pinus brutia, Platanus orientalis* and *Quercus alnifolia*; under *Pinus nigra* ssp. *pallasiana* at higher elevations and also in shady places beside streams in garigue, on igneous formations; 2,000–6,000 ft. alt.; fl. (late April–) late May–June (–early July).

DISTR.: Division 2. Also Turkey, N. Syria and Israel.

2. Locally frequent on the Troödos Range: Prodhromos, Platres, Stavros tis Psokas, Livadhi tou Pasha, Kryos Potamos, Troödos, etc. *Kotschy* 755! *Holmboe* 867! *Syngrassides* 1593! *Kennedy* 271! 273–277! 1511–1513! *Lindberg f.* s.n.! *Davis* 3504! *N. Macdonald* 168! *D. P. Young* 7254! 7448! *Meikle* 2700! 2842! *P. Laukkonen* 400! etc.

NOTES: This distinctive plant differs from the typical subspecies in its rather smaller, entirely green flowers which have a slender, attenuate spur. It is a plant of Pine forests on igneous formations and is consequently restricted to the Troödos massif.

P. chlorantha ssp. *holmboei* can no longer be regarded as a Cyprus endemic. Feinbrun (1960) and Dafni (1979) record it from the Upper Galilee region of Palestine. Recent examination of an alcohol specimen at Kew collected from a limestone area north of Latakia has confirmed its occurrence in Syria; it is also cited from Turkey by Davies & Huxley (1983).

7. OPHRYS *L.*

Sp. Plant., ed. 1, 945 (1753).
Gen. Plant., ed. 5, 406 (1754).

Tubers 2 (–3), globose or ovoid, sessile or stipitate; stem glabrous; leaves basal, rosulate, and/or cauline, unspotted; perianth segments free, ± spreading, the 3 outer sepals larger, usually glabrous, the 2 inner petals smaller, often hirsute; labellum without a spur, entire or 3-lobed, glabrous or velutinous, often strongly convex, sometimes with 2 conical protuberances at the base, or at the base of the lateral lobes, and with a variously shaped, often shining, glabrous central area or *speculum*, above which, in most species, is an often brightly coloured, shield-shaped area or *basalfield*, at the base and to either side of which may be 2 small swellings or callosities, apical sinus with either a distinct, deflexed, horizontal or upturned, glabrous appendage, or a less distinct mucro; column having a minute rostellum with the position of the staminodes often being indicated by coloured points, one on either side of the stigmatic cavity, its apex drawn out into a short or rostrate, obtuse, acute or acuminate structure or *anther-connective*; viscidia 2, placed in 2 simple pouches or *bursiculae.*

About 40–50 species and subspecies distributed throughout Europe, North Africa and western Asia, attaining their greatest diversity in the Mediterranean Basin. Pollination is effected by male Aculeate Hymenoptera.

Basalfield and apical appendage of lip absent; anther-connective obtuse:
  Lip with a deflexed margin, marginal zone narrow, velvety, reddish or yellowish **1. O. fusca**
  Lip with a flat margin, marginal zone broad, glabrous or hirsute, yellow -   **2. O. lutea**
Basalfield and apical appendage of lip present; anther-connective acute or acuminate, short or rostrate:
  Lip normally entire:
    Speculum inconspicuous, reduced to a small patch or a short-branched H which is broader than long and adjacent to the apex of the basalfield; rudimentary lateral lobes of lip always absent; dorsal sepal curving forward; petals pubescent; staminodal points conspicuous - - - - - - - - - - | **3. O. bornmuelleri**
    Speculum conspicuous, H or pi (Π)-shaped, sometimes lacking the cross-bar, much longer than broad, or rarely scutelliform; rudimentary lateral lobes of lip sometimes present; dorsal sepal erect or slightly reflexed; petals glabrous, rarely papillose or pubescent on upper surface; staminodal points inconspicuous or apparently absent:
      Anther-connective short, to 1 mm. long; lateral protuberances of lip usually prominent:

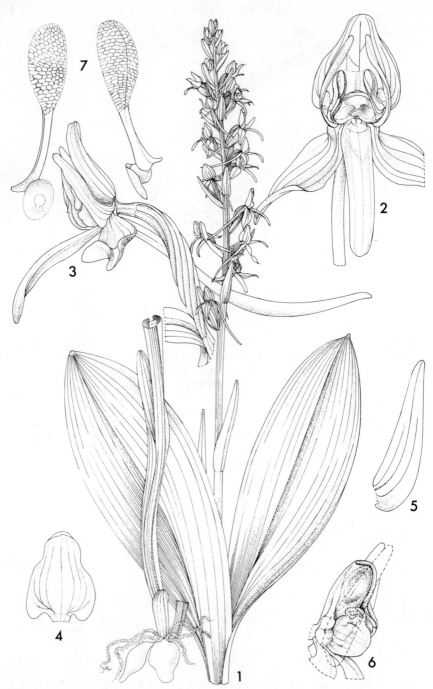

Plate 87. PLATANTHERA CHLORANTHA (Custer) Reichb. ssp. HOLMBOEI (Lindberg f.) J. J. Wood
1, habit, ×½; **2,** flower, front view, ×6; **3,** flower, side view, ×6; **4,** dorsal sepal ×7·2; **5,** petal ×7·2; **6,** column, side view, ×6·5; **7,** pollinia, ×14. (**1** from *Meikle* 2700 & *Davis* 3504; **2–7** from *Kennedy* 1511.)

Sepals yellowish- to olive-green; lip (8–) 10–12 mm. long, pale to dark brown; basalfield bright yellowish-green or dark brown **4. O. sphegodes** ssp. **sphegodes**
Sepals, particularly the laterals, ± stained dull red; lip 9–15 (–17) mm. long, blackish-brown to maroon; basalfield reddish-brown, rarely yellowish-brown
4. O. sphegodes ssp. **mammosa**
Anther-connective long, rostrate, 1·5–3·5 mm. long; lateral protuberances of lip usually inconspicuous or absent; sepals dull olive-green, the laterals stained reddish-purple on their lower halves; basalfield dark brown- **4. O. sphegodes** ssp. **transhyrcana**
Lip normally 3-lobed:
Anther-connective 1·5–2 mm. long, rostrate or acuminate, flexuous; apical appendage of lip longer than broad, deflexed; speculum of lip reduced, collar-shaped or scutelliform, usually with 2 distal rudimentary spots or blotches - - - - **5. O. apifera**
Anther-connective usually less than 1·5 mm. long, acute; apical appendage of lip often broader than long and variously tridentate, often upturned or otherwise reduced to a small, narrow, deflexed mucro; speculum not as above:
Speculum horseshoe-shaped or ± spectacle-shaped, sometimes reduced to 2 crescents or flecks; apical appendage of lip reduced to a small, narrow, deflexed mucro; staminodal points apparently absent - - - - **6. O. argolica** ssp. **elegans**
Speculum variable in form, either irregular and separated by 3 or 4 scutelliform patches, annular or occasionally H-shaped, covering the greater part of the lip; apical appendage of lip often broader than long and variously tridentate, often upturned; staminodal points present, red:
Lip 12–15 mm. long; border of speculum pure white; basal callosities of lip weakly developed, whitish; sepals pure green, only occasionally flushed pale pink
7. O. kotschyi
Lip 6–12 mm., rarely to 15 mm. long; border of speculum creamy or pale yellow; basal callosities of lip strongly developed, green, brown, black or reddish; sepals pure green to purple-violet:
Mid-lobe of lip broadest at or below the middle; sepals pale pink to purple-violet; lip (8–) 10–12 mm. long - - - - - - - - **8. O. scolopax**
Mid-lobe of lip broadest towards the apex:
Sepals greenish-white to pale pink; lip 6–8 (–10) mm. long
9. O. umbilicata ssp. **umbilicata**
Sepals pure green; lip (6–) 8–15 mm. long - - **9. O. umbilicata** ssp. **attica**

1. O. fusca *Link* in Schrader, Journ. Bot., 2: 324 (1800); Unger et Kotschy, Die Insel Cypern, 208 (1865); Boiss., Fl. Orient., 5: 75 (1884); Soó in Botanisch. Archiv. 23: 22, 130 (1928); Renz in Feddes Repert., 27: 211 (1929); Post, Fl. Pal., ed. 2, 2: 572 (1933); Seligman in Quart. Bull. Alpine Gard. Soc., 20 233 (1952); R. S. Davis in Orch. Journ., 3: 164 (1954); Gumprecht in Jahresb. Naturwiss. Ver. Wuppertal, 19: 38 (1964); Sundermann in Die Orchidee, 20: 79 (1969); Soó in Acta Bot. Acad. Sci. Hung., 16: 380 (1970); Osorio-Tafall et Seraphim, List Vasc. Plants Cyprus, 25 (1973); J. et P. Davies in Quart. Bull. Alpine Gard. Soc., 48: 121, 123 (1980).

Plant (8–) 10–40 cm. high; tubers 2, ovoid, sessile or with either one or both stipitate; stem bearing 1 or 2 amplexicaul, ovate-elliptic, acute sheaths; leaves 3–6, rosulate, narrowly to broadly elliptic, obtuse, acute or mucronate, 2·5–9 cm. long, 1–3 cm. wide; inflorescence 1–5 (–10)-flowered, lax; floral bracts ovate-elliptic, acute, as long as or longer than the ovary; sepals green, rarely pinkish, oblong or ovate, obtuse, subglabrous or papillose at the margin, 6–11 mm. long; dorsal sepal 3–6 mm. wide, curving forward; lateral sepals 3–8 mm. wide, somewhat spreading; petals green, yellowish-green or brownish-green, linear to linear-oblong, obtuse, subglabrous or papillose, 5–8 mm. long, c. 2 mm. wide; labellum dark brown, reddish-brown or purplish-brown, the mid-lobe often with a narrow yellow or red marginal zone, obovate in outline, 9–23 mm. long, 7–21 mm. wide, narrowed and obliquely or horizontally extended at the base, sometimes geniculately deflexed, margins deflexed, distinctly 3-lobed, lateral lobes oblong-ovate, obtuse, mid-lobe obcordate, ovate to reniform, emarginate, velutinous, lateral protuberances absent, basal callosities absent or sometimes represented by rudimentary swellings, basalfield absent, speculum greyish, greyish-mauve, pale silvery-blue, iridescent blue, red or brown, with a white, yellow or red border which is sometimes omega-shaped, 2-lobed; column yellow or greenish, stigmatic cavity often forming a V-shaped indentation into the base of the speculum, staminodal points absent; anther-connective yellow or green, obtuse, hidden by the dorsal sepal.

ssp. **fusca**; Gumprecht in Jahresb. Naturwiss. Ver. Wuppertal, 19: 38 (1964); Soó in Acta Bot. Acad. Sci. Hung., 16: 380 (1970); Osorio-Tafall et Seraphim, List Vasc. Plants Cyprus, 25 (1973).
TYPE: Portugal; "Häufig um Lissabon".

Inflorescence 1–10-flowered; petals subglabrous; labellum 9–15 mm. long, 7–21 mm. wide, base longitudinally furrowed and V-shaped towards the stigmatic cavity, pale green beneath, marginal zone narrow, yellow or red, speculum greyish, greyish-mauve or pale silvery-blue, occasionally pale yellow mottled red, never iridescent blue, border white, yellow or red, often rather hazy.

HAB.: In garigue, on dry turf on limestone hillsides, in Pine forest, under Olives and *Cistus*; sea-level–4,000 ft. alt.; fl. Febr.–April.

DISTR.: Divisions 1–3, 5–8. Also Portugal, Spain, Balearic Islands, southern France, Corsica, Sardinia, Italy, Sicily, Malta, Yugoslavia, Albania, Greece, Crete, Rhodes, western Turkey, Lebanon, Israel, Palestine; western North Africa.

1. Toxeftera (Akamas), 1962, *Meikle* 2001! 2985! Psindron (Akamas), 1962, *Meikle* 2124b! Ayios Minas near Smyies, 1962, *Meikle* 2180! Between Khrysokhou and Polis, 1977, *J. J. Wood* 68!
2. Kakopetria, 1929, *Renz* 1438! Platres, 1939, *Kennedy* 1514!
3. Near Omodhos, 1859, *Kotschy* 570a; Alethriko, 1905, *Holmboe* 212a; Near Anglisidhes, 1929, *Renz*; Near Lefkara, 1929, *Renz*; Paramytha, 1962, *G. Nikolaides* s.n.! Episkopi, 1964, *J. B. Suart* 185! Between Axylou and Nata, 1978, *Chesterman* 59!
5. Between Nicosia and Eylenja, 1929, *Renz*; Delikipo, 1929, *Renz* 1440! Near Trikomo, 1929, *Renz*; Nicosia, 1929, *Renz*; Kythrea, 1937, *Miss Godman* s.n.!
6. Skouriotissa, 1927, *Rev. A. Huddle* 91! Dhiorios Forest Station, 1930, *Norman* 397! Kormakiti, 1937, *Syngrassides* 1417! 1430! Orga, 1941, *Davis* 2099!
7. Kyrenia Pass, 1929, *Renz* 1427! 1428! Lapithos, 1929, *Renz* 1429! Akanthou Pass, 1929, *Renz* 1430! 1433! Kremama Kamilou, 1929, *Renz* 1431! Mandres, 1929, *Renz* 1437! Myrtou, 1929, *Renz* 1432! Buffavento, 1929, *Renz* 437! Kantara, 1929, *Renz* 1434!; Melounda, 1929, *Renz*; Agridhaki, 1929, *Renz* 1436! Phlamoudhi, 1929, *Renz* 1435! Olymbos, 1929, *Renz*; Ayios Nikolaos, 1929, *Renz* 1439! Halevga, 1937, *Roger-Smith* s.n.! also 1952, *F. M. Probyn* 77! and, 1956, *G. E. Atherton* 1045! Kyrenia, 1948, *Duke-Woolley* 62! 64b! also, 1929, *Casey* 328! and, 1956, *G. E. Atherton* 901! 962! St. Hilarion, 1949, *Casey* 549! and, 1959, *C. E. H. Sparrow* 24! between St. Hilarion and Aghirda, 1959, *P. H. Oswald* 103!
8. Between Komi Kebir and Ephtakomi, 1929, *Renz*; Rizokarpaso, 1941, *Davis* 2390!

NOTES: On Cyprus *O. fusca* ssp. *fusca* is often very abundant in *Pinus brutia* forest, where it is commonly associated with *Orchis morio* ssp. *picta*. Plants blooming early in the season very often bear rather larger flowers than those produced later. The lip is also narrower and slightly geniculate in the earlier flowers but broader and straight in those appearing later. Pollination is effected by bees of the genus *Andrena*.

O. fusca *Link* ssp. **fusca** × **O. lutea** *Cav.* ssp. **galilaea** (*H. Fleischm. et Bornm.*) *Soó*

Sepals green; petals yellowish-green; labellum almost entirely reddish-brown, marginal zone narrow, yellow, speculum bluish-mauve, sometimes mottled reddish-brown.

HAB.: In garigue, maquis and in *Pinus-Cupressus* woodland.

DISTR.: Divisions 7, 8. Recorded from mixed populations in many parts of the Mediterranean.

7. Ardhana-Phlamoudhi, 1929, *Renz* 1202! Phlamoudhi, 1929, *Renz* 446! Myrtou, 1929, *Renz* 1203!
8. Ephtakomi, 1929, *Renz* 440!

ssp. **fleischmannii** (*Hayek*) *Soó* in Notizbl. Bot. Gart. Berlin, 9: 905 (1926); Davies et Huxley, Wild Orch. Brit. Europe, 161 (1983).
     O. fleischmannii Hayek in Feddes Repert., 22: 388 (1926); Baumann et Künkele, Wild-wachsenden Orch. Europ., 224 (1982).
     O. heldreichii (non Schltr.) H. Fleischm. in Oesterr. Bot. Zeitschr., 74: 182, t. 2, fig. 6 (1925).

O. omegaifera H. Fleischm. var. *fleischmannii* (Hayek) Soó in Acta Bot. Acad. Sci. Hung., 25: 361 (1979).

[*O. omegaifera* (non H. Fleischm.) Gumprecht in Jahresb. Naturwiss. Ver. Wuppertal, 19: 38 (1964); Sundermann, Europ. Medit. Orch., ed. 1: 63 (1970).]

[*O. fusca* Link ssp. *omegaifera* (non (H. Fleischm.) Erich Nelson) Sundermann in Die Orchidee, 20: 79 (1969); Soó in Acta Bot. Acad. Sci. Hung., 16: 381 (1970); Hermjakob, Orchids of Greece Cyprus, 1: 27 (1974); Landwehr, Wilde Orch. Europa, 2: 382 (1977); J. et P. Davies in Quart. Bull. Alpine Gard. Soc., 48: 121, 124 (1980).]

TYPE: Crete; "Distr. Khania, Akrotiri, Karstboden bei Hagia Triadha, 3. März 1904, leg. I. Dörfler (Nr. 154)" (W).

Inflorescence 1–10-flowered; petals papillose; labellum 10–16 mm. long, 8–11 mm. wide, base flat, without longitudinal V-shaped furrow, marginal zone very narrow and yellowish or absent, speculum brownish-mauve or greyish-violet, border conspicuous, W-shaped, white.

HAB.: In garigue on limestone hillsides; 1,500–2,500 ft. alt.; fl. Febr.–April.

DISTR.: Divisions 1–3, 7. Also Greece, Crete, Rhodes, Turkey (S. Anatolia), Lebanon, Israel, Syria.

1. One mile S. of Kathikas, 1977, *J. J. Wood* 113 ! S. of Stroumbi, 1978, *Kalteisen* s.n.
2. Kakopetria, 1970, *Baumann* s.n.; between Ayios Amvrosios and Kissousa, 1977, *J. J. Wood* 26 ! 29 ! Gorge below Mallia, 1977, *J. J. Wood* 45 ! 125 ! Between Pano Lefkara and Vavatsinia, 1977, *J. J. Wood* 54 ! Between Mallia and Omodhos, 1977, *J. J. Wood* 126 !
3. Amargeti, 1978, *Kalteisen* s.n.; between Eledhiou and Pendalia, 1978, *Breiner* s.n.; W. of Mari, 1978, *Kalteisen* s.n.; S. of Kato Lefkara, 1978, *Kalteisen* s.n.
7. Pentadaktylos, 1880, *Sintenis & Rigo* 158 ! Between Vasilia and Larnaka tis Lapithou, 1970, *Baumann* s.n.; N.E. of Kythrea, 1970, *Baumann* s.n.; Mandres, 1980, *Baumann* s.n.

NOTES: This subspecies has an omega- or w-shaped speculum-border like that of *O. fusca* ssp. *omegaifera* (H. Fleischm.) Erich Nelson, with which it has been confused in the past. All reports reputed to be of ssp. *omegaifera* are referable to ssp. *fleischmannii*. *O. fusca* ssp. *omegaifera* is unrecorded from Cyprus, but may yet be discovered with diligent searching. It is quite distinct from ssp. *fleischmannii*, differing by its larger flowers having a reddish-brown, geniculate lip. The latter has been reported from Mt. Olympus and between Larnaca and Vasilia. It is quite frequent on the limestone foothills of the Troödos Range.

In the past various authors have puzzled as to where exactly *O. fleischmannii* should be placed within the *O. fusca* Link complex. Nelson (1962) states that, in his opinion, it is certainly not conspecific with ssp. *omegaifera*, but could represent either ssp. *iricolor* or possibly ssp. *fusca*. Soó's (1928) treatment is the most reasonable and it is the one followed here. He cites several collections from Crete and Cyprus. Of these, two are represented at Kew, viz. *Heldreich* s.n., from "Cydonia", Crete and *Sintenis & Rigo* 158, from "Pentedactylon", Cyprus. The remaining Cyprus collections cited by Soó, viz. *Kotschy* 233, 269, 270, (1859) and 230 (1862) are all from Lefkara, Larnaca or Limassol, and are not represented at Kew.

nothossp. **bayeri** (*H. Baumann*,) *J. J. Wood* **stat. nov.**
O. × bayeri H. Baumann (*O. fusca* Link ssp. *fleischmannii* (Hayek) Soó × *O. fusca* Link ssp. *fusca*), in Beih. Veröff. Naturschutz Landschaftspflege Bad.-Württ., 19: 142, 147, t. 15 (1981).

Sepals green; petals yellowish-green; labellum brown, with greenish-yellow margin and reddish or purple-brown hairs, speculum bilobed, bluish-grey, lateral lobes distinctly shorter than mid-lobe; stigmatic cavity only slightly V-shaped.

HAB.: In garigue.

DISTR.: Division 4.

4. Larnaca, 1980, *Bayer* s.n.

ssp. **iricolor** (*Desf.*) *Holmboe*, Veg. Cypr., 56 (1914); Erich Nelson, Gestaltwandel, 208 (1962); Gumprecht in Jahresb. Naturwiss. Ver. Wuppertal, 19: 38 (1964); Sundermann in Die Orchidee, 20: 79 (1969); Soó in Acta Bot. Acad. Sci. Hung., 16: 380 (1970); Osorio-Tafall et Seraphim, List Vasc. Plants Cyprus, 25 (1970); Hermjakob, Orchids of Greece Cyprus, 1: 26 (1974); Williams et al., Field Guide, 54 (1978); J. et P. Davies in Quart. Bull. Alpine Gard. Soc., 48: 123, 124 (1980); Sundermann, Europ. Medit. Orch., ed. 3, 89 (1980); Davies et Huxley, Wild Orch. Brit. Europe, 161 (1983).

O. iricolor Desf. in Ann. Mus. Nat. Hist. Paris, 10: 224, t. 13 (1807); Post, Fl. Pal., ed. 2, 2:

572 (1933); Renz in Feddes Repert., 27: 212 (1929); Seligman in Quart. Bull. Alpine Gard. Soc., 20: 233 (1952); R. S. Davis in Orch. Journ., 3: 164 (1954); Sundermann, Europ. Medit. Orch., ed. 1, 61 (1970).

O. fusca Link var. *iricolor* (Desf.) Reichb. f., Icones Fl. Germ. et Helv., XIII–XIV: 73–75 (1851); Boiss., Fl. Orient., 5: 75 (1884); Elektra Megaw, Wild Flowers of Cyprus, 13, t. 23 (1973).

TYPE: Based on Tournefort's *Orchis Orientalis, fucum referens, flore maximo, scuto azureo* (Coroll. Inst. Rei. Herb., 30; 1700), figured by Aubriet, and reproduced in Desfontaines' paper (*loc. cit.*).

Inflorescence 1–4 (–5)-flowered; petals often papillose; labellum 15–23 mm. long, c. 21 mm. wide, base longitudinally furrowed and V-shaped towards the stigmatic cavity, brownish-red beneath, marginal zone narrow, yellow or red, speculum iridescent blue, often divided into 2 symmetrical halves, border usually absent, rarely pale blue.

HAB.: In garigue and Pine forest, on limestone hillsides, under Carob trees in association with *Neotinea maculata*, or under *Cistus* spp. and *Calycotome villosa*; sea-level–2,000 ft. alt.; fl. Febr.–April.

DISTR.: Divisions 1–3, 5, 7, 8. Also Spain, southern France, Corsica, Sardinia, Italy (Riviera, Monte Argentario), Yugoslavia, Greece, Crete, Rhodes, Turkey, Lebanon, Israel, Palestine, Syria.

1. Baths of Aphrodite, 1962, *Meikle* 2222!
2. Between Pano Lefkara and Vavatsinia, 1977, *J. J. Wood* 61! Between Mallia and Omodhos, 1977, *J. J. Wood* 127!
3. Mazotos, 1905, *Holmboe* 179; Alethriko, 1905, *Holmboe* 209; Ypsonas, 1970, *A. Genneou* in ARI 1395! Between Apsiou and Phasoula, 1978, *Chesterman* 76!
5. Near Trikomo, 1929, *Renz*; Kythrea, 1937, *Miss Godman* s.n.!
7. Kyrenia Pass, 1929, *Renz* 1467! Alakati, 1929, *Renz* 1470! Lapithos, 1929, *Renz* 1468! Akanthou Pass, 1929, *Renz* 1469! 1471! Olymbos, 1929, *Renz*; Melounda, 1929, *Renz*; Kantara, 1929, *Renz*; Near Pentadaktylos, 1929, *Renz*; Ayios Nikolaos, 1929, *Renz* 1474! Ardhana — Phlamoudhi, 1929, *Renz* 1475! W. end of Northern Range, 1930, *C. Norman* 395! Between Halevga and Kharcha, 1941, *Davis* 2157! Halevga, 1952, *F. M. Probyn* 81! and, 1956, *G. E. Atherton* 1019! Kyrenia, 1956, *G. E. Atherton* 1022! and, 1973, *P. Laukkonen* 146! Between Palaeosophos and Karavas, 1968, *Economides* in ARI 1098!
8. Between Komi Kebir and Ephtakomi, 1929, *Renz* 1472! 1473!

NOTES: This subspecies often flowers a little later than ssp. *fusca*.

2. O. lutea *Cav.*, Icones, 2: 46, t. 160 (1793); Unger et Kotschy, Die Insel Cypern, 208 (1865); Boiss., Fl. Orient., 5: 75 (1884); Holmboe, Veg. Cypr., 56 (1914); Soó in Botanisch. Archiv, 23: 22, 130 (1928); Renz in Feddes Repert., 27: 212 (1929); Post, Fl. Pal., ed. 2, 2: 573 (1933); Seligman in Quart. Bull. Alpine Gard. Soc., 20: 233 (1952); R. S. Davis in Orch. Journ., 3: 164 (1954); Gumprecht in Jahresb. Naturwiss. Ver. Wuppertal, 19: 38 (1964); Sundermann in Die Orchidee, 20: 79 (1969); Soó in Acta Bot. Acad. Sci. Hung., 16: 379 (1970); Sundermann, Europ. Medit. Orch., ed. 1, 65 (1970), ed. 2, 87 (1975), ed. 3, 91 (1980); Hermjakob, Orch. Greece Cyprus, 1: 28 (1974); Landwehr, Wilde Orch. Europa, 2: 390 (1977); Williams et al., Field Guide, 56 (1978).

Plant 7–30 (–40) cm. high; tubers 2, ovoid, 1 or both stipitate; stem bearing 1, occasionally 2, amplexicaul, narrowly elliptic, acute sheaths; leaves (3–) 4–8, rosulate, narrowly to broadly elliptic, narrowly ovate or narrowly obovate, acute, sometimes obtuse and mucronate, 2–6 (–9) cm. long, 1–2 (–2·5) cm. wide; inflorescence 2–7-flowered, lax; floral bracts ovate-elliptic, obtuse to acute, as long as or longer than the ovary; flowers slightly scented; sepals green, ovate-oblong, obtuse, subglabrous or papillose at the margin, 6–10 mm. long, 4–6 mm. wide; dorsal sepal curving forward; lateral sepals spreading; petals green, yellowish-green or yellowish, linear to linear-oblong, ± truncate, subglabrous or papillose at the margin, 3–7 mm. long, 2–3 mm. wide; labellum dark reddish-brown or blackish-purple, with a 1–3 mm. wide, yellow, orange or greenish marginal zone, suborbicular or oblong in outline, 9–18 mm. long, 6–20 mm. wide, distinctly 3-lobed, the lobes either subequal or with a larger or smaller, often bilobed mid-lobe, the lateral lobes sometimes overlapping the mid-lobe, papillose to hairy, marginal zone glabrous, lateral protuberances absent, basal callosities absent, basalfield absent, speculum lead- or mauve-grey, 2-lobed or ± entire, narrow or broad; column yellow or greenish, stigmatic cavity forming a V-shaped indentation into the base of the speculum, staminodal points absent; anther-connective yellow or green, obtuse, hidden by the dorsal sepal.

ssp. **galilaea** (*H. Fleischm. et Bornm.*) Soó in Notizbl. Bot. Gart. Berlin, 9: 906 (1926); P. Mouterde, Nouv. Fl. Liban Syrie, 1: 330, t. CVIII, fig. 4 (1966).

 Ophrys galilaea H. Fleischm. et Bornm. in Ann. Naturh. Mus. Wien, 36: 12 (1923); Post, Fl. Pal., ed. 2, 2: 573 (1933).

 Arachnites lutea (Cav.) Tod.var. *minor* Tod., Orchid. Sic., 97, t. 2, figs. 9, 10 (1842).

 Ophrys lutea Cav. var. *minor* (Tod.) Guss., Fl. Sic. Syn., 2 (2): 550 (1844); Erich Nelson, Gestaltwandel, 214 (1962); Sundermann, Europ. Medit. Orch., ed. 1, 65 (1970); Hermjakob, Orch. Greece Cyprus, 1: 29 (1974).

 O. sicula Tineo, Plant. Rar. Sic., 13 (1846).

 O. lutea Cav. ssp. *murbeckii* (Fl.) Soó in Feddes Repert., 24: 25 (1927); J. et P. Davies in Quart. Bull. Alpine Gard. Soc., 48: 121, 123 (1980); Davies et Huxley, Wild Orch. Brit. Europe, 160 (1983).

 O. lutea Cav. ssp. *minor* (Tod.) O. et E. Danesch, Orch. Europ. Ophrys-Hybr., 43 (1972) nom. invalid.; Osorio-Tafall et Seraphim, List Vasc. Plants Cyprus, 25 (1973) nom. invalid.

 TYPE: Palestine; "Galilaea: ad Hunin, 990 m. s.m.; 21.IV.1897, leg. J. Bornmüller (Iter Syriacum 1897, Nr. 1489)" (B).

Labellum 9–12 mm. long, 6–8 (–10) mm. wide, the mid-lobe subequalling or smaller than the lateral lobes, marginal zone of the lobes 1–2 mm. wide, the reddish-brown pigment often extending into the yellow marginal zone of the midlobe, sometimes in the form of an inverted letter V, speculum relatively broad.

HAB.: In garigue, on dry turf, in rocky places; Pine forest; Olive groves; on calcareous soils; sea-level–2,000 ft. alt.; fl. March–April (or often late Febr. in Cyprus).

DISTR.: Divisions: 1–8. Also in central and southern Portugal, Spain, Balearic Islands, southern France, Corsica, Sardinia, Italy, Sicily, Malta, Lampedusa, Yugoslavia, Albania, Greece, Crete, Rhodes, Turkey, Lebanon, Israel, Palestine; North Africa. This subspecies replaces *O. lutea* ssp. *lutea* in much of the eastern Mediterranean.

1. Ayios Neophytos, 1959, *C. E. H. Sparrow* 27! Near Ayios Yeoryios (Akamas), 1962, *Meikle* 2019! Psindron (Akamas), 1962, *Meikle* 2124a!
2. Between Ayios Amvrosios and Kissousa, 1977, *J. J. Wood* 22! Gorge below Mallia, 1977, *J. J. Wood* 46! Between Pano Lefkara and Vavatsinia, 1977, *J. J. Wood* 59!
3. Stavrovouni, 1787, *Labillardière*; Between Limassol and Omodhos, 1859, *Kotschy* 450; Lefkara, 1862, *Kotschy* 229; also, 1929, *Renz* 458! Alethriko, 1905, *Holmboe* 212; Between Anglisidhes and Skarinou, 1929, *Renz* 456! Episkopi, 1964, *J. B. Suart* 12! Ypsonas, 1970, *A. Genneou* in ARI 1397! Khirokitia, 1970, *A. Genneou* in ARI 1694!
4. Xylophago, 1905, *Holmboe* 76.
5. Between Eylenja and Nicosia, 1929, *Renz* 423! Near Trikomo, 1929, *Renz* 434! Nicosia, 1929, *Renz*; Delikipo, 1929, *Renz* 459! Lapathos, 1962, *Meikle* 2986!
6. Kykko metokhi, 1900, *A. G. & M. E. Lascelles* s.n.!
7. Pentadaktylos, 1880, *Sintenis & Rigo* 159! and, 1929, *Renz*; Kyrenia Pass 1929, *Renz* 425! Lapithos, 1929, *Renz* 427! Akanthou Pass, 1929, *Renz* 428! Kremama Kamilou, 1929, *Renz* 431! Myrtou, 1929, *Renz* 432! Melounda, 1929, *Renz* 442! Between Kantara and Mandres, 1929, *Renz*; Ayios Epiktitos, 1929, *Renz* 429! Klepini, 1929, *Renz* 450! Ardhana–Phlamoudhi, 1929, *Renz* 447! Halevga, 1937, *Roger-Smith* s.n.! also, 1952, *F. M. Probyn* 78! Tjiklos, 1973, *P. Laukkonen* 165! Kyrenia, 1948, *Collingwood Ingram* 20! also, 1948, *Duke-Woolley* 64a!
8. Between Komi Kebir and Ephtakomi, 1929, *Renz*; Between Ayios Philon and Rizokarpaso, 1941, *Davis* 2244!

NOTES: Pollination is effected by bees of the genus *Andrena*.

3. O. bornmuelleri *M. Schulze ex Bornm.* in Verh. Zool.-Bot. Ges. Wien, 48: 635 (1898); Soó in Botanisch. Archiv. 23: 25, 131 (1928); Renz in Feddes Repert., 27: 208, 214 (1929); Post, Fl. Pal., ed. 2, 2: 574 (1933); Seligman in Quart. Bull. Alpine Gard. Soc., 20: 233 (1952); R. S. Davis in Orch. Journal, 3: 164 (1954); Erich Nelson, Gestaltwandel, 171 (1962); Gumprecht in Jahresb. Naturwiss. Ver. Wuppertal, 19: 38 (1964); Sunderman in Die Orchidee, 20: 79 (1969); Soó in Acta Bot. Acad. Sci. Hung., 16: 391 (1970); Sundermann, Europ. Medit. Orch., ed. 1: 85 (1970); Osorio-Tafall et Seraphim, List Vasc. Plants Cyprus, 25 (1973); Hermjakob, Orch. Greece Cyprus, 13 (1974); B. et E. Willing in Die Orchidee, 26: 74–78 (1975); Renz in Rechinger f., Fl. Iran., no. 126: 73 (1978); Williams et al., Field Guide, 42 (1978); J. & P. Davies in Quart. Bull. Alpine Gard. Soc., 48: 122, 229 (1980).

O. tenthredinifera Willd. var. – Bornm. in Verh. Zool.-Bot. Ges. Wien, 48: 635 (1898); Renz in Rechinger f., Fl. Iran., 126: 73 (1978).

O. fuciflora (F. W. Schmidt) Moench ssp. *bornmuelleri* (M. Schulze ex Bornm.) B. et E. Willing in Die Orchidee, 26: 74 (1975); Williams et al., Field Guide, 42 (1978); Sundermann, Europ. Medit. Orch., ed. 3, 109 (1980); Davies et Huxley, Wild Orch. Brit. Europe, 175 (1983).

O. holoserica (Burm. f.) Greuter ssp. *bornmuelleri* (M. Schulze ex Bornm.) [Osorio-Tafall et Seraphim, List Vasc. Plants Cyprus, 25 (1973) nom. invalid.; Elektra Megaw, Wild Flowers of Cyprus, 12, t. 22 (1973) nom. invalid.; Landwehr, Wilde Orch. Europa, 2: 472 (1977) comb. invalid.]

Plant 10–30 (–40) cm. high; tubers 2, sessile or with one stipitate; stem often bearing 1 cauline leaf and 1 or sometimes 2 amplexicaul, acute sheaths; leaves 3–4, grouped at the base, narrowly elliptic to narrowly obovacte, obtuse to acute, 6–11 (–13) cm. long, 1·3–3 cm. wide; inflorescence (2–) 7–10 (–15)-flowered, usually rather lax; floral bracts ovate-elliptic, acute, usually longer than the ovary, the lowermost up to 3·5 cm. long; sepals pale green or whitish-green, sometimes suffused pink particularly at the base, ovate-oblong or ovate-elliptic, obtuse, glabrous, dorsal sepal 6–12 mm. long, up to 4 mm. wide, curving forward or reflexed; lateral sepals up to 5 mm. wide, spreading; petals yellowish-green, pale yellow or pale pink, triangular-ovate, ± acute, pubescent, the margins somewhat villous, 1–2 mm. long, up to 1·5 mm. wide at the base; labellum brown to purplish-brown, becoming paler towards the margin, broadly ovate, suborbicular, quadrate or trapeziform-flabellate and emarginate in outline, 7–12 mm. long, 10–15 mm. wide, flattened or slightly to strongly convex, velutinous, entire, margins flattened, sometimes deflexed or patent, basal protuberances pale brown, prominent, conical, hairy, up to 3 mm. long, basal callosities dark green or blackish, linked by a blackish-violet or dark green band, basalfield reddish or brown, often having a black border, passing over into the stigmatic cavity, speculum small, mostly restricted to the area at the apex of the basalfield, steel-blue or mauve, with a creamy border widening laterally, often resembling a short-branched letter H, sometimes represented by two small spots only, hair tuft present or replaced by a thickened area, apical appendage yellowish-green, subquadrate, often shallowly tridentate, upcurved; column yellowish or greenish-cream; staminodal points reddish; anther-connective green, short or very short, apiculate.

ssp. **bornmuelleri**

TYPE: Israel; "Galiläa: Buschige Abhänge bei Hunin, 900 m.", *Bornmueller* 1493 (B).

Habit rather slender; inflorescence (2–) 7–10 (–15)-flowered, usually rather elongated, lax; sepals sometimes suffused pink; dorsal sepal curving forward; labellum usually rather flat, rarely slightly convex, velutinous, 7–12 mm. long, c. 10 mm. wide, patent or deflexed, basalfield reddish, speculum rarely reduced to two small spots, hair tuft sometimes present, but never very prominent; anther-connective short.

HAB.: In garigue, on dry turf, beneath Pine trees, also in grassy places beneath Carob trees (particularly on the Northern Range), on both calcareous and igneous formations; sea-level–3,000 ft. alt.; fl. (late Febr.–) mid March–mid April.

DISTR.: Divisions 1–3, 7, 8. Also southern Turkey (southern Anatolia), Lebanon, Syria, Palestine, Israel, Iraq.

1. Tala, 1974, *Meikle* 4014! Between Khrysokhou and Polis, 1977, *J. J. Wood* 70! Between Baths of Aphrodite and Fontana Amorosa, 1977, *J. J. Wood* 76! Between Lachi and Neokhorio, 1977, *J. J. Wood* 98! and, 1978, *Chesterman* 49! Near Smyies, 1977, *J. J. Wood* 106! Between Dhrousha and Kritou Terra, 1977, *J. J. Wood* 111! 112!

2. Between Pano Lefkara and Vavatsinia, 1977, *J. J. Wood* 64!
3. Between Lefkara and Kornos, 1929, *Renz*; Mesayitonia, 1948, *Mavromoustakis* s.n.!
   Paramali, 1960, *C. E. H. Sparrow* 50! and, 1964, *J. B. Suart* 207! 4 miles N. of Kandou, 1963,
   J. B. Suart 101! 128! Episkopi, 1963–4, *J. B. Suart* 43! 48! 186!
7. Ayios Amvrosios, 1937, *Miss Godman* s.n.! Dhikomo, 1941, *Davis* 2171! Yaïla, 1941, *Davis*
   2949! Kyrenia, 1956, *G. E. Atherton* 963! 965! 1316! Between St. Hilarion and Aghirda,
   1959, *P. H. Oswald* 102! S. of Kyrenia, 1962, *Casey* 1743! Halevga, 1963, *J. B. Suart* 120!
   Klepini, 1970, *Landwehr*; and, *B. & E. Willing* s.n.
8. Leonarisso, 1968, *Economides* in ARI 1116!

NOTES: A widespread plant on Cyprus, where it replaces the closely related *O. holoserica*, and
tending to be more frequent on the Northern Range particularly around Kyrenia and Halevga.
It often flowers two or three weeks later than ssp. *grandiflora*. Pollination is probably effected
by bees of the genus *Eucera*.

ssp. **grandiflora** (*Fleischm. et Soó*) *Renz et Taubenheim* in Notes RBG Edinb., 41: 276 (1983).
    O. bornmuelleri M. Schulze ex Bornm. f. *grandiflora* Fleischm. et Soó in Feddes Repert.,
24: 26 (1927); Post, Fl. Pal., ed. 2, 2: 574 (1933).
    O. fuciflora (F. W. Schmidt) Moench ssp. *bornmuelleri* (M. Schulze ex Bornm.) B. et E.
Willing var. *grandiflora* (Fleischm. et Soó) B. et E. Willing in Die Orchidee, 26: 78
(1975); Sundermann, Europ. Medit. Orch., ed. 3: 109 (1980); Davies et Huxley, Wild Orch.
Brit. Europe, 175 (1983).
    O. holoserica (Burm. f.) Greuter ssp. *bornmuelleri* (M. Schulze ex Bornm.) [Landwehr
var. *grandiflora* (Fleischm. et Soó) Landwehr, Wilde Orch. Europa, 2: 472 (1977) comb.
invalid].
TYPE: Syria, "Cum typo in Syria, sic Svedia (Kotschy) Jerusalem (Dinsmore)", (B).

Plant rather more robust, but generally less tall; inflorescence 3–5 (–7)-
flowered, generally less elongated, rather more compact; sepals usually pure
green rarely suffused pink or white suffused green; dorsal sepal reflexed;
labellum convex, particularly in the area around the basalfield, basal
protuberances often with more pronounced "shoulders", marginal zone
and lateral protuberances densely hairy, 10–12 mm. long, 10–15 mm. wide,
usually strongly deflexed with the apex sometimes almost touching the
stem, basalfield brown, a little paler than the remainder of the labellum,
speculum noticeably reduced towards the middle, often only represented by
two small patches or spots, hair tuft often prominent; anther-connective
very short.

HAB.: In garigue, amongst scrub, in Pine forest, on damp grassy ground beside streams, also
amongst Juniper scrub; 200–2,000 ft. alt.; fl. mid March–early April.

DISTR.: Divisions 1, 3, 5–8. Also Turkey (southern and south east Anatolia), Palestine,
western Syria and no doubt elsewhere in the eastern Mediterranean. Apparently commoner in
Cyprus than elsewhere.

1. Peyia Forest, 1962, *Meikle* 2043! Smyies, 1962, *Meikle* 2184! Between Khrysokhou and
   Polis, 1977, *J. J. Wood* 66! Between Lachi and Neokhorio, 1977, *J. J. Wood* 99–101! S.W. of
   Polis, 1978, *Chesterman* 54!
3. Paramytha, 1962, *G. Nikolaides* s.n.! Episkopi, 1964, *J. B. Suart* 6! Paramali, 1964, *J. B.
   Suart* 206! Moni, 1964, *J. B. Suart* 219!
5. Delikipo, 1929, *Renz* 1283!
6. Dhiorios Forest Station, 1930, *C. Norman* 392! Kormakiti, 1937, *Syngrassides* 1428! 1431!
   Orga, *B. & E. Willing* s.n.
7. Alakati, 1929, *Renz* 430! Kyrenia Pass, 1929, *Renz* 1284! 1285! Klepini, 1929, *Renz* 1282!
   also, 1970, *Landwehr*; Agridhaki, 1929, *Renz* 445! Kantara, 1929, *Renz* 439! Halevga, 1937,
   Roger-Smith s.n.! and, 1956, *G. E. Atherton* 1048! 1050! Kyrenia, 1948, *Duke-Woolley* 63!
   Tjiklos, 1973, *P. Laukkonen* 166!
8. Ephtakomi, 1929, *Renz* 441! Komi Kebir, 1929, *Renz* 435! Ronnas Bay near Rizokarpaso,
   1941, *Davis* 2335!

NOTES: *O. bornmuelleri* ssp. *grandiflora* has a more convex lip and often flowers a little earlier
than ssp. *bornmuelleri*. Both plants frequently grow together in the same locality but are very
often separated by a difference in flowering period of from two to three weeks. Only one will
generally be found to be in full flower at any given period. Although not rare on Cyprus, it seems
to be less frequent than the typical plant. Intermediate forms between the two have not been
seen.

O. bornmuelleri *M. Schulze ex Bornm.* ssp. **grandiflora** (*Fleischm. et Soó*) *Renz et Tau-*

benheim × **O. sphegodes** *Mill.* ssp. **mammosa** *(Desf.) Soó* apud B. et E. Willing in Die Orchidee, 26: 77 (1975).

Plant 10 cm. high; inflorescence up to 10-flowered; sepals green, stained red in places; petals rather long; labellum having a yellowish margin, subquadrate, densely hairy on the shoulders and marginal zone, speculum shaped like an inverted letter U; anther-connective very short.

HAB.: On calcareous rocks; 2,500 ft. alt.; fl. early April.

DISTR.: Division 2, 12 plants observed.

2. S. of Omodhos, 1974, *B. & E. Willing.*

4. O. sphegodes *Mill.*, Gard. Dict., ed. 8, no. 8 (1768); Soó in Acta Bot. Acad. Sci. Hung., 16: 381 (1970); Sundermann, Europ. Medit. Orch., ed. 3, 93 (1980); Davies et Huxley, Wild Orch. Brit. Europe, 162 (1983).

Plant (10–) 20–60 (–80) cm. high; tubers 2, oblong or ovoid, shortly stipitate or sessile; stem erect or slightly flexuous, bearing 1–4 cauline leaves grading into 1–3 amplexicaul, acute sheaths in the upper part of the stem; leaves 3–6, mostly grouped at the base, narrowly elliptic to broadly elliptic, sometimes narrowly obovate, obtuse, usually mucronate, often glaucescent, spreading, 6–15 (–20) cm. long, (1·2–) 1·5–2·5 (–3·5) cm. wide; inflorescence 2–10-flowered, lax; floral bracts ovate-elliptic, concave, obtuse, mucronate or acute, as long as or longer than the ovary, the lowermost up to 5 cm. long; sepals yellowish- to dull olive-green, sometimes suffused dull red or reddish-purple, oblong, ovate-oblong to elliptic, obtuse, revolute, glabrous, rarely papillose, 6–15 (–17) mm. long, 3–7 mm. wide; dorsal sepal slightly curving forward, erect or slightly reflexed; lateral sepals spreading or slightly reflexed; petals yellowish-green, dark olive-green, reddish-green, brownish-green or dull red, oblong-triangular or linear-ligulate, rarely ovate to narrowly-elliptic, obtuse or acute, margin usually undulate, sometimes revolute, glabrous or very shortly pubescent on the upper surface, spreading to strongly reflexed, 4–8 (–11) mm. long, 2–3 (–5) mm. wide; labellum brown, dark brown, blackish-brown, purplish-brown or maroon, with an often yellowish-green, orange-yellow or red marginal zone, particularly towards the apex, ovate-oblong or ovate-triangular, nearly orbicular when flattened, (8–) 10–14 (–17) mm. long, 8–14 (–17) mm. wide, entire or subentire, i.e. with a small incision each side, convex, margin deflexed or flattened, velutinous, the sides sometimes villose, papillose at the centre, marginal zone glabrous, lateral protuberances present or absent, basal callosities green, bluish-mauve or dull bluish-violet, linked by a dull violet to bluish band, basalfield bright yellowish-green, reddish-brown or dark brown, rarely yellowish-brown, abbreviated or elongate and ± cordate, speculum sky-blue, dull violet, bluish-mauve or bluish-violet, with a white border H–Π-shaped, sometimes lacking the cross-bar, hair tuft either absent or poorly developed, apical appendage conspicuous or reduced to a small mucro; column green, olive-green, greenish-cream or yellowish-cream, sometimes flushed dull red; stigmatic cavity long or short, traversed by a prominent brownish–blackish-purple or bluish-mauve transverse band, staminodal points absent; anther-connective short to rostrate, acute.

ssp. **sphegodes** *Erich Nelson*, Gestaltwandel, 189 (1962) — as *O. sphecodes*; Gumprecht in Jahresb. Naturwiss. Ver. Wuppertal, 19: 38 (1964); Soó in Acta Bot. Acad. Sci. Hung., 16: 381 (1970); Hermjakob, Orch. Greece Cyprus, 17 (1974); Landwehr, Wilde Orch. Europa, 2: 396 (1977); J. J. Wood in Die Orchidee, 31: 229 (1980), et in Orch. Rev., 89: 293 (1981).
    O. aranifera Huds., Fl. Angl., ed. 2, 2: 392 (1778); Boiss., Fl. Orient., 5: 77 (1884); Post, Fl. Pal., ed 2, 2: 574 (1933).
    TYPE: England; "grows naturally in dry pastures in several parts of England" *Miller* (? B.M.).

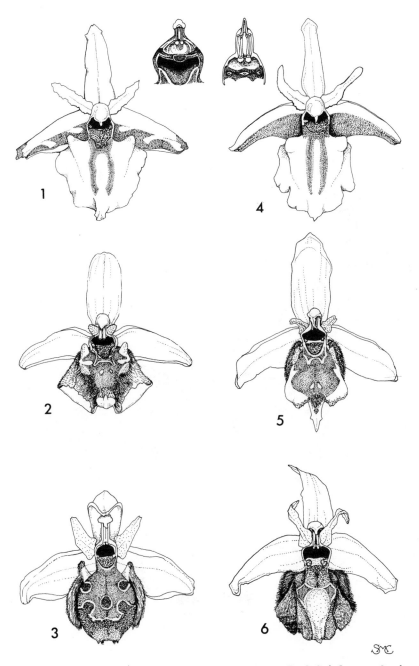

Plate 88. OPHRYS flowers: **1,** O. SPHEGODES Mill. ssp. MAMMOSA (Desf.) Soó, flower, × 2; column showing stigmatic cavity, × 6 (from *J. J. Wood* 124); **2,** O. BORNMUELLERI M. Schulze ex Bornm. × 2 (from *J. J. Wood* 106); **3,** O. UMBILICATA Desf. ssp. ATTICA (Boiss. et Orph.) J. J. Wood, × 2 (from *J. J. Wood* 8); **4,** O. SPHEGODES Mill. ssp. TRANSHYRCANA (Czernjak.) Soó, flower, × 2; column showing stigmatic cavity, × 6 (after E. Nelson, Gestaltwandel; 1962); **5,** O. APIFERA Huds., × 2 (from *J. J. Wood* 97); **6,** O. ARGOLICA H. Fleischm. ssp. ELEGANS (Renz) E. Nelson, × 2 (from *R. S. Davis* s.n.)

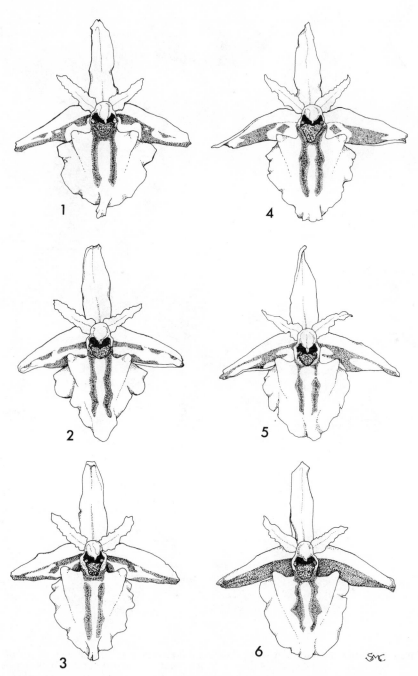

Plate 89. Speculum variation in OPHRYS SPHEGODES Mill. ssp. MAMMOSA (Desf.) Soó, × 2 (**1** from *J. J. Wood* 124; **2** from *J. J. Wood* 24; **3** from *J. J. Wood* 33; **4, 5** from *J. J. Wood* 49; **6** from *J. J. Wood* 17.)

Plant rather slender; inflorescence (2–4–) 5–9 (–10)-flowered; sepals yellowish- to olive-green, 6–10 (–12) mm. long, 3–4 mm. wide; petals obtuse; labellum pale to dark brown, ovate-oblong to ovate-triangular, entire, rarely obscurely 3-lobed, sometimes weakly emarginate, in eastern Mediterranean forms usually longer than broad and broadest towards the apex when spread out, (8–) 10–12 mm. long, 8–12 mm. wide, lateral protuberances usually present in eastern Mediterranean forms, basal callosities green, basalfield bright yellowish-green or dark brown, abbreviated or sometimes elongate and cordate, speculum H-shaped, apical appendage conspicuous in eastern Mediterranean forms; stigmatic cavity rather short; anther-connective short, up to 1 mm. long.

HAB.: In garigue, on dry turf, on calcareous formations; alt. unknown; fl. March–April.

DISTR.: Division 3. Noted from Cyprus by Nelson, Gumprecht, Soó and Hermjakob, but without precise localisation. Widespread in western, central and southern Europe, northwards to southern England and central Germany, southeastwards to the Crimea, Caucasus area, northern Turkey and northern Iran.

3. Between Vavla and Kato Dhrys, 1981, *Mrs. G. Phillips*, transparency only!

NOTES: Characters such as petal and lip shape, the conspicuous lateral protuberances and apical appendage of the lip indicate that much of the eastern Mediterranean population of ssp. *sphegodes* is of a somewhat transitional nature intergrading with ssp. *mammosa*. Pollination is effected by bees of the genera *Andrena* and *Colletes*.
The specimens observed by Mrs. Phillips were all apochromic and lacking in both speculum and basalfield.

ssp. **mammosa** (*Desf.*) *Soó* in Izv. Bot. Inst. Akad. Bulg., 6: 370 (1958); Erich Nelson, Gestaltwandel, 184 (1962); Gumprecht in Jahresb. Naturwiss. Ver. Wuppertal, 19: 38 (1964); Sundermann in Die Orchidee, 20: 79 (1969); Soó in Acta Bot. Acad. Sci. Hung., 16: 383 (1970); Osorio-Tafall et Seraphim, List Vasc. Plants Cyprus, 25 (1973); Elektra Megaw, Wild Flowers of Cyprus, 13, t. 23 (1973); Hermjakob, Orch. Greece Cyprus, 20 (1974); B. et E. Willing in Die Orchidee, 26: 74 (1975); 27: 114 (1976); Williams et al., Field Guide, 48 (1978); J. et P. Davies in Quart. Bull. Alpine Gard. Soc., 48: 123, 124 (1980); Sundermann, Europ. Medit. Orch., ed. 3, 95 (1980); J. J. Wood in Die Orchidee, 31: 229 (1980), Orch. Review, 89: 293 (1981); Davies et Huxley, Wild Orch. Brit. Europe, 163 (1983).
  O. mammosa Desf. in Ann. Mus. Paris, 10: 222, t. 12 (1807); Holmboe, Veg. Cypr., 56 (1914); Renz in Feddes Repert., 27: 212 (1929); Post, Fl. Pal., ed. 2, 2: 574 (1933); Seligman in Quart. Bull. Alpine Gard. Soc., 20: 233 (1952); R. S. Davis in Orch. Journ., 3: 164 (1954); Sundermann, Europ. Medit. Orch., ed. 1, 69 (1970); Landwehr, Wilde Orch. Europa, 2: 400 (1977).
  O. aranifera Huds. var. *mammosa* Reichb. f., Icones Fl. Germ., 13–14: 89 (1851).
  O. aranifera Huds. ssp. *mammosa* Soó in Notizbl. Bot. Gart. Berlin, 9: 907 (1926).
  O. mammosa Desf. f. *subtriloba* Renz in Feddes Repert., 25: 249 (1928); 27: 212 (1929); Erich Nelson, Gestaltwandel, 186 (1962); B. et E. Willing in Die Orchidee, 27: 115 (1976).
  [*O. atrata* (non Lindl.) Kotschy in Unger et Kotschy, Die Insel Cypern, 209 (1865); Boiss., Fl. Orient., 5: 78 (1884); H. Stuart Thompson in Journ. Bot., 44: 339 (1906).]
  TYPE: Based on Tournefort's *Orchis orientalis, fucum referens, flore mammoso* (Coroll. Inst. Rei. Herb., 30; 1700, figured by Aubriet and reproduced in Desfontaines' paper, *loc. cit.*).

Plant rather robust; inflorescence (2–) 4–10-flowered; sepals, particularly the lateral, often stained dull red; dorsal sepal 7–10 (–12) mm. long, 3–4 mm. wide; lateral sepals 7–13 mm. long, 4–5 (–6) mm. wide; petals ± acute; labellum blackish-brown, purplish-brown or maroon, orbicular to ovate, 9–15 (–17) mm. long, 8–14 (–17) mm. wide, entire, rarely 3-lobed, lateral protuberances prominent, up to 4 mm. long, basal callosities bluish-mauve, occasionally dark green, prominent, basalfield reddish-brown, rarely yellowish-brown, elongate and ± cordate, broadest towards stigmatic cavity, speculum H-shaped, apical appendage reduced to a small mucro; stigmatic cavity long; anther-connective short to semi-rostrate, 1–2 mm. long. *Plate 88, fig. 1; Plate 89.*

HAB.: In garigue, on dry turf; in Pine forest; on calcareous formations; sea-level–2,200 ft. alt.; fl. March–May (–early June), or often February in Cyprus.

DISTR.: Divisions 1–8. Also Bulgaria, Yugoslavia, Albania, Greece, Aegean Islands, Crete, Rhodes, western Turkey.

1. Psindron (Akamas), 1962, *Meikle* 2124a! S. of Kathikas, 1977, *J. J. Wood* 115!
2. Between Ayios Amvrosios and Kissousa, 1977, *J. J. Wood* 17! 23–25! 28! Gorge below Mallia, 1977, *J. J. Wood* 33–35! 37! 124! Between Pano Lefkara and Vavatsinia, 1977, *J. J. Wood* 49! 51! 52! 55! 58!
3. Lefkara, 1862, *Kotschy* 230! Mazotos, 1905, *Holmboe* 180; Alethriko, 1905, *Holmboe* 210; Near Anglisidhes, 1929, *Renz*; Khapotomi, 1960, *C. E. H. Sparrow* 51b! Episkopi, 1964, *J. B. Suart* 2–5! Between Pakhna and Mallia, 1979, *Chesterman* 130!
4. W. of Larnaca, 1977, *J. J. Wood* 2!
5. Between Nicosia and Eylenja, 1929, *Renz* 1569! Near Trikomo, 1929, *Renz* 1572! Delikipo, 1929, *Renz* 1578! Kythrea, 1937, *Miss Godman* s.n.!
6. W. of Nicosia, 1929, *Renz* 1571!; Nicosia, 1960, *C. E. H. Sparrow* 51a!
7. Pentadaktylos, 1880, *Sintenis & Rigo* 157! Kyrenia Pass, 1929, *Renz* 1570! Akanthou Pass, 1929, *Renz* 1573! Kremama Kamilou, 1929, *Renz*; Melounda, 1929, *Renz*; Kantara, 1929, *Renz* 1574! Between Kantara and Mandres, 1929, *Renz*; Near Klepini, 1929, *Renz*; Ardhana-Phlamoudhi, 1929, *Renz* 1577! Halevga, 1937, *Roger-Smith* s.n.! also, 1952, *F. M. Probyn* 75! Kyrenia, 1929, *Renz* 426! also 1948, *Duke-Woolley* 46! 1949, *Casey* 418! and, 1956, *G. E. Atherton* 900! 1013! 1014! Ayios Nikolaos, 1929, *Renz* 1576!
8. Komi Kebir, 1929, *Renz*; Ephtakomi, 1929, *Renz* 1575! Ayios Andronikos, 1938, *Syngrassides* 1787! and, 1948, *Mavromoustakis* s.n.! Leonarisso, 1968, *Economides* in ARI 1115!

NOTES: Cyprian populations of ssp. *mammosa* often have a greater percentage of red anthocyanin pigment on the lateral sepals, particularly on their lower halves; this is a feature also shared by ssp. *transhyrcana* (Czernjak.) Soó. Forms from the Northern Range exhibiting rudimentary lobing of the lip have been distinguished by Renz as f. *subtriloba*. Pollination is effected by bees of the genus *Andrena*.

ssp. **transhyrcana** (*Czernjak.*) *Soó* in Acta Bot. Acad. Sci. Hung., 5: 444 (1959); Renz in Rechinger f., Fl. Iran., no. 126: 70 (1978); Sundermann, Europ. Medit. Orch., ed. 3, 95 (1980); J. J. Wood in Die Orchidee, 31: 230 (1980), in Orch.Review, 89: 294 (1981).

    O. transhyrcana Czernjak. in Not. Syst. Herb. Hort. Bot. Petrop., 4: 1 (Jan., 1923); Erich Nelson, Gestaltwandel, 156 (1962); Baumann et Künkele, Wildwachsenden Orch. Europas, 280 (1982).

    O. sintenisii H. Fleischm. et Bornm. in Ann. Nat. Mus. Wien, 36: 10 (Febr., 1923); Soó in Acta Bot. Acad. Sci. Hung., 16: 391 (1970); Sundermann, Europ. Medit. Orch., ed. 1, 69 (1970) Landwehr, Wilde Orch. Europa, 2: 412 (1977).

    O. spruneri Nym. var. *orientalis* Schltr. in Keller et Schltr., Mon. Orch., 1: 112 (1926); Post, Fl. Pal., ed. 2, 2: 555 (1933).

    O. sphegodes Mill. ssp. *sintenisii* (H. Fleischm. et Bornm.) Erich Nelson, Gestaltwandel, 181 (1962); Gumprecht in Jahresb. Naturwiss. Ver. Wuppertal, 19: 38 (1964); Sundermann in Die Orchidee, 20: 79 (1969); Osorio-Tafall et Seraphim, List Vasc. Plants Cyprus, 25 (1973); B. et E. Willing in Die Orchidee, 27: 114 (1976); Williams et al., Field Guide, 50 (1978); J. et P. Davies in Quart. Bull. Alpine Gard. Soc., 48: 124 (1980); Davies et Huxley, Wild Orch. Brit. Europe, 164 (1983).

TYPE: U.S.S.R.: Turkmenistan; "Regio Transcaspica, distr. Krasnowodsk, in promont. Kopet-dag, prope Karakala: Alty-Waj, 3400′ in fruticetis, in declivibus herbosis, 7.iv. 1916.", *Czernjakovska* 451 (LE).

Plant usually robust; inflorescence 4–10-flowered; sepals dull olive -green, the lower half of the laterals stained reddish-purple; dorsal sepal appearing distinctly narrow because of the strongly revolute margin, 10–15 (–17) mm. long, 4–5 mm. wide when flattened; lateral sepals 10–17 mm. long, 4–7 mm. wide when flattened; petals narrowly ligulate or linear, acute, margins somewhat revolute, rarely flat, ± glabrous or very shortly pubescent, often strongly reflexed, 7–11 mm. long, 2–3 mm. wide; labellum dark brown, broadly obovate, 10–16 mm. long, 13 mm. wide when flattened, entire or sometimes obscurely 3-lobed, strongly convex, margin flat and spreading or strongly deflexed, densely covered with short, dark purple, rather rigid hairs towards the periphery, lateral protuberances poorly developed or absent, basal callosities dull bluish-violet, insignificant, flat, basalfield dark brown, abbreviated and broad or elongate and ± cordate, speculum ± Π-shaped, i.e. composed of two parallel vertical bars joined above by a transverse cross-bar running along the apex of the basalfield and extending laterally, the two vertical bars sometimes becoming isolated from the cross-bar, otherwise H-shaped, apical appendage reduced to a small, deflexed mucro; stigmatic cavity short; anther-connective rostrate, margin involute, 1·5–3·5 mm. long. *Plate 88, fig. 4.*

HAB.: In garigue or Pine forest, often in moist situations, on calcareous formations; 200–2,500 ft. alt.; fl. mid March–mid April.

DISTR.: Divisions 1, 2, 4, 7, 8. Also U.S.S.R. (Turkmenistan), Turkey (Anatolia [Cilicia]), Lebanon, Syria, Palestine, Israel, Iraq, western and northern Iran.

1. Neokhorio, 1962, *Meikle* 2206! Between Khrysokhou and Polis, 1977, *J. J. Wood* 74!
2. Gorge below Mallia, 1977, *J. J. Wood* 36! 38–43! Between Pano Lefkara and Vavatsinia, 1977, *J. J. Wood* 53! 56! 60!
4. Dhekelia, 1977, *A. Della* in ARI 1675!
7. Halevga, 1956, *G. E. Atherton* 1046! 1047! Antiphonitis, 1959, *C. E. H. Sparrow* 25!
8. Komi Kebir, 1936, *Syngrassides* 1338! Between Ayios Philon and Rizokarpaso, 1941, *Davis* 2242!

NOTES: In Cyprus *O. sphegodes* ssp. *transhyrcana* grades into large, subentire-lipped forms of ssp. *mammosa*. This situation makes accurate identification rather difficult at times. In pure populations it may be recognised by the absence or inconspicuousness of the protuberances of the lip and by the long, rostrate anther-connective.

The high percentage of red anthocyanin pigment often present in the sepals and the misconception that this subspecies has a distinctly 3-lobed lip, has led to the belief that ssp. *transhyrcana* may be closely related to ssp. *amanensis* from S. Turkey. The lip is never deeply 3-lobed and being convex is easily split during pressing and drying. Thus many herbarium specimens often appear to be strongly 3-lobed. Nelson's *O. sphegodes* ssp. *amanensis* is now thought, and quite correctly in my opinion, to be conspecific with the Turkish populations of ssp. *spruneri*. It is obvious that ssp. *transhyrcana* is not closely allied to ssp. *spruneri*, differing markedly in lip-shape and the shorter stigmatic cavity. The sepals are never entirely pink or reddish-purple as are those of the latter. Recent observations have shown that the areas of distribution of both subspecies overlap in southern Turkey and consequently hybrids may occur. Pollination is probably effected by bees of the genus *Andrena*.

5. O. apifera *Huds.*, Fl. Angl., ed. 1, 340 (1762); Boiss., Fl. Orient., 5: 79 (1884); Post, Fl. Pal., ed. 2, 2: 578 (1933); Erich Nelson, Gestaltwandel, 175 (1962); Gumprecht in Jahresb. Naturwiss. Ver. Wuppertal, 19: 38 (1964); Sundermann in Die Orchidee, 20: 79 (1969); Hermjakob, Orch. Greece Cyprus, 15 (1974); Sundermann, Europ. Medit. Orch., ed. 2, 107 (1975), ed. 3, 117 (1980); Landwehr, Wilde Orch. Europa, 2: 474 (1977); Renz in Rechinger f., Fl. Iran., no. 126: 74 (1978); Davies et Huxley, Wild Orch. Brit. Europe, 177–178 (1983).

Plant (15–) 25–45 (–50) cm. high; tubers 2, globose, sessile or with 1 stipitate; stem bearing 2–3 cauline leaves and 1 or 2 amplexicaul, acute sheaths; leaves 3–6 (–9), mostly rosulate, narrowly elliptic to ovate, sometimes oblanceolate, the lowermost obtuse, sometimes acute, the uppermost acute, 4–10 (–16) cm. long, 1–3 cm. wide; inflorescence 2–6 (–17)-flowered, up to 25 cm. long, lax or subdense; floral bracts elliptic or ovate-elliptic, acute, equal to or longer than the ovary; sepals bright purple-violet, violet-rose, pink, greenish-white or white, with green nerves, oblong or ovate-oblong, obtuse, glabrous, spreading or reflexed, (8–) 10–16 mm. long, 5–9 mm. wide; petals green, yellowish-green or pink, triangular to linear, apex ± cucullate, margin revolute, densely whitish velutinous, sometimes glabrous, 3–4 mm. long, ± 3 mm. wide; labellum chestnut-brown or blackish-purple, rarely yellowish-green or bicoloured, broadly ovate or oval in outline, 10–13 mm. long, 5–8 mm. wide, convex, strongly 3-lobed, partly papillose and partly villous, mid-lobe obovate, lateral lobes oblique, triangular-ovate, deflexed, lateral protuberances brown, with creamy or yellow hairs, to 3 mm. long, basal callosities greenish-yellow or brownish-green, basalfield orange, orange-brown or reddish-brown, speculum pale violet, mauve or brown, with a cream or pale yellow border, small, collar-shaped or scutelliform, usually with two distal rudimentary yellow spots or blotches positioned along the apex of the basalfield, hair tuft yellowish, prominent, apical appendage yellow or greenish-yellow, lanceolate, rarely tridentate, glabrous, deflexed,

2–3 mm. long; column green, c. 9 mm. long, staminodal points absent; anther-connective green, rostrate or acuminate, flexuous, 1·5–2 mm. long.

var. **apifera**; Erich Nelson, Gestaltwandel, 175 (1962).
TYPE: England; "in pratis et pascuis siccioribus", *Hudson*.

Sepals bright purple-violet, violet-rose or pink, rarely white; labellum chestnut-brown or blackish-purple, never bicoloured or yellowish-green, basalfield prominent, orange, orange-brown or reddish-brown, speculum prominent, pale violet, mauve or brown, with a cream or pale yellow border. *Plate 88, fig. 5.*

HAB.: In garigue, on dry turf, in Pine forest or in marshy meadows that become seasonally dry; sea-level–3,000 ft. alt.; fl. mid April–early May.

DISTR.: Divisions 1, 3, 4, 6. Widespread in western, central and southern Europe, eastwards to Turkey (Anatolia), Lebanon, Syria, Palestine, Israel, Caucasus area, western U.S.S.R. and northern Iran; North Africa.

1. Between Lachi and Neokhorio, 1977, *J. J. Wood* 97! Near Neokhorio, 1977, *J. J. Wood* 109!
3. Kakoradjia, 1937, *Syngrassides* 1527! Zagala near Trimiklini, 1941, *Davis* 3172! 4 m. N. of Kandou, 1963, *J. B. Suart* 121! 6 m. S. of Trimiklini, 1963, *J. B. Suart* 126! Above Pano Lefkara, 1974, *Meikle* 5006!
4. Near Tekke, W. of Larnaca Salt Lake, 1979, *St. White* s.n.
6. Syrianokhori, 1938, *Syngrassides* 1814!

NOTES: Specimens having dark purple-violet or very dark pink sepals occur sporadically among a large population of var. *chlorantha* near Neokhorio in the Akamas Peninsula. Unlike other members of this genus, *O. apifera* is almost entirely autogamous.

var. **bicolor** (*Naegeli*) *Erich Nelson*, Gestaltwandel, 178 (1962).
O. *bicolor* Naegeli in Ber. Schweiz. Bot. Ges., 23: 64 (1914).
TYPE: Switzerland; "Geneva".

Sepals pink; labellum bicoloured, consisting of a pale yellow or greenish-yellow basal zone extending to or just beyond the middle and a reddish-brown apical zone, both being separated by a brown median line, lateral lobes less prominent, basalfield very small, greenish-yellow to apricot, speculum absent.

HAB.: In garigue, bushy places, or on grassy slopes between shrubs; 250 ft. alt.; fl. early April.

DISTR.: Division 1. Also France, Switzerland, Italy and Albania.

1. Near Loukrounou, 1977, *J. J. Wood* 117!

var. **chlorantha** (*Hegetschw.*) *Richter*, Pl. Eur., 1: 264 (1890); Erich Nelson, Gestaltwandel, 178 (1962).
O. *chlorantha* Hegetschw., Fl. Schweiz, 876 (1840).
TYPE: Switzerland; not precisely localized, "Hin und wieder mit den Vorigen [*O. apifera* Huds.], doch mehr an beschatteten Stellen".

Sepals white, greenish-white or yellowish-green; petals yellowish-green; labellum lemon-yellow shading into green, usually becoming bright yellowish-green towards the base and darker green towards the apex, basalfield often very small, yellowish-green, speculum pale greenish-cream, often barely discernible or virtually absent.

HAB.: Calcareous banks amongst garigue containing *Sarcopoterium*, *Prasium*, *Pistacia* and *Cistus*, etc.; 600 ft. alt.; fl. early April.

DISTR.: Divisions 1, 3. Also France, Switzerland, West Germany and Rhodes.

1. Near Neokhorio, 1977, *J. J. Wood* 108!
3. Near Trimiklini, 1982, *Tennant* s.n.!

NOTES: A large colony of this apochromic variety occurs near Neokhorio. Some of the individuals in this colony have flowers with a normal-sized basalfield on the lip. Klein (in Die Orchidee, 29: 29; 1978) points out that after anthesis flowers of the almost identical and obviously conspecific var. *flavescens* Rosb. may change their colour and resemble those of var. *chlorantha* on the same plant. This confirms how equivocal such colour variations very often are.

6. O. argolica *H. Fleischm.* in Verh. Zool. Bot. Ges. Wien, 69: 295 (1919).

O. ferrum-equinum Desf. ssp. *argolica* (H. Fleischm.) Soó in Notizbl. Bot. Gart. Berlin, 9: 907 (1926).

Plant (10–) 15–25 (–35) cm. high; tubers 2, small, globose or ovoid, 1 sessile, 1 stipitate; stem bearing 2–3 cauline leaves and 1 or 2 obtuse, mucronate or acute sheaths; leaves 3–4 (–6), mostly rosulate, narrowly elliptic, obtuse and mucronate or acute, 3–6 (–18) cm. long, 1–1·8 (–2·5) cm. wide; inflorescence (1–) 4–6-flowered, usually lax; floral bracts elliptic, acute, mostly longer than the ovary; sepals purple-violet, lilac-pink, whitish or greenish, with a green mid-nerve, oblong, ovate or ovate-elliptic, 8–12 mm. long, 4–6 mm. wide, obtuse, revolute, glabrous; dorsal sepal sometimes appearing linear-lanceolate due to the strongly revolute margin, usually reflexed, sometimes curving forward; lateral sepals spreading or reflexed; petals pale pinkish-violet, lilac or purple, sometimes brownish-violet, lanceolate, ovate, elliptic or triangular, obtuse or acute, sometimes slightly revolute, rather undulate, papillose or velutinous, spreading or reflexed, $\frac{1}{2}$ $\frac{2}{3}$ as long as the sepals; labellum reddish-brown, brown or blackish-violet, with whitish or yellowish hairs on the shoulders and often with a narrow, yellowish marginal zone, suborbicular to ovate in outline, 7–12 mm. long, 8–10 mm. wide, entire or 3-lobed, slightly convex or flat, papillose, marginal zone glabrous, lateral protuberances absent, basal callosities conspicuous, often long, greenish and black or dark olive-green, linked by an inconspicuous, mauve or orange band, basalfield reddish- or orange-brown, often lacking a clearly defined margin, speculum dull violet to brown, with a white border, variable in shape, horseshoe- or spectacle-shaped, transversely linear, or often reduced to two spots, flecks or crescents, occasionally consisting of one spot or line only, hair tuft poorly developed or absent, apical appendage reduced to a small greenish or brownish, narrow, deflexed mucro; column cream or yellowish-green; stigmatic cavity rather small, traversed by a broad or narrow brownish-purple transverse band, staminodal points reddish-brown or absent; anther loculi long; anther-connective short, $\pm$ acute; caudicles of the pollinarium long.

ssp. **elegans** *(Renz) Erich Nelson,* Gestaltwandel, 153 (1962); Gumprecht in Jahresb. Naturwiss. Ver. Wuppertal, 19: 38 (1964); Sundermann in Die Orchidee, 20: 79 (1969); Soó in Acta Bot. Acad. Sci. Hung., 16: 391 (1970); Sundermann, Europ. Medit. Orch., ed. 1, 89 (1970); Osorio-Tafall et Seraphim, List Vasc. Plants Cyprus, 25 (1973); Hermjakob, Orch. Greece Cyprus, 3 (1974); Sundermann, Europ. Medit. Orch., ed. 2, 109 (1975), ed. 3, 119 (1980); B. et E. Willing in Die Orchidee, 26: 74 (1975); Landwehr, Wilde Orch. Europa, 2: 434, 435, figs. 1, 2 (1977); Williams et al., Field Guide, 32 (1978); J. et P. Davies in Quart. Bull. Alpine Gard. Soc., 48: 230 (1980); Davies et Huxley, Wild Orch. Brit. Europe, 169 (1983).
  O. gottfriediana Renz ssp. *elegans* Renz in Feddes Repert., 27: 206, 213, t. 1, figs. 1, 4 (1929); Post, Fl. Pal., ed. 2, 2: 576 (1933).
  O. elegans (Renz) H. Baumann et Künkele in Mitt. Arbeitskr. Heim. Orch. Baden-Württ., 13: 350 (1981), Die Wildwachsenden Orch. Europas, 216 (1982).
  [*O. scolopax* Cav. var. *picta* (non (Link) Reichb. f.) Unger et Kotschy, Die Insel Cypern, 209 (1865).]
  [*O. gottfriediana* (non Renz) R. S. Davis in Orch. Journ., 3: 164 (1954).]
  TYPE: Cyprus; "Bogas tis Kerynias, am südfuss des Passes, östlich von Agridia. Cupressus — Pinus Wäldchen, Schiefer, 12. März 1929, *Renz* 424" (RENZ).

Plant (7–) 10–15 (–25) cm. high; inflorescence (1–) 2–3-flowered; sepals pale pinkish-violet, strongly reflexed; petals slightly papillose, strongly reflexed; labellum 7–10 mm. long, c. 8 mm. wide, velutinous, 3-lobed, occasionally subentire, rather flat when subentire, the mid-lobe becoming convex when the labellum is distinctly 3-lobed, margins often strongly deflexed and revolute, sometimes overlapping on the undersurface, so that the apex is more attenuate than in ssp. *argolica*, lateral lobes usually pale reddish-brown at the base, greenish at the apex, often fused with the mid-lobe for the greater part of their length, with a forward-pointing, glabrous, free apical portion, margins slightly deflexed, basal callosities long, dark olive-green, basalfield shiny, often surrounded by the rudiments of the speculum border, speculum normally spectacle-shaped, composed of two diamond-shaped patches each extending towards the centre of the labellum and linked by a narrower isthmus; staminodal points absent. *Plate 88, fig. 6*.

HAB.: Pine and Cypress forests particularly on the Northern Range, in garigue, Juniper scrub, on sand dunes, etc.; most frequent near the coast and on low hills; on chalk and tertiary marls; sea-level–4,000 ft. alt.; fl. Febr.–early March (–April).

DISTR.: Divisions 1–3, 5–8. Southern Turkey.

1. Peyia Forest, 1962, *Meikle* 2055! and, 1978, *Chesterman* 53! Smyies, 1962, *Meikle* 2196!
2. Yironas R. below Madhari, 1962, *Meikle* 2832!
3. Stavrovouni, 1948, *Mavromoustakis* s.n.! Evdhimou, 1960, *M. Hawkins* 3! Episkopi, 1964, *J. B. Suart* 7! 16! 13–15! Between Kouklia and Pano Arkhimandrita, 1979, *Chesterman* 119!
5. Between Nicosia and Eylenja, 1929, *Renz* 1499!
6. Between Kormakiti and Liveras, 1937, *Syngrassides* 1416! Nicosia, 1957, *P. H. Oswald* 79!
7. Ardhana-Phlamoudhi, 1929, *Renz* 1508! Kyrenia Pass, 1929, *Renz* 424! Akanthou Pass, 1929, *Renz* 1502! Kremama Kamilou, 1929, *Renz* 1504! Myrtou, 1929, *Renz* 1500! Olymbos, 1929, *Renz* 1503! Melounda, 1929, *Renz* 1506! between Kantara and Mandres, 1929, *Renz*; Klepini, 1929; *Renz* 449! Near Ayios Amvrosios, 1941, *Davis* 2188! Halevga, 1952, *F. M. Probyn* 76! Kyrenia, 1956, *G. E. Atherton* 964! Ayios Nikolaos, 1929, *Renz* 1509!
8. Komi Kebir, 1929, *Renz* 1501! Between Komi Kebir and Ephtakomi, 1929, *Renz*; Ephtakomi, 1929, *Renz* 1505! 1507! Between Ayios Philon and Rizokarpaso, 1941, *Davis* 2243! Near Khelones, 1941, *Davis* 2296! Ronnas Bay, 1941, *Davis* 2334!

NOTES: Renz (1929) cited a doubtful record for this subspecies from Arisch-Kebir, near Alexandretta in the Amanus region of south-east Turkey. Recent observations made by Sundermann and Taubenheim (1978) have confirmed its occurrence in Turkey. Specimens collected by Vöth in southern Turkey during April 1967 are shown to be indistinguishable from those of Cyprian populations.
Some individuals within a population may, on first inspection, bear a superficial resemblance to the Greek *O. ferrum-equinum* Desf. ssp. *gottfriediana* (Renz) Erich Nelson.
Very small plants are to be found growing in dry, exposed localities or at high elevations.

O. argolica *Fleischm.* ssp. **elegans** (*Renz*) *Erich Nelson* × **O. bornmuelleri** *M. Schulze ex Bornm.* ssp. **grandiflora** (*Fleischm. et Soó*) *Renz et Taubenheim*; B. et E. Willing in Die Orchidee, 26: 77 (1975).

Sepals pale pink, reflexed; petals lilac-pink, rather long; labellum entire, lateral protuberances present, basalfield reddish-brown, speculum dull brown with a white border, spectacle-shaped; staminodal points present, reddish; anther-connective short, ± acute.

HAB.: In *Pinus halepensis* forest; 2,000 ft. alt.; fl. early April.

DISTR.: Division 7, two plants observed.

7. Above Halevga Forest Station, Northern Range, 8.4.1974, *B. et E. Willing*.

NOTES: Demut (in Die Orchidee, 32: 37; 1981) records *O. argolica* Fleischm. ssp. *elegans* (Renz) Erich Nelson × *O. umbilicata* Desf. ssp. *umbilicata* from Cyprus, but without habitat or locality.

7. **O. kotschyi** *H. Fleischm. et Soó* in Notizbl. Bot. Gart. Berlin, 9:908(1926); Soó in Botanisch. Archiv, 23: 135 (1928); R. S. Davis in Orch. Journ., 3: 164 (1954); Erich Nelson, Gestaltwandel, 144 (1962); Gumprecht in Jahresb. Naturwiss. Ver. Wuppertal, 19: 38 (1964); Sundermann in Die Orchidee, 20: 79 (1969); Soó in Acta Bot. Acad. Sci. Hung., 16: 391 (1970); Sundermann, Europ. Medit. Orch., ed. 1, 93 (1970); Osorio-Tafall et Seraphim, List Vasc. Plants Cyprus, 26 (1973); Hermjakob, Orch. Greece Cyprus, 5 (1974); Sundermann, Europ. Medit. Orch., ed. 2, 111 (1975), ed. 3, 123 (1980); B. et E. Willing in Die Orchidee, 26: 75 (1975); Landwehr, Wilde Orch. Europa, 2: 426, 427, figs. 1, 2 (1977); Williams et al., Field Guide, 30 (1978); J. et P. Davies in Quart. Bull. Alpine Gard. Soc., 48: 122, 123 (1980); Baumann et Künkele, Wildwachsenden Orch. Europas, 246 (1982); Davies et Huxley, Wild Orch. Brit. Europe, 170 (1983).
O. sintenisii H. Fleischm. et Bornm. ssp. *kotschyi* (H. Fleischm. et Soó) Soó in Botanisch. Archiv, 23: 31 (1928).
O. cypria Renz in Feddes Repert., 27: 202, 213 (1929); Osorio-Tafall et Seraphim, List Vasc. Plants Cyprus, 26 (1973).

TYPE: Cyprus; "Mt. Pentedactylos (Sintenis et Rigo no. 157, 159 p.p. !), Lefkara, Larnaca (Kotschy no. 269, 270/1859 p.p. !)".

Plant (8–) 20–30 (–35) cm. high; tubers 2, large, globose or ovoid, usually stipitate; stem often bearing 1 cauline leaf and 1 amplexicaul, acute sheath; leaves dark bluish-green, shining, (3–) 4–8 (–10), all except 1 grouped at the base, narrowly elliptic, obtuse and mucronate or acute, 4–8 cm. long, 1·5–2·5 cm. wide; inflorescence (3–) 4–7 (–10)-flowered, lax; floral bracts ovate to elliptic, concave, acute, longer than the ovary; sepals 6–15 mm. long, 5–7 mm. wide, green, occasionally flushed pale pink, rarely entirely pale pink or brownish, median nerve green, broadly ovate, obtuse, revolute, glabrous, dorsal sepal curving forward, lateral sepals spreading or slightly reflexed; petals dark olive-green, mostly suffused purple at the base, broadly elliptic or ovate, or triangular-ovate, 4–7 mm. long, 2–3 mm. wide at the base, acute, shortly hairy, usually reflexed; labellum dark brown to blackish-purple, ovate in outline, 12–15 mm. long, 5–7 mm. wide, distinctly 3-lobed, strongly convex, velutinous, mid-lobe broadest at the middle, the margin revolute, particularly towards the apex, lateral lobes somewhat paler than the mid-lobe, narrow, convex, strongly deflexed and revolute, sometimes almost touching on the undersurface of the labellum, lateral protuberances to 2 mm. long, curving forward or ± erect, covered with dark brown to blackish-purple villose hairs above, lower surface glabrous, basal callosities whitish, marked black, linked by an irregular blackish-purple band, inconspicuous, basalfield dark brown to blackish-purple, cordate, speculum dark brown to dull violet, sometimes almost blackish, with a snowy-white border, large, irregular, the border isolating 3 blackish-purple scutelliform patches, otherwise rarely ± H-shaped, hair tuft blackish-purple, distinct, apical appendage green to light brown, small, unarticulated, horizontal or upturned; column whitish to green, to 6 mm. long, staminodal points reddish; stigmatic cavity marked by deep purple transverse bands; anther-connective green, acute; caudicles of the pollinarium long.

HAB.: Grassy, shady places under Pine, Cypress, also under *Acacia* and *Eucalyptus*, in garigue or in sand dunes, or amongst lush vegetation in damp situations, on calcareous soils; sea-level–3,000 ft. alt.; fl. early March–early April.

DISTR.: Divisions 3–8. Probably endemic; Sundermann (1975) states that it may possibly occur in southern Turkey.

3. Ypsonas, 1970, *A. Genneou* in ARI 1393! 1·5 m. E. of Kophinou, 1978, *Chesterman* 40! and, 1979, *P. Davies* s.n.!
4. Near Larnaca, 1959, *Kotschy* 269.
5. Between Nicosia and Eylenja, 1929, *Renz* 1064! Near Trikomo, 1929, *Renz* 1299! Lapathos, 1962, *Meikle* 2987!
6. W. of Nicosia, 1929, *Renz*; Near Kormakiti, 1937, *Syngrassides* 1429!
7. Pentadaktylos, 1880, *Sintenis & Rigo* 157; 159 p.p. Olymbos, 1929, *Renz* 1288! Myrtou, 1929, *Renz* 1287! and, 1962, *Meikle* 2420! Klepini, 1929, *Renz* 1286! Between Halevga and Kharcha, 1941, *Davis* 2770! Kyrenia, 1956, *G. E. Atherton* 1315! Above Ayios Amvrosios, 1969, *D. Poole* s.n.!
8. Ayios Andronikos, 1938, *Syngrassides* 1788! Ronnas Bay, 1941, *Davis* 2333! Yialousa, 1941, *Davis* 2402!

NOTES: This large-flowered and distinctive species belongs, together with *O. cretica* (Vierh.) Erich Nelson, within the subsection *Aegaeae* of the section *Orientales*. The largest populations occur in the central and northern part of the island.

8. O. scolopax *Cav.*, Icones, 2: 46, t. 161 (1793); Boiss., Fl. Orient., 5: 80 (1884); Renz in Feddes Repert., 27: 213 (1929); R. S. Davis in Orch. Journ., 3: 164 (1954); Erich Nelson, Gestaltwandel, 159 (1962); Gumprecht in Jahresb. Naturwiss. Ver. Wuppertal, 19: 38 (1964); Soó in Acta Bot. Acad. Sci. Hung., 16: 386 (1970); Sundermann, Europ. Medit. Orch., ed. 1, 83 (1970); Osorio-Tafall et Seraphim, List Vasc. Plants Cyprus, 26 (1973); Hermjakob, Orch. Greece Cyprus, 8 (1974); Landwehr, Wilde Orch. Europa, 2: 436 (1977); Renz in Rechinger f., Fl. Iran., no. 126: 79 (1978); Williams et al., Field Guide, 34 (1978); J. et P.

Davies in Quart. Bull. Alpine Gard. Soc., 48: 124 (1980); Davies et Huxley, Wild Orch. Brit. Europe, 172 (1983).
    O. picta Link in Schrader, Journ. Bot., 2: 325 (1799); Boiss., Fl. Orient., 5: 80 (1884); Post, Fl. Pal., ed. 2, 2: 576 (1933).
    O. scolopax Cav. var. *picta* (Link) Reichb. f., Icones Fl. Germ., 13 & 14: 98, 99 (1851); Boiss., Fl. Orient., 5: 80 (1884); Holmboe, Veg. Cypr., 57 (1914).
    O. fuciflora (F. W. Schmidt) Moench ssp. *scolopax* (Cav.) Sundermann, Europ. Medit. Orch., ed. 3, 39, 109 (1980).
    TYPE: Spain; "Habitat floretque cum praecedenti [*O. lutea*]", i.e. "in humidis umbrosis prope Albaydae [Albaida] oppidum, praesertim in collibus radicem iuxta fontis originem". (MA).

Plant 10–45 (–60) cm. high; tubers 2, elliptic or subglobose, sessile or with one tuber stipitate; stem slender, bearing 1–2 (–3) cauline leaves and 1–2 amplexicaul, acute sheaths, or 1 obtuse, mucronate or acute sheath only; leaves (3–) 4–7 (–10), mostly grouped at the base, narrowly elliptic or oblong-ligulate, somewhat concave, obtuse, mucronate, acute or acuminate, 5–12 cm. long, 1–2 (–2·5) cm. wide; inflorescence (2–) 3–7 (–12)-flowered, lax; floral bracts ovate to elliptic, concave, obtuse or acute to acuminate, the lowermost longer than the ovary; sepals pale pink to purple-violet, seldom whitish or greenish, obtuse to subacute, revolute, glabrous, median nerve green, prominent; dorsal sepal oblong or ovate-oblong, slightly concave, erect or slightly reflexed, very rarely curving forward, 6–12 (–15) mm. long, 4–7 mm. wide; lateral sepals obliquely ovate-elliptic, spreading or slightly reflexed, 6–13 (–15) mm. long, 4–7 mm. wide; petals pale pink, purplish-pink, apricot or greenish-white, ligulate, triangular or ovate-elliptic, auriculate at the base, acute or obtuse, sometimes revolute, velutinous or densely villous, spreading or slightly reflexed, 4–5 mm. long, 2–3 mm. wide at the base; labellum dark brown to blackish- or purplish-brown or reddish-brown, suborbicular, ovate, obovate or oblong in outline, (8–) 10–12 mm. long, c. 8 mm. wide, distinctly 3-lobed, strongly convex, mid-lobe broadest at or just below the middle, velutinous, shortly hairy towards the periphery, marginal zone very narrow, margin recurved and revolute, particularly towards the apex, making the labellum appear narrow from the front, lateral lobes clothed with brown or reddish-brown villose hairs on the outer surface, inner surface yellowish, greenish or brownish, ovate-oblong to triangular, strongly deflexed or sometimes laterally spreading, lateral protuberances 4–6 (–8) mm. long, with brown, villose hairs in the lower part, apex yellowish, greenish or brownish, glabrous, usually small, broad, obtuse, horizontal or decurved, rarely erect, basal callosities green, marked black, dark olive-green or brownish-green, spherical to long, linked by a purplish-brown band or by a dull greenish-brown area, basalfield yellowish- to bright orange-brown, with a blackish- to bluish-violet or purple border, speculum dull violet to dull steel-blue, reddish- or purplish-brown, with a creamy yellow or whitish border, variable in shape, often large, extending from the base of the labellum towards the middle of the mid-lobe, irregular, annular or H-shaped, often enclosing 1–3 dark brown, velutinous, oval, scutelliform patches, the two vertical bars usually becoming reduced so that the often larger, central patch is not enclosed, hair tuft insignificant, apical appendage yellowish-green or brownish, large, relatively long, tridentate, horizontal or deflexed; column cream, yellowish or green, c. 6 mm. long; stigmatic cavity marked with a purplish-brown transverse band, not forming a V-shaped indentation into the base of the basalfield, staminodal points reddish; anther-connective green, acute, short; caudicles of the pollinarium long.

    HAB.: In garigue, on dry turf, or in woodland; 200–2,000 ft. alt.; fl. early March–April, or to June at higher elevations.

    DISTR.: Divisions 1–3, 5–8. Also Portugal, southern Spain, southern France, Balearic Islands,

Corsica, Sardinia, Italy (Monte Gargano), Greece, Rhodes, western and southern Turkey, Caucasus, Transcaucasia, northern Iran; North Africa.

1. Between Baths of Aphrodite and Fontana Amorosa, 1977, *J. J. Wood* 77! 1 m. S. of Kathikas, 1977, *J. J. Wood* 114! Near Smyies, 1977, *J. J. Wood* 105! 107!
2. Near Mallia, 1977, *J. J. Wood* 19! Between Pano Lefkara and Vavatsinia, 1977, *J. J.Wood* 50!
3. Alethriko, 1905, *Holmboe* 211! Skarinou, 1929, *Renz*; Between Lefkara and Kornos, 1929, *Renz*.
5. Between Nicosia and Eylenja, 1929, *Renz*.
6. W. of Nicosia, 1929, *Renz*.
7. St. Hilarion, 1929, *Renz*; Kantara, 1929, *Renz*; Melounda, 1929, *Renz*; Akanthou, 1929, *Renz*; Kyrenia, 1929, *Renz*; Olymbos, 1929, *Renz*; Between Pentadaktylos and Klepini, 1929, *Renz*; Ayios Amvrosios, 1938, *Lady Loch* 56!
8. Between Komi Kebir and Ephtakomi, 1929, *Renz*; Between Ayios Philon and Rizokarpaso, 1941, *Davis* 2239!

NOTES: Pollination is effected by bees of the genus *Eucera*.

9. O. umbilicata *Desf.* in Ann. Mus. Hist. Nat. Paris, 10: 227, t. 15 (1807); Baumann and Künkele in Mitt. Arbeitskr. Heim. Orch. Baden-Württ., 13: 285–301 (1981); Wildwachsenden Orch. Europas, 282 (1982).

Plant (8–) 15–45 cm. high; tubers 2, oblong or subglobose, shortly stipitate or sessile; stem slender or robust, erect or slightly flexuous, usually bearing 1 or 2 cauline leaves and 1 or 2 amplexicaul, acute sheaths; leaves 3–6 (–8), mostly rosulate, narrowly elliptic or oblanceolate, obtuse, mucronate or acute, erect to spreading, 4–15 cm. long, 1·5–3·5 cm. wide; inflorescence (3–) 5–12-flowered, short to elongated, narrowly or broadly cylindrical, 4–15 cm. long, 5–6 cm. wide, subdense; floral bracts ovate, narrowly elliptic or oblanceolate, concave, acute, the lowermost longer than the ovary, the uppermost as long as the ovary; sepals pale to olive-green or greenish-white to pale pink, rarely whitish, ovate-oblong or oblong, obtuse, revolute, glabrous, dorsal sepal curving forward, 6–10 (–13) mm. long, 4–6 mm. wide; lateral sepals oblique, spreading or reflexed, 6–12 (–13) mm. long, 5–7 mm. wide; petals yellowish- to olive-green, rarely whitish or purplish, triangular or triangular-ligulate, auriculate at the base, obtuse or ± acute, villous, erect or slightly reflexed, 4–5 mm. long, 2–3 mm. wide at the base; labellum brownish or brownish-purple, ovate, oblong or flabelliform in outline, 6–7 (–15) mm. long, 6–11 mm. wide when expanded, distinctly 3-lobed, with or without a distinctly excised sinus between the mid and lateral lobes, mid-lobe flabelliform when expanded, cuneate at the base, broadest towards the middle or ± truncate apex, strongly convex, margin revolute, papillose to velutinous, marginal zone membranous, glabrous, lateral lobes small, elliptic or obliquely ligulate, deflexed or spreading, clothed with dark brown villose hairs on the upper surface, lower surface yellowish-green, lateral protuberances with dense, brown villose-velutinous hairs on the lower part, apex yellowish-green or creamy, glabrous, 1–4 mm. long, basal callosities greenish, brownish, blackish or reddish, linked by a steel-blue to blackish-purple band, basalfield dark brown or bright orange-brown, with a blackish, bluish-violet or purple border, speculum reddish-brown, dull steel-blue or brownish-violet, with a creamy-white or yellowish border, extending from the base of the labellum to the middle of the mid-lobe, irregular, enclosing 3–4 brownish or brownish-purple, velutinous, scutelliform or oblong patches, the lateral rudiments of the speculum often connected to the central portion, hair-tuft sometimes prominent, apical appendage yellowish-green, small, broad, sometimes obscurely tridentate, upturned or horizontal; column green or whitish, 5–6 mm. long; staminodal points reddish; anther-connective green, acute, usually short, c. 1 mm. long.

ssp. **umbilicata**

O. dinsmorei Schltr. in Feddes Repert., 19: 46 (1923); Post, Fl. Pal., ed. 2, 2: 576 (1933).
O. cornuta Stev. ssp. *orientalis* Renz in Feddes Repert., 27: 205, 213 (1929); Rechinger f. in
Oesterr. Bot. Zeitschr., 94: 194 (1947).

O. oestrifera M. Bieb. ssp. *orientalis* (Renz) Soó in Keller et Schltr., Mon. et Icon. Orch., 2:
65 (1931).

O. scolopax Cav. ssp. *orientalis* (Renz) Erich Nelson, Gestaltwandel, 158 (1962);
Gumprecht in Jahresb. Naturwiss. Ver. Wuppertal, 19: 38 (1964); Sundermann in Die
Orchidee, 20: 79 (1969); Soó in Acta Bot. Acad. Sci. Hung., 16: 391 (1970); Sundermann,
Europ. Medit. Orch., ed. 1, 83 (1970); Osorio-Tafall et Seraphim, List Vasc. Plants Cyprus,
26 (1973); Hermjakob, Orch. Greece Cyprus, 7 (1974); B. et E. Willing in Die Orchidee, 26:
78 (1975); Landwehr, Wilde Orch. Europa, 2: 438, 439, fig. 4 (1977); Renz in Rechinger f.,
Fl. Iran., no. 126: 81 (1978); Williams et al., Field Guide, 36 (1978); J. et P. Davies in Quart.
Bull. Alpine Gard. Soc., 48: 121, 124 (1980); Davies et Huxley, Wild Orch. Brit. Europ.,
173 (1983).

O. carmeli H. Fleischm. et Bornm. ssp. *orientalis* (Renz) Soó in Acta Bot. Acad. Sci.
Hung., 18: 381 (1973); Sundermann, Europ. Medit. Orch., ed. 3, 113 (1980).

[*O. oestrifera* (non M. Bieb.) Holmboe, Veg. Cypr., 57 (1914).]

[*O. oestrifera* M. Bieb. var. *cornuta* (non Boiss.) Seligman in Quart. Bull. Alpine Gard.
Soc., 20: 233 (1952).]

[*O. cornuta* (non Stev.) R. S. Davis in Orch. Journ., 3: 164 (1954).]

[*O. scolopax* Cav. ssp. *cornuta* (non (Stev.) Camus) Elektra Megaw, Wild Flowers of
Cyprus, 13, t. 24 (1973); Osorio-Tafall et Seraphim, List Vasc. Plants Cyprus, 26 (1973).]

TYPE: Probably Greece, Khios or Samos, 1701 or 1702, based on Tournefort's *Orchis
orientalis, fucum referens, flore parvo umbilicato* (Coroll. Inst. Rei. Herb., 30; 1700, figured by
Aubriet and reproduced in Desfontaines' paper, *loc. cit.*), *Tournefort* (P, missing).

Plant 8–15 (–30) cm. high; stem bearing 1 sheath only; leaves 4–6, basal
only; inflorescence 5–12-flowered; sepals greenish-white to pale pink; dorsal
sepal strongly curving forward, 6–10 mm. long, 4–6 mm. wide, lateral sepals
spreading, 6–11 mm. long, 5–6 mm. wide; labellum 6–8 (–10) mm. long, c. 6
mm. wide, mid-lobe broadening sharply from a narrow base and broadest
towards the middle and apex, lateral protuberances 1–3 mm. long, speculum
frequently covering a large area of the labellum, but locally reduced in size,
the border enclosing 3–4 oval patches, hair tuft prominent, apical
appendage broad, short, sometimes obscurely tridentate, horizontal;
stigmatic cavity forming a V-shaped indentation into the base of the
basalfield.

HAB.: Widespread on chalk, tertiary marls and sandstones, in Pine and Cypress forest on the
Northern Range, in *Cistus* garigue on the southern foothills of the Troödos, in grassy areas
among Pines, *Eucalyptus* and *Acacia*, also under Juniper and *Sarcopoterium*; sea-level–2,000 ft.
alt.; fl. mid March–late April (–early May).

DISTR.: Divisions 2–8. Also Kos, ?Khios, ?Samos, western and southern Turkey, Lebanon,
Syria, Palestine and Israel.

2. Between Ayios Amvrosios and Kissousa, 1977, *J. J. Wood* 18 ! 21 ! 27 !
3. Skarinou, 1929, *Renz* 457 ! Between Lefkara and Kornos, 1929, *Renz*; Episkopi, 1964, *J. B.
Suart* 10 ! Ypsonas, 1970, *A. Genneou* in ARI 1392 !
4. By Salt Lake W. of Larnaca, 1977, *J. J. Wood* 5 ! 9 !
5. Between Nicosia and Eylenja, 1929, *Renz* 1300 ! Trikomo, 1929, *Renz* 1307 ! Kythrea, 1937,
Miss Godman s.n. !
6. W. of Nicosia, 1929, *Renz* 436 ! Kremama Kamilou, 1929, *Renz* 1304 !
7. St. Hilarion, 1929, *Renz*; Kantara, 1929, *Renz* 1308 ! Mandres, 1929, *Renz* 448 ! Melounda,
1929, *Renz*; Ayios Nikolaos, 1929, *Renz* 1310 ! Akanthou, 1929, *Renz* 444 ! Ardhana —
Phlamoudhi, 1929, *Renz* 1309 ! Kyrenia, 1929, *Renz*; also, 1932, *Syngrassides* 299 ! and, 1973,
P. Laukkonen 145 ! Between Kantara and Mandres, 1929, *Renz*; Myrtou, 1929, *Renz* 1305 !
Olymbos, 1929, *Renz* 1306 ! Pentadaktylos, Klepini, 1929, *Renz* 1311 !, Kyrenia Pass, 1929,
Renz 1301 ! Ayios Amvrosios, 1938, *Lady Loch* 59 ! Halevga, 1952, *F. M. Probyn* 93 !
8. N. of Komi Kebir, 1929, *Renz* 1302a ! 1302b ! 1302c ! Ephtakomi, 1929, *Renz* 1313 ! Between
Ayios Philon and Rizokarpaso, 1941, *Davis* 2241 ! 2296 ! Ronnas Bay, 1941, *Davis* 2337 !

ssp. **attica** *(Boiss. et Orph.) J. J. Wood* in Kew Bull., 38: 136 (1983).

O. arachnites Hoffm. var. *attica* Boiss. et Orph. in Boiss., Diagn., 2, 4: 91 (1859); Boiss.,
Fl. Orient., 5: 77 (1884).

O. arachnites Lam. ssp. *attica* (Boiss. et Orph.) Richter, Plant. Eur., 1: 262 (1890).

O. attica (Boiss. et Orph.) Soó in Notizbl. Bot. Gart. Berlin, 9: 909 (1926); Renz in Feddes

Repert., 27: 214 (1929); R. S. Davis in Orch. Journ., 3: 164 (1954); Sundermann in Die Orchidee, 20: 79 (1969), Europ. Medit. Orch., ed. 1, 83 (1970); Hermjakob, Orch. Greece Cyprus, 6 (1974); Landwehr, Wilde Orch. Europa, 2: 444 (1977); Sundermann et Taubenheim in Die Orchidee, 29: 177 (1978); Williams et al., Field Guide, 34 (1978).

O. attica (Boiss. et Orph.) Soó f. *flavo-marginata* Renz in Feddes Repert., 27: 214 (1929).

O. attica (Boiss. et Orph.) Soó f. *holocheila* Renz in Feddes Repert., 27: 214 (1929).

O. scolopax Cav. ssp. *attica* (Boiss. et Orph.) Erich Nelson, Gestaltwandel, 156 (1962); Elektra Megaw, Wild Flowers of Cyprus, 13, t. 24 (1973); Osorio-Tafall et Seraphim, List Vasc. Plants Cyprus, 26 (1973); B. et E. Willing in Die Orchidee, 26: 75–78 (1975); Williams et al., Field Guide, 34 (1978).

O. carmeli H. Fleischm. et Bornm. ssp. *attica* (Boiss. et Orph.) Renz in Rechinger f., Fl. Iran., no. 126: 79 (1978); Sundermann, Europ. Medit. Orch., ed. 3, 113 (1980).

O. flavomarginata (Renz) H. Baumann et Künkele in Mitt. Arbeitskr. Heim. Orch. Baden-Württ., 13: 301, t. 4 (1981); Wildwachsenden Orch. Europas, 222 (1982).

[*O. carmeli* (non H. Fleischm. et Bornm.) Erich Nelson, Gestaltwandel, 156 (1962); Williams et al., Field Guide, 34 (1978); J. et P. Davies in Quart. Bull. Alpine Gard. Soc., 48: 121, 124 (1980); Davies et Huxley, Wild Orch. Brit. Europe, 173 (1983).]

TYPE: Greece; "in *Atticâ* propé *Stadium* rarissima cl. Prof. *Orphanides*" (ATHU).

Plant (10–) 15–25 (–35) cm. high; inflorescence (3–) 5–10-flowered, usually short, broadly cylindrical, up to 6 cm. wide; petals triangular; labellum without a distinctly excised sinus between the mid and lateral lobes, (6–) 8–15 mm. long. *Plate 88, fig. 3.*

HAB.: In garigue, on dry turf, in Pine forest, on limestone; sea-level–1,000 ft. alt.; fl. (mid Febr.–) mid March–mid April (–mid May at higher elevations).

DISTR.: Divisions 1–8. Also Greece (Atticâ, Peloponnese and Ionian Islands), Rhodes, western and southern Turkey, Lebanon, Syria, Palestine and Israel.

1. Psindron (Akamas), 1962, *Meikle* 2124b! Smyies, 1962, *Meikle* 2197! Between Lachi and Neokhorio, 1977, *J. J. Wood* 103!
2. Between Ayios Amvrosios and Kissousa, 1977, *J. J. Wood* 20!
3. Mazotos, 1905, *Holmboe* 181! Skarinou, 1929, *Renz*; Mesayitonia, 1948, *Mavromoustakis* s.n.! Paramali, 1960, *C. E. H. Sparrow* 53! Paramytha, 1962, *Nicolaides* s.n.! Episkopi, 1964, *J. B. Suart* 1! 184!
4. Dhekelia, 1973, *P. Laukkonen* 164! Salt Lake W. of Larnaca, 1977, *J. J. Wood* 1! 6! 8! 10!
5. Between Nicosia and Eylenja, 1929, *Renz* 1239! 1247! Near Trikomo, 1929, *Renz* 1242! Salamis, 1929, *Renz*; Athalassa, 1933, *Syngrassides* 46!
6. W. of Nicosia, 1929, *Renz*; and, 1936, *Syngrassides* 936! Kormakiti, 1937, *Syngrassides* 1664–1666! Kremama Kamilou, 1929, *Renz* 1245! Orga, 1970, *Baumann & Künkele* s.n.
7. Kyrenia Pass, 1929, *Renz* 1240! 1241! Akanthou Pass, 1929, *Renz* 1224! 1244! Myrtou, 1929, *Renz* 1246! and, 1962, *Meikle* 2420a! Olymbos, 1929, *Renz* 433! Melounda, 1929, *Renz*; Pentadaktylos, 1929, *Renz*; Between Kantara and Mandres, 1929, *Renz*; Kyrenia, 1932, *Syngrassides* 300! and, 1956, *G. E. Atherton* 977! 978! Halevga, 1937, *Roger-Smith* s.n.!
8. Between Komi Kebir and Ephtakomi, 1929, *Renz*; Komi Kebir, 1929, *Renz* 1243! Between Ayios Philon and Rizokarpaso, 1941, *Davis* 2240! Ronnas Bay, 1941, *Davis* 2338!

NOTES: Individuals having rather large flowers appear to predominate in Cyprus. In both Cyprus and Anatolia in Turkey, the line of demarcation separating *O. umbilicata* ssp. *attica* and ssp. *umbilicata* is rather blurred.

Renz (1929) recognised two Cyprian forms of ssp. *attica*, each differing slightly in lip characters. The first, *f. flavomarginata* Renz, recently treated as a species by Baumann & Künkele (1981), bears flowers having a flat, broad yellowish marginal zone to the mid-lobe. It has been found in grassy places around Nicosia, on Pentadaktylos in the Northern Range and in the northerly mountains to the west of Mt. Olympus. The second form, *f. holocheila* Renz, produces one or two flowers with a subentire or ± entire lip on an otherwise normal inflorescence and may superficially resemble *O. bornmuelleri*. It has been found in Pine forest in the Northern Range and growing amongst *Sarcopoterium* west of Trikomo in the Mesaoria.

O. umbilicata Desf. ssp. **attica** (*Boiss. et Orph.*) *J. J. Wood* × **O. bornmuelleri** *M. Schulze ex Bornm.* ssp. **grandiflora** (*Fleischm. et Soó*) *Renz et Taubenheim*; B. et E. Willing in Die Orchidee, 26: 78 (1975).

Dorsal sepal erect; petals flushed lilac-pink, auriculate at the base; labellum entire, margins somewhat recurved, densely hairy on the marginal zone, speculum brown with a yellowish-green border, irregular.

HAB.: In *Pinus halepensis* forest; 2,000 ft. alt.; fl. late March.

DISTR.: Division 7, one plant observed.

7. Halevga Forest Station, Northern Range, 26.3.1974, *B. & E. Willing.*

[O. BOMBYLIFLORA *Link* in Schrader, Journ. für die Bot., 2: 325 (1799).
Young tubers long stipitate, with stolons up to 10 cm. long; sepals green; labellum deeply 3-lobed, brown, lateral protuberances prominent, speculum scutelliform or bipartite; anther-connective obtuse. This species is widespread in the Mediterranean Basin and in the east occurs in Greece, Crete, Rhodes, southern Turkey and Lebanon, but it has not so far been recorded for Cyprus.]

[O. SPECULUM *Link* in Schrader, Journ. für die Bot., 2: 324 (1799).
Sepals green or yellowish, often brown-striate; labellum deeply 3-lobed, dark purple, margin long-villose, speculum large, iridescent blue with a yellow border; anther-connective obtuse. Although widespread in Greece, Rhodes, Turkey and Lebanon, this species is, curiously, absent from both Cyprus and Crete.]

O. SPHEGODES *Mill.* ssp. AESCULAPII *(Renz) Soó* in Acta Bot. Acad. Sci. Hung., 16: 383 (1970); De Langhe and D'Hose in Bull. Soc. Roy. Bot. Belg., 115: 300, 303 (1982).
Sepals pale to dark olive-green; petals greenish-yellow; labellum entire or subentire, 10–14 mm. long, (12–) 13–15 (–18) mm. wide, blackish-brown or purplish with a broad, glabrous, yellowish margin, speculum greyish-mauve with a white border, H-shaped, the lower arms sometimes joined, lateral protuberances small or absent. This subspecies occurs in Greece and has recently been reported from S.W. Cyprus by De Langhe and D'Hose (1982). I have never personally seen any material of ssp. *aesculapii* from Cyprus and De Langhe's report may refer to one of the forms intermediate between ssp. *mammosa* and ssp. *transhyrcana,* having a yellowish margin to the lip. The authenticity of this report must therefore remain unsubstantiated.

[O. SPHEGODES *Mill.* ssp. SPRUNERI *(Nym.) Erich Nelson,* Gestaltwandel, 183 (1962); Elektra Megaw, Wild Flowers of Cyprus, 12, 13, t. 22 (1973), as "*O. sphecodes* ssp. *spruneri*"; B. et E. Willing in Die Orchidee, 27: 115 (1976); Sundermann, Europ. Medit. Orch., ed. 3, 95 (1980); J. J. Wood in Die Orchidee, 31: 231 (1980), Orch. Review, 89: 295 (1981).
O. hiulca Sprun. ex Reichb., Icones Fl. Germ. et Helv., 13–14: 93 (1851); Post in Mém. Herb. Boiss., 18: 100 (1900); H. Stuart Thompson in Journ. Bot., 44: 338 (1906); Holmboe, Veg. Cypr., 57 (1914), "In the plains of Cyprus, according to Post."
Dorsal and lateral sepals usually heavily stained purplish-pink; labellum distinctly 3-lobed, very rarely subentire, lateral protuberances insignificant or absent, apical appendage present, lateral lobes deflexed. This subspecies occurs in Greece, Crete, the Aegean Islands and southern Turkey (Adana Province) but is absent from Cyprus. Cyprian material identified as this in the Kew herbarium by Mr V. S. Summerhayes has proved to be *O. argolica* Fleischm. ssp. *elegans* (Renz) Erich Nelson. Post's record ("Plaines de Chypre, mars" [1894]) may refer to forms of *O. sphegodes* Mill. ssp. *mammosa* (Desf.) Soó or *O. sphegodes* Mill. ssp. *transhyrcana* (Czernjak.) Soó. The plant illustrated by Elektra Megaw (1973) has an entire lip and is clearly not ssp. *spruneri.*]

[O. TENTHREDINIFERA *Willd.,* Sp. Plant., 4: 67 (1805); Sibth. et Sm., Fl. Graec. Prodr., 2: 217 (1816), Fl. Graec., 10: 22, t. 929 (1840); Poech, Enum. Plant. Ins. Cypr., 11 (1842); Unger et Kotschy, Die Insel Cypern, 208 (1865), as "O. tentredinifera Willd."; H. Stuart Thompson in Journ. Bot., 44: 338 (1906); Holmboe, Veg. Cypr., 56 (1914); Soó in Botanisch. Archiv, 23: 25, 132 (1928); Elektra Megaw, Wild Flowers of Cyprus, 12, t. 21 (1973); B. et E. Willing in Die Orchidee, 26: 75–77 (1975); Sundermann, Europ. Medit. Orch., ed. 3, 105 (1980).

Sepals purplish-pink to violet, broadly ovate; labellum entire, rarely 3-lobed, quadrate or flabelliform, lateral protuberances inconspicuous, speculum reduced, hair tuft pronounced; anther-connective obtuse. Although mentioned frequently in the literature, there is no satisfactory evidence for the natural occurrence of *O. tenthredinifera* in Cyprus. Some of the records probably refer to forms of *O. bornmuelleri* M. Schulze ex Bornm. ssp. *grandiflora* (Fleischm. et Soó) Renz et Taubenheim with varying degrees of pink pigmentation on the sepals.]

8. ANACAMPTIS *L. C. M. Rich.*
Orch. Eur. Annot., 25 & 33 (1817), preprint from
Mém. Mus. Hist. Nat. Paris, 4: 47 & 55 (1818).

Tubers entire; stem leafy, glabrous; leaves rosulate and cauline, unspotted; inflorescence dense; floral bracts membranous; perianth segments concave, the dorsal sepal and petals connivent to form a galea, the lateral sepals spreading, glabrous; labellum glabrous, spurred, flat, deeply 3-lobed, with 2 longitudinal calli at the base; column short; stigmas 2, clearly separated by and lying on either side of the anther on the column, above the mouth of the spur; rostellum small; viscidium solitary, narrow, transversely ligulate, placed in a simple bursicle.

A monotypic genus distributed in Europe and North Africa, eastwards to Iran.

1. **A. pyramidalis** (*L.*) *L. C. M. Rich.*, Orch. Eur. Annot., 33 (1817) preprint from Mém. Mus. Hist. Nat. Paris, 4: 55 (1818); Boiss., Fl. Orient., 5: 57 (1884); Holmboe, Veg. Cypr., 58 (1914); Soó in Botanisch. Archiv, 23: 93, 171 (1928); Renz in Feddes Repert., 27: 217 (1929); Post, Fl. Pal., ed. 2, 2: 563 (1933); Rechinger f. in Arkiv. för Bot., ser. 2, 1: 419 (1950); R. S. Davis in Orch. Journ., 3: 164 (1954); Sundermann, Europ. Medit. Orch., ed. 1, 105 (1970), ed. 3, 131 (1980); Osorio-Tafall et Seraphim, List Vasc. Plants Cyprus, 27 (1973); Davies et Huxley, Wild Orch. Brit. Europe, 143 (1983).
    Orchis pyramidalis L., Sp. Plant., ed. 1, 940 (1753).
    Aceras pyramidalis (L.) Reichb. f., Icon. Fl. Germ., 13: 6 (1851); Unger et Kotschy, Die Insel Cypern, 206 (1865).
    TYPE: "*in* Helvetiae, Belgii, Angliae, Galliae *arenosis, cretaceis.*"

Plant (16–) 25–60 (–75) cm. high; tubers usually 2, ovoid or subglobose, entire, sessile or shortly stipitate; stem erect, often becoming arcuate, slender, smooth, bearing 2–3 brownish basal sheaths and leaves towards the base and several ovate to ovate-elliptic, acute, sheathing bracts above; leaves 3–7, rosulate (during winter), cauline (during summer), the winter leaves withering before flowering commences, linear to linear-lanceolate, rarely narrowly elliptic, obtuse and mucronate to acute, canaliculate, 8–25 cm. long, 0·7–1·2 (–2) cm. wide, the lowermost often becoming recurved; inflorescence many-flowered, conical (pyramidal) at first, becoming shortly cylindrical, 2–10 (–12) cm. long, 3 (–4) cm. wide, dense; floral bracts linear to linear-lanceolate or narrowly ovate-elliptic, apex acute to acuminate, often reflexed, membranous, slightly exceeding or a little shorter than the ovary, usually purplish-pink; flowers pale pink to dark reddish-pink, rarely bright blood-red or white, the labellum having a cream or white basal and central portion; sepals concave, acute, (4–) 6–8 mm. long, c. 3 mm. wide; dorsal sepal and petals loosely connivent, forming a galea; dorsal sepal oblong-ovate to ovate-elliptic; lateral sepals obliquely narrowly elliptic to oblong-ovate; petals obliquely ovate to ovate-elliptic, acuminate, 4–7 mm. long, c. 2·5 mm. wide at the base; labellum usually broader than long, flabellate-cuneate, flat, 6–9 mm. long, 8–10 (–12) mm. wide, distinctly and ± equally 3-lobed, with 2 prominent, cream or white, narrow, erect, often spreading longitudinal calli at its base, extending from the mouth of the spur towards the centre of the labellum and acting as guide plates for insects, mid-lobe linear to oblong, obtuse, usually narrower than the laterals, lateral lobes oblong-rhombic, obliquely triangular or linear, obtuse, sometimes a little longer than the mid-lobe; spur filiform, ± acute, horizontal to strongly downturned, equal to or exceeding the ovary, those of the lowermost flowers often projecting below the base of the inflorescence; column short, obtuse.

HAB.: In garigue, in dry grassy places, in light *Pinus brutia* forest, on waste ground; on calcareous soils; sea-level–3,000 ft. alt.; fl. March–May.

DISTR.: Divisions 1–5, 7, 8. Widely distributed in western, central and southern Europe, extending northwards to Scandinavia and eastwards to the Caucasus, Iraq and northern and western Iran; North Africa.

1. Paphos, 1862, *Kotschy* 653; Fontana Amorosa, 1962, *Meikle* 2293! Between Neokhorio and Prodhromi, 1977, *J. J. Wood* 110!
2. Pano Panayia, 1941, *Davis* 3344!
3. Limassol, 1913, *Haradjian* 617; Polemidhia, 1948 and 1952, *Mavromoustakis* 43! Cape Gata, 1948, *Mavromoustakis* s.n.! Akrotiri, 1955, *Merton* 2062! and, 1963, *J. B. Suart* 100! Episkopi, 1963, *J. B. Suart* A5! Evdhimou — Pakhna, 1978, *Holub* s.n.!
4. Cape Greco, 1862, *Kotschy* 183; Ormidhia, 1929, *Renz* 454! Cape Pyla, 1929, *Renz* 455!
5. Near Apostolos Varnavas, 1948, *Mavromoustakis* s.n.! and, 1962, *Meikle* 2637!
7. Kremama Kamilou, 1929, *Renz*; St. Hilarion, 1931, *J. A. Tracey* 8! Kornos, 1941, *Davis* 3051! Kyrenia, 1954, *Casey* 733! Bellapais, 1956, *G. E. Atherton* 1317! Tjiklos, 1973, *P. Laukkonen* 184! 337!
8. Aphendrika, 1880, *Sintenis & Rigo* 559! Near Rizokarpaso, 1912, *Haradjian* 227; Ayios Andronikos, 1948, *Mavromoustakis* s.n.!

NOTES: Under adverse conditions flower-resupination sometimes fails to occur with the result that the lip is borne uppermost. The flowers, which are delicately scented, are visited by butterflies and both day- and night-flying moths. When a Lepidopteran visits a flower the end of the ligulate viscidium curls round and clasps the insect's proboscis.

In many southern populations the mid- and lateral lobes of the lip may be equal in length and width.

9. BARLIA *Parl.*

Nuov. Gen. Piant. Monocot., 5 (1858).

Tubers large, entire; stem erect, robust, leafy below, glabrous; basal leaves large, broad, unspotted, shiny, cauline leaves grading into amplexicaul bracts; inflorescence many-flowered, dense; floral bracts membranous; flowers large, olive-green to purple-violet; dorsal sepal connivent with the petals to form a galea; lateral sepals spreading; labellum deeply 3-lobed, with 2 parallel median ridges running from the mouth of the spur to the middle of the mid-lobe, mid-lobe bilobed, lateral lobes falcate, undulate; spur conical, saccate, with a globular, papillose to hairy nectary within; column broad; stigmatic cavity elongate, ± cordate; staminodes prominent; viscidium placed in a large, broad bursicle; caudicles of the pollinarium short.

A monotypic genus widespread in southern Europe and North Africa, extending from the Canary Islands to Turkey.

1. **B. robertiana** (*Loisel.*) *Greuter* in Boissiera, 13: 192 (1967); Erich Nelson, Mon. Icon. Serapias, Aceras, Loroglossum, Barlia, 65 (1968); Osorio-Tafall et Seraphim, List Vasc. Plants Cyprus, 26 (1973); Sundermann, Europ. Medit. Orch., ed. 2, 123 (1975), ed. 3, 135 (1980); Landwehr, Wilde Orch. Europa, 2: 344 (1977); J. et P. Davies in Quart. Bull. Alpine Gard. Soc., 48: 121, 123 (1980); Davies et Huxley, Wild Orch. Brit. Europe, 141 (1983).
   Orchis robertiana Loisel., Fl. Gall., ed. 1, 2: 606 (1806).

O. longibracteata Biv., Sic. Plant. Cent., 1: 57 (1806); Sintenis in Oesterr. Bot. Zeitschr., 31: 255 (1881).

Aceras longibracteata (Biv.) Reichb. f., Icon. Fl. Germ., 13 & 14: 3 (1851); Unger et Kotschy, Die Insel Cypern, 206 (1865); Boiss., Fl. Orient., 5: 55 (1884); Holmboe, Veg. Cypr., 58 (1914); R. S. Davis in Orch. Journ., 3: 164 (1954).

Barlia longibracteata (Biv.) Parl., Fl. Ital., 3: 447 (1858).

Himantoglossum longibracteatum (Biv.) Schltr., Die Orchideen, 52 (1914); Soó in Botanisch. Archiv, 23: 91, 169 (1928); Renz in Fedde Repert., 27: 217 (1929).

TYPE: France; Toulon, "in collibus siccis petrosis circa Telonem, invenit D. ROBERT".

Plant (19–) 30–60 (–80) cm. high; tubers 2 (–3), large, ovoid to spherical, sessile, the youngest long-stipitate, roots numerous, long, thick; stem erect, robust, straight, angled, with 2 membranous sheaths at the base, leafy above, dull greyish-green below, suffused brownish-red to purple above; leaves 5–6 (–8), the lowermost grouped together towards the base of the stem, often very large, broad, elliptic, ovate-elliptic, oblong-elliptic, occasionally narrowly elliptic, obtuse and mucronate, fleshy, with prominent nerves on the upper surface, 6–25 (–30) cm. long, 4–10 cm. wide, spreading to erect, shining dark green or grey-green, often suffused violet towards the apex, the uppermost grading into narrowly elliptic to lanceolate, acute, amplexicaul bracts, sometimes reaching the base of the inflorescence, often suffused dull brownish-red or violet; inflorescence many-flowered, 6–23 cm. long, dense, rarely subdense; floral bracts linear-lanceolate, acute, membranous, longer than the galea, pale olive-green, often suffused brownish-red or purple-violet, particularly towards the apex; flowers large, with a scent like an Iris; sepals pale olive-green, brownish-red to purple-violet, the inner surface, particularly of the laterals, spotted purple, ovate, obtuse, concave; dorsal sepal curving forward with the petals to form a galea, 10–12 mm. long, 6–8 mm. wide; lateral sepals slightly oblique, 10–15 mm. long, 7–9 mm. wide, spreading; petals pale olive-green, unspotted or only lightly spotted pale purple, narrowly ovate to linear, obtuse, 7–10 mm. long, 3 mm. wide; labellum pale purple-violet or olive-green with a central white area usually bearing irregular purple-violet flecks, lateral lobes and lobules of the mid-lobe dull olive-green, 3-lobed, 15–20 mm. long, 14–16 mm. wide across lateral lobes, with 2 parallel median ridges running from the mouth of the spur to the middle of the mid-lobe, mid-lobe bilobed, 8–10 mm. long, sometimes with a small mucro in the sinus between the oblong, rounded or truncate, slightly undulate, often upturned lobules, lateral lobes shorter than the mid-lobe, falcate, attenuate, obtuse to ± acute, strongly undulate, often touching the lobules of the mid-lobe; spur 4–6 mm. long, white to pale purple-violet, conical, saccate, obtuse, downturned, with a globular papillose or hairy nectary on the lower surface inside; column greenish to purple-violet, broad; stigmatic cavity elongate, ± cordate, edged with dark purple; staminodes prominent; anther-connective short, broad; caudicles of the pollinarium yellowish; pollinia dark green.

HAB.: In garigue and dry grassy places; roadside banks; woodland; on calcareous and neutral soils; sea-level–1,500 ft. alt.; fl. (Jan.–) Febr.–March.

DISTR.: Divisions 1–3, 4, 6. Also Canary Islands, Portugal, southern and eastern Spain, Balearic Islands, southern France, Italy, Corsica, Sardinia, Sicily, Yugoslavia (southern Dalmatia), Albania, Greece, Aegean Islands, Crete, Rhodes, southern Turkey; North Africa.

1. Below Neokhorio, 1962, *Meikle* 2214!
2. Evrykhou, 1929, *Renz.*
3. Near Vouni, 1859, *Kotschy* s.n.; Moni, 1964, *J. B. Suart* 218! Asgata, 1970, *A. Genneou* in ARI 1388!
4. Near Larnaca, 1880, *Sintenis* teste Boiss.
6. W. of Nicosia, 1929, *Renz* 2291!

10. ACERAS *R. Br.*
in Aiton f., Hort. Kew., ed. 2, 5: 191 (1813).

Tubers entire; stem leafy, glabrous; leaves rosulate and cauline, unspotted; inflorescence dense; floral bracts membranous; perianth segments connivent to form a galea, glabrous; labellum without a spur, deeply 3-lobed, resembling a man in outline, slightly concave at the base, otherwise flat, pendent, with 2 thick, converging calli at the base, the upper and lower surfaces densely papillose; column very short; stigmatic cavity obliquely oval, low; staminodes rudimentary, small, rounded, free at the apex; rostellum insignificant; anther short, obtuse, connective absent; viscidia 2, elliptic, often ± adherent to one another, placed in a triangular bursicle; caudicles of the pollinarium short.

A monotypic genus distributed in western and southern Europe and North Africa, extending eastwards as far as Lebanon.

1. **A. anthropophorum** *(L.) R. Br.* in Aiton f., Hort. Kew., ed. 2, 5: 191 (1813); Unger et Kotschy, Die Insel Cypern, 205 (1865); Boiss., Fl. Orient., 5: 55 (1884); H. Stuart Thompson in Journ. Bot., 44: 338 (1906); Holmboe, Veg. Cypr., 58 (1914); Soó in Botanisch. Archiv, 23: 88, 168 (1928); Post, Fl. Pal., ed. 2, 2: 561 (1933); Sundermann, Europ. Medit. Orch., ed. 1, 101 (1970), ed. 2, 124 (1975), ed. 3, 137 (1980); Osorio-Tafall et Seraphim, List Vasc. Plants Cyprus, 26 (1973); Landwehr, Wilde Orch. Europa, 2: 338 (1977); Williams et al., Field Guide, 66 (1978); Davies et Huxley, Wild Orch. Brit. Europ., 135 (1983).

Ophrys anthropophora L., Sp. Plant., ed. 1, 948 (1753).
TYPE: "*in* Italia, Lusitania, Gallia."

Plant (10–) 20–40 (–60) cm. high; tubers 2, ovoid or subglobose, entire, sessile or stipitate; stem erect or flexuous, bearing several whitish or brownish, membranous sheaths towards the base, with leaves, and several ovate to ovate-elliptic, acute, sheathing bracts above; leaves (3–) 5–8, rosulate and cauline, oblong to oblong-elliptic, occasionally obovate, obtuse and mucronate to acute, canaliculate, 5–12 (–15) cm. long 1–2 (–2·5) cm. wide, erect, becoming porrect to spreading later, dark to greyish-green or bluish-green; inflorescence many-flowered (flowers up to 90), narrowly cylindrical, 3–22 cm. long, 1·5–2·5 cm. wide, dense, the lowermost flowers becoming spaced out as the spike develops; floral bracts linear to triangular, 1-nerved, acute, membranous, shorter than the ovary, bluish-green; sepals pale to dull apple-green, with a narrow violet, brick-red or brownish-purple marginal zone and pale red nerves; sepals and petals connivent, forming a ± horizontal to downturned semi-globose galea, the sepals often free at the apex; sepals ovate to ovate-elliptic, concave, 1-nerved, obtuse, 5–7 mm. long, 3–4 mm. wide; dorsal sepal symmetrical; lateral sepals asymmetrical; petals pale green, narrowly linear, 1-nerved, obtuse, 5–6 mm. long, 1 mm. wide; labellum dull yellow, orange, brick-red or reddish-brown, usually yellowish-green at the centre of the mid-lobe, 12–15 mm. long, deeply 3-lobed, longer than the galea, slightly concave in the greenish zone lying between the stigmatic cavity and its base, upper and lower surface densely papillose, with 2 thick, white or yellowish, shiny, converging calli at the base, developed laterally from the lower edges of the stigmatic cavity, mid-lobe linear, longer than the lateral lobes, c. 2 mm. wide, distally bifurcate, the lobules narrowly linear, obtuse, divergent, separated by a mucro, lateral lobes narrowly linear, $\frac{1}{2}$–$\frac{3}{4}$ as long as the mid-lobe, 0·5–1 mm. wide; spur absent, replaced by a shallow, nectar-secreting depression; stigmatic cavity with a pale violet or purple margin; rostellum whitish; anther dull violet-purple; caudicle and bursicle whitish; pollinia yellow.

HAB.: In garigue, in short grass or light woodland; on calcareous and neutral soils; 1,500 ft. alt.; fl. late March–May.

DISTR.: Division 3, rare. Widespread in western Europe and the Mediterranean Basin, extending north-westwards to England and eastwards to south-west Turkey and Lebanon; North Africa.

3. Between Omodhos and Limassol, near Ayios Therapon, 1859, *Kotschy* 55.

NOTES: The flowers emit a rather unpleasant smell with a hint of coumarin (new-mown hay) which may attract hover-flies and ants. Hybrids are frequently reported, particularly involving *Orchis simia* (× *Orchiaceras bergonii* (Nant.) Camus and *O. italica* (× *Orchiaceras bivonae* Soó).

11. ORCHIS *L.*

Sp. Plant., ed. 1, 939 (1753).
Gen. Plant., ed. 5, 405 (1754).

Tubers 2 (–3), globose, ovoid or ellipsoid, entire, sessile or stipitate; stem glabrous; leaves basal, rosulate and/or cauline, spotted or unspotted; floral bracts thin, membranous; perianth-segments usually concave and connivent to form a galea, or with the 3 outer sepals more flattened and spreading to deflexed, the sepals and the 2 inner petals either equal in length or the latter shorter, glabrous; lip (or labellum) spurred, entire to 3-lobed, porrect to deflexed, the surface either flat, undulate or deflexed, sometimes with subfalcate lobes, with or without calli at the base, glabrous or papillose; column with a 3-lobed rostellum, the median lobe of which is short and lamelliform; anther-connective absent; viscidia 2, placed in a simple bursicle.

About 35 species distributed over Europe, the Mediterranean Basin and temperate Asia.

Lateral sepals spreading or deflexed, not connivent with the petals; dorsal sepal sometimes connivent with the petals to form a galea:
  Leaves cauline, linear to lanceolate, attenuate towards the apex, acute, unspotted:
    Mid-lobe of the lip usually shorter than the lateral lobes; lip 7 (–10) mm. long, 8 (–12) mm. wide, its narrow white central area lacking violet-pink markings; spur broadened and often bilobed at the apex - - - - - - - **1. O. laxiflora**
    Mid-lobe of the lip usually longer than the lateral lobes; lip (8–) 9–15 mm. long, (10–) 11–15 (–20) mm. wide, its narrow white central area with violet-pink markings; spur of equal width or attenuate at the apex - - - - - - - **2. O. palustris**
  Leaves basal, rosulate, narrowly elliptic or oblong, to ovate or obovate, attenuate towards the base, obtuse and mucronate, spotted or unspotted:
    Lip distinctly 3-lobed:
      Spur broadened, conical at the mouth, linear towards the apex, longer than the ovary; lip with many spots:
        Leaves 1·5–2 cm. wide; lip (8–) 10–14 (–17) mm. long, 10–14 (–17) mm. wide; spur 1·5–2·2 cm. long, horizontal or upturned - - - - - **3. O. anatolica**
        Leaves 2–2·5 (–3) cm. wide; lip 13–17 mm. long, 13–22 mm. wide; spur 2–2·5 cm. long, sharply upturned - - - - - - - **3. O. anatolica** var. **troodi**
      Spur filiform, not broadened at the mouth, downturned, as long as or a little shorter than the ovary; lip usually with 4 spots - - - - - **4. O. quadripunctata**
    Lip entire; spur conical-saccate, shorter than the ovary; lip unspotted - **5. O. collina**
Lateral and dorsal sepals connivent with the petals to form a galea:
  Lip entire, cuneate, constricted and clawed at the base - - - **6. O. papilionacea**
  Lip distinctly 3-lobed, not cuneate nor constricted and clawed at the base:
    Galea obtuse at the apex:
      Flowers mostly pink to pale purple; plant (10–) 15–25 (–32) cm. high; inflorescence lax; sepals 6–8 mm. long; spur spreading to upturned, as long as or longer than the ovary
7. O. morio ssp. **picta**
      Flowers mostly greenish-yellow, yellow or brownish-yellow; plant 22–60 (–70) cm. high; inflorescence dense; sepals 8–10 mm. long; spur downturned, shorter than the ovary
8. O. punctulata
    Galea long-acuminate, acuminate or occasionally acute at the apex:
      Spur conical, often incurved towards the underside of the lip; margin of the lateral lobes of the lip not entire; floral bracts as long as or longer than the ovary:
        Lip spotted, lateral lobes having a crenate or denticulate margin; galea acuminate, occasionally acute, 5–10 (–12) mm. long; spur not abruptly incurved toward underside of lip - - - - - - **9. O. coriophora** ssp. **fragrans**

Lip unspotted, lateral lobes having an acutely dentate to sinuately serrate margin; galea long-acuminate, 9–12 (–15) mm. long; spur abruptly incurved toward underside of lip - - - - - - - - - **10. O. sancta**

Spur cylindrical, not incurved towards the underside of the lip; margin of the lateral lobes of the lip entire; floral bracts much shorter than the ovary:

Lobes of the lip acute, flat, never subfalcate and upcurved; two calli present at the base of the lip; leaves often spotted, margin undulate - - **11. O. italica**

Lobes of the lip obtuse, subfalcate, upcurved; calli absent; leaves never spotted, margin not undulate - - - - - - - - **12. O. simia**

1. O. laxiflora *Lam.*, Fl. Franç., ed. 1, 3: 504 (1778); Boiss., Fl. Orient., 5: 71 (1884); Renz in Feddes Repert., 27: 216 (1929); Post, Fl. Pal., ed. 2, 2: 570 (1933); Mrs Frank Tracey in Journ. R.H.S., 58: 304 (1933); B. et E. Willing in Die Orchidee, 27: 113–114 (1976); Sundermann, Europ. Medit. Orch., ed. 3, 167 (1980); Davies et Huxley, Wild Orch. Brit. Europe, 138 (1983).

TYPE: France; "on la trouve dans les prés montagneux".

Plant 30–80 (–100) cm. high; tubers subglobose or ellipsoid, subsessile or shortly stipitate; stem erect or slightly flexuous, terete, sometimes fistulose in robust specimens, angular, green below, suffused reddish-purple above; leaves 3–8, cauline, linear to lanceolate, somewhat conduplicate, sheathing at the base, acute, erect or slightly spreading, unspotted, 10–15 cm. long, 1 (–2) cm. wide; inflorescence 6–20-flowered, sometimes elongated, 6–25 (–45) cm. long, lax; floral bracts narrowly elliptic, 3–7-veined, acute, shorter than, and enclosing the ovary, suffused reddish-purple; flowers reddish-purple to violet with a pure white central area usually extending from the base to the apex of the labellum, rarely wholly white; sepals with dark purple veins, ovate to oblong, obtuse, 7 (–10) mm. long, 2–4 mm. wide, dorsal sepal somewhat concave, usually connivent with the petals, erect or curving forwards, lateral sepals reflexed upwards or spreading, often bending towards the centre of the flower with age; petals yellowish or reddish-purple to violet, obliquely oblong, obtuse, curving forwards over the column, 6–8 mm. long, 2·5–4 mm. wide; labellum variable in shape, suborbicular, obovate or triangular-obovate in outline, cuneate at the base, 7 (–10) mm. long, 8 (–12) mm. wide, distinctly 3-lobed, mid-lobe subquadrate or obcordate, entire, emarginate or shallowly bilobed, margin crenulate or undulate, usually much shorter than the lateral lobes, lateral lobes broadly oblong or semi-oblong, oblong-orbicular or slightly rectangular, entire or shallowly bilobed, margin undulate, deflexed after anthesis; spur reddish-purple to violet, cylindrical, broadened and usually bilobed at the apex, gently upturned or horizontal, $\frac{1}{2}\frac{2}{3}$ as long as the ovary, 10–12 mm. long; column pinkish-white, short, 4 mm. long; stigmatic cavity creamy-white, broad, with a swelling above, on either side.

HAB.: In marshes, by streamsides or in damp sandy places; 200–2,900 ft. alt.; fl. April–May (–June).

DISTR.: Divisions 1–3, 6. Western, central and southern Europe, westwards to the Channel Islands, eastwards to southern U.S.S.R., eastern Turkey and Lebanon; North Africa.

1. Between Khrysokhou and Polis, 1977, *J. J. Wood* 67! Between Lachi and Neokhorio, 1978, *Chesterman* 62!
2. Mandria, 1939, *Kennedy* 1516! and, 1952, *F. M. Probyn* 93! Aphamis, 1941, *Davis* 3171!
3. Near Perapedhi, 1962, *Merton* s.n.!
6. Near Nicosia, 1929, *Renz*.

NOTES: *O. laxiflora* is found in the western part of Cyprus, and was once frequent in coastal marshes before widespread drainage. The hybrid ×*O. lloydiana* Rouy, (*O. laxiflora*×*O. palustris*), with a dense-flowered inflorescence and a deeply lobed lip, is known to occur in mixed populations. It has not been recorded from Cyprus.

2. O. palustris *Jacq.*, Collect. Bot., 1: 75 (1787); Unger et Kotschy, Die Insel Cypern, 207 (1865); Boiss., Fl. Orient., 5: 70 (1884); Post, Fl. Pal., ed. 2, 2: 570 (1933); Sundermann, Europ. Medit. Orch., ed. 1, 141 (1970); B. et E. Willing in Die Orchidee, 27: 114 (1976); Renz in Rechinger f., Fl. Iran., no. 126: 99 (1978).

O. elegans Heuff. in Flora, 18: 250 (1835); Renz in Rechinger f., Fl. Iran., no. 126: 102 (1978).

O. laxiflora Lam. ssp. *palustris* (Jacq.) Aschers. et Graebn., Syn. Mitteleur. Fl., 3: 712 (1907); Holmboe, Veg. Cypr., 58 (1914).

O. laxiflora Lam. ssp. *elegans* (Heuff.) Soó in Notizbl. Bot. Gart. Berlin, 9: 910 (1926); B. et E. Willing in Die Orchidee, 27: 114 (1976).

O. laxiflora Lam. ssp. *dielsiana* Soó in Notizbl. Bot. Gart. Berlin, 9: 910 (1926); Soó in Botanisch. Archiv, 23: 61, 154 (1928); Post, Fl. Pal., ed. 2, 2: 570 (1933); Sundermann, Europ. Medit. Orch., ed. 1, 141 (1970); Osorio-Tafall et Seraphim, List Vasc. Plants Cyprus, 27 (1973); B. et E. Willing in Die Orchidee, 27: 114 (1976).

TYPE: Austria; "in pratis Austriae palustribus, imprimis circa Himberg", *Jacquin*.

Plant 20–80 (–100) cm. high; tubers subglobose or ellipsoid, subsessile or shortly stipitate; stem green below, often suffused reddish-purple above, erect or slightly flexuous, terete, sometimes fistulose in robust specimens, angular; leaves 3–8, cauline, linear to lanceolate, somewhat conduplicate, sheathing at the base, apex acute, 10–25 (–30) cm. long, 1·5–2·5 (–3) cm. wide, erect or slightly spreading, unspotted; inflorescence 6–50-flowered, often elongated, 7–25 (–45) cm. long, often rather dense; floral bracts narrowly elliptic, acute, enclosing and usually equalling the ovary, the lowermost exceeding the ovary, green, suffused pale reddish-purple; flowers dark purple, sometimes pink, rarely pure white, labellum with a central white area, flecked violet-pink; sepals ovate to oblong, obtuse, 7–12 mm. long, 3·5–5 mm. wide, dorsal sepal slightly concave, usually connivent with the petals, erect or curving forward, lateral sepals reflexed upwards or spreading, often bending towards the centre of the flower with age; petals sometimes yellowish, obliquely oblong, obtuse, 6–10 mm. long, 2·5–4 mm. wide, curving forward over the column; labellum variable in size and shape, suborbicular, obovate, triangular-obovate, transversely oblong or obcordate in outline, cuneate at the base, 9–15 mm. long, 11–15 (–20) mm. wide, 3-lobed, sometimes entire or subentire, flat or slightly convex, mid-lobe subquadrate or obcordate, entire or somewhat emarginate at the apex, sometimes shallowly bilobed, usually longer than, or sometimes equal to, the lateral lobes, usually crenulate, lateral lobes (when present) broadly oblong or semi-oblong, oblong-orbicular or slightly rectangular, entire or shallowly bilobed, rarely deflexed after anthesis; spur reddish-purple to violet, cylindrical, attenuate at the apex or of uniform width, rarely bilobed, equalling or only slightly shorter than the ovary, 10–15 mm. long, strongly upturned, horizontal or slightly downturned; column pinkish-white, short, 4 mm. long; stigmatic cavity creamy-white, broad, with a swelling above, on either side.

HAB.: In marshes, by streamsides, or on damp turf; sea-level–100 ft. alt.; fl. (late April) late May–June.

DISTR.: Divisions 2, 3, 7. Western, central and southern Europe, northwards to Sweden (Gotland), eastwards to the U.S.S.R. and the Caucasus, Turkey (Anatolia, Smyrna, Cilicia, Pontus and Armenia), Syria, Jordan, Palestine, Iraq, Iran, Afghanistan and central Asia (Balkhash, Pamir-Alai); North Africa.

2. Phini, 1862, *Kotschy*.
3. Limassol, 1905, *Holmboe* 656! N. of Limassol, 1931, *J. A. Tracey* 5! Asomatos, 1939, 1948, *Mavromoustakis* 134! s.n.! Akrotiri, 1960, *C. E. H. Sparrow* 55!
7. Near Dhikomo, 1862, *Kotschy* 495 and, 1932, *Syngrassides* 498!

NOTES: Populations commonly contain intermediate individuals displaying a variability in lip form ranging from 3-lobed to entire. Several of these intermediates have been described as varieties or forms of which only two need concern us here.

Plants with rather small flowers with sepals to 8 mm. long, a subentire or obscurely 3-lobed, triangular-obovate lip 8–10 mm. long, 10–12 mm. wide and a spur of either uniform width or very slightly attenuate at the apex, have been described as *O. laxiflora* Lam. ssp. *dielsiana* by Soó (1926). Robust plants with broad leaves, elongated inflorescences and very large flowers with a broad, entire or subentire lip have been described at both specific, varietal and formal level as *O. elegans* Heuff. (1835), *O. palustris* Jacq. var. *elegans* (Heuff.) Beck (1903) and *O.*

laxiflora Lam. ssp. *palustris* (Jacq.) Aschers. et Graebn. f. *elegans* (Heuff.) Aschers. et Graebn. (1907), [as "*O. laxiflorus* β. *paluster* c. *elegans*"].

These extremes of variation are of little significance and the two mentioned above fall within the range of variants to be found alongside normal individuals in many eastern populations.

3. O. anatolica *Boiss.*, Diagn., 1, 5: 56 (1844); Kotschy in Oesterr. Bot. Zeitschr., 12: 275 (1862); Unger et Kotschy, Die Insel Cypern, 207 (1865); Sintenis in Oesterr. Bot. Zeitschr., 31: 229, 392 (1881); Boiss., Fl. Orient., 5: 70 (1884); Holmboe, Veg. Cypr., 57 (1914); Soó in Botanisch. Archiv, 23: 56, 153 (1928); Renz in Feddes Repert., 27: 215 (1929); Post, Fl. Pal., ed. 2, 2: 570 (1933); Seligman in Quart. Bull. Alpine Gard. Soc., 20: 233 (1952); R. S. Davis in Orch. Journ., 3: 164 (1954); Sundermann in Die Orchidee, 20: 79 (1969); Sundermann, Europ. Medit. Orch., ed. 1, 147 (1970), ed. 2, 133 (1975), ed. 3, 161 (1980); Elektra Megaw, Wild Flowers of Cyprus, 12, t. 20 (1973); Osorio-Tafall et Seraphim, List Vasc. Plants Cyprus, 27 (1973); B. et E. Willing in Die Orchidee, 27: 114 (1976); Landwehr, Wilde Orch. Europa, 1: 282 (1977); Renz in Rechinger f., Fl. Iran., no. 126: 96 (1978); J. et P. Davies in Quart. Bull. Alpine Gard. Soc., 48: 124 (1980); Davies et Huxley, Wild Orch. Brit. Europe, 132–133 (1983).

Plant (6–) 15–40 (–50) cm. high; tubers subglobose or ovoid, sessile or shortly stipitate; stem slender or robust, erect or slightly flexuous, bearing 1 or 2 ovate, acute sheaths below, naked above, dark pink to purplish-red, pale green in albino forms; leaves 2–5, rosulate, narrowly elliptic or oblong to obovate, attenuate towards the base, obtuse and mucronate, usually blotched or spotted purple to purplish-brown, (3–) 8–13 cm. long, (0·8–) 1·5–2·5 (–3) cm. wide; inflorescence lax, up to 20 cm. long, (2–) 8–14-flowered, sometimes secund and evenly spaced along the stem, floral bracts narrowly elliptic to linear, 1–3-nerved, acute, usually a little shorter than the ovary, dark pink to purplish-red, greenish-white in albino forms; sepals and petals pale pink to purple, the lateral sepals sometimes flushed greenish-brown on the inner surface, occasionally white; sepals oblong, ovate or narrowly elliptic, obtuse, 6–10 mm. long, 4–5 mm. wide, dorsal sepal oblong, somewhat cucullate, ± obtuse at the apex, 5 mm. wide, often reflexed, lateral sepals rather curved, acute, spreading, 4 mm. wide; petals oblong to obliquely ovate, obtuse, with 3 green nerves, curving forward over the column, 6–8 mm. long, 3 mm. wide; labellum pale pink to purple, occasionally white, with a white central area spotted and flecked purplish-red, obovate or suborbicular in outline, (8–) 10–14 (–17) mm. long, (8–) 10–14 (–22) mm. wide, cuneate at the base, 3-lobed, flat or with a slightly reflexed margin, mid-lobe cuneate-oblong or somewhat quadrate, truncate or slightly excised, more or less velutinous, longer than the lateral lobes, lateral lobes rhombic, truncate; spur pale pink to purple, linear-cylindrical, dilated and often paler at the mouth, narrowed at the apex, horizontal or curved upwards, longer than the ovary, 1·5–2·5 cm. long; column 4–5 mm. long; anther-cap obtuse or shortly apiculate.

var. **anatolica**

TYPE: Turkey; "Hab. in Caria, Pinard, Cilicia, *Aucher* no. 2236, insula Chio, *Olivier* in DC. herb".

Plant (5–) 10–40 (rarely to 50) cm. high; stem often rather slender; leaves (0·8–) 1·5–2 cm. wide; inflorescence up to 15 cm. long; labellum (8–) 10–14 mm. long, (8–) 10–12 mm. wide; spur horizontal or upturned, 1·5–2·2 cm. long.

HAB.: In garigue, on dry turf, in stony places, in light woodland, mostly on calcareous soil; 700–3,000 ft. alt.; fl. (late Febr.–) March–May.

DISTR.: Divisions 3, 6–8. Also Aegean Islands (Cyclades), Crete, Rhodes, Turkey (Anatolia), Lebanon, Syria, Palestine, Iraq and western Iran.

3. Pano Lefkara, 1929, *Renz* ; Mosphiloti, 1962, *Meikle* 2977 !
6. Dhiorios Forest Station, 1930, *C. Norman* 398 !
7. Pentadaktylos, 1862, *Kotschy* 377, also, 1880, *Sintenis & Rigo* 151 ! 1929, *Renz* 451 ! 452 !

and, 1956, *G. E. Atherton* 1038! Ayios Nikolaos, 1929, *Renz* 1878! Kyrenia Pass, 1929, *Renz* 1871! Lapithos, 1929, *Renz* 1872! Akanthou Pass, 1929, *Renz* 1873! Ardhana-Phlamoudhi, 1929, *Renz* 1880! St. Hilarion, 1929, *Renz*, also, 1941, *Davis* 2169! 1949, *Casey* 295! 1959, *C. E. H. Sparrow* 23! and 1963, *J. B. Suart* 125! Alakati, 1929, *Renz* 1874! Olymbos, 1929, *Renz* 1875! and, 1957, *Merton* 2895! Near Mandres, 1929, *Renz*; Buffavento, 1929, *Renz* 1876! and, 1939, *Syngrassides* 1247! Kantara, 1929, *Renz* 1877! Melounda, 1929, *Renz*; Agridaki, 1929, *Renz* 1879! Kyrenia, 1927, *Rev. A. Huddle* 90! Halevga, 1932, *Syngrassides* 58! and, 1956, *G. E. Atherton* 1039! 1040! Above Kharcha, 1948, *Duke-Woolley* 1a! 66! Kyparissovouno, 1970, *I. M. Hecker* 49!
8. N. of Komi Kebir, 1929, *Renz.*

NOTES: The typical variety prefers rather open, sunny positions and is widespread in northern and central Cyprus. Individuals, perhaps of hybrid origin and somewhat intermediate between *O. anatolica* and *O. quadripunctata* Cyr. ex Ten., are sometimes found in the south of the island. Their flowers are a little smaller, like *O. quadripunctata*, but retain the general morphological features of *O. anatolica*. An albino variety occurs which has pale green nerves on the tepals and a few purplish-red spots arranged in two parallel rows along the centre of the lip.

var. **troodi** (*Renz*) *Soó* in Feddes Repert., Sonderbeiheft A, 2: 177 (1933); R. S. Davis in Orch. Journ., 3: 164 (1954); Sundermann, Europ. Medit. Orch., ed. 2, 147 (1975), ed. 3, 161 (1980); Landwehr, Wilde Orch. Europa, 1: 286, 287 (1977); Robatsch in Die Orchidee, 31: 195–200 (1980); Davies et Huxley, Wild Orch. Brit. Europe, 132–133 (1983).
   O. anatolica Boiss. ssp. *troodi* Renz in Feddes Repert., 27: 209, 215, t. 3, fig. 3–4 (1929); Rechinger f. in Oesterr. Bot. Zeitschr., 94: 194 (1947); Sundermann, Europ. Medit. Orch., ed. 1, 133 (1970); Osorio-Tafall et Seraphim, List Vasc. Plants Cyprus, 27 (1973); B. et E. Willing in Die Orchidee, 27: 114 (1976); J. et P. Davies in Quart. Bull. Alpine Gard. Soc., 48: 124 (1980).
   [*O. anatolica* (non Boiss.) Matthews, Lilies of the Field, 39 (1968).]
   TYPE: Cyprus; "Troödos Gebirge, berghang ob Kakopetria, gegen die Quelle Vryssi Platania, Nordseite des Troödos-Massivs, vulkanisches Gestein, 9. April 1929, *Renz* 453" (RENZ).

Often attaining 50 cm. high; stem robust; leaves 2–2·5 (–3) cm. wide; inflorescence up to 20 cm. long, flowers often secund and evenly spaced along the stem; labellum 13–17 mm. long, up to 22 mm. wide; spur sharply upturned, 2–2·5 cm. long. *Plate 90.*

HAB.: Bare ground in *Pinus brutia/Quercus alnifolia* forest, often associated with *Dactylorhiza romana* and *Thymus integer*; 1,000–5,300 ft. alt.; fl. late March–early April.

DISTR.: Divisions 2, 3. Endemic.

2. Locally common. Near Galata, Kakopetria, Saïttas, Platania, Kryos Potamos, Platres, Pyrgos, Stavros tis Psokas, Prodhromos, Kykko, Ayia valley, Zakharou, etc. *Kotschy* 927; *Renz*; *Syngrassides* 344! *Lady Loch* 44! *Kennedy* 265! 266! 1515! *Mavromoustakis* 51! *Merton* 2870! *Meikle* 2260! 2843! *J. J. Wood* 90! *Chesterman* 55! 57! 121! 124! *Edmondson & McClintock* E2806! etc.
3. Stavrovouni, 1862, *Kotschy* 185!

NOTES: This robust, large flowered variety is restricted to shady pine forests and may represent a local ecotype. Albino variants are sometimes found.

4. **O. quadripunctata** Cyr. ex Ten., Prodr. Fl. Nap., 1: LIII (1811); Boiss., Fl. Orient., 5: 69 (1884); Holmboe, Veg. Cypr., 57 (1914); Soó in Botanisch. Archiv, 23: 55, 153 (1928); Post, Fl. Pal., ed. 2, 2: 569 (1933); Burtt et Davis in Kew Bull., 4: 114 (1949); R. S. Davis in Orch. Journ., 3: 164 (1954); Sundermann, Europ. Medit. Orch., ed. 1, 131 (1970), ed. 2, 145 (1975), ed. 3, 159 (1980); Osorio-Tafall et Seraphim, List Vasc. Plants Cyprus, 27 (1973); B. et E. Willing in Die Orchidee, 26: 74 & 75 (1975), 27: 114 (1976); Landwehr, Wilde Orch. Europa, 2: 296 (1977); Williams et al., Field Guide, 98 (1978); J. et P. Davies in Quart. Bull. Alpine Gard. Soc., 48: 230 (1980); Davies et Huxley, Wild Orch. Brit. Europe, 132 (1983).
   TYPE: Italy; not indicated.

Plant (10–) 20–30 cm. high; tubers subglobose or ovoid, sessile or shortly stipitate; stem slender, erect or slightly flexuous, bearing 1 or 2 ovate, acute or mucronate sheaths below, naked above, pale to purplish-red; leaves 2–6, rosulate, narrowly elliptic, oblong or oblong-linear, attenuate towards the base, obtuse and mucronate, with or without purplish-red spots and blotches, 4–10 cm. long, 1–1·7 (–3) cm. wide; inflorescence many-flowered, ovoid to cylindrical, lax to dense; floral bracts narrowly elliptic to linear, 1–3-veined, acute, shorter than or equalling the ovary, pale to purplish-red;

Plate 90. ORCHIS ANATOLICA Boiss. var. TROODI (Renz) Soó   **1,** habit, × 1; **2,** habit, × ½; **3,** flower, front view, × 1·75; **4,** flower, side view, × 1·75; **5,** dorsal sepal, × 2·75; **6,** petal, × 2·75; **7,** lateral sepal, × 2·75; **8,** column, side view, × 3; **9,** column, front view, × 3; **10,** pollinia, × 9. (**1** from *J. J. Wood* 90 & *Meikle* 2260; **2** from *Merton* 2870; **3–9** from *J. J. Wood* 90; **10** from *Chesterman* 57.)

sepals and petals pink to purple-violet, sometimes white; sepals oblong to ovate, obtuse, 3–6 mm. long, dorsal sepal recurved, lateral sepals spreading; petals obliquely ovate, obtuse, curving forward over the column, 2–3·5 (–5) mm. long; labellum pink to purple-violet, sometimes white, with a white basal patch carrying four or more pink to purple-violet spots, two of which are visible and two usually hidden inside at the junction with the stigmatic cavity, orbicular or rhombic in outline, 3-lobed, rarely entire, 4–5 (–7) mm. long, 6–7 (–10) mm. wide, mid-lobe oblong to quadrate, truncate, lateral lobes oblong, ovate or rhombic, convergent or divergent, equal to or sometimes slightly shorter than the mid-lobe; spur pink to purple-violet, filiform, not dilated at the mouth, ·downturned or horizontal, equal to or slightly shorter than the ovary, 1–1·2 cm. long; column 2 mm. long; anther obovate, obtuse.

HAB.: In garigue, on stony slopes and dry turf, on calcareous or igneous soils; 800–3,000 ft. alt.; fl. (March–) April–May (–June).

DISTR.: Divisions 1–3, ?7. Also southern Italy, Sardinia, Sicily, Jugoslavia, Albania, Greece, Crete, southern Turkey and Lebanon.

1. Ayios Minas near Smyies, 1962, *Meikle* 2178!
2. Kaminaria, 1941, *Kennedy* in *Davis* 2628! Above Khrysorroyiatissa Monastery, 1981, *Hewer* 4776!
3. Stavrovouni, 1934, *Syngrassides* 1390! also, 1938, *Lady Loch* 43! Lefkara, 1941, *Davis* 2769! N. of Kandou, 1963, *J. B. Suart* 124!
7. Buffavento, 1862, *Kotschy* 417 teste Boiss. See Notes.

NOTES: This species is most frequent on the limestone foothills of the Troödos Range. Individuals from Cyprus often have a rather broad lip and noticeably downward-pointing spur. Albino plants often occur and have been called var. *albiflora* Raulin [as *O. brancifortii* Biv. var. *albiflora* Raulin, L'Ile de Crete, 862 (1858)]. Hybrids between *O. quadripunctata* and *O. anatolica* Boiss. may occur in mixed populations. The Kotschy record from Buffavento cited by Boissier must be regarded as doubtful in view of the fact that no one else has found *O. quadripunctata* in this area.

5. O. collina *Banks et Soland.* in Russell, Nat. Hist. Aleppo, ed. 2, 2: 264 (1794); Osorio-Tafall et Seraphim, List Vasc. Plants Cyprus, 27 (1973); Sundermann, Europ. Medit. Orch., ed. 2, 141 (1975), ed. 3, 155 (1980); Landwehr, Wilde Orch. Europa, 1: 240 (1977); Renz in Rechinger f., Fl. Iran., no. 126: 98 (1978).

  O. saccata Ten., Prodr. Fl. Nap.: LIII (1811); Boiss., Fl. Orient., 5: 67 (1884); Renz in Feddes Repert., 27: 216 (1929); Post, Fl. Pal., ed. 2, 2: 569 (1933); R. S. Davis in Orch. Journ., 3: 164 (1954); Sundermann in Die Orchidee, 20: 79 (1969); J. et P. Davies in Quart. Bull. Alpine Gard. Soc., 48: 121 (1980); Davies et Huxley, Wild Orch. Brit. Europe, 127 (1983).

  O. sparsiflora Ten. ex Boiss., Fl. Orient., 5: 67 (1884).

  [*O. papilionacea* (non L.) H. Stuart Thompson in Journ. Bot., 44: 338 (1906); Holmboe, Veg. Cypr., 57 (1914) quoad spec. cypr.]

  TYPE: Syria; Aleppo.

Plant (8–) 10–35 (–40) cm. high; tubers ovate-oblong or oblong, shortly stipitate; stem erect, robust, rather fleshy, often red or purplish above, bearing several cauline leaves and narrowly elliptic, acute sheaths; leaves 2–4, mostly rosulate, the lowermost ovate to oblong-ligulate, attenuate towards the base, spreading, the uppermost smaller, amplexicaul, obtuse and mucronate or acute, dark green, shining, sometimes spotted brown, 3–10 (–12) cm. long, 1–3 cm. wide; inflorescence 4–20-flowered, oblong or cylindrical, 3–15 (–20) cm. long, lax; floral bracts narrowly elliptic or ovate-oblong, 5–7-nerved, acute or acuminate, concave, the lowermost longer than the ovary, the uppermost mostly equal to the ovary, often brownish-purple; flowers variable in colour, ranging from purple-violet, vinaceous, brownish- or greenish-purple to dull olive-green, greenish-cream, rarely yellowish-green or dirty white; sepals oblong to ovate, obtuse, 9–12 mm. long, 3–4 mm. wide, margins irregularly undulate, dorsal sepal slightly concave, connivent with the petals to form a loose galea, lateral sepals oblique, spreading; petals oblong-lanceolate, narrowly ligulate or narrowly ovate, obliquely sub-falcate, obtuse, 7–10 (–12) mm. long, 2–3 mm. wide at the base; labellum entire, ovate, obovate or orbicular, 9–12 mm. long, 9–11 mm. wide, margin sometimes slightly uneven, minutely crenulate or emarginate at the apex; spur creamy-white or rose, green at the apex, saccate, conical or broadly cylindrical, obtuse, slightly curved, downturned, half as long as the ovary, 5–7 (–10) mm. long; column 4–5 mm. long; anther shortly apiculate.

HAB.: In garigue, in dry sandy and stony places, on margins of cultivated ground, in light woodland, mostly calcareous soils; sea-level–1,200 ft. alt.; fl. (mid Jan.–) Febr.–March (–April).

DISTR.: Divisions 2, 4–6. Also Portugal, southern Spain, Balearic Islands, southern France, southern Italy, Corsica, Sardinia, Sicily, Malta; Greece, Crete, Rhodes, Turkey, Caucasus, Syria, Lebanon, Israel, Iraq, western and northen Iran, Turkestan; North Africa.

2. Aphamis, 1979, *Chesterman* 129!
4. Larnaca, 1951, *Mrs. Grove* in *Casey* 1101!
5. Between Nicosia and Eylenja, 1929, *Renz* 1089! Athalassa, 1936, *Lady Loch* s.n.! Trikomo, 1939, *Lady Loch* 64! Kythrea, Near Kephalovryso, 1941, *Davis* 2161! Gypsos, 1969, *A. Genneou* in ARI 1314!
6. Government House, Nicosia, 1903, *A. G. & M. E. Lascelles* s.n.! and, *Miss Haig* s.n.! W. of Nicosia, 1929, *Renz* 2212! Also around Nicosia, 1959, 1960, *P. H. Oswald* 47! *C. E. H. Sparrow* 52!

6. O. papilionacea *L.*, Syst. Nat., ed. 10, 1242 (1759); Boiss., Fl. Orient., 5: 60 (1884); Post, Fl. Pal., ed. 2, 2: 565 (1933); Sundermann, Europ. Medit. Orch., ed. 1, 127 (1970), ed. 2, 139 (1975); Osorio-Tafall et Seraphim, List Vasc. Plants Cyprus, 26 (1973); Williams et al., Field Guide, 94 (1978); Davies et Huxley, Wild Orch. Brit. Europe, 118 (1983).

O. papilionacea L. var. *rubra* (Jacq.) Reichb. f., Icon. Fl. Germ., 13 & 14: 16 (1851); Boiss., Fl. Orient., 5: 60 (1884); Post, Fl. Pal., ed. 2, 2: 565 (1933).
TYPE: Not indicated.

Plant (12–) 20–40 (–50) cm. high; tubers subglobose or ellipsoid, subsessile or shortly stipitate; stem erect, straight or slightly flexuous, bearing several ovate or narrowly elliptic, prominently nerved, acute sheaths, upper part of stem and sheaths greenish-red, red or purplish-red; leaves (3–) 4–8 (–10), mostly rosulate, a few cauline, linear, lanceolate-ligulate or narrowly elliptic, obtuse and mucronate or acute, erect to spreading, sometimes slightly recurved, unspotted, (4–) 10–18 cm. long, 0·5–1·5 cm. wide; inflorescence 2–14-flowered, ovoid or shortly oblong, rarely cylindrical, lax or rather dense; floral bracts narrowly elliptic, 3–5-nerved, acute or apiculate, equal to or longer than the ovary, usually red to purplish-red with darker nerves, sepals and petals dark pink, red to purple, with darker nerves, connivent, forming a loose galea when the flower first opens and parting at a later stage; sepals oblong to ovate, obtuse or acute, 9–15 (–18) mm. long, 4–8 mm. wide, dorsal sepal becoming porrect, lateral sepals oblique, becoming spreading; petals obliquely oblong-ligulate or ovate, obtuse, sometimes acute, usually slightly shorter than the sepals, 8–13 (–16) mm. long, 3–5 mm. wide; labellum either cream, flushed pink or sometimes completely pale pink or pale to dark red or cream, flushed pink with darker pink or red spots and streaks, 10–16 mm. long, 7–12 (–15) mm. wide, entire, rarely obscurely 3-lobed, suborbicular, obcordate or dilated and flabellate to ± ovate in outline, becoming cuneate, constricted and clawed at the base, margin irregular, crenulate; spur pink to purple, cylindrical, usually attenuate towards the apex, obtuse, often slightly bifid, decurved, 8–12 mm. long, shorter than the ovary; column 4–5 mm. long; anther apiculate.

HAB.: In garigue, in dry grassy places, light woodland, preferring calcareous soils; 30 ft. alt.; fl. (Febr.–) March–April (–May).

DISTR.: Division 3. Very rare. Also Portugal, southern Spain, ?Balearic Islands, southern France, Italy, Corsica, Sardinia, Sicily, Malta, Jugoslavia, Romania, Bulgaria, Albania, Greece, Crete, Rhodes, Aegean Islands, Caucasus, Turkey, Syria, Lebanon, Israel, Palestine, Jordan; North Africa.

3. Akrotiri, by N. side of Limassol Salt Lake, 1960, *C. E. H. Sparrow* 55a!

NOTES: Small-flowered individuals predominate in the eastern Mediterranean. Populations further west together with those from North Africa are usually larger-flowered, and some have been distinguished as var. *grandiflora* Boiss. *O. papilionacea* has been found only once in Cyprus, and may no longer grow there.

7. O. morio *L.*, Sp. Plant., ed. 1: 940 (1753).

Plant (8–) 15–40 cm. high; tubers subglobose or ellipsoid, shortly stipitate; stem erect, bearing 2–4 ovate or narrowly elliptic, acute sheaths towards the base, becoming naked above, usually flushed purplish-red; leaves (3–) 5–8 (–15), rosulate, narrowly elliptic, obovate or oblong-ligulate, obtuse and mucronate, unspotted, (4–) 7–12 cm. long, 1–1·5 (–2) cm. wide; inflorescence few- to many-flowered, shortly oblong or pyramidal, rarely cylindrical, 3–8 (–13) cm. long, dense or lax; floral bracts narrowly elliptic, acuminate, equal to or sometimes shorter than or slightly longer than the ovary, usually flushed purple-violet; flowers reddish-purple, lilac, pale mauve, rarely pink, white or greenish-white; sepals connivent with the petals to form a galea, often bearing prominent green or dark purple nerves; dorsal sepal oblong, obtuse, concave, 5–9 mm. long, 4–5 mm. wide; lateral sepals obliquely oblong or ovate, obtuse, 5–12 mm. long, 4–6 mm. wide; petals oblong, obtuse, 5–7 mm. long, 3 mm. wide; labellum with a usually paler creamy central area spotted dark pink to pale purple, rarely unspotted, shallowly 3-lobed or indistinctly lobed or subentire, usually broader than long, transversely oblong, transversely rectangular, suborbicular or reniform in outline, 6–10 mm. long, 10–20 mm. wide, mid-lobe triangular-ovate or quadrate, dilated, truncate or emarginate, serrate or crenate at the apex, usually rather small and shorter or sometimes longer than the lateral lobes, lateral lobes broadly rounded, truncate, margins simple, crenate or serrate, usually reflexed; spur cylindrical, dilated or slightly attenuated towards the apex, obtuse, horizontal or slightly upturned, 8–10 mm. long, shorter than or equal to, rarely longer than the ovary; column short, 4 mm. long; anther obtuse or shortly apiculate.

ssp. **picta** (*Loisel.*) *Aschers. et Graebn.*, Syn. Mitteleur. Fl. 3: 667 (1907); Soó in Botanisch. Archiv, 23: 38, 140 (1928); Sundermann, Europ. Medit. Orch., ed. 1, 125 (1970); Osorio-Tafall et Seraphim, List Vasc. Plants Cyprus, 26 (1975); Renz in Rechinger f., Fl. Iran., no. 126: 108 (1978); J. et P. Davies in Quart. Bull. Alpine Gard. Soc., 48: 121 (1980).
    O. picta Loisel. in Mém. Soc. Linn. Paris, 6: 431 (1827); Post, Fl. Pal., ed. 2, 2: 565 (1933); R. S. Davis in Orch. Journ., 3: 164 (1954).
    [*O. morio* (non L.) Sibth. et Sm., Fl. Graec. Prodr., 2: 212 (1816); Poech, Enum. Plant. Ins. Cypr., 11 (1842); Unger et Kotschy, Die Insel Cypern, 206 (1865); Sintenis in Oesterr. Bot. Zeitschr., 31: 229, 257, 392 (1881); Boiss., Fl. Orient., 5: 60 (1884); Holmboe, Veg. Cypr., 57 (1914).]

Plant (8–) 15–25 (–32) cm. high; stem slender; inflorescence often few-flowered, short, lax; flowers small; sepals 5–8 mm. long; labellum broadly oblong, slightly attenuate at the base, 5–8 mm. long, 5–10 (–12) mm. wide, mid-lobe quadrate, shorter than or longer than the lateral lobes; spur slender, about as long as the ovary.

var. **picta** (*Loisel.*) *Reichb.f.*, Icon. Fl. Germ., 13 & 14: 17 (1851); Boiss., Fl. Orient., 5: 60 (1884).
    O. morio L. var. *albiflora* Boiss., Fl. Orient., 5: 60 (1884); Post, Fl. Pal., ed. 2, 2: 565 (1933).
    O. syriaca Boiss. et Blanche ex Boiss., Fl. Orient., 5: 60 (1884) nom. invalid.
    O. morio L. ssp. *syriaca* Camus in E. G. Camus et Bergon, Mon. Orch., 106 (1908); Sundermann, Europ. Medit. Orch., ed. 3, 151 (1980).
    O. picta Loisel. var. *albiflora* (Boiss.) Renz in Feddes Repert., 25: 243 (1928), 27: 215 (1929).
    TYPE: France; "environs de Toulon", *M. Robert.*

Labellum distinctly 3-lobed, with dark pink to pale purple spots.

HAB.: In garigue, woodland, *Pinus brutia/Quercus alnifolia* forest, mostly on calcareous or, less often, igneous soils; near sea-level–3,000 ft. alt.; fl. Febr.–April.

DISTR.: Divisions 1–4, 6–8. Also Spain, ?Balearic Islands, southern France, Corsica, Sardinia, Italy, Sicily, Yugoslavia, Albania, Bulgaria, Romania, Greece, Rhodes, Turkey, Crimea, Caucasus area, Syria, Lebanon, Israel, southern Iran; North Africa.

1. Peyia Forest, 1962, *Meikle* 2042! Between Khrysokhou and Polis, 1977, *J. J. Wood* 71!
2. Troödos area, 1951, *F. M. Probyn* s.n.! Ayia valley, 1977, *J. J. Wood* 85! 88!
3. Stavrovouni 1948, *Mavromoustakis* s.n.! Episkopi, 1964, *J. B. Suart* 168! Near Kophinou, 1978, *Chesterman* 46!
4. Xylophagou, 1905, *Holmboe* 77! Larnaca Salt Lake, 1978, *Chesterman* 37!
6. Dhiorios Forest Station, 1930, *C. Norman* 393! Between Kormakiti and Myrtou, 1937, *Syngrassides* 1415!
7. Pentadaktylos, 1880, *Sintenis & Rigo* 154! and, 1900, *A. G. & M. E. Lascelles* s.n.! Kyrenia, 1932, *Syngrassides* 697! also 1948, *Duke-Woolley* 65! and, 1956, *G. E. Atherton* 889! 899! 903! 905! 906! 923! 1026! Halevga, 1952, *F. M. Probyn* 74! Near Myrtou, 1956, *Merton* 2517! Palaeosophos–Karavas road, 1968, *Economides* in ARI 1097! W. of Bellapais, 1973, *P. Laukkonen* 147!
8. Yialousa, 1938, *Lady Loch* 29! Between Ayios Philon and Rizokarpaso, *Davis* 2245!

NOTES: This subspecies is widely distributed on Cyprus, being found up in the mountains and also locally in light woodland in lowland districts. On Troödos colonies can be found growing under *Pinus brutia* in association with *Dactylorhiza romana*, *Orchis anatolica* var. *troodi* and *Thymus integer*.

Individuals having greenish sepals with darker green nerves and a white, unspotted lip were given varietal status by Boissier as *O. morio* var. *albiflora* (syn. *O. syriaca* Boiss. et Blanche ined.). This albino was subsequently treated at subspecific or varietal rank as *O. morio* ssp. *syriaca* Camus (1908) and *O. morio* ssp. *pictus* var. *syriacus* (Camus) Soó (1933) respectively. Soó states that, as one travels further east, ssp. *picta* grades into both var. *libani*, described below, and var. *syriacus*. He admits that what he recognises as true var. *syriacus* may only be a colour variant of ssp. *picta*. This albino can be found growing amongst typical ssp. *picta* in Paphos Forest.

var. **libani** (*Renz*) Soó in Keller et Schltr., Mon. et Icon. Orch. Europ., 2: 139 (1933); Williams et al., Field Guide, 92 (1978).

O. picta Loisel. ssp. *libani* Renz in Feddes Repert., 27: 209, 215 (1929); Post, Fl. Pal., ed. 2, 2: 565 (1933); Soó in Keller et Schltr., Mon. et Icon. Orch. Europ., 2: 139 (1933); Seligman in Quart. Bull. Alpine Gard. Soc., 20: 233 (1952); Sundermann in Die Orchidee, 20: 79 (1969), as *O. morio* ssp. *libani* Renz.

O. morio L. ssp. *libani* (Renz) Sundermann, Europ. Medit. Orch., ed. 1, 125 (1970); Landwehr, Wilde Orch. Europa, 1: 250 (1977); Davies et Huxley, Wild Orch. Brit. Europe, 119 (1983) nom. invalid.

[*O. syriaca* Baumann et Künkele, Wildwachsenden Orch. Europas, 354 (1982) quoad descr. nec *O. morio* L. ssp. *syriaca* Camus (1908).]

TYPE: Cyprus; "auf Kalk im Nordgebirge, sowohl in Wäldern wie auch in der Gestrüppformation; in der Ebene an einigen Stellen in Wäldchen."

Habit often rather delicate, 10–25 cm. high; inflorescence very lax; galea pale green, suffused pink to pale red or completely pale pink to pale red; labellum subentire or only indistinctly 3-lobed, white, cream or rose-violet, without spotting.

HAB.: In garigue, light woodland; most frequent on calcareous soils in the Northern Range; sea-level–2,000 ft. alt.; fl. March–April.

DISTR.: Divisions 5–8. Also south-east Turkey, Israel, Syria and Lebanon.

5. Near Trikomo, 1929, *Renz*.
6. W. of Nicosia, 1929, *Renz*; Kremama Kamilou, 1929, *Renz* 2131!
7. Kyrenia Pass, 1929, *Renz* 2128! 2129! Aghirda, 1929, *Renz*; Lapithos, 1929, *Renz*; Akanthou Pass, 1929, *Renz* 2130! Ayios Nikolaos, 1929, *Renz* 2134! Myrtou, 1929, *Renz* 1085! and, 1956, *Merton* 2517a! Olympbos, 1929, *Renz*; Melounda, 1929, *Renz*; Between Kantara and Mandres, 1929, *Renz*; Near Klepini, 1929, *Renz*; Ardhana–Phlamoudhi, 1929, *Renz* 2133! Between Halevga and Kharcha, 1941, *Davis* 2155! Halevga, 1956, *G. E. Atherton* 1041! Near Keurmurju, 1959, *P. H. Oswald* 39! Kyrenia, 1956, *G. E. Atherton* 890! and, 1973, *P. Laukkonen* 147a!
8. Ephtakomi, 1929, *Renz* 2132!

NOTES: Renz (1929) states that the characters given above remain constant throughout eastern populations. Soó (1933) states that in the south-east Mediterranean, e.g. Cyprus and Syria, ssp. *picta* var. *picta* grades into var. *libani*, which is quite likely. He appears uncertain of the true status of var. *libani* and questions whether it can really be upheld even as a variety. Sundermann (1975) and Landwehr (1977), on the other hand, treat var. *libani* at subspecific level and agree that it replaces ssp. *picta* in the east. Landwehr defines western and central Turkey as the easternmost limit of true ssp. *picta*.

After having examined both living and preserved plants from Cyprus I find that they correspond with true ssp. *picta* var. *picta*. These plants have a typical distinctly 3-lobed lip ranging in colour from purple (frequent), through pink to white (rare). Examination of preserved material from further east in eastern Turkey and Syria has also shown the same degree of variation. This upsets the generally held view that ssp. *picta* var. *picta* is gradually replaced in the east by the pale-flowered var. *libani*.

In recent literature var. *libani* is always described as having a white lip. Renz (1929) in his original description described the lip as rose-violet. Examination of collector's notes of material referable to var. *libani* indicates lip colour as being either white or cream. Whether or not the proportion of individuals having a green, pink-suffused galea and a white lip is any greater in the east is hard to ascertain without extensive field observation. Typical ssp. *picta* certainly extends into the east and my views are upheld by Renz (1978) who cites northern Iran as being the easternmost extension of its geographical range.

Individuals of the albino variant of ssp. *picta*, variously described as var. *albiflora* or var. *syriaca*, mentioned above, seem to have been confused with var. *libani*. However, the most noticeable distinguishing feature of the latter is the subentire or scarcely 3-lobed lip. This taxon appears insufficiently distinct from ssp. *picta* to warrant subspecific status. However, it would seem justifiable to recognise it at varietal level in order to define this atypical eastern population.

8. O. punctulata *Stev. ex Lindl.*, Gen. et Sp. Orch., 273 (1835); Boiss., Fl. Orient., 5: 64 (1884); Soó in Botanisch. Archiv, 23: 45, 147 (1928); Post, Fl. Pal., ed. 2, 2: 568 (1933); Burtt et Davis in Kew Bull., 4, 1: 114 (1949); Sundermann, Europ. Medit. Orch., ed. 1, 117 (1970); Osorio-Tafall et Seraphim, List Vasc. Plants Cyprus, 27 (1973); Sundermann, Europ. Medit. Orch., ed. 2, 131 (1975), ed. 3, 145 (1980); Landwehr, Wilde Orch. Europa, 2: 326 (1977); Renz in Rechinger f., Fl. Iran., no. 126: 113 (1978); Williams et al., Field Guide, 86 (1978); J. et P. Davies in Quart. Bull. Alpine Gard. Soc., 48: 229 (1980); Davies et Huxley, Wild Orch. Brit. Europ., 126 (1983).

TYPE: U.S.S.R., Crimea; "in Tauria meridionali", *Steven*, (H).

Plant 22–60 (–70) cm. high; tubers oblong or ellipsoid, sessile; stem erect, stout, leafy below, with 1–3 narrowly elliptic, amplexicaul, acute sheaths above, naked and somewhat keeled below the inflorescence; leaves 3–7 (–9), mostly rosulate, except for 1 or 2 just above the main rosette, oblong, oblong-ligulate or narrowly elliptic, obtuse, obtuse and mucronate or acute, 8–15 (–30) cm. long, 2·5–5 cm. wide, unspotted; inflorescence many-flowered, cylindrical, 6–15 (–28) cm. long, dense; floral bracts ovate or narrowly elliptic, acute, membranous, much shorter than the ovary, 2·5–4·5 mm. long, whitish; flowers smelling of Lily-of-the-valley (*Convallaria majalis*); sepals and petals yellowish-green, yellow or brownish-yellow, rarely flushed pink, the former with blackish-violet dotted lines along the nerves, connivent, forming a galea; dorsal and lateral sepals oblong-ovate or oblong-ligulate, obtuse, 8–10 (–13) mm. long, 3·5–4 mm. wide at the base; petals yellowish-green, linear or oblong, obtuse to acute, margin minutely papillose, 8–9 mm. long, 1·5 mm. wide; labellum yellowish-green at the base, ochre to brown or brownish-violet towards the ends of the lobes, with numerous tufts of violet-red papillae, 6–11 mm. long, 8–11 mm. wide, mid-lobe broadly linear, 6–8 mm. long, 2–3 mm. wide at the base, becoming flabellate, bilobulate, lobules obliquely quadrate-oblong, truncate, slightly divaricate, 3 mm. long, 2 mm. wide, with a small mucro 1 mm. long in the sinus, lateral lobes linear-falcate, slightly dilated towards the obtuse apex, 4–5 mm. long, 1 mm. wide; spur green or yellowish-green, broadly cylindrical, downturned, 6 mm. long, about half as long as the ovary; column short, 3–4 mm. long; anther obtuse or slightly apiculate.

HAB.: In garigue, on dry stony slopes, in open maquis, beneath Pines, or on streamsides, preferring calcareous soils; 400–1,000 ft. alt.; fl. Febr.

DISTR.: Divisions 1, 3, 7. Also Greece (Thrace), Rhodes, Turkey (Anatolia), Crimea, Caucasus, Syria, Palestine, Lebanon, Israel, Jordan, Iraq, western and northern Iran.

1. Near Paphos, *Mrs Gunnis* s.n.! Between Lachi and Neokhorio, 1977, *J. J. Wood* 102! also, 1978 and 1979, *Chesterman* 50! 120!
3. Asgata, 1970, *A. Genneou* in ARI 1389! E. of Alaminos, 1981, *G. & P. Phillips*!
7. Near Myrtou, 1939, *Lady Loch* 67!

NOTES: Davies & Huxley (1983) report that extremely robust, large-flowered forms are to be found growing in the same location as rather small-flowered slender plants. The large-flowered forms are as much as three or four times the height of the others.

9. O. coriophora *L.*, Sp. Plant., ed. 1, 2: 940 (1753); Unger et Kotschy, Die Insel Cypern, 207 (1865); Boiss., Fl. Orient., 5: 61 (1884); Holmboe, Veg. Cypr., 57 (1914); Post, Fl. Pal., ed. 2, 2: 565 (1933); Renz in Rechinger f., Fl. Iran., no. 126: 108 (1978).

Plant (12–) 15–35 (–60) cm. high; tubers subglobose or ellipsoid, shortly stipitate; stem slender or stout and robust, straight, leafy below, with several leafy, ovate to narrowly elliptic, amplexicaul, acute sheaths above; leaves 4–7 (–10), rosulate, linear or lanceolate, occasionally oblong, often plicate, acute to acuminate, rarely obtuse and mucronate, erect to spreading, (5–) 8–15 (–25) cm. long, (0·8–) 1–2 (–3·5) cm. wide, unspotted; inflorescence dense, many-flowered, cylindrical or oblong, sometimes elongated, (3–) 5–12 (–25) cm. long; floral bracts narrowly elliptic, acute to acuminate, 1-nerved, as long as or longer than the ovary; flowers dark purple, brownish-purple, reddish-purple, cerise, purplish olive-green, greenish-pink, rarely pure green, the labellum spotted or blotched dark purple, foetid, or having a rather sickly sweet scent reminiscent of bitter almonds or vanilla, rarely inodorous; sepals and petals connivent, forming a porrect to ± erect galea, the former usually free at the apex; dorsal sepal ovate-elliptic to oblong, acute or acuminate, 5–9 mm. long, 2·5–3 mm. wide; lateral sepals whitish centrally with a dark green mid-nerve, obliquely ovate-elliptic, acute, rostrate or acuminate, 7–10 mm. long, 3–4 mm. wide; petals linear-lanceolate, acute or acuminate, 1-nerved, 4–6 mm. long, 1–2 mm. wide; labellum 3-lobed, deflexed, longer than broad, 5–7 (–11) mm. long, 7 (–10) mm. wide, mid-lobe ligulate-oblong, obtuse to acute, up to 4 mm. long, a little or distinctly longer than the lateral lobes, lateral lobes with green nerves, subquadrate, subrhombic or semi-ovate, margin entire, crenate or denticulate; spur conical, attenuate towards the subacute or ± obtuse apex, downturned, apex often slightly incurved towards the underside of the labellum, shorter than, to as long as, or rarely longer than the ovary; column short, 3 mm. long; anther minutely apiculate.

ssp. **fragrans** (*Poll.*) *Camus*, Mon. Orch. Eur., 136 (1908); Holmboe, Veg. Cypr., 57 (1914); Osorio-Tafall et Seraphim, List Vasc. Plants Cyprus, 26 (1973); Sundermann, Europ. Medit. Orch., ed. 2, 127 (1975), ed. 3, 139 (1980); B. et E. Willing in Die Orchidee, 27: 113 (1976); Landwehr, Wilde Orch. Europa, 2: 302 (1977); Williams et al., Field Guide, 82 (1978); J. et P. Davies in Quart. Bull. Alpine Gard. Soc., 48: 122, 229 (1980); Davies et Huxley, Wild Orch. Brit. Europe, 121 (1983).
    O. fragrans Poll., Element. Bot., 2: 155, t. 9, fig. 2 (1811); Renz in Feddes Repert., 27: 216 (1929); Sundermann in Die Orchidee, 20: 79 (1969); Europ. Medit. Orch., ed. 1, 111 (1970).
    O. polliniana Spreng., Plant. Pugill., 2: 58 (1815); Sintenis in Oesterr. Bot. Zeitschr., 32: 364, 397, 398 (1882); Soó in Botanisch. Archiv, 23: 39 (1928).
    O. coriophora L. var. *polliniana* (Spreng.) Poll., Fl. Veron., 3: 7 (1824); Reichb. f., Icones Fl. Germ., 13, 14: 21 (1851); Unger et Kotschy, Die Insel Cypern, 207 (1865).
    O. coriophora L. var. *fragrans* (Poll.) Boiss., Voy. Bot. Esp., 2: 593 (1841), Fl. Orient., 5: 61 (1884); H. Stuart Thompson in Journ. Bot., 44: 338 (1906); Soó in Botanisch. Archiv, 23: 39, 40, 142 (1928); Post, Fl. Pal., ed. 2, 2: 566 (1933).
    [*O. coriophora* (non L.) Post in Mém. Herb. Boiss., 18: 100 (1900); H. Stuart Thompson in Journ. Bot., 44: 338 (1906); Mrs Frank Tracey in Journ. R.H.S., 58: 305 (1933); R. S. Davis in Orch. Journ., 3: 164 (1954).]
    TYPE: Italy; "in pratis siccis secus viam qua itur ex pago *Villafranca* ad pagum *Valleggio* dictum, et copiose in illis secus flumen Mincium" *Pollini*.

Galea acuminate; labellum (6–) 8–11 mm. long, mid-lobe distinctly longer than the lateral lobes; flowers with a sweet scent reminiscent of bitter almonds or vanilla, rarely inodorous.

HAB.: In garigue and dry grassy places; by roadsides; in Pine forest, or in marshy places, on calcareous soils; sea-level–2,000 ft. alt.; fl. April–May (–June).

DISTR.: Divisions 1–4, 6–8. Also Spain, Balearic Islands, southern France, Italy, Corsica, Sardinia, Sicily, Malta, Yugoslavia, Albania, Greece, Crete, Rhodes, Turkey, Syria, Lebanon, Crimea, Caucasus, Iraq and northern and western Iran; North Africa.

1. Near Smyies, 1941, *Davis* 3337!
2. Stavros tis Psokas, 1979, *Hewer* 4646!
3. Kophinou, 1905, *Holmboe* 601! Between Anglisidhes and Skarinou, 1929, *Renz* 1932! N. of Skarinou, 1929, *Renz*; Cape Gata, 1948, *Mavromoustakis* s.n.! Akrotiri, 1955, *Merton* 2071! and, 1963, *J. B. Suart* 99! Paramali, 1959, *C. E. H. Sparrow* 20! 20a! Near Souni, 1960, *N. Macdonald* 128! Episkopi, 1963, *J. B. Suart* 71! Between Nikoklia and Phasoula, 1977, *J. J. Wood* 123! Between Evdhimou and Pakhna, 1978, *Holub* s.n.!
4. Larnaca, 1905, *Holmboe* 565! W. of Larnaca, 1977, *J. J. Wood* 4! Athna Forest, 1967, *Merton* in ARI 432!
6. Peristerona, 1937, *Syngrassides* 1567! 1813! Dhiorios, 1952, *F. M. Probyn* 88! Orga, 1974, *B. & E. Willing*.
7. Between Kyrenia and Lapithos, 1862, *Kotschy* 497! Kremama Kamilou, 1929, *Renz*; Kyrenia, 1931, *J. A. Tracey* 6! also, 1956, *G. E. Atherton* 1330! and, 1970, *A. Genneou* in ARI 1529! Kyrenia Pass, 1952, *F. M. Probyn* s.n.!
8. Aphendrika, 1880, *Sintenis & Rigo* 156! N. of Komi Kebir, 1929, *Renz* 1931! Ayios Andronikos, 1948, *Mavromoustakis* s.n.! Matakos near Galinoporni, 1962, *Meikle* 2448!

10. O. sancta L., Sp. Plant., ed. 2, 2: 1330 (1763); Boiss., Fl. Orient., 5: 62 (1884); Soó in Botanisch. Archiv, 23: 40, 143 (1928); Post, Fl. Pal., ed. 2, 2: 566 (1933); Mrs Frank Tracey in Journ. R.H.S., 58: 305 (1933); Rechinger f. in Archiv för Bot., ser. 2, 1: 419 (1950); R. S. Davis in Orch. Journ., 3: 164 (1954); Osorio-Tafall et Seraphim, List Vasc. Plants Cyprus, 26 (1973); Sundermann, Europ. Medit. Orch., ed. 2, 127 (1975), ed. 3, 141 (1980); B. et E. Willing in Die Orchidee, 26: 74 (1975), 27: 113 (1976); Landwehr, Wilde Orch. Europa, 2: 308 (1977); J. et P. Davies in Quart. Bull. Alpine Gard. Soc., 48: 122, 229 (1980); Davies et Huxley, Wild Orch. Brit. Europe, 121 (1983).

O. coriophora L. var. *sancta* (L.) Reichb. f., Icones Fl. Germ., 13, 14: 173 (1851); Unger et Kotschy, Die Insel Cypern, 207 (1865); Holmboe, Veg. Cypr., 57 (1914).

O. coriophora L. ssp. *sancta* (L.) Hayek in Beih. Feddes Repert., 30, 3: 386 (1933); B. et E. Willing in Die Orchidee, 26: 74 (1975), 27: 113 (1976).

TYPE: "*in* Palaestina."

Plant 13–45 cm. high; tubers subglobose or ellipsoid, stipitate; stem erect, leafy at the base, bearing numerous ovate to narrowly elliptic, acute sheaths above, usually flushed pink to red below the inflorescence; leaves 5–12, rosulate, linear, narrowly elliptic or oblong, acute or obtuse and mucronate, (4–) 6–12 cm. long, 0·8–1·3 cm. wide, unspotted; inflorescence 6- to many-flowered, cylindrical, dense; floral bracts ovate, acute or long-acuminate, the lowermost 3–5-nerved, equal to or longer than the ovary; flowers pale lilac, rose or carmine-red, the sepals often flushed green, the labellum unspotted; sepals and petals connivent, forming a porrect to ± erect galea, the apex of which is usually curved down giving a rostrate appearance; sepals ovate, long-acuminate, 9–12 (–15) mm. long, 3–4 mm. wide; petals ovate, acute to acuminate, 8–10 (–13) mm. long, 1–3 mm. wide; labellum 8–12 mm. long, 8–10 mm. wide, with prominent venation, 3-lobed, mid-lobe oblong, entire or denticulate, acute, longer than the lateral lobes, lateral lobes broadly semi-ovate or subrhombic, margin acutely dentate or sinuately serrate; spur conical, abruptly attenuate towards the apex, incurved towards the underside of the labellum, 8–10 mm. long, equal in length to the ovary; column short, 3 mm. long.

HAB.: In garigue, on dry hillsides, often amongst *Juniperus* and *Pistacia*; in *Pinus brutia* forest, or on stabilised sand dunes, preferring calcareous soils; sea-level–600 ft. alt.; fl. mid April–May (–June).

DISTR.: Divisions 1, 3, 4, 5, 7, 8. Also Aegean Islands (Chios, Lesbos, Samos), Rhodes, southern and western Turkey, Syria, Lebanon, Palestine and Israel.

1. Lyso, 1913, *Haradjian* 887; Fontana Amorosa, 1978, *Holub* s.n.!
3. Khoulou, 1905, *Holmboe* 746! Near Episkopi, *Mavromoustakis* 42! and, 1963, *J. B. Suart* 122! Paramali, 1959, *C. E. H. Sparrow* 30! Cape Gata, 1962, *Meikle* 2906!
4. Near Ayia Napa, 1862, *Kotschy* 97–99.
5. 1 m. S.W. of Boghaz, near Gastria, 1962, *Meikle* 2629!
7. Between Kyrenia and Lapithos, 1862, *Kotschy* 497a! Kyrenia, 1931, *J. A. Tracey* 7! also, 1948, *Duke-Woolley* 97! 1956, *G. E. Atherton* 1332! and, 1973, *P. Laukkonen* 307!
8. Near Rizokarpaso, 1912, *Haradjian* 228! Near Ayios Andronikos, 1948, *Mavromoustakis* s.n.! and, 1952, *F. M. Probyn* s.n.! Yioti Forest, 1957, *Poore* 63! Koma tou Yialou, 1970, *A. Genneou* in ARI 1486!

NOTES: *O. sancta* is most frequent near the coast in the north-eastern part of Cyprus. Intermediate individuals probably of hybrid origin involving *O. coriophora* ssp. *fragrans* have been observed near Orga in N.W. Cyprus. They are said to have a greenish-red lip similar in shape to *O. sancta*, but having the characteristic spotting of *O. coriophora* ssp. *fragrans*. The author has not seen such plants and cannot therefore confirm their identity. Such hybrids no doubt do occur in mixed populations, such as those near Ayios Andronikos in the Karpas Peninsula.

11. O. italica *Poir.* in Lam., Encycl. Méth., 4: 600 (1798); Soó in Botanisch. Archiv 44, 147 (1928); Sundermann in Die Orchidee, 20: 79 (1969), Europ. Medit. Orch., ed. 1, 121 (1970); Osorio-Tafall et Seraphim, List Vasc. Plants Cyprus, 27 (1973); Sundermann, Europ. Medit. Orch., ed. 2, 135 (1975), ed. 3, 149 (1980); Landwehr, Wilde Orch. Europa, 2: 320 (1977); J. et P. Davies in Quart. Bull. Alpine Gard. Soc., 48: 229 (1980); Davies et Huxley, Wild Orch. Brit. Europ., 123 (1983).

    O. longicruris Link in Schrader, Journ. Bot., 2: 323 (1799); Unger et Kotschy, Die Insel Cypern, 207 (1865); Sintenis in Oesterr. Bot. Zeitschr., 31: 392 (1881); Boiss., Fl. Orient., 5: 65 (1884); Holmboe, Veg. Cypr., 57 (1914); Renz in Feddes Repert., 27: 215 (1929); Post, Fl. Pal., ed. 2, 2: 568 (1933); R. S. Davis in Orch. Journ., 3: 164 (1954).

    O. undulatifolia Biv., Pl. Sic. Cent., 2: 44, t. 6 (1807); Sibth. et Sm., Fl. Graec. Prodr., 2: 213 (1816), Fl. Graec., 10: 20, t. 927 (1840).

TYPE: Italy; "observée en Italie par M. Vahl."

Plant (15–) 20–40 (–50) cm. high; tubers ovoid or ellipsoid, sometimes divided into secondary lobes, stipitate; stem erect or slightly flexuous, leafy at the base, bearing above 2–4 ovate or narrowly elliptic, acute sheaths, often with undulate margins; leaves 5–8 (–10), mostly rosulate, some often cauline, oblong-elliptic or narrowly elliptic, obtuse and mucronate or ± acute, margin usually undulate, (3–) 7–12 cm. long, (1–) 1·5–2 (–2·5) cm. wide, often spotted or flecked purplish-brown; inflorescence many-flowered, conical at first, becoming globose or ovoid, occasionally shortly cylindrical, dense; floral bracts ovate, acute, membranous, 1-nerved, much shorter than the ovary; flowers rose-pink or lilac, rarely white, the sepals with darker pink to purple nerves, the labellum usually spotted darker pink; sepals and petals connivent, forming a galea, usually becoming free at a later stage; sepals ovate to ovate-elliptic, acute to acuminate, (8–) 10–12 (–15) mm. long, 4–5 mm. wide; petals ovate to ovate-elliptic, obtuse, 3–12 mm. long, 2–3 mm. wide; labellum longer than broad, 12–16 (–18) mm. long, 3–8 mm. wide below lobules of mid-lobe, distinctly 3-lobed, flat, with 2 oblong to triangular, glabrous to minutely papillose calli at the base, mid-lobe longer than the lateral lobes, divided into 2 obliquely oblong to linear, acute lobules separated by a narrowly linear mucro, lateral lobes linear, acute; spur 4–7 mm. long, cylindrical, emarginate to shallowly bilobed at the apex, downturned, shorter than the ovary; column short, 2 mm. long.

HAB.: In garigue or light woodland, especially of Pine; in dry stony places, also sometimes in lusher, damp habitats beside streams in association with *Serapias* spp. and *Asphodelus aestivus* on calcareous and neutral soils; 300–2,500 ft. alt.; fl. March–April (–May).

DISTR.: Divisions 2, 3, 6–8. Also Portugal, Spain, Balearic Islands, Italy, Sicily, Yugoslavia, Albania, Greece, Crete, Rhodes, Aegean Islands, Turkey, Lebanon, Palestine; North Africa.

2. Between Ayios Amvrosios and Kissousa, 1977, *J. J. Wood* 31!
3. Near Lefkara, 1862, *Kotschy*; Stavrovouni, 1862, *Kotschy* 232; Between Anglisidhes and Skarinou, 1929, *Renz* 2039! Kouklia, 1960, *N. Macdonald* 56! Khirokitia 1970, *A. Genneou* in ARI 1441!

6. English School, Nicosia, 1959, *P. H. Oswald* 101!
7. Pentadaktylos, 1862, *Kotschy* 350! also 1929, *Renz* 2038! and, 1880, *Sintenis & Rigo* 152!
Olymbos, 1929, *Renz* 2033! also, 1957, *Merton* 2894! Ayios Nikolaos, 1929, *Renz* 2036!
Buffavento, 1929, *Renz*; Melounda, 1929, *Renz* 2034! Akanthou, 1929, *Renz* 2037! Between
Kantara and Mandres, 1929, *Renz*; near Klepini, 1929, *Renz*; Ardhana-Phlamoudhi, 1929,
Renz 2035! Dhikomo, 1932, *Syngrassides* 497! St. Hilarion, 1949, *Casey* 547! Halevga, 1937,
Syngrassides 1467! also, 1941–1974, *Davis* 2767! *F. M. Probyn* 72! *G. E. Atherton* 1051!
C. E. H. Sparrow 22! *Meikle* 4034! Above Kharcha, 1948, *Duke-Woolley* 1! Templos, 1956,
G. E. Atherton 1200! Near Vasilia, 1970, *I. M. Hecker* 37!
8. Between Komi Kebir and Ephtakomi, 1929, *Renz*.

NOTES: Hybrids between *O. italica* and *O. simia* have been reported from the north-eastern
part of the Kyrenia Range and also south of the main Troödos massif.

12. O. simia *Lam.*, Fl. Franç., ed. 1, 3: 507 (1778); Sintenis in Oesterr. Bot. Zeitschr,, 32: 192
(1882); Boiss., Fl. Orient., 5: 63 (1884); Holmboe, Veg. Cypr., 57 (1914); Soó in Botanisch.
Archiv, 23: 44, 146 (1928); Renz in Feddes Repert., 27: 216 (1929); Post, Fl. Pal., ed. 2, 2:
567 (1933); Rechinger f. in Archiv för Bot., ser. 2, 1: 419 (1950); R. S. Davis in Orch. Journ.,
3: 164 (1954); Sundermann in Die Orchidee, 20: 79 (1969), Europ. Medit. Orch., ed. 1, 121
(1970), ed. 2, 135 (1975), ed. 3, 149 (1980); Osorio-Tafall et Seraphim, List Vasc. Plants
Cyprus, 27 (1973); Landwehr, Wilde Orch. Europa, 2: 322 (1977); Williams et al., Field
Guide, 90 (1978); Davies et Huxley, Wild Orch. Brit. Europe, 124 (1983).
TYPE: France; not indicated.

Plant (18–) 20–45 cm. high; tubers ovoid or ellipsoid, stipitate; stem erect
or slightly flexuous, leafy at the base, bearing 1–3 ovate or narrowly elliptic,
acute sheaths above, naked or with 1 inconspicuous scaly sheath below the
inflorescence; leaves 3–6, oblong, oblong-elliptic, oblong-ovate or narrowly
elliptic, attenuate towards the base, obtuse or acute, (6–) 10–15 (–20) cm.
long, 2–3·5 (–4·5) cm. wide, erect to spreading, glossy green, unspotted;
inflorescence many-flowered, ovoid or broadly cylindrical, dense, the
flowers opening from the top downwards; floral bracts triangular-ovate to
elliptic, membranous, whitish, hyaline, 1–4 mm. long; sepals and petals
partly connivent, forming a galea, the sepals free at the apex, pale greyish-
pink or creamy-lilac, flecked violet-red, with dark pink to violet-red nerves;
sepals ovate to narrowly elliptic, sometimes oblong-elliptic, acuminate, the
lateral oblique, 10–12 (–14) mm. long, 3–4 mm. wide; petals linear,
acuminate, 1-nerved, 9–10 (–12) mm. long, c. 1 mm. wide; labellum pale pink
or white with tufts of purple papillae at the base and centre, 14–16 (–20) mm.
long, 1·5–2 mm. wide below lobules of mid-lobe, distinctly 3-lobed, lobes
dark purple or pinkish-violet, rarely entirely white or pink, mid-lobe longer
than the lateral lobes, ligulate, divided into 2 narrowly linear, obtuse,
subfalcate-upcurved, divergent lobules separated by a mucro or lacinule,
lateral lobes narrowly linear, obtuse, subfalcate-upcurved, c. 9 mm. long;
spur cream or pale greyish-pink, narrowly cylindrical, 5 mm. long,
emarginate at the apex, downturned, about half as long as the ovary;
column short, 3 mm. long; anther obtuse, sometimes minutely apiculate.

HAB.: In garigue and dry grassy places, on stony slopes, in Pine woods, on calcareous soils;
1,000–2,000 ft. alt.; fl. (late March–) April–May.

DISTR.: Divisions 3, 7, 8. Also south-east England, France, Belgium, Netherlands (found
once), Luxembourg, Germany, south-west Switzerland, Spain, Italy, Jugoslavia, Albania,
Hungary, Romania, Bulgaria, Greece, Crete, Rhodes, Crimea, Caucasus, Turkey (Anatolia),
Syria, Lebanon, Palestine, Jordan, Iraq, northern and western Iran; North Africa.

3. Near Monagri, 1977, *J. J. Wood* 132! Between Apsiou and Phasoula, 1978 and 1979,
Chesterman 77! 122!
7. Near Kantara, 1880, *Sintenis & Rigo* 153! also, 1929, *Renz*; 1955, *J. Hughes* s.n.! and, 1970,
A. Hansen 610; Olymbos, 1929, *Renz*; Ardhana-Phlamoudhi, 1929, *Renz* 2282! Phlamoudhi,
1929, *Renz* 2283! Near Mandres, 1929, *Renz*.
8. Near Kantara Castle, 1905, *Holmboe* 546.

[O. LACTEA *Poir.* in Lam., Encycl. Méth. Bot., 4: 594 (1798).

O. acuminata Desf., Fl. Atl., 2: 318 (1800); Kotschy in Oesterr. Bot. Zeitschr., 12: 276 (1862).

Leaves unspotted; flowers white or greenish-pink with a pink-spotted labellum; sepals acuminate; mid-lobe of labellum longer than the lateral lobes, denticulate at the apex; spur half as long as or longer than the ovary. This species is widespread in the Mediterranean Basin and Balkan Peninsula, extending eastwards as far as Lebanon, but absent from Cyprus. Records of this species from Cyprus may refer to either *O. coriophora* L. ssp. *fragrans* (Poll.) Camus or to *O. italica* Poir.]

[O. MASCULA *L.* ssp. PINETORUM (*Boiss. et Ky.*) *Camus*, Mon. Orch. Europ., 156 (1908).

Leaves glossy green, unspotted; flowers bright rose-purple; mid-lobe of labellum longer than the lateral lobes; spur shorter than or equal to the ovary. This subspecies is characteristically found in Pine and Oak forest and is widespread in Turkey, Transcaucasia, Iraq and Iran. It would appear to fill the same ecological niche in those countries as does *O. anatolica* Boiss. var. *troodi* (Renz) Soó in Cyprus.]

[O. PROVINCIALIS *Balbis*, Misc. Bot., 2: 33, t. 2 (1806).

Leaves spotted reddish-brown; flowers pale yellow or white with a darker yellow, red-spotted labellum; mid-lobe of labellum small, only a little longer than the lateral lobes; spur equal to or longer than the ovary. Although this species is found in Greece, Crete, Rhodes and Turkey it appears to be absent from Cyprus.]

[O. TRIDENTATA *Scop.*, Fl. Carn., ed. 2, 2: 190 (1772); Holmboe, Veg. Cypr., 57 (1914), as "*O. tridentatus* Scop."

O. variegata All., Fl. Pedem., 2: 147 (1785); Sintenis in Oesterr. Bot. Zeitschr., 32: 291, 364, 397, 398 (1882).

Leaves unspotted; flowers lilac-pink to pale violet with a purplish-pink-spotted labellum; sepals acute; mid-lobe of labellum longer than the lateral lobes, entire or emarginate at the apex; spur shorter than the ovary. This species is found as far east as Turkey (Anatolia), Caucasus, Lebanon and Iraq. Modern authorities do not regard it as occurring naturally in Cyprus, nor have I seen any correctly named specimens of *O. tridentata* from Cyprus in the Kew herbarium. One sheet named as such proved to be *Anacamptis pyramidalis* (L.) L. C. M. Rich. Records of this species from Cyprus are most likely to refer to poor specimens of *O. italica* Poir.]

12. **DACTYLORHIZA** *Necker ex Nevski*

Acta Inst. Bot. Acad. Sci. URSS, ser. 1 (Fl. & Syst. Plant. Vasc.), fasc. 4: 332 (1937).

Tubers 2 (–3), ovoid, oblong-cylindrical to napiform, often attenuated, usually palmately bifid to 5-fid, rarely entire or shallowly bifid or trifid at the apex (in our area), normally sessile; stem erect or flexuous, fistular or solid, glabrous; leaves basal, rosulate and/or cauline, most commonly the latter, often cucullate at the apex, spotted or unspotted; floral bracts rather fleshy, leafy; lateral sepals free, spreading or deflexed, the dorsal sepal and petals often connivent to form a galea, very rarely all connivent, glabrous; labellum entire to 3-lobed, spurred, porrect to deflexed, the surface either flat, convex, undulate or deflexed, without calli at the base, glabrous or papillose; column short; anther-connective absent; viscidia 2, placed in a simple bursicle.

About 42 species distributed throughout Europe, the Mediterranean Basin, extending eastwards through Afghanistan to Pakistan and Japan, including 2 species in North America.

Leaves mostly rosulate; stem without stolons; flowers yellow (in Cyprus); lateral sepals strongly reflexed; lip unspotted; spur upturned or horizontal, longer than the ovary
1. D. romana

Leaves cauline; stem with stolons; flowers whitish or pink to greyish-violet; lateral sepals not strongly reflexed; lip spotted and flecked; spur downturned, shorter than the ovary
2. D. iberica

1. **D. romana** (*Seb.*) *Soó* in Ann. Univ. Sci. Budapest, Sect. Biol., 5: 3 (1962); Sundermann in Die Orchidee, 20: 79 (1969), Europ. Medit. Orch., ed. 1, 147 (1970); Osorio-Tafall et Seraphim, List Vasc. Plants Cyprus, 27 (1973); Erich Nelson, Mon. Ikon. Dact., 103 (1976); Landwehr, Wilde Orch. Europa, 1: 62 (1977): Williams et al., Field Guide, 108 (1978); J. et P. Davies in Quart. Bull. Alpine Gard. Soc., 48: 124, 125 (1980).

Orchis romana Seb., Roman. Plant., fasc. 1: 12 (1813); Seb. et Mauri, Fl. Roman. Prodr., 308 (1818); Holmboe, Veg. Cypr., 58 (1914); Soó in Botanisch. Archiv, 23: 63, 158 (1928); Renz in Fedde Repert., 27: 216 (1929); Post, Fl. Pal., ed. 2, 2: 571 (1933); R. S. Davis in Orch. Journ., 3: 164 (1954).

O. pseudosambucina Ten., Syn. Nov. Plant., ed. 1, 64 (1815); Unger et Kotschy, Die Insel Cypern, 208 (1865); Boiss., Fl. Orient., 5: 72 (1884); H. Stuart Thompson in Journ. Bot., 44: 338 (1906).

Dactyorhiza sulphurea (Link) Franco ssp. *pseudosambucina* (Ten.) Franco in Bot. Journ. Linn. Soc., 76: 336 (1978); Davies et Huxley, Wild Orch. Brit. Europe, 99 (1983).

D. sambucina (L.) Soó ssp. *pseudosambucina* (Ten.) Sundermann, Europ. Medit. Orch., ed. 3, 40, 177 (1980).

TYPE: Italy; "in ericeto vulgo dicto Pigneto di Sacchetti", *Sebastiani*.

Plant (10–) 20–30 (–35) cm. high; tubers ovoid to oblong-cylindrical, fusiform, entire or shortly bifid at the apex, sessile; stem erect, often slender, ridged, fistular, with 2–3 large, brownish, membranous sheaths at the base which clasp the basal leaves; leaves 4–7 (–10), mostly rosulate, narrowly oblong to linear-lanceolate, broadest above the middle, attenuate towards the base, with a short sheathing base, acute, not cucullate, 4–15 (–20) cm. long, 0·7–1·5 (–3·5) cm. wide, weakly conduplicate, lowermost porrect to spreading, uppermost bract-like, ± erect, unspotted; inflorescence narrowly cylindrical, short, occasionally ovoid, few- to many-flowered, lax or dense; floral bracts herbaceous, ovate to narrowly elliptic, usually acute, sometimes obtuse, longer than the ovary, the lowermost often reaching the lower and middle flowers, green; flowers clear lemon or sulphur-yellow, dull rose-pink or purple-violet; sepals free or the dorsal connivent with the petals to form a galea, oblong, ovate or narrowly elliptic, obtuse, 6–10 (–13) mm. long, 4–5 mm. wide, spreading; dorsal sepal 3-nerved, sometimes slightly shorter than the laterals; lateral sepals 3–5-nerved, usually strongly reflexed; petals obliquely ovate, obtuse, usually a little shorter than the sepals, 5–10 mm. long, 3·5–7 mm. wide; labellum with a central creamy or yellow area in pink or purple-violet-flowered individuals, suborbicular or obovate in outline, distinctly or indistinctly 3-lobed towards the apex, often somewhat conduplicate, porrect, 7–8 (–12) mm. long, 7–10 (–15) mm. wide, glabrous or minutely papillose, mid-lobe subquadrate or suborbicular, sometimes ovate-triangular, entire or emarginate, obtuse, shorter than the lateral lobes, lateral lobes oblong-ovate, rounded, margins entire or crenulate; spur cylindrical, terete, rarely conical or saccate-cylindrical, shallowly curved, slightly dilated at the apex, horizontal or strongly upturned, distinctly longer than the ovary, (13–) 17–20 (–25) mm. long; column short, 3–4 mm. long; stigmatic cavity ± circular, the stigma with a pale mauve-pink to purple margin; staminodes insignificant; anther pale mauve-pink; pollinia pale bluish-green; ovary terete, slightly twisted, glabrous, 1–1·5 cm. long.

HAB.: Dry rocky places in the mountains, also widespread in *Pinus brutia* forest, often in association with *Orchis morio* ssp. *picta* and *Thymus integer*; 500–3,000 ft. alt.; fl. March–May (–June).

DISTR.: Divisions 1–3. Also Portugal, Spain, Italy, Sicily, Jugoslavia, Corcula Island, Romania, Bulgaria, ?European Russia, Albania, Greece, Crete, ?Rhodes, ?Crimea, ?Caucasus area, Turkey, Syria, Lebanon; North Africa.

1. Smyies, 1962, *Meikle* 2201!
2. Kakopetria, 1929, *Renz* 2225! Phileyia, 1937, *Lady Loch* 20! and, 1957, *Merton* 2869! Stavros tis Psokas, 1952, *F. M. Probyn* 110! also, 1967, *D. N. Paton* s.n.! Ayia valley, 1977, *J. J. Wood* 87!
3. Stavrovouni, 1938, *Lady Loch* 42!

NOTES: Only the yellow-flowered form occurs in Cyprus. The author has not seen any Cyprus specimens of the pink-flowered plant, either in the field or the herbarium, despite the fact that it is quite common in neighbouring Turkey.

2. D. iberica (*M. Bieb. ex Willd.*) Soó in Ann. Univ. Sci. Budapest, 5: 3 (1962); Sundermann, Europ. Medit. Orch., ed. 1, 147 (1970); Osorio-Tafall et Seraphim, List Vasc. Plants Cyprus, 27 (1973); Sundermann, Europ. Medit. Orch., ed. 2, 161 (1975), ed. 3, 175 (1980); Erich Nelson, Mon. Ikon. Dact., 106 (1976); Landwehr, Wilde Orch. Europa, 1: 56 (1977); Williams et al., Field Guide, 108 (1978); J. et P. Davies in Quart. Bull. Alpine Gard. Soc., 48: 125 (1980); Robatsch in Die Orchidee, 31: 195–200 (1980); Davies et Huxley, Wild Orch. Brit. Europe, 98 (1983).

    Orchis iberica M. Bieb. ex Willd., Sp. Plant., 4: 25 (1805); Boiss., Fl. Orient., 5: 66 (1884); Holmboe, Veg. Cypr., 57 (1914); Soó in Botanisch. Archiv, 23: 64, 157 (1982); Post, Fl. Pal., ed. 2, 2: 569 (1933); Lindberg f., Iter Cypr., 11 (1946); R. S. Davis in Orch. Journ., 3: 164 (1954).

    O. angustifolia M. Bieb., Fl. Taur.-Cauc., 2: 368 (1808); Boiss., Fl. Orient., 5: 65 (1884).
TYPE: "Habitat in Iberia."

Plant (20–) 35–42 (–60) cm. high; tubers cylindrical to napiform, attenuate, entire or sometimes bifid or trifid at the apex, sessile; stem erect, thin, weakly ridged, fistular, with several membranous sheaths below, bearing subterranean stolons at the base just above the tubers; leaves 3–5 (–6), cauline, linear or narrowly lanceolate, broadest at or above the middle, with a long sheathing base, acuminate, sometimes obtuse, (8–) 10–20 cm. long, (1–) 1·5–2 cm. wide, conduplicate, remote, the longest placed towards the middle of the stem, erect to spreading, grading into acute bracts above, unspotted; inflorescence narrowly cylindrical, often elongated, occasionally ovoid, few- to many-flowered, 6–15 (–20) cm. long, lax or dense; floral bracts herbaceous, narrowly elliptic, acute, the lowermost longer than, and the uppermost ± equalling the ovary; flowers pink, sometimes whitish or light greyish-violet, labellum spotted and flecked dark purple; sepals and petals connivent, forming a galea; sepals oblong to ovate-elliptic, acute, 3-nerved; dorsal sepal 8–10 mm. long, 4 mm. wide; lateral sepals oblique, 9–11 mm. long, 3–4 mm. wide, often becoming free at the apex; petals obliquely lanceolate, acuminate, 6–8 mm. long, 2 mm. wide; labellum obovate or suborbicular, cuneate to flabellate at the base, shallowly 3-lobed distally, rarely entire, ± flat, 6–11 mm. long, 7–10 mm. wide, minutely papillose-pubescent, mid-lobe narrowly triangular, linear or tooth-like, acute, equalling or slightly exceeding the lateral lobes, lateral lobes broadly triangular or semi-ovate, falcately rounded, obtuse, outer margins crenulate; spur narrowly cylindrical, slender, ± acute, 5–7 mm. long, shorter than the ovary, downturned, slightly curved towards the underside of the labellum; column short, 4 mm. long; stigmatic cavity narrowly cordate; staminodes small; rostellum very short; anther dull purple-violet; pollinia dark bluish-green; ovary cylindrical, slightly curved, twisted, glabrous.

HAB.: In marshy meadows, or damp ground around springs and streamsides, calcicole; 3,300–5,500 ft. alt.; fl. (May–) June–July (–August).

DISTR.: Division 2. Also Greece, Turkey, Caucasus area, Crimea, Lebanon, Iraq, western and northern Iran.

2. Common in central Troödos: Prodhromos, Platania, Kryos Potamos, Xerokolymbos, Troödos, also in the Cedar Valley, Paphos Forest. *Sintenis & Rigo* 870 ! *Haradjian* 469 ! *C. B. Ussher* 9 ! *Syngrassides* 1096 ! *Kennedy* 267–270 ! *Lindberg f.* s.n. ! *D. P. Young* 7319 ! 7375 ! *Economides* in ARI 1199 ! *P. Laukkonen* 399 ! etc.

NOTES: *D. iberica* is unique within the genus on account of its habit of producing stolons. This method of vegetative reproduction enables it to form large, uniform populations. The specific name refers to Caucasian Iberia, roughly corresponding with modern Georgia (U.S.S.R.).

13. NEOTINEA *Reichb. f.*
Poll. Orch. Gen., 29 (1852).

Tubers 2, globose or ellipsoid, entire; stem rather glaucous; leaves basal, rosulate and cauline, glaucous, spotted or unspotted; inflorescence dense; floral bracts thin, membranous; perianth-segments $\pm$ equal, connivent to form a galea, glabrous; labellum spurred, 3-lobed, as long as or scarcely longer than the perianth-segments, without calli, minutely papillose, spur very short; column very small, with 2 separate stigmas, borne one on each side of the column at the mouth of the spur; anther-connective absent; viscidia 2, distinct, enclosed in a simple bursicle; caudicles of the pollinarium very short.

A monotypic genus distributed from western Ireland and the Isle of Man eastwards across southern Europe from Portugal and the Canary Islands to western Turkey and the Lebanon; North Africa.

1. **N. maculata** (*Desf.*) *Stearn* in Ann. Mus. Goulandris, 2: 79 (1975); Landwehr, Wilde Orch. Europa, 1: 44 (1977); Robatsch in Die Orchidee, 31: 195–200 (1980); Sundermann, Europ. Medit. Orch., ed. 3, 171 (1980); Davies et Huxley, Wild Orch. Brit. Europe, 114 (1983).
   Satyrium maculatum Desf., Fl. Atlant., 2: 319 (1799).
   Orchis secundiflora Bertol., Rar. Ital. Pl. Decas 2: 42 (1806); Sintenis in Oesterr. Bot. Zeitschr., 32: 192 (1882).
   Aceras densiflora (Brot.) Boiss., Voy. Bot. Espagne, 2: 595 (1841); Kotschy in Oesterr. Bot. Zeitschr., 12: 275 (1862).
   A. intacta (Link) Reichb. f., Icones Fl. Germ., 13: 2, t. 148 (1850); Unger et Kotschy, Die Insel Cypern, 205 (1865).
   Neotinea intacta (Link) Reichb. f., Poll. Orch. Gen., 29 (1852); Holmboe, Veg. Cypr., 58 (1914); Soó in Botanisch. Archiv, 23: 91, 170 (1928); Renz in Fedde Repert., 27: 217 (1929); Post, Fl. Pal., ed. 2, 2: 563 (1973); R. S. Davis in Orch. Journ., 3: 164 (1954); Sundermann, Europ. Medit. Orch., ed. 1, 107 (1970); Osorio-Tafall et Seraphim, List Vasc. Plants Cyprus, 25 (1973); Sundermann, Europ. Medit. Orch., ed. 2, 155 (1975); B. et E. Willing in Die Orchidee, 27: 114 (1976); J. et P. Davies in Quart. Bull. Alpine Gard. Soc., 48: 227 (1980).
   Tinaea intacta (Link) Boiss., Fl. Orient., 5: 58 (1884).
   Neotinea intacta (Link) Reichb. f. 1. *luteola* Renz in Fedde Repert., 27: 210, 217 (1929); Sundermann in Die Orchidee, 20: 79 (1969); Stearn in Ann. Mus. Goulandris, 2: 81 (1975); Landwehr, Wilde Orch. Europa, 1: 46, 47, fig. 5 (1978).
   TYPE: Algeria; "in Atlante prope Belide" [Blida].

Plant 8–25 (–40) cm. high; tubers 2, globose or ellipsoid, sessile, or with 1 or both stipitate; stem slender, often pinkish-purple and rather glaucous, with 1 or 2 membranous sheaths at the base, leafy below, with 1–3 ovate-elliptic, subacute to acute sheaths above; leaves 2–4 (–6), narrowly elliptic to oblong, obtuse and mucronate, the lowermost rosulate, 3–12 cm. long, 1–4·5 cm. wide, spreading, the uppermost cauline, smaller, erect, glaucous bluish-green, with small, purplish-brown spots arranged in longitudinal rows or completely green; inflorescence many-flowered, 2–6 cm. long, sometimes secund, dense; floral bracts ovate, acute, shorter than the ovary, 4 mm. long; sepals and petals connivent, forming a galea, dull pink, pinkish-purple, often darker purple at the base, greenish-white, yellowish or straw-coloured; sepals ovate, obtuse to acuminate, 1-nerved, the laterals concave at the base, 3–4 mm. long, 2 mm. wide; petals ligulate, 1-nerved, 3 mm. long, 1–2 mm. wide, apex truncate or shallowly bilobed; labellum dull pink, pinkish-purple with darker purple nerves, greenish-white, yellowish or straw-coloured, 3–4 mm. long, 3-lobed, minutely papillose, mid-lobe ligulate, bifurcate or trifurcate, truncate, lateral lobes linear, shorter than the mid-lobe; spur broadly conical, obtuse, 1–2 mm. long; column short.

HAB.: In garigue, *Pinus brutia* forest or under *Cupressus sempervirens*, on rough grassland, usually calcicole; 300–5,000 ft. alt.; fl. March–April.

DISTR.: Divisions 1–3, 5–8. Distribution that of the genus.

1. Near Ayios Yeoryios (Akamas), 1962, *Meikle* 2069! Smyies, 1962, *Meikle* 2183! Ayios Neophytos, 1971, *Guichard* CYP 7!
2. Near Prodhromos, 1862, *Kotschy* 755a; also, 1961, *D. P. Young* 7273! 7286! Between Kakopetria and Platania, 1929, *Renz*; Stavros tis Psokas, 1952, *F. M. Probyn* 106! Moutoullas, 1969, *A. Genneou* in ARI 1355a! Between Ayios Amvrosios and Kissousa, 1977, *J. J. Wood* 32! Between Pano Lefkara and Vavatsinia, 1977, *J. J. Wood* 57! Ayia valley, 1977, *J. J. Wood* 86! 89! and, 1978, *Chesterman* 60! Zakharou, 1979, *Edmondson & McClintock* E2807!
3. Between Limassol and Omodhos, 1859, *Kotschy* 414! Stavrovouni, 1862, *Kotschy* 186! Between Lefkara and Kornos, 1929, *Renz*; Lefkara, 1941, *Davis* 2764!
5. Kythrea, 1937, *Miss Godman* s.n.! Delikipo, 1929, *Renz* 2427!
6. Kremama Kamilou, 1929, *Renz* 2419!
7. Olymbos, 1929, *Renz* 2420! Melounda, 1929, *Renz* 2422! Akanthou, 1929, *Renz* 2423! Phlamoudhi, 1929, *Renz* 2424! Ardhana–Phlamoudhi, 1929, *Renz* 2425! Near Mandres, 1929, *Renz*; Near Klepini, 1929, *Renz* 2426! Panagra, 1936, *Syngrassides* 1022! Near Halevga, 1937, *Syngrassides* 1468! also, 1952, *F. M. Probyn* 80! and, 1959, *C. E. H. Sparrow* 21! Ayios Amvrosios, 1937, *Miss Godman* s.n.! Kornos, 1941, *Davis* 3052! St. Hilarion, 1949, *Casey* 548! Kyrenia, 1956, *G. E. Atherton* 118! 1187!
8. Between Komi Kebir and Ephtakomi, 1929, *Renz*; Komi Kebir, 1929, *Renz* 2421!

NOTES: In the Mediterranean region the leaves of most populations are spotted and the flowers are pink to purple. However, specimens with greenish-white or straw-coloured flowers very often have faintly spotted leaves. Renz (1930) reported and described *N. intacta* lusus *luteola* from the Troödos range which differs in having pale yellow to greenish bracts and flowers. It has also been recorded from Attica and the Ionian islands of Corfu and Cefalonia. Both white- and pink-flowered forms are found near the coast, but only the former occurs in Troödos.

93a. MUSACEAE

Very large, often tree-like herbs; leaves alternate or basal, the leaf-sheaths rolled around one another to form what appears to be a stem or trunk; leaves large, oblong, entire with a prominent midrib and parallel lateral nerves often splitting or tearing between the nerves; inflorescences spicate, pendent, conspicuously bracteate; flowers generally unisexual, in whorls, the males in the axils of the upper bracts, the females in the axils of the middle and lower bracts; perianth of 5 segments connate to form a narrow, 5-lobed tube split down the adaxial side, the sixth segment free; male flowers with 5 stamens (sometimes with an additional staminode); filaments filiform; anthers 2-thecous; female flowers with an inferior, 3-locular ovary; ovules numerous, axile; style filiform; stigma 3-lobed; staminodes present. Fruit an elongate berry; seeds (when present) with a hard testa, copious endosperm and a straight embryo.

The family includes one genus (*Musa* L.) or possibly 2 genera (*Musa* L. and *Ensete* Bruce) widely distributed in the tropics of the Old World, with estimates of the number of species ranging from 30–45. No members of the family are native or naturalized in Cyprus, but the Banana, probably of Malaysian origin, consisting of named cultivars (mostly sterile triploids) derived from *M. acuminata* Colla (Mem. Gen. Musa, 66; 1820) or hybrids between this and the closely allied *M. balbisiana* Colla (Mem. Gen. Musa, 56; 1820). (*M.* × *paradisiaca* L., *M.* × *sapientum* L.) is now extensively cultivated around Paphos, and was noted as growing on the island as long ago as 1458: "some other trees called banana, which produce fruit very much like small cucumbers; when it is ripe it is yellow and very sweet of savour" (Gabriele Capodilista, Itinerario della Terra Santa nel 1458; Perugia ?1485, quoted in Cobham, Excerpta Cypria, 35; 1908). While successful in favoured coastal areas, the Banana is too tender for general cultivation in Cyprus, and, even near the coast, can be damaged by exceptionally cold winters.

94. IRIDACEAE

Perennial herbs with rhizomes, bulbs or corms; leaves often flattened laterally, equitant and distichous; inflorescence terminal, cymose or flowers solitary; flowers hermaphrodite; perianth petaloid, actinomorphic or zygomorphic, segments 6 in 2 similar or dissimilar series, often united basally to form a long or short tube; stamens 3, opposite the outer perianth-segments; filaments free or partly connate; anthers 2-thecous, extrorse, dehiscing longitudinally; ovary inferior (or very rarely superior), 3-locular with axile placentation (or rarely 1-locular with parietal placentation); style slender, 3-lobed above, the lobes sometimes petaloid, the stigma apical or sub-apical; ovules numerous or few, anatropous. Fruit a loculicidally dehiscent capsule; seeds with copious endosperm and a small embryo.

About 65 genera widely distributed in temperate and tropical regions of the world.

Style-branches petaloid with a stigmatiferous flap on their undersurface; outer perianth-segments with a deflexed apex (or "fall"):
    Perianth-tube present; plants growing from a rhizome or bulb - - - IRIS (p. 1559)
    Perianth-tube absent; plants growing from a corm - - - - - **1. Gynandriris**
Style-branches not petaloid; outer perianth-segments without a deflexed apex (or "fall"):
    Flowers actinomorphic; inflorescence not spiciform:
        Scape absent; ovary below ground-level; leaves all basal; perianth-tube long and slender; bracts membranous - - - - - - - - - - **2. Crocus**
        Scape present; ovary not below ground level; leaves basal and cauline; perianth-tube short; bracts herbaceous, at least in part - - - - - - - **3. Romulea**
    Flowers zygomorphic; inflorescence spiciform - - - - - - **4. Gladiolus**

IRIS *L.*

Sp. Plant., ed. 1, 38 (1753).

Gen. Plant., ed. 5, 24 (1754).

W. R. Dykes, The Genus Iris, 245 pp., 48 pls. (1913).

Perennial bulbous or rhizomatous herbs; stems commonly developed; leaves all basal, or basal and cauline, usually equitant, often folded isobilaterally, sometimes flat or canaliculate; flowers solitary or in few-

flowered cymes, commonly subtended by spathaceous bracts, actinomorphic, sessile or pedicellate; perianth-segments connate at the base,
sometimes with a distinct tube, 3 exterior segments (*"falls"*) patent or
spreading, with an expanded, deflexed apical limb, and usually with a
narrower, often canaliculate base (the *"haft"* or *"claw"*), the 3 interior
segments (*"standards"*) usually erect, sometimes very small or obsolete;
style divided above into 3 conspicuous petaloid branches, overlying the
claws of the outer perianth-segments and covering the 3 stamens, each style-
branch with a transverse stigmatiferous flap near the apex of its underside;
ovary 3-locular; ovules numerous. Fruit a fusiform or ellipsoid, 3-angled
capsule, sometimes with a sterile apical beak; seeds usually numerous,
sometimes arillate, subglobose, ovoid, angled or flattened.

About 300 species widely distributed in temperate regions of the northern
hemisphere. All the Cyprus representatives are cultivated or escapes from
cultivation, and belong to subgenus *Iris* sect. *Iris*, a group of rhizomatous
perennials, with equitant leaves, and generally with bearded outer perianth-
segments or falls.

Flowers blue, purple or violet; stems usually with well-developed branches; cauline leaves not
    closely investing the stems -   -   -   -   -   -   -   I. GERMANICA (p. 1560)
Flowers white; stems usually unbranched or with very short lateral branches; cauline leaves
    almost tubular, closely investing the stems -   -   -   -   I. ALBICANS (p. 1561)

I. GERMANICA *L.*, Sp. Plant., ed. 1, 38 (1753); Sibth. et Sm., Fl. Graec. Prodr., 1: 26 (1806), Fl.
    Graec., 1: 29, t. 40 (1806); Boiss., Fl. Orient., 5: 137 (1882); Dykes, The Genus Iris, 162
    (1913).
        I. cypriana Baker et Foster in Gard. Chron., ser. 3, 4: 182 (1888), Dykes, The Genus Iris,
    177 (1913); Randolph in The Iris Year Book, 1955: 38 (1955); Osorio-Tafall et Seraphim,
    List Vasc. Plants Cyprus, 24 (1973).
        I. mesopotamica Dykes, The Genus Iris, 176 (1913); Post, Fl. Pal., ed. 2, 2: 601 (1933).

Robust rhizomatous perennial, up to 1 m. high, with a stout, horizontal
rhizome and tough roots; stems glabrous, usually branched, the branches
spreading, much exceeding the bracts or cauline leaves; basal leaves
equitant, isobilaterally folded, ensiform, glaucous-grey, up to 57 cm. long
and 2·5–3·5 cm. wide, with numerous longitudinal, parallel nerves; cauline
leaves usually much reduced, not tubular or closely investing the stems;
bracts oblong-navicular, acute, about 4 cm. long, 2 cm. wide, herbaceous
and greenish towards the base with a scarious apex and margins;
inflorescence rather open, with a 2–3-flowered terminal head of flowers and
2–3 lateral, 2 (sometimes 3)-flowered heads; flowers fragrant, shortly
pedicellate; perianth-tube 2–3 cm. long, greenish with purplish streaks
extending from the bases of the standards; outer perianth-segments (falls)
broadly obovate, 9–10 cm. long, 4–5 cm. wide, reddish-lilac or violet, apex
rounded, base obscurely clawed, beard conspicuous, about 4 cm. long, hairs
long, white tipped orange; inner perianth-segments usually broader than
outer, up to 7 cm. wide, paler, with plicate-undulate margins; style-branches
about 5 cm. long, 1·5 cm. wide, paler than the perianth-segments, apex
deeply 2-lobed, with acute, irregularly dentate-lacerate lobes; filaments
glabrous, whitish, about 1–1·5 cm. long; anthers creamy, linear, to about
2 cm. long, 0·2 cm. wide, with a shortly sagittate base; ovary glabrous,
narrowly ellipsoid, about 2·5 cm. long, 0·5 cm. wide. Fruit not seen, said to
be elongate-ellipsoid.

HAB.: Gardens, roadsides, waste ground; from below sea-level to ? 4,000 ft. alt.; fl.
April–June.

DISTR.: Divisions 3, 7; planted or an escape from cultivation; probably to be found in S.
Turkey, Syria and Palestine, but see Notes.

3. Kato Lefkara, 1937, *Syngrassides* 1543 partly !

7. Bellapais, 1885, *E. C. Kenyon* s.n. (cult. Kew)! also, 1954, *L. F. Randolph* s.n.! Trimithi, 1885, *E. C. Kenyon* s.n., also, 1954, *L. F. Randolph* s.n.! Vasilia, 1937, *Syngrassides* 1543 partly!

NOTES: Probably widespread in the island as a cultivated plant or escape from cultivation, but very seldom collected. In addition to the above, there is an unlocalized specimen in herb. Kew, collected by C. B. Ussher in 1934, and records for *Iris florentina* L. in Nattrass, First List Cyprus Fungi, 35, 46 (1937) from Nicosia, 1933 & 1935, *Nattrass* 306, 531 and Kato Lefkara, 1937, *Nattrass* 841, probably belong here too.

I have deliberately used the name *Iris germanica* in the wide sense to include *I. cypriana, I. mesopotamica, I. trojana* and other cultivated Irises of garden origin, now widely distributed or escapes, or occasionally naturalized, over areas of Europe, the Mediterranean region and western Asia. L. F. Randolph (Iris Year Book, 1955: 39) remarks: "Comparison of living specimens of *I. cypriana* from Cyprus and of *I. mesopotamica* from several localities in Lebanon near Beirut did not reveal any significant morphological differences". Little is known as to the origin of these cultivated Irises of the *I. germanica-pallida-albicans* complex, nor, allowing that they may be of hybrid origin, is it easy to suggest a parentage. One thing is, however, clear: that *I. cypriana* cannot be considered indigenous or endemic in our area.

The plants described as *I. cypriana* by Baker and Foster were originally sent to Kew by Miss Eliza C. Kenyon. Writing to Sir Joseph Hooker (letter dated Jan. 11 1885) she says: "they grow wild in the neighbourhood of Bellapais and Tremithi and probably in other localities".

I. ALBICANS *Lange* in Vid. Meddel. Dansk Naturh. Foren. Kjoeben., 1860: 76 (1861), Descr. Plant. Nov. Hisp., fasc. 3: 19, t. 33 (1866); Dykes, The Genus Iris, 161 (1913); Post, Fl. Pal., ed. 2, 2: 602 (1933).

[*I. florentina* (vix L.) Sm. in Sibth. et Sm., Fl. Graec. Prodr., 1: 26 (1806), Fl. Graec., 1: 28, t. 39 (1806); Unger et Kotschy, Die Insel Cypern, 202 (1865); Boiss., Fl. Orient., 5: 138 (1882); Gennadius, Rep. Agric. Cyprus, Pt. 2: 33 (1896); ? Holmboe, Veg. Cypr., 54 (1914); Osorio-Tafall et Seraphim, List Vasc. Plants Cyprus, 24 (1973).]

Similar to *I. germanica* L., but seldom more than 60 cm. high, with relatively short, broad ensiform leaves, the cauline often closely investing the stout, unbranched or sparingly branched stem; inflorescences commonly terminal, consisting of 2–3 shortly pedicellate flowers, partly enveloped by large, navicular bracts 4–5 cm. long, 2–3 cm. wide, with an herbaceous base and broadly scarious margins; lateral inflorescences few, subsessile or on very short branches; flowers fragrant, typically pure white, sometimes tinged blue or purple; exterior perianth-segments 7–8 cm. long, 3–4 cm. wide, obovate with a rounded apex, beard about 3 cm. long, hairs whitish, tipped yellow; interior perianth-segments broadly obovate or suborbicular, about 7 cm. long and almost as wide; style-branches about 4 cm. long, 1–1·5 cm. wide, the apex bifid with acute, deltoid lobes; filaments glabrous, about 1·5 cm. long; anthers creamy, linear, about 1·5 cm. long, 0·2 cm. wide; ovary fusiform, glabrous, about 2 cm. long, 0·4 cm. wide. Fruit and seeds not seen.

HAB.: Gardens, waste ground, roadsides; 4,500–5,000 ft. alt.; fl. May–June.

DISTR.: Division 2. Planted and sometimes naturalized in S. Europe and the Mediterranean region east to Iran and Arabia.

2. Garden at Prodhromos, 1862, *Kotschy* 888; Marathasa valley, 1896, *Gennadius*; below Troödos on road to President's house, 1963, *D. P. Young* 7851 (see Notes)!

NOTES: The identity of *Iris florentina* L. is somewhat obscure (and certainly not at all clear to Linnaeus!), but it is currently regarded as little more than a whitish-flowered variant of *I. germanica* L. *Iris albicans* is, however, tolerably distinct from *I. germanica*, and recognizable by its stocky habit, stout, often unbranched, or very shortly branched stems, and convolute, sheathing cauline leaves. In many areas of its distribution the flowers are uniformly white, but blue and purplish variants are recorded, though perhaps incorrectly. Dykes (The Genus Iris, 161; 1913) refers *Kotschy* 888 to this species, and *D. P. Young* 7851 is probably also *I. albicans*, but the specimen is without flowers and unsatisfactory. *Holmboe* 827, from St. Hilarion, also without flowers, is more probably a form of *I. germanica* (*I. cypriana*). In addition to these specimens, Baker and Foster, in their original description of *I. cypriana*, refer to "a white Iris, which proved to be identical with *I. albicans* (Lange)", which was sent alive to Kew by Miss E. C. Kenyon in 1885: unfortunately no herbarium specimens were made from the living material nor is it known where Miss Kenyon collected it though very probably from a garden in Kyrenia (where she lived) or from the neighbourhood of Kyrenia.

Iris albicans is apparently wild, or at least thoroughly naturalized, in Yemen and adjacent parts of S. Arabia, and might there be regarded as an indigenous species, but Randolph (The Iris Year Book, 1955: 40) refers it to the hybrid complex that includes *I. germanica*. In the eastern Mediterranean region it is frequently planted in Moslem cemeteries.

I. XIPHIUM *L.*, Sp. Plant., ed. 1, 40 (1753), the Spanish Iris, a slender, bulbous plant, usually 40–50 cm. high, with narrow, spreading canaliculate leaves, violet or yellow, scentless flowers, and a relatively short, small, beardless exterior perianth-limb, has been collected at (7) Tjiklos, 1973 (*P. Laukkonen* 305), where, like some other plants from this locality, it is clearly a relic of cultivation. *I. xiphium* is a native of S.W. Europe and western North Africa.

1. GYNANDRIRIS *Parl.*

Nuovi Gen. Sp. Monocot., 49 (1854).

Perennial herbs arising from reticulately fibrous, tunicate corms; leaves few, spreading, canaliculate; stem developed, bracts spathaceous, tubular-amplexicaul towards base of inflorescence; membranous; flowers resembling those of *Iris* L., but without a perianth-tube, the perianth-segments inserted on the slender rostrate apex of the ovary; outer segments with an ascending claw and deflexed limb; inner segments erect; stamens 3; filaments agglutinate to lower surface of style-branches, but not fused with them; anthers extrorse; style divided into 3 petaloid style-branches with a stigmatiferous flap as in *Iris*. Capsule 3-locular, membranous; seeds numerous, pyriform or subglobose, scarcely compressed.

One (or two) species in the Mediterranean region eastwards to Central Asia, otherwise South African, where there are about 20 species. Although superficially resembling *Iris*, *Gynandriris* is nowadays regarded as more closely allied to the African genus *Moraea* Mill.

1. **G. sisyrinchium** (*L.*) *Parl.*, Nuovi Gen. Sp. Monocot., 49 (1854); Unger et Kotschy, Die Insel Cypern, 202 (1865); Sintenis in Oesterr. Bot. Zeitschr., 31: 392 (1801); Osorio-Tafall et Seraphim, List Vasc. Plants Cyprus, 24 (1973); Elektra Megaw, Wild Flowers of Cyprus, 13, t. 25 (1973).
 Iris sisyrinchium L., Sp. Plant., ed. 1, 40 (1853); Sibth. et Sm., Fl. Graec. Prodr., 1: 28 (1806), Fl. Graec., 1: 30, t. 42 (1806); Poech, Enum. Plant. Ins. Cypr., 10 (1842); Boiss., Fl. Orient., 5: 120 (1882); Holmboe, Veg. Cypr., 54 (1914); Post, Fl. Pal., ed. 2, 2: 589 (1933).
 Helixyra sisyrinchium (L.) N.E. Br. in Trans. Roy. Soc. S. Africa, 17: 349 (1929); R. Seligman in Quart. Bull. Alpine Gard. Soc., 20: 230 (1952).
 TYPE: "in Hispania, Lusitania".

Erect perennial 10–50 cm. high; corm broadly ovoid or subglobose, 1–2·5 cm. diam., clothed with a tunic of brown, reticulate fibres; stem glabrous, bluntly angled; leaves 2, arising at soil-level, 14–60 cm. long, 0·2–0·8 cm. wide, usually recurved, glabrous, canaliculate, with 5–13 prominent nerves and a tubular basal sheath; inflorescences terminal or terminal and lateral, the lowermost sometimes terminal on a well-developed, erect branch; flowers in clusters of 2–4, with shortly lanuginose pedicels about 1 cm. long, enveloped by several spathaceous, thinly scarious, prominently nerved, acuminate bracts, tubular at the base and tapering to a slender acumen, 2–7 cm. long, 0·8–1·5 cm. wide; perianth sessile on the apex of the elongate ovary-beak, without a perianth-tube; exterior segments rich lavender-blue with a white median patch at the base of the fall, claw erect, 1·8–2 cm. long, 1–1·3 cm. wide, fall broadly ovate, subacute, deflexed, about 1–1·5 cm. long and almost as wide, upper surface shortly papillose but without a beard, inner segments erect, narrowly oblong, shortly acute, lavender-blue, about 2·5–3 cm. long, 0·8 cm. wide; filaments glabrous, 4–5 mm. long; anthers linear, pallid, 9–10 mm. long, about 1 mm. wide; ovary membranous, narrowly fusiform, 15–20 mm. long, about 2 mm. wide, the apex produced into a slender, almost filiform, beak 20–25 mm. long; styles united for about 5–6 mm., then divided into 3 petaloid, lavender-blue, oblong branches 2–2·5 cm. long, 0·8 cm. wide, bifid at the apex into narrowly deltoid-acuminate lobes 8–10 mm. long; stigmas prominent, gibbous, 2-lobed. Fruit a membranous, fusiform capsule about 20 mm. long, 2·5–3 mm. wide; seeds subglobose or pyriform, about 2 mm. long, 1·8 mm. wide, at most slightly compressed, often with a distinct lateral ridge or narrow wing; testa fuscous-brown, asperulous-rugulose.

HAB.: Cultivated and fallow fields, roadsides, waste ground, stony hillsides; sea-level to 2,000 ft. alt.; fl. Febr.–April.

DISTR.: Divisions 1, 3–7. Widespread in the Mediterranean region and eastwards to Central Asia.

1. Threshing floor near Neokhorio, 1962, *Meikle* 2189!
3. Near Limassol, 1956, *Maj. Gen. G. E. R. Bastin* s.n.
4. Near Dhekelia ["Redgelia"], 1880, *Sintenis & Rigo* 163b!
5. Kythrea, 1927, *Rev. A. Huddle* 94! also, 1950, *Chapman* 88!
6. Nicosia, 1927, *Rev. A. Huddle* 93! and, 1930, *F. A. Rogers* 0657! Geunyeli, 1934 and 1936, *Nattrass* 432; 774.
7. Kyrenia, 1949, *Casey* 314! also, 1956, *G. E. Atherton* 985! Dhavlos, 1957, *H. Painter* 34! 35! Tjiklos, 1973, *P. Laukkonen* 70!

NOTES: So common in lowland Cyprus that it is generally ignored by collectors, or recorded without precise localization. It is probably to be found in all Divisions. The flowers are exceptionally fugacious, expanding towards dusk and shrivelling at dawn. The seeds ripen very soon after anthesis.

2. CROCUS L.*

Sp. Plant., ed. 1, 36 (1753).
Gen. Plant., ed. 5, 23 (1754).

Perennial herbs with tunicated corms; tunic membranous, coriaceous or fibrous; cataphylls (or basal sheaths) up to 5, sheathing the aerial shoot; leaves appearing with the flowers (synanthous) or after them (hysteranthous), all basal, flat or canaliculate on the upper surface, usually with a whitish median stripe; lower surface with 2 deep grooves on either side of a flattish keel; scape absent; flowers infundibuliform, autumnal or vernal, 1–several, each on a short subterranean pedicel which is sometimes subtended by a membranous sheathing prophyll ("basal spathe" of various authors); bract membranous; bracteole similar, or reduced, or absent; perianth actinomorphic; tube long and slender; segments 6, in 2 equal or subequal whorls, inner whorl sometimes distinctly smaller than outer; style 3-branched to multifid; ovary subterranean. Capsule cylindrical to ellipsoid, maturing at or just above ground level by elongation of the pedicel; seeds numerous, globose to ellipsoid, very variable in surface architecture of the testa and degree of development of the raphe and caruncle.

About 90 species, chiefly in S. Europe, the Mediterranean region, and eastwards to Central Asia.

Style divided into at least six branches; flowering time November (–January)   **3. C. veneris**
Style divided into 3 branches; flowering time January–April:
  Anthers dark purplish-maroon; leaves equalling or exceeding the flower at anthesis; corm
    tunic splitting into parallel fibres for most of its length   -   **2. C. hartmannianus**
  Anthers yellow; leaves only shortly developed at anthesis, sometimes reaching the base of the
    flower; corm tunic largely membranous, splitting into parallel fibres only near the base,
    and basally circumscissile, producing horizontal rings-   -   -   -   **1. C. cyprius**

* By B. F. Mathew.

1. C. cyprius *Boiss. et Kotschy* in Unger et Kotschy, Die Insel Cypern, 203 (1865); Boiss., Fl. Orient., 5: 114 (1882); Holmboe, Veg. Cypr., 55 (1914); Druce in Proc. Linn. Soc., 141st Session, 52 (1930); Mountfort in Quart. Bull. Alpine Gard. Soc., 16: 115 (1948); Osorio-Tafall et Seraphim, List Vasc. Plants Cyprus, 24 (1973); Elektra Megaw, Wild Flowers of Cyprus, 14, t. 27 (1973).

    C. vernus Sibth. et Sm. (non Hill), Fl. Graec. Prodr., 1: 24 (1806); Poech, Enum. Plant. Ins. Cypr., 10 (1842); Unger et Kotschy, Die Insel Cypern, 203 (1865).

    C. aerius Herb. var. *cyprius* (Boiss. et Kotschy) Baker in Journ. Linn. Soc. Bot., 16: 83 (1877).

    TYPE: Cyprus; "Bei Prodromo im Aufsteigen gegen den Troodos am Schnee 5000′, 5 April 1859" *Kotschy* 257 (K!).

Cormous perennial herb up to 10 cm. in overall height at anthesis; corm ovoid, flattened at the base, about 0·7–1 cm. in diameter; tunics rather papery, splitting lengthways into strips and circumscissile at the base forming fragile rings of tissue; cataphylls 3, papery, white; leaves 3–4, synanthous, equalling the flower or with the tips only just visible at anthesis, 1–2 mm. wide, grey-green with a narrow median white stripe on the upper surface, glabrous; flowers 1–2, fragrant, ground colour white or lilac, heavily stained violet on the exterior, especially towards the base of the segments; throat deep yellow, pubescent; prophyll absent; bract and bracteole present, subequal, membranous, silvery-white or brownish; perianth-tube (2·5–) 3–7 cm. long; segments more or less equal, 1·5–3·5 cm. long, (0·4–) 0·6–1 cm. wide, oblanceolate or elliptic, obtuse; filaments 4–7 mm. long, yellow-orange, papillose; anthers 7–12 mm. long, yellow; style shorter than, equalling or exceeding the anthers, divided into 3 apically expanded reddish-orange branches; capsule ellipsoid c. 1–1·5 cm. long, 0·7 cm. wide, carried at ground level at maturity. 2n = 10. *Plate 91, figs. 1–5.*

    HAB.: Open rocky slopes, flowering near melting snow, sometimes in *Berberis* or *Juniperus* scrub or beneath *Pinus*; 3,500–6,400 ft. alt.; fl. (Jan.–) Febr.–April.

    DISTR.: Division 2, locally abundant. Endemic.

    2. Khionistra, 3,500–6,400 ft. alt., locally abundant near the summit, also about Troödos, Kannoures Springs, Prodhromos, Amiandos, down to Platania, 1787–1978, *Sibthorp*; *Kotschy* 257! 772; *Druce* s.n.! *C. B. Ussher* 99! *Syngrassides* 899! *Kennedy* 242–250! 252–258! *Lady Loch* 61! *F. M. Probyn* 100! *Mavromoustakis* 65! *P. H. Oswald* 66! *C. E. H. Sparrow* 54! *N. Macdonald* 47! *P. Synge* s.n.! *Osorio-Tafall et Seraphim* s.n.! *J. J. Wood* 130! *J. Holub* s.n.!

2. C. hartmannianus *Holmboe*, Veg. Cypr., 54 (1914); Mountfort in Quart. Bull. Alpine Gard. Soc., 16: 115 (1948); Osorio-Tafall et Seraphim, List Vasc. Plants Cyprus, 24 (1973); Elektra Megaw, Wild Flowers of Cyprus, 14, t. 27 (1973).

    TYPE: Cyprus; "northern slopes of Kionia on open places in the forest above the monastery of Makhaeras, ca. 880 m." 16 Febr. 1905, *E. Hartmann* (O).

Cormous perennial herb up to 15 cm. in overall height at anthesis; corm ovoid, flattened at the base, about 1 cm. in diameter; tunics papery, splitting into many parallel fibres; cataphylls 3, papery, white; leaves 3–4 (–5), synanthous, equalling or overtopping the flower at anthesis, 1–1·5 mm. wide, grey-green with a narrow median white stripe on the upper surface, glabrous; flowers 1–2, with a lilac or white ground colour, heavily stained or sometimes striped violet on the exterior; throat orange-yellow, finely pubescent or glabrous; prophyll absent; bract and bracteole present, subequal, membranous, brownish, usually speckled darker; perianth-tube c. 4–7·5 cm. long; segments more or less equal, 2·2–2·7 cm. long, 0·7–0·9 cm. wide, oblanceolate, obtuse; filaments 5–7 mm. long, orange-yellow, papillose or glabrous; anthers 7–11 mm. long, dark purplish-maroon; style equalling or exceeding the anthers, divided into 3 reddish-orange branches. Capsule not seen. 2n = 20. *Plate 91, fig. 6.*

    HAB.: Dry stony slopes in *Cistus* scrub or sparse *Pinus brutia* woodland; 2,400–3,000 ft. alt.; fl. Jan.–Febr.

Plate 91. Figs. **1–5.** CROCUS CYPRIUS Kotschy **1,** habit, × ⅔; **2,** leaves and fruit, × ⅔; **3,** flower opened out, × ⅔; **4,** fruit, × 2; **5,** seed, × 2; fig. **6.** CROCUS HARTMANNIANUS Holmboe habit, × ⅔; fig. **7.** CROCUS VENERIS Tappeiner flower opened out, × ⅔. (**1, 3** from *Kennedy* 246; **2, 4, 5** from *Kennedy* 255; **6** from *Lady Loch* 60; **7** from *Kennedy* 1259.)

DISTR.: Divisions 2, 7, rare. Endemic.

2. Kionia, above Makheras Monastery, 1905, *Hartmann* s.n.; also, 1939, *Lady Loch* 60! and, 1940, *Kennedy* 1573! Noted from three localities in this area, 1972, *Osorio-Tafall & Seraphim.*
7. "Peak to the west of Trypa Vouno, Kyrenia Range; 2,400 ft. alt.; on north-facing slopes below the summit: in pockets of stiff, red clay overlying hard limestone under light *Pinus brutia* woodland, growing through thin fine litter. Very local and uncommon'', 30 Jan. 1956, *Merton* 2475!

NOTES: The plant from the Kyrenia Range differs from the rest of the material in having a glabrous throat and filaments, and striped flowers. Its occurrence on this limestone mountain, far from its type area, is most surprising, but is very precisely recorded. Further investigation into the Northern Range populations is clearly desirable.

3. C. veneris *Tappeiner* in Poech, Enum. Plant. Ins. Cypr., 10 (1842); Kotschy in Oesterr. Bot. Zeitschr., 12: 278 (1862); Unger et Kotschy, Die Insel Cypern, 203 (1865); Boiss., Fl. Orient., 5: 109 (1882) pro parte; Holmboe, Veg. Cypr.: 55 (1914); Mountfort in Quart. Bull. Alpine Gard. Soc., 16: 115 (1948); Seligman in Quart. Bull. Alpine Gard. Soc., 20: 233 (1952); B. L. Burtt in Kew Bull., 1954: 71 (1954); Osorio-Tafall et Seraphim, List Vasc. Plants Cyprus, 24 (1973); Elektra Megaw, Wild Flowers of Cyprus, 14, t. 27 (1973).
TYPE: Cyprus; "in collibus calcareis prope Paphos", 1840, *Kotschy* (?W).

Cormous perennial herb about 4–8 cm. in overall height at anthesis; corm ovoid, about 6–10 mm. in diameter; tunics membranous with parallel fibres; cataphylls usually 3, papery, white; leaves 3–4, synanthous, usually equalling the flower at anthesis but occasionally with only the tips showing, 0·5–1 mm. wide, dark green with a narrow silvery median stripe on the upper surface, glabrous or slightly scabrid on the margins; flowers 1–2 (–3), fragrant, white, usually with a violet stripe or feathering on the outside of the outer three segments; throat yellow pubescent; prophyll absent; bract and bracteole present, subequal, white; perianth-tube c. 3–4 (–5) cm. long, white; segments more or less equal, 1·4–2·5 (–3) cm. long, 0·3–0·5 cm. wide, oblanceolate, acute; filaments 3–5 mm. long, yellow, pubescent; anthers 6–8 mm. long, yellow; style usually slightly exceeding or sometimes equalling the stamens, divided into at least 6, and usually many more, yellow or orange thread-like branches. Capsule ellipsoid-subglobose, c. 5 mm. long, 4 mm. wide, carried above ground on a short pedicel at maturity. 2n = 16. *Plate 91, fig. 7.*

HAB.: Stony and grassy places, in garigue, maquis or open conifer woods; 350–3,000 ft. alt.; fl. Nov. (–Jan.).

DISTR.: Divisions 1, 2, 4, 5, 7, 8. Endemic.

1. Between Ktima and the sea, 1840, *Kotschy*; Polemi, 1937, *Lady Loch* 11! Also noted from about Ayios Neophytos Monastery and Stroumbi, 1972, *Osorio-Tafall & Seraphim.*
2. Aphamis, 1938, *Kennedy* 1259! Also from near Sarandi, 1972, *Osorio-Tafall & Seraphim*; Ayia Moni near Statos, 1982, *Elektra Megaw* s.n.!
4. Pergamos, 1954, *Merton* 1820!
5. Near Trikomo, 1972, *Osorio-Tafall & Seraphim.*
7. Frequent, especially on north-facing slopes, along the Northern Range from Kornos Peak to Antiphonitis Monastery, 1940–1973, *Davis* 2009! *Casey* 1061! *Osorio-Tafall & Seraphim* 8014! 8217!
8. On gypsum near Gastria, 1972, *Osorio-Tafall & Seraphim.*

3. ROMULEA *Maratti**
Plant. Rom. Sat., 13 (1772) nom. cons.

Small winter- and spring-flowering plants; corm asymmetrical with a basal ridge on one side; tunic brown, hard, brittle, smooth, with apical teeth and a basal fringe of fibres; basal leaves 2 per corm, distichous, compressed-terete, 4-grooved; cauline leaves similar usually shorter; scape not normally

---

* By W. Marais.

produced above ground at anthesis; pedicels semiterete; bract navicular, herbaceous; bracteole herbaceous or membranous; flowers actinomorphic, gamopetalous; style filiform with 3 deeply bifid branches. Capsule borne on a ± elongated scape, enclosed in the persistent bract and bracteole; seeds globose or angled, testa brown.

About 70 species in South Africa where two subgenera are distinguished on flower-shape, and nine sections on the type of corm. About 4 species from high altitudes in tropical East Africa and Arabia. No review of the Mediterranean and European species has been carried out since A. Béguinot, 1907–09, Revisione Monografica del genere *Romulea* Maratti, in Malpighia, 21: 49–122, 364–478; 22: 377–469; 23: 55–117, 185–239, 257–296, but there are certainly far fewer species than the 29 recognized in this work.

Perianth deep violet (rarely some pure white individuals in a population), 2·1–3·7 cm. long; tube 8·5–17 mm. long - - - - - - - - - - **1. R. tempskyana**
Perianth white with dark veins to deep lilac-blue, or greenish-yellow flushed violet, 0·9–1·8 cm. long; tube 2·5–4 mm. long:
Bract and bracteole largely herbaceous, fairly rigid, closely and conspicuously veined
2. R. ramiflora ssp. **ramiflora**
Bract herbaceous; bracteole with wide membranous margins, not conspicuously veined
3. R. columnae ssp. **columnae**

1. R. tempskyana *Freyn* in Bull. Herb. Boiss., 5: 798 (1897); H. Stuart Thompson in Journ. Bot., 44: 339 (1906); Holmboe, Veg. Cypr., 55 (1914); Osorio-Tafall et Seraphim, List Vasc. Plants Cyprus, 24 (1973); Elektra Megaw, Wild Flowers of Cyprus, 14, t. 27 (1973).
TYPE: Cyprus; "in peninsula Karpas die 7. januario 1894 leg. Deschamps (exs. 465)" (? BRNM).

Leaves 3–6, 8–16 cm. long, 1–1·25 mm. wide, often recurved and adpressed to the ground; scape not produced at anthesis, 1–6-flowered; pedicels 2–5 cm. long; bract 16–22 mm. long, herbaceous with narrow membranous margins which are speckled with reddish-brown, the whole bract sometimes tinged reddish or violet; bracteole almost entirely membranous, brownish or whitish, speckled with reddish-brown; perianth deep violet, rarely pure white, always with a yellow throat, 2·1–3·7 cm. long; tube 8·5–17 mm. long, narrow; segments lanceolate to ovate, acute or obtuse, the outer often greenish on the back; filaments glabrous; top of anthers reaching about half way up the perianth-segments; style purple; stigmas paler, held just above or much above the tops of the anthers. Ripe capsule not seen.

HAB.: In garigue, grassland, coastal maquis, open glades in Pine forests or on limestone or igneous hillsides; sea-level to 4,000 ft. alt.; fl. Jan. –Febr. (–March).

DISTR.: Divisions 2–8. Also Aegean Islands, S. Turkey, Palestine.

2. Between Saïttas and Trimiklini, 1932, *Syngrassides* 1343! Kryos Potamos, 3,000–4,000 ft. alt., 1937, *Kennedy* 260! 261! 262! Platres, 1937, *Kennedy* 259!
3. Curium, 1964, *J. B. Suart* 136! 137!
4. Near Larnaca, 1880, *Sintenis & Rigo* 163!
5. Athalassa, 1967–1969, *Economides* in ARI 1157! 1158 (white flowers)! 1304! 1305 (white flowers)!
6. Between Orga and Kormakiti, 1941, *Davis* 2103! Makhedonitissa Monastery, near Nicosia, 1941, *Davis* 2128! 2129 (white flowers)! Nicosia, 1973, *P. Laukkonen* 11a!
7. Mountains above Kythrea, 1880, *Sintenis & Rigo* 163a! Halevga, 1932, *Syngrassides* 59! Kyrenia Pass, 1935–55, *Syngrassides* 809! *Lady Loch* 21! *G. E. Atherton* 792! Kyrenia, 1938–1973, *C. H. Wyatt* 18! *G. E. Atherton* 808! 837! 975! *O. Huovila* 62! *P. Laukkonen* 11! Above Sisklipos, 1940, *Davis* 2118! Drako Vrysi near Karmi, 1949, *Casey* 213! Near Kantara, 1939, *Lady Loch* 65! 66! Pentadaktylos, 1950, *Casey* 981! and, 1956, *G. E. Atherton* 1061! Near Myrtou, 1956, *Merton* 2464!
8. Karpas peninsula, 1894–1937, *Deschamps* 465! *H. Roger-Smith* s.n.! *Mrs Stagg* s.n.!

2. R. ramiflora *Ten.*, App. Ind. Sem. Hort. Neap., 3 (1827) et in Atti Reale Accad. Sci. Napoli, 3: 113–119, t. 7 (1832); Boiss., Fl. Orient., 5: 117 (1882); Holmboe, Veg. Cypr., 55 (1914); Osorio-Tafall et Seraphim, List Vasc. Plants Cyprus, 24 (1973).

Leaves 3–6 (–8), 6–30 cm. long, 1–2 mm. wide, often recurved, 2 basal, 1 cauline low on the scape, the others apical, scape elongating to 30 cm. after anthesis, 1–4 (–6)-flowered; pedicels to 30 cm. long in fruit; bract 12–20 mm. long, sometimes longer than perianth, herbaceous, strongly veined; bracteole with narrow or wide hyaline margins, mostly herbaceous; perianth pale to deep lilac-blue, pink or greenish-yellow flushed violet, throat orange, 1–3 cm. long; tube 2·5–4·5 mm. long; segments oblanceolate, acute; filaments glabrous or papillose at the base; top of anthers reaching about two-thirds way up the perianth-segments; stigmas held just below or just above the top of the anthers, whitish. Capsule cylindrical to clavate, very variable in size.

ssp. ramiflora

TYPE: Italy; cult. Naples Bot. Gard. from material collected *"in pascuis sterilibus prope Neapolim, in loco vulgo* Strada del Campo *prope Aediculam divae Mariae del Pianto"* (NAP).

Leaves 6–30 cm. long, 1–1·5 mm. wide; bracteole with a narrow hyaline margin; perianth 10–18 mm. long, pale to deep lilac-blue or greenish-yellow, flushed violet, throat orange; stigmas held just below the tops of the anthers.

HAB.: In grassland, along water channels, in wet and marshy places; water meadows; 250–3,300 ft. alt.; fl. (Febr.–) March.

DISTR.: Divisions 2, 5–7. Widespread in the Mediterranean region.

2. Kryos Potamos, 3,000–3,300 ft. alt., 1937, 1939, *Kennedy* 263 ! 1339 partly !
5. By the bridge before Pyroi, 1936, *Syngrassides* 1077 ! Dhikomo marshes, 1936, *Syngrassides* 1190 !
6. Peristerona (Morphou), 1937, *Syngrassides* 1563 !
7. Near Ayios Khrysostomos Monastery, 1880, *Sintenis & Rigo* 162 ! Kambyli, 1956–1957, *Merton* 2524 ! 2979 !

3. R. columnae *Seb. et Mauri*, Fl. Rom. Prodr., 18 (1818); Boiss., Fl. Orient., 5: 112 (1882); Holmboe, Veg. Cypr., 55 (1914); Post, Fl. Pal., ed. 2, 2: 587 (1933); Osorio-Tafall et Seraphim, List Vasc. Plants Cypr., 24 (1973).
    Trichonema columnae (Seb. et Mauri) Reichb., Fl. Germ. Excurs., 83 (1830); Unger et Kotschy, Die Insel Cypern, 203 (1865); Sintenis in Oesterr. Bot. Zeitschr., 32: 54 (1882) quoad nomen.

Leaves 3–8, up to 30 cm. long, 0·3–1 mm. wide, erect or recurved; scape not produced above ground at anthesis; pedicels short; bract 6–13 mm. long, herbaceous, often tinged purple or speckled red-brown; bracteole almost wholly membranous, usually densely speckled; perianth pale lilac or pale violet to white, usually dark-veined, 9–15 mm. long; tube 2·5–4 mm. long, the throat usually yellow; segments lanceolate to oblanceolate, acute; filaments papillose or glabrous; top of anthers reaching about half way up the segments; stigmas held below the top of the anthers. Fruiting scape to 5 cm. above ground. Capsule subglobose, triquetrous.

ssp. columnae

TYPE: Italy; "In collibus etiam urbis. *Sul Testaccio e sul Gianicolo in copia, presso Tor-di-Quinto".*

Leaves up to 7 cm. long, 0·6–1 mm. wide; filaments frequently glabrous.

HAB.: On dry banks amongst short grass or on grassy ground overlying rocks; 800–3,300 ft. alt.; fl. late Jan. to early April.

DISTR.: Divisions 2, 7. Mediterranean region and Atlantic coasts as far north as Great Britain; Atlantic Islands.

2. Platres, 1939, *Kennedy* 1339, partly ! Asinou, 1953, *Kennedy* 1773 !
7. Above Karmi, 2,600 ft. alt., *Casey* 377 !

NOTES: Also recorded by Kotschy (Die Insel Cypern, 203; 1865) from the neighbourhood of (2) Prodhromos, 1862, no. 813, and from above (7) Ayios Khrysostomos Monastery, 1859, no. 982. The identity of the plants recorded must be questioned, in view of the fact that the much commoner *R. tempskyana* Freyn was not known to Kotschy or his contemporaries, who frequently attached the epithet *columnae* to plants of this latter species, or to *R. ramiflora* Ten.

4. GLADIOLUS *L.*

Sp. Plant., ed. 1, 36 (1753).
Gen. Plant., ed. 5, 23 (1754).

Perennial herbs arising from corms, the latter usually clothed with a fibrous tunic; stem generally well developed, mostly erect with 2–4 basal sheaths and 1 or more ligulate or ensiform, sessile, sheathing cauline leaves, the upper often much reduced and bract-like; inflorescence terminal, spiciform, usually simple; flowers often showy, frequently secund, sessile or shortly pedicellate; bracts and bracteoles generally well developed and persistent; perianth zygomorphic, hooded or open and funnel-shaped, usually curved, segments 6, in 2 series, united basally into a short tube, free and porrect above, generally with a narrow basal claw and an expanded limb, the abaxial segments often blotched or streaked; stamens 3, usually lying under the adaxial perianth-segments, inserted at apex of perianth-tube; anthers extrorse, linear, sharply sagittate at base; style filiform, terete, divided above into 3 relatively short, apically expanded or subspathulate, stigmatiferous branches. Capsule oblong-subglobose or cylindrical, loculicidally 3-locular, often membranous or scarious; seeds numerous, often flattened and/or winged.

About 300 species, mostly in tropical and South Africa, with about 12 species in S. Europe, western Asia, North Africa and Atlantic Islands. Several species (and hybrids) are popular as cultivated ornamentals.

Anthers longer than filaments or equalling them; plant 30–130 cm. high; flowers vivid magenta-pink; lowermost bract often 6 cm. long or longer; perianth-segments 3–4 cm. long
1. G. italicus

Anthers shorter than filaments; plant 15–50 cm. high; flowers pale or dark rose-pink; lowermost bract not more than 3 cm. long; perianth-segments 2·5–3 cm. long - **2. G. triphyllus**

1. **G. italicus** *Mill.*, Gard. Dict., ed. 8, no. 2 (1768); Osorio-Tafall et Seraphim, List Vasc. Plants Cyprus, 24 (1973).
G. segetum Ker Gawler in Bot. Mag., 19: t. 719 (1804); Unger et Kotschy, Die Insel Cypern, 202 (1865); Boiss., Fl. Orient., 5: 139 (1882); Holmboe, Veg. Cypr., 84 (1914); Post, Fl. Pal., ed. 2, 2: 603 (1933).
TYPE: not indicated.

Erect perennial 30–130 cm. high; corm broadly ovoid, 2–3 cm. diam., tunic brownish, reticulately fibrous; basal sheath 10–18 cm. long, tubular for the greater part of the length, apex subacute, 1–1·5 cm. long, slightly reflexed, leaves usually 4–5, linear, the lowermost 10–30 cm. long, 0·6–1·4 cm. wide, distinctly 3-nerved, with rather obscure secondary nerves, apex shortly acute, base sheathing for 5 cm. or more, middle leaves much longer, sometimes exceeding 30 cm., with a finely tapering acuminate apex, uppermost leaf often less than 15 cm. long, 0·5 cm. wide, with a tapering apex and long-sheathing base; inflorescence a lax spike at first about 15 cm. long, lengthening to 30 cm. or more with age; rhachis subterete, glabrous; flowers 6–12 (–17), rather remote, sessile; bracts linear-subulate, herbaceous, the lowermost frequently 6 cm. or more in length, about 0·5 cm. wide at the base, the upper becoming progressively shorter; bracteoles similar to bracts but distinctly smaller, seldom more than 2·5 cm. long; perianth bright magenta-pink, the 3 abaxial segments with paler, creamy median stripes, tube narrowly infundibuliform, generally less than 1 cm. long, segments oblanceolate, 3–4 cm. long, 0·6–1 cm. wide, shortly acute, or obtuse, with a tapering, shortly clawed base, the adaxial segment generally longer than the laterals or abaxial, and generally rather remote from them; filaments 5–8 mm. long, glabrous; anthers linear, yellow, 14–18 mm. long, about 2 mm. wide; ovary shortly ovoid-oblong, about 5 mm. long, 4 mm. wide, glabrous; style slender, about 3 cm. long; stigmas about 2·5 mm. long, apex dilated-subspathulate. Capsule subglobose or very broadly ellipsoid, 0·8–1·5 cm. long, 0·8–1·2 cm. wide, rigidly scarious, distinctly furrowed along the septa, pale brown, subtended by the persistent bracts and bracteoles; seeds subglobose or irregularly angled, about 2 mm. diam.; testa dull reddish-brown, rugulose.

HAB.: In cereal fields; occasionally in garigue on dry rocky hillsides or in clefts of limestone rocks; sea-level to 4,500 ft. alt.; fl. March–April.

DISTR.: Divisions 1–8, locally abundant; S. Europe and Mediterranean region eastwards to Central Asia.

1. Near Polis, 1962, *Meikle* 2269!
2. Near Prodhromos, 1862, *Kotschy* 748! Lefka, 1932, *Syngrassides* 253! Platres, 1937, *Kennedy* 241!
3. Foot of Stavrovouni, 1787, *Sibthorp* teste *Kotschy*; Mazotos, 1905, *Holmboe* 187; 2 m. S. of Pyrgos (Limassol), 1963, *J. B. Suart* 42! 67!
4. Near Larnaca, 1880, *Sintenis & Rigo* 948! Kondea, 1933, *Nattrass* 301; Between Ayia Napa and Paralimni, 1958, *N. Macdonald* 46! Cape Greco, 1962, *Meikle* 2595!
5. Kythrea, 1880, *Sintenis & Rigo*; Lakatamia, 1932, *Syngrassides* 368! Geunyeli, 1956, *G. E. Atherton* 1277! Strongylos, 1967, *Merton* in ARI 440!
6. Nicosia, 1927, *Rev. A. Huddle* 95! Ayia Irini, 1941, *Davis* 2586! Dhiorios, 1962, *Meikle*! Near Liveras, 1974, *Meikle*!
7. Myrtou, 1934, *Nattrass* 493; also, 1970, *I. Hecker* 52! Kantara, 1955, *Mrs J. Hughes* s.n. (partly)!
8. Near Akradhes, 1962, *Meikle*!

2. G. triphyllus (*Sm.*) *Ker-Gawler* in Bot. Mag., 25: t. 992 (1807); Unger et Kotschy, Die Insel Cypern, 203 (1865); Boiss., Fl. Orient., 5: 141 (1882); Holmboe, Veg. Cypr., 54 (1914); Osorio-Tafall et Seraphim, List Vasc. Plants Cyprus, 24 (1973); Elektra Megaw, Wild Flowers of Cyprus, 14, t. 26 (1973).

G. *communis* L. var. *triphyllus* Sm. in Sibth. et Sm., Fl. Graec. Prodr., 1: 25 (1806); Poech, Enum. Plant. Ins. Cypr., 10 (1842).

G. *communis* L. *varietas* Sm. in Sibth. et Sm., Fl. Graec., 1: 28, t. 38 (1806).

G. *trichophyllus* Sintenis in Oesterr. Bot. Zeitschr., 32: 365, 398 (1882) nomen.

TYPE: "In Cypri campestribus ad meridiem montis *Troodos* dicti, solo fertiliore. *D. F. Bauer*". (OXF, K!)

Slender erect perennial 15–50 cm. high; corm broadly ovoid, 1·5–2·5 (–3) cm. long, 0·8–1·5 (–2) cm. wide, tunic brown, coarsely reticulate-fibrous; basal sheath 2·5–7 cm. long, tubular for the greater part of its length, apex subacute, 0·3–1 cm. long, erect or slightly reflexed; leaves usually 3–4, linear, the 2 lowermost 10–30 cm. long, less than 0·5 cm. wide, distinctly 3–4-nerved, with obscure secondary nerves, grey-green or dull green, tapering to an acute apex, the upper leaves generally much smaller, the uppermost often bract-like, about 5 cm. long and less than 0·3 cm. wide, with a tapering apex and a long sheathing base; inflorescence a lax spike, at first 2–10 cm. long, lengthening to 15 cm. or more with age; rhachis subterete, glaucous or purplish, glabrous; flowers 1–7, rather remote, sessile, secund, sweet-smelling towards dusk; bracts linear-acuminate, 1·5–3 cm. long, usually less than 0·5 cm. wide, glabrous, often glaucous, sometimes with a membranous margin; bracteoles similar to bracts, but usually less than 1·5 cm. long; perianth pale or dark rose-pink, the 3 abaxial segments paler than the adaxial, with pale, whitish, median stripes, tube narrowly infundibuliform, 0·7–1·3 cm. long, segments oblanceolate, 2·5–3 cm. long, 0·7–1 cm. wide, acute, obtuse or rounded, with a tapering, distinctly or indistinctly clawed base, the adaxial segment usually rather longer than the laterals or abaxial, and generally (but not always) remote from them; filaments 10–15 mm. long, glabrous; anthers linear, yellow, 6–8 mm. long, about 1–1·5 mm. wide;

LRM

Plate 92. GLADIOLUS TRIPHYLLUS Sm. **1,** habit, ×1; **2,** fruiting stem, ×1; **3,** seed, ×12; **4,** androecium and gynoecium, ×3; **5,** flower, front view, ×1½. (**1, 4, 5** from *Meikle* 2440 and *Merton* 621; **2, 3** from *Davis* 1767.)

ovary broadly ovoid or subglobose, 3–4 mm. long, 2·5–3 mm. wide, glabrous; style slender, 2–2·5 cm. long; stigmas about 2–2·5 mm. long, apex dilated-subspathulate. Capsule oblong-obovate, about 1 cm. long, 0·8 cm. wide, rigidly scarious, rugulose, distinctly furrowed along the septa, pale brown, subtended by the persistent bracts and bracteoles; seeds subglobose, about 2 mm. diam.; testa reddish-brown, rugulose. *Plate 92.*

HAB.: In open Pine forest or in garigue or maquis on dry, rocky limestone or igneous hillsides; sea-level to 4,000 ft. alt.; fl. March–May.

DISTR.: Divisions 1–4, 7, 8. Endemic.

1. Frequent; Polis, Paphos, Ayios Yeoryios (Akamas), Baths of Aphrodite, Fontana Amorosa, etc. 1905–1978, *Holmboe; Druce* s.n. ! *E. Megaw* s.n. ! *Davis* 3339 ! *Merton* 3045 ! *Meikle* 2016 ! 2145 ! 2292 ! *J. J. Wood* 81 ! *Edmondson & McClintock* E2738 ! *Chesterman* 61 ! etc.
2. Southern slopes of Troödos range, 1787, *Bauer* ! Between Ayia and Stavros tis Psokas, 1937, *Syngrassides* 1597 ! Platres, 1937–38, *Kennedy* 239 ! 240 ! Near Dhodheka Anemi Pass, Tripylos, 1940, *Davis* 1767 ! and between Tripylos and Stavros tis Psokas, 1941, *Davis* 3482 ! Kambos, 1952, *F. M. Probyn* 109 ! Yialia, 1955, *Merton* 2120 ! Road from Kykko Monastery to Pomos, 1979, *Hewer* 4661 !
3. Between Pissouri and Kouklia, 1862, *Kotschy*; Cape Gata, 1905, *Holmboe* 671; Episkopi, 1954, *Merton* 1890 ! Paramali, 1959, *C. E. H. Sparrow* 10 ! Kouklia, 1960, *N. Macdonald* 53 ! Kellaki-Limassol road, 1967, *Merton* in ARI 621 !
4. About 3 m. S.E. of Goshi, 1974, *Meikle* 4092 !
7. Between Melandrina and Antiphonitis Church, 1862, *Kotschy* 531 ! St. Hilarion, 1931, *J. A. Tracey* 4 ! Kornos, above Larnaka tis Lapithou, 1941, *Davis* 3036 !
8. Near Aphendrika (Epiotissa), 1880, *Sintenis & Rigo* 164 ! Ayios Bakos near Yialousa, 1962, *Meikle* 2440 !

NOTES: On April 13, 1787, Sibthorp noted in his diary (Walpole, Travels, 15; 1820) "At eight we left the convent [Stavrovouni Monastery]; the Pinus pinea was less frequent as we advanced in our descent. I observed a new species of Gladiolus, G. montanus". The "new species" is usually referred to *G. italicus* (*G. segetum*), but this was surely a familiar species to Sibthorp, even if he knew it by a different name, and "*G. montanus*" is more likely to be an unpublished synonym of *G. triphyllus*, later drawn by Bauer from a plant found on the southern slopes of Troödos. The name *Gladiolus montanus* had already been given to a South African species by Linné f. (Suppl. Plant., 95; 1781).

Despite frequent rumours, there has, as yet, been no satisfactory evidence for the occurrence of *G. triphyllus* outside Cyprus. It is clearly related to *G. italicus*, but in the living state is readily identified by its slender habit, paler flowers, and, towards evening, by its sweet Carnation-like scent.

95. AMARYLLIDACEAE

Bulbous or rhizomatous perennials; leaves all basal, often distichous, linear or lorate, frequently rather fleshy; inflorescence scapose, the flowers frequently in umbels, or sometimes solitary, generally subtended by 1–2 (or more) membranous bracts or spathes, hermaphrodite, actinomorphic or occasionally zygomorphic; perianth inserted on the apex of the ovary, with or without a basal tube; segments 6, the 3 outer and 3 inner similar or distinct; corona sometimes present; stamens 6, opposite the perianth segments, inserted at the apex of the perianth-tube or near the base of the perianth-segments; filaments free or connate at the base; anthers basifixed or versatile, usually introrse; ovary inferior, 3-locular (or rarely 1-locular by abortion), with axile (or rarely parietal) placentation; style usually slender; stigma capitate or 3-lobed; ovules generally numerous, anatropous. Fruit a loculicidal capsule, or irregularly dehiscent and berry-like; seeds usually numerous, sometimes compressed or flattened; endosperm fleshy; embryo small, straight; vegetative reproduction by bulbils not uncommon.

About 85 genera and more than 1,000 species, chiefly in tropical and subtropical areas. The family includes many showy species, popular as garden or house plants.

Filaments free from corona; anthers included in corona, or at most very shortly exserted
                                                                1. Narcissus
Filaments adnate to corona; anthers distinctly exserted from corona   -   -   **2. Pancratium**

1. NARCISSUS *L.*

Sp. Plant., ed. 1, 289 (1753).
Gen. Plant., ed. 5, 141 (1754).

Perennials with tunicate bulbs; leaves linear or lorate, appearing with or after the flowers; scape frequently compressed; bract (or spathe) solitary, split laterally, papery or membranous; flowers solitary or several in irregular, frequently subsecund, umbels; perianth-tube short or long, often greenish, segments patent or reflexed, subequal, white or yellow; corona generally well developed, erect or spreading, with an entire, crenate-undulate or shortly lobulate, sometimes reflexed, margin; stamens inserted at the apex, middle or base of the perianth-tube, usually included in corona, sometimes exserted and declinate; filaments short or long, free from corona; anthers mostly linear or narrowly oblong, medifixed and versatile or subbasifixed; ovary 3-locular; style often long and slender; stigma obscurely or conspicuously 3-lobed. Capsule trigonous, 3-locular; pericarp dry, subcoriaceous; seeds subglobose or variously angled; testa generally blackish.

About 30 species, mostly in western Europe, especially the Iberian peninsula. Many of the species are ornamental, and some sweetly scented, and the longstanding popularity of the genus in horticulture has undoubtedly added very considerably to its taxonomic complexity.

Corona minute, of 6 semicircular lobes; leaves very slender, filiform, usually developing after anthesis; scape with transverse articulations; flowers appearing in autumn
                                                                2. N. serotinus
Corona cup-shaped, subentire; leaves linear or lorate, developing with the flowers in winter or spring; scape without articulations:
  Leaves glaucescent, 4–8 mm. wide:
    Umbels 2–5 (–6)-flowered; perianth white, corona yellow   -   -   -   **1. N. tazetta**
    Umbels usually more than 6-flowered; perianth and corona white
                                                                N. PAPYRACEUS (p. 1574)
  Leaves dark green, 3–4 mm. wide; flowers uniformly bright yellow   N. JONQUILLA (p. 1576)

1. N. tazetta *L.*, Sp. Plant., ed. 1, 290 (1753); Sibth. et Sm., Fl. Graec., 4: 7, t. 307 (1823); Boiss., Fl. Orient., 5: 150 (1882); Post, Fl. Pal., ed. 2, 2: 608 (1933); Elektra Megaw, Wild Flowers of Cyprus, 15, t. 29 (1973).
    Hermione cypri Haworth in Phil. Mag., n.s., 9: 184 (March 1831); Unger et Kotschy, Die Insel Cypern, 205 (1865).
    Narcissus cypri (Haworth) Sweet, Brit. Fl. Gard., 4: t. 92 (April 1831).
    N. tazetta L. var. *cypri* (Haworth) Boiss., Fl. Orient., 5: 151 (1882).
    N. tazetta L. ssp. *cypri* (Haworth) Holmboe, Veg. Cypr., 53 (1914); Osorio-Tafall et Seraphim, List Vasc. Plants Cyprus, 23 (1973).
    TYPE: "*in* Galliae Narbonensis, Lusitaniae, Hispaniae *maritimis*".

Bulb ovoid, 3–4 cm. long, 2–3·5 cm. wide, covered with a thin, papery, dark brown, dull or shining tunic; basal sheaths cylindrical, 5–7 cm. long, 0·8–1 cm. wide, membranous with a truncate or oblique apex; leaves linear or lorate, rather fleshy, glaucescent, 13–30 cm. long, 0·4–0·8 cm. wide at anthesis, elongating, sometimes to 50 cm. or more after anthesis, apex obtuse; nervation obscure; scapes 12–30 cm. long at anthesis, fistulose or solid, often distinctly compressed, spathe papery, 3–3·5 cm. long, 0·6–1 cm.

wide, pale brownish, usually tubular towards base, unilaterally split above, apex acute; flowers 2–5 (–6) in our area, sometimes 12–17 or more in cultivation, sweet-scented, forming a loose, subsecund umbel; pedicels angular, glabrous, 1·5–3·5 cm. long at anthesis, lengthening to 5–7 cm. in fruit; ovary oblong-ovoid, about 5–6 mm. long, 3–4 mm. wide, perianth-tube greenish, narrow, about 2 cm. long, 0·5 cm. wide at apex; perianth-segments white, spreading or reflexed, ovate-oblong, about 1–1·7 cm. long, 0·6–1 cm. wide, obtuse or subacute with a mucronate or shortly cuspidate apex; corona yellow or tinged orange, cup-shaped, about 3–4 mm. long, 7–8 mm. diam., entire or irregularly lobulate, sometimes shallowly plicate; stamens inserted in 2 series near the apex of the perianth-tube; filaments very short, not exceeding 2 mm.; anthers included, medifixed, linear, 7–8 mm. long, about 1·5 mm. wide; style about 2 cm. long, included in corona, apex shortly 3-lobed. Capsule oblong-trigonous, 1–1·5 cm. long, 0·8 cm. wide, pericarp rather papery, pale brown, transversely rugulose; seeds irregularly angular and compressed, about 3 mm. long, 2 mm. wide; testa black, minutely rugulose. *Plate 93, figs. 1–3a.*

HAB.: In clefts of limestone rocks, or in garigue on limestone; formerly also common as a weed in cereal fields but now rarer; sea-level to 2,500 ft. alt.; fl. Nov.–Febr.

DISTR.: Divisions 3–7. Widespread in the Mediterranean region, and, as a cultivated plant, throughout temperate regions.

3. Limassol, 1801, *Hume.*
4. Larnaca, 1801, *Hume,* and, 1880, *Sintenis & Rigo* 857!
5. Dhikomo, 1938, and Trakhonas, 1938 ("Common in several fields"). *Syngrassides* 1745! Angastina, Exo Metokhi, 1969, *A. Genneou* in ARI 1348!
6. Between Orga and Liveras, 1956–1957, *Merton* 2609! 2852!
7. Glykyotissa (Snake) Island near Kyrenia, 1930, *Druce* s.n.! also, 1955, *N. Chiotellis* 778! Halevga, 1932, *Syngrassides* 60! and, 1936, *Lady Loch* 9! Akanthou, 1940, *Davis* 2028! Kyrenia, 1949, *Casey* 202! and, 1955, *G. E. Atherton* 780! 789! also, 1956, *G. E. Atherton* 316! Panagra, 1962, *Meikle!* Tjiklos, 1973, *P. Laukkonen* 10!

NOTES: The Cyprus plant agrees closely with *Hermione cypri* and with *Narcissus cypri* as figured by Sweet; there is, however, no positive evidence that either came from Cyprus. The earliest reference I have traced to a "Cyprus Narcissus" is in Parkinson's *Paradisus,* 86 (1656) where mention is made of *Narcissus Cyprius flore pleno luteo polyanthos,* "very double and of a fine pale yellow colour". Just over a century later, Miller (Gard. Dict., ed. 7; 1759) says, "There is also one [Narcissus] with very double Flowers, whose Outer Petals are white, and those in the Middle are some white, and others are of an Orange Colour, which have a very agreeable Scent and is the earliest in flowering; it is generally called the *Cyprus Narcissus.*" The two authors are clearly describing different variants, of which the latter, or a "single"-flowered derivative thereof, ultimately became *Hermione cypri* Haworth. I agree with Baker (in Burbidge, The Narcissus, 78; 1875) that *H. cypri* is best united with typical *Narcissus tazetta,* though fewer-flowered than most of the cultivated strains of this variable species. It is unquestionably wild in Cyprus, and seemingly always with a white perianth and yellow corona there.

N. PAPYRACEUS *Ker-Gawler* in Bot. Mag., 24: t. 947 (1806); Boiss., Fl. Orient., 5: 515 (1882); Holmboe, Veg. Cypr., 53 (1914); Osorio-Tafall et Seraphim, List Vasc. Plants Cyprus, 23 (1973).

Hermione papyracea (Ker-Gawler) Haworth, Suppl. Plant. Succ. (Narciss. Rev.), 143 (1819), Narciss. Monogr., 12 (1831) as *H. "papyratia"*; Kunth, Enum. Plant., 5: 746 (1850); Unger et Kotschy, Die Insel Cypern, 204 (1865); Sintenis in Oesterr. Bot. Zeitschr., 31: 191 (1881).

The Paperwhite Narcissus, with many-flowered umbels of uniformly white flowers, regarded by many as yet another variant of *N. tazetta* L., and very commonly grown for its early, sweet-scented flowers, is incorrectly listed as indigenous in Cyprus. Kunth's record of *N. papyraceus* "in Insula Cypri" (Enum. Plant., 5: 747; 1850) is based on an erroneous transcription of M. Roemer, Fam. Nat. Regn. Veg. Syn. Mon., 4: 230 (1847), where the plant is recorded "Ad Vesuvium et in insulâ *Capri*". Sintenis recorded *Hermione papyracea* as growing in fields about Larnaca, 17 Febr., 1880, but

Plate 93. Figs. **1–3a.** NARCISSUS TAZETTA L. **1,** habit, × 1; **2,** flower, longitudinal section, × 3; **3,** infructescence, × 1; **3a,** seed, × 24; figs. **4, 5.** NARCISSUS SEROTINUS L. **4,** habit, × 1; **5,** flower, longitudinal section, × 3. (**1** from *Syngrassides* 1745; **2** from *Genneou* in ARI 1348; **3** from *Merton* 2609; **3a** from *Merton* 2852; **4, 5,** from *Lady Loch* 6.)

specimens collected on this date, and distributed as *Narcissus tazetta* L. var. *syriacus* (Boiss. et Gaill.) Boiss., are referable to the common form of *N. tazetta*, with a white perianth and yellow corona.

It is quite likely that the Paperwhite Narcissus is grown in gardens in Cyprus, but it cannot be admitted to the flora as a native or naturalized plant.

2. N. serotinus *L.*, Sp. Plant., ed. 1, 290 (1753); Poech, Enum. Plant. Ins. Cypr., 11 (1842); Boiss., Fl. Orient., 5: 152 (1882); Holmboe, Veg. Cypr., 53 (1914); Post, Fl. Pal., ed. 2, 2: 608 (1933); Osorio-Tafall et Seraphim, List Vasc. Plants Cyprus, 23 (1973); Elektra Megaw, Wild Flowers of Cyprus, 15, t. 29 (1973).
    Hermione serotina (L.) Haworth, Narciss. Monogr., 13 (1831); Unger et Kotschy, Die Insel Cypern, 205 (1865).
    TYPE: "*in* Hispania".

Bulb ovoid, 1·5–3 cm. long, 1–2 cm. wide, with a rather thick, papery, dark fuscous-brown tunic usually produced into a slender apical neck; basal sheaths membranous, truncate or oblique at apex, 3–7 cm. long, 0·4 cm. wide; leaves 1–2, generally developing after anthesis, filiform, 10–20 cm. long, 1–1·5 mm. wide, apex acute or subacute; scape 1–2 (–3)-flowered, slender, terete, 10–25 cm. long, with a few distinct transverse articulations or nodes especially near the base; spathe 1–3 cm. long, 0·3–0·5 cm. wide, tubular towards base, split unilaterally above, membranous, pale brownish, with a simple or bifid, acuminate apex; flowers fragrant; pedicels angular, up to 3 cm. long, elongating slightly in fruit; ovary oblong-ovoid, 3–4 mm. long, 2–2·5 mm. wide; perianth-tube slender, greenish, 1·5–2 cm. long, about 0·4 cm. diam. at apex; segments patent, oblong, 1–1·5 cm. long, 0·4–0·8 cm. wide, white, apex rounded or subacute, the 3 outer often shortly cuspidate; corona very small, consisting of six, semicircular, orange lobes about 1 mm. long, 1·5–2 mm. wide at base; stamens inserted at apex of, and about ⅓ way down perianth-tube; filaments very short; anthers included or the upper very shortly exserted, narrowly oblong, about 1·5 mm. long, 0·8 mm. wide; style slender, about 1·5 cm. long, apex shortly and obscurely 3-lobed. Capsule broadly oblong-trigonous, about 0·8–1 cm. long, 0·6–1 cm. wide; pericarp pale, papery; seeds irregularly angular, about 2 mm. long, 1–1·5 mm. wide; testa rugulose, blackish. *Plate 93, figs. 4, 5*.

HAB.: On shallow soil over rocks, or in small pockets of soil on rocks; sea-level to 600 ft. alt.; fl. end of Sept.–Dec.

DISTR.: Divisions 1, 3, 5–8. Widespread in the Mediterranean region.

1. Between Paphos and Ktima, 1840, *Kotschy* 54; common about Paphos, 1981, *Meikle*!
3. Episkopi, 1963, *J. B. Suart* 130!
5. Athalassa, 1965, *Merton* in ARI 45! also, 1967, *Economides* in ARI 1034!
6. Government House and Ayia Paraskevi, Nicosia, 1935, *Syngrassides* 860!
7. Kyrenia 1936, *Kennedy* 126! also, 1948, *Casey* 153! 1955, *G. E. Atherton* 751! and, 1968, *I. M. Hecker* 14! Ayios Epiktitos, 1936, *Lady Loch* 10! Vasilia, 1937, *Lady Loch* 6! and, 1940, *Davis* 2004!
8. Karpas peninsula, 1929, *C. B. Ussher* 30!

NOTES: Probably widespread in the lowlands, but overlooked by collectors because of its very late (or very early!) flowering period.

N. JONQUILLA *L.*, Sp. Plant., ed. 1, 290 (1753), the Jonquil, a spring-flowering species, with 1–3, slender, dark green leaves 20–30 cm. long, 0·3–0·4 cm. wide, and uniformly bright yellow, sweet-scented flowers, sometimes solitary, but usually in loose umbels of 2–5. The individual flowers resemble those of *N. tazetta* L., with a long, slender, greenish perianth-tube and a short, cup-shaped corona.

Recorded in 1937 and 1939 as growing by the seashore near (7) Kyrenia (*Lady Loch* 19! 75!), but almost certainly a garden outcast there, which may or may not have persisted. *N. jonquilla* is not otherwise recorded, though Lady Loch notes on the label of one of the specimens: "I also have been shown some picked up in the hills far from a village". It is a native of Spain and Portugal, occasionally naturalized in other parts of S. Europe.

2. PANCRATIUM *L.*

Sp. Plant., ed. 1, 290 (1753).
Gen. Plant., ed. 5, 141 (1754).

Perennials with tunicate bulbs; leaves linear or lorate, frequently developing after anthesis; scape compressed; bracts 2 (or spathe 2-valved), membranous or scarious; flowers 2–15 or more in loose umbels, sessile or pedicellate, often fragrant; perianth white or greenish, tube often elongate, slender, narrowly infundibuliform, segments subequal, generally patent; corona conspicuous, cup-shaped, with 6 or 12 apical lobes; stamens inserted at apex of perianth-tube; filaments adnate to corona for the greater part of their length; anthers narrowly oblong or linear, dorsifixed, versatile, often conspicuously exserted; ovary ellipsoid, 3-locular; style elongate, slender; stigma capitate or obscurely 3-lobed. Capsule broadly ellipsoid-trigonous or subglobose, loculicidally 3-valved; seeds numerous, compressed-angular; testa blackish.

About 15 species in the Mediterranean region, Africa and tropical Asia.

1. **P. maritimum** *L.*, Sp. Plant., ed. 1, 291 (1753); Sibth. et Sm., Fl. Graec. Prodr., 1: 220 (1809), Fl. Graec., 4: 8, t. 309 (1823); Poech, Enum. Plant. Ins. Cypr., 11 (1842); Unger et Kotschy, Die Insel Cypern, 204 (1865); Boiss., Fl. Orient., 5: 152 (1882); Holmboe, Veg. Cypr., 53 (1914); Post, Fl. Pal., ed. 2, 2: 609 (1933); Osorio-Tafall et Seraphim, List Vasc. Plants Cyprus, 23 (1973); Elektra Megaw, Wild Flowers of Cyprus, 14, t. 28 (1973).
   TYPE: "*in* Hispaniae *maritimis circa* Valentiam & *infra* Monspelium".

Bulb large, ovoid, 8–12 cm. long, 5–8 cm. wide, usually deeply buried and produced above into a long, cylindrical neck; tunic papery or membranous, pallid or brownish; leaves 5–6, appearing after anthesis, linear or lorate, glaucous, up to 50 cm. long, 1·5–2 cm. wide, often twisted or undulate, apex obtuse or subacute; scape usually solitary, stout, compressed, 7–30 cm. long; bracts 5–8 cm. long, about 1–2·5 cm. wide, united at the base, free above and gradually acuminate; flowers 2–6 (sometimes more), loosely umbellate, fragrant; pedicels short, stout, seldom exceeding 1 cm.; perianth-tube greenish, 8–10 cm. long, about 2 cm. diam. at apex, segments white, spreading, narrowly oblong, acute, about 4–5 cm. long, 0·5–1 cm. wide, corona broadly cupular-infundibuliform, about 4 cm. long, apex about 4 cm. diam., with 12, broadly deltoid, acute, somewhat reflexed lobes about 8 mm. long; filaments slender, glabrous, about 4–5 cm. long, adnate for most of their length to the corona, the apical part free for about 1 cm. and exserted from corona; anthers yellow, narrowly oblong, 4–5 mm. long, 1–1·5 mm. wide; ovary glabrous, glaucous, at first narrowly ellipsoid, 1–1·5 cm. long, 0·5 cm. wide, swelling rapidly after fertilisation, style slender, 10–12 cm. long; stigma exserted, capitate or obscurely 3-lobed. Capsule broadly ellipsoid or subglobose, 2–3 cm. diam., commonly with a conspicuous, peg-like apical projection; seeds irregularly angular and compressed (often resembling an oyster shell in shape), about 1–1·3 cm. long and wide; testa black, sublustrous, minutely and closely foveolate.

HAB.: Sandy seashores and sand-dunes; about sea-level; fl. Aug.–Oct.

DISTR.: Divisions 1–8, probably in suitable sites all round the coast; widespread in the Mediterranean region.

1. Coast E. of Paphos, 1981, *Meikle*!

2. Coast at Pakhyammos, 1959, *P. H. Oswald.*
3. Coast near Amathus, 1959, *P. H. Oswald* 152! Episkopi, 1959, *C. E. H. Sparrow* 45!
4. Famagusta, 1900, *A. G. & M. E. Lascelles* s.n.! also, 1930, *C. B. Ussher* 88! and, 1959, *P. H. Oswald.*
5. Salamis, 1962, *Meikle*!
6. Ayia Irini, 1952, *F. M. Probyn* 143!
7. "Auf der Nordküste in Sand von Melandrina gegen Acathu [Akanthou] bei Eurosi [Ayios Amvrosios]", 1862, *Kotschy.*
8. Boghaz, 1955, *G. E. Atherton* 430!

NOTES: Not infrequently dug up and transplanted into gardens, and collected, as a garden plant, at (7) Myrtou, 1955, *G. E. Atherton* 422! and (7) Kyrenia, 1955, *G. E. Atherton* 428!

95a. AGAVACEAE

Mostly robust, often monocarpic, herbs, or woody perennials; leaves generally crowded towards the base of the stem, commonly fibrous and frequently thick and fleshy, with entire or spinous-toothed margins; flowers hermaphrodite, often very numerous in spikes, racemes or panicles; perianth tubular, 6-lobed, or divided to the base, or almost to the base, into 6 segments; stamens 6, inserted on the perianth-tube or at the base of the perianth-segments; anthers linear, introrse, dehiscing by longitudinal slits; ovary superior or inferior, 3-locular with axile placentation; style slender; stigma capitate or 3-lobed; ovules generally numerous. Fruit a loculicidal capsule; seeds with fleshy endosperm and a small embryo.

A heterogeneous group forming an ill-defined family, parts of which are variously assigned to *Amaryllidaceae* and *Liliaceae* by Bentham & Hooker (1883) and Engler & Prantl (1930) or united in a separate family (Hutchinson, 1959) which, with the exclusion of tribes *Dracaeneae, Phormieae* and *Nolineae,* is the classification adopted here. With this modification the family includes about 12 genera, mostly from the New World. It is not represented in Cyprus by any indigenous plants, but the Century Plant, *Agave americana* L., Sp. Plant., ed. 1, 323 (1753), with a large rosette, 2–4 m. diam., of thick, recurved, spiny-margined glaucous-green leaves, and huge panicles of yellow-green flowers crowded at the tips of horizontal inflorescence-branches, is not infrequently planted, and has been recorded by Bevan (Notes on Agriculture in Cyprus, 91; 1919) and by Osorio-Tafall et Seraphim (List Vasc. Plants Cyprus, 23; 1973). The popular name, Century Plant, records the legend that *A. americana* does not flower until it is 100 years old, and dies after ripening its fruit. In fact the plant can flower after 10 years, and commonly reproduces itself by stolons. It is generally believed to be a native of Mexico, but is planted or naturalized all round the shores of the Mediterranean.

Agave sisalana Perrine (in U.S. House of Reps. 25th Congress, 2nd session, Rep. no. 564: 87; 1838) from Yucatan, a smaller plant than *A. americana,* with stiff, erect, spine-pointed, entire or shortly denticulate (var. *armata* Trelease), bright green leaves, and a shorter panicle of glaucous-green flowers intermixed with numerous viviparous bulbils, is also noted by Bevan (1919) and has been seen naturalized on sandy ground by the shore near (5) Salamis. It is grown commercially in many tropical countries as the source of Sisal Hemp.

The genus *Furcraea* Vent. resembles *Agave* in habit, but has lax inflorescences of whitish, pendulous flowers with a short perianth-tube and relatively large, rotately spreading perianth-lobes; the stamens, conspicuously exserted in *Agave*, are included or almost included in *Furcraea*, and the filaments are distinctly swollen in the lower half. One species, *Furcraea foetida* (L.) Haworth (*Agave foetida* L., *Furcraea gigantea* Vent.), Mauritius Hemp, a native of Central America, is widely cultivated in the tropics and subtropics and has become naturalized (and sometimes persistently invasive) in some areas. It is noted by Bevan (Notes on Agriculture in Cyprus, 91: 1919) as grown on the island, but is scarcely established.

In *Agave* and *Furcraea* the ovary is inferior; in the woody, suffruticose or arborescent genus *Yucca* L. the ovary is superior, and the large, creamy-white or greenish, lantern-shaped flowers have the perianth divided to the base into 6 broad segments. In some *Yucca* species the crowded leaves are rigid and spine-tipped (as in *Agave sisalana* Perrine), in others they are pliable and reflexed. The genus is certainly represented in Cyprus gardens, but not naturalized, nor, in the absence of specimens, can the cultivated species be identified.

The other member of the family *Agavaceae*, namely the Tuberose, *Polianthes tuberosa* L. (Sp. Plant., ed. 1, 316; 1753), unknown as a wild plant, but probably of Central American origin, is widely cultivated for its long racemes of richly perfumed, white, funnel-shaped flowers. It is a rhizomatous perennial, with a bulb-like base, and numerous, long, lax, linear, basal leaves. A "double-flowered" variant is more common in cultivation than the normal, "single" form.

96. DIOSCOREACEAE

I. H. Burkill in Journ. Linn. Soc. (Bot.), 56: 317–412 (1960).

Herbs with tuberous rhizomes or thick, woody rootstocks; stems usually climbing; leaves generally alternate, stipulate or exstipulate, entire, often sagittate or cordate, sometimes lobed or digitately compound; nervation reticulate; petiole often twisting, sometimes articulated at the base; flowers small, greenish, unisexual (dioecious) or hermaphrodite, actinomorphic, mostly in axillary spikes or racemes, occasionally clustered or solitary; perianth campanulate or spreading, 6-lobed or 6-partite, the lobes subequal in 2 series; stamens 6, or 3, with or without 3 staminodes; ovary inferior, 3-locular; styles 3, free or connate; ovules generally 2, superimposed in each loculus, anatropous; placentation axile. Fruit a 3-valved, often winged, capsule, or a berry; seed sometimes winged, with horny endosperm and a small embryo.

Six genera, widely distributed, but chiefly in tropical and subtropical regions.

1. TAMUS *L.*

Sp. Plant., ed. 1, 1028 (1753).
Gen. Plant., ed. 5, 454 (1754).

Perennial dioecious herbs with large, cylindrical, ovoid or irregular stem-tubers; stems twining; leaves alternate, sagittate, cordate or lobed; flowers in axillary racemes, the male inflorescences usually more elongate than the female; male flowers campanulate or subrotate, deeply divided into 6 subequal lobes; stamens 6, inserted at base of perianth around the base of rudimentary styles; filaments filiform; anthers small; female flowers with 6 small, narrow perianth-lobes, with or without minute staminodes; ovary 3-locular; styles connate and columnar below, divided above into 3, recurved, 2-lobed stigmas. Fruit a succulent berry, becoming red when ripe; seeds 1–6, ovoid or spherical.

1. **T. communis** L., Sp. Plant., ed. 1, 1028 (1753); Sibth. et Sm., Fl. Graec. Prodr., 2: 258 (1816); Poech, Enum. Plant. Ins. Cypr., 9 (1842); Unger et Kotschy, Die Insel Cypern, 201 (1865); Boiss., Fl. Orient., 5: 344 (1882); Holmboe, Veg. Cypr., 44 (1914); Post, Fl. Pal., ed. 2, 2: 664 (1933); I. H. Burkill in Journ. Ecol., 32: 121 (1944); Lindberg f., Iter Cypr., 10 (1946); Osorio-Tafall et Seraphim, List Vasc. Plants Cyprus, 24 (1973).

   T. cretica L., Sp. Plant., ed. 1, 1028 (1753); Sibth. et Sm., Fl. Graec. Prodr., 2: 258 (1816), Fl. Graec., 10: 48, t. 958 (1840); Poech, Enum. Plant. Ins. Cypr., 10 (1842); Unger et Kotschy, Die Insel Cypern, 202 (1865).

   T. communis L. var. *subtriloba* Guss., Fl. Sic. Syn., 2: 880 (1845).

   T. communis L. var. *cretica* (L.) Parl., Fl. Ital., 3: 64 (1858); Boiss., Fl. Orient., 5: 345 (1882); Holmboe, Veg. Cypr., 44 (1914).

   T. communis L. ssp. *cretica* (L.) Nyman, Consp. Fl. Europ., 718 (1882); Osorio-Tafall et Seraphim, List Vasc. Plants Cyprus, 24 (1973).

   TYPE: *"in* Europa *australi"*.

Perennial herb 1–4 m. high, growing from a large, irregular, dark-coloured subterranean tuber; stems scandent, twisting to the left, glabrous, leaves variable, ovate-cordate to strongly 3-lobed, the median lobe narrow or broad, the laterals conspicuous or inconspicuous, 1–8 (–10) cm. long, 1–8 (–10) cm. wide, glabrous, dark green, glossy, margins entire, apex often shortly cuspidate; nervation inconspicuous, nerves 3–9, spreading palmately from the base of the lamina; petiole often long and slender, glabrous, sometimes 10 cm. long; stipules reduced to 2 short, horn-like prominences; flowers in lax, axillary racemes, the male racemes often 15 cm. or more long, the females usually much shorter, or sometimes reduced to clusters or a solitary flower; bracts minute, subulate, scarious, about 1·5 mm. long, 0·3 mm. wide; pedicels slender, to 3 mm. long; male flowers with a campanulate, pale green, glabrous perianth, divided more than half-way into 6 oblong lobes about 2–3 mm. long, 0·8–1 mm. wide; stamens 6; filaments about 0·6 mm. long, glabrous; anthers pale yellow, about 0·3 mm. long and almost as wide; female flowers with a conspicuous ovoid or ellipsoid, glabrous ovary, 2–3 mm. long, 1·5–2 mm. wide; perianth divided almost to base into 6 narrow, oblong, recurved lobes; styles united for about 0·3 mm., forming a short column; stigmas about 0·5 mm. long. Fruit a globose or subglobose red berry, about 1 cm. diam., surmounted by the remains of the flower; seeds globose, about 3 mm. diam.; testa brown, smooth or very obscurely rugulose.

HAB.: Climbing over bushes or tall herbs in thickets, by roadsides or in Pine forest, usually in moist places near streams; sea-level to 4,500 ft. alt.; fl. March–June.

DISTR.: Divisions 1–4, 7, 8. Southern Europe and Mediterranean region eastwards to Caucasus and Caspian.

1. Lyso, 1913, *Haradjian* 863! Near Ayios Yeoryios (Akamas), 1962, *Meikle* 2147!
2. Near Prodhromos, 1880, *Sintenis* 1071! Kykko Monastery, 1913, *Haradjian* 933! Platania, 1935, *Syngrassides* 739! and, 1939, *Lindberg f.* s.n.; Kryos Potamos, 1937, *Kennedy* 124! 125!
3. Mazotos, 1905, *Holmboe* 172.
4. Cape Greco, 1862, *Kotschy* 115!
7. Pentadaktylos, 1862, *Kotschy*; Dhavlos, 1880, *Sintenis*; St. Hilarion, 1901, *A. G. & M. E. Lascelles* s.n.! Mylous Forest Station, 1933, *Syngrassides* 1365! Kyrenia, 1949, *Casey* 498! Trypimeni, 1957, *Merton* 3092! Akanthou, 1967, *Merton* in ARI 781! Tjiklos, 1973, *P. Laukkonen* 352!

8. Valia, 1905, *Holmboe* 499! Koma tou Yialou, 1950, *Chapman* 235!

NOTES: It is impossible to distinguish *Tamus communis* L. and *T. cretica* L. (*T. communis* L. var. *subtriloba* Guss.) on leaf shape alone. A higher percentage of Cyprus specimens are slender, with conspicuously trilobed leaves, when compared with W. European material, but the distinction, though perhaps statistically significant, scarcely warrants nomenclatural recognition.

97. LILIACEAE
K. Krause in Engl. et Prantl, Pflanzenfam., ed. 2, 15a: 227–386 (1930).

Perennial herbs with bulbs, tubers or rhizomes, sometimes shrubby or arborescent, rarely annuals; stems sometimes climbing; leaves usually alternate, sometimes whorled or opposite, or all basal, mostly simple and entire with parallel or rarely reticulate nervation; flowers usually in racemes or panicles, less often in cymes or umbels, or solitary, mostly hermaphrodite, actinomorphic or occasionally somewhat zygomorphic, frequently showy; perianth of 2 whorls, each of (2–) 3 (–5) free or more or less united segments; stamens generally as many as perianth-segments, free or connate; anthers commonly introrse, mostly 2-thecous and dehiscing longitudinally; ovary superior (rarely semi-inferior), generally 3-locular with axile placentation, rarely 1-locular with parietal placentation; ovules usually numerous, in 2 ranks in each loculus; styles often well developed, free or connate; stigmas simple or bifid. Fruit generally a loculicidal or septicidal capsule, or a berry; seeds with abundant endosperm, and a straight or curved embryo.

About 250 genera with a cosmopolitan distribution, including many decorative species popular in gardens.

Flowers lateral or axillary, arising from elongate aerial stems:
  Leaves cordate or sagittate, reticulately veined; stems scrambling, usually armed with prickles - - - - - - - - - - - - - **1. Smilax**
  Leaves (or leaf-like cladodes) not cordate or sagittate:
    Cladodes (or apparent leaves) ovate or lanceolate; filaments of male flowers united into a tube; fruit a scarlet berry - - - - - - - - - **3. Ruscus**
    Cladodes linear or subulate; filaments free; fruit a black berry - - - **2. Asparagus**
Flowers terminal:
  Leaves thick, succulent, with rigid, spinose marginal teeth; perianth tubular ALOE (p. 1592)
  Leaves not thick or succulent, nor armed with spinose teeth:
    Leaves extending up aerial stems:
      Plants rhizomatous; perianth-segments free to base or almost to base, spreading **5. Asphodeline**
    Plants bulbous:
      Flowers in umbels or capitula, enveloped in bud by a spathaceous bract or bracts; plant usually smelling of garlic when crushed - - - - **10. Allium**
      Flowers not in umbels or capitula, nor enveloped by a spathaceous bract or bracts:
        Flowers pendulous or nodding; perianth-segments with a well-marked basal nectary-pit - - - - - - - - - **8. Fritillaria**
        Flowers erect; perianth-segments without a basal nectary-pit:
          Flowers large, cup-shaped, showy, often solitary; perianth-segments frequently red with a conspicuous basal blotch - - - - - **7. Tulipa**
          Flowers small, star-shaped, not showy; perianth-segments yellow, greenish or white, without a basal blotch - - - - - - **9. Gagea**
  Leaves all basal; flowers on a leafless scape or apparently arising directly from the ground:
    Flowers in umbels or capitula, enveloped in bud by a spathaceous bract or bracts:
      Perianth segments free or almost free, style gynobasic; plant usually smelling of garlic when crushed - - - - - - - - - **10. Allium**
      Perianth-segments shortly connate; style terminal; plant not smelling of garlic when crushed - - - - - - - - - **10a.** NOTHOSCORDUM

Flowers not in umbels or capitula, nor enveloped by a spathaceous bract or bracts:
Plants rhizomatous; inflorescences commonly branched -   -   - **4. Asphodelus**
Plants bulbous; inflorescences not branched:
  Flowers arising directly from the ground; ovary subterranean at anthesis; styles
    usually 3 -   -   -   -   -   -   -   -   -   - **6. Colchicum**
  Flowers borne above ground on leafless scapes; style 1:
    Flowers appearing in advance of the leaves in summer or autumn:
      Flowers white or whitish, leaves broadly lorate, 2–10 cm. wide; seeds flattened
                                       12. Urginea
      Flowers blue or bluish; leaves linear-filiform, about 2 mm. wide; seeds pyriform
                                       13. Scilla
    Flowers appearing with the leaves in spring:
      Flowers without a distinct perianth-tube:
        Perianth-segments whitish, with a greenish dorsal stripe; bracts conspicuous
                                    11. Ornithogalum
        Perianth-segments blue or tinged blue, without a dorsal stripe; bracts small,
          inconspicuous -   -   -   -   -   -   -   -   - **13. Scilla**
      Flowers with a distinct perianth-tube:
        Perianth-segments much longer than tube; flowers few (1–4); capsule
          bluntly trigonous -   -   -   -   -   - **14. Chionodoxa**
        Perianth-segments shorter than tube; flowers usually more than 4:
          Perianth-segments 1 cm. long (or longer), not much shorter than tube;
            capsule bluntly trigonous -   -   - HYACINTHUS (p. 1643)
          Perianth-segments (or lobes) small, much less than 1 cm. long, and much
            shorter than the perianth-tube:
            Capsule depressed-globose, bluntly trigonous; leaves with conspicuous
              elevated nervation -   -   -   -   - **15. Hyacinthella**
            Capsule sharply triquetrous; leaves without conspicuous elevated
              nervation:
              Mouth of perianth-tube distinctly flared; stamens inserted in one
                series at base of perianth-lobes; seeds pruinose - **16. Bellevalia**
              Mouth of perianth-tube often more or less constricted, not flared;
                stamens inserted in 2 series usually towards the middle of the
                perianth-tube; seeds black, not pruinose -   -   **17. Muscari**

SUBFAMILY 1. **Smilacoideae** Climbing or scrambling shrubs with reticulately veined leaves, flowers mostly unisexual (dioecious) in axillary or terminal racemes, panicles or umbels; ovules 1–2 in each loculus, pendulous, orthotropous or semi-anatropous. Fruit a berry; seeds with hard endosperm.

1. SMILAX *L.*

Sp. Plant., ed. 1, 1028 (1753).
Gen. Plant., ed. 5, 455 (1754).

Scrambling shrubs or subshrubs; stems sometimes armed with prickles; leaves alternate or rarely opposite, often coriaceous and persistent, with 3–5 conspicuous parallel primary nerves and reticulate venation; flowers unisexual (dioecious) in umbels or umbellate fascicles along an elongate axis, rarely subsolitary; bracts usually very small; perianth-segments 6, free to base, subequal, generally spreading or recurved at anthesis; stamens 6, inserted around base of perianth; filaments free, slender; anthers introrse, longitudinally dehiscent; rudimentary ovary absent in male flowers; female flowers with a sessile, 3-locular ovary; stigmas 3, sessile or subsessile, soon deciduous; stamens usually 6, filiform, deciduous. Fruit a globose berry.

About 350 species widely distributed in the tropics and subtropics, less abundantly represented in temperate regions.

1. **S. aspera** *L.*, Sp. Plant., ed. 1, 1028 (1753); Sibth. et Sm., Fl. Graec. Prodr., 2: 259 (1816), Fl. Graec., 10: 49, t. 959 (1840); Poech, Enum. Plant. Ins. Cypr., 9 (1842); Unger et Kotschy, Die Insel Cypern, 201 (1865); Boiss., Fl. Orient., 5: 343 (1882); Holmboe, Veg. Cypr., 53 (1914); Post, Fl. Pal., ed. 2, 2: 663 (1933); Lindberg f., Iter Cypr., 10 (1946); Chapman,

Cyprus Trees and Shrubs, 30 (1949); P. Vernet in Bull. Soc. Bot. France, Mém., 1966: 140–146 (1967), Osorio-Tafall et Seraphim, List Vasc. Plants Cyprus, 23 (1923).

S. mauritanica Desf., Fl. Atl., 2: 367 (1800); Gaudry, Recherches Sci. en Orient, 197 (1854); Unger et Kotschy, Die Insel Cypern, 201 (1865).

S. aspera L. var. *altissima* Moris et De Not., Florula Capr., 127 (1837).

S. aspera L. var. *mauritanica* (Desf.) Gren. et Godr., Fl. France, 3: 234 (1955); Boiss., Fl. Orient., 5: 343 (1882); H. Stuart Thompson in Journ. Bot., 44: 339 (1906); Holmboe, Veg. Cypr., 53 (1914); Post, Fl. Pal., ed. 2, 2: 663 (1933); Lindberg f., Iter Cypr., 11 (1946); Rechinger f. in Arkiv för Bot., ser. 2, 1: 419 (1950).

S. aspera L. ssp. *mauritanica* (Desf.) Aschers. et Graebn., Syn. Mitteleurop. Fl., 3: 323 (1906); Osorio-Tafall et Seraphim, List Vasc. Plants Cyprus, 23 (1973).

[*Aristolochia altissima* (non Desf.) Lindberg f., Iter Cypr., 12 (1946).]

TYPE: "*in* Hispaniae, Italiae, Siciliae *sepibus*".

Evergreen, dioecious climber with a woody base; stems glabrous, sometimes attaining 2 m. or more, irregularly branched, distinctly ribbed or angled, more or less densely armed with patent or recurved prickles or sometimes unarmed, distinctly flexuous on the younger growths; leaves very variable, narrowly or broadly sagittate or cordate, coriaceous, sometimes blotched with white, 1·5–10 cm. long, 1–6 cm. wide, margins entire, apex sharply apiculate, base with a wide sinus; primary nerves 3–5, conspicuous, sometimes armed with prickles on the underside of the leaf; venation inconspicuously reticulate; petiole 5–20 mm. long, sometimes prickly; stipules (or stipular sheath) adnate to petiole, produced some distance above apparent apex into 2 long, twisting tendrils, or occasionally ecirrhose; inflorescence axillary and/or terminal, consisting of 5–8 or more flowers-clusters strung along an elongate, flexuous axis 2–10 cm. long; clusters (1–) 3–8-flowered, umbellate; bracts minute, suborbicular, scarious, dark brown, imbricate, forming a shallow cup about 1 mm. long; pedicels slender, up to 8 mm. long, those of male flowers generally longer than those of females; perianth creamy-white, fragrant, segments narrowly lanceolate or strap-shaped, those of the male flowers up to 5 mm. long, 1 mm. wide, patent or reflexed at anthesis, those of the female flowers smaller, about 2–3 mm. long, the 3 exterior about 0·6 mm. wide, the 3 interior rather narrower, all erect or spreading at anthesis; stamens 6; filaments glabrous, about 2–3 mm. long; anthers whitish, narrowly oblong, about 1 mm. long, 0·5 mm. wide; staminodes in female flower reduced to sterile threads about 1 mm. long; ovary glabrous, oblong, trigonous, about 1·5 mm. long, 0·8 mm. wide, stigmas recurved, subsessile, about 0·2 mm. long. Fruit a red berry 6–8 mm. diam., the berries usually numerous and clustered like small bunches of grapes; seeds broadly ovoid or subglobose, 4–5 mm. long, 4 mm. wide; testa dark red-brown, shining, minutely and closely rugulose.

HAB.: In maquis and garigue, generally in moist thickets; sometimes climbing over *Quercus alnifolia*; sea-level to 4,500 ft. alt.; fl. Aug.–Nov.

DISTR.: Divisions 1–8, locally abundant; widespread in S. Europe and Mediterranean region.

1. Ayios Yeoryios (Akamas), 1962, *Meikle*! Polis area 1962, *Meikle*!
2. Galata, 1862, *Kotschy*; and between Sina Oros and Galata, 1973, *P. Laukkonen* 87! Troödos, 1900, *A. G. & M. E. Lascelles* s.n.! Makheras Monastery, 1905, *Holmboe* 1121; Platres, 1929, *C. B. Ussher* 14! also, 1937, *Kennedy* 167–169! Saïttas, 1932, *Syngrassides* 346! Stavros tis Psokas, 1938, *Chapman* 319–321! Mesapotamos, 1939, *Lindberg f.* s.n.! Kakopetria, 1955, *N. Chiotellis* 436! Prodhromos, 1955, *G. E. Atherton* 573! 635!
3. Kalavasos, 1905, *Holmboe* 607; Kato Lefkara, 1937, *Syngrassides* 1522! Kakoradjia, 1937, *Syngrassides* 1709! Khalassa, 1956, *Poore* 33! Episkopi, 1964, *J. B. Suart* 166! Phasouri, 1970, *A. Genneou* in ARI 1938! N.E. Of Apsiou, 1979, *Edmondson &|McClintock* E2696!
4. Cape Greco, 1862, *Kotschy*; Ormidhia, 1880, *Sintenis & Rigo* 867! Ayia Napa, 1905, *Holmboe* 57!
5. Kythrea, 1880, *Sintenis*; also 1931, *Nattrass* 40.
6. Dhiorios area, 1962, *Meikle*!
7. Pentadaktylos, 1862, *Kotschy*; Ayios Elias, 1862, *Kotschy*; Armenian Monastery, 1939, *Lindberg f.* s.n.; Myrtou, 1940, *Davis* 1972! Kyrenia, 1947, *Casey* 525! Below Bellapais, 1948,

Casey 921! Lapithos, 1955, *Miss Mapple* 94! also, 1955, *G. E. Atherton* 686! St. Hilarion, 1955, *G. E. Atherton* 741!

8. Komi Kebir, 1912, *Haradjian* 273; Akradhes area, 1962, *Meikle*!

NOTES: P. Vernet (Bull. Soc. Bot. France, Mém., 1966: 140–146; 1967) notes that in S. France the form with cordate leaves (var. *altissima* Moris et De Not.) is characteristic of riverine (ripicolous) communities; cordate-triangular leaves are prevalent in forest, while forms with narrow sagittate-hastate leaves are found in degraded garigue. The same is broadly true in Cyprus, but the different leaf-forms are there connected by every possible intermediate, clearly indicating the uselessness of infraspecific analysis in this complex.

SUBFAMILY 2. **Asparagoideae** Rhizomatous perennials, stems erect or climbing, sometimes woody or woody at base; leaves generally reduced to small scales, bearing leaf-like modified stems (cladodes, phylloclades) in their axils; flowers solitary or in clusters, hermaphrodite or unisexual (dioecious); perianth-segments free or partly united; stamens 6 or 3, the filaments free or united into a column. Fruit a berry; seeds solitary or few, with hard endosperm.

2. ASPARAGUS *L.*

Sp. Plant., ed. 1, 313 (1753).
Gen. Plant., ed. 5, 147 (1754).

Rhizomatous perennials commonly with fusiform tubers; stems erect or climbing, herbaceous or woody and persistent, generally much branched; leaves reduced to small, scarious, often sharply spurred scales, subtending one or more leaf-like, subulate or filiform cladodes; flowers hermaphrodite or unisexual (dioecious), solitary or in fascicles, umbels or racemes at the base of the cladodes; pedicels usually slender, articulated, commonly bracteolate at the base; perianth small, campanulate or rotate, with 6 subequal segments connate at or for some distance above base; stamens 6, inserted at base of perianth-segments; anthers dorsifixed, introrse; ovary sessile, 3-locular; styles connate or occasionally free to base; stigmas 3, generally recurved, ovules 2–8 in each loculus. Fruit a berry; seeds usually few (1–6) by abortion, generally subglobose, testa thin; endosperm hard; embryo small, cylindrical.

About 300 species widely distributed in the Old World.

Cladodes in fascicles of (2–) 5–10 (–15), slender, 2–8 (–15) mm. long, dark green
1. A. acutifolius
Cladodes solitary or sometimes in fascicles of 3, robust, rigid, 5–40 mm. long, glaucous
2. A. stipularis

A. OFFICINALIS *L.*, Sp. Plant., ed. 1, 313 (1753), the Asparagus of gardens, with erect, herbaceous, much-branched, smooth stems, often 1 m. or more in height; erecto-patent, filiform cladodes 1–3 cm. long in fascicles of 4–15 or more; unisexual flowers usually in pairs, with slender pedicels articulate at or near the middle, and red globose berries, is grown as a vegetable in Cyprus, and sometimes occurs as a self-sown garden escape, or as a relic in neglected or abandoned gardens. It has been recorded as an escape at (7) Tchingen, Kyrenia (1955, *G. E. Atherton* 359!).

1. **A. acutifolius** *L.*, Sp. Plant., ed. 1, 314 (1753); Unger et Kotschy, Die Insel Cypern, 200 (1865); Boiss., Fl. Orient., 5: 337 (1882); Holmboe, Veg. Cypr., 53 (1914); Post, Fl. Pal., ed. 2, 2: 661 (1933); Lindberg f., Iter Cypr., 20 (1946); Chapman, Cyprus Trees and Shrubs, 29 (1949); Osorio-Tafall et Seraphim, List Vasc. Plants Cyprus, 22 (1973).
   [*A. verticillatus* (non L.) Kotschy in Unger et Kotschy, Die Insel Cypern, 200 (1865).]
   TYPE: "*in* Lusitania, Hispania".

Erect, scandent or sprawling perennial 30–200 cm. high, arising from a dense, compact rhizome with fleshy roots; stems conspicuously angular, papillose-scabridulous, often purplish, generally with numerous, short spreading branches; reduced leaves of main stem mostly with a strongly recurved, rigid, pungent spur; cladodes dark green in fascicles of (2–) 5–10 (–15), very variable in size, linear-subulate, 2–8 (–15) mm. long, 0·5–1 mm. wide, subequal, spreading, distinctly keeled, with a sharp, spinulose apex and smooth or papillose margins; flowers mostly solitary, creamy white, pedicel slender, 3–8 mm. long, glabrous, articulated at or rather below the middle; bracteoles minute, membranous; male flowers with a widely infundibuliform, whitish perianth, segments oblong-ovate, obtuse, concave, about 3–3·5 mm. long, 1·5 mm. wide, united only at the base, the 3 exterior a little longer and more acute than the 3 interior; filaments glabrous, 2–2·5 mm. long, tapering to apex; anthers pallid, about 1·5 mm. long, 0·8 mm. wide, the thecae with distinctly divaricate bases; rudimentary ovary present; female flowers similar but smaller, perianth-segments 2–2·5 mm. long; ovary obovoid, about 1·5 mm. long, 1 mm. wide; stigmas sessile, recurved, about 0·6 mm. long; staminodes represented by reduced filaments. Fruit a black globose berry about 8–10 mm. diam., generally with the persistent remains of the perianth attached at the apex of the fruiting pedicel; seeds 1–4, irregularly subglobose, about 4 mm. diam.; testa black, rugose.

HAB.: Dry rocky hillsides in garigue or maquis, by roadsides or field-margins, sometimes in Pine forest; on igneous rocks or limestone; sea-level to 5,000 ft. alt.; fl. July–Oct.

DISTR.: Divisions 1–4, 7, 8. Europe and Mediterranean region.

1. Toxeftera near Ayios Yeoryios (Akamas), 1962, *Meikle* 2077!
2. Tris Elies, 1862, *Kotschy* 836; between Platres and Perapedhi, 1900, *A. G. & M. E. Lascelles* s.n.! Saïttas, 1932, *Syngrassides* 342! Platania, 1932, *Chapman* 115! Ayia (Paphos Forest), 1934, *Chapman* 177! Platres, 1937, *Kennedy* 170! Tembria, 1937, *Syngrassides* 1670! Stavros, 1938, *Chapman* 318! 322–325! Troodos area, 1950, *F. M. Probyn* s.n.! Prodhromos, 1955, *G. E. Atherton* 634! and, 1968, *Economides* in ARI 1253!
3. Mazotos, 1862, *Kotschy* 567, also, 1905, *Holmboe* 169.
4. Larnaca, 1862, *Kotschy*; Ayia Napa, 1862, *Kotschy* 381; Ayios Antonios, Sotira, 1950, *Chapman* 586!
7. Armenian Monastery, Pentadaktylos, 1939, *Lindberg f.* s.n.! St. Hilarion, 1939, *Lindberg f.* s.n.; Bellapais, 1948, *Casey* 100! Kyrenia, 1955, *G. E. Atherton* 242! 715! 740! 742! Lapithos, 1955, *G. E. Atherton* 688! East of Kyrenia, 1968, *Barclay* 1055!
8. Ayios Theodhoros, 1939, *Lindberg f.* s.n.; Yialousa, 1970, *A. Genneou* in ARI 145!

NOTES: Widespread and rather variable in Cyprus; often used, and offered for sale, in place of *A. officinalis* L., but the shoots are relatively spindly, with a slightly bitter taste.

2. **A. stipularis** *Forssk.*, Fl. Aegypt.-Arab., (1775); Boiss., Fl. Orient., 5: 338 (1882); Holmboe, Veg. Cypr., 52 (1914); Post, Fl. Pal., ed. 2, 2: 662 (1933); Lindberg f., Iter Cypr., 10 (1946); Chapman, Cyprus Trees and Shrubs, 29 (1949); Osorio-Tafall et Seraphim, List Vasc. Plants Cyprus, 22 (1973).
 A. horridus L. f., Suppl. Plant., 203 (1781); Sibth. et Sm., Fl. Graec. Prodr., 1: 236 (1809), Fl. Graec., 4: 33, t. 339 (1823), Poech, Enum. Plant. Ins. Cypr., 9 (1842); Unger et Kotschy, Die Insel Cypern, 200 (1865).
 A. aphyllus L. var. *stipularis* (Forssk.) Bak. in Journ. Linn. Soc., 14: 600 (1875).
 A. stipularis Forssk. var. *brachyclados* Boiss., Fl. Orient., 5: 338 (1882); Holmboe, Veg. Cypr., 52 (1914); Chapman, Cyprus Trees and Shrubs, 29 (1949).
 A. stipularis Forssk. var. *tenuispinus* Holmboe, Veg. Cypr., 52 (1914); Lindberg f., Iter Cypr., 10 (1946); Chapman, Cyprus Trees and Shrubs, 29 (1949).
 [*A. aphyllus* (non L.) Kotschy in Unger et Kotschy, Die Insel Cypern, 200 (1865); Post in Mém. Herb. Boiss., 18: 101 (1900); H. Stuart Thompson in Journ. Bot., 44: 339 (1906).]
 TYPE: Egypt; "Alexandriae; inculta habitans, rarior", *Forsskaal* (C).

Erect or sprawling shrub to 1 m. high; roots with large, fusiform tubers; stems woody, persistent, angular, striate, glaucous-green, smooth or minutely papillose, generally much branched, with short spreading branches; leaves reduced to papery or membranous, persistent, acuminate, pale brown scales, up to 12 mm. long on the main stem, with a conspicuous, deflexed, pointed spur; cladodes usually solitary, sometimes 3 together, very robust and rigid, patent, 3–4-angled, glaucous, smooth or papillose, 5–40 mm. long, 1·5–2·5 mm. wide, apex spinulose-mucronate; flowers unisexual, 1–6 (occasionally more) in clusters at the base of the cladodes; bracteoles 2–3, membranous, semicircular, about 2 mm. diam.; pedicels rarely exceeding 3 mm., glabrous, angular, articulated a little below the middle; perianth greenish-white, campanulate; male flowers with 6 subequal, oblong, obtuse, somewhat concave segments, about 2·5 mm. long, 1 mm. wide, free almost to base; stamens 6; filaments glabrous, about 2·5 mm. long, strongly dilated and 0·8 mm. wide at base; anthers pallid, about 1·2 mm. long, 1 mm. wide, with markedly divaricate thecae; rudimentary ovary relatively large; female flowers similar to male but much smaller; perianth-segments usually less than 2 mm. long; ovary subglobose, about 1 mm. diam., narrowed above to a stylar column about 0·5 mm. long; stigmas 3, spreading, about 0·3 mm. long. Fruit a depressed-globose, obscurely trigonous, blue-black, pruinose berry about 7 mm. diam.; seeds usually 3, depressed-globose or almost lenticular, about 3·5 mm. diam.; testa black, minutely and regularly punctulate.

HAB.: Dry rocky or sandy ground by the sea, or on rocky slopes inland; occasionally in thin Pine forest; sea-level to ? 2,000 ft. alt., usually lowland; fl. March–June.

DISTR.: Divisions 1–8. Widespread in the Mediterranean region, Atlantic Islands.

1. Paphos, 1905, *Holmboe* 722; Skali near Ayios Yeoryios (Akamas), 1962, *Meikle* 2084! Summit of Ayios Yeoryios Island, 1962, *Meikle* 2152!
2. Yialia, 1938, *Chapman* 315! Near Vouni Palace, 1963, *Townsend* 63/57!
3. Between Moni and Amathus, 1862, *Kotschy* 578!
4. Near Larnaca, 1962, *Kotschy*; also, 1877, *herb. Post* 561! 562! and 1939, *Lindberg f.* s.n.; Ayia Napa, 1905, *Holmboe* 44! Perivolia, 1939, *Lindberg f.* s.n.; Athna Forest, 1956, *Merton* 2661! Cape Greco, 1958, *N. Macdonald* 58! Also noted as forming impenetrable hedges in Famagusta (1949, *Chapman*).
5. Athalassa, 1932, *Syngrassides* 401! 407! also, 1939, *Lindberg f.* s.n.; Salamis, 1939, *Lindberg f.* s.n.!
6. Nicosia, 1905, *Holmboe* 287, and, 1939, *Lindberg f.* s.n.
7. Mountains above Kythrea, 1880, *Sintenis & Rigo* 169! Kyrenia, 1900, *A. G. & M. E. Lascelles* s.n.! Opposite Glykyotissa (Snake) Island, 1968, *I. M. Hecker* 17!
8. Klidhes Islands, 1933, *Chapman* 22! Near Boghaz, 1966, *Economides* in ARI s.n.

3. RUSCUS L.

Sp. Plant., ed. 1, 1041 (1753).
Gen. Plant., ed. 5, 463 (1754).
P. F. Yeo in Notes Roy. Bot. Gard. Edinb., 28: 237–264 (1968).

Rhizomatous shrubs or perennial herbs; rhizomes compact; stems simple or branched; leaves reduced to membranous or scarious scales, subtending leaf-like, whorled, opposite or alternate, evergreen cladodes; inflorescence a cluster or condensed raceme borne in the axil of a small or large membranous or herbaceous bract generally situated near the middle of the cladode; flowers unisexual (monoecious or dioecious) or rarely hermaphrodite, developing one at a time; pedicel articulated; perianth-segments 6, free to base, the 3 exterior generally a little broader than the 3 interior; filaments united into a hollow, violet or violet-striped column; anthers 3, sessile or sunk into the apex of the staminal column; ovary 1–3-locular; stigma sessile or subsessile or pileate. Fruit a red berry; seeds 1–4, globose or angled; endosperm hard; embryo small.

Four or five species in Europe, the Mediterranean region east to Transcaucasia; Atlantic Islands.

1. R. aculeatus *L.*, Sp. Plant., ed. 1, 1041 (1753); Boiss., Fl. Orient., 5: 340 (1882); Holmboe, Veg. Cypr., 53 (1914); Post, Fl. Pal., ed. 2, 2: 662 (1933); Lindberg f., Iter Cypr., 10 (1946); Chapman, Cyprus Trees and Shrubs, 30 (1948); Rechinger f. in Arkiv för Bot., ser. 2, 1: 419 (1950); Osorio-Tafall et Seraphim, List Vasc. Plant. Cypr., 23 (1973).

Erect or sprawling, shrubby perennial; rhizome short; stems terete, dark green, glabrous, closely striate; leaves reduced to scarious scales at the base of the stems, or subtending the cladodes as subulate, membranous scales 2–5 mm. long, 1–2·5 mm. wide at base; cladodes 12–30 mm. long, 3–12 (–25) mm. wide, ovate or lanceolate, dark green, glabrous, rigid, with numerous parallel nerves, sessile or subsessile, apex acuminate, spinulose (rarely broadly acute), margins entire; flowers generally unisexual (dioecious), in small clusters about two thirds way down the cladodes, shortly pedicellate (pedicels 1–2·5 mm. long) or subsessile, subtended by minute, broadly ovate, membranous bract-scales; perianth whitish, broadly infundibuliform, segments 6, recurved, the 3 outer 2·5–3 mm. long, about 1·5 mm. wide, ovate-elliptic, the 3 inner much smaller and narrower; about 1·5 mm. long, 0·6 mm. wide; male flowers with 3 stamens; filaments about 1 mm. long, united in a column around the abortive ovary, glabrous; anthers pallid, broadly ovate or suborbicular, about 0·8 mm. long and almost as wide; female flowers with the sterile, anther-less staminodes united into a tube around a sessile, narrowly ovoid, glabrous ovary about 1·3 mm. long, 0·8 mm. wide; stigma sessile or subsessile, pileate, about 1 mm. diam. Fruit a globose, scarlet berry about 1–1·5 cm. diam.; seeds usually 2, semi-globose, translucent, about 7 mm. diam.; testa very thin, minutely granulose.

var. **angustifolius** *Boiss.*, Fl. Orient., 5: 341 (1882); H. Stuart Thompson in Journ. Bot., 44: 339 (1906).
 R. aculeatus L. var. vel forma *pumilus* Druce in Rep. B.E.C., 9: 471 (1931).
 TYPE: Turkey; "in fauce Gülek Boghas Tauri Cilicici" *Kotschy* (G).

Cladodes mostly four or more times as long as broad, lanceolate or narrowly ovate, long-acuminate.

HAB.: In hedges and thickets, sometimes on rocky slopes, usually in shade; near sea-level to 3,500 ft. alt.; fl. Febr.–March.

DISTR.: Divisions 2, 7, 8. Var. *angustifolius* in the eastern Mediterranean (Italy and eastwards) to Transcaucasia; typical *R. aculeatus* widespread in western, central and southern Europe.

2. Platres, 1900, *A. G. & M. E. Lascelles* s.n.! also, 1955, *Kennedy* 1868! Mylos [? date], *Chapman* 210; Saïttas, 1932, *Syngrassides* 356! Perapedhi, 1962, *Meikle* 2804!
7. Lapithos, 1905, *Holmboe* 839; also 1928 or 1930, *Druce*; St. Hilarion, 1937, *Syngrassides* 1740! Bellapais, 1939, *Lindberg f.* s.n.; between Bellapais and Buffavento, 1941, *Davis* 2175! Yaïla, 1941, *Davis* 3604! Three-quarters of a mile E. of St. Hilarion, 1948, *Casey* 148!
8. Rizokarpaso, 1912, *Haradjian* 155; Koronia, 1937, *Chapman* 244! Ronnas Valley on N.E.-facing marl slopes, 1957, *Merton* 2929!

NOTES: The distinction between *R. aculeatus* var. *aculeatus* and var. *angustifolius* appears to rest solely on the size and shape of the cladode; intergradations between the two varieties are not uncommon, even in Cyprus, and the distinction cannot be reckoned of much significance, though marked enough when a range of specimens from Europe and the eastern Mediterranean regions is compared.

SUBFAMILY 3. **Asphodeloideae** Rhizomatous (rarely bulbous) perennials, or rarely annuals; leaves often all basal, parallel-nerved, generally long and linear; inflorescence racemose or paniculate; flowers hermaphrodite; perianth-segments 6, free to base or ± united, mostly spreading; stamens 6 (rarely 3); anthers introrse, basifixed or versatile; ovary 3-locular. Fruit a loculicidal capsule.

4. ASPHODELUS L.

Sp. Plant., ed. 1, 308 (1753).
Gen. Plant., ed. 5, 146 (1754).

Rhizomatous perennials, or rarely annuals; leaves all basal; scape simple or branched, solid or fistulose; bracts scarious or membranous; flowers numerous, white or pinkish; pedicels articulated; perianth-segments subequal, free or united only at base, usually patent; stamens 6; filaments often dilated at base and enveloping the ovary; anthers oblong or almost linear, versatile; ovary sessile, 3-locular; style filiform; stigma capitate or 3-lobed; ovules 2 in each loculus. Capsule trigonous, coriaceous; seeds 6 or fewer by abortion, angular; testa black, often rugose; endosperm cartilaginous.

About 12 species in S. Europe, the Mediterranean region and eastwards to the Himalayas.

Plant robust, often 1 m. or more high, leaves flat, strap-shaped; roots with fusiform tubers
                                                                    1. A. aestivus
Plant slender, generally less than 70 cm. high; leaves grass-like, fistulose; roots without tubers:
    Plant perennial; perianth-segments 12–14 mm. long; capsule about 5 mm. diam.
                                                                    2. A. fistulosus
    Plant annual; perianth-segments 4–7 mm. long; capsule 3–4 mm. diam.   3. A. tenuifolius

1. A. aestivus Brot., Fl. Lusit., 1: 525 (1804); Post, Fl. Pal., ed. 2, 2: 655 (1933); Nattrass, First List Cyprus Fungi, 2, 14 (1937); Osorio-Tafall et Seraphim, List Vasc. Plants Cyprus, 19 (1973).
    A. microcarpus Viv., Fl. Cors. Diagn., 5 (1824), App. Altera Fl. Cors. Prodr., 6 (1830); Boiss., Fl. Orient., 5: 313, (1882); Elektra Megaw, Wild Flowers of Cyprus, 15, t. 30 (1973).
    A. ramosus L. ssp. microcarpus (Viv.) Baker in Journ. Linn. Soc., 15: 270 (1876); Holmboe, Veg. Cypr., 44 (1914); Lindberg f., Iter Cypr., 10 (1946).
    [A. ramosus (non L.) Sibth. et Sm., Fl. Graec., 3: 28, t. 334 (1823); Kotschy in Oesterr. Bot. Zeitschr., 12: 277 (1862); Unger et Kotschy, Die Insel Cypern, 199 (1865); Sintenis in Oesterr. Bot. Zeitschr., 31: 191, 229, 326 (1881) et 32: 123, 192, 365 (1882); C. E. Gresham in New Flora & Silva, 5: 271 (1933).]
    TYPE: Portugal; "in Transtagana et Extremadura, ad vias praesertim in fossulis, soloque depresso, et in Beiro prope Fundão", Brotero (? LISU).

Robust erect perennial to 1 m. high or more; rhizome compact, the clustered fleshy roots bearing large, fusiform tubers; leaves strap-shaped, flat, slightly keeled, glabrous, glaucous, 20–60 (or more) cm. long, 1·2–3 (or more) cm. wide, crowded at the base of the scape, and surrounded by a collar of fibres formed from the decayed remains of old leaf-rosettes, tapering gradually above to an acute (generally damaged) apex; cauline leaves absent; scape terete, solid, commonly more than 12 mm. diam., glabrous, unbranched below, divided above into 5–8 or more erecto-patent inflorescence-branches 12–30 (–60) cm. long, the terminal branch generally somewhat longer than the laterals; bracts scarious, brownish, linear-subulate, soon deciduous, the lower often 2–3 (or more) cm. long, 0·5–0·8 cm. wide, the upper much smaller, rarely exceeding 1 cm. in length, often slenderly caudate-acuminate; flowers very numerous, rather crowded along the inflorescence-branches, mostly solitary in the axil of each bract; pedicel slender, about 1 cm. long, articulated near the middle; perianth whitish, the segments with a distinct pinkish-brown median nerve, subequal, spreading, narrowly oblong, subacute or obtuse, about 15 mm. long, 4 mm. wide; filaments almost as long as perianth-segments, filiform above, dilated at base and enveloping ovary; anthers yellow, narrowly oblong, ·2–2·5 mm. long, 0·6 mm. wide; style a little longer than stamens; stigma obscure, subcapitate. Fruit a small coriaceous, sharply trigonous, transversely rugose, brownish, obovoid capsule, about 8–10 mm. long, 6–8 mm. wide, encircled at the base by the remnants of the perianth; seeds generally 6, sharply angled, shaped like the segments of an orange, 5–6 mm. long, 2 mm. wide; testa blackish-brown, minutely and closely verruculose.

HAB.: Roadsides, waste ground, garigue, clearings in Pine forest, seashores, etc.; sea-level to 6,000 ft. alt.; fl. Jan.–June.

DISTR.: Divisions 1–8, common. Widespread in the Mediterranean region, also Portugal, and eastwards to Iraq; Atlantic Islands.

1. Ayios Yeoryios (Akamas), 1962, *Meikle*! Polis, 1962, *Meikle*!
2. Prodhromos, 1862, *Kotschy*; also, 1939, *Lindberg f.* s.n.! Troödos, 1914, *Feilden* s.n.!
3. Near Stavrovouni, 1880, *Sintenis*.
4. Between Ormidhia and Famagusta, 1862, *Kotschy*; Cape Greco, 1862, *Kotschy*; Phaneromene near Larnaca, 1880, *Sintenis & Rigo* 858!
5. Kythrea, 1880, *Sintenis*; Lefkoniko, 1880, *Sintenis*; Salamis, 1962, *Meikle*!
6. Nicosia, 1932–1936, *Nattrass* 183, 472, 755; Dhiorios, 1962, *Meikle*! 15 miles W. of Nicosia, 1973, *P. Laukkonen* 9!
7. Pentadaktylos, 1862, *Kotschy*; Kantara, 1880, *Sintenis*; Kornos, 1936, *Syngrassides* 974! Kyrenia, 1956, *G. E. Atherton* 846! Kharcha, 1956, *G. E. Atherton* 1077! Tjiklos, 1973, *P. Laukkonen* 53!
8. Rizokarpaso; 1880, *Sintenis*; Akradhes, 1962, *Meikle*!

NOTES: One of the commonest plants on the island, and for this reason, and also because of its ungainly size, rarely collected.

2. A. fistulosus *L.*, Sp. Plant., ed. 1, 309 (1753); Unger et Kotschy, Die Insel Cypern, 200, (1865); Boiss., Fl. Orient., 5: 314 (1882); Holmboe, Veg. Cypr., 45 (1914); Post, Fl. Pal., ed. 2, 2: 655 (1933); Lindberg f., Iter Cypr., 10 (1946); Rechinger f. in Arkiv för Bot., ser. 2, 1: 419 (1950); Osorio-Tafall et Seraphim, List Vasc. Plants Cyprus, 19 (1973).
TYPE: "*in* Gallo-Provincia, Hispania, Creta".

Erect perennial 40–70 cm. high; rhizome compact, emitting numerous fleshy roots, but without tubers; leaves numerous, all basal, narrowly linear, fistulose, 13–30 cm. long, 0·2–0·5 cm. wide above the abruptly dilated, sheathing, submembranous base, tapering gradually to a slender acumen, glabrous or minutely asperulous along the prominent nerves; scapes generally several from each leaf-rosette, terete, fistulose, glabrous, generally unbranched below, with 2–8 lax, elongate, erecto-patent branches in the region of the inflorescence; bracts ovate-caudate, the lower up to 10 cm. long, 5 mm. wide, the upper gradually diminishing, all membranous with a conspicuous dark midrib; flowers ephemeral, numerous, in slender, rather lax racemes; pedicels very short at anthesis, generally shorter than the subtending bract, articulated towards the base, elongating to 8–10 mm. in fruit, then erecto-patent and articulated near the middle; perianth white or pinkish with a darker brownish midrib running the length of each segment; segments spreading, narrowly oblong, obtuse or subacute, 12–14 mm. long, 2–2·5 mm. wide; stamens 6; filaments about 8 mm. long at anthesis, flattened and puberulous but not dilated at base nor enveloping the ovary; anthers oblong, about 2·5 mm. long, 1·5 mm. wide, ovary sessile, subglobose, trigonous, glabrous; style slender, about 7 mm. long, stigma rather large, subcapitate, distinctly 3-lobed. Capsule subglobose, about 5 mm. diam., strongly rugose transversely; seeds usually 2 in each loculus, shaped like the segments of an orange, about 3·5 mm. long, 2 mm. wide; testa blackish with 3–5 conspicuous transverse depressions, minutely and closely verruculose.

HAB.: Roadsides, edges of fields; sea-level to 2,200 ft. alt.; fl. Febr.–June.

DISTR.: Divisions 1, 2, 4–8, local. Mediterranean region; Atlantic Islands.

1. Khrysokhou, 1862, *Kotschy*; Ktima, 1913, *Haradjian* 652! Inia, 1941, *Davis* 3292!
2. Kakopetria, 1949, *Casey* 854!
4. Ayia Napa, 1904, *Holmboe* 56; also, 1939, *Lindberg f.* s.n.! Famagusta, 1912, *Haradjian* 103.
5. Nisou, 1936, *Syngrassides* 1256!
6. Strovolos, 1902, *A. G. & M. E. Lascelles* s.n.! Skylloura-Philia road, 1954, *Merton* 1873!

7. Kalogrea, 1934, *Syngrassides* 4944! Akanthou, 1940, *Davis* 2022! Lapithos, 1967, *Merton* in ARI 553!
8. Rizokarpaso, 1880, *Sintenis & Rigo* 560! also, 1941, *Davis* 2375!

3. A. tenuifolius *Cav.* in Anal. Cienc. Nat., 3: 46, t. 27, fig. 2 (1801), Icones, 6: 63, t. 587, fig. 2 (1801); Boiss., Fl. Orient., 5: 314 (1882); Post, Fl. Pal., ed. 2, 2: 655 (1933).
    A. fistulosus L. var. *tenuifolius* (Cav.) Baker in Journ. Linn. Soc., 15: 272 (1876).
    A. fistulosus L. ssp. *tenuifolius* (Cav.) Arcang., Comp. Fl. Ital., ed. 2, 140 (1894).
    TYPE: Morocco; "El Sr. Broussonet la encontró junto á Mogador". (? MPU).

Erect annual (7–) 10–50 cm. high; roots fibrous, fused above into what appears to be a short taproot; leaves very similar to those of *A. fistulosus* L., slender, linear, fistulose, 3–10 cm. long, 0·1–0·3 cm. wide above the conspicuously dilated, membranous base, glaucous green, usually papillose-asperulous, sometimes conspicuously so, along the nerves; scapes often solitary, sometimes simple, but usually 2–5-branched in the region of the inflorescence, terete, fistulose, frequently papillose-asperulous towards the base, branches erecto-patent, generally less than 20 cm. long; bracts ovate-caudate, white-membranous with a dark midrib, generally less than 5 mm. long, 4 mm. wide, often minute; flowers ephemeral, solitary in the axils of the bracts, forming very slender, lax, many-flowered racemes; pedicels 2–3 mm. long at anthesis, elongating to 5–8 mm. in fruit, erect, glabrous, articulated about one-third way up; perianth white or pinkish, segments spreading, oblong, obtuse, 4–7 mm. long, 2–2·5 mm. wide, with a dark midrib; stamens 6; filaments about 2·5 mm. long, scarcely dilated, but flattened and puberulous at base, not enveloping the ovary; anthers oblong, about 1·3 mm. long, 0·8 mm. wide; ovary subglobose, trigonous, about 0·8 mm. diam., glabrous; style about 2·5 mm. long; stigma distinctly 3-lobed. Capsule subglobose, about 3–4 mm. diam., distinctly rugose transversely; seeds usually 6, shaped like the segments of an orange, about 3 mm. long, 1·5 mm. wide, faces with deep transverse depressions; testa fuscous-black, minutely and closely verruculose.

HAB.: On dry limestone or sandstone slopes; near sea-level; fl. March–April.

DISTR.: Divisions 1, 2, rare. Mediterranean region (chiefly N. Africa and eastern area), Atlantic Islands and Asia eastwards to Baluchistan and India.

1. Ayios Yeoryios Island (Akamas), 1962, *Meikle* 2162!
2. Vouni Palace, on steep, chalky slopes, 1974, *Meikle* 4823!

NOTES: Closely allied to *A. fistulosus*, but distinct and replacing the latter over much of western Asia. It may be commoner in lowland areas of Cyprus than would appear, but seems considerably less frequent than *A. fistulosus*.

5. ASPHODELINE *Reichb.*
Fl. Germ. Excurs., 116 (1830).

Erect perennials with fleshy roots; stems often leafy from base to apex, leaves alternate, linear with conspicuously dilated membranous bases; inflorescences simple or branched, racemose; bracts membranous, caudate; flowers solitary or in pairs (sometimes more) in the axil of each bract; pedicels articulated; perianth-segments 6, narrow, free to the base or almost to the base, spreading; stamens 6, or the outer whorl sometimes sterile; filaments unequal, declinate, dilated at base and wholly or partly enveloping ovary; anthers versatile, oblong or linear, introrse; ovary sessile, subglobose, 3-locular; style filiform; stigma small, sometimes 3-lobed. Fruit a subglobose, coriaceous, loculicidally dehiscent capsule; seeds 2 or 1 (by abortion) in each loculus, angular; testa usually black, rugose; endosperm present.

Inflorescence dense, spiciform, unbranched, with crowded flowers; bracts 20–30 mm. long

1. A. lutea

Inflorescence lax, branched, paniculate, flowers not crowded; bracts generally less than 20 mm. long, often very small - - - - - - - - - - **2. A. brevicaulis**

1. A. lutea (*L.*) *Reichb.*, Fl. Germ. Excurs., 116 (1830); Boiss., Fl. Orient., 5: 316 (1882); Post, Fl. Pal., ed. 2, 2: 656 (1933).

> *Asphodelus luteus* L., Sp. Plant., ed. 1, 309 (1753).
>
> [*Asphodeline liburnica* (non (Scop.) Reichb.) Seligman in Quart. Bull. Alpine Gard. Soc., 20: 235 (1952).]
>
> TYPE: "*in* Sicilia".

Robust erect perennial 60–100 cm. high; roots fleshy but not tuberous; stem unbranched, terete, 1–1·5 cm. diam., covered with broad, membranous, transparent leaf-bases; leaves rather crowded at base of stem, more spaced above, linear, the basal 15–30 cm. or more long, 0·2–0·4 cm. wide, glabrous, distinctly nerved, the upper progressively smaller and narrower, with broad membranous bases and a filiform-subulate lamina; inflorescence dense, unbranched, 10–15 cm. long at anthesis, elongating to 30 cm. or more in fruit; bracts ovate-caudate, membranous, 20–30 mm. long, 10–15 mm. wide, with a clearly defined midrib; rhachis glabrous, longitudinally ridged; flowers solitary or paired (sometimes more) in the axil of each bract, marcescent and persistent; pedicels about 1 cm. long, angular, glabrous, articulated about the middle, elongating to 2 cm. or more in fruit; perianth yellow, segments narrowly oblong, obtuse, about 20 mm. long, 4 mm. wide, with a dark midrib, free from slightly above top of ovary; stamens 6 (or sometimes 3); filaments of exterior stamens about 12 mm. long, those of inner stamens about 20 mm. long; anthers of outer stamens about 3 mm. long, 1·5 mm. wide, those of inner stamens about 6 mm. long, almost 2 mm. wide; ovary ovoid-subglobose, trigonous, about 4 mm. long, 3·5 mm. wide; style about 25 mm. long; stigmas small, very obscurely 3-lobed. Capsule leathery, subglobose, about 12 mm. diam.; seeds generally solitary in each loculus, about 5 mm. long and almost as wide, sharply 3- or 4-angled; testa fuscous-black, obscurely verruculose.

HAB.: Rocky limestone slopes under *Cupressus sempervirens*; about 2,000–2,500 ft. alt.; fl. April.

DISTR.: Division 7, rare. Balkan peninsula and eastern Mediterranean region.

7. Kornos, 2,000–2,500 ft. alt. on north-facing slopes, 1950, *Casey* 972! and, same area, 1954, *Seligman* in *Casey* 1334!

2. A. brevicaulis (*Bertol.*) *Gay* in Balansa, Pl. d'Orient, 1855, no. 316, et ex Baker in Journ. Linn. Soc., 15: 276 (1876); Boiss., Fl. Orient., 5: 317 (1882); Post, Fl. Pal., ed. 2, 2: 657 (1933).

> *Asphodelus brevicaulis* Bertol. in Nov. Comm. Acad. Bonon., 5: 430 (1842), Misc. Bot., 1: 20 (1842).
>
> [*Asphodelus liburnicus* (non Scop.) Sintenis in Oesterr. Bot. Zeitschr., 32: 192 (1882).]
>
> [*Asphodeline liburnica* (non (Scop.) Reichb.) Boiss., Fl. Orient., 5: 316 (1881) quoad spec. cypr.; Holmboe, Veg. Cypr., 45 (1914); Rechinger f. in Arkiv för Bot., ser. 2, 1: 419 (1950); Osorio-Tafall et Seraphim, List Vasc. Plants Cyprus, 19 (1973).]
>
> TYPE: Turkey or Iraq; "ex oris Euphratis [1836, *Chesney*] Pl. sicc. Euphr. n. 163" (? BOLO ? G).

Erect, slender or (less often) robust perennial 60–100 cm. high; roots rather fleshy but not at all tuberous; stems subterete, fistulose, glabrous; leaves crowded towards base of stem, but extending rather sparsely upwards almost to inflorescence, linear, 10–30 cm. long, 0·2–0·4 cm. wide, decreasing in size towards apex of stem, tapering from a conspicuously dilated, membranous, sheathing base; nervation rather prominent, minutely asperulous; inflorescence with 2–6 (or more) spreading, glabrous, angular branches; bracts membranous with a well-defined midrib, ovate-caudate, sometimes 20 mm. long, 5 mm. wide at base of inflorescence, becoming progressively smaller, broader and more shortly cuspidate upwards; flowers lax, solitary or in pairs in the axil of the bracts; pedicels glabrous, 5–10 mm. long at anthesis, elongating considerably in fruit, usually articulated near the base; perianth widely infundibuliform or spreading, segments yellow, the 3 exterior stained orange externally, subequal, narrowly oblong, obtuse or subacute, 15–20 mm. long, 3–4 mm. wide, free almost to base, each segment with a conspicuous dark midrib; stamens 6 in 2 series; filaments of the outer series slender, 10–13 mm. long, not much dilated at the papillose base, inner series similar but about 13–16 mm. long; anthers oblong, the 3 outer about 3·5 mm. long, 1·5 mm. wide, the 3 inner about 4·5 mm. long, 1·5 mm. wide; ovary sessile, subglobose, trigonous, glabrous, about 1·5 mm. diam.; style slender, glabrous, 12–18 mm. long; stigma small, obscurely 3-lobed. Capsule subglobose, about 12 mm. diam., coriaceous, transversely rugose; fully developed seeds not seen, probably similar to those of *A. lutea* (L.) Reichb.

HAB.: Open limestone mountainside; c. 2,000 ft. alt.; fl. April–May.

DISTR.: Division 7 only, rare. S. Turkey, Syria, Palestine, Iraq.

7. Ridge E. of Olymbos Peak, above Akanthou, 1880, *Sintenis & Rigo* 561 ! also, 1950, *Casey* 1042 ! and 1957, 1967, *Merton* 2896 ! and in ARI 777 !

NOTES: Also noted by Seligman (Quart. Bull. Alpine Gard. Soc., 20: 235; 1952) "in great quantities from about 2,200 ft. to the top of Mt. Kornos", but the record should be referred to *A. lutea*. Sintenis was the first to misidentify the plant from the eastern end of the Northern Range as *Asphodeline liburnica*, and has been followed by most writers on the flora. True *A. liburnica* differs in its tuberous roots, foliage largely restricted to base of stems, and simple or rarely few-branched inflorescences.

ALOË *L.*

Sp. Plant., ed. 1, 319 (1753).
Gen. Plant., ed. 5, 150 (1754).
G. W. Reynolds, The Aloes of Tropical Africa and Madagascar,
537 pp. (1966).

Succulent, acaulescent, shrubby or arborescent perennials, mostly with rosettes of thick, amplexicaul, lanceolate or linear, spinose or spinose-margined leaves; inflorescence scapose, simple or branched; flowers hermaphrodite, generally pedicellate, in lax or dense racemes; perianth reddish, yellowish or whitish, tubular, straight or curved, segments 6, subequal, free to the base or more or less connate; stamens 6, the 3 inner usually rather longer, with narrower filaments, than the 3 outer; anthers linear or oblong; ovary sessile, ovoid, trigonous, 3-locular; style filiform; stigma small, capitate; ovules numerous in 2 ranks in each loculus. Fruit a papery or coriaceous, globose, ovoid or shortly cylindrical capsule, dehiscing loculicidally; seeds numerous, angled or flattened; testa grey or black.

About 350 species, chiefly in tropical and South Africa, Madagascar and Arabia.

A. VERA (*L.*) *Burm. f.*, Fl. Ind., 83 (1768); Boiss., Fl. Orient., 5: 329 (1882); Holmboe, Veg. Cypr., 45 (1914); Post, Fl. Pal., ed. 2, 2: 659 (1933); Osorio-Tafall et Seraphim, List Vasc. Plants Cyprus, 19 (1973).
    A. perfoliata L. var. II *vera* L., Sp. Plant., ed. 1, 320 (1753).
    A. barbadensis Mill., Gard. Dict., ed. 8, no. 2 (1768); Poech, Enum. Plant. Ins. Cypr., 8 (1842); Reynolds, Aloes Trop. Afr., 144 (1966).
    A. vulgaris Lam., Encycl. Méth., 1: 86 (1783); Sibth. et Sm., Fl. Graec. Prodr., 1: 238 (1809), Fl. Graec., 4: 34, t. 341 (1823); Unger et Kotschy, Die Insel Cypern, 192 (1865).
    TYPE: "*in* Indiis".

Acaulescent, suckering perennial, leaves in dense rosettes, suberect, fleshy, lanceolate, about 40–50 cm. long, 6–7 cm. wide at base, grey-green, sometimes tinged or spotted red, margins armed with remote, deltoid, shortly spinulose teeth; scape 60–100 cm. high, simple or very sparingly branched; flowers in dense, cylindrical racemes 20–40 cm. long, 5–6 cm. diam., erect or spreading in bud, pendulous at anthesis, bracts pallid, papery, reflexed, oblong- or ovate-acuminate, 10–15 mm. long, 5–6 mm. wide; pedicels stout, 4–5 mm. long, recurved near apex; perianth yellow, cylindrical or narrowly ventricose, 20–30 mm. long, 5–7 mm. wide; segments erect, acute, free for about one-third of their length, the exterior connate marginally to the back of the rather broader interior segments; filaments slender, about as long as perianth; anthers linear-oblong, shortly exserted, about 4 mm. long, 1 mm. wide; ovary oblong, 5–6 mm. long, 3 mm. wide; style slender, glabrous, about 25 mm. long; stigma truncate, exserted. Fruit not seen.

HAB.: Waste ground; seashores; sea-level to 300 ft. alt.; fl. April–May.

DISTR.: Divisions 1, 4, 6, an old introduction. Widely naturalized in the Mediterranean region; Atlantic Islands; possibly a native of S. Africa.

1. Between Paphos and Ktima, 11 May, 1787, *Sibthorp*.
4. Famagusta, 18 April, 1787, *Sibthorp*; Larnaca, 1862, *Kotschy*; Ayia Napa, 1931, *J. A. Tracey* 75!
6. Khrysiliou, 1936, *Syngrassides* 1224!

NOTES: Perhaps less common than formerly, but often ignored by collectors unwilling to cope with the fleshy, prickly growths. It was at one time highly valued as a medicinal plant, and, in Cyprus at least, is always a garden escape or outcast.

SUBFAMILY 4. **Melanthoideae** Erect or rarely climbing, sometimes acaulescent perennials arising from rhizomes, bulbs or corms; scape developed, or flowers arising from within subterranean leaf-sheaths; perianth-segments equal or subequal; anthers introrse or extrorse, basifixed or versatile. Fruit a loculicidally or septicidally dehiscent capsule.

6. COLCHICUM *L.*

Sp. Plant., ed. 1, 341 (1753).
Gen. Plant., ed. 5, 159 (1754).
B. Stefanov, Mon. *Colchicum* in Sborn. Bulg. Akad. Nauk, 22: 1–99 (1926).
N. Feinbrun in Pal. Journ. Bot., J. series, 6: 71–95 (1953).

Perennial, acaulescent herbs arising from tunicated corms or rarely from short rhizomes; tunics often produced apically into a tubular "neck" or collar; leaves all basal, linear, strap-shaped or elliptical, parallel-nerved, appearing with the flowers (synanthous) or after anthesis (hysteranthous); flowers hermaphrodite, solitary or in 2–30-flowered fascicles, subsessile; bracts much reduced; perianth infundibuliform, pink, white or purple, sometimes tessellated; segments 6, in 2 subequal series, united basally into a long or short tube; stamens 6, inserted at base of perianth-segments; filaments sometimes swollen at base, or with lateral swellings or *lamellae*; anthers introrse, dorsifixed, versatile; ovary subterranean; styles 3, free, stigmas punctiform, truncate or decurrent along the style. Fruit a septicidal capsule; seeds numerous, subglobose; testa brownish; endosperm hard; embryo small.

About 60 species in Europe, the Mediterranean region and eastwards to Central Asia.

Leaves appearing after anthesis (hysteranthous), strap-shaped, 13–22 (–30) cm. long, 1·5–4·5 cm. wide; perianth-segments 2·5–4·5 cm. long, 0·5–1·2 cm. wide, white or pale pink
1. C. troodi

Leaves appearing with the flowers (synanthous), linear, 3–15 cm. long, 0·1–0·4 cm. wide;
    perianth-segments 1–3 cm. long, 0·2–0·6 cm. wide:
Perianth-segments white, lilac or mauve, 10–20 mm. long; anthers about 2·5 mm. long,
    usually purplish, greyish or brownish before dehiscence -   -   - **2. C. pusillum**
Perianth-segments pink (rarely white), about 30 mm. long, anthers about 3 mm. long, yellow
    before dehiscence - - - - - - - - - **3. C. stevenii**

1. **C. troodi** *Kotschy* in Unger et Kotschy, Die Insel Cypern, 190 (1862); Baker in Journ. Linn.
    Soc., 17: 430 (1879); Boiss., Fl. Orient., 5: 161 (1882); Hooker f. in Bot. Mag., 112, t. 6901
    (1886); Stefanov, Mon. *Colchicum*, 56 (1926); Osorio-Tafall et Seraphim, List Vasc. Plants
    Cyprus, 19 (1973); Elektra Megaw, Wild Flowers of Cyprus, 17, t. 37 (1973).
    C. decaisnei Boiss., Fl. Orient., 5: 157 (1882); Stefanov, Mon. *Colchicum*, 67 (1926); Post,
    Fl. Pal., ed. 2, 2: 612 (1933); Feinbrun in Pal. Journ. Bot., J series, 6: 83 (1953).
    [*C. autumnale* (non L.) Poech, Enum. Plant. Ins. Cypr., 8 (1842).]
    TYPE: Cyprus; "Im Gebirge um Prodromo [Prodhromos] zerstreut [*Kotschy*] n. 904" (K !).

Hysteranthous perennial; corm ovoid, 3–6 cm. long, 2·5–6 cm. wide,
densely covered with several layers of thin, papery, rather lustrous, dark
brown tunics produced apically into a subtruncate neck 5–9 cm. long; leaves
2–5, suberect or spreading, strap-shaped, 13–22 (–30) cm. long, 1·5–4·5 cm.
wide, glabrous, ciliate or sometimes thinly pilose all over, dark green or
glaucescent, apex obtuse or shortly acute; flowers generally several, 2–6
(–12) from each corm, the earlier ones often shrivelled but persistent as the
later ones expand; perianth-tube slender, 12 cm. or more long, exceeding the
tunic-neck by 1·5–7 cm., segments erecto-patent, narrowly oblong-
lanceolate, white or pale pink, about 2·5–4·5 cm. long, 0·5–1·2 cm. wide,
tapering to base, rather abruptly narrowed at apex; filaments 1·5–2 cm.
long, very slender, glabrous or sometimes thinly pilose near point of
insertion; anthers yellow, linear, 6–7 mm. long, 0·8 mm. wide; styles 14 cm.
or more long, equalling or shortly exceeding the stamens; stigmas a little
recurved, punctiform or very shortly decurrent. Fruit ellipsoid, glabrous, or
minutely puberulous 1·5–2·5 (–3) cm. long, 0·8–1·5 cm. wide; seeds
subglobose, 3 mm. diam.; testa light brown, granulose.

HAB.: In garigue on dry, rocky, calcareous or non-calcareous slopes; in Pine forest or Hazel
groves; near sea-level to 6,400 ft. alt.; fl. Sept.–Nov.

DISTR.: Divisions 2, 3, 7, 8, locally abundant. S.E. Turkey, Syria, Palestine.

2. Troödhitissa, Prodhromos, Troödos, Evrykhou, Platania, Moniatis R., Xerokolymbos,
    Kryos Potamos, Asprokremnos, Platres, Ayia valley, Tripylos, Papoutsa, Lagoudhera,
    Amiandos, etc. often very abundant, 1840–1977, *Kotschy* 904 ! 1086; *Sintenis & Rigo* 866 !
    A. G. & M. E. Lascelles s.n. ! *C. B. Ussher* 103 ! *Syngrassides* 830 ! 1672 ! *Kennedy* 127–155 !
    Lady Loch 1 ! *Davis* 1974 ! 1990 ! 3497 ! *S. G. Cowper* s.n. ! *Casey* in K. 3132 ! 3133 !
    3134 ! *Mavromoustakis* 20 ! *Merton* 1956 ! 3147 ! in ARI 47 ! *G. E. Atherton* 509 ! *P. H.
    Oswald* 148 ! *E. Pasche jr.* L74/6 ! *Barclay* 1907 ! *J. J. Wood* s.n. ! etc.
3. Episkopi, 1963, *J. B. Suart* 131 ! Between Episkopi and Paramali, 1964, *J. B. Suart* s.n. !
    Temple of Apollo near Episkopi, 1964, *J. B. Suart* s.n. !
7. St. Hilarion, 1929, *C. B. Ussher* 53 ! Near Vasilia and on N. slopes of Kornos, 1940–1948,
    Davis 1996 ! 2193 ! *Kennedy* 1699 ! Between Halevga and Kharcha, 1941, *Davis* 2167 !
    Kyrenia Pass, 1948, *Casey* 126 !
8. Between Rizokarpaso and Apostolos Andreas, 1941, *Davis* 2297 !

NOTES: Having examined a very wide range of specimens, I am convinced that the supposed
differences between *Colchicum troodi* and *C. decaisnei* are illusory, and that, far from being an
endemic as generally supposed, the former has a wide distribution in the eastern Mediterranean
area. The flowers of *C. troodi* vary enormously in overall size and in the relative width of the
perianth-segments, this variation is largely due to the vigour of individual plants and is to be
found even within a single population, as is also variation in the colour of the perianths, from
white to pink. The leaves are decidedly less variable, retaining much the same proportions in
lush and depauperate plants. Likewise there is little noteworthy variation in stamens and
styles. As in other *Colchicum* species, glabrousness and hairiness of leaves is of trivial
significance and may vary from one extreme to the other in adjacent plants.

2. **C. pusillum** *Sieber* in Flora, 5: 248 (1822); Stefanov, Mon. *Colchicum*, 55 (1926); Rechinger f.,
    Fl. Aegaea, 709 (1943).

C. hiemale Freyn in Bull. Herb. Boiss., 5: 802 (1987); Stefanov, Mon. *Colchicum*, 56 (1926); A. K. Jackson et Turrill in Kew Bull., 1938: 468 (1938); Seligman in Quart. Bull. Alpine Gard. Soc., 20: 233 (1952); Meikle in R.H.S. Lily Year Book 1968: 98 (1967); Osorio-Tafall et Seraphim, List Vasc. Plants Cyprus, 19 (1973).

[*C. bertolonii* (non Steven) Kotschy in Unger et Kotschy, Die Insel Cypern, 191 (1865); Boiss., Fl. Orient., 5: 165 (1882) pro parte; Post in Mém. Herb. Boiss., 18: 102 (1900); H. Stuart Thompson in Journ. Bot., 44: 339 (1906); Holmboe, Veg. Cypr., 44 (1914).]

[*C. cupanii* (non Guss.) Osorio-Tafall et Seraphim, List Vasc. Plants Cyprus, 19 (1973).]

TYPE: Crete; "am Cap Meleca im Kreta auf dürren trocknen Stellen" (? G).

Synanthous perennial; corm narrowly ovoid, 1·5–2 cm. long, 1–1·4 cm. wide, rather densely covered with several layers of thin papery tunics, the inner usually rather lustrous and rich chestnut-brown, the outer fuscous, neck of tunic subtruncate, 3–8 cm. long; leaves 2–6 (or more), developing with the flowers, arcuate-recurved, linear, 3–8 cm. long (and probably much longer in fruiting specimens), 0·1–0·4 cm. wide, glabrous, apex obtuse or subacute; flowers generally several, 2–6 (or more) from each corm; perianth-tube very slender, 5–10 cm. or more long, exceeding the tunic-neck by 1·5–6 cm.; segments erecto-patent or spreading, narrowly oblong-lanceolate or oblanceolate, 10–20 mm. long, 2–5 mm. wide, pale mauve, lilac or whitish, tapering to base, rounded or acute at apex; filaments white, about 7 mm. long, glabrous, coloured orange-yellow at point of insertion; anthers fuscous, purplish, greyish or brownish, rarely yellow, linear, about 2·5 mm. long, 0·7 mm. wide, styles 6–11 cm. long, more or less equalling the stamens; stigmas punctiform, scarcely recurved. Fruit and seeds not seen.

HAB.: Dry, rocky hillsides, dry pastures and cultivated land, sea-level to 3,000 ft. alt.; fl. Nov.–Dec.

DISTR.: Divisions 2, 4–8. Also in Crete and Aegean Islands.

2. Aphamis, 1938, *Kennedy* 1260!
4. Between Larnaca and Varosha, 1892, *Deschamps* 469; between Avgorou and Famagusta, 1862, *Kotschy* 179.
5. Athalassa, 1936, *Lady Loch* 15! and, 1967, *Economides* in ARI 1033! Mile 5, Nicosia-Famagusta road, 1953, *Merton* 1093!
6. Nicosia 1898–1973, *Post* s.n., *Syngrassides* 858! *Davis* 2070! *Kennedy* 156! *P. H. Oswald* 20! *P. Laukkonen* 39! Dhiorios, 1953–54, *Kennedy* 1807! 1818!
7. Ayios Khrysostomos Monastery, 1936, *Lady Loch* 14! 18! Halevga, 1936, *Lady Loch* 8! Kornos, 1937, *Lady Loch* 14a! Kyrenia Pass, 1955, *G. E. Atherton* 765! Aghirda, 1955, *G. E. Atherton* 774!
8. Cape Elea, 1937, *Lady Loch* 16! S. of Neta, 1956, *Poore* 41!

NOTES: In the absence of fully mature specimens the leaf dimensions may be inaccurate.

3. C. stevenii *Kunth*, Enum. Plant., 4: 144 (1843); Boiss., Fl. Orient., 5: 165 (1882); Stefanov, Mon. *Colchicum*, 57 (1926); Post, Fl. Pal., ed. 2, 2: 615 (1933); A. K. Jackson et Turrill in Kew Bull., 1938; 467 (1938); Osorio-Tafall et Seraphim, List Vasc. Plants Cyprus, 19 (1973).

TYPE: "Mons Atlas (Desf.), Syria (Labill., Bove), Persia (Olivier) et ? Hispania (Loefl.)".

Synanthous perennial; corm ovoid, about 3 cm. long, 1·5–2·5 cm. wide, rather densely covered with several layers of thin, papery tunics, the inner usually rather lustrous, dark brown, the outer fuscescent, neck of tunic subtruncate, 6–7 cm. long; leaves 4–6 (or more), developing with the flowers, becoming recurved and often sinuous after anthesis, linear, 7–15 cm. long, 0·3–0·4 cm. wide, glabrous, apex obtuse or subacute; flowers sometimes solitary, more often 2–4 (occasionally more) from each corm; perianth-tube very slender, 10–12 cm. long, exceeding the tunic-neck by 4–8 cm.; segments erecto-patent, narrowly oblong-lanceolate or oblanceolate, about 30 mm. long, 4–6 mm. wide, bright pink (or sometimes white), tapering to base, subacute at apex; filaments white, about 14 mm. long, glabrous, coloured orange-yellow at point of insertion; anthers yellow, linear-oblong, about 3 mm. long, 0·8 mm. wide; styles 7–8 mm. long, often rather shorter than stamens; stigmas punctiform, not recurved. Fruit oblong-ovoid, about 12 mm. long, 9 mm. wide; seeds globose, about 3 mm. diam.; testa brown, granulose.

HAB.: On dry banks between fields; 50 ft. alt.; fl. Oct.–Dec.

DISTR.: Division 7, very rare. W. & S. Turkey, Syria, Palestine.

7. Below Vasilia, 1937, *Lady Loch* 9 !

NOTES: Very closely akin to *C. pusillum*, from which it differs most obviously in its larger, brighter flowers with larger, yellow anthers.

SUBFAMILY 5. **Lilioideae** Bulbous, caulescent perennials; inflorescence terminal, racemose or subumbellate, or flowers solitary; perianth-segments free to base, generally subequal; stamens 6; anthers introrse, basifixed or versatile. Fruit a loculicidal, or rarely a septicidal capsule; seeds generally compressed, discoid.

7. TULIPA *L.*

Sp. Plant., ed. 1, 305 (1753).
Gen. Plant., ed. 1, 145 (1754).
A. D. Hall, The Genus Tulipa, 171 pp., 40 pls. (1940).

Bulbs tunicate, the tunics usually lanate within, sometimes glabrous; stems simple or rarely branched above, solid, terete; leaves mostly basal, alternate, rather fleshy, diminishing markedly upwards; flowers terminal, generally erect, showy; bracts wanting or much reduced; perianth cup-shaped or infundibuliform, segments erect or spreading, subequal, often blotched at the base, but without a nectary-pit; stamens included; filaments often dilated at base, glabrous or pilose near the base; anthers basifixed, commonly purplish or violet; ovary sessile or subsessile, oblong; style very short or wanting; stigmas 3, often large and recurved; ovules numerous. Fruit an oblong, ellipsoid or subglobose, loculicidal capsule; seeds numerous, flattened, discoid or irregular, sometimes winged; testa brown or pallid, endosperm fleshy; embryo often small.

About 100 species chiefly in western and central Asia. Both Cyprus species belong to the type section *Tulipa*, with the filaments and bases of the perianth-segments glabrous.

Perianth-segments bright red with a conspicuous, acute, yellow-bordered basal blotch
　　　　　　　　　　　　　　　　　　　　　　　　　1. T. agenensis
Perianth-segments purplish-crimson; basal blotch usually small, subtruncate or rounded, sometimes obscure or obsolescent - - - - - - - - **2. T. cypria**

1. T. agenensis *DC.* in Redouté, Liliacées, 1: 60 (1804); Osorio-Tafall et Seraphim, List Vasc. Plants Cyprus, 20 (1973).

T. oculus-solis St. Amans in Rec. Trav. Soc. Agric. Sci. Arts Agen, 1: 75 (1804); Boiss., Fl. Orient., 5: 192 (1882); Holmboe, Veg. Cypr., 48 (1914); Post, Fl. Pal., ed. 2, 2: 620 (1933); A. D. Hall, The Genus Tulipa, 108, t. XXIII (1940); Elektra Megaw, Wild Flowers of Cyprus, 16, t. 32 (1973).

T. veneris A. D. Hall in Journ. Bot., 76: 312 (1939), The Genus Tulipa, 116 t. XXVII (1940).

[*T. praecox* (vix Tenore) Osorio-Tafall et Seraphim, List Vasc. Plants Cyprus, 20 (1973).]

TYPE: France; Lot-et-Garonne, "M. Lamouroux a découvert dans les environs d'Agen une nouvelle Tulipe".

Bulb ovoid, 2·5–3 cm. long, 2–2·5 cm. wide, tunic dark brown, rather lustrous, papery, lined internally with thick, woolly, pale fawn-coloured indumentum; stem unbranched, up to 30 cm. high; leaves usually 4, lanceolate, long-acuminate, the basal sometimes 30 cm. (or more) long, 2–4 cm. wide, the upper smaller and narrower, often almost linear and less than 1 cm. wide, all glaucescent, the margins sometimes undulated with a few scattered, short hairs fringing the cartilaginous edge; flower solitary, loosely cup-shaped or subinfundibuliform, segments subequal, oblong, 2·5–9 cm. long, 0·8–3·5 cm. wide, bright scarlet with a conspicuous, black, subacute, basal blotch 1–2·5 cm. long, bordered with a narrow yellow zone, apex of segments obtuse, acute or sometimes caudate-acuminate; filaments stout, violet-black, sometimes with a yellow apex, glabrous, 5–10 mm. long, tapering to apex from a wide base; anthers linear-oblong, 8–12 mm. long, 2 mm. wide, blackish before dehiscence; pollen yellowish; ovary oblong-ellipsoid, 10–12 mm. long, 3 mm. wide, tapering to an inconspicuous style; stigmas conspicuous, recurved. Capsule oblong-ellipsoid, 3–4 cm. long, 1·5–2 cm. wide, with 6 conspicuous ribs; seeds flattened, irregularly orbicular-cuneiform, winged, 0·8–1 cm. long, 0·5–0·7 cm. wide; testa bright brown, distinctly and regularly reticulate.

HAB.: Vineyards, orchards and cereal fields; 300–1,500 ft. alt.; fl. March–April.

DISTR.: Divisions 1, 3, 5, 6. Naturalized in S. France and Italy, also in Aegean Islands, Turkey, Syria, Palestine eastwards to Iran, but perhaps not strictly indigenous in any of these areas, and generally growing near habitations or as a weed of cultivated fields.

1. Between Stroumbi and Polemi, 1977, *Elektra Megaw* s.n. ! and, 1978, *Chesterman* 63 !
3. Monagri village, near Amaskou Church, 1954, *Scott-Moncrieff* in *Casey* 1652 !
5. Strongylos, 1905, *Holmboe* 358; Mora, in barley fields, 1956, *Merton* 2528 !
6. Kykko metokhi near Nicosia, 1904, *Miss Samson* s.n. ! also, 1927, *Rev. A. Huddle* 103 ! 1936, *Syngrassides* 896 ! and, 1950, *F. M. Probyn* s.n. !

NOTES: Some confusion exists as to the type-locality of *T. veneris* A. D. Hall: the author (Journ. Bot., 76: 319; 1939, The Genus Tulipa, 116) cites his cultivated material as coming originally from (7) Myrtou, which has long been known as one of the localities for *T. cypria* Stapf (infra). But W. B. Turrill (in Bot. Mag., 157: t. 9363) says: "Sir Daniel Hall ... received two consignments of bulbs from Mr. M. T. Dawe, at that time Director of Agriculture in Cyprus. The one consignment came from Myrtou, the other from the Kykko property near Nicosia. The two groups of plants are stated to be identical". The last sentence may be questioned, for while the Myrtou plants were almost certainly *T. cypria*, the only Tulip occurring at Kykko metokhi is *T. agenensis* (*T. oculus-solis*) and this, apart possibly from its chromosome number, is indistinguishable from *T. veneris*. In the circumstances one may reasonably assume that *T. veneris* came from Kykko metokhi near Nicosia. Likewise one may assume that Sir Daniel Hall erred in recording *T. cypria* from "Nicosia" (The Genus Tulipa, 112; 1940).

2. T. cypria *Stapf* in Bot. Mag., 157: t. 9363 (1934); A. D. Hall, The Genus Tulipa, 112 (1940); Seligman in Quart. Bull. Alpine Gard. Soc., 20: 230 (1952); Osorio-Tafall et Seraphim, List Vasc. Plants Cyprus, 20 (1973); Elektra Megaw, Wild Flowers of Cyprus, 16, t. 33 (1973).
[*T. montana* (non Lindl.) Kotschy in Unger et Kotschy, Die Insel Cypern, 191 (1865); H. Stuart Thompson in Journ. Bot., 44: 339 (1906); Holmboe, Veg. Cypr., 48 (1914).]
[*T. systola* (non Stapf) Osorio-Tafall et Seraphim, List Vasc. Plants Cyprus, 20 (1973).]
TYPE: Cyprus; "Myrtou; received through *M. T. Dawe*, cult. & comm. John Innes Hortic. Inst., 5th May 1931" (K).

Very similar to *T. agenensis* DC. in habit, bulb, leaves and inflorescence; perianth-segments oblong-obovate, 3–6 cm. long, 1·5–3·5 cm. wide, dark purplish-crimson (or rarely bright crimson), apex rounded or shortly cuspidate, base with a rather small truncate or rounded black blotch, usually about 1 cm. long, bordered with a creamy or yellowish zone, or blotch sometimes very obscure or obsolescent; stamens and pistils as in *T. agenensis*. Capsule broadly oblong, prominently 6-ribbed, about 3 cm. long, 2 cm. wide; seeds not seen. *Plate 94.*

HAB.: In cereal fields; sometimes in pastures or in Juniper forest, on deep, marly soils; 400–900 ft. alt.; fl. March–April.

DISTR.: Divisions 1, 6, 7. Endemic.

1. Kako Skala, in Juniper forest above Ayios Nikolaos, E. of the Akamas, 1948, *Kennedy* 1611 !
6. Kormakiti, 1941, *Davis* 2539 ! and, 1956, *Merton* 2614 ! 2615 ! Dhiorios, 1953, *Kennedy* s.n. ! and, 1962, *Meikle* 2331 !
7. Myrtou, Ayios Panteleimon, 1862, *Kotschy*; 1927–1970, *Rev. A. Huddle* 102 ! *Druce* s.n. ! *M. T. Dawe* s.n. ! *Syngrassides* 699 ! 1341 ! *Lady Loch* 74 ! *Kennedy* 1653 ! *Merton* 2578 !

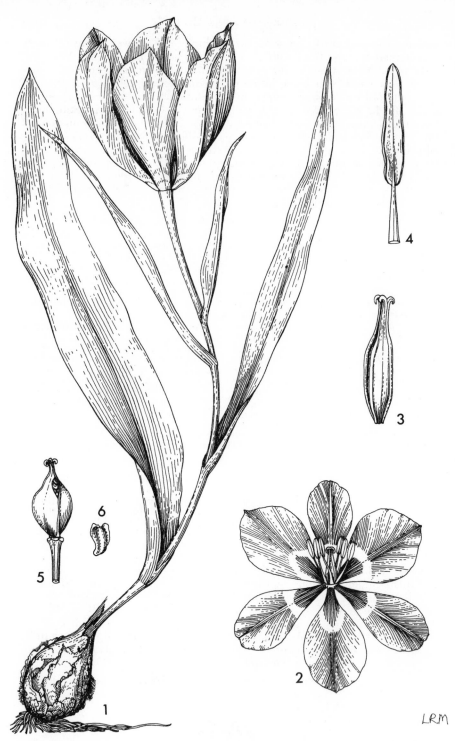

Plate 94. TULIPA CYPRIA Stapf **1**, habit, ×⅔; **2**, flower, ×⅔; **3**, gynoecium, ×2½; **4**, stamen, ×2½; **5**, fruit, ×⅔; **6**, seed, ×4. (**1–4** from *Kennedy* 1611; **5, 6** from *Merton* 2615.)

Economides in ARI 1162! *I. M. Hecker* 42! Dhikomo, 1932, *Syngrassides* 700! Panagra, 1950, *Seligman*; also, 1962, *Meikle* 2368! Kyrenia (planted), 1956, *G. E. Atherton* 1190! and, 1973, *P. Laukkonen* 67, 347, 386; near Lapithos, 1967, *Polunin* 8550!

NOTES: Very closely allied to, and probably a local mutant of *T. agenensis* DC., perhaps of relatively recent origin, and with the exception of the Division 1 record, largely restricted to cereal fields about Myrtou, Kormakiti and Dhiorios. The absence of references to Tulips in the literature of the Ancient World, and their tendency to survive as segetals in the Mediterranean region, suggests that with possibly a few exceptions, they may have arrived, from further east, within the last three or four hundred years.

8. FRITILLARIA *L.*

Sp. Plant., ed. 1, 304 (1753).
Gen. Plant., ed. 5, 144 (1754).
W. B. Turrill et J. R. Sealy in Hooker's Icones, 39 (1, 2):
280 pp. (1980).

Bulbous perennials; bulb usually rather small with a few, fleshy scales; tunics membranous, tending to disappear as the bulb develops; stems generally simple; leaves opposite, verticillate or alternate; flowers generally solitary, nodding or pendulous, occasionally in racemes; perianth campanulate, segments free to base, connivent or occasionally spreading; sometimes conspicuously tessellated, with a well-developed basal nectary-pit; stamens 6, hypogynous, or adherent to base of perianth-segments, linear-filiform; anthers linear, attached to filament a little above the base; ovary sessile or subsessile, styles slender, caducous; stigmas introrsely 3-lobed or entire. Capsule erect, subtruncate, shortly cylindrical or sub-globose, 6-angled or sometimes 6-winged, loculicidally dehiscent; seeds numerous, flattened, irregularly suborbicular-cuneate, sometimes narrowly winged; testa pallid or brownish; embryo usually small.

About 100 species in Europe, temperate Asia and W. North America.

Flowers solitary (or rarely in pairs); leaves few (5–9), rather remote - **1. F. acmopetala**
Flowers numerous (3–34 or more); leaves numerous, rather crowded - - **2. F. persica**

1. **F. acmopetala** *Boiss.*, Diagn., 1 (7): 164 (1846), Fl. Orient., 5: 180 (1882); Post, Fl. Pal., ed. 2, 2: 618 (1933); A. K. Jackson et Turrill in Kew Bull., 1938: 467 (1938); Osorio-Tafall et Seraphim, List Vasc. Plants Cyprus, 19 (1973); Elektra Megaw, Wild Flowers of Cyprus, 15, t. 31 (1973); Sealy et Turrill in Hooker's Icones, 39 (1, 2): t. 3819 (1980).
TYPE: Turkey, "in monte *Amano* Syriae supra *Baylan* Aucher No. 2181". (G, K!).

Bulb depressed-globose, 1–1·8 cm. diam., consisting of 2 fleshy scales, with a basal collar of old withered scales; stem erect, 25–30 cm. high, very slender in the subterranean part, 2–4 mm. diam. in the upper part, glabrous, often suffused purple; basal leaves 1–2, elliptic or obovate-elliptic, 4–8 cm. long, 1·5–3·5 cm. wide, glaucous, apex acute, base tapering to a slender, subterranean petiole 9–20 cm. long; cauline leaves 4–7, alternate, or the 2 lowest subopposite, linear, 5–10 cm. long, 0·2–0·5 cm. wide, the lower broader than the upper and less acuminate, or sometimes obtuse, all sessile, green or glaucous; flowers solitary (rarely 2), nutant, campanulate, 2–3 cm. diam. at mouth; perianth-segments recurved at apex, the exterior oblong, 3–4 cm. long, 0·8–1 cm. wide, obtuse or subacute, greenish-yellow suffused greyish-brown, the interior obovate, subacute, 1·5–2·5 cm. wide, greenish-yellow suffused reddish-purple with a brownish apex; nectary-pits narrowly elliptic, about 4·5 mm. long, 2 mm. wide, filaments 9–10 mm. long, slender, thinly papillose-puberulous; anthers narrowly oblong, distinctly apiculate, about 5 mm. long, 1·5 mm. wide, ovary cylindrical, about 10 mm. long, 3 mm. wide, glabrous, pale green; styles about 8 mm. long, divided to more than half way into 3 recurving lobes. Capsule oblong-cylindrical, truncate, longitudinally sulcate, not winged, 2·5–3 cm. long, 1–1·5 cm. wide; seeds flattened, orbicular-cuneate, about 8 mm. long, 6 mm. wide; testa pale brown, foveolate.

HAB.: In cereal fields; sea-level to 800 ft. alt.; fl. March–April.

DISTR.: Division 7, rare. Also S. Turkey, Syria, Palestine.

7. Karmi, 1937, *Lady Loch* 17! Temblos, 1938, *Lady Loch* 55! Karmi to Trimithi, 1938 *Syngrassides* 1784! Ayios Yeoryios near Kyrenia, 1941, *Davis* 2765! and, 1948, *Kennedy* 1652.

2. F. persica *L.*, Sp. Plant., ed. 1, 304 (1753).
    Theresia libanotica Boiss., Diagn., 1 (13): 20 (1853).
    Fritillaria libanotica (Boiss.) Baker in Journ. Linn. Soc., 14: 270 (1874); Boiss., Fl. Orient., 5: 189 (1882); Post in Mém. Herb. Boiss., 18: 102 (1900); Holmboe, Veg. Cypr., 48 (1914); Post, Fl. Pal., ed. 2, 2: 260 (1933); Osorio-Tafall et Seraphim, List Vasc. Plants Cyprus, 19 (1973); Elektra Megaw, Wild Flowers of Cyprus, 15, t. 31 (1973); Sealy et Turrill in Hooker's Icones, 39 (1, 2): t. 3839 (1980).
    TYPE: "*in* Persia? *e Susis venit in Europam*".

Bulb large, subglobose, 4–5 cm. diam., consisting of a single massive scale invested by the remnants of old, withered scales; stem stout, erect, 50–80 cm. high; basal leaves (when present) obovate-elliptic, 10–15 cm. long, 1·5–3·5 cm. wide, bright green, apex obtuse or acute, base tapering to a slender petiole 8–12 cm. long; cauline leaves very numerous (to 30 or more), rather crowded, narrowly lanceolate or linear-lanceolate, 5–17 cm. long, 0·5–2·5 cm. wide, mostly alternate, glaucous-green, apex acuminate or acute, base sessile, slightly amplexicaul; flowers 3–34 (or more) in a rather lax, bracteate, terminal raceme; bracts herbaceous, linear-subulate, 0·5–2·5 (–4) cm. long, 0·1–0·5 cm. wide; pedicels 1–4 cm. long, decurved near apex; perianth nutant, widely campanulate, 2–3 cm. diam.; segments similar, oblong, subacute, about 2 cm. long, 0·8–1 cm. wide, dark purplish-brown or paler greenish-purple, distinctly, longitudinally 8–10-veined, nectary-pit broadly ovate, about 2–3 mm. long, 2 mm. wide; filaments 6–9 mm. long, glabrous; anthers ovate-oblong, 2–2·5 mm. long, 1·5 mm. wide, brownish or purplish; ovary oblong, glabrous, about 4 mm. long, 2 mm. wide; style 8–10 mm. long; stigma truncate or very obscurely 3-lobulate. Capsule depressed-globose, about 1·5 cm. long, 2–2·5 cm. wide, deeply 6-sulcate, not winged; seeds flattened, obovate, 6–9 mm. long, 4–5 mm. wide; testa brown, reticulate-foveolate.

HAB.: In cereal fields and field borders, sometimes on chalky hillsides; sea-level to 2,000 ft. alt.; fl. Febr.–March.

DISTR.: Divisions 3, 4. Also Turkey, Syria, Palestine.

3. Kato Lefkara, 1934, *Syngrassides* 1384! Also, 1939, *Lady Loch* 72! and, 1941, *Davis* 2731! 1955, *Merton* 2055!
4. Kiti, 1888, *Post*; Perivolia, 1905, *Holmboe* 161; Kondea, 1954, *Merton* 1847! Also recorded (no exact data) from Engomi near Famagusta.

SUBFAMILY 6. **Allioideae** Perennials with tunicate bulbs or rhizomes; leaves all basal, or basal and cauline, sometimes fistulose; inflorescence racemose or cymose-subumbellate, subtended by one or more conspicuous bracts, or flowers rarely solitary; perianth-segments free, spreading, or united only towards base; stamens 6 or 3; anthers basifixed or dorsifixed, introrse. Fruit a loculicidal capsule; seeds subglobose, angular or flattened.

9. GAGEA *Salisb.*

in Konig et Sims, Ann. Bot., 2: 555 (1806).
A. Terracciano in Bull. Herb. Boiss., ser. 2, 5: 1061–1076, 1113–1128 (1905); ser. 2, 6: 105–120 (1906).
A. A. Pascher in Bull. Soc. Imp. Nat. Moscou, n.s., 19: 353–375 (1907).
G. Stroh in Beih. Bot. Centralbl., 57B: 485–520 (1937).
C. C. Heyn et A. Dafni in Israel Journ. Bot., 20: 214–233 (1971); 26: 11–22 (1977).

Bulbs small; basal leaves 1–3, cauline leaves 1–3 or wanting; inflorescence corymbose or subumbellate or flowers sometimes solitary; bracts solitary or in subopposite pairs, generally conspicuous, foliaceous; pedicels usually long; perianth yellowish, greenish or rarely white, persistent, often accrescent, segments 6, subequal, spreading or rarely erecto-patent, free to base, nectary-pit absent; stamens inserted at base of perianth-segments; filaments slender, often flattened; anthers usually shortly ovate or oblong; ovary trigonous or triquetrous, sessile or subsessile; ovules numerous; style usually simple, slender; stigma subcapitate or rarely divided into 3 short, stigmatiferous lobes. Capsule membranous; seeds globose, ovoid, angular, or flattened; testa pale or dark, endosperm fleshy; embryo small.

About 100 species in temperate Europe and Asia. A difficult genus taxonomically, the species superficially alike and often growing in mixed populations, where there may be some hybridization. The Cyprus species call for further investigation in the field.

Perianth white with greenish or purplish veins - - - - - - **6. G. graeca**
Perianth yellow, stained or striped green externally:
  Perianth-segments sharply acuminate; bulb-tunic of interwoven fibres, often produced apically into a long collar - - - - - - - - - **4. G. fibrosa**
  Perianth-segments blunt or subacute:
    Cauline leaves filiform, very similar to basal; pedicels usually glabrous, at least at apex, bulb-tunic fibrous - - - - - - - - - - **5. G. chlorantha**
    Cauline leaves much dilated towards base, distinct from basal leaves; bulb-tunic reticulate-foveolate:
      Cauline leaves subopposite at apex of naked stem; flowers usually numerous, rather crowded; pedicels villose-lanuginose - - - - - - - **1. G. villosa**
      Cauline leaves generally alternate, spaced, the lower often attached near middle of stem:
        Flowers 1–3 (–7); perianth-segments 10–16 mm. long; pedicels erect, rather stout, white-lanuginose, especially towards apex - - - - **3. G. peduncularis**
        Flowers (2–) 3–15 (–25), perianth-segments 8 (–12) mm. long; pedicels slender, glabrous or thinly villose, often bearded at apex - - - - **2. G. juliae**

SUBGENUS 1. **Gagea** Seeds globose, ovoid or angular, not flattened.

1. **G. villosa** (*M. Bieb.*) *Duby*, Bot. Gall., 1: 467 (1828); Osorio-Tafall et Seraphim, List Vasc. Plants Cyprus, 20 (1973); Heyn et Dafni in Israel Journ. Bot., 26: 13 (1977).
    Ornithogalum arvense Pers. in Usteri, Ann. Bot., 11: 8, t. 1, fig. 2 (1794) nom. illeg.; Sibth. et Sm., Fl. Graec. Prodr., 1: 230 (1809), Fl. Graec., 4: 27, t. 332 (1823).
    O. villosum M. Bieb., Fl. Taur.-Cauc., 1: 274 (1808).
    Gagea arvensis Dumort., Florula Belg., 140 (1827); Unger et Kotschy, Die Insel Cypern, 192 (1865) pro parte; Boiss., Fl. Orient., 5: 205 (1882) pro parte; Holmboe, Veg. Cypr., 47 (1914) pro parte; Post, Fl. Pal., ed. 2, 2: 624 (1933) pro parte; Stroh in Beih. Bot. Centralbl., 57B: 490 (1937).
    G. villosa (M. Bieb.) Duby var. *hermonis* Dafni et Heyn in Israel Journ. Bot., 26: 16 (1917) quoad plant. cypr.
    [*G. billardieri* (non Kunth) Kotschy in Unger et Kotschy, Die Insel Cypern, 192 (1865) pro parte.]
    TYPE: U.S.S.R., "Copiosum in Tauriae agris et campis, in promontorio Caucasico quoque reperitur" (LE).

Bulbs 2, enveloped by a common, fuscous-brown, minutely reticulate-foveolate tunic, subglobose or broadly ovoid, 1–1·5 cm. diam.; roots slender, fibrous; basal leaves 2, narrowly linear, 8–20 cm. long, 0·1–0·3 cm. wide, flat, glabrous or shortly pilose, faintly nerved, apex obtuse or subacute; stem erect, rather robust, 3–10 cm. long, shortly pilose especially towards apex, with spreading white hairs; cauline leaves subopposite, crowded with the bracts at apex of stem, 4·5–8 cm. long, 0·3–0·6 cm. wide, attenuate from a broad semi-amplexicaul base to a slender apex; bracts similar but shorter and narrower, often sparsely villose with crispate white hairs; inflorescences (3–) 5–15 (–20)-flowered, irregularly branched, often rather condensed; pedicels very unequal, 1–5 cm. long at anthesis, elongating in fruit, rather conspicuously villose-lanuginose; perianth infundibuliform, segments greenish-yellow, narrowly lanceolate, 10–13 mm. long, 1·5–2·5 mm. wide, acute or subacute, obscurely 3-nerved, generally shortly villose externally towards base and along the nerves; filaments 5–7 mm. long, glabrous; anthers yellow, shortly oblong, about 1·8 mm. long, 1·5 mm. wide; ovary narrowly triquetrous-obcordate, about 3 mm. long, 2 mm. wide towards apex, often sparsely pilose along angles; style stout, about 5 mm. long; stigma subcapitate, obscurely 3-lobed. Capsule about 12 mm. long, 7 mm. wide near apex, tapering to base; seeds subglobose-pyriform, about 2·2 mm. long, 1·5 mm. wide, testa brown, distinctly reticulate-foveolate.

HAB.: Moist shaded situations in Pine forest or under *Juniperus*; 3,600–6,400 ft. alt.; fl. April–May.

DISTR.: Division 2, apparently rare. Widespread in Europe and eastwards to Iran.

2. Platres, 1937, *Kennedy* 162! Khionistra, 5,200–6,400 ft. alt., *Kennedy* 163! 164! 166! *Edmondson & McClintock* E2849!

2. G. juliae *Pascher* in Sitzungsb. Deutsch Naturw. Ver. Böhmen "Lotus", n.s., 14: 125 (1904); Holmboe, Veg. Cypr., 47 (1914); Stroh in Beih. Bot. Centralbl., 57B: 492 (1937); Osorio-Tafall et Seraphim, List Vasc. Plants Cyprus, 20 (1973).
  [? *Ornithogalum spathaceum* (non Hayne) Sibth. et Sm., Fl. Graec. Prodr., 1: 229 (1809), Fl. Graec., 4: 26, t. 331 (1823).]
  [? *Gagea spathacea* (non (Hayne) Salisb.) Poech, Enum. Plant. Ins. Cypr., 8 (1842); Unger et Kotschy, Die Insel Cypern, 192 (1865).]
  [*G. billardieri* (non Kunth) Kotschy in Unger et Kotschy, Die Insel Cypern, 192 (1865) pro parte]
TYPE: Cyprus; "in monte Pentedactylos 28 Martio 1880", *Sintenis & Rigo* 180 (? PRC, K!). See Notes.

Bulbs 2, enveloped by a common, fuscous-brown, minutely reticulate-foveolate tunic, subglobose or broadly ovoid, 0·7–1·3 cm. diam.; roots very slender, fibrous; basal leaves 2, narrowly linear or filiform, 9–25 cm. long, 0·1–0·25 cm. wide, flat, glabrous, faintly nerved, tapering to a slender, obtuse or subacute acumen; stem erect, 2–11 cm. long, glabrous or very thinly lanuginose with long, crispate hairs, often purplish, cauline leaves alternate, commonly remote, with one near the middle of the stem, the other just below the inflorescence, narrowly lanceolate-acuminate, semi-amplexicaul, subglabrous, 1·5–10 cm. long, 0·3–0·8 cm. wide near base, distinct from, and much wider than the basal leaves; bracts linear-filiform, 1–2 cm. long, 0·1–0·2 cm. wide, sparsely white-villose; inflorescences (2–) 3–15 (–25)-flowered, irregularly branched, lax; pedicels very unequal, 1–5 cm. long at anthesis, elongating in fruit, very slender, glabrous, subglabrous or thinly villose with long, spreading woolly hairs, frequently rather densely bearded with long hairs at the point of the junction with the perianth; perianth widely infundibuliform, segments yellow internally, with a broad greenish stripe externally, oblong-lanceolate, 8 (–12) mm. long, 2 (–3) mm. wide, acute or obtuse, obscurely 3-nerved, generally glabrous except for the tuft of long hairs at the base of the segments externally; filaments 5–6 mm. long, glabrous; anthers yellow, shortly oblong, 1–1·5 mm. long, 0·8–1 mm. wide; ovary broadly triquetrous-obcordate, about 3–5 mm. long, 2–3 mm.

wide near apex, glabrous; style stout, 3–5 mm. long; stigma subcapitate, obscurely 3-lobed. Capsule 6–7 mm. long, 5–6 mm. wide, broadly oblong-oblanceolate, not tapering much to base, pale brown; seeds subglobose-pyriform, about 1 mm. diam.; testa brown, reticulate-foveolate. *Plate 95, figs. 1–2.*

HAB.: Usually on damp, shaded hillsides in garigue, sometimes by roadsides or in moist rock-crevices, or in Pine forest; 100–5,400 ft. alt.; fl. Febr.–April.

DISTR.: Divisions 1–3, 7, 8. Endemic.

1. Erimidhes near Ayios Yeoryios (Akamas), 1962, *Meikle* 2136! Smyies, 1962, *Meikle* 2182!
2. Platres, 1937, *Kennedy* 157 in part! also, 1978, *Chesterman* 68! Kryos Potamos, 2,800–5,400 ft.alt., 1937, *Kennedy* 159! 160! 161! S.W. of Khionistra, 4,300 ft. alt., 1937, *Kennedy* 165! Kionia, 1941, *Davis* 2704! and, 1967, *Merton* in ARI 517! Vavatsinia, 1941, *Kennedy & Davis* 2740! Makheras road, 1949, *Casey* 283! Above Kakopetria, 1951, *Merton* 86! Above Klirou, 1956, *Merton* 2489! Below Government Cottage, Troödos, 1957, *Merton* 3188! N. of Stavros tis Psokas, 1959, *P. H. Oswald* 91! and, 1962, *Meikle* 2704!
3. Near Paramytha, 1947, *Mavromoustakis* s.n.! Limassol–Nicosia road, 1956, *Merton* 2480! Above Kellaki, 1956, *Merton* 2696! Above Mosphiloti, 1962, *Meikle* 2978! Below Mallia, 1977, *J. J. Wood* 62! N. of Kophinou, 1978, *Chesterman* 42! Between Kellaki and Parekklisha, 1980, *De Langhe* 35/81!
7. Pentadaktylos, 1862, *Kotschy* 368, 416; also, 1880, *Sintenis & Rigo* 180 in part! and, 1956, *G. E. Atherton* 1060! St. Hilarion, 1941, *Davis* 2514! Kantara Forest road, 1951, *Merton* 246! Kyparissovouno, 1954, *Merton* 1867! Near Kyrenia, 1956, *G. E. Atherton* 1029! 1285! N. side of Kornos, 1956, *Poore* in *Merton* 2696!
8. Yialousa, 1938, *Lady Loch* 30!

NOTES: Pascher (1904) records that the type material of *G. juliae* is a mixture, including some plants of *G. peduncularis* (J. et C. Presl) Pascher: isotypes of *Sintenis & Rigo* 180 at Kew are likewise mixed gatherings, some consisting solely of *G. peduncularis*, others of *G. juliae* and *G. chlorantha*, or all three intermingled on the sheet. This confusion is perhaps due to the fact that *Gagea* species commonly grow in mixed populations in Cyprus, or perhaps because *Sintenis & Rigo* pursued the regrettable policy of amalgamating specimens of what was thought to be a single species under one name, locality number and date. All this type material is starved and depauperate, resembling small-flowered *G. peduncularis*; well-grown specimens of *G. juliae* are many-flowered, and are frequently mistaken for *G. villosa*.

The identity of the plant figured as *Ornithogalum spathaceum* in Flora Graeca is uncertain, the plate shows a few-flowered specimen, like *G. peduncularis* (with which Holmboe united *O. spathaceum*), but Smith describes it as having glabrous stems and very glabrous pedicels, characters not associated with *G. peduncularis*, but sometimes found in *G. juliae*, which also has the smallish flowers of *Ornithogalum spathaceum* Fl. Graec. t. 331.

3. G. peduncularis (*J. et C. Presl*) *Pascher* in Sitzungsb. Deutsch Naturw. Ver. Böhmen "Lotus", 14: 114 (1904); Holmboe, Veg. Cypr., 47 (1914); Stroh in Beih. Bot. Centralbl. 57B: 491 (1937); Osorio-Tafall et Seraphim, List Vasc. Plants Cyprus, 20 (1973); Elektra Megaw, Wild Flowers of Cyprus, 14, t. 26 (1973).
  Ornithogalum pedunculare J. et C. Presl, Delic. Prag., 150 (1822).
  [? *Gagea foliosa* (non (J. et C. Presl) J. A. et J. H. Schultes) Boiss., Fl. Orient., 5: 205 (1882) pro parte; Post in Mém. Herb. Boiss., 18: 102 (1900).]
TYPE: Crete; "in nemorosis umbrosis insulae Cretae. Sieber".

Bulbs 2 enveloped by a common, fuscous-brown, minutely reticulate-foveolate tunic, subglobose or broadly ovoid, 7–10 mm. diam.; roots very slender, fibrous; basal leaves 2, filiform, 3–15 cm. long, generally less than 0·1 cm. wide, flat, glabrous, faintly nerved, tapering to a very slender, subacute acumen; stem erect, 1–10 cm. long, usually thinly lanuginose with long, white, crispate hairs, occasionally glabrous; cauline leaves 1–3, the lowermost towards the middle or upper part of the stem, the upper solitary or sometimes paired and subopposite at the base of the inflorescence, all narrowly lanceolate-caudate 1·5–10 (–12) cm. long, 0·2–0·7 cm. wide near the semi-amplexicaul base, glabrous or thinly ciliate, distinct from the basal leaves; bracts similar but much smaller, often subulate, 1–3·5 cm. long, 0·1–0·3 cm. wide, glabrous or ciliate; inflorescence 2–3 (rarely to 7)-flowered, or flowers commonly solitary; pedicels 2–3 cm. long at anthesis, elongating a little in fruit, erect, often rather robust and usually white-lanuginose

Plate 95. Figs. **1–2.** GAGEA JULIAE Pascher   **1,** habit, × 1; **2,** flower, longitudinal section, × 3; figs. **3–4.** GAGEA PEDUNCULARIS (J. et C. Presl) Pascher   **3,** habit, × 1; **4,** flower, longitudinal section, × 3; figs. **5–6.** GAGEA FIBROSA (Desf.) J. A. et J. H. Schultes   **5,** habit, × 1; **6,** flower, longitudinal section, × 6; figs. **7–8.** GAGEA CHLORANTHA (M. Bieb.) J. A. et J. H. Schultes   **7,** habit, × 1; **8,** flower, longitudinal section, × 6. (**1, 2** from *Meikle* 2704; **3, 4** from *P. H. Oswald* 42; **5, 6** from *Merton* 2526; **7, 8** from *Poore* 47.)

especially towards apex; perianth widely infundibuliform, segments golden-yellow internally, stained green externally, oblong-lanceolate, 10–16 mm. long, 2–3·5 mm. wide, obtuse or subacute, obscurely (3–) 5-nerved, generally glabrous except for a lanuginose beard along the lower half of the median nerve externally; filaments 7–8 mm. long, glabrous; anthers yellow, oblong, about 1·8 mm. long, 1·2 mm. wide; ovary triquetrous-oblong, 3–4 mm. long, 2–3 mm. wide, glabrous; style stout, about 4 mm. long; stigma subcapitate or obscurely 3-lobed. Capsule rather narrowly oblong, 10–12 mm. long, 5–6 mm. wide, tapering somewhat to base, pale brown; seeds subglobose, often shortly beaked, about 1·3 mm. long and almost as wide; testa rich brown, conspicuously reticulate-foveolate. *Plate 95, figs. 3–4.*

HAB.: On rocky hillsides and in short turf on rocky pastures; sea-level to 2,600 ft. alt., usually lowland; fl. Jan.–March.

DISTR.: Divisions 2–7. Also Balkan peninsula and Aegean region.

2. Saïttas, 1932, *Syngrassides* 352 ! also, 1978, *Chesterman* 72 ! Valley of Kryos Potamos, 2,600 ft. alt., 1937, *Kennedy* 158 ! Above Klirou, 1956, *Merton* 2493 ! 3 m. E. of Pano Platres, 1973, *P. Laukkonen* 116 !
3. Mazotos, 1905, *Holmboe* 190; Plains near Polemidhia, 1948, *Mavromoustakis* s.n. ! Between Apsiou and Phasoula, 1978, *Chesterman* 73 !
4. 4 m. W. of Famagusta, 1951, *Merton* 185 ! Mile 10, Nicosia–Famagusta road, 1954, *Merton* 1822 ! Ayios Andronikos, Phrenaros, 1959, *P. H. Oswald* 42 !
5. Nicosia–Larnaca road, 1940, *Davis* 2132 ! Lapathos, 1962, *Meikle* 2988 !
6. Makhedonitissa Monastery, 1941, *Davis* 2130 partly ! S. of Ayia Irini, 1957, *Merton* 2847 ! Orga–Liveras road, 1957, *Merton* 2856 !
7. Pentadaktylos, 1880, *Sintenis & Rigo* 180 partly ! 181 partly ! also, 1956, *G. E. Atherton* 1056 ! Bellapais, 1936, *Syngrassides* 1165 ! Near Kyrenia, 1937–1956, *Mrs. Stagg* s.n. ! *Casey* 269 ! *Merton* s.n. ! *G. E. Atherton* 886 ! 197 ! Vavilas, 1938, *Lady Loch* 22 ! Halevga, 1941, *Davis* 2148 ! Above Kharcha, 1941, *Davis* 2151 ! Kazaphani, 1949, *Casey* 249 ! and, 1956, *G. E. Atherton* 927 ! Panagra, 1971, *E. Hodgkin* s.n. !

SUBGENUS 2. **Hornungia** *(Bernh.) Pascher* Seeds irregularly deltoid, flattened.

4. **G. fibrosa** *(Desf.) J. A. et J. H. Schultes* in Roemer et Schultes, Syst. Veg., 7: 552 (1829); Post, Fl. Pal., ed. 2, 2: 626 (1933); Stroh in Beih. Bot. Centralbl., 57B: 491 (1937); Heyn et Dafni in Israel Journ. Bot., 20: 277 (1971).
   Ornithogalum fibrosum Desf., Fl. Atlant., 1: 294, t. 84 (1798).
   Gagea rigida Boiss. et Spruner in Boiss., Diagn., 1, 7: 108 (1846) pro parte.
   G. reticulata (Pall.) J. A. et J. H. Schultes var. *fibrosa* Boiss., Fl. Orient., 5: 208 (1882) pro parte.
   G. reticulata (Pall.) J. A. et J. H. Schultes f. *rigida* (Boiss. et Spruner) Pascher in Bull. Soc. Imp. Nat. Moscou, n.s., 19: 367 (1907) pro parte.
   G. reticulata (Pall.) J. A. et J. H. Schultes ssp. *fibrosa* (Desf.) Maire et Weiller, Fl. Afr. Nord, 5: 126 (1958).
   [*G. reticulata* (Pall.) J. A. et J. H. Schultes var. *tenuifolia* (non Boiss.) Holmboe, Veg. Cypr., 48 (1914); Seligman in Quart. Bull. Alpine Gard. Soc., 20: 233 (1952).]
   [*G. reticulata* (Pall.) J. A. et J. H. Schultes ssp. *circinata* (non *Ornithogalum circinatum* L.f.) Osorio-Tafall et Seraphim, List Vasc. Plants Cyprus, 20 (1973) nom. invalid.]
   TYPE: Tunisia; "in arenis prope Kerwan [Qairwan]" (P).

Bulb solitary, ovoid, 3–10 mm. diam., enveloped by a dull brown reticulate-fibrous tunic often produced apically into an elongate, cylindrical collar; roots numerous, fibrous, tangled around the bulb (or aggregated bulbs) occasionally distinctly thickened and vermiform; basal leaves 1–2, filiform or narrowly linear, semiterete or flattened, canaliculate above or ribbon-like, glabrous, often recurved, 3–10 cm. long, 0·1–0·3 cm. wide; stems usually very short, less than 2 cm. long, often totally enveloped by bulb-tunic; cauline leaves almost indistinguishable from basal, sometimes sparsely ciliate; bracts foliaceous, narrowly lanceolate-filiform, 1–2 cm. long; inflorescence subumbellate 1–3-flowered in our area, sometimes up to 16-flowered elsewhere, or flowers solitary; pedicels 1–3 (–6) cm. long, shortly white-pilose with adpressed or spreading, often crispate hairs; perianth spreading, stellate at anthesis, segments linear-oblong, 10–15 mm. long, 1–2 mm. wide, yellow internally, greenish and glabrous externally, attenuate to a sharp, purplish acumen, nerves numerous, obscure; filaments about 5 mm. long, glabrous, tapering from base to apex; anthers rather narrowly oblong, yellow, about 1·8 mm. long, less than 1 mm. wide; ovary oblong-triquetrous, glabrous, 2·5–3 mm. long, 2 mm. wide; style stout, about 5 mm. long; stigma subcapitate. Capsule ellipsoid or obcordate, 5–7 mm. long, 4–5 mm. wide, pale brown; seeds deltoid, flattened, about 1·5 mm. long, 1·2 mm. wide; testa brown, reticulate with a raised margin. *Plate 95, figs. 5–6.*

HAB.: Rocky pastures, dry grassy banks, sometimes in areas of bare, trodden soil; sea-level to 2,900 ft. alt.; fl. Febr.–March.

DISTR.: Divisions 2–8. Eastern Mediterranean region and eastwards to Iraq, Iran and Transcaspia. North Africa.

2. Platres, 1937, *Kennedy* 157 partly! Vavatsinia, 1941, *Kennedy & Davis* 2740 partly!
3. Curium, 1964, *J. B. Suart* 140!
4. Ayia Napa, 1905, *Holmboe* 65; 4 m. W. of Famagusta, 1951, *Merton* 185 partly! Famagusta, 1959, *P. H. Oswald* 44 partly!
5. Athalassa, 1933, *Syngrassides* 49! also, 1962, *Meikle* 2974! Syngrasis, 1936, *Syngrassides* 932! Mia Milea, 1951, *Merton* 44 partly! 51! and, 1954, *Merton* 1826! 1827!
6. Kokkini Trimithia, 1935, *Syngrassides* 780! Mandres, 1951, *Merton* 112 partly! Ayia Irini, 1957, *Merton* 2845! English School, Nicosia, 1959, *P. H. Oswald* 82! Dhiorios, 1972, *G. Joscht* 6709!
7. Pentadaktylos, 1880, *Sintenis & Rigo* 179! Above Kharcha, 1941, *Davis* 2152! Kyrenia, 1949, *Casey* 312! Kambyli, 1956, *Merton* 2526!
8. Between Khelones and Rizokarpaso, 1941, *Davis* 2300!

NOTES: As noted by Holmboe, all the Cyprus specimens are very dwarf, with unusually small flowers. There is considerable variation in leaf-width, even within single populations, and I am not wholly satisfied that, in Cyprus at least, any specific distinction can be drawn between *G. fibrosa* and the closely allied *G. reticulata* (Pall.) J. A. et J. H. Schultes.

5. G. chlorantha (*M. Bieb.*) *J. A. et J. H. Schultes* in Roemer et Schultes, Syst. Veg., 7: 551 (1829); Boiss., Fl. Orient., 5: 209 (1882); Holmboe, Veg. Cypr., 47 (1914); Post, Fl. Pal., ed. 2, 2: 626 (1933); Stroh in Beih. Bot. Centralbl., 57B: 502 (1937); Rechinger f. in Arkiv för Bot., ser. 2, 1: 419 (1950); Heyn et Dafni in Israel Journ. Bot., 20: 222 (1971); Osorio-Tafall et Seraphim, List Vasc. Plants Cyprus, 20 (1973).

　　Ornithogalum chloranthum M. Bieb., Fl. Taur.-Cauc., 3: 264 (1918).

　　G. chlorantha (M. Bieb.) J. A. et J. H. Schultes var. *cyprica* Pascher in Bull. Soc. Imp. Nat. Moscou, n.s., 19: 370 (1907).

　　TYPE: U.S.S.R.; "in promontorii Caucasici septentrionalis campestribus apricis, in hortis circa oppidum Kisljar" (LE).

Bulb ovoid or subglobose, 3–7 mm. long and almost as wide, enveloped by a brownish, fibrous tunic, sometimes with a short apical collar; roots numerous, slender, fibrous; basal leaves 1–2, filiform, 3–11 cm. long, usually less than 0·1 cm. wide, glabrous or shortly pilose (especially towards base), sometimes distinctly ribbed, canaliculate above with a slender acute apex; stem very slender, 1–3 cm. long, glabrous or often retrorse- or crispate-pilose; cauline leaves 1–2, very similar to basal, filiform, 1–3 cm. long, with a slightly dilated, ciliate base; bracts similar to cauline leaves but smaller, often thinly villose; flowers usually solitary in our area, occasionally paired; pedicels 1–3 cm. long, very slender, usually glabrous and reddish towards apex, sometimes shortly pilose with spreading hairs; perianth infundibuliform, the segments closing together soon after anthesis; segments 8–17 mm. long, 2–3·5 mm. wide, narrowly oblong, obtuse, glabrous, yellow internally, greenish externally; filaments very slender, glabrous, 6–8 mm. long; anthers yellow, rather narrowly oblong, 1–1·5 mm. long, 0·6–1 mm. wide; ovary glabrous, narrowly oblong, about 3 mm. long, 1–1·5 mm. wide; style about 5 mm. long; stigma subcapitate, obscurely 3-lobed. Capsule oblong, 8–10 mm. long, 5 mm. wide, dark brown; seeds irregularly deltoid,

1·5–2 mm. long, 1 mm. wide, flattened; testa brown, reticulate with a raised margin. *Plate 95, figs. 7–8.*

HAB.: Rocky pastures, dry grassy banks and bare ground, often with *G. fibrosa*; sea-level to 3,000 ft alt.; fl. Jan.–March.

DISTR.: Divisions 4–7. Also Caucasus, Turkey, Syria, Palestine, Iraq, Iran.
4. Near Famagusta, 1959, *P. H. Oswald* 44 partly !
5. Kythrea, 1927, *Rev. A. Huddle* 104 ! By Larnaca road near Nicosia, 1941, *Davis* 2134 ! Hill above Syngrasis, 1951, *Merton* 79 ! Mia Milea, 1951, *Merton* 44 partly ! Athalassa, 1960, *P. H. Oswald* 51 ! and, 1969, *Economides* in ARI 1303 !
6. Makhedonitissa Mon. near Engomi, 1941, *Davis* 2130 ! Nicosia, 1951, *Merton* s.n. ! Mandres, 1951, *Merton* 112 ! Mitsero, 1957, *Poore* 47 !
7. Pentadaktylos, 1880, *Sintenis & Rigo* 180 partly !.Kephalovryso, Kythrea, 1941, *Davis* 2162 !

NOTES: The Cyprus specimens, like those of *G. reticulata*, are on the whole small and rather depauperate, but scarcely deserve nomenclatural recognition, even at varietal rank.

6. G. graeca (*L.*) *A. Terracc.* in Bull. Soc. Bot. Fr., 52, Mém., 2: 25 (1905); Stroh in Beih. Bot. Centralbl., 57B: 500 (1937); Greuter in Israel Journ. Bot., 19: 155 (1970).
   Anthericum graecum L., Sp. Plant., ed. 2, 444 (1762); Sibth. et Sm., Fl. Graec. Prodr., 1: 234 (1809), Fl. Graec., 4: 30, t. 336 (1823); Poech, Enum. Plant. Ins. Cypr., 9 (1842).
   Lloydia graeca (L.) Endl. ex Kunth, Enum. Plant., 4: 245 (1843); Unger et Kotschy, Die Insel Cypern, 192 (1865); Boiss., Fl. Orient., 5: 203 (1882); Holmboe, Veg. Cypr., 48 (1914); Post, Fl. Pal., ed. 2, 2: 623 (1933); Osorio-Tafall et Seraphim, List Vasc. Plants Cyprus, 20 (1973).
   TYPE: "*in* Oriente *Burmannus*".

Bulb ovoid, 7–12 mm. long and almost as wide, tunic fuscous, coarsely lacerate at apex, but scarcely fibrous; basal leaves, 2–3, narrowly linear, 3–10 cm. long, 0·1–0·2 cm. wide, glabrous, distinctly nerved, apex very slender, subacute, commonly broken off at anthesis; cauline leaves 4–6, linear-filiform, 0·5–5 cm. long, shortly sheathing at base, gradually tapering to apex; inflorescence 1–5-flowered, loosely racemose, flowers slightly nodding; bracts subulate-filiform, glabrous, about 5 mm. long; pedicels slender, glabrous, 5–40 mm. long; perianth infundibuliform, segments white, veined green or purplish externally, erect, connivent, narrowly oblong, 12–17 mm. long, 3–4 mm. wide, often somewhat carinate with a well-marked midrib, and an acute or obtuse, somewhat recurved apex; filaments very slender; glabrous, 5–6 mm. long; anthers shortly oblong, about 1·5 mm. long, 1 mm. wide; ovary narrowly oblong, 4–7 mm. long, 1·5–2·5 mm. wide; style stout, about 2 mm. long; stigma capitate. Capsule oblong, pale brown, about 10 mm. long, 5 mm. wide; seeds deltoid, flattened, about 1·3 mm. long, 1 mm. wide; testa bright yellow-brown, reticulate, with a narrowly raised margin.

HAB.: In garigue on grassy or rocky hillsides; sometimes in open *Pinus brutia* forest; sea-level to 2,000 ft. alt.; fl. March–April.

DISTR.: Divisions 1, 3–5, 7, 8. Eastern Mediterranean region from Greece to Palestine.
1. Smyies, 1948, *Kennedy* 1604 ! Kako Skala (Akamas), 1948, *Kennedy* 1677 ! Ayios Neophytos, 1959, *C. E. H. Sparrow* 15 ! also, 1971, *K. M. Guichard* 4a ! Aprilokambos near Ayios Yeoryios (Akamas), 1962, *Meikle* 2015 ! Mirillis R. near Ayios Yeoryios (Akamas), 1962, *Meikle* 2068 !
3. Stavrovouni, 1862, *Kotschy* 194; Mazotos, 1905, *Holmboe* 175; near Limassol, 1956, *Maj. Gen. Basten* s.n. ! Kouklia, 1960, *N. Macdonald* 54 ! Polemidhia, 1964, *J. B. Suart* 191 ! Pissouri, 1981, *Hewer* 4768 ! Below Kilani, 1981, *Hewer* 4786 !
4. Near Famagusta, 1862, *Kotschy.*
5. Near Kythrea, 1880, *Sintenis & Rigo* 182 ! Dennarga Forest, 1955, *Merton* 1993 !
7. Between Melandrina and Antiphonitis, 1862, *Kotschy* 515; Kyrenia Pass, 1902, *A. G. & M. E. Lascelles* s.n. ! Aghirda, 1936, *Syngrassides* 1161 ! Ayios Amvrosios, 1937, *Miss Godman* 38 ! Kyrenia, 1938, *C. H. Wyatt* 56 ! also, 1949, *Casey* 379 ! and, 1956, *G. E. Atherton* 1245 ! Kharcha, 1941, *Davis* 2776 ! and, 1956, *G. E. Atherton* 1058 ! Lakovounara Forest, 1950, *Chapman* 361 ! St. Hilarion, 1952, *F. M. Probyn* 58 ! Between Kyrenia and Bellapais, 1970, *I. M. Hecker* 32 !
8. Komi Kebir, 1934, *Syngrassides* 1392 ! Koma tou Yialou, 1950, *Chapman* 212 !

10. ALLIUM L.

Sp. Plant., ed. 1, 294 (1753).
Gen. Plant., ed. 5, 143 (1754).
E. Regel in Acta Hort. Petrop., 3: 1–266 (1875).
W. T. Stearn in Ann. Mus. Goulandris, 4: 83–198 (1978).

Perennial bulbous herbs, often with a characteristic, heavy odour; bulbs usually tunicate, leaves filiform, linear or ovate-elliptic, with a sheathing base, lamina flat, canaliculate, triquetrous or fistulose; inflorescence scapose; flowers usually in a terminal, cymose umbel, at first enveloped in a membranous or herbaceous spathe; pedicels commonly with 2 basal bracteoles; perianth campanulate or subglobose, segments 1–3-veined, free or shortly connate at base; stamens included or exserted; filaments inserted near base of perianth-segments, free or shortly connate at base, simple or with lateral teeth or appendages; anthers introrse, dorsifixed; ovary sessile, 3-locular; ovules mostly 2 (occasionally up to 10) in each loculus; style gynobasic, short, straight; stigma truncate or capitate, rarely 3-lobed. Fruit a membranous capsule; seeds angled, triquetrous or compressed, rarely subglobose; testa generally black.

About 500 species, widely distributed in temperate regions of the northern hemisphere.

The term *bulblets* here refers to offsets or divisions of the bulb; *bulbils* are formed in the inflorescence and wholly or partly replace the flowers (cf. Stearn in Tutin et al., Fl. Europaea, 5: 49; 1980).

Leaves all basal or arising from ground level, the sheaths not extending up the aerial parts of the stem (or scape):
  Leaves pilose, at least along the margins and sheaths:
    Leaves 0·9–1 cm. wide; perianth widely cup-shaped, with spreading segments
          2. A. trifoliatum
    Leaves 0·1–0·4 (–0·7) cm. wide, perianth cup-shaped with connivent segments
          3. A. cassium var. **hirtellum**
  Leaves glabrous:
    Perianth cup-shaped in fruit, with connivent segments; ovary 2·5–3·5 mm. diam.; ovules 2 in each loculus:
      Stems triquetrous (3-angled), flowers white; inflorescence with bulbils
          1. A. neapolitanum
      Stems terete; flowers pink; inflorescence sometimes with bulbils -   - **4. A. roseum**
    Perianth spreading or reflexed in fruit; ovary 3·5–4 mm. diam.; ovules 4–8 in each loculus:
      Stem 60–90 cm. high; leaves up to 9 cm. wide, broadly acuminate - **18. A. nigrum**
      Stem 10–40 cm. high; leaves up to 2·5 cm. wide, tapering to a very slender acumen
          19. A. orientale
Leaf sheaths, or leaves, extending up the aerial part of the stem (or scape):
  Plants flowering in late autumn, just in advance of the winter rains - **6. A. autumnale**
  Plants flowering in spring or summer:
    Bulb-tunic of interwoven fulvous or fuscous fibres; flowers few with very unequal pedicels
          5. A. cupani
    Bulb-tunics membranous, papery, coriaceous or crustaceous, sometimes lacerate, but not of interwoven fibres:
      Inflorescence consisting entirely of bulbils; bulb of numerous bulblets enveloped in a common tunic - - - - - - - A. SATIVUM (p. 1617)
      Inflorescence consisting entirely of flowers, or of flowers mixed with bulbils:
        Leaves flat (sometimes keeled) with clearly defined margins:
          All the stamens similar, with simple, linear or subulate filaments:
            Stamens included or at most a little exserted; stem 20–70 cm. high
              7. A. paniculatum
            Stamens distinctly exserted, stem 5–30 (–35) cm. high - **8. A. stamineum**
          Stamens of 2 kinds, the outer 3 with simple filaments, the inner 3 with 3-cuspidate filaments:
            Perianth 4–6 mm. long, 3–4 mm. wide, segments connivent, overlapping:
              Bulblets usually numerous; umbels 3–10 cm. diam.; stamens included or very shortly exserted:

Bulblets yellowish-brown - - - - - **9. A. ampeloprasum**
Bulblets blackish-crimson or vinous - - - **10. A. scorodoprasum**
Bulblets few or absent; stamens exserted; umbel often 10–20 cm. diam.
A. PORRUM (p. 1618)
Perianth 2–3 mm. long, 2–2·5 mm. wide, segments loosely sub-connivent, but not
overlapping - - - - - - - - **17. A. willeanum**
Leaves fistulose (hollow internally) without clearly defined margins:
Leaves and stems distinctly swollen or inflated some distance above base
A. CEPA (p. 1609)
Leaves and stems not swollen or inflated above base:
Perianth 4–7 mm. long, 1·5–4 mm. wide, commonly reddish or purple:
Leaves about 8 mm. wide; pedicels very unequal, the median up to 40 mm.
long, the outer seldom more than 15 mm. long - **15. A. amethystinum**
Leaves 0·5–4 mm. wide; pedicels subequal, less than 30 mm. long:
Stamens clearly exserted; pedicels 5–20 (–30) mm. long; leaf-sheaths usually
concealing stipitate bulblets - - - **11. A. sphaerocephalon**
Stamens included; pedicels generally less than 4 mm. long; bulblets generally
few or absent - - - - - - - - **12. A. junceum**
Perianth 2–4 mm. long, 1·5–4 mm. wide:
Leaf-sheaths extending beyond half the length of the stem; perianth segments
2–3 mm. long, 1–1·2 mm. wide - - - **16. A. margaritaceum**
Leaf-sheaths not extending beyond one-third the length of the stem; perianth
segments 3·5–4 mm. long, 2–2·5 mm. wide:
Pedicels 5–40 mm. long; stamens shortly exserted - - **14. A. curtum**
Pedicels 3–6 mm. long; stamens included or anthers shortly exserted
13. A. rubrovittatum

SECT. 1. **Cepa** (*Mill.*) *Prokh.* Bulbs subcylindrical or globose, usually clustered on a short rhizome; leaves sheathing lower part of stem, distichous, fistulose; stem terete, fistulose; spathe equalling or shorter than pedicels; stamens simple or with the bases of the inner filaments shortly toothed; ovary with distinct nectariferous pores; stigma entire; seeds angled.

A. CEPA *L.*, Sp. Plant., ed. 1, 301 (1753), the Onion, with globose or subglobose bulbs; tunics membranous, brownish, glossy; stem to 1 m. high, conspicuously inflated towards base; leaves green, fistulose, to 40 cm. long, 2 cm. diam.; flowers numerous, crowded into a subglobose or hemispherical umbel 4–9 cm. diam.; perianth-segments whitish with a greenish stripe; stamens exserted, the inner filaments with shortly toothed bases. Capsule about 5 mm. diam.; seeds black, angular.

An important culinary vegetable, thought to be of Central Asiatic origin, but now known only in cultivation, or as an escape from cultivation. It is widely grown in Cyprus, as elsewhere in the Mediterranean region, but does not appear to have become a naturalized element in the island's flora.

SECT. 2. **Molium** *G. Don ex Koch* Bulbs ovoid or subglobose, not attached to a rhizome; leaves flat, sheathing at base; stem terete or angled; umbel crowded or lax; spathe equalling or shorter than pedicels; perianth-segments spreading or connivent, often conspicuous; stamens usually included; ovary with nectariferous pores; ovules 2 in each loculus; stigma entire; seeds angled.

1. **A. neapolitanum** *Cyr.*, Plant. Rar. Neap., 1: XIII, t. 4 (1788); Unger et Kotschy, Die Insel Cypern, 198 (1865); Regel in Acta Hort. Petrop., 3: 224 (1875); Boiss., Fl. Orient., 5: 274 (1882); Post in Mém. Herb. Boiss., 18: 101 (1900); H. Stuart Thompson in Journ. Bot., 44: 339 (1906); Holmboe, Veg. Cypr., 47 (1914); Post, Fl. Pal., ed. 2, 664 (1933); Osorio-Tafall et Seraphim, List Vasc. Plants Cyprus, 22 (1973).
TYPE: Italy, "Colitur in Hortis Neapolitanis, ob florem pulchritudinem, & modo in apricis circa urbem sponte crescere incipit" (? NAP).

Bulb subglobose, 1–2 cm. diam., outer tunic crustaceous, grey-brown, inner tunic white, smooth; stem 25–50 (–70) cm. high, glabrous, triquetrous, the angles sometimes narrowly winged; leaves 2–4, linear, 6–50 cm. long, 1·5–4 cm. wide, glabrous, somewhat keeled, base sheathing the stem for 6–10 cm. or more, apex slender-acuminate (often broken off at anthesis), margins minutely papillose; spathe 1, membranous, shortly cuspidate, about 1·5–2 cm. long, 1–1·5 cm. wide; umbels hemispherical, many flowered, rather lax; pedicels glabrous, 1·5–3·5 cm. long, angular; perianth cup-shaped, segments spreading, ovate, obtuse, white, translucent, about 10–12 mm. long, 3–4 mm. wide; stamens included; filaments about 5–6 mm. long, subulate, expanded towards base, glabrous; anthers oblong, grey-green, about 1·5 mm. long, 0·5 mm. wide; ovary trigonous-subglobose, about 2·5 mm. diam., pallid; style about 4 mm. long, tapering to apex; stigma truncate. Capsule membranous, pale brown, about 6 mm. diam.; seeds irregularly angular, about 2·5 mm. long and wide; testa black, closely papillose with rows of blunt papillae.

HAB.: Cultivated and fallow fields; roadsides; Hazel groves and moist places in Pine forest; sea-level to 5,500 ft. alt., but usually lowland; fl. Febr.–May.

DISTR.: Divisions 1–8, locally common. Mediterranean region.

1. Ayios Yeoryios (Akamas), 1962, *Meikle*! Polis area, 1962, *Meikle*!
2. Troödos area, 1948, *Mavromoustakis* s.n.! Above Askas, W. of Palekhori, 1979, *Edmondson & McClintock* E2903!
3. Stavrovouni, 1880, *Sintenis*; Kouklia, 1905, *Holmboe* 364.
4. Larnaca, 1862, *Kotschy* 304; also, 1880, *Sintenis*, and 1884, *Deflers* 1278; Cape Greco, 1958, *N. Macdonald* 41! Dhekelia, 1973, *P. Laukkonen* 175! Near Troulli, 1981, *Hewer* 4734!
5. Nisou, ? 1898, *Post* s.n.; Kythrea, 1933, *Nattrass* 584.
6. Kokkini Trimithia, 1862, *Kotschy* 481; Nicosia, 1927, *Rev. A. Huddle* 97! also, 1930, *F. A. Rogers* 0666! and, 1934, *Nattrass* 429; Morphou, 1941, *Davis* 2623!
7. Locally common, 1862–1973; Pentadaktylos, Buffavento, Kyrenia, Kantara, Dhavlos, Thermia, Myrtou, St. Hilarion, etc. *Kotschy* 412; *Sintenis & Rigo* 989 pro parte! *Syngrassides* 86! *G. E. Atherton* 873! 958! *Polunin* 6603! *Merton* in ARI 210! *Economides* in ARI 1045! etc.
8. Cape Andreas, 1880, *Sintenis*; Gastria, 1936, *Syngrassides* 927! Koma tou Yialou, 1950, *Chapman* 257! Ronnas Valley, 1957, *Merton* 2939!

2. A. trifoliatum *Cyr.*, Plant. Rar. Neap., 2: xi, t. 3 (1792); Boiss., Fl. Orient., 5: 270 (1882) pro parte; Post, Fl. Pal., ed. 2, 2: 642 (1933); Osorio-Tafall et Seraphim, List Vasc. Plants Cyprus, 22 (1973).

A. graecum Urv., Enum. Plant. Ins. Archip., 37 (1833); Kotschy, Die Insel Cypern, 197 (1865).

A. subhirsutum L. var. *graecum* (Urv.) Regel in Acta Hort. Petrop., 3: 271 (1875).

A. subhirsutum L. ssp. *trifoliatum* (Cyr.) Holmboe, Veg. Cypr., 46 (1914).

[*A. subhirsutum* (non L.) Sibth. et Sm., Fl. Graec. Prodr., 1: 223 (1809), Fl. Graec., 4: 13, t. 313 (1823) pro parte quoad plant. cypr.; Poech, Enum. Plant. Ins. Cypr., 9 (1842); Boiss., Fl. Orient., 5: 270 (1882) pro parte quoad plant. cypr.]

[*A. hirsutum* (non Zucc.) Post in Mém. Herb. Boiss., 18: 101 (1900).]

[*A. subhirsutum* L. ssp. *ciliatum* (non *A. ciliatum* Cyr.) Holmboe, Veg. Cypr., 46 (1914).]

[*A. ciliatum* (non Cyr.) Osorio-Tafall et Seraphim, List Vasc. Plants Cyprus, 22 (1973).]

TYPE: Italy, "*in Insula Capreorum*" [Capri], *Nicodemus* [Gaetano Nicodemo] (? NAP).

Bulb subglobose, 1–1·5 cm. diam., outer tunic grey-brown, crustaceous, slightly exasperate, inner tunic whitish, smooth; stems slender, terete, glabrous, 7–30 cm. long, leaves 1–3, linear, 6–25 cm. long, 0·2–0·8 cm. wide, ciliate or thinly pilose all over, base sheathing the stem for 5 cm. or more, often densely retrorse-pilose, apex tapering to a slender acumen (often broken off at anthesis), margins not papillose; spathe 1, membranous, shortly cuspidate, about 1–1·5 cm. long, 0·8–1 cm. wide; umbels lax, hemispherical, many flowered, pedicels slender, glabrous, angled, 1–2 (–3) cm. long; perianth cup-shaped, segments spreading, ovate, acute, 5–8 mm. long, 2–3 mm. wide, white with a pink or purple midrib or flushed pink, rarely pure white; stamens included; filaments about 3 mm. long, subulate, expanded towards base, glabrous; anthers oblong, greenish- or greyish-yellow, about 1·3 mm. long, 0·6 mm. wide, ovary trigonous-subglobose, yellowish, about 1·5 mm. diam.; style 3 mm. long; stigma truncate. Capsule subglobose with an excavate apex, about 4 mm. diam.; seeds angular, about 2 mm. long, 1·5 mm. wide; testa black, papillose with short, broad-based papillae.

HAB.: Pastures; salt-steppe; in garigue on rocky hillsides; occasionally in cultivated or fallow fields; sea-level to 1,000 ft. alt.; fl. March–May.

DISTR.: Divisions 1, 3–8, locally common. Mediterranean region from S. Italy and Sicily eastwards.

1. Cape Arnauti (Cape Akamas), 1884, *Deflers* 1227; Polis area, 1962, *Meikle*! Fontana Amorosa, 1962, *Meikle* 2289!
3. Limassol, 1898, *Post* s.n., Polemidhia hills, 1948, *Mavromoustakis* s.n.! Cherkez, 1954, *Mavromoustakis* 4! Kolossi, 1961, *Polunin* 6679! and, 1964, *J. B. Suart* 181! Near Moni, 1964, *J. B. Suart* 211!
4. Larnaca aerodrome, 1950, *Chapman* 419! Larnaca Salt Lake, 1955, *Merton* 2011! Between Larnaca and Kiti, 1972, *G. Joscht* 6643! Ayia Napa, 1981, *Hewer* 4722!
5. Kythrea, 1880, *Sintenis & Rigo* 165! 167! Sotira, 1905, *Holmboe* 407a; 407b; Kythrea, 1932, *Syngrassides* 261! Syngrasis, 1936, *Syngrassides* 931! Salamis, 1962, *Meikle*!
6. Nicosia, 1927, *Rev. A. Huddle* 96! Ayia Irini, 1941, *Davis* 2593! Syrianokhori, 1941, *Davis* 2509! Dhiorios, 1962, *Meikle*!
7. Ayios Khrysostomos Monastery, 1862, *Kotschy*; Kyrenia, 1931–56, *J. A. Tracey* 3! *Syngrassides* 696! *C. H. Wyatt* 44! *Casey* 587! *G. E. Atherton* 1020! 1197! Myrtou, 1932, *Nattrass* 211 and, 1936, *Syngrassides* 1205! Larnaka tis Lapithou, 1941, *Davis* 2993! Kornos, 1941, *Davis* 3041! St. Hilarion, 1949, *Casey* 690! Orga, 1956, *Poore* 2654! Lapithos, *Merton* in ARI 308!
8. Akradhes, 1962, *Meikle*!

NOTES: Closely allied to *A. subhirsutum* L. (*A. ciliatum* Cyr.) which has long-pedicellate, pure white flowers with bluntish perianth-segments, and longer filaments with violet (or brown) anthers. It has been several times recorded from Cyprus, but I think through confusion with white-flowered *A. trifoliatum*, which is not rare, though generally to be found growing with the typical, pink-flushed species. The Cyprus populations of *A. trifoliatum* are fairly uniform.

3. A. cassium *Boiss.*, Diagn., 1, 13: 28 (1853), Fl. Orient., 5: 271 (1882); Post, Fl. Pal., ed. 2, 2: 643 (1933); Osorio-Tafall et Seraphim, List Vasc. Plants Cyprus, 22 (1973).
    A. roseum L. var. *cassium* (Boiss.) Regel in Acta Hort. Petrop., 3: 229 (1875).
    [*A. hirsutum* (non Zucc.) Kotschy in Unger et Kotschy, Die Insel Cypern, 198 (1865).]

Bulb broadly ovoid or subglobose, 1–2 cm. diam., exterior tunic dark grey, minutely pitted, interior tunics whitish, smooth; stems slender, 8–30 cm. long, terete, glabrous or subglabrous; leaves 2–4, linear, flat, 8–20 (–30) cm. long, 0·1–0·4 (–0·7) cm. wide, glabrous, subglabrous or pilose with rather sparse, long hairs, base long sheathing, apex obtuse or subacute, often withered at anthesis, margins smooth, nervation often rather prominent; spathe membranous, about 10 mm. long, 6 mm. wide, short, acute, simple or 2-lobed; inflorescence a lax, 4–15 (–26)-flowered umbel; pedicels slender, 1–2 cm. long, glabrous; perianth rather narrowly campanulate with connivent, narrowly oblong, whitish or pink-tinged, acuminate segments 8–10 mm. long, 1·5–2·5 mm. wide; filaments subulate, simple, about 5 mm. long, tapering upwards from a wide base; anthers yellow, oblong, about 1·5 mm. long, 0·7 mm. wide; ovary trigonous-subglobose, about 2 mm. diam., glabrous; style slender, about 4 mm. long; stigma shortly clavate. Capsule globose, pale brown, about 5 mm. diam.; seeds angular, about 2 mm. long, 1·5 mm. wide; testa black, minutely and bluntly papillose.

var. **hirtellum** *Boiss.*, Fl. Orient., 5: 272 (1882); Holmboe, Veg. Cypr., 47 (1914); Post, Fl. Pal., ed. 2, 2: 643 (1933).
    A. troödi Lindberg f., Iter Cypr., 10 (1946); Osorio-Tafall et Seraphim, List Vasc. Plants Cyprus, 22 (1973).
TYPE: [Turkey] "in vallibus jagi [sic] Bulghardagh Tauri Cilicici 6500'–7000' (Ky 13! et 31!) in Libano ad Cedros (Bl. 656 bis) prope Balbeck (Bl!)" (G).

Leaves, and sometimes base of stem, thinly, but rather conspicuously, pilose with long hairs.

HAB.: Stony slopes and banks at high altitudes on igneous mountainsides; 4,300–5,500 ft. alt.; fl. May–June.

DISTR.: Division 2. Also S.E. Turkey, Syria, Lebanon.

2. Between Prodhromos and Tris Elies ["Trisedies"], 1862, *Kotschy* 528, 768! Troödos area, 1880, *Sintenis* 1073 teste *Lindberg f.* also, 1912, *Haradjian* 448!, 1939, *Lindberg f.* s.n.! and, 1948, *Mavromoustakis* s.n.! Near Government Cottage, Troödos, 1905, *Holmboe* 864! Kalogeros, S.W. of Khionistra, 1937, *Kennedy* 191! Xerokolymbos, 1937, *Kennedy* 192! Kryos Potamos, 5,100 ft. alt., 1937, *Kennedy* 193! 194! and, 1941, *Davis* 3506! Seven Sisters, below Troödos, 1960, *N. Macdonald* 195!

NOTES: Although the Cyprus plant is generally rather smaller than Turkish or Lebanese specimens, I do not think it is specifically distinct; contrary to what Lindberg f. says, the outermost tunics of the bulbs are distinctly alveolate, even in the type material of his *A. troödi* (K!).

4. A. roseum *L.*, Sp. Plant., ed. 1, 296 (1753); Regel in Acta Hort. Petrop., 3: 228 (1875); Boiss., Fl. Orient., 5: 273 (1882); Osorio-Tafall et Seraphim, List Vasc. Plants Cyprus, 22 (1973).

Bulb broadly ovoid or subglobose, 1–1·5 cm. diam., exterior tunic crustaceous, grey-brown, deeply and conspicuously pocked with circular or hexagonal pits, interior tunic smooth, whitish; bulblets usually numerous, ovoid; stem terete, 10–60 cm. high; leaves 2–6, linear, flat, 10–30 cm. long, 0·5–1 cm. wide, glabrous, distinctly nerved, base long-sheathing, apex slender, acuminate, margins smooth or papillose; spathe papery, 1-valved, 1–1·5 cm. long, deeply 3–4-lobed at anthesis; inflorescences with or without bulbils, usually many-flowered, often rather lax; pedicels slender, glabrous, 1–4 cm. long; perianth campanulate or cup-shaped with a rounded base, segments commonly pink, sometimes white, free almost to base, oblong, 9–10 mm. long, 4·5–5·5 mm. wide, thin, translucent, apex rounded or obtuse, midrib often prominent, dark pink; filaments subulate, tapering from a broad base, alternately about 5 mm. and 5·5 mm. long; anthers yellow, oblong, about 1·5 mm. long, 1 mm. wide; ovary trigonous-subglobose, glabrous, about 3·5 mm. diam.; style about 5 mm. long; stigma clavate. Capsule 4–4·5 mm. diam., seeds angular, 3 mm. long, 2–2·5 mm. wide; testa black, bluntly papillose.

var. **bulbiferum** *Desf. ex DC.* in Lam. et DC., Fl. Franç., 3: 221 (1805); Kunth, Enum. Plant., 4: 439 (1843); Boiss., Fl. Orient., 5: 274 (1882).
A. carneum Bertol., Rar. Ligur. Plant. Dec., 1: 7 (1803), Plant. Genuens., 51 (1804); Savi, Due Cent. Plant., 87 (1804); Santi, Viaggio Terzo Prov. Senesi, 315, t. 6 (1806).
A. ambiguum Sm. in Sibth. et Sm., Fl. Graec. Prodr., 1: 227 (1809), Fl. Graec., 4: 23, t. 327 (1823) non DC. in Lam. et DC., Fl. Franç., 3: 220 (1805).
A. roseum L. var. *carneum* (Bertol.) Reichb., Icones Fl. Germ., 10: 28, t. 1103 (1848); Holmboe, Veg. Cypr., 47 (1914).
TYPE: Cultivated in the Jardin des Plantes, Paris (? P, ? G).

Inflorescence bulbilliferous; bulbils ovoid, often purple; flowers rather few, commonly sterile.

HAB.: Dry fields; c. 1,500 ft. alt.; fl. March–May.

DISTR.: Division 1, rare. Widespread in S. Europe and the Mediterranean region.

1. Tsadha, 1905, *Holmboe* 740!

SECT. 3. **Briseis** (*Salisb.*) *Stearn* Bulbs subglobose, not rhizomatous, leaves and stems triquetrous; spathe shorter than pedicels, 2-valved, persistent; perianth-segments spreading or connivent; stamens included, filaments without appendages, ovules 2 in each loculus of ovary; stigma 3-lobed. Seeds angular with a white elaiosome.

A. TRIQUETRUM *L.*, Sp. Plant., ed. 1, 300 (1753) with triquetrous leaves and stems and secund inflorescences of rather large white, drooping flowers, is recorded from the mountains above Kythrea, 1880, by Sintenis (in Oesterr.

Bot. Zeitschr., 31: 392; 1881), but no specimens can be traced, nor is this western Mediterranean species likely to occur in our area, unless planted. The record may be a slip of the pen.

SECT. 4. **Scorodon** *Koch* Bulbs ovoid or subglobose, not rhizomatous; stems terete; leaves sheathing at the base, usually filiform; spathe generally shorter than pedicels; perianth-segments spreading or connivent; stamens included or sometimes shortly exserted; filaments simple or those of inner stamens occasionally with 2 short basal teeth; ovules 2 in each loculus of ovary; stigma entire. Seeds angular.

5. **A. cupani** *Rafin.*, Caratt., 86 (1810); Regel in Acta Hort. Petrop., 3: 123 (1875); Boiss., Fl. Orient., 5: 265 (1882) pro parte; Holmboe, Veg. Cypr., 46 (1914); Lindberg f., Iter Cypr., 10 (1946); Osorio-Tafall et Seraphim, List Vasc. Plants Cyprus, 22 (1973).

Bulb ovoid, often 2–3 cm. long, 0·7–1·5 cm. diam., tunic of interwoven fulvous or fuscous fibres, sometimes breaking away from the base of the bulb; stems slender, terete, 6–50 cm. high; leaves 3–4, running far up the stem, filiform, canaliculate above, up to 8 cm. long, 0·4–0·5 mm. wide, glabrous or shortly pilose, especially on the sheaths, usually withered before anthesis; spathe 1–3 cm. long, usually less than 0·5 cm. wide, papery, persistent, distinctly nerved, with a narrowly tubular base and an acuminate or cuspidate-acuminate apex; inflorescence rather fastigiate, 3–15-flowered; pedicels very unequal, 1–4 cm. long, glabrous; perianth narrowly tubular-campanulate, segments connivent, whitish or flushed pink with a darker midrib, narrowly oblong, 3–9 mm. long, 1–2 mm. wide, acute or obtuse; stamens included; filaments subulate, the outer 1–4 mm. long, the inner 0·8–3 mm. long; anthers yellow, oblong, 0·8–1 mm. long; 0·5–0·8 mm. wide; ovary trigonous-subglobose, 1·5–3·5 mm. diam., glabrous, style 0·7–3 mm. long; stigma truncate. Capsule depressed-globose, 3–4 mm. diam., pale brownish; seeds angular or often flattened, 2·5–3·5 mm. long, 1·5–2 mm. wide; testa black, closely shagreened-verruculose.

ssp. **cyprium** *Meikle* in Ann. Musei Goulandris, 6: 94 (1983). See App. V, p. 1898.
[*A. cupani* (non Rafin.) Boiss., Fl. Orient., 5: 265 (1882) pro parte quoad plant. cypr.; Holmboe, Veg. Cypr., 46 (1914); Lindberg f., Iter Cypr., 10 (1946).]
TYPE: Cyprus; "In montibus inter Potami et Evriku, 14 Junio 1880", *Sintenis & Rigo* 860 (holotype K!).

Bulb-tunics generally breaking away from base of bulb, leaf-sheaths shortly pilose; flowers small; perianth-segments 3–4 mm. long, 1–1·5 mm. wide; spathe usually short, less than 1·5 cm. long, with an abruptly narrowed, cuspidate apex.

HAB.: Dry hillsides and sterile fields; 500–1,500 ft. alt.; fl. May–June.

DISTR.: Divisions 5, 6, rare. Endemic.

5. Athalassa, 1939, *Lindberg f.* s.n.!
6. Between Potami and Evrykhou, 1880, *Sintenis & Rigo* 860! Akaki ["Akacha"] near Peristerona, 1905, *Holmboe* 845.

NOTES: Apparently related to *A. cupani* Rafin. ssp. *hirtovaginatum* (Kunth) Stearn, but consistently distinguished by its small flowers. Evidently rare on the island, and still inadequately known.

6. **A. autumnale** *P. H. Davis* in Kew Bull., 1949: 114 (1949); Osorio-Tafall et Seraphim, List Vasc. Plants Cyprus, 23 (1973).
TYPE: Cyprus; near Myrtou, 14 Oct. 1940, *Davis* 1967 (K!).

Bulb ovoid, 2–2·5 cm. long, 1–2 cm. diam., tunics membranous, the outermost fuscous, the inner white; stems erect, slender, terete, glabrous, 12–60 cm. high; leaves 2–3, sheaths glabrous, conspicuously sulcate, often extending almost to apex of stem; lamina narrowly linear, glabrous, glaucescent, fistulose, conspicuously sulcate, up to 25 cm. long, 0·2 cm. wide, generally withered and broken off well before anthesis, spathe 1-valved, subulate, 2–3 cm. long, 0·4–0·5 cm. wide at base, tapering to a slender acumen; inflorescence a rather lax few- to many-flowered umbel; pedicels subequal, 1–3 cm. long, glabrous, angled, at first spreading or erect, becoming erect and sometimes subsecund in fruit, purplish, subglaucous; perianth subcylindrical, about 2 mm. diam., segments oblong, 4–5 mm. long, 2 mm. wide, whitish or purplish with a greenish-brown median stripe, concave or carinate, apex rounded or obtuse; stamens included or almost included; filaments slender, simple, about 3·5 mm. long; anthers yellow, oblong, about 1·2 mm. long, 0·8 mm. wide; ovary trigonous-oblong, about 2–2·5 mm. long, 1–1·3 mm. wide, glabrous; style slender about 2·5 mm. long; stigma subcapitate. Capsule depressed-globose, about 5 mm. diam., pale brown, glabrous, reticulate-alveolate externally, the loculi rounded at the apex; seeds angular or compressed, about 3·5 mm. long, 2 mm. wide; testa black, minutely and closely verruculose.

HAB.: In garigue on limestone, sandstone or igneous hillsides, sometimes under Pines or in vineyards or hedgerows; 500–4,000 ft. alt.; fl. Oct.–Nov.

DISTR.: Divisions 2, 7. Endemic.

2. East of Platres, 1936, *Kennedy* 1587! 1588! Saïttas, 1970, *A. Genneou* in ARI 1603!
7. Near Myrtou, 1940, *Davis* 1967! Vasilia, 1940, *Davis* 2015! Akanthou, 1940, *Davis* 2023! Karavas, 1944, *Casey* 965!

NOTES: Clearly allied to *A. callimischon* Link from Greece and Crete, but uniformly distinct in stature and inflorescence.

SECT. 5. **Codonoprasum** *Reichb.* Bulbs ovoid, not rhizomatous; stems terete; leaf-sheaths extending far up the stem; spathes generally 2-valved, the valves dilated at the base, produced apically into unequal caudate appendages generally equalling or often exceeding the pedicels; perianth-segments connivent, forming a narrowly cup-shaped, infundibuliform or cylindrical perianth; filaments simple, subulate, without lateral appendages; ovary with inconspicuous nectary-pores; ovules 2 in each loculus. Seeds angular or compressed.

7. **A. paniculatum** *L.*, Syst. Nat. ed. 10, 2: 978 (1759), Sp. Plant., ed. 2, 1: 428 (1762); Regel in Acta Hort. Petrop., 2: 191 (1875); Boiss., Fl. Orient., 5: 259 (1882); Post, Fl. Pal., ed. 2, 2: 640 (1933); Osorio-Tafall et Seraphim, List Vasc. Plants Cyprus, 22 (1973).

Bulbs ovoid, 1–2·5 cm. long, 0·8–2 cm. wide, tunics membranous, pallid or the outermost greyish-fuscous; stems terete, 20–70 cm. high; leaves 3–5, the sheaths extending ⅓–½ way up stem, glabrous or occasionally shortly pilose, lamina linear, canaliculate, smooth, glabrous, up to 30 cm. long, less than 0·5 cm. wide, often withered before anthesis; spathe with 2 unequal valves, spreading or lightly reflexed at anthesis, base dilated, strongly nerved, 5–12 mm. wide, apex produced into a linear-caudate, herbaceous appendage (1–) 3–14 cm. long; inflorescence usually many-flowered; pedicels sometimes conspicuously unequal, the longest up to 4 cm. long, the shortest often less than 1 cm. long; perianth-segments erect, pink, white, greenish or brownish, connivent to form a narrowly cup-shaped flower, individual segments oblong, about 3·5–7 mm. long, 1·5–2·5 mm. wide, apex obtuse or rounded, rarely subacute; stamens included or very shortly exserted; filaments slender, subulate, about as long as perianth-segments, united basally into a narrow annulus; anthers yellowish or greenish, oblong, about 1·2 mm. long, 0·8 mm. wide; ovary ellipsoid or subglobose, about 3–5 mm. long, 2–3 mm. wide, sometimes papillose-veruculose; style 1–2 mm. long; stigma subclavate.

Capsule trigonous-ellipsoid or trigonous-subglobose, about 5 mm. long, 2·5–5 mm. wide; seeds angular, about 4 mm. long, 1·5 mm. wide; testa black, closely verruculose.

ssp. **fuscum** (*Waldst. et Kit.*) *Arcang.*, Comp. Fl. Ital., ed. 2, 136 (1894).
    A. fuscum Waldst. et Kit., Descr. et Icon. Plant. Rar. Hung., 3: 267, t. 241 (1808–09); Regel in Acta Hort. Petrop., 3: 190 (1875).
    A. paniculatum L. var. *fuscum* (Waldst. et Kit.) Boiss., Fl. Orient., 5: 260 (1882).
    [*A. paniculatum* (non L.) Sm. in Sibth. et Sm., Fl. Graec. Prodr., 1: 225 (1809), Fl. Graec., 4: 16, t. 318 (1823).]
    TYPE: Hungary: "in rupibus calcareis Banatus, frequens ed thermas Herculis" (PR).

Plant tall, often 50–70 cm. high; leaf-sheaths glabrous, spathe-appendages often 8–14 cm. long, much exceeding the inflorescence; pedicels very unequal, the longest often 3–4 cm. long; perianth-segments brownish or greenish, not pink or white; ovary commonly tapering to apex and base, sometimes distinctly papillose-verruculose.

    HAB.: In Olive groves; c. 100 ft. alt.; fl. May–July.

    DISTR.: Division 7, rare. S.E. Europe, Balkans.
7. Kyrenia, 1949 and 1956, *Casey* 816! 1728!

ssp. **pallens** (*L.*) *Arcang.*, Comp. Fl. Ital., ed. 2, 136 (1894).
    A. pallens L., Sp. Plant., ed. 2, 1: 427 (1762); Lindberg f., Iter Cypr., 10 (1946).
    A. coppoleri Tineo, Cat. Plant. Hort. Panorm., 275 (1827); Osorio-Tafall et Seraphim, List Vasc. Plants Cyprus, 23 (1973).
    A. paniculatum L. var. *pallens* (L.) Gren. et Godr., Fl. France, 3: 209 (1855); Regel in Acta Hort. Petrop., 3: 193 (1875); Boiss., Fl. Orient., 5: 260 (1882); Post, Fl. Pal., ed. 2, 2: 640 (1933).
    TYPE: "*in* Italia, Hispania, Monspelii, Pannonia".

Plant 40–55 cm. high; leaf-sheaths glabrous; spathe-appendages generally less than 6 cm. long, exceeding the inflorescence; pedicels not markedly unequal, forming a rather compact umbel, usually less than 2 cm. long; perianth-segments whitish with a green or purplish midrib; ovary usually trigonous-subglobose, not tapering noticeably to apex and base, generally smooth.

    HAB.: Fields, vineyards, roadsides, sometimes on dry stony hillsides or on seashores; sea-level to 2,800 ft. alt.; fl. May–July.

    DISTR.: Divisions 1–6. Southern Europe and Mediterranean region.
1. Ayios Neophytos Monastery, 1939, *Lindberg f.* s.n.
2. Near Evrykhou, 1880, *Sintenis & Rigo* 863! Between Perapedhi and Mandria, 1937, *Kennedy* 185!
3. Lefkara, 1940, *Davis* 1916!
4. Cape Pyla, 1934, *Syngrassides* 432! Larnaca, 1939, *Lindberg f.* s.n. Famagusta, 1939, *Lindberg f.* s.n.; between Famagusta and Salamis, 1939, *Lindberg f.* s.n.! Athna Forest, 1952, *Merton* 839!
5. Athalassa, 1939, *Lindberg f.* s.n.!
6. Between Peristerona and Potami, 1880, *Sintenis & Rigo* 864! Nicosia, 1973, *P. Laukkonen* 336!
    NOTES: Sintenis has labelled the Kew specimen of *Sintenis & Rigo* 864 "*Allium myrianthum* Boiss.*" but the flowers are too few, and the spathe-valves too little reflexed for this species.

ssp. **exaltatum** *Meikle* in Ann. Musei Goulandris, 6: 94 (1983). See App. V, p. 1898.
    TYPE: Cyprus; Xerokolymbos, S.W. of Khionistra, 4930 ft. alt., 1937, *Kennedy* 181 (K, holotype!).

Plant slender, generally about 20–40 cm. high; spathe-appendages short, generally less than 3 cm. long, equalling or exceeding the inflorescence; pedicels not markedly unequal, usually less than 1·5 cm. long, often very short, forming a compact subglobose or ovoid umbel; perianth-segments whitish or brownish-green; ovary usually trigonous-subglobose, smooth.

Capsule indented at apex, the top of the loculi running smoothly into the sinus and without a subapical protuberance.

HAB.: Rocky or stony mountainsides, or crevices of igneous rocks; sometimes under Pines; 4,000–5,450 ft. alt.; fl. July–Aug.

DISTR.: Division 2. Possibly endemic.

2. West of Asprokremnos on S. slope of Khionistra, 1937, *Kennedy* 180! Xerokolmybos, S.W. of Khionistra, 1937, *Kennedy* 181–183! Kryos Potamos, 4,000 ft. alt., 1939, *Kennedy* 184! Platres, 4,000 ft. alt., 1940, *Davis* 1849!

NOTES: *Davis* 1849 has been determined as *A. bassitense* Thiéb. by M. Koyuncu (1979), and this may well be correct, though there are some discrepancies between our plant and *A. bassitense* (from Syria, near Akra Dağ) to judge from Thiébaut's brief description. I do not think the Cyprus plant can be considered specifically distinct from *A. paniculatum* (incl. *A. pallens*).

8. A. stamineum *Boiss.*, Diagn., 2, 4: 119 (1859); Regel in Acta Hort. Petrop., 3: 195 (1875); Boiss., Fl. Orient., 5: 256 (1882); Post, Fl. Pal., ed. 2, 2: 639 (1933); A. K. Jackson in Kew Bull., 1936: 16 (1936); Lindberg f., Iter Cypr., 10 (1946); Osorio-Tafall et Seraphim, List Vasc. Plants Cyprus, 23 (1973).

TYPE: Turkey; "Caria 1843" *Pinard* s.n. (G, K!).

Bulb ovoid, 1·5–2 cm. long, 1 cm. diam., tunic papery, finely striate, whitish or yellow-brown; stem 5–30 (–35) cm. long, terete, commonly somewhat curved, greenish or purplish and often glaucous, especially towards apex, leaves 2–5, semiterete, 6–15 (–20) cm. long, 0·1–0·2 cm. wide, sometimes fistulose, canaliculate above, convex and distinctly nerved below, sheath 5–20 cm. long, finely striate, glabrous or shortly pilose; inflorescence a lax, few- to many-flowered umbel; valves of spathe unequal, spreading, 1–6 cm. long, dilated and 5–8 mm. wide at base, abruptly narrowed above to a long subulate appendage; pedicels unequal, 0·5–2 (–5) cm. long, slender, glabrous, erect, spreading and pendulous at anthesis, sometimes becoming more uniformly erect in fruit; perianth cup-shaped or subglobose, segments connivent, greenish, lilac or purplish, oblong, 3–3·5 mm. long, 2 mm. wide, concave or bluntly keeled, apex rounded or very obtuse; filaments subulate, purplish or violet, glabrous, exceeding the perianth-segments by 0·5 mm. or more; anthers oblong, yellow, about 0·6 mm. long, 0·5 mm. wide; ovary sessile or subsessile, glabrous, trigonous-ellipsoid, tapering slightly to apex and base, about 2 mm. long, 1·5 mm. wide; style slender, 2–2·5 mm. long; stigma truncate. Capsule trigonous-subglobose, about 4 mm. diam., pale brown, tops of the loculi rounded, running smoothly into apical sinus; seeds compressed, D-shaped, about 4 mm. long, 2 mm. wide; testa black, minutely and regularly rugulose.

HAB.: On eroded sandstone slopes, or on limestone or igneous mountainsides under Pines; 1,000–5,700 ft. alt.; fl. May–Aug.

DISTR.: Divisions 2, 5, 7. Greece and eastern Mediterranean region eastwards to Iran.

2. Near Prodhromos towards Lemithou, 1880, *Sintenis & Rigo* 859! Asprokremnos, S. of Khionistra, 1937, *Kennedy* 171! 173! Khionistra, 1937, *Kennedy* 172! 175! 177! Xerokolymbos, 5,300 ft. alt., 1937, *Kennedy* 174! Kryos Potamos, 4,800–5,600 ft. alt., 1937–38, *Kennedy* 176! 179! 1146! Platres, 1937–38, *Kennedy* 178! 1143! 1144! 1145! Kyperounda, 1940, *Davis* 1881!
5. Athalassa, 1939, *Lindberg f.* s.n.
7. Halevga, 1940, *Davis* 1731! Yaïla, 1941, *Davis* 3607! Kythrea, 1941, *Davis* 3642!

NOTES: All the Cyprus specimens examined have glabrous leaf-sheaths and few (10–30)-flowered umbels of smallish flowers with shortly exserted filaments. The plant from the Northern Range appears to be less suffused with purple than those from high altitudes on Khionistra, where, however, many plants in a wide range of families depart from the norm in their heavily anthocyanin-pigmented leaves and stems.

In view of the known polymorphism of *A. stamineum*, I have not attempted to distinguish the Cyprus plants infraspecifically, though future research may show that this is warranted. M. Koyuncu has determined *Davis* 3642 as *A. amphipulchellum* Zahariadi.

SECT. 6. **Allium** Bulbs ovoid or subglobose, not rhizomatous, leaves linear, flat or fistulose, sheathing the basal part of the stem; spathe usually caducous, 1- or 2-valved, the valves generally produced into a subulate appendage; perianth-segments mostly connivent; stamens usually dimorphic, the outer simple, the inner 3-cuspidate, the median, antheriferous lobe flanked by lateral, sterile cusps; ovary with distinct nectariferous pores; ovules 2 in each loculus; stigma entire. Seeds angular.

A. SATIVUM *L.*, Sp. Plant., ed. 1, 296 (1753), Garlic, with clustered, ovoid, strong-smelling bulbs, stems commonly 1 m. or more in height, clad in the lower half with long, flat, keeled leaves, and with a bulbiferous umbel enveloped by a single, long-rostrate spathe, is cultivated in Cyprus, as elsewhere in S. Europe, and recorded for the island by Kotschy (in Unger et Kotschy, Die Insel Cypern, 196; 1865). It is thought to be a cultivated derivative of the Central Asiatic *A. longicuspis* Regel.

9. A. ampeloprasum *L.*, Sp. Plant., ed. 1, 294 (1753); Regel in Acta Hort. Petrop., 3: 52 (1875); Boiss., Fl. Orient., 5: 232 (1882); Holmboe, Veg. Cypr., 45 (1914); Post, Fl. Pal., ed. 2, 2: 634 (1933); Lindberg f., Iter Cypr., 10 (1946); Rechinger f. in Arkiv för Bot., ser. 2, 1: 418 (1950); v. Bothmer in Bot. Notiser, 125: 62 (1972), Opera Bot., 34: 1–104 (1974); Osorio-Tafall et Seraphim, List Vasc. Plants Cyprus, 22 (1973).

    A. ampeloprasum L. var. *holmense* Aschers. et Graebn., Syn. Mitteleurop. Fl., 3: 105 (1905).

    A. ampeloprasum L. f. *holmense* (Aschers. et Graebn.) Holmboe, Veg. Cypr., 45 (1914).

    [*A. ampeloprasum* L. var. *leucanthum* (non (K. Koch) Ledeb.) Lindberg f., Iter Cypr., 10 (1946).]

TYPE: "*in* Oriente, *inque insula* Holms *Angliae*".

Bulb broadly ovoid, 3–6 cm. long, 2–6 cm. diam., tunic whitish, membranous, bulblets generally numerous, shining yellowish-brown, ovoid, shortly acute, with a narrow marginal wing; stem stout, erect, 40–180 cm. high; leaves 4–10, mostly withered by anthesis, enveloping the basal quarter of the stem, sheaths glabrous, striatulate, lamina linear, 20–50 cm. long, 0·5–4 cm. wide, flat, carinate, glaucous, tapering from base to apex, margins minutely scabridulous or almost smooth; spathe 1-valved with a compressed beak, caducous; umbel globose, 3–10 cm. diam., many-flowered; pedicels subequal, 1·5–5 cm. long, sharply angled; perianth cup-shaped, 4–5 mm. long, 3–4 mm. wide, segments generally reddish, purple or lilac, occasionally white, connivent, concave, somewhat carinate, the keels smooth or obscurely and irregularly scabridulous, outer segments 4·5–5·5 mm. long, 2·5–3 mm. wide, oblong, subacute, the inner more membranous, oblong-obovate, obtuse, a little shorter and proportionately wider; stamens slightly exserted; filaments of outer stamens simple, narrowly deltoid-subulate, 4–5 mm. long, 1–1·3 mm. wide at base, papillose-ciliate, the inner oblong, about 1·5 mm. wide, 3-cuspidate, the antheriferous cusp about 1·5 mm. long, the flanking sterile cusps filiform, commonly more than 3 mm. long; anthers oblong, violet, 1·5–2 mm. long, 1 mm. wide; ovary shortly stipitate on a thick torus, broadly triquetrous-ovoid, about 2·5–3 mm. long, 3 mm. wide, glabrous; style 2·5–3 mm. long; stigma subtruncate. Capsule about 4 mm. diam., pale brown; seeds angular, about 3·5 mm. long, 2 mm. wide, testa black, closely papillose-verruculose.

HAB.: Roadsides, cultivated and waste ground, sometimes on dry, stony hillsides; sea-level to 2,800 ft. alt.; fl. May–June.

DISTR.: Divisions 1, 2, 4–8. S. and W. Europe, Mediterranean region and eastwards to Iraq and Caucasus.

1. Yeroskipos, 1905, *Holmboe* 715.
2. Pano Panayia, 1905, *Holmboe* 1054; between Mandria and Perapedhi, 1937, *Kennedy* 188! Kambos, 1968, *Economides* in ARI 1210! Between Pano Platres and Moniatis, 1973, *P. Laukkonen* 438!

4. Larnaca, 1939, *Lindberg f.* s.n.! Near Paralimni, 1947–48, *Mavromoustakis* s.n.! Cape Greco, 1958, *N. Macdonald* 78! Between Sotira and Ayia Napa, 1979, *Chesterman* 110! also, 1981, *Hewer* 4726!
5. Salamis, 1939, *Lindberg f.* s.n.; Athalassa, 1939, *Lindberg f.* s.n.!
6. Morphou, 1972, *W. R. Price* 1057!
7. Between Kyrenia and Lapithos, 1880, *Sintenis & Rigo* 861! Kyrenia, 1939, *Lindberg f.* s.n.! also, 1948, *M. Casey* in *Casey* 190! 1955, *Miss Mapple* 23! and, 1956, *Casey* 1729! St. Hilarion, 1939, *Lindberg f.* s.n.
8. Komi Kebir, 1912, *Haradjian* 311, 320, 321.

A. PORRUM *L.*, Sp. Plant., ed. 1, 295 (1753), the Leek, generally considered to be a cultivated derivative of *A. ampeloprasum* L., and differing in its mild flavour, general absence of bulblets, long-rostrate spathe, very large umbels (often 10–20 cm. diam.) of greenish or pale purplish flowers, and distinctly exserted stamens, is found in Cyprus gardens, and noted by Kotschy (Die Insel Cypern, 196; 1865) and by Nattrass (First List Cyprus Fungi, 14; 1937). It is not naturalized on the island.

10. A. scorodoprasum *L.*, Sp. Plant., ed. 1, 297 (1753); Sibth. et Sm., Fl. Graec. Prodr., 1: 224 (1809); Poech, Enum. Plant. Ins. Cypr., 8 (1842); Unger et Kotschy, Die Insel Cypern, 196 (1865); Regel in Acta Hort. Petrop., 3: 42 (1875); Boiss., Fl. Orient., 5: 232 (1882); Holmboe, Veg. Cypr., 45 (1914); Post, Fl. Pal., ed. 2, 2: 634 (1933); Osorio-Tafall et Seraphim, List Vasc. Plants Cyprus, 22 (1973); W. T. Stearn in Ann. Mus. Goulandris, 4: 178 (1978).

Bulb ovoid, 1–2 cm. long, 1–1·5 cm. wide, tunics membranous, or the outermost subcoriaceous and yellowish-brown, sometimes coarsely lacerated at the apex; bulblets often numerous, ovoid, or subglobose, acute, shining blackish-crimson or vinous; stems erect, terete, 30–100 cm. long; leaves 2–5, enveloping $\frac{1}{3}$ to $\frac{1}{2}$ the stem, linear flat, carinate, glaucescent, 10–30 cm. long, 0·3–1·5 cm. wide, apex acute, margins entire; spathe caducous, chartaceous, shortly acute, about 1·5 cm. long and about as wide; umbels compact, dense, 1–5 cm. diam., consisting of numerous dark vinous bulbils mixed with relatively few flowers, or of numerous flowers without bulbils; pedicels subequal, 1–2 cm. long, sharply angled; perianth ovoid, about 4–6 mm. long, 3 mm. wide, crimson-purple or lilac, tapering from a rounded base to a subacute apex, segments ovate-oblong, rather coarsely papillose-verruculose externally, concave but scarcely keeled, the outer 5–6 mm. long, 2·5 mm. wide, acute, the inner 7 mm. long, 2·5 mm. wide, obtuse or subacute; stamens included, the outer with simple, subulate filaments 3–4 mm. long, 1 mm. wide at base, the inner 3-cuspidate, with an oblong, papillose basal part 4–4·5 mm. long, 1 mm. wide and a median, antheriferous cusp 1 mm. long, flanked by 2 filiform, sterile cusps about 2·5 mm. long; anthers oblong, yellowish, 2 mm. long, 0·8 mm. wide; ovary triquetrous-globose, sessile, about 3 mm. diam., style 3·5 mm. long; stigma truncate. Capsule globose, about 4 mm. diam., pale brown, enveloped by the persistent perianth; seeds sharply angled, about 3·5 mm. long, 2 mm. wide; testa black, closely and minutely papillose-verruculose.

[ssp. SCORODOPRASUM
    Type: "*in* Oelandia, Dania, Pannonia".

Leaves commonly 0·8–1·5 cm. wide; umbel consisting chiefly of dark shining bulbils intermixed with a few, scattered, sterile, long-pedicellate flowers.

HAB. and DISTR. unknown. Recorded for Cyprus by Sibthorp (Fl. Graec. Prodr., 1: 224; 1809), but almost certainly in error, since the plant has not been re-collected, and is generally absent from this part of the Mediterranean. The Sibthorp reference is the basis of all subsequent records from the island. *A. scorodoprasum* ssp. *scorodoprasum* is believed to be a derivative of *A.*

scorodoprasum ssp. *rotundum* (infra), and was formerly grown as a potherb (rocambole) particularly in N. Europe. It may have been cultivated in Cyprus, but this seems unlikely].

ssp. **rotundum** (*L.*) *Stearn* in Ann. Mus. Goulandris, 4: 128 (1978).
  A. rotundum L., Sp. Plant., ed. 2, 1: 423 (1762); Sibth. et Sm., Fl. Graec. Prodr., 1: 222 (1809); Poech, Enum. Plant. Ins. Cypr., 9 (1842); Unger et Kotschy, Die Insel Cypern, 196 (1865); Regel in Acta Hort. Petrop., 3: 57 (1875); Boiss., Fl. Orient., 5: 233 (1882); Post in Mém. Herb. Boiss., 18: 101 (1900); Holmboe, Veg. Cypr., 45 (1914); Post, Fl. Pal., ed. 2, 2: 635 (1933); v. Bothmer in Bot. Notiser, 125: 63 (1972); Osorio-Tafall et Seraphim, List Vasc. Plants Cyprus, 22 (1973).
  TYPE: "*in* Europa *australiori*".

Leaves 3–10 mm. wide; umbel consisting entirely of crowded, fertile flowers, without bulbils.

HAB.: In garigue on the borders of vineyards or on dry rocky, calcareous hillsides; 100–3,600 ft. alt.; fl. May–July.

DISTR.: Divisions 2, 3, rare. S. Europe and Mediterranean region eastwards to Iran.

2. Aphamis, 3,600 ft. alt., 1937, *Kennedy* 186a (intermixed with *A. sphaerocephalon* L., *Kennedy* 186b)!
3. Pissouri, ? 1898, *Post* s.n.

NOTES: Also recorded by Sibthorp (Fl. Graec. Prodr., 1: 222; 1807) but without precise locality.

11. A. sphaerocephalon *L.*, Sp. Plant., ed. 1, 297 (1753); Regel in Acta Hort. Petrop., 3: 45 (1875); Boiss., Fl. Orient., 5: 236 (1882); Post in Mém. Herb. Boiss., 18: 101 (1900); H. Stuart Thompson in Journ. Bot., 44: 339 (1906); Holmboe, Veg. Cypr., 46 (1914); Post, Fl. Pal., ed. 2, 2: 636 (1933); Rechinger f. in Arkiv för Bot., ser. 2, 1: 419 (1950); v. Bothmer in Bot. Notiser, 125: 64 (1972).
  TYPE: "*in* Italia, Sibirica, Helvetia".

Bulb broadly ovoid or subglobose, 2–3 cm. long, 2–2·5 cm. diam.; tunics whitish, membranous, or the outermost subcoriaceous, yellowish-brown, coarsely lacerate at apex and base, bulblets usually present, compressed, acuminate, about 1 cm. long, 0·5–0·8 cm. wide, stipitate, concealed under the base of the leaf-sheaths, pallid or yellow-brown and lustrous; stems 20–90 cm. high, slender, terete, leaves 2–6, sheathing the lower $\frac{1}{4}$–$\frac{1}{2}$ of the stem, sheaths glabrous, distinctly sulcate, lamina fistulose, compressed-canaliculate, glabrous, narrowly linear, up to 30 cm. long, 0·1–0·4 cm. wide, tapering to apex, spathe persistent, 2–4-valved, shortly beaked, papyraceous, 1–2 cm. long, 1–1·5 cm. wide, reflexed at anthesis, and much shorter than the inflorescence; umbel dense, many-flowered, globose, or broadly ovoid, 2–4 (–6) cm. diam.; bulbils occasionally present (not seen in our area); pedicels subequal, angular, glabrous, 0·5–2 (–3) cm. long; perianth ovoid, constricted towards apex, reddish-purple, 4–6 mm. long, 2·5–3 mm. wide near base, segments subequal, ovate, concave, somewhat carinate, about 4–5·5 mm. long, 2–2·5 mm. wide, apex obtuse or mucronate- subacute, smooth or minutely papillose externally especially along keel; stamens conspicuously exserted; outer filaments simple, subulate, thinly papillose, 5–6 mm. long, about 1 mm. wide at base, inner filaments 3-cuspidate, with an oblong, papillose basal part, about 2·5 mm. long, 1 mm. wide, and a median antheriferous cusp about 2·5 mm. long, flanked by 2 filiform sterile cusps about 4–5 mm. long; anthers oblong, purplish 1–1·2 mm. long, about 0·5 mm. wide; ovary glabrous, triquetrous-ovoid, sessile, about 4 mm. long, 2·5 mm. wide; style about 4 mm. long; stigma truncate. Capsule broadly ovoid or subglobose, about 3·5 mm. diam., pale yellowish-brown; seeds angled, about 2·5 mm. long, 0·5 mm. wide; testa black, minutely and closely papillose-verruculose.

HAB.: In vineyards or in garigue on rocky limestone or igneous mountainsides; waste ground; roadsides; sea-level to 4,000 ft. alt.; fl. May–July.

DISTR.: Divisions 1, 2, rare. Europe and Mediterranean region eastwards to Turkey and Palestine.

1. Ktima, 1913, *Haradjian* 710.
2. Near Omodhos, 1880; *Sintenis & Rigo* 862! Above Kykko Monastery, 1898, *Post* s.n.; Aphamis, 1937, *Kennedy* 186b! 187!

NOTES: As noted under *A. scorodoprasum* L. ssp. *rotundum* (L.) Stearn, *Kennedy* 186 is a mixture, consisting mainly of *A. sphaerocephalon* (186b) but including one plant of *A. scorodoprasum* ssp. *rotundum* (186a).

Many authors note that in *A. sphaerocephalon* the median cusp of the inner filaments is as long as, or longer than, the lateral cusps; in the few Cyprus specimens examined this is not so, the median cusp being distinctly shorter than the laterals.

12. A. junceum *Sm.* in Sibth. et Sm., Fl. Graec. Prodr., 1: 226 (1809), Fl. Graec., 4: 19, t. 322 (1823); Poech, Enum. Plant. Ins. Cypr., 9 (1842); Unger et Kotschy, Die Insel Cypern, 197 (1865); Regel in Acta Hort. Petrop., 3: 71 (1875); Boiss., Fl. Orient., 5: 238 (1882); Holmboe, Veg. Cypr., 46 (1914); Lindberg f., Iter Cypr., 10 (1946); Osorio-Tafall et Seraphim, List Vasc. Plants Cyprus, 22 (1973).
TYPE: "In insulâ Cypro" (OXF).

Bulb ovoid, 1–3 cm. long, 0·5–1·5 cm. wide, tunics rather fibrous, splitting irregularly at apex and base, the outermost fuscous, the inner brownish or pallid; bulblets usually absent; stems 10–45 cm. high, glabrous, slender, striate or angled; leaves 2–4, sheaths glabrous, striate, extending up ¼–½ of the stem, lamina fistulose, linear-subulate, 7–30 cm. long, 0·5–0·2 mm. diam., glaucous-green, tapering to a slender acumen; spathe persistent, membranous, 2-valved, shortly apiculate, about 1–1·5 cm. long, 0·6–1 cm. wide, shorter than the inflorescence; umbel compact, dense, many-flowered, globose or broadly ovoid, 1–2·5 cm. diam.; pedicels short, subequal, angled, glabrous, generally less than 4 mm. long; perianth narrowly campanulate, purple (or greenish-white), 5–7 mm. long, 2–3 mm. wide, segments erect, connivent, subequal, ovate, concave, the exterior about 6·5 mm. long, 3–3·5 mm. wide, acute, with a sharply defined, papillose keel, the interior about 7 mm. long, less acute and less markedly keeled; stamens included; filaments of outer stamens lanceolate, 4–4·5 mm. long, 0·8 mm. wide, glabrous, filaments of inner stamens 3-cuspidate, glabrous, the basal part oblong, about 4 mm. long, 1·8 mm. wide, the median antheriferous cusp narrowly deltoid, about 1 mm. long, the lateral cusps narrower, up to 1·8 mm. long, their outer margins often sharply laciniate-lobulate; anthers oblong, purplish, 1·5 mm. long, 0·8 mm. wide; ovary narrowly triquetrous-oblong, about 3–3·5 mm. long, 1·5 mm. wide, glabrous, the loculi with shortly rostrate apices; style 1–1·5 mm. long; stigma truncate-subcapitate. Capsule ovoid, acute, about 4–5 mm. long, 3 mm. diam., pallid; seeds angular, about 3 mm. long, 1–1·2 mm. wide, testa black, closely and densely papillose-verruculose.

HAB.: In dry pastures or garigue on limestone or igneous hillsides, occasionally by roadsides; sea-level to 2,900 ft. alt.; fl. March–May.

DISTR.: Divisions 1–3, 7. Also Turkey.

1. Stroumbi, 1913, *Haradjian* 745! Dhrousha, 1941, *Davis* 3240! Aprilokambos near Ayios Yeoryios, 1962, *Meikle* 2016a!
2. Near Prodhromos, 1862, *Kotschy* 767 (see note); between Mandria and Perapedhi, 1937, *Kennedy* 189! 190! Platres, 1938, *Kennedy* 1139! Mile 42, Xeros–Yialia road, 1957, *Merton* 2954! Moummouros, 1962, *Meikle* 2964!
3. Between Episkopi and Pissouri, 1862, *Kotschy* 624! Between Episkopi and Evdhimou, 1905, *Holmboe* 709; Polemidhia, 1948, *Mavromoustakis* s.n.! Episkopi, 1954, *Merton* 1886! also, 1963, *J. B. Suart* 69! and, 1967, *Merton* in ARI 707! Between Alassa and Lania, 1970, *A. Genneou* in ARI 1524! Between Nikoklia and Phasoula, 1977, *J. J. Wood* 237! Between Evdhimou and Pakhna, 1978, *J. Holub* s.n.; Pissouri, 1981, *Hewer* 4763!
7. St. Hilarion, 1880, *Sintenis & Rigo* 617! also, 1934, *Syngrassides* 698! Kyrenia, 1932–1973, *Nattrass* 578; *Syngrassides* 7! *Lindberg f.* s.n.! *G. E. Atherton* 155! 1299! *P. Laukkonen* 378!

NOTES: Kotschy's record from Prodhromos is questionable, as no other collector has found *A. junceum* at such an altitude.

A. junceum is apparently related to *A. rubrovittatum* Boiss. et Heldr. (infra) but is readily distinguishable by its larger flowers, and by the lateral cusps of the inner filaments being normally distinctly lobed or laciniate; the statement by Boissier (Fl. Orient., 5: 238) that these inner filaments are 5-cuspidate at the apex is not strictly true, and is based on a misinterpretation of Flora Graeca t. 322; the filaments are in fact 3-cuspidate, and the lateral cusps may vary from being quite entire to deeply laciniate-lobed.

13. **A. rubrovittatum** *Boiss. et Heldr.* in Boiss., Diagn., 1, 13: 29 (1853), Fl. Orient., 5: 234 (1882); Regel in Acta Hort. Petrop., 3: 68 (1875); Rechinger f., Fl. Aegaea, 716 (1943).
[*A. curtum* (non Boiss. et Gaill.) Sintenis in Oesterr. Bot. Zeitschr., 32: 397 (1882); Holmboe, Veg. Cypr., 46 (1914).]
TYPE: "in Cretâ, in saxosis faucis *Kordaliotiko* in Eparchiâ *Agio Vasili* (Heldr. [pl. Cret. exsicc. 1847])".

Bulb ovoid, 1–2 cm. long, 0·7–1 cm. diam., tunics membranous, pallid, the outermost thicker, yellow-brown, coarsely lacerated at apex and base; bulblets absent; stem erect or curved, 10–23 (–35) cm. long, glabrous, often purplish, distinctly striate or almost angled near apex; leaves 2–4, mostly withered by anthesis, the sheaths striate-sulcate, glabrous, enveloping the lower ¼ of the stem, 4–11 cm. long, lamina linear, fistulose, 4–20 cm. long, 0·5–2·5 mm. wide, glabrous, striate; spathe membranous, 2-valved, acute, 4–8 mm. long, 3–6 mm. diam., much shorter than the inflorescence, persisting and becoming reflexed at anthesis; umbels broadly ovoid or subglobose, 0·2–1·5 (–2) cm. diam., many-flowered, compact; pedicels subequal, 3–6 mm. long, glabrous, angular; perianth cup-shaped or subglobose, 2·5–3 mm. long and about as wide, brownish-pink or purplish, the segments with darker median stripes, segments erect, connivent, concave, somewhat keeled, the outer ovate, acute, about 3·5 mm. long, 2 mm. wide, the inner rather shorter and broader, about 3 mm. long, 2·5 mm. wide, obtuse or subacute; stamens included or anthers slightly exserted, the outer with simple, lanceolate-subulate filaments about 3 mm. long, the inner filaments 3-cuspidate, with an oblong basal part about 2·5 mm. long, 1 mm. wide, median cusp 1 mm. long, lateral cusps filiform, 2 mm. long, basal part of inner filaments shortly papillose-ciliate; anthers oblong, purplish, about 2–2·5 mm. long, 1–2 mm. wide; ovary sessile, subglobose, 4–5 mm. diam., glabrous; style 2–3 mm. long, bright purple; stigma truncate-subcapitate. Capsule triquetrous-subglobose, about 4–5 mm. diam., yellowish-brown; seeds angled, rugose, about 2·5 mm. long, 1·3 mm. wide; testa black, closely verruculose. *Plate 96.*

HAB.: In dry, rocky, sometimes saline, pastures, or in clearings in Pine forest; sea-level to 3,000 ft. alt.; fl. May–June.

DISTR.: Divisions 4, 5, 7, rather rare. Also Crete.

4. Between Famagusta and Dherinia, 1948, *Mavromoustakis* s.n.! Larnaca airport, 1952, *Merton 823*! Near Famagusta, 1972, *W. R. Price 1010*!
5. Near Kythrea, 1880, *Sintenis & Rigo 357*! Mia Milea, 1950, *Chapman 679*! also, 1952, *Merton 828*!
7. Near Kyrenia, 1880, *Sintenis & Rigo 855*! Summit of Yaïla, 1940, *Davis 1742*! Lakovounara, 1950, *Chapman 515*!

NOTES: Specimens collected by *Lindberg f.* at (5) Athalassa near Nicosia (Iter Cypr., 10; 1946) appear to belong here rather than to *A. curtum* Boiss. et Gaill. as determined; the material is, however, overripe and difficult to name with certainty.

14. **A. curtum** *Boiss. et Gaill.* in Boiss., Diagn., 2, 4: 116 (1859), Fl. Orient., 5: 245 (1882); Regel in Acta Hort. Petrop., 65 (1875); Post, Fl. Pal., ed. 2, 2: 638 (1933); Feinbrun in Pal. Journ. Bot., J. Series, 3: 14 (1943); Lindberg f., Iter Cypr., 10 (1946).
A. curtum Boiss. et Gaill. ssp. *palaestinum* Feinbrun in Pal. Journ. Bot., J. Series, 3: 14 (1943); Osorio-Tafall et Seraphim, List Vasc. Plants Cyprus, 22 (1973).
[*A. arvense* (non Guss.) Osorio-Tafall et Seraphim, List Vasc. Plants Cyprus, 22 (1973).]
TYPE: Lebanon, "in arenosis ultrà *Abarouh* in vicinitis *Sidonis Syriae* cl. Gaillardot" (G).

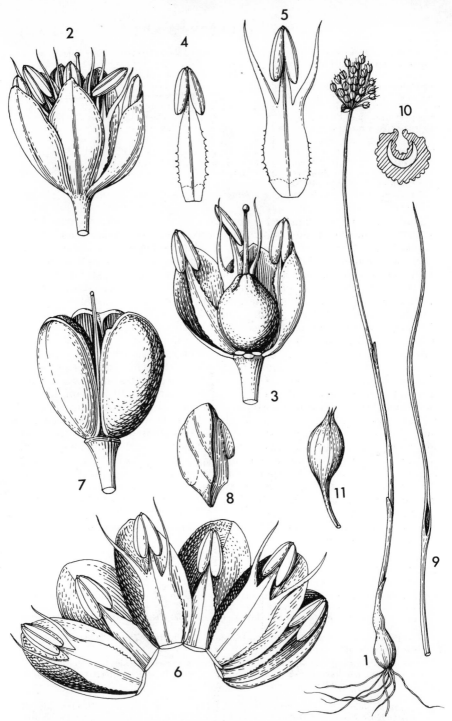

Plate 96. ALLIUM RUBROVITTATUM Boiss. et Heldr. **1,** habit, × ⅔; **2,** flower, × 12; **3,** flower, longitudinal section, × 12; **4,** outer stamen, × 12; **5,** inner stamen, × 12;. **6,** perianth-segments and stamens, × 12; **7,** capsule, × 12; **8,** seed, × 12; **9,** leaf, × ⅔; **10,** leaf, transverse section, × 12; **11,** bulblet, × 2. (**1–6, 11** from *Merton* 828; **7, 8** from *Mavromoustakis* s.n.; **9, 10** from *Chapman* 679.)

Bulb broadly ovoid or subglobose, 1·5–2·5 cm. long and almost as wide; tunics membranous, whitish, or the outermost greyish; bulblets usually present, ovoid, acute, 0·5–1 cm. long, 0·3–0·8 cm. wide; stem erect or curved towards base, 20–60 cm. high, terete, glabrous, longitudinally striate; leaves 2–3, sheathing the lower third of the stem, sheaths glabrous, striate, lamina narrowly linear, fistulose-canaliculate, up to 43 cm. long, 0·2 cm. wide, striate, glabrous or thinly papillose-scaberulous, tapering to a filiform apex; spathe 2-valved, membranous, 5–12 mm. long, 5–15 mm. wide, much shorter than inflorescence, apex acute-cuspidate; umbel subglobose or ovoid, about 1·5–3 cm. diam., rather dense, many-flowered; pedicels subequal, 5–40 mm. long, glabrous, angled; perianth cup-shaped, 3–4 mm. long and about as wide, segments ovate-oblong, concave obtuse, glabrous, greenish or purplish with a whitish margin, about 4 mm. long, the outer distinctly keeled, about 2 mm. wide, the inner about 2·5 mm. wide, scarcely keeled; stamens shortly exserted; filaments of outer stamens simple, linear-lanceolate, about 4·5 mm. long, 0·5 mm. wide at base, papillose-ciliate, filaments of inner stamens 3-cuspidate, basal part oblong, about 2 mm. long, 1 mm. wide, sparsely papillose-ciliate, median cusp 1·5–2 mm. long, a little longer than (sometimes equalling or shorter than) the lateral cusps; anthers yellowish, oblong, about 1·5 mm. long, 0·2 mm. wide; ovary trigonous-ovoid, glabrous, sessile, about 2·5–2·8 mm. long, 2 mm. wide; style 2·3 mm. long; stigma truncate-subcapitate. Capsule subglobose, greenish, 3–3·5 mm. diam.; seeds angular, about 3 mm. long, 1·5 mm. wide; testa black, closely verruculose.

HAB.: Sandy seashores and sandy fields near the sea; sea-level to 100 ft. alt.; fl. May–June.
DISTR.: Divisions 4, 5, rare. Also Syria, Palestine, Egypt.
4. Near Famagusta, 1938, *Syngrassides* 1827! and, 1939, *Lindberg f.* s.n.!
5. Trikomo, 1939, *Lindberg f.* s.n.! and, 1970, *A. Hansen* s.n. cult. Copenhagen!

NOTES: *Sintenis & Rigo* 557, from Kythrea, referred to *A. curtum* by Boissier and Holmboe, is correctly *A. rubrovittatum* Boiss. et Heldr., a close ally of *A. curtum*, but smaller in every respect, with denser, purplish umbels. *Syngrassides* 1827 has been misidentified as *A. sphaerocephalon* L. var. *viridialbum* Tineo, which Stearn (in Ann. Mus. Goulandris, 4: 181; 1978) treats as synonymous with *A. sphaerocephalon* L. ssp. *arvense* (Guss.) Arcang. (*A. arvense* Guss.).

15. **A. amethystinum** *Tausch* in Syll. Plant. Nov. Ratisbon., 2: 255 (1828); v. Bothmer in Bot. Notiser, 125: 70 (1972), Garbari et Corsi in Inform. Bot. Ital., 4: 125 (1972).
[*A. descendens* (non L.) Sibth. et Sm., Fl. Graec. Prodr., 1: 224 (1809), Fl. Graec., 4: 15, t. 816 (1823); Poech, Enum. Plant. Ins. Cypr., 9 (1842); Unger et Kotschy, Die Insel Cypern, 197 (1865); Boiss., Fl. Orient., 5: 236 (1882); Osorio-Tafall et Seraphim, List Vasc. Plants Cyprus, 22 (1973).]
TYPE: Yugoslavia; "in Dalmatia, unde herbarius Sieberi summitates absque foliis attulit" (PRC).

Bulb ovoid 2–4 cm. long, 1·5–3 cm. diam., tunics pallid, membranous or papery, the outermost tinged greyish-brown, rugulose; bulblets absent or sometimes stipitate and enveloped by leaf-sheaths 2–3 cm. above bulb, yellowish, acute or acuminate, with a coarsely reticulate surface; stem robust, erect, 80–100 cm. high, terete, striatulate, sometimes purplish towards apex; leaves 3–5 (–7), extending to ⅓ the total length of stem, the sheaths pallid, glabrous, striate, lamina linear, fistulose-canaliculate, up to 50 cm. long, 0·8 cm. wide, glabrous, sulcate, tapering to apex, margins often remotely scaberulous; spathe 1-valved, caducous, rostrate, 1–5 cm. long, 0·8–2 cm. wide; umbel ovoid or subglobose 2·5–6·5 cm. diam., many-flowered, pedicels unequal, the median erect or suberect, up to 40 mm. long, glabrous, sulcate or angular, often purplish, the outer seldom more than 15 mm. long, frequently recurved, bracteoles conspicuous, white, membranous, forming an apparent involucre; perianth purplish or greenish, shortly cylindrical, 4–5 mm. long, 1·5–4 mm. wide, those of the outer, short-pedicellate flowers generally much smaller than those of the median, erect flowers, segments erect, connivent, oblong, concave, or the outer slightly keeled, obtuse, glabrous, about 4–5 mm. long, the outer about 2·5 mm. wide, the inner 1·5–2 mm. wide; stamens shortly exserted, filaments of outer stamens simple, linear-lanceolate, 4–5 mm. long, 0·8 mm. wide at base, tapering to apex, glabrous or margins papillose-ciliate, filaments of inner stamens 3-cuspidate, the basal part oblong, about 2·5 mm. long, 2 mm. wide, papillose-ciliate, the median cusp filiform-subulate, about 3 mm. long, the lateral cusps filiform, very slender, often 5–7 mm. long, commonly twisted and curled in the upper part; anthers oblong, yellowish or purplish, 1·8 mm. long, 0·8 mm. wide; ovary subglobose-trigonous, 4–5 mm. diam., glabrous, loculi with a distinct apical prominence; style 3 mm. long; stigma subcapitate. Capsule about 5 mm. diam., pale brown; seeds angular, about 3·5 mm. long, 1·5 mm. wide; testa black, rugulose, obscurely verruculose.

HAB.: In cultivated fields; sea-level to 500 ft. alt.; fl. May–June.

DISTR.: Division 5, rare. Central and eastern Mediterranean region from Italy to Turkey and Cyprus.

5. Near Kythrea on the way from Nicosia, in cereal fields, 1934, *Syngrassides* 440 !

NOTES: Easily passed by as a slender *A. ampeloprasum* L. and perhaps commoner than records indicate. The plant figured in Fl. Graeca, t. 316 appears to be this, and may be based on Cyprus material, but, unfortunately, Sibthorp cites no localities.

A. amethystinum is most obviously distinguished from *A. ampeloprasum* by its fistulose-canaliculate leaves. These are often withered at anthesis, but normally persistent at least in part.

Kollmann (in Taxon 19: 789–792 and in Not. Roy. Bot. Gard. Edinb., 31: 119–124; 1971) has dismissed the name *A. descendens* L. as a *nomen ambiguum*; she names the present species *A. segetum* Jan ex J. A. et J. H. Schultes (1830), but this name is antedated by *A. amethystinum* Tausch.

16. A. margaritaceum *Sm.* in Sibth. et Sm., Fl. Graec. Prodr., 1: 224 (1809), Fl. Graec., 4: 14, t. 315 (1823); Poech, Enum. Plant. Ins. Cypr., 9 (1842); Unger et Kotschy, Die Insel Cypern, 197 (1865); Regel in Acta Hort. Petrop., 3: 49 (1875); Boiss., Fl. Orient., 5: 239 (1882); Holmboe, Veg. Cypr., 46 (1914); Osorio-Tafall et Seraphim, List Vasc. Plants Cyprus, 22 (1973).

Bulb ovoid, 1·5–2·5 cm. long, 1–2 cm. diam., tunics membranous, the outermost yellowish-brown or fuscous, coarsely lacerate-fibrous at apex and base; bulblets generally absent; stems slender, erect, terete, striate, glabrous, 10–90 cm. long, leaves 2–5, subfiliform, up to 30 cm. long, seldom more than 0·2 cm. wide, fistulose-canaliculate, glabrous or sometimes papillose-asperulous, tapering to a very slender apex, commonly withered and wanting at anthesis, sheaths extending up $\frac{1}{2}$–$\frac{2}{3}$ of the stem; spathe 1-valved, caducous, 1–5 cm. long, 1–2 cm. wide, rostrate; umbel subglobose, 1–3·5 cm. diam., rather dense, many-flowered; pedicels slender, angular, subequal, about 5–15 mm. long; bracteoles membranous, pallid, conspicuous, forming an apparent involucre, perianth small, greenish with a median green or red stripe or blotch along each segment, cup-shaped, 2–3 mm. long, 1·5–2 mm. wide, segments erect, oblong, obtuse or subacute, subequal, glabrous, concave or the outer slightly keeled, about 2–3 mm. long, 1–1·2 mm. wide; stamens shortly exserted, filaments of outer stamens simple, subulate, about 2 mm. long, 0·8 mm. wide at base, filaments of inner stamens 3-cuspidate, the basal part oblong, about 1·3 mm. long, 0·8 mm. wide, sometimes sparsely papillose-ciliate, median cusp subulate, about 1·5 mm. long, the lateral cusps to 2 mm. or more, subulate-filiform; anthers oblong, yellow, about 0·8 mm. long, 0·4 mm. wide; ovary trigonous-ovoid; glabrous, 1·3 mm. long, 1 mm. wide; style 1–1·3 mm. long; stigma subcapitate. Capsule ovoid-triquetrous, about 3 mm. long, 2 mm. wide, pale brown; seeds angular, about 3 mm. long, 1·3 mm. wide; testa black, conspicuously depressed-verruculose.

var. **margaritaceum**
> *A. sardoum* Moris, Stirp. Sard., 2: 10 (1828).
> *A. sphaerocephalon* L. var. *sardoum* (Moris) Regel in Acta Hort. Petrop., 3: 47 (1875).
> *A. guttatum* Stev. ssp. *sardoum* (Moris) Stearn in Ann. Mus. Goulandris, 4: 184 (1978), et in Tutin et al., Fl. Europ., 5: 67 (1980).
> TYPE: "Circa Bursam Bithyniæ; etiam in monte Athô, insulisque Naxo, Cypro et Cimalo" (OXF). Stearn (Ann. Mus. Goulandris, 4: 185; 1978) notes that Bauer's original drawing in *Fl. Graeca* t. 315 is labelled "Cimolos".

Perianth-segments whitish or greenish with a green median stripe.

HAB. and DISTR. not known, recorded without precise localization in Fl. Graec. Prodr. and Fl. Graec., and subsequently cited by Kotschy, Holmboe and others as a Cyprus plant, largely on the basis of Sibthorp's supposed discovery. In fact it is much more likely that Sibthorp's plant and *Kotschy* 292a from Ayios Khrysostomos, were the superficially similar *A. willeanum* Holmboe (infra).

var. **guttatum** (*Steven*) Gay in Ann. Sci. Nat., ser. 3, 8: 223 (1847).
> *A. guttatum* Steven in Mém. Soc. Nat. Moscou, 2: 173 (1809); de Wilde-Duyfjes in Meded. Laundbouwh. Wagen., 76 (11): 34 (1976); Stearn in Ann. Mus. Goulandris, 4: 184 (1978) et in Tutin et al., Fl. Europ., 5: 67 (1980).
> TYPE: U.S.S.R.; "Circa colonias Svevicas districtus Odessani haud infrequens" (LE, K !).

Perianth-segments whitish or greenish with a median purple (or sometimes dark green) blotch.

HAB.: In garigue on igneous hillsides; 2,500–3,000 ft. alt.; fl. June–July.

DISTR.: Division 2, rare. Also Balkan peninsula, S.W. European Russia, Turkey.

2. Between Apliki and Gourri ("Vouri"), 1940, *Davis* 1896 !

NOTES: Evidently rare on the island, where it has been much confused with the distinct *A. willeanum* Holmboe.

De Wilde-Duyfjes (1976) and Stearn (1978) both regard the name *Allium margaritaceum* Sm. as an illegitimate later homonym of *A. margaritaceum* Moench, Suppl. Meth. Plant., 80 (1802). It will, however, be found that Moench used the epithet at varietal rank, leaving it available for later use at a different rank. At species level the plant named *A. ampeloprasum* L. var. *margaritaceum* Moench is correctly *A. neglectum* Wenderoth (= *A. scorodoprasum* L. teste Regel).

I do not regard the distinctions between *A. margaritaceum* Sm. var. *margaritaceum* and *A. margaritaceum* Sm. var. *guttatum* (Stev.) Gay as of any great significance.

17. A. willeanum *Holmboe*, Veg. Cypr., 45, Fig. 9 (1914); Lindberg f., Iter Cypr., 10 (1946); Osorio-Tafall et Seraphim, List Vasc. Plants Cyprus, 22 (1973).
> TYPE: Cyprus; "in the valley beyond Kaminaria", 1905, *Holmboe* 1070 (G).

Bulb ovoid, 1·5–3 cm. long, 1·3–2·5 cm. diam., tunics membranous, the outermost fuscous, coarsely lacerate-fibrous, bulblets generally present, long-stipitate, concealed under the leaf-sheaths 2–6 cm. above apex of bulb, yellowish-brown, ellipsoid-acuminate, compressed, striatulate, 1–2 cm. long, 0·5–0·7 cm. wide, stem erect, terete, striatulate, sometimes bluntly and sparsely scabridulous, often purplish, 40–70 (–100) cm. high, leaves 4–6, withering before anthesis, sheaths shallowly sulcate, extending up ¼–⅓ of the stem, lamina narrowly linear, flat or canaliculate, keeled dorsally, up to 30 cm. long, 0·3–0·5 cm. wide, sulcate, often papillose along the margins and keel, spathe 1-valved, membranous, 1–2 cm. wide at base, abruptly constricted to a subulate beak up to 6 cm. long, caducous; umbel globose 2–5 cm. diam., rather crowded, many-flowered, pedicels subequal, slender, angular, glabrous, 1–2·3 cm. long, or sometimes the median flowers fertile and long-pedicellate, the external apparently sterile and more shortly pedicellate; bracteoles inconspicuous; perianth whitish, cup-shaped, each segment with a green keel, 2–3 mm. long, 2–2·5 mm. wide, segments oblong, erect, sub-connivent, 3–3·5 mm. long, 1·5 mm. wide, distinctly keeled, the keel shortly papillose, apex obtuse, truncate or shallowly emarginate, often somewhat cucullate; stamens exserted; filaments of outer stamens simple, subulate, about 3·5 mm. long, 0·6 mm. wide at base, margins distinctly papillose-ciliate, filaments of inner stamens 3-cuspidate, the basal part oblong, about 2 mm. long, 1·3 mm. wide, margins papillose-ciliate, median cusp about 2–2·5 mm. long, lateral cusps filiform, about 3 mm. long, often curled; anthers oblong, yellowish, about 1 mm. long, 0·4 mm. wide; ovary ovoid-triquetrous, glabrous, about 1·8 mm. long, 1·2 mm. wide; style about 1·3 mm. long; stigma subcapitate. Capsule subglobose, pale brown, about 3 mm. diam.; seeds angled, about 2·5 mm. long, 1·8 mm. wide; testa black, closely and bluntly verruculose.

HAB.: In garigue on dry, rocky, limestone or igneous hillsides; occasionally in crevices on cliffs or rocks, or in dried-up river-beds; by roadsides, waste ground or edges of cultivated fields; sea-level to 3,100 ft. alt.; fl. June–July.

DISTR.: Divisions 1–4, 6, 7. Endemic.

1. Near Ayios Neophytos Monastery, 1939, *Lindberg f.* s.n.!
2. Between Prodhromos and Galata, 1880, *Sintenis & Rigo* 914! Near Kaminaria, 1905, *Holmboe* 1070; Kakopetria, 1936, *Syngrassides* 1260! Platres, 1938, *Kennedy* 1140! Between Ambelikou and Kambos, 1939, *Lindberg f.* s.n.; Kambos 1968, *Economides* in ARI 1209!
3. Lefkara, 1940, *Davis* 1902! Skarinou, 1940, *Davis* 2034!
4. Cape Greco, 1958, *N. Macdonald* 78A!
6. Orga, 1967, *Economides* in ARI 1030!
7. St. Hilarion, 1938, *Kennedy* 1141! Buffavento, 1938, *Kennedy* 1142! Kyrenia, 1939, *Lindberg f.* s.n.! also, 1955, *G. E. Atherton* 79! 188! 738! Bellapais, 1939, *Lindberg f.* s.n.; Lapithos, 1940, *Davis* 1736! Halevga, 1940, *Davis* 1924! Near Panagra, 1970, *Chicken* 152!

NOTES: Probably first collected on the island, but without precise localization, in 1854, by Gaudry (herb. Gay, K!) unless, as is probable, Sibthorp's record for *A. margaritaceum* belongs here. *Kotschy* 292a, collected at Ayios Khrysostomos on 15 April 1862 probably also belongs here, though at this early date the specimen can hardly have been in flower.

A. willeanum is evidently a fairly common plant in Cyprus; it is, as Holmboe notes, allied to *A. cappadocium* Boiss., but probably more closely to *A. margaritaceum* Sm., from which it is readily distinguished by its flat (not fistulose) leaves. It is also a more robust plant than *A. margaritaceum*.

SECT. 7. **Melanocrommyum** *Webb et Berth*. Bulbs subglobose or ovoid, not rhizomatous, leaves all basal, the sheaths not extending up the stem; spathe persistent, 2–4-lobed, usually shorter than the umbel; perianth-segments spreading at anthesis, often becoming reflexed in fruit; stamens all simple; ovary with 4–8 ovules in each loculus; stigma entire or shortly lobed. Seeds angular. Plants not smelling of Garlic.

18. **A. nigrum** *L.*, Sp. Plant., ed. 2, 430 (1762); Sibth. et Sm., Fl. Graec. Prodr., 1: 226 (1809), Fl. Graec., 4: 20, t. 323 (1823); Poech, Enum. Plant. Ins. Cypr., 9 (1842); Unger et Kotschy, Die Insel Cypern, 199 (1865); Regel in Acta Hort. Petrop., 225 (1875); Boiss., Fl. Orient., 5: 279 (1882); Post, Fl. Pal., ed. 2: 645 (1933); Stearn in Ann. Mus. Goulandris, 4: 189 (1978).
　　A. multibulbosum Jacq., Fl. Austr., 1: 9, t. 10 (1773); Regel in Acta Hort. Petrop., 3: 226 (1875).
　　A. bauerianum Baker in Gard. Chron., 1874 (2): 131 (1874), et in Journ. Bot., 12: 289 (1874).
　　A. nigrum L. ssp. *multibulbosum* (Jacq.) Holmboe, Veg. Cypr., 47 (1914); Osorio-Tafall et Seraphim, List Vasc. Plants Cyprus, 22 (1973); De Wilde-Duyfjes in Meded. Landbouwh. Wagen., 76 (11): 195 (1976).
TYPE: "Algiriae *inque* Galloprovincia".

Bulb ovoid, 2–3 cm. long and almost as wide, tunics membranous, the outermost greyish-brown, smooth or rugulose, bulblets sometimes numerous (ssp. *multibulbosum* (Jacq.) Holmboe) or represented by a subglobose, fleshy gemma enveloped in a cucullate, gemmiferous leaf (ssp. *nigrum*); leaves (2–) 3–6, all basal, broadly linear or linear-lanceolate, fleshy, up to 50 cm. long and 9 cm. wide, acuminate, glabrous, with smooth or papillose-denticulate margins; stem terete, striate, robust, 60–90 cm. long, 0·5–1 cm.

diam.; spathe persistent, papery, 2–4 valved, ultimately reflexed, the valves ovate-concave, up to 3 cm. long, 2·5 cm. wide, shortly acute; umbel hemispherical, dense, many-flowered, 5–10 cm. diam., generally without bulbils; pedicels 2–5 cm. long, glabrous, bluntly angular, perianth widely cup-shaped or stellate, 1·5–2·5 cm. diam.; segments oblong-elliptic, 10–13 mm. long, 3–4 mm. wide, obtuse, whitish or pinkish with a greenish or purplish, papillose midrib; stamens included; filaments subulate, glabrous, fleshy, 3–4 (–5) mm. long, connate at the base into a narrow annulus; anthers oblong, yellowish, about 3·5 mm. long, 1·4 mm. wide; ovary trigonous-subglobose, about 4 mm. diam., greenish or blackish, the loculi sulcate dorsally, minutely papillose, style about 1·5 mm. long; stigma very shortly and obscurely 3-lobed. Capsule coriaceous, trigonous, subglobose-ovoid, about 10 mm. long, 8 mm. wide; seeds angular, about 4 mm. long, 3 mm. wide, testa black, minutely verruculose.

HAB.: Generally in cereal fields, sometimes on waste ground or dry hillsides; sea-level to 1,600 ft. alt.; fl. March–April.

DISTR.: Divisions 1, 3–5, 7. S. Europe and Mediterranean region.

1. 10 m. S.E. of Polis, near Lyso, 1979, *Edmondson & McClintock* E2822!
3. Kato Lefkara, 1955, *Merton* 2041!
4. Near Larnaca, 1862, *Kotschy* 235! Kalokhorio, 1862, *Kotschy*; between Paralimni and Famagusta, 1905, *Holmboe* 430.
5. Near Pera, 1862, *Kotschy*; near Kythrea, 1880, *Sintenis & Rigo* 168! also, 1933, *Nattrass* 297, and, 1934, *Syngrassides* 438! 439!
7. Lapithos, 1862, *Kotschy*; Myrtou, 1932, *Nattrass* 210; Panagra, 1938, *Mrs. Dray* s.n.! Above Vasilia, 1940, *Davis* s.n.!

NOTES: As Holmboe remarks, the Cyprus specimens have exceptionally large, broad leaves, and have been referred to *A. multibulbosum* Jacq. by various investigators; the distinctions, however, between this and *A. nigrum* L. are unsatisfactory. The specimen collected at Panagra by Mrs. Dray has been recently (1977) determined as *A. multibulbosum*, though it has the gemmiferous leaf said (by B. De Wilde-Duyfjes) to be characteristic of *A. nigrum*.

19. A. orientale *Boiss.*, Diagn., 1, 13: 25 (1854), Fl. Orient., 5: 232 (1882); Regel in Acta Hort. Petrop., 3: 229 (1875); Holmboe, Veg. Cypr., 47 (1914); Post, Fl. Pal., ed. 2, 2: 645 (1933); Osorio-Tafall et Seraphim, List Vasc. Plants Cyprus, 22 (1973); Stearn in Ann. Mus. Goulandris, 4: 190 (1978).

A. macrospermum Boiss. et Kotschy ex Kotschy in Unger et Kotschy, Die Insel Cypern, 198 (1865).

[*A. decipiens* (non Fisch. ex J. A. et J. H. Schultes) Sintenis in Oesterr. Bot. Zeitschr., 31: 193, 327 (1881); C. et W. Barbey, Herborisations au Levant, 99 (1882).]

TYPE: "in *Oriente, Asia minori* in *Caria* (Pinard), *Sipylo* monte (Aucher No. 2200) ex Ciliciâ (No. 2188), monte *Solyma Lyciae* (Heldr.) in agris *Syriae* ad *Aleppum* Kotschy No. 71, in desertis *Arabicis* ad fines *Palaestinae* (Boiss.), in *Mesopotamia* (Aucher No. 2215) (G, K!).

Bulb subglobose, 1·5–3 cm. diam.; tunics papery, pallid, or the outermost greyish-brown; bulblets commonly present, subglobose, 5–10 mm. diam.; stem robust, terete, striate, glabrous, 10–40 cm. high; leaves 2–6, all basal, persistent until anthesis, linear-lanceolate, long-acuminate, 10–50 cm. long, 0·5–2·5 cm. wide, glabrous, or papillose along the margins, striate; spathe 2–4-lobed, membranous, the lobes shortly acute, broadly ovate, 1–2 cm. long, 1·5–2 cm. wide, persistent and ultimately reflexed; umbel hemispher-ical, 3–7 cm. diam., rather dense, many-flowered; pedicels 1–2·5 cm. long, glabrous, angular, rather swollen at apex, 1–2·5 cm. long; perianth widely cup-shaped, 6–10 mm. long and almost as wide, segments white or pink, shining, translucent, oblong-lanceolate, 8–10 mm. long, 1·5–3 mm. wide, obtuse or shortly acute; stamens included; filaments stout, subulate, concave adaxially, alternately 3–4 mm. and 4–5 mm. long, connate basally into a narrow annulus; anthers yellowish or purplish, oblong, 3–3·5 mm. long, 1–1·5 mm. wide; ovary trigonous-subglobose, about 3·5 mm. diam., minutely papillose, brownish, the loculi sharply sulcate dorsally, with distinctly gibbous apices; style stout, about 6 mm. long; stigma obscurely 3-

lobed. Capsule coriaceous, trigonous-subglobose, pale brown, about 8 mm. diam.; seed broadly ovoid or subglobose, about 4 mm. diam., angled; testa black, rugulose, minutely papillose.

HAB.: In cereal fields; sea-level to 500 ft. alt.; fl. Febr.–April.

DISTR.: Divisions 4–8. Eastern Mediterranean region from Turkey to Egypt.

4. Larnaca, 1881, *C. & W. Barbey* s.n.; also, 1905, *Holmboe* 150; between Paralimni and Ayia Napa, 1936, *Syngrassides* 1159! Famagusta, 1939, *Lady Loch* 73! Pyroi, 1981, *Hewer* 4790!
5. Between Nicosia and Kythrea, 1880, *Sintenis & Rigo* 166! Lakatamia, 1932, *Syngrassides* 372! Mora, 1956, *Merton* 2529!
6. Nicosia, 1862, *Kotschy*; also, 1905, *Holmboe* 430, and, 1959, *P. H. Oswald* 74!
7. Kyrenia, 1862, *Kotschy*; Kantara, 1880, *Sintenis & Rigo* 939 pro parte (mixed with *A. neapolitanum*).
8. Leonarisso, 1968, *Economides* in ARI 1118!

A. DIOSCORIDIS *Sibth. ex Sibth. et Sm.*, Fl. Graec. Prodr., 1: 222 (1809) an imperfectly described species, said to have been seen "In frutetis densis umbrosis Cariae, Mysiae, et insulae Cypri", cannot, in the absence of an illustration or specimen, be identified with any certainty. Kotschy (in Unger et Kotschy, Die Insel Cypern, 199; 1865) supposed it to be *Nectaroscordum siculum* (Ucria) Lindl., but there is no evidence that this occurs in Cyprus, though the identification may be correct for the other Sibthorp records. Most probably the Cyprus plant was the relatively common *A. nigrum* L., as suggested by Smith and G. Don.

10a. NOTHOSCORDUM *Kunth*

Enum. Plant., 4: 457 (1843) nom. cons.

Bulbous perennials resembling *Allium* but without the characteristic smell; leaves basal, linear, sheathing; spathe 2-valved, the valves connate basally, inflorescence a lax umbel; perianth persistent, campanulate, the segments shortly connate at the base, subequal, 1-nerved, stamens 6, included, filaments simple, inserted near base of perianth; anthers oblong, introrse, dorsifixed; ovary sessile, 3-locular; style terminal; stigma truncate or subcapitate; ovules 4–12 in each loculus of the ovary. Capsule membranous; seeds angular, with a black testa.

About 35 species, all American; *N. inodorum* is widely naturalised in Europe and the Mediterranean region.

1. N. INODORUM (*Ait.*) *Nicholson*, Ill. Dict. Gard., 2: 457 (1885); Maire, Fl. Afr. Nord., 5: 243, fig. 887 (1958); Stearn in Tutin et al., Fl. Europ., 5: 69 (1980).
   Allium inodorum Ait., Hort. Kew., ed. 1, 1: 427 (1789).
   Allium fragrans Vent., Hort. Cels., t. 26 (1801).
   Nothoscordum fragrans (Vent.) Kunth, Enum. Plant., 4: 441 (1843).
TYPE: "*Nat.* of Carolina", cult. Kew.

Bulb ovoid, about 2 cm. long, 1·5 cm. wide, tunics pale brown, papery; bulblets usually numerous, small, ovoid, less than 5 mm. long; stems erect, slender, striate, glabrous, 30–50 cm. long; leaves 4–6, basal, linear, glabrous, striate, 30–40 cm. long, 0·2–0·8 cm. wide; spathe membranous, pallid, 2-valved, the valves 8–10 mm. long, 5–7 mm. wide, acute or acuminate; umbel lax, 6–12-flowered; pedicels erect, glabrous, striate, 1–4 cm. long; perianth cup-shaped, about 1 cm. long, 1–1·5 cm. wide, base shortly tubular-infundibuliform, segments whitish with a brownish or pinkish midrib, oblong, 5–8 mm. long, 2·5–3 mm. wide, apex obtuse or subacute; stamens included; filaments linear-subulate, 5–6 mm. long, 1·5 mm. wide at base; anthers oblong, about 1 mm. long, 0·5 mm. wide, greenish-yellow; ovary trigonous-ovoid, about 3 mm. long, 2·5 mm. wide; style 6 mm. long; stigma subcapitate. Capsule ovoid, about 7 mm. long, 5 mm. wide, valves separating widely on dehiscence, papery, pale brown; seeds subovoid, bluntly angular and shortly rostrate, about 3 mm. long, 2 mm. wide; testa shining black, rugulose-cerebriform.

HAB.: Dry pastures; c. 500 ft. alt.; fl. April–May.

DISTR.: Division 6. A native of temperate South America, widely naturalised in the Mediterranean region and in other warm-temperate areas.

6. Nicosia, June 25, 1930, *David Fairchild* 3588!

NOTES: Grown by the U.S. Department of Agriculture from seed said to have been collected in Cyprus; the specimens submitted to Kew for identification in 1935 were erroneously determined as *Allium subhirsutum* and, have since lain unnamed amongst a miscellany of "spp. indet." at the end of the *Allium* collections. They are unquestionably *Nothoscordum inodorum*. It is a little odd that no one else should have collected the plant in Cyprus, but, at the same time, there is no reason to doubt the record, since *N. inodorum* is well established in the Mediterranean region and Atlantic Islands. It spreads very rapidly by seeds and bulblets, and is said to become a persistent weed in gardens.

SUBFAMILY 7. **Scilloideae** Perennials with tunicate bulbs; leaves all basal, not fistulose; inflorescence racemose or racemose-subumbellate, flowers subtended by membranous, conspicuous or inconspicuous bracts; perianth-segments free or united, spreading or connivent, or perianth sometimes tubular or globose; stamens usually 6; filaments free or rarely united; anthers introrse, dorsifixed; ovary trigonous, 3-locular; ovules 2–many. Capsule loculicidally dehiscent; seeds globose, angular or compressed.

11. ORNITHOGALUM L.

Sp. Plant., ed. 1, 306 (1753).
Gen. Plant., ed. 5, 145 (1754).

Bulbous perennials; bulbs renewed annually or over a period of 2–4 years, scales free or concrescent; leaves all basal, often with a pale median stripe; flowers in an elongate or corymbiform raceme, the lower pedicels often markedly accrescent after anthesis; bracts often conspicuous; perianth usually white with a greenish dorsal stripe along each segment, sometimes yellow or orange; segments 6, free to the base or almost to the base; stamens 6, free; filaments simple or tricuspidate at apex; anthers dorsifixed, introrse; ovary sessile; stigma truncate or subcapitate. Fruit a loculicidally dehiscent, angled capsule; seeds usually numerous, globose or angled, sometimes compressed or flattened.

About 150 species in Europe, the Mediterranean region and western Asia; also in Africa south to the Cape; introduced elsewhere.

Inflorescence an elongate many (30 or more)-flowered cylindrical raceme; pedicels subequal; testa of seeds rugulose but not reticulate - - - - - **5. O. narbonense**
Inflorescence corymbiform or loosely subspiciform; flowers generally fewer than 30; lower pedicels distinctly longer than upper; testa of seeds reticulate:
Leaves 3–12 mm. wide, linear or lorate:
Inflorescence lax, corymbiform-subumbellate; lower pedicels 5–8 cm. long; leaves with a pale median stripe - - - - - - **1. O. umbellatum ssp. divergens**
Inflorescence compact, crowded; lower pedicels less than 2 cm. long; leaves without a pale median stripe - - - - - - - - **4. O. chionophilum**
Leaves linear-filiform, 1–3 mm. wide:
Inflorescence lax, corymbiform, lower pedicels 2–7 cm. long; leaves usually 5–12 or more **2. O. pedicellare**
Inflorescence usually crowded, subspiciform; lower pedicels 1–1·5 cm. long; leaves usually 4–8 - - - - - - - - - - **3. O. trichophyllum**

SECT. 1. **Heliocharmos** (*Baker*) *Benth.* Inflorescence subcorymbose, the lower pedicels usually much longer than the upper; leaves generally 4 or more, linear or lanceolate.

1. O. umbellatum *L.*, Sp. Plant., ed. 1, 307 (1753); Boiss., Fl. Orient., 5: 218 (1882); Holmboe, Veg. Cypr., 50 (1914); Post, Fl. Pal., ed. 2, 2: 629 (1933); Osorio-Tafall et Seraphim, List Vasc. Plants Cyprus, 21 (1973).

    O. minus L. f., Mantissa Altera, 364 (1771); Osorio-Tafall et Seraphim, List Vasc. Plants Cyprus, 21 (1973).

    O. umbellatum L. var. *minus* (L. f.) Aschers. et Graebn., Syn. Mitteleurop. Fl., 3: 245 (1905); Holmboe, Veg. Cypr., 50 (1914).

Bulb ovoid, 2–3 cm. long, 1·5–2·5 cm. wide, tunic membranous, pale brown, scales concrescent, with few or numerous leafy or leafless offsets or bulbils; leaves 2–6, linear, 20–30 cm. or more long, 0·4–0·6 cm. wide, usually longer than the scape, tapering to apex and base, glabrous, bright green with a whitish median stripe; scape stout, erect, 12–20 cm. long; inflorescence 3–15 (–30)-flowered, lax, subcorymbose; bracts linear-subulate, 1·5–5 cm. long, greenish; lower pedicels 5–8 cm. long, elongating in fruit, erecto-patent, divaricate or sometimes a little reflexed; upper pedicels progressively shorter, though still exceeding the bracts; perianth-segments spreading, narrowly oblong, 15–20 mm. long, 4–6 mm. wide, white, the 3 outer with a very broad green dorsal stripe, the 3 inner rather smaller, with a narrower dorsal stripe; filaments oblong-acuminate, about 6 mm. long, 2–3 mm. wide at base; anthers yellow, oblong, about 3 mm. long, 1·5 mm. wide, base bluntly sagittate; ovary green or yellow, oblong, glabrous, about 5 mm. long, 3 mm. wide; style about 3 mm. long; stigma subtruncate. Capsule oblong or oblong-obovoid, 12–15 mm. long, 8–10 mm. wide, pale brown, venulose, sharply 6-angled or narrowly winged, apex emarginate; seeds broadly ovoid, about 1·8 mm. long, 1 mm. wide, distinctly flattened on one side, with a short tongue-like appendage; testa black, reticulate.

ssp. **divergens** (*Boreau*) *Aschers. et Graebn.*, Syn. Mitteleurop. Fl., 3: 246 (1905); Holmboe, Veg. Cypr., 50 (1914).

    O. divergens Boreau, Fl. Centre France, ed. 2, 507 (1849); Boiss., Fl. Orient., 5: 218 (1882); Feinbrun in Pal. Journ. Bot., J. series, 2: 140 (1941); Osorio-Tafall et Seraphim, List Vasc. Plants Cyprus, 21 (1973).

    TYPE: France; "*Limoges* (Lamy) — *Vallée de la Loire*, etc." (ANG).

Bulb with basal, leafless bulbils; pedicels commonly patent or arcuate-reflexed after anthesis; flowers frequently large.

    HAB.: Cereal fields and cultivated or fallow land; sea-level to 3,000 ft. alt.; fl.March–April.

    DISTR.: Divisions 1–6, 8. Widespread in S. Europe and the Mediterranean region.

1. Tsadha, 1981, *Hewer* 4754!
2. Aphamis, 1939, *Kennedy* 1541!
3. Episkopi, 1964, *J. B. Suart* 202! Anaphotia, 1978, *Chesterman* 41! Near Kophinou, 1978, *Chesterman* 44!
4. Kouklia (Mesaoria), 1905, *Holmboe* 387.
5. Near Kythrea, 1880, *Sintenis & Rigo* 177! Lakatamia, 1932, *Syngrassides* 370! Dhikomo, 1936, *Syngrassides* 1185!
6. Syrianokhori, 1935, *Syngrassides* 766! also, 1956, *Merton* 2625! Morphou, 1956, *Merton* 2485!
8. Ronnas Valley, Rizokarpaso, 1957, *Merton* 2935!

    NOTES: Although often regarded as specifically distinct, *O. umbellatum* ssp. *divergens* is questionably deserving of subspecific status and may be no more than an ecological form of the typical species. Small plants tend to merge with the following, *O. pedicellare* Boiss. et Kotschy, and *O. umbellatum* L. var. *minus* (L.f.) Aschers. et Graebn., cited from (4) Ayia Napa by Holmboe (H.34), may be such a plant; the specimen cited by Holmboe has not been examined.

2. O. pedicellare *Boiss. et Kotschy* in Unger et Kotschy, Die Insel Cypern, 195 (1865); Boiss., Fl. Orient., 5: 219 (1882); Holmboe, Veg. Cypr., 50 (1914); Osorio-Tafall et Seraphim, List Vasc. Plants Cyprus, 22 (1973).

    O. pedicellare Boiss. et Kotschy ssp. *cylindrocarpum* Zahariadi in Ann. Mus. Goulandris, 4: 254 (1978).

    [*O. nanum* (non Sm.) Sintenis in Oesterr. Bot. Zeitschr., 32: 22 (1882).]

TYPE: Cyprus; "Auf dem Wege zwischen Larnaca und Athienu entdeckt 1859", *Kotschy* 412 (G).

Bulb ovoid, 1·5–2·5 cm. long, 1–2 cm. wide, tunic membranous, greyish or brownish, scales free, leaves 5–12 or more, linear-filiform, 4–20 cm. long, 0·1–0·3 cm. wide, flat, glabrous, erect or spreading-recurved with or without a pallid median stripe; scape 1–9 (–12) cm. long, sometimes almost wanting, erect; bracts conspicuous, whitish-membranous, linear-subulate, 8–30 mm. long, 2–6 mm. wide at base, generally much shorter than the lower pedicels; inflorescence lax, subcorymbose, 2–8 (–15)-flowered, lower pedicels 2–7 cm. long, at first erecto-patent, often becoming strongly divaricate after anthesis, upper pedicels usually shorter; perianth-segments spreading, oblong, acute, 10–15 mm. long, 3–4 mm. wide, white internally with a broad green stripe externally; filaments narrowly oblong-acuminate, 5–7 mm. long, about 1·5–2 mm. wide at base; anthers yellow, oblong, about 2 mm. long, 1 mm. wide, base bluntly subsagittate; ovary greenish, subglobose or oblong, 2–4 mm. long, 1·5–3 mm. wide; style slender 1·5–2 mm. long; stigma subcapitate. Capsule oblong, 7–11 mm. long, 5–8 mm. wide, brown, papery, apex shallowly emarginate, angles not very prominent; seeds ovoid or subglobose, about 1·5 mm. long and almost as wide, with a small, protruding appendage; testa black, reticulate.

HAB.: Pastures; in garigue on rocky hillsides, or in shallow, terra rossa grasslands; sea-level to 5,400 ft. alt.; fl. March–April.

DISTR.: Divisions 1–8, locally abundant. Believed to be endemic, but possibly also occurring in S. Greece, Aegean area and S. Turkey.

1. Erimidhes near Ayios Yeoryios (Akamas), 1962, *Meikle* 2137! Near the Baths of Aphrodite, 1977, *E. Georgiadou* s.n.; between Lachi and Neokhorio, 1978, *Chesterman* 51!
2. Kryos Potamos, 2,800 ft. alt., 1937, *Kennedy* 222! Kryos Potamos, 4,000 ft. alt., 1937, *Kennedy* 223! Platres, 1937, *Kennedy* 224! Livadhi tou Pasha, 1937, *Kennedy* 225!
3. 4 miles N. of Kandou, 1963, *J. B. Suart* 52! Episkopi, 1964, *J. B. Suart* 174!
4. Between Larnaca and Athienou, 1859, *Kotschy* 412; Famagusta, 1862, *Kotschy* 93a; near Dhekelia ["Redgelia"], 1880, *Sintenis* teste *Zahariadi*; Famagusta, 1936, *Syngrassides* 1154! Cape Greco, 1958, *N. Macdonald* 38!
5. Ayia Varvara near Nisou, 1936, *Syngrassides* 1180! Mile 3, Nicosia-Famagusta road, 1950, *Chapman* 388!
6. Nicosia, 1927, *Rev. A. Huddle* 101!
7. Near Ayios Khrysostomos Monastery, 1880, *Sintenis et Rigo* 175! Kyrenia, 1949, *Casey* 300! also, 1956, *G. E. Atherton* 911! Templos, 1956, *G. E. Atherton* 1201! 7½ miles N.E. of Halevga, 1959, *P. H. Oswald* 114! Tjiklos, 1973, *P. Laukkonen* 92!
8. Lythrangomi, 1935, *Syngrassides* 704!

NOTES: Commonly misidentified as *O. tenuifolium* Guss. (non Delaroche) ≡ *O. gussonii* Tenore, but readily distinguished by its very numerous, generally reflexed, leaves, short scapes, and long, divergent lower pedicels, the latter often twice or three times longer than the scape. The name *O. tenuifolium* has, however, been applied to a wide range of distinct taxa.

3. O. trichophyllum *Boiss. et Heldr.* in Boiss., Diagn., 2, 4: 108 (1859); Feinbrun in Pal. Journ. Bot., J. series, 2: 142 (1941); Osorio-Tafall et Seraphim, List Vasc. Plants Cyprus, 22 (1973).

   O. tenuifolium Guss. var. *trichophyllum* (Boiss. et Heldr.) Boiss., Fl. Orient., 5: 219 (1882).
   O. tenuifolium Guss. ssp. *trichophyllum* (Boiss. et Heldr.) Holmboe, Veg. Cypr., 50 (1914).
   [? *O. tenuifolium* (non Guss.) Kotschy in Unger et Kotschy, Die Insel Cypern, 195 (1865); Boiss., Fl. Orient., 5: 219 (1882); Holmboe, Veg. Cypr., 49 (1914).]
TYPE: Egypt; "in herbosis propé *Alexandrium* ad Castrum *Maxi*", *Samaritani* s.n. (G).

Bulb ovoid, 2–2·5 cm. long, 1·3–2 cm. wide, tunic membranous, brownish externally, silvery white and pustulate internally; leaves 4–8 or more, filiform, 7–17 cm. long, 0·1–0·15 cm. wide, glabrous, uniformly green, prostrate (in living plants) often apparently erect in dried specimens and longer than the scape, not tapering much to apex or base; scape 2·5–9 cm. long; inflorescence 2–5 (–10)-flowered, often rather crowded; bracts membranous, longitudinally nerved, lanceolate-acuminate, 8–15 mm. long,

3–4 mm. wide at base, mostly equalling or exceeding the pedicels; lower pedicels erect or ascending, 1–1·5 cm. long, relatively stout, not elongating much in fruit, upper pedicels often very short or almost wanting; perianth-segments oblong, acute or subacute, 12–15 (–20) mm. long, 3–4 mm. wide, white internally, with a broad, rather ill-defined, green stripe externally; filaments lanceolate-acuminate, about 6–7 mm. long, 2 mm. wide at base; anthers yellow, oblong, about 2·5 mm. long, 1 mm. wide, base subsagittate; ovary sessile, oblong, longitudinally sulcate, 4·5–5 mm. long, 3 mm. wide, minutely squamose-papillose; style 4 mm. long; stigma subcapitate. Capsule broadly oblong, 7–10 mm. long, 5–8 mm. wide, angled, apex subtruncate; seeds oblong or subglobose, 1·5–2·5 mm. long, 1·5–2 mm. wide; testa black, reticulate.

HAB.: Rocky pastures; 400–600 ft. alt.; fl. March–April.

DISTR.: Divisions 5, 6, rare. Also Egypt and Palestine.

5. Strongylos, 1905, *Holmboe* 357; Petra tou Dhiyeni, 1951, *Merton* 793!
6. Nicosia, 1880, *Sintenis & Rigo* 176! Kokkinotrimithia, 1933, *Syngrassides* 320!

NOTES: Here also may belong Holmboe's record for *O. tenuifolium* Guss. from Ephtakomi (*Holmboe* 529) and perhaps the plant collected by Kotschy between Omodhos and Limassol, 1859 (*Kotschy* 498) though the specimen (G) is robust, many-flowered and very different from typical *O. trichophyllum*. Kotschy's second record for *O. tenuifolium* ("in Vorbergen um Prodromo") may be referable to *O. pedicellare* or *O. chionophilum*; it is most unlikely that *O. trichophyllum* would be found outside the dry steppe areas of the Mesaoria.

4. O. chionophilum *Holmboe*, Veg. Cypr., 49 (1914); Lindberg f., Iter Cypr., 10 (1946); Osorio-Tafall et Seraphim, List Vasc. Plants Cyprus, 21 (1973).
  O. cyprium Zahariadi in Ann. Mus. Goulandris, 6: 186 (1983).
  [*O. montanum* (non Cyr.) Kotschy in Unger et Kotschy, Die Insel Cypern, 196 (1865); Boiss., Fl. Orient., 5: 217 (1882) pro parte; Holmboe, Veg. Cypr., 50 (1914); Osorio-Tafall et Seraphim, List Vasc. Plants Cyprus, 21 (1973).]
  [*O. lanceolatum* (non Labill.) Kotschy in Unger et Kotschy, Die Insel Cypern, 196 (1865); Boiss., Fl. Orient., 5: 216 (1882) pro parte; Holmboe, Veg. Cypr., 50 (1914); Osorio-Tafall et Seraphim, List Vasc. Plants Cyprus, 21 (1973).]
  [*O. huetii* (non Boiss.) Kotschy in Oesterr. Bot. Zeitschr., 12: 278 (1862).]
  [*O. brevipedicellatum* (non Boiss.) Boiss., Fl. Orient., 5: 216 (1882) pro parte quoad plant. cypr.]
  TYPE: Cyprus; "near the top of Chionistra, ca. 1900 m. above the sea", 1905, *Holmboe* 991 (O).

Bulbs broadly ovoid, 2–4·5 cm. long, 2–3·5 cm. wide, tunic fuscous brown externally, membranous and translucent internally, scales apparently concrescent; leaves 2–5, linear-lorate, up to 30 cm. long, 0·3–1·2 cm. wide, glabrous, green, without a pallid median stripe, sometimes absent at anthesis; scape erect, stout, 2–9 cm. long, usually much shorter than the leaves; inflorescences often subcylindrical, crowded, many-flowered, or densely corymbiform, rarely flowers solitary or 2–3 together; bracts membranous, lanceolate-caudate, 2–3 cm. long, 3–4 mm. wide at base, whitish or greenish; lower pedicels erecto-patent, to about 2 cm. long, not lengthening much in fruit, upper pedicels often very short or almost wanting; perianth-segments narrowly oblong, 10–15 mm. long, 3–4 mm. wide, the 3 outer often shortly mucronate, the 3 inner obtuse, all white internally with a broad green external stripe; filaments lanceolate, 6–7 mm. long, tapering upwards from a base about 2–3 mm. wide; anthers yellow, oblong, subsagittate, about 2·5–3 mm. long, 1 mm. wide; ovary oblong, about 5 mm. long, 3 mm. wide, sharply ribbed longitudinally. Capsule oblong-obovate, pale brown, papery, venulose, about 12 mm. long, 7 mm. wide, sharply angled but not winged; seeds subglobose, about 2–2·5 mm. diam.; testa black, reticulate.

HAB.: Damp, gravelly mountainsides; 2,500–6,400 ft. alt.; fl. March–June.

DISTR.: Division 2. Endemic.

2. Troödos Range (Khionistra, Tripylos, Palekhori, Lagoudhera, Prodhromos, Kryos Potamos, Moniatis, Xerokolymbos, etc.) 1859–1979, *Kotschy* 37, 707! 749; *Sintenis & Rigo* 868! *Haradjian* 442! *Feilden* s.n.! *Tracey* 1! *Kennedy* 220! 221! 226–238! *Lindberg f.* s.n.! *Davis* 3459! *Merton* 2452! 2999! 3185! 3197! *Meikle* 2678! *J. B. Suart* 110! *Economides* in ARI 1327! *Osorio-Tafall & Meikle* s.n.! *Edmondson & McClintock* E 2853! *Hewer* 4623! etc.

NOTES: Obviously allied to the Asiatic *O. lanceolatum* Labill., *O. cuspidatum* Bertol. and the eastern Mediterranean *O. montanum* Cyr., but distinguished by its compact, often spiciform inflorescences; small, shortly pedicellate flowers, and narrow leaves. Specimens growing in very wet ground can develop very long, flaccid, grass-like leaves, but revert to more typical *O. chionophilum* when grown under drier conditions.

SECT. 2. **Beryllis** (*Salisb.*) *Benth.* Inflorescence racemose, the pedicels all equal or subequal, generally at least 10 mm. long; leaves usually 4 or more.

5. **O. narbonense** *L.*, Cent. Plant., 2, 15 (1756); Boiss., Fl. Orient., 5: 214 (1882); Holmboe, Veg. Cypr., 49 (1914); Post, Fl. Pal., ed. 2, 2: 627 (1933); Feinbrun in Pal. Journ. Bot., J. series, 2: 134 (1941); Rechinger f. in Arkiv för Bot., ser. 2, 1: 419 (1950); Osorio-Tafall et Seraphim, List Vasc. Plants Cyprus, 21 (1973).

[*O. pyrenaicum* (non L.) Kotschy in Unger et Kotschy, Die Insel Cypern, 196 (1865); H. Stuart Thompson in Journ. Bot., 44: 339 (1906); Osorio-Tafall et Seraphim, List Vasc. Plants Cyprus, 21 (1973).]

TYPE: "*in* Galliae *australis*, Italiae *agris*".

Bulb ovoid, 2–3·5 cm. long, 2–3 cm. wide, tunic membranous, brownish externally, silvery-white within, scales free; leaves 2–4, linear, commonly 30–60 cm. long, 0·3–0·7 cm. wide, bright or pale green, sometimes glaucescent, flat or keeled, tapering to apex from a long sheathing base; scape 20–60 cm. (or more) long, erect, terete; inflorescence a rather dense, many-flowered, cylindrical raceme, 12–20 cm. long, 3–5 cm. diam. at anthesis, elongating to 30 cm. or more in fruit; bracts subulate, 15–20 mm. long, 1·5–2 mm. wide at base, with a slender, filiform apex, membranous, with faint greenish nerves; pedicels 1·5–2 cm. long, erecto-patent at anthesis, suberect and adpressed to rhachis in fruit; perianth-segments narrowly oblong, 12–13 mm. long, 2–3 mm. wide, white internally with a well-marked green median stripe externally, outer segments bluntly mucronate, inner segments obtuse; filaments membranous, oblong, about 6 mm. long, 2 mm. wide, abruptly constricted to a short, filiform apex; anthers yellow, oblong, subsagittate, about 3·5 mm. long, 1·5 mm. wide; ovary oblong, about 3–4 mm. long, 2·5–3 mm. wide, greenish, costate, minutely papillose, apex subtruncate; style 3 mm. long; stigma subcapitate. Capsule oblong, angular, pale brown, about 12 mm. long, 5–7 mm. wide; seeds oblong, about 4–5 mm. long, 1·5–2 mm. wide, with a short, tongue-shaped appendage; testa black, rugulose and minutely papillose, but not reticulate.

HAB.: Cultivated fields, vineyards, pastures, or in garigue; sea-level to 4,500 ft. alt.; fl. April–May.

DISTR.: Divisions 1–8, locally frequent. Widespread in S. Europe and Mediterranean region.
1. Dhrousha, 1941, *Davis* 3247! Kato Paphos, 1973, *Holub* s.n.! Between Lachi and Baths of Aphrodite, 1979, *Edmondson & McClintock* E 2750!
2. Prodhromos, 1862, *Kotschy* s.n.! (cited as 910 in Die Insel Cypern); Between Perapedhi and Mandria, 1937, *Kennedy* 219! Aphamis, 1979, *Hewer* 4657!
3. Limassol, 1913, *Haradjian* 600; below Stavrovouni, 1959, *C. E. H. Sparrow* 36! Pyrgos, 1963, *J. B. Suart* 18! Amathus, 1978, *Holub* s.n.! Pissouri, 1981, *Hewer* 4761!
4. Cape Greco, 1958, *N. Macdonald* 72! and 1962, *Meikle* 2607!
5. Near Kythrea, 1880, *Sintenis & Rigo* 178! Dheftera, 1901, *A. G. & M. E. Lascelles* s.n.! Near Aphania, 1954, *Merton* 1878!
6. Between Koutraphas and Vizakia, 1936, *Syngrassides* 1226! Dhiorios, 1971, *G. Joscht* 6402!
7. Kyrenia, 1948, *Casey* 780! Karavas, 1955, *Merton* 2187!
8. Rizokarpaso, 1912, *Haradjian* 226; Peristerges near Kilanemos, 1962, *Meikle* 2436! Galouni, Apostolos Andreas Forest, 1962, *Meikle* 2483!

O. ARABICUM *L.*, Sp. Plant., ed. 1, 307 (1753), with dense corymbose or thyrsoid racemes of exceptionally large milky-white flowers (without the usual dorsal green stripe) and shining blackish ovaries, is included in Osorio-Tafall et Seraphim, List Vasc. Plants Cyprus, 21 (1973), but without any reference to the source of the record. *O. arabicum* is not infrequently cultivated as a garden plant, but I have no evidence of its occurrence as a wild or naturalized plant in Cyprus. It has a scattered distribution in the Mediterranean region and Atlantic Islands.

12. URGINEA *Steinh.*
Ann. Sci. Nat., series 2, 1: 322 (1834).

Bulbous perennials; bulb tunicate; leaves all basal, linear or oblong-elliptic; scape unbranched; inflorescence racemose; flowers usually numerous, rather small; pedicels articulated; bracts membranous-scarious, sometimes spurred basally; perianth usually whitish, sometimes tinged pink or purple; segments 6, free, subequal, patent or somewhat connivent, often with 1 or more pink or purplish median nerves; stamens 6, inserted at base of perianth-segments and generally shorter than segments; filaments simple, filiform or dilated towards the base; anthers dorsifixed, introrse; ovary sessile, 3-locular, generally triquetrous; ovules numerous in each loculus; style simple, filiform; stigma truncate or subcapitate. Capsule papyraceous, dehiscing loculicidally; seeds flattened or strongly compressed, often with a marginal wing; testa black.

About 80 species in Europe, the Mediterranean region, Africa and India. W. T. Stearn (in Ann. Mus. Goulandris, 4: 199–210; 1978) follows J. P. Jessopp (in Journ. South Afr. Bot., 43: 265–319; 1977) and unites *Urginea* Steinh. (1834) with *Drimia* Jacq. (1797); other authors have preferred to maintain the two as distinct, which latter course has been followed here, without necessarily rejecting the possibility that examination of all the species in both genera would dictate a different policy.

1. **U. maritima** (*L.*) *Baker* in Journ. Linn. Soc., 13: 221 (1872); Boiss., Fl. Orient., 5: 224 (1882); Holmboe, Veg. Cypr., 48 (1914); Post, Fl. Pal., ed. 2, 2: 631 (1933); Lindberg f., Iter Cypr., 11 (1946); Osorio-Tafall et Seraphim, List Vasc. Plants Cyprus, 20 (1973); Elektra Megaw, Wild Flowers of Cyprus, 16, t. 34 (1973).
   Scilla maritima L., Sp. Plant., ed. 1, 308 (1753); Gaudry, Recherches Sci. en Orient, 187 (1853–54); L. J. S. Littlejohn, Island of Cyprus, 159 (1947).
   Urginea scilla Steinh. in Ann. Sci. Nat., ser. 2, 1: 330 (1884); Kotschy in Unger et Kotschy, Die Insel Cypern, 195 (1865).
   [*U. undulata* (non (Desf.) Steinh.) Boiss., Fl. Orient., 5: 223 (1882) pro parte quoad plant. cypr.; Holmboe, Veg. Cypr., 48 (1914); Osorio-Tafall et Seraphim, List Vasc. Plants Cyprus, 20 (1923).]
   TYPE: "*ad* Hispaniae, Siciliae, Syriae littoria arenosa".

Bulb subglobose or broadly ovoid, 5–20 cm. diam., tunic papery, pale brown; leaves usually numerous, broadly lorate, shining green, 20–50 cm. long, 2–10 cm. wide, withering and disappearing before anthesis; scape robust, terete, erect, 60–150 cm. high; inflorescence an elongate, narrowly cylindrical raceme, 25–50 cm. or more long, 4–6 cm. diam.; flowers 50–150 or more, not very crowded; basal bracts narrowly lingulate, strongly reflexed, 5–8 mm. long, 3–4 mm. wide, upper bracts small, filiform, spreading, about 3 mm. long, 0·5 mm. wide at base; pedicels slender, patent, up to 2 cm. long, articulated at base; perianth saucer-shaped or widely campanulate, segments whitish, often with a purplish median nerve, oblong, obtuse or subacute, about 8 mm. long, 2·5 mm. wide, becoming erect after anthesis and enveloping the developing fruit; filaments spreading, subulate, about 5 mm. long, 2 mm. wide at base; anthers oblong, greenish, about 2·5–3 mm.

long, 1·5 mm. wide; ovary sessile, oblong, about 4 mm. long, 3 mm. wide; style about 3 mm. long; stigma truncate. Capsule oblong, sharply triquetrous, 8–13 mm. long, 5–8 mm. wide, coriaceous, pale brown, apex truncate or shortly emarginate; seeds oblong, flattened, 7–8 mm. long, 3–3·5 mm. wide; testa black, shining, minutely rugulose with narrowly rectangular raised areas.

HAB.: Sandy seashores; roadsides; or amongst garigue on dry hillsides; sea-level to 4,500 ft. alt.; fl. July–Sept.

DISTR.: Divisions 1–8, common. Mediterranean region and Atlantic Islands.

1. Ayios Yeoryios (Akamas), 1962, *Meikle*! Near Polis, 1962, *Meikle*!
2. Makheras Monastery, 1862, *Kotschy*; Platres, 1937, *Kennedy* 213! 214! 215! Prodhromos, 1955, *G. E. Atherton* 658!
3. Lefkara, 1862, *Kotschy*; and, 1905, *Holmboe* 1102.
4. Larnaca, 1862, *Kotschy*; Kalokhorio, 1862, *Kotschy*; Larnaca, 1880, *Sintenis & Rigo* 939! 998! Cape Greco, 1958, *N. Macdonald* 96!
5. Kythrea, 1939, *Lindberg f.* s.n.! Athalassa, 1970, *A. Genneou* in ARI 1582! Salamis, 1962, *Meikle*!
6. Dhiorios, 1962, *Meikle*!
7. Bellapais, 1939, *Lindberg f.* s.n.; Kyrenia, 1955, *G. E. Atherton* 402!
8. Akradhes, 1962, *Meikle*!

NOTES: Widespread both inland and by the sea in Cyprus, but, partly because of its fleshy texture, and partly because of its separate leafing and flowering periods, very largely ignored by herbarium collectors.

Boissier's record for *Urginea undulata*, repeated by Holmboe, is apparently based upon *Sintenis & Rigo* 939, "Larnaka: in maritimis, 12.ix.1880", a duplicate of which is labelled *Urginea undulata* Desf. at Kew. The specimen so labelled (one of the last to be collected by Sintenis in Cyprus) is, however, typical *U. maritima*. What may possibly be part of the same collection, but numbered *Sintenis & Rigo* 998, is correctly labelled *Scilla maritima* L.

13. SCILLA *L.*

Sp. Plant., ed. 1, 308 (1753).
Gen. Plant., ed. 5, 146 (1754).
F. Speta in Naturk. Jahrb. Stadt Linz, 25: 19–198 (1981).

Bulbous perennials; bulb perennial, tunicate, scales free; leaves all basal, rarely with a pallid median stripe, synanthous or hysteranthous; scape simple; inflorescence an elongate or corymbiform raceme, or flowers solitary; bracts membranous or scarious, sometimes spurred basally, occasionally absent; perianth campanulate or shallowly concave, segments 6, free to the base, sometimes reflexing with age, generally blue, purplish or whitish, subequal, 1-nerved; stamens 6, free, inserted at base of perianth segments, included or exserted; filaments usually slender, subulate, often spreading, anthers dorsifixed, introrse, yellow or violet; ovary 3-locular, sessile or very shortly stipitate, trigonous; ovules 2–10 in each loculus; style short, straight or bent near the base; stigma truncate or subcapitate. Capsule subglobose-trigonous, sometimes spongy, dehiscing loculicidally; seeds subglobose or pyriform, not flattened; testa black; caruncle (or elaiosome) sometimes present.

About 80 species in temperate Europe and Asia, the Mediterranean region, Africa, Atlantic Islands.

Plants flowering in autumn, generally in advance of the development of the leaves
                                                   1. S. autumnalis

Plants flowering in spring, with or after the development of the leaves:
    Flowers small, less than 1 cm. diam., very numerous (50 or more) in a long cylindrical raceme, leaves 1·5–2·5 cm. wide; scape often more than 1 m. tall   S. HYACINTHOIDES (p. 1636)
    Flowers large, more than 1 cm. diam., solitary or 2–6 in a lax raceme; leaves 0·5–1·3 cm. wide:
        Flowers bright purplish-blue, segments spreading or often somewhat recurved; bracts conspicuous, lacerate - - - - - - - - - **2. S. cilicica**
        Flowers milky-white tinged lilac, segments generally subconnivent; bracts very small, truncate or emarginate - - - - - - - - **3. S. morrisii**

1. S. autumnalis *L.*, Sp. Plant., ed. 1, 309 (1753); Poech, Enum. Plant. Ins. Cypr., 8 (1842); Unger et Kotschy, Die Insel Cypern, 194 (1865); Boiss., Fl. Orient., 5: 224 (1882); Holmboe, Veg. Cypr., 48 (1914); Post, Fl. Pal., ed. 2, 2: 631(1933); Osorio-Tafall et Seraphim, List Vasc. Plants Cyprus, 21 (1973).
TYPE: "*in* Hispania, Gallia, Verona *solo glareoso*".

Bulb ovoid or subglobose, 1·3–5 cm. long, 1–2 cm. wide, tunics membranous, the outermost dull brown, the inner whitish; leaves usually numerous, 8–10 or more, appearing after anthesis, linear-filiform, 7–12 cm. long, 0·2 cm. wide, glabrous, rather prominently nerved, apex acute or subacute; scapes usually 2 or more from each bulb, erect, slender, purplish, 7–20 cm. long; elongating after anthesis; inflorescence a lax or dense raceme 1·5–10 cm. long, 1–2 cm. diam.; pedicels filiform, 5–10 mm. long, spreading, becoming arcuately erect in fruit; flowers 6–15 or more; perianth widely campanulate, segments mauve-blue or purplish, oblong, 3–4 mm. long, 1–1·5 mm. wide, with a prominent median nerve and an obtuse or subacute apex; filaments filiform, purplish, about 2 mm. long; anthers oblong, blue-purple, about 1 mm. long, 0·4 mm. wide; ovary ovoid-trigonous, about 2 mm. long, 1·6 mm. wide; ovules 2 in each loculus; style straight, slender, about 2 mm. long; stigma truncate. Capsule subglobose, about 3 mm. diam. with an indented apex, papery, pale brown, usually enveloped by the persistent perianth; seeds pyriform, irregularly compressed, about 3 mm. long, 1·8 mm. wide; testa lustrous brown or blackish, minutely rugulose.

HAB.: Rocky hillsides and seashores or in dry pastures overlying rocks; sea-level to 3,000 ft. alt.; fl. Sept.–Nov.

DISTR.: Divisions 1–3, 5–8. Southern and western Europe, the Mediterranean region.

1. Paphos, 1840, *Kotschy* 56; and, 1981, *Meikle*!
2. Platanistasa, 1905, *Holmboe* 1163; Saïttas, 1934, *Syngrassides* 523! and, 1970, *A. Genneou* in ARI 1608! Kryos Potamos, 2,900 ft. alt.; *Kennedy* 216! 217! Evrykhou, 1937, *Syngrassides* 1677! Kalopanayiotis, 1968, *A. Genneou* in ARI 1258!
3. Episkopi, 1963, *J. B. Suart* 129!
5. Between Vouno and Mia Milea, 1937, *Syngrassides* 1732! Athalassa, 1967, 1969, *Economides* in ARI 1032! *A. Genneou* in ARI 1302!
6. Government House, Nicosia, 1935, *Syngrassides* 859!
7. Kyrenia area, 1936–1970, *Kennedy* 218! *Lady Loch* 8! 13! *Davis* 2010! *Casey* 149! *G. E. Atherton* 748! 769! 784! *I. M. Hecker* 15!
8. Karpas, 1937, *Lady Loch* 5!

S. HYACINTHOIDES *L.*, Syst. Nat., ed. 12, 2: 243 (1767); Osorio-Tafall et Seraphim, List Vasc. Plants Cyprus, 21 (1973); F. Speta in Naturk. Jahrb. Stadt Linz, 25: 186 (1981).

A robust plant, superficially somewhat resembling *Urginea maritima*. Scape commonly more than 1 m. high; leaves present with the inflorescences, ligulate-lorate, 20–50 cm. long, 1·5–2·5 cm. wide, bright green, canaliculate, racemes many-flowered, 20–30 or more cm. long, 4–6 cm. diam.; bracts minute, semicircular, membranous; pedicels slender, 2–3 cm. long, patent; perianth widely campanulate, segments pale or bright blue, oblong, about 5 mm. long, 2·5 mm. wide, obtuse or subacute; stamens included; anthers blue. Capsules subglobose-trigonous, pale brown, 5–6 mm. diam.; seeds broadly pyriform, 6–7 mm. long, 5–6 mm. wide; testa blackish, rugulose.

Listed by Osorio-Tafall et Seraphim (List Vasc. Plants Cyprus, 21), and said by Speta (*loc. cit.*) to have been collected at Cape Greco (Div. 4), but no specimens have been seen, nor am I aware that the plant has been found in Cyprus by any other collector. Pending confirmation of the record, it is here admitted tentatively to the Flora. *S. hyacinthoides* has a wide distribution in the Mediterranean region and eastwards to Iraq. It is recorded from shores adjacent to Cyprus, and is a very likely occurrence on the island. It commonly grows on rocky calcareous hillsides, and flowers during March and April.

2. S. cilicica Siehe in Gard. Chron., ser. 3, 44: 194 (1908); Meikle in Lily Year Book 1962: 125 (1961); Osorio-Tafall et Seraphim, List Vasc. Plants Cyprus, 21 (1973); Meikle in Bot. Mag., 181: t. 745 (1977); Speta in Naturk. Jahrb. Stadt Linz, 25: 131 (1981).
S. veneris Speta in Naturk. Jahrb. Stadt Linz, 22: 69 (1977).
[S. amoena (non L.) Kotschy in Unger et Kotschy, Die Insel Cypern, 194 (1865); Seligman in Quart. Bull. Alpine Gard. Soc., 19: 56 (1981), 20: 234 (1952).]
[S. siberica (non Haw.) Holmboe, Veg. Cypr., 48 (1914); Druce in Proc. Linn. Soc., 141st sess.: 51 (1930); Seligman in Quart. Bull. Alpine Gard. Soc., 20: 234 (1952); Osorio-Tafall et Seraphim, List Vasc. Plants Cyprus, 21 (1973) as S. "sibirica".]
[S. cernua (non Delaroche) Sintenis in Oesterr. Bot. Zeitschr., 32: 21 (1882); Boiss., Fl. Orient., 5: 227 (1882) pro parte quoad plant. cypr.; Rechinger f. in Arkiv för Bot., ser. 2, 1: 419 (1950).]
TYPE: Turkey, "in half-shaded districts of Cilicia, in the lower mountain regions", cult. Mersin, Siehe (BM! E).

Bulb broadly ovoid, 1·5–2 cm. long and almost as wide, sometimes producing thick retractile roots, tunic glossy purple-black, leaves 2–4 (–6), linear, 15–30 cm. long, 0·7–1·3 cm. wide, spreading, canaliculate, bright green, developing through the winter and often withered at the cucullate, subacute or shortly acute apex before anthesis, translucent with distinct nervation when dried; scape 10–30 cm. long, subterete, often purplish, erect at anthesis, arcuate in fruit; inflorescence a lax 2–6-flowered raceme or flowers sometimes solitary; rhachis slender, purple, up to 8 cm. long; bracts membranous, 2–4 mm. long and almost as wide, irregularly lacerate, white or purplish, sometimes shortly spurred; pedicels slender, patent or decurved, 3–10 mm. long, bronze-purple, scarcely accrescent; perianth slightly concave, or flat, or with the segments sometimes recurved, purplish-blue, segments oblong-lanceolate, 12–15 mm. long, 3–4 mm. wide, apex subacute or obtuse, minutely papillose externally; stamens included; filaments subulate, about 8 mm. long, flattened dorsiventrally, whitish or tinged blue; anthers narrowly oblong, dark blue, about 1·8 mm. long, 1 mm. wide; ovary greenish or stained violet, subglobose, trigonous, about 3 mm. diam.; ovules 4 in each loculus; style straight or bent obliquely, slender, 6–7 mm. long, pallid or tinged violet; stigma truncate. Capsule broadly and bluntly ellipsoid or subglobose, 1–1·4 cm. long, about 1 cm. wide, at first spongy, becoming papery and rugose on dehiscence; seeds 3–4 in each loculus, subglobose, ecarunculate, about 2 mm. diam.; testa dull black, minutely rugulose. Plate 97, figs 3–6.

HAB.: In crevices of shaded limestone rocks; 100–3,300 ft. alt.; fl. Jan.–March.

DISTR.: Divisions 1, 7. Also S. Turkey, Syria, Lebanon.

1. Baths of Aphrodite, 1953, Kennedy 1778! Between Androlikou and Prodhromi, 1957, Merton 3032!
7. N. side of Buffavento, 1862, Kotschy 414; same area, 1880, Sintenis & Rigo 170! also, 1937, Syngrassides 1426! and, 1959, C. E. H. Sparrow 8! Dramia, 1938, Lady Loch 25! St. Hilarion, 1937, Mrs. Stagg s.n.! also, 1941, Davis 2170! 1949, Casey 342! and, 1961, Polunin 6615! Kyrenia, 1953, Kennedy 1777! also, 1956, G. E. Atherton 895! Kyparissovouno, 1954, Merton 1853! and, 1962, Meikle s.n.! Above Keurmurju, 1959, P. H. Oswald 37!

3. S. morrisii Meikle in Kew Bull., 30: 537 (1975); Speta in Naturk. Jahrb. Stadt Linz, 25: 129 (1981).
TYPE: Cyprus; Paphos Distr., near Khrysorroyiatissa Monastery, 1974, H. M. Morris & R. D. Meikle in Meikle 4015 (K!).

Bulb broadly ovoid, 2–2·5 cm. long, 1·8–2·3 cm. wide, tunic purple-black; leaves generally 3, flaccid, sprawling on the ground, developed during the winter and beginning to wither at the tip by anthesis, linear, up to 70 cm.

Plate 97. Figs. **1–2.** SCILLA MORRISII Meikle **1,** habit, × 1; **1a,** infructescence, × 1; **1b,** seed, × 6; **2,** flower, longitudinal section, × 3; figs. **3–6.** SCILLA CILICICA Siehe **3,** habit, × 1; **3a,** infructescence, × 1; **4,** capsule, × 3; **5,** seed, × 6; **6,** flower, longitudinal section, × 3. (**1–2** from *Meikle* 4015; **3** from *Syngrassides* 1426; **4, 5** from *Merton* 3032; **6,** from *Mrs Stagg* s.n.)

long, 0·5–1·2 (–1·4) cm. wide, carinate, bright green, with a pallid stripe along the upper surface of the keel; apex acute, shortly cucullate, scapes 1–3, erect or arcuate (or almost prostrate in fruit), 10–35 cm. long, greenish or tinged bronze-purple; inflorescence a lax 2–3 (–5)-flowered raceme, or flowers sometimes solitary; rhachis 2–6 cm. long; bracts small, membranous, oblong, about 1·5–2 mm. long and almost as wide, without a basal spur; pedicels generally less than 1 cm. long, patent or suberect at anthesis becoming recurved in fruit; perianth campanulate, milky-white tinged lilac or blue, segments subconnivent (or briefly patent), subequal, oblong, 10–15 mm. long, 2–4 mm. wide, obscurely nerved, apex obtuse or subacute, sometimes a little cucullate; stamens included; filaments subulate, white, about 6 mm. wide at base; anthers pale blue, oblong, about 1·2 mm. long, 0·8 mm. wide; ovary sessile, subglobose, yellowish, about 4 mm. diam.; ovules (2–) 4 (–5) in each loculus; style straight, subulate, 4–5 mm. long; stigma truncate or sub-capitate. Capsules globose or subglobose, about 1–1·3 cm. diam., spongy at first, becoming papery and rugose on dehiscence; seeds subglobose, ecarunculate, about 2 mm. diam.; testa black, minutely rugulose. *Plate 97, figs. 1–2.*

HAB.: Moist shaded crevices and banks under *Quercus infectoria* ssp. *veneris*; c. 2,700 ft. alt.; fl. March–April.

DISTR.: Division 2, very rare. Endemic.

2. Near Khrysorroyiatissa Monastery, 1974, *H. M. Morris & R. D. Meikle* in *Meikle* 4015!

S. BIFOLIA *L.* Sp. Plant., ed. 1, 309 (1753), generally with 2 erect, concave leaves, developing in spring with the inflorescence, the latter usually less than 15 cm. high, consisting of 2–6 (–10) small, bright blue, starry flowers in a short, subsecund raceme; lower pedicels usually well developed, upper often very short; bracts minute or absent.

Recorded from a churchyard at Famagusta by Kotschy (Die Insel Cypern, 194: 1865) but no doubt planted there. Holmboe's inclusion of the plant in Studies on the Vegetation of Cyprus, 48 (1914) is based on Kotschy's record. There is no evidence that *S. bifolia* occurs naturally in Cyprus, though it is widely distributed in S. Europe and Western Asia.

14. CHIONODOXA *Boiss.*

Diagn., 1, 5: 61 (1844).

Scilla L. series *Chionodoxa* (Boiss.) Speta in Naturk. Jahrb. Stadt Linz, 21: 12 (1976).

Bulbous perennials; bulb perennial, tunicate, scales free, leaves generally 2, basal, synanthous, without a median stripe; scape simple, inflorescence a lax, subsecund raceme, or flowers solitary; bracts membranous, very small, not spurred basally; perianth hypocrateriform with a short basal tube, segments subequal, spreading, bright or pale blue, often white or whitish at the base; stamens 6, free, erect, connivent, exserted from perianth-tube; filaments strongly flattened dorsiventrally, truncate or emarginate, alternatively longer and shorter, often white or pallid; anthers dorsifixed, introrse, oblong, blue or yellow; ovary 3-locular, sessile, trigonous; ovules 4–6 or more in each loculus; style short, straight; stigma truncate. Capsule subglobose-trigonous, at first spongy, dehiscing loculicidally; seeds subglobose, not flattened; testa black; caruncle (or elaiosome) present.

Six species in Crete, W. Turkey, Cyprus.

1. **C. lochiae** *Meikle* in Kew Bull., 1954: 495 (1954) et in Bot. Mag., 171: t. 281 (1956); Osorio-

Tafall et Seraphim, List Vasc. Plants Cyprus, 21 (1973); Elektra Megaw, Wild Flowers of Cyprus, 17, t. 35 (1973).
  Scilla lochiae (Meikle) Speta in Naturk. Jahrb. Stadt Linz, 21: 49 (1976), 25: 67 (1981).
TYPE: Cyprus; Troödos, northern flank, near Spilia, 3,000–4,000 ft. alt.; 13 March 1953, *E. W. Kennedy* 1776 (K!).

Bulb subglobose, 1·5–2 cm. diam., tunic papery, rough and fuscous-brown externally, whitish internally; leaves 2 (–3), linear, concave and sheathing the lower half of the scape, 6–25 cm. long, 0·4–0·8 cm. wide, at first suberect, later recurved or flaccid, dark lustrous green tinged bronze, apex shortly acute or apiculate, often cucullate; scape solitary, simple, 7–30 cm. long, erect, tinged bronze; inflorescence a lax 2–3 (–4)-flowered raceme, or flowers often solitary, bracts minute, 1–2 mm. long, membranous, subtruncate, whitish or purplish; pedicels 0·5–1·8 cm. long, slender, erecto-patent or decurved; flowers often cernuous; perianth uniformly bright blue, tube 5–7 mm. long, about 4 mm. wide, rugulose, segments at first patent, later distinctly recurved, the 3 exterior oblong-lanceolate, 12–13 mm. long, 4 mm. wide, the median nerve distinctly cristate above internally and produced into a small, papillose apiculus, the 3 inner segments as long as the exterior, 5–6 mm. wide, the median nerve less prominent internally, and not produced into an apiculus; filaments unequal, alternately about 3 mm. and 4 mm. long, erect, oblong, tapering to a subtruncate or slightly emarginate apex, whitish or pale blue; anthers blue, narrowly oblong, about 4·5 mm. long, 1 mm. wide, shortly bifid at apex and base; ovary subglobose or broadly ovoid, about 4 mm. long, 3–3·5 mm. wide; ovules 4–5 in each loculus. Capsule globose, trigonous, about 1·5 cm. diam.; seeds subglobose, about 2·5 mm. diam; testa black.

HAB.: Moist humus-rich ground under Pines; 3,000–4,500 ft. alt.; March–April.

DISTR.: Division 2, rare. Endemic.

2. Here and there around the central part of the Troödos Range; near Spilia, Pedhoulas, Prodhromos, Platania, etc. *Kennedy* 1776! 1871! *Merton* 1849! *Polunin* 8532!

15. HYACINTHELLA *Schur*
Oesterr. Bot. Wochenbl., 6: 227 (1856).
N. Feinbrun in Bull. Res. Council Israel, 10D: 324–347 (1961).

Bulbous perennials; bulb small, tunicate; leaves generally 2, usually with distinct, elevated nervation; scapes simple; inflorescence a lax or dense raceme; bracts minute; pedicels erecto-patent, sometimes very short or wanting; perianth generally blue, campanulate, tubular or infundibuliform, not constricted above, divided into 6 erect lobes for about ⅓ of the total length; stamens inserted below base of perianth-lobes in 2 (often ill-defined) series; anthers mostly oblong, dorsifixed, introrse; ovary subglobose or ovoid; style short, straight; stigma truncate or subcapitate. Capsule depressed-globose, bluntly trigonous, coriaceous; seeds 1–2 in each loculus, subglobose or broadly ovoid, not flattened overall, but sometimes compressed on one side; testa black, glossy, reticulate-rugose, sometimes pitted.
About 10 species in S.E. Europe and western Asia.

1. **H. millingenii** (Post) Feinbrun in Bull. Res. Council Israel, 10D: 335 (1961); K. Persson et P. Wendelbo in Candollea, 36: 531 (1981).
  Bellevalia millingeni Post in Mém. Herb. Boiss., 18: 101 (1900); H. Stuart Thompson in Journ. Bot., 44: 339 (1906).
  Hyacinthus nervosus Bertol. ssp. *millingeni* (Post) Holmboe, Veg. Cypr., 52 (1914).
  Hyacinthella nervosa (Bertol.) Chouard ssp. *millingeni* (Post) Osorio-Tafall et Seraphim, List Vasc. Plants Cyprus, 21 (1973) nom. invalid.

[*Bellevalia nervosa* (non *Hyacinthus nervosus* Bertol.) Sintenis in Oesterr. Bot. Zeitschr., 31: 226, 255, 326 (1881); Boiss., Fl. Orient., 5: 306 (1882) pro parte quoad plant. cypr.; Seligman in Quart. Bull. Alpine Gard. Soc., 20: 233 (1952).]
TYPE: Cyprus; "prope Nicosia, Cypri; floret decembro. [*Post*] No. 920" (BEI).

Bulb ovoid, about 2 cm. long, 0·8–1·4 cm. wide, tunics membranous, the outermost pale greyish-brown; leaves 2, linear, erect or recurved, 5–20 cm. long, 0·2–0·8 cm. wide, glabrous, glaucous, with numerous, prominent parallel nerves, base sheathing the scape often for more than half its length, tinged reddish-purple, apex shortly acute, cucullate, margins often minutely scabridulous; scape slender, erect, 4–15 cm. long; inflorescence spiciform usually less than 2·5 cm. long, 0·8 cm. wide, flowers 3–15 (–20), sessile, pale blue; bracts minute, purplish; perianth campanulate or shortly tubular, about 4–6 mm. long, 2·5–3 mm. diam., lobes ovate, acute, about 2 mm. long, 1·8 mm. wide at base, apex incurved; stamens inserted in middle part of corolla-tube; filaments subequal, about 1·5 mm. long, the lower part adnate to perianth-tube, glabrous, tinged reddish; anthers violet, ovate, about 1·3 mm. long, 0·8 mm. wide; ovary subglobose-trigonous, 1·5 mm. diam.; style 2 mm. long; stigma subcapitate. Capsule strongly depressed-globose, bluntly trigonous, about 2·5 mm. long, 5 mm. wide, pale brown, coriaceous; seed usually solitary, subglobose, compressed unilaterally, about 2 mm. diam., testa blackish-brown, minutely pitted.

HAB.: Rocky, calcareous hillsides; 200–2,500 ft. alt.; fl. Nov.–Febr.

DISTR.: Divisions 1–8. Also S. Turkey.

1. Smyies (Akamas), 1948, *Kennedy* 1700!
2. Above Lythrodhonda, 1938, *Lady Loch* 28!
3. Above Kandou, 1966, *A. Matthews* 5!
4. Near Larnaca, 1859, *Samaritani*; also, 1880, *Sintenis & Rigo* 886!
5. Mia Milea, 195–, *Merton* s.n.; *Athalassa*, 1967, *Economides* in ARI 1174!
6. Nicosia, ?1898, *Post* 920; also, 1944, *Evenari*; between Kormakiti and Orga, 1941, *Davis* 2102!
7. Frequent; Buffavento, Kyrenia Pass, St. Hilarion, Bellapais, Ayios Epiktitos, Myrtou, Keurmurju, Panagra, etc. 1935–1971, *Syngrassides* 810! *Davis* 2038! *Casey* 203! 208! 398! *G. E. Atherton* 784a! *Merton* 2461! *P. H. Oswald* 40! *E. Hodgkin* s.n.! etc.
8. Karpas, rocks by the sea near 'Macheriona' [*sic*], 1894, *Deschamps* s.n.; also noted from the Karpas by *Lady Loch*.

NOTES: Feinbrun (Bull. Res. Council Israel, 10D: 336; 1961) also notes a specimen said to have been collected by Merton at "Mia Malea, Chionistra". As the species appears to be absent from Khionistra, I have assumed that Mia Milea near Nicosia is intended, but the specimen may possibly have come from Mallia on the southern edge of the Troödos Range.

16. BELLEVALIA Lapeyr.

Journ. Phys. Chim. Hist. Nat. Arts, 67: 425 (1808) nom. cons. prop.
N. Feinbrun in Pal. Journ. Bot., J. series, 1: 42–54, 131–142, 336–409 (1938–39).

Bulbous perennials, bulbs tunicate, leaves linear, lorate or lanceolate, with a hyaline, membranous, smooth or scabridulous-ciliate margin; nervation not conspicuously elevated; scape simple; inflorescence a conical, cylindrical or spiciform raceme; bracts very small, sometimes absent; upper flowers occasionally sterile; perianth campanulate, tubular or infundibuliform, at first often bluish or purplish, later whitish, yellowish or livid, tube well developed, lobes 6, spreading or erect, tube not constricted at their base; stamens inserted in one series at base of tube; anthers introrse; ovary sessile, 3-locular; ovules 2–6 in each loculus; style straight; stigma truncate or subcapitate. Capsule sharply triquetrous, seeds subglobose, minutely rugulose-pitted with a blackish, pruinose testa.

Raceme lax; pedicels 3–13 (–16) mm. long; flowers at first purplish-blue, becoming dirty yellowish-green or brownish with age; perianth-tube 5–10 (–13) mm. long **1. B. trifoliata**

Raceme condensed, cylindrical-spiciform; pedicels very short or almost wanting, rarely up to 4 mm. long; flowers whitish or putty-coloured, sometimes at first tinged mauve or pink; perianth-tubes 2–5 mm. long- - - - - - - - - **2. B. nivalis**

1. **B. trifoliata** (*Ten.*) *Kunth*, Enum. Plant., 4: 308 (1843); Unger et Kotschy, Die Insel Cypern, 193 (1865); Boiss., Fl. Orient., 5: 303 (1882); Post, Fl. Pal., ed. 2, 2: 651 (1933); Feinbrun in Pal. Journ. Bot., J. ser., 1: 343 (1940); Osorio-Tafall et Seraphim, List Vasc. Plants Cyprus, 21 (1973).
   Hyacinthus trifoliatus Ten., Fl. Nap., 3: 376, t. 136 (1824); Holmboe, Veg. Cypr., 51 (1914).
   TYPE: Italy; "Nasce in *Puglia*, al *Tavoliere*, tre le biade" (NAP).

Bulb broadly ovoid, 2–5 cm. long, 1·5–3 cm. wide, tunics thinly papery, brownish; leaves (2–) 3–4 (–6), linear, 20–50 cm. long, 0·8–3 cm. wide, base long-sheathing, apex acuminate, margins almost smooth, scabridulous or shortly ciliate; scape 15–30 cm. long; inflorescence cylindrical, usually rather lax, 5–12 cm. long, 3–4 cm. wide at anthesis, elongating to 25 cm. or more in fruit; flowers 12–60, at first suberect, soon patent or slightly nodding; rhachis angular, often purplish; bracts minute, often purplish; pedicels slender, 3–13 (–16) mm. long; perianth tubular-campanulate, 8–13 (–16) mm. long, 3–4 mm. wide at the distinctly flared mouth, at first tinged purplish-blue, becoming dirty yellowish-green or brownish with age, lobes suberect, concave, subacute, somewhat overlapping, about 3 mm. long, 2·5 mm. wide; stamens inserted at base of perianth-tube, in one series; filaments adnate to the perianth-tube for the greater part of their length, glabrous, the free part 2–4 mm. long; anthers oblong, violet-purple, about 2·8 mm. long, 1·3 mm. wide, almost as long as the perianth-lobes; ovary ovoid- or ellipsoid-triquetrous, about 4 mm. long, 2 mm. wide; style 8–9 mm. long; stigma subtruncate. Fruiting pedicels patent or slightly incurved; capsule orbicular-triquetrous or broadly obcordate-triquetrous, about 8–13 mm. diam.; valves pale brown, distinctly veined; seeds subglobose, about 3 mm. diam.; testa minutely rugulose-pitted, sublustrous, black, pruinose.

HAB.: Cereal fields; roadsides; moist banks; sea-level to 1,900 ft. alt.; fl. Febr.–May.

DISTR.: Divisions 1, 3–8, locally common. Widespread in the Mediterranean region.

1. Smyies (Akamas), 1948, *Kennedy* 1602! Baths of Aphrodite, 1953, *Kennedy* 1775! Ayios Yeoryios (Akamas), 1962, *Meikle*! Polis, 1962, *Meikle*!
3. Amathus, 1964, *J. B. Suart* 148; Kolossi and Akrotiri, 1966, *Ann Matthews* 12! Below Mallia, 1977, *J. Wood* 44! Between Ayios Yeoryios and Kithasi, 1978, *Chesterman* 65! 66!
4. Larnaca, 1880, *Sintenis & Rigo* 173! Sotira, 1952, *Merton* 621! Between Cape Greco and Paralimni, 1958, *N. Macdonald* 4! Between Mazotos and Alaminos, 1978, *Chesterman* 35!
5. Kythrea, 1880, *Sintenis & Rigo*; also 1931–1934, *Nattrass* 193, 201, 300, 428, 737; Lakatamia, 1932, *Syngrassides* 701! Dhikomo, 1936, *Syngrassides* 1035!
6. Nicosia, 1927, *Rev. A. Huddle* 98! Also, 1930, *F. A. Rogers* 0674! and 1959, *P. H. Oswald* 73! Kormakiti peninsula, 1972, *P. M. Synge* s.n.!
7. Above Vasilia, 1941, *Davis* 2202! Also, 1959, *P. H. Oswald* 59! Kyrenia, 1949, *Casey* 265! 313! Also, 1956, *G. E. Atherton* 989! 1006! 1271! Thermia, 1956, *G. E. Atherton* 880! Tjiklos, 1973, *P. Laukkonen*! Ayios Epiktitos, 1967, *Merton* in ARI 585! 587!
8. Rizokarpaso, 1941, *Davis* 2374!

2. **B. nivalis** *Boiss. et Kotschy* in Boiss., Diagn., 2, 4: 110 (1858); Unger et Kotschy, Die Insel Cypern, 194 (1865); Sintenis in Oesterr. Bot. Zeitschr., 31: 255, 326 (1881); Boiss., Fl. Orient., 5: 304 (1882); Post in Mém. Herb. Boiss., 18: 102 (1900); Post, Fl. Pal., ed. 2, 2: 652 (1933); Feinbrun in Pal. Journ. Bot., J. ser., 1: 377 (1940); Seligman in Quart. Bull. Alpine Gard. Soc., 20: 233 (1952); Osorio-Tafall et Seraphim, List Vasc. Plants Cyprus, 21 (1973).
   Hyacinthus nivalis (Boiss. et Kotschy) Baker in Journ. Linn. Soc., 11: 430 (1870); Holmboe, Veg. Cypr., 51 (1914).
   H. pieridis Holmboe, Veg. Cypr., 51 (1914).
   [*H. romanus* (non L.) Sm. in Sibth. et Sm., Fl. Graec. Prodr., 1: 237 (1809), Fl. Graec., 4: 34 (1823); Poech, Enum. Plant. Ins. Cypr., 8 (1842); Holmboe, Veg. Cypr., 51 (1914) pro parte quoad plant. cypr.]
   [*Bellevalia romana* (non (L.) Reichb.) Kotschy in Unger et Kotschy, Die Insel Cypern,

193 (1865); Boiss., Fl. Orient., 5: 301 (1882); Osorio-Tafall et Seraphim, List Vasc. Plants Cyprus, 21 (1973) pro parte quoad plant cypr.]
TYPE: Syria; "in humo argilloso ad nives Alpium *Mandschura Syriae* alt. 6500′ cl. Kotschy". (G).

Bulb ovoid, 2–3 cm. long, 1·3–2·5 cm. wide, tunics thinly papery, the outermost fuscous-brown; leaves (2–) 3–4 (–5), spreading, linear, 5–25 (–35) cm. long, 0·3–1·2 cm. wide, base not long-sheathing, apex slenderly acuminate, margins shortly scabridulous-ciliate; scapes 1–3, erect, 5–12 (–24) cm. long, usually much shorter than the leaves; raceme cylindrical-spiciform, 2–4 (–7) cm. long, 1·5–2 cm. wide at anthesis, somewhat accrescent in fruit; bracts lingulate, membranous, obtuse, generally less than 2 mm. long, 0·5 mm. wide; flowers 7–14 (–27), erecto patent; pedicels usually very short or almost wanting, occasionally up to 4 mm. long; perianth tubular-campanulate with a distinctly flared mouth, dirty white or putty-coloured, often stained livid towards base externally, the buds and topmost flowers sometimes tinged mauve or pink, tube 2–5 mm. long, 2–3 mm. diam., lobes oblong-ovate, subacute, 2–2·5 mm. long, 1·5–2 mm. wide, erect or slightly spreading; stamens inserted near base of perianth-tube; filaments adnate for the greater part of their length to perianth-tube, free part 1·5–2 mm. long; anthers violet, oblong-ovate, subacute, about 1·8 mm. long, 1 mm. wide; ovary sessile, ovoid- or ellipsoid-triquetrous, about 3 mm. long, 2 mm. wide, style 1·5–4 mm. long; stigma subtruncate. Capsule subsessile or on a short fruiting pedicel up to 5 mm. long, subglobose-triquetrous, about 0·8 mm. diam., valves brownish, distinctly veined; seeds subglobose, about 2 mm. diam.; testa blackish, pruinose, minutely rugulose-pitted.

HAB.: In coastal garigue or on rocky hillsides inland, sometimes in gardens or cultivated or fallow fields; sea-level to 3,000 (–4,500) ft. alt.; fl. Jan.–April.

DISTR.: Divisions 1–8, locally common. Also Syria, ?S. Turkey.

1. Ayios Nikolaos (Akamas), 1948, *Kennedy* 1610! Smyies, 1948, *Kennedy* 1701!
2. Above Prodhromos towards the summit of Khionistra, 1859, *Kotschy* 411; Makheras Monastery, 1862, *Kotschy* 216! Kryos Potamos, 2,900 ft. alt., 1937, *Kennedy* 195! 196! 197! 198! Moniatis River, 3,500 ft. alt., 1937, *Kennedy* 199! Asinou, 1967, *Polunin* 8538!
3. Akrotiri Bay, 1948, *Mavromoustakis* s.n.! Yermasoyia River, 1948, *Mavromoustakis* s.n.! 1·5 miles E. of Kophinou, 1978, *Chesterman* 39! Between Ayios Yeoryios and Kithasi, 1978, *Chesterman* 67!
4. Near Avgorou, 1862, *Kotschy* 93! Dhekelia, 1880, *Sintenis & Rigo* 174! Ayia Napa, 1905, *Holmboe* 71; Famagusta, 1936, *Syngrassides* 1067! Cape Greco, 1958, *N. Macdonald* 27!
5. Kythrea, 1880, *Sintenis & Rigo*; also, 1956, *G. E. Atherton* 1062! Athalassa, 1962, *Meikle* 2984! also, 1969, *Economides* in ARI 1306! Between Eylenja and Athalassa, 1972, *G. Joscht* 6617!
6. Nicosia, 1927, *Rev. A. Huddle* 99! Kykko metokhi, 1952, *F. M. Probyn* 39! English School, Nicosia, 1959, *P. H. Oswald* 78! Cape Kormakiti, 1972, *P. M. Synge* s.n.!
7. St. Hilarion, 1900, *A. G. & M. E. Lascelles* s.n.! Dramia, 1938, *Lady Loch* 27! West of St. Hilarion, 1949, *Casey* 532! Kyrenia, 1956, *G. E. Atherton* 925! 972! Ayios Khrysostomos, 1941, *Davis* 2180! Vasilia, 1941, *Davis* 2192! also, 1956, *G. E. Atherton* 1101! and, 1971, *E. Hodgkin* s.n.! Above Kephalovryso, Kythrea, 1967, *Merton* in ARI 344! Near Klepini, 1972, *G. Joscht* 6701!
8. Apostolos Andreas, 1934, *Syngrassides* 1393! Koma tou Yialou, 1935, *Syngrassides* 705!

NOTES: Feinbrun (Pal. Journ. Bot., J. ser., 1: 378; 1940) has remarked on the puzzling ecology of this species, it being apparently a high alpine in Syria, while predominantly lowland in Cyprus, where only Kotschy seems to have seen it at altitudes above 3,500 feet. Despite this discrepancy, I can find no satisfactory morphological distinctions between Cyprus and Syrian plants; though additional material and information from Syrian localities is much needed.

HYACINTHUS ORIENTALIS *L.*, Sp. Plant., ed. 1, 316 (1753), the ancestor of the cultivated Hyacinth, a bulbous perennial with linear-lorate leaves and stout scapes bearing lax racemes of sweetly scented blue (sometimes pink or white) flowers, is grown in Cyprus gardens, and sometimes survives as a garden relict. It has been collected at Neapolis, Nicosia (1973, *P. Laukkonen*

98), and, probably on the strength of similar occurrences, is cited in Osorio-Tafall et Seraphim, List Vasc. Plants Cyprus, 20; 1973. It cannot, however, be regarded as an established member of the Cyprus flora, though indigenous in adjacent regions of Turkey, Syria and Palestine. *Hyacinthus* differs from *Hyacinthella* and *Bellevalia* in its large perianth with deeply cut, recurved lobes, and in its large subglobose, bluntly trigonous capsules; the fleshy leaves do not have the conspicuous, raised nervation of *Hyacinthella*. From *Muscari* (including *Leopoldia*) it is most obviously distinguished by its flared (not constricted) perianth-tube with deeply cut, recurved lobes.

17. MUSCARI *Mill.*

Gard. Dict. Abridg., ed. 4 (1754).
D. C. Stuart in The Lily Year Book, 29: 125–138 (1965), et in Notes Roy. Bot. Gard. Edinb., 30: 189–196 (1970).
F. Garbari et W. Greuter in Taxon, 19: 329–335 (1970).

Bulbous perennials, bulbs tunicate; leaves all basal, linear or lorate, with or without a hyaline, membranous margin, nervation not conspicuously elevated, scape simple; inflorescence a cylindrical, ovoid or thyrsoid raceme or spike; bracts very small, sometimes absent; apical flowers often sterile and distinct from the lower, fertile flowers; perianth globose, urceolate or shortly cylindrical, the tube generally constricted at its apex, with small, recurved lobes, uniformly blue or violet, or the fertile flowers sometimes brownish or yellowish; stamens in 1 series or in 2 obscure or well-defined series, inserted towards middle of corolla tube; anthers included, introrse; ovary sessile; ovules 2 in each of the 3 loculi. Capsule conspicuously triquetrous; seeds subglobose with a rugose-reticulate, lustrous black testa.

About 40 species in S. Europe, the Mediterranean region and western Asia.

Fertile flowers brownish or yellowish, distinct from the blue or purple sterile flowers; raceme
    8–30 cm. long -   -   -   -   -   -   -   -   - **1. M. comosum**
Fertile flowers blue or violet, similar to the sterile but usually darker in colour; raceme less than
    4 cm. long:
    Flowers appearing in autumn (Oct.–Dec.); perianth sky-blue, oblong-campanulate, scarcely
        constricted at mouth - - - - - - - - - **4. M. parviflorum**
    Flowers appearing in winter or spring (Dec.–March); perianth dark blue or violet:
        Perianth-lobes white or whitish; perianth-tube constricted at mouth; leaves 1·5–6 mm.
            wide - - - - - - - - - - **2. M. neglectum**
        Perianth-lobes concolorous with tube, dark indigo or violet; tube not (or indistinctly)
            constricted at apex; leaves 1–3 mm. wide - - - - **3. M. inconstrictum**

SUBGENUS 1. **Leopoldia** *(Parl.) Rouy* Fertile flowers tubular or elongate-urceolate, brownish or yellowish, sterile flowers distinct, generally blue, purple or violet, often in a terminal coma; stamens in 2 series; seeds with a strongly reticulate-rugose testa.

1. **M. comosum** *(L.) Mill.*, Gard. Dict., ed. 8, no. 2 (1768); Poech, Enum. Plant. Ins. Cypr., 8 (1842); Boiss., Fl. Orient., 5: 291 (1884); Holmboe, Veg. Cypr., 52 (1914); Post, Fl. Pal., ed. 2, 2: 648 (1933); Lindberg f., Iter Cypr., 10 (1946); Osorio-Tafall et Seraphim, List Vasc. Plants Cyprus, 20 (1973).
    Hyacinthus comosus L., Sp. Plant., ed. 1, 318 (1753); Sibth. et Sm., Fl. Graec. Prodr., 1: 238 (1809).
    Bellevalia comosa (L.) Kunth, Enum. Plant., 4: 306 (1843); Unger et Kotschy, Die Insel Cypern, 194 (1865).
    Leopoldia comosa (L.) Parl., Fl. Palerm., 438 (1845); B. Bentzer in Bot. Notiser, 126: 75 (1973).
    [*Muscari pinardi* (non Boiss.) Post in Mém. Herb. Boiss., 18: 101 (1900); H. Stuart Thompson in Journ. Bot., 44: 839 (1906); Holmboe, Veg. Cypr., 52 (1914); Osorio-Tafall et Seraphim, List Vasc. Plants Cyprus, 20 (1973).]

[*M. creticum* (non Vierh.) Vierh. in Oesterr. Bot. Zeitschr., 66: 169 (1916) pro parte quoad plant. cypr.]

TYPE: "*in* Galliae & Europae *australis agris*".

Bulb ovoid, 4–5 cm. long, 3–4 cm. wide, tunic papery, fuscous externally, pinkish internally, leaves (1–) 2–4, linear, glaucous-green, 15–40 cm. long, 0·8–2 cm. wide, base long-sheathing, apex acuminate, margins hyaline-membranous, scabridulous or shortly ciliate; scape solitary, 15–33 cm. long; inflorescence lax, cylindrical, 8–30 cm. long, 2–4 cm. diam., bearing 10–50 (or more) fertile flowers; bracts minute, whitish, membranous, less than 1·5 mm. long; pedicels of fertile flowers patent, 5–12 mm. long; sterile flowers violet-blue, forming a conspicuous terminal tuft or coma, pedicels often 2 cm. long, perianth subglobose or oblong-globose; fertile flowers brownish, or yellowish at apex and base, perianth-tube shortly cylindrical, 5–6 mm. long, 4 mm. diam., distinctly constricted at apex, lobes very short, recurved, pallid, ovate-deltoid, about 1·5 mm. long and almost as wide at base; stamens inserted in 2 series near the middle of the perianth-tube; free part of filaments about 1·5 mm. long; anthers shortly oblong, 1–2 mm. long, 1–1·2 mm. wide, violet-blue; ovary sessile, ovate-triquetrous, about 2 mm. long, 2 mm. wide at base; style about 2 mm. long; stigma truncate. Capsule broadly ovoid-triquetrous, about 8 mm. long, 3 mm. wide, valves pale brown, distinctly nerved, apex rounded or slightly emarginate; seeds subglobose, about 2·5 mm. diam., testa black, rugulose-reticulate, minutely pitted.

HAB.: Cultivated fields; occasionally in garigue on rocky coastal pastures or on hillsides; sea-level to 6,400 ft. alt.; fl. March–May (to Aug.).

DISTR.: Divisions 1–8, often common. Central and southern Europe and Mediterranean region east to Transcaucasia and Iran.

1. Ayios Yeoryios (Akamas), 1962, *Meikle*! Polis area, 1962, *Meikle*! Between Prodhromi and Androlikou, 1978, *Chesterman* 52!
2. Prodhromos, 1862, *Kotschy* 892! Khionistra, 1905, *Holmboe* 988; also, 1937, *Kennedy* 205! 1939, *Lindberg f.* s.n.! and, 1941, *Davis* 3500! Kryos Potamos, 2,900 ft. alt., 1937, *Kennedy* 200! 203! Evrykhou, 1935, *Syngrassides* 823! Between Saïttas and Kalokhorio, 1978, *Chesterman* 71! Platres, 1937, *Kennedy* 202! 206! Vouni, 1970, *I. M. Hecker* 53!
3. Kouklia, 1905, *Holmboe* 371; Lefkara, 1941, *Davis* 2737! Limassol, 1963, *J. B. Suart* 74! Between Apsiou and Phasoula, 1978, *Chesterman* 70! also, near Apsiou, 1979, *Edmondson & McClintock* E2703!
4. Larnaca, 1862, *Kotschy* 184; Cape Greco, 1958, *N. Macdonald* 3!
5. Kythrea, 1862, *Kotschy* 351; Pera, 1949, *Casey* 301!
6. Nicosia, 1927, *Rev. A. Huddle* 100! Dhiorios, 1962, *Meikle*!
7. N. side Pentadaktylos, 1862, *Kotschy*; Melandrina, 1862, *Kotschy* 510; Kyrenia, 1941, *Casey* 443! also, 1956, *G. E. Atherton* 1236!
8. Akradhes, 1962, *Meikle*!

NOTES: One specimen (*Casey* 301, from Pera; mycol. herb. K!) is heavily infected with *Ustilago vaillantii*, and the deformed flowers bear a superficial resemblance to those of *Bellevalia trifoliata*.

SUBGENUS 2. **Botryanthus** (*Kunth*) *Rouy* Fertile flowers globose or oblong-urceolate, bright or dark blue or purplish-blue, sterile flowers few, smaller and paler than the fertile flowers, sessile or shortly pedicellate; seeds almost smooth or with a weakly reticulate-rugose testa.

2. **M. neglectum** *Guss. ex Ten.*, Syll. Fl. Neap., App. 5, 13 (1842); Boiss., Fl. Orient., 5: 296 (1882); Post, Fl. Pal., ed. 2, 2: 650 (1933); Osorio-Tafall et Seraphim, List Vasc. Plants Cyprus, 20 (1973).

M. atlanticum Boiss. et Reuter, Pugill. Plant. Nov., 114 (1852); Osorio-Tafall et Seraphim, List Vasc. Plants Cyprus, 20 (1973).

[*M. racemosum* (non Mill.) Sintenis in Oesterr. Bot. Zeitschr., 31: 226, 229, 255 (1881); Boiss., Fl. Orient., 5: 295 (1882); Post, Fl. Pal., ed. 2, 2: 649 (1933); Osorio-Tafall et Seraphim, List Vasc. Plants Cyprus, 20 (1973).]

[*M. pulchellum* (vix Heldr. et Sart. ex Boiss.) A. K. Jackson in Kew Bull., 1937: 345

(1937); Feinbrun in Bull. Res. Council Israel, 8D: 170 (1960); Osorio-Tafall et Seraphim, List Vasc. Plants Cyprus, 20 (1973).]

TYPE: Italy; "In cultis passim, tam in elatioribus quam in demissis, *da Castel di Sangro a Reggio*" (? NAP).

Bulb ovoid, 1·5–2·5 cm. long, 1·2–2 cm. wide, tunic dark brown, membranous; roots slender; bulbils sometimes present; leaves 3–6 or more, linear or slightly dilated towards apex, concave, dark green, 4–20 (–30) cm. long, 1·5–6 mm. wide, tapering to a non-sheathing base, apex somewhat cucullate, obtuse or subacute; scapes generally solitary, erect, 6–20 cm. long, often tinged bronze or purplish in the upper part; inflorescence ovoid or shortly cylindrical, 1·5–4 cm. long, 0·8–1·8 cm. wide, rather dense at first, becoming lax with age, 12–40 (or more)-flowered; pedicels slender, bluish, sometimes very short, or up to 4–5 mm. long, patent, often recurved with age; bracts white, minute, lacerate, seldom more than 1 mm. long; perianth of fertile flowers oblong-ovoid, dark blue, 3–5 mm. long, 2–4 mm. wide, tube constricted towards apex, lobes white, semi-circular, less than 1 mm. long, 1–1·3 mm. wide at base; stamens included, inserted in 2 series near the middle of the perianth-tube; free part of filaments about 0·6 mm. long; anthers oblong, dark violet, about 1·5 mm. long, 1 mm. wide; ovary sessile, subglobose-triquetrous, about 2 mm. long and almost as wide; style 2 mm. long; stigma truncate; sterile flowers pale violet-blue, often narrowly oblong and closed at the apex. Capsule suborbicular-triquetrous, 5–7 mm. diam., valves pale brown, distinctly nerved, apex often conspicuously emarginate; seeds irregularly compressed-subglobose; about 2 mm. diam.; testa black, irregularly rugulose.

HAB.: Cultivated fields, roadsides, grassy banks; occasionally in garigue on calcareous (? and igneous) hillsides; 150–2,000 ft. alt.; fl. Febr.–March.

DISTR.: Divisions 1–6, locally common. Widespread in central and southern Europe and the Mediterranean region and east to Iran.

1. Between Polemi and junction of Paphos–Polis road, 1978, *Chesterman* 56!
2. Saïttas, 1932, *Syngrassides* 347!
3. Kouklia, 1953, *Kennedy* 1774! Lefkara, 1953, *Kennedy* 1772! Above Ypsonas, 1966, *Ann Matthews* 2! 1·5 miles east of Kophinou, 1978, *Chesterman* 45!
4. Stavrovouni (mixed with *M. inconstrictum* Rechinger f.), 1880, *Sintenis & Rigo* 941! also, 1948, *Mavromoustakis* s.n.! E. of Dhekelia, 1981, *Hewer* 4717!
5. Near Athienou, 1880, *Sintenis & Rigo* 171!
6. Nicosia, 1901, *A. G. & M. E. Lascelles* s.n.! also, 1959, *P. H. Oswald* 60! Yerolakkos, 1932, *Syngrassides* 67! Morphou, 1951, *Merton* 144!

3. M. inconstrictum *Rechinger f.* in Arkiv för Bot., ser. 2, 2: 314 (1952); Stuart in The Lily Year Book, 29: 127 (1966); Osorio-Tafall et Seraphim, List Vasc. Plants Cyprus, 20 (1973).
    M. racemosum Mill. var. *brachyanthum* Boiss., Fl. Orient., 5: 295 (1882) pro parte quoad Kotschy Suppl. 405; Holmboe, Veg. Cypr., 52 (1914); Seligman in Quart. Bull. Alpine Gard. Soc., 20: 234 (1952).
    [*Botryanthus parviflorus* (non (Desf.) Kunth) Kotschy in Unger et Kotschy, Die Insel Cypern, 193 (1865).]
    [*Muscari heldreichii* (non Boiss.) Sintenis in Oesterr. Bot. Zeitschr., 31: 226, 229 (1881).]
    [*M. commutatum* (non Guss.) Meikle in Elektra Megaw, Wild Flowers of Cyprus, 17 (1973) excl. icon.]

TYPE: Arabia Petraea, Petra, Felsen, 1,000 m., 27.II.1937, *Dinsmore* 10371 (S).

Bulb about 1·5 cm. long, 1–1·3 cm. wide, tunic membranous, fuscous; leaves 2–4, very narrow, almost filiform, 5–22 cm. long, 0·1–0·3 cm. wide, dull green above, slightly reddish towards base, tapering slightly downwards from a subacute apex; inflorescences 1–2 from each bulb; scape erect, 5–17 cm. long; raceme rather lax, 1–8 cm. long, 1–1·5 cm. diam., fertile flowers 3–15 (or more); bracts whitish, membranous, lacerate, less than 1 mm. long; pedicels slender, up to 2 mm. long, often recurved; perianth oblong-campanulate, 3–4 mm. long, 2–2·5 mm. wide, dark indigo-blue, the tube and lobes concolorous, lobes semicircular, mucronulate, erect, about

0·4 mm. long, 1 mm. wide; sterile flowers smaller, pale blue; stamens inserted about middle of perianth-tube; filaments purplish, about 0·4 mm. long; anthers violet, broadly oblong, about 1 mm. long, 0·8 mm. wide; ovary subglobose-triquetrous, about 6·5 mm. diam., minutely papillose; style straight, about 1·3 mm. long; stigma subtruncate. Capsule sharply triquetrous, valves suborbicular, about 5·5 mm. diam., pale rosy-brown, apex shortly emarginate; seeds compressed-subglobose, about 2 mm. diam., testa shining black, almost smooth.

HAB.: In dry rocky pastures, or in batha or garigue on calcareous or igneous hillsides; sea-level to 4,500 ft. alt.; fl. Dec.–March.

DISTR.: Divisions 1–7. S. Turkey, Syria, Palestine east to Iraq and Iran.

1. Smyies, 1962, *Meikle* s.n.!
2. Troodhitissa Monastery, 1862, *Kotschy* 405! also, 1971, *E. Hodgkin* K.72.10247! Kryos Potamos, 2,800–3,000 ft. alt., 1937, *Kennedy* 208! 209! 210! Xerokolymbos, 4,300 ft. alt., 1937, *Kennedy* 211! Platres, 3,800 ft. alt., *Kennedy* 212! Above Lythrodhonda, 1938, *Lady Loch* 26!
3. Stavrovouni, 1880, *Sintenis & Rigo* 941 partly! Yerasa, 1947, *Mavromoustakis* s.n.! Episkopi, 1964, *J. B. Suart* 173!
4. Near Larnaca, 1880, *Sintenis & Rigo* 941 partly!
5. Near Kythrea, 1880, *Sintenis & Rigo* 172! Athalassa Forest, 1959, *P. H. Oswald* 52!
6. Nicosia airport, 1951, 1954, *Merton* 57! 1381! S. of Ayia Irini, 1957, *Merton* 2846! Mitsero, 1957, *Poore* 46! Nicosia, 1973, *P. Laukkonen* 96a!
7. Kyrenia Pass, 1934, *Syngrassides* 969! Orga, 1941, *Davis* 2114! Between Halevga and Kharcha, 1941, *Davis* 2154! Near Karmi, 1949, *Casey* 216! Pentadaktylos, 1950, *Casey* 980! Near Aghirdha, 1954, *Merton* 1824! Kyrenia, 1956, *G. E. Atherton* 996! 1031! Keurmurju, 1959, *P. H. Oswald* 38! Klepini, 1971, *E. Hodgkin* s.n.! W. of Bellapais, 1973, *P. Laukkonen* 96!

NOTES: The Cyprus plants are, on the whole, smaller in all respects than those from mainland Asia; it is questionable, however, if the difference can be used as the basis for a taxonomic distinction.

SUBGENUS 3. **Pseudomuscari** *Stuart* Fertile flowers oblong-campanulate, the perianth at most slightly constricted above; perianth-lobes with a darker median stripe; pedicels ascending; seeds with a rugulose testa.

4. **M. parviflorum** *Desf.*, Fl. Atlant., 6: 309 (1798); Boiss., Fl. Orient., 5: 299 (1882); Post in Mém. Herb. Boiss., 18: 101 (1900); H. Stuart Thompson in Journ. Bot., 44: 339 (1906); Holmboe, Veg. Cypr., 52 (1914); Post, Fl. Pal., ed. 2, 2: 650 (1933); Osorio-Tafall et Seraphim, List Vasc. Plants Cyprus, 20 (1973).

? Botryanthus parviflorus Kotschy in Unger et Kotschy, Die Insel Cypern, 193 (1868).
TYPE: Tunis; "ad maris litora prope Carthaginem eversam" (P).

Bulb ovoid, 10–25 mm. long, 8–20 mm. wide, tunic membranous, rather pale fuscous-brown, leaves 3–4, narrowly linear, 7–21 cm. long, 1·5–2·5 mm. wide, tapering from an acute or subacute apex to a slender base; scape usually solitary, erect, 7–23 cm. long; flowers 8–15 in a lax raceme rarely much more than 3·5 cm. long, 1·5 cm. diam., bracts whitish membranous, subulate, less than 1 mm. long, often with a larger median division sited some distance below the pedicel-base, accompanied by 1 or 2 minute divisions sited laterally at the base of the pedicel; pedicels ascending, 2·5–5 mm. long; perianth oblong-campanulate, sky-blue, 3–4 mm. long, 2·5–3 mm. wide, lobes deltoid, about 1·2 mm. long, and about as wide at the base, apex shortly recurved; stamens inserted in 2 distinct series near the middle and apex of the perianth-tube; filaments about 1 mm. long; anthers blue, shortly oblong, about 1 mm. long, 0·8 mm. wide; ovary sessile, subglobose-triquetrous, about 2·5–3 mm. diam., minutely papillose; style 2 mm. long; stigma truncate or subcapitate. Capsule sharply triquetrous, valves pale greyish-brown, suborbicular, 4–6 mm. diam., indistinctly nerved and minutely reticulate; seeds compressed-subglobose, about 2 mm. diam., testa black, distinctly rugulose.

HAB.: Cultivated and fallow fields; grassy hillsides in garigue; sea-level to 600 ft. alt.; fl. Oct.–Dec.

DISTR.: Divisions 1, 3, 6–8. Mediterranean region from Balearics and Italy eastwards.

1. Common about Paphos, 1981, *Meikle*!
3. Limassol, 1899, *Post*.
6. Makhedonitissa Monastery near Nicosia, 1935, *Syngrassides* 369! English School, Nicosia, 1958, *P. H. Oswald* 27!
7. Kyrenia, 1936–1968, *Kennedy* 207! *Lady Loch* 11! *Casey* 1060! *G. E. Atherton* 758! *I. M. Hecker* 16! Akanthou, 1940, *Davis* 2021! and, 1970, *A. Genneou* in ARI 1621! Vasilia, 1940, *Davis* 2006! Lapithos, 1955, *G. E. Atherton* 673!
8. Karpas, 1929, *C. B. Ussher* 32!

NOTES: Kotschy's record (Die Insel Cypern, 193) for *Botryanthus parviflorus* (Desf.) Kunth has been referred to *M. racemosum* Mill. var. *brachyanthum* Boiss., which, so far as concerns Cyprus references, is here regarded as synonymous with *M. inconstrictum* Rechinger f.

98. JUNCACEAE

F. Buchenau in Engl., Pflanzenr., 25 (IV.36): 1–284 (1906).
S. Snogerup in Rechinger f., Fl. Iranica, 75: 1–35 (1971).

Perennial rhizomatous herbs or annuals; leaves linear, flat, canaliculate, compressed or terete, all basal or extending up the stems, sometimes reduced to cataphylls; inflorescence usually compound, with bracts and bracteoles; flowers small, actinomorphic, hermaphrodite or rarely unisexual (dioecious); perianth-segments 6 in 2 series, often scarious and persistent, generally greenish, brownish or fuscous; aestivation imbricate; stamens 6, in 2 series, or the inner series sometimes wanting, inserted at the base of the perianth-segments; filaments free; anthers erect, basifixed, introrse; ovary superior, 3-carpellate, 1- or 3-locular; ovules 3 to many, ascending, anatropous; style simple, long or short; stigmas 3, opposite the outer perianth-segments. Fruit a loculicidal capsule; seeds generally small, embryo straight, small, situated at the base of the endosperm near the micropyle.

Nine genera with a cosmopolitan distribution, but mostly confined to high altitudes in the tropics.

1. JUNCUS *L.*

Sp. Plant., ed. 1, 325 (1753).
Gen. Plant., ed. 5, 152 (1754).

Glabrous perennial or annual herbs; leaves various, often with auriculate sheaths, lamina linear, flat or canaliculate, or resembling the stems and compressed or terete; stems leafy or leafless, sometimes filled with spongy, continuous or interrupted pith; inflorescence terminal or apparently lateral, cymose, or flowers occasionally few or solitary; bracteoles often present; perianth-segments persistent, scarious; stamens generally 6, sometimes fewer. Capsule 1- or 3-locular; seeds small, numerous, sometimes with an appendage; testa often conspicuously sculptured, frequently becoming mucilaginous when wetted.

About 200 species distributed throughout the world, but mostly in cold or temperate regions, usually in marshy or moist situations.

Flowering stems without cauline leaves; inflorescence apparently lateral, exceeded by a subulate bract:

Small annuals; leaves and flowering stems distinct; inflorescence consisting of one or more
 dense heads - - - - - - - - - - - **12. J. capitatus**
Robust rhizomatous or caespitose perennials; leaves and flowering stems closely similar:
 Stems with interrupted pith, glaucous, with 12–18 prominent longitudinal ridges, not very
  rigid or pungent; perianth-segments narrow, lanceolate-acuminate - **6. J. inflexus**
 Stems with continuous pith, not glaucous or prominently ridged, very rigid and pungent:
  Flowers greenish or straw-coloured; perianth-segments narrowly ovate- or elliptic-
   acuminate, the inner without hyaline apical auricles:
   Flowers in dense glomerules; anthers a little longer than filaments; capsule equalling
    or shorter than perianth-segments - - - - - - **1. J. maritimus**
   Flowers not in dense glomerules; anthers much longer than filaments; capsule
    distinctly exceeding perianth-segments - - - - - **2. J. rigidus**
  Flowers brownish or fuscous; perianth-segments ovate-apiculate, the inner with hyaline
   apical auricles:
   Capsule 3 mm. or more wide, clearly exceeding the persistent perianth; stems very
    rigid and pungent, 2–5 mm. diam. - - - - - - **3. J. acutus**
   Capsule less than 3 mm. wide, scarcely exceeding the perianth:
    Capsule acute, 1·5–2 mm. wide; stems slender, 1·5–2 mm. diam.
                5. J. heldreichianus
    Capsule obtuse, apiculate, 2–2·5 mm. wide; stems 2·5–3 mm. diam. **4. J. littoralis**
Flowering stems with cauline leaves; inflorescence apparently terminal, not exceeded by
 bracts:
Leaves without transverse walls (not septate):
 Robust rhizomatous perennials; stems 30–120 cm. high - - - - **7. J. subulatus**
 Slender annuals; stems generally less than 30 cm. high:
  Capsule broadly oblong-ovoid or subglobose, 2 mm. long, 1·5 mm. wide; perianth-
   segments 2–2·5 mm. long - - - - - - - **8. J. sphaerocarpus**
  Capsule oblong-ellipsoid, 3–4·5 mm. long; perianth-segments 3·5–6 mm. long:
   Inner perianth-segments obtuse or shortly acute, generally shorter than the capsule
                   11. J. ambiguus
   Inner perianth-segments acuminate, exceeding the capsule:
    Flowers solitary (or occasionally 2–3 together) spaced along the inflorescence-
     branches; seeds distinctly striate-reticulate - - - - **9. J. bufonius**
    Flowers generally in dense, fan-shaped glomerules; seeds almost smooth
                 10. J. hybridus
Leaves with transverse walls (septate):
 Perianth-segments 3–4 mm. long; flowers in dense echinate glomerules; capsule clearly
  exceeding perianth-segments, distinctly rostrate **13. J. fontanesii** ssp. **pyramidatus**
 Perianth-segments about 2·5 mm. long; flowers in dense, rounded glomerules; capsule
  slightly exceeding perianth-segments, not rostrate - - - **14. J. articulatus**

SUBGENUS 1. **Juncus** Robust rhizomatous perennials; flowering stems
leafy at the base; leaves terete, rigid, pungent, pith-filled, not transversely
septate, resembling the stems but dilated and sheathing at the base; 2 lowest
bracts foliaceous, often pungent; bracteoles wanting. Seeds with
appendages.

1. J. maritimus *Lam.*, Encycl. Méth., 3: 264 (1789); Boiss., Fl. Orient., 5: 354 (1882) pro parte;
 Post, Fl. Pal., ed. 2, 2: 666 (1933) pro parte; Osorio-Tafall et Seraphim, List Vasc. Plants
 Cyprus, 18 (1973).
 J. acutus L. var. *β* L., Sp. Plant., ed. 1, 325 (1753).
 TYPE: "croit naturellement en Angleterre, en France, etc. aux lieux maritimes et maré-
cageux" (P).

Robust caespitose perennial; rhizomes usually rather crowded, often
much branched; intravaginal shoots absent; stems 50–100 cm. high,
generally 1·5–2 mm. diam., erect, rigid, with 40–50 or more longitudinal
ridges; pith not interrupted; cataphylls 3–4, acute, dull brownish or
purplish-brown; leaves 2–4, sub-basal, similar to the stems but usually
shorter, pungent, dilated and sheathing towards the base; inflorescence
usually lax, much branched, subsecund, 5–14 cm. long, 2–8 cm. wide;
lowermost bracts foliaceous, terete, pungent, the outer up to 50 cm. long, the
inner seldom exceeding 5 cm.; upper bracts becoming progressively smaller
towards tips of inflorescence-branches, scarious, carinate, lanceolate-
acuminate, usually less than 10 cm. long, 3 mm. wide; flowers greenish or

straw-coloured, in dense, terminal, 3–6 (sometimes more)-flowered glomerules; perianth-segments narrowly elliptic-acuminate, carinate, about 3
mm. long, 1 mm. wide; filaments flattened, about 0·8 mm. long; anthers
narrowly oblong, about 1 mm. long, 0·5 mm. wide; style about 0·8 mm. long,
deciduous, leaving a short, apiculate style-base; stigmas 0·8 mm. long.
Capsule ovoid-prismatic, about 2·5 mm. long, 2 mm. wide, shining light
brown; seeds irregularly fusiform, about 0·8 mm. long, 0·2 mm. wide, shortly
membranous-appendiculate at either extremity; testa dark brown,
minutely rugulose-reticulate.

HAB.: Marshy ground by the sea; near sea-level; fl. July–Aug.

DISTR.: Division 3. Western and Central Europe, Mediterranean region and eastwards to W.
Iran.

3. Asomatos marshes near Limassol Salt Lake, 1939, *Mavromoustakis* 138 !

NOTES: The only reliable record, identified by Snogerup in 1968. Most authors (Kotschy,
Sintenis, Holmboe, etc.) have failed to distinguish *J. maritimus* from the following (*J. rigidus*
Desf.), which is evidently much commoner in Cyprus, and the great majority of records for *J.
maritimus* are referable to *J. rigidus*. It is not unlikely that *J. maritimus* is, however, commoner
than would appear; because of their late flowering and fruiting, Rushes in this group are
inadequately collected.

2. J. rigidus *Desf.*, Fl. Atlant., 1: 312 (1798): Snogerup in Rechinger f., Fl. Iran., 75: 4 (1971).
    J. maritimus Lam. var. *arabicus* Aschers. et Buchenau in Boiss., Fl. Orient., 5: 354
    (1882); Post, Fl. Pal., ed. 2, 2: 666 (1933).
    [*J. maritimus* (non Lam.) Kotschy in Unger et Kotschy, Die Insel Cypern, 190 (1865);
    Sintenis in Oesterr. Bot. Zeitschr., 31: 184, 256 (1881); Holmboe, Veg. Cypr., 44 (1914);
    Lindberg f., Iter Cypr., 10, et auct. mult.]
TYPE: Not localized ("in arenis ad maris littora") (P).

Robust caespitose perennial, rhizomes crowded, often much branched;
intravaginal shoots absent; stems 50–150 cm. high, 2·5–5 mm. diam., erect,
rigid, with 40–50 or more longitudinal ridges; pith not interrupted;
cataphylls 3–4, acute, shining mid-brown, the uppermost commonly
produced into a terete lamina; leaves 2–4, sub-basal, similar to the stems but
usually shorter, pungent, dilated and sheathing towards the base;
inflorescence subsecund, much-branched, commonly rather fastigiate, with
a long central axis and relatively short, erect lateral branches, up to 22 cm.
long, usually less than 5 cm. wide, lowermost bracts foliaceous, terete,
pungent, the outer up to 30 cm. long, the inner seldom exceeding 5 cm.;
upper bracts lanceolate-cuspidate, carinate, scarious, usually less than 5
mm. long and 3 mm. wide; flowers greenish or straw-coloured, solitary or 2–3
together, spaced along the branches, but seldom forming obvious
glomerules; perianth-segments narrowly ovate-acuminate, shortly mucronate, about 3 mm. long, 1 mm. wide; filaments very short, less than 0·2 mm.
long; anthers large, narrowly oblong, 2·5–3 mm. long, 0·8 mm. wide; ovary
narrowly ovoid-trigonous, about 1·5 mm. long, 0·8 mm. wide; style about 0·5
mm. long, deciduous, leaving a very short, apiculate style-base; stigmas
spirally twisted, about 0·8–1 mm. long. Capsule narrowly ovoid-prismatic,
about 4·5 mm. long, 2·5 mm. wide, shining light-brown; seeds irregularly
fusiform, about 0·6–0·8 mm. long, 0·2 mm. wide, with long, membranous,
tail-like appendages at either extremity; testa rich brown, minutely
rugulose-reticulate.

HAB.: Marshy, saline ground near the sea or occasionally inland; dried-up river beds; sea-level
to 500 ft. alt.; fl. June–Aug.

DISTR.: Divisions 4–8. Mediterranean region from Sardinia eastwards, extending further east
to Afghanistan and Pakistan.

4. Larnaca, 1862, *Kotschy*; also, 1880, *Sintenis* 603 ! and, 1905, *Holmboe* 99; Famagusta, 1929,
    C. B. *Ussher* 51 ! Cape Pyla, 1934, *Syngrassides* 426 ! Prastio, 1935, *Nattrass* 670; Prastio to
    Kouklia, 1939, *Lindberg f.* s.n.; north of Yialias R., 1968, *Economides* in ARI 1195 !

5. By rivulet N. of Lefkoniko, 1970, *A. Hansen* 349!
6. Dhiorios Forest, 1954, *Merton* 1910!
7. Panagra–Orga road, 1967, *Economides* in ARI 1023!
8. Boghaz, 1937, *Syngrassides* 1649!

3. J. acutus *L.*, Sp. Plant., ed. 1, 325 (1753); Boiss., Fl. Orient., 5: 353 (1882) pro parte;
Holmboe, Veg. Cypr., 44 (1914); Post, Fl. Pal., ed. 2, 2: 666 (1833) pro parte; Osorio-Tafall
et Seraphim, List Vasc. Plants Cyprus, 19 (1973).
TYPE: "*in* Angliae, Galliae, Italiae *maritimis paludosis*".

Robust, caespitose perennial; rhizomes crowded, often branched;
intravaginal shoots often present; stems 50–180 cm. high, 2–5 mm. diam.,
erect, very rigid, almost smooth or with numerous minute longitudinal
ridges; pith not interrupted; cataphylls few, acute, shining dark brown,
leaves 2–5, sub-basal, very similar to, but often longer than the stems,
pungent and dilated towards the base into long, dark brown, lustrous
sheaths; inflorescence much branched, often rather compact, the main axis
scarcely longer than the lateral branches, not noticeably secund or
fastigiate; lowermost bracts foliaceous, terete, sharply pungent, the outer
3–30 cm. long, the inner usually less than 5 cm. long; upper bracts ovate-
acuminate, those towards the base of the inflorescence 5–6 mm. long, 3–4
mm. wide, concave, scarious, pale brown with a fuscous-brown base, those
towards the apex of the inflorescence becoming progressively smaller;
flowers not in distinct glomerules; perianth-segments broadly ovate,
concave, about 3 mm. long, 2·5 mm. wide, the exterior shortly apiculate,
hyaline towards base, fuscous-brown near apex, the interior obtuse, pallid,
with short rounded, hyaline auricles on either side of the apex; filaments
very short, 0·2–0·3 mm. long; anthers oblong, about 1·8 mm. long, 0·8 mm.
wide; ovary narrowly ovoid, about 1·5 mm. long, 0·8 mm. wide; style stout,
about 1 mm. long, deciduous, leaving a very short truncate-apiculate style-
base; stigmas spirally twisted, about 1 mm. long. Capsule broadly ovoid-
ellipsoid, obscurely prismatic or almost rounded, lustrous brown, about 5
mm. long, 3–3·5 mm. wide, clearly exceeding the persistent perianth; seeds
fusiform, about 0·8 mm. long, 0·3 mm. wide, with long, membranous tail-like
appendages at either extremity; testa rich brown, minutely rugulose-
reticulate.

HAB.: Moist ground or ditches by the sea, or occasionally inland; sea-level to 500 ft. alt.; fl.
March–May.

DISTR.: Divisions 1–4, 6–8, W. Europe, Mediterranean region east to the Caspian; Atlantic
Islands, and with allied forms in South Africa, N. & S. America, Australia, New Zealand.

1. Seashore below Baths of Aphrodite, 1962, *Meikle* 2232!
2. Lefka, 1951, *Merton* 466!
3. Limassol, 1905, *Holmboe* 661; see Notes.
4. Kouklia, 1905, *Holmboe* 394; see Notes.
6. Syrianokhori, 1935, *Syngrassides* 1330! Dhiorios Forest, 1954, *Merton* 1916!
7. Kyrenia, 1955, *G. E. Atherton* 81! 698!
8. Koma tou Yialou, 1950, *Chapman* 218!

NOTES: Probably not rare in Cyprus, but its spiky growths deter collectors. *Holmboe* 394 and
661 may be more correctly referable to *J. heldreichianus* Marsson ex Parl. with which *J. acutus*
has been repeatedly (and surprisingly) confused.

4. J. littoralis *C. A. Mey.*, Pflanz. Cauc., 34 (1831); Snogerup in Rechinger f., Fl. Iran., 75: 6
(1971).
J. tommasinii Parl., Fl. Ital., 2: 315 (1852).
J. acutus L. ssp. *tommasinii* (Parl.) Arcang., Comp. Fl. Ital., 715 (1882).
J. acutus L. var. *tommasinii* (Parl.) Buchenau in Engl., Bot. Jahrb., 12: 421 (1890).
J. acutus L. × *J. maritimus* Lam. ["*J. acuto-maritimus*"] Ledeb., Fl. Ross., 4: 234
(1852); Boiss., Fl. Orient., 5: 362 (1882).
TYPE: U.S.S.R.; Caspian, "In insula Sara" (LEN).

Very similar to *J. acutus* L. in habit and appearance, but generally less robust; stems usually less than 100 cm. high and 3 mm. diam.; leaves 2–6, erect, rigid, sharply pungent; pith not interrupted; inflorescence much branched, rather dense and many-flowered, but usually less compact than in *J. acutus*; lowermost bracts foliaceous, pungent, broadly sheathing towards the base, the outer 2–10 (–15) cm. long, often shorter than inflorescence, the inner frequently very short; perianth-segments dark fuscous-brown, the outer ovate-apiculate, concave, about 2·8–3 mm. long, 1·5 mm. wide, the inner about 2 mm. wide, with conspicuous, rounded, hyaline auricles on either side of the apex; filaments very short; anthers about 2 mm. long, 0·8 mm. wide; ovary and style as in *J. acutus*; stigmas spirally twisted, up to 2 mm. long. Capsule broadly ovoid-ellipsoid, about 3 mm. long, 2–2·5 mm. wide, obscurely trigonous, dark brown, lustrous, not much exceeding the persistent perianth, apex obtuse or somewhat pyramidal, crowned with a conspicuous, apiculate style-base; seeds as in *J. acutus*.

HAB.: Edge of salt-marsh; near sea-level; fl. April–May.

DISTR.: Division 3, rare. Mediterranean region, Black Sea and Caspian.

3. Akrotiri, at edge of salt-marsh wth *Juncus heldreichianus*, 1967, *Merton* in ARI 730!

5. J. heldreichianus *Marsson ex Parl.*, Fl. Ital., 2: 315 (1852); Snogerup in Rechinger f., Fl. Iran., 75: 6 (1971).

    J. acutus L. var. *heldreichianus* (Marsson ex Parl.) Hal., Consp. Fl. Graec., 13: 280 (1904).

    [*J. acutus* (non L.) Boiss., Fl. Orient., 5: 353 (1882) pro parte; H. Stuart Thompson in Journ. Bot., 43: 332 (1905).]

    TYPE: Greece; "In planitie maritima Atticae ad Phalerum, 1851, *Heldreich* 1993 (FI, G, K!).

Densely caespitose perennial; rhizomes crowded; intravaginal shoots often numerous; stems 20–50 cm. high, 1·5–2 mm. diam., erect, rigid, almost smooth or with very obscure longitudinal ridges; pith not interrupted; cataphylls few, shining dark brown; leaves 2–4, basal, very similar to the stems, but often rather shorter, sharply pungent and dilated towards the base into long, brown sheaths; inflorescence lax, irregularly branched, 3–10 cm. long, usually less than 3 cm. wide, often with an elongate axis, with several long-pedunculate basal branches arising from the lowermost bracts, the latter foliaceous, pungent, the outer 3–15 (rarely to 30) cm. long, the inner usually less than 3 cm. long, often shorter than the inflorescence but sometimes exceeding it, rich reddish-brown towards the sheathing base; upper bracts ovate-acuminate, carinate, reddish brown; those towards the base of the inflorescence branches 5–8 mm. long, 2–3 mm. wide, the upper progressively smaller; flowers 3–6 (or more) in distinct glomerules; perianth-segments reddish-brown with a paler midrib, the outer 3–3·5 mm. long, 1 mm. wide, sharply apiculate, carinate, the inner about 3 mm. long, 1 mm. wide, less strongly carinate, with conspicuous, rounded hyaline apical auricles; filaments very short; anthers oblong, 1·8 mm. long, 0·8 mm. wide; ovary broadly ovoid, about 1·3 mm. long, 1 mm. wide; style 1·3 mm. long; stigmas spirally twisted, about 2 mm. long. Capsule ovoid-prismatic, acute, 2–2·5 mm. long, 1·5–2 mm. wide, dark shining brown, minutely verruculose, equalling or shorter than the persistent, erect perianth-segments; seeds broadly fusiform, about 0·8 mm. long, 0·4 mm. wide, with short membranous appendages at one or both extremities; testa dark shining brown, minutely rugulose-reticulate.

HAB.: In salt-marshes or brackish swamps; in seasonally inundated ground or by springs and streams; sea-level to 5,500 ft. alt.; fl. April–Aug.

DISTR.: Divisions 2–7. Greece, Crete, Aegean area, Turkey.

2. Moniatis river, 3,600 ft. alt.; 1938, *Kennedy* 1083! Troödos area, 1950, *F. M. Probyn* s.n.!

Prodhromos, 1961, *D. P. Young* 2356! Kryos Potamos, 5,000 ft. alt., 1963, *D. P. Young* 7839! Livadhi tou Pasha, 1966, *Merton* in ARI 63!
3. Akrotiri, 1967, *Merton* in ARI 731!
4. Larnaca aerodrome, 1950, *Chapman* 693! Akhyritou, 1962, *Meikle* 2613!
5. Kythrea, 1880, *Sintenis & Rigo* 558! Dhikomo marshes, 1936, *Syngrassides* 1183! Mia Milea to Lakovounara, 1950, *Chapman* 474!
6. Near Myrtou, 1932, *Syngrassides* 84!
7. Kyrenia, 1949, *Casey* 772! also, 1955, *G. E. Atherton* 190! Bellapais, 1956, *G. E. Atherton* 1321!

SUBGENUS 2. **Genuini** *Buchenau* Rhizomatous perennials; flowering shoots leafless; sterile shoots with a single leaf similar to the flowering stem; leaves terete, pith-filled, not transversely septate; lowermost bracts foliaceous; bracteoles present. Seeds usually without appendages.

6. J. inflexus *L.*, Sp. Plant., ed. 1, 326 (1753); Snogerup in Rechinger f., Fl. Iran., 75: 8 (1971); Osorio-Tafall et Seraphim, List. Vasc. Plants Cyprus, 18 (1973).
   J. glaucus [Ehrh., Beitr. Naturk., 6: 83 (1791) nom. nud.] Sibth., Fl. Oxon., 113 (1794); Boiss., Fl. Orient., 5: 353 (1882); Holmboe, Veg. Cypr., 44 (1914); Post, Fl. Pal., ed. 2, 2: 666 (1933); Lindberg f., Iter Cypr., 10 (1946); Rechinger f. in Arkiv för Bot., ser. 2, 1: 419 (1950).
   J. longicornis Bast. in Journ. de Bot., 3: 21 (1814).
   J. glaucus Ehrh. var. *longicornis* (Bast.) Grognot in Mém. Hist. Nat. Soc. Éduenne, 1: 198 (1865); Buchenau in Engl., Pflanzenr., 25 (IV.36): 134 (1906).
   J. glaucus Ehrh. f. *longicornis* (Bast.) Aschers. et Græbn., Syn. Mitteleurop. Fl., 2, Abth. 2: 449 (1904); Holmboe, Veg. Cypr., 44 (1914).
   J. inflexus L. var. *longicornis* (Bast.) Briq., Prodr. Fl. Corse, 1: 256 (1910).
   J. cyprius Lindberg f., Iter Cypr., 9 (1946); Osorio-Tafall et Seraphim, List Vasc. Plants Cyprus, 18 (1973).
   TYPE: "in Europa australi".

Robust, erect, rhizomatous perennial; rhizomes compact; intravaginal shoots absent; stems 50–100 (–150) cm. high, 1·5–2·5 mm. diam., rigid, glaucous, with 12–18 prominent longitudinal ridges; pith interrupted; cataphylls 3–4, acuminate or acute, shining dark brown; leaves solitary, very similar to the stems and arising from similar cataphylls, apex less rigid and pungent than in *Juncus* subgenus *Juncus*; inflorescence lax, secund, 3·5–12 cm. long, 1·5–5 cm. wide, effuse and widest towards apex, much branched, generally much shorter than the lowermost bract; flowers not in glomerules; bracts foliaceous, somewhat pungent, the outer often 20–30 cm. long, the inner often very short or almost wanting, with a filiform lamina, both with greenish-glaucous, sheathing bases; upper bracts narrowly ovate-lanceolate, acuminate-caudate, shortly carinate, greenish or brownish, 2–8 mm. long, 1–3 mm. wide; bracteoles ovate-acuminate, 1–1·5 mm. long, 0·5–0·8 mm. wide, membranous-scarious forming an involucre at the apex of the short (c. 2 mm. long) pedicels; perianth greenish or brownish; perianth-segments connivent, lanceolate-acuminate, concave or keeled, the outer about 3·5 mm. long, 1 mm. wide near base, the inner slightly shorter, broader and more obtuse, about 3 mm. long, 1·5 mm. wide; filaments about 0·8 mm. long; anthers narrowly oblong, about 0·8 mm. long, 0·2 mm. wide; ovary ovoid-ellipsoid, about 1 mm. long, 0·7 mm. wide, trigonous; style short, thick, about 0·5 mm. long, stigmas 0·8 mm. long, connivent and forming a cone. Capsule ovoid- or ellipsoid-prismatic, about 2 mm. long, 1·2–1·5 mm. wide, shining dark brown, crowned with an apiculate style-base, equalling or shorter than the persistent perianth-segments; seeds irregularly oblong, about 0·3 mm. long, 0·2 mm. wide, not, or scarcely, appendiculate; testa dark brown, finely reticulate, with close transverse ridges.

HAB.: In ditches and by streams; 500–5,100 ft. alt.; fl. June–Sept.

DISTR.: Divisions 2, 5, 7. Widespread in Europe, Africa and Asia; introduced elsewhere.

2. Troödos, 1912, *Haradjian* 475; Pano Panayia, 1913, *Haradjian* 917! Kannaviou, 1939, *Lindberg f.* s.n.! Platania, 1939, *Lindberg f.* s.n.! also, 1966, *Merton* in ARI 64! and, 1970, *A. Genneou* in ARI 1626! Kryos Potamos, 5,100 ft. alt., 1952, *Kennedy* 1767! Prodhromos, 1955, *G. E. Atherton* 609! and, 1961, *D. P. Young* 7855!

5. Kythrea, 1880, *Sintenis & Rigo* 674!

7. Between Pano Dhikomo and Ayios Khrysostomos Monastery, 1880, *Sintenis & Rigo* 603.

NOTES: To judge from type material, *Juncus cyprius* Lindberg f. has thicker stems and paler, less shining cataphylls than is usual in Cyprian *Juncus inflexus*; the Sintenis & Rigo specimen also cited as *J. cyprius* by Lindberg f. is, however, transitional to *J. inflexus*, and it is questionable if the two can be distinguished even as varieties.

SUBGENUS 3. **Subulati** *Buchenau* Rhizomatous perennials; leaves extending far up the flowering shoots, terete, pith-filled, not transversely septate; lowermost bracts inconspicuous; bracteoles present; seeds without obvious appendages.

7. J. subulatus *Forssk.*, Fl. Aegypt.–Arab., 75 (1775); Boiss., Fl. Orient., 5: 354 (1882); Holmboe, Veg. Cypr., 43 (1914); Post, Fl. Pal., ed. 2, 2: 666 (1933); Lindberg f., Iter Cypr., 10 (1946); Osorio-Tafall et Seraphim, List Vasc. Plants Cyprus, 19 (1973).

TYPE: Egypt; "Alexandriae" (C).

Erect, rhizomatous perennial; rhizomes elongate, branched, scaly; stems 30–120 cm. high, terete, with numerous, rather obscure longitudinal ridges; cataphylls 2–4, rather loose, shining reddish-brown; leaves terete, 30 cm. long or more, 5–6 mm. diam., rather soft with open, spongy pith, sheaths commonly extending almost to apex of stem, apex subulate-acuminate but scarcely pungent; inflorescence often rather fastigiate, with numerous erect, unequal branches, 5–20 cm. long, 2–6 cm. wide; lowermost bracts subulate-filiform, much shorter than the inflorescence, the outer 2–3 cm. long, with a long, sheathing base, the inner rather smaller, upper bracts scarious, pale brown, lanceolate-acuminate, carinate, about 4–8 mm. long, 2–3 mm. wide; bracteoles scarious, acuminate, about 2 mm. long, 1 mm. wide, forming a distinct involucre at the base of each flower; flowers sessile or shortly pedicellate, not in glomerules; perianth-segments erect, pale brown, scarious, the outer lanceolate, long-acuminate, about 3·2 mm. long, 0·8 mm. wide, strongly concave, the inner more shortly acuminate and less concave, about 2·8–3 mm. long, 0·8 mm. wide; filaments very short, scarcely 0·2 mm. long; anthers oblong, about 1·3 mm. long, 0·6 mm. wide; ovary ovoid-ellipsoid, reddish, about 1·3 mm. long, 0·8 mm. wide; style slender, about 1 mm. long; stigmas spirally twisted. Capsule ovoid-ellipsoid, dark, shining, reddish-brown, about 2·5 mm. long, 1·8 mm. wide, obscurely trigonous, surmounted by a very short, apiculate style-base; seeds bluntly oblong, about 0·6 mm. long, 0·2 mm. wide; testa shining brown, minutely rugulose; appendages obscure or wanting.

HAB.: Salt-marshes; sometimes in (? brackish) marshes inland; sea-level to 2,500 ft. alt.; fl. May–June.

DISTR.: Divisions 1, 3, 4, 7. Mediterranean region and eastwards to the Caspian.

1. Cape Arnauti, 1913, *Haradjian* 816!

3. Limassol Salt Lake, 1905, *Holmboe* 660.

4. Larnaca, 1939, *Lindberg f.* s.n.; Larnaca airport, 1952, *Merton* 1552! Larnaca Salt Lake, 1967, *Merton* in ARI 396! Famagusta, 1939, *Lindberg f.* s.n.!

7. Marshes near Pano Dhikomo, 1880, *Sintenis & Rigo* 603! Near Kyrenia, 1939, *Lindberg f.* s.n.; also, 1955, *Miss Mapple* 109! and *G. E. Atherton* 149! Kalogrea, 1932, *Syngrassides* 285! Near Lefkonico Pass, 1967, *Merton* in ARI 798!

SUBGENUS 4. **Poiophylli** *Buchenau* Annuals; leaves flat or compressed, cauline; inflorescence terminal; bracteoles present. Seeds without appendages.

8. J. sphaerocarpus *Nees* in Flora, 1: 521 (1818); Boiss., Fl. Orient., 5: 759 (1884); Holmboe, Veg. Cypr., 43 (1914); Lindberg f., Iter Cypr., 10 (1946); Osorio-Tafall et Seraphim, List Vasc. Plants Cyprus, 19 (1973).
TYPE: "auf der Röhn", *F. X. Heller*. Ascherson & Graebner note that Heller's original specimens came from W. Germany; Retzbach (on the R. Main).

Slender annual 3–15 cm. high; stems much branched at the base, erect, less than 1 mm. diam. leaves filiform, usually less than 30 mm. long, 0·5 mm. wide, subacute, with inconspicuous, submembranous basal sheaths; inflorescence very lax and slender, dichotomously branched; branches erect or erecto-patent, sometimes flexuous; lowermost bracts sometimes resembling the leaves, with a filiform lamina, or sometimes lanceolate, concave-carinate, membranous, about 4 mm. long, 0·5 mm. wide, upper bracts similar but smaller, often long-acuminate; flowers solitary, sessile or subsessile, subtended by an involucre of pallid, membranous bracts, each about 1 mm. long, 0·8 mm. wide, ovate-acuminate, concave; perianth-segments greenish, the outer narrowly lanceolate-acuminate, about 2·5 mm. long, 0·8 mm. wide, with a broad midrib and narrow hyaline-membranous margins, the inner less acuminate, about 2 mm. long, 0·8 mm. wide, with broader hyaline margins; stamens 6; filaments slender, about 0·5 mm. long, anthers narrowly oblong, about 0·5 mm. long, 0·1 mm. wide; ovary brown, ovoid, about 1·5 mm. long, 1 mm. wide, noticeably trigonous; style very short, 0·1–0·2 mm. long; stigmas deflexed, twisted, 0·5–0·8 mm. long. Capsule broadly oblong-ovoid, about 2 mm. long, 1·5 mm. wide, shining brown; seeds oblong, about 0·3 mm. long, 0·1 mm. wide; testa brown, minutely rugulose-reticulate; appendages wanting.

HAB.: Moist, trodden ground and marshy places; 600–5,000 ft. alt.; fl. April–July.

DISTR.: Divisions 2, 3. Central Europe and Mediterranean region east to Afghanistan.

2. Prodhromos, 1880, *Sintenis & Rigo* 874 (as *J. tenageia* L.f.)! also, 1905, *Holmboe* 895; Xerokolymbos R. ravine, 1961, *D. P. Young* 7442!
3. Kakoradjia near Stavrovouni, 1937, *Syngrassides* 1514!

9. J. bufonius *L.*, Sp. Plant., ed. 1, 328 (1753); Boiss., Fl. Orient., 5: 361 (1882) pro parte; Buchenau in Engl., Pflanzenr., 25 (IV. 36): 105 (1906) pro parte; Holmboe, Veg. Cypr., 43 (1914) pro parte; Post, Fl. Pal., ed. 2, 2: 668 (1933) pro parte; Lindberg f., Iter Cypr., 9 (1946); Snogerup in Rechinger f., Fl. Iran., 75: 16 (1971); Osorio-Tafall et Seraphim, List Vasc. Plants Cyprus, 19 (1973) pro parte; T. A. Cope et C. A. Stace in Watsonia, 12: 121 (1978).
TYPE: "*in* Europae *inundatis*".

Erect annual 7–25 cm. high; stems slender, usually much branched from the base, sometimes solitary; leaves 1–5, linear-filiform, 2–13 cm. long, 0·5–2 mm. wide, basal sheaths inconspicuous, with narrow membranous margins; inflorescence usually lax, repeatedly but irregularly dichotomous, with long, erecto-patent branches; flowers solitary or occasionally in clusters of 2–3, but not conspicuously glomerulate, rather widely spaced along the inflorescence branches, sessile or shortly pedicellate; lowermost bracts foliaceous, the outer to 5 cm. long or more, the inner rather shorter; upper bracts membranous-scarious, pallid, ovate, acute, 3–5 mm. long, 2–3 mm. wide; bracteoles 2 (–3) ovate, concave, acute, membranous, about 2 mm. long, 1·5 mm. wide; perianth greenish, segments suberect, the outer lanceolate-acuminate, about 4·5 mm. long, 1·2 mm. wide at base with a slender acumen, and rather wide membranous-hyaline margins, inner segments about 4 mm. long, 1·2 mm. wide, more shortly acuminate with a narrower midrib and wider hyaline margins; stamens 6; filaments very slender, 1–1·5 mm. long; anthers narrowly oblong, about 0·5 mm. long, 0·1 mm. wide; ovary reddish, oblong, ellipsoid, about 3 mm. long, 2 mm. wide, not noticeably trigonous; style very short, barely 0·5 mm. long; stigmas about 1 mm. long, deflexed, twisted. Capsule oblong-ellipsoid, 3–4 mm. long, about 1·8 mm. wide, shorter than the persistent perianth; shining greenish-brown, apex rounded or obtuse, very shortly mucronulate; seeds bluntly oblong-ellipsoid, about 0·4 mm. long, 0·2 mm. wide; testa rich brown, minutely reticulate with rather prominent longitudinal ridges; appendages wanting.

HAB.: Moist paths, roadsides, ditches and other damp open ground; 3,600–5,000 ft. alt.; fl. May–July.

DISTR.: Division 2, rather rare. Found as a weed throughout the world.

2. Prodhromos, 1880, *Sintenis & Rigo* 871 partly! also, 1905, *Holmboe* 894; Platania, 1939, *Lindberg f.* s.n.; also, 1961, *D. P. Young* 7411! Xerokolymbos, 4,300 ft. alt., 1961, *D. P. Young* 7443!

NOTES: Evidently restricted, in Cyprus, to altitudes above 3,000 ft., but no deliberate effort has been made to plot its distribution on the island, and it may be commoner than would appear.

10. J. hybridus *Brot.*, Fl. Lusit., 1: 513 (1804); Lindberg f., Iter Cypr., 10 (1946); Snogerup in Rechinger f., Fl. Iran., 75: 17 (1971); Osorio-Tafall et Seraphim, List Vasc. Plants Cyprus, 19 (1973); T. A. Cope et C. A. Stace in Watsonia, 12: 123 (1978).

   J. mutabilis Savi, Fl. Pisana, 1: 364 (1798) non Lam. (1789) nom. illeg.

   J. insulanus Viv., Fl. Cors., 5 (1824).

   J. bufonius L. var. *hybridus* (Brot.) Parl., Fl. Ital., 2: 353 (1852); Post, Fl. Pal., ed. 2, 2: 669 (1933).

   J. bufonius L. ssp. *hybridus* (Brot.) Arcang., Comp. Fl. Ital., 218 (1882).

   J. bufonius L. f. *mutabilis* (Savi) Aschers. et Graebn., Syn. Mitteleurop. Fl., 2, Abth. 2: 422 (1904); Holmboe, Veg. Cypr., 43 (1914).

   [*J. pygmaeus* (non L. Rich.) Boiss., Fl. Orient., 5: 360 (1882) pro parte quoad plant. cypr.]

   [*J. bufonius* (non L.) Kotschy in Unger et Kotschy, Die Insel Cypern, 190 (1865); Boiss., Fl. Orient., 5: 361 (1882) pro parte; H. Stuart Thompson in Journ. Bot., 44: 339 (1906) et auctt. mult.]

   TYPE: Portugal; "in udis cum sequenti [*J. bufonius* L.], circa Conimbricam [Coimbra], et alibi in Beira" (? destroyed).

Slender annual (3–) 7–20 (–40) cm. high; stems erect, usually much branched from the base, sometimes solitary; leaves 1–5, linear-filiform, 3–15 (–30) cm. long, 0·5–1·5 mm. wide, basal sheaths inconspicuous, with narrow, membranous margins; inflorescence usually lax, repeatedly but irregularly dichotomous, with long erecto-patent branches; flowers usually 3–5 (–10) in distinct, fan-shaped glomerules, sometimes solitary, usually spaced along the inflorescence-branches, sessile or shortly pedicellate; lowermost bracts foliaceous, filiform-subulate, the outer to 5 cm. long or more, the inner rather shorter; upper bracts membranous-scarious, pallid, ovate, acute, 3–5 mm. long, 2–3 mm. wide; bracteoles 2 (–3) ovate, concave, acute, membranous, about 2 mm. long, 1·5 mm. wide; perianth greenish, segments suberect, concave, the outer narrowly lanceolate-acuminate, about 6 mm. long, 1·5 mm. wide, with a thick midrib, slender acumen and wide hyaline margins, the inner about 4–4·5 mm. long, 1·5 mm. wide, with a narrower midrib and broader hyaline margins; stamens 6; filaments slender, about 1 mm. long; anthers narrowly oblong, about 0·8 mm. long, 0·3 mm. wide; ovary oblong, about 3·5 mm. long, 1·8 mm. wide; style slender, 0·4–0·5 mm. long; stigmas deflexed, twisted, about 0·8–1 mm. long. Capsule oblong, about 4–4·5 mm. long, 1·8 mm. wide, distinctly shorter than the persistent perianth, shining dark reddish-brown, apex obtuse or subacute, very shortly mucronulate; seeds broadly and bluntly oblong-ellipsoid, about 0·3 mm. long, 0·2 mm. wide; testa yellowish-brown, obscurely and minutely rugulose or almost smooth; appendages wanting.

HAB.: Moist paths, ditches, streamsides and other damp places; sea-level to 5,400 ft. alt.; fl. April–July.

DISTR.: Divisions 1–5, 7. South-west Europe, Mediterranean region east to Iraq, Atlantic Islands, and probably as an introduced weed elsewhere.

1. Fontana Amorosa, 1962, *Meikle* 2306! Paphos harbour, 1978, *J. Holub* s.n.!
2. Livadhi tou Pasha, 1937, *Kennedy* 107! Kryos Potamos, 5,400 ft. alt., 1938, *Kennedy* 1084! Mesapotamos, 1939, *Lindberg f.* s.n.; Xerokolymbos, 4,300 ft. alt., 1961, *D. P. Young* 7441!
3. Akrotiri, 1967, *Merton* in ARI 691!
4. Near Larnaca and Mazotos, 1862, *Kotschy* 63! 559a; Larnaca, 1877, *J. Ball* 2436! and same area, 1877, *herb. Post*! also, 1950, *Chapman* 453! Kouklia, 1952, *Merton* 1570! Chali, near Famagusta, 1962, *Meikle* 2619!
5. Mile 14, Nicosia–Limassol road, 1950, *Chapman* 538! Sha, 1955, *Merton* 2071! Dheftera, 1967, *Merton* in ARI 301!
7. Myrtou, 1932, *Syngrassides* 98! Larnaka tis Lapithou, 1941, *Davis* 3013! Kyrenia, 1955, *Miss Mapple* 26! also, 1956, *G. E. Atherton* 1221! 1336! Kambyli, 1956, *Merton* 2731!

11. J. ambiguus *Guss.*, Fl. Sic. Prodr., 1: 435 (1827); T. A. Cope et C. A. Stace in Watsonia, 12: 113–128 (1978).

J. ranarius Song. et Perr. in Billot, Annot., 192 (1860); Snogerup in Tutin et al., Fl. Europ., 5: 107 (1980).

J. mutabilis Savi ssp. *ambiguus* (Guss.) Nyman, Consp. Fl. Europ., 749 (1882).

TYPE: Sicily: "In udis arenosis maritimis; *Spaccaforno, Trapani*" (? NAP).

Slender annual; stems branched at the base or solitary, 6–20 cm. high, erect or sometimes ascending from a procumbent base; leaves 1–4, linear-filiform, 3–7 cm. or more long, 0·5–1 mm. wide, basal sheaths inconspicuous, with narrow membranous margins, inflorescence usually lax, irregularly dichotomous, branches erecto-patent, 1–5 (or more) cm. long, flowers usually spaced along the branches, sessile or subsessile, solitary, or sometimes 2–4 (–5) together in dense glomerules; lowermost bracts foliaceous, filiform-subulate, the outer to 5 cm. long, the inner generally much shorter; upper bracts membranous-scarious, ovate, acute, 3–5 mm. long, 2–3 mm. wide; bracteoles 2 (–3), ovate, acute, concave, membranous, 2 mm. long, 1·5 mm. wide; perianth greenish, segments suberect, concave, the outer narrowly lanceolate-acuminate, about 4·5 mm. long, 1·2 mm. wide, with a thick midrib, slender acumen and wide hyaline margins, the inner about 3·5 mm. long, 1–1·2 mm. wide, apex shortly acute, midrib stout but narrower than in the outer segments, hyaline margins very broad; stamens 6; filaments slender, about 1·2 mm. long; anthers narrowly oblong, about 0·8 mm. long, 0·1–0·2 mm. wide; ovary oblong, about 1·2 mm. long, 0·5 mm. wide; style short, about 0·3 mm. long; stigmas recurved. Capsule oblong-obovate, about 4 mm. long, 2 mm. wide, distinctly longer than the inner perianth-segments, shining dark brown; apex rounded or subtruncate, very shortly mucronulate; seeds broadly and bluntly oblong-ellipsoid or barrel-shaped, about 3·5–4·5 mm. long, 3 mm. wide; testa shining brown, almost smooth or very minutely striate-reticulate.

HAB.: Damp margin of salt-lake, amongst *Salicornia fruticosa*; near sea-level; fl. April–May.

DISTR.: Division 3, apparently rare. North, central and east Europe, seemingly reaching the eastern limit of its distribution in Cyprus; also in N. America.

3. Limassol Salt Lake E. of Akrotiri, 1941, *Davis* 3515! teste T. A. Cope.

NOTES: Only once collected, and the specimens are a little over-mature. To be looked for in the neighbourhood of Larnaca and Limassol Salt Lakes.

SUBGENUS 5. **Juncinella** *V. Krecz. et Gontsch.* Small annuals with solitary or caespitose stems; leaves basal, linear-subulate, flat or semi-terete, long-sheathing; flowers solitary or in clusters, ebracteolate. Seeds conspicuously reticulate.

12. J. capitatus *Weigel*, Obs. Bot., 28, t. 2, fig. 5 (1772); Boiss., Fl. Orient., 5: 361 (1882);

Buchenau in Engl., Pflanzenr., 25 (IV. 36): 256 (1906); Post, Fl. Pal., ed. 2, 2: 668 (1933).
TYPE: Germany; "in Pomerania" *Rev. P. Wilke.*

Erect annual 6–10 cm. high; stems branched at the base or sometimes
solitary, slender; leaves all basal, bright green, very smooth, filiform-
subulate, 10–50 mm. long, 0·4–1·5 mm. wide, gradually dilated basally into
reddish, membranous-margined sheaths; inflorescences capitate, secund,
generally solitary, sometimes 2–3; lowermost bracts foliaceous, subulate,
the outer sometimes 3 cm. long, the inner usually less than 1 cm. long;
flowers 5–6 or more, sessile; perianth pale brownish-green, segments erect,
connivent, the outer narrowly lanceolate-acuminate, 6·5–7 mm. long, 1·8
mm. wide, strongly carinate, with a thickened midrib and wide hyaline
margins, apex tapering to a slender acumen, inner segments almost wholly
hyaline, lanceolate, shortly acuminate, about 4 mm. long, 1·3 mm. wide;
stamens 3, opposite the outer segments; filaments slender, 1·5 mm. long;
anthers narrowly oblong, about 0·8 mm. long, 0·2 mm. wide; ovary reddish,
oblong, shortly stipitate, about 1·8 mm. long, 1·2 mm. wide; style very
slender, about 0·7 mm. long; stigmas deflexed, twisted, 1–1·5 mm. long.
Capsule reddish-brown, ovoid-trigonous, about 2 mm. long, 1·5 mm. wide,
apex subacute, shortly mucronulate; seeds oblong, about 0·3 mm. long, 0·2
mm. wide; testa pale yellowish-brown, glossy, minutely and closely
reticulate with close, transverse ridges; appendages wanting.

HAB.: Moist cart-rut in Pine forest; c. 900 ft. alt.; fl. March–April.

DISTR.: Division 6, apparently rare. Most of Europe, Mediterranean region eastwards to
Turkey and Palestine; tropical Africa; Australia; N. America.

6. Near Dhiorios, in sandy Pine forest, 1962, *Meikle* 2342!

NOTES: The Cyprus specimens have larger flowers than is usual in western European material;
but *J. capitatus* is variable in this respect.

SUBGENUS 6. **Septati** *Buchenau* Rhizomatous or occasionally caespitose
perennials, rarely annuals; leaves cauline, terete, compressed or rarely
canaliculate, unitubulose or sometimes divided longitudinally by internal
walls and pluritubulose, with coincident or non-coincident transverse septa;
flowers in glomerules; bracteoles wanting. Seeds commonly reticulate;
appendages rarely present.

13. J. fontanesii *Gay ex Laharpe* in Mém. Soc. Hist. Nat. Par., 3: 130 (1827); Duval-Jouve in
Rev. Sci. Nat. Montpellier, 1: 117–150 (1872); Buchenau in Engl., Pflanzenr., 25 (IV. 36):
190 (1906); Post, Fl. Pal., ed. 2, 2: 668 (1933); Snogerup in Rechinger f., Fl. Iran., 75: 24
(1971).

Erect or decumbent perennial 10–50 cm. high; rhizome scarcely
developed, the plants arising from the nodes of elongate, rooting stolons;
stems solitary or in tufts; cataphylls inconspicuous; cauline leaves 3–5,
unitubulose, linear, compressed, distinctly septate, 5–30 cm. long, 1–3 mm.
wide, tapering to a slender acumen from a long, sheathing, auriculate base,
auricles subacute, membranous, 3–5 mm. long, 2–3 mm. wide; inflorescence
lax or congested, irregularly branched, 2–13 cm. long, 1·5–10 cm. wide,
consisting of 3–20 (or more) globose, sub-echinate, 15–30-flowered
glomerules, 6–15 mm. diam.; lowermost bracts foliaceous, 1–5 cm. long, the
outer usually much longer than the inner; inflorescence-branches and
partial inflorescences subtended by membranous-scarious, ovate, acute
bracts 2–6 mm. long, 2–3 mm. wide at base; perianth greenish or brownish-
red, segments erect, subequal, concave, lanceolate-acuminate, about 3–4
mm. long, 0·8–1 mm. wide, with a narrow hyaline margin; stamens 6;
filaments slender, about 0·2 mm. long; anthers narrowly oblong, about 0·8
mm. long, 0·2 mm. wide; ovary narrowly oblong-prismatic, sharply angled,

about 4 mm. long, 0·8 mm. wide, apex often produced into a short beak; style slender, about 0·5 mm. long; stigmas divergent, about 1 mm. long. Capsule oblong-prismatic or sub-pyramidal, about 4 mm. long, 1 mm. wide, usually tapering to a short rostrate apex and equalling or exceeding the perianth; valves shining red-brown, margins involute on dehiscence; seeds broadly oblong-ellipsoid, about 0·4 mm. long, 0·3 mm. wide; testa yellowish-brown, minutely rugulose-reticulate; appendages absent.

ssp. **pyramidatus** (*Laharpe*) *Snogerup* in Rechinger f., Fl. Iran., 75: 25 (1971); Osorio-Tafall et Seraphim, List Vasc. Plants Cyprus, 19 (1973).

    J. pyramidatus Laharpe in Mém. Soc. Hist. Nat. Par., 3: 128 (1827); Boiss., Fl. Orient., 5: 359 (1882).

    J. fontanesii Gay var. *pyramidatus* (Laharpe) Buchenau in Engl., Bot. Jahrb., 1: 140 (1880); Post, Fl. Pal., ed. 2, 2: 668 (1933).

    J. fontanesii Gay f. *pyramidatus* (Laharpe) Holmboe, Veg. Cypr., 44 (1914); Lindberg f., Iter Cypr., 9 (1946).

    TYPE: "Cette espèce a été rapportée de Syrie par M. Labillardière, et d'Egypte par M. Savigny" (FI, K!).

Capsule narrowly pyramidal above, tapering from below the middle to a short rostrate apex and clearly exceeding the erect, persistent perianth-segments, thereby giving the glomerules an echinate appearance; plants said to be more compact and less stoloniferous than in *J. fontanesii* ssp. *fontanesii*, but this is to a great degree dependent on habitat.

    HAB.: In running or stagnant water, or in mud by springs, streams and ponds; 700–5,000 ft. alt.; fl. June–Aug.

    DISTR.: Divisions 1–3, 5, 7. Eastern Mediterranean region and eastwards to Iraq.

1. Ayios Neophytos Monastery, 1939, *Lindberg f.* s.n.!
2. Near Galata, 1880, *Sintenis & Rigo* 875! Trikoukkia, 1935, *Syngrassides* 232! Kakopetria, 1949, *Casey* 826! also, 1970, *E. Chicken* 46!
3. Episkopi, 1905, *Holmboe* 306; Mamonia, 1960, *N. Macdonald* 161!
5. Near Kythrea, 1880, *Sintenis & Rigo* 562!
7. Near Myrtou, 1932, *Syngrassides* 31! 175 (mixed with *J. articulatus* L.)! Boghazi, 1939, *Lindberg f.* s.n.!

14. J. articulatus *L.*, Sp. Plant., ed. 1, 327 (1753); Snogerup in Rechinger f., Fl. Iran., 75: 22 (1971); Osorio-Tafall et Seraphim, List Vasc. Plants Cyprus, 19 (1973).

    J. lampocarpus Ehrh. ex Hoffm., Deutschl. Fl., 125 (1791); Boiss., Fl. Orient., 5: 358 (1882); Buchenau in Engl., Pflanzenr., 25 (IV. 36): 217 (1906); Holmboe, Veg. Cypr., 44 (1914); Post, Fl. Pal., ed. 2, 2: 667 (1933); Lindberg f., Iter Cypr., 10 (1946).

    TYPE: "*in* Europae *aquosis*".

Erect or ascending perennial; rhizomes usually short; stems rather crowded, 5–50 cm. high; cataphylls rather few, 5–18 mm. long, about 4 mm. wide, sheathing the base of the stems, reddish, with membranous margins produced into short auricles at apex; cauline leaves 3–6, narrowly linear, 3–20 cm. long, 1–3 mm. wide, unitubulose, distinctly septate, apex shortly acute, base long-sheathing, the sheaths with scarious margins produced into short blunt auricles; inflorescence generally lax and spreading, 3–10 cm. long and almost as wide, much branched, consisting of well-spaced, 4–15 (–30)-flowered glomerules on erect or divaricate branches; glomerules generally compact and dense, not echinate in appearance; lowermost bracts foliaceous, subulate, the outer up to 5 cm. long, the inner generally much shorter; upper bracts ovate-acuminate, scarious, concave, 2–5 mm. long, 1·5–3 mm. wide; perianth greenish or brownish at anthesis, segments erect, subequal, lanceolate-acuminate, about 2·5 mm. long, 0·5 mm. wide, concave, with a thick midrib and scarious-membranous margins; stamens 6; filaments slender, about 0·5 mm. long; anthers narrowly oblong, about 0·5 mm. long, 0·2 mm. wide; ovary narrowly ellipsoid-prismatic, sharply angled, about 2·5 mm. long, 1 mm. wide; style very short or almost wanting;

stigmas deflexed, twisted, 1–1·5 mm. long. Capsule ovoid-triquetrous, acuminate, shining reddish-brown or fuscous, about 3 mm. long, 1·3 mm. wide, slightly exceeding the erect, persistent perianth-segments; seeds bluntly oblong-ellipsoid, 0·4–0·5 mm. long, 0·3 mm. wide; testa honey-brown, conspicuously reticulate.

HAB.: By springs, streams or irrigation channels; in roadside ditches, or by ponds or in other marshy places; sea-level to 5,500 ft. alt.; fl. April–Sept.

DISTR.: Divisions 2, 3, 6, 7. Widespread in Europe, Asia and temperate North America; N. and S. Africa, S. Australia, New Zealand.

2. Evrykhou, 1880, *Sintenis* 876! Prodhromos, 1905, *Holmboe* 948; also, 1949, *Casey* 905! and, 1955, *G. E. Atherton* 562! 575! Xerokolymbos, 4,300 ft. alt., *Kennedy* 106! Kryos Potamos, 4,000 ft. alt., 1937, *Kennedy* 109! Mesapotamos, 1939, *Lindberg f.* s.n.; Platania, 1939, *Lindberg f.* s.n.! also, 1961, *D. P. Young* 7412! and, 1966, *Merton* in ARI 71! Trikoukkia, 1961, *D. P. Young* 7381!
3. Asomatos Marshes, 1939, *Mavromoustakis* 130! Arminou, 1941, *Davis* 3407!
6. Syrianokhori, 1952, *Merton* 1504!
7. St. Hilarion, 1880, *Sintenis & Rigo* 556! Near Myrtou, 1932, *Syngrassides* 175 partly! Above Thermia, 1948, *Casey* 24! Kyrenia, 1955, *G. E. Atherton* 109!

J. articulatus *L.* × **J. fontanesii** *Gay ex Laharpe* ssp. **pyramidatus** (*Laharpe*) *Snogerup*

Intermediate in habit and inflorescence between the presumed parents, with rather large, dense, globose, red-brown glomerules, and with the sterile, shortly rostrate capsules slightly exceeding the acuminate perianth-segments.

Determined by S. Snogerup from material collected in (6) Dhiorios Forest, 1954, *Merton* 1918! The collector notes that the hybrid was found "in a small marshy valley with *Juncus* spp. etc.", but there are no collections of *Juncus fontanesii* or *J. articulatus* from the locality.

98a. PALMAE

Shrubs, trees or rhizomatous perennials, sometimes climbing; stems branched or unbranched, or almost wanting; primary roots disappearing and replaced by roots from the base of the stem; leaves crowded in a terminal cluster or scattered, entire, palmately or pinnately divided, folded induplicately or reduplicately in bud; rhachis or petiole often expanded basally into a fibrous sheath; inflorescence axillary or less often terminal; flowers unisexual (monoecious or dioecious), polygamous or hermaphrodite, small, actinomorphic, usually numerous in a branched panicle or spadix, enveloped by one or more membranous, papyraceous or coriaceous spathes; perianth double or occasionally rudimentary; sepals 3, free or connate, imbricate, or open in bud; petals 3, free or connate, usually valvate in male flowers, imbricate in female flowers; stamens generally 6 in 2 series; anthers 2-thecous, dehiscing by longitudinal slits; pollen-grains often smooth; ovary superior, 1–3 (–7)-locular, or carpels 3, connate only at base; ovule solitary, erect or pendulous in each loculus or carpel. Fruit a berry or drupe, exocarp often fibrous or scaly; seeds free or adherent to endocarp; endosperm present, frequently hard or bony, sometimes ruminate; embryo small.

A large, predominantly tropical family said to include more than 200 genera and 2,500 species.

Although it is likely that several distinct Palms are grown in Cyprus, there are no indigenous representatives of the family, nor in the absence of specimens or reliable records, is it possible to say much about those in cultivation.

The Date, *Phoenix dactylifera* L. (Sp. Plant., ed. 1, 1188; 1753), often a tall tree up to 30 m. high, with a terminal cluster of long, plumose, pinnatisect leaves; dioecious inflorescences; 3 free carpels, and large drooping clusters of oblong-ellipsoid sweet, sticky fruits, is commonly cultivated, usually in the lowlands, though Holmboe (Veg. Cypr., 42; 1914) records it at c. 1,500 ft., at Lythrodhonda. The Date is thought to be a native of Africa and S.W. Asia.

Washingtonia filifera (J. A. Linden ex André) H. Wendl. ex S. Watson, Bot. Calif., 2: 211 (1880) (*Pritchardia filifera* J. A. Linden ex André) a native of California and Arizona, is a conspicuous feature in the Municipal Gardens, Nicosia, where there is a fine avenue of mature specimens. It has a massive trunk and terminal cluster of persistent, flabelliform, palmatisect leaves with fibrous threads in the sinuses.

Washingtonia robusta H. Wendl., from Mexico, a taller, more slender palm, without threads in the sinuses of mature leaves, has been recorded by G. Frangos (Cypr. Agric. Journ., 18: 86; 1923) from a garden at Dhali, but the record may be an error for the more commonly cultivated *W. filifera*.

Readers wishing to know more about Palms grown in the Mediterranean region should consult: B. Chabaud, Les Palmes de Côtes d'Azur, 208 pp. (1915), and J. de Carvalho-e-Vasconcellos et J. do Amaral-Franco in Portug. Act. Biol., ser. 2, 2: 289–425 (1948).

99. TYPHACEAE

P. Graebner in Engl., Pflanzenr., 2 (IV. 8): 1–26 (1900).

Perennial, rhizomatous aquatic and marsh plants; flowering stems simple, farctate, clothed basally with distichous cataphylls; leaves sub-basal, alternate, long-sheathing, distichous, lamina linear, entire, flat, concave or carinate, parallel-nerved, sheaths sometimes auriculate; flowers unisexual, minute, in dense cylindric, contiguous or remote, monoecious spikes, often subtended by deciduous, leafy spathes; male inflorescences above female; male flowers consisting of 1–3 (–8) stamens subtended by filiform or linear, simple or lacerate scales; filaments free or partly fused; anthers 4-thecous, often with a prominent, convex or conical connective; pollen-grains single or in tetrads; female flowers frequently associated with sterile pistillodes, borne on long, slender gynophores clothed with simple or clavate-tipped hairs, ebracteolate or with a spathulate, entire or lacerate bracteole or scale; ovary clavate or fusiform, 1-locular; ovule solitary, pendulous; style slender, persistent; stigma unilateral, linear or sub-spathulate. Fruit dry, tardily dehiscent, with a membranous or coriaceous pericarp; seed yellowish, with a membranous testa and copious endosperm.

One genus and about 10 species with a cosmopolitan distribution.

1. TYPHA *L*.

Sp. Plant., ed. 1, 971 (1753).
Gen. Plant., ed. 5, 418 (1754).

Description, etc., that of the family.

1. **T. domingensis** *Pers.*, Syn. Plant., 2: 352 (1807); J. B. Gèze, Etudes Bot. Agron. Typha, 118 (1912); D. M. Napper in Fl. Trop. E. Africa, Typhaceae, 2 (1971).

T. australis Schumacher, Beskr. Guin. Plant., 401, (1827); Osorio-Tafall et Seraphim, List Vasc. Plants Cyprus, 5 (1973).

T. angustata Bory et Chaub., Exped. Sci. Morée Bot., 3 (2): 338 (1832); Boiss., Fl. Orient., 5: 50 (1882); Holmboe, Veg. Cypr., 30 (1914); Post, Fl. Pal., ed. 2, 2: 559 (1933); Lindberg f., Iter Cypr., 6 (1946).

[*T. latifolia* (non L.) Gaudry, Recherches Sci. en Orient, 190 (1855); Unger et Kotschy, Die Insel Cypern, 212 (1865); Boiss., Fl. Orient., 5: 49 (1884) pro parte quoad plant. cypr.; Holmboe, Veg. Cypr., 30 (1914); Osorio-Tafall et Seraphim, List Vasc. Plants Cyprus, 5 (1973).]

TYPE: Dominican Republic; "ad St. Domingo" (? L).

Robust, glaucous, rhizomatous perennial to 3·5 m. high; leaf-sheaths glabrous, closely striate, sometimes embracing the stem for 1·5 m. or more, apex with a somewhat auriculate, narrowly membranous margin; lamina linear, flattish or flat above and slightly convex below, 1 m. long or more, 4–14 mm. wide, apex obtuse or subacute; peduncle terete, finely striate, generally exserted from the uppermost sheath for 30 cm. or more; male and female spikes remote; male spikes slender, cylindrical, about 10 cm. long, 0·4–0·5 cm. wide, scales membranous, whitish, linear or narrowly oblanceolate, about 0·8 mm. long, 0·2 mm. wide, apex irregularly lacerate or subentire; filaments about 1·5 mm. long; anthers usually in groups of 3–4, linear-oblong, about 2 mm. long, 0·3 mm. wide, connective distinct, blunt, brownish; female spikes narrowly cylindrical, brown, velvety, sometimes 30 cm. long or more, usually less than 1·4 cm. diam., hairs whitish, scabridulous, to 5 mm. long; ovary narrowly ellipsoid, about 1 mm. long 0·3 mm. wide, blotched brown, style filiform, about 1 mm. long; bracteole to 4 mm. long, apex subspathulate. Ripe fruit and seed not seen.

HAB.: Margins of ponds; sides of streams and water channels; ditches; swamps; sea-level to 600 ft. alt.; fl. June–July.

DISTR.: Divisions 1, 3–5, 7, 8. Throughout the world in warm-temperate and tropical regions.
1. Yeroskipos, 1934, *Syngrassides* 505! Akhelia, 1939, *Lindberg f.* s.n.
3. Kakoradjia, between Nicosia and Limassol, 1932, *Syngrassides* 393!
4. Near Larnaca, 1862, *Kotschy*; Perivolia, 1939, *Lindberg f.* s.n.!
5. Near Vatili, 1862, *Kotschy*.
7. Between Dhikomo and Aghirda, 1956, *Merton* 2836!
8. Near Gastria, 1905, *Holmboe* 473; between Yialousa and Rizokarpaso, 1937, *Syngrassides* 1640!

NOTES: Neither Gaudry nor Kotschy recognized *T. domingensis*, so it may be assumed that their records for *T. latifolia* L. are erroneous. The occurrence of the latter in Cyprus is questionable though, considering its cosmopolitan distribution, it may yet be found, perhaps at higher altitudes than *T. domingensis*. It is easily recognized by its *contiguous* male and female spikes, and by its broader leaves.

100. SPARGANIACEAE

P. Graebner in Engl., Pflanzenr., 2 (IV. 10): 1–24 (1900).

Perennial, rhizomatous, aquatic and marsh plants; stems erect or floating, farctate, usually unbranched except in region of inflorescence, clothed basally with distichous cataphylls; leaves basal, sheathing, distichous, lamina linear, entire, flat, concave or carinate, parallel-nerved; inflorescence branched or unbranched; flowers monoecious, in dense globose, sessile or pedunculate, bracteate or ebracteate, contiguous or remote, spikes, the upper spikes generally male, the lower female; male flowers with 1–6 membranous or thick, scale-like perianth-segments; stamens 1–8; filaments free or shortly connate; anthers oblong, 4-thecous, with the connective dilated apically; female flowers with (1–) 3–6 thick or membranous perianth-segments; ovary ovoid or obovoid, superior, 1–3-locular; style filiform, persistent; stigma unilateral, linear, spathulate or capitate. Fruit indehiscent, with a spongy exocarp and hard endocarp, terete or angled; seeds with a membranous testa and copious endosperm.

One genus and about 15 species, in temperate regions of both hemispheres, but chiefly in Europe, Asia and North America.

1. SPARGANIUM L.

Sp. Plant., ed. 1, 971 (1753).
Gen. Plant., ed. 5, 418 (1754).

Description etc., that of the family.

1. S. erectum *L.*, Sp. Plant., ed. 1, 971 (1753); Post, Fl. Pal., ed. 2, 2: 558 (1933).
  S. ramosum Huds., Fl. Angl., ed. 2, 401 (1778); Boiss., Fl. Orient., 5: 48 (1882).

Erect, rhizomatous perennial, 30–150 (–200) cm. high; leaves linear, carinate towards base, glabrous, shining green, 60–100 cm. or more long, 0·8–1·8 cm. wide, apex obtuse or subacute; inflorescence branched, with erecto-patent branches 7–12 cm. long arising from the axils of leaf-like bracts 2–30 cm. long, or the uppermost branches sometimes ebracteate; female spikes 1–3 (–6), usually remote, sessile or shortly pedunculate, about 1 cm. diam. at anthesis, 2 cm. diam. or more in fruit; perianth-segments 4–6, spathulate or obcuneate, carinate, about 3 mm. long, 1 mm. wide, rather fleshy, with a dark brown, subtruncate apex; ovary narrowly ellipsoid, somewhat beaked, about 4 mm. long, 1·3 mm. wide; style about 2 mm. long, caducous; stigma linear, unilateral; male spikes sometimes 20 or more, terminal and lateral, contiguous or the lower remote, 4–6 mm. diam.; perianth-segments 4–6, obcuneate, about 1·8–2 mm. long, 0·5 mm. wide, apex subtruncate, dark brown; stamens 4–6, filaments slender, about 3 mm. long, free to base; anthers oblong, about 0·7 mm. long, 0·4 mm. wide. Fruit ellipsoid or obpyramidal, angled or subterete, shortly beaked, about 5–9 mm. long, 2–7 mm. wide, uniformly shining pale brown, or the upper part darker, 1–3-locular; seeds narrowly ellipsoid-fusiform, about 2 mm. long, 0·5 mm. wide.

ssp. **neglectum** (*Beeby*) *Schinz et Keller*, Fl. Suisse (Ed. Franç.) , 1: 26 (1908); Osorio-Tafall et Seraphim, List Vasc. Plants Cyprus, 5 (1973).
  S. neglectum Beeby in Journ. Bot., 23: 193, t. 258 (1885).
  S. ramosum Huds. ssp. *neglectum* (Beeby) Aschers. et Graebn., Syn. Mitteleurop. Fl., 1: 281 (1897); Rechinger f. in Arkiv för Bot., ser. 2, 1: 417 (1950).

TYPE: Probably the plate (t. 258) in Journ. Bot., 23 (1885)! drawn from a plant collected in Reigate, Surrey.

Fruit uniformly pale glossy brown when ripe, subterete, generally 1-locular, 7–9 mm. long, 2–3·5 mm. wide, apex acuminate.

HAB.: Marshes and swamps; c. 3,000 ft. alt.; fl. July–Aug.

DISTR.: Division 2, rare. Widespread in Europe, the Mediterranean region and east to Central Asia.

2. Pano Panayia, 1913, *Haradjian* 919.

NOTES: Only once recorded; the subspecies is found in most countries adjacent to Cyprus.

101. ARACEAE

Engler et Krause in Engler, Pflanzenr., 21 (IV. 23B.) (1905); 37 (IV. 23B.) (1908); 48 (IV. 23C.) (1911); 55 (IV. 23Da.) (1912); 60 (IV. 23 Db.) (1913); 64 (IV. 23Dc.) (1915); 71 (IV. 23E.) (1920); 73 (IV. 23F.) (1920); 74 (IV. 23A.) (1920).

by S. J. Mayo & R. D. Meikle

Perennial herbs with acrid or milky sap; stem tuberous or rhizomatous (also climbing, epiphytic, rarely arborescent, or floating aquatics); leaves alternate, petiolate, basally sheathing (sometimes pulvinate at apex of petiole); lamina variable in size and shape, often sagittate or hastate or cordate, simple or lobed; venation pinnate, often forming inframarginal veins (rarely parallel), finer veins reticulate (or striate); inflorescence pedunculate, consisting of a more or less cylindric *spadix* bearing numerous, sessile, ebracteate flowers crowded on a fleshy axis and sometimes with a sterile terminal appendix, the spadix subtended by an herbaceous, sometimes brightly-coloured, spreading or convolute, bract-like *spathe*; flowers naked (or with a perigonium), unisexual (or bisexual), female (pistillate) flowers at base of spadix, male (staminate) flowers higher up, often separated by a sterile zone; stamens free or connate; anthers sessile or with filaments, dehiscing by pores or slits; ovary superior, unilocular (or plurilocular), ovules several to many (or solitary) in each loculus, placentation parietal or basal (axile or apical); stigma sessile (or with a more or less distinct style). Fruit a fleshy, often brightly coloured berry (rarely forming a syncarp); seeds 1 to several, minute or relatively large, variable in shape, with or without endosperm.

A largely tropical family of about 110 genera and more than 2,000 species. Apart from the introduced *Colocasia esculenta* (L.) Schott, all Cyprus species belong to the tribe *Areae* with naked unisexual flowers, sterile terminal spadix appendix and unilocular ovaries with orthotropous ovules. Family characters falling outside those of the Cyprus representatives are cited in parentheses in the above family description.

Leaf blade peltate, pendent on erect petiole - - - - - - COLOCASIA (p. 1670)
Leaf blade not peltate, suberect or held more or less perpendicular to petiole, never pendent:
    Spathe-tube with connate margins; stamens distant; spadix lacking sterile filaments
                                           2. Arisarum
    Spathe-tube with free, convolute margins; stamens densely congested; sterile filaments
      present on spadix (in Cyprus species) - - - - - - - - - **1. Arum**

1. ARUM *L.*

Sp. Plant., ed. 1, 964 (1753).
Gen. Plant., ed. 5, 413 (1754).

Terrestrial perennial herbs; stem tuberous, subglobose to subcylindric, growing point vertical or oblique; leaves several, radical, sagittate or hastate, sometimes cordate, fine venation reticulate; inflorescence solitary, terminal; spathe with lower part (tube) convolute, subcylindric to subglobose or ovoid, more or less inflated, usually somewhat constricted towards apex, upper part (limb) ovate to oblong, acuminate, erect, expanded, variously coloured; spadix with basal female part usually separated from upper male part by sterile zone of subulate filaments, terminal, subcylindric, sterile appendix usually separated from male part by zone of subulate filaments; flowers unisexual, naked; stamens densely congested; anthers subsessile, dehiscing by oblique apical pores; ovary unilocular with several to many ovules on a single parietal placenta; stigma subglobose, sessile. Berries 1–several-seeded, glossy scarlet or orange, borne in a cylindric spike; seeds subglobose, foveolate-rugose, with abundant endosperm.

About 12 species found chiefly in the Mediterranean region but extending from the Azores to the Himalayas. In the Cyprus species the leaves appear in late autumn and are fully developed by the time of flowering in spring; they are generally quite withered by the time the fruit ripens in late spring and early summer.

Spathe-limb conspicuously spotted or blotched purple internally   -   -   **1. A. dioscoridis**
Spathe-limb not spotted or blotched internally:
  Spadix-appendix pale yellow; spathe-limb yellowish- or whitish-green internally
                                                    5. A. italicum
  Spadix-appendix pinkish-purple or purple; spathe-limb purple or margined or tinged purple
    internally:
    Spathe-tube elongate-ovoid, about as long as or sometimes longer than limb; spadix-
      appendix very slender, less than 0·3 cm. diam.   -   -   -   **3. A. hygrophilum**
    Spathe-tube shortly cylindric, much shorter than limb; spadix-appendix subcylindric to
      fusiform, 0·3 cm. or more:
      Spadix-appendix fusiform, 1–3·5 cm. diam., pinkish-purple; peduncle usually much
        longer than petiole -   -   -   -   -   -   -   - **2. A. conophalloides**
      Spadix-appendix subcylindric, 0·3–0·6 cm. diam., purple; peduncle usually shorter than
        petiole   -   -   -   -   -   -   -   -   -   - **4. A. orientale**

1. A. dioscoridis *Sm.* in Sibth. et Sm., Fl. Graec. Prodr., 2: 245 (1816), Fl. Graec., 10: 37, t. 947 (1840); Poech, Enum. Plant. Ins. Cypr., 12 (1842); Unger et Kotschy, Die Insel Cypern, 211 (1865); Boiss., Fl. Orient., 5: 35 (1882); Holmboe, Veg. Cypr., 42 (1914); Post, Fl. Pal., ed. 2, 2: 552 (1933); Osorio-Tafall et Seraphim, List Vasc. Plants Cyprus, 18 (1973); Elektra Megaw, Wild Flowers of Cyprus, 18, t. 38 (1973).
    A. cyprium Schott in Bonplandia, 9: 368 (1861); Unger et Kotschy, Die Insel Cypern, 211 (1865).
    A. dioscoridis Sm. var. *cyprium* (Schott) Engl., Pflanzenr., 73 (IV. 23F.): 73 (1920).
    A. dioscoridis Sm. var. *smithii* Engl. in DC., Mon. Phan., 2: 583 (1879), Pflanzenr., 73 (IV. 23F.): 73 (1920).
    [*A. detruncatum* (non C. A. Mey. ex Schott) H. Stuart Thompson in Journ. Bot., 44: 338 (1906).]
    TYPE: "In insulae Cypri cultis, inter segetes, vulgaris", Fl. Graec. t. 947!

Tuber subglobose or shortly cylindric, growing point vertical; leaves hastate-sagittate, up to 30 cm. long, 22 cm. wide across basal lobes, dark green, basal lobes often strongly divergent, petiole as long as or longer than the lamina; peduncle stout, up to about 20 cm. long, usually shorter than the petioles; inflorescence strongly malodorous; spathe up to 30 cm. long; tube 4–5 cm. long, 2·5–3·5 cm. diam., green externally, constricted at apex; limb narrowly ovate, acuminate, 4·5–6·5 cm. wide, inner surface dark purple in basal half, shading to greenish towards apex and heavily marked with purple spots and larger blotches; spadix shorter than the spathe, 12–20 cm. long, appendix narrowly clavate or subcylindric, dark purple, 7–15 cm. long, 0·6–1 cm. diam., distinctly but not very strongly constricted towards base into a stipe 2–5 cm. long; sterile filaments purplish, present above and below male zone.

HAB.: Margins of cultivated ground; rocky hillsides; margins of streams; sea-level to 3,400 ft. alt.; fl. March–May.

DISTR.: Divisions 2–5, 7. Also southern Turkey eastwards to Iraq.

2. Platres, 1937, *Kennedy* 121! Limnitis valley, 1954, *Mrs. G. Scott-Moncrieff* in *Casey* 1664!
3. Kolossi, 1862, *Kotschy*; 5 miles E. of Amathus, 1954, *Mrs. G. Scott-Moncrieff* in *Casey* 1662! Kalokhorio to Athrakos, 1979, *Edmondson & McClintock* E 2978!
4. Ormidhia, 1787, *Sibthorp*; also, 1862, *Kotschy* s.n.; Cape Greco, 1862, *Kotschy* 181! and, 1905, *Holmboe* 422; also, 1958, *N. Macdonald* 55!; Ayia Napa, 1905, *Holmboe* 62; Ayios Antonios, Sotira, 1981, *T. F. Hewer* 4747!
5. Mile 3, Nicosia–Famagusta road, 1950, *Chapman* 378!
7. Kephalovryso, near Kythrea, 1941, *Davis* 2957! Kyrenia, 1929, *Mrs. C. B. Ussher* 22! 6 miles E. of Kyrenia, 1954, *Mrs. G. Scott-Moncrieff* in *Casey* 1663!

NOTES: Fruiting specimens from 2. Aphamis, 1937, *Kennedy* 122! 123! are also probably referable here, but cannot be identified with complete certainty in the absence of flowers.

2. **A. conophalloides** *Kotschy ex Schott*, Prodr. Syst. Aroid., 97 (1860); Engl., Pflanzenr., 73 (IV. 23F.): 75 (1920); Post, Fl. Pal., ed. 2, 2: 552 (1933).
    A. detruncatum C. A. Mey. ex Schott var. *conophalloides* (Kotschy ex Schott) Boiss., Fl. Orient., 5: 36 (1882).
TYPE: Turkey; "Cataonia, Gorumse. Kotschy" (W? destroyed, K!).

Tuber subglobose or shortly cylindric, growing point vertical; leaves hastate-sagittate, up to 30 cm. long, 20 cm. wide across basal lobes, dark green, basal lobes divergent but not usually as strongly as in *A. dioscoridis* Sm.; petiole as long as or longer than the lamina; peduncle stout, rigidly erect, up to about 50 cm. long, usually much longer than the petioles; inflorescence smelling of horse dung but not offensively so; spathe up to 50 cm. long; tube up to about 8 cm. long, and about 3 cm. diam., shortly cylindric, green externally, constricted at apex; limb narrowly oblong-lanceolate or ovate-lanceolate, acuminate, 5–10 cm. wide at widest part, inner surface dull purple, becoming greenish towards apex and base, not spotted or blotched; spadix a little shorter than spathe, 17–35 cm. long, appendix fusiform, pinkish-purple, 12–28 cm. long, 1–3·5 cm. diam., tapering basally to a short, ill-defined stipe; sterile filaments crimson-purple, present above and below male zone.

HAB.: Rocky igneous mountainsides; 2,000–4,000 ft. alt.; fl. April–May.

DISTR.: Division 2, rare. Also Turkey, Iraq, Iran.

2. Limnitis valley, c. 2,000 ft. alt., 1954, *Mrs. G. Scott-Moncrieff* in *Casey* 1659! 1660! 1661! Dhodeka Anemi, 1962, *Meikle* s.n.! Side of Tripylos, by road from Stavros tis Psokas to Kykko Monastery, 1979, *Edmondson & McClintock* E 2838!

NOTES: All records are from a limited area of the north slopes of Paphos Forest, at high altitudes.

3. **A. hygrophilum** *Boiss.*, Diagn., 1, 13: 8 (1853); Sintenis in Oesterr. Bot. Zeitschr., 31: 325, 326 (1881); 32: 121 (1882); Boiss., Fl. Orient., 5: 37 (1882); Holmboe, Veg. Cypr., 43 (1914); Engl., Pflanzenr., 73 (IV. 23F.): 77 (1920); Post, Fl. Pal., ed. 2, 2: 553 (1933); Lindberg f., Iter Cypr., 9 (1946); Osorio-Tafall et Seraphim, List Vasc. Plants Cyprus, 18 (1973).
    TYPE: Syria; "ad fossas humidas et rivulos vallis *Antilibani* infrâ pagum *Zebdani* frequentissimè. Legi floriferum Maio 1846". Boissier in Plant. Syr. ex. 1846. (G, K!).

Tuber subglobose or shortly cylindric, up to 5 cm. diam., growing point vertical; leaves hastate-sagittate, 11–20 cm. long, 8–20 cm. wide across the patently divergent basal lobes, dark glossy green, median lobe oblong-lanceolate, apex shortly acute to obtuse; petiole much exceeding lamina, often three times as long; peduncle slender, up to 45–50 cm. long, generally much shorter than petiole; spathe 8–14 cm. long; tube 2·5–6 cm. long, about as long as or sometimes even longer than limb, 1–2·5 cm. diam., elongate-ovoid, greenish or stained purple externally, constricted at apex; limb narrowly ovate or lanceolate, acuminate, to about 3·5 cm. wide, pale to whitish green, tinged purple externally and often margined dark purple, not spotted; spadix about ⅔ length of spathe, 7–8 cm. long, appendix very slender, narrowly cylindric or almost caudate, 4–5 cm. long, less than 0·3 cm. diam., dark purple, scarcely stipitate at base; sterile filaments dark purple, very long, present above and below male zone. Berries orange. *Plate 98.*

HAB.: Moist, shaded places; roadside ditches and irrigation channels; 500–600 ft. alt.; fl. March–April.

DISTR.: Divisions 1, 5, 7. Also Israel, Lebanon, Syria.

1. Tala, 1982, *Mrs. A. A. Butcher* s.n.!
5. Kythrea, 1880, *Sintenis & Rigo* 129! Also, 1939–1967, *Lindberg f.* s.n.! *Meikle* 2552! *Merton* in ARI 355!
7. Kazaphani, 1937, *Syngrassides* 1462! Also, 1954–1956, *Mrs. G. Scott-Moncrieff* in *Casey* 1658! *G. E. Atherton* 1147! Between Melounda and Halevga, 1963, *J. B. Suart* 63!

4. A. orientale *M. Bieb.*, Fl. Taur.-Cauc., 2: 407 (1808); Boiss., Fl. Orient., 5: 39 (1882); Engl., Pflanzenr., 73 (IV. 23F.): 78 (1920); Post, Fl. Pal., ed. 2, 2: 553 (1933).
    A. gratum Schott, Syn. Aroid., 11 (1856), Prodr. Syst. Aroid., 89 (1860); Osorio-Tafall et Seraphim, List Vasc. Plants Cyprus, 18 (1973).
    A. orientale M. Bieb. var. *gratum* (Schott) Engl. in DC., Mon. Phan., 2: 588 (1879); Boiss., Fl. Orient., 5: 39 (1882); Sintenis in Oesterr. Bot. Zeitschr., 32: 121 (1882); Holmboe, Veg. Cypr., 43 (1914); Engl., Pflanzenr., 73 (IV. 23F.): 79 (1920).
    A. orientale M. Bieb. var. *sintenisii* Engl., Pflanzenr., 73 (IV. 23F.): 80 (1920).
TYPE: U.S.S.R., Crimea and Caucasus; "Frequens in umbrosis sylvaticis tam Tauriae, quam Caucasi. Floret Majo".

Tuber stout, subglobose to shortly cylindric, 2·5–4 cm. diam., growing point vertical; leaves hastate-sagittate, 15–25 cm. long, 7–22 cm. wide across basal lobes, dark green; petiole longer than the lamina; peduncle 25–35 cm. long, usually slightly shorter than petiole; spathe 13–25 cm. long; tube 3–5 cm. long, about 2 cm. diam., shortly cylindric, green externally (?), constricted at apex; limb ovate-lanceolate, acuminate, 3–6 cm. wide at widest part, inner surface strongly tinged brown-purple, not spotted or blotched; spadix 8–14 cm. long, more than half as long as spathe, appendix subcylindric, only slightly narrowed towards base, purple, 7–10 cm. long, 0·3–0·6 cm. diam.; sterile filaments present above and below male zone.

HAB.: Moist ground by irrigation channels in cultivated ground; 500–600 ft. alt.; fl. April–May.

DISTR.: Divisions 5, 7. Also Eastern Greece, Turkey, Crimea, Caucasus, Armenia, Iran.

5. Kythrea, 1880–1950, *Sintenis & Rigo* 130! *Syngrassides* 376! 1790! *Miss Godman* 39! *Chapman* 82!
7. Near Kalogrea, 1880, *Sintenis & Rigo* 130a!

NOTES: A taxonomically difficult species, which seems to be closely related to *A. italicum* Miller.

5. A. italicum *Mill.*, Gard. Dict., ed. 8, no. 2 (1768); Boiss., Fl. Orient., 5: 40 (1882); Holmboe, Veg. Cypr., 43 (1914); Engl., Pflanzenr., 73 (IV. 23F.): 82 (1920); Osorio-Tafall et Seraphim, List Vasc. Plants Cyprus, 18 (1973).
    [*A. ponticum* (non Schott) Unger et Kotschy, Die Insel Cypern, 211 (1865).]
    [*A. italicum* Mill. subvar. *nickellii* (? Schott) Engl., Pflanzenr., 73 (IV. 23F.): 85 (1920), pro parte quoad *Kotschy plant. cypr.* 1862 no. 739.]
TYPE: "grows naturally in Italy, Spain and Portugal", cultivated, Chelsea Physic Garden (BM ?).

Tuber shortly cylindric, rhizomatous, about 5 cm. long, 2·5 cm. diam., growing point obliquely-lateral; leaves hastate-sagittate, 15–35 cm. long, 15–20 cm. wide across basal lobes, dark green, main veins often yellowish-

LRM

Plate 98. ARUM HYGROPHILUM Boiss. **1,** habit, × $\frac{2}{3}$; **2,** inflorescence with spathe removed, × $\frac{2}{3}$; **3,** fruit, × $\frac{2}{3}$. (**1–3** from *G. E. Atherton* 1147.)

green on upper surface, basal lobes strongly divergent; petiole longer than lamina; peduncle stout, usually about half as long as petiole, or less, 8–16 cm. long; spathe 15–30 cm. long; tube 3–4 cm. long, 1·5–2 cm. diam., subcylindric, green and sometimes tinged purplish externally, somewhat constricted at apex; limb ovate, acuminate, 5–13 cm. wide at the widest point, inner surface yellowish- to whitish-green, not spotted or blotched; spadix usually less than half as long as spathe, 6–13 cm. long, appendix clavate or cylindric, distinctly stipitate towards base, 3·5–10 cm. long, 0·5–0·8 cm. diam. or more, pale yellow; sterile filaments present above and below male zone.

HAB.: Moist, shaded places and grassy banks, on chalk and limestone; 700–3,000 ft. alt.; fl. May.

DISTR.: Divisions 2, 7, rare. Also Azores eastwards to Turkey.

2. On chalk near Phini, 1862, *Kotschy* 739!
7. Lapithos, 1955, *Merton* 2226!

[A. CRETICUM *Boiss. et Heldr.* (in Boiss., Diagn., 2, 13: 9; 1853), tentatively reported from Cyprus by Kotschy (in Oesterr. Bot. Zeitschr., 12: 279; 1862) is unquestionably an error, nor is the record repeated in Die Insel Cypern (1865).]

2. ARISARUM *Mill.*

Gard. Dict. Abridg., ed. 4, (1754).

Terrestrial herbs; stem tuberous to rhizomatous, ovoid to cylindric; leaves 1 to 3, radical, cordate-sagittate, sagittate or sagittate-hastate, fine venation reticulate; inflorescence solitary, terminal; spathe tubular in lower part with connate margins, erect, upper part cucullate, procurved, shorter than tube or very elongated and tapering to an erect, subulate appendix; spadix shorter than spathe, erect, ± procurved towards apex, basal pistillate part very short and few-flowered, contiguous with upper staminate part, terminal half of spadix a cylindric to clavate, sterile appendix; flowers unisexual, naked; stamens ± scattered and distant from one another; filaments distinct; anthers with thecae connate at apex, dehiscing by a single continuous slit; ovary unilocular, with several to many ovules on a basal placenta, style attenuate, stigma capitate. Berries few, obconic, truncate at apex, dull greenish-brown; seeds with endosperm.

Two or three species, occurring throughout the Mediterranean region and west to the Canaries and Azores.

1. **A. vulgare** *Targ.-Tozz.* in Ann. Mus. Fis. Firenze, 2 (2): 67 (1810); Boiss., Fl. Orient., 5: 44 (1882); Engl., Pflanzenr., 73 (IV. 23F.): 145 (1920); Post, Fl. Pal., ed. 2: 555 (1933); Osorio-Tafall et Seraphim, List Vasc. Plants Cyprus, 18 (1973); Elektra Megaw, Wild Flowers of Cyprus, 18, t. 39 (1973).
    Arum arisarum L., Sp. Plant., ed. 1, 966 (1753); Sm. in Sibth. et Sm., Fl. Graec. Prodr., 2: 246 (1816), Fl. Graec., 10: 38, t. 948 (1840); Poech, Enum. Plant. Ins. Cypr., 11 (1842).
    A. incurvatum Lam., Fl. Fr., 3: 538 (1778).
    Arisarum veslingii Schott, Syn. Aroid., 4 (1856).
    A. libani Schott, Prodr. Syst. Aroid., 21 (1860); Unger et Kotschy, Die Insel Cypern, 210 (1865).
    A. sibthorpii Schott, Prodr. Syst. Aroid., 21 (1860); Unger et Kotschy, Die Insel Cypern, 210 (1865).
    A. crassifolium Schott in Bonplandia, 9: 369 (1861); Unger et Kotschy, Die Insel Cypern, 211 (1865).
    A. vulgare Targ.-Tozz. var. *veslingii* (Schott) Engl. in DC., Mon. Phan., 2: 563 (1879); Boiss., Fl. Orient., 5: 44 (1882).
    A. vulgare Targ.-Tozz. ssp. *incurvatum* (Lam.) Holmboe, Veg. Cypr., 43 (1914); Osorio-Tafall et Seraphim, List Vasc. Plants Cyprus, 18 (1973).

A. *vulgare* Targ.-Tozz. ssp. *veslingii* (Schott) Holmboe, Veg. Cypr., 43 (1914); Osorio-Tafall et Seraphim, List Vasc. Plants Cypr., 18 (1973).

TYPE: "*in* Mauritaniae, Italiae, Lusitaniae, Hispanicae, Galloprovinciae *nemoribus*".

Tuber subcylindric, 1–2 cm. diam., with vertical growing point, often producing short, more slender, horizontal rhizomes and forming dense stands; leaves 1–2, cordate- to hastate-sagittate, 3–17 cm. long, 2·5–10 cm. wide across retrorse to divergent, usually rounded basal lobes, glossy green, median lobe acute to obtuse and shortly cuspidate; petiole much longer than lamina, 10–40 cm. long, brownish-green speckled; peduncle equalling or exceeding petiole, 10–40 cm. long, slender, purplish-green speckled; spathe 3·5–7 cm. long; tube erect, cylindric, equalling or sometimes longer than limb, white with longitudinal green, purplish-green or brownish-green stripes, 2·5–4·0 cm. long, 1–1·5 cm. diam.; limb green, purplish- or reddish-brown, about 2 cm. wide, cucullate, procurved, ovate, shortly acuminate; spadix equalling or slightly exceeding spathe, appendix very slender, narrowly clavate to narrowly cylindric, 2–4 cm. long, usually less than 2 mm. diam., upper part slightly thicker, exserted, procurved and pendent, greenish- to brownish-purple.

HAB.: Shaded rock-crevices, or shaded banks on limestone or igneous hillsides; sea-level to 1,000 ft. alt.; fl. Dec.–May.

DISTR.: Divisions 1, 2, 4, 6–8. Widespread in the Mediterranean region.

1. Near Ktima, 1859, *Kotschy* 59; Toxeftera near Ayios Yeoryios (Akamas), 1962, *Meikle* 2092! Near Polis, 1962, *Meikle* s.n.!
2. Skouriotissa, 1927, *Rev. A. Huddle* 105!
4. Larnaca, 1862, *Kotschy* 280a; also, 1880, *Sintenis & Rigo* 131! 865! between Larnaca Salt Lake and the Phaneromene, 1862, *Kotschy*; Ayia Napa, 1905, *Holmboe* 40; Ormidhia, 1905, *Holmboe* 82; Dhekelia, 1973, *P. Laukkonen* 103!
6. Dhiorios, 1962, *Meikle*!
7. Pentadaktylos, 1880, *Sintenis & Rigo* 131a! Lapithos, 1934, *Syngrassides* 492! Ayios Amvrosios, 1937, *Syngrassides* 1445! 1729! Near Kyrenia, 1937–1956, *Mrs. Stagg* s.n.! *Casey* 150! 200! *G. E. Atherton* 753! 818! Vasilia, 1940, *Davis* 2007! Above Keurmurju, 1959, *P. H. Oswald* 36! Pakhyammos, 1959, *P. H. Oswald* 64! Ayios Epiktitos–Kyrenia road, 1968, *S. Economides* in ARI 1155!
8. Karpas, 1937, *Roger-Smith* s.n.!

COLOCASIA ESCULENTA (*L.*) *Schott* in Schott et Endl., Melet. Bot., 18 (1832); Nicolson in A. C. Smith, Fl. Vit. Nov., 1: 456 (1979).

Arum esculentum L., Sp. Plant., ed. 1, 965 (1753).

A. colocasia L., Sp. Plant., ed. 1, 965 (1753); Sibth. et Sm., Fl. Graec. Prodr., 2: 245 (1816); Poech, Enum. Plant. Ins. Cypr., 12 (1842).

Colocasia antiquorum Schott in Schott et Endl., Melet. Bot., 18 (1832); Unger et Kotschy, Die Insel Cypern, 211 (1865); Boiss., Fl. Orient., 5: 45 (1882); Holmboe, Veg. Cypr., 43 (1914); Engler et Krause in Engl., Pflanzenr., 71 (IV. 23E.): 65 (1920); Post, Fl. Pal., 2: 556 (1933).

Colocasia antiquorum Schott var. *esculenta* (L.) Schott ex Seem., Fl. Vit., 284 (1868); Engler et Krause in Engl., Pflanzenr., 71 (IV. 23E.): 67 (1920).

Colocasia esculenta (L.) Schott var. *antiquorum* (Schott) Hubbard et Rehder in Bot. Mus. Leafl., 1 (1): 5 (1932).

Colocasia antiquorum Schott ssp. *esculenta* (L.) Haudricourt in Rev. Bot. Appl. Agr. Trop., 21: 62 (1941).

Robust, tuberous perennial; leaves all radical; petioles up to 85 cm. long, exceeding the lamina, sheathing for ⅓ of base, lamina pendent, peltate, cordate-sagittate, about 30 cm. long, or more, 20 cm. wide or more, glaucescent on lower surface; inflorescences rarely if ever produced in Cyprus; peduncle up to 50 cm. long; expanded part of spathe yellowish, about 20 cm. long, 2·5 cm. wide, spadix much shorter than spathe, with white, cylindric to clavate staminodes scattered among the pistils, intermediate sterile part about 1 cm. long, staminate part 4–4·5 cm. long, naked terminal appendix under 3 cm. long, conical, rugulose.

Native of tropical Southeast Asia and cultivated throughout the tropics and widely grown in the lowlands of Cyprus. According to Holmboe it was first recorded for the island by Etienne de Lusignan in 1573, though he mentions that it is represented on 13th and 14th century capitals. It is also noted for Cyprus by Linnaeus (1753) and Sibthorp (1789). Gaudry (Recherches Sci. en Orient, 168; 1855) says it is particularly grown on the south coast about Paphos, and Kotschy (Die Insel Cypern, 211; 1865) records it from Lefkara, Kiti, Morphou, Panteleimon, Kyrenia and Rizokarpaso; he says the best tubers come from Lefkara. *Colocasia esculenta* is still fairly popular in cultivation and occasionally survives in moist situations as an escape.

Records for *Colocasia* in Dioscorides and Pliny appear to refer to a different plant, *Nelumbo nucifera*, but the term Colocasi had a wider application, being used, it would appear, for a range of edible root crops (see Portères in Journ. Agric. Trop. Bot. Appl., 7: 169–192; 1962).

102. ALISMATACEAE
F. Buchenau in Engl., Pflanzenr., 16 (IV. 15): 1–66 (1903).

Aquatic and marsh plants, usually perennial; leaves generally basal, sometimes submerged or floating; petioles sheathing; inflorescence scapose; flowers hermaphrodite or unisexual, generally in bracteate umbels or branched verticels; sepals 3, persistent; petals 3, imbricate, often caducous; stamens 3–many, inserted spirally or in whorls; filaments usually elongate; anthers 2-thecous, dehiscing extrorsely or by lateral, longitudinal slits; ovaries 3–many, in spirals, whorls or scattered over a capitate receptacle, generally 1-locular; ovules 1–many; style terminal or subventral; stigma terminal or lateral. Fruit usually a cluster or head of achenes or follicles; seeds mostly small, without endosperm; embryo curved or horseshoe-shaped.

About 13 genera widely distributed in temperate and tropical regions.

Fruits oblong, blunt; pedicels scarcely incrassate in fruiting specimens - - - **1. Alisma**
Fruits narrowly triangular, acute or shortly rostrate, spreading stellately when ripe; pedicels distinctly incrassate in fruiting specimens - - - - - **2. Damasonium**

1. ALISMA *L.*
Sp. Plant., ed. 1, 342 (1753).
Gen. Plant., ed. 5, 160 (1754).
I. Bjorkquist in Opera Bot. (Lund), 17: 1–128 (1967); 19: 1–138 (1968).

Leaves sometimes floating or submerged, generally lanceolate or narrowly ovate, sometimes subcordate; flowers hermaphrodite, usually in branched verticels, occasionally in racemes or umbels; stamens 6, inserted in pairs opposite the petals; ovaries numerous, free, 1-locular, arranged in a single whorl; style subventral. Fruit a laterally compressed, ventrally beaked achene.

About 10 species in the northern hemisphere.

1. A. lanceolatum *With.*, Arrang. Brit. Plants, ed. 3, 2: 362 (1796); Lindberg f., Iter Cypr., 6 (1946).

A. *plantago-aquatica* L. var. *lanceolata* (With.) Sm., Fl. Brit., 1: 401 (1800); Post, Fl. Pal., ed. 2, 2: 538 (1933).

TYPE: not indicated; probably the plate in Gerard's Herball, ed. 1, 337, fig. 2 (1597)!

Erect, glabrous perennial to 1 m. high; leaves all basal, lamina narrowly elliptic, tapering to apex and base, 8–27 cm. long, 1·5–6 cm. wide, entire, generally with 3–5 prominent primary nerves, and numerous, close, parallel, divaricate, secondary nerves, apex sharply acuminate; petioles 6–20 cm. long, base shortly sheathing; inflorescence a lax, diffuse, branched panicle of verticels, 20–50 cm. or more long, 11–30 cm. or more wide, axis and primary branches often somewhat angled; bracts scarious, ovate or lanceolate-acuminate, 0·4–2·5 cm. long, 0·2–0·5 cm. wide; pedicels slender, straight, 1–3 cm. long; sepals greenish, persistent, subcoriaceous, concave, obtuse or shortly apiculate, 2·5–3 mm. long, 2–2·5 mm. wide; petals lilac-pink, marcescent, suborbicular, about 3·5 mm. diam.; filaments subulate, 1·3 mm. long; anthers oblong, about 1·3 mm. long, 0·8 mm. wide; ovary strongly compressed laterally, ovate-deltoid, glabrous, about 0·6 mm. long, 0·5 mm. wide, style about 1 mm. long; stigma decurrent along one side of style for about 0·3 mm. Achenes numerous, forming a single whorl about 4 mm. diam., oblong, compressed, about 2 mm. long, 1·5 mm. wide, pericarp whitish, spongy, translucent; seeds oblong, compressed, about 1·3 mm. long, 0·8 mm. wide; testa dark, shining, red-brown, translucent, irregularly rugulose.

HAB.: Marshy ground by the sea; near sea-level; fl. April–July.

DISTR.: Division 4, rare. Widely distributed in Europe, the Mediterranean region, Africa and Asia.

4. Marshy shore near Famagusta, 1939, *Lindberg f.* s.n.!

2. DAMASONIUM *Mill.*

Gard. Dict. Abridg., ed. 4 (1754).

Aquatic and marsh plants, similar to *Alisma* in habit and general appearance, but with 6–9 carpels, radiating stellately in a single whorl, each containing 2-many ovules. Fruit a star-shaped whorl of compressed, acute or shortly rostrate follicles.

About 5 species in Europe, the Mediterranean region, western and Central Asia, Australia and California.

1. **D. alisma** *Mill.*, Gard. Dict., ed. 8, no. 1 (1768); Holmboe, Veg. Cypr., 31 (1914).

Alisma damasonium L., Sp. Plant., ed. 1, 343 (1753).

Damasonium bourgaei Coss., Notes Plant. Nouv. Crit. Midi Espagne, 2: 47 (1849); Boiss., Fl. Orient., 5: 10 (1882); Post, Fl. Pal., ed. 2, 2: 538 (1933).

D. alisma Mill. var. *compactum* M. Micheli in DC., Mon. Phan., 3: 43 (1881); Buchenau in Engl., Pflanzenr., 16 (IV. '15): 19 (1903).

D. alisma Mill. ssp. *bourgaei* (Coss.) Osorio-Tafall et Seraphim, List Vasc. Plants Cyprus, 5 (1953) nom. invalid.

TYPE: "a native of England", *Miller* (? BM).

Erect or spreading, glabrous annual, 8–40 cm. high; leaves all basal; lamina elliptic-oblong, 2–7 cm. long, 0·5–2 cm. wide, entire, generally with 3 rather prominent primary nerves, and indistinct divaricate secondary nerves, apex sharply acute or obtuse, base subcordate, petiole up to 19 cm. long, sometimes tinged pink; inflorescence a rather rigid, branched panicle of verticels, branches erecto-patent, 8–10 cm. or more long, or sometimes very reduced or almost wanting; bracts membranous, ovate-lanceolate, up to 10 mm. long, 4 mm. wide, pedicels rigid, 8–15 mm. long, thickening in fruit; sepals papyraceous, persistent but soon shrivelling, broadly and bluntly ovate, concave, about 2 mm. long, 1·5 mm. wide, petals pinkish, inconspicuous, marcescent, suborbicular, about 2·5 mm. long, 2 mm. wide; filaments subulate, about 2 mm. long; anthers oblong, about 0·5 mm. long, 0·4 mm. wide; ovaries at first erect, fusiform, acuminate, laterally compressed, glabrous, about 2 mm. long, 0·6 mm. wide at base; style very short; stigma unilaterally decurrent for about 0·3 mm. Follicles 6, patent, narrowly triangular, longitudinally ribbed, about 6–7 mm. long, 2·5 mm. wide at base; seed oblong, about 1·8–2 mm. long, 1 mm. wide; testa dark brown, closely transversely rugulose, with minutely and closely furrowed ridges.

HAB.: Wet sandy ground, or on mud by the edges of pools and lakes; sea-level to 100 ft. alt.; fl. April.

DISTR.: Divisions 4, 8, locally abundant. W. and S. Europe, Mediterranean region and eastwards to Central Asia.

4. Near Famagusta, about the mouth of the Pedieos R., 1905–1962, *Holmboe* 435; *Lindberg f.* s.n.! *Meikle* 2624! *F. J. F. Barrington* s.n.!
8. Galatia, 1962, *Meikle* 2535!

NOTES: Like many aquatics and paludals, very variable in size, the Barrington specimens are abnormally small, some less than 2 cm. high, but *D. alisma* can be as large as *Alisma lanceolatum* With. Varieties and subspecies based on differences in overall robustness are taxonomically worthless.

103. JUNCAGINACEAE
F. Buchenau in Engl., Pflanzenr., 16 (IV. 14): 1–19 (1903).

Perennial or annual marsh or aquatic plants; leaves basal, alternate, distichous, linear, sheathing at base; inflorescence a terminal spike or raceme; flowers small, greenish, hermaphrodite or occasionally unisexual; perianth-segments (2–) 6, sepaloid; stamens (2–) 3–6, inserted at base of perianth-segments; filaments very short or wanting; anthers 2-thecous, extrorse, dehiscing by longitudinal slits; carpels 3 or 6, free or connate, 1-ovulate; style wanting; stigma oblique. Fruit apocarpous or syncarpous; seed basal, erect, generally without endosperm; embryo straight.

Three genera with a wide distribution in temperate regions of both hemispheres.

1. TRIGLOCHIN *L.*
Sp. Plant., ed. 1, 338 (1753).
Gen. Plant., ed. 5, 157 (1754).
H. Horn af Rantzien in Svensk Bot. Tidskr., 55: 81–117 (1961).

Flowers hermaphrodite; perianth-segments 6, concave or somewhat saccate; stamens 6 or fewer by abortion; anthers sessile or subsessile; carpels 6 (often only the inner 3 fertile), each 1-ovulate, free or connate, separating from the central axils at maturity; seeds without endosperm.

About 12 species chiefly in temperate regions of both hemispheres.

1. **T. bulbosa** *L.*, Mantissa Altera, 226 (1771); Buchenau in Engl., Pflanzenr., 16 (IV. 14): 11 (1903); Holmboe, Veg. Cypr., 31 (1914); Lindberg f., Iter Cypr., 6 (1946); Osorio-Tafall et Seraphim, List Vasc. Plants Cyprus, 6 (1973).
   T. barrelieri Loisel., Fl. Gall., ed. 1, 705 (1807); Boiss., Fl. Orient., 5: 13 (1882).

T. bulbosa L. ssp. *barrelieri* (Loisel.) Rouy in Rouy et Fouc., Fl. France, 13: 271 (1912).
TYPE: "*ad* Cap. b. spei". (Cape of Good Hope).

Erect perennial 7–30 cm. high; rootstock bulbous, covered with a brown or fuscous tunic of fine or coarse, interwoven fibres; leaves linear, tapering to apex, glabrous, up to 23 cm. long, 0·1–0·3 cm. wide, base membranous whitish, sheathing, apex of sheath obscurely auriculate; scape glabrous; raceme 2–10 cm. long at anthesis, lengthening to 15 cm. or more in fruit; flowers remote or crowded, subsessile or with arcuate-ascending pedicels 2–3 mm. long at anthesis, lengthening to 5 mm. or more in fruit; perianth-segments submembranous, oblong, deeply concave, 1·8–2 mm. long, 1·5–1·8 mm. wide, apex rounded; anthers sessile, broadly oblong or suborbicular, yellow, about 1·2 mm. long and almost as wide; fertile carpels 3, about 2–3 mm. long, 0·8–1 mm. wide, connate almost to apex; stigmas oblique along adaxial surface of carpel apex, broadly ovate or suborbicular, about 0·3 mm. long. Fruit fusiform, trigonous, 5–6 mm. long, 1–1·5 mm. wide, shortly 3-lobed at apex, dehiscing upwards from base into 3 fruitlets; seed narrowly fusiform, about 4 mm. long, 0·8 mm. wide; testa membranous, striatulate, adhering to the dark brown, fibrilliferous carpel-wall. *Plate 99.*

HAB.: Salt-marshes and margins of brackish pools and lakes; near sea-level; fl. Febr.–April.

DISTR.: Divisions 3–5. S.W. Europe and Mediterranean region; also Angola, Zambia, Malawi and South Africa.

3. Limassol, 1905, *Holmboe* 658.
4. Larnaca, 1905, *Holmboe* 98; Famagusta, 1939, *Lindberg f.* s.n.; near Paralimni, 1953, *Kennedy* 1782!
5. Ayios Seryios near Famagusta, 1936, *Syngrassides* 1156!

NOTES: Probably not uncommon in salt-marshes, but rarely collected.

104. POTAMOGETONACEAE

P. Graebner in Engl., Pflanzenr., 31 (IV. 11): 39–142 (1907).

Submerged or floating aquatic perennial or annual herbs, rarely subterrestrial; stems elongate, generally much branched; leaves alternate, rarely opposite or whorled, sessile or petiolate, generally with a membranous, stipuloid sheath or scale, free or partly adnate to the leaf-base; inflorescence a pedunculate, ebracteate spike; flowers hermaphrodite; perianth-segments 4, often shortly clawed, sepaloid; stamens 4, inserted at base of perianth-segments; anthers sessile, 2-thecous, opening by longitudinal slits; carpels (1–) 4, free or basally connate, 1-ovulate; style short; stigma truncate or subtruncate. Fruit a small drupe or achene; seed without endosperm; embryo hooked or spirally coiled.

Two genera (*Potamogeton* L. and *Groenlandia* Gay) with a cosmopolitan distribution.

1. POTAMOGETON *L.*

Sp. Plant., ed. 1, 126 (1753).
Gen. Plant., ed. 5, 61 (1754).

Leaves (apart from those subtending the inflorescences) alternate. Fruits drupaceous, with a hard endocarp. Otherwise characters those of the family.
About 90 species, with a cosmopolitan distribution.

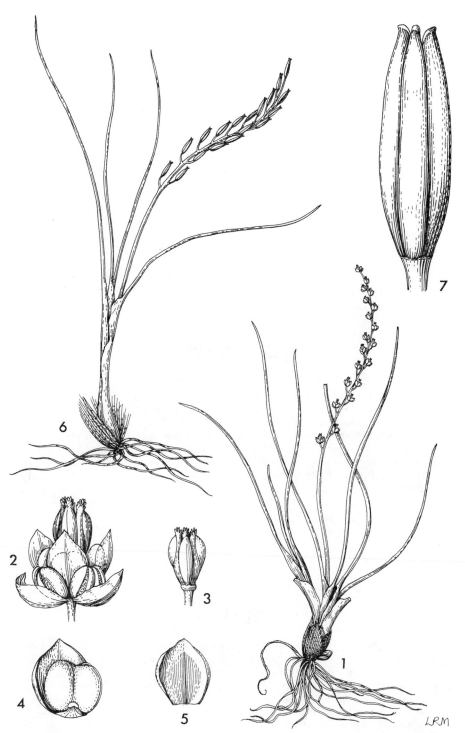

Plate 99. TRIGLOCHIN BULBOSA L. **1,** habit, ×$\frac{2}{3}$; **2,** flower, ×10; **3,** gynoecium, ×10; **4,** perianth-segment and stamen, ×10; **5,** perianth-segment, ×10; **6,** fruiting specimen, habit, ×$\frac{2}{3}$; **7,** fruit, ×10. (**1–5** from *Kennedy* 1782; **6, 7** from *Bornmueller* 10018.)

1. P. nodosus *Poir.* in Lam., Encycl. Méth., Suppl. 4: 535 (1816); Osorio-Tafall et Seraphim, List Vasc. Plants Cyprus, 6 (1973).

    P. fluitans Roth, Tent. Fl. Germ., 1: 72 (1788), 2: 202 (1789) pro parte; Boiss., Fl. Orient., 5: 16 (1882); Post, Fl. Pal., ed. 2, 2: 541 (1933).

    [*P. natans* (non L.) A. K. Jackson in Kew Bull., 1937, 345 (1937).]

    TYPE: Canaries; "Cette plante a été recueillie aux îles Canaries par M. Broussonet" (P).

Stems elongate; leaves submerged and floating, the submerged thin, translucent, lanceolate or elliptic, up to 20 cm. long, 4 cm. wide, apex obtuse, base narrowing to a well-developed petiole, the floating subcoriaceous, opaque, elliptic or ovate-elliptic, to 16 cm. long, 6 cm. wide, apex obtuse, base broadly cuneate or almost rounded; petioles of floating leaves often very long, equalling or up to twice as long as lamina; sheaths conspicuous, membranous, up to 5 cm. long, rather persistent; peduncle stout, up to 20 cm. long, generally much longer than subtending leaves; spike cylindrical, compact, about 4 cm. long, 0·6 cm. wide; perianth-segments suborbicular, 1·8–2 mm. diam.; anthers suborbicular, 0·8 mm. diam.; carpel strongly compressed laterally, about 1·5 mm. diam.; style about 0·5 mm. long, rather thick; stigma small, suborbicular. Fruit compressed, broadly ovoid, about 3 mm. long, 2·5 mm. wide, brownish-red, rugulose, keeled dorsally; seed hooked, about 2·5 mm. long; testa membranous, pale brown.

    HAB.: In slow-running streams, pools and water-tanks; sea-level to 3,500 ft. alt.; fl. May–July.

    DISTR.: Divisions 1, 2, 6. Cosmopolitan distribution.

1. Khrysokhou, 1934, *Syngrassides* 842!
2. Platania, Forestry nursery, in an old water-tank, 1955, *Merton* 2330!
6. Syrianokhori, pond near village, 1952, *Merton* 793!

P. PERFOLIATUS *L.*, Sp. Plant., ed. 1, 126 (1753) is listed by Osorio-Tafall & Seraphim (List Vasc. Plants Cyprus, 6; 1973), but I have not seen any specimens from Cyprus. It has no floating leaves and the submerged leaves are sessile, cordate and amplexicaul at the base, with a blunt or rounded, somewhat cucullate apex; the leaf-sheaths are small and evanescent.

    P. perfoliatus has a wide distribution in the northern hemisphere, but is generally rare in the eastern Mediterranean area, probably through lack of suitable habitats. In the absence of localized material it cannot be fully admitted to the flora.

105. RUPPIACEAE

Submerged, aquatic perennials; stems usually elongate, much branched; leaves alternate, or the involucral subopposite, linear or filiform, sheathing at the base, with membranous, auriculate sheaths; inflorescence a pedunculate, ebracteate, few-flowered spike; flowers 2, hermaphrodite; perianth wanting; stamens 2; anthers relatively large, sessile, dehiscing extrorsely by a longitudinal slit; carpels usually 4 (sometimes more), free, 1-ovulate, at first sessile, but becoming long-stipitate in fruit; stigma sessile, peltate. Fruit drupaceous, indehiscent; seeds slightly curved, without endosperm.

One genus with a cosmopolitan distribution.

1. RUPPIA L.

Sp. Plant., ed. 1, 127 (1753).
Gen. Plant., ed. 5, 61 (1754).

W. A. Setchell in Proc. Calif. Acad. Sci., ser. 4, 25: 469–478 (1946).
G. Reese in Zeitschr. für Bot., 50: 237–264 (1962).
G. C. Gamarro in Darwiniana, 14: 575–608 (1968).

Description that of the family.
About 6 species with a cosmopolitan distribution.

Leaves capilliform, subterete, up to 0·5 mm. wide; fruiting peduncles often twisted, but not
　spirally coiled　-　-　-　-　-　-　-　-　-　-　-　-　- **1. R. maritima**
Leaves filiform, flattened, 0·5–1 mm. wide; fruiting peduncles spirally coiled　- **2. R. cirrhosa**

1. R. maritima L., Sp. Plant., ed. 1, 127 (1753); Sibth. et Sm., Fl. Graec. Prodr., 1: 110 (1806);
　Poech, Enum. Plant. Ins. Cypr., 11 (1842); Unger et Kotschy, Die Insel Cypern, 210 (1865);
　Osorio-Tafall et Seraphim, List Vasc. Plants Cyprus, 6 (1973).
　　R. maritima L. var. rostrata Agardh in Physiogr. Sällsk. Årsb., 6: 37 (1823); Post, Fl.
　Pal., ed. 2, 2: 543 (1933).
　　R. rostellata Koch in Reichb., Icones Plant. Crit., 2: 66, t. 174, fig. 306 (1824); Boiss., Fl.
　Orient., 5: 20 (1882).
　　R. maritima L. ssp. rostellata (Koch) P. Graebner in Engl., Pflanzenr., 31 (IV. 11): 144
　(1906); Holmboe, Veg. Cypr., 31 (1914).
　TYPE: "in Europeae maritimis".

Stems very slender, often much branched and forming loose or dense
submerged mats; leaves capilliform, extremely slender and subterete, up to
10 cm. long, about 0·5 mm. wide, apex (often broken off in dried specimens)
acute, minutely but sharply serrulate, base sheathing, the sheaths not very
conspicuous, to about 10 mm. long, those of the upper, subopposite,
involucral leaves about 3·5 mm. wide, the others narrower; peduncles
slender, very short at anthesis, lengthening to 6–7 cm. in fruit, curved or
twisted, but not spirally coiled; anthers yellow, thecae broadly ellipsoid or
subglobose, about 1 mm. long and almost as wide; carpels usually 4 together,
clavate, about 0·5 mm. long, 0·3 mm. wide, at first sessile, becoming stipitate
later, with stipes up to 13 mm. long; stigma sessile, peltate, about 0·3 mm.
diam. Fruit ovoid, compressed, about 2·5 mm. long, 1·8 mm. wide, apex
shortly and bluntly rostellate; seed with a very thin, membranous, whitish
testa.

HAB.: In brackish pools and ditches; near sea-level; fl. April–July.

DISTR.: Divisions 3, 4, 8. Cosmopolitan.

3. Akrotiri, near Limassol Salt Lake, 1979, Edmondson & McClintock E 2971!
4. Larnaca, 1862, Kotschy 267! also, 1905, Holmboe 567! Famagusta, 1970, A. Hansen 715!
8. Boghaz, 1970, A. Hansen 747!

2. R. cirrhosa (Petagna) Grande in Bull. Ort. Bot. Napoli, 5: 58 (1918).
　　Buccaferrea cirrhosa Petagna, Inst. Bot., 5: 1826 (1787).
　　Ruppia spiralis L. ex Dumort., Florula Belg., 164 (1827); Boiss., Fl. Orient., 5: 19 (1882)
　pro parte; Lindberg f., Iter Cypr., 6 (1946); Osorio-Tafall et Seraphim, List Vasc. Plants
　Cyprus, 6 (1973).
　　R. maritima L. var. spiralis (L. ex Dumort.) Moris, Stirp. Sard. Elench., 1: 43 (1827);
　Post, Fl. Pal., ed. 2, 2: 543 (1933).
　　R. maritima L. ssp. spiralis (L. ex Dumort.) P. Graebner in Engl., Pflanzenr., 31 (IV. 11):
　142 (1907); Holmboe, Veg. Cypr., 31 (1914).
　TYPE: Italy, near Naples; "In Lacu vulgo di Licola".

Superficially very similar to R. maritima L., but altogether more robust;
leaves filiform, distinctly flattened, up to 10 cm. long, 0·5–1 mm. wide, apex
subacute, entire or minutely serrulate, basal sheaths, especially those of
the upper, subopposite, involucral leaves, conspicuous, up to 5 mm. wide;

peduncles slender, generally shorter than involucral sheaths at anthesis, elongating to 8 cm. or more, and becoming spirally coiled in fruit; anthers yellow, thecae broadly oblong, about 2 mm. long, 1 mm. wide; carpels usually 5 or more together, broadly clavate, about 0·5 mm. long, 0·3 mm. wide, at first sessile, becoming stipitate later, with stipes up to 15 mm. long; stigma sessile, peltate, about 0·3 mm. diam. Fruit ovoid, compressed, distinctly but bluntly rostellate, about 3 mm. long, 1·8 mm. wide; seed with a very thin, whitish, membranous testa.

HAB.: In brackish pools and ditches; near sea-level; fl. April–July.

DISTR.: Division 4. Cosmopolitan.

4. Larnaca, 1928, *Druce* s.n.! also, 1939, *Lindberg f.* s.n.; Famagusta, 1939, *Lindberg f.* s.n.!

105a. NAJADACEAE

Submerged annual or perennial aquatics; stems slender, much branched; leaves subopposite or apparently whorled; linear, entire or spinose-dentate, with a sheathing base; flowers unisexual (monoecious or dioecious), minute, axillary; male flower generally enveloped by a membranous, tubular spathe, and an inner, membranous, 2-lobed perianth, comprising a single, subsessile, 1–4-thecous anther, dehiscing by an apical slit; pollen-grains subglobose or ellipsoid; female flower naked or enveloped in a spathe; carpel 1, 1-locular, 1-ovulate; stigmas (or stigmatic arms) 2–4, elongate, filiform. Fruit a small, ellipsoid or fusiform nut; exocarp reticulate or alveolate; seed without endosperm.

NAJAS MINOR *All.*, Auct. Syn. Stirp. Hort. Taur., 3 (1773) has been listed by Osorio-Tafall & Seraphim (List Vasc. Plants Cyprus, 7; 1973), but I have no other evidence of its occurrence on the island. It is a slender annual, with crowded, narrowly linear, translucent, spinose-dentate leaves, up to 30 mm. long, 1 mm. wide; sheaths with rounded or truncate auricles; flowers monoecious, the male flowers with a spathe, the female naked. *N. minor* has been found elsewhere in the eastern Mediterranean region but is uncommon there. Its occurrence in Cyprus requires confirmation by more substantial evidence.

106. POSIDONIACEAE

Submerged, marine, rhizomatous perennials; roots tough; rhizome densely covered with the intact or frayed leaf-sheaths; leaves sessile, distichous, flat, grass-like, with an obtuse apex and distinct, articulated basal sheath; inflorescence terminal, borne on a flattened scape; flowers hermaphrodite or the uppermost functionally male, in short, congested, axillary spikes subtended by leaf-like bracts; perianth absent; stamens 3;

anthers sessile or subsessile, rather large, dehiscing extrorsely by longitudinal slits, connective dilated and produced into an aristate or muticous acumen; pollen filiform; carpel 1, ovoid or ellipsoid, 1-ovulate; stigma sessile, irregularly laciniate. Fruit drupaceous, resembling a small Olive, dehiscing irregularly near apex, the juvenile growths appearing as a small tuft of sheathing leaves.

One genus only.

1. POSIDONIA *Koenig*

in Koenig et Sims, Ann. Bot., 2: 95, t. 6 (1805) nom. cons.

Description, etc. that of the family.

Four species, one in the Mediterranean and adjacent parts of the Atlantic, the others in Australian waters.

1. **P. oceanica** (*L.*) *Del.*, Descr. Égypte Hist. Nat., 2: 78 (1813); Boiss., Fl. Orient., 5: 26 (1882); P. Graebner in Engl., Pflanzenr., 31 (IV. 11): 38 (1907); Holmboe, Veg. Cypr., 31 (1914); Post, Fl. Pal., ed. 2, 2: 546 (1933); Osorio-Tafall et Seraphim, List Vasc. Plants Cyprus, 6 (1973).

    Zostera oceanica L., Mantissa 1, 123 (1767).

    Posidonia caulini Koenig in Koenig et Sims, Ann. Bot., 2: 96 (1805).

    TYPE: "*in* Oceano".

Rhizomatous perennial forming a dense turf, usually on sand in deep water; rhizomes covered with densely imbricating pale, glossy brown fibrous leaf-sheaths; leaves 30–35 cm. long, about 0·7 cm. wide, entire, glabrous, apex rounded, base abruptly articulated, convex, joined to a pallid, concave-tipped sheath up to 5 cm. long, 1–1·5 cm. wide; scape 15–25 cm. long, flattened; bracts 2–5 cm. long, 0·3–0·6 cm. wide, blunt or rounded, distichous; flower-spikes with a stout axis 15–20 mm. long; flowers usually 3, sessile; anther-connective broadly oblong-obovate, about 5 mm. long, 4 mm. wide, scarious, brown spotted, apex subtruncate with a median aristate acumen 2·5–3 mm. long; anther-thecae attached to the dorsal surface of the connective, about 3·5 mm. long, 1·5 mm. wide; carpel narrowly ellipsoid, about 3 mm. long, 1 mm. wide; stigma sessile, deeply and narrowly laciniate, some of the lobes up to 1 mm. long. Fruit ellipsoid, about 1·5 cm. long, 1 cm. wide; pericarp thick, whitish and chalky internally.

    HAB.: In 2–25 ft. of water, or deeper, usually on sand; fl. ? Oct.

    DISTR.: Divisions 1–4. Throughout the Mediterranean.

1. Paphos, 1929, *Druce*; also 1956, *Merton* 2778!
2. Pomos, 1934, *Syngrassides* 961!
3. Petra tou Romiou, 1978, *J. Holub* s.n.!
4. Larnaca, 1905, *Holmboe* 212.

    NOTES: Probably all around the coast, but usually growing in deep water and seldom collected.

107. ZANNICHELLIACEAE

Submerged aquatic perennials (or annuals) with slender, rooting rhizomes; leaves alternate or subopposite, linear, sheathing at the base; flowers unisexual (monoecious), the two sexes often closely approximate, solitary: perianth present or absent; stamens 1–3, sometimes connate;

anthers 1–2-thecous, dehiscing by longitudinal slits; pollen-grains globose; carpels 1–9; free, sessile or stipitate, 1-ovulate; style simple or lobed; stigmas often dilated, peltate. Fruit achaenial, indehiscent; seed without endosperm.

Three genera, one with a cosmopolitan distribution, the others in Europe and Australia.

1. ZANNICHELLIA L.

Sp. Plant., ed. 1, 969 (1753).
Gen. Plant., ed. 5, 416 (1754).
G. Reese in Biol. Zentralbl., 86, suppl.: 277–306 (1967).

Slender rhizomatous perennials (or annuals); flowers axillary, solitary, monoecious, the male and female closely approximate; male flower with an elongated peduncle (like a filament); anthers 1 or 2, sometimes connate, 2-thecous (or apparently 4-thecous where connate); female flowers with a small, cupular perianth; carpels (1–) 4 (–9), often keeled and crenate dorsally; styles simple; stigmas peltate.

Two species, one cosmopolitan, the other endemic in S. Africa.

1. **Z. palustris** L., Sp. Plant., ed. 1, 969 (1753); Boiss., Fl. Orient., 5: 14 (1882); Holmboe, Veg. Cypr., 31 (1914); Post, Fl. Pal., ed. 2, 2: 541 (1933); Osorio-Tafall et Seraphim, List Vasc. Plants Cyprus, 6 (1973).
   Z. *palustris* L. var. *pedicellata* Wahl. et Rosén in Nov. Act. Sci. Upsal., 8: 254 (1821).
   Z. *repens* Boenningh., Prodr. Fl. Monast., 272 (1824); Druce in Rep. B.E.C., 9: 471 (1931).
   Z. *pedicellata* (Wahl. et Rosén) Fries, Novit. Fl. Suec. Mantissa 3, 9: 135 (1845); Lindberg f., Iter Cypr., 6 (1946).
   Z. *marina* Nielsen in Bot. Tidsskr., ser. 2, 1: 204 (1872).
   Z. *palustris* L. ssp. *pedicellata* (Wahl. et Rosén) Holmboe, Veg. Cypr., 31 (1914); Osorio-Tafall et Seraphim, List Vasc. Plants Cyprus, 6 (1973).
   Z. *palustris* L. ssp. *marina* (Nielsen) Holmboe, Veg. Cypr., 31 (1914).
   TYPE: "*in Europae, Virginiae fossis, fluviis*".

Slender, much-branched perennial (or annual sec. A. A. Obermeyer in *Fl. Southern Afr.*, 1: 77; 1966) forming loose, submerged mats; leaves alternate on sterile growths, usually opposite or ternate on fertile branches, linear, 20–60 mm. long, 0·3–0·5 mm. wide, tapering to an acute apex, base surrounded by a tubular, unilaterally fissured, membranous sheath; flowers axillary; male flower consisting of a single, naked stamen (or 2 fused stamens) borne at the side of the female flower; anther oblong, about 0·8 mm. long, 0·4 mm. wide, at first subsessile or shortly stipitate, the stipe (or filament) lengthening to 8 mm. or more at anthesis; thecae 2 (or 4 in fused anthers), connective produced as a blunt, brownish apiculus; female flower a shortly stipitate, membranous cupule (or perianth) containing (1–) 4 (–9), narrowly oblong-ovoid, yellowish, glabrous, sessile or shortly stipitate carpels about 0·4 mm. long, 0·2 mm. wide; style about 0·2 mm. long; stigma oblique, cochleariform, peltate, marcescent, about 0·6 mm. long, 0·4 mm. wide. Fruit oblong, curved, somewhat compressed laterally, about 2 mm. long, 1 mm. wide, pale brown when ripe, the dorsal margin keeled, and generally crenate or widely dentate, the apex rostrate with the persistent style; seed about 1·5 mm. long, 0·8 mm. wide; testa very thin, membranous, hyaline.

HAB.: Rock-pools, ponds and lakes; sea-level to 4,500 ft. alt.; fl. March–June.

DISTR.: Divisions 1, 2, 4, 6–8. Cosmopolitan.

1. Between Peyia and Ayios Yeoryios (Akamas), 1962, *Meikle* 2054!
2. Livadhia, 1939, *Lindberg f.* s.n.!
4. Larnaca, 1877, *Ball ex herb. Post* 636! also, 1928 or 1930, *Druce* s.n.; Kouklia, 1905, *Holmboe*

368, 381; Paralimni, 1905, *Holmboe* 424; Famagusta, 1939, *Lindberg f.* s.n.; Ayios Antonios, Sotira, 1952, *Merton* 730!
6. Myrtou, 1932, *Syngrassides* 96! Kokkini Trimithia, 1937, *Syngrassides* 1561! Between Skylloura and Philia, 1956, *Merton* 2766! Near Prophitis Elias Monastery beyond Ayia Marina, 1957, *Merton* 2886!
7. Kambyli, 1954, *Merton* 1870!
8. Cape Elea, 1905, *Holmboe* 513; Galatia, 1962, *Meikle* 2531!

NOTES: Presumably annual in the seasonal rock-pools where it commonly grows.

107A. ZOSTERACEAE

Submerged marine perennials with creeping, rooting rhizomes; leaves linear, grass-like, distichous, with compressed, sheathing bases; inflorescence enveloped by a sheathing, leaf-like spathe; flowers unisexual, monoecious or dioecious, arranged along a flattened axis (or spadix); perianth wanting; male flower consisting of 1 dorsifixed anther; pollen filiform; female flower sessile or shortly stipitate, consisting of a 1-locular carpel bearing 2 long, filiform stigmas. Fruit indehiscent or dehiscing irregularly; seed without endosperm.

Two species of the monoecious genus *Zostera* L., namely *Z. marina* L., Sp. Plant., ed. 1, 968 (1753) and *Z. nana* Roth, Enum. Plant. Phaen. Germ., 1: 8 (1827), are listed by Osorio-Tafall & Seraphim (List Vasc. Plants Cyprus, 6; 1973). I have, however, seen only one specimen from Cyprus purporting to belong to the genus, and this was a misidentification of *Cymodocea nodosa* (Ucria) Ascherson. Although both *Zostera marina* and *Z. nana* have been reported from adjacent coasts, neither species can be admitted to the flora of Cyprus without positive evidence of its occurrence on the island.

108. CYMODOCEACEAE

Submerged marine perennials with creeping, rooting rhizomes; leaves alternate or subopposite, sheathing at the base, often in distichous tufts at the ends of short unbranched stems; flowers unisexual (dioecious), generally subsessile or on very short lateral branches at the base of the leaf-tufts, without perianths; male flower consisting of 2 fused anthers borne on an elongating stipe (or filament), dehiscence extrorse by longitudinal slits; pollen filiform; female flowers of 2, collateral, 1-locular, 1-ovulate carpels; stigmas 2, long, subulate. Fruit drupaceous, indehiscent; seed without endosperm.

Four genera and about 18 species widely distributed in tropical and subtropical waters.

1. CYMODOCEA *Koenig*

in Koenig et Sims, Ann. Bot., 2: 96, t. 7 (1805) nom. cons.

Marine perennials with tough, jointed, creeping rhizomes; stems erect, with conspicuous leaf-scars; leaves usually 2–7 in rather congested terminal tufts, grass-like or lorate, thin or subcoriaceous, basal sheath adnate to lamina, often conspicuous and pallid, auriculate, folded around the stem; flowers on abbreviated shoots at the base of the leaf-tufts, subtended by reduced, bract-like leaves. Fruit drupaceous, with a bony pericarp.

Seven species widely distributed in tropical and subtropical waters; one species extending into the Mediterranean.

1. **C. nodosa** (*Ucria*) *Ascherson* in Sitz.-Ber. Ges. Naturf. Freunde Berlin, 1869: 4 (1869); Boiss., Fl. Orient., 5: 21 (1882); P. Graebner in Engl., Pflanzenr., 31 (IV. 11): 147 (1907); Post, Fl. Pal., ed. 2, 2: 544 (1933); Lindberg f., Iter Cypr., 6 (1946).
    Phucagrostis major Theophrasti Cavolini, Phucagrost. Theophrast. Anthesis, XIII, t. 1 (1792) nom. invalid.
    Zostera nodosa Ucria in Nuov. Racc. Opusc. Aut. Sicil., 6: 12 (1793).
    Cymodocea aequorea Koenig in Koenig et Sims, Ann. Bot., 2: 96 (1805).
    Phucagrostis major Cavolini ex Willd., Sp. Plant., 4 (2): 649 (1806); Bornet in Ann. Sci. Nat., ser. 5, 1: 5–51 (1864).
    Cymodocea major (Cavolini ex Willd.) Grande in Nuov. Giorn. Bot. Ital., n.s., 27: 238 (1920); Osorio-Tafall et Seraphim, List Vasc. Plants Cyprus, 7 (1973).
    TYPE: Not indicated; possibly an unpublished plate in Cupani, *Pamphyton Siculum.*

Creeping rhizomatous perennial; rhizomes often 30 cm. or more long, rooting at the nodes in mud or sand; stems short, erect, with conspicuous transverse leaf-scars; leaves linear, grass-like, up to 30 cm. or more long, 0·2–0·4 cm. wide, apex rounded, sharply serrulate, margins flat, entire or remotely serrulate towards apex, base enveloped by whitish membranous sheaths up to 5 cm. long, with truncate, auriculate apices; flowers developing at the base of the leaf-sheaths; male flower a naked stamen (or 2 fused stamens), stipe, or apparent filament, up to 3·5 cm. long at anthesis; anther linear, 10–12 mm. long, 1·5 mm. wide, 4-thecous; female flower of 2 collateral, broadly ovoid, acute carpels, 5–7 mm. long, 4–5 mm. wide; stigmas filiform, 5–7 mm. long. Fruit a broadly ovoid, acute drupe or nut about 8–10 mm. long, 6–8 mm. wide.

HAB.: On sand under 2 feet or more of water; fl. June–Aug.

DISTR.: Divisions 1, 4. Mediterranean and Atlantic coasts.

1. Paphos, in the harbour, 1956, *Merton* 2777!
4. Larnaca, Perivolia, 1939, *Lindberg f.* s.n.!

NOTES: Probably much more frequent around the coasts of Cyprus than the records suggest, but, like other thalassophytes, seldom collected, and even more rarely collected in flower.

109. CYPERACEAE

A. L. Juss., Gen. Plant., 26 (1789).

C. B. Clarke in Bull. Misc. Inf. (Kew), add. ser. 8 (1908), Ill. Cyp. (1912).

by Sheila S. Hooper

Rhizomatous perennials or, less often, fibrous-rooted annuals; rhizome-scales grading into sheathing cauline leaves; ligular appendages sometimes present; lamina without articulation between sheath and blade, generally linear and bifacial; sheath closed, or rarely open; cauline stem solid, often trigonous, usually with nodes only at the base; branches usually bearing a 2-keeled or 2-lobed prophyll at the base; inflorescence generally bracteate, various, but generally capitate or irregularly umbellate (anthelate); flowers very small, each in the axil of an often chaffy bracteole (glume), unisexual or more often bisexual, generally aggregated either in 2 ranks (distichously) into flattened spikelets, or spirally into rounded (terete) spikelets; perianthal structures absent or in some bisexual flowers bristle-like (in *Carex* a sac-like prophyll or utricle envelops the pistil); stamens free, predominantly 1–3; anthers basifixed, longitudinally dehiscent, generally with a crested connective; ovary superior, 1-locular; ovule solitary, basal, erect, anatropous; style sometimes enlarged at base, continuous or articulated with the ovary, generally with 2–3 papillose branches (stigmas). Fruit (nut, achene) indehiscent, hard-walled, often trigonous or lenticular; endosperm present. Anatomy with (chlorocyperoid), or without (eucyperoid) an inner chlorenchymatous layer to the bundle sheath and radiate chlorenchyma; silica bodies almost always present. For embryo types see Van der Veken, P. A. J. B. in Bull. Jard. Bot. Brux., 35: 285–354 (1965).

A cosmopolitan family of about 90 genera and 4,000 species, with a preponderance of tropical genera; predominantly monoecious herbs with a preference for wet, open habitats.

Pistils and stamens never within the same glume (flowers unisexual); ovary concealed within a
    flask- or bottle-shaped covering (utricle) - - - - - - - **11. Carex**
Pistils and stamens within the same glume (flowers bisexual); pistil exposed, though sometimes
    surrounded by bristles:
    Only the central glumes of the spikelet fertile, generally markedly larger than the sterile basal
        and apical glumes:
        Internodes of the rhachilla very short between the lower empty glumes, enlarged and zig-
            zag between the fertile ones; leaves subterete and smooth; stamens 3   **9. Schoenus**
        Internodes of the rhachilla all similar in length; leaves pseudo-dorsiventral and very
            scabrid; stamens 2; plants very large - - - - - **10. Cladium**
    All glumes of the spikelets, except occasionally the lowermost, fertile and similar in size:
        Glumes in 2 ranks and generally sharply V-shaped, so that the spikelets are flattened;
            rhachilla quadrangular in section:
            Nut trigonous or rounded (occasionally flattened with a wide face against the rhachilla);
                spikelets more or less elliptical in section; stigmas 3 (rarely 2) - **1. Cyperus**
            Nut lenticular with the narrow edge against the rhachilla; spikelets much compressed;
                stigmas 2 - - - - - - - - - - - **2. Pycreus**
        Glumes in several ranks, and either U-shaped or bluntly V-shaped; rhachilla cylindrical or
            nearly so, making spikelets terete:
            Spikelets solitary and terminal with no obvious bract, nut crowned by the persistent,
                enlarged style-base; leaf-blades reduced to an apiculus on the sheath **4. Eleocharis**
            Spikelets generally several, accompanied by stem-like or leaf-like bracts, or, if one only,
              then nut without a persistent style-base and plant leafy:
                Spikelets densely clustered in spherical heads; glumes few, very scabrid
                                  8. Scirpoides
                Spikelets loosely arranged in apparently lateral clusters, with or without rays; glumes
                    many, glabrous or pubescent:
                    Culms not leafy above base; bract stem-like, erect at least in the young
                      inflorescence:
                      Spikelets large, more than 5 mm. long; large perennials - **6. Schoenoplectus**
                      Spikelets small, not more than 4 mm. long, annuals or small perennials
                                      5. Isolepis
                Culms leafy; bracts leaf-like, flat, some at least spreading:
                  Glumes thin, apically retuse or lacerate, awned - - **7. Bolboschoenus**
                  Glumes firm, apically rounded, entire, apiculate - - **3. Fimbristylis**

SUBFAMILY 1. **Cyperoideae** At least some, often most, of the glumes in the spikelet enclosing both pistil and stamens (i.e. flowers bisexual). Includes various embryo types.

TRIBE 1. **Cypereae** (sens. lat. incl. *Scirpeae*). Glumes containing bisexual flowers occupying all the spikelet except, often, for the prophyll and reduced apical glumes; rhachilla internodes uniform in length.

1. CYPERUS *L.*

Sp. Plant., ed. 1, 44 (1753).
Gen. Plant., ed. 5, 26 (1754).
Juncellus (Griseb.) C. B. Cl. in Hook. f., Fl. Br. Ind., 6: 394 (1893).
Chlorocyperus Rikli in Jahrb. Wiss. Bot., 27: 563 (1895).
Cyperus subgen. *Eu-Cyperus* et *Juncellus* (Griseb.) C. B. Clarke; Kükenthal
in Engl., Pflanzenr., 101 (IV. 20): 42–326 (1935–1936).

Annuals or rhizomatous perennials generally with trigonous culms; flowers bisexual (or distal one male by reduction), without perianthal structures, distichously arranged in prophyllate spikelets; spikelets flattened, elliptic in section, typically lanceolate in outline, digitately or spicately arranged in rayed or sessile, simple or branched spikes, subtended by secondary bracts, the whole forming a once, twice or thrice anthelate (irregularly umbellate) or capitate inflorescence, subtended by one or more whorls of 3 primary bracts; glumes falling before the rhachilla and more or less at the same time as the ripe nut; stigmas 3, and nut trigonous or triquetrous, or occasionally stigmas 2, and then nut lenticular with its flat face against the widened side of the rhachilla. Embryo cyperoid. Anatomy chloro- or eu-cyperoid.

About 300 species (when *Pycreus, Mariscus*, etc. are excluded) predominantly in the tropics and subtropics of the Old World, extending to north central Europe and South Africa, and with about 150 species in the New World tropics.

Inflorescence a head of spikelets without rays:
Head obviously terminal, up to 6 cm. wide; far-creeping sand-dune plants -   **6. C. capitatus**
Head apparently lateral, flanked by an erect bract:
Bracts 2 or 3, leaf-like; leaves well developed, flat; nut dark, trigonous   -   **3. C. cyprius**
Bracts apparently solitary, terete and culm-like; leaf-blades reduced to an apiculus or when developed, thick, crescentiform or triangular; nut pale, compressed
7. C. laevigatus
Inflorescence spreading, anthelate, with short to long rays:
Plants with thick, far-creeping rhizomes:
Rays few, not more than 5 cm. long; slender, montane plants -   -   **3. C. cyprius**
Rays numerous, 8–12 (–20) cm. long; tall, lowland plants   -   -   **2. C. longus**
Plants with thin, or short, rhizomes, or rhizomes absent:
Spikelets not more than 5 mm. long, in small round, dense spikes   -   **5. C. fuscus**
Spikelets 7–40 mm. long:
Plant densely caespitose, without elongated rhizomes; glumes regularly and closely imbricate -   -   -   -   -   -   -   -   -   -   **4. C. glaber**
Plant not densely caespitose, with slender tuber-bearing rhizomes; glumes loosely imbricate -   -   -   -   -   -   -   -   -   **1. C. rotundus**

SUBGENUS 1. **Cyperus** Spikelets on an elongated rhachis, forming distinct spikes; rhachilla not or only slightly deeper than wide. Nut generally trigonous; stigmas generally 3. Anatomy chlorocyperoid.

1. **C. rotundus** *L.*, Sp. Plant., ed. 1, 45 (1753); Boiss., Fl. Orient., 5: 376 (1882); H. Stuart Thompson in Journ. Bot., 44: 340 (1906); Holmboe, Veg. Cypr., 40 (1914); Post, Fl. Pal., ed. 2, 2: 675 (1933); Lindberg f., Iter Cypr., 9 (1946); Osorio-Tafall et Seraphim, List Vasc. Plants Cyprus, 17 (1973).
   Chlorocyperus rotundus (L.) Palla in Allg. Bot. Zeitschr., 6: 61 (1900).
   TYPE: "*in* India".

Rhizomatous perennial; rhizomes numerous, slender, 1–2 mm. wide, wiry, purplish-brown, bearing purple-veined, red-brown scales, or their fibrous remains, either short and upwardly curved or long, directed downwards and ending in a swollen, finally woody, ellipsoid, obscurely ridged tuber (or "nut"); culms rather rigid, 15–60 cm. long, 1·5–2·5 (–6) mm. wide, triquetrous, ridged; leaves several, basal, crowded, shorter or longer than the culms; sheaths short, obscurely ridged dorsally, ventrally membranous, red-veined and dotted, soon splitting; ligule very small or none; blade flat, 2–5 (–10) mm. wide, gradually acuminate, minutely scabridulous marginally in the upper part, dark green above, pale or greyish-green beneath; inflorescence once or twice anthelate; bracts 3, leafy, unequal, the longest generally exceeding the rays; rays 3–9, compressed, smooth, 0·5–5 (–11) cm. long; spikes broadly ovoid, lax, rhachis glabrous or feebly scabrid, flexuous, slender, sometimes with short to long secondary rays; spikelets linear-lanceolate, 10–40 mm. long, about 2 mm. wide (sometimes accrescent and up to 60 mm. long), acute, becoming obtuse in fruit, pale to dark purple-red; glumes ovate, 3–3·5 mm. long, 2 mm. wide, obtuse, thin except for the green, acute, minutely excurrent, slightly curved, 3-nerved keel, sides brown to red-purple, each with 2 parallel nerves and a narrow to broad, hyaline margin decurrent in a transparent or red-dotted wing to the node below; stamens 3; anthers 1·5–2·5 mm. long, connective-tip conical, red; style about 1 mm. long; stigmas 3, long, much exserted. Nut (rather rarely produced) ellipsoid, about 1·5 mm. long, 0·6–1 mm. wide, red to blackish-brown, minutely apiculate, unequally trigonous with flat faces and sharp angles; surface smooth.

HAB.: A weed of moist, cultivated ground, ditches and waste land; sea-level to 700 ft. alt.; fl. July–Nov.

DISTR.: Divisions 1–3, 6, 7. A widespread weed in the warmer regions of the world.

1. Between Evretou and Sarama, 1941, *Davis* 3346!
2. Yialia, 1905, *Holmboe* 800; Mesapotamos Monastery, 1939, *Lindberg f.* s.n.!
3. Kalavasos, 1905, *Holmboe* 627; Kannaviou, 1939, *Lindberg f.* s.n.; Kolossi, 1939, *Lindberg f.* s.n.; Akhelia, 1939, *Lindberg f.* s.n.! also, 1967, *Economides* in ARI 1004! Asomatos, 1939, *Mavromoustakis* 90! Limassol, 1979, *Holub* s.n.! Phasouri, 1969, *A. Genneou* in ARI 1360!
6. Nicosia, 1939, *Lindberg f.* s.n.; and, 1973, *P. Laukkonen* 332, 388a.
7. Lapithos, 1932, *Syngrassides* 236! Kyrenia, 1936, *Kennedy* 118! also, 1955, *G. E. Atherton* 11! 244! 302A! 335! 364! 706! Akanthou, 1940, *Davis* 2027!

NOTES: Two luxuriant specimens from western Cyprus (Kannaviou, 1939, *Lindberg f.* and between Evretou and Sarama, 1941, *Davis* 3346) account for the exceptionally large culm and leaf-blade dimensions given above in brackets.

The only Cyprus record (Rechinger f. in Arkiv för Bot., ser. 2, 1: 418; 1950) for *Cyperus esculentus* L., a similar species with larger, edible tubers and golden, distinctly nerved glumes, remains doubtful because of the possibility of confusion with *C. rotundus*. It may, however be correct, since *C. esculentus* is widely cultivated in Mediterranean climatic zones for its edible tubers ("Tiger nuts"), the source of "*horchata de chufas*". The species is recorded from Kykko ("Ticco"), 3,500–4,300 ft. alt., 28–30 June, 1913, *Haradjian* 942.

2. C. longus L., Sp. Plant., ed. 1, 45 (1753); Boiss., Fl. Orient., 5: 375 (1882); Holmboe, Veg. Cypr., 40 (1914); Post, Fl. Pal., ed. 2, 2: 674 (1933); Lindberg f., Iter Cypr., 9 (1946); Osorio-Tafall et Seraphim, List Vasc. Plants Cyprus, 17 (1973).
Chlorocyperus longus (L.) Palla in Allg. Bot. Zeitschr., 6: 201 (1901).

Rhizomatous, loosely tufted perennial, roots pectinate with many rootlets; rhizome robust, horizontal, flexuous, about 5 mm. wide, thickened at the stem-bases, covered with purple-nerved scales which become fibrous with age; culms erect, 20–100 cm. long, 3–5 mm. wide, acutely triquetrous, smooth; leaves basal and cauline; sheaths long, herbaceous and distinctly ribbed dorsally, hyaline above on the ventral side, outermost sheaths dingy red-purple; ligule none; blade to 60 cm. long, 3–10 mm. wide, scabrid, especially towards the long, acuminate tip; inflorescence once to thrice anthelate; bracts leafy, 3–6, up to 60 cm. long, often exceeding the inflorescence, gradually acuminate; rays 8–12 (–20) cm. long, flattened, smooth, secondary rays very slender; spikes few to numerous, broadly ovate-cuneate, loosely or densely spikeleted; spikelets digitately arranged, erecto-patent to erect, narrowly oblong to oblong-ellipsoid, (2–) 5–20 mm. long, 1–2·5 mm. wide, flattened, acute; glumes regularly but rather loosely imbricate, ovate, 2–2·5 mm. long, 1–1·75 mm. wide, obtuse to subacute, sides red-brown or dark brown, indistinctly 2-nerved, margin hyaline, decurrent into a persistent wing, keel acute, green, 3-nerved, excurrent in a minute apiculus; stamens 3; anthers 1·25–2 mm. long, connective-tip conical; style 1 mm. long; stigmas 3, long. Nuts (rarely produced) oblong-ellipsoid, 0·9–1 mm. long, 0·3–0·4 mm. wide, trigonous, very shortly apiculate, almost smooth, shining dark brown to iridescent black.

var. **longus**
    TYPE: "*in* Italiae, Galliae *paludibus*".

Spikes well developed; spikelets 7–20 mm. long, usually many per spike.

    HAB.: Wet ground by streams and irrigation channels; sea-level to 1,000 ft. alt.; fl. May–Sept.

    DISTR.: Division 3, also Mediterranean region into Central Europe and British Isles; Atlantic Islands, and, with closely allied species, widespread in Africa and Asia.

    3. Khoulou, 1905, *Holmboe* 761; Akhelia, 1939, *Lindberg f.* s.n.!

var. **heldreichianus** (*Boiss.*) *Boiss.*, Fl. Orient., 5: 375 (1882); Post, Fl. Pal., ed. 2, 2: 674 (1933); Rechinger f. in Arkiv för Bot., ser. 2, 1: 418 (1950).
    C. heldreichianus Boiss., Diagn., 1, 13: 39 (1853).
    C. longus L. f. *heldreichianus* (Boiss.) Holmboe, Veg. Cypr., 40 (1914); Lindberg f., Iter Cypr., 9 (1946).
    TYPE: Turkey; "ad aquas calidas *Ciliciae* in faucibus *Tauri* prope *Tchiftèckan*", *Heldreich* "pl. Anat. exs. 1847".

Spikes depauperate; spikelets 2–7 mm. long, usually rather few (up to 7) per spike.

    HAB.: By streams and irrigation channels; about 700 ft. alt.; fl. May–Sept.

    DISTR.: Division 5; also eastern Mediterranean region from Yugoslavia to Palestine.

    5. Kythrea, 1880–1941, *Sintenis & Rigo* 671! *Syngrassides* 419! *Nattrass* 534; *Lindberg f.* s.n.! *Davis* 3633!

    NOTES: Apparently confined to this one locality, where it may well have been introduced, perhaps from adjacent areas of Turkey.

3. C. cyprius *Post* in Mém. Herb. Boiss., 18: 102 (1900); H. Stuart Thompson in Journ. Bot., 44: 340 (1906); Holmboe, Veg. Cypr., 40 (1914); Osorio-Tafall et Seraphim, List Vasc. Plants Cyprus, 17 (1973).
    C. longus L. var. *cypricus* C. B. Clarke in Journ. Linn. Soc., 21: 165 (1884).
    C. longus L. ssp. *badius* (Desf.) Aschers. et Graebn. f. *depauperatus* Holmboe, Veg. Cypr., 40 (1914).
    C. longus L. ssp. *badius* Aschers. et Graebn. f. *cypricus* (C. B. Clarke) Kükenth. in Engl., Pflanzenr., 101 (IV. 20): 102 (1935).
    [*C. olivaris* (non Targ.-Tozz.) Kotschy in Unger et Kotschy, Die Insel Cypern, 190 (1865).]
    TYPE: Cyprus; "inter lapides ad ripas rivulorum montis Troodi", 1898, *Post* 923 (BEI).

Loosely tufted perennial with a creeping rhizome about 2 mm. wide, bearing purple-red, thick roots and purplish, eventually fibrous, scales; culms erect, 4–30 (–45) cm. long, 1·5–2·5 mm. wide, sharply trigonous, smooth; leaves basal, shorter than culms; sheaths papery, purple-ribbed and spotted; ligule none; blade flat or keeled, 2–4 mm. wide, tapering to a scabrid, trigonous acumen; inflorescence once-anthelate, or subcapitate in small specimens; bracts 3, leaf-like, very unequal, the longest several times exceeding inflorescence; rays 1–3 (–6), up to 5 cm. long, flattened, smooth; spikes few, hemispherical or broadly cuneate, rather lax, with a very short rhachis; spikelets few (2–7 per spike), narrowly elliptic or oblong, 4–14 mm. long, 1·5–2 mm. wide, acute, flattened, reddish- or purplish-brown; glumes ovate, 2–2·75 mm. long, 1·8–2 mm. wide, obtuse or subacute, keel green, 3-nerved, produced into a very short apiculus, sides red-brown, indistinctly nerved with narrow, hyaline margins decurrent on the rhachilla; stamens 3; anthers 1–1·5 mm. long, connective-tip conical, reddish; stigmas 3, long. Nut obovoid-ellipsoid, about 1·25 mm. long, 0·6 mm. wide, sharply trigonous, obtuse and shortly apiculate, blackish, pruinose through the reticulate areolation.

HAB.: By streams and irrigation channels on igneous mountainsides; 2,000–5,150 ft. alt.; fl. May–Oct.

DISTR.: Division 2; endemic, but see Notes.

2. Prodhromos, 1862–1955, *Kotschy* 771! *Casey* 952! *G. E. Atherton* 572! Near Podithou Monastery, Galata, 1880, *Sintenis & Rigo* 872a! 872b! Kryos Potamos, 3,200–5,150 ft. alt., 1938, *Kennedy* 1137! 1138! Roudhkias valley, 1941, *Davis* 3387!

NOTES: There is much resemblance between this high altitude taxon and the normally much taller *C. longus* L. ssp. *badius* (Desf.) Aschers. et Graebn. (*C. badius* Desf., Fl. Atlant., 1: 45, t. 7, fig. 2; 1798), as Holmboe has indicated. However, it seems unlikely to be a depauperate form of ssp. *badius*, which is not known to occur in Cyprus. There are, in addition, small morphological differences, such as the strictly digitate arrangement of the spikelets, which differentiate between the more luxuriant plants of *C. cyprius* and small plants of *C. longus* ssp. *badius*.

4. C. glaber *L.*, Mantissa Altera, 179 (1771); Boiss., Fl. Orient., 5: 371 (1882); Post, Fl. Pal., ed. 2, 2: 673 (1933); Lindberg f., Iter Cypr., 9 (1946); Osorio-Tafall et Seraphim, List Vasc. Plants Cyprus, 17 (1973).
   Chlorocyperus glaber (L.) Palla in Allg. Bot. Zeitschr., 6: 201 (1901).
   TYPE: Italy; "*in* Veronae *humentibus*".

Robust annual or perhaps perennial with fibrous roots; culms clustered, 10–50 cm. high, 1–3 mm. wide, triquetrous or trigonous, smooth, thickened at base; leaves basal and cauline; sheaths long, membranous on the ventral side, the outer tinged and spotted pink-purple, ligule none; blade flat, 3–4 mm. wide, the upper exceeding the culms, gradually acuminate, scabrid on the midrib and margins towards tip; inflorescence simply anthelate; bracts 3, leaf-like, the longest up to 20 cm., erect, or spreading; rays generally less than 5, short, 0·3–3 (–6) cm. long, flattened, spikes hemispherical; rhachis very short; spikelets crowded, oblong, 7–15 mm. long, 2 mm. wide, acute, flattened; rhachilla with narrow, white, persistent wings; glumes neatly imbricate, ovate, 2·5–3 mm. long, 2 mm. wide, keel broad, green, subterminally apiculate, scabridulous, sides spotted purple-brown, 3–4-nerved, with a distinct, white, membranous margin above; stamens 3; anthers 0·5 mm. long, connective-tip rounded; style very short, conical; stigmas 3, long. Nut obovoid, about 1·25 mm. long, 0·75 mm. wide, obtuse, shortly beaked, trigonous with concave sides, iridescent black, regularly punctulate.

HAB.: By streams, springs and irrigation channels; 600–1,000 ft. alt.; fl. June–Oct.

DISTR.: Divisions 1–3, 7. Mediterranean region extending into southern Europe and eastwards to Afghanistan.

1. Near Paphos, 1840, *Kotschy* s.n.! Maa Beach, 1967, *Economides* in ARI 934!
2. Kambos, 1939, *Lindberg f.* s.n.; Milikouri, 1939, *Lindberg f.* s.n.
3. Kannaviou, 1939, *Lindberg f.* s.n.
7. Bellapais, 1939, *Lindberg f.* s.n.! also, 1948, *Casey* 107! Kyrenia, Tchingen, 1955, *G. E. Atherton* 302B!

SUBGENUS 2. **Anosporum** (*Nees*) *C. B. Clarke* Spikelets digitate or fasciculate on a short rhachis. Nut trigonous or flattened with 3 or 2 stigmas respectively. Anatomy eucyperoid.

5. C. fuscus *L.*, Sp. Plant., ed. 1, 46 (1753); Hume in Walpole, Mem. Europ. Asiatic Turkey, 254 (1817); Boiss., Fl. Orient., 5: 370 (1882); Post, Fl. Pal., ed. 2, 2: 673 (1933); A. K. Jackson in

Kew Bull., 1937: 345 (1937); Osorio-Tafall et Seraphim, List Vasc. Plants Cyprus, 17 (1973).

[*C. fuscus* L. var. *virescens* (non (Hoffm.) Vahl) Lindberg f., Iter Cypr., 9 (1946).]

TYPE: "*in* Galliae, Germaniae, Helvetiae *pratis humidis*".

Glabrous tufted annual with reddish, fibrous roots; culms clustered, 10–45 cm. high, 1–3 mm. wide, soft, triquetrous with concave sides; leaves equalling or exceeding culms; sheaths loose, membranous, red-veined and spotted; ligule membranous, obtuse, about 1 mm. long; blade flat, 1·5–5 mm. wide, acuminate, tip trigonous, slender, very rough; inflorescence once, or, in robust specimens, twice anthelate, often rather compact; bracts 3, leaf-like, unequal, at least the lowest much exceeding the rays; rays 3–6 (–8), up to 4 cm. long, somewhat flattened, smooth; spikes dense, subspherical; rhachis short or almost none; branches (if present) short, patent; spikelets lanceolate to oblong, small, 2·5–4·5 mm. long, 1·5 mm. wide, subcompressed, obtuse, generally dark-coloured, rhachilla elongating in fruit; glumes ovate, 1–1·5 mm. long, 0·8–1 mm. wide, obtuse, closely imbricate when young, spreading later, keel broad and flattened below, ± sharp above, green, 3-nerved, minutely excurrent in a recurved mucro; sides nerveless, thin, generally tinged with dark purple, margin hyaline, very shortly decurrent; stamens 2; anthers 0·2–0·3 mm. long, connective-tip minute; style very short; stigmas (2–) 3. Nut narrowly obovoid, 0·75 mm. long, 0·3–0·4 mm. wide, sharply trigonous or occasionally lenticular, distinctly apiculate, shortly stipitate, greenish to golden-brown with pale angles, microscopically reticulated by the quadrate surface-cells.

HAB.: Moist ground by streams and irrigation channels; sea-level to 4,500 ft. alt.; fl. June–Oct.

DISTR.: Divisions 1–4, 7, 8. Also in southern Europe, Mediterranean region, Atlantic Islands, and Asia east to China and Siam; introduced elsewhere.

1. Maa Beach, 1967, *Economides* in ARI 983!
2. Prodhromos, 1955, *G. E. Atherton* 632! Kalopanayiotis, 1968, *A. Genneou* in ARI 1290!
3. Limassol, 1801, *Hume*; Kannaviou, 1939, *Lindberg f.* s.n.!
4. Larnaca, 1801, *Hume*.
7. Kyrenia Pass, 1935, *Syngrassides* 955! Kyrenia, 1936, *Kennedy* 117! Thermia, 1948, *Casey* 75! and, 1955, *G. E. Atherton* 496!
8. Ayios Andronikos, 1934, *Syngrassides* 529!

SUBGENUS 3. **Juncellus** (*Griseb.*) *C. B. Clarke* (1884) emend. Inflorescence capitate; rhachilla deeper than wide; achene much flattened dorsiventrally; stigmas 2 or 3. Anatomy chlorocyperoid.

6. C. capitatus *Vand.*, Fasc. Plant., 5 (1771); Kükenthal, Pflanzenr., 101 (IV.20): 267 (1935); Hooper in Israel Journ. Bot., 26: 98 (1977).

Schoenus mucronatus L., Sp. Plant., ed. 1, 42 (1753).
Cyperus kalli Forssk., Fl. Aegypt.-Arab., 15 (1775); Holmboe, Veg. Cypr., 40 (1914); Lindberg f., Iter Cypr., 9 (1946).
C. schoenoides Griseb., Spic. Fl. Rum. et Bith., 2: 421 (1844); Boiss., Fl. Orient., 5: 368 (1882).
Galilea mucronata (L.) Parl., Fl. Palermo, 1: 299 (1845).
Cyperus mucronatus (L.) Briquet, Prodr. Fl. Corse, 1: 225 (1910); Post, Fl. Pal., ed. 2, 2: 672 (1933); Osorio-Tafall et Seraphim, List Vasc. Plants Cyprus, 17 (1973) non Rottb. (1772) nom. illeg.

TYPE: "*in* Lusitaniae *transtagum maritimis*".

Extensively creeping, tough, glaucous perennial with a much elongated, horizontal rhizome, 2–3 mm. wide, bearing acute, red-brown, purple-veined scales and fleshy, sand-binding roots; culms up to 30 cm. long and 3–4 mm. wide, wiry, subterete; leaves tough, yellowish (when dried), flexuous or recurved, about equalling the culms; sheath papery dorsally, hyaline ventrally, becoming fibrous, red-brown or tinged purple-brown, with an oblique mouth and no ligule; blade flat or shallowly concave, up to 7 mm. wide, tapering to an indurated, almost pungent tip, margins often inrolled, scabrid; inflorescence capitate, 1·5–4 cm. across, of one to several, more or less confluent spikes; bracts 3 or 4, leaf-like, erect, spreading or reflexed, much exceeding the inflorescence; spikelets numerous, densely crowded, lanceolate, compressed (but less flattened than in other *Cyperus* species), 10–15 mm. long, 3·5–5 mm. wide, acute; rhachilla indurate, quadrangular, fragile, conspicuously barred transversely by the glume-scars; glumes loosely imbricate, irregularly distichous, broadly ovate, up to 8 mm. long, 4–6 mm. wide, with a green keel, indistinct below but excurrent in a rigid cusp 0·5–2 mm. long above, sides papery, tinged reddish-purple with numerous, slender nerves only conspicuous towards apex, margin smooth, narrowly hyaline; stamens 3, filaments membranous, strap-shaped, 0·4 mm. wide; anthers 1·25 mm. long, connective tip small, acute; style 1·5 mm. long; stigmas 3, long. Nut narrowly obovoid-ellipsoid, compressed-trigonous, shortly apiculate, 3·5–4 mm. long, 2 mm. wide, blackish-brown, densely and minutely foveolate, sometimes minutely glandular near apex.

HAB.: Sandy shores and sand-dunes; sea-level; fl. Febr.–May.

DISTR.: Divisions 3, 5, 6, 8. Also Mediterranean region and Atlantic Islands.

3. Between Limassol and Amathus; 1941–1978, *Davis* 3591! *Mavromoustakis* s.n.! *J. J. Wood* 16! *J. Holub* s.n.!
5. Salamis, 1905–1962, *Holmboe* 462; *Lindberg f.* s.n.! *Merton* 3216! *Meikle* 2564!
6. Ayia Irini, 1953, *Kennedy* 1794! and, 1967, *Economides* in ARI 929!
8. Ayios Philon near Rizokarpaso, 1941, *Davis* 2276!

NOTES: The generic position of this species is doubtful; apart from superficial differences it departs from most *Cyperus* species in the irregularly distichous glume arrangement, which has the appearance of being derived; it also departs in the sublomentoid structure of the rhachilla. These peculiarities are the basis for referring it to the monotypic genus *Galilea* Parl.

7. C. laevigatus *L.*, Mantissa Altera, 179 (1771); Boiss., Fl. Orient., 5: 366 (1882); Post, Fl. Pal., ed. 2, 2: 671 (1933).
   C. mucronatus Rottb., Descr. Plant. Rar., 17 (1772).
   Juncellus laevigatus (L.) C. B. Clarke in Fl. Brit. Ind., 6: 596 (1893).
   Acorellus laevigatus (L.) Palla in Koch, Syn. Deutsch. Schweiz. Fl., ed. 3, 2: 2558 (1905).
   TYPE: S. Africa; "*ad* Cap. b. spei I. G. KOENIG".

Rhizomatous, mat-forming perennial with reddish, fibrous roots; rhizome horizontally creeping with very short or elongated, purple-red internodes; culms tufted or solitary, 8–40 cm. high, about 1 mm. wide, terete or obscurely trigonous, wiry, often curved; leaves few, much reduced; sheaths papery, reddish, the lower ending in an apiculus, the uppermost in a very short to short, thickly crescentiform or triangular, channelled blade, flattened towards the obtuse tip; inflorescence pseudo-lateral; bracts unequal, the lowermost erect, stem-like, often greatly exceeding the spikelets, the upper small, glumaceous; spikelets 1–40, sessile, clustered, oblong lanceolate, 4–20 mm. long, 2 mm. wide, slightly compressed or rather turgid; rhachilla persistent, often becoming curved, quadrangular, with deep sides, wingless, pitted, with prominent transverse glume scars; glumes closely imbricate, broadly ovate, about 2 mm. long, 1·5 mm. wide, obtuse, sometimes apiculate, rounded above, flattened at the base on the back, sides faintly nerved, pale greenish or stramineous, or often dark purple, sometimes with a clear hyaline edge; stamens 3, filaments wide, flat, rather persistent; anther 1 mm. long, connective-tip lanceolate; style long; stigmas 2. Nut ellipsoid, ovoid or obovoid, 1·25–1·5 mm. long, 0·75–1 mm. wide, apiculate with a small persistent style-base, dorsiventrally concave-convex to lenticular, stramineous to brown, shining, minutely reticulated by the quadrate surface cell-walls.

var. **distachyos** (*All.*) *Coss. et Dur.*, Explor. Algér. Bot., 2: 251 (1868); Kükenthal in Engl., Pflanzenr., 101 (IV.20): 324 (1935).
   C. distachyos All., Auct. Fl. Pedem., 48, t. 2, fig. 5 (1789); Sintenis in Oesterr. Bot. Zeitschr., 32: 121, 290 (1882); Holmboe, Veg. Cypr., 40 (1914); Post, Fl. Pal., ed. 2, 2: 671 (1933); Rechinger f. in Arkiv för Bot., ser. 2, 1: 418 (1950); Osorio-Tafall et Seraphim, List Vasc. Plants Cyprus, 17 (1973).
   C. junciformis Cav., Icones, 3, 1: t. 204, fig. 1 (1795); Boiss., Fl. Orient., 5: 367 (1882). *Chlorocyperus junciformis* (Cav.) Rikli in Jahrb. Wiss. Bot., 27: 563 (1895).
   TYPE: France; "Locis paludosis inter *Nicaeam*, & flumen *le Var*" (TO).

Spikelets 1–7, up to 20 mm. long, often curved; glumes dark purple-black, apiculate. Nut ellipsoid.

   HAB.: Marshy ground by streams and irrigation channels; sea-level to 1,000 ft. alt.; fl. Nov.–May.

   DISTR.: Divisions 1, 5–8. Widespread in the Mediterranean region and eastwards to Pakistan; Atlantic Islands; tropical and South Africa; Lower California, Mexico, Bolivia, S. Australia.
1. Neokhorio (Akamas), 1957, *Merton* 3038! also, 1962, *Meikle* 2210!
5. N. of Lefkoniko, 1970, *A. Hansen* 350.
6. Between Kokkinotrimithia and Nicosia, 1937, *Syngrassides* 1566! Syrianokhori marshes, 1952, *Merton* 783!
7. Between Kythrea and Ayios Khrysostomos Monastery, 1880, *Sintenis & Rigo* 353! Kyrenia, 1936, *Kennedy* 119! also, 1955, *G. E. Atherton* 349!
8. Near Komi Kebir, 1912, *Haradjian* 313!

2. PYCREUS *P. Beauv.*

Fl. Oware, 2: 48 (1816).

Cyperus L. subgen. *Pycreus* (P. Beauv.) C. B. Clarke in Journ. Linn. Soc., 21: 33 (1884); Kükenthal in Engl., Pflanzenr., 101 (IV. 20): 326–402 (1936).

   Annuals or perennials; flowers bisexual, without hypogynous structures, distichously arranged on a flattened, usually winged rhachilla, in much flattened, many-flowered, typically lanceolate spikelets; spikelets digitately or spicately arranged in rayed or sessile spikes accompanied by leafy bracts; glumes shed together with, or a little before, the ripe nut, from the more persistent rhachilla; stigmas 2. Nut laterally flattened with 2 angles, one against the rhachilla. Embryo cyperoid. Anatomy chlorocyperoid.

   About 90 species, commonest in Africa, but occurring in the warmer parts of both Old and New Worlds.

1. **P. flavidus** (*Retz.*) *Koyama* in Journ. Jap. Bot., 51: 316 (1976).
   Cyperus flavidus Retz., Obs. Bot., 5: 13 (1788).
   C. globosus All., Auct. Fl. Pedem., 49 (1789): Boiss., Fl. Orient., 5: 364 (1882); Post, Fl. Pal., ed. 2, 2: 671 (1933); Lindberg f., Iter Cypr., 9 (1946); Osorio-Tafall et Seraphim, List Vasc. Plants Cyprus, 17 (1973).
   Chlorocyperus globosus (All.) Palla in Allg. Bot. Zeitschr., 6: 60 (1900).
   Pycreus globosus (All.) Reichb., Fl. Germ. Excurs., 140[10] (1830).
   TYPE: India; "E Tranquebaria misit *honor*. KÖNIG" (LD).

   Glabrous, somewhat glaucous annual or tufted perennial with slender, fibrous roots; culms clustered, (10–) 15–40 (–60) cm. long, 1–2 mm. wide, trigonous, ridged; leaves erect, linear to setaceous, shorter than or equalling the culms; sheaths tinged purple-pink, the outermost becoming fibrous; blade gradually narrowed to a slender, scabrid tip; inflorescence loosely capitate or compactly anthelate; bracts 2–4, leaf-like, erect to patent, the lowest much exceeding the rays and erect, at least in the young inflorescences; rays few, to 3 (–5) cm. long, often curved; spikes 5–20, hemispherical to subspherical, shortly spicate; spikelets elliptic to oblong, 10–15 (–30) mm. long, 2–3 mm. wide, acute, pale or silvery brown; rhachilla red-spotted, persistent, elongating, continuously winged; glumes ovate-oblong, 2–2·8 mm. long, 1·2–1·8 mm. wide, obtuse, midrib broad, flattish, 3-nerved, sides nerveless or 1-nerved, straw-coloured or red-streaked, margin broad, hyaline, often becoming incurved apically; stamens 2; anthers 0·25–0·5 mm. long, connective-tip minute, red, rounded; style 0·5–1 mm. long; stigmas 2. Nut obovoid or oblong with a flattened corner, 1–1·2 mm. long, 0·5–0·7 mm. wide, strongly compressed, apiculate, yellow to brown, shining, minutely and regularly punctulate, with squarish surface cells.

HAB.: Moist ground by roadsides and irrigation channels; 300–1,000 ft. alt.; fl. June–Sept.

DISTR.: Division 7, rare; widespread in Old World tropics and subtropics.

7. Lapithos, 1939, *Lindberg f.* s.n.! Upper Thermia, 1955, *G. E. Atherton* 497!

NOTES: This species has a similarity, in general appearance, to slender specimens of *Cyperus glaber* L., but the 2-stigmatic, flattened nut easily distinguishes it when fruiting.

3. FIMBRISTYLIS *Vahl*
Enum. Plant., 2: 285 (1805) nom. cons.

Annuals or shortly rhizomatous perennials, generally leafy and with leaf-like bracts; ligule present or absent; flowers bisexual, without hypogynous structures, generally spirally imbricate in terete few- to many-flowered spikelets; style-base bulbously thickened, deciduous with the 2–3-fid style. Embryo fimbristyloid. Anatomy chlorocyperoid.

About 300 species, chiefly in the Old World tropics.

1. F. ferruginea (*L.*) *Vahl*, Enum. Plant., 2: 291 (1805).
    Scirpus ferrugineus L., Sp. Plant., ed. 1, 50 (1753).
    Fimbristylis mauritiana Tausch ex Schultes in Roemer et Schultes, Syst. Veg., ed. 15, Mant., 475 (1824).
    F. sieberiana Kunth, Enum. Plant., 2: 237 (1837); Kern in Blumea, 8: 131 (1955).
    TYPE: "*in* Jamaicae *paludibus maritimis*".

Tufted, glaucous, slender perennial with a more or less woody base; culms 25–55 cm. high, about 1 mm. wide, triquetrous, ridged, leafy in the lower part; leaves shorter than culms, often much reduced; sheaths loose, sparsely hispidulous above, membranous on the ventral side, red-dotted with a concave mouth; ligule a dense fringe of short hairs; blade up to 25 cm. long, 1–2 mm. wide, flat, not keeled, scabridulous on the thickened margins towards the suddenly acute apex; inflorescence of (2–) 3–5 spikelets; bracts 2–3, leaf-like, with brown pubescent auricles, the lowermost bract often stiffly erect and much exceeding inflorescence; rays up to 2 cm. long, much flattened; spikelets ovoid to oblong-ovoid, 8–10 mm. long, 3–4 mm. wide, acute with a dark brown rhachilla; glumes broadly ovate or suborbicular, 2·75–3·75 mm. long, 2·35–3 mm. wide, deciduous, red-brown but densely grey-tomentose apically, rounded on the back, with a single median nerve excurrent in a short apiculus; stamens 3; anthers about 1 mm. long, filaments broad; style flattened, slightly expanded below, densely ciliate above; stigmas 2. Nut broadly obovate, 1·5–1·75 mm. long, about 1·25 mm. wide, biconvex, stipitate, umbonate, pale to dark brown with a dark stipe, almost smooth with a subquadrate surface areolation.

HAB.: Damp ground by springs and irrigation channels, etc., to 300 ft. alt.; fl. Sept.–Oct.

DISTR.: Divisions 1, 7. Scattered in the Mediterranean region and eastwards to the Philippines and Queensland; tropical and S. Africa; tropical America.

1. Baths of Aphrodite, 1966, *Merton* in ARI 82!
7. Upper Thermia, 1955, *G. E. Atherton* 498!

NOTES: A widespread, but scattered, species, nowhere common, which seems to be spreading into the Mediterranean region, having recently been recorded in Spain and Crete. The Mediterranean form has a tufted habit, slender stems and well-developed leaf-blades and so would be named *F. sieberiana* Kunth (= *F. mauritiana* Schultes) if Kern's key is followed. However I do not find that Kern's division of this pantropic taxon is satisfactory when the whole range of its variation is considered, and doubt whether this leafy form is more than

ecotypically distinct from the halophilous, rhizomatous, stout-stemmed, short-bladed type
found in coastal vegetation both in the West Indies and elsewhere.

4. ELEOCHARIS *R. Br.*

Prodr. Fl. Nov. Holl., 224 (1810).
Heleocharis Lestib., Ess. Cyp., 41 (1819) et auct. mult.

Annuals or perennials, often with terete culms; leaf-blades reduced to an
apiculus on the sheath; ligule absent; flowers bisexual, with up to 10 or
without bristles, generally numerous, spirally imbricate in a solitary,
terminal, globose, ellipsoid or cylindric head; bracts glume-like; style-base
bulbously thickened, persistent on the ripe nut; stigmas 2 or 3. Nut generally lenticular, sometimes trigonous. Embryo fimbristyloid. Anatomy eucy-
period.
About 160 species, found in damp situations worldwide.

1. **E. palustris** (*L.*) *Roem. et Schult.*, Syst. Veg., 2: 151 (1817); Boiss., Fl. Orient., 5: 386 (1882);
   Post, Fl. Pal., ed. 2, 2: 679 (1933); H. K. Svenson in Rhodora, 41: 43–77 (1939); S. M.
   Walters in Journ. Ecol., 37: 192–206 (1949); Osorio-Tafall et Seraphim, List Vasc. Plants
   Cyprus, 16 (1973).
   Scirpus palustris L., Sp. Plant., ed. 1, 47 (1753); Holmboe, Veg. Cypr., 40 (1914) as *S.
   paluster.*
   TYPE: "*in* Europae *fossis & inundatis*".

Creeping or tufted, somewhat glaucous, rhizomatous perennial; rhizome
2–3 mm. thick with short to very long internodes; culms solitary or crowded,
(10–) 30–60 (–90) cm. long, 1·5–4·5 mm. wide, terete, hard or spongy, finely
ridged; sheaths thin, papery, straw-coloured to red-purple, purple-spotted
towards the transverse mouth; bracts 1 or 2, similar to the glumes but
without keels; spikelets elliptic, 10–25 mm. long, 3–5 mm. wide, subacute,
straw-coloured or purple; glumes numerous, appressed, becoming patent,
ovate-oblong, 3–5 mm. long, 1·5–1·8 mm. wide, acute or obtuse, with a low,
green midrib, sides purple-streaked with wide hyaline margins; hypogynous
bristles usually 4, retrorsely spinulose, equalling or exceeding the achene
and style-base; stamens 3; anthers 2–3 mm. long, minutely apiculate;
stigmas 2; style-base ⅔ width of nut, pyramidal or mitriform, 0·3–0·6 mm.
high, more or less obtuse, whitish to brown. Nut ellipsoid to obovoid, 1·1–1·3
mm. long, 1·1–1·2 mm. wide, yellow or red-brown, minutely punctulate.

HAB.: Marshes and margins of fresh-water and brackish lakes and ponds; sea-level to 4,500 ft.
alt.; fl. March–June.

DISTR.: Divisions 2, 4, 6, 8; almost cosmopolitan, but mainly in temperate regions.
2. Prodhromos, 1905, *Holmboe* 906.
4. Kouklia, 1905, *Holmboe* 390.
6. Syrianokhori, 1935, *Syngrassides* 789! and, 1953, *Merton* 1035!
8. Galatia, 1962, *Meikle* 2539!

NOTES: *Eleocharis palustris* is here interpreted in a wide sense. *Syngrassides* 789 has the single
bract and long glumes characteristic of *E. uniglumis* (Link) Schult. and has been so determined
by S. M. Walters. However it lacks the typically slender habit and few glumes of that species
and is immature, whilst it comes from the same locality (Syrianokhori) as typical *E. palustris*
(*Merton* 1035). More collecting of this genetically and ecologically variable complex is needed
before deciding if more than one species of the *palustris* group is present in Cyprus.

5. ISOLEPIS *R. Br.*

Prodr. Fl. Nov. Holl., 221 (1810).
Scirpus L., Sp. Plant., ed. 1, 47 (1753) pro parte
E. Palla in Engl., Bot. Jahrb., 10: 293–301 (1888).

Usually small, caespitose annuals, occasionally perennials; flowers
bisexual, without hypogynous bristles, numerous and spirally imbricate in
1–several, terete, often sessile, more or less clustered spikelets subtended at

least when young, by a more or less erect and stem-like bract; stamens 1–3; style-base not enlarged. Nut trigonous, papillose or longitudinally ridged (in Cyprus). Embryo cyperoid. Anatomy eucyperoid.

About 30 species, hardly distinguishable morphologically from *Schoenoplectus*, but with a cyperoid embryo; usually — unlike *Schoenoplectus* — small plants.

Mature nut clearly longitudinally ribbed; bract often exceeding spikelets in length
1. I. setacea
Mature nut minutely reticulate; bract seldom exceeding spikelets in length - **2. I. cernua**

1. I. setacea (*L.*) *R. Br.*, Prodr. Fl. Nov. Holl., 221 (1810); Osorio-Tafall et Seraphim, List Vasc. Plants Cyprus, 17 (1973).
  Scirpus setaceus L., Sp. Plant., ed. 1, 49 (1753); Boiss., Fl. Orient., 5: 379 (1882); Post, Fl. Pal., ed. 2, 2: 676 (1933).
  S. setaceus L. var. *pseudoclathratus* Schramm in Flora, 41: 709 (1858); Holmboe, Veg. Cypr., 40 (1914).
  TYPE: "*in* Europae *litoribus maritimis*".

Small tufted annual or perennial with fine, fibrous roots and, sometimes, a short, knotted rhizome; culms 20–100 (–300) mm. high, 0·3–0·5 mm. wide, trigonous, ridged; leaves capillary, shorter than culms; sheaths loose, membranous, purple-tinged; blade crescentiform (in section), blunt-tipped, the lowermost reduced to setaceous points; inflorescence of 1–4 (–10) clustered, sessile spikelets, subtended by an erect, stem-like bract, variable in length, up to 3 cm. long, but often much shorter; spikelets ovoid, 1·5–2·5 mm. long, 1–2 mm. wide; glumes ovate, 1–1·5 mm. long, 0·8–1 mm. wide, rounded with the green, thickened, keeled midrib exserted as a blunt mucro, sides membranous, dark purple-brown, clearly nerved; stamens 1–2 (–3); anthers 0·3 mm. long, connective-tip blunt; stigmas 3. Nut broadly obovoid, 0·5–0·8 mm. long, 0·4–0·5 mm. wide, dark brown, angles obtuse, apex rounded, apiculate, conspicuously longitudinally ridged by the short ends of transversely elongate, rectangular cells in vertical rows.

HAB.: Moist ground in Pine forest; 4,500–5,500 ft. alt.; fl. June–July.

DISTR.: Division 2, very rare. Also widespread in Europe, Asia and eastwards to China; Australia, Atlantic Islands, N., E. and S. Africa; introduced into N. America.
2. Moist ground in forest (? Damaskinari) near Prodhromos, 1880, *Sintenis & Rigo* 975! also, 1905, *Holmboe* 897.

2. I. cernua (*Vahl*) *Roem. et Schult.*, Syst. Veg., 2: 106 (1817); Osorio-Tafall et Seraphim, List Vasc. Plants Cyprus, 17 (1973).
  Scirpus cernuus Vahl, Enum. Plant., 2: 245 (1805); Holmboe, Veg. Cypr., 41 (1914).
  S. savii Seb. et Maur., Fl. Rom. Prodr., 22 (1808); Boiss., Fl. Orient., 5: 380 (1882).
  TYPE: Portugal; "in Lusitania. Rathke" (C).

A densely tufted, slender annual with fine fibrous roots; culms setaceous, up to 20 cm. high, 0·5 mm. wide, triquetrous, leaves inconspicuous, much shorter than culms; sheaths membranous, loose, tinged purple; blades usually reduced to a short, blunt apiculus, or the uppermost up to 2 cm. long, crescentiform (in section), ridged; inflorescence of 1 or 2, sessile spikelets; bracts up to 1 cm. long, but usually not exceeding the spikelets, erect when young, but becoming deflexed; spikelet ovoid, acute or subacute, 2–4 mm. long, 1·5–2·5 mm. wide, pale, or (in our area) often fuscous brown; glumes ovate, 1·5–2 mm. long, 1–1·5 mm. wide, distinctly keeled, with a thickened green midrib ending in a minute apiculus, sides lightly nerved; stamens 1–2; anthers about 0·3 mm. long; connective-tip blunt or subacute; styles 3. Nut broadly obovoid or suborbicular, 1 mm. long, 0·7–0·9 mm. wide, unequally trigonous with one rounded angle, apically blunt with a minute apiculus, black, minutely white-papillose.

HAB.: Muddy ground by pools and marshes; sea-level to 800 ft. alt.; fl. April–May.

DISTR.: Divisions 3, 6, 7. Also, Mediterranean region and Atlantic Europe from the British Isles southward; Atlantic Islands, S. Africa; N. & S. America.

3. Asomatos, 1962, *Meikle* 2924! Akrotiri, 1967, *Merton* in ARI 686!
6. Syrianokhori marshes, 1952, *Merton* 800!
7. Between Myrtou and Panagra, 1934, *Syngrassides* 1355!

NOTES: The nuts of Cyprus specimens are consistently larger than those of specimens from western Europe, and the spikelets are very dark compared to those of many European collectings. However, some Sicilian and southern Italian gatherings closely approach the Cyprus plants, while differing slightly. These gatherings may be *Isolepis sicula* Presl or *Scirpus minaae* Todaro (*Scirpus savii* Seb. et Maur. ssp. *minaae* (Tod.) Arcang.), but such an identification must be tentative in the absence of relevant type material.

6. SCHOENOPLECTUS *(Reichb.) Palla*

in Verh. Zool.-Bot. Ges. Wien, 38 Sitzb.: 49 (1888) nom. cons.

Scirpus L. sect. *Schoenoplectus* Reichb., Icones Fl. Germ., 8: 40 (1846); A. A. Beetle in Am. Journ. Bot., 28: 691–700 (1941); 29: 653–656 (1942).

Annuals or perennials with terete or trigonous, nodeless culms; leaf-blades usually reduced; ligule present (in Cyprus species); flowers bisexual without or with up to 6 bristles, numerous, spirally imbricate in 1–several, generally large, ovoid or ellipsoid spikelets; spikelets sessile or shortly rayed in an apparently lateral cluster overtopped by the erect, stem-like lowermost bract; style not bulbously thickened, continuous with the ovary, generally 3-fid. Nut beaked, with a smooth or often transversely rugulose surface. Embryo schoenoplectoid. Anatomy eucyperoid.

About 80 species, mostly in the tropics of the Old World.

Hypogynous bristles expanded, plumose; culms sharply trigonous below inflorescence
　　　　　　　　　　　　　　　　　　　　　　　　　　　　　　　　1. S. litoralis
Hypogynous bristles retrorsely scabrid; culms terete or subterete below inflorescence
　　　　　　　　　　　　　　　　　　　　　　　　　　　　　　　　2. S. lacustris

1. S. litoralis *(Schrader) Palla* in Engl., Bot. Jahrb., 10: 299 (1888).
　　　Scirpus litoralis Schrader, Fl. Germ., 1: 142 (1806); Unger et Kotschy, Die Insel Cypern, 189 (1865); Boiss., Fl. Orient., 5: 383 (1882); Holmboe, Veg. Cypr., 41 (1914); Post, Fl. Pal., ed. 2, 2: 677 (1933); Lindberg f., Iter Cypr., 9 (1946); Osorio-Tafall et Seraphim, List Vasc. Plants Cyprus, 16 (1973).
　　　TYPE: Germany; "In lacustribus, inundatis maritimis Duini et Monfalconii (*Wulfen*)".

Rhizomatous, sometimes tufted, glabrous perennial with wiry roots; rhizome either short, with clustered stems, or much elongated and creeping, about 2·5 mm. thick, producing leafy, submerged shoots (? only in running water); culms erect, 30–200 cm. high, smooth, matt, sharply trigonous above, obtusely trigonous below, thickened by loose sheaths at the base; sheaths long, membranous, often splitting and decaying on the flowering shoots; ligule oblique, membranous, with an obtuse, free edge about 1 mm. long; blades of the submerged leaves (when present) linear-lanceolate, long-acuminate, flat, soft, with a single nerve and transverse anastomoses, sometimes much elongated and probably partly floating, those of emergent leaves much reduced, about 3–10 cm. long, 0·3–0·5 cm. wide, more or less marcescent; inflorescence lax, once- or twice-anthelate; rays 0·5–1 mm. wide, flattened apically; bract trigonous and stem-like or flattened, generally slightly exceeding inflorescence; spikelets generally many, solitary or in clusters, oblong-ovoid, sometimes becoming cylindric, acute, 6–12 (–22) mm. long, 3–5 mm. wide, chestnut-brown, becoming paler with age; glumes broadly elliptic or suborbicular, about 3 mm. long, 2–3 mm. wide, scarious, tightly appressed, dorsally rounded with a thin raised midrib excurrent in a slightly scabrid to smooth, sharp mucro; sides red-brown, nerveless, often transversely wrinkled, margin broadly membranous, often splitting at the mucro or irregularly lacerate, and sometimes ciliolate; hypogynous bristles 4 (–5), as long as, or somewhat exceeding the nut, expanded, finely and densely fringed with long, multicellular reddish hairs, especially above; stamens 3; filaments strap-shaped; anthers 1·5–2 mm. long, including the conspicuous, rounded, white, apiculate or fimbriate connective-tip; stigmas 2. Nut broadly obovoid, 2–2·5 mm. long, 1·25–1·5 mm. wide, olive-brown to black, lenticular or plano-convex, distinctly beaked, almost smooth.

HAB.: Brackish pools, marshes, ditches and wet ground; near sea-level; fl. April–July.

DISTR.: Divisions 3–6, 8. Also in S. Europe and Mediterranean region eastwards to Afghanistan and central Asia; Africa; Australia.

3. Asomatos, 1939, *Mavromoustakis* 77 ! Limassol, 1939, *Lindberg f.* s.n.
4. Larnaca, 1862, *Kotschy* 289 ! also, 1880, *Sintenis*; Cape Pyla, 1880, *Sintenis*; Famagusta, 1939, *Lindberg f.* s.n. ! Mouth of Pedieos R., 1955, *Merton* 2255 !
5. Vatili, 1905, *Holmboe* 351; near Salamis, 1962, *Meikle* 2634 !
6. Syrianokhori, 1953, *Merton* 1765 !
8. Cape Andreas, 1937, *Syngrassides* 1661 !

2. S. lacustris (*L.*) *Palla* in Engl., Bot. Jahrb., 10: 299 (1888).
   Scirpus lacustris L., Sp. Plant., ed. 1, 48 (1753); Boiss., Fl. Orient., 5: 383 (1882); Post, Fl. Pal., ed. 2, 2: 677 (1933).

Robust, erect, glabrous perennial with a thick, creeping rhizome; culms up to 350 cm. high and 1·5 cm. wide, thickened at base by sheaths, terete, spongy, green or glaucous; leaves reduced or sometimes well developed when submerged; sheaths long, membranous, splitting early, brownish or purplish; ligular arc unequally deltoid, ligule brown, membranous; blade of submerged leaves ribbon-like, those of emergent leaves linear-subulate, up to 15 cm. long, but often much less, tapering to a blunt apex; inflorescence of numerous spikelets, lax and once or twice anthelate or congested; rays up to 10 cm. long, flattened, scabridulous; bracts subulate, the lowermost erect, shorter to longer than the inflorescence, the upper with a membranous base and subulate, scabridulous tip, up to 1 cm. long; spikelets solitary or in small clusters, narrowly ovoid, 6–15 mm. long, 3–5 mm. wide, acute or subacute, rufous brown; glumes broadly ovate, 2·5–4 mm. long, 1·5–3 mm. wide, scarious, tightly appressed, dorsally rounded with a thin raised midrib sometimes excurrent in a narrow, scabrid apiculus; sides red-brown, nerveless, smooth or glandular and red-spotted, transversely wrinkled, margins broadly hyaline-membranous, often ciliolate and becoming lacerate apically; hypogynous bristles 4–6, about as long as the nut, conspicuously retrorse strigose, brownish; stamens 3; filaments strap-shaped; anthers about 2–2·5 mm. long, connective-tip blunt, pilose or glabrous; stigmas 2 or 3. Nut broadly obovoid, lenticular or bluntly trigonous, 2–2·5 mm. long, 1·5–2 mm. wide, grey or fuscous, smooth or minutely punctulate, shining.

ssp. **tabernaemontani** (*C. Gmel.*) *A. & D. Löve* in Folia Geobot. Phytotax., 10: 275 (1975).
   Scirpus tabernaemontani C. Gmel., Fl. Bad., 1: 101 (1805); Rechinger f. in Arkiv för Bot., ser. 2, 1: 418 (1950); Osorio-Tafall et Seraphim, List Vasc. Plants Cyprus, 16 (1973).
   S. glaucus Sm., English Botany, 33: t. 2321 (1812).
   S. lacustris L. var. *digynus* Godr., Fl. Lorraine, 3: 90 (1844); Boiss., Fl. Orient., 5: 383 (1882).
   S. lacustris L. ? ssp. *glaucus* (Sm.) Hartm., Svensk Norsk Excurs. Fl., 10 (1846) taxon incerti gradus.
   S. lacustris L. ssp. *tabernaemontani* (C. Gmel.) Syme, English Botany, ed. 3, 10: 64 (1870).
   TYPE: Germany; "Utrinque in stagnis et paludibus praesertim sylvaticis Rheno vicinis frequens".

Culms glaucous, generally shorter and more slender than in ssp. *lacustris*; glumes beset with red, glandular spots; connective-tip of anthers glabrous, or shortly and obscurely spiculate; stigmas 2. Nut compressed-lenticular (not trigonous).

HAB.: Swampy ground by margins of open water, often in brackish situations; sea-level to 4,000 ft. alt.; fl. March–June.

DISTR.: Divisions 2, 3, rare. Also throughout Europe, N. Africa, S. Africa, E. Asia.

2. Panayia, 3,000–4,000 ft. alt., 1913, *Haradjian* 916; Pomos, 1934, *Syngrassides* 962!
3. Asomatos, 1939, *Mavromoustakis* 39!

7. BOLBOSCHOENUS *Aschers. ex Palla*

in Koch, Syn. Deutsch. Schweiz. Fl. (ed. Hallier et Brand), 2531 (1905).
Scirpus L. α *Bolboschoenus* Aschers., Fl. Brandenb., 1: 753 (1864) sine grad. tax.
Scirpus L. sect. *Bolboschoenus* (Aschers.) Beetle in Amer. Journ. Bot., 29: 82 (1942).

Perennials; culms with swollen tuberous bases; leaves cauline, laminate; flowers bisexual, spirally arranged in large, pedunculate spikelets; bracts leaf-like, several; hypogynous bristles 0–6; style-base persistent, not enlarged. Nut smooth. Embryo schoenoplectoid. Anatomy eucyperoid.

About 8 species, mainly in eastern Asia and eastern N. America.

1. **B. maritimus** (*L.*) *Palla* in Koch, Syn. Deutsch. Schweiz. Fl. (ed. Hallier et Brand), 2532 (1905).
    Scirpus maritimus L., Sp. Plant., ed. 1, 51 (1753); Sibth. et Sm., Fl. Graec. Prodr., 1: 34 (1806); Poech, Enum. Plant. Ins. Cypr., 8 (1842); Unger et Kotschy, Die Insel Cypern, 190 (1865); Boiss., Fl. Orient., 5: 384 (1882); Holmboe, Veg. Cypr., 41 (1914); Post, Fl. Pal., ed. 2, 2: 678 (1933); Lindberg f., Iter Cypr., 9 (1946); T. Norlindh in Bot. Notiser, 125: 397–405 (1972).
    TYPE: "*in* Europae *litoribus maritimis*".

Glabrous rhizomatous perennial with fleshy roots; rhizome often long-creeping, about 2 mm. wide, branching, sometimes tuberiferous, covered with sheath-like scales becoming hard and black; tubers subglobose or ellipsoid, up to 15 mm. across; culms solitary, sometimes curved, (10–) 25–100 cm. high, leafy in the lower half, triquetrous with the angles sharp or narrowly winged and the sides concave above, flat below and ribbed; leaves often exceeding the culms; sheaths commonly long, tight, membranous and red-dotted on the ventral side below the shallowly convex, eligulate mouth, lowermost sheaths without blades; blades linear, 10–35 cm. long, 0·5–0·7 cm. wide, gradually long-acuminate, flat, many-nerved, margins scaberulous, thickened, midrib channelled above and prominent beneath; inflorescence capitate or subumbellate, of 1–10 solitary or glomerate, sessile or rayed spikelets; bracts scabrid, the lower leaf-like, with the lowermost generally about twice as long as the rays; rays flattened, up to 4 cm. long or almost wanting; spikelets terete, ovoid becoming oblong, acute, 10–20 (–40) mm. long, 5–6 mm. wide; glumes ovate or elliptic, 3·5–4 mm. long, 2·5 mm. wide, membranous, stramineous with a dark border, or dark red-brown, sparsely hispid, apically retuse or lacerate with a narrow, raised midrib excurrent in a long, recurved, scabrid awn; hypogynous bristles (0–) 3–6, white, minutely retrorsely scabrid, unequal, shorter than, to slightly exceeding, the nut; stamens 3; anthers 3 mm. long, with a scabrid, conical connective-tip; stigmas (2–) 3. Nut obovoid, trigonous below, plano-convex above, or occasionally lenticular, about 2·25 mm. long, 1·3 mm. wide, shortly beaked, brown, smooth.

HAB.: Freshwater and brackish marshes and ditches; sea-level to 4,500 ft. alt.; fl. April-Sept.

DISTR.: Divisions 1–5, 7, 8. Cosmopolitan, except Polar regions.

1. Istinjo, 1905, *Holmboe* 776.
2. Prodhromos, 1862, *Kotschy* 848.

3. Limassol, 1905, *Holmboe* 669, also, 1939, *Lindberg f.* s.n.; Asomatos, 1939, *Mavromoustakis* 69! 70! Zakaki, 1939, *Mavromoustakis* 101! Akrotiri, 1967, *Merton* in ARI 687!
4. Larnaca, 1862, *Kotschy* 310; also, 1939, *Lindberg f.* s.n.! Kouklia, 1905, *Holmboe* 389; also, 1967, *Merton* in ARI 447! Akhyritou, 1951, *Merton* 1059!
5. Kythrea, 1934, *Syngrassides* 862A! Dhikomo, 1936, *Syngrassides* 1192!
7. Between Myrtou and Panagra, 1934, *Syngrassides* 1358! Larnaka tis Lapithou, 1941, *Davis* 3001!
8. Galatia, 1962, *Meikle* 2540! N. of Boghaz, 1970, *A. Hansen* 744!

NOTES: The Lindberg f. specimen from Larnaca Salt Lake has the very long spikelets (up to 40 mm. long) characteristic of plants growing in very brackish locations, and sometimes named *Scirpus maritimus* L. forma *macrostachys* (Willd.) Junge.

8. SCIRPOIDES *Ség.*

Plant. Veron., 3: 73 (1754).
Holoschoenus Link, Hort. Reg. Bot. Berol., 1: 293 (1827).
A. Becherer in Candollea, 4: 130–145 (1929).

Rush-like perennials with terete stems and reduced leaves; flowers bisexual, spirally arranged in small spikelets, clustered into sessile or unequally pedunculate, globose spikes, without hypogynous bristles; bract culm-like, usually erect; glumes spiculate-scabrid in the upper half; stamens 3; style very short. Nut broadly ellipsoid. Embryo cyperoid. Anatomy eucyperoid.

A single species (or species complex) widely distributed in tropical and warm-temperate zones of the Old World.

1. **S. holoschoenus** (*L.*) *Soják* in Čas, Nár. Muz. (Prague), 140: 127 (1972).
    Scirpus holoschoenus L., Sp. Plant., ed. 1, 49 (1753) as *S. "holoscoenus"*; Boiss., Fl. Orient., 5: 382 (1882); Post in Mém. Herb. Boiss., 18: 102 (1900); H. Stuart Thompson in Journ. Bot. 44: 340 (1906); Holmboe, Veg. Cypr., 41 (1914); Lindberg f., Iter Cypr., 9 (1946); Rechinger f. in Arkiv för Bot., ser. 2, 1: 418 (1950).
    S. romanus L., Sp. Plant., ed. 1, 49 (1753).
    S. australis Murr. in L., Syst. Veg., ed. 13, 85 (1774).
    Holoschoenus vulgaris Link, Hort. Reg. Bot. Berol., 1: 293 (1827); Osorio-Tafall et Seraphim, List Vasc. Plants Cyprus, 17 (1973).
    Scirpus holoschoenus L. var. *australis* (Murr.) Koch, Syn. Fl. Germ. Helv., ed. 2, 857 (1844).
    S. holoschoenus L. var. *romanus* (L.) Koch, Syn. Fl. Germ. Helv., ed. 2, 857 (1844).
    TYPE: "*in* Europa *australi*".

Densely tufted, glabrous, glaucous perennial with sand-binding roots and a short, woody, creeping rhizome; culms densely clustered, 30–90 cm. high, 2–3 mm. wide, though much thickened at the base by sheaths, soft, flattened below, smooth or ridged, harsh above, branched intravaginally; leaves basal, mostly reduced to sheaths; sheaths long, dorsally herbaceous, becoming brown and sometimes hard, ventrally membranous, pale, red-spotted, prolonged into a rounded anti-ligule, soon splitting into fibres (often in a herring-bone pattern); ligule none; blade short, hard, terete, channelled, becoming flat at the apex, scabrid on the margins at least below, with an obtuse, scabrid, often brown, flattened tip; inflorescence pseudo-lateral, shortly once or twice anthelate; bracts leaf-like, the upper erect and much exceeding the inflorescence, 10–40 cm. long, the lower spreading or deflexed, much shorter than the upper but very variable in length, generally flat; rays short, usually less than 5 cm. long, firm, flattened, sometimes marginally scabrid; spikes 1–4 (–10), sessile or pedunculate, spherical, 2–12 mm. diam., composed of many spikelets; spikelets glomerate, apparently confluent when young, ovoid, rather obtuse, 2–3·5 mm. long, 1–2 mm. wide; glumes obovate, 1·25 mm. long, 0·8–1 mm. wide, obtuse, rounded on the back with a green, acute keel, prominent above and excurrent in a short, obtuse mucro, sides membranous, nerved, red-tinged, scabrid with white spiculae in the upper half; anthers about 0·8 mm. long; connective-tip small, rounded; style about 0·3 mm. long; stigmas 3, about 0·7 mm. long, thick, crystalline-papillose. Nut broadly ellipsoid, greyish, 1 mm. long, 0·65 mm. wide, shortly beaked, unequally trigonous, thickened on the angles, with a minutely reticulate surface.

HAB.: Marshes and moist ground by streams, springs and irrigation channels, occasionally on damp grass steppe and field margins or muddy roadsides; sea-level to 5,400 ft. alt.; fl. April–Sept.

DISTR.: Divisions 1–8. General distribution that of genus.

1. Ktima, 1913, *Haradjian* 649!
2. Troödos, 1880, *Sintenis & Rigo* 877! and, 1912, *Haradjian* 496! Livadhi tou Pasha, 1938, *Kennedy* 1085! and, 1952, *Merton* 898! Near Amiandos, 1940, *Davis* 1889! Prodhromos, 1944, *Casey* 951! Loumadho valley, 1952, *Kennedy* 1770! Above Trikoukkia, 1961, *D. P. Young* T385!
3. Mazotos, 1905, *Holmboe* 188; Stavrovouni, 1937, *Syngrassides* 1483! and, 1939, *Lindberg f.* s.n.! Arminou, 1941, *Davis* 3407A! Near Mamonia, 1960, *N. Macdonald* 160!
4. Mile 3, Larnaca–Famagusta road, 1950, *Chapman* 446! Larnaca aerodrome, 1950, *Chapman* 695!
5. Tymbou, 1932, *Syngrassides* 386!
6. Nicosia, 1929 or 1930; *C. B. Ussher* 98! Myrtou, 1932, *Syngrassides* 204! Syrianokhori, 1941, *Davis* 2622! Pano Dheftera, 1979, *Edmondson & McClintock* E 2939!
7. Kalogrea, 1932, *Syngrassides* 284! Thermia, 1948, *Casey* 76! and, 1955, *G. E. Atherton* 494! Kyrenia, 1927, *Rev. A. Huddle* 106! and, 1955, *Miss Mapple* 16! also, 1955, *G. E. Atherton* 36! 126!
8. Eleousa, 1962, *Meikle* 2542! Yialousa, 1970, *A. Genneou* in ARI 1458!

NOTES: The Cyprus population of this variable species consists chiefly of plants with three heads in the inflorescence, one sessile and two usually rayed (subsp. *australis* (Murr.) Soják). On the upper slopes of Troödos an extreme variant, generally with a single head and approximating to *S. romanus* (L.) Soják, is apparently common; in the lowlands specimens with many-headed inflorescences resembling those of *S. holoschoenus* var. *holoschoenus*, but rarely quite so luxuriant are occasionally collected. The varieties are not, however, clear-cut and it seems best to leave the aggregate undivided.

TRIBE 2. **Rhynchosporeae** Glumes containing bisexual flowers few, central or apical in the spikelet with smaller, sterile or male, glumes below and sometimes also above; rhachilla internodes between bisexual flowers elongated and flexuose (zig-zag).

9. SCHOENUS *L.*

Sp. Plant., ed. 1, 42 (1753).
Gen. Plant., ed. 5, 26 (1754).
G. Kükenthal in Fedde Repert., 44 (1938).

Perennials, occasionally annuals; leaf-blades narrow, occasionally wanting, basal sheaths open; flowers bisexual, or the uppermost male only, distichously arranged in 1–4-flowered spikelets, with 0–6 plumose or ciliate hypogynous bristles; fertile glumes larger than basal, sterile glumes; rhachilla elongated between fertile flowers, flexuous or zig-zag; stigmas 3. Nut often 3-ribbed.

About 90 species, mainly in S.E. Asia and Australia.

1. **S. nigricans** *L.*, Sp. Plant., ed. 1, 43 (1753); Boiss., Fl. Orient., 5: 393 (1882); Holmboe, Veg. Cypr., 40 (1914); Post, Fl. Pal., ed. 2, 2: 681 (1933); Lindberg f., Iter Cypr., 9 (1946); Osorio-Tafall et Seraphim, List Vasc. Plants Cyprus, 17 (1973).
    [*Chaetospora ferruginea* (non (L.) Reichb.) Kotschy in Unger et Kotschy, Die Insel Cypern, 189 (1865)].
    [*Schoenus ferrugineus* (non L.) H. Stuart Thompson in Journ. Bot., 44: 340 (1906)].
    TYPE: "*in Europae paludibus aestate, exsiccatis*".

Densely tufted, glabrous perennial with fleshy, often dark purple roots, often growing up through the sheaths; culms 15–80 cm. long, 1–1·5 mm. wide at base, smooth, wiry, obtusely trigonous below, somewhat flattened above, ribbed; leaves all basal, half as long as the culm or more; sheaths open with overlapping membranous edges, stiff, ribbed, keeled at least above, all except the innermost dark reddish-brown and shining inside and out; ligule very short, minutely ciliate; blade linear, wiry, subterete or laterally compressed, adaxially channelled, apically trigonous; inflorescence capitate, ovoid or subglobose, 1–1·5 cm. long and as wide or a little wider, made up of crowded bracteate fascicles, each including about 5 spikelets; lowest 2 bracts with ovate, sheathing bases and leaf-like tips, the lowermost 2–5 times the length of the inflorescence; spikelets lanceolate, 10–12 mm. long, 2–4 mm. wide, laterally flattened, often curved; two lowermost glumes about 5 mm. long, lanceolate, apiculate, strongly ciliate on the keel, the 2 lower fertile glumes about 6 mm. long, subacute, keel smooth or apically rough, upper 3–4 fertile glumes ciliate on the keel; rhachilla robust, zig-zag, winged; hypogynous bristles 3–4, very short, lanceolate, antrorsely ciliate to spiculate; stamens 3, filaments 10–12 mm. long; anthers 4 mm. long, connective-tip yellow, conical or bifurcate; style 3–5 mm. long; stigmas long-papillose. Nut ovoid-ellipsoid, 1·5–1·9 mm. long, 1·2 mm. wide, obtusely trigonous with convex sides, ivory to grey, smooth, shining.

HAB.: Marshy ground by streams and springs, or on the drier parts of freshwater marshes or salt-marshes; sea-level to 4,500 ft. alt.; fl. March–July.

DISTR.: Divisions 1–5, 7. Europe and Mediterranean region E. to Central Asia; tropical and S. Africa, N. America.

1. Baths of Aphrodite, 1905, *Holmboe* 789; also, 1962, *Meikle* 2216! Kapsala near Ayios Yeoryios (Akamas), 1962, *Meikle* 2075!
2. Prodhromos, 1862, *Kotschy* 600 or 890! also, 1905, *Holmboe* 947; Mesapotamos, 1939, *Lindberg f.* s.n.; above Trikoukkia, 1961, *D. P. Young* 7384!
3. Cape Gata, 1862, *Kotschy* 600 or 890! Limassol, 1905, *Holmboe* 663; also, 1939, *Lindberg f.* s.n.! Alethriko, 1905, *Holmboe* 234; Kakoradjia, 1937, *Syngrassides* 1506! Akrotiri, 1941, *Davis* 3582! and, 1967, *Merton* in ARI 684!
4. Larnaca aerodrome, 1950, *Chapman* 413! 457!
5. Mile 3, Larnaca–Famagusta road, 1950, *Chapman* 429!
7. Kalogrea, 1932, *Syngrassides* 286! Kyrenia, 1949, *Casey* 539! Kyrenia Pass, 1955, *G. E. Atherton* 5!

10. CLADIUM *P. Browne*

Nat. Hist. Jamaica, 114 (1756).
J. H. Kern in Act. Bot. Neerl., 8: 263–268 (1959).

Perennial; culm hollow, leafy throughout; leaves pseudo-dorsiventral, very scabrid, eligulate; flowers bisexual or male by reduction, spirally arranged in 2–4-flowered, small spikelets in capitate clusters in a compound panicle; hypogynous structures represented only by a basal disk; rhachilla straight. Nut with a hard endocarp.

Two critical species; *C. mariscus* almost cosmopolitan.

1. **C. mariscus** (*L.*) *Pohl*, Tent. Fl. Bohem., 1: 32 (1809); Boiss., Fl. Orient., 5: 392 (1882); Holmboe, Veg. Cypr., 40 (1914); Post, Fl. Pal., ed. 2, 2: 681 (1933); Lindberg f., Iter Cypr., 9 (1946); Osorio-Tafall et Seraphim, List Vasc. Plants Cyprus, 17 (1973).
Schoenus mariscus L., Sp. Plant., ed. 1, 42 (1753).
Cladium germanicum Schrader, Fl. Germ., 1: 75 (1806).
TYPE: "*in* Europae *paludibus*".

Harsh, densely tufted, stout, somewhat glaucous perennial with thick, fleshy roots; rhizome more or less elongated and branched, about 8 mm. wide, covered with dark, purplish-brown, triangular to lanceolate scales; culms clustered or remote, up to 2 m. tall, 4–8 mm. wide, thickened basally by sheaths, terete, finely ridged, with hollow internodes; leaves several, mostly basal but 2–3 cauline, long-sheathing, harsh; outermost scale-like with short, open, or soon splitting sheaths, shining within, successive leaves much elongated, inner leaves with the sheaths terete ridged, closed except at the top; ligule absent; blade erect, linear, gradually narrowed to a solid triangular tip, rounded on the back, thick, hard, finely nerved with transverse anastomoses, scabrid-serrate (lacerating) on the margin and midrib, especially towards the tip, pseudo-dorsiventral; inflorescence an elongate leafy panicle, up to 50 cm. long, with 1–2 lateral, compound, bracteose, more or less corymbose partial panicles at each node; bracts long, sheathing, the lower leaf-like, the upper with reduced blades; peduncles unequal, flattened, with acute-serrulate margins, twice or thrice terminally branched, branches prophyllate, subtended by slender, non-sheathing bracts, unequal, those of the last order bearing subspherical, subdigitate clusters of few spikelets; spikelets ovoid, terete, 2–4-flowered, light red-brown, 2–4 mm. long, 0·75–2 mm. wide; glumes papery, acute with one thin median nerve, the lowermost 2 (–4) empty, small, fertile glumes ovate, about 4 mm. long, 3 mm. wide; stamens 2; anthers 3 mm. long, with a conical red, densely papillose connective-tip; stigmas 3. Nut ellipsoid, about 3 mm. long, 1·5 mm. wide, dark brown, shining, brittle, terete but flattened basally, with a conical beak, frequently shed with an obscurely 3-lobed, saucer-like basal disk; exocarp spongy; endocarp stony.

HAB.: Marshy ground; near sea-level; fl. May–June.

DISTR.: Divisions 3, 6. Europe and Mediterranean area east to China and Japan; Malay Islands, Philippines, New Guinea, Australia, tropical and S. Africa, Atlantic Islands and tropical America.

3. Between Limassol and Akrotiri, 1905–1962, *Holmboe* 655; *Lindberg f.* s.n.! *Davis* 3544! *Meikle* 2920! *Merton* in ARI 685!
6. Syrianokhori marshes, 1952, *Merton*.

NOTES: Not collected by Merton at Syrianokhori, but repeatedly mentioned as an associate of other marsh plants, and evidently common (or locally dominant) in the area.

SUBFAMILY 2. **Caricoideae** No glumes enclosing both pistil and stamens (i.e. flowers unisexual); female flowers partly to entirely enclosed in a sac-like prophyll (utricle). Embryo caricoid.

11. CAREX L.

Sp. Plant., ed. 1, 972 (1753).
Gen. Plant., ed. 5, 420 (1754).
G. Kükenthal in Engl., Pflanzenr., 38 (IV. 20): 67–824 (1909).

Caespitose or creeping perennials with more or less well-developed woody rhizomes; leaves 3-ranked, blade usually linear, sheath ligulate, the basal often becoming fibrous; flowers (actually single-flowered spikelets) uni-sexual, spirally arranged and aggregated, often in large numbers, into bracteate, often spherical to cylindrical, bisexual or unisexual spikes, the males usually consisting of 3 stamens, and the females of a pistil ensheathed within a bottle- or flask-shaped utricle, both subtended by a generally acute to acuminate glume; stigmas 2 or 3. Embryo caricoid; anatomy eucyperoid.

About 2,000 species widespread in temperate and subarctic regions, and at high altitudes in the tropics.

Female spikes pendulous, green, cylindrical, up to 18 cm. long; plant very robust   **4. C. pendula**
Female spikes either absent or not pendulous, much less than 18 cm. long:
  Spikes ovoid or rounded with male flowers at the top and female below (androgynous); stigmas 2:
    Plants with a long, creeping rhizome and dense, brown inflorescence   -   **1. C. divisa**
    Plants tufted:
      Inflorescence interrupted, lower spikes ± distant, green except when old; culms 2–3 mm. wide at base   -   -   -   -   -   -   -   -   **2. C. divulsa**

Inflorescence dense, lower spikes approximate, often branched, soon becoming brown;
culms more than 3 mm. wide at base - - - - - - - **3. C. otrubae**
Spikes either all male or all female (occasionally the uppermost female partly male at the
top); stigmas 3:
  Inflorescence short, of one terminal male spike and one female below; small tufted plants;
    utricle clearly ribbed, inflated-trigonous, scabridulous-pubescent   **10. C. halleriana**
  Inflorescence of more than 2 spikes; utricle not both clearly ribbed and scabridulous-
    pubescent:
    Female spikes very few-flowered (flowers 1–6); utricles substipitate   -**11. C. illegitima**
    Female spikes many-flowered; utricles not stipitate:
      Utricles beakless, somewhat inflated - - - - **5. C. flacca** ssp. **serrulata**
      Utricles with a more or less pronounced, bifid beak:
        Male spikes 3–4; utricle hispid-scabridulous, not ribbed - - **6. C. hispida**
        Male spikes 1–2; utricle glabrous:
          Leaves narrow, inrolled; female spikes approximate; utricle nervose-ribbed
                9. C. extensa
          Leaves flat; female spikes remote:
            Male spikes fulvous or purplish-brown, up to 50 mm. long; culms up to 90 cm.
            long - - - - - - - - - - **7. C. distans**
            Male spikes dark purple, up to 18 mm. long; culms rarely exceeding 25 cm.
                8. C. troodi

SUBGENUS 1. **Vignea** (*P. Beauv.*) *Nees* Spikes single or several, all more or
less similar and bisexual; stigmas usually 2.

1. **C. divisa** *Huds.*, Fl. Angl., ed. 1, 348 (1762); Unger et Kotschy, Die Insel Cypern, 188 (1865);
Boiss., Fl. Orient., 5: 401 (1882); H. Stuart Thompson in Journ. Bot., 44: 340 (1906);
Holmboe, Veg. Cypr., 41 (1914); Post, Fl. Pal., ed. 2, 2: 683 (1933); Lindberg f., Iter Cypr., 9
(1946); Osorio-Tafall et Seraphim, List Vasc. Plants Cyprus, 17 (1973).
TYPE: British Isles; "*In the meadows near the* Hithe *at* Colchester *in* Essex, Mr. Newton; *by*
Hithe *in* Kent, Mr. J. Sherard. *R. Syn.*"

Creeping, glaucous perennial; rhizomes often long, 2–3 mm. wide,
sometimes branched, covered with fuscous-brown scales which become
fibrous with age; culms clustered, paired or distant, (6–) 15–30 (–50) cm.
long, 0·8–3 mm. wide, triquetrous, scabridulous above, thickened by
sheaths below; leaves rather distant, generally shorter than culms, sheaths
pale, outer sometimes tinged purple, long, papery; ligular arc 2–3 mm. high,
obtuse, ligule distinct, about 1 mm. wide; blade 1–4 (–5) mm. wide, flat or
involute, distinctly nerved, gradually tapering to a slender, trigonous,
scabridulous apex; bracts glumaceous with a scabrid, setaceous tip, that of
the lowermost bract variable in length, from shorter than to several times
longer than the inflorescence; inflorescence spicate, 1·5–4·5 cm. long, ovoid
or ellipsoid, of (3–) 5–10 androgynous spikes; spikes rusty- to purplish-
brown, ovoid, all crowded or the lowermost occasionally distant, 6–10 mm.
long, 3–6 mm. wide, sessile; glumes ovate, 3·5–4 mm. long, 2–2·5 mm. wide,
with membranous, brown sides and a very slender, raised midrib excurrent
into a narrow, awn-like acumen, male narrower and less acuminate than
female; stamens 3; anthers about 3 mm. long, connective-tip shortly
apiculate; utricle broadly compressed-ovoid with a rounded base, 3·5–4 mm.
long, about 2·5 mm. wide, becoming corky or coriaceous and dark brown
when ripe, abruptly narrowed above into a broad, sometimes serrulate-
winged, distinctly bifid beak, ventral face indistinctly veined, dorsal face
prominently ribbed; stigmas 2. Nut lenticular, 1·5 mm. diam., brown,
shining.

HAB.: Marshy ground by streams and lakes, often in brackish situations; sea-level to 5,200 ft.
alt.; fl. March–May (–August).

DISTR.: Divisions 1–4, 6–8. Mainly coastal areas of Europe, from Ireland and Holland
southwards to Spain; Mediterranean region and eastwards to Central Asia.

1. Paphos harbour, 1978, *J. Holub* s.n.!

2. Livadhia, 1939, *Lindberg f.* s.n.! Lefka, 1951, *Merton* 464! Livadhi tou Pasha, 1954, *Merton* 1928! Perapedhi, 1970, *A. Genneou* in ARI 1522!
3. Kolossi, 1862, *Kotschy* 620; Limassol, 1905, *Holmboe* 659! Akrotiri, 1951, *Merton* 346! 347!
4. Kouklia, 1905, *Holmboe* 400; also, 1967, *Merton* in ARI 446! Larnaca, 1930, *Druce* s.n.! also, 1938, *Syngrassides* 1831! Famagusta, 1936, *Syngrassides* 879!
6. Syrianokhori, 1935, *Syngrassides* 772! and, 1941, *Davis* 2616! 2625!
7. Kephalovryso, Kythrea, 1941, *Davis* 2918! Kyrenia, 1934, *Syngrassides* 647! also, 1949, *Casey* 417! Larnaka tis Lapithou, 1941, *Davis* 3000!
8. Galatia, 1962, *Meikle* 2537! 2541!

2. C. divulsa *Stokes* in With., Arr. Brit. Pl., ed. 2, 2: 1035 (1787); Unger et Kotschy, Die Insel Cypern, 188 (1865); H. Stuart Thompson in Journ. Bot., 44: 340 (1906); Lindberg f., Iter Cypr., 9 (1946); Osorio-Tafall et Seraphim, List Vasc. Plants Cyprus, 17 (1973).
[*C. muricata* (non L.) Kotschy in Unger et Kotschy, Die Insel Cypern, 188 (1865); H. Stuart Thompson in Journ. Bot., 44: 340 (1906).]
TYPE: British Isles; "in Norfolk and Suff. Mr. Woodw. — Also in meadows. St."

Tufted perennial with short, knotted rhizomes; culms 20–70 cm. long, 0·2–0·3 cm. wide at base, erect or slightly curved, ribbed, trigonous, scabridulous above; sheaths dull brown becoming black-fibrous with age, not infrequently corrugated on the membranous ventral face, ligular arc about 1 mm. high, ligule acute; blade flat, shorter than or equalling culm, 2–4 mm. wide, gradually tapering to a slender, trigonous-filiform apex; bracts ovate-deltoid with a filiform tip, generally very short, seldom exceeding 1·5 cm., frequently absent; inflorescence spiciform, 2–8 cm. long, of 6–12 androgynous spikes; spikes ovoid, 0·5–1 cm. long, 0·4–0·8 cm. wide, sessile, upper crowded, lower 3–4 usually discrete (but not remote), glumes ovate, 2·5–3·5 mm. long, 1·5–2 mm. wide, shorter than the utricle, membranous except for a narrow green midrib excurrent in a short, scabridulous cusp or awn; male glumes commonly persistent, scarious; stamens 3; anthers 1·5–2 mm. long, connective minutely cuspidate; utricle broadly ovoid or suborbicular, 3·5–5 mm. long, 1·6–3 mm. wide, greenish becoming brown, suddenly narrowed into a beak 0·7–1·3 mm. long, flat, sometimes narrowly winged, generally serrulate, distinctly and sharply bifid; nerves on ventral face of utricle indistinct, dorsal face nerved or sulcate at the base only; stigmas 2, long. Nut depressed-subglobose, 1·5–2 mm. long, dark brown, minutely punctulate.

HAB.: By streams and irrigation channels, or occasionally in garigue on dry calcareous or non-calcareous hillsides; sea-level to 4,600 ft. alt.; fl. April–June.

DISTR.: Divisions 1–3, 5, 7. Also W. and S. Turkey, Syria, Lebanon, Palestine.

1. Near Inia, 1957, *Merton* 3021!
2. Prodhromos, 1862, *Kotschy* 855! also, 1905, *Holmboe* 905; and, 1961, *D. P. Young* 7357! Milikouri, 1939, *Lindberg f.*; Mandria, 1939, *Kennedy* 1738! Aphamis, 1939, *Kennedy* 1739! Between Platres and Aphamis, 1941, *Davis* 3192! Áyia valley, 1956, *Merton* 2831! Askas, 1957, *Merton* 3122! also, 1979, *Edmondson & McClintock* E2902!
3. Episkopi, 1862, *Kotschy* 620.
5. Kythrea, 1962, *Meikle* 2556!
7. Lapithos, 1880, *Sintenis & Rigo* 523! also, 1939, *Lindberg f.* s.n.!

NOTES: Problems exist in the taxonomy of the *Carex divulsa* complex which require further field and laboratory study for their elucidation. The Cyprus population of this variable species shows more or less small differences which exclude it from the described variants from neighbouring areas (*C. coriogyne* Nelmes, *C. polyphylla* Kar. et Kir., *C. leersiana* Rauschert); but the temptation to describe another closely allied taxon is resisted because much remains to be discovered of the cytology and crossing behaviour of this group. The above description applies to the Cyprus population rather than the whole complex.

C. REMOTA *L.*, Fl. Angl., 24 (1754); Unger et Kotschy, Die Insel Cypern, 189 (1865); H. Stuart Thompson in Journ. Bot., 44: 340 (1906); Holmboe, Veg. Cypr., 41 (1914); Osorio-Tafall et Seraphim, List Vasc. Plants Cyprus, 17 (1973).
This slender, tufted sedge with small, green, distant spikes, subtended by

long, leaf-like bracts, has been recorded "an Quellenabflussen bei Prodromos [Kotschy] n. 826", but has not been found since on the island. Although unlikely, in view of the amount of collecting done in the vicinity of Prodhromos, its occurrence on the Amanus Range (Turkey–Syria frontier) raises the possibility of Kotschy's record being confirmed.

3. C. otrubae *Podp.* in Publ. Fac. Sci. Univ. Masaryk, 12: 15 (1922).
    C. nemorosa Rabentisch, Fl. Neomarch., 21 (1804) non Schrank (1789).
    C. vulpina L. var. *nemorosa* Koch, Syn. Fl. Germ. et Helv., ed. 2, 866 (1844); Holmboe, Veg. Cypr., 41 (1914).
    [*C. vulpina* (non L.) Kotschy in Unger et Kotschy, Die Insel Cypern, 189 (1865); Osorio-Tafall et Seraphim, List Vasc. Plants Cyprus, 17 (1973).]
    TYPE: Czechoslovakia; "in pratis pr. Grygov ad Olomouc urbem (*J. Otruba*)" (BRNM).

Robust, tufted perennial with short knotted rhizomes; culms (20–) 30–80 cm. long, 0·4–0·6 cm. wide at base, erect or slightly curved, scabridulous above, very sharply trigonous with concave sides; sheaths dull brownish, becoming blackish-fibrous with age; ventral face of the sheath hyaline, tearing easily; ligular arc about 10 mm. high, subobtuse, ligule wide; blade flat, keeled, up to 10 mm. wide, mostly shorter than culm, tapering to a slender scabridulous apex; bracts setaceous, usually inconspicuous; inflorescence spiciform, cylindrical or narrowly ovoid, simple or branched, 2–6 (–8) cm. long, of 8–20 (–40) androgynous spikes; spikes ovoid to spherical, about 10 mm. long, 7 mm. wide, echinate, the upper crowded, the lower approximate, rarely distant; female glumes ovate, twice as long as the utricle, acute, pale or rufous-tinged with a green midrib excurrent in a scabrid apiculus; male glumes oblong, thin, pale, about 3–3·5 mm. long, 1–1·5 mm. wide; stamens 3; anthers about 2 mm. long, with a long, sharp connective-tip; utricle broadly ovoid, plano-convex, 3·5–4·5 mm. long, about 2 mm. wide, rounded below, narrowed to a serrulate, narrowly winged, deeply bifid beak about 1·5 mm. long, clearly nerved on both faces, brown, shining and spreading at maturity; stigmas 2. Nut ovoid, flattened, mid-brown, about 2 mm. long, minutely punctulate.

HAB.: Ditches and moist, grassy ground; sea-level to 4,500 ft. alt.; fl. April–June.

DISTR.: Divisions 2, 3. Throughout Europe to W. Russia; Syria, Lebanon, Palestine, Turkey east to Afghanistan; N. Africa.

2. Prodhromos, 1862, *Kotschy* 712.
3. Episkopi, 1905, *Holmboe* 695!

SUBGENUS 2. **Carex** Spikes several, the uppermost male, the others female or mixed; stigmas generally 3.

4. C. pendula *Huds.*, Fl. Angl., ed. 1, 352 (1762); Boiss., Fl. Orient., 5: 418 (1884); Druce in Rep. B.E.C., 9: 471 (1931); Post, Fl. Pal., ed. 2, 2: 685 (1938); Osorio-Tafall et Seraphim, List Vasc. Plants Cyprus, 18 (1973).
    TYPE: British Isles; "in sepibus inter Hampstead et Highgate copiose".

Massively tufted, glaucous green, glabrous perennial with thick, red-brown, somewhat woolly roots; rhizome thick, shortly branched, with red-brown, persistent scales; culms 60–80 cm. high, stout below, slender and somewhat nodding above, trigonous, ridged, leafy; lower leaves merging with red-brown sheaths, the upper numerous, long-sheathing; ligular arc 4–8 cm. long, ligule membranous, dark brown, up to 1 mm. wide; blade coriaceous, linear, plicate, up to 100 cm. long, 1–2 cm. wide, rather suddenly acuminate, scabrid on the nerves and margins towards the apex, nerves fine, with numerous transverse striae; inflorescence about $\frac{1}{3}$ culm length, with 1 (2) terminal male spike(s) and 4–6 remote, pendulous female spikes; bracts foliaceous, the lower long-sheathing, as long as or shorter than the inflorescence; male spike(s) linear-cylindric, up to 11 cm. long, dense-flowered, tawny-brown; glumes lanceolate, 5–8 mm. long, with a strongly nerved keel excurrent in a ciliate-tipped mucro, margins pallid, hyaline-membranous; stamens 3: anthers 4–5 mm. long, with a flat, obtuse connective-tip; female spikes dense-flowered, linear-cylindric, up to 18 cm. long, 0·5 cm. wide, the uppermost sometimes apically male, glumes ovate, 2–2·5 mm. long, about 1 mm. wide, dark brown with a green, 3-nerved keel excurrent in a smooth or scabridulous mucro or short awn, margin narrowly hyaline; utricle ellipsoid, 2–3 mm. long, about 1 mm. wide, inflated, trigonous, pale green or brownish, abruptly narrowing to a short, entire or shortly bidentate, ciliate beak, marginal nerves distinct, the others thin or faint. Nut small, oval, trigonous, stipitate; style-base cylindrically thickened; stigmas 3.

HAB.: Streamsides in Pine forest; c. 1,500 ft. alt.; fl. April–May.

DISTR.: Division 2. Also west, central and south Europe E. to Russia (Crimea), Turkey, Lebanon and N. Africa.

2. Stavros tis Psokas, 1930, *Druce* s.n.! Ayia valley, 1956, *Merton* 2830! and, 1962, *Meikle* 2660!

5. C. flacca *Schreb.*, Spic. Fl. Lips. App. no. 669 (1771); Osorio-Tafall et Seraphim, List Vasc. Plants Cyprus, 18 (1973).
   C. glauca Scop., Fl. Carn., ed. 2, 2: 223 (1772); Unger et Kotschy, Die Insel Cypern, 189 (1865); Boiss., Fl. Orient., 5: 417 (1882); H. Stuart Thompson in Journ. Bot., 44: 340 (1906); Holmboe, Veg. Cypr., 41 (1914).
   [*C. panicea* (non L.) Sintenis in Oesterr. Bot. Zeitschr., 32: 192 (1882).]

A very variable, glaucous, creeping perennial, generally with a long rhizome, about 1·5 mm. wide, clothed with the remains of fuscous-brown scales; culms variable, commonly about 40 cm. long, 1–2 mm. wide, sharply trigonous, smooth; leaves glaucous, firm, shorter than the culms, sheaths pale brown or red-spotted, ligular arc about 3 mm. long, ligule about 0·5 mm. wide, blade 3–6 mm. wide, weakly plicate, strongly scabridulous, rather suddenly narrowed to a trigonous acumen; bracts leaf-like, with a short, dark-auricled sheath, lowermost bracts exceeding inflorescence; inflorescence ⅕–¼ culm length, usually with 4–5 (–8) ellipsoid to cylindric, dark, often approximate spikes, the uppermost (1–) 2 (–4) male, the rest female, though the upper may be male-tipped; terminal male spike 20–40 (–80) mm. long, 2–3 (–4) mm. wide, others shorter, purple-brown; glumes ovate, about 4 mm. long, 1–1·5 mm. wide, lower acute, upper obtuse, with a flat, green midrib disappearing below the tip and weakly fimbriate margins; stamens 3; anthers 5 mm. long, connective-tip acute; female spikes usually dense-flowered, cylindric or ovoid-cylindric, 20–30 mm. (or more) long, 3–5 mm. wide, the upper subsessile or shortly pedunculate, the lowermost frequently long-pedunculate and nodding; glumes ovate to lanceolate-ovate, acute or sometimes cuspidate-aristate, shorter or longer than the utricle, 2–5 mm. long, about 1·5 mm. wide, dark purplish-brown with an included or excurrent green midrib; utricle ovoid or obovoid, somewhat inflated, obscurely trigonous, 2–4 mm. long, 1–2 mm. wide, pale greenish-brown, often becoming dark purplish, glabrous or scabridulous, with distinct submarginal nerves, erostrate with a round fuscous orifice; stigmas (2–) 3. Nut obovoid, brown, 3-angled, 1·5 mm. long, 1 mm. wide.

ssp. **serrulata** (*Biv.*) *Greuter* in Boissiera, 13: 167 (1967).
   C. cuspidata Host, Gram. Austr., 1: 71 (1801).
   C. serrulata Biv., Stirp. Rar. Sic. Decr., [manipulus] 4: 9 (1816).
   C. glauca Scop. ssp. *serrulata* (Biv.) Arcang., Comp. Fl. Ital., ed. 2, 92 (1894).
   C. glauca Scop. var. *cuspidata* (Host) Aschers. et Graebn., Syn. Mitteleurop. Fl., 2 (2): 138 (1903); Lindberg f., Iter Cypr., 9 (1946).

C. flacca Schreb. ssp. *cuspidata* (Host) C. Vicioso, Estud. Monogr. Gen. "Carex" Esp., 100 (1959).

TYPE: Sicily; "Panormi in pascuis montosis frequens" (BASSA, FI, LIV).

Spikes large (female up to 70 mm. long; male up to 90 mm. long and 4 mm. wide), usually approximate, erect, shortly-pedunculate, the lowermost spikes occasionally distant and long-pedunculate; glumes often markedly cuspidate or aristate, with a scabridulous tip; utricles large, usually 4 mm. long, often suberect.

HAB.: Sides of streams, springs or irrigation channels, or often on dry, stony, calcareous or igneous slopes in garigue or in open Pine forest; sea-level to 5,500 ft. alt.; fl. March–Sept.

DISTR.: Divisions 1–3, 6–8. S.W. Europe, Mediterranean region to Caspian.

1. Dhrousha, 1941, *Davis* 3249! Terra, near Khrysokhou, 1955, *Merton* 2127! Near Inia (Akamas), 1957, *Merton* 3037! Erimidhes, 1962, *Meikle* 2132!
2. Prodhromos, 1862, *Kotschy* 826a; also, 1880, *Sintenis & Rigo* 946! 1955, *G. E. Atherton* 608! and, 1961, *D. P. Young* 7354! Moniatis R., 3,000–3,400 ft. alt., 1937, *Kennedy* 100! 111! Kryos Potamos, 1937, *Kennedy* 112! Platres, 1937, *Kennedy* 116! also, 1941, *Davis* 3152! and, 1977, *J. J. Wood* 131! Platania, 1939, *Lindberg f.* s.n.; Khrysorroyiatissa Monastery, 1941, *Davis* 3440! Yialia, 1962, *Meikle* 2317! Saittas, 1962, *Meikle* 2794! Between Mandria and Perapedhi, 1962, *Meikle* 2871! Between Troödos and Pano Amiandos, 1979, *Edmondson & McClintock* E 2858!
3. 4 m. N. of Kandou, 1963, *J. B. Suart* 60!
6. Dhiorios Forest, 1954, *Merton* 1918! Myrtou–Kormakiti road, 1967, *Merton* in ARI 121!
7. Above Kyrenia, 1880, *Sintenis & Rigo* 355! Near Kalogrea, 1932, *Syngrassides* 283! Yaïla, 1941, *Davis* 2895! 3632! Larnaka tis Lapithou, 1941, *Davis* 2968! Kornos, 1941, *Davis* 3039! Above Akanthou, 1951, *Merton* 235! Near Bellapais, 1952, *Casey* 1271! Kambyli, 1954, *Merton* 1856! Karavas, 1955, *Merton* 2192! Kyrenia, 1955, *G. E. Atherton* 35! 278! 1237! N. side Sina Oros, 1967, *Merton* in ARI 782!
8. Yialousa, 1930, *Druce* s.n.!

NOTES: The application of infraspecific names in this polymorphic species is unsatisfactory. Cyprus specimens come nearest to the Mediterranean ssp. *serrulata*, with large, suberect utricles and aristate female glumes; they tend to differ in having very large male spikes. It would be interesting to know if the chromosome number differs from that of the accepted number of *C. flacca* ssp. *flacca* (2n = 76).

6. C. hispida *Willd. ex Schkuhr*, Beschr. Abbild. Riedgr., 63, t. 5, fig. 64 (1801), nachtr., 29 (1806); Holmboe, Veg. Cypr., 41 (1914); Post, Fl. Pal., ed. 2, 2: 685 (1933); Lindberg f., Iter Cypr., 9 (1946); Rechinger f. in Arkiv för Bot., ser. 2, 1: 418 (1950); Osorio-Tafall et Seraphim, List Vasc. Plants Cyprus, 18 (1973).

C. echinata Desf., Fl. Atlant., 1: 338 (1798); Boiss., Fl. Orient., 5: 417 (1882) non Murr. (1770) nom. illeg.

[*C. paludosa* (non Good.) Sintenis in Oesterr. Bot. Zeitschr., 32: 121 (1882).]

[*C. acutiformis* (non Ehrh.) Holmboe, Veg. Cypr., 42 (1914); Osorio-Tafall et Seraphim, List Vasc. Plants Cyprus, 18 (1973).]

TYPE: "in der Barbaren", ? *Vahl* (B– WILLD).

Robust, loosely tufted perennial from a stout rhizome, 5–8 mm. wide and covered with dark brown fibrous scales; culms robust up to 150 cm. long, 1 cm. wide at base, erect, sharply trigonous, smooth; leaves harsh, as long as or longer than culms, sheaths persistent, dark purple brown, sharply keeled, ventral face becoming fibrous in herring-bone pattern, ligular arc sharp, about 1 cm. long with a dark ligule about 1·5 mm. wide, blade up to 12 mm. wide below, gradually tapering to a very slender, coarsely scabridulous, trigonous acumen; lower bracts leaf-like, exceeding inflorescence, with conspicuous, dark, rounded, hyaline-margined auricles, upper bracts (to male spikes) abruptly reduced, filiform or glumaceous; inflorescence $\frac{1}{4}-\frac{1}{3}$ culm length, with 6–8 distant, erect, cylindric spikes, male spikes usually 3–4, (3·5–) 5·5–7·5 (–10) cm. long, 0·3–0·6 cm. wide, dark purple-brown, glumes lingulate to oblanceolate, 6–8 mm. long, 1·5–2 mm. wide, with an included pale midrib, rounded or subtruncate apex and conspicuous erose-membranous margins; stamens 3, anthers about 3·5 mm. with a small sharp apiculus; female spikes dense-flowered, (2–) 3–11 (–12) cm. long, 0·6–0·9 cm.

wide, the upper sessile, the lower sometimes shortly pedunculate; glumes oblong, about 5 mm. long, 1–1·5 mm. wide, shortly to slightly longer than the utricle, with a blunt or conspicuously scabridulous-aristate apex and hyaline membranous margins; utricle broadly elliptic or obovate, 3–4 mm. long, 2·5–3 mm. wide, flattened, pale brown, dark-punctate, thinly hispid-scabridulous on surface and margins with 2 distinct submarginal nerves, beak short, cylindrical, obscurely bifid; stigmas 3. Nut ellipsoid, 2–2·5 mm. long, 1–1·5 mm. wide, strongly trigonous, stipitate, beaked, fuscous.

HAB.: Marshes and sides of streams or springs; sea-level to 5,500 ft. alt.; fl. March–July.

DISTR.: Divisions 2, 3, 6–8. Also Portugal, Spain and widespread in the Mediterranean region.
2. Troödos, 1913, *Haradjian* 994; Moniatis River, 3,400 ft. alt., 1937, *Kennedy* 114! Mesapotamos Monastery, 1939, *Lindberg f.* s.n.! Khrysorroyiatissa Monastery, 1941, *Davis* 3438! Platania, 1955, *Merton* 2339! Psilondendron (Kryos Potamos), 1962, *Meikle* 2788!
3. Episkopi, 1905, *Holmboe* 694; Asomatos, 1939, *Mavromoustakis* 126! Akrotiri, 1955, *Merton* 2051! S. of Kellaki, 1956, *Merton* 2697!
6. Myrtou, 1937, *Syngrassides* 1421! Dhiorios Forest, 1954, *Merton* 1918!
7. Kephalovrysi, Kythrea, 1880, *Sintenis & Rigo* 354! also, 1941, *Davis* 2944! Mylous Forest Station, 1933, *Syngrassides* 1378! Kyrenia Pass, 1939, *Lindberg f.* s.n.! 3 miles E. of Kyrenia, 1948, *Casey* 93! Kyrenia, 1955, *Miss Mapple* 15! Upper Thermia, 1955, *G. E. Atherton* 495! Ayios Demetrios, 1956, *G. E. Atherton* 1273!
8. Yialousa, 1941, *Davis* 2379! Ronnas R. valley, 1957, *Merton* 2941! and 1962, *Meikle* 2548!

NOTES: Specimens with short female spikes and blunt or retuse female glumes have been called *C. hispida* var. (or forma) *soleirolii* (*C. soleirolii* Duby, Bot. Gall., 2: 498; 1828), but in *Merton* 2941, from the Ronnas R. valley, both forma *soleirolii* and typical *hispida* are to be found growing from one rhizome!

7. **C. distans** *L.*, Syst. Nat., ed. 10, 1263 (1759); Unger et Kotschy, Die Insel Cypern, 188 (1865); Boiss., Fl. Orient., 5: 425 (1882); Holmboe, Veg. Cypr., 41 (1914); Post, Fl. Pal., ed. 2, 2: 686 (1933); Lindberg f., Iter Cypr., 9 (1946); Osorio-Tafall et Seraphim, List Vasc. Plants Cyprus, 18 (1973).
    C. sinaica Nees ex Steud., Syn. Plant. Cyper. (Syn. Plant. Glum., pars 2), 233 (1855) nom. illeg. superfl.
    C. distans L. f. *sinaica* [Boeck. in Linnaea, 41: 269; 1887] Kük. in Engl., Pflanzenr., 38 (IV.20): 664 (1909).
    C. distans L. f. *major* Kneucker in Allg. Bot. Zeitschr., 5: 127 (1899).
    C. distans L. ssp. *binerviformis* Holmboe, Veg. Cypr., 42 (1914); Osorio-Tafall et Seraphim, List Vasc. Plants Cyprus, 18 (1973).
    TYPE: Not indicated; perhaps a specimen in the Linnaean herbarium, or based on Morison, Plant. Hist., 3: 243, t. 12, fig. 18 (1699).

Tufted, shortly rhizomatous, polymorphic perennial, with a knotted rhizome 2–3 mm. wide, covered with red-brown scales and coarse roots; culms slender, erect or somewhat nodding, up to 90 cm. long, seldom more than 0·2 cm. wide at base, trigonous, smooth; leaves basal, much shorter than culms, sheaths pale, becoming purple-brown with age, often with rusty spots and with a rounded, rusty-spotted antiligule, ligular arc about 3 mm. long, ligule membranous, dark; blade firm, usually flat, 2–5 mm. wide, rather suddenly tapering to a trigonous acumen; bracts leaf-like, shorter than inflorescence, with rounded, brownish auricles; inflorescence $\frac{1}{2}$–$\frac{2}{3}$ culm length with (2–) 3–4 (–5) distant, more or less erect spikes, males 1 (–2), fusiform, 15–50 mm. long, 3–5 mm. wide, fulvous or purplish-brown, glumes narrowly oblong, obtuse, 3–5 mm. long, 1–1·5 mm. wide, midrib pallid, included or shortly exserted, margins erose; stamens 3; anthers 2·5 mm. long, with a distinct apiculate connective-tip; female spikes 2–3, cylindric or ovoid, 10–30 (–50) mm. long, 3–6 mm. wide, subsessile or shortly pedunculate, glumes ovate or oblong-ovate, 2–3 mm. long, 1·5 mm. wide, rusty-brown or fuscous with a paler, included or exserted, apically scabridulous midrib; utricle broadly ovate, trigonous, about 4 mm. long, 1·5–2 mm. wide, often brownish or reddish-brown or red-spotted, with 2 conspicuous, generally green submarginal nerves, and more or less well-developed facial nerves, beak distinct, about 1 mm. long, bifid, with distinct, scabridulous, sharp teeth; stigmas 3. Nut ellipsoid, sharply trigonous, about 2 mm. long, 1·5 mm. wide, brown, minutely punctulate. ·

HAB.: Marshes; wet ground by streams, springs and irrigation channels; sea-level to 3,000 ft. alt.; fl. March–May.

DISTR.: Divisions 1–4, 6–8. Widespread in Europe (except arctic regions); Mediterranean region and east to Central Asia; Atlantic Islands.

1. Akamas, 1950, *Casey* 1019! Ayios Minas, near Smyies, 1962, *Meikle* 2174! Fontana Amorosa, 1978, *J. Holub* s.n.! between Inia and Smyies, 1979, *Edmondson & McClintock* E 2769!
2. Evrykhou, 1935, *Syngrassides* 818! Arminou, 1941, *Davis* 3406! Khrysorroyiatissa, 1941, *Davis* 3436! Maliades, between Mandria and Perapedhi, 1962, *Meikle* 2870!
3. Episkopi, 1862, *Kotschy* 612! Akrotiri marshes, 1902–1967, *Hartmann* 385! *Syngrassides* 1805! *Merton* 344! 2058! *Merton* in ARI 688! Limassol, 1930, *Druce* s.n.!
4. Kouklia, 1905, *Holmboe* 367; Larnaca, 1928, *Druce* s.n.!
6. Dhiorios, 1930, *Druce* s.n.! Myrtou, 1932, *Syngrassides* 32! Syrianokhori, 1935, *Syngrassides* 642! and, 1941, *Davis* 2615! 2619!
7. Ayios Khrysostomos Monastery, 1862, *Kotschy* 394! Larnaka tis Lapithou, 1941, *Davis* 3002! Kyrenia, 1949, *Casey* 262! Karakoumi, 1956, *G. E. Atherton* 1234!
8. Ronnas valley, 1957, *Merton* 2942! 3068! and, 1962, *Meikle* 2547! Near Leonarisso, 1957, *Merton* 2949!

NOTES: Holmboe's record for *C. diluta* M. Bieb. (Veg. Cypr., 42; 1914) from "Vrysi tou Eroton" (Baths of Aphrodite) is almost certainly a misidentification of *C. distans* var. *distans*, which has been collected in the same area.

An abnormal form, in which the female spikes are branched, whilst the male are reduced or absent, has been collected (2) by the Kryos Potamos at 5,200 ft. alt., below Government Cottage, Troödos, 1961, *D. P. Young* 7446! A gathering (*Merton* 2057) from (3) Akrotiri Marshes, 1955, has exceptionally dense-flowering spikes with well-developed scabridulous, female glume-tips; the nuts appear shrivelled so that hybridisation may be suspected.

8. C. troodi *Turrill* in Kew Bull., 1930: 125 (1930); Druce in Rep. B.E.C., 9: 471 (1931); Osorio-Tafall et Seraphim, List Vasc. Plants Cyprus, 18 (1973).

C. distans L. f. *minor* (? Post, Fl. Syr., 838; 1896) Holmboe, Veg. Cypr., 42 (1914).

TYPE: Cyprus, "Troodos, Chionistra, 1,900 m., sent by the *Forestry Department Cyprus*, July, 1928, to Dr G. C. Druce F.R.S." (OXF!)

Similar to the preceding species but generally smaller with culms usually about 20–25 cm. long, but up to 50 cm. at lower altitudes; leaves about 10 cm. long; spikes shorter than those of *C. distans*, males distant, oblong-ellipsoid, 10–18 mm. long, 3–4 mm. wide, with dark purple glumes; females ovoid or shortly oblong, 5–15 mm. long, 4–6 mm. wide, the lowermost sometimes remote and sub-basal, glumes dark purple with a green, included or shortly exserted midrib.

HAB.: By swamps, streams and springs at high altitudes; 2,200–6,300 ft. alt.; fl. March–July.

DISTR.: Division 2. Possibly also on Crete and ? elsewhere (see Notes).

2. Prodhromos, 1880–1961, *Sintenis & Rigo* 878! *D. P. Young* 7378! 7387! 7438! Khionistra, below Government Cottage, 1905, *Holmboe* 873! and, 6,300 ft. alt., 1928, *Forestry Dept.* comm. *Druce* s.n.! Platania, 1930, *Druce* s.n.! also, 1939, *Lindberg f.* s.n.! Kryos Potamos, 5,150 ft. alt., 1937, *Kennedy* 108! Moniatis River, 3,400 ft. alt., 1937, *Kennedy* 113! Xerokolymbos, 4,300 ft. alt., 1937, *Kennedy* 115! Tripylos, 1941, *Davis* 3457! and, 1962, *Meikle* 2675! 2679! Above Amiandos, 1951, *Merton* 376! Kyperounda, 1952, *Merton* 892! Pumping station below Troödos, 1955, *Merton* 2376! Saïttas, 1962, *Meikle* 2790! Between Troödos and Pano Platres, 1973, *P. Laukkonen* 412!

NOTES: The status of this taxon in relation to *C. distans* L. on the one hand and *C. diluta* M.Bieb. on the other, requires further study, as does the Cretan population at high altitudes, which is closely similar, if not conspecific.

9. C. extensa Good. in Trans. Linn. Soc., 2: 175, t. 21, fig. 7 (1794); Boiss., Fl. Orient., 5: 424 (1882); Druce in Rep. B.E.C., 9: 471 (1931); Post, Fl. Pal., ed. 2, 2: 686 (1933).

TYPE: British Isles; "in palustribus prope Harwich — On the marshy part of Braunton Burrows in Devonshire".

Tufted, glaucous perennial with short rhizomes; culms rigid, curved, up to

35 cm. long, seldom more than 1·5 mm. wide, smooth, ribbed; leaves shorter or longer than culms, sheaths dull brown or purplish, becoming somewhat fibrous with age, ligular arc flat or retuse, ligule small, dark, blade firm, inrolled, generally appearing to be not more than 1·5–2 mm. wide, with an acute subulate apex; bracts leaf-like, erecto-patent, the lower often much exceeding inflorescence with a sheathing, membranous-margined base; inflorescence generally less than ⅙ of culm; spikes (2–) 3–4, approximate or sometimes with a remote, sub-basal female spike; male spike 1, very shortly pedunculate, narrowly cylindrical, 7–20 mm. long, 2–3 mm. wide, red-brown, glumes ovate, 3–4 mm. long, 1·5 mm. wide, midrib conspicuously 3-nerved, excurrent in a short apiculus; stamens 3; anthers 2 mm. long, with a short apiculate connective-tip; female spikes 2–3, approximate or the lowest remote, sessile or shortly (rarely long-)pedunculate, ovoid or shortly cylindrical, 5–15 mm. long, 4–6 mm. wide; glume ovate, 2–2·5 mm. long, 1–1·5 mm. wide, reddish-brown with an excurrent greenish midrib; utricle broadly ovoid, about 3 mm. long, 2 mm. wide, grey-green or spotted red-brown, conspicuously nervose-ribbed, abruptly narrowed above to a shortly bifid beak, 0·5 mm. long; stigmas 3. Nut broadly ellipsoid, sharply trigonous, about 2 mm. long, 1·5 mm. wide, fuscous with a pale furfuraceous surface.

HAB.: Brackish marshes near salt-lakes; just above sea-level; fl. April–May.

DISTR.: Divisions 3, 4. Throughout coastal Europe except Arctic; N. African coast; Black Sea; Caspian Sea; Atlantic Islands.

3. Limassol Salt Lake, E. of Akrotiri, 1941, *Davis* 3566! Pano Livadhia, Asomatos, 1962, *Meikle* 2923!
4. Larnaca, 1930, *Druce* s.n.!

10. C. halleriana *Asso*, Syn. Stirp. Arag., 133, t. 9, fig. 2 (1779); Holmboe, Veg. Cypr., 41 (1914); Post, Fl. Pal., ed. 2, 2: 684 (1933); Lindberg f., Iter Cypr., 9 (1946); Rechinger f. in Arkiv för Bot., ser. 2, 1: 418 (1950); Osorio-Tafall et Seraphim, List Vasc. Plants Cyprus, 17 (1973).
C. gynobasis Vill., Hist. Plant. Dauph., 2: 206 (1787); Boiss., Fl. Orient., 5: 409 (1882); Sintenis in Oesterr. Bot. Zeitschr., 32: 192 (1882).
TYPE: Spain; "*en el monte de Herrera, en la Sierra de Villaroya*, circa *Calcena*" (? MA).

A loosely caespitose perennial with short or somewhat elongated rhizomes; old leaf-sheaths fibrous, light- or fuscous-brown; leaves numerous, basal, crowded, bright green, 4–30 cm. long, 0·1–0·2 cm. wide, tapering gradually to a slender often curled apex, erect or recurved, margins recurved, scabridulous; culms 8–30 cm. long, slender, trigonous, striate; inflorescence usually short, with 1 terminal, male spike and 1 or 2, approximate, female spikes, sometimes with a basal, remote, female spike on a filiform peduncle almost as long as the culm; bracts foliaceous, lowermost generally exceeding inflorescence; male spike narrowly fusiform, 10–15 mm. long, 1·5–2 mm. wide, glumes narrowly ovate, acute or acuminate, 3–6 mm. long, 1·5–2 mm. wide, chestnut-brown with a greenish scabridulous midrib and conspicuous, lacerate, membranous margins; stamens 3; anthers 3·5 mm. long with an acuminate connective-tip; female spikes ellipsoid becoming obovoid at maturity, 6–13 mm. long, 4–7 mm. wide, with 3–9 flowers, glumes conspicuous, erecto-patent, generally exceeding utricles, ovate, acute or cuspidate, 4–7 mm. long, 3–4 mm. wide, brown with green scabridulous midrib and conspicuous, membranous margin; utricle ellipsoid, 3·5–6 mm. long, 2–2·5 mm. wide, trigonous, minutely scabridulous-pubescent, with numerous prominent nerves, beak about 0·5 mm. long, apically subtruncate or irregularly retuse. Nut obtuse, stipitate, brown.

HAB.: Dry rocky slopes in garigue; sea-level to 3,700 ft. alt.; fl. Febr.–May.

DISTR.: Divisions 2–4, 6–8. Portugal, Spain, south-central Europe, and Mediterranean region E. to N. Iran and Afghanistan.

2. Pano Platres, 1930, *E. Wall* s.n.
3. Yerasa, 1941, *Davis* 3076! Amathus, 1978, *J. Holub* s.n.! Kakomallis, 1962, *Meikle* 2902!
4. Cape Kiti, 1905, *Holmboe* 159; Athna, 1930, *Druce* s.n.! Akhyritou, 1962, *Meikle* 2966!
6. Between Kormakiti and Liveras, 1937, *Syngrassides* 1420! Áyia Irini, 1951, *Merton* 150!
7. Pentadaktylos, 1862, *Kotschy* 394! Kantara, 1880, *Sintenis & Rigo* 356! Yaïla, 1941, *Davis* 2794! 2864! St Hilarion, 1949, *Casey* 395! Kyrenia, 1950, *Casey* 989! Karavas, 1955, *Merton* 2189! Antiphonitis Monastery, 1955, *Merton* 2285!
8. Between Ayios Philon and Rizokarpaso, 1941, *Davis* 2251! Ronnas Bay, 1941, *Davis* 2382! Koma tou Yialou, 1950, *Chapman* 238! Eleousa, 1962, *Meikle* 2994!

NOTES: Host to an ovary-infesting smut, *Anthracoidea caricis* (Pers.) Bref. (*Cintractia caricis* (Pers.) Magn. sens. lat.).

11. C. illegitima *Cesati* in Friedrichsthal, Reise, 271 (1888); Boiss., Fl. Orient., 5: 407 (1882); A. K. Jackson et Turrill in Kew Bull., 1938: 468 (1938); Osorio-Tafall et Seraphim, List Vasc. Plants Cyprus, 18 (1973).

TYPE: Aegean; Poros, *Friedrichsthal.*

Creeping perennial with a knotted rhizome about 3 mm. wide producing slender shoots which are purple and sheath-covered below; leaves from the upper part of the culm erect, harsh; ligule flat, short, fringed like the margin of the orifice, blades up to 30 cm. long, exceeding culms, 2·5–4 mm. wide, gradually narrowed to a long, filiform, often curled acumen, margins scabridulous, incurved; culms 15–25 cm. long, about 1 mm. wide, trigonous; inflorescence sometimes simple with a single terminal spike, or sometimes with up to 4 sessile or slenderly pedunculate lateral spikes, the lowermost very remote and enveloped by the leaf-cluster; spikes androgynous, narrowly fusiform, 1·5–2·2 cm. long; bracts sometimes much exceeding spikes, the lower leaf-like, diminishing upwards, the upper glume-like, shortly connate basally, dark fuscous-brown with a green midrib excurrent in a scabridulous arista; male glumes dark fuscous- or chestnut-brown, numerous, closely imbricate, oblong-ovate, acute or shortly acuminate, the lowermost scabrid-aristate, 3–4 mm. long, 1–1·5 mm. wide, with or without a narrow membranous margin; stamens 3; anthers 3 mm. long, cuspidate; female glumes about 3, rather distant, lanceolate, about 6 mm. long, 1 mm. wide, purple-brownish with a green midrib, shortly sheathing at the base with a long, but easily detached flagellum, opposed by a flattened, winged, ciliate, brown rhachilla section; utricle inflated, ellipsoid or obovoid, about 5·5 mm. long, 2 mm. wide, slightly curved, pale green becoming brown, with 2 clear lateral nerves, base distinctly stipitate, beak very short with a round, fringed orifice; stigmas 3; rudimentary rhachilla glume-like, acuminate, as long as the nut, slenderly 2-nerved, ciliate-edged. Nut inflated-trigonous, with an acute, trigonous tip, shortly stipitate. *Plate 100.*

HAB.: On rocky, limestone hillsides; 1,100–1,650 ft. alt.; fl. March–April.

DISTR.: Division 7, rare. Also coasts of Adriatic, Ionian and Aegean seas.

7. At foot of Northern Range near Kyrenia, 1936, *Kennedy* 120! Kantara forest road above Akanthou, 1951, *Merton* 236!

[C. DISTACHYA Desf., Fl. Atlant., 2: 336, t. 118 (1800); Osorio-Tafall et Seraphim, List Vasc. Plants Cyprus, 17 (1973), closely resembles *C. illegitima* but differs in the yellowish-brown colour of the spikes and in the presence of a short, but distinct, bifid beak to the utricle. It has a wide distribution in the Mediterranean region, and occurs on the coast of Turkey adjacent to Cyprus. While it might well be found in Cyprus, the unlocalized record in Osorio-Tafall et Seraphim, *loc. cit.* is more likely to have been based on a misidentification of *C. illegitima.*]

Plate 100. CAREX ILLEGITIMA Cesati **1**, habit, $\times \frac{2}{3}$; **2**, flowering spike, $\times 4$; **3**, fruiting spike, $\times 4$; **4**, female flower, $\times 4$; **5**, male flower, $\times 4$; **6**, utricle, $\times 4$. (**1, 2, 4, 5** from *Kennedy* 120; **3, 6** from *Merton* 236.)

110. GRAMINEAE

by N. L. Bor

with key to genera by T. A. Cope

C. E. Hubbard in Hutch., Fam. Flow. Plants, 2: 710–741 (1934); R. Pilger in Engl. et Prantl, Pflanzenfam., ed. 2, 14c: 1–205 (1940); R. Pilger in Engl. Bot. Jahrb., 76: 281–383 (1954); R. Pilger et Eva Potzdal in Engl. et Prantl, Pflanzenfam., ed. 2, 14d: 1–168 (1956); Eva Potzdal in Engl. et Prantl, Pflanzenfam., ed. 2, 14e, Nachtrag: 171–220 (1956); W. D. Clayton in Arber, Gramineae, reprint, I–XXXII (1965).

Annual or perennial herbs, rarely (some *Bambuseae*) shrubs or trees with woody stems often reaching a considerable height; stems jointed, branched or simple, erect or prostrate and creeping at the base and then rooting at the nodes, less often erect upon stilt roots (the latter usually developed in forest or shade-loving species and those living in swamps), cylindrical, compressed, flattened on one side or rarely quadrangular, hollow in the internodes, solid at the nodes, rarely completely solid; underground root system often well developed in perennials, extending sometimes to considerable distances by means of rhizomes; stolons, which are sometimes very tough and woody, are developed by some species and radiate over the soil to a distance of up to several metres. In branched culms, at the point where the branching occurs, a two-keeled organ, called the *prophyllum*, is developed with its back to the culm. The leaves are arranged alternately in two ranks and consist of sheath, ligule and blade; sheaths surrounding the culm with overlapping, free or sometimes connate margins, often swollen at the base, ribbed or smooth, often with small auricles at the mouth; the free margin sometimes ciliate; ligule situated at the junction of sheath and blade, membranous, chartaceous, or a fringe of bristles or hairs, rarely absent; blades usually long and narrow, rarely broad, ovate, sometimes filiform or setaceous, joined to or jointed upon the sheath with a narrowed, rounded or clasping base, rarely petiolate.

Inflorescence various, rarely consisting of one spikelet, more often in spikes, few- to many-spiculate panicles, racemes or false racemes in which a sessile spikelet is accompanied by a pedicellate spikelet or its rudiment. Flowers mostly hermaphrodite, sometimes unisexual, usually consisting of stamens and pistil and two or three minute, hyaline scales (*lodicules*) representing the perianth enclosed and sessile between two bracts (*lemma* and *palea*), the whole called a floret. Florets, if more than one, alternate in one plane on the opposite sides of a jointed or tough axis (*rhachilla*), rarely subspically arranged, with two empty bracts (*glumes*) at the base of the spikelet, rarely one or both of the glumes absent. Stamens hypogynous, in Cyprus species 1, 3 or 6, consisting of long delicate filaments and 2-locular anthers, opening usually by longitudinal slits, occasionally by pores; ovary 1-celled, sometimes with a terminal hairy appendage (*Bromus, Triticeae*); ovules solitary, anatropous, attached to a point or line on the adaxial side of the carpel, the point or line of attachment being visible on the grain as the hilum; styles usually 2, rarely 1 or 3, sometimes connate at the base, and their remains crowning the grain as a *stylopodium* (*Sclerochloa*); stigmas mostly plumose; caryopsis 1-seeded, indehiscent, rarely with a mucilage-forming pericarp; embryo abaxial, large or small.

One of the largest families of flowering plants, represented in almost every part of the world, and including about 620 genera and 10,000 species.

Spikelets 1–many-flowered, breaking up at maturity above the ± persistent glumes, or if
  falling entire then not 2-flowered with the upper floret bisexual and the lower male or
  barren; spikelets usually laterally compressed or terete:
  Inflorescence a panicle, or a raceme of spikelets with long or short (1–1·5 mm.) pedicels;
    spikelets solitary at each node of the axis:
    Spikelets with 2 or more bisexual florets, rarely the lower floret male:
      Spikelets with 2 bisexual florets surmounted by a series of empty lemmas reduced to a
        clavate, knob-shaped mass (*Meliceae*) -   -   -   -   -   - **19. Melica**
      Spikelets with 2 or more bisexual florets (rarely the lowest floret male), succeeding
        lemmas not reduced to a knob-shaped mass:
        Tall reed-like grasses with plumose panicles and broad leafblades (*Arundineae*):
          Lemmas hairy on the back; rhachilla glabrous   -   -   -   - **54. Arundo**
          Lemmas glabrous on the back; rhachilla long-villous -   -   - **55. Phragmites**
        Not tall reed-like grasses:
          Glumes longer than the lowest floret:
            Lemmas with a geniculate awn twisted at the base, or sometimes the awn
              straight, simple; ligule membranous (*Aveneae*):
              Awn articulated in the middle, the upper part clavate, the lower part twisted,
                bearded at the articulation   -   -   -   -   - **23. Corynephorus**
              Awn not articulated in the middle, straight or geniculate:
                Lower floret male with a geniculate awn; upper floret bisexual with a short,
                  straight awn   -   -   -   -   -   -   - **26. Arrhenatherum**
                Lowest floret always bisexual:
                  Spikelets 2-flowered, small; rhachilla not produced   -   - **22. Aira**
                  Spikelets 2–4-flowered; rhachilla produced:
                    Inflorescence a large, effuse, nodding panicle   -   - **-25. Avena**
                    Inflorescence a ± dense, erect panicle:
                      Glumes very unequal   -   -   -   -   -   - **27. Avellinia**
                      Glumes subequal:
                        Lemma lobes produced as bristles ± as long as the lemma; awn
                          geniculate, twisted below the knee   -   - **24. Trisetaria**
                        Lemma lobes acute or obtuse; awn straight, not twisted below
                          28. Lophochloa
            Lemmas awnless or with a very short awn between the teeth of the bilobed
              lemma; ligule a row of hairs (*Danthonieae*) -   -   -   - **53. Schismus**
          Glumes shorter than the lowest floret:
            Ovary with a hairy apical appendage:
              Spikelets in an open panicle (*Bromeae*) -   -   -   -   - **43. Bromus**
              Spikelets borne upon a pedicel about 1·5 mm. long (*Brachypodieae*):
                Plant perennial   -   -   -   -   -   -   - **44. Brachypodium**
                Plant annual -   -   -   -   -   -   -   - **45. Trachynia**
            Ovary without a hairy apical appendage although sometimes hairy at the tip:
              Lemma (7–) 9–11-nerved, rounded on the back, the nerves prominent
                (*Aeluropodeae*, in part) -   -   -   -   -   -   - **61. Aeluropus**
              Lemmas 3–5(–7)-nerved, rounded or keeled on the back, the nerves not
                prominent:
                Lemmas 3-nerved, keeled, glabrous (*Eragrostideae*, in part):
                  Spikelets in slender, ± secund, racemosely arranged racemes
                    57. Leptochloa
                  Spikelets in open panicles   -   -   -   -   - **58. Eragrostis**
                Lemmas 5-nerved, if 3-nerved, panicle dichotomously branched (5.
                  Cutandia), if 7-nerved, grain with a stylopodium (14. *Sclerochloa*);
                  lemmas keeled or rounded on the back, glabrous or hairy on the nerves
                  (*Poeae*, in part):
                  Inflorescence very dense, globular and prickly, long-exserted on a slender
                    peduncle   -   -   -   -   -   -   -   - **9. Echinaria**
                  Inflorescence dense or loosely paniculate, not globular or, if so, loosely
                    long-awned (6. *Cynosurus*):
                    Spikelets all alike:
                      Lemmas firmly keeled on the back:
                        Panicle dichotomously branched; lemmas 3-nerved   **5. Cutandia**
                        Panicle not dichotomously branched:
                          Spikelets broadly ovate, erect or pendulous; lemmas inflated,
                            awnless, cordate at the base, closely imbricate, boat-
                            shaped, horizontally spreading -   -   -   - **2. Briza**
                          Spikelets and lemmas not as above:
                            Lemmas strongly 7-nerved, subcartilaginous at the base,
                              hyaline or herbaceous above, obtuse; grain crowned with
                              a stylopodium   -   -   -   -   - **14. Sclerochloa**

Lemmas not with the above combination of characters; grain without a stylopodium:
Spikelets borne in dense 1-sided clusters on the branches of a racemose panicle - - - - **8. Dactylis**
Spikelets in loose or dense panicles - - - **16. Poa**
Lemmas rounded on the back:
Florets with 1 stamen; panicle rather dense, erect, of long-awned spikelets - - - - - - - -**11. Vulpia**
Florets with 3 stamens; panicle effuse or contracted:
Lemmas obtuse, hyaline at the apex - - **18. Puccinellia**
Lemmas acute, awned or awnless:
Plants annual:
Panicle effuse:
Branches whorled; pedicels filiform, not inflated towards the tip - - - - - **17. Lindbergella**
Branches not whorled, not more than 2 at each node; pedicels inflated towards the tip - **15. Sphenopus**
Panicle strict, often reduced to a raceme:
Glumes ± strongly keeled; grain with punctiform hilum
3. Catapodium
Glumes not keeled; grain with linear hilum
4. Micropyrum
Plants perennial - - - - - - **10. Festuca**
Spikelets of 2 kinds, fertile and sterile:
Fertile spikelets 2–3-flowered; sterile spikelets comprising numerous awned, empty lemmas - - - - - **6. Cynosurus**
Fertile spikelets 1-flowered; sterile spikelets comprising numerous obtuse awnless empty lemmas - - - - **7. Lamarckia**
Spikelets with 1 fertile floret, this either solitary or with empty lemmas below it:
Fertile floret with empty lemmas below it (these short and scale-like):
Glumes minute, represented by 2 semicircular inconspicuous lips; lemma and palea similar in texture, firmly compressed; sterile lemmas 2 (*Oryzeae*) - **1. Oryza**
Glumes well-developed, as long as the spikelet; sterile lemmas 1 or 2 (*Phalarideae*)
29. Phalaris
Fertile floret solitary, without supporting lemmas:
Spikelets falling entire (*Agrostideae*, in part) - - - - - **36. Polypogon**
Spikelets not falling entire, breaking up above the glumes:
Lemmas hyaline or membranous at maturity, awned or awnless, hairy or glabrous:
Lemmas 5-nerved, awned or awnless; glumes acute or acuminate (rarely awned); grains with an adherent pericarp (*Agrostideae*):
Panicle enclosed by the uppermost leaf-sheaths; lemma awnless; keel of glumes winged (if not, see 62. *Crypsis*) - - - - **34. Maillea**
Panicle ± exserted from the uppermost leaf-sheaths on a short or long peduncle, if enclosed then lemma awned:
Panicle a short dense ovoid, fluffy white head with linear densely pectinate-ciliate glumes and long-awned lemmas - - - **39. Lagurus**
Panicle lax or dense; glumes not linear nor densely pectinate-ciliate:
Glumes with dilated bases forming a bulb-like swelling - **37. Gastridium**
Glumes not swollen at the base:
Inflorescence very dense, spike-like, cylindrical, ovoid or ellipsoid; spikelets firmly compressed laterally:
Spikelets 12–14 mm. long - - - - -**40. Ammophila**
Spikelets less than 10 mm. long:
Lemma with a geniculate awn from near the base and 2 long apical bristles - - - - - - **38. Triplachne**
Lemma without 2 long apical bristles:
Inflorescence ovoid or ellipsoid; spikelets with an awn visible beyond the glumes - - - - - **32. Alopecurus**
Inflorescence cylindrical; awn, if present, not visible outside the spikelets - - - - - - **33. Phleum**
Inflorescence not spike-like nor very dense; spikelets fusiform, terete or compressed:
Glumes long-awned - - - - - **36. Polypogon**
Glumes awnless:
Glumes acuminate; callus long-villous - - **31. Calamagrostis**
Glumes acute; callus not villous - - - **35. Agrostis**
Lemmas 1–3-nerved, awnless; glumes and lemmas similar in texture, truncate or acute; grain with a loose (mucilaginous when wet) pericarp (*Sporoboleae*):
Inflorescence a dense ovoid or ellipsoid head, embraced below by the inflated

uppermost leaf-sheaths; spikelets laterally compressed (if keel of glumes
winged, see 34. *Maillea*) - - - - - - **62. Crypsis**
Inflorescence a contracted panicle, linear and spike-like, well exserted from the
uppermost leaf-sheaths; spikelets fusiform - - - **63. Sporobolus**
Lemmas indurated at maturity, glabrous, or, if hairy, the hairs clavate or simple:
Spikelets awnless; hairs on the lemma, if present, clavate (*Milieae*) **30. Milium**
Spikelets awned; hairs on the lemma, if present, simple, not clavate:
Awn tripartite, the branches glabrous (*Aristideae*) - - **64. Aristida**
Awn simple, glabrous or hairy (*Stipeae*):
Lemma partially terete, wrapped around the grain; awns geniculate, rarely
straight, usually twisted below, glabrous or hairy; callus long and sharp
or very short and blunt - - - - - - - **41. Stipa**
Lemma dorsally flattened, not wrapped around the grain; awn short,
straight, scabrid and caducous; callus short and blunt- **42. Oryzopsis**
Inflorescence a simple spike of sessile spikelets seated on a central fragile or tough axis (if
spikelets 3 at a node, the laterals often pedicelled), or of digitately arranged spikes, or the
spikes panicled:
Spikes panicled; lemmas prominently (7–) 9–11-nerved (*Aeluropodeae*, in part)
61. Aeluropus
Spikes solitary and terminal, or digitately arranged, if panicled, then lemma 1–3-nerved:
Inflorescence of digitately arranged or panicled spikes:
Spikelets 3–5-flowered, all florets fertile; spike axis terminating in a point
(*Eragrostideae*, in part) - - - - - - **56. Dactyloctenium**
Spikelets with 1 fertile floret, sometimes the rhachilla produced and bearing a
rudimentary spikelet at its summit (*Chlorideae*):
Spikes digitately arranged at the tip of the peduncle - - - **59. Cynodon**
Spikes not digitate, arranged along an elongated axis - - - **60. Spartina**
Inflorescence a simple spike with 1–3 spikelets at each node:
Ovary with a hairy apical appendage; hilum linear; mouth of sheath with auricles;
spikelets 1–3 at each node, distichous; lateral spikelets of a triad often pedicelled
(*Triticeae*):
Spikelets solitary at each node of the axis:
Spikelets 2-flowered; keels of the lemma prominently pectinate-ciliate; glumes
subulate - - - - - - - - - - **49. Secale**
Spikelets more than 2-flowered; lemma not pectinate-ciliate on the keel; glumes
not subulate:
Plants perennial; nerves of the lemma confluent at the tip - **46. Agropyron**
Plants annual; nerves of the lemma not confluent at the tip:
Glumes and lemmas rounded on the back; grain tightly enclosed between the
lemma and palea and adhering to the latter, or free - **47. Aegilops**
Glumes and lemmas keeled on the back, at least in the upper half; grain free
between lemma and palea - - - - - - **48. Triticum**
Spikelets (including the sterile where present) more than one at each node of the
axis:
Spikelets 2 at each node:
Spikelets with long (6–10 cm.), recurved awns - - **51. Taeniatherum**
Spikelets with awns less than 3 cm. long - - - - **50. Crithopsis**
Spikelets 3 at each node, the 2 lateral male or barren, or both fertile in cultivated
barley - - - - - - - - - - **52. Hordeum**
Ovary without a hairy apical appendage; hilum basal, punctiform or linear; mouth of
sheath without auricles; spikelets solitary at each node of the axis:
Spikelets sunk in cavities in the axis; glumes 2, placed side by side, or the upper
covering the floret while the lower is suppressed or minute (*Hainardieae*):
Both glumes present, placed side by side and entirely covering the floret
20. Parapholis
Lower glume suppressed (except in the terminal spikelet), the upper entirely
covering the floret - - - - - - - **21. Hainardia**
Spikelets not sunk in cavities in the axis; glume 1, the lower suppressed (except in
the terminal spikelet) (*Poëae*, in part):
Spike long and flexuous, slender; upper glume very short - - **13. Psilurus**
Spike rather rigid; upper glume well developed - - - **12. Lolium**
Spikelets 2-flowered, falling entire at maturity, with the upper floret bisexual and the lower
male or barren and in the latter case often much reduced; spikelets usually dorsally
compressed:
Spikelets all bisexual, or each bisexual spikelet paired with a male or barren one:
Spikelets solitary; glumes usually membranous, the lower usually smaller or sometimes
suppressed; upper lemma papery to polished and stony, awnless (*Paniceae*):
Spikelets subtended by numerous bristles which remain on the axis after the spikelets
have fallen - - - - - - - - - - **70. Setaria**

Spikelets not subtended by bristles:
  Inflorescence an open panicle - - - - - - - - **68. Panicum**
  Inflorescence consisting of 1-sided spikes or racemes, these either digitate or scattered
    along a central axis:
    Upper lemma coriaceous to crustaceous, with narrow inrolled margins clasping only
      the edge of the palea:
      Lower glume present:
        Racemes mostly 4-rowed, the spikelets in clusters of 2 or more; spikelets
          gibbously plano-convex, cuspidate to awned; upper palea acute with
          reflexed tip - - - - - - - - - **67. Echinochloa**
        Racemes 2-rowed, the spikelets borne singly or in pairs; upper palea obtuse, its
          tip not reflexed:
          Spikelets pubescent, 2–2·5 mm. long; upper lemma muticous
            65. Brachiaria
          Spikelets usually glabrous (sometimes pubescent or setosely fringed), 3·5–5
            mm. long; upper lemma with a mucro 0·3–1 mm. long
              UROCHLOA (p. 000)
      Lower glume absent - - - - - - - - **69. Paspalum**
    Upper lemma chartaceous to cartilaginous, with thin flat margins covering most of
      the palea and often overlapping - - - - - - **66. Digitaria**
  Spikelets typically paired, with 1 sessile and the other pedicelled, those of a pair usually
    dissimilar (the pedicelled sometimes much reduced) or the spikelets all alike; glumes as
    long as the spikelet and enclosing the florets, ± rigid and firmer than the hyaline or
    membranous lemmas; upper lemma often with a geniculate awn (*Andropogoneae*):
    Spikelets of each pair alike, one of them pedicelled; inflorescence a plumose panicle:
      Panicle spike-like, silvery, apparently not composed of racemes - **74. Imperata**
      Panicle open, the component racemes distinct- - - - - **75. Saccharum**
    Spikelets of each pair different, sometimes the pedicelled reduced:
      Racemes arranged in a panicle with its common axis longer than the racemes, not
        supported by spathes:
        Sessile spikelets dorsally compressed - - - - - **76. Sorghum**
        Sessile spikelets laterally compressed - - - - - **71. Chrysopogon**
      Racemes paired, supported by spathes:
        Lower glume of sessile spikelet convexly rounded on the back, without marginal
          keels; awn hairy - - - - - - - - **73. Hyparrhenia**
        Lower glume of sessile spikelet 2-keeled; awn glabrous - - **72. Andropogon**
  Spikelets unisexual, the sexes in separate inflorescences (*Maydeae*) - - - **77. Zea**

TRIBE 1. **Oryzeae** *Dumort.* Annual or perennial grasses, usually aquatic; leaf-blades narrow, linear or lanceolate, with festucoid anatomy; silica-bodies dumb-bell-shaped, placed transversely; micro-hairs present, 2-celled, elongate; ligule membranous, lacerate, often long; spikelets all alike, hermaphrodite (or very rarely unisexual or monoecious), laterally compressed, 1-flowered or rarely 3-flowered, with the terminal floret hermaphrodite, and the 2 lower, if present, empty and scale-like; inflorescence an open or contracted panicle; glumes very minute or suppressed, sometimes represented by 2 semicircular or shallowly concave, flattened lips; lemmas 3 or 1, when 3, the lower 2 scale-like, when solitary, membranous to coriaceous, even crustaceous, 3–9-nerved, awned from the tip or awnless; palea of the same texture; stamens 6, 3 or 1; lodicules 2, 2-lobed or entire; ovary glabrous; styles 2; stigmas plumose. Grain with a linear hilum; starch-grains compound. Chromosomes small; basic number 12.

1. ORYZA *L.*

Sp. Plant., ed. 1, 333 (1753).
Gen. Plant., ed. 5, 155 (1754).

Annual or perennial grasses with erect, unbranched culms and flat leaf blades; inflorescence an open or contracted panicle of short pedicellate spikelets; spikelets firmly laterally compressed, 3-flowered, disarticulating above the glumes; glumes equal, each reduced to an annular semicircular rim at the tip of the pedicel; two lower lemmas sterile, narrow or bristle-like, without a palea; the third hermaphrodite, keeled, coriaceous, awned or awnless, strongly 5-nerved; palea as long as the lemma and of the same texture; stamens 6; lodicules 2, 2-lobed or entire. Embryo about one-quarter the length of the laterally compressed elliptical grain; hilum linear.

This genus comprises some 20–25 species found in tropical and subtropical areas of the Old World.

Rice, *Oryza sativa* L., forms the staple diet of millions of people and has been introduced and cultivated in suitable places in the New World.

1. O. SATIVA *L.*, Sp. Plant., ed. 1, 333 (1753); Kotschy in Unger et Kotschy, Die Insel Cypern, 176 (1865); Post, Fl. Pal., ed. 2, 2: 712 (1933); N. L. Bor in C. C. Townsend et al., Fl. Iraq, 9: 47, t. 14 (1968).
    TYPE: "*in* Ethiopia, *colitur in* Indiae *paludosis*" (LINN!).

Annual grass; culms up to 1 m. or more tall, erect or somewhat decumbent at the base and then rooting at the lower nodes, leafy almost to the panicle, smooth and glabrous, simple; leaf-blades linear-acuminate, up to 60 cm. long, 6–8 mm. wide, scabrid on the upper surface and on the margins particularly towards the tip, scaberulous below, sometimes smooth, glabrous; lower sheaths papery or spongy, disintegrating, smooth and glabrous, sometimes sparsely pilose, the upper sheaths clasping, loose, striate, smooth and glabrous; ligule up to 2 cm. long, membranous, split; inflorescence a panicle, contracted at first then spreading, nodding, up to 30 cm. long, the base often enclosed in the uppermost leaf-sheath; rhachis rather robust, angled, scabrid; branches scabrid; spikelets laterally compressed, oblong-elliptic in outline, 8–12 mm. long, seated on pedicels which are two-lipped at the apex, the lips representing reduced glumes; rhachilla tough; empty lemmas lanceolate-acute, narrow, 1–2 mm. long; fertile lemma crustaceous in texture, navicular, finely tessellate, 5-nerved, sparsely hairy, straw-coloured, or even purplish, awned or not; palea of the same texture, acute or shortly mucronate; awn, if present, up to 4 mm. long; anthers 3 mm. long.

HAB.: Cultivated ground; not native or naturalized.

DISTR.: Division 3; cultivated near Kolossi, 1865, *Kotschy*. Found throughout the warmer regions of the world.

TRIBE 2. **Poëae** Annual or perennial grasses; leaf-blades linear, flat, filiform or setaceous with festucoid anatomy; silica-bodies circular or oblong; sheaths sometimes tubular; 2-celled micro-hairs absent; ligule membranous; inflorescence of pedicellate spikelets in open or contracted panicles or sessile in panicles or solitary spikes; spikelets all alike, laterally compressed, 2–many-flowered, hermaphrodite, or some hermaphrodite and some sterile, the latter consisting of empty lemmas and glumes (*Cynosurus, Lamarckia*), rarely lower glume absent in lateral florets (*Lolium, Psilurus*); rhachilla disarticulating above the upper glume and between the florets; glumes usually persistent, the lower more or less equal to or shorter than the upper, or absent altogether, rarely both minute, the lower 1-nerved, the upper 3-nerved; lemmas herbaceous, 5–13-nerved, awned or awnless; palea 2-keeled; stamens 3, rarely 1 (*Vulpia*); lodicules 2, usually lanceolate-acute, entire, lacerate, denticulate or lobed, sometimes absent. Grain tightly enclosed between lemma and palea; hilum basal, punctiform or linear; starch-grains compound. Chromosomes large, basic number 7.

2. BRIZA *L.*

Sp. Plant., ed. 1, 70 (1753).
Gen. Plant., ed. 5, 32 (1754).

Annual or perennial herbs with flat leaf-blades; inflorescence a loose, delicate panicle with filiform branches or rather dense and narrow; spikelets broad to very broad, firmly laterally compressed or somewhat plump, several-flowered, hermaphrodite, pendulous or erect; rhachilla disarticulating above the glumes and between the florets; glumes subequal, broad, almost orbicular when flattened, keeled; lemmas papyraceous to chartaceous, very broad, almost cordate to rounded at the base, many-nerved, imbricate; palea smaller than the lemma, orbicular-obovate, 2-keeled; stamens 3; lodicules 2, lanceolate-acute or acuminate. Grain rounded on the back, flattened adaxially, enclosed in the lemma and palea; embryo obovate, about one-sixth the length of the grain; hilum small, elliptic, suprabasal.

About 20 species in temperate parts of the northern hemisphere and in South America; naturalized elsewhere.

Spikelets erect, grey, in a narrow erect panicle - - - - - - **1. B. humilis**
Spikelets pendulous; branches widely spreading:
  Spikelets large, 10–20 mm. long, 8–15 mm. wide - - - - - **2. B. maxima**
  Spikelets much smaller:
    Annual; uppermost ligule up to 8 mm. long; anthers 0·5 mm. long - - **3. B. minor**
    Perennial; uppermost ligule 1–2 mm. long; anthers 2 mm. long - - **4. B. media**

1. B. humilis *M. Bieb.*, Fl. Taur.-Cauc., 1: 66 (1808); N. L. Bor in C. C. Townsend et al., Fl. Iraq, 9: 56, t. 17 (1968); Osorio-Tafall et Seraphim, List Vasc. Plants Cyprus, 7 (1973).
    B. spicata Sm. in Sibth. et Sm., Fl. Graec. Prodr., 1: 57 (1806), Fl. Graec., 1: 60, t. 77 (1808); Boiss., Fl. Orient., 5: 593 (1884); Holmboe, Veg. Cypr., 35 (1914); Lindberg f., Iter Cypr., 7 (1946) non *B. spicata* Burm. f. (1768).
    Brizochloa spicata (Sm.) Jirásek et Chrtek, Nov. Bot. Inst. Bot. Univ. Carol. Prag., 39–41 (1967).
    Brizochloa humilis (M. Bieb.) Chrtek et Hadač in Candollea, 24: 70 (1969).
    TYPE: U.S.S.R., "In Tauriae meridionalis et Iberiae mediae collibus" (LE).

An annual grass; culms erect or loosely fascicled, rarely densely tufted with the outer stems geniculate at the base, slender, simple or branched at the base, smooth and glabrous, with purplish nodes, up to 40 cm. tall; leaf-blades flat, finally involute or rolled, linear-acuminate, up to 15 cm. long, 1–1·5 mm. wide, smooth and glabrous on the lower surface, scabrid above and on the margins; sheaths tightly clasping or rather loose, striate, smooth and glabrous; ligule membranous, 2–4 mm. long, acute, lacerate; inflorescence a narrow, dense, or sometimes lax, panicle up to 12 cm. long, usually not more than 1·5 cm. wide; spikelets plump, ovate-acute in outline, usually grey, finally straw-coloured, 6–8-flowered, 4–6 mm. long, short-pedicellate; glumes more or less equal, broadly obovate-acute when flattened, gibbous in the spikelet, 3–3·5 mm. long, both keeled, 3-nerved, glabrous; lemmas almost orbicular if flattened, apiculate, overlapping, smooth and glabrous or minutely adpressed-hairy in the lower half, scarious on the margins; palea broadly elliptic to almost orbicular, 2-nerved; anthers 1·5 mm. long. *Plate 101.*

HAB.: On bare, rocky mountainsides or in Pine forest on igneous formations; 2,200–3,800 ft. alt.; fl. April–May.

DISTR.: Division 2. Also S.E. Europe, Turkey south to Palestine and eastwards to Caucasus, Iran, Iraq.

2. Common on Troödos Range, Kykko, Platres, Chakistra, Perapedhi, Pedhoulas, Tripylos, Phini, etc. *Hartmann* 507 ! *Kennedy* 61 ! 1118 ! 1119 ! *Davis* 3182 ! 3494 ! *Meikle* 2742 ! 2806 ! etc.

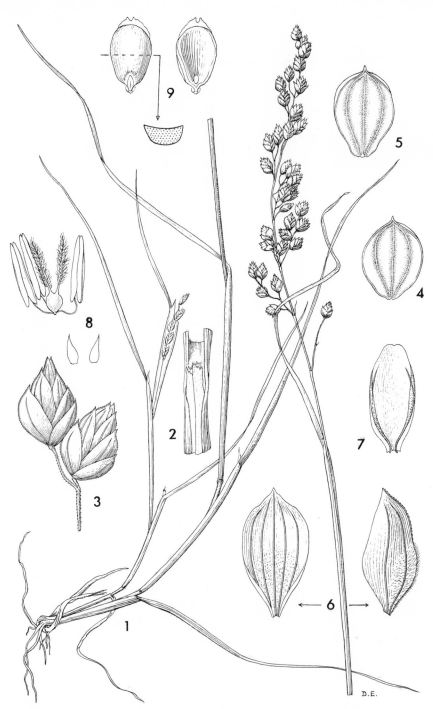

Plate 101. BRIZA HUMILIS M. Bieb. **1,** habit, × 1; **2,** ligule, × 4; **3,** pair of spikelets, × 4; **4,** lower glume, × 10; 5, upper glume, × 10; **6,** lemma from back and side, × 10; **7,** palea, × 10; **8,** flower, × 6; **9,** grain, × 8. (**1–9** from *Rawi* 8966.)

2. B. maxima *L.*, Sp. Plant., ed. 1, 70 (1753); Sibth. et Sm., Fl. Graec. Prodr., 1: 57 (1806), Fl. Graec., 9: 60, t. 76 (1808); Kotschy in Unger et Kotschy, Die Insel Cypern, 181 (1865); Sintenis in Oesterr. Bot. Zeitschr., 31: 392 (1881), 32: 398 (1882); Boiss., Fl. Orient., 5: 593 (1884); Holmboe, Veg. Cypr., 35 (1914); Post, Fl. Pal., ed. 2, 2: 755 (1933); Osorio-Tafall et Seraphim, List Vasc. Plants Cyprus, 7 (1973).

TYPE: "*In* Italia, Lusitania" (LINN!).

An annual grass; culms very often solitary or forming loose tufts, up to 50 cm. tall, slender, smooth and glabrous; leaf-blades linear-acuminate, up to 25 cm. long, 2–4 mm. wide, flat, minutely scabrid on the upper surface and on the margins; sheaths tightly clasping, or the lower slipping from the culms, smooth and glabrous; panicle rather loose, often of only a few spikelets, in robust plants more or less pyramidal in outline; spikelets drooping, very broadly ovate- or oblong-obtuse, 12–20 mm. long, 7–15 mm. wide, 6–18-flowered, pale green or more or less suffused with purple; glumes more or less equal, 6–7 mm. long, 5–9-nerved, glabrous, deeply navicular and round on the back, hooded, very often deeply stained with blackish purple; lemma very broadly ovate, obtuse when flattened, wider than long, 6–8 mm. long, imbricate, obliquely gibbous when viewed from the side, glabrous on the somewhat chartaceous back, or very sparsely covered with minute glandular hairs, 7–9-nerved; palea broadly elliptic in outline, rounded at the top, 2-nerved; anthers about 2 mm. long.

HAB.: Cultivated and fallow land, roadsides, or in garigue on stony hillsides; sea-level to 3,200 ft. alt.; fl. March–May.

DISTR.: Divisions 1–8, common in Division 7. Native of Mediterranean region, now well established in most warm-temperate countries.

1. Dhrousha, 1941, *Davis* 3222!
2. Platres-Mandria, 1937, *Kennedy* 62! Platres, 1938, *Kennedy* 1120! Between Platres and Aphamis, 1941, *Davis* 3184! Pomos, near shore, 1962, *Meikle* 2245!
3. Limassol, 1859, *Kotschy* 431; Mazotos, 1905, *Holmboe* 168; Akrotiri, 1927, *Rev. A. Huddle* 110!
4. Larnaca, 1862, *Kotschy* 309; Cape Greco, S.W. arm, 1958, *N. Macdonald* 35! Dhekelia, 1973, *P. Laukkonen* 218!
5. Athalassa, 1933, *Syngrassides* 45!
6. Skarinou, 1941, *Davis* 2679! Dhiorios, 1962, *Meikle* 2328!
7. Common; *Sintenis & Rigo* 384! *Syngrassides* 54! *Davis* 2875! *Casey* 382! etc.
8. Karpas, 1880, *Sintenis & Rigo*.

NOTES: Holmboe mentions what he calls a gracile form with only one spikelet (*Holmboe* 168). This is a phenomenon met with in many species of grass, particularly those of the genus *Bromus*, when growing in a particularly arid habitat.

3. B. minor *L.*, Sp. Plant., ed. 1, 70 (1753); Sibth. et Sm., Fl. Graec. Prodr., 1: 56 (1806), Fl. Graec., 1: 58, t. 74 (1808); Kotschy in Unger et Kotschy, Die Insel Cypern, 181 (1865); Sintenis in Oesterr. Bot. Zeitschr., 31: 392 (1881), 32: 398 (1882); Boiss., Fl. Orient., 5: 593 (1884); Holmboe, Veg. Cypr., 35 (1914); Post, Fl. Pal., ed. 2, 2: 755 (1933); N. L. Bor in C. C. Townsend et al., Fl. Iraq, 9: 56 (1968); Osorio-Tafall et Seraphim, List Vasc. Plants Cyprus, 7 (1973).

TYPE: "*In* Helvetia, Italia" (LINN!).

An annual grass; culms up to 50 cm. tall, somewhat loosely fascicled, rarely solitary, slender, erect or geniculately ascending, smooth and glabrous; leaf-blades linear- or lanceolate-acuminate, up to 15 cm. long, 2–6 mm. wide, flat, rather thin, rolled when dry, glabrous, minutely scabrid on the upper surface and on the margins; sheaths rather loose, the lower slipping from the culm, smooth and glabrous; ligule membranous, obtuse or truncate, 2–4 mm. long; panicle lax, obovate in outline with widely spreading branches, up to 15 cm. long, 12 cm. wide, or occasionally the spikelets bunched in a compact head 4 cm. long, 1–1·5 cm. wide; branches filiform; spikelets all nodding, very broadly ovate, almost triangular to orbicular in outline, compressed, 4–8-flowered, pale green or suffused with purple; glumes spreading horizontally, equal, 2–3·5 mm. long, membranous,

3–5-nerved; lemmas closely imbricate, kidney-shaped when flattened, 2–3 mm. long, gibbous when viewed from the side, almost cordate at their insertion on the rhachilla, broadly scarious on the margins, glabrous, 7–9-nerved; palea elliptic, 2-nerved; anthers small, about 0·5 mm. long.

HAB.: Pastures, cultivated ground, sometimes in fresh-water or brackish marshes, or on damp ground in forests; sea-level to 3,050 ft. alt.; fl. April–June.

DISTR.: Divisions 1–4, 6–8, common in Division 7 on the plains and the lower slopes of the Northern Range. Most of Europe, Eastern Mediterranean region through Iraq to Iran; India; naturalised throughout the warm-temperate region.

1. Below Stroumbi, 1941, *Davis* 3374!
2. Platres, 1938, *Kennedy* 1117! Yialia, near the sea, *Merton* 2121! Ayios Merkourios, 1962, *Meikle* 2276!
3. Skarinou, 1941, *Davis* 2690!
4. Kouklia, 1905, *Holmboe* 392; Larnaca aerodrome, 1950, *Chapman* 417!
6. Ayia Irini (Morphou), 1941, *Davis* 2596!
7. Common; *Sintenis & Rigo* 383! *Syngrassides* 1319! 1360! *Davis* 2972! *Miss Mapple* 96! etc.
8. Karpas, 1880, *Sintenis & Rigo*.

4. B. media *L.*, Sp. Plant., ed. 1, 70 (1753); Sibth. et Sm., Fl. Graec. Prodr., 1: 57 (1806); Kotschy in Unger et Kotschy, Die Insel Cypern, 181 (1865); Boiss., Fl. Orient., 5: 592 (1884); H. Stuart Thompson in Journ. Bot., 44: 340 (1906).
    TYPE: "*In* Europae *partis siccioribus*" (LINN!).

A perennial grass with short rhizomes; culms loosely caespitose, often stout, erect or shortly decumbent at the base, simple, smooth and glabrous, up to 80 cm. tall; leaf-blades linear-acute or -acuminate, flat, up to 20 cm. long, 3–5 mm. wide, glabrous, minutely rough on the upper surface and on the margins; sheaths not tightly clasping, smooth below, rough above, glabrous; ligule membranous, obtuse or truncate, 1–2 mm. long; inflorescence a loose panicle, usually pyramidal in outline but may be obovate or elliptic, up to 15 cm. long, 15 cm. wide but in youth much narrower; spikelets drooping, loosely scattered, in shape almost triangular to ovate, 4–8-flowered; glumes equal in length, 2·5–3·5 mm. long, narrowly navicular, hooded, rounded on the back, 3–5-nerved, membranous, smooth and glabrous, pale green or suffused with purple; lemmas cordate at the base, oblate when flattened, obliquely gibbous when seen from the side, smooth and glabrous, 4 mm. long, 7–9-nerved; palea elliptic, 2-nerved; anthers 2–2·5 mm. long.

DISTR.: There are no Cyprus specimens of this widespread European and Asiatic species at Kew. The name is included on the authority of Sibthorp et Smith, Fl. Graec. Prodr., 1: 57 (1806). Holmboe (Veg. Cypr., 35; 1914) suggests that Sibthorp's plant may be *B. minor* L. Sibthorp does not indicate a precise locality.

3. CATAPODIUM *Link*

Hort. Berol., 1: 44 (1827).
Scleropoa Griseb., Spic. Fl. Rum. et Bith., 2: 431 (1844).
Paunero in Ann. Inst. Bot. Cav., 25: 207–241 (1976).

Annual grasses with flat leaf-blades; inflorescence a rather stiff, linear, spicate or ovate, one-sided panicle, shortly branched below, bearing pedicellate spikelets above; spikelets 3–10-flowered, hermaphrodite, or upper florets imperfect, compressed, adpressed to the axis or branches, with rather stout grooved pedicels; glumes more or less equal, keeled; lemmas rather firm, rounded on the back, glabrous or hairy, spaced on the scabrid rhachilla, longer than the glumes; palea as long as the lemma, 2-keeled, scabrid on the keels; stamens 3; lodicules 2, oblong, divided above into two teeth of which one is acute and much longer than the other. Grain grooved on the adaxial surface, oblong-fusiform in shape; embryo small, sunken, about one-fifth the length of the grain; hilum a short linear spot, suprabasal.

Two species in Europe and western Asia.

Inflorescence a one-sided raceme of shortly pedicellate distichous spikelets - **1. C. marinum**
Inflorescence a panicle - - - - - - - - - - **2. C. rigidum**

1. C. marinum (*L.*) *C. E. Hubb.* in Kew Bull., 9: 375 (1954); Osorio-Tafall et Seraphim, List Vasc. Plants Cyprus, 7 (1973).

> *Festuca marina* L., Amoen. Acad., ed. 1, 4: 96 (1759).
> *Poa loliacea* Huds., Fl. Angl., 35 (1762).
> *Triticum loliaceum* (Huds.) Sm. in Sowerby et Sm., Eng. Bot., 4: 221 (1795).
> *T. rottbolla* DC. in Lam. et DC., Fl. Franç., 3: 86 (1805).
> *Brachypodium loliaceum* (Huds.) Roem. et Schult., Syst. Veg., 2: 652 (1817).
> *Catapodium loliaceum* (Huds.) Link, Hort. Bot. Berol., 1: 45 (1827); Boiss., Fl. Orient., 5: 634 (1884); Post, Fl. Pal., ed. 2, 2: 767 (1933); A. K. Jackson in Kew Bull., 8: 407 (1933).
> *Desmazeria loliacea* (Huds.) Nym., Sylloge, 426 (1854–55).
> *Scleropoa loliacea* (Huds.) Gren. et Godr., Fl. France, 3: 557 (1856).

TYPE: British Isles; "Found by Mr *Newton* at *Bare* about a Mile from Lancaster; as also nigh the *Salt-Pans*, about a Mile from *Whitehaven, Cumberland*; and at Bright Helmston in Sussex, and elsewhere on the Sea-Coasts." (Ray, Synopsis, ed. 3, 395; 1724).

Annual herb; culms fascicled, simple, rarely branched, erect or spreading, covered by the sheaths almost to the inflorescence, up to 10 cm. tall (including the inflorescence), smooth and glabrous; leaf-blades linear-acute, flat at first, then rolled or involute, glabrous on both surfaces, smooth below, somewhat rough on the nerves above, up to 10 cm. long, 2–3 mm. wide; sheaths rather lax, becoming scarious, striate, smooth and glabrous; ligule membranous, truncate, up to 2 mm. long; inflorescence spicate, 2·5–5 cm. long, straight or curved, green or purplish; axis white with flat, narrow, green margins, smooth and glabrous, scabrid on the margin; pedicels cuneate; spikelets up to 20 in the raceme, 5–6 mm. long, elliptic-acute in outline, laterally compressed, 5–7-flowered, rarely more; glumes somewhat unequal or equal, coriaceous, keeled, scarious on the margins, the lower 1–3-nerved, the upper 3-nerved, smooth and glabrous or scaberulous on the keels, oblong-acute when flattened, 3·5 mm. long; lemma narrowly elliptic-oblong, acute, 3·5 mm. long, 3-nerved, scarious at the tip, smooth and glabrous, rounded, not keeled on the back, the lateral nerves green and sometimes prominent particularly towards the tip; palea a little shorter than the lemma, 2-keeled; densely ciliate-scabrid on the keels; anthers 0·5 mm. long.

HAB.: On rocky ground and walls, usually near the sea; sea-level to 300 ft. alt.; fl. April.

DISTR.: Divisions 2, 7, 8. Coasts of Atlantic and Mediterranean.

2. Three miles above Lefka on the road to Kykko, 1927, *Rev. A. Huddle* 113!
7. Kyrenia, 1932, *Syngrassides* 3!
8. Cape Andreas, 1880, *Sintenis & Rigo* 390! Apostolos Andreas, near the monastery, 1962, *Meikle* 2467!

NOTES: In Linnaeus' Flora Anglica (in Amoen. Acad., 4: 96; 1759 not in the 1754 Dissertation) the name *Festuca marina* is cited with the number 395–4. This is a reference to the third edition (ed. J. J. Dillenius, 1724) of Ray's Synopsis Methodica Stirpium Britannicarum, where the species is described (no. 4) on page 395. Oddly enough Linnaeus does not mention *Festuca marina* in Sp. Plant., ed. 2, 1762.

2. C. rigidum (*L.*) *C. E. Hubb.* in Dony, Fl. Bedford, 437 (1953); N. L. Bor in C. C. Townsend et al., Fl. Iraq, 9: 62, t. 20 (1968); Osorio-Tafall et Seraphim, List Vasc. Plants Cyprus, 7 (1973).

> *Poa rigida* L., Cent. 1 Plant., 5 (1755), Amoen. Acad., 4: 265 (1759); Sibth. et Sm., Fl. Graec. Prodr., 1: 54 (1806); Sintenis in Oesterr. Bot. Zeitschr., 32: 194 (1882).
> *Festuca rigida* (L.) Kunth, Enum., 1: 392 (1833); Kotschy in Unger et Kotschy, Die Insel Cypern, 183 (1865); H. Stuart Thompson in Journ. Bot., 44: 340 (1906).
> *Scleropoa rigida* (L.) Griseb., Spic. Fl. Rum. et Bith., 2: 431 (1845); Sintenis in Oesterr.

Bot. Zeitschr., 32: 54 (1882); Boiss., Fl. Orient., 5: 638 (1884); Holmboe, Veg. Cypr., 37 (1914); Post, Fl. Pal., ed. 2, 2: 769 (1933); Lindberg f., Iter Cypr., 8 (1946).
TYPE: *"in* Gallia, Anglia" (LINN!).

Annual; culms fasciculate, very rarely solitary, erect or the outermost geniculate at the base, finally erect, smooth and glabrous, often somewhat glaucous, up to 45 cm. tall, usually much shorter; leaf-blades flat at first then folded or rolled, linear, attenuate to a stout-acuminate tip, up to 15 cm. long, 1·5–2 mm. wide, scabrid on the upper surface and on the margins, smooth on the lower surface; sheaths clasping, striate, smooth and glabrous; ligules membranous, truncate-lacerate, 2–3 mm. long; inflorescence a more or less unilateral panicle, elliptic- or oblong-acute in outline, consisting of solitary, compound or simple branches, bearing few spikelets, up to 10 cm. long, 1–2·5 cm. wide; spikelets 5–7 mm. long, seated on trigonous pedicels, 6–9-flowered; florets spaced on the fragile rhachilla; glumes more or less equal in length, 1·8–2·2 mm. long, green with scarious margins, the lower 1-nerved, the upper 3-nerved; lemmas elliptic or oblong, acute or obtuse when flattened, with narrow scarious margins, rounded on the back, 3–5-nerved, 2–2·5 mm. long, minutely punctulate; palea 2-nerved; anthers less than 1 mm. long.

HAB.: Walls, roadsides, cultivated and fallow ground, dry slopes and banks by the sea, sometimes on moist ground by ditches and streams; sea-level to 4,600 ft. alt.; fl. March–Sept.

DISTR.: Divisions 1, 2, 4–8, common in Division 7. W. and S. Europe, Mediterranean region and eastwards to Iran; Atlantic Islands. Naturalized in Australia, New Zealand, Tasmania, S. Africa, N. and S. temperate America.

1. Dhrousha, 1941, *Davis* 3220!
2. Stavros tis Psokas Forest, Kambos, 1937, *Syngrassides* 1588! Platres, 1938, *Kennedy* 1122! Polystipos, 1955, *Merton* 2423! Prodhromos, 1961, *D. P. Young* 7313!
4. Mile 3, Larnaca–Famagusta road, 1950, *Chapman* 433!
5. Nicosia–Limassol road, 1967, *Merton* in ARI 598!
6. Nicosia, Department of Agriculture, 1950, *Chapman* 529! 619! Near Dhiorios, 1962, *Meikle* 2335! Dheftera, 1967, *Merton* in ARI 499!
7. Common; *Sintenis & Rigo* 379! *Syngrassides* 262! *Casey* 8! 493! *Chapman* 648!, etc.
8. Koma tou Yialou, Karpas, 1950, *Chapman* 223!

4. MICROPYRUM *(Gaudin) Link*
in Linnaea, 17: 397 (1843).

Triticum L. subgenus *Micropyrum* Gaudin, Fl. Helvet., 1: 366 (1828).

Annual grasses with narrowly linear leaf-blades; inflorescence spike-like, of shortly pedicellate, distichous, alternate, laterally compressed spikelets arranged with the broad side against the shallow excavations of a tough axis; spikelets ovate-acute in outline, finally cuneate, 3–9-flowered; rhachilla fragile below each floret; glumes somewhat unequal, oblong-elliptic, acute when flattened, faintly 5-nerved, rounded on the back, muticous or shortly awned, with incurved margins; palea as long as the lemma, oblong-lanceolate, 2-keeled, scabrid on the keels; lodicules 2, 2–3-toothed at the top, glabrous; ovary glabrous; styles 2, long; stigmas plumose; stamens 3. Caryopsis firmly adherent to lemma and palea, somewhat compressed; hilum linear, as long as the grain; embryo about a quarter as long.

One (or possibly two) species in S. Europe, the Mediterranean region and Atlantic Islands.

1. **M. tenellum** *(L.) Link* in Linnaea, 17: 398 (1843).
   Triticum tenellum L., Syst. Nat., ed. 10, 880 (1759).
   T. poa Lam. ex Lam. et DC., Fl. Franç., 3: 86 (1805).
   T. lachenalii Gmel., Fl. Bad., 1: 291 (1806).
   T. halleri Viv., Fl. Ital. Fragm., 24 (1808).

Brachypodium poa (Lam. ex Lam. et DC.) Roem. et Schult., Syst. Veg., 2: 746 (1817).
Festuca poa (Lam. ex Lam. et DC.) Raspaol in Ann. Sci. Nat., ser. 1, 5: 445 (1825);
Kunth, Rév. Gram., 1: 129 (1829).
Agropyrum halleri (Viv.) Reichb., Fl. Germ. Excurs., 20 (1830).
Nardurus lachenalii (Gmel.) Godr., Fl. Lorr., 3: 187 (1844).
N. tenellus (L.) Duval-Jouve in Bull. Bot. Soc. Fr., 13: 932 (1866).
N. poa (Lam. ex Lam. et DC.) Boiss., Fl. Orient., 5: 633 (1884).
Catapodium tenellum (L.) Trabut in Batt. et Trab., Fl. Alg. Monocots., 232 (1895).

TYPE: Without locality (France; "Monspelii? Sauvages" in L., Sp. Plant., ed. 2, 127; 1762)
(LINN!).

An annual grass; culms mostly solitary, sometimes more or less fascicled, geniculately ascending or erect, smooth and glabrous or minutely scaberulous below the spike, up to 20 cm. tall in Cyprus plants, angled towards the top, with purplish nodes; leaf-blades up to 7 cm. long, 1–1·5 mm. wide when flattened, linear-acuminate, flat at first then folded, involute or convolute, becoming filiform, smooth and glabrous on the lower surface, scaberulous on the margins; sheaths tightly clasping or loose at the base, striate, scabrid; ligule membranous, about 1 mm. long, truncate, lacerate; inflorescence spiciform, up to 9 cm. long, 0·5–1 cm. broad, erect; axis tough; spikelets greenish, shortly pedicellate, 4–5 mm. long, 3–9-flowered; rhachis fragile; glumes somewhat unequal, not strongly keeled, the lower 3–3·5 mm. long, 1–3-nerved, oblong, acute, the upper oblong-elliptic, acute, rounded on the dorsal surface, faintly 5-nerved, smooth and glabrous but scaberulous between the lateral nerves and the margins, muticous or shortly awned; palea 2-keeled; anthers 3, 0·5–0·6 mm. long; awn, if present, 5 mm. long or less.

HAB.: Rocky slopes and mountainsides on igneous formations; 2,500–4,700 ft. alt.; fl. April–June.

DISTR.: Division 2, locally frequent. General distribution that of genus.

2. Locally frequent, Prodhromos, Tripylos, Pedhoulas, Stavros tis Psokas, etc. *Kennedy* 57!
1131! *Syngrassides* 1610! *Davis* 3496! *Meikle* 2759! *A. Hansen* 801! etc.

NOTES: The form with shortly awned lemmas is alone found in Cyprus; it has been separated as a variety as follows: *Micropyrum tenellum* (L.) Link var. *aristatum* (Tausch) Pilger in Bot. Jahrb., 74: 567 (1949) (*Triticum lolioides* Pers. var. *aristatum* Tausch in Flora, 20: 116; 1837).

5. CUTANDIA *Willk.*
in Bot. Zeit., 18: 130 (1860).

Annual, much-branched grasses; leaf-blades much narrower than the sheaths; panicles numerous, dichotomously branched, often from each node, partially enclosed in the spathe-like sheaths, with short, often divaricate, 3-angled branches; spikelets 3–8-flowered, laterally compressed, hermaphrodite, in the ultimate dichotomy seated on stout, 3-angled pedicels of unequal length; glumes unequal, membranous, keeled, 1-nerved; lemmas of similar texture, muticous or with a short awn, 3-nerved; paleas 2-keeled; stamens 3; lodicules 2, obliquely truncate with one margin produced into an acuminate tooth. Grain shallowly concave on the adaxial surface; embryo about one-eighth the length of the grain; hilum punctiform.

The silica-bodies in the leaf-blade epidermis are frequently dumb-bell-shaped, a circumstance which renders its inclusion in the tribe Poëae doubtful (see Metcalfe, Anat. Monocots., 1: 117; 1960).

Six species in the Mediterranean region.

Spikelets oblong or elliptic-oblong in outline, 6–9-flowered, about 4 mm. broad
                      1. C. maritima
Spikelets linear in outline, 3–6-flowered, about 1–2 mm. broad - - **2. C. dichotoma**

1. **C. maritima** (*L.*) *Richter*, Plant. Europ., 1: 78 (1890); Osorio-Tafall et Seraphim, List Vasc. Plants Cyprus, 7 (1973).

Triticum maritimum L., Sp. Plant., ed. 2, 1: 128 (1762).

Scleropoa maritima (L.) Parl., Fl. Ital., 468 (1848); Sintenis in Oesterr. Bot. Zeitschr., 32: 398 (1882); Boiss., Fl. Orient., 5: 637 (1884); Holmboe, Veg. Cypr., 37 (1914); Post, Fl. Pal., ed. 2, 2: 768 (1933).

[*Cutandia memphitica* (non (Sprengel) Richter) A. K. Jackson in Kew Bull., 1934: 273 (1934); Osorio-Tafall et Seraphim, List Vasc. Plants Cyprus, 7 (1973).]

TYPE: "*in* Galliae et Angliae *maritimis*" (LINN!).

An annual grass, green, glaucous or reddish at maturity; culms densely fascicled, the inner erect, the outer geniculate at the base and widely spreading, rarely rooting at the purplish nodes, smooth and glabrous, shining, up to 45 cm. tall; leaf-blades linear-acuminate, flat or becoming convolute or involute towards the stout tip, smooth below, scabrid on the upper surface and on the margins, up to 20 cm. long, 2–6 mm. wide; sheaths somewhat loose, striate, glabrous; ligule membranous, truncate-lacerate, 2–3 mm. long; inflorescence a dense or open panicle; axis angled, smooth; pedicels gradually widened towards the tips; branches shortly dichotomous, each division with a pedicellate spikelet and the branch again dividing in similar fashion and finally terminated by a long-pedicellate spikelet; spikelets oblong or elliptic-oblong in outline, 6–9-flowered, up to 14 mm. long, 4–5 mm. wide; glumes unequal, keeled, oblong or lanceolate, obtuse, or acute when flattened, coriaceous, scarious on the margins, the lower 3·5–4·5 (–5) mm. long, 3-nerved, the upper a little longer and broader, 3-nerved; lemma about 5 mm. long, coriaceous, strongly keeled, oblong-acute, mucronate, strongly 3-nerved; palea 2-keeled; anthers about 2 mm. long.

HAB.: Coastal sand-dunes; about sea-level; fl. April.

DISTR.: Divisions 5, 7, 8. Mediterranean region.

5. Salamis, 1905, *Holmboe* 449!
7. Dhavlos, 1905, *Holmboe* 541; between Ayios Epiktitos and Ayios Amvrosios, *Davis* 3614! Mile 6, Kyrenia-Akanthou road, 1950, *Chapman* 617! 6 miles E. of Kyrenia, 1951, *Casey* 1185!
8. On the North coast of the Karpas (a few miles N. of Aphendrika), 1880, *Sintenis & Rigo*; S. of Galinoporni, 1962, *Meikle* 2499! Koma tou Yialou, 1970, *A. Genneou* in ARI 1478!

NOTES: *Holmboe* 449, which was described as a small form with sterile spikelets, is simply a starved state of *C. maritima*.

2. **C. dichotoma** (*Forssk.*) *Trabut* in Batt. et Trab., Fl. Alg. Monocots., 237 (1895); Osorio-Tafall et Seraphim, List Vasc. Plants Cyprus, 7 (1973).

Festuca dichotoma Forssk., Fl. Aegypt.-Arab., 22 (1775).

Scleropoa dichotoma (Forssk.) Boiss., Fl. Orient., 5: 639 (1884).

Scleropoa memphitica (Spreng.) Parl. var. *dichotoma* (Forssk.) Bonn. et Barr., Cat. Plant. Tunis., 483 (1896); Post, Fl. Pal., ed. 2, 2: 770 (1933).

TYPE: Egypt; "Alexandria" *Forsskål* (C!).

Annual grass; culms usually many, fascicled, the central erect, the lateral geniculate at the base, finally erect, smooth and glabrous, up to 30 cm. tall, usually much shorter, with dark purple nodes; leaf-blades involute or rolled, linear-acute, up to 8 cm. long, 1–1·5 mm. wide when flattened, scabrid on the nerves on the upper surface and on the margins, smooth and glabrous on the lower surface; sheaths inflated, striate, smooth and glabrous, scarious on the margins; ligule membranous, truncate, lacerate, 2–3 mm. long; inflorescence an elongate panicle partly sunk in the uppermost leaf-sheath; branches dichotomous in the following way: a sessile spikelet and a branchlet forming the first dichotomy, subsequent divisions similarly a sessile spikelet and a short branch; axis, branches, branchlets and pedicels triquetrous, scabrid; spikelets linear, 8–10 mm. long, 2–6-flowered; glumes unequal, chartaceous to coriaceous, broadly scarious on the margins, 1-nerved, keeled, the lower about 3–3·5 mm. long, lanceolate, the upper 4–5 mm. long, linear-lanceolate; lemmas 5–5·5 mm. long, linear, acute or acuminate, conspicuously 3-nerved, scarious on the margins; palea 2-nerved; anthers 1·5 mm. long.

HAB.: On sandy soils and stabilized dunes; sea-level to c. 150 ft. alt.; fl. March.

DISTR.: Divisions 4, 6. Eastern Mediterranean region, Iraq, Iran, C. Asia, N. Africa, Egypt.

4. Famagusta, in the Nursery Garden, 1933, *Syngrassides* 42!
6. Ayia Irini, 1941, *Davis* 2542! and, 1962, *Meikle* 2383! 2 miles N. of Syrianokhori, 1941, *Davis* 2603!

6. CYNOSURUS *L.*

Sp. Plant., ed. 1, 72 (1753).
Gen. Plant., ed. 5, 33 (1754).

Perennial or annual grasses with flat leaf-blades; inflorescence a one-sided panicle, sometimes ovoid or globular, bristly with long-awned spikelets, or oblong, stiff, with unawned spikelets; spikelets of two kinds, some fertile, others sterile, the latter enclosing or concealing the former; sterile spikelets consisting of glumes and empty lemmas, distichous and alternate on a tough rhachilla; fertile spikelets laterally compressed, with 1–5 fertile florets or the uppermost reduced; rhachilla disarticulating above the glumes and between the florets; glumes more or less equal, narrow, thin; lemmas rounded on the dorsal surface, membranous, 3–5-nerved, mucronate or awned at the tip; palea as long as the lemma, 2-nerved; stamens 3; lodicules 2, oblong, 2-toothed, one tooth much longer than the other. Grain elliptic or oblong-elliptic in outline, somewhat dorsally compressed, flattened and grooved on the adaxial surface, adherent to the palea; embryo very small, about one-twelfth the length of the grain; hilum punctiform, basal.

Five or six species in the N. temperate region of the Old World.

Glumes and lemmas of the sterile spikelets produced into awns which are purple at the base and pale above, 15–20 mm. long; anthers 0·4–0·6 mm. long - - - - **1. C. coloratus**
Glumes and lemmas of the sterile spikelets produced into pale awns up to 15 mm. long; anthers 1–4 mm. long:
Peduncle of the inflorescence erect; glumes of the fertile spikelets as wide as the lemmas
2. C. echinatus
Peduncle curved immediately below the inflorescence; glumes of the fertile spikelets much narrower than the lemmas - - - - - - - - - **3. C. elegans**

1. **C. coloratus** *Lehm. ex Nees*, Fl. Afr. Austr., 439 (1841); Ascher. et Schweinf., Ill. Fl. Egypte, 172 (1887); Osorio-Tafall et Seraphim, List Vasc. Plants Cyprus, 7 (1973).
C. callitrichus C. et W. Barbey, Herborisations au Levant, 165 (1882) t. 10, figs. 1–8, sine descr.; Boiss., Fl. Orient., 5: 571 (1884); Holmboe, Veg. Cypr., 36 (1914).
C. echinatus var. *callitrichus* (C. et W. Barbey) Bornm. in Beihefte Bot. Centralbl., 31: 269 (1914); Post, Fl. Pal., ed. 2, 2: 748 (1933).
TYPE: S. Africa; "In districtu Zwellendam legit Mundt".

An annual grass; culms loosely fasciculate, rarely solitary, slender, simple, smooth and glabrous, up to 20 cm. tall, usually much shorter; leaf-blades linear, tapering to an acute or obtuse, stout tip, up to 5 cm. long, 2–3 mm. wide, flat or twisted, quite glabrous, smooth below but becoming very scabrid towards the tip and on the margins; sheaths rounded on the back, somewhat loose, striate, glabrous, sometimes scaberulous; ligule membranous, whitish, up to 15 mm. long, becoming lacerate; inflorescence an ovoid, ellipsoid or almost globose, very dense, one-sided panicle, 2·5 cm. long, 1 cm. wide excluding the awns; spikelets of two kinds, sterile and fertile, the latter hidden by the former; sterile spikelets a series of distichous, empty, spaced scales on a straight, tough rhachilla, the lower 6–9 narrow, each ending in a purplish awn about 12 mm. long, the upper 3–4 broader, imbricate, ending in similar awns; fertile spikelets always 1-flowered; glumes equal, hyaline, narrowly lanceolate, keeled, 4–7 mm. long, with an awn as long as or shorter than themselves, 1-nerved; lemma broadly elliptic-acute 3·5–4 mm. long, rounded on the back, smooth in the lower half, scabrid above, minutely 2-

toothed at the tip, awned; awn 12–20 mm. long, scabrid, purple to deep purple; palea 2-nerved; anthers about 0·5 mm. long.

HAB.: Pastures, fallow fields and stony hillsides, usually on calcareous soils; 100–2,400 ft. alt. ; fl. March–April.

DISTR.: Divisions 2, 4–7. Crete, Syria, Palestine, Egypt, Libya, S. Africa (? introduced).

2. Above Pano Lefkara, 1974, *Meikle* 5009!
4. Opposite Pyla marshes, 1936, *Syngrassides* 1027! Site of Akhyritou Reservoir, 1970, *A. Hansen* 545! S.E. of Goshi, 1974, *Meikle* 4093!
5. Vatili, 1905, *Holmboe* 337; Malounda, 1950, *Chapman* 288!
6. Kokkini Trimithia, 1966, *Ioannou* in ARI s.n.!
7. Lakovounara Forest, 1950, *Chapman* 73!

2. C. echinatus *L.*, Sp. Plant., ed. 1, 72 (1753); Sibth. et Sm., Fl. Graec. Prodr., 1: 58 (1806), Fl. Graec. 1: 61 t. 78 (1808); Boiss., Fl. Orient., 5: 571 (1884); Holmboe, Veg. Cypr., 36 (1914); Post, Fl. Pal., ed. 2, 2: 748 (1933); Lindberg f., Iter Cypr., 7 (1946); N. L. Bor in C. C. Townsend et al., Fl. Iraq, 9: 74 (1968); Osorio-Tafall et Seraphim, List Vasc. Plants Cyprus, 7 (1973).
TYPE: "*in* Europa *australiori*" (LINN!).

An annual grass; culms loosely fasciculate rarely solitary, erect or shortly decumbent at the base, spreading, finally erect, smooth and glabrous; leaf-blades usually flat, up to 15 cm. long, 5 mm. wide, linear, acute or acuminate from a rounded base, scabrid on the upper surface particularly at the tip and on the margins, scaberulous on the lower surface or smooth, glabrous, not spreading because of the arched, ligular ring; sheaths tight or slightly loose; ligule up to 10 mm. long, truncate or rounded; inflorescence a dense, ovoid, oblong or almost globose, one-sided panicle, up to 4 cm. long, 1–2 cm. wide, green, shining, often tinged with purple; spikelets of two kinds, sterile and fertile, the former shielding the latter; sterile spikelets compressed, broadly ovate in outline, consisting of a number of distichous, alternate, awned bracts or scales, somewhat distant on a straight axis (rhachilla); fertile spikelets cuneate, 8–12 mm. long, 2–3-flowered, compressed; glumes equal, narrowly lanceolate-acuminate, 7–12 mm. long, 1-nerved; lemmas ovate-acute when flattened, 5–7 mm. long, 5-nerved, 2-toothed at the tip, awned, glabrous, scaberulous towards the tip; palea 2-nerved; awn scabrid, about 12 mm. long; anthers 3–4 mm. long.

HAB.: Dry stony hillsides, usually on igneous formations; 2,500–4,000 ft. alt.; fl. March–June.

DISTR.: Division 2, local. Also S. Europe, Mediterranean region and Atlantic Islands eastwards to India.

2. Locally common about Prodhromos, Platres, Kykko, etc. *Syngrassides* 1573! *Kennedy* 1110! 1111! *Lindberg f.* s.n.; *Merton* 2397!, etc.

3. C. elegans *Desf.*, Fl. Atlant., 1: 82, t. 17 (1798); Kotschy in Unger et Kotschy, Die Insel Cypern, 182 (1865); Sintenis in Oesterr. Bot. Zeitschr., 31: 393 (1881), 32: 124, 194 (1882); Boiss., Fl. Orient., 5: 571 (1884); Holmboe, Veg. Cypr., 36 (1914); Post, Fl. Pal., ed. 2, 2: 746 (1933); N. L. Bor in C. C. Townsend et al., Fl. Iraq, 9: 73, t. 25 (1968); Osorio-Tafall et Seraphim, List Vasc. Plants Cyprus, 7 (1973).
TYPE: "in Atlante prope Mayane Algeriae" (P).

An annual; culms up to 45 cm. tall, loosely fascicled or solitary, erect or geniculately ascending, smooth, shining, glabrous, slender; leaf-blades greyish-green, linear, acuminate, soft, flat, softly hairy on the upper surface, smooth and glabrous below, scaberulous on the margins, up to 12 cm. long, 4 mm. broad; sheaths tight or slightly loose, the lower slipping from the culms, striate, smooth and glabrous; ligules up to 5 mm. long, lanceolate, acute or obtuse; inflorescence a dense or somewhat loose, greyish-green, one-sided panicle seated at the top of a peduncle which is abruptly curved just below the panicle, up to 4 cm. long, 1–1·5 cm. wide, excluding the awns; spikelets of two kinds, fertile and sterile, the latter concealing the former; sterile spikelets consisting of empty hairy or glabrous bracts, distichous on a straight, tough rhachilla; bracts subulate, lanceolate or ovate, acute, awned, 6–12 in number; fertile spikelets 1–3-flowered, 3–3·5 mm. long; glumes equal, filiform, about as long as the spikelet or a little shorter; lemma 2·75–3 mm. long, elliptic, acute, rounded on the back, smooth below, scabrid above, mid-nerve strong, with 2 faint laterals on each side, all joining at the tip and carried out as a scabrid awn 10 mm. long; palea 2-keeled; anthers 1–2 mm. long.

HAB.: Stony hillsides, sometimes in Pine forest, on igneous or calcareous formations; 1,200–5,700 ft. alt.; fl. March–May.

DISTR.: Divisions 2, 7, locally common. Mediterranean region, Atlantic Islands, eastwards to Iran and Central Asia.

2. Locally common; Khionistra, Tripylos, Perapedhi, Papoutsa, Kakopetria, etc. *Kennedy* 87 ! *Davis* 3095 ! *Merton* 2303 ! and in ARI 524 ! *Meikle* 2747 ! 2795 ! etc.
7. Locally common; Buffavento, Pentadaktylos, above Kyrenia, Halevga, etc. *Kotschy* 337 ! *Sintenis & Rigo* 381 ! *Davis* 2817 ! *Casey* 665 ! etc.

7. **LAMARCKIA** *Moench*

Meth., 201 (1794) nom. cons.
Achyrodes Boehm. in Ludwig, Def. Gen. Plant., 420 (1760).

Annual grasses with flat leaf-blades; spikelets in fascicles of 3–4 sterile surrounding 1 fertile, gathered together into a dense, one-sided panicle; sterile and fertile fascicles falling together at maturity by disarticulation of the basal pedicel; sterile spikelets linear with subulate glumes, supporting a number of muticous, ovate, empty scales; fertile spikelets 1–2-flowered, the second floret when present represented by a rudimentary, awned lemma seated on a prolongation of the rhachilla; glumes narrow, membranous, 1-nerved; lemma faintly 5-nerved, awned below the tip; palea as long as the lemma, 2-nerved, adherent to the grain; stamens 3, small; lodicules 2, ovate-acute, entire, glabrous, small; caryopsis adhering to the palea and lemma, obovate, dorsally compressed, grooved adaxially with a short, linear hilum in the groove; embryo one quarter to one third the length of the grain. Sterile spikelets seated on a hairy pedicel; glumes similar to those of the fertile spikelets; lemmas 8–12, reduced to obovate, emarginate scales.

One species, widespread in the Mediterranean region; Atlantic Islands; Ethiopia. Sometimes as an introduction elsewhere.

1. **L. aurea** (*L.*) *Moench*, Meth., 201 (1794); Poech, Enum. Plant. Ins. Cypr., 7 (1842); Kotschy in Unger et Kotschy, Die Insel Cypern, 182 (1865); Sintenis in Oesterr. Bot. Zeitschr., 31: 393 (1881), 32: 124 (1882); Boiss., Fl. Orient., 5: 570 (1884); H. Stuart Thompson in Journ. Bot., 44: 340 (1906); Holmboe, Veg. Cypr., 36 (1914); Osorio-Tafall et Seraphim, List Vasc. Plants Cyprus, 7 (1973).
   Cynosurus aureus L., Sp. Plant., ed. 1, 73 (1753); Sibth. et Sm., Fl. Graec. Prodr., 1: 59 (1806), Fl. Graec., 1: 62, t. 79 (1808).
   Poa aurea (L.) Sibth. ex Walpole, Travels, 23 (1820).
   TYPE: "*in* Europa australi" (LINN !).

An annual grass; culms up to 30 cm. tall, usually much shorter, rarely solitary, more often loosely or densely fasciculate, erect or shortly decumbent at the base and finally erect, smooth and glabrous but scaberulous below the panicle; leaf-blades up to 10 cm. long, 1 mm. wide, linear-acuminate, flaccid, flat, scabrid on the margins, smooth and glabrous on the surfaces, or the upper scaberulous; sheaths rather loose, markedly striate, smooth and glabrous, the lower slipping from the culms; ligules up to 10 mm. long, membranous, rounded or pointed; inflorescence up to 6–8 cm. long, 2–3 cm. wide, dense, one-sided, erect; spikelets of two kinds, the outer sterile, shielding the fertile inner spikelets; sterile spikelets fascicled, 3–4, surrounding one fertile, each seated on a hairy pedicel, linear, 6–7 (–8) mm. long, with two lanceolate narrow glumes 3–4 mm. long, followed by 6–8 empty, distichous, obovate, emarginate scales, 1·5–2 mm. long, pale, often tinged with purple; fertile spikelets 1–2-flowered; glumes equal, 3 mm. long narrow, acuminate, 1-nerved, membranous; lower lemma broadly elliptic-truncate when flattened, 2·5 mm. long, smooth on the back below, scabrid towards the tip; awned below the tip; palea 2-nerved; rhachilla produced and crowned by a rudimentary lemma which is awned; awns straight, scaberulous, 10–12 mm. long; anthers 0·5 mm. long.

HAB.: Cultivated fields, waste ground, roadsides; occasionally on stony hillsides; usually on calcareous soils; 200–400 ft. alt.; fl. April.

DISTR.: Divisions 3, 4, 6, 7, rare. Distribution that of the genus.

3. Limassol, 1862, *Kotschy*.
4. Ayios Antonios, Sotira, 1950, *Chapman* 603 ! and, 1951, *Merton* 649 !
6. On the road to Peristerona, 1787, *Sibthorp*.
7. Near Pentadaktylos, 1880, *Sintenis & Rigo* s.n.; E. end of the Kyrenia range, on the S. side, 1880, *Sintenis & Rigo*.

8. DACTYLIS *L.*

Sp. Plant., ed. 1, 71 (1753).
Gen. Plant., ed. 5, 32 (1754).

Perennial grasses with flat leaf-blades; inflorescence an effuse or linear, contracted panicle with the spikelets pedicellate and crowded in fascicles; spikelets laterally compressed, with 2–5 bisexual flowers; rhachilla disarticulating above the glumes and between the florets; glumes unequal, narrow, keeled, somewhat curved, ciliate on the keels; lemma chartaceous to coriaceous, 5-nerved, keeled, ciliate on the keel, mucronate at the apex; palea 2-keeled; stamens 3; lodicules 2, oblong-cuneate, 2-lobed at the top. Grain loosely enclosed by the lemma and palea, ellipsoid or lanceolate-ellipsoid in shape, flattened on the adaxial surface, rounded abaxially; embryo small, about one-sixth the length of the grain; hilum a suprabasal dot.

Five species in temperate Europe and Asia; one in Cyprus.

1. **D. glomerata** *L.*, Sp. Plant., ed. 1, 71 (1753); Sibth. et Sm., Fl. Graec. Prodr., 1: 58 (1806); Kotschy in Unger et Kotschy, Die Insel Cypern, 182 (1865); Boiss., Fl. Orient., 5: 596 (1884); H. Stuart Thompson in Journ. Bot., 44: 340 (1906); Holmboe, Veg. Cypr., 35 (1914); Post, Fl. Pal., ed. 2, 2: 756 (1933); Rechinger f. in Arkiv för Bot., ser. 2, 1: 417 (1950); T. Photiades in The Countryman, Sept. 1958: 22–23 (1958); Osorio-Tafall et Seraphim, List Vasc. Plants Cyprus, 8 (1973).
    D. hispanica Roth, Catalect. Bot., 1: 8 (1797).
    D. glomerata L. var. *hispanica* (Roth) Koch, Syn. Fl. Germ. Helv., 808 (1837); Post, Fl. Pal., ed. 2, 2: 756 (1933).
    D. glomerata L. var. *australis* Willk. in Willk. et Lange, Prodr. Fl. Hisp., 1: 88 (1861).
    D. glomerata L. ssp. *hispanica* (Roth) Nyman, Consp. Fl. Europ., 819 (1882); Lindberg f., Iter Cypr., 7 (1946); Osorio-Tafall et Seraphim, List Vasc. Plants Cyprus, 8 (1973).
    D. australis (Willk.) Sintenis in Oesterr. Bot. Zeitschr., 32: 399 (1882).
    TYPE: "*in* Europae *cultis ruderatis*" (LINN!).

A perennial grass; culms densely tufted, slender to stout, the outer stems somewhat geniculate at the base and spreading, up to 60 cm. tall, but in Cyprus usually very much shorter, smooth and glabrous; leaf-blades narrow, linear, abruptly acute, up to 25 cm. long, 1–3 mm. wide, greyish-green, folded, very rarely flat, pilose on the upper surface, glabrous, shining and smooth below, scabrid on the margins; upper sheaths rounded, closely clasping, the lower scarious, slipping from the culms, densely pilose to glabrous, somewhat rough or smooth; ligule membranous, 2–10 mm. long; inflorescence a one-sided dense panicle consisting of closely packed spikes of spikelets terminating the peduncle, or very rarely a similar group of spikelets terminating 2 or more short branches; spikelets firmly compressed, 6–7 mm. long, 3–5-flowered, elliptic, oblong or wedge-shaped; glumes unequal, the lower 3·5 mm. long, oblong-acuminate, 1-nerved, the upper 4–4·5 mm. long, elliptic-acuminate, 3-nerved, both keeled and scabrid on the keels; lemmas closely imbricate, oblong-elliptic when flattened, entire or notched at the apex with a stout mucro in the cleft, long ciliate on the keel or merely scabrid, 5-nerved; palea 2-keeled, scabrid on the keels; anthers 2–4 mm. long.

HAB.: In dry places on hillsides, in garigue; by roadsides, or in salt steppe; sea-level to 4,300 ft. alt.; fl. April–Sept.

DISTR.: Divisions 1–4, 7, 8. Throughout temperate Europe, Asia and North Africa; ssp. *hispanica* (Roth) Nyman widespread in the Mediterranean region and eastwards to Central Asia.

1. Dhrousha, 1941, *Davis* 3265!
2. Prodhromos, Platres, Kakopetria, etc., frequent; *Kotschy* 877a; *Kennedy* 1116! *D. P. Young* 7302! *Economides* in ARI 957! etc.
3. Frequent; Kophinou, Kakomallis, Cape Gata, etc. *Holmboe* 578; *Chapman* 556! *Meikle* 2881! 2907! *A. Genneou* in ARI 1491! 1493! etc.
4. Larnaca aerodrome, 1950, *Chapman* 414! Ayios Antonios, Sotira, 1950, *Chapman* 585; Akhyritou, 1962, *Meikle* 2614!
7. Frequent; Kyrenia, Lapithos, Karavas, St. Hilarion, Lakovounara, etc. *Lindberg f.*; *Casey* 57! 73! *Syngrassides* 645! *Chapman* 659! etc.
8. Between Rizokarpaso and Ayios Symeon (Elisis), 1880, *Sintenis & Rigo* 380! Rizokarpaso, 1912, *Haradjian* 207!

NOTES: Numerous subspecies and varieties of this polymorphic species have been described, but in the absence of satisfactory discriminatory characters, the aggregate is left intact here. All the indigenous material of *D. glomerata* L. belongs to ssp. *hispanica* (Roth) Nyman (*D. glomerata* L. var. *australis* Willk.), this is normally less tall than typical *D. glomerata*, with narrower, glaucous leaves and narrow interrupted, spike-like inflorescences. In ssp. *hispanica* the tip of the lemma is notched, with the mucro inserted in the cleft.

9. ECHINARIA *Desf.*

Fl. Atlant., 2: 385 (1799) nom. cons.

Annual grasses with flat, folded or convolute, scabrid, linear-acute leaf-blades; inflorescence a dense, globular or ovoid, capitate, prickly panicle seated at the tip of a long-exserted peduncle; spikelets all alike, 3–4-flowered, seated on very short, scabrid pedicels; florets hermaphrodite, or the upper reduced; glumes equal in length, the lower with two strong nerves produced as firm teeth, the upper emarginate with the median nerve produced as a mucro; lemma broadly elliptic, 5-nerved, scabrid and shortly hairy on the rounded dorsal surface, the five (sometimes seven) nerves produced as very firm, flattened, scabrid awns which are reflexed at maturity; palea as broad as the lemma, 2-nerved, the nerves being produced as awns; rhachilla fragile; lodicules 2, lanceolate-acuminate with one or two small teeth on the sides; stamens 3; styles joined at the base forming a short stylopodium. Grain oblong-truncate, pubescent at the apex, laterally compressed, flattened adaxially; embryo one-third the length of the grain; hilum punctiform, suprabasal.

One species only, widely distributed in the Mediterranean region, also Caucasus, Iraq, Iran and Central Asia.

1. **E. capitata** (*L.*) *Desf.*, Fl. Atlant., 2: 385 (1799); Kotschy in Unger et Kotschy, Die Insel Cypern, 179 (1865); Boiss., Fl. Orient., 5: 565 (1884); H. Stuart Thompson in Journ. Bot., 44: 340 (1906); Holmboe, Veg. Cypr., 34 (1914); Post, Fl. Pal., ed. 2, 2: 746 (1933); Osorio-Tafall et Seraphim, List Vasc. Plants Cyprus, 8 (1973).
   Cenchrus capitatus L., Sp. Plant., ed. 1, 1049 (1753); Sibth. et Sm., Fl. Graec. Prodr., 1: 75 (1806), Fl. Graec., 1: 81, t. 100 (1808).
   TYPE: "*in* G. Narbonensi, Italia" (LINN!).

An annual; culms sometimes solitary, more often loosely, sometimes densely, fasciculate, erect or geniculately ascending, up to 20 cm. tall, striate, smooth and glabrous, or pubescent at the nodes, more or less scabrid below the panicle; leaf-blades rather stiff, linear, abruptly narrowed to an acute or obtuse, hard point, green or glaucous, flat, very shortly hairy and scabrid on both surfaces, scabrid on the margins, up to 6 cm. long, 2–3 mm. wide; upper sheaths clasping, striate, green or greyish green, the lower scarious or falling from the culms; ligules about 1 mm. long, membranous, lacerate; panicle globose, prickly, seated on a long-exserted peduncle, green at first then pale, straw-coloured; spikelets shortly pedicellate on short branches, 3–4-flowered, slightly compressed; glumes membranous, oblong-ovate, more or less equal in length, 2–2·5 mm. long, the lower 2-keeled, obtuse, with the two keels shortly produced as firm awns, the upper with the keel-nerve produced as an awn; lemma broadly elliptic when flattened, shortly hairy on the dorsal surface, 5-nerved, divided at the top into 5 stout, awned lobes which become recurved at maturity, 3 mm. long, with the lobes 3·5 mm. long; palea 2-nerved, 2-awned; anthers 1·5 mm. long.

HAB.: On igneous mountainsides, in the open or in Pine forest; c. 4,000 ft. alt.; fl. April–May.

DISTR.: Division 2, rare. Distribution that of the genus.

2. About Prodhromos, 1862, *Kotschy* 833; Platres, 1938, *Kennedy* 1108 ! 1109 !

10. FESTUCA *L.*

Sp. Plant., ed. 1, 73 (1753).
Gen. Plant., ed. 5, 33 (1754).

Caespitose perennials with erect, green or glaucous culms; leaf-blades flat, folded or rolled, occasionally auriculate; sheaths split to the base or more or less tubular; inflorescence a panicle, dense or, in the Cyprus species, effuse and spreading; spikelets several-flowered, all alike, hermaphrodite; rhachilla fragile; glumes subequal, narrow, subulate; lemmas rounded on the back, membranous or papyraceous, 5-nerved, awned or not; palea 2-keeled; stamens 3; lodicules 2, free or very shortly connate at the base, lanceolate or ovate, 2-toothed at the apex, one tooth sometimes longer than the other; ovary glabrous or with a few hairs at the apex. Grain fusiform, grooved on the adaxial surface; hilum the length of the grain, occupying the groove; embryo obovate, one-fifth the length of the grain.

About 450 species, distributed throughout the world.

1. **F. arundinacea** *Schreb.*, Spicil. Fl. Lips., 57 (1771); Post, Fl. Pal., ed. 2, 2: 764 (1933); Osorio-Tafall et Seraphim, List Vasc. Plants Cyprus, 8 (1973).
      F. elatior L. var. *arundinacea* (Schreb.) Wimm., Fl. Schles., ed. 3, 59 (1875); Boiss., Fl. Orient., 5: 622 (1884).
    TYPE: Germany; Leipzig, "in prato acclivi *hinter dem Biniz*, loco humido" *Schreber* (M).

A tufted perennial grass without rhizomes; culms usually erect, robust, up to 150 cm. tall, simple, smooth and glabrous but scabrid below the panicle; leaf-blades green, linear, tapering to a long, acuminate tip, flat, scabrid on the upper surface, and on the margins, smooth and shining below, up to 45 cm. long, 3–10 mm. wide; sheaths furnished at the mouth with ciliate falcate auricles, scaberulous or smooth, glabrous, striate; ligules membranous, 2 mm. long; inflorescence an effuse panicle, nodding; branches usually in pairs of unequal length, the shorter 2–3-spiculate; spikelets elliptic to oblong, compressed, 10–18 mm. long, 3–10-flowered; glumes more or less equal in length, the lower narrowly lanceolate, 1-nerved, 4–6 mm. long, the upper oblong-acute or lanceolate-oblong-acute, 4·5–6 mm. long, 3-nerved; lemmas elliptic-acute when flattened, awnless, rarely with a very short awn, smooth and glabrous; palea membranous, scabrid on the keel; anthers 3–4 mm. long.

HAB.: Moist roadsides at high altitudes on igneous formation; 5,200 ft. alt.; fl. July.

DISTR.: Division 2, very rare. Most of Europe and Mediterranean region; Atlantic Islands; western and Central Asia.

2. Near Kryos Potamos bridge, 1963, *D. P. Young* 7842!

11. VULPIA *C. C. Gmel.*

Fl. Bad., 1: 8 (1806).
Hackel in Flora, 63: 467–477 (1880).
Herrard in Blumea, 2: 299–326 (1937).

Annual, slender herbs with linear, narrow, folded, convolute or involute leaf-blades; sheaths not tubular; inflorescence a narrow panicle very shortly branched below, often a raceme of shortly pedicellate spikelets turned to one side; spikelets with 3–9 florets, which are hermaphrodite or more usually the uppermost reduced to empty lemmas; rhachilla disarticulating above the glumes and below each floret; glumes as a rule very unequal, linear-acuminate or rarely more or less equal, long-awned; lemmas membranous, lanceolate in shape, rounded on the back with involute margins at maturity, narrowed upwards into a fine, straight, scabrid awn, somewhat spaced on the scabrid rhachilla; stamens 1, rarely 3; lodicules 2, hyaline, 2-toothed, one tooth being larger than the other. Grain narrow, oblong, pointed at the base, deeply grooved on the adaxial face, tightly embraced by lemma and palea; embryo very small; hilum a short, vertical line just above the base.

About 25–30 species in temperate regions of Europe, the Mediterranean, Pacific and N. and S. America.

Glumes and lemmas glabrous or shortly scaberulous:
  Stamens 3; lemmas (7–) 10–12 (–15) mm. long -   -   -   -   -   **1. V. membranacea**
  Stamens 1; lemmas usually less than 7 mm. long:
    Lower glume awned, up to 10 mm. long (including awn) -   -   -   **2. V. brevis**
    Lower glume not awned, less than 10 mm. long:
      Inflorescence usually well exserted from uppermost leaf sheath, 3–15 cm. long
                                                3. V. muralis
      Inflorescence usually not fully exserted from uppermost leaf sheath, 5–25 (–35) cm. long
                                              4. V. myuros
Glumes glabrous; lemmas long-pilose -   -   -   -   -   -   -   -   **5. V. ciliata**

1. V. membranacea (*L.*) *Dumort.*, Obs. Gram. Belg., 100 (1823) in adnot; Sintenis in Oesterr. Bot. Zeitschr., 32: 364 (1882); Post, Fl. Pal., ed. 2, 2: 765 (1933); Osorio-Tafall et Seraphim, List Vasc. Plants Cyprus, 8 (1973).
    Stipa membranacea L., Sp. Plant., ed. 1, 560 (1753).
    Festuca uniglumis Sol. in Ait., Hort. Kew., ed. 1, 1: 108 (1789); Boiss., Fl. Orient., 5: 629 (1884); Holmboe, Veg. Cypr., 37 (1914).
    TYPE: Spain; "*in* Hispania", *Loefling* (LINN!).

An annual grass; culms densely fascicled, rarely solitary; central stems erect, the lateral slightly geniculate, finally erect, slender, smooth and glabrous, rarely scaberulous below the panicle, up to 15 cm. tall, usually much less; leaf-blades flat at first, finally involute, linear-acute, up to 4 cm. long, 1–2 mm. wide, smooth and glabrous on the lower surface, puberulous or hirsute on the upper, scabrid on the margins; sheaths rather loose, smooth and glabrous, striate, those enclosing the base of the inflorescence almost spathe-like; ligule a very narrow, membranous ring; inflorescence a very dense panicle, excluding the awns 3–5 cm. long, 1–1·5 cm. wide, one-sided; branches very short, antrorsely scabrid or sometimes smooth; spikelets, without the awns, 1–1·5 cm. long, 3–9-flowered; florets spaced on the tough rhachilla, the uppermost sterile, all falling together; lower glume almost absent in the lateral spikelets, up to 3 mm. long towards the top of the panicle, the upper glume 8–10 mm. long, linear-acuminate, with broad hyaline margins, 3-nerved, rather rough towards the tip, produced into a scabrid awn 6–8 mm. long; lemma 12 mm. long, awned, 3-nerved, smooth and glabrous in the lower two thirds, scabrid at the tip; awn scabrid, 16–18 mm. long, straight; palea shorter, with two scabrid keels; stamens 3; anthers 1·5–2 mm. long.

HAB.: On seashores and sand-dunes, or in garigue near the sea; at or near sea-level; fl. Febr.–May.

DISTR.: Divisions 2–8. Widespread in the Mediterranean region, extending to western Europe. Introduced into Australia.

2. Yialia, 1955, *Merton* 2014!
3. W. of Limassol, 1902, *Hartmann* 285a!
4. Cape Kiti, 1955, *Merton* 2014!
5. Salamis, 1957, *Merton* 3208!
6. Common about Ayia Irini, *Syngrassides* 762! *Davis* 2544! *Merton* 1099! and in ARI 132!
7. Mile 6, Kyrenia–Akanthou road, 1950, *Chapman* 618!
8. Near Yialousa, 1880, *Sintenis & Rigo* 372! Near Apostolos Andreas, 1941, *Davis* 2328! Cape Andreas Forest, 1957, *Merton* 3069!

2. V. brevis *Boiss. et Kotschy* in Boiss., Diagn., 2, 4: 139 (1859); Post, Fl. Pal., ed. 2, 2: 765 (1933); Osorio-Tafall et Seraphim, List Vasc. Plants Cyprus, 8 (1973).
  Festuca inops Del., Fl. d'Egypte, 52 (1813) nomen.
  Vulpia inops Hackel in Flora, 63: 476 (1880); Boiss., Fl. Orient., 5: 630 (1884).
  Festuca brevis (Boiss. et Kotschy) Druce in Rep. B.E.C., 9: 47 (1931).
  TYPE: Lebanon; "in graminosis *Mar Tserkis Libani* prope Bscherre, alt. 4,800 ft. Fl. Jul. Cl. Kotschy" (G).

An annual grass; culms fasciculate, erect or the outer culms bent at the base and spreading, up to 6 cm. tall, leafy to the panicle, smooth and glabrous, or scaberulous, or puberulent below the panicle; leaf-blades up to 5 cm. long, linear-acute, folded, filiform and twisted, very narrow (1 mm. wide), pubescent on the upper surface, smooth and glabrous or hairy below, scaberulous on the margins; sheaths very loose, the uppermost almost spathe-like, striate, smooth and glabrous, membranous on the margins; ligule membranous, lacerate, about 1 mm. long; inflorescence a very dense panicle, when well developed trichotomously branched, usually the trichotomy obscure; branches short; spikelets compressed, green or suffused with purple, with one fertile floret surmounted by a number of empty awned scales; lower glume lanceolate-subulate, awned, with the awn 10 mm. long, the upper oblong-acute, 5 mm. long, 3-nerved, awned with a bristle 7 mm. long, both scabrid or shortly hairy on the back; fertile lemma oblong-obtuse, 4 mm. long, awned, 5-nerved, scabrid on the dorsal surface; awn 8–9 mm. long, scabrid; stamens 1; anthers 0·5–0·75 mm. long.

HAB.: In dry pans on margin of salt-marshes; on sand-dunes; near sea-level; fl. March–April.

DISTR.: Divisions 4, 6, rare. Egypt, Sinai, Palestine, Syria.

4. Larnaca airport, 1951, *Merton* 326!
6. Myrtou, 1928 or 1930, *Druce* s.n.; Cape Kormakiti, 1956, *Merton* 290!

NOTES: The name *Festuca inops* first appeared in Delile, Flore d'Egypte, 52 (1813), without a description. At the same time as the text was written a plate was prepared figuring this plant among others. Strangely enough this plate was not published at that time with the 62 other plates which illustrated Delile's work. When C. & W. Barbey published "Herborisations au Levant" in 1882, a plate reproduced from photographs of the missing plates 63 and 64, which were at Montpellier, was inserted among others in that work. They are of very poor quality and the sketch purporting to be *Festuca inops*, exhibits little that could identify this species. Barbey (loc. cit., p. 175) gives an account of the missing illustration. Apparently the first description of *Vulpia inops* to be validly published is that of Hackel, but unfortunately this is antedated by *Vulpia brevis* Boiss. et Kotschy.

A new subgenus, *Spirachne*, was created in *Vulpia*, by Hackel (Flora, 63: 467–477; 1880) to distinguish this species from its congeners. He found that the glumes and lower lemma were not distichously arranged, as in the very great majority of grass species, but form a spiral. The empty lemmas terminating the rhachilla, follow usually, but not always, the same plan.

3. V. muralis *(Kunth) Nees* in Linnaea, 19: 694 (1847).
  Festuca muralis Kunth in H.B.K., Nov. Gen. et Sp. Plant., 7: 485, t. 691 (1826).
  Vulpia myuros (L.) C. C. Gmel. var. *tenella* Boiss., Voy. Bot. Esp., 2: 668 (1841).

V. broteri Boiss. et Reut., Pug. Plant. Nov., 128 (1852).
V. sciuroides (Roth) C. C. Gmel. γ *longe-aristata* Willk. in Willk. et Lange, Prodr. Fl. Hisp., 1: 91 (1861).
[*Festuca myuros* (non L.) Kunth in H.B.K., Nov. Gen. et Sp. Plant., 1: 155 (1816) as *F. myurus*]
[*Vulpia sciuroides* (non (Roth) C. C. Gmel.) Sintenis in Oesterr. Bot. Zeitschr., 32: 194 (1882) pro parte]
[*V. bromoides* (non (L.) S. F. Gray) Osorio-Tafall et Seraphim, List Vasc. Plants Cyprus, 8 (1973)]
TYPE: Ecuador; "in muris quibus horti circumsepti sunt prope Conocoto Quitensium, alt. 1350 hexap." *Humboldt* (?P, B).

An annual grass; culms solitary or loosely tufted, up to 60 cm. tall, when solitary erect, when tufted the lateral culms geniculate, finally erect, slender, smooth and glabrous; leaf-blades up to 15 cm. long, 1–3 mm. broad, flat or rolled, minutely hairy on the upper surface, scaberulous on the margins; sheaths clasping; ligules short, membranous; panicle long-exserted from the uppermost leaf-sheath, oblong, usually dense, up to 15 cm. long, green or purplish; spikelets oblong or wedge-shaped, 5–10-flowered; rhachilla fragile below each floret; glumes persistent, the lower 1–3 mm. long, 1-nerved, lanceolate, the upper 3–10 mm. long, 3-nerved, broader; lemmas 4–7 mm. long, rounded on the back, 5-nerved, tipped with a scabrid straight awn, up to 13 mm. long; stamens 1; anthers 0·3–0·6 mm. long.

HAB.: Dry, calcareous hillsides and limestone screes; c. 2,500 ft. alt.; fl. April–May.

DISTR.: Division 7, rare. Mediterranean region; introduced into the New World.

7. Kantara, 1880, *Sintenis & Rigo* 400, in part ! Same locality, 1952, *Merton* 770B !

NOTES: A good example of the dangers of amalgamating collectings: *Sintenis & Rigo* 400 includes three distinct species, *V. muralis*, *V. ciliata* and *V. myuros*.

4. **V. myuros** (*L.*) *C. C. Gmel.*, Fl. Bad., 1: 8 (1806); Boiss., Fl. Orient., 5: 628 (1884); Post, Fl. Pal., ed. 2, 2: 764 (1933); N. L. Bor in C. C. Townsend et al., Fl. Iraq, 9: 88, (1968); Osorio-Tafall et Seraphim, List Vasc. Plants Cyprus, 8 (1973).
Festuca myuros L., Sp. Plant., ed. 1, 74 (1753); Sibth. et Sm., Fl. Graec. Prodr., 1: 60 (1806); Lindberg f., Iter Cypr., 7 (1946).
[*Vulpia sciuroides* (non (Roth) Gmel.) Sintenis in Oesterr. Bot. Zeitschr., 32: 194 (1882) pro parte]
TYPE: "*in* Anglia, Italia" (LINN !).

An annual grass; culms usually densely fascicled, the outer stems geniculate at the base, finally erect, up to 45 cm. tall, smooth and glabrous, leafy to the panicle; leaf-blades linear-acute, flat at first then involute, up to 15 cm. long, 1–2 mm. wide, shortly pilose on the upper surface or glabrous and scaberulous, smooth and shining below, scabrid on the margins; sheaths somewhat loose, the uppermost opening out, smooth and glabrous; inflorescence a linear, dense or lax, curved or nodding panicle, sometimes racemose in the terminal part, 5–35 cm. long, green or suffused with purple; spikelets oblong at first, then cuneate, excluding the awns 7–10 mm. long, 3–7-flowered; rhachilla fragile; glumes persistent, extremely unequal, the lower often a mere scale or as long as one sixth the length of the upper lemma, the upper linear-subulate, 3-nerved, the two lateral nerves short, 3–8 mm. long; lemmas oblong-lanceolate, acute, 5–7 mm. long, awned, 5-nerved, scabrid on the back or smooth, glabrous; awn up to 15 mm. long; palea with 2 scabrid keels; stamens 1 (2); anthers 0·3–0·6 mm. long.

HAB.: Hillsides, roadsides; on sandy or stony ground; 700–4,500 ft. alt.; fl. (Jan.–) April–June.

DISTR.: Divisions 2, 7. Europe eastwards to Central Asia and India; North Africa; Atlantic Islands.

2. Common in central Troödos; Platres, Platania, Troödhitissa, Kakopetria, Phini, etc. *Lindberg f.* s.n. ! *Kennedy* 1538 ! *Merton* 2207 ! 2301 ! 3135 ! 3180 ! *Meikle* 2739 ! 2873a ! etc.
7. Bellapais, 1956, *G. E. Atherton* 856 ! Kantara road near Kantara, 1952, *Merton* 770a (partly) !

5. V. ciliata *Link*, Hort. Berol., 1: 147 (1827); Boiss., Fl. Orient., 5: 628 (1884); Post, Fl. Pal., ed. 2, 2: 765 (1933); N. L. Bor in C. C. Townsend et al., Fl. Iraq, 9: 87 (1968); Osorio-Tafall et Seraphim, List Vasc. Plants Cyprus, 8 (1973).

   Festuca ciliata Danth. in Lam. et DC., Fl. Franç., 3: 55 (1805).

   F. danthonii Aschers. et Graebn. Syn. Mitteleur. Fl., 2: 550 (1901).

   Vulpia danthonii (Aschers. et Graebn.) Volk in Schinz et Keller, Fl. Schweiz, ed. 2, 57 (1905).

   [*V. sciuroides* (non (Roth) Gmel.) Sintenis in Oesterr. Bot. Zeitschr., 32: 194 (1882) pro parte]

   TYPE: "in Europa australe," *Link* (B).

An annual grass; culms rarely solitary, more often densely fasciculate with the outermost stems geniculately ascending, up to 30 cm. tall, smooth and glabrous, often leafy right up to the panicle; leaf-blades up to 15 cm. long, 1·5 mm. wide, linear-acuminate, involute, setaceous, curved, scabrid on both surfaces and on the margins, or smooth on the lower surface, glabrous; sheaths firmly clasping or the uppermost spathe-like, surrounding the base of the panicle, striate, smooth and glabrous; ligule very short; panicle very narrow, up to 15 cm. long, one-sided, branched towards the base; spikelets cuneate, gaping, 3–7-flowered, of which only the two lowest are fertile, very shortly pedicellate; lower glume minute, triangular or ovate, about 0·5 mm. long, the upper 3–4 mm. long, lanceolate-acuminate, both hyaline; lowest lemma linear-lanceolate, acute, awned, covered all over the dorsal surface with long white hairs; palea as long as the lemma, scabrid on the two keels; awn 10–12 mm. long; stamens 1; anthers 0·4–0·5 mm. long.

HAB.: On waste ground and in garigue, sometimes on sandy ground near the sea; sea-level to 680 ft. alt.; fl. April–May.

DISTR.: Divisions 2, 4, 5, 7. Mediterranean region and eastwards to Crimea, Transcaucasia, N. W. Pakistan and Central Asia.

2. Yialia, 1955, *Merton* 2157 !
4. Mile 17, Nicosia–Larnaca road, 1950, *Chapman* 404 ! near Famagusta, 1951, *Merton* 353 ! Larnaca Salt Lake, 1967, *Merton* in ARI 407 !
5. Mile 3, Nicosia–Famagusta road, 1950, *Chapman* 394 !
7. Near Kantara, 1880, *Sintenis & Rigo* 375 ! 400 ! Lakovounara Forest, 1950, *Chapman* 146 ! Lapithos, 1955, *Merton* 2222 !

12. LOLIUM *L.*

Sp. Plant., ed. 1, 83 (1753).
Gen. Plant., ed. 5, 36 (1754).
E. E. Terrell in U.S. Dept. Agric. Techn. Bull., 1392: 1–65 (1968).

Annual or perennial grasses with flat leaf-blades; inflorescence a terminal spike of spikelets, distichously and alternately arranged in the same plane and at the nodes in the cavities of a more or less tough axis; rhachilla disarticulating above the glumes and between the florets; spikelets sessile, more or less compressed, 3–11- or even more-flowered, hermaphrodite or the upper florets more or less reduced; glumes of the terminal spikelet equal and similar, the lower suppressed in the lateral spikelets; lemmas 5–7-nerved, overlapping, awned or awnless; palea 2-keeled, scabrid on the keels; stamens 3; lodicules 2, ovate-lanceolate, fleshy below, glabrous. Grain tightly enclosed by the hardened lemma and palea; embryo small, elliptic; hilum in a furrow and as long as the grain.

Six species widely distributed in the Old World and introduced elsewhere.

Florets turgid at maturity -   -   -   -   -   -   -   -   **1. L. temulentum**
Florets not turgid at maturity:
   Upper glume of the lateral spikelets much shorter than (about half as long as) the spikelet; florets oblong or oblong-lanceolate:
      Perennial; leaf-blades folded about the midrib when young; lemmas awnless, muticous or blunt -   -   -   -   -   -   -   -   -   -   -   **2. L. perenne**

Annual; leaf-blades rolled when young; lemmas awned from near the tip

3. L. multiflorum

Upper glume reaching to or exceeding the tip of the spikelet or at least more than half as long; florets elliptic to ovate, awned or awnless:

Axis of the spikelets slender; spike straight, rigid - - - - - - **4. L. rigidum**

Axis of the spike stout, almost terete - - - - - - - **5. L. loliaceum**

1. **L. temulentum** *L.*, Sp. Plant., ed. 1, 83 (1753); Sibth. et Sm., Fl. Graec. Prodr., 1: 70 (1806); Kotschy in Unger et Kotschy, Die Insel Cypern, 184 (1865); Sintenis in Oesterr. Bot. Zeitschr., 31: 193 (1881); Boiss., Fl. Orient., 5: 681 (1884); Holmboe, Veg. Cypr., 39 (1914); Post, Fl. Pal., ed. 2, 2: 790 (1933); Nattrass, First List Cypr. Fungi, 16, 45 (1934); Rechinger f. in Arkiv för Bot., ser. 2, 1: 418 (1950); N. L. Bor in C. C. Townsend et al., Fl. Iraq, 9: 96 (1968); Osorio-Tafall et Seraphim, List Vasc. Plants Cyprus, 8 (1973).

Lolium temulentum var. *arvense* (With.) Lilj., Utkast Sv. Fl., 80 (1816); Holmboe, Veg. Cypr., 39 (1914); Post, Fl. Pal., ed. 2, 2: 79 (1933); Rechinger f. in Arkiv för Bot., ser. 2, 1: 418 (1950).

L. arvense With., Arr. Brit. Pl., ed. 3, 2: 168 (1796); Sibth. et Sm., Fl. Graec. Prodr., 70 (1806).

L. temulentum var. *muticum* Boiss., Fl. Orient., 5: 681 (1884); Rechinger f. in Arkiv för Bot., ser. 2: 1, 418 (1950).

Lolium temulentum var. *leptochaeton* A. Braun in Flora, 17: 252 (1834).

TYPE: "*in* Europae *agris inter Hordeum, Linum.*" (LINN !).

An annual grass; culms up to 60 cm. tall, rarely solitary and erect, usually loosely to densely tufted and then the lateral stems geniculate at the base and spreading, slender or stout, simple, smooth and glabrous or scaberulous below the spike; leaf-blades linear, tapering to a very acute, fine tip, up to 30 mm. long, 3–10 mm. wide, flat, glabrous, smooth or scabrid on the surfaces, scabrid on the margins; sheaths tightly clasping, loose below, glabrous, smooth or scaberulous; ligule membranous, 1·5–2 mm. long, obtuse or rounded; spike erect, up to 30 cm. long, rigid; spikelets oblong, 12–20 mm. long, 4–6 mm. wide, 4–10-flowered; upper glume exceeding the length of the spikelet, linear-obtuse, rigid, flat on the dorsal surface, 7–9-nerved; lemmas elliptic to ovate when flattened, obtuse-emarginate at the tip, 6–8 mm. long, swollen and hard when mature, 5–9-nerved; long- or short-awned just below the tip or awnless; palea 2-keeled, finely scabrid on the keels; anthers 2·5 mm. long; awn straight, 0·5–2 cm. long, scabrid or almost absent.

HAB.: Cultivated fields and fallows, sometimes in Pine forest at high altitudes; sea-level to 5,400 ft. alt.; fl. April–May.

DISTR.: Divisions 1–8. Europe and Mediterranean region eastwards to Siberia and India.

1. Lyso ["Lisso"], 1913, *Haradjian* 846; Polis, 1962, *Meikle* 2311 !
2. Pano Platres, 1930, *E. Wall* s.n.; Platres, 1938, *Kennedy* 1125 ! Kryos Potamos, 4,100 ft. alt., 1938, *Kennedy* 1129 ! Khionistra, 1938, *Kennedy* 1130 !
3. Moni, 1862, *Kotschy* 588; Limassol, 1953, *Merton* 958 ! Asomatos, 1956, *Merton* 2740 !
4. Larnaca, 1862, *Kotschy* 28; Old Larnaca, near the aqueduct, 1880, *Sintenis*; The Phaneromene near Larnaca, 1905, *Holmboe* 153.
5. Syngrasis, 1935, *Nattrass* 509.
6. Morphou, 1931, *Nattrass* 22; Nicosia, 1950, *Chapman* 627 ! 704 !
7. Above Kythrea, 1880, *Sintenis & Rigo* 654 ! Kyrenia, 1949, *Casey* 540 !, also, 1956, *G. E. Atherton* 1085 ! 1161 ! Halevga, 1950, *Chapman* 644 ! Karavas, 1955, *Merton* 2185 !
8. Tavros, 1935, *Nattrass* 507; Boghaz, 1936, *Nattrass* 769.

NOTES: The variability in the length of the awns of this species has led botanists into describing a very large number of named varieties, subvarieties and forms, several of which are listed above.

If the awnless form (*L. arvense* With.) is compared with the typical long-awned form, one gets the impression of two distinct species, but, in fact, there is a complete range from awnless to long-awned spikelets. In these circumstances, no useful purpose would be served by maintaining a large number of varietal names.

2. **L. perenne** *L.*, Sp. Plant., ed. 1, 83 (1753); Sibth. et Sm., Fl. Graec. Prodr., 1: 70 (1806); Kotschy in Unger et Kotschy, Die Insel Cypern, 184 (1865); Sintenis in Oesterr. Bot. Zeitschr., 31: 193 (1881); Boiss., Fl. Orient., 5: 679 (1884); Holmboe, Veg. Cypr., 39 (1914); Lindberg f., Iter Cypr., 8 (1946); Rechinger f. in Arkiv för Bot., ser. 2, 1: 418 (1950); N. L.

Bor in C. C. Townsend et al., Fl. Iraq, 9: 93 (1968); Osorio-Tafall et Seraphim, List Vasc. Plants Cyprus, 8 (1973).

TYPE: "*in* Europae *ad agrorum versuras solo fertili.*" (LINN!).

A perennial grass; culms tufted or solitary, erect or, when decumbent at the base, the outermost spreading, stout or slender, usually unbranched, smooth and glabrous; leaf-blades linear-acute, glabrous, folded in the young shoots, up to 20 cm. long, 4–10 mm. wide, scaberulous on the upper surface and on the margins, or smooth on both surfaces, glossy below; sheaths rounded on the back, clasping the culms, scabrid or smooth; ligule 1–2 mm. long, membranous; spikes stout or slender, up to 30 cm. long, erect or curved; spikelets oblong to elliptic in outline, 5–14-flowered, 8–20 mm. long; upper glume much shorter than the spikelet, oblong-lanceolate when flattened, obtuse, 5–7-nerved, smooth; lemma oblong-acute or -obtuse when flattened, 5–7 mm. long, unawned, rounded on the back, smooth and glabrous, 5-nerved.

HAB.: Cultivated fields, fallows, roadsides and streamsides; sea-level to 5,600 ft. alt.; fl. March–June.

DISTR.: Divisions 1, 2, 4, 6, 7. Europe and Mediterranean region eastwards to Central Asia; introduced elsewhere.

1. Dhrousha, 1941, *Davis* 3269!
2. Troödos, 1912, *Haradjian* 463; by the Loumadho, E. of Khionistra, 1937, *Kennedy* 56! Troödos, "Olympus Camp Hotel," 1939, *Lindberg f.* s.n.! Prodhromos, 1939, *Lindberg f.* s.n.; above Trikoukkia near Prodhromos, 1961, *D. P. Young* 7376!
4. Old Larnaca, near the aqueduct, 1880, *Sintenis & Rigo.*
6. Syrianokhori, 1935, *Syngrassides* 807!
7. Around Ayios Khrysostomos Mon., 1862, *Kotschy* 451; St. Hilarion, 1934, *Syngrassides* 646! Kyrenia, 1952, *Casey* 1254! Kambyli, 1956, *Merton* 2730!

NOTES: *Lolium perenne* L. var. *tenue* (L.) Schrad., Fl. Germ., 397 (1806), based on *L. tenue* L., Sp. Plant., ed. 2, 1: 122 (1762) and cited by Lindberg f. from Troödos, is only a depauperate plant of *L. perenne* L.

3. L. multiflorum *Lam.*, Fl. Franç., 3: 621 (1778); Boiss., Fl. Orient., 5: 679 (1884); N. L. Bor in C. C. Townsend et al., Fl. Iraq, 9: 93 (1968); Osorio-Tafall et Seraphim, List Vasc. Plants Cyprus, 8 (1973).

TYPE: France; "sur le bord des prés & des champs, dans les environs de Péronne." (P).

An annual or biennial grass; culms up to 60 cm. tall, erect or geniculately ascending, more usually loosely or densely tufted, rarely solitary, smooth and glabrous or somewhat scaberulous below the spike; leaf-blades linear-acute or acuminate, glabrous, rolled when young, up to 30 cm. long, 10 mm. wide, scabrid or smooth above, scabrid on the margins, smooth and glossy beneath, furnished at the base with small falcate auricles; sheaths tightly clasping the culms or somewhat loose, smooth, glabrous or scaberulous; ligule 1–2 mm. long, rounded, membranous; inflorescence a true spike, erect or more usually nodding, up to 30 cm. long, compressed, green or suffused with purple; spikelets 5–15-flowered, up to 25 mm. long, oblong; both glumes present in the terminal spikelet only; the lower suppressed in the lateral spikelets, the upper varying in length but shorter than the spikelet, oblong or lanceolate-oblong, obtuse or acute, 4–7-nerved, smooth; lemmas oblong or lanceolate-oblong, obtuse, rarely minutely 2-toothed, 5–8 mm. long, scarious on the margins, 5-nerved, awned from the tip; palea as long as the lemma, 2-keeled, scabrid on the keels; awn straight, scabrid, up to 10 mm. long; anthers 3–4 mm. long.

HAB.: Cultivated fields and waste ground; introduced.

DISTR.: Division? Widespread in Europe and the Mediterranean region eastwards to Iran; introduced into most temperate countries.

?Div.: "in citrus groves in Anatoliko" (ARI Techn. Bull., 9, app. 2; 1972).

L. multiflorum *Lam.* x **L. rigidum** *Gaud.*

A robust plant with stiff vertical spikes (sometimes feebly branched); sheaths and culms very slightly rough; spikelets about 10 mm. long; lateral glume about half as long as spikelet; lemmas shortly awned.

HAB.: Weed of cultivation; sea-level to c. 500 ft. alt.; fl. April–May.

DISTR.: Division 5. Found wherever the parents grow together, or sometimes elsewhere as a weed or casual.

5. Athalassa, Agricultural Research Institute, 1970, *A. Genneou* in ARI 1421!

4. L. rigidum *Gaud.*, Agrost. Helvet., 1: 334 (1811); Boiss., Fl. Orient., 5: 680 (1884); H. Stuart Thompson in Journ. Bot., 44: 341 (1906); Druce in Rep. B.E.C., 9: 471 (1931); Post, Fl. Pal., ed. 2, 2: 790 (1933); Rechinger f. in Arkiv för Bot., ser. 2, 1: 418 (1950); Osorio-Tafall et Seraphim, List Vasc. Plants Cyprus, 8 (1973).

L. *strictum* Presl, Cypr. Gram. Sicul., 49 (1820); Lindberg f., Iter Cypr., 8 (1946).

L. *strictum* Presl var. *compressum* Boiss., Diagn., 2, 3: 144 (1859); Kotschy in Unger et Kotschy, Die Insel Cypern, 184 (1865); Holmboe, Veg. Cypr., 39 (1914).

TYPE: Italy; "Augustae Praetoriae [Aosta] ad vias apricas anno 1809 inveni", *Gaudin* (G).

An annual grass; culms usually tufted with the outer stems very shortly decumbent, eventually erect, rarely solitary, up to 45 cm. long, usually less, smooth and glabrous, simple; leaf-blades linear-acuminate, flat, glabrous, very scabrid on the nerves on the upper surface, scaberulous or almost smooth on the margins, up to 15 cm. long, 5 mm. broad; sheath rounded on the dorsal surface, markedly striate, smooth and glabrous, rarely slightly scaberulous; ligule membranous, 1–2 mm. long, truncate or rounded; spike 4–15 cm. long, rigid, straight, rarely curved; spikelets 10–13 mm. long, 6–8-flowered, oblong, gaping, finally cuneate; glume lanceolate-acute or obtuse, firm, 5–7-nerved, smooth and glabrous, 8–13 mm. long, scarious on the margins; lemma 5–9 mm. long, oblong, obtuse, 5-nerved, smooth and glabrous, slightly rough, scarious on the margins; palea as long as the lemma, 2-keeled, finely scabrid on the keels; anthers 3 mm. long.

HAB.: Cultivated fields and fallows, roadsides, waste ground; dry stream beds; occasionally on sandy or salty ground near the sea; sea-level to 4,900 ft. alt.; fl. March–June.

DISTR.: Divisions 2–7, common in divisions 2 and 7. S. Europe and Mediterranean region eastwards to N.W. India and Central Asia; introduced elsewhere.

2. Platres, Khionistra, Kakopetria, etc. *Syngrassides* 1264! *Lindberg f.* s.n.! *Kennedy* 1126!–1128! *Chapman* 293!
3. Limassol, 1928 or 1930, *Druce* s.n.; Limassol, 1930, *Wall* s.n.; 2 miles S. of Pyrgos, 1963, *Suart* 84!
4. Larnaca, 1862, *Kotschy* 262! Pergamos, 1935, *Nattrass* 510; Larnaca aerodrome, 1950, *Chapman* 451!
5. Near Athalassa, 1933, *Syngrassides* 7! Mile 3, Nicosia–Myrtou road, 1950, *Chapman* 569! Mia Milea, 1967, *Merton* in ARI 192!
6. Syrianokhori, 1935, *Syngrassides* 643! Morphou, 1935, *Nattrass* 487; Nicosia, 1955, *Baker* 160!
7. Kyrenia, Bellapais, Kharcha, etc. *Chapman* 112! 264! 355! 665! *Casey* 1255! *Miss Mapple* 76!

L. rigidum *Gaud.* x **L. temulentum** *L.*

A robust plant with very scabrid sheaths, and with the culms rough below the erect, stiff inflorescence; the lateral glume is more than three-quarters the length of the spikelet (which is 20–25 mm. long) and the lemmas are shortly awned.

HAB.: Growing with the parent species in cultivated and waste land; c. 600 ft. alt.; fl. April–May.

DISTR.: Divisions 5–7. Found wherever the parent species grow together.

5. Athalassa, Agricultural Research Institute, 1970, *A. Genneou* in ARI 1422!
6. Mile 3, Larnaca–Myrtou road, 1950, *Chapman* 569a!
7. Lakovounara Forest, 1950, *Chapman* 353! 666!

5. L. loliaceum (*Bory et Chaub.*) *Hand.-Mazz.* in Ann. Naturh. Mus. Wien, 28: 32 (1941).

Rottboellia loliacea Bory et Chaub., Exped. Sci. Mor., 3 (2): 46, t. 3, fig. 2 (1832).
   L. rigidum Gaud. var. *rottboellioides* Heldr. ex Boiss., Fl. Orient., 5: 680 (1884).
   L. rigidum Gaud. var. *loliaceum* (Bory et Chaub.) Halacsy, Consp. Fl. Graec., 446 (1904);
Post, Fl. Pal., ed. 2, 2: 790 (1933).
   L. subulatum Vis., Fl. Dal., 1: 90, t. 3 (1842); T. Photiades in The Countryman 8 (10): 12
(1945).
   TYPE: Greece; "Les environs de Modon [Methone], où elle est assez commune," (P).

A stout annual; culms very rarely solitary and then erect, most often
fasciculate and then geniculate at the base, finally erect, up to 40 cm. tall,
2–4 mm. in diameter, smooth and glabrous; leaf-blades linear-acute or
acuminate, rounded at the base, up to 15 cm. long, 2–4 mm. wide, flat, green,
smooth and glabrous on both surfaces; inflorescence a terete, stiff, thick
spike up to 15 cm. long, usually straight but sometimes curved; axis very
stout, smooth and glabrous, tough, excavated to take the spikelets;
internodes 8–15 mm. long; spikelets at first flush with the internodes of the
spike, eventually gaping; upper glume oblong-acute or acuminate, 10–25
mm. long, flat or curved, strongly nerved, much longer than the florets,
smooth and glabrous; florets 4–6; lemmas oblong-acute, coriaceous, smooth
and glabrous, 3–8 mm. long, sometimes the tip produced into a short arista
or awn; palea as long as the lemma, 2-keeled.

   HAB.: Cultivated ground or borders of fields; sea-level to 4,500 ft. alt.; fl. March–June.

   DISTR.: Divisions 2, 5, 6. Mediterranean region E. of Sicily and eastwards to Iraq and
Caucasus. Introduced elsewhere.

2.   Prodhromos, 1939, *Lindberg f.* s.n. ! Near Klirou, 1967, *Merton* in ARI 281 !
5.   Malounda, 1950, *Chapman* 292 ! Kythrea, 1957, *Merton* 2964 !
6.   Syrianokhori, 1935, *Syngrassides* 804 ! 807a (mixed with *L. rigidum*) ! Dhiorios, 1962, *Meikle*
      2338 !

   NOTES: Terrell (1968) prefers to call this species *Lolium subulatum* Vis., reducing *L. loliaceum*
to synonymy under *L. rigidum* var. *rottboellioides*. This is a matter of opinion, and the writer
prefers the synonymy given above.

13. PSILURUS *Trin.*
Fund. Agrost., 93 (1820).

Annual, slender herbs with weak culms and filiform leaves; inflorescence a
terminal, very slender, curved or flexuous spike of spaced, distichous
spikelets seated in the shallow cavities of the joints of the tardily
disarticulating, angled rhachis; spikelets 1-flowered, with occasionally a
terminal, vestigial, sterile floret seated on a long, flexuous rhachilla; lower
glume of the terminal floret very small, those of the lower spikelets absent,
upper glume also small, somewhat laterally placed; lemma linear-lanceolate,
membranous, prolonged into an awn; palea equal in length, membranous, 2-
keeled; stamen one only; lodicules 2, oblong, 2-toothed, one tooth long,
slender, acuminate, the other acute, much shorter. Grain long, slender, the
length of the lemma; embryo very small; hilum linear.

A monotypic genus, widely distributed in the Mediterranean region and
eastwards to Central Asia; introduced elsewhere.

1.   **P. incurvus** *(Gouan) Schinz et Thell.* in Vierteljahrschr. Naturf. Ges. Zurich, 58: 40 (1913); N.
      L. Bor in C. C. Townsend et al., Fl. Iraq, 9: 102, t. 35 (1968); Osorio-Tafall et Seraphim, List
      Vasc. Plants Cyprus, 9 (1973).
         Nardus incurva Gouan, Hort. Monsp., 33 (April 1762).
         Nardus aristatus L., Sp. Plant., ed. 2, 1: 78 (Sept. 1762); Sibth. et Sm., Fl. Graec. Prodr.,
      1: 35 (1806).
         Psilurus nardoides Trin., Fund. Agrost., 73 (1820); Kotschy in Unger et Kotschy, Die
      Insel Cypern, 186 (1865); Sintenis in Oesterr. Bot. Zeitschr., 32: 194 (1882); Boiss., Fl.
      Orient., 5: 682 (1884); H. Stuart Thompson in Journ. Bot., 44: 341 (1906).
         P. aristatus (L.) Duval-Jouve in Bull. Soc. Bot. Fr., 13: 132 (1866); Holmboe, Veg. Cypr.,
      39 (1914).
      TYPE: France; "Luxuriabat in aridis circa Caunelles" (K !).

An annual grass; culms very slender, solitary or loosely fasciculate, erect or geniculately ascending, up to 30 cm. tall, smooth and glabrous; leaf-blades usually extremely narrow and twisted when old, very narrowly linear-acute, more or less flat or folded, 1–1·5 mm. wide, up to 5 cm. long, somewhat scaberulous at the tip; sheaths smooth and glabrous; ligules membranous, 1–2 mm. long, truncate or obtuse; spike up to 15 cm. long, curved, usually embraced at the base by the uppermost leaf-sheath, with spaced, distichous spikelets; axis of the spike scabrid on the angles and sometimes on the surface; spikelets sessile, linear-lanceolate, about 5 mm. long (awn excluded) 2-flowered, the second being sterile and reduced, seated on the produced rhachilla; lower glume absent or rudimentary in the lateral spikelets, developed but short in the terminal, upper glume about 1·5 mm. long, triangular in shape, 1-nerved, laterally orientated; lemma of the fertile floret linear-lanceolate, cylindrical, glabrous or somewhat scaberulous, 1-nerved, 3–6 mm. long, awned; palea with 2 ciliolate keels; anther 0·4–2·5 mm. long; awn 2 mm. long.

HAB.: Cultivated and fallow fields, roadsides; hillsides in garigue or Pine forest on igneous or non-igneous formations; 150–4,600 ft. alt.; fl. March–June.

DISTR.: Divisions 2, 5–8, common in Division 2. Distribution that of the genus.

2. Common; Prodhromos, Kakopetria, Spilia, Platres, etc. *Kotschy* 796 ! *Kennedy* 1132 !–1134 ! *Davis* 3191 ! *Meikle* 2809 ! etc.
5. Vatili, 1905, *Holmboe* 340; Nicosia–Limassol road, Mile 10, 1967, *Merton* in ARI 596 !
6. Nicosia, 1905, *Holmboe* 310; Skylloura, 1952, *Merton* 814 !
7. N.W. of Bellapais, 1952, *Casey* 1263 !
8. Around Kantara Monastery, 1880, *Sintenis & Rigo*; Cape Andreas, 1938, *Syngrassides* 1778 ! Koma tou Yialou, 1950, *Chapman* 225 !

NOTES: Kennedy 1133 consists of 10 plants all gathered in the same area. Several of these have inflorescences which are covered with short, stiff, white bristles and conform to the description of *Psilurus hirtellus* Simonkai in Oesterr. Bot. Zeitschr., 38: 344 (1888) (*P. aristatus* var. *hirtellus* (Simonkai) Aschers. et Graebn., Syn. Mitteleur. Fl., 2: 767; 1902). In view of the fact that this minor variant is found with the glabrous form there does not seem to be any good reason for making a variety, much less a species, of it.

14. **SCLEROCHLOA** *P. Beauv.*

Ess. Agrost., 97 (1812).

Annual grasses with flat or folded leaf-blades; inflorescence a one-sided compact panicle in which the main axis is somewhat flattened, with the spikelets subsessile or shortly pedicellate on the main axis in the upper part and often on short branches below; spikelets 3–5-flowered, the uppermost sterile, the lower hermaphrodite; rhachilla not fragile, thick; glumes very unequal, membranous with broad scarious margins; lemmas membranous to coriaceous with broad scarious margins, prominently 5–7-nerved, smooth and glabrous; palea 2-keeled, chartaceous; stamens 3; lodicules 2, oblong-cuneate from a broad base, irregularly undulate on the upper margin. Grain lanceolate-oblong in outline, flattened adaxially, angled abaxially, with a prominent beak at the apex (stylopodium), loosely enclosed by the lemma and palea; embryo basal, one-sixth the length of the grain; hilum short, oblong, suprabasal.

Two (or three) species in Europe and western Asia.

1. **S. dura** (*L.*) *P. Beauv.*, Ess. Agrost., 97, 177, t. 19, fig. 4 (1812); Boiss., Fl. Orient., 5: 635 (1884); Holmboe, Veg. Cypr., 36 (1914); Post, Fl. Pal., ed. 2, 2: 768 (1933); N. L. Bor in C. C. Townsend et al., Fl. Iraq, 9: 194, t. 36 (1968); Osorio-Tafall et Seraphim, List Vasc. Plants Cyprus, 9 (1973).
   Cynosurus durus L., Sp. Plant., ed. 1, 72 (1753); Walpole, Travels, 23 (1820).
   Sesleria dura (L.) Schrank, Baier. Fl., 1: 351 (1789); Poech, Enum. Plant. Ins. Cypr., 7 (1842).

Festuca dura (L.) Vill., Hist. Plant. Dauph., 2: 94 (1787); Kotschy in Unger et Kotschy, Die Insel Cypern, 183 (1865); H. Stuart Thompson in Journ. Bot., 44: 340 (1906).
Poa dura (L.) Scop., Fl. Carn., 1: 70 (1772); Sibth. et Sm., Fl. Graec. Prodr., 1: 53 (1806).
TYPE: *"in* Europa *australi"* (LINN !).

An annual, glaucous herb; culms fasciculate, often prostrate and spreading radially, erect at the tips, smooth and glabrous, 5–15 cm. long; leaf-blades up to 7 cm. long, 1–4 mm. wide, flat or folded, the lower often rolled or twisted, smooth and glabrous on both surfaces, smooth or very slightly scaberulous on the margins, glaucous, rarely green; sheaths rather loose, keeled, striate, smooth and glabrous; ligule membranous, obtuse or truncate, 2 mm. long; inflorescence a very dense, one-sided panicle, very rarely rather loose, ovoid, ovoid-oblong or oblong in outline; branches very short, carrying one or two spikelets; spikelets 7–10 mm. long, 2–4-flowered, oblong in shape; glumes unequal, the lower about 2 mm. long, broader than long, ovate-obtuse, somewhat firmer about the 3 median nerves, glabrous, the upper 4–5 mm. long, oblong, broad, 5–7-nerved, hyaline on the margins, almost coriaceous around the nerves, glabrous; lemmas 5–6 mm. long, 5-nerved, oblong, broadly hyaline on the margins; palea equal in length, with two ciliate-scabrid keels; anthers about 1·5 mm. long.

HAB.: Roadsides, waste ground; 400–4,500 ft. alt.; fl. April–May.

DISTR.: Divisions 2, 4, 6, rare. S. Europe and Mediterranean region eastwards to N.W. India and Central Asia; introduced elsewhere.

2. Prodhromos, 1859, *Kotschy.*
4. Sotira, 1950, *Chapman* 583 !
6. Near Peristerona, 1787, *Sibthorp* s.n.

15. SPHENOPUS *Trin.*

Fund. Agrost., 135 (1820).

Weak, slender, annual herbs; inflorescence an ample panicle of small spikelets; spikelets compressed laterally, 2–7-flowered; rachilla disarticulating above the glumes and below the florets; seated on rather stout grooved pedicels; glumes very unequal, the lower minute, the upper somewhat longer, nerveless, thin, keeled; lemmas membranous, 3-nerved, rounded on the back below, keeled above; paleas 2-keeled, 2-lobed; stamens 3, very small; lodicules 2, lanceolate-acute or cuneate, 2-toothed, much smaller than the ovary. Grain oblong-oblanceolate, depressed on the adaxial surface; hilum punctiform; embryo one-quarter to one-fifth the length of the grain.
Two (or three) species, chiefly in the Mediterranean region.

1. **S. divaricatus** *(Gouan) Reichb.*, Fl. Germ. Excurs., 45 (1830); Sintenis in Oesterr. Bot. Zeitschr., 32: 122 (1882); Boiss., Fl. Orient., 5: 575 (1884); Holmboe, Veg. Cypr., 35 (1914); Post, Fl. Pal., ed. 2, 2: 750 (1933); N. L. Bor in C. C. Townsend et al., Fl. Iraq, 9: 108 (1968); Osorio-Tafall et Seraphim, List Vasc. Plants Cyprus, 9 (1973).
Poa divaricata Gouan, Ill. Obs. Bot., 4, t. 2, fig. 1 (1773); Sibth. et Sm., Fl. Graec. Prodr., 1: 54 (1806).
P. expansa J. F. Gmel., Syst. Nat., 2: 181 (1791).
Festuca expansa (J. F. Gmel.) Kunth, Rev. Gram., 1: 129 (1829).
Glyceria sphenopus Steud., Syn. Plant. Glum., 1: 287 (1854); Kotschy in Unger et Kotschy, Die Insel Cypern, 181 (1865).
TYPE: France; Gramond (Aveyron), "Cum Scirpo Micheliano primo Vere & in pratulis gramuntionis oritur." (K !).

An annual grass; culms slender, up to 30 cm. tall, often densely fasciculate, with the outer stems geniculate at the base and spreading, smooth and glabrous, shining, often purplish, or pale but with dark purple nodes, simple; leaf-blades setaceous, rolled, up to 7 cm. long, less than 1 mm. wide when flattened, linear, tapering to a very firm point; sheaths clasping, striate, smooth and glabrous; ligule up to 4 mm. long, membranous; inflorescence a panicle, contracted at first, finally very effuse; branches usually geminate at the nodes of the axis, di- and trichotomously branching and rebranching, the branchlets divergent, the latter carrying the pedicellate spikelets; pedicels 3-angled, wider below the spikelet; spikelets 2–3 mm. long, 2–5-flowered, oblong; lower glume a minute scale, the upper about 1 mm. long, ovate-acute or obtuse, both nerveless; lemmas ovate- or elliptic-oblong, obtuse or rounded at the apex, rarely mucronate, 1·25–1·5 mm. long, 3-nerved; lateral nerves near the margin, shortly ciliate; palea hyaline, 2-nerved; anthers 0·3 mm. long.

HAB.: By roadsides; in grass steppe; in salt marshes or by streams, in dry or moist (often brackish) situations; sea level to c. 4,000 ft. alt.; fl. March–June.

DISTR.: Divisions 2, 4, 5, 7. S.W. Europe, Mediterranean region, eastwards to Central Asia; Atlantic Islands. Introduced into S. Africa, Australia.

2. Around Prodhromos, 1862, *Kotschy* 291 !
4. Larnaca, 1939, *Lindberg f.* s.n. ! Larnaca aerodrome, 1950, *Chapman* 452 ! Larnaca Salt Lake, 1967, *Merton* in ARI 408 !
5. Lefkoniko, 1880, *Sintenis & Rigo* 377 ! Vatili, 1905, *Holmboe* 356; Near Pyroi, 1934, *Syngrassides* 460 ! Pyroi Bridge, 1938, *Syngrassides* 1835 ! Salamis, 1962, *Meikle* 2585 !
7. Lakovounara Forest, 1950, *Chapman* 70 ! 222 ! Mia Milea–Lakovounara road, 1950, *Chapman* 475 !

16. POA *L.*

Sp. Plant., ed. 1, 67 (1753).
Gen. Plant., ed. 5, 31 (1754).

Annual or perennial grasses with flat leaf-blades hooded at the apex; inflorescence an effuse or contracted panicle; spikelets compressed, with few to many florets seated distichously on the joints of a fragile rhachilla which disarticulates above the glumes and at the base of the lemmas; florets hermaphrodite with the upper rudimentary or occasionally female; glumes somewhat unequal, the lower 1-, the upper 3-nerved; lemmas keeled, 5-nerved, hyaline at the tip and on the margins, sometimes pubescent on the nerves, and sometimes woolly on the callus; palea hyaline, 2-keeled, keels scabrid or pilose for all or part of their length; stamens 3; lodicules 2, lanceolate-acuminate, sometimes with an additional tooth. Grain oblong or elliptic-oblong, triangular in section, slightly depressed on the adaxial surface; embryo about one-eighth the length of the grain; hilum basal, punctiform.

About 300 species, found throughout the world, but chiefly in temperate regions, or on mountains in the tropics.

Plants bulbous at the base - - - - - - - - - - - **1. P. bulbosa**
Plants not bulbous:
  Plants with moniliform rhizomes; spikelets usually 2-flowered - - - **2. P. silvicola**
  Plants without moniliform roots or rhizomes; spikelets always 3- or more-flowered:
    Annual or biennial grasses:
      Anthers minute, not more than 0·5 mm. long; lemmas 2–2·5 mm. long - **8. P. infirma**
      Anthers 0·8–1·3 mm. long; lemmas 3–4 mm. long - - - - **7. P. annua**
    Perennial grasses:
      Ligule pointed; lower sheaths scabrid; stolons present but no rhizomes **3. P. trivialis**
      Ligule truncate or rounded; lower sheaths smooth; plants rhizomatous:
        Culms terete; lower branches of the panicle 4–5-nate; lateral nerves of the lemmas prominent:
          Basal leaves very narrow, about 1 mm. wide; lemmas 2–3 mm. long
              5. P. angustifolia
          Basal leaves broad, 2–4 mm. wide; lemmas 3–4 mm. long - - **4. P. pratensis**
        Culms strongly compressed; lateral nerves of the lemmas obscure - **6. P. compressa**

1. P. bulbosa *L.*, Sp. Plant., ed. 1, 70 (1753); Kotschy in Unger et Kotschy, Die Insel Cypern, 180 (1865); Boiss., Fl. Orient., 5: 605 (1884); H. Stuart Thompson in Journ. Bot., 44: 340

(1906); Holmboe, Veg. Cypr., 36 (1914); Post, Fl. Pal., ed. 2, 2: 759 (1933); Lindberg f., Iter Cypr., 8 (1946); N. L. Bor in C. C. Townsend et al., Fl. Iraq, 9: 114 (1968); Osorio-Tafall et Seraphim, List Vasc. Plants Cyprus, 9 (1973).

A perennial grass with fibrous roots; culms tufted, bulbous at the base, up to 50 cm. tall, clothed at the base with the scarious remains of old sheaths, slender, simple, smooth and glabrous; leaf-blades usually flat, flaccid, green or greyish green, sometimes folded, filiform, linear, abruptly contracted at the tip to a hooded point, minutely scabrid on the upper surface and on the margins, otherwise smooth and glabrous; sheaths closely clasping, with scarious margins, looser below, the lower swollen; ligules membranous, pale, 4–6 mm. long; inflorescence an oblong-ovoid panicle, 2–9 cm. long, often one-sided, rather densely spiculate; branches 2- to 3-nate; spikelets ovate-oblong, 3–6-flowered, 4–6 mm. long; nearly always viviparous in which state the parts are much distorted and often enlarged, shedding all or part of their indumentum; glumes, in normal spikelets, more or less equal, 2·5–3 mm. long, the lower oblong-lanceolate, 1-nerved, the upper broader, 3-nerved; lemma oblong-lanceolate, 3 mm. long, hyaline at the tip and along the margins, sometimes suffused with purple, 5-nerved, ciliate on keel and nerves; palea 2-nerved, 2-keeled, scabrid on the keels; anthers 1–1·5 mm. long; wool present on the callus in non-viviparous spikelets.

var. **bulbosa**
TYPE: France; "*in* Gallia" (LINN!).

Spikelets not viviparous.

HAB.: On dry rocky or gravelly ground; by roadsides or borders of fields; sea-level to 6,000 ft. alt.; fl. March–June.

DISTR.: Divisions 2, 4–7; common in Divisions 2 and 7. Southern and Central Europe, Mediterranean region eastwards to N.W. India and Central Asia; Atlantic Islands; introduced elsewhere.
2. Common about Troödos, Platres, Pedhoulas, Chakistra, Makheras Mon., etc. *Kennedy* 65–69! 86! *Chapman* 290! *Merton* 2308! 2334! *D. P. Young* 7334! *A. Hansen* 800! etc.
4. Near Larnaca, 1862, *Kotschy* 71; Larnaca aerodrome, 1950, *Chapman* 26!
5. Mile 3, Nicosia–Famagusta road, 1950, *Chapman* 30! Athalassa, 1967, *Merton* in ARI 102!
6. Nicosia, 1905, *Holmboe* 298; Kokkini Trimithia, 1935, *Syngrassides* 771! Dhiorios, 1962, *Meikle* 2330! 2335!
7. Common; *Sintenis & Rigo* 392b! *Casey* 448! *Chapman* 113! 148! 149! 182! 188! 193! *G. E. Atherton* 1162!

var. **vivipara** *Koeler*, Descr. Gram., 189 (1802); Kotschy in Unger et Kotschy, Die Insel Cypern, 181 (1865); Holmboe, Veg. Cypr., 36 (1914); Lindberg f., Iter Cypr., 8 (1946).
    Poa bulbosa L. f. *vivipara* (Koeler) Dinsmore in Post, Fl. Pal., ed. 2, 2: 759 (1933).
TYPE: Germany; "prope Moguntiam [Mainz] in arenosis" (G).

Spikelets viviparous.

HAB.: In the same habitats as var. *bulbosa*, and at similar altitudes; fl. March–June.

DISTR.: Divisions 2–4, 6, 7; common in Divisions 2 and 7. General distribution similar to that of var. *bulbosa*.
2. Common on the Troödos Range; *Kennedy* 70–72! 75! 85! *Lindberg f.* s.n.! *Chapman* 287! *Merton* 2200! *P. Laukkonen* 913! etc.
3. Kophinou, 1905, *Holmboe* 591; Kalavasos, 1905, *Holmboe* 638; Above Kellaki near Panayia Glossa, 1956, *Merton* 2549!
4. Mile 17, Nicosia–Larnaca road, 1950, *Chapman* 407!
6. Nicosia, 1955, *Baker* 162! Dhiorios, 1962, *Meikle* 2319!
7. Common; *Sintenis & Rigo* 392! *Syngrassides* 56! *Davis* 2878! *Casey* 354!

2. **P. sylvicola** *Guss.*, Enum. Plant. Vasc. Inarime, 371, t. 18 (1854).
    P. trivialis L. var. *sylvicola* (Guss.) Hackel, Verh. Zool.-Bot. Ges. Wien, 40: 127 (1890).
    P. trivialis L. ssp. *sylvicola* (Guss.) Lindberg f. in Finska Vet.-Soc. Forhandl., 38 (13): 9 (1906), Iter Cypr., 8 (1946).
    [*P. trivialis* L. ssp. *attica* (non *P. attica* Boiss. et Heldr.) Holmboe, Veg. Cypr., 36 (1914).]

TYPE: Italy; "in sylvaticis apricis ubique vulgatissima Inarime [Ischia]; nec non prope Neapolim et Stabias," *Gussone* (NAP).

A perennial grass with conspicuously moniliform, spreading rhizomes; culms solitary or loosely tufted, erect or briefly decumbent at the base, slender, simple, smooth and glabrous, up to 60 cm. tall; leaf-blades linear-acuminate, green or greyish-green, flat, flaccid, up to 20 cm. long, 2–6 mm. wide, scaberulous on the upper surface and on the margins, smooth and glabrous or scabrid on the mid-nerve beneath; sheaths rather loose, keeled, striate, the uppermost smooth and glabrous, the lower scabrid with retrorse, minute scabridities; ligule membranous, up to 10 mm. long on uppermost leaves, acute; inflorescence a dense narrow panicle when young, rather loose when flowering, up to 20 cm. long, 3–6 cm. wide, green or greyish-green; branches whorled at the somewhat distant nodes (up to 5 cm. apart), 5-nate, filiform, flexuous; spikelets ovate-acute, compressed, usually 2-flowered, 2·5–3 mm. long, gaping at maturity; lower glume narrow, 1-nerved, curved, 2·5 mm. long, the upper much broader 3 mm. long, 3-nerved, both scarious on the margins; lemma 2·5 mm. long, rather firm, lateral nerves distinct, ciliate on the keel in the lower half and also on the lateral nerves; palea 2-keeled, scabrid on the keels; wool copious; anthers 1·5–2 mm. long.

HAB.: Usually in damp situations by streams or in moist meadows; 1,800–5,500 ft. alt.; fl. April–July.

DISTR.: Division 2, locally common. Also S. Europe and western Asia.

2. Locally common about Kyperounda, Platania, Prodhromos, Kannoures Springs, Polystipos, etc. *Holmboe* 904! *Syngrassides* 490! 1099! *Lindberg f.* s.n.! *Merton* 904! 2355! 2424! *Meikle* 2738!

NOTES: The inflorescence of this species is remarkably similar to that of *P. trivialis* L., with which it may have been confused. In *P. sylvicola* the spikelets are usually 2-flowered, rarely 3-flowered, and the plant has distinctive, moniliform rhizomes, while *P. trivialis* is stoloniferous, but not rhizomatous.

3. P. trivialis *L.*, Sp. Plant., ed. 1, 67 (1753); Sibth. et Sm., Fl. Graec. Prodr., 1: 55 (1806); Boiss., Fl. Orient., 5: 602 (1884); Post, Fl. Pal., ed. 2, 2: 759 (1933); N. L. Bor in C. C. Townsend et al., Fl. Iraq, 9: 113, t. 39 (1968); Osorio-Tafall et Seraphim, List Vasc. Plants Cyprus, 9 (1973).

TYPE: "*in* Europae *pascuis*". (LINN!).

A perennial grass with creeping leafy stolons; culms up to 90 cm. tall, loosely tufted, erect or more often spreading from a decumbent base, finally erect, smooth and glabrous; leaf-blades up to 15 cm. long, not more than 5 mm. wide, linear, tapering to an acuminate tip, scabrid on both surfaces and on the margins, glabrous; sheaths somewhat keeled, usually harshly rough with retrorse scabridities, occasionally smooth (var. *glabra* Doell, Rhein Fl., 92; 1943); ligule ovate- or oblong-acute, 4–6 mm. long, membranous; inflorescence a variable erect or nodding, ovate or pyramidal panicle; 7·5–20 cm. long, compact or loose; branches whorled, 4–5-nate, scaberulous, filiform; spikelets oblong- to ovate-acute, 4–5 mm. long, usually crowded, 3–4-flowered, green, sometimes tinged with purple; lower glume 2–2·5 mm. long, very narrowly lanceolate-acuminate, 1-nerved, the upper 3–3·5 mm. long, broadly lanceolate- or ovate-acute when flattened, 3-nerved; lemma 2·5–3 mm. long, oblong-obtuse or -acute when flattened, conspicuously 5-nerved, ciliate on the keel below, ciliate or glabrous on the side nerves; palea 2-keeled, scabrid on the keels; wool copious; anthers 1·5 mm. long.

HAB.: In moist meadows; c. 800 ft. alt.; fl. April.

DISTR.: Division 7 only, very rare. Also Europe, Mediterranean region, Atlantic Islands, temperate Asia. Introduced elsewhere.

7. Kambyli, 1956, *Merton* 2733!

NOTES: Apart from an unlocalized Sintenis specimen, cited by Boissier (Fl. Orient., 5: 602), this is the only authentic record of *P. trivialis* from Cyprus.

4. **P. pratensis** *L.*, Sp. Plant., ed. 1, 67 (1753); Sibth. et Sm., Fl. Graec. Prodr., 1: 55 (1806); Boiss., Fl. Orient., 5: 601 (1884); Post, Fl. Pal., ed. 2, 2: 758 (1933); N. L. Bor in C. C. Townsend et al., Fl. Iraq, 9: 112 (1968); Osorio-Tafall et Seraphim, List Vasc. Plants Cyprus, 9 (1973).
TYPE: "*in* Europae *pratis fertilissimis*". (LINN!).

A closely or loosely tufted, perennial grass with both intravaginal shoots and long wiry rhizomes; culms erect, up to 80 cm. tall, stout, smooth and glabrous; leaf-blades linear, always flat, shortly tapering to a stout point, firm, or in shade flaccid, the basal up to 25 cm. long, 5 mm. wide, those of the culms shorter and narrower, green or greyish-green, scabrid on the margins, glabrous or rarely hairy; sheaths smooth and glabrous; ligules about 2 mm. long, membranous, truncate; inflorescence an ovate or oblong panicle, very compact and dense at first, widely spreading at flowering time, up to 20 cm. long, 12 cm. wide; lowest branches 5-nate; spikelets oblong, elliptic or ovate in outline, up to 5·5 mm. long, 2–5-flowered; lower glume 2–3 mm. long, 1-nerved, lanceolate-acute, the upper 2·5–3·5 mm. long, 3-nerved, elliptic-lanceolate or even oblong, obtuse, both hyaline on the margins; lemmas 3–3·5 (–4) mm. long, oblong, subobtuse when flattened, 5-nerved, hyaline at the tip and on the margins, ciliate on the keel and on the lower portion of the lateral nerves; palea with scabrid keels; wool copious; anthers 1·5–2 mm. long.

HAB.: Moist ground by streams and springs at high altitudes on igneous formations; c. 5,000 ft. alt.; fl. April–July.

DISTR.: Division 2, rare. Widespread in Europe, Mediterranean region, Atlantic Islands, and eastwards to Central Asia. Introduced elsewhere.

2. Amiandos, 1951, *Merton* 374! Livadhi tou Pasha, near Amiandos, 1952, *Merton* 893!

5. **P. angustifolia** *L.*, Sp. Plant., ed. 1, 67 (1753); Sibth. et Sm., Fl. Graec. Prodr., 1: 55 (1806); N. L. Bor in C. C. Townsend et al., Fl. Iraq, 9: 112 (1968); Osorio-Tafall et Seraphim, List Vasc. Plants Cyprus, 9 (1973).
P. pratensis L. var. *angustifolia* (L.) Sm., Fl. Brit., 1: 105 (1800); Holmboe, Veg. Cypr., 37 (1914).
P. pratensis L. ssp. *angustifolia* (L.) Hayek, Prodr. Fl. Pen. Balc., 3: 269 (1932); Lindberg f., Iter Cypr., 8 (1946).
TYPE: "*in* Europa *ad agrorum versuros*" (LINN!).

A perennial tufted grass with hard, widely spreading rhizomes; culms usually erect, less often geniculately ascending, up to 90 cm. tall, smooth and glabrous; leaf-blades of two kinds, the basal 10–30 cm. long, conduplicate and filiform, flexuous or straight, not more than 1 mm. broad when flattened, linear, smooth and glabrous, smooth or scaberulous on the margins; those of the culms shorter but broader than the basal, slightly rough on the upper surface and scabrid on the margins, smooth below; sheaths tight, smooth and glabrous, striate; ligule membranous, truncate, up to 2 mm. long, usually much shorter; inflorescence an oblong-pyramidal panicle, sometimes nodding, 5–10 cm. long, up to 5 cm. broad; branches 5-nate at the lowest node; spikelets ovate to oblong, 2·5–5 mm. long, 2–5-flowered, compressed; lower glume 2 mm. long, lanceolate-acuminate when flattened, 1-nerved, the upper 2·5 mm. long, 3-nerved, oblong-acute when flattened; lemmas 2·5–3 mm. long, keeled, 5-nerved, oblong-obtuse when flattened, hyaline at the tip and on the margins, ciliate on the keel and on the lower part of the lateral nerves; palea 2-keeled, scabrid on the keels; wool copious; anthers 1·5 mm. long.

HAB.: In moist situations and by streams, sometimes in *Pinus* woodland or in vineyards, often at high altitudes; 800–6,400 ft. alt.; fl. April–July.

DISTR.: Divisions 2, 7, common locally in central Troödos. Widespread in Europe and eastwards to Central Asia.

2. Locally common; Troödos, Khionistra, Platres, Prodhromos, etc.; *Sintenis & Rigo* 1006! *Hartmann* 276a! *Kennedy* 73! 74! 1092! 1093! 1094! *Lindberg f.* s.n., etc.
7. Kambyli, 1957, *Merton* 2991!

6. P. compressa *L.*, Sp. Plant., ed. 1, 69 (1753); Sibth. et Sm., Fl. Graec. Prodr., 1: 55 (1806); Poech, Enum. Plant. Ins. Cypr., 7 (1842); Kotschy in Unger et Kotschy, Die Insel Cypern, 181 (1865); Boiss., Fl. Orient., 5: 602 (1884); H. Stuart Thompson in Journ. Bot., 44: 340 (1906); Holmboe, Veg. Cypr., 36 (1914); Post, Fl. Pal., ed. 2, 2: 759 (1933); Osorio-Tafall et Seraphim, List Vasc. Plants Cyprus, 9 (1973).
TYPE: "*in* Europae et Americae *septentrionalis siccis, muris, tectis*" (LINN!).

A perennial grass with extensively creeping rhizomes; culms 20–50 cm. tall, strongly compressed, rather stiff, decumbent and rooting at the basal nodes, smooth and glabrous; leaf-blades up to 10 cm. long, 1–4 mm. wide, linear, abruptly tapering to a rather stout tip, somewhat stiff, scabrid on the upper surface and on the margins, particularly at the tip, otherwise glabrous; sheaths strongly compressed with a sharp keel, smooth and glabrous, striate, rather loose; ligules membranous, rounded at the top, 1·5 mm. long; inflorescence a panicle, dense and compact before flowering, then somewhat loose but never widely spreading; lower branches 2- to 5-nate; spikelets about 4·5 mm. long, 3–7-flowered; glumes more or less equal in length, 2–2·5 mm. long, 3-nerved, hyaline on the margins, the lower oblong, acute, the upper oblong-elliptic, acute; lemmas 2·25–2·75 (–3) mm. long, oblong-obtuse when flattened, obscurely 5-nerved, with the lateral nerves very close to the margin, ciliate on the keel and on the lateral nerves; palea 2-keeled, scabrid on the keels; wool absent or scanty; anthers 1·5 mm. long.

HAB.: Moist ground by streams at high altitudes on igneous formations, or as a weed in gardens; 5,000–5,600 ft. alt.; fl. July.

DISTR.: Division 2, rare. Europe and S.W. Asia; introduced elsewhere.

2. Kryos Potamos, 5,150 ft. alt., 1938, *Kennedy* 1095! Weed at Troödos Hotel, Troödos, 1963, *D. P. Young* 7856! Weed in garden of President's House, Troödos, 1963, *D. P. Young* 7857!

P. NEMORALIS *L.* (Sp. Plant., ed. 1, 69; 1753) a slender, laxly caespitose perennial, up to 80 cm. tall, with narrow, flaccid leaves (held at a right angle to the culm) and lax, often slightly nodding, panicles of small, elliptic-acute or lanceolate-acute spikelets, may yet be found in Cyprus, possibly in moist, shaded situations at high altitudes. It is not uncommon in adjacent countries, and in the Aegean islands. In *P. nemoralis* the ligule is often almost absent, or reduced to a very narrow, membranous ring.

7. P. annua *L.*, Sp. Plant., ed. 1, 68 (1753); Sibth. et Sm., Fl. Graec. Prodr., 1: 56 (1806); Kotschy in Unger et Kotschy, Die Insel Cypern, 180 (1865); Boiss., Fl. Orient., 5: 601 (1884); Holmboe, Veg. Cypr., 36 (1914); Post, Fl. Pal., ed. 2, 2: 758 (1933); Lindberg f., Iter Cypr., 8 (1946); N. L. Bor in C. C. Townsend et al., Fl. Iraq, 9: 122 (1968); Osorio-Tafall et Seraphim, List Vasc. Plants Cyprus, 9 (1973).
TYPE: "*in* Europa *ad vias*" (LINN!).

An annual, sometimes biennial, very rarely a short-lived perennial; culms usually geniculate at the base, finally erect, up to 30 cm. tall but usually shorter, smooth and glabrous; leaf-blades 2–3·5 cm. long, up to 5 cm. in favourable habitats, 2–5 mm. wide, linear, suddenly contracted to an acute, hooded tip, usually flat, dark green, glabrous, scaberulous on the margins; sheaths smooth and glabrous, somewhat compressed, striate; ligules membranous, 1·5–3 mm. long; inflorescence a loose, roughly pyramidal panicle, very variable, up to 12 cm. long and often as wide; branches 2-nate or solitary; spikelets more or less crowded at the tips of the branches, green or sometimes suffused with violet, 3–5-flowered, ovate or elliptic-oblong in shape, 4–6 mm. long; glumes unequal, the lower lanceolate, acute or acuminate, 1·5–2 mm. long, 1-nerved, hyaline on the margins, the upper 2–2·5 mm. long, broader, elliptic-acute when flattened, 3-nerved, hyaline on the margins; lemmas 3 mm. long, oblong-obtuse when flattened, 5-nerved, silky-ciliate on the keel and on the lateral nerves in the lower half; palea elliptic-truncate, long ciliate on the keels, 2-nerved; anthers 0·6–0·8 mm. long; wool absent.

HAB.: Roadsides, gardens, waste ground, sometimes by springs or streams, or in moist hollows in Pine forest at high altitudes; sea-level to 5,200 ft. alt.; fl. Jan.–June.

DISTR.: Divisions 2–4, 7. Europe, Mediterranean region, Atlantic Islands eastwards to India and Central Asia; introduced elsewhere.

2. Common; Khionistra, Prodhromos, Troödhitissa, Platres, etc. *Kennedy* 60! 1121! *Merton* 2366! 2442! 3198! *D. P. Young* 7454!
3. Skarinou, 1941, *Davis* 2784!
4. Near Larnaca, 1862, *Kotschy*; Ayia Napa, 1862, *Kotschy* 106.
7. Kyrenia, 1949, *Casey* 421! Bellapais, 1956, *G. E. Atherton* 857!

8. **P. infirma** *Kunth* in H.B.K., Nov. Gen. et Sp., 1: 158 (1816); N. L. Bor in C. C. Townsend et al., Fl. Iraq, 9: 124, t. 42 (1968); Osorio-Tafall et Seraphim, List Vasc. Plants Cyprus, 9 (1973).

    P. exilis (Tomm.) Murb. in Aschers. et Graebn., Syn. Mitteleur, Fl., 2, 1, abt. 1: 389 (1900).
    P. remotiflora Murb., Contrib. N.-Ouest Afr., 3: 22 (1900).
    TYPE: Colombia; "Crescit in frigidis . . . Bogota" (P).

An annual grass; culms rather slender, loosely tufted, erect or geniculately ascending, up to 20 cm. tall, usually very much less, smooth and glabrous; leaf-blades soft, thin, flaccid, yellowish green, up to 6 cm. long, 5 mm. wide, linear, tapering abruptly to a stout tip, scaberulous on the upper surface, on the margins and on the midrib below, otherwise smooth and glabrous; sheaths rather loose, smooth and glabrous; ligules membranous, 1–2 mm. long, rounded; inflorescence an open panicle with ascending branches; branches in pairs, often a longer accompanied by a shorter; spikelets oblong-obtuse in shape, 3–5-flowered, 4–4·5 mm. long; rhachilla-joints often visible; glumes somewhat unequal, the lower ovate-acute, 1–1·25 mm. long, 1-nerved, the upper elliptic-acute, 1·5 mm. long, 3-nerved, both glabrous, scabrid on the keel; lemmas 2·5 mm. long, oblong-obovate, 5-nerved, hyaline at the tip and along the margins, densely ciliate on the keel and on the nerves; palea 2-keeled, long ciliate on the keels; no wool; anthers minute 0·2–0·3 mm. long.

HAB.: Gardens, roadsides, waste ground, often in damp situations; 250–2,500 ft. alt.; fl. Febr.–March.

DISTR.: Divisions 1, 2, 5, 6. S.W. England, Mediterranean region eastwards to Tibet and N.W. India; Atlantic Islands; introduced elsewhere.

1. Gorge between Androlikou and Prodhromi, 1957, *Merton* 3036! Neokhorio, Phrangos (Akamas), 1962, *Meikle* 2211!
2. Evrykhou, 1935, *Syngrassides* 817! Kalokhorio (Makheras), 1950, *Chapman* 286! Gourri, 1950, *Chapman* 302! Above Palekhori, 1957, *Merton* 3132!
5. Kythrea, 1936, *Syngrassides* 942!
6. Nicosia moat, nr. Famagusta Gate, 1952, *Merton* 715!

17. LINDBERGELLA *Bor*

Svensk. Bot. Tidskr., 63: 368 (1969).
Lindbergia Bor in Svensk. Bot. Tidskr., 62: 467 (1968) non Kindb. (1897).

Annual grasses; culms usually fascicled, erect or the lateral somewhat geniculate at the base, slender, smooth and glabrous, usually suffused with purple; leaf-blades flat at first, eventually complicate or involute, linear-acute; sheaths of the culm clasping, the basal shorter and scarious; ligules membranous; inflorescence a panicle, narrow when young, very lax at maturity with an erect axis at the nodes of which arise binate or whorls of branches spreading widely, or even slightly declinate; branches and branchlets few-spiculate; pedicels longer than the spikelets; spikelets 2–5-flowered, laterally compressed, at first elliptic in outline, eventually gaping; glumes unequal, 3-nerved, elliptic-acute; lemmas elliptic-oblong, 3-nerved, apiculate, covered with silky appressed hairs in the lower half, scaberulous on nerves and keel; palea equal in length to the lemma, elliptic-acute, 2-keeled, densely ciliate in the lower $\frac{7}{8}$ of the keel, scabrid above; stamens 3; stigmas plumose; lodicules broadly ovate-acute, toothed on one margin. Caryopsis grooved adaxially; hilum punctiform, at the base of the groove; embryo small, $\frac{1}{8}$ to $\frac{1}{10}$ as long as the grain.

One species, endemic in Cyprus.

1. **L. sintenisii** (*Lindberg f.*) *Bor* [in Svensk. Bot. Tidskr., 63: 369 (1969) non rite publ.] **comb. nov.**

  Poa sintenisii Lindberg f. in Årsbok Soc. Sci. Fenn., 20B (7): 5 (1942), Iter Cypr., 8 (1946).

  P. persica Trin. var. *alpina* Boiss., Fl. Orient., 5: 610 (1884) pro parte; Holmboe, Veg. Cypr., 37 (1914).

  P. persica Trin. ssp. *cypria* Samuells. ap. Rechinger f. in Arkiv för Bot., ser. 2, 1: 417 (1950).

  [*P. persica* (non Trin.) Kotschy in Unger et Kotschy, Die Insel Cypern, 181 (1865); Holmboe, Veg. Cypr., 37 (1914).]

  TYPE: Cyprus; Troödos, June 20, 1880, *Sintenis* 881 (LD, K !).

An annual grass; culms 8–21 cm. tall, simple, more or less fascicled, the central erect, the lateral shortly decumbent at the base; leaf-blades up to 6 cm. long, 1–2 mm. wide, linear-acuminate when flattened, scabrid on both surfaces and on the margins; sheaths smooth and glabrous; ligules lacerate at the top, membranous, 2–3 mm. long; panicle at first nodding, narrow with ascending branches finally erect, 2–11 cm. long, 1–9 cm. wide, with horizontal or slightly declinate, smooth and glabrous branches; branches 2–5 at each node, unequal in length; pedicels longer than the spikelets; spikelets 2–5-flowered, 3–4·5 mm. long, elliptic-acute in outline, finally gaping; rhachilla disarticulating above the glumes and between the florets, crowned with a vestigial floret; lower glume 3 mm. long, elliptic-acute, the upper 3·5 mm. long, rather broader than the lower, both 3-nerved, scabrid on the nerves and sometimes between the nerves, keeled, hyaline on the margins, the central part suffused with purple; lowest lemma 4 mm. long, 3-nerved, keeled, oblong-elliptic when flattened, acute, mucronate or apiculate, covered with appressed silky-white hairs in the lower half of the dorsal surface; palea as long as the lemma, 2-keeled, densely ciliate in the lower $\frac{7}{8}$, scabrid towards the tip; anthers 0·5 mm. long; grain 2·5–2·75 mm. long. *Plate 102, figs. 9–13.*

HAB.: Rocky mountainsides on igneous formations at high altitudes, often in moist places or by streams; 3,300–6,000 ft. alt.; fl. May–July.

DISTR.: Division 2, locally common. Endemic.

2. Locally common in central Troödos area; Prodhromos, Khionistra, Mesapotamos, Platres, Kakopetria, Pedhoulas, etc. *Sintenis* 881 ! *Kennedy* 93–96 ! 1100–1104 ! *Lindberg f.* s.n. ! *Davis* 3454A ! *Meikle* 2844 ! 2853 ! etc.

18. PUCCINELLIA *Parl.*

Fl. Ital., 1: 366 (1848) nom. cons.

E. Paunero in Anal. Inst. Bot. Cavanilles, 17 (2): 31–55 (1959).

Perennial, caespitose grasses or rarely annuals, with flat leaf-blades; spikelets panicled, several- to many-flowered, terete or slightly laterally compressed; rhachilla disarticulating above the glumes and below each

Plate 102. Figs. 1–8. BRACHYPODIUM FIRMIFOLIUM Lindberg f. **1**, habit, ×⅔; **2**, spikelet, ×2; **3**, floret, ×6; **4**, gynoecium, ×10; **5**, stamen, ×10; **6**, palea, ×10; **7**, lower lemma, ×10; **8**, upper lemma, ×10; figs. 9–13. LINDBERGELLA SINTENISII (Lindberg f.) Bor **9**, habit, ×⅔; **10**, glume, ×10; **11**, lemma, ×10; **12**, gynoecium, ×10; **13**, stamen, ×10. (**1–8** from *Lindberg f.* s.n. isotype; **9–13** from *Lindberg f.* s.n. isotype.)

lemma; glumes unequal, shorter than the lowest lemma, the lower 1–3-nerved, the upper 3–7-nerved, broadly obtuse or somewhat rounded at the apex, rather thin in texture; lemmas firmer, rounded on the back, pointed or rounded at the scarious apex, rarely acuminate, 5-nerved; palea as long as the lemma or slightly shorter; lodicules 2, cuneate, with 2 teeth, one of which is larger than the other; stamens 3; styles 2, short; stigma plumose. Grain oblong, more or less dorsally compressed; embryo one sixth to one fifth as long as the grain; hilum punctiform.

About 100 species widely distributed in North temperate regions; also in South Africa.

Panicle large, up to 22 cm. long, 12 cm. wide; keels of the palea scabrid - - **1. P. gigantea**
Panicle smaller, up to 9 cm. long, 7 cm. wide; keels of the palea long-ciliate below
2. P. ciliata

1. P. gigantea *(Grossh.) Grossh.*, Fl. Kavk., 1: 114 (1928).
  Atropis gigantea Grossh. in Monit. Jard. Bot. Tiflis, 46: 35, t. 2 (1919).
  TYPE: U.S.S.R.; "in Talysch, Kumbashi" (LE).

A non-rhizomatous (?) perennial grass; culms closely tufted, erect, up to 70 cm. tall (including the inflorescence), smooth and glabrous, simple; leaf-blades folded or involute, rather firm, erect, smooth and rounded on the lower surface, costate on the upper, minutely scabrid on the margins, up to 15 cm. long, 3 mm. broad; sheaths smooth and glabrous, short, broad and rather lax below, firmly clasping above; ligule acute, 3 mm. long, carried on the margins of the sheath; inflorescence a panicle, up to 22 cm. long, 12 cm. broad, of distantly spaced, fascicled, long and short branches, the lowest being up to six in number, branching and rebranching; branches minutely scabrid; spikelets very shortly pedicellate, 5–8 mm. long, linear, 5–8-flowered, deep purple in colour; lower glume elliptic-acute, 1·75 mm. long, 1-nerved, glabrous; upper glume 2·2–2·5 mm. long, narrowly elliptic-acute, 3-nerved, glabrous; lowest lemma 2·5 mm. long, oblong to slightly obovate, rounded at the tip, very shortly apiculate, thickly hairy at the very base, more or less hairy in the lower quarter; stamens 1·5 mm. long; keels of palea scabrid above, scaberulous below.

HAB.: Brackish marshes; c. 50 ft. alt.; fl. May.

DISTR.: Division 4, rare. Southern U.S.S.R., Turkey, Iran.

4. Akhyritou reservoir, 1953, *Merton* 900a! (1095).

2. P. ciliata *Bor* in Notes Roy. Bot. Gard. Edinb., 28: 299 (1968).
  TYPE: Turkey; "Ismir, Meremen", 1951, *Miles & Donald* s.n. (E!, K!).

A densely tufted non-rhizomatous grass; culms erect, smooth and glabrous, simple, up to 45 cm. tall; leaf-blades folded or involute, up to 15 cm. long, 2–3 mm. wide, smooth and glabrous on the lower surface, scabrid and costate on the upper surface, scaberulous on the margins; sheaths smooth and glabrous, the lower short, lax, the upper closely clasping; ligule 2·5–4 mm. long, acute; panicle 9 cm. long, 7 cm. wide; branches binate or solitary at the nodes, eventually wide-spreading; branches and branchlets scabrid; spikelets 5–7 mm. long, 5–6-flowered, linear in outline, seated on pedicels up to 5 mm. long; lower glume 2·5 mm. long, 1-nerved or with two very short lateral nerves, elliptic-acute when flattened; upper glume 3 mm. long, 3-nerved, narrowly elliptic-acute, both smooth and glabrous; lowest lemma oblong, 2–2·75 mm. long, rounded-acute at the tip, slightly wider upwards, 5-nerved, hairy in the lower part of the dorsal surface, rounded on the back; palea elliptic-obtuse, 2-keeled, ciliate on the keels in the lower half, scabrid above; anthers 1·5 mm. long.

HAB.: Cultivated ground, probably introduced; 450 ft. alt.; fl. June.

DISTR.: Division 6, rare, probably not indigenous.

6. Nicosia, nursery garden, 1955, *Merton* 2426!

NOTES: I know of only two species of *Puccinellia* with the palea long-ciliate in the lower half, namely *P. festuciformis* (Host) Parl. and *P. ciliata* Bor. The type of the former (Herb. Vindob. !) shows that the lemmas are longer (3·5–4 mm.) and wider than in *P. ciliata*. *P. festuciformis* is also a much more robust plant.

TRIBE 3. **Meliceae** *Reichb.* Perennial grasses with narrow, flat leaf-blades and tubular sheaths; anatomy festucoid; silica-bodies rounded; ligules membranous or rarely absent; inflorescence a panicle, effuse with filiform branches, or spike-like with subsessile or shortly pedicellate spikelets; spikelets with 2-several bisexual florets; rhachilla disarticulating above the glumes and between the florets, produced above the uppermost floret and crowned with several sterile lemmas wrapped together and forming a compact, obconical knob; glumes more or less unequal, papyraceous with scarious margins; lemmas rounded on the dorsal surface, prominently 5- to many-nerved, membranous, more or less pilose or glabrous; palea 2-keeled; stamens 3; lodicules 2, laterally connate into an oblong band. Grain ellipsoid, loosely or tightly enclosed in the lemma and palea, grooved on the adaxial surface; embryo small, basal, about one-fifth the length of the grain; hilum linear, short, suprabasal in the groove. Chromosomes large; basic number 9.

19. MELICA *L.*

Sp. Plant., ed. 1, 66 (1753).

Gen. Plant., ed. 5, 31 (1754).

C. Papp, Monographie der europäischen Arten der Gattung Melica L., in Engl., Bot. Jahrb., 65: 275–348 (1932).

W. Hempel, Taxonomische und chorologische Untersuchungen an Arten von Melica, in Feddes Repert., 81: 131–145 (1970).

Perennial grasses with erect culms, often shortly decumbent at the base, with flat leaf-blades and tubular sheaths; spikelets with 1–2 hermaphrodite florets; rhachilla disarticulating above the glumes and between the florets, produced into a clavate mass consisting of empty sterile lemmas; glumes membranous, the lower usually shorter than the upper; lemmas rounded on the dorsal surface, hairy all over or only on the marginal nerves or quite glabrous; stamens 3; styles 2; stigmas plumose; lodicules fused. Grain, embryo and hilum as above.

About 70 species widely distributed in temperate regions, excluding Australia.

1. **M. minuta** *L.*, Mantissa 1, 32 (1767); Sibth. et Sm., Fl. Graec. Prodr., 1: 50 (1806); Poech, Enum. Plant. Ins. Cypr., 7 (1842); Kotschy in Unger et Kotschy, Die Insel Cypern, 182 (1865); Sintenis in Oesterr. Bot. Zeitschr., 32: 192 (1882); Post, Fl. Pal., ed. 2, 2: 753 (1933); Lindberg f., Iter Cypr., 8 (1946); Osorio-Tafall et Seraphim, List Vasc. Plants Cyprus, 12 (1973).

   M. ramosa Vill., Hist. Pl. Dauph., 2: 91 (1787); Boiss., Fl. Orient., 5: 585 (1884).

   M. saxatilis Sm. in Sibth. et Sm., Fl. Graec. Prodr., 1: 51 (1806).

   M. minuta L. var. *vulgaris* Cosson, Not. Plant. Crit., 11 (1848); Holmboe, Veg. Cypr., 35 (1914).

   M. ramosa Vill. var. *vulgaris* (Coss.) Boiss., Fl. Orient., 5: 585 (1884).

   M. ramosa Vill. var. *parviflora* Boiss., Fl. Orient., 5: 585 (1884).

   M. ramosa Vill. var. *saxatilis* (Sm.) Boiss., Fl. Orient., 5: 585 (1884).

   M. minuta L. var. *parviflora* (Boiss.) Holmboe, Veg. Cypr., 35 (1914).

   TYPE: "in Italia" (LINN!).

A perennial grass; culms slender, more or less densely caespitose from a knotty underground root-stock, erect, or somewhat decumbent at the base, smooth, glabrous, shining, simple, up to 45 cm. tall; leaf-blades greyish-green, narrowly linear-acuminate, flat at first, rapidly becoming involute and almost setaceous, up to 15 cm. long, 1–1·5 mm. wide, smooth and glabrous on the lower surface, minutely hairy on the upper surface, more or less smooth on the margins, scaberulous near the tip; sheaths smooth and glabrous, sometimes scaberulous, particularly on the lower sheaths; ligule membranous, 1–3 mm. long, lacerate; inflorescence a rather narrow or open panicle, especially after flowering, up to 10 cm. long, 2–5 cm. wide; branches solitary, scabrid, bare in the lower half, often branched; spikelets erect at first then pendulous, 2-flowered, ellipsoid, 7–10 mm. long, pale, purple or variegated brown and purple; glumes unequal, broadly scarious on the margins, the lower 5 mm. long, 5-nerved, elliptic-acute when flattened, the upper 6·5 mm. long, 7-nerved, ovate-acuminate; lower lemma 8 mm. long, 7-nerved, rounded on the back, smooth or scaberulous and glabrous; upper lemma 5 mm. long, 7-nerved; paleas a little shorter than the lemmas, scabrid on the keels; anthers 2·5 mm. long; sterile lemmas an ellipsoid knob, rounded or truncate, 3·5 mm. long.

HAB.: Dry rocky ground on igneous or calcareous hills; on roadsides, walls, or sometimes in forests; sea-level to 4,600 ft. alt.; fl. March–June.

DISTR.: Divisions 2, 3, 6–8. Widely distributed in the Mediterranean region.

2. Common; Prodhromos, Platres, Khrysorroyiatissa Mon., Stavros tis Psokas, etc. *Kotschy* 682! 687! *Syngrassides* 1606! *Kennedy* 1115! *N. Macdonald* 174!
3. Alethriko, 1905, *Holmboe* 226; Kophinou, 1905, *Holmboe* 579; Skarinou, 1941, *Davis* 2759! 2783!
6. Forest of Kormakiti, 1936, *Syngrassides* 1198! West of Orga, 1956, *Merton* 2601! Near Panagra, 1962, *Meikle* 2400!
7. Common; *Kotschy* 520; *Syngrassides* 59! *Mavromoustakis* 25! *Casey* 565! *Chapman* 643! etc.
8. Common; *Sintenis & Rigo* 385! *Holmboe* 528! *Chapman* 237! *Meikle* 2437! 2518!

NOTES: All botanists who have dealt with this plant have described it as very variable, and with good reason. A considerable number of varieties has been recognized because of the luxuriance or otherwise of the panicle, the amount of indumentum, stature, colour and so forth. Although Papp in his monograph confines himself to six varieties, at least seventeen others have appeared in the literature in addition to seven other names under *M. ramosa*, itself a synonym of *M. minuta*. It is quite obvious that the establishment of varieties on slender deviations from what is considered to be typical *M. minuta* is quite irrational, and, until cytological evidence is forthcoming, the writer has decided not to recognize any varieties.

TRIBE 4. **Hainardieae** *Greuter* Annual grasses; leaf-blades linear to linear-lanceolate with festucoid anatomy; silica-bodies circular, oblong or elliptic; no 2-celled micro-hairs; ligules membranous; spikelets all alike, sessile, solitary, alternate, sunk in the cavities of an erect or curved, tough or fragile rhachis, forming a spicate inflorescence; glumes 2, and these placed side by side, or the lower suppressed and the upper covering the cavity completely; lemma hyaline, awnless, nearly as long as the glumes, 1–3-nerved; palea 2-nerved; stamens 3; ovary lobed and appendaged at the top; styles 2; stigmas plumose; lodicules 2, entire, glabrous. Grain free with a small, oblong hilum; starch grains compound. Chromosomes large; basic numbers 7, 13.

20. **PARAPHOLIS** *C. E. Hubb.*

Blumea, Suppl. 3: 14 (1946).
E. Paunero, Notas sobre gramineas III, Parapholis
in Anal. Inst. Bot. Cav., 22: 186–220 (1964).

Annual grasses with characteristically curved or rarely strictly erect culms; leaf-blades flat; inflorescence a terminal curved or straight spike with a tough or fragile rhachis at the nodes of which are seated alternately and distichously sessile spikelets; spikelets oblong-acute, with one hermaphrodite floret, sunk in the cavities of the rhachis; rhachilla disarticulating above the glumes, not produced; glumes equal, seated side by side at the node, 5-nerved, flat or keeled; lemma oblique to the rhachis, navicular or rounded on the back, membranous, 3-nerved with lateral nerves very short; palea as long as the lemma, 2-nerved; stamens 3; lodicules 2, ovate-acute, glabrous, fleshy below. Grain oblong, truncate, with a short apical appendage; embryo small, about one-sixth to one-eighth the length of the grain; hilum small, basal, punctiform.

Six species in Europe and the Mediterranean region eastwards to Central Asia and N.W.India.

Outer keel of the glumes distinctly winged - - - - - - -	**1. P. marginata**
Outer keel of the glumes not winged - - - - - - -	**2. P. incurva**

1. P. marginata *Runemark* in Bot. Notiser, 115: 8 (1962).
    TYPE: Greece; "Kiklades, Sirina, Dio Adelfi, the W-island 14.5.1960 Runemark and Nordenstam" (LD).

An annual with erect, prostrate, or curved culms, 5–15 cm. tall, smooth and glabrous, covered by the sheaths which become reddish with age; leaf-blades plicate, involute or rolled, 1–3 cm. long, 0·5–2·5 mm. broad, smooth and glabrous on the lower surface, markedly ridged on the upper surface and scabrid on the ridges, linear-acute, somewhat pungent, sometimes filiform and flexuous; sheaths smooth and glabrous, somewhat striate, with hyaline margins, more or less inflated, overlapping; ligule a very narrow membrane; spikes always supported at the base by the uppermost leaf-sheath, straight or more often curved, sometimes into a half circle, often a spike arising from each node, up to 5 cm. long, green or purplish; spikelets 4–6 mm. long, ovate or lanceolate, acute; glumes equal, with the outer keel conspicuously winged; anthers 0·5–0·8 mm. long. 2n = 14.

    HAB.: Sandy shores and salt marshes, occasionally on saline ground inland; sea-level to 650 ft. alt.; fl. April–May.

    DISTR.: Divisions 3–8. S. Europe and Mediterranean region eastwards to Central Asia and N.W. India; introduced elsewhere.
3. Akrotiri, 1905, *Holmboe* 688; Phasouri, 1962, *Meikle* 2925!
4. Kouklia, 1905, *Holmboe* 386; Larnaca Salt Lake, *Merton* in ARI 381! 403a!
5. Near Pyroi, 1934, *Syngrassides* 462! Vatili, 1862, *Kotschy* 536.
6. Asomatos, 1956, *Merton* 2746!
7. Kyrenia, 1880, *Sintenis & Rigo* 655! also *Syngrassides* 5! *Merton* 1128! Lakovounara Forest, 1950, *Chapman* 263! 354!
8. Cape Andreas, 1880, *Sintenis & Rigo* 389! Apostolos Andreas, 1962, *Meikle* 2468! Elisis, near Galinoporni, 1962, *Meikle* 2490!

2. P. incurva (*L.*) *C. E. Hubb.* in Blumea, Suppl. 3: 14 (1946); N. L. Bor in C. C. Townsend et al., Fl. Iraq, 9: 266 (1968); Osorio-Tafall et Seraphim, List Vasc. Plants Cyprus, 11 (1973).
    Aegilops incurva L., Sp. Plant., ed. 1, 1051 (1753).
    A. incurvata L., Sp. Plant., ed. 2, 1490 (1763).
    Rottboellia incurvata L. f., Suppl. Plant., 114 (1781); Sibth. et Sm., Fl. Graec. Prodr., 1: 71 (1806), Fl. Graec., 1: 72, t. 91 (1808).
    Lepturus incurvatus (L.) Trin., Fund. Agrost., 123 (1820); Poech, Enum. Plant. Ins. Cypr., 7 (1842); Kotschy in Unger et Kotschy, Die Insel Cypern, 186 (1865); Boiss., Fl. Orient., 5: 684 (1884); Holmboe, Veg. Cypr., 39 (1914); Post, Fl. Pal., ed. 2, 2: 792 (1933).
    [*L. filiformis* (non Roth) Trin., Fund. Agrost., 123 (1820); Kotschy in Unger et Kotschy, Die Insel Cypern, 187 (1865); Boiss., Fl. Orient., 5: 684 (1884).]
    TYPE: "*in* Oriente" (LINN!).

An annual grass; culms very rarely solitary, more usually densely or loosely tufted in such a fashion that they are prostrate at first and then curved into a half circle, rigid, smooth and glabrous, up to 15 cm. tall; leaf-blades linear-acute, up to 3 cm. long, 1–2 mm. wide, flat, folded or rolled, scabrid on the upper surface and on the margins, smooth on the lower surface, glabrous; sheaths rather loose, the lower scarious, the upper clasping, striate, smooth and glabrous; ligule less than 1 mm. long, membranous; inflorescence a rigid, curved, cylindrical spike up to 6 cm. long; axis fragile, jointed, hollowed out distichously to take the spikelets; spikelets seated in the concavities, longer than the joints, 4–6 mm. long, 1-flowered; glumes as long as the spikelet, narrowly oblong-acute, coriaceous, rigid, 3–4-nerved, glabrous, seated side by side covering the lemma; keels not winged; lemma elliptic-acute when flattened, hyaline, 3-nerved, lateral nerves short; palea of the same texture, 2-nerved; anthers 0·5–1 mm. long. 2n = 38.

HAB.: Salt marshes and moist ground by the sea; sometimes on saline ground inland; sea-level or a little above sea-level; fl. April–May.

DISTR.: Divisions 3, 4, rare. Mediterranean region eastwards to Central Asia; introduced elsewhere.

3. Mazotos, 1862, *Kotschy* 239!
4. Near Larnaca, 1862, *Kotschy* 248! Larnaca Salt Lake, 1967, *Merton* in ARI 403!

NOTES: *Kotschy* 248 from Larnaca may be a mixed collecting, consisting at least in part of *P. marginata* Runemark. The Kew sheet is wholly *P. incurva*, but Runemark (Bot. Not., 115: 14) cites what is probably the same number (Larnaca ["Lacarna"] 1862 Kotschy) from the Vienna herbarium, under his *P. marginata*.

21. HAINARDIA *W. Greuter*
in Boissiera, 13: 178 (1967).

Annual, erect grasses with flat, involute or folded leaf-blades; inflorescence a straight, vertical, terminal spike with a tough rhachis on which, alternately and distichously arranged, are seated the sessile spikelets, sunk in the cavities of the internodes; spikelets 1-flowered, hermaphrodite; glumes 2, equal in the terminal spikelets, only one, the upper, in the lateral spikelets, covering the cavity in the internode, leathery, strongly nerved; lemma with its back to the rhachis; membranous, hyaline, 3-nerved; palea membranous, 2-nerved; stamens 3; lodicules 2, broad, thick and fleshy at the base, tapering to an acute or acuminate tip, oblique, glabrous. Grain oblong, somewhat dorsally compressed with an appendage at the apex; embryo small, about one-fifth the length of the grain; hilum short, linear, suprabasal.

One species in western Europe, the Mediterranean region, Atlantic Islands and eastwards to Iran. Introduced in Australia, S. Africa, N. & S. America.

1. **H. cylindrica** (*Willd.*) *Greuter* in Boissiera, 13: 177 (1967).
    Rottboellia cylindrica Willd., Sp. Plant., 1: 464 (1797).
    Monerma cylindrica (Willd.) Coss. et Dur., Expl. Sci. Alg., 214 (1858); Boiss., Fl. Orient., 5: 683 (1884); Holmboe, Veg. Cypr., 39 (1914); Post, Fl. Pal., ed. 2, 2: [791](1933); N. L. Bor in C. C. Towsend et al., Fl. Iraq, 9: 268 (1968); Osorio-Tafall et Seraphim, List Vasc. Plants Cyprus, 11 (1973).
    Lepturus cylindricus (Willd.) Trin., Fund. Agrost., 123 (1820).
    TYPE: "*in* Europa *australi*" (B).

An annual grass; culms up to 12 cm., exceptionally to 20 cm. tall, straight or somewhat curved, erect or spreading, solitary or loosely fascicled, smooth and glabrous; leaf-blades up to 6 cm. long, 1–2·5 mm. wide, flat at first, folded or rolled in older plants, linear-acuminate, scabrid on the upper surface and on the margins, smooth below, glabrous; sheaths rather loose below, tighter above, markedly striate, smooth and glabrous; ligule membranous, 1–1·5 mm. long, lacerate; inflorescence a true spike, up to 10 cm. long, usually partly enveloped at the base by the uppermost leaf-sheath, consisting of a noded fragile axis at each node of which in alternate, distichous excavations, are seated the solitary spikelets; spikelets fitting the excavations in the axis but longer than the internodes, 6–7 mm. long; glumes two in the terminal spikelet, solitary in the lateral, as long as the spikelet, gaping at flowering time, coriaceous, 5–7-nerved, the central nerve particularly strong, linear-acute or acuminate, smooth and glabrous; lemma about 5 mm. long, hyaline, lanceolate-acute, with the dorsal surface appressed against the excavation wall; palea 2-nerved, hyaline; anthers about 4 mm. long.

HAB.: Moist brackish ground, and as a weed on a garden path; c. 600 ft. alt.; fl. May.

DISTR.: Division 7, rare. General distribution that of genus.

7. Lakovounara Forest, 1950, *Chapman* 550 ! Kyrenia, 1952, *Casey* 1269 !

TRIBE 5.   **Aveneae** *Dumort.* Spikelets all alike, 2–7-flowered, more or less laterally compressed, with all the florets hermaphrodite or the uppermost more or less reduced, or when 2 only, the lower sometimes male and the upper hermaphrodite; rhachilla not disarticulating in cultivated oats, disarticulating above the glumes and between the florets in wild species; spikelets pedicellate and arranged in open, ample or contracted panicles, rarely in racemes; glumes persistent, equal or unequal, usually as long as or longer than the spikelets, often with silvery membranous margins; lemmas membranous to crustaceous, 5–7-nerved, hyaline or scarious on the margins, entire or 2-lobed at the apex, awned; lobes often produced as bristles; awn usually consisting of a twisted column and a straight bristle, rarely in two parts (the lower twisted), articulated at the joint, sometimes a short straight bristle only; palea hyaline, 2-nerved; stamens 3, rarely 2; ovary hairy or glabrous; styles 2; stigmas plumose; lodicules 2, lanceolate or 2-toothed, fleshy at the base. Grain tightly or loosely enclosed by the lemma and palea; embryo one-sixth to one-third the length of the grain; hilum punctiform, basal or linear and as long as the grain; starch grains compound.

Annual or perennial grasses; leaf-blades narrow, linear, with festucoid anatomy; silica bodies elliptic-oblong; 2-celled hairs absent; ligules well developed, membranous. Chromosomes large; basic number 7.

E. Paunero, Las Aveneas españolas, 1. Anal. Inst. Bot. Cav., 13: 149–229 (1955); 14: 187–251 (1956); 15: 377–415 (1967); 17 (2): 257–376 (1959).

22. AIRA *L.*

.Sp. Plant., ed. 1, 63 (1753).
Gen. Plant., ed. 5, 31 (1754).

Annual, very slender grasses with flat leaf-blades; panicles very effuse or the spikelets somewhat crowded; spikelets small with 2 bisexual florets; rhachilla disarticulating above the glumes and between the florets, not produced beyond the second floret; glumes keeled, subequal, with hyaline margins, 1-nerved, longer than the lemmas; lemmas rounded on the back, glabrous, shining, brownish, muticous or 2-toothed at the tip, awned below the middle, or only the lower lemma awned, and the upper unawned, or both unawned; stamens 3; styles 2; stigmas plumose; lodicules 2, lanceolate. Grain oblong in outline, flattened or slightly concave on the inner surface, firmly adherent to the lemma and palea at maturity; embryo one-sixth to one-eighth the length of the grain; hilum basal, punctiform.

About 12 species widely distributed in temperate regions, and on mountains in the tropics.

Pedicels of the spikelets 4–8 times as long as the spikelets; panicle very lax; spikelets loose, not
    crowded   -   -   -   -   -   -   -   -   -   -   -   -   **1. A. elegans**

Pedicels 2–4 times as long as the spikelets; panicle not lax; spikelets often gathered in groups at the tips of the branches - - - - - - - - A. CARYOPHYLLEA (p. 1755)

1. **A. elegans** *Willd. ex Gaud.*, Agrost. Helvet., 1: 130 (1811); Osorio-Tafall et Seraphim, List Vasc. Plants Cyprus, 13 (1973).
  A. capillaris Host, Gram. Austr., 4: 20 (1809); Sintenis in Oesterr. Bot. Zeitschr., 32: 194 (1882); Boiss., Fl. Orient., 5: 529 (1884); Holmboe, Veg. Cypr., 34 (1914); Post, Fl. Pal., ed. 2, 2: 732 (1933) non Savi (1798).
  [*A. caryophyllea* (non L.) Kotschy in Unger et Kotschy, Die Insel Cypern, 179 (1865); Holmboe, Veg. Cypr., 34 (1914); H. Stuart Thompson in Journ. Bot., 44: 340 (1906).]
  TYPE: Italy; "Circa *Papiam* [Pavia] invenit L. THOMAS" (LAU).

An annual grass; culms slender, smooth and glabrous, solitary or fascicled, erect or somewhat decumbent at the base, up to 40 cm. tall; leaf-blades very narrow, linear, acute or obtuse, up to 6 cm. long, 0·5–0·8 mm. wide, flat, flaccid, rolled or twisted when dry, scabrid on both surfaces and on the margins; sheaths clasping, scaberulous, striate, glabrous; ligule lanceolate, pointed, 2–3 mm. long; inflorescence a very diffuse panicle, 4–15 cm. long, broadly elliptic, ovate or even cuneate-orbicular in outline with erect or widely spreading, capillary branches, bare in the lower half, branching towards the tips, carrying the pedicellate spikelets which are usually lax, sometimes crowded; spikelets 1·5–2 mm. long, 2-flowered, lanceolate in shape, soon gaping; glumes membranous, equal, ovate-acute when flattened, 1-nerved; lower lemma lanceolate-acute when flattened, obscurely 5-nerved, smooth on the dorsal surface below but scabrid towards the tip, 1–1·5 mm. long; paleas as long, oblong-obtuse, 2-nerved; upper lemma somewhat longer, split at the apex into 2 slender lobes, 5-nerved, the central nerve issuing from the dorsal surface as a perfect awn, 3 mm. long, twisted in the lower half; palea similar to the lower; anthers 0·3–0·5 mm. long.

HAB.: Dry ground amongst garigue or under Pines on rocky calcareous or igneous hillsides; 100–c. 4,000 ft. alt.; fl. March–June.

DISTR.: Divisions 2, 3, 5–8; common in Division 2. S. Europe, Mediterranean region.

2. Common; Mesopotamos, Platres, Perapedhi, Xeros, etc.; *Kotschy* 841! *Davis* 3190! 3455! *Kennedy* 92! 1105! *Merton* 3000! *D. R. Harris* 3104! etc.
3. Episkopi, 1954, *Merton* 1888!
5. Vatili, 1905, *Holmboe* 482.
6. Dhiorios, 1962, *Meikle* 2323!
7. Yaïla, 1941, *Davis* 2883! Halevga, 1970, *A. Hansen* 698!
8. Near Kantara, 1880, *Sintenis & Rigo* 386! Ayios Thyrsos near Rizokarpaso, 1938, *Syngrassides* 1769!

[A. CARYOPHYLLEA *L.*, Sp. Plant., ed. 1, 66 (1753) is similar to *A. elegans* Willd. ex Gaud. but has a more condensed inflorescence, with the spikelets often in groups at the tips of the inflorescence-branches, and with the pedicels conspicuously shorter than those of *A. elegans*. It is a common plant in many parts of Europe and the Mediterranean region, and is very likely to occur in Cyprus, though it has not yet been recorded.]

23. **CORYNEPHORUS** *P. Beauv.*

Essai Agrost., 90 (1812), nom. cons.
Weingartneria Bernh., Syst. Verz. Pfl. Erfurt, 23, 51 (1800).
J. Jirásek et J. Chrtek, Systematische Studie über die Arten der Gattung *Corynephorus* P. Beauv. in Preslia, 34: 374–386 (1962).
M. Breistroffer in Procés-verbaux de la Soc. dauphinoise, Ser. 3, no. 11: 3–4 (1950).

Annual or perennial herbs with conduplicate, setaceous leaf-blades; inflorescence a lax or contracted panicle; spikelets firmly laterally compressed, each with two hermaphrodite florets; rhachilla disarticulating above the glumes, prolonged beyond the upper floret; glumes slightly unequal, longer than the lemmas, the lower 1-, the upper 3-nerved; lemmas ovate-acute, 2-toothed at the apex, awned from the base; awn in two parts, the lower about the length of the lemma, twisted, brown, articulated with an upper somewhat shorter and more slender club-shaped part; point of articulation decorated with a crown of hairs; palea 2-keeled, notched or 4-lobed at the tip; stamens 3; lodicules 2, lanceolate, sometimes 2-toothed. Grain oblong, more or less dorsally compressed; embryo about one-sixth the length of the grain, at the base of a groove, as long as the grain; hilum punctiform, suprabasal.

About six species in Europe and the Mediterranean region.

Jirásek and Chrtek, *loc. cit.*, published an account of their exhaustive studies on the genus *Corynephorus*. The result of their investigation was that only one of the species described, *C. canescens* (L.) P. Beauv., a perennial, was retained in the genus *Corynephorus*, while the remainder, all annuals, were relegated to a new genus, *Anachortus* Jir. et Chr. The authors bring together an impressive list of characters, both morphological and anatomical, whereby *Anachortus* can be separated from *Corynephorus*. The differences, however, in the opinion of this writer, are not such as would justify the separation of *Anachortus* from *Corynephorus*. This opinion is based upon the illustrations in the paper under consideration, as well as examination of the specimens, in which the various differences in glume and lemma-shape are only minimal. The only real difference, which the writer considers of importance, between the two genera is that the species of *Corynephorus* is a perennial, while those relegated to *Anachortus* are annuals. The chromosome numbers are, in records available up to date, identical.

1. C. divaricatus (*Pourr.*) *Breistr.* in Procés-verb. Soc. dauph., 3 (11): 3 (1950).
  Aira caryophyllea L. var. β Lam., Encycl. Meth., 1: 600 (1785).
  A. divaricata Pourr. in Mem. Acad. Roy. Soc. Toulouse, 3: 307 (1788).
  Aira articulata Desf., Fl. Atlant., 1: 70, t. 13 (1798).
  Corynephorus articulatus (Desf.) P. Beauv., Essai Agrost., 90, 159 (1812); Boiss., Fl. Orient., 5: 530 (1884); Osorio-Tafall et Seraphim, List Vasc. Plants Cyprus, 13 (1973).
  Weingartneria articulata (Desf.) Aschers. et Graebn., Syn. Fl. Mitteleurop., 2: 301 (1899); Holmboe, Veg. Cypr., 34 (1914) (as ssp. *euarticulata*); Post, Fl. Pal., ed. 2, 2: 732 (1933).
  Anachortus articulatus (Desf.) Jirásek et Chrtek in Preslia, 34: 383 (1962).
  Anachortus divaricatus (Pourr.) Lainz in Inst. Forestal Invest. Exper., 35 (1971).
  TYPE: France, "Environs de Narbonne", *Pourret* (P!).

An annual, green or glaucescent grass; culms solitary or more often fascicled, sometimes branched from the base, up to 40 cm. tall, smooth and glabrous, slender; leaf-blades folded, up to 6 cm. long, 0·5–1·5 mm. wide when flattened, linear-acute, scabrid on both surfaces and on the margins; sheaths tightly clasping, striate, the uppermost slightly inflated, smooth and glabrous; ligule pointed, 4–6 mm. long, membranous. Inflorescence an effuse or contracted panicle 6–20 cm. long and as wide; branches capillary, in pairs, smooth and glabrous, bare in the lower half; spikelets 4 mm. long, shortly pedicellate at the tips of the branchlets, 2-flowered; glumes more or less equal, elliptic-acute when flattened, the lower 1-nerved, the upper 3-nerved, scabrid on the keels, otherwise smooth and glabrous, suffused with violet below, hyaline above; lower lemma elliptic or ovate, acuminate, 2·5 mm. long, hyaline, awned from the base; palea much smaller, hyaline; awn in two parts, the basal part lying in a groove in the lemma on the abaxial side, brown, rigid, slightly shorter than the lemma, articulated to the equally long, clubshaped, hyaline, distal part; point of articulation decorated with a crown of hairs; anthers minute.

HAB.: In alluvial fields and fallows; in open, rocky pastures; on sand-dunes; c. 50–1,000 ft. alt.; fl. May.

DISTR.: Divisions 3–5, 8. Mediterranean region, western Asia.

3. Near Michael Arkhangelos Monastery, W. of Stavrovouni Monastery, 1957, *Merton* 3153!
4. Phrenaros, 1955, *Merton* 2254!
5. Salamis, 1905, *Holmboe* 464.
8. Valia Minor State Forest, 1953, *Merton* 1096!

NOTES: Breistroffer (*loc. cit.*) made the new combination *C. divaricatus* (Pourr.) Breistr. purely on the statement of Timbal-Lagrave in Bull. Soc. Sci. Phys. Nat. Toulouse 2: 112 (1875) that Dr. Bubani (author of Flora Pyrenaica, 1897–1901) had informed him "in litt." that on a visit to the Madrid herbarium, he had seen the *Aira divaricata* of Pourr. and that this plant was none other than *Corynephorus articulatus* (Desf.) P. Beauv.

No authentic sheet could recently be found in Madrid, but Lainz discovered in the Paris herbarium what appears to be a plant collected in the locus classicus by Pourret and presented to the Muséum by Dr. Barbier, into whose hands Pourret's herbarium had fallen. The plant is named *Aira divaricata* Pourr. and is undoubtedly what has been known for so many years as *Corynephorus articulatus* (Desf.) P. Beauv. An acceptance of Breistroffer's new combination seems unavoidable.

24. TRISETARIA *Forssk.*

Fl. Aegypt.-Arab., 27 (1775).

E. Paunero, Las especies españolas del género *Trisetaria* Forssk. in Anal. Jard. Bot. Madrid, 9: 503–582 (1950).

Annual grasses with solitary or fasciculate culms, with flat or rolled, soft leaf-blades; inflorescence dense, cylindrical, spicate or somewhat loose, enclosed at the base in the topmost leaf-sheath, in which the laterally compressed spikelets are seated on short scabrid branches attached to the scabrid central axis; spikelets with 2 flowers, the upper seated on an elongate, hairy rhachilla-joint; glumes equal or unequal, the lower the smaller and narrower, 1-nerved, the upper 3-nerved; lower lemma membranous, 5-nerved, terete with 2 apical setae and armed with a perfect awn from the back; upper lemma similar but smaller; stamens 3; anthers minute; lodicules 2, cuneate-truncate, minute. Grain elongate, compressed, grooved on the adaxial surface; embryo about one-eighth the length of the grain; hilum punctiform.

Three species in the Mediterranean region and eastwards to Central Asia and Pakistan; one species in Cyprus.

1. **T. linearis** *Forssk.*, Fl. Aegypt.-Arab., LX & 27 (1775); Osorio-Tafall et Seraphim, List Vasc. Plants Cyprus, 13 (1973).
   Trisetum lineare (Forssk.) Boiss., Diagn., 1, 13: 49 (1853), Fl. Orient., 5: 533 (1884); Post, Fl. Pal., ed. 2, 2: 735 (1933).
   TYPE: Egypt; "*Alexandriae*, in peninsula *Râs-ettin.*" (C).

An annual grass; culm slender, fasciculate or more usually solitary, erect or somewhat geniculate at the base or rarely even with prostrate bases, then rooting at the nodes, glabrous or puberulous below the nodes, up to 40 cm. tall, usually much shorter; leaf-blades flat, soft, linear-acuminate, 2 mm. wide, up to 8 cm. long, eventually involute, the uppermost glabrescent, the lower sparsely villous on the upper surface, pubescent on both surfaces; sheaths rounded, the uppermost glabrous and embracing the base of the inflorescence, the lower villous and pubescent; hairs retrorse; ligule membranous, 1–1·5 mm. long; inflorescence a dense, cylindrical, spike-like panicle up to 12 cm. long, 1 cm. wide, with numerous erect spikelets shortly pedicellate on short branches; spikelets 2-flowered, cuneate; glumes more or less equal, narrowly lanceolate, silvery-hyaline on the margins, 5–6·5 mm. long, 3-nerved, scabrid on the keel-nerve, glabrous; lower lemma 3–4 mm. long, membranous, 5-nerved, scaberulous on the dorsal surface, 2-lobed, the lobes produced as aristulae 5 mm. long; awned in the sinus with a geniculate awn 9 mm. long, with a chestnut, twisted column; stamens 3; anthers 1–1·5 mm. long; second floret often rudimentary.

HAB.: Maritime sands; at or near sea-level; fl. March.

DISTR.: Division 5. Also N. Africa, Palestine, Egypt, Caucasus, Iraq, Iran.

5. East of Trikomo, 1970, *A. Hansen* 641!

NOTES: In addition to the above, what appears to be old and rather decayed material of the same grass was collected at (8) Valia, May, 1953, *Merton* s.n.

25. AVENA *L.*

Sp. Plant., ed. 1, 79 (1753).
Gen. Plant., ed. 5, 34 (1754).
A. I. Malzer in Bull. Appl. Bot. Gen. Plant-Breed., Suppl., 38 (1930).
G. Ladizinsky et D. Zohary in Canad. Journ. Gen. Cyt., 10 (1): 68–81 (1968).
G. Ladizinsky in Israel Journ. Bot., 20: 133–151 (1971).

Annual grasses with flat leaf-blades; inflorescence an open, effuse or contracted panicle or rarely a raceme of pedicellate spikelets; spikelets large, 1- to several-flowered, hermaphrodite; rhachilla fragile at least below the lowest floret, in the wild species, tough in cultivated forms; glumes equal or markedly unequal, chaffy, the lower 5- to 7-, the upper 7- to 9-nerved; lemmas coriaceous to crustaceous, 7-nerved, 2-lobed or entire with a stout geniculate awn issuing from the dorsal surface; lobes awned, with or without a lateral tooth, or unawned; callus of the articulated florets cushion-shaped, in non-articulated florets with a horizontal or oblique fracture; palea 2-keeled, hairy on the keels; stamens 3; lodicules 2, ovate or lanceolate-acuminate, fleshy below. Grain oblong, hairy, adherent to the lemma and palea or more or less free; embryo about one-eighth the length of the grain; hilum basal, punctiform.

About 70 species in temperate regions, and on mountains in the tropics.

Callus short, inarticulate or articulate, if the latter, callus cushion-like:
  Glumes very unequal:
    Spikelets 3–5-flowered; rhachilla articulate below all the florets   -   A. CLAUDA (p. 1760)
    Spikelets 2–3-flowered; rhachilla articulated below the lowest floret only, but not between
      the florets - - - - - - - - - - **3. A. eriantha**
  Glumes more or less equal in length:
    Lemmas with two fine bristles or awns up to 12 mm. long from the apical lobes, in addition
      to the dorsal awn:
      Bristles exceeding the length of the glumes - - - - - **6. A. hirtula**
      Bristles shorter than the tips of the glumes:
        Lobes of the lemma gradually tapering into the bristle; glumes 20–30 mm. long
          1. A. barbata
        Lobes of the lemma toothed below the base of the bristle; glumes 12–20 mm. long
          9. A. wiestii
    Lemmas 2-toothed at the apex, rarely produced into 2 bristles and if so, the bristles not
      more than 2 mm. long:
      Rhachilla articulated between the florets or continuous in cultivated forms and then
        with the false callus horizontal:
        Rhachilla articulated below all the florets - - - - **4. A. fatua**
        Rhachilla not articulated - - - - - - A. SATIVA (p. 1763)
      Rhachilla not articulated between the florets; false callus oblique:
        Lowest lemmas articulate with the rhachilla:
          Spikelets 3–5-flowered; glumes 30–50 mm. long; lowest lemma 25–40 mm. long
            8. A. sterilis
          Spikelets 2–3-flowered; glumes 25–30 mm. long; lowest lemma 20–25 mm. long
            7. A. ludoviciana
        Lowest lemma not articulated with the rhachilla; caryopsis corticate (i.e. adherent to
         lemma and palea) - - - - - - - **5. A. byzantina**
Callus 5–7 mm. long, with a sharp articulation - - - - **2. A. ventricosa**

1. A. barbata *Pott ex Link* in Schrad., Journ. Bot., 2: 315 (1799); C. et W. Barbey,
Herborisations au Levant, 99, 169 (1880); Boiss., Fl. Orient., 5: 543 (1884); Holmboe, Veg.
Cypr., 34 (1914); Post, Fl. Pal., ed. 2, 2: 738 (1933); R. M. Nattrass, First List Cypr. Fungi,
17, 18 (1937); Lindberg f., Iter Cypr., 7 (1946); N. L. Bor in C. C. Townsend et al., Fl. Iraq,
9: 328, t. 120 (1968); Osorio-Tafall et Seraphim, List Vasc. Plants Cyprus, 13 (1973).
    A. hirsuta Roth, Catalect. Bot., 3: 19 (1806); Kotschy in Unger et Kotschy, Die Insel
Cypern, 180 (1865).
    ? *A. cypria* Sibth. in Walpole, Travels, 23 (1820).
    A. strigosa Schreb. ssp. *barbata* (Pott ex Link) Thell. in Viertelj. Naturf. Ges. Zürich, 56:
330 (1911).
    TYPE: Portugal; "wächst häufig wild um Lissabon" (B).

An annual grass; culms solitary or fascicled, erect or geniculately
ascending, up to 1 m. tall, smooth and glabrous even at the nodes; leaf-
blades linear, tapering to a long acuminate tip, up to 30 cm. long, 7 mm.
wide, scabrid on the upper surface and on the margins, glabrous or sparsely
covered with hairs; sheaths clasping the culms, striate, glabrous or sparsely
hairy, more or less rough towards the top; ligule membranous, 1–3 mm. long,
scabrid on the outer surface, truncate; panicle effuse with spreading
branches or somewhat one-sided; branches fascicled; spikelets 2–3-flowered;
glumes more or less equal, 20–30 mm. long, the lower 3- to 5-, the upper 5- to
7-nerved, lanceolate or elliptic, acute; all lemmas articulated to the rhachilla
and caducous, the lowest up to 25 mm. long, cleft at the apex into two
aristulate lobes, very hairy on the dorsal surface; aristulae not exceeding the
glumes; awns of the lemmas dorsal, twisted below the knee; paleas shorter,
ciliate on the keels; anthers 2·5 mm. long.

HAB.: Cultivated and waste ground, roadsides; sometimes on stony hillsides or in open Pine
forest; sea-level to 4,500 ft. alt.; fl. Febr.–June.

DISTR.: Divisions 2, 4–8. Europe, Mediterranean region and eastwards to N.W. India;
Atlantic Islands; introduced into S. Africa, Australia and N. America.

2. Common (Prodhromos, Kakopetria, Platres, etc.); *Kennedy* 78! 1106! 1107! *Merton* 363!
   D. P. Young 7312!
4. Common around Larnaca, 1862, *Kotschy* 68! Kondea, 1938, *Syngrassides* 1832! Kouklia,
   1951, *Merton* 443! Larnaca Salt Lake Plantation, 1955, *Merton* 2002!
5. Kythrea, 1932, *Syngrassides* 279! Near Athalassa, 1933, *Syngrassides* 6! Between Nisou and
   Stavrovouni, 1933, *Syngrassides* 34a! Mile 10, Nicosia-Famagusta old road, 1967, *Merton* in
   ARI 488!
6. Nursery Garden, Nicosia, 1932, *Syngrassides* 229! Syrianokhori, 1935, *Syngrassides* 644!
   Nicosia, 1935, *Nattrass* 497; Nicosia airport, 1951, *Merton* 310b!
7. Common; *Syngrassides* 1162a! *Chapman* 664! *Merton* 1683! *Economides* in ARI 1138! etc.
8. Rizokarpaso, 1941, *Davis* 2370!

NOTES: Kotschy (Die Insel Cypern, 180) identifies Sibthorp's *Avena cypria* with *A. hirsuta*
Roth, a synonym of *A. barbata*. Sibthorp records that *A. cypria* was collected near Peristerona
(6).

2. A. ventricosa *Bal. ex Cosson* in Bull. Soc. Bot. Fr., 1: 14 (1854); N. L. Bor in C. C. Townsend et
al., Fl. Iraq, 9: 330, t. 121 (1968); Osorio-Tafall et Seraphim, List Vasc. Plants Cyprus, 13
(1973).
    A. bruhnsiana Gruner in Bull. Soc. Nat. Mosc., 40 (4): 458 (1867).
    [*A. clauda* (non Dur.) Druce in Rep. B.E.C., 9: 471 (1931).]
    TYPE: Algeria; "lieux incultes de l'Algérie occidentale", *Balansa* 557 (P, K!). The Kew sheet
is labelled "*Oran, dans les champs sabbloneux avoisinant la Batterie espagnole*. 5 April, 1852."

An annual grass; culms reaching 60 cm. in height, erect or geniculately
ascending, solitary or fascicled, smooth and glabrous or somewhat pilose on
the lower nodes; leaf-blades linear-acuminate, flat, scaberulous on both
surfaces, scabrid on the margins, up to 15 cm. long, 3–4 mm. wide,
sometimes pilose on the upper surface towards the base and ciliate on the
margins; sheaths clasping the culms, smooth and glabrous, or the lower
loosely pilose; ligule membranous, 2–3 mm. long, lacerate; panicle most
often a unilateral, simple raceme of unispiculate branches; spikelets more or
less pendulous, 2-flowered; rhachilla disarticulating below the lowest floret;

glumes somewhat unequal, the lower 22–24 mm. long, 5–7-nerved, the upper 25–30 mm. long, 9-nerved, lanceolate or elliptic, acute, papery with membranous margins; lowest lemma indurated, 15–20 mm. long, glabrous on the dorsal surface, hairy towards the top, bifid with the lobes produced as short bristles, 7-nerved, awned; palea narrow, 2-keeled; callus 5–6 mm. long, densely hairy; anthers about 1·5 mm. long.

HAB.: On dry, stony soil; in grazed and ungrazed grasslands; 450–1,500 ft. alt.; fl. March–April.

DISTR.: Divisions 2, 5, 6. Transcaucasia, Iraq, Iran, N. Africa.

2. Kambos, [Gambo], 1928, *Druce* 34! (as *A. clauda* Dur.).
5. Between Nisou and Stavrovouni, 1933, *Syngrassides* 34b! Athalassa, 1967, *Merton* in ARI 104! Athalassa Farm, 1967, *Merton* in ARI 160!
6. Nicosia airport, 1951, *Merton* 308! 310a!

[A. CLAUDA *Durieu* (in Duchartre, Rev. Bot., 1: 360; 1845) recorded by Druce (Rep. B.E.C., 9: 471; 1931), from Kambos ["Gambo"], is an error. The plant collected by Druce was *A. ventricosa* Bal. ex Coss. (q.v.). True *A. clauda*, with very unequal glumes, 3–5-flowered spikelets, and with the rhachilla articulated below all the florets, is widely distributed in the eastern Mediterranean and eastwards to Central Asia. It is very likely to occur in Cyprus, but has yet to be collected.]

3. A. eriantha *Durieu* in Duchartre, Rev. Bot., 1: 360 (1845); N. L. Bor in C. C. Townsend et al., Fl. Iraq, 9: 334 (1968); Osorio-Tafall et Seraphim, List Vasc. Plants Cyprus, 13 (1973).
    Trisetum pilosum Roem. et Schult., Syst. Veg., 2: 662 (1817).
    Avena pilosa (Roem. et Schult.) M. Bieb., Fl. Taur.-Cauc., 3, suppl.: 84 (1819); Boiss., Fl. Orient., 5: 542 (1884); Post, Fl. Pal., ed. 2, 2: 737 (1933) non Scopoli (1772).
    TYPE: Algeria; "Les lieux secs et montueux des terrains calcaires de l'ouest de l'Algérie: *Oran*, *Mascara*, etc."

An annual grass; culms up to 60 cm. tall, solitary or fasciculate, often shortly decumbent at the base, somewhat glaucous, smooth and glabrous; leaf-blades linear-acute or lanceolate-acute, up to 10 cm. long, 2–5 mm. wide, green, scabrid on both surfaces, glabrous or pilose; sheaths rather tight, glabrous or furnished with longish hairs; inflorescence a unilateral panicle; branches often 1-spiculate; spikelets more often 2-flowered, sometimes 2–3-flowered, widely gaping; glumes very unequal, the lower 10–15 mm. long, 3–5-nerved, the upper 20–25 mm. long, 7-nerved, membranous, glabrous, hyaline on the margins; rhachilla fragile below the lowest floret, tough between the florets; lowest lemma lanceolate, more or less 20 mm. long, shortly and sparsely pilose near the tip or glabrescent; lobes produced as two aristulae; lemmas awned; awn perfect, geniculate with a twisted column; stamens about 1 mm. long.

HAB.: Cultivated and waste ground; roadsides; bare stony hillsides, sometimes forming a turf with other annuals; 100–1,500 ft. alt.; fl. March–May.

DISTR.: Divisions 6, 7. Eastern Mediterranean region and eastwards to Central Asia.

6. West of Orga, 1956, *Merton* 2598!
7. Lakovounara Forest, 1950, *Chapman* 147! 190! and *Merton* 315! Ayios Demetrianos, Kythrea, 1950, *Chapman* 85! Mia Milea, 1950, *Chapman* 486! Aghirda, 1936, *Syngrassides* 1162! Kambyli, 1956, *Merton* 2736!

NOTES: This species has been persistently named *A. pilosa* (Roem. et Schult.) M. Bieb. despite the fact that this name is antedated by the distinct *A. pilosa* Scopoli.

4. A. fatua L., Sp. Plant., ed. 1, 80 (1753); Boiss., Fl. Orient., 5: 545 (1884); H. Stuart Thompson in Journ. Bot., 44: 340 (1906); Holmboe, Veg. Cypr., 34 (1914); Post, Fl. Pal., ed. 2, 2: 738 (1933); N. L. Bor in C. C. Townsend et al., Fl. Iraq, 9: 334 (1968); Osorio-Tafall et Seraphim, List Vasc. Plants Cyprus, 13 (1973).
    TYPE: "*in* Europae *agris inter segetes*" (LINN!).

An annual grass; culms erect or geniculately ascending, solitary or tufted, rather stout, smooth and glabrous, up to 1·5 m. tall; leaf-blades linear, acute or acuminate, flat, up to 30 cm. long, 4–10 mm. wide, rough on both surfaces and on the margins; sheaths clasping the culms, striate, more or less loosely pilose at the base; ligules membranous, up to 6 mm. long; inflorescence a broad or narrow pyramidal panicle, 10–40 cm. long, up to 20 cm. wide with pendulous spikelets; spikelets gaping, 18–25 (rarely 30) mm. long, 2–3-flowered; rhachilla articulate below the lipped basal scars of all the lemmas; glumes more or less equal, as long as the spikelets, smooth, glabrous, 7–11-nerved, papery with silvery margins; lemmas elliptic-acute when flattened, bifid at the apex, glabrous or hairy on the dorsal surface, 14–20 mm. long, 7–9-nerved, yellow, finally brown, smooth below, rough above, awned; awn from the middle of the back, stout, geniculate, twisted below the knee; palea shorter, 2-keeled, ciliate on the keels; anthers 3 mm. long.

HAB.: ? an adventive in cereal fields.

DISTR.: Not known (see Notes).

NOTES: Recorded, but without locality, by H. Stuart Thompson (Journ. Bot., 44: 340; 1906) from material said to have been collected in Cyprus in 1904 by Miss E. A. Samson. No specimens are to be found at Kew, so it is impossible to check the identification. Dr. D. Zohary (personal communication to Dr. N. L. Bor) suggests that its occurrence may have been sporadic, as an adventive in cereal crops. The fact that it has not since been re-found indicates that, even as an adventive, it must be unusual.

5. A. byzantina *C. Koch* in Linnaea, 21: 392 (1848); Osorio-Tafall et Seraphim, List Vasc. Plants Cyprus, 13 (1973).

    A. fatua L. var. *glabrescens* Coss. et Durieu, Fl. d'Algérie, II, Phan., Glumacées [in Expl. Sci. Algérie]: 113 (1855); Boiss., Fl. Orient., 5: 544 (1884); Post, Fl. Pal., ed. 2, 2: 738 (1933).

    A. sterilis L. ssp. *byzantina* (C. Koch) Thell. in Viertelj. Naturf. Ges. Zürich, 56: 316 (1911).

    [? *A. orientalis* (non Schreb.) R. M. Nattrass, First List Cyprus Fungi, 4, 11, 12, 18, 34, 44, (1937).]

TYPE: Turkey; neighbourhood of Istanbul, *C. Koch* (B).

An annual grass, rather greyish in colour; culms up to 150 cm. tall, rather stout, fascicled or solitary, erect or geniculately ascending, smooth and glabrous; leaf-blades linear-acuminate, tapering to a fine tip, up to 40 cm. long, 4–5 mm. wide, rough on both surfaces and on the margins; sheaths clasping, striate, glabrous and smooth or very slightly scaberulous towards the top; ligule membranous, short; panicle pyramidal or unilateral, up to 30 cm. long, 5–15 cm. wide; spikelets ± 30 mm. long, lanceolate in outline, finally gaping, 2–3-flowered; glumes 25–30 mm. long, papery, elliptic-acuminate, 7–9-nerved, membranous on the margins, more or less equal in length, smooth and glabrous; lowest lemma usually awned, ± 25 mm. long, fracturing from the rhachilla with a circular, oblique pseudo-callus, hirsute up to the middle on the dorsal surface, glabrescent or even glabrous; palea much narrower, 2-keeled, scabrid on the keels; awn straight or flexuous, sometimes kneed but not powerfully twisted below the knee; anthers 2–3 mm. long.

HAB.: Cultivated and waste ground; c. 800 ft. alt.; fl. April–June.

DISTR.: Division 5. Mediterranean region, and introduced elsewhere.

5. Nisou, 1938, *Nattrass and Papaiannou* in *Nattrass* 928!

NOTES: The cited specimen was collected as a host for *Puccinia graminis* (see Cyprus Agric. Journ., 34: 26; 1939) and is now in the Commonwealth Mycological Institute. It was originally identified as *Avena orientalis* Schreb., and raises the possibility that other specimens so listed by Nattrass (First List Cyprus Fungi, *loc. cit.*) are also *A. byzantina*. The two Oats most widely cultivated in Cyprus are, according to J. Parisinos (Countryman, Sept. 1959, p. 22), "White Algerian" and "Red Algerian"; both of which are considered to be cultivars of *A. byzantina*.

6. A. hirtula *Lag.*, Gen. et Sp. Nov., 4 (1816); Boiss., Fl. Orient., 5: 543 (1884); N. L. Bor in C. C. Townsend et al., Fl. Iraq, 9: 336 (1968); Osorio-Tafall et Seraphim, List Vasc. Plants Cyprus, 13 (1973).

TYPE: Spain; "locis ruderatis, incultis, ad agrorum versuras et juxta vias Matriti, Murciae, Orcelis, Gadibus, alibique", *Lagasca* (?MA).

An annual grass; culms often fascicled, sometimes solitary, erect or geniculately ascending, up to 50 cm. tall, smooth and glabrous; nodes reddish; leaves mostly collected at the base of the plants; blades of the culm-sheaths very short, at the most 5 cm. long, 2–3 mm. broad, the lower up to 15 cm. long, 2–4 mm. wide, sparsely covered with soft, white hairs or glabrous; sheaths rather loose, striate, covered with soft, white hairs, the lowest becoming scarious; ligule membranous, 2–3 mm. long, truncate, lacerate; inflorescence sometimes a spreading panicle, often reduced to a loose raceme which is spreading or secund; branches solitary or in pairs, sparsely hairy, flexuous, usually carrying a solitary spikelet; spikelets 2–3-flowered, 15–25 mm. long; florets all articulate to the rhachilla and caducous; rhachilla very hairy below each floret; scar oblong-linear; glumes as long as the spikelet, elliptic-acuminate in shape with hyaline tips, 5–7-nerved; lowest lemma 12–15 mm. long, becoming coriaceous, very hairy on the lower half, awned with a geniculate awn, twisted at the base, bifid at the tip, with the lobes produced as 2 fine awns which overtop the glumes and are furnished with a tooth on one side.

HAB.: Roadsides, waste ground; in Pine forest; 500–4,500 ft. alt.; fl. April–May.

DISTR.: Division 2. Widely dispersed in S. Europe and Mediterranean region eastwards to Iraq and Arabia; Madeira.

2. Prodhromos, 1953, *Merton* 1068! Yialia, 1955, *Merton* 2153! Kakopetria, 1955, *Merton* 2302!

7. A. ludoviciana *Durieu* in Act. Soc. Linn. Bordeaux, 20: 41 (1855); Sintenis in Oesterr. Bot. Zeitschr., 32: 398 (1882); Holmboe, Veg. Cypr., 34 (1950); Rechinger f. in Arkiv för Bot., ser. 2, 1: 418 (1950); N. L. Bor in C. C. Townsend et al., Fl. Iraq, 9: 336, t. 125, 126 (1968); Osorio-Tafall et Seraphim, List Vasc. Plants Cyprus, 13 (1973).

A. sterilis L. *var. ludoviciana* (Durieu) Husnot, Gram., 2: 39 (1897).

TYPE: France; "assez fréquente dans les environs de Bordeaux", *Durieu* (P).

An annual grass with tufted or solitary culms up to 1·5 m. tall, erect or somewhat decumbent at the base, stout, smooth and glabrous; leaf-blades flat, firm, rough on both surfaces and on the margins, glabrous, up to 60 cm. long, 6–12 mm. wide, linear, acuminate or acute; sheaths clasping the culms, pilose below; ligules membranous, truncate or obtuse, up to 8 mm. long; inflorescence an erect or nodding, pyramidal panicle, up to 40 cm. long, 12–20 cm. wide; spikelets pendulous, gaping, 23–32 mm. long, 2-awned, 2–3-flowered; rhachilla fragile above the glumes, but tough between the florets; glumes more or less equal, lanceolate-oblong, acute when flattened, hyaline on the margins, 9–11-nerved, papery in texture; lemma when flattened elliptic-acute, shortly 2-toothed at the tip, 2-nerved, hairy on the dorsal surface in the lower half, with a lipped elliptic or ovate scar to the lower articulation, stoutly awned from the middle of the dorsal surface; awn geniculate, twisted below the knee; palea narrow, 2-keeled, scabrid or shortly ciliate on the keels; anthers 2·5–3 mm. long.

HAB.: By tracks and roadsides, on waste or cultivated land; 150–5,600 ft. alt.; fl. March–July.

DISTR.: Divisions 2–5, 7, 8, common in Division 7. Widely distributed in the Mediterranean region and eastwards to Central Asia; Atlantic Islands. Introduced elsewhere.

2. Ayios Merkourios, 1962, *Meikle* 2278! Apliki, 1967, *Merton* in ARI 291! Troödos, 1963, *D. P. Young* 7829!
3. Two miles S. of Pyrgos (Limassol), 1963, *J. B. Suart* 80!
4. Near Old Larnaca, 1880, *Sintenis & Rigo* 883! Paralimni, 1969, *A. Genneou* in ARI 1366!

5. Near Athalassa, 1933, *Syngrassides* 6a! Between Nisou and Stavrovouni, 1933, *Syngrassides* 34! Athalassa, Agric. Research Inst., *Merton* in ARI 159! Mia Milea, 1967, *Merton* in ARI 190! Ayios Seryios, 1973, *P. Laukkonen* 30! Mile 10, Nicosia–Famagusta (old) Road, 1967, *Merton* in ARI 489!

7. Common; *Sintenis & Rigo* 371! *Chapman* 191! 352! *Economides* in ARI 1139! 1181! etc.

.8. Karpas, near the tip of Cape Andreas, 1880, *Sintenis & Rigo*; Leonarisso, 1938, *Syngrassides* 1786!

A. SATIVA *L.*, Sp. Plant., ed. 1, 79 (1753); Unger et Kotschy, Die Insel Cypern, 180 (1865); Boiss., Fl. Orient., 5: 541 (1884); Post, Fl. Pal., ed. 2, 2: 737 (1933); N. L. Bor in C. C. Townsend et al., Fl. Iraq, 9: 338, t. 127 (1968).
TYPE: Not localized (LINN!).

A more or less glaucescent annual; culms up to 1·5 m. tall, solitary or fascicled, erect, smooth and glabrous; leaf-blades flat, linear-acuminate, up to 45 cm. long, 4–15 mm. wide, rough on both surfaces and on the margins, glabrous, linear-acuminate, firm, rounded at the base; sheaths glabrous, clasping or rather loose; ligule short, membranous, toothed; inflorescence a pyramidal nodding panicle or more or less unilateral; spikelets lanceolate, finally gaping, pendulous, 2–3-flowered; rhachilla below and between the florets tough, breaking with a horizontal, irregular fracture; glumes more or less equal, 17–28 mm. long, broadly elliptic-acuminate in shape when flattened, 7–9-nerved, papery, membranous on the margins; lemmas awned or unawned, chartaceous, elliptic-oblong, acute, shining, smooth and glabrous, markedly 7-nerved in the upper half; awn, if present, on the lowest lemma only, rather slender, not twisted; palea enclosed by the margins of the lemma, narrow, 2-keeled, scabrid on the keels; anthers 2–3 mm. long.

Though a valuable crop in many cool-temperate regions, the Oat is of minor importance in Cyprus, where it would seem that the two most popular varieties are derived from *A. byzantina* C. Koch and not from *A. sativa* (see Notes under *A. byzantina* p. 1761).

8. **A. sterilis** *L.*, Sp. Plant., ed. 2, 118 (1762); Sibth. et Sm., Fl. Graec. Prodr., 1: 66 (1806); Unger et Kotschy, Die Insel Cypern, 180 (1865); Boiss., Fl. Orient., 5: 542 (1884); H. Stuart Thompson in Journ. Bot., 44: 340 (1906); Holmboe, Veg. Cypr., 34 (1914); Post, Fl. Pal., ed. 2, 2: 737 (1933); N. L. Bor in C. C. Townsend et al., Fl. Iraq, 9: 340 (1968); Osorio-Tafall et Seraphim, List Vasc. Plants Cyprus, 13 (1973).
TYPE: Spain; "*in* Hispania. ☉. Alströmer" (LINN!).

An annual green or glaucous grass; culms over 1 m. tall, solitary or fascicled, spreading, erect or geniculately ascending, smooth and glabrous, glabrous or pubescent at the nodes; leaf-blades linear or linear-lanceolate, acuminate, up to 35 cm. long, 1·5 cm. wide, scaberulous on both surfaces, scabrid on the margins, glabrous or sometimes ciliate on the margins below; sheaths clasping, striate, glabrous, or the lower more or less pilose; ligules membranous, truncate, 4–6 mm. long; panicle spreading, pyramidal or unilateral, up to 40 cm. long; branches fascicled; spikelets 30–50 mm. long, widely gaping, 2–5-flowered, the lower two usually awned, the others awnless; rhachilla tough throughout; glumes usually equal, as long as the spikelet, elliptic-acuminate, 9–11-nerved, papery, membranous on the margins; lemma 7-nerved, lanceolate-acute when flattened, 20–40 mm. long, very hairy in the lower half, rarely glabrescent or glabrous; callus very short, densely hairy, scabrid towards the 2-toothed tip, awned; palea narrow, 2-keeled; awn 3–6 cm. long, geniculate, twisted below the knee; anthers 4 mm. long.

HAB.: A weed of cultivated and waste land; sea-level to 1,200 ft. alt.; fl. March–April.

DISTR.: Divisions 4, 6, 7, common in Division 7. Mediterranean region extending (? as an introduction) into Central Europe; western Asia to Afghanistan; Atlantic Islands.

4. Near Larnaca, rare, 1862, *Kotschy* 3; Pergamos, 1956, *Merton* 2668!

6. Asomatos, 1956, *Merton* 2742 !
7. Common; *Rev. A. Huddle* 112 ! *Syngrassides* 1757 ! *Casey* 423 ! *Merton* 2709 ! 2988 !

NOTES: Much commoner than records suggest; it is widespread at lower altitudes particularly on rich soils, along the margins of cultivation.

9. A. wiestii *Steud.*, Syn. Plant. Glum., 1: 231 (1854); Druce in Rep. B.E.C., 9: 471 (1931); Boiss., Fl. Orient., 5: 543 (1884); Post, Fl. Pal., ed. 2, 2: 738 (1933); N. L. Bor in C. C. Townsend et al., Fl. Iraq, 9: 340, t. 129 (1968); Osorio-Tafall et Seraphim, List Vasc. Plants Cyprus, 13 (1973).
    A. barbata Brot. ssp. *wiestii* (Steud.) Holmboe, Veg. Cypr., 34 (1914); Lindberg f., Iter Cypr., 7 (1946).
    TYPE: Egypt; "In Aegypto pauca specimina legit *Wiest*" (P).

An annual with slender, solitary or fascicled culms, up to 70 cm. tall, smooth and glabrous, erect or geniculately ascending; leaf-blades up to 20 cm. long, 1–3 (rarely up to 8) mm. wide, linear-acuminate, scabrid on both surfaces, glabrous or sparsely pilose; sheaths also glabrous or sparsely pilose, occasionally somewhat rough towards the top; ligule membranous, 2–4 mm. long; panicle pyramidal, narrow, 10–20 cm. long; spikelets most often 2-flowered with florets all articulate; glumes membranous, more or less equal, the lower 3-nerved, the upper 5–7-nerved, glabrous; lower lemma hairy on the back almost to the base of the two lobes which are produced as two aristulae 3–6 mm. long, not exceeding the tip of the glumes, awned from the back, with a stout perfect awn; margins of the lobes denticulate.

HAB.: On rocky ungrazed land; in forests; on gullies and banks of ditches; (170–) 350–c. 3,000 ft. alt.; fl. March–May.

DISTR.: Divisions 2–7. Eastern Mediterranean and eastwards to Afghanistan; Caucasus; N. Africa.
2. Stavros tis Psokas forests, 1937, *Syngrassides* 1607 ! Milikouri near Kykko Monastery, 1939, *Lindberg f.* s.n.; Yialia, 1955, *Merton* 2143 ! Palekhori, 1957, *Merton* 3003 !
3. Mazotos, 1905, *Holmboe* 174.
4. Pergamos, 1955, *Merton* 1967 !
5. Vatili, 1905, *Holmboe* 345; Kythrea, 1939, *Lindberg f.* s.n.
6. Nicosia, 1928 or 1930, *Druce* s.n.; Mile 3, Nicosia–Myrtou Road, 1950, *Chapman* 577 ! Nicosia airport, 1951, *Merton* 314; Argaki, 1967, *Economides* in ARI 222 !
7. Pentadaktylos, 1928 or 1930, *Druce* s.n.; Lakovounara Forest, 1950, *Chapman* 351 ! 1¼ miles W. of Larnaka tis Lapithou, 1956, *Merton* 2701 !

26. ARRHENATHERUM *P. Beauv.*

Essai Agrost., 55 (1812).
T. J. Jenkin in Welsh Plant Breeding Station, Ser. H. No. 12, Seasons 1921–1931: 126–147 (1931).

Tall perennial grasses with erect, stout culms and flat leaf-blades which roll up on drying; roots yellow; lower internodes sometimes swollen and bulb-like, or the base surrounded with a cluster of bulbils; panicles rather strict or somewhat effuse, nodding, branched; spikelets 2- or rarely 3-flowered; rhachilla disarticulating above the glumes and between the florets, produced beyond the upper floret; lower floret male, upper female or hermaphrodite; glumes persistent, unequal, the lower only half the length of the upper; lower lemma membranous, 7-nerved, with a hairy callus, awned on the back at the base with a long, geniculate awn with a twisted column; upper lemma glabrous or hairy on the back, awned with a short, straight awn just below the tip; stamens 3; styles 2; stigmas plumose; lodicules 2, lanceolate-acuminate. Grain oblong-oblanceolate, hairy, grooved on the inner surface; embryo about one-sixth the length of the grain; hilum linear, about half as long as the grain.

About 4 species in Europe, the Mediterranean region and western Asia.

1. A. album (*Vahl*) *W. D. Clayton* in Kew Bull., 16: 250 (1962).

 Avena alba Vahl, Symb. Bot., 2: 24 (1791).

 Arrhenatherum erianthum Boiss. et Reut., Pugill. Plant. Nov., 121 (1852).

 ? *A. palaestinum* Boiss., Diagn., 1, 13: 51 (1853).

 [*Avena elatior* L. var. *bulbosa* (non Gaudin) Kotschy in Unger et Kotschy, Die Insel Cypern, 180 (1865).]

 [*Arrhenatherum elatius* (non (L.) P. Beauv.) Boiss., Fl. Orient., 5: 550 (1884) pro parte quoad spec. cypr.; Osorio-Tafall et Seraphim, List Vasc. Plants Cyprus, 13 (1973).]

 [*Avena elatior* L. f. *tuberosa* (non Aschers.) Holmboe, Veg. Cypr., 34 (1914).]

 TYPE: Tunis, *Vahl* (C).

A perennial grass with yellowish roots; culms loosely tufted, erect or very shortly decumbent at the base, up to 100 cm. tall, smooth and glabrous or hairy at the nodes, the lower internodes usually enlarged into bulb-like swellings; leaf-blades linear-acuminate, up to 35 cm. long, 2–8 mm. wide, flat, glabrous or sparsely pilose on the upper surface, rough on the upper surface and on the margins, smooth below; sheaths clasping the culms, usually glabrous, sometimes sparsely hairy; ligule membranous, up to 3 mm. long; inflorescence a lanceolate to oblong, erect or nodding panicle, loose or rather dense, 5–15 cm. long, 2–3 cm. broad; spikelets oblong or gaping, 7–11 mm. long, 2-flowered, rarely 3- or 4-flowered, the lower male, the upper hermaphrodite, florets falling together at maturity leaving the persistent glumes; glumes unequal, the lower the shorter, lanceolate, 1-nerved, 5·5–7 mm. long, the upper narrowly ovate-acute, 3-nerved, 8·5–9 mm. long, both hyaline on the margins, shining; lower lemma ovate or oblong-elliptic, acute, 8–10 mm. long, 7-nerved, glabrous or sparsely hairy on the dorsal surface below, the central nerve issuing from the base of the lemma as a perfect awn 15–22 mm. long, twisted at the base; palea narrowly oblong-elliptic, acute, 2-nerved, scabrid on the keels; upper lemma similar to the lower, covered in the lower two-thirds with golden hairs, 7·5–9 mm. long, but with a straight short awn just below the tip, very rarely with a geniculate awn; anthers 4–5 mm. long.

HAB.: In Pine and Cedar forests; 1,700–4,600 ft. alt.; fl. May–June.

DISTR.: Division 2, locally common. S. Europe and Mediterranean region, possibly extending to Palestine and Iraq.

2. Kykko, Ambelitzia, 1937, *Syngrassides* 1583! Tripylos, 1941, *Davis* 3495! Adelphi Forest, 1957, *Merton* 3139! Prodhromos, 1961, *D. P. Young* 7359!

NOTES: In the synonymy of *A. album* I have included *A. palaestinum* Boiss., albeit with a query. The reason for this is that although the structure of the spikelets in both is fundamentally identical, the size of the glumes, lemmas and lower awns over a range of specimens is larger in *A. palaestinum*. This, however, is not clear-cut, and there is an overlap, at the lower end, between *A. album* and *A. palaestinum*, and at the upper between *A. palaestinum* and *A. kotschyi*. It is just possible that *A. palaestinum* arose as a result of intragression between *A. album* and *A. kotschyi*.

27. AVELLINIA *Parl.*

Plant. Nov., 59 (1842).

Annual grasses with flat leaf-blades, convolute when dry; panicle at first lax, then contracted, erect; spikelets 2–4-flowered, pedicellate on the short branchlets of the panicle; glumes very unequal, the lower the shorter, the upper as long as the lowest lemma; lemmas convolute or compressed, very narrow, bifid at the apex, awned; awn slender, scabrid, from between the lobes; palea hyaline, shortly cleft at the apex, scabrid on the keels; stamens 3; lodicules 2, cuneate, truncate or somewhat rounded; grain fusiform; embryo one-seventh to one-sixth the length of the grain; hilum linear, about half as long as the grain.

One species widely distributed in the Mediterranean region, and as an introduction elsewhere.

1. **A. michelii** (*Savi*) *Parl.*, Plant. Nov., 61 (1842); Holmboe, Veg. Cypr., 34 (1914).
   Bromus michelii Savi, Bot. Etrusc., 1: 78 (1808).
   Vulpia michelii (Savi) Reichb., Fl. Germ. Excurs., 140³ (1830); Sintenis in Oesterr. Bot.
   Zeitschr., 32: 291 (1882).
   Festuca michelii (Savi) Kunth, Enum. Plant., 1: 397 (1833).
   Koeleria michelii (Savi) Coss. et Durieu in Fl. d'Algérie II, Phan., Glumacées [in Expl.
   Sci. Algérie]: 120 (1855); Boiss., Fl. Orient., 5: 574 (1884).
   Colobanthium michelii (Savi) Osorio-Tafall et Seraphim, List Vasc. Plants Cyprus, 9
   (1973) nom. invalid.
   TYPE: Italy; "In Agro Florentino. Ex Herb. Micheliano" (FI).

An annual grass; culms very slender, rarely reaching 30 cm. in height,
usually fascicled, rarely solitary, shortly and sparsely retrorsely pubescent;
leaf-blades very narrow, linear-acute, involute or convolute when dry, up to
5 cm. long, 1–1·5 mm. wide, more or less shortly pubescent on both surfaces,
sometimes glabrous below, scabrid and pubescent on the upper surface;
sheaths clasping or slightly inflated, striate, retrorsely and shortly hispid;
ligule about 1 mm. long, truncate, denticulate or lacerate; inflorescence a
lanceolate or linear-lanceolate panicle, sometimes somewhat lax, dense after
flowering, up to 7 cm. long, 5–15 mm. broad; spikelets shining, greenish,
sometimes tinged with purple, laterally compressed, oblong-lanceolate in
outline, 4–5 mm. long, 2–4-flowered; glumes very unequal, the lower
lanceolate to subulate, 1·5–1·8 mm. long, glabrous, 1-nerved, the upper
lanceolate, as long as the spikelet, 3-nerved, scabrid on the keel, otherwise
smooth and glabrous, with broad hyaline margins; lowest lemma about 4
mm. long, narrow, linear-acute, bifid at the apex, awned in the cleft, smooth
and glabrous; palea about half as long as the lemma, hyaline, 2-nerved,
scabrid on the keels; awn 2–3 mm. long, straight, scabrid; anthers 0·4–0·5
mm. long.

HAB.: In garigue, or on stony slopes or seashores; sea-level to 900 ft. alt.; fl. March–April.

DISTR.: Divisions 4, 6, 8. Distribution that of genus.

4. Mile 3, Larnaca-Famagusta road, 1950, *Chapman* 430!
6. Karpasha Forest near Kapouti, 1956, *Merton* 2596!
8. On a ridge near Dhavlos, 1880, *Sintenis & Rigo* 365!

28. **LOPHOCHLOA** *Reichb.**

Fl. Germ. Excurs., 42 (1830).
K. Domin in Bibliotheca Bot., 65: 254–296 (1907).

Slender annual grasses with flat leaf-blades; inflorescence a compact,
dense or somewhat lobed panicle; spikelets firmly compressed with 3–5
bisexual florets or the uppermost reduced; rhachilla disarticulating above
the glumes and between the florets, glabrous or hairy; glumes equal or
unequal, the upper broader and usually longer than the lower, 3-nerved;
lemmas keeled, bifid at the tip, with the tips obtuse or produced as short
awns, glabrous or hairy, awned just below the tip with a straight awn, or
with a perfect awn with twisted column from the apical cleft; palea as long as
the lemma, 2-keeled; keel nerves produced as short aristae or not; stamens 3;
anthers minute; lodicules 2, oblong-obtuse, fleshy below, glabrous. Grain
narrowly oblong; embryo small, one-eighth to one-sixth the length of the
grain; hilum basal, short.

About 6 species in Europe, the Mediterranean region and Asia.

Panicle ovoid, ellipsoid or shortly cylindrical, very bristly; awns of the lemma 4 mm. long,
    scabrid - - - - - - - - - - - - - **5. L. hispida**
Panicle usually oblong, not at all bristly; awns if present much shorter than 4 mm.:
    Palea conspicuously longer than lemma and ending in 2 aristulae - - **1. L. berythea**

* Recently reduced to synonymy under *Rostraria* Trin. (1820) [T.A.C. & R.D.M.].

Palea shorter than lemma and not ending in aristulae:
  Glumes and lemmas obtuse at the tips:
    Lemmas 4–4·5 mm. long, smooth, very shortly awned-   - -   **3. L. obtusiflora**
    Lemmas 2 mm. long, more or less covered with minute warts -   **4. L. amblyantha**
  Glumes and lemmas acute or acuminate - - - - - -   **2. L. cristata**

1. **L. berythea** (*Boiss. et Blanche*) *Bor* in Taxon, 16: 68 (1967) et in C. C. Townsend et al., Fl. Iraq, 9: 348, t. 132 (1968); Osorio-Tafall et Seraphim, List Vasc. Plants Cyprus, 13 (1973).
  Koeleria berythea Boiss. et Blanche in Boiss., Diagn., 2, 4: 135 (1859); Druce in Rep. B.E.C., 9: 471 (1931).
  K. phleoides (Vill.) Pers. var. *grandiflora* Boiss., Fl. Orient., 5: 573 (1884).
  K. phleoides (Vill.) Pers. ssp. *berythea* (Boiss. et Blanche) Domin sec. Dinsmore in Post, Fl. Pal., ed. 2, 2: 749 (1933).
  TYPE: Lebanon; "circâ *Beyrouth* cl. Blanche" (G).

Culms erect or shortly decumbent at the base, up to 45 cm. tall, smooth and glabrous; leaf-blades linear-acuminate, flat, somewhat rigid or flaccid, covered on both surfaces with sparse or dense, long or short hairs, scabrid on the margins, up to 15 cm. long, 3–4 mm. broad; sheaths clasping or the lower falling from the culms, long-ciliate on the free margin, glabrous or loosely pilose on the dorsal surface; ligule membranous, truncate, lacerate, 1–2 mm. long; inflorescence a dense, oblong panicle, cylindrical, sometimes lobed, up to 8 cm. long, 2 cm. broad; spikelets 4–5-flowered, elliptic, then gaping; glumes unequal, the lower narrowly oblong-acuminate, 1-nerved, hyaline on the margins, the upper much broader, 3-nerved, oblong-elliptic, acute or acuminate, hyaline on the margins; lowest lemma 4·5–5 mm. long, broadly elliptic-acute when flattened, 5-nerved, hyaline on the margins, awned with a straight awn 1–3 mm. long just below the tip, minutely hairy on the back; palea narrowly linear, 2-nerved, the nerves excurrent at the tip into two short aristulae; anthers 0·5–0·75 mm. long.

HAB.: Dry, rocky hillsides; margins of cultivated fields; sea-level to 1,800 ft. alt.; fl. Febr.–April.

DISTR.: Divisions 1, 4–8, common in Division 7. Syria, Lebanon, Palestine, Iraq, Iran.

1. Paphos, 1928 or 1930, *Druce*; above Peyia, 1962, *Meikle* 2031!
4. Larnaca, 1928, *Druce*!
5. Salamis, 1928, *Druce*!
6. Nicosia, *P. Laukkonen* 119! 276! *O. Huovila* 64!
7. Common; Kyrenia, Larnaka tis Lapithou, Kantara, Myrtou, etc. *Davis* 3014! *Casey* 349! *Meikle* 2419! *P. Laukkonen* 235! etc.
8. Between Komi Kebir and Dhavlos, 1941, *Davis* 2451!

2. **L. cristata** (*L.*) *Hyl.* in Bot. Notiser, 1953: 355 (1953), Nord. Kärlväxtfl., 1: 283 (1953).
  Festuca cristata L., Sp. Plant., ed. 1, 76 (1753).
  F. phleoides Vill., Fl. Delph. (in Gilibert, Syst. Plant. Europ., I), 7 (1786), Hist. Plant. Dauph., 2: 95, t. II, fig. 7 (1787).
  Koeleria phleoides (Vill.) Pers., Syn. Plant., 1: 97 (1805); C. et W. Barbey, Herborisations au Levant, 99, 169 (1882); Boiss., Fl. Orient., 5: 572 (1884); Holmboe, Veg. Cypr., 35 (1914); Post, Fl. Pal., ed. 2, 2: 748 (1933).
  K. cristata (L.) Bertol., Amoen. Ital., 67 (1819) non *K. cristata* (L.) Pers. (1805) nom. illeg.
  Lophochloa phleoides (Vill.) Reichb., Fl. Germ. Excurs., 42 (1830); N. L. Bor in C. C. Townsend et al., Fl. Iraq, 9: 351 (1968); Osorio-Tafall et Seraphim, List Vasc. Plants Cyprus, 13 (1973).
  [*Koeleria pubescens* (non P. Beauv.) Druce in Rep. B.E.C., 9: 471 (1931).]
  [*Trisetum paniceum* (non Pers.) Druce in Rep. B.E.C., 9: 471 (1931).]
  TYPE: Portugal; "*in Lusitaniae collibus sterilibus*", Loefling.

Culms fasciculate, rarely solitary, simple, smooth and glabrous, leafy almost to the panicle, up to 50 cm. tall; leaf-blades flat, flaccid, loosely hairy on both surfaces, scabrid on the margins, linear-acuminate, up to 15 cm. long, 3–5 mm. wide; sheaths clasping, covered with loose spreading hairs, densely hairy in the throat; ligule truncate, toothed, lacerate, about 2 mm.

long. Inflorescence a cylindrical panicle, sometimes lobed, 0·5–6 cm. long, 0·4–1·5 cm. wide; spikelets 3–7-flowered, compressed, elliptic, finally gaping; rhachilla disarticulating above the glumes and underneath the florets; glumes unequal in length and breadth, the lower 1-nerved, oblong, acute or acuminate, 2·5–4 mm. long, the upper 3-nerved, 3·5–5·5 mm. long, broader, elliptic-acute, both silvery-hyaline on the margins, scaberulous on the keels; lowest lemma elliptic-acute when flattened, about 4–5 mm. long, with 5 prominent nerves between which the surface is reticulate or scrobiculate, bidentate at the tip, awned in the sinus; awn short, straight; palea shorter than the lemma, 2-nerved; anthers 0·5–0·6 mm. long.

HAB.: Stony pastures and waste ground; sea-level to 1,200 ft. alt.; fl. Febr.–April.

DISTR.: Divisions 1, 3–8, common in Divisions 5, 6 and 7. S. Europe and Mediterranean region eastwards to Central Asia; introduced elsewhere.

1. Karavopetres near Ayios Yeoryios (Akamas), 1962, *Meikle* 2062! Fontana Amorosa, 1962, *Meikle* 2291!
3. Skarinou, 1941, *Davis* 2682! Curium, 1961, *Polunin* 6691!
4. Ayia Napa, 1905, *Holmboe* 54; Phaneromene near Larnaca, 1905, *Holmboe* 259; Larnaca, 1928, *Druce* s.n.! Athna Plantation, 1930, *Druce* s.n.! Ayios Antonios, Sotira, 1952, *Merton* 661!
5. Common; *Chapman* 33! 37! *Merton* in ARI 148! 298! 466!
6. Common about Nicosia, Dhiorios and Ayia Irini; *Rev. A. Huddle* 107! *Druce* 351! *Nattrass* 24; 473; *Meikle* 2350! 2387! Pendayia, 1969, *A. Genneou* in ARI 1630!
7. Common about Kyrenia; *Casey* 447! 590! 1148! 1253! *Chapman* 109! 152! etc.
8. Gastria, 1941, *Davis* 2427! 2432! Platanisso, 1950, *Chapman* 211! Eleousa near Rizokarpaso, 1962, *Meikle* 2991!

NOTES: The specimen collected by Druce at Larnaca in April 1928 and cited (Rep. B.E.C., 9: 471) as *Koeleria pubescens* Beauv. is a variety of the present species (*Koeleria phleoides* (Vill.) Pers. var. *submutica* Ball) differing from typical *L. cristata* in having shorter, hairier and very short-awned lemmas. The distinctions are not such as to warrant formal nomenclatural recognition.

3. L. obtusiflora (*Boiss.*) *Gontsch.* in Komarov et al., Fl. U.R.S.S., 2: 338 (1934); N. L. Bor in C. C. Townsend et al., Fl. Iraq, 9: 348, t. 133 (1968); Osorio-Tafall et Seraphim, List Vasc. Plants Cyprus, 13 (1973).

    Koeleria obtusiflora Boiss., Diagn., 1, 7: 121 (1846).

    K. obtusiflora Boiss. var. *condensata* Boiss., Diagn., 2, 4: 134 (1859), Fl. Orient., 5: 573 (1884); Domin in Bibl. Bot., 65: 274 (1907).

    K. phleoides (Vill.) Reichb. var. *obtusiflora* (Boiss.) Boiss., Fl. Orient., 5: 573 (1884).

    K. phleoides (Vill.) Reichb. ssp. *obtusiflora* (Boiss.) Domin sec. Dinsmore in Post, Fl. Pal., ed. 2, 2: 749 (1933).

    TYPE: Iran; "prope pagum *Radar* provinciae *Schiraz* Kotschy No. 131" (G).

Culms fasciculate, rarely solitary, erect or geniculately ascending, often spreading, smooth and glabrous, shining, up to 45 cm. tall, green or greyish; leaf-blades lanceolate or linear, acuminate, flat, up to 15 cm. long, 3 mm. wide, covered on both surfaces with rather sparse long hairs, scabrid on the margins; sheaths clasping or the lower loose, striate, smooth and glabrous or sparsely covered with longish hairs; ligule 1–1·5 mm. long, membranous, toothed or lacerate, truncate; inflorescence a very dense, cylindrical panicle rounded at the apex, sometimes lobed, 1–8 cm. long, 0·4–1·5 cm. broad, very pale or golden; spikelets, in the type of the species, 4–5 mm. long, 5-flowered, oblong, then gaping; glumes very unequal, the lower 2 mm. long, very narrow, acuminate, 1-nerved, the upper much broader, elliptic or oblong, acute when flattened, 3-nerved, 3·5 mm. long or somewhat longer, glabrous; lowest lemma 4·5 mm. long or sometimes 5·5 mm. long, oblong when flattened, glabrous, prominently nerved, ending above in two rounded lobes, slightly emarginate, mucronate or not in the notch; palea much narrower, 2-nerved; anthers less than 0·5 mm. long.

HAB.: On sandy or rocky pastures, usually near the sea; sea-level to c. 500 ft. alt.; fl. March–May.

DISTR.: Divisions 4, 5, 7. S.W. Europe (probably introduced), Syria, Lebanon and Palestine eastwards to Central Asia.

4. Larnaca, 1862, *Kotschy* 259! Larnaca aerodrome, 1950, *Chapman* 29! 708! Cape Greco, by lighthouse, 1962, *Meikle* 2599!
5. Near Athalassa, 1933, *Syngrassides* 12!
7. Ayios Yeoryios, Kyrenia, 1951, *Merton* 302!

NOTES: Domin (*loc. cit.*) has identified *Kotschy* 259 as *Koeleria obtusiflora* var. *condensata*; but the differences between this and typical *Lophochloa* (*Koeleria*) *obtusiflora* are so unsatisfactory that the distinction has not been maintained here.

4. L. amblyantha (*Boiss.*) *Bor* **comb. nov.**
    Koeleria phleoides (Vill.) Pers. var. *amblyantha* Boiss., Diagn., 2, 4: 134 (1859), Fl. Orient., 5: 573 (1884); Holmboe, Veg. Cypr., 35 (1914); Post, Fl. Pal., ed. 2, 2: 749 (1933).
    K. obtusiflora Boiss. var. *amblyantha* (Boiss.) Domin in Bibl. Bot., 14: 273 (1907).
    TYPE: Lebanon; "ad *Beyrouth* cl. Blanche" (G).

Culms erect, very slender, up to 40 cm. tall, often much shorter, smooth and glabrous, shining, simple; leaf-blades flat or folded, linear-acuminate, up to 20 cm. long, 1–4 mm. broad, sparsely covered on both surfaces with spaced hairs, those on the upper surface longer than those on the lower, smooth on the margins; upper sheaths tightly clasping, the lower rather loose or falling from the culms, somewhat inflated, smooth and glabrous; ligules about 1 mm. long, truncate, membranous; inflorescence a terminal, dense panicle, cylindrical or somewhat lobed, 2–7 cm. long, 5–10 mm. wide; branches very short; spikelets 2·5–3 mm. long, widely gaping, 4–5-flowered; lower glume very narrow, linear-acuminate, 2 mm. long, 1-nerved, the upper broadly elliptic-acute, 2–2·5 mm. long, 2 mm. wide when flattened, 3-nerved, with hyaline margins; lemmas oblong or obovate-oblong, rounded at the apex, very slightly emarginate, keeled, but somewhat inflated, very faintly nerved or lateral nerves quite obscure, more or less tuberculate on the dorsal surface, 2 mm. long, with a minute mucro at the tip or without a mucro; anthers 0·5 mm. long.

HAB.: In *Cistus* garigue and dry rocky pastures near the sea; sea-level to 200 ft. alt.; fl. April.

DISTR.: Divisions 4, 8. Also Lebanon, Jordan.

4. Larnaca, 1880, *W. Barbey* sec. Boiss.
8. Rizokarpaso, 1930, *Druce* s.n.! Eleousa near Rizokarpaso, 1962, *Meikle* 2544!

5. L. hispida (*Savi*) *Jonsell* in Bot. Journ. Linn. Soc., 76: 321 (1978).
    Festuca hispida Savi, Fl. Pis., 1: 117 (1798).
    Koeleria hispida (Savi) DC., Hort. Monsp., 119 (1813).
    Bromus hispidus (Savi) Savi, Bot. Etrusc., 2: 62 (1815).
    TYPE: Italy; "Monte Pisano vicino a Calci", *Savi* (PI).

Culms fasciculate or solitary, erect or geniculately ascending, sometimes spreading, up to 30 cm. tall, shining, smooth and glabrous; leaf-blades flat, lanceolate or linear, acuminate, up to 10 cm. long, 3–5 mm. wide, glabrous on the lower surface, more or less densely ciliate on the scabrid margins, covered on the upper surface with spaced, spreading hairs; sheaths clasping the culms, sparsely pilose, striate; ligule truncate, about 1 mm. long; inflorescence a very dense, ovoid, ellipsoid or shortly cylindrical panicle rounded at the apex, 1–3 cm. long, 1–1·5 cm. wide, bristly; branches very short; spikelets 3–4-flowered, about 4 mm. long excluding the awns; rhachilla disarticulating above the glumes and below the florets, glabrous; glumes very unequal, the lower 2·5–3 mm. long, 1-nerved, subulate, the upper much broader, hyaline on the margins, 3–3·5 mm. long, prominently 3-nerved, lemma 3–3·5 mm. long, prominently 5-nerved, with a few long hairs on the dorsal surface, very shortly 2-toothed at the tip, awned in the sinus; awn 4 mm. long, straight, scabrid; palea shorter than the lemma, 2-nerved, ciliate on the nerves; anthers 1 mm. long.

HAB.: Damp ground by side of river; near sea-level; fl. April.

DISTR.: Division 4, rare. Also Italy, Sardinia, Sicily, Greece, Lampedusa, ? Morocco, Algeria, Tunisia.

4. Chali, near Famagusta, 1962, *Meikle* 2620 !

TRIBE 6. **Phalarideae** *Kunth* Annual or perennial grasses; leaf-blades narrow, linear, with festucoid anatomy; silica-bodies oblong with smooth or sinuous outlines, rarely rounded or cube-shaped; micro-hairs absent; ligules membranous; spikelets all alike, hermaphrodite, or some sterile and reduced, club-shaped; fertile spikes strongly compressed, (1–) 3-flowered, the uppermost flower hermaphrodite, the lower two reduced to scales or one or both absent; rhachilla disarticulating above the glumes but not between the florets; inflorescence a dense, contracted, cylindrical or ovoid panicle; glumes persistent, equal, and as long as the spikelet, winged on the keels or not winged; lower 2 lemmas small, scale-like, empty, or one or both missing; fertile lemma becoming indurated when mature, awnless, glossy; stamens 3; ovary glabrous; styles 2; stigmas plumose; lodicules 2. Grain closely invested by lemma and palea; hilum oblong, short; starch grains compound. Chromosomes large, basic numbers 5, 6, 7. [Tribe probably best included in *Aveneae* Dumort. T.A.C. & R.D.M.]

29. PHALARIS *L.*
Sp. Plant., ed. 1, 54 (1753).
Gen. Plant., ed. 5, 29 (1754).
E. Paunero in Anal. Jard. Bot. Madrid, 8: 475–522 (1948).
D. E. Anderson in Iowa State Journ. Sci., 36: 1–96 (1961).

Annual or perennial grasses with linear, flat leaf-blades; inflorescence an ellipsoid or oblong panicle, sometimes lobed, consisting of many hermaphrodite spikelets, or rarely the basal spikelets rudimentary or reduced; spikelets strongly compressed, 3-flowered, the lower two florets reduced to the lemmas and the upper hermaphrodite; glumes regularly or irregularly winged on the keels or wingless, ± equal in size; lower lemmas reduced to small linear or lanceolate scales; upper lemma coriaceous, compressed, rounded on the back, glabrous or very hairy; palea of the same texture; stamens 3; lodicules 2 or absent, when present lanceolate-acute. Grain compressed, ellipsoid, with a groove the length of, or shorter than the grain; embryo about one-third the length of the grain; hilum oblong, at the base of a groove which is as long as, or shorter than the grain.

About 20 species widely distributed in temperate regions of both hemispheres.

Perennial, robust grass, culms bulbous at the base - - - - - **5. P. aquatica**
Annual grasses:
  Spikelets of 2 kinds, the basal very much reduced; the upper in groups, with sterile spikelets
    surrounding a single fertile one - - - - - - - - **4. P. paradoxa**
  Spikelets all similar:
    Wings of the glumes toothed or undulate - - - - - - **3. P. minor**
    Wings of the glumes entire:
      Empty lemmas at the base of the fertile floret oblong-acute, about 3 mm. long
                                       2. P. canariensis
      Empty lemmas at the base of the fertile floret ovate-acute, about 1 mm. long
                                       1. P. brachystachys

1. **P. brachystachys** *Link* in Schrader, Neues Journ., 1 (3): 134 (1806); Boiss., Fl. Orient., 5: 471 (1884); Post, Fl. Pal., ed. 2, 2: 713 (1933); Lindberg f., Iter Cypr., 8 (1946); Rechinger f. in Arkiv för Bot., ser. 2, 1: 418 (1950); N. L. Bor in C. C. Townsend et al., Fl. Iraq, 9: 362, t. 138 (1968); Osorio-Tafall et Seraphim, List Vasc. Plants Cyprus, 14 (1973).

[*P. canariensis* (non L.) Brot., Fl. Lusit., 1: 79 (1804).]

TYPE: Portugal; "inter segetes, locisque incultis spontanea; colitur etiam, sed parcè", *Brotero* (?LISU, GOET).

An annual grass; culms loosely tufted, rarely solitary, erect or geniculately ascending, in robust specimens up to 60 cm. tall, usually much shorter, smooth and glabrous; leaf-blades flat, linear-acuminate, flaccid, up to 30 cm. long, 3–12 mm. wide, dark green, definitely scaberulous on both surfaces and on the margins, glabrous; sheaths rather loose, the uppermost somewhat inflated, smooth and glabrous; ligule up to 3 mm. long, hyaline or milky white; inflorescence a dense cylindrical or ellipsoid panicle rounded at the top, 1·5–4 cm. long, 10–15 mm. wide; branches and pedicels very short; spikelets firmly compressed, 5–8·5 mm. long, rather glaucous; glumes equal, as long as the spikelet, hyaline on the margins, keeled, 3-nerved, winged in the upper half or two-thirds, together obovate-cuneate in the spikelet; wings entire on the margin; empty lemmas two small, ovate-acute scales c. 1 mm. long, lanceolate-acute; fertile lemma 5 mm. long, compressed, keeled, shortly appressed-hairy; palea of the same texture; anthers 3 mm. long.

HAB.: Cultivated and fallow fields; by marshes or on temporarily flooded ground; occasionally on dry hillsides; sea-level to 2,000 ft. alt.; fl. March–July.

DISTR.: Divisions 1, 3, 5–7. Mediterranean region eastwards to Iran; Atlantic Islands; occasionally as a weed elsewhere.

1. Stroumbi, 1913, *Haradjian* 728!
3. Two miles S. of Pyrgos, 1963, *J. B. Suart* 85!
5. Near Aphania, 1952, *Merton* 1558! Mora, 1956, *Merton* 2559!
6. Asomatos, 1956, *Merton* 2744!
7. Kythrea, 1939, *Lindberg f.* s.n.; Ayios Dhimitrianos, Kythrea, 1950, *Chapman* 163! One mile E. of Kyrenia, 1956, *Merton* 2770! Kambyli, 1957, *Merton* 2984!

2. **P. canariensis** *L.*, Sp. Plant., ed. 1, 54 (1753); Sibth. et Sm., Fl. Graec. Prodr., 1: 36 (1806), Fl. Graec., 1: 40, t. 55 (1808); Kotschy in Oesterr. Bot. Zeitschr., 12: 277 (1862); Boiss., Fl. Orient., 5: 471 (1884); Post, Fl. Pal., ed. 2, 2: 713 (1933); R. M. Nattrass, First List Cyprus Fungi, 4 (1937).

TYPE: "*in* Europa *australi*" (LINN!).

An annual grass; culms usually fascicled, erect or geniculately ascending, up to 60 cm. tall, smooth and glabrous; leaf-blades linear-acuminate, flat, 10–26 cm. long, 3–12 mm. wide, more or less scaberulous on both surfaces, smooth or minutely scaberulous on the margins, glabrous; sheaths loose, the uppermost inflated, glabrous, striate, obscurely scaberulous; ligule membranous, 3–5 mm. long; panicle very dense, cylindrical or ovate-cylindrical; axis and branches scaberulous; spikelets 7–9 mm. long, 4 mm. wide, obovate in outline, much compressed; glumes as long as the spikelet, 3-nerved, very sparsely covered with soft white hairs, scabrid, winged on the keel in the upper two-thirds; wings entire on the margin; sterile lemmas 2, oblong-acute, slightly curved on the back, 3–4 mm. long; fertile lemma papyraceous, velvety, lanceolate or ovate-lanceolate, acute when flattened, 5–6 mm. long, eventually smooth and glossy; palea of the same texture, compressed-hairy on the keels; anthers about 3 mm. long.

HAB.: Waste and cultivated ground, probably introduced; sea-level to 3,750 ft. alt.; fl. June–Sept.

DISTR.: Divisions 2, 6, rare. Widely naturalized in S. Europe and the Mediterranean region, perhaps a native of N.W. Africa and the Canary Islands.

2. Platres, 1938, *Kennedy* 1087!
6. Morphou, 1932, *Nattrass* 176.

NOTES: Kotschy recorded it as "common about the margins of wheat fields around Kiti" (Oesterr. Bot. Zeitschr., 12: 277; 1862), but did not repeat the record in Die Insel Cypern, where (p. 177) *P. minor* is listed from this locality.

3. **P. minor** *Retz.*, Obs. Bot., 3: 8 (1783); Unger et Kotschy, Die Insel Cypern, 177 (1865); Boiss., Fl. Orient., 5: 472 (1884); Holmboe, Veg. Cypr., 31 (1914); Post, Fl. Pal., ed. 2, 2: 714 (1933); N. L. Bor in C. C. Townsend et al., Fl. Iraq, 9: 364, t. 139 (1968); Osorio-Tafall et Seraphim, List Vasc. Plants Cyprus, 14 (1973).

   P. gracilis Parl., Plant. Nov., 36 (1842).
   P. minor Retz. var. *gracilis* (Parl.) Parl., Fl. Ital., 1: 70 (1848); Boiss., Fl. Orient., 5: 472 (1884).

   TYPE: Described from cultivated specimens of unstated origin (LD).

An annual grass; culms solitary or more often loosely tufted, erect or geniculately ascending, smooth and glabrous, slender or stout and up to 100 cm. tall; leaf-blades up to 25 cm. long, 10 mm. wide, rounded at the base, linear, tapering to a long acuminate point, dark green, more or less scaberulous on both surfaces and on the margins, glabrous; sheaths somewhat loose, the uppermost definitely inflated, minutely scabrid, glabrous; ligule membranous, 6–8 mm. long; inflorescence a very dense cylindrical, ovate- or lanceolate-cylindrical panicle, 2–5 cm. long, 1–1·5 cm. broad, of closely packed spikelets, seated on short branches and pedicels, axis scaberulous; spikelets elliptic-acute in outline when young, eventually gaping, 6–6·5 mm. long, firmly compressed; glumes equal, as long as the spikelet, rather pale, with 3 green nerves, keeled, smooth and glabrous, winged in the upper two-thirds; wing denticulate-undulate; sterile lemma one, subulate, c. 1 mm. long, curved; fertile lemma 3 mm. long, appressed-pilose, indurated, finally smooth and glabrous, shining, broadly ovate-acute when flattened; palea of the same texture; anthers 2 mm. long.

HAB.: Cultivated and fallow land, waste ground, roadsides, often in damp situations; sea-level to 3,800 ft. alt.; fl. Febr.–June.

DISTR.: Divisions 1, 2, 4, 6–8, often common. S. Europe, Mediterranean region eastwards to Central Asia; Atlantic Islands, occurring as a weed almost throughout temperate regions and the tropics.

1. Baths of Aphrodite, 1952, *Merton* 669! Ayios Merkourios, 1962, *Meikle* 2277!
2. Perapedhi, 1938, *Kennedy* 1086! Platres, 1938, *Kennedy* 1088!
4. Larnaca, 1862, *Kotschy*; Kiti, 1862, *Kotschy*; near Old Larnaca, 1880, *Sintenis and Rigo*; Larnaca, 1905, *Holmboe* 116; Perivolia, 1955, *Merton* 2016!
6. Nicosia, 1862, *Kotschy* 454, and, 1905, *Holmboe* 265 and, 1950, *Chapman* 622! Syrianokhori, 1935, *Syngrassides* 806! Asomatos, 1956, *Merton* 2745.
7. Common; *Sintenis & Rigo* 373! *Casey* 528! *G.E. Atherton* 1218! 2712! 2728!
8. Near Patriki, 1934, *Syngrassides* 859A! Ovgoros, 1941, *Davis* 2479! S. of Galinoporni, 1962, *Meikle* 2500! Galatia, 1962, *Meikle* 2534!

NOTES: Boissier (Fl. Orient., 5: 472; 1884) refers *Kotschy* 304, "in insulâ Cypro" to *P. minor* var. *gracilis*. Like many varieties of Mediterranean plants, it is distinguished by its slender habit only. There is no spikelet difference from typical *P. minor*, and there appears to be no valid reason for maintaining the name.

4. **P. paradoxa** *L.*, Sp. Plant., ed. 2, 1665 (1763); Sibth. et Sm., Fl. Graec. Prodr., 1: 39 (1806); Fl. Graec., 1: 43, t. 58 (1808); Unger et Kotschy, Die Insel Cypern, 177 (1865); Boiss., Fl. Orient., 5: 472 (1884); Holmboe, Veg. Cypr., 31 (1914); Post, Fl. Pal., ed. 2, 2: 714 (1933); Rechinger f. in Arkiv för Bot., ser. 2, 1: 418 (1950); N. L. Bor in C. C. Townsend et al., Fl. Iraq, 9: 366 (1968); Osorio-Tafall et Seraphim, List Vasc. Plants Cyprus, 14 (1973).

   P. praemorsa Lam., Fl. Franç., 3: 566 (1778).

Annual grass; culms fascicled, those in the centre of a tuft erect, the outer geniculately ascending, very leafy, smooth and glabrous, up to 60 cm. tall; leaf-blades linear-acuminate, up to 30 cm. long, 2–6 mm. broad, glabrous, scaberulous on the upper surface and on the margins, smooth or scaberulous below; sheaths somewhat loose, the uppermost inflated and often partially clasping the base of the inflorescence, glabrous, scaberulous; inflorescence a very dense, cylindrical-obovoid or oblanceolate panicle, 2–7 cm. long, 1–2·5 cm. wide, partially enclosed at the base or free; spikelets of three kinds; those at the base of the panicle which are neuter, deformed and consisting of empty, shapeless, club-like scales; those that resemble fertile spikelets but which are actually sterile, and the fertile spikelets, so arranged that in a fascicle of spikelets the central is fertile seated on a scabrid pedicel, or sessile and surrounded by sterile spikelets; sterile spikelets consisting of two empty club-shaped glumes seated on a smooth and glabrous pedicel; glumes 4 mm. long, narrow, keeled; fertile spikelets 6–8 mm. long seated on a coarsely scabrid pedicel, firmly compressed; glumes produced into an awn about 3 mm. long, winged on the keel with an erect tooth-like wing, 7-nerved, smooth and glabrous; sterile lemmas two, very minute, filiform or absent; fertile lemma about 3 mm. long, ovate-acute, indurated, with a few long soft hairs towards the tip, finally smooth, glabrous and shining; palea of the same texture; anthers 2 mm. long.

var. **paradoxa**
TYPE: "*in* Oriente *P. Forskåhl*" — cultivated at Uppsala (LINN!).

All the sterile spikelets uniform, or only a few reduced to club-like structures.

HAB.: Moist ground by margins of rivers, ponds or irrigation channels, or on seasonally flooded cultivated land; sea-level to 2,400 ft. alt.; fl. March–June.

DISTR.: Divisions 3–8. Widespread in the Mediterranean region and eastwards to Afghanistan; Atlantic Islands; Ethiopia; S. Africa. Introduced and naturalized in many temperate regions of the world.

3. Limassol, 1913, *Haradjian* 601; Perapedhi, 1938, *Kennedy* 1090!
4. Perivolia, 1955, *Merton* 2017! Chali, near Famagusta, 1962, *Meikle* 2625!
5. Syngrasis, 1862, *Kotschy* 539.
6. Peristerona, 1936, *Syngrassides* 985! Xeros, 1937, *Syngrassides* 1565! Asomatos, 1956, *Merton* 2743!
7. Around Kythrea, 1880, *Sintenis & Rigo*; Kyrenia, 1949, *Casey* 617!
8. Galatia, 1962, *Meikle* 2533!

var. **praemorsa** *Coss. et Durieu* in Fl. d'Algérie II, Phan., Glumacées [in Expl. Sci. Algérie]: 25 (1855); Boiss., Fl. Orient., 5: 472 (1884); Post, Fl. Pal., ed. 2, 2: 714 (1933); N. L. Bor in C. C. Townsend et al., Fl. Iraq, 9: 368, t. 140 (1968).
P. paradoxa L. f. monstr. *praemorsa* Lindberg f., Iter Cypr., 8 (1946).
TYPE: France; "On trouve cette plante en Provence" (P).

Many of the sterile spikelets reduced to club-like structures.

HAB.: In the same habitats as var. *paradoxa*; sea-level to 4,100 ft. alt.; fl. March–July.

DISTR.: Divisions 1–3, 7. General distribution as in var. *paradoxa*.

1. Between Dhrousha and Kato Arodhes, 1941, *Davis* 3280!
2. Kryos Potamos, 1938, *Kennedy* 1089! Platres, 1938, *Kennedy* 1091!
3. Near Limassol, 1939, *Lindberg f.* s.n.! 2 miles S. of Pyrgos, 1963, *J. B. Suart* 88!
7. Myrtou, 1932, *Syngrassides* 37! Lakovounara Forest, 1950, *Chapman* 368!

NOTES: There is no clear-cut distinction between the two varieties, which are of minimal taxonomic significance.

5. P. aquatica *L.*, Amoen. Acad., 4: 264 (1754); C. E. Hubbard in Milne-Redhead et Polhill, Fl. Trop. E. Afr., Gramineae, 1: 97 (1970).
P. tuberosa L., Mant. Alt., 557 (1771); Post, Fl. Pal., ed. 2, 2: 714 (1933); N. L. Bor in C. C. Townsend et al., Fl. Iraq, 9: 368 (1968); Osorio-Tafall et Seraphim, List Vasc. Plants Cyprus, 14 (1973).
P. nodosa Murr., Syst. Veg., 88 (1774); Sibth. et Sm., Fl. Graec. Prodr., 1: 37 (1806), Fl. Graec., 1: 41, t. 56 (1808); Boiss., Fl. Orient., 5: 473 (1884).
TYPE: "*in* Aegypto" [*Hasselquist*] (LINN-lectotype!).

Perennial grass; culms prominently swollen at the base, loosely tufted, the lateral stems geniculately ascending, smooth and glabrous, reaching a height of 1·5 m.; leaf-blades green or greyish, linear-acuminate, rounded-truncate at the base, up to 35 cm. long, 2–15 mm. broad, scaberulous to almost smooth on both surfaces, scaberulous on the margins; sheaths clasping, except the uppermost which is somewhat inflated, smooth and glabrous; ligule 4–6 mm. long, entire; inflorescence a dense panicle, lanceolate or oblong in outline, 3–12 cm. long, 1–3 cm. wide; branches very short or exceptionally 2·5 cm. long, carrying closely packed spikelets; spikelets firmly compressed, about 5 mm. long, elliptic-acute when young, finally widely gaping, pale with green nerves; glumes scarious on the margins, keeled, as long as the spikelets, winged along the whole length of the keel; wing entire on the margin; sterile lemmas one, sometimes two, minute or up to 2 mm. long; fertile lemma 3–4·5 mm. long, ovate-acute when flattened, densely pubescent; palea as long, and of the same texture; anthers 3–3·5 mm. long.

HAB.: Wet ground by streams and channels, or as a weed in moist fields.

DISTR.: Division 7, rare. Widespread in the Mediterranean region and eastwards to Iraq; Atlantic Islands. Introduced and cultivated in India, Africa, Australia and N. America.

7. Myrtou-Panagra, 1934, *Syngrassides* 1383! Kambyli, 1957, *Merton* 2985!

NOTES: Photiades (Countryman 8 (10): 12; 1954 and Sept. 1958: 22) reports the experimental planting of *P. aquatica* (*P. tuberosa*) at Nicosia and at Dennarga near Gouphes. It was said to be "a drought-resistant perennial grass which can also give excellent yields of forage under irrigation". Experimental plantings at Dennarga became established and persisted for three years.

TRIBE 7. **Milieae** *Endl.* Spikelets 1-flowered, slightly dorsally compressed; rhachilla disarticulating above the glumes, not produced; inflorescence an effuse or narrow panicle; glumes herbaceous or membranous, 3-nerved, equal to or longer than the lemma, subequal, persistent, the upper wrapped round the lower; lemma smooth and glossy, obscurely 5-nerved, chartaceous, becoming strongly indurated in the fruit or hyaline, covered with clavate hairs, eventually becoming chartaceous; palea of the same texture also becoming indurated, 2-nerved; lodicules 2, acute, entire; stamens 3; ovary glabrous; styles 2; stigmas plumose. Grain with a linear hilum; embryo one quarter to one-fifth the length of the grain; starch grains compound.

Annual or perennial grasses; leaf-blades linear to linear-lanceolate with festucoid anatomy; silica bodies roundish; no 2-celled micro-hairs; ligules membranous, well developed. Chromosomes large; basic numbers 4, 7, 9.

30. MILIUM *L.*
Sp. Plant., ed. 1, 61 (1753).
Gen. Plant., ed. 5, 30 (1754).

Annual or perennial grasses with linear, flat leaf-blades; inflorescence an effuse or narrow panicle, with whorled branches; spikelets 1-flowered, hermaphrodite, ellipsoid, acute; glumes subequal, 3-nerved, longer than the lemma, persistent; lemma elliptic-acute in outline, coriaceous, very smooth, becoming indurated in fruit, folded round the palea; lodicules 2, linear, acute or acuminate; stamens 3. Grain elliptic in outline, dorsally somewhat compressed; embryo about one-fifth the length of the grain; hilum linear, one-fifth to half as long as the grain.

Species 4–6, widely distributed in N. temperate regions.

Annuals, up to 60 cm. tall:
    Panicles constricted, up to 11 cm. long, 6 cm. wide; pedicels mostly less than 5 mm. long
1. M. vernale

Panicles very effuse, up to 15 cm. or more wide with widely spreading branches; pedicels
rarely less than 10 mm. long - - - - - - - **2. M. pedicellare**
Perennial up to 150 cm. tall - - - - - - - - - 3. M. EFFUSUM

1. **M. vernale** *M. Bieb.*, Fl. Taur.-Cauc., 1: 53 (1808); Boiss., Fl. Orient., 5: 510 (1884); Holmboe,
Veg. Cypr., 33 (1914); Post, Fl. Pal., ed. 2, 2: 727 (1933); N. L. Bor in C. C. Townsend et al.,
Fl. Iraq, 9: 278, t. 97 (1968); Osorio-Tafall et Seraphim, List Vasc. Plants Cyprus, 12
(1973).
[*M. vernale* M. Bieb. var. *montianum* (non (Parl.) Coss.) Boiss., Fl. Orient., 5: 510 (1884)
quoad spec. cypr.]
[*M. vernale* M. Bieb. ssp. *montianum* (non *M. montianum* Parl.) Holmboe, Veg. Cypr., 33
(1914); Osorio-Tafall et Seraphim, List Vasc. Plants Cypr., 12 (1973).]
TYPE: U.S.S.R.; "in Tauriae et Caucasi subalpini collibus" (LE).

Annual; culms erect, solitary or loosely fasciculate, leafy almost to the
panicle, glabrous, up to 45 cm. tall, smooth or scaberulous below the nodes;
leaf-blades flat, flaccid, green or yellowish-green, up to 10 cm. long, 1–3 mm.
wide, linear-lanceolate, acuminate, scabrid on the upper surface and on the
margins, glabrous on both surfaces, smooth below; ligule membranous, 2–3
mm. long; inflorescence an open panicle, up to 10 cm. long, 4–6 cm. broad;
branches fascicled or whorled, bare in the lower half or two-thirds, smooth,
ascending, sometimes horizontally divaricate, the upper part carrying the
shortly pedicellate spikelets; spikelets green or more usually tinged with
purple, elliptic-acute in shape, 2·5–3 mm. long; glumes equal, as long as the
spikelets, glabrous, scaberulous, broadly elliptic-acute when flattened, 3-
nerved, or with an extra very short lateral pair, the lower glume embracing
the upper; lemma slightly shorter than the glume, becoming chartaceous,
very smooth, shining, elliptic-obtuse in outline, cream-coloured or white;
palea of the same texture, narrow; anthers 2 mm. long.

HAB.: Shady ground by streams or roadsides at high elevations; 1,800–5,000 ft. alt.; fl.
April–May.

DISTR.: Division 2, [? 7]. S.E. Europe and eastern Mediterranean region from Turkey and
Palestine eastwards to Central Asia.

2. Xerokolymbos, S.W. of Khionistra, 1937, *Kennedy* 91 ! Papoutsa, above Palekhori, 1941,
Davis 3096 ! Tripylos, Paleokhori, 1962, *Meikle* 2666 ! Mavres Sykies, 1962, *Meikle* 2737 !

NOTES: The plant recorded as *M. vernale* from near Kantara Monastery (Sintenis in Oesterr.
Bot. Zeitschr., 32: 192; 1882) was most probably *M. pedicellare* (Bornm.) Roschev. ex Meld.,
which is known from this locality. There is no Sintenis material of *M. vernale* at Kew.

2. **M. pedicellare** (*Bornm.*) *Roschev. ex Meld.* in Rechinger f., Arkiv för Bot., ser. 2, 2: 291 (1952);
N. L. Bor in C. C. Townsend et al., Fl. Iraq, 9: 278, t. 98 (1968); Osorio-Tafall et Seraphim,
List Vasc. Plants Cypr., 12 (1973).
M. vernale M. Bieb. var. *pedicellare* Bornm. in Beih. Bot. Centralbl., 31 (2): 267 (1913);
Post, Fl. Pal., ed. 2, 2: 727 (1933).
[? *M. vernale* (non M. Bieb.) Sintenis in Oesterr. Bot. Zeitschr., 32: 192 (1882).]
TYPE: Lebanon; "Antilibanon, westl. Abhänge bei Baalbek, 12–1300 m. (Nr. 12925)"
Bornmüller (W !).

Annual grass; culms usually solitary, sometimes loosely tufted, very
slender, smooth and glabrous, simple, shining, up to 60 cm. tall; leaf-blades
linear-acuminate, flat, flaccid, dark-green, up to 8 cm. long, 2–4 mm. wide,
scaberulous on the nerves on the upper surface and on the margins, smooth
below; sheaths at first closely clasping, eventually loose, distinctly rough,
glabrous; ligule membranous, closely applied to the culm, 2–4 mm. long,
truncate, lacerate; panicle when fully developed very effuse, up to 15 cm.
long, 8–16 cm. wide; branches in pairs widely spreading at right-angles, bare
at the base, they and the branchlets and pedicels scabrid; pedicels up to 1
cm. long or less; spikelets elliptic-acute in outline, 3 mm. long, greyish-green
or pale green in colour; glumes subequal, the lower the broader and almost

enveloping the upper, 3-nerved, scaberulous on the nerves, hyaline towards the margins and at the tips. Grain 1·5–1·75 mm. long, elliptic-obtuse, very smooth.

HAB.: Shaded north-facing slopes on limestone mountains; c. 1,000–2,500 ft. alt.; fl. April–May.

DISTR.: Division 7, rare. Eastern Mediterranean region from Turkey to Palestine and eastwards to Iran.

7. Near Kalogrea, 1880, *Sintenis & Rigo* 394! Kantara road, near Kantara, 1952, *Merton* 759!

[3. M. EFFUSUM *L.*, Sp. Plant., ed. 1, 61 (1753); Sibth. et Sm., Fl. Graec. Prodr., 1: 44 (1806); Poech, Enum. Plant. Ins. Cypr., 7 (1842); Unger et Kotschy, Die Insel Cypern, 177 (1865); Boiss., Fl. Orient., 5: 510 (1884); H. Stuart Thompson in Journ. Bot., 44: 340 (1906); Holmboe, Veg. Cypr., 33 (1914); Osorio-Tafall et Seraphim, List Vasc. Plants Cyprus, 12 (1973).

TYPE: "*in* Europae *nemoribus umbrosis*" (LINN!).

A loosely tufted perennial; culms up to 150 cm. tall, erect, or somewhat decumbent at the base, smooth and glabrous; leaf-blades linear-acute, up to 30 cm. long, 5–15 mm. wide, glabrous, scaberulous on the margins; sheaths clasping, smooth and glabrous; ligules up to 10 mm. long, membranous; panicle very loose, of widely spreading branches, whorled at the nodes, few-spiculate; spikelets ovate to elliptic, 3–4 mm. long, on pedicels 1–3 mm. long; glumes persistent, greenish with white margins, scaberulous, 3-nerved, equal; lemma lanceolate to elliptic, as long as the glumes, finally indurated, very smooth and shining; palea as long as the lemma and of similar texture.

HAB.: "In insulae Cypri nemorosis" *Sibthorp* (see Notes).

DISTR.: ? Division. Widespread in Europe, temperate Asia and North America.

NOTES: Recorded only by Sibthorp, without precise indication of locality, and almost certainly an error, either of identification or localization. The species does not occur in any areas adjacent to Cyprus, and while it might possibly be re-found in moist, shaded places high up on the Troödos Range, its occurrence on the island must be reckoned very unlikely.]

TRIBE 8. **Agrostideae** *Dumort.* Spikelets 1-flowered, all alike, hermaphrodite, breaking up at maturity, rarely falling entire or attached to the pedicel; rhachilla disarticulating above the glumes, rarely tough, or rarely produced as a naked or hairy bristle; inflorescence effuse or dense, contracted and spike-like; glumes usually persistent, equal or more or less unequal, sometimes much compressed, awnless or awned, very rarely winged; lemma more delicate in texture than the glumes, glabrous or pilose, awned from a 2-lobed or entire tip, or from the back, or awnless; palea present or absent, 2-nerved or nerveless; lodicules 2, rarely absent; stamens 3, rarely 2, or very rarely 1. Grain often tightly enclosed between lemma and palea; embyo small; hilum punctiform; starch grains compound.

Slender annual or perennial grasses; leaf-blades linear-acuminate, very narrow or broad, flat or folded, with festucoid anatomy; silica-bodies elliptic oblong; 2-celled hairs absent. Chromosomes large; basic number 7. [Tribe probably best included in *Aveneae* Dumort., T.A.C. & R.D.M.]

E. Paunero, Las Agrostideas españolas, in Anal. Inst. Bot. Cav., 11 (1): 319–417 (1952).

31. CALAMAGROSTIS *Adans.*
Fam. Plant., 2: 31 (1763).

Perennial grasses, often tall, with long, flat leaf-blades; inflorescence an effuse or dense panicle; spikelets 1-flowered, pedicellate, hermaphrodite; rhachilla jointed below the lemma, produced or not; callus short- or long-bearded; glumes lanceolate-acuminate, equal or unequal, the lower 1-nerved, the upper 3-nerved; lemma hyaline, much shorter than the glumes, cleft or entire at the apex, 3–5-nerved, awned from the tip or just below or from the base; palea a hyaline scale, 2-nerved; stamens 3; lodicules 2, lanceolate, acuminate, fleshy at the base. Grain ellipsoid in outline, enclosed by the delicate lemma and palea; embryo small, about one quarter the length of the grain; hilum a punctiform dot, sub-basal.

About 80 species in temperate regions of both hemispheres, or on mountains in the tropics.

1. **C. epigejos** (*L.*) *Roth*, Tent. Fl. Germ., 1: 34 (1788); Boiss., Fl. Orient., 5: 525 (1884); A. K. Jackson in Kew Bull., 1937: 346 (1937); R. M. Nattrass, First List Cypr. Fungi, 20 (1937); Rechinger f. in Arkiv för Bot., ser. 2, 1: 418 (1950); N. L. Bor in C. C. Townsend et al., Fl. Iraq, 9: 299 (1968); Osorio-Tafall et Seraphim, List Vasc. Plants Cyprus, 12 (1973).

Arundo epigejos L., Sp. Plant., ed. 1, 81 (1753).

TYPE: "*in* Europae *collibus aridis*" (LINN!).

A tall, rhizomatous grass forming more or less dense tussocks; culms up to 180 cm. tall, usually stout, smooth and glabrous, except below the panicle which is scabrid; leaf-blades linear-acuminate, involute, rarely flat, 20–60 cm. long, 3–10 mm. wide, rather tough and rigid, rough on both surfaces particularly the upper, glabrous; sheaths closely clasping the culms, striate, glabrous, minutely scaberulous; ligules up to 10 mm. long, membranous above, somewhat chartaceous below; inflorescence a dense, lobed, lanceolate, linear or oblong panicle up to 45 cm. long, 3–4 cm. wide, pale, greenish or purple; axis and branches more or less tough; spikelets lanceolate, 5–7 mm. long, densely packed; glumes lanceolate-acuminate, the lower 1-nerved, slightly narrower than the 3-nerved upper glume, scabrid on the keels, hyaline; lemma hyaline, elliptic-oblong, 3-nerved, awned from the middle of the dorsal surface, 3–3·5 mm. long, surrounded by the fine callus hairs up to 6 mm. long, the two lateral nerves projecting above as two points; palea hyaline, 2-nerved, shorter than the lemma; awn 2 mm. long, straight, emerging from the back or emarginate tip of the lemma; anthers 2 mm. long.

HAB.: By sides of streams and roadside ditches at high altitudes on igneous formation; 4,500–5,400 ft. alt.; fl. July–Sept.

DISTR.: Division 2, locally frequent on Troödos Range. Widespread in Europe and Asia, south to N.W. India, with a distinct variety in tropical Africa and South Africa; introduced elsewhere.

2. Troödos, 5,000–6,400 ft. alt., 1912, *Haradjian* 495! Prodhromos, 1935, *Syngrassides* 800! Livadhi tou Pasha, 1936–1938, *Kennedy* 80! Trikoukkia, 1937, *Kennedy* 79! and, 1970, *A. Genneou* in ARI 1150! Between Amiandos and Troödos, 1940, *Davis* 1934! Lunata valley, 1951, *Chapman* s.n.! Livadhi tou Pasha, 1952, *Merton* 894!

32. ALOPECURUS *L.*

Sp. Plant., ed. 1, 60 (1753).

Gen. Plant., ed. 5, 30 (1754).

E. Paunero in Anal. Inst. Bot. Cav., 10 (2): 301–345 (1952).

Annual or perennial grasses with flat leaf-blades; inflorescence an oblong, elliptic or terete, spike-like panicle of many closely packed spikelets seated on and deciduous from the cupuliform tips of short pedicels; spikelets firmly compressed, 1-flowered, hermaphrodite; glumes subequal, compressed, often partly connate at the margins below, more or less ciliate on the keels; lemma hyaline, truncate, with a median or basal awn usually visible beyond the glumes, or rarely enclosed; palea present or absent; stamens 3. Grain compressed laterally, wedge-shaped in section, rounded on the dorsal (abaxial) surface, loosely enclosed in the lemma and glumes; no lodicules;

embryo about one-third the length of the grain; hilum basal, short, oblong or punctiform.

About 20 species widely distributed in temperate regions.

Annual grasses:
Panicle narrowly cylindrical, 3–10 cm. long, 5–8 mm. wide, often purplish; uppermost
    sheaths not inflated - - - - - - - - - **1. A. myosuroides**
Panicle ellipsoid or ovoid in outline, 2–4 cm. long, 1–1·5 cm. wide, green; uppermost leaf-
    sheath conspicuously inflated - - - - - - - **2. A. utriculatus**
Perennial grass; panicle cylindrical - - - - - - - 3. A. PRATENSIS

1. **A. myosuroides** *Huds.*, Fl. Angl., 23 (1762); A. K. Jackson in Kew Bull., 1937: 345 (1937); N.
    L. Bor in C. C. Townsend et al., Fl. Iraq, 9: 288, t. 102 (1968); Osorio-Tafall et Seraphim,
    List Vasc. Plants Cyprus, 12 (1973).
      A. agrestis L., Sp. Plant., ed. 2, 89 (1762); Boiss., Fl. Orient., 5: 485 (1884); Post, Fl. Pal.,
    ed. 2, 2: 719 (1933).
      A. agrestis L. var. *tonsus* Blanche ex Boiss., Fl. Orient., 5: 485 (1884).
      A. myosuroides Huds. var. *tonsus* (Blanche ex Boiss.) Lindberg f., Iter Cypr., 6 (1946).
    TYPE: England; *"in arvis, et ad vias"* Hudson.

An annual grass; culms loosely or somewhat densely tufted, up to 60 cm. tall, erect or decumbent at the base, smooth and glabrous; leaf-blades linear-acuminate, up to 15 cm. long, 2–6 mm. wide, glabrous, rough on both surfaces and on the margins; sheaths clasping the culm, except for the uppermost which is slightly inflated; ligules membranous, minutely toothed, 2–4 mm. long; inflorescence a dense, spike-like panicle up to 10 cm. long, 4–6 mm. wide, cylindrical, somewhat tapering towards the tip; spikelets very firmly laterally compressed, 1-flowered, falling entire from the very short pedicels, imbricate in the spike, 4·5–7 mm. long, green to purplish; glumes subequal, connate in the lower third by their margins, oblong or elliptic, acute, 3-nerved, narrowly winged on the keels, ciliate on the wing and keel and on the lateral nerves at the base; lemma as long as or slightly longer than the glumes, smooth and glabrous, 5-nerved, but the central nerve issuing from the dorsal surface just above the base; awn scabrid, 9–15 mm. long or rarely not exserted; anthers 3–4 mm. long.

HAB.: Cultivated and waste ground; margins of fields; 100–3,000 ft. alt.; fl. April.

DISTR.: Divisions 2, 6, rare. Widespread in Europe and temperate Asia, and introduced into N. America and other temperate regions.
2. Platres, 1936, *Syngrassides* 946! Kakopetria, 1939, *Lindberg f.* s.n.
6. Xeros (Lefka), 1937, *Syngrassides* 1564!

NOTES: The variety with a very short, inconspicuous awn (var. *tonsus* (Blanche ex Boiss.) Lindberg f.; var. *breviaristatus* Merch. ex Aschers. et Graebn.) is not uncommon in Europe. Lindberg's specimen has not been examined.

2. **A. utriculatus** *Banks et Sol.* in Russell, Nat. Hist. Aleppo, ed. 2, 2: 243 (1794); Sibth. et Sm.,
    Fl. Graec. Prodr., 1: 43 (1806), Fl. Graec., 1: 47, t. 63 (1808); Post, Fl. Pal., ed. 2, 2: 720
    (1933); N. L. Bor in C. C. Townsend et al., Fl. Iraq, 9: 292, t. 104 (1968); Osorio-Tafall et
    Seraphim, List Vasc. Plants Cyprus, 12 (1973).
      A. anthoxanthoides Boiss., Diagn., 1, 13: 42 (1853), Fl. Orient., 5: 486 (1884); Unger et
    Kotschy, Die Insel Cypern, 177 (1865); Holmboe, Veg. Cypr., 33 (1914).
    TYPE: Syria; near Aleppo, *Russell* (BM).

An annual grass; culms laxly fasciculate, slender, smooth and glabrous, erect or decumbent at the base, leafy, attaining 50 cm.; leaf-blades flat, flaccid, linear-acuminate, up to 10 cm. long, 2–8 mm. wide, minutely scabrid on both surfaces and on the margins; sheaths rather loose, the uppermost often considerably inflated; ligule membranous, 2–4 mm. long, rounded; panicle ovate or oblong in outline, 1–4 cm. long, 1–1·5 cm. broad, of densely imbricating spikelets; spikelets firmly compressed, 6–6·5 mm. long, narrowly elliptic-acute in shape or slightly gaping; glumes equal, keeled, green or purplish, connate by the margins in the lower third or half, narrowly winged on the keel in the upper third, scabrid or ciliate on keel, ciliate on the connate portion or rarely pilose on the outer sides; lemma as long as the glumes, narrowly oblong-acute, smooth and glabrous, faintly 4-nerved, the fifth nerve issuing from the dorsal surface at the base as an awn reaching 12 mm. in length, of which the lower half is a twisted column; palea absent; anthers 4 mm. long.

HAB.: Crevices of limestone rocks, 250–400 ft. alt.; fl. March–April.

DISTR.: Division 4, rare. Turkey, Syria, Palestine, Iraq, Iran.

4. Ayia Napa, 1862, *Kotschy* 130; same locality, 1905, *Holmboe* 55; Ayios Antonios, Sotira, 1950, *Chapman* 593!; same locality, 1951, *Merton* 165! 651! 973!

NOTES: *Alopecurus utriculatus* (L.) Pers. (1805) is a later homonym of the above-named plant. It was given a new name (*A. rendlei* Eig) in Journ. Bot., 75: 187 (1937).

[3. A. PRATENSIS *L.*, Sp. Plant., ed. 1, 60 (1753); Sibth. et Sm., Fl. Graec. Prodr., 1: 42 (1806); Unger et Kotschy, Die Insel Cypern, 177 (1865); H. Stuart Thompson in Journ. Bot., 44: 340 (1906); Post, Fl. Pal., ed. 2, 2: 720 (1933).
TYPE: "*in* Europae *pratis*" (LINN!).

A perennial with widely creeping stolons; culms loosely or densely tufted, erect or decumbent at the base, up to 120 cm. tall, smooth and glabrous, slender to moderately stout; leaf-blades linear-acuminate, flat, smooth and glabrous, scabrid on the margins, up to 20 cm. long, 4–8 mm. broad; sheaths smooth, cylindrical, the upper slightly inflated, smooth and glabrous; ligule up to 2·5 mm. long, membranous; panicle cylindrical, spike-like, obtuse at the tip, very dense, soft, 2–9 cm. long, 5–10 mm. wide, long-exserted; branches very short, erect; spikelets very shortly pedicellate, lanceolate-oblong or elliptic, 4–6 mm. long, firmly laterally compressed, 1-flowered, deciduous from the pedicels; glumes, seen from the side, narrowly lanceolate-acute with the margins connate towards the base, 3-nerved, finely ciliate on the keels, hairy on the lateral nerves and sometimes between the nerves and towards the margins; lemma as long as or slightly longer than the glumes, elliptic-acute when flattened, keeled with margins connate below, membranous, smooth, 5-nerved, the central nerve issuing from the back as a scabrid awn 7–9 mm. long; palea absent; 2–3·5 mm. long.

Recorded by Sibthorp "Circa Athenas, et in insulâ Cypro", but as Holmboe (Veg. Cypr., 33) remarks, its occurrence in Cyprus seems very dubious, especially as it has not been noted or collected since. The species is widely distributed in Europe and the eastern Mediterranean region, as far east as Greece, but is not known to occur in any of the areas adjacent to Cyprus.]

33. PHLEUM *L.*

Sp. Plant., ed. 1, 59 (1753).
Gen. Plant., ed. 5, 29 (1754).

Annual or perennial grasses with linear, flat leaf-blades; inflorescence a terete, spike-like panicle of many, closely imbricate, shortly pedicellate spikelets; spikelets firmly compressed, 1-flowered, hermaphrodite, falling entire from the pedicels; rhachilla disarticulating above the glumes, rarely produced; glumes equal, abruptly awned or mucronate or gradually acute, firm but hyaline on the margins, often coarsely ciliate on the keels, rarely quite glabrous; lemma shorter than the glumes, truncate or bluntly acute, awnless; palea slightly shorter than the lemma, 2-nerved; stamens 3; lodicules 2, cuneate or oblong and 2-lobed or irregularly toothed on the truncate margin. Grain ellipsoid, enclosed by the indurated lemma and palea; embryo one-fifth to one-sixth the length of the grain; hilum basal, punctiform.

About 12 species in temperate regions of Europe, Asia and both Americas.

Spikelets elliptic in outline, closely imbricated, giving the inflorescence a smooth appearance
1. P. subulatum

Spikelets oblong-obovate in outline, less closely imbricated, giving the inflorescence a rough
appearance - - - - - - - - - - - - - 2. P. PANICULATUM

1. **P. subulatum** *(Savi) Aschers. et Graebn.*, Syn. Mitteleur. Fl., 2: 154 (1899); Post, Fl. Pal., ed.
      2, 2: 718 (1933); Osorio-Tafall et Seraphim, List Vasc. Plants Cyprus, 12 (1973).
      Phalaris subulata Savi, Fl. Pisana, 1: 57 (1798).
      Phleum tenue Schrad., Fl. Germ., 1: 191 (1806); Boiss., Fl. Orient., 5: 480 (1884).
   TYPE: Italy; "Nel Monte Pisano", *Savi* (PI).

An annual grass with fascicled, erect or geniculately ascending culms up to 30 cm. tall, smooth and glabrous, often with purple nodes; leaf-blades flat, up to 10 cm. long, 2–3 mm. wide, smooth and glabrous on both surfaces, scaberulous on the margins, flat, green; sheaths rounded on the back, clasping the culms, except the uppermost which is somewhat inflated, smooth and glabrous or minutely scaberulous; ligule up to 4 mm. long, somewhat pointed; inflorescence a dense, cylindrical, spike-like panicle tapering at both ends, of closely packed, shortly pedicellate spikelets, 3–6 cm. long, 4–5 mm. wide; spikelets 2·5–3 mm. long, elliptic or oblong, acute in shape seen from the side; glumes as long as the spikelets, firmly compressed, mucronate, joined below, 3-nerved, ciliate or scabrid on the keels; lateral nerves raised and close to the keel nerve; margins membranous; lemma about half as long as the glumes, when flattened almost orbicular, 5-nerved, denticulate on the almost truncate upper margin, somewhat scaberulous; palea almost as long, 2-nerved, much narrower; anthers 1·8 mm. long.

   HAB.: In garigue, or in pastures on limestone; sea-level to 150 ft. alt.; fl. Febr.–April.

   DISTR.: Division 6, rare. Eastern Mediterranean region from Aegean Islands to Egypt; sometimes as a casual around ports in other parts of the Mediterranean region or in western Europe.

   6. S. of Liveras, 1956, *Merton* 2616! Near Orga, 1956, *Merton* 2648! and, 1957, *Merton* 2842.

2. P. PANICULATUM *Huds.*, Fl. Angl., 23 (1762).
      P. asperum Jacq., Collectanea, 1: 110 (1786); Unger et Kotschy, Die Insel Cypern, 177
      (1865); Boiss., Fl. Orient., 5: 481 (1884); H. Stuart Thompson in Journ. Bot., 44: 340
      (1906).
   TYPE: England; "in pratis infra King's Weston prop. Bristolium" *Hudson*.

An annual with geniculately ascending, fascicled, rarely solitary, erect culms, up to 30 cm. tall, smooth and glabrous, simple; leaf-blades flat, lanceolate-acuminate or linear-acuminate, flaccid, green, almost or entirely smooth on both surfaces, minutely scaberulous on the margins, up to 10 cm. long, 3–4 mm. wide; sheaths short, clasping, or the uppermost slightly inflated, smooth and glabrous; ligule membranous, 2–4 mm. long; inflorescence a very dense, cylindrical, spike-like, terminal panicle 2–8 cm. long, of very shortly pedicellate spikelets, seated on very short branches; spikelets 2–3 mm. long, laterally compressed, oblong-obovate seen from the side, often inflated in the upper half; glumes equal, obovate when flattened, mucronate, keeled, ciliate or scabrid on the keels, 3-nerved; lemma very much smaller, only 1–1·5 mm. long, hyaline, oblong-obtuse when flattened, faintly 5-nerved; palea much narrower, 2-nerved; stamens very small, less than 1 mm. long.

   HAB.: In cultivated fields near irrigation ditches; near sea-level; fl. April–May.

   DISTR.: Division 3. Widespread in S. Europe and temperate Asia; and as a casual elsewhere.

   3. Episkopi, 1862, *Kotschy* s.n.; Kouklia, 1862, *Kotschy* 616a.

NOTES: Not listed by any other collector, and perhaps, as Holmboe (Veg. Cypr., 33) suggests, an error. The specimens have not been examined. *P. paniculatum*, though recorded (questionably) from Crete, is generally absent from the eastern Mediterranean region.

34. MAILLEA *Parl.*
Plant. Nov., 31 (1842).

Dwarf annual with very narrow, folded leaf-blades; inflorescence a very dense, ellipsoid panicle of closely packed spikelets; spikelets all alike, hermaphrodite, 1-flowered, very firmly laterally compressed; glumes equal, semi-elliptic in outline, navicular, keeled, with a firm cartilaginous scabrid wing on the keel, equal in width to one half the glume; lemma very much smaller than the glumes, when flattened broader than long, truncate, faintly 5-nerved; palea shorter and much narrower, ovate-obtuse, 2-nerved; lodicules absent; stamens 3; styles 2; stigmas plumose. Caryopsis globose in outline, somewhat laterally compressed; embryo a little less than half the length of the grain; hilum punctiform, basal.

One species in Sardinia, Greece, Crete, Aegean Islands and Cyprus.

1. **M. crypsoides** (*Urv.*) *Boiss.*, Fl. Orient., 5: 479 (1884).
   Phalaris crypsoides Urv. in Mém. Soc. Linn. Par., 1: 263 (1822).
   Maillea urvillei Parl., Plant. Nov., 32 (1842).
   Phleum crypsoides (Urv.) Hackel in Bull. Soc. Bot. Fr., 39: 274 (1892).
   TYPE: Greece; "*Cl. D'Urville* legit in scopulo Raphti ad littus atticum, ubi copiose invenitur, ut ipse cl. auctor me monuit" (FI, K!).

An annual grass; culms in robust plants not more than 2·5 cm. tall, usually shorter, covered with leaf-sheaths to the base of the panicle, smooth and glabrous, with short internodes; leaf-blades linear-acute, 2 cm. long or less, 1–1·5 mm. wide, folded, recurved, scaberulous on both surfaces and on the margins; leaf-sheaths lax, smooth or scaberulous, scarious on the margins, the uppermost much inflated and surrounding the base of the inflorescence; ligule membranous, elliptic, 1 mm. long; panicle very dense, ellipsoid, up to 1 cm. long, 8 mm. wide, the lower part sunk in the uppermost leaf-sheaths; axis thick; pedicels very short, thick; spikelets imbricate, pale, flattened laterally, 3·5 mm. long, elliptic in outline; glumes equal, as long as the spikelet, broadly winged on the keel, apiculate at the apex, coarsely scabrid on the keel, smooth and glabrous or minutely scabrid on the sides; lemma c. 1 mm. long, membranous, twice as wide as long, faintly 5-nerved, truncate; palea shorter, hyaline, 2-nerved; anthers 1·5 mm. long.

HAB.: In rock-clefts by the sea; sea-level; fl. March.

DISTR.: Division 1, rare. General distribution that of the genus.

1. Tip of Cape Arnauti, 1962, *Meikle* 2305!

35. AGROSTIS *L.*
Sp. Plant., ed. 1, 61 (1753).
Gen. Plant., ed. 5, 30 (1754).
E. Paunero in Anal. Inst. Bot. Cav., 7: 561–644 (1948).

Annual or perennial grasses with flat, involute or convolute leaf-blades; inflorescence an effuse or contracted panicle with numerous small spikelets seated on short pedicels on branches often arranged verticillately; spikelets 1-flowered, hermaphrodite; rhachilla sometimes produced as a naked or penicillate bristle; glumes more or less equal, 1-nerved; lemma more delicate, hyaline, elliptic-oblong, truncate, 5-nerved, the lateral nerves sometimes excurrent, awnless or awned from the middle or base of the dorsal surface, glabrous or hairy; palea hyaline, 2-nerved or nerveless, sometimes absent altogether; stamens 3; lodicules 2, lanceolate. Grain free, more or less dorsally compressed or somewhat terete; embryo small, about one quarter the length of the grain; hilum punctiform or shortly oblong, basal.

About 100 species widely distributed in temperate regions.

Plants stoloniferous, without rhizomes -   -   -   -   -   -   -   **1. A. stolonifera**
Plants rhizomatous, without stolons   -   -   -   -   -   -   -   **2. A. cypricola**

1. **A. stolonifera** *L.*, Sp.Plant., ed. 1, 62 (1753); Sibth. et Sm., Fl. Graec. Prodr., 1: 46 (1806); N. L. Bor in C. C. Townsend et al., Fl. Iraq, 9: 286 (1968); Osorio-Tafall et Seraphim, List Vasc. Plants Cyprus, 12 (1973).
    [*A. alba* (non L.) Sowerby et Sm., Eng. Bot., 17: t. 1189 (1803); Boiss., Fl. Orient., 5: 514 (1884); Post, Fl. Pal., ed. 2, 2: 729 (1933).]

A perennial grass up to 40 cm. tall, emitting numerous leafy stolons at the base; culms tufted, usually decumbent at the base, finally erect, rooting at the lower nodes, slender, simple, smooth and glabrous; leaf-blades linear-acuminate, convolute when young, finally flat, attaining 10 cm. in length, 1–4 mm. wide, rough on both surfaces and on the margins; sheaths tight, striate, smooth or slightly rough; ligule rounded, membranous, 2–5 mm. long; panicle pyramidal or cylindrical, up to 30 cm. long, with verticillate branches spreading horizontally at flowering-time but erect when fruiting; spikelets fusiform or lanceolate in outline, 2–3 mm. long, 1-flowered; rhachilla disarticulating above the glumes; glumes more or less equal, the lower 1-, the upper 3-nerved, as long as the spikelet, the lower usually the longer, broadly or narrowly lanceolate when flattened, minutely scabrid on the keels towards the tips or sometimes all over; lemma very broadly elliptic-truncate, 5-nerved, each nerve carried out as a minute lobe at the upper margin, about two-thirds as long as the glumes; palea shorter than the lemma, 2-nerved, hyaline; anthers 1–1·5 mm. long.

var. **scabriglumis** (*Boiss. et Reut.*) *C. E. Hubb.* in A. W. Hill (ed.), Fl. Trop. Afr., 10 (1): 172 (1937).
    A. scabriglumis Boiss. et Reut., Pugill. Plant. Nov., 125 (1852).
    A. alba L. var. *scabriglumis* (Boiss. et Reut.) Boiss., Fl. Orient., 5: 514 (1884).
    ? *A. alba* L. var. *densiflora* Guss., Suppl. Fl. Sic. Prodr., 15 (1832).
    TYPE: Spain; "ad rivulos et aquas montis *Sierra Nevada* (Boiss.)" (G).

Glumes scabrid on the sides and keel.

HAB.: Wet ground by springs and streams at high altitudes on igneous formations; 5,000–5,600 ft. alt.; fl. July–Aug.

DISTR.: Division 2. Widespread in the Mediterranean region.

2. Asprokremnos, S. of Khionistra, 1937, *Kennedy* 97 ! Kryos Potamos, 5,600 ft. alt., 1938, *Kennedy* 1099 ! Kannoures Springs, 1955, *Kennedy* 1869 ! and, 1959, *Casey* 1669 !

NOTES: *Agrostis stolonifera* was also recorded by Sintenis (Oesterr. Bot. Zeitschr., 31: 326; 1881) from the neighbourhood of Kythrea, but the plant collected there was almost certainly *Polypogon semiverticillatus* (Forssk.) Hyl. (*Agrostis verticillata* Vill.), to which the record is here referred.
    The rare intergeneric hybrid, × *Agropogon littoralis* (Sm.) C. E. Hubb. in Journ. Ecol., 33: 333 (1946), and in Hutchinson, British Flowering Plants, 327 (1948), a sterile plant resulting from the crossing of *Agrostis stolonifera* L. and *Polypogon monspeliensis* (L.) Desf., has been found on several occasions in the British Isles. Since the two parents grow in Cyprus, collectors should look out for this interesting hybrid. It is usually a definite perennial which differs from the *Polypogon* parent by its persistent spikelets, and from *Agrostis* by the emarginate, awned glumes.

2. **A. cypricola** *Lindberg f.* in Soc. Sci. Fenn. Årsb., B., no. 7: 5 (1942), Iter Cypr., 6, fig. 2 (1946).
    TYPE: Cyprus; "Boghazi, in ripa rivuli in valle supra opp. Kyrenia" *Lindberg f.* (H !).

A perennial grass with a strong rhizome; culms erect, up to 70 cm. tall, smooth and glabrous, simple; leaf-blades linear-acuminate, flat, glaucous, 15–20 cm. long, 2–4 mm. wide, scabrid on both surfaces and on the margins, ascending not spreading; sheaths clasping, striate, smooth and glabrous; ligule 4–5 mm. long, rounded or lacerate at the tip; panicle rather dense, pale, 5 cm. long, 1·5–2 cm. wide; branches more or less fascicled, 1–1·5 cm. long, branched and rebranched, scaberulous; spikelets fusiform, finally gaping, 2 mm. long; glumes as long as the spikelets, elliptic-acute, 1-nerved, scabrid on the keel; lemma 1·5 mm. long, elliptic-truncate when flattened, faintly 5-nerved, glabrous; middle nerve minutely produced; palea 2-nerved, 1 mm. long; anthers 1 mm. long.

HAB.: Moist streamside, c. 1,250 ft. alt.; fl. May–June.

DISTR.: Division 7, very rare. Endemic.

7. Boghazi above Kyrenia [Kyrenia Pass], 1939, Lindberg f. s.n.!

NOTES: The base of the plant emerges from what appears to be a strong rhizome. Apart from this feature it resembles *A. stolonifera* very closely in respect of foliage, ligule, panicle and spikelets. So much so indeed, that one suspects that the rhizome may be a stolon which has become buried in sand. Oddly enough, the species has only been collected on one occasion. I suspect this gathering to be *A. stolonifera* but maintain it as a species on account of the obvious rhizome. Should this, indeed, prove to be a buried stolon, the name will merge in that of *A. stolonifera*.

36. POLYPOGON *Desf.*

Fl. Atlant., 1: 66 (1798).

Annual or perennial grasses with flat leaf-blades; inflorescence a very dense, bristly panicle of very small, awned, rarely awnless, spikelets; spikelets with one hermaphrodite floret, when mature falling with the pedicel or a portion thereof; callus below the glumes slightly swollen; glumes subequal, rounded on the back, narrowly oblong, entire at the apex and awnless, or emarginate and then with the median nerve carried out as a scabrid awn; lemma about half as long as the glumes, elliptic in outline, truncate at the top, 5-nerved, hyaline and shining at maturity, awnless or the midnerve produced as a very short awn; palea hyaline, 2-nerved; stamens 3; lodicules 2, oblong, ending in 2 teeth, one of which is longer than the other. Grain obliquely ellipsoid in shape, concave on the adaxial surface; embryo less than half the length of the grain; hilum basal, punctiform.

About 8–10 species in the Mediterranean region and S.W. Asia; naturalized elsewhere.

Glumes acute or obtuse, awnless - - - - - - - **3. P. semiverticillatus**
Glumes emarginate, long-awned:
  Margins of glumes long-ciliate at the tip - - - - - **1. P. maritimus**
  Margins of the glumes not long-ciliate at the tip- - - - - **2. P. monspeliensis**

1. **P. maritimus** *Willd.* in Ges. Naturf. Fr. Berlin, Neue Schrift, 3: 442 (1801); Sintenis in Oesterr. Bot. Zeitschr., 32: 121 (1882); Boiss., Fl. Orient., 5: 520 (1884); Holmboe, Veg. Cypr., 33 (1914); Post, Fl. Pal., ed. 2, 2: 731 (1933); Lindberg f., Iter Cypr., 8 (1946); N. L. Bor in C. C. Townsend et al., Fl. Iraq, 9: 314, t. 114 (1968); Osorio-Tafall et Seraphim, List Vasc. Plants Cyprus, 12 (1973).
    P. subspathaceus Req. in Ann. Sci. Nat., ser. 1, 5: 386 (1825).
    P. maritimus Willd. var. *subspathaceus* (Req.) Duby in DC., Bot. Gall., 1: 508 (1828); Lindberg f., Iter Cypr., 8 (1946).
    TYPE: France; La Rochelle, "Ich erhielt diese Art vom Herrn. Dr. Bonpland, der sie zu La Rochelle am Meerestrade gefunden hat" (B).

A dwarf, annual grass with fascicled, erect or geniculately ascending, slender, smooth and glabrous culms, up to 45 cm. tall; leaf-blades linear-acute, up to 15 cm. long, 1–2·5 mm. wide, glabrous, scabrid on both surfaces and on the margins, flat, twisted, rolled or folded when dry; sheaths short, smooth and glabrous, rather lax, the uppermost sometimes grotesquely swollen and containing the base of the inflorescence; ligule membranous, up to 5 mm. long, often fimbriate; inflorescence an oblong-cylindrical, sometimes lobed, bristly panicle of closely packed spikelets, 1–6 cm. long; spikelets 1·5–3 mm. long, at first oblong in outline, finally gaping; glumes equal, shortly connate below, hyaline and shining in the upper two-thirds, somewhat firmer in the lower third which is usually armed with more or less stiff bristles, 2-lobed at the tip; lobes ciliate; awned from the fissure with a straight awn up to 7 mm. long, which is sometimes purplish in colour; lemma one-third the length of the glumes, hyaline, faintly 5-nerved, awnless, rarely with a short straight awn; anthers 0·3–0·4 mm. long.

HAB.: On stony or sandy seashores; in salt-marshes, or on cultivated ground, usually in damp situations, by ditches or muddy pools; sea-level to 5,500 ft. alt.; fl. April–July.

DISTR.: Divisions 2–4, 6, 7. Eastern Mediterranean region and eastwards to Central Asia; Atlantic Islands; naturalized in N. America, S. Africa and Australasia.

2. Livadhi tou Pasha, 1938, *Kennedy* 1097! 1939, *Lindberg f.* s.n. and, 1955, *Merton* 2373! Kannoures Springs, 1939, *Lindberg f.* s.n.

3. Akrotiri, 1905, *Holmboe* 686; Perapedhi, 1938, *Kennedy* 1096!

4. Near Famagusta, 1939, *Lindberg f.* s.n.; Mile 3, Larnaca-Famagusta road, 1950; *Chapman* 431! Larnaca Salt Lake, 1967, *Merton* in ARI 411a!

6. Nicosia, 1950, *Chapman* 686! Karpasha Forest, 1956, *Merton* 2574!

7. Near Kythrea, 1880, *Sintenis & Rigo*; Kambyli, 1956, *Merton* 2732! and, 1957, *Merton* 2980.

NOTES: Some authors have considered *P. subspathaceus* specifically distinct from *P. maritimus*, differing in its laxer, ovate panicle, partly immersed in the uppermost leaf-sheath, and in its glumes, more deeply notched at the apex, and scaly at the base. The author does not accept these as valid distinctions and would, at most, regard *P. subspathaceus* as a form of *P. maritimus*. Apart from the *Lindberg f.* specimen from near Famagusta, three others: *Merton* 2732, 2574 and *Merton* in ARI 411a (cited above) might also be referred to *P. subspathaceus*.

2. P. monspeliensis (*L.*) *Desf.*, Fl. Atlant., 1: 67 (1798); Unger et Kotschy, Die Insel Cypern, 178 (1865); Sintenis in Oesterr. Bot. Zeitschr., 32: 293 (1882); H. Stuart Thompson in Journ. Bot., 44: 340 (1906); Holmboe, Veg. Cypr., 33 (1914); Post, Fl. Pal., ed. 2, 2: 730 (1933); Lindberg f., Iter Cypr., 8 (1946); N. L. Bor in C. C. Townsend et al., Fl. Iraq, 9: 318, t. 115 (1968); Osorio-Tafall et Seraphim, List Vasc. Plants Cyprus, 12 (1973).

Alopecurus monspeliensis L., Sp. Plant., ed. 1, 61 (1753).

Phleum crinitum Schreb., Beschr. Graes., 1: 151 (1769); Sibth. et Sm., Fl. Graec. Prodr., 1: 42 (1806), Fl. Graec., 1: 46, t. 62 (1808).

TYPE: France; "Monspelii" (LINN!).

An annual grass forming small tufts; culms erect or more usually geniculately ascending, up to 80 cm. tall, simple or sometimes branched at the base, smooth and glabrous but scabrid below the panicle; leaf-blades linear-acuminate, flat, up to 20 cm. long, 2–8 mm. wide, scabrid on the upper surface and on the margins, smooth below, glabrous; sheaths clasping, the uppermost somewhat inflated, smooth and glabrous or slightly scaberulous towards the top; ligule membranous, 5–12 mm. long, denticulate on the upper margin; inflorescence 2–15 cm. long, an oblong or elliptic, occasionally lobed, bristly panicle of closely packed, subsessile spikelets; spikelets deciduous from the denticulately cup-tipped, short pedicels, narrowly oblong, finally gaping, 2–3 mm. long; glumes equal, oblong-elliptic, emarginate at the apex, 1-nerved, minutely ciliate on the margins, scaberulous on the dorsal surface, awned from the notch; awn straight, fine, 4–7 mm. long; lemma 1–1·5 mm. long, broadly elliptic-truncate, 5-nerved, the central nerve produced as a short straight awn or awnless, hyaline, smooth and glabrous; palea much narrower, 2-nerved; anthers minute, 0·4–0·5 mm. long.

HAB.: Damp, muddy ground by streams, pools or roadsides; sea-level to 800 (– 2,000) ft. alt.; fl. April–June.

DISTR.: Divisions 3–8; W. and S. Europe, Mediterranean region, Atlantic Islands and throughout temperate Asia; occurring as a weed in warmer areas almost throughout the world.

3. Mazotos, 1862, *Kotschy* 560; near Stavrovouni Mon., 1939, *Lindberg f.* s.n.; Kolossi Castle, 1941, *Davis* 3532k!
4. Ayia Napa, 1862, *Kotschy* 107; Larnaca, 1877, *Post* 671! 672! Kouklia, 1905, *Holmboe* 384; Famagusta, 1912, *Haradjian* 113! and, 1939, *Lindberg f.* s.n.; Chali near Famagusta, 1962, *Meikle* 2621!
5. Near Kythrea, 1880, *Sintenis & Rigo* 378! Mile 14, Nicosia-Limassol road, 1950, *Chapman* 540!
6. Skylloura, 1936, *Syngrassides* 1200! West of Orga, 1970, *Chicken* 179!
7. Panagra, 1934, *Syngrassides* 1354! Kyrenia, 1939, *Lindberg f.* s.n.!
8. Kamaeres ["Camares"] 1862, *Kotschy*; Ayios Andronikos, 1880, *Sintenis & Rigo*; Galatia, 1962, *Meikle* 2536!

NOTES: Druce (Rep. B.E.C., 9: 471; 1931) mentions a var. *"purpurascens"* of this species, but there is no diagnosis for the variety, nor have I been able to trace any publication of the name.

3. **P. semiverticillatus** (*Forssk.*) *Hyl.* in Uppsala Univ. Årsbok, no. 7: 74 (1945); N. L. Bor in C. C. Townsend et al., Fl. Iraq, 9: 318 (1968); Osorio-Tafall et Seraphim, List Vasc. Plants Cyprus, 12 (1973).
   Phalaris semiverticillatus Forssk., Fl. Aegypt.-Arab., 17 (1775).
   Agrostis verticillata Vill., Prosp. Plant. Dauph., 16 (1779); Boiss., Fl. Orient., 5: 513 (1884); Holmboe, Veg. Cypr., 33 (1914); Post, Fl. Pal., ed. 2, 2: 729 (1933); Lindberg f., Iter Cypr., 6 (1946); Rechinger f. in Arkiv för Bot., ser. 2, 1: 418 (1950).
   A. semiverticillata (*Forssk.*) Christens. in Dansk Bot. Archiv., 4: 12 (1922).
   [*A. stolonifera* (non L.) Sintenis in Oesterr. Bot. Zeitschr., 31: 326 (1881).]
   TYPE: Egypt; *"Rosettae & Kahirae* frequens, Aprile ineunte florens" *Forsskål* (C!).

An annual, sometimes perennial, grass, with long spreading stolons rooting at the nodes; culms prostrate at the base, finally erect, up to 50 cm. tall, smooth and glabrous, simple; leaf-blades linear-acute, up to 15 cm. long, 3–10 mm. wide, scabrid on both surfaces and on the margins; sheaths clasping, smooth and glabrous; ligule membranous, rounded, 2–4 mm. long; panicle erect, ovoid or ellipsoid, mostly lobed, rather dense, up to 15 cm. long, 1–4 cm. wide; spikelets 1-flowered, oblong or fusiform, 1·5–2·2 mm. long, slightly gaping, falling with the pedicel attached; glumes equal, elliptic, acute or obtuse when flattened, the lower 1-, the upper 3-nerved, membranous, scabrid all over; lemma broadly elliptic-truncate, 5-nerved, hyaline; palea nearly as long as the lemma, 2-nerved; anthers 0·5–0·7 mm. long. Grain pale brown, 1 mm. long.

HAB.: Damp situations, by ditches and irrigation channels; occasionally on waste ground, or by roadsides or in gardens; sea-level to 4,600 ft. alt.; fl. April–Nov.

DISTR.: Divisions 1, 2, 6–8, locally abundant. S. Europe, Mediterranean region eastwards to Central Asia; Atlantic Islands; Ethiopia, Eritrea. Introduced into S. Africa, Australia, N. & S. America, and parts of Central Europe.
1. Stroumbi, 1913, *Haradjian* 775!
2. Locally abundant about Troödos, Prodhromos and Kakopetria; *Sintenis & Rigo* 916! *Chiotellis* 462! *G. E. Atherton* 528! 665! *D. P. Young* 7300!
6. Nicosia, 1950, *Chapman* 620!
7. Locally abundant about Kythrea, Lapithos, Kyrenia, Myrtou, Kambyli, Larnaka tis Lapithou; *Sintenis & Rigo* 387! *Syngrassides* 29! *Meikle* 2558! *Merton* in ARI 33!
8. Ayios Theodhoros, 1970, *A. Genneou* in ARI 1473!

NOTES: In Bull. Soc. Bot. Fr., 89, Session Extraordinaire, 110: 56–58 (1963), Breistroffer published *Polypogon viridis* (Gouan) Breistr. as a new combination based on the name *Agrostis viridis* Gouan (Hort. Reg. Monsp., 546; 1762), claiming that this is an earlier name for the plant currently known as *Polypogon semiverticillatus* (Forssk.) Hyl. and based on *Phalaris semiverticillata* Forssk. (Fl. Aegypt.-Arab., 17; 1775). This assumption is founded entirely on the literature written 200–250 years ago, since the author of the new combination admits (in litt.) that he has not seen the type specimen of *Agrostis viridis* Gouan, nor does he know where it is to be found. The type may well be in Herb. Kew., since Gouan's herbarium and papers were bought by Sir W. Hooker, through the intermediacy of Bentham in 1821, the year of Gouan's death. Gouan's specimens are disseminated throughout the herbarium, but the type of *Agrostis viridis* Gouan has not, after intensive search, been traced. Gouan gives the briefest of descriptions for *Agrostis viridis* but refers to Scheuchzer (Agrostographia 130; 1719) which Breistroffer says is generally accepted as a description of *Agrostis semiverticillata* (Forssk.) Christens.

In Fl. Suecica, 22 (1755) Linnaeus cites the Scheuchzer phrase-name as a synonym of *Agrostis stolonifera* L. var. β. In Sp. Plant., ed. 2, 93 (1762) the Scheuchzer name becomes the sole basis of *A. stolonifera* L. var. β. Breistroffer concludes that two distinct plants are involved, that of Fl. Suecica being a variety of *A. stolonifera*, that of Sp. Plant., ed. 2 being *Polypogon viridis* (or *P. semiverticillatus*).

Three years after Gouan had described *Agrostis viridis* he reduced it to a variety of *Agrostis stolonifera* (Fl. Monsp., 118; 1765). In 1796 (Herborisations Env. Monp., 7) he seems to have identified his species with *A. stolonifera* L. Breistroffer says that *Agrostis viridis* is common in Languedoc, but Bentham, who toured there in 1825, has not listed this species in his catalogue of 1826. This is all the more strange since Bentham and Gouan were well acquainted with one another. Moreover Bentham does not seem to have collected this plant if, indeed, it is different from *A. stolonifera* L., athough he collected in the neighbourhood of Montpellier for 5 or 6 years.

In view of these uncertainties, and in the absence of the type of *Agrostis viridis* Gouan, the author prefers to regard the latter name as a *nomen dubium*. It is clear that neither Gouan nor Linnaeus was sure of the identity of *Agrostis viridis*.

37. GASTRIDIUM *P. Beauv.*

Ess. Agrost., 21 (1812).

Annual grasses with flat leaf-blades. Inflorescence a narrow, dense, oblong panicle, tapering towards the tip, of many spikelets shortly pedicellate on the short, erect, scabrid branches; pedicels slightly widened at the tip; spikelets with one bisexual floret; rhachilla breaking up above the glumes, very shortly produced and pilose; glumes somewhat unequal with firm, globular, shining bases, the rest keeled, awl-shaped, membranous, both 1-nerved, very rarely the upper 3-nerved and then only faintly so; lemma much shorter, broadly elliptic-truncate when flattened, 5-nerved, with the median nerve issuing from the dorsal surface as a perfect awn twisted at the base, or occasionally awnless, sparsely or densely pilose; palea hyaline, 2-keeled; stamens 3; styles 2; stigmas plumose; lodicules 2, extremely delicate, oblong, truncate at the base, tapering above the middle to an acute or obtuse tip. Grain ellipsoid, tightly enclosed between the lemma and palea, shallowly grooved on the adaxial surface; palea somewhat adherent; embryo orbicular, basal, small; hilum sub-basal, punctiform.

Two (or three) species in the Mediterranean region and western Asia.

1. **G. phleoides** (*Nees et Mey.*) *C. E. Hubb.* in Kew Bull., 9: 375 (1954); N. L. Bor in C. C. Townsend et al., Fl. Iraq, 9: 302, t. 110 (1968); Osorio-Tafall et Seraphim, List Vasc. Plants Cyprus, 12 (1973).
   Lachnagrostis phleoides Nees et Mey. in Nov. Act. Acad. Caes. Leop. Carol., 19, Suppl. 1: 146 (1843).
   [*Milium lendigerum* (non L.) Sm. in Sibth. et Sm., Fl. Graec. Prodr., 1: 44 (1806), Fl. Graec., 1: 45, t. 65 (1808).]
   [*Gastridium ventricosum* (non (Gouan) Schinz et Thell.) A. K. Jackson in Kew Bull., 1937: 346 (1937); Osorio-Tafall et Seraphim, List Vasc. Plants Cyprus, 12 (1973).]
   [*Gastridium lendigerum* (non (L.) Desv.) Lindberg f., Iter Cypr., 7 (1946).]
   TYPE: Chile; "Circa Valparaiso in republica Chilensi", *Nees* (B).

An annual grass; culms up to 60 cm. tall, erect when solitary, but usually more or less densely fascicled and then the outermost stems prominently decumbent at the base, nodes few, smooth and glabrous; leaf-blades linear, long-acuminate from a rounded base, up to 10 cm. long, 1–3 mm. wide, scabrid on both surfaces and on the margins, often flaccid; sheaths glabrous, scaberulous, striate; ligules 2–6 mm. long, membranous, pointed; panicles terminal, erect, dense, silvery-green, lanceolate or cylindrical and then slightly tapering at tip and base, with short densely spiculate branches, 5–18 cm. long, 5–15 mm. wide; spikelets 6–7·5 mm. long, lanceolate, shortly pedicellate, erect; glumes unequal, with a prominent bulge at the base, the lower as long as the spikelet, linear-acuminate, keeled, 1-nerved, hyaline on the margins, scabrid on the keel, 1-nerved, the upper about 1–1·5 mm.

shorter, lanceolate-acuminate, keeled above, scabrid on the keel, 1-nerved; lemma broadly elliptic-truncate when flattened, 5-nerved, hyaline, loosely hairy on the dorsal surface, 1–1·5 mm. long, the central nerve produced from the dorsal surface above the middle as an awn; awn 7 mm. long, the lower half twisted; palea hyaline, 2-nerved, much narrower than the lemma; anthers 0·5 mm. long.

HAB.: On stony or sandy soil by roadsides, on cultivated land or on hillsides in garigue, sometimes in open Pine forest; on limestone or igneous formations; sea-level to 5,200 ft. alt.; fl. April–June.

DISTR.: Divisions 2–4, 6–8, common in Divisions 2 and 7. S. Europe, Mediterranean region eastwards to Iran; Atlantic Islands. Introduced into Australia, S. Africa, West Indies, N. and temperate S. America.

2. Common; Platania, Platres, Prodhromos, Mesapotamos, etc.; *Syngrassides* 1090! *Kennedy* 1112! 1113! *Lindberg f.* s.n.; *D. P. Young* 7451! etc.
3. Mile 24, Nicosia-Limassol road, 1962, *Meikle* 2956! 2957!
4. Larnaca aerodrome, 1950, *Chapman* 691! Athna (Akhna) Forest, 1952, *Merton* 798!
6. Ayia Irini, 1967, *Economides* in ARI 936!
7. Common; Kyrenia, Kythrea, Halevga, Kantara, etc.; *Sintenis & Rigo* 364! *Casey* 805! *Chapman* 642! *Meikle* 2951! etc.
8. Eleousa Experimental Station, 1962, *Meikle* 2992!

38. TRIPLACHNE *Link*
Hort. Reg. Bot. Berol., 2: 241 (1833).

Annual grasses with linear-acuminate, flat leaf-blades; inflorescence a dense panicle, lanceolate, ovate or oblong in outline, of closely packed, shortly pedicellate spikelets; spikelets with one bisexual floret; rhachilla disarticulating above the glumes, produced beyond the floret as a hairy stipe; glumes unequal, the lower longer than the upper, hyaline on the margins, somewhat firmer at the base; lemma very much smaller than the glumes, broadly elliptic when flattened, 5-nerved, the two lateral produced above as 2 aristae as long as the lemma, the median nerve carried out as a perfect awn just above the base with a twisted column; palea as long as the lemma, 2-keeled; stamens 3; lodicules absent. Grain elliptic in outline, slightly obovate, grooved on the adaxial surface; embryo about one-fifth the length of the grain; hilum sub-basal, punctiform.

One species in the Mediterranean region and Atlantic Islands.

1. **T. nitens** (*Guss.*) *Link*, Hort. Reg. Bot. Berol., 2: 241 (1833); Holmboe, Veg. Cypr., 33 (1914); Post, Fl. Pal., ed. 2, 2: 730 (1933); Osorio-Tafall et Seraphim, List Vasc. Plants Cyprus, 13 (1973).
 Agrostis nitens Guss., Ind. Sem. Hort. Boccadifalco, 1825: 1 (1825), Fl. Sic. Prodr., 1: 59 (1827); Sintenis in Oesterr. Bot. Zeitschr., 32: 264 (1882).
 Gastridium nitens (Guss.) Coss. et Durieu, Fl. d'Algérie, II, Phan., Glumacées [in Expl. Sci. Algérie]: 68 (1855); Boiss., Fl. Orient., 5: 519 (1884).
 TYPE: Sicily; "In herbidis, et arenosis maritimis, *Trapani tra l'Acquedotto* ed il mare, S. Croce alla spiaggia del Braccetto" *Gussone* (NAP).

An annual with fasciculate or solitary, erect or geniculately ascending, smooth and glabrous culms up to 30 cm. tall with purplish nodes; leaf-blades flat, up to 8 cm. long, 4 mm. wide, linear-acuminate, rough on both surfaces, very rough on the margins; sheaths rounded, clasping, striate, somewhat rough; ligule up to 4 mm. long, membranous; panicle pale, ovoid or fusiform, up to 6 cm. long, 1 cm. wide, of densely packed spikelets; spikelets linear-lanceolate; glumes unequal, the lower 4 mm. long, the upper 3·5 mm. long, both 1-nerved, shining, keeled above, scabrid on the keel; lemma ovate or elliptic, truncate when flattened, hyaline, membranous, appressed hairy on the dorsal surface, faintly 5-nerved, the median issuing just above the base as a perfect awn, the outer lateral nerves prolonged at the tip as two aristae;

awn about 4 mm. long, of which the lower half is twisted; palea as long as the lemma but narrower, hyaline, 2-nerved; anthers 0·5 mm. long.

HAB.: Sandy seashores; sea-level; fl. April.

DISTR.: Division 8, rare. General distribution that of genus.

8. Cape Andreas, 1880, *Sintenis & Rigo* 358! Dhavlos, 1880, *Sintenis & Rigo*.

39. LAGURUS L.

Sp. Plant., ed. 1, 81 (1753).
Gen. Plant., ed. 5, 34 (1754).

Annual grasses with flat, linear leaf-blades; inflorescence an ovoid, cylindrical or globular, soft, silky panicle of densely packed spikelets shortly pedicellate on very short branchlets; spikelets containing one herma-phrodite floret or consisting of glumes only at the base of the panicle; rhachilla disarticulating at maturity above the glumes; glumes persistent, equal, narrowly lanceolate, subulate, 1-nerved; lemma shorter than the glumes, elliptic when flattened, produced at the apex into two aristulae, awned from the back with a perfect awn with twisted column and straight bristle; palea hyaline, 2-nerved, expanded at the base; stamens 3; lodicules 2, hyaline, very delicate, elliptic-acuminate in shape. Grain terete, apiculate; embryo about one-sixth the length of the grain; hilum basal, punctiform.

One species, widely distributed in the Mediterranean region, and along the coast of S.W. Europe as far north as the Channel Islands; Atlantic Islands. Introduced into N. and S. America, Australia, S. Africa, etc.

1. **L. ovatus** *L.*, Sp. Plant., ed. 1, 81 (1753); Sibth. et Sm., Fl. Graec. Prodr., 1: 68 (1806), Fl. Graec., 1: 71, t. 90 (1808); Unger et Kotschy, Die Insel Cypern, 179 (1865); Boiss., Fl. Orient., 5: 521 (1884); Holmboe, Veg. Cypr., 33 (1914); Post, Fl. Pal., ed. 2, 2: 731 (1933); Osorio-Tafall et Seraphim, List Vasc. Plants Cyprus, 12 (1973).
    TYPE: "*in* Italia, Gallia, Sicilia, Lusitania" (LINN!).

An annual grass; culms solitary or fascicled, and then decumbent at the base, finally erect, up to 60 cm. tall, markedly striate, covered with short, patent hairs; leaf-blades linear or lanceolate, acuminate or acute, up to 15 cm. long, 4–10 mm. wide, rounded or almost cordate at the base, flat, softly pilose on both surfaces, greyish-green; sheaths also softly hairy, the uppermost somewhat inflated; ligule hairy, membranous, about 3 mm. long; inflorescence a globular, ovoid, oblong-ellipsoid or cylindrical, dense, hairy panicle, 1–4 cm. long, 1–1·5 cm. broad, pale or greyish rarely tinged with purple; spikelets very shortly pedicellate on the short branches, 8–10 mm. long; glumes equal, very narrowly lanceolate, tapering at the tip into a fine bristle, 1-nerved, covered with fine spreading hairs; lemma hyaline, elliptic-acute in shape when flattened, slightly hairy at the base, or appressed hairy on the dorsal surface, 5-nerved, the central nerve emerging as a perfect awn in the upper third, the two lateral on each side joining towards the apex and carried out as two aristae, 3–5 mm. long; median awn up to 15 mm. long, twisted in the lower half or third; callus bristly; anthers 1·5–2 mm. long.

HAB.: On dry stony or sandy soil, in garigue on hillsides, by roadsides or on waste ground; sea-level to 3,000 ft. alt.; fl. March–July.

DISTR.: Divisions 3–7, often abundant. General distribution that of genus.

3. Ayios Nikolaos, Cape Gata, 1859, *Kotschy*; between Kalokhorio and Stavrovouni, 1862, *Kotschy*; Episkopi Bay, 1950, without collector's name [no. 34!].
4. Common about Larnaca; *Kotschy* 313; *Holmboe* 85; *Haradjian* 5! *Chapman* 12! *Merton* in ARI 400!
5. Vatili, 1905, *Holmboe* 327; Laxia ["Latchia"], 1932, *Syngrassides* 74! Athalassa, 1936, *Syngrassides* 1178!

6. Between Ayia Irini and Kormakiti, 1941, *Davis* 2523! Ayia Irini, 1941, *Davis* 2547! Near Dhiorios, 1962, *Meikle* 2349!
7. Common; Kyrenia, Kythrea, Buffavento, Yaïla, etc.; *Sintenis & Rigo* 388! *Davis* 2859! *Casey* 411! *Chapman* 77! *G. E. Atherton* 156! 1284! *P. Laukkonen* 284! etc.

NOTES: Numerous subspecies and varieties have been described, differing in characters such as size of plant or inflorescence, hairiness of leaves or lemmas, etc. Most of the supposed differences are taxonomically valueless, and the species is here left as an intact unit.

40. AMMOPHILA *Host*
Gram. Austr., 4: 24, t. 41 (1809).

Perennial grasses with creeping scaly rhizomes, and with rigid, long, involute leaf-blades; inflorescence a very dense, erect, linear, terminal panicle of many closely packed, erect spikelets; spikelets 1-flowered, firmly compressed; rhachilla disarticulating above the glumes, produced beyond the floret as a penicillate stipe; glumes more or less equal, chartaceous to coriaceous, the lower 1-nerved, the upper 3-nerved; lemma with a prominent bearded callus, 5-nerved, papyraceous, shortly bifid at the apex with a very short rigid awn in the sinus; callus long- or short-hairy; stamens 3; styles 2; stigmas plumose; lodicules 2, narrowly elliptic, long acuminate, more or less shortly toothed on the margins. Grain oblong, deeply sulcate adaxially; embryo about one quarter the length of the grain; hilum linear, about half the length of the grain.

Two species in Europe, the Mediterranean region and the Atlantic coasts of N. America.

1. **A. arenaria** (*L.*) *Link*, Hort. Reg. Berol., 1: 105 (1827); Boiss., Fl. Orient., 5: 526 (1884); Post, Fl. Pal., ed. 2, 2: 731 (1933); Osorio-Tafall et Seraphim, List Vasc. Plants Cyprus, 12 (1973).
   A. arundinacea Host, Gram. Austr., 4: 24, t. 41 (1809).
   Psamma australis Mabille, Recherches Plant. Corse, 1: 33 (1867).
   Ammophila arenaria (L.) Link var. *arundinacea* (Host) Husnot, Gram., 19 (1896); Post, Fl. Pal., ed. 2, 2: 732 (1933).
   A. arenaria (L.) Link var. *australis* (Mab.) Aschers. et Graebn., Syn. Mitteleurop. Fl., 2: 221 (1899); Holmboe, Veg. Cypr., 33 (1914); Lindberg f., Iter Cypr., 6 (1964).
   TYPE: "*in Europa ad maris litora arenosa*" (LINN!).

A perennial grass with stout, scaly, branched rhizomes forming dense tufts; culms erect, up to 100 cm. tall, terete, rigid, stout, smooth and glabrous; leaf-blades up to 60 cm. long, 4–5 mm. wide when flattened, tightly involute, rigid, markedly ribbed above, densely and minutely hairy on the ribs, smooth below; sheaths leathery, overlapping; ligules narrow, pointed, up to 2·5 cm. long, minutely scabrid on the dorsal surface; panicle linear, spike-like, 7–30 cm. long, 1–3 cm. wide, pale; branches short; spikelets gaping, 10–16 mm. long, closely packed and erect, firmly laterally compressed; glumes elliptic-lanceolate, acute, more or less equal or the upper somewhat the longer, the lower 1-, the upper 3-nerved, smooth and glabrous, slightly scabrid on the keels, 10–13 mm. long; lemma 9–12 mm. long, lanceolate, 5–7-nerved, bifid at the apex, minutely scabrid, surrounded at the base with a circle of hairs up to 5 mm. long; awn in the sinus very short; palea as long as the lemma, narrow, 2-nerved; anthers 4–7 mm. long. Grain enclosed by the indurated lemma and palea.

HAB.: Sandy seashores and sand-dunes; sea-level; fl. May–July.

DISTR.: Division 4, rare. Widespread on the coasts of W. Europe and the Mediterranean.

4. Ayia Napa, 1905, *Holmboe* 1, and, 1939, *Lindberg f.*! Between Cape Greco and Ayia Napa, 1957, *Poore* 51!

NOTES: Both Holmboe and Lindberg f. refer the Cyprus plant to *A. arenaria* (L.) Link var. *australis* (Mab.) Aschers. et Graebn., but the differences between this and typical *A. arenaria*, based upon the pilosity of the rhachis, are not convincing, and scarcely deserving of recognition even at varietal rank.

TRIBE 9. **Stipeae** *Dumort.* Spikelets all alike, hermaphrodite, terete or dorsally compressed, 1-flowered; rhachilla disarticulating above the glumes, not produced; inflorescence a spreading or contracted panicle; glumes persistent, more delicate in texture than lemma or palea, often translucent, 1–3(–5)-nerved, lemma awned, terete, enveloping the palea, or somewhat compressed dorsally and convex, with a sharp, curved or obtuse callus, clothed with caducous hairs, or hairy on the nerves, or dorsally glabrous; palea hyaline or of the same texture as the lemma; awn short, straight, or long and straight or once or twice geniculate with a twisted column and a straight or flexuous bristle, glabrous or variously villose; stamens 3; styles 2; lodicules normally 3, often 2. Grain tightly enclosed by lemma and palea; hilum linear, as long as the grain; embryo small, one-eighth to one quarter the length of the grain.

Annual or perennial grasses; leaf-blades very narrow, rarely flat, most often convolute or plicate; chlorenchyma non-radiate; silica bodies narrowly elliptic, or oblong, cross-shaped or dumb-bell-shaped; 2-celled hairs present, elongate; ligules membranous or scarious. Chromosomes small, basic numbers 9, 11, 12, 13 or 17.

41. STIPA *L.*

Sp. Plant., ed. 1, 78 (1753).
Gen. Plant., ed. 5, 34 (1754).

Perennial, rarely annual, grasses with convolute, sometimes flat, leaf-blades; inflorescence a diffuse or contracted panicle of pedicellate spikelets; spikelets very narrowly ellipsoid or terete, 1-flowered, hermaphrodite; rhachilla disarticulating above the glumes, not produced above the floret; glumes hyaline or membranous, often several centimetres long, 1–3-nerved, muticous or mucronate; lemma very firm, with a bearded, pungent or rarely obtuse callus, convolute, terete, tapering at the tip or minutely 2-lobed, awned at the tip or between the lobes; awn usually with a twisted column and a bristle, glabrous, plumose or hairy in various ways; palea hyaline, 2-nerved; stamens 3, often with anthers hairy at the tips; lodicules 3, rarely 2, hyaline, elongate, tapering to an obtuse tip. Grain closely invested by the lemma and palea, fusiform-terete; embryo about one-fifth the length of the grain; hilum linear, nearly as long as the grain.

About 300 species in temperate and tropical regions, mostly xerophytic.

Awn not more than 2 cm. long, straight, glabrous, callus blunt - - - **2. S. bromoides**
Awn much longer; callus of lemma sharp-pointed:
    Panicle very dense, erect; awns eventually twisted into a characteristic tail **3. S. capensis**
    Panicle and awns not as above:
        Bristle of the awn glabrous, scabrid or with very short hairs; hairs on column dense, 2 mm.
           long; leaf-blades glabrous; lemma 14–16 mm. long - - - - **4. S. lagascae**
        Bristle of awn hairy:
           Column of awn hairy; hairs on the bristle up to 2 mm. long; lemma 9 mm. long, rarely
             longer - - - - - - - - - - - **1. S. barbata**
           Column of the awn glabrous; hairs on the bristle up to 5 mm. long; lemma 1·5–2 cm. long
                            5. S. pennata

1. **S. barbata** *Desf.*, Fl. Atlant., 1: 97, t. 27 (1798); Boiss., Fl. Orient., 5: 503 (1884); Holmboe, Veg. Cypr., 32 (1914); Post, Fl. Pal., ed. 2, 2: 725 (1933); N. L. Bor in C. C. Townsend et al., Fl. Iraq, 9: 399, t. 150 (1968); Osorio-Tafall et Seraphim, List Vasc. Plants Cyprus, 14 (1973).
    S. szowitsiana Trin. ex Griseb. in Ledeb., Fl. Rossica, 4: 450 (1853).
    TYPE: Algeria; "in collibus incultis circa Mascar et Tlemsen" (P).

A perennial grass; culms often densely caespitose, surrounded at the base with dead leaf-sheaths, smooth and glabrous or puberulous below the nodes, covered with the leaf-sheaths almost to the panicle, up to 100 cm. tall, slender to stout, simple; leaf-blades conduplicate, up to 30 cm. long, 1 mm. wide, smooth and glabrous on the outer surface, circular or elliptic in cross section, densely but minutely tomentose on the markedly nerved upper surface, more or less scaberulous on the margins; sheaths clasping, glabrous, scaberulous; ligule lanceolate, membranous, finely ciliate; inflorescence a rather narrow panicle enveloped at the base by the uppermost leaf-sheath, lax, few-spiculate; spikelets fusiform, 2·5–3·5 cm. long, gaping at maturity; glumes more or less equal, lanceolate-acuminate, hyaline, the lower 7-nerved, the upper 5-nerved; lemma convolute, fusiform, indurated, 5-nerved, 9–10 mm. long, hairy between the nerves in the lower half, with a crown of hairs at the tip, awned; callus pointed; awn 14–20 cm. long, twice geniculate, the column 4–5 cm. long, twisted; hairs 2 mm. long; anthers 5 mm. long with bearded tips.

HAB.: Dry, rocky hillsides and pastures; 250–750 ft. alt.; fl. March–May.

DISTR.: Divisions 3, 5, 6. Mediterranean region and eastwards to Pakistan and Central Asia.

3. Khirokitia, 1905, *Holmboe* 603; Episkopi, 1905, *Holmboe* 708.
5. Athalassa, 1936, *Syngrassides* 1170! and, 1951, *Merton* 1027! Kythrea, 1941, *Davis* 2954! Mile 10, Nicosia-Famagusta (old) road, 1967, *Merton* in ARI 486!
6. Ayia Marina (Myrtou), 1952, *Merton* 810!

2. S. bromoides (*L.*) *Doerfl.*, Herb. Norm. Cent., 34: 129, no. 3386 (1897); Post, Fl. Pal., ed. 2, 2: 726 (1933); N. L. Bor in C. C. Townsend et al., Fl. Iraq, 9: 400 (1968); Osorio-Tafall et Seraphim, List Vasc. Plants Cyprus, 14 (1973).
  Agrostis bromoides L., Mantissa Plant., 1, 30 (1767); Gouan, Illustr. et Obs. Bot., 3, t. 1, fig. 3 (1773).
  Stipa aristella L., Syst. Nat., ed. 12, 3: 229 (1768); Sibth. et Sm., Fl. Graec. Prodr., 1: 66 (1806), Fl. Graec., 1: 69, t. 87 (1808); Holmboe, Veg. Cypr., 32 (1914); Lindberg f., Iter Cypr., 9 (1946).
  Aristella bromoides (L.) Bertol., Fl. Ital., 1: 690 (1833); Boiss., Fl. Orient., 5: 504 (1884).
  TYPE: France; Monspelii. D. *Gouan* (LINN!) ("Primum in Ericetis legeram circa *Lamalou* Anno 1765" *Gouan loc. cit.*).

A perennial grass, with a shortly creeping, woody rhizome; culms solitary or few-fascicled, erect, up to 1 m. tall, slender, smooth and glabrous; leaf-blades up to 30 cm. long, 1–2 mm. broad, linear-acuminate, involute or convolute, glabrous and smooth on the lower surface, scabrid on the upper surface, and on the margins, or the latter smooth; sheaths clasping, rounded, striate, smooth and glabrous; ligule membranous, very short; inflorescence a very narrow, erect, sometimes lax, panicle up to 20 cm. long; axis scaberulous, angled; spikelets greenish, eventually straw-coloured, fusi-form, 8–9 mm. long; glumes membranous, equal, 3-nerved (nerves green); elliptic-oblong, acuminate when flattened, smooth and glabrous; lemma up to 6 mm. long, somewhat indurated, pilose below, scabrid towards the tip, pale turning brown at maturity, awned; palea of the same texture; callus rounded; awn straight, scabrid, 1–1·3 cm. long; anthers 3–5·5 mm. long.

HAB.: Dry rocky slopes, in garigue or open forest; 1,000–4,200 ft. alt.; fl. May–Aug.

DISTR.: Divisions 2, 3, 7; locally common. Mediterranean region eastwards to Iran.

2. Locally common about Prodhromos, Platres, Galata, etc., *Sintenis & Rigo* 660a! 660b! *Holmboe* 939; *Kennedy* 76! 77! 1123! 1124! *Lindberg f.* s.n.! *D. P. Young* 7421!
3. Oritaes Forest, 1956, *Merton* 2722!
7. Locally common about Kyrenia, Karavas, St. Hilarion, Lefkoniko Pass, Halevga, etc. *Sintenis & Rigo* 660! *Lindberg f.* s.n.; *Casey* 787! *Syngrassides* 442! *Merton* in ARI 41! etc.

3. S. capensis *Thunb.*, Prodr. Plant. Cap., 19 (1794); N. L. Bor in C. C. Townsend et al., Fl. Iraq, 9: 402 (1968); Osorio-Tafall et Seraphim, List Vasc. Plants Cyprus, 14 (1973).
  S. retorta Cav., Obs. Reyno Valencia, 1: 119 (1795).

S. tortilis Desf., Fl. Atlant., 1: 99, t. 31 (1798); Unger et Kotschy, Die Insel Cypern, 178 (1865); C. et W. Barbey, Herborisations au Levant, 99 (1880); Boiss., Fl. Orient., 5: 500 (1884); Post, Fl. Pal., ed. 23, 2: 725 (1933).

[*S. paleacea* (non Poir.) Sm. in Sibth. et Sm., Fl. Graec. Prodr., 1: 65 (1806); Fl. Graec., 1: 68, t. 86 (1808).]

TYPE: Not localized (UPS!).

An annual grass; culms densely fascicled, rarely solitary, erect or the outermost decumbent at the base, geniculately ascending, covered by the sheaths up to the panicle, up to 50 cm. tall, usually much shorter, smooth and glabrous or somewhat scaberulous below the panicle; leaf-blades linear-acuminate, flat, rapidly becoming involute, up to 15 cm. long, 1–1·5 mm. wide, the lower narrow, filiform, puberulous or glabrous on the upper surface, smooth and glabrous on the lower surface; sheaths clasping, except the uppermost which is much inflated and supports the base of the panicle, hairy at the collar; ligule very short, truncate, minutely ciliate; inflorescence a very dense panicle, 4–6 cm. long; axis smooth and glabrous; spikelets 15–16 mm. long, linear at first; glumes unequal, membranous, hyaline, linear-acuminate, 3-nerved, the lower 15–16 mm. long, the upper about 12 mm. long, or both more or less equal, rarely the lower shorter than the upper; lemma chartaceous, 4–7 mm. long, 5-nerved, loosely hairy on the dorsal surface, awned; callus pointed, 2 mm. long; awns bigeniculate, up to 10 cm. long, often much shorter; column about 3 cm. long, shortly hirsute, strongly twisted when mature, all the awns twisted together, forming a kind of tail; anthers 3–3·5 mm. long, bearded at the tips.

HAB.: Waste and fallow land, sometimes in impoverished batha and garigue, abundant and sometimes dominant in overgrazed areas; sea-level to 2,000 ft. alt.; fl. March–May.

DISTR.: All divisions, often locally common. Mediterranean region eastwards to N.W. India and Central Asia; Atlantic Islands; S. Africa.

1. Yeroskipos, 1862, *Kotschy*; Akamas, between Neokhorio and Smyies forest station, 1941, *Davis* 3304A! Ayios Yeoryios Island, Akamas, 1962, *Meikle* 2159!
2. Philani, 1862, *Kotschy* 222; Vouni, near the palace ruins, 1963, *Townsend* 63/59!
3. 2 miles S. of Pyrgos, 1963, *J. B. Suart* 82! Episkopi, 1981, *De Langhe* 58/81!
4. Around Larnaca, 1862, *Kotschy* 78! also, 1881, *C. & W. Barbey* s.n. and, 1950, *Chapman* 10!
5. Between Kythrea and Lefkoniko, 1880, *Sintenis & Rigo* s.n.; between Athalassa and Eylenja, 1933, *Syngrassides* 1! Mile 5, Nicosia-Famagusta road, 1950, *Chapman* 34! 44! Mile 7, Nicosia-Famagusta road, 1950, *Chapman* 370!
6. Nicosia, 1905, *Holmboe* 288; Kokkini Trimithia, 1933, *Syngrassides* 337! Dhiorios, 1962, *Meikle* 2320!
7. Common; *Kotschy* (5– or 7–) 25; *Sintenis & Rigo* 391! *Casey* 521! 1160! Lakovounara Forest, *Chapman* 69! 150! 151!
8. Between Cape Andreas and Rizokarpaso, 1880, *Sintenis & Rigo*.

NOTES: This species appears in Maire (Fl. d'Afr. Nord, 2: 68; 1953) as *Stipa retorta* Cav. (1795) which antedates *S. tortilis* Desf. (1798) the name by which the Mediterranean plant had been generally known.

Recently, however, some botanists have come to the conclusion that *Stipa capensis* Thunb. is the correct name for the species, a conclusion which has been challenged by Chrtek and Hadač (in Candollea, 24: 180; 1969). These authors say: "a comparison of plants from S. Africa and S.W. Asia clearly shows them to be distinct. The South African plants have shorter awns and lemmas as compared with the saharo-sindian material." The statement is not borne out by examination. A large series of specimens of *S. retorta* Cav. from the Mediterranean region has been examined and awn length has been measured as well as length of lemma. The result of this survey of Mediterranean plants shows that the length of the awn varies from 5 cm. to 10 cm., and that of the lemma varies from 4 mm. to 8 mm. For South African plants the length of the awn varies from 6 cm. to 10 cm., and that of the lemma from 4 mm. to 7 mm. Between these limits there is every gradation, and the same applies to the length of the glumes. There may be firm differences to distinguish *S. retorta* and *S. capensis*, but the author has been unable to find them!

4. S. lagascae *Roem. et Schult.*, Syst. Veg., 2: 333 (1817); Boiss., Fl. Orient., 5: 500 (1884); N. L. Bor in C. C. Townsend et al., Fl. Iraq, 9: 406 (1968); Osorio-Tafall et Seraphim, List Vasc. Plants Cyprus, 14 (1973).

S. holosericea Trin. et Rupr., Sp. Gram. Stip. (1842), et in Mém. Acad. Sci. Petersb., ser. 8, 7: 71 (1842).
S. fontanesii Parl., Fl. Ital., 1: 187 (1848); Post, Fl. Pal., ed. 2, 2: 725 (1933); Lindberg f., Iter Cypr., 9 (1946).

A perennial grass; culms densely caespitose, woody, stout, erect, smooth and glabrous, or the lower internodes pilose, up to 60 cm. tall, filled with pith below; leaf-blades stiff, rigid, curved, conduplicate, up to 30 cm. long, 1–1·5 mm. wide, circular or elliptic in cross-section, glabrous or hairy below, shortly hairy and scabrid on the inner surface, when flattened narrowly linear tapering to an acuminate tip; sheaths of the upper nodes smooth and glabrous, those of the lower nodes very shortly hairy; ligule about 2 mm. long, truncate or erose, ciliate; inflorescence a very narrow, loose, nodding panicle, the base of which is supported by and included in the uppermost leaf-sheath, up to 30 cm. long; axis smooth; branches up to 10 cm. long, few-spiculate; spikelets silvery, about 4 cm. long; glumes more or less equal, as long as the spikelets, lanceolate-acuminate, very delicate, the lower 5–7-nerved, the upper 5-nerved; lemma 10–15 mm. long (with a callus densely bearded, 3 mm. long), silky hairy in the lower half, smooth and glabrous above, scabrid below the articulation, convolute, awned; palea of the same texture; awn twice geniculate, 10–25 cm. long, very minutely ciliate; anthers about 7 mm. long.

HAB.: On dry, rocky hillsides in garigue or on bare chalk; 550–700 ft. alt.; fl. Apr.–June.

DISTR.: Division 5, locally common. Mediterranean region and eastwards to Central Asia.

5. Around Athalassa; *Lindberg f.* s.n. ! *Davis* 3197 ! *Chapman* 638 ! *M. E. D. Poore* 9 !

5. S. pennata *L.*, Sp. Plant., ed. 1, 78 (1753); Sibth. et Sm., Fl. Graec. Prodr., 1: 65 (1806); Poech, Enum. Plant. Ins. Cypr., 7 (1842); Unger et Kotschy, Die Insel Cypern, 178 (1865); Boiss., Fl. Orient., 5: 502 (1884); H. Stuart Thompson in Journ. Bot., 44: 430 (1906); Holmboe, Veg. Cypr., 32 (1914); Post, Fl. Pal., ed. 2, 2: 725 (1933); N. L. Bor in C. C. Townsend et al., Fl. Iraq, 9: 408, t. 154 (1968); Osorio-Tafall et Seraphim, List Vasc. Plants Cyprus, 14 (1973).
TYPE: "*in* Austria, Gallia" (LINN !).

A perennial, caespitose grass; culms erect, rigid, up to 60 cm. tall, simple, clothed with long sheaths; leaf-blades linear, acute, up to 50 cm. long, 1 mm. wide, convolute, glaucous, rigid, erect, smooth or scaberulous, glabrous on the lower surface, somewhat hairy on the upper surface, scabrid on the margins; panicle few-spiculate, enclosed at the base in the uppermost leaf-sheath; branches ascending, pedicels shorter than the spikelets; spikelets fusiform; glumes linear-acuminate, smooth and glabrous, 3–6 cm. long; lemma 1·5–2 cm. long, 5-nerved, indurated, hairy at the base with a very sharp hairy callus; awn up to 30 cm. long, geniculate, twisted and glabrous on the 10 cm. column; bristle plumose almost to the tip.

HAB.: Not known.

DISTR.: Recorded by Sibthorp (*loc. cit. supra*) "et in Cypro insulâ", but without further localization. Mediterranean region east to Iran.

NOTES: Not collected in Cyprus since, and probably an error, as it is unlikely that such a distinctive grass would be overlooked.

S. TENACISSIMA *L.*, Cent. Plant., 1, 6 (1753), Esparto Grass, a native of Spain and Portugal, was introduced into Salamis Plantation early in the present century (Bovill, Rep. Plant. Work Cyprus, 14; 1915), but does not seem to have survived in the area. It is a robust perennial, forming large tufts, with culms 1–2 m. high, spikelets 3 cm. long, and geniculate awns densely hairy below the knee. Apart from its commercial value as a source of paper-pulp, *S. tenacissima* is sometimes cultivated in gardens as an ornamental.

42. ORYZOPSIS *Michx.*
Fl. Bor.-Amer., 1: 51, t. 9 (1803).

Perennial grasses with long, linear, flat leaf-blades; inflorescence an effuse panicle of loosely aggregated or solitary, pedicellate spikelets; spikelets ovate or elliptic, acute or lanceolate in outline, dorsally compressed, 1-flowered; rhachilla articulated above the glumes, not produced beyond the floret; glumes equal, acute or acuminate, the lower 5-nerved, the upper 3-nerved; lemma coriaceous, lanceolate, ovate or elliptic in outline, acute, awned with a short, caducous awn at the tip, turning dark brown or black at maturity, more or less hairy; palea of the same texture; stamens 3; lodicules 2 or 3, roughly lanceolate-acuminate. Grain elliptic in outline, somewhat compressed, flattened and grooved on the adaxial surface; embryo orbicular, about half the length of the grain; hilum linear, as long as the grain.

About 50 species in temperate and subtropical regions of the northern hemisphere.

Branches of the panicle fascicled, often very numerous; spikelets narrowly elliptic-acuminate, 2–3 (–3·5) mm. long - - - - - - - - - - - **1. O. miliacea**
Branches of the panicle in pairs; spikelets 6–8 mm. long, elliptic-acute - **2. O. coerulescens**

1. O. miliacea (*L.*) *Aschers. et Schweinf.* in Mém. Inst. Égypt., 2: 169 (1887); Holmboe, Veg. Cypr., 33 (1914); Post, Fl. Pal., ed. 2, 2: 726 (1933); Lindberg f., Iter Cypr., 8 (1946); Rechinger f. in Arkiv för Bot., ser. 2, 1: 418 (1950); N. L. Bor in C. C. Townsend et al., Fl. Iraq, 9: 414, t. 156 (1968); Osorio-Tafall et Seraphim, List Vasc. Plants Cyprus, 14 (1973).
  Agrostis miliacea L., Sp. Plant., ed. 1, 61 (1753).
  Milium arundinaceum Sm. in Sibth. et Sm., Fl. Graec. Prodr., 1: 45 (1806), Fl. Graec., 1: 50, t. 66 (1808).
  Urachne parviflora Trin., Fund. Agrost., 110 (1820); Unger et Kotschy, Die Insel Cypern, 178 (1865).
  Piptatherum miliaceum (L.) Coss., Not. Crit., 129 (1851); Boiss., Fl. Orient., 5: 507 (1884).
  TYPE: "*in* Europa" (LINN!).

A perennial grass from a knotty, woody base or short rhizome; culms erect or geniculately ascending, slender, branched at the base or at the nodes, up to 1·5 m. tall, more or less scabrid, glabrous; leaf-blades linear-acuminate, up to 30 cm. long, 2–4 mm. wide, flat, finally involute or rolled, scabrid on the upper surface and on the margins, pilose above, smooth (or scaberulous) and glabrous below; sheaths much shorter than the internodes, rounded, clasping, smooth and glabrous; ligules membranous, 1–1·5 mm. long, minutely pubescent, truncate; panicle up to 30 cm. long, lax or dense, erect or nodding; axis smooth and glabrous; branches fascicled, often very numerous, ascending, unequally long, slender, bare for half their length, the lower sometimes without spikelets, the upper carrying numerous spikelets; spikelets narrowly elliptic-acute or -acuminate, 2–3 (–3·5) mm. long, green or flushed with violet; glumes more or less equal, membranous, elliptic-acute when flattened, the lower a little longer than the upper, 3-nerved; lemma about 2·5 mm. long or less, chartaceous, smooth and glabrous, shining, 3-nerved, awned; palea of the same texture; awn 3 mm. long, scabrid; anthers 1·5 mm. long, bearded.

HAB.: Rocky hillsides amongst garigue, occasionally on sandy or marshy ground; sea-level to 2,500 ft. alt.; fl. March–Aug.

DISTR.: Divisions 1–3, 5, 7; common in Division 7. Mediterranean region eastwards to Iran; Atlantic Islands; introduced elsewhere.

1. Stroumbi, 1913, *Haradjian* 753! Ayios Merkourios, 1962, *Meikle* 2279! Kissonerga to Maa Beach, 1967, *Economides* in ARI 993!
2. Khrysorroyiatissa Mon., 1862, *Kotschy* 586!

3. Kolossi, 1862, *Kotschy*; Kophinou, 1905, *Holmboe* 597; Limassol, 1930, *Wall*; Asomatos, 1939, *Mavromoustakis* 131! Near Kithasi, 1970, *Chicken* 79 (var. *thomasii* (Duby) Richter)!
5. Athalassa, 1939, *Lindberg f.* s.n.
7. Common on the Northern Range; *Sintenis & Rigo* 658! *Holmboe* 542; *Miss Mapple* 7! *G. E. Atherton* 314! *Economides* in ARI 1099!

NOTES: In var. *thomasii* (Duby) Richter, Plant. Europ., 33 (1890), based on *Milium thomasii* Duby, Bot. Gall., 1: 505 (1828) the lower verticels of the panicle branches are bare, without spikelets. While it is true that in other forms all the lower branches are spiculate, these two extremes are connected by intermediates, with some of the branches in a verticel spiculate, while others (in the same verticel) are bare. In consequence, it is virtually impossible to draw the line between one variant and the other. Until more is known about the reasons for the abortion of spikelets on the lower verticels, it seems best to disregard the varietal name.

Oryzopsis miliacea was sown experimentally at Dennarga in 1954 (Photiades in The Countryman, Sept. 22, 1958). The results were good, and it was reported that the grass had persisted for three years.

2. O. coerulescens (*Desf.*) *Richter*, Plant. Europ., 1: 34 (1890); Holmboe, Veg. Cypr., 33 (1914); Post, Fl. Pal., ed. 2, 2: 726 (1933); Lindberg f., Iter Cypr., 8 (1946); Osorio-Tafall et Seraphim, List Vasc. Plants Cyprus, 14 (1973).

    Milium coerulescens Desf., Fl. Atlant., 1: 66, t. 12 (1798); Sibth. et Sm., Fl. Graec. Prodr., 1: 45 (1806).

    Piptatherum coerulescens (Desf.) P. Beauv., Essai Agrost., 18 (1812); Boiss., Fl. Orient., 5: 507 (1884); Sintenis in Oesterr. Bot. Zeitschr., 32: 192 (1882).

    Urachne coerulescens (Desf.) Trin., Fund Agrost., 110 (1820); Unger et Kotschy, Die Insel Cypern, 178 (1865).

TYPE: North Africa; "in fissuris rupium Atlantis" (P).

A perennial grass; culms loosely tufted, up to 60 cm. tall, erect, rather wiry, smooth and glabrous, simple; leaf-blades linear-acuminate, up to 40 cm. long, 1–5 mm. wide, scaberulous on both faces and on the margins, or smooth on the lower face only, flat at first then involute or rolled, glabrous, strongly nerved on the upper surface; sheaths closely clasping or the lower slipping from the culms, glabrous, scabrid; ligule 4–6 mm. long; panicle erect or somewhat nodding, up to 20 cm. long, with widely spreading, smooth branches up to 20 cm. long, in pairs, one branch much shorter than the other, carrying few spikelets; spikelets 6–8 mm. long, elliptic-acute in outline, green, tinged with purple; glumes more or less equal, or the lower slightly longer, ovate- or lanceolate-acute when flattened, 3–5-nerved, somewhat keeled towards the apex, scaberulous on the keel; lemma elliptic-acute in outline, brown to brownish black, 3·5–4 mm. long, awned, obscurely 5-nerved, shining, smooth, entirely glabrous except for a pencil of hairs on the sides at the base; callus rounded; palea of the same texture as the lemma; awn caducous, scabrid, straight, 3 mm. long; anthers 3 mm. long, with one or two hairs at the tip.

HAB.: Dry, rocky ground and rock fissures on calcareous or igneous formations; near sea-level to 4,000 ft. alt.; fl. March–June.

DISTR.: Divisions 2, 3, 6–8. W. and S. Europe and Mediterranean region eastwards to Pakistan.

2. Khrysorroyiatissa Mon., 1862, *Kotschy* 694a! Kryos Potamos, 3,100 ft. alt., 1937, *Kennedy* 63! Platres, 1938, *Kennedy* 1114! Mesapotamos, 1939, *Lindberg f.* s.n.! Stavros tis Psokas, 1962, *Meikle* 2745! Between Lefkara and Vavatsinia, 1967, *Merton* in ARI 726!
3. Amathus, 1862, *Kotschy* 586; near Akrotiri, 1905, *Holmboe* 676; Kornos-Lefkara road, 1967, *Merton* in ARI 589!
6. Ayia Irini, 1941, *Davis* 2528!
7. Kyrenia, Mylous Forest Station, 1933, *Syngrassides* 52! Larnaka tis Lapithou, 1941, *Davis* 2992! and, 1956, *Merton* 2705!
8. Panayia tou Kantara, 1880, *Sintenis & Rigo*.

TRIBE 10. **Bromeae** *Dumort.* Spikelets all alike, several- to many-flowered, terete or laterally compressed, hermaphrodite; rhachilla disarticulating above the glumes and below each floret; spikelets arranged in large, ample, often nodding panicles or in dense, erect panicles of crowded spikelets; glumes unequal, the lower 1–3-nerved, the upper 3–7 (–9)-nerved, glabrous or hairy, often scarious on the margins; lemmas rounded or keeled on the dorsal surface, 5–9-nerved, 2-cleft or -lobed at the hyaline tip, awnless or with 1, 3, 7 or 9 straight or recurved awns; paleas hyaline, 2-nerved; stamens 3 or 2; ovary with a terminal, fleshy, hairy appendage; styles 2; stigmas pilose; lodicules 2. Grain adherent to the lemma and palea; hilum linear; starch grains simple.

Annual or perennial grasses; leaf-blades narrow, linear-acuminate, flat with festucoid anatomy; silica-bodies oblong; 2-celled hairs absent; sheaths tubular, soon splitting; ligules often well developed, membranous. Chromosomes large; basic number 7.

43. BROMUS L.

Sp. Plant., ed. 1, 76 (1753).
Gen. Plant., ed. 5, 33 (1754).
Z. Ovadiahu-Yavin in Israel Journ. Bot., 18: 195–216 (1969).
T. Kowal et W. Rudnicka-Sternowa in Mon. Bot., 29: 1–39 (1969).
H. Scholz in Willdenowia, 6: 139–159 (1970).
P. M. Smith in Annals of Bot., n.s., 36 (144): 1–30 (1972).

Annual or perennial grasses with tubular sheaths and flat leaf-blades; inflorescence a panicle, contracted, erect or effuse and nodding; spikelets wedge-shaped or lanceolate in outline, very large or small, long- or short-pedicellate, several-flowered, hermaphrodite; rhachilla disarticulating above the glumes and between the florets; glumes awl-shaped or lanceolate, the lower 1–3-nerved, the upper 3- to 9-nerved; lemmas rounded or keeled on the back, 5–9-nerved, bifid at the apex, 1- to 3-awned from below the tip or awnless and muticous; palea 2-keeled, rounded at the apex, ciliate with spaced hairs on the keels; stamens 3; lodicules 2, oblong, obovate or lanceolate, net-veined, glabrous. Grain flattened, oblong, tapering at the base, rounded and appendaged at the apex, adherent to the lemma and palea; embryo one-eighth the length of the grain; hilum linear, as long as the grain.

About 50 species chiefly in temperate regions, a few on mountains in the tropics.

Spikelets oblong or wedge-shaped, not elliptic or lanceolate, nor tapering towards the tip; lower glume 1-nerved, the upper 3-nerved:
  Panicle contracted or rather loose, but with the spikelets clustered in groups:
    Spikelets 7–9 cm. long, including the awns:
      Callus of the lemmas rounded at the tip - - - - - - **5. B. diandrus**
      Callus of the lemmas markedly pointed at the tip - - - - **6. B. rigidus**
    Spikelets less than 7 cm. long, often much less:
      Culms hairy below the inflorescence:
        Inflorescence broadly obovate, not acutely cuneate at the base; ligule up to 5 mm. long, lacerate, milky-white; spikelets compressed; lemmas 13–15 mm. long, 2–2·5 mm. wide, with a definite scarious margin; anthers 0·6–1 mm. long; awns and lemmas straight - - - - - - - - **12. B. rubens**
        Inflorescence obovate, sharply cuneate at the base; ligule 3 mm. long, hyaline; lemmas 12–14 mm. long, 1–1·5 mm. wide, rounded on the back, with involute margins, which are hardly scarious; anthers 0·3–0·4 mm. long; awns and lemmas curved outwards - - - - - - - - **7. B. fasciculatus**
      Culms glabrous or very sparsely hairy:
        Spikelets 3·5–6 cm. long, erect, often dense, obovate; lemmas 12–19 mm. long, 3 mm. wide; anthers 0·5–1 mm. long - - - - - **11. B. madritensis**
        Spikelets up to 3 cm. long (including the awns), if slightly longer (*B. tectorum*) then the spikelets strongly compressed, with a tendency to turn to one side in clusters of parallel spikelets:

Panicle very dense, flabellate, cuneate at the base; lemmas 1–1·5 mm. wide with involute margins, slightly recurved from the rhachilla; anthers 0·3–0·4 mm. long - - - - - - - - - - - - **7. B. fasciculatus**

Panicle dense or somewhat loose, not flabellate or cuneate at the base; spikelets in parallel or roughly parallel clusters, erect or turned to one side; lemmas with silvery margins; anthers more than 0·3–0·4 mm. long - **17. B. tectorum**

Panicle loose, branches widely spreading, up to 10 cm. long, carrying 1 or 2 spikelets, or sometimes more:

Spikelets (including the awns) 7–9 cm. long, several on each branch:

Callus of the lemmas blunt, rounded - - - - - - - **5. B. diandrus**

Callus of the lemmas sharply pointed - - - - - - - **6. B. rigidus**

Spikelets (including the awn) much shorter, usually solitary at the tip of each branch
16. B. sterilis

Spikelets ovate, lanceolate, elliptic or oblong, tapering towards the tip; lower glume 3–7-nerved, upper 5–9-nerved:

Glumes and lemmas strongly keeled:

Lemmas without awns - - - - - - - - - B. CATHARTICUS (p. 1812)

Lemmas shortly awned - - - - - - - - B. CARINATUS (p. 1812)

Glumes and lemmas rounded on the dorsal surface:

Lemmas, at least the uppermost, with 3 awns - - - - **2. B. danthoniae**

Lemmas with one awn only:

Spikelets 3–5 cm. long (often shorter in starved specimens of *B. squarrosus*):

Panicle oblanceolate or oblong in outline; spikelets terete, short-pedicellate, ascending; awns recurved at maturity - - - - **14. B. alopecuros**

Panicle with spreading branches; spikelets laterally compressed:

Lemmas 8–10 mm. long, 6 mm. wide, with a broad hyaline margin, conspicuously widened above the middle, with the margins forming an obtuse angle; awns of the lower lemmas frequently reduced - - - - - **15. B. squarrosus**

Lemmas 11–14 mm. long, 5–6 mm. wide with a broad hyaline margin, not as above; lower lemmas with perfect awns, and 2 acuminate apical teeth; awns recurved at maturity - - - - - - - - **10. B. lanceolatus**

Spikelets less than 3 cm. long:

Lemmas rhomboidal when flattened, margins conspicuously hyaline
15. B. squarrosus

Lemmas elliptic or oblanceolate when flattened; margins not conspicuously hyaline:

Palea shorter than the lemma; anthers 3–4·5 mm. long - - **3. B. arvensis**

Palea as long as, or longer than the lemma; anthers up to 2 mm. long:

Spikelets hairy, hairs patent, long and short, curved - **4. B. molliformis**

Spikelets glabrous, or, if hairy, then the hairs usually equal and adpressed:

Inflorescence a rather loose, nodding panicle; spikelets long-pedicellate, narrowly lanceolate; awns of the uppermost lemmas much longer than those of the lower - - - - - - - **9. B. japonicus**

Inflorescence and spikelets not as above; the inflorescence erect, dense or rather loose:

Inflorescence very dense, obovate or oblong; spikelets crowded on short pedicels, sometimes arranged in pseudowhorls and panicle somewhat interrupted; upper and lower glumes 5–6 mm. and 6–7 mm. long respectively; lemmas 6–8 mm. long - - - **13. B. scoparius**

Inflorescence not as above:

Spikelets less than 10 mm. long - - B. BRACHYSTACHYS (p. 1797)

Spikelets more than 10 mm. long:

Spikelets 18–22 mm. long; lemma 8–11 mm. long, glabrous or hairy, rather firm; awn straight or slightly bent
1. B. COMMUTATUS (p. 1798)

Spikelets about 15 mm. long; lemmas hairy, about 9 mm. long; glumes 6–8 mm. long; awns strongly recurved at maturity
8. B. intermedius

[B. BRACHYSTACHYS *Hornung* in Flora 16: 417 (1833) is recorded from Cyprus by Osorio-Tafall and Seraphim (List Vasc. Plants Cyprus, 9; 1973), but the source of the record has not been traced. H. Scholz (in Bot. Jahrb., 91: 462; 1972) distinguishes the mid-European *B. brachystachys* from the grass so named in the eastern Mediterranean area and western Asia; the latter he re-names *B. pseudobrachystachys* H. Scholz. It is to *B. pseudobrachystachys* rather than to *B. brachystachys* that the Cyprus record should probably be referred, though, in the absence of specimens, this can only be surmised. *B. pseudobrachystachys* is reported from S. Turkey, Syria, Lebanon and Palestine, and is quite likely to be found in Cyprus.]

[1. B. COMMUTATUS *Schrad.*, Fl. Germ., 1: 354 (1806); Boiss., Fl. Orient., 5: 654 (1884); N. L. Bor in C. C. Townsend et al., Fl. Iraq, 9: 136 (1968); Osorio-Tafall et Seraphim, List Vasc. Plants Cyprus, 9 (1973).
   TYPE: Germany; "Inter segetes, ad vias, sepes, alibique" (? GOET).

An annual or biennial grass; culms slender or somewhat stout, loosely tufted or rarely solitary, simple, smooth and glabrous, up to 100 cm. tall; leaf-blades linear-acuminate, rounded at the base, flat, loosely hairy on both surfaces, rough on the upper surface and on the margins, up to 30 cm. long, 4–8 mm. wide; sheaths on the upper part of the culm closely fitting, glabrous, the lower looser and hairy; ligules lacerate, membranous, up to 4 mm. long; inflorescence a rather loose, nodding panicle with the lower branches fasciculate, up to 24 cm. long; spikelets at first lanceolate in outline then oblong, 1·5–2·5 cm. long, 4–10-flowered; rhachilla disarticulating at maturity below each floret; glumes more or less equal in length, 5–9 mm. long, the lower the narrower, 5-nerved, elliptic-acute, the upper broader, 7–9-nerved, smooth and glabrous; lemmas 8–11 mm. long, broadly elliptic when flattened, finely but obscurely 7–11-nerved, notched at the apex, awned; awn emerging just below the tip, straight, scabrid, 4–10 mm. long; palea shorter than the lemma; anthers 1·5–2 mm. long.

   HAB.: Pastures and hillsides, on marl or limestone; sea-level to 900 ft. alt.; fl. March–April.

   DISTR.: Divisions 4, 6, 7. Throughout Europe, Turkey, Caucasus, Iraq, Iran; introduced from Europe into S. Africa and N. America.

   4. Larnaca aerodrome, 1950, *Chapman* 19!
   6. Near Myrtou, 1956, *Merton* 2583!
   7. Lakovounara Forest, 1950, *Chapman* 66! 145! 192!

   NOTES: All the cited specimens are depauperate and untypical, and are questionably distinct from forms of *B. japonicus* Thunb. R.D.M.]

2. **B. danthoniae** *Trin.* in C. A. Mey., Verz. Plant. Cauc., 24 (1831); Kotschy in Oesterr. Bot. Zeitschr., 12: 275 (1862); N. L. Bor in C. C. Townsend et al., Fl. Iraq, 9: 136, t. 47 (1968); Osorio-Tafall et Seraphim, List Vasc. Plants Cyprus, 9 (1973).
   Triniusa danthoniae (Trin.) Steud., Syn. Plant. Glum., 1: 328 (1854); Hackel in Flora, 62: 153–158 (1879).
   Boissiera danthoniae (Trin.) A. Braun, Ind. Sem. Hort. Berol., 3 (1857).
   Bromus macrostachys Desf. var. *triaristatus* Hackel in Flora, 62: 155 (1879); Boiss., Fl. Orient., 5: 652 (1884).
   B. lanceolatus Roth var. *danthoniae* (Trin.) Dinsm. in Post, Fl. Pal., ed. 2, 2: 775 (1933).
   TYPE: U.S.S.R., Azerbaijan; "In locis lapidosis aridis montium Talüsch prope pagum Swant (alt. 670 hexapl.)" *C. A. Meyer* (LE).

An annual grass; culms up to 40 cm. tall, loosely tufted, very rarely solitary and then only on very arid poor soils, smooth and glabrous or puberulous below the panicle and at the nodes; leaf-blades linear-acuminate, up to 15 cm. long, 2–4 mm. wide, covered on both surfaces with dense or sparse, soft, white hairs or glabrescent, rough on the margins; sheaths tightly clasping, shortly keeled upward, the uppermost glabrescent, the lower sparsely to densely, retrorsely pilose; ligules up to 4 mm. long, truncate, lacerate; inflorescence a dense or somewhat loose panicle, erect, rarely nodding; branches usually short, often unispiculate; spikelets oblong or oblong-elliptic, exceptionally 5 cm. long, 5 mm. broad, 28-flowered, usually much shorter, 1·5–2·5 mm. long, 6–7 mm. wide, 10–15-flowered; lower glume oblong-acute, 5-nerved, glabrous, 6 mm. long, the upper broader, oblong or elliptic-oblong, acute, 7–9-nerved, 7·5 mm. long, chartaceous, glabrous; lemma varying in texture from membranous to chartaceous, when flattened almost trapeziform, bifid at the apex, glabrous and smooth, sparsely hairy or velutinous on the dorsal surface, awned; palea shorter than the lemma; awns usually 3 from the upper lemmas, of which the central is stout, often purple, recurved and twisted, the side awns of the lower lemmas shorter or sometimes absent; anthers 1·5–2 mm. long.

HAB.: Open, rocky ground near the sea; fl. March–May.

DISTR.: Division 4. Turkey, Syria, Lebanon, Palestine eastwards to Pakistan.

4. Cape Greco, 1862, Kotschy.

NOTES: The only record, and possibly an error, as it is not repeated by Kotschy in Die Insel Cypern, generally an indication of second thoughts. Since the grass occurs in all adjacent countries, it may yet be re-found in Cyprus.

3. B. arvensis L., Sp. Plant., ed. 1, 77 (1753); Boiss., Fl. Orient., 5: 655 (1884); Lindberg f., Iter Cypr., 7 (1946); Osorio-Tafall et Seraphim, List Vasc. Plants Cyprus, 9 (1973).
  Serrafalcus arvensis (L.) Godr., Fl. Lorr., 3: 185 (1844); A. K. Jackson in Kew Bull., 1934: 273 (1934).
  S. arvensis (L.) Godr. var. pubescens Caldesi in Nuov. Giorn. Bot. Ital., 12: 279 (1880).
  Bromus arvensis L. f. pubescens (Caldesi) Lindberg f., Iter Cypr., 7 (1946).
  B. arvensis L. f. puberulus Lindberg f., Iter Cypr., 7 (1946).
  B. arvensis L. var. recurvatus Lindberg f., Iter Cypr., 7 (1946).

TYPE: "in Europa ad versuras agrorum".

An annual grass; culms slender, simple, loosely fasciculate, rarely solitary, erect or geniculately ascending, smooth and glabrous; leaf-blades linear-acuminate, up to 20 cm. long, 1–4 mm. wide, flat, loosely and sparsely covered with soft, white hairs, scabrid on the margins; sheaths tightly clasping toward the top of the culm, rather loose below, covered with similar hairs; ligule membranous, lacerate, up to 4 mm. long; panicle rather open, nodding, ovate in outline, up to 20 cm. long and as broad, sometimes reduced to a few spikelets; branches in pairs, unequal or occasionally the lower fascicled; spikelets lanceolate in outline when young, oblong or lanceolate-oblong, compressed when older, 1–2 cm. long, 3–4 mm. wide, 4–8-flowered; rhachilla disarticulating below each floret; glumes unequal, the lower 3-nerved, elliptic-acute, 4–6 mm. long, the upper 5–7-nerved, broadly elliptic-acute, often somewhat broader above the middle, 6–8 mm. long; lemma 7–9 mm. long, oblong-obovate when flattened, notched at the apex, 7-nerved, awned, minutely scabrid; awn straight, rough, 6–10 mm. long; anthers 3–4 mm. long; palea as long as the lemma or slightly shorter.

HAB.: Roadsides, field borders, sometimes on dry rocky slopes; sea-level to 5,500 ft. alt.; fl. March–July.

DISTR.: Divisions 1, 2, 4–7. Widespread in Europe and temperate Asia; introduced into N. America and elsewhere.

1. Erimidhes, near Ayios Yeoryios (Akamas), 1962, Meikle 2146!
2. Prodhromos, Perapedhi, Galata, Troödos, etc. Lindberg f.! Meikle 2807! D. P. Young 7822! etc.
4. Chali, near Famagusta, 1962, Meikle 2623!
5. Between Nisou and Stavrovouni, 1933, Syngrassides 18!
6. Near Dhiorios, 1962, Meikle 2351!
7. Milestone 7, Nicosia-Kyrenia road, 1952, Casey 1235! Bank of nullah, N.W. of Bellapais, 1952, Casey 1266!

NOTES: Lindberg f. recognizes three variants of this polymorphic species, viz.: f. pubescens (Caldesi) Lindberg f., f. puberulus Lindberg f. and var. recurvatus Lindberg f., the first two are distinguished by the pubescence of the culms, leaves or inflorescences, the third by its recurved awns. The distinctions are of trifling importance, and scarcely worth names. A fourth variant, var. fragilis Aschers. et Graebn. (Syn. Mitteleur. Fl. 2: 611; 1901) is no more than a starved state, with the inflorescence reduced to a raceme. It, too, cannot be considered deserving of recognition at any rank.

4. B. molliformis Lloyd, Fl. Ouest Fr., ed. 1, 536 (1854).
  Serrafalcus lloydianus Godr. ex Gren. et Godr., Fl. de France, 3: 591 (1856).

Bromus lloydianus (Godr. ex Gren. et Godr.) Nyman, Suppl. Syll. Fl. Europ., 73 (1865).
B. intermedius Guss. ssp. *divaricatus* Douin in Bonnier, Fl. Complète de France, Suisse et Belg., 12: 56 (1934).
B. hordeaceus L. ssp. *molliformis* Maire et Weiller ex Maire, Fl. Afr. Nord, 3: 255 (1955).
TYPE: France; "sables, friches, talus des fossés dans la région maritime. C. jusqu'à *Belle-ile*. — Au midi de *la Loire*, on le retrouve dans l'intérieur, à *Nancras, Pons, Saintes, Surgères, Cognac, Fontenay*, etc." *Lloyd* (NTM).

An annual grass; culms loosely tufted, rarely solitary, geniculate at the base, finally erect, smooth and glabrous, sparsely hairy at the nodes; leaf-blades linear-lanceolate, flat, covered on both surfaces with rather sparse, whitish hairs, scaberulous on the margins, up to 25 cm. long, 2–5 mm. wide; sheaths tightly clasping, the uppermost glabrous or glabrescent, the lower with sparse or dense, white, retrorse hairs; ligule membranous, 2–3 mm. long, truncate-lacerate; inflorescence a rather dense head of spikelets, seated on short hairy pedicels; spikelets elliptic in shape, erect, 1–2 cm. long, 6–9-flowered; glumes and lemmas covered with a peculiar kind of pubescence consisting of hairs which stand out, some longer, 0·75 mm., some shorter, 0·3 mm., from the dorsal surface; glumes slightly unequal in length, the lower 7 mm. long, 3 mm. wide, 3–5-nerved, elliptic-acute when flattened, the upper 7·5 mm. long, 4 mm. wide when flattened, broadly elliptic-acute, 7–9-nerved; lemmas when flattened oblong-obtuse, shortly notched at the apex, 7·5 mm. long, 3 mm. wide, 7-nerved, awned, hyaline on the margins, the hyaline portion covered with antrorse scabridities; palea much shorter than the lemma; awn scabrid, as long as the lemma; anthers 0·3–0·5 mm. long.

HAB.: On rocky ground in Cypress forest; 1,500 ft. alt.; fl. April.

DISTR.: Division 7, rare. W. Europe and Mediterranean region.

7. Halevga Forest Station, 1950, *Chapman* 115!

5. **B. diandrus** *Roth* in Bot. Abhandl., 44 (1787); N. L. Bor in C. C. Townsend et al., Fl. Iraq, 9: 141, t. 48 (1968); Osorio-Tafall et Seraphim, List Vasc. Plants Cyprus, 9 (1973).
B. gussonei Parl., Plant. Rar., 2: 8 (1840); R. M. Nattrass, First List Cyprus Fungi, 11 (1973); Rechinger f. in Arkiv för Bot., ser. 2, 1: 417 (1950).
B. rigidus Roth var. *gussonei* (Parl.) Boiss., Fl. Orient., 5: 649 (1884).
B. villosus Forssk. ssp. *gussonei* (Parl.) Holmboe, Veg. Cypr., 37 (1914).
B. villosus Forssk. var. *gussonei* (Parl.) Dinsm. in Post, Fl. Pal., ed. 2, 2: 773 (1933).
TYPE: Cultivated from "semina inter Passulas majores [raisins!] lecta" (? B).

An annual grass; culms loosely tufted, rarely solitary, erect or geniculately ascending from a decumbent base, slender or somewhat robust, simple, up to 60 cm. tall, scabrid and hairy below the panicle and nodes, or glabrous; leaf-blades linear-acuminate, loosely hairy on both surfaces, rough on the upper surface and on the margins, flat, up to 25 cm. long, 4–8 mm. wide; sheaths tightly clasping above, looser below, sparsely covered with retrorse white hairs; ligule 3–6 mm. long, membranous, lacerate; inflorescence a very loose panicle of few to many large spikelets; branches in pairs, threes or fours, up to 10 cm. long, scabrid, carrying one spikelet, rarely two; spikelets at first oblong, becoming cuneate and gaping, including the awns 7–9 cm. long, 5–8-flowered, compressed; rhachilla disarticulating below each floret; callus of the lemma obtuse; glumes unequal, narrowly lanceolate-acuminate, the lower 15–23 mm. long, 1–3-nerved, the upper 20–32 mm. long, 3–5-nerved, rather broader; lemma lanceolate-acute, bifid at the apex, scabrid on the dorsal surface, 22–36 mm. long, 7-nerved, awned; palea shorter than the lemma; awn straight, scabrid, 4–6 cm. long; anthers 0·8–1·5 mm. long. *Plate 103*.

HAB.: Cultivated and waste ground, roadsides; occasionally in garigue or in open woodland.

DISTR.: Divisions 1, 2, 4–8. Mediterranean region eastwards to Central Asia; introduced into W. Europe, N. & S. America, S. Africa, etc.

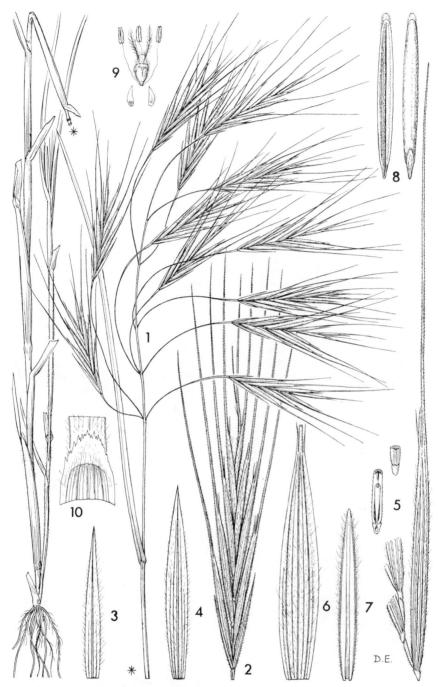

Plate 103. BROMUS DIANDRUS Roth **1,** habit, × $\frac{2}{5}$; **2,** spikelet, × 1$\frac{1}{2}$; **3,** lower glume, × 3; **4,** upper glume, × 3; **5,** rhachilla and callus, × 3; **6,** lowest lemma, × 3; **7,** its palea, × 3; **8,** grain, × 3; **9,** flower, × 3; **10,** ligule, × 2. (**1–10** from *Rawi* 9866X.)

1. Neokhorio, Phrangos, 1962, *Meikle* 2202!
2. Near Prodhromos, 1880, *Sintenis & Rigo* 888a!
4. Larnaca Salt Lake plantation, 1955, *Merton* 2004!
5. Kythrea, 1967, *Merton* 359!
6. Nicosia, nursery garden, 1932, *Syngrassides* 233! Nicosia, 1934, *Nattrass* 434; Karpasha, 1956, *Merton* 2577!
7. Yaïla, 1941, *Davis* 2824! Kyrenia, 1956, *G. E. Atherton* 1125! 1238!
8. Leonarisso, 1968, *Economides* in ARI 1066! 1137!

6. B. rigidus *Roth* in Roem. et Usteri, Bot. Mag., 4: 21 (1790); Boiss., Fl. Orient., 5: 649 (1884); Druce in Rep. B.E.C., 9: 471 (1931); R. M. Nattrass, First List Cyprus Fungi, 14 (1937).

 B. villosus Forssk., Fl. Aegypt.-Arab., 23 (1775); Holmboe, Veg. Cypr., 37 (1914); Post, Fl. Pal., ed. 2, 2: 772 (1933) non *B. villosus* Scop. (1772) nom. illeg.

 B. maximus Desf., Fl. Atlant., 1: 95, t. 26 (1798); Unger et Kotschy, Die Insel Cypern, 183 (1865).

 TYPE: Cultivated from seed of undisclosed origin (B).

An annual grass; culms loosely tufted, erect or decumbent at the base, stout, smooth and glabrous, but crisply tomentose to retrorsely hairy below the panicle, up to 60 cm. tall, often much shorter; leaf-blades flat, linear-acuminate, up to 25 cm. long, 8 mm. wide, but shorter and narrower in arid habitats, sparsely villous on both surfaces, scaberulous on the margins; inflorescence an open or compact nodding panicle 6–12 cm. long, the lower branches straight, fasciculate, up to 2 cm. long; spikelets cuneate, gaping, including the awns 6–9 cm. long, 4–9-flowered; callus of the lemmas about 1 mm. long, sharp, with retrorse hairs, stout; glumes somewhat unequal, smooth and glabrous, the lower 1·5–2 cm. long, 1-nerved, lanceolate-acuminate, the upper 2·5–3 cm. long, 3-nerved, broader, both hyaline on the margins; lemmas 2·5–3 cm. long, narrowly oblong-lanceolate, broadly silvery hyaline on the margins, bifid at the apex, with the lobes 4 mm. long, conspicuously 7-nerved, awned; palea much shorter than the lemma; awn stout, straight, scabrid, 3·5–5 cm. long; anthers up to 5 mm. long.

 HAB.: Cultivated and waste ground, often locally common on sandy ground or slacks in sand-dunes; sea-level to 4,500 ft. alt.; fl. Jan.–June.

 DISTR.: Divisions 2, 4–8. Mediterranean region, Atlantic Islands.

2. Near Prodhromos, 1862, *Kotschy* 837; and, 1880, *Sintenis & Rigo* 888b!
4. Larnaca, 1862, *Kotschy* s.n.; and, 1928 or 1930, *Druce*; Famagusta, 1936, *Syngrassides* 875! and, 1969, *Kierans* s.n.!
5. Salamis, 1905, *Holmboe* 448.
6. Nicosia, 1936, *Nattrass* 753! and, 1950, *Chapman* 526! Ayia Irini, 1951, *Merton* 291!
7. Kyrenia, 1952, *Casey* 1214!
8. Valia Minor State Forest, 1953, *Merton* 1097!

7. B. fasciculatus *Presl*, Cyper. et Gram. Sic., 39 (1820); Boiss., Fl. Orient., 5: 650 (1884); Holmboe, Veg. Cypr., 37 (1914); Post, Fl. Pal., ed. 2, 2: 773 (1933); N. L. Bor in C. C. Townsend et al., Fl. Iraq, 9: 142, t. 49 (1968); Osorio-Tafall et Seraphim, List Vasc. Plants Cyprus, 9 (1973).

 TYPE: Sicily; "in arvis arenosis, Panormi in planitie della Consulazione" (PR).

An annual grass; culms very slender, fascicled, very rarely solitary, erect or decumbent at the base, finally ascending, up to 15 cm. tall, smooth and glabrous, but shortly puberulous below the panicle; leaf-blades linear-acuminate, up to 4 cm. long, 1–2 mm. broad, covered on both surfaces with short white hairs, scaberulous on the margins; sheaths closely clasping the culms, covered with similar hairs; ligule up to 4 mm. long, lacerate; inflorescence a dense, flabellate panicle, not more than 5 cm. long including the awns, 2–3 cm. wide; spikelets, including the awns, 2–3 cm. long; glumes unequal, the lower 7–8 mm. long, subulate, 1-nerved, the upper lanceolate, acuminate, 3-nerved, 12–14 mm. long, both more or less hairy; lemmas 13–15 mm. long, 1–1·5 mm. wide, with strongly involute margins, rounded on the back, slightly arcuate outwards, glabrous or covered with very short hairs, oblong when flattened, produced at the tip into two acuminate lobes 2 mm. long, awned below the apex; awn scabrid, 13–18 mm. long, widely divaricate when mature; palea shorter than the lemma; anthers 0·3–0·4 mm. long.

HAB.: Dry banks and hillsides on Kythrean formation; 600–650 ft. alt.; fl. April.

DISTR.: Division 7. Mediterranean region eastwards to Iran.

7. Lakovounara Forest, 1950 and 1962, *Chapman* 183! 187! 358! *Merton* 962!

NOTES: Boissier (Fl. Orient., 5: 650) cites Kotschy as the authority for an unlocalized Cyprus record, but no Kotschy specimen has been seen, nor does Kotschy mention the plant in any of his writings.
The species is usually described as having two varieties, viz.: var. *fasciculatus*, in which the parts of the spikelet are glabrous, and var. *alexandrinus* Thell., in which the parts of the spikelet are hairy. The latter variety has not yet been collected in Cyprus.

8. B. intermedius *Guss.*, Fl. Sic. Prodr., 1: 114 (1827); Boiss., Fl. Orient., 5: 653 (1884); Holmboe, Veg. Cypr., 38 (1914); Post, Fl. Pal., ed. 2, 2: 775 (1933); Lindberg f., Iter Cypr., 7 (1946); N. L. Bor in C. C. Townsend et al., Fl. Iraq, 9: 146 (1968); Osorio-Tafall et Seraphim, List Vasc. Plants Cyprus, 9 (1973).
[*B. squarrosus* (non L.) Kotschy in Unger et Kotschy, Die Insel Cypern, 184 (1865); Holmboe, Veg. Cypr., 37 (1914).]
[*B. squarrosus* L. var. *villosus* (non (Gmel.) Koch) Holmboe, Veg. Cypr., 37 (1914).]
TYPE: Sicily; "in pascuis apricis collium; *Palermo a Baida, Ficuzza, Piana" Gussone* (NAP).

An annual grass; culms slender, fasciculate, erect or geniculately ascending, smooth and glabrous or sometimes puberulous below the nodes, simple, up to 80 cm. tall; leaf-blades flat, linear, tapering at the tip into a long acuminate point, more or less sparsely hairy on the upper surface, glabrous below, ciliate on the margins, up to 20 cm. long, 2–4 mm. wide; sheaths clasping the stems, the upper striate, glabrous, the lower more or less covered with sparse, white, retrorse hairs; ligule short, membranous; inflorescence a long-exserted panicle of few spikelets, loose or somewhat dense, ovate-oblong or oblong; lower branches 2–5-fasciculate, flexuous, filiform, spreading or reflexed; spikelets elliptic, finally oblong, 1·5–2·5 mm. long (including the awns), 6–10-flowered, green, rarely purplish; rhachilla disarticulating below the florets; lower glume 4–6 mm. long, narrowly elliptic-acute, 3–5-nerved, superior elliptic-acute, 6–8-nerved, both covered with crisped hairs; lemmas 8–9 mm. long, two-toothed at the tip, 7-nerved, covered on the dorsal surface with crisped hairs, awned below the apex; awns about as long as the lemmas, strongly recurved; palea shorter than the lemma, anthers 1–1·5 mm. long.

HAB.: Pastures, roadsides, fallow fields, often on sandy ground near the sea; occasionally in open woodland; sea-level to 3,900 ft. alt.; fl. April–July.

DISTR.: Divisions 2, 3, 5, 7, 8. Mediterranean region eastwards to Iran.

2. Ayios Nikolaos Mon., 1902, *Hartmann* 525; Platania, 1939, *Lindberg f.* s.n.
3. Kophinou, 1905, *Holmboe* 575.
5. Mia Milea, 1950, *Chapman* 483!
7. Halevga, 1950, *Chapman* 647! Buffavento Castle, 1970, *E. Chicken* 132!
8. Valia, Ayios Theodhoros, 1905, *Holmboe* 477!

9. B. japonicus *Thunb.*, Fl. Jap., 51, t. 11 (1784); Post, Fl. Pal., ed. 2, 2: 776 (1933); N. L. Bor in C. C. Townsend et al., Fl. Iraq, 9: 146 (1968); Osorio-Tafall et Seraphim, List Vasc. Plants Cyprus, 9 (1973).
B. patulus Mert. et Koch in Roehl., Deutsch. Fl., 1: 685 (1823); Boiss., Fl. Orient., 5: 655 (1884).
[*B. alopecuroides* (non Poir.) Sintenis teste Boiss. sec. Holmboe, Veg. Cypr., 38 (1914) quoad *Sintenis & Rigo* 1003.]

An annual grass; culms up to 60 cm. tall, erect or decumbent at the base, finally erect, solitary or more usually loosely tufted, unbranched, smooth and glabrous; leaf-blades linear-acuminate, up to 20 cm. long, 2–4 mm. broad, covered on both surfaces with more or less sparse or somewhat dense, soft, white hairs, rough on the margins; sheaths tightly clasping, the uppermost glabrous or sparsely hairy, the lower with soft white hairs; ligule 2–3 mm. long, truncate; inflorescence a somewhat compact panicle, sometimes very effuse, up to 12 cm. long, 2–9 cm. broad; lower branches 2–5-fasciculate, flexuous, filiform, carrying 1–2 spikelets; spikelets ovate-lanceolate, elliptic or oblong-elliptic in outline, compressed, 2·5–3·5 cm. long (awn included), 4–6 mm. wide, 8–12-flowered; rhachilla fragile; glumes unequal, oblong-lanceolate, acute, the upper the broader, the lower 4·5–5 mm. long, 3–5-nerved, the upper 6–7 mm. long, 5–7-nerved; lemma broadly elliptic-obtuse, 7–9-nerved, shortly notched at the tip, awned below the tip, 6–9 mm. long, 4–5 mm. wide, glabrous or hairy on the dorsal surface; palea shorter than the lemma; awns of the lower lemmas short, 1–3 mm. long, sometimes the lowest without an awn, the upper awns 8–9 mm. long, slightly curved; anthers about 1 mm. long.

var. **japonicus**

TYPE: Japan; "in Kosido et iuxta Nagasaki", *Thunberg.*

Spikelets glabrous.

HAB.: Roadsides, waste ground, grassy banks and pastures, occasionally in forest clearings; 350–4,500 ft. alt.; fl. April–June.

DISTR.: Divisions 2–6. Mediterranean region to Central Asia; introduced into many parts of the world.

2. Frequent; Pedhoulas, Yialia, Kakopetria, Prodhromos, Chakistra, etc.; *Sintenis & Rigo* 1003! *Merton* 377! 2158! 2299! *D. P. Young* 7430! *Meikle* 2727! etc.
3. Pissouri, 1967, *Merton* in ARI 670!
4. Larnaca, 1862, *Kotschy* s.n.! Akhna Forest, 1952, *Merton* 726! Larnaca Salt Lake, 1967, *Merton* in ARI 365!
5. Athalassa, 1967, *Merton* in ARI 162!
6. Nicosia Airport, 1951, *Merton* 319!

var. **velutinus** (*Koch*) *Aschers. et Graebn.*, Syn. Mitteleurop. Fl., 2 (1): 619 (1901); Bornm. in Engl. Bot. Jahrb., 61, Beibl. 140: 176 (1927).
    B. patulus Mert. et Koch var. *velutinus* Koch, Syn. Fl. Germ. Helv., ed. 1, 821 (1837).
TYPE: Not localized.

Spikelets pubescent.

HAB.: With var. *japonicus* in similar habitats.

DISTR.: Divisions 2, 3, 4, 7. General distribution as for var. *japonicus.*

2. Platania, 1955, *Merton* 2340! Kakopetria, 1955, *Merton* 2342! Spilia, 1967, *Merton* in ARI 822!
3. Lania, 1970, *A. Genneou* in ARI 1532!
4. Skarinou, 1941, *Davis* 2685! Ayios Antonios, Sotira, 1951, *Merton* 265! Dhekelia, 1972, *P. Laukkonen* 219!
7. Above Kythrea, 1880, *Sintenis & Rigo* 656!

NOTES: The two varieties are often found in the same population, and the distinction is of trifling significance.

10. **B. lanceolatus** *Roth,* Catalect. Bot., 1: 18 (1797); Sintenis in Oesterr. Bot. Zeitschr., 32: 398 (1882); Post, Fl. Pal., ed. 2, 2: 774 (1933); N. L. Bor in C. C. Townsend et al., Fl. Iraq, 9: 147 (1968); Osorio-Tafall et Seraphim, List Vasc. Plants Cyprus, 9 (1973).
    B. macrostachys Desf., Fl. Atlant., 1: 96, t. 19, fig. 2 (1798); Boiss., Fl. Orient., 5: 652 (1884); Holmboe, Veg. Cypr., 38 (1914); R. M. Nattrass, First List Cyprus Fungi, 14 (1937).
    B. divaricatus Rohde ex Loisel. in Desv., Journ. de Bot., 2: 214 (1809); Unger et Kotschy, Die Insel Cypern, 184 (1865); H. Stuart Thompson in Journ. Bot., 44: 340 (1906).

An annual grass; culms solitary or loosely tufted, erect or geniculately ascending from a decumbent base, smooth and glabrous but retrorsely bearded at the nodes, up to 80 cm. tall; leaf-blades linear, tapering to an acuminate tip, loosely covered with soft white hairs on both surfaces, flat, scabrid towards tip, up to 30 cm. long, 5 mm. wide; sheaths tightly clasping, glabrous in the upper part of the culms, covered with retrorse hairs below; ligule truncate, 2 mm. long, lacerate; inflorescence usually dense, less often loose, oblong or lanceolate in outline, up to 20 cm. long, usually erect; spikelets 2–3 cm. long, 8–16-flowered, glabrous or hairy; glumes unequal, oblong or oblong-elliptic, acute, the lower 9–15 mm. long, 3–5-nerved, the upper 12–20 mm. long, 7–9-nerved, both hyaline on the margins; lemmas oblong-obovate, acute, 12–15 mm. long, bifid at the apex, 9–11-nerved, awned; palea shorter than the lemma; awn divaricate at maturity, inserted 2–3 mm. below tip, 10–12 mm. long; anthers 1·5–2 mm. long.

var. **lanceolatus**
TYPE: Cultivated from seed communicated by Roemer (B).

Spikelets glabrous.

HAB.: Fields near the coasts and at high altitudes; up to 4,000 ft. alt.; fl. April–May (see Notes).

DISTR.: Division 2 (?) Eastern Mediterranean region and North Africa eastwards to Central Asia; introduced elsewhere.

2. "Frequens in campestribus maritimis et montanis circa Prodhromos altitud. 1–4000 ped. mense Aprilis et Maii, 1862, *Kotschy* 798" (see Notes)!

NOTES: The only record for typical *B. lanceolatus* from Cyprus. The specimen was identified by Kotschy as *B. madritensis* L., and it is clear that his notes are to be read as referring to the general distribution within the island of what he considered to be a frequent grass; whether the Kew specimen numbered *Kotschy* 798 came from Prodhromos is open to doubt, for it is not positively stated to have been collected there. One other specimen from Prodhromos (*Sintenis & Rigo* 1089!) may possibly be the same as *Kotschy* 798, but the Kew material is over-mature and difficult to identify with certainty. The frequent occurrence of *B. lanceolatus* Roth var. *lanuginosus* (Poir.) Dinsm. (infra) in the same area suggests that further search might lead to the re-finding of var. *lanceolatus*.

var. **lanuginosus** (*Poir.*) *Dinsm.* in Post, Fl. Pal., ed. 2, 2: 774 (1933); N. L. Bor in C. C. Townsend et al., Fl. Iraq, 9: 148 (1968).
    B. lanuginosus Poir. in Lam., Encycl. Méth. Suppl., 1: 703 (1810).
    B. macrostachys Desf. var. *lanuginosus* (Poir.) Coss. et Durieu, Fl. d'Algérie II, Phan., Glumacées [in Expl. Sci. Algérie]: 162 (1855); Boiss., Fl. Orient., 5: 652 (1884); Holmboe, Veg. Cypr., 38 (1914); Rechinger f. in Arkiv för Bot., ser. 2, 1: 417 (1950).
TYPE: France; "en Provence, aux environs de Toulon et aux îles d'Hières" (P).

Spikelets hairy.

HAB.: Roadsides, waste ground, fallows and pastures, often on ground that is flooded in winter; sea-level to 4,600 ft. alt.; fl. April–July.

DISTR.: Divisions 1–7, locally common. General distribution as for *B. lanceolatus* Roth var. *lanceolatus*.

1. Dhrousha, 1941, *Davis* 3221!
2. Near Prodhromos, 1880, *Sintenis & Rigo* 1004!, also, 1961, *D. P. Young* 7290! Platres, 1939, *Kennedy* 1536!
3. Zakaki, 1902, *Hartmann* 527! Limassol aerodrome, 1950, *Chapman* 558! Akrotiri marshes, 1952, *Merton* 752.
4. Kouklia near Lysi, 1962. *Meikle* 2616!
5. Strongylos, 1967, *Merton* in ARI 460!
6. Myrtou, 1932, *Syngrassides* 24! 206! Asomatos, 1956, *Merton* 2735! Near Morphou, 1972, *W. R. Price* 1055!
7. Common about Kyrenia, Larnaka tis Lapithou, Ayios Epiktitos; *Casey* 1251! *Merton* 2082! 2093! 2702! ARI 570! *G. E. Atherton* 1343! *P. Laukkonen* 241!

11. **B. madritensis** *L.*, Cent. I Plant., 5 (1755), Amoen. Acad., 4: 265 (1759); Unger et Kotschy, Die Insel Cypern, 183 (1865); Boiss., Fl. Orient., 5: 649 (1884); Holmboe, Veg. Cypr., 37 (1914); Post, Fl. Pal., ed. 2, 2: 772 (1933); Lindberg f., Iter Cypr., 7 (1946); N. L. Bor in C. C. Townsend et al., Fl. Iraq, 9: 148 (1968); Osorio-Tafall et Seraphim, List Vasc. Plants Cyprus, 10 (1973).

An annual grass; culms mostly loosely tufted, rarely solitary, erect or geniculately ascending from a decumbent base, simple, smooth and glabrous or minutely puberulous below the panicle, up to 60 cm. tall; leaf-blades linear-acuminate, up to 20 cm. long, 5 mm. wide, sparsely covered on both surfaces and ciliate on the margins with soft, white hairs, rarely completely glabrous or ciliate on the margins only, rough on the margins; inflorescence a dense or somewhat loose panicle, often cuneate in shape, sometimes nodding, up to 12 cm. long or longer; branches up to 3 cm. long, ascending or slightly spreading, scaberulous; spikelets 1–2 on a branch, cuneate, compressed, including the awns 3–6 cm. long, 6–13-flowered; glumes unequal, the lower subulate, 1-nerved, 6–11 mm. long, the upper oblong-elliptic, acute when flattened, 3-nerved, 10–16 mm. long, both membranous on the margins; lemmas oblong-elliptic, acute, 12–19 mm. long, notched at the tip, awned; palea shorter than the lemma; awn straight, scabrid, 12–16 mm. long; anthers 0·5–1 mm. long.

var. **madritensis**
TYPE: "*in* Hispania" (LINN!).

Lemmas glabrous.

HAB.: Cultivated and waste ground, roadsides, a common weed; sea-level to c. 5,700 ft. alt.; fl. March–June.

DISTR.: Divisions 1, 2, 4–7. Mediterranean region eastwards to Afghanistan; Atlantic Islands. Widely naturalized in Europe, Australia, N. & S. America, etc.

1. Dhrousha, 1941, *Davis* 3224!
2. Platres, 1937, *Kennedy* 59B! Troödos, "Olympus Camp Hotel", 1939, *Lindberg f.* s.n.! Prodhromos, 1961, *D. P. Young* 7307!
4. Ayios Antonios, Sotira, 1951, *Merton* 256!
5. Lapathos, 1962, *Meikle* 2963! Athalassa, 1967, *Merton* in ARI 98!
6. Monastery farm 1 m. S.W. of Nicosia, 1927, *Rev. A. Huddle* 109! Nicosia, 1950, *Chapman* 531! and, 1973, *P. Laukkonen* 212! 257! Near Morphou, 1972, *W. R. Price* 1056!
7. Tchiklos, 1956, *G. E. Atherton* 1136! 1½ miles W. of Larnaka tis Lapithou, 1956, *Merton* 2699!

var. **ciliatus** *Guss.*, Fl. Sic. Syn., 1: 78 (1843); Post, Fl. Pal., ed. 2, 2: 772 (1933); N. L. Bor in C. C. Townsend et al., Fl. Iraq, 9: 150 (1968).
TYPE: Sicily; "*Palermo, Terranova*", *Gussone* (NAP).

Lemmas hairy.

HAB.: Cultivated and waste ground, roadsides; sometimes on stabilized sand-dunes; sea-level to 600 ft. alt.; fl. Febr.–April.

DISTR.: Divisions 1, 3, 4, 6–8. General distribution that of var. *madritensis*.

1. Above Peyia, 1962, *Meikle* 2030!
3. Curium, 1961, *Polunin* 6695! 2 m. S. of Pyrgos, 1963, *J. B. Suart* 86!
4. Larnaca aerodrome, 1950, *Chapman* 447!
6. Nicosia airport, 1951, *Merton* 320!
7. Kyrenia, 1949, *Casey* 374! 415! and, 1956, *G. E. Atherton* 1347!
8. Ayios Philon near Rizokarpaso, 1941, *Davis* 2270!

NOTES: In addition to the above records, *B. madritensis* L. in the aggregate sense has been noted from:

2. Kakopetria, 1939, *Lindberg f.* s.n.
3. Kophinou, 1905, *Holmboe* 576.
4. Larnaca, 1880, *C. & W. Barbey* 946.
5. Kythrea, 1880, *Sintenis & Rigo* 369.
7. (or 8). Kantara, 1880, *Sintenis & Rigo* 368.

In the absence of specimens, these records cannot be determined at varietal level.

B. madritensis *L.* × **B. rubens** *L.*

Intermediate in vegetative and floral characters between the presumed parents.

HAB.: Poor pastures and rocky hillsides; sea-level to 2,000 ft. alt.; fl. March–April.

DISTR.: Divisions 3, 4, 7; to be found wherever the two parent species grow together.

3. Yerasa, 1941, *Davis* 3079 !
4. Athna, 1956, *Merton* 2666 !
7. Kyrenia, Mylos plantation, 1949, *Casey* 419 !

12. **B. rubens** *L.*, Cent. I Plant., 5 (1755), Amoen. Acad., 4: 265 (1759); Sibth. et Sm., Fl. Graec. Prodr., 1: 63 (1806), Fl. Graec., 1: 66, t. 83 (1808); Unger et Kotschy, Die Insel Cypern, 183 (1865); C. et W. Barbey, Herborisations au Levant, 99 (1882); Boiss., Fl. Orient., 5: 650 (1884); Holmboe, Veg. Cypr., 37 (1914); Post, Fl. Pal., ed. 2, 2: 773 (1933); R. M. Nattrass, First List Cyprus Fungi, 14 (1937); N. L. Bor in C. C. Townsend et al., Fl. Iraq, 9: 151, t. 51 (1968); Osorio-Tafall et Seraphim, List Vasc. Plants Cyprus, 10 (1973).

An annual grass; culms loosely tufted or solitary, up to 45 cm. tall, erect or decumbent at the base, smooth and glabrous but densely and shortly hairy below the panicle, up to 45 cm. tall, simple; leaf-blades flat, up to 12 cm. long, 5 mm. wide, linear-acuminate, loosely hairy on both surfaces, scaberulous and ciliate on the margins; sheaths closely clasping, the upper slightly keeled and sometimes glabrous, the lower densely covered with white, short, retrorse hairs; ligule up to 5 mm. long, lacerate; inflorescence a very dense, erect panicle, more or less long-exserted, obovate or obovate-oblong, cuneate at the base, up to 10 cm. long, usually much shorter; branches short, the lower in fascicles, hairy; spikelets somewhat wedge-shaped, 4–10-flowered, 25–35 cm. long (including the awns), often reddish in colour; glumes unequal, hyaline on the margins, hairy or glabrous, the lower subulate, 1-nerved, 5–7 mm. long, the upper narrowly lanceolate, acuminate, 3-nerved, 8–10 mm. long, both glabrous or hispid; lemmas 10–13 mm. long, 2–2·5 mm. wide, oblong-elliptic, 5–7-nerved, rounded on the back, bifid at the acute tip, glabrous or hairy, awned below the tip; palea shorter than the lemma; awn 10–12 mm. long; anthers 2, 0·3–1 mm. long.

var. **rubens**
TYPE: Spain; "*in* Hispania. *Loefling.*"

Spikelets hairy.

HAB.: Poor pastures and waste ground; sea-level to 550 ft. alt.; fl. March–April.

DISTR.: Divisions 4, 6, 7. Mediterranean region eastwards to Central Asia; Atlantic Islands; introduced into N. America.

4. Larnaca aerodrome, 1950, *Chapman* 14 !
6. Nicosia airport, 1951, *Merton* 321 !
7. Kyrenia, 1952, *Casey* 1241 !

var. **glabriglumis** *Maire* in Emberger et Maire, Cat. Plant. Maroc., 4: 943 (1941); N. L. Bor in C. C. Townsend et al., Fl. Iraq, 9: 152 (1968).
TYPE: Morocco; not localized.

Spikelets glabrous.

HAB.: Poor pastures and rocky hillsides, frequent on Kythrean formation; sea-level to 1,000 ft. alt.; fl. March–May.

DISTR.: Divisions 4, 5–7. General distribution as for var. *rubens*.

4. Dhekelia, 1880, *Sintenis & Rigo* 891 !
5. Between Nisou and Stavrovouni, 1933, *Syngrassides* 33 !
6. Near Kythrea, 1880, *Sintenis & Rigo* 366 !
7. Kyrenia, 1927, *Rev. A. Huddle* 108 ! Lakovounara Forest, 1950, *Chapman* 71 !

NOTES: The following records have not been examined, and cannot be determined varietally:
2. Mandria, 1936, *Nattrass* 823.
4. Near Larnaca, 1862, *Kotschy* 307a and, 1880, *C. & W. Barbey* 944; Phaneromene near Larnaca, 1905, *Holmboe* 155; Ayia Napa, 1905, *Holmboe* 58.
8. Between Apostolos Andreas Mon. and Aphendrika, 1880, *Sintenis & Rigo*.

13. B. scoparius *L.*, Cent. I Plant., 6 (1755), Amoen. Acad., 4: 266 (1759); Sibth. et Sm., Fl. Graec. Prodr., 1: 63 (1806); Boiss., Fl. Orient., 5: 650 (1884); Holmboe, Veg. Cypr., 38 (1914); Post, Fl. Pal., ed. 2, 2: 773 (1933); R. M. Nattrass, First List Cyprus Fungi, 14 (1937); N. L. Bor in C. C. Townsend et al., Fl. Iraq, 9: 152 (1968); Osorio-Tafall et Seraphim, List Vasc. Plants Cyprus, 10 (1973).

An annual grass; culms loosely tufted, rarely solitary, erect or decumbent at the base and finally erect, smooth and glabrous, simple, slender, up to 50 cm. tall; leaf-blades flat, linear-acuminate, up to 20 cm. long, 4 mm. wide, usually much shorter and narrower, glabrous on the undersurface, loosely hairy on the upper surface with long white hairs, scabrid on the margins; inflorescence a dense panicle, cylindrical, rounded at the top, somewhat wedge-shaped at the base, interrupted at times, up to 7 cm. long, 2·5 cm. wide; branches 2–6-fasciculate, very short; spikelets oblong-acute, up to 3 cm. long (awns included), gaping a little at flowering time, 8–12-flowered; glumes somewhat unequal, hyaline on the margins, the lower 3–4 mm. long, 3-nerved, elliptic-acute, the upper 5–7 mm. long, 3-nerved, oblong-elliptic, acute; lemmas oblong-oblanceolate when flattened, 8–10 mm. long, hyaline on the margins, glabrous or hairy on the back, 7-nerved, bifid, awned; palea shorter than the lemma; awn up to 14 mm. long, slightly twisted, recurved at maturity; anthers 0·3–1 mm. long.

var. **scoparius**
    B. scoparius L. var. *stenanthus* Stapf in Kew Bull., 1907: 369 (1907).
TYPE: Spain; "*in* Hispania" (LINN !).

Spikelets glabrous.

HAB.: Cultivated and waste ground, roadsides; dry rocky hillsides; sea-level to 2,000 ft. alt.; fl. Febr.–April.

DISTR.: Divisions 1, 3–8, locally common; Mediterranean region eastwards to Central Asia; introduced into N. America.
1. Above Peyia, 1962, *Meikle* 2029 ! Ayios Yeoryios Island (Akamas), 1962, *Meikle* 2158 !
3. Zakaki, 1902, *Hartmann* 526 ! 2 miles S of Pyrgos, 1963, *J. B. Suart* 83 !
4. Near Old Larnaca, 1880, *Sintenis & Rigo* 886 ! Ayios Antonios, Sotira, 1951, *Merton* 254 ! 264 ! Cape Greco, 1958, *N. Macdonald* 45 ! Famagusta, 1969, *Kierans* s.n. !
5. Common; Kythrea, Knodhara, Laxia, etc. *Sintenis & Rigo* 367 ! 376b ! *Syngrassides* 77 ! 378 ! 941 ! *Merton* 204 ! etc.
6. Nicosia, 1950, *Chapman* 527 !
7. Lakovounara Forest, 1950, *Chapman* 153 ! Ayios Epiktitos, 1951, *Casey* 1116 ! 1½ miles W. of Larnaka tis Lapithou, 1956, *Merton* 2700 !
8. Gastria, 1941, *Davis* 2426 ! Platanisso, 1950, *Chapman* 247 !

NOTES: *B. scoparius* L. var. *stenanthus* Stapf, allegedly distinguished from the type by its longer and narrower lemmas, is recorded from Cyprus by Stapf (*loc. cit.*, 369) on the strength of specimens collected on the hills above Kythrea (*Sintenis & Rigo* 367 !). It was subsequently noted from Limassol, 1928 or 1930, by Druce (Rep. B.E.C., 9: 471; 1931). The variety is hardly more than a favourable habitat-state of typical *B. scoparius*.

var. **hirtulus** *Regel* in Act. Hort. Petrop., 7: 602 (1881).
    var. *villiglumis* Maire et Weiller in Maire, Fl. Afr. Nord, 3: 259 (1955) quoad descr.
TYPE: Central Asia: "Taschkent (Krause), Katti-kurgan, prope Samarkand (O. Fedtschenko), ad fluvium Syr-Darja (Golike)" (LE).

Spikelets hairy.

HAB.: Roadsides, waste ground, dry hillsides; sea-level to 2,000 ft. alt.; fl. March–April.

DISTR.: Divisions 2, 7. General distribution that of var. *scoparius*.
2. Gourri, 1950, *Chapman* 289 !
7. Kyrenia, 1956, *G. E. Atherton* 1184 ! 1270 !

14. B. alopecuros *Poir.*, Voy. en Barbarie, 2: 100 (1789); Boiss., Fl. Orient., 5: 651 (1884); Post, Fl. Pal., ed. 2, 2: 774 (1933); Osorio-Tafall et Seraphim, List Vasc. Plants Cyprus, 10 (1973).

B. alopecuroides Poir., in Lam., Encycl. Méth., Suppl. 1: 703 (1810); Holmboe, Veg. Cypr., 38 (1914).

An annual grass; culms mostly fasciculate, rarely solitary, erect or geniculately ascending, smooth and glabrous, or retrorsely pilose on the nodes and often on the stems below the nodes, up to 60 cm. tall; leaf-blades linear-acuminate, flat, rather thin, flaccid, up to 20 cm. long, 2–4 mm. wide, sparsely pilose on the upper surface, glabrous below, ciliate and scaberulous on the upper surface and on the margins; sheaths striate, closely clasping, retrorsely hairy below; ligule membranous, 1–2 mm. long, lacerate; inflorescence long exserted, of more or less closely packed, shortly pedicellate, ascending spikelets on a stout, scabrid axis; spikelets 8–15-flowered, cylindrical, 2–4 cm. long; glumes unequal, hyaline on the margins, oblong-elliptic, acute, the lower 6–8 mm. long, 3-nerved, the upper 10–12 mm. long, 6–9-nerved; lemmas oblong-obovate, acute, 10–14 mm. long, glabrous or shortly hairy on the dorsal surface, bifid at the apex, hyaline on the margins, 9-nerved, awned 3–4 mm. below the tip; awn twisted at the base and divaricate at maturity, up to 18 mm. long; anthers about 1·5 mm. long.

var. **alopecuros**
TYPE: Algeria; "dans les prairies aux environs de la Calle" (P).

Spikelets hairy.

HAB.: Roadsides, waste ground, dry hillsides in garigue; 200–1,500 ft. alt.; fl. March–May.

DISTR.: Divisions 3, 4, 6, 7. Mediterranean region and western Asia.

3. Mile 27, Nicosia-Limassol road, 1950, *Chapman* 557.
4. Athna Forest, 1952, *Merton* 726B.
6. Myrtou, 1932, *Syngrassides* 200! Near Panagra, 1962, *Meikle* 2401!
7. Kyrenia, 1949–55, *Casey* 486! 1231! 1306 (partly)! *Merton* 2081! Halevga Forest, 1950, *Chapman* 645!

var. **calvus** *Hal.*, Consp. Fl. Graec., 3: 400 (1904).
TYPE: Greece; Heldreich pl. fl. hellen. a. 1880.

Spikelets glabrous.

HAB.: Roadsides, waste ground, hillsides in garigue, vineyards, etc.

DISTR.: Divisions 6, 7; general distribution that of *B. alopecuros* Poir. var. *alopecuros*.

6. Asomatos, 1956, *Merton* 2734!
7. Larnaka tis Lapithou, 1936, *Syngrassides* 937! and, 1956, *Merton* 2698! Kornos, 1956, *Poore* s.n.! Panagra, 1962, *Meikle* 2401!

NOTES: [Four of the specimens cited above (*Merton* 2081, 2698, 2734 and *Casey* 1231) are referred to a new species, *Bromus caroli-henrici* Greuter by its author (in Ann. Naturhist. Mus. Wien, 75: 83–89; 1971). The new species differs from *B. alopecuros* chiefly in having a narrower inflorescence, with the spikelets usually solitary at each node, in its acuminate lemmas with deeply and narrowly emarginate, aristulate apices, and in longer anthers. *B. caroli-henrici* is said to have a wide distribution in the eastern Mediterranean, from Crete and the southern Aegean islands through S. Turkey and Syria to S. Palestine. Since all the Cyprus material determined above as *B. alopecuros* is clearly conspecific, it may reasonably be assumed that it should all be re-named *B. caroli-henrici*, if the distinctions between this and *B. alopecuros* prove valid. R.D.M.]

15. **B. squarrosus** *L.*, Sp. Plant., ed. 1, 76 (1753); Boiss., Fl. Orient., 5: 651 (1884); Post, Fl. Pal., ed. 2, 2: 774 (1933); N. L. Bor in C. C. Townsend et al., Fl. Iraq, 9: 158 (1968); Osorio-Tafall et Seraphim, List Vasc. Plants Cyprus, 10 (1973).
TYPE: "*in* Gallia, Helvetia, Sibiria" (LINN!).

An annual or biennial grass; culms loosely tufted, rarely solitary, erect or spreading, smooth and glabrous or shortly hairy below and at the nodes, simple, up to 40 cm. tall; leaf-blades greyish-green, up to 15 cm. long, 4 mm. broad, flat or rolled when dry, linear-acuminate, glabrous below, covered loosely on the upper surface with soft white hairs, rough on the margins; sheaths closely appressed, striate, the lower covered with thick retrorse hairs, the upper glabrescent; ligule about 2 mm. long, lacerate, villous on the outer surface; inflorescence few-spiculate, up to 15 cm. long, usually nodding; branches slender, the lowest solitary or in fascicles of 2–4, somewhat thickened at the tips, each carrying 1–2 spikelets; spikelets large, often 5 cm. long, elliptic or oblong-elliptic in outline, compressed, up to 1·5 cm. broad, 8–30-flowered; rhachilla hairy, disarticulating below the mature florets; glumes unequal, ovate-acute, broadly scarious on the margins, the lower 6–9 mm. long, 5-nerved, the upper 8–12 mm. long, 9-nerved; lemmas when flattened very broadly elliptic, rhomboidal or oblong-obovate, 10–11 mm. long, 5–6 mm. broad, with very broadly scarious margins forming an obtuse angle, 9–11-nerved, slightly notched at the apex, awned below the tip; palea shorter than the lemma; awn reflexed almost at right angles, slightly twisted; anthers 1–1·5 mm. long.

HAB.: Roadsides and dry slopes; 250–2,500 ft. alt.; fl. March–May.

DISTR.: Divisions 2, 5, rare; Mediterranean region eastwards to Central Asia; introduced into W. Europe, and into other parts of Asia.

2. Stavros tis Psokas, 1951, *Merton* 458!
5. Athalassa, 1936, *Syngrassides* 1172!

NOTES: Holmboe misunderstood this species: his records (Veg.Cypr.) from Vatili (*Holmboe* 319) and Valia (*Holmboe* 477) are misidentifications; no. 477 is *B. intermedius* Guss.; no. 319 is a very starved plant, and not identifiable with certainty. The specimen from Kophinou (*Holmboe* 575) named *B. squarrosus* L. var. *villosus* Koch, is likewise a form of *B. intermedius*. One of Kotschy's unlocalized records for *B. squarrosus* L. (*Kotschy* 270) is similarly an error for *B. intermedius*, and the same is probably true for his no. 214, but this specimen has not been examined.

16. B. sterilis *L.*, Sp. Plant., ed. 1, 77 (1753); Sibth. et Sm., Fl. Graec. Prodr., 1: 62 (1806); Boiss., Fl. Orient., 5: 648 (1884); Druce in Rep. B.E.C., 9: 471 (1931); Post, Fl. Pal., ed. 2, 2: 771 (1933); Lindberg f., Iter Cypr., 7 (1946); N. L. Bor in C. C. Townsend et al., Fl. Iraq, 9: 158, t. 54 (1968); Osorio-Tafall et Seraphim, List Vasc. Plants Cyprus, 10 (1973).

An annual plant; culms solitary or loosely fasciculate, erect or shortly decumbent at the base and finally erect, slender, smooth and glabrous, up to 80 cm. tall; leaf-blades flat, flaccid, linear-acuminate, softly hairy on both surfaces, scabrid on the margins; inflorescence a very loose open panicle, erect at first then nodding, really a raceme in that the long filiform branches carry a solitary spikelet, very rarely two; spikelets cuneate, including the awns 4–6 cm. long, 4–10-flowered; rhachilla very fragile below each floret; glumes unequal, the lower subulate, 1-nerved, 6–14 mm. long, the upper lanceolate-acuminate or oblong-lanceolate, 16–20 mm. long, 3-nerved; lemmas oblong or lanceolate, acuminate, 7-nerved, bifid at the apex, awned below the tip, scabrid on the nerves, narrowly hyaline on the margins; palea shorter than the lemma; awn straight, 15–30 mm. long; anthers 1–1·8 mm. long.

var. **sterilis**
TYPE: "*in* Europae *australioris agris, sylvis*" (LINN!).

Lemmas glabrous.

HAB.: Roadsides, waste or cultivated ground, or in maritime or montane garigue; sea-level to 4,500 ft. alt.; fl. March–June.

DISTR.: Divisions 2, 4, 7, 8; Atlantic Europe and Mediterranean region eastwards to Central Asia; introduced elsewhere.

2. Prodhromos, 1880, *Sintenis & Rigo* 888! Platania, 1939, *Lindberg f.* s.n.! and, 1955, *Merton* 2353! Kakopetria, 1955, *Merton* 2300!
4. Larnaca Salt Lake, 1967, *Merton* in ARI 421!

7. Near Kythrea, 1880, *Sintenis & Rigo* 369! Kyrenia, 1949, *Casey* 420! and, 1956, *G. E. Atherton* 1195!
8. Near Kantara, 1880, *Sintenis & Rigo* 368!

var. **velutinus** *Volkart* in Hegi, Illustr. Fl. Mitteleur., 1: 362 (1908).
TYPE: Switzerland, without precise locality.

Lemmas hairy.

HAB.: On brackish ground; near sea-level; fl. April–July.

DISTR.: Division 4. General distribution that of *B. sterilis* L. var. *sterilis*.

4. Larnaca, 1862, *Kotschy* 256a!

NOTES: The following records for *B. sterilis*, which have not been checked, cannot be assigned with certainty to a variety: (2) Prodhromos, 1939, *Lindberg f.*; (3) Limassol, 1928 or 1930, *Druce*; (8) between Apostolos Andreas Monastery and Aphendrika, 1880, *Sintenis*.

17. B. tectorum *L.*, Sp. Plant., ed. 1, 77 (1753); Sibth. et Sm., Fl. Graec. Prodr., 1: 63 (1806), Fl. Graec., 1: 65, t. 82 (1808); Unger et Kotschy, Die Insel Cypern, 183 (1865); Boiss., Fl. Orient., 5: 647 (1884); H. Stuart Thompson in Journ. Bot., 44: 340 (1906); Holmboe, Veg. Cypr., 37 (1914); Post, Fl. Pal., ed. 2, 2: 771 (1933); Lindberg f., Iter Cypr., 7 (1946); N. L. Bor in C. C. Townsend et al., Fl. Iraq, 9: 160 (1968); Osorio-Tafall et Seraphim, List Vasc. Plants Cyprus, 10 (1973).

An annual grass; culms loosely fascicled, sometimes solitary, erect or, when several, slightly geniculate at the base, simple, smooth and glabrous or minutely hairy below the inflorescence; leaf-blades linear-acuminate, flat, up to 15 cm. long, 2–3 mm. wide, the lowest often involute-filiform; softly and shortly hairy on both surfaces; sheaths tubular, the upper smooth and glabrous, shining, the lower softly hairy; ligules membranous, lacerate, 3–5 mm. long; panicle often dense, characteristic; the spikelets shortly pedicellate on the branches or turned to one side in parallel groups; spikelets wedge-shaped, finally gaping, 3–9-flowered, excluding the awns up to 3 cm. long; glumes narrowly elliptic or oblong, acute or acuminate, the lower 1-nerved, 5–7 mm. long, the upper 3-nerved, 8–12 mm. long, glabrous; lemmas elliptic-acute, 2-lobed at the apex, 9–15 mm. long, glabrous or hairy, 7-nerved, membranous on the margins, awned just below the tip; awn scabrid, straight, 10–15 mm. long, anthers small, 1 mm. long or less.

var. **tectorum**
TYPE: "*in* Europae *collibus siccis & tectis terrestribus*" (LINN!).

Lemmas glabrous.

HAB.: Roadsides, waste ground and stony slopes at high altitudes on igneous formations; 3,800–4,630 ft. alt.; fl. April–June.

DISTR.: Division 2. Central Europe and Mediterranean region eastwards to China; Atlantic Islands; introduced elsewhere.

2. Prodhromos, 1961, *D. P. Young* 7299! 7368! N. side of Tripylos, 1962, *Meikle* 2685!

var. **hirsutus** *Regel* in Act. Hort. Petrop., 7: 600 (1880), Descr. Plant. Nov., fasc. 8: 61 (1881); N. L. Bor in C. C. Townsend et al., Fl. Iraq, 9: 161 (1968).
TYPE: U.S.S.R.; "in montibus karatavicis in promontoriis Mogal-tau, Kcharli-tau et Dschinbulak (Sewerzow)" (LE).

Lemmas hairy.

HAB.: Rocky slopes and summits, or open Pine forest on igneous formations at high altitudes; sometimes a component of dwarf, snow-patch vegetation; 2,400–6,400 ft. alt.; fl. April–July.

DISTR.: Division 2. General distribution that of *B. tectorum* var. *tectorum*.

2. Summit of Khionistra, 1880, *Sintenis & Rigo* 887! also, 1937, *Kennedy* 58! and, 1961, *D. P. Young* 7328!, and, 1973, *P. Laukkonen* 951! Platres, 1937, *Kennedy* 59! 59A! Troödos, 1939, *Lindberg f.* s.n.! Livadhi tou Pasha, 1951, *Merton* 389! 588! Amiandos, 1952, *Merton* 818! Kakopetria, 1955, *Merton* 2300A! Selladhi tou Mavrou Dasous, above Spilia, 1962, *Meikle* 2812! Palekhori, 1969, *Economides* in ARI 1342!

NOTES: Kotschy's records for *B. tectorum* from Larnaca and Cape Greco (Die Insel Cypern, 183) are almost certainly errors, since the species is apparently confined, in Cyprus, to high altitudes. In the absence of specimens, it is difficult to judge what species was intended. The "gracile form" mentioned by Holmboe (Veg. Cypr., 37) is simply a starved state of *B. tectorum*; such forms are common on Khionistra.

B. CARINATUS *Hook. et Arn.*, Bot. Beechey Voy., 403 (1940). A robust annual or biennial with erect culms up to 1 m. tall, and scaberulous, often ciliate, leaves up to 45 cm. long and 3–12 mm. wide; inflorescence an effuse, nodding panicle up to 40 cm. long and 30 cm. wide, with 2–3 scabrid branches, up to 10 cm. long, at each node; spikelets oblong or ovate-oblong, 2–4 cm. long, 0·6–1 cm. wide, greyish or glaucous, finally gaping; glumes unequal, the lower 6–9 mm. long, 3–5-nerved, hyaline at the margins, the upper 10–13 mm. long, acuminate, 5–7-nerved; lemmas oblong-acuminate, 10–15 mm. long, minutely pilose or glabrous, awned from the tip, awn 7–15 mm. long; palea shorter than or nearly as long as the lemma; anthers 4–5 mm. long.

A native of western North America, cultivated and naturalized in many parts of Europe, and elsewhere. There is evidence (comm. Merton) that it was introduced into Cyprus in 1954 as a possibly useful forage grass, but there are no records of its subsequent performance, or of its present status on the island.

Another robust Brome, B. CATHARTICUS *Vahl*, Symb. Bot., 2: 22 (1791) (*B. unioloides* H.B.K.; *B. willdenowii* Kunth) is reported (Jones & Merton, Rep. Pasture Res. Survey and Dev. Cyprus, 40; 1958) as being produced, on an experimental scale, in the Plant Introduction Nursery. It resembles *B. carinatus* in habit, but the glumes are blunter, and the lemmas mucronate, or at most with a very short awn, up to 2 mm. long; the palea is conspicuously shorter than the lemma.

B. catharticus is a native of South America, now widely naturalized in many parts of the world, and popularly known as "Rescue Grass", on account of its value as a forage plant. Nothing is known of its history and behaviour in Cyprus.

TRIBE 11. **Brachypodieae** *Harz* Spikelets all alike, very shortly pedicellate, terete, several- to many-flowered; rhachilla disarticulating above the glumes and between the florets; inflorescence terminal, spike-like with one side of the lemmas adjacent to the rhachis; glumes unequal, persistent, the lower 1–5-nerved, the upper 7–9-nerved; lemmas 7–9-nerved, rounded on the back, awned from the tip; palea 2-keeled, hyaline, pectinately ciliate; awn straight, scabrid; stamens 3; ovary with a terminal, hairy appendage; styles 2; stigmas plumose; lodicules 2. Grain adherent to the palea and lemma; hilum linear; starch grains simple. Annual or perennial grasses; leaf-blades linear-acuminate, flat, rounded at the base, with festucoid anatomy; silica-bodies oblong; hairs 1-celled; ligules short, membranous. Chromosomes small; basic number 5, 7, 9.

44. BRACHYPODIUM *P. Beauv.*
Essai Agrost., 100 (1812).
H. Mühlberg in Feddes Repert., 81: 119–130 (1970).

Slender annual or perennial herbs with narrow, flat, acuminate leaf-blades; inflorescence a terminal raceme with the spikelets seated on very short pedicels, nodding or erect; spikelets terete, tapering towards the apex, several- to many-flowered; florets bisexual or the uppermost reduced;

rhachilla disarticulating above the glumes and between the florets; glumes unequal, persistent; lemmas membranous, rounded on the back, lanceolate-acuminate, awned from the tip; palea as long as the lemma, 2-keeled; stamens 3; lodicules 2, lanceolate-acuminate, oblique, ciliate on the margins. Grain with an apical, hairy appendage, narrowly elliptic or oblong, tightly invested by the lemma and palea; embryo very small, about one-eighth the length of the grain; hilum linear, the length of the grain.

Eight or nine species in temperate regions, or on mountains in the tropics.

Plants rhizomatous; awns of the lemmas shorter than the lemmas - - - **1. B. pinnatum**
Plants without rhizomes; awns as long as, or longer than the lemmas:
  Leaf-sheaths hairy; plants green - - - - - - - - **2. B. sylvaticum**
  Leaf-sheaths glabrous; plants yellowish or glaucous-green - - - **3. B. firmifolium**

1. B. pinnatum (*L.*) *P. Beauv.*, Essai Agrost., 101, 155, t. 19, fig. 3 (1812); Boiss., Fl. Orient., 5: 658 (1884); Post, Fl. Pal., ed. 2, 2: 778 (1933); R. M. Nattrass, First List Cyprus Fungi, 14 (1937); Lindberg f., Iter Cypr., 7 (1946); N. L. Bor in C. C. Townsend et al., Fl. Iraq, 9: 167, t. 57 (1968); Osorio-Tafall et Seraphim, List Vasc. Plants Cyprus, 10 (1973).
    Bromus pinnatus L., Sp. Plant., ed. 1, 78 (1753); Sibth. et Sm., Fl. Graec. Prodr., 1: 63 (1806).
    TYPE: "*in* Europae *sylvis montosis asperis*" (LINN!).

A perennial grass with widely spreading, scaly rhizomes, loosely or densely caespitose; culms erect, simple, smooth and glabrous, up to 100 cm. tall; leaf-blades linear-acuminate, flat or rolled, flaccid or somewhat rigid, glabrous on both surfaces or sparsely hairy on the upper surface; sheaths tightly clasping, glabrous and smooth, or the lower somewhat hairy; ligule membranous, up to 2 mm. long; inflorescence a simple raceme of 6–12 very shortly pedicellate spikelets; spikelets oblong or lanceolate in outline, straight or curved, 10–20-flowered, 2–3 cm. long; rhachilla disarticulating below each lemma; glumes unequal, persistent, lanceolate-acute, the lower 3–5 mm. long, 3–6-nerved, the upper 5–7 mm. long, 7-nerved, glabrous; lemmas oblong-acute, 8–10 mm. long, 7-nerved, smooth and glabrous, awned; awn shorter than the lemma; anthers 4–4·5 mm. long.

HAB.: Roadsides, waste ground, gardens; by streams and irrigation channels; in marshy places and moist thickets; sea-level to 4,500 ft. alt.; fl. June–Sept.

DISTR.: Divisions 2, 3, 6, 7. Europe and Mediterranean region eastwards to Siberia; introduced elsewhere.

2. Prodhromos, 1955, *G. E. Atherton* 541! S.W. of Troödos, 1970, *E. Chicken* 69!
3. Salt Lake near Limassol, 1939, *Lindberg f.* s.n.; Asomatos marshes, 1939, *Mavromoustakis* 133!
6. Myrtou, 1935, *Nattrass* 706.
7. Kythrea, 1939, *Lindberg f.* s.n.; Bellapais, 1939, *Lindberg f.* s.n.; Boghazi, 1939, *Lindberg f.* s.n.

2. B. sylvaticum (*Huds.*) *P. Beauv.*, Essai Agrost., 101, 155 (1812); Boiss., Fl. Orient., 5: 657 (1884); Holmboe, Veg. Cypr., 38 (1914); Post, Fl. Pal., ed. 2, 2: 777 (1933); Lindberg f., Iter Cypr., 7 (1946); N. L. Bor in C. C. Townsend et al., Fl. Iraq, 9: 167 (1968); Osorio-Tafall et Seraphim, List Vasc. Plants Cyprus, ed. 1, 10 (1973).
    Festuca sylvatica Huds., Fl. Angl., ed. 1, 38 (1762).
    Bromus sylvaticus (Huds.) Sm., Fl. Brit., 136 (1800); Sibth. et Sm., Fl. Graec. Prodr., 1: 63 (1806).
    [*Brachypodium pinnatum* (non (L.) P. Beauv.) Post in Mém. Herb. Boiss., 18: 102 (1900); H. Stuart Thompson in Journ. Bot., 44: 340 (1906); Holmboe, Veg. Cypr., 38 (1914).]
    TYPE: British Isles; "*in sylvis et sepibus frequens*". Hudson.

A caespitose perennial grass; culms erect or somewhat decumbent at the base, up to 80 cm. tall, smooth and glabrous but bearded at the nodes; leaf-blades linear or lanceolate, acuminate, up to 30 cm. long, 8 mm. broad, flaccid, drooping, flat, loosely pilose on the upper and lower surfaces or glabrous, scaberulous on the margins and on the upper surface; sheaths clasping, more or less pilose; ligule membranous, up to 5 mm. long, truncate; inflorescence a simple nodding raceme of shortly pedicellate spikelets; spikelets cylindrical, oblong-acute or lanceolate-acute in outline, 6–12-flowered; rhachilla disarticulating below each lemma at maturity; glumes unequal, the lower lanceolate, 6–8 mm. long, 5–7-nerved, the upper oblong-acute, 7·5–10 mm. long, 7–9-nerved, shortly hairy; lemmas oblong-elliptic, acute, 7-nerved, 8–11 mm. long, rounded on the back, covered with short hairs, awned from the tip; awn straight, scabrid, up to 10 mm. long; anthers 3·5–4 mm. long.

HAB.: Rocky mountain slopes; roadsides, field margins, gardens, waste ground; sides of irrigation channels; 500–4,500 ft. alt.; fl. March–Aug.

DISTR.: Divisions 2, 5, 6. Europe, Mediterranean region, Atlantic Islands, temperate Asia, Australia; introduced elsewhere.

2. Prodhromos, 1880, *Sintenis & Rigo* 889! Kyperounda ["Kippalunga"], 1898, *Post* s.n.! Above Kakopetria, 1936, *Syngrassides* 1265! Platania, 1939, *Lindberg f.* s.n.! Below Galata, 1956, *Merton* 2808!
5. Kythrea, 1932, *Syngrassides* 228! and, 1935, *Syngrassides* 838!
6. Near Dhiorios, 1962, *Meikle* 2352!

3. **B. firmifolium** *Lindberg f.*, Iter Cypr., 7, fig. 1 (1946); Osorio-Tafall et Seraphim, List Vasc. Plants Cyprus, 10 (1973).

TYPE: Cyprus; "M. Troodos, Mesopotamos, in siccis juxta cataractam, 21.6.1939." *Lindberg f.* (K!).

A caespitose, perennial grass; culms slender, up to 45 cm. tall, erect or decumbent at the base, smooth and glabrous, shortly bearded at the nodes; leaf-blades linear-acuminate, tapered to the base, up to 10 cm. long, 5 mm. wide, glaucous green in colour, very sparsely and shortly stiffly hairy on the upper surface, scaberulous above and on the margins; sheaths closely clasping, smooth and glabrous, striate; ligule membranous, short, truncate; panicle a simple raceme of somewhat crowded, shortly pedicellate spikelets; spikelets 2–5, 1·5–2 cm. long, 4–10-flowered; glumes slightly unequal, the lower 6 mm. long, 5–7-nerved, the upper 6·5–7 mm. long, 7–9-nerved, glabrous and smooth; lemma oblong-elliptic, acute, conspicuously 7-nerved, awned, minutely hairy in the lower half on the dorsal durface; awns up to 8 mm. on the upper lemmas, much shorter on the lower; palea as long as the lemmas, rounded at the apex, pectinate-ciliate; anthers 1·5 mm. long. *Plate 102, figs. 1–8.*

HAB.: Igneous mountainsides, usually in damp, shaded situations by streams and cataracts, sometimes in open forest or meadows; 3,300–5,500 ft. alt.; fl. June–July.

DISTR.: Division 2, locally common. Endemic.

2. Mesapotamos, 1939, *Lindberg f.* s.n.! Kryos Potamos, 5,500 ft. alt., 1951, *Merton* 538! and, same locality, 5,000 ft. alt., 1963, *D. P. Young* 7840! 7848! Livadhi tou Pasha, 1952, *Merton* 897!

45. TRACHYNIA *Link*
Hort. Reg. Bot. Berol., 1: 42 (1827).

Annual glaucous grasses with flat, linear-acuminate leaf-blades; spikelets all alike, normally 1–3, sometimes, but rarely, as many as 7; spikelets 8–12-flowered, crowded at the tip of the long-exserted peduncle, at first terete, eventually somewhat laterally compressed; rhachilla disarticulating above the glumes and between the florets; glumes unequal, rounded on the dorsal surface, coriaceous; lemmas oblong or elliptic, acute, rounded on the dorsal surface, awned from the tip, prominently nerved, awn straight, scabrid; stamens 3; styles 2; lodicules 2, lanceolate, acuminate, more or less ciliate towards the tip; palea as long as the lemma. Grain adnate to the palea, free from the lemma, terete or oblong-ellipsoid, somewhat dorsally compressed; embryo about one quarter as long as the grain; hilum linear, as long as the grain. 2n = 30.

One species widely distributed in the Mediterranean region and eastwards to Central Asia; Atlantic Islands; Ethiopia; S. Africa; introduced elsewhere.

1. **T. distachya** (*L.*) *Link*, Hort. Reg. Bot. Berol., 1: 43 (1827); N. L. Bor in C. C. Townsend et al., Fl. Iraq, 9: 168, t. 58 (1968); Osorio-Tafall et Seraphim, List Vasc. Plants Cyprus, 10 (1973).
  Bromus distachyos L., Cent. II Plant., 8 (1756); Sibth. et Sm., Fl. Graec. Prodr., 1: 64 (1806).
  Festuca distachya (L.) Roth, Catalecta Bot., 1: 11 (1797); Unger et Kotschy, Die Insel Cypern, 183 (1865); Sintenis in Oesterr. Bot. Zeitschr., 31: 257, 398 (1881).
  Brachypodium distachyum (L.) P. Beauv., Essai Agrost., 101, 155 (1812); Boiss., Fl. Orient., 5: 657 (1884); Holmboe, Veg. Cypr., 38 (1914); Post, Fl. Pal., ed. 2, 2: 777 (1933); Lindberg f., Iter Cypr., 7 (1946).
  TYPE: "*in* Europa *australi*, Oriente" (LINN!).

An annual, green or, more usually, glaucous grass; culms erect or geniculately ascending, fasciculate and spreading, rarely solitary, up to 45 cm. tall but in arid habitats only 2 cm. tall, upwards smooth and glabrous or more often retrorsely scabrid under the panicle, shortly pubescent or glabrous; leaf-blades flat, lanceolate, ending in a rigid, acute tip, with thickened, scabrid margins, 1–6 cm. long, 2–5 mm. wide, dark green or glaucescent, strongly nerved on the upper surface, sparsely pilose on the upper and lower surfaces or glabrescent; sheaths striate; inflorescence consisting of 1–3 (–7) subsessile spikelets often closely aggregate at the tip of the peduncle; spikelets somewhat laterally compressed, lanceolate or oblong, 8–12-flowered, 1–2·5 cm. long; glumes unequal, coriaceous, the lower 5–6 mm. long, 5-nerved, the upper 7–9 mm. long, rather broader, 7-nerved, mucronate; lowest lemma 8–9 mm. long, oblong-elliptic, acute, mucronate or shortly awned, glabrous or covered on the dorsal surface with minute scabridities, 7-nerved; nerves confluent at the apex; upper lemmas with an awn about 9 mm. long; palea as long as the lemma, bristly-pectinate on the keels at the tip; anthers 0·75–1 mm. long.

HAB.: In garigue on hillsides; by roadsides and in Pine forest; in cultivated fields or damp sandy ground; sea-level to 4,500 ft. alt.; fl. March–May.

DISTR.: Divisions 1–8, often common. Distribution that of genus.

1. Ayia Nikola (Akamas), 1941, *Davis* 3330! Polis, 1955, *Merton* 2169!
2. Common; Lefka, Makheras, Platres, Ayios Nikolaos, etc. *Kennedy* 1112A! *Casey* 278! *Meikle* 2643! 2847! etc.
3. Stavrovouni, 1862, *Kotschy* 215; Alethriko, 1905, *Holmboe* 227; between Nisou and Stavrovouni, 1933, *Syngrassides* 37! 38! Khalassa-Lania, 1970, *A. Genneou* in ARI 1523!
4. S. of Xylotymbou ["Timbo"], 1880, *Sintenis & Rigo* s.n.; Mile 3, Larnaca-Famagusta road, 1950, *Chapman* 428!
5. Vatili, 1905, *Holmboe* 318; Athalassa, 1950, *Chapman* 637 and, 1967, *Merton* in ARI 95!
6. Kokkimi Trimithia, 1933, *Syngrassides* 319! Mile 3, Nicosia-Myrtou road, 1950, *Chapman* 574! Nicosia airport, 1951, *Merton* 334!
7. Near Pentadaktylos, 1880, *Sintenis & Rigo*; Lakovounara Forest, 1950, *Chapman* 68! 189! Kyrenia, 1952, *Casey* 1250!
8. Eleousa Experimental Station near Rizokarpaso, 1962, *Meikle* 2993!

TRIBE 12. **Triticeae** *Dumort.* Spikelets solitary or in groups of 2–3 at each node of the spike-axis, 1- to many-flowered, sessile and alike or the lateral spikelets of a triad pedicellate, male or barren or much reduced, alternating on the opposite sides of a fragile or tough spike-axis; rhachilla disarticulating above the glumes or tough in cultivated species; inflorescence a spike or pseudoraceme; glumes coriaceous or membranous, often strongly nerved, sometimes reduced, awl-shaped, rarely absent; lemmas 5- to many-nerved, awned or awnless; awn short or long, straight or recurved, not strongly twisted or kneed; stamens 3; ovary with a lobed, terminal, hairy appendage; styles 2, stigmas plumose; lodicules 2, normally fimbriate on the upper margin. Grain often adhering to lemma and palea; hilum, long, linear.

Annual or perennial grasses with linear, narrow leaf-blades and festucoid anatomy; silica-bodies rounded, elliptic or oblong; 2-celled micro-hairs absent; ligules membranous. Chromosomes large with a basic number of 7.

The tribe *Triticeae*, more than any other, has been the subject of intensive research from many angles perhaps because of the fact that it includes so many important food and forage plants. Because of the readiness with which many of the genera hybridize, it has been suggested that the whole tribe should be considered as one huge genus. On the other hand those who take a narrow view of genera are inclined to make further generic divisions, while those who follow the middle road reject some of these, while accepting others. A recent study (H. Runemark and W. K. Heneen in Bot. Notiser, 121: 51–79; 1968) suggests the relegation of all the perennial species of *Agropyron* to the genus *Elymus*; the genus *Agropyron* is retained here in its wider and more generally accepted sense. Additional comments on the taxonomy of wheat, barley and rye have been published by W. M. Bowden in Canad. Journ. Bot., 37: 657–684 (1959).

46. AGROPYRON *Gaertn.*

Nov. Comm. Acad. Petrop., 14: 539 (1770).
H. Runemark & W. K. Heneen in Bot. Notiser, 121: 51–79 (1968).
W. K. Heneen & H. Runemark in Hereditas, 70 (2): 155–164 (1972).

Perennial herbs with erect culms and flat leaf-blades; inflorescence an erect terminal spike with a tough, persistent or fragile rhachis; spikelets several-flowered, solitary at each node, so orientated that the inner margins of the glumes and lemmas lie against the axis, erect or slightly curved outwards; glumes more or less equal, several-nerved, usually shorter than the lowest lemma, truncate, obtuse, emarginate, or acute and awned; lemmas rounded on the back, truncate, obtuse, or acute, awned or awnless; stamens 3; lodicules 2, acute, obovate-lanceolate or ovate in outline, fleshy at the base, ciliate in the upper half, occasionally with a lateral tooth. Grain fusiform or oblong in outline with a hairy apical appendage, concave on the adaxial surface; hilum linear, as long as the grain, embryo one-sixth to one-eighth the length of the grain.

More than 100 species, widely distributed in temperate regions of the world.

Lemmas awned; awns up to 30 mm. long　-　-　-　-　-　- **1. A. panormitanum**
Lemmas muticous:
　Rhachis of the spike-axis tough:
　　Nerves on the upper surface of the leaf-blade densely covered with a short, stiff pubescence
　　　　　　　　　　　　　　　　　　　　　　　　　5. A. haifense
　　Nerves on the upper surface of the leaf-blade glabrous, rough, or with long, sparsely scattered hairs:
　　　Plant with long rhizomes　-　-　-　-　-　-　-　- **3. A. repens**
　　　Plant densely tufted, without rhizomes　-　-　-　- **4. A. elongatum**
　Rhachis of the spike-axis very fragile　-　-　-　-　-　- **2. A. junceum**

1. **A. panormitanum** *Parl.*, Rar. Plant. Sic., 2: 20 (1840); Boiss., Fl. Orient., 5: 663 (1884); Post, Fl. Pal., ed. 2, 2: 778 (1933); N. L. Bor in C. C. Townsend et al., Fl. Iraq, 9: 216, t. 72 (1968); Osorio-Tafall et Seraphim, List Vasc. Plants Cyprus, 11 (1973).
　　TYPE: Sicily; "In sterilibus montosis calcareis. *Palermo alla Pizzuta, alla Moarta, e a Monte Cuccio*" *Parlatore* (FI).

A perennial grass with shortly creeping rhizomes, forming loose tufts; culms erect or geniculately ascending, reaching 1 m. in height, smooth and glabrous with brown nodes; leaf-blades flat at first becoming involute or rarely revolute or folded, up to 30 cm. long, 2–8 mm. wide, linear-acuminate, scabrid on the upper surface and on the margins, occasionally sparsely hairy; sheaths auricled, clasping, smooth and glabrous, the lower sometimes minutely pubescent; ligule very short, less than 1 mm. long, truncate; spike up to 20 cm. long, erect or more usually nodding; axis flat facing the spikelets, convex on the opposite side, glabrous; spikelets imbricate, fusiform, pressed to the axis, 3–5-flowered; glumes more or less equal, oblong-elliptic, acute, slightly wider above the middle, 18–25 mm. long, with 5–11 strong nerves, coriaceous, very scabrid on the nerves, mucronate or shortly aristate; lemma oblong-elliptic, acute, 15–18 mm. long, with 5 faint nerves in the lower half which become very prominent and scabrid towards the tip, the latter prolonged into a scabrid straight or slightly recurved awn up to 3 cm. long; palea as long as the lemma.

HAB.: Igneous mountainsides at high altitudes, in Pine forest or in orchards and vineyards; 4,000–4,700 ft. alt.; fl. May–Sept.

DISTR.: Division 2. S. Europe and Mediterranean region eastwards to Iraq and Iran.

2. Near Prodhromos, 1880, *Sintenis & Rigo* 880! Pedhoulas-Prodhromos road, 1955, *Merton* 2384! Troödhitissa Monastery, 1956, *Merton* 2810! Trikoukkia, 1961, *D. P. Young* 7457!

2. A. junceum (*L.*) *P. Beauv.*, Essai Agrost., 146 (1812); Boiss., Fl. Orient., 5: 665 (1884); H. Stuart Thompson in Journ. Bot., 44: 341 (1906); Post, Fl. Pal., ed. 2, 2: 779 (1933); Lindberg f., Iter Cypr., 6 (1946); Osorio-Tafall et Seraphim, List Vasc. Plants Cyprus, 11 (1973).

Triticum junceum L., Cent. 1 Plant., 6 (1755), Amoen. Acad., 4: 266 (1759), Sp. Plant., ed. 2, 128 (1762); Unger et Kotschy, Die Insel Cypern, 185 (1865); Holmboe, Veg. Cypr., 39 (1914).

TYPE: "*in* Helvetia, Oriente".

A perennial with widely spreading rhizomes; culms more or less glaucous, fascicled at the nodes of stolons or rhizomes, erect, covered with the sheaths almost to the base of the spike, 20 cm.–1 m. tall (*teste* Maire), brittle, smooth and glabrous below the inflorescence, glabrous at the nodes; leaf-blades linear-acuminate, usually involute, up to 15 cm. long in Cyprus specimens, 2–3 mm. wide, smooth and glabrous below, strongly nerved on the upper surface; nerves densely covered with a short, white pubescence; sheaths rather loose, long, rounded on the back, usually covered with very short, crisped or retrorse pubescence; ligule very short, truncate; spike up to 20 cm. long; axis smooth and glabrous, fragile, the spikelets falling with the joint next below; spikelets 1·5–2 cm. long (in Cyprus specimens), 6–8-flowered; glumes 10–13 mm. long (in Cyprus specimens), oblong-acute, keeled, 7–9-nerved, smooth and glabrous; lemmas 15 mm. long, coriaceous, lanceolate, blunt, keeled towards the tip, apiculate; palea broader than and shorter than the lemma.

HAB.: On sandy seashores and sand-dunes; near sea-level; fl. May–July.

DISTR.: Divisions 1, 4, 5. Maritime Europe and Mediterranean region; introduced elsewhere.
1. N. of Paphos, 1862, *Kotschy* 671a.
4. Near Famagusta, 1936, *Iacovou* 88! and, 1938, *Syngrassides* 1837! Perivolia, 1939, *Lindberg f.* s.n.! Larnaca airport, 1952, *Merton* 830a!
5. Salamis, 1905, *Holmboe* 461! and, 1939, *Lindberg f.* s.n.! Trikomo, 1939, *Lindberg f.* s.n.

NOTES: According to W. Simonet (in Bull. Soc. Bot. Fr., 82: 624–628; 1936), this species exists in two forms, the tetraploid 2n = 28, and the hexaploid 2n = 42. The former is said to be confined to western Europe, while the latter is a Mediterranean plant. Simonet established a new subspecies (ssp. *mediterraneum*) for the latter. The differences between the two, from the morphological angle, seem to be minor; in any case the Cyprus specimens seem to be typical *A. junceum*. The Latin diagnosis of the subspecies has not been traced.

3. A. repens (*L.*) *P. Beauv.*, Essai Agrost., 102, 146, 180, t. 20, fig. 2 (1812); Boiss., Fl. Orient., 5:

663 (1884); Post, Fl. Pal., ed. 2, 2: 778 (1933); N. L. Bor in C. C. Townsend et al., Fl. Iraq, 9: 220, t. 74 (1968); Osorio-Tafall et Seraphim, List Vasc. Plants Cyprus, 11 (1973).

　　Triticum repens L., Sp. Plant., ed. 1, 86 (1753).

TYPE: *"in* Europae *cultis"* (LINN).

A perennial grass with long, creeping rhizomes; culms erect or geniculately ascending, up to 120 cm. high, smooth and glabrous; leaf-blades linear-acuminate, up to 24 cm. long, 1 cm. wide, usually flat, often sparsely clothed with long hairs on the nerves above; sheaths rounded on the back with short, spreading auricles, smooth, glabrous or hairy; ligules short, truncate, membranous; spike erect, straight, 10–20 (–30) cm. long, green or occasionally somewhat glaucous; spike-axis tough, scabridulous on the margins, glabrous or hairy; spikelets usually overlapping, oblong-elliptic or narrowly cuneate, 3–8-flowered, 10–20 mm. long; glumes equal or subequal, lanceolate or lanceolate-oblong, 7–12 mm. long, blunt or acute; lemmas overlapping, lanceolate or lanceolate-oblong, keeled towards apex, 8–13 mm. long, blunt or sometimes sharply pointed; paleas nearly as long as lemmas, with 2 rough keels.

HAB.: Moist, shaded igneous mountainsides at high elevations; c. 5,400 ft. alt.; fl. July–Aug.

DISTR.: Division 2, apparently rare. Widespread in Europe, Mediterranean area, Atlantic Islands and temperate Asia; introduced elsewhere.

2. Kannoures Springs, 1954, *Casey* 1668!

NOTES: A bad weed of cultivated land in many temperate countries, but evidently rare in Cyprus, and perhaps absent from the lowlands. While there is no evidence of introduction in the locality cited above, it may possibly be a relatively recent arrival in Cyprus.

4. A. elongatum (*Host*) *P. Beauv.*, Essai Agrost., 146 (1812); Boiss., Fl. Orient., 5: 665 (1884); Post, Fl. Pal., 2, 2: 779 (1933); Lindberg f., Iter Cypr., 6 (1946); Osorio-Tafall et Seraphim, List Vasc. Plants Cyprus, 11 (1973).

　　Triticum elongatum Host, Icon. et Descr. Gram. Austr., 2: 18, t. 23 (1802).

TYPE: Yugoslavia; "In siccis, & locis aqua marina inundatis Tergesti *alle Saule*; *alle Saline di Capo d'Istria, Pirano*, alibique". *Host* (? W).

A perennial, more or less glaucous, densely tufted grass; culms erect, capable of reaching 120 cm. in height but usually much less, smooth and glabrous, with brownish nodes; leaf-blades usually flat, involute with age, up to 25 cm. long, 2–4 mm. wide, linear-acuminate, scabrid on the upper surface and on the margins, smooth below, glabrous; sheaths auricled, the upper smooth and glabrous, the lower glabrous or very shortly pubescent; ligule very short, truncate; spike stiff, linear, up to 20 cm. long; axis flattened on the face towards the spikelet, rounded on the other side, glabrous, scabrid or smooth on the angles; spikelets ovate or lanceolate, laterally compressed, elliptic-acute in shape, somewhat diverging from the axis, 14–18 mm. long, 6–11-flowered; glumes slightly unequal, coriaceous, 9–11 mm. long, linear-oblong, obtuse or truncate, scarious on the margins, 5–11-nerved; lemmas coriaceous, with a prominent median nerve, oblong, obtuse or emarginate at the apex, smooth and glabrous, or slightly scaberulous on the median nerve; margins ciliate with minute clavate hairs; palea as long as the lemma.

HAB.: Margins of salt-marsh, and sandy ground near the sea; near sea-level; fl. May.

DISTR.: Divisions 3, 4; Mediterranean region, western Asia.

3. Near Limassol, 1939, *Lindberg f.* s.n.

4. Larnaca Salt Lake, 1939, *Lindberg f.* s.n.; Larnaca airport, 1951, *Merton* 454!

5. A. haifense (*Meld.*) *Bor* **comb. nov.**

　　Elytrigia elongata (Host) Nevski var. *haifensis* Meld. in Arkiv för Bot., ser. 2, 2: 304 (1952).

Elymus elongatus (Host) Runemark ssp. *haifensis* (Meld.) Runemark in Hereditas, 70 (2): 156 (1972).
[*Agropyron elongatum* (non (Host) P. Beauv.) A. K. Jackson in Kew Bull., 1936: 16 (1936).]
TYPE: Palestine; "Haifa, Sandstrand", *Samuelsson* 1025 (W).

A caespitose perennial; culms erect or very slightly curved at the swollen base, smooth and glabrous, more or less glaucous, up to 45 cm. tall; leaf-blades linear-acuminate, up to 25 cm. long, 2–3 mm. wide, involute, smooth and glabrous on the lower surface, covered with short crisped hairs on the upper surface, among which are occasional longer hairs, strongly nerved; sheaths smooth and glabrous (in Cyprus specimens), striate, clasping; ligule very short; spike 17–30 cm. long (*teste* Melderis): axis tough; joints smooth on the margins, up to 11-spiculate in Cyprus specimens; spikelets 1·5–2 cm. long, rhomboidal in outline, 7–14-flowered; glumes 5–8 mm. long, mostly 5–7-nerved, oblong, obtuse, truncate or obliquely truncate at the tip; lemma 7–10 mm. long, oblong, obtuse or truncate, or concave at the tip, smooth and glabrous; palea as long as the lemma, 2-keeled.

HAB.: Salt-marshes; near sea-level; fl. June–July.

DISTR.: Division 4. Eastern Mediterranean region.

4. Pyla, 1934, *Syngrassides* 427 !

NOTES: Melderis (*loc. cit.*) remarks that this species takes, in many respects, an intermediate position between *A. elongatum* (Host) P. Beauv. and *A. junceum* (L.) P. Beauv., and indeed, might be taken for a hybrid between these two, were it not for the fact that the pollen grains are fertile. [It appears to me that all the material from Cyprus identified as *A. elongatum* and the above are conspecific, and that one or other name should be deleted. R.D.M.]

47. AEGILOPS *L.*

Sp. Plant., ed. 1, 1050 (1753).
Gen. Plant., ed. 5, 470 (1754).
P. Zhukovsky in Bull. Appl. Bot., Gen. & Plant Breed., 181: 417–609 (1928).
A. Eig in Fedde, Repert. Sp. Nov., Beiheft 55: 1–228 (1929).
K. Hammer in Feddes Repert., 91: 225–258 (1980).

Annual grasses with narrow, flat, rarely rolled leaf-blades; inflorescence a spike in two parts, the lower much the smaller, with a tough, continuous, flexuous rhachis terminating the peduncle, at each node of which are 1–3 alternate, distichously arranged, vestigial spikelets (or rarely spikelets wanting) ending in an abscission layer, with which the main, and larger, part of the spike is articulated; terminal portion of the spike with a fragile or partially tough rhachis, at each node of which is attached a solitary, hermaphrodite spikelet, all these spikelets alike, or the upper rudimentary, arranged alternately and distichously, falling together or separately with the adjacent joint or the one next below; spikelets terete or more or less ovoid or obovoid, 2–8-flowered, the upper florets male or neuter, the others usually hermaphrodite; glumes rather thick, leathery, truncate above and awnless or with one or more teeth or awns, keeled or not keeled; lemmas chartaceous or membranous below, becoming firmer and definitely nerved towards the tips, toothed or awned above; paleas 2-keeled; stamens 3; lodicules 2, cuneate, fimbriate on the upper margin. Grain compressed, mostly oblong, tapering towards the base, with a hairy apical appendage; embryo about one-fifth the length of the grain; hilum linear, as long as the grain.

About 25 species in the Mediterranean region and western Asia.

Base of the spike without vestigial spikelets; spike slender, narrow, linear, with the compressed spikelets in 2 rows; glumes 2-toothed, not awned - - - - - **5. Ae. bicornis**

Base of the spike with 1–3 vestigial spikelets; spikes lanceolate or ovate in outline; glumes
    usually awned:
Nerves of the lower glume narrow, equally wide, approximately parallel, inter-nerve spaces
    distinct:
    Glumes 5–7 mm. long, 3–4 mm. wide; lemmas with well-developed awns; glumes with equal
        awns, or the middle awn, where 3 are present, reduced to a tooth or absent, and then
        the 2 lateral awns separated by a space -   -   -   -   -   - **4. Ae. kotschyi**
    Glumes 6–8 mm. long, 4–6 mm. wide, wider below the middle; lemma usually without
        awns, and those of the glumes 2 or 3, mostly irregular, or their place taken by irregular
        teeth -   -   -   -   -   -   -   -   -   -   -   - **3. Ae. peregrina**
Nerves of the lower glume unequally wide, flattened, with shallow or indistinct inter-nerve
    spaces:
    Spikes linear-lanceolate, gradually tapering upwards; fertile spikelets (3–) 4–5 (–7)
                                             6. Ae. triuncialis
    Spikes suddenly contracted above the lowest fertile spikelet; fertile spikelets 2–3 (–4):
        Uppermost spikelet fertile; fertile spikelets 2, rarely 3, not crowded, ellipsoid; lower
           glume of uppermost spikelet 3-awned, that of the lowermost spikelet 2–3-awned,
           the awns shorter than those of the uppermost; awns of the lemma distinctly shorter
           than those of the glumes; vestigial spikelets 1 or 2   -   - **1. Ae. biuncialis**
        Uppermost spikelet sterile; spikelets 3, rarely 2 or 4, ovoid (wider below the middle);
           lower glume usually more than 3-awned (very rarely 2-awned); awns of the lemma
           strongly developed, a little shorter than the awns of the glume; vestigial spikelet 1
                                             2. Ae. geniculata

1. **Ae. biuncialis** *Vis.*, Fl. Dalmat., 1: t. 1, fig. 2 (1842), 3: 344 (1852); Post, Fl. Pal., ed. 2, 2: 784
    (1933).
    Ae. lorentii Hochst. in Lorent, Wander. Morgenl., 326 (1845); Post, Fl. Pal., ed. 2, 2: 784
    (1933); N. L. Bor in C. C. Townsend et al., Fl. Iraq, 9: 176 (1968); Osorio-Tafall et
    Seraphim, List Vasc. Plants Cyprus, 10 (1973).
    Ae. ovata L. var. *lorentii* (Hochst.) Boiss., Fl. Orient., 5: 674 (1884).
    Triticum ovatum (L.) Rasp. var. *biunciale* (Vis.) Aschers. et Graebn., Syn. Mitteleur. Fl.,
    2: 706 (1902).
    T. ovatum (L.) Rasp. ssp. *biunciale* (Vis.) Holmboe, Veg. Cypr., 38 (1914).
    Ae. biuncialis Vis. var. *vulgaris* Zhuk. in Bull. Appl. Bot., Gen. et Pl. Breed., 18: 483
    (1928).
    TYPE: Yugoslavia; Lesina [Hvar], *Stalio* (PAD).

    An annual grass; culms often densely fascicled, more or less geniculate at
the base, finally erect, up to 40 cm. tall, smooth and glabrous, striate; leaf-
blades linear-acuminate, up to 10 cm. long, 1–2 mm. broad, flaccid, flat,
sparsely covered on both surfaces with long white hairs, scabrid on the
margins and towards the tip, auriculate at the base; sheaths rather loose,
striate, sparsely pilose; ligule membranous, very short, lacerate; spike
supported at the base by 1 or 2 vestigial spikelets, 2-, rarely 3-spiculate,
lanceolate or elliptic-lanceolate, excluding the awns 2–3 cm. long, falling
entire at maturity; spikelets more or less urn-shaped (in Cyprus specimens),
wider below the middle, 4–5-flowered, of which 1–2 are fertile; glumes 6–9
mm. long, 5–6 mm. wide, elliptic-truncate, with 6–7 flattened, unequally
wide nerves, 2–3-awned; awns of the terminal spikelet 3–5 cm. long, those of
the lowest shorter; lemma of the lowest spikelet as long as the glumes, with
much shorter awns.

    HAB.: Roadsides, rocky hillsides, or on sandy ground near the sea; to 3,100 ft. alt.; fl.
April–May.

    DISTR.: Divisions 2–4, 7. Mediterranean region eastwards to Iran.

2. Platres, 1937, *Kennedy* 55 !
3. Kalavasos, 1905, *Holmboe* 631; Limassol road near turning to Mosphiloti, 1967, *Merton* in
    ARI 600 !
4. Larnaca airport, 1950, *Chapman* 450 !
7. Pentadaktylos, 1880, *Sintenis & Rigo* 657 ! Kyrenia, 1932, *Syngrassides* 14 ! Karavas, 1955,
    Merton 2182 !

2. **Ae. geniculata** *Roth*, Bot. Abh. Beob., 45 (1787).
    Ae. ovata L., Sp. Plant., ed. 1, 1050 (1753) pro parte; Sintenis in Oesterr. Bot. Zeitschr.,

31: 257, 393 (1881); Boiss., Fl. Orient., 5: 673 (1884); Post, Fl. Pal., ed. 2, 2: 783 (1933); N. L. Bor in C. C. Townsend et al., Fl. Iraq, 9: 186 (1968); Datta, Evenari et Gutterman in Israel Journ. Bot., 19: 463 (1970); Osorio-Tafall et Seraphim, List Vasc. Plants Cyprus, 10 (1973).

Triticum ovatum (L.) Gren. et Godr., Fl. France, 3: 601 (1856); Holmboe, Veg. Cypr., 38 (1914) pro parte.

TYPE: "Semina hujus plantae et *Aegil. ovatae* Linn. inter passulas majores legi", *Roth.*

An annual grass; culms mostly densely fascicled, often prostrate at the base and geniculately ascending, up to 40 cm. tall, smooth and glabrous, with purplish nodes; leaf-blades linear or lanceolate, acute, sparsely covered on both surfaces with bulbous-based whitish hairs, striate and scabrid on the upper surface and on the margins, up to 10 cm. long, 3–4 mm. wide; sheaths of the upper leaves somewhat inflated, smooth and glabrous, the lower scarious, loose, sparsely covered with spreading hairs; ligule membranous, very short, denticulate; spike, exclusive of the awns, 1–2·5 cm. long, falling entire, 2–4-spiculate; vestigial spikelet at the base one, rarely two; spikelets 2–4, usually 3, urn-shaped, that is, contracted at base and apex and rounded about the middle or below, sometimes ellipsoid, the lower spikelets 5-flowered of which the lowest two flowers are fertile; glumes 7–8 mm. long, 5–6 mm. wide; nerves irregular, unequally broad, flattened, glabrous or covered with short spines or hirsute, awned; awns 15–30 mm. long, scabrid, 3–4 on the lowest glumes, 2–5 on the upper; lowest lemma as long as the glume, chartaceous below, more rigid above, where the 5 nerves are strongly marked, truncate at the top where several of the nerves emerge as teeth or short awns.

HAB.: Amongst garigue on dry hillsides; by roadsides and margins of cultivated fields; often common near coast; sea-level to 3,100 ft. alt.; fl. March–May.

DISTR.: Divisions 2–8, common in Division 7. Mediterranean region and Atlantic Islands eastwards to Iran.

2. Platres, 1937, *Kennedy* 55a!
3. Kophinou, 1905, *Holmboe* 577! Kalavasos, 1905, *Holmboe* 632! Akrotiri, 1927, *Rev. A. Huddle* 111! between Nisou and Stavrovouni, 1933, *Syngrassides* 32!
4. Near Xylotymbou ["Timbo"], 1880, *Sintenis & Rigo* s.n.; Akhyritou, 1962, *Meikle* 2162!
5. Athalassa, 1967, *Merton* in ARI 99! 155! Nicosia, 1905, *Holmboe* 294!
6. Dhiorios, 1938, *Nattrass & Papaioannou* 90B; Syrianokhori, 1952, *Merton* 777!
7. Common; *Sintenis & Rigo* 362! *Syngrassides* 14a! *Casey* 497! *Chapman* 67! 105! 656! etc.
8. Eleousa Experimental Station near Rizokarpaso, 1962, *Meikle* 2990! Yialousa, 1970, *A. Genneou* in ARI 1488!

NOTES: The "variety with only two very short awns — of unequal length" collected by Holmboe at (8) Valia near Ayios Theodhoros (*Holmboe* 506!) proves on examination to be *Ae. kotschyi* Boiss. The Kotschy specimen from Larnaca (*Kotschy* 274) also cited by Holmboe (Veg. Cypr., 38) may be *Ae. kotschyi* too, but the specimen has not been examined.

3. Ae. peregrina (*Hackel*) *Maire et Weiller*, Fl. Afr. Nord, 3: 358 (1955); N. L. Bor in C. C. Townsend et al., Fl. Iraq, 9: 186 (1968); Osorio-Tafall et Seraphim, List Vasc. Plants Cyprus, 10 (1973).

Triticum peregrinum Hackel in Ann. Scott. Nat. Hist., 1907, no. 62: 102 (1907), et J. Fraser in Ann. Scott. Nat. Hist., 1908, no. 66: 105, t. 3 (1908).

Aegilops variabilis Eig in Fedde Repert., Beiheft, 55: 121 (1929); Post, Fl. Pal., ed. 2, 2: 785 (1933).

[*Ae. caudata* (non L.) N. L. Bor in C. C. Townsend et al., Fl. Iraq, 9: 178 (1968) quoad cit. cypr.; ? Osorio-Tafall et Seraphim, List Vasc. Plants Cyprus, 10 (1973).]

TYPE: Scotland; "introductam in Scotia prope Edinburgh (Slateford et Leith Docks), invenit J. Fraser" [1906] (? E; cultivated derivatives of type in K!).

An annual grass with many geniculately ascending, smooth and glabrous culms up to 40 cm. tall with purplish nodes; leaf-blades linear-acuminate, up to 7 cm. long, 1–3 mm. wide, flat, finally twisted or convolute, sparsely covered on both surfaces with soft, white hairs, scabrid on the margins; sheaths closely appressed, rounded on the back, auriculate at the mouth, sparsely covered with soft white hairs or sometimes glabrous; ligule very short, membranous, denticulate; spike cylindrical or lanceolate-cylindrical in outline, up to 7 cm. long excluding the awns, falling entire at maturity, 2–7-spiculate, with 3 rudimentary spikelets at the tip of the peduncle; spikelets ellipsoid, cylindrical or somewhat gaping, 3–6-flowered, of which the lower are fertile and the remainder sterile; glumes 6–8 mm. long, with 7–9 raised, parallel, scabrid nerves on the dorsal surface, awned or unawned or the awns reduced to irregular teeth which are sometimes recurved; awns if present 2–3; lowest lemma membranous, 5-nerved, usually without awns but occasionally 1–3 short, scabrid awns are developed.

HAB.: Amongst garigue on dry hillsides, or on coastal sand-dunes; sea-level to 800 ft. alt.; fl. April–May.

DISTR.: Divisions 1, 4, 5, 7, 8. Mediterranean region eastwards to Iran.

1. Between Neokhorio and Smyies, 1941, *Davis* 3315!
4. Near Ayios Memnon, 1948, *Mavromoustakis* s.n.! Ayios Antonios, Sotira, 1950, *Chapman* 597! Athna [Akhna] Forest, 1952, *Merton* 758!
5. Near Athalassa, 1933, *Syngrassides* 11! also, 1967, *Merton* in ARI, 155! 169!
7. Mile 6, Kyrenia-Akanthou road, 1950, *Chapman* 616! Lapithos, 1955, *Merton* 2228! Akanthou, 1967, *Merton* in ARI 775!
8. Valia minor State Forest, 1953, *Merton* 1098! Yioti, 1962, *Meikle* 2523!

Ae. geniculata *Roth* × **Ae. peregrina** (*Hackel*) *Maire et Weiller*

Two specimens, *Syngrassides* 335 (Div. 6, between Kokkini Trimithia and Kafkalou, 1933) and *Meikle* 2347 (Div. 6, Dhiorios, 1962) are intermediate between *Ae. geniculata* and *Ae. peregrina*, with only one vestigial spikelet, and with the awns irregular and reduced. They may, in the circumstances, be reasonably assumed to be hybrids between these two species.

4. Ae. kotschyi *Boiss.*, Diagn., 1, 7: 129 (1846); Post, Fl. Pal., ed. 2, 2: 785 (1933); N. L. Bor in C. C. Townsend et al., Fl. Iraq, 9: 184 (1968); Osorio-Tafall et Seraphim, List Vasc. Plants Cyprus, 10 (1973).
Ae. triuncialis L. var. *kotschyi* (Boiss.) Boiss., Fl. Orient., 5: 674 (1884).
[*Ae. ovata* (non L.) Kotschy in Unger et Kotschy, Die Insel Cypern, 186 (1865).]
[*Triticum ovatum* (non (L.) Gren. et Godr.) Holmboe, Veg. Cypr., 38 (1914) pro parte quoad *Kotschy* 274 et *Holmboe* 506.]
TYPE: Iran; "in Persia australi prope *Schiras* ad radices monti *Sabst-Buschom* Kotschy No. 366a et 1003" (W, G, K!).

An annual grass; culms usually densely fasciculate, rarely solitary, up to 25 cm. tall, erect or geniculately ascending, smooth and glabrous with purplish brown nodes; leaf-blades linear or lanceolate, acuminate, flat, striate and scabrid on the upper surface as well as on the margins, smooth below, glabrous or with a few stiff hairs above and/or below, up to 7 cm. long, 2–4 mm. wide; sheaths somewhat loose, sparsely covered with whitish hairs, auriculate at the mouth; ligule membranous, truncate, toothed, very short; spike lanceolate in outline, 2–6- (usually 4-) spiculate, excluding the awns 2–3 cm. long, falling entire at maturity; spikelets linear-oblong, 3–4-flowered, of which the lower 2 or 3 are fertile; glumes 5–7 mm. long, with 7–9 parallel scabrid nerves, usually 3-awned; awns 2–3 cm. long; lemma membranous below, somewhat indurated above, 5-nerved, 1–3-awned.

HAB.: Dry fields, and seashores; sea-level to c. 100 ft. alt.; fl. April–May.

DISTR.: Divisions 4, 8. Eastern Mediterranean region eastwards to Iran.

4. Near Larnaca, 1862, *Kotschy* 274; near Dhekelia, 1956, *Merton* 2664!
8. Valia, near Ayios Theodhoros, 1905, *Holmboe* 506!

5. Ae. bicornis (*Forssk.*) *Jaub. et Spach*, Illustr. Fl. Orient., 4: 10, t. 309 (1850); Boiss., Fl. Orient., 5: 677 (1884) pro parte; Post, Fl. Pal., ed. 2, 2: 788 (1933); A. K. Jackson in Kew Bull., 1937: 346 (1937).

Triticum bicorne Forssk., Fl. Aegypt.-Arab., 26 (1775); Delile, Fl. d'Egypte, 35, t. 15, fig. 1 (1813); Druce in Rep. B.E.C., 9: 471 (1930).

[*Ae. ligustica* (non (Savignone) Coss.) Osorio-Tafall et Seraphim, List Vasc. Plants Cyprus, 10 (1973).]

TYPE: Egypt; "Alexandriae", *Forsskål* (? C).

An annual grass; culms loosely fascicled, branched from the base, geniculately ascending, slender, up to 45 cm. tall; leaf-blades flat, up to 8 cm. long, 2–3 mm. wide, linear-acute or -acuminate, more or less pilose on both surfaces or glabrescent below, scaberulous on the margins and very scabrid at the tip, involute with age; auricles developed; lower sheaths often villous with spreading hairs, scarious; the upper clasping, smooth and glabrous, long-ciliate on the margins; ligule membranous, short, ciliate; spike 5–8 cm. long (excluding the awns), 5 mm. wide, laterally compressed; spikelets in two rows, alternate, distichous, 6–8 mm. long, 3-flowered (of which 1–2 florets are sterile); lowest 1–2 spikelets often unawned and rudimentary; glumes equal, oblong-elliptic, 2-toothed at the apex, 5–6-nerved, 5·5 mm. long, scaberulous on the nerves; lowest lemma elliptic-acute, 7-nerved, awned; palea hyaline, 2-keeled, as long as the lemma, scabrid on the keels; awn straight, scabrid, 3·5–7 cm. long; anthers 2–3 mm. long.

HAB.: Sand-dunes and sandy seashores; about sea-level; fl. March–April.

DISTR.: Divisions 5, 6. Eastern Mediterranean region.

5. Salamis, 1928 or 1930, *Druce* (see Notes); also, 1957, *Merton* 3217!
6. Ayia Irini, 1936, *Syngrassides* 1206!, also, 1951, *Merton* 292!, and, 1962, *Meikle* 2384! Syrianokhori, 1952, *Merton* 796!

NOTES: The Druce record from Salamis (Rep. B.E.C., 9: 417; 1930) is cited under the name "*Triticum bicorne* Forsk. var. *nana*". No description is given for the variety, nor is there a reference to a published description. In the circumstances, it must be assumed that *T. bicorne* var. *nana* is simply an invalid name for a dwarfed or starved condition of the typical plant, which is recorded independently from the same locality.

6. Ae. triuncialis *L.*, Sp. Plant., ed. 1, 1051 (1753); Unger et Kotschy, Die Insel Cypern, 186 (1865); Boiss., Fl. Orient., 5: 674 (1884); H. Stuart Thompson in Journ. Bot., 44: 341 (1906); Post, Fl. Pal., ed. 2, 2: 784 (1933); Rechinger f. in Arkiv för Bot., ser. 2, 1: 417 (1950); N. L. Bor in C. C. Townsend et al., Fl. Iraq, 9: 190, t. 65 (1968); Osorio-Tafall et Seraphim, List Vasc. Plants Cyprus, 10 (1973).
Triticum triunciale (L.) Gren. et Godr., Fl. France, 3: 602 (1855); Holmboe, Veg. Cypr., 38 (1914); Lindberg f., Iter Cypr., 9 (1946).
TYPE: "*in* Monspelii, Massiliae, Smyrnae *aridis*".

A glaucous annual with fasciculate, erect or geniculately ascending culms, up to 45 cm. tall, smooth and glabrous; leaf-blades linear-acuminate, up to 10 cm. long, 2–3 mm. wide, scaberulous on the upper surface, scabrid on the margins, smooth below, covered on both surfaces with whitish hairs or glabrous below, sheaths closely clasping or somewhat inflated, auricled at the mouth, the uppermost smooth and glabrous, the lower often scarious, sparsely covered with spreading white hairs; ligule membranous, very short, denticulate; spike lanceolate in outline, 3–6 cm. long, erect, 4–5-spiculate, gradually tapering towards the tip; lowest spikelet 4-flowered (the lower two florets fertile); glumes elliptic-oblong, mostly 8–9 (–10) mm. long, with flattened, unequally broad nerves, awned, glabrous or covered with stiff hairs; awns usually three, scabrid, or the central missing; awns of the uppermost spikelets 5–6 cm. long, those of the lower spikelets often much shorter or reduced; lowest lemma as long as the glumes, hyaline below, indurated above, 5-nerved, 3-toothed or shortly 3-awned.

HAB.: Mountainsides; waste ground, roadsides, sometimes in open Pine forest, usually on igneous formations; 200–5,600 ft. alt.; fl. April–July.

DISTR.: Divisions 2–4, 7; common in Division 2. Mediterranean region and eastwards to Central Asia; introduced elsewhere.

2. Troödos, 1912, *Haradjian* 433; Kryos Potamos, 5,600 ft. alt., 1937, *Kennedy* 103!
Prodhromos, 1939, *Lindberg f.* s.n., also, 1961, *D. P. Young* 7369! Troödos, "Olympus Camp
Hotel", 1939, *Lindberg f.* s.n.!, also, from Troödos, 1940, *Davis* 1814! and 1963, *D. P. Young*
7796! Platania, 1939, *Lindberg f.* s.n., and 1955, *Merton* 2352! Amiandos, 1952, *Merton* 817!
Near Phini, 1962, *Meikle* 2873!
3. Between Yerasa and Kalokhorio, on serpentine, 1962, *Meikle* 2889!
4. Kiti, 1862, *Kotschy*.
7. Pentadaktylos, 1880, *Sintenis & Rigo* 657! Kyrenia, 1949, *Casey* 445! and 1950, *Casey* 1023!

NOTES: In the absence of a specimen, Kotschy's record from near Kiti must be regarded as
rather questionable; unlike other *Aegilops* species in Cyprus, *Ae. triuncialis* would appear to be
uncommon except at high altitudes on the igneous Troödos range.

Ae. triaristata Willd. (Sp. Plant., 4: 943; 1806) is noted for Cyprus in Osorio-Tafall et
Seraphim, List Vasc. Plants Cyprus, 10 (1973), but no correctly named material of this species
has been seen, though the name has, on occasion been incorrectly attached to Cyprus
specimens. *Ae. triaristata* has the spike abruptly contracted above, 3 (rarely 2) vestigial
spikelets at the tip of the peduncle, 4 spikelets, the lower 2 fertile, the upper 2 sterile and remote;
the glumes are 8–11 mm. long, 5–7 mm. wide, with 3 (rarely 2) subequal awns 3·5–4·5 cm. long.

Ae. triaristata has a wide distribution in the Mediterranean region eastwards to Iran, and
may yet be satisfactorily recorded from Cyprus.

48. TRITICUM *L.*

Sp. Plant., ed. 1, 85 (1753).
Gen. Plant., ed. 5, 37 (1754).

Annual grasses with flat leaf-blades; inflorescence a distichous, compound
spike consisting of two lateral rows of spikelets seated alternately at the
nodes of a sinuous, tough or fragile rhachis which ends in a perfect or
vestigial, terminal spikelet; spikelets laterally compressed with 3–9 bisexual
florets or one or several of the uppermost sterile; glumes subequal, shorter
than the spikelet or longer (*T. polonicum*), asymmetrical, more or less
keeled, toothed, apiculate or awned; lemmas navicular, keeled from the base
or towards the apex only, awned or unawned; palea 2-nerved, 2-keeled,
ciliate on the keels, sometimes split; stamens 3; lodicules 2, ovate, ciliate on
the upper margin. Grain free between lemma and palea, or adherent, oblong-
elliptic, hairy at the apex, deeply grooved adaxially; embryo about one-fifth
the length of the grain; hilum linear, as long as the grain.

About 15 species (many cultivated) in the Mediterranean region and
eastwards to Central Asia.

Axis of the spike visible from the side, fragile; spikelets standing out from the axis **1. T. spelta**
Axis of the spike not visible from the side, tough; spikelets closely adpressed:
    Glumes firmly compressed and keeled from base to apex:
        Leaf-blades glabrous; spike linear, axis of spike ciliate; endosperm vitreous **2. T. durum**
        Leaf-blades velutinous; axis of spike densely ciliate with white hairs; endosperm mealy
                3. T. turgidum
    Glumes umbonate below, keeled above - - - - - - - **4. T. aestivum**

1. T. spelta *L.*, Sp. Plant., ed. 1, 86 (1753); Unger et Kotschy, Die Insel Cypern, 184 (1865);
Post, Fl. Pal., ed. 2, 2: 782 (1933).
TYPE: Not localized.

An annual plant; culms up to 150 cm. tall, smooth and glabrous, erect or
somewhat geniculate at the base, with thin walls; leaf-blades linear-
acuminate, up to 20 cm. long, 3–8 mm. wide, usually smooth and glabrous or
with a few long hairs on the upper surface, or with a row of hairs on the
midrib in young blades; sheaths clasping; ligule membranous, short; spike
10–15 cm. long, variously coloured from red, grey, white to black; very loose,
so that the brittle spike-axis is plainly visible from the side; spikelets awned
or unawned, laterally compressed, 3–4-flowered, of which 2–3 are fertile;
glumes about 10 mm. long, sharply keeled, truncate at the top, but the keel
ending in a sharp or blunt tooth, glabrous on the back but somewhat ciliate

on the margins; lowest lemma somewhat similar but not keeled, glabrous, coriaceous, 9-nerved, sometimes toothed or rarely awned; palea with two ciliolate keels.

HAB.: Cultivated fields, or as a casual on waste ground; fl. March–April.

DISTR.: Probably Division 2. Cultivated in Central and northern Europe, chiefly in hilly areas.

? 2. "am Gebirge gebaut", 1862, *Kotschy*.

NOTES: No recent records, but Spelt Wheat may still survive here and there in mountain country.

2. T. durum *Desf.*, Fl. Atlant., 1: 114 (1798); Post, Fl. Pal., ed. 2, 2: 783 (1933); N. L. Bor in C. C. Townsend et al., Fl. Iraq, 9: 205 (1968); Osorio-Tafall et Seraphim, List Vasc. Plants Cyprus, 10 (1973).
TYPE: North Africa; "colitur in Barbaria" (P).

Culms up to 1·5 m. tall, glabrous, smooth or somewhat scaberulous, leaf-blades linear-acuminate, up to 60 cm. long, 1–2 cm. wide, glabrous on the upper surface, puberulous on the lower; spike 4–11 cm. long (awns excluded), quadrangular in section, 1·2–1·75 cm. wide; axis normally tough, sometimes brittle, ciliate on the margins, bearded below the insertion of the spikelets; spikelets 5–7-flowered, 10–15 mm. long, 8–15 mm. wide, with 2–4 fertile florets; glumes 8–12 mm. long, firmly keeled from base to apex, toothed at the apex, hairy or glabrous; lemmas thin, pale, 9–15-nerved, the lowest long-awned; awns white, red or black; endosperm vitreous. .

HAB.: Cultivated; fl. March–April.

DISTR.: Division 5, cultivated. Widely distributed as a cultivated plant (Durum Wheat) in the Mediterranean region and eastwards to Central Asia; also cultivated in America (North, Central and South) and elsewhere.

5. Kythrea, 1936, *Syngrassides* 1216!

NOTES: Durum Wheat is widely cultivated in Cyprus, but rarely collected, hence the above record has little relevance to the distribution of the cereal on the island. This is the Wheat principally used in the manufacture of Macaroni and Spaghetti; it has a hard grain, and bread made from it is inferior to that made from *T. aestivum* L.

3. T. turgidum *L.*, Sp. Plant., ed. 1, 86 (1753); Unger et Kotschy, Die Insel Cypern, 184 (1865); Post, Fl. Pal., ed. 2, 2: 782 (1933); N. L. Bor in C. C. Townsend et al., Fl. Iraq, 9: 208 (1968).
TYPE: Not localized.

Culms up to 160 cm. tall, smooth and glabrous, the upper nodes solid or the walls thick, leaf-blades when young covered on both surfaces with soft, white hairs which are velvety to the touch, up to 45 cm. long, 1–2 cm. wide; spike 7–10 cm. long, often compound and branched below, 1–1·5 cm. broad, nodding, quadrangular or oblong in section, axis tough, bearded on the joints below the spikelets, ciliate on the margins; spikelets 10–13 mm. long, 8–15 mm. wide, 5–7-flowered, of which 3–5 are fertile; glumes white, yellow, red or blackish, sometimes glaucous, glabrous or pubescent, 8–11 mm. long, oblique, keeled to the strongly toothed tip; lemma rounded on the dorsal surface, awned; awn 8–16 cm. long. Endosperm mealy. 2n = 28.

HAB.: Cultivated; fl. April–May.

DISTR.: Recorded as a cultivated Wheat by Kotschy (1865), and frequently mentioned in agricultural reports on Cyprus cereals in cultivation at the present time. Rivet, Egyptian or Poulard Wheat is a winter wheat, said to have a high yield where conditions suit it, and to be resistant to rust disease.

4. T. aestivum *L.*, Sp. Plant., ed. 1, 85 (1753) emend. Fiori et Paoletti, Fl. Ital., 1: 107 (1896); Post, Fl. Pal., ed. 2, 2: 782 (1933); N. L. Bor in C. C. Townsend et al., Fl. Iraq, 9: 197, t. 67 (1968).
T. hybernum L., Sp. Plant., ed. 1, 86 (1753).
T. sativum Lam., Encycl. Méth., 2: 554 (1786).

T. vulgare Vill., Hist. Plant. Dauph., 2: 153 (1787); Unger et Kotschy, Die Insel Cypern, 184 (1865).

TYPE: Not localized.

Culms up to 75 cm. tall, smooth and glabrous, sometimes pubescent at the nodes; leaf-blades when young usually hairy, eventually glabrous, up to 60 cm. long, 1–1·5 cm. wide, linear-acuminate in shape, flat, scaberulous on the margins; sheaths usually glabrous, sometimes hairy; ligule truncate, membranous; spike 6–18 cm. long, quadrangular in section; axis tough, continuous, not bearded at the nodes; spikelets 3–6-flowered, 10–15 mm. long, 9–18 mm. wide; glumes glabrous, hairy or villous, reddish, yellowish or pale, oblique, usually keeled in the upper three quarters only, toothed at the tip; lemma rounded on the dorsal surface in the lower quarter, keeled above, muticous or long-awned. 2n = 42.

HAB.: Cultivated; fl. March–April.

DISTR.: Noted as frequent by Kotschy, and still widely cultivated on the island. It is grown in temperate regions throughout the world, and is the wheat normally used for making bread.

Bor (Fl. Iraq, 9: 200; 1968) notes that *T. aestivum* is an allohexaploid with one set of 14 chromosomes derived from *Aegilops speltoides* Tausch, one from *Ae. tauschii* Coss. and one from *Triticum monococcum* L.

49. SECALE L.

Sp. Plant., ed. 1, 84 (1753).
Gen. Plant., ed. 5, 36 (1754).

Annual or perennial grasses with flat leaf-blades; inflorescence a dense spike of distichous, alternately arranged spikelets, each seated at a node of the tough or fragile rhachis; spikelets firmly laterally compressed, appressed tangentially to the rhachis, with 2 hermaphrodite florets; rhachilla articulate below the florets, produced and crowned by a rudimentary floret; glumes subequal, coriaceous, narrow, acute or ending in an awn; lemma coriaceous, firmly keeled, spinously ciliate on the keel, 5-nerved, produced at the tip into a fine scabrid awn; palea membranous, 2-keeled, finely scabrid on the keels; stamens 3; lodicules 2, large, cuneate, rounded above and fimbriate on the rounded margin. Grain free between lemma and palea, oblong-obtuse, hairy at the summit; embryo about one quarter the length of the grain; hilum linear, the length of the grain.

Three or four species in Europe, Asia and S. Africa.

1. **S. cereale** L., Sp. Plant., ed. 1, 84 (1753); Boiss., Fl. Orient., 5: 671 (1884); N. L. Bor in C. C. Townsend et al., Fl. Iraq, 9: 260 (1968).

TYPE: Not localized.

A glaucous annual; culms up to 150 cm. tall, solitary or loosely fasciculate, erect or very shortly geniculate at the base, very shortly pilose below the spike, otherwise smooth and glabrous; leaf-blades flat, linear-acuminate, flaccid, scaberulous above, smooth below, up to 30 cm. long, 2–5 mm. wide, sometimes sparsely pilose on the upper surface of the lower blades, scabrid on the margins; sheaths closely appressed, smooth and glabrous, the lower becoming scarious; ligule very short, truncate; spike erect or slightly bent, dense, up to 20 cm. long, laterally compressed; axis tough, flat, densely ciliate on the margins and sometimes on the two surfaces; spikelets compressed, 2-flowered; glumes linear or narrowly oblong-elliptic, gradually passing into a short awn, including the latter 15 mm. long; lemmas firmly compressed, about 1·5 cm. long, awned, 5-nerved, pectinate-spinulose on the keel, otherwise smooth and glabrous, passing gradually into a straight, scabrid awn 3–5 cm. long, palea rather shorter than the lemma, 2-nerved, scabrid on the keels.

HAB.: Cultivated; fl. March–May.

DISTR.: No specimens have been seen, but Rye is frequently referred to in the annual reports of the Department of Agriculture of Cyprus; it is, however, a cereal of minor economic importance, grown in hilly areas on soils unsuitable for Wheat and Barley.

50. CRITHOPSIS *Jaub. et Spach*
Illustr. Plant. Orient., 4: 30, t. 321 (1851).

Annual grasses usually with geniculate culms and flat leaf-blades auriculate at the base; inflorescence a dense spike consisting of a fragile rhachis at the nodes of which are seated, alternately and distichously, the spikelets in pairs; spikelets with 2 florets, the lower bisexual, the upper rudimentary, seated at the nodes of the flattened rhachilla; rhachilla disarticulating below the floret; glumes similar, equal, coriaceous, strap-like, scabrid; lower lemma becoming coriaceous, flat or depressed on the abaxial surface, 5-nerved, attenuate at the apex into a flattened awn; palea 2-keeled, ciliate on the keels; lemma and palea firmly adherent to the grain; stamens 3; lodicules 2, lanceolate or ovate, acute, entire, or with 1 or 2 very short lobes on the margins, fimbriate at the apex and on the margins. Grain dorsally compressed, oblong-elliptic, appendaged at the apex; embryo basal, about one-fifth the length of the grain; hilum linear, the length of the grain.

A monotypic genus restricted to the eastern Mediterranean Region, Iraq and Iran.

1. **C. delileana** (*Schult.*) *Roschevicz*, Gräser, 319 (1937) in obs.; N. L. Bor in C. C. Townsend et al., Fl. Iraq, 9: 226, t. 77 (1968); Osorio-Tafall et Seraphim, List Vasc. Plants Cyprus, 11 (1973).

   Elymus geniculatus Del., Fl. Egypte, 30, t. 13, fig. 1 (1813); Post, Fl. Pal., ed. 2, 2: 795 (1933) non Curtis ex Smith (1800).

   Elymus delileanus Schult. in Roem. et Schult., Syst. Veg., 2, Mantissa: 424 (1824); Boiss., Fl. Orient., 5: 692 (1884).

   Elymus aegyptiacus Spreng., Syst. Veg., 1: 328 (1824).

   Crithopsis rhachitricha Jaub. et Spach, Illustr. Plant. Orient., 4: 30, t. 321 (1851).

   TYPE: Egypt; "à Alexandrie dans les champs d'orge, entre le lac *Mareotis* et la mer, au mois de mars 1800", *Delile* (MPU).

An annual grass; culms geniculate at the base, eventually erect, fasciculate, glaucous, up to 30 cm. tall, smooth and glabrous; leaf-blades linear, acute or acuminate, up to 4 cm. long, 1–3 mm. wide, flat or somewhat involute at the margins, rather rigid with a pungent tip, smooth and glabrous below, sparsely pilose on the upper surface, scabrid on the sometimes ciliate margins; sheaths somewhat inflated towards the panicle, smooth and glabrous, auricled at the mouth, the lower shorter, scarious, loose; ligule very short; inflorescence a somewhat dense spike, oblong in outline, 2–3·5 cm. long (excluding the awns) with a very fragile, hairy axis, at each node of which are seated the spikelets in pairs, spikelets 2-flowered, the lower fertile, the upper rudimentary; glumes equal, linear, coriaceous, about 12 mm. long, 3-nerved, very scabrid, imperceptibly merging into a scabrid bristle 5 mm. long; lower lemma about 7 mm. long, chartaceous to coriaceous, oblong-lanceolate, passing gradually into a scabrid awn 5–7 mm. long, flat on the back, minutely scabrid, with a few scattered spinules toward the top, 5-nerved; palea of the same texture.

HAB.: Dry, calcareous banks; 600–900 ft. alt.; fl. April.

DISTR.: Division 6. General distribution that of genus.
6. Near Ayia Marina, 1956, *Merton* 2751 ! Between Skylloura and Philia, 1956, *Merton* 2757 !

51. TAENIATHERUM *Nevski*

Acta Univ. Asiae Med., ser. 8b, Bot., fasc. 17: 38 (1934).

Annual herbs with flat leaf-blades which roll up when dry, shortly auricled at the base; inflorescence a bristly spike consisting of a tough, angled rhachis at the nodes of which are seated the long-awned spikelets in pairs; spikelets with 2 florets, of which the lower is hermaphrodite and the upper rudimentary; rhachilla disarticulating above the glumes with an oblique plane of fracture, continuous, linear and flat between the florets; glumes equal, coriaceous, subulate, dilated and indurated at the base, scabrid; lemma papyraceous, 5-nerved, attenuate above into a long, scabrid, flat, erect or recurved awn; palea equal in length to the lemma, 2-keeled, scabrid on the keels; stamens 3; lodicules 2, obliquely lanceolate, humped on the outer margin, fimbriate on the margins and also on the surface in the upper quarter. Grain narrowly elliptic-oblong, deeply grooved adaxially, dorsally compressed, with a hairy appendage at the apex; embryo small, about one-eighth the length of the grain; hilum linear, as long as the grain.

Two (or possibly 1) species in the Mediterranean region and eastward to Central Asia.

Robust; awns of the lemma stout, erect, somewhat recurved at maturity, over 0·8 mm. wide at
   the base; glumes erect -   -   -   -   -   -   -   -   -   **1. T. crinitum**
Slender, weak; awns of the lemmas flexuous, arcuate, not more than 0·5 mm. wide at the base;
   glumes spreading -   -   -   -   -   -   -   -   -   **2. T. asperum**

1. **T. crinitum** (*Schreb.*) *Nevski* in Acta Univ. Asiae Med., ser. 8b, Bot., fasc. 17: 38 (1934); N. L. Bor in C. C. Townsend et al., Fl. Iraq, 9: 264 (1968); Osorio-Tafall et Seraphim, List Vasc. Plants Cyprus, 11 (1973).
   Elymus crinitus Schreb., Beschr. Gräser, 2: 15, t. 24, fig. 1 (1772); Sibth. et Sm., Fl. Graec. Prodr., 1: 72 (1806), Fl. Graec., 1: 76, t. 96 (1808); Post, Fl. Pal., ed. 2, 2: 795 (1933).
   Hordeum caput-medusae (L.) Coss. et Durieu ssp. *crinitum* (Schreb.) Aschers. et Graebn., Syn. Mitteleur. Fl., 2 (1): 744 (1901); Holmboe, Veg. Cypr., 39 (1914).
   [*Elymus caput-medusae* (non L.) Boiss., Fl. Orient., 5: 691 (1884).]
   [*Hordeum caput-medusae* (non L.) Coss. et Durieu) Lindberg f., Iter Cypr., 7 (1946).]
   TYPE: Presumably grown at Erlangen, and figured from cultivated specimens (? M).

An annual, glaucous grass; culms loosely fasciculate, rarely solitary, mostly geniculate at the base then erect, up to 45 cm. tall, smooth and glabrous; leaf-blades flat at first then involute, linear-acuminate, up to 12 cm. long, 2–4 mm. wide, smooth and glabrous on the lower surface, hairy on the upper surface and scabrid on the margins; sheaths tightly clasping on the upper internodes, rather loose below and slipping from the culms, smooth and glabrous, or the lower shortly and retrorsely hairy; ligule membranous, truncate, less than 1 mm. long; inflorescence an erect, bristly spike, without the awns up to 5 cm. long; spikelets in pairs at each node of the tough spike-axis, 2-flowered; glumes equal in shape and length, indurate at the base, 1–2·5 cm. long, straight, very scabrid; lemma 8–12·5 mm. long, flattened on the back, lanceolate in outline, minutely scabrid on the back, scabrid on the rounded margins towards the tip where it is narrowed into a firm, stiff, scabrid, straight (eventually slightly curved) awn up to 12 cm. long; palea of the same texture, 2-keeled. *Plate 104.*

HAB.: On rocky hillsides, in open garigue, generally on igneous formations; 600–6,400 ft. alt.; fl. April–July.

DISTR.: Divisions 2, 3, 5. Central and S.E. Europe eastwards to Central Asia.

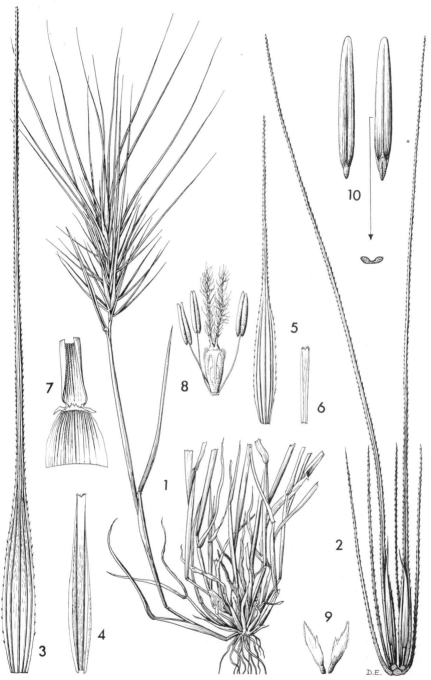

Plate 104. TAENIATHERUM CRINITUM (Schreb.) Nevski **1,** habit ×⅔; **2,** pair of spikelets showing subulate glumes, ×2; **3,** lower lemma, ×4; **4,** its palea, ×4; **5,** upper lemma, ×4; **6,** its palea, ×4; **7,** ligule, ×3; **8,** floret, ×10; **9,** lodicules, ×10; **10,** grain, ×4. (**1–10** from *Davis & Hedge* D.28120 Turkish material.)

2. Locally common (see Notes); Prodhromos, Mavrovouni, Agros, Kakopetria, Troödos, Evrykhou, Platania, Jemali Bridge, Perapedhi, Kykko, Phini, etc. *Sintenis & Rigo* 879! *Syngrassides* 1089! *Kennedy* 101! *Davis* 1794! *Lindberg f.* s.n.! *Merton* 1087! 2788! *Meikle* 2642! *Economides* in ARI 951! *A. Genneou* in ARI 1339!, etc.

3. Mile 19, Nicosia-Limassol road, 1950, *Chapman* 547! Kakomallis, 1962, *Meikle*!

5. Mile 10, Nicosia-Limassol road, 1967, *Merton* in ARI 599!

NOTES: In Cyprus frequently an indicator of metalliferous rocks, particularly copper ores. It would be interesting to know if its distribution elsewhere is restricted to a similar habitat.

2. T. asperum (*Simonkai*) *Nevski* in Acta Univ. Asiae Med., ser. 8b, Bot., fasc. 17: 39 (1934); N. L. Bor in C. C. Townsend et al., Fl. Iraq, 9: 265 (1968); Osorio-Tafall et Seraphim, List Vasc. Plants Cyprus, 11 (1973).

    Cuviera caput-medusae (L.) Simonkai var. *aspera* Simonkai in Term. Közl., 29, Pótfüz., 230 (1897).

    Hordeum caput-medusae (L.) Coss. et Durieu ssp. *asperum* (Simonkai) Degen. in Aschers. et Graebn., Syn. Mitteleur. Fl., 2 (1): 744 (1901); Holmboe, Veg. Cypr., 39 (1914).

    Cuviera aspera (Simonkai) Simonkai in Magyar Bot., Lapok, 3: 86 (1904).

    Elymus caput-medusae L. var. *aspera* (Simonkai) Hal., Consp. Fl. Graec., 3: 426 (1904).

TYPE: Hungary.

A slender annual grass; culms rather loosely fasciculate, usually procumbent at the base, finally erect, up to 30 cm. tall, smooth and glabrous; leaf-blades linear-acuminate, flat, becoming folded, involute or rolled, up to 6 cm. long, 1–2 mm. wide, smooth and glabrous on the lower surface, scabrid on the upper surface and on the margins; sheaths rather loose, the uppermost often spathaceous, very shortly auricled at the mouth, somewhat purplish below, smooth and glabrous; ligule membranous, very short, lacerate; inflorescence a rather weak, bristly spike, 3–4 cm. long (excluding the awns), axis tough; spikelets in pairs at each node, 2-flowered; glumes subulate, 25–30 mm. long, spreading or curved at the base, scabrid; lemma 7–8 mm. long, elliptic in outline, scaberulous, green or tinged with purple, attenuate at the tip and forming a thin, weak awn up to 7 cm. long which is usually flexuous or irregularly curved; palea 2-keeled.

HAB.: On rocky hillsides in open garigue on igneous formations; 4,600–6,400 ft. alt.; fl. May–June.

DISTR.: Division 2. Central and S.E. Europe eastwards to Central Asia.

2. Prodhromos, Khionistra, Kryos Potamos (5,600 ft. alt.); *Sintenis & Rigo* 879 a! *Kennedy* 99! 100! 1021! *D. P. Young* 7315!

NOTES: [In Cyprus, scarcely distinct from *T. crinitum* with which it is sometimes collected, and of which it appears to be a depauperate or juvenile state; both have been collected on the summit of Khionistra. R.D.M.]

52. HORDEUM *L.*

Sp. Plant., ed. 1, 84 (1753).
Gen. Plant., ed. 5, 37 (1754).

Annual or perennial herbs with flat leaf-blades auricled at the base; inflorescence an erect or nodding, distichous, compressed spike in which the spikelets are in triplets, each triplet consisting of a sessile spikelet flanked by two sessile (cultivated species) or pedicellate lateral spikelets, seated alternately and distichously at the nodes of a tough or fragile rhachis; spikelets 1-flowered, that of the central spikelet hermaphrodite, those of the lateral spikelets male or barren or all fertile in cultivated barley, the three spikelets falling together; rhachilla produced in the central spikelet; glumes linear, mostly awn-like, sometimes expanded at the base; lemma 5-nerved, more or less long-awned or awnless, or the tip produced into a trifid appendage; palea 2-nerved; lemma and palea adherent to the grain or the latter free, stamens 3; lodicules 2, obliquely wedge-shaped or spathulate, densely ciliate in the cultivated species, or lanceolate, or with an additional lateral tooth slightly ciliate above in the wild species. Grain elliptic, deeply furrowed adaxially, with an apical hairy appendage; embryo about one-fifth to one quarter the length of the grain; hilum linear, the length of the grain. About 20 species widely distributed in temperate regions.

Culms bulbous at the base - - - - - - - - - - - **1. H. bulbosum**
Culms not bulbous at the base:
  Wild grasses; axis readily disarticulating at maturity:
    Awns of the spikelets up to 12 cm. long - - - - - - **6. H. spontaneum**
    Awns much shorter:
      Lowest floret of the central spikelet sessile:
        Both glumes of the lateral spikelets subulate, or one slightly swollen but not winged
                                        4. H. geniculatum
        One glume of the lateral spikelet with a wing at the base - - - **5. H. marinum**
      Lowest floret of the central spikelet stipitate:
        Inflorescence 2–9 cm. long, 4–8 (–10) mm. wide, rather dense; hairs on the rhachis
        0·25–0·75 mm. long; anthers 0·2–0·5 mm. long (very rarely longer)
                                          2. H. glaucum
        Inflorescence to 7 cm. long, 1–2·5 cm. wide, rather loose; hairs on the rhachis very
        short or absent; anthers 0·8–1·8 mm. long - - - **3. H. leporinum**
  Cultivated grasses; axis tough:
    All spikelets fertile, forming a 6-rowed spike - - - - - - **7. H. vulgare**
    Lateral spikelets sterile, forming a 2-rowed spike - - - - - **8. H. distichon**

1. H. bulbosum L., Cent. Plant. II, 8 (1756), Amoen. Acad., 4: 304 (1759); Sibth. et Sm., Fl. Graec. Prodr., 1: 73 (1806); Fl. Graec., 1: 79, t. 98 (1806); Poech, Enum. Plant. Ins. Cypr., 7 (1842); Unger et Kotschy, Die Insel Cypern, 185 (1865); Boiss., Fl. Orient., 5: 688 (1884); Holmboe, Veg. Cypr., 39 (1914); Post, Fl. Pal., ed. 2, 2: 794 (1933); Lindberg f., Iter Cypr., 7 (1946); Nattrass, First List Cyprus Fungi, 14, 17, 45 (1973); N. L. Bor in C. C. Townsend et al., Fl. Iraq, 9: 245 (1968); Osorio-Tafall et Seraphim, List Vasc. Plants Cyprus, 11 (1973).
    TYPE: "*in* Italia, Oriente" (LINN!).

A densely caespitose perennial; culms erect or geniculately ascending, the lower 1–2 nodes swollen and forming a fleshy bulb or bulbs which are covered with remains of the membranous or fibrous, withered sheaths, smooth and glabrous; leaf-blades up to 30 cm. long, 2–6 mm. wide, linear-acuminate, the lower more or less villous with spaced hairs on both surfaces, the uppermost glabrescent, very scabrid on the upper surface and on the margins; upper sheaths smooth and glabrous, auricled, clasping, the lower loose, more or less villous; ligule very short, truncate; spike straw-coloured, sometimes pale-purplish, 6–11 cm. long, 1 cm. wide, straight or slightly curved; axis very fragile when mature, flattened, minutely ciliate on the margins; spikelets in threes falling with the joint next below; central spikelet fertile, 1-flowered; glumes 3 mm. long, narrowly elliptic, flattened and ciliate on the margins, with an awn up to 8 mm. long; lemma sessile, 9–10 mm. long, oblong, elliptic-acute, faintly 5-nerved, produced at the apex into an awn up to 3·5 cm. long; palea as long as the lemma; lateral spikelets: glumes unequal, the upper subulate, 2 mm. long, the lower slightly wider at the base, both awned; awns about 2 cm. long; lemma elliptic-acute, 8 mm. long, awnless.

HAB.: Roadsides, waste ground, margins of fields; sometimes in coastal garigue, or on stony hillsides inland; sea-level to 5,100 ft. alt.; fl. April–June.

DISTR.: Divisions 2–8, locally common. Mediterranean region eastwards to Central Asia; tropical Africa.

2. Lefka, 1932, *Syngrassides* 105! Kryos Potamos, 1938, *Kennedy* 1135! Platres, 1938, *Kennedy* 1136! Prodhromos, 1961, *D. P. Young* 7394!
3. Cape Gata, 1862, *Kotschy* 620a; between Yerasa and Kalokhorio, 1962, *Meikle* 2890!
4. Freshwater Lake, Famagusta, 1930, *Druce* s.n.! Near Kalopsidha, 1967, *Merton* in ARI 471!
5. Mile 14, Nicosia-Limassol Rd., 1950, *Chapman* 545! Salamis, 1962, *Meikle* 2574!
6. Myrtou, 1932, *Syngrassides* 199! Nicosia, 1935, *Nattrass* 503.

7. Common about Kyrenia; *Wyatt* 36! *Kennedy* 1685! *Casey* 593! *G. E. Atherton* 1323! etc.; Ayios Demetrianos, Kythrea, 1950, *Chapman* 165!
8. Near Dhavlos, 1880, *Sintenis & Rigo* 393! S. of Galinoporni, 1962, *Meikle* 2501!

2. H. glaucum *Steudel*, Syn. Plant. Glum., 1: 352 (1854); N. L. Bor in C. C. Townsend et al., Fl. Iraq, 9: 248 (1968); Osorio-Tafall et Seraphim, List Vasc. Plants Cyprus, 11 (1973).

H. stebbinsii Covas in Madroño, 10: 18 (1949).

[*H. murinum* (non L.) Kotschy in Unger et Kotschy, Die Insel Cypern, 185 (1865); Holmboe, Veg. Cypr., 39 (1914).]

TYPE: "Mons Sinai" Hrbr. [*Schimper*] un. it. arab. nr. 383.

An annual grass; culms usually densely fascicled, rarely solitary, strongly geniculate at the base, the upper part erect, smooth and glabrous, 6–40 cm. tall; leaf-blades linear-acuminate or acute, 3–15 cm. long, 4·7 mm. wide, sometimes glabrous on both surfaces, usually more or less sparsely hairy, scaberulous on the margins; sheaths tight or rather lax and slipping from the culms, smooth and glabrous, shining, finally scarious; ligule membranous, 1–3 mm. long; inflorescence linear, 2–9 cm. long, 4–8 mm. wide when young, with closely packed spikelets, the lateral of which spread widely when mature; spikelets in threes at the nodes of a very fragile, flat rhachis, the margins of which are ciliate with hairs 0·25–0·75 mm. long; lateral spikelets on a slender stipe; glumes subulate; lemma 7–10 mm. long, male or neuter; central spikelet sessile; glumes subulate; lemma oblong-elliptic, stipitate, 5–6 mm. long, tipped with an awn 10–15 mm. long; anthers 0·2–0·5 mm. long.

HAB.: Roadsides, waste ground; sea-level to c. 1,000 ft. alt.; fl. March–May.

DISTR.: Divisions 4–7; Mediterranean region eastwards to N.W. India. Frequently as an adventive elsewhere.

4. Phaneromene, Larnaca, 1905, *Holmboe* 154!
5. Laxia, 1932, *Syngrassides* 76! Athalassa, 1967, *Merton* in ARI 97!
6. Nicosia, 1950, *Chapman* 528!
7. Kythrea, 1862, *Kotschy* 357!

3. H. leporinum *Link* in Linnaea, 9: 133 (1835); Post, Fl. Pal., ed. 2, 2: 793 (1933); N. L. Bor in C. C. Townsend et al., Fl. Iraq, 9: 249 (1968); Osorio-Tafall et Seraphim, List Vasc. Plants Cyprus, 11 (1973).

H. murinum L. var. *leporinum* (Link) Bory et Chaub., Nouv. Fl. Pélop., 8 (1838).

H. murinum L., ssp. *leporinum* (Link) Arcang., Comp. Fl. Ital., 805 (1882).

[*H. murinum* (non L.) Sm. in Sibth. et Sm., Fl. Graec. Prodr., 1: 74 (1806) teste Link, loc. cit. (1835).]

TYPE: Greece; "Frequens in Graecia", *Link* (B).

An annual grass; culms loosely or densely fascicled, geniculately ascending or erect, up to 50 cm. tall, smooth and glabrous; leaf-blades linear-acuminate, up to 20 cm. long, 2–4 mm. wide, laxly pilose on both surfaces, scabrid on the margins; sheaths auricled at the mouth, somewhat but only slightly inflated above, or tight, smooth and glabrous, the lowest often scarious and slipping from the culms, sparsely pilose or glabrous, shining; ligules rather short, less than 1 mm. long, membranous; inflorescence a dense, erect or slightly inclined spike-like panicle up to 7 cm. long, 1–2·5 cm. wide, often glaucous; spikelets in threes at the nodes of a very fragile rhachis, falling together, the two lateral pedicellate; lateral spikelets: glumes unequal, the lower 5 mm. long, narrowly elliptic-acute, ciliate on the margins, narrowed into an awn 1–2 cm. long, the upper reduced to a setaceous scabrid bristle, as long as or longer than the lower; lemma oblong, 5-nerved, 14 mm. long, tapering into an awn 2·5 cm. long; palea narrow, 2-nerved; anthers 1·5 mm. long; median spikelet sessile; glumes equal, very narrowly elliptic, 10 mm. long, densely ciliate on both margins, narrowed into a scabrid awn 14 mm. long; lemma stipitate, 10 mm. long, elliptic-acute in shape, chartaceous, 5-nerved, produced as a scabrid awn 2–3 cm. long; anthers 0·8–1·5 mm. long.

HAB.: Waste ground, roadsides, often a weed of cultivation, occasionally in open garigue; sea-level to 4,600 ft. alt.; fl. March–June.

DISTR.: Divisions 2–4, 7. Mediterranean region and eastwards to China; introduced into Australia, S. Africa, N. America, etc.

2. Prodhromos, 1961, *D. P. Young* 7310!
3. Asomatos, 1956, *Merton* 2738!
4. Larnaca aerodrome, 1950, *Chapman* 13! Famagusta, 1951, *Merton* 262!
7. Kythrea, 1932, *Syngrassides* 290! Kyrenia, 1949, *Casey* 449! and, 1956, *G. E. Atherton* 1145! St. Hilarion, 1941, *Davis* 2497!

NOTES: Rajhathy and Morrison (in Canad. Journ. Gen. Cyt., 4: 245; 1962) declare that cytogenetic evidence such as interfertility, identity of karyotypes and regular chromosome pairing, has shown that *Hordeum murinum* L. and *H. leporinum* Link should not be considered separate species. One does not question the cytogenetic results obtained by the two authors from the plants investigated by them, but the correctness of the identifications of the plants investigated may be legitimately questioned. The writer is convinced that *H. murinum* and *H. leporinum* are distinct. Apart from the differences in the lemma of the lower floret (stipitate in *H. leporinum*, sessile in *H. murinum*), the lateral spikelets in *H. leporinum* are longer and wider. *H. murinum* has not yet been found in Cyprus.

4. H. geniculatum *All.*, Fl. Pedem., 2: 259, t. 91, fig. 3 (1785); N. L. Bor in C. C. Townsend et al., Fl. Iraq, 9: 246, t. 85 (1968).
  H. hystrix Roth, Catalect. Bot., 1: 23 (1797).
  H. winkleri Hackel in Oesterr. Bot. Zeitschr., 27: 49 (1977).
  H. gussoneanum Parl. ex Nyman, Consp. Fl. Europ., 838 (1882); Lojac-Poj., Fl. Sic., 3: 364 (1909) —based on *H. gussoneanum* [*gussonianum*] Parl., Fl. Palerm., 1: 246 (1845) nom. invalid. provis.
  H. maritimum With. ssp. *gussoneanum* (Parl. ex Nyman) Aschers. et Graebn., Syn. Mitteleurop. Fl., 2 (1): 737 (1901).
  [*H. secalinum* (non Schreb.) Gussone, Plant Rar., 58, t. 11 (1826).]
  [*H. maritimum* (non With.) Lindberg f., Iter Cypr., 8 (1946).]
  TYPE: France; "*Cl.* Bellardi legit ad litora maris *Nicaeensis* prope *Portum di Limpia*" (TO).

An annual grass with loosely tufted or solitary culms up to 40 cm. tall, erect or geniculate at the base, rather slender, simple, smooth and glabrous; leaf-blades linear-acuminate, up to 6 cm. long, 2–3 mm. wide, flat, pilose or glabrous, minutely scabrid on the margins; upper sheaths slightly inflated, with the uppermost sometimes embracing the base of the inflorescence, smooth and glabrous, the lower more or less scarious, glabrous or covered with a minute pubescence; ligules very short, membranous, less than 1 mm. long; inflorescence a very dense, cylindrical panicle, 2–4 cm. long excluding the awns, 1–1·5 cm. wide; spikelets in threes, the two lateral pedicellate, the central sessile; awns of all the spikelets setaceous, up to 2 cm. long, scabrid; lateral spikelets frequently sterile, reduced to a series of awnlike scales; fertile spikelets: glumes setaceous, scabrid; lemma sessile, papyraceous, 5-nerved, elliptic-acute, 6–7 mm. long, produced into an awn 1·5–2 cm. long; palea as long as lemma, 2-nerved, 2-keeled.

HAB.: Roadsides; fallows and sandy flats near the sea; sea-level to 5,500 ft. alt.; fl. April–June.

DISTR.: Divisions 2, 4, 6. W. Central and S. Europe, eastern Mediterranean region and eastwards to Central Asia; tropical S. Africa; introduced elsewhere.

2. Between "Olympus Camp Hotel" and Khionistra, 1939, *Lindberg f.* s.n.! Livadhi tou Pasha, 1955, *Merton* 2333! Troödos, 1956, *Merton* 2797! Prodhromos, 1961, *D. P. Young* 7404!
4. Larnaca, 1967, *Merton* in ARI 370!
6. Asomatos, 1956, *Merton* 2737!

5. H. marinum *Huds.*, Fl. Angl., ed. 2, 57 (1778); Druce in Rep. B.E.C., 9: 471 (1931); Post, Fl. Pal., ed. 2, 2: 794 (1933); N. L. Bor in C. C. Townsend et al., Fl. Iraq, 9: 250 (1968); Osorio-Tafall et Seraphim, List Vasc. Plants Cyprus, 11 (1973).
  H. maritimum With., Bot. Arr., ed. 2, 1: 27 (1787); Unger et Kotschy, Die Insel Cypern,

185 (1865); Boiss., Fl. Orient., 5: 687 (1884); Holmboe, Veg. Cypr., 39 (1914); Lindberg f., Iter Cypr., 8 (1946).

TYPE: England; "*in pratis et pascuis*", *Hudson*.

An annual grass with loosely tufted or solitary culms, erect or geniculately ascending, up to 40 cm. tall, smooth and glabrous; leaf-blades linear-acuminate, up to 8 cm. long, 3 mm. wide, glaucous, minutely hairy on one or both surfaces or glabrous, scaberulous to smooth on the margins; sheaths auricled at the mouth, smooth and glabrous on the upper part of the stems, scarious and often covered with sparse, spreading hairs below; ligule membranous, less than 1 mm. long; inflorescence a very dense spike-like panicle, cylindrical, 2–5 cm. long, 1·5–2·5 cm. wide, greenish or purplish; spikelets in threes at each fragile node, the two lateral pedicellate and neuter, the central sessile; glumes of the lateral spikelets unequal, the lower with an awn up to 25 mm. long, scabrid, the upper with a broad wing at the base and with the awn about 22 mm. long; lemma about 5 mm. long, shortly awned; glumes of the sessile spikelet equal, scabrid, setaceous, 10–25 mm. long; lemma when flattened, elliptic or oblong, acute, sessile, tipped with an awn up to 25 mm. long, 5-nerved; palea narrow, as long as the lemma, 2-nerved.

HAB.: Roadsides, cultivated land; damp saline ground and seashores; sea-level to 900 ft. alt.; fl. March–May.

DISTR.: Divisions 3–5, 7, 8. Western and southern Europe and Mediterranean region eastwards to Afghanistan; Atlantic Islands; tropical S. Africa; introduced into N. America.

3.   Akrotiri marshes, 1938, *Syngrassides* 1806! Phasouri (Asomatos), 1962, *Meikle* 2926!
4.   Near Larnaca, 1862, *Kotschy* 534! and, 1928 or 1930, *Druce* s.n.; Famagusta, 1939, *Lindberg f.* s.n.!
5.   Dhikomo, 1936, *Syngrassides* 1193! Mia Milea-Koutsovendis road, 1950, *Chapman* 476! Mile 3, Nicosia-Myrtou road, 1950, *Chapman* 571!
7.   Myrtou, 1932, *Syngrassides* 28!
8.   "bei Camares" (near Gastria), 1862, *Kotschy* 535.

6. H. spontaneum *K. Koch* in Linnaea, 21: 430 (1848); Post, Fl. Pal., ed. 2, 2: 792 (1933); A. K. Jackson in Kew Bull., 1937: 346 (1937); Nattrass et Papaioannou in Cypr. Agric. Journ., 34: 26 (1939); N. L. Bor in C. C. Townsend et al., Fl. Iraq, 9: 252 (1968); Osorio-Tafall et Seraphim, List Vasc. Plants Cyprus, 11 (1973).

H. ithaburense Boiss., Diagn., 1, 13: 70 (1853), Fl. Orient., 5: 686 (1884).

TYPE: Caucasus; "Auf den steppenartigen Matten des schirwan'schen Theiles des Kaukasus, 500–1000' hoch", *K. Koch* (? B, LE).

A robust, annual grass; culms usually solitary and erect, or loosely fasciculate and geniculately ascending, up to 70 cm. tall, smooth and glabrous, with brown nodes; leaf-blades linear-acuminate, flat, somewhat flaccid, up to 15 cm. long, 4–8 mm. wide, very sparsely pilose and scaberulous on the upper surface or glabrous, smooth and glabrous on the lower surface, very scabrid on the margins; sheaths auricled at the mouth, all rather loose, the lower particularly so, smooth and glabrous; ligule about 1·5 mm. long, truncate, lacerate; inflorescence a spike, excluding the awns 3–5 cm. long, dense, erect, straw-coloured; spikelets in threes at the nodes of the flat, very fragile axis, which is densely ciliate on the margins, the median fertile, the two lateral sterile; glumes of the fertile spikelet equal, very narrowly elliptic, 6 mm. long, shortly appressed-hirsute, each passing into a scabrid awn 12 mm. long; lemma coriaceous, 10–13 mm. long, oblong, flattened on the back, 5-nerved, scabrid on the nerves, attenuate into a straight scabrid awn 6–12 cm. long; palea of the same texture; glumes of the sterile spikelets similar to those of the fertile; lemma oblong, obtuse or acute, flat, scabrid towards the tip, coriaceous, 3-nerved, 13 mm. long, awnless or very shortly awned.

HAB.: Roadsides, cultivated land and on hillsides amongst garigue; near sea-level to 1,500 ft. alt.; fl. March–April.

DISTR.: Divisions 2, 5–7. S.E. Europe and eastern Mediterranean region eastwards to Central Asia.

2. Between Lefka and Evrykhou, 1957, *D. R. Harris* 2909 !
5. Near Athalassa, 1936, *Syngrassides* 1173 ! and, 1938, *Nattrass & Papaioannou* 911.
6. Asomatos, 1956, *Merton* 2739 !
7. Kyrenia, 1949, *Casey* 422 !

7. H. vulgare *L.*, Sp. Plant., ed. 1, 84 (1753); Unger et Kotschy, Die Insel Cypern, 185 (1865); Post, Fl. Pal., ed. 2, 2: 793 (1933); N. L. Bor in C. C. Townsend et al., Fl. Iraq, 9: 254, t. 88 (1968); Osorio-Tafall et Seraphim, List Vasc. Plants Cyprus, 11 (1973).

H. hexastichon L., Sp. Plant., ed. 1, 85 (1753); Unger et Kotschy, Die Insel Cypern, 185 (1865).

TYPE: Not localized (LINN !).

An annual grass; culms solitary or loosely fasciculate, up to 80 cm. tall, erect, smooth and glabrous; leaf-blades flat, linear-acute, glabrous, scaberulous on both surfaces, scabrid on the margins, up to 60 cm. long, 10–15 mm. broad; sheaths falcate-auriculate, rather loose, striate, smooth and glabrous, the lower scarious, very loose; ligule very short, membranous; inflorescence a four- or six-sided spike, excluding the awns 6–10 cm. long; axis tough, not articulated, flattened, ciliate on the margins; spikelets in threes at each node of the axis, all hermaphrodite, 1-flowered, with the rhachilla produced as a glabrous or hairy filament; glumes very narrow, hairy on the dorsal surface, attenuate into a scabrid awn, awn and glume 2–3 cm. long; lemma papyraceous or chartaceous, broadly elliptic-acute, distinctly 5-nerved towards the tip, carried out above into a scabrid awn up to 15 cm. long; palea as long as the lemma, of the same texture, 2-keeled.

HAB.: Cultivated ground, or occasionally as an escape by roadsides or on open hillsides; 3,500–3,900 ft. alt.; fl. April–June.

DISTR.: Division 2. Cultivated throughout Europe and in most temperate parts of the world, or on mountains in the tropics.

2. Moniatis River, 1937, *Kennedy* 104 ! Kryos Potamos, 1937, *Kennedy* 105 !

NOTES: Kotschy (loc. cit., 185) notes that *H. vulgare* was commonly grown all over Cyprus in 1862 as a fodder plant, and that it was mown three times in the spring.

8. H. distichon *L.*, Sp. Plant., ed. 1, 85 (1753); Unger et Kotschy, Die Insel Cypern, 185 (1865); Post, Fl. Pal., ed. 2, 2: 793 (1933); N. L. Bor in C. C. Townsend et al., Fl. Iraq, 9: 246 (1968).

H. zeocriton L., Sp. Plant., ed. 1, 85 (1753); Unger et Kotschy, Die Insel Cypern, 185 (1865).

TYPE: Not localized.

An annual; culms solitary or loosely fasciculate, up to 90 cm. tall, smooth and glabrous, erect, green; leaf-blades linear-acuminate, flat, scaberulous on both surfaces, glabrous, very scabrid on the margins, up to 60 cm. long, 1–1·5 cm. broad; upper leaf-sheaths somewhat loose, falcately auriculate at the mouth, smooth and glabrous, the lower very loose, scarious, disintegrating, slipping from the culms; ligule membranous, very short, truncate; inflorescence an erect or somewhat curved spike, excluding the awns 6–12 cm. long, laterally compressed; rhachis tough, briefly ciliate on the margins; spikelets in threes at the nodes, the central fertile and sessile, the lateral shortly pedicellate and sterile; fertile spikelet: glumes equal, narrowly elliptic, glabrous or slightly hairy, prolonged into an awn, awn and glume 9–10 mm. long; lemma coriaceous, elliptic or oblong-elliptic, acute, 5-nerved, glabrous, scaberulous on the nerves towards the tip, 10 mm. long, awned; awn flattened, scabrid, up to 12 cm. long. Grain falling easily from the glumes. Sterile spikelets: glumes similar to those of the fertile spikelet;

lemma narrowly oblong-obtuse, 8–9 mm. long, glabrous, flat; palea shorter than the lemma.

HAB.: Cultivated fields; fl. March–May.

DISTR.: Said, by Kotschy, to be cultivated in the neighbourhood of Larnaca, but no specimens from Cyprus have been examined. *H. distichon* is, however, in cultivation in most temperate regions of the world, and has been grown since prehistoric times.

TRIBE 13. **Danthonieae** (*Beck*) *C. E. Hubb.* Annual or perennial grasses; leaf-blades narrowly linear or setaceous, rolled or flat, with bambusoid anatomy; silica-bodies rounded or dumb-bell-shaped; micro-hairs 2-celled, slender; ligule a fringe of hairs; spikelets all alike, hermaphrodite, 2–10-flowered, rhachilla disarticulating above the glumes and between the florets; inflorescences effuse or contracted panicles; glumes more or less equal, as long as or longer than the lowest lemma, often hyaline on the margins, and often strongly 3–11-nerved; lemmas membranous or coriaceous, 5–11-nerved, 2-lobed, awned in the sinus; awn very short or perfect; palea 2-keeled, hyaline, 2-lobed or obtuse; stamens 3; ovary glabrous; styles 2; stigma plumose; lodicules 2. Grain loosely held between lemma and palea; starch grains compound. Chromosomes small; basic numbers 6, 9, 12.

53. SCHISMUS *P. Beauv.*
Essai Agrost., 73 (1812).

Annual grasses with flat or involute, setaceous leaf-blades; inflorescence a panicle, usually dense, with spikelets on short pedicels crowded on and articulated to the branches; spikelets firmly laterally compressed with 5–10 bisexual florets or the uppermost reduced; rhachilla disarticulating above the glumes and between the florets; glumes subequal, strongly nerved with scarious margins, much longer than the lowest lemma, lemma rounded on the back, membranous, hyaline on the margins and at the bifid or bilobed tip, 9-nerved, awnless or with a very short awn in the sinus; palea as long as or nearly as long as the lemma, 2-keeled; stamens 3; lodicules 2, cuneate, truncate, glabrous, small, fleshy. Grain elliptic, obovate-obtuse, triangular in section, flat on the abaxial, convex on the adaxial surface, transparent, shining; embryo elliptic, one-third to one half the length of the grain; hilum suprabasal, small, elliptic.

Five species in Africa, the Mediterranean region and eastwards to N.W. India.

1. **S. arabicus** *Nees*, Fl. Afr. Austr., 422 (1841); Sintenis in Oesterr. Bot. Zeitschr., 32: 122 (1882); Boiss., Fl. Orient., 5: 597 (1884); Holmboe, Veg. Cypr., 36 (1914); Druce in Rep. B.E.C., 9: 47 (1931); Post, Fl. Pal., ed. 2, 2: 757 (1933); Rechinger f. in Arkiv för Bot., ser. 2, 1: 417 (1950); N. L. Bor in C. C. Townsend et al., Fl. Iraq, 9: 377 (1968); Osorio-Tafall et Seraphim, List Vasc. Plants Cyprus, 14 (1973).
TYPE: Sinai; "in valle Hamme Arabiae petraeae, (Schimper)" (B, destroyed).

A glaucous or rarely green, annual grass; culms very slender, densely fasciculate, erect, sometimes geniculately ascending or spreading, smooth and glabrous, shining; leaf-blades narrowly linear-acuminate, flat or convolute, filiform, flexuous, loosely covered with long white hairs, rarely glabrous, scaberulous on the surfaces and scabrid on the margins; sheaths rather loose, broadly hyaline, particularly the lower, on the margins, markedly striate on the rounded dorsal surface, smooth and glabrous; ligule a fringe of hairs; inflorescence a rather dense panicle, elliptic-ovate or oblong in outline, of shortly pedicellate spikelets, 1–5 cm. long, 1–1·5 cm. wide; axis erect, smooth below, scabrid above; branches short, slender, ascending, scaberulous; spikelets cuneate, gaping, 5–7 mm. long, 5–10-flowered; rhachilla disarticulating below each floret; glumes as long as the spikelets, persistent, lanceolate or oblong, acute, more or less equal in length, with broad scarious margins, glabrous, 3–7-nerved, green; lower lemmas elliptic or obovate when flattened, cleft at the apex into two acuminate lobes, 2–3 mm. long, 7–9-nerved, hairy in the lower half, aristulate in the sinus or not; palea shorter than the lemma, just reaching the base of the cleft; anthers minute, about 0·3–0·4 mm. long.

HAB.: Sandy seashores and gravelly waste ground; sea-level to c. 600 ft. alt.; fl. April–May.

DISTR.: Divisions 1, 3, 5, 6. uncommon. Mediterranean region and eastwards to N.W. India and Central Asia.

1. Ktima, 1970, *A. Hansen* 833!
3. Limassol, 1930, *E. Wall*.
5. Lefkoniko, 1880, *Sintenis & Rigo* 376!
6. Nicosia, Government House Grounds, 1935, *Syngrassides* 791! Nicosia, Dept. of Agriculture, 1950, *Chapman* 681! Between Astromeritis and Koutraphas, 1955, *Merton* 2296!

S. BARBATUS *(L.) Thell.* (in Bull. Herb. Boiss., ser. 2, 7: 391; 1907), differing from *S. arabicus* Nees chiefly in its shortly notched lower lemma, with acute or obtuse lobes, and in its relatively longer palea, often as long as the lemma, is listed for Cyprus in Osorio-Tafall and Seraphim, List Vasc. Plants Cyprus, 14 (1973). This record, probably based on the inclusion of Cyprus in the general distribution of *S. barbatus* cited in Fl. Iraq, 9: 380 (1968), must be expunged; both references to Cyprus can be traced to the misidentification of *Merton* 2296 from "between Astromeritis and Koutraphas" (cited above).

TRIBE 14. **Arundineae** *Dumort.* Perennial, reed-like grasses with broad linear leaf-blades, cordate or rounded at the base, with panicoid anatomy; silica-bodies cross-shaped; hairs 2-celled, long, narrow; ligule a row of stiff hairs; spikelets 2–10-flowered, laterally compressed, hermaphrodite or dioecious (in *Cortaderia* Stapf); rhachilla disarticulating above the glumes and between the florets; sometimes penicillate with long hairs; inflorescence a large, plumose panicle; glumes hyaline or membranous, acuminate, persistent; lemmas 3–7-nerved, all alike or the lowest neuter, glabrous or hairy all over the dorsal surface; palea hyaline, 2-nerved; stamens 3; ovary glabrous; styles 2; stigmas plumose; lodicules 2 or 3. Grain loosely enclosed; hilum basal; starch grains compound. Chromosomes small; basic number 12.

54. ARUNDO *L*.

Sp. Plant., ed. 1, 81 (1753).
Gen. Plant., ed. 5, 35 (1754).

Tall, stout, perennial grasses with hollow culms and flat, broad leaf-blades; inflorescence a large, plume-like, decompound panicle with the lower branches fascicled; spikelets laterally compressed, few-flowered (florets mostly bisexual); rhachilla articulated above the glumes and between the florets; glumes more or less equal, membranous; lemmas 2-fid, with a short awn in the sinus, 3–5-nerved, covered all over the back below the middle with long soft hairs; palea shorter, 2-keeled, thickly ciliate on the keels; stamens 3; lodicules 2, cuneate, 3-nerved, glabrous, with 2 or 3 irregular teeth on the upper margin. Grain oblanceolate-terete, depressed on the adaxial surface, striate; embryo about one-third the length of the grain; hilum oblong, suprabasal.

About 12 species widely distributed in tropical and warm-temperate regions.

1. **A. donax** *L.*, Sp. Plant., ed. 1, 81 (1753); Sibth. et Sm., Fl. Graec. Prodr., 1: 68 (1806); Unger et Kotschy, Die Insel Cypern, 179 (1865); Boiss., Fl. Orient., 5: 564 (1884); Holmboe, Veg. Cypr., 341 (1914); Post, Fl. Pal., ed. 2, 2: 745 (1933); Lindberg f., Iter Cypr., 7 (1946); N. L. Bor in C. C. Townsend et al., Fl. Iraq, 9: 372, t. 142 (1968); Osorio-Tafall et Seraphim, List Vasc. Plants Cyprus, 14 (1973).
    TYPE: "*In* Hispania, Galloprovincia" (LINN!).

A tall grass spreading by stout, knotty rhizomes; culms erect, stout, up to 6 m. tall, usually shorter, smooth and glabrous, 1–2 cm. in diameter, rather glaucous; leaf-blades cordate at the base, linear, up to 60 cm. long, 4–8 cm. broad, tapering to a long, acuminate tip, smooth and glabrous on both surfaces, scabrid on the margins; sheaths rounded, closely applied to the culms, hairy in the throat, smooth and glabrous; ligule short, membranous, 1–2 mm. long; panicle very dense, plumose, erect, rarely nodding, 30–60 cm. long, greyish or white; spikelets lanceolate-acuminate in outline, 2–5-flowered, finally gaping; glumes lanceolate-acuminate, glabrous, the lower 1–3-nerved, the upper 5-nerved, 11–13 mm. long; lowest lemma about 10–11 mm. long, 5–7-nerved, narrowly elliptic-acuminate, pilose in the lower third; palea elliptic-acute, 2-nerved, hyaline, ciliate on the margins; rhachilla glabrous; anthers 2·5–3 mm. long.

HAB.: Damp ground by the sides of rivers and drains; sea-level to c. 2,000 ft. alt.; fl. June–Oct.

DISTR.: Divisions 2, 4–6. Southern Europe, Mediterranean region, Atlantic Islands and eastwards through warm-temperate and tropical Asia to Polynesia and New Zealand; introduced into most warm-temperate and tropical parts of the world.

2. Galata, on the banks of the Karyotes Potamos, 1939, *Lindberg f.* s.n.
4. Famagusta, 1933, *Nattrass* 315.
5. Around Kythrea, 1880, *Sintenis & Rigo*, and, same area, 1931, *Nattrass* 138.
6. Nicosia, 1931, *Nattrass* 125, and, 1935, *Nattrass* 600.

NOTES: Common in many parts of lowland Cyprus, but rarely collected.

55. PHRAGMITES *Trin.*
Fund. Agrost., 134 (1820).

Tall, stout, perennial grasses with creeping rhizomes, flat leaf-blades, and tall, hollow culms; panicle a very large, decompound, plumose panicle; spikelets slightly laterally compressed, 2–10-flowered (florets bisexual); rhachilla disarticulating above the glumes and between the lemmas, long pilose with silky hairs; glumes equal or unequal, rather thin; lowest lemma neuter or male, similar to the upper glume but longer, the remainder more or less caudate-acuminate, 3-nerved; callus long, covered with silky hairs; palea much shorter; stamens 3; lodicules 2, obovate from a narrow base. Grain oblanceolate-ellipsoid, enclosed by the lemma and palea; embryo less than half the length of the grain; hilum short, oblong.

Three species with a cosmopolitan distribution.

1. **P. australis** (*Cav.*) *Trin. ex Steud.*, Nomenclator Bot., ed. 2, 2: 324 (1841); N. L. Bor in C. C. Townsend et al., Fl. Iraq, 9: 374, t. 143 (1968); Osorio-Tafall et Seraphim, List Vasc. Plants Cyprus, 14 (1973).
    Arundo australis Cav., Anal. Hist. Nat., 1: 100 (1799).
    Arundo phragmites L., Sp. Plant., ed. 1, 181 (1753); Sibth. et Sm., Fl. Graec. Prodr., 1: 68 (1806).
    Phragmites communis Trin., Fund. Agrost., 134 (1820); Unger et Kotschy, Die Insel Cypern, 179, (1865); Boiss., Fl. Orient., 5: 563 (1884); H. Stuart Thompson in Journ. Bot., 44: 340 (1906); Holmboe, Veg. Cypr., 34 (1914); Post, Fl. Pal., ed. 2, 2: 745 (1933); Lindberg f., Iter Cypr., 8 (1946).

TYPE: Australia; "á la Bahía Botánica [Botany Bay] viniendo de Jackson [Port Jackson]. Alli la cogio en April Don Luis Née" (MA !).

A tall, perennial reed, spreading principally by stout rhizomes and occasionally by stolons; culms erect, strong, stout, smooth and glabrous, up to 3 m. tall; leaf-blades linear, long-acuminate, flat, greyish-green, rather stiff, smooth and glabrous, scabrid on the margins, finally deciduous from the sheaths; sheaths closely clasping the culms, overlapping; ligule a dense ring of short hairs; panicle dense, erect, plumose, finally nodding, 15–30 cm. long, 10–20 cm. broad, brownish or purplish; spikelets shortly pedicellate, 2–6-flowered, at first lanceolate in outline, finally widely gaping, with the lowest floret male or empty and the others hermaphrodite; rhachilla long-pilose, breaking up below each fertile lemma; glumes unequal, persistent, the lower elliptic, acute, 3-nerved, 4–5 mm. long, the upper elliptic-oblong, acute, 5-nerved, 7–8 mm. long, membranous, glabrous; lowest lemma lanceolate-oblong, acute, 10–12 mm. long, 3-nerved, glabrous; fertile lemmas elliptic-caudate, 3-nerved, 9–10 mm. long; palea 3–14 mm. long, narrow; anthers 1·5–2 mm. long.

HAB.: Wet ground by rivers, streams and ditches; sea-level to 600 ft. alt.; fl. June–Oct.

DISTR.: Divisions 3, 4, 6–8. Widespread in temperate regions of the world extending into the tropics.

3. Kolossi, 1862, *Kotschy*; between Yeroskipos and Limassol, 1935, *Syngrassides* 504 !
4. Larnaca 1935, *Nattrass* 615; Famagusta (Fresh Water Lake), 1935, *Nattrass* 675; Perivolia, 1939, *Lindberg f.* s.n.
6. Near Nicosia, 1955, *Pallis* s.n. !
7. Akhiropiitos, near Karavas, 1936, *Syngrassides* 1292! Kyrenia, 1955, *G. E. Atherton* 84 ! 699 !
8. Ayios Andronikos, 1935, *Nattrass* 541.

CORTADERIA SELLOANA *(Schult. et Schult. f.) Aschers. et Graebn.*, Syn. Mitteleur. Fl., 2 (1): 325 (1900). (*Arundo selloana* Schult. et Schult. f., Syst. Veg., Mantissa III, 605 (1827); *Cortaderia argentea* (Nees) Stapf in Gard. Chron., ser. 3, 22 (2): 396; 1897), Pampas Grass, a robust, densely caespitose perennial, 2–3 m. high, with long, arching, sharp-cutting leaves, and silvery plumose panicles of dioecious spikelets at the apex of long, erect culms, is grown for ornament in many temperate regions, and is noted for Cyprus by Osorio-Tafall and Seraphim (List Vasc. Plants Cyprus, 14; 1973). No specimens have been seen. Pampas Grass is a native of S. America, extending from southern Brazil to northern Patagonia; Stapf comments that it is absent from the greatest part of those vast grassy plains which we generally call "Pampas", so that the popular name, now probably fixed, is inappropriate.

TRIBE 15. **Eragrostideae** *Stapf* Annual or perennial, often glandular, grasses; glands taking the form of sunken pits, patches or dots; leaf-blades narrow, often setaceous, with panicoid anatomy; silica-bodies saddle-shaped; micro-hairs present, 2-celled, swollen, club-shaped; ligule a fringe of hairs, rarely membranous; spikelets 2- to many-flowered, hermaphrodite, usually firmly laterally compressed; rhachilla disarticulating above the glumes, and the spikelets breaking up from above downwards, or somewhat tough and the glumes and lemmas falling from below upwards in succession together with the paleas, or the latter falling later; spikelets pedicellate in open or contracted panicles or sessile in panicled spikes; glumes shorter than the lemmas; lemmas membranous to coriaceous, 3-nerved, entire or emarginate at the apex, sometimes mucronate; palea hyaline, 2-keeled, scabrid or ciliate; stamens 2 or 3; ovary glabrous; styles 2; stigmas plumose;

lodicules 2. Grain sometimes with a free pericarp; hilum punctiform; starch grains compound. Chromosomes small; basic numbers 9, 10.

56. DACTYLOCTENIUM *Willd.*

Enum. Plant. Hort. Berol., 1029 (1809).

Annual or perennial grasses, the latter stoloniferous and/or rhizomatous, with narrowly linear leaf-blades; inflorescence a simple umbel of 2–5 digitately arranged spikes seated at the tip of a peduncle; spikelets compressed, 2-ranked, secund on the undersurface of a straight or curved rhachis, 2–5-flowered, broader than long, hermaphrodite; rhachilla disarticulating above the glumes and between the lemmas; glumes unequal, the upper mucronate, cuspidate, sometimes awned, the lower 1-, the upper 3-nerved; lemmas acuminate, 5-nerved, mucronate or occasionally even shortly awned; stamens 3; styles 2; stigmas plumose; lodicules 2, hyaline, 2-toothed. Grain enclosed in a thin pericarp; embryo about two-thirds the length of the grain; hilum basal, punctiform.

About 10 species in the warmer regions of the world.

1. **D. aegyptium** (*L.*) *P. Beauv.*, Essai Agrost., Explic. Planches, 10 [Pl. XV.] (1812); Post, Fl. Pal., ed. 2, 2: 741 (1933); N. L. Bor in C. C. Townsend et al., Fl. Iraq, 9: 426, t. 161 (1968). *Cynosurus aegyptius* L., Sp. Plant., ed. 1, 72 (1753).
    TYPE: "*in* Africa, Asia, America" (LINN!).

An annual or perennial, stoloniferous grass; culms erect or decumbent at the base and rooting at the nodes, simple, smooth and glabrous, up to 40 cm. long; leaf-blades flat, linear-acute or linear-obtuse, up to 10 cm. long, 5 mm. wide, glabrous on both surfaces or covered with tubercle-based hairs on the upper surface, scabrid on the margins; sheaths rather lax and scarious below, tighter above, glabrous, rarely sparsely hairy, smooth; ligule membranous, fimbriate; panicle of 2–5 digitate spikes seated on a long-exserted peduncle; spikes up to 4 cm. long, with a straight or incurved rhachis ending in a sharp point; spikelets 2-ranked, secund, crowded, 3–5-flowered, firmly compressed, about 3 mm. long, 4 mm. broad, pallid; lower glume 1-nerved, keeled, narrowly winged on the keel, ovate-acute when flattened, 2·5 mm. long; upper glume broader, similar, 3-nerved, furnished at the tip with a mucro or short awn; lowest lemma 2–3·5 mm. long, ovate-acuminate when flattened, scabrid towards the tip; palea 2-keeled; anthers 0·5–0·7 mm. long.

HAB.: In *Citrus* groves, an introduced weed; near sea-level; fl. June–Oct.

DISTR.: Divisions 4, 7. Widespread in the tropics of the Old World, extending (as a weed) into the Mediterranean region; introduced into warm-temperate and tropical America.

4. Ayios Memnon, Famagusta, 1969, *A. Genneou* in ARI 1357!
7. Kyrenia, 1968, *Hecker* 23!

57. LEPTOCHLOA *P. Beauv.*

Essai Agrost., 71 (1812).

Annual or perennial grasses; leaf-blades elliptic, lanceolate or linear, acuminate, flaccid or rather firm, flat, filiform or setaceous; inflorescence a panicle consisting of a long or short axis upon which are alternately arranged, patent, long or short racemes of spikelets, each 2–several-flowered, often minute, crowded or somewhat distant on the slender rhachis; rhachilla disarticulating above the glumes and between the florets; glumes equal or unequal, 1-nerved, persistent; lemmas acute or obtuse, 3-nerved, frequently pubescent on the nerves; palea 2-nerved, hairy or ciliate on the keels; stamens 3, minute; styles 2; stigmas plumose; lodicules 2. Grain, seen from the side, semi-circular in outline, triangular in section, grooved on the back; embryo half the length of the grain; hilum broad, punctiform. About 27 species with a wide tropical and subtropical distribution.

1. **L. filiformis** (*Lam.*) *P. Beauv.*, Essai Agrost., 71, 166 (1812).
    Festuca filiformis Lam., Tabl. Encycl. Méth., 1: 191 (1791).
    TYPE: "Ex Amer. merid. *Comm. D. Richard*" (P).

An annual grass; culms erect, slender, simple, smooth and glabrous even at the nodes, up to 90 cm. tall; leaf-blades linear-acuminate, flaccid, dark green, glabrous and smooth on both surfaces, minutely scaberulous on the margins, up to 15 cm. long, 8 mm. wide; sheath smooth, but with a sparse covering of tubercle-based hairs becoming denser towards the mouth; ligule 1·5 mm. long, readily splitting into hairs; inflorescence a panicle of spikes; spikes slender, nodding, up to 15 cm. long, numerous; spikelets 2–3-flowered, green turning purplish, flattened; glumes unequal, keeled, lanceolate-acute, scabrid on the keels, the lower 1·25 mm., the upper 2·5 mm. long; lowest lemma oblong, obtuse, 3-nerved, keeled, long-ciliate on keel and lateral nerves, which are close to the margin, 1·75 mm. long; palea 2-keeled, long-ciliate on the keels.

HAB.: In *Citrus* groves, an introduced weed; near sea-level; fl. June–Oct.

DISTR.: Division 7. Western tropical America and West Africa, occurring as a weed elsewhere.
7. Karavas, 1969, and 1970, *A. Genneou* in ARI 1356! 1579!

58. ERAGROSTIS *P. Beauv.*
Essai Agrost., 70 (1812) emend. Reichb. (1828).

Annual or perennial, glandular or eglandular grasses with narrow, linear leaf-blades; inflorescences effuse or rarely spike-like panicles of 2- to many-flowered, laterally compressed, long- or short-pedicelled spikelets; rhachilla disarticulating from above downwards, or tough, or persistent and parts falling off from below upwards; florets hermaphrodite; glumes unequal or subequal; lemmas 3-nerved, more or less overlapping, keeled, ciliate on the margins and/or keel or quite glabrous, membranous to coriaceous; paleas hyaline, 2-keeled, long-ciliate or not on the keels; stamens 3 or 2; lodicules 2, minute, cuneiform-truncate or rarely absent. Grain globose, ellipsoid or obovoid, slightly laterally compressed, striate or tessellately ridged; embryo one-third to one half the length of the grain; hilum basal, punctiform.

About 300 species with a cosmopolitan distribution, but mostly tropical or subtropical.

1. **E. cilianensis** (*All.*) *Vign.-Lut.* [in Malpighia, 18: 386 (1904) nom. invalid.] *ex Janchen* in
    Mitt. Naturw. Ver. Univ. Wien, 5 (9): 110 (1907); N. L. Bor in C. C. Townsend et al., Fl.
    Iraq, 9: 440, t. 167 (1968); Osorio-Tafall et Seraphim, List Vasc. Plants Cyprus, 15 (1973).
      Briza eragrostis L., Sp. Plant., ed. 1, 70 (1753).
      Poa cilianensis All., Fl. Pedem., 2: 246, t. 91, fig. 2 (1785).
      Poa megastachya Koeler, Descript. Gram., 181 (1802).
      Poa eragrostis (L.) Brot., Fl. Lusit., 1: 103 (1804); Sibth. et Sm., Fl. Graec. Prodr., 1: 54
    (1806), Fl. Graec., 1: 57, t. 73 (1808).
      Eragrostis megastachya (Koeler) Link, Hort. Berol., 1: 187 (1827); Boiss., Fl. Orient., 5:
    580 (1884); Holmboe, Veg. Cypr., 34 (1914); Post, Fl. Pal., ed. 2, 2: 751 (1933); Lindberg f.,
    Iter Cypr., 7 (1946).
      E. megastachya (Koeler) Link f. *cilianensis* (All.) Aschers. et Graebn., Syn. Mitteleurop.
    Fl., 2 (1): 371 (1900); Holmboe, Veg. Cypr., 34 (1914).
    TYPE: Italy; "in agro patrio *Ciliani*, legit cl. Bellardi" (TO).

An annual grass; culms erect or geniculate at the base, smooth and glabrous, simple or branched towards the base, with purple nodes which are often decorated with a row of depressed glands on the lower margin; leaf-blade linear-acuminate, up to 20 cm. long, 8 mm. wide, flat, rigid, minutely rough on both surfaces, glabrous or very sparsely pilose with long white hairs, margins with raised crateriform glands, nerves on the lower surface often with depressed glands; sheaths clasping or somewhat inflated, often with depressed oval or circular glandular patches; ligule membranous, 1–2 mm. long, ciliate; panicle lax or contracted, ovate or elliptic in outline, up to 20 cm. long, 12 cm. wide; spikelets oblong or elliptic in outline, compressed, 6–40-flowered, seated on pedicels which often bear 1 to several crateriform glands; glumes unequal, keeled, lanceolate or ovate-lanceolate, the lower 1-, the upper 3-nerved, with 1 to several glands on the keels, 1·5–2 mm. long; lemma broadly elliptic-acute or retuse when flattened, 3-nerved, with lateral nerves close to the margin, 2–2·5 mm. long, whitish-greyish or tinged with purple, sometimes with glandular keels; palea 2-nerved; anthers minute.

HAB.: Cultivated, irrigated land, and by margins of streams and other moist situations; sea-level to 3,400 ft. alt.; fl. June–Oct.

DISTR.: Divisions 1–4, 6, 7. Mediterranean region, Atlantic Islands and eastwards throughout the warmer parts of Asia; N.E. Tropical Africa; introduced into S. Africa; Australia, N. & S. America.

1. Ayios Neophytos, 1939, *Lindberg f.* s.n.; Khlorakas, 1967, *Economides* in ARI 966! Between Paphos and Yeroskipos, 1981, *Meikle* 5046!
2. Vavatsinia, 1905, *Holmboe* 1111; Platres, 1938, *Kennedy* 1533!
3. Near Stavrovouni monastery, 1939, *Lindberg f.* s.n.! Kannaviou, by the R. Ezuza, 1939, *Lindberg f.* s.n.; Asomatos marshes, 1939, *Mavromoustakis* 135!
4. Styllos, 1968, *A. Genneou* in ARI 1261!
6. Nicosia, by R. Pedieos, 1939, *Lindberg f.* s.n.; Syrianokhori, 1970, *E. Chicken* 168!
7. Lapithos, 1932, *Syngrassides* 239! and 1955, *G. E. Atherton* 678! Kyrenia, 1936, *Kennedy* 98! and, 1955, *G. E. Atherton* 15! 295! Ayios Epiktitos, 1955, *G. E. Atherton* 229!

E. CURVULA (*Schrad.*) *Nees* (Fl. Afr. Austr., 397; 1841), a native of S. Africa, has been planted experimentally at Dennarga, N. of Gouphes (Div. 5) according to Photiades (Countryman, Sept. 22, 1958). It is not known if it survived, nor have any specimens been seen. It is readily distinguished from *E. cilianensis* by its long, flexuous, filiform leaf-blades.

TRIBE 16. **Chlorideae** *Agardh* Annual or perennial grasses; leaf-blades narrow, linear, with panicoid anatomy; silica-bodies saddle-shaped; micro-hairs present, 2-celled, swollen; ligule a row of hairs or membranous; first seedling leaf broad, flat, horizontal; spikelets 1- to several-flowered, compressed laterally, with one floret hermaphrodite and with or without imperfect florets above or below it; spikelets arranged in one or two rows on the continuous rhachis, forming solitary, digitate, or scattered spikes or spike-like racemes; rhachilla disarticulating above the glumes and between the florets, produced, or crowned with one to several rudimentary florets; glumes unequal; lemmas firmly membranous to chartaceous, 1–3-nerved, more or less hairy, awned or awnless; palea often obovate, 2-keeled; stamens 3; ovary glabrous; styles 2; stigmas plumose. Grain loosely enclosed; hilum basal; starch grains compound. Chromosomes small; basic numbers 9, 10.

59. CYNODON L. Rich.

in Pers., Syn. Plant., 1: 85 (1805) nom. cons.

Perennial grasses with creeping stolons and/or rhizomes and short narrow leaf-blades; inflorescence a collection of spikes digitately arranged at the tip of the culms; spikelets all alike, small, laterally compressed, 2-seriate and secund on the rhachis, 1-flowered; glumes more or less equal, thin, keeled;

lemma longer and broader than the glumes, firmly membranous to chartaceous, boat-shaped, containing a bisexual floret; paleas 2-keeled; stamens 3; lodicules 2, minute, cuneate, concave on the upper margin. Grain oblong or elliptic-oblong in outline, free within the lemma and palea, triangular in section with rounded angles, shallowly grooved on the adaxial surface, striate on the surface; embryo one-third the length of the grain; hilum punctiform, basal.

About 10 species, widely distributed in the tropics and subtropics.

1. **C. dactylon** (*L.*) *Pers.*, Syn. Plant., 1: 85 (1805); Unger et Kotschy, Die Insel Cypern, 179 (1865); Boiss., Fl. Orient., 5: 553 (1884); H. Stuart Thompson in Journ. Bot., 44: 340 (1906); Holmboe, Veg. Cypr., 34 (1914); Post, Fl. Pal., ed. 2, 2: 740 (1933); Lindberg f., Iter Cypr., 7 (1946); N. L. Bor in C. C. Townsend et al., Fl. Iraq, 9: 454 (1968); Osorio-Tafall et Seraphim, List Vasc. Plants Cyprus, 15 (1973).

Panicum dactylon L., Sp. Plant., ed. 1, 58 (1753); Sibth. et Sm., Fl. Graec. Prodr., 1: 40 (1806), Fl. Graec., 1: 45, t. 60 (1808).

A perennial grass forming a close turf by means of widely spreading, branching stolons which root at the nodes and send up vertical flowering culms; underground scaly rhizomes present; flowering culms erect or very shortly decumbent, slender, smooth and glabrous, up to 30 cm. tall; leaf-blades linear-acute, flat, rarely folded, up to 10 cm. long, 2–4 mm. wide, minutely rough on the upper surface, and on the margins, otherwise smooth, glabrous or loosely pilose; sheaths clasping, glabrous or loosely hairy; ligule a row of dense hairs; inflorescence of 3–6 digitate spikes radiating from the tip of the culm; spikes spreading, straight or sometimes curved, 2–5 cm. long; spikelets sessile, in two rows secund to the axis and appressed to it, firmly compressed, 2–2·8 mm. long, 1-flowered; glumes more or less equal in length, 1·5–2·5 mm. long, 1-nerved, the lower oblong-acute, the upper broader, acuminate, both persistent, hyaline; lemma as long as the spikelet, very firmly compressed, chartaceous, keeled, 3-nerved, minutely hairy on the keel and on the nerves in the lower half; palea as long as the lemma, 2-keeled; rhachilla sometimes produced and carrying a rudimentary floret; anthers up to 1·5 mm. long.

var. **dactylon**
  TYPE: "*in* Europa *australi*" (LINN!).

Racemes more or less glabrous; leaves not crowded.

  HAB.: Gardens, roadsides, waste ground, sometimes abundant in saline situations near the sea; sea-level to about 600 ft. alt.; fl. March–Nov.

  DISTR.: Divisions 1, 3, 4, 6, 7, locally common in Divisions 3 and 6. Found throughout the warmer regions of the world, probably originated in tropical Africa.

1. Near Polis, 1962, *Meikle* 2316! Kissonerga, 1967, *Economides* in ARI 991!
3. Locally common at Limassol; *Kotschy*; *Holmboe* 650; *Davis* 3539! Phasouri (Asomatos), 1962, *Meikle* 2930!
4. Larnaca, 1880, *Sintenis & Rigo* 661! Athienou, 1934, *Nattrass*162.
6. Locally common at Nicosia, 1934–1935, *Nattrass* 199, 514, 539; *Syngrassides* 1617! *Lindberg f.* s.n.! Morphou, *Hiotellis* 482!
7. Kythrea, 1935, *Nattrass* 669; Kyrenia, 1955, *G. E. Atherton* 28!

var. **villosus** *Regel* in Bull. Soc. Imp. Nat. Mosc., 42: 305 (1869); N. L. Bor in C. C. Townsend et al., Fl. Iraq, 9: 456, t. 174 (1968).
  TYPE: W. Asia (U.S.S.R.): "Am Kaspischen Meere bei Sarepta", *Becker*.

Racemes densely villous; leaves conspicuously crowded.

  HAB.: On sandy ground; c. 500 ft. alt.; fl. March–Nov.

  DISTR.: Division 6, evidently rare. Throughout the range of the species.

6. Nicosia, 1956, *C. E. Shaw*!

60. SPARTINA *Schreb.*

Gen. Plant., ed. 8, 43 (1789).

Perennial grasses with scaly rhizomes; leaf-blades flat, involute or rolled; ligule a ring of hairs; inflorescence a panicle of spikes arranged along a short, 3-angled axis; spikes usually ascending, consisting of a 3-angled rhachis upon which are arranged 2 rows of sessile, 1-flowered, laterally compressed spikelets which fall entire; glumes keeled, more or less unequal; lemma firmer than the glumes; palea 2-nerved; styles 2, stigmas plumose; stamens 3; lodicules absent.

Sixteen species, chiefly in temperate America, a few in Europe and Asia; Tristan da Cunha.

1. **S. anglica** *C. E. Hubb.*, Grasses, ed. 2, 359 (1968) nomen, Bot. Journ. Linn. Soc., 76: 364 (1978).
   [*S. stricta* (non (Ait.) Roth) Lindberg f., Iter Cypr., 8 (1946).]
   [*S. maritima* (non (Curt.) Fernald) Osorio-Tafall et Seraphim, List Vasc. Plants Cyprus, 15 (1973).]
   TYPE: England: W. Sussex; Bosham, 1968, *C. E. Hubbard* D17868A (K!).

A perennial grass with widely spreading rhizomes; culms up to 100 cm. tall, erect or very shortly curved at the base, smooth and glabrous; leaf-blades linear-acuminate, green or greyish-green, flat or involute towards the tips, abruptly narrowed to the sheath, smooth and glabrous, up to 45 cm. long, 6–15 mm. wide; sheaths smooth and glabrous, overlapping; ligule a rim of hairs 2–3 mm. long; panicles contracted, comprising 2–12 erect spikes of spikelets, 12–40 cm. long; spikes stiff, up to 25 cm. long; spikelets biseriate, adpressed, compressed, 1–2-flowered, 14–21 mm. long; glumes elliptic-acute, more or less unequal, the upper broader, 1-nerved, sparsely hairy on the dorsal surface, keeled; lemma shorter than the upper glume, oblong-acute, sparsely hairy, with broad hyaline margins; palea hyaline, 2-nerved; anthers 8–13 mm. long, sagittate. Embryo green, broad, as long as the grain.

HAB.: Maritime mud-flats; sea-level; fl. July–Nov.

DISTR.: Division 4, introduced. Arose in S. England as a hybrid between *S. alterniflora* Lois. and *S. maritima* (Curt.) Fernald.

4. Near Famagusta, 1939, *Lindberg f.* s.n.!

NOTES: No doubt deliberately planted; *S. anglica* has been introduced into many countries, to consolidate and accelerate the reclamation of saline mud-flats.

TRIBE 17. **Aeluropodeae** *Nevski ex Bor* Spikelets laterally compressed, all alike, hermaphrodite, 6- to many-flowered, spaced in spikes along a vertical axis or more or less gathered together in a dense oblong or globular terminal head; rhachilla disarticulating above the glumes and between the florets, produced beyond the terminal floret; glumes unequal, persistent, membranous, the lower the smaller, 1–3-nerved, the upper 5–7-nerved, both rounded on the dorsal surface, mucronate, cuspidate or apiculate; lemmas longer than the glumes, 9–11-nerved, apiculate or mucronate; paleas hyaline, 2-keeled with broad flaps; stamens 3; ovary glabrous; styles 2; stigmas plumose. Grain free; hilum punctiform; lodicules 2.

Perennial grasses; leaf-blades often coriaceous, convolute, pungent, sometimes flat, linear or lanceolate, acuminate; anatomy panicoid; silica-bodies dumb-bell- or cross-shaped; micro-hairs swollen, 2-celled; ligule a rim of short hairs. Chromosomes small; basic number 10.

61. AELUROPUS *Trin.*
Fund. Agrost., 143 (1820).

Perennial grasses with folded, involute or convolute, rarely flat, leaf-blades; inflorescence a dense capitate terminal panicle or spiciform unilateral panicle in which the spikelets are closely packed or rhachides arranged on one side of the vertical axis; spikelets with 4–11 bisexual florets, slightly laterally compressed; glumes unequal, keeled, mucronate, membranous; lemmas similar in texture, 9–11-nerved, keeled, mucronate; palea thinner, hyaline, ciliate on the keels; stamens 3; lodicules 2, oblong-truncate, slightly cuneate. Grain oblong-elliptic or oblong-obovate in outline, flat adaxially, rounded abaxially, loosely enclosed by the lemma and palea; embryo elliptic, about half the length of the grain; hilum punctiform, basal.

Three or four species distributed from the Mediterranean region eastwards to India.

Spikelets glomerate in a globular or dense, oblong, pilose head - - **2. Ae. lagopoides**
Spikelets forming a narrow, elongate inflorescence, not glomerate:
   Plants glabrous - - - - - - - - - - - **1. Ae. littoralis**
   Plants ± pilose - - - - - - - **2. × 1. Ae. lagopoides × Ae. littoralis**

1. **Ae. littoralis** *(Gouan) Parl.*, Fl. Ital., 1: 461 (1848); Boiss., Fl. Orient., 5: 594 (1884); Post in Mém. Herb. Boiss., 18: 102 (1900), Fl. Pal., ed. 2, 2: 756 (1933); H. Stuart Thompson in Journ. Bot., 44: 340 (1906); N. L. Bor in C. C. Townsend et al., Fl. Iraq, 9: 422, t. 159 (1968); Osorio-Tafall et Seraphim, List Vasc. Plants Cyprus, 14 (1973).
   Poa littoralis Gouan, Fl. Monsp., 470 (1765).
   Festuca littoralis (Gouan) Sm. in Sibth. et Sm., Fl. Graec. Prodr., 1: 61 (1806), Fl. Graec., 1: 63, t. 80 (1808).
   TYPE: France; "in maritimis arenosis à *Villeneuve, Maguelone*", Gouan (K !).

A perennial, rhizomatous grass with many prostrate stoloniferous stems eventually erect or sending up vertical culms from the nodes; culms often forming small tufts up to 30 cm. tall, glabrous, scabrid below the panicle; leaf-blades lanceolate, rounded at the base to the sheath, up to 5 cm. long, 2 mm. wide, acuminate, strongly nerved on the upper surface, scabrid on the nerves and on the margins, sometimes sparsely pilose; sheaths tight or loose, striate, pilose at the mouth; ligule a rim of hairs; inflorescence a spike of spikes spaced upon a vertical axis; spikes 5–10 mm. long, of sessile spikelets, alternate, distichous, appressed to the axis; spikelets ovate-oblong, 6–9-flowered; glumes more or less equal, 1·25–1·5 mm. long, lower 1–3-nerved, the upper 5-nerved, both glabrous; lemmas glabrous, 1·75–2·5 mm. long, elliptic-oblong, acute, 8–9-nerved; palea hyaline, 2-nerved; flaps of the glabrous keels equally broad from apex to base or wider below; anthers 2 mm. long.

HAB.: Salt marshes and sandy grasslands near the sea, about sea-level; fl. April–July.

DISTR.: Divisions 4, 6; Mediterranean region eastwards to N. China, Mongolia.

4. Larnaca, 1894, *Post.*
6. Kondemenos, 1935, *R. M. Nattrass* 659.

NOTES: Also collected in 1959 by P. H. Oswald (no. 125), but without precise localization; probably more common on the island than records suggest.

2. **Ae. lagopoides** *(L.) Trin. ex Thwaites*, Enum. Plant. Zeyl., 374 (1864); N. L. Bor in C. C. Townsend et al., Fl. Iraq, 9: 423, t. 160 (1968); Osorio-Tafall et Seraphim, List Vasc. Plants Cyprus, 14 (1973).
   Dactylis lagopoides L., Mantissa 1, 33 (1767).
   Dactylis repens Desf., Fl. Atlant., 1: 79, t. 15 (1798).
   Aeluropus repens (Desf.) Parl., Fl. Ital., 1: 462 (1848); R. M. Nattrass, First List Cypr. Fungi, 21 (1937).

Ae. littoralis (Gouan) Parl. var. *repens* (Desf.) Coss. et Durieu, Fl. d'Algérie, II, Phan., Glumacées [in Expl. Sci. Algérie]: 155 (1855); Boiss., Fl. Orient., 5: 594 (1884).

Ae. littoralis (Gouan) Parl. ssp. *repens* (Desf.) Holmboe, Veg. Cypr., 35 (1914).

TYPE: "*in* India *Burmannus*". (LINN!).

A perennial stoloniferous and/or rhizomatous grass, at times densely tufted, at others with widely spreading prostrate stems not rooting at the nodes, and eventually sending up one or several vertical culms from each node; culms up to 15 cm. tall, smooth and glabrous or more usually densely scabrid below the inflorescence, branched; leaf-blades lanceolate or linear-lanceolate, 2–8 cm. long, 2–4 mm. wide, flat or folded, involute-subulate, pilose above and below or very minutely hairy below; sheaths pilose or glabrescent, ciliate on the margins; ligule a rim of hairs; inflorescence a glomerate, terminal head of solitary spikelets or of spikes of spikelets, densely aggregated; rhachis of the spikes short, hairy; spikelets sessile, ovate in shape, 4–8-flowered; lower glume 1·5 mm. long, oblong-obtuse, 1–3-nerved, with hyaline margins, villous; upper glume 2 mm. long, similar in shape, 3–5-nerved, villous; lemmas 2·5–2·75 mm. long, elliptic-apiculate, 8–11-nerved, villous from tubercles between the nerves; lower lemmas often empty; keels of palea hairy; flaps of the palea narrowed from apex to base; anthers 1–1·25 mm. long.

HAB.: Salt marshes and sandy grassland near the sea; sea-level to 500 ft. alt.; fl. April–July.

DISTR.: Divisions 3–6. Also Sicily and eastern Mediterranean region eastwards to India and Central Asia; N. Africa, N. Sudan, Eritrea, Somalia, Mauritania.

3. Akrotiri Bay, 1939, *Mavromoustakis* 92! and, 1949, *Mavromoustakis* 32! Limassol Salt Lake, between Akrotiri and Phasouri, 1941, *Davis* 3564!
4. Locally common around Larnaca Salt Lake, *Sintenis & Rigo* 882! *Lindbergf.* s.n.! *Chapman* 421! 688! *Merton* 1176! Famagusta, 1905, *Holmboe* 1095, and, 1939, *Lindbergf.* s.n.
5. Near Pyroi, by the bridge, 1935, *Syngrassides* 656!
6. Nicosia, 1935, *R. M. Nattrass* 650.

NOTES: *Aeluropus lagopoides* (L.) Trin. is based upon *Dactylis lagopoides* L. of which there is a specimen in the Linnean Herbarium in London. The specimen has been examined and, in the writer's opinion, this species occurs and is remarkably constant over a very wide range, namely from India, through Iran, Iraq to Egypt and further to West Africa by way of the North African States.

2. × 1. Ae. lagopoides (*L.*) *Trin. ex Thwaites* × **Ae. littoralis** (*Gouan*) *Parl.*

More or less intermediate between *Ae. lagopoides* and *Ae. littoralis*, but all parts of the plant are covered with long, soft, white hairs. The inflorescence is a spike of spikes, similar to that of *Ae. littoralis* but the spikelets are softly hairy. An examination of the anthers shows them to be devoid of pollen.

HAB.: Salt marsh; near sea-level; fl. April–July.

DISTR.: Division 5.

5. Salamis, in a salt marsh, 1962, *Meikle* 2590!

NOTES: This specimen has a spicate inflorescence, as opposed to the globular or cylindrical inflorescence of *Ae. lagopoides*, but the spikelets and most of the plant are covered with long soft hairs. Typical *Ae. littoralis* (Gouan) Parl. as understood by me, has glabrous spikelets, but many specimens are more or less hairy, in other words this hairiness is an extremely variable character. In Flora USSR., 2: 358 (1934) Roschevicz, in his description of *Ae. littoralis*, mentions a variety, var. *dasyphylla* Trautv., which has hairy sheaths and leaves. The variety *dasyphylla* Trautv. appears in Fedtsch. in Bull. Jard. Pierre Grand, 14: 74 (1915), without a description, but *Ae. intermedius* Regel (Bull. Soc. Nat. Mosc., 41: 292; 1869) is cited as being the same plant. *Aeluropus intermedius* Regel is considered to be a good species by Tzvelev (Not. Syst. Acad. Sc. URSS., 25; 1966) and an identical plant with hairy spikelets has been given specific rank as *Aeluropus korshinskyi* Tzvelev. This view however is not held by all botanists. Some consider that *Ae. repens* (Desf.) Parl. collected in Egypt and based on *Dactylis repens* of Desfontaines, is specifically distinct from *Ae. lagopoides*, based on *Dactylis lagopoides* L. of South India. There is however no gap in the distribution of *Ae. lagopoides* from South India to West Africa, and no differences which could be called specific have been detected. Others, e.g. Maire, Fl. Afr. Nord, 3: 71 (1955), consider *Ae. lagopoides* to be a variety of *Ae. littoralis* (Gouan)

Parl. which is called var. *repens* (Desf.) Coss. et Dur. This is, in the writer's opinion an extreme instance of "lumping" with which he cannot agree.

TRIBE 18. **Sporoboleae** *Stapf* Annual or perennial grasses; leaf-blades narrow, linear, with panicoid anatomy; silica-bodies saddle-shaped; micro-hairs 1- or 2-celled, distal cell swollen; ligule a rim of hairs sometimes connate at the base; spikelets all alike, hermaphrodite, terete or gaping, 1-flowered, small; rhachilla disarticulating above or below the glumes; inflorescence an open or contracted, sometimes spike-like panicle; glumes hyaline or membranous, equal or unequal, truncate or acute, more or less persistent, nerveless or 1-nerved; lemmas 1–3-nerved, membranous, not becoming indurated, usually shining; palea hyaline, 1–3-nerved; stamens 3 or 2; ovary glabrous; styles 2; stigmas plumose; lodicules 2, broadly cuneate. Grain enclosed in a delicate, mucilaginous pericarp; hilum punctiform, basal; starch grains compound. Chromosomes small; basic numbers 9, 10, 12 or 6 (Kai Larsen in Dansk Bot. Archiv., 20: 211–275; 1963).

62. CRYPSIS *Ait.*

Hort. Kew., ed. 1, 1: 48 (1789) nom. cons.
C. E. Hubbard in Hooker's Icones Plant., 35: t. 3457 (1947).
Pallasia Scop., Introd., 72 (1777).
Heleochloa Host, Gram. Austr., 1: 23 (1801).

Annual grasses with prostrate, finally erect, culms and flat leaf-blades; inflorescence a spiciform panicle consisting of a dense, globose, ovoid or cylindrical mass of closely packed spikelets on very short pedicels and branches, often partly embraced at the base by the topmost, subtending sheaths; spikelets with one bisexual flower, strongly laterally compressed; rhachilla not produced, disarticulating above or sometimes below the glumes; glumes persistent, subequal, complicate, markedly compressed; lemma acutely keeled, 1-nerved, membranous; palea as long as the lemma, 1–3-nerved; stamens 2–3; lodicules none. Grain oblong or elliptic-oblong, laterally compressed, wrinkled or smooth, surrounded by a membranous, free pericarp; embryo elliptic, as long as, or three-quarters the length of the grain; hilum punctiform, basal.

About 12 species distributed from the Mediterranean region to N. China; also in tropical Africa.

Inflorescence an elongate, narrow, oblanceolate, spike-like panicle	-	**1. C. alopecuroides**
Inflorescence an ellipsoid panicle or a capitulum of spikes:		
Inflorescence ellipsoid, longer than broad - - - - -		**4. C. schoenoides**
Inflorescence a capitulum, broader than long:		
Spikelets 4–6 mm. long, 1 mm. wide; stamens 3 - - -		**3. C. factorovskyi**
Spikelets 3–3·5 mm. long, 1–1·5 mm. wide; stamens 2 - - - -		**2. C. aculeata**

1. C. alopecuroides (*Pill. et Mitterp.*) *Schrad.*, Fl. Germ., 1: 167 (1806); N. L. Bor in C. C. Townsend et al., Fl. Iraq, 9: 462, t. 177 (1968); Osorio-Tafall et Seraphim, List Vasc. Plants Cyprus, 15 (1973).
   Phleum alopecuroides Pill. et Mitterp., Iter Poseg. Sclav., 147, t. 16 (1783).
   Heleochloa alopecuroides (Pill. et Mitterp.) Host, Gram. Austr., 1: 23, t. 29 (1801); Boiss., Fl. Orient., 5: 476 (1884); Post, Fl. Pal., ed. 2, 2: 717 (1933).
   Phalaris geniculata Sm. in Sibth. et Sm., Fl. Graec. Prodr., 1: 38 (1806).
   TYPE: Hungary; "... ad *Adony* transivimus. In udis circa hoc oppidum & *Budae* quoque, sed rarius reperitur", *Piller & Mitterpacher* (BP).

An annual grass; culms usually numerous, often widely spreading and prostrate but not rooting at the nodes, finally erect, smooth and glabrous, a few centimetres tall in the Cyprus plant, up to 30 cm. tall in favourable habitats elsewhere; leaf-blades glaucous or green, lanceolate or linear, acute,

2–6 cm. long, 1–3 mm. wide, strongly nerved on both surfaces, glabrous, smooth, scabrid or loosely to densely pilose on both surfaces, scabrid on the margins; sheaths rather loose, the uppermost inflated and often surrounding the base of the panicle, smooth and glabrous, or pilose above or throughout, striate, ciliate on the margins; ligule a densely ciliate rim; inflorescence a very dense, narrow panicle, 1–5 cm. long, 3–5 mm. wide, cylindrical, narrowly ellipsoid or narrowly oblanceolate in outline; branches very short; pedicels minute; spikelets firmly compressed, green or tinged with purple, mostly 2·5 mm. long, gaping; glumes unequal, keeled, scarious on the margins, ciliate on the keels, the lower 2 mm., the upper 2·5 mm. long, keel-nerve green; lemma 2–2·5 mm. long, slightly overtopping the glumes, oblong-acute, keeled, 1-nerved, scabrid on the keel; palea 2-nerved; anthers yellow or purple, 1·5 mm. long.

HAB.: On the dried mud of peaty pools; 5,200 ft. alt.; fl. June–Sept.

DISTR.: Division 2, rare. Central Europe and Mediterranean region eastwards to Central Asia.

2. Livadhi tou Pasha, 1954, *Merton* 1950 !

2. C. aculeata (*L.*) *Ait.*, Hort. Kew., ed. 1, 1: 48 (1789); Boiss., Fl. Orient., 5: 475 (1884); Post, Fl. Pal., ed. 2, 2: 716 (1933); Lindberg f., Iter Cypr., 7 (1946); N. L. Bor in C. C. Townsend et al., Fl. Iraq, 9: 460, t. 176 (1968); Osorio-Tafall et Seraphim, List Vasc. Plants Cyprus, 15 (1973).

Schoenus aculeatus L., Sp. Plant., ed. 1, 42 (1753).

TYPE: "*in* Italia, Narbona, Lusitania, Archipelagi *insulis*" (LINN !).

An annual grass; culms sometimes very short, branched from the base, 3–4 cm. tall, or sometimes prostrate and widely spreading with the aspect of *C. factorovskyi* (infra), smooth and glabrous, green or somewhat glaucous; leaf-blades lanceolate-acuminate, flat or partly involute, up to 5 cm. long, usually shorter, 2–3 mm. wide, glabrous or sparsely hirsute on the lower surface, striate above, often pilose with soft hairs towards the base, scabrid on the margins; sheaths rather loose, striate, those subtending the inflorescence much inflated and with reduced blades; inflorescence a capitulum, wider than long, of closely packed spikelets immersed in the uppermost inflated leaf-sheaths; spikelets firmly compressed, 3–3·5 mm. long, 1–1·5 mm. wide; glumes very narrow, scabrid on the keels, the lower 2–2·5 mm. long, the upper 3 mm. long; lemma broader, scabrid on the keel; palea 1–3-nerved, usually 1-nerved; anthers 2, 1–2 mm. long.

HAB.: Damp sandy ground by the seashore; near sea-level; fl. Sept.–Nov.

DISTR.: Division 4, rare. Central Europe and Mediterranean region eastwards to Central Asia; introduced into S. Africa.

4. Famagusta, on the seashore, 1939, *Lindberg f.* s.n. !

3. C. factorovskyi *Eig* in Zionist Org. Inst. Agr. Nat. Hist. Bulletin 6: 58 (1972); Post, Fl. Pal., ed. 2, 2: 716 (1933); A. K. Jackson in Kew Bull., 6: 345 (1937); Osorio-Tafall et Seraphim, List Vasc. Plants Cyprus, 15 (1973).

TYPE: Israel; Ramath-Gan, near Tel-Aviv, *Eig* (HUJ, K !).

An annual grass; culms many from the base, prostrate, much branched, up to 90 cm. long (fide Eig) finally geniculately ascending, sending up flowering shoots from each node, glaucous, smooth and glabrous; leaf-blades linear-acute, glaucous, up to 6 cm. long, 1–3 mm. wide, involute or flat, pilose on the upper surface, with white hairs from tubercle-bases, smooth and glabrous on the lower surface, scabrid on the margins; upper 2–3 sheaths very broad with reduced blades, supporting the inflorescence as an involucre, the remainder rounded on the back, rather firm, slipping from the culm, smooth and glabrous; ligule a dense row of hairs; inflorescence a capitulum of spikelets terminating the culms and the branches; spikelets 4–6 mm. long, 1 mm. wide, firmly compressed, closely packed together; glumes more or less equal in length, 1-nerved, very narrow, oblong, acute or acuminate, the upper somewhat broader than the lower, keeled, ciliate on the keels and puberulous on the surface in the upper third; lemma delicate, compressed, keeled, 1-nerved, ciliate on the keel in the upper part; palea 2-nerved; stamens 3, 2–3 mm. long.

HAB.: Dried muddy ground, usually in saline or brackish situations; sea-level to c. 500 ft. alt.; fl. June–Oct.

DISTR.: Divisions 3, 4, 6, rare. Also Palestine and Syria.

3. Near Akrotiri Salt Lake, 1970, *E. Chicken* 112 !
4. Kouklia, below the reservoir, 1952, *Merton* 913 ! Larnaca, 1970, *A. Genneou* in ARI 1589 !
6. Makhedonitissa Mon., 1935, *Syngrassides* 793 !

NOTES: An unlocalized specimen was also collected in Cyprus, in 1860, by *J. D. Hooker & D. Hanbury* (K !). It came almost certainly from the neighbourhood of Larnaca, probably from near the Salt Lake (Div. 4).

4. **C. schoenoides** (*L.*) *Lam.*, Encycl. Méth., 1: 166, t. 42, fig. 1 (1791); Lindberg f., Iter Cypr., 7 (1946); N. L. Bor in C. C. Townsend et al., Fl. Iraq, 9: 464, t. 178 (1968); Osorio-Tafall et Seraphim, List Vasc. Plants Cyprus, 15 (1973).
  Phleum schoenoides L., Sp. Plant., ed. 1, 60 (1753).
  Heleochloa schoenoides (L.) Host, Gram. Austr., 1: 23 (1801); Boiss., Fl. Orient., 5: 476 (1884); Post, Fl. Pal., ed. 2, 2: 717 (1933).
  TYPE: "*in* Italia, Smyrna, *inque* Hispania. *Loefling*" (LINN !).

An annual grass with many, spreading, prostrate stems, not rooting at the nodes, finally ascending and terminating in the inflorescence, smooth and glabrous, somewhat glaucous, more or less branched; leaf-blades lanceolate-acuminate, flat or involute, sometimes up to 6 cm. long, but usually much shorter, markedly ribbed on the upper surface, hairy above, glabrous below; sheath closely clasping, smooth and glabrous; ligule a very narrow, ciliate rim; inflorescence an ellipsoid, very dense panicle of spikelets, or rarely ovate in outline, subtended by one or two inflated leaf-sheaths with reduced blades; spikelets firmly compressed, 3–3·5 mm. long, elliptic-acute in outline, very shortly pedicellate; glumes slightly unequal, linear-acute, scabrid on the keel, 1-nerved, 2·5–3 mm. long; lemma elliptic-acute, 3·25–3·5 mm. long, 1-nerved; palea almost as broad, 2-nerved; anthers 3, yellow, 1 mm. long.

HAB.: Damp ground, often in brackish situations; sea-level to 100 ft. alt.; fl. April–Oct.

DISTR.: Divisions 4, 7. West, Central and southern Europe and Mediterranean region eastwards to Central Asia and N.W. India; introduced elsewhere.

4. Near the Salt Lake, Larnaca, 1939, *Lindberg f.* s.n. ! Kouklia, below the reservoir, 1952, *Merton* 912 !
7. Kyrenia, 1939, *Lindberg f.* s.n.; Panagra, 1970, *E. Chicken* 153 !

NOTES: *Merton* 502 (Livadhi·tou Pasha, peaty margin of dried-up pool) a depauperate plant only 2 cm. high, looks like *C. schoenoides*, but may be a new species.

63. SPOROBOLUS *R. Br.*
Prodr. Fl. Nov. Holl., 169 (1810).

Perennial, rarely annual grasses with flat or convolute leaf-blades; spikelets very small, 1-flowered, pale or greyish, loosely or closely panicled, articulate or not on the pedicels, fusiform in outline; rhachilla disarticulating above the glumes; glumes equal or unequal, usually hyaline, nerveless, the lower usually much shorter than the spikelet, often truncate, the upper as long as or shorter than the spikelet, truncate or acute; lemma elliptic-acute, 1-nerved or more or less indistinctly 3-nerved; palea almost as long as the lemma, 2-nerved, hyaline, folded between the nerves; lodicules 2, broadly cuneate, truncate or crenulate; stamens 2 or 3; styles 2; stigmas plumose. Caryopsis cuneate, elliptic or globose, compressed or not, enclosed in a thin envelope which becomes mucilaginous when moistened; embryo rather large; hilum basal, punctiform.

About 150 species widely distributed in warm-temperate and tropical regions of the world.

1. **S. virginicus** (*L.*) *Kunth*, Rev. Gram., 1: 67 (1829).
   Agrostis virginica L., Sp. Plant., ed. 1, 63 (1753).
   Agrostis arenaria Gouan, Illustr. Obs. Bot., 3 (1773).
   Agrostis pungens Schreb., Descr. Gräs., 2: 46 (1779).
   Sporobolus pungens (Schreb.) Kunth, Rev. Gram., 1: 67 (1829); Boiss., Fl. Orient., 5: 512 (1884); Druce in Rep. B.E.C., 9: 47 (1931).
   S. arenarius (Gouan) Duval-Jouve in Bull. Soc. Bot. Fr., 16: 294 (1869); Post, Fl. Pal., ed. 2, 2: 728 (1933); Lindberg f., Iter Cypr., 9 (1946).
   TYPE: U.S.A. "*in* Virginia" *Clayton* (BM!).

A perennial grass with widely creeping scaly rhizomes, sending up flowering shoots from the nodes; culms loosely tufted, smooth and glabrous, simple, covered with sheaths almost to the panicle; leaf-blades 2–8 cm. long, 2–5 mm. wide, involute, stiff, linear-acuminate with a pungent tip, smooth and glabrous on the lower, more or less hairy on the upper surface, scaberulous or almost smooth on the margins, distichously arranged in one plane and often close together; sheaths overlapping, smooth and glabrous or hairy at the collar; ligule a dense row of hairs; panicle rather dense, ovate in outline, axis smooth; branches solitary, ascending; spikelets lanceolate or fusiform, pale, 2–2·5 (–3) mm. long; lower glume 2 mm. long, nerveless or faintly 1-nerved, oblong-acute, the upper as long as the spikelet, oblong-acute, 1-nerved; lemma 2·5 (–3) mm. long, lanceolate-acute, glabrous, smooth, 1-nerved; palea as long as the lemma, 2-nerved; anthers 1–1·5 mm. long.

HAB.: Sandy shores and sand-dunes near sea-level; fl. June–Sept.

DISTR.: Divisions 3–5, 7. Widespread in tropical and subtropical regions.
3. Limassol Salt Lake E. of Akrotiri, 1941, *Davis* 3578!
4. Ayia Napa, 1939, *Lindberg f.* s.n.; Larnaca airport, 1952, *Merton* 846! Famagusta, 1956, *Merton* 2839! Coast S.E. of Varosha, 1959, *P. H. Oswald* 173! Larnaca, 1970, *E. Chicken* 197!
5. Salamis, 1928 or 1930, *Druce*; and, 1939, *Lindberg f.* s.n.; Trikomo, 1939, *Lindberg f.* s.n.
7. Ayios Trikas ["Trikaz"], 1968, *Economides* in ARI 1130!

TRIBE 19. **Aristideae** *C. E. Hubb. ex Bor* Tufted perennial or rarely annual grasses; leaf-blades narrow, often convolute, pungent, rarely flat, with panicoid anatomy; silica-bodies dumb-bell-shaped or elliptic-oblong; micro-hairs 2-celled, elongate; ligule a fringe of hairs; spikelets all alike, hermaphrodite, 1-flowered; rhachilla disarticulating above the glumes, not produced; inflorescence an effuse or narrow, contracted panicle; glumes membranous, usually unequal, often markedly so, acute, mucronate or shortly awned, entire or bifid at the tip; lemma convolute, at first chartaceous, finally firmer, cylindric, entire or 2-fid at the tip, awned; awn tripartite from the base or at the tip of a column, rarely the column attached to the upper half of the lemma which falls with it; column straight or twisted; branches of the awn glabrous or hairy; palea oblong, hyaline, 2-nerved; stamens 3 or 1; ovary glabrous; styles 2; stigmas plumose; lodicules 2. Grain tightly enclosed between lemma and palea; hilum linear; starch grains compound. Chromosomes small; basic numbers 11, 12.

64. ARISTIDA *L.*

Sp. Plant., ed. 1, 82 (1753).
Gen. Plant., ed. 5, 35 (1754).

Perennial, tufted grasses, rarely annual, with narrow, often convolute leaf-blades; inflorescence an effuse or contracted panicle of pedicellate awned spikelets; spikelets 1-flowered, hermaphrodite; rhachilla disarticulating above the glumes, not produced beyond the floret; glumes persistent, equal or unequal, rarely the lower longer than the upper, sometimes shortly awned; lemma convolute, chartaceous to coriaceous, terete, with a bearded, pungent, obtuse or rarely forked callus with a tripartite glabrous awn which is continuous with the lemma or articulated to its tip or to the tip of a column, the latter straight or twisted; palea a 2-nerved or nerveless scale; stamens 3 or 1; lodicules 2 or 3; oblong-truncate (*A. plumosa*) or oblanceolate, glabrous, strongly nerved. Grain terete, tightly embraced by the lemma; embryo elliptic, up to two-thirds the length of the grain; hilum linear, nearly as long as the grain.

About 260 species, widespread especially in tropical and subtropical regions.

1. **A. caerulescens** *Desf.*, Fl. Atlant., 1: 109, t. 21, fig. 2 (1798); Boiss., Fl. Orient., 5: 491 (1884); N. L. Bor in C. C. Townsend et al., Fl. Iraq, 9: 384, t. 146 (1968).

   A. adscensionis L. var. *caerulescens* (Desf.) Hackel in Stuck. Anal. Nac. Buenos Aires, 11: 90 (1904).

   [*A. adscensionis* (non L.) Dinsmore in Post, Fl. Pal., ed. 2, 2: 721 (1933).]

   TYPE: Tunisia; "in arvis prope Kerwan [Kairouan]", *Desfontaines* (P).

A perennial grass with numerous sterile innovation shoots; culms usually erect, sometimes geniculate at the base, simple or branched, glabrous, scabrid, up to 40 cm. tall; leaf-blades flat at first, folded, involute or convolute, filiform, up to 15 cm. long, 0·5–1·5 mm. wide, linear-acuminate, smooth and glabrous on the lower surface, scabrid and striate on the upper surface; sheaths rounded on the dorsal surface, scarious on the margins, smooth and glabrous or scaberulous; ligule a very narrow ciliate ring; inflorescence a narrow, dense or lax, purplish panicle about 10 cm. long, 1–1·5 cm. wide, erect or nodding; axis angled, scabrid; branches ascending, binate or solitary; spikelets almost always suffused with violet, fusiform, 9–10 mm. long, seated on short pedicels; glumes unequal, oblong-acute, the lower 4·5–6·5 mm. long, 1-nerved, the upper 7–8·5 mm. long, 3-nerved, thin, bifid at the tip, with a short mucro in the sinus; lemma 8–10 mm. long, convolute, smooth and glabrous or scabrid towards the trifurcation, awned; awn with three scaberulous, capillary branches, more or less equal, or the central the longest, up to 2·5 cm. long.

HAB.: Stony hillsides and roadsides on igneous formations; 2,800 ft. alt.; fl. May–June.

DISTR.: Division 2, rare. Mediterranean region, Atlantic Islands eastwards to Afghanistan; Ethiopia.

2. Between Vavatsinia and Ayii Vavatsinias, 1967, *Merton* in ARI 715!

NOTES: The distinctions between the annual *A. adscensionis* L. and *A. caerulescens* Desf. are tenuous, and there are strong arguments for uniting the two, at least at species level.

TRIBE 20. **Paniceae** *R. Br.* Annual or perennial grasses, or sometimes woody outside our area (*Panicum turgidum, Pennisetum divisum*); leaf-blades linear or sometimes lanceolate, acute or acuminate, tapering to or rounded at the base, usually flat, with panicoid anatomy; silica-bodies nodular, dumb-bell-shaped or cruciform; micro-hairs 2-celled, filiform or club-shaped (in some *Digitaria* spp. not found in Cyprus); first foliate leaf flat and horizontal; ligule membranous, a row of hairs, or rarely absent; spikelets usually similar, hermaphrodite, rarely unisexual (but not in Cyprus species), usually deciduous from the pedicels at maturity, 2-flowered, with the lower male or barren and the upper hermaphrodite; glumes usually unequal, membranous or herbaceous, the lower the smaller, rarely absent, the upper as long as, or shorter, rarely much shorter than the spikelet; lower lemma the same texture as the glumes, paleate or not, the upper much firmer in texture, smooth or rugose, awnless or with a short mucro at the tip; rhachilla not produced, rarely the lower joint swollen (*Eriochloa*); stamens 3; ovary glabrous; styles 2; stigmas plumose; lodicules usually 2. Caryopsis firmly enclosed between lemma and palea; hilum basal, punctiform; embryo half the length of the grain; starch grains simple. Chromosomes small; basic numbers 7 (rarely), 9, 10, 15, 17, 19.

65. BRACHIARIA *Griseb.*

in Ledebour, Fl. Ross., 4: 469 (1853).

Annual or perennial grasses with narrow, flat leaf-blades and upright culms often decumbent at the base; inflorescence a panicle made up of adaxially orientated spikelets, arranged in closely packed spikes or racemes or more loosely arranged in pairs on the branches; spikelets somewhat dorsally compressed, 2-flowered; glumes unequal, the lower very much smaller than the upper which is almost the length of the spikelet; lower lemma of the same texture and shape as the upper glume, containing a male floret; palea hyaline; upper lemma coriaceous to crustaceous with involute margins, obtuse or mucronate, transversely rugose or smooth, containing a bisexual floret; palea of the same texture; stamens 3; lodicules 2, broadly cuneate, irregularly undulate on the upper margin, glabrous. Grain oblong-elliptic, dorsally compressed or plano-convex; embryo half as long as the grain or longer; hilum punctiform, just above the base.

Over 50 species widely distributed in the warmer regions of both hemispheres.

1. **B. eruciformis** (*Sm.*) *Griseb.* in Ledebour, Fl. Ross., 4: 469 (1853); A. K. Jackson in Kew Bull., 1934: 273 (1934); N. L. Bor in C. C. Townsend et al., Fl. Iraq, 9: 472, t. 181 (1968); Osorio-Tafall et Seraphim, List Vasc. Plants Cyprus, 15 (1973).

   Panicum eruciforme Sm. in Sibth. et Sm., Fl. Graec. Prodr., 1: 40 (1806), Fl. Graec., 1: 44, t. 59 (1808); Boiss., Fl. Orient., 5: 437 (1884); Post, Fl. Pal., ed. 2, 2: 696 (1933); Lindberg f., Iter Cypr., 8 (1946).

   TYPE: Samos; "in arvis circa Junonis templum in insulâ Samo" *Sibthorp* (OXF); figured in Fl. Graec., 1: t. 59 (1808)!

An annual; culms loosely tufted from a decumbent and geniculately ascending base, rooting at the nodes, up to 60 cm. tall, smooth and glabrous, very shortly pilose at the nodes or glabrous; leaf-blades linear-lanceolate, somewhat rigid, softly hairy on both surfaces or more or less glabrous above, scabrid on both surfaces and on the margins, up to 9 cm. long, 3–7 mm. wide, greyish-green; inflorescence of 2–10 spike-like racemes, curved, ascending, and appressed to the central axis; racemes of very shortly pedicellate, adaxial spikelets, 12–15 mm. long; pedicels hairy; spikelets dorsally compressed, elliptic in outline, 2·5 mm. long, more or less hairy; lower glume a hyaline nerveless scale 0·25 mm. long, truncate, the upper as long as the spikelet, elliptic-obtuse when flattened, 5-nerved; lemma of the lower floret similar to the upper glume, 5-nerved, hairy, neuter or enclosing a male flower; palea hyaline; upper lemma elliptic-oblong, obtuse, very firm, shining, white, 1·25–1·75 mm. long; palea of the same texture; anthers about 1 mm. long.

HAB.: Cultivated ground, often in damp situations on clay; sea-level to c. 1,500 ft. alt.; fl. June–Oct.

DISTR.: Divisions 2, 5–7. S. Europe and eastern Mediterranean region eastwards to India; tropical and S. Africa; introduced elsewhere.

2. Kannaviou, by R. Ezuza, 1939, *Lindberg f.* s.n.
5. Near Kythrea, 1939, *Lindberg f.* s.n.!
6. Makhedonitissa Monastery, 1935, *Syngrassides* 792!
7. Lapithos, Experimental Lime Plantation, 1932, *Syngrassides* 217!

UROCHLOA PANICOIDES *P. Beauv.*, Essai Agrost., 52, t. 11, fig. 1 (1812), a tropical grass, indigenous in eastern Tropical Africa, Yemen and India, and introduced into Australia and elsewhere, has been collected as a weed in Citrus groves at Morphou (6) (1970, *A. Genneou* in ARI 1613), and may occur as an alien elsewhere in Cyprus. It is a tufted annual, with culms 10–100 cm. high, often ascending from a prostrate rooting base; the leaf blades are linear to narrowly lanceolate, 2–5 cm. long, 0·5 to 1·5 cm. wide, subamplexicaul, coarse, glabrous or pubescent, the margins tuberculate-ciliate at least near the base; the inflorescence comprises 2–7 (–10 or rarely more) racemes on a common axis 1–6 cm. long, each raceme about 1–6 cm. long, bearing single or sometimes paired spikelets on a narrowly winged rhachis, the pedicels (sometimes also the rhachis) clothed with white hairs; the spikelets are elliptic (2·5–) 3·5–5 mm. long, acute; the lower glume ovate, a quarter to almost half as long as the spikelet, very rarely more, 3–5-nerved, obtuse to subacute; the upper glume is glabrous or pubescent, often with cross-veins; the lower lemma sometimes has a setose fringe; the upper lemma has a mucro 0·3–1 mm. long.

66. DIGITARIA *Haller*

Stirp. Helv., 2: 244 (1768) nom. cons.

Annual or perennial grasses with narrow, flat, often soft, leaf-blades and erect culms, mostly decumbent at the base; inflorescence rarely a panicle, mostly of racemes in pairs or digitately arranged or more or less whorled or alternate; rhachis flat or triquetrous, often winged; spikelets, when racemose on a rhachis, abaxial, 2- to 3-nate, pedicellate, 2-flowered, all alike or occasionally dimorphic; glumes unequal, the lower minute, hyaline or absent, the upper rarely as long as the spikelet, narrow, rarely absent; lower lemma flat, shape of the spikelet, 5–7-nerved, membranous; palea absent or, if present, minute; upper lemma chartaceous, with hyaline, flat, not involute, margins, 3-nerved, containing a hermaphrodite flower; palea of same texture; stamens 3; lodicules 2, broadly cuneate. Grain plano-convex in section, oblong-elliptic in outline; embryo one-third the length of the grain; hilum suprabasal, small, elliptic in outline.

Over 300 species widely distributed in the warmer parts of both hemispheres, especially abundant in Africa.

1. **D. sanguinalis** (*L.*) *Scop.*, Fl. Carn., ed. 2, 1: 52 (1772); N. L. Bor in C. C. Townsend et al., Fl. Iraq, 9: 478, t. 183 (1968); Osorio-Tafall et Seraphim, List Vasc. Plants Cyprus, 15 (1973). *Panicum sanguinale* L., Sp. Plant., ed. 1, 57 (1753); Sibth. et Sm., Fl. Graec. Prodr., 1: 40 (1806); Boiss., Fl. Orient., 5: 433 (1884); Holmboe, Veg. Cypr., 32 (1914); Post, Fl. Pal., ed. 2, 2: 695 (1933); Lindberg f., Iter Cypr., 8 (1946).
   TYPE: "*in* America, Europa *australi*" (LINN!).

An annual grass; culms prostrate at the base, rooting at the lower nodes, spreading, often branched, smooth and glabrous or pilose at the nodes, green or purplish, up to 45 cm. tall; leaf-blades linear or lanceolate, acuminate, rounded at the base, flat, flaccid, up to 18 cm. long, 3–10 mm. wide, usually much shorter, pilose on both surfaces or glabrous, scabrid on the margins; sheaths densely or sparsely covered with tubercle-based hairs, rarely glabrous, rather loose; ligule 1–2 mm. long, membranous; inflorescence consisting of a number (4–10) of racemes of spikelets more or less whorled at the tip of the peduncle, erect or spreading; racemes of shortly pedicellate spikelets, secund on a 3-angled axis, 3–15 cm. long; spikelets in pairs, elliptic in outline, 2·5–3 mm. long, 1-flowered; lower glume a minute, hyaline, nerveless scale, the upper narrowly triangular or lanceolate, 1–1·5 mm. long, 3-nerved, shortly hairy; lower lemma flat, elliptic, 7-nerved, scabrid on the nerves in the upper half, minutely hairy or not; palea a hyaline scale; upper lemma ovate-acute, firm, smooth; palea of the same texture, with wide flaps below; anthers 0·6 mm. long.

HAB.: Cultivated, often irrigated, land; sea-level to 3,000 ft. alt.; fl. March–Sept.

DISTR.: Divisions 2–4, 6–8. Europe, Mediterranean region eastwards to Central Asia and India, N. America; cosmopolitan as a weed in warm and temperate regions of the world.

2. Palekhori, 1905, *Holmboe* 1139; Kambos, 1939, *Lindberg f.* s.n.; Kakopetria, on the road to Troödos, 1939, *Lindberg f.* s.n.; Kakopetria, 1955, *Hiotellis* 444!
3. Asomatos marshes, 1939, *Mavromoustakis* 136!
4. Famagusta, 195–, *Merton* 2820!
6. Nicosia, Government House Gardens, 1935, *Syngrassides* 794! Kokkini Trimithia, 1966, *Merton* in ARI 20! Syrianokhori, 1970, *E. Chicken* 166!
7. Kythrea, 1934, *Syngrassides* 533! Kyrenia, 1935, *Syngrassides* 650! Kyrenia, 1948, *Casey* 5!
8. Ayios Andronikos, 1934, *Syngrassides* 513!

67. ECHINOCHLOA *P. Beauv.*
Essai Agrost., 53 (1812).

Annual (both Cyprus species) or perennial grasses with flat, often flaccid, leaf-blades, ligule usually absent; inflorescence a panicle made up of racemes of spikelets in false spikes; spikelets sessile or pedicellate, closely packed, usually awned or cuspidate, rounded on the back, flat on the ventral surface, 2-flowered; glumes unequal, membranous, the lower much the smaller, mucronate, the upper the length of the spikelet, mucronate or short-awned; lower lemma membranous, enclosing a male flower or barren, similar to the upper glume, mucronate or awned; palea hyaline; upper lemma chartaceous to coriaceous, containing a bisexual floret, with involute margins clasping a palea of the same texture; stamens 3; lodicules 2, broadly cuneate-truncate. Grain broadly elliptic in outline, plano-convex in profile; embryo half to three-quarters the length of the grain; hilum basal, circular.

About 30 species widely distributed in the warmer regions of the world.

Spikelets 2–3 mm. long; glumes and lemmas mucronate, not awned or cuspidate

1. E. colonum

Spikelets larger, 3–4 mm. long; glumes often awned and lower lemma cuspidate, acuminate or awned -   -   -   -   -   -   -   -   -   -   -   - 2. E. crusgalli

1. E. colonum (*L.*) *Link*, Hort. Berol., 2: 209 (1833); N. L. Bor in C. C. Townsend et al., Fl. Iraq, 9: 479, t. 184 (1968); Osorio-Tafall et Seraphim, List Vasc. Plants Cyprus, 15 (1973).
Panicum colonum L., Syst. Nat., ed. 10, 2: 870 (1759); Boiss., Fl. Orient., 5: 435 (1884); Post in Mém. Herb. Boiss., 18: 102 (1900); H. Stuart Thompson in Journ. Bot., 44: 340 (1906); Holmboe, Veg. Cypr., 32 (1914); Post, Fl. Pal., ed. 2, 2: 696 (1933); Lindberg f., Iter Cypr., 8 (1946).
TYPE: Based on Jamaican specimens figured in Sloane, Nat. Hist. Jam., t. 64, fig. 3 (1707)! and Ehret, Plant. et Pap. Rar., t. 3, fig. 3 (1742)!

An annual grass; culms loosely tufted, usually decumbent at the base and rooting at the nodes, finally erect and up to 60 cm. tall, smooth and glabrous or rarely sparsely pilose at the lower nodes; leaf-blades linear or lanceolate, acuminate, rounded at the base, completely glabrous on both surfaces, up to

10 cm. long, 3–8 mm. wide, flat, flaccid, thin, scaberulous on the margins; lower sheaths very lax, the upper clasping, smooth and glabrous, striate; ligule absent; inflorescence consisting of a number of ascending racemes of closely packed spikelets, spaced upon a central axis; racemes of very shortly pedicellate spikelets secund upon a scabrid rhachis; spikelets in four rows, ovate-acute in outline, 2·5–3 mm. long, green or tinged with purple; glumes very unequal, the lower broadly ovate-cuspidate, 1 mm. long, 3-nerved, hispid on the nerves, the upper elliptic-obtuse, as long as the spikelet, 5–7-nerved, hispid or scabrid on the nerves; lower floret male; lemma similar to the upper glume, but flat; palea oblong-acute, hyaline; upper floret hermaphrodite; lemma 2 mm. long, ovate-acute, more or less cuspidate, whitish; palea of the same texture; anthers 1 mm. long.

HAB.: Cultivated and waste land, roadsides, ditches; sea-level to c. 700 ft. alt.; fl. June–Dec.

DISTR.: Divisions 1, 3–7. Found as a weed throughout the warmer regions of the world.

1. Kissonerga, 1967, *Economides* in ARI 979 !
3. Akhelia, 1939, *Lindberg f.* s.n.
4. Larnaca, 1930, *Ussher* 122 ! Perivolia, 1939, *Lindberg f.* s.n. !
5. Kythrea, 1935, *Syngrassides* 658 ! and, 1939, *Lindberg f.* s.n.
6. Nicosia, 1894, *Post* s.n.; Morphou, 1951, *Merton* 539 ! and, 1955, *Hiotellis* 481 ! Kokkini Trimithia, 1966, *Merton* s.n. !
7. Kyrenia, 1936, *Kennedy* 81 ! and, 1948, *Casey* 20 !

2. E. crusgalli (*L.*) *P. Beauv.*, Essai Agrost., 53, 161, t. 11, fig. 2 (1812); N. L. Bor in C. C. Townsend et al., Fl. Iraq, 9: 480, t. 185 (1968); Osorio-Tafall et Seraphim, List Vasc. Plants Cyprus, 15 (1973).
      Panicum crusgalli L., Sp. Plant., ed. 1, 56 (1753); Sibth. et Sm., Fl. Graec. Prodr., 1: 40 (1806); Boiss., Fl. Orient., 5: 435 (1884); Post, Fl. Pal., ed. 2, 2: 696 (1933); A. K. Jackson in Kew Bull., 1936: 345 (1936).

An annual grass; culms often robust, up to 1 m. tall, branched below, loosely tufted, geniculately ascending, smooth and glabrous; leaf-blades linear-acuminate, up to 25 cm. long, flat, often flaccid, dark green, scabrid on both surfaces towards the tip, smooth and glabrous on the lower surface towards the base, glabrous above, scabrid on the margins; lower leaf-sheaths papery, keeled, upper leaf-sheaths rather loose, glabrous; ligule completely absent; panicle consisting of a series of racemes of secund, closely packed spikelets, shortly pedicellate on a scabrid rhachis; racemes spaced on a scabrid axis; spikelets 3–3·7 mm. long, ovate-elliptic in outline, acute, cuspidate or awned; lower glume membranous, broadly ovate, clasping, obtuse to pointed, 1·25 mm. long, 5-nerved, scaberulous on the nerves, the upper herbaceous, very broadly oblong-ovate, acute, when flattened, as long as the spikelet, 5–7-nerved, spinulose on the nerves, appressed pubescent between the nerves or glabrous; lemma of the lower floret empty, similar to the upper glume, flat on the back, cuspidate or produced into a scabrid, often long, flexuous awn, 7-nerved; palea hyaline; lemma of the upper floret elliptic-ovate in outline, cuspidate, chartaceous to coriaceous, white, polished; palea of the same texture; awn up to 3 cm. long; anthers 1 mm. long.

var. **crusgalli**
      TYPE: "*in* Europae, Virginiae *cultis* (LINN !).

Lower lemma awned.

HAB.: Cultivated ground, ditches, marshes; sea-level to 4,500 ft. alt.; fl. Aug.–Nov.

DISTR.: Divisions 2, 6. Widespread in the warmer parts of the world.

2. Platani above Kakopetria, 1935, *Syngrassides* 748 ! Prodhromos, 1955, *G. E. Atherton* 581 ! Trikoukkia, 1970, *A. Genneou* in ARI 1558 !
6. Syrianokhori, 1935, *Syngrassides* 747 !

var. **submutica** Neilr., Fl. Nied.-Oesterr., 31 (1859).

TYPE: Not indicated.

Lower lemma without awns.

HAB.: Cultivated ground, ditches and marshes; near sea-level; fl. July–Nov.

DISTR.: Division 6. Distribution that of var. *crusgalli*.

6. Syrianokhori, 1970, *E. Chicken* 167!

68. PANICUM *L.*

Sp. Plant., ed. 1, 55 (1753).

Gen. Plant., ed. 5, 29 (1754).

Annual or perennial grasses with linear, lanceolate or ovate leaf-blades; inflorescence a contracted or effuse panicle of small, pedicellate, 2-flowered, awnless spikelets falling entire from the pedicels at maturity or breaking up into parts; glumes membranous, the lower smaller than the upper which equals the spikelet in length; lower lemma similar to the upper glume in texture, with or without a palea, containing a male floret or empty; upper lemma coriaceous, smooth with involute margins, containing a bisexual flower; palea of similar texture; stamens 3; lodicules 2, cuneate-truncate, upper margin undulate. Grain elliptic in outline, enclosed by the indurated lemma and palea; embryo about half the length of the grain; hilum suprabasal, circular.

About 500 species widely distributed in tropical and warm-temperate countries.

1. **P. repens** *L.*, Sp. Plant., ed. 2, 87 (1762); Sibth. et Sm., Fl. Graec. Prodr., 1: 41 (1806), Fl. Graec., 1: 45, t. 61 (1808); Boiss., Fl. Orient., 5: 440 (1884); Holmboe, Veg. Cypr., 32 (1914); Post, Fl. Pal., ed. 2, 2: 607 (1933); Lindberg f., Iter Cypr., 8 (1946); Rechinger f. in Arkiv för Bot., ser. 2, 1: 418 (1950); N. L. Bor in C. C. Townsend et al., Fl. Iraq, 9: 488, t. 187 (1968); Osorio-Tafall et Seraphim, List Vasc. Plants Cyprus, 15 (1973).

TYPE: Spain; "*in* Hispania" (LINN!).

A perennial grass with widely spreading, scaly, sharp-pointed rhizomes; culms up to 1 m. tall, usually much shorter, decumbent at the base and rooting at the nodes, simple or branched at the base, smooth and glabrous; leaf-blades linear-lanceolate, long-acuminate, up to 30 cm. long, 2–5 mm. wide, flat, folded or involute, glabrous or sparsely hairy on the upper surface, scabrid on the margins, often decorated on the margins at the base with tubercle-based hairs; sheaths rather loose, smooth and glabrous; ligule a narrow, chartaceous, ciliate ring; inflorescence an open panicle, with binate or irregularly attached, ascending, angular, scabrid branches and branchlets up to 20 cm. long, 8 cm. wide; spikelets ovate-lanceolate, acute in outline, rather pale, 2·5–3·5 mm. long; glumes very unequal, the lower membranous, 0·5–0·75 mm. long, clasping, broader than long, rounded or usually apiculate, 3-nerved, the upper as long as the spikelet, elliptic-acute, 7–9-nerved; lemma of the lower floret similar to the upper glume, containing a male flower; palea as long as lemma, hyaline; lemma of the upper floret elliptic-oblong, acute, indurated, 2 mm. long, shining, whitish; palea of the same texture; anthers about 2 mm. long.

HAB.: Wet places, banks of rivers, irrigated vegetable gardens, damp sandy ground near the sea, etc.; sea-level to 900 ft. alt.; fl. June–Oct.

DISTR.: Divisions 1, 3–7. Widely distributed in warm-temperate and tropical regions of the Old World, also in the southern states of U.S.A.

1. Prodhromi near Polis, 1937, *Syngrassides* 1713! Akhelias Chiftlik, Paphos, 1956, *Poore* 21!
3. Near Limassol, 1939, *Lindberg f.* s.n.; Asomatos marshes, 1939, *Mavromoustakis* 132! Cherkez, 1939–40, *Mavromoustakis* 4! 57! 67! Akrotiri Salt Lake, 1970, *E. Chicken* 114!

4. Karaolos Plantation, N. of Famagusta, 1930, *Papaiacovou* 125! Near Famagusta, 1939, *Lindberg f.* s.n.
5. Near Ayios Andronikos (Kythrea), 1880, *Sintenis & Rigo* 681! Kythrea, 1935, *Syngrassides* 657! also, same area, 1935, *Nattrass* 663; 1937, *Nattrass & Papaioannou* 886; 1939, *Lindberg f.* s.n.!
6. Myrtou, 1935, *Nattrass* 681; Syrianokhori, 1936, *Syngrassides* 1244! and, 1953, *Merton* 1039!
7. Kyrenia, 1935, *Nattrass* 680!

NOTES: A very troublesome weed, almost impossible to eradicate because of its widely spreading rhizomes.

P. MILIACEUM *L.*, Sp. Plant., ed. 1, 58 (1753), Millet, is sometimes cultivated on a small scale in Cyprus, and has been collected in a melon field at (Div. 5 or 6) Peristerona (1936, *Syngrassides* 1081!).

P. miliaceum is a robust, erect annual up to 120 cm. high, with leaf-blades up to 30 cm. long and 2 cm. wide, and roughly hairy leaf-sheaths; the inflorescence is a large, much-branched terminal panicle often 25–30 cm. long and 10–15 cm. wide, bearing numerous ovate-acute spikelets 4·5–5 mm. long. The seeds are valued as a cereal, and as a food for cage-birds, and the green plant is an excellent fodder.

Millet is believed to have been originally cultivated in India, but is now grown or naturalized in most warm-temperate and tropical regions.

P. DECOMPOSITUM *R. Br.*, Prodr. Fl. Nov. Holl., 191 (1810), from Australia, is said by Bovill (Rep. Plant. Work Cyprus, 14; 1915) to have been cultivated in Salamis Plantation, but no specimens have been seen. It is a robust annual or short-lived perennial up to 100 cm. tall, with long, narrow, glabrous leaves and glabrous leaf-sheaths; the panicle is 30–40 cm. long, at first strict with a dense sheaf of verticillate, semi-verticillate or alternate branches issuing from the uppermost leaf-sheath, finally emerging and spreading widely; the spikelets are 3–3·5 mm. long, ovate-acuminate in outline, borne on pedicels longer than the spikelets themselves.

It is just possible that the grass may have survived in the Salamis area (Div. 5).

P. TURGIDUM *Forssk.*, Fl. Aegypt.-Arab., 18 (1775); N. L. Bor in C. C. Townsend et al., Fl. Iraq, 9: 490, t. 188 (1968) is listed in Osorio-Tafall et Seraphim, List Vasc. Plants Cyprus, 15 (1973), presumably on the evidence of the external distribution cited in Fl. Iraq, *loc. cit.* It is a grass of sandy, desert areas in Africa and S.W. Asia, and its presence in Cyprus is unlikely. No specimen has been seen, and the error may have arisen through supposing that *Sintenis & Rigo* 1040 came from Cyprus; it is actually from Sarona, Syria (coll. 16 Sept. 1880).

69. PASPALUM *L.*

Syst. Nat., ed. 10, 855 (1759).

Annual or perennial herbs with linear, flat leaf-blades; inflorescence of solitary, digitate or racemosely arranged spikes; rhachis of the racemes flat, sometimes winged; spikelets secund, 2- or 4-seriate, solitary or paired, orbicular, orbicular-obovate or oblong in outline, plano-convex in profile, 2-flowered; glumes dissimilar, the lower, if present, a minute scale, the upper membranous, covering the dorsal surface; the lower lemma similar to the upper glume in texture, flat, empty; the upper chartaceous to coriaceous, with involute margins containing a bisexual floret; palea of the same texture with very broad flaps; stamens 3; lodicules 2, roughly cuneate, glabrous, upper margin rounded, coarsely lobed or toothed. Grain orbicular in outline, plano-convex in profile; embryo orbicular, less than half the length of the grain; hilum sub-basal, punctiform.

About 150–250 species, chiefly in tropical and temperate America, but also represented in the Old World.

Spikelets orbicular, fringed on the margin with long white hairs; upper glume glabrous
                                               1. P. dilatatum

Spikelets without white hairs on the margin; upper glume adpressed-pubescent
                                               2. P. distichum

1. P. dilatatum *Poir.* in Lam., Encycl. Méth., 5: 35 (1804); N. L. Bor in C. C. Townsend et al., Fl. Iraq, 9: 492, t. 189 (1968); Osorio-Tafall et Seraphim, List Vasc. Plants Cyprus, 15 (1973).
TYPE: Argentina; "Cette plante a été recueillie à Buenos-Ayres par Commerson" (P).

A robust perennial with a stout, woody rhizome; culms erect or somewhat decumbent at the base, simple or branched below, smooth and glabrous, or the lower nodes pilose, up to 1·5 m. tall; leaf-blades flat, linear, long-acuminate, up to 60 cm. long, 5–15 mm. wide, with cartilaginous, often finely undulate, scabrid margins, glabrous except for the long, white hairs behind the ligule; lower sheaths keeled, loose, softly hairy, the upper clasping with very short blades; ligule short, membranous to chartaceous; inflorescence of 4–6 (rarely more) alternately arranged, false spikes, sessile or shortly pedicellate on a slender axis; rhachis of the racemes flat; racemes 5–10 cm. long, drooping, curved; spikelets secund in two to four rows on the undersurface of the rhachis, orbicular- or broadly ovate-apiculate, hemispherical, 3–3·5 mm. long, lower glume absent, the upper convex, as long as and the shape of the spikelet, membranous, 5-nerved, glabrous on the back, very pilose on the margins; lemma of the lower floret membranous, flat, a little shorter than the spikelet, 5-nerved, shortly pilose on the margins; palea absent; lemma and palea of the upper floret suborbicular, obtuse, chartaceous, white, polished, 2·5 mm. long, faintly 5-nerved; stamens 1·5 mm. long.

HAB.: Ditches, marshes and irrigation channels; introduced; near sea-level; fl. June–Oct.

DISTR.: Division 6. A native of S. America, but now well established as a weed in many warm countries.

6. Morphou, 1951, *Merton* 1261 ! and, 1970, *A. Genneou* in ARI 1610 ! Syrianokhori, 1970, *E. Chicken* 165 !

NOTES: An excellent fodder grass.

2. P. distichum *L.*, Syst. Nat., ed. 10, 2: 855 (1759).
P. paspalodes (Michx.) Scribn. in Torr. Bot. Club Mem., 5: 29 (1894); N. L. Bor in C. C. Townsend et al., Fl. Iraq, 9: 494, t. 190 (1968); Osorio-Tafall et Seraphim, List Vasc. Plants Cyprus, 15 (1973).
Digitaria paspalodes Michx., Fl. Bor. Amer., 1: 46 (1803).
TYPE: U.S.A.; S. Carolina, "in pascuis aridis, juxta *Charlston* [Charleston]", *Michaux* (P).

A perennial plant with widely spreading rhizomes; culms decumbent at the base, rooting at the nodes, finally erect, up to 40 cm. tall, smooth and glabrous except for the stiffly hairy nodes; leaf-blades flat, usually more or less erect, linear-acute, up to 12 cm. long, 2–4 mm. wide, glabrous but ciliate towards the base of the scaberulous margins; lower sheaths scarious, loose, the upper green, keeled, ciliate on the margin, otherwise smooth and glabrous; ligule very narrow, rarely up to 1·5 mm. long; inflorescence a pair of spike-like racemes, each consisting of a flattened rhachis on the under surface of which are situated two rows of shortly pedicellate, closely packed spikelets; spikelets oblong- or elliptic-oblong, acute, flattened on one surface, convex on the other, 2·5–3·5mm. long; lower glume absent or represented by a small triangular scale, the upper as long as the spikelet, 3–5-nerved, appressed-pubescent; lemma of the lower, neuter, floret flat, shape of and as long as the spikelet, 3-nerved; lemma of the upper, hermaphrodite, spikelet indurated, smooth, glabrous, apiculate; tip minutely hirsute.

HAB.: By an irrigation channel; introduced; c. 100 ft. alt.; fl. June–Oct.

DISTR.: Division 7. A native of America, now widespread as an introduction in the warmer parts of the world.

7. Kyrenia, 1955, *Casey* 1679!

PENNISETUM VILLOSUM *R. Br.* in Fresen., Mus. Senckenb., 2: 134 (1837) has been collected in Cyprus by Merton (no. 2936) who records (letter 23/11/1970) that it was introduced by the Department of Agriculture, presumably as an ornamental. The specimen is not localized, nor is there any record of the subsequent performance of *P. villosum* on the island. It is a perennial, rhizomatous grass, up to 1 m. high in favourable habitats, but often very much smaller; the leaves are linear, acute, 15–20 cm. long (up to 60 cm. long in cultivation), 2–4 mm. wide, flat or folded, green or glaucous, glabrous or sparsely covered on the upper surface with white, tubercle-based hairs; the inflorescence is a very dense, cylindrical, false spike up to 10 cm. long and 2 cm. wide consisting of numerous solitary or grouped, awnless spikelets seated on a pedicellate involucre of numerous free bristles which are glabrous or covered with white hairs in the lower half. *P. villosum* is a native of Ethiopia and of adjacent regions in Africa and Arabia; it is cultivated as an ornamental in many warm parts of the world, where it is often recorded as an escape or naturalized.

70. SETARIA *P. Beauv.*
Essai Agrost., 51 (1812) nom. cons.

Annual or perennial grasses with flat, or pleated, fan-like leaf-blades; inflorescence spike-like, dense or somewhat loose and lobed, in which the small, awnless, pedicellate spikelets are supported by one to several, yellowish or reddish, scabrid, persistent bristles; spikelets with two florets, the lower male or barren, the upper hermaphrodite, rounded on the back, flat on the ventral surface; glumes unequal, membranous, the lower smaller than the upper, which may be as long as the spikelet or shorter; lower lemma membranous, with or without a palea; upper coriaceous to crustaceous, obscurely keeled on the back, transversely rugose or smooth, with involute margins; palea similar in texture; stamens 3; lodicules 2, oblong-cuneate, truncate above. Grain oblong or ellipsoid, tightly enclosed in the hardened lemma and palea; embryo somewhat longer than half the grain; hilum basal, obovate, small.

About 140 species in tropical and warm-temperate parts of the world.

Spikelets 3 mm. long; upper glume shorter than the upper lemma; upper lemma very coarsely
rugose - - - - - - - - - - - - - - - **1. S. pumila**
Spikelets 2–3·5 mm. long; upper glume as long as or almost as long as lemma; upper lemma
finely rugose:
   Bristles retrorsely scabrid - - - - - - - - **3. S. verticillata**
   Bristles antrorsely scabrid:
      Inflorescence 10 cm. or more long, 2–3 cm. wide, dense, often lobulate; grains persistent
                          2. S. italica
      Inflorescence much shorter and narrower, loose; grains caducous - - **4. S. viridis**

1. S. pumila (*Poir.*) *Roem. et Schult.*, Syst. Veg., 2: 891 (1817); Rauschert in Feddes Repert., 83: 661 (1973).
   Panicum pumilum Poir. in Lam., Encycl. Méth., Suppl., 4: 273 (1816).
   [*Panicum glaucum* (non L.) Sm. in Sibth. et Sm., Fl. Graec. Prodr., 1: 39 (1806).]

[*Setaria glauca* (non (L.) P. Beauv.) Boiss., Fl. Orient., 5: 442 (1884); Post, Fl. Pal., ed. 2, 2: 699 (1933); Lindberg f., Iter Cypr., 8 (1946); N. L. Bor in C. C. Townsend et al., Fl. Iraq, 9: 500, t. 192 (1968); Osorio-Tafall et Seraphim, List Vasc. Plants Cyprus, 16 (1973).]

TYPE: Not localized, probably France or N. Africa, *Desfontaines* (P ?).

An annual grass; culms solitary or loosely tufted, erect or geniculately ascending, slender, smooth and glabrous, scaberulous below the panicle; leaf-blades linear-acute, flat, up to 30 cm. long, 4–10 mm. broad, glabrous or hairy towards the base on the upper surface, scaberulous on the margins; sheaths keeled at the base of the culms, clasping above, smooth and glabrous; ligule a fringe of hairs; panicle dense, cylindrical, rounded at the top, bristly, erect, up to 10 cm. long, 4–8 mm. wide excluding the bristles; branches very short; bristles antrorsely scabrid, yellow to reddish, up to 10 mm. long; spikelets broadly elliptic-obtuse in outline, semi-elliptic from the side, 3 mm. long; lower glume broadly ovate, obtusely acute, the lower about 1·5 mm. long, 3-nerved, the upper 2–2·5 mm. long, 5-nerved; lemma of the lower, male or barren, floret as long as the spikelet, flat, elliptic-acute, 5-nerved; palea short; lemma of the upper, hermaphrodite, floret elliptic in shape, chartaceous, rounded on the back, transversely rugose, yellow; palea of the same texture; anthers 1·5 mm. long.

HAB.: Cultivated and waste ground, roadsides, often in shaded, wet situations by the sides of ditches and irrigation channels; near sea-level to 4,500 ft. alt.; fl. July–Oct.

DISTR.: Divisions 2, 7. Europe, Mediterranean region, Atlantic Islands and eastwards across Asia to China and Japan; introduced elsewhere.

2. Trikoukkia (Prodhromos), 1937, *Syngrassides* 1686 ! Milikouri near Kykko, 1939, *Lindberg f.* s.n. ! Prodhromos, 1955, *G. E. Atherton* 525 ! S.E. of Kakopetria, 1970, *E. Chicken* 40 ! 107 !
7. Karavas, 1970, *A. Genneou* in ARI 1624 !

NOTES: The identity of the *Panicum glaucum* L. recorded by Hume (in Walpole, Mem. Europ. Asiatic Turkey, ed. 1, 254; 1817) from (Div. 4) "Larnica and Limosol" (June–July, 1801) remains uncertain in the absence of a specimen; it may well be *S. pumila*.

2. S. italica (*L.*) *P. Beauv.*, Essai Agrost., 51, 178 (1812); Post, Fl. Pal., ed. 2, 2: 699 (1933); N. L. Bor in C. C. Townsend et al., Fl. Iraq, 9: 502 (1968); Osorio-Tafall et Seraphim, List Vasc. Plants Cyprus, 16 (1973).

    Panicum italicum L., Sp. Plant., ed. 1, 56 (1753).

TYPE: "*in* Indiis" (LINN !).

An annual grass; culms erect or geniculately ascending, slender or usually stout, up to 1 m. tall, simple or more or less branched from the base, smooth and glabrous, scaberulous below the panicle; leaf-blades linear from a contracted base, tapering to an acuminate tip, flat, up to 45 cm. long, 6–20 mm. wide, scabrid on the upper surface and on the margins; sheaths rather loose, smooth and glabrous, ciliate on the margin, striate; ligule a densely ciliate rim; inflorescence an erect or nodding, continuous and cylindric, or more or less lobed, false spike, up to 30 cm. long, 2–3 cm. wide; axis rather stout, angled, scabrid; spikelets persistent, 2·5–3·5 mm. long, closely packed in small groups each supported by 2–4, antrorsely scabrid bristles up to 10 mm. long; lower glume broadly ovate-acute, 1 mm. long, 1–3-nerved, the upper elliptic-acute or -obtuse, 5–7-nerved, 1·5–2 mm. long; lemma of the neuter, lower floret similar to the upper glume, flat on the back, 5-nerved; palea, if present, a minute hyaline scale; lemma of the hermaphrodite upper floret as long as the spikelet, yellowish or reddish, crustaceous, smooth or almost so; palea of the same texture; anthers 1 mm. long.

HAB.: Cultivated ground; 550 ft. alt.; fl. June–Oct.

DISTR.: Division 7, introduced. Widely cultivated in the Mediterranean, and throughout the tropics and subtropics.

7. Kythrea, in the school garden, 1936, *Syngrassides* 1285 !

NOTES: Foxtail or Italian Millet, grown to provide seed for cage-birds, or occasionally as a quick-growing cereal for human consumption.

3. S. verticillata (*L.*) *P. Beauv.*, Essai Agrost., 51, 178 (1812); Boiss., Fl. Orient., 5: 443 (1884); H. Stuart Thompson in Journ. Bot., 44: 340 (1906); Holmboe, Veg. Cypr., 32 (1914); Post, Fl. Pal., ed. 2, 2: 699 (1933); Lindberg f., Iter Cypr., 8 (1946); N. L. Bor in C. C. Townsend et al., Fl. Iraq, 9: 503, t. 193 (1968); Osorio-Tafall et Seraphim, List Vasc. Plants Cyprus, 16 (1973).

   Panicum verticillatum L., Sp. Plant., ed. 2, 82 (1762); Sibth. et Sm., Fl. Graec. Prodr., 1: 39 (1806); Unger et Kotschy, Die Insel Cypern, 178 (1865).

   TYPE: "*in* Europa *australi & Oriente*" (LINN!).

An annual grass; culms loosely tufted, erect or somewhat decumbent at the base and geniculately ascending, up to 50 cm. tall, smooth and glabrous but scaberulous below the panicle; leaf-blades linear-acuminate, up to 30 cm. long, 4–16 mm. wide, flat, glabrous, scaberulous on the upper surface and on the margins, flaccid, sparsely pilose with tubercle-based hairs on both surfaces or almost glabrous; sheaths somewhat compressed and loose, ciliate on the margins; ligule very short, densely ciliate; inflorescence a narrow erect panicle, cylindrical in outline, rarely very dense, usually somewhat open and often interrupted towards the base, very bristly, up to 10 cm. long, 10–15 mm. wide; axis angled, scabrid; branches short; spikelets 1·9–2·2 mm. long, elliptic in outline, rounded on the back, flat on the face, supported by retrorsely scabrid bristles; glumes very unequal, the lower ovate, rounded or acute, 0·6–0·7 mm. long, 1–3-nerved, the upper as long as and the shape of the spikelet, convex, membranous, 5–7-nerved; lemma of the neuter lower floret flat, as long as the spikelet; palea short; lemma of the hermaphrodite floret indurated, a little shorter than the spikelet, finely granular; palea of the same texture, flat; anthers 0·3 mm. long.

HAB.: Cultivated, often irrigated land, roadsides, waste ground, etc.; sea-level to 4,000 ft. alt.; fl. June–Oct.

DISTR.: Divisions 1–4, 6, 7. Central and S. Europe and Mediterranean region eastwards to Indonesia; Atlantic Islands; introduced and naturalized in most of the warmer parts of the world.

1. Ayios Neophytos, below the Monastery, 1939, *Lindberg f.* s.n.; Kissonerga, 1967, *Economides* in ARI 978!
2. Platres, 1937, *Kennedy* 90! and, 1939, *Kennedy* 1539! 1540! Milikouri, near Kykko Monastery, 1939, *Lindberg f.* s.n.!
3. Limassol, 1862, *Kotschy* 606a.
4. Kouklia (Famagusta) 1931, *Nattrass* 112; Famagusta, 1935, *Nattrass* 671; Perivolia, 1939, *Lindberg f.* s.n.
6. Government House Gardens, Nicosia, 1935, *Syngrassides* 796!
7. Kyrenia, 1948, *Casey* 25! Ayios Yeoryios, Kyrenia, 1955, *G. E. Atherton* 53! Myrtou, 1955, *G. E. Atherton* 415! Lapithos Reformatory School, 1955, *G. E. Atherton* 676!

4. S. viridis (*L.*) *P. Beauv.*, Essai Agrost., 51, 178, t. 13, fig. 3 (1812); Boiss., Fl. Orient., 5: 443 (1884); Post, Fl. Pal., ed. 2, 2: 699 (1933); A. K. Jackson in Kew Bull., 1934: 273 (1934); Lindberg f., Iter Cypr., 8 (1946); N. L. Bor in C. C. Townsend et al., Fl. Iraq, 9: 504 (1968); Osorio-Tafall et Seraphim, List Vasc. Plants Cyprus, 16 (1973).

   Panicum viride L., Syst. Nat., ed. 10, 2: 870 (1759); Sibth. et Sm., Fl. Graec. Prodr., 1: 39 (1806).

   TYPE: Not localized (LINN!).

An annual; culms loosely tufted, geniculately ascending, finally erect, smooth and glabrous, but scaberulous below the panicle, up to 50 cm. tall; leaf-blades up to 25 cm. long, 4–10 mm. wide, linear-acuminate, flat, flaccid, glabrous, scaberulous on the upper surface and on the margins; sheaths rounded on the back, smooth and glabrous, ciliate on the margins; ligule a dense fringe of silky hairs; panicle cylindrical, very dense, usually erect, rounded at the top, 1–10 cm. long, 4–10 mm. wide, excluding the bristles; axis scabrid; branches short; spikelets elliptic-oblong, green or purplish,

each supported by a solitary, antrorsely scabrid bristle 2–3 mm. long;
glumes unequal, the lower 0·6–1 mm. long, membranous, 1–3-nerved; the
upper as long as the spikelet, convex on the back, 5-nerved; lemma of the
lower floret resembling the upper glume and as long, but flat, 5–7-nerved;
palea hyaline, short; lemma of the hermaphrodite floret elliptic or elliptic-
oblong, obtuse, finely punctulate, becoming indurated; palea of the same
texture; anthers 0·8 mm. long.

   HAB.: Cultivated ground, roadsides, often in moist situations; 450–4,500 ft. alt.; fl. June–Oct.

   DISTR.: Divisions 2, 5, 7. Europe, Mediterranean region, and most of Asia. Naturalized in
many of the warmer parts of the world.
   2. Prodhromos, 1939, *Lindberg f.* s.n.! Kyperounda, 1954, *Merton* 1932! Trikoukkia (near
       Prodhromos), 1961, *D. P. Young*, 7458!
   5. Athalassa, 1966, *Merton* in ARI 15!
   7. Lapithos, Experimental Lime Plantation, 1932, *Syngrassides* 237! Kythrea, 1939, *Lindberg
       f.* s.n.; Lapithos, 1939, *Lindberg f.* s.n.; Bellapais, 1948, *Casey* 108!

   TRIBE 21. **Andropogoneae** *Dumort.* Annual or perennial grasses with flat
leaf-blades which may be linear-lanceolate in shape, rounded at the base, or
rarely cordate or sagittate, with panicoid anatomy, rarely petiolate; silica-
bodies dumb-bell-shaped; micro-hairs present, 2-celled, filiform; first foliage
leaf of the seedling flat and horizontal; ligules membranous, or reduced to a
ciliate rim, or absent; spikelets rarely solitary or in threes, usually paired,
one of each pair sessile, the other pedicellate; the sessile spikelet most often
hermaphrodite, 2-flowered, with the lower floret male or barren or
sometimes missing, the upper hermaphrodite or rarely female; the
pedicellate spikelet also 2-flowered, with the lower male or barren, the upper
male or empty, or both absent or rarely the spikelet itself reduced to the
pedicel or the latter reduced or entirely absent; pedicellate spikelets
deciduous from the pedicels at maturity; rhachis usually breaking up (tough
in *Imperata*) and the sessile spikelets falling with the adjacent joint and
pedicel; rhachilla not produced; spikelets paniculate, racemose or spicate on
tough or fragile rhachides; glumes firmer than the lemmas, both as long as
the spikelets; lemmas more delicate, the upper of the sessile sometimes
reduced to the hyaline base of the awn or broader and two-lobed or -cleft,
with a stout geniculate awn from the tip, sinus or cleft, or awnless; paleas
shorter than the lemma, frequently one or both absent; ovary glabrous;
styles 2; stigmas plumose; lodicules 2, cuneate-truncate. Grain loosely
enclosed between lemma and palea; hilum sub-basal, punctiform; embryo
large; starch grains simple or compound. Chromosomes small; basic
numbers 5, 9–15, 17, 19.

71. CHRYSOPOGON *Trin.*
Fund. Agrost., 187 (1820) nom. cons.

   Perennial herbs, usually decumbent at the base, with narrow leaf-blades,
and sheaths which are often sharply keeled; inflorescence a terminal panicle
with the awned spikelets in threes, one sessile and two pedicellate,
articulated to the bearded tips of whorled branches; spikelets 2-flowered,
with a distinct callus (rarely long and decurrent on the pedicel), falling
entire, the lower floret reduced to an empty lemma, the upper hermaphrodite in the sessile, male in the pedicellate, spikelets; glumes
coriaceous, the lower of the sessile 2-keeled and muriculate towards the tip,
the upper rounded below, keeled above, awned, the lower of the pedicellate
2-keeled, the upper flattened, awnless or awned; the upper lemma in the
sessile spikelet minutely 2-lobed or entire, with the mid-nerve passing out

into a perfect awn; paleas very small or absent; stamens 3; lodicules 2, cuneate-truncate, glabrous. Embryo half the length of the laterally compressed grain; hilum oblong-punctiform, basal.

About 16 species in the tropics or subtropics of the Old World.

1. C. gryllus (*L.*) *Trin.*, Fund. Agrost., 188 (1820); Boiss., Fl. Orient., 5: 458 (1884); N. L. Bor in C. C. Townsend et al., Fl. Iraq, 9: 512, t. 196 (1968); Osorio-Tafall et Seraphim, List Vasc. Plants Cyprus, 16 (1973).

Andropogon gryllus L., Cent. Plant. II, 33 (1756); Sibth. et Sm., Fl. Graec. Prodr., 1: 46 (1806), Fl. Graec., 1: 52, t. 67 (1808); Poech, Enum. Plant. Ins. Cypr., 7 (1842); Unger et Kotschy, Die Insel Cypern, 188 (1865); H. Stuart Thompson in Journ. Bot., 44: 340 (1906); Holmboe, Veg. Cypr., 32 (1914); Post, Fl. Pal., ed. 2, 2: 706 (1933).

TYPE: "*in* Rhaetia, Helvetia, Veronae, *Sauvages, Seguier*". (LINN!).

A densely caespitose perennial; culms usually erect, rarely decumbent below, sometimes spreading, up to 1·5 m. tall, terete or compressed, striate with a compact, woody rootstock, smooth and glabrous; leaf-blades somewhat glaucous, up to 20 cm. long, 1–2 mm. wide, linear-acute or acuminate, flat, rarely folded or twisted, covered on both surfaces with tubercle-based hairs, scaberulous on the margins; basal sheaths short, scarious, glabrous, compressed, keeled, those on the culms tight or loose, striate, glabrous; ligule a very narrow ciliate rim; panicle very effuse, 15–30 cm. long, 8–15 cm. wide, with whorled, capillary, flexuous, scabrid branches bearing a triplet or very short false raceme of spikelets at the rusty-haired, oblique tips; sessile spikelet with a bearded callus 1–2 mm. long; lower glume rounded on the back, smooth or more often with a row of tubercles on the sides, which are pectinate-ciliate at the apex, coriaceous in texture; upper glume as long, oblong, bifid at the apex, rounded on the dorsal surface below, keeled above, coriaceous with hyaline margins, glabrous below, shortly ciliate towards the apex, obscurely nerved, the central nerve carried out just below the apex as a slender scabrid awn; lower floret empty; lemma oblong-acute, hyaline, ciliate on the margins; upper floret hermaphrodite; lemma 6–7 mm. long, forming the hyaline base of an awn; awn up to 3·5 cm. long, of which 1·8 cm. is a hairy, chestnut-coloured, twisted column; anthers 4 mm. long; pedicellate spikelet: pedicels slender, 6 mm. long; lower glume lanceolate-acuminate, 8 mm. long, with an arista 8 mm. long, straight, 5-nerved; upper glume 8 mm. long, 3-nerved; anthers 4·5 mm. long.

HAB.: Unknown; fl. April–Aug.

DISTR.: ? Division. Widely distributed in S. Europe, the Mediterranean region and eastwards to Malaysia. Recorded by Sibthorp & Smith (Fl. Graec. Prodr., 1: 46) "In asperis et petrosis Cretae et Cypri", but not seen since, though there is no reason to doubt the accuracy of the record.

72. ANDROPOGON *L.*

Sp. Plant., ed. 1, 1045 (1753).
Gen. Plant., ed. 5, 468 (1754).

Perennial, caespitose grasses with flat leaf-blades; inflorescence of 2 to several false, fragile racemes, usually spatheate, terminating the culms or the branches; spikelets in pairs, differing in size, a sessile 2-flowered hermaphrodite, together with a pedicellate, 2-flowered, male or neuter, the sessile falling with the adjacent pedicel, the pedicellate deciduous from the pedicel; joints of the racemes linear, widened upwards or turbinate; sessile spikelets: glumes dorsally or laterally compressed, equal or subequal; lower floret usually reduced to an empty lemma without a palea; the upper floret hermaphrodite, its lemma 2-toothed or cleft, awned in the cleft with a perfect awn twisted at the base; palea a hyaline scale or absent; lodicules 2, cuneate, minute; stamens 3. Grain fusiform; embryo half as long as the grain; hilum punctiform. Pedicellate spikelet usually compressed dorsally; glumes subequal; florets 2, male or empty; lemmas if present hyaline, ciliate, not awned.

About 113 species widely distributed in the tropics and subtropics.

1. **A. distachyos** *L.*, Sp. Plant., ed. 1, 1046 (1753); Sibth. et Sm., Fl. Graec. Prodr., 1: 48 (1806), Fl. Graec., 1: 53, t. 69 (1808); Unger et Kotschy, Die Insel Cypern, 187 (1865); Sintenis in Oesterr. Bot. Zeitschr., 32: 19 (1882); H. Stuart Thompson in Journ. Bot., 44: 340 (1906); Holmboe, Veg. Cypr., 32 (1914); Lindberg f., Iter Cypr., 7 (1946); Osorio-Tafall et Seraphim, List Vasc. Plants Cyprus, 16 (1973).

    Pollinia distachya (L.) Spreng., Plant. Pugill. Secund., 12 (1815); Post, Fl. Pal., ed. 2, 2: 705 (1933).

    TYPE: Switzerland; *"in* Helvetia" (LINN!).

A perennial caespitose grass; culms erect, simple or branched at the base, up to 60 cm. tall, smooth and glabrous; leaf-blades linear, tapering to a long acuminate tip, more or less setaceous with age, pilose on the upper surface, glabrous on the lower, scabrid on the margins, up to 18 cm. long, 1–3 mm. broad; sheaths clasping the culms, striate, glabrous and smooth, or the lowest more or less silky-hairy; ligules truncate, ciliate, about 1 mm. long; inflorescence a pair of false racemes, 6–10 cm. long, one shortly pedicellate, the other sessile; rhachis fragile at maturity, 3-angled, ciliate on the inner angles; sessile spikelets with a bearded callus 1·5 mm. long; lower glume 10 mm. long, elliptic-acuminate, ending in 2 aristulae, winged on both margins in the upper third or two-thirds, 7–9-nerved; upper glume navicular, 3-nerved, 7 mm. long, compressed towards the tip, ciliate on the margins, scabrid on the keel above, otherwise glabrous; central nerve produced as a straight, capillary awn 9–10 mm. long; lower floret empty; lemma 8 mm. long, 2-nerved, ciliate on the margins; palea absent; upper floret hermaphrodite; lemma 8 mm. long, humped on the back, cleft to the middle, hyaline, glabrous; lobes long, acuminate; awn perfect, inserted in the cleft, 30 mm. long; column brown, twisted, 10 mm. long; anthers 4 mm. long; pedicellate spikelet male; pedicel 3·5 mm. long, flattened, ciliate and toothed on one margin; lower glume 8–9 mm. long, elliptic-acuminate, winged in the upper two-thirds, 2-toothed at the apex, awned between the teeth with a fine straight awn 6–7 mm. long; upper glume somewhat shorter, acute; lemmas elliptic-acute, hyaline, both, or one, male, or both empty.

    HAB.: Dry rocky slopes, on calcareous and non-calcareous formations; sea-level to c. 2,000 ft. alt.; fl. Febr.–Aug.

    DISTR.: Divisions 1, 3, 7, 8, locally common in Division 7. Widespread in the Mediterranean region.

    1. Ayios Neophytos, 1939, *Lindberg f.* s.n.; Dhrousha, 1941, *Davis* 3246! Below Stroumbi, 1941, *Davis* 3376!
    3. Kolossi, 1939, *Lindberg f.* s.n.; also, 1953, *Merton* 953!
    7. Common; *Sintenis & Rigo* 398! *Lindberg f.* s.n.! *Merton* 785! 2183! 2706! *Meikle* 2402! etc.
    8. Yioti ["Juti"] 1905, *Holmboe* 533! between Rizokarpaso and Ronnas Bay, 1941, *Davis* 2361! Kantara Castle, 1941, *Davis* 2455!

    NOTES: *Merton* 785, from near Lefkoniko Pass (Division 7) is infected with a smut *Sorosporium polliniae* P. Magnus (*teste* R. W. G. Dennis).

73. HYPARRHENIA *Anderss. ex Fourn.*

Mex. Plant. Gram., 51, 67 (1886).
Anderss. in Nov. Act. Soc. Sci. Upsala, ser. 3 (2): 254 (1855) nomen.
W. D. Clayton in Kew Bull., Additional Series II: 1–196 (1969).

Annual or perennial coarse grasses with flat leaf-blades; inflorescence a spatheate panicle of a very large or small number of few- to many-jointed raceme-pairs, each terminating a peduncle included in a spatheole; spikelets arranged on a fragile rhachis, a sessile accompanied by a pedicellate, 2-flowered, the sessile hermaphrodite, awned, the pedicellate male or neuter, unawned, the lower pairs often homogamous, male or neuter; sessile spikelet falling with the adjacent joint as well as with the pedicel of the pedicellate spikelet, the latter falling from the pedicel; glumes equal in length, the lower dorsally flat, the upper shallowly boat-shaped; lower lemma a hyaline scale enclosing a male flower, with or without a minute hyaline palea; upper lemma of the sessile spikelet minutely 2-lobed at the apex, very narrow, forming the stipitate base of a stout, geniculate awn which is hairy below the knee; palea of the upper floret absent or a minute hyaline scale; lodicules two, very small, cuneate-truncate, with a minute tooth on the outer edge, glabrous; stamens 3. Grain oblanceolate in outline; embryo about half its length; hilum obovate or round, basal. Pedicellate spikelets acute or acuminate; lemmas of both florets developed as hyaline scales; paleas usually absent.

About 75 species in the Mediterranean region, Africa and Arabia.

1. **H. hirta** (*L.*) *Stapf* in Prain, Fl. Trop. Afr., 9: 315 (1919); R. M. Nattrass, First List Cyprus Fungi, 6, 7, 14 (1937); N. L. Bor in C. C. Townsend et al., Fl. Iraq, 9: 530, t. 204 (1968); Osorio-Tafall et Seraphim, List Vasc. Plants Cyprus, 16 (1973).

Andropogon hirtus L., Sp. Plant., ed. 1, 1046 (1753); Sibth. et Sm., Fl. Graec. Prodr., 1: 48 (1806); Unger et Kotschy, Die Insel Cypern, 188 (1865); Boiss., Fl. Orient., 5: 465 (1884); Holmboe, Veg. Cypr., 32 (1914); Post, Fl. Pal., ed. 2, 2: 710 (1933); Lindberg f., Iter Cypr., 7 (1946); Rechinger f. in Arkiv för Bot., ser. 2, 1: 418 (1950).

TYPE: "*in* Lusitania, Sicilia, Smyrnae" (LINN!).

A more or less caespitose perennial grass; culms up to 1 m. tall or even taller, smooth and glabrous; leaf-blades narrowly linear, acute or acuminate, flat, up to 10 cm. long, 1–3 mm. wide, rolled or folded when dry, scabrid on the margins, more or less hairy at the base on the upper surface, often ciliate on the margins below, glabrous on the undersurface; lower culm-sheaths clasping, those from the upper nodes spathe-like with reduced blades, the uppermost spathaceous; ligule membranous, 3–4 mm. long, lacerate; inflorescence consisting of pairs of false racemes enclosed in spathes, terminating the culms and branches; racemes divergent, not declinate; racemes 3–4 cm. long, one long-pedicellate the other shortly pedicellate or sessile, seated on a hairy, flexuous or straight peduncle; joints and pedicels profusely covered with white hairs; sessile spikelets with a cuneate bearded callus, hermaphrodite; lower glume 4–5 mm. long, membranous, 9–11-nerved, villous on the dorsal surface, upper glume as long, obtuse at the tip, 3-nerved, keeled, apiculate, ciliate on the margins; lower lemma hyaline, elliptic-acute, empty; palea absent; upper lemma very narrow, 4·5 mm. long, hyaline, minutely bifid at the apex with a stout perfect awn in the sinus; awn up to 2 cm. long, of which the lower 8 mm. is a twisted, hairy column; palea absent or a minute hyaline scale; anthers 1·5–2·5 mm. long; pedicellate spikelet: pedicel 3 mm. long, with a single tooth, slender, ciliate on the margins; spikelets about 6 mm. long; lower glume oblong-elliptic, acute, 7–9-nerved, pilose on the dorsal surface; upper glume glabrous, 3-nerved; lemmas hyaline, empty or male, awnless.

HAB.: Dry rocky banks and hillsides; roadsides; sometimes in open *Quercus* or *Pinus* forest, on calcareous or igneous formations; sea-level to c. 4,500 ft. alt.; fl. Febr.–Nov.

DISTR.: Divisions 1–8, locally common. S. Europe and Mediterranean region eastwards to N.W. India; Atlantic Islands; tropical Africa; S. Africa; Australia.

1. Stroumbi, 1913, *Haradjian* 774! Polis, 1934, *Nattrass* 440; Akamas, W. of Ayios Yeoryios, 1941, *Davis* 3298! Skala, above Peyia, 1962, *Meikle* 2039!

2. Locally common around Platres, Perapedhi, Mesapotamos, etc.; *Sintenis & Rigo* 399a!
   Kennedy 1258! 1534! 1535! *Davis* 1842!
3. Near Episkopi, 1862, *Kotschy* 617! Kalavasos, 1905, *Holmboe* 637! Limassol, 1935, *Nattrass*
   628; Kolossi, 1939, *Lindberg f.* s.n.; near Mamonia, 1960, *N. Macdonald* 157!
4. Near Dhekelia, 1935, *Nattrass* 478; Larnaca aerodrome, 1950, *Chapman* 692! Famagusta,
   1955, *Miss Mapple* 8!
5. Athalassa, 1934, *Syngrassides* 954, and, 1939, *Lindberg f.* s.n.! also, 1950, *Chapman* 629!
6. Dhiorios, 1932, *Syngrassides* 287!
7. Common; *Sintenis & Rigo* 399! *Davis* 1740! *Casey* 79! *G. E. Atherton* 247! 465!, etc.
8. Rizokarpaso, 1912, *Haradjian* 253; and, 1941, *Davis* 2371! Komi Kebir, 1912, *Haradjian*
   315!

74. IMPERATA *Cyr.*

Plant. Rar. Neap., 2: 26, t. 11 (1792).

Perennial rhizomatous herbs with erect, robust or slender, solid culms and narrow, flat, long, linear leaf-blades; inflorescence a terminal silky, erect spike-like panicle; spikelets all alike, surrounded by hairs from the callus and base of the glume, in pairs at the joints of the false racemes, one short, the other long-pedicellate, falling entire, 2-flowered, the lower often reduced to an empty lemma, rarely male, the upper hermaphrodite; rhachis of the racemes tough; glumes subequal, membranous, enveloped in long silky hairs from the lower half of the glume and callus; lemmas hyaline, short; paleas shorter, the lower narrow, the upper broader; stamens 1–2; lodicules wanting. Grain oblong-ellipsoid or ellipsoid; embryo elliptic, half the length of the grain; hilum basal, punctiform.

Five or six species widely distributed in the tropics and subtropics.

1. **I. cylindrica** (*L.*) *Raeuschel*, Nom. Bot., ed. 3, 10 (1797); Boiss., Fl. Orient., 5: 452 (1884); Post
   in Mém. Herb. Boiss., 18: 102 (1900); H. Stuart Thompson in Journ. Bot., 44: 340 (1906);
   Holmboe, Veg. Cypr., 31 (1914); Post, Fl. Pal., ed. 2, 2: 703 (1933); Lindberg f., Iter Cypr.,
   8 (1946); N. L. Bor in C. C. Townsend et al., Fl. Iraq, 9: 532, t. 205 (1968); Osorio-Tafall et
   Seraphim, List Vasc. Plants Cyprus, 16 (1973).
   Lagurus cylindricus L., Syst. Nat., ed. 10, 878 (1759).
   Saccharum cylindricum (L.) Lam., Encycl. Méth., 1: 594 (1785); Sibth. et Sm., Fl. Graec.
   Prodr., 1: 36 (1806), Fl. Graec., 1: 40, t. 54 (1808).
   Imperata arundinacea Cyr., Plant. Rar. Neap., 2: 26 (1792); Unger et Kotschy, Die Insel
   Cypern, 187 (1865); Sintenis in Oesterr. Bot. Zeitschr., 32: 19 (1882).
   TYPE: Not indicated, but a reference to "*D. Gerard*" implies that the type material was sent
   to Linnaeus by Louis Gerard (1733–1819), author of *Flora Gallo-Provincialis* (1761), and that it
   came from S. France (LINN!).

A perennial with a long, creeping, white, succulent rhizome; culms from 10 cm. to 1 m. tall, solid, glabrous and smooth but with densely bearded nodes; leaf-blades linear-acuminate, canaliculate with a broad midrib on the upper surface, extremely scabrid on the margins, up to 100 cm. long, 5 mm. wide, covered on the upper surface towards the base with white hairs, glabrous elsewhere; sheaths rather loose, overlapping at the base, glabrous or hairy; ligules membranous, lacerate, narrow; inflorescence a plumose, erect, silky-white, dense panicle up to 20 cm. long, 2 cm. wide; pedicels and joints slightly dilated towards the tips, silky-hairy; spikelets in pairs, one short-pedicellate, the other long-pedicellate, otherwise similar, fusiform or lanceolate in outline, 3·5–4·5 mm. long; callus hairs slender, white, about three times as long as the spikelet; short-pedicellate spikelet with membranous glumes 3·5–4·5 mm. long, the lower 7–9-nerved, covered on the dorsal surface with long white hairs, upper glume 3-nerved, covered below with similar hairs; lower lemma hyaline, empty, lanceolate-acute, epaleate; upper lemma hermaphrodite, ovate-lanceolate, nerveless, acute; palea a small hyaline scale; anthers 2, c. 2·5–3 mm. long; stigmas long, plumose conspicuously purple.

HAB.: Sandy ground near the sea or inland, sometimes on river banks or by the sides of ditches; sea-level to 900 ft. alt.; fl. April–June.

DISTR.: Divisions 1, 3–8, common in Division 7. W. and S. Europe, the Mediterranean region and eastwards to Central Asia; Atlantic Islands.

1. Polis, 1932, *Nattrass* 222.
3. Limassol, 1905, *Holmboe* 654; Kakoradjia, 1934, *Syngrassides* 488!
4. Pyla, 1934, *Syngrassides* 428!
5. Mesaoria, 1894, *Post* s.n.; Trikomo, by the seashore, 1939, *Lindberg f.* s.n.!
6. Morphou, 1972, *W. R. Price* 1052!
7. Common; Myrtou, Larnaka tis Lapithou, above Kythrea, Lefkoniko Pass, etc. *Kotschy* 370! *Sintenis & Rigo* 988! *Syngrassides* 443! *Davis* 3015! *Merton* in ARI 801! etc.
8. S. of Galinoporni, 1962, *Meikle* 2505!

NOTES: A very persistent and troublesome weed. A small piece of the rhizome left in the ground can develop into a new plant.

75. SACCHARUM *L.*

Sp. Plant., ed. 1, 54 (1753).
Gen. Plant., ed. 5, 28 (1754).

Erect, perennial, often very tall herbs with polished usually solid culms, and linear, flat, often broad leaf-blades; inflorescence a terminal plume-like panicle of many spikelets, seated in pairs, a sessile accompanied by a pedicellate, at the joints of articulate, very fragile racemes; spikelets 2-flowered, usually surrounded by hairs from the lower part of the glume, the pedicellate falling from its pedicel, the sessile deciduous, accompanied by the adjoining pedicel; glumes subequal, membranous; lemmas hyaline, the lower empty, the upper awnless or awned from the tip, containing a hermaphrodite flower; paleas very small, hyaline or absent; lodicules 2, cuneate with a blunt tooth on one or each margin; stamens 3. Grain terete; embryo up to half the length of the grain; hilum punctiform, basal.

Five species widely distributed in the tropics and subtropics.

Spikelets awnless - - - - - - - - - | S. OFFICINARUM (p. 1868)
Spikelets with a conspicuous awn - - - - - - - - 1. **S. ravennae**

1. **S. ravennae** (*L.*) *Murr.*, Syst. Veg., ed. 13, 88 (1774); Sibth. et Sm., Fl. Graec. Prodr., 1: 36 (1806), Fl. Graec., 1: 38, t. 52 (1808); Unger et Kotschy, Die Insel Cypern, 187 (1865); N. L. Bor in C. C. Townsend et al., Fl. Iraq, 9: 539, t. 207 (1968).
   Andropogon ravennae L., Sp. Plant., ed. 2, 1481 (1763).
   Erianthus ravennae (L.) P. Beauv., Essai Agrost., 14 (1812); Boiss., Fl. Orient., 5: 454 (1884); Holmboe, Veg. Cypr., 32 (1914); Post, Fl. Pal., ed. 2, 2: 704 (1933); Lindberg f., Iter Cypr., 7 (1946); Osorio-Tafall et Seraphim, List Vasc. Plants Cyprus, 16 (1973).
   TYPE: "*in* Italia".

A rhizomatous, caespitose, perennial grass; culms erect, up to 2 m. tall, forming great tufts of densely packed stems, smooth and glabrous; leaf-blades up to 1 m. long, 5–15 mm. wide, rough on both surfaces, linear-acuminate in shape, flat, with a very conspicuous midrib which in some blades occupies a large portion of the surface, sparsely to densely hairy towards the base on the upper surface; sheaths clasping, tight, glabrous or sparsely hairy with hairs from tubercle-bases; ligule a narrow, membranous ring, long-ciliate; inflorescence a dense, plumose, reddish, purplish or white panicle up to 60 cm. long, 15 cm. broad, ellipsoid in shape, more or less lobed; spikelets in pairs at the joints of the racemes, a sessile and a pedicellate; joints and pedicels slender, slightly dilated at the tips, long ciliate; sessile spikelet lanceolate in outline, 4–6 mm. long, surrounded at the base by hairs about the same length; glumes subequal, the lower membranous, 2-toothed, 2-keeled, smooth and glabrous, scabrid on the keels, the upper more or less boat-shaped, 3-nerved, ciliate on the margins; lower floret male or empty; lemma lanceolate-acute, 3 mm. long; upper floret hermaphrodite; lemma hyaline, ovate-lanceolate, acute, awned; awn 4–6 mm. long; anthers 3 mm. long; pedicellate spikelet rather similar to the sessile; glumes hairy.

HAB.: Moist ground by streams and ditches; sea-level to 3,500 ft. alt.; fl. June–Oct.

DISTR.: Divisions 2–4, 7. Mediterranean region and eastwards to N. India.

2. Near Saïttas, 1934, *Syngrassides* 507! Platania, 1940, *Davis* 1980!
3. Alaminos, 1936, *Nattrass* 831.
4. Varosha, Famagusta, 1862, *Kotschy*; Famagusta, 1933, *Nattrass* 288, 298 and, 1934, *Nattrass* 441; Kondea, *Nattrass* 811.
7. Myrtou, 1935, *Nattrass* 657; Boghazi, 1939, *Lindberg f.* s.n.! Kyrenia, 1955, *M. Pallis* s.n.!

S. OFFICINARUM *L.*, Sp. Plant., ed. 1, 54 (1753), Sugar-cane, a perennial cultivated grass with very tall (up to 6 m.) stout, many-noded culms, smooth and glabrous or pubescent below the panicle, more or less with a waxy bloom below the nodes; leaf-blades linear or lanceolate, acuminate, narrowed towards the base, up to 1·5 m. long, 1–5 cm. wide, rigid, with a very broad, white midrib, scabrid on the upper surface and on the margins; sheaths clasping, striate, smooth and glabrous when old, hairy when young; ligule a very short membranous ring, ciliate; inflorescence a very large, pyramidal, dense, silvery or golden panicle, 30–60 cm. long, with the lower branches more or less verticillate; spikelets lanceolate in outline, the sessile and pedicellate similar, 3·5–4 mm. long, surrounded at the base by long, silky hairs from the callus; glumes more or less equal, membranous, the lower glabrous, 2–4-nerved, the upper 3-nerved, hairy; lower lemma empty, hyaline, almost as long as the spikelet; palea absent; fertile lemma long or short, broad or narrow, hyaline, ciliate on the margins or not, awnless; anthers nearly 2 mm. long. Thought to have originated in New Guinea, and to have spread from there through tropical Asia to Europe; sugar-cane is now cultivated in all the warmer regions of the world.

Gaudry (Recherches Sci. en Orient, 156; 1855) quoting L. de Mas Latrie (Histoire de l'île de Chypre sous le Regne des Princes de la Maison de Lusignan, 2: 88; 1852) records that vast plantations of sugar-cane existed in the Middle Ages around Episkopi, Kouklia and the Gulf of Pentagia (Morphou Bay) and provided the island with one of its principal revenues at that time. By the end of the 16th century the cultivation of sugar-cane had ceased, possibly because of the disordered state of the island, but more probably for economic reasons, and because (Porcacchi, 1576) cotton had become a more profitable crop. By the 18th century (Heyman, 1758) even the techniques of sugar manufacture had been forgotten. Kotschy (Die Insel Cypern, 187; 1865) notes that, in 1862, *Saccharum officinarum* still survived in gardens at Episkopi. It is not mentioned by subsequent writers on the flora of Cyprus.

76. SORGHUM *Moench*

Meth., 207 (1794) nom. cons.

Annual or perennial herbs, with stout, erect culms, flat leaf-blades, and large, often congested, panicles, made up of few to many pairs of spikelets, one sessile, one pedicellate and much narrower, at the joints of tough (in cultivated forms) or fragile axes terminating the branches; florets 2, lower reduced to an empty lemma, the upper hermaphrodite in the sessile, male or neuter in the pedicellate spikelet; glumes coriaceous, equal, the lower flattened and keeled, the upper shallowly boat-shaped; lemmas hyaline, the upper 2-lobed, awnless or with a geniculate awn; paleas minute or absent; stamens 3; lodicules 2, truncate-cuneate, ciliate on the upper margin, embryo half the length of the obovoid or ellipsoid grain; hilum basal, a circular or obovate spot.

About 60 species widely distributed in the tropics and subtropics.

Inflorescence a loose pyramidal panicle of lanceolate or elliptic-oblong spikelets:
    Perennial with widely spreading rhizomes; nodes silky-pubescent - - **1. S. halepense**
    Annual with glabrous or glabrescent nodes - - - - S. SUDANENSE (p. 1870)
Inflorescence dense, glomerate, often coloured; spikelets obovoid or globose
    S. BICOLOR (p. 1869)

1. S. halepense (*L.*) *Pers.*, Syn. Plant., 1: 101 (1805); Boiss., Fl. Orient., 5: 459 (1884); Post, Fl. Pal., ed. 2, 2: 707 (1933); N. L. Bor in C. C. Townsend et al., Fl. Iraq, 9: 548, t. 211 (1968); Osorio-Tafall et Seraphim, List Vasc. Plants Cyprus, 16 (1973).
    Holcus halepensis L., Sp. Plant., ed. 1, 1047 (1753).
    Andropogon halepensis (L.) Brot., Fl. Lus., 1: 89 (1804); Sibth. et Sm., Fl. Graec. Prodr., 1: 47 (1806), Fl. Graec., 1: 52, t. 68 (1808); Unger et Kotschy, Die Insel Cypern, 187 (1865); H. Stuart Thompson in Journ. Bot., 44: 340 (1906).
    Andropogon sorghum (L.) Brot. ssp. *halepensis* Hackel in DC., Mon. Phan., 6: 501 (1889); Holmboe, Veg. Cypr., 32 (1914); Lindberg f., Iter Cypr., 7 (1946).
    TYPE: "*in* Syria, Mauritania" (LINN!).

A perennial with long, widely spreading rhizomes; culms up to 2·5 m. tall, erect or somewhat decumbent at the base and rooting at the nodes, smooth and glabrous but with finely silky-pubescent nodes; leaf-blades broadly linear-lanceolate, tapering to a fine acuminate tip, rounded at the base, glabrous, smooth, very scabrid on the margins; sheaths striate, glabrous; ligules short, membranous, erose, hairy on the back; inflorescence a decompound panicle, 15–30 cm. long; branches terminating in few-noded, false racemes; joints and pedicels similar, nearly as long as the spikelets; sessile spikelet ovate or elliptic, acute, 4–5 mm. long; lower glume flat on the back, keeled on the margins above, more or less hairy on the dorsal surface; upper glume 3-nerved, somewhat keeled towards the apex; lower lemma sterile, hyaline, as long as the glumes; palea absent; fertile lemma short, cleft at the apex, hyaline, awned in the cleft; palea absent or a minute, hyaline scale; awn perfect, 12 mm. long, of which 5 mm. form a chestnut, twisted column; anthers 3 mm. long; pedicellate spikelets male, as long as the sessile, but narrower; lower glume narrowly lanceolate, the upper navicular; lemmas hyaline; paleas usually absent; anthers 3 mm. long.

HAB.: Cultivated and fallow fields, usually in moist situations; sea-level to 3,750 ft. alt.; fl. April–Nov.

DISTR.: Divisions 1–7, common in Divisions 6, 7. Mediterranean region eastwards to Central Asia; Atlantic Islands, but now found wild or cultivated in most of the warmer regions of the world.

1. Polis-Ktima road, 1952, *Merton* 921!
2. Pharmakas, 1905, *Holmboe* 1126; Palekhori, 1905, *Holmboe* 1140! Platres-Perapedhi, 1937, *Kennedy* 83! 84!, also, 1940, *Davis* 1851! Platres, 1960, *N. Macdonald* 175!
3. Between Limassol and Kolossi, 1862, *Kotschy*; Limassol-Yermasoyia, 1925, *Syngrassides* 717!
4. Larnaca, 1930, *Ussher* 124! Kouklia, 1931, *Nattrass* 111.
5. Tymbou, 1932, *Syngrassides* 389! Athienou, 1967, *Economides* in ARI 1037!
6. Nicosia, 1932 and 1935, *Nattrass* 231, 575, 642, 674; Kokkinotrimithia, ? 1967, *Merton* in ARI 22!
7. Near Kythrea, 1880, *Sintenis & Rigo* 679! Karavas, 1933, *Nattrass* 631; Kyrenia, 1936, *Kennedy* 82! also, 1948, *Casey* 32! and, 1955, *G. E. Atherton* 288! etc.

NOTES: The leaves of this grass are occasionally poisonous to stock, particularly when wilted, owing to the presence of cyano-genetic glycosides in the tissues of the blade.

S. BICOLOR (*L.*) *Moench*, Meth., 207 (1794) (*Holcus bicolor* L., Mantissa Alt., 301; 1771), an annual or perennial grass; culms erect 1·5–2·5 m. tall, solitary or fasciculate, about 1 cm. diam., solid, pubescent at the nodes or not; leaf-blades rounded at the base, linear or lanceolate, acuminate, up to 60 cm.

long by 6 cm. broad, hairy on the upper surface at the base, otherwise glabrous, scabrid on the margins; sheaths rather loose or the upper tightly clasping; ligule very short, scarious; panicle very variable, dense or loose, erect or the peduncle curved, up to 25 cm. long; branches short; racemes stout, compact, 2–4-noded; sessile spikelet 4–6 mm. long, 3–4 mm. broad, elliptic or obovate in shape; rhachilla not disarticulating above the glumes; glumes more or less equal, shape of the spikelet, variously coloured from pale yellow to chestnut-brown or black; lower lemma empty, hyaline; palea absent; upper lemma broadly ovate, 3 mm. long, 2-lobed at the apex, awned in the notch; palea a minute, hyaline scale or absent; awn up to 10 mm. long, feebly twisted at the base; anthers 3 mm. long. Grain broadly obovoid or globular. Pedicellate spikelets neuter, lanceolate, persistent, 3–4 mm. long.

"Sorgho" is frequently cultivated in Mediterranean countries and almost certainly occurs in Cyprus. No specimens have been seen but the plant is included here so that it can be identified if met with; it is noted from Morphou (Div. 6), 1935, by *Nattrass* (576 in First List Cyprus Fungi, 10; 1937) under the name *S. technicum* (Koern.) Fiori.

S. SUDANENSE (*Piper*) *Stapf* in Prain, Fl. Trop. Afr., 9: 113 (1917) (*Andropogon sorghum* Brot. var. *sudanensis* Piper in Proc. Biol. Soc. Wash., 28: 33 (1915); *S.* × *drummondii* (Nees ex Steud.) Millspaugh et Chase in Publ. Field Columb. Mus. Bot., 3: 21; 1903). An annual grass with slender glabrous culms, up to 3 m. high, and dull, green leaves 30 cm. long, 2 cm. wide, often mottled with purple; panicle loose, ovate-pyramidal, with ascending branches. A native of the Sudan, but sometimes cultivated elsewhere as a summer fodder crop. It is recorded from Kyrenia (Div. 7), 1931, *Nattrass* 26, and from Agros (Div. 2), 1933, *Nattrass* 639.

TRIBE 22. **Maydeae** *Dumort.* Annual or perennial grasses with linear or lanceolate, often broad, leaf-blades, rounded at the base, with panicoid anatomy; silica-bodies dumb-bell-shaped or cruciform; micro-hairs 2-celled, slender, elongate; ligules membranous, well-developed; spikelets unisexual, dissimilar, monoecious in different inflorescences (*Zea*) or in the same inflorescence; male spikelets 2-flowered, paired, one sessile, the other pedicellate; glumes more or less equal, enclosing the florets, membranous or chartaceous; lemmas and paleas (if present) hyaline; stamens 3; anthers linear; lodicules 0 or 2, fleshy; female spikelets 2-flowered (lower floret barren) solitary or sometimes paired, crowded in rows on a thick spongy rhachis (*Zea*) or sunk in the cavities of a fragile rhachis, or female spikelets three-flowered (one floret fertile, 2 sterile) and then enclosed in a bony involucre derived from a metamorphosed sheath (*Coix* L.); glumes firm or thin, emarginate or lobed; lemmas membranous; paleas similar or the lower absent. Grain in *Zea* wedge-shaped, rectangular in section, in other genera variously shaped; hilum basal, punctiform; embryo as long as the grain; starch grains simple. Chromosomes medium; basic numbers 5, 9. [Tribe now generally regarded as falling within the *Andropogoneae* Dumort. T.A.C. & R.D.M.]

77. ZEA *L.*

Sp. Plant., ed. 1, 971 (1753).
Gen. Plant., ed. 5, 419 (1754).

Tall, annual, erect grasses with broad drooping leaf-blades; inflorescences monoecious; male in numerous spike-like racemes forming spreading panicles, terminating the shoots; spikelets 2-flowered, in pairs, one subsessile, the other pedicellate; glumes equal, membranous, lemmas and palea hyaline; stamens 3; female inflorescence axillary, sheathed, presumably of several spike-like racemes fused into a thick spongy central axis with the spikelets arranged on its periphery; female spikelets in pairs in longitudinal rows, 2-flowered, the lower floret fertile, the upper sterile; glumes subequal, broad, fleshy below, hyaline above; lemmas shorter, hyaline; paleas similar; lodicules wanting. Grain usually cuneate, flattened above; embryo nearly as long as the grain; hilum as broad as long, or broader, basal, small.

1. Z. MAYS *L*., Sp. Plant., ed. 1, 971 (1753); Unger et Kotschy, Die Insel Cypern, 176 (1865); N. L. Bor in C. C. Townsend et al., Fl. Iraq, 9: 558, t. 215 (1968).
   TYPE: "*in* America" (LINN !).

A robust annual herb; culms up to 3 m. tall, smooth and glabrous, shining, solid, rooting from the lower nodes, erect, leaf-blades linear-acuminate, up to 75 cm. long, 2·5–12 cm. wide, flat, hairy or glabrous on the upper surface, ciliate on the margins, smooth and glossy below, drooping; sheaths rather loose, rounded on the back, striate, often overlapping, smooth and glabrous; ligule short, 1–2 mm. long, membranous; male inflorescence a panicle of spreading or drooping false racemes on a stout, terete, smooth and glabrous axis up to 45 cm. long, 20 cm. wide; racemes up to 30 cm. long; spikelets all alike, 2-flowered; glumes membranous, elliptic-obtuse, many-nerved, smooth, glabrous or hirsute with short hairs, 6–10 mm. long; lemmas and palea hyaline; female inflorescence consisting of a thick fleshy axis, seated in a number of papery, spathe-like sheaths; spikelets sessile, half-sunk in the fleshy or fibrous axis, conical in shape, arranged in longitudinal rows; lower glume emarginate, ciliate, the upper acute or 2-lobed, ciliate; lemma of the lower floret empty, hyaline; palea shorter or absent; lemma of the upper floret containing a female flower, membranous; palea hyaline; stigmas very long, drooping, often tinged with purple, 2-lobed at the tip.

Noted as a cultivated plant in Cyprus by Kotschy, and sometimes occurring as an escape from cultivation. Maize was cultivated by the American Indians long before the time of Columbus, and reached Europe about 500 years ago (see P. C. Mangelsdorf, Corn; its origin, evolution and improvement. Cambridge, Mass., 262 pp.; 1975).

111. PTERIDOPHYTA

by B. S. Parris.

Stems leafless, ± cylindrical, green or brown, with sporangia borne in terminal strobili
        2. Equisetum
Stems with leaves; the sporangia borne in strobili, or plant with fronds bearing sporangia on their undersurface:
  Plant moss-like with creeping stems; sporangia in strobili - - - - **1. Selaginella**
  Plant not as above:
    Fertile fronds or fertile portions of fronds different from sterile:
      Sterile fronds or parts of fronds entire, fleshy - - - - **3. Ophioglossum**
      Sterile fronds dichotomously branched or 1–2-pinnate, herbaceous **5. Anogramma**
    Fertile and sterile fronds or parts of frond similar:
      Sporangia not protected by indusia, scales or reflexed frond-margins:
        Sporangia scattered along veins, not in discrete sori - - **5. Anogramma**

Sporangia in discrete ± circular sori   -   -   -   -   -   **12. Polypodium**
Sporangia protected by indusia, scales or reflexed frond-margins:
  Sori linear, oblique to main veins of pinnae or pinnules, not near frond-margins
                                                               11. Asplenium
  Sori ± circular in outline, or ± marginal, not oblique to main veins of pinnae or
    pinnules:
    Sori ± marginal, linear, protected by reflexed frond-margin:
      Frond 1-pinnate; sori protected by reflexed frond-margin   -   -   **4. Pteris**
      Frond several times pinnate; sori protected by reflexed frond-margin and
        indusium   -   -   -   -   -   -   -   -   **8. Pteridium**
    Sori either not marginal or not linear:
      Sori not near margin, not protected by reflexed frond-margin:
        Sori with ovate to ovate-lanceolate basally attached indusium **9. Cystopteris**
        Sori with reniform centrally attached indusium   -   -   **10. Dryopteris**
      Sori ± marginal, sometimes protected by reflexed frond-margin:
        Sori protected by scales or hairs, or by the inrolled lamina and its modified
        margin; sterile pinnules without lobes divided again into dentate lobes
        with each vein ending in a tooth   --   -   -   -   -   **6. Cheilanthes**
      Sori borne on modified, reflexed marginal flaps (or false indusia), the sterile
      pinnules with lobes divided into dentate lobes, with each vein terminating
      in a tooth   -   -   -   -   -   -   -   -   **7. Adiantum**

LYCOPSIDA

Sporophytes with simple or usually dichotomously branched stems; leaves usually small and spirally arranged; stele without leaf-gap; homosporous or heterosporous; sporangia thick-walled, borne singly in the axil of a leaf (sporophyll), or on its upper surface near its base.

SELAGINELLALES

Terrestrial or epiphytic; ligule present; heterosporous. Spermatozoids biciliate.
One family.

111a. SELAGINELLACEAE

Milde, Höher Sporenpfl. Deutschl., 4: 136 (1865).
Hieronymus et Sadebeck in Engl. et Prantl, Pflanzenfam., 1 (4): 621–715 (1900–1901); Reimers in Engl., Syllabus, ed. 12, 1: 276 (1954).

Herbs with elongate stems producing leafless branches (rhizophores) which bear roots; leaves small, 1-veined, ligulate, either spirally arranged and monomorphic or 4-ranked and dimorphic; sporangia grouped in terminal strobili, solitary near the base of the upper surface of sporophylls, unilocular; megasporangia in lower, microsporangia in upper part of strobilus; megaspores (1–) 4 (–42); microspores numerous. Female prothallus many-celled, retained in megaspore and protruding from its apex; archegonia several at top of prothallus; male prothallus retained in microspore until near maturity, consisting of one prothallial cell and an antheridium producing 128 or 256 spermatozoids. Fertilization sometimes taking place before megaspore shed.

One extant genus with about 700 species; mainly tropical with some temperate.

1. SELAGINELLA *P. Beauv.*

Prodr. Aethéog., 101 (1805).

Description as for family.

1. **S. denticulata** (*L.*) *Link*, Fil. Sp., 159 (1841); Unger et Kotschy, Die Insel Cypern, 176 (1865); Sintenis in Oesterr. Bot. Zeitschr., 31: 226, 229, 393 (1881); Boiss., Fl. Orient., 5: 746 (1884); Holmboe, Veg. Cypr., 28 (1914); Post, Fl. Pal., ed. 2, 2: 817 (1933); Lindberg f., Iter Cypr., 5 (1946); Davis, Fl. Turkey, 1: 36 (1965); Osorio-Tafall et Seraphim, List Vasc. Plants Cyprus, 1 (1973).

   Lycopodium denticulatum L., Sp. Plant., ed. 1, 1106 (1753).
   TYPE: "Lusitania, Hispania, Iberia."

Perennial, stems green, up to 15 cm. long, creeping, much-branched and often producing rhizophores at the nodes; leaves arranged in four rows, up to 2·5 mm. long, ovate to narrowly ovate, finely toothed, obtuse to acuminate at apex, sessile, dimorphic, the lateral leaves spreading and the dorsal leaves appressed to the stem, the former sometimes rather larger than the latter; strobili terminal on side branches and not sharply differentiated from them, c. 1 cm. long; sporophylls lanceolate to broadly lanceolate, acuminate at apex, finely toothed, each with a single axillary sporangium.

HAB.: Shaded, moist rocks, walls and earthy banks; sea-level to 2,000 ft. alt.; fr. Febr.–May.

DISTR.: Divisions 1–3, 5, 7. Mediterranean region.

1. Stroumbi, 1913, *Haradjian* 759! Toxeftera near Ayios Yeoryios (Akamas), 1962, *Meikle* 2100!
2. Near Lefka on road to Kykko, 1927, *Rev. A. Huddle* 123!
3. Limassol, 1840, *Kotschy*; Stavrovouni, 1862, *Kotschy* 190 a; also, 1880, *Sintenis*; Klavdhia, 1880, *Sintenis & Rigo* 709! Alethriko, 1905, *Holmboe* 238.
5. Syngrasis, 1936, *Syngrassides* 1279!
7. Pentadaktylos, 1880, *Sintenis*; Kyrenia Pass, 1939, *Lindberg f.* s.n.; Akanthou, 1940, *Davis* 2033! Above Antiphonitis Monastery, 1955, *Merton* 2268! Lapithos, 1967, *Merton* in ARI 729! 3 m. W. of Bellapais, 1973, *P. Laukkonen* 172!

SPHENOPSIDA

Sporophyte with stem simple or branches in whorls; leaves simple, in whorls, joined at bases to form a sheath around the stem; stele without leaf-gap; homosporous or with incipient heterospory; sporangia borne on peltate sporangiophores arranged in strobili.

One extant order.

EQUISETALES

Terrestrial or subaquatic perennial herbs; leaves in whorls united into a sheath at the base, without ligules. Spermatozoids multiciliate.

One family.

111b. EQUISETACEAE

Sadebeck in Engl. et Prantl, Pflanzenfam., 1 (4): 520–548 (1900–1901).
Reimers in Engl., Syllabus, ed. 12, 1: 285–286 (1954).

Rhizome creeping, giving rise to aërial stems at intervals; stems either all photosynthetic or of two kinds, sterile and photosynthetic and fertile without chlorophyll; stem grooved, simple or branched from the base, or with whorls of branches at nodes; traversed by a central cavity, a ring of smaller cavities (vallecular canals) in the cortex and a ring of smaller carinal canals alternating with the vallecular canals; leaves small, usually the same number as the grooves on the stem, united at base to form a sheath, with those on the branches smaller than those of the stem; strobili solitary, terminal on main stem and occasionally on branches; spores numerous, more or less spherical, with 2 centrally attached elaters which are spathulate at each end, spirally coiled when moist and springing outwards when dry. Prothallus green, more or less cushion-shaped, lobed or branched, with rhizoids on the lower surface, archegonia and antheridia borne on separate prothalli (the female larger) or successively on the same prothallus.

One extant genus of 23 species; cosmopolitan, except for Australasia.

2. EQUISETUM *L.*

Sp. Plant., ed. 1, 1061 (1753).
Gen. Plant., ed. 5, 484 (1754).

Description as for family.

Sterile stems whitish (brown towards base), smooth, with (14–) 20–40 branches at nodes
2. E. telmateia

Sterile stems greenish, minutely tuberculate, with 0–6 branches at nodes
1. E. ramosissimum

1. **E. ramosissimum** *Desf.*, Fl. Atlant., 2: 398 (1799); Post, Fl. Pal., ed. 2, 2: 816 (1933); Lindberg f., Iter Cypr., 5 (1946); Davis, Fl. Turkey, 1: 32 (1965); Zohary, Fl. Palaest., 1: 4, t. 2 (1966); Osorio-Tafall et Seraphim, List Vasc. Plants Cyprus, 1 (1973).
    E. ramosum [Schleicher ex] DC., Syn. Plant. Fl. Gall., 118 (1806); Schleicher, Cat. Plant. Helv., ed. 2, 27 (1808); Boiss., Fl. Orient., 5: 742 (1884).
    E. pannonicum Waldst. et Kit. ex Willd., Sp. Plant., 5: 6 (1810).
    E. ramosissimum Desf. var. *subverticillatum* [A. Braun ex] Milde, Höher Sporenpfl. Deutschl., 117 (1865), Mon. Equiset., 436 (1867); Holmboe, Veg. Cypr., 27 (1914).
    E. ramosissimum Desf. var. *pannonicum* (Waldst. et Kit. ex Willd.) Aschers. in Aschers. et Graebn., Syn. Mitteleurop. Fl., 1: 140 (1896); Holmboe, Veg. Cypr., 27 (1914).
    TYPE: Tunisia; "ad radices montis Zowan apud Tunetanos" (P).

Perennial; stems usually dying in autumn, not dimorphic, 12–130 cm. high, hollow for $\frac{1}{2}$–$\frac{2}{3}$ diam., greenish, minutely tuberculate, markedly ridged, unbranched or with up to 6 branches at each node; sheaths 4–20 mm. long, green at first, then becoming pale brown, darker brown at base and with dark brown teeth, the latter as many as the ridges on the stem; strobilus 0·4–2 cm. long, 0·2–0·5 cm. wide, ± cylindrical, apiculate, terminal on the main stem, sometimes also terminal on the side branches, occasionally with subsidiary strobili at the node immediately below the main strobilus; branches up to at least 30 cm. long, green, simple, minutely tuberculate, ridged, the lowest branch-sheaths dark brown to dark chestnut-brown, the next light brown with teeth tipped dark brown, all others greenish with teeth tipped dark brown.

HAB.: Marshy ground, and damp places by springs and streams; sea-level to 4,500 ft. alt.; fr. April–Sept.

DISTR.: Divisions 1–5, 7, 8. South and Central Europe, Asia and Africa.

1. Ayios Minas near Smyies, 1962, *Meikle* 2176! Maa beach, 1967, *Economides* in ARI 992! By rivulet S. of Khrysokhou, 1970, *A. Hansen* 829! Between Inia and Smyies, 1979, *Edmondson & McClintock* E 2770!
2. Prodhromos, 1905, *Holmboe* 890; and, 1961, *D. P. Young* 7352! Platres, 1929, *C. B. Ussher* 40! Above Platania Forest Station, 1959, *P. H. Oswald* 163! and below Platania, 1966, *Merton* in ARI 73!
3. Alethriko, 1905, *Holmboe* 225; Episkopi, 1905, *Holmboe* 696; Akrotiri marshes, 1956, *Merton* 750! Kakoradjia, 1937, *Syngrassides* 1507!
4. Kouklia, 1905, *Holmboe* 391!
5. Kythrea, 1880, *Sintenis & Rigo* 97! also, 1932, *Syngrassides* 288! and, 1939, *Lindberg f.* s.n. !

7. Near Lapithos, 1880, *Sintenis & Rigo* 633! Mylous Forest Station, Kyrenia, 1933, *Syngrassides* 1364! Kyrenia Pass, 1939, *Lindberg f.* s.n.! Kyrenia, 1949, *Casey* 508! also, 1955, *G. E. Atherton* 2! 75!
8. Yialousa, 1970, *A. Genneou* in ARI 1459!

2. E. telmateia *Ehrh.*, Hannover Mag., 21: 287 (1783); Unger et Kotschy, Die Insel Cypern, 173 (1865) as *"telmateja"*; Boiss., Fl. Orient., 5: 741 (1884); Davis, Fl. Turkey, 1: 34 (1965); Zohary, Fl. Palaest., 1: 3, t. 1 (1966); Osorio-Tafall et Seraphim, List Vasc. Plants Cyprus, 1 (1973).

[*E. maximum* (non Lam.) Holmboe, Veg. Cypr., 27 (1914); Post, Fl. Pal., ed. 2, 2: 816 (1933); Lindberg f., Iter Cypr., 5 (1946).]

TYPE: Not indicated; probably best typified by the illustration of *Equisetum primum* in Matthiolus, *Commentarii*, cited by Ehrhart.

Herbaceous perennial; fertile stems appearing in spring before the sterile, and dying after the spores are shed, the sterile dying down in autumn; sterile stems to 60 cm. long, hollow for at least $\frac{2}{3}$ diam., whitish or brown towards base, smooth, with inconspicuous ridges, sheaths 1–1·5 cm. long, greenish, with long-attenuate, brown teeth, the latter as many as the ridges on the stem, branches (14–) 20–40 at all except the lowest nodes, up to 25 cm. long, green, simple, minutely tuberculate, 4-ridged, the lowest branch-sheaths 4-toothed, chestnut-brown, scarious, the other branch-sheaths 4-toothed, green, the lowest internode shorter than adjoining stem-sheath; fertile stems to 30 cm. long, brownish, unbranched, with 6–9 sheaths up to 3·5 cm. long (the longest at the top of the stem), sheaths pale basally, brown distally; strobilus solitary, c. 4·5–7 cm. long, 1–1·5 cm. wide, ± cylindrical, obtuse at apex.

HAB.: Wet ground by streams and springs; 600–4,500 ft. alt.; fr. April.

DISTR.: Divisions 2, 7. North Temperate regions.

2. Near Prodhromos on the way to Lemithou and Tris Elies, 1862, *Kotschy* 864a.
7. Kazaphani, 1937, *Syngrassides* 1448! Lapithos, 1939, *Lindberg f.* s.n.! Below Bellapais, 1948, *Casey* 101! 102!

NOTES: No fertile material has been seen from Cyprus and the description above has been drawn up from fertile Turkish material.

No material has been seen of the following, nor, on the basis of existing evidence, can they be accepted as part of the Cyprus flora:

E. ARVENSE *L.*, Sp. Plant., ed. 1, 1061 (1753); Osorio-Tafall et Seraphim, List Vasc. Plants Cyprus, 1 (1973).

E. SYLVATICUM *L.*, Sp. Plant., ed. 1, 1061 (1753); Sibth. et Sm., Fl. Graec. Prodr., 2: 269 (1816); Unger et Kotschy, Die Insel Cypern, 173 (1865); Boiss., Fl. Orient., 5: 742 (1884); Holmboe, Veg. Cypr., 27 (1914) as *"silvaticum"*; Davis, Fl. Turkey, 1: 33 (1965); Osorio-Tafall et Seraphim, List Vasc. Plants Cyprus, 1 (1973). Recorded, without locality, by Sibthorp & Smith, and, independently, by Kotschy, from (2) between Vrecha and "Yophyri", 9 May, 1862 (*Kotschy* 697a). A very unlikely occurrence in the absence of *E. sylvaticum* from all areas adjacent to Cyprus, and almost certainly a misidentification.

E. PALUSTRE *L.*, Sp. Plant., ed. 1, 1061 (1753); Boiss., Fl. Orient., 5: 742 (1884). Cited by Boissier from Cyprus, but the specimen so labelled is a misidentification of the relatively common *E. ramosissimum* Desf.

PTEROPSIDA

Sporophytes with spirally arranged fronds, usually large and often compound; stele with leaf-gaps; homosporous or heterosporous; sporangia thick- or thin-walled, often grouped in sori, borne on margin or underside of lamina of the frond, or on specialized frond-segments.

Six extant orders.

OPHIOGLOSSALES

Terrestrial (occasionally epiphytic, but not in Cyprus); vernation not circinnate, but fronds folded or bent; homosporous; sporangia maturing simultaneously. Spermatozoids multiciliate.

One family.

111c. OPHIOGLOSSACEAE

(R. Br.) Agardh, Aphor. Bot., 8: 113 (1822).
Bitter in Engl. et Prantl, Pflanzenfam., ed. 1, 1 (4): 449–472 (1900).
Clausen in Mem. Torrey Bot. Club, 19 (2): 1–177 (1938).
Reimers in Engl., Syllabus, ed. 12, 1: 290 (1954).

Stock usually short, fleshy, usually erect, without scales; rootlets fleshy; fronds usually solitary, rarely 2, the stipe protected by the sheathing base of the previous year's stipe; fertile lamina divided into sterile and fertile parts; sporangia in 2 rows, on margin of fertile part of lamina, each developed from a group of cells, sessile or nearly so, with a wall several cells deep, annulus absent. Prothallus usually subterranean, tuber-like, without chlorophyll, but with endotrophic mycorrhiza bearing both archegonia and antheridia on the same prothallus, the latter sunk into the tissues.

Three (or seven) genera comprising 70–90 species; with a cosmopolitan distribution.

3. OPHIOGLOSSUM L.

Sp. Plant., ed. 1, 1062 (1753).
Gen. Plant., ed. 5, 484 (1754).

Stock usually unbranched and usually subterranean, bearing one or several fronds annually; vegetative reproduction by adventitious buds or roots; sterile lamina usually simple, linear to ovate, entire, sometimes forked or palmately lobed; fertile lamina usually simple, sometimes forked or pinnately branched; venation reticulate; sporangia sunken, close-set on each side of the midrib opening by a transverse slit.

Thirty to fifty species; with a cosmopolitan distribution.

Winter-green; sterile lamina less than 1 cm. broad   -   -   -   -   **1. O. lusitanicum**
Summer-green; sterile lamina more than 3 cm. broad   -   -   -   -   **2. O. vulgatum**

1. O. lusitanicum L., Sp. Plant., ed. 1, 1063 (1753); Boiss., Fl. Orient., 5: 720 (1884); Holmboe, Veg. Cypr., 27 (1914); Post, Fl. Pal., ed. 2, 2: 805 (1933); Davis, Fl. Turkey, 1: 40 (1965); Zohary, Fl. Palaest., 1: 5, t. 3 (1966); Osorio-Tafall et Seraphim, List Vasc. Plants Cyprus, 1 (1973).

TYPE: "*in* Lusitania".

Herbaceous perennial; stock short, erect, subterranean, covered with bases of old fronds and producing stout fleshy roots, some of which may give rise to new plants; fronds 1–2, produced annually in winter and dying down in spring, consisting of a stipe with a sterile lamina and usually also an erect fertile spike; common stipe 0·7–4 cm. long, $\frac{1}{2}$ to $1\frac{1}{2}$ times as long as sterile lamina, straw-coloured, ± completely underground, glabrous; sterile lamina 1–4·5 cm. long, 0·1–0·5 cm. wide, linear to narrowly lanceolate, acute at apex, entire, fleshy, glabrous; fertile spike 0·4–1·4 cm. long, 0·2 cm. wide, glabrous, with 4–12 sunken sporangia on each side, the apex ± acute and sterile, on stipe 1–4·5 cm. long from its junction with the common stipe.

HAB.: Short, grassy sward on terra rossa, or on grassy banks in *Pinus brutia* forest; sea-level to 1,700 ft. alt.; fr. Jan.–March.

DISTR.: Divisions 1, 2, 4, 6, 8. Mediterranean region and W. Europe.

1. Near Fontana Amorosa, 1962, *Meikle* 2294!
2. Above Klirou, Kanopetra Forest, 1956, *Merton* 2490!
4. Below the monastery at Ayia Napa, and between Ayia Napa and Xylophagou, 1905, *Holmboe* 22.
6. Ayia Irini Forest, 1956, *Merton* 2459! Near Dhiorios, 1962, *Meikle* 2339!
8. Valia, near Ayios Theodhoros, 1905, *Holmboe*.

2. O. vulgatum *L.*, Sp. Plant., ed. 1, 1062 (1753); Boiss., Fl. Orient., 5: 720 (1884); Holmboe, Veg. Cypr., 27 (1914); Post, Fl. Pal., ed. 2, 2: 805 (1933); Lindberg f., Iter Cypr., 5 (1946); Davis, Fl. Turkey, 1: 40 (1965); Osorio-Tafall et Seraphim, List Vasc. Plants Cyprus, 1 (1973).

TYPE: "*in* Europae *pratis sylvaticis*".

Herbaceous perennial; stock short, erect, subterranean, covered with bases of old fronds and producing stout fleshy roots, some of which give rise to new plants; fronds solitary, produced annually in spring and dying in autumn, consisting of a stipe with a broad sterile lamina and usually also an erect fertile spike; common stipe 10–15 cm. long, 2–3 times as long as the sterile lamina, straw-coloured to pale chestnut at base and green above, underground for c. $\frac{1}{2}$ its length; sterile lamina 4–9 cm. long, 3–4 cm. wide, ovate to ovate-lanceolate, obtuse at apex, entire, fleshy, glabrous; fertile spike 1–2 cm. long, 0·2–0·3 cm. wide, glabrous, with c. 10–30 sunken sporangia on each side from its junction with the common stipe.

HAB.: Moist grassy ground by streams and springs at high elevations on igneous mountainsides; 3,000–4,500 ft. alt.; fr. June.

DISTR.: Division 2, rare. North Temperate regions.

2. Panayia Trikoukkia near Prodhromos, 1905, *Holmboe* 891; Platania, 1939, *Lindberg f.* s.n.!

FILICALES

Terrestrial or epiphytic, rarely aquatic; vernation circinnate; homosporous; sporangia usually not maturing simultaneously. Spermatozoids multiciliate.

About 34 families; with a cosmopolitan distribution.

111d. PTERIDACEAE

Ching in Webbia, 35 (2): 239 (1982).

Terrestrial; rhizome-scales opaque; fronds not jointed to rhizome; veins free or anastomosing; stipe with U-shaped vascular strand; sori without true indusium, submarginal, more or less continuous, protected by the scarious, reflexed frond margin, spores tetrahedral.

Six genera; with a cosmopolitan distribution.

4. PTERIS L.

Sp. Plant., ed. 1, 1073 (1753).
Gen. Plant., ed. 5, 484 (1754).

Terrestrial; rhizome dictyostelic, short; stipe with single U-shaped vascular strand; fronds monomorphic, simple to several times pinnate; sori continuous along a sub-marginal vein connecting the other veins, not on apices of segments nor on the sinuses connecting them, protected by the scarious deflexed margin; annulus of 16–34 thickened cells; spores tetrahedral, occasionally bilateral.

About 250 species; with a cosmopolitan distribution, but mainly tropical.

1. **P. vittata** L., Sp. Plant., ed. 1, 1074 (1753); Davis, Fl. Turkey, 1: 44 (1965); Zohary, Fl. Palaest., 1: 8, t. 7 (1966); Osorio-Tafall et Seraphim, List Vasc. Plants Cyprus, 2 (1973).
   [*P. longifolia* (non L.) Boiss., Fl. Orient., 5: 728 (1884); Holmboe, Veg. Cypr., 26 (1914); Post, Fl. Pal., ed. 2, 2: 809 (1933); Lindberg f., Iter Cypr., 5 (1946).]
   TYPE: "*in* China. *Osbeck*".

Perennial; rhizome ± erect; rhizome-scales linear-lanceolate, long-acuminate and often crisped at apices, light brown; stipe c. 8–14 cm. long, $\frac{1}{3}$ –$\frac{1}{5}$ as long as lamina, abaxially grooved, mid- to chestnut-brown at base, becoming pale yellowish to reddish-brown above, with scattered scales similar to those of the rhizome, and occasional ovate-lanceolate or lanceolate but otherwise similar scales; lamina to at least 90 cm. long, 20 cm. wide, oblanceolate to linear-oblanceolate in outline, herbaceous to thinly coriaceous, rhachis pale yellowish to reddish-brown, with occasional scales similar to those of the rhizome but smaller, 1-pinnate, the pinnae acuminate (or acute in small pinnae) at apex and broadly, unequally cuneate to cordate at base, the edges dentate when not fertile; sori linear, submarginal, covered when young by an indusium formed from the reflexed lamina-margin, in fully fertile plants continuous from near the base to near the apex of nearly all pinnae; in less fully fertile fronds continuous in the basal $\frac{1}{2}$ of the pinnae in the apical part of the frond.

HAB.: Moist, shaded ground by sides of streams; 400–1,000 ft. alt.; fr. June–Oct.

DISTR.: Divisions 1, 7, rare. Old World tropics and subtropics.

1. Near the Baths of Aphrodite, 1905, *Holmboe* 788.
7. Larnaka tis Lapithou, 1936, *Syngrassides* 1300! Ayios Amvrosios, 1937, *Syngrassides* 1444! Kyrenia Pass (Boghazi), 1939, *Lindberg f.* s.n.; Mylous Plantation, Kyrenia 1949, *Casey* 530! Kyrenia, 1955, *G. E. Atherton* 319! 727!

NOTES: The unlocalized Cyprus record cited by Boissier (Fl. Orient., 5: 728) is erroneous; Sintenis & Rigo collected their specimens of the fern in Syria, not in Cyprus.

111e. ADIANTACEAE

(C. Presl) Ching in Sunyatsenia, 5: 229 (1940).

Terrestrial; rhizome sclerostelic, creeping or ascending; rhizome-scales opaque; fronds not jointed to rhizome, undivided to several times pinnate, veins usually free; stipe with one or two vascular strands at base, uniting to form a single 4-angled strand distally, or dividing to form several strands; sori submarginal on veinlets at ends of veins, occasionally on surface between veinlets, sometimes along veins, without indusium, protected by reflexed frond-margins; spores tetrahedral, occasionally bilateral.

Thirty-two genera in tropical and temperate regions.

5. ANOGRAMMA *Link*

Filic. Sp., 1: 137 (1841).

Terrestrial; rhizome short; sporophyte annual, but gametophyte apparently perennial; stipe with 2 vascular strands; fronds slightly dimorphic, 2–3-pinnate, veins free, margin flat; sori spread along veins, confluent when mature; annulus of c. 22 thickened cells; spores tetrahedral.

Seven species in north and south temperate zones.

1. **A. leptophylla** (*L.*) *Link*, Fil. Sp., 1: 137 (1841); Davis, Fl. Turkey, 1, 45 (1965); Zohary, Fl. Palaest., 1: 8, t. 8 (1966); Osorio-Tafall et Seraphim, List Vasc. Plants Cyprus, 3 (1973).

 Polypodium leptophyllum L., Sp. Plant., ed. 1, 1092 (1753).

 Gymnogramma leptophylla (L.) Desv. in Ges. Naturf. Freunde Berlin Mag. Neuesten Entdeck. Gesammten Naturk., 5: 305 (1811); Unger et Kotschy, Die Insel Cypern, 173 (1865); Sintenis in Oesterr. Bot. Zeitschr., 31: 231 (1881); Boiss., Fl. Orient., 5: 721 (1884); Holmboe, Veg. Cypr., 26 (1914); Post, Fl. Pal., ed. 2, 2: 807 (1933); Lindberg f., Iter Cypr., 5 (1946).

 TYPE: "*in* Hispania, Lusitania, Galloprovincia".

Annual; gametophyte said to be perennial, tuberous and producing new sporophytes annually; rhizome erect; rhizome-hairs scattered, c. 1 mm. long, whitish to pale brown, multicellular; fronds weakly dimorphic, the sterile ones produced first and often dead when the fertile ones are mature, the later-produced fertile fronds larger and on longer stipes than the earlier ones; lamina pale yellowish-green, herbaceous, glabrous; sterile fronds 1–3, stipe c. 0·5–1·5 cm. long, ± as long to c. ½ as long as the lamina, chestnut-brown at base, shading to yellowish brown above, with scattered hairs similar to those of the rhizome basally; first sterile frond ± semicircular in outline, usually dichotomously branched into 2 major lobes and c. 8 ultimate lobes, sometimes only one degree of lobing present; later-formed sterile fronds 1–2·5 cm. long, 1–2 cm. wide, broadly triangular to triangular in outline, rhachis greenish, 1–2-pinnate, the ultimate segments lobed, obovate, obtuse at apex and cuneate at base; sterile fronds often grading into fully fertile fronds; fertile fronds 1–4, stipe (1–) 3–8 (–12) cm. long, ± as long to 1½ times as long as lamina, coloured as the stipes of sterile fronds or sometimes greenish above, hairs the same as for the stipes of sterile fronds, lamina (1–) 3–10 (–12) cm. long, (0·5–) 1–3 cm. wide, ovate, narrowly triangular or lanceolate in outline, 2–3-pinnate, rhachis yellowish to greenish, the ultimate segments lobed, ovate to obovate, obtuse at apex and cuneate at base; sori without indusium, produced along the veins in the ultimate segments and sometimes covering their entire undersurface.

HAB.: Shady walls, rocks and earthy banks; sea-level to 3,300 ft. alt.; fr. Febr.–May.

DISTR.: Divisions 1–4, 7. Scattered in temperate and tropical regions of both hemispheres.

1. Toxeftera, near Ayios Yeoryios (Akamas), 1962, *Meikle* 2091!
2. Near Lefka, 1927, *Rev. A. Huddle* 118! Platres, 1937, *Kennedy* 211! Milikouri, 1939, *Lindberg f.* s.n.; Ayia valley, 1982, *D. Tennant* s.n.!
3. Stavrovouni, 1862, *Kotschy* 186a; Akrotiri, 1982, *D. Tennant* s.n.!
4. Ayia Napa, 1905, *Holmboe* 64.
7. Pentadaktylos, 1862, *Kotschy* 337a! Lapithos, 1939, *Lindberg f.* s.n.; above Vasilia, 1941, *Davis* 2203! Karini, 1941, *Davis* 2778! Kyrenia, 1952, *Casey* 1230! Above Antiphonitis Monastery, 1955, *Merton* 2267! Kazaphani, 1956, *G. E. Atherton* 1157!

6. CHEILANTHES *Swartz*

Syn. Fil., 5 (1806) nom. cons.

Terrestrial; rhizome short; stipe with a single vascular strand at base which divides into 2 or 3 strands distally; fronds not dimorphic, once-pinnate to several times pinnate; veins free; sorus submarginal sometimes covered by the reflexed frond margins; annulus of 14–24 thickened cells; spores tetrahedral.

About 180 species; in tropical and temperate regions.

Undersurface of lamina covered with scales - - - - - - **1. C. marantae**
Undersurface of lamina not covered with scales:
  Undersurface of lamina covered with hairs - - - - - - **2. C. vellea**
  Undersurface of lamina glabrous - - - - - - - **3. C. pteridioides**

1. C. marantae (*L.*) *Domin* in Biblioth. Bot., 85: 133 (1915); Davis, Fl. Turkey, 1: 42 (1965); Osorio-Tafall et Seraphim, List Vasc. Plants Cyprus, 2 (1973).
  Acrostichum marantae L., Sp. Plant., ed. 1, 1071 (1753); Sibth. et Sm., Fl. Graec. Prodr., 2: 271 (1816), Fl. Graec., 10: 54 (1840).
  Notholaena marantae (L.) Desv. in Journ. Bot. Agric., 1: 92 (1813); Unger et Kotschy, Die Insel Cypern, 174 (1865) as "*Notochlaena*"; Boiss., Fl. Orient., 5: 725 (1884) as "*Notochlaena*"; Holmboe, Veg. Cypr., 210 (1914) as "*Notochlaena*"; Post, Fl. Pal., ed. 2, 2: 808 (1933).
  TYPE: "*in* Europa *australi*".

Perennial; rhizome short-creeping, rhizome-scales linear-lanceolate, long-acuminate, chestnut-brown; stipe 5–18 cm., ± as long as lamina to twice as long as lamina, ± terete, dark chestnut-brown, dark purplish-brown or blackish, with scattered scales similar to those of the rhizome but smaller; lamina 3–9 cm. long, 1·5–3 cm. wide, lanceolate to ovate-lanceolate in outline, coriaceous, rhachis the same colour as the stipe and with similar scales, 2-pinnate, the pinnules obtuse at apex and truncate or sessile at base, narrowly triangular or oblong, entire, the undersurface of the lamina and pinna-rhachis densely covered with narrowly lanceolate, whitish to chestnut-brown scales, the upper surface of the lamina glabrous or with scattered very small glands, the upper surface of the pinna-rhachis with scattered scales like those of the lamina and pinna-rhachis undersurfaces; sori submarginal, continuous on each pinnule, without an indusium, and not covered by the frond-margins, but partly protected by the scales of the lamina undersurface. *Plate 105.*

HAB.: In crevices of igneous rocks, or under loose rock debris in dryish situations at high altitudes; 4,900–5,700 ft. alt.; fr. Febr.–Dec.

DISTR.: Division 2, rare. Mediterranean region, N.W. Africa, Himalayas to Southwest China. 2. Upper slopes of Khionistra, 1880–1940, *Sintenis & Rigo* 710! *Kennedy* 22! 23! 24! *Davis* 2075!

NOTES: Records for *Notholaena marantae* in Die Insel Cypern, 174 (1865) are mis-identifications of *Cheilanthes pteridioides*, likewise the Sibthorp record, said (by Holmboe) to have been based on material collected at (3) Stavrovouni.
  C. marantae has been transferred to *Gymnopteris* by Ching because of the arrangement of the sporangia along the veins and because the closely related Chinese species *G. delavayi* is placed in that genus. The arrangement of the sporangia, however, is rather variable in the material of *C. marantae* examined; they may be distributed along each vein for a distance of up to 0·5 mm. or arise ± from one point on the vein. Although it is evidently not closely related to the other species of *Cheilanthes* treated here it seems best to retain *C. marantae* in this genus pending a world-wide review of generic limits of the Cheilanthoid ferns and their relatives.

2. C. vellea (*Aiton*) *F. Mueller*, Fragm., 5: 123 (1866).
  Acrostichum velleum Aiton, Hort. Kew., ed. 1, 3: 457 (1789).
  Notholaena vellea (Aiton) R. Br., Prodr., 146 (1810); A. K. Jackson in Kew Bull., 1937: 346 (1937).

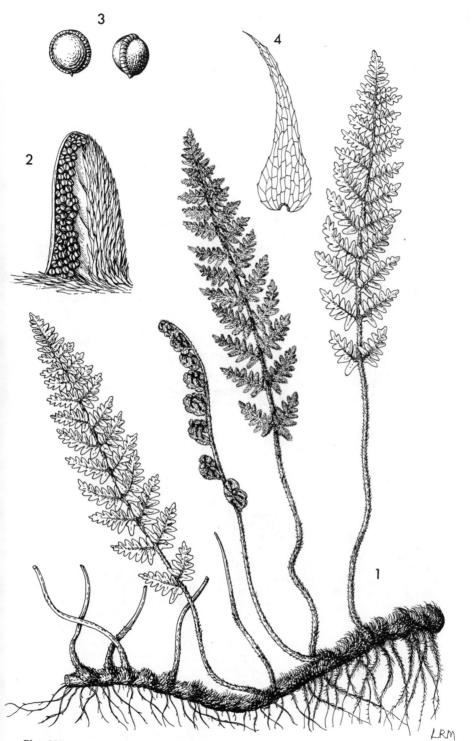

Plate 105. CHEILANTHES MARANTAE (L.) Domin **1**, habit, ×⅔; **2**, pinnule with some of the scales removed to show sori, ×10; **3**, sori, ×32; **4**, single scale, ×32. (**1–4** from *Balansa* 760, Turkish material.)

Notochlaena lanuginosa (Desf.) Poiret in Lam., Encycl. Méth., Suppl., 4: 110 (1816); Sintenis in Oesterr. Bot. Zeitschr., 31: 393 (1881); Boiss., Fl. Orient., 5: 725 (1884).
Acrostichum catanense Cosent. in Atti Accad. Gioenia, ser. 1, 2: 207, 218 (1827).
Cheilanthes catanensis (Cosent.) H. P. Fuchs in Brit. Fern Gaz., 9: 45 (1961); Davis, Fl. Turkey, 1: 43 (1965); Zohary, Fl. Palaest., 1: 6, t. 5 (1966); Osorio-Tafall et Seraphim, List Vasc. Plants Cyprus, 2 (1973).
TYPE: "*Nat.* of Madeira. Mr. *Francis Masson*".

Perennial; rhizome ± erect to short-creeping, sometimes branched; rhizome-scales linear-lanceolate, long-acuminate, crisped at apex, whitish to pale chestnut brown; stipe 1–2 cm., $\frac{1}{2}$ to $\frac{1}{10}$ as long as the lamina, dark chestnut brown to purplish brown, with scattered scales similar to those of the rhizome but rather smaller, and many crisped whitish multicellular hairs which may darken with age to pale yellow-brown or pale chestnut-brown; lamina 5–10 cm. long, 1–2 cm. wide, lanceolate to oblanceolate, linear-lanceolate or linear-oblanceolate in outline; coriaceous, rhachis the same colour as the stipe and with the same type of hair, 2-pinnate, the pinnules obtuse at apex and truncate to subcordate at base, suborbicular, ovate or oblong, with few blunt lobes, the undersurface densely covered, the upper less so, with the same type of hair as on the stipe and rhachis; sori near the margin on the lower surface of the lamina, ± confluent within each lobe, without an indusium and not covered by the frond-margin, but partly protected by the hairs on the lamina undersurface.

HAB.: Crevices of dry, exposed limestone, or rarely igneous, rocks; 1,000–3,000 ft. alt.; fr. Jan.–Nov.

DISTR.: Divisions 2, 7. Mediterranean region to N.W. Africa and W. Himalayas.

2. Skouriotissa, 1927, *Rev. A. Huddle* 119!
7. Pentadaktylos, 1880, *Sintenis & Rigo* 100! Above Dhikomo, 1936, *Syngrassides* 990! Above Aghirda, 1940, *Davis* 2037! also, 1941, *Davis* 2491! and, 1949, *Casey* 206! Above Larnaka tis Lapithou, 1941, *Davis* 2095! Yaïla, 1941, *Davis* 2868! Buffavento, 1974, *Meikle*!

3. C. pteridioides (*Reichard*) *C. Chr.*, Index Fil., 178 (1905).
Polypodium pteridioides Reichard in L., Syst. Plant., ed. 4, 4: 424 (1780) nom. nov. pro *Polypodium fragrans* L. (1771) non L. (1753).
Cheilanthes fragrans Swartz, Syn. Fil., 127 (1806); Unger et Kotschy, Die Insel Cypern, 174 (1865); Sintenis in Oesterr. Bot. Zeitschr., 31: 229 (1881); Boiss., Fl. Orient., 5: 725 (1884); Holmboe, Veg. Cypr., 26 (1914); Druce in Proc. Linn. Soc., 141st Sess., 51 (1930), Rep. B.E.C., 9: 471 (1931); Post, Fl. Pal., ed. 2, 2: 809 (1933); Lindberg f., Iter Cypr., 5 (1946); Davis, Fl. Turkey, 1: 41 (1965); Zohary, Fl. Palaest., 1: 6, t. 4 (1966); Osorio-Tafall et Seraphim, List Vasc. Plants Cyprus, 1 (1973).
C. suaveolens Swartz, Syn. Fil., 127 (1806); Sibth. et Sm., Fl. Graec. Prodr., 2: 278 (1816), Fl. Graec., 10: 56 (1840).
TYPE: "*in* Gallia *australi* Baro Capucinus; *ad muros* Funschal. *Koenig*".

Perennial; rhizome ± erect to short-creeping, sometimes branched; rhizome-scales linear-lanceolate, long-acuminate, medium brown to chestnut-brown; stipe 1–11 cm. long, usually ± as long as lamina, but sometimes $\frac{1}{2}$ as long or twice as long, dark chestnut-brown, dark purplish-brown or blackish, with scattered scales similar to those of the rhizome but smaller; lamina smelling of newly dried hay, (1·5–) 3–11 cm. long, (0·7–) 1–3·5 cm. wide, lanceolate to ovate-lanceolate in outline, coriaceous, rhachis the same colour as the stipe and with similar scales, (2–) 3-pinnate, the pinnules obtuse at apex and broadly cuneate to truncate or cordate at base, suborbicular to ovate and oblong, entire or with a few obtuse lobes, lamina glabrous on both surfaces or occasionally with scattered very small glands on the upper surface, the pinna-rhachis and pinnule-rhachis with scales similar to those of stipe and main rhachis but rather smaller; sori submarginal, protected by the inrolled lamina and also by an indusium formed by the modified lamina-margin, solitary at apex of pinnule-lobes, and several along margin of entire pinnules.

HAB.: In clefts of igneous or calcareous rocks; sea-level to 4,300 ft. alt.; fr. Jan.–Dec.

DISTR.: Divisions 1–3, 5, 7. Mediterranean region to W. Himalayas.

1. Smyies, 1962, *Meikle* 2200!
2. Three miles S. of Lefka, 1927, *Rev. A. Huddle* 122! Troödhitissa, 1936, *Kennedy* 25! 27! Kryos Potamos, 2,600–4,300 ft. alt., 1936, *Kennedy* 26! 29! 30! 31! 33! E. of Kokkinovouno, 1937, *Kennedy* 27! Between Platres and Mandria, 1937, *Kennedy* 32! Platres, 1937, *Kennedy* 34! Mesapotamos, 1939, *Lindberg f.* s.n.; Ayios Theodhoros, 1941, *Davis* 3056! Phlevas Sawmill, 1957, *Merton* 2871! and, between Phlevas and Vroisha, 1962, *Meikle* 2720! Lorovouno, 1979, *Edmondson & McClintock* E 2800! Ayia valley 1982, *D. Tennant* s.n.!
3. Stavrovouni, 1862, *Kotschy* 192! Lania, 1928 or 1930, *Druce* s.n.; near Silikou, 1940, *Davis* 2074! N.E. of Apsiou, 1979, *Edmondson & McClintock* E 2700! Pissouri, 1982, *D. Tennant* s.n.!
5. Kythrea, 1941, *Davis* 2955!
7. Between Dhikomo and Kyrenia, 1862, *Kotschy* 464! Pentadaktylos, 1880, *Sintenis & Rigo* 98! and, 1928 or 1930, *Druce* s.n.; Ayios Khrysostomos Monastery, 1927, *Rev. A. Huddle* 120! Mylous Forest Station 1935, *Syngrassides* 808! Aghirda, 1936, *Lady Loch* 16! and, 1955, *G. E. Atherton* 776! Buffavento Castle, 1937, *Syngrassides* 1735! St. Hilarion, 1939, *Lindberg f.* s.n.; Kyrenia Pass, 1939, *Lindberg f.* s.n.; Vasilia, 1940, *Davis* 1997! Larnaka tis Lapithou, 1941, *Davis* 2089!

7. ADIANTUM *L.*

Sp. Plant., ed. 1, 1094 (1753).
Gen. Plant., ed. 5, 485 (1754).

Terrestrial; rhizome short; stipe with 2 vascular strands at base, uniting to form a single 4-angled strand distally; fronds not dimorphic, once to several times pinnate; veins free; sorus borne on reflexed marginal lobes; annulus of c. 18 thickened cells; spores tetrahedral.

About 200 species; cosmopolitan but chiefly in South America.

1. **A. capillus-veneris** *L.*, Sp. Plant., ed. 1, 1096 (1753); Hume in Walpole, Mem. Europ. Asiatic Turkey, 254 (1817) as "*Adianthum*"; Gaudry, Recherches Sci. en Orient, 198 (1855); Unger et Kotschy, Die Insel Cypern, 174 (1865); Sintenis in Oesterr. Bot. Zeitschr., 31: 229, 326 (1881); Boiss., Fl. Orient., 5: 730 (1884); H. Stuart Thompson in Journ. Bot., 44: 341 (1906); Holmboe, Veg. Cypr., 26 (1914); Post, Fl. Pal., ed. 2, 2: 811 (1933); Lindberg f., Iter Cypr., 5 (1946); Davis, Fl. Turkey, 1: 43 (1965); Zohary, Fl. Palaest., 1: 7, t. 6 (1966); Osorio-Tafall et Seraphim, List Vasc. Plants Cyprus, 2 (1973).
TYPE: "*in* Europa *australi*".

Perennial; rhizome short-creeping; rhizome-scales linear-lanceolate, yellowish-brown, chestnut-brown, or dark brown; stipe (1–) 4–14 (–19) cm. long, ± as long as, to ½ as long as lamina, angled, dark chestnut-brown to blackish-brown at base, rather lighter at apex, the lower part with scales similar to those of rhizome, elsewhere glabrous; lamina (3–) 5–20 (–25) cm. long, (1–) 2–10 (–12) cm. wide, broadly ovate to lanceolate in outline, thinly herbaceous, rhachis dark chestnut-brown to blackish-brown, 2–3-pinnate, pinnules obtuse at apex and cuneate at base, obovate, oblong or semicircular, glabrous, usually divided rather deeply into 2–3 main lobes, each of which is divided less deeply 2–3 times to form lobes which in fertile pinnae are truncate and have a sorus at their apex, and which in sterile pinnae are again divided into dentate lobes, with each vein terminating in a tooth; sori reniform to oblong or linear, on a modified, reflexed marginal flap or false indusium, each sorus extending across several vein endings, on the upper and outer edge of the pinnules.

HAB.: Wet walls and damp, shaded rocks by springs, streams and irrigation channels; sea-level to 4,500 ft. alt.; fr. Apr.–Sept.

DISTR.: Divisions 1–5, 7, 8. Widespread in temperate and subtropical regions.

1. Baths of Aphrodite, 1905, *Holmboe* 787; also 1979, *Edmondson & McClintock* E 2743! Ayios Minas near Smyies, 1962, *Meikle* 2170!

2. Phini, 1862, *Kotschy* 883; near Lefka, 1927, *Rev. A. Huddle* 114! Palekhori, 1940, *Davis*
   2020! Between Perapedhi and Trimiklini, 1941, *Davis* 3447! Kakopetria, 1955, *Chiotellis*
   449! Prodhromos, 1955, *G. E. Atherton* 639!
3. Limassol, 1801, *Hume*; Stavrovouni, 1880, *Sintenis*; Pissouri, 1967, *Merton* in ARI 704!
4. Larnaca, 1801, *Hume*; Ayia Napa, 1905, *Holmboe* 42.
5. Kythrea, 1862, *Kotschy*; also, 1880, *Sintenis*.
7. Kyrenia, 1927, *Rev. A. Huddle* 115! also, 1948, *Casey* 50! and, 1955, *Miss Mapple* 29! *G. E.
   Atherton* 318! 672! Lapithos, 1939, *Lindberg f.* s.n.; also, 1955, *Merton* 2243! Boghazi, 1939,
   Lindberg f. s.n.!
8. Ayios Andronikos, 1934, *Syngrassides* 512!

111f. DENNSTAEDTIACEAE
Pichi-Sermolli in Webbia, 24: 704 (1970).

Terrestrial; rhizome covered with hairs; fronds not jointed to the rhizome;
veins free (anastomosing in some species outside our area); stipe with several
vascular strands uniting to form a U-shaped strand; sporangia marginal,
more or less continuous, protected by an indusium and the reflexed frond-
margins; spores tetrahedral.
Seventeen genera; distribution cosmopolitan.

8. **PTERIDIUM** *Scop.*
Fl. Carniol., 169 (1760) nom. cons.

Terrestrial; rhizome solenostelic, long-creeping; stipe with numerous
vascular strands below, fusing to form a single U-shaped strand above;
fronds monomorphic, several times pinnate; sorus continuous along the
frond margin, borne on the connecting vein; annulus of c. 13 thickened cells.
Spores tetrahedral.
Variously treated as containing from 1 to at least 6 species; with a
cosmopolitan distribution.

1. **P. aquilinum** (*L.*) *Kuhn* in Decken, Reise in Ost-Afrika, 3 (3), Bot., 11 (1879); Holmboe, Veg.
   Cypr., 26 (1914); Ward et Gresham in Journ. Roy. Hort. Soc., 59: 72 (1934); Davis, Fl.
   Turkey, 1: 46 (1965); Osorio-Tafall et Seraphim, List Vasc. Plants Cyprus, 2 (1973).
      Pteris aquilina L., Sp. Plant., ed. 1, 1075 (1753); Sibth. et Sm., Fl. Graec. Prodr., 2: 278
   (1816); Gaudry, Recherches Sci. en Orient, 199 (1855); Unger et Kotschy, Die Insel
   Cypern, 175 (1865); H. Stuart Thompson in Journ. Bot., 44: 31 (1906); Post, Fl. Pal., ed. 2,
   2: 809 (1933).
      [*Pteridium aquilinum* (L.) Kuhn f. *lanuginosum* (vix *Pteris lanuginosa* Bory ex Willd.)
   Holmboe, Veg. Cypr., 26 (1914).]
      [*P. aquilinum* (L.) Kuhn var. *lanuginosum* (vix *Pteris lanuginosa* Bory ex Willd.)
   Lindberg f., Iter Cypr., 5 (1946).]
   TYPE: "*in* Europae *sylvis, praesertim caeduis*" (Hb. Cliff. BM).

Perennial; rhizome long-creeping and much-branched, deeply buried,
with short shoots arising from it which bear the fronds; rhizome-hairs
multicellular, chestnut-brown to dark brown, shining; stipe c. 12–15 cm.
long, ± ½ as long as lamina, adaxially grooved, dark brown at base and there
bearing hairs similar to those of the rhizome but somewhat smaller and pale
chestnut-brown, straw-coloured to pale chestnut-brown above and ±
glabrous; lamina c. 22–36 cm. long, 31–36 cm. wide, broadly triangular to
broadly ovate in outline, coriaceous, rhachis straw-coloured to pale
chestnut-brown with occasional hairs like those of the stipe base, tripinnate,

the pinnules sessile, entire or with a few large lobes at base, acute at apex, the undersurface of the lamina, especially the pinnule midvein and other veins, with frequent whitish to pale yellowish hairs and the undersurface of the pinna- and pinnule-rhachis with sparse similar hairs which may be chestnut-brown or have chestnut-brown crosswalls; hairs on the upper surface of lamina, veins and pinna- and pinnule-rhachis like the hairs of their lower surfaces but much fewer; sori marginal, linear, along both sides of the pinnules but not in their apices, protected by an outer indusium formed by the reflexed frond margin and by an inner indusium hidden beneath the outer.

HAB.: In Pine forest on igneous mountains; 2,000–6,000 ft. alt.; fr. August.

DISTR.: Division 2. Europe, S.W. Asia and Africa.

2. Troödos, 1862, *Kotschy*; Prodhromos, 1905, *Holmboe* 956; also, 1949, *Casey* 901! and, 1955, *G. E. Atherton* 504! Stavros tis Psokas, 1939, *Lindberg f.* s.n.; Platania, 1939, *Lindberg f.* s.n.; and, 1973, *P. Laukkonen* 416! Between Troödos and Pano Platres, 1973, *P. Laukkonen*, 404! Kakopetria, 1955, *Chiotellis* 434! "Very common in the Pine-forests of the southern mountain-range", *Holmboe*.

111g. ASPIDIACEAE
Mett. ex Frank in Leunis, Syn. Pflanzenk., ed. 2, 3: 1469 (1977).

Terrestrial; rhizome dictyostelic, scales opaque; fronds not jointed to rhizome; veins free (or sometimes anastomosing outside our area); stipe with 2–7 vascular strands; sori superficial, usually protected by a round or reniform indusium; spores bilateral.

About 54 genera with a cosmopolitan distribution.

9. CYSTOPTERIS *Bernh.*
in Schrader, Neue Journ. Bot., 1 (2): 26 (1805) nom. cons.

Terrestrial, rhizome short; stipe with 2 vascular strands at base, uniting upwards to form a single U-shaped strand; fronds monomorphic, bi- or tri-pinnate; veins free; sori dorsal on veinlets, round, indusium attached below the sorus on its inner side; annulus of 14–16 thickened cells; spores bilateral.

About 18 species; with a cosmopolitan distribution, but mainly in the north temperate zone.

1. **C. fragilis** (*L.*) *Bernh.* in Schrader, Neue Journ. Bot., 1 (2): 27 (1805); Kotschy in Oesterr. Bot. Zeitschr., 12: 279 (1862) as "*Cistopteris*"; Unger et Kotschy, Die Insel Cypern, 175 (1865); Boiss., Fl. Orient., 5: 740 (1884); Post, Fl. Pal., ed. 2, 2: 815 (1933); Davis, Fl. Turkey, 6: 54 (1965); Osorio-Tafall et Seraphim, List Vasc. Plants Cyprus, 3 (1973).
Polypodium fragile L., Sp. Plant., ed. 1, 1091 (1753).
Cystopteris fragilis (L.) Bernh. var. *genuina* Boiss., Fl. Orient., 5: 740 (1884).
C. fragilis (L.) Bernh. ssp. *eu-fragilis* Holmboe, Veg. Cypr., 25 (1914); Lindberg f., Iter Cypr., 5 (1946).
TYPE: "*in collibus* Europae *frigidioris*".

Perennial; rhizome short-creeping, occasionally branched; rhizome-scales broadly lanceolate to linear-lanceolate, long-acuminate at apex, chestnut-brown; stipe 4–14 cm. long, ± as long as lamina to $\frac{1}{3}$ as long as lamina, angled to shallowly grooved adaxially, chestnut-brown to straw-coloured at base, chestnut-brown, straw-coloured or greenish towards apex, with a few scales similar to those of the rhizome, but smaller especially towards the base; frond 5–16 cm. long, 2–6 cm. wide, broadly lanceolate to oblong-lanceolate in outline, herbaceous, rhachis straw-coloured to greenish, sometimes chestnut-brown towards base, lamina 2-pinnate, the pinnules all ± decurrent, lobed rather deeply, the lobes often with blunt teeth, the pinnules obtuse to acute at apex and broadly cuneate at base or sessile to the pinna-rhachis wing, a few very small translucent hairs with red crosswalls and sometimes a red terminal cell present on the margin of rhachis, pinna-rhachis or pinnule-rhachis of young fronds, but these usually shed with age, sori ± orbicular, covered by a thin pale ovate to ovate-lanceolate indusium, acuminate at apex, which is attached basally and is reflexed when the sporangia are mature, spores spiny.

HAB.: Moist, shaded rock-crevices; 3,000–6,300 ft. alt.; fr. May.

DISTR.: Division 2 only. Cosmopolitan.

2. N. side Khionistra summit, 1862, *Kotschy* 914; also, 1905, *Holmboe* 967; 1937, *Kennedy* 6! and 1941, *Davis* 3167! Platres, 1937, *Kennedy* 1! 2! Asprokremnos, 1937, *Kennedy* 3! Xerokolymbos, 5,400 ft. alt., 1937, *Kennedy* 4! 7! 8! Kalogeros, 1937, *Kennedy* 7! Kannoures springs, 1939, *Lindberg f.* s.n.! Agros, 1941, *Davis* 3062! Lagoudhera, 1955, *Merton* 1964! Askas, 1957, *Merton* 3123! Khorteri, near Stavros tis Psokas, 1962, *Meikle* 2714! Yironas River, below Madhari, 1962, *Meikle* 2826! Trikoukkia, S. of Prodhromos, 1963, *D. P. Young* s.n.!

10. DRYOPTERIS *Adanson*
Fam. Plant., 2: 20 (1763) nom. cons.

Terrestrial; rhizome short; stipe with several vascular strands; fronds monomorphic, once to several times pinnate; veins free; sori dorsal on veinlets, round, with reniform indusia; annulus of c. 14 thickened cells; spores bilateral.

About 150 species; near cosmopolitan, but mainly in north temperate zone.

1. **D. pallida** (*Bory*) *Fomin* in Věstn. Tifflisk. Bot. Sada, 20: 49 (1911); Davis, Fl. Turkey, 1: 59 (1965) pro parte; Fraser-Jenkins in Candollea, 32: 312 (1977).

   Nephrodium pallidum Bory, Expéd. Sci. Morée, Bot., 287 (1832); Unger et Kotschy, Die Insel Cypern, 175 (1865) quoad nomen.

   [*Dryopteris rigida* (non A. Gray) Post, Fl. Pal., ed. 2, 2: 814 (1933).]

   [*D. villarii* (non (Bellardi) Woynar ex Schinz et Thell.) Zohary, Fl. Palaest., 1: 12, t. 13 (1966).]

Perennial; rhizome erect to short-creeping; rhizome-scales lanceolate to ovate, long-acuminate and crisped at apex, pale brown to chestnut-brown, stipe 11–34 (–50) cm. long, $\frac{1}{4}-\frac{2}{3}$ as long as lamina, adaxially grooved, chestnut-brown to straw-coloured, occasionally greenish towards apex, with scales similar to those of the rhizome but smaller, especially common towards the stipe base; lamina 25–76 cm. long, 11–28 cm. wide, broadly lanceolate to ovate or narrowly triangular in outline, thinly coriaceous, rhachis pale chestnut-brown or straw-coloured, occasionally greenish, 2-(3-) pinnate, the pinnules lobed, sometimes very deeply, the lobes obtuse at apex and toothed, the pinnules obtuse to acute at apex, and broadly unequally cuneate to truncate at base, the rhachis and pinna-rhachis with scattered scales similar to those of the stipe but smaller, which intergrade with multicellular hairs, both surfaces of the lamina sometimes with scattered pale glands; sori orbicular, covered with a reniform indusium, usually near the margin, but in some plants midway between the pinnule midvein and the margin (these may be intermediates between ssp. *libanotica* and ssp. *pallida*, which are known to intergrade in southern Turkey).

ssp. **libanotica** (*Rosenst.*) *Nardi* in Webbia, 32: 97 (1977); Fraser-Jenkins in Candollea, 32: 315 (1977).

Aspidium libanoticum Rosenst. in Mém. Herb. Boiss., 9: 1 (1900).

Dryopteris libanotica (Rosenst.) C. Chr., Index Fil., 275 (1905); Rechinger f. in Arkiv för Bot., ser. 2, 1: 417 (1950).

[*Nephrodium pallidum* (non Bory) Unger et Kotschy, Die Insel Cypern, 175 (1865) quoad plant.]

[*Aspidium filix-mas* (non (L.) Sw.) Gaudry, Recherches Sci. en Orient, 199 (1855); Sintenis in Oesterr. Bot. Zeitschr., 32: 194 (1882); Holmboe, Veg. Cypr., 26 (1914).]

[*Nephrodium filix-mas* (non (L.) Strempel) Boiss., Fl. Orient., 5: 737 (1884) pro parte quoad plant. cypr.]

[*N. rigidum* (non (Sw.) Desv.) Boiss., Fl. Orient., 5: 738 (1884) pro parte quoad plant. cypr.]

[*Dryopteris rigida* (non A. Gray) Rechinger f. in Arkiv för Bot., ser. 2, 1: 417 (1950).]

[*D. villarii* (non (Bellardi) Woynar ex Schinz et Thell.) Zohary, Fl. Palaest., 12, t. 13 (1966); Osorio-Tafall et Seraphim, List Vasc. Plants Cyprus, 5 (1973).]

[*D. filix-mas* (non (L.) Schott) Osorio-Tafall et Seraphim, List Vasc. Plants Cyprus, 5 (1973).]

TYPE: Lebanon; "in regione inferiore prope Brummana, 700–800 m" July, 1897, *J. Bornmüller 1764* (B ? K !).

Sori marginal; otherwise as *D. pallida* (Bory) Fomin ssp. *pallida*.

HAB.: Moist, shaded mountainsides, often under Pines; 800–4,500 ft. alt.; fr. May–Nov.

DISTR.: Divisions 2, 7. S.W. Asia.

2. Phini, 1862, *Kotschy*; Prodhromos, 1862, *Kotschy* 884; Kykko Monastery, 1880–1939, *Sintenis & Rigo* 693 ! *Holmboe* 1013; *Haradjian* 968 ! *Lindberg f.* s.n. ! Stavros tis Psokas, 1928–1962, *Druce* s.n.; *Lindberg f.* s.n. ! *Meikle* 2761 ! Kalogeros, 1937, *Kennedy* 36 ! Milikouri, 1939, *Lindberg f.* s.n.; Kambos, 1939, *Lindberg f.* s.n.; Lagoudhera, 1955, *Merton* 1964A ! Zakharou, 1979, *Edmondson & McClintock* E 2802 !

7. Pentadaktylos, 1862, *Kotschy* 375 ! Bellapais, 1880, *Sintenis & Rigo* s.n. ! and, 1939, *Lindberg f.* s.n.; Lapithos, 1905, *Holmboe* 838; Kyrenia, 1927–1955, *Rev. A. Huddle* 121 ! *C. B. Ussher* 59 ! *Kennedy* 36 ! *Casey* 65 ! *Miss Mapple* 48 ! Ayios Amvrosios, 1937, *Syngrassides* 1443 ! St. Hilarion, 1939, *Lindberg f.* s.n.; and, 1956, *G. E. Atherton* 1345 ! Trypa Vouno, 1956, *G. E. Atherton* 1384 ! Sina Oros, 1967, *Merton* in ARI 750 !

111h. ASPLENIACEAE

Mett. ex Frank in Leunis, Syn. Pflanzenk., ed. 2, 3: 1465 (1877).

Terrestrial (or sometimes epiphytic outside our area); rhizome-scales clathrate; fronds not jointed to rhizome; veins free or anastomosing; stipe with 2 vascular strands often forming a single x-shaped strand; sori superficial (not marginal or sub-marginal in Cyprus species), ± linear, on one or both sides of the fertile vein, usually with an indusium, sometimes exindusiate; spores bilateral.

Five to eight genera; with a cosmopolitan distribution.

11. ASPLENIUM *L.*

Sp. Plant., ed. 1, 1078 (1753).
Gen. Plant., ed. 5, 485 (1754).
Ceterach DC. in Lam. et DC., Fl. France, 2: 566 (1805).

Terrestrial (or sometimes epiphytic outside our area); rhizome dictyostelic, erect or long-creeping; stipe with 2 vascular strands at base, uniting above to give a single x-shaped strand; fronds monomorphic, undivided to several times pinnate (sometimes proliferous but not in Cyprus), venation usually free, sometimes anastomosing; sori oval to linear, elongate, usually along only one side of the fertile veinlets, oblique to the midrib of the pinna or pinnule, the indusium usually opening inwards towards the midrib; annulus of 20–28 thickened cells; spores bilateral.

About 650 species; with a cosmopolitan distribution.

Fronds 2–3-pinnate:
Lower pinnae acute to obtuse at apex, not attenuate - - - **3. A. adiantum-nigrum**
Lower pinnae attenuate at apex - - - - - - - **4. A. onopteris**
Fronds 1-pinnate:
Undersurface of lamina densely covered with scales - - - - **5. A. ceterach**
Undersurface of lamina without a dense covering of scales:
Rhachis blackish - - - - - - - - - - **1. A. trichomanes**
Rhachis green - - - - - - - - - - **2. A. viride**

1. **A. trichomanes** *L.*, Sp. Plant., ed. 1, 1080 (1753); Boiss., Fl. Orient., 5: 731 (1884); Holmboe, Veg. Cypr., 26 (1914); Post, Fl. Pal., ed. 2, 2: 811 (1933); Davis, Fl. Turkey, 1: 48 (1965); Osorio-Tafall et Seraphim, List Vasc. Plants Cyprus, 2 (1973).

Perennial; rhizome ± erect, sometimes branched and forming small clumps; rhizome-scales linear-lanceolate, long-acuminate and frequently crisped at apex, dark brown with pale brown margins, clathrate; stipe 1–8 cm. long, ¼ to ⅙ as long as lamina, adaxially flattened, the angles with a narrow, membranous, brown wing, dark shining chestnut- to purplish-brown, glabrous; lamina 4–21 cm. long, 1–2 cm. wide, linear in outline, membranous, 1-pinnate, the pinnae broadly ovate or suborbicular at base of frond, commonly oblong above, obtuse at apex, shallowly crenate, particularly at apex and acroscopic margin, eventually deciduous; rhachis concolorous with stipe and winged like it, glabrous; sori linear, indusiate, the indusium entire or crenate, sometimes irregularly so.

ssp. **quadrivalens** *D. E. Meyer*, Ber. Deutsch. Bot. Ges., 74: 456 (1962).
TYPE: Germany; "Kienburg bei Ruhpolding in Bayern, leg. D. E. Meyer, 22–24, 7. 1957" (B).

Pinnae often rather crowded, mostly opposite, transversely (not obliquely) inserted, almost sessile, symmetrical, oblong, rarely auriculate, up to 11 mm. long, convex or sometimes flat with unrolled margins; sori rather numerous, 4–9 (–12), up to 3 mm. long; stipe thick, often dark brown or blackish. [Tetraploid (n = 72).]

HAB.: Crevices of igneous rocks by streams and rivers; 3,500–4,000 ft. alt.; April–Dec.

DISTR.: Division 2. Northern and southern hemispheres.

2. Makheras, below monastery, 1905, *Holmboe* 415; Kryos Potamos, 4,000 ft. alt., 1936, *Kennedy* 19! Palekhori, 1940, *Davis* 2019! Yironas River, Madhari, 4,000 ft. alt., 1962, *Meikle* 2834!

2. **A. viride** *Hudson*, Fl. Angl., ed. 1, 385 (1762); Unger et Kotschy, Die Insel Cypern, 175 (1865); Boiss., Fl. Orient., 5: 731 (1884); H. Stuart Thompson in Journ. Bot., 44: 341 (1906); Holmboe, Veg. Cypr., 26 (1914); Davis, Fl. Turkey, 1: 49 (1965); Osorio-Tafall et Seraphim, List Vasc. Plants Cyprus, 2 (1973).
TYPE: England; "*in rupibus humidis in comitatibus* Eboracensi *et* Westmorlandico, *passim*" (? lost).

Perennial; rhizome erect to short-creeping; rhizome-scales linear-lanceolate, long-acuminate and frequently crisped at apex, dark brown, clathrate, shining; stipe (0·7–) 1–4 (–8) cm. long, (⅐–) ⅓–½ (–1) times as long as the lamina, adaxially grooved, dark chestnut- to purplish-brown basally, shading through chestnut-brown, to green distally, covered when young with scales similar to those of the rhizome, but smaller, which intergrade with multicellular hairs, both scales and hairs being mostly shed with age except at the base of the stipe; lamina (1–) 3–7 (–8) cm. long, (0·4–) 0·5–0·9 (–1·1) cm. wide, linear in outline, membranous, 1-pinnate, the pinnae flabellate, ovate or broadly ovate, obtuse at apex, with shallow obtuse lobes; rhachis concolorous with lamina, adaxially grooved, sometimes with small scales and multicellular hairs similar to those of the stipe, the hairs occasionally present also on the undersurface of the pinnae; sori linear, indusiate, the indusium entire or irregularly toothed.

HAB.: Shady ravines at high altitudes; c. 5,000–6,400 ft. alt.

DISTR.: Division 2, very rare. Widely distributed in N. temperate regions.

2. "An schattigen Stellen in Schluchten des Troodos, Mai, [1862, *Kotschy*] n. 864".

NOTES: Not seen since, and possibly a misidentification of *A. trichomanes* L., which Kotschy does not record. *A. viride* is normally found in fissures of limestone rocks, and is an unlikely occurrence on the igneous Troödos range. The description has been drawn up from Turkish material.

3. A. adiantum-nigrum *L.*, Sp. Plant., ed. 1, 1081 (1753); Kotschy in Oesterr. Bot. Zeitschr., 12: 278 (1862); Unger et Kotschy, Die Insel Cypern, 175 (1865); Boiss., Fl. Orient., 5: 784 (1884); Post, Fl. Pal., ed. 2, 2: 812 (1933); Lindberg f., Iter Cypr., 5 (1946); Davis, Fl. Turkey, 1: 50 (1965); Zohary, Fl. Palaest., 1: 10, t. 10 (1966); Osorio-Tafall et Seraphim, List Vasc. Plants Cyprus, 2 (1973).

    A. adiantum-nigrum L. ssp. *nigrum* Heufler in Verh. Zool-bot. Ver. Wien, 6: 310 (1856); Holmboe, Veg. Cypr., 26 (1914) nom. illeg.

TYPE: "*in* Europa *australiore*" (LINN).

Perennial; rhizome short-creeping, occasionally erect, sometimes branched and forming small clumps; rhizome-scales linear-lanceolate, long-acuminate and frequently crisped at apex, dark brown, clathrate, often shining; stipe (2–) 4–12 (–19) cm. long, ± as long as lamina or occasionally up to 1½ the lamina length, adaxially grooved, sometimes dark chestnut- to purplish-brown throughout, sometimes dark chestnut- to purplish-brown at base, chestnut towards the middle and green at the apex, covered when young with scales similar to those of the rhizome but smaller, which intergrade with multicellular hairs, both scales and hairs being shed with age except at the stipe base; lamina (2–) 4–13 cm. long, (1–) 2–6 cm. wide, broadly to narrowly triangular or ovate in outline, coriaceous, rhachis concolorous with lamina or brown basally and grooved adaxially, 2–3-pinnate, the pinnae acute to obtuse at apex, lobed, the lobes with acute to acuminate, rarely obtuse, apices, the rhachis, main axes of pinnae and pinnules, and occasionally both lamina surfaces, sometimes with dark multicellular hairs like those on the stipe; sori linear, indusiate, the indusium with an entire or erose margin; spores dark brown, oblong in outline, (33–) 37–53 (–62) μm. long including perispore.

HAB.: Crevices of igneous rocks by mountain streams at high altitudes; 4,000–6,200 ft. alt.; fr. March–Nov.

DISTR.: Division 2. North Temperate region; mountains of Tropical Africa, Mascarenes, S. Africa; Hawaii.

2. Confined to the upper slopes of Khionistra: Asprokremnos, Kryos Potamos, Xerokolymbos, etc. *Kennedy* 9–17! *Lindberg f.* s.n.! *Davis* 1862! 1938! *Meikle* 2784! etc.

NOTES: Holmboe's record from St. Hilarion (Veg. Cypr., 26 (1914) — as *A. adiantum-nigrum* L. ssp. *nigrum* f. *lancifolium* Heufl.) is almost certainly referable to *A. onopteris* L., which has been collected in this locality. The most common form of *A. adiantum-nigrum* in Cyprus is that on serpentine rocks, which approaches *A. cuneifolium* Viv. in frond morphology. The spores of all Cyprus material examined are similar in size, those of the serpentine plants not being of the smaller size expected of the diploid *A. cuneifolium*. Although central and eastern European plants of *A. cuneifolium* are diploid, a tetraploid of this species is known from Corsica, and cytological study of the Cyprus serpentine populations is desirable to establish whether they are *A. adiantum-nigrum* (as treated here by analogy with morphologically similar British serpentine forms of *A. adiantum-nigrum*, which have been studied cytologically) or tetraploid *A. cuneifolium*.

4. A. onopteris *L.*, Sp. Plant., ed. 1, 1081 (1753); Davis, Fl. Turkey, 1: 50 (1965); Osorio-Tafall et Seraphim, List Vasc. Plants Cyprus, 2 (1973).

A. adiantum-nigrum L. ssp. *onopteris* (L.) Heufler, Verh. Zool.-bot. Ver. Wien, 6: 310 (1856); Holmboe, Veg. Cypr., 26 (1914); Lindberg f., Iter Cypr., 5 (1946); Zohary, Fl. Palaest., 1: 10 (1966).

A. adiantum-nigrum L. var. *onopteris* (L.) Fiori, Nuov. Fl. Anal. Ital., 1: 27 (1923); Post, Fl. Pal., ed. 2, 2: 812 (1933).

TYPE: "*in* Italia, Gallia".

Perennial; rhizome erect to short-creeping; rhizome-scales linear-lanceolate, long-acuminate and frequently crisped at apex, dark brown, clathrate, often shining; stipe (7–) 10–25 (–30) cm. long, ± as long as lamina, occasionally up to 1½ times the lamina length and rarely twice the lamina length, adaxially grooved, dark chestnut- to purplish-brown throughout or sometimes pale chestnut-brown or green towards the apex, covered when young with scales similar to those of the rhizome, but smaller, which intergrade with multicellular hairs, both scales and hairs being mostly shed with age except at the base of the stipe; lamina (7–) 10–20 (–24) cm. long, (4–) 5–13 (–16) cm. wide, broadly to narrowly triangular to ovate in outline, rather coriaceous, rhachis concolorous with lamina or brown basally and grooved adaxially, 3-pinnate to 4-pinnatifid, the apices of at least the lower pinnae attenuate (except in very small fronds which have the lower pinnae acute at apices), the ultimate segments usually acute to acuminate at apex, lobed, the lobes with acuminate apices, the rhachis, main axes of pinnae and pinnules and veins on the lower surface occasionally with dark multicellular hairs like those on the stipe; sori linear, (25–) 31–44 (–48) mm. long including perispore.

HAB.: Shady crevices of calcareous or igneous rocks, often near streams; 1,800–4,500 ft. alt.; fr. Febr.–Nov.

DISTR.: Divisions 2, 7. S.W. Europe (N. to Ireland), Mediterranean region.

2. Near Phini, 1862, *Kotschy* 737! Kykko, 1880–1939, *Sintenis & Rigo* 694! *Holmboe* 1018; *Haradjian* 974! *Lindberg f.* s.n. Mesapotamos, 1937, *Kennedy* 15! Kokkinovouno Ravine, Platres, 1937, *Kennedy* 18! Saïttas, 1962, *Meikle* 2789! Between Pano Panayia and Peravasa, 1977, *J. J. Wood* 121!

7. St. Hilarion, 1937, *Syngrassides* 1741! also, 1961, *Polunin* 6600! Between Bellapais and Buffavento, 1941, *Davis* 2178! 3½ miles W. of Kantara, 1952, *Merton* 769! N. side Sina Oros, 1967, *Merton* in ARI 787!

5. A. ceterach *L.*, Sp. Plant., ed. 1, 1080 (1753).

Ceterach officinarum DC. in Lam. et DC., Fl. France, 2: 566 (1805); Sintenis in Oesterr. Bot. Zeitschr., 31: 229, 393 (1881); Boiss., Fl. Orient., 5: 722 (1884); H. Stuart Thompson in Journ. Bot., 44: 34 (1906); Holmboe, Veg. Cypr., 26 (1914); Post, Fl. Pal., ed. 2, 2: 807 (1933); Lindberg f., Iter Cypr., 5 (1946); Davis, Fl. Turkey, 1: 52 (1965); Zohary, Fl. Palaest., 1: 11, t. 11 (1966); Osorio-Tafall et Seraphim, List Vasc. Plants Cyprus, 2 (1973).

TYPE: "*in* Walliae, Italiae *fissuris rupium*".

Perennial; rhizome ± erect, sometimes branched and forming small clumps; rhizome-scales lanceolate to linear-lanceolate, acuminate at apex, dark brown with a narrow margin of smaller pale brown cells, clathrate; stipe 1–4 (–6) cm., ± as long as lamina to ⅓ as long as lamina, ± densely covered with scales, those at the base of the stipe similar to those of the rhizome but lanceolate, mixed with and grading into ovate to ovate-lanceolate pale brown shining clathrate scales, the dark scales most frequent basally and the pale scales most frequent apically; lamina (2–) 4–9 (–11) cm. long, 1–2 cm. wide, linear-elliptic to linear-oblanceolate, rarely lanceolate, in outline, coriaceous, pinnate to deeply pinnatifid, pinnae or lobes obtuse at apex, the undersurface of the lamina and rhachis densely covered with pale brown scales like those on the stipe, which are less common on the upper surface of the rhachis; sori linear, with indusium absent or rudimentary, the indusium with an entire margin.

HAB.: Crevices of calcareous or igneous rocks; 300–5,500 ft. alt.; fr. Jan.–Dec.

DISTR.: Divisions 1–3, 7, 8. W. Europe, Mediterranean region east to Himalayas.

1. Cliffs at Smyies, 1962, *Meikle* 2186!
2. Marathasa valley, 1905, *Holmboe*; Platres, 1914–1977, *Feilden* s.n.! *Kennedy* 20! *Davis* 2076! *P. Laukkonen* 129! *J. J. Wood* 119! 3 miles from Lefka, 1927, *Rev. A. Huddle* 116! Milikouri, 1939, *Lindberg f.* s.n.; Palekhori, 1940, *Davis* 2017! Kionia, 1967, *Merton* in ARI 504! Above Askas, 1979, *Edmondson & McClintock* E 2910!
3. Stavrovouni, 1880, *Sintenis*; Dhoros, 1940, *Davis* 2073!
7. Pentadaktylos, 1880, *Sintenis*; Kyrenia, 1927, *Rev. A. Huddle* 117! also, 1956, *G. E. Atherton* 1281! St. Hilarion, 1929, *C. B. Ussher* 58! and, 1961, *Polunin* 6602! Ayios Epiktitos, 1936. Syngrassides 1432! Vasilia, 1940, *Davis* 1998! Larnaka tis Lapithou, 1941, *Davis* 2090! Above Bellapais, 1948, *Casey* 195! Above Antiphonitis Monastery, 1955, *Merton* 2277! Aghirda, 1955, *G. E. Atherton* 777!
8. Ephtakomi, 1905, *Holmboe* 522.

111i. POLYPODIACEAE

Berchtold et J. S. Presl, Prirozen Rostl.; 1: 272 (1820).
Diels in Engl. et Prantl, Pflanzenfam., 1 (4): 139–339 (1899–1900).
Reimers in Engl., Syllabus, ed. 12, 1: 300–307 (1954).

Epiphytic or terrestrial; rhizome-scales opaque or clathrate; fronds usually jointed to rhizome; veins free or anastomosing; sori on undersurface of lamina, without indusium, usually discrete; spores bilateral.

About 50 genera; cosmopolitan.

12. POLYPODIUM *L*.

Sp. Plant., ed. 1, 1082 (1753).
Gen Plant., ed. 5, 485 (1754).

Epiphytic or terrestrial; rhizome dictyostelic, creeping. Stipe with 1–3 major vascular strands; fronds not dimorphic, pinnatifid to pinnate (rarely more divided); veins free or anastomosing to form regular areoles each with one central excurrent included veinlet; sori terminal or nearly so on the lowest acroscopic veinlet, forming usually one row on each side of the midrib; annulus of 4–18 thickened cells; spores bilateral.

About 75 species; mainly northern hemisphere, especially tropical America.

1. **P. cambricum** *L.*, Sp. Plant., ed. 1, 1086 (1753).
    P. australe Fée, Mém. Fam. Foug., 5: 236 (1852); Davis, Fl. Turkey, 1: 62 (1956); Osorio-Tafall et Seraphim, List Vasc. Plants Cyprus, 3 (1973).
    [*P. vulgare* (non L.) Kotschy in Unger et Kotschy, Die Insel Cypern, 174 (1865); Sintenis in Oesterr. Bot. Zeitschr., 32: 194 (1882); Boiss., Fl. Orient., 5: 723 (1884); Holmboe, Veg. Cypr., 27 (1914); Post, Fl. Pal., ed. 2, 2: 808 (1933); Rechinger f. in Arkiv för Bot., ser. 2, 1: 417 (1950); Zohary, Fl. Palaest., 1: 13, t. 14 (1966); Osorio-Tafall et Seraphim, List Vasc. Plants Cyprus, 3 (1973).]
    [*P. vulgare* L. var. *brevipes* (vix *P. vulgare* L. f. *brevipes* Milde) Druce in Rep. B.E.C., 9: 47 (1931).]
    TYPE: "*in* Anglia".

Perennial; rhizome short- to long-creeping, sometimes branched, with a chalky covering; rhizome-scales narrowly lanceolate to ovate-lanceolate, toothed, long-acuminate and often crisped at apex, chestnut-brown to medium or dark brown; stipe 1–14 cm. long, ± as long as to ⅓ as long as the lamina, adaxially grooved, straw-coloured to greenish, with occasional scales similar to those of the rhizome or linear-lanceolate; lamina 2·5–19 cm. long, 2·5–11 cm. wide, triangular to narrowly ovate in outline, slightly coriaceous, 1-pinnate to very deeply pinnately lobed, the pinnae or lobes obtuse to acute at apex, serrate or crenate, the teeth not usually obvious in small specimens except towards the pinna or lobe apex, the rhachis or midvein of the lamina sometimes with occasional scales like those of the stipe on the undersurface; sori orbicular to broadly elliptic or broadly oblong, without an indusium but with branched paraphyses, in one row on each side of the pinna or lobe midvein and nearer to it than to the margin; indurated cells of annulus (4–) 5–7 (–9).

HAB.: In crevices of limestone or igneous rocks and cliffs; 1,000–2,500 ft. alt.; fr. March–May.

DISTR.: Divisions 1, 2, 7, 8. W. Europe, Mediterranean region.

1. Smyies, 1948, *Kennedy* 1603! Yiolou, above Baths of Aphrodite, 1962, *Meikle* 2229! Between Inia and Smyies, 1979, *Edmondson & McClintock* E 2753!
2. Between Pano Panayia and Peravasa, 1977, *J. J. Wood* 120!
7. Pentadaktylos, 1862, *Kotschy* 338a; also, 1928 or 1930, *Druce* s.n.; Kantara, 1880, *Sintenis & Rigo* 986! also, 1905, *Holmboe* 556; St. Hilarion, 1937, *Syngrassides* 1742! Vasilia, 1940, *Davis* 1999! 3¼ miles W. of Kantara, 1952, *Merton* 769! Sina Oros, 1970, *A. Hansen* 402!
8. Komi Kebir, 1912, *Haradjian* 303.

APPENDIX IV

COLLECTORS CITED IN VOLS. 1 & 2 OF THE FLORA

	Collected
Antoniades	1939
Arnold, Mrs. N.	1981
Atherton, Miss G. E.	1955–1956
Aucher-Eloy, Pierre Martin Rémi (1793–1838)	1831
Austin-Harrison	1953
Baker, E. W.	1955
Baker, Sir Samuel White (1821–1893)	1879
Ball, John (1818–1889)	1877
Barbey, Caroline (1847–1918)	1880
Barbey, William (1842–1914)	1880
Barclay, Sir Colville (b. 1913)	1968
Barrington, F. J. F.	1950
Bastin, Maj. Gen. G. E. R.	1956
Bauer, Ferdinand Lucas (1760–1826)	1787
Bayer, M.	1980
Biddulph, Sir Robert (1835–1918)	1882
Bovill, Alfred Karslake	1882–1920
Breiner, R.	1979
Butcher, Mrs. A. A.	1982
Campbell, H.	1928
Carletti, Dr.	? 1862
Casey, Edwards C.	1947–1958
Casey, Mercy M.	1947–1958
Chapman, Mrs. E. F.	1933–1955
Chapman, G. W.	1933–1955
Chatterton, Mrs. I.	1967
Chesterman, D. D.	1977, 1979
Chicken, Eric	1970, 1973
Chiotellis, N.	1955
Chrtek, Jindřich (b. 1930)	1978
Clarke, Edward Daniel (1769–1822)	1801
Cowper, S. G.	1948
Croker, Mrs. F.	1880
Davidson, D. F.	1948
Davies, Paul	1979
Davis, Peter Hadland (b. 1918)	1940–1941
Dawe, Morley Thomas (1880–1943)	1931
Deflers, Albert (1841–1921)	1884
De Langhe, J. E.	1980
Della (née Genneou) Mrs. A.	1976
Deschamps, E.	1892–1894
Dickson, Mrs. H. R. P.	1951
Dray, Mrs. M. E.	1936
Druce, George Claridge (1850–1932)	1928, 1930
Drummond, Alexander	1745, 1750
Duke-Woolley, H.	1948
Economides	1966–1970
Edmondson, J. R. (b. 1948)	1979
Evenari, M.	1944
Faik, N.	1948

Fairchild, David	1930
Feilden, G. T.	1914
Fell, Adrienne	1982
Foggie, A.	1933
Frangos, G.	1923
Gaudry, Jean Albert (1827–1908)	1853, 1854
Gennadius, Panagiotes G.	1895–1901
Genneou, A.	1969–1970
Georgiadou, Elizabeth	1977–1980
Giuseppi, Paul Leon (d. 1947)	1932
Godfrey, G.	1982
Godman, Miss C. Edith	1937
Gordon, W. A.	1947
Graves, Mrs.	1954
Gresham, Major C. E.	1933
Grove, Mrs.	1954
Guichard, Kenneth M.	1971
Gunnis, Mrs.	c. 1936
Hadjichristodoulou	1970
Haig, Miss Elizabeth	1934
Hanbury, Daniel (1825–1875)	1860
Hansen, Alfred (b. 1925)	1970
Haradjian, Manoog (d. ? 1914)	1912–1913
Harris, D. R.	1937
Hartmann, Ernst	1904–1905
Hasselquist, Fredrik (1722–1752)	1751
Haussknecht, Heinrich Carl (1838–1903)	1866
Hawkins, John (? 1758–1841)	1787
Hecker, Mrs. I. M.	1968
Hewer, Thomas Frederick (b. 1903)	1979, 1981
Hiotellis, N. = Chiotellis, N.	1955
Hodgkin, Eliot (1906–1973)	1971
Holmboe, Jens (1880–1943)	1905
Holub, Josef (b. 1930)	1976
Holubova, A.	1976
Hooker, Joseph Dalton (1817–1911)	1860
Houstoun, Mrs.	1922
Huddle, Rev. A.	1927
Hughes, Mrs. Jean	1955
Hume, John, *not* Sir Abraham	1801
Huovila, Olli	1965–1966
Ingram, Collingwood (1880–1982)	1948
Ioannou, Y.	1966
Joscht, Mrs. Gerda	1971–1974
Julin, Erik R.	1971
Kalteisen, M.	1978
Kennedy, Mrs. Elizabeth Warren (d. 1964)	1936–1955
Kenyon, Miss Eliza C.	1885
Kierans, M.	1969
Kotschy, Karl Georg Theodor (1813–1866)	1840, 1859, 1862
Koutsoftas, C.	1948
Kridhiotis	1938
Labillardière, Jacques-Julien Houtton de (1755–1834)	1787
Larsen, H. Breyntroj	1972
Lascelles, A. G.	1900–1902
Lascelles, Miss M. E.	1900–1902
Laukkonen, P.	1973
Lefèvre, ? Louis Victor (b. 1810)	c. 1840
Leontiades, Leontios I.	1964–1982
Lindberg, Harald (1871–1963)	1939
Loch, Margaret, Lady	1936–1952

McClintock, M. A. S.	1979
Macdonald, Sq. Ldr. and Mrs. N.	1958
Mapple, Miss Beatrice M.	1955
Mariti, Giovanni	1760
Marshall, Mrs.	1948
Mason, P. H.	1964
Matthews, Miss Ann	1965–1974
Mavromoustakis, G. A.	1939–1948, 1957
Megaw, Mrs. Elektra	1977–1982
Meikle, Robert Desmond (b. 1923)	1962, 1974, 1981
Merton, Lionel Francis Herbert (1919–1974)	1951–1957, 1967
Michaelides, H.	1948
Michaelides, Michael G.	1901–1905
Morris, Hubert Meredydd (d. 1978)	1930–1951, 1974
Mouillefert, Pierre (1846–1903)	1893
Nattrass, Roland Marshall (b. 1895)	1935
Nicolaides, George	1961, 1962
Noel, Lady	1936
Norman, Cecil (1872–1947)	1930
Osorio-Tafall, Bibiano F. (b. 1903)	1967–1974
Oswald, Philip	1959, 1960
Painter, H.	1957
Pallis, M.	1954
Paolides, P.	1954–1955
Papadakis	1964
Papaiacovou, M.	1930–1936
Papaioannou, P.	1938
Pasche, E.	1974
Paton, D. N.	1967
Pavlides, P.	1954–1955
Payatsos, K.	1948
Phillips, Mrs. G.	1981
Pichler, Thomas (1828–1903)	1889
Pierides, Luke	1905
Pococke, Rev. Richard (1704–1765)	1738
Polunin, Oleg Vladimirovitsch	1961
Poole, D.	1969
Poore, Martin Edward Duncan (b. 1925)	1956
Post, George Edward (1838–1909)	1894–1898
Price, William Robert (1886–1975)	1972
Probyn, Mrs. F. M.	1947–1952
Reid, Clement (1853–1916)	1908
Renz, Jany (b. 1907)	1929
Rigo, Gregorio (1841–1922)	1880
Rogers, Rev. Frederick Arundel (1876–1944)	1930
Roger-Smith, H.	1936–1937
Samaritani	1859
Samson, Miss E. A.	1904
Scott-Moncrieff, Mrs. Gundred	1950, 1954, 1955
Seligman, Richard Joseph Simon	1954
Seraphim, George M.	1960–1974
Sestini, Domenico (1750–1832)	1782
Shaw, C. E.	1956
Sibthorp, John (1758–1796)	1787
Sintenis, Paul Ernst Emil (1847–1907)	1880
Slavik, B.	1978
Soteriadou, Mrs. A. C.	1951–1974
Sparrow, Brigadier C. H. E.	1959
Spitzenberger	1973
Spooner, Mrs. B.	1982
Spyrou, J.	1948

Stagg, Mrs.	1937
Suart, J. B.	1963, 1964
Synge, Patrick Millington (1910–1982)	1972
Syngrassides, Anastasios	1932–1939
Tennant, David	1982
Topali, Mlle. S. P.	1931
Townsend, Clifford Charles (b. 1926)	1971
Tracey, J. A.	1931
Trenbath, Brian Rahr	1963
Tsiakouris, S.	1937
Unger, Franz Joseph Andreas Nicolaus (1800–1870)	1862
Ussher, C. B.	1929–1930
Voskarides, J.	1948
Wall, Erik (1871–1959)	1930
Waterer, Ralph Ronald (1902–1971)	1929–1951
White, Roger St.	1979
Wieringen, G. J. van	1981
Wood, Jeffrey J. (b. 1952)	1977
Wyatt, C. H.	1937, 1938
Yakoumi, V.	1948
Young, Donald Peter (1917–1972)	1961, 1963

Collectors' numbers are usually cited immediately after their names, except where the number is an institutional one, as with the Agricultural Research Institute (near Nicosia); here the specimens are cited as, for example, "*Economides* in ARI 960". The letters "K", "G", etc. following specimen citations are herbarium abbreviations listed in *Index Herbariorum* Part I, ed. 6, compiled by P. K. Holmgren and W. Keuken, and published in *Regnum Vegetabile* vol. 92: 303–354 (1974).

APPENDIX V

LIST OF NEW NAMES, NEW COMBINATIONS, ETC. INCLUDED IN VOL. 2

Page 838   **Centranthus calcitrapa** (*L.*) *Dufr.* ssp. **orbiculatus** (*Sm.*) *Meikle* **stat. nov.** BASIONYM: *Valeriana orbiculata* Sm. in Sibth. et Sm., Fl. Graec. Prodr., 1: 21 (1806).

Page 869   **Bellis annua** *L.* ssp. **minuta** (*DC.*) *Meikle* **stat. nov.** BASIONYM: *B. annua* L. var. *minuta* DC., Prodr., 5: 304 (1836).

Page 890   **Inula conyzae** (*Griesselich*) *Meikle* **comb. nov.** BASIONYM: *Aster conyzae* Griesselich, Kleine Bot. Schrift., 122 (before July 1836).

Page 949   **Carduus argentatus** *L.* ssp. **acicularis** (*Bert.*) *Meikle* **stat. nov.** BASIONYM: *C. acicularis* Bert. in Ann. Stor. Nat. Bologna, 1: 274 (1829).

Page 971   **Centaurea calcitrapa** *L.* ssp. **angusticeps** (*Lindberg f.*) *Meikle* **stat. nov.** BASIONYM: *C. angusticeps* Lindberg f. in Act. Soc. Sci. Fenn., n.s. B., 2: 34 (1946).

Page 1055   **Legousia falcata** (*Ten.*) *Fritsch ex Janchen* var. **scabra** (*Lowe*) *Meikle* **comb. nov.** BASIONYM: *Prismatocarpus scaber* Lowe in Trans. Cambridge Phil. Soc., 6 (3): 16 (1838).

Page 1070   **Limonium ocymifolium** (*Poir.*) *O. Kuntze* ssp. **bellidifolium** (*Sm.*) *Meikle* **comb. nov.** BASIONYM: *Statice bellidifolia* Sm. in Sibth. et Sm., Fl. Graec. Prodr., 1: 211 (1806).

Page 1242   **Orobanche minor** *Sm.* var. **pubescens** (*Urv.*) *Meikle* **stat. nov.** BASIONYM: *O. pubescens* Urv. in Mém. Soc. Linn. Par., 1: 332 (1822).

Page 1249   **Phyla filiformis** (*Schrader*) *Meikle* **comb. nov.** BASIONYM: *Lippia filiformis* Schrader, Ind. Sem. Hort. Gotting. (1834) et in Ann. Sci. Nat., ser. 2, 6: 99 (1836).

Page 1260   **Mentha longifolia** (*L.*) *L.* ssp. **cyprica** (*H. Braun*) *R. Harley* **stat. nov.** BASIONYM: *M. cyprica* H. Braun in Verhl. Zool. Bot. Ges. Wien, 39: 217 (1889).

Page 1281   **Acinos exiguus** (*Sm.*) *Meikle* **comb. nov.** BASIONYM: *Thymus exiguus* Sm. in Sibth. et Sm., Fl. Graec. Prodr., 1: 421 (1809).

Page 1294   **Salvia veneris** *Hedge* **nom. nov.** REPLACED SYNONYM: *S. crassifolia* Sm. in Sibth. et Sm., Fl. Graec. Prodr., 1: 17 (1806) non Jacq. (1800–1809) nom. illeg.

Page 1427   **Thymelaea passerina** (*L.*) *Coss. et Germ.* ssp. **pubescens** (*Guss.*) *Meikle* **comb. nov.** BASIONYM: *Stellera pubescens* Guss., Fl. Sic. Prodr., 1: 466 (1827).

Page 1479   **Quercus infectoria** *Olivier* ssp. **veneris** (*A. Kerner*) *Meikle* **comb. nov.** BASIONYM: *Q. veneris* A. Kerner in Schneider, Ill. Handb. Laubholzk., 1: 191 (1904).

Page 1769   **Lophochloa amblyantha** (*Boiss.*) *Bor* **comb. et stat. nov.** BASIONYM: *Koeleria phleoides* (Vill.) Pers. var. *amblyantha* Boiss., Diagn., 2, 4: 134 (1859).

Page 1818   **Agropyron haifense** (*Meld.*) *Bor* **comb. et stat. nov.** BASIONYM: *Elytrigia elongata* (Host) Nevski var. *haifensis* Meld. in Arkiv för Bot., ser. 2, 2: 304 (1952).

Page 1747   **Lindbergella sintenisii** (*Lindberg f.*) *Bor* [in Svensk Bot. Tidskr., 63: 369 (1969) without direct reference to basionym] **comb. nov.** BASIONYM: *Poa sintenisii* Lindberg f. in Årsbok Soc. Sci. Fenn., 20B (7): 5 (1942), Act. Soc. Sci. Fenn., n.s. B, 2 (7): 8 (1946).

NAMES OF NEW TAXA

To ensure valid publication under the Code, the following names are here accompanied by brief Latin diagnoses and type-citations. Detailed descriptions should have appeared in *Annales Musei Goulandris* in spring of 1983, but at the date of writing (Dec. 1983) are still awaited. [Published in *Ann. Mus. Goulandris*, 6: 87-94, 29 Dec., 1983.]

Page 913　　**Anthemis plutonia** *Meikle* **sp. nov.** *A. tricolori* Boiss. affinis, sed foliis acute bipinnatisectis, achaeniis magnis, conspicue coronatis, indumento densius sericeo-piloso vel lanuginoso facile distinguitur.
　　　　　　TYPUS: Cyprus; Platres, 1,200 m. alt., 7 April, 1937, *Kennedy* 950 (holo-K).

Page 934　　**Senecio glaucus** *L.* **ssp. cyprius** *Meikle* **ssp. nov.** a *S. glauco* ssp. *glauco* foliis grosse dentato-lobatis; involucro maturitate ad basin indurato, calyculo deficiente, satis differt.
　　　　　　TYPUS: Cyprus; Xeros, on Kafkalla by the sea; 19 March, 1970, *A. Genneou* in ARI 1379 (holo-K, iso-ARI Cyprus).

Page 1021　　**Taraxacum aphrogenes** *Meikle* **sp. nov.** Sectioni *Scariosae* Hand.-Mazz. emend. Dahlst. pertinens, forsan *T. hellenico* Dahlst. vel *T. bithynico* DC. affine sed foliis linearibus, lobulis parvis obtusis numerosis dissectis, plane differt.
　　　　　　TYPUS: Cyprus; in rock-fissures close to the sea at Kato Paphos, 30 Oct. 1981, *Meikle* 5044 (holo-K).

Page 1070　　**Limonium albidum** (*Guss.*) *Pignatti* ssp. **cyprium** *Meikle* **ssp. nov.** a subspecie typica differt bracteolis interioribus rectis (nec curvatis), floribus minoribus, vix bracteolas excedentibus, calycis tubo recto.
　　　　　　TYPUS: Cyprus; Cape Andreas, on rocks by the sea; c. 6 m. alt.; 11 Nov. 1951, *Merton* 606 (holo-K).

Page 1146　　**Lithodora hispidula** (*Sm.*) *Griseb.* ssp. **versicolor** *Meikle* **ssp. nov.** a *L. hispidula* ssp. *hispidula* foliis supra persistenter adpresse-strigosis, setulisque crassioribus, basi tuberculatis sparse indutis; inflorescentiis plerumque trifloris, corollis infundibuliformibus plerumque roseis vel salmoneo-rubris, interdum purpureis, rarissime (vel nunquam) caeruleis distinguenda.
　　　　　　TYPUS: Cyprus; "Prope Klafdea [Klavdhia]. 29 Februario 1880." *Sintenis & Rigo* 102 (holo-K).

Page 1277　　**Micromeria chionistrae** *Meikle* **sp. nov.** *M. microphyllae* (Urv.) Benth. affinis sed floribus subsessilibus verticillastros condensatos formantibus, bracteolae pedicellos perbreves valde excedentes.
　　　　　　TYPUS: Cyprus; Phini, 1,000 m. alt., in cracks of bare rock; 6 June, 1939, *Kennedy* 1495 (holo-K).

Page 1323　　**Phlomis cypria** *Post* var. **occidentalis** *Meikle* **var. nov.** a *P. cypria* Post var. *cypria* foliis oblongis subacuminatis, 3–7 cm. longis, 1–2·3 cm. latis, saepe flavido-tomentosis, ad apicem obtusum vel subacutum sensim angustatis.
　　　　　　TYPUS: Cyprus; between Lyso and Abdoulinas Junction, 360 m. alt.; 29 April, 1962, *Meikle* 2770 (holo-K).

Page 1331　　**Scutellaria cypria** *Rechinger f.* var. **elatior** *Meikle* **var. nov.** a *S. cypria* Rechinger f. var. *cypria* caulibus erectis elatioribus 20–30 cm. longis, pilis albis recurvatis, crispatis, pilis longioribus patentibus (praecipue ad caulium basin) interdum mixtis; foliis 2–3·5 cm. longis, 1–2·5 cm. latis, utrinque dense crispato-pubescentibus; inflorescentiis elongatis, laxioribus, usque 25 cm. longis differt.
　　　　　　TYPUS: Cyprus; 1 km. N.E. of Apsiou, 15 km. N. of Limassol. Macchia on loose, stony serpentine slopes; 450 m. alt.; 8 April, 1979, *J. R. Edmondson* and *M. A. S. McClintock* E 2692 (holo-K).

Page 1613　　**Allium cupani** *Rafin.* ssp. **cyprium** *Meikle* **ssp. nov.** ab *A. cupani* typico floribus parvis, perianthii-segmentis 3–4 mm. longis, 1–1·5 mm. latis spatha plerumque brevi, minus quam 1·5 cm. longa, apice cuspidata abrupte coarctata differt.
　　　　　　TYPUS: Cyprus; "In montibus inter Potami et Evriku [Evrykhou]; 14 Junio 1880", *Sintenis & Rigo* 860 (holo-K).

Page 1615　　**Allium paniculatum** *L.* ssp. **exaltatum** *Meikle* **ssp. nov.** a specie typica differt habitu tenui, spathae appendicibus brevibus vix 3 cm. attingentibus et inflorescentiam aequantibus vel paullo superantibus, pedicellis subaequalibus brevibus, plerumque minus quam 1·5 cm. longis, umbellam compactam, globosam vel ovoideam formantibus.
　　　　　　TYPUS: Cyprus; Xerokolymbos, S.W. of Khionistra, 1,480 m. alt.; 25 July 1937, *Kennedy* 181 (holo-K).

APPENDIX VI

KEY TO FAMILIES INCLUDED IN THE FLORA OF CYPRUS

The Key applies only to families as represented in Cyprus, and is not intended for general use.

Plants without flowers or seeds - - - - - - **111. Pteridophyta** (p. 1871)
Plants with flowers and seeds:
  Plants without chlorophyll, variously coloured, but not green, parasites and saprophytes
           GROUP A (p. 1899)
  Plants green, with chlorophyll:
    Flowers crowded into heads surrounded by an involucre - - GROUP B (p. 1899)
    Flowers not crowded into involucrate heads:
      Leaves reduced to small scales or absent - - - - GROUP C (p. 1899)
      Leaves (or apparent leaves) with a distinct lamina:
        Trees, shrubs and woody climbers:
          Leaves opposite or whorled - - - - - GROUP D (p. 1900)
          Leaves alternate or irregularly inserted - - - - GROUP F (p. 1902)
        Herbs:
          Leaves opposite or whorled - - - - - GROUP E (p. 1901)
          Leaves alternate or all basal:
            Leaves compound or deeply divided - - - - GROUP G (p. 1904)
            Leaves simple - - - - - - - - GROUP H (p. 1905)

GROUP A. PLANTS WITHOUT CHLOROPHYLL, VARIOUSLY COLOURED BUT NOT GREEN; PARASITES AND SAPROPHYTES

Trailing or climbing plants with thread-like stems - - - - **65. Convolvulaceae**
Erect herbs:
  Flowers in sessile or subsessile clusters; plant without any obvious stem **77. Rafflesiaceae**
  Flowers in stalked spikes or racemes:
    Flowers actinomorphic; corolla polypetalous; ovary superior - - **56. Monotropaceae**
    Flowers zygomorphic:
      Ovary superior; corolla tubular - - - - - - **68. Orobanchaceae**
      Ovary inferior; corolla of free petals - - - - - **93. Orchidaceae**

GROUP B. FLOWERS CROWDED INTO HEADS SURROUNDED BY AN INVOLUCRE

Petals free:
  Leaves opposite, simple - - - - - - - - **16. Caryophyllaceae**
  Leaves alternate, generally compound or deeply lobed:
    Ovary superior - - - - - - - - - **33. Leguminosae**
    Ovary inferior - - - - - - - - - **45. Umbelliferae**
Petals united into a tubular or ligulate corolla:
  Anthers united into a tube around the style - - - - - **53. Compositae**
  Anthers free:
    Leaves in whorls - - - - - - - - - **49. Rubiaceae**
    Leaves not in whorls:
      Ovary superior - - - - - - - - - **71. Labiatae**
      Ovary inferior - - - - - - - - - **52. Dipsacaceae**

GROUP C. LEAVES REDUCED TO SMALL SCALES, OR ABSENT

Leaves reduced to small scales adpressed to stems:
  Flowers with a coloured perianth:
    Perianth pink or pinkish - - - - - - - **19. Tamaricaceae**
    Perianth yellow - - - - - - - - - **80. Thymelaeaceae**
  Flowers without a perianth - - - - - - - **2. Cupressaceae**

Leaves absent:
  Stems transversely articulated or jointed:
    Flowers with a conspicuous, coloured corolla - - - - - - **43a. Cactaceae**
    Flowers inconspicuous, without a corolla:
      Stems succulent - - - - - - - - **74. Chenopodiaceae**
      Stems not succulent:
        Fruit fleshy; shrubs and woody climbers - - - - **3. Ephedraceae**
        Fruit not fleshy, cone-like; tall shrubs and trees - - - **88a. Casuarinaceae**
  Stems not transversely articulated or jointed:
    Stems branched, often climbing or sprawling - - - - - - **97. Liliaceae**
    Stems not branched, erect - - - - - - - - **98. Juncaceae**

GROUP D. TREES, SHRUBS AND WOODY CLIMBERS WITH UNIFORMLY OPPOSITE OR WHORLED LEAVES

Flowers inconspicuous, without any obvious perianth or coloured bracts:
  Leaves simple:
    Leaves subulate or scale-like - - - - - - - **2. Cupressaceae**
    Leaves oblong-elliptic - - - - - - - - **74. Chenopodiaceae**
  Leaves imparipinnate - - - - - - - - - **60. Oleaceae**
Flowers with a coloured perianth or corolla, or conspicuous coloured bracts:
  Bracts conspicuous, coloured; perianth inconspicuous - - - **72a. Nyctaginaceae**
  Bracts inconspicuous; perianth or corolla conspicuous:
    Petals (or coloured sepals) free:
      Stamens very numerous:
        Woody climbers, leaves with twisting petioles - - - **4. Ranunculaceae**
        Erect shrubs or trees; petioles not twisting:
          Leaves gland-dotted:
            Petals yellow - - - - - - - - **21. Guttiferae**
            Petals not yellow - - - - - - - **39. Myrtaceae**
          Leaves not gland-dotted:
            Sepals red, fleshy, persistent - - - - - **40a. Punicaceae**
            Sepals not red or fleshy - - - - - - **12. Cistaceae**
      Stamens 10 or fewer:
        Leaves fleshy - - - - - - - - **25. Zygophyllaceae**
        Leaves not fleshy:
          Sepals united into a tube - - - - - **16. Caryophyllaceae**
          Sepals free, not forming a tube - - - - - **3. Aceraceae**
    Petals united:
      Ovary inferior:
        Leaves imparipinnate - - - - - - - **48. Sambucaceae**
        Leaves simple or irregularly lobed:
          Leaves exstipulate, opposite; flowers zygomorphic - - **47. Caprifoliaceae**
          Leaves stipulate, opposite or apparently whorled; flowers actinomorphic
                                  49. Rubiaceae

      Ovary superior:
        Stamens 2:
          Flowers actinomorphic - - - - - - - **60. Oleaceae**
          Flowers zygomorphic - - - - - - - **71. Labiatae**
        Stamens 4–10:
          Leaves compound:
            Leaves palmately compound - - - - - **70. Verbenaceae**
            Leaves pinnately compound - - - - - **69a. Bignoniaceae**
          Leaves simple:
            Flowers actinomorphic:
              Stamens 8–10, twice as many as corolla-lobes - - - **55. Ericaceae**
              Stamens 4–5, as many as corolla-lobes:
                Leaves toothed - - - - - - **62a. Loganiaceae**
                Leaves entire:
                  Corolla with a central corona - - - - **62. Asclepiadaceae**
                  Corolla without a central corona - - - - **61. Apocynaceae**
            Flowers zygomorphic:
              Style gynobasic; fruit a nutlet - - - - - **71. Labiatae**
              Style terminal:
                Fruit fleshy, drupaceous - - - - - **70. Verbenaceae**
                Fruit a capsule or follicle:
                  Inflorescence spicate, conspicuously bracteate; seeds not winged
                                  69c. Acanthaceae

Inflorescence paniculate, not conspicuously bracteate; seeds winged:
Corolla white, blotched yellow; fruit linear, very elongate and narrow
             69a. Bignoniaceae
Corolla blue or mauve; fruit an ovoid, woody capsule
             67. Scrophulariaceae

GROUP E. HERBS WITH OPPOSITE OR WHORLED LEAVES

Leaves compound or lobed almost to base:
  Petals free:
    Flowers zygomorphic, spurred - - - - - - - **8. Fumariaceae**
    Flowers actinomorphic, not spurred:
      Fruit beaked - - - - - - - - **26. Geraniaceae**
      Fruit not beaked - - - - - - - - **25. Zygophyllaceae**
  Petals united into a tube:
    Inflorescence a spike - - - - - - - - **70. Verbenaceae**
    Inflorescence a cyme, corymb or umbel:
      Flowers zygomorphic, corolla-tube longer than limb - - - **51. Valerianaceae**
      Flowers actinomorphic, subrotate, corolla-tube shorter than limb **48. Sambucaceae**
Leaves simple:
  Flowers without petals or coloured perianth:
    Aquatic herbs, floating or rooting in mud at the bottom of ponds or pools
             38. Callitrichaceae
    Terrestrial herbs:
      Fruit a capsule:
        Fruit 2 (–3)-lobed, with 1 seed in each lobe - - - - **83. Euphorbiaceae**
        Fruit without lobes, containing numerous small seeds:
          Ovary 1-locular; placentation free-central - - - - **16. Caryophyllaceae**
          Ovary 3–5-locular; placentation axile - - - - - - **44. Aizoaceae**
      Fruit an indehiscent nut, drupe or achene:
        Leaves linear, subulate or narrowly oblong, without a distinct lamina and petiole
             17. Illecebraceae
        Leaves with a distinct lamina and petiole:
          Leaves glabrous or subglabrous, somewhat fleshy - - **50. Theligonaceae**
          Leaves pubescent, strigose or hispid, not fleshy - - - **84. Urticaceae**
  Flowers with petals or coloured perianth:
    Petals (or perianth-segments) free:
      Leaves dotted with translucent glands - - - - - - **21. Guttiferae**
      Leaves not gland-dotted:
        Leaves fleshy - - - - - - - - **36. Crassulaceae**
        Leaves not fleshy:
          Sepals united into a tube:
            Style 1, simple, undivided - - - - - - **40. Lythraceae**
            Styles 2 or more, free or connate towards base:
              Styles connate below - - - - - - **15. Frankeniaceae**
              Styles free to base - - - - - **16. Caryophyllaceae**
          Sepals not united into a tube:
            Style 1, simple; stamens usually numerous - - - **12. Cistaceae**
            Styles 2 or more, free or connate towards base:
              Flowers in cymes or pseudo-racemes - - - **16. Caryophyllaceae**
              Flowers solitary, axillary - - - - - **20. Elatinaceae**
    Petals (or perianth) united into a tube:
      Ovary inferior:
        Leaves in apparent whorls - - - - - - - **49. Rubiaceae**
        Leaves opposite in pairs - - - - - - - **51. Valerianaceae**
      Ovary superior or apparently superior:
        Corolla zygomorphic:
          Fruit a capsule:
            Capsule at least twice as long as wide - - - **69b. Pedaliaceae**
            Capsule not twice as long as wide - - - **67. Scrophulariaceae**
          Fruit a nutlet or berry:
            Ovary deeply 4-lobed; style usually gynobasic, corolla mostly strongly
              zygomorphic - - - - - - - - - **71. Labiatae**
            Ovary not, or shallowly 4-lobed; style terminal; corolla often weakly
              zygomorphic - - - - - - - - **70. Verbenaceae**
        Corolla actinomorphic:
          Corolla-lobes asymmetric - - - - - - - **61. Apocynaceae**
          Corolla-lobes symmetric or nearly so:

Fruit a capsule or follicle containing 2 or more seeds:
   Inflorescence a dense spike; capsule circumscissile   -   **72. Plantaginaceae**
   Inflorescence not a spike; capsule not circumscissile:
     Corolla with a central corona   -   -   -   -   **62. Asclepiadaceae**
     Corolla without a central corona:
       Fruit a follicle; seeds crowned with silky hairs   -   **61. Apocynaceae**
       Fruit a capsule:
         Stamens opposite corolla-lobes; flowers solitary, axillary
                                   58. Primulaceae
         Stamens alternating with corolla-lobes; flowers in terminal cymes
                                   63. Gentianaceae
   Fruit indehiscent, 1-seeded   -   -   -   -   -   -   **72a. Nyctaginaceae**

GROUP F. TREES, SHRUBS AND WOODY CLIMBERS WITH ALTERNATE OR IRREGULARLY INSERTED LEAVES

Leaves needle-like:
   Fruit a woody cone, with overlapping scales; flowers without a perianth   -   **1. Pinaceae**
   Fruit not a woody cone; flowers with a perianth   -   -   -   **79a. Proteaceae**
Leaves not needle-like:
   Leaves compound or lobed almost to base:
     Woody climbers -   -   -   -   -   -   -   -   -   **30. Vitaceae**
     Shrubs or trees:
       Leaves harsh, leathery or fibrous, crowded at the apex of a bare trunk   **98a. Palmae**
       Leaves not as above:
         Leaves pleasantly or unpleasantly aromatic when crushed:
           Plant armed with prickles -   -   -   -   -   -   -   **34. Rosaceae**
           Plant unarmed:
             Shrub with glandular-punctate leaves   -   -   -   -   **28. Rutaceae**
             Tree; leaves not glandular-punctate   -   -   -   **88. Juglandaceae**
         Leaves not aromatic or malodorous when crushed:
           Leaves stipulate:
             Fruit a pod -   -   -   -   -   -   -   -   **33. Leguminosae**
             Fruit indehiscent, fleshy or spongy -   -   -   -   **34. Rosaceae**
           Leaves exstipulate:
             Leaves pinnate:
               Leaves with 13–40 leaflets:
                 Leaflets ovate, coarsely toothed near base; fruit a samara
                                   28a. Simaroubaceae
                 Leaflets linear-lanceolate, entire or serrulate; fruit a drupe
                                     32. Anacardiaceae
               Leaves with fewer than 13 leaflets:
                 Stamens with filaments united into a column; fruit a drupe   **28b. Meliaceae**
                 Stamens free:
                   Fruit an inflated capsule or large, fleshy, yellow or orange drupe
                                     30a. Sapindaceae
                   Fruit a small, brownish or reddish resinous drupe   -   **32. Anacardiaceae**
             Leaves irregularly dissected, not pinnate -   -   -   -   **79a. Proteaceae**
   Leaves simple:
     Woody climbers:
       Stems prickly -   -   -   -   -   -   -   -   -   **97. Liliaceae**
       Stems not prickly:
         Leaves persistent, evergreen; plant without tendrils:
           Flowers solitary, axillary; perianth tubular -   -   -   -   **78. Aristolochiaceae**
           Flowers in umbels; perianth not tubular   -   -   -   -   **46. Araliaceae**
        · Leaves deciduous; plant with tendrils   -   -   -   -   **30. Vitaceae**
     Shrubs or trees:
       Flowers with distinct sepals and petals:
         Petals free:
           Ovary borne on a distinct stipe (or gynophore); filaments tinged mauve or pink
                                   10. Capparaceae
         Ovary sessile or subsessile:
           Leaves clothed with stellate hairs:
             Stamens connate, forming a staminal column; epicalyx generally present
                               22. Malvaceae
             Stamens not forming a staminal column; epicalyx absent   -   **23. Tiliaceae**
           Leaves with simple hairs or glabrous:
             Fruit a pod -   -   -   -   -   -   -   -   **33. Leguminosae**

Fruit fleshy or spongy, indehiscent:
 Flowers very small; petals often absent; stamens 4–5 - **29. Rhamnaceae**
 Flowers conspicuous; petals well developed; stamens usually numerous
              34. Rosaceae
Petals united:
 Corolla-tube distinctly narrowed above (urceolate or sub-urceolate):
  Flowers unisexual, solitary or in small axillary clusters - -**58a. Ebenaceae**
  Flowers hermaphrodite in branched terminal panicles - - **55. Ericaceae**
 Corolla-tube cylindrical or widened towards apex:
  Inflorescence racemose - - - - - - **59. Styracaceae**
  Inflorescence not racemose; flowers solitary or in cymes or clusters:
   Fruit a capsule or many-seeded berry - - - - **66. Solanaceae**
   Fruit a 1-seeded drupe- - - - - **64. Boraginaceae**
Flowers apetalous or with an undifferentiated calyx and corolla:
 Leaves palmately lobed:
  Fruits crowded into pedunculate, globose clusters:
   Base of petiole calyptrate, covering the bud like a cap; fruit immersed in a basal
    tuft of bristles - - - - - - - **87. Platanaceae**
   Base of petiole not calyptrate; fruit without a basal tuft of bristles
           37. Hamamelidaceae
  Fruits not in pedunculate, globose clusters:
   Fruit fleshy, indehiscent - - - - - - - **86. Moraceae**
   Fruit a capsule - - - - - - **83. Euphorbiaceae**
 Leaves not palmately lobed:
  Leaves strongly aromatic when crushed:
   Stamens very numerous, conspicuous - - - **39. Myrtaceae**
   Stamens few, inconspicuous - - - - - **79. Lauraceae**
  Leaves not (or very indistinctly) aromatic when crushed:
   Fruit a fleshy berry, drupe or cluster of drupes:
    Leaves covered with silvery, flat, medifixed scale-hairs - **81. Elaeagnaceae**
   Leaves not covered with medifixed scale-hairs:
    Branches spinose or prickly:
     Flowers bright yellow - - - - - **6. Berberidaceae**
     Flowers creamy-white - - - - - **97. Liliaceae**
    Branches not spinose or prickly:
     Leaves (or apparent leaves) rigid, pungent - - **97. Liliaceae**
     Leaves not rigid or pungent:
      Fruit a cluster of drupelets or a fleshy, hollow *syconium* with an apical
       aperture or ostiole - - - - **86. Moraceae**
      Fruit not as above:
       Perianth brightly coloured - - - **80. Thymelaeaceae**
       Perianth greenish, inconspicuous:
        Fruit bright red or orange when ripe:
         Leaves sparse, linear; flowers in short axillary racemes
             82. Santalaceae
         Leaves copious, broadly lanceolate, ovate or elliptic; flowers in
          terminal spikes or panicles - - **73. Amaranthaceae**
        Fruit dull greenish, brownish or blackish when ripe:
         Leaves coarsely toothed or lobed - - - **85. Ulmaceae**
         Leaves shortly toothed or entire - - **29. Rhamnaceae**
 Fruit a nut, capsule, samara or achene, not fleshy:
  Fruit papery, conspicuously winged:
   Leaves entire, glutinous when young; fruit often tinged pink-purple
           30a. Sapindaceae
   Leaves toothed; fruit pale green - - - - **85. Ulmaceae**
  Fruit not papery or winged:
   Flowers unisexual, the males forming deciduous catkins:
    Plants dioecious; seeds enveloped in a silky tuft of hairs - **92. Salicaceae**
    Plants monoecious; seeds without a silky tuft of hairs:
     Fruit enveloped in a spiny husk or seated in a scaly cupule
           91. Fagaceae
     Fruit without a spiny husk or scaly cupule:
      Fruit an ovoid nut, enveloped by a foliaceous involucre
          90. Corylaceae
      Fruit without a foliaceous involucre - - - **89. Betulaceae**
   Flowers hermaphrodite or, if unisexual, males not forming catkins:
    Flowers greenish, inconspicuous - - - **74. Chenopodiaceae**
    Flowers coloured, conspicuous:

Fruit woody, persistent, conspicuous -   -   -   **79a. Proteaceae**
Fruit not woody, inconspicuous, often hidden in the persistent perianth
            80. Thymelaeaceae

GROUP G.   HERBS WITH ALTERNATE OR BASAL, COMPOUND OR DEEPLY DIVIDED LEAVES

Leaves all basal:
   Flowers crowded into a dense spike -   -   -   -   -   -   **72. Plantaginaceae**
   Flowers not in a spike:
     Leaves 3-foliolate; leaflets obcordate   -   -   -   -   **27. Oxalidaceae**
     Leaves not 3-foliolate; leaflets not obcordate:
       Ovary inferior; flowers in umbels   -   -   -   -   **45. Umbelliferae**
       Ovary superior; flowers not in umbels:
         Carpels united until ripe, with a central beak -   -   -   **26. Geraniaceae**
         Carpels free, sometimes individually beaked but without a central beak
            4. Ranunculaceae
Leaves not all basal:
   Flowers with a distinct calyx and corolla:
     Petals free:
       Petals 4 or fewer:
         Stamens numerous:
           Flowers actinomorphic; stems exuding latex when broken   -   **7. Papaveraceae**
           Flowers zygomorphic; stem not exuding latex when broken -   **4. Ranunculaceae**
         Stamens 6 or fewer:
           Petals all similar, equal or subequal:
             Leaves palmately compound   -   -   -   -   **10. Capparaceae**
             Leaves pinnately or irregularly dissected or lobed   -   -   **9. Cruciferae**
           Petals in dissimilar, unequal pairs:
             Flowers spurred or saccate, in racemes or spikes   -   -   **8. Fumariaceae**
             Flowers not spurred, solitary or in cymes -   -   -   -   **7. Papaveraceae**
       Petals 5 or more:
         Flowers zygomorphic:
           Ovary superior:
             Sepals united into a basal tube; fruit a pod   -   -   **33. Leguminosae**
             Sepals free or almost free; fruit a capsule -   -   -   **11. Resedaceae**
           Ovary inferior   -   -   -   -   -   -   -   **45. Umbelliferae**
         Flowers actinomorphic:
           Leaves and stems stellate-pilose -   -   -   -   **23. Malvaceae**
           Leaves and stems not stellate-pilose:
             Leaves dotted over with translucent glands   -   -   **28. Rutaceae**
             Leaves not gland-dotted:
               Leaves trifoliolate with obcordate leaflets   -   -   **27. Oxalidaceae**
               Leaves not trifoliolate:
                 Leaves paripinnate -   -   -   -   -   **25. Zygophyllaceae**
                 Leaves not paripinnate:
                   Fruit of 2 or more free achenes or follicles (apocarpous):
                     Land plants:
                       Cauline leaves stipulate -   -   -   -   -   **34. Rosaceae**
                       Cauline leaves exstipulate:
                         Sepals unequal, grading into petals; fruit consisting of 2 or more
                           leathery or fleshy follicles   -   -   -   **5. Paeoniaceae**
                         Sepals equal or subequal, not grading into petals; fruit not leathery
                           or fleshy -   -   -   -   -   **4. Ranunculaceae**
                     Water plants; leaves with capillary divisions   -   **4. Ranunculaceae**
                   Fruit not apocarpous; carpels fused at least until maturity:
                     Fruit with a central beak -   -   -   -   **26. Geraniaceae**
                     Fruit without a central beak   -   -   -   **35. Saxifragaceae**
     Petals united into a tube:
       Plants with tendrils; fruit large, fleshy, indehiscent   -   -   **43. Cucurbitaceae**
       Plants without tendrils; fruit not as above:
         Flowers zygomorphic   -   -   -   -   -   **67. Scrophulariaceae**
         Flowers actinomorphic:
           Flowers in scorpioid cymes -   -   -   -   **63a. Hydrophyllaceae**
           Flowers solitary or paired, axillary   -   -   -   **65. Convolvulaceae**
   Flowers without a distinct calyx and corolla:
     Leaves palmate; shoots and young leaves resinous, aromatic -   **85a. Cannabaceae**
     Leaves pinnate, pinnatisect or irregularly laciniate:

Leaves imparipinnate; fruit a pendulous capsule- - - - **42. Datiscaceae**
Leaves not imparipinnate; fruit a small nut or achene:
  Perianth-segments 5; plants erect, glandular-aromatic - - **74. Chenopodiaceae**
  Perianth-segments 4; plants prostrate, not aromatic - - - **34. Rosaceae**

GROUP H. HERBS WITH SIMPLE, ALTERNATE LEAVES, OR LEAVES ALL BASAL

Flowers with a distinct calyx and corolla:
Petals free:
  Indumentum of stellate hairs:
    Petals 4, cruciately arranged - - - - - - - **9. Cruciferae**
    Petals 5 or more:
      Flowers conspicuous; fruit a schizocarp or 5-locular capsule - - **22. Malvaceae**
      Flowers inconspicuous; fruit 3-locular - - - - - **83. Euphorbiaceae**
  Indumentum of simple hairs or plants glabrous:
    Flowers actinomorphic:
      Petals 3; sepals 3; fruit apocarpous - - - - - - **102. Alismataceae**
      Petals and sepals 4 or more:
        Ovary inferior or semi-inferior:
          Flowers in simple or compound umbels - - - - **45. Umbelliferae**
          Flowers not in umbels:
            Seeds with a coma of silky hairs - - - - **41. Onagraceae**
            Seeds without a coma of silky hairs:
              Style 1,bifid, the arms united below - - - - **53. Compositae**
              Styles 2 or more:
                Styles 2, divergent, like minute horns - - - **35. Saxifragaceae**
                Styles 3 or more - - - - - - - **44. Aizoaceae**
        Ovary superior:
          Fruit apocarpous, comprising a group of achenes or follicles:
            Leaves fleshy - - - - - - - - **36. Crassulaceae**
            Leaves not fleshy - - - - - - - **4. Ranunculaceae**
          Fruits not apocarpous; carpels united at least until maturity:
            Petals and sepals 4, cruciately arranged - - - - **9. Cruciferae**
            Petals and sepals usually 5 or more:
              Fruits with a central beak - - - - - **26. Geraniaceae**
              Fruits without a central beak:
                Leaves glandular-punctate, strong-smelling when crushed **28. Rutaceae**
                Leaves not glandular-punctate or strong-smelling:
                  Fruit woody, spinose - - - - - - **34. Rosaceae**
                  Fruit not woody or spinose:
                    Fruit a circumscissile capsule - - - - **18. Portulacaceae**
                    Fruit not circumscissile:
                      Fruit dehiscing into 10, 1-seeded valves - - **24. Linaceae**
                      Fruit a 1–6-locular capsule:
                        Style simple, filiform:
                          Sepals free or almost free; stamens numerous **23. Tiliaceae**
                          Sepals united into a tube; stamens 12 or fewer
                              40. Lythraceae
                        Styles 3–5, free or partly connate - - - **44. Aizoaceae**
    Flowers zygomorphic:
      Lowermost petal spurred or saccate at base; flowers solitary - - **13. Violaceae**
      Lowermost petal not spurred or saccate at base; flowers in spikes or racemes:
        Lowermost petal with a conspicuous fimbriate apex, fruit flattened, 2-locular
                            14. Polygalaceae
        Lowermost petal without a fimbriate apex:
          Fruit a pod - - - - - - - - **33. Leguminosae**
          Fruit a capsule, open and gaping at apex - - - **11. Resedaceae**
Petals united:
  Flowers distinctly zygomorphic:
    Leaves all basal; flowers solitary:
      Leaves densely viscid-glandular above - - - - **69. Lentibulariaceae**
      Leaves not viscid-glandular - - - - - - **67. Scrophulariaceae**
    Leaves not all basal:
      Flowers solitary - - - - - - - - **54. Campanulaceae**
      Flowers in spikes, cymes, racemes or panicles:
        Fruit a nutlet; leaves strigose or bristly - - - **64. Boraginaceae**
        Fruit a capsule; leaves not strigose or bristly:
          Leaves clammy-glandular - - - - - - **66. Solanaceae**

Leaves not clammy-glandular:
  Capsule oblong-quadrangular   -   -   -   -   **69b. Pedaliaceae**
  Capsule not oblong-quadrangular   -   -   -   **67. Scrophulariaceae**
Flowers actinomorphic or almost so:
  Ovary inferior:
    Flowers unisexual; fruit fleshy   -   -   -   -   -   **43. Cucurbitaceae**
    Flowers hermaphrodite; fruit a capsule -   -   -   -   **54. Campanulaceae**
  Ovary superior:
    Styles 5, free or partly connate   -   -   -   -   **57. Plumbaginaceae**
    Style 1, simple or shortly bifid at apex:
      Corolla scarious, brownish or greenish; capsule circumscissile   **72. Plantaginaceae**
      Corolla not scarious, not brownish or greenish:
        Corolla funnel-shaped, tapering from a wide apex:
          Fruit woody, usually spinose   -   -   -   -   -   **66. Solanaceae**
          Fruit not woody or spinose -   -   -   -   -   **65. Convolvulaceae**
        Corolla not funnel-shaped:
          Corolla narrowly tubular   -   -   -   -   -   **66. Solanaceae**
          Corolla not narrowly tubular:
            Fruit a berry -   -   -   -   -   -   -   **66. Solanaceae**
            Fruit a capsule:
              Capsule circumscissile -   -   -   -   -   **66. Solanaceae**
              Capsule not circumscissile:
                Leaves hairy:
                  Flowers in spikes or racemes -   -   -   **67. Scrophulariaceae**
                  Flowers not in spikes or racemes   -   -   **65. Convolvulaceae**
                Leaves glabrous or very sparsely pubescent   -   **58. Primulaceae**
Flowers without a distinct calyx and corolla:
  Submerged or floating aquatics:
    Plants growing in the sea:
      Rhizomes covered with dense fibres from persistent, decayed leaf-sheaths
                                                106. Posidoniaceae
      Rhizomes not densely fibrous:
        Leaves in congested, terminal tufts; apex sharply serrulate; basal sheaths conspicuous
                                                108. Cymodoceaceae
        [Leaves not in terminal tufts, apex not serrulate   -   -   **107a. Zosteraceae**]
    Plants growing in fresh water or brackish pools and lakes:
      Perianth well developed, of 4 segments; flowers often numerous in pedunculate spikes
                                                104. Potamogetonaceae
      Perianth absent or much reduced:
        Fruits borne at the apex of a long, slender (often spirally twisted) peduncle
                                                105. Ruppiaceae
        Fruits sessile or shortly stipitate:
          Leaves entire or subentire; fruits keeled, usually crenate or dentate dorsally
                                                107. Zannichelliaceae
          [Leaves serrulate; fruits not keeled, not crenate or dentate dorsally
                                                105a. Najadaceae]
  Terrestrial plants, or, if aquatic, not floating or submerged:
    Inflorescence consisting of a fleshy cylindrical spadix, surrounded by a conspicuous,
      convolute, leafy spathe; leaves usually all basal, sagittate   -   -   **101. Araceae**
    Inflorescence not consisting of a spadix and spathe:
      Flowers without any obvious perianth:
        Flowers unisexual forming readily distinguishable inflorescences on the same plant
          (monoecious):
          Inflorescences dense, cylindrical, terminal on stout erect, unbranched stems;
            female inflorescences velutinous, dark brown   -   -   -   **99. Typhaceae**
          Inflorescences not as above:
            Inflorescences globose; stems terete -   -   -   -   -   **100. Sparganiaceae**
            Inflorescences not globose:
              Stems terete; female flowers enveloped by papery, spathaceous sheaths
                                                110. Gramineae
              Stems trigonous or triquetrous; female flowers not enveloped by spathaceous
                sheaths   -   -   -   -   -   -   **109. Cyperaceae**
      Flowers hermaphrodite, or, if unisexual, not forming distinct inflorescences:
        Stems mostly hollow except at the well-marked nodes, terete or subterete; leaf-
          sheaths open, separated from blade by a distinct articulation; styles 2, distinct
                                                110. Gramineae
        Stems pith-filled, without obvious nodes, often trigonous or triquetrous; leaf-
          sheath usually tubular, closed, not clearly separated from blade; style 1, often
          divided into branches above   -   -   -   -   -   **109. Cyperaceae**

Flowers with a well-developed perianth:
  Ovary inferior:
    Plant tree-like in habit with very large oblong leaves -   -   - **93a. Musaceae**
    Plant not tree-like in habit:
      Leaves thick, fleshy, with spinose margins   -   -   - **95a. Agavaceae**
      Leaves not spinose-margined:
        Leaves cordate, reniform or lobed:
          Perianth tubular, zygomorphic-   -   -   - **78. Aristolochiaceae**
          Perianth not tubular, actinomorphic   -   - **96. Dioscoreaceae**
        Leaves not cordate, reniform or lobed:
          Flowers zygomorphic:
            Stamens 3 with filaments and anthers -   -   -   - **94. Iridaceae**
            Stamens 1–2, sessile, pollen generally agglutinated into masses (pollinia)
                                          93. Orchidaceae
          Flowers actinomorphic or almost actinomorphic:
            Perianth small, inconspicuous, greenish; fruit indehiscent
                                        82. Santalaceae
            Perianth conspicuous, white or coloured:
              Stamens 3   -   -   -   -   -   -   - **94. Iridaceae**
              Stamens 6   -   -   -   -   -   -   - **95. Amaryllidaceae**
  Ovary superior:
    Leaves lanceolate, linear or subulate with parallel nerves (monocotyledons):
      Perianth conspicuous, white or brightly coloured -   -   - **97. Liliaceae**
      Perianth inconspicuous, membranous or glumaceous:
        Inflorescence spicate; carpels 3, free; styles wanting - **103. Juncaginaceae**
        Inflorescence not spicate; carpels united; styles present -   - **98. Juncaceae**
    Leaves not lanceolate, linear or subulate; nervation not parallel (dicotyledons):
      Stems climbing; flowers in tail-like spikes-   -   -   - **74a. Basellaceae**
      Stems not climbing:
        Fruit a black juicy berry -   -   -   -   -   - **75. Phytolaccaceae**
        Fruit not a juicy berry:
          Stipules united into a membranous, tubular, often fringed or laciniate
            sheath (ochrea) around the stem   -   -   - **76. Polygonaceae**
          Stipules not united into a membranous, tubular sheath, or plants
            exstipulate:
            Fruit 2–3-locular, septicidally dehiscent   -   - **83. Euphorbiaceae**
            Fruit 1-locular, circumscissile or indehiscent:
              Perianth membranous or scarious   -   -   - **73. Amaranthaceae**
              Perianth herbaceous or fleshy:
                Stigma penicillate; filaments inflexed, springing out elastically at
                  anthesis -   -   -   -   -   -   - **84. Urticaceae**
                Stigma not penicillate; filaments not springing out elastically at
                  anthesis -   -   -   -   -   - **74. Chenopodiaceae**

INDEX to vols. 1 & 2

Accepted names are in roman type; synonyms and misidentifications in italic. Page numbers in bold face are references to detailed treatment of the taxa concerned; subsidiary references are in roman type.

Abelicea Reichb. 1463
 cretica (Spach) Holmboe 8, 10, 11, 1463
Abelmoschus esculentus (L.) Moench 313
Acacia Mill. 17, **596**
 albida Del. 603
 armata G. Frangos 601
 armata R. Br. 602
 constricta Benth. 602
 cultriformis A. Cunn. ex G. Don 599
 cunninghamii Hook. 600
 cyanophylla Lindl. 597
 cyclops A. Cunn. ex G. Don 599
 erioloba E. Mey. 601
 farnesiana (L.) Willd. 601
 giraffae auct. 601
 greggii A. Gray 602
 julibrissin Willd. 604
 karroo Hayne 601
 ligulata A. Cunn. ex Benth. 598
 longifolia (Andr.) Willd. 600
  var. *sophorae* (Labill.) F. Muell. 600
 lophantha Willd. 604
 melanoxylon R. Br. 600
 pravissima F. Muell. 599
 pycnantha Benth. 598
 retinodes Schlechtend. 597
 ? *saligna* auct. 598
 sophorae (Labill.) R. Br. 600
 verticillata (L'Hérit.) Willd. 597
Acanthaceae 1247
Acanthus L. 1247
 mollis L. 1247
 spinosus L. 1247
 syriacus Boiss. 1247
Acarna cancellata (L.) Willd. 946
Acer L. **363**
 creticum auct. 363
 negundo L. 364
 obtusifolium Sibth. et Sm. **363**
 orientale L.
  var. *obtusifolium* (Sibth. et Sm.) Spach 363
 syriacum Boiss. et Gaillardot 363
  var. *cyprium* Kotschy 363
Aceraceae 19, **362**
Aceras R. Br. **1538**
 anthropophorum (L.) R. Br. **1538**
 densiflora (Brot.) Boiss. 1557
 intacta (Link) Reichb. f. 1557
 longibracteata Reichb. f. 1537
 pyramidalis (L.) Reichb. f. 1535
Achillea L. **903**
 aegyptiaca auct. 903
 biebersteinii Afan. **903**
 cretica auct. 904
 cretica L. **906**
 erioclada DC. 906

Achillea L. (*cont.*)
 micrantha Willd. 903
 santolina auct. 906
 santolina L. 7, 904
 tournefortii auct. 903
 wilhelmsii C. Koch 904
Acinos Mill. **1280**
 exiguus (Sm.) Meikle **1281**
 graveolens (M. Bieb.) Link 1281
  var. integrifolius Raulin 1281
 rotundifolius Pers. 1282
 troodi (Post) Leblebici **1282**
Acorellus laevigatus (L.) Palla 1689
Acrostichum catanense Cosent. 1882
 marantae L. 1880
 velleum Ait. 1880
Adhatoda vasica Nees 1247
Adiantaceae **1878**
Adiantum L. **1883**
 capillus-veneris L. **1883**
Adonis L. **40**
 aestivalis L. **41**
  ssp. *dentata* (Del.) Holmboe 42
  ssp. *microcarpa* (DC.) Holmboe 42
 annua L. **41**
 autumnalis L. 41
 cupaniana Guss. 42
 dentata Del. **42**
 microcarpa DC. **42**
Aegialophila cretica Boiss. et Heldr. 973
  var. *alpina* Post 973
 pumila auct. 973
Aegilops L. **1819**
 bicornis (Forssk.) Jaub. et Spach **1822**
 biuncialis Vis. **1820**
  var. *vulgaris* Zhuk. 1820
 caudata auct. 1821
 geniculata Roth **1820**
  × Ae. peregrina (Hackel) Maire et Weiller **1822**
 incurva L. 1752
 incurvata L. 1752
 kotschyi Boiss. **1822**
 ligustica auct. 1823
 lorentii Hochst. 1820
 ovata auct. 1822
 ovata L. 1820
  var. *lorentii* (Hochst.) Boiss. 1820
 peregrina (Hackel) Maire et Weiller **1821**
 speltoides Tausch 1826
 tauschii Coss. 1826
 triaristata Willd. 1824
 triuncialis L. **1823**
  var. *kotschyi* (Boiss.) Boiss. 1822
 variabilis Eig 1821
Aeluropus Trin. **1845**
 intermedius Regel 1846

Aeluropus Trin. (*cont.*)
  korshinskyi Tzvelev 1846
  lagopoides (L.) Trin. **1845**
    × Ae. littoralis (Gouan) Parl. **1846**
  littoralis (Gouan) Parl. **1845**
    var. dasyphylla Trautv. 1846
    ssp. *repens* (Desf.) Holmboe 1846
    var. *repens* (Desf.) Coss. et Durieu
      1846
  repens (Desf.) Parl. 1845
Aeonium arboreum (L.) Webb et Berth.
  654
Aeschynomene sesban L. 490
Aethiorhiza Cass. **1013**
  bulbosa (L.) Cass. **1014**
    ssp. *microcephala* Rechinger f. 1014
Agavaceae **1578**
Agave americana L. 1578
  foetida L. 1579
  sisalana Perrine 1578
    var. armata Trelease 1578
Agrimonia L. **619**
  eupatoria L. **619**
× Agropogon littoralis (Sm.) C. E. Hubb.
  1782
Agropyron Gaertn. **1816**
  elongatum auct. 1819
  elongatum (Host) P. Beauv. **1818**
  haifense (Meld.) Bor **1818**
  halleri (Viv.) Reichb. 1723
  junceum (L.) P. Beauv. **1817**
    ssp. mediterraneum Simonet 1817
  panormitanum Parl. **1816**
  repens (L.) P. Beauv. **1817**
Agrostis L. **1781**
  alba auct. 1782
  alba L. 1782
    var. *densiflora* Guss. 1782
    var. *scabriglumis* (Boiss. et Reut.)
      Boiss. 1782
  arenaria Gouan 1850
  bromoides L. 1791
  cypricola Lindberg f. **1782**
  miliacea L. 1794
  nitens Guss. 1787
  pungens Schreb. 1850
  scabriglumis Boiss. et Reut. 1782
  semiverticillata (Forssk.) Christens. 1785
  stolonifera auct. 1785
  stolonifera L. **1782**
    var. scabriglumis (Boiss. et Reut.) C. E.
      Hubb. **1782**
  verticillata Vill. 1785
  virginica L. 1850
  viridis Gouan 1785
Ailanthus altissima (Mill.) Swingle 351
  glandulosa Desf. 351
Ainsworthia Boiss. **765**
  cordata Boiss. 765
  trachycarpa Boiss. **765**
Aira L. **1754**
  articulata Desf. 1756
  capillaris Host 1755
  caryophyllea auct. 1755
  caryophyllea L. 1755
    var. β Lam. 1756
  divaricata Pourr. 1756
  elegans Willd. ex Gaud. **1755**

Aïzoaceae 19, **681**
  tribe Aïzoeae **683**
  tribe Mesembryanthemeae Reichb. **681**
  tribe Molligineae (Fenzl) Benth. et
    Hook. f. **684**
  tribe Telephieae (Bartl.) DC. **685**
Aïzoön L. **683**
  hispanicum L. 8, **683**
Ajuga L. **1342**
  chamaepitys (L.) Schreber **1345**
    ssp. *chia* auct. 1345
    ssp. cypria P. H. Davis **1346**
    ssp. palaestina (Boiss.) Bornm. **1345**
  chia auct. 1345
  chia Schreber 1345
    ssp. *chia* auct. 1345
    ssp. *tridactylites* auct. 1345
    var. *tridactylites* auct. 1345
  iva (L.) Schreber **1344**
    ssp. *pseudo-iva* (Robill. et Cast. ex DC.)
      Holmboe 1344
    var. *pseudo-iva* (Robill. et Cast. ex DC.)
      Steud. 1344
    var. *spatulifolia* Mutel 1344
  orientalis L. **1343**
  palaestina Boiss. 1345
  pseudo-iva Robill. et Cast. ex DC. 1344
  tridactylites auct. 1345
  tridactylites Gingins ex Benth. 1346
    var. *integrifolia* Sintenis 1346
Albizia Durazz. 603
  julibrissin Durazz. 603
  lebbeck (L.) Benth. 604
  lophantha (Willd.) Benth. 604
Alcea L. 302
  pontica Janka 303
  setosa (Boiss.) Alef. 303
Alectorolophus minor (L.) Wimm. et Grab.
  1232
Alhagi Adans. **532**
  camelorum Fisch. 534
    var. *turcorum* (Boiss.) Boiss. 535
  graecorum Boiss. **532**
  mannifera Desv. 532
  maurorum auct. 535
  maurorum DC. 532
    var. *karduchorum* Boiss. 532
  maurorum Medik. **534**
    var. maurorum **534**
    var. turcorum (Boiss.) Meikle **535**, 807
  pseudalhagi Desv. 534
  turcorum Boiss. 535
Alisma L. **1671**
  damasonium L. 1672
  lanceolatum With. **1671**
  plantago-aquatica L. 1672
    var. *lanceolata* (With.) Sm. 1672
Alismataceae **1671**
Alkanna Tausch **1139**
  lehmanii (Tineo) A.DC. **1139**
  matthioli Tausch 1139
  tinctoria Tausch 1139
  tuberculata (Forssk.) Meikle 1139
  tuberculata Greuter 1139
Alliaria Scop. **163**
  officinalis Andrz. ex M. Bieb. 164
  petiolata (M. Bieb.) Cavara et Grande
    164

Allium L. **1608**
  sect. Allium **1616**
  sect. Briseis (Salisb.) Stearn **1612**
  sect. Cepa (Mill.) Prokh. **1609**
  sect. Codonoprasum Reichb. **1614**
  sect. Melanocrommyum Webb et
    Berth. **1626**
  sect. Molium G. Don ex Koch **1609**
  sect. Scorodon Koch **1613**
  ambiguum Sm. 1612
  amethystinum Tausch **1623**
  ampeloprasum L. **1617**
    f. *holmense* (Aschers. et Graebn.)
      Holmboe 1617
    var. *holmense* Aschers. et Graebn.
      1617
    var. *leucanthum* auct. 1617
    var. margaritaceum Moench 1625
  amphipulchellum Zahariadi 1616
  arvense auct. 1621
  arvense Guss. 1623
  autumnale P. H. Davis **1613**
  bassitense Thiébaut 1616
  bauerianum Baker 1626
  cappadocicum Boiss. 1626
  carneum Bertol. 1612
  cassium Boiss. 1611
    var. hirtellum Boiss. **1611**
  cepa L. 1609
  ciliatum auct. 1610
  coppoleri Tineo 1615
  cupani auct. 1613
  cupani Rafin. **1613**
    ssp. cyprium Meikle **1613**
  curtum auct. 1621
  curtum Boiss. et Gaill. **1621**
    ssp. *palaestinum* Feinbrun 1621
  decipiens auct. 1627
  descendens auct. 1623
  dioscoridis Sibth. ex Sibth. et Sm. 1628
  fragrans Vent. 1628
  fuscum Waldst. et Kit. 1615
  graecum Urv. 1610
  guttatum Stev. 1625
    ssp. *sardoum* (Moris) Stearn 1625
  hirsutum auct. 1610
  inodorum Ait. 1628
  junceum Sm. **1620**
  longicuspis Regel 1617
  macrospermum Boiss. et Kotschy 1627
  margaritaceum Sm. **1624**
    var. guttatum (Steven) Gay **1625**
    var. margaritaceum **1625**
  myrianthum Boiss. 1615
  neapolitanum Cyr. **1609**
  neglectum Wenderoth 1625
  nigrum L. **1626**
    ssp. *multibulbosum* (Jacq.) Holmboe
      1626
  orientale Boiss. **1627**
  pallens L. 1615
  paniculatum auct. 1615
  paniculatum L. **1614**
    ssp. exaltatum Meikle **1615**
    ssp. fuscum (Waldst. et Kit.) Arcang.
      1615
    var. *fuscum* (Waldst. et Kit.) Boiss.
      1615

Allium L. (*cont.*)
    ssp. pallens (L.) Arcang. **1615**
    var. pallens (L.) Gren. et Godr. 1615
  porrum L. 1618
  roseum L. **1612**
    var. bulbiferum Desf. ex DC. **1612**
    var. carneum (Bertol.) Reichb. 1612
    var. cassium (Boiss.) Regel 1611
  rotundum L. 1619
  rubrovittatum Boiss. et Heldr. **1621**
  sardoum Moris 1625
  sativum L. 1617
  scorodoprasum L. **1618**
    ssp. rotundum (L.) Stearn **1619**
    ssp. scorodoprasum 1618
  segetum Jan ex J. A. et J. H. Schultes
    1624
  sphaerocephalon L. **1619**
    ssp. arvense (Guss.) Arcang. 1623
    var. *sardoum* (Moris) Regel 1625
    var. viridialbum Tineo 1623
  stamineum Boiss. **1616**
  subhirsutum auct. 1610
  subhirsutum L. 1610
    ssp. *ciliatum* auct. 1610
    var. graecum (Urv.) Regel 1610
    ssp. *trifoliatum* (Cyr.) Holmboe 1610
  trifoliatum Cyr. **1610**
  triquetrum L. 1612
  troödi Lindberg f. 1611
  willeanum Holmboe **1625**
Alnus Mill. **1473**
  oblongata auct. 1473
  orientalis Decne. **1473**
    var. *longifolia* H. Winkler 1473
    var. *ovalifolia* H. Winkler 1473
    f. *puberula* Callier 1473
    var. *weissii* H. Winkler 1473
    f. *winkleri* Callier 1473
Aloë L. **1592**
  barbadensis Mill. 1592
  perfoliata L. 1592
    var. *vera* L. 1592
  vera (L.) Burm. f. 1592
  vulgaris Lam. 1592
Alopecurus L. **1777**
  agrestis L. 1778
    var. *tonsus* Blanche ex Boiss. 1778
  anthoxanthoides Boiss. 1778
  monspeliensis L. 1784
  myosuroides Huds. **1778**
    var. *breviaristatus* Merch. ex Aschers. et
      Graebn. 1778
    var. *tonsus* (Blanche ex Boiss.) Lind-
      berg f. 1778
  pratensis L. **1779**
  rendlei Eig 1779
  utriculatus Banks et Sol. **1778**
  utriculatus (L.) Pers. 1779
Alsine Gaertn. 265
  bocconii Scheele 275
  brevis auct. 268
  cupaniana Jord. et Fourr. 259
  filicaulis Lindberg f. 271
  globulosa (Labill.) Hal. 268
  intermedia Boiss. 269
  lydia Boiss. 270
    var. *kotschyana* Boiss. 270

Alsine Gaertn. (*cont.*)
  media L. 259
  mediterranea (Ledeb.) J. Malý 272
  meyeri auct. 268
  montana auct. 269
  picta (Sibth. et Sm.) Fenzl 267
  procumbens (Vahl) Fenzl 266
  sintenisii Lindberg f. 267
  smithii Fenzl 268
  subtilis Fenzl 270
  tenuifolia (L.) Crantz 271
    var. *hispida* Boiss. ex Kotschy 271
    ssp. *kotschyana* auct. 271
    var. *mucronata* Boiss. 272
    var. *subtilis* Fenzl 270
    var. *viscosa* Boiss. 271
  thymifolia (Sibth. et Sm.) Fenzl 269
Althaea L. **302**
  hirsuta L. **303**
  lavateriflora auct. 303
  rosea auct. 303
  setosa Boiss. **303**
Altingiaceae Lindl. 654
  subfam. *Liquidambaroideae* 654
Alyssum L. **134**
  akamasicum B. L. Burtt 2, 6, **138**
  alpestre L. 135
    var. *obtusifolium* auct. 135
    var. *suffrutescens* auct. 136
  argenteum auct. 136
  campestre L. 138
    ssp. *hirsutum* auct. 139
    var. *micropetalum* (Fisch. ex DC.) Koch 139
  cedrorum Schott et Kotschy 139
  chondrogynum B. L. Burtt 2, 7, **136**
  condensatum auct. 136
  coriaceum Nyárády 135
  cypricum Nyárády 2, **136**
  foliosum Bory et Chaub. **141**
  fulvescens auct. 140
  hirsutum auct. 139
  maritimum (L.) Lam. 142
  micropetalum Fisch. ex DC. 139
  minus (L.) Rothm. **138**
  minutum Schlechtend. ex DC. **140**
  murale Boiss. 136
  sativum auct. 170
  strigosum Banks et Sol. **139**
    ssp. *cedrorum* (Schott et Kotschy) Dudley 139
  troodi Boiss. 2, **135**
  umbellatum Desv. **140**
  virgatum Nyárády 136
    var. *mutabile* Nyárády 136
Amaracus cordifolius Aucher-Eloy et Montbret ex Benth. 1262
Amaranthaceae **1363**
Amaranthus L. **1365**
  albus L. 1369
  angustifolius Lam. 1368
  blitum auct. 1369
  blitum L. 1368
    ssp. *graecizans* (L.) Lindberg f. 1369
    var. *graecizans* (L.) Moq. 1368
  caudatus auct. 1366
  chlorostachys Willd. 1366
  cruentus L. 1366

Amaranthus L. (*cont.*)
  delilei Richter et Loret 1367
  gracilis Desf. ex Poir. 1367
  graecizans L. **1368**
    ssp. *graecizans* **1368**
    ssp. *silvestris* (Vill.) Brenan **1369**
    var. *silvestris* (Vill.) Aschers. 1369
  hybridus L. **1365**
    ssp. *cruentus* (L.) Thell. **1366**
    var. *erythrostachys* Moq. 1366
    ssp. *hybridus* **1366**
    var. *hybridus* 1366
  paniculatus L. 1366
  patulus Bertol. 1366
  retroflexus L. 1367
    var. *delilei* (Richter et Loret) Thell. 1367
  silvestris Vill. 1369
    var. *graecizans* (L.) Boiss. 1369
  viridis L. **1367**
Amaryllidaceae **1572**
Ambrosia L. **899**
  maritima L. **899**
Ammi L. **740**
  majus L. **741**
  visnaga (L.) Lam. **740**
Ammophila Host **1789**
  arenaria (L.) Link **1789**
    var. *arundinacea* (Host) Husnot 1789
    var. *australis* (Mab.) Aschers. et Graebn. 1789
  arundinacea Host 1789
Amygdalus communis L. 609
  persica L. 609
Anacamptis L. C. M. Rich. **1535**
  pyramidalis (L.) L. C. M. Rich. **1535**
Anacardiaceae 19, **364**
Anachortus Jir. et Chr. 1756
  articulatus (Desf.) Jir. et Chr. 1756
  divaricatus (Pourr.) Lainz 1756
Anacyclus creticus L. 914
  orientalis auct. 925
Anagallis L. **1083**
  arvensis L. **1083**
    var. *albiflora* Druce 1085
    ssp. *arvensis* **1084**
    var. *arvensis* **1084**
    f. *azurea* Hyl. 1084
    f. *caerulea* Lüdi 1084
    ssp. *caerulea* (Schreb.) Hartm. 1085
    var. *caerulea* Gouan **1084**
    var. *caerulea* (Schreb.) Gren. et Godr. 1085
    ssp. *foemina* (Mill.) Schinz et Thell. **1085**
    var. *gentianea* Beck 1085
    var. *pallida* Hook. f. **1085**
    var. *phoenicea* Gouan 1084
    ssp. *phoenicea* (Scop.) Vollmann 1084
  caerulea L. 1084
  caerulea Schreb. 1085
  foemina Mill. 1085
  phoenicea Scop. 1084
Anagyris L. **376**
  foetida L. **376**
Anchusa L. **1131**
  subg. *Anchusa* **1131**
  subg. *Buglossoides* Guşul. 1135

Anchusa L. (*cont.*)
  subg. *Buglossum* (Gaertn.)Guşul. 1131
  subg. Lycopsis (L.) Guşul. **1135**
  aegyptiaca (L.) DC. **1135**
  aggregata Lehm. 1136
  azurea Mill. **1133**
  bracteolata Viv. 1139
  humilis (Desf.) I. M. Johnst. **1136**
  italica Retz. 1133
  paniculata Ait. 1133
  parviflora auct. 1136
  rhizochroa Viv. 1139
  strigosa Labill. **1134**
  tinctoria (L.) L. 1139
  tuberculata Forssk. 1139
  undulata L. **1132**
    ssp. hybrida (Ten.) Bég. **1132**
    ssp. undulata 1132
  ventricosa Sm. 1138
Andropogon L. **1863**
  distachyos L. **1864**
  gryllus L. 1863
  halepensis (L.) Brot. 1869
  hirtus L. 1865
  ravennae L. 1867
  sorghum (L.) Brot. 1869, 1870
    ssp. *halepensis* Hackel 1869
    ssp. *sudanensis* Piper 1870
Androsace L. **1076**
  maxima L. **1077**
Androsaemum foetidum Willk. et Lange 295
  hircinum (L.) Spach 295
Anemone L. **38**
  blanda Schott et Kotschy **40**
  coronaria L. **38**
    var. alba (Goaty et Pons) Burnat **39**
    var. coronaria **39**
    var. cyanea (Risso) Arduino **39**
    var. parviflora Boiss. **39**
    var. rosea (Hanry) Rouy et Fouc. **39**
  hortensis L. **7, 10, 39**
  stellata Lam. 39
Anethum L. **750**
  foeniculum L. 749
  graveolens L. **751**
  piperitum Ucria 750
Anogramma Link **1879**
  leptophylla (L.) Link **1879**
Anredera cordifolia (Ten.) Steenis 1394
Anthemis L. **907**
  sect. Anthemis **910**
  sect. Chia Yavin **915**
  sect. Cota (Gay) Rupr. **908**
  sect. Maruta (Cass.) Reichb. **915**
  altissima auct. 908, 910
  amblyolepis Eig **908**
  arvensis auct. 918
  australis Sm. 919
  bornmuelleri Stoy. et Acht. 916
  chia L. **915**
  complanata (Sm.) Hal. 925
  cota auct. 910
    f. *apiculata* Holmboe 909
    ssp. *melanolepis* (Boiss.) Holmboe 909
    ssp. *palaestina* auct. 908
  cotula L. **915**
    var. *hierosolymitana* Eig 916

Anthemis L. (*cont.*)
  ssp. *paleacea* Eig 916
  cretica (L.) Nyman 914
  cypricola Lindberg f. 912
  galilaea Eig 916
  melanolepis Boiss. 909
    var. *cypria* Eig 910
    var. *genuina* Eig 910
  montana L.
    var. *tenuiloba* auct. 925
  palaestina auct. 908
  palaestina (Reuter ex Kotschy) Boiss. **909**
    ssp. *amblyolepis* (Eig) Feinbrun 908
  pamphylica auct. 918
  parvifolia Eig **917**
  peregrina L. 919
  plutonia Meikle **913**
  pontica auct. 912, 923
  pseudocotula auct. 917
  pseudocotula Boiss. **917**
    ssp. rotata (Boiss.) Eig **918**
  pusilla Greuter 914
  rigida Boiss. ex Heldr. **914**
  rosea auct. 912
  rosea Sm. 919
  rotata Boiss. 918
  tinctoria L. 919
  tomentosa L. 919
  topaliana Beauverd 913
  tricolor Boiss. **910**
    f. *artemisioides* (Holmboe) Lindberg f. 913
    var. *artemisioides* Holmboe 913
Anthericum graecum L. 1607
Anthriscus Pers. **695**
  caucalis M. Bieb. **695**
  scandix Beck 695
  vulgaris Pers. 695
Anthyllis L. 471
  tetraphylla L. 471
Antirrhinum L. **1212**
  albifrons Sm. 1204
  aparinoides Willd. 1203
  chalepense L. 1204
  cymbalaria auct. 1205
  elatine L. 1207
  graecum Bory et Chaub. 1207
  majus L. **1212**
    ssp. *angustifolium* (Chav.) Lindberg f. 1213
    var. angustifolium Chav. 1213
    f. *glandulosum* Rothm. 1213
    ssp. *majus* 1213
    ssp. *tortuosum* (Bosc ex Lam.) Rouy 1213
  micranthum Cav. 1202
  orontium L. 1211
  parviflorum Jacq. 1202
  pelisserianum L. 1201
  siculum auct. 1213
  simplex Willd. 1202
  spurium L. 1209
  strictum Sm. 1203
Apargia tuberosa (L.) Willd. 1017
Aphanes L. **618**
  arvensis L. **618**
  floribunda auct. 618

Apium L. **736**
  graveolens L. **737**
  nodiflorum (L.) Lag. **737**
Apocynaceae **1096**
Apocynum L. 1100
  venetum L. 1100
Arabidopsis (DC.) Heynh. **169**
  thaliana (L.) Heynh. **169**
Arabis L. **148**
  albida auct. 149
  albida Stev. 148
    var. *billiardieri* auct. 148
  auriculata auct. 151
  billiardieri auct. 148
  cremocarpa (Boiss. et Bal.) Kotschy 152
  cypria Holmboe **148**
  deflexa auct. 148
  glabra auct. 152
  kennedyae Meikle **151**
  laxa Sibth. et Sm. 152
    var. *cremocarpa* (Boiss. et Bal.) Boiss. 152
  montbretiana auct. 151
  petiolata M. Bieb. 164
  purpurea Sibth. et Sm. **149**
  thaliana L. 169
  verna R. Br. **150**
    var. dasycarpa Godr. ex Rouy et Fouc. **150**
    var. *liocarpa* Lindberg f. 150
    var. verna **150**
Araceae **1664**
Arachnites lutea (Cav.) Tod. 1517
  var. *minor* Tod. 1517
Araliaceae 19, **767**
Arbutus L.
  andrachne L. 1, 4, **1063**
    f. *arguteserrata* Lindberg f. 1063
    × A. unedo L. 1063
  × andrachnoides Link 1063
  unedo L. 6, **1062**
Arctium L. **947**
  lappa L. **947**
  majus (Gaertn.) Bernh. 948
  minus (Hill) Bernh. 948
  tomentosum Mill. 948
  vulgare (Hill) Druce 948
Arenaria L. **261**
  sect. *Spergularia* Pers. 274
  boissieri Pax 264
  cerignensis Sibth. 267
  ciliata auct. 265
  cretica Spreng. 265
  cypria Holmboe 262
  diandra Guss. 275
  filiformis Labill. 266
  geniculata Poir. 266
  globulosa Labill. 268
  hybrida Vill. 271
  leptoclados (Reichb.) Guss. **263**
    var. viscidula (Rouy et Fouc.) F. N. Williams **264**
  leptophylla auct. 264
  macrosepala auct. 264
  macrosepala Boiss. 264
    var. *minor* Boiss. 264
    ssp. *saponarioides* (Boiss. et Bal.) Holmboe 264

Arenaria L. (*cont.*)
  marina (L.) All. 276
  mediterranea Ledeb. 272
  muralis auct. 263
  nana Boiss. 264
  oxypetala auct. 262, 263
  pamphylica auct. 262
  pamphylica Boiss. et Heldr. **262**
    ssp. kyrenica McNeill **263**
  picta Sibth. et Sm. 266
  procumbens Vahl 266
  rhodia Boiss. **262**
    ssp. cypria (Holmboe) McNeill 6, **262**
  rubra auct. 275
  rubra L. 276
    var. *marina* L. 276
  saponarioides Boiss. et Bal. **264**
    ssp. *boissieri* (Pax) McNeill 264
  serpyllifolia auct. 264
  serpyllifolia L. 264
    ssp. *leptoclados* auct. 264
    var. *leptoclados* Reichb. 263
    var. *viscida* auct. 264
    var. *viscidula* Rouy et Fouc. 264
  tenuifolia L. 271
  thymifolia Sibth. et Sm. 269
Argyrolobium Eckl. et Zeyh. **380**
  uniflorum (Decne.) Jaub. et Spach 7, **380**
Arisarum Mill. **1669**
  crassifolium Schott 1669
  libani Schott 1669
  sibthorpii Schott 1669
  veslingii Schott 1669
  vulgare Targ.-Tozz. **1669**
    var. *incurvatum* (Lam.) Holmboe 1669
    ssp. *veslingii* (Schott) Holmboe 1670
    var. *veslingii* (Schott) Engl. 1669
Aristella bromoides (L.) Bertol. 1791
Aristida L. **1851**
  adscensionis auct. 1851
  adscensionis L. 1851
    var. *caerulescens* (Desf.) Hackel 1851
  caerulescens Desf. **1851**
Aristolochia L. **1416**
  altissima auct. 1583
  altissima Desf. 1417
  baetica L. 1419
  hirta L. 1419
  parvifolia Sm. **1419**
  sempervirens L. **1417**
Aristolochiaceae **1416**
Arnoseris pusilla auct. 1015
Arrhenatherum P. Beauv. **1764**
  album (Vahl) W. D. Clayton **1765**
  elatius auct. 1765
  erianthum Boiss. et Reut. 1765
  palaestinum Boiss. 1765
Artedia L. **716**
  squamata L. 7, **717**
Artemisia L. **927**
  annua L. **928**
  arborescens L. **927**
  campestris auct. 928
Arthrocnemum Moq. **1385**
  fruticosum (L.) Moq. 1386
  glaucum Ungern-Sternb. 1385
  macrostachyum (Moric.) Moris et Delponte **1385**

Arthrolobium scorpioides (L.) Desf. 516
Arum L. **1665**
  arisarum L. 1669
  colocasia L. 1670
  conophalloides Kotschy ex Schott **1666**
  creticum Boiss. et Heldr. 1669
  cyprium Schott 1665
  detruncatum auct. 1665
  detruncatum C. A. Mey. ex Schott 1666
    var. *conophalloides* (Kotschy ex
    Schott) Boiss. 1666
  dioscoridis Sm. **1665**
    var. *cyprium* (Schott) Engl. 1665
    var. *smithii* Engl. 1665
  esculentum L. 1670
  gratum Schott 1667
  hygrophilum Boiss. **1666**
  incurvatum Lam. 1669
  italicum Mill. **1667**
    subvar. *nickellii* (Schott) Engl. 1667
  orientale M. Bieb. **1667**
    var. *gratum* (Schott) Engl. 1667
    var. *sintenisii* Engl. 1667
  ponticum auct. 1667
Arundo L. **1837**
  australis Cav. 1838
  donax L. **1838**
  epigejos L. 1777
  phragmites L. 1838
  selloana Schult. et Schult. f. 1839
Asarum hypocistis L. 1415
Asclepiadaceae **1102**
  subfam. Asclepiadoideae **1103**
  subfam. Periplocoideae **1102**
Asclepias L. 1104
  canescens Willd. 1105
  curassavica L. 1104
  fruticosa L. 1104
Aspalthium Medik. **488**
  bituminosum (L.) Fourr. **489**
Asparagus L. **1584**
  acutifolius L. **1584**
  aphyllus auct. 1585
  aphyllus L. 1585
    var. *stipularis* (Forssk.) Bak. 1585
  horridus L. f. 1585
  officinalis L. 1584
  stipularis Forssk. **1585**
    var. *brachyclados* Boiss. 1585
    var. *tenuispinus* Holmboe 1585
  verticillatus auct. 1584
Asperugo L. **1129**
  procumbens L. **1129**
Asperula L. **791**
  arvensis L. **793**
  calabrica L.f. 773
  cypria Ehrend. **792**, 807
  humifusa (M. Bieb.) Besser 781
    var. *pycnantha* (Boiss.) Boiss. 782
  pycnantha Boiss. 782
    var. *lasiocarpa* Boiss. 782
  setosa Jaub. et Spach 794
  stricta Boiss. **793**
Asphodeline Reichb. **1590**
  brevicaulis (Bertol.) Gay **1591**
  liburnica auct. 1591
  lutea (L.) Reichb. **1591**
Asphodelus L. **1588**

Asphodelus L. (*cont.*)
  aestivus Brot. **1588**
  brevicaulis Bertol. 1591
  fistulosus L. **1589**
    ssp. *tenuifolius* (Cav.) Arcang. 1590
    var. *tenuifolius* (Cav.) Baker 1590
  liburnicus auct. 1591
  luteus L. 1591
  microcarpus Viv. 1588
  ramosus auct. 1588
  ramosus L. 1588
    ssp. *microcarpus* (Viv.) Baker 1588
  tenuifolius Cav. **1590**
Aspidiaceae **1885**
Aspidium filix-mas auct. 1887
  libanoticum Rosenst. 1887
Aspleniaceae **1887**
Asplenium L. **1887**
  adiantum-nigrum L. **1889**
    f. lancifolium Heufler 1889
    ssp. *nigrum* Heufler 1889
    ssp. *onopteris* (L.) Heufler 1890
    var. *onopteris* (L.) Fiori 1890
  ceterach L. **1890**
  cuneifolium Viv. 1889
  onopteris L. **1889**
  trichomanes L. **1888**
    ssp. quadrivalens D. E. Meyer **1888**
  viride Huds. **1888**
Aster *conyzae* Griesselich 890
  squamatus (Spreng.) Hieron. 871
Asteriscus Mill. **896**
  aquaticus (L.) Less. **896**
  spinosus Sch. Bip. 898
Asterolinon Hoffsgg. et Link **1082**
  linum-stellatum (L.) Duby **1082**
Astragalus L. **490**
  alexandrinus Boiss. 501
  angustifolius auct. 503
  asterias Stev. ex Ledeb. **495**
  boeticus L. **496**
  caprinus L. **501**
    ssp. laniger (Desf.) Maire **501**
  contortuplicatus L. 495
  corrugatus Bertol. 495
  cruciatus auct. 495
  cyprius Boiss. **503**
  depressus L. 504
  dyctiocarpus auct. 503
  echinus DC. **502**
    ssp. *chionistrae* Lindberg f. 503
    var. chionistrae (Lindberg f.) Meikle
      503, 806
    var. *echinus* **503**
  epiglottis L. **493**
  glaux L. 502
  hamosus L. **496**
  incanus auct. 503
  laniger Desf. 501
  lusitanicus Lam. **498**
    ssp. orientalis Chater et Meikle **499**
  macrocarpus DC. **500**
    ssp. lefkarensis Agerer-Kirchoff et
      Meikle 7, **500**
  pelecinus (L.) Barneby **492**
  pentaglottis L. 495
  pseudostella auct. 494
  sesameus auct. 494

Astragalus L. (*cont.*)
  sinaicus Boiss. **494**
  spruneri auct. 503
  stella auct. 495
  suberosus Banks et Sol. **497**
    var. hartmannii (Holmboe) Meikle **498, 806**
    var. suberosus **498**
  tribuloides auct. 494
  tuberculosus DC. 497
    var. *hartmannii* Holmboe 498
    ssp. *mersinensis* Širj. 497
Athamanta multiflora Sibth. et Sm. 742
Athanasia maritima (L.) L. 907
Atractylis L. **946**
  cancellata L. **946**
Atriplex L. **1378**
  halimus L. **1378**
  hastata auct. 1380
  hastata L. 1380
    var. *salina* auct. 1380
  laciniata auct. 1379
  leucoclada Boiss. 1379
  nummularia Lindl. 1379
  patula L. **1381**
  portulacoides L. 1382
  prostrata Boucher ex DC. **1380**
  rosea L. **1379**
  semibaccata R. Br. **1381**
  tatarica auct. 1379
  triangularis Willd. 1380
Atropis gigantea Grossh. 1749
Avellinia Parl. **1765**
  michelii (Savi) Parl. **1766**
Avena L. **1758**
  alba Vahl 1765
  barbata Pott ex Link **1759**
    ssp. *wiestii* (Steud.) Holmboe 1764
  bruhnsiana Gruner 1759
  byzantina C. Koch **1761**
  clauda auct. 1759
  clauda Durieu 1760
  cypria Sibth. 1759
  elatior L. 1765
    var. *bulbosa* auct. 1765
    f. *tuberosa* auct. 1765
  eriantha Durieu **1760**
  fatua L. **1760**
    var. *glabrescens* Coss. et Durieu 1761
  hirsuta Roth 1759
  hirtula Lag. **1762**
  ludoviciana Durieu **1762**
  orientalis auct. 1761
  pilosa (Roem. et Schult.) M. Bieb. 1760
  sativa L. 1763
  sterilis L. **1763**
    ssp. *byzantina* (C. Koch) Thell. 1761
    var. *ludoviciana* (Durieu) Husnot 1762
  strigosa Schreb. 1758
    ssp. *barbata* (Pott ex Link) Thell. 1759
  ventricosa Bal. ex Cosson **1759**
  wiestii Steud. **1764**

Baccharis dioscoridis L. 873
Ballota L. **1315**
  sect. Acanthoprasium Benth. **1317**
  sect. Ballota **1316**
  foetida Lam. 1316

Ballota L. (*cont.*)
  integrifolia Benth. **1317**
  nigra L. **1316**
    ssp. *foetida* auct. 1317
    ssp. foetida Hayek 1316
    var. *meridionalis* Bég. 1317
    f. *uncinata* Fiori et Bég. 1317
    ssp. uncinata (Fiori et Bég.) Patzak **1317**
  wettsteinii Rechinger 1317
Balsamita tridentata Del. 926
Barkhausia foetida (L.) F. W. Schmidt 1008
Barlia Parl. **1536**
  longibracteata (Biv.) Parl. 1537
  robertiana (Loisel.) Greuter **1536**
Bartsia latifolia (L.) Sm. 1226
  trixago L. 1228
  viscosa L. 1227
Basellaceae **1394**
Bassia hirsuta auct. 1392
  pulverulenta Lindberg f. 1392
Bauhinia L. 591
  acuminata L. 591
  galpinii N. E. Br. 591
  purpurea L. 591
  tomentosa L. 591
  variegata L. 591
Behen vulgaris Moench 231
Bellardia All. **1228**
  trixago (L.) All. **1228**
Bellevalia Lapeyr. **1641**
  comosa (L.) Kunth 1644
  millingenii Post 1640
  nervosa auct. 1641
  nivalis Boiss. et Kotschy **1642**
  romana auct. 1642
  trifoliata (Ten.) Kunth **1642**
Bellis L. **867**
  annua L. **869**
    var. microcephala (Lange) Nyman 871
    ssp. minuta (DC.) Meikle **869**
    var. *minuta* DC. 869
  microcephala Lange 871
  perennis L. **868**
  sylvestris Cyr. **868**
Berberidaceae 19, **69**
Berberis L. **70**
  cretica L. **70**
Beta L. **1372**
  maritima L. 1373
  vulgaris L. **1372**
    var. cicla L. 1373
    ssp. maritima (L.) Arcang. **1373**
    var. *maritima* (L.) Moq. 1373
    var. rubra (L.) Moq. 1373
Betulaceae **1473**
Bifora Hoffm. **718**
  testiculata (L.) DC. **718**
Biforis testiculata Spreng. ex Schultes 719
Bignonia tomentosa Thunb. 1216
Bignoniaceae **1244**
Bilderdykia convolvulus (L.) Dumort. 1404
Biscutella L. **123**
  ciliata DC. 124
  columnae Ten. 124
  didyma L. **123**
    var. *ciliata* (DC.) Vis. 124
    ssp. *columnae* (Ten.) Holmboe 124

Biscutella L. (*cont.*)
   var. columnae (Ten.) Hal. **124**
   var. didyma **124**
   var. lejocarpa Vis. **124**
   laevigata auct. 124
   leiocarpa DC. 124
Biserrula pelecinus L. 492
Blackstonia Huds. **1108**
   acuminata (Koch et Ziz) Domin **1109**
   ssp. acuminata 1110
   ssp. aestiva (K. Maly) Zeltner 1110
   perfoliata (L.) Huds. **1108**
   ssp. intermedia (Ten.) Zeltner **1109**
   ssp. perfoliata 1109
Blitum virgatum L. 1374
Boissiera danthoniae (Trin.) A. Braun 1798
Bolboschoenus Aschers. ex Palla **1696**
   maritimus (L.) Palla **1696**
Bombycilaena (DC.) Smolj. **876**
   bombycina (Lag.) Smolj. 876
   discolor (Pers.) Lainz **876**
   erecta (L.) Smolj. 877
Bonaveria securidaca (L.) Scop. 514
Bongardia C. A. Mey. **70**
   chrysogonum (L.) Endl. **72**
Bonjeania recta (L.) Reichb. 476
Boraginaceae **1119**
   subfam. Boraginoideae **1124**
   subfam. Cordioideae **1120**
   subfam. Heliotropoideae **1120**
Borago L. **1130**
   officinalis L. **1130**
Bosea L. **1363**
   cypria Boiss. ex Schinz et Autran v, **1363**
   yervamora auct. 1364
Botryanthus parviflorus auct. 1646, 1647
Bougainvillea Commerson ex Juss. **1362**
   × buttiana Holttum et Standl. **1362**
   glabra Choisy **1362**
   peruviana Kunth 1362
   spectabilis Willd. **1362**
Boussingaultia baselloides hort. 1394
Brachiaria Griseb. **1852**
   eruciformis (Sm.) Griseb. **1852**
Brachychiton populneus R. Br. 314
Brachypodium P. Beauv. **1812**
   distachyum (L.) P. Beauv. **1815**
   firmifolium Lindberg f. **1814**
   loliaceum (Huds.) Roem. et Schult. 1721
   pinnatum auct. 1813
   pinnatum (L.) P. Beauv. **1813**
   poa (Lam. ex Lam. et DC.) Roem. et
     Schult. 1723
   sylvaticum (Huds.) P. Beauv. **1813**
Brassica L. **97**
   cretica Lam. 97
   ssp. *hilarionis* Lindberg f. 97
   var. *hilarionis* (Post) O. E. Schulz 97
   cretica auct. 97
   eruca L. 104
   hilarionis Post **97**
   nigra (L.) Koch **99**
   oleracea L. **100**
   orientalis L. 115
   tournefortii Gouan **99**
Briza L. **1717**
   eragrostis L. 1841
   humilis M. Bieb. **1717**

Briza L. (*cont.*)
   maxima L. **1719**
   media L. **1720**
   minor L. **1719**
   spicata Sm. 1717
Brizochloa humilis (M. Bieb.) Chrtek et
   Hadač 1717
   spicata (Sm.) Jirásek et Chrtek 1717
Bromus L. **1796**
   alopecuroides auct. 1803
   alopecuroides Poir. 1808
   alopecuros Poir. **1808**
   var. alopecuros **1808**
   var. calvus Hal. **1808**
   arvensis L. **1799**
   var. *fragilis* Aschers. et Graebn.
     1799
   f. *puberulus* Lindberg f. 1799
   f. *pubescens* (Caldesi) Lindberg f. 1799
   var. *recurvatus* Lindberg f. 1799
   brachystachys Hornung 1797
   carinatus Hook. et Arn. 1812
   caroli-henrici Greuter 1809
   catharticus Vahl 1812
   commutatus Schrad. 1798
   danthonii Trin. **1798**
   diandrus Roth **1800**
   distachyos L. 1815
   divaricatus Rohde ex Loisel. 1804
   fasciculatus Presl **1802**
   var. alexandrinus Thell. 1803
   var. fasciculatus 1803
   gussonei Parl. 1800
   hispidus (Savi) Savi 1769
   hordeaceus L. 1800
   ssp. *molliformis* Maire et Weiller 1800
   intermedius Guss. **1803**
   ssp.*divaricatus* Douin 1800
   japonicus Thunb. **1803**
   var. japonicus **1804**
   var. velutinus (Koch) Aschers. et
     Graebn. **1804**
   lanceolatus Roth **1804**
   var. *danthoniae* (Trin.) Dinsm. 1798
   var. lanceolatus **1805**
   var. lanuginosus (Poir.) Dinsm. **1805**
   lanuginosus Poir. 1805
   lloydianus (Godr. ex Gren. et Godr.)
     Nyman 1800
   macrostachys Desf. 1804
   var. *lanuginosus* (Poir.) Coss. et
     Durieu 1805
   var. *triaristatus* Hackel 1798
   madritensis L. **1805**
   var. ciliatus Guss. **1806**
   var. madritensis **1806**
   × B. rubens L. **1806**
   maximus Desf. 1802
   michelii Savi 1766
   molliformis Lloyd 1799
   patulus Mert. et Koch 1803
   var. *velutinus* Koch 1804
   pinnatus L. 1813
   pseudobrachystachys H. Scholz 1797
   rigidus Roth **1802**
   var. *gussonei* (Parl.) Boiss. 1800
   rubens L. **1807**
   var. glabriglumis Maire **1807**

Bromus L. (*cont.*)
    var. rubens **1807**
  scoparius L. **1808**
    var. hirtulus Regel **1808**
    var. scoparius **1808**
    var. *stenanthus* Stapf 1808
    var. *villiglumis* Maire et Weiller 1808
  squarrosus auct. 1803
  squarrosus L. **1809**
    var. *villosus* auct. 1803
  sterilis L. **1810**
    var. sterilis **1810**
    var. velutinus Volkart **1811**
  sylvaticus (Huds.) Sm. 1813
  tectorum L. **1811**
    var. hirsutus Regel **1811**
    var. tectorum **1811**
  unioloides H.B.K. 1812
  villosus Forssk. 1802
    ssp. *gussonei* (Parl.) Holmboe 1800
    var. *gussonei* (Parl.) Dinsm. 1800
  willdenowii Kunth 1812
Brotera corymbosa Willd. 940
Broussonetia papyrifera (L.) Vent. 1465
Brugmansia Pers. 1191
  aurea Lagerh. 1192
  × candida Pers. 1191
  sanguinea (Ruiz et Pavon) D. Don 1191
  suaveolens (Humb. et Bonpl. ex Willd.)
    Bercht. et J. Presl 1191
Bryonia L. **678**
  cretica L. **678**
  dioica auct. 678
  multiflora auct. 678
Buccaferrea cirrhosa Petagna 1677
Buddleja davidii Franch. 1107
  globosa Hope 1108
  madagascariensis Lam. 1108
Buglossoides Moench **1147**
  arvensis (L.) I. M. Johnston **1148**
    ssp. arvensis **1149**
    ssp. gasparrinii (Heldr. ex Guss.) R.
      Fernandes **1150**
    ssp. sibthorpiana (Griseb.) R.
      Fernandes **1149**
  incrassata (Guss.) I. M. Johnston 1150
  tenuiflora (L.f.) I. M. Johnston **1147**
Bulliarda DC. 641
  vaillantii DC. 642
Bunias aegyptiaca L. 131
  cakile L. 115
  myagroides L. 114
  tenuifolia Sibth. et Sm. 113
Bunium L. **743**
  creticum Mill. 720
  ferulaceum Sibth. et Sm. **743**
  ferulifolium Desf. 743
  napiforme Willd. ex Spreng. 720
Buphthalmum aquaticum L. 896
  asteroideum Viv. 898
  spinosum L. 898
Bupleurum L. **729**
  fontanesii Guss. 732
  gerardii All. **735**
  glaucum Robill. et Cast. 733
  glumaceum Sibth. et Sm. 732
  gracile auct. 734
  gracile Urv. **732**

Bupleurum L. (*cont.*)
  heterophyllum Link 729
  intermedium (Lois. ex DC.) Steudel 730
  lancifolium Hornem. **729**
  nodiflorum Sibth. et Sm. **731**
  odontites L. **732**
  orientale Snogerup **734**
  protractum Hoffsgg. et Link 730
    var. *heterophyllum* (Link) Boiss. 729
  rigoi Huter ex David 735
  rotundifolium L. 730
    var. *intermedium* Lois. ex DC. 730
  rotundifolium Sibth. et Sm. 730
  semicompositum L. **733**
    var. *glaucum* (Robill. et Cast.) H.
      Wolff 733
    var. pseudodontites (Rouy et Camus)
      H. Wolff 734
  sintenisii Aschers. et Urban ex Huter **735**
  subovatum Link ex Spreng. **730**
    var. *heterophyllum* (Link) H. Wolff 730
  tenuissimum L. 734
    ssp. *gracile* auct. 734
  trichopodum Boiss. et Spruner **736**
    var. *depauperatum* Boiss. 736
Bursa concava Druce 129
  rubella auct. 129

Cachrys L. **726**
  crassiloba (Boiss.) Meikle **728**, 807
  cretica Lam. 726
  pterochlaena auct. 727
  scabra (Fenzl) Meikle **727**, 807
  sicula auct. 727
Cactaceae 19, **680**
Cactus ficus-indica L. 680
Caesalpinia coriaria (Jacq.) Willd. 585
  gilliesii (Hook.) Dietr. 585
Cakile Mill. **114**
  maritima Scop. **115**
Calamagrostis Adans. **1776**
  epigejos (L.) Roth **1777**
Calamintha L. **1279**
  cretica auct. 1279
  clinopodium Spenner 1283
  crassinervis (Lindberg f.) Osorio-Tafall et
    Seraphim 1281
  exigua (Sm.) Hal. 1281
  graveolens auct. 1281
  incana (Sm.) Boiss. ex Benth. **1279**
  nepeta (L.) Savi 1279
  troodi Post 1282
  vulgaris (L.) Karsten 1283
Calendula L. **935**
  aegyptiaca Pers. 936
  arvensis L. **936**
    ssp. *micrantha* (Tineo et Guss.)
      Holmboe 936
  gracilis DC. 936
  micrantha Tineo et Guss. 936
  persica C. A. Mey. 936
    ssp. *gracilis* (DC.) Holmboe 936
    var. *gracilis* (DC.) Boiss. 936
Calepina Adans. **109**
  corvini (All.) Desv. 109
  irregularis (Asso) Thellung **109**
Calicotome Link 381
Callistemma (Mertens et Koch) Boiss. 850

Callistemma (Mertens et Koch) Boiss. (*cont.*)
  brachiatum (Sm.) Boiss. 856
  palaestinum (L.) Heldr. 856
Callitrichaceae 19, **657**
Callitriche L. **657**
  brutia Petagna **657**
  hamulata Kuetz. 657
    ssp. *pedunculata* (DC.) Syme 658
    var. *pedunculata* (DC.) Bab. 657
  intermedia Hoffm. 658
    ssp. *pedunculata* (DC.) Clapham 658
    var. *pedunculata* (DC.) Druce 658
  pedunculata DC. 657
Calycotome Link **381**
  villosa (Poir.) Link 4, **382**
Calystegia R. Br. **1165**
  sepium (L.) R. Br. **1165**
Camelina Crantz **170**
  microcarpa Andrz. 170
    var. *albiflora* (Boiss.) Bornm. 170
  rumelica Velenovský **170**
  sativa auct. 170
  sylvestris auct. 170
  sylvestris Wallr. 170
    var. *albiflora* Boiss. 170
Campanula L.
  attica Boiss. et Heldr. 1050
  creutzburgii Greuter 1050
  cypria Rechinger f. 1049
  delicatula Boiss. **1050**
  drabifolia Sm. **1049**
  erinus L. **1048**
  hybrida L. 1054
  pentagonia L. 1053
  peregrina L. **1052**
  pinatzii Greuter 1050
  podocarpa Boiss. **1049**
  raveyi Boiss. 1050
  rhodensis A. DC. 1050
  speculum-veneris L. 1053
Campanulaceae **1047**
  subfam. Campanuloideae **1048**
  subfam. Lobelioideae **1056**
Camphorosma monspeliaca L. 1383
  pteranthus L. 284, 1383
Campsis grandiflora (Thunb.) K. Schum.
    1246
  radicans (L.) Seem. 1246
Cannabaceae **1464**
Cannabis L. 1464
  sativa L. 1464
Capnophyllum peregrinum (L.) Lange 752
Capparaceae 19, **172**
Capparis L. **172**
  ovata Desf. 173
    var. *canescens* (Coss.) Heywood 173
  sicula Duham. 173
  spinosa L. **172**
    ssp. *sicula* (Duham.) Holmboe 173
    var. canescens Cosson **173**
Caprifoliaceae 19, **769**
Capsella Medik. **128**
  batavorum E. Almq. 130
  bursa-pastoris (L.) Medik. **129**
    var. brachycarpa Heldr. **129**
    var. bursa-pastoris **129**
  concava E. Almq. 130
  mediterranea E. Almq. 130

Capsella Medik. (*cont.*)
  patagonica E. Almq. 130
  procumbens (L.) Fries 130
  rubella auct. 129
  treviorum E. Almq. 130
  turoniensis E. Almq. 130
Capsicum annuum L. 1185
  baccatum L. 1185
Cardamine L. **146**
  fontanum Lam. 153
  graeca L. **147**
  hirsuta L. **147**
    var. *glabra* Sintenis 147
Cardaria Desv. **120**
  draba (L.) Desv. **120**
    ssp. chalepensis (L.) O. E. Schulz **121**
    ssp. draba **120**
Cardiospermum halicacabum L. 362
Cardopatium A. L. Juss. **940**
  corymbosum (L.) Pers. **940**
  orientale Spach 940
Carduncellus Adanson **983**
  caeruleus (L.) C. Presl **983**
    var. dentatus DC. 984
Carduus L. **948**
  acanthoides L. 949
  acarna L. 953
  acicularis Bert. 949
  albidus M. Bieb. 951
  arabicus Jacq.
    ssp. marmoratus (Boiss. et Heldr.)
      Kazmi 951
  argentatus L. **949**
    ssp. acicularis (Bert.) Meikle **949**
    × C. pycnocephalus L. 950
  argenteus Sintenis 949
  lanceolatus L. 954
  leucographus L. 951
  marianus L. 963
  marmoratus Boiss. et Heldr. 951
  pycnocephalus L. **950**
    ssp. albidus (M. Bieb.) Kazmi **951**
    var. *albidus* (M. Bieb.) Boiss. 951
  syriacus L. 952
  vulgaris Savi 954
Carex L. **1700**
  subgen. Carex **1703**
  subgen. Vignea (P. Beauv.) Nees **1701**
  acutiformis auct. 1705
  coriogyne Nelmes 1702
  cuspidata Host 1704
  diluta M. Bieb. 1707
  distachya Desf. 1709
  distans L. **1706**
    ssp. *binerviformis* Holmboe 1706
    f. *major* Kneucker 1706
    f. *minor* Holmboe 1707
    f. *sinaica* Kük. 1706
  divisa Huds. **1701**
  divulsa Stokes **1702**
  echinata Desf. 1705
  extensa Good. **1707**
  flacca Schreb. **1704**
    ssp. *cuspidata* (Host) C. Vicioso 1704
    ssp. serrulata (Biv.) Greuter **1704**
  glauca Scop. 1704
    var. *cuspidata* (Host) Aschers. et
      Graebn. 1704

Carex L. (*cont.*)
 ssp. *serrulata* (Biv.) Arcang. 1704
 gynobasis Vill. 1708
 halleriana Asso **1708**
 hispida Willd. ex Schkuhr **1705**
 illegitima Cesati **1709**
 leersiana Rauschert 1702
 muricata auct. 1702
 nemorosa Rabentisch 1703
 otrubae Podp. **1703**
 paludosa auct. 1705
 panicea auct. 1704
 pendula Huds. **1703**
 polyphylla Kar. et Kir. 1702
 remota L. 1702
 serrulata Biv. 1704
 sinaica Nees ex Steud. 1706
 soleirolii Duby 1706
 troodi Turrill **1707**
 vulpina auct. 1703
 vulpina L. 1703
  var. *nemorosa* Koch 1703
Carlina L. 943
 corymbosa L.
  ssp. *graeca* auct. 944
  ssp. *involucrata* (Poir.) Holmboe 944
  var. *involucrata* (Poir.) Boiss. 944
 curetum auct. 944
 graeca auct. 944
 involucrata Poir. 944
  ssp. cyprica Meusel et Kästner **944**
  × C. pygmaea (Post) Holmboe 944
 lanata L. **945**
  var. *pygmaea* Post 945
 pygmaea (Post) Holmboe **945**
 vulgaris auct. 944
Carrichtera DC. **105**
 annua (L.) DC. **105**
 vellae DC. 105
Carthamus L. **978**
 sect. Atractylis Reichb. **981**
 sect. Carthamus **978**
 sect. Lepidopappus Hanelt **980**
 sect. Odontagnathius (DC.) Hanelt **979**
 boissieri Hal. **980**
 caeruleus L. 983
 creticus L. 982
 dentatus Vahl **979**
  ssp. ruber (Link) Hanelt **980**
 glaucus auct. 980
 glaucus M. Bieb.
  ssp. *boissieri* (Hal.) Holmboe 980
  var. *tenuis* (Boiss. et Blanche) Boiss. 981
 lanatus auct. 982
 lanatus L. **982**
  ssp. boeticus (Boiss. et Reut.) Nyman **982**
  ssp. *creticus* (L.) Holmboe 982
 ruber auct. 981
 ruber Link 980
 tenuis (Boiss. et Blanche) Bornm. **981**
 tinctorius L. **978**
Carum L. **742**
 carvi L. 743
 ferulifolium (Desf.) Boiss. 743
 multiflorum (Sibth. et Sm.) Boiss. **742**

Caryophyllaceae 19, **212**
 subfam. *Paronychioideae* 279
 tribe Alsineae DC. **252**
 tribe Caryophylleae **213**
 tribe Polycarpeae (DC.) Benth. **277**
Cassia artemisioides Gaudich. ex DC. 589
 didymobotrya Fres. 589
 fistula L. 589
 senna L. 589
Castanea *castanea* Karsten 1478
 sativa Mill. 1478
 vesca Gaertn. 1478
 vulgaris Lam. 1478
Casuarina L. 1472
 cunninghamiana Miq. 1472
 equisetifolia L. 1472
 glauca Sieb. ex Sprengel 1472
 quadrivalvis Labill. 1472
 stricta Ait. 1472
 verticillata Lam. 1472
Casuarinaceae 1472
Catalpa bignonioides Walt. 1244
Catananche L. **986**
 lutea L. **986**
Catapodium Link **1720**
 loliaceum (Huds.) Link 1721
 marinum (L.) C. E. Hubb. **1721**
 rigidum (L.) C. E. Hubb. **1721**
 tenellum (L.) Trabut 1723
Caucalis L. **704**
 subgen. *Turgenia* (Hoffm.) Benth. 705
 arvensis Huds. 700
 daucoides L. 704, 707
 erythrotricha (Reichb.) Boiss. et Hausskn. 703
 fallax Boiss. et Bl. 700
  var. *brevipes* Boiss. 700
 glabra Forssk. 713
 latifolia L. 706
 leptophylla L. 702
  var. *erythrotricha* (Reichb.) Post 703
 maritima Gouan 708
 nodosa (L.) Scop. 699
 platycarpos L. **704**
 pumila L. 708
 purpurea Ten. 701
 tenella Del. 703
Cedrus Trew **21**
 brevifolia (Hook. f.) Henry 22
 libani auct. 22
 libani A. Rich. **21**
  ssp. brevifolia (Hook. f.) Meikle 2, 6, **22**
   var. *brevifolia* Hook. f. 22, 806
 libanitica Trew ex Pilger 21
  ssp. *brevifolia* (Hook. f.) O. Schwarz 22
  ssp. stenocoma O. Schwarz 22
 libanotica Link 21
  ssp. *brevifolia* (Hook. f.) Holmboe 22
Celsia L. **1197**
 arcturus auct. 1198
 glandulosa Bouché 1198
 horizontalis auct. 1198
 orientalis L. 1199
Celtis L. **1461**
 aspera (Audib. ex Spach) Steven 1462
 asperrima Lojac. 1462
 australis L. **1461**

Celtis L. (*cont.*)
  var. *lutescens* Guss. 1462
  caucasica Willd. 1462
  glabrata Steven ex Planch. 1462
  kotschyana Steven 1462
  tournefortii Lam. **1462**
  var. *aetnensis* Torn. 1462
  var. *aspera* Audib. ex Spach 1462
  var. *glabrata* Boiss. 1462
  var. *laevis* Spach 1462
  var. tournefortii 1462
Cenchrus capitatus L. 1729
Centaurea L. **967**
  sect. Acrolophus (Cass.) DC. **968**
  sect. Aegialophila (Boiss. et Heldr.) O. Hoffm. **973**
  sect. Calcitrapa DC. **969**
  sect. *Crupina* Pers. 966
  sect. Cyanus (Mill.) DC. **974**
  sect. Mesocentron (Cass.) DC. **969**
  sect. Microlophus (Cass.) DC. **968**
  aegialophila Wagenitz **973**
  angusticeps Lindberg f. 971
  f. *albiflora* Lindberg f. 971
  behen auct. 965
  behen L. 969
  benedicta (L.) L. 977
  calcitrapa L. **970**
  ssp. angusticeps (Lindberg f.) Meikle **971**
  ssp. cilicia (Boiss. et Bal.) Wagenitz 971
  cerinthifolia Sm. 965
  cretica (Boiss et Heldr.) Nyman 973
  crupinastrum Moris 966
  cyanoides Berggren et Wahlenb. **975**
  hyalolepis Boiss. **971**
  iberica Trev. ex Spreng. **970**
  monacantha Clarke 972
  pallescens auct. 972
  ssp. *hyalolepis* (Boiss.) Holmboe 971
  var. *hyalolepis* (Boiss.) Boiss. 971
  pumila auct. 973
  pumilio L. 974
  reuteriana Boiss. 975
  salmantica L. 976
  solstitialis L. **969**
  veneris B. L. Burtt et Davis 7, **968**
Centaurium Hill **1111**
  erythraea Rafn **1111**
  ssp. *grandiflorum* Osorio-Tafall et Seraphim 1112
  ssp. rhodense (Boiss. et Reuter) Melderis **1112**
  grandiflorum (Pers.) Ronniger 1112
  majus (Hoffmsgg. et Link) Ronniger 1112
  ssp. *rhodense* (Boiss. et Reuter) Zeltner 1112
  maritimum (L.) Fritsch **1117**
  pulchellum (Swartz) Druce **1114**
  ssp. *ramosissimum* (Vill.) P. Fourn. 1114
  ramosissimum (Vill.) Druce 1114
  spicatum (L.) Fritsch **1116**
  tenuiflorum (Hoffmsgg. et Link) Fritsch **1115**
  ssp. acutiflorum (Schott) Zeltner 1116

Centaurium Hill (*cont.*)
  var. hermanii (Sennen) Zeltner 1116
Centranthus Necker ex Lam. et DC. **836**
  calcitrapa (L.) Dufr. **838**
  ssp. orbiculatus (Sm.) Meikle **838**
  var. *orbiculatus* (Sm.) DC. 838
  ruber (L.) DC. **837**
  ssp. *sibthorpii* (Heldr. et Sart. ex Boiss.) Hal. 837
  var. sibthorpii (Heldr. et Sart. ex Boiss.) Baldacci **837**
  sibthorpii Heldr. et Sart. ex Boiss. 837
Cephalanthera L. C. M. Rich. **1494**
  alba (Crantz) Simonk. 1495
  damasonium (Mill.) Druce **1495**
  ensifolia (Murray) L. C. M. Rich. 1496
  grandiflora S. F. Gray 1495
  latifolia Janch. 1495
  longifolia (L.) Fritsch **1496**
  pallens (Jundz.) L. C. M. Rich. 1495
  rubra (L.) L. C. M. Rich. **1496**
Cephalaria Schrader ex R. & S. **848**
  syriaca (L.) Schrader **848**
  f. *pedunculata* (DC.) Boiss. 848
  var. *pedunculata* DC. 848
  ssp. phoeniciaca Bobr. **849**
  ssp. syriaca **849**
Cephalorrhynchus Boiss. **1024**
  candolleanus auct. 1025
  cypricus (Beauverd) Rechinger f. **1024**
  tuberosus auct. 1025
Cerastium L. **253**
  anomalum Waldst. et Kit. 253
  brachypetalum Pers. **256**
  ssp. roeseri (Boiss. et Heldr.) Nyman **257**
  var. *roeseri* (Boiss. et Heldr.) Boiss. 257
  comatum Desv. 254
  dichotomum L. **255**
  dubium (Bast.) Schwarz **253**
  fontanum Baumg. 258
  fragillimum Boiss. 258
  glomeratum Thuill. **255**
  holosteoides Fr. 258
  illyricum auct. 254
  illyricum Ard. **254**
  ssp. *comatum* (Desv.) Sell et Whitehead 254
  ssp. pilosum Rouy et Fouc. **254**
  pilosum Sibth. et Sm. 254
  roeseri Boiss. et Heldr. 257
  semidecandrum L. **257**
  viscosum Sibth. et Sm. 255
  vulgatum auct. 258
Cerasus avium Moench 610
Ceratocapnos Durieu **86**
  palaestinus Boiss. **86**
Ceratocephala Moench **43**
  falcata (L.) Pers. **43**
  var. *exscapa* Boiss. 43
Ceratonia L. **589**
  siliqua L. **589**
Cercidium floridum Benth. 587
Cercis siliquastrum L. 591
  var. hebecarpa Bornm. 592
Cerinthe orientalis L. 1162
Ceterach DC. 1887

1922 INDEX

Ceterach DC. (*cont.*)
   officinarum DC. 1890
Chaenorhinum (DC.) Reichb. **1210**
   rubrifolium (Robill. et Cast. ex DC.)
     Fourr. **1210**
Chaetospora ferruginea auct. 1698
Chamaenerion angustifolium (L.) Scop. 670
   parviflorum Schreber 672
   spicatum (Lam.) S. F. Gray 670
Chamaepeuce alpini Jaub. et Spach 956
   var. *camptolepis* auct. 956
   fruticosa auct. 956
   mutica DC. 956
Chamerion angustifolium (L.) Holub 670
Cheilanthes Swartz **1880**
   catanensis (Cosent.) H. P. Fuchs 1882
   fragrans Swartz 1882
   marantae (L.) Domin **1880**
   pteridioides (Reichard) C. Chr. **1882**
   suaveolens Swartz 1882
   vellea (Ait.) F. Mueller **1880**
Cheiranthus bicornis Sibth. et Sm. 157
   chius L. 161
   coronopifolius Sibth. et Sm. 156
   flexuosus Sibth. et Sm. 160
   fruticulosus L. 156
   incanus L. 155
   longipetalus Vent. 157
   lyratus (L.) Sibth. et Sm. 162
   sinuatus L. 155
   tricuspidatus L. 158
   tristis L. 156
Chelidonium hybridum L. 82
   violaceum Lam. 82
Chenopodiaceae **1370**
      tribe Atripliceae C. A. Meyer **1378**
      tribe Beteae Moq. **1372**
      tribe Camphorosmeae Moq. **1383**
      tribe Chenopodieae **1373**
      tribe Polycnemeae (Dumort.) Moq.
       1371
      tribe Salicornieae (Dumort.) Moq.
       1383
Chenopodium L. **1373**
   aegyptiacum Hasselq. 1389
   album L. **1377**
   botrys L. **1374**
   foliosum (Moench) Aschers. **1374**
   fruticosum L. 1388
   hircinum Schrader 1377
   maritimum L. 1389
   murale L. **1376**
   opulifolium Schrader ex Koch et Ziz
     1376
   rubrum L. 1375
   scoparia L. 1383
   virgatum (L.) Ambrosi 1375
   vulvaria L. **1375**
Chilopsis linearis (Cav.) Sweet 1244
Chionodoxa Boiss. **1639**
   lochiae Meikle **1639**
Chironia maritima (L.) Willd. 1117
Chlamydophora Ehrenb. ex Less. **926**
   tridentata (Del.) Ehrenb. ex Less. **926**
Chlora acuminata Koch et Ziz 1110
   intermedia Ten. 1109
   perfoliata (L.) L. 1109
     ssp. *intermedia* (Ten.) Nyman 1109

Chlora Adans. (*cont.*)
   var. *minor* Ten. 1109
     ssp. *serotina* (Koch ex Reichb.)
      Arcang. 1110
     var. *sessilifolia* Griseb. 1110
   serotina Koch ex Reichb. 1110
Chlorocyperus Rikli 1684
   glaber (L.) Palla 1687
   globosus (All.) Palla 1690
   junciformis (Cav.) Rikli 1690
   longus (L.) Palla 1685
   rotundus (L.) Palla 1684
Chondrilla L. **1023**
   juncea L. **1023**
Chrozophora A. H. L. Juss. **1449**
   hierosolymitana Spreng. 1450
   obliqua (Vahl) A. H. L. Juss. ex Spreng.
     1450
   tinctoria (L.) Raf. **1450**
     var. *hierosolymitana* (Spreng.) Muell.-
      Arg. 1450
     var. *verbascifolia* (Willd.) Muell-Arg.
      1450
   verbascifolia (Willd.) A. H. L. Juss. ex
     Spreng. 1450
Chrysanthemum L. **920**
   balsamita auct. 922
   coronarium L. **921**
     var. coronarium **921**
     var. discolor Urv. **922**
   myconis L. 922
   parthenium (L.) Bernh. 923
   segetum L. **920**
Chrysopogon Trin. **1862**
   gryllus (L.) Trin. **1863**
Cicer L. **535**
   arietinum L. **536**
   ervoides Brign. 565
   incisum (Willd.) K. Malý 536
Cicerbita cyprica Beauverd 1024
Cichorium L. **988**
   divaricatum Schousboe 990
   endivia L. **989**
     f. divaricatum (Schousboe) Webb **989**
     ssp. *divaricatum* (Schousboe) P. D. Sell
      990
     ssp. *pumilum* (Jacq.) Hegi 990
   intybus L. **988**
     ssp. *divaricatum* (Schousboe) Bonnier
      et Layens 990
     var. *divaricatum* (Schousboe) DC. 990
   pumilum Jacq. 989
   spinosum L. 990
Cionura Griseb. **1106**
   erecta (L.) Griseb. **1106**
Circinnus circinnatus (L.) O. Ktze. 472
Cirsium Mill. **954**
   acarna (L.) Moench 953
   chamaepeuce (L.) Ten. 956
     ssp. *camptolepis* auct. 956
     ssp. *fruticosum* auct. 956
   lanceolatum (L.) Scop. 954
   syriacum (L.) Gaertn. 952
   vulgare (Savi) Ten. **954**
Cistaceae 19, **181**
Cistanche Hoffmsgg. et Link **1232**
   lutea (Desf.) Hoffmsgg. et Link 1233
   phelypaea (L.) P. Cout. **1233**

Cistanche Hoffmsgg. et Link (*cont.*)
tinctoria (Forssk.) G. Beck-Mannagetta 1233
Cistus L. 4, **182**
aegyptiacus L. 196
arabicus L. 197
creticus L. 3, **182**
var. creticus **183**
var. tauricus (Presl) Dunal **183**
crispus L. 187
× cyprius Lam. 187
guttatus L. 188
hirtus Sibth. et Sm. 191
incanus auct. 184
incanus L. 182, 184
ladanifer L. 6, **187**
ladanifer L. × laurifolius L. 187
laevis Cav. 200
ledifolius L. 195
monspeliensis L. 3, 6, **186**
var. *minor* Willk. 186
monspeliensis L. × parviflorus Lam. **186**
niloticus L. 195
parviflorus Lam. **184**
parviflorus Lam. × salviifolius L. **185**
pilosus L. 182
polymorphus Willk. 183
racemosus auct. 190
salicifolius L. 194
salviifolius L. 3, **185**
var. *biflorus* Willk. 185
f. *brevipedunculatus* Willk. 185
f. *longipedunculatus* Willk. 185
var. *undulatifolius* Kotschy 185
serratus Cav. 189
× skanbergii Lojac. **186**
stipulatus Forssk. 192
syriacus Jacq. 190
tauricus Presl 183
thymifolius L. 199
villosus L. 182
var. *creticus* (L.) Boiss. 183
var. *tauricus* (Presl) Grosser 183
Citrullus Eckl. et Zeyh. **676**
colocynthis (L.) Schrad. **676**
edulis Spach 677
lanatus (Thunb.) Mats. et Nakai 677
vulgaris Eckl. et Zeyh. 677
Citrus aurantifolia (Christm.) Swingle 351
aurantium L. 350
bergamia Risso et Poit. 351
deliciosa Ten. 350
grandis (L.) Osbeck 351
limetta Risso 350
limon (L.) Burm. f. 350
medica L. 350
paradisi Macf. 350
sinensis (L.) Osbeck 350
Cladium P. Browne **1699**
germanicum Schrader 1699
mariscus (L.) Pohl 7, **1699**
Clematis L. **35**
cirrhosa L. **36**
vitalba L. **38**
viticella L. **36**
Cleome L. **173**
cypria Čelak. 174
iberica DC. **174**

Cleome L. (*cont.*)
ornithopodioides auct. 174
ornithopodioides L. **174**
ssp. *cypria* (Čelak.) Holmboe 174
Clinopodium L. **1283**
vulgare L. **1283**
ssp. orientale v. Bothmer 1284
Closterandra minor Boiv. 76
Clypeola L. **143**
glabra Boiss. 144
hispida Presl 144
jonthlaspi L. **143**
var. glabra Boiss. **144**
var. hispida (Presl) Reynier **144**
var. *intermedia* Hal. 144
var. jonthlaspi **144**
var. lasiocarpa Guss. **144**
maritima L. 142
minor L. 138
Cnicus L. **977**
benedictus L. **977**
syriacus Willd. 952
Cochlearia coronopus L. 121
Cola acuminata (P. Beauv.) Schott et Endl. 314
nitida (Vent.) Schott et Endl. 314
Colchicum L. **1593**
autumnale auct. 1594
bertolonii auct. 1595
cupanii auct. 1595
decaisnei Boiss. 1594
hiemale Freyn 1595
pusillum Sieber **1594**
stevenii Kunth **1595**
troodi Kotschy **1594**
Colobanthium michelii (Savi) Osorio-Tafall et Seraphim 1766
Colocasia *antiquorum* Schott 1670
ssp. *esculenta* (L.) Haudricourt 1670
var. *esculenta* (L.) Schott ex Seem. 1670
esculenta (L.) Schott 1670
var. *antiquorum* (Schott) Hubbard et Rehder 1670
Colocynthis Mill. 676
vulgaris Schrad. 676
Compositae **861**
subfam. Asteroideae **866**
subfam. Lactucoideae **984**
tribe Anthemideae Cass. **903**
tribe Arctoteae Cass. **937**
tribe Astereae **867**
tribe Calenduleae Cass. **935**
tribe Cardueae Cass. **938**
tribe Eupatorieae Cass. **866**
tribe Helenieae Benth. **902**
tribe Heliantheae Cass. **898**
tribe Lactuceae Cass. **984**
tribe Senecionideae Cass. **928**
Conium L. **724**
maculatum L. **725**
Conringia Adans. **115**
orientalis (L.) Dum. **115**
Consolida ambigua (L.) P. W. Ball 65
phrygia (Boiss.) Soó 65
Convolvulaceae **1163**
Convolvulus L. **1166**
althaeoides L. **1172**

Convolvulus L. (*cont.*)
  arvensis L. **1172**
    var. arvensis **1172**
    var. *cherleri* (Agardh) Hal. 1173
    var. linearifolius Choisy **1173**
  betonicifolius Mill. 1171
  cantabrica L. 1168
  cherleri Agardh 1173
  coelesyriacus Boiss. **1173**
  cyprius auct. 1168
  × cyprius Boiss. 7, **1171**
  dorycnium L. **1166**
    ssp. *oxysepalus* (Boiss.) Rechinger f.
      1166
    var. *oxysepalus* Boiss. 1166
  evolvuloides Desf. 1175
  hirsutus M. Bieb. 1171
  humilis Jacq. **1175**
  lineatus L. **1170**
    var. *angustifolius* Kotschy ex Sa'ad
      1171
  littoralis L. 1164
  oleifolius Desr. 7, **1167**
    var. deserti Pamp. **1168**
    var. oleifolius **1167**
    var. pumilus Pamp. 7, **1168**
  pentapetaloides L. **1174**
  purpureus L. 1165
  sepium L. 1165
  siculus L. **1174**
  sintenisii Boiss. 1173
  stolonifer Cyr. 1164
  undulatus Cav. 1175
Conyza Less. **871**
  bonariensis (L.) Cronq. **872**
  canadensis (L.) Cronq. **872**
  geminiflora Ten. 885
  squamata Spreng. 871
  squarrosa L. 890
  tenorii Spreng. 885
  tomentosa Forssk. 885
  vulgaris Lam. 890
Corchorus L. **315**
  capsularis L. 315
  olitorius L. 315
  trilocularis auct. 315
  trilocularis L. **316**
Cordia myxa L. 1120
Coriandrum L. **717**
  aquilegifolium All. 724
  sativum L. **717**
  testiculatum L. 719
Coridothymus capitatus (L.) Reichb. f. 1271
Coronilla L. **511**
    sect. Coronilla **513**
    sect. Emerus Desv. **511**
    sect. Scorpioides Benth. et Hook. f.
      516
  cretica auct. 513
  emeroides Boiss. et Sprun. 512
  emerus L. **511**
    ssp. emeroides (Boiss. et Sprun.)
      Holmboe **512**
  minima L. 513
  parviflora Willd. 513
    var. *rubriflora* Cand. 514
  repanda (Poir.) Guss. **517**
    ssp. dura (Cav.) Maire 519

Coronilla L. (*cont.*)
    ssp. repanda 519
  rostrata Boiss. et Sprun. **513**
  scorpioides (L.) Koch **516**
  securidaca L. **514**
  varia L. 513
Coronopus J. G. Zinn **121**
  procumbens Gilib. 121
  ruellii All. 121
  squamatus (Forssk.) Aschers. **121**
  verrucarius Muschler et Thell. 121
Cortaderia *argentea* (Nees) Stapf 1839
  selloana (Schult. et Schult. f.) Aschers. et
    Graebn. 1839
Corydalis Vent. **85**
  rutifolia (Sibth. et Sm.) DC. **86**
Corylaceae **1474**
Corylus L. **1476**
  avellana auct. 1477
  avellana L. **1476**
  maxima Mill. 1477
  tubulosa Willd. 1477
Corynephorus P. Beauv. **1755**
  articulatus (Desf.) P. Beauv. 1756
  canescens (L.) P. Beauv. 1756
  divaricatus (Pourr.) Breistr. **1756**
Cosmos bipinnatus Cav. 902
Cota altissima auct. 910
  palaestina Reuter ex Kotschy 909
Cotinus coggygria Scop. 372
Cotoneaster Medik. **637**
  nummularia Fisch. et Mey. 638
  racemiflorus (Desf.) C. Koch **638**
    var. nummularius (Fisch. et Mey.)
      Dippel **638**
Cotula aurea Loefl. 925
  complanata Sm. 925
  coronopifolia auct. 926
  tridentata (Del.) Dinsm. 2, 926
Cotyledon horizontalis Guss. 643
  rupestris Salisb. 643
  tuberosa (L.) Hal. 643
Crambe L. **110**
  corvini All. 109
  hispanica L. 7, **110**
Crassula L. **641**
  alata (Viv.) Berger **641**
  cespitosa Cav. 652
  microcarpa Sm. 653
  rubens (L.) L. 651
  tillaea auct. 641
  vaillantii (Willd.) Roth **642**
Crassulaceae 19, **640**
Crataegus L. **635**
  aria L. 634
  aronia (L.) Bosc. ex DC. 635
  azarolus L. **635**
    var. *aronia* L. 635
  calycina auct. 636
  graeca Spach 634
  monogyna Jacq. **636**
  orientalis auct. 635
  × *ruscinonensis* Grenier et Blanc 637
  × sinaica Boiss. **637**
Crepis L. **1003**
  aculeata auct. 1012
  altissima Balb. 993
  aspera L. 1012

Crepis L. (cont.)
　　var. inermis (Cass.) Boiss. 1012
　　barbata L. 992
　　bulbosa (L.) Tausch 1014
　　　　var. polycephala Boiss. 1014
　　bureniana Boiss. 1005
　　commutata (Spreng.) Greuter 1009
　　dioscoridis auct. 1004
　　fallax Boiss. 1008
　　foetida L. 1007
　　　　ssp. commutata (Spreng.) Babcock
　　　　　　1009
　　　　ssp. foetida 1008
　　　　var. glandulosa Lindberg f. 1008
　　　　ssp. vulgaris Babcock 1008
　　　　　　f. fallax (Boiss.) Babcock 1008
　　　　var. zacynthia (Marg. et Reut.) Lind-
　　　　　　berg f. 1008
　　fraasii Sch. Bip. 1004
　　glandulosa Guss. 1008
　　hierosolymitana auct. 1004, 1005
　　kotschyana Boiss. 1005
　　micrantha Czerep. 1011
　　montana Urv. 1004
　　muricata Sm. 1011
　　nemausensis Gouan 1010
　　palaestina (Boiss.) Bornm. 1006, 1007
　　parviflora Desf. ex Pers. 1011
　　pulchra L. 1006
　　raulinii auct. 1004
　　reuteriana Boiss. 1005
　　rhoeadifolia M. Bieb. 1008
　　sancta (L.) Bornm. 1009
　　　　ssp. bifida (Vis.) Thell. 1010
　　　　ssp. nemausensis (Gouan) Thell. 1010
　　sieberi Boiss. 1004
　　virgata Desf. 993
　　zacintha (L.) Babcock 1011
Cressa L. 1176
　　cretica L. 1176
Crithmum L. 748
　　maritimum L. 748
Crithopsis Jaub. et Spach 1827
　　delileana (Schult.) Roschevicz 1827
　　rhachitricha Jaub. et Spach 1827
Crocus L. 1563
　　aerius Herb. 1564
　　　　var. cyprius (Boiss. et Kotschy) Baker
　　　　　　1564
　　cyprius Boiss. et Kotschy 1564
　　hartmannianus Holmboe 1564
　　veneris Tappeiner 1566
　　vernus Sibth. et Sm. 1564
Croton obliquus Vahl 1450
　　tinctorius L. 1450
　　verbascifolius Willd. 1450
　　villosus Sm. 1450
Crucianella L. 794
　　aegyptiaca L. 796
　　angustifolia auct. 795
　　herbacea Forssk. 796
　　imbricata Boiss. 796
　　latifolia auct. 796
　　latifolia L. 794
　　macrostachya auct. 796
　　macrostachya Boiss. 795
Cruciata Mill. 790
　　articulata (L.) Ehrend. 790

Cruciata Mill. (cont.)
　　pedemontana (Bell.) Ehrend. 791
Cruciferae 19, 94
　　tribe Alysseae Gren. et Godr. 133
　　tribe Arabideae DC. 146
　　tribe Brassiceae 96
　　tribe Drabeae O. E. Schulz 144
　　tribe Euclideae DC. 131
　　tribe Hesperideae Prantl 158
　　tribe Lepidieae DC. 116
　　tribe Matthioleae O. E. Schulz 154
　　tribe Sisymbrieae DC. 163
Crupina (Pers.) DC. 966
　　crupinastrum (Moris) Vis. 966
　　vulgaris auct. 966
Crypsis Ait. 1847
　　aculeata (L.) Ait. 1848
　　alopecuroides (Pill. et Mitterp.) Schrad.
　　　　1847
　　factorovskyi Eig 1848
　　schoenoides (L.) Lam. 1849
Cryptoceras rutifolium Schott et Kotschy
　　86
Cucubalus aegyptiacus L. 235
　　behen L. 231
　　inflatus Salisb. 231
　　italicus L. 228
　　venosus Gilib. 231
Cucumis L. 679
　　callosus (Rottl.) Cogn. 679
　　colocynthis L. 676
　　dudaim L. 679
　　ficifolia Bouché 679
　　maxima Lam. 679
　　melo L. 679
　　mixta Pangalo 679
　　moschata (Lam.) Poiret 679
　　pepo L. 679
　　sativus L. 679
Cucurbitaceae 19, 675
Cupressaceae 19, 27
Cupressus L. 27
　　arizonica Greene 29
　　fastigiata DC. 28
　　guadalupensis S. Wats. 29
　　horizontalis Mill. 28
　　macrocarpa Hartw. 29
　　pyramidalis Targ.-Tozz. 28
　　sempervirens L. 1, 28
　　　　var. horizontalis (Mill.) Aiton 28
　　　　var. pyramidalis (Targ.-Tozz.) Nyman
　　　　　　28
　　　　f. pyramidalis (Targ.-Tozz.) Holmboe
　　　　　　28
　　　　var. sempervirens 28
Cuscuta L. 1177
　　campestris Yuncker 1178
　　epilinum Weihe 1179
　　epithymum auct. 1178, 1179
　　globularis Bert. 1178
　　major auct. 1177
　　minor auct. 1179
　　monogyna Vahl 1177
　　palaestina Boiss. 1178
　　planiflora Ten. 1179
Cutandia Willk. 1723
　　dichotoma (Forssk.) Trabut 1724
　　maritima (L.) Richter 1723

Cutandia Willk. (*cont.*)
  memphitica auct. 1724
Cuviera aspera (Simonkai) Simonkai 1830
  caput-medusae (L.) Simonkai 1830
  var. *aspera* Simonkai 1830
Cyclamen L. **1077**
  cilicium Boiss. et Heldr. 1080
  coum Mill. 1079
  creticum Hildebr. 1079
  cyprium Kotschy **1079**
  cyprium Sibth. 1078
  cypro-graecum E. & N. Mutch 1080
  graecum Link 7, **1080**
  hederifolium auct. 1078
  latifolium Sm. 1078
  neapolitanum auct. 1079
  persicum Mill. **1078**
  repandum Sm. 1079
Cydonia oblonga Mill. 633
  vulgaris Pers. 633
Cymbalaria Hill **1205**
  longipes (Boiss. et Heldr.) Cheval. **1205**
  muralis auct. 1206
Cymodocea Koenig **1682**
  aequorea Koenig 1682
  major (Cavolini ex Willd.) Grande 1682
  nodosa (Ucria) Aschers. **1682**
Cymodoceaceae **1681**
Cynanchum erectum L. 1106
Cynara L. **960**
  cardunculus L. **961**
  cornigera Lindley **962**
  horrida Aiton 961
  humilis auct. 962
  scolymus L. 962
  sibthorpiana Boiss. et Heldr. 962
Cynocrambe prostrata Gaertn. 799
Cynodon L. Rich. **1842**
  dactylon (L.) Pers. **1843**
    var. dactylon **1843**
    var. villosus Regel **1843**
Cynoglossum L. **1124**
  apenninum L. 1129
  creticum Mill. **1128**
  germanicum Jacq. 1125
  hungaricum Simonkai 1125
  montanum L. **1124**
    var. *asiaticum* auct. 1126
    var. *asiaticum* Brand 1125
    ssp. extraeuropaeum Brand **1125**
  nebrodense Guss. 1125
  pictum Ait. 1128
  pustulatum Boiss. 1125
  teheranicum H. Riedl 1125
  troodi Lindberg f. **1126**
Cynosurus L. **1725**
  aegyptius L. 1840
  aureus L. 1727
  callitrichus C. et W. Barbey 1725
  coloratus Lehm. ex Nees **1725**
  durus L. 1739
  echinatus L. **1726**
    var. *callitrichus* (C. et W. Barbey) Bornm. 1725
  elegans Desf. **1726**
Cyperaceae **1682**
  subfam. Caricoideae **1700**
  subfam. Cyperoideae **1683**

Cyperaceae (*cont.*)
  tribe Cypereae **1683**
  tribe Rhynchosporeae **1698**
  tribe *Scirpeae* 1683
Cyperus L. **1684**
  subgenus Anosporum (Nees) C. B. Clarke **1687**
  subgenus Cyperus **1684**
  subgenus *Eu-Cyperus* C. B. Clarke 1684
  subgenus *Juncellus* (Griseb.) C. B. Clarke **1688**
  subgenus *Pycreus* (P. Beauv.) C. B. Clarke 1690
  badius Desf. 1687
  capitatus Vand. **1688**
  cyprius Post **1686**
  distachyos All. 1690
  esculentus L. 1685
  flavidus Retz. 1690
  fuscus L. **1687**
    var. *virescens* auct. 1688
  glaber L. **1687**
  globosus All. 1690
  heldreichianus Boiss. 1686
  junciformis Cav. 1690
  kalli Forssk. 1688
  laevigatus L. **1689**
    var. distachyos (All.) Coss. et Dur. **1690**
  longus L. **1685**
    ssp. *badius* (Desf.) Aschers. et Graebn. 1686
    f. *cypricus* (C. B. Clarke) Kükenth. 1686
    var. *cypricus* C. B. Clarke 1686
    f. *depauperatus* Holmboe 1686
    f. *heldreichianus* (Boiss.) Holmboe **1686**
    var. heldreichianus (Boiss.) Boiss. **1686**
    var. longus **1686**
  mucronatus (L.) Briq. 1688
  mucronatus Rottb. 1689
  olivaris auct. 1686
  rotundus L. **1684**
  schoenoides Griseb. 1688
Cyprinia Browicz **1102**
  gracilis (Boiss.) Browicz **1103**
Cystopteris Bernh. **1885**
  fragilis (L.) Bernh. **1885**
    ssp. *eu-fragilis* Holmboe 1885
    var. *genuina* Boiss. 1885
Cytinus L. **1415**
  hypocistis (L.) L. **1415**
    ssp. *clusii* Nyman 1415
    ssp. *kermesinus* (Guss.) Arcang. 1415
    var. *kermesinus* Guss. 1415
    ssp. *orientalis* Wettst. 1415
  ruber (Fourr.) Komarov 1415
Cytisopsis dorycniifolia Jaub. et Spach 473
Cytisus laniger DC. 382
  uniflorus Decne. 381

Dactylis L. **1728**
  australis (Willk.) Sintenis 1728
  glomerata L. **1728**
    var. *australis* Willk. 1728

Dactylis L. (*cont.*)
  ssp. hispanica (Roth) Nyman 1728
    var. *hispanica* (Roth) Koch 1728
  hispanica Roth 1728
  lagopoides L. 1845
  repens Desf. 1845
Dactyloctenium Willd. **1840**
  aegyptium (L.) P. Beauv. **1840**
Dactylorhiza Necker ex Nevski **1554**
  iberica (M. Bieb. ex Willd.) Soó **1556**
  romana (Seb.) Soó **1555**
  sambucina (L.) Soó 1555
    var. *pseudosambucina* (Ten.) Sundermann 1555
  sulphurea (Link) Franco 1555
    ssp. *pseudosambucina* (Ten.) Franco 1555
Dahlia Cav. 901
  coccinea Cav. 901
  × cultorum Thorsr. et Reis. 901
  pinnata Cav. 901
Dalbergia sissoo Roxb. 583
Damasonium Mill. **1672**
  alisma Mill. **1672**
    ssp. *bourgaei* (Coss.) Osorio-Tafall et Seraphim 1672
    var. *compactum* M. Micheli 1672
  bourgaei Coss. 1672
Danaä aquilegifolia (All.) All. 724
Daphne L. 1423
  argentea Sm. 1426
  oleoides Schreber 1423
    var. brachyloba Meissner 1424
    var. glandulosa (Bertol.) Keissler 1424
  tartonraira L. 1425
Datisca L. **674**
  cannabina L. **674**
  glomerata (Presl) Baill. 674
Datiscaceae 19, **674**
Datura L. **1190**
  innoxia Mill. **1191**
  metel auct. 1191
  stramonium L. **1190**
    var. tatula (L.) Torr. 1190
Daucus L. **709**
  aureus Desf. **711**
  bicolor Sibth. et Sm. 712
  broteri Ten. **711**
  carota L. **709**
    ssp. carota **710**
    ssp. maximus (Desf.) Ball **710**
  durieua Lange 2, **715**
  glaber (Forssk.) Thell. **713**
  guttatus Sibth. et Sm. **713**
    var. *brachylaenus* (Boiss.) Hayek 713
  involucratus Sibth. et Sm. **715**
  littoralis Sibth. et Sm. 714
    var. *forskahlei* Boiss. 714
    var. *genuinus* Boiss. 714
  maritimus auct. 714
  maximus Desf. 710
  muricatus auct. 714
  muricatus L. 714
    var. *littoralis* (Sibth. et Sm.) H. Stuart Thompson 714
  setulosus Guss. 713
    var. *brachylaenus* Boiss. 713
  subsessilis Boiss. 715

Daucus L. (*cont.*)
  visnaga L. 740
Delphinium L. **63**
  subgen. Consolida (DC.) Dalla Torre et Harms **65**
  subgen. Delphinium **66**
  ajacis auct. 65
  ambiguum L. **65**
  caseyi B. L. Burtt **67**
  eriocarpum (Boiss.) Hal. 66
  peregrinum L. **66**
    ssp. *eriocarpum* (Boiss.) Holmboe 66
    var. eriocarpum Boiss. **66**
  phrygium Boiss. **65**
  staphisagria L. **67**
Dennstaedtiaceae **1884**
Desmazeria loliacea (Huds.) Nym. 1721
Dianthus L. **214**
  cinnamomeus Sibth. et Sm. 218
  crinitus Sm. 218
  cyprius A. K. Jackson et Turrill **216**
  diffusus Sibth. et Sm. 218
  multipunctatus Ser. 215
    var. *troodi* Post 215
  pendulus Boiss. et Bl. 218
    ssp. *cyprius* (A. K. Jackson et Turrill) Osorio-Tafall et Seraphim **216**
  quadrilobus auct. 215
  strictus Banks et Sol. 215
    ssp. *axilliflorus* (Fenzl) Eig **216**
    ssp. *troodi* (Post) Osorio-Tafall et Seraphim 215, 806
    var. troodi (Post) S. S. Hooper **215**
  sulcatus auct. 215
  tripunctatus Sibth. et Sm. **214**
  velutinus Guss. 218
Dichondra micrantha Urban 1176
  repens auct. 1176
Didesmus Desv. **112**
  aegyptius (L.) Desv. **112**
    var. aegyptius **112**
    var. tenuifolius (Sibth. et Sm.) Heldr. **113**
  bipinnatus auct. 113
  tenuifolius (Sibth. et Sm.) DC. 113
Digitaria Haller **1853**
  paspalodes Michx. 1858
  sanguinalis (L.) Scop. **1853**
Dioscoreaceae **1579**
Diospyros ebenum J. G. Koenig 1087
  kaki Thunb. 1087
  lotus L. 1087
Diotis Desf. 907
  maritima (L.) Desf. 907
Diplotaxis DC. **103**
  viminea (L.) DC. **103**
Dipsacaceae **848**
Dodonaea viscosa (L.) Jacq. 362
Dolichos lablab L. 582
  purpureus L. 582
Dorycnium Mill. **475**
  graecum (L.) Ser. **477**
  hirsutum (L.) Ser. 475
  kotschyi Boiss. et Reut. ex Boiss. 477
  latifolium Willd. 477
  rectum (L.) Ser. **476**
Doxantha unguis-cati (L.) Miers 1246
Draba verna L. 145

Draba L. (*cont.*)
  ssp. *vulgaris* (DC.) Holmboe 145
Dryopteris Adanson **1886**
  filix-mas auct. 1887
  libanotica (Rosenst.) C. Chr. 1887
  pallida (Bory) Fomin **1886**
    ssp. libanotica (Rosenst.) Nardi **1887**
    ssp. pallida 1886
  rigida auct. 1886
  villarii auct. 1886
Duranta repens L. 1252

Ebenaceae **1086**
Ecballium A. Rich. **677**
  elaterium (L.) A. Rich. **677**
Echinaria Desf. **1729**
  capitata (L.) Desf. **1729**
Echinochloa P. Beauv. **1854**
  colonum (L.) Link **1854**
  crusgalli (L.) P. Beauv. **1855**
    var. crusgalli **1855**
    var. submutica Neilr. **1856**
Echinophora L. **693**
  sibthorpiana Guss. 694
  tenuifolia L. **694**
    ssp. sibthorpiana (Guss.) Holmboe **694**
  tenuifolia auct. 694
Echinops L. **938**
  corymbosus L. 940
  creticus Boiss. et Heldr. 939
  glandulosus Weiss 939
  spinosissimus Turra **939**
    ssp. bithynicus (Boiss.) Kož. 940
  spinosus auct. 939
  viscosus DC. 939
    ssp. *creticus* (Boiss. et Heldr.) Rech. f.
      939
    ssp. *glandulosus* (Weiss) Rech. f. 939
Echinopsilon hirsutus auct. 1392
Echites antidysenterica Roth 1101
Echium L. 1152
  angustifolium Mill. **1154**
  arenarium Guss. **1156**
  boissieri Steud. 1153
  creticum auct. 1154
  diffusum auct. 1156
  elegans Lehm. 1154
  glomeratum Poir. **1152**
  hispidum Sm. 1154
  humile Desf. 1136
  italicum auct. 1152
  italicum L. **1153**
  lycopsis auct. 1155
  plantagineum L. **1155**
  pomponium Boiss. 1153
  pyramidatum DC. 1153
  rauwolfi auct. 1154
  rubrum Forssk. 1154
    ssp. *halacsyi* (Holmboe) Osorio-Tafall
      et Seraphim 1154
  sericeum auct. 1154
  sericeum Vahl 1154
    var. *diffusum* Boiss. 1154
    ssp. *elegans* (Lehm.) Holmboe 1154
    ssp. *halacsyi* Holmboe 1154
    var. *hispidum* Boiss. 1154
  sibthorpii Roemer et Schultes 1154
  vulgare L. 1156

Elaeagnaceae **1427**
Elaeagnus L. **1428**
  angustifolia L. **1428**
    var. angustifolia 1429
    var. inermis Lam. et DC. 1429
    var. *orientalis* (L.) O. Kuntze 1428
  hortensis M. Bieb. 1428
  orientalis L. 1428
Elatinaceae 19, **292**
Elatine L. **292**
  campylosperma Seub. 292
  hydropiper auct. 292
  macropoda Guss. **292**
Eleocharis R. Br. **1692**
  palustris (L.) Roem. et Schult. **1692**
  uniglumis (Link) Schult. 1692
Elymus aegyptiacus Spreng. 1827
  caput-medusae auct. 1828
  caput-medusae L. 1830
    var. *aspera* (Simonkai) Hal. 1830
  crinitus Schreb. 1828
  delileanus Schult. 1827
  elongatus (Host) Runemark 1819
    ssp. *haifensis* (Meld.) Runemark 1819
  geniculatus L. 1827
Elytrigia elongata (Host) Nevski 1818
    var. *haifensis* Meld. 1818
Emex Campderá **1413**
  spinosa (L.) Campderá **1414**
Enarthrocarpus Labill. **107**
  arcuatus Labill. 8, **108**
  lyratus (Forssk.) DC. **108**
  strangulatus auct. 109
Ensete Bruce 1558
Ephedra L. **33**
  alata Decne. 35
  alte C. A. Mey. 35
  fragilis Desf. **33**
    ssp. campylopoda (C. A. Mey.) Aschers.
      et Graebn. **33**
    var. *campylopoda* (C. A. Mey.) Stapf 33
    ssp. fragilis 34
  major Host **34**
    ssp. *procera* (Fisch. et Mey.) Markgraf 34
  nebrodensis Tineo 34
  procera (Fisch. et Mey.) Chapman 34
Ephedraceae 19, **32**
Epilobium L. **669**
  sect. Chamaenerion Tausch **670**
  sect. Epilobium **671**
  adnatum Griseb. 673
  angustifolium L. **670**
  × aschersonianum Hausskn. **673**
  hirsutum L. **671**
  lanceolatum Seb. et Mauri **672**
  lanceolatum Seb. et Mauri × parviflorum
    Schreb. **673**
  montanum L. 672
    var. *lanceolatum* (Seb. et Mauri) Koch
      672
  parviflorum Schreb. **672**
  spicatum Lam. 670
  tetragonum L. **673**
    var. lamyi F. W. Schultz 674
Epipactis Zinn **1497**
  alba Crantz 1495
  condensata Boiss. ex D. P. Young **1502**
  consimilis Boiss. ex Hook. f. 1498

Epipactis Zinn (*cont.*)
  helleborine (L.) Crantz **1503**
    ssp. *condensata* (Boiss. ex D. P. Young)
      Sundermann 1502
    ssp. *troodi* (Lindberg f.) Sundermann
      1499
  latifolia (L.) All. 1499, 1503
    var. *parvifolia* auct. 1499
  microphylla (Ehrh.) Swartz **1501**
    var. *congesta* Boiss. 1502
    var. firmior Schur 1501
  palustris (L.) Crantz 1503
  persica (Soó) Nannfeldt 1499, 1501
    ssp. *troodi* (Lindberg f.) Landwehr
      1499
  purpurata Sm. 1502
  troodi Lindberg f. **1499**
  veratrifolia Boiss. et Hohen. **1498**
Equisetaceae **1873**
Equisetales **1873**
Equisetum L. **1874**
  arvense L. 1875
  maximum auct. 1875
  palustre L. 1875
  pannonicum Waldst. et Kit. ex Willd.
    1874
  ramosissimum Desf. **1874**
    var. *pannonicum* (Waldst. et Kit. ex
      Willd.) Aschers. 1874
    var. *subverticillatum* Milde 1874
  ramosum DC. 1874
  sylvaticum L. 1875
  telmateia Ehrh. **1875**
Eragrostis P. Beauv. **1841**
  cilianensis (All.) Vign.-Lut. **1841**
  curvula (Schrad.) Nees 1842
  megastachya (Koeler) Link 1841
    f. *cilianensis* (All.) Aschers. et Graebn.
      1841
Erianthus ravennae (L.) P. Beauv. 1867
Erica L. **1059**
  manipuliflora Salisb. **1061**
  sicula Guss. **1060**
    ssp. *libanotica* (C. et W. Barbey) P. F.
      Stevens 1060
    var. *libanotica* (C. et W. Barbey)
      Holmboe 1060
  verticillata Forssk. 1061
Ericaceae **1059**
  subfam. Ericoideae **1059**
  subfam. *Monotropoideae* 1064
  subfam. Vaccinioideae **1061**
Erigeron bonariense L. 872
  canadense L. 872
  canadensis auct. 872
  crispus Pourr. 872
  graveolens L. 892
  linifolius Willd. 872
  siculum L. 895
  viscosum L. 891
Ernodea montana Sibth. et Sm. 773
Erodium L'Hérit. **336**
  botrys (Cav.) Bert. **341**
  cavanillesii Willk. 343
  chium auct. 342
  ciconium (L.) L'Hérit. **339**
  cicutarium (L.) L'Hérit. **337**
  crassifolium L'Hérit. 7. **344**

Erodium L'Hérit. (*çont.*)
  var. *salinarium* Sibth. ex DC. 344
  gruinum (L.) L'Hérit. **340**
  hirtum auct. 342
  hirtum Willd. 344
  laciniatum (Cav.) Willd. **341**
    var. *cavanillesii* auct. 342
    f. glanduloso-pilosum Vierh. **343**
    f. laciniatum **342**
  malacoides (L.) Willd. **343**
  moschatum (L.) L'Hérit. **338**
  pulverulentum (Cav.) Willd. 343
  triangulare (Forssk.) Muschler 341
Erophaca baetica (L.) Boiss. 498
Erophila DC. **145**
  minima C. A. Mey. **146**
  praecox (Steven) DC. 145
    var. *virescens* (Jord.) O. E. Schulz 145
  verna (L.) Chevall. **145**
  vulgaris DC. 145
Eruca Mill. **104**
  cappadocica Reut. 104
    var. *eriocarpa* Boiss. 105
  sativa Mill. **104**
    var. eriocarpa (Boiss.) Post **105**
    var. *longirostris* (Uechtritz) Rouy 105
    var. sativa **104**
Erucaria Gaertn. **113**
  aleppica Gaertn. 114
  hispanica (L.) Druce **113**
  myagroides (L.) Hal. 113
Erucastrum incanum (L.) Koch 102
Ervum L. 561
  ervilia L. 558
  gracile auct. 560
  gracile (Loisel.) DC. 559
  hirsutum L. 558
  lenticula Schreb. ex Hoppe 565
  lentoides Ten. 564
  lunatum Boiss. et Bal. 557
  nigricans M. Bieb. 562
  orientale Boiss. 564
  pubescens DC. 560
Eryngium L. **689**
  campestre L. **691**
  creticum Lam. **690**
  cyaneum Sibth. et Sm. 690
  glomeratum Lam. **691**
  maritimum L. **692**
  ? *pusillum* auct. 690
Erysimum L. **162**
  alliaria L. 164
  repandum L. **162**
Erythraea centaurium Borkh. 1111
  var. *laxa* Boiss. 1112
  latifolia auct. 1115
  maritima (L.) Pers. 1117
  pulchella (Swartz) Fries 1114
    ssp. *ramosissima* (Pers.) Lindberg f.
      1114
  ramosissima (Vill.) Pers. 1114
    var. *pulchella* (Swartz) Griseb. 1114
  rhodensis Boiss. et Reuter 1112
  sanguinea Mabille 1112
  spicata (L.) Pers. 1116
    f. *albiflora* Lindberg f. 1116
  tenuiflora Hoffmsgg. et Link 1115
Eschscholzia californica Chamisso 85

Eucalyptus L'Hérit. 17, **660**
  camaldulensis Dehnhardt 662
  cornuta Labill. 661
  gomphocephala DC. 661
  melliodora A. Cunn. ex Schauer 663
  occidentalis Endl. 662
  rostrata Schlecht. 662
  rudis Endl. 663
  tereticornis Sm. 662
  torquata Luehmann 663
  umbellata Chapman 662
Eufragia latifolia (L.) Griseb. 1226
  var. *flaviflora* Boiss. 1226
  viscosa (L.) Benth. 1227
Eupatorium L. **866**
  cannabinum L. **866**
    ssp. *syriacum* (Jacq.) Lindberg f. 867
    var. syriacum (Jacq.) Boiss. **867**
  syriacum Jacq. 867
Euphorbia L. **1433**
  aleppica L. **1442**
  alexandrina Del. 1447
  altissima Boiss. 1437
  amygdaloides auct. 1448
  amygdaloides L. 1448
  apios auct. 1437
  arguta Banks et Sol. **1442**
  biglandulosa auct. 1446
  calendulifolia Del. 1442
  canescens L. 1434
  cassia Boiss. **1439**
    ssp. cassia **1439**
    ssp. rigoi (Boiss. ex Freyn) Holmboe **1440**
  chamaepeplus Boiss. et Gaill. **1445**
  chamaesyce L. **1434**
    ssp. *canescens* (L.) Holmboe 1434
    var. *canescens* (L.) Boiss. 1434
    ssp. *massiliensis* (DC.) Thellung 1435
  characias auct. 1448
  cybirensis Boiss. 1440
    var. *acutifolia* Boiss. 1440
    var. *dehiscens* Boiss. 1440
  cypria Boiss. 1440
  dimorphocaulon P. H. Davis **1437**
  diversifolia Poir. 1447
  dumosa Boiss. 1436
  exigua L. **1443**
    var. *acuta* L. 1443
    var. exigua **1443**
    var. retusa L. **1443**
    var. *tricuspidata* (La Peyr.) Koch 1443
    var. *truncata* Koch 1443
  falcata L. **1445**
    var. *rubra* (Cav.) Boiss. 1445
  geniculata Ortega 1436
  helioscopia auct. 1441
  helioscopia L. **1441**
  helioscopioides Loscos et Pardo 1441
  herniariifolia Willd. **1446**
  heterophylla L. 1436
  hierosolymitana Boiss. **1436**
  hyberna L. 1439
  kotschyana auct. 1440, 1448
  lanata Sieb. ex Spreng. 1435
  leiosperma Sm. 1447
  macrostegia Boiss. 1448
  maculata auct. 1435

Euphorbia L. (*cont.*)
  malacophylla Clarke 1435
  massiliensis DC. 1434
  myrsinites auct. 1446
  nutans Lag. **1435**
  obliquata Forssk. 1447
  obtusifolia Lam. 1447
  paralias L. **1446**
  peplis L. **1434**
  peploides auct. 1444
  peploides Gouan 1444
  peplus L. **1443**
    var. *maritima* Boiss. 1444
    var. minima DC. **1444**
    f. *peploides* (Gouan) Knoche 1444
    ssp. *peploides* (Gouan) Rouy 1444
    var. *peploides* (Gouan) Vis. 1444
    var. peplus **1444**
  petiolata Banks et Sol. **1435**
  portlandica auct. 1447
  preslii Guss. 1435
  provincialis Willd. 1447
  prunifolia Jacq. 1436
  pubescens Vahl **1437**
  pumila Sm. 1446
  rigida M. Bieb. 1446
  rigoi Boiss. 1440
  rubra auct. 1443
  rubra Cav. 1445
  seticornis Poir. 1447
  sintenisii Boiss. ex Freyn **1441**
  sylvatica auct. 1448
  syriaca Spreng. 1435
  terracina L. **1447**
  thamnoides Boiss. 1436
    var. *hierosolymitana* (Boiss.) Boiss. 1436
  thompsonii Holmboe 7, **1448**
  tricuspidata La Peyr. 1443
  troodii Post 1440
  valerianifolia Lam. **1440**
  veneris M. S. Khan **1446**
  verrucosa L. 1437
  zahnii Heldr. ex Hal. 1440
Euphorbiaceae **1432**
Euphrasia latifolia L. 1226
  viscosa Sieb. 1227
Evax Gaertn. **873**
  asterisciflora auct. 874
  asterisciflora (Lam.) Pers. 875
  contracta Boiss. **874**
  eriosphaera Boiss. et Heldr. **875**
  pygmaea (L.) Brot. **873**

Factorovskya Eig **430**
  aschersoniana (Urb.) Eig **430**
Fagaceae **1477**
Fagonia L. **326**
  cretica L. 7, **326**
Farsetia eriocarpa DC. 134
Fedia cornucopiae (L.) Gaertn. 840
  lasiocarpa Stev. 843
  muricata Stev. 846
  orientalis Schlecht. 846
  tridentata Stev. 844
  truncata Reichb. 846
Ferula L. **752**
  amani Zohary et Davis 754

Ferula L. (*cont.*)
 anatriches (Kotschy) Sintenis 753
 cassii Zohary et Davis 754
 communis L. **753**
  var. *anatriches* Kotschy 753
  ssp. communis **753**
  ssp. glauca (L.) Rouy et Camus **753**
 cypria Post **754**
 elaeochytris Korovin 754
 graeca Sibth. 753
 hispida Friv. 757
 nodiflora L. 753
 scabra Fenzl 727
Ferulago Koch **754**
 cypria (Post) Lindberg f. 754
 cypria H. Wolff **755**
 syriaca Boiss. **755**
Festuca L. **1730**
 arundinacea Schreb. **1730**
 brevis (Boiss. et Kotschy) Druce 1732
 ciliata Danth. 1734
 cristata L. 1767
 danthonii Aschers. et Graebn. 1734
 dichotoma Forssk. 1724
 distachya (L.) Roth 1815
 dura (L.) Vill. 1740
 elatior L. 1730
  var. *arundinacea* (Schreb.) Wimm. 1730
 expansa (J. F. Gmel.) Kunth 1740
 filiformis Lam. 1841
 hispida Savi 1769
 inops Del. 1732
 littoralis (Gouan) Sm. 1845
 marina L. 1721
 michelii (Savi) Kunth 1766
 muralis Kunth 1732
 myuros auct. 1733
 myuros L. 1733
 phleoides Vill. 1767
 poa (Lam. ex Lam. et DC.) Raspail 1723
 rigida (L.) Kunth 1721
 sylvatica Huds. 1813
 uniglumis Sol. 1731
Fibigia Medik. **133**
 eriocarpa (DC.) Boiss. **134**
Ficaria grandiflora Rob. 47
Ficus L. **1467**
 benjamina auct. 1468
 carica L. **1467**
  f. *genuina* Boiss. 1467
 drupacea Thunb. 1468
 elastica Roxb. ex Hornem. 1469
 macrophylla Desf. ex Pers. 1469
 microcarpa L. f. 1468
 nitida auct. 1468
 sycomorus L. 1468
 vegalensis Frangos 1468
 vidaliana Warb. 1468
Filago L. **877**
 sect. Filago **877**
 sect. Gifolaria Coss. et Kralik **880**
 sect. Oglifa (Coss.) DC. **880**
 aegaea Wagenitz **879**
  ssp. aristata Wagenitz **879**
 arvensis L. **880**
  ssp. *lagopus* (Stephan ex Willd.) Nyman 880

Filago L. (*cont.*)
  var. *lagopus* (Stephan ex Willd.) DC. 880
 contracta (Boiss.) Chrtek et Holub 874
 eriocephala Guss. 877
 gallica L. 881
  ssp. *tenuifolia* auct. 881
 germanica L. 878, 879
  ssp. *decumbens* Holmboe 878
  ssp. *eriocephala* (Guss.) Holmboe 877
  var. *eriocephala* (Guss.) Parl. 877
  ssp. *spathulata* (C. Presl) Holmboe 878
 lagopus (Stephan ex Willd.) Parl. 880
 mareotica Del. **880**
  var. floribunda (Pomel) Maire 880
 maritima L. 907
 prostrata Parl. 878
 pygmaea L. 873
 pyramidata L. **878**
 spathulata C. Presl 878
  var. *prostrata* Heldr. 878
 tenuifolia C. Presl 882
 vulgaris Lam. 879
  ssp. *eriocephala* (Guss.) Osorio-Tafall et Seraphim 877
  ssp. *prostrata* (Heldr.) Osorio-Tafall et Seraphim 878
Filicales **1877**
Fimbristylis Vahl **1691**
 ferruginea (L.) Vahl **1691**
 mauritiana Tausch ex Schultes 1691
 sieberiana Kunth 1691
Foeniculum Mill. **749**
 capillaceum Gilib. 750
 divaricatum Griseb. 750
 officinale All. 749
 piperitum (Ucria) Presl 750
  ssp. *divaricatum* (Griseb.) Nym. 750
  var. *pluriradiatum* Boiss. 750
 vulgare Mill. **749**
  ssp. *capillaceum* (Gilib.) Holmboe 750
  ssp. piperitum (Ucria) Coutinho **750**
  ssp. vulgare **750**
Frankenia L. **210**
 hirsuta L. **210**
  var. brevipes Hausskn. **211**
  ssp. *hispida* (DC.) Holmboe 211
  var. hispida (DC.) Boiss. **211**
  var. *intermedia* Boiss. 211
  ? ssp. *laevis* auct. 211
 ? *laevis* auct. 211
 pulverulenta L. **210**
Frankeniaceae 19, **209**
Fraxinus L. 1091
 angustifolia Vahl 1092
  ssp. angustifolia 1092
  ssp. oxycarpa (Willd.) Franco et Rocha Afonso 1092
 excelsior L. 1092
 ornus L. 1092
 velutina Torrey 1093
Fritillaria L. **1599**
 acmopetala Boiss. **1599**
 libanotica (Boiss.) Baker 1600
 persica L. **1600**
Fumana (Dunal) Spach 4, **197**
 arabica (L.) Spach **197**

Fumana (Dunal) Spach (*cont.*)
  var. arabica **198**
  var. incanescens Hausskn. **198**
  glutinosa (L.) Boiss. 199
    var. *viridis* (Ten.) Boiss. 200
  laevis (Cav.) Sennen 200
  spachii auct. 199
  thymifolia (L.) Verlot **199**
    var. laevis (Cav.) Grosser **200**
    var. thymifolia **199**
Fumaria L. **88**
  asepala A. K. Jackson et Turrill 93
  bracteosa Pomel **92**
  capreolata L. **90**
  densiflora DC. **92**
  gaillardotii Boiss. 8, **88**
  judaica Boiss. **89**
  macrocarpa Parl. **89**
    var. *laxa* Sintenis 90
    var. *oxyloba* (Boiss.) Hammar 90
  micrantha Lag. 92
    var. *parlatoriana* Boiss. 92
  officinalis L. **91**
  oxyloba Boiss. 90
  parlatoriana Kralik 92
  parviflora Lam. **93**
  petteri Reichb. **90**
    ssp. thuretii (Boiss.) Pugsl. **91**
  rutifolia Sibth. et Sm. 86
  spicata L. 93
  thuretii Boiss. 91
Fumariaceae 19, **85**
Furcraea Vent. 1578
  foetida (L.) Haworth 1579
  gigantea Vent. 1579

Gagea Salisb. **1601**
  subgenus Gagea **1601**
  subgenus Hornungia (Bernh.) Pascher **1605**
  arvensis Dumort. 1601
  billardieri auct. 1601, 1602
  chlorantha (M. Bieb.) J. A. et J. H. Schultes **1606**
    var. *cyprica* Pascher 1606
  fibrosa (Desf.) J. A. et J. H. Schultes **1605**
  foliosa auct. 1603
  graeca (L.) Terracc. **1607**
  juliae Pascher **1602**
  peduncularis (J. et C. Presl) Pascher **1603**
  reticulata (Pall.) J. A. et J. H. Schultes 1606
    ssp. *circinata* auct. 1605
    ssp. *fibrosa* (Desf.) Maire et Weiller 1605
    var. *fibrosa* Boiss. 1605
    f. *rigida* (Boiss. et Spruner) Pascher 1605
    var. *tenuifolia* auct. 1605
  rigida Boiss. et Spruner 1605
  spathacea auct. 1602
  villosa (M. Bieb.) Duby **1601**
    var. *hermonis* Dafni et Heyn 1601
Galilea mucronata (L.) Parl. 1688
Galium L. **780**
  aparine L. **784**

Galium L. (*cont.*)
  var. *macrocarpum* Boiss. et Kotschy 784
  articulatum (L.) Roem. et Schultes 790
  canum Req. **780**
    ssp. *hilarionis* Lindberg f. 781
    var. *musciforme* (Boiss.) Boiss. 780
  cyprium Sibth. 787
  decaisnei Boiss. 783
  divaricatum Pourret ex Lam. **788**
  floribundum Sibth. et Sm. **786**
  humifusum M. Bieb. **781**
    var. humifusum 782
    var. lasiocarpum (Boiss.) Meikle **782,** 807
  laurae Holmboe 779
  murale (L.) All. **789**
    var. *troödi* Lindberg f. 790
  musciforme Boiss. 780
  parisiense L. **787**
    var. leiocarpum Tausch **787**
    var. parisiense **787**
  pauciflorum Kotschy 789
  pedemontanum (Bell.) All. 791
  peplidifolium Boiss. **783**
    f. *pygmaea* Kotschy 783
  pisiferum Boiss. 7, **785**
  recurvum Req. **789**
    var. *glabratum* Holmboe 789
    var. *pauciflorum* Boiss. 789
  rubioides L. 780
  saccharatum All. 786
  setaceum Lam. **782**
    f. *leiocarpum* Bornm. 783
    var. setaceum **782**
    var. urvillei (Req.) Hal. **783**
  spurium L. 784
    ssp. *tenerum* auct. 784
    var. *tenerum* auct. 784
    ssp. *vaillantii* auct. 784
    var. *vaillantii* auct. 784
  suberosum Sibth. et Sm. 792
  tenuissimum M. Bieb. **788**
  tricorne Stokes 785
  tricornutum Dandy **785**
  urvillei Req. 783
  vaillantii auct. 784
  valantia Weber 786
  verrucosum Huds. **786**
Garidella nigellastrum L. 60
  unguicularis Poir. 61
Gastridium P. Beauv. **1786**
  lendigerum auct. 1786
  nitens (Guss.) Coss. et Durieu 1787
  phleoides (Nees et Mey.) C. E. Hubb. **1786**
  ventricosum auct. 1786
Genista L. **382**
  acanthoclada auct. 383
  sphacelata Decne. 4, **383**
    var. *bovilliana* Holmboe 383
    var. crudelis Meikle **384,** 806
    var. sphacelata **383**
  tinctoria L. 384
  uniflora (Decne.) Briq. 381
Gentiana maritima L. 1117
  perfoliata L. 1108
  pulchella Swartz 1114

Gentiana L. (*cont.*)
  ramosissima Vill. 1114
  spicata L. 1116
Gentianaceae **1108**
Geraniaceae 19, **329**
Geranium L. **329**
  botrys Cav. 341
  ciconium L. 339
  cicutarium L. 337
    var. *moschatum* L. 338
  columbinum L. **334**
  dissectum L. **333**
  gruinum L. 340
  hirtum Forssk. 344
  laciniatum Cav. 341
  malacoides L. 343
  modestum Jord. 334
  molle L. **332**
  purpureum (L.) Vill. **334**
  pusillum Burm. f. **331**
  robertianum auct. 335
  robertianum L. 334
    ssp. *purpureum* (Vill.) Murb. 335
    var. *purpureum* (Vill.) DC. 334
  rotundifolium L. **331**
  triangulare Forssk. 341
  tuberosum L. **330**
Geropogon L. **1041**
  glabrum L. 1041
  hirsutum L. 1042
  hybridus (L.) Sch. Bip. **1041**
Geum L. 612
  sect. *Orthurus* Boiss. 612
  heterocarpum Boiss. 612
Gifola eriocephala (Guss.) Chrtek et Holub 877
  spathulata (C. Presl) Reichb. 878
Gladiolus L. **1569**
  communis L. 1570
    var. *triphyllus* Sm. 1570
  italicus Mill. **1569**
  segetum Ker-Gawler 1569
  trichophyllus Sintenis 1570
  triphyllus (Sm.) Ker-Gawler **1570**
Glandularia G. F. Gmelin 1252
  aristigera (S. Moore) Troncoso 1252
Glaucium Mill. **80**
  corniculatum (L.) J. H. Rudolph **80**
    var. corniculatum **80**
    var. flaviflorum DC. **81**
    ssp. *phoeniceum* (Crantz) Holmboe 80
    var. *phoeniceum* (Crantz) DC. 80
    ssp. *tricolor* (Bernh. ex Spreng.) Holmboe 81
    var. tricolor (Bernh. ex Spreng.) Ledeb. **81**
  flavum Crantz **81**
    var. leiocarpum (Boiss.) Stoj. et Stef. **81**
  leiocarpum Boiss. 81
  luteum Scop. 81
  phoeniceum auct. 81
  phoeniceum Crantz 80
  tricolor Bernh. ex Spreng. 81
  violaceum (Lam.) Juss. 82
Glaucosciadium Burtt et Davis **766**
  cordifolium (Boiss.) Burtt et Davis **766**
Gleditsia triacanthos L. 587

Gleditsia L. (*cont.*)
  f. inermis (L.) Zabel 587
Glinus Loefl. ex L. **684**
  lotoides L. **684**
Glochidopleurum sintenisii (Aschers. et Urb. ex Huter) Kozo-Polj. 735
Glyceria sphenopus Steud. 1740
Glycyrrhiza L. **505**
  glabra L. **505**
    var. glabra **506**
    var. glandulifera (W. et K.) Reg. et Herd. **506**
Gnaphalium arvense L. 880
  asterisciflorum Lam. 875
  conglobatum Viv. 887
  gallicum L. 881
  italicum Roth 888
  lagopus Stephan ex Willd. 880
  luteo-album L. 886
  siculum Spreng. 887
Gomphocarpus R. Br. 1104
  fruticosus (L.) Ait. f. 1104
Goniolimon tataricum (L.) Boiss. 1073
Gossypium L. 313
  barbadense L. 313
  herbaceum L. 313
  hirsutum L. 313
Gramineae **1711**
  tribe Aeluropodeae Nevski ex Bor **1844**
  tribe Agrostideae Dumort. **1776**
  tribe Andropogoneae Dumort. **1862**
  tribe Aristideae C. E. Hubb. ex Bor **1850**
  tribe Arundineae Dumort. **1837**
  tribe Aveneae Dumort. **1754**
  tribe Brachypodeae Harz **1812**
  tribe Bromeae Dumort. **1795**
  tribe Chlorideae Agardh **1842**
  tribe Danthonieae (Beck) C. E. Hubb. **1836**
  tribe Eragrostideae Stapf **1839**
  tribe Hainardieae Greuter **1751**
  tribe Maydeae Dumort. **1870**
  tribe Meliceae Reichb. **1750**
  tribe Milieae Endl. **1774**
  tribe Oryzeae Dumort. **1715**
  tribe Paniceae R. Br. **1851**
  tribe Phalarideae Kunth **1770**
  tribe Poëae **1716**
  tribe Sporoboleae Stapf **1847**
  tribe Stipeae Dumort. **1790**
  tribe Triticeae Dumort. **1815**
Grevillea robusta A. Cunn. 1422
Gundelia L. **937**
  tournefortii L. **937**
Guttiferae 19, **293**
Gymnogramma leptophylla (L.) Desv. 1879
Gymnopteris delavayi 1880
Gynandriris Parl. **1562**
  sisyrinchium (L.) Parl. **1562**
Gypsophila L. **221**
  sect. *Petrorhagia* Ser. 219
  pilosa Huds. **221**
  porrigens (L.) Boiss. 221

Hainardia W. Greuter **1753**
  cylindrica (Willd.) Greuter **1753**

Hakea gibbosa (Sm.) Cav. 1422
　suaveolens R. Br. 1422
Halimione Aellen **1382**
　portulacoides (L.) Aellen **1382**
Halocnemum M. Bieb. **1384**
　strobilaceum (Pall.) M. Bieb. **1384**
Halopeplis Bunge ex Ungern-Sternb. **1383**
　amplexicaulis (Vahl) Ungern-Sternb.
　　1384
Hamamelidaceae 19, **654**
Haplophyllum Adr. Juss. **349**
　buxbaumii (Poir.) G. Don **349**
Hasselquistia aegyptiaca L. 764
Hedera L. **768**
　helix L. **768**
　　ssp. helix **769**
　　ssp. poetarum (Bertol.) Nym. **768**
Hedypnois Mill. **997**
　cretica (L.) Dum.-Cours. 998
　monspeliensis Willd. 998
　polymorpha DC. 998
　rhagadioloides (L.) F. W. Schmidt **997**
　　ssp. *cretica* (L.) Holmboe 998
Hedysarum L. **523**
　aequidentatum Sibth. et Sm. 528
　alhagi auct. 532
　alhagi L. 534
　atomarium auct. 524
　capitatum Desf. 523
　　var. *pallens* Moris 523
　caput-galli L. 527
　coronarium L. 526
　crista-galli Murr. 529
　cyprium Boiss. **524**
　flexuosum L. 524
　pallens (Moris) Hal. 523
　pseud-alhagi M. Bieb. 534
　saxatile auct. 524
　saxatile L. 531
　spinosissimum L. **523**
　　ssp. *pallens* (Moris) Holmboe 523
　syriacum auct. 524
　venosum Desf. 530
Heleocharis Lestib. 1692
Heleochloa Host 1847
　alopecuroides (Pill. et Mitterp.) Host 1847
　schoenoides (L.) Host 1849
Helianthemum Mill. **189**
　sect. Argyrolepis Spach **189**
　sect. Brachypetalum Dunal **194**
　sect. Eriocarpum Dunal **192**
　sect. *Fumana* Dunal 197
　sect. Helianthemum **190**
　sect. *Tuberaria* Dunal 187
　aegyptiacum (L.) Mill. **196**
　appenninum auct. 191
　arabicum (L.) Pers. 197
　chamaecistus Mill. 7, **190**
　ellipticum auct. 192
　fasciculi Greuter 190
　glutinosum (L.) Benth. 199
　　var. *laeve* (Cav.) Benth. 200
　guttatum (L.) Mill. 189
　　var. *cavanillesii* Dunal 189
　　ssp. *eriocaulon* (Dunal) Holmboe 189
　hirtum auct. 191
　lasiocarpum Desf. ex Jacq. et Hérincq
　　196

Helianthemum Mill. (*cont.*)
　lavandulifolium auct. 190
　ledifolium (L.) Mill. 8, **195**
　　f. *dissitiflorum* Willk. 196
　　ssp. lasiocarpum (Desf. ex Jacq. et
　　　Hérincq) Nyman **196**
　　var. *lasiocarpum* (Desf. ex Jacq. et
　　　Hérincq) Bornm. 196
　　f. laxiflorum Grosser **196**
　　ssp. ledifolium **195**
　　var. microcarpum Coss. ex Willk. **196**
　lippii (L.) Dum.-Cours. 192
　　var. *ehrenbergii* (Willk.) Boiss. 192
　niloticum (L.) Pers. 195
　nummularium auct. 190
　obovatum auct. 191
　obtusifolium Dunal **191**
　pulverulentum auct. 192
　racemosum auct. 190
　salicifolium Mill. **194**
　　var. glabrum Meikle **195**, 806
　stipulatum (Forssk.) C. Christens. **192**
　syriacum (Jacq.) Dum.-Cours. **189**
　thibaudii Pers. 190
　thymifolium (L.) Pers. 199
　　var. *glutinosum* (L.) Lindberg f.
　　　199
　viride Ten. 200
　vulgare Gaertn. 190
　　var. *microphyllum* 191
Helianthus L. 901
　annuus L. 901
　tuberosus L. 901
Helichrysum Mill. **886**
　conglobatum (Viv.) Steudel **887**
　italicum (Roth) Don **888**
　　var. *microphyllum* auct. 888
　microphyllum auct. 888
　rupicola Pomel 887
　　ssp. *brachyphyllum* (Boiss.) Holmboe
　　　887
　siculum Boiss. 887
　　ssp. *brachyphyllum* (Boiss.) Lindberg
　　　f. 887
　　var. *brachyphyllum* Boiss. 887
　stoechas (L.) Moench ssp. *barrelieri* (Ten.)
　　Nyman 887
Heliotropium L. **1120**
　dolosum De Not. **1122**
　eichwaldii auct. 1122
　europaeum L. **1121**
　　ssp. *tenuiflorum* auct. 1122
　　ssp. *tenuiflorum* Guss. 1121
　　var. *tenuiflorum* auct. 1122
　hirsutissimum Grauer **1122**
　supinum L. **1123**
　undulatum auct. 1122
　villosum Willd. 1122
Helixyra sisyrinchium (L.) N. E. Br. 1562
Helleborine longipetala Ten. 1505
　palustris Hill 1503
　pseudo-cordigera Sebast. 1505
Hellenocarum H. Wolff 742
　multiflorum H. Wolff 742
Helosciadium Koch 736
　nodiflorum (L.) Koch 738
Heracleum absinthifolium Vent. 760
　carmeli Labill. 759

Heracleum L. (*cont.*)
  tomentosum Sibth. et Sm. 760
Hermione cypria Haworth 1573
  papyracea (Ker-Gawler) Haworth 1574
  serotina (L.) Haworth 1576
Herniaria L. **281**
  cinerea DC. **282**
  glabra L. **284**
  hirsuta L. **282**
  incana auct. 282
  micrantha A. K. Jackson et Turrill **283**
  parnassica auct. 283
Hesperis verna L. 150
Hibiscus L. **313**
  esculentus L. 313
  trionum L. **313**
Hieracium sanctum L. 1010
  sprengerianum L. 1001
Himantoglossum longibracteatum (Biv.)
    Schltr. 1537
Hippocrepis L. **519**
  biflora Boiss. 521
  bisiliqua Forssk. 521
  ciliata Willd. **522**
  multisiliquosa L. **521**
  unisiliquosa L. **519**
    ssp. bisiliqua (Forssk.) Bornm. **520**
    ssp. unisiliquosa **520**
Hippomarathrum Hoffmsgg. et Link 726
  boissieri auct. 727
  crassilobum Boiss. 728
  cristatum auct. 727
  scabrum (Fenzl) Boiss. 727
Hirschfeldia Moench **102**
  adpressa Moench 102
  incana (L.) Lagrèze-Fossat **102**
Holarrhena antidysenterica (Roth) A.DC.
    1101
Holcus bicolor L. 1869
  halepensis L. 1869
Holoschoenus Link 1697
  vulgaris Link 1697
Holosteum L. **252**
  umbellatum L. **252**
    var. glandulosum Vis. **252**
Hordeum L. **1830**
  bulbosum L. **1831**
  caput-medusae auct. 1828
  caput-medusae Coss. et Durieu 1828
    ssp. *asperum* (Simonkai) Degen 1830
    ssp. *crinitum* (Schreb.) Aschers. et
      Graebn. 1828
  distichon L. **1835**
  geniculatum All. **1833**
  glaucum Steudel **1832**
  gussoneanum Parl. ex Nyman 1833
  hexastichon L. 1835
  hystrix Roth 1833
  ithaburense Boiss. 1834
  leporinum Link **1832**
  marinum Huds. **1833**
  maritimum auct. 1833
  maritimum With. 1833
    ssp. *gussoneanum* (Parl. ex Nyman)
      Aschers. et Graebn. 1833
  murinum auct. 1832
  murinum L. 1832
    ssp. *leporinum* (Link) Arcang. 1832

Hordeum L. (*cont.*)
  var. *leporinum* (Link) Bory et Chaub.
    1832
  secalinum auct. 1833
  spontaneum K. Koch **1834**
  stebbinsii Covas 1832
  vulgare L. **1835**
  winkleri Hackel 1833
  zeocriton L. 1835
Hormuzakia aggregata (Lehm.) Guşul. 1136
Hornungia procumbens (L.) Hayek 130
Humulus L. 1464
Hutchinsia procumbens (L.) Desv. 130
Hyacinthella Schur **1640**
  millingenii (Post) Feinbrun **1640**
  nervosa (Bertol.) Chouard 1640
    ssp. *millingenii* (Post) Osorio-Tafall et
      Seraphim 1640
Hyacinthus L. **1643**
  comosus L. 1644
  nervosus Bertol. 1640
    ssp. *millingenii* (Post) Holmboe 1640
  nivalis (Boiss. et Kotschy) Baker 1642
  orientalis L. 1643
  pieridis Holmboe 1642
  romanus auct. 1642
  trifoliatus Ten. 1642
Hydrophyllaceae **1118**
Hymenocarpos Savi **472**
  circinnatus (L.) Savi **472**
Hymenolobus Nutt. **130**
  procumbens (L.) Schinz et Thell. 7, **130**
Hyoscyamus L. **1192**
  albus L. **1192**
  aureus L. **1193**
Hyoseris L. **991**
  cretica L. 998
  hedypnois L. 998
  microcephala Coss. 991
  minima auct. 1015
  rhagadiolus L. 998
  scabra L. **991**
Hyparrhenia Anderss. ex Fourn. **1864**
  hirta (L.) Stapf **1865**
Hypecoum L. **83**
  aegyptiacum (Forssk.) Aschers. et
    Schweinf. 84
  grandiflorum Benth. 84
  imberbe Sibth. et Sm. **84**
  pendulum L. **84**
  procumbens L. **83**
Hypericum L. **294**
    sect. Adenosepalum Spach **296**
    sect. Androsaemum (Duhamel)
      Godron **294**
    sect. Coridium Spach **295**
    sect. Drosocarpium Spach **297**
    sect. Hypericum **299**
    sect. Oligostema (Boiss.) Stef. **298**
    sect. Taeniocarpium Jaub. et Spach
      296
  atomarium auct. 297
  ciliatum Lam. 298
  confertum Choisy **296**
    ssp. stenobotrys (Boiss.) Holmboe **296**
    var. *stenobotrys* (Boiss.) Boiss. 296
  coris auct. 295
  crispum L. 300

Hypericum L. (*cont.*)
  dentatum Loisel. 298
  empetrifolium Willd. **295**
  gracile Boiss. 297
  hircinum L. **294**
  hyssopifolium Chaix 294
  lanuginosum Lam. **297**
    ssp. *gracile* (Boiss.) Holmboe 297
    var. *gracile* (Boiss.) Boiss. 297
    var. lanuginosum **297**
    ssp. *millepunctatum* Holmboe 297
  leprosum auct. 298
  modestum Boiss. 298
  myrtifolium Spach 298
  nummularium L. 294
  perfoliatum L. **298**
  perforatum L. **299**
    var. *angustifolium* DC. 299
  repens L. **298**
  stenobotrys Boiss. 296
  tenellum Clarke 298
  triquetrifolium Turra **300**
Hypochaeris L. **1015**
  achyrophorus L. **1016**
  aethnensis (L.) Ball 1016
  glabra L. **1015**
    var. *erostris* Boiss. 1015
    var. glabra **1015**, 1016
    var. .oiseleuriana Godr. 1015
    ssp. minima (Cyrill.) Holmboe 1015
  minima Cyrill. 1015
Hypocistis rubra Fourr. 1415
Hyssopus officinalis L. 1273

Iberis L. **122**
  acutiloba Bertol. 122
  odorata L. 2, **122**
Illecebraceae 19, 212, **279**
Illecebrum paronychia L. 280
Imperata Cyr. **1866**
  arundinacea Cyr. 1866
  cylindrica (L.) Raeuschel **1866**
Imperatoria ostruthium L. 758
Inula L. **889**
  sect. Bubonium DC. **889**
  sect. Cupularia Willk. **891**
  sect. Limbarda (Adans.) DC. **891**
  arabica L. 895
  britannica L. 890
  conyza DC. 890
  conyzae (Griesselich) Meikle **890**
  crithmoides L. **851**
  dentata Sm. 894
  dysenterica L. 893
  graveolens (L.) Desf. **892**
  pulicaria auct. 895
  viscosa (L.) Ait. **891**
  vulgaris Trevisan 890
Ioxylon pomiferum Rafin. 1466
Ipomoea L. **1163**
  littoralis (L.) Boiss. 1164
  purpurea (L.) Roth **1165**
  sagittata Poir. 7, **1164**
  stolonifera (Cyr.) J. F. Gmel. 7, **1164**
Iridaceae **1559**
Iris L. 1559
  sect. Iris 1560
  albicans Lange 1561

Iris L. (*cont.*)
  cypriana Baker et Foster 1560
  florentina auct. 1561
  germanica L. 1560
  mesopotamica Dykes 1560
  sisyrinchium L. 1562
  xiphium L. 1562
Isolepis R. Br. **1692**
  cernua (Vahl) Roem. et Schult. **1693**
  setacea (L.) R. Br. **1693**
  sicula Presl 1694

Jacaranda mimosifolia D. Don 1244
Jasminum L. 1089
  angulare Vahl 1091
  azoricum L. 1091
  fruticans L. 1091
  grandiflorum L. 1090
  mesnyi Hance 1091
  nudiflorum Lindl. 1091
  officinale L. 1090
  polyanthum Franch. 1091
  primulinum Hemsl. 1091
Jasonia sicula (L.) DC. 895
Juglandaceae **1471**
Juglans regia L. 1471
Juncaceae **1648**
Juncaginaceae **1673**
Juncellus (Griseb.) C. B. Clarke 1684
  laevigatus (L.) C. B. Clarke 1689
Juncus L. **1648**
  subgenus Genuini Buchenau **1653**
  subgenus Juncinella V. Krecz. et
    Gontsch. **1657**
  subgenus Juncus **1649**
  subgenus Poiophylli Buchenau **1654**
  subgenus Septati Buchenau **1658**
  subgenus Subulati Buchenau **1654**
  acuto-maritimus Ledeb. 1651
  acutus auct. 1652
  acutus L. **1651**
    var. β L. 1649
    var. *heldreichianus* (Marsson ex Parl.)
      Hal. 1652
    ssp. *tommasinii* (Parl.) Arcang. 1651
    var. *tommasinii* (Parl.) Buchenau
      1651
  ambiguus Guss. **1657**
  articulatus L. **1659**
    × fontanesii Gay ex Laharpe
    ssp. pyramidatus (Laharpe) Snog.
      1660
  bufonius auct. 1656
  bufonius L. **1655**
    ssp. *hybridus* (Brot.) Arcang. 1656
    var. *hybridus* (Brot.) Parl. 1656
    f. *mutabilis* (Savi) Aschers. et Graebn.
      1656
  capitatus Weigel **1657**
  cyprius Lindberg f. 1653
  fontanesii Gay ex Laharpe **1658**
    f. *pyramidatus* (Laharpe) Holmboe
      1659
    ssp. pyramidatus (Laharpe) Snogerup
      1659
    var. *pyramidatus* (Laharpe) Buchenau
      1659
  glaucus Ehrh. 1653

Juncus L. (*cont.*)
  f. *longicornis* (Bast.) Aschers. et Graebn. 1653
  var. *longicornis* (Bast.) Grognot 1653
  heldreichianus Marsson ex Parl. **1652**
  hybridus Brot. **1656**
  inflexus L. **1653**
  var. *longicornis* (Bast.) Briq. 1653
  insulanus Viv. 1656
  lampocarpus Ehrh. ex Hoffm. 1659
  littoralis C. A. Mey. **1651**
  longicornis Bast. 1653
  maritimus auct. 1650
  maritimus Lam. **1649**
  var. *arabicus* Aschers. et Buchenau 1650
  mutabilis Savi 1656
  ssp. *ambiguus* (Guss.) Nyman 1657
  pygmaeus auct. 1656
  pyramidatus Laharpe 1659
  ranarius Song. et Perr. 1657
  rigidus Desf. **1650**
  sphaerocarpus Nees **1655**
  subulatus Forssk. **1654**
  tenageia auct. 1655
  tommasinii Parl. 1651
Juniperus L. **29**
  excelsa M. Bieb. 2, **31**
  foetidissima Willd. 2, **31**
  occidentalis Hook. 32
  oxycedrus L. **29**
  ssp. macrocarpa (Sibth. et Sm.) Ball 30
  ssp. *rufescens* (Link ex Endl.) Holmboe 29
  phoenicea L. 8, **30**
  rufescens Links ex Endl. 29
  virginiana L. 32
Jurinea Coss. **957**
  aucheriana DC. 958
  brevicaulis Boiss. 958
  cypria Boiss. **957**
Justicia adhatoda L. 1247

Kentranthus Necker 836
Kentrophyllum baeticum Boiss. et Reut. 982
  lanatum auct. 982
  lanatum (L.) DC. 982
  tenue Boiss. 981
Kickxia Dumort. **1206**
  commutata (Bernh. ex Reichb.) Fritsch **1206**
  ssp. graeca (Bory et Chaub.) R. Fernandes **1207**
  elatine (L.) Dumort. **1207**
  ssp. *crinita* (Mabille) Greuter 1208
  ssp. sieberi (Arcang.) Hayek **1208**
  lanigera (Desf.) Hand.-Mazz. **1209**
  sieberi H. H. Allan 1208
  spuria (L.) Dumort. **1209**
  ssp. integrifolia (Brot.) R. Fernandes 1210
  ssp. spuria 1210
Knautia palaestina L. 856
  plumosa L. 856
Kochia *hirsuta* auct. 1392
  scoparia (L.) Schrader 1383
Koeleria berythea Boiss. et Blanche 1767
  cristata (L.) Bertol. 1767

Koeleria Pers. (*cont.*)
  cristata (L.) Pers. 1767
  hispida (Savi) DC. 1769
  michelii (Savi) Coss. et Durieu 1766
  obtusiflora Boiss. 1768
  var. *amblyantha* (Boiss.) Domin 1769
  var. *condensata* Boiss. 1768
  phleoides (Vill.) Pers. 1767
  var. *amblyantha* Boiss. 1769
  ssp. *berythea* (Boiss. et Blanche) Domin 1767
  var. *grandiflora* Boiss. 1767
  ssp. *obtusiflora* (Boiss.) Domin 1768
  var. *obtusiflora* (Boiss.) Boiss. 1768
  pubescens auct. 1767
Koelpinia Pallas **994**
  linearis Pallas **994**
Koelreuteria paniculata Laxm. 362
Kohlrauschia Kunth **218**
  velutina (Guss.) Reichb. **218**
Koniga libyca (Viv.) R. Br. 142
  maritima (L.) R. Br. 142
Krubera Hoffm. **751**
  leptophylla Hoffm. 752
  peregrina (L.) Hoffm. **751**

Labiatae **1254**
  tribe Lamieae **1303**
  tribe Lavanduleae Boiss. **1256**
  tribe Nepeteae Dumort. **1300**
  tribe Ocimeae Dumort. **1256**
  tribe Prasieae Benth. **1327**
  tribe Rosmarineae Briq. **1299**
  tribe Salvieae Dumort. **1287**
  tribe Saturejeae Benth. **1257**
  tribe Scutellarieae Benth. **1329**
  tribe Teucrieae Dumort. **1332**
Lablab niger Medik. 582
  purpureus (L.) Sweet 582
  vulgaris Savi 582
Lachnagrostis phleoides Nees et Mey. 1786
Lactuca L. **1027**
  aculeata auct. 1028
  cretica Desf. 1026
  hispida auct. 1024
  ssp. *candolleana* auct. 1025
  leucophaea Sm. 1026
  saligna L. **1027**
  sativa L. 1028
  var. capitata L. 1029
  var. crispa L. 1029
  var. longifolia Lam. 1029
  scariola L. 1028
  serriola L. **1028**
  tetrantha B. L. Burtt et P. Davis 1030
  triquetra (Labill.) Boiss. 1031
  tuberosa Jacq. 1026
  viminea (L.) J. & C. Presl 1029
Lagenaria *leucantha* Rusby 679
  siceraria (Molina) Standley 679
  vulgaris Ser. 679
Lagoecia L. **692**
  cuminoides L. **692**
Lagonychium stephanianum (M. Bieb.) M. Bieb. 592
Lagoseris bifida (Vis.) Koch 1010
  sancta (L.) K. Malý 1010
Lagurus L. **1788**

Lagurus L. (*cont.*)
　cylindricus L. 1866
　ovatus L. **1788**
Lamarckia Moench **1727**
　aurea (L.) Moench **1727**
Lamium L. **1310**
　amplexicaule L. **1311**
　garganicum L. **1310**
　　ssp. *garganicum* 1311
　　ssp. striatum (Sm.) Hayek **1311**
　maculatum auct. 1311
　moschatum Mill. **1312**
　　var. *micranthum* Boiss. 1313
　purpureum auct. 1311
　striatum Sm. 1311
Lamyra (Cass.) Cass. subgenus *Notobasis*
　　Cass. 952
Lantana camara L. 1248
　var. hybrida (Neub.) Mold. 1248
　hybrida Neub. 1248
Lappa major Gaertn. 947
　vulgaris Hill 947
Lapsana koelpinia L. f. 994
　rhagadiolus L. 995
　stellata L. 996
　zacintha L. 1011
Laser cordifolium (Boiss.) Thell. 766
Lathraea phelypaea L. 1233
Lathyrus L. **566**
　sect. Aphaca (Adans.) Reichb. **568**
　sect. Cicercula (Medik.) Gren. et Godr.
　　569
　sect. Clymenum (Adans.) DC. **567**
　sect. Lathyrus **575**
　sect. Orobastrum Boiss. **576**
　amoenus Fenzl 573
　amphicarpos auct. 575
　annuus L. **569**
　　ssp. *cassius* (Boiss.) Holmboe 570
　　var. *hierosolymitanus* (Boiss.) Post 569
　aphaca L. **568**
　bithynicus L. 546
　blepharicarpos Boiss. **574**
　　var. cyprius Meikle **575, 806**
　cassius Boiss. **570**
　cicera L. **572**
　cyprius Rechinger f. 575
　gorgonei Parl. **573**
　hierosolymitanus Boiss. 569
　latifolius L. 575
　monanthos (L.) Willd. 555
　nigricans auct. 564
　ochrus (L.) DC. **567**
　odoratus L. 575
　pseudoaphaca Boiss. 569
　sativus L. **573**
　saxatilis (Vent.) Vis. **577**
　setifolius L. **576**
　sphaericus Retz. **576**
　　var. *pilosulus* Murb. 576
Launaea Cass. **1039**
　mucronata auct. 1039
　resedifolia (L.) O. Kuntze **1039**
　　var. aegyptiaca Amin 1040
Lauraceae **1420**
Laurentia Adans. 1056
　minuta (L.) A. DC. 1056
　　f. *nobilis* E. Wimmer 1057

Laurentia Adans. (*cont.*)
　tenella A. DC. 1057
Laurus L. **1420**
　nobilis L. **1421**
　persea L. 1421
Lavandula L. **1256**
　augustifolia Mill. 1257
　officinalis Chaix 1257
　spica L. 1257
　stoechas L. 2, **1256**
　vera DC. 1257
Lavatera L. **304**
　bryoniifolia Mill. **306**
　cretica L. **304**
　punctata All. **306**
　tomentosa Dum.-Cours. 306
　unguiculata Desf. 306
Lawsonia inermis L. 668
Lecokia DC. **725**
　cretica (Lam.) DC. **726**
Legousia J. F. Durande **1052**
　castellana (Lange) Samp. 1056
　falcata (Ten.) Fritsch ex Janchen **1054**
　　var. falcata **1055**
　　var. pusilla Boiss. 1055
　　var. scabra (Lowe) Meikle **1055**
　hybrida (L.) Delarbre **1054**
　pentagonia (L.) Thell. 1053
　perfoliata (L.) Britton 1056
　speculum-veneris (L.) Chaix **1053**
Leguminosae 19, **372**
　subfam. Caesalpinioideae **585**
　subfam. Mimosoideae **592**
　subfam. Papilionoideae **376**
　tribe Caesalpinieae Endl. **585**
　tribe Cassieae Bronn **589**
　tribe Cerceae Bronn **591**
　tribe Dalbergieae DC. **582**
　tribe Galegeae (Bronn) Benth. **488**
　tribe Genisteae (Adans.) Benth. **378**
　tribe Hedysareae DC. **506**
　tribe Loteae DC. **471**
　tribe Phaseoleae (Bronn) DC. **581**
　tribe Podalyrieae Benth. **376**
　tribe Sophoreae (Bronn) DC. **583**
　tribe Trifolieae (Bronn) Benth. **384**
　tribe Vicieae (Bronn) DC. **535**
Lens Mill. **561**
　culinaris Medik. **562**
　ervoides (Brign.) Grande **565**
　esculenta Moench 562
　lenticula (Schreb. ex Hoppe) Alef. 565
　nigricans (M. Bieb.) Godr. 562
　orientalis (Boiss.) Hand.-Mazz. **564**
Lentibulariaceae **1242**
Leontice L. **72**
　chrysogonum L. 72
　leontopetalum L. **72**
Leontodon L. **1017**
　bulbosus L. 1014
　laevigatus auct. 1020
　tuberosus L. **1017**
　　ssp. *olivieri* (DC.) Holmboe 1017
Leopoldia comosa (L.) Parl. 1644
Lepidium L. **116**
　chalepense L. 121
　cornutum Sibth. et Sm. 117
　draba L. 120

Lepidium L. (*cont.*)
    ssp. *chalepense* (L.) Holmboe 121
    latifolium L. **118**
    perfoliatum L. **118**
    procumbens L. 130
    sativum L. **117**
    spinosum Ard. **117**
    squamatum Forssk. 121
Leptochloa P. Beauv. **1840**
    filiformis (Lam.) P. Beauv. **1841**
Lepturus cylindricus (Willd.) Trin. 1753
    filiformis auct. 1752
    incurvatus (L.) Trin. 1752
Leucanthemum myconis (L.) Giraud 922
Ligia passerina (L.) Fasano 1426
    ssp. *pubescens* (Guss.) Arcang. 1427
Ligusticum cornubiense L. 724
    cyprium Spreng. 742
Ligustrum vulgare L. 668
Liliaceae **1581**
    subfam. Allioideae **1600**
    subfam. Asparagoideae **1584**
    subfam. Asphodeloideae **1587**
    subfam. Lilioideae **1596**
    subfam. Melanthoideae **1593**
    subfam. Scilloideae **1629**
    subfam. Smilacoideae **1582**
Limodorum Boehm. **1493**
    abortivum (L.) Swartz **1493**
Limonium Mill. 7, **1066**
    albidum (Guss.) Pignatti **1069**
        var. cyprium Meikle **1070**
    delicatulum (Girard) O. Kuntze 1068
    echioides (L.) Mill. **1072**
        ssp. echioides **1072**
        ssp. exaristatum (Murb.) Maire **1073**
    exaristatum (Murb.) P. Fournier 1073
    gmelini auct. 1068
    meyeri (Boiss.) O. Kuntze **1068**
    mucronulatum (Lindberg f.) Osorio-Tafall
        et Seraphim 1069
    narbonense Mill. **1068**
    ocymifolium (Poir.) O. Kuntze **1070**
        ssp. bellidifolium (Sm.) Meikle **1070**
        var. *bellidifolium* Rechinger f. 1071
    oleifolium auct. 1071
    sinuatum (L.) Mill. **1067**
    virgatum (Willd.) Fourr. **1071**
    vulgare auct. 1068
Limosella L. 1216
    aquatica L. 1216
Linaceae 19, **317**
Linaria Mill. **1201**
    sect. *Chaenorhinum* DC. 1210
    albifrons (Sm.) Spreng. **1203**
    aparinoides Chav. 1203
    arvensis (L.) Desf. 1202
        var. *flaviflora* Boiss. 1202
        ssp. *simplex* Fourn. 1202
    bombycina Boiss. et Bl. 1208
    chalepensis (L.) Mill. **1204**
        var. brevicalyx Davis 1205
    commutata Bernh. ex Reichb. 1207
        var. *polygonoides* Hal. 1207
    crinita Mabille 1208
    cymbalaria auct. 1205
    elatine auct. 1208
    elatine (L.) Mill. 1207

Linaria Mill. (*cont.*)
    ssp. *sieberi* Arcang. 1208
        var. *villosa* Boiss. 1208
    lanigera Desf. 1209
    longipes Boiss. et Heldr. 1205
    micrantha (Cav.) Hoffmsgg. et Link
        1202
    parviflora Desf. 1202
    parviflora (Jacq.) Hal. 1202
    pelisseriana (L.) Mill. **1201**
    peloponnesiaca Boiss. et Heldr. 1203
    prestandreae Tineo 1208
    rubrifolia Robill. et Cast. ex DC. 1210
    sibthorpiana Boiss. et Heldr. ex Boiss.
        1203
        var. *parnassica* (Boiss. et Heldr.)
            Boiss. 1203
        var. *peloponnesiaca* (Boiss. et Heldr.)
            Boiss. 1203
    sieberi Reichb. ex Heldr. 1208
        var. *bombycina* (Boiss. et Bl.) Hal.
            1208
        ssp. *crinita* (Mabille) Nyman 1208
    simplex Desf. **1202**
    spuria (L.) Mill. 1209
        ssp. *spuria* 1209
    stricta (Sm.) Guss. 1203
Lindbergella Bor **1746**
    sintenisii (Lindberg f.) Bor **1747**
Lindbergia Bor 1746
Linum L. **317**
    sect. Dasylinum **319**
    sect. Halolinum (Planch.) Meikle **323**,
        806
    sect. Limoniopsis (Planch.) Juz. **323**
    sect. Linastrum (Planch.) Benth. **320**
    sect. Linum **318**
    angustifolium Huds. 318
    bienne Mill. **318**
    corymbulosum Reichb. **321**
    crepitans Dumort. 319
    cribrosum Reichb. 318
    gallicum auct. 321
    gallicum L. 320
    hirsutum auct. 320
    humile Mill. 319
    maritimum L. 7, **323**
    nodiflorum L. **323**
    pubescens Banks et Sol. **319**
    sibthorpianum Marg. et Reut. 320
    strictum auct. 322
    strictum L. **322**
        ssp. spicatum (Pers.) Lindberg f. **322**
        var. *spicatum* Pers. 322
    trigynum L. **320**
    usitatissimum L. **319**
        var. bienne Dierb. **319**
        var. crepitans Boenningh. **319**
        var. *hybernum* Dierb. 319
        var. *hyemale* A. DC. 319
        var. *romanum* Heer 319
        var. usitatissimum **319**
        var. *vulgare* Boenningh. 319
    viscosum auct. 320
Lippia canescens auct. 1249
    filiformis Schrader 1249
    nodiflora auct. 1249
    nodiflora (L.) Michx. 1249

Lippia auct. (*cont.*)
 var. *rosea* (D. Don) Macbride 1249
Liquidambar L. **655**
 imberbe auct. 655
 orientalis auct. 655
 styraciflua L. **655**
Lithodora Griseb. **1145**
 hispidula (Sm.) Griseb. **1145**
 ssp. cyrenaica (Pamp.) Brullo et
  Furnari 1146
 ssp. hispidula 1146
 ssp. versicolor Meikle **1146**
Lithospermum L. 1145
 apulum (L.) Vahl 1151
 arvense L. 1148
  var. *coerulescens* DC. 1150
  var. *minus* Ten. 1150
  ssp. *sibthorpianum* (Griseb.) Holmboe
   1149
  var. *sibthorpianum* (Griseb.) Hal. 1149
  var. *splitgerberi* (Guss.) Fiori 1149
 hispidulum Sm. 4, 1145
  var. *cyrenaicum* Pamp. 1146
 incrassatum auct. 1147, 1149
 incrassatum Guss. 1150
  ssp. *gasparrinii* (Heldr. ex Guss.)
   Nyman 1150
  var. *gasparrinii* (Heldr. ex Guss.)
   Cesati 1150
 lehmanii Tineo 1139
 sibthorpianum Griseb. 1149
 splitgerberi Guss. 1149
 tenuiflorum auct. 1149
 tenuiflorum L. f. 1147
 tinctorium L. 1139
Lloydia graeca (L.) Endl. ex Kunth 1607
Lobelia bivonae Tineo 1057
 minuta L. 1056
 setacea Sm. 1056
 tenella Biv. 1057
Lobularia Desv. **141**
 libyca (Viv.) Webb et Berth. **142**
 maritima (L.) Desv. **142**
Loganiaceae **1107**
Logfia arvensis (L.) Holub 881
 davisii Holub 881
 gallica (L.) Coss. et Germ. 881
Lolium L. **1734**
 arvense With. 1735
 loliaceum (Bory et Chaub.) Hand.-Mazz.
  1737
 multiflorum Lam. **1736**
  × rigidum Gaud. **1736**
 perenne L. **1735**
  var. *tenue* (L.) Schrad. 1736
 rigidum Gaud. **1737**
  var. *loliaceum* (Bory et Chaub.) Hal.
   1738
  var. *rottboellioides* Heldr. ex Boiss.
   1738
  × temulentum L. **1737**
 subulatum Vis. 1738
 temulentum L. **1735**
  var. *arvense* (With.) Lilj. 1735
  var. *leptochaeton* A. Braun 1735
  var. *muticum* Boiss. 1735
 tenue L. 1736
Lonicera L. **769**

Lonicera L. (*cont.*)
 etrusca Santi **769**
  ssp. *roeseri* (Heldr. ex Boiss.) Osorio-
   Tafall et Seraphim 770
  var. *roeseri* Heldr. ex Boiss. 769
 implexa auct. 770
 japonica Thunb. 770
 periclymenum auct. 770
Lophochloa Reichb. **1766**
 amblyantha (Boiss.) Bor **1769**
 berythea (Boiss. et Blanche) Bor **1767**
 cristata (L.) Hyl. **1767**
 hispida (Savi) Jonsell **1769**
 obtusiflora (Boiss.) Gontsch. **1768**
 phleoides (Vill.) Reichb. 1767
Lotus L. **477**
 sect. Krokeria (Moench) Ser. **486**
 sect. Lotus **478**
 angustissimus L. **483**
 collinus (Boiss.) Heldr. **479**
 commutatus Guss. 480
  var. *collinus* (Boiss.) Brand 480
 corniculatus L. **480**
  var. *alpinus* auct. 481
  ssp. *tenuifolius* (L.) Hartm. 481
  var. tenuifolius L. **481**
 creticus auct. 478
 creticus L. 478
  ssp. *collinus* (Boiss.) Holmboe 479
  var. *collinus* Boiss. 479
  var. *cytisoides* (L.) Boiss. 478
  var. *genuinus* Boiss. 478
 cytisoides L. **478**
  ssp. *collinus* (Boiss.) Murb. 479
 diffusus auct. 481, 483, 485
 edulis L. **486**
 graecus L. 477
 halophilus Boiss. et Spruner **484**
 hirsutus L. 475
 judaicus Boiss. ex Bornm. 480
 lamprocarpus Boiss. 482
 ornithopodioides L. **485**
 palustris Willd. **482**
  var. *villosissimus* Lindberg f. 482
 peregrinus L. **483**
 perpusillus Sintenis 484
 pusillus Viv. 484
 rectus L. 476
 tenuifolius (L.) Reichb. 481
 tenuis Waldst. et Kit. ex Willd. 481
 tetragonolobus L. 487
  ssp. *palaestinus* Holmboe 487
 villosus Forssk. 484
Luffa *aegyptiaca* Mill. 679
 cylindrica (L.) M. J. Roem. 679
Lunaria libyca Viv. 142
Lupinus L. **378**
 angustifolius L. **379**
 hirsutus auct. 379
 hirsutus L. 379
  ssp. *micranthus* (Guss.) Holmboe 379
  var. *micranthus* (Guss.) Boiss. 379
 micranthus Guss. **379**
Lycium L. **1186**
 afrum L. **1188**
 chinense auct. 1188
 europaeum auct. 1187
 ferocissimum Miers **1188**

Lycium L. (*cont.*)
  halimifolium auct. 1188
  horridum auct. 1188
  intricatum Boiss. 1187
  schweinfurthii U. Dammer **1187**
  vulgare auct. 1187
Lycopersicon *esculentum* Mill. 1180
  lycopersicum (L.) Farw. 1180
Lycopodium denticulatum L. 1873
Lycopsida **1872**
Lycopsis *aegyptiaca* L. 1135
  echioides L. 1139
Lygia passerina (L.) Fasano 1426
  ssp. *pubescens* (Guss.) Lindberg f. 1427
  pubescens (Guss.) C. A. Mey. 1427
Lyonnetia abrotanifolia auct. 914
  pusilla Cass. 914
  rigida DC. 914
Lysimachia linum-stellatum L. 1082
Lythraceae 19, **664**
Lythrum L. **665**
  bibracteatum Boiss. 667
  flexuosum auct. 665
  graefferi Ten. 665
  hyssopifolia L. **666**
    ssp. thymifolia (L.) Batt. 667
  junceum Banks et Sol. **665**
  salzmannii Jord. 668
  thymifolia L. 667
    var. major DC. 668
  tribracteatum Salzm. ex Spreng. **667**
    var. candollei E. Koehne 668
    var. salzmannii (Jord.) E. Koehne 668

Macadamia integrifolia Maiden et Betcke 1422
  ternifolia F. Muell. 1422
Machaerium tipu Benth. 583
Maclura *aurantiaca* Nuttall 1466
  pomifera (Rafin.) C. K. Schneider 1466
Maillea Parl. **1781**
  crypsoides (Urv.) Boiss. 6, **1781**
  urvillei Parl. 1781
Majorana bevanii (Holmes) A. Cheval. 1267
  crassifolia Benth. 1267
  dubia (Boiss.) Briq. 1267
  hortensis Moench 1268
  syriaca (L.) Rafin. 1266
Malcolmia R. Br. **159**
  sect. Malcolmia **160**
  sect. Sisymbrioideae Boiss. **159**
  chia (L.) DC. **161**
    var. chia **162**
    var. lyrata (Sibth. et Sm.) Boiss. **162**
  confusa Boiss. 159
  flexuosa (Sibth. et Sm.) Sibth. et Sm. **160**
  lyrata (Sibth. et Sm.) Sibth. et Sm. 162
  micrantha Boiss. et Reut. 161
  nana (DC.) Boiss. **159**
    var. glabra Meikle **160,** 806
    var. nana **160**
  torulosa (Desf.) Boiss. 168
Malva L. **307**
  aegyptia L. **308**
  alcea L. 311
  althaeoides Cav. 308
  althaeoides auct. 303

Malva L. (*cont.*)
  cretica Cav. **307**
    var. althaeoides (Cav.) Gavioli **308**
    var. cretica **308**
  cypriana D. Don 312
  flexuosa Hornem. 311
  montana auct. 310
  montana Forssk. 310
  neglecta Wallr. **310**
  nicaeênsis All. **310**
  oxyloba Boiss. 311
  parviflora L. **311**
    var. oxyloba (Boiss.) Kristofferson **311**
    var. parviflora **311**
  pusilla auct. 310
  sherardiana L. 312
  sylvestris L. **309**
    var. eriocarpa Boiss. 309
    var. oxyloba Post **309**
    var. sylvestris **309**
  verticillata L. 310
Malvaceae 19, **302**
Malvella Jaub. et Spach **312**
  sherardiana (L.) Jaub. et Spach **312**
Mandragora L. **1188**
  autumnalis Bert. 1189
  haussknechtii Heldr. 1189
  officinarum L. **1189**
    ssp. *haussknechtii* (Heldr.) Vierh. 1189
  vernalis Bert. 1189
Mantisalca Cass. **975**
  salmantica (L.) Briq. et Cavill. **976**
Maresia nana (DC.) Batt. 159
Marrubium L. **1307**
  apulum Ten. 1308
  vulgare L. **1308**
    ssp. *apulum* (Ten.) Osorio-Tafall et Seraphim 1308
    var. *opulum* (Ten.) Heldr. 1308
    var. *lanatum* Benth. 1308
Marsdenia erecta (L.) R. Br. 7, 1106
Maruta cotula (L.) DC. 915
Matricaria L. **924**
  aurea (Loefl.) Sch. Bip. **925**
  chamomilla L. 924
    var. coronata Gay ex Boiss. 925
    var. pappulosa Margot et Reuter 925
    ssp. pusilla (Willd.) Holmboe 925
    var. pusilla (Willd.) Fiori et Paol. 925
  coronata (Gay ex Boiss.) Gay ex Koch 925
  parthenium L. 923
  pusilla Willd. 925
  recutita L. **924**
    var. coronata (Gay ex Boiss.) Gruenberg-Fertig **925**
    var. *recutita* 925
Matthiola R. Br. **154**
  bicornis (Sibth. et Sm.) DC. 157
    var. *oxyceras* (DC.) Bornm. 157
  coronopifolia (Sibth. et Sm.) DC. 156
  fruticulosa (L.) Maire **156**
  incana (L.) R. Br. **154**
  longipetala (Vent.) DC. **157**
    var. *bicornis* (Sibth. et Sm.) Zohary 157
    var. *oxycera* (DC.) Zohary 157

Matthiola R. Br. (*cont.*)
sinuata (L.) R. Br. **155**
tenella DC. 157
tricuspidata (L.) R. Br. **158**
var. *integrifolia* G. Strobl 158
tristis (L.) R. Br. 156
Mattiastrum lithospermifolium auct. 1126
Medicago L. **409**
sect. Medicago **411**
sect. Orbiculares Urb. **413**
sect. Spirocarpos Ser. **414**
aculeata Willd. **427**
apiculata Bast. 428
arabica (L.) Huds. **422**
blancheana Boiss. **416**
var. bonarotiana (Arcang.) Arcang.
416
var. *inermis* Post 416
bonarotiana Arcang. 416
ciliaris (L.) All. 429
circinnata L. 472
constricta Dur. **426**
coronata (L.) Bartal. **418**
cylindracea auct. 424
denticulata Willd. 422
var. *genuina* Boiss. 422
var. *vulgaris* Benth. 422
disciformis DC. **420**
var. apiculata Urban **421**
var. disciformis **421**
elegans Jacq. ex Willd. 415
falcata L. 411
gerardi Waldst. et Kit. ex Willd. 425
globosa Urban 426
hispida Gaertn. 422
ssp. *denticulata* (Willd.) Holmboe 422
var. *denticulata* (Willd.) Burnat 422
ssp. *lappacea* auct. 422
intertexta (L.) Mill. **428**
var. ciliaris (L.) Heyn **429**
lappacea auct. 422
littoralis Rohde ex Lois.-Deslong. **423**
var. inermis Moris **424**
var. littoralis **424**
var. *subinermis* Boiss. 424
var. *tricycla* (DC.) Holmboe 424
lupulina L. **429**
var. *willdenowiana* Koch 429
maculata Willd. 422
marina L. **412**
minima (L.) Bartal. **419**
oliviformis Guss. 427
orbicularis (L.) Bartal. **413**
polymorpha L. **421**
var. *arabica* L. 422
var. *ciliaris* L. 429
var. *coronata* L. 418
var. *minima* L. 419
var. *rigidula* L. 425
var. *scutellata* L. 414
var. *turbinata* L. 427
var. vulgaris (Benth.) Shinners **422**
praecox DC. **418**
rigidula (L.) All. **425**
rotata Boiss. **417**
rugosa Desr. **415**
sativa auct. 412
sativa L. **411**

Medicago L. (*cont.*)
ssp. *falcata* (L.) Naeg. et Thell. 411
ssp. *sativa* 411
ssp. *varia* (Martyn) Naeg. et Thell. 412
scutellata (L.) Mill. **414**
tribuloides Desr. 424
truncatula Gaertn. **424**
tuberculata Willd. 427
var. *apiculata* Urban 428
var. *spinosa* H. Stuart Thompson 428
turbinata (L.) All. **427**
var. apiculata (Urban) Heyn **428**
var. turbinata **428**
× varia Martyn **411**
willdenowii Boenn. 430
Melia azedarach L. 352
Meliaceae 19, **352**
Melica L. **1750**
minuta L. **1750**
var. *parviflora* (Boiss.) Holmboe 1750
var. *vulgaris* Cosson 1750
ramosa Vill. 1750
var. *parviflora* Boiss. 1750
var. *saxatilis* (Sm.) Boiss. 1750
var. *vulgaris* (Coss.) Boiss. 1750
saxatilis Sm. 1750
Melilotus Mill. **432**
albus Medik. **433**
ssp. *parviflora* (Boiss.) O. E. Schulz
433
var. parviflorus Boiss. **433**
arborea Castagne ex Ser. 433
gracilis DC. 434
indicus (L.) All. **434**
messanensis (L.) All. **436**
neapolitanus Ten. 434
parviflora Desf. 434
sicula (Turra ex Vitm.) B. D. Jackson
436
sulcatus Desf. **435**
Melissa L. **1284**
altissima Sm. 1285
clinopodium Benth. 1283
graveolens Benth. 1281
nepeta L. 1279
officinalis L. **1285**
var. officinalis 1285
var. *romana* (Mill.) Woodv. 1285
var. *villosa* Benth. 1285
romana Mill. 1285
Mentha L. **1258**
sect. Mentha **1259**
sect. Pulegium (Mill.) DC. **1258**
aquatica L. **1259**
crispa L. 1261
cyprica H. Braun 1260
var. *galatae* H. Braun 1260
hirsuta Huds. 1259
longifolia auct. 1260
longifolia (L.) L. **1259**
ssp. cyprica (H. Braun) R. Harley
1260
var. *cyprica* (H. Braun) Briq. 1260
var. *galatae* (H. Braun) Briq. 1260
ssp. hymalaiensis Briq. 1261
ssp. noeana (Boiss. ex Briq.) Briq.
1261
ssp. *royleana* (Benth.) Briq. 1261

Mentha L. (*cont.*)
ssp.typhoides (Briq.) R. Harley 1260
microphylla auct. 1260, 1261
pulegium L. **1258**
sieberi C. Koch 1261
spicata L. 1261
var. *longifolia* L. 1259
ssp. spicata **1261**
ssp. tomentosa (Briq.) R. Harley **1261**
sylvestris L. 1259
var. *glabra* W. Koch 1261
var. *nemorosa* auct. 1260
var. *stenostachya* Boiss. 1261
var. *viridis* L. 1261
tomentosa auct. 1260
tomentosa Urv. 1261
viridis (L.) L. 1261
Mercurialis L. **1450**
annua L. **1451**
tomentosa L. 1451
Mesembryanthemum L. **682**
crystallinum L. **682**
nodiflorum L. **682**
Meum segetum Guss. 739
Microlonchus salmanticus (L.) DC. 976
Micromeria Benth. 4, **1274**
chionistrae Meikle **1277**
cypria Kotschy 1276
var. *villosissima* Lindberg f. 1276
graeca auct. 1275
graeca (L.) Benth. 1276
ssp. *cypria* (Kotschy) Chapman 1276
var. *latifolia* auct. 1276
juliana auct. 1275
juliana (L.) Benth. 1275
var. *myrtifolia* (Boiss. et Hoh.) Boiss. 1275
microphylla (Urv.) Benth. **1276**
myrtifolia Boiss. et Hoh. **1275**
nervosa (Desf.) Benth. **1274**
Micropus L. sect. *Bombycilaena* DC. 876
bombicinus Lag. 876
discolor Pers. 876
erectus auct. 876
Micropyrum (Gaudin) Link **1722**
tenellum (L.) Link **1722**
Milium L. **1774**
arundinaceum Sm. 1794
coerulescens Desf. 1795
effusum L. 1776
lendigerum auct. 1786
pedicellare (Bornm.) Roschev. ex Meld. **1775**
thomasii Duby 1795
vernale auct. 1775
vernale M. Bieb. **1775**
ssp. *montianum* auct. 1775
var. *montianum* auct. 1775
var. *pedicellare* Bornm. 1775
Mimosa farcta Banks et Sol. 592
farnesiana L. 601
lebbeck L. 604
longifolia Andr. 600
sophorae Labill. 600
stephaniana M. Bieb. 592
verticillata L'Hérit. 597
Minuartia L. **265**
subgen. Minuartia **267**

Minuartia L. (*cont.*)
subgen. Rhodalsine (J. Gay) McNeill **266**
subgen. Spergella (Fenzl) McNeill **266**
decipiens (Fenzl) Bornm. 269
ssp. *damascena* McNeill 269
filicaulis (Lindberg f.) Rechinger f. 271
geniculata (Poir.) Thell. **266**
globulosa (Labill.) Schinz et Thell. **268**
hybrida (Vill.) Schischk. **271**
intermedia (Boiss.) Hand.-Mazz. **269**
mediterranea (Ledeb.) K. Malý **272**
meyeri (Boiss.) Bornm. 268
var. *cypricola* Mattf. 268
picta (Sibth. et Sm.) Bornm. **266**
sintenisii (Lindberg f.) Rechinger f. **267**
subtilis (Fenzl) Hand.-Mazz. **270**
ssp. filicaulis (Lindberg f.) McNeill **271**
tenuifolia (L.) Hiern 271
thymifolia (Sibth. et Sm.) Bornm. **269**
Mirabilis L. 1362
jalapa L. 1362
Misopates Rafin. **1211**
orontium (L.) Rafin. **1211**
Mollugo tetraphylla L. 277
Moluccella L. **1313**
frutescens auct. 1317
laevis L. **1314**
spinosa L. **1315**
Momordica elaterium L. 677
Monerma cylindrica (Willd.) Coss. et Dur. 1753
Monotropa L. **1064**
hypophegea Wallr. 1065
hypopithys L. **1065**
var. glabra Roth 1065
var. hypopithys 1065
Monotropaceae **1064**
Moraceae **1464**
Morocarpus foliosus Moench 1374
Morus L. 1465
alba L. 1465
nigra L. 1466
papyrifera L. 1465
rubra auct. 1466
rubra L. 1466
Musa L. 1558
acuminata Colla 1559
balbisiana Colla 1559
× paradisiaca L. 1559
× sapientum L. 1559
Musaceae **1558**
Muscari Mill. **1644**
subgenus Botryanthus (Kunth) Rouy **1645**
subgenus Leopoldia (Parl.) Rouy **1644**
subgenus Pseudomuscari Stuart **1647**
atlanticum Boiss. et Reuter 1645
commutatum auct. 1646
comosum (L.) Mill. **1644**
creticum auct. 1645
heldreichii auct. 1646
inconstrictum Rechinger f. **1646**
neglectum Guss. ex Ten. **1645**
parviflorum Desf. **1647**
pinardi auct. 1644
pulchellum auct. 1645
racemosum auct. 1645

Muscari Mill. (*cont.*)
  racemosum Mill. 1646
    var. *brachyanthum* Boiss. 1646
Myagrum aegyptium L. 112
  irregulare Asso 109
  paniculatum L. 133
Myosotis L. **1141**
  apula L. 1151
  collina auct. 1142
  cretica auct. 1141
  discolor auct. 1142
  hispida Schlechtendal 1142
  idaea auct. 1141
  incrassata auct. 1141
  minutiflora Boiss. et Reuter **1143**
  pusilla Loisel. **1141**
  ramosissima Rochel **1142**
  refracta Boiss. **1144**
  stricta auct. 1142, 1143
  sylvatica Hoffm. **1144**
Myosurus L. **43**
  minimus L. **43**
Myrtaceae 19, **658**
Myrtus L. 3, **659**
  communis L. **659**
    var. leucocarpa DC. 660

Najadaceae **1678**
Najas minor All. 1678
Narcissus L. **1573**
  cypri (Haworth) Sweet 1573
  jonquilla L. 1576
  papyraceus Ker-Gawler 1574
  serotinus L. **1576**
  tazetta L. **1573**
    ssp. cypri (Haworth) Holmboe 1573
    var. cypri (Haworth) Boiss. 1573
    var. syriacus (Boiss. et Gaill.) Boiss.
      1576
Nardurus lachenalii (Gmel.) Godr. 1723
  poa (Lam. ex Lam. et DC.) Boiss. 1723
  tenellus (L.) Duval-Jouve 1723
Nardus aristatus L. 1738
  incurva Gouan 1738
Nasturtium R. Br. **153**
  fontanum (Lam.) Aschers. 153
    var. *simplicifolium* (Neum.) Holmboe
      153
  microphyllum (Boenn.) Reichb. 154
  officinale R. Br. **153**
    var. *asarifolium* Kralik ex Rouy et
      Fouc. 153
    var. *simplicifolium* Neum. 153
Neatostema I. M. Johnston **1151**
  apulum (L.) I. M. Johnston **1151**
Nectaroscordum siculum (Ucria) Lindl.
  1628
Nemauchenes aspera (L.) Steudel 1012
  inermis Cass. 1012
Neotinea Reichb. f. **1557**
  intacta (Link) Reichb. f. 1557
    f. *luteola* Renz 1557
  maculata (Desf.) Stearn **1557**
Nepeta L. **1300**
  cataria L. 1300
  mussinii auct. 1301
  orientalis auct. 1301
  sibthorpii auct. 1301

Nepeta L. (*cont.*)
  troodi Holmboe **1301**
Nephrodium filix-mas auct. 1887
  pallidum auct. 1886
  rigidum auct. 1887
Nerium L. **1098**
  indicum Mill. 1099
  kotschyi Boiss. 1099
  mascatense A. DC. 1099
  odorum Aiton 1099
  oleander L. **1098**
Neslia Desv. **132**
  apiculata C. A. Mey. **132**
  paniculata auct. 132
  paniculata (L.) Desv. **133**
Neurada L. **629**
  procumbens L. 7, **629**
    var. *orbicularis* Del. 629
Nicotiana L. **1194**
  glauca Graham **1194**
  pusilla L. 1195
  rustica L. 1195
  tabacum L. 1195
Nigella L. **60**
  subgenus Garidella (L.) Brand **60**
  subgenus Nigella **61**
  arvensis L. **61**
    ssp. *cretica* auct. 62
    var. glauca Boiss. **62**
    var. *microcarpa* auct. 62
    var. *microcarpa* Boiss. 62
  ciliaris DC. **64**
  damascena L. **63**
  elata auct. 63
  fumariifolia Kotschy **62**
    var. *normalis* Terracino 62
  nigellastrum (L.) Willk. **60**
  sativa L. **63**
  stellaris auct. 62
  tuberculata Griseb. 62
  unguicularis (Poir.) Spenner **61**
Noaea Moq. **1392**
  mucronata (Forssk.) Aschers. et
    Schweinf. **1393**
  spinosissima Moq. 1393
Nonea Medik. **1137**
  echioides auct. 1138
  philistaea Boiss. **1137**
  ventricosa (Sm.) Griseb. **1138**
Nothochlaena lanuginosa (Desf.) Poiret
  1882
Notholaena marantae (L.) Desv. 1880
  vellea (Ait.) R. Br. 1880
Nothoscordum Kunth 1628
  fragrans (Vent.) Kunth 1628
  inodorum (Ait.) Nicholson 1628
Notobasis (Cass.) Cass. **952**
  syriaca (L.) Cass. **952**
Notoceras cardaminifolium DC. 117
Nyctaginaceae **1362**

Obione portulacoides (L.) Moq. 1382
Ochrosia elliptica Labill. 1101
Ocimum *americanum* L. 1256
  basilicum L. 1256
  canum Sims 1256
Octhodium DC. **131**
  aegyptiacum (L.) DC. **131**

Odontites Ludwig **1230**
  bocconei auct. 1230
  cypria Boiss. **1230**
  frutescens auct. 1230
  lutea auct. 1230
Odontospermum Necker ex Sch. Bip. 896
  aquaticum Sch. Bip. 897
Oenothera hirta Link 674
  micrantha Hornem. 674
Olea L. **1094**
  africana Mill. 1096
  chrysophylla Lam. 1096
  europaea L. 4, **1094**
    ssp. *oleaster* (Hoffmsgg. et Link)
      Negodi 1095
    var. *oleaster* (Hoffmsgg. et Link) DC.
      1095
    var. *sativa* (Hoffmsgg. et Link) DC.
      1095
    ssp. *sylvestris* (Mill.) Hegi 1095
    var. *sylvestris* (Mill.) Lehr 1095
  ferruginea Royle 1096
  indica Burm. f. 1096
  oleaster Hoffmsgg. et Link 1095
  sativa Hoffmsgg. et Link 1095
  sylvestris Mill. 1095
Oleaceae **1089**
Onagraceae 19, **669**
Onobrychis Mill. **527**
  subgenus Onobrychis **527**
  subgenus Sisyrosema (Bunge ex Boiss.)
    Širjaev **530**
  aequidentata (Sibth. et Sm.) Urv. **528**
  caput-galli (L.) Lam. **527**
  crista-galli (Murr.) Lam. **529**
  gaertneriana Boiss. 529
  saxatilis (L.) Lam. **531**
  squarrosa Viv. 529
  venosa (Desf.) Desv. **530**
Ononis L. **384**
  sect. Natrix DC. emend. Griseb. **385**
    subsect. Biflorae Širjaev **387**
    subsect. Natrix Willk. emend. Širjaev
      385
    subsect. Pubescentes (Širjaev) Meikle
      392, 806
    subsect. Reclinatae Širjaev emend.
      Meikle **391**
    subsect. Torulosae Širjaev **386**
    subsect. Viscosae Širjaev emend.
      Meikle **388**
  sect. Ononis **393**
    subsect. Bugranoides Willk. emend.
      Širjaev **393**
    subsect. Diffusae Širjaev **397**
    subsect. Mitissimae Širjaev **398**
    subsect. Ononis **394**
    subsect. Salzmannianae Širjaev **400**
    subsect. Variegatae Širjaev **396**
  alopecuroides L. **400**
  antiquorum auct. 395
  antiquorum L. 395
    var. *leiosperma* (Boiss.) Post 395
  biflora Desf. **387**
    var. *maroccana* (Batt. et Pitard) Jah.
      et Maire 387
  breviflora DC. 388
  cherleri auct. 391, 393

Ononis L. (*cont.*)
  cherleri L. 393
  columnae All. 393
    var. *orientalis* Širjaev 393
  crispa L. 386
  dentata Sol. ex Lowe 392
  diffusa Ten. **397**
  leiosperma Boiss. 395
    var. *tomentosa* Boiss. 395
  macracantha Clarke 395
  mitissima L. **398**
  mollis Savi 391
  natrix auct. 392
  natrix L. 385, 392
    ssp. hispanica (L. f.) P. Cout. 385
  ornithopodioides L. **386**
  pubescens L. **392**
  pusilla L. **393**
  reclinata L. **391**
    var. minor Moris **391**
    ssp. mollis (Savi) Bég. 391
    var. *mollis* (Savi) Heldr. 391
    ssp. *monophylla* Bég. 392
    var. monophylla (Bég.) Pamp. **392**
    var. *tridentata* (Lowe) Lowe 392
  serrata Forssk. **397**
    subsp. *diffusa* Rouy et Fouc. 397
    var. *major* Lange 397
  sicula Guss. **390**
  spinosa L. **394**
    ssp. *antiquorum* auct. 395
    ssp. leiosperma (Boiss.) Širjaev **395**
    var. leiosperma 395
    var. tomentosa (Boiss.) Širjaev **395**
  variegata L. **396**
  viscosa L. **388**
    ssp. breviflora (DC.) Nyman **388**
    ssp. *sicula* (Guss.) Huber-Mor. **390**
Onopordum L. **958**
  boissieri Freyn et Sint. 960
  boissieri Willk. 959
  bracteatum Boiss. et Heldr. **960**
  carduiforme Boiss. 959
  cyprium Eig **958**
  graecum auct. 958
  illyricum auct. 959
  insigne Holmboe 960
    f. *pallida* Lindberg f. 958
  sibthorpianum auct. 959
    ssp. *anatolicum* auct. 959
  tauricum Willd.
    ssp. *elatum* auct. 959
  virens auct. 958
Onosma L. **1157**
  caespitosum Kotschy **1160**
  frutescens auct. 1160
  frutescens Lam. 1160
  fruticosum Labill. 1158
  fruticosum Sm. **1158**
  giganteum Lam. **1161**
    var. giganteum 1162
    var. hispidum Boiss. **1162**
  mite Boiss. et Heldr. **1160**
  orientale (L.) L. 1162
  troodi Kotschy **1159**
Ophioglossaceae **1876**
Ophioglossales **1876**
Ophioglossum L. **1876**

Ophioglossum L. (*cont.*)
  lusitanicum L. **1876**
  vulgatum L. **1877**
Ophrys L. **1511**
  anthropophora L. 1538
  apifera Huds. **1525**
    var. apifera **1526**
    var. bicolor (Naegeli) E. Nelson **1526**
    var. chlorantha (Hegetschw.) Richter **1526**
    var. flavescens Rosb. 1526
  arachnites Hoffm. 1532
    var. *attica* Boiss. et Orph. 1532
  arachnites Lam. 1532
    ssp. *attica* (Boiss. et Orph.) Richter 1532
  aranifera Huds. 1523
    ssp. *mammosa* Soó 1523
    var. *mammosa* Reichb. f. 1523
  argolica H. Fleischm. **1527**
    ssp. elegans (Renz) E. Nelson **1527**
    × O. bornmuelleri M. Schulze ssp. grandiflora (Fleischm. et Soó) Renz et Taubenheim **1528**
  atrata auct. 1523
  attica (Boiss. et Orph.) Soó 1532
    f. *flavomarginata* Renz 1533
    f. *holocheila* Renz 1533
  × *bayeri* H. Baumann 1515
  bicolor Naegeli 1526
  bombyliflora Link 1534
  bornmuelleri M. Schulze ex Bornm. **1517**
    ssp. bornmuelleri **1518**
    f. *grandiflora* Fleischm. et Soó 1519
    ssp. grandiflora (Fleischm. et Soó) Renz et Taubenheim **1519**
    × O. sphegodes Mill. ssp. mammosa (Desf.) Soó **1520**
  carmeli auct. 1533
  carmeli H. Fleischm. et Bornm. 1533
    ssp. *attica* (Boiss. et Orph.) Renz 1533
    ssp. *orientalis* (Renz) Soó 1532
  chlorantha Hegetschw. 1526
  cornuta auct. 1532
  cornuta Stev. 1532
    ssp. *orientalis* Renz 1532
  cretica (Vierh.) E. Nelson 1529
  cypria Renz 1528
  dinsmorei Schltr. 1532
  elegans (Renz) H. Baumann et Künkele 1527
  ferrum-equinum Desf. 1527
    ssp. *argolica* (H. Fleischm.) Soó 1527
    ssp. *gottfriediana* (Renz) E. Nelson 1528
  flavomarginata (Renz) H. Baumann et Künkele 1533
  fleischmannii Hayek 1514
  fuciflora (F. W. Schmidt) Moench 1518
    ssp. *bornmuelleri* (M. Schulze) B. et E. Willing 1518
    var. *grandiflora* (Fleischm. et Soó) B. et E. Willing 1519
    ssp. *scolopax* (Cav.) Sunderm. 1530
  fusca Link **1513**
    nothosubsp. bayeri (H. Baumann) J. J. Wood **1515**
    ssp. fleischmannii (Hayek) Soó **1514**

Ophrys L. (*cont.*)
    ssp. fusca **1514**
    ssp. iricolor (Desf.) Holmboe **1515**
    var. *iricolor* (Desf.) Reichb. f. 1516
    ssp. omegaifera auct. 1515
    ssp. omegaifera (H. Fleischm.) E. Nelson 1515
    × O. lutea Cav. ssp. galilaea (H. Fleischm. et Bornm.) Soó **1514**
  galilaea H. Fleischm. et Bornm. 1517
  gottfriediana auct. 1527
  gottfriediana Renz 1527
    ssp. *elegans* Renz 1527
  heldreichii H. Fleischm. 1514
  hiulca Sprun. ex Reichb. 1534
  holoserica (Burm. f.) W. Greuter 1518
    ssp. bornmuelleri (M. Schulze) Landwehr 1518
    var. grandiflora (H. Fleischm. et Soó) Landwehr 1519
  iricolor Desf. 1515
  kotschyi H. Fleischm. et Soó **1528**
  lutea Cav. **1517**
    ssp. galilaea (H. Fleischm. et Bornm.) Soó **1517**
    ssp. *minor* (Tod.) O. et E. Dänesch 1517
    var. *minor* (Tod.) Guss. 1517
    ssp. *murbeckii* (Fl.) Soó 1517
  mammosa Desf. 1523
    f. *subtriloba* Renz 1523
  oestrifera auct. 1532
  oestrifera M. Bieb. 1532
    ssp. *orientalis* (Renz) Soó 1532
    var. *cornuta* auct. 1532
  omegaifera auct. 1515
  omegaifera H. Fleischm. 1515
    var. *fleischmannii* (Hayek) Soó 1515
  picta Link 1530
  scolopax Cav. **1529**
    ssp. attica (Boiss. et Orph.) E. Nelson 1533
    ssp. *cornuta* auct. 1532
    ssp. *orientalis* (Renz) E. Nelson 1532
    var. *picta* auct. 1527
    var. *picta* (Link) Reichb. f. 1530
  sicula Tineo 1517
  sintenisii H. Fleischm. et Bornm. 1524
    ssp. *kotschyi* (H. Fleischm. et Soó) Soó 1528
  speculum Link 1534
  sphegodes Mill. **1520**
    ssp. aesculapii (Renz) Soó 1534
    ssp. amanensis E. Nelson 1525
    ssp. mammosa (Desf.) Soó **1523**
    ssp. *sintenisii* (H. Fleischm. et Bornm.) E. Nelson 1524
    ssp. sphegodes **1520**
    ssp. spruneri (Nym.) E. Nelson 1534
    ssp. transhyrcana (Czernjak.) Soó **1524**
  spiralis L. 1504
  spruneri Nym. 1524, 1534
    var. *orientalis* Schltr. 1524
  tenthredinifera Willd. 1535
    var. – 1518
  transhyrcana Czernjak. 1524
  umbilicata Desf. **1531**

Ophrys L. (*cont.*)
  ssp. attica (Boiss. et Orph.) J. J. Wood **1532**
  ssp. umbilicata **1532**
  × bornmuelleri M. Schulze ssp. grandiflora (Fleischm. et Soó) Renz et Taubenheim **1533**
Opopanax Koch **757**
  hispidus (Friv.) Griseb. **757**
  orientale Boiss. 757
Opuntia Mill. **680**
  ficus-indica (L.) Mill. 680
  vulgaris Mill. 681
  vulgaris Ten. 681
× Orchiaceras bergenii (Nant.) Camus 1539
  bivonae Soó 1539
Orchidaceae **1491**
Orchis L. **1539**
  abortiva L. 1493
  acuminata Desf. 1554
  anatolica auct. 1543
  anatolica Boiss. **1542**
    var. anatolica **1542**
    ssp. *troodi* Renz 1543
    var. troodi (Renz) Soó **1543**
  angustifolia M. Bieb. 1556
  brancifortii Biv. 1545
    var. albiflora Raulin 1545
  chlorantha Custer 1509
  collina Banks et Sol. **1545**
  coriophora auct. 1550
  coriophora L. **1550**
    ssp. fragrans (Poll.) Camus **1550**
    var. *fragrans* (Poll.) Boiss. 1550
    var. *polliniana* (Spreng.) Poll. 1550
    ssp. *sancta* (L.) Hayek 1551
    var. *sancta* (L.) Reichb. f. 1551
  elegans Heuff. 1541
  fragrans Poll. 1550
  iberica M. Bieb. ex Willd. 1556
  italica Poir. **1552**
  lactea Poir. 1553
  laxiflora Lam. **1540**
    ssp. *dielsiana* Soó 1541
    f. *elegans* (Heuff.) Aschers. et Graebn. 1542
    ssp. *elegans* (Heuff.) Soó 1541
    var. *elegans* (Heuff.) Beck 1541
    ssp. *palustris* (Jacq.) Aschers. et Graebn. 1541
  × lloydiana Rouy 1540
  longibracteata Biv. 1537
  longicruris Link 1552
  mascula L. 1554
    ssp. pinetorum (Boiss. et Kotschy) Camus 1554
  morio auct. 1547
  morio L. **1547**
    var. *albiflora* Boiss. 1547
    ssp. libani (Renz) Sundermann 1548
    var. libani (Renz) Soó **1548**
    ssp. picta (Loisel.) Aschers. et Graebn. **1547**
    var. picta (Loisel.) Reichb. f. **1547**
    ssp. *syriaca* Camus 1547
  palustris Jacq. **1540**
  papilionacea auct. 1545
  papilionacea L. **1546**

Orchis L. (*cont.*)
  var. grandiflora Boiss. 1547
  var. *rubra* (Jacq.) Reichb. f. 1546
  picta Loisel. 1547
    ssp. *libani* Renz 1548
  polliniana Spreng. 1550
  provincialis Balbis 1554
  pseudosambucina Ten. 1555
  punctulata Stev. ex Lindl. **1549**
  pyramidalis L. 1535
  quadripunctata Cyr. ex Ten. **1543**
  robertiana Loisel. 1536
  romana Seb. 1555
  saccata Ten. 1545
  sancta L. **1551**
  secundiflora Bertol. 1557
  simia Lam. **1553**
  sparsiflora Ten. ex Boiss. 1545
  syriaca auct. 1548
  syriaca Boiss. et Blanche ex Boiss. 1547
  tridentata Scop. 1554
  undulatifolia Biv. 1552
  variegata All. 1554
  vomeracea Burm. f. 1505
Origanum L. **1262**
  sect. Amaracus Benth. **1262**
  sect. Majorana (Mill.) DC. **1266**
  sect. Origanum **1265**
  sect. Prolaticorolla Ietswaart **1264**
  bevanii Holmes 1267
  cordifolium (Aucher-Eloy et Montbret ex Benth.) Vogel **1262**
  dubium Boiss. **1267**
  heracleoticum auct. 1266
  hirtum Link 1266
  laevigatum Boiss. **1264**
  majorana auct. 1267, 1269
  majorana L. **1268**
    var. tenuifolium Weston **1269**
  majoranoides auct. 1267
  majoranoides Willd. 1269
  maru auct. 1269
  maru L. 1266
    var. viridulum H. S. Thompson 1266
  microphyllum auct. 1269
  pseudo-onites Lindberg f. 1266
  syriacum auct. 1267
  syriacum L. **1266**
    var. bevanii (Holmes) Ietswaart **1267**
    ssp. *dubium* (Boiss.) Holmboe 1267
    var. syriacum **1267**
  vestitum Clarke 1267
  vulgare L. **1265**
    ssp. *heracleoticum* auct. 1266
    ssp. hirtum (Link) Ietswaart **1266**
Orlaya Hoffm. **706**
  daucoides (L.) Greuter **707**
  kochii Heywood 707
  maritima (Gouan) Koch 708
  platycarpos Koch 707
  pumila (L.) Hal. 708
Ornithogalum L. **1629**
  sect. Beryllis (Salisb.) Benth. **1633**
  sect. Heliocharmos (Baker) Benth. **1629**
  arabicum L. 1633
  arvense Pers. 1601
  brevipedicellatum auct. 1632

Ornithogalum L. (*cont.*)
  chionophilum Holmboe **1632**
  chloranthum M. Bieb. 1606
  cuspidatum Bertol. 1633
  cyprium Zahariadi 1632
  divergens Boreau 1630
  fibrosum Desf. 1605
  gussonii Ten. 1631
  huetii auct. 1632
  lanceolatum auct. 1632
  lanceolatum Labill. 1633
  minus L. f. 1630
  montanum auct. 1632
  montanum Cyr. 1633
  nanum auct. 1630
  narbonense L. **1633**
  pedicellare Boiss. et Kotschy **1630**
    ssp. *cylindrocarpum* Zahariadi 1630
  pedunculare J. et C. Presl 1603
  pyrenaicum auct. 1633
  spathaceum auct. 1602
  tenuifolium auct. 1631
  tenuifolium Guss. 1631
    ssp. *trichophyllum* (Boiss. et Heldr.) Holmboe 1631
    var. *trichophyllum* (Boiss. et Heldr.) Boiss. 1631
  trichophyllum Boiss. et Heldr. **1631**
  umbellatum L. **1630**
    ssp. divergens (Boreau) Aschers. et Graebn. **1630**
    var. *minus* (L. f.) Aschers. et Graebn. 1630
  villosum M. Bieb. 1601
Ornithopus L. **510**
  compressus L. **510**
  repandus Poir. 517
  scorpioides L. 516
Orobanchaceae **1232**
Orobanche L. **1233**
  sect. Orobanche **1237**
  sect. Trionychon Wallr. **1234**
  aegyptiaca Pers. **1236**
  alba Stephan 1238
  crenata Forssk. **1240**
  cypria Reuter **1238**
    var. pterocephali Beck-Mannagetta 1238
  lavandulacea auct. 1236
  longiflora auct. 1236
  loricata Reichb. 1241
  minor Sm. **1241**
    var. minor **1241**
    var. pubescens (Urv.) Meikle **1242**
  mutelii F. Schultz 1235
  nana (Reuter) Noë ex Beck-Mannagetta 1234, 1237
  orientalis Beck-Mannagetta **1237**
  ovata Blakelock 1238
  oxyloba (Reuter) Beck-Mannagetta 1234, 1237
  picridis F. Schultz 1241
  pruinosa Lapeyr. 1240
  pubescens Urv. 1242
  ramosa auct. 1236
  ramosa L. **1234**
    var. brevispicata (Ledeb.) R. A. Graham **1235**

Orobanche L. (*cont.*)
  var. *minor* Loret et Barr. 1235
  ssp. *mutelii* (F. Schultz) Coutinho 1235
  ssp. *nana* (Reuter) Coutinho 1234
  var. ramosa **1235**
  schultzii Beck-Mannagetta 1235
  speciosa DC. 1240
  tinctoria Forssk. 1233
  versicolor F. Schultz 1242
Orobus saxatilis Vent. 577
Orthurus Juzepczuk **612**
  heterocarpus (Boiss.) Juzepczuk **612**
Oryza L. 1715
  sativa L. 1716
Oryzopsis Michx. **1794**
  coerulescens (Desf.) Richter **1795**
  miliacea (L.) Aschers. et Schweinf. **1794**
    var. thomasii (Duby) Richter 1795
Osyris L. **1429**
  alba L. **1429**
Otanthus Hoffmsgg. et Link **907**
  maritimus (L.) Hoffmsgg. et Link **907**
Oxalidaceae 19, **345**
Oxalis L. **345**
  acetosella L. 347
  cernua Thunb. 346
  corniculata L. **346**
  pes-caprae L. **346**

Paeonia L. **68**
  arietina Anders. 69
    var. *orientalis* (Thiébaut) F. C. Stern 69
  corallina Retz. 69
    var. *orientalis* Thiébaut 69
    var. *triternata* auct. 69
  decora auct. 69
  mascula (L.) Mill. **68**
  officinalis L. 69
    var. *mascula* L. 69
Paeoniaceae 19, **68**
Paliurus Mill. **356**
  aculeatus Lam. 356
  australis Gaertn. 356
  spina-Christi Mill. **356**
Pallasia Scop. 1847
Pallenis Cass. **897**
  spinosa (L.) Cass. **898**
Palmae **1660**
Pancratium L. **1577**
  maritimum L. **1577**
Pandorea jasminoides (Lindl.) K. Schum. 1246
Panicum L. **1856**
  colonum L. 1854
  crusgalli L. 1855
  dactylon L. 1843
  decompositum R. Br. 1857
  eruciforme Sm. 1852
  glaucum auct. 1859
  italicum L. 1860
  miliaceum L. 1857
  pumilum Poir. 1859
  repens L. **1856**
  sanguinale L. 1853
  turgidum Forssk. 1857
  verticillatum L. 1861

Panicum L. (*cont.*)
 viride L. 1861
Papaver L. **74**
 argemone auct. 76
 argemone L. 76
  var. *glabrum* auct. 76
 belangeri Boiss. 76
 desertorum Grossh. 76
 dubium L. **75**
 gracile Auch. ex Boiss. **79**
 humile Fedde 75
 hybridum L. **78**
 minus (Boiv.) Meikle **76**
 postii Fedde **75**
 rhoeas L. **74**
  ssp. *humile* Holmboe 75
  var. oblongatum Boiss. **75**
  var. *pryorii* auct. 75
 setigerum DC. **78**
 somniferum auct. 78
 somniferum L. 78
  ssp. *setigerum* (DC.) Corb. 78
  var. *setigerum* (DC.) Elk. 78
 subpiriforme Fedde 75
Papaveraceae 19, **73**
Paracaryum myosotoides auct. 1126
Parapholis C. E. Hubb. **1751**
 incurva (L.) C. E. Hubb. **1752**
 marginata Runemark **1752**
Parentucellia Viv. **1226**
 flaviflora (Boiss.) Nevski 1226
 latifolia (L.) Caruel **1226**
  var. *albiflora* Hal. 1226
  ssp. flaviflora (Boiss.) Hand.-Mazz.
   1226
  ssp. latifolia 1227
 viscosa (L.) Caruel **1227**
Parietaria L. **1457**
 cretica L. **1458**
 diffusa Mert. et Koch 1457
 erecta auct. 1457
 judaica L. **1457**
 lusitanica L. **1459**
 officinalis auct. 1457
 officinalis L. 1457
  ssp. *diffusa* (Mert. et Koch) Arcang.
   1457
  ssp. *erecta* auct. 1457
  ssp. *judaica* (L.) Bég. 1457
 vulgaris Hill 1457
Parkinsonia aculeata L. 587
 torreyana Wats. 587
Paronychia Mill. **279**
 argentea Lam. **279**
  var. *rotundata* (DC.) Chaudhri 280
 capitata L. 280
  ssp. *macrosepala* (Boiss.) Maire et Weiller 280
 capitata Sibth. et Sm. 280
 hispanica DC. 280
  var. *rotundata* DC. 280
 macrosepala Boiss. **280**
Paspalum L. **1857**
 dilatatum Poir. **1858**
 distichum L. **1858**
 paspalodes (Michx.) Scribn. 1858
Passerina hirsuta L. 1424
Pastinaca opopanax L. 757

Paulownia *imperialis* Sieb. et Zucc. 1216
 tomentosa (Thunb.) Steud. 1216
Pedaliaceae 1246
Peganum L. **325**
 harmala L. **325**
Pennisetum divisum (Gmel.) Henr. 1851
 villosum R. Br. 1859
Pentapera sicula (Guss.) Klotzsch 1060
  ssp. *libanotica* (C. et W. Barbey) Yalt-
   irik 1060
  var. *libanotica* C. et W. Barbey 1060
Periploca gracilis Boiss. 1103
Persea americana Mill. 1421
Persicaria maculata S. F. Gray 1400
 salicifolia (Brouss. ex Willd.) Assenov
  1399
Petrorhagia (Ser.) Link **219**
 cretica (L.) P. W. Ball et Heywood **220**
 kennedyae (A. K. Jackson et Turrill)
  Meikle **220**, 806
Petroselinum crispum (Mill.) A. W. Hill
 738
Peucedanum ostruthium (L.) Koch 758
 veneris Kotschy 728, 758
Phaca boetica L. 498
Phacelia tanacetifolia Benth. 1118
Phagnalon Cass. **882**
 graecum Boiss. et Heldr. 883
 rupestre (L.) DC. 4, **883**
  ssp. graecum (Boiss. et Heldr.) Hayek
   883
  ssp. *illyricum* (Lindberg f.) Ginzberger
   885
  var. *illyricum* Lindberg f. 885
  ssp. rupestre **883**
Phalaris L. **1770**
 aquatica L. **1773**
 brachystachys Link **1771**
 canariensis auct. 1771
 canariensis L. **1771**
 crypsoides Urv. 1781
 geniculata Sm. 1847
 gracilis Parl. **1772**
 minor Retz. **1772**
  var. *gracilis* (Parl.) Parl. 1772
 nodosa Murr. 1773
 paradoxa L. **1772**
  var. paradoxa **1773**
  f. monstr. *praemorsa* Lindberg f. 1773
  var. praemorsa Coss. et Durieu 1773
 praemorsa Lam. 1772
 semiverticillatus Forssk. 1785
 subulata Savi 1780
 tuberosa L. 1773
Phaseolus aureus Roxb. 582
 coccineus L. 582
 lunatus L. 582
  var. macrocarpus Benth. 582
 mungo L. 582
 vulgaris L. 581
  var. nanus (L.) Aschers. 581
Phelipaea aegyptiaca (Pers.) Walpers 1236
 lutea Desf. 1233
 mutelii (F. Schultz) Reuter 1235
  var. *nana* (Reuter) Boiss. 1234
 oxyloba Reuter 1234
 ramosa (L.) C. A. Mey. 1234
  var. *brevispicata* Ledeb. 1235

Phelipaea Pers. (*cont.*)
  var. *mutelii* (F. Schultz) Boiss. 1235
  var. *nana* (Reuter) Boiss. 1234
  var. *simplex* Vis. 1234
Phillyrea L. **1093**
  angustifolia L. 1094
  latifolia L. **1093**
  media L. 1093
Phleum L. **1779**
  alopecuroides Pill. et Mitterp. 1847
  asperum Jacq. 1780
  crinitum Schreb. 1784
  crypsoides (Urv.) Hackel 1781
  paniculatum Huds. 1780
  schoenoides L. 1849
  subulatum (Savi) Aschers. et Graebn.
    1780
  tenue Schrad. 1780
Phlomis L. **1319**
  bailanica Vierh. 1326
  bertramii Post 1326
  brevibracteata Turrill **1323**
  cypria Post **1321**
    var. *cypria* **1321**
    var. *occidentalis* Meikle **1323**
  fruticosa auct. 1321
  fruticosa L. **1320**
  longifolia Boiss. et Bl. **1325**
    var. bailanica (Vierh.) Huber-Morath
      1326
  lunariifolia Sm. 6, **1324**
  viscosa auct. 1326
  viscosa Poir. 1326
    var. *angustifolia* Boiss. 1325
Phoenix dactylifera L. 1661
Phragmites Trin. **1838**
  australis (Cav.) Trin. ex Steud. **1838**
  communis Trin. 1838
Phucagrostis major Cavolini ex Willd. 1682
  major Theophrasti Cavolini 1682
Phyla Lour. **1249**
  canescens auct. 1249
  filiformis (Schrader) Meikle **1249**
  nodiflora (L.) Greene **1249**
    var. *rosea* (D. Don) Mold. 1249
Physalis somnifera L. 1186
Physanthyllis Boiss. **471**
  tetraphylla (L.) Boiss. **471**
Physospermum Cusson ex Jussieu **723**
  aquilegifolium (All.) Koch 724
  cornubiense (L.) DC. **724**
Phytolacca L. **1394**
  americana L. **1395**
  decandra L. 1395
  dioica L. **1395**
  pruinosa Fenzl **1396**
  stricta auct. 1396
Phytolaccaceae **1394**
Picnomon Adans. **953**
  acarna (L.) Cass. **953**
Picridium hispanicum Poir. 1034
  intermedium Sch. Bip. 1033
  orientale (L.) DC. 1034
  tingitanum (L.) Desf. 1034
    var. *minus* Boiss. 1034
  vulgare auct. 1033
  vulgare Desf. 1032
Picris L. **999**

Picris L. (*cont.*)
  altissima Del. **1001**
  cyprica Lack **1000**
  kotschyi auct. 1002, 1043
  longirostris Sch. Bip. 1043
    var. *kotschyi* Sch. Bip. 1002, 1043
  pauciflora auct. 1000
  pauciflora Willd. **999**
  sprengeriana auct. 999, 1001
Pimpinella L. **744**
  anisum L. **745**
  cretica Poir. **745**
  cypria Boiss. **747**
  peregrina L. **746**
    var. hispida (Lois.) Thell. 746
    var. peregrina 746
  tragium auct. 747
  tragium Vill. 747
    var. *cypria* (Boiss.) H. Wolff 747
    var. *pseudotragium* auct. 747
Pinaceae 19, **21**
Pinguicula L. **1243**
  crystallina Sm. **1243**
Pinus L. **22**
  brutia Tenore 1, 2, 3, **25**, 26
  canariensis C. Smith 27
  caramana Gaudry 24
  cedrus L. 21
  halepensis auct. 25
  halepensis Mill. **26**
    ssp. *brutia* (Ten.) Holmboe 25
  laricio Poir. 24
    var. *caramanica* Loudon 24
  maritima auct. 25
  nigra Arnold **24**
    var. *caramanica* (Loudon) Rehder 2,
      24
    ssp. pallasiana (D. Don) Holmboe 2,
      24
  pallasiana D. Don 24
  pinea L. **26**
  radiata D. Don 27
Piptatherum coerulescens (Desf.) P. Beauv.
    1795
  miliaceum (L.) Coss. 1794
Pircunia dioica (L.) Moq. 1395
Pistacia L. 7, **365**
  sect. Butmela Zohary **367**
  sect. Eu-lentiscus Zohary **365**
  sect. Pistacia **368**
  atlantica Desf. **367**
    ssp. *cypricola* Lindberg f. 367
    var. *latifolia* DC. 367
  lentiscus L. 4, **365**
  mutica Fisch. et Mey. 367
    ssp. *cypricola* (Lindberg f.) Lindberg f.
      367
  palaestina (Boiss.) Kotschy 368
    × saportae Burnat 6, **366**
  terebinthus auct. 367
  terebinthus L. 4, **368**
    ssp. *palaestina* (Boiss.) Engl. 368
  vera auct. 368
  vera L. 369
Pisum L. **578**
  arvense auct. 580
  arvense L. 580
    ssp. *elatius* auct. 580

Pisum L. (*cont.*)
  ssp. *humile* (Boiss. et Noê) Holmboe 580
  var. *variegatum* Guss. 580
  elatius auct. 580
  elatius M. Bieb. 580
  fulvum auct. 580
  fulvum Sibth. et Sm. 580, 581
  humile Boiss. et Noê 580
  ochrus L. 567
  pumilio (Meikle) Greuter 580
  sativum L. 579
    var. brevipedunculatum Davis et Meikle 580
    ssp. elatius (M. Bieb.) Aschers. et Graebn. 580
    var. elatius (M. Bieb.) Alef. 580
    var. pumilio Meikle 580
    var. sativum 579
    ssp. *syriacum* Berger 580
Planera abelicea (Lam.) J. A. Schultes 1463
  cretica (Spach) Kotschy 1463
Plantaginaceae 1347
Plantago L. 1347
  subgenus Plantago 1348
  subgenus Psyllium Harms 1360
  afra L. 1361
  albicans L. 1355
  amplexicaulis Cav. 1350
    var. bauphula (Edgew.) Pilger 1351
  bellardii All. 1359
    var. deflexa Pilger 1360
  commutata Guss. 1349
  coronopus L. 1348
    ssp. commutata (Guss.) Pilger 1349
    ssp. *simplex* Decne. 1349
  crassifolia Forssk. 1350
  cretica L. 1358
  dubia L. 1354
  indica auct. 1361
  lagopus L. 1354
    f. *minor* (Ten.) Pilger 1354
    var. *minor* Ten. 1354
  lanceolata L. 1351
    var. bakeri C. E. Salmon 1353
    var. dubia (L.) Liljeblad 1354
    subvar. euryphylla Pilger 1354
    var. lanceolata 1353
    var. *mediterranea* Pilger 1353
    var. *timbali* auct. 1353
  loeflingii L. 1357
  major L. 1348
  maritima auct. 1350
  maritima L. 1349
    ssp. crassifolia (Forssk.) Holmboe 1350
  notata Lag. 1357
  ovata auct. 1357
  ovata Forssk. 1356
  psyllium auct. 1361
  squarrosa Murr. 1360
    var. *brachystachys* Boiss. 1360
  stricta auct. 1361
Platanaceae 1469
Platanthera L. C. M. Rich. 1509
  bifolia (L.) L. C. M. Rich. 1510
    ssp. *montana* Holmboe 1510
  chlorantha auct. 1510

Platanthera L. C. M. Rich. (*cont.*)
  chlorantha (Custer) Reichb. 1509
    ssp. chlorantha 1510
    ssp. holmboei (Lindberg f.) J. J. Wood 1510
  holmboei Lindberg f. 1510
  montana Reichb. f. 1509
Platanus L. 1470
  × acerifolia (Ait.) Willd. 1471
  hispanica Muenchh. 1471
  hybrida Brot. 1471
  occidentalis L. 1471
  orientalis L. 1470
    var. *insularis* A. DC. 1470
Platycapnos spicata (L.) Bernh. 93
Pluchea dioscoridis (L.) DC. 873
Plumbaginaceae 1066
Plumbago L. 1074
  auriculata Lam. 1074
  capensis Thunb. 1074
  europaea L. 1074
Poa L. 1741
  angustifolia L. 1744
  annua L. 1745
  attica Boiss. et Heldr. 1742
  aurea (L.) Sibth. ex Walpole 1727
  bulbosa L. 1741
    var. bulbosa 1742
    f. *vivipara* (Koeler) Dinsm. 1742
    var. vivipara Koeler 1742
  cilianensis All. 1841
  compressa L. 1745
  divaricata Gouan 1740
  dura (L.) Scop. 1740
  eragrostis (L.) Brot. 1841
  exilis (Tomm.) Murb. 1746
  expansa J. F. Gmel. 1740
  infirma Kunth 1746
  littoralis Gouan 1845
  loliacea Huds. 1721
  megastachya Koeler 1841
  nemoralis L. 1745
  persica auct. 1747
  persica Trin. 1747
    var. *alpina* Boiss. 1747
    ssp. *cypria* Samuels. 1747
  pratensis L. 1744
    ssp. angustifolia (L.) Hayek 1744
    var. angustifolia (L.) Sm. 1744
  remotiflora Murb. 1746
  rigida L. 1721
  sintenisii Lindberg f. 1747
  sylvicola Guss. 1742
  trivialis L. 1743
    ssp. attica auct. 1742
    ssp. *sylvicola* (Guss.) Lindberg f. 1742
    var. *sylvicola* (Guss.) Hackel 1742
Podospermum canum C. A. Meyer 1044
  jacquinianum Koch 1044
  villosum auct. 1046
Podranea ricasoliana (Tanf.) Sprague 1246
Poinciana coriaria Jacq. 585
  gilliesii Hook. 585
Poinsettia heterophylla (L.) Klotsch et Garcke 1436
Pollinia distachya (L.) Spreng. 1864
Polycarpon L. 277

Polycarpon L. (*cont.*)
  alsinefolium auct. 278
  alsinefolium DC. 278
  diphyllum Cav. 278
  tetraphyllum (L.) L. **277**
Polycnemum L. **1371**
  arvense L. **1371**
Polygala L. **207**
  apopetala T. S. Brandeg. 209
  glumacea Sibth. et Sm. 208
  monspeliaca L. **207**
  venulosa Sibth. et Sm. **208**
Polygalaceae 19, **207**
Polygonaceae **1398**
Polygonum L. **1398**
    sect. Persicaria (L.) DC. **1399**
    sect. Polygonum **1399**
    sect. Tiniaria Meissner **1404**
  arenastrum auct. 1401
  aviculare L. **1401**
    var. *heterophyllum* Druce 1401
  bellardii auct. 1403
  convolvulus L. **1404**
  equisetiforme Sm. **1402**
  heterophyllum Lindm. 1401
  lapathifolium auct. 1399
  lapathifolium L. **1400**
    ssp. maculatum (S. F. Gray) Dyer et
      Trimen **1400**
  maritimum L. **1401**
  nodosum Pers. 1400
  patulum M. Bieb. **1403**
  salicifolium Brouss. ex Willd. **1399**
  scabrum Poir. 1399
    var. *salicifolium* (Boiss.) Dinsmore
      1399
  serrulatum Lag. 1399
    var. *salicifolium* Boiss. 1399
  tenuiflorum J. et C. Presl 1400
Polypodiaceae **1891**
Polypodium L. **1891**
  australe Fée 1891
  cambricum L. **1891**
  fragile L. 1885
  fragrans L. 1882
  leptophyllum L. 1879
  pteridioides Reichard 1882
  vulgare auct. 1891
    f. brevipes Milde 1891
    var. *brevipes* auct. 1891
Polypogon Desf. **1783**
  maritimus Willd. **1783**
    var. *subspathaceus* (Req.) Duby 1783
  monspeliensis (L.) Desf. **1784**
  semiverticillatus (Forssk.) Hyl. **1785**
  subspathaceus Req. 1783
  viridis (Gouan) Breistr. 1785
Populus L. **1489**
  alba L. 1490
  dilatata auct. 1490
  nigra L. **1489**
    var. afghanica Aitch. et Hemsley **1490**
    var. *italica* auct. 1490
    var. italica Muenchh. 1490
    var. *pyramidalis* auct. 1490
    var. *thevestina* (Dode) Bean 1490
  pyramidalis auct. 1490
  thevestina Dode 1490

Populus L. (*cont.*)
  usbekistanica Kom. 1490
    cv. *Afghanica* W. Bugaļa 1490
Portulaca L. **287**
  oleracea L. **287**
    ssp. sativa (Haw.) Thell. 288
    ssp. *silvestris* (DC.) Thell. 287
    var. *sylvestris* DC. 287
Portulacaceae 19, **287**
Posidonia Koenig **1679**
  caulini Koenig 1679
  oceanica (L.) Del. **1679**
Posidoniaceae **1678**
Potamogeton L. **1674**
  fluitans Roth 1676
  natans auct. 1676
  nodosus Poir. **1676**
  perfoliatus L. 1676
Potamogetonaceae **1674**
Potentilla L. **614**
  hirta auct. 616
  hirta L. 616
    ssp. *pedata* (Willd. ex Spreng.)
      Holmboe 616
  pedata Willd. ex Spreng. 616
  recta L. **614**
  reptans L. **616**
Poterium L. **620**
  dictyocarpum Spach 622
  gaillardotii Boiss. 622
  magnolii Spach 622
  obtusum (Maxim.) Franch. et Savat. 623
  sanguisorba L. **620**, 623
    ssp. dictyocarpum (Spach) Rouy et
      Camus **622**
    ssp. muricatum (Spach) Rouy et
      Camus 622
  spinosum L. 623
  verrucosum Link ex G. Don **622**
Prasium L. **1327**
  majus L. **1328**
Prenanthes L. **1031**
  triquetra Labill. **1031**
  viminea L. 1029
Primula L. **1075**
  acaulis (L.) Hill 1076
  veris L. 1076
    var. *acaulis* L. 1076
  vulgaris Huds. **1076**
Primulaceae **1075**
Prismatocarpus falcatus Ten. 1054
  scaber Lowe 1055
Pritchardia filifera J. A. Linden ex André
  1661
Prosopis L. **592**
  farcta (Banks et Sol.) Macbride **592**
  juliflora (Sw.) DC. 594
  pubescens Benth. 596
  stephaniana (M. Bieb.) Kunth ex Spreng.
    592
Protea sp. 1422
Proteaceae **1422**
Prunella L. **1326**
  vulgaris L. **1326**
Prunus L. **607**
  subgenus Amygdalus (L.) Focke **609**
  subgenus Cerasus Pers. **609**
  subgenus Prunus **608**

Prunus L. (*cont.*)
 amygdalus Batsch 609
 armeniaca L. 608
 avium L. **610**
 cerasus Osorio-Tafall et Seraphim 610
 domestica L. **608**
  var. aubertiana DC. 608
  var. juliana DC. 608
 dulcis (Mill.) D. A. Webb 609
 persica (L.) Batsch 609
  var. nectarina (Ait.) Maxim. 609
Psamma australis Mabille 1789
Pseudognaphalium Kirpiczn. **885**
 luteo-album (L.) Hilliard et B. L. Burtt **886**
Pseudorlaya Murb. **707**
 pumila (L.) Grande **708**
Psidium guajava L. 665
Psilurus Trin. **1738**
 aristatus (L.) Duv.-Jouve 1738
  var. *hirtellus* (Simonkai) Aschers. et Graebn. 1739
 hirtellus Simonkai 1739
 incurvus (Gouan) Schinz et Thell. 1738
 nardoides Trin. 1738
Psoralea L. 488
 bituminosa L. 489
  var. *palaestina* (Gouan) Hal. 490
  var. *plumosa* (Reichb.) Reichb. 490
Psylliostachys spicata (Willd.) Nevski 1073
Pteranthus Forssk. **284**
 dichotomus Forssk. 2, **284**
 echinatus Desf. 284
Pteridaceae **1877**
Pteridium Scop. **1884**
 aquilinum (L.) Kuhn **1884**
  f. *lanuginosum* auct. 1884
  var. *lanuginosum* auct. 1884
Pteridophyta **1871**
Pteris L. **1878**
 aquilina L. 1884
 lanuginosa Bory ex Willd. 1884
 longifolia auct. 1878
 vittata L. **1878**
Pterocephalus Adanson **858**
 brevis Coult. **859**
 involucratus Spreng. 859
 multiflorus Poech **860**
  ssp. multiflorus **860**
  ssp. obtusifolius Holmboe **861**
 obtusifolius (Holmboe) C. E. Gresham 861
 palaestinus (L.) Coult. 856
 papposus auct. 859
 papposus (L.) Coult. var. *luteiflorus* Lindberg f. 859
 plumosus auct. 859
 plumosus (L.) Coult. 858
Pteropsida **1875**
Pterotheca bifida (Vis.) Fisch. et Mey. 1010
Ptilostemon Cass. **955**
 chamaepeuce (L.) Less. **956**
  ssp. *camptolepis* auct. 956
  var. cyprius Greuter **956**
  ssp. *polycephalus* auct. 956
 fruticosus auct. 956
 muticus Cass. 956

Puccinellia Parl. **1747**
 ciliata Bor **1749**
 festuciformis (Host) Parl. 1750
 gigantea (Grossh.) Grossh. **1749**
Pulegium vulgare Mill. 1258
Pulicaria Gaertn. 893
 dentata (Sm.) DC. 894
 dysenterica (L.) Bernh. **893**
  ssp. *dentata* (Sm.) Holmboe 894
  var. *microcephala* Boiss. 894
  ssp. uliginosa Nyman **894**
 uliginosa Stev. ex DC. 894
Punica granatum L. 669
 protopunica Balf. f. 669
Punicaceae 19, **668**
Putoria Pers. **773**
 calabrica (L.f.) DC. **773**
Pycreus P. Beauv. **1690**
 flavidus (Retz.) Koyama **1690**
 globosus (All.) Reichb. 1690
Pyrethrum balsamita auct. 922
 balsamita (L.) Willd. var. *tanacetoides* Boiss. 922
 parthenium (L.) Sm. 923
Pyrostegia ignea (Vell.) C. Presl 1246
Pyrus L. **631**
 aria Ehrh. 634
  var. *cretica* Lindl. 634
 communis L. 631
 malus L. 633
 syriaca Boiss. **631**

Quercus L. **1478**
 sect. Cerris Spach 1482
  subsect. Cypriotes J. Gay 1482
 sect. Gallifera Spach 1481
 abelicea Lam. 1463
 aegilops auct. 1480
 aegilops L. 1482
 alnifolia Poech 6, **1481**
 boissieri Reuter 1479
  ssp. *latifolia* (Boiss.) O. Schwarz 1480
  ssp. *petiolaris* (Boiss. et Heldr.) O. Schwarz 1480
  var. *petiolaris* (Boiss. et Heldr.) Zohary 1480
  ssp. *tauricola* (Kotschy) O. Schwarz 1480
 calliprinos Webb 1485
  var. *arcuata* Kotschy ex A. DC. 1485
  var. *eucalliprinos* A. DC. 1485
  var. *palaestina* (Kotschy) Zohary 1485
 canariensis Willd. 1481
 coccifera auct. 1485
 coccifera L. 4, **1484**
  ssp. calliprinos (Webb) Holmboe **1485**
  var. *calliprinos* (Webb) Boiss. 1485
  ssp. coccifera 1486
  ssp. *palaestina* (Kotschy) Holmboe 1485
  var. *palaestina* (Kotschy) Boiss. 1485
  ssp. *pseudococcifera* Boiss. 1485
 cypria Hochst. et Kotschy 1480
 cypria Jaubert et Spach 1481
 faginea Lam. 1481
 fruticosa Brot. 1481
 ilex auct. 1481
 ilex L. 1482

Quercus L. (*cont.*)
inermis Ehrenb. ex Kotschy 1480
infectoria auct. 1480, 1481
infectoria Olivier **1479**
 ssp. *boissieri* (Reuter) O. Schwarz 1480
 var. *insularis* A. Camus 1480
 ssp. *petiolaris* (Boiss. et Heldr.) O.
  Schwarz 1480
 var. *petiolaris* (Boiss. et Heldr.) Hand.-
  Mazz. 1480
 ssp. veneris (A. Kerner) Meikle **1479**
 var. *veneris* (A. Kerner) Lindberg f.
  1480
lusitanica auct. 1479, 1480
lusitanica Lam. 1480
 var. *genuina* auct. 1480
 ssp. *infectoria* auct. 1480
 var. *latifolia* Boiss. 1480
 ssp. *orientalis* A. DC. 1479
 var. *petiolaris* (Boiss. et Heldr.) A. DC.
  1480
 var. *syriaca* (Kotschy) A. DC. 1480
 ssp. *veneris* (A. Kerner) Holmboe 1480
macrocarpa Michaux 1484
macrolepis Kotschy 1482
palaestina Kotschy 1485
petiolaris Boiss. et Heldr. 1480
petraea (Mattuschka) Liebl. 1484
pfaeffingeri Kotschy 1480
 var. *cypria* Kotschy 1480
pseudo-coccifera Labill. 1485
robur L. 1484
 Spielart *petraea* Mattuschka 1484
sessiliflora Salisb. 1484
suber L. 1484
syriaca Kotschy 1480
tauricola Kotschy 1480
veneris A. Kerner 1480

Rafflesiaceae **1414**
Ranunculaceae 19, **35**
Ranunculus L. **44**
 subgenus Batrachium (DC.) A. Gray
  45
 subgenus Ficaria (Huds.) L. Benson
  47
 subgenus Ranunculus **48**
aquatilis auct. 45
aquatilis L. 45
 ssp. *circinatus* auct. 46
 ssp. *circinatus* Dinsmore 46
 ssp. *confusus* auct. 46
 var. *sphaerospermus* (Boiss. et
  Blanche) Boiss. 46
 var. *submersus* Gren. et Godr. 45
arvensis L. **59**
asiaticus L. **50**
 var. albus Hayek **50**
 var. flavus Dörfl. **50**
 var. sanguineus (Mill.) DC. **50**
 var. tenuilobus DC. **50**
bullatus L. **48**
 f. *cuneifolius* (Coust. et Gand.) Coust. et
  Gand. 48
 var. *cuneifolius* Coust. et Gand. 48
 var. *cyrenaicus* Pamp. 48
 ssp. cytheraeus (Hal.) Vierhapper et
  Rechinger f. **48**

Ranunculus L. (*cont.*)
 var. *cytheraeus* Hal. 48
cadmicus auct. 52
cadmicus Boiss. **51**
 ssp. *cyprius* (Boiss.) Vierh. 52
 var. cyprius Boiss. **52**
calthifolius Jord. 47
chaerophyllos auct. 50
chius DC. **54**
 var. chius **54**
 var. leiocarpus P. H. Davis **55**
cicutarius auct. 49
circinatus auct. 46
constantinopolitanus (DC.) Urv. **55**
cornutus DC. **58**
falcatus L. 43
ficaria L. **47**
 ssp. *calthifolius* (Jord.) Holmboe 47
 var. *calthifolius* Guss. 47
 ssp. ficariiformis (F. Schultz) Rouy et
  Fouc. **47**
 var. *grandiflorus* (Rob.) Strobl 47
ficariiformis F. Schultz 47
flabellatus Desf. 50
incrassatus Guss. 54
isthmicus Boiss. **49**
kykkoënsis Meikle **52**
lanuginosus L. **55**
 var. *constantinopolitanus* DC. 55
leptaleus DC. 53
lomatocarpus auct. 56
lomatocarpus Fisch. et Mey. 56, 58
marginatus Urv. **56**
 var. trachycarpus (Fisch. et Mey.)
  Azn. **56**
millefoliatus auct. 53
millefoliatus Vahl **52**
 ssp. leptaleus (DC.) Meikle **53**
millefolius Banks et Soland. **53**
muricatus L. **58**
myriophyllus DC. 53
neapolitanus Ten. **54**
 var. adpresse-pilosus Freyn emend.
  Vierh. **54**
 subvar. *brevirostris* Freyn 54
 ssp. *tommasinii* (Reichb.) Vierh. 54
orientalis L. 53
paludosus Poir. **50**
? *pantothrix* Brot. 46
? *parviflorus* auct. 54
? *paucistamineus* Tausch 46
peltatus Schrank **45**
 var. microcarpus Meikle **46**
 ssp. peltatus **46**
 ssp. sphaerospermus (Boiss. et
  Blanche) Meikle **46**
rumelicus Griseb. **49**
saniculifolius auct. 46
sphaerospermus Boiss. et Blanche 46
trachycarpus auct. 58
trachycarpus Fisch. et Mey. 56, 58
trilobus auct. 56
trilobus Desf. 56
 var. *tripetalus* Holmboe 56
troodi Lindberg f. 56
ulmifolius auct. 611
sp. indet. **59**
Raphanus L. **106**

Raphanus L. (*cont.*)
 lyratus Forssk. 109
 × micranthus (Uechtr.) O. E. Schulz 107
 raphanistrum L. **106**
 sativus L. **107**
Rapistrum Crantz **111**
 aegyptium (L.) Crantz 112
 perenne (L.) All. 112
 rugosum (L.) All. **111**
  var. *orientale* auct. 111
  var. venosum (Pers.) DC. **111**
Reichardia Roth **1032**
 intermedia (Sch. Bip.) Coutinho **1033**
 orientalis (L.) Hochr. 1034
 picroides (L.) Roth **1032**
 tingitana (L.) Roth **1034**
  var. *orientalis* (L.) Aschers. et
   Schweinf. 1034
Reseda L. **176**
 alba L. **177**
 lutea L. **178**
  var. crispa West. **178**
 luteola L. **177**
  var. crispata (Link) Muell. Arg. **178**
 macrosperma Reichb. 179
  var. *orientalis* Muell. Arg. 179
 mediterranea auct. 179
 odorata L. **181**
 orientalis (Muell. Arg.) Boiss. ex
   Kotschy **179**
 phyteuma auct. 179
 truncata auct. 178
 undata auct. 177
Resedaceae 19, **176**
Rhagadiolus Juss. **995**
 edulis Gaertn. **995**
  koelpinia (L.f.) Willd. 994
 stellatus (L.) Gaertn. **996**
  ssp. *edulis* (Gaertn.) DC. 995
  var. *edulis* (Gaertn.) Holmboe 995
  var. *leiocarpus* DC. 996
  ssp. *stellatus* 996
  var. *stellatus* 996
Rhamnaceae 19, **352**
Rhamnus L. **353**
 sect. Alaternus DC. **353**
 sect. Cervispina (Moench) DC. **354**
 alaternus L. **353**
 graecus Boiss. et Reut. 355
 heldreichii auct. 355
 lotus L. 359
 oleoides L. **354**
  ssp. graecus (Boiss. et Reut.) Holmboe
   355
  ssp. microphyllus (Hal.) P. H. Davis
   355
  var. microphyllus Hal. 355
  var. *parvifolia* Kotschy 355
 paliurus L. 356
 punctatus auct. 355
 spina-Christi L. 358
 zizyphus L. 358
Rhinanthus minor (L.) L. 1232
 trixago (L.) L. 1228
Rhodalsine geniculata (Poir.) F. N.
   Williams 266
Rhus L. **371**
 coriaria L. **371**

Rhus L. (*cont.*)
 var. *humilior* Pojero 371
 cotinus L. 372
Ricinus L. **1452**
 communis L. **1452**
  var. microcarpus Muell. Arg. 1453
  f. viridis (Willd.) Muell. Arg. 1453
Ridolfia Moris **739**
 segetum (Guss.) Moris **739**
Robinia hispida L. 490
 pseudoacacia L. 490
 spinosa Hume 490
Rodigia commutata Spreng. 1009
Roemeria Medik. **82**
 hybrida (L.) DC. **82**
  var. eriocarpa DC. **82**
  var. hybrida **82**
Romulea Maratti **1566**
 columnae Seb. et Mauri **1568**
  ssp. columnae **1568**
 ramiflora Ten. **1567**
  ssp. ramiflora **1568**
 tempskyana Freyn **1567**
Rosa L. **624**
 canina L. **626**
  var. canina **626**
  var. *corymbifera* (Borkh.) Rouy 628
  ssp. *dumalis* Holmboe 628
  var. dumetorum (Thuill.) Desv. **628**
  var. *glabra* auct. 628
  var. *vulgaris* Koch 626
 chionistrae Lindberg f. **625**
 corymbifera Borkh. 628
 damascena Mill. 625
 dumalis Chapman 628
 dumetorum Thuill. 628
  var. *thuillieri* Christ 628
  f. *trichoneura* Christ 628
 gallica L. 625
 noisettiana Thory 625
 sempervirens L. 628
Rosaceae 19, **606**
 subfam. Neuradoideae **628**
 subfam. Pomoideae **629**
 subfam. Prunoideae **607**
 subfam. Rosoideae **610**
 tribe Potentilleae Spreng. **612**
 tribe Poterieae (Reichb.) Hook. f. **618**
 tribe Roseae **624**
 tribe Rubeae Dum. **610**
Rosmarinus L. **1299**
 officinalis L. 8, **1299**
Rosularia (DC.) Stapf **644**
 cypria (Holmboe) Meikle **644**, 807
 pallidiflora (Holmboe) Meikle **645**, 807
Rottboellia cylindrica Willd. 1753
 incurvata L.f. 1752
 loliacea Bory et Chaub. 1738
Rubia L. **776**
 brachypoda Boiss. 777
 doniettii Griseb. 777
 laurae (Holmboe) Airy Shaw **779**
 lucida auct. 777
 olivieri A. Rich. 777
  ssp. *brachypoda* (Boiss.) Holmboe 777
  ssp. *doniettii* (Griseb.) Holmboe 777
  var. *elliptica* Boiss. 777
  var. *stenophylla* Boiss. 777

Rubia L. (*cont.*)
  tenuifolia Urv. **777**
  tinctorum L. **776**
Rubiaceae 19, **772**
  tribe *Theligoneae* Wunderlich 798
Rubus L. **611**
  anatolicus (Focke) Hausskn. 611
  candicans auct. 611
  discolor auct. 611
  sanctus Schreb. **611**
  ulmifolius Schott 611
    ssp. *anatolicus* Focke 611
    var. *anatolicus* Focke ex Heldr. 611
Rumex L. **1404**
  subgenus Acetosa (Mill.) Rechinger f. **1405**
  subgenus Platypodium (Willk.) Rechinger f. **1412**
  subgenus Rumex **1406**
  aquaticus L. 1406
  bucephalophorus L. **1412**
    ssp. aegaeus Rechinger f. **1413**
    ssp. *bucephalophorus* auct. 1412
    ssp. *gallicus* auct. 1412
    ssp. graecus (Steinh.) Rechinger f. **1412**
    var. *graecus* Steinh. 1412
  cassius Boiss. 1409
  conglomeratus Murr. **1409**
  cristatus DC. **1408**
  cyprius Murb. **1405**
    ssp. *disciformis* Samuelsson 1405
    ssp. *eucyprius* Samuelsson 1405
  dentatus L. **1411**
    ssp. mesopotamicus Rechinger f. **1412**
  divaricatus L. 1410
  graecus Boiss. et Heldr. 1408
  hymenocephalus auct. 1413
  hymenosepalus Torrey 1413
  orientalis auct. 1408
  orientalis Bernh. 1408
    var. *graecus* (Boiss. et Heldr.) Boiss. 1408
  patientia auct. 1408
  patientia L. 1408
    ssp. *graecus* (Boiss. et Heldr.) Holmboe 1408
  pulcher L. **1409**
    ssp. *anodontus* auct. 1410
    ssp. cassius (Boiss.) Rechinger f. **1409**
    ssp. *divaricatus* auct. 1411
    ssp. divaricatus (L.) Arcang. **1410**
    ssp. pulcher 1410
  roseus auct. 1405
  spinosus L. 1414
  tingitanus auct. 1405
  tuberosus L. 1406
  vesicarius L. 1405
    ssp. *cyprius* (Murb.) Holmboe 1405
    ssp. *roseus* auct. 1405
Ruppia L. **1677**
  cirrhosa (Petagna) Grande **1677**
  maritima L. **1677**
    ssp. *rostellata* (Koch) P. Graebner 1677
    var. *rostrata* Agardh 1677
    ssp. *spiralis* (L. ex Dumort.) P. Graebner 1677
    var. *spiralis* (L. ex Dumort.) Moris 1677

Ruppia L. (*cont.*)
  rostellata Koch 1677
  spiralis L. ex Dumort. 1677
Ruppiaceae **1676**
Ruscus L. **1586**
  aculeatus L. **1586**
    var. aculeatus 1587
    var. angustifolius Boiss. **1587**
    var. *pumilus* Druce 1587
Ruta L. **348**
  angustifolia auct. 349
  bracteosa DC. 348
  buxbaumii Poir. 349
  chalepensis L. **348**
    ssp. *bracteosa* (DC.) Batt. 348
    var. *bracteosa* (DC.) Boiss. 348
    var. *latifolia* (Salisb.) Fiori 348
  graveolens auct. 348
  linifolia auct. 350
  spathulata Sibth. et Sm. 350
Rutaceae 19, **348**
  subfam. Aurantioideae 350
  tribe Citreae 350
  subtribe Citrinae 350
  tribe Ruteae 350

Saccharum L. **1867**
  cylindricum (L.) Lam. 1866
  officinarum L. 1868
  ravennae (L.) Murr. **1867**
Sagina L. **273**
  apetala Ard. **273**
  maritima G. Don **273**
  melitensis Gulia ex Duthie 273
  procumbens auct. 273, 274
Salicaceae **1486**
Salicornia L. **1386**
  amplexicaulis Vahl 1384
  europaea L. **1387**
    var. *fruticosa* L. 1386
  fruticosa (L.) L. **1386**
  glauca Del. 1385
  herbacea L. 1387
  macrostachya Moric. 1385
  strobilacea Pall. 1384
Salix L. **1486**
  alba L. **1487**
    × babylonica L. 1488
  australior auct. 1487
  babylonica L. 1488
    × fragilis L. 1488
  excelsa S. G. Gmelin 1488
  fragilis auct. 1487
  micans Anderss. 1488
    f. kassanogluensis auct. 1488
    × pendulina Wend. 1488
    × sepulcralis Simonk. 1488
Salsola L. **1390**
  echinus Labill. 1393
  fruticosa (L.) L. 1388
  hirsuta auct. 1392
  inermis Forssk. **1391**
  kali L. **1390**
  laniflora auct. 1392
  mucronata Forssk. 1393
  soda L. **1391**
Salvia L. **1287**

Salvia L. (*cont.*)
  aethiopis L. **1293**
  argentea Sibth. 1295
  candidissima auct. 1294
  candidissima Vahl 1295
  cassia Rechinger f. 1295
  cerignensis Sibth. 1298
  cilicica Boiss. et Kotschy 1295
  clandestina auct. 1297
  clandestina L. 1296
    var. *multifida* (Sm.) DC. 1296
  controversa auct. 1297
  controversa Ten. 1296
  crassifolia Sm. 2, 1294
  cyanescens Boiss. et Bal. 1295
  cypria Kotschy 1289
  dominica L. **1293**
  fruticosa Mill. **1289**
  grandiflora auct. 1288
  grandiflora Etl. 1289
    f. *albiflora* Lindberg f. 1288
    var. *brachyodonta* Boiss. 1288
    ssp. *willeana* Holmboe 1288
  graveolens Vahl 1294
  hierosolymitana Boiss. **1295**
  horminum L. 1291
  lanigera Poir. **1297**
  libanotica Boiss. et Gaill. 1289
  multicaulis Vahl 1299
  multifida Sm. 1296
  pinnata L. **1291**
  sibthorpii Sm. 1298
  tomentosa Mill. 1289
  triloba L.f. 1289
    ssp. *cypria* (Kotschy) Holmboe 1289
    var. *cypria* (Kotschy) Lindberg f. 1289
    ssp. *libanotica* (Boiss. et Gaill.)
      Holmboe 1289
  veneris Hedge **1294**
  verbenaca L. **1296**
    ssp. *clandestina* (L.) Briq. 1296
    var. *clandestina* (L.) Hal. 1296
    var. *serotina* Boiss. 1296
    var. *vernalis* Boiss. 1296
  virgata Jacq. 1298
  viridis L. **1291**
    ssp. *horminum* (L.) Holmboe 1291
  willeana (Holmboe) Hedge **1288**
Sambucaceae 19, **771**
Sambucus L. **771**
  ebulus L. **771**
  nigra L. **772**
Samolus L. **1085**
  valerandi L. **1086**
Sanguisorba minor Scop. 620
  ssp. *magnolii* (Spach) Briq. 622
  ssp. *verrucosa* (Link ex G. Don)
    Holmboe 622
  spinosa (L.) Bertol. 623
Santalaceae **1429**
Santalum album L. 1429
Santolina rigida Sm. 914
Sapindaceae 19, **362**
Sapindus mukorossi Gaertn. 362
Sapium *thomsonii* God.-Leb. ex Jum. 1453
  verum Hemsley 1453
Saponaria L. **222**, 223
  cretica L. 220

Saponaria L. (*cont.*)
  cypria Boiss. **222**
  cyprica Post et Autran 222
  depressa Biv. 222
    ssp. *cypria* (Boiss.) Holmboe 222
  orientalis L. **223**
  segetalis Neck. 224
  vaccaria L. 224
Sarcopoterium Spach **623**
  spinosum (L.) Spach 4, **623**
Satureja L. **1272**
  capitata L. 1271
  crassinervis Lindberg f. 1281
  exigua (Sm.) Holmboe 1281
    var. *integrifolia* Holmboe 1281
  graeca L. 1276
    ssp. *cypria* (Kotschy) Holmboe 1276
  incana (L.) Briq. 1279
  nepeta (L.) Scheele 1279
  nervosa Desf. 1274
  spinosa auct. 1271
  thymbra auct. 1271
  thymbra L. **1273**
  troodi (Post) Holmboe 1282
  vulgaris (L.) Fritsch 1283
Satyrium maculatum Desf. 1557
Saxifraga L. **639**
  hederacea L. **639**
  tridactylites L. **639**
Saxifragaceae 19, **638**
Scabiosa L. **850**
  albocincta Greuter 856
  argentea L. **851**
  atropurpurea L. 850
    var. atropurpurea 850
    ssp. *maritima* (L.) Arcang. 850
    var. setifera (Lam.) DC. 850
  brachiata Sm. **856**
  cerignensis Sibth. 856
  coronopifolia Sm. 851
  crenata Cyr. 850
  cretica L. **856**
  cyprica Post **854**
  divaricata Jacq. 852
  involucrata Sm. 859
  kurdica Post 854
  maritima L. 850
  minoana Greuter 856
  papposa L. 858
  paucidentata Hub.-Mor. 854
  plumosa (L.) Sm. 858
  prolifera L. **853**
  sicula L. **852**
  syriaca L. 848
  ucranica L. 851
    var. *sicula* (L.) Coult. 852
Scaligeria DC. **719**
  cretica (Mill.) Boiss. **719**
  napiformis (Willd. ex Spreng.) Grande
    720
  tournefortii Boiss. 721
Scandix L. **696**
  anthriscus L. 695
  australis L. **697**
    ssp. *grandiflora* (L.) Thell. 698
    var. lasiactina Boiss. 698
  grandiflora L. **698**
  latifolia Sibth. et Sm. 726

Scandix L. (*cont.*)
　macrorrhyncha auct. 697
　pecten-veneris L. **697**
　　ssp. *macrorrhyncha* auct. 697
　　ssp. *macrorrhyncha* Rouy et Camus
　　　697
　pinnatifida Vent. 696
　stellata Banks et Sol. **696**
Scariola F. W. Schmidt **1029**
　tetrantha (B. L. Burtt et P. Davis) Soják
　　1030
　viminea (L.) F. W. Schmidt **1029**
Schanginia baccata (Forssk.) Moq. 1389
Schinus L. **370**
　molle L. **370**
　terebinthifolius Raddi 371
Schismus P. Beauv. **1836**
　arabicus Nees **1836**
　barbatus (L.) Thell. 1837
Schoenoplectus (Reichb.) Palla **1694**
　lacustris (L.) Palla 1695
　　ssp. tabernaemontani (C. Gmel.) A. &
　　　D. Löve **1695**
　litoralis (Schrader) Palla **1694**
Schoenus L. **1698**
　aculeatus 1848
　ferrugineus auct. 1698
　mariscus L. 1699
　mucronatus L. 1688
　nigricans L. **1698**
Scilla L. **1635**
　amoena auct. 1637
　autumnalis L. **1636**
　bifolia L. 1639
　cernua auct. 1637
　cilicica Siehe 6, **1637**
　hyacinthoides L. 7, 1636
　lochiae (Meikle) Speta 1640
　maritima L. 1634
　morrisii Meikle **1637**
　siberica auct. 1637
　veneris Speta 1637
Scirpoides Ség. **1697**
　holoschoenus (L.) Soják **1697**
Scirpus L.
　α *Bolboschoenus* Aschers. 1696
　　sect. *Bolboschoenus* (Aschers.) Beetle
　　　1696
　　sect. *Schoenoplectus* Reichb. 1694
　australis Murr. 1697
　cernuus Vahl 1693
　ferrugineus L. 1691
　glaucus Sm. 1695
　holoschoenus L. 1697
　　ssp. *australis* (Murr.) Soják 1698
　　var. *australis* (Murr.) Koch 1697
　　var. *holoschoenus* 1698
　　var. *romanus* (L.) Koch 1697
　lacustris L. 1695
　　var. *digynus* Godr. 1695
　　ssp. *glaucus* (Sm.) Hartm. 1695
　　ssp. *tabernaemontani* (C. Gmel.) Syme
　　　1695
　litoralis Schrader 1694
　maritimus L. 1696
　　f. *macrostachys* (Willd.) Junge 1697
　minaae Tod. 1694
　palustris L. 1692

Scirpus L. (*cont.*)
　romanus L. 1697
　savii Seb. et Maur. 1693
　　ssp. *minaae* (Tod.) Arcang. 1694
　setaceus L. 1693
　　var. *pseudoclathratus* Schramm 1693
　tabernaemontani C. Gmel. 1695
Scleranthus L. **285**
　annuus L. **285**
　　ssp. delortii (Gren.) Meikle **286**, 806
　　ssp. *polycarpos* Thell. 286
　　ssp. *ruscinonensis* (Gillot et Coste)
　　　P. D. Sell 286
　　ssp. *verticillatus* (Tausch) Arcang. 286
　collinus Hornung et Opiz 286
　delortii Gren. 286
　polycarpos Sibth. et Sm. 286
　ruscinonensis Gillot et Coste 286
　verticillatus Tausch 286
　　ssp. *delortii* (Gren.) Nyman 286
Sclerochloa P. Beauv. **1739**
　dura (L.) P. Beauv. **1739**
Scleropoa Griseb. 1720
　dichotoma (Forssk.) Boiss. 1724
　loliacea (Huds.) Gren. et Godr. 1721
　maritima (L.) Parl. 1724
　memphitica (Spreng.) Parl. 1724
　　var. *dichotoma* (Forssk.) Bonn. et
　　　Barr. 1724
　rigida (L.) Griseb. 1721
Scolymus L. **984**
　hispanicus L. **985**
　maculatus L. **984**
Scorpiurus L. **506**
　muricatus L. **506**
　　var. subvillosus (L.) Lam. **508**
　　var. *sulcatus* (L.) Lam. 508
　subvillosus L. 508
　　var. *breviaculeatus* Batt. et Trab. 508
　sulcatus L. 508
Scorzonera L. 1043
　araneosa Sm. 1047
　buphthalmoides DC. 1041
　cana Hoffm. 1045
　cypria Kotschy 1047
　jacquiniana (Koch) Čelak. **1044**
　　var. subintegra Boiss. **1045**
　laciniata L. **1044**
　mollis auct. 1046
　orientalis L. 1034
　papposa auct. 1046
　parviflora Jacq. 1047
　picroides L. 1032
　resedifolia L. 1039
　subintegra (Boiss.) Thiébaut 1045
　tingitana L. 1034
　troodea Boiss. **1046**
Scrophularia L. **1214**
　canina auct. 1215
　hypericifolia Wydl. 1215
　peregrina L. **1214**
　peyronii Post **1215**
　sphaerocarpa auct. 1215
　xanthoglossa auct. 1215
Scrophulariaceae **1196**
　　tribe Antirrhineae Chav. **1200**
　　tribe Digitaleae Benth. **1217**
　　tribe Euphrasieae Benth. **1226**

Scrophulariaceae (cont.)
  tribe Gratioleae Benth. **1216**
  tribe Scrophularieae **1213**
  tribe Verbasceae Benth. **1197**
Scutellaria L. **1329**
  albida auct. 1331
  columnae All. 1329
    var. sibthorpii auct. 1331
    var. sibthorpii Benth. 1329
  columnae auct. 1330
  cypria Rechinger f. **1330**
    var. cypria **1331**
    var. elatior Meikle **1331**
  fl. rubro nana Kotschy 1330
  hirta auct. 1330
  peregrina auct. 1330
    ssp. sibthorpii auct. 1331
    ssp. sibthorpii (Benth.) Holmboe 1329
    var. sibthorpii (Benth.) Boiss. et Reut. 1329
  rubicunda Hornem. 1329
    ssp. sibthorpii (Benth.) Osorio-Tafall et Seraphim 1329
  sibthorpii (Benth.) Hal. **1329**
  utriculata auct. 1330
Secale L. **1826**
  cereale L. **1826**
Securigera coronilla DC. 514
  securidaca (L.) Deg. et Doerfl. 514
Sedum L. **647**
  sect. Cepaea Koch **648**
  sect. Epeteium Boiss. **650**
  sect. Sedum **647**
  sect. Telmissa (Fenzl) Schönland **653**
  altissimum Poir. 648
  caespitosum (Cav.) DC. **652**
  cyprium A. K. Jackson et Turrill **649**
  lampusae (Kotschy) Boiss. **648**
    var. microstachyum (Kotschy) Fröderström 649
  litoreum Guss. **652**
  microcarpum (Sm.) Schönland 653
  microstachyum (Kotschy) Boiss. **649**
  palaestinum auct. 652
  pallidum auct. 652
  porphyreum Kotschy **650**
    var. parviglandulosum Lindberg f. 650
  rubens L. **651**
  rubrum (L.) Thell. 652
  sediforme (Jacq.) Pau **647**
Selaginella P. Beauv. **1873**
  denticulata (L.) Link **1873**
Selaginellaceae **1872**
Selaginellales **1872**
Sempervivum arboreum L. 654
  globiferum auct. 648
  sediforme Jacq. 648
Senebiera coronopus (L.) Poir. 121
Senecio L. **929**
  aegyptius L. **931**
    var. arabicus (L.) Holmboe 931
    var. discoideus Boiss. **931**
  arabicus L. 931
  bicolor (Willd.) Tod. ssp. cineraria (DC.) Chater 934
  cineraria DC. 934
  crassifolius auct. 932
  glaucus L. **934**

Senecio L. (cont.)
  ssp. coronopifolius (Maire) J. Alex. 934
  ssp. cyprius Meikle **934**
  ssp. joppensis (Dinsmore) Feinbrun 934
  joppensis Dinsmore 934
  leucanthemifolius Poir. **931**
    var. vernalis (Waldst. et Kit.) J. Alex. **932**
  squalidus L. 935
  vernalis Waldst. et Kit. 932
  vulgaris L. **930**
Serapias L. **1505**
  columnae (Aschers. et Graebn.) Fleischm. 1508
  cordigera L. 1506
    ssp. laxiflora (Soó) Sundermann 1508
    ssp. orientalis (W. Greuter) Sundermann 1507
    ssp. vomeracea (Burm. f.) Sundermann 1506
  damasonium Mill. 1495
  ensifolia Murray 1496
  helleborine L. 1503
    var. longifolia L. 1496
  latifolia Mill. 1495
  laxiflora Chaub. 1508
    var. columnae Reichb. f. 1508
  longifolia Huds. 1496
  longipetala (Ten.) Poll. 1505
  microphylla Ehrh. 1501
  neglecta De Not. 1508
  pallens Jundz. 1495
  parviflora Parl. 1508, 1509
    var. columnae Aschers. et Graebn. 1508
    ssp. laxiflora Soó 1508
  pseudo-cordigera (Sebast.) Moric. 1505
  rubra L. 1496
  vomeracea (Burm. f.) Briq. **1505**
    l. heldreichii Soó 1506
    ssp. laxiflora (Soó) Gölz et Reinhard **1508**
    rsskr. orientalis E. Nelson 1507
    ssp. orientalis W. Greuter **1507**
    ssp. vomeracea **1506**
Serrafalcus arvensis (L.) Godr. 1799
    var. pubescens Caldesi 1799
  lloydianus Godr. ex Gren. et Godr. 1799
Serratula L. **964**
  cerinthifolia (Sm.) Boiss. **965**
  chamaepeuce L. 956
  cordata Cass. 965
Sesamum indicum L. 1246
Sesbania grandiflora (L.) Pers. 490
  sesban (L.) Merrill 490
Seseli annuum L. 749
  coloratum Ehrh. 749
Sesleria dura (L.) Schrank 1739
Setaria P. Beauv. **1859**
  glauca auct. 1860
  italica (L.) P. Beauv. **1860**
  pumila (Poir.) Roem. et Schult. **1859**
  verticillata (L.) P. Beauv. **1861**
  viridis (L.) P. Beauv. **1861**
Sherardia L. **797**
  arvensis L. **797**

Sherardia L. (*cont.*)
  muralis L. 789
Sideritis L. **1303**
  cilicica Boiss. 1304
    var. *cypria* (Post) Lindberg f. 1304
  curvidens Stapf **1303**
  cypria Post **1304**
  incana L. 1307
  perfoliata L. **1306**
  pullulans auct. 1304
  romana auct. 1303
  romana L. 1303
    ssp. *curvidens* (Stapf) Holmboe 1303
Silene L. **224**
  sect. Atocion Otth **234**
  sect. Behenantha Otth **238**
  sect. Conoïmorphae Otth **250**
  sect. Dichotomae (Rohrb.) Chowdhuri **242**
  sect. Dipterospermae (Rohrb.) Chowdhuri **247**
  sect. Erecto-refractae Chowdhuri **242**
  sect. Inflatae Boiss. **230**
  sect. Lasiocalycinae Boiss. **241**
  sect. Lasiostemones Boiss. **229**
  sect. Paniculatae Boiss. **226**
  sect. Rigidulae Boiss. **233**
  sect. Silene **245**
  aegyptiaca (L.) L.f. **235**
  apetala Willd. **249**
    var. alexandrina Aschers. **250**
    var. apetala **249**
    var. grandiflora Boiss. **249**
  atocion auct. 236
  atocion Juss. ex Jacq. 235, 236
  behen L. **239**
  bipartita Desf. 247
    var. *eriocaulon* Boiss. 248
    var. *stenophylla* Boiss. 248
  brachypetala Rob. et Cast. 245
  cerastoides Lindberg f. 246
  colorata Poir. **247**
    var. colorata **248**
    ssp. *decumbens* (Biv.) Holmboe 248
    var. decumbens (Biv.) Rohrb. **248**
    var. distachya (Brot.) Rohrb. 248
    ssp. *oliveriana* (Otth) Rohrb. 248
    var. oliveriana (Otth) Muschler **248**
    var. *stenophylla* (Boiss.) Dinsm. 248
  commutata Guss. 232
  conica Sibth. et Sm. 251
  conoidea L. **250**
  cretica L. **237**
  cucubalus Wibel 231
  damascena auct. 247
  damascena Boiss. et Gaill. 235, 247
  decumbens Biv. 248
  dichotoma Ehrh. **242**
    ssp. racemosa (Otth) Hayek **243**
    var. racemosa **243**
  discolor Sibth. et Sm. **243**
  fraudatrix Meikle **236**
  fruticosa L. **226**
  fuscata Link **234**
  galataea Boiss. **229**
  gallica L. **245**
    var. quinquevulnera (L.) Mert. et Koch **246**

Silene L. (*cont.*)
  gemmata Meikle **238**
  gigantea L. **227**
  inflata Sm. 231
    var. *colorata* Hampe 232
    var. *major* Kotschy 232
    var. *rubriflora* Boiss. 232
  italica (L.) Pers. **228**
  kotschyi Boiss. **233**
    var. maritima Boiss. **233**
    var. *stenocalyx* (Lindberg f.) Chowdhuri 233
  laevigata Sibth. et Sm. **240**
  leucophaea Sibth. et Sm. 247
  longipetala Vent. **229**
  macrodonta Boiss. **251**
  nicaeënsis auct. 243
  nicaeënsis All. 243
    var. *latifolia* Boiss. 243
  nocturna L. **244**
    var. brachypetala (Rob. et Cast.) Benth. **245**
    var. nocturna **245**
  oliveriana auct. 248
  oliveriana Otth 248
  orchidea L.f. 235
  otites (L.) Wibel 230
  palestina Boiss. 247
    var. *damascena* auct. 247
  papillosa Boiss. **241**
  paradoxa L. **228**
  pendula L. 242
  pontica Brandza 233
  porrigens Gouan ex L. 221
  pseudo-atocion auct. 229
  racemosa Otth 243
    var. *sibthorpiana* (Reichb.) Boiss. 243
  rigoi Sintenis 249
  rubella L. **234**
  sedoides Poir. **237**
  stenocalyx Lindberg f. 233
  thymifolia Sibth. et Sm. 232
  tridentata Desf. **246**
  venosa Aschers. 231
    var. *commutata* (Guss.) Holmboe 232
    var. *rubriflora* (Boiss.) Holmboe 232
  vespertina auct. 247
  vulgaris (Moench) Garcke **230**
    f. colorata (Hampe) Hayek **232**
    ssp. commutata (Guss.) Hayek **232**
    ssp. macrocarpa Turrill **232**
Siler cordifolium Boiss. 766
Silybum Adans. **962**
  marianum (L.) Gaertn. **963**
Simaroubaceae 19, **351**
Sinapis L. **100**
  alba L. **101**
  arvensis L. **100**
    var. arvensis **101**
    var. orientalis (L.) Koch et Ziz **101**
    var. schkuhriana (Reichb.) Hagenb. **101**
  hispanica L. 113
  incana L. 102
  nigra L. 99
  orientalis L. 101
  schkuhriana Reichb. 101
Sisymbrium L. **164**

Sisymbrium L. (*cont.*)
    sect. *Arabidopsis* DC. 169
    sect. *Torularia* Cosson 168
    alliaria (L.) Scop. 164
    columnae Jacq. 165
    irio L. **165**
    nanum DC. 159
    nasturtium-aquaticum L. 153
    officinale (L.) Scop. **167**
        var. leiocarpum DC. **167**
        var. officinale **167**
    orientale L. **165**
    polyceratium L. **166**
    thalianum (L.) J. Gay 169
    torulosum Desf. 168
    vimineum L. 103
Sium nodiflorum L. 738
Smilax L. **1582**
    aspera L. **1582**
        var. *altissima* Moris et De Not. 1582
        ssp. *mauritanica* (Desf.) Aschers. et Graebn. 1582
        var. *mauritanica* (Desf.) Gren. et Godr. 1582
    mauritanica Desf. 1582
Smyrnium L. **721**
    cognatum Kotschy 723
    connatum Boiss. et Kotschy **723**
    olusatrum L. **721**
    perfoliatum Sibth. et Sm. 722
    rotundifolium Mill. **722**
Solanaceae **1180**
Solanum L. **1181**
    alatum auct. 1183
    elaeagnifolium Cav. 1184
    glaucum Dunal 1185
    luteum Mill. 1183
    lycopersicum L. 1180
    melongena L. 1184
    miniatum Bernh. 1183
    nigrum L. **1181**
        ssp. *villosum* (L.) Ehrh. 1183
        var. *villosum* L. 1182
    rubrum Mill. 1183
    seaforthianum Andrews 1184
    tuberosum L. 1183
    villosum auct. 1181
    villosum (L.) Lam. 1183
    villosum Mill. **1182**
        ssp. alatum (Moench) Edmonds 1183
        ssp. villosum 1183
Solenanthus apenninus (L.) Fisch. et Mey. 1129
Solenopsis C. Presl **1056**
    bivonaeana C. Presl 1057
    minuta (L.) C. Presl **1056**
        ssp. nobilis (E. Wimmer) Meikle **1057**
Sonchus L. **1035**
    arvensis L. 1039
    asper (L.) Hill **1037**
        ssp. asper **1037**
        ssp. glaucescens (Jordan) Ball **1037**
        × S. oleraceus L. **1038**
    chondrilloides auct. 1032
    ciliatus Lam. 1036
    glaucescens Jordan 1037
    nymanii Tineo et Guss. 1037
    oleraceus L. **1035**

Sonchus L. (*cont.*)
    f. *albescens* Neumann 1036
    var. *albescens* (Neumann) Druce 1036
    var. *asper* L. 1037
    var. *ciliatus* (Lam.) Druce 1036
    var. *triangularis* Wallr. 1036
    picroides (L.) Lam. 1032
    tenerrimus L. **1038**
    tingitanus (L.) Lam. 1034
Sophora japonica L. 585
Sorbus L. **633**
    aria auct. 634
    aria (L.) Crantz **634**
        ssp. cretica (Lindl.) Holmboe **634**
        var. *graeca* Boiss. 634
    cretica (Lindl.) Fritsch et Rechinger 634
    domestica L. **633**
    graeca (Spach) Kotschy 634
    umbellata auct. 635
    umbellata (Desf.) Fritsch 634
        var. *cretica* (Lindl.) Schneid. 634
Sorghum Moench **1868**
    bicolor (L.) Moench 1869
        × *drummondii* (Nees ex Steud.) Millspaugh et Chase 1870
    halepense (L.) Pers. **1869**
    sudanense (Piper) Stapf 1870
Sparganiaceae **1663**
Sparganium L. **1663**
    erectum L. **1663**
        ssp. neglectum (Beeby) Schinz et Keller **1663**
    neglectum Beeby 1663
    ramosum Huds. 1663
        ssp. *neglectum* (Beeby) Aschers. et Graebn. 1663
Sparmannia africana L.f. 316
Spartina Schreb. **1844**
    alterniflora Lois. 1844
    anglica C. E. Hubb. **1844**
    maritima auct. 1844
    maritima (Curt.) Fernald 1844
    stricta auct. 1844
Spartium villosum Poir. 382
Specularia falcata (Ten.) A. DC. 1054
        var. *scabra* (Lowe) A. DC. 1055
    hybrida (L.) A. DC. 1054
    speculum (L.) A. DC. 1053
Spergularia (Pers.) J. & C. Presl **274**
    atheniensis Aschers. 275
    bocconii (Scheele) Aschers. et Graebn. **275**
    campestris auct. 275
    diandra auct. 276
    diandra (Guss.) Heldr. et Sart. **275**
    marina (L.) Griseb. **276**
    media auct. 276
    rubra auct. 275
    rubra (L.) J. & C. Presl 275
        var. *atheniensis* Heldr. et Sart. 275
    salina J. & C. Presl 276
        f. *heterosperma* (Fenzl) Holmboe 277
        f. *leiosperma* (Kindb.) Holmboe 277
Sphenopsida **1873**
Sphenopus Trin. **1740**
    divaricatus (Gouan) Reichb. **1740**
Spinacia oleracea L. 1383
Spiranthes L. C. M. Rich. **1504**

Spiranthes L. C. M. Rich. (*cont.*)
  autumnalis L. C. M. Rich. 1504
  spiralis (L.) Chevall. **1504**
Sporobolus R. Br. **1849**
  arenarius (Gouan) Duval-Jouve 1850
  pungens (Schreb.) Kunth 1850
  virginicus (L.) Kunth **1850**
Stachys L. **1309**
  cretica L. **1309**
  italica Mill. 1309
Staehelina L. **964**
  apiculata Labill. 964
  chamaepeuce (L.) L. 956
  lobelii DC. **964**
Statice L. 1066
  albida Guss. 1069
  aristata Sm. 1072
  aucheri Girard 1070
  bellidifolia Sm. 1070
  delicatula Girard 1068
  echioides auct. 1071
  echioides L. 1072
    ssp. *exaristata* Murb. 1073
    var. *exaristata* (Murb.) Lindberg f. 1073
  globulariifolia Desf. 1068
    var. *glauca* Boiss. 1068
  gmelini auct. 1068
  gmelini Willd. 1068
    var. *laxiflora* Boiss. 1068
  graeca auct. 1071
    var. *microphylla* auct. 1071
  limonium auct. 1068
  limonium L. 1068
    var. *macroclada* auct. 1068
  meyeri Boiss. 1068
  mucronulata Lindberg f. 1069
  ocymifolia Poir. 1070
    ssp. *bellidifolia* (Sm.) Nyman 1070
    var. *bellidifolia* (Sm.) Boiss. 1070
  oleifolia auct. 1071
  psiloclada Boiss. 1069
    var. *albida* (Guss.) Boiss. 1069
  raddiana Boiss. 1068
  rorida Sm. 1071
  sinuata L. 1067
  spicata Willd. 1073
  tatarica L. 1073
  virgata Willd. 1071
Stellaria L. **258**
  apetala Murb. 260
  apetala Ucria 260
  cilicica Boiss. et Bal. **261**
  dubia Bast. 253
  glabella Jord. et Fourr. 260
  media (L.) Vill. **259**
    ssp. apetala Celak. **260**
    ssp. cupaniana (Jord. et Fourr.) Nyman **259**
    var. *glabella* (Jord. et Fourr.) Briq. 260
    var. *major* Sintenis 259
    ssp. media **259**
    ssp. *neglecta* auct. 260
    ssp. *postii* Holmboe 259
    var. *pubescens* Post 259
  neglecta Weihe 259
    ssp. *postii* (Holmboe) Rech. f. 259
  stellera passerina L. 1426

Stellera L. (*cont.*)
  pubescens Guss. 1427
Stenophragma thalianum (L.) Celak. 169
Steptorhamphus Bunge **1025**
  tuberosus (Jacq.) Grossh. **1026**
Sterculiaceae 19, **314**
Stipa L. **1790**
  aristella L. 1791
  barbata Desf. **1790**
  bromoides (L.) Doerfl. **1791**
  capensis Thunb. **1791**
  fontanesii Parl. 1793
  holosericea Trin. ex Rupr. 1793
  lagascae Roem. et Schult. **1792**
  membranacea L. 1731
  paleacea auct. 1792
  pennata L. **1793**
  retorta Cav. 1791
  szowitsiana Trin. ex Griseb. 1790
  tenacissima L. 1793
  tortilis Desf. 1792
Styracaceae **1087**
Styrax L. **1087**
  benzoin Dryand. 1089
  officinalis L. 4, **1088**
    var. californicus (Torr.) Rehd. 1088
  officinarum Sintenis 1088
Suaeda Forssk. ex Scop. **1387**
  aegyptiaca (Hasselq.) Zohary 1389
  baccata Forssk. 1389
  fruticosa Forssk. 1388
  fruticosa (L.) Dumort. 1388
  maritima (L.) Dumort. 1389
  vera Forssk. **1388**
Sycomorus antiquorum Gasparrini 1468
Synelcosciadium Boiss. **759**
  carmeli (Labill.) Boiss. **759**

Taeniatherum Nevski **1828**
  asperum (Simonkai) Nevski **1830**
  crinitum (Schreb.) Nevski **1828**
Tagetes L. **902**
  erecta L. 902
  minuta L. **902**
  patula L. 902
Tamaricaceae 19, **288**
Tamarix L. **288**
  aphylla (L.) Karst. 291
  gallica L. 291
    var. mannifera Ehrenb. 291
  hampeana Boiss. et Heldr. **290**
  mannifera auct. 289
  meyeri Boiss. 289
  pallasii auct. 291
  parviflora auct. 290
  parviflora DC. 290
    var. *cypria* Kotschy 290
  smyrnensis Bunge **291**
  tetragyna Ehrenb. **289**
    ssp. *meyeri* (Boiss.) Holmboe 289
    var. *meyeri* (Boiss.) Boiss. 289
  tetrandra Pall. ex M. Bieb. **289**
Tamus L. **1579**
  communis L. **1580**
    ssp. *cretica* (L.) Nyman 1580
    var. *cretica* (L.) Parl. 1580
    var. *subtriloba* Guss. 1580
  cretica L. 1580

Tanacetum L. **922**
  balsamita L. **922**
  parthenium (L.) Sch. Bip. **923**
  uliginosum Sm. 926
Taraxacum Weber **1018**
  sect. Erythrosperma Dahlst. 1023
  sect. Scariosae Hand.-Mazz. 1021
  aleppicum auct. 1019
  aphrogenes Meikle **1021**
  bithynicum DC. 1021
  calocephalum auct. 1019
  cyprium Lindberg f. **1019**
  gracilens Dahlst. 1023
  gymnanthum auct. 1019
  hellenicum Dahlst. **1020**
  holmboei Lindberg f. **1021**
  laevigatum auct. 1020
  megalorrhizon auct. 1019, 1020
  minimum auct. 1019
  officinale auct. 1019, 1020
    var. *genuinum* auct. 1019
  pseudonigricans auct. 1020
Tecoma stans (L.) Kunth 1246
Tecomaria capensis (Thunb.) Spach 1246
Teesdalia R. Br. **119**
  coronopifolia (Bergeret) Thell. **119**
  lepidium DC. 119
  regularis Sm. 119
Telephium L. **685**
  imperati L. **685**
    ssp. orientale (Boiss.) Nym. **686**
    var. *orientale* (Boiss.) Boiss. 686
  orientale Boiss. 686
Telmissa Fenzl **653**
  microcarpa (Sm.) Boiss. **653**
  sedoides Fenzl 653
Tetragonolobus Scop. **487**
  palaestinus (? Boiss. et Bl.) H. Stuart
    Thompson 487, 488
  pseudopurpureus Uechtr. 488
  purpureus Moench **487**
Teucrium L. **1332**
  sect. Chamaedrys Schreber **1336**
  sect. Polium Schreber **1338**
  sect. Scordium Reichb. **1335**
  sect. Scorodonia Schreber **1334**
  sect. Teucrium **1333**
  brevifolium Schreber 1342
  chamaepitys L. 1345
  creticum L. **1333**
  cypricum Post 1338
  cyprium Boiss. **1338**
    ssp. cyprium **1339**
    ssp. kyreniae P. H. Davis **1339**
  divaricatum Kotschy **1336**
    ssp. canescens (Čelak.) Holmboe **1337**
    var. *canescens* Čelak. 1337
    f. *kotschyanum* (Bornm.) Rechinger f.
      1337
    var. *kotschyanum* Bornm. 1337
    ssp. *sieberi* (Čelak.) Holmboe 1337
  flavum L. 1336
    f. *divaricatum* Kotschy 1336
    var. *purpureum* Benth. 1336
  iva L. 1344
  kotschyanum Poech **1334**
  lucidum auct. 1337
  micropodioides Rouy **1341**

Teucrium L. (*cont.*)
  polium auct. 1341
  polium L. 1342
    ssp. *micropodioides* (Rouy) Holmboe
      1341
    var. *purpurascens* auct. 1341
    var. *roseum* auct. 1341
  pseudo-chamaedrys Sibth. 1337
  pseudo-polium Sibth. 1341
  rosmarinifolium Lam. 1333
  scordioides Schreber 1336
  scordium L. **1335**
    ssp. scordioides (Schreber) Arcang.
      1336
    var. *scordioides* (Schreber) Sm. 1336
    ssp. scordium **1335**
  sieberi Čelak. 1337
    var. *canescens* Čelak. 1337
  smyrnaeum Boiss. 1334
Thapsia foetida auct. 755
Theligonaceae 19, **798**
Theligonum L. **799**
  cynocrambe L. **799**
    var. minor Ulbrich 800
Theobroma cacao L. 314
Theresia libanotica Boiss. 1600
Thesium L. **1430**
  bergeri Zucc. 1431
  divaricatum Jan ex Mert. 1432
  humile Vahl **1430**
Thlaspi L. **125**
  annuum auct. 128
  bursa-pastoris L. 129
  coronopifolium Bergeret 119
  cyprium Bornm. **125**
  densiflorum auct. 125
  microstylum Boiss. 126
  natolicum auct. 128
  perfoliatum auct. 128
  perfoliatum L. **126**
    ssp. *annuum* auct. 128
    var. perfoliatum **128**
    var. stylatum Post **128**
  violascens auct. 125
Thrincia tuberosa (L.) DC. 1017
    var. *olivieri* DC. 1017
Thymbra L. 1271, **1278**
  spicata auct. 1271
  spicata L. **1278**
Thymelaea Mill. **1424**
  sect. Chlamydanthus (C. A. Mey.)
    Endl. **1425**
  sect. Ligia (Fasano) Meissner **1426**
  sect. Piptochlamys (C. A. Mey.) Endl.
    1424
  argentea (Sm.) Endl. 1426
  arvensis Lam. ssp. *pubescens* (Guss.)
    Arcang. 1427
  gussonei Boreau 1427
  hirsuta (L.) Endl. **1424**
  mesopotamica (Jeffrey) Peterson 1427
  passerina (L.) Coss. et Germ. **1426**
    ssp. pubescens (Guss.) Meikle **1427**
  pubescens auct. 1427
  tartonraira (L.) All. **1425**
    ssp. argentea (Sm.) Holmboe **1426**
    var. linearifolia K. Tan **1426**
Thymelaeaceae **1423**

Thymus L. **1270**
billardieri Boiss. 1270
capitatus L. 4, **1271**
exiguus Sm. 1281
graveolens M. Bieb. 1281
incanus Sm. 1279
integer Griseb. **1270**
microphyllus Urv. 1276
tragoriganum auct. 1271
tragoriganum L. 1273
villosus auct. 1270
Tilia argentea Desf. 316
tomentosa Moench 316
Tiliaceae 19, **314**
Tillaea L. 641
alata Viv. 641
muscosa auct. 641
rubra L. 652
vaillantii Willd. 642
Tinaea intacta (Link) Boiss. 1557
Tipuana (Benth.) O. Ktze. 583
tipu (Benth.) O. Ktze. 583
Tolpis Adans. **992**
altissima (Balb.) Pers. 993
barbata (L.) Gaertn. **992**
quadriradiata Biv. 993
umbellata auct. 993
umbellata Bertol. 992
virgata (Desf.) Bertol. **993**
Tordylium L. **761**
aegyptiacum (L.) Poir. **763**
apulum L. **761**
cordatum Holmboe 765
cordatum (Jacq.) Poir. 765
ssp. *trachycarpum* (Boiss.) Holmboe 765
maximum L. **762**
nodosum L. 699
peregrinum L. 752
syriacum L. **763**
Torilis Adanson **699**
arvensis (Huds.) Link **700**
ssp. arvensis **700**
ssp. divaricata (Moench) Thell. **700**
ssp. *elongata* (Hoffmsgg. et Link) Cannon 701
var. *heterocarpa* (Batt.) Maire 701
ssp. neglecta (Spreng.) Thell. **701**
ssp. *purpurea* (Ten.) Hayek 701
erythrotricha Reichb. **703**
heterophylla Guss. **702**
homophylla Stapf et Wettst. 701
leptophylla (L.) Reichb. **702**
nodosa (L.) Gaertn. **699**
var. *bracteosa* Murb. 700
f. homoeocarpa Thellung **700**
f. nodosa **700**
purpurea (Ten.) Guss. **701**
tenella (Del.) Reichb. **703**
triradiata auct. 702
triradiata Post 701
Torularia (Cosson) O. E. Schulz **168**
torulosa (Desf.) O. E. Schulz 168
var. scorpiuroides (Boiss.) O. E. Schulz 168
var. torulosa **169**
Toxicodendron altissima Mill. 351
Trachomitum Woodson **1100**

Trachomitum Woodson (*cont.*)
sarmatiense Woodson 1100
venetum (L.) Woodson 1100
ssp. *sarmatiense* (Woodson) Avetisian 1100
ssp. scabrum (Russan.) Rechinger f. 1101
Trachynia Link **1814**
distachya (L.) Link **1815**
Tragopogon L. **1040**
australis Jord. 1040
buphthalmoides (DC.) Boiss. 1041
coelesyriacus Boiss. 1040
crocifolius auct. 1040
eriospermus auct. 1040
glaber (L.) Hoffm. 1042
hirsutum (L.) Kotschy 1042
hirsutus Gouan 1042
hybridum L. 1041
longirostris Bisch. ex Sch. Bip. 1040
picroides L. 1043
porrifolius auct. 1040
porrifolius L. ssp. *longirostris* (Bisch. ex Sch. Bip.) Holmboe 1040
sinuatus Avé-Lall. **1040**
Tremastelma Raf. 850
palaestinum (L.) Janchen 856
Tribulus L. **328**
terrestris L. **328**
Trichocrepis bifida Vis. 1010
Trichonema columnae (Seb. et Mauri) Reichb. 1568
Trifolium L. **437**
sect. Chronosemium Ser. **467**
sect. Fragifera Koch **455**
sect. Mistyllus (Presl) Godr. **462**
sect. Trichocephalum Koch **453**
sect. Trifoliastrum Ser. **464**
sect. Trifolium **439**
agrarium auct. 469
angustifolium auct. 445
angustifolium L. **444**
argutum Banks et Sol. 6, **463**
arvense L. **443**
var. *longisetum* (Boiss. et Bal.) Boiss. 443
bicorne Forssk. 460
blancheanum Boiss. 446
boissieri Guss. ex Soyer-Willemet et Godron **468**
campestre Schreb. **469**
ssp. campestre **469**
ssp. paphium Meikle **470**
cherleri L. **440**
clusii Gren. et Godr. 460
clypeatum L. **450**
ssp. *scutatum* (Boiss.) Holmboe 451
dasyurum C. Presl **450**
desvauxii auct. 445
desvauxii Boiss. et Bl. 446
var. *blancheanum* auct. 445
var. *blancheanum* (Boiss.) Boiss. 446
dichroanthum auct. 446
dubium Sibth. **470**
echinatum M. Bieb. **446**
filiforme auct. 470
formosum Urv. 450
fragiferum L. **457**

Trifolium L. (*cont.*)
  glaucescens Hausskn. 469
  globosum L. 8, **454**
  glomeratum L. **466**
  hirtum All. **440**
  lagrangei Boiss. 470
  lappaceum L. **442**
    var. *brachyodon* Hausskn. 442
    ssp. *brevidens* (Hal.) Lindberg f. 442
    var. *brevidens* Hal. 442
  leucanthum auct. 446
  leucanthum M. Bieb. **447**
  longisetum Boiss. et Bal. 443
  Melilotus indica L. 434
  Melilotus sicula Turra ex Vitm. 436
  messanense L. 436
  nidificum Griseb. 454
  nigrescens Viv. **465**
    ssp. petrisavii (Clem.) Holmboe **465**
  ovatifolium Bory et Chaub. 458
  pamphylicum Boiss. et Heldr. **444**
    var. blancheanum (Boiss.) Meikle **446**, 806
    var. dolichodontium Hossain **445**
    var. pamphylicum **445**
  petrisavii Clem. 465
  physodes Stev. ex M. Bieb. **457**
  pilulare Boiss. **455**
  pratense L. **439**
  procumbens auct. 469, 470
    var. erythranthum Griseb. 470
  purpureum Loisel. 445
    var. *laxiusculum* (Boiss. et Bl.) Hossain 446
    ssp. *pamphylicum* (Boiss.- et Heldr.) Holmboe 445
    var. *pamphylicum* (Boiss. et Heldr.) Zohary 445
    var. *roussaeanum* auct. 446
  repens L. **464**
  resupinatum L. **458**
    var. *clusii* (Gren. et Godr.) Dinsmore 460
    var. minus Boiss. **460**
    var. resupinatum **460**
  scabrum L. **448**
  scutatum Boiss. **451**
  speciosum Willd. 467
  spicatum Sibth. 434
  spumosum L. **462**
  stellatum L. **449**
    var. adpressum Turrill **449**
    var. stellatum **449**
  striatum auct. 448
  striatum L. **452**
    var. macrodontum Boiss. 453
    var. spinescens Lange 453
  subterraneum L. **453**
    var. *brachycladum* Gib. et Belli 453
  suffocatum L. **466**
  supinum Savi 446
  tomentosum L. **461**
    var. *curvisepalum* (V. Täckh.) Hossain 462
  xerocephalum Fenzl 463
Triglochin L. **1673**
  barrelieri Loisel. 1673
  bulbosa L. **1673**

Triglochin L. (*cont.*)
  ssp. *barrelieri* (Loisel.) Rouy 1673
Trigonella L. **401**
  aschersoniana Urb. 430
  balansae Boiss. et Reut. **401**
  berythea Boiss. et Bl. **408**
  cariensis Boiss. **406**
  corniculata auct. 402
  elatior auct. 402
  foenum-graecum auct. 408
  foenum-graecum L. 406, **407**
  hamosa auct. 402
  monspeliaca L. **406**
  sibthorpii Boiss. 402
  spicata Sibth. et Sm. **405**
  spinosa L. **404**
  sprunerana Boiss. **402**
  strangulata Boiss. **403**
  torulosa Griseb. 402
Triniusa danthoniae (Trin.) Steud. 1798
Triplachne Link **1787**
  nitens (Guss.) Link **1787**
Trisetaria Forssk. **1757**
  linearis Forssk. **1757**
Trisetum lineare (Forssk.) Boiss. 1757
  paniceum auct. 1767
  pilosum Roem. et Schult. 1760
Triticum L. **1824**
  subgenus *Micropyrum* Gaudin 1722
  aestivum L. **1825**
  bicorne Forssk. 1823
    var. *nana* Druce 1823
  durum Desf. **1825**
  elongatum Host 1818
  halleri Viv. 1722
  hybernum L. 1825
  junceum L. 1817
  lachenalii Gmel. 1722
  loliaceum (Huds.) Sm. 1721
  maritimum L. 1724
  monococcum L. **1826**
  ovatum auct. 1822
  ovatum (L.) Rasp. 1820
    ssp. *biunciale* (Vis.) Holmboe 1820
    var. *biunciale* (Vis.) Aschers. et Graebn. 1820
  peregrinum Hackel 1821
  poa Lam. ex Lam. et DC. 1722
  repens L. 1818
  rottbolla DC. 1721
  sativum Lam. 1825
  spelta L. **1824**
  tenellum L. 1722
  triunciale (L.) Gren. et Godr. 1823
  turgidum L. **1825**
  vulgare Vill. 1826
Trixago apula Steven 1228
Tuberaria (Dunal) Spach **188**
  guttata (L.) Fourr. **188**
    var. *eriocaulon* (Dunal) Grosser 189
    var. *plantaginea* (Willd.) Grosser 189
  variabilis Willk. 188
Tulipa L. **1596**
  agenensis DC. **1596**
  cypria Stapf 6, 7, **1597**
  montana auct. 1597
  oculus-solis St. Amans 1596
  praecox auct. 1596

Tulipa L. (*cont.*)
  systola auct. 1597
  veneris A. D. Hall 1596
Tunica auct. 219
  kennedyae A. K. Jackson et Turrill 220
  pachygona Fisch. et Mey. 220
  prolifera auct. 219
  velutina (Guss.) Fisch. et Mey. 219
Turgenia Hoffm. **705**
  latifolia (L.) Hoffm. **705**
Turritis L. **151**
  glabra auct. 152
  laxa (Sibth. et Sm.) Hayek **152**
Tussilago L. **929**
  farfara L. **929**
Typha L. **1662**
  angustata Bory et Chaub. 1662
  australis Schumacher 1662
  domingensis Pers. **1662**
  latifolia auct. 1662
Typhaceae **1661**
Tyrimnus (Cass.) Cass. **951**
  leucographus (L.) Cass. **951**

Ulex europaeus auct. 383
Ulmaceae **1459**
Ulmus L. **1460**
  abelicea Sm. 1463
  campestris auct. 1460
  canescens Melville **1460**
  glabra Huds. 1461
  minor auct. 1460
  minor Mill. 1460
    ssp. *canescens* (Melville) Browicz et
      Zieliński 1460
  montana With. 1461
Umbelliferae 19, **686**
    subfam. Apioideae **693**
    subfam. Saniculoideae **689**
Umbilicus DC. **642**
    sect. *Rosularia* DC. 644
  cyprius Holmboe 644
  globulariifolius auct. 644, 645
  horizontalis (Guss.) DC. **643**
  ? intermedius auct. 643
  lampusae Kotschy 648
  microstachyus Kotschy 649
  pallidiflorus Holmboe 645
  pendulinus DC. 643
  pestalozzae auct. 645
  pestalozzae Kotschy 644
  rupestris (Salisb.) Dandy **643**
Urachne coerulescens (Desf.) Trin. 1795
  parviflora Trin. 1794
Urginea Steinh. **1634**
  maritima (L.) Baker **1634**
  scilla Steinh. 1634
  undulata auct. 7, 1634
Urochloa panicoides P. Beauv. 1853
Urospermum Scop. **1042**
  picroides (L.) F. W. Schmidt **1043**
Urtica L. **1453**
  caudata Vahl 1456
  dioica L. **1454**
    ssp. cypria Lindberg f. **1455**
    ssp. dioica **1455**
    ssp. gracilis (Aiton) Selander 1455
  dubia Forssk. 1456

Urtica L. (*cont.*)
  gracilis Aiton 1455
  membranacea Poir. **1456**
  pilulifera L. **1456**
  urens L. **1454**
Urticaceae **1453**

Vaccaria Wolf **223**
  pyramidata Medik. **224**
  segetalis (Neck.) Garcke 224
Vaillantia Hoffm. 774
Valantia L. **774**
  aparine L. 786
  articulata L. 790
  columella (Ehrenb. ex Boiss.) Meikle
      776, **807**
  hispida L. **775**
  lanata Del. ex Coss. 776
  muralis auct. 775
  muralis L. **774**
  pedemontana Bell. 791
Valeriana L. **835**
  angustifolia Sm. 837
  calcitrapa L. 838
  cornucopiae L. 840
  dioscoridis Sm. 835
  echinata L. 845
  italica Lam. **835**
  locusta L. 844
    var. *coronata* L. 842
    var. *discoidea* L. 841
    var. *multifida* Gouan 844
    var. *olitoria* L. 844
    var. *pumila* L. 844
    var. *vesicaria* L. 843
  orbiculata Sm. 838
  pumila Willd. 844
  rotundifolia Sm. 838
  rubra L. 837
  sisymbriifolia auct. 835, 836
  tuberosa auct. 835
Valerianaceae **835**
Valerianella Moench **840**
    sect. Cornigerae Soy.-Willem. **841**
    sect. Coronatae Boiss. **841**
    sect. Platycoele DC. **841**
    sect. Siphonocoele Soy.-Willem. **841**
  carinata Lois. **844**
  chlorostephana auct. 841
  coronata (L.) DC. **842**
  dentata (L.) Poll. 847
  discoidea (L.) Lois. **841**
  echinata (L.) DC. **845**
  eriocarpa auct. 847
  eriocarpa Desv. 847
  lasiocarpa (Stev.) Betcke **843**
  locusta (L.) Laterrade 844
  muricata (Stev.) Baxt. **846**
  olitoria (L.) Poll. **844**
  orientalis (Schlecht.) Boiss. et Bal. **846**
  pumila (Willd.) DC. **844**
  triceras Bornm. **846**
  tridentata (Stev.) Reichb. 845
  truncata (Reichb.) Betcke 846
    var. *muricata* (Stev.) Boiss. 847
Velezia L. **213**
  rigida L. **213**
Vella annua L. 105

Verbascum L. **1197**
  blattaria L. **1197**
  glandulosum (Bouché) O. Kuntze 1198
  levanticum I. K. Ferguson **1198**
  orientale (L.) All. 1199
  sinuatum L. **1200**
Verbena L. **1250**
  aristigera S. Moore 1252
  nodiflora L. 1249
  officinalis L. **1251**
  supina L. **1251**
    f. petiolulata Lindberg f. **1252**
    f. supina **1252**
  tenuisecta Briquet 1252
Verbenaceae **1247**
  subfam. Verbenoideae **1248**
  subfam. Viticoideae **1252**
Veronica L. 1217
  acinifolia auct. 1218
  agrestis L. 1221
  anagallis-aquatica L. **1224**
    f. anagallidiformis Boreau **1225**
    var. *anagallidiformis* (Boreau) Franchet 1225
    f. anagallis-aquatica **1224**
    ssp. *divaricata* Krösche 1225
    var. *montioides* (Boiss.) Boiss. 1225
    var. *pseudo-anagalloides* Gren. 1225
  arvensis L. **1218**
  beccabunga auct. 1224, 1225
  buxbaumii Ten. 1221
  caespitosa Boiss. 1225
    var. leiophylla Boiss. 1226
  cymbalaria Bodard **1222**
  didyma Ten. 1220
    var. *thellungiana* (E. Lehm.) Druce 1220
  hederifolia L. 1223
    var. *triloba* auct. 1223
  hispidula Boiss. et Huet 1218
    ssp. *ixodes* (Boiss. et Bal.) M. Fischer 1218
  ixodes Boiss. et Bal. **1218**
  montioides Boiss. 1225
  persica Poir. **1221**
  polita Fries **1220**
    ssp. *thellungiana* E. Lehm. 1220
  pusilla auct. 1218
  thellungiana (E. Lehm.) Dalla Torre et Sarnth. 1220
  tournefortii C. C. Gmel. 1221
  trichadena Jord. 1223
  triphyllos L. **1219**
Vicia L. **536**
  sect. Cracca S. F. Gray **548**
  sect. Ervum (L.) S. F. Gray **558**
  sect. Faba (Adans.) Aschers. et Graebn. **546**
  sect. Vicia **538**
  amphicarpa auct. 543
  angustifolia (L.) L. 542
  articulata Hornem. 554
  assyriaca Boiss. **541**
  bithynica (L.) L. **545**
  calcarata Desf. 552
  carnea Kotschy 544
  cassia Boiss. **550**
  cordata auct. 542

Vicia L. (*cont.*)
  cracca auct. 550
  cracca L. 550
    ssp. *tenuifolia* (Roth) Gaudin 549
  cretica Boiss. et Heldr. **551**
  cypria Kotschy **555**
  dasycarpa auct. 553
  dumetorum L. 548
  eriocarpa (Hausskn.) Hal. 553
  ervilia (L.) Willd. **558**
  faba L. 548
  gracilis Loisel. 559
  hirsuta (L.) S. F. Gray **558**
  hybrida L. **539**
  johannis Tamamsch. 547
  lathyroides L. **544**
  laxiflora Brot. **559**
  lens (L.) Coss. et Germ. 562
  lenticula (Schreb. ex Hoppe) Janka 565
  lentoides (Ten.) Holmboe 564
  lunata (Boiss. et Bal.) Boiss. **557**
    var. *grandiflora* U. Plitm. 557
  michauxii auct. 544
  microphylla auct. 553
  monantha Retz. **551**
    ssp. monantha 552
    ssp. triflora (Ten.) Burtt et Lewis 552
  monanthos (L.) Desf. 554
  narbonensis auct. 548
  narbonensis L. **546**
    var. narbonensis **547**
    var. *salmonea* (Mout.) Schäfer 547
    ssp. *serratifolia* (Jacq.) Nym. 547
    var. serratifolia (Jacq.) Ser. **547**
  noëana Reut. ex Boiss. 541
  onobrychoides L. 548
  palaestina Boiss. **554**
  pannonica Crantz **538**
  peregrina L. **554**
  pinardii Boiss. 550
  pubescens (DC.) Link **560**
  salaminia Heldr. et Sart. 553
    ssp. *macrophyllaria* Cand. 553
  sativa L. **541**
    ssp. *amphicarpa* (Boiss.) Aschers. et Graebn. 543
    var. amphicarpa Boiss. **543**
    ssp. *angustifolia* Aschers. et Graebn. 543
    var. angustifolia L. **542**
    ssp. *nigra* (L.) Ehrh. 542
    var. *nigra* L. 542
    var. *obovata* Ser. ex DC. 542
    var. sativa **542**
    var. *segetalis* (Thuill.) Ser. ex DC. 543
  sepium L. 538
  sericocarpa Fenzl 540
    var. *microphylla* auct. 539
  serratifolia Jacq. 547
  subterranea Gérard ex Dorthes 543
  tenuifolia Roth **549**
    ssp. *elegans* (Guss.) Holmboe 549
    var. *elegans* (Guss.) Dinsm. 550
    var. laxiflora Griseb. **549**
    ssp. *stenophylla* Vel. 549
    var. *stenophylla* Boiss. 549
  tenuissima Schinz et Thell. 559
  varia Host 553

Vicia L. (*cont.*)
  var. *eriocarpa* Hausskn. 553
  villosa Roth **552**
  ssp. eriocarpa (Hausskn.) P. W. Ball
    553
  ssp. *microphylla* auct. 553
Vigna Savi 582
  mungo (L.) Hepper 582
  radiata (L.) Wilczek 582
Vinca L. 1097
  herbacea Waldst. et Kit. 1098
  major L. 1097
  minor L. 1098
Vincetoxicum Wolf **1105**
  canescens (Willd.) Decne. **1105**
  hirundinaria Medik. 1107
  officinale auct. 1106, 1107
Viola L. **201**
  sect. Melanium Ging. **203**
  sect. Viola **201**
  alba Bess. **202**
  ssp. *dehnhardtii* (Ten.) W. Becker 202
  ssp. *thessala* (Boiss. et Sprun.) Hayek
    202
  arvensis Murr. **203**
  ssp. kitaibeliana (Roem. et Schult.) W.
    Becker **205**
  canina auct. 202
  cypria Lindberg f. 202
  dehnhardtii Ten. 202
  heldreichiana Boiss. **205**
  kitaibeliana Roem. et Schult. 205
  mirabilis auct. 203
  occulta auct. 205
  odorata L. **201**
  parvula Tineo **206**
    var. *subarachnoidea* Lindberg f. 206
  riviniana auct. 203
  riviniana Reichb. 203
  ssp. *neglecta* auct. 203
  sieheana W. Becker **202**
    var. *oblongifolia* W. Becker 202
    f. *grandistipulata* W. Becker 202
  silvestris auct. 203
  silvestris Lam. 203
  ssp. *riviniana* auct. 203
  suavis auct. 201
  sylvatica auct. 203
  sylvatica Fries 203
    var. *riviniana* auct. 203
  thessala Boiss. et Sprun. 202
  tricolor L. 205
    var. *arvensis* auct. 205
Violaceae 19, **200**
Vitaceae 19, **360**
Vitex L. **1253**
  agnus-castus L. **1253**
Vitis L. **360**
  sylvestris Gmel. 361
  vinifera L. **360**
  ssp. sylvestris (Gmel.) Berger 361
  ssp. vinifera 361
Vogelia paniculata auct. 132
  paniculata (L.) Hornem. 133
Vulpia C. C. Gmel. **1731**
  subgenus Spirachne Hackel 1732
  brevis Boiss. et Kotschy **1732**
  bromoides auct. 1733

Vulpia C. C. Gmel. (*cont.*)
  broteri Boiss. et Reut. 1733
  ciliata Link **1734**
  danthonii (Aschers. et Graebn.) Volk
    1734
  inops Hackel 1732
  membranacea (L.) Dumort. **1731**
  michelii (Savi) Reichb. 1766
  muralis (Kunth) Nees 1732
  myuros (L.) C. C. Gmel. **1733**
    var. *tenella* Boiss. 1732
  sciuroides auct. 1733, 1734
  sciuroides (Roth) C. C. Gmel. 1733
    var. *longe-aristata* Willk. 1733

Washingtonia filifera (J. A. Linden ex
    André) H. Wendl. ex S. Watson
    1661
  robusta H. Wendl. 1661
Weingartneria Bernh. 1755
  articulata (Desf.) Aschers. et Graebn.
    1756
Withania Pauquy **1185**
  somnifera (L.) Dunal **1186**

Xanthium L. **899**
  brasilicum Vell. 900
  spinosum L. **900**
  strumarium L. **900**
Xeranthemum L. **942**
  annuum L. 942
    ssp. *annettae* auct. 942
    ssp. *inapertum* L. 942
  inapertum (L.) Mill. **942**
  squarrosum auct. 942

Zacintha verrucosa Gaertn. 1011
Zannichellia L. **1680**
  marina Nielsen 1680
  palustris L. **1680**
    ssp. *marina* (Nielsen) Holmboe 1680
    ssp. *pedicellata* (Wahl. et Rosén)
      Holmboe 1680
    var. *pedicellata* Wahl. et Rosén 1680
  pedicellata (Wahl. et Rosén) Fries 1680
  repens Boenningh. 1680
Zannichelliaceae **1679**
Zappania nodiflora (L.) Lam. 1249
    var. *rosea* D. Don 1249
Zea L. **1870**
  mays L. **1871**
Zelkova Spach **1463**
  abelicea (Lam.) Boiss. 8, 10, **1463**
  cretica Spach 1463
Zinnia elegans Jacq. 901
Ziziphora L. **1286**
  capitata L. **1286**
Ziziphus Mill. **357**
  jujuba Mill. 358
  lotus (L.) Lam. 357, **359**
  officinarum Medik. 358
  paliurus (L.) Willd. 356
  sativa Gaertn. 358
  spina-Christi (L.) Willd. **357**
  vulgaris Lam. 358
  zizyphus (L.) Meikle **358**, 806
Zosima Hoffm. **759**
  absinthiifolia (Vent.) Link **760**

Zostera L. 1681
    marina L. 1681
    nana Roth 1681
    nodosa Ucria 1682
    oceanica L. 1679
Zosteraceae **1681**

Zygophyllaceae 19, **324**
    tribe Peganeae Engl. **324**
    tribe Tribuleae Engl. **327**
    tribe Zygophylleae Engl. **325**
Zygophyllum L. **326**
    album L.f. **327**

RIZOKARPASO

8

FAMAGUSTA

LEFKONIKO

LARNACA

5

KYRENIA

NICOSIA

7

MORPHOU

6

LIMASSOL

TROÖDOS

3

2

1

PAPHOS

0   7   14 miles